Definitions, units and conversion fa

In the SI, units are given for both defining relations of the magnetization, $\boldsymbol{B}=\mu_0(\boldsymbol{H}+\boldsymbol{M})$ and $\boldsymbol{B}=\mu_0\boldsymbol{H}+\boldsymbol{M}$, respectively. $\mu_0=4\pi\cdot 10^{-7}\,\mathrm{Vs\,A^{-1}\,m^{-1}}$, A: molar mass, ϱ: mass density, $\boldsymbol{P}$: magnetic moment, $\boldsymbol{M}$: magnetic moment per unit volume (magnetization, magnetic polarization).

Quantity	cgs/emu	SI	
$\boldsymbol{B}$	$\mathrm{G}=(\mathrm{erg\,cm^{-3}})^{1/2}$	$\mathrm{T}=\mathrm{Vs\,m^{-2}}$	
	$1\,\mathrm{G}\triangleq$	$10^{-4}\,\mathrm{T}$	
$\boldsymbol{H}$	$\mathrm{Oe}=(\mathrm{erg\,cm^{-3}})^{1/2}$	$\mathrm{Am^{-1}}$	
	$1\,\mathrm{Oe}\triangleq$	$10^3/4\pi\,\mathrm{Am^{-1}}$	
$\boldsymbol{M}$	$\boldsymbol{B}=\boldsymbol{H}+4\pi\boldsymbol{M}$	$\boldsymbol{B}=\mu_0(\boldsymbol{H}+\boldsymbol{M})$	$\boldsymbol{B}=\mu_0\boldsymbol{H}+\boldsymbol{M}$
	G	$\mathrm{Am^{-1}}$	T
	$1\,\mathrm{G}\triangleq$	$10^3\,\mathrm{Am^{-1}}$	$4\pi\cdot 10^{-4}\,\mathrm{T}$
$\boldsymbol{P}$	$\boldsymbol{P}=\boldsymbol{M}V$	$\boldsymbol{P}=\boldsymbol{M}V$	$\boldsymbol{P}=\boldsymbol{M}V$
	$\mathrm{G\,cm^3}$	$\mathrm{Am^2}$	$\mathrm{Vs\,m}$
	$1\,\mathrm{G\,cm^3}\triangleq$	$10^{-3}\,\mathrm{Am^2}$	$4\pi\cdot 10^{-10}\,\mathrm{Vs\,m}$
σ	$\sigma=M/\varrho$	$\sigma=M/\varrho$	$\sigma=M/\varrho$
	$\mathrm{G\,cm^3\,g^{-1}}$	$\mathrm{Am^2\,kg^{-1}}$	$\mathrm{Vs\,m\,kg^{-1}}$
	$1\,\mathrm{G\,cm^3\,g^{-1}}\triangleq$	$1\,\mathrm{Am^2\,kg^{-1}}$	$4\pi\cdot 10^{-7}\,\mathrm{Vs\,m\,kg^{-1}}$
σ_m	$\sigma_\mathrm{m}=\sigma A$	$\sigma_\mathrm{m}=\sigma A$	$\sigma_\mathrm{m}=\sigma A$
	$\mathrm{G\,cm^3\,mol^{-1}}$	$\mathrm{Am^2\,mol^{-1}}$	$\mathrm{Vs\,m\,mol^{-1}}$
	$1\,\mathrm{G\,cm^3\,mol^{-1}}\triangleq$	$10^{-3}\,\mathrm{Am^2\,mol^{-1}}$	$4\pi\cdot 10^{-10}\,\mathrm{Vs\,m\,mol^{-1}}$
χ	$\boldsymbol{P}=\chi\boldsymbol{H}$	$\boldsymbol{P}=\chi\boldsymbol{H}$	$\boldsymbol{P}=\chi\mu_0\boldsymbol{H}$
	$\mathrm{cm^3}$	$\mathrm{m^3}$	$\mathrm{m^3}$
	$1\,\mathrm{cm^3}\triangleq$	$4\pi\cdot 10^{-6}\,\mathrm{m^3}$	$4\pi\cdot 10^{-6}\,\mathrm{m^3}$
χ_v	$\chi_v=\chi/V$	$\chi_v=\chi/V$	$\chi_v=\chi/V$
	$\mathrm{cm^3\,cm^{-3}}$	$\mathrm{m^3\,m^{-3}}$	$\mathrm{m^3\,m^{-3}}$
	$1\,\mathrm{cm^3\,cm^{-3}}\triangleq$	$4\pi\,\mathrm{m^3\,m^{-3}}$	$4\pi\,\mathrm{m^3\,m^{-3}}$
χ_g	$\chi_\mathrm{g}=\chi_v/\varrho$	$\chi_\mathrm{g}=\chi_v/\varrho$	$\chi_\mathrm{g}=\chi_v/\varrho$
	$\mathrm{cm^3\,g^{-1}}$	$\mathrm{m^3\,kg^{-1}}$	$\mathrm{m^3\,kg^{-1}}$
	$1\,\mathrm{cm^3\,g^{-1}}\triangleq$	$4\pi\cdot 10^{-3}\,\mathrm{m^3\,kg^{-1}}$	$4\pi\cdot 10^{-3}\,\mathrm{m^3\,kg^{-1}}$
χ_m	$\chi_\mathrm{m}=\chi_\mathrm{g}A$	$\chi_\mathrm{m}=\chi_\mathrm{g}A$	$\chi_\mathrm{m}=\chi_\mathrm{g}A$
	$\mathrm{cm^3\,mol^{-1}}$	$\mathrm{m^3\,mol^{-1}}$	$\mathrm{m^3\,mol^{-1}}$
	$1\,\mathrm{cm^3\,mol^{-1}}\triangleq$	$4\pi\cdot 10^{-6}\,\mathrm{m^3\,mol^{-1}}$	$4\pi\cdot 10^{-6}\,\mathrm{m^3\,mol^{-1}}$

Experimental errors

In this volume, experimental errors are given in parentheses referring to the last decimal places. For example, 1.352 (12) stands for 1.352 ± 0.012, and 342.5 (21) stands for 342.5 ± 2.1.

Landolt-Börnstein / New Series

Landolt-Börnstein

Numerical Data and Functional Relationships
in Science and Technology

New Series
Editor in Chief: O. Madelung

Units and Fundamental Constants in Physics and Chemistry

Nuclear and Particle Physics (Group I)

Atomic and Molecular Physics (Group II)

Solid State Physics (Group III)

Macroscopic Properties of Matter (Group IV)

Geophysics (Group V)

Astronomy and Astrophysics (Group VI)

Biophysics (Group VII)

Landolt-Börnstein

Numerical Data and Functional Relationships in Science and Technology

New Series / Editor in Chief: O. Madelung

Group III: Solid State Physics

Volume 27

Magnetic Properties of Non-Metallic Inorganic Compounds Based on Transition Elements

Subvolume j1

Halides I

Editor: H.P.J. Wijn

Contributors:

K. Katsumata, M. Matsuura, H.P.J. Wijn

Springer-Verlag

Berlin Heidelberg NewYork

London Paris Tokyo

Hong Kong Barcelona Budapest

ISSN 0942-7988 (Solid State Physics)

ISBN 3-540-56068-8 Springer-Verlag Berlin Heidelberg New York
ISBN 0-387-56068-8 Springer-Verlag New York Berlin Heidelberg

Library of Congress Cataloging in Publication Data
Zahlenwerte und Funktionen aus Naturwissenschaften und Technik, Neue Serie
Editor in Chief: O. Madelung
Vol. III/27j1 : Edited by H.P.J. Wijn
At head of title: Landolt-Börnstein. Added t.p.: Numerical data and functional relationships in science and technology.
Tables chiefly in English.
Intended to supersede the Physikalisch-chemische Tabellen by H.H. Landolt and R. Börnstein of which the 6th ed.
began publication in 1950 under title: Zahlenwerte und Funktionen aus Physik, Chemie, Astronomie,
Geophysik und Technik.
Vols. published after v. 1 of group I have imprint: Berlin, New York, Springer-Verlag
Includes bibliographies.
1. Physics--Tables. 2. Chemistry--Tables. 3. Engineering--Tables.
I. Börnstein, R.(Richard), 1852-1913. II. Landolt, H. (Hans), 1831-1910.
Physikalisch-chemische Tabellen. IV. Title: Numerical data and functional relationships in science and
technology.
QC61.23 502`.12 62-53136

Typesetting: Macmillan India Ltd.
Printing: Druckhaus Langenscheidt KG, Berlin
Binding: Lüderitz & Bauer, Berlin

SPIN: 10057716 63/3020 - 5 4 3 2 1 0 - Printed on acid-free paper

Editor

H.P.J. Wijn, Institut für Werkstoffe der Elektrotechnik der Rheinisch-Westfälischen Technischen Hochschule Aachen, Templergraben, 52062 Aachen, FRG

Authors

K. Katsumata, Institute of Physical and Chemical Research (RIKEN), Wako, Saitama, 351-01, Japan

M. Matsuura, Department of Electronics and Information Science, Kyoto Institute of Technology, Matsugasaki Sakyoku, Kyoto 606, Japan

H.P.J. Wijn, Institut für Werkstoffe der Elektrotechnik der Rheinisch-Westfälischen Technischen Hochschule Aachen, Templergraben, 52062 Aachen, FRG

Preface

In Volume III/27 of the New Series of Landolt-Börnstein, entitled "Magnetic Properties of Non-Metallic Inorganic Compounds Based on Transition Elements" the magnetic properties are given of inorganic compounds which contain, apart from other elements, at least one type of a transition element. In a way this volume can be considered as a supplement to the Volumes III/4 and III/12 "Magnetic and Other Properties of Magnetic Oxides and Related Compounds", published in 1970 and in the period 1978 – 1982, respectively. However it seemed appropriate to change the title of the volume, since not only magnetic properties of oxides, but also those of pnictides, chalcogenides and halides have been a major field for research in the last decades.

The magnetic properties of halides have been compiled in Subvolume III/27j, the first part of which is presented herewith. It deals with the halides having the general chemical formula MX_n, where M represents a 3d element and X a halogen element. Also graphite compounds intercalated by these transition metal chlorides have been included.

A survey of the contents of all subvolumes planned for LB III/27 is printed on the inside of the front cover.

The use of two systems of units illustrates the situation in magnetic research today, where conversion from CGS units to SI units and vice versa has to be frequently performed. The user of the tables and figures of this subvolume is helped to convert the units by the list of definitions, units and conversion factors for the magnetic quantities occurring most frequently. This list is reproduced in the front cover.

Many thanks are due to the authors for the agreeable co-operation, the Landolt-Börnstein editorial office in Darmstadt, especially Dr.W.Polzin, Frau R.Brangs and Frau I.Lenhart for the great help with the editorial work, and to Springer Verlag for their thoughtful help in the final preparation of this volume.

Aachen, December 1993 **The Editor**

Table of contents

Magnetic properties of non-metallic inorganic compounds based on transition elements
Subvolume j1: Halides I

9 Magnetic properties of halides

9.1 MX_2 and $MX_2 \cdot nH_2O$ compounds (M = 3d element, X = halogen element)

9.1.1 Introduction

9.1.1.1 General remarks

This chapter is devoted to description of macroscopic as well as microscopic magnetic and related properties of MX_2 and $MX_2 \cdot nH_2O$ compounds. Here, M stands for 3d transition metal ions, whereas X represents halogen ions.

The macroscopic magnetic properties are the Curie constant, paramagnetic Curie temperature and the magnetic ordering temperature of MX_2 and $MX_2 \cdot nH_2O$ compounds. Most of these compounds show antiferromagnetic ordering and the remaining ones helimagnetic ordering at low temperatures. The magnetic susceptibility per mole (χ_m) of antiferromagnets and helimagnets well above the Néel temperature is expressed by

$$\chi_m = C_m/(T - \Theta)\,,$$

where T is the absolute temperature, C_m the Curie constant per mole and Θ is the paramagnetic Curie temperature. By fitting the susceptibility data with this formula, one obtains the values of C_m and Θ.

The magnetic phase transition from the paramagnetic to antiferromagnetic or helimagnetic state is of second order. Specific heat measurement on antiferromagnetic and helimagnetic substances gives the most accurate value for the Néel temperature. Susceptibility and neutron scattering measurements also give the Néel temperature.

The spin structures in the ordered states of antiferromagnets and helimagnets are directly determined from neutron scattering study. Electron spin resonance (ESR) and nuclear magnetic resonance (NMR) experiments also give information on the spin structure.

The macroscopic data are presented in separate tables for each of the MF_2 (Table 1), MCl_2 (Table 2), MBr_2 (Table 3), MI_2 (Table 4) and $MX_2 \cdot nH_2O$ (Tables 5, 5a) compounds followed by the relevant figures. The figures are generally arranged in the following order: first the crystal and spin structures, then the specific heat and magnetic susceptibility data as a function of temperature. Next, the magnetization is given as a function of applied magnetic field as well as the magnetic field vs. temperature phase diagram.

For the microscopic magnetic and related properties of MX_2 and $MX_2 \cdot nH_2O$ compounds we concentrate here to the Hamiltonians which are the basis of the microscopic magnetism of these compounds. The magnetic excitations in the ordered states of antiferromagnets and helimagnets are well described by spin-wave picture at low temperatures. The spin-wave spectra are calculated with the Hamiltonians. The magnetic excitation spectra are directly observed by neutron inelastic scattering technique from which the parameters of the Hamiltonians can be derived. Some of the parameters of the Hamiltonian are determined from ESR and far-infrared absorption experiments. The Hamiltonians and their parameters for each of the compounds of MF_2, MCl_2, MBr_2 and $MCl_2 \cdot 2H_2O$ are given in the Tables 6 and 7. Finally the experimental spin-wave spectra are presented from which these parameters have been obtained.

In Landolt-Börnstein New Series Group II, Subvolumes 2, 8, 10, 11, 12a and 12b, which appeared in the years 1966 and 1976–1984, more data have been compiled, specifically on the paramagnetic susceptibility and on electron spin resonance experiments. Some of these data have been included in the present compilation with the additional reference LB II/n, where n stands for subvolume n. For more detailed data on the crystal structures and on the lattice parameters see Landolt-Börnstein, New Series, Group III, Subvolume 7a.

For further data see also edge compositions (x = 0 and x = 1) in section 9.2 ($M_{1-x}M'_xX_2$ and $M_{1-x}M'_xX_2 \cdot nH_2O$).

9.1.1.2 List of frequently used symbols and abbreviations

Symbols

a, b, c	Å	lattice parameters
C_m	emu K mol^{-1} = cm^3 K mol^{-1}	Curie constant per mole
C_p (c_p)	J K^{-1} mol^{-1}, cal K^{-1} mol^{-1}, (J K^{-1} g^{-1})	specific heat (1 cal = 4.184 J)
D	cm^{-1}	single ion anisotropy constant with the corresponding Hamiltonian of the form DS_z^2
E	cm^{-1}	single ion anisotropy constant with the corresponding Hamiltonian of the form $E\{S_x^2 - S_y^2\}$
E_n	meV, cm^{-1}, THz	excitation energy (1 meV = 8.06554 cm^{-1} = 0.241799 THz)
g_i		spectroscopic splitting factor (i = x, y, z)
$\boldsymbol{H}$	Oe, A m^{-1}	applied magnetic field (often given as $\mu_0 H$ in tesla (T))
H_A		anisotropy field
h, k, l		reciprocal lattice indices
i, j, k		site indices in real space
J_{ij}	cm^{-1}	exchange interaction constant between the spins i and j (J/k in K)
K_{ij}	cm^{-1}	anisotropic parts of the exchange interaction constant
$\boldsymbol{k}$	Å^{-1}	wave vector
k		Boltzmann constant
M_v	G	magnetic moment per unit volume = magnetization
p_A	μ_B	magnetic moment per atom (ion)
p_{eff}	μ_B	effective magnetic moment per molecule (formula unit)
$\boldsymbol{S}$		spin operator
T	K	temperature
T_N	K	Néel temperature
x, y, z		coordinate axes
z_i		number of interacting neighbors

α		crystal axis (for special definition see corresponding captions)
α, β, γ	deg	lattice parameters
ζ		reduced wavevector coordinate
Θ	K	paramagnetic Curie temperature
θ	deg	angle
λ	Å	neutron wavelength
μ_B	$J T^{-1}$	Bohr magneton
σ	$emu g^{-1} = G cm^3 g^{-1}$	magnetic moment per gram (specific magnetization)
σ_m	$emu mol^{-1} = G cm^3 mol^{-1}$	magnetic moment per mole
χ_g	$emu g^{-1} = cm^3 g^{-1}$	magnetic susceptibility per gram
χ_m	$emu mol^{-1} = cm^3 mol^{-1}$	magnetic susceptibility per mole
$\chi_{\parallel}, \chi_{\perp}$		magnetic susceptibility parallel and perpendicular to a principal crystallographic direction
χ', χ''		real, imaginary part of ac magnetic susceptibility

Abbreviations

A.F.	antiferromagnetic phase
ESR	electron spin resonance
Heli.	helimagnetic phase
NMR	nuclear magnetic resonance
P.	paramagnetic phase
S.F.	spin flop phase

9.1.2 Macroscopic magnetic and related properties of MX_2 compounds

9.1.2.1 MF_2 compounds

Table 1. MF_2 compounds.

Substance (Figs.)	Crystal type	Lattice constants*)	Molar Curie constant (in cm^3K/mol)	Paramagnetic Curie temperature	Magnetic order	Critical temperature	Spin structure	Ordered moment
VF_2 (Figs. 1 ··· 3)	Rutile Fig. 1	$a = 4.80$ Å $c = 3.24$ Å [66S]	1.82 [67S2]	− 80 K [67S2]	Heli. ← P.	7.0 K [66S]	Spiral along c with a turn angle of 96.0°; ⊥ c [69L]	
CrF_2	Distorted rutile ($P2_1/c$)	$a = 4.732$ Å $b = 4.718$ Å $c = 3.505$ Å $\beta = 96.52°$ [57J]			A.F. ← P.	53 K [60C]	The moments at the corner sites are directed oppositely to those at the center [60C]	$3.96\mu_B$ [60C]
MnF_2 (Figs. 4 ··· 9)	Rutile	$a = 4.8734$ Å $c = 3.3099$ Å [54S]	4.47 [50C]	− 97.0 K [50C]	A.F. ← P.	67.33 K [65T1] 67.29 K [73B1, LBII/12a, p. 124]	Fig. 4 [53E]	
FeF_2 (Figs. 10 ··· 12)	Rutile	$a = 4.6966$ Å $c = 3.3091$ Å [54S]	3.88 [41B]	− 117 K [41B]	A.F. ← P.	78.35 K [55C2]	The same as in MnF_2 [53E]	
CoF_2 (Figs. 13 ··· 16)	Rutile	$a = 4.6951$ Å $c = 3.1796$ Å [54S]			A.F. ← P.	37.70 K [55C2]	The same as in MnF_2 [53E]	

NiF_2 (Figs. 17 ⋯ 24)	Rutile	$a = 4.6506$ Å $c = 3.0836$ Å [54S]	1.36 [65C2]	− 128 K [65C2]	A.F. ← P.**)	73.22 K [55C1] 73.3(1) K [73 B2; LBII/12a, p. 320]	The moments at the corner sites are directed oppositely to those at the center [53E] Spin ⊥ c ≈ 1.3° away from [100] [61S]

*) Room temperature unless otherwise specified.
**) Antiferromagnetic with weak ferromagnetism. Transverse weak ferromagnetism $\sigma_{D\perp} = 169(2)\,G\,cm^3/mol$; longitudinal weak ferromagnetism $\sigma_{D\parallel} = 57(4)\,G\,cm^3/mol$ [73B2; LBII/12a, p. 320]. Also reported $\sigma_{D\perp} = 169(2)\,G\,cm^3/mol$ and $\sigma_{D\parallel} = 73(4)\,G\,cm^3/mol$ [73B3; LBII/12a, p. 320].

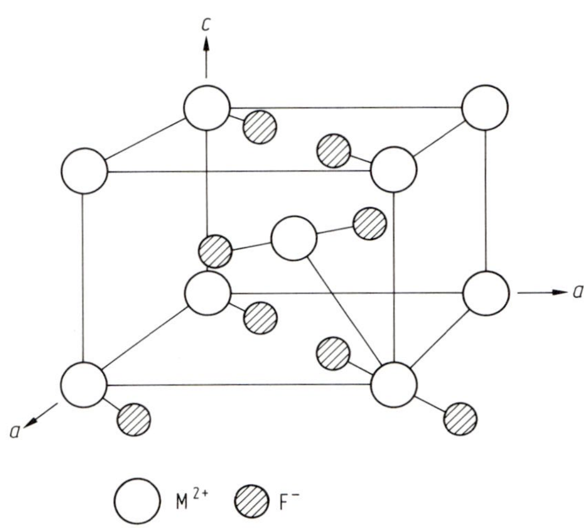

Fig. 1. Rutile type crystal structure (D_{4h}^{14}-$P4_2/mnm$).

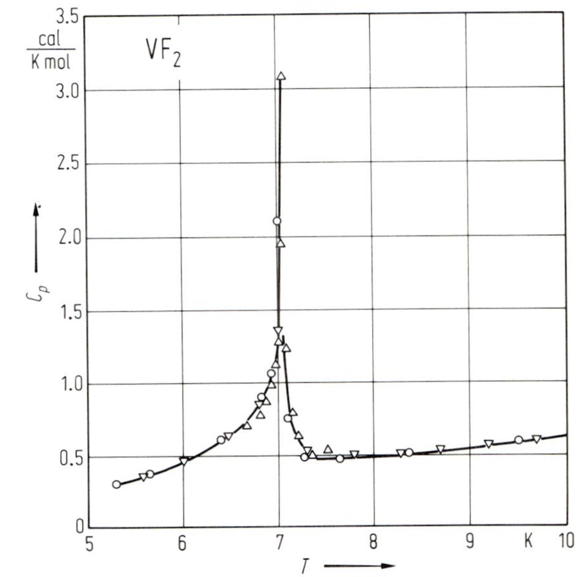

Fig. 2. VF_2. Molar specific heat vs. temperature. Different symbols indicate different series [66S].

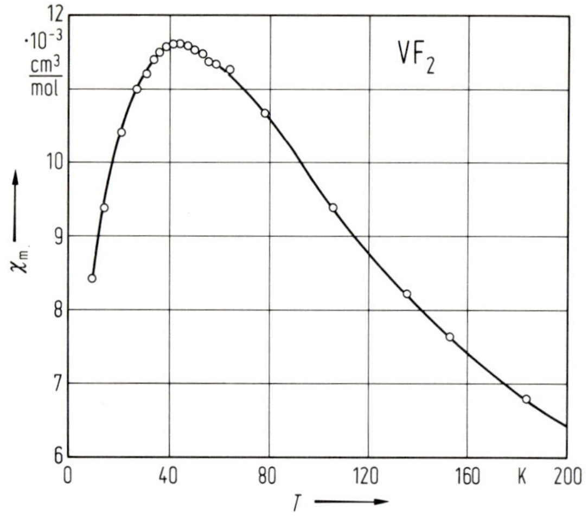

Fig. 3. VF_2. Molar magnetic susceptibility of randomly oriented powder vs. temperature [67S2].

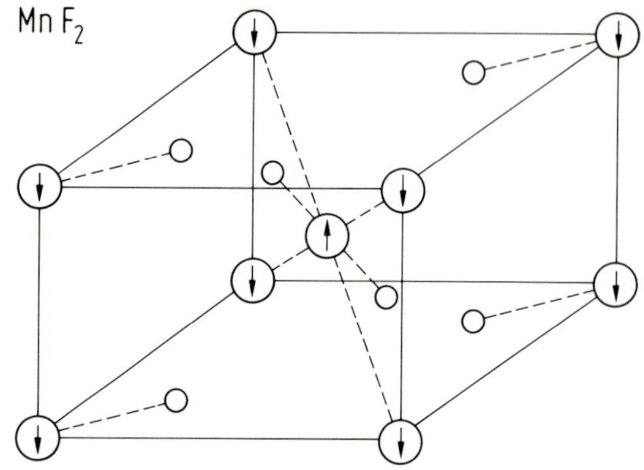

Fig. 4. MnF_2. Spin structure. The arrows show the Mn^{2+} magnetic moments. The small circles correspond to fluorine sites [53E].

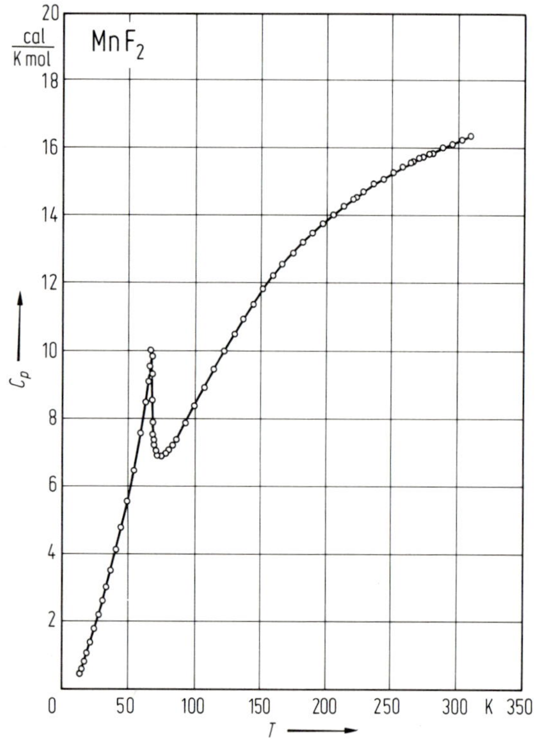

Fig. 5. MnF_2. Molar specific heat vs. temperature [42S].

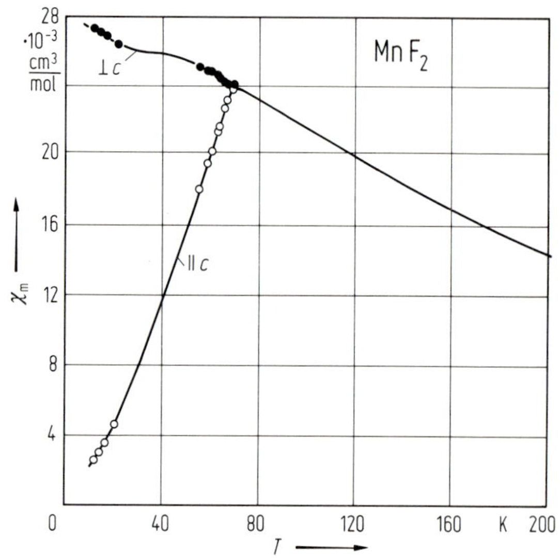

Fig. 6. MnF_2. Molar magnetic susceptibility parallel and perpendicular to the c axis vs. temperature [50G].

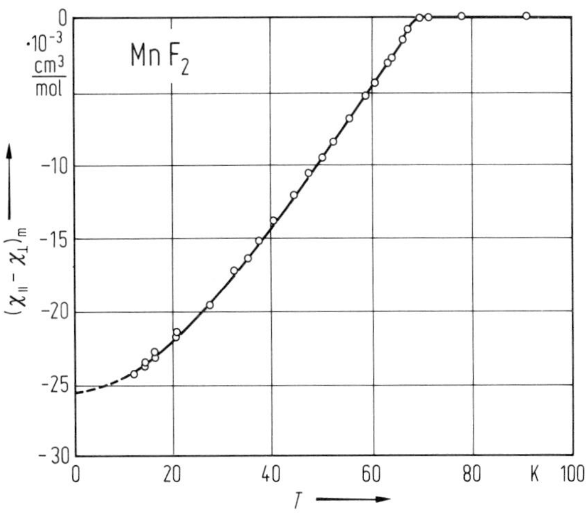

Fig. 7. MnF_2. Anisotropy of molar magnetic susceptibility vs. temperature [50G].

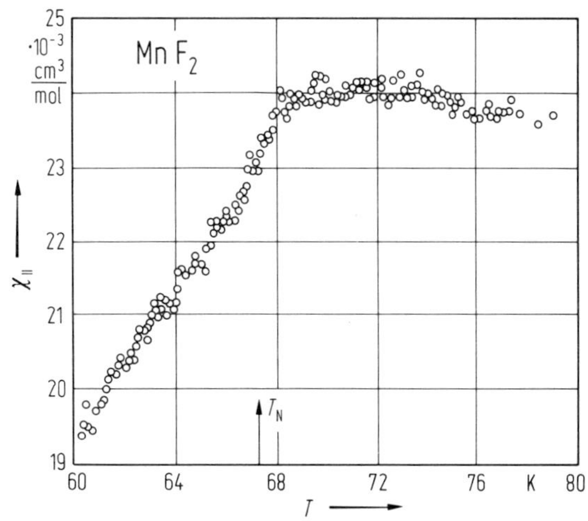

Fig. 8. MnF_2. Temperature dependence of longitudinal molar magnetic susceptibility $\chi_{||}$. The Néel point defined as the maximum of the temperature for the maximum slope of the curve, $T_N = 67.29(5)$ K [73B1; LBII/12a, p. 589].

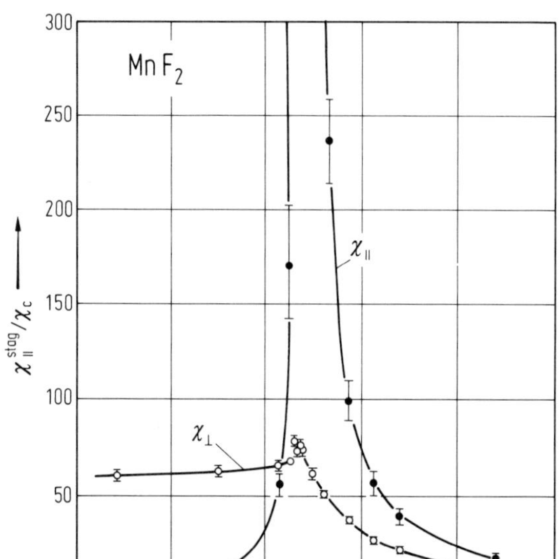

Fig. 9. MnF_2. Transverse ($\chi_\perp$) and longitudinal ($\chi_{||}$) staggered susceptibilities vs. temperature as derived from inelastic neutron scattering. The quantity $\chi_C \equiv C/T_N$ is the susceptibility of an ideal Curie-law spin-5/2 paramagnet at $T_N = 67.46$ K [71S].

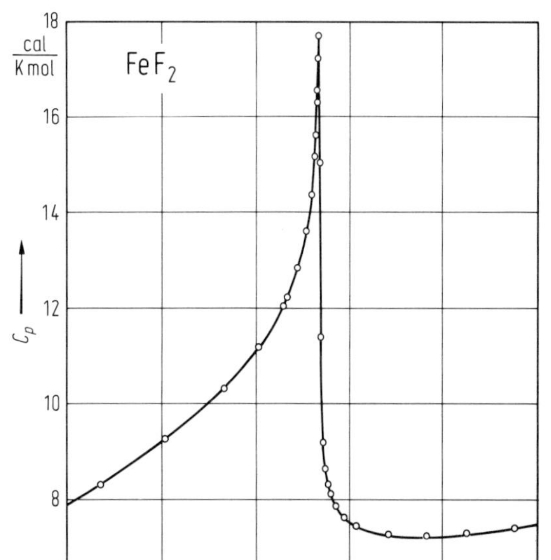

Fig. 10. FeF_2. Molar specific heat vs. temperature [55C2].

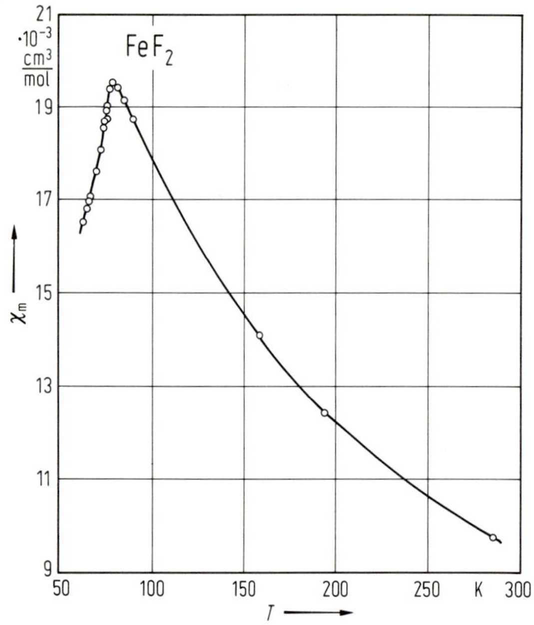

Fig. 11. FeF_2. Molar magnetic susceptibility vs. temperature [41B].

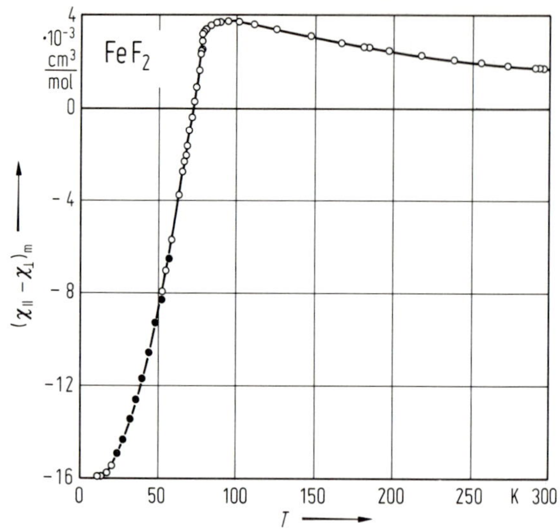

Fig. 12. FeF_2. Anisotropy of molar magnetic susceptibility vs. temperature [53S].

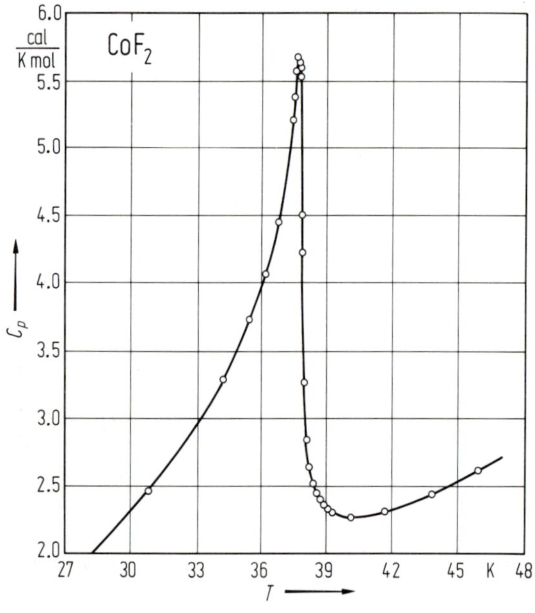

Fig. 13. CoF_2. Molar specific heat vs. temperature [55C2].

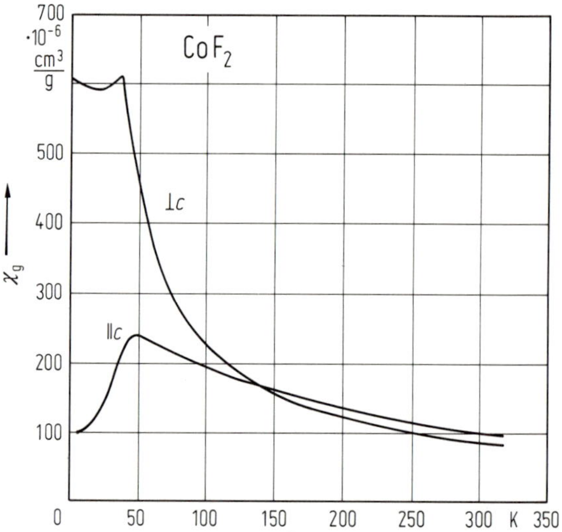

Fig. 14. CoF_2. Susceptibility per gram parallel and perpendicular to the c axis vs. temperature [65L].

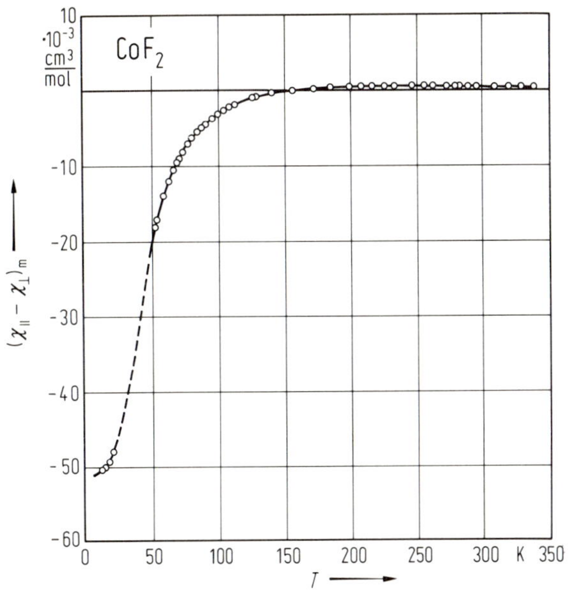

Fig. 15. CoF_2. Anisotropy of molar magnetic susceptibility vs. temperature [53S].

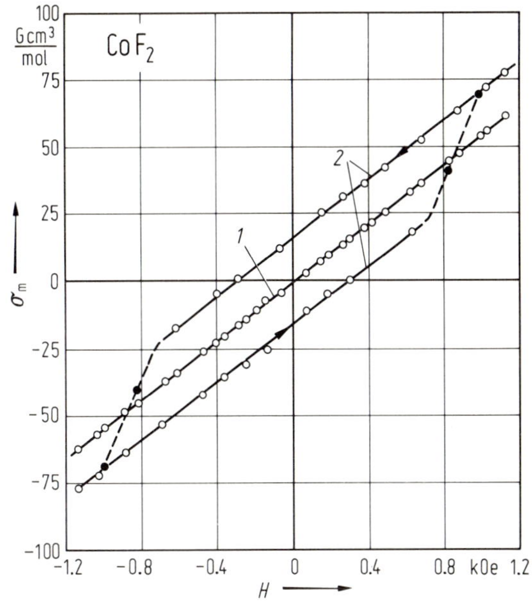

Fig. 16. CoF_2. Molar magnetic moment along y axis vs. applied magnetic field. $T = 20.4$ K. *1*: in the absence of stress, *2*: under a shear stress ($\sigma_{xz} = 340$ kg/cm^2) showing piezomagnetism [60B].

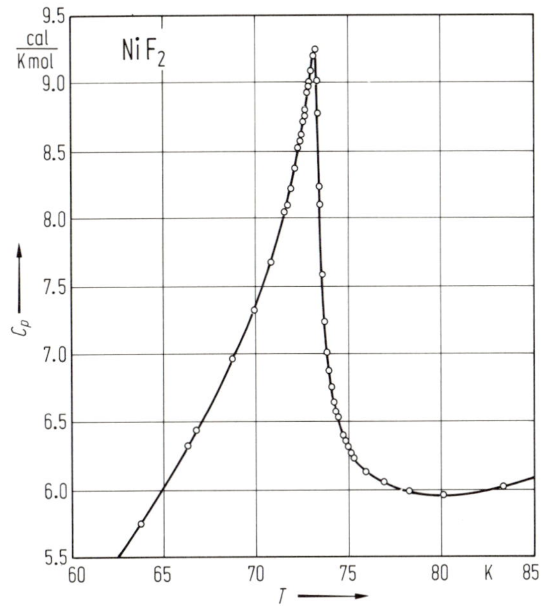

Fig. 17. NiF_2. Molar specific heat vs. temperature [55C1].

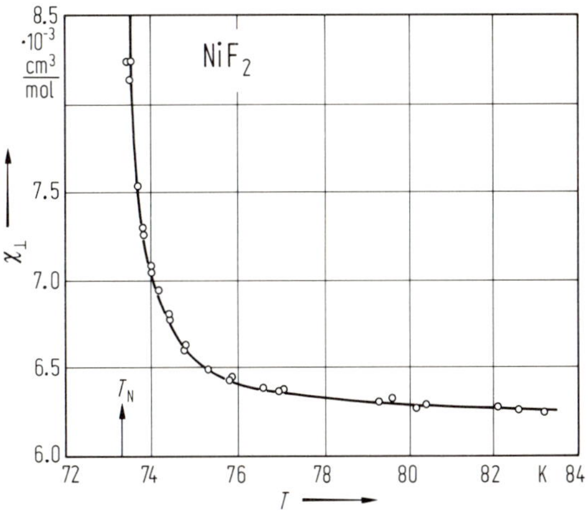

Fig. 18. NiF_2. Molar magnetic susceptibility in the basal plane, $\chi_{\perp}$, immediately above the Néel point [65C2].

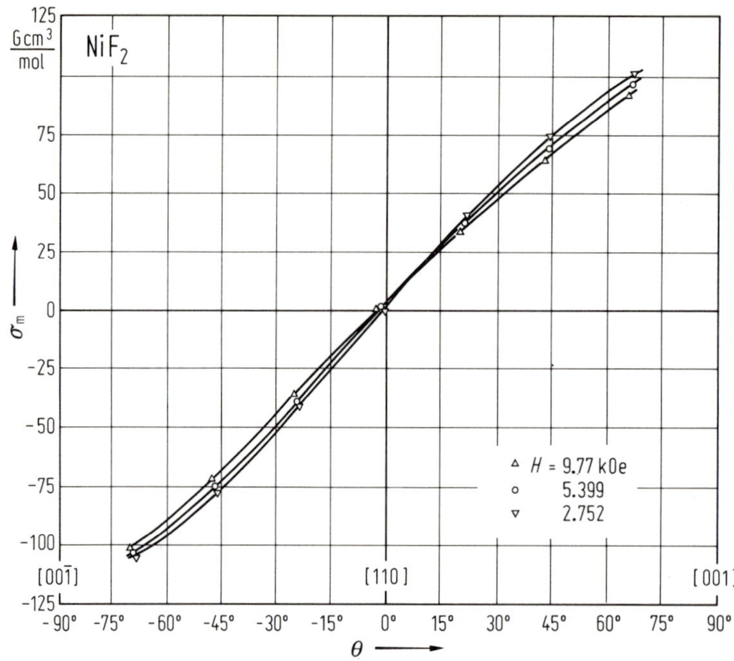

Fig. 19. NiF_2. Molar perpendicular magnetic moment vs. angle between field and the [110] direction. Rotation in the (110) plane. $T = 20.4$ K [54M].

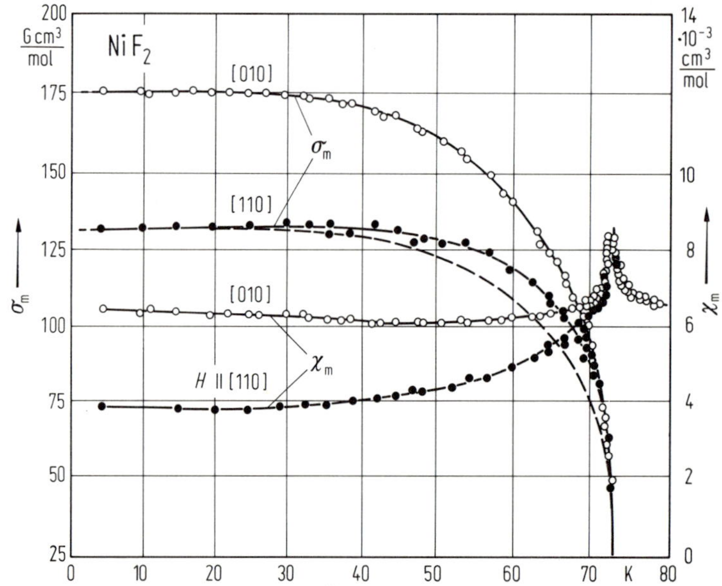

Fig. 20. NiF_2 single crystal. Temperature dependence of $\sigma_{D\perp}$ and $\chi_\perp(T)$ at $\boldsymbol{H} \parallel [010]$ and $\sigma'(T)$ and $\chi'_m(T)$ at $\boldsymbol{H} \parallel [110]$. $\sigma'(T)$ and $\chi'_m(T)$ obtained by reducing the linear sections of the $\sigma_{\parallel}(T)$ curves for $\psi = 45°$ at temperatures higher than $T = 50$ K. The values of σ' for temperatures below $T = 50$ K are directly the values of $\sigma_{\parallel}(H)$ for $\psi = 45°$ at $H = 0$. The dashed curve shows the dependence of $\sigma_{D\perp}(T)\sqrt{2}/2$, corresponding to a model wherein the spontaneous magnetic moment $\sigma_{D\perp}(T)$ is fixed on the [010] axis [73B2; LBII/12a, p. 626]. $T_N = 73.3$ K.

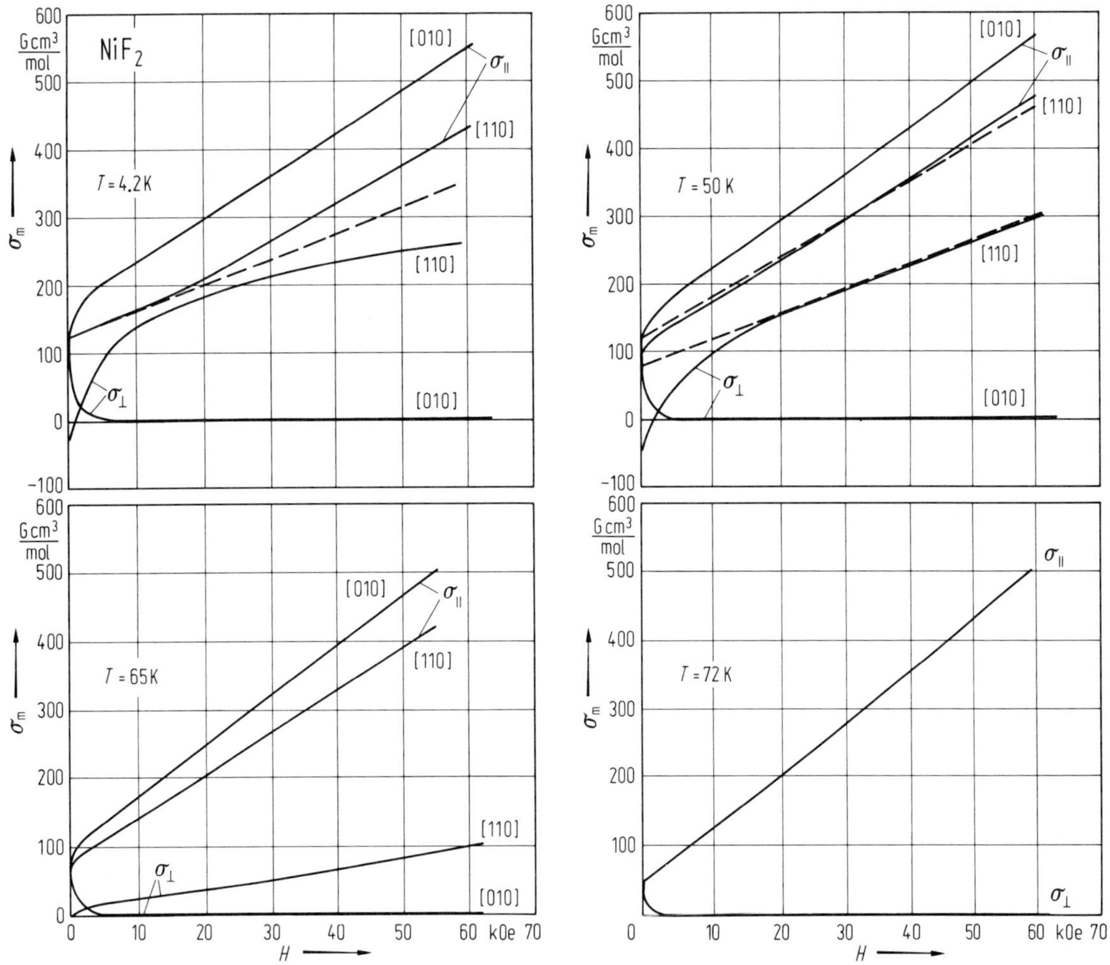

Fig. 21. NiF_2 single crystal. Plots of $\sigma_{\|}(H)$ and $\sigma_{\perp}(H)$ against the temperature for two directions of the field, $\boldsymbol{H} \| [010]$ and $\boldsymbol{H} \| [110]$. The corresponding axes are marked on the curves [73B2; LBII/12a, p. 626].

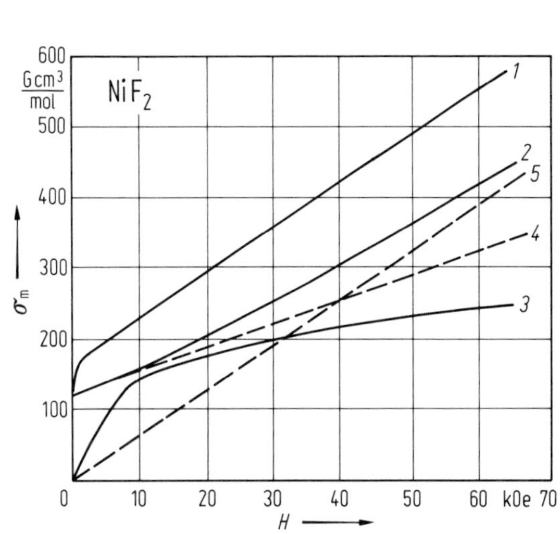

Fig. 22. NiF_2 single crystal. Plots of $\sigma_{\|}(H)$ for $\boldsymbol{H} \| [010]$ (curve *1*), for $\boldsymbol{H} \| [110]$ (curve *2*) and of $\sigma_{\perp}(H)$ for $\boldsymbol{H} \| [110]$ (curve *3*); the dashed lines are plots of $\sigma_{\|}(H)$ in the limiting states $\sigma_{\|}(H) = \sigma_{D\perp}/\sqrt{2} + \chi_{\perp}H/2$, when the antiferromagnetic vector $\boldsymbol{L}$ is fixed on the [100] axis (curve *4*), and $\sigma_{\|}(H) = \chi_{\perp}H$, when $\boldsymbol{L} \perp \boldsymbol{H}$ and $\boldsymbol{L} \| [1\bar{1}0]$ (curve *5*) [73B3; LBII/12a, p. 625].

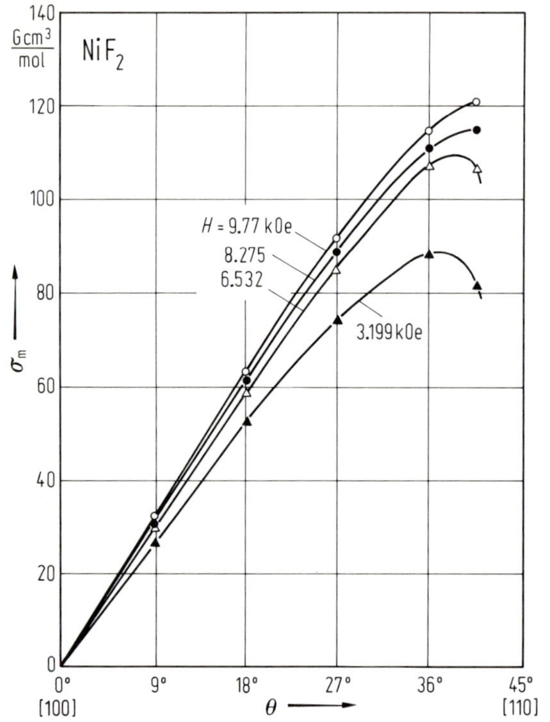

Fig. 23. NiF_2. Molar perpendicular magnetic moment vs. angle between field and the [100] direction. Rotation about the *c* axis. $T = 20.4$ K [54M].

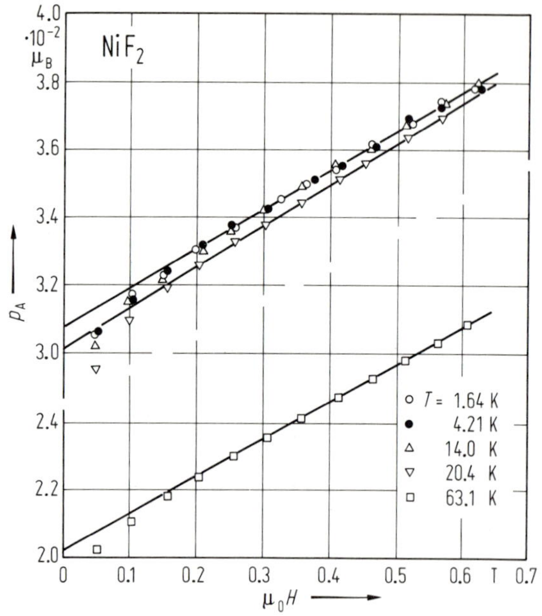

Fig. 24. NiF_2. Magnetic moment per Ni atom measured along the *a* axis vs. applied magnetic field [65C2].

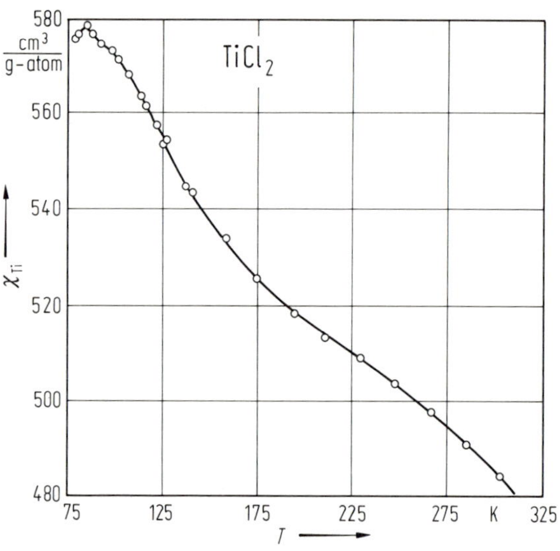

Fig. 25. $TiCl_2$. Temperature dependence of the atomic magnetic susceptibility χ_{Ti} [62L; LBII/2, p. 2–380].

9.1.2.2 MCl_2 compounds

Table 2. MCl_2 compounds.

Substance (Figs.)	Crystal type	Lattice constants*)	Molar Curie constant (in cm^3 K/mol)	Paramagnetic Curie temperature	Magnetic order	Critical temperature	Spin structure	Ordered moment
$TiCl_2$ (Fig. 25)	CdI_2	$a = 3.43(1)$ Å $c = 6.10(1)$ Å [62G]			A.F. ← P.	85K [62L; LBII/2, p. 2–1]		
VCl_2 (Fig. 27)	CdI_2 Fig. 26	$a = 3.60$ Å $c = 5.84$ Å [77N]	1.96 [77N]	− 437 K [77N]	A.F. ← P.	36.0 K [83H1]	3-sublattice A.F. [83H1]	
$CrCl_2$ (Fig. 28)	Distorted TiO_2	$a = 6.638$ Å $b = 5.984$ Å $c = 3.476$ Å [60C]	3.26 [40S]	− 149 K [40S]	A.F. ← P.	16.06 K [62S]	Ferromagn. (011) planes antiparallel to adjacent planes [60C]	$3.86\mu_B$ [60C]
$MnCl_2$ (Figs. 30, 31)	$CdCl_2$ Fig. 29	$a = 3.709$ Å $c = 17.59$ Å [63F]	4.07 [40S]	− 3.3 K [40S]	A.F.I ← P. A.F.II ← A.F.I	1.96 K 1.81K [62M]		
$FeCl_2$ (Figs. 32 ··· 36)	$CdCl_2$	$a = 3.593$ Å $c = 17.60$ Å [63F]	5.20 ($\parallel c$) 4.0 ($\perp c$) [74B]	21 K ($\parallel c$) 9 K ($\perp c$) [74B]	A.F. ← P.	23.8 K [72L]	Fig. 32 [59W]	$4.5\mu_B$ [59W]
$CoCl_2$ (Figs. 37, 38)	$CdCl_2$	$a = 3.553$ Å $c = 17.41$ Å [63F]	3.46 [40S]	38.1 K [40S]	A.F. ← P.	24.71 K [62C1]	The same as in $FeCl_2$ except the spin direction; spin $\perp c$ [59W]	$3.0\mu_B$ [59W]

continued

Table 2 (*continued*)

Substance (Figs.)	Crystal type	Lattice constants*)	Molar Curie constant (in cm^3 K/mol)	Paramagnetic Curie temperature	Magnetic order	Critical temperature	Spin structure	Ordered moment
$NiCl_2$ (Figs. 39 ··· 41)	$CdCl_2$	a = 3.481 Å c = 17.30 Å [63F]	1.30 [71D]	67 K [71D]	A.F. ← P.	52 K [52B]	The same as in $FeCl_2$ except the spin direction; spin ⊥ c [73K]	2.11μ_B [71D]
$CuCl_2$ (Figs. 42 ··· 44)	Monoclinic (C2/m)	a = 6.85 Å b = 3.30 Å c = 6.70 Å β = 121° [47W]	0.536 [40S]	− 109 K [40S]	A.F. ← P.	23.91 K [62S]		

*) Room temperature unless otherwise specified.

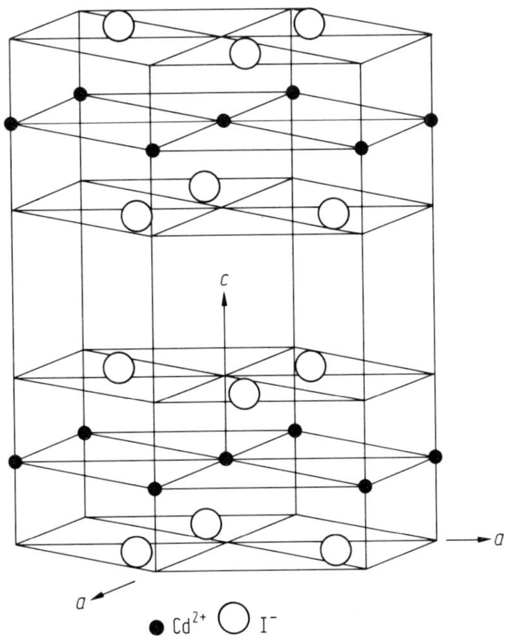

Fig. 26. CdI_2 type crystal structure (D^3_{3d} ($C\bar{3}m$)).

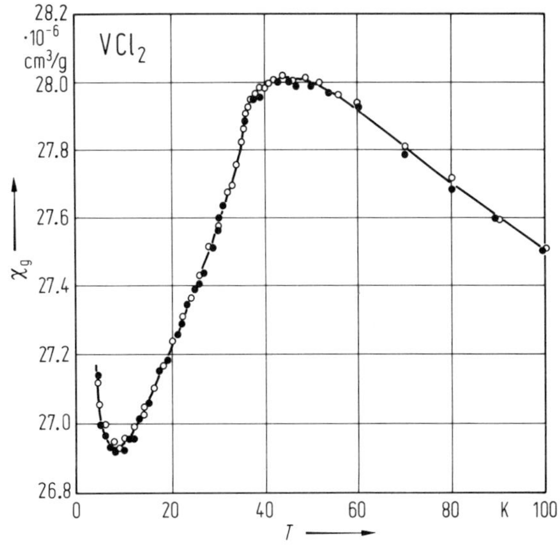

Fig. 27. VCl_2. Magnetic susceptibility per gram vs. temperature. $H = 0.2$ kOe. ●: $\parallel c$, ○: $\perp c$ [83H2].

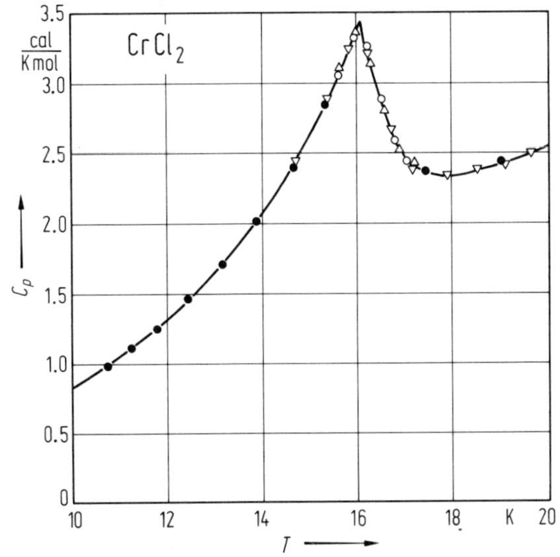

Fig. 28. $CrCl_2$. Molar specific heat vs. temperature. Different symbols indicate different series [62S].

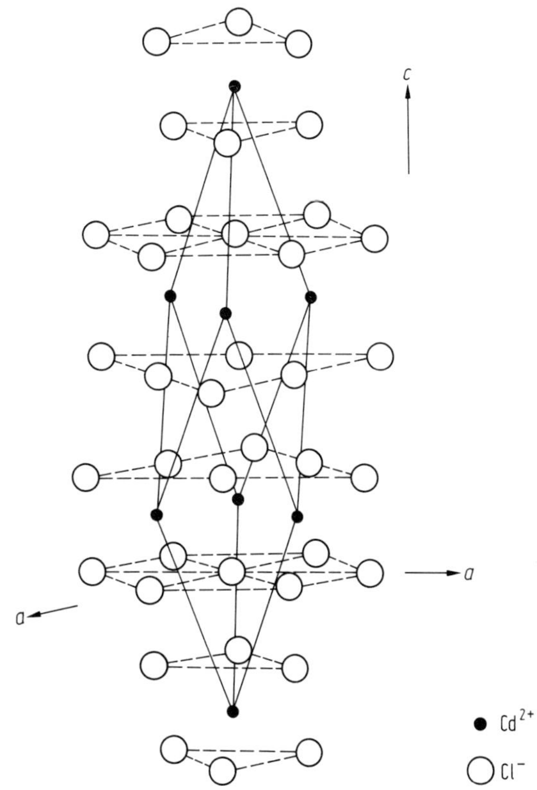

Fig. 29. $CdCl_2$ type crystal structure (D^5_{3d}($R\bar{3}m$)).

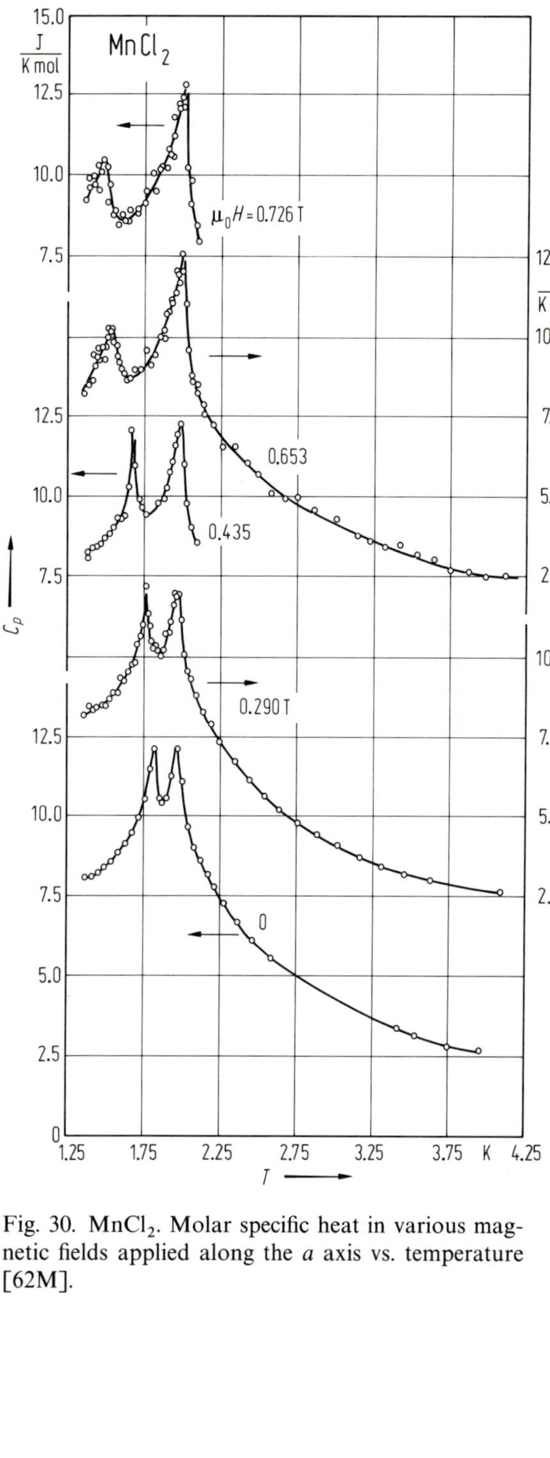

Fig. 30. $MnCl_2$. Molar specific heat in various magnetic fields applied along the *a* axis vs. temperature [62M].

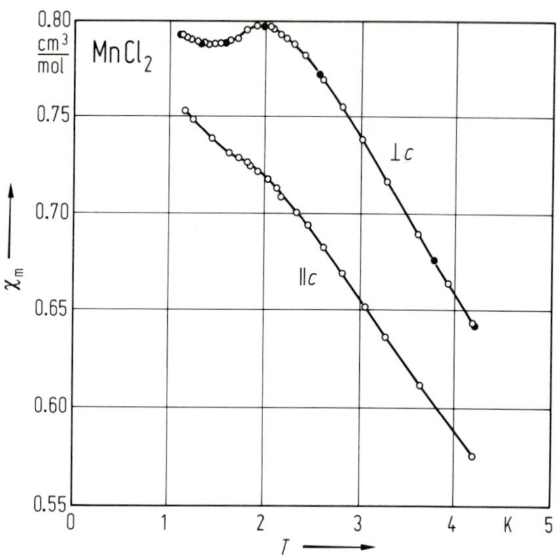

Fig. 31. $MnCl_2$. Molar magnetic susceptibility parallel and perpendicular to the *c* axis vs. temperature. Open and full circles represent different experiments [55M].

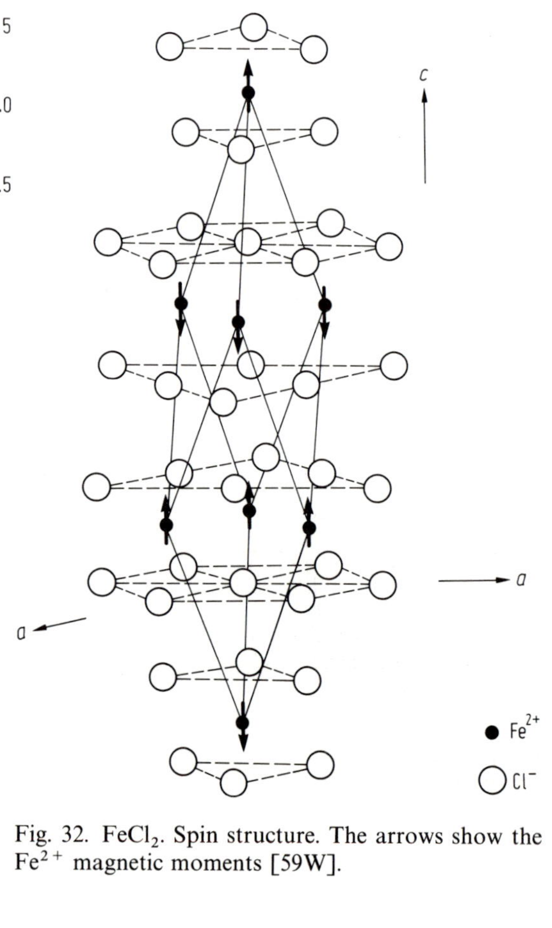

Fig. 32. $FeCl_2$. Spin structure. The arrows show the Fe^{2+} magnetic moments [59W].

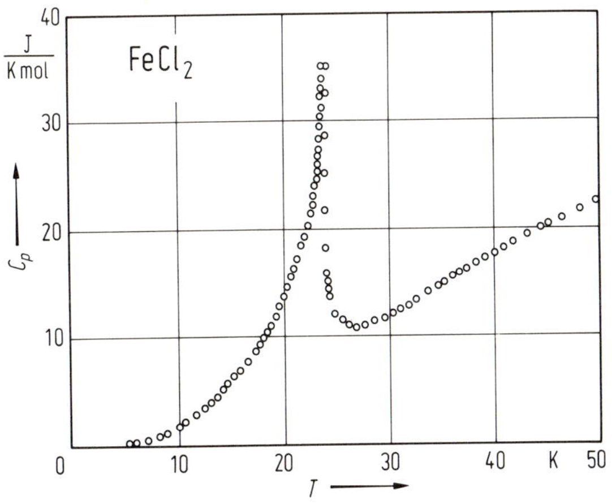

Fig. 33. $FeCl_2$. Molar specific heat vs. temperature [72L].

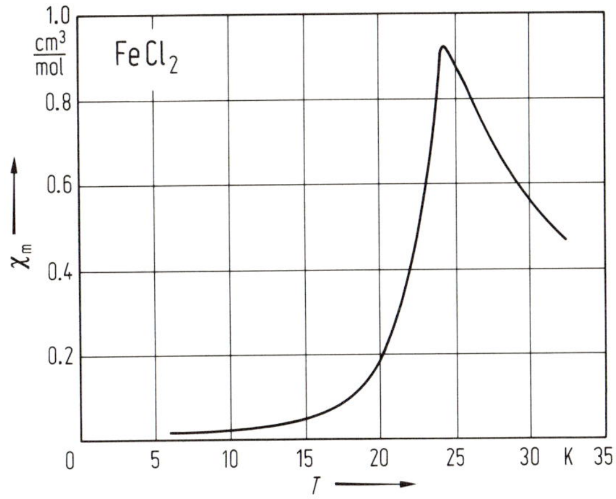

Fig. 34. $FeCl_2$. Molar magnetic susceptibility along the c axis vs. temperature [74B].

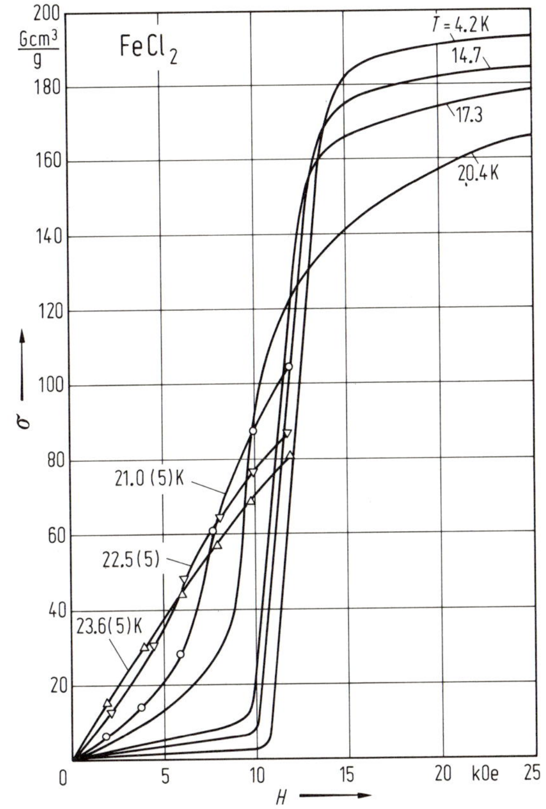

Fig. 35. $FeCl_2$. Magnetic moment per gram (specific magnetization) at various temperatures vs. magnetic field applied along the c axis [67J].

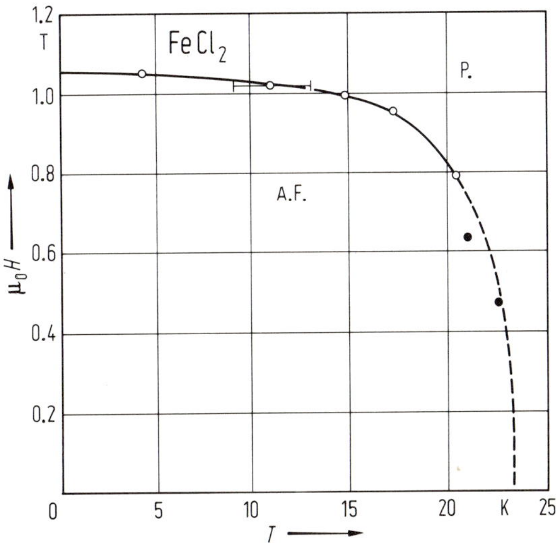

Fig. 36. $FeCl_2$. Magnetic field vs. temperature phase diagram. Solid line: first-order transition, dashed line: higher-order transition [67J].

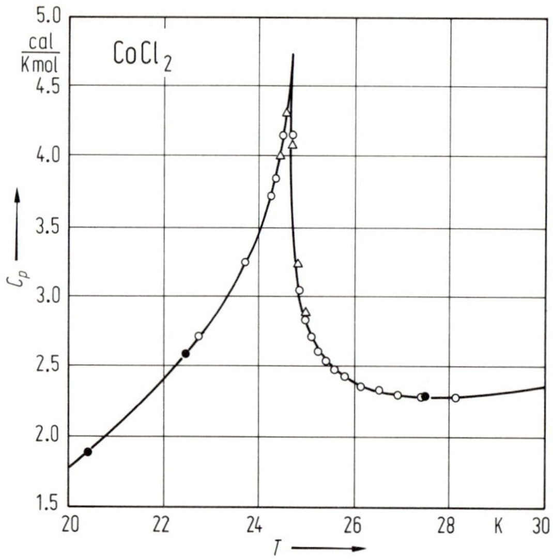

Fig. 37. $CoCl_2$. Molar specific heat vs. temperature. Different symbols indicate different series [62C1].

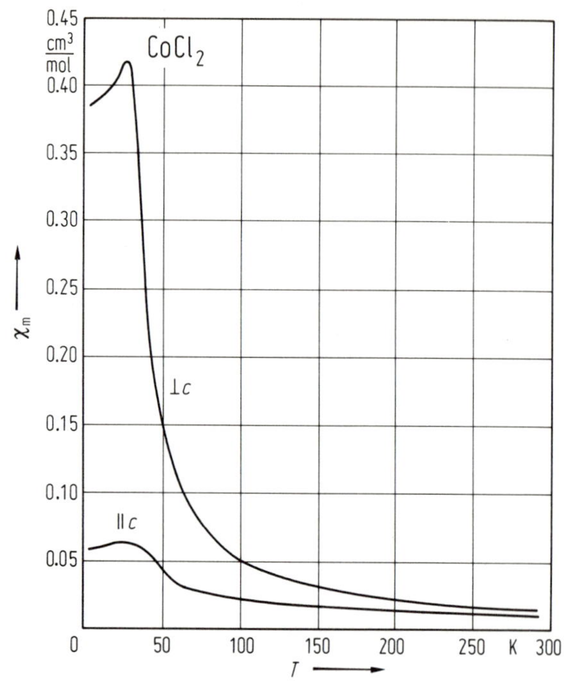

Fig. 38. $CoCl_2$. Molar magnetic susceptibility parallel and perpendicular to the c axis vs. temperature [56B].

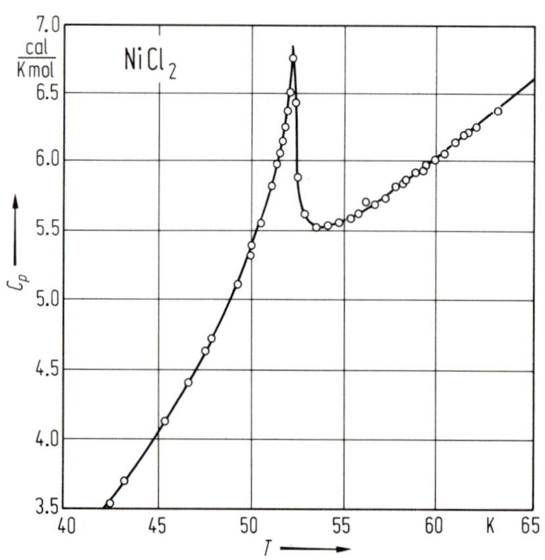

Fig. 39. $NiCl_2$. Molar specific heat vs. temperature [52B].

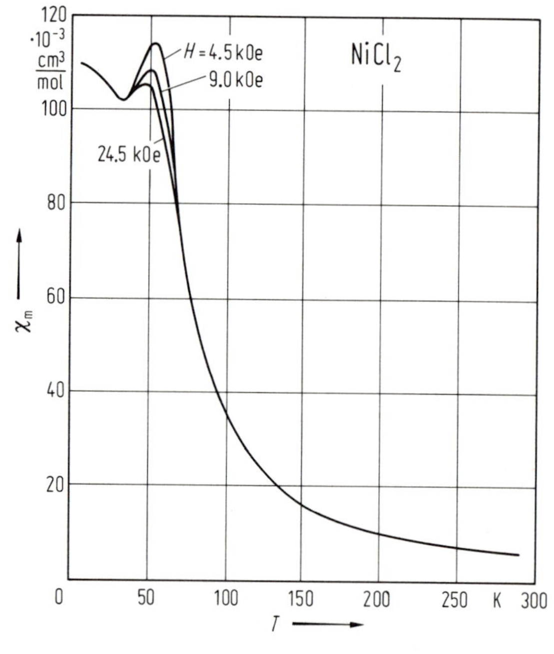

Fig. 40. $NiCl_2$. Molar magnetic susceptibility in various magnetic fields vs. temperature [56B].

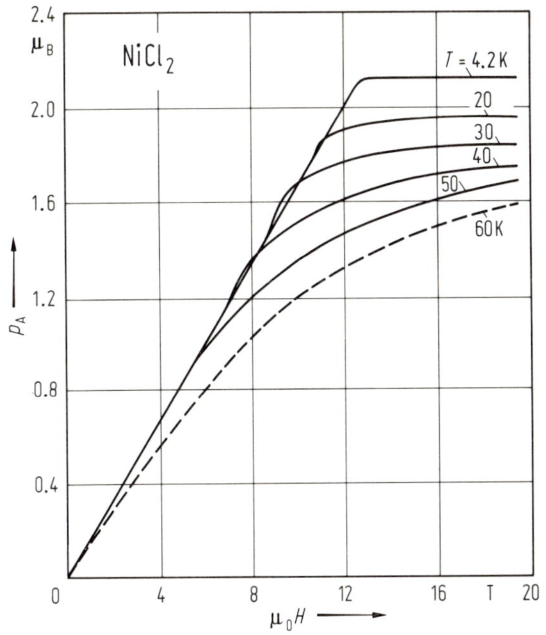

Fig. 41. $NiCl_2$. Magnetic moment per atom at various temperatures vs. magnetic field applied parallel to the *c* axis [71D].

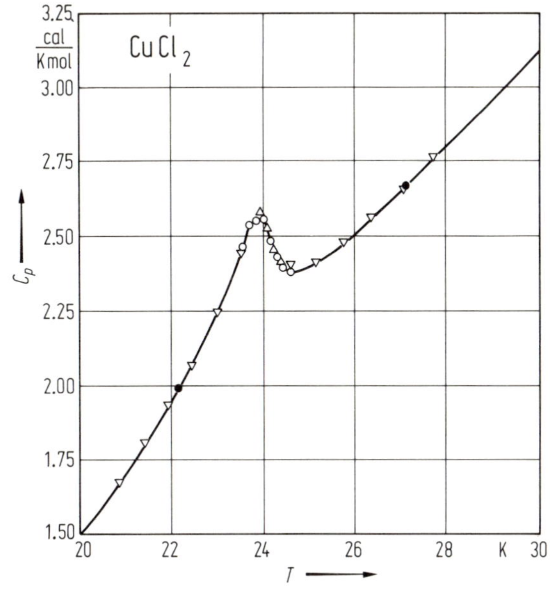

Fig. 42. $CuCl_2$. Molar specific heat vs. temperature. Different symbols indicate different series [62S].

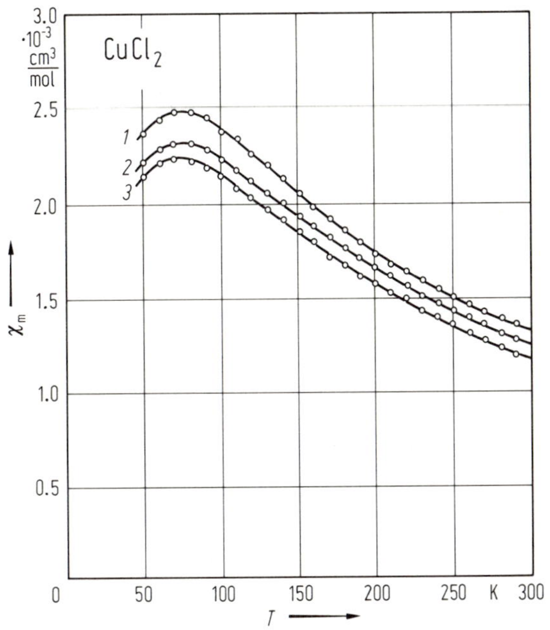

Fig. 43. $CuCl_2$. Molar magnetic susceptibility vs. temperature. *1*: ∥*c*, *2*: ∥*a*, *3*: ⊥(*b*, *c*) [78B1].

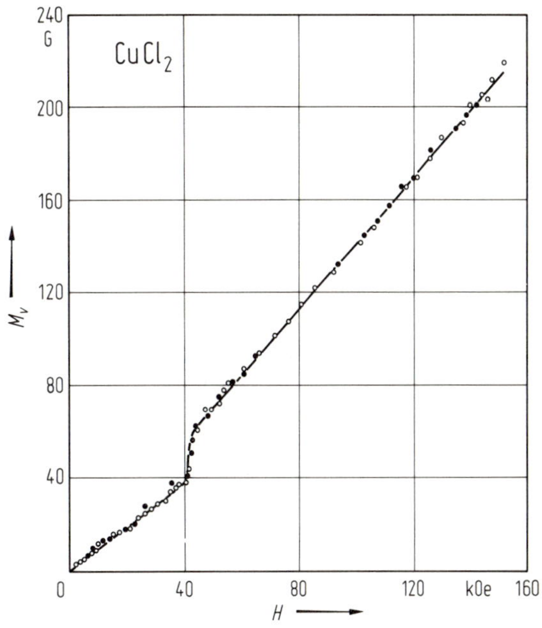

Fig. 44. $CuCl_2$. Magnetization vs. magnetic field applied along the *b* axis measured in an increasing (●) and decreasing (○) fields. $T = 4.2$ K [78B2].

9.1.2.3 MBr_2 compounds

Table 3. MBr_2 compounds.

Substance (Figs.)	Crystal type	Lattice constants*)	Molar Curie constant (in cm^3 K/mol)	Paramagnetic Curie temperature	Magnetic order	Critical temperature	Spin structure	Ordered moment
VBr_2 (Fig. 45)	CdI_2 Fig. 26	$a = 3.78$ Å $c = 6.23$ Å [77N]	2.07 [77N]	− 335 K [77N]	A.F. ← P.	29.5 K [83H1]	3-sublattice A.F. [83H1]	
$MnBr_2$ (Fig. 46)	CdI_2	$a = 3.868$ Å $c = 6.272$ Å [58W]			A.F. ← P.	2.16 K [58W]	Fig. 46	
$FeBr_2$ (Figs. 47 ··· 53)	CdI_2	$a = 3.772$ Å $c = 6.223$ Å [59W]	5.17 ($\parallel c$) 3.70 ($\perp c$) [74B]	3.5 K ($\parallel c$) − 3.0 K ($\perp c$) [74B]	A.F. ← P.	14.2 K [72L]	Fig. 47 [59W]	$3.9\mu_B$ [73F1]
$CoBr_2$ (Fig. 54)	CdI_2	$a = 3.728$ Å $c = 6.169$ Å [59W]			A.F. ← P.	19 K [80Y]	The same as in $FeBr_2$ except the spin direction spin $\perp c$ [59W]	$2.8\mu_B$ [59W]
$NiBr_2$ (Figs. 55 ··· 58)	$CdCl_2$ Fig. 29	$a = 3.71$ Å $c = 18.30$ Å [76D]			A.F. ← P. Heli. ← A.F.	52 K 22.8 K [80A]	High T phase The same as in $FeBr_2$ except the spin direction; spin $\perp c$ Low T phase Fig. 55; spin $\perp c$ [80A]	$2.0\mu_B$ [80A]

*) Room temperature unless otherwise specified.

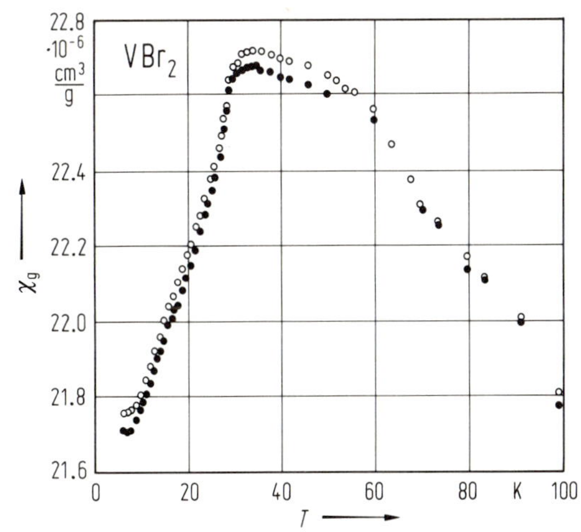

Fig. 45. VBr_2. Magnetic susceptibility per gram vs. temperature. $H = 0.2$ kOe. ●: $\parallel c$, ○: $\perp c$ [83H2].

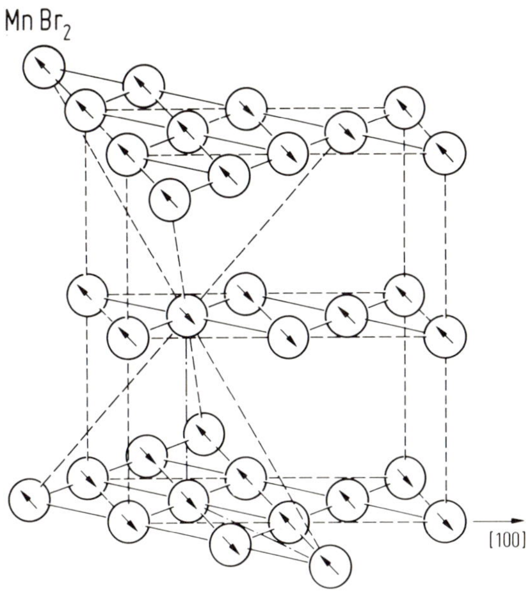

Fig. 46. $MnBr_2$. Spin structure. This figure shows half of the orthorhombic antiferromagnetic cell [58W].

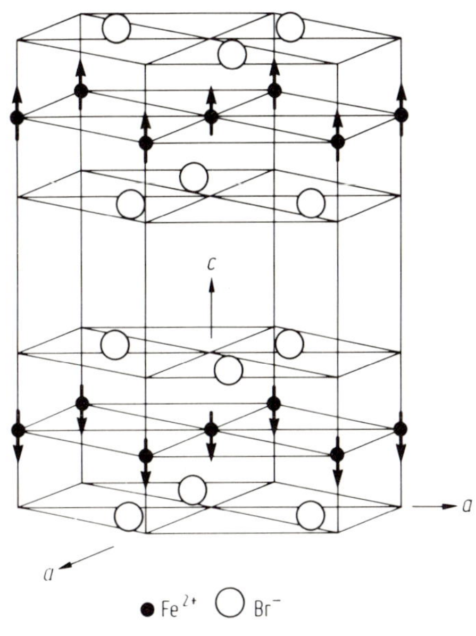

Fig. 47. $FeBr_2$. Spin structure.

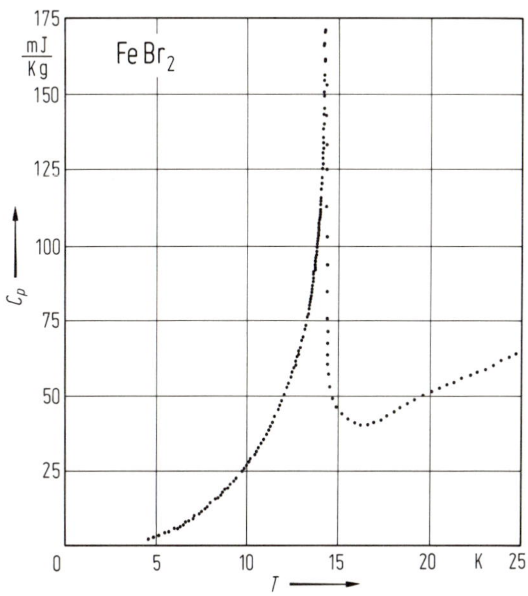

Fig. 48. $FeBr_2$. Specific heat per gram vs. temperature. One mole ($FeBr_2$) = 215.7 g [72L].

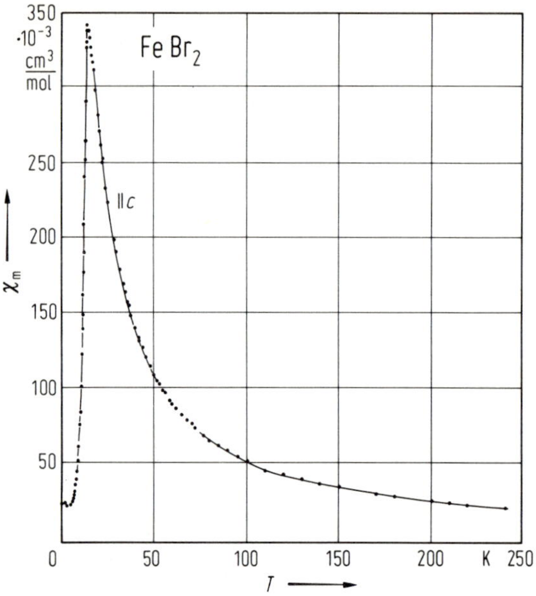

Fig. 49. $FeBr_2$. Molar magnetic susceptibility along the *c* axis vs. temperature [74B].

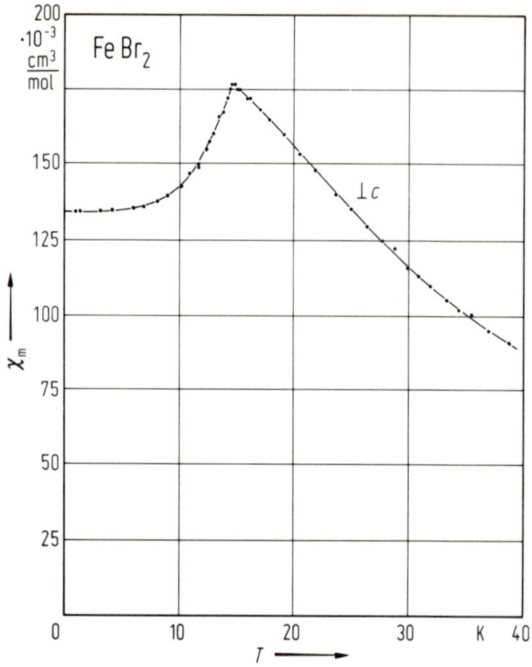

Fig. 50. $FeBr_2$. Molar magnetic susceptibility perpendicular to the *c* axis vs. temperature [74B].

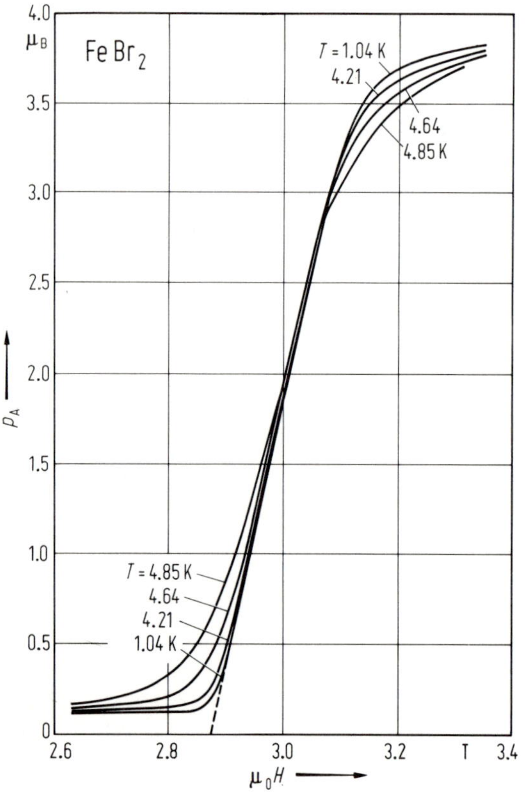

Fig. 51. $FeBr_2$. Magnetic moment per atom at low temperatures vs. magnetic field applied along the *c* axis showing first-order transition [73F1].

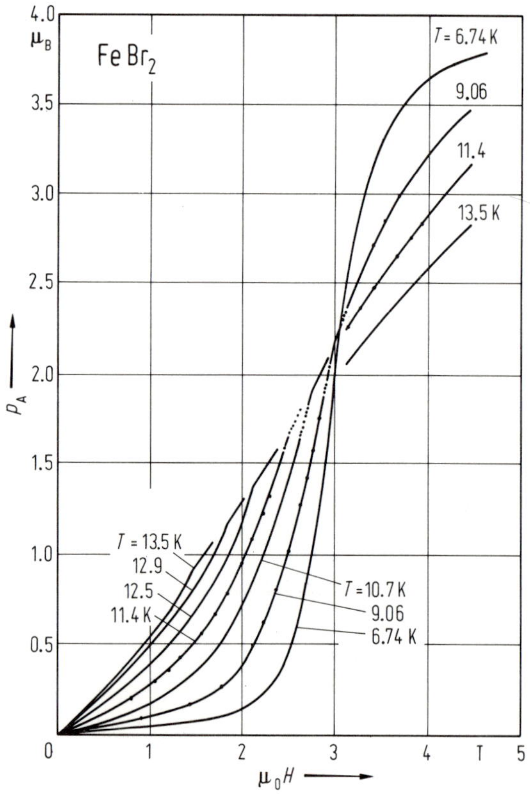

Fig. 52. $FeBr_2$. Magnetic moment per atom at high temperatures vs. magnetic field applied along the *c* axis showing second-order transition [73F1].

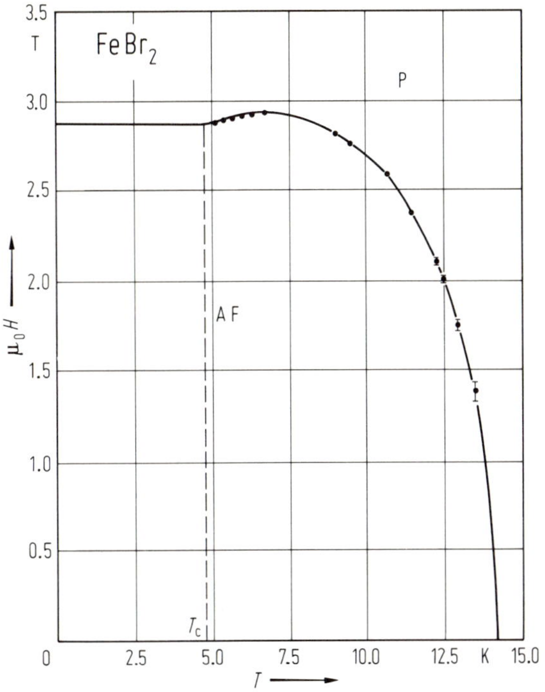

Fig. 53. $FeBr_2$. Magnetic field vs. temperature phase diagram. Below T_c the transition is of first-kind. Above T_c second-order transition takes place [73F1].

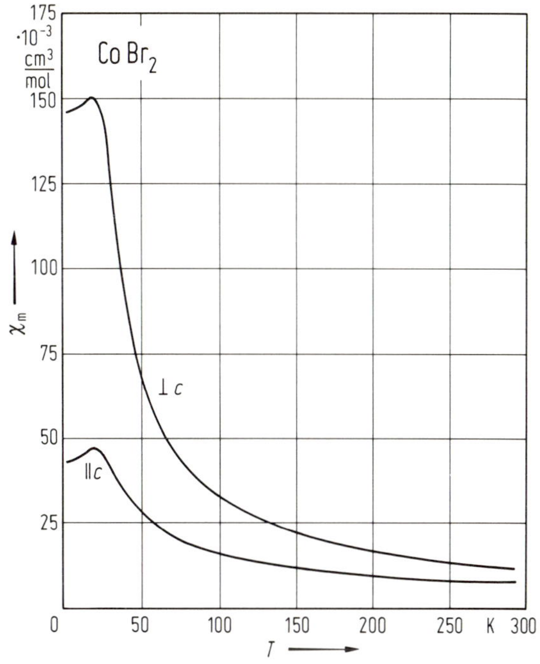

Fig. 54. $CoBr_2$. Molar magnetic susceptibility parallel and perpendicular to the c axis vs. temperature [58B2].

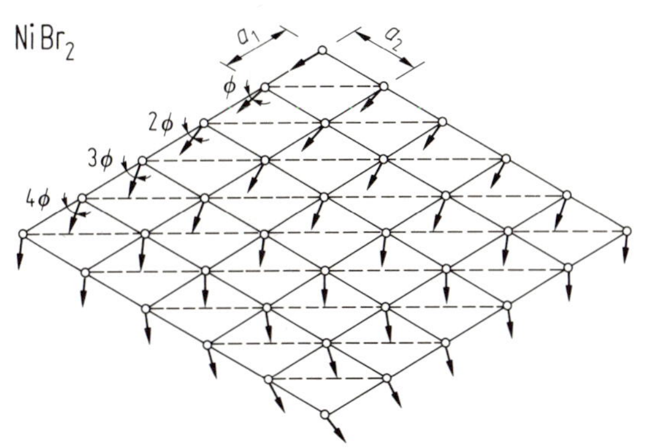

Fig. 55. $NiBr_2$. Spin structure within the c plane of the helimagnetic phase. The propagation vector $\boldsymbol{k} = [0.027, 0.027, 3/2]$ at $T = 4.2$ K [80A].

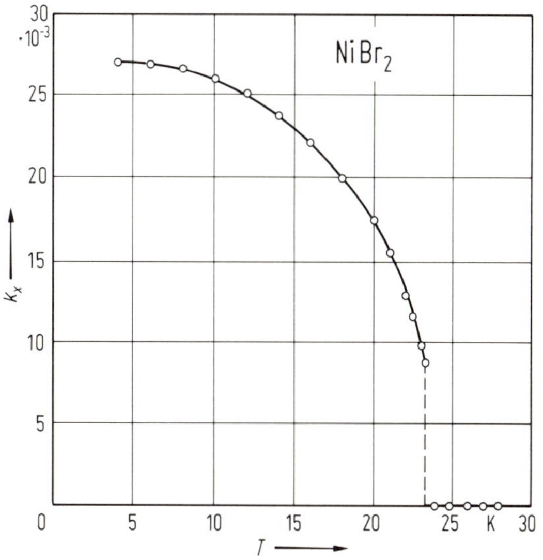

Fig. 56. $NiBr_2$. The x component of the propagation vector ($\boldsymbol{k}$) vs. temperature. The data are taken with increasing temperature. k_x becomes zero at 22.8 K [80A].

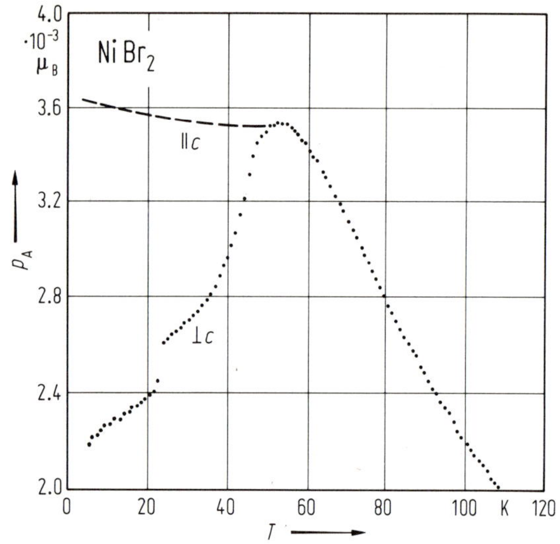

Fig. 57. $NiBr_2$. Magnetic moment per atom in a magnetic field of 0.06 T applied parallel and perpendicular to the c axis vs. temperature [80A].

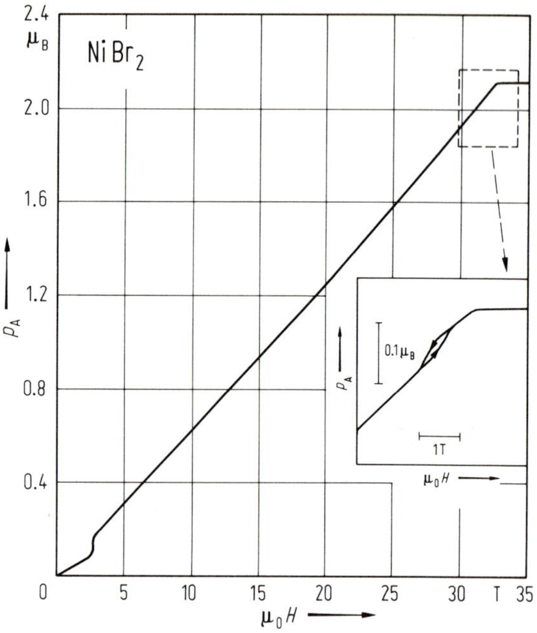

Fig. 58. $NiBr_2$. Magnetic moment per atom vs. magnetic field applied perpendicular to the c axis showing the successive transition: Heli → A.F. → Fan → P. $T = 1.3$ K [83K].

9.1.2.4 MI_2 compounds

Table 4. MI_2 compounds.

Substance (Figs.)	Crystal type	Lattice constants*)	Molar Curie constant (in cm^3 K/mol)	Paramagnetic Curie temperature	Magnetic order	Critical temperature	Spin structure	Ordered moment
VI_2 (Fig. 59)	CdI_2 Fig. 26	$a = 4.06$ Å $c = 6.75$ Å [77N]	2.07 [77N]	− 143 K [77N]	A.F.I ← P. A.F.II ← A.F.I	16.3 K 14.4 K [83H1]	High T phase 3-sublattice A.F. [83H1]	
MnI_2	CdI_2	$a = 4.146$ Å $c = 6.829$ Å [62C2]			Heli. ← P.	3.40 K [62C2]	Ferromagn. (307) planes rotate by $2\pi/16$ in successive planes [62C2]	$4.6\mu_B$ [62C2]
FeI_2 (Figs. 60 ··· 62)	CdI_2	$a = 4.03$ Å $c = 6.75$ Å [74G]	4.00 ($\parallel c$) 3.26 ($\perp c$) [74B]	− 21.5 K ($\parallel c$) − 24 K ($\perp c$) [74B]	A.F. ← P.	9.3 K [74B]	Fig. 60	$3.7\mu_B$ [74G]
CoI_2 (Fig. 63)	α-form CdI_2	$a = 3.96$ Å $c = 6.65$ Å [59B; LBII/2, p. 2–148]			A.F. ← P.	3 K [59B; LBII/2, p. 2–148]		

*) Room temperature unless otherwise specified.

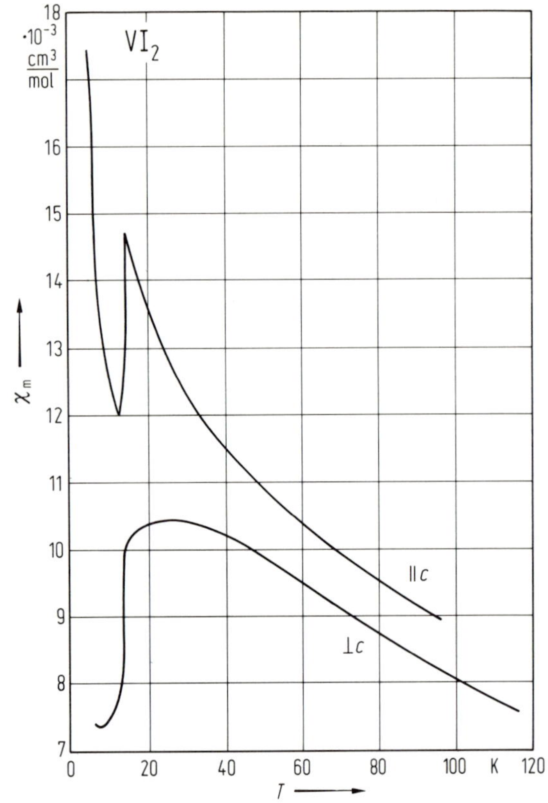

Fig. 59. VI_2. Molar magnetic susceptibility parallel and perpendicular to the *c* axis vs. temperature [79K].

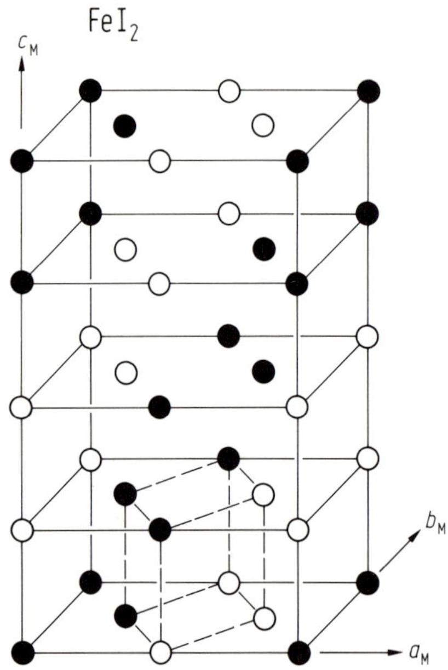

Fig. 60. FeI_2. Spin structure. The circles show Fe^{2+} moments. Filled circles: parallel to the *c* axis, open circles: antiparallel to the *c* axis. The dashed lines represent the chemical unit cell [74G]. Index M: magnetic.

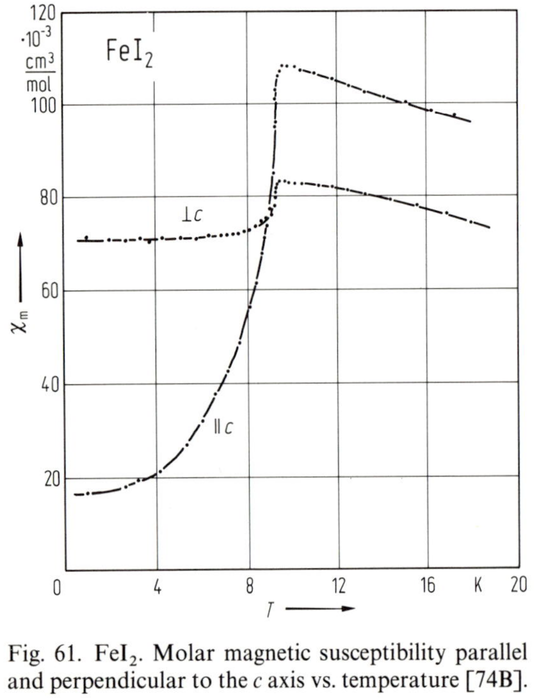

Fig. 61. FeI_2. Molar magnetic susceptibility parallel and perpendicular to the *c* axis vs. temperature [74B].

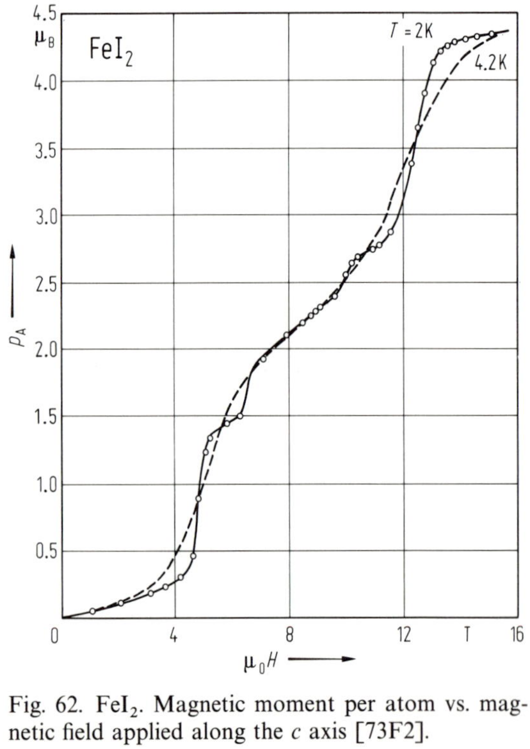

Fig. 62. FeI_2. Magnetic moment per atom vs. magnetic field applied along the *c* axis [73F2].

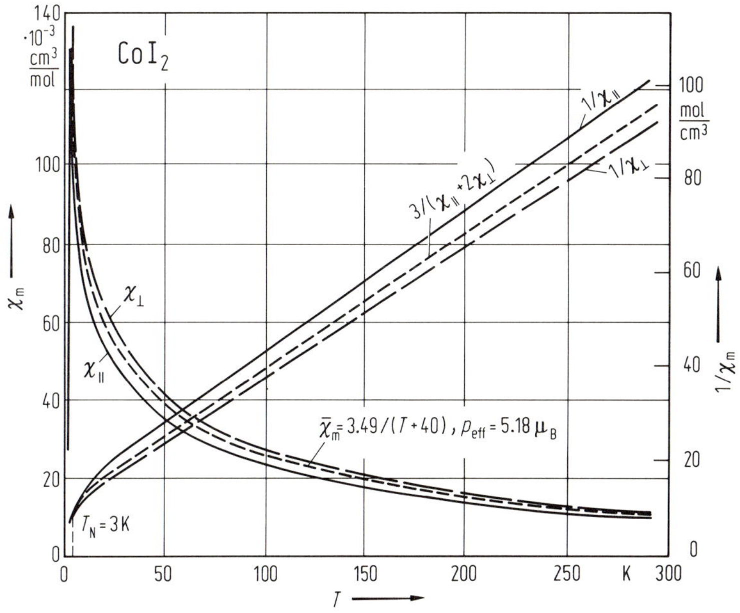

Fig. 63. CoI_2. Molar magnetic susceptibility as a function of temperature [58B2; LBII/2, p. 2–396].

9.1.3 Macroscopic magnetic and related properties of $MX_2 \cdot nH_2O$ compounds

Table 5. $MX_2 \cdot nH_2O$ compounds.

Substance (Figs.)	Crystal type	Lattice constants*)	Molar Curie constant (in cm^3 K/mol)	Paramagnetic Curie temperature	Magnetic order	Critical temperature	Spin structure	Ordered moment
$VF_2 \cdot 4H_2O$ $VCl_2 \cdot 2H_2O$ $VCl_2 \cdot 4H_2O$ $VBr_2 \cdot 4H_2O$ $VI_2 \cdot 6H_2O$ $CrCl_2 \cdot 4H_2O$ $CrBr_2 \cdot 6H_2O$ $CrI_2 \cdot 5H_2O$ (Fig. 64) $CrI_2 \cdot 6H_2O$	Table 5a							
$MnCl_2 \cdot 2H_2O$ (Fig. 65)	monoclinic (C2/m)	$a = 7.409$ Å $b = 8.800$ Å $c = 3.691$ Å $\beta = 98.67(2)°$ [65M]		− 7.5 K [64N3]	A.F. ← P.	1.65 K [64N3]		
$MnCl_2 \cdot 4H_2O$ (Figs. 66 ··· 69)	Monoclinic ($P2_1/n$)	$a = 11.186$ Å $b = 9.513$ Å $c = 6.186$ Å $\beta = 99.74°$ [64 Z]	4.41 [59G]	− 1.9 K [59G]	A.F. ← P.	1.62 K [69R]		
$MnBr_2 \cdot 4H_2O$ (Fig. 70)	Monoclinic	$\beta = 99.6°$ [67S1; LBII/8, p. 187]				2.1 K [59G] 2.13 K [67S1; LBII/8, p. 187]		

$FeCl_2 \cdot 2H_2O$ (Figs. 71 ··· 74)	$MnCl_2 \cdot 2H_2O$ Fig. 65	$a = 7.355$ Å $b = 8.548$ Å $c = 3.637$ Å $\beta = 98.18°$ [65M]	4.3 ($\parallel \alpha$)**) 2.7 ($\perp \alpha$) [65N]	12 K ($\parallel \alpha$) 5 K ($\perp \alpha$) [65N]	A.F. ← P.	23 K [65N]	Ferromagn. chain along c parallel within (100) planes, antiparallel along a; spin $\parallel \alpha$ [73S]	4.25 μ_B [65N]
$FeCl_2 \cdot 4H_2O$ (Figs. 75, 76)	Monoclinic ($P2_1/c$)	$a = 5.91$ Å $b = 7.17$ Å $c = 8.44$ Å $\beta = 112°10'$ [59P]	3.61 [62F]	− 2 K [62F]	A.F. ← P.	1.097 K [72R]		
$CoF_2 \cdot 4H_2O$	see Table 5a							
$CoCl_2 \cdot 2H_2O$ (Figs. 77 ··· 80)	$MnCl_2 \cdot 2H_2O$	$a = 7.279$ Å $b = 8.553$ Å $c = 3.569$ Å $\beta = 97.58°$ [66M]	5.0 ($\parallel b$) 0.79 ($\perp b$) [64N2]	1 K [64N2]	A.F. ← P.	17.20 K [64S1]	Ferromagn. chain along c parallel within (100) planes, antiparallel along a; spin $\parallel b$ [66C1]	3.3 μ_B [66C1]
$CoCl_2 \cdot 2D_2O$	Monoclinic (C2/m) [65C1]	$a = 7.282$ Å $b = 8.555$ Å $c = 3.570$ Å $\beta = 97°3.3'$ [65C1]			A.F. ← P.		At 4.2 K AF moment of face centered Co ion antiparallel to that of corner ions. Both spins parallel b axis [65C1]	2.84 μ_B [65C1]
$CoCl_2 \cdot 6H_2O$ (Figs. 81 ··· 86)	$CoCl_2 \cdot 6H_2O$ Fig. 81	$a = 10.34$ Å $b = 7.06$ Å $c = 6.67$ Å $\beta = 122°20'$ [60M]		− 3.3 K [60H]	A.F. ← P.	2.29 K [60R]	Fig. 82 [70K]	3.8 μ_B [70K]

continued

Table 5 (*continued*)

Substance (Figs.)	Crystal type	Lattice constants*)	Molar Curie constant (in cm^3 K/mol)	Paramagnetic Curie temperature	Magnetic order	Critical temperature	Spin structure	Ordered moment
$CoCl_2 \cdot 6D_2O$ (Fig. 85, Table 5a)					A.F. ← P.	≈ 3.0 K [60H]		
$CoBr_2 \cdot 2H_2O$ (Fig. 87)	$MnCl_2 \cdot 2H_2O$	a = 7.6296 Å b = 8.7700 Å c = 3.7648 Å β = 97.539° [67M1]	4.3 (∥ b) 1.4 (⊥ b) [64N1]	≈ 0 K [64N1]	A.F. ← P.	9.5 K [64N1]		
$CoBr_2 \cdot 6H_2O$ (Figs. 88 ⋯ 90)	$CoCl_2 \cdot 6H_2O$	a = 11.029 Å b = 7.178 Å c = 6.908 Å β = 124.71° [73M2]			A.F. ← P.	3.07 K [59F]	spin ∥ c [60G]	
$NiF_2 \cdot 4H_2O$	see Table 5a							
$NiCl_2 \cdot 2H_2O$ (Figs. 91, 92)	Monoclinic***) (I2/m)	a = 6.9093 Å b = 6.8858 Å c = 8.8298 Å β = 92.246° [67M2]			AF.I ← P. A.F.II ← A.F.I	7.258 K 6.309 K [72P]		
$NiCl_2 \cdot 4H_2O$ (Figs. 93 ⋯ 95)	Monoclinic $MnCl_2 \cdot H_2O$	a = 10.90 Å b = 9.35 Å c = 6.00 Å β = 100.5° [73M1; LBII/12a, p. 321]		Figs. 93, 94	A.F. ← P. A.F. spin flop transition Fig. 95 [74P; LBII/12a, p. 627]	2.99(1) K [73M1; LBII/12a, p. 321] [74P; LBII/12a, p. 627]		

$NiCl_2 \cdot 6H_2O$ (Figs. 96 ··· 103)	$CoCl_2 \cdot 6H_2O$	$a = 10.23$ Å $b = 7.05$ Å $c = 6.57$ Å $\beta = 122°10'$ [61M1]	1.18 [60R]	− 10 K [59H] see also Table 5a	A.F. ← P. A.F. spin flop in magnetic field	5.34 K [60R]	The same as in $CoCl_2 \cdot 6H_2O$ except the spin direction; spin direction 22.5° from the a^* axis towards the a axis [67K]
$NiBr_2 \cdot 6H_2O$ (Fig. 104)					A.F. ← P.	6.50 K [59S]	
$CuF_2 \cdot 2H_2O$ (Figs. 105, 106)	Monoclinic (C^3_{2h}-I2/m)	$a = 6.416$ Å $b = 7.397$ Å $c = 3.301$ Å $\beta = 99.6°$ [58G]	0.46 [58B1]	− 37 K [58B1]	A.F. ← P.	10.90 K [66C2]	The moments at the corner sites are directed oppositely to those at the center [62A]. Spin 3.5° away from c [62P]
$CuCl_2 \cdot 2H_2O$ (Figs.107 ··· 110)	Ortho-rhombic (D^7_{2h}-Pbmn)	$a = 7.38$ Å $b = 8.04$ Å $c = 3.72$ Å [36H]	0.446 ($\parallel a$) 0.385 ($\parallel b$) [52V]	− 5 K [52V]	A.F. ← P.	4.357 K [69B]	Ferromagn. (001) plane antiparallel along c; spin $\parallel a$****) [65S]

*) Room temperature unless otherwise specified.
**) The α axis lies in the ac plane making an angle 32° with the c axis.
***) A crystallographic phase transition takes place at 230 ± 20 K [72B2].
****) A canting of the spin moments is observed in $CuCl_2 \cdot 2D_2O$. The weak ferromagnetic moment is parallel to c and amounts to $\approx 0.1\,\mu_B$ [67U].

Table 5a. Paramagnetic properties of $MX_2 \cdot nH_2O$ compounds. Data earlier compiled in the Landolt-Börnstein subvolumes of group II as given under "Ref." p_{eff}: effective magnetic moment per molecule.

Compound	T [K]	χ_g [$\cdot 10^{-6}$ cm^3/g]	χ_m [$\cdot 10^{-6}$ cm^3/mol]	p_{eff} [μ_B]	Θ [K]	Ref.
$VF_2 \cdot 4H_2O$	290	38.78		3.85	0	[68S1; LBII/8, p. 68]
	196	57.44		3.84		
	82	132.9		3.78		
$VCl_2 \cdot 2H_2O$	290	27.82		3.24		[68S2; LBII/8, p. 68]
	196	33.55		2.92		
	82	41.22		2.09		
$VCl_2 \cdot 4H_2O$	290	32.47		3.86	0	[68S2; LBII/8, p. 68]
	196	47.59		3.84		
	82	112.5		3.83		
$VBr_2 \cdot 4H_2O$	290	22.34		3.88	0	[68S2; LBII/8, p. 69]
	196	32.61		3.85		
	82	76.45		3.80		
$VI_2 \cdot 6H_2O$	290	15.35		3.90	0	[68S2; LBII/8, p. 69]
	196	22.77		3.89		
	82	53.97		3.85		
$CrCl_2 \cdot 4H_2O$	323.8		9453	4.97	0	[65F; LBII/8, p. 127]
	206.1		14710	4.94		[65E; LBII/8, p. 127]
	94.0		32680	4.98		
$CrBr_2 \cdot 6H_2O$	324.8		9376	4.96	0	[65F; LBII/8, p. 127]
	206.5		14750	4.96		[65E; LBII/8, p. 127]
	96.0		31500	4.94		
$CrI_2 \cdot 5H_2O$	320 ··· 100		Fig. 64		0	[65E; LBII/8, p. 127]
$CrI_2 \cdot 6H_2O$	293	25.0		4.98		[65F; LBII/8, p. 127]
$MnBr_2 \cdot 4H_2O$	294 ··· 15			5.93	− 1.3[a])	[59G; LBII/2, p. 2–67]
				5.93	− 2.5[b])	[59T; LBII/2, p. 2–67]
$CoF_2 \cdot 4H_2O$	323		9670	5.2	− 28	[63R; LBII/2, p. 2–146]
	293		10470			
	197		15300			
	77		32100			
$CoCl_2 \cdot 6D_2O$	4.2 ··· 1.42		Fig. 85			[60H; LBII/2, p. 2–148]
$NiF_2 \cdot 4H_2O$	293		4230	3.26	− 19	[62R; LBII/2, p. 2–220]
	195		6220			
	77		13650			
$NiCl_2 \cdot 6H_2O$	20.4 ··· 13.8		$\parallel a$ 44158		− 7.50	[73H; LBII/12a, p. 322]
			$\perp c$ 43494		− 8.11	
			$\chi_b = 42649$		− 8.37	

[a]) For $H \parallel c$, χ is nearly isotropic.
[b]) For $H \parallel b$.

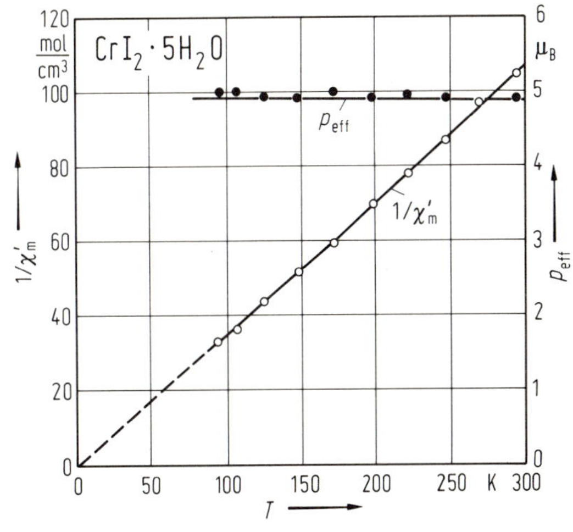

Fig. 64. $CrI_2 \cdot 5H_2O$. Reciprocal molar magnetic susceptibility and effective magnetic moment per molecule as a function of temperature [65E; LBII/8, p. 867].

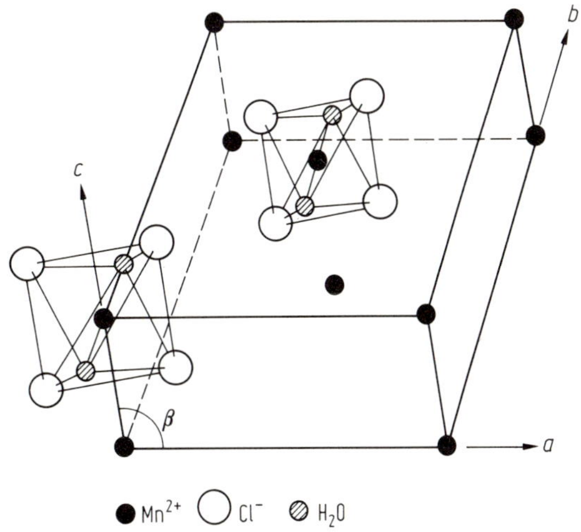

Fig. 65. $MnCl_2 \cdot 2H_2O$. Crystal structure (C2/m) [65M].

For Fig. 66 see next page.

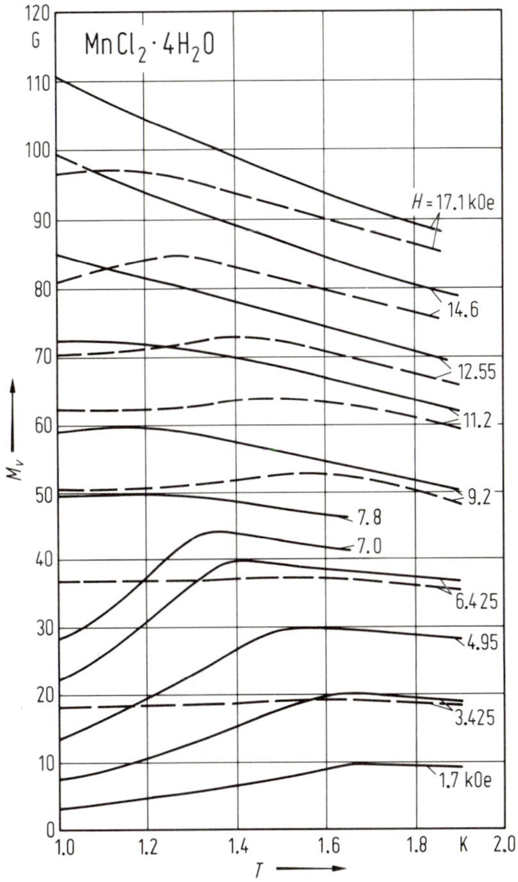

Fig. 67. $MnCl_2 \cdot 4H_2O$. Magnetization along the *b* (- - -) and *c* (—) axes in various magnetic fields vs. temperature [59G].

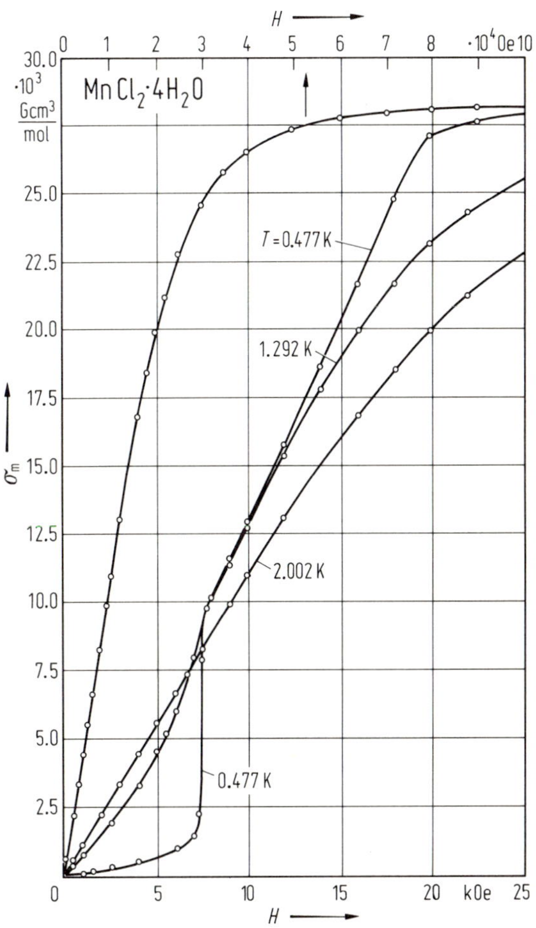

Fig. 68. $MnCl_2 \cdot 4H_2O$. Molar magnetic moment at various temperatures vs. magnetic field applied along the *c* axis. The upper curve is the entire set of measurements at $T = 2.002$ K [69R].

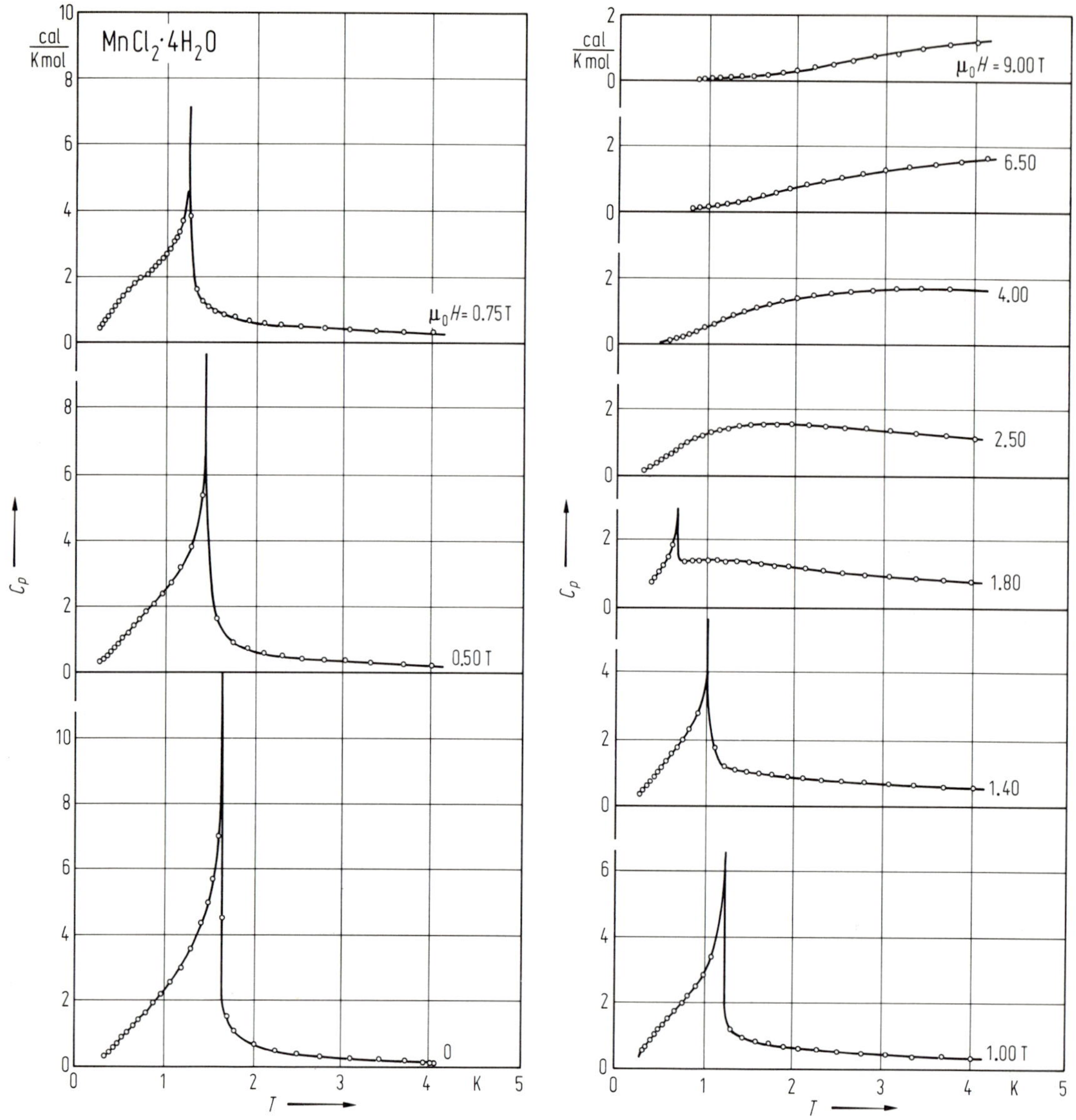

Fig. 66. $MnCl_2 \cdot 4H_2O$. Molar specific heat in various magnetic fields applied along the *c* axis vs. temperature [69R].

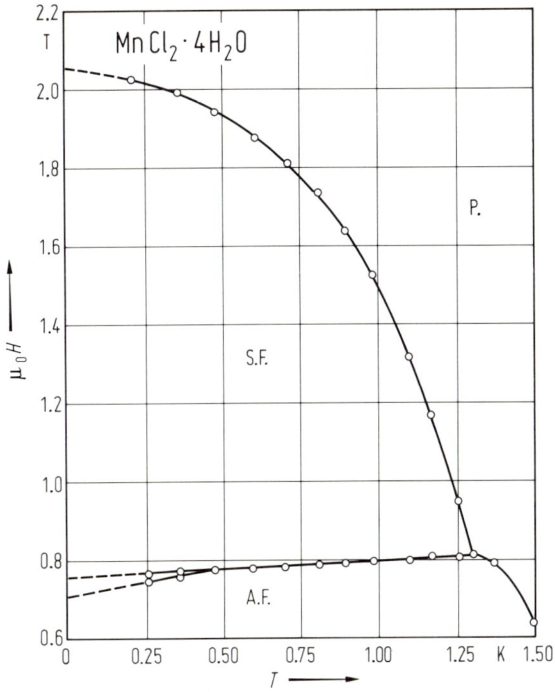

Fig. 69. $MnCl_2 \cdot 4H_2O$. Magnetic field vs. temperature phase diagram. The magnetic field is applied along the c' axis [67R]. c' axis is perpendicular to *ab* plane.

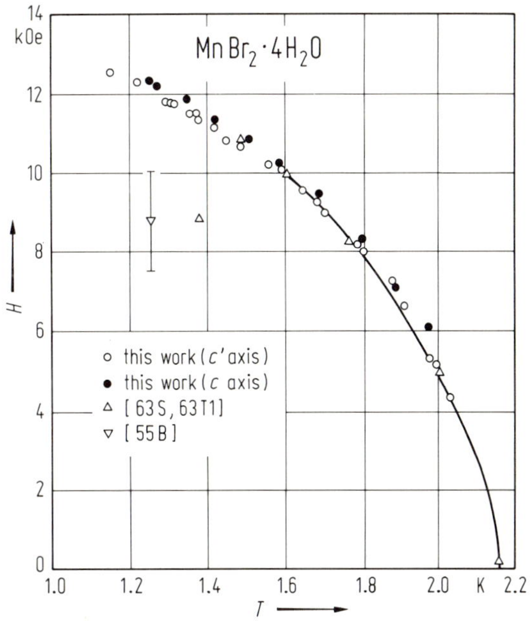

Fig. 70. $MnBr_2 \cdot 4H_2O$. Phase diagram, transition inducing field H vs. temperature T, with $\boldsymbol{H} \parallel c'$ axis, where c' is orthogonal to a and b axes and 9.6° away from the crystallographic axis in the *ac* plane [67S1; LBII/8, p. 880].

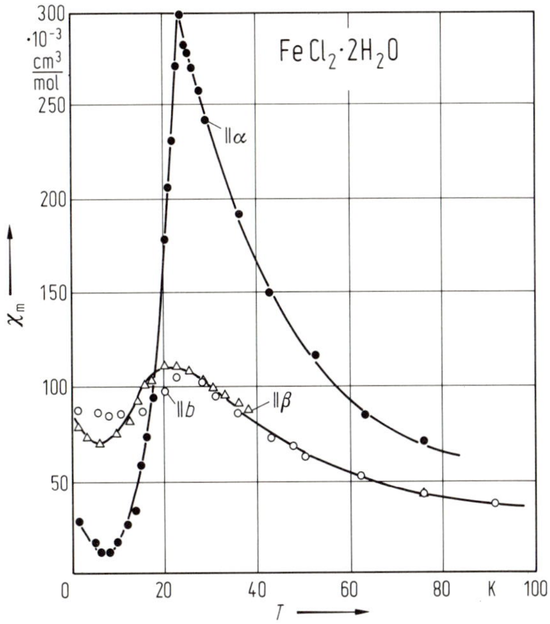

Fig. 71. $FeCl_2 \cdot 2H_2O$. Molar magnetic susceptibility along the α, β and b axes vs. temperature. The α axis lies in the *ac* plane and makes an angle of 32° with the c axis. The β axis lies in the *ac* plane being orthogonal to the α axis [65N].

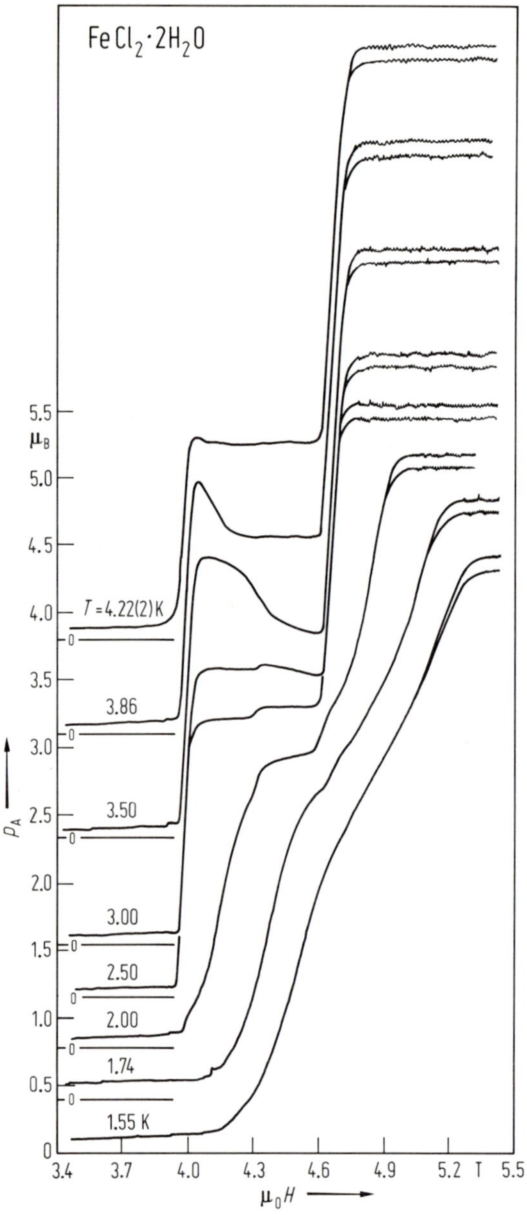

Fig. 72. $FeCl_2 \cdot 2H_2O$. Magnetic moment per atom at various temperatures vs. magnetic field applied along the α axis showing a large hysteresis and appearance of a metastable state. The data are taken in an increasing field [75K].

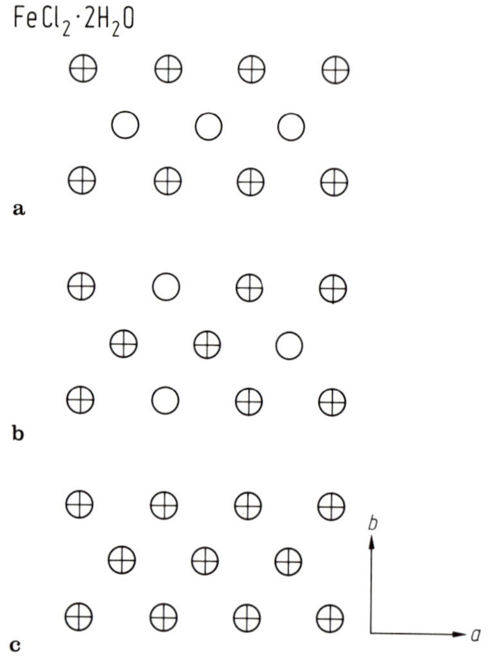

Fig. 73. $FeCl_2 \cdot 2H_2O$. The spin structures appearing in magnetic field. This figure shows the projection of the spin arrangements on the *ab* plane. (a) A.F. phase, (b) ferrimagnetic phase with the magnetization one-third of that in the saturated phase, and (c) saturated phase [75K].

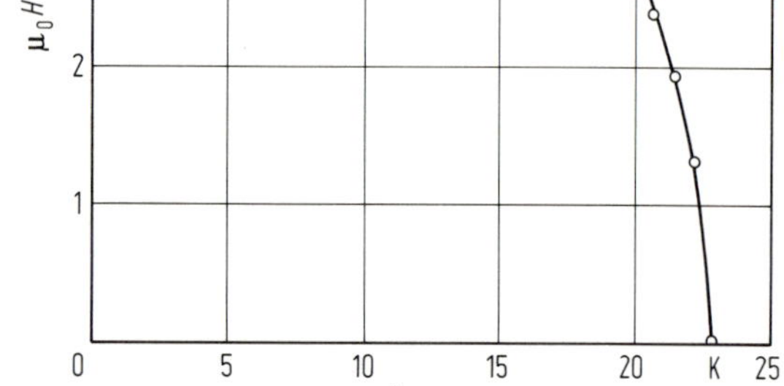

Fig. 74. $FeCl_2 \cdot 2H_2O$. Magnetic field vs. temperature phase diagram [73S]. For the spin structures in the three phases, see Fig. 73.

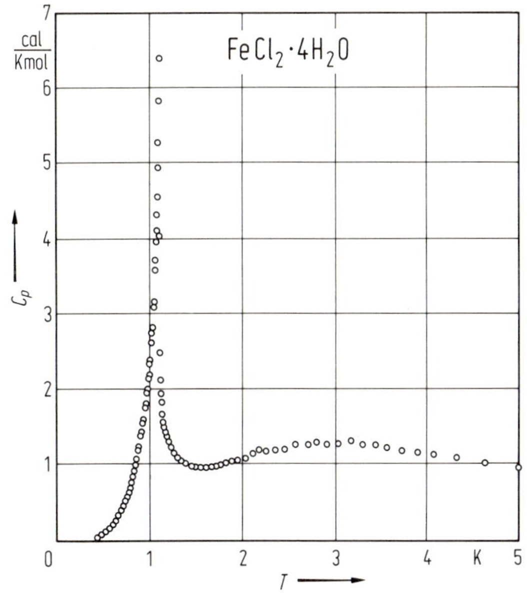

Fig. 75. $FeCl_2 \cdot 4H_2O$. Molar specific heat vs. temperature [72R].

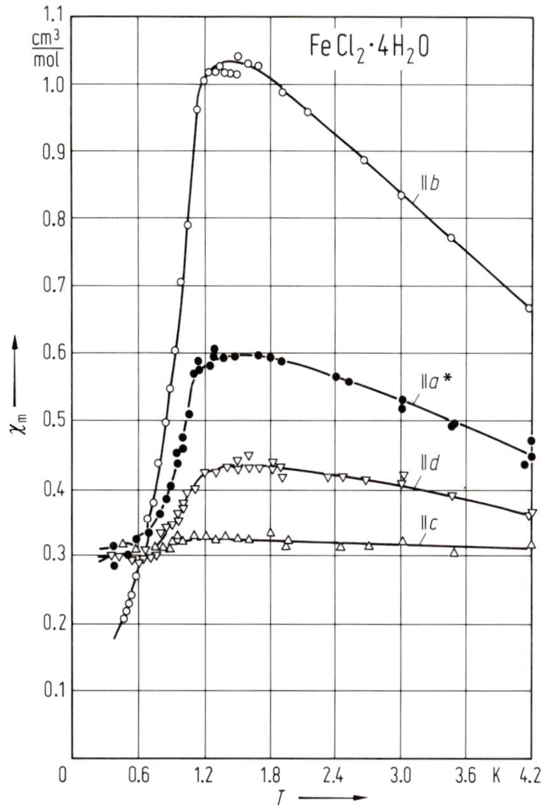

Fig. 76. $FeCl_2 \cdot 4H_2O$. Molar magnetic susceptibility ▶ along the a^*, b, c and d axes vs. temperature. Here, the b and c are conventional axes of monoclinic crystal, a^* is perpendicular to both b and c, and d is at 45° between and in the plane of a and c [64S2].

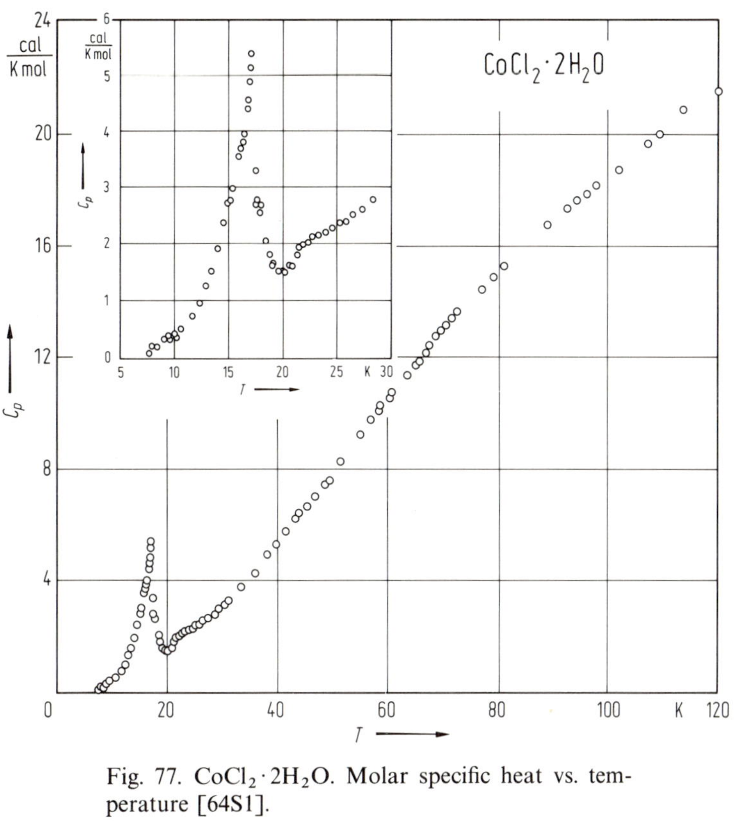

Fig. 77. $CoCl_2 \cdot 2H_2O$. Molar specific heat vs. temperature [64S1].

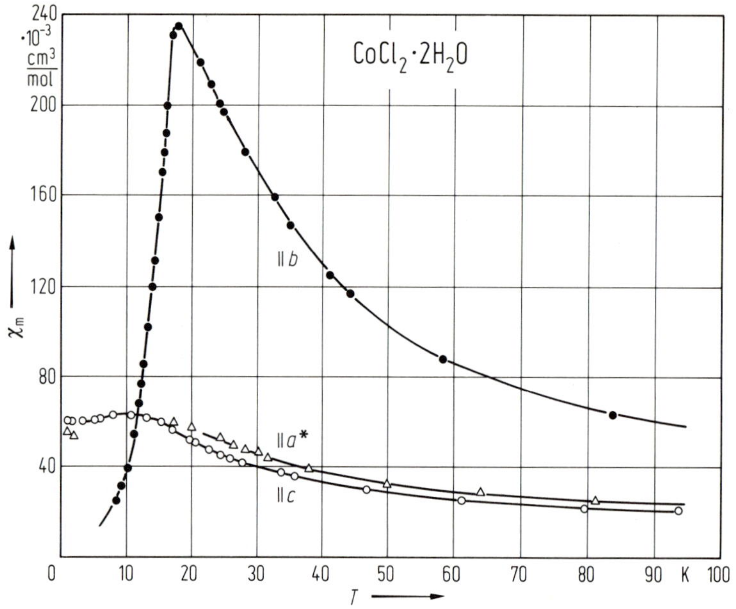

Fig. 78. $CoCl_2 \cdot 2H_2O$. Molar magnetic susceptibility along the a^*, b and c axes vs. temperature. Here, a^* is perpendicular to both b and c [64N2].

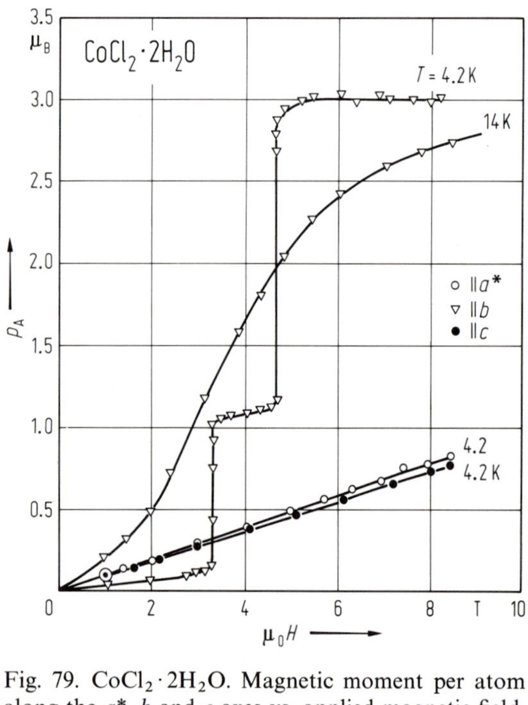

Fig. 79. $CoCl_2 \cdot 2H_2O$. Magnetic moment per atom along the a^*, b and c axes vs. applied magnetic field. Here, a^* is perpendicular to both b and c [64K].

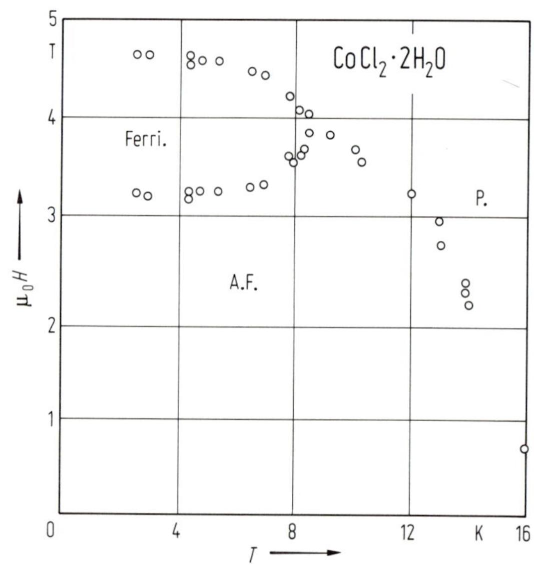

Fig. 80. $CoCl_2 \cdot 2H_2O$. Magnetic field vs. temperature phase diagram [71L]. For the spin structures in the three phases, see Fig. 73.

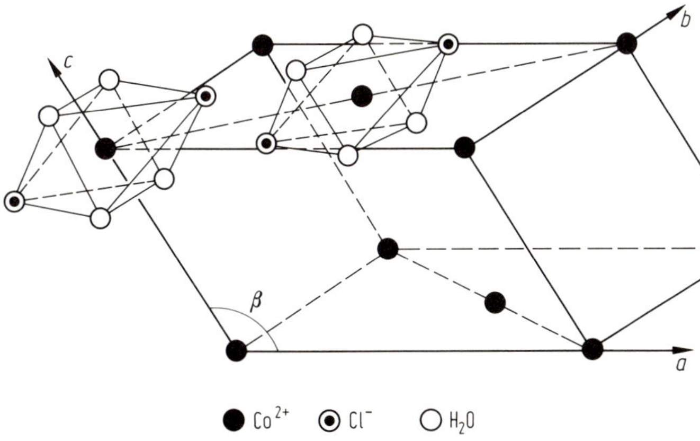

Fig. 81. $CoCl_2 \cdot 6H_2O$. Crystal structure (C_{2h}^3-C2/m).

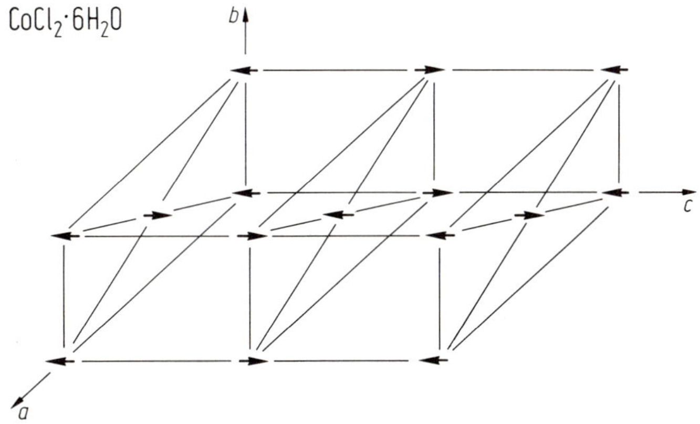

Fig. 82. $CoCl_2 \cdot 6H_2O$. Spin structure [70K].

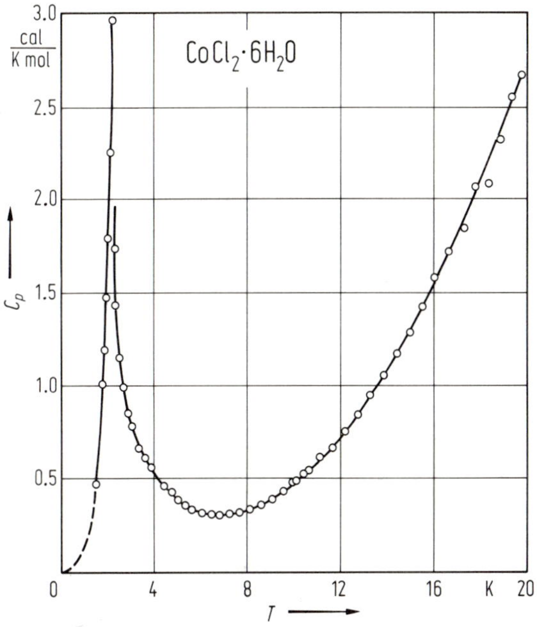

Fig. 83. $CoCl_2 \cdot 6H_2O$. Molar specific heat vs. temperature [60R].

For Fig. 84 see next page.

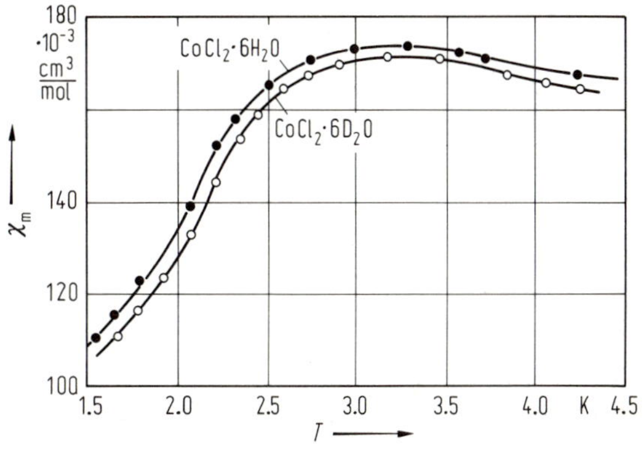

Fig. 85. $CoCl_2 \cdot 6H_2O$; $CoCl_2 \cdot 6D_2O$. Temperature dependence of the molar magnetic susceptibility [60H; LBII/2, p. 2–395].

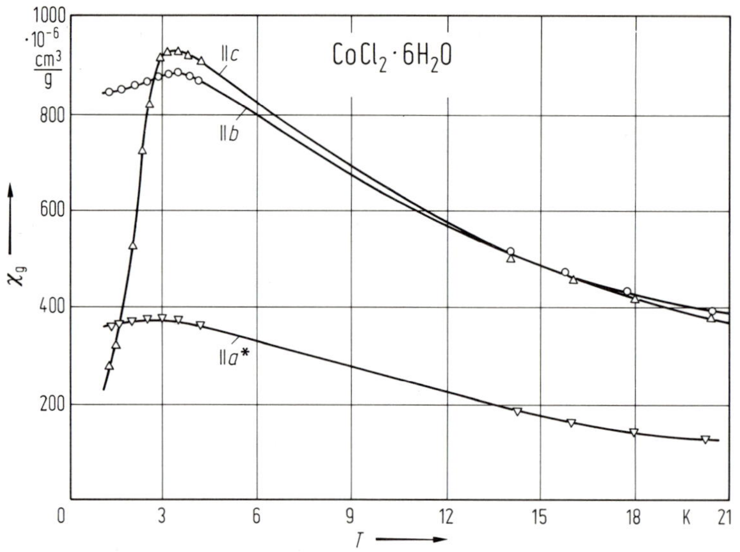

Fig. 84. $CoCl_2 \cdot 6H_2O$. Magnetic susceptibility per gram along the a^*, b and c axes vs. temperature. Here, a^* is perpendicular to both b and c [60H].

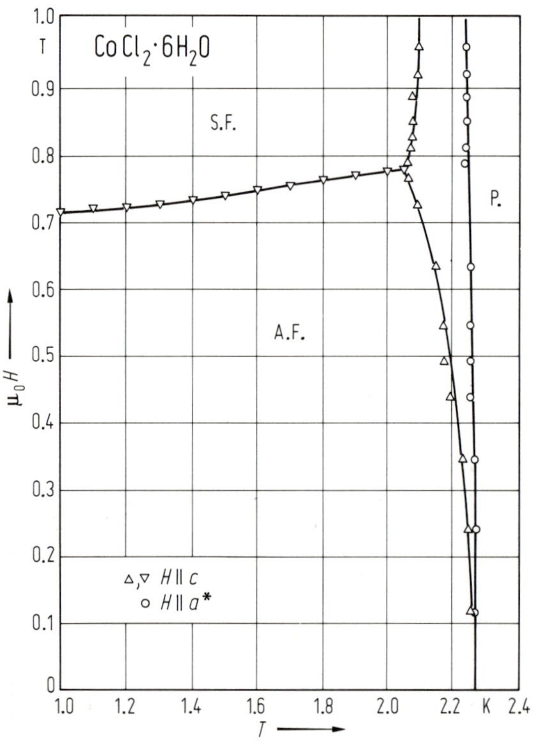

Fig. 86. $CoCl_2 \cdot 6H_2O$. Magnetic field vs. temperature phase diagram. Here, a^* is perpendicular to both b and c [60V].

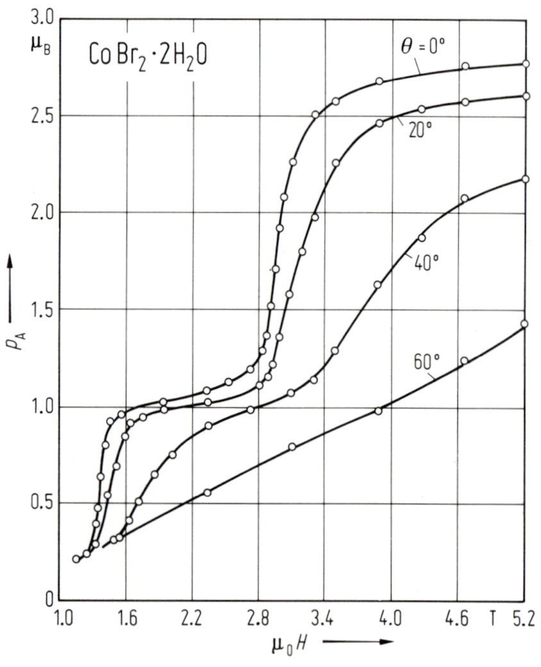

Fig. 87. $CoBr_2 \cdot 2H_2O$. Magnetic moment per atom as a function of angle between the magnetic field and the b axis vs. applied magnetic field. $T = 4.0$ K [64N1].

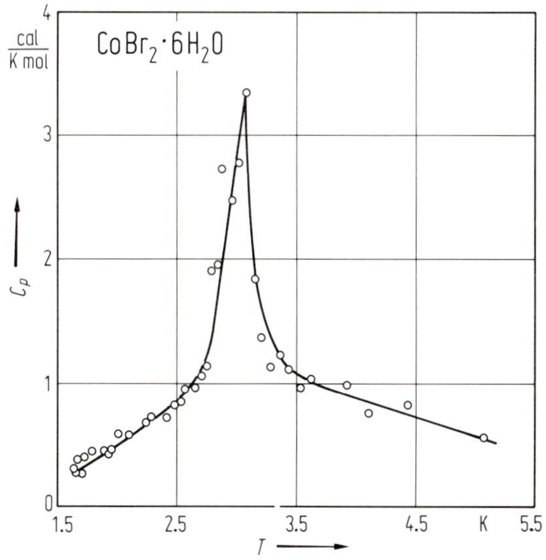

Fig. 88. $CoBr_2 \cdot 6H_2O$. Molar specific heat vs. temperature [59F].

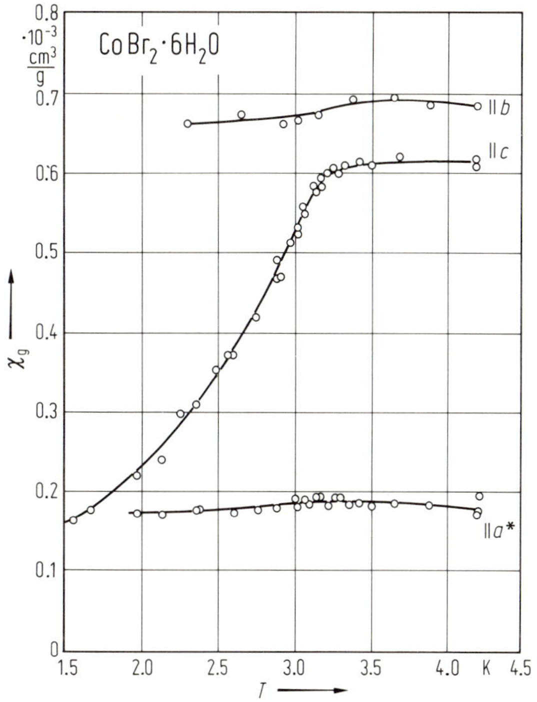

Fig. 89. $CoBr_2 \cdot 6H_2O$. Magnetic susceptibility per gram along the a^*, b and c axes vs. temperature. Here, a^* is perpendicular to both b and c [60G].

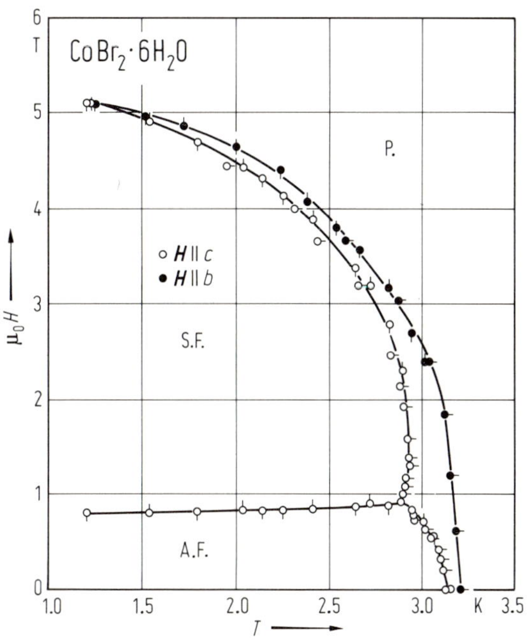

Fig. 90. $CoBr_2 \cdot 6H_2O$. Magnetic field vs. temperature phase diagram. The dash indicates the direction in which the phase boundary is crossed [73M2].

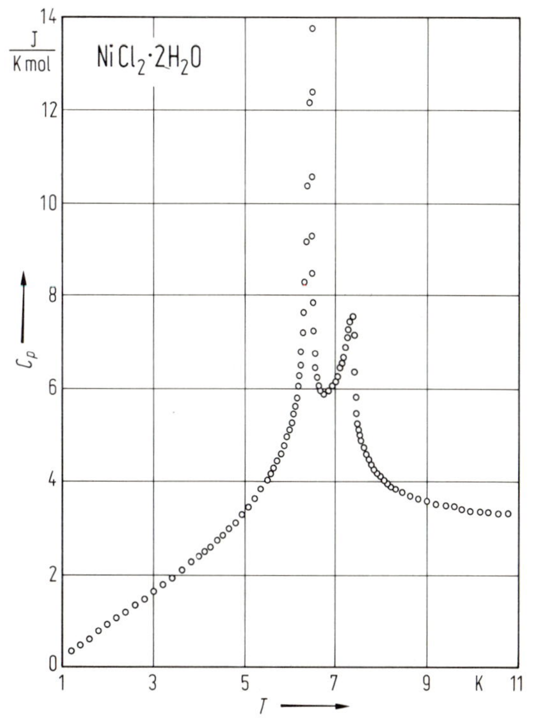

Fig. 91. $NiCl_2 \cdot 2H_2O$. Molar specific heat vs. temperature [72P].

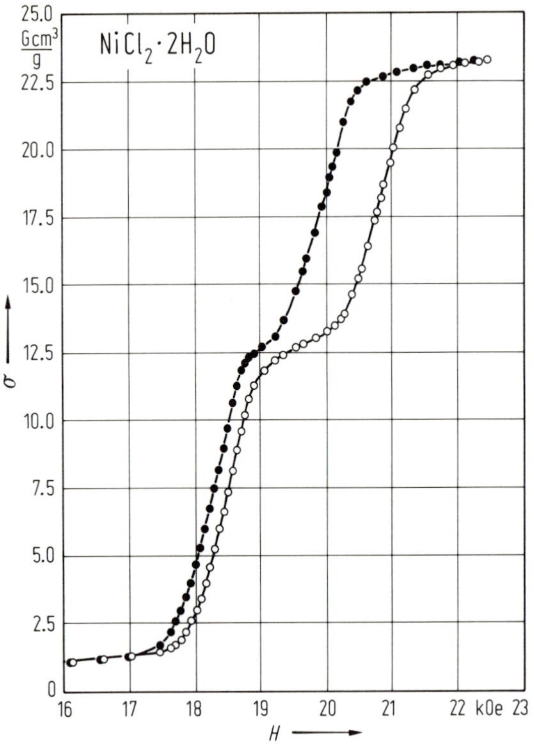

Fig. 92. $NiCl_2 \cdot 2H_2O$. Magnetic moment per gram (specific magnetization) vs. applied magnetic field. $T = 1.1$ K. ○: increasing field, ●: decreasing field [72B2].

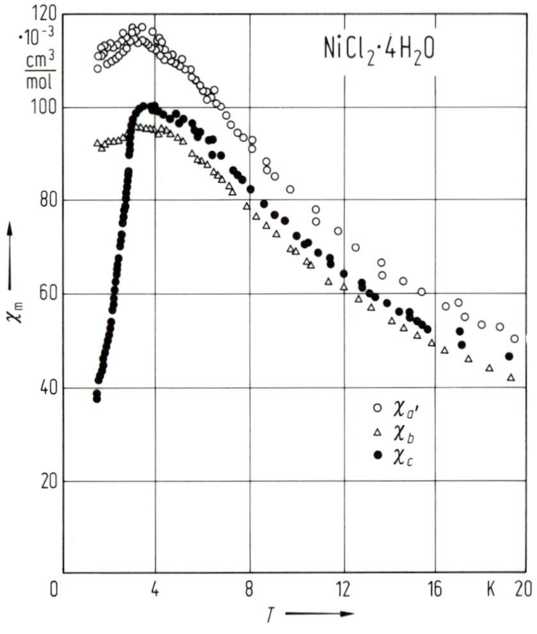

Fig. 93. $NiCl_2 \cdot 4H_2O$ single crystal. Temperature dependence of molar magnetic susceptibility along the a', b and c axes [73M1; LBII/12a, p. 627].

For Fig. 94 see next page.

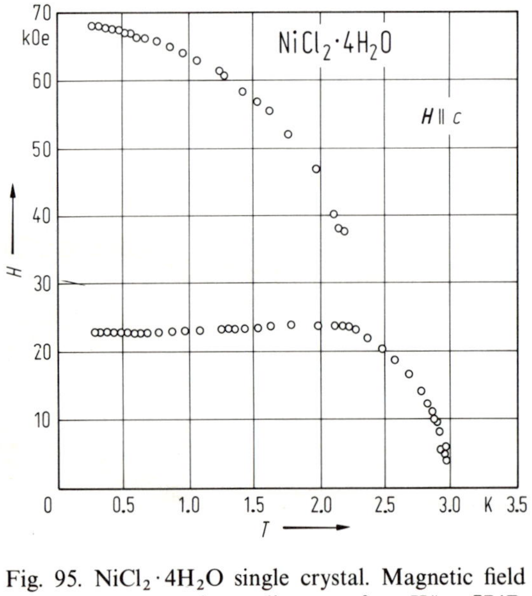

Fig. 95. $NiCl_2 \cdot 4H_2O$ single crystal. Magnetic field vs. temperature phase diagram for $\boldsymbol{H} \parallel c$ [74P, LBII/12a, p. 627].

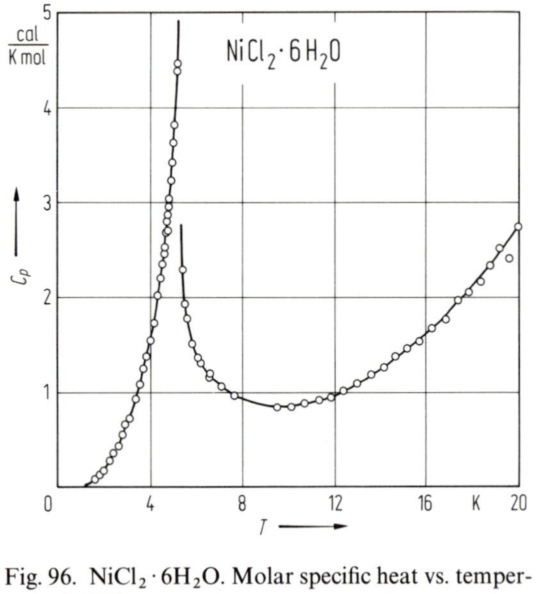

Fig. 96. $NiCl_2 \cdot 6H_2O$. Molar specific heat vs. temperature [60R].

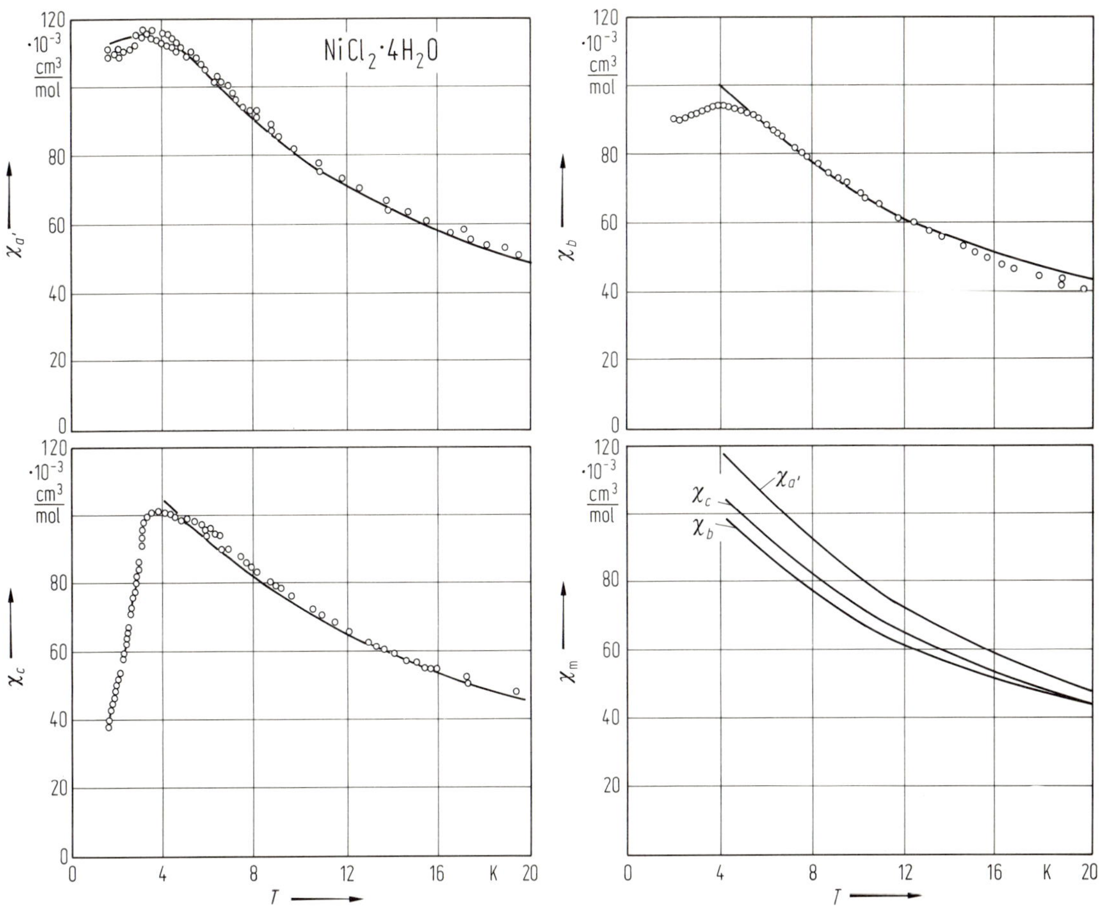

Fig. 94. $NiCl_2 \cdot 4H_2O$ single crystal. Comparison of the fitted molar magnetic susceptibility curves to the corresponding experimental data and the resulting anisotropy [73M1; LBII/12a, p. 627].

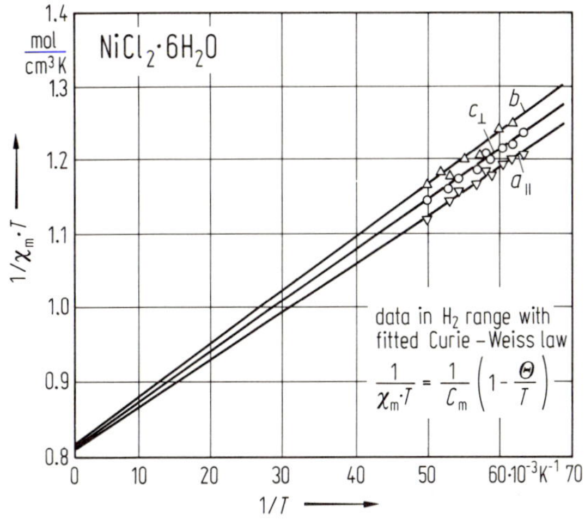

Fig. 98. $NiCl_2 \cdot 6H_2O$ single crystal. Temperature dependence of $1/(\chi_m \cdot T)$ ($T > T_N = 5.34$ K) measured along the $a_{||}$, b and $c_{\perp}$ directions [73H; LBII/12a, p. 629].

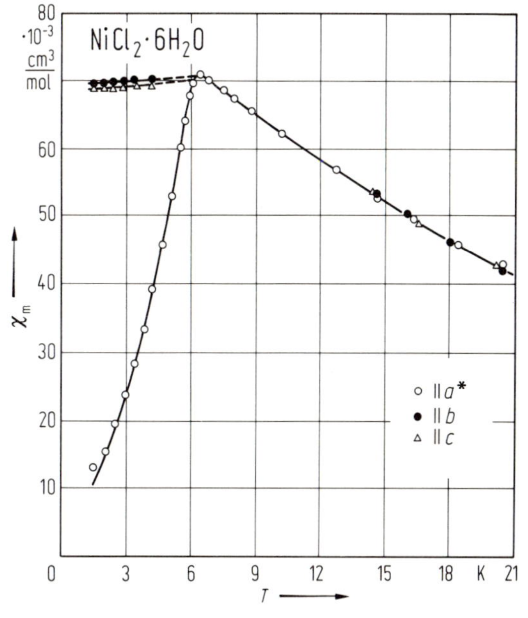

Fig. 97. $NiCl_2 \cdot 6H_2O$. Molar magnetic susceptibility along the a^*, b and c axes vs. temperature. Here, a^* is perpendicular to both b and c [59H].

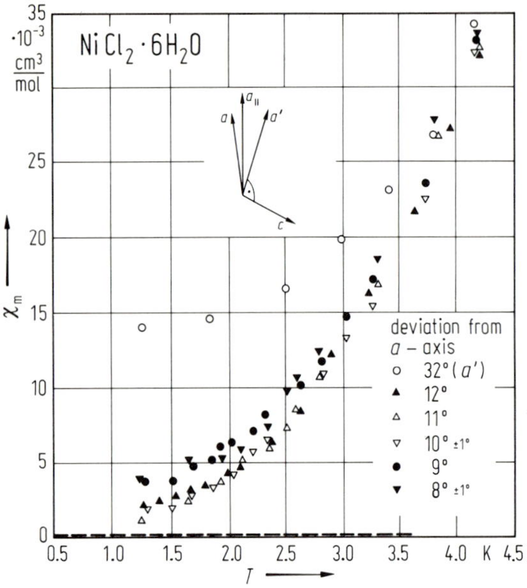

Fig. 99. $NiCl_2 \cdot 6H_2O$ single crystal. Temperature dependence of χ_m below $T = 4.2$ K measured in the *ac* plane at various angles from the *a* axis. The preferred direction of spin alignment is found to be $a - 10.5 \pm 1.5° \equiv a_{||}$. The direction $a - 32° \equiv a'$ was originally thought to be preferred. The temperature-independent susceptibility is indicated by the dashed line [73H; LBII/12a, p. 628].

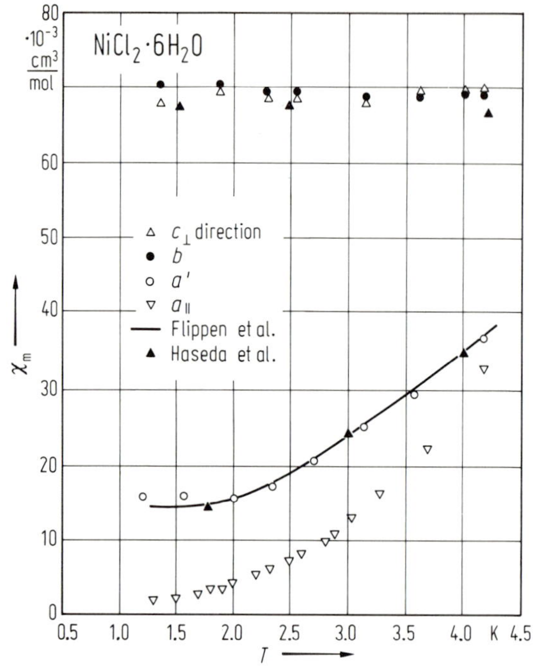

Fig. 100. $NiCl_2 \cdot 6H_2O$ single crystal. Temperature dependence of χ_m below $T = 4.2$ K measured along the principal axes $a_{||}$, b and $c_\perp$ as well as a'. Typical values from Flippen et al. and Haseda et al. (for references see original lit.) are shown for comparison. $\chi_{a||}$ is an average over measurements nominally at 10°, 11° and 12° from the *a* axis in the *ac* plane [73H; LBII/12a, p. 628].

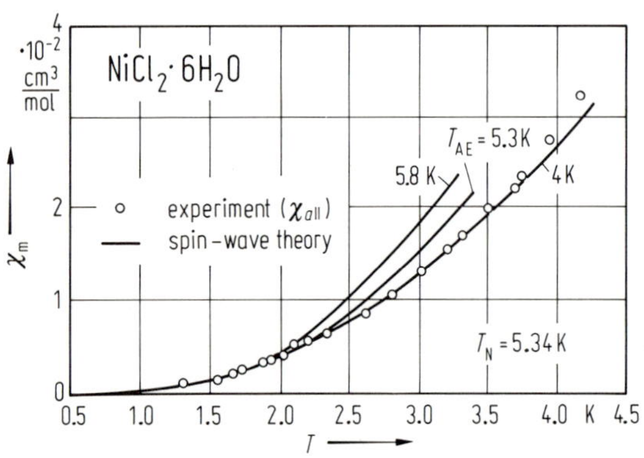

Fig. 101. $NiCl_2 \cdot 6H_2O$ single crystal. Temperature dependence of $\chi_{a||}$. Curves fitted for a simple spin-wave model with several values of the gap temperature, T_{AE} [73H; LBII/12a, p. 628].

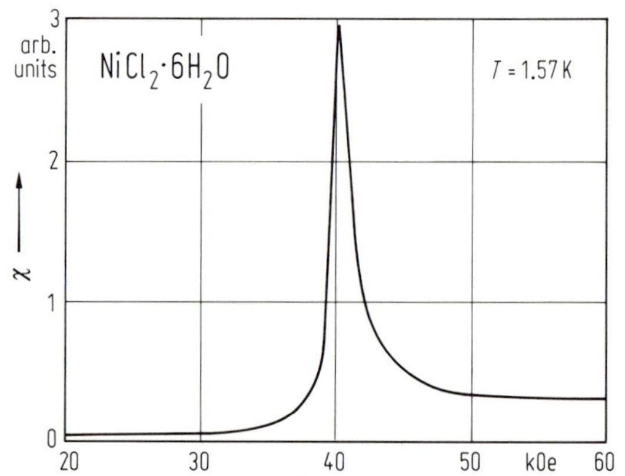

Fig. 102. $NiCl_2 \cdot 6H_2O$ single crystal. Susceptibility χ vs. H at constant temperature showing the antiferromagnetic spin-flop transition [73C; LBII/12a, p. 628].

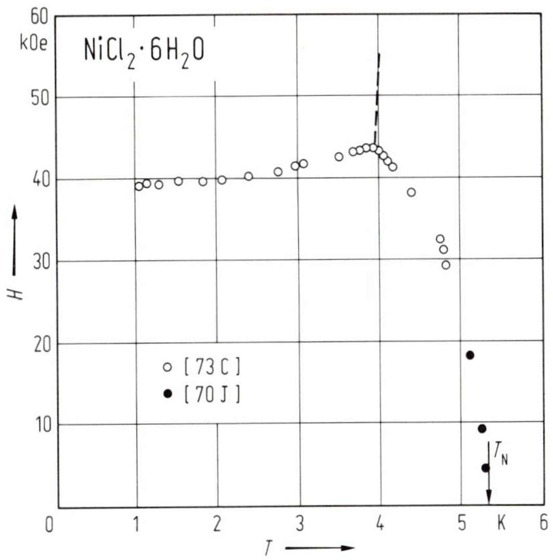

Fig. 103. $NiCl_2 \cdot 6H_2O$ single crystal. Magnetic field vs. temperature phase diagram [73C; LBII/12a, p. 628].

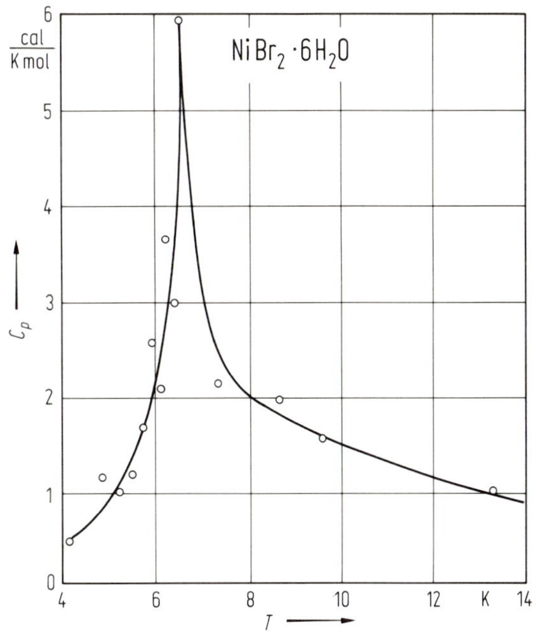

Fig. 104. $NiBr_2 \cdot 6H_2O$. Molar specific heat vs. temperature. $T_N = 6.50$ K [59S].

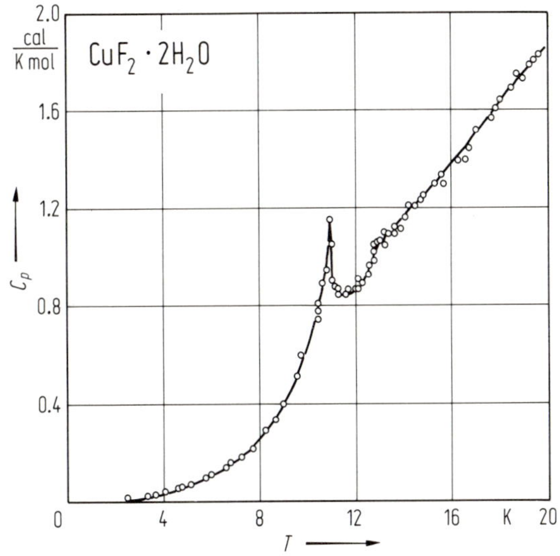

Fig. 105. $CuF_2 \cdot 2H_2O$. Molar specific heat vs. temperature [66C2].

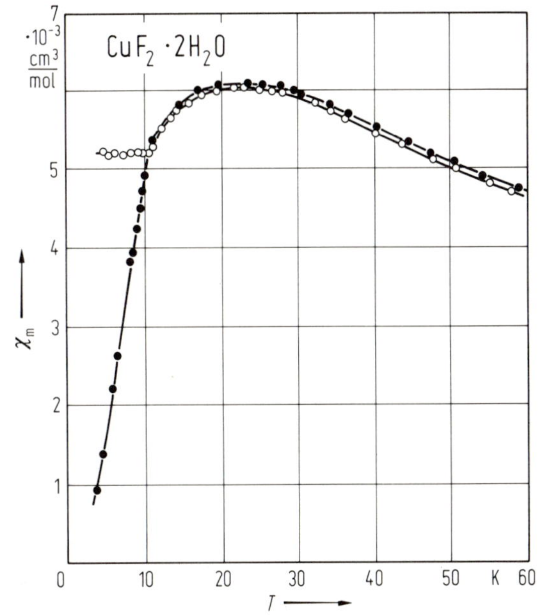

Fig. 106. $CuF_2 \cdot 2H_2O$. Molar magnetic susceptibility parallel (●) and perpendicular (○) to the c axis vs. temperature [65T2].

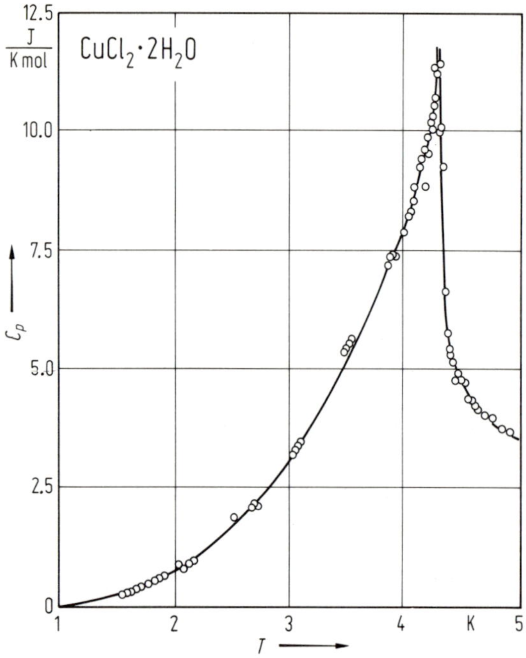

Fig. 107. $CuCl_2 \cdot 2H_2O$. Molar specific heat vs. temperature [52F].

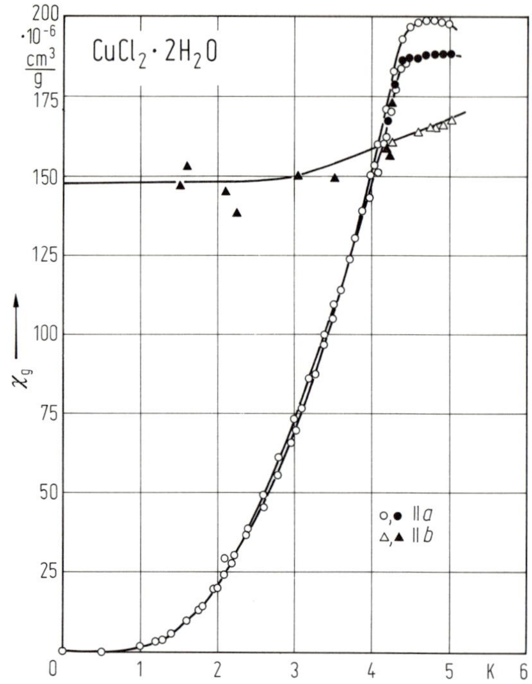

Fig. 108. $CuCl_2 \cdot 2H_2O$. Magnetic susceptibility per gram along the *a* and *b* axes vs. temperature [55V].

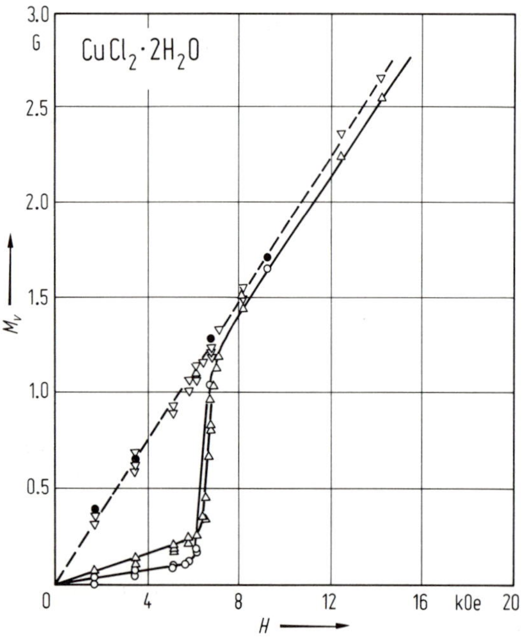

Fig. 109. $CuCl_2 \cdot 2H_2O$. Magnetization along the *a* and *c* axes vs. applied magnetic field. △: ‖*a* ($T = 2.1$ K), ▽: ‖*c* ($T = 2.1$ K), ○: ‖*a* ($T = 1.5$ K), ●: ‖*c* ($T = 1.5$ K) [52V].

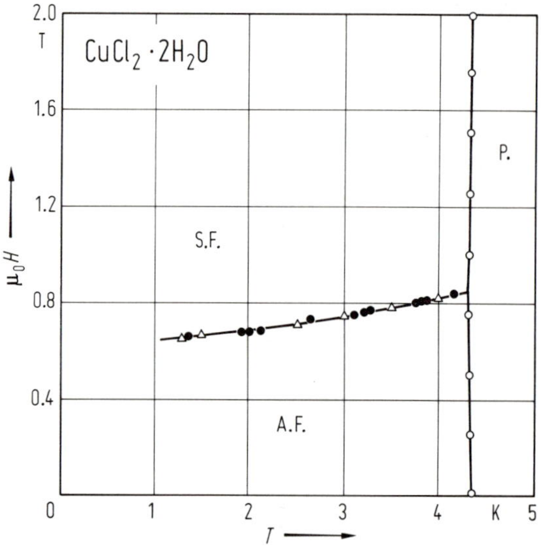

Fig. 110. $CuCl_2 \cdot 2H_2O$. Magnetic field vs. temperature phase diagram. Triangles indicate NMR data, other symbols from magnetothermal measurements [69B].

9.1.4 Microscopic magnetic and related properties of MX_2 and $MX_2 \cdot nH_2O$ compounds

9.1.4.1 MX_2 compounds

Table 6. MX_2 compounds. For the definition of the exchange paths in the compounds with rutile structure see Fig. 111, with $CdCl_2$ and CdI_2 structure Fig. 116.

Substance (Fig.)	Hamiltonian	Parameters
VF_2 (Fig. 111)	$g\mu_B \boldsymbol{H}\cdot\boldsymbol{S} + D\,[S_z^2 - S(S+1)/3] + E(S_x^2 - S_y^2)$ $S = 3/2$, $z \parallel c$	$g = 1.97$ $D = 0.425\ \mathrm{cm}^{-1}$ $\lvert E\rvert = 0.153\ \mathrm{cm}^{-1}$ [60P]
MnF_2 (Fig. 112)	$-\sum_{ij} J_{ij}\boldsymbol{S}_i\cdot\boldsymbol{S}_j + \mu_B\sum_i \boldsymbol{H}\tilde{\boldsymbol{g}}\boldsymbol{S}_i + \sum_i \boldsymbol{S}_i\tilde{\boldsymbol{D}}\boldsymbol{S}_i$ $S = 5/2$	$J_1 = 0.23\ \mathrm{cm}^{-1}$ $J_2 = -1.23\ \mathrm{cm}^{-1}$ $J_3 = -0.03\ \mathrm{cm}^{-1}$ [69N] $g_x = 2.002$ $g_y = 2.002$ $g_z = 2.002$ $D_x = 2.1\cdot 10^{-3}\ \mathrm{cm}^{-1}$ $D_y = 1.03\cdot 10^{-2}\ \mathrm{cm}^{-1}$ $D_z = -1.24\cdot 10^{-2}\ \mathrm{cm}^{-1}$ [56T]
FeF_2 (Fig. 113)	$-\sum_{ij} J_{ij}\boldsymbol{S}_i\cdot\boldsymbol{S}_j + \mu_B\sum_i \boldsymbol{H}\tilde{\boldsymbol{g}}\boldsymbol{S}_i + \sum_i D(S_i^z)^2$ $S = 2$	$J_1 = 0.048\ \mathrm{cm}^{-1}$ $J_2 = -3.64\ \mathrm{cm}^{-1}$ $J_3 = -0.194\ \mathrm{cm}^{-1}$ $D = -6.46\ \mathrm{cm}^{-1}$ [70H1] $g_{\parallel c} = 2.25$ [61O]
CoF_2 (Fig. 114)	$-2\sum_{ij} J_{ij}\boldsymbol{S}_i\cdot\boldsymbol{S}_j + \sum_i D(S_i^z)^2$ $S = 3/2$	$J_1 = 0.323\ \mathrm{cm}^{-1}$ $J_2 = -2.00\ \mathrm{cm}^{-1}$ $D = -12.2\ \mathrm{cm}^{-1}$ [68M]
NiF_2 (Fig. 115)	$-\sum_{ij} J_{ij}\boldsymbol{S}_i\cdot\boldsymbol{S}_j + \sum_{i\,(\mathrm{corner})}[D(S_i^z)^2 + E\{(S_i^x)^2 - (S_i^y)^2\}] + \sum_{l\,(\mathrm{center})}[D(S_l^z)^2 - E\{(S_l^x)^2 - (S_l^y)^2\}] + \mu_B\sum_k \boldsymbol{S}_k g\boldsymbol{H}$ $S = 1$	$J_1 = 0.22\ \mathrm{cm}^{-1}$ $J_2 = -13.87\ \mathrm{cm}^{-1}$ $J_3 = -0.79\ \mathrm{cm}^{-1}$ [70H3] $g = 2.33$ $D = 4.183\ \mathrm{cm}^{-1}$ $\lvert E\rvert = 2.67\ \mathrm{cm}^{-1}$ [60P]

continued

Table 6 (*continued*)

Substance (Fig.)	Hamiltonian	Parameters
$FeCl_2$ (Fig. 117)	$-2\sum_{ij} J_{ij}\boldsymbol{S}_i\cdot\boldsymbol{S}_j - 2\sum_{ij} K_{ij} S_i^z S_j^z + \sum_i D[(S_i^z)^2 - 2/3]$ $S = 1$	$g\mu_B H_A = -2D(S-1/2) - 2S\sum_m K_{lm} + 2S\sum_{l'} K_{ll'} = 16\ \mathrm{cm}^{-1}$ $J_1 = 2.7\ \mathrm{cm}^{-1}$ $J_2 = -0.36\ \mathrm{cm}^{-1}$ $J'_1 = -0.12\ \mathrm{cm}^{-1}$ [72B1][a]) $\lvert D\rvert = 6.3\ \mathrm{cm}^{-1}$ $g_{\parallel c} = 4.0$ [90T]
$CoCl_2$ (Fig. 118)	$-2\sum_{ij} J_{ij}(\delta S_i^x S_j^x + S_i^y S_j^y + S_i^z S_j^z)$ $S = 1/2$ $x \parallel c$	$J_1 = 18.7\ \mathrm{cm}^{-1}$ $J_2 = -9.0\ \mathrm{cm}^{-1}$ $\delta = 0.4$ [77Z] $g_{\parallel c} = 3.04$ $g_{\perp c} = 4.95$ [61M2]
$NiCl_2$ (Fig. 119)	$-2\sum_{ij} J_{ij}\boldsymbol{S}_i\cdot\boldsymbol{S}_j + D\sum_i (S_i^z)^2$ $S = 1$ $z \parallel c$	$J_1 = 15.1\ \mathrm{cm}^{-1}$ $J_2 = -3.37\ \mathrm{cm}^{-1}$ [75L] $J' = -0.53\ \mathrm{cm}^{-1}$ $D = 0.55\ \mathrm{cm}^{-1}$ [73K][a])
$FeBr_2$		Intralayer interactions $J_1/k = 6.35$ K, $J_2/k = -7.6$ K [74B]
FeI_2		Intralayer interactions $J_1/k = -8.3$ K, $J_2/k = -26.4$ K [74B]
$CoBr_2$ (Fig. 120)	$-\sum_{ij}(J_{ij}\boldsymbol{S}_i\cdot\boldsymbol{S}_j - K_{ij} S_i^x S_j^x)$ $S = 1/2$	$J_1 = 16.2\ \mathrm{cm}^{-1}$ $J_2 = 0.15\ \mathrm{cm}^{-1}$ $K_1 = 9.20\ \mathrm{cm}^{-1}$ $J_3 = -2.73\ \mathrm{cm}^{-1}$ $J' = -7.8\ \mathrm{cm}^{-1}$ $K_2 = K_3 = 0$ $K' = -0.48\ \mathrm{cm}^{-1}$ [80Y][b]) $g_{\parallel c} = 2.71$ $g_{\perp c} = 4.99$ [61M2]
$NiBr_2$ (Fig. 121)	$-\sum_{ij} J_{ij}\boldsymbol{S}_i\cdot\boldsymbol{S}_j + D\sum_i (S_i^z)^2$ $S = 1$	$J_1 = 12.6\ \mathrm{cm}^{-1}$ $J_2 = -0.15\ \mathrm{cm}^{-1}$ $J_3 = -3.3\ \mathrm{cm}^{-1}$ $J' = -1.47\ \mathrm{cm}^{-1}$ $D = 1.2\ \mathrm{cm}^{-1}$ [82R][b])

[a]) J'_1, J': Exchange interaction constants between *c* planes.
[b]) J', K': Exchange interaction constants between *c* planes.

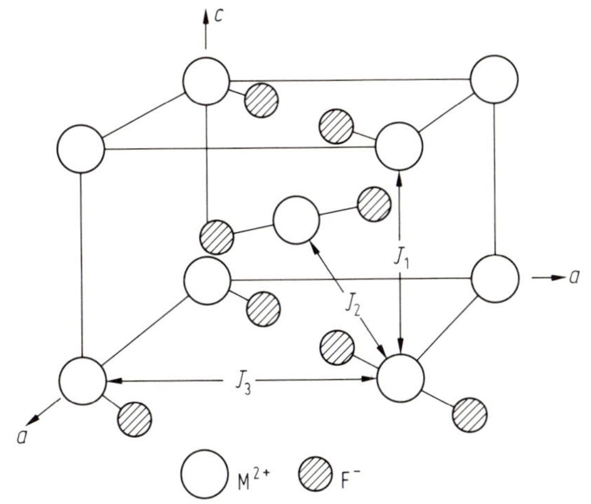

Fig. 111. The exchange interaction paths in the compounds with the rutile structure.

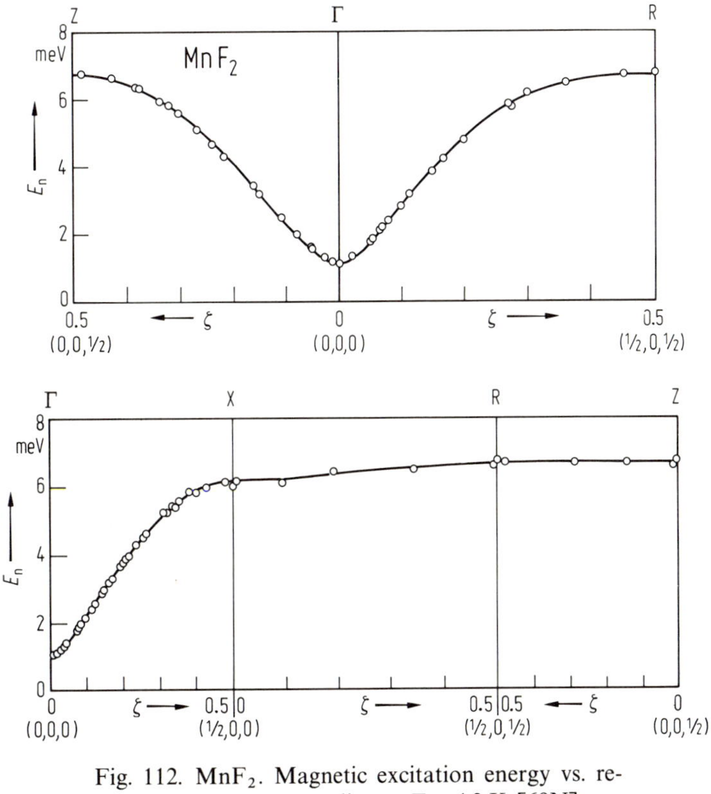

Fig. 112. MnF_2. Magnetic excitation energy vs. reduced wave vector coordinate, $T = 4.2$ K [69N].

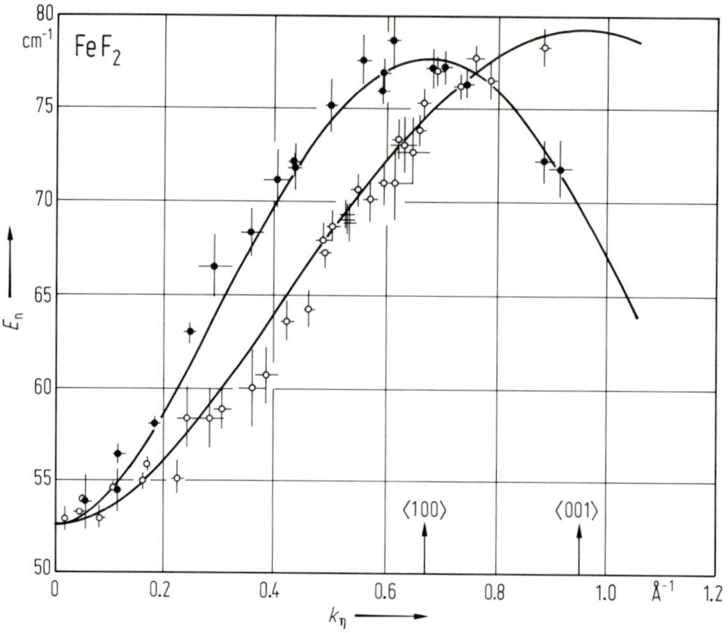

Fig. 113. FeF_2. Magnetic excitation energy vs. wave vector k. $T = 4.2$ K. ● $\langle\eta 0 0\rangle$, ○ $\langle 0 0 \eta\rangle$ [70H1].

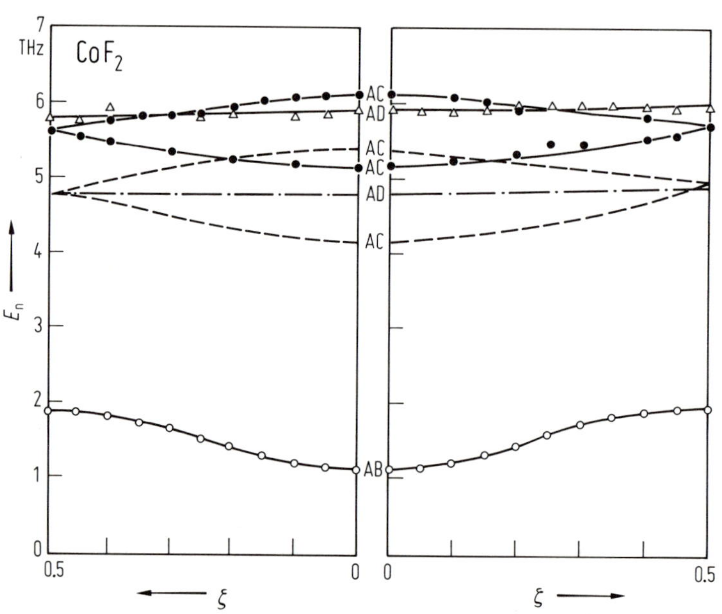

Fig. 114. CoF_2. Magnetic excitation energy vs. reduced wave vector coordinate. Left: along $\langle 100\rangle$, right: along $\langle 001\rangle$. A, B, C, D are the four lowest energy levels. $T = 4.2$ K transitions: △ (A D), ● (A C), ○ (A B). $T = 80$ K transitions –·–· (A D), - - - - - (A C) [68M].

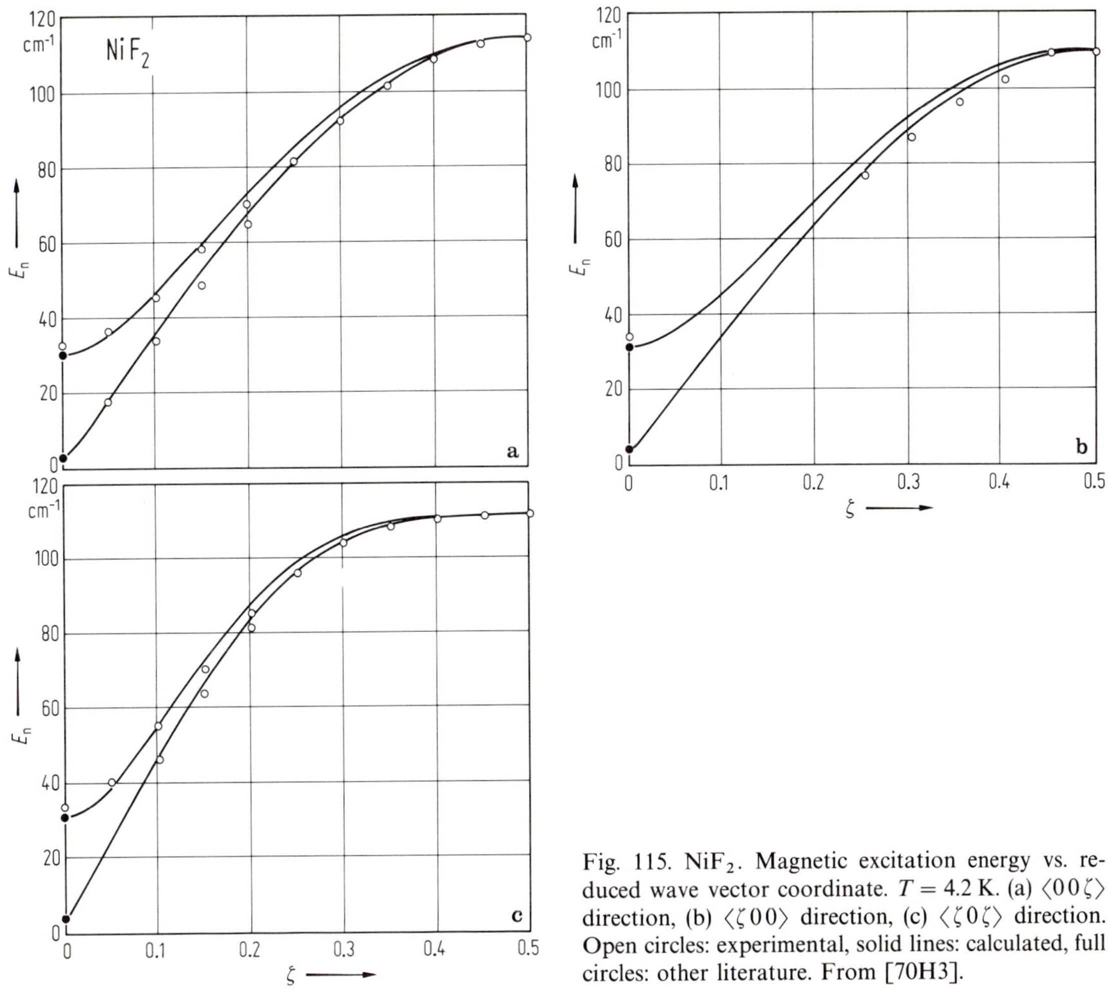

Fig. 115. NiF_2. Magnetic excitation energy vs. reduced wave vector coordinate. $T = 4.2$ K. (a) $\langle 0 0 \zeta \rangle$ direction, (b) $\langle \zeta 0 0 \rangle$ direction, (c) $\langle \zeta 0 \zeta \rangle$ direction. Open circles: experimental, solid lines: calculated, full circles: other literature. From [70H3].

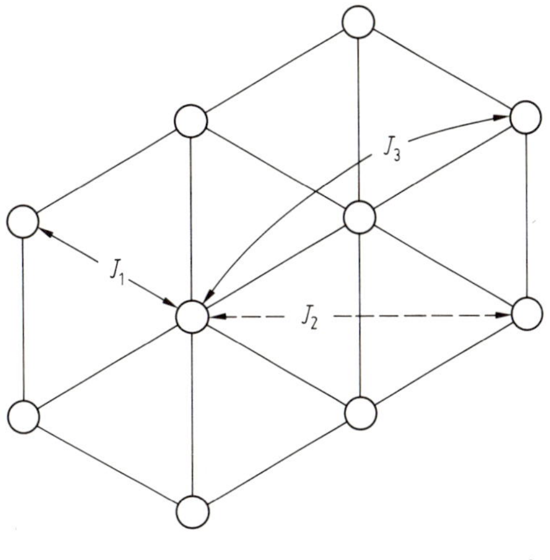

Fig. 116. The exchange paths in the c plane of the compounds with the $CdCl_2$ and CdI_2 structure.

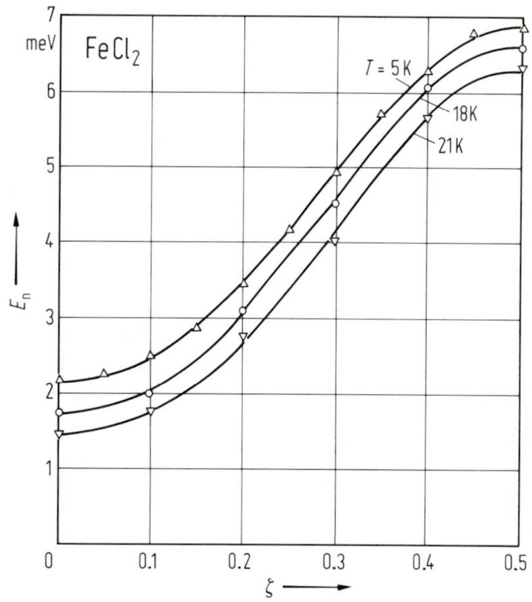

Fig. 117. $FeCl_2$. Magnetic excitation energy vs. reduced wave vector coordinate along the ⟨ζ 0 0⟩ direction [72B1].

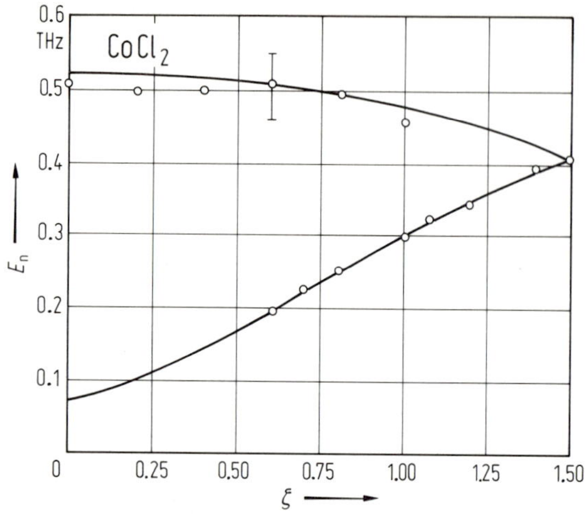

Fig. 118. $CoCl_2$. Magnetic excitation energy vs. reduced wave vector coordinate along the ⟨0 0 ζ⟩ direction. $T = 4.2$ K [77Z].

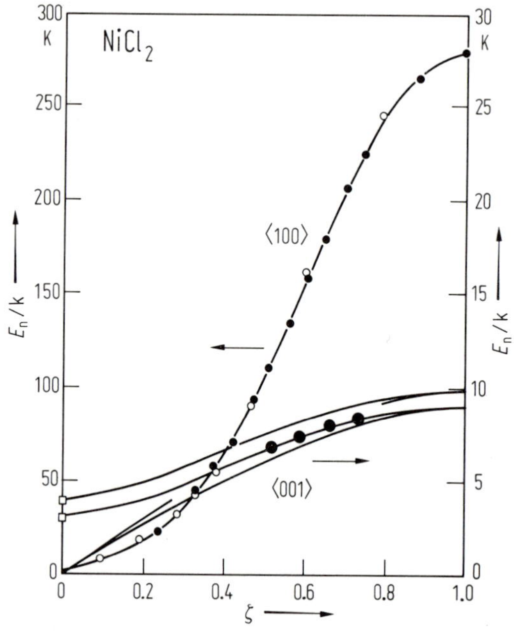

Fig. 119. $NiCl_2$. Magnetic excitation energy vs. reduced wave vector coordinate along the ⟨100⟩ and ⟨001⟩ directions. Curves: theory, ○: from neutron scattering, □: from NMR [75L]. 1 eV ≙ $1.16 \cdot 10^4$ K.

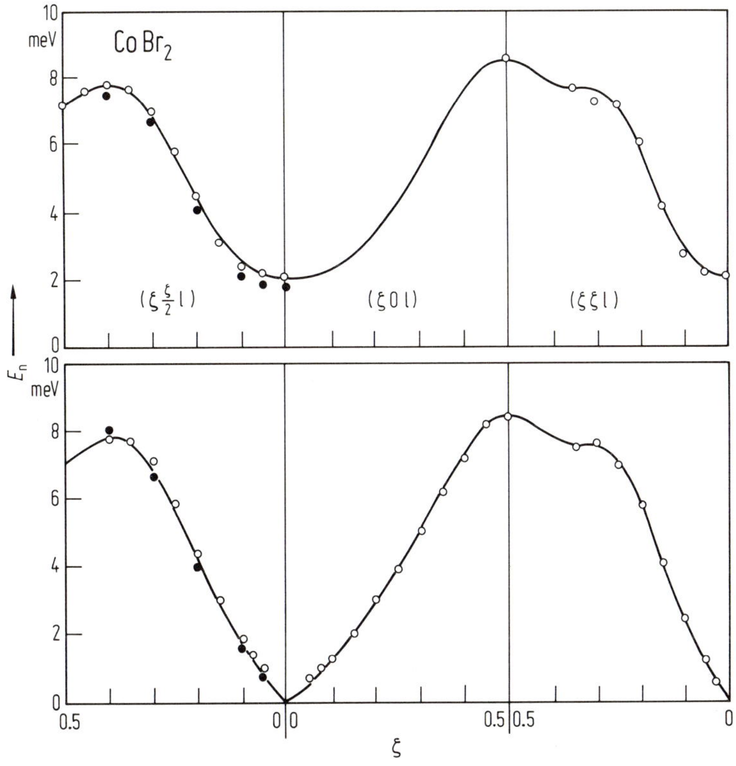

Fig. 120. $CoBr_2$. Magnetic excitation energy vs. reduced wave vector coordinate. Upper panel: $l = 0.75$, lower panel: $l = 0.5$. ○: $T = 5$ K, ●: $T = 15$ K [80Y].

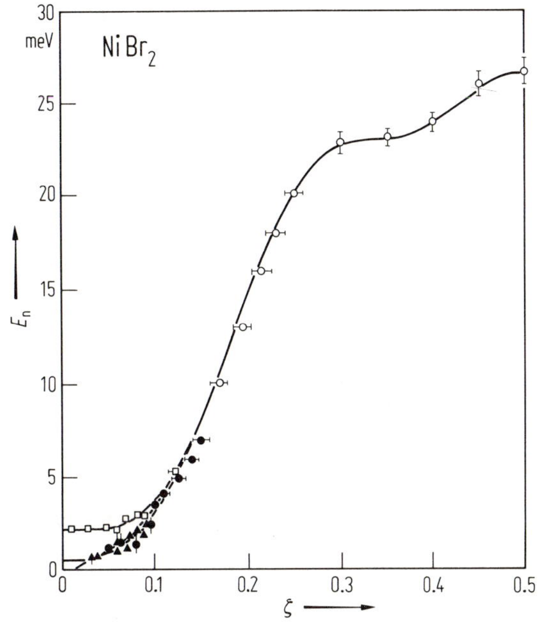

Fig. 121. $NiBr_2$. Magnetic excitation energy vs. reduced wave vector coordinate along the $\langle \zeta \zeta l \rangle$ direction. $T = 4.2$ K. □: $l = 0.75$ ($\lambda = 2.2$ Å), ▲: $l = 1.5$ ($\lambda = 2.2$ Å), ●: $l = 0$ ($\lambda = 2.2$ Å), ○: $l = 0$ ($\lambda = 1.2$ Å) [82R].

9.1.4.2 $MX_2 \cdot nH_2O$ compounds

Table 7. $MX_2 \cdot nH_2O$ compounds. For the definition of the exchange interaction paths in the compounds with $MnCl_2 \cdot 2H_2O$ structure see Fig. 122.

Substance (Figs.)	Hamiltonian	Parameters
$FeCl_2 \cdot 2H_2O$	$-2\sum_{ij} J_{ij} \boldsymbol{S}_i \cdot \boldsymbol{S}_j - \mu_B \sum_i \boldsymbol{S}_i \tilde{\boldsymbol{g}} \boldsymbol{H} - D \sum_i (S_i^z)^2$ $S = 2$	$g_\parallel = 2.23 \quad D = 9.58\ cm^{-1}$ [70H2] $J_1 = -0.281\ cm^{-1} \quad J_2 = -0.046\ cm^{-1}$ [75K]
$CoCl_2 \cdot 2H_2O$ (Fig. 123)	$-\sum_{ij}(J_{ij}^{zz} S_i^z S_j^z + J_{ij}^{xx} S_i^x S_j^x + J_{ij}^{yy} S_i^y S_j^y) - \sum_i g_\parallel \mu_B H_z S_i^z$ $S = 1/2 \quad z \parallel b$	$g_\parallel = 6.81 \quad J_0^{zz} = 12.66\ cm^{-1} \quad J_1^{zz} = -3.25\ cm^{-1}$ $J_2^{zz} = -0.7\ cm^{-1} \quad J_3^{zz} = 0.09\ cm^{-1}$ $J_0^{\perp}/J_0^{zz} = 0.155 \quad J_1^{\perp}/J_1^{zz} = J_2^{\perp}/J_2^{zz} = 0.280$ $J_0^{a}/J_0^{zz} = 0.10 \quad J_1^{a}/J_1^{zz} = J_2^{a}/J_2^{zz} = 0.16$ [69T]*)
$CoCl_2 \cdot 6H_2O$		Intralayer exchange $J/k = -2.05(10)$ K [74D; LB II/12a, p. 227]
$CoBr_2 \cdot 6H_2O$		Intralayer exchange from high temperature susceptibility $J/k = -2.45(10)$ K [74D; LBII/12a, p.227]
$NiCl_2 \cdot 4H_2O$	$-2\sum_{ij} J_{ij} \boldsymbol{S}_i \cdot \boldsymbol{S}_j + D(S_z^2 - \frac{2}{3}) + E(S_x^2 - S_y^2) + \mu_B \sum_i g_i H_i S_i$	Intralayer exchange $zJ/k = -2.9$ K [74P; LBII/12a, p. 321] $D/k = -11.5(1)$ K; $E/k = 0.1(1)$ K
$NiCl_2 \cdot 6H_2O$	$-2\sum_{ij} J_{ij} \boldsymbol{S}_i \cdot \boldsymbol{S}_j + D(S_z^2 - \frac{2}{3}) + E(S_x^2 - S_y^2) + \mu_B \sum_i g_i H_i S_i$	Intralayer exchange $zJ/k = -6.0$ K [73H; LBII/12a, p. 322] $D/k = -1.5(5)$ K; $E/k = 0.26(40)$ K

*) $J_{ij}^{\perp} \equiv (J_{ij}^{xx} + J_{ij}^{yy})/2$, $J_{ij}^{a} \equiv (J_{ij}^{xx} - J_{ij}^{yy})/2$.

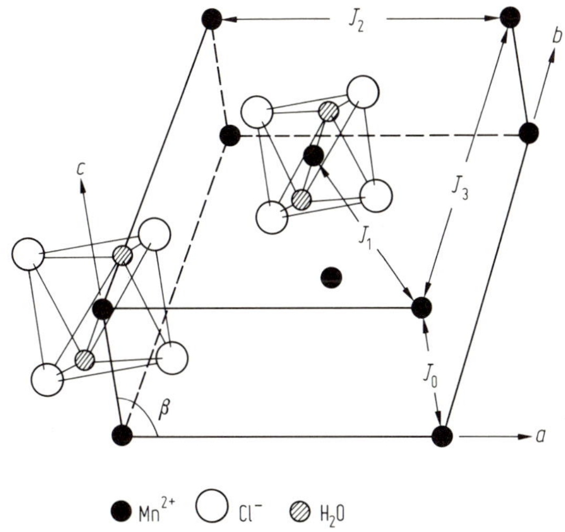

Fig. 122. The exchange paths of the compounds with the $MnCl_2 \cdot 2H_2O$ structure.

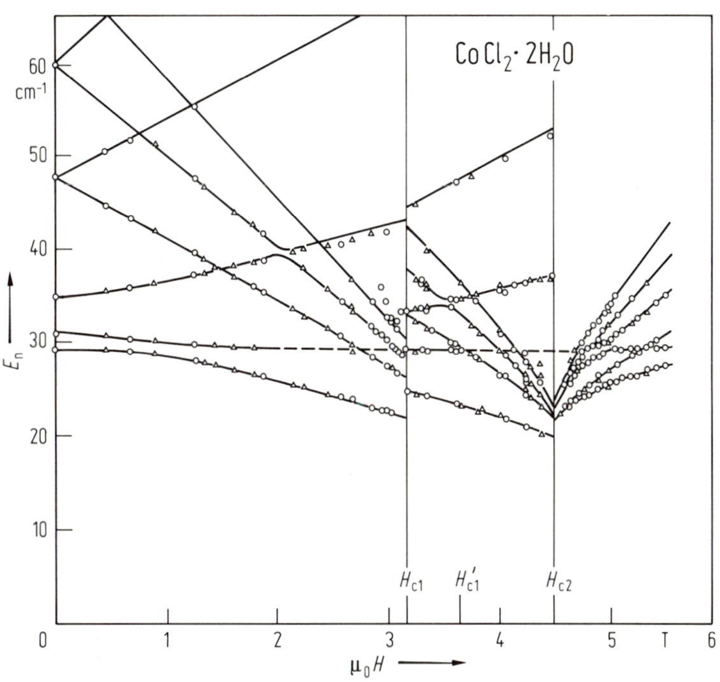

Fig. 123. $CoCl_2 \cdot 2H_2O$. Magnetic excitation energy vs. applied magnetic field. The dashed line represents phonon excitation. The external field is directed parallel to the *b* axis. △: $T \lesssim 1.6$ K, ○: $T = 6$ K [69T].

9.1.5 References for 9.1

36H Harker, D.: Z. Kristallogr. **93** (1936) 136.
40S Starr, C., Bitter, F., Kaufmann, A.R.: Phys. Rev. **58** (1940) 977.
41B Bizette, H., Tsaï, B.: C. R. Acad. Sci. **212** (1941) 119.
42S Stout, J.W., Adams, H.E.: J. Am. Chem. Soc. **64** (1942) 1535.
47W Wells, A.F.: J. Chem. Soc. (1947) 1670.
50C Corliss, L., Delabarre, Y., Elliott, N.: J. Chem. Phys. **18** (1950) 1256.
50G Griffel, M., Stout, J.W.: J. Chem. Phys. **18** (1950) 1455.
52B Busey, R.H., Giauque, W.F.: J. Am. Chem. Soc. **74** (1952) 4443.
52F Friedberg, S.A.: Physica **18** (1952) 714.
52V Van den Handel, J., Gijsman, H.M., Poulis, N.J.: Physica **18** (1952) 862.
53E Erickson, R.A.: Phys. Rev. **90** (1953) 779.
53S Stout, J.W., Matarrese, L.M.: Rev. Mod. Phys. **25** (1953) 338.
54M Matarrese, L.M., Stout, J.W.: Phys. Rev. **94** (1954) 1792.
54S Stout, J.W., Reed, S.A.: J. Am. Chem. Soc. **76** (1954) 5279.
55B Bölger, B.: Communie., Conference de Phys. des Basses Temp., Paris (1955) 244.
55C1 Catalano, E., Stout, J.W.: J. Chem. Phys. **23** (1955) 1284.
55C2 Catalano, E., Stout, J.W.: J. Chem. Phys. **23** (1955) 1803.
55G Gerritsen, H.J., Okkes, R., Bölger, B., Gorter, C.J.: Physica **21** (1955) 629.
55M Murray, R.B., Roberts, L.D.: Phys. Rev. **100** (1955) 1067.
55V Van der Marel, L.C., Van den Broek, J., Wasscher, J.D., Gorter, C.J.: Physica **21** (1955) 685.
56B Bizette, H., Terrier, C., Tsaï, B.: C. R. Acad. Sci. **243** (1956) 1295.
56K Kapadnis, D.G., Hartmans, R.: Physica **22** (1956) 181.
56T Tinkham, M.: Proc. R. Soc. A**236** (1956) 535.
57J Jack, K.H., Maitland, R.: Proc. Chem. Soc. (1957) 232.
58B1 Bozorth, R.M., Nielsen, J.W.: Phys. Rev. **110** (1958) 879.
58B2 Bizette, H., Terrier, C., Tsaï, B.: C. R. Acad. Sci. **246** (1958) 250.
58G Geller, S., Bond, W.L.: J. Chem. Phys. **29** (1958) 925.
58W Wollan, E.O., Koehler, W.C., Wilkinson, M.K.: Phys. Rev. **110** (1958) 638.
59B Bizette, H.C., Terrier, C., Tsaï, B.: J. Phys. Radium **20** (1959) 421.
59F Forstat, H., Taylor, G., Spence, R.D.: Phys Rev. **116** (1959) 897.
59G Gijsman, H.M., Poulis, N.J., Van den Handel, J.: Physica **25** (1959) 954.
59H Haseda, T., Kobayashi, H., Date, M.: J. Phys. Soc. Jpn. **14** (1959) 1724.
59P Penfold, B.R., Grigor, J.A.: Acta Crystallogr. **12** (1959) 850.
59S Spence, R.D., Forstat, H., Khan, G.A., Taylor, G.: J. Chem. Phys. **31** (1959) 555.
59T Tsujikawa, I.: Nippon Kagaku Zasshi **80** (1959) 1391.
59W Wilkinson, M.K., Cable, J.W., Wollan, E.O., Koehler, W.C.: Phys. Rev. **113** (1959) 497.
60B Borovik-Romanov, A.S.: Sov. Phys. JETP **11** (1960) 786.
60C Cable, J.W., Wilkinson, M.K., Wollan, E.O.: Phys. Rev. **118** (1960) 950.
60G Garber, M.: J. Phys. Soc. Jpn. **15** (1960) 734.
60H Haseda, T.: J. Phys. Soc. Jpn. **15** (1960) 483.
60M Mizuno, J.: J. Phys. Soc. Jpn. **15** (1960) 1412.
60P Peter, M., Mock, J.B.: Phys. Rev. **118** (1960) 137.
60R Robinson, W.K., Friedberg, S.A.: Phys. Rev. **117** (1960) 402.
60V Van der Lugt, W., Poulis, N.J.: Physica **26** (1960) 917.
61M1 Mizuno, J.: J. Phys. Soc. Jpn. **16** (1961) 1574.
61M2 Morigaki, K.: J. Phys. Soc. Jpn. **16** (1961) 1639.
61O Ohlmann, R.C., Tinkham, M.: Phys. Rev. **123** (1961) 425.
61S Shulman, R.G.: Phys. Rev. **121** (1961) 125.
62A Abrahams, S.C.: J. Chem. Phys. **36** (1962) 56.

62C1 Chisholm, R.C., Stout, J.W.: J. Chem. Phys. **36** (1962) 972.
62C2 Cable, J.W., Wilkinson, M.K., Wollan, E.O., Koehler, W.C.: Phys. Rev. **125** (1962) 1860.
62F Friedberg, S.A., Cohen, A.F., Schelleng, J.H.: J. Phys. Soc. Jpn. Suppl. **17** (1962) 515.
62G Gal'perin, E.L., Sandler, R.A.: Kristallografiya **7** (1962) 217; Sov. Phys. Crystallogr. **7** (1962) 169.
62L Lewis, J., Machin, D.J., Newnham, I.E., Nyholm, R.S.: J. Chem. Soc. (1962) 2036.
62M Murray, R.B.: Phys. Rev. **128** (1962) 1570.
62P Peter, M., Moriya, T.: J. Appl. Phys. Suppl. **33** (1962) 1304.
62R Rüdorff, W., Kändler, J., Babel, D.: Z. Anorg. Allgem. Chem. **317** (1962) 261.
62S Stout, J.W., Chisholm, R.C.: J. Chem. Phys. **36** (1962) 979.
63F Ferrari, A., Braibanti, A., Bigliardi, G.: Acta Crystallogr. **16** (1963) 846.
63R Rüdorff, W., Lincke, G., Babel, D.: Z. Anorg. Allgem. Chem. **320** (1963) 150.
63S Schelleng, J.H., Friedberg, S.A.: J. Appl. Phys. **34** (1963) 1017.
63T1 Tsujikawa, I., Kanda, E.: J. Phys. Radium **20** (1959) 352; J. Phys. Soc. Jpn. **18** (1963) 1382.
63T2 Trapp, C., Stout, J.W.: Phys. Rev. Lett. **10** (1963) 157.
64K Kobayashi, H., Haseda, T.: J. Phys. Soc. Jpn. **19** (1964) 765.
64N1 Narath, A.: J. Phys. Soc. Jpn. **19** (1964) 2244.
64N2 Narath, A.: Phys. Rev. **136** (1964) A766.
64N3 Narath, A., Barham, D.C.: Bull. Amer. Phys. Soc. **9** (1964) 112.
64S1 Shinoda, T., Chihara, H., Seki, S.: J. Phys. Soc. Jpn. **19** (1964) 1637.
64S2 Schriempf, J.T., Friedberg, S.A.: Phys. Rev. **136** (1964) A518.
64Z Zalkin, A., Forrester, J.D., Templeton, D.H.: Inorg. Chem. **3** (1964) 529.
65C1 Cox, D.E., Frazer, B.C., Shirane, G.: Phys. Lett. **17** (1965) 103.
65C2 Cooke, A.H., Gehring, K.A., Lazenby, R.: Proc. Phys. Soc. **85** (1965) 967.
65E Earnshaw, A., Larkworthy, L.F., Patel, K.S.: J. Chem. Soc. (1965) 3267.
65F Fackler, J.P., Holah, D.G.: Inorg. Chem. **4** (1965) 954.
65L Lines, M.E.: Phys. Rev. **137** (1965) A982.
65M Morosin, B., Graeber, E.J.: J. Chem. Phys. **42** (1965) 898.
65N Narath, A.: Phys. Rev. **139** (1965) A1221.
65S Shirane, G., Frazer, B.C., Friedberg, S.A.: Phys. Lett. **17** (1965) 95.
65T1 Teaney, D.T.: Phys. Rev. Lett. **14** (1965) 898.
65T2 Tazawa, S., Nagata, K., Date, M.: J. Phys. Soc. Jpn. **20** (1965) 181.
66C1 Cox, D.E., Shirane, G., Frazer, B.C., Narath, A.: J. Appl. Phys. **37** (1966) 1126.
66C2 Clay, R.M., Staveley, L.A.K.: Proc. Low Temp. Calorimetry Conf. Helsinki, 1966 (Ann. Acad. Sci. Fennicae A, VI, 210, p. 194).
66M Morosin, B.: J. Chem. Phys. **44** (1966) 252.
66S Stout, J.W., Boo, W.O.J.: J. Appl. Phys. **37** (1966) 966.
67J Jacobs, I.S., Lawrence, P.E.: Phys. Rev. **164** (1967) 866.
67K Kleinberg, R.: J. Appl. Phys. **38** (1967) 1453.
67M1 Morosin, B.: J. Chem. Phys. **47** (1967) 417.
67M2 Morosin, B.: Acta Crystallogr. **23** (1967) 630.
67R Rives, J.E.: Phys. Rev. **162** (1967) 491.
67S1 Schmidt, V.A., Friedberg, S.A.: J. Appl. Phys. **38** (1967) 5319.
67S2 Stout, J.W., Lau, H.Y.: J. Appl. Phys. **38** (1967) 1472.
67U Umebayashi, H., Shirane, G., Frazer, B.C., Cox, D.E.: J. Appl. Phys. **38** (1967) 1461.
68M Martel, P., Cowley, R.A., Stevenson, R.W.H.: Can. J. Phys. **46** (1968) 1355.
68S1 Seifert, H.J., Loh, H.W., Jungniekel, K.: Z. Anorg. Allgem. Chem. **360** (1968) 62.
68S2 Seifert, H.J., Auel, T.: Z. Anorg. Allgem. Chem. **360** (1968) 50.
69B Butterworth, G.J., Zidell, V.S.: J. Appl. Phys. **40** (1969) 1033.
69L Lau, H.Y., Stout, J.W., Koehler, W.C., Child, H.R.: J. Appl. Phys. **40** (1969) 1136.
69N Nikotin, O., Lindgård, P.A., Dietrich, O.W.: J. Phys. **C2** (1969) 1168.
69R Reichert, T.A., Giauque, W.F.: J. Chem. Phys. **50** (1969) 4205.

69T Torrance, Jr. J.B., Tinkham, M.: Phys. Rev. **187** (1969) 595.

70H1 Hutchings, M.T., Rainford, B.D., Guggenheim, H.J.: J. Phys. C**3** (1970) 307.

70H2 Hay, K.A., Torrance, Jr. J.B.: Phys. Rev. B**2** (1970) 746.

70H3 Hutchings, M.T., Thorpe, M.F., Birgeneau, R.J., Fleury, P.A., Guggenheim, H.J.: Phys. Rev. B**2** (1970) 1362.

70J Johnson, W.L., Reese, W.: Phys. Rev. B**2** (1970) 1355.

70K Kleinberg, R.: J. Chem. Phys. **53** (1970) 2660.

71D De Gunzbourg, J., Papassimacopoulos, S., Miedan-Gros, A., Allain, Y.: J. Phys. Colloq. C1 **32** (1971) C1-125.

71L Lowe, M.A., Abeledo, C.R., Misetich, A.A.: Phys. Lett. **37**A (1971) 274.

71S Schulhof, M.P., Nathans, R., Heller, P., Linz, A.: Phys. Rev. B**4** (1971) 2254.

72B1 Birgeneau, R.J., Yelon, W.B., Cohen, E., Makovsky, J.: Phys. Rev. B**5** (1972) 2607.

72B2 Bongaarts, A.L.M., Van Laar, B., Botterman, A.C., De Jonge, W.J.M.: Phys. Lett. **41**A (1972) 411.

72L Lanusse, M.C., Carrara, P., Fert, A.R., Mischler, G., Redoulès, J.P.: J. Phys. **33** (1972) 429.

72P Polgar, L.G., Herweijer, A., De Jonge, W.J.M.: Phys. Rev. B**5** (1972) 1957.

72R Raquet, C.A., Friedberg, S.A.: Phys. Rev. B**6** (1972) 4301.

73B1 Bragg, E.E., Seehra, M.S.: Phys. Rev. **B7** (1973) 4197.

73B2 Borovik-Romanov, A.S., Bazhan, A.N., Kreines, N.M.: Zh. Eksp. Teor. Fiz. **64** (1973) 1367; Sov. Phys. JETP **37** (1973) 695.

73B3 Bazahn, A.N.: Zh. Eksp. Teor. Fiz. **65** (1973) 2479; Sov. Phys. JETP **38** (1974) 1238.

73C Castilla Becerra, C., Paduan Filho, A.: Phys. Lett. **44**A (1973) 13.

73F1 Fert, A.R., Carrara, P., Lanusse, M.C., Mischler, G., Redoules, J.P.: J. Phys. Chem. Solids **34** (1973) 223.

73F2 Fert, A.R., Gelard, J., Carrara, P.: Solid State Commun. **13** (1973) 1219.

73H Hamburger, A.I., Friedberg, S.A.: Physica **69** (1973) 67.

73K Katsumata, K., Yamasaka, K.: J. Phys. Soc. Jpn. **34** (1973) 346.

73M1 McElearny, J.N., Losee, D.B., Merchant, S., Carlin, R.L.: Phys. Rev. **B7** (1973) 3314.

73M2 Metselaar, J.W., De Klerk, D.: Physica **65** (1973) 208; and **63** (1973) 191.

73S Schneider, W., Weitzel, H.: Solid State Commun. **13** (1973) 303.

74B Bertrand, Y., Fert, A.R., Gélard, J.: J. Phys. **35** (1974) 385.

74D De Jongh, L.J., Betts, D.D., Austen, D.J.: Solid State Commun. **15** (1974) 1711.

74G Gelard, J., Fert, A.R., Meriel, P., Allain, Y.: Solid State Commun. **14** (1974) 187.

74P Paduan Filho, A., Becerra, C.C., Oliveira, Jr., N.F.: Phys. Lett. A**50** (1974) 51.

75K Katsumata, K.: J. Phys. Soc. Jpn. **39** (1975) 42.

75L Lindgård, P.A., Birgeneau, R.J., Als-Nielsen, J., Guggenheim, H.J.: J. Phys. C**8** (1975) 1059.

76D Day, P., Dinsdale, A., Krausz, E.R., Robbins, D.J.: J. Phys. C**9** (1976) 2481.

77N Niel, M., Cros, C., Le Flem, G., Pouchard, M., Hagenmuller, P.: Physica **86**–**88**B (1977) 702.

77Z Ziebeck, K.R.A., Escribe, C., Redoules, J.P., Gelard, J.: Solid State Commun. **23** (1977) 867.

78B1 Billerey, D., Terrier, C., Mainard, R., Perrin, M., Hubsch, J.: Phys. Lett. **68**A (1978) 275.

78B2 Billerey, D., Terrier, C.: Phys. Lett. **68**A (1978) 278.

79K Kuindersma, S.R., Haas, C., Sanchez, J.P., AI, R.: Solid State Commun. **30** (1979) 403.

80A Adam, A., Billerey, D., Terrier, C., Mainard, R., Regnault. L.P., Rossat-Mignod, J., Mériel, P.: Solid State Commun. **35** (1980) 1.

80Y Yoshizawa, H., Ubukoshi, K., Hirakawa, K.: J. Phys. Soc. Jpn. **48** (1980) 42.

82R Régnault, L.P., Rossat-Mignod, J., Adam, A., Billerey, D., Terrier, C.: J. Phys. **43** (1982) 1283.

83H1 Hirakawa, K., Kadowaki, H., Ubukoshi, K.: J. Phys. Soc. Jpn. **52** (1983) 1814.

83H2 Hirakawa, K., Ikeda, H., Kadowaki, H., Ubukoshi, K.: J. Phys. Soc. Jpn. **52** (1983) 2882.

83K Katsumata, K., Sugiyama, K., Date, M.: J. Phys. Soc. Jpn. **52** (1983) 3312.

90T Tuchendler, J., Katsumata, K.: Solid State Commun. **74** (1990) 1159.

9.2 $M_{1-x}M'_xX_2$ and $M_{1-x}M'_xX_2 \cdot nH_2O$ compounds (M, M' = 3d element, X = halogen element)

9.2.1 Introduction

9.2.1.1 General remarks

This chapter is devoted to description of static as well as dynamical magnetic and related properties of the random systems $M_{1-x}M'_xX_2$ and $M_{1-x}M'_xX_2 \cdot nH_2O$ compounds. Here, M and M' stand for 3d transition metal ions, whereas X represents halogen ions. When M or M' is non-magnetic ions such as Mg^{2+}, Zn^{2+} and Cd^{2+}, we have randomly diluted magnetic systems.

In section 9.2.2 we present the static magnetic properties of the alloy systems. Generally, first the concentration vs. transition temperature phase diagrams are given, followed by the magnetic data from which the phase diagrams were constructed. Random magnetic systems exhibit novel magnetic phases such as spin-glass and mixed ordering (or oblique antiferromagnetic) phases.

The magnetic susceptibility per mole (χ_m) of the random magnets well above the transition temperature is expressed by

$$\chi_m = C_m/(T - \Theta) ,$$

where T is the absolute temperature, C_m the Curie constant per mole and Θ is the paramagnetic Curie temperature. By fitting the susceptibility data with this formula, one obtains the values of C_m and Θ.

The magnetic phase transition from the paramagnetic to antiferromagnetic or mixed ordering state is of second order. Specific heat measurement on the substances showing these transitions gives the most accurate value for the Néel temperature. Susceptibility and neutron scattering measurements also give the Néel temperature. Spin-glass transition is widely studied with susceptibility and magnetization measurements. The magnetization of spin-glasses strongly depends on the history of the temperature variation, i.e., whether the sample is cooled from the paramagnetic phase in an applied magnetic field or in zero field. Similar behavior is observed in randomly diluted uniaxial antiferromagnets.

The spin structures in the ordered states, except spin-glasses, are directly determined from neutron scattering study. Electron spin resonance (ESR) and nuclear magnetic resonance (NMR) experiments also give information on the spin structure.

In section 9.2.3 we present the dynamical magnetic properties of $M_{1-x}M'_xX_2$ and $M_{1-x}M'_xX_2 \cdot nH_2O$ compounds. The dynamical properties are directly measured by inelastic neutron scattering and ESR. The latter gives information at wave vector zero, though the energy resolution is better than the former. The magnetic excitation energy of the random systems change with the alloying. From this change in the energy, one can obtain information on the exchange interaction and the anisotropy energy in these random systems. One also finds novel magnetic excitations such as localized excitation and fractons in the random magnets.

9.2.1.2 List of frequently used symbols and abbreviations

Symbols

a, b, c	Å	lattice parameters
C_m	$emu\,K\,mol^{-1} = cm^3\,K\,mol^{-1}$	Curie constant per mole
C_{mag}	$J\,K^{-1}\,mol^{-1}, cal\,K^{-1}\,mol^{-1}$	magnetic specific heat (1 cal ≙ 4.184 J)
C_p	$J\,K^{-1}\,mol^{-1}, cal\,K^{-1}\,mol^{-1}$	specific heat (1 cal ≙ 4.184 J)
E_n	meV, cm^{-1}, THz	excitation energy (1 meV ≙ 8.06554 cm^{-1} ≙ 0.241799 THz)
E_0	meV	energy of incident neutron
g_i		spectroscopic splitting factor (i = x, y, z)
$\boldsymbol{H}$	Oe, Am^{-1}	applied magnetic field (mostly given as $\mu_0 H$ in Tesla)
H_A		anisotropy field
H_E		exchange field
H_{hf}		hyperfine field
H_S		saturation field
h, k, 1		reciprocal lattice indices
i, j, k		site indices in real space
I_n		neutron scattering intensity (peak or integrated value)
I_N		NMR intensity
I_X		X-ray scattering intensity
I_γ		γ-ray absorption intensity
k		Boltzmann constant
l		length
M_v	G	magnetic moment per unit volume = magnetization
Δn		birefringence
p_A	μ_B	magnetic moment per atom (ion)
$\boldsymbol{S}$		spin operator
T	K	temperature
$T_{H,L}$	K	upper, lower transition temperature
T_N	K	Néel temperature
v_D	$mm\,s^{-1}, cm\,s^{-1}$	Doppler velocity
x		concentration
α, β, γ	deg	lattice parameters
ζ		reduced wavevector coordinate
Θ	K	paramagnetic Curie temperature
θ	$deg\,cm^{-1}$	Faraday rotation
$\kappa_{T,L}$		transverse, longitudinal inverse correlation length in reciprocal lattice units (r.l.u.)
λ	Å	neutron wavelength
μ_B	$J\,T^{-1}$	Bohr magneton
ν	Hz	frequency
ν_E	GHz	ESR frequency
ν_N	MHz	NMR frequency

σ	$\text{emu}\,g^{-1} = G\,cm^3\,g^{-1}$	magnetic moment per gram (specific magnetization)
σ_m	$\text{emu}\,mol^{-1} = G\,cm^3\,mol^{-1}$	magnetic moment per mole
χ_g	$\text{emu}\,g^{-1} = cm^3\,g^{-1}$	magnetic susceptibility per gram
χ_m	$\text{emu}\,mol^{-1} = cm^3\,mol^{-1}$	magnetic susceptibility per mole
$\chi_{\parallel}, \chi_{\perp}$		magnetic susceptibility parallel and perpendicular to a principal crystallographic direction
χ', χ''		real, imaginary part of ac magnetic susceptibility

Abbreviations

AF	antiferromagnetic phase
ESR	electron spin resonance
F	ferromagnetic phase
FC	field cooled state (or sample)
FW	field warmed state (or sample)
FWHM	full width at half maximum
Heli	helimagnetic phase
LRZ	Lorentzian line shape
LSQ	Lorentzian squared line shape
NMR	nuclear magnetic resonance
P	paramagnetic phase
SF	spin flop phase
SG	spin glass phase
ZFC	zero field cooled state (or sample)

9.2.2 Static magnetic and related properties of $M_{1-x}M'_xX_2$ and $M_{1-x}M'_xX_2 \cdot nH_2O$ compounds

9.2.2.1 Static magnetic and related properties of $M_{1-x}M'_xF_2$ compounds

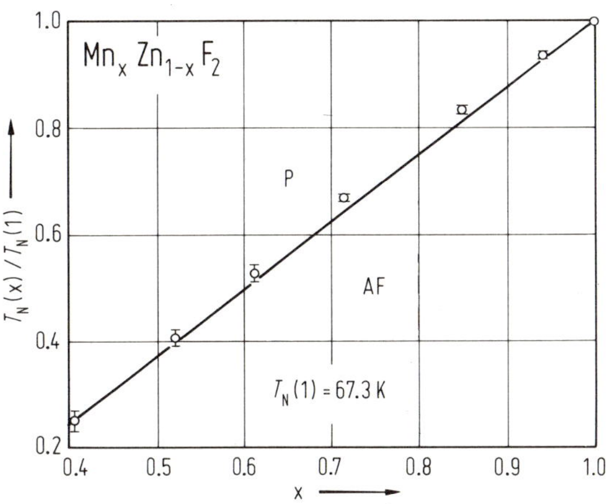

Fig. 1. $Mn_xZn_{1-x}F_2$. Concentration vs. transition temperature phase diagram determined from the linear magnetic birefringence measurement [80B1].

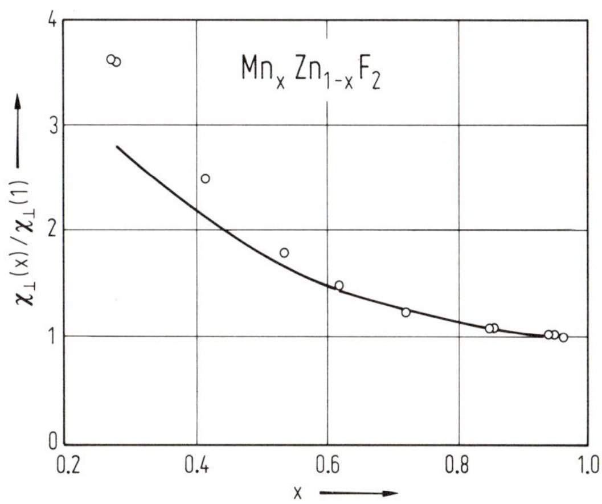

Fig. 2. $Mn_xZn_{1-x}F_2$. Concentration vs. perpendicular susceptibility measured at 4.2 K by a Faraday balance in fields between 0.5 and 1.0 T. The solid curve represents the result of computer simulation [81K2].

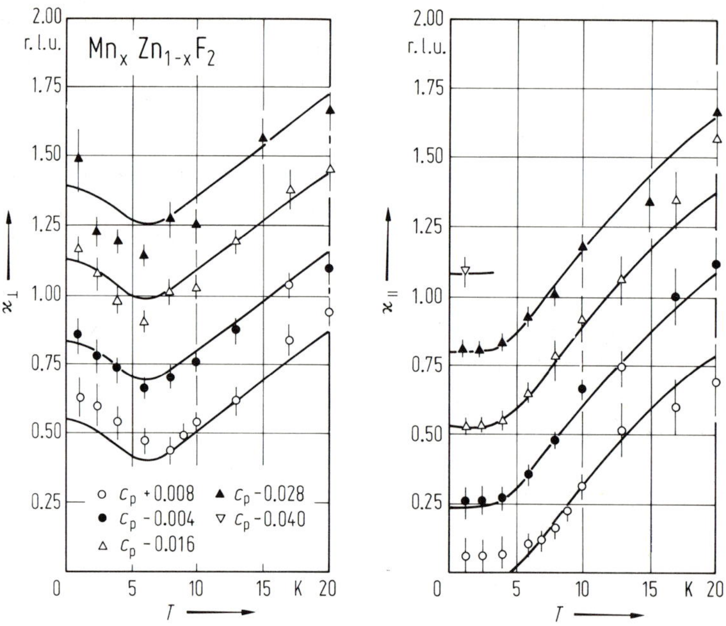

Fig. 3. $Mn_xZn_{1-x}F_2$. Temperature vs. perpendicular (left) and parallel (right) inverse correlation lengths near the percolation concentration c_p (x = 0.245) [80C].

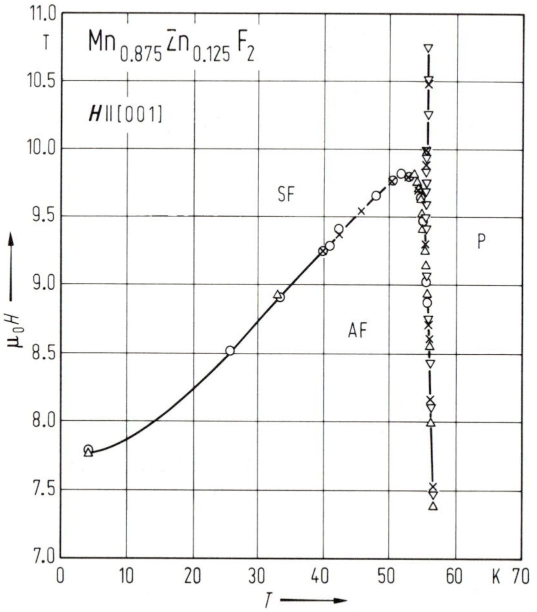

Fig. 4. $Mn_{0.875}Zn_{0.125}F_2$. Temperature vs. applied magnetic field phase diagram. Different symbols are from different types of measurement [83S2].

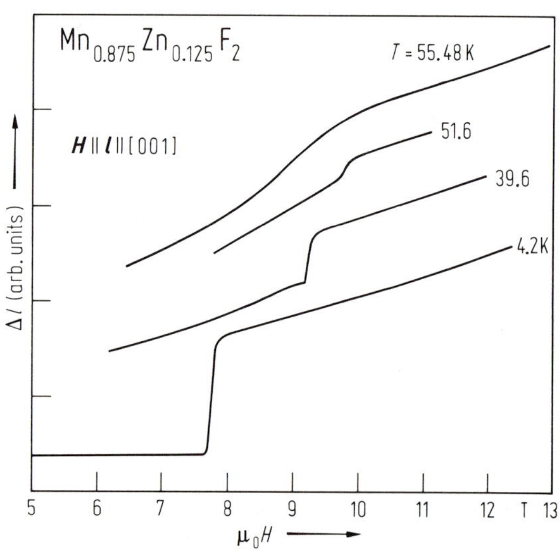

Fig. 5. $Mn_{0.875}Zn_{0.125}F_2$. Magnetostriction vs. applied magnetic field at several temperatures [83S2].

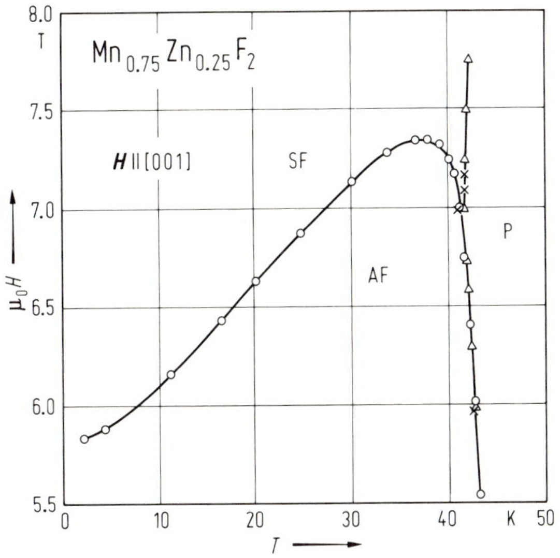

Fig. 6. $Mn_{0.75}Zn_{0.25}F_2$. Temperature vs. applied magnetic field phase diagram. ○: from magnetostriction, △, ×: from thermal expansion [84S].

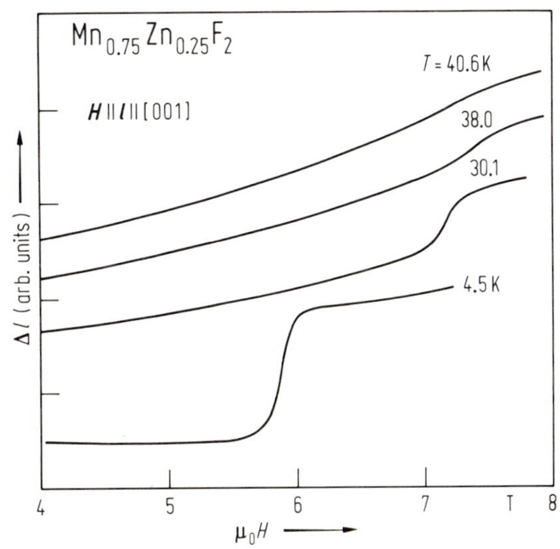

Fig. 7. $Mn_{0.75}Zn_{0.25}F_2$. Magnetostriction vs. applied magnetic field at several temperatures [84S].

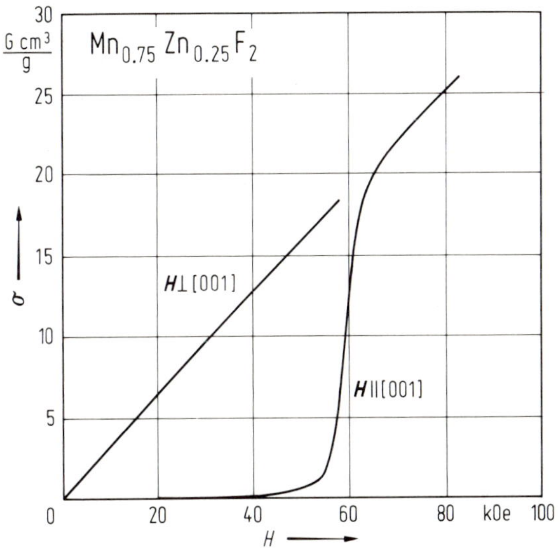

Fig. 8. $Mn_{0.75}Zn_{0.25}F_2$. Magnetic moment per gram vs. applied magnetic field at 4.2 K [84S].

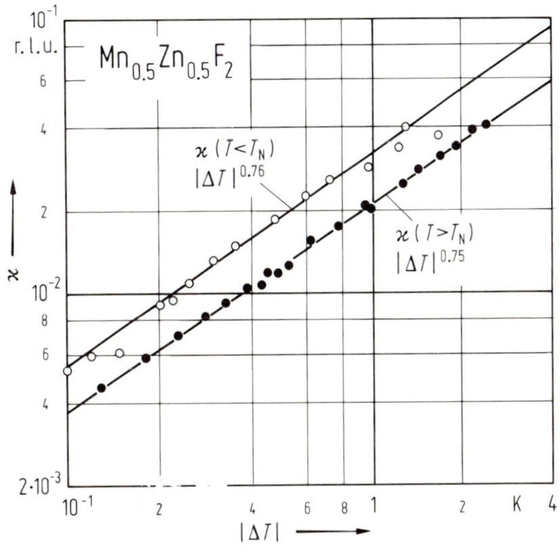

Fig. 10. $Mn_{0.5}Zn_{0.5}F_2$. Inverse correlation length vs. temperature difference from the Néel temperature ($T_N = 21.0$ K) [86M2].

For Fig. 9 see next page.

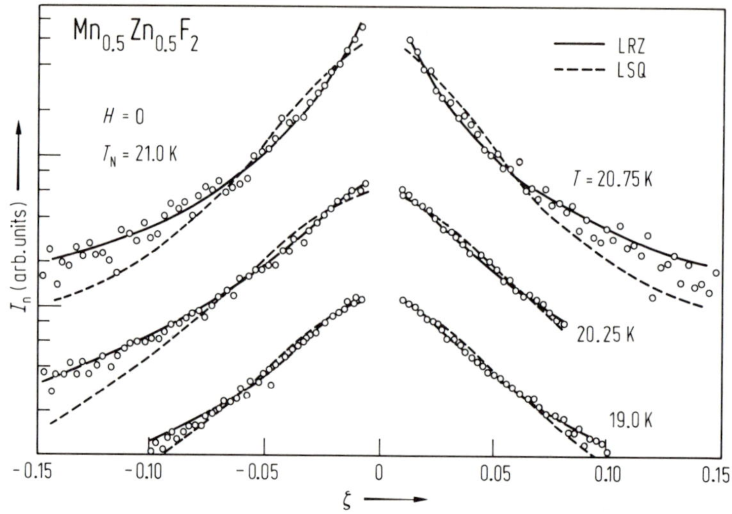

Fig. 9. $Mn_{0.5}Zn_{0.5}F_2$. Transverse scans through (1, 0, 0) Bragg reflection. LRZ: Lorentzian line shape, LSQ: Lorentzian squared line shape [86M2].

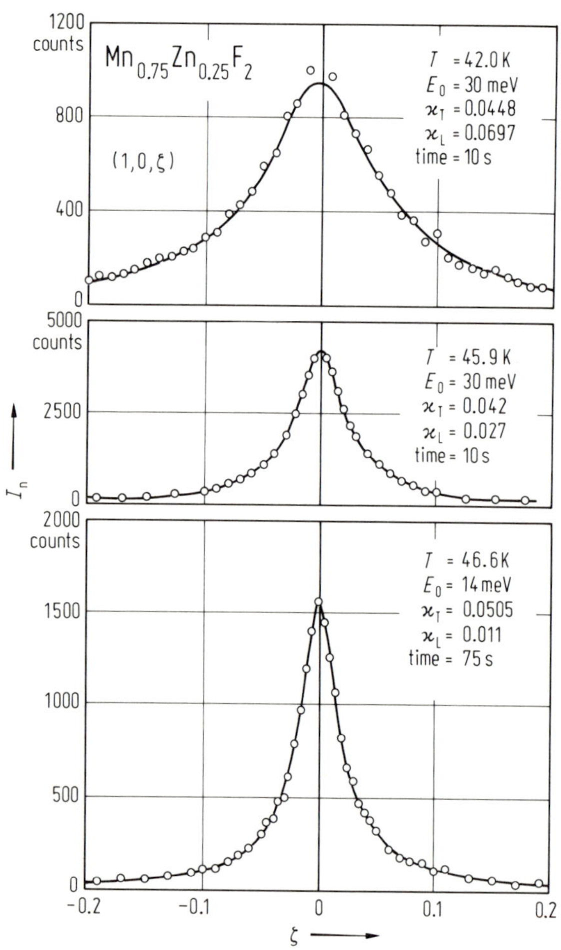

Fig. 11. $Mn_{0.75}Zn_{0.25}F_2$. Transverse scans through the (1,0,0) Bragg reflection. E_0: energy of incident neutrons, $\kappa_{T,L}$ in r.l.u. [86M2].

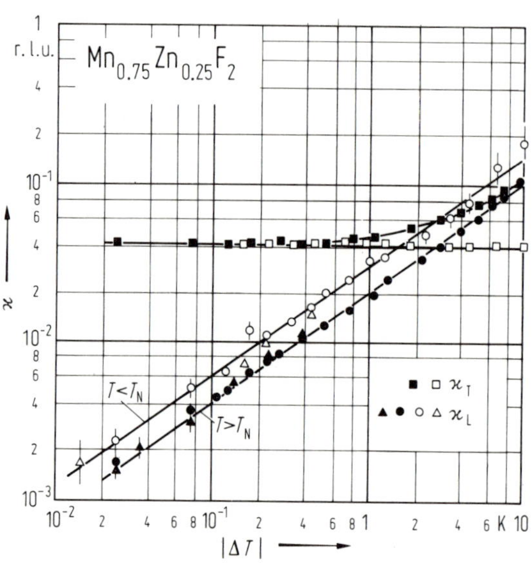

Fig. 12. $Mn_{0.75}Zn_{0.25}F_2$. Transverse and longitudinal inverse correlation lengths (in r.l.u.) vs. temperature difference from the Néel temperature ($T_N = 46.2(1)$ K). Solid points above T_N, open symbols below T_N. △ and ▲ are points from $E_0 = 14$ meV, others from 30.5 meV [86M2].

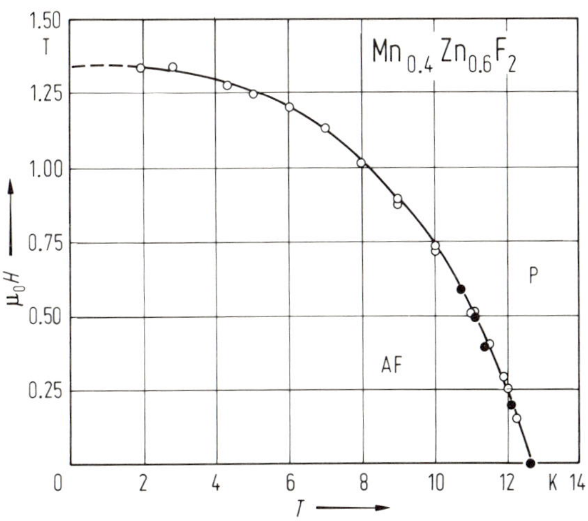

Fig. 13. $Mn_{0.4}Zn_{0.6}F_2$. Temperature vs. applied magnetic field phase diagram obtained from the birefringence measurements. ● from $\frac{d(\Delta n)}{dT}$ vs. T; ○ from $\frac{d(\Delta n)}{dH}$ vs. H [88R].

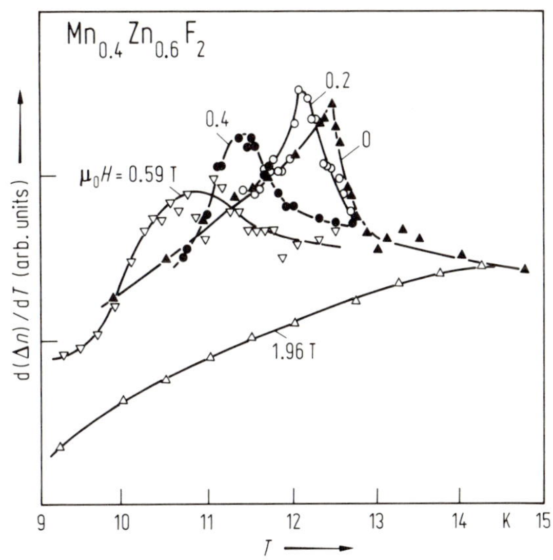

Fig. 14. $Mn_{0.4}Zn_{0.6}F_2$. Derivative of the optical birefringence vs. temperature in various applied magnetic fields [88R].

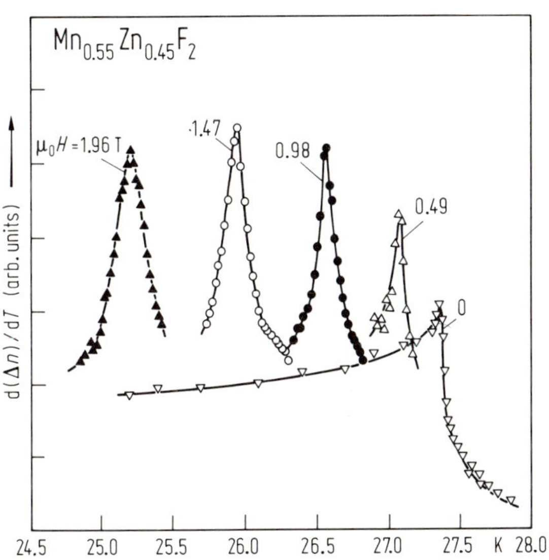

Fig. 15. $Mn_{0.55}Zn_{0.45}F_2$. Derivative of the optical birefringence vs. temperature in various applied magnetic fields [88R].

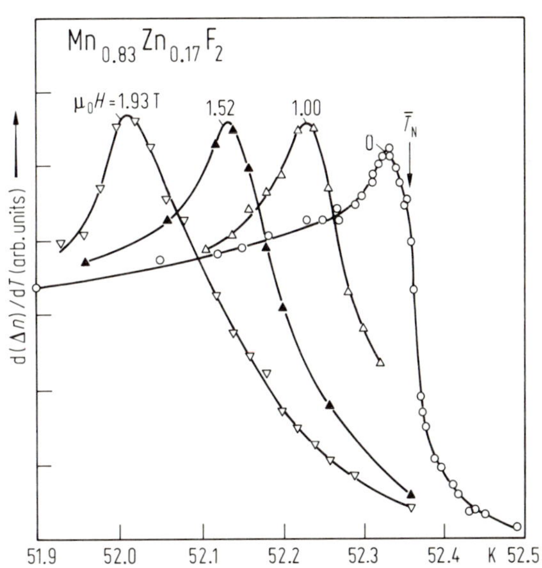

Fig. 16. $Mn_{0.83}Zn_{0.17}F_2$. Derivative of the optical birefringence vs. temperature in various applied magnetic fields. $\bar{T}_N$: average transition temperature [88R].

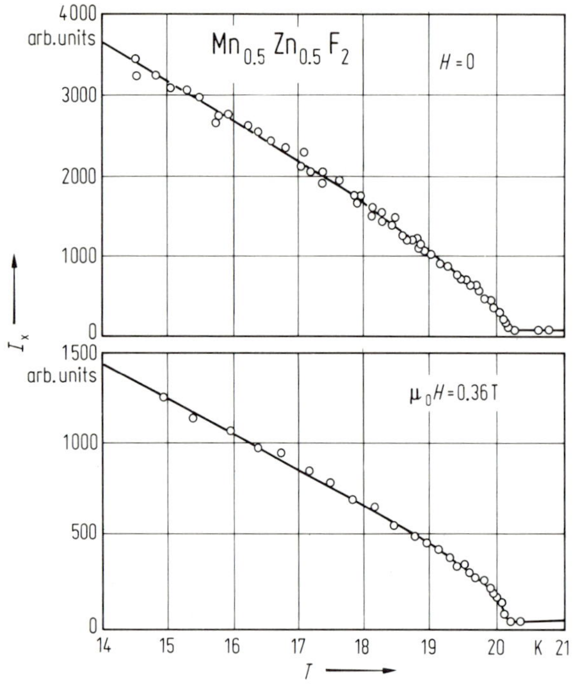

Fig. 17. $Mn_{0.5}Zn_{0.5}F_2$. Integrated intensity of the (1, 0, 0) Bragg point vs. temperature measured by the synchrotron magnetic X-ray scattering technique [88T].

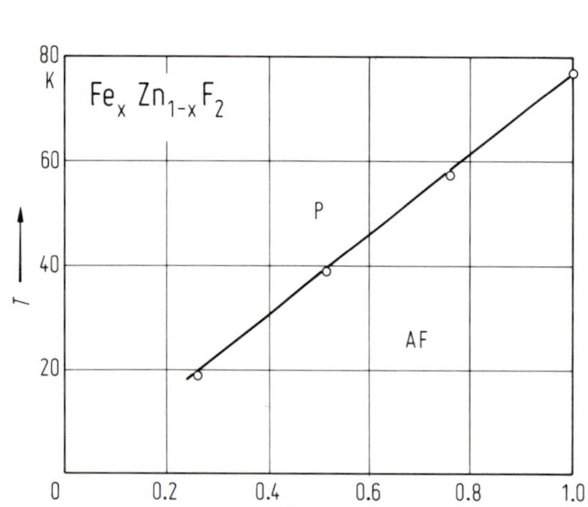

Fig. 18. $Fe_xZn_{1-x}F_2$. Concentration vs. transition temperature phase diagram determined from the Mössbauer spectroscopy [66W].

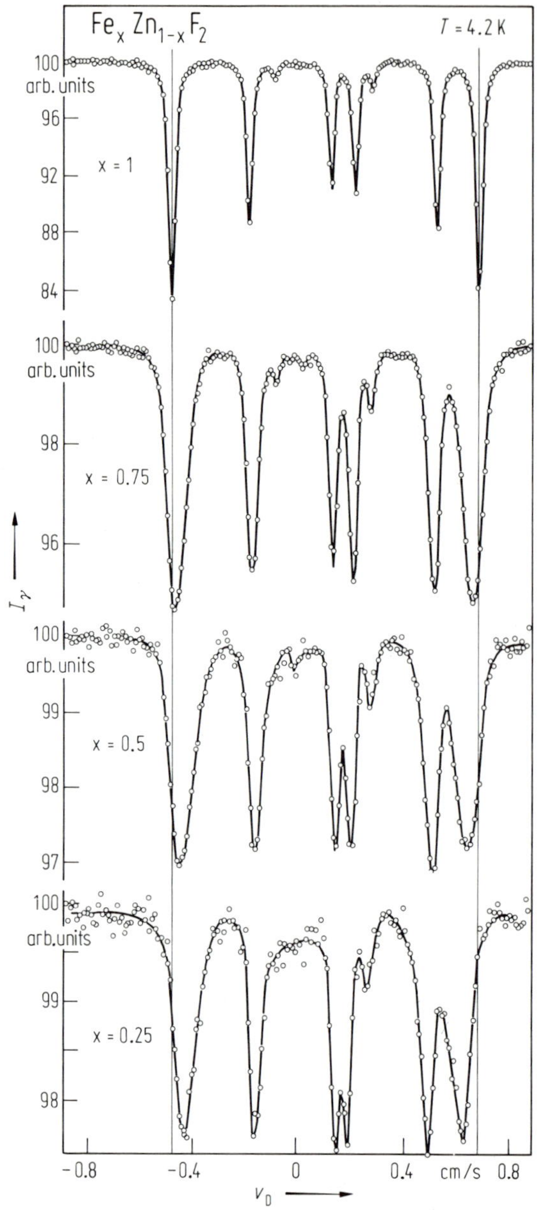

Fig. 19. $Fe_xZn_{1-x}F_2$. Mössbauer-effect absorption spectrum of ^{57}Fe at 4.2 K [66W].

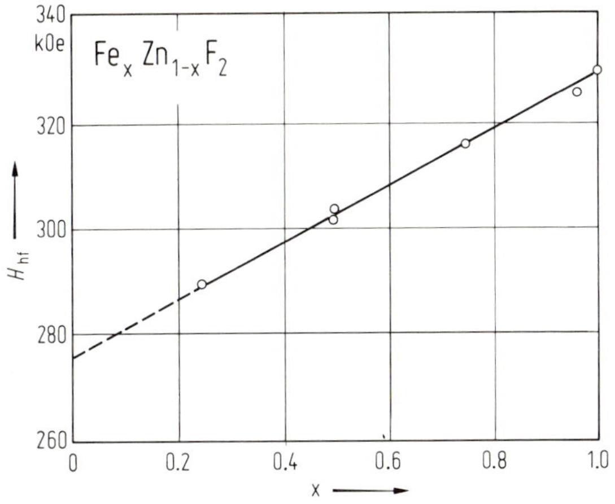

Fig. 20. $Fe_xZn_{1-x}F_2$. The saturation value of the effective magnetic field at ^{57}Fe [66W].

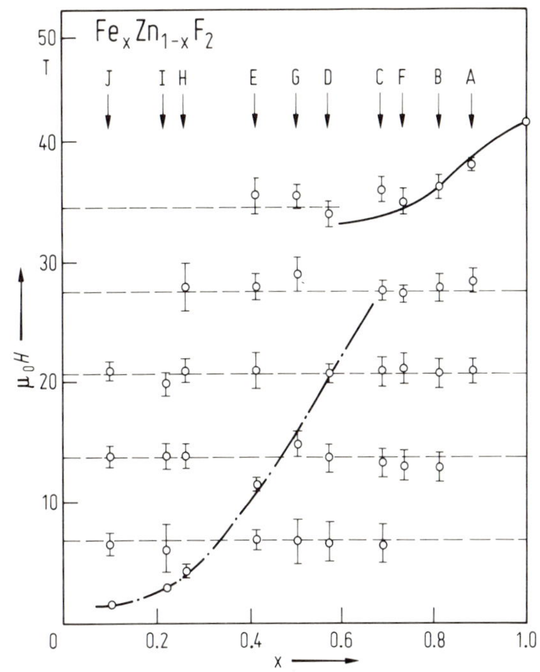

Fig. 21. $Fe_xZn_{1-x}F_2$. dM/dH peak fields vs. concentration obtained from the pulsed field experiment. Horizontal dashed lines are theoretical exchange-flip fields. The solid line is theoretical spin-flop field. The dash-dotted line marks the anomalous peak in (dM/dH) vs. H [81K1].

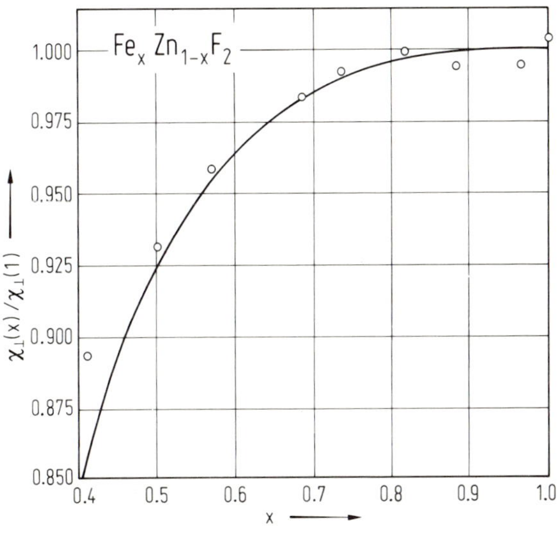

Fig. 22. $Fe_xZn_{1-x}F_2$. Concentration vs. perpendicular susceptibility measured at 4.2 K by a Faraday balance in fields between 0.5 and 1.0 T. The solid curve represents the result of computer simulation [81K2].

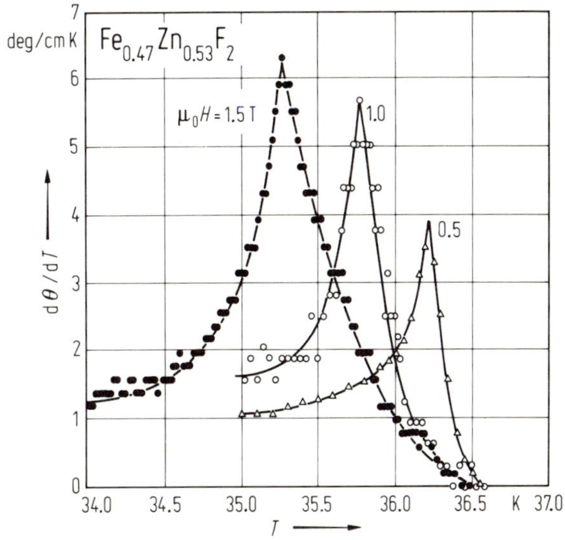

Fig. 23. $Fe_{0.47}Zn_{0.53}F_2$. Faraday rotation $(d\theta/dT)$ vs. temperature for zero field cooled sample measured at the designated applied magnetic fields [86K].

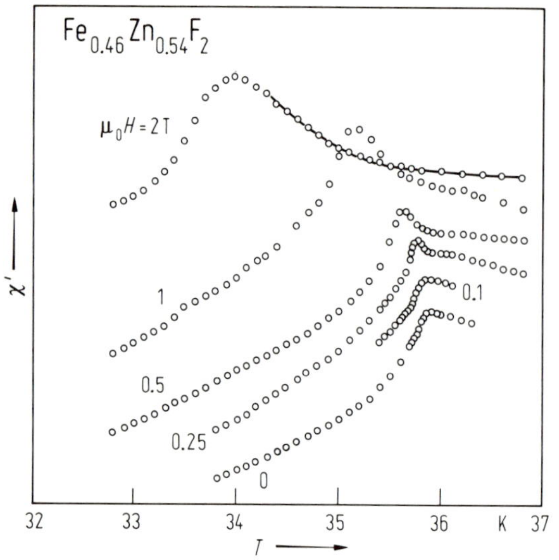

Fig. 24. $Fe_{0.46}Zn_{0.54}F_2$. Real part of ac susceptibility vs. temperature for zero field cooled sample measured at various applied magnetic fields. $\boldsymbol{H} \parallel c$ axis. The curves were displaced vertically to separate them. Solid line is guide for the eye [86M1].

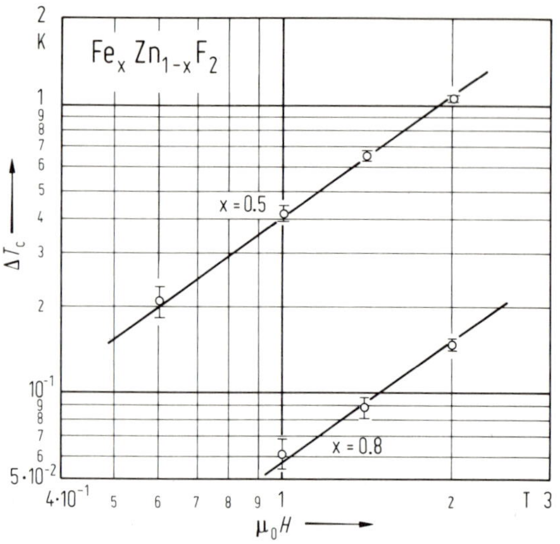

Fig. 26. $Fe_xZn_{1-x}F_2$. Applied magnetic field vs. shift in the ordering temperature obtained from the birefringence measurement. $\Delta T_c = T_N - T_c$ where T_N is Néel temperature of zero field antiferromagnet [82B].

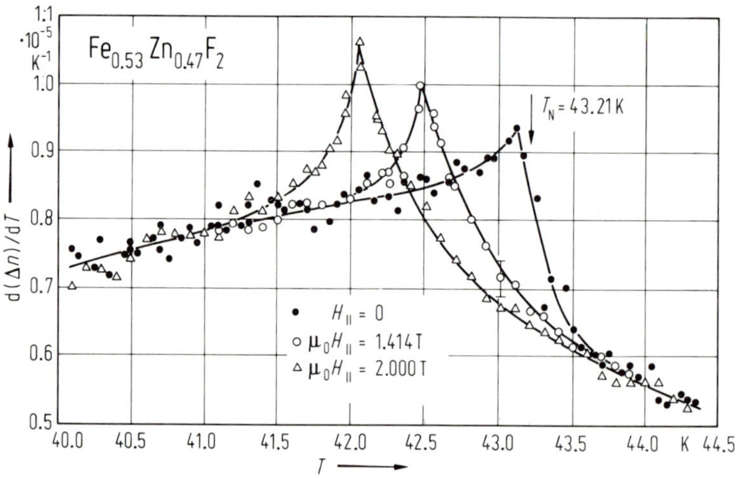

Fig. 25. $Fe_{0.53}Zn_{0.47}F_2$. Temperature derivative of the linear birefringence vs. temperature. $\boldsymbol{H} \parallel c$ axis [82B].

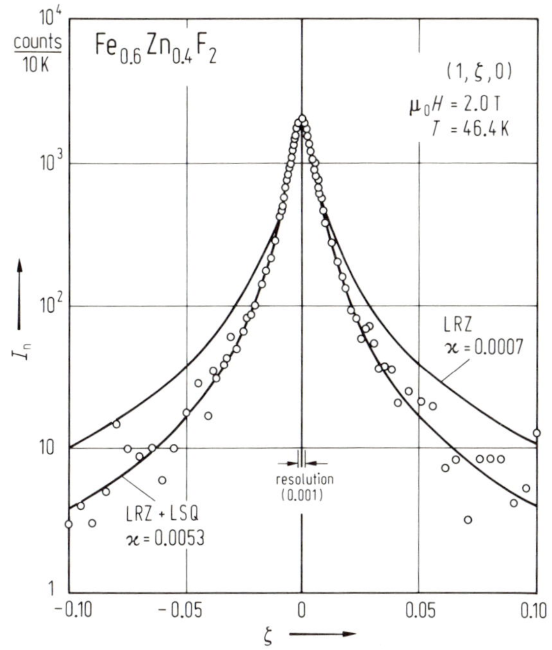

Fig. 27. $Fe_{0.6}Zn_{0.4}F_2$. Transverse scans through the (1,0,0) Bragg reflection with $\mu_0 H = 2.0$ T. LRZ: Lorentzian line shape, LSQ; Lorentzian squared line shape; κ in r.l.u. [85Y].

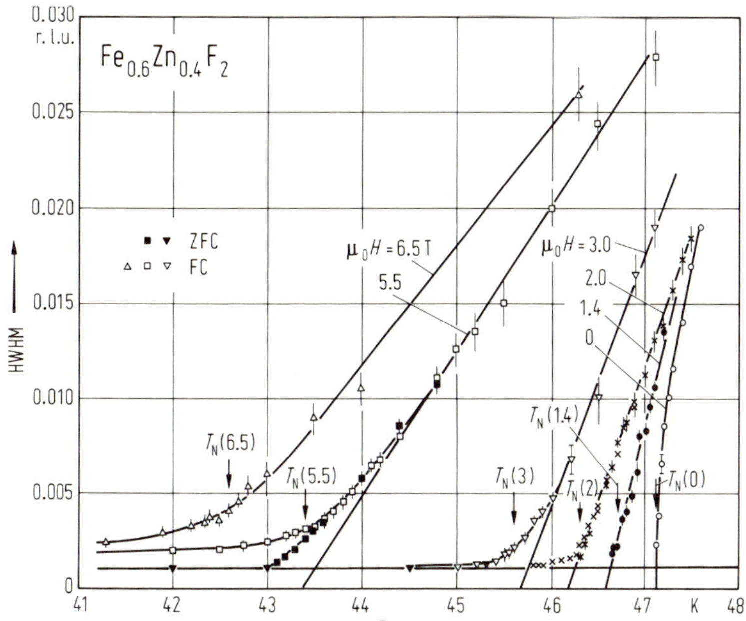

Fig. 28. $Fe_{0.6}Zn_{0.4}F_2$. Measured half width at half maximum (HWHM) of the neutron scattering profile vs. temperature for the designated applied magnetic fields [85Y].

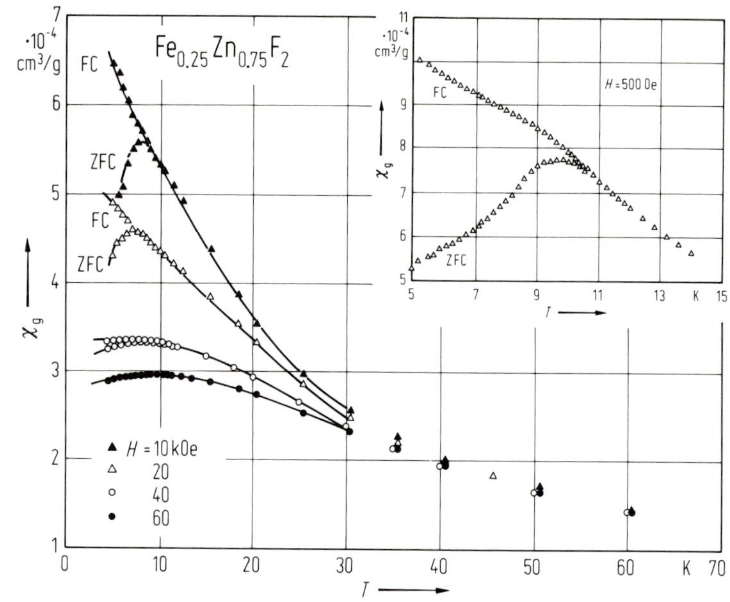

Fig. 29. $Fe_{0.25}Zn_{0.75}F_2$. Zero field cooled and field cooled susceptibilities vs. temperature measured for several applied magnetic fields. Inset shows the zero field and field cooled susceptibilities for H = 500 Oe. $\boldsymbol{H} \parallel c$ axis [88M].

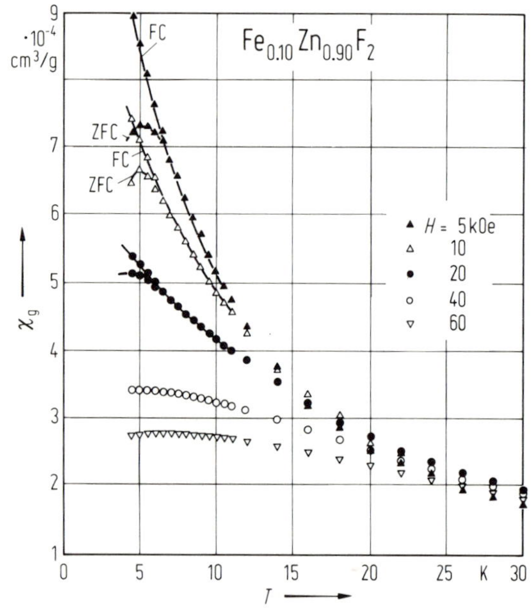

Fig. 30. $Fe_{0.10}Zn_{0.90}F_2$. Zero field cooled and field cooled susceptibilities vs. temperature for several applied magnetic fields. $\boldsymbol{H} \parallel c$ axis [88M].

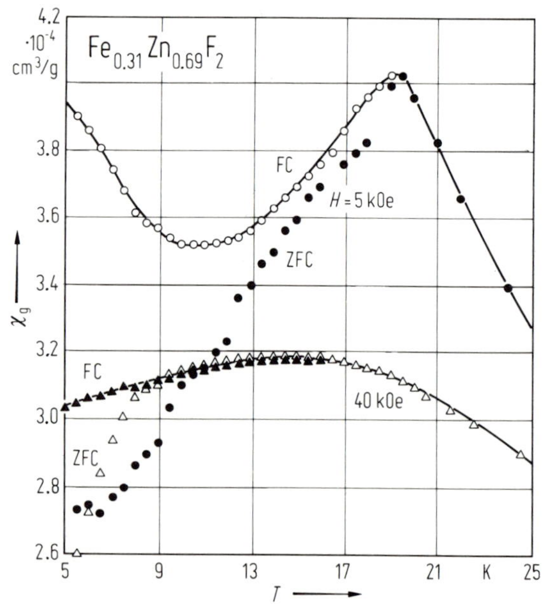

Fig. 31. $Fe_{0.31}Zn_{0.69}F_2$. Zero field cooled and field cooled susceptibilities vs. temperature for two values of the applied magnetic field. $\boldsymbol{H} \parallel c$ axis [88M].

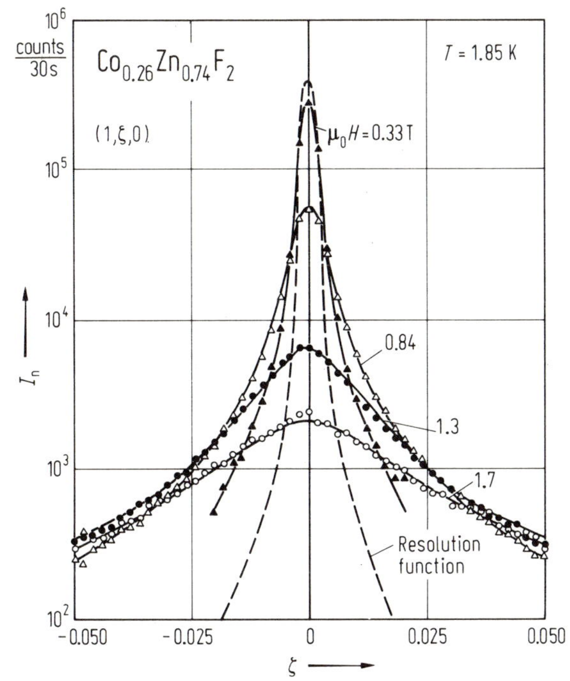

Fig. 32. $Co_{0.26}Zn_{0.74}F_2$. Transverse scans through the (1, 0, 0) Bragg reflection for various applied magnetic fields [83H].

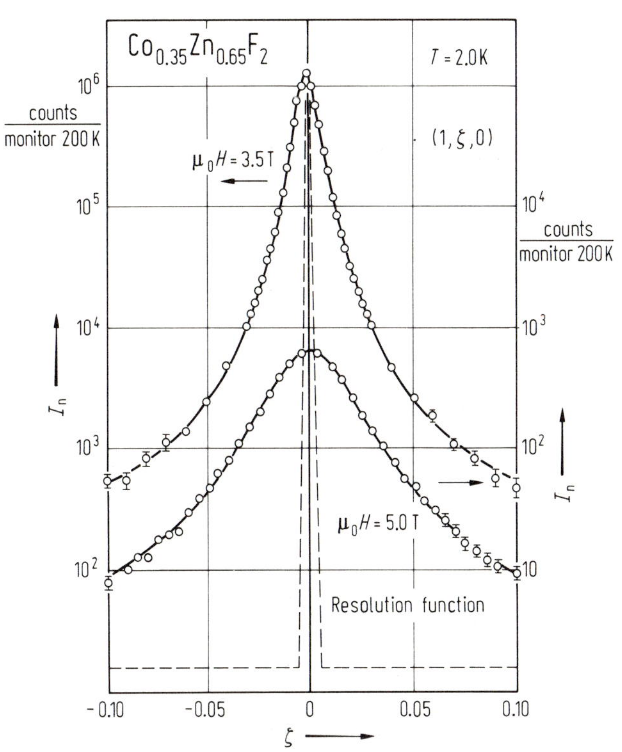

Fig. 33. $Co_{0.35}Zn_{0.65}F_2$. Transverse scans through the (1, 0, 0) Bragg reflection for the designated applied magnetic fields [83H].

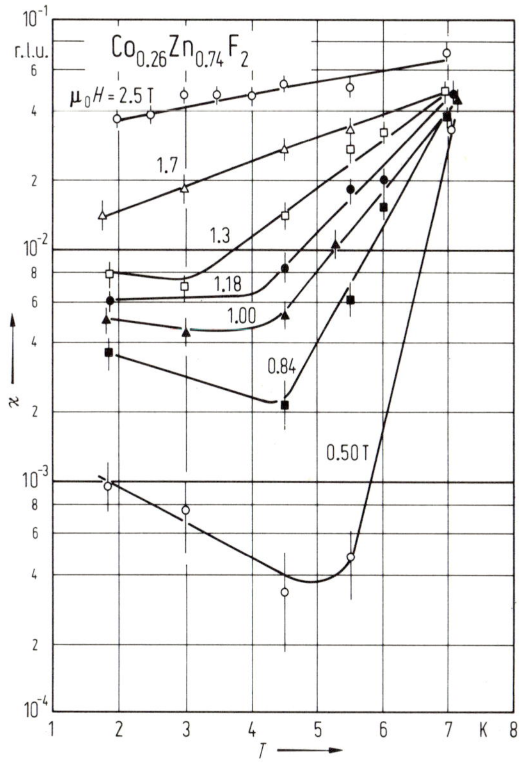

Fig. 34. $Co_{0.26}Zn_{0.74}F_2$. Inverse correlation length vs. temperature for various applied magnetic fields (field cooled case) [83H].

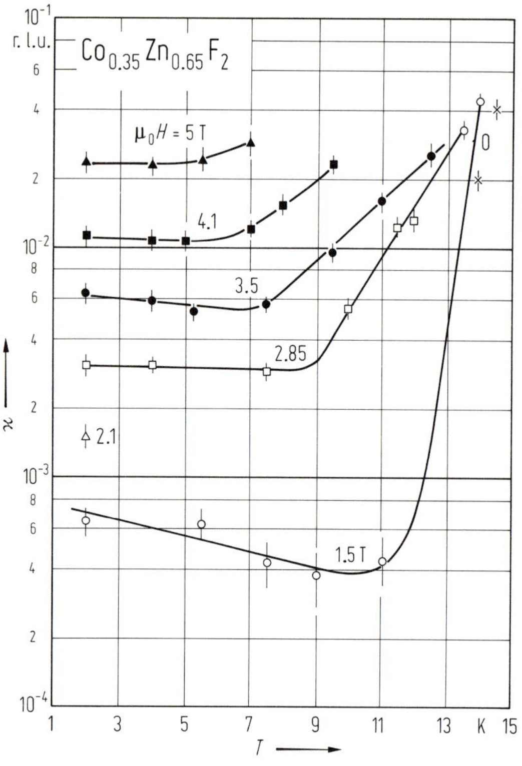

Fig. 35. $Co_{0.35}Zn_{0.65}F_2$. Inverse correlation length vs. temperature for various applied magnetic fields (field cooled case) [83H].

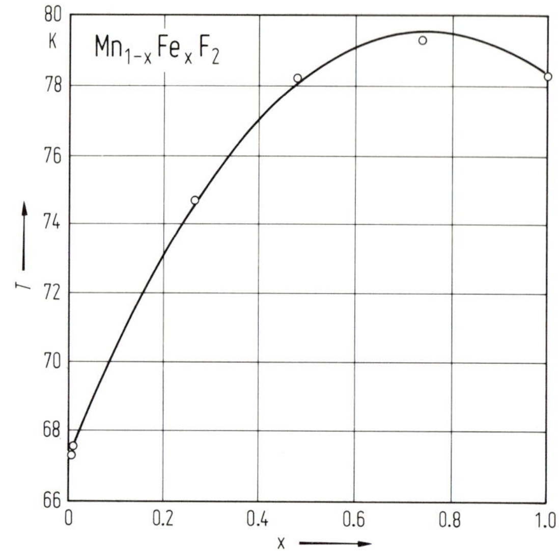

Fig. 36. $Mn_{1-x}Fe_xF_2$. Concentration vs. transition temperature phase diagram determined from the Mössbauer spectroscopy [69W].

9.2.2.2 Static magnetic and related properties of $M_{1-x}M'_xCl_2$ compounds

Table 1. Critical temperatures in $Fe_xMg_{1-x}Cl_2$ [84B].

x	T_N [K]	T_f [K]
1	24.0	
0.964	22.5	
0.900	20.5	
0.807	17.6	
0.716	13.6	
0.610	10.0	2.1
0.552	8.5	3.0
0.495		4.0
0.457		3.5
0.413		3.0
0.286		1.8
0.184		—

The Néel temperatures (T_N) are determined as the position of the susceptibility peak. The freezing temperatures (T_f) correspond to the reversibility temperatures.

Table 2. Néel temperatures of $Fe_{1-x}Cd_xCl_2$ [81B1].

x	T_N [K]
0	24
0.101(10)	20.1
0.213(15)	13
0.322(15)	7
0.369(15)	4
0.522(25)	< 1

The Néel temperatures (T_N) are determined as the position of susceptibility peak.

Table 3. Néel temperatures of $Fe_{1-x}Mn_xCl_2$ [81B1].

x	T_N [K]
0.0015(5)	24
0.039(5)	22.9
0.105(10)	19.7
0.151(10)	17.9
0.251(15)	13.4
0.348(15)	10.3
0.578(30)	2.1
0.757(15)	2
1	1.9

The Néel temperatures(T_N) are determined as the position of susceptibility peak.

Table 4. Tricritical temperatures T_t(x) and fields H_t (x) for $Fe_{1-x}Mn_xCl_2$ [81H].

x (± 0.005)	T_t(x) [K] (± 0.1)	$\mu_0 H_t$(x) [T] (± 0.02)
0.079	16.25	0.80
0.135	13.70	0.795
0.159	13.20	0.755
0.195	12.5	0.76
0.249	9.95	0.725
0.293	< 1.5	

T_t(x) and H_t(x) are determined from field-induced light scattering and Faraday rotation.

Table 5. Macroscopic magnetic properties of $Fe_{1-x}Co_xCl_2$ (from susceptibility measurements) [83W1].

Co concentration	Transition temperature		Paramagnetic Curie temperature		Molar Curie constant [cm^3 K mol^{-1}]	
x (± 0.001)	upper T_H [K] (± 0.1 K)	lower T_L [K] (± 1.0 K)	$\Theta_{\parallel}$ [K]	$\Theta_{\perp}$ [K]	$C_m^{\parallel}$	$C_m^{\perp}$
0.0977	21.4		21.85(34)	9.51(19)	5.91(6)	5.72(3)
0.1625	20.0		19.75(34)	9.66(20)	6.40(6)	6.49(3)
0.2641	16.8	6.0	18.33(35)	11.40(19)	5.72(6)	6.35(3)
0.2863	16.2	12.0	19.30(47)	11.68(21)	5.55(7)	6.81(4)
0.3995	17.2	6.0	15.51(32)	13.69(16)	5.09(4)	6.61(3)
0.4591	18.1		13.31(20)	14.84(19)	4.75(2)	6.25(4)
0.6035	20.1		9.64(24)	15.41(11)	4.07(2)	6.46(2)
0.8011	22.4		1.02(14)	17.20(10)	3.25(1)	6.62(2)

Table 6. Transition temperatures of $Fe_{1-x}Co_xCl_2$ (from neutron scattering) [83W1]. Results from 16 measurements of 6 different samples. Compositions are estimated from atomic absorption analyses made on different parts of each sample.

x	T_H [K]	T_L [K]
0.275	16.40(5)	8.0(5)
0.266	16.80(10)	7.0(5)
0.255	17.35(10)	
0.400	17.20(5)	
0.295	15.60(5)	12.5(5)
0.340	15.70(10)	12.0(10)
0.360	16.10(5)	11.0(5)
0.369	16.35(10)	9.5(5)
0.350	16.15(15)	11.5(5)
0.335	15.75(5)	13.0(5)
0.315	15.30(10)	
0.291	15.95(5)	11.5(5)
0.293	15.70(10)	
0.258	17.10(10)	< 5.0
0.275	16.50(10)	7.5(10)
0.286	15.90(10)	11.0(5)

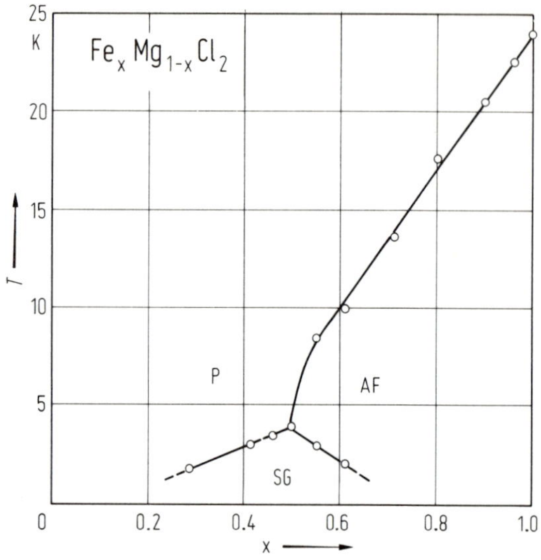

Fig. 37. $Fe_xMg_{1-x}Cl_2$. Concentration vs. transition temperature phase diagram determined from the susceptibility measurements [84B].

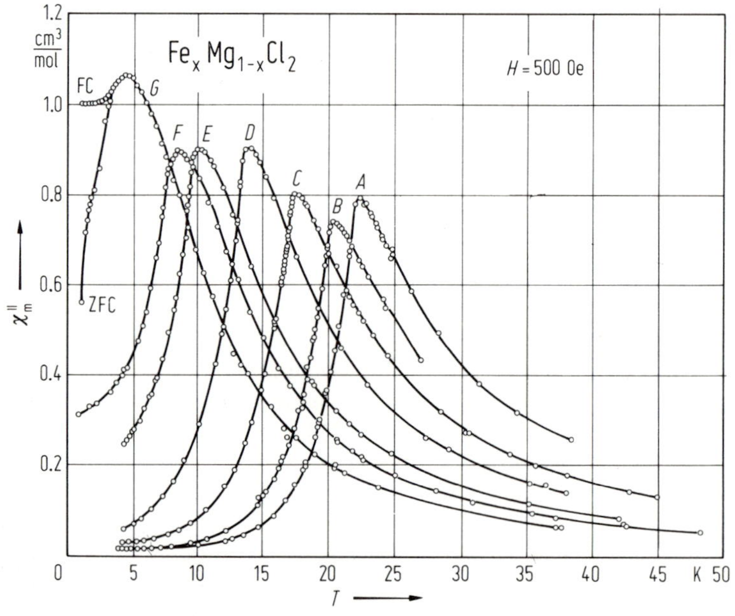

Fig. 38. $Fe_xMg_{1-x}Cl_2$. Susceptibility parallel to the *c* axis vs. temperature for various concentrations. The values of x are *A*: 0.964; *B*: 0.900; *C*: 0.807; *D*: 0.716; *E*: 0.610; *F*: 0.552; *G*: 0.495 [84B].

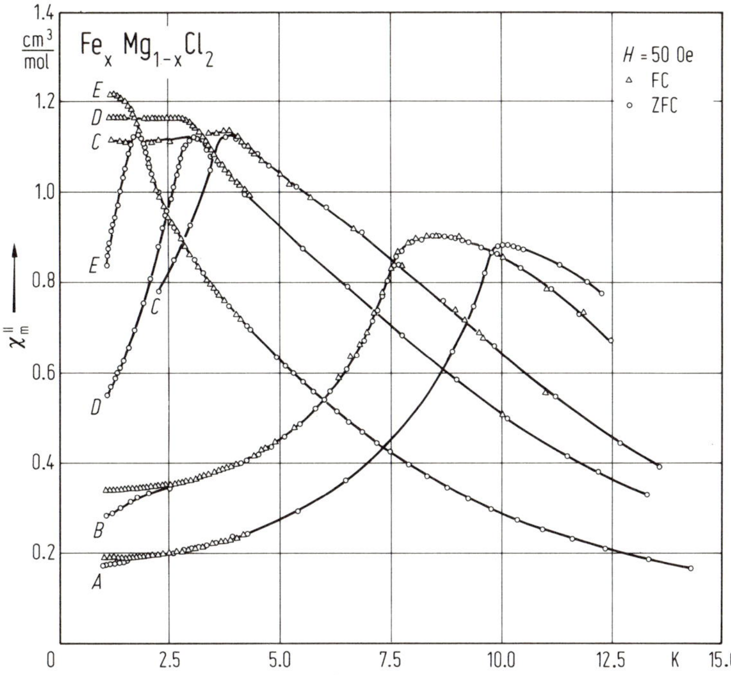

Fig. 39. $Fe_xMg_{1-x}Cl_2$. Susceptibility parallel to the *c* axis vs. temperature for various concentrations. The values of x are *A*: 0.610; *B*: 0.552; *C*: 0.495; *D*: 0.413; *E*: 0.286 [84B].

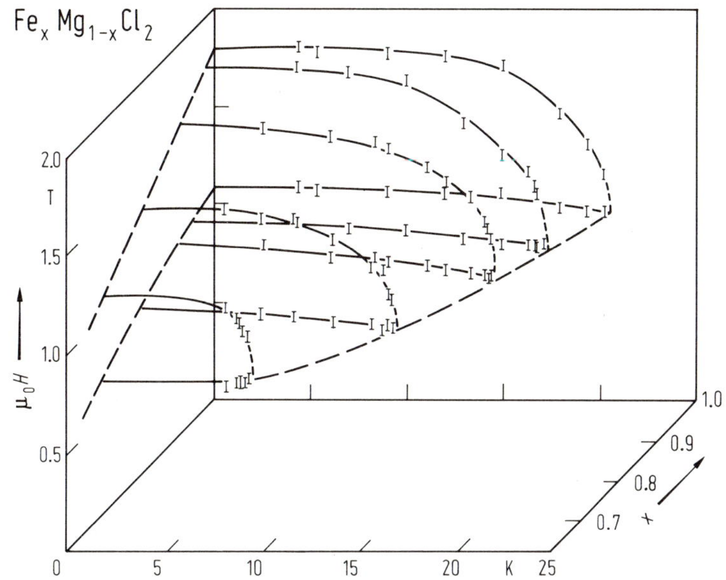

Fig. 40. $Fe_xMg_{1-x}Cl_2$. The mixed phase region of the magnetic phase diagram determined from the optical measurements [77W].

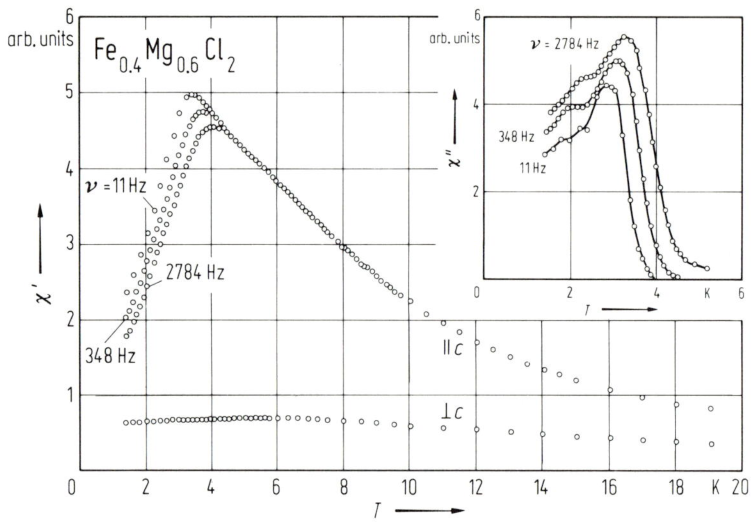

Fig. 41. $Fe_{0.4}Mg_{0.6}Cl_2$. Real part of ac susceptibility vs. temperature parallel and perpendicular to the c axis. The inset shows imaginary part of the ac susceptibility [87W].

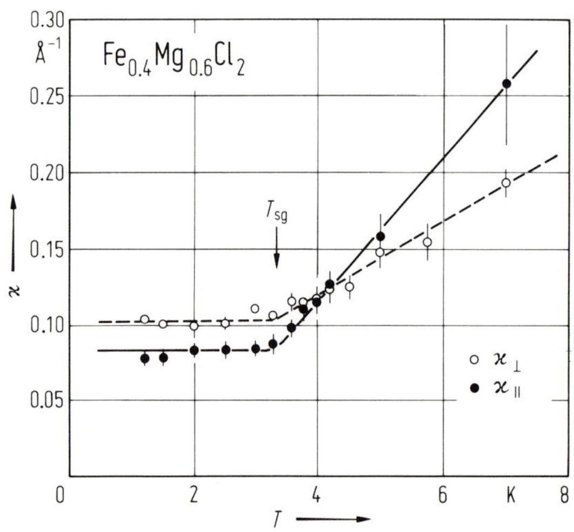

Fig. 42. $Fe_{0.4}Mg_{0.6}Cl_2$. Temperature vs. inverse correlation lengths parallel and perpendicular to the c axis. T_{sg}: spin glass transition temperature [87W].

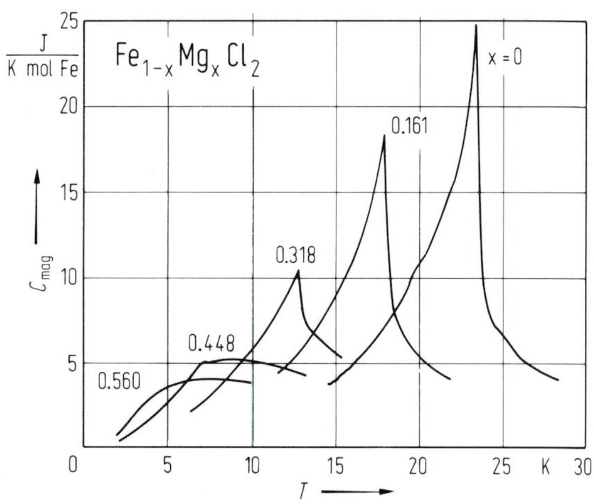

Fig. 43. $Fe_{1-x}Mg_xCl_2$. Magnetic specific heat vs. temperature for five samples with different concentrations [85W].

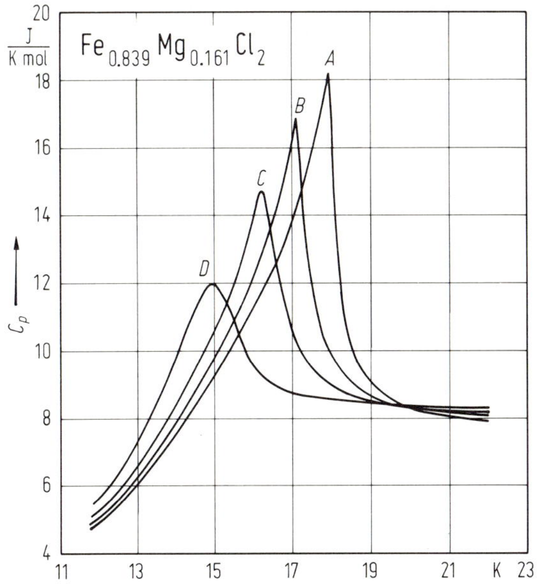

Fig. 44. $Fe_{0.839}Mg_{0.161}Cl_2$. Specific heat vs. temperature at constant applied magnetic fields. *A*: $\mu_0 H = 0$; *B*: 0.480 T; *C*: 0.674 T; *D*: 0.866 T [83W2].

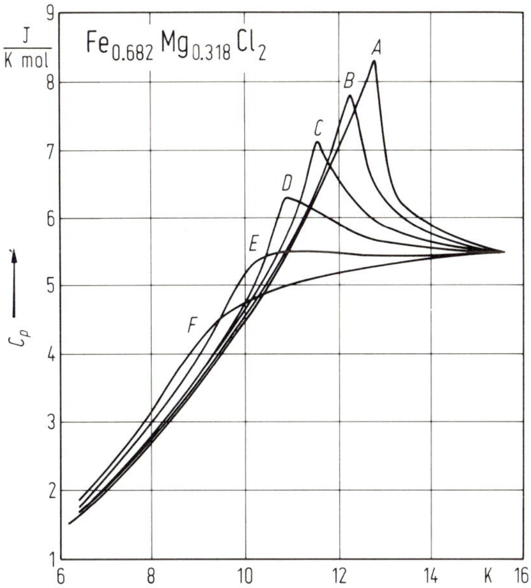

Fig. 45. $Fe_{0.682}Mg_{0.318}Cl_2$. Specific heat vs. temperature at constant applied magnetic fields. *A*: $\mu_0 H = 0$; *B*: 0.211 T; *C*: 0.352 T; *D*: 0.448 T; *E*: 0.544 T; *F*: 0.641 T [83W2].

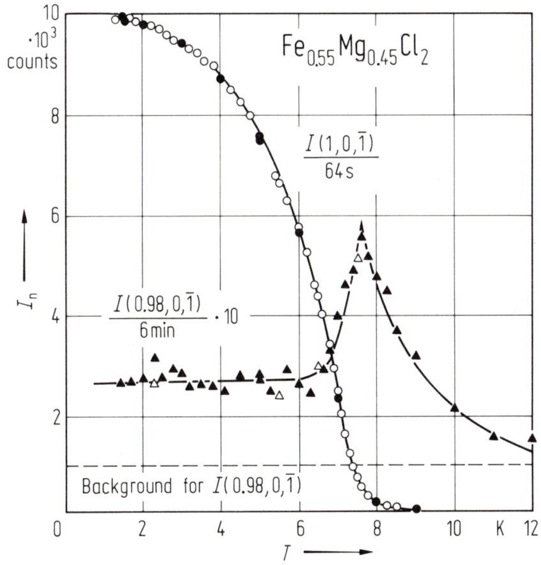

Fig. 46. $Fe_{0.55}Mg_{0.45}Cl_2$. Magnetic Bragg scattering intensity at $(1, 0, \bar{1})$ and diffuse scattering intensity at $(0.98, 0, \bar{1})$ vs. temperature. ● ▲: warming, ○ △: cooling [85W].

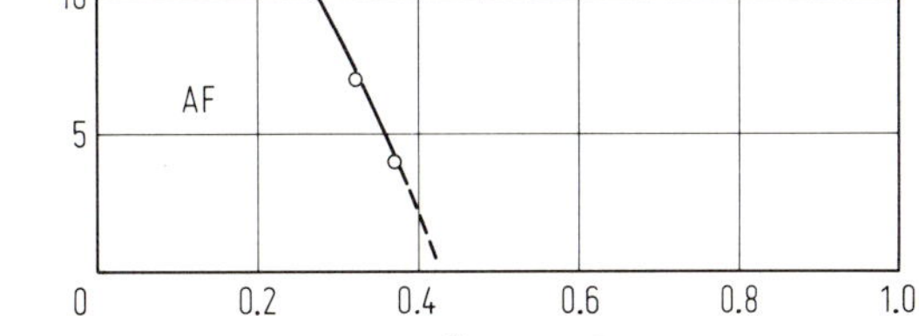

Fig. 47. $Fe_{1-x}Cd_xCl_2$. Concentration vs. transition temperature phase diagram determined from the susceptibility measurements [81B1].

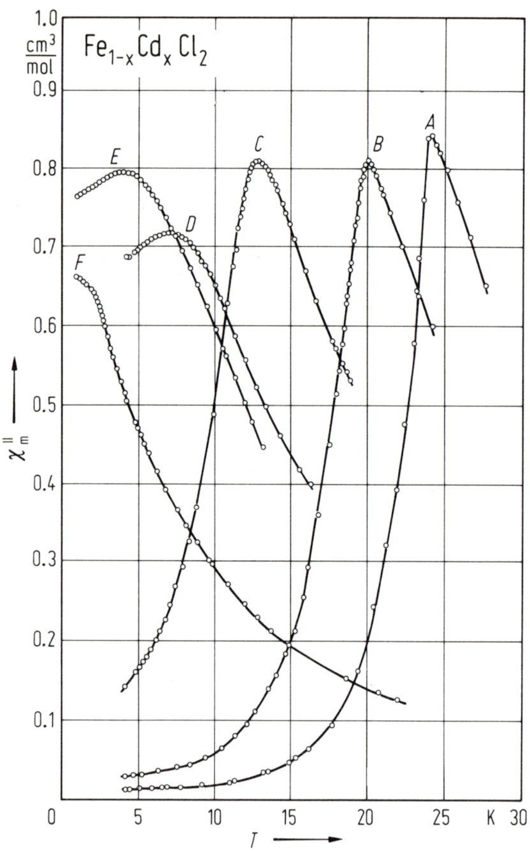

Fig. 48. $Fe_{1-x}Cd_xCl_2$. Susceptibility parallel to the c axis vs. temperature. The values of x are *A*: 0; *B*: 0.101; *C*: 0.213; *D*: 0.322; *E*: 0.369; *F*: 0.522 [81B1].

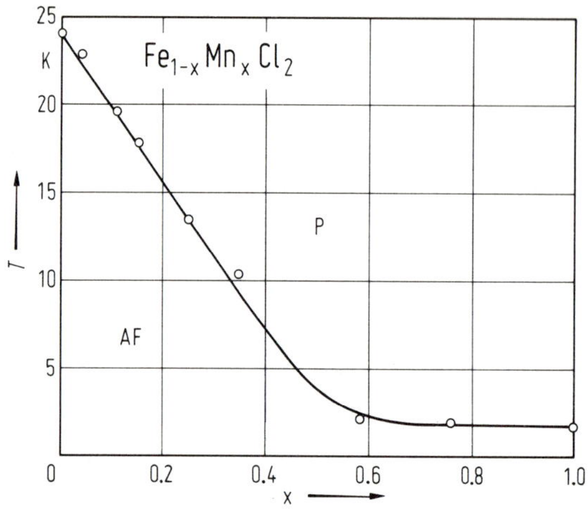

Fig. 49. $Fe_{1-x}Mn_xCl_2$. Concentration vs. transition temperature phase diagram determined from the susceptibility measurements [81B1].

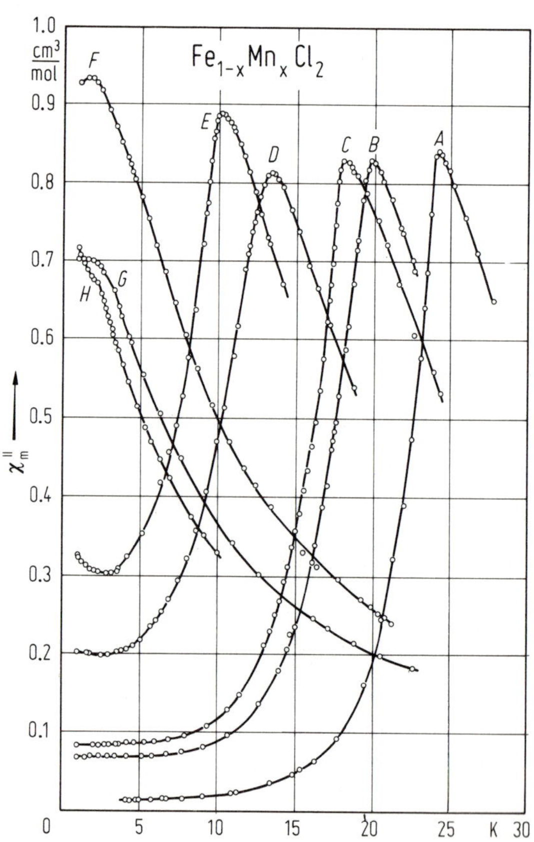

Fig. 50. $Fe_{1-x}Mn_xCl_2$. Susceptibility parallel to the c axis vs. temperature. The values of x are *A*: 0.001; *B*: 0.105; *C*: 0.151; *D*: 0.251; *E*: 0.348; *F*: 0.578: *G*: 0.757; *H*: 1 [81B1].

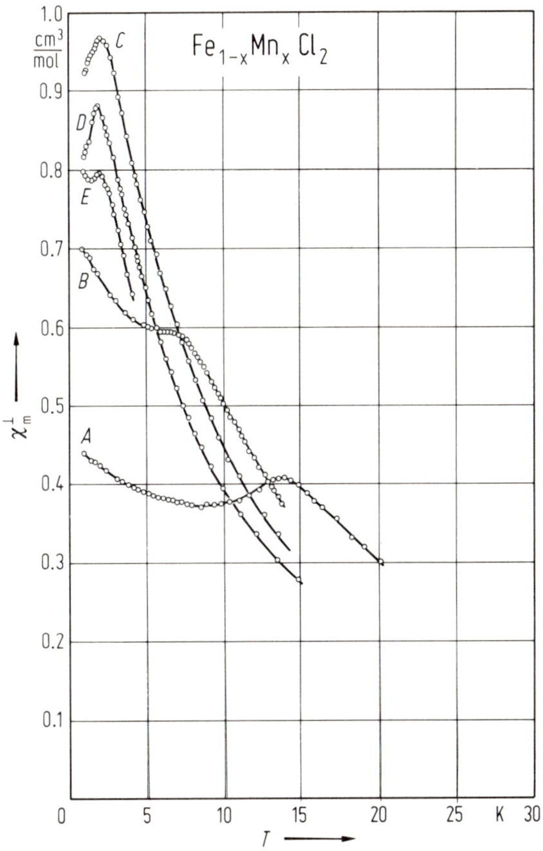

Fig. 51. $Fe_{1-x}Mn_xCl_2$. Susceptibility perpendicular to the c axis vs. temperature. The values of x are *A*: 0.251; *B*: 0.348; *C*: 0.578; *D*: 0.757; *E*: 1 [81B1].

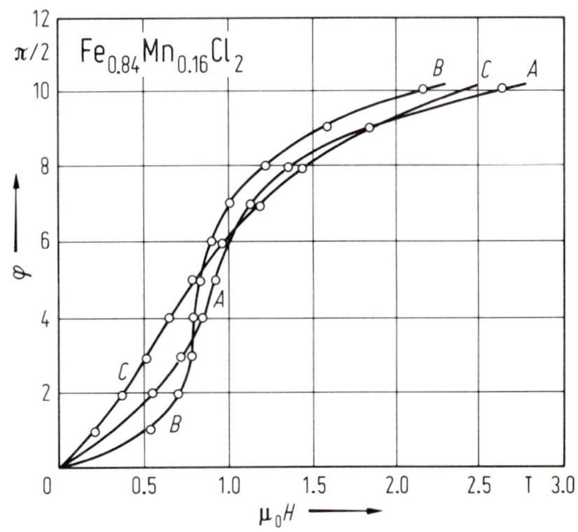

Fig. 52. $Fe_{0.84}Mn_{0.16}Cl_2$. Faraday rotation angle in units of $\pi/2$ vs. applied magnetic field for several temperatures. *A*: $T = 16.2$ K, *B*: 21.0 K, *C*: 13.9 K [81H].

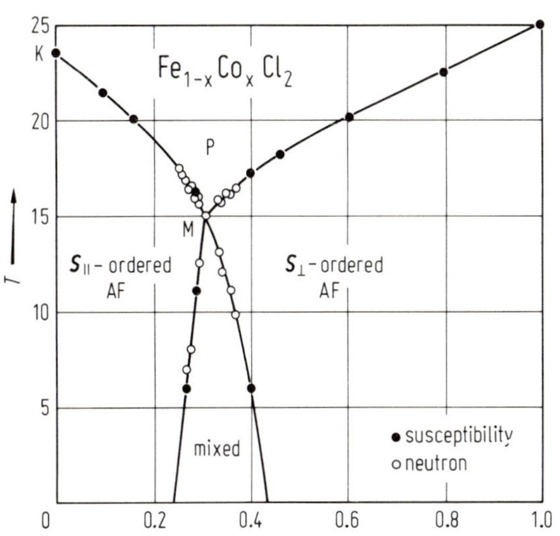

Fig. 53. $Fe_{1-x}Co_xCl_2$. Concentration vs. transition temperature phase diagram determined from the susceptibility and neutron scattering measurements [80W1].

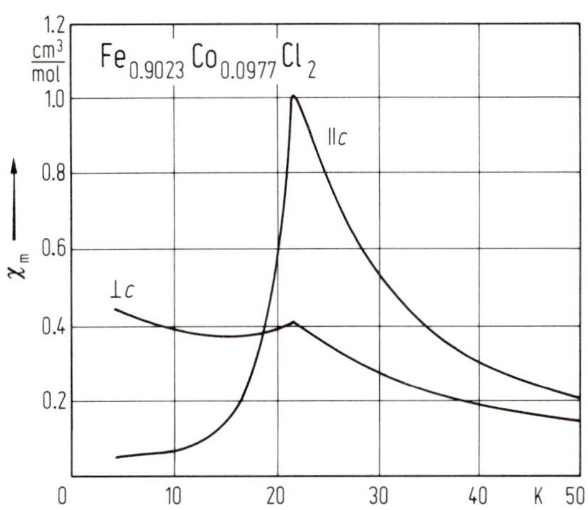

Fig. 54. $Fe_{0.9023}Co_{0.0977}Cl_2$. Susceptibility parallel and perpendicular to the c axis vs. temperature [83W1].

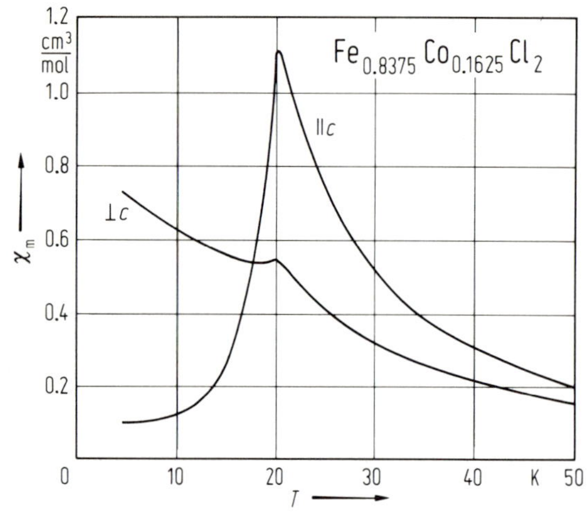

Fig. 55. $Fe_{0.8375}Co_{0.1625}Cl_2$. Susceptibility parallel and perpendicular to the c axis vs. temperature [83W1].

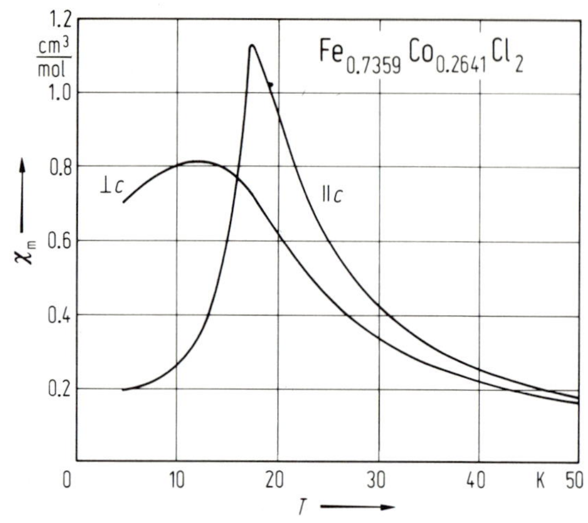

Fig. 56. $Fe_{0.7359}Co_{0.2641}Cl_2$. Susceptibility parallel and perpendicular to the c axis vs. temperature [83W1].

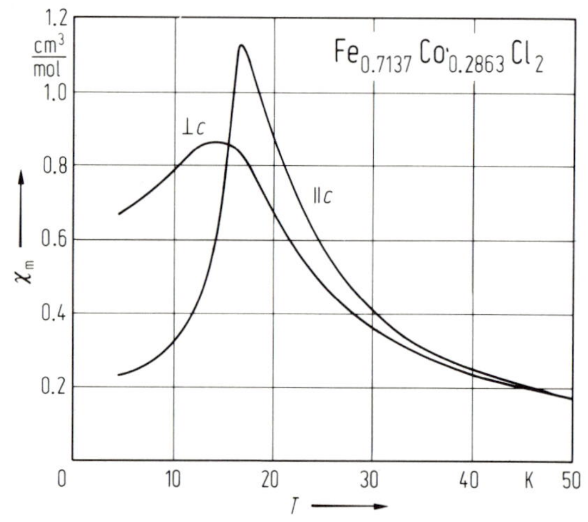

Fig. 57. $Fe_{0.7137}Co_{0.2863}Cl_2$. Susceptibility parallel and perpendicular to the c axis vs. temperature [83W1].

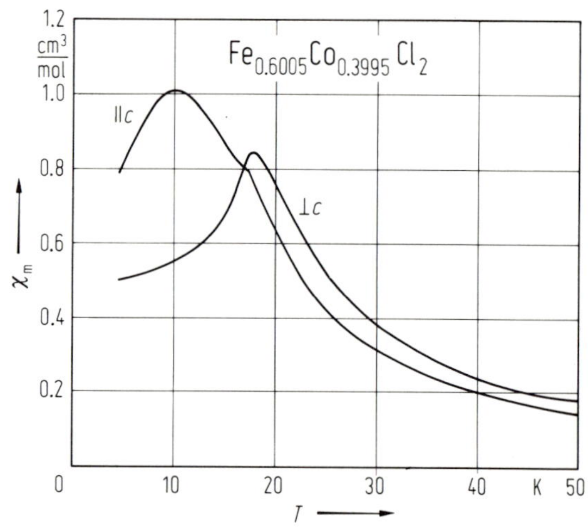

Fig. 58. $Fe_{0.6005}Co_{0.3995}Cl_2$. Susceptibility parallel and perpendicular to the c axis vs. temperature [83W1].

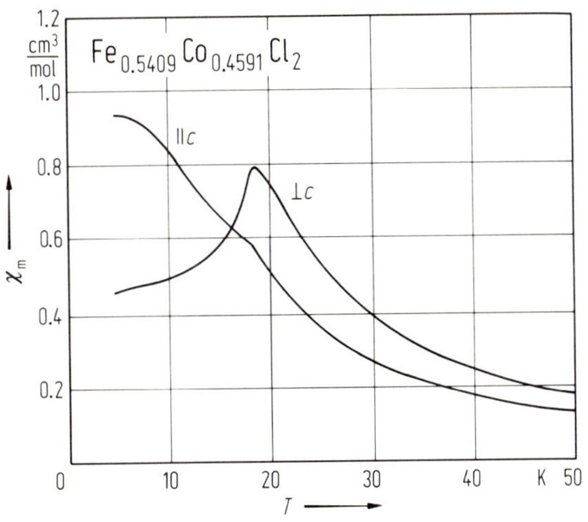

Fig. 59. $Fe_{0.5409}Co_{0.4591}Cl_2$. Susceptibility parallel and perpendicular to the c axis vs. temperature [83W1].

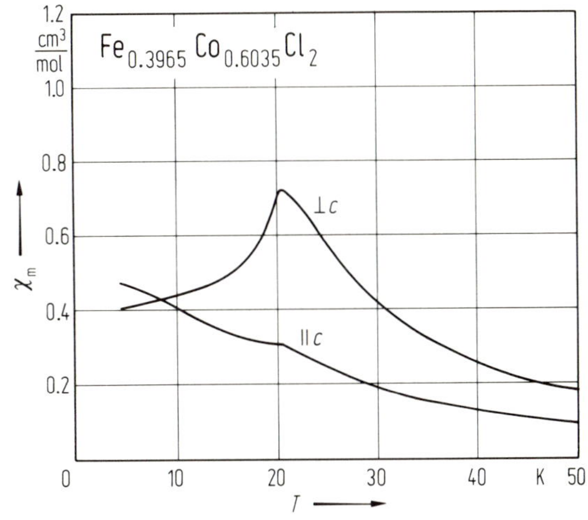

Fig. 60. $Fe_{0.3965}Co_{0.6035}Cl_2$. Susceptibility parallel and perpendicular to the c axis vs. temperature [83W1].

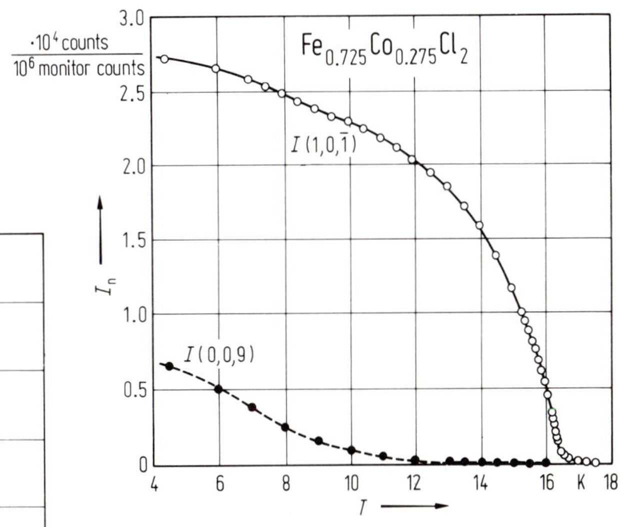

Fig. 62. $Fe_{0.725}Co_{0.275}Cl_2$. Peak intensity of scattered neutrons at the (1, 0, $\bar{1}$) and (0, 0, 9) Bragg points vs. temperature [83W1].

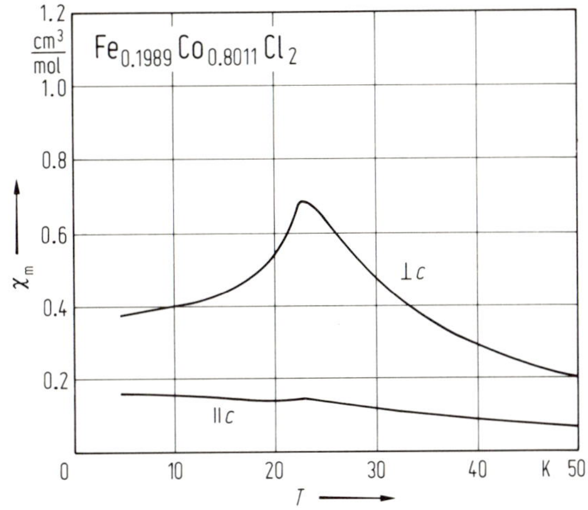

Fig. 61. $Fe_{0.1989}Co_{0.8011}Cl_2$. Susceptibility parallel and perpendicular to the c axis vs. temperature [83W1].

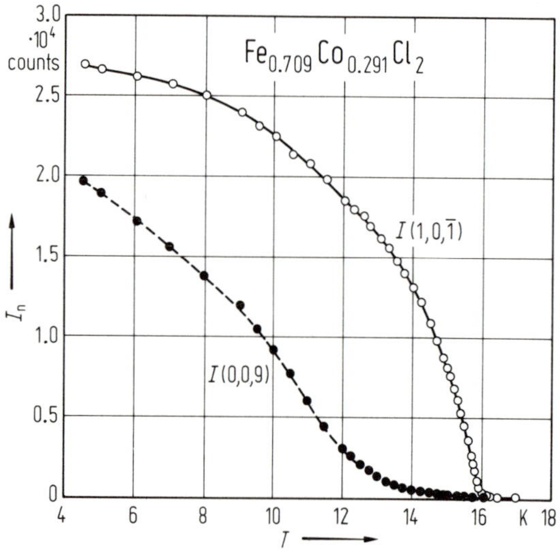

Fig. 63. $Fe_{0.709}Co_{0.291}Cl_2$. Integrated intensity of scattered neutrons at the $(1, 0, \bar{1})$ and $(0,0,9)$ Bragg points vs. temperature [83W1].

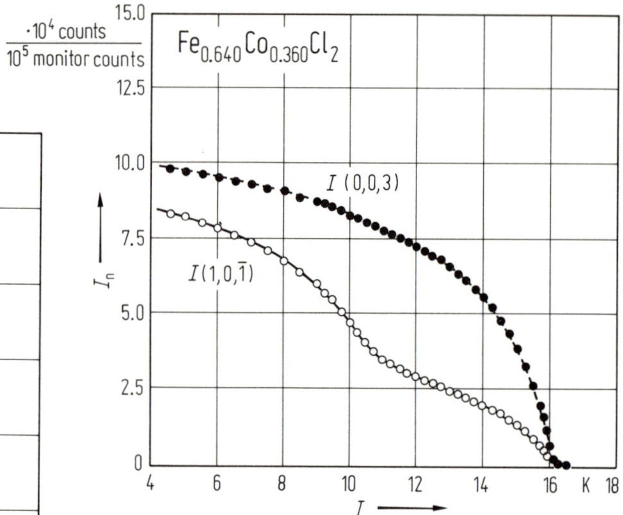

Fig. 65. $Fe_{0.640}Co_{0.360}Cl_2$. Peak intensity of scattered neutrons at the $(1, 0, \bar{1})$ and $(0, 0, 3)$ Bragg points vs. temperature [83W1].

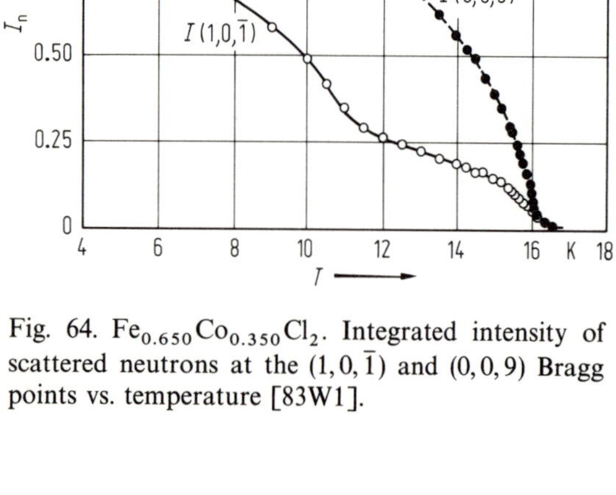

Fig. 64. $Fe_{0.650}Co_{0.350}Cl_2$. Integrated intensity of scattered neutrons at the $(1,0,\bar{1})$ and $(0,0,9)$ Bragg points vs. temperature [83W1].

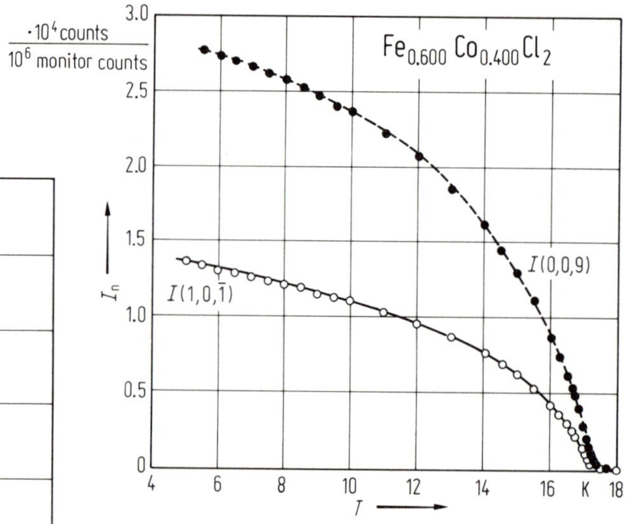

Fig. 66. $Fe_{0.600}Co_{0.400}Cl_2$. Peak intensity of scattered neutrons at the $(1, 0, \bar{1})$ and $(0, 0, 9)$ Bragg points vs. temperature [83W1].

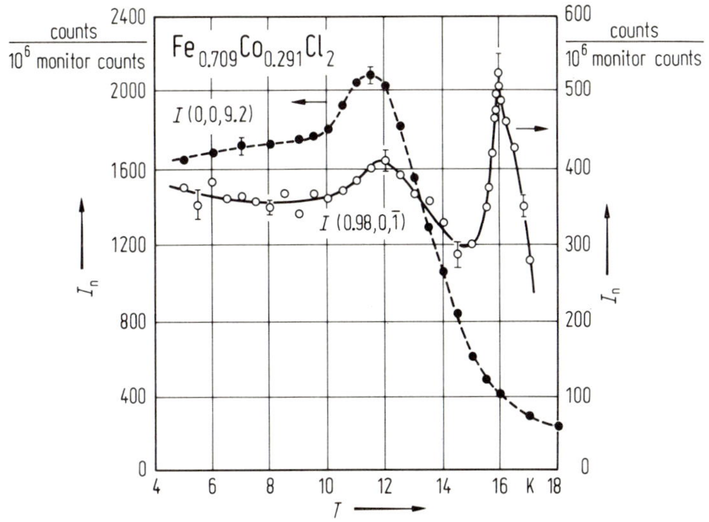

Fig. 67. $Fe_{0.709}Co_{0.291}Cl_2$. Intensity of scattered neutrons at the (0, 0, 9.2) and (0.98, 0, $\bar{1}$) reciprocal lattice points vs. temperature [83W1].

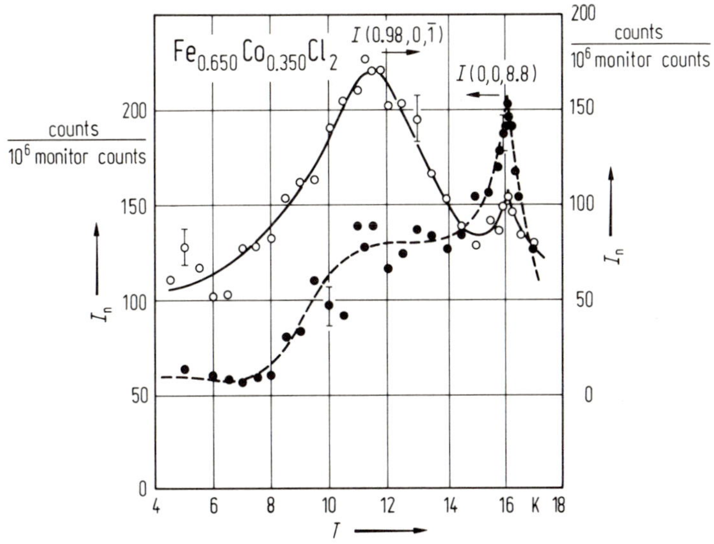

Fig. 68. $Fe_{0.650}Co_{0.350}Cl_2$. Intensity of scattered neutrons at the (0, 0, 8.8) and (0.98, 0, $\bar{1}$) reciprocal lattice points vs. temperature [83W1].

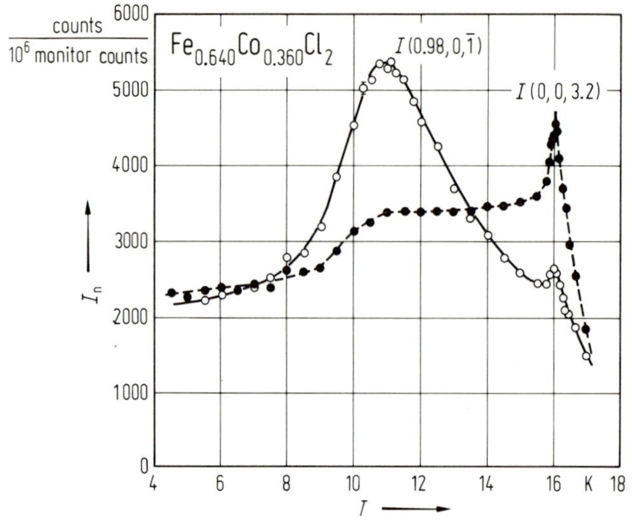

Fig. 69. $Fe_{0.640}Co_{0.360}Cl_2$. Intensity of scattered neutrons at the (0, 0, 3.2) and (0.98, 0, $\bar{1}$) reciprocal lattice points vs. temperature [83W1].

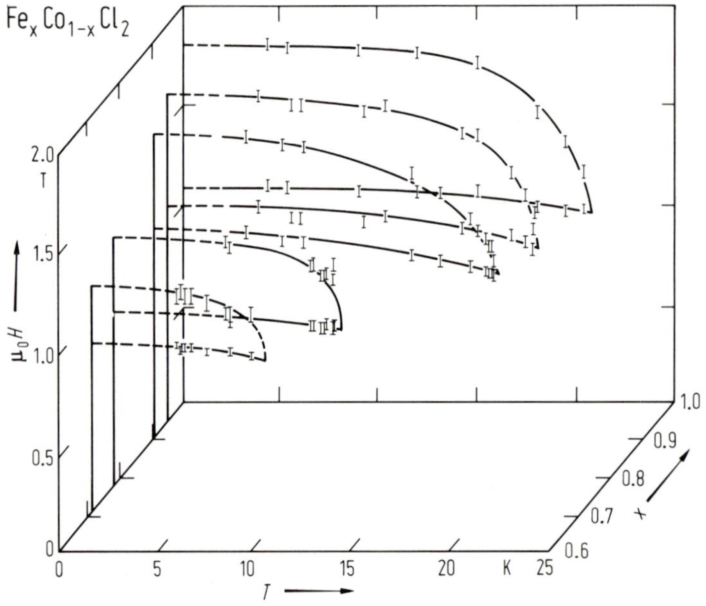

Fig. 70. $Fe_xCo_{1-x}Cl_2$. The mixed phase region of the magnetic phase diagram determined from the optical measurements [80W2].

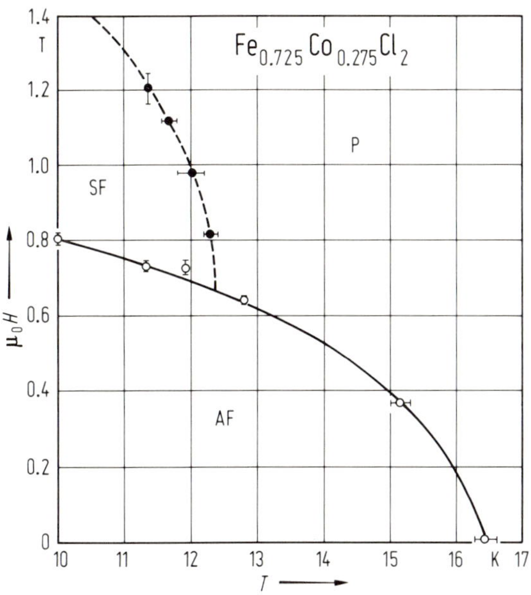

Fig. 71. $Fe_{0.725}Co_{0.275}Cl_2$. Temperature vs. applied magnetic field phase diagram determined from the neutron scattering measurements. The spin-flop phase is disordered [84W].

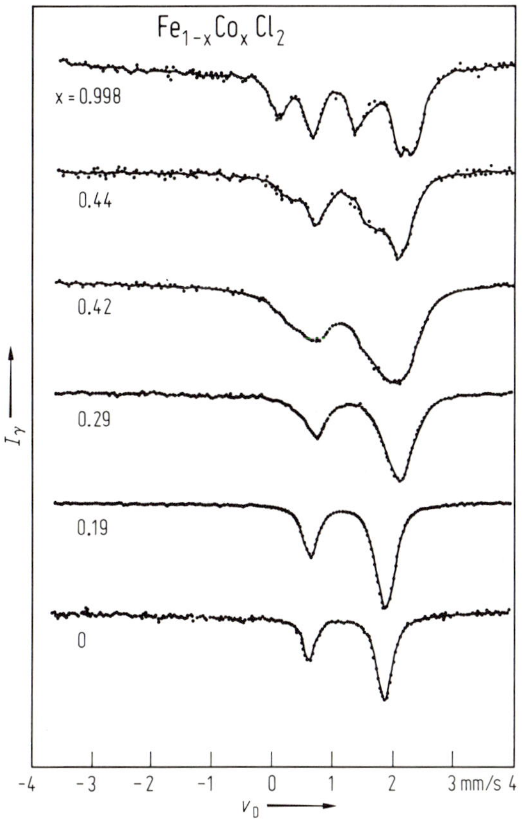

Fig. 72. $Fe_{1-x}Co_xCl_2$. Mössbauer spectra at 4.2 K for various concentrations [84H].

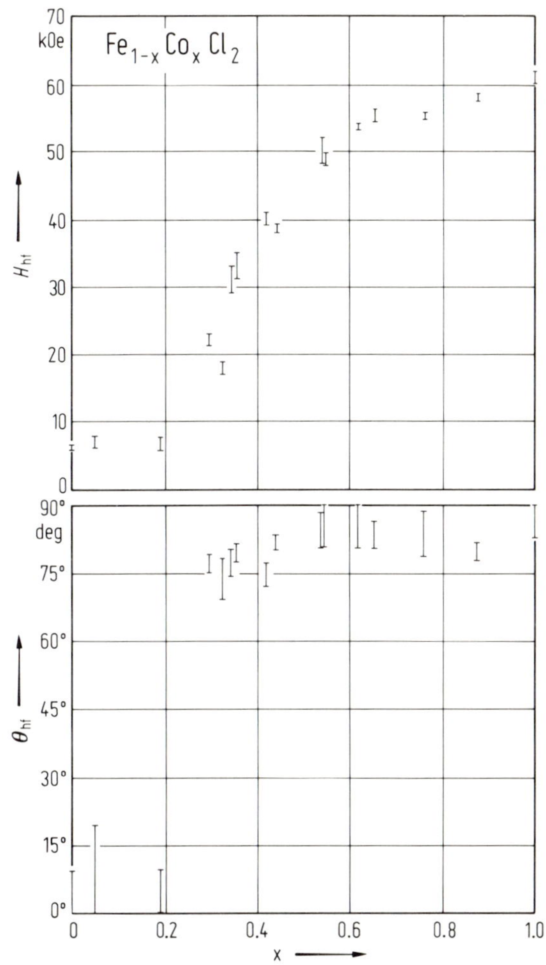

Fig. 73. $Fe_{1-x}Co_xCl_2$. Hyperfine field vs. concentration (upper) and the direction of H_{hf} vs. concentration (lower) deduced from the analysis of the Mössbauer spectra at 4.2 K [84H]. θ_{hf} is angle between the direction of hyperfine field and the principal axis of the electric field gradient tensor.

For Fig. 74 see p. 87.

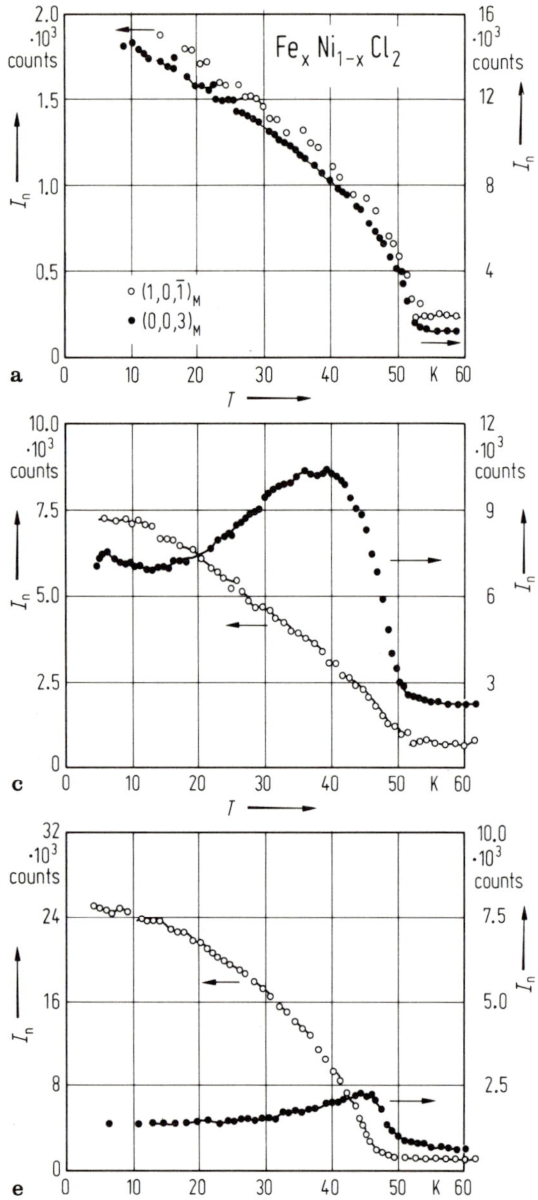

Fig. 75. $Fe_xNi_{1-x}Cl_2$. Neutron scattering intensity at the (0, 0, 3) and (1, 0, $\bar{1}$) Bragg points vs. temperature. The values of x are (a) ~ 0.005, (b) ~ 0.03, (c) ~ 0.06, (d) 0.11, (e) ~ 0.12 [83I]. Index M: magnetic.

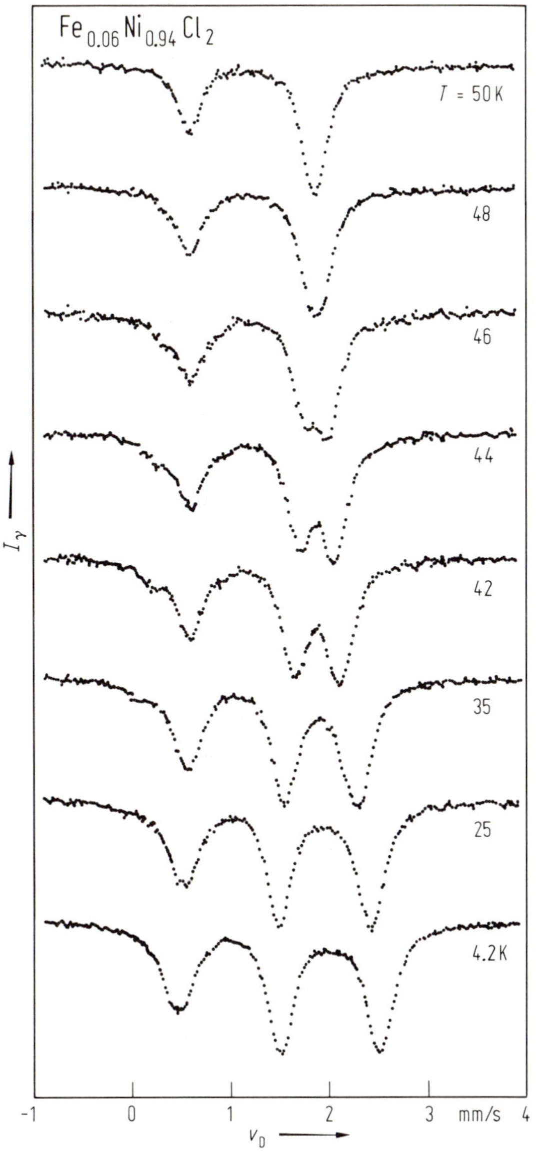

Fig. 76. $Fe_{0.06}Ni_{0.94}Cl_2$. Mössbauer spectrum at various temperatures [83I].

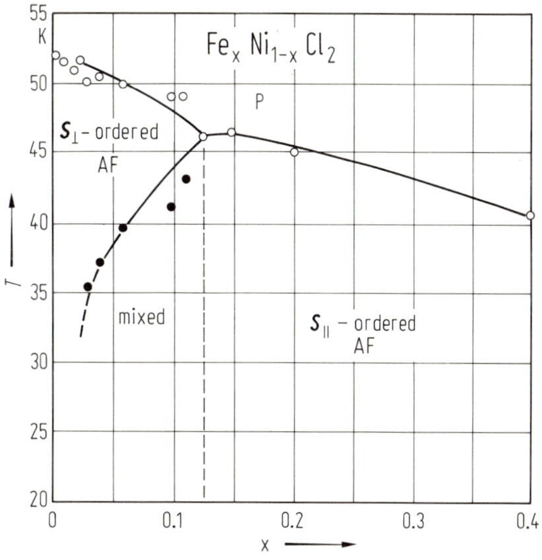

Fig. 74. $Fe_xNi_{1-x}Cl_2$. Concentration vs. transition temperature phase diagram determined from the neutron scattering measurements [83I].

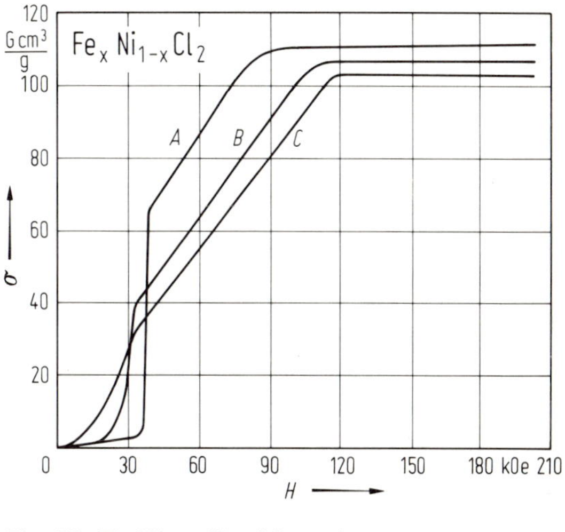

Fig. 77. $Fe_xNi_{1-x}Cl_2$. Magnetic moment per gram vs. applied magnetic field parallel to the *c* axis at 4.2 K. The values of x are, *A*: 0.15; *B*: 0.10; *C*: 0.06 [86I].

9.2.2.3 Static magnetic and related properties of $Fe_{1-x}Co_xBr_2$ compound

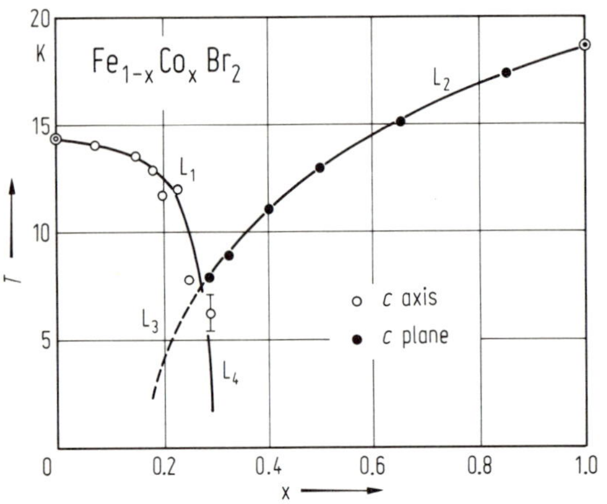

Fig. 78. $Fe_{1-x}Co_xBr_2$. Concentration vs. transition temperature phase diagram determined from susceptibility and magnetization measurements parallel and perpendicular to the c axis. Included are T_N for x = 0 from C_p measurements and for x = 1 from neutron scattering [84K].

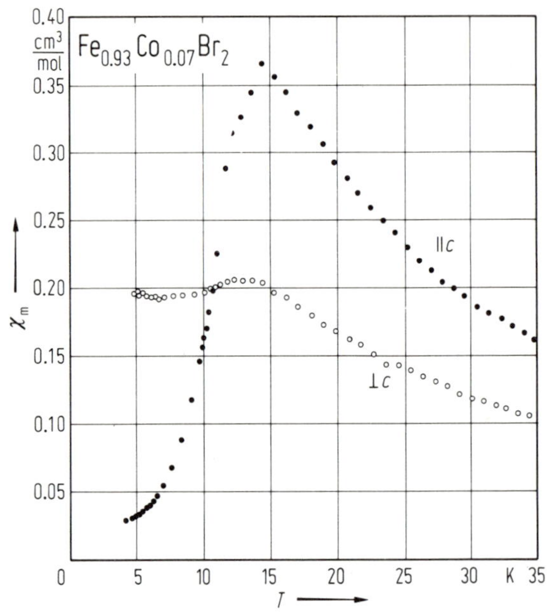

Fig. 79. $Fe_{0.93}Co_{0.07}Br_2$. Susceptibility parallel and perpendicular to the c axis vs. temperature [84K].

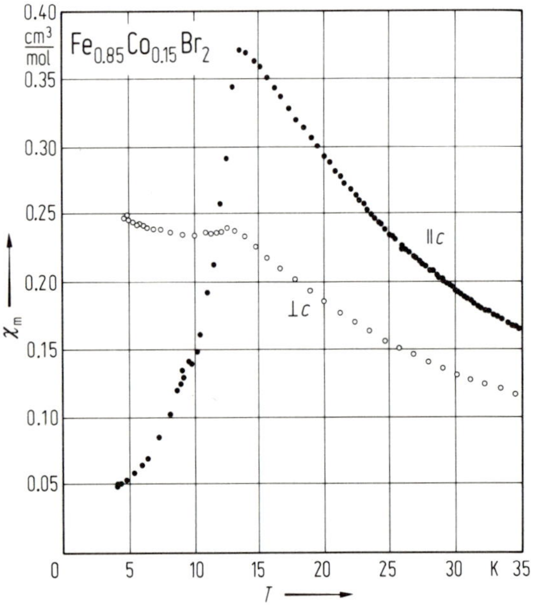

Fig. 80. $Fe_{0.85}Co_{0.15}Br_2$. Susceptibility parallel and perpendicular to the c axis vs. temperature [84K].

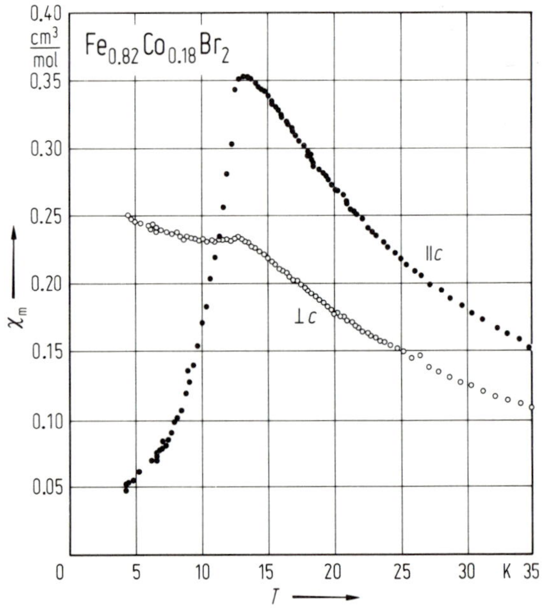

Fig. 81. $Fe_{0.82}Co_{0.18}Br_2$. Susceptibility parallel and perpendicular to the c axis vs. temperature [84K].

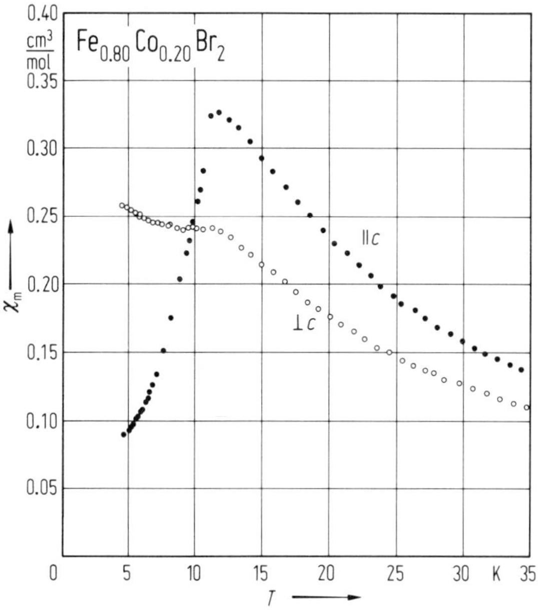

Fig. 82. $Fe_{0.80}Co_{0.20}Br_2$. Susceptibility parallel and perpendicular to the c axis vs. temperature [84K].

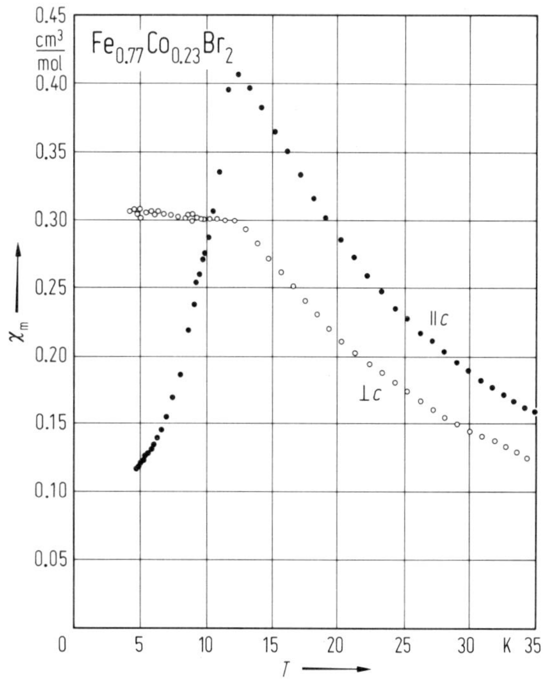

Fig. 83. $Fe_{0.77}Co_{0.23}Br_2$. Susceptibility parallel and perpendicular to the c axis vs. temperature [84K].

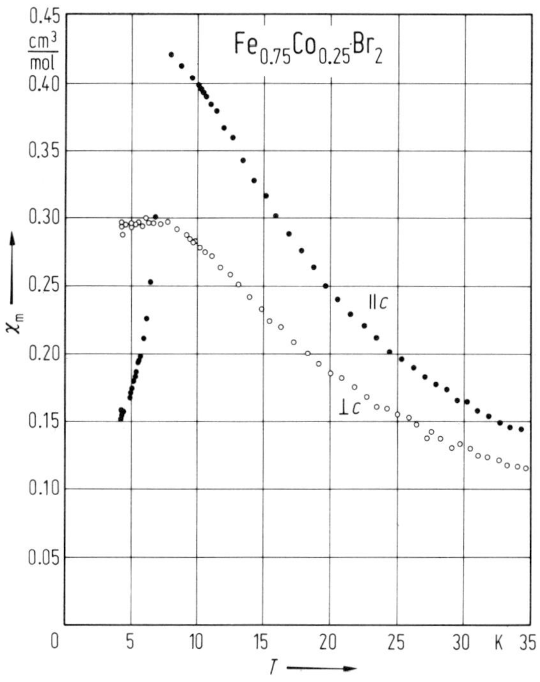

Fig. 84. $Fe_{0.75}Co_{0.25}Br_2$. Susceptibility parallel and perpendicular to the c axis vs. temperature [84K].

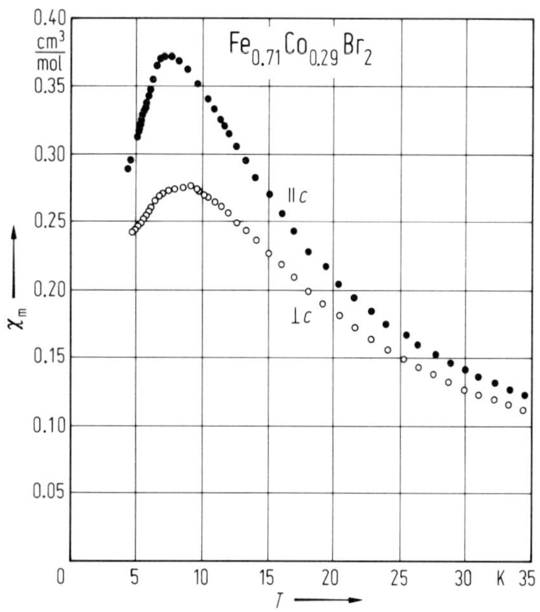

Fig. 85. $Fe_{0.71}Co_{0.29}Br_2$. Susceptibility parallel and perpendicular to the c axis vs. temperature [84K].

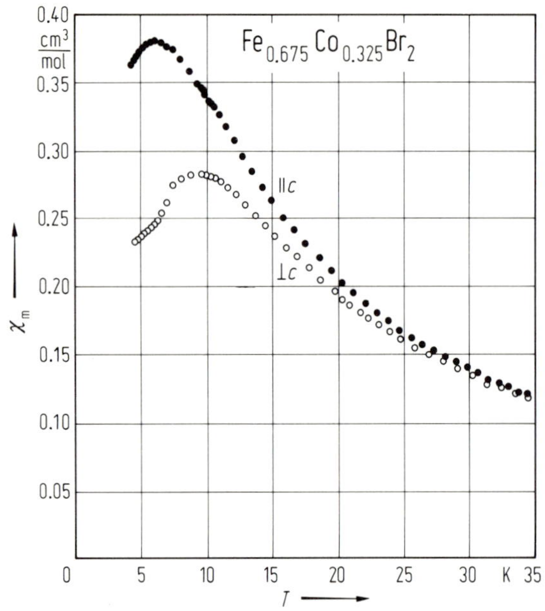

Fig. 86. $Fe_{0.675}Co_{0.325}Br_2$. Susceptibility parallel and perpendicular to the *c* axis vs. temperature [84K].

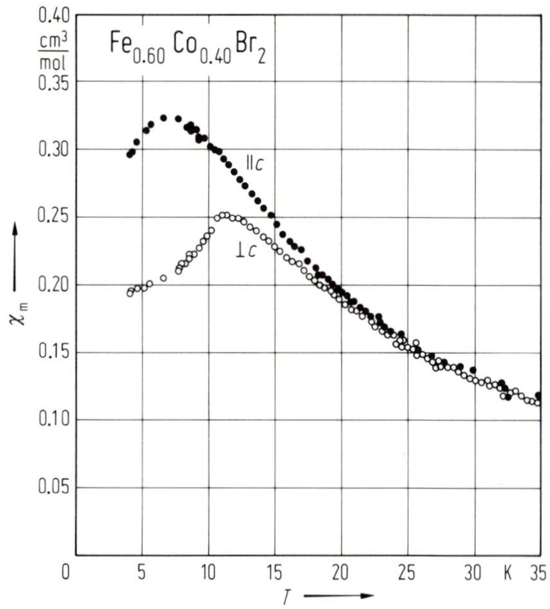

Fig. 87. $Fe_{0.60}Co_{0.40}Br_2$. Susceptibility parallel and perpendicular to the *c* axis vs. temperature [84K].

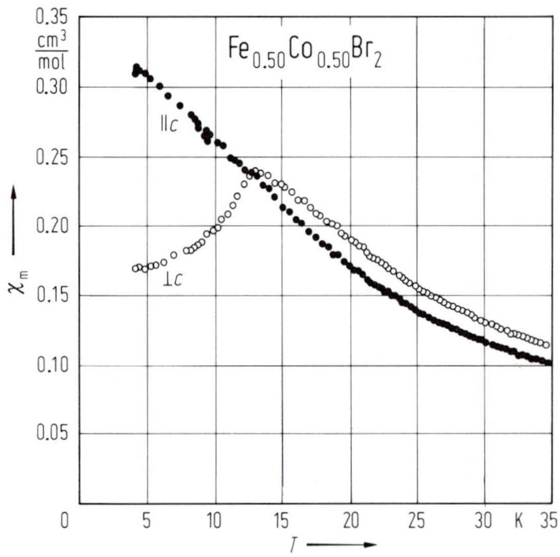

Fig. 88. $Fe_{0.50}Co_{0.50}Br_2$. Susceptibility parallel and perpendicular to the *c* axis vs. temperature [84K].

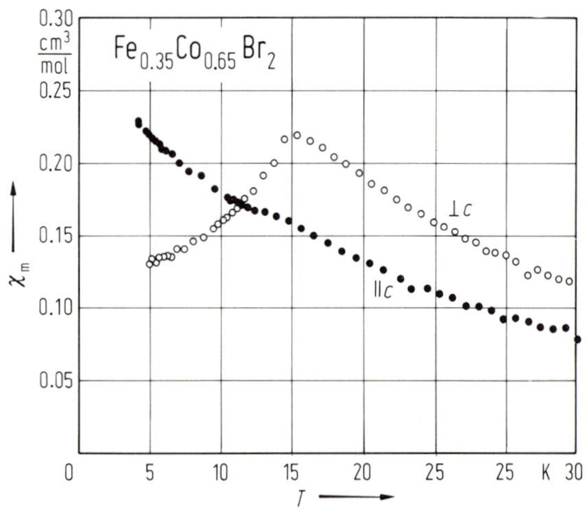

Fig. 89. $Fe_{0.35}Co_{0.65}Br_2$. Susceptibility parallel and perpendicular to the *c* axis vs. temperature [84K].

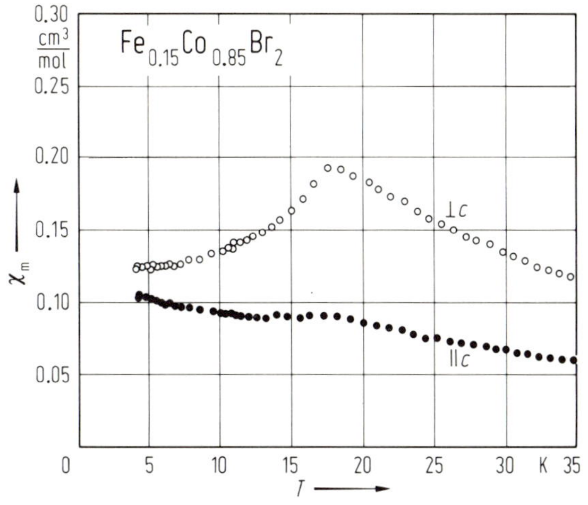

Fig. 90. $Fe_{0.15}Co_{0.85}Br_2$. Susceptibility parallel and perpendicular to the c axis vs. temperature [84K].

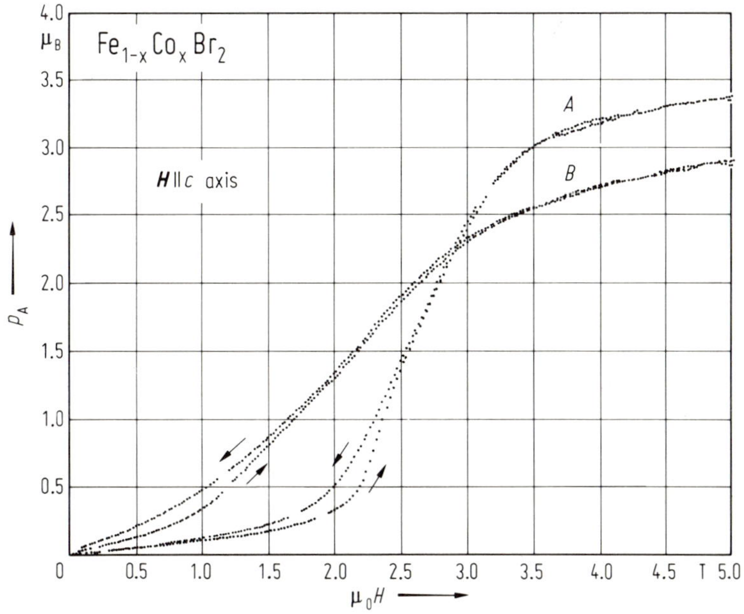

Fig. 91. $Fe_{1-x}Co_xBr_2$. Magnetic moment per atom vs. applied magnetic field. A: x = 0.15 and T = 2.9 K, B: x = 0.25 and T = 2.4 K [84K].

9.2.2.4 Static magnetic and related properties of $M_{1-x}M'_xX_2 \cdot nH_2O$ compounds

Table 7. Mössbauer parameters of $Fe_{1-x}Co_xCl_2 \cdot 2H_2O$ at 4.2 K [83 S1]. θ_H: angle between direction of H_{hf} and b axis.

x		Relative intensity [%]	θ_H [°]	H_{hf} [kOe]	$e^2qQ/2$ [mm/s]	T_N [K]
0.069		100	90	251(3)	2.4(1)	21
0.412	I	72(3)	82(1)	254(3)	2.5(2)	13
	II	28(3)	87···90	~ 250	~ 2.5	
0.474	I	78(3)	79(1)	260(3)	2.3(2)	—
	II	22(3)	84···87	~ 250	~ 2.5	
0.544	I	83(2)	78(1)	253(3)	2.4(2)	11
	II	17(2)	0···15	~ 135	~ 2.6	
0.569	I	80(2)	78(1)	257(3)	2.4(2)	12
	II	20(2)	0···15	~ 140	~ 2.6	
0.65	I	67(3)	78(1)	262(3)	2.1(2)	12
	II	33(3)	0···15	~ 140	~ 2.6	
0.813	I	50(5)	0···15	138(3)	2.6(2)	15
	II	50(5)	77(1)	255(6)	2.6(2)	

Table 8. Transition temperatures of $Fe_{1-x}Co_xCl_2 \cdot 2H_2O$ [79K2].

x	$T_N(=T_H)$ [K]	T_L [K]
0.000	22.3(5)	—
0.330	15.4(5)	—
0.389	13.10(5)	5.1(2)
0.452	11.80(5)	6.80(5)
0.499	9.7(1)	9.2(1)
0.568	10.30(5)	7.5(2)
0.657	12.10(5)	4.8(2)
0.813	14.8(5)	—
0.924	16.5(5)	—
1.000	17.30(5)	—

Table 9. Transition temperatures of $Co_{1-x}Ni_xCl_2 \cdot 6H_2O$ [70T].

x	$T_N(x)$
0	2.29
0.1	2.35
0.208	2.58
0.231	2.80
0.25	2.55
0.473	3.08
0.5	2.75
0.627	3.49
0.708	4.26
0.857	4.71
1	5.34

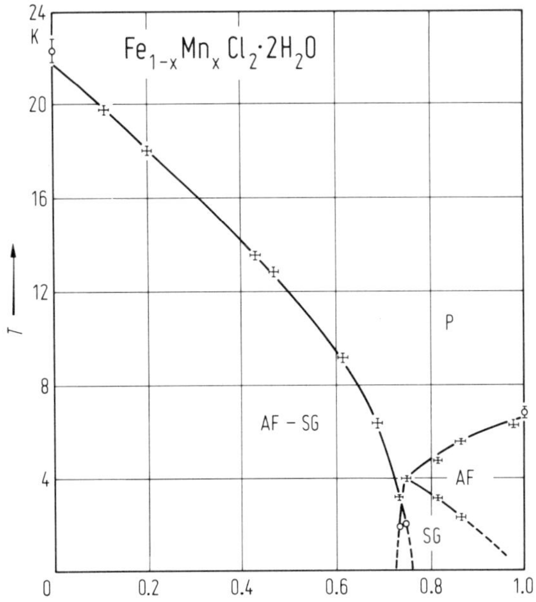

Fig. 92. $Fe_{1-x}Mn_xCl_2 \cdot 2H_2O$. Concentration vs. transition temperature phase diagram determined from the susceptibility measurements [84D].

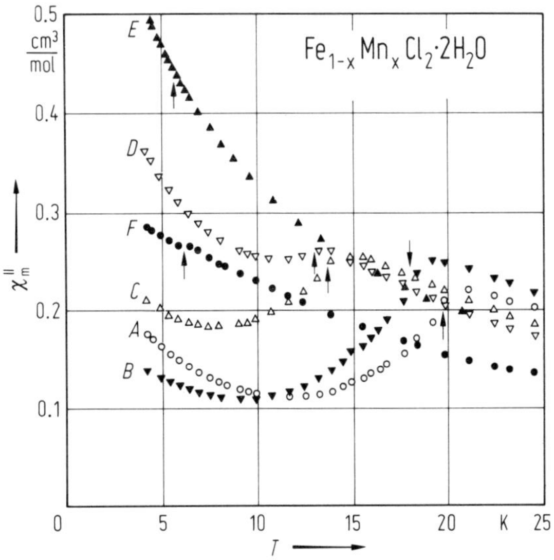

Fig. 93. $Fe_{1-x}Mn_xCl_2 \cdot 2H_2O$. Susceptibility along the *c* axis vs. temperature. The values of x are *A*: 0.108, *B*: 0.201, *C*: 0.429, *D*: 0.468, *E*: 0.864, *F*: 0.977. Arrows indicate estimated magnetic transition temperatures [84D].

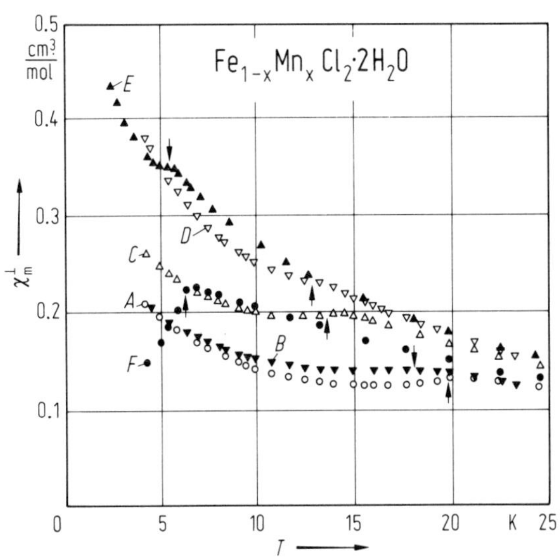

Fig. 94. $Fe_{1-x}Mn_xCl_2 \cdot 2H_2O$. Susceptibility perpendicular to the *c* axis vs. temperature. The values of x are *A*: 0.108, *B*: 0.201, *C*: 0.429, *D*: 0.468, *E*: 0.864, *F*: 0.977 [84D]. Arrows: see Fig. 93.

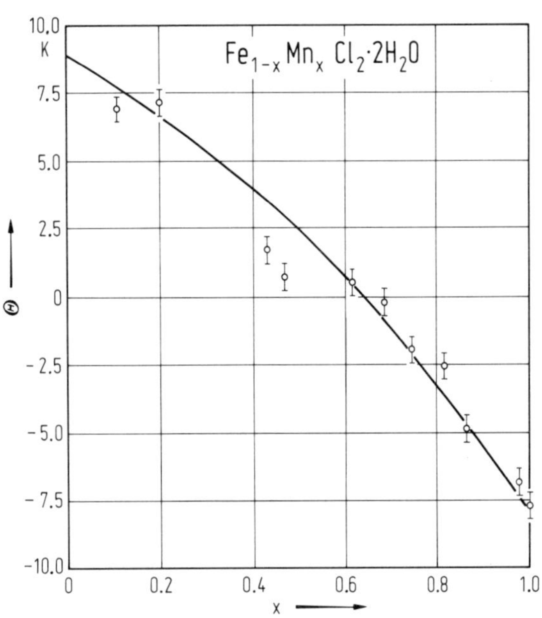

Fig. 95. $Fe_{1-x}Mn_xCl_2 \cdot 2H_2O$. Paramagnetic Curie temperature vs. concentration [84D].

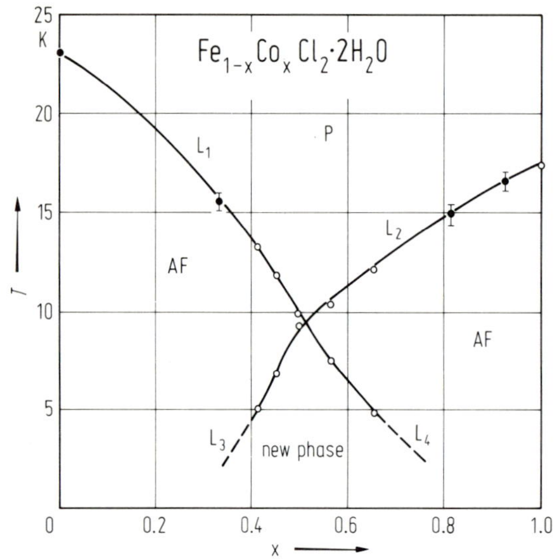

Fig. 96. $Fe_{1-x}Co_xCl_2 \cdot 2H_2O$. Concentration vs. transition temperature phase diagram determined from the specific heat (○) and susceptibility (●) measurements [79K1].

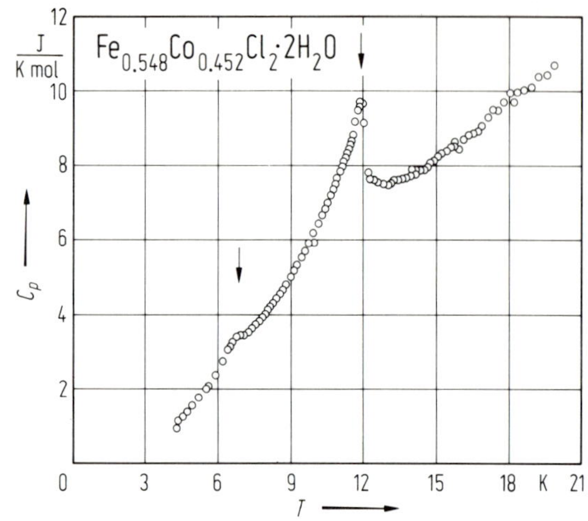

Fig. 97. $Fe_{0.548}Co_{0.452}Cl_2 \cdot 2H_2O$. Specific heat vs. temperature. Arrows: T_N, T_L, see Table 8 [79K2].

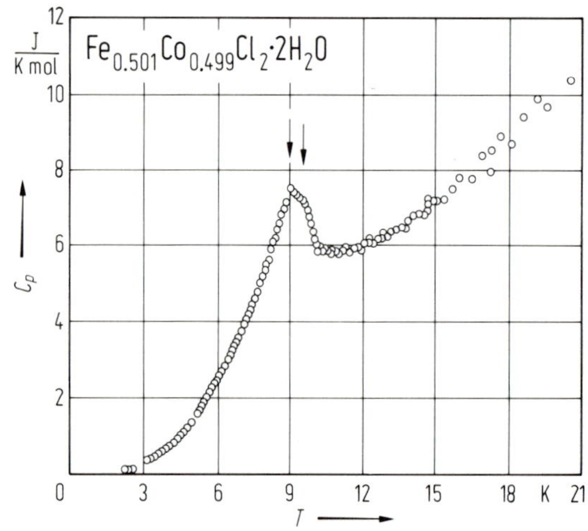

Fig. 98. $Fe_{0.501}Co_{0.499}Cl_2 \cdot 2H_2O$. Specific heat vs. temperature. Arrows: see Table 8 [79K2].

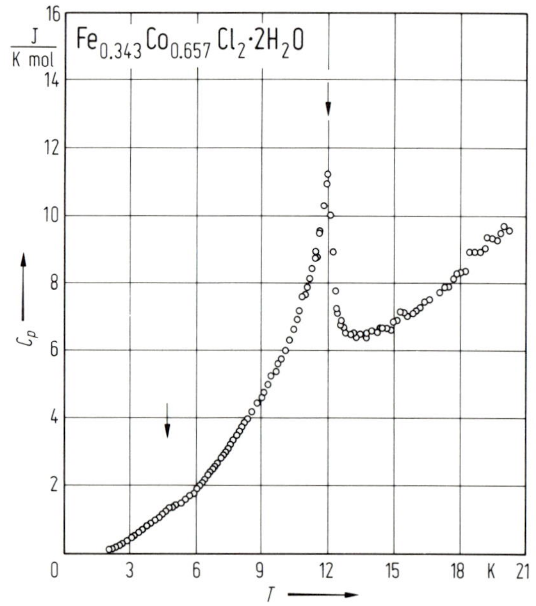

Fig. 99. $Fe_{0.343}Co_{0.657}Cl_2 \cdot 2H_2O$. Specific heat vs. temperature. Arrows: see Table 8 [79K2].

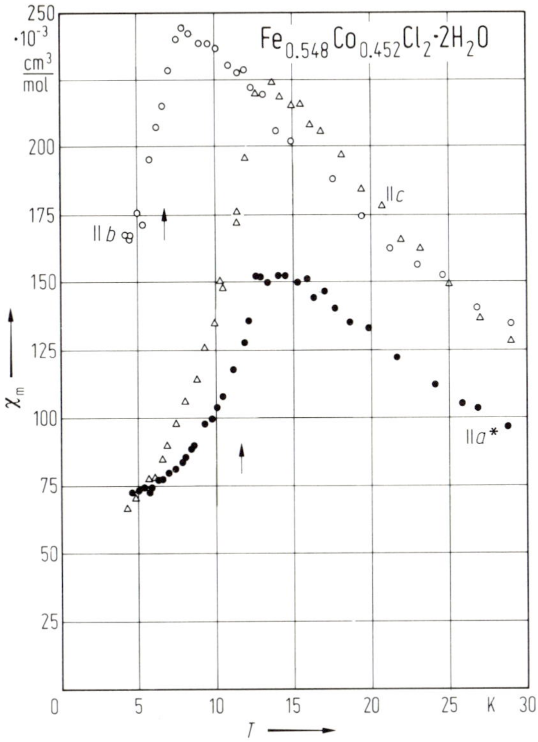

Fig. 100. $Fe_{0.548}Co_{0.452}Cl_2 \cdot 2H_2O$. Susceptibility along the a^*, b and c axes vs. temperature. Arrows: magnetic transition temperatures from specific heat measurements. a^* is perpendicular to both b and c [79K1].

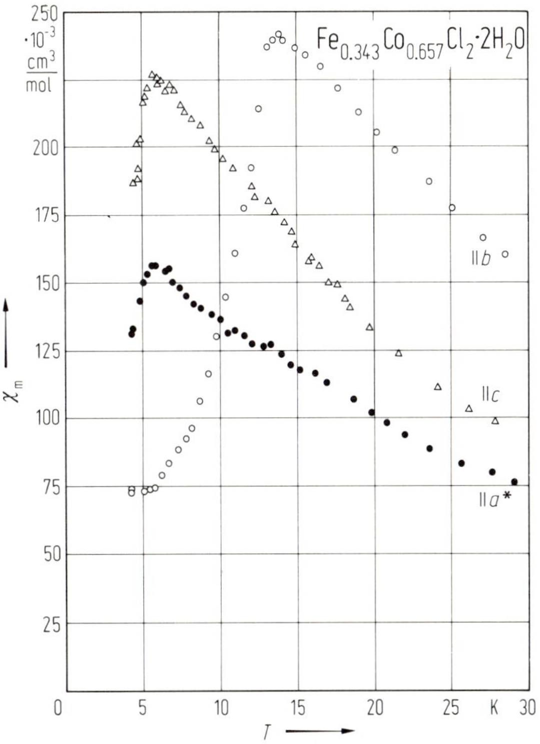

Fig. 101. $Fe_{0.343}Co_{0.657}Cl_2 \cdot 2H_2O$. Susceptibility along the a^*, b and c axes vs. temperature. a^* is perpendicular to both b and c [79K1].

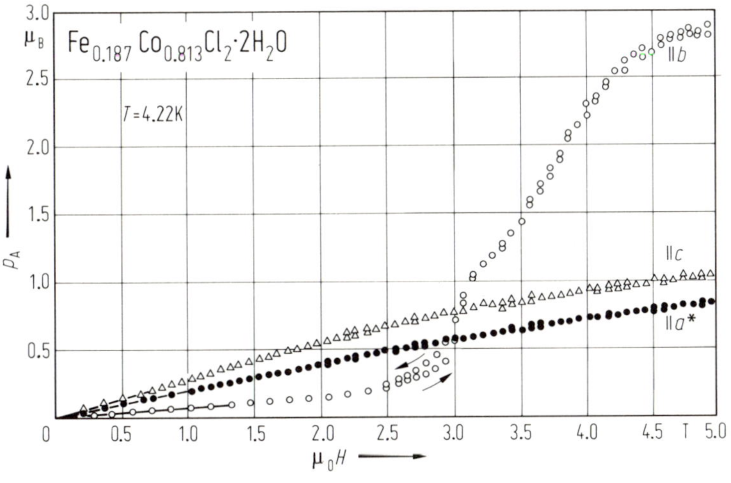

Fig. 102. $Fe_{0.187}Co_{0.813}Cl_2 \cdot 2H_2O$. Magnetic moment per atom along the a^*, b and c axes vs. applied magnetic field. a^* is perpendicular to both b and c [79K1].

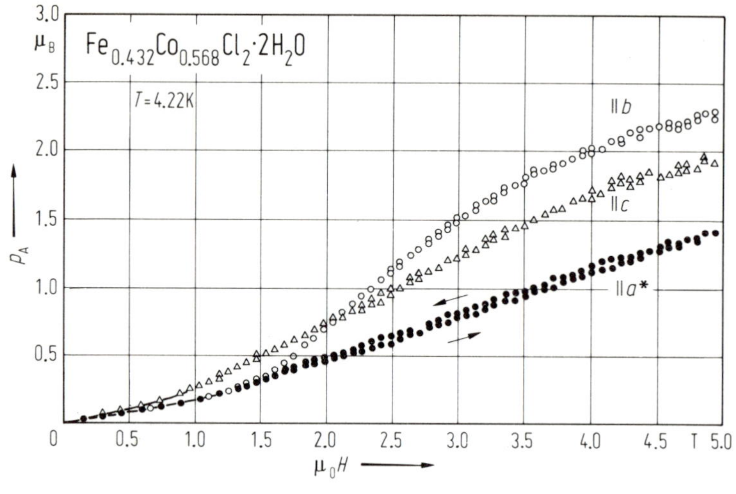

Fig. 103. $Fe_{0.432}Co_{0.568}Cl_2 \cdot 2H_2O$. Magnetic moment per atom along the a^*, b and c axes vs. applied magnetic field. a^* is perpendicular to both b and c [79K1].

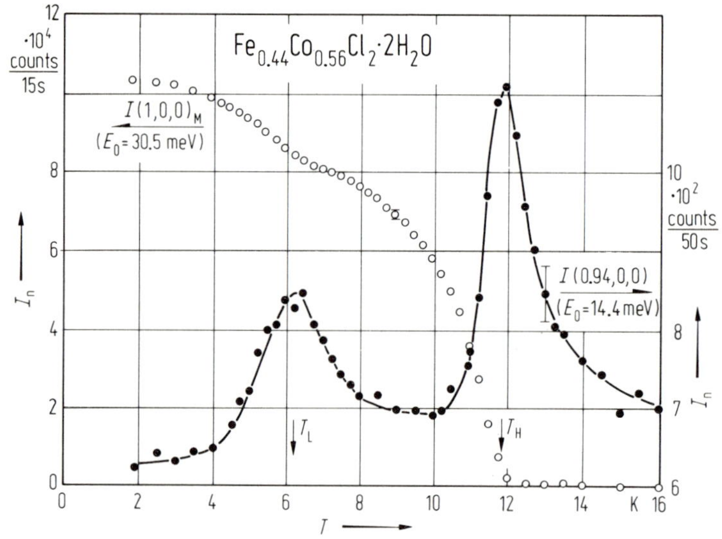

Fig. 104. $Fe_{0.44}Co_{0.56}Cl_2 \cdot 2H_2O$. Intensity of scattered neutrons at the (1, 0, 0) Bragg point and that at the (0.94, 0, 0) reciprocal lattice point vs. temperature [85K1].

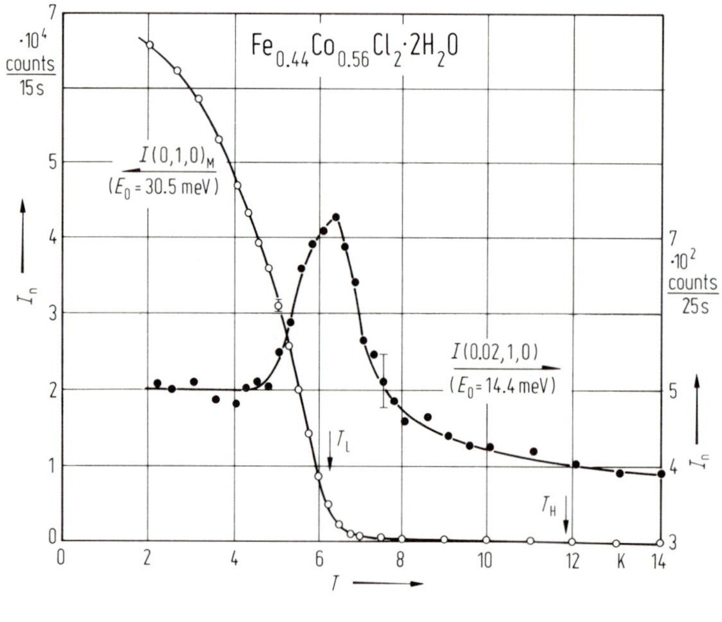

Fig. 105. $Fe_{0.44}Co_{0.56}Cl_2 \cdot 2H_2O$. Intensity of scattered neutrons at the (0, 1, 0) Bragg point and that at the (0.02, 1, 0) reciprocal lattice point vs. temperature [85K1].

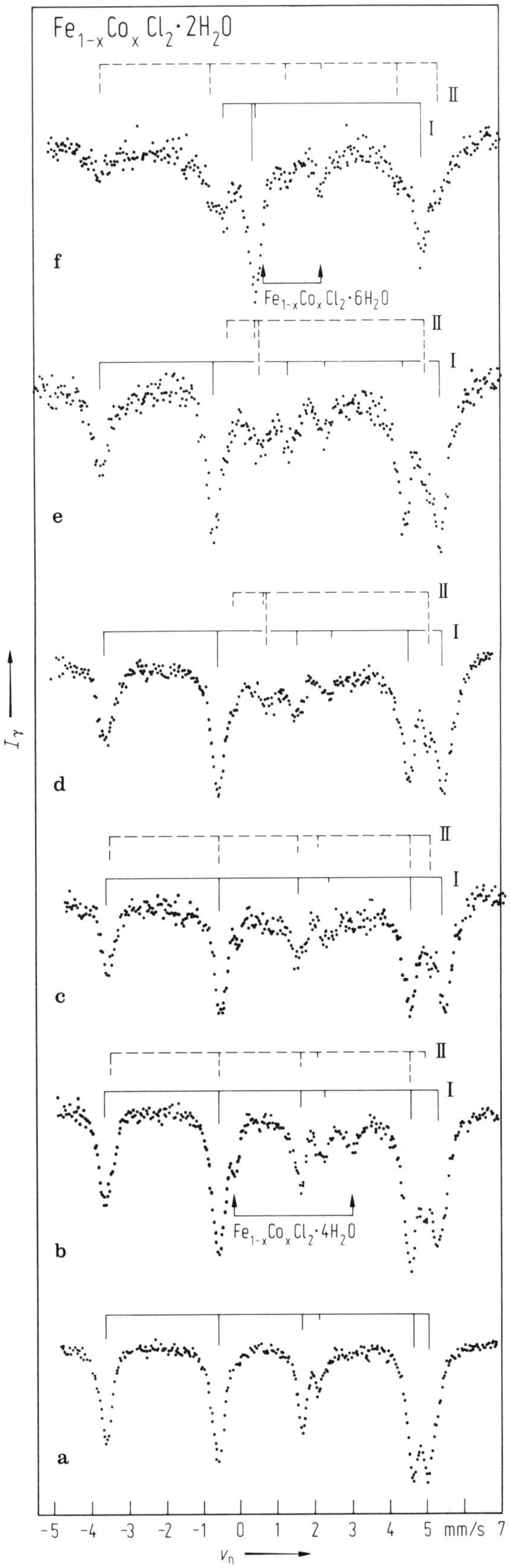

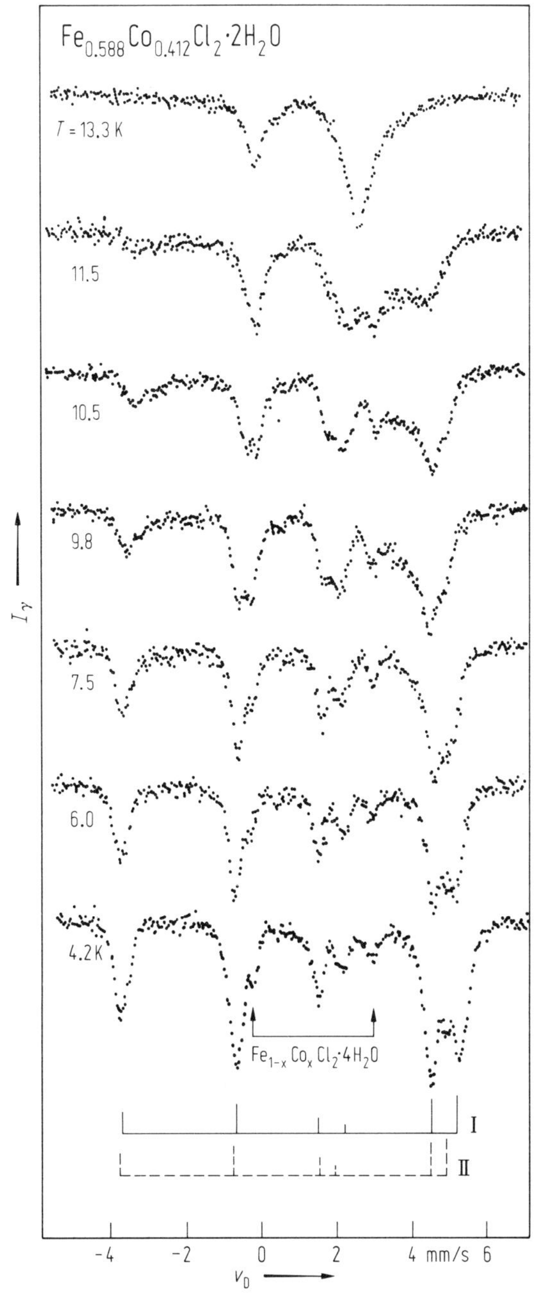

Fig. 107. $Fe_{0.588}Co_{0.412}Cl_2 \cdot 2H_2O$. Mössbauer spectrum obtained at the designated temperatures [83S1].

◀

Fig. 106. $Fe_{1-x}Co_xCl_2 \cdot 2H_2O$. Mössbauer spectrum at 4.2 K for various concentrations. The values of x are (a) 0.069, (b) 0.412, (c) 0.474, (d) 0.544, (e) 0.569, (f) 0.813 [80I].

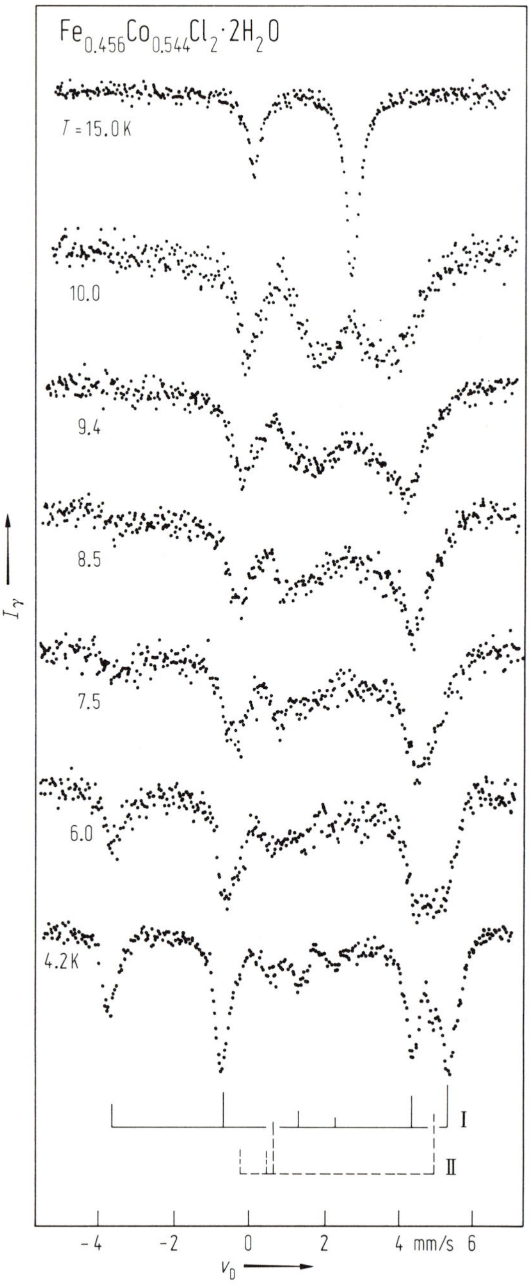

Fig. 108. $Fe_{0.456}Co_{0.544}Cl_2 \cdot 2H_2O$. Mössbauer spectrum obtained at the designated temperatures [83S1].

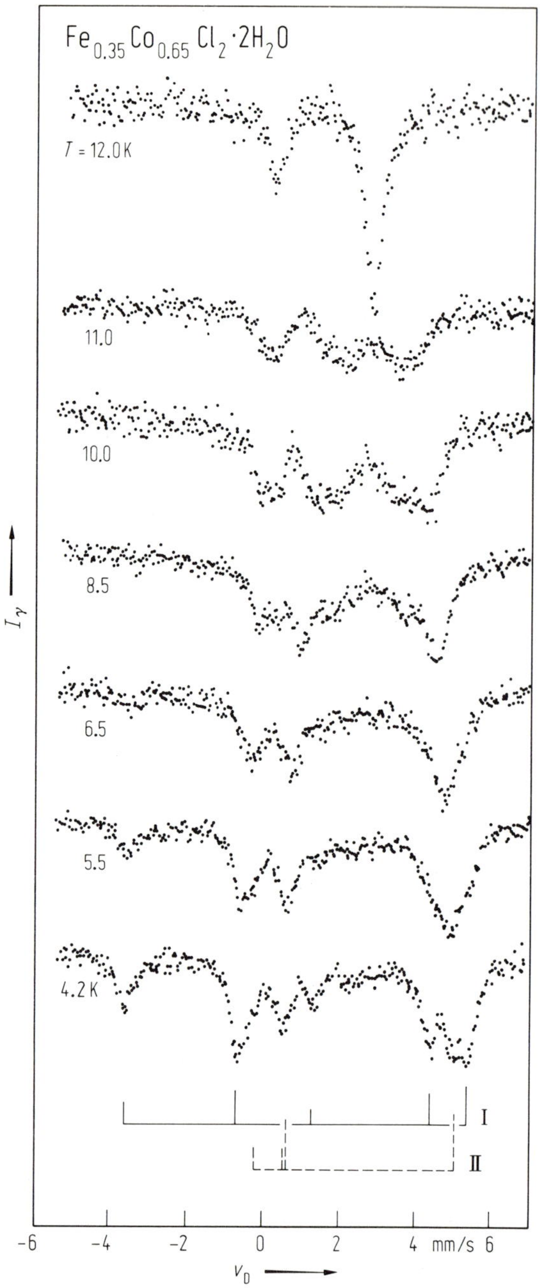

Fig. 109. $Fe_{0.35}Co_{0.65}Cl_2 \cdot 2H_2O$. Mössbauer spectrum obtained at the designated temperatures [83S1].

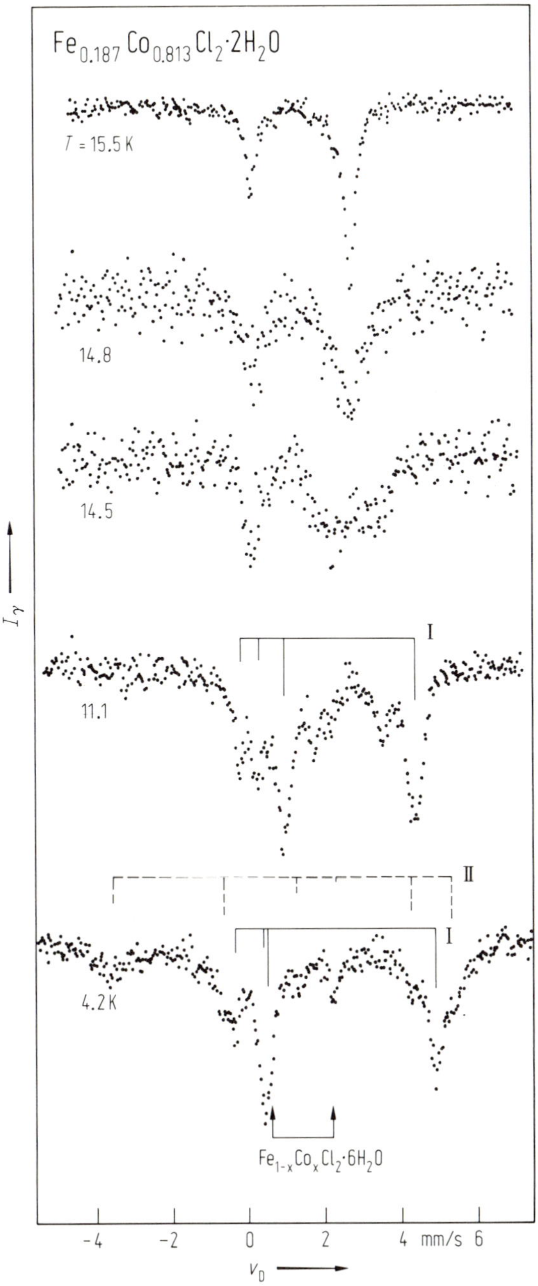

Fig. 110. $Fe_{0.187}Co_{0.813}Cl_2 \cdot 2H_2O$. Mössbauer spectrum obtained at the designated temperatures [83S1].

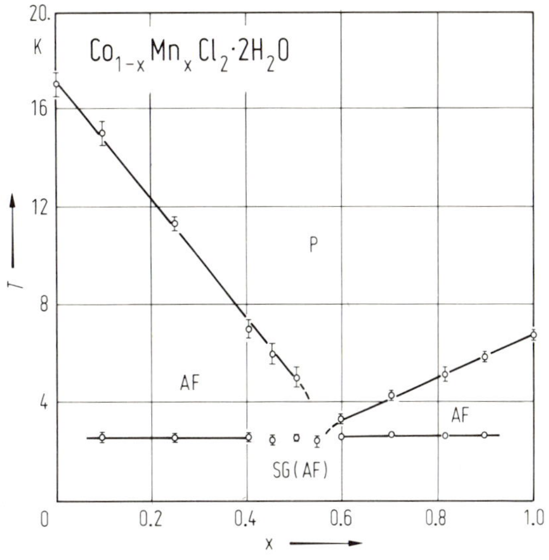

Fig. 111. $Co_{1-x}Mn_xCl_2 \cdot 2H_2O$. Concentration vs. transition temperature phase diagram obtained from the susceptibility and magnetization measurements [88D].

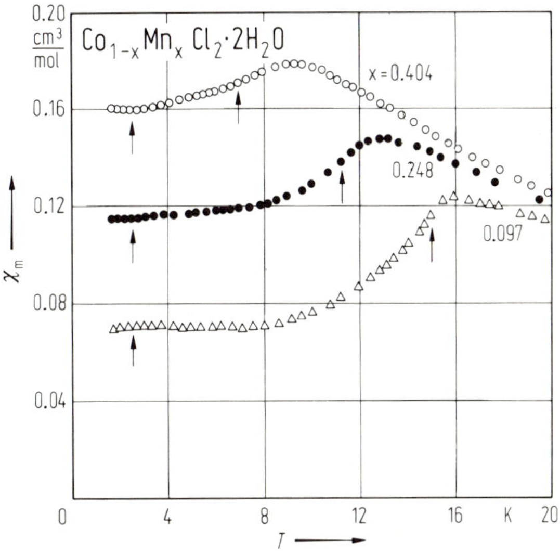

Fig. 112. $Co_{1-x}Mn_xCl_2 \cdot 2H_2O$. Susceptibility vs. temperature for polycrystalline samples with x = 0.404, 0.248 and 0.097. Arrows indicate estimated magnetic transition temperatures (see original paper) [88D].

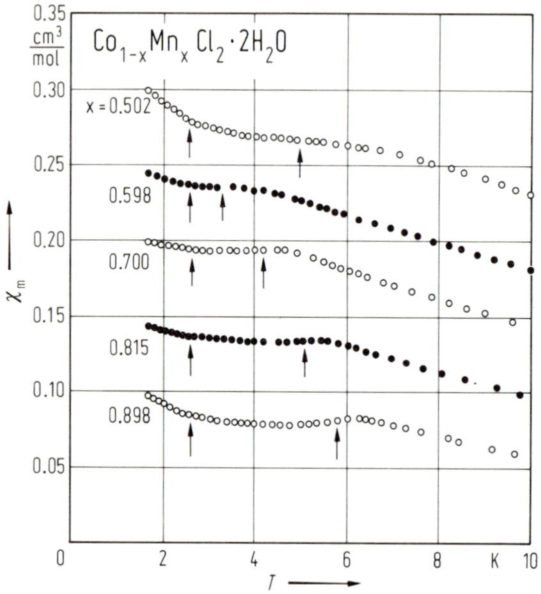

Fig. 113. $Co_{1-x}Mn_xCl_2 \cdot 2H_2O$. Susceptibility vs. temperature for polycrystalline samples with concentrations shown in the figure. Arrows indicate estimated magnetic transition temperatures (see original paper) [88D].

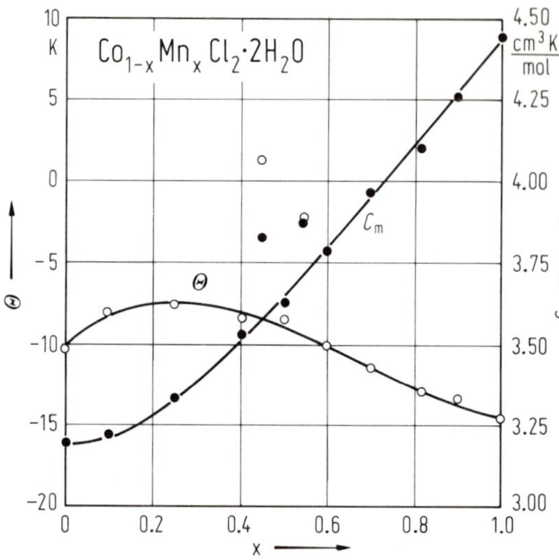

Fig. 114. $Co_{1-x}Mn_xCl_2 \cdot 2H_2O$. Concentration vs. paramagnetic Curie temperature (○) and Curie constant (●) [88D].

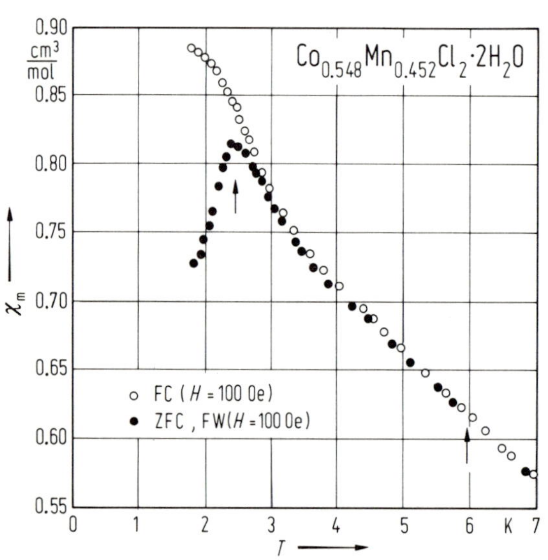

Fig. 115. $Co_{0.548}Mn_{0.452}Cl_2 \cdot 2H_2O$. Field cooled (FC) and zero field cooled (ZFC) or field warmed (FW) susceptibilities vs. temperature. Arrows indicate estimated magnetic transition temperature [88D].

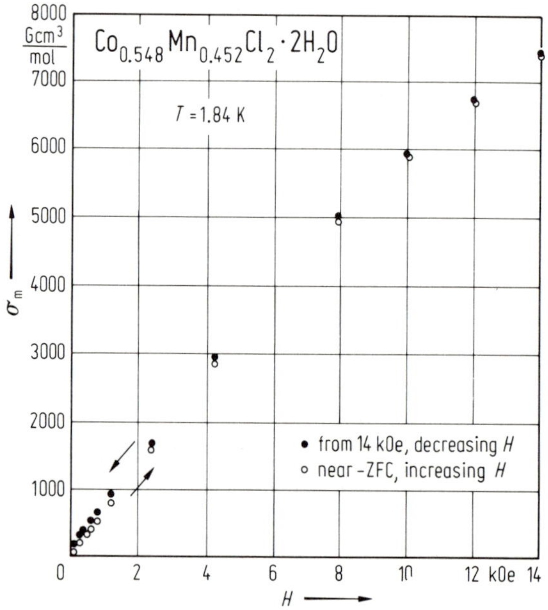

Fig. 116. $Co_{0.548}Mn_{0.452}Cl_2 \cdot 2H_2O$. Magnetic moment per mole vs. applied magnetic field [88D].

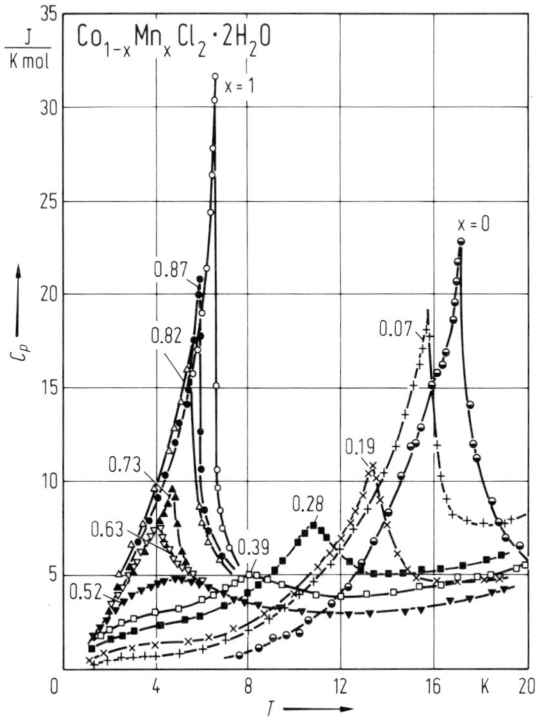

Fig. 117. $Co_{1-x}Mn_xCl_2 \cdot 2H_2O$. Specific heat vs. temperature for single crystals with concentrations shown in the figure [90D].

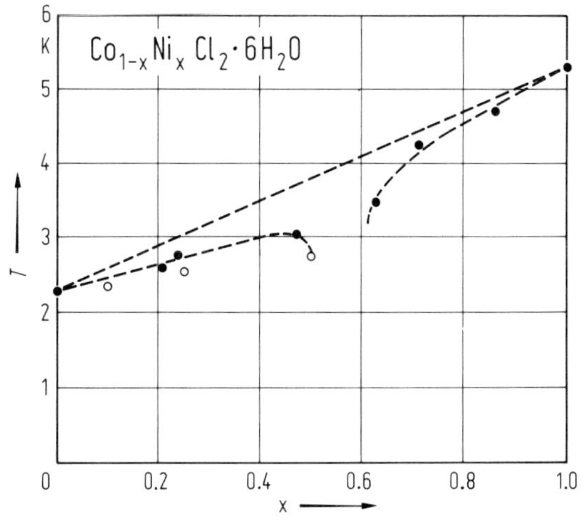

Fig. 119. $Co_{1-x}Ni_xCl_2 \cdot 6H_2O$. Concentration vs. transition temperature phase diagram obtained from specific heat measurements [70T]. Open circles are from other literature.

▶

Fig. 118. $Co_{1-x}Mn_xCl_2 \cdot 2H_2O$. NMR spectrum for single crystals with concentrations shown in the figure at 1.45(5) K [92Z].

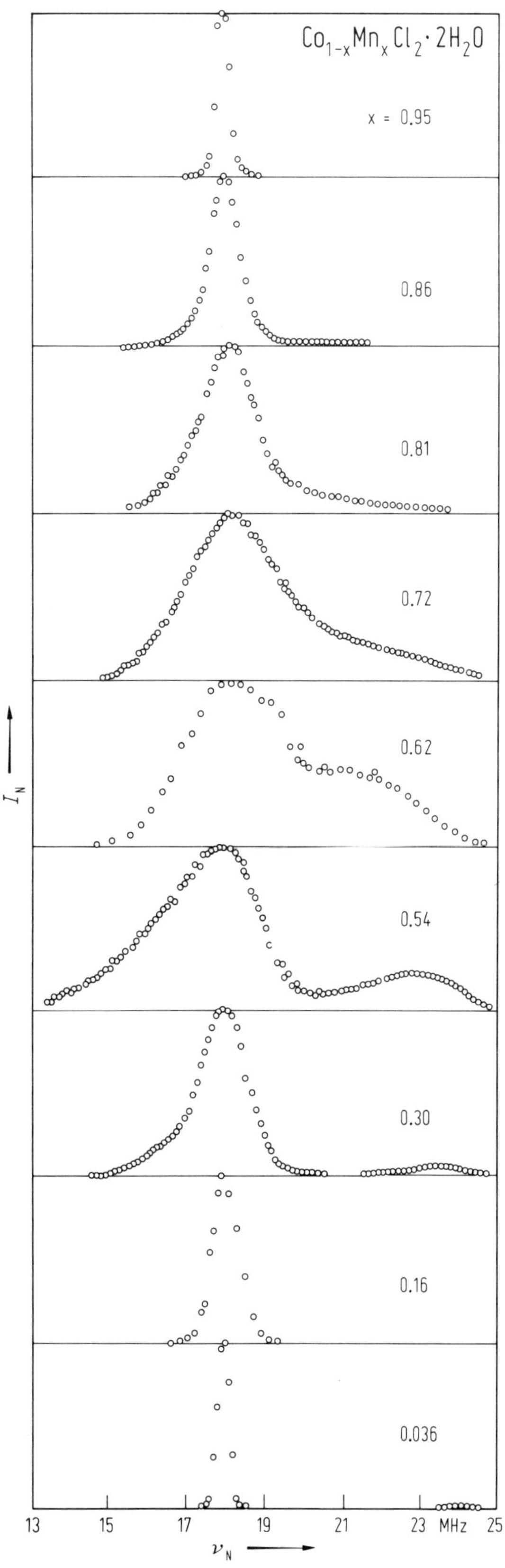

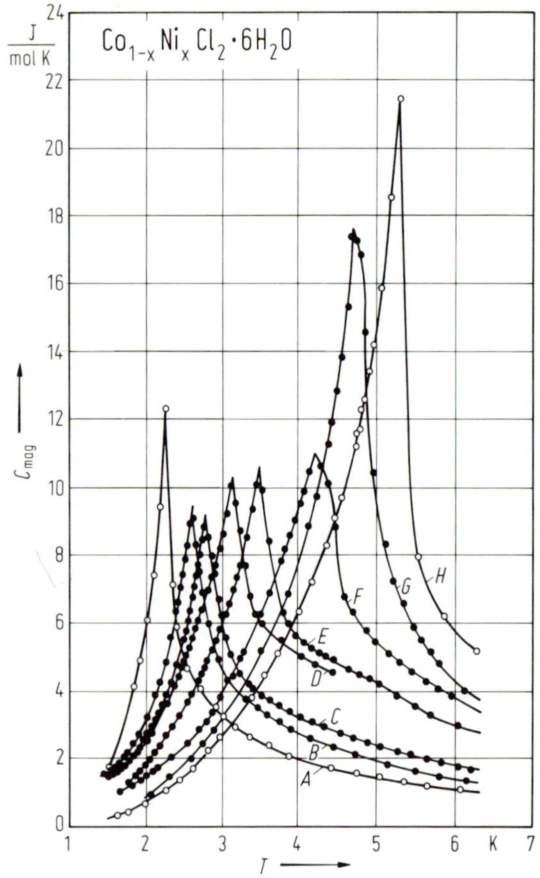

Fig. 120. $Co_{1-x}Ni_xCl_2 \cdot 6H_2O$. Magnetic specific heat vs. temperature. The values of x are *A*: 0; *B*: 0.208; *C*: 0.231; *D*: 0.473; *E*: 0.627; *F*: 0.708; *G*: 0.857; *H*: 1 [70T].

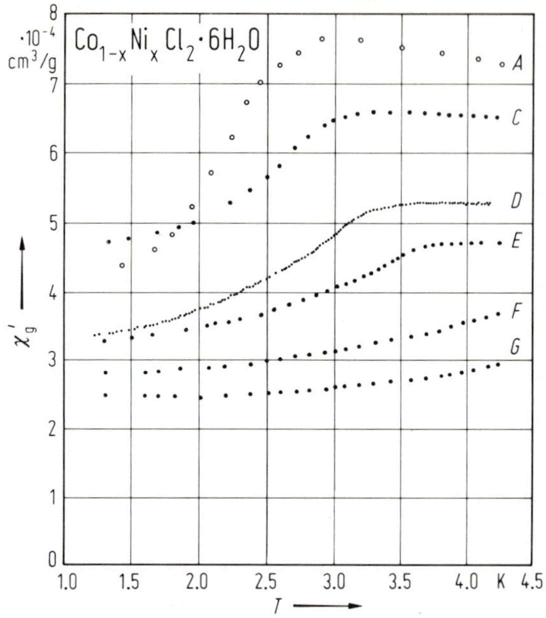

Fig. 121. $Co_{1-x}Ni_xCl_2 \cdot 6H_2O$. Real part of ac susceptibility vs. temperature. The values of x are *A*: 0; *C*: 0.231; *D*: 0.473; *E*: 0.627; *F*: 0.708; *G*: 0.857 [70T].

9.2.3 Dynamical magnetic and related properties of $M_{1-x}M'_xX_2$ and $M_{1-x}M'_xX_2 \cdot nH_2O$ compounds

9.2.3.1 Dynamical magnetic and related properties of $M_{1-x}M'_xF_2$ compounds

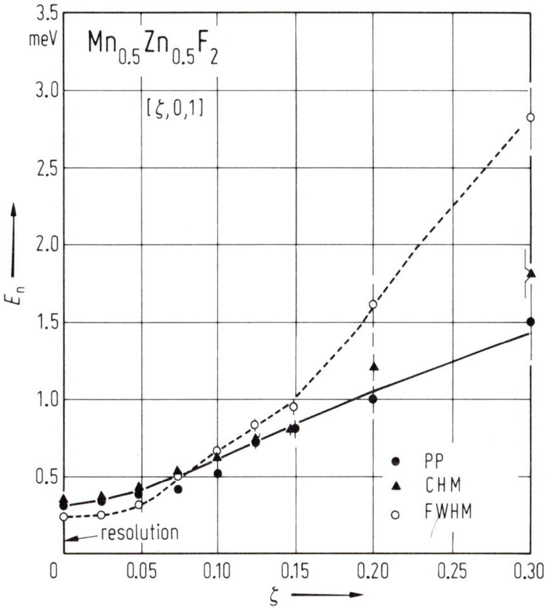

Fig. 122. $Mn_{0.5}Zn_{0.5}F_2$. Magnetic excitation energy vs. reduced wave vector coordinate along the [ζ, 0, 1] direction measured at 5 K. Peak position (PP), center of half maximum (CHM), and the full width at half maximum (FWHM) read off from the energy spectra are plotted. Solid line is calculated [87U].

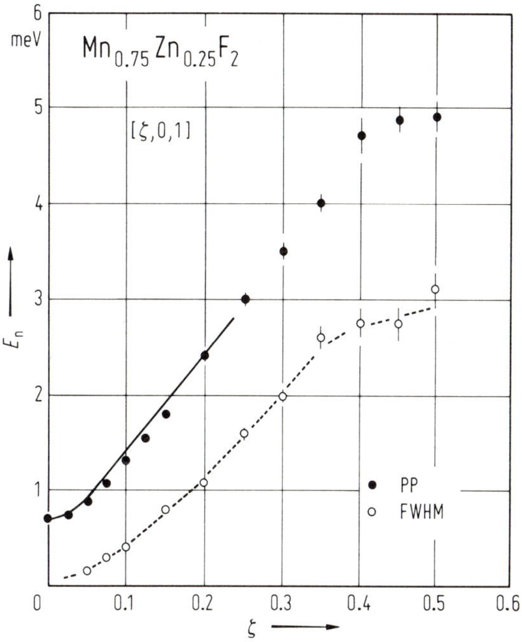

Fig. 123. $Mn_{0.75}Zn_{0.25}F_2$. Magnetic excitation energy vs. reduced wave vector coordinate along the [ζ, 0, 1] direction measured at 5 K. Peak position (PP) and the full width at half maximum (FWHM) of the energy spectra are plotted. Solid line is calculated [87U].

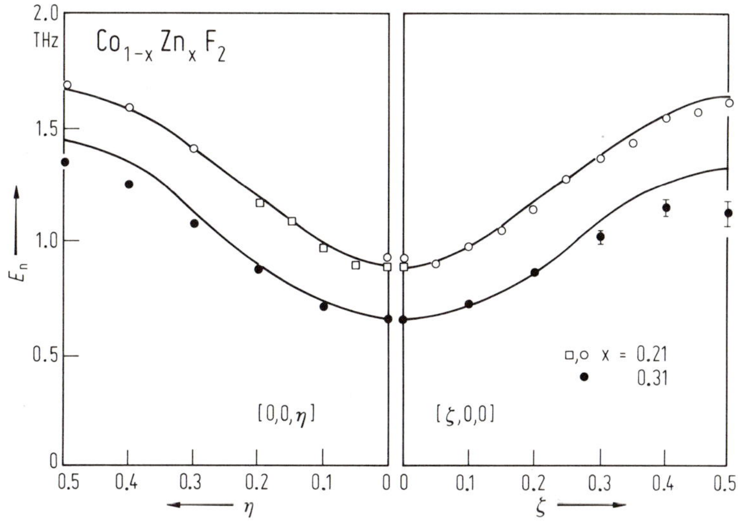

Fig. 124. $Co_{1-x}Zn_xF_2$. Magnetic excitation energy vs. reduced wave vector coordinate along the [0, 0, η] and [ζ, 0, 0] directions measured at 5 K. The values of x are 0.21 and 0.31 [75C].

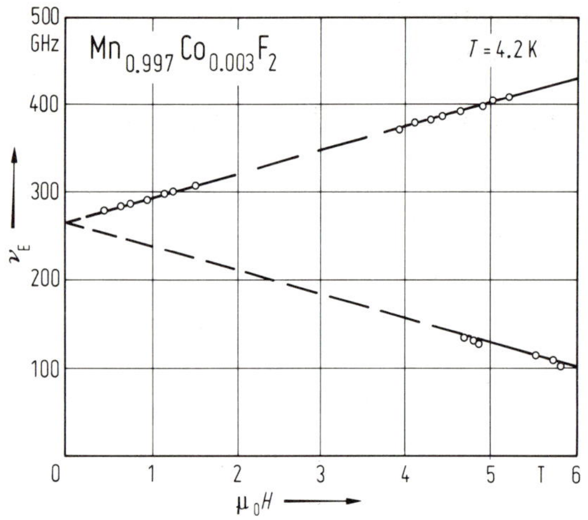

Fig. 125. $Mn_{0.997}Co_{0.003}F_2$. Frequency vs. applied magnetic field for the ESR position. $\boldsymbol{H} \parallel c$ axis [85T].

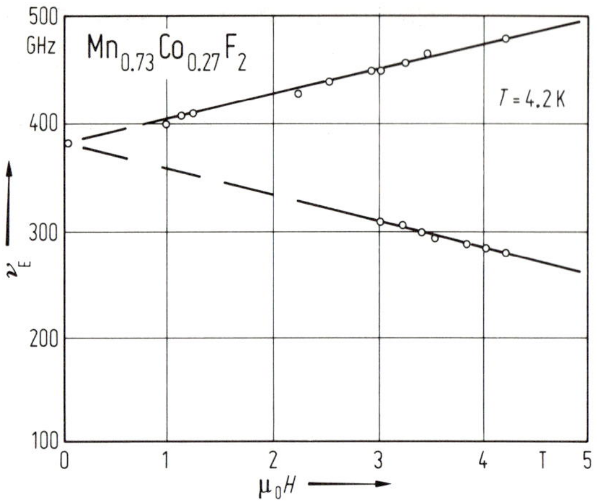

Fig. 126. $Mn_{0.73}Co_{0.27}F_2$. Frequency vs. applied magnetic field for the ESR position. $\boldsymbol{H} \parallel c$ axis [85T].

9.2.3.2 Dynamical magnetic and related properties of $M_{1-x}M'_xCl_2$ compounds

Table 10. Magnon gap energies and g value of $Fe_{1-x}Mn_xCl_2$ at 4.2 K [80B2].

x	Antiferromagnetic gap ν [GHz]	Ferromagnetic gap ν [GHz]	g value $g_{\parallel}$
0.0015(5)	498(10)	404(5)	4.00
0.039(5)	473(10)	379(6)	3.94
0.105(10)	393(10)	315(5)	3.89
0.151(10)	363(10)	282(10)	4.09
0.251(15)	295(20)	239(10)	4.00
0.578(30)		156(20)	3.8

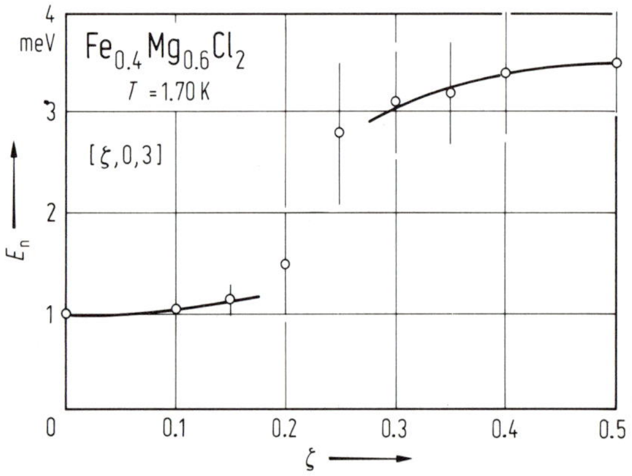

Fig. 127. $Fe_{0.4}Mg_{0.6}Cl_2$. Magnetic excitation energy vs. reduced wave vector coordinate along the [ζ, 0, 3] direction [87W].

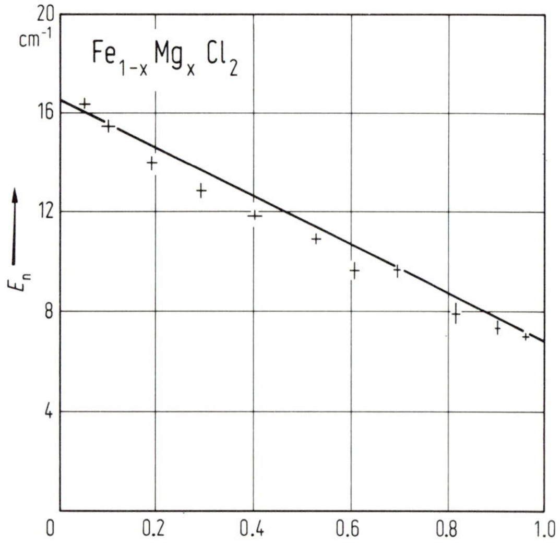

Fig. 128. $Fe_{1-x}Mg_xCl_2$. Peak position of far-infrared absorption vs. concentration at 1.5 K [76H].

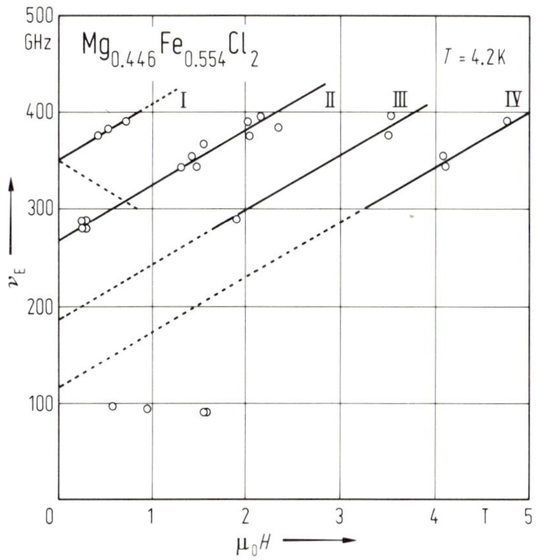

Fig. 129. $Mg_{0.446}Fe_{0.554}Cl_2$. Frequency vs. applied magnetic field for the ESR positions. $\boldsymbol{H} \parallel c$ axis. Straight lines I, II, III are theoretical, IV is only guide to the eye [90T].

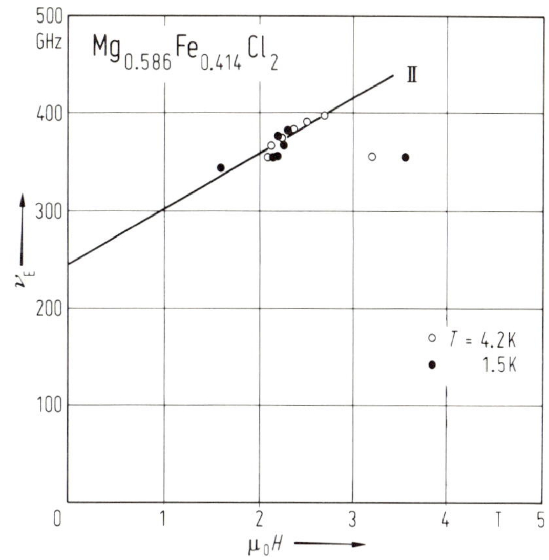

Fig. 130. $Mg_{0.586}Fe_{0.414}Cl_2$. Frequency vs. applied magnetic field for the ESR position II. $\boldsymbol{H} \parallel c$ axis. Line II is theoretical [90T].

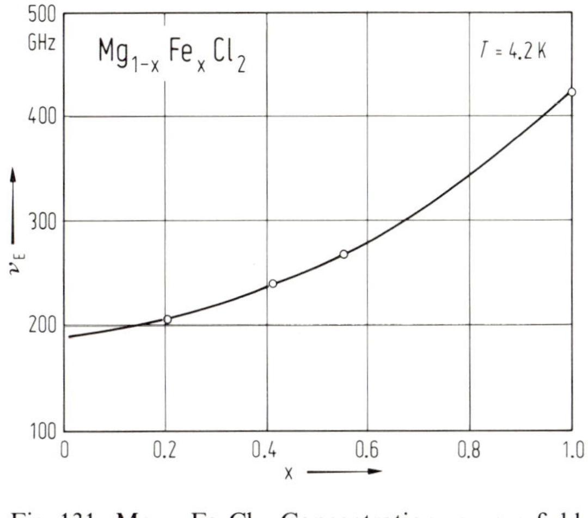

Fig. 131. $Mg_{1-x}Fe_xCl_2$. Concentration vs. zero field gap energy deduced from the ESR measurements [90T].

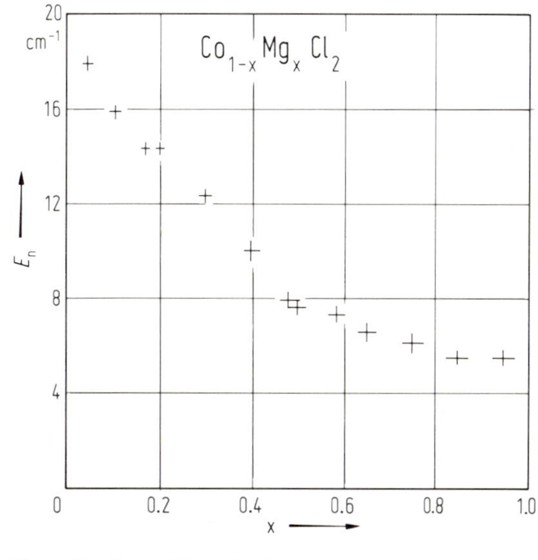

Fig. 132. $Co_{1-x}Mg_xCl_2$. Peak position of far-infrared absorption at 1.5 K vs. concentration [79W].

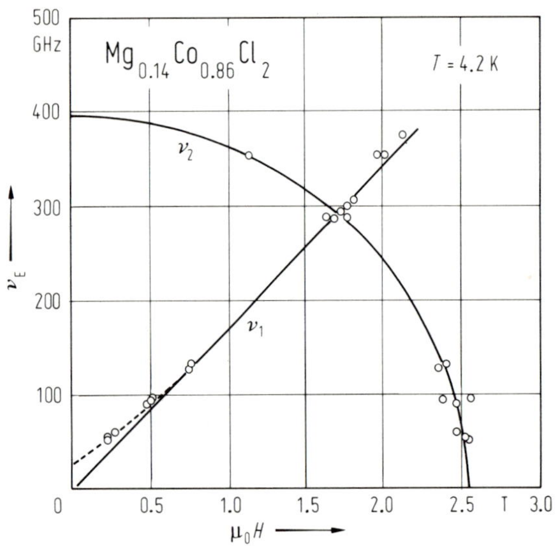

Fig. 133. $Mg_{0.14}Co_{0.86}Cl_2$. Frequency vs. applied magnetic field for the ESR positions. $\boldsymbol{H} \perp c$ axis. Full lines are theoretical [88K].

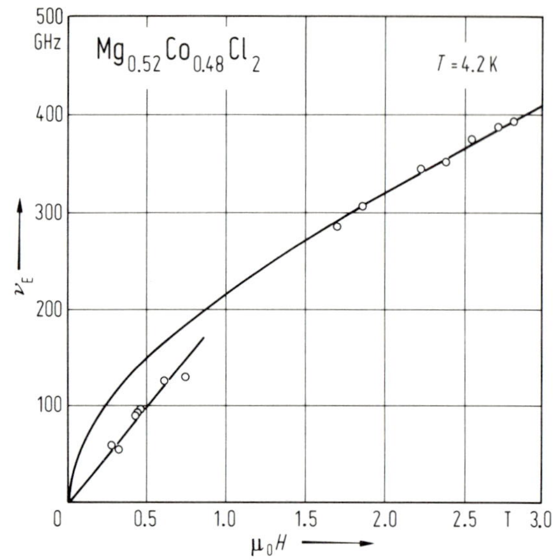

Fig. 134. $Mg_{0.52}Co_{0.48}Cl_2$. Frequency vs. applied magnetic field for the ESR positions. $\boldsymbol{H} \perp c$ axis. Full lines are theoretical [88K].

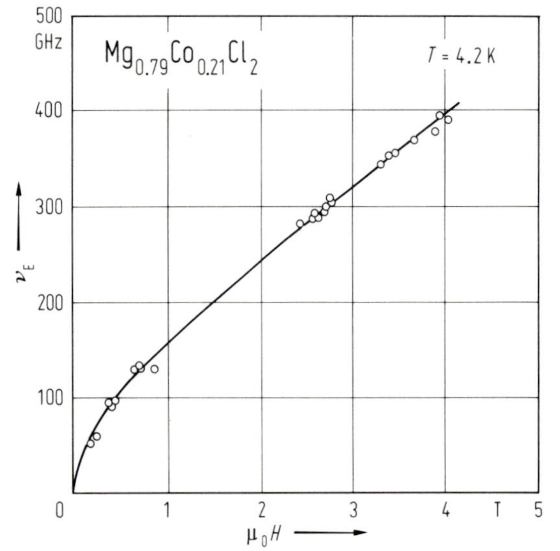

Fig. 135. $Mg_{0.79}Co_{0.21}Cl_2$. Frequency vs. applied magnetic field for the ESR positions. $\boldsymbol{H} \perp c$ axis. Full line is theoretical [88K].

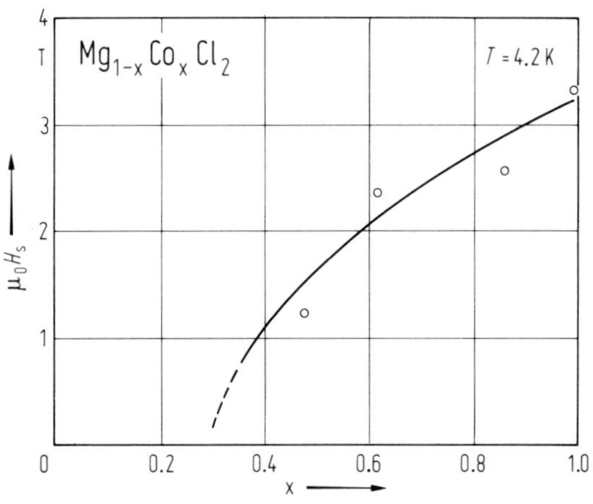

Fig. 138. $Mg_{1-x}Co_xCl_2$. Saturation field from the antiferromagnetic to ferromagnetic phase vs. concentration deduced from the ESR measurements. $\boldsymbol{H} \perp c$ axis. Full line is guide to the eye [88K].

For Figs. 136, 137 see next page.

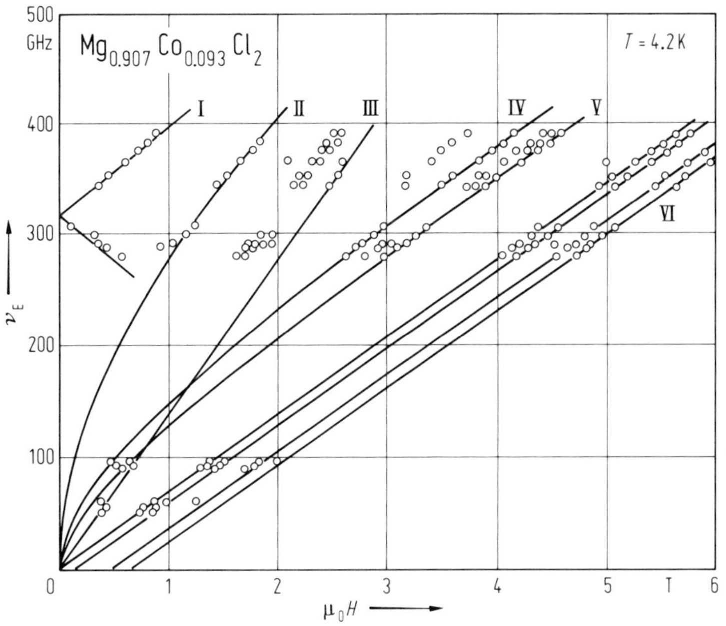

Fig. 136. $Mg_{0.907}Co_{0.093}Cl_2$. Frequency vs. applied magnetic field for the ESR positions. $H \perp c$ axis. Full lines I through V are theoretical, line VI is guide to the eye [89K].

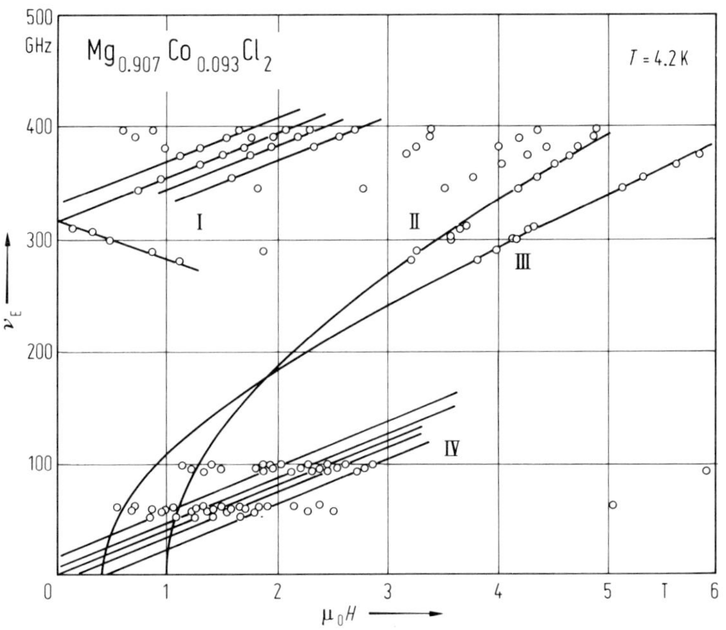

Fig. 137. $Mg_{0.907}Co_{0.093}Cl_2$. Frequency vs. applied magnetic field for the ESR positions. $H \parallel c$ axis. Full lines are theoretical [89K].

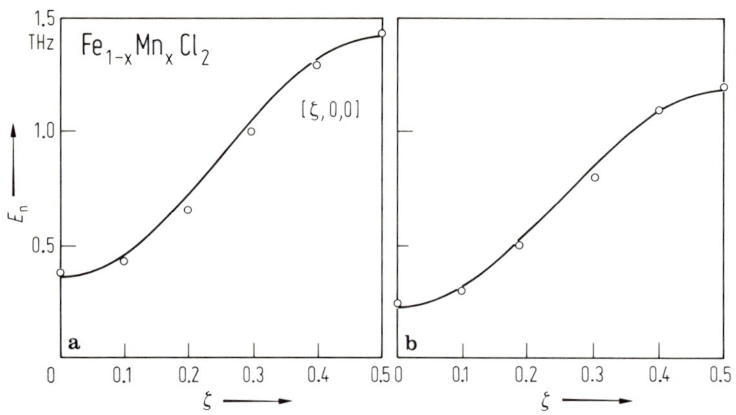

Fig. 139. $Fe_{1-x}Mn_xCl_2$. Magnetic excitation energy vs. reduced wave vector coordinate along the [ζ, 0, 0] direction at 3 K. (a) x = 0.105, (b) x = 0.251 [81B2].

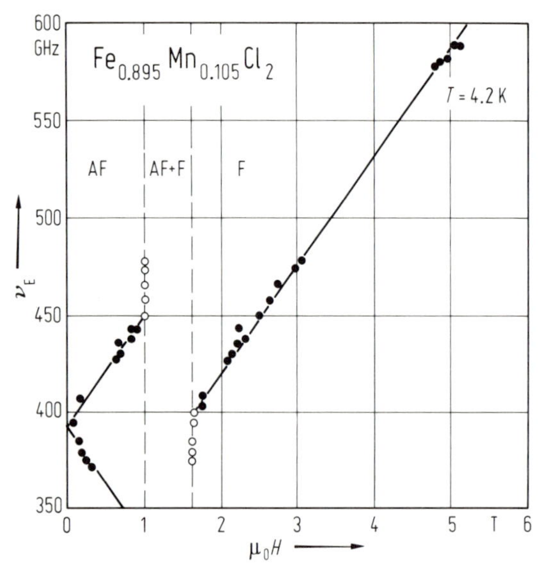

Fig. 140. $Fe_{0.895}Mn_{0.105}Cl_2$. Frequency vs. applied magnetic field for the ESR positions. **H** ∥ *c* axis [80B2].

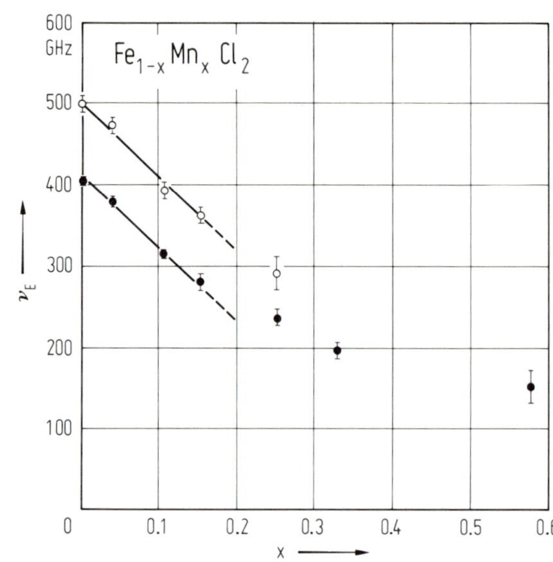

Fig. 141. $Fe_{1-x}Mn_xCl_2$. Concentration vs. zero field gap energy deduced from the ESR measurements at 4.2 K. (○) AF phase, (●) F phase [80B2].

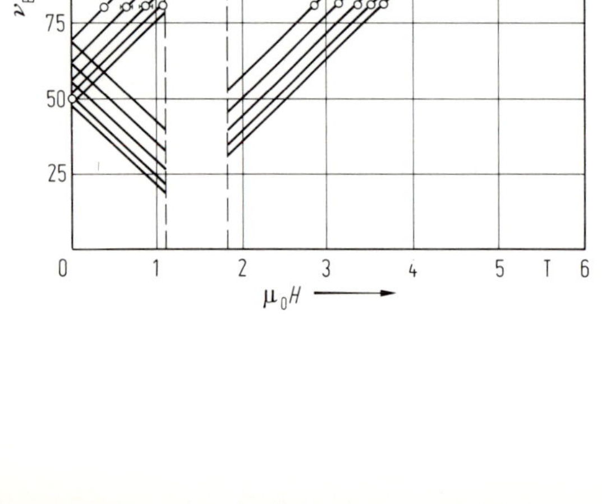

Fig. 142. $Fe_{0.995}Mn_{0.005}Cl_2$. Frequency vs. applied magnetic field for the ESR positions. **H** ∥ *c* axis [80T].

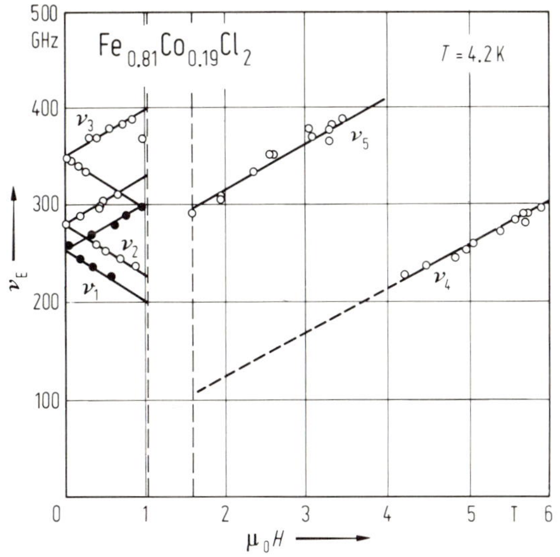

Fig. 143. $Fe_{0.81}Co_{0.19}Cl_2$. Frequency vs. applied magnetic field for the ESR positions. $\boldsymbol{H} \parallel c$ axis [84T1].

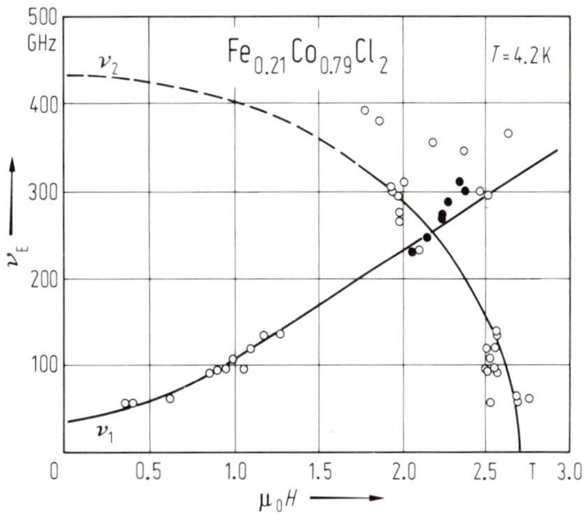

Fig. 144. $Fe_{0.21}Co_{0.79}Cl_2$. Frequency vs. applied magnetic field for the ESR positions. $\boldsymbol{H} \perp c$ axis. Full line ν_2 is theoretical; open circles: uniform mode, full circles: localized mode [84T1].

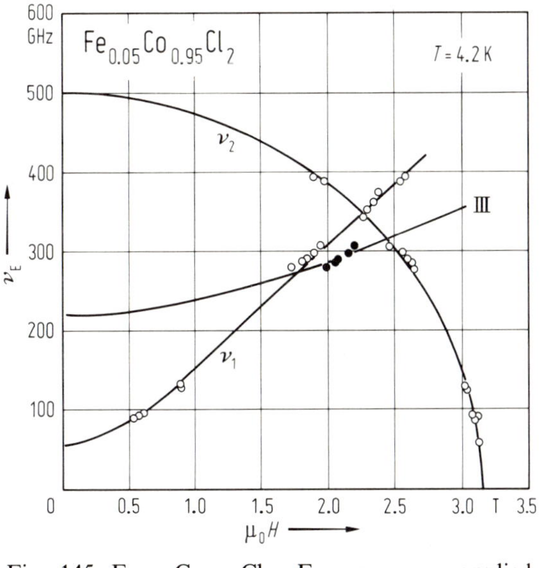

Fig. 145. $Fe_{0.05}Co_{0.95}Cl_2$. Frequency vs. applied magnetic field for the ESR positions. $\boldsymbol{H} \perp c$ axis. (○) antiferromagnetic resonance, (●) Fe^{2+} localized mode. Full lines ν_2 and III are theoretical [87K].

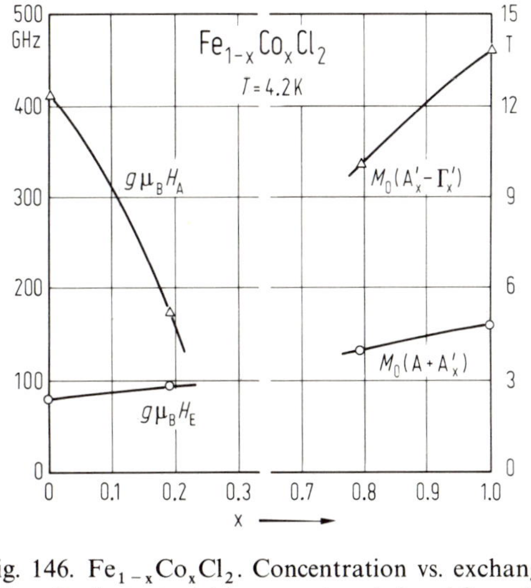

Fig. 146. $Fe_{1-x}Co_xCl_2$. Concentration vs. exchange (O) and anisotropy (△) parameters deduced from the ESR measurements [84T1].

9.2.3.3 Dynamical magnetic and related properties of $Fe_{1-x}Co_xBr_2$ compound

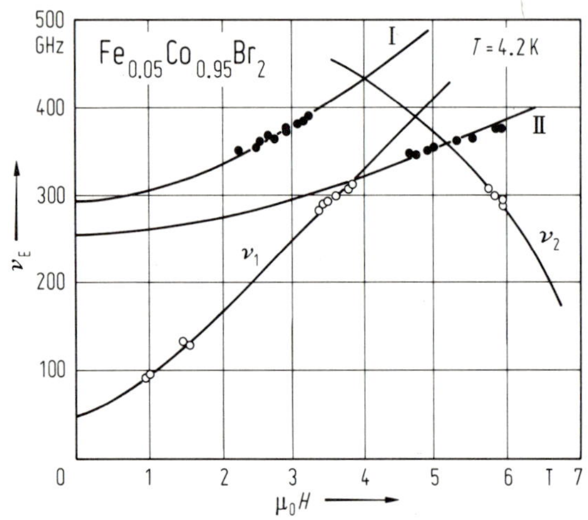

Fig. 147. $Fe_{0.05}Co_{0.95}Br_2$. Frequency vs. applied magnetic field for the ESR positions. $\boldsymbol{H} \perp c$ axis. (○) antiferromagnetic resonance, (●) Fe^{2+} localized mode. Full lines I and II are theoretical [85K3].

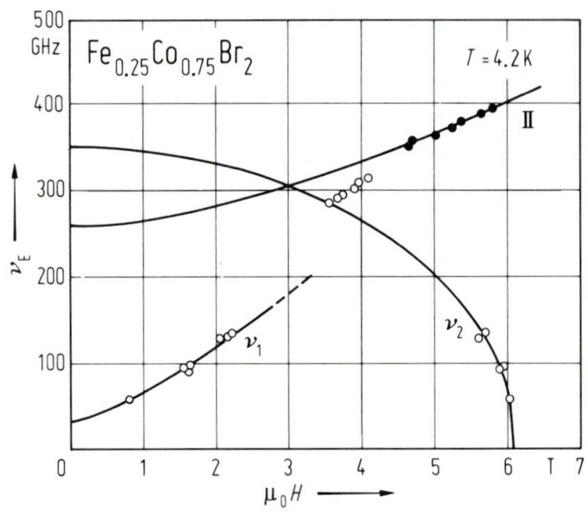

Fig. 148. $Fe_{0.25}Co_{0.75}Br_2$. Frequency vs. applied magnetic field for the ESR positions. $\boldsymbol{H} \perp c$ axis. (○) antiferromagnetic resonance, (●) Fe^{2+} localized mode. Full line II is theoretical [85K3].

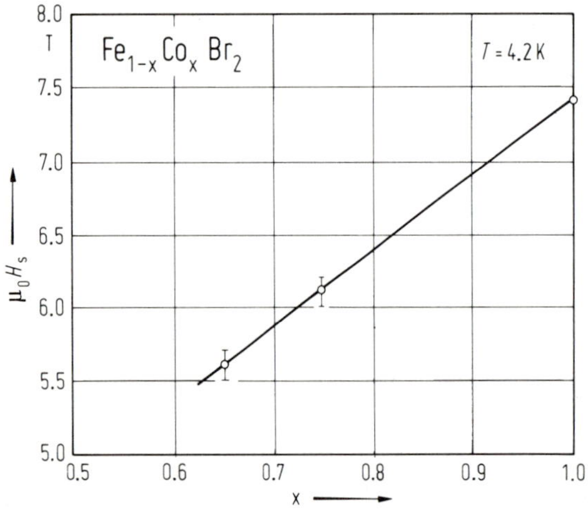

Fig. 149. $Fe_{1-x}Co_xBr_2$. Saturation field from the antiferromagnetic to ferromagnetic phase vs. concentration deduced from the ESR measurements. $\boldsymbol{H} \perp c$ axis [84T2].

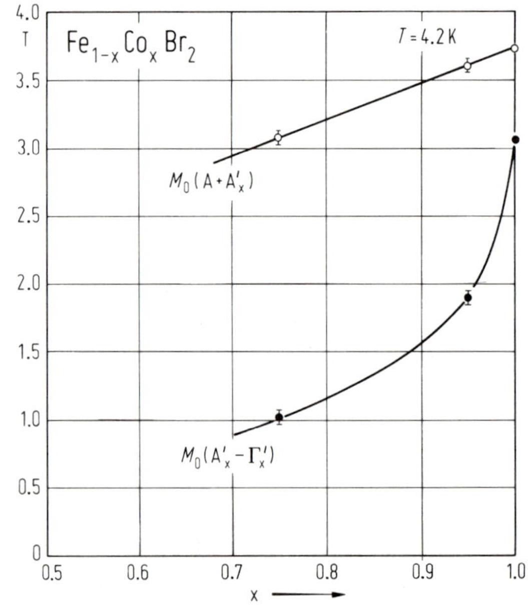

Fig. 150. $Fe_{1-x}Co_xBr_2$. Concentration vs. exchange (○) and anisotropy (●) parameters deduced from the ESR measurements [84T2].

9.2.3.4 Dynamical magnetic and related properties of $Fe_{1-x}Co_xCl_2 \cdot 2H_2O$ compound

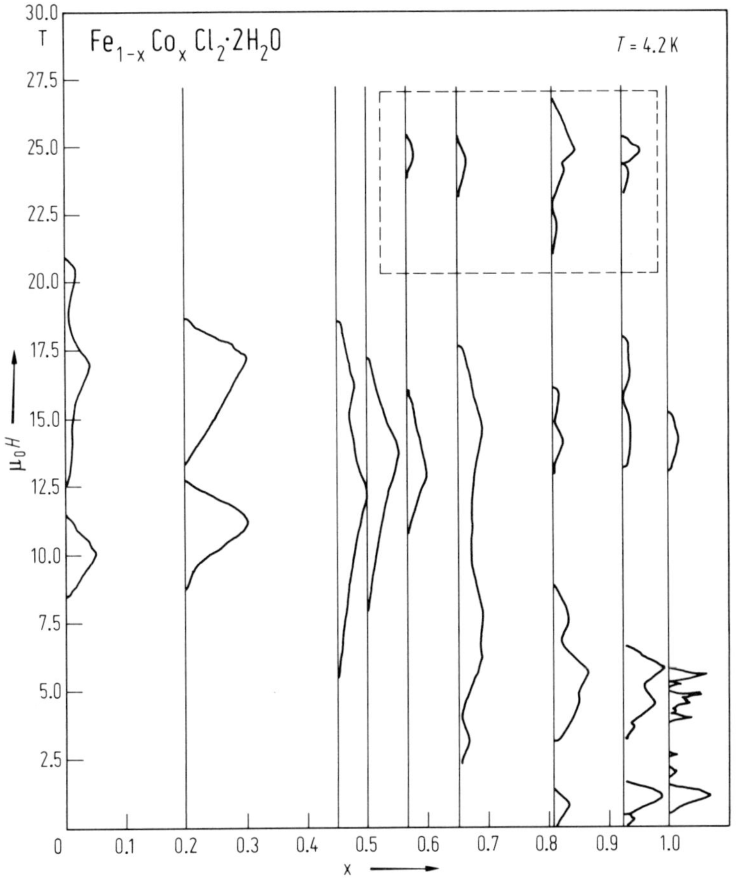

Fig. 151. $Fe_{1-x}Co_xCl_2 \cdot 2H_2O$. ESR spectra in single crystals with different concentrations observed at 4.2 K with electromagnetic waves at 0.337 mm. $\boldsymbol{H} \parallel b$ axis. The spectra in the dotted rectangle show an Fe^{2+} localized excitation [85K2].

9.2.4 References for 9.2

66W Wertheim, G.K., Buchanan, D.N.E., Guggenheim, H.J.: Phys. Rev. **152** (1966) 527.
69W Wertheim, G.K., Guggenheim, H.J.: Phys. Rev. **178** (1969) 804.
70T Takeda, K., Matsuura, M., Haseda, T.: J. Phys. Soc. Jpn. **29** (1970) 885.
75C Cowley, R.A., Dietrich, O.W., Jones, D.A.: J. Phys. C **8** (1975) 3023.
76H Hayes, W., Walker, P.J., Wiltshire, M.C.K.: J. Phys. C **9** (1976) L255.
77W Wood, T.E., Day, P.: J. Phys. C **10** (1977) L333.
79K1 Katsumata, K., Kobayashi, M., Satō, T., Miyako, Y.: Phys. Rev. B **19** (1979) 2700.
79K2 Kobayashi, M., Katsumata, K., Satō, T., Miyako, Y.: J. Phys. Soc. Jpn. **46** (1979) 1467.
79W Wiltshire, M.C.K.: J. Phys. C **12** (1979) 3571.
80B1 Belanger, D.P., Borsa, F., King, A.R., Jaccarino, V.: J. Magn. Magn. Mater. **15–18** (1980) 807.
80B2 Bertrand, D., Fert, A.R., Magariño, J., Tuchendler, J., Legrand, S.: J. Phys. C **13** (1980) 5165.
80C Cowley, R.A., Shirane, G., Birgeneau, R.J., Svensson, E.C., Guggenheim, H.J.: Phys. Rev. B **22** (1980) 4412.
80I Ito, A., Someya, Y., Katsumata, K.: Solid State Commun. **36** (1980) 681.
80T Tuchendler, J., Magariño, J., Bertrand, D., Fert, A.R.: J. Phys. C **13** (1980) 233.
80W1 Wong, Po-zen, Horn, P.M., Birgeneau, R.J., Safinya, C.R., Shirane, G.: Phys. Rev. Lett. **45** (1980) 1974.
80W2 Wood, T.E., Day, P.: J. Magn. Magn. Mater. **15–18** (1980) 782.
81B1 Bertrand, D., Fert, A.R., Legrand, S., Redoulès, J. P., Schmidt, M.C.: J. Phys. C **14** (1981) 1789.
81B2 Bertrand, D., Petitgrand, D., Hennion, B., Legrand, S.: J. Phys. C **14** (1981) 1975.
81H Haywood, S.K., Wood, T.E., Day, P.: J. Phys. C **14** (1981) 2697.
81K1 King, A.R., Jaccarino, V., Sakakibara, T., Motokawa, M., Date, M.: Phys. Rev. Lett. **47** (1981) 117.
81K2 King, A.R. Jaccarino, V.: J. Appl. Phys. **52** (1981) 1785.
82B Belanger, D.P., King, A.R., Jaccarino, V.: Phys. Rev. Lett. **48** (1982) 1050.
83H Hagen, M., Cowley, R.A., Satija, S.K., Yoshizawa, H., Shirane, G., Birgeneau, R.J., Guggenheim, H.J.: Phys. Rev. B **28** (1983) 2602.
83I Ito, A., Tamaki, T., Someya, Y., Ikeda, H.: Physica **120B** (1983) 207.
83S1 Someya, Y., Ito, A., Katsumata, K.: J. Phys. Soc. Jpn. **52** (1983) 254.
83S2 Shapira, Y., Oliveira jr., N.F.: Phys. Rev. B **27** (1983) 4336.
83W1 Wong, P., Horn, P.M., Birgeneau, R.J., Shirane, G.: Phys. Rev. B **27** (1983) 428.
83W2 Wong, Po-zen, von Molnar, S., Dimon, P.: Solid State Commun. **48** (1983) 573.
84B Bertrand, D., Bensamka, F., Fert, A.R., Gelard, J., Redoulès, J.P., Legrand, S.: J. Phys. C **17** (1984) 1725.
84D DeFotis, G.C., Pohl, C., Pugh, S.A.: J. Chem. Phys. **80** (1984) 2079.
84H Howes, B.D., Price, D.C., Wiltshire, M.C.K.: J. Phys. C **17** (1984) 3669.
84K Katsumata, K., Tuchendler, J., Legrand, S.: Phys. Rev. B **30** (1984) 1377.
84S Shapira, Y., Oliveira jr., N.F., Foner, S.: Phys. Rev. B **30** (1984) 6639.
84T1 Tuchendler, J., Katsumata, K.: Solid State Commun. **50** (1984) 849.
84T2 Tuchendler, J., Katsumata, K., Legrand, S.: Solid State Commun. **51** (1984) 557.
84W Wong, Po-zen, Cable, J.W.: Phys. Rev. B **30** (1984) 485.
85K1 Katsumata, K., Yoshizawa, H., Shirane, G., Birgeneau, R.J.: Phys. Rev. B **31** (1985) 316.
85K2 Katsumata, K., Uyeda, C., Motokawa, M., Date, M.: J. Phys. Soc. Jpn. **54** (1985) 1244.
85K3 Katsumata, K., Tuchendler, J.: Solid State Commun. **54** (1985) 429.
85T Tuchendler, J.: Solid State Commun. **55** (1985) 1017.
85W Wong, Po-zen, von Molnar, S., Palstra, T.T.M., Mydosh, J.A., Yoshizawa, H., Shapiro, S.M., Ito, A.: Phys. Rev. Lett. **55** (1985) 2043.

85Y Yoshizawa, H., Cowley, R.A., Shirane, G., Birgeneau, R.J.: Phys. Rev. B **31** (1985) 4548.

86I Ito, A., Torikai, E., Kitazawa, M., Tamaki, T., Goto, T., Sakakibara, T., Todo, S., Oguro, I.: J. Magn. Magn. Mater. **54–57** (1986) 39.

86K Kleemann, W., King, A.R., Jaccarino, V.: Phys. Rev. B **34** (1986) 479.

86M1 Mydosh, J.A., King, A.R., Jaccarino, V.: J. Magn. Magn. Mater. **54–57** (1986) 47.

86M2 Mitchell, P.W., Cowley, R.A., Yoshizawa, H., Böni, P., Uemura, Y.J., Birgeneau, R.J.: Phys. Rev. B **34** (1986) 4719.

87K Katsumata, K., Tuchendler, J.: Solid State Commun. **61** (1987) 659.

87U Uemura, Y.J., Birgeneau, R.J.: Phys. Rev. B **36** (1987) 7024.

87W Wong, Po-zen, Yoshizawa, H., Shapiro, S.M., Mydosh, J.A., van den Berg, J., Palstra, T.T.M.: Phys. Rev. Lett. **58** (1987) 1276.

88D DeFotis, G.C., Mantus, D.S., McGhee, E.M., Echols, K.R., Wiese, R.S.: Phys. Rev. B **38** (1988) 11486.

88K Katsumata, K., Kawai, S., Tuchendler, J.: Solid State Commun. **65** (1988) 1211.

88M Montenegro, F.C., Rezende, S.M., Coutinho-Filho, M.D.: J. Appl. Phys. **63** (1988) 3755.

88R Ramos, C.A., King, A.R., Jaccarino, V.: Phys. Rev. B **37** (1988) 5483.

88T Thurston, T.R., Peters, C.J., Birgeneau, R.J., Horn, P.M.: Phys. Rev. B **37** (1988) 9559.

89K Katsumata, K., Tuchendler, J.: J. Phys.: Condens. Matter **1** (1989) 10507.

90D Deguchi, H., Takahashi, K., Kubo, H., Takeda, K.: J. Phys. Soc. Jpn. **59** (1990) 1071.

90T Tuchendler, J., Katsumata, K.: Solid State Commun. **74** (1990) 1159.

92Z Zenmyo, K., Kubo, H.: J. Phys. Soc. Jpn. **61** (1992) 1025.

9.3 MX_3 compounds (M = 3d element, X = halogen element)

9.3.1 Introduction

9.3.1.1 General remarks

This chapter is devoted to description of macroscopic as well as microscopic magnetic and related properties of MX_3 compounds. Here, M stands for 3d transition metal ions, whereas X represents halogen ions.

The macroscopic magnetic properties such as Curie constant, paramagnetic Curie temperature and magnetic ordering temperature are presented in section 9.3.2. Most of these compounds show antiferromagnetic ordering and some helimagnetic or ferromagnetic ordering at low temperatures. The magnetic susceptibility per mole (χ_m) of these magnets well above the Néel or Curie temperature is expressed by

$$\chi_m = C_m/(T - \Theta) ,$$

where T is the absolute temperature, C_m the Curie constant per mole and Θ is the paramagnetic Curie temperature. By fitting the susceptibility data with this formula, one obtains the values of C_m and Θ.

The magnetic phase transition from the paramagnetic to antiferromagnetic, helimagnetic or ferromagnetic state is of second order. Specific heat measurement on magnetic substances gives the most accurate value for the transition temperature. Susceptibility and neutron scattering measurements also give the transition temperature.

The spin structures in the ordered states of antiferromagnets and helimagnets are directly determined from neutron scattering study. Electron spin resonance (ESR) and nuclear magnetic resonance (NMR) experiments also give information on the spin structure.

Table 1 lists the paramagnetic susceptibilities for some MX_3 compounds.

The other data are presented in a table for each of the MF_3 (Table 2), MCl_3 (Table 3), MBr_3 (Table 4), and MI_3 (Table 5) compounds, followed by the relevant figures. These figures are arranged in the following order: first the crystal and spin structures are given. Then, the specific heat and magnetic susceptibility data are presented as a function of temperature. Finally, the magnetization data is given as a function of applied magnetic field.

In section 9.3.3 we summarize the microscopic magnetic and related properties of MX_3 compounds. We concentrate here to the Hamiltonians which are the basis of the microscopic magnetism of these compounds. The magnetic excitations in the ordered states of antiferromagnets, helimagnets and ferromagnets are well described by spin-wave picture at low temperatures. The spin-wave spectra are calculated with the Hamiltonians. The magnetic excitation spectra are directly observed by neutron inelastic scattering technique. Thus, one can determine the parameters of the Hamiltonians experimentally. Some of the parameters of the Hamiltonian are determined from NMR experiments. The Hamiltonians and their parameters of MX_3 compounds are given in Table 6. The experimental spin-wave spectra and NMR data from which these parameters have been obtained are presented.

9.3.1.2 List of frequently used symbols and abbreviations

Symbols

a, b, c	Å	lattice parameters
C_m	$emu\,K\,mol^{-1} = cm^3\,K\,mol^{-1}$	Curie constant per mole
C_{mag}	$J\,K^{-1}\,mol^{-1}$	magnetic specific heat
C_p	$cal\,K^{-1}\,mol^{-1}$	specific heat (1 cal = 4.184 J)

D	cm^{-1}	single ion anisotropy constant with the corresponding Hamiltonian of the form DS_z^2
E_n	meV, cm^{-1}, THz	excitation energy (1 meV = 8.06554 cm^{-1} = 0.241799 THz)
g_i		spectroscopic splitting factor (i = x, y, z)
$\boldsymbol{H}$	Oe, $A\,m^{-1}$	applied magnetic field (sometimes given as $\mu_0 H$ in tesla (T))
H_A		anisotropy field
H_D		Hamiltonian with dipole-dipole interaction only
h, k, l		reciprocal lattice indices
i, j, k		site indices in real space
J_{ij}	cm^{-1}	exchange interaction constant between the spins i and j (J/k in K)
k		Boltzmann constant
p_A	μ_B	magnetic moment per atom (ion)
p_{eff}	μ_B	effective magnetic moment per molecule (formula unit)
$p_{Fe}^{\parallel}$	μ_B	magnetic moment per Fe atom parallel to net ferromagnetic moment
$\boldsymbol{S}$		spin operator
T	K	temperature
T_C	K	Curie temperature
T_N	K	Néel temperature
x, y, z		coordinate axes
z_i		number of interacting neighbors
α, β, γ	deg	lattice parameters
ζ		reduced wavevector coordinate
Θ	K	paramagnetic Curie temperature
μ_B	$J\,T^{-1}$	Bohr magneton
ν	Hz	frequency
σ	$emu\,g^{-1} = G\,cm^3\,g^{-1}$	magnetic moment per gram (specific magnetization)
σ_m	$emu\,mol^{-1} = G\,cm^3\,mol^{-1}$	magnetic moment per mole
χ_g	$emu\,g^{-1} = cm^3\,g^{-1}$	magnetic susceptibility per gram
χ_m	$emu\,mol^{-1} = cm^3\,mol^{-1}$	magnetic susceptibility per mole
$\chi_{\parallel}, \chi_{\perp}$		magnetic susceptibility parallel and perpendicular to a principal crystallographic direction

Abbreviations

AF	antiferromagnetic phase
ESR	electron spin resonance
F	ferromagnetic phase
Heli	helimagnetic phase
NMR	nuclear magnetic resonance
P	paramagnetic phase

9.3.2 Macroscopic magnetic and related properties of MX_3 compounds

Table 1. MX_3 compounds. Paramagnetic susceptibilities derived from Gouy measurements and already earlier given in Landolt-Börnstein subvolumes II/2 and 10.

Compound	T [K]	χ_m [$\cdot 10^{-6}$ cm^3/mol]	Ref.
TiF_3	0···300	Fig. 2	74A
$TiCl_3$	80···450	Fig. 10	60O
$TiBr_3$	441	520	47K2
	273	660	
	195	680	
	90	220	
VBr_3	293	2910	36K
	195	4230	
	90	8660	
VI_3	295	3610	69J2
	195	5420	
	90	16990	

9.3.2.1 MF_3 compounds

Table 2. MF_3 compounds.

Substance (Figs.)	Crystal type	Lattice constants*)	Molar Curie constant (in $cm^3\,K\,mol^{-1}$)	Paramagnetic Curie temperature	Magnetic order	Critical temperature	Spin structure
TiF_3 (Fig. 2)	VF_3 Fig. 1 $R\bar{3}c$	$a = 5.521$ Å $\alpha = 58.94°$ [61M; LBIII/7a, p. 51]		− 10 K [54E; LBIII/4a, p. 220] − 75 K [74A; LBII/12a, p. 573]	48 K [74A]		
VF_3**) (Fig. 1)	VF_3	$a = 5.373$ Å $\alpha = 57.52°$ [51J]			P down to 4.2 K [58 W]		
CrF_3***) (Figs. 3 ··· 5)	VF_3	$a = 5.2643$ Å $\alpha = 56.563°$ [57J]	1.85 [59H1]	− 124 K [59H1]	Canted ← P AF [58W, 59H2]	69.8 K [58H]	Spins couple antiferro-magnetically with their six nearest neighbors. Spins ∥ (111) planes [58W]
MnF_3 (Figs. 6, 7)	Monoclinic ($C2/c$-C^6_{2h}) Fig. 6	$a = 8.904$ Å $b = 5.037$ Å $c = 13.448$ Å $\beta = 92.74°$ [57H2]	3.10 [58B]	8 K [58B]	AF ← P [58B]	magnetic measurement: 47 K [58B]; neutron diffraction: 43 K [58W]	(101) ferromagnetic sheets which couple antiferromagnetically with the adjacent planes. Spins in the (101) planes [58W]
FeF_3 (Figs. 8, 9)	VF_3	$a = 5.362$ Å $\alpha = 57.99°$ [57H1]			Canted ← P AF [58W, 68L] canting angle Fig, 9	magnetic measurement: 362.2 K [68L]; neutron diffraction: 394 K [58W]	Spins couple antiferro-magnetically with their six nearest neighbors. Spins ∥ (111) planes [58W]

Table 2 (*continued*)

Substance (Figs.)	Crystal type	Lattice constants*)	Molar Curie constant (in $cm^3 K mol^{-1}$)	Paramagnetic Curie temperature	Magnetic order	Critical temperature	Spin structure
					weak ferro-magnetism: $p^{\parallel}_{Fe}(295 K)$ = $0.005\mu_B$ [66S1, LBIII/4a, p. 221]		$p_{Fe}(4.2 K) = 5\ \mu_B$ [58W]
FeF_3	Hexagonal tungsten bronze (P6/mmm) [86L]	a = 7.413 Å c = 3.7949 Å [86L]			AF ← P [86L]	97 K [86L]	Three sublattices at 120° one from each other in the (001) plane. Antiferromagnetic along the c axis [86L]
FeF_3	Cubic pyrochlore (Fd3m) [86D]	a = 10.325 Å [86D]			AF ← P [86F]	20 K [86F]	Four ferromagnetic sublattices oriented at 109° from each other, so that the resulting moment is zero [86F]
CoF_3	VF_3	a = 5.279 Å α = 56.97° [57H1]			AF ← P	460 K [58W, 69F]	Spins couple antiferro-magnetically with their six nearest neighbors. Spins ⊥ (111) planes [58W]

*) Room temperature unless otherwise specified.
**) VF_3 has an orthorhombic structure (Cmcm) with a = 7.425 Å, b = 12.835 Å and c = 7.563 Å when prepared from a solution [87D].
***) CrF_3 has an orthorhombic structure (Cmcm) with a = 7.276 Å, b = 12.480 Å and c = 7.364 Å when prepared from a solution [87D].

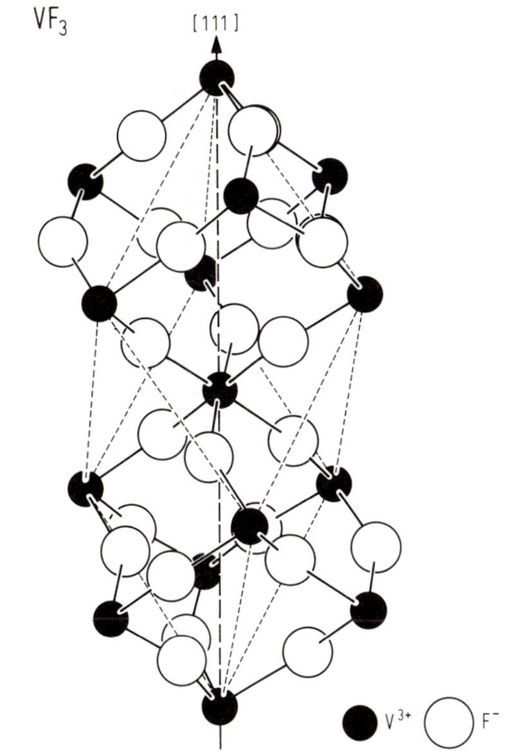

Fig. 1. VF_3. Crystal structure (space group R$\bar{3}$c). Filled circles: V^{3+} ions; open circles: F^- ions. The arrow indicates the [111] direction [57H1].

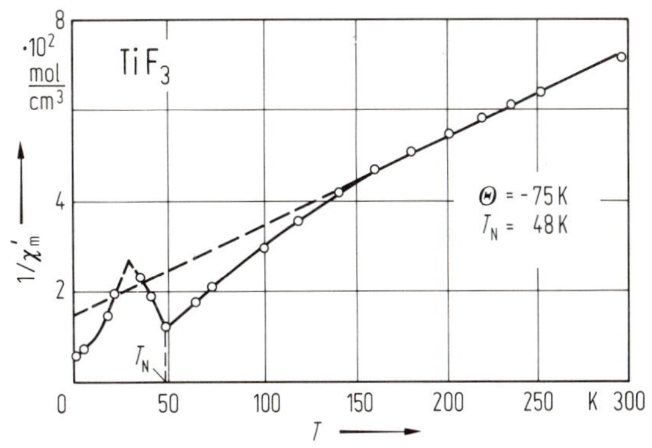

Fig. 2. TiF_3. Temperature dependence of reciprocal molar susceptibility. [74A, LBII/12a, p. 573].

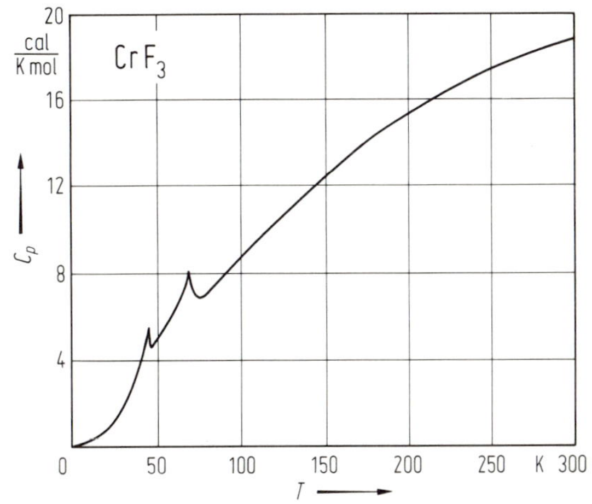

Fig. 3. CrF_3. Molar specific heat vs. temperature [58H].

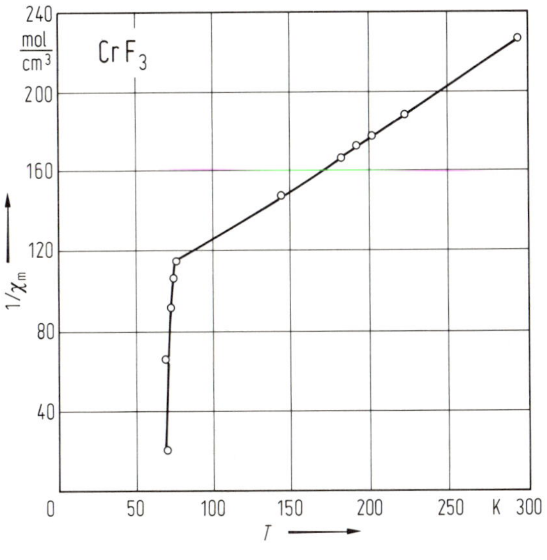

Fig. 4. CrF_3. The reciprocal susceptibility per mole vs. temperature [59H1].

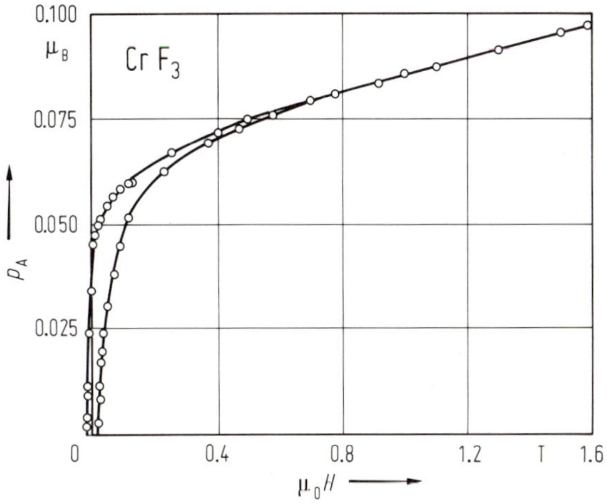

Fig. 5. CrF_3. Magnetic moment per atom vs. magnetic field at 4.2 K [59H2].

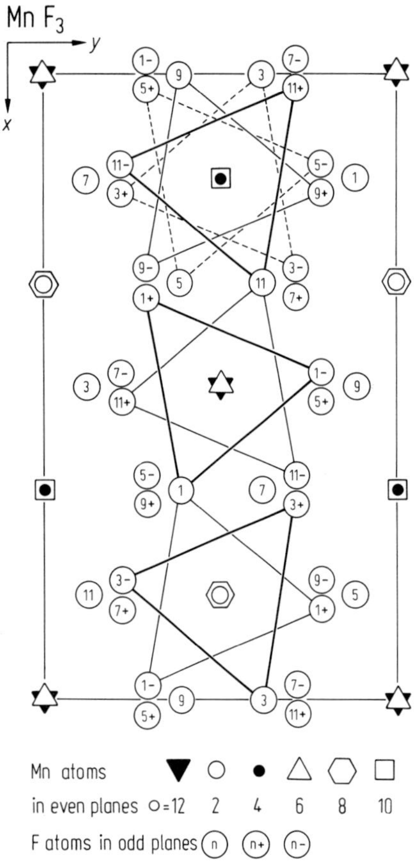

Fig. 6. MnF_3. Projection along [001] on (001) of the MnF_3 unit cell [57H2].

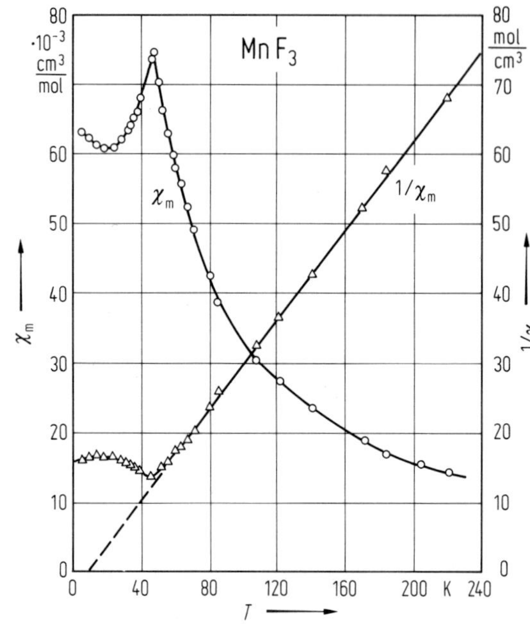

Fig. 7. MnF_3. Molar magnetic susceptibility and its reciprocal vs. temperature [58B].

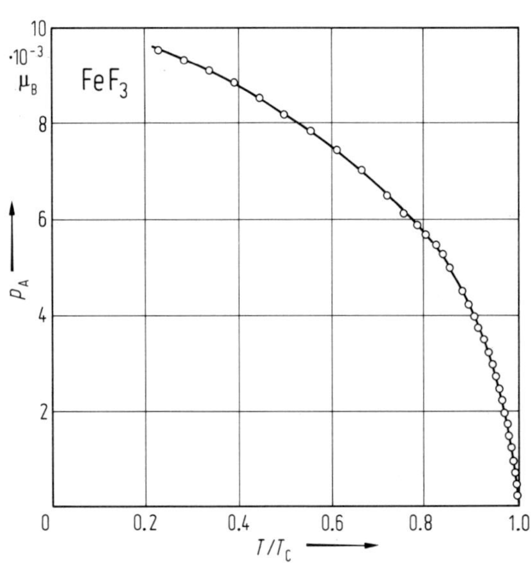

Fig. 8. FeF_3. Macroscopic magnetic moment (weak ferromagnetic moment) per atom vs. reduced temperature [68L]. $T_C = 362$ K from $\sigma(T)/\sigma(0) \propto (T_C - T)^\beta$.

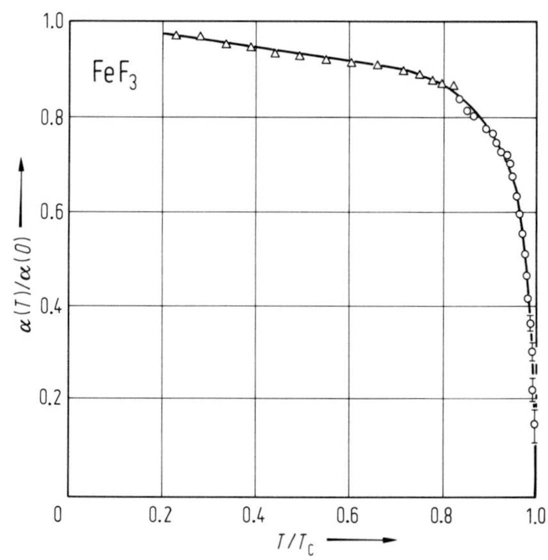

Fig. 9. FeF_3. Reduced canting angle $\alpha(T)/\alpha(0)$ as a function of T/T_C. T_C: Curie temperature for weak ferromagnetism [68L].

9.3.2.2 MCl_3 compounds

Table 3. MCl_3 compounds.

Substance (Figs.)	Crystal type	Lattice constants*)	Molar Curie constant (in cm^3 K mol^{-1})	Paramagnetic Curie temperature	Magnetic order	Critical temperature	Spin structure
α-$TiCl_3$ (Figs. 10, 11)	Hexagonal	a = 6.12 Å c = 17.5 Å [47K1]**)	p_{eff} = 1.31μ_B [62L; LBII/2,	p. 2–2]		265 K [62L; LBII/2, p. 2–2]	
VCl_3 (Fig. 12)	Hexagonal	a = 6.045(5) Å c = 17.45(2) Å [64M2; LBIII/7a, p. 386]	1.005 [40S]	− 30.1 K [40S]			
$CrCl_3$ (Fig. 13···16)	Monoclinic (C2/m) Fig. 13	a = 5.9588 Å b = 10.3206 Å c = 6.1138 Å β = 108.495° [64M1]***)	1.89 [40S]	27.0 K [40S]	AF ← P [61C]	16.8 K [58H]	Spins in each hexagonal layer are parallel and anti-parallel to those in the adjacent layers. Spin $\perp c$ [61C]
$FeCl_3$ (Figs. 17···20)	Hexagonal, see also LBIII/7a, p. 394	a = 6.065 Å c = 17.44 Å [62C] a = 6.0649(5) Å c = 17.443(2) Å [66B; LBIII/7a, p. 394]	4.07 [40S]	− 11.5 K [40S] Fig. 17	Heli ← P [62C]	8.83 K [81J]	Spins lie within $(14\bar{5}0)$ planes and rotate by $2\pi/15$ in successive planes along the $[14\bar{5}0]$ direction [62C]

For footnotes see next page.

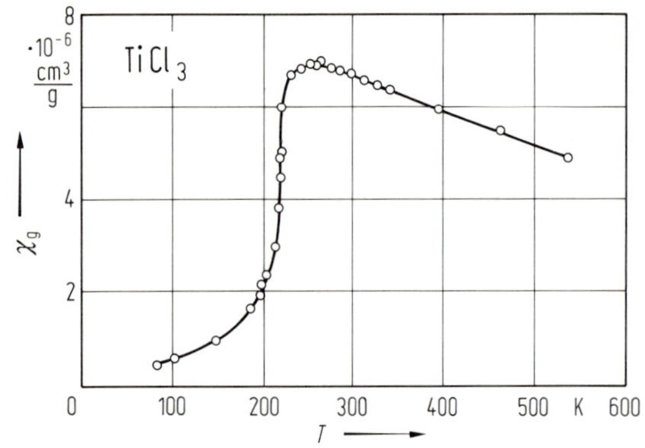

Fig. 10. $TiCl_3$. Magnetic susceptibility per gram vs. temperature [60O], see also [62L].

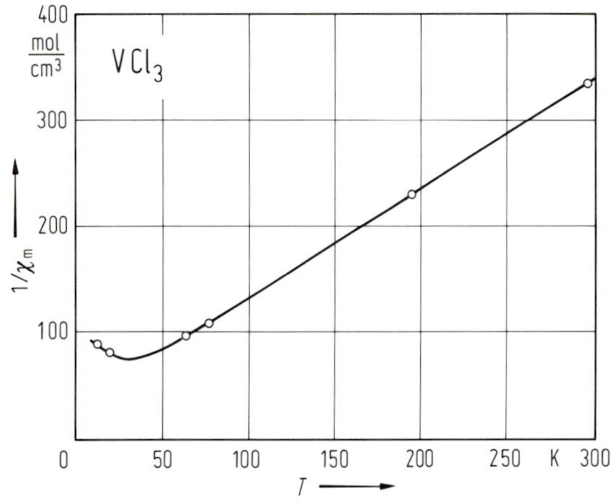

Fig. 12. VCl_3. The reciprocal susceptibility per mole vs. temperature [40S].

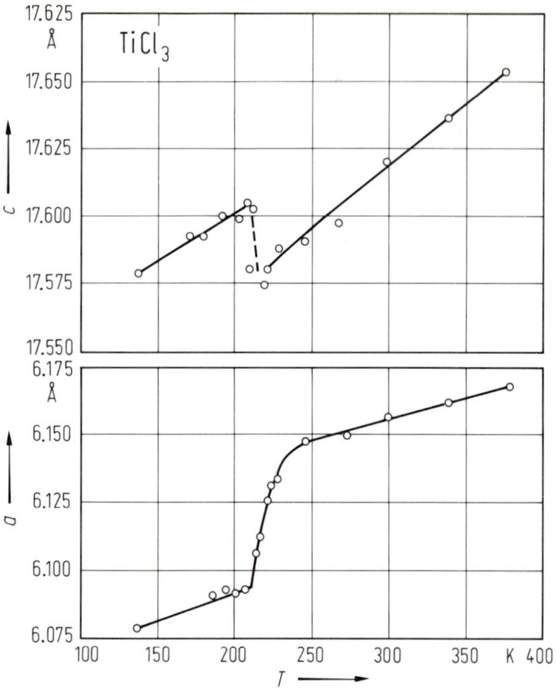

Fig. 11. $TiCl_3$. Temperature dependence of lattice constants a and c [60O].

Footnotes for Table 3

*) Room temperature unless otherwise specified.

**) Further data [LBIII/7a, p. 382]

a [Å]	c [Å]	Ref.
6.1581	17.617	67T
6.1573(3)	17.623(3)	67Z

The lattice constants change abruptly at 217 K where the susceptibility decreases sharply, Fig. 11 [60O].

***) First-order phase transition near 240 K. At 225 K the crystal structure is rhombohedral ($R\bar{3}$) with $a = 5.942$ Å and $c = 17.333$ Å [64M1].

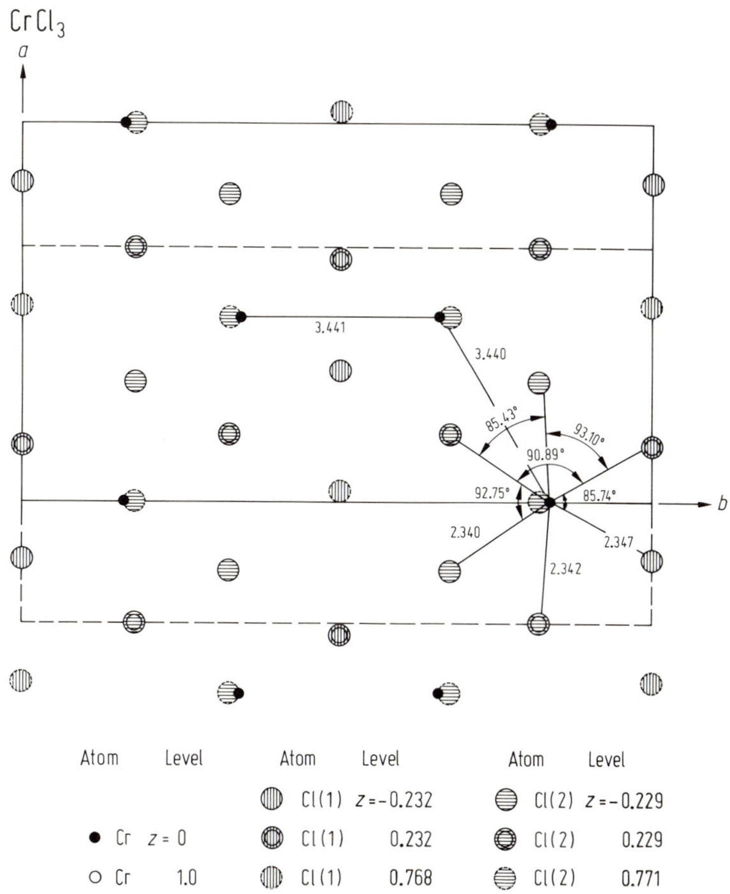

Fig. 13. $CrCl_3$. Projection of the crystal structure on (001) plane [64M]. Distances in Å.

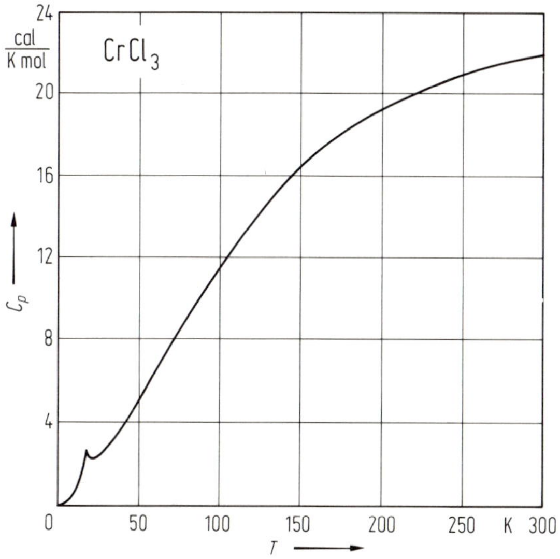

Fig. 14. $CrCl_3$. Molar specific heat vs. temperature [58H].

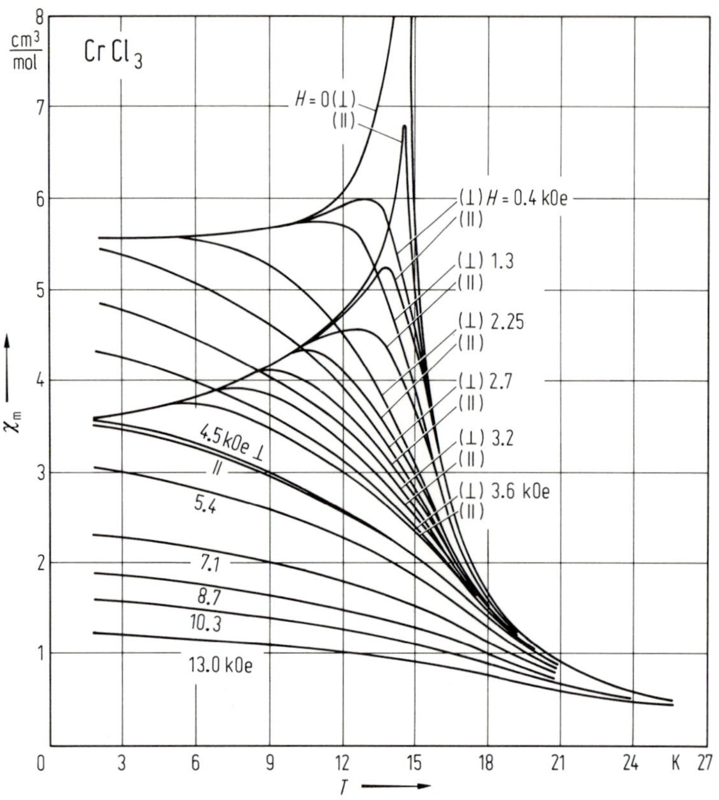

Fig. 15. $CrCl_3$. Molar magnetic susceptibility along the three fold axis (∥) and perpendicular to it (⊥) vs. temperature for various magnetic fields [61B].

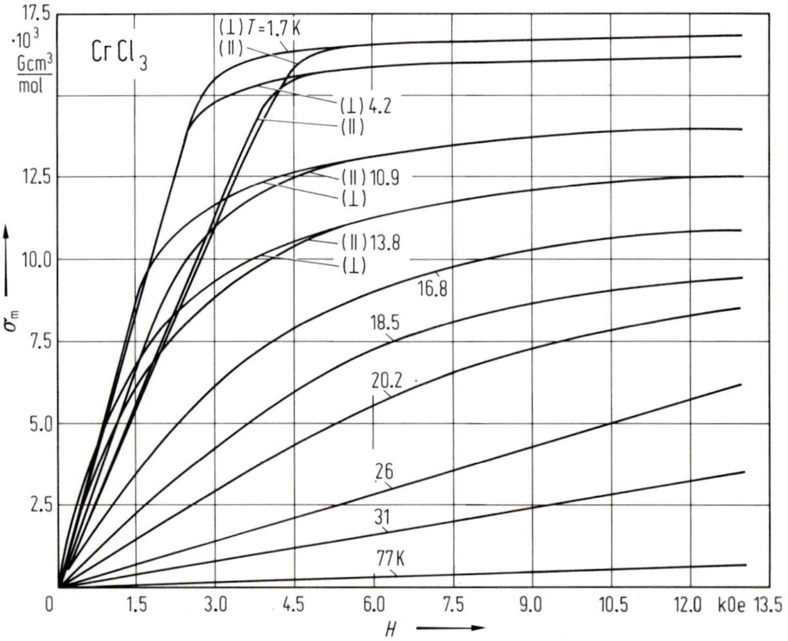

Fig. 16. $CrCl_3$. Magnetic moment per mole along the three fold axis (∥) and perpendicular to it (⊥) vs. applied magnetic field [61B].

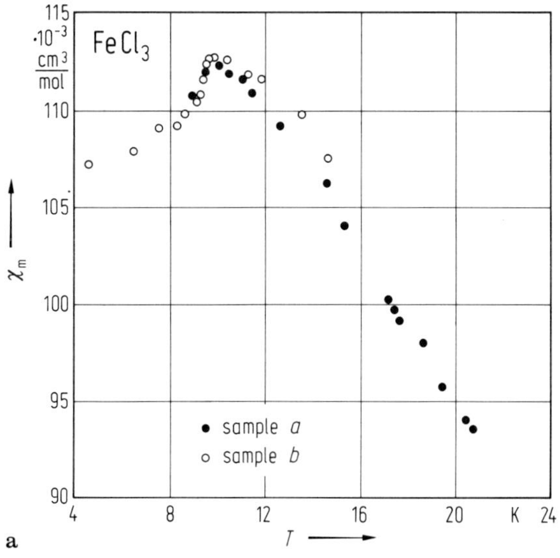

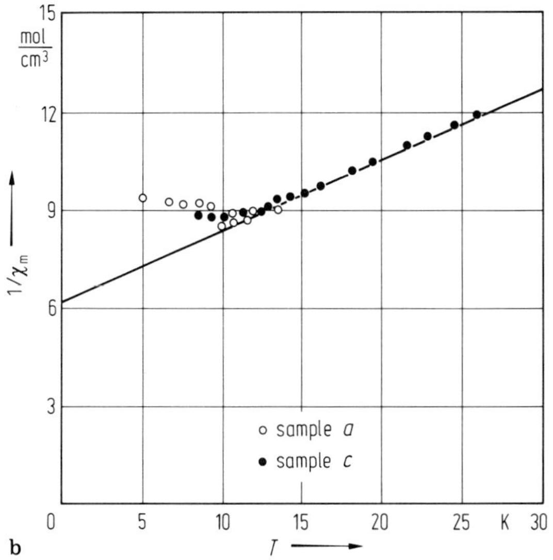

Fig. 17. $FeCl_3$. (a) Temperature dependence of molar susceptibility. (b) Temperature dependence of reciprocal molar susceptibility [69J1, LBII/10, p. 689].

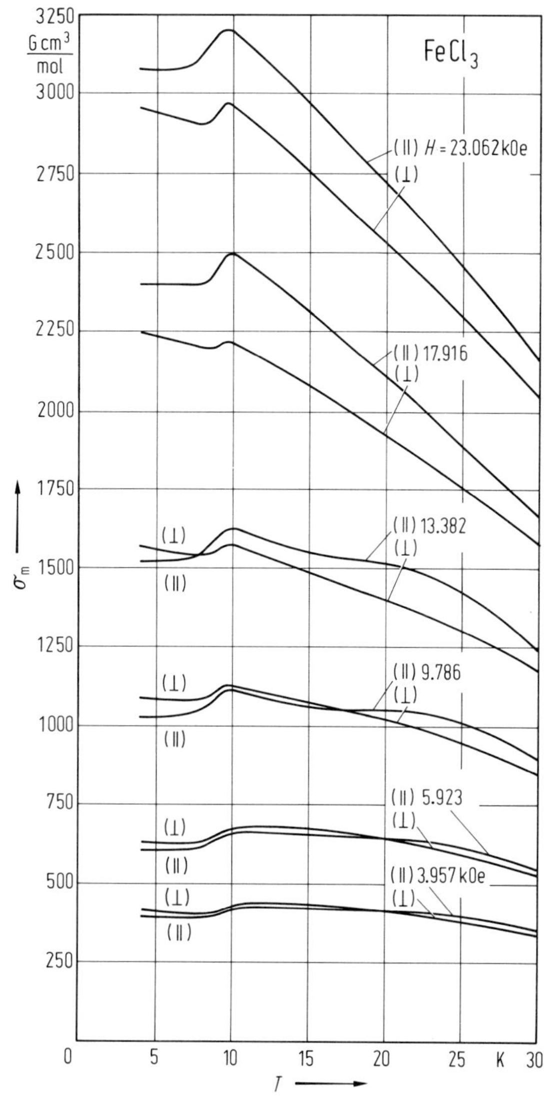

Fig. 18. $FeCl_3$. Magnetic moment per mole along the c axis (∥) and perpendicular to it (⊥) vs. temperature for various magnetic fields [70B].

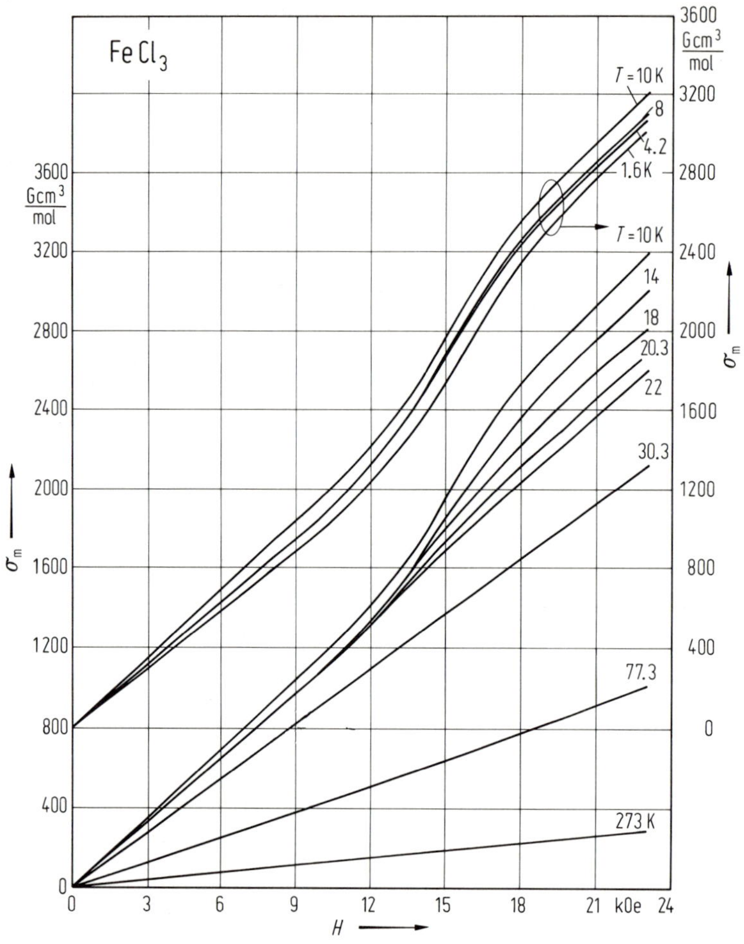

Fig. 19. $FeCl_3$. Magnetic moment per mole along the *c* axis vs. applied magnetic field for various temperatures [70B].

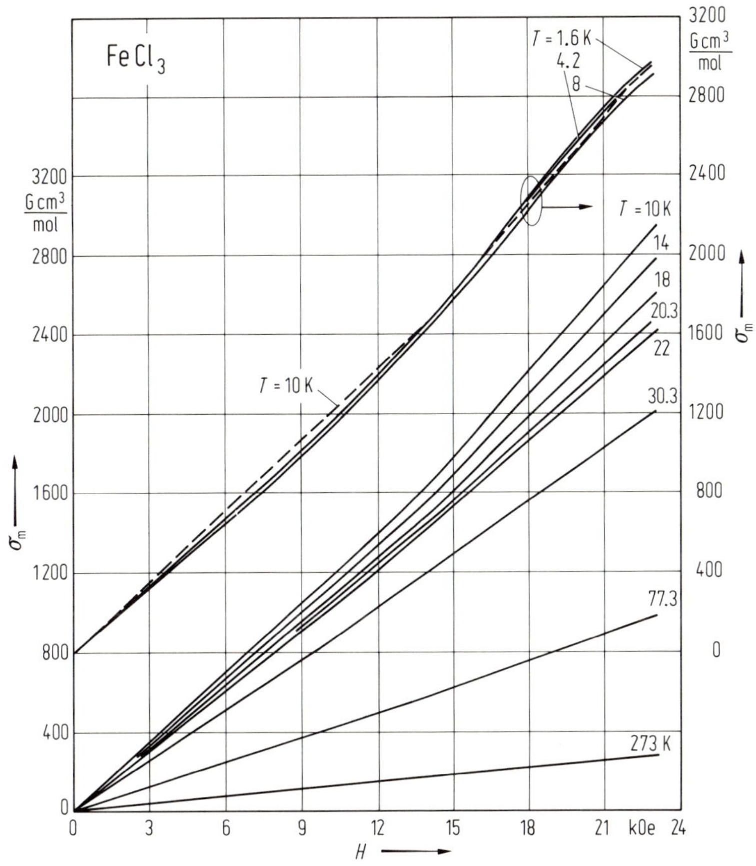

Fig. 20. $FeCl_3$. Magnetic moment per mole perpendicular to the *c* axis vs. applied magnetic field for various temperatures [70B].

9.3.2.3 MBr_3 compounds

Table 4. MBr_3 compounds.

Substance (Figs.)	Crystal type	Lattice constants*)	Molar Curie constant (in cm^3 K mol^{-1})	Paramagnetic Curie temperature	Magnetic order	Critical temperature	Spin structure	Ordered moment
α-$TiBr_3$		$a = 6.47$ Å $c = 18.65$ Å [70N; LBIII/7a, p. 532]	Table 1			230 K [47K2; LBII/2, p. 2–3]		
VBr_3	$FeCl_3$	$a = 6.400(5)$ Å $c = 18.53(2)$ Å [64M2; LBIII/7a, p. 534]	$p_{eff} = 2.67\mu_B$ [36K] Table 1	-20 K [36K; LBII/2, p. 2–12]				
$CrBr_3$ (Figs. 21 ⋯ 24)	Hexagonal Fig. 21	$a = 3.66$ Å $c = 18.30$ Å [60T]	1.84 [60T]	47 K [60T]	F ← P [60T]	32.55 K [65J]	Spin ∥ c [60T]	$3\mu_B$ [60T]

*) Room temperature unless otherwise specified.

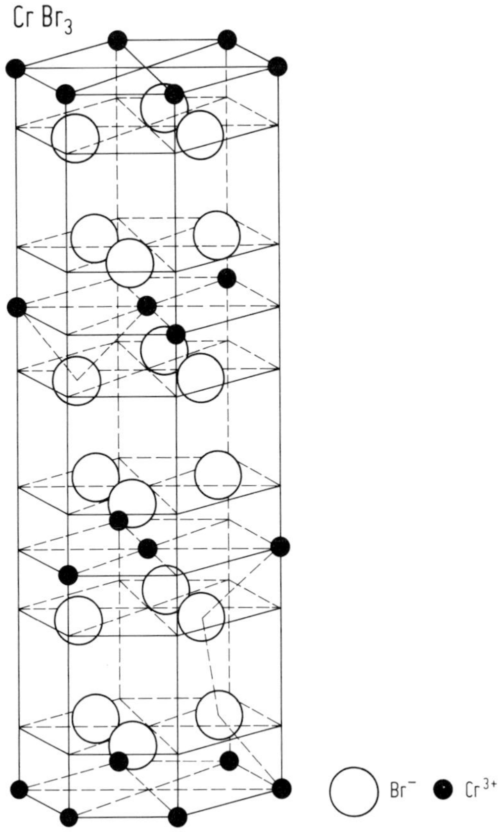

Fig. 21. $CrBr_3$. Crystal structure [60T].

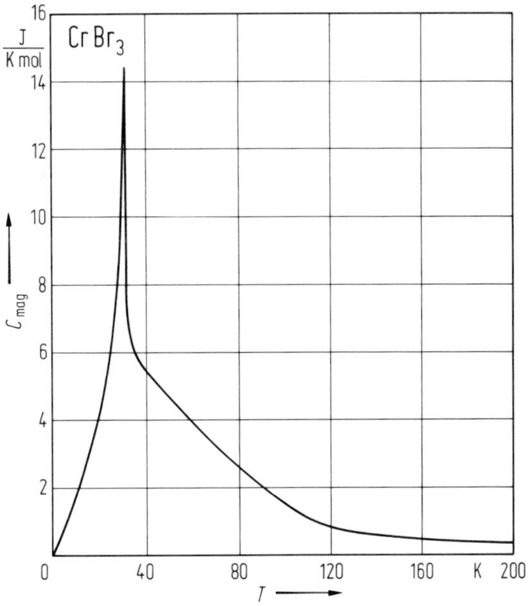

Fig. 22. $CrBr_3$. Molar magnetic specific heat after subtracting the contribution of the lattice vs. temperature [65J].

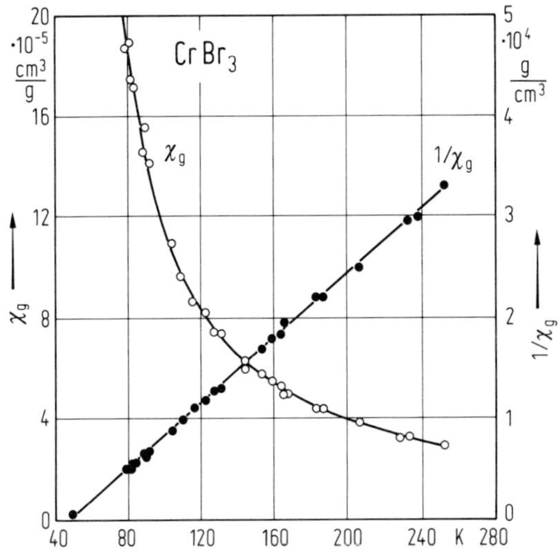

Fig. 23. $CrBr_3$. Magnetic susceptibility per gram and its reciprocal vs. temperature [60T].

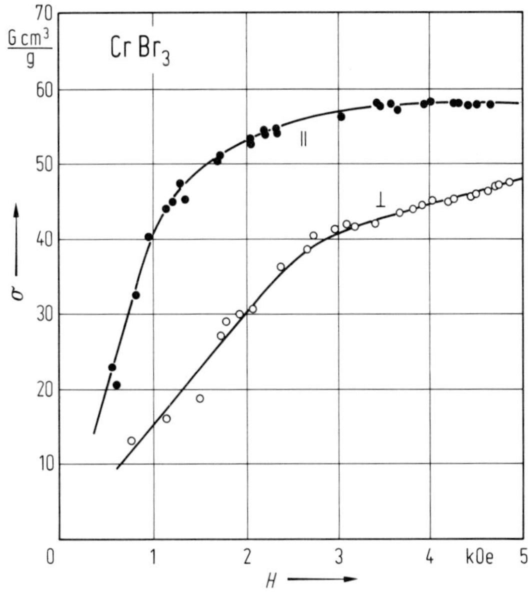

Fig. 24. $CrBr_3$. Magnetic moment per gram parallel to the c axis (||) and perpendicular to it (⊥) vs. applied magnetic field at $T = 4.2$ K [60T].

9.3.2.4 MI_3 compounds

Table 5. MI_3 compounds.

Substance (Figs.)	Crystal type	Lattice constants*)	Molar Curie constant (in $cm^3\,K\,mol^{-1}$)	Paramagnetic Curie temperature	Magnetic order	Critical temperature	Spin structure	Ordered moment
TiI_3	Hexagonal	$a = 7.146(5)$ Å $c = 6.502(5)$ Å [66S2; LBIII/7a, p. 616]				> 434 K [47K2; LBII/2, p. 2–3]		
VI_3	BiI_3	$a = 6.925(5)$ Å $c = 19.91(2)$ Å [69J2; LBIII/7a, p. 619]	Table 1					
CrI_3 (Figs. 25, 26)	Hexagonal	$a = 6.86$ Å $c = 19.88$ Å [52H]	2.03 [59H1]	70 K [59H1]	F ← P [65D]	68 K [65D]	Spin ∥ c [65D]	$3.10\mu_B$ [65D]

*) Room temperature unless otherwise specified.

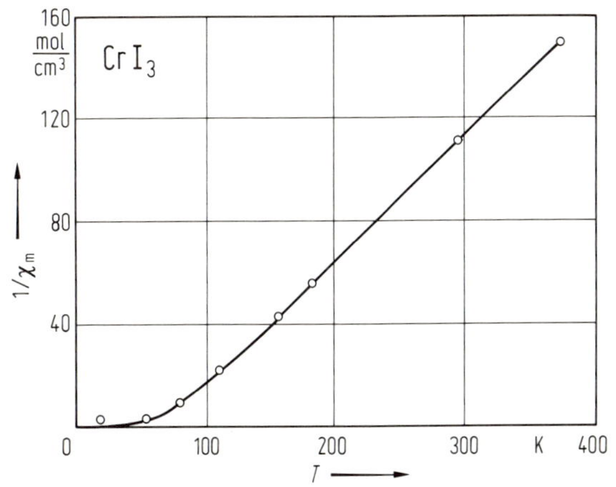

Fig. 25. CrI_3. The reciprocal susceptibility per mole vs. temperature [59H1].

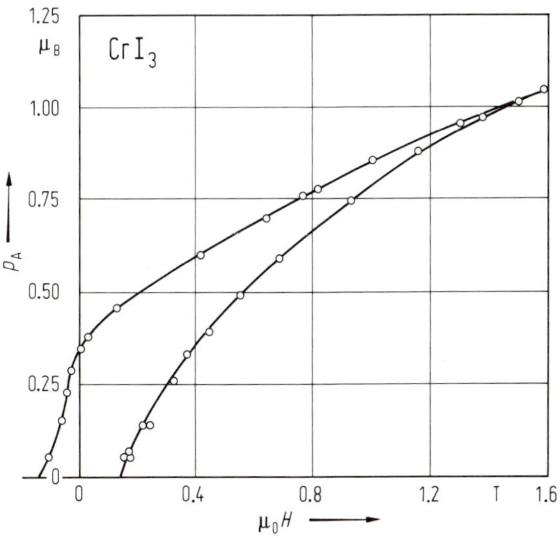

Fig. 26. CrI_3. Magnetic moment per atom vs. magnetic field at 4.2 K [59H2].

9.3.3 Microscopic magnetic and related properties of CrX_3 compounds

Table 6. CrX_3 compounds.

Substance (Figs.)	Hamiltonian	Parameters
CrF_3	Isotropic exchange constant for nearest-neighbour interaction $J_{nn}/k = -7.2$ K (from neutron scattering) [68R; LBIII/4a, p. 221]	
$CrCl_3$ (Fig. 27)	$-\sum_{ii',pp'} J_{ii',pp'} \boldsymbol{S}_{ip} \cdot \boldsymbol{S}_{i'p'} + H_D + D\sum_{ip}(S^y_{ip})^2$ $S = 3/2$	Jz (in plane) $= 9.42\ cm^{-1}$ $J_L z_L$ (between planes) $= -0.045\ cm^{-1}$ $DS/g\mu_B = -2000$ Oe Dipole field (001) = 2254 Oe [63N]
$CrBr_3$ (Fig. 28, 29)	$-\sum_{nn',jj'} J_{nn',jj'} \boldsymbol{S}_{nj} \cdot \boldsymbol{S}_{n'j'} - g\mu_B H_A \sum_{nj} S^z_{nj}$ $S = 3/2$	$J_1 = 6.85\ cm^{-1}$ $J_2 = -0.194\ cm^{-1}$ $J_3 = -0.115\ cm^{-1}$ $J_4 = 0.055\ cm^{-1}$ $J_5 = 0.105\ cm^{-1}$ $g\mu_B H_A = 0.65\ cm^{-1}$ [71S]
CrI_3 (Fig. 30)	$-\sum_{ii',pp'} J_{ii',pp'} \boldsymbol{S}_{ip} \cdot \boldsymbol{S}_{i'p'} - g\mu_B H_A \sum_{i,p} S^z_{ip}$ $S = 3/2$	J_T (in plane) $= 9.38\ cm^{-1}$ J_L (between planes) $= 1.20\ cm^{-1}$ H_A ($T = 0$ K) $= 28.6$ kOe $g = 2.07$ [65N]

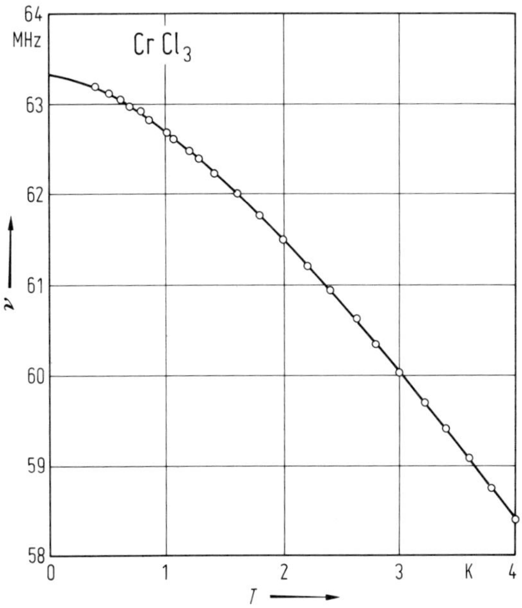

Fig. 27. $CrCl_3$. ^{53}Cr NMR (central component of quadrupole triplet) vs. temperature. Solid curve shows the best fit of the two-dimensional spin-wave model to the data [63N]. The corresponding Hamiltonian and the values of the parameters are given in Table 6.

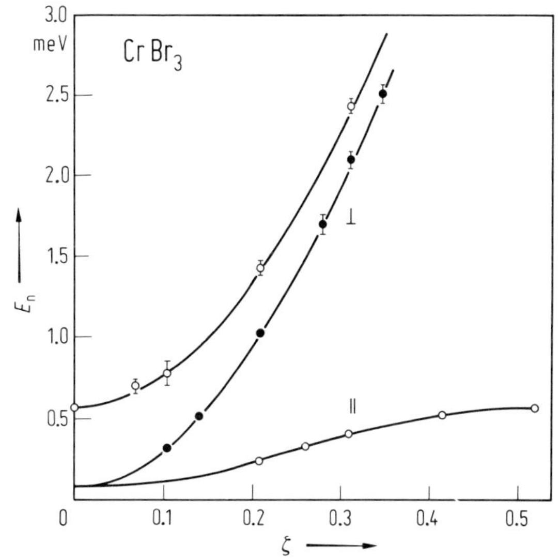

Fig. 28. $CrBr_3$. Magnetic excitation energy vs. reduced wave vector coordinate parallel and perpendicular to the c axis at 6 K. Solid curves show the best fit of the spin-wave model to the data [71S]. The corresponding Hamiltonian and the values of the parameters are given in Table 6.

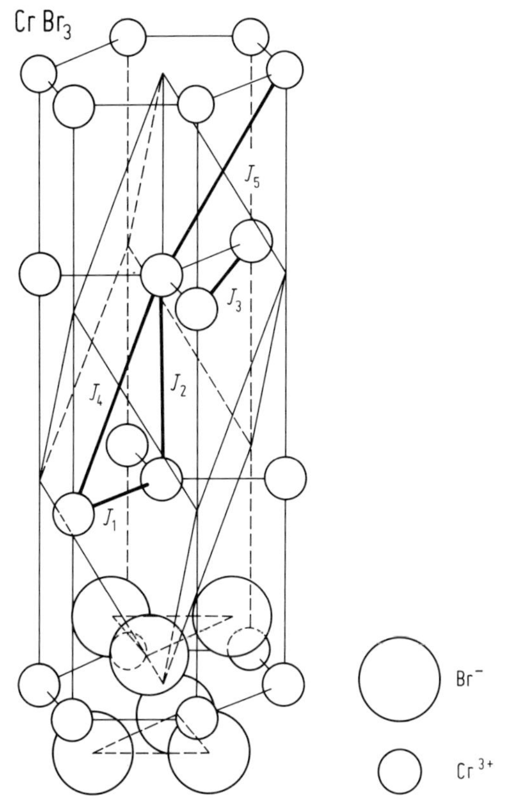

Fig. 29. $CrBr_3$. The exchange paths. Small balls: Cr^{3+} ions. Large balls: Br^- ions. Each Cr^{3+} layer has a layer of Br^- ions on either side, but for clarity only one such "sandwich" is shown [71S].

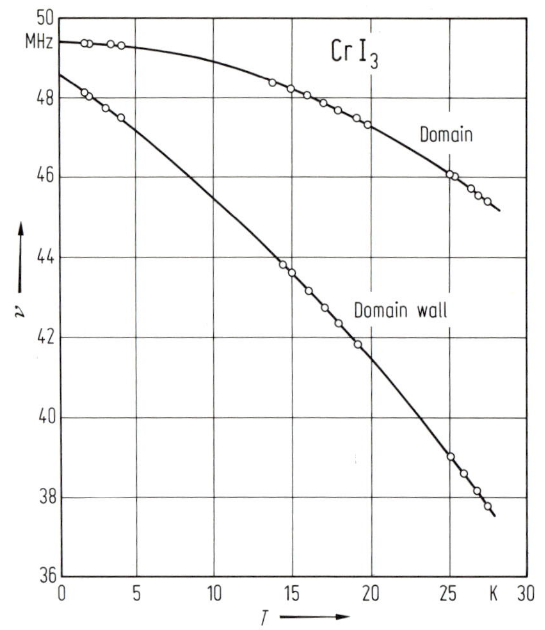

Fig. 30. CrI_3. Domain and domain-wall ^{53}Cr NMR vs. temperature. For the domain resonance only the central (1/2↔ − 1/2) transition frequency is shown [65N]. The values of the parameters in the Hamiltonian obtained from fit to the data are given in Table 6.

9.3.4 References for 9.3

36K Klemm, W., Hoschek, E.: Z. Anorg. Allg. Chem. **226** (1936) 359.
40S Starr, C., Bitter, F., Kaufmann, A.R.: Phys. Rev. **58** (1940) 977.
47K1 Klemm, W., Krose, E.: Z. Anorg. Allg. Chem. **253** (1947) 218.
47K2 Klemm, W., Krose, E.: Z. Anorg. Allg. Chem. **253** (1947) 209.
51J Jack, K.H., Gutmann, V.: Acta Crystallogr. **4** (1951) 246.
52H Handy, L.L., Gregory, N.W.: J. Am. Chem. Soc. **74** (1952) 891.
54E Ehrlick, P., Pietzka, G.: Z. Anorg. Allgem. Chem. **275** (1954) 121.
57H1 Hepworth, M.A., Jack, K.H., Peacock, R.D., Westland, G.J.: Acta Crystallogr. **10** (1957) 63.
57H2 Hepworth, M.A., Jack, K.H.: Acta Crystallogr. **10** (1957) 345.
57J Jack, K.H., Miltland, R.: Proc. Chem. Soc. (1957) 232.
58B Bozorth, R.M., Nielsen, J.W.: Phys. Rev. **110** (1958) 879.
58H Hansen, W.N., Griffel, M.: J. Chem. Phys. **28** (1958) 902.
58K Klemm, W., Holze, E., Basualdo, W.: Congr. Inter. Chim. Pure Appl. 16^e, Paris **1957**, 837 (1958).
58N Nevitt, M.V., Schwartz, L.H.: Trans. Met. Soc. AIME **212** (1958) 700.
58W Wollan, E.O., Child, H.R., Koehler, W.C., Wilkinson, M.K.: Phys. Rev. **112** (1958) 1132.
59H1 Hansen, W.N., Griffel, M.: J. Chem. Phys. **30** (1959) 913.
59H2 Hansen, W.N.: J. Appl. Phys. Suppl. **30** (1959) 304S.
60O Ogawa, S.: J. Phys. Soc. Jpn. **15** (1960) 1901.

60T Tsubokawa, I.: J. Phys. Soc. Jpn. **15** (1960) 1664.
61B Bizette, H., Terrier, C., Adam, A.: C. R. Acad. Sci. **252** (1961) 1571.
61C Cable, J.W., Wilkinson, M.K., Wollan, E.O.: J. Phys. Chem. Solids **19** (1961) 29.
61M Muetterties, E.L., Castle, J.E.: J. Inorg. Nucl. Chem. **18** (1961) 148.
62C Cable, J.W., Wilkinson, M.K., Wollan, E.O., Koehler, W.C.: Phys. Rev. **127** (1962) 714.
62L Lewis, J., Machin, D.J., Newnham, I.E., Nyholm, R.S.: J. Chem. Soc. (1962) 2036.
63N Narath, A.: Phys. Rev. **131** (1963) 1929.
64M1 Morosin, B., Narath, A.: J. Chem. Phys. **40** (1964) 1958.
64M2 McCarley, R.E., Roddy, J.W., Berry, K.O.: Inorg. Chem. **3** (1964) 50.
65D Dillon, Jr., J.F., Olson, C.E.: J. Appl. Phys. **36** (1965) 1259.
65J Jennings, L.D., Hansen, W.N.: Phys. Rev. **139** (1956) A1694.
65N Narath, A.: Phys. Rev. **140** (1965) A854.
66B Blairs, S., Shelton, R.A.J.: J. Inorg. Nucl. Chem. **28** (1966) 1855.
66S1 Shane, J.R., Kedzie, R.W., Kestigian, M., Lyons, D.H., Wang, F.F.Y.: Rep. Air Force Cambridge Res. Lab. AF **19** (628) (1966) 5128.
66S2 Schnering, H.G. von: Naturwissenschaften **53** (1966) 359.
67K Kocher, C.W.: Phys. Lett. **24**A (1967) 93.
67T Tornqvist, E.G.M., Richardson, J.T., Wilchinsky, Z.W., Looney, R.W.: J. Catalysis **8** (1967) 189.
67W Wertheim, G.K., Guggenheim, H.J., Buchanan, D.N.E.: Solid State Commun. **5** (1967) 537.
67Z Zannetti, R.: Ric. Sci. **37** (1967) 291.
68C Canty, A.J., Coutts, R.S.P., Wailes, P.C.: Aust. J. Chem. **21** (1968) 807.
68L Levinson, L.M.: J. Phys. Chem. Solids **29** (1968) 1331.
68R Rao, L.M., Murthy, N.S.S., Venkataraman, G., Iyengar, P.K.: Phys. Letters **26**A (1968) 108.
69F Friedt, J.M., Adloff, J.P.: Inorg. Nucl. Chem. Lett. **5** (1969) 163.
69J1 Jones, E.R., Morton, O.B., Cathey, L., Auel, T., Amma, E.L.: J. Chem. Phys. **50** (1969) 4755.
69J2 Juza, D., Giegling, D., Schäfer, H.: Z. Anorg. Allgem. Chem. **366** (1969) 121.
70B Bizette, H., Adam, A., Ciret, N.: C. R. Acad. Sci. **271** (1970) B608.
70N Newland, B.G., Shelton, R.A.J.: J. Less-Common Met. **22** (1970) 369.
71S Samuelsen, E.J., Silberglitt, R., Shirane, G., Remeika, J.P.: Phys. Rev. B**3** (1971) 157.
74A Alter, E., Hoppe, R.: Z. Anorg. Allg. Chem. **403** (1974) 127.
81J Johnson, P.B., Friedberg, S.A., Rayne, J.A.: J. Appl. Phys. **52** (1981) 1932.
86D De Pape, R., Ferey, G.: Mater. Res. Bull. **21** (1986) 971.
86F Ferey, G., De Pape, R., Leblanc, M., Pannetier, J.: Rev. Chim. Miner. **23** (1986) 474.
86L Leblane, M., De Pape, R., Ferey, G.: Solid State Commun. **58** (1986) 171.
87D De Pape, R., Le Bail, A., Lubin, F., Ferey, G.: Rev. Chim. Miner. **24** (1987) 545.

The author wishes to express his sincere thanks to Ms. Masayo Ajiro for her help in compiling the data and preparing the manuscripts of chapters 9.1, 9.2 and 9.3.

9.4 MCl_2–GIC, MCl_3–GIC (M = 3d element)

9.4.1 Introduction

9.4.1.1 General remarks

In this chapter, the magnetic phase transitions and the ordering properties of MCl_2–GICs, MCl_3–GICs, where M stands for a 3d transition metal atom i.e. Cr, Mn, Fe, Co, Ni or Cu, and the related intercalation compounds are presented.

GICs (graphite intercalation compounds) form a family of complex physical systems of layer structure [81D]. There has been found a characteristic lattice structure, in which "stage structure" and "island structure" coexist as in the following. Various substances e.g. MCl_2 or MCl_3 in this case, are introduced between the carbon layers of graphite making a two-dimensional (2D) lattice and each 2D lattice of the intercalated substance (intercalant) is separated from the adjacent ones regularly by a certain number of carbon layers called "stage number". Each intercalant lattice is not generally extended infinitely and divided into many island-like 2D clusters of a certain finite size.

Such a characteristic lattice structure is schematically shown in Figs. 1.1 ··· 4, and Table 1.1. It is formed cooperatively in the intercalation process of graphite [80S1, 84H, 85K3] not only by direct interactions among the intercalants but by various indirect ones caused by the interactions between the intercalant and the intermediary carbon layers. The lattice structure of GICs is thus heterogeneous and qualitatively different from usual layer structure compounds of a simple regular lattice.

Transition metal chlorides MCl_2 and MCl_3 have a layer structure of a hexagonal symmetry in which the transition metal ions form a triangular lattice in MCl_2 and a honeycomb lattice in MCl_3, respectively (see Fig. 1.3). When these compounds are intercalated in graphite, the lattice structure of each intercalant layer is found almost the same as that of the corresponding layer in the pristine compound and has a hexagonal symmetry, although it is not usually commensurable to graphite lattice.

The interlayer distance between the adjacent intercalant lattices is much longer than that in the pristine compound, especially in high stage ones as seen in Table 1.1. As the result, the intralayer exchange interaction between the transition metal ions in a GIC is almost the same as that in the pristine compound but the interlayer magnetic interaction in the former is much weaker than that in the latter. Thus, MCl_2– and MCl_3–GICs present us a variety of model systems for 2D magnets.

Spin ordering on a 2D lattice has long been an essential problem of fundamental physics, especially from the view point of cooperativity. A variety of possibilities have been predicted so far theoretically. It has exactly been proved that a 2D Ising (I)-type ferromagnet goes into a long range order (LRO) at a finite temperature [44O] but that a 2D *XY*- or Heisenberg (H)-type one does not down to 0 K at all [66M], although a characteristic ordered phase without spontaneous magnetization but accompanied by a divergent susceptibility was conjectured to appear [66S]. Actually, a topological order characterized by bound vortices of opposite helicity was predicted for a 2D *XY*-type ferromagnet [73K2, 74K4], even with a hexagonal symmetry breaking field [77J]. A topological order has been predicted also for 2D triangular antiferromagnets [84K2, 84M] in which local frustration effect makes an important role in the ordering process. The ordering characteristics of 2D magnets are summarized in Table 1.2.

Experimentally, neither true 2D spin LRO has ever been found nor topological spin order. A quasi-2D phase transition in real magnets is always accompanied by a spontaneous 3D LRO although in some cases the critical phenomena or the dynamical aspects could be interpreted by an ideal 2D model. It can be explained as a result of so called "dimensionality crossover" due to the weak interlayer interaction. For a complex physical system like a random or a heterogeneous magnet, however, the situation is quite different and the crossover phenomena may not take place, because the spin correlation cannot always be extended infinitely in each layer, hindered by various origins in the system.

In the case of GICs, where spin correlation length should be limited by island size, a true 2D spin LRO could, thus, be realized in a finite temperature range [91M]. In such a case, another phase transition may happen among the already ordered 2D islands by some possible inter-island interaction. As the result, 2-step successive magnetic transition of a hierarchical type may be suspected to be found in some magnetic GICs, with an intermediate state of a true 2D (intra-island) LRO with totally (inter-island) disorder over the system.

In this chapter 9.4 figures and tables are numbered separately for each subchapter containing three digits, i.e. Table 1.2 is the second table of subchapter 9.4.1, Fig. 2.2 is the second figure of subchapter 9.4.2.

Table 1.1. Lattice parameters.

Compounds	(Stage)	In-plane lattice structure	a [Å]	d [Å]	In-plane island size R [Å]	References
$CoCl_2$		Triangle	3.55	5.79	∞	59W
$CoCl_2$–GIC	(S1)	Triangle	3.55	9.40		87Y1
	(S2)	Triangle	3.56	12.70	150 ··· 900*)	87W1, 85M1*), 88W2*)
$NiCl_2$		Triangle	3.47	5.77	∞	77B
$NiCl_2$–GIC	(S1)	Triangle	3.50	9.37		89H2
	(S2)	Triangle	3.46	12.78	100 ··· 200*)	87W1, 83F*)
$MnCl_2$		Triangle	3.68	5.83	∞	82B, 80E
$MnCl_2$–GIC	(S1)	Triangle	3.69	9.48		82B
	(S2)	Tirangle	3.69	12.76	130 ··· 200*)	82B, 84G*)
$CuCl_2$		Triangle	3.3	6.70	∞	47W
$CuCl_2$–GIC	(S1)	Triangle		9.3		83H
	(S2)	Triangle		12.7	300*)	83H, 85N*)
$CrCl_3$		Honeycomb	3.43	5.78	∞	82K
$CrCl_3$–GIC	(S1)	Honeycomb				
	(S3)	Honeycomb		16.0		90R2
$FeCl_3$		Honeycomb	3.49	5.81	∞	73S
$FeCl_3$–GIC	(S1)	Honeycomb		9.4		82M4
	(S2)	Honeycomb		12.6		82M4

Table 1.2. Ordering characteristics in 2D lattices.

Type of interaction*)	Ferromagnets and quadratic antiferromagnets				Triangular antiferromagnets		
	I	*XY*	*XY'*	H	I	*XY*	H
Frustration	no	no	no	no	yes	yes	yes
Spontaneous magnetization	yes	no	yes	no	no	no	no
Phase transition	Single	Single	2-step	no	no	Single	Single
Ordered state**)	S-LRO	T1 (KT)	T1 & S-LRO	no	no	C-LRO	T2
Spin configuration	Colin. (F, AF)	Vortex	Vort. & Colin.	—	—	120°	Vortex (Z_2)
References	44O	66M 73K2	77J	66M 80S2	50H 50W	84M	84K2

*) I: Ising model, *XY*: *XY* model, *XY'*: *XY* model with 6-fold symmetry breaking field and H: Heisenberg model.

**) S and C represent "Spin" and "Chirality", respectively. T1 is a topological order of KT-type, characterized by bound vortices. T2 is a another topological order, which is characterized by bound different (Z_2) vortices.

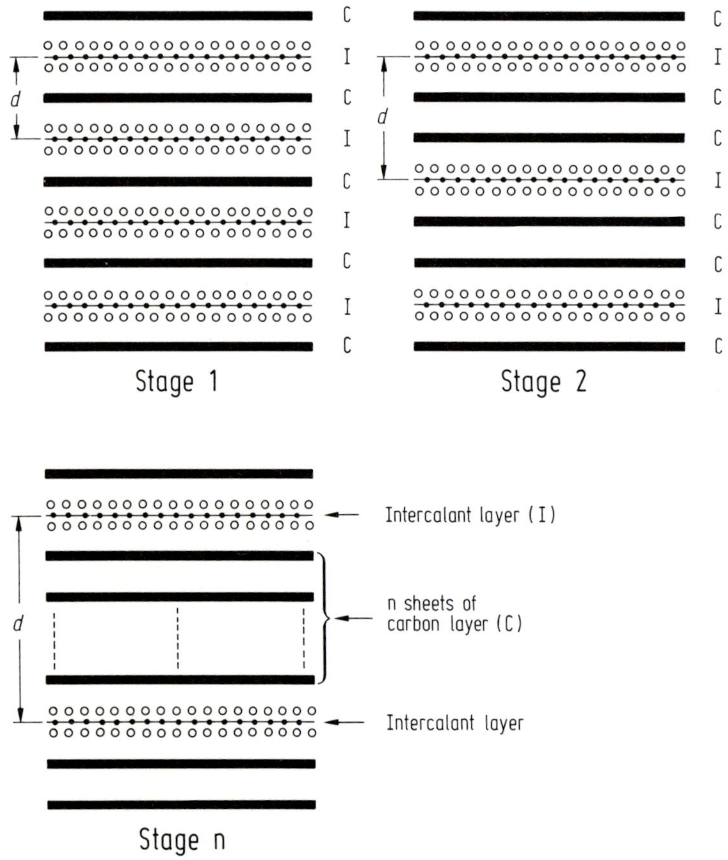

Fig. 1.1. Stage structure of GICs.

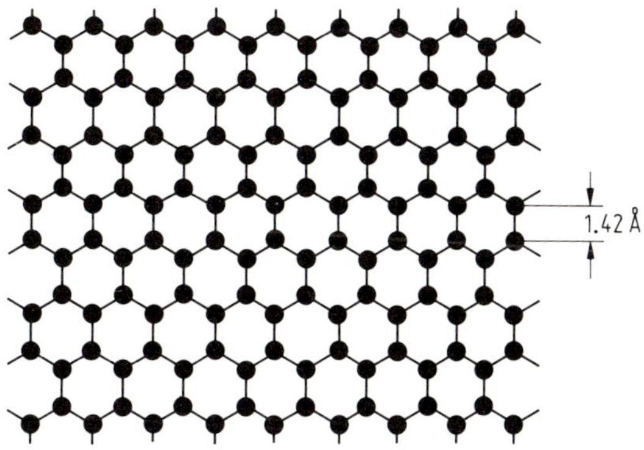

Fig. 1.2. Carbon layer structure.

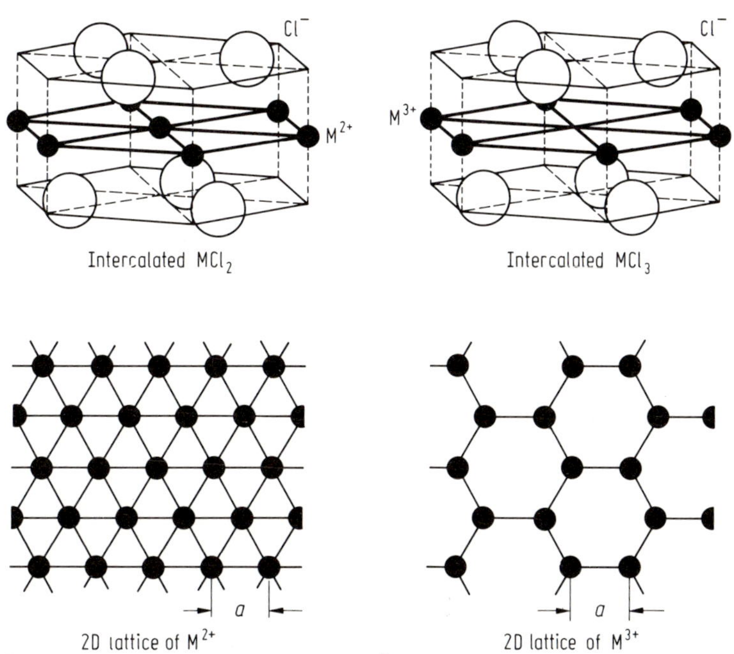

a b

Fig. 1.3. Lattice structure of intercalant layer in MCl_2–GIC (a) and MCl_3–GIC (b).

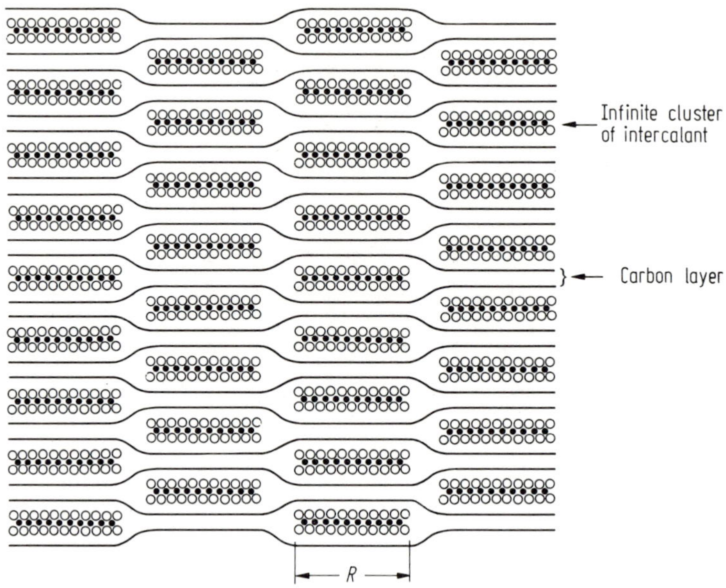

Fig. 1.4. Schematic island structure of stage 2 MCl_n–GIC (sectional view).

9.4.1.2 List of frequently used symbols

In the following list, the symbols which are used frequently in this chapter are presented. The units given in the second column are the most frequently used ones in the tables or the figures in this chapter. In most cases, magnetic quantities are presented by CGS-emu units.

Symbol	Unit	Quantity
a, b, c	Å	lattice parameters, unit vectors for crystallographic axes
a^*, b^*, c^*	$Å^{-1}$	reciprocal lattice parameters
C	$cm^3\,mol^{-1}\,K$	molar Curie constant
$C_\perp$	$cm^3\,mol^{-1}\,K$	C observed in the direction perpendicular to c-axis
$C_\parallel$	$cm^3\,mol^{-1}\,K$	C observed in the direction parallel to c-axis
C_p	$J\,mol^{-1}\,K^{-1}$, $J\,g^{-1}\,K^{-1}$	heat capacity at constant pressure
C_M	$J\,mol^{-1}\,K^{-1}$	magnetic contribution of heat capacity
d	Å	c-axis repeat distance for layer structure lattice
D	eV	single ion anisotropy energy (D/k in K)
E	eV	energy
E_{sw}	eV	spinwave energy
g	—	g-value, spectroscopic g-factor
$g_\perp$	—	g along the direction perpendicular to c-axis
$g_\parallel$	—	g along the direction parallel to c-axis

Symbol	Unit	Quantity
h	Oe	amplitude of magnetic AC field
H	Oe	magnetic field (strength)
H_c	Oe	cooling field
H_r	Oe	resonance field
H_t	Oe	transition field
ΔH	Oe	linewidth of magnetic resonance spectrum
ΔH_{pp}	Oe	peak to peak linewidth for magnetic resonance spectrum
I	—	intensity, count number
I_i	—	integrated intensity
J	eV	exchange energy, intra-layer exchange energy (J/k in K)
J'	eV	inter-layer exchange energy
J_A	eV	anisotropy of exchange energy
k	JK^{-1}	Boltzmann constant
M	G	magnetization
M_r	G	thermoremanent magnetization
M'_ω	G	fundamental in-phase component of AC magnetic response $M(t)$
$M'_{3\omega}$	G	3rd harmonic in-phase component of AC magnetic response $M(t)$
n	—	stage number
p	bar	pressure
p_A	μ_B	magnetic moment per atom (ion)
p_{eff}	μ_B	effective magnetic moment
q	$Å^{-1}$	wavevector, scattering vector
Q		reciprocal vector component for neutron scattering (in reciprocal lattice units)
R	$JK^{-1} mol^{-1}$	gas constant
R	Å	island size, cluster size
t	s	time
T	K	temperature
T_c	K	critical temperature of magnetic phase transition
T_{cu}	K	upper critical temperature in successive phase transition
T_{cl}	K	lower critical temperature in successive phase transition
T_g	K	paramagnetic spin glass transition temperature
T_{KT}	K	Kosterlitz–Thouless transition temperature
T_N	K	Néel temperature
$T_{\chi max}$	K	critical temperature characterized by maximum point of χ
v	$mm\,s^{-1}$	velocity for Mössbauer spectroscopy
x	—	concentration index of a composition element of sample
θ	deg	angle
Θ	K	Curie–Weiss temperature
$\Theta_\parallel$	K	Θ observed in the direction parallel to c-axis
$\Theta_\perp$	K	Θ observed in the direction perpendicular to c-axis

Symbol	Unit	Quantity
κ	$Å^{-1}$	inverse correlation length
ν	Hz	frequency
ν_r	Hz	magnetic resonance frequency
ξ	Å	correlation length
ξ_c	Å	correlation length along c-axis
ρ	Ω cm	resistivity
σ	$G cm^3 g^{-1} = emu g^{-1}$	magnetic moment per gram (specific magnetization)
σ_m	$G cm^3 mol^{-1} = emu mol^{-1}$	magnetic moment per mole (molar magnetization)
σ_r	$G cm^3 g^{-1}$	thermoremanent specific magnetization
σ_0	$G cm^3 g^{-1}$	specific magnetization under ZFC
$\sigma_\perp$	$G cm^3 g^{-1}$	specific magnetization in the direction perpendicular to c-axis
$\sigma_\parallel$	$G cm^3 g^{-1}$	specific magnetization in the direction parallel to c-axis
χ	—	volume susceptibility
χ_m	$cm^3 mol^{-1}$	susceptibility per mole
χ_g	$cm^3 g^{-1}$	susceptibility per gram
χ_{AC}	—, $cm^3 mol^{-1}$, $cm^3 g^{-1}$	AC susceptibility
χ'	—, $cm^3 mol^{-1}$, $cm^3 g^{-1}$	real part of complex AC susceptibility
χ''	—, $cm^3 mol^{-1}$, $cm^3 g^{-1}$	imaginary part of complex AC susceptibility
χ'_r	—, $cm^3 mol^{-1}$, $cm^3 g^{-1}$	AC susceptibility in ZFHAFC
χ'_0	—, $cm^3 mol^{-1}$, $cm^3 g^{-1}$	AC susceptibility in ZFC
$\chi_\parallel$	—, $cm^3 mol^{-1}$, $cm^3 g^{-1}$	susceptibility observed in the direction parallel to c-axis
$\chi_\perp$	—, $cm^3 mol^{-1}$, $cm^3 g^{-1}$	susceptibility observed in the direction perpendicular to c-axis
$\Delta\chi$	—, $cm^3 mol^{-1}$, $cm^3 g^{-1}$	anisotropy of susceptibility ($\equiv \chi_\perp - \chi_\parallel$)
ω	s^{-1}	angular frequency

— in the Unit column means "non-dimension"

9.4.1.3 List of abbreviations

Through this chapter, some abbreviations are used, which are listed below with the full explanations.

AC	alternating current
AES	Auger electron spectroscopy
AF	antiferromagnet(ic)
arb	arbitrary
D	dimension(al)
DC	direct current
DHvA	De Haas van Alphen
EM	electron microscope(py)
EPR	electron paramagnetic resonance

ER	electric resistance(ivity)
ESR	electron spin resonance
F	ferromagnet(ic)
FC	field cooling
FCM	field cooled magnetization
FH	field heating
FWHM	full width half maximum
GBIC	graphite bi-intercalation compound
GIC	graphite intercalation compound
H	Heisenberg
HOPG	highly oriented pyrolytic graphite
I	Ising
KT	Kosterlitz and Thouless
LRO	long range order
M	metal (mostly used for transition metal) atom
max	maximum
min	minute
MR	magnetic resonance
ND	neutron diffraction
NMR	nuclear magnetic resonance
P	paramagnet(ic)
S	stage
SCKG	single crystal kish graphite
SdH	Shubnikov de Haas
SIM	scanning ion microscope(py)
SG	spin glass
SP	super-paramagnet(ic)
SRO	short range order
STM	scanning tunneling microscope(py)
SW	spin wave
TRM	thermo-remanent magnetization
XPS	X-ray photo-spectroscopy
ZFC	zero field cooling
ZFCM	zero field cooled magnetization
ZFH	zero field heating
ZFHAFC	zero field heating after field cooling
$\perp$, $\parallel$	perpendicular, parallel

9.4.2 Singly intercalated compounds

9.4.2.1 General survey

In this section, magnetic transitions in singly intercalated graphite compounds are presented. As mentioned in 9.4.1.1, MCl_2– and MCl_3–GICs, especially the high stage compounds are taken as excellent model systems of a 2D magnet, although the influence of the characteristic lattice structure, including finite size effect due to the island structure, would bring some modifications which do not exist in infinitely extended homogeneous systems. A number of experimental studies have, therefore, been carried out extensively, intending to look after some new aspects of cooperative phenomena.

The transition metal ions, Mn^{2+}, Co^{2+}, Ni^{2+} and Cu^{2+} in MCl_2–GIC form a triangular lattice and the characteristics of magnetic ordering are presented in 9.4.2.2–9.4.2.4. The intralayer exchange interaction of $CoCl_2$– and $NiCl_2$–GICs is found ferromagnetic. For stage 2 compounds or higher stage ones, a two-step magnetic ordering occurs through a 2D ordered state between two successive transition temperatures T_{cu} and T_{c1} ($< T_{cu}$).

The intralayer exchange interaction of $MnCl_2$– and $CuCl_2$–GICs is antiferromagnetic. For $MnCl_2$–GIC, a single phase transition is found at T_c and the characteristic is very similar to a typical spin glass. For $CuCl_2$–GIC, the magnetic ordering is characterized by a broad maximum around 75 K and a Curie-like increase of susceptibility at low temperatures. Any clear evidence of phase transition has not been found without some correction of the obtained result.

The transition metal ions, Cr^{3+} and Fe^{3+} in MCl_3–GICs form a honeycomb lattice and the characteristics of magnetic ordering are presented in 9.4.2.5. The intralayer exchange interaction of $CrCl_3$– and $FeCl_3$–GIC is ferromagnetic and antiferromagnetic, respectively. A phase transition is clearly found for both these compounds. The ordering characteristic, however, is a little complicated.

In Table 2.1, the magnetic properties of MCl_2– and MCl_3–GICs are summarized and in Table 2.2, the references for these compounds are classified, according to the investigated ways.

Table 2.1. Magnetic properties.

Compounds	(Stage)	Intra-plane interaction J/k [K]	Inter-plane interaction J'/J	Final ordered state	Phase transition T_c [K]	References
$CoCl_2$		F(XY)	AF	AF(3D)	Single	59W
		12.7	$\sim -10^{-3}$		24.9(T_N)	
$CoCl_2$–GIC	(S1)	F(XY)	AF	AF(3D)	Single (or 2-step?)	85I, 86W1
		10.4	$\sim -10^{-3}$		9.9(T_N)	
	(S2)	F(XY)	AF		2-step	86W1, 88W1,
				AF-like	9.0(T_{cu}, 2D)	90M2
		7.75	$\sim -10^{-3\cdots4}$	SG-like	7.1(T_{cl})	
	(S3)	F(XY)	AF		2-step	86W1, 90M4
					9.2(T_{cu})	
		6.75	$\sim -10^{-4}$		7.1(T_{cl})	

continued

Table 2.1 (*continued*)

Compounds	(Stage)	Intra-plane interaction J/k [K]	Inter-plane interaction J'/J	Final ordered state	Phase transition T_c [K]	References
$NiCl_2$		F(*XY*)	AF	AF(3D)	Single	75L
		21.7	$\sim -10^{-2}$		52.3(T_N)	
$NiCl_2$–GIC	(S1)	F(*XY*)	AF	AF(3D)	Single	89N1
			$\sim -10^{-3}$		22.0	
	(S2)	F(*XY*)	AF		2-step	87M3, 89H1,
		8.75		AF-like	21(T_{cu})	83S1
			$\sim -10^{-4}$	SG-like	17(T_{cl})	
$MnCl_2$		AF(H)	AF	AF(3D)	2-step	84G, 55M
					1.81(T_{cu})	
		$-3.3(\Theta_\perp)$			1.96(T_{cl})	
$MnCl_2$–GIC	(S1)	AF(H)			Single	86K2, 86W1
		$-5.4(\Theta_\perp)$		SG-like	1.2	
	(S2)	AF(H)	F	$2\sqrt{3}\times 2\sqrt{3}$	Single	86S3, 86W1,
				(3D)		87M1
		$-5.9(\Theta_\perp)$		SG-like	1.2	
$CuCl_2$		AF(H)	AF	AF(3D)	Single	85K2, 62S,
					23.9(T_N)	78B
		−36			75($T_{\chi_{max}}$)	
$CuCl_2$–GIC	(S1)	AF(H)			2-step?	85N
		−40			75($T_{\chi_{max}}$)	
	(S2)	AF(H)			No?	85N
		−40			75($T_{\chi_{max}}$)	
$CrCl_3$		F	AF	AF(3D)	Single	91C, 65N,
						82K
		$29(\Theta_\perp)$	$\sim -10^{-3}$		16.8(T_N)	
$CrCl_3$–GIC	(S1)	F				91C
		$25(\Theta_\perp)$				
	(S3)	F	F		Single	91C, 90H
		$26(\Theta_\perp)$		SG-like	11.4	
$FeCl_3$		AF(H)		AF (3D)	Single	73S
		$-20(\Theta_\perp)$			8.76(T_N)	
$FeCl_3$–GIC	(S1)	AF(H)			Single	82M4
		$-3.8(\Theta_\perp)$			4.3	
	(S2)	AF(H)			Single	82M4
		$-7.6(\Theta_\perp)$			1.3	
	(S3)	AF(H)			Single	87I2
					1.7 ··· 1.8	

Table 2.2. List of references.

Compounds	Crystal properties	Magnetic properties	Neutron diffraction	ESR, NMR, Mössbauer	Heat capacity	Others	Theory
$CoCl_2$–GIC (S2)	85M1 85Y 87W1 89S3 90S1 90S2 91S2	74K1 74K2 76K 82E 83E1 83M1 83S5 84C 84E 85M1 85M2 85O 85Y 86M 86R1 86R3 86S4 86W1 87M2 87M3 88M2 88M3 88N 89H1 89Y1 90M2 90M3 90N1 90N3 90S2 90Y 91M 91S2	83S5 84S1 85I 86S3 86W1 86W2 87W1 87W2 87Z1 88W1 88W2 89W1 89W2 90M3 91M	85T2	76K 83S3 83S4 85M1 85T1 86A 86M 86S4	88K 89K 89S1 89Y1 89Y3 90M1	85D 85R1 85S1 85S2 85S3 86K1 86K3 86R1 86S1 86S2 87D 89M
$CoCl_2$–GIC (S3–5)		82E 83E1 83E2 84E 85O 86M 86W1 88M2 90M4		83S4	83B		
$CoCl_2$–GIC (S1)	85Y 89S3 90N1 90N2 91S2	85O 85Y 86M 86S1 86W1 87C 87Y1 88C 88M1 88M2 88N 90N1 90N2 90N3 91S2	85I 86S4 89R	85T2	86A 86S4	87Y2 88K 89K 89S1 89S4 89Y1 89Y3 90N2	
$NiCl_2$–GIC (S2)	81F 83S1 85S3 87W1 89N1	74K1 74K2 74K3 76K 81S 83S1 83S2 83S6 85M1 85O 86M 86R1 87M3 88M3 88R 89H1 89N1 89Y1 90R2 90S1 90Y 91H	83F 87W1	80F 83S6 84S3	74K2 76K 82O		

$NiCl_2$–GIC (S3–5)		82E		
$NiCl_2$–GIC (S1)	74K1 85S3 89F 89N1 89H2	74K1 89F 89H2 89N1 89N2 90N3 91H	89F	87Y2
$MnCl_2$–GIC (S1, 2)	82B 84G 84K1 86K2 88L 90S2 91S2	84G 84K1 85K1 86S3 86K2 86S3 86W1 87M1 88M3 90S2 91S2	84G 84K1 85K2 84K1 87M1 89S2	86L 87Y2 88L
$CuCl_2$–GIC (S1–3)	83H 86R2 89Y2	85N 86R2	85K2	89Y2 90O
$CrCl_3$–GIC (S1–3)	90R2 91C	90H 90R2	91C	
$FeCl_3$–GIC (S2–10)	80T 82M3 90M1	82E 82M4 83M2 84S1 85R2 86I 87I2 90M1	82M1 82M2 82M3 82O 86S4 82M4 83M2	87I1 87I3 88K 87Z2 88Z 88P 89A
$FeCl_3$–GIC (S1)		82M4 86I 87I2 84S1 88R 90M1	82M1 82M2 83M3 82O 86S4	87Y2 88K 89K

9.4.2.2 $CoCl_2$–GIC

9.4.2.2.1 Stage 2 $CoCl_2$–GIC
Survey of the properties

I. Lattice parameters

$a = 3.56$ Å [87W1]
$a = 3.55$ Å [90S2], [90S1], [90Y], [91S2]
$d = 12.70$ Å [87W1], [90N3], [90S1], [90S2], [90Y]
$d = 12.79$ Å [91S2]

II. Magnetic data

$T_{cu} = 9.5$ K (ND) [87W1], [88W2]
$T_{cu} = 9.4$ K (ND) [87W2], [88W1], [89W1]
$T_{cu} = 9.0$ K ($M'_{3\omega}$) [90M2]
$T_{cu} = 9.1$ K [90S1], [90S2], [90Y], [91S2]
$T_{cu} = 9.2$ K (χ') [88M2]
$T_{cl} = 8.8$ K (ND) [87W1], [88W1], [88W2], [89W1]
$T_{cl} = 8.1$ K (χ') [88M2]
$T_{cl} = 8.0$ K [90S2], [90S1], [90Y], [91S2]
$T_{cl} = 7.1$ K ($M'_{3\omega}$) [90M2]
$\Theta_{\perp} = 23.2$ K [86W1], [90S1], [90S2], [90Y], [91S2]
$C_{\perp} = 3.84$ cm^3 K/mol [86W1], [90S1], [90Y]
$p_{eff} = 5.54\ \mu_B$ [90S1], [90S2], [90Y], [91S2]
$J/k = 7.3$ K (ND) [87Z1]
$J/k = 7.75$ K [86W1], [90S1], [90S2], [90Y], [91S2]
$J_A/k = 3.72$ K [90S1], [90S2], [90Y], [91S2]
$J'/J = -2.5 \cdot 10^{-2}$ (ND) [87Z1]
$g_{\perp} = 6.4$ [86W1], [90S1], [90S2], [90Y], [91S2]

III. Other physical data

Thermal conductivity: [88K], [89K]
Thermoelectric power: [88K], [89K]
Electric resistivity: Fig. 2.30, 31 [89K], [89S1], [89Y3]
^{35}Cl and ^{37}Cl NMR: Fig. 2.32 [85T2]
DHvA: Fig. 2.29 [88N]
C_p: [83S4], Fig. 2.25–28 [86S4], [85T1]
SdH: [83S4]
ND: Fig. 2.11–18 [87W1], [88W1], [89W2]
Electron microscopy: [85M1]

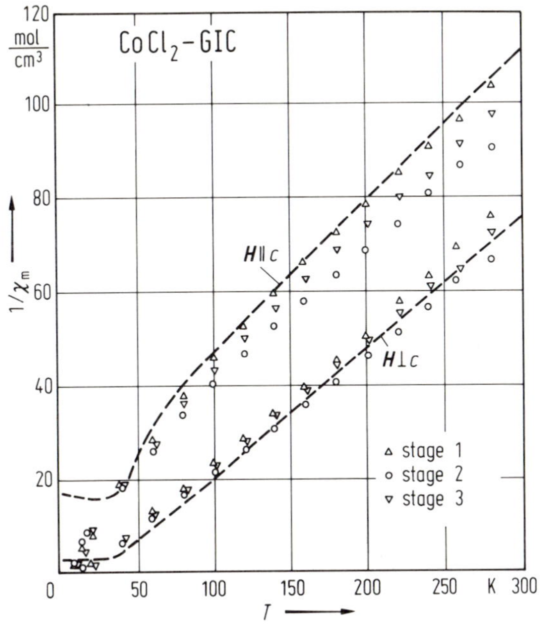

Fig. 2.1. $CoCl_2$–GIC. Temperature dependence of inverse susceptibilities in two field directions for stage 1, 2 and 3 samples. $H = 1.0$ kOe. The dashed line refers to data for pristine $CoCl_2$ [86W1].

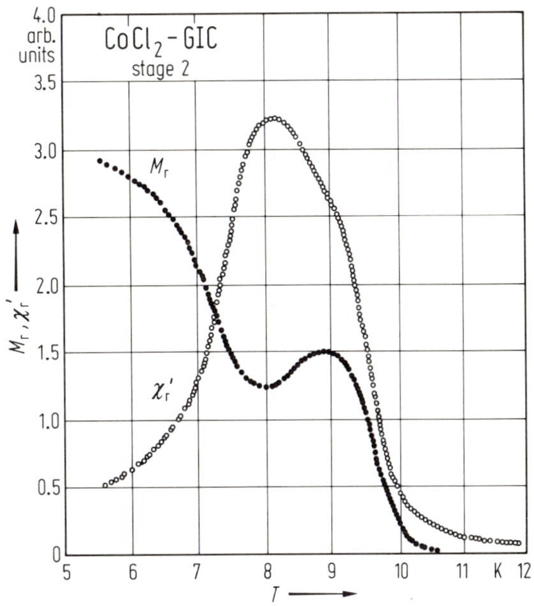

Fig. 2.2. $CoCl_2$–GIC, stage 2. Temperature dependence of TRM and AC susceptibility measured simultaneously. $H = 0$, $H_c = 120$ mOe, $h = 10$ mOe, $\nu = 10$ Hz, $\boldsymbol{H}_c \perp c$ and $\boldsymbol{h} \perp c$ [83M1].

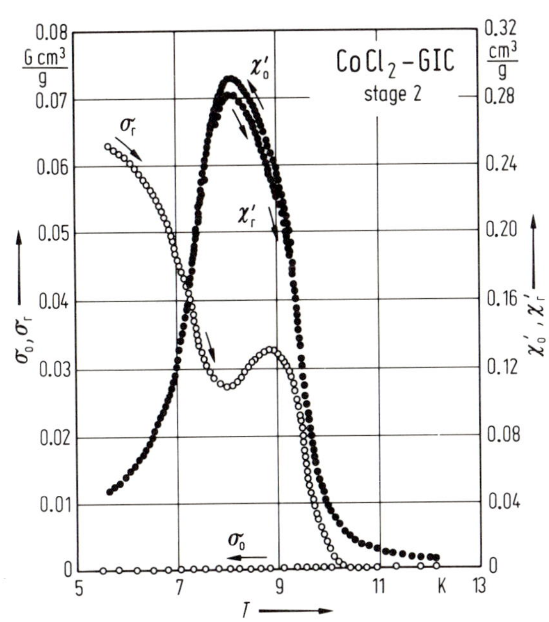

Fig. 2.3. $CoCl_2$–GIC, stage 2. Temperature dependence of AC susceptibilities and magnetizations under zero field in two different processes, cooling (ZFC) and heating after field cooling (ZFHAFC). $h(\perp c) = 1.0$ mOe, $\nu = 10$ Hz, $H_c(\perp c) = 0.12$ Oe, $H = 0$ [88M2].

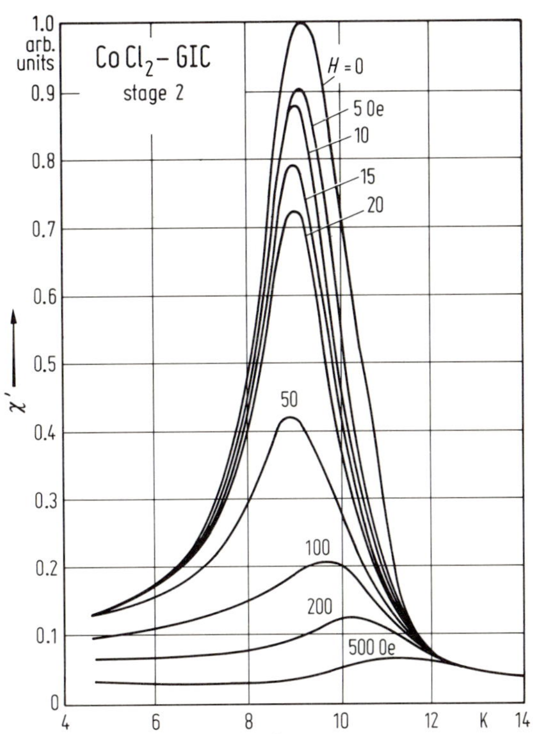

Fig. 2.4. $CoCl_2$–GIC, stage 2. Temperature dependence of AC susceptibilities in DC magnetic field. $\boldsymbol{H} \perp c$. DC field is indicated in the figure [84C]. See also Fig. 2.35.

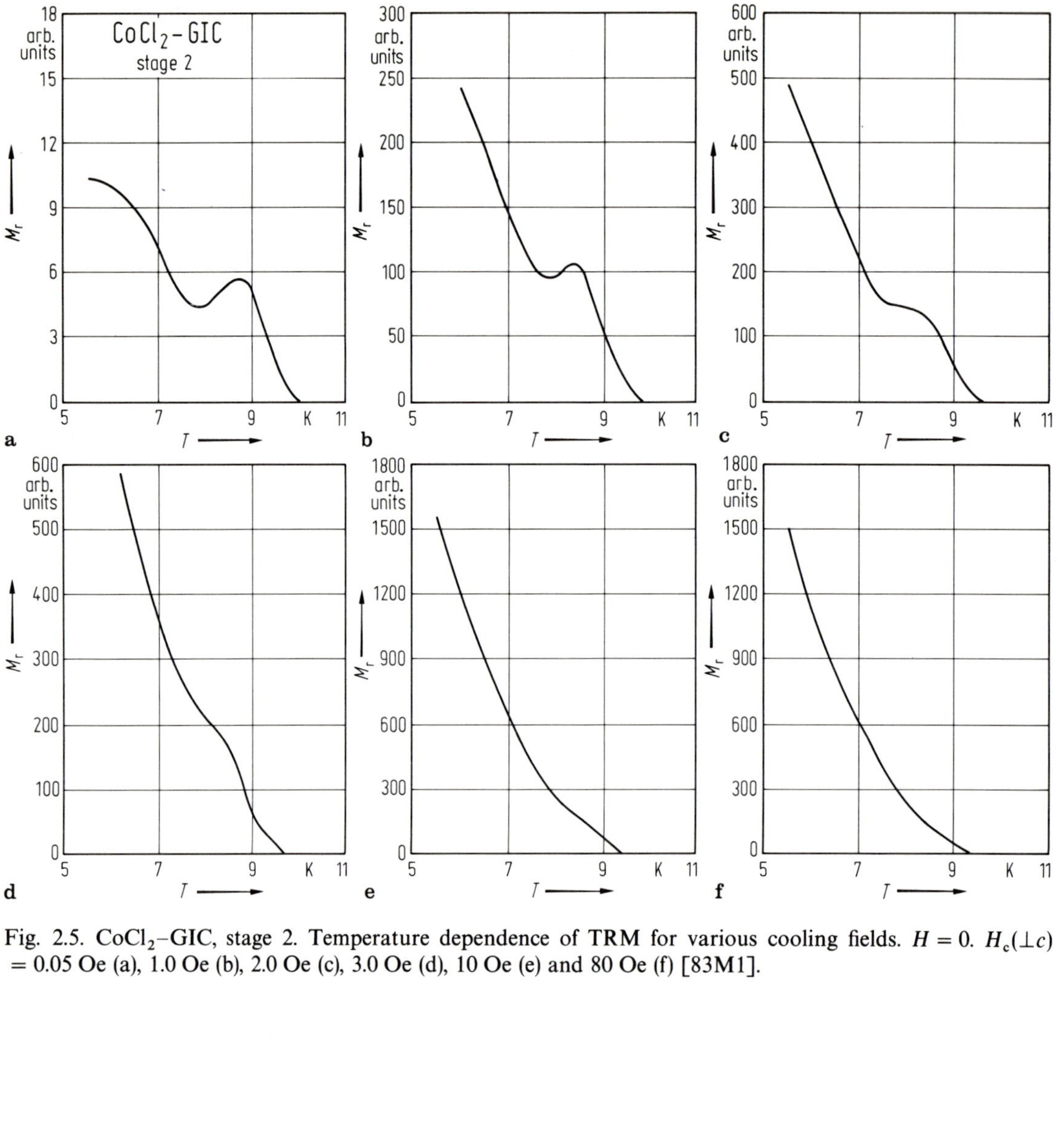

Fig. 2.5. $CoCl_2$–GIC, stage 2. Temperature dependence of TRM for various cooling fields. $H = 0$. $H_c(\perp c) = 0.05$ Oe (a), 1.0 Oe (b), 2.0 Oe (c), 3.0 Oe (d), 10 Oe (e) and 80 Oe (f) [83M1].

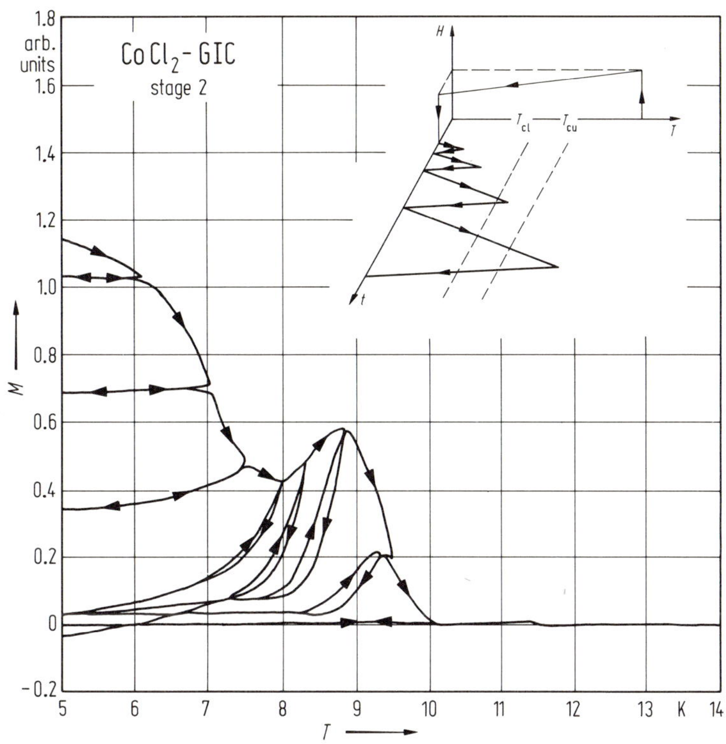

Fig. 2.6. $CoCl_2$–GIC, stage 2. Temperature dependence of magnetization measured in the process shown in the inset. $H = 0$, $H_c = 120$ mOe ($\perp c$) [85M2].

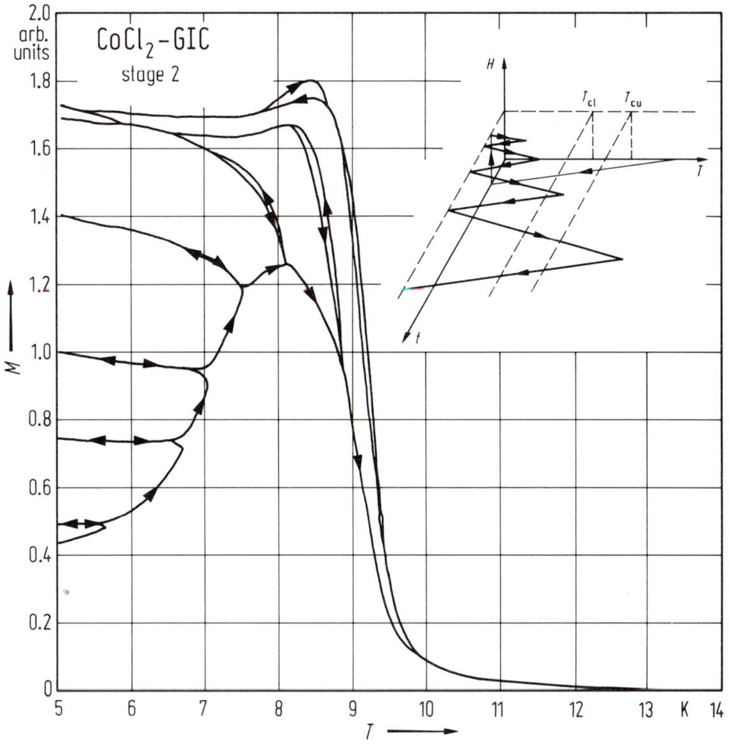

Fig. 2.7. $CoCl_2$–GIC, stage 2. Temperature dependence of magnetization measured in the process shown in the inset. $H = 120$ mOe ($\perp c$) [88M3].

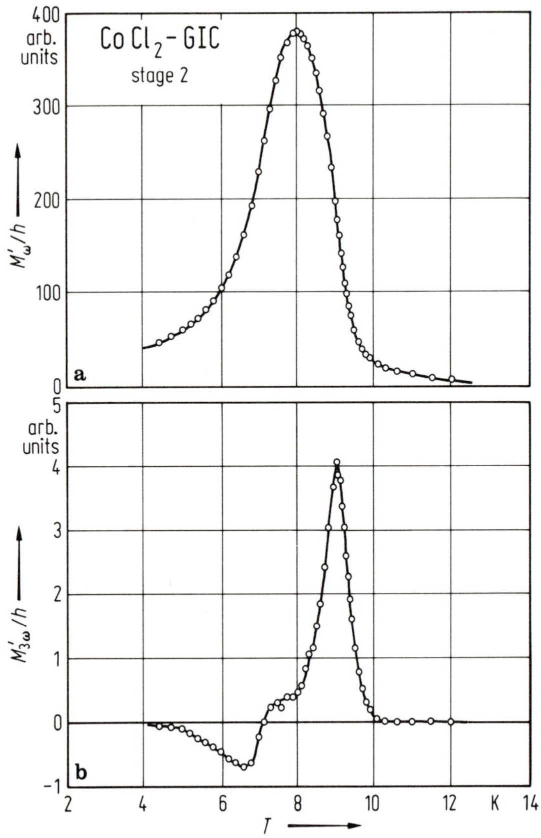

Fig. 2.8. $CoCl_2$–GIC, stage 2. Temperature dependence of (a) linear and (b) nonlinear magnetic responses to AC field measured at $h = 0.8$ Oe simultaneously. $\boldsymbol{h} \perp c$, $\nu = 0.1$ Hz [90M2].

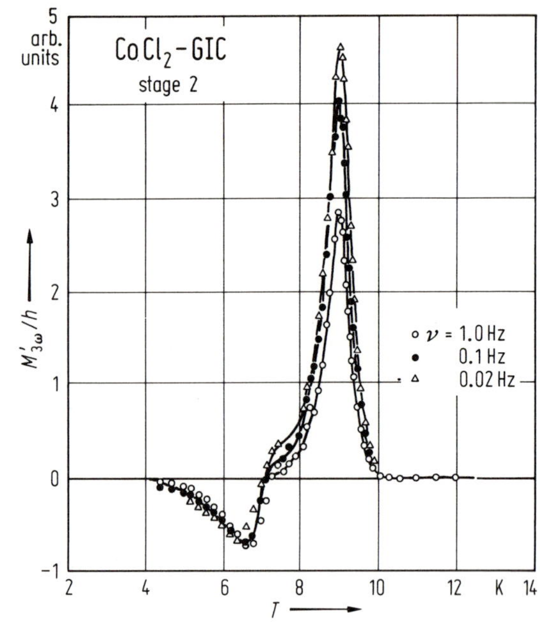

Fig. 2.9. $CoCl_2$–GIC, stage 2. Temperature dependences of AC nonlinear responses at various frequencies. $h = 0.8$ Oe ($\perp c$) [90M2].

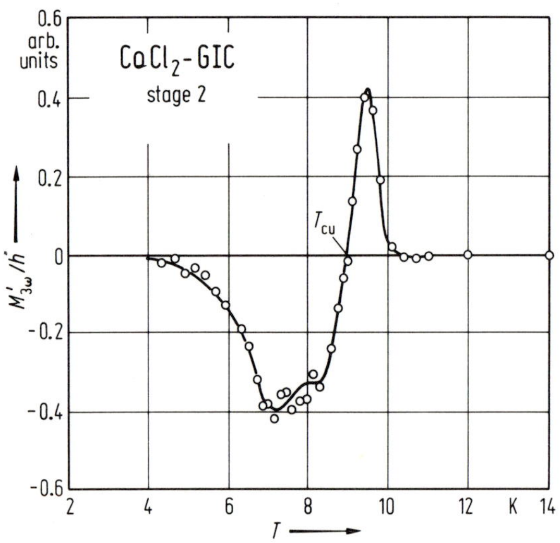

Fig. 2.10. $CoCl_2$–GIC, stage 2. Temperature dependence of AC nonlinear responses. $h = 0.2$ Oe ($\perp c$), $\nu = 0.1$ Hz [90M2].

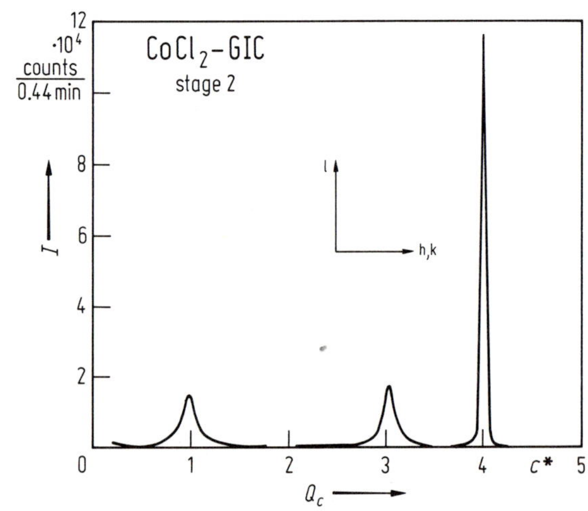

Fig. 2.11. $CoCl_2$–GIC, stage 2. Neutron scattering (nuclear part) along (0, 0, l). The inset shows the direction of the scan in reciprocal space. $T = 16$ K [87W1].

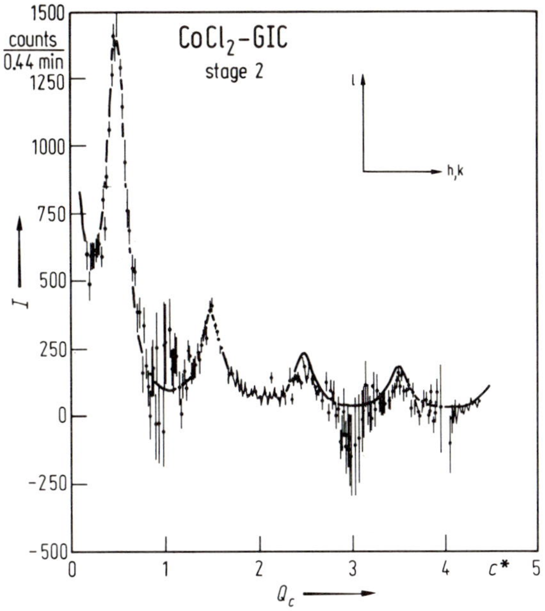

Fig. 2.12. $CoCl_2$–GIC, stage 2. Neutron scattering (magnetic part) along (0, 0, l). $T = 4.4$ K [87W1]. The solid line is a fit after Eq. (16) of [87W1].

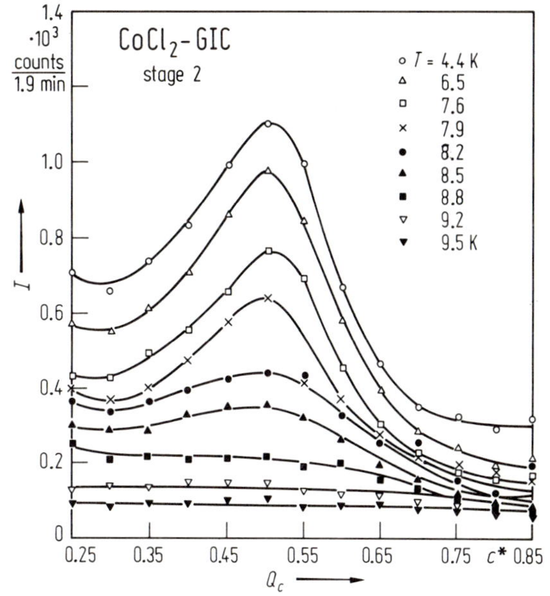

Fig. 2.13. $CoCl_2$–GIC, stage 2. Neutron scattering (magnetic part) along (0, 0, l) for various temperatures [87W1].

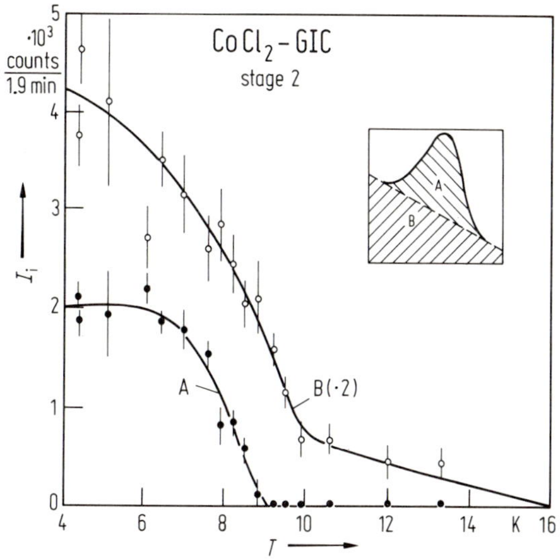

Fig. 2.14. $CoCl_2$–GIC, stage 2. Temperature dependence of the integrated neutron scattering intensity under Lorentzian AF component A and the 2D magnetic ridge B [87W1].

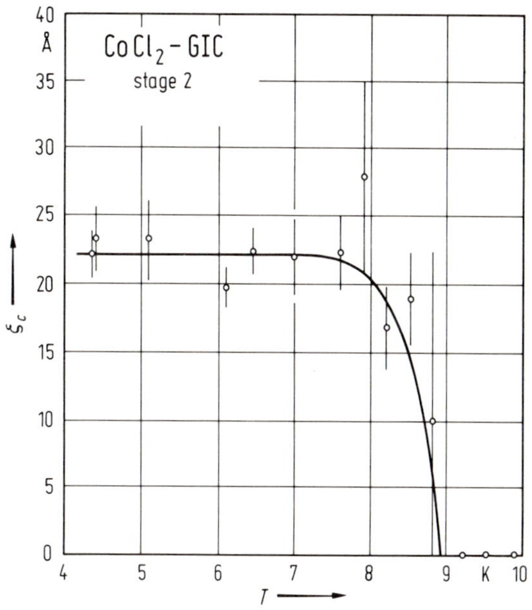

Fig. 2.15. $CoCl_2$–GIC, stage 2. Temperature dependence of the spin correlation length along the c-axis. The solid line is a guide to the eye [87W1].

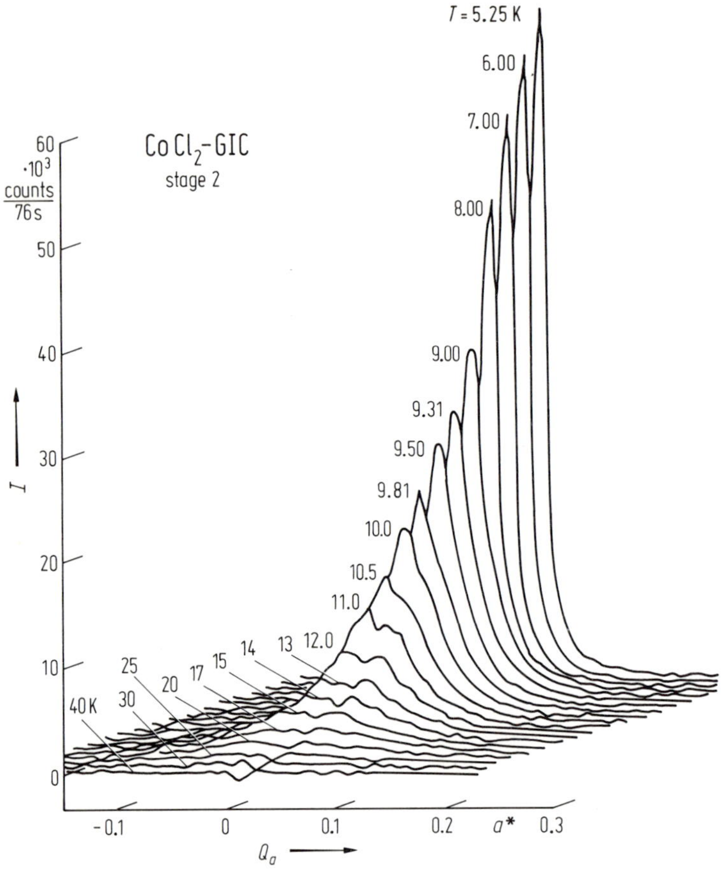

Fig. 2.16. CoCl$_2$–GIC, stage 2. The 2D critical neutron scattering along (Q_a, 0, 0.2). Nuclear scattering intensity has been subtracted from these data. The numbers associated with each scan show the temperatures in Kelvin [89W2].

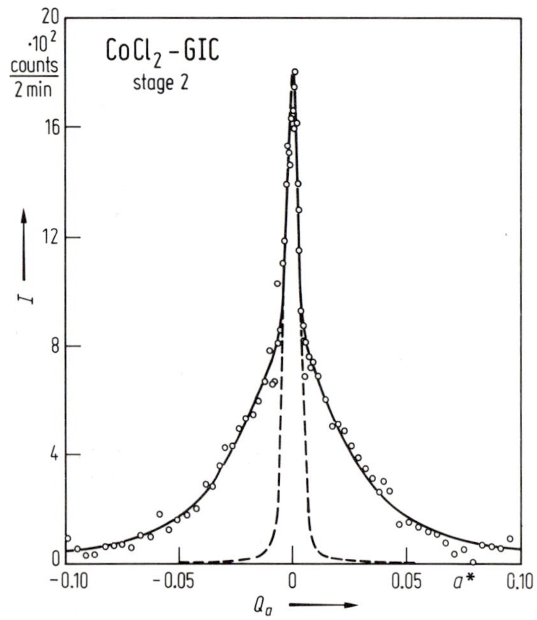

Fig. 2.17. CoCl$_2$–GIC, stage 2. Fits of structure factors based on Kummer function (dashed line) and on a sum of SRO and LRO terms (solid line) with experimental data at $T = 9.0$ K [88W1].

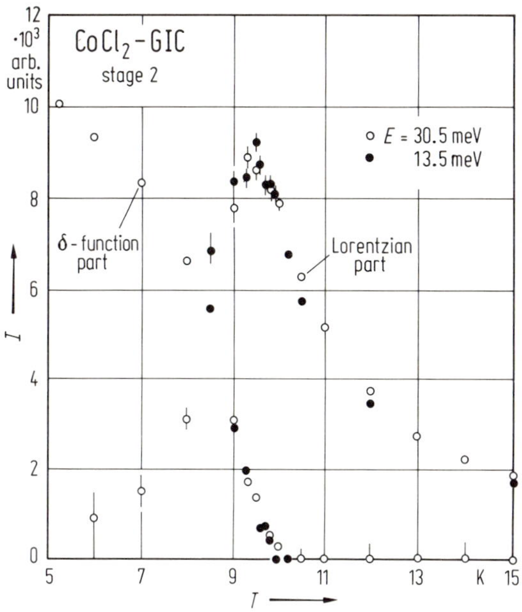

Fig. 2.18. $CoCl_2$–GIC, stage 2. Intensities of Lorentzian and δ-function parts of 2D magnetic scattering as a function of temperature for different energies of incident neutrons [89W2].

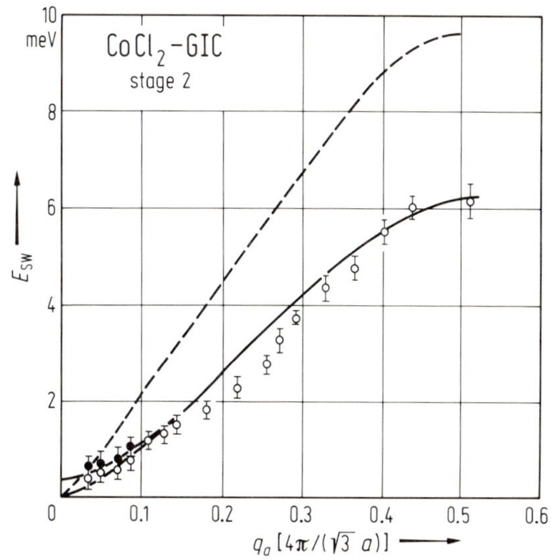

Fig. 2.19. $CoCl_2$–GIC, stage 2. Spin wave energy dispersion at $T = 4.5$ K. The solid lines are calculated spin wave energies according to Eqs. (5) and (6) in [87Z1]. The dashed line indicates the spin wave dispersion of pristine $CoCl_2$ [87Z1].

For Fig. 2.21 see next page.

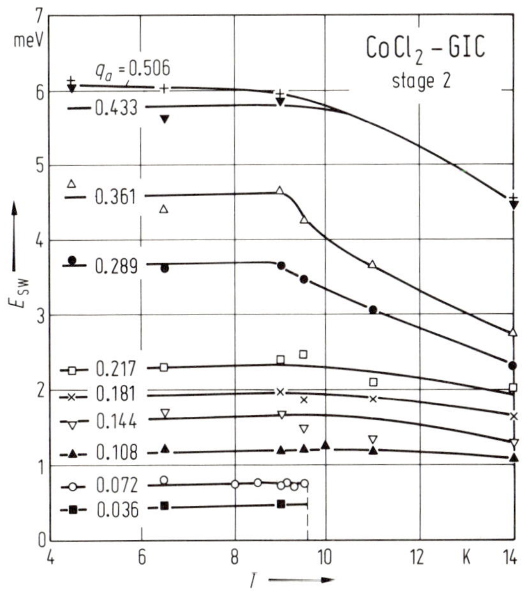

Fig. 2.20. $CoCl_2$–GIC, stage 2. Magnon renormalization with temperature. The numbers refer to the magnon wave vector in reciprocal lattice units. Lines are guides to the eye [89W1].

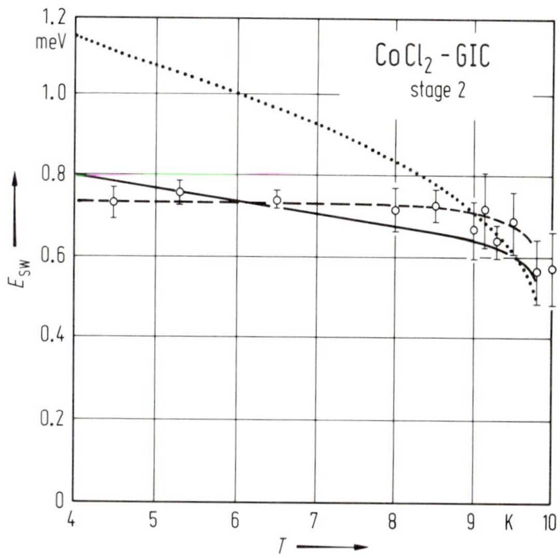

Fig. 2.22. $CoCl_2$–GIC, stage 2. Spin wave energies at $Q_a = 0.072a^*$ as a function of temperature. The solid, dashed and dotted lines are theoretical predictions for spin wave energies close to T_{KT} by Ohta–Jasnow, Nelson–Kosterlitz and Pokrovskii–Uimin, respectively [89W2].

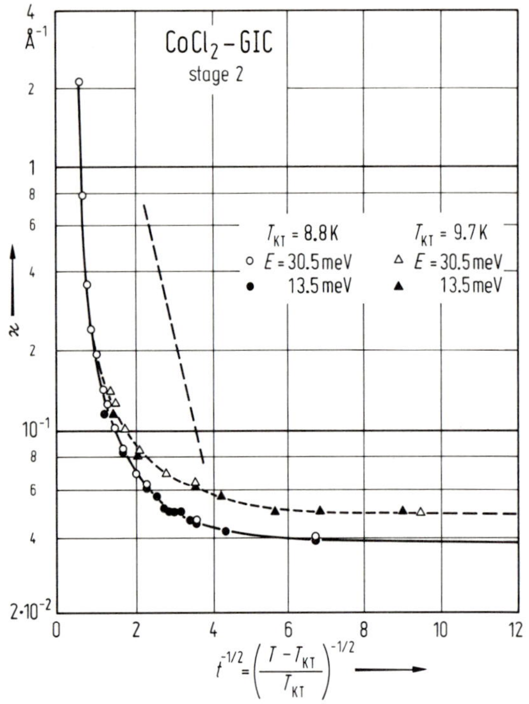

Fig. 2.21. $CoCl_2$–GIC, stage 2. Temperature dependence of inverse correlation length. Experimental data are plotted for $[(T-T_{KT})/T_{KT}]^{-1/2}$ with two trial values of T_{KT} as indicated in the figure. The straight dashed line shows the theoretical slope for $b = 1.5$ by Kosterlitz–Thouless ($\kappa = \kappa_0 \exp(-bt^{-1/2})$). *E*: energy of incident neutrons [89W2].

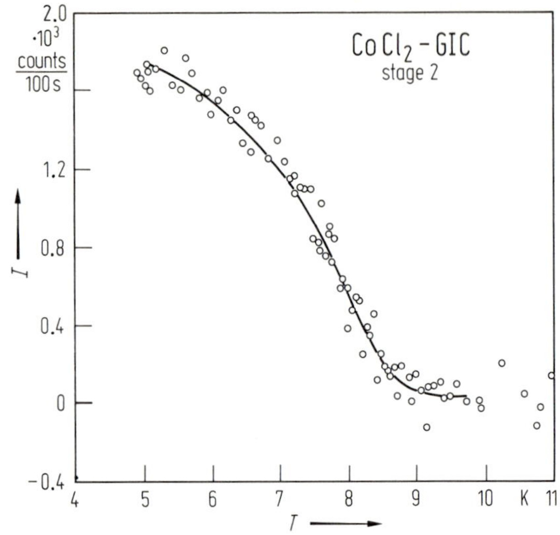

Fig. 2.23. $CoCl_2$–GIC, stage 2. Temperature dependence of AF Bragg peak intensity at (0, 0, 0.5). $H = 0$. Solid line is a guide to the eye [83S5].

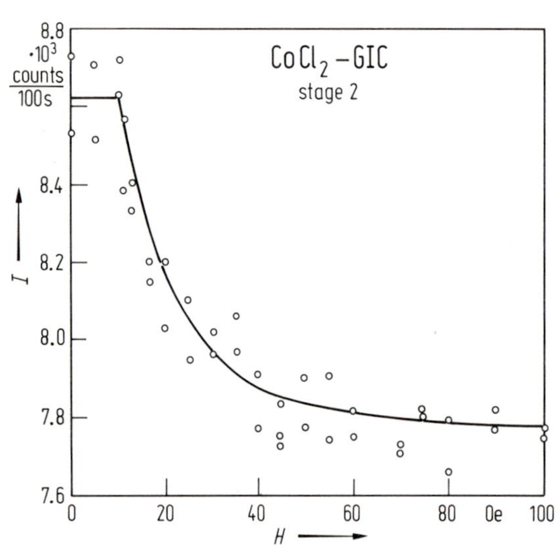

Fig. 2.24. $CoCl_2$–GIC, stage 2. Field dependence of AF Bragg peak intensity at (0, 0, 0.5) at $T = 6.0$ K. $\boldsymbol{H} \perp c$. Solid line is a guide to the eye [83S5].

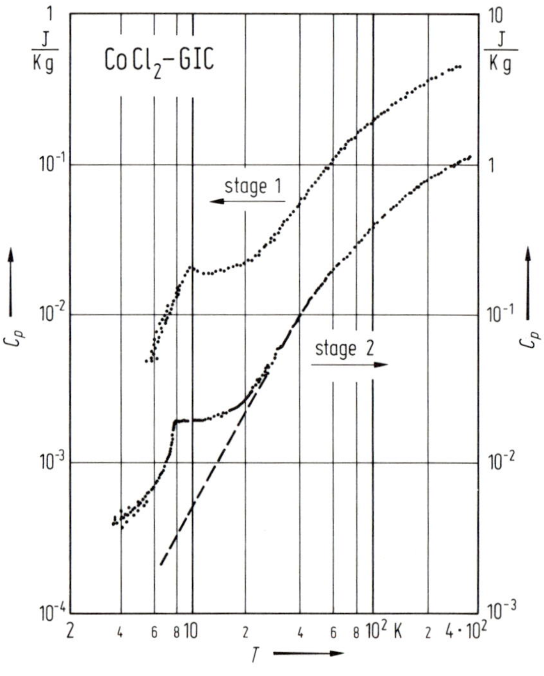

Fig. 2.25. $CoCl_2$–GIC. Temperature dependence of heat capacity for stage 1 and 2 samples. Dashed line is the estimated lattice contribution [86S4].

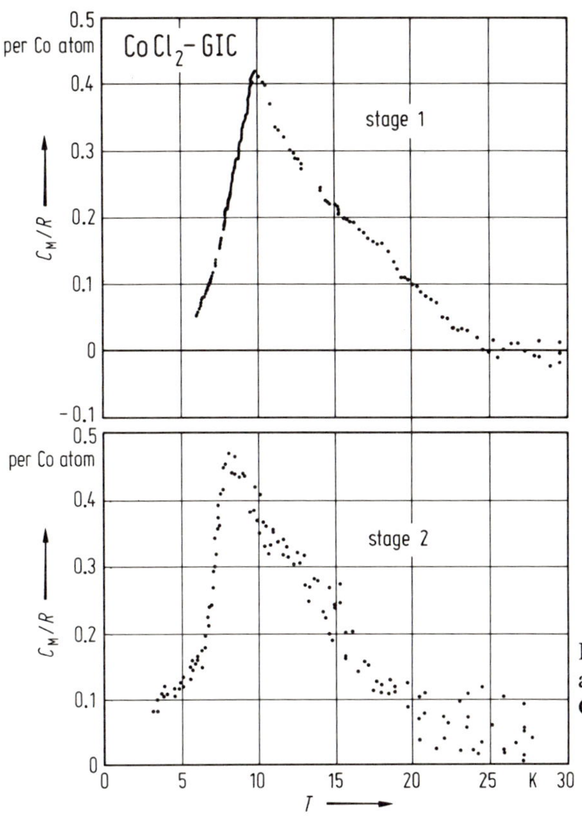

Fig. 2.26. $CoCl_2$–GIC. Temperature dependence of magnetic contribution of heat capacity for stage 1 and 2 samples [86S4].

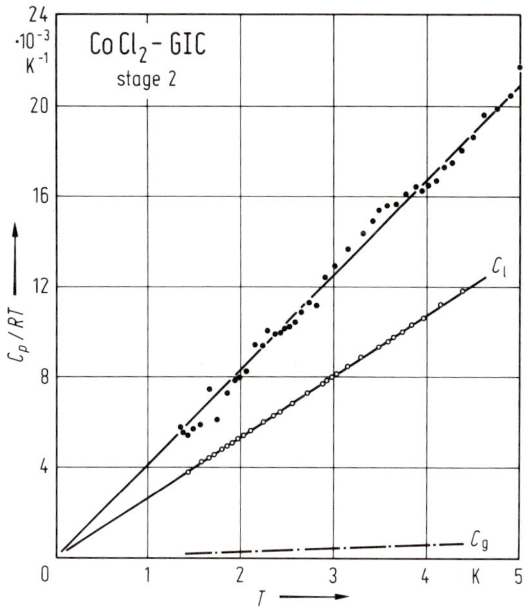

Fig. 2.28. $CoCl_2$–GIC, stage 2. C_p/RT vs. T plot. C_l and C_g are estimated lattice contributions of $CoCl_2$–GIC and of graphite, respectively [85T1].

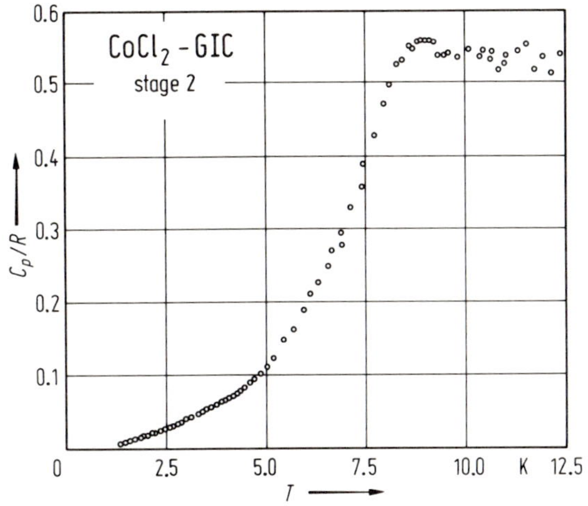

Fig. 2.27. $CoCl_2$–GIC, stage 2. Temperature dependence of heat capacity [85T1].

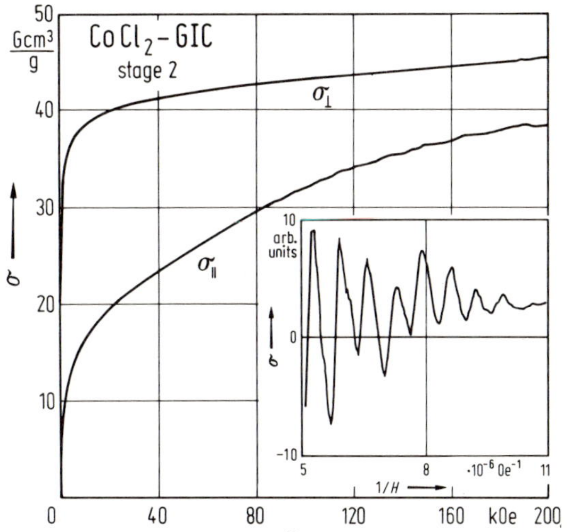

Fig. 2.29. $CoCl_2$–GIC, stage 2. Magnetization curves in two field directions relative to the c axis. $T = 4.2$ K. The figure inset shows the DHvA oscillations vs. $1/H$ that have been separated from the other magnetization contributions [88N].

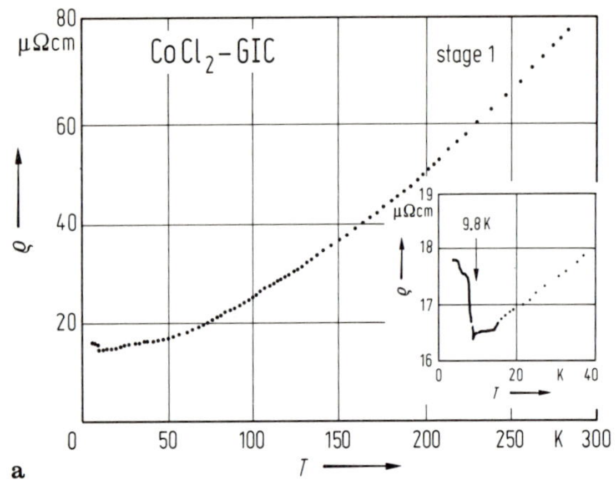

Fig. 2.30. $CoCl_2$–GIC. Experimental zero-field resistivities for stage 1 (a) and stage 2 (b) samples as a function of temperature [89S1].

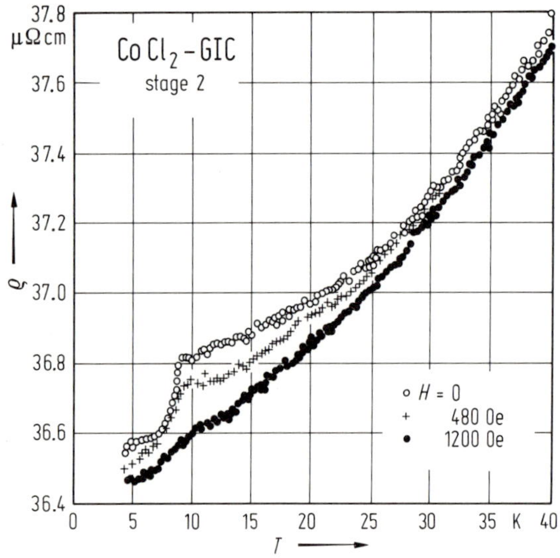

Fig. 2.31. $CoCl_2$–GIC, stage 2. Experimental resistivities in various external fields as a function of temperature [89Y3].

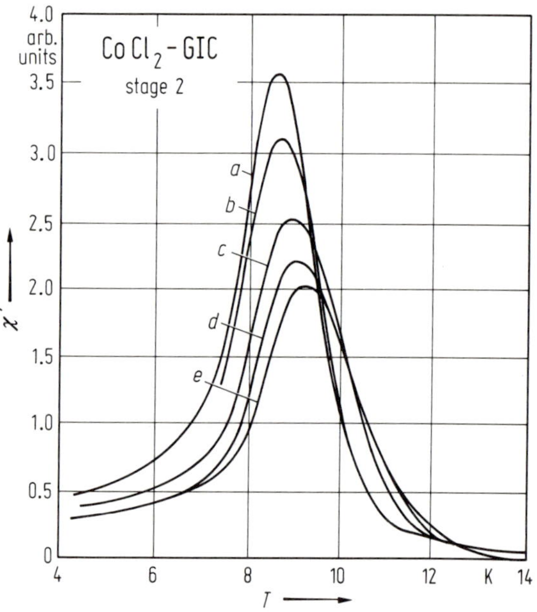

Fig. 2.33. $CoCl_2$–GIC, stage 2. AC susceptibilities as a function of temperature under pressures of 0 kbar (curve *a*), 3.1 kbar (*b*), 8.6 kbar (*c*), 12.5 kbar (*d*) and 15 kbar (*e*). $H = 0$, $h \leq 0.1$ Oe [90N1].

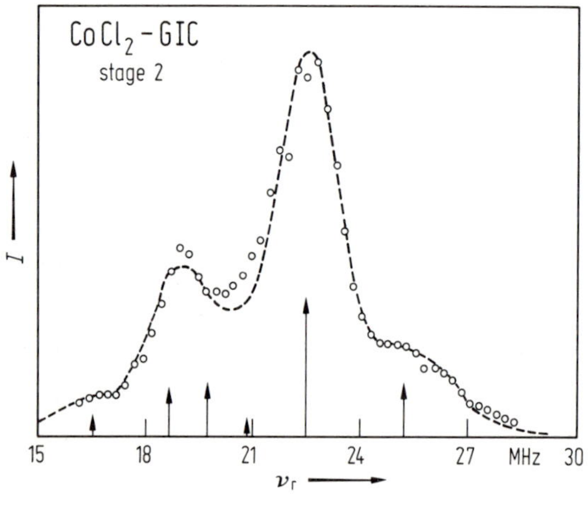

◀ Fig. 2.32. $CoCl_2$–GIC, stage 2. Spin echo spectra of ^{35}Cl and ^{37}Cl NMR signals. Upward arrows indicate resonance positions and relative intensities. The dotted curve is the calculated spectrum [85T2].

9.4.2.2.2 Stage n $CoCl_2$–GIC ($n \geqq 3$)
Survey of the properties

II. Magnetic data

$T_{cu} = 9.2$ K (χ', $M'_{3\omega}$) for S3 [88M2], [90M4]
$T_{cl} = 8.0$ K (χ') for S3 [88M2]
$T_{cl} = 7.1$ K ($M'_{3\omega}$) for S3 [90M4]
$\Theta_{\perp} = 20.3$ K for S3 [86W1]
$C_{\perp} = 3.64$ cm^{3}K/mol for S3 [86W1]
$J/k = 6.75$ K for S3 [86W1]
$g_{\perp} = 6.23$ for S3 [86W1]

III. Other physical data

Thermal conductivity: [83B]
Thermoelectric power: [83B]
C_p: [83S4]
SdH: [83S4]

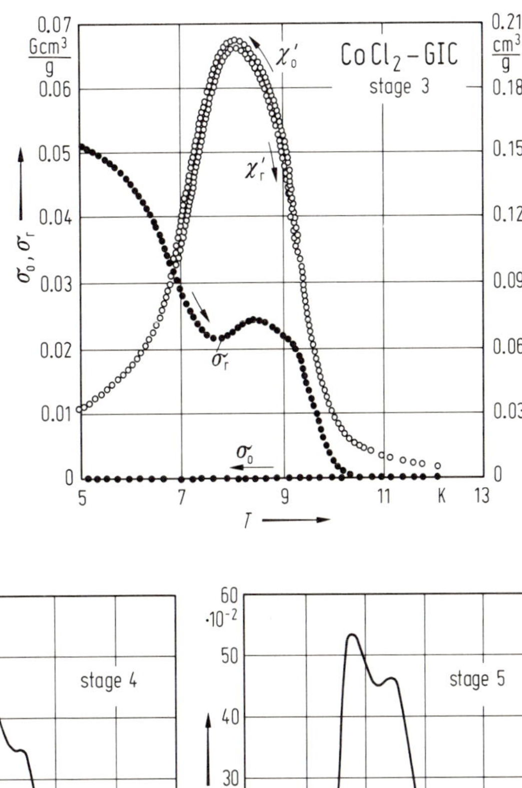

Fig. 2.34. $CoCl_2$–GIC, stage 3. Temperature dependence of AC susceptibilities and magnetizations measured in two different processes ZFC and ZFHAFC. $h(\perp c) = 1.0$ mOe, $\nu = 10$ Hz, $H = 0$, $H_c(\perp c) = 0.12$ Oe [88M2].

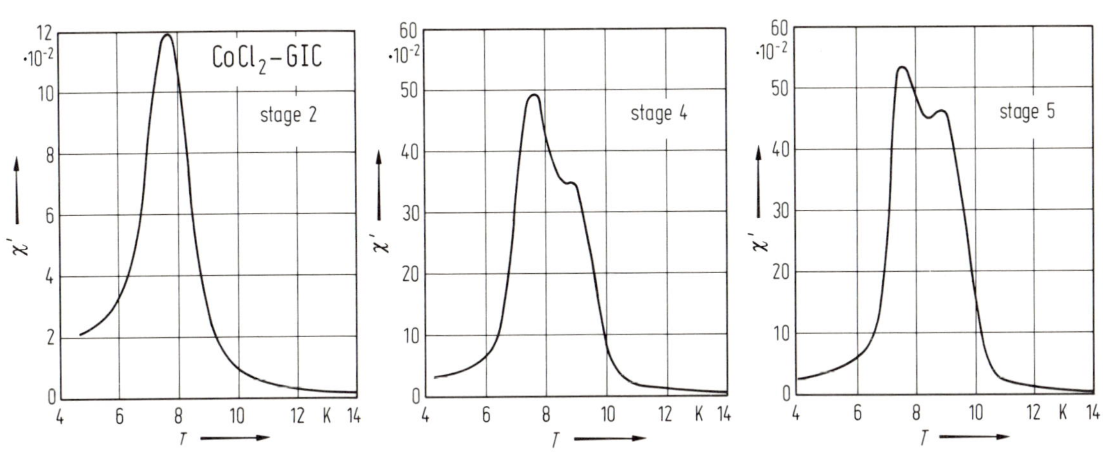

Fig. 2.35. $CoCl_2$–GIC. Temperature dependence of AC volume susceptibilities for stage 2, 4 and 5 samples. $H = 0$, $h(\perp c) = 0.6$ Oe, $\nu = 300$ Hz [84E]. For susceptibilities of stages 2, 3, see also Fig. 2.1.

9.4.2.2.3 Stage 1 $CoCl_2$–GIC
Survey of the properties

I. Lattice parameters

$a = 3.55$ Å [87C], [87Y1]
$d = 9.40$ Å [87C], [87Y1]
$d = 9.38$ Å [90N3]

II. Magnetic data

$T_N = 9.9$ K (ND) [85I]
$T_N = 9.8$ K (ρ) [89S1]
$T_N = 9.7$ K (χ') [90N1]
$T_{cu} = 9.77$ K (χ'') [87Y1]
$T_{cu} = 10.2$ K (χ') [88M2]
$T_{cl} = 8.22$ K (χ'') [87Y1]
$T_{cl} = 9.9$ K (χ') [88M2]
$\Theta_\perp = 31.1$ K [86W1]
$C_\perp = 3.30$ cm^{3}K/mol [86W1]
$J/k = 10.38$ K [86W1]
$g_\perp = 5.94$ [86W1]

III. Other physical data

Thermal conductivity: [88K], [89K]
Thermoelectric power: [88K], [89K]
Electric resistivity: Fig. 2.43 [89K], [89S1], [89S4], [89Y3]; see also Fig. 2.30
ND: Fig. 2.38, 39 [85I], [86S4], [89R]
C_p: Fig. 2.25, 26 [86S4]

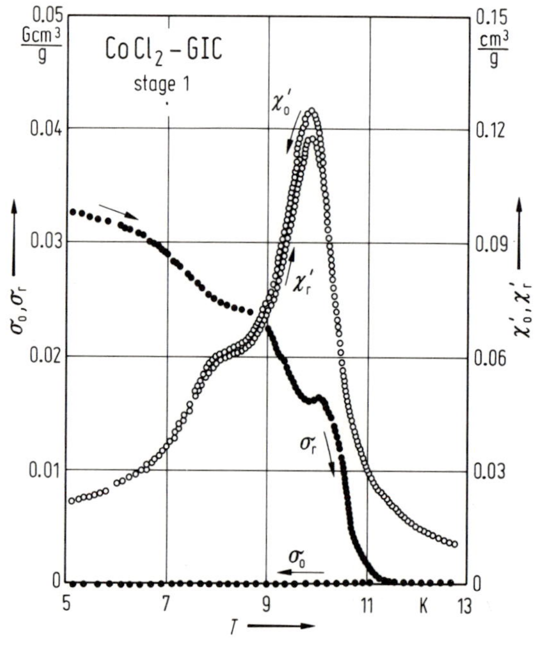

Fig. 2.36. $CoCl_2$–GIC, stage 1. Temperature dependence of AC susceptibilities and magnetizations measured in two different processes ZFC and ZFHAFC. $h(\perp c) = 1.0$ mOe, $\nu = 10$ Hz, $H = 0$, $H_c(\perp c) = 0.12$ Oe [88M2].

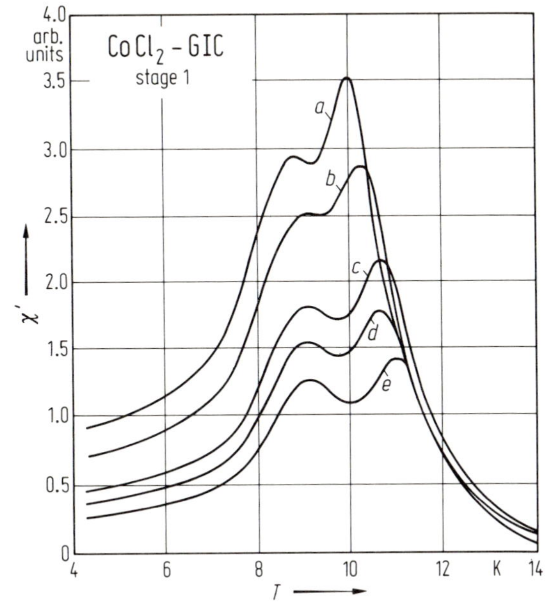

Fig. 2.37. $CoCl_2$–GIC, stage 1. AC susceptibilities as a function of temperature under pressures of 0 kbar (curve *a*), 3.1 kbar (*b*), 8.6 kbar (*c*), 12.5 kbar (*d*) and 15 kbar (*e*). $H = 0$, $h \leqq 0.1$ Oe [90N1].

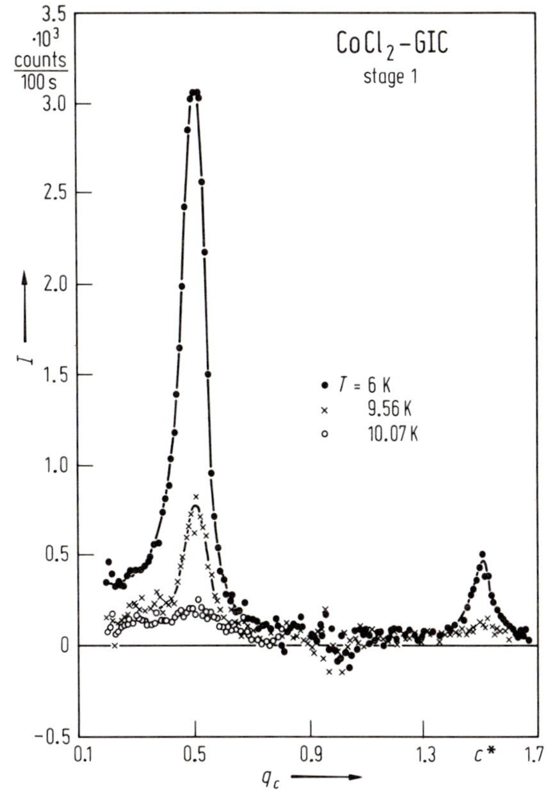

Fig. 2.38. $CoCl_2$–GIC, stage 1. Intensity distribution of the magnetic neutron scattering along (0, 0, *l*) direction at various temperatures [85I].

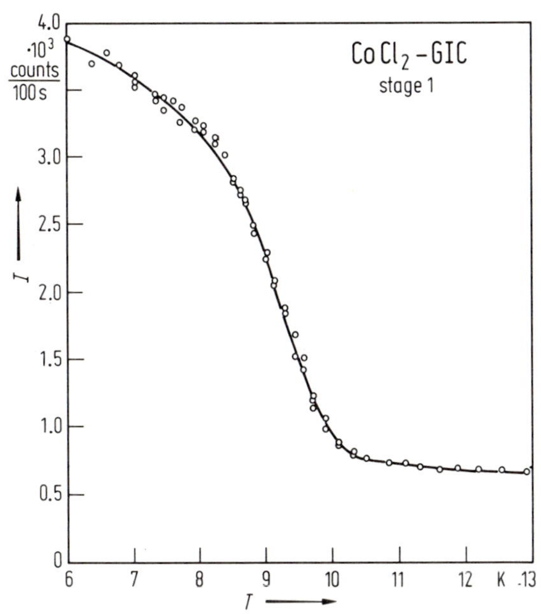

Fig. 2.39. $CoCl_2$–GIC, stage 1. The neutron scattering intensities at the AF (0, 0, 0.5) position as a function of temperature [85I].

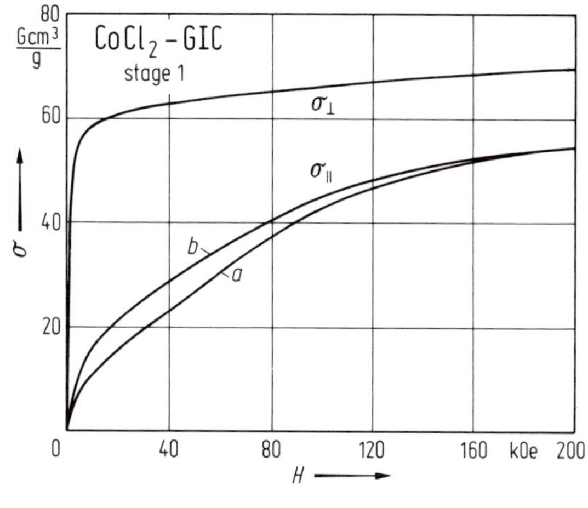

Fig. 2.40. $CoCl_2$–GIC, stage 1. Magnetization curves in two field directions relative to *c* axis. $T = 4.2$ K. The virgin trace $\sigma_{\parallel}$ is labelled (*a*), on reducing the magnetic field, the curve (*b*) shows a hysteresis [88N].

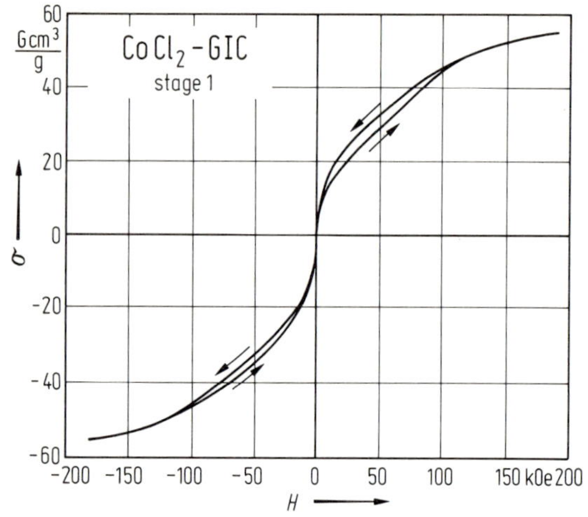

Fig. 2.41. $CoCl_2$–GIC, stage 1. A full magnetization hysteresis loop. $T = 4.2$ K. The field $H(\|c)$ is swept in the sequence $0 \rightarrow 180$ kOe $\rightarrow 0 \rightarrow -180$ kOe $\rightarrow 0$ [90N3].

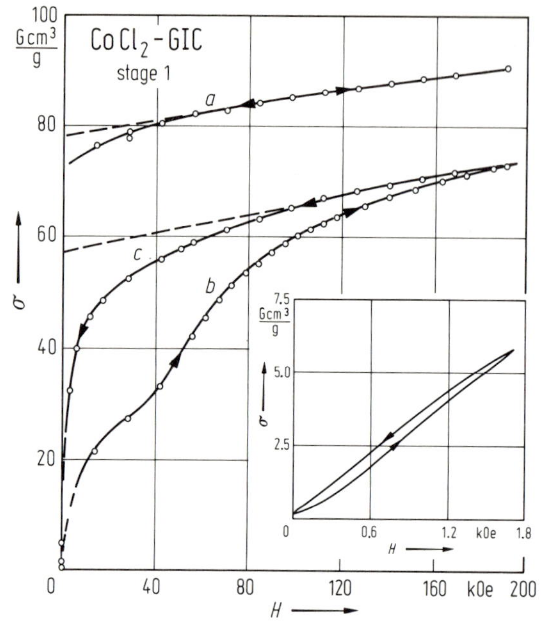

Fig. 2.42. $CoCl_2$–GIC, stage 1. Magnetization curves up to 190 kOe in two field directions. $T = 4.2$ K. (Curve *a*) for $\boldsymbol{H} \perp c$, (*b*) and (*c*) for $\boldsymbol{H} \| c$ first as increasing field and then as decreasing field, respectively. Inset shows the magnetization hysteresis in the field up to 1.6 kOe of $\boldsymbol{H} \| c$ at 4.2 K [87Y1].

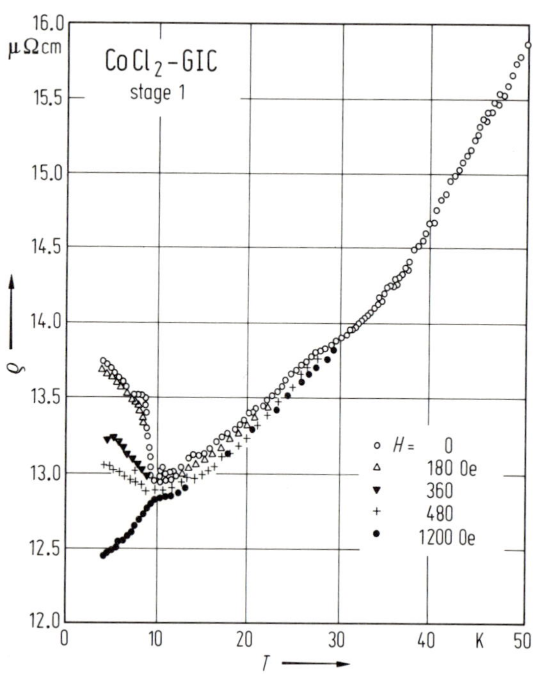

Fig. 2.43. $CoCl_2$–GIC, stage 1. Experimental resistivities in various external fields as a function of temperature [89Y3].

9.4.2.3 $NiCl_2$–GIC

9.4.2.3.1 Stage 2 $NiCl_2$–GIC
Survey of the properties

I. Lattice parameters

$a = 3.46$ Å [87W1], [90S1], [90Y]
$a = 3.478$ Å [80F]
$a = 3.48$ Å [83F]
$d = 12.65$ Å [80F], [83F]
$d = 12.82$ Å [85S3]
$d = 12.7$ Å [83S1], [90S1], [90Y]
$d = 12.67$ Å [89N1]
$d = 12.78$ Å [91H], [87W1]

II. Magnetic data

$T_{cu} = 21$ K (χ', $M'_{3\omega}$) [89H1]
$T_{cu} = 20.3$ K (M_{ZFC}) [76K]
$T_{cu} = 20.2$ K (M_{FC}) [74K2]
$T_{cu} = 19.4$ K [81S]
$T_{cu} = 22.0$ K [83S1], [90S1], [90Y]
$T_{cu} = 21.3$ K (χ_{AC}) [83S2]
$T_{cl} = 17$ K ($M'_{3\omega}$) [89H1]
$T_{cl} = 18.1$ K [76K]
$T_{cl} = 18$ K [74K2]
$T_{cl} = 17.3$ K [81S]
$T_{cl} = 17.5$ K [83S1], [90S1], [90Y]
$T_{cl} = 18.0$ K [83S2]
$T_{cl} = 19.5$ K [89N1]
$\Theta_{\perp} = 62$ K [83S6], [83S2]
$\Theta_{\perp} = 70$ K [83S1], [90S1], [90Y]
$C_{\perp} = 1.35$ cm^{3}K/mol [83S1], [90S1], [90Y]
$p_{eff} = 3.29\mu_B$ [83S1], [90S1], [90Y]
$J/k = 8.75$ K [83S1], [90S1], [90Y]
$J'/J = -8.1 \cdot 10^{-4}$ (ESR) [83S6]
$J'/J = -(6 \cdots 8) \cdot 10^{-4}$ [81S]
$D/k = 0.8$ K [90S1], [90Y]
$g_{\parallel} = 2.096$ [84S3]
$g_{\perp} = 2.156$ [84S3]
$g_{\perp} = 2.33$ [90S1], [90Y]

III. Other physical data

C_p: Fig. 2.56, 57 [82O]
ESR: [80F], [83S6], [84S3]
ND: Fig. 2.51–55 [87W1]

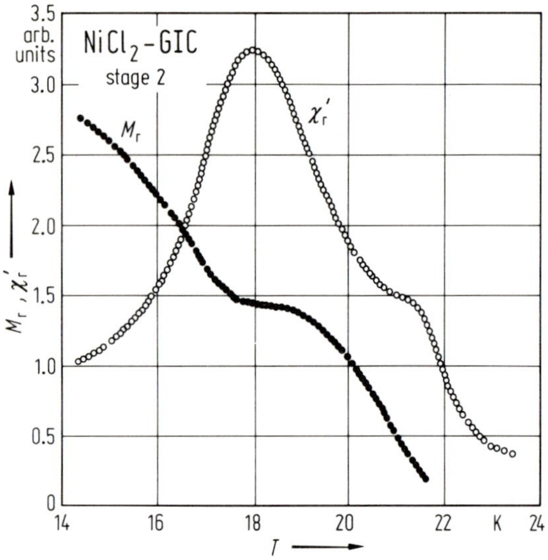

Fig. 2.44. $NiCl_2$–GIC, stage 2. Temperature dependence of TRM and AC susceptibility measured simultaneously. $H = 0$, $H_c = 120$ mOe, $h = 10$ mOe, $\nu = 10$ Hz, $\boldsymbol{H}_c \perp c$ and $\boldsymbol{h} \perp c$ [83S2].

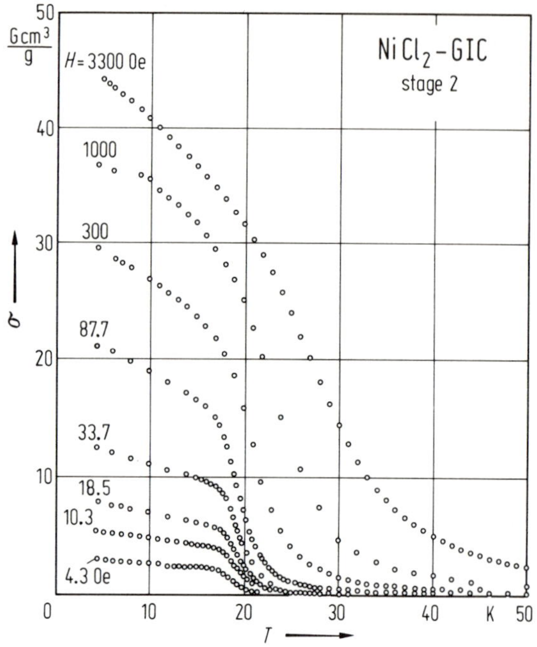

Fig. 2.46. $NiCl_2$–GIC, stage 2. Temperature dependence of FCM as a function of external field. The values of $H(\perp c)$ are indicated [83S1]. For field dependence of magnetization see also Fig. 2.62. See also Fig. 3.7.

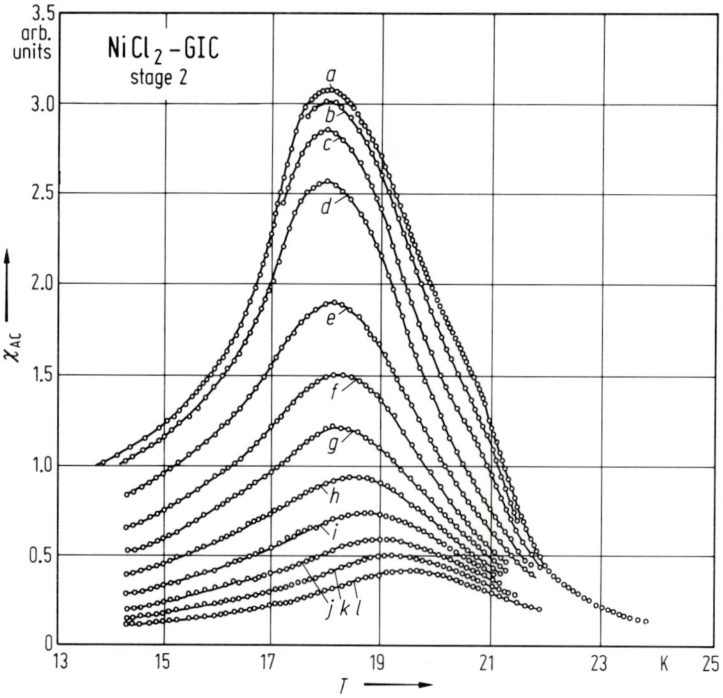

Fig. 2.45. $NiCl_2$–GIC, stage 2. Temperature dependence of AC susceptibilities in various external fields. $h = 100$ mOe, $\nu = 530$ Hz, $\boldsymbol{H} \perp c$ and $\boldsymbol{h} \perp c$. $H = 0.5$ Oe (curve *a*), 3.0 Oe (*b*), 5.5 Oe (*c*), 10 Oe (*d*), 20 Oe (*e*), 30 Oe (*f*), 40 Oe (*g*), 50 Oe (*h*), 60 Oe (*i*), 70 Oe (*j*), 80 Oe (*k*) and 90 Oe (*l*) [83S2].

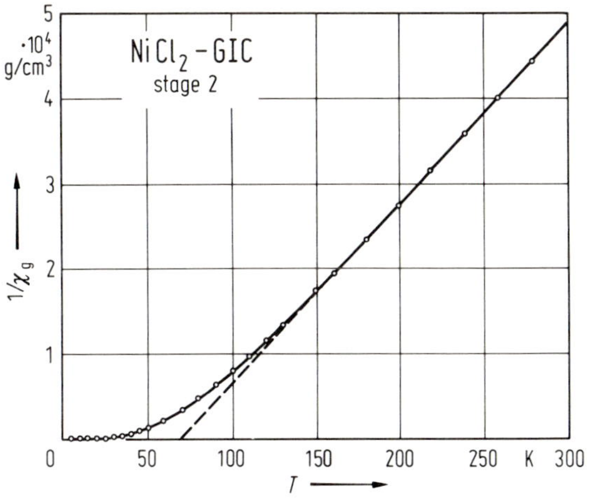

Fig. 2.47. $NiCl_2$–GIC, stage 2. The inverse DC susceptibility vs. temperature. $H(\perp c) = 1000$ Oe [83S1].

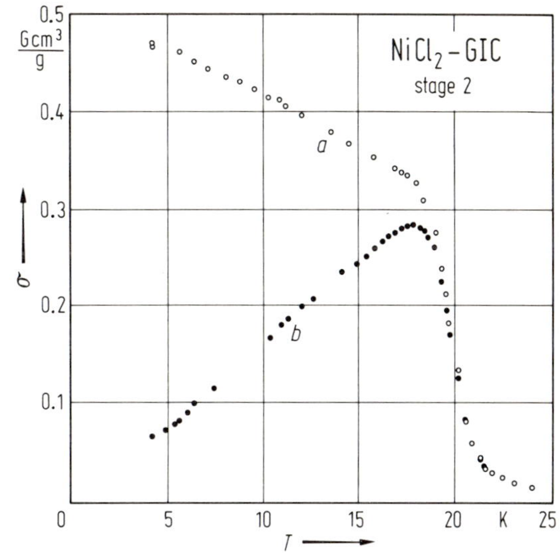

Fig. 2.48. $NiCl_2$–GIC, stage 2. Temperature dependence of FCM (curve *a*) and ZFCM (*b*). $H(\perp c) = 5$ Oe and $H_c(\perp c) = 5$ Oe [90R2].

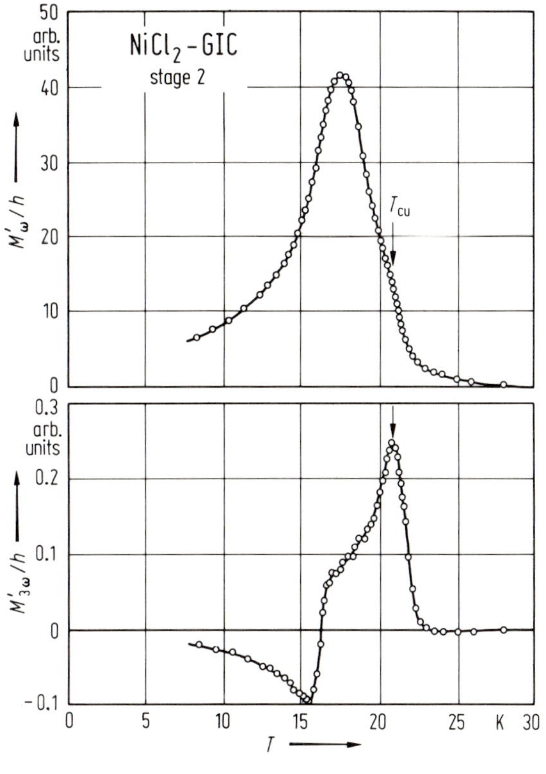

Fig. 2.49. $NiCl_2$–GIC, stage 2. Temperature dependences of linear and nonlinear magnetic responses to AC field measured simultaneously at $h = 0.8$ Oe. $\boldsymbol{h} \perp c$, $\nu = 0.01$ Hz [89H1].

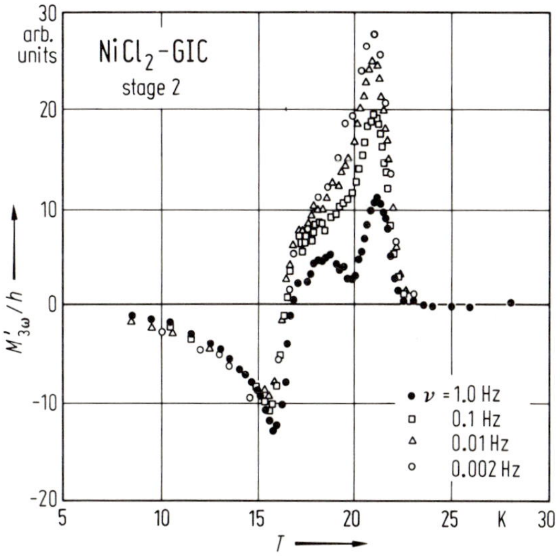

Fig. 2.50. $NiCl_2$–GIC, stage 2. Temperature dependences of nonlinear magnetic responses to AC field measured at $h = 0.8$ Oe at various frequencies. $\boldsymbol{h} \perp c$ [89H1].

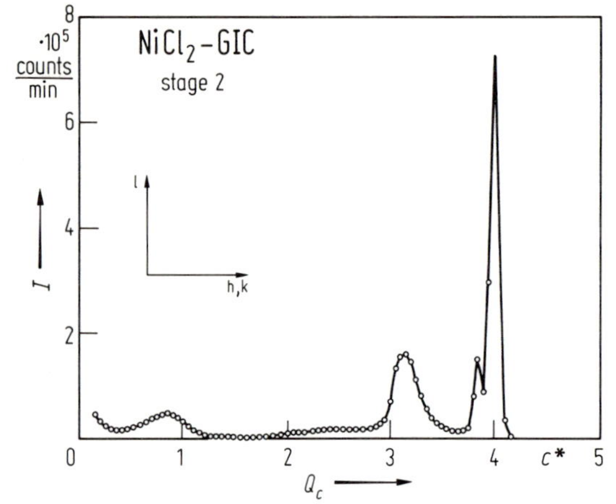

Fig. 2.51. $NiCl_2$–GIC, stage 2. Neutron scattering (nuclear part) along (0, 0, l). The inset shows the direction of the scan in reciprocal space. $T = 36.0$ K. The peak at $l = 3.85$ is due to pristine $NiCl_2$ [87W1].

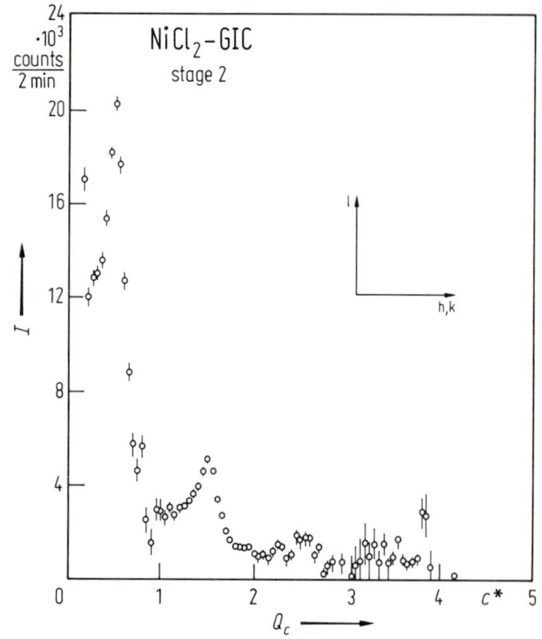

Fig. 2.52. $NiCl_2$–GIC, stage 2. Neutron scattering (magnetic part) along (0, 0, l). $T = 8.22$ K [87W1].

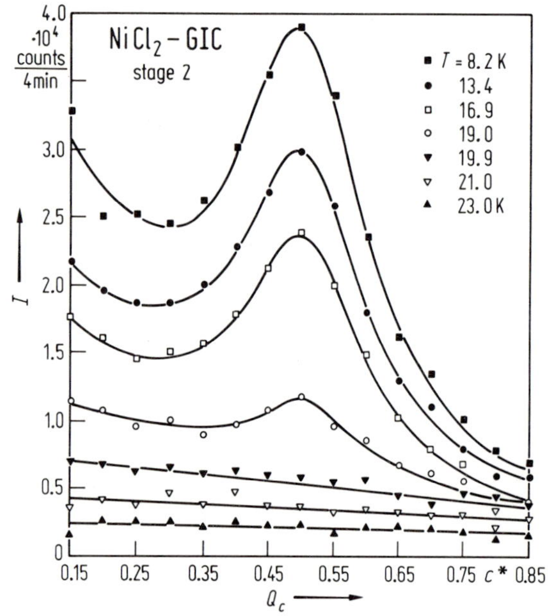

Fig. 2.53. $NiCl_2$–GIC, stage 2. Neutron scattering (magnetic part) along (0, 0, l) for various temperatures [87W1].

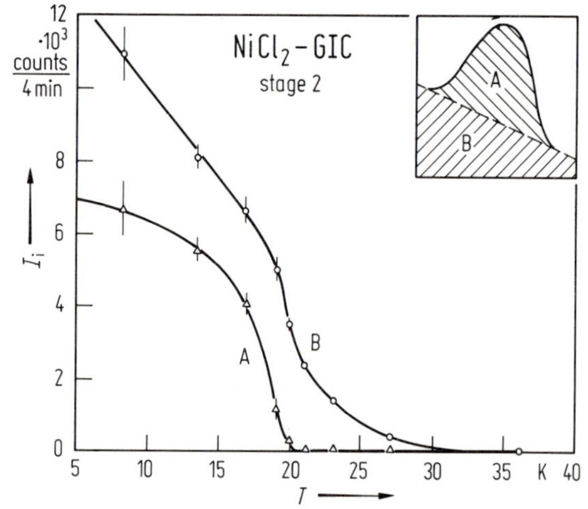

Fig. 2.54. $NiCl_2$–GIC, stage 2. Temperature dependence of the integrated neutron scattering intensity under Lorentzian AF component A and the 2D magnetic ridge B of neutron scattering [87W1].

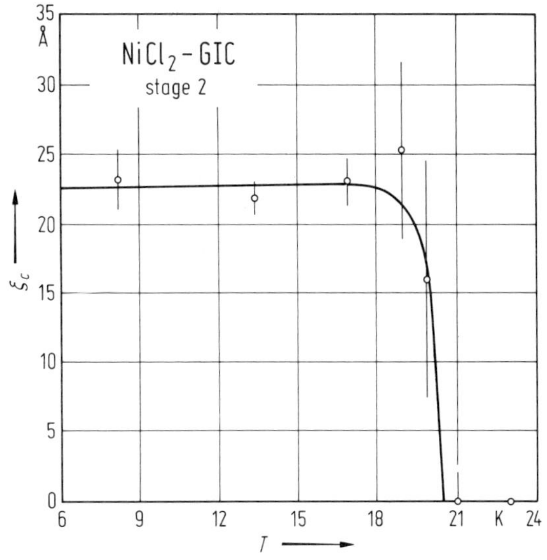

Fig. 2.55. $NiCl_2$–GIC, stage 2. Temperature dependence of the spin correlation length along the *c*-axis. The solid line is a guide to the eye [87W1].

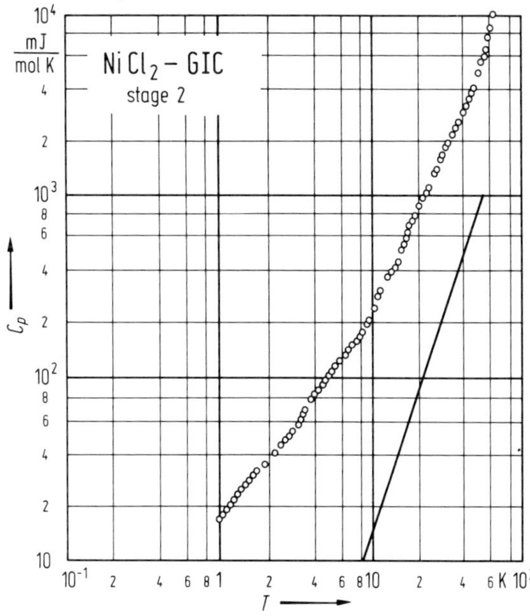

Fig. 2.56. $NiCl_2$–GIC, stage 2. Temperature dependence of heat capacity of $C_{11.3}NiCl_{2.13}$. Solid line is the heat capacity of HOPG [82O].

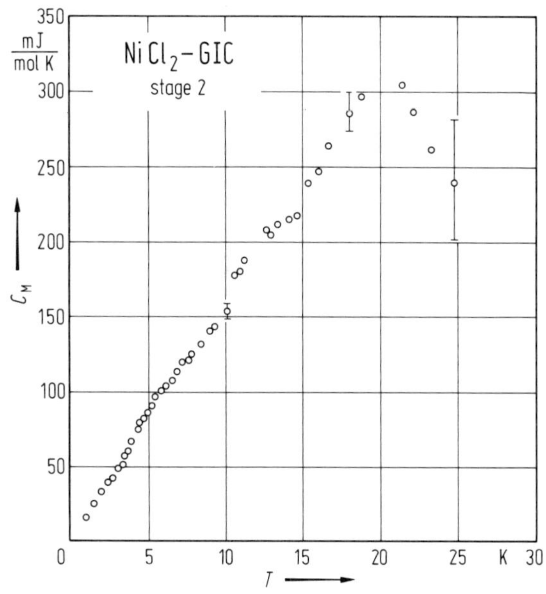

Fig. 2.57. $NiCl_2$–GIC, stage 2. Temperature dependence of magnetic heat capacity of $C_{11.3}NiCl_{2.13}$ [82O].

9.4.2.3.2 Stage n $NiCl_2$–GIC (n ≧ 3)
Survey of the properties

I. Magnetic data

T_{cu} = 18.2 K (χ_{AC}) for S3 and S5 [82E]
T_{cl} = 17.8 K (χ_{AC}) for S3 and S5 [82E]

9.4.2.3.3 Stage 1 $NiCl_2$–GIC
Survey of the properties

I. Lattice parameters

a = 3.50 Å [89H2]
a = 3.53 Å [74K1]
d = 9.37 Å [89H2]
d = 9.36 Å [89N1], [89N2]
d = 9.43 Å [85S3]

II. Magnetic data

T_N = 21 K (M_{ZFC}) [89F]
T_N = 22.0 K (χ_{AC}) [89N1]
T_N = 17.85 K (M_{FC}) [89H2]
T_N = 18.1 K [74K1]
$\Theta_\perp$ = 60 K [89F]
p_{eff} = 3.7 μ_B [89H2]
p_{eff} = 3.3 μ_B [89F]
$J'/J = -2.5\cdot 10^{-4}$ [89F]
$J'/J = -5.7\cdot 10^{-4}$ [89N2]
$J'/J = -1.5\cdot 10^{-3}$ [89N1]

III. Other physical data

Electric resistivity: Fig. 2.63 [89N1]
ESR: [89F]

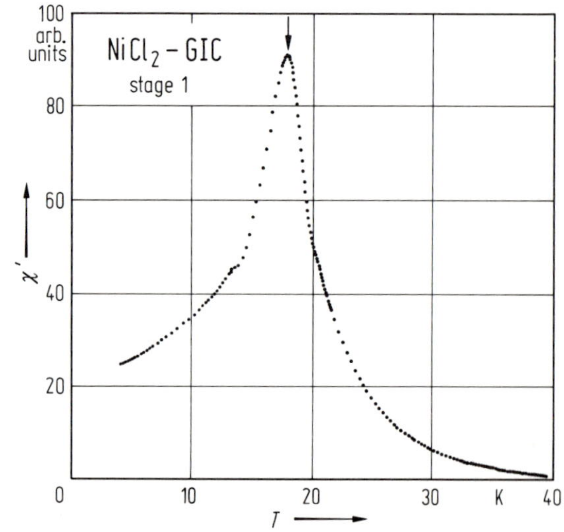

Fig. 2.58. $NiCl_2$–GIC, stage 1. Temperature dependence of AC susceptibility. $h(\perp c)$ = 0.3 Oe, ν = 1000 Hz, H = 0 [91H].

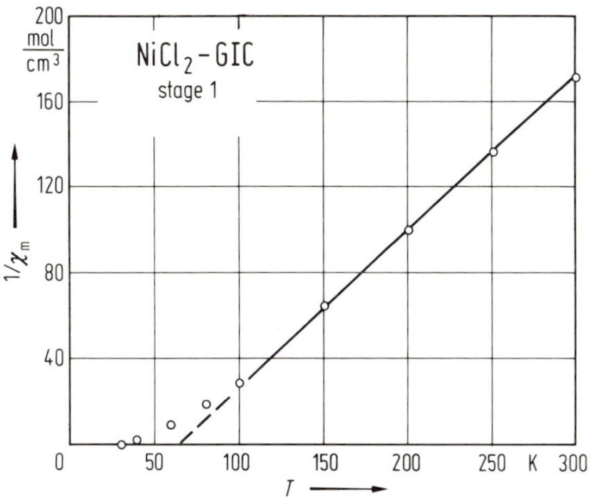

Fig. 2.59. $NiCl_2$–GIC, stage 1. The inverse DC susceptibility vs. temperature. $H(\perp c) = 1000$ Oe [89H2].

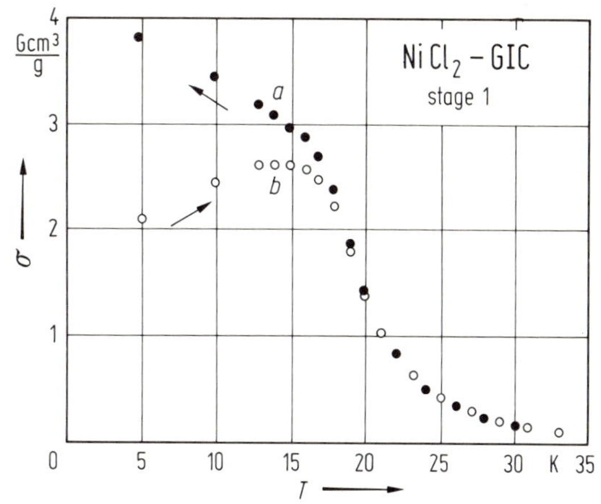

Fig. 2.60. $NiCl_2$–GIC, stage 1. Temperature dependence of FCM (curve *a*) and ZFCM (*b*). $H(\perp c) = 3$ Oe and $H_c(\perp c) = 3$ Oe [89F].

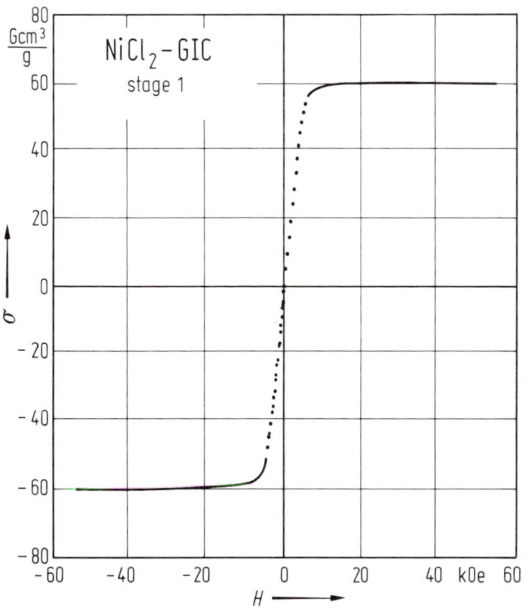

Fig. 2.61. $NiCl_2$–GIC, stage 1. Magnetization curve. $T = 4.2$ K. The field $H(\parallel c)$ is swept in the sequence $0 \rightarrow 50$ kOe $\rightarrow 0 \rightarrow -50$ kOe $\rightarrow 0$ [90N3].

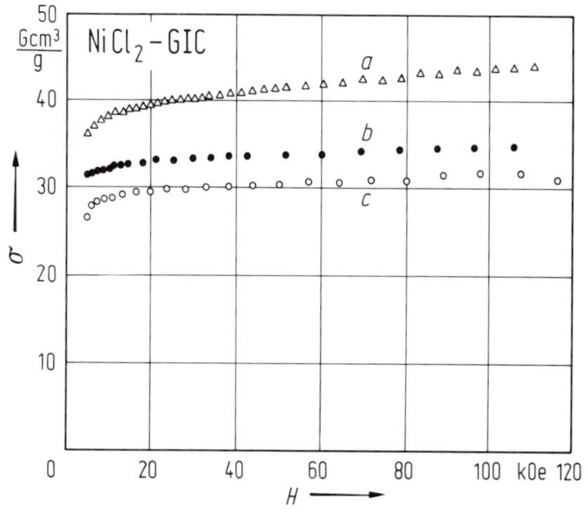

Fig. 2.62. $NiCl_2$–GIC. Field dependence of magnetization for stage 1 and 2 samples. $T = 4.2$ K. (Curve *a*) for stage 1, $\boldsymbol{H} \perp c$, (*b*) and (*c*) for stage 2, $\boldsymbol{H} \perp c$ and $\boldsymbol{H} \parallel c$, respectively [91H].

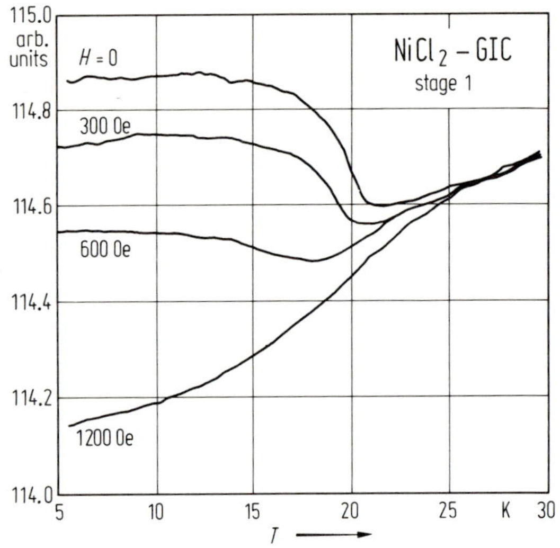

Fig. 2.63. $NiCl_2$–GIC, stage 1. Resistivity vs. temperature of a single flake sample based on natural graphite in the external field. The values of $H(\perp c)$ are indicated [89N1].

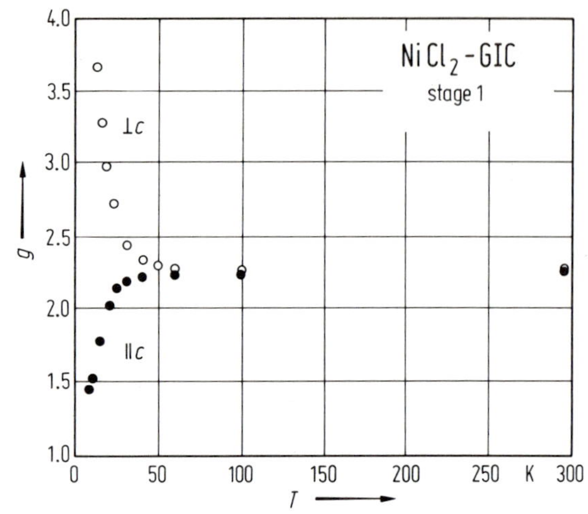

Fig. 2.64. $NiCl_2$–GIC, stage 1. Temperature dependence of g-values of Ni^{2+} ion along two external field directions [89F].

9.4.2.4 MCl_2–GIC (M = Mn, Cu)

9.4.2.4.1 Stage n $MnCl_2$–GIC
Survey of the properties

I. Lattice parameters

a = 3.67 Å for S2 [90S2], [91S2]
a = 3.69 Å for S2 and S1 [82B]
a = 3.71 Å for S2 [89S2]
d = 9.48 Å for S1 [82B]
d = 12.76 Å for S2 [82B], [91S2]
d = 12.65 Å for S2 [89S2]
d = 12.89 Å for S2 [91S2]

II. Magnetic data

T_c = 1.2 K (ND) for S2 [86S3]
T_c = 1.1 K (χ max) for S2 [87M1], [90S2], [91S2]
T_c = 1.2 K (χ max) for S1 [86K2]
$\Theta_\perp$ = − 5.94 K for S2 [86S3]
$\Theta_\perp$ = − 8 K for S2 and S1 [84G]
$\Theta_\perp$ = − 5.9 K for S2 [86W1], [90S2]
$\Theta_\perp$ = − 5.4 K for S1 [86W1]

$\Theta_{\parallel} = -9.06$ K for S2 [86S3]
$\Theta_{\parallel} = -10.99$ K for S2 [91S2]
$\Theta_{\parallel} = -9$ K for S2 and S1 [84G]
$\Theta_{\parallel} = -9.1$ K for S2 and -7.2 K for S1 [86W1]
$C_{\perp} = 4.30$ cm^{3}K/mol for S2 [84G]
$C_{\perp} = 4.26$ cm^{3}K/mol for S2 [86W1]
$C_{\perp} = 4.40$ cm^{3}K/mol for S1 [84G]
$C_{\perp} = 4.12$ cm^{3}K/mol for S1 [86W1]
$C_{\perp} = C_{\parallel} = 4.25$ cm^{3}K/mol for S2 [86S3]
$C_{\parallel} = 4.25$ cm^{3}K/mol for S2 [86W1]
$C_{\parallel} = 4.01$ cm^{3}K/mol for S1 [86W1]
$p_{eff} = 5.83\ \mu_B$ for S2 [86S3]
$p_{eff} = 5.84\ \mu_B$ for S2 [90S2]
$p_{eff} = 6.04\ \mu_B$ for S2 [91S2]
$p_{eff} = 5.87\ \mu_B$ for S2 and 5.93 for S1 [84G]
$J/k = -0.20$ K for S2 [86W1], [90S2], [91S2]
$J/k = -0.17$ K for S1 [86W1]
$D/k = 0.97$ K for S2 [86W1], [90S2], [91S2]
$D/k = 0.56$ K for S1 [86W1]
$g_{\perp} = g_{\parallel} = 1.97$ for S2 [86W1], [90S2]
$g_{\perp} = 1.94$ for S1 [86W1]
$g_{\perp} = 2.04$ for S2 [91S2]
$g_{\parallel} = 1.91$ for S1 [86W1]

III. Other physical data

AES: [88L], [86L]
XPS: [88L], [86L]
C_p: [87M1]
ND: Fig. 2.75 [86S3]
EPR: Fig. 2.76, 77 [86K2]

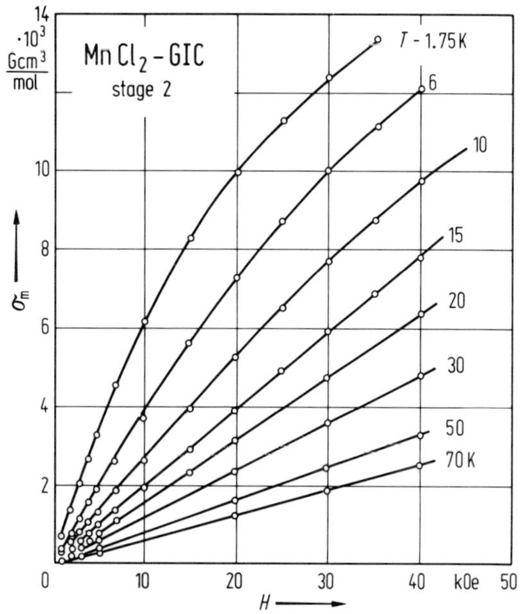

Fig. 2.65. $MnCl_2$–GIC, stage 2. Field dependence of magnetizations at various temperatures. $\boldsymbol{H} \parallel c$. Measured temperature is indicated [86W1].

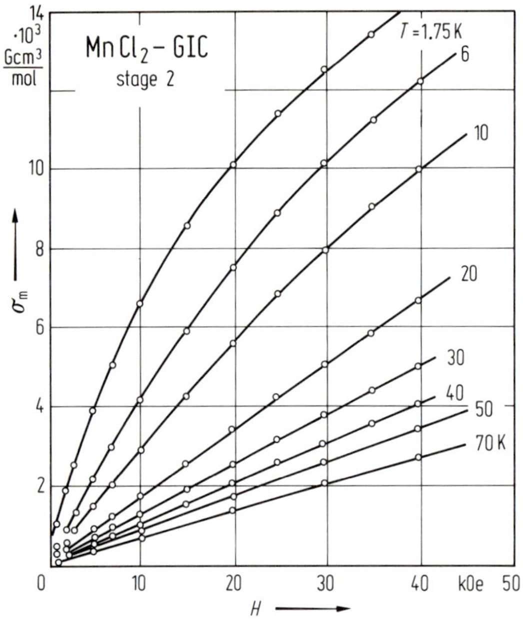

Fig. 2.66. $MnCl_2$–GIC, stage 2. Field dependence of magnetizations at various temperatures. ***H***⊥*c*. Measured temperature is indicated [86W1].

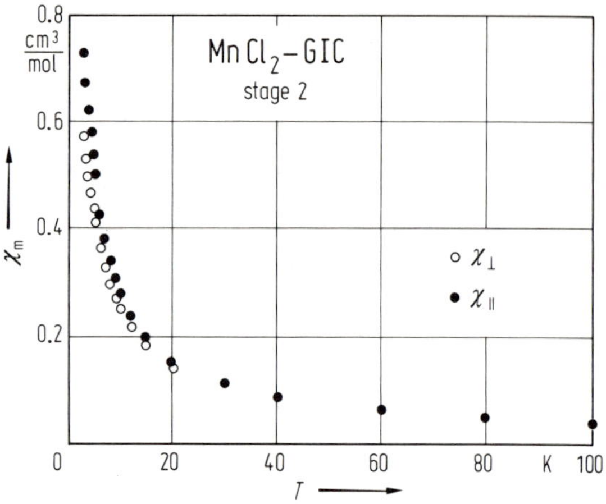

Fig. 2.67. $MnCl_2$–GIC, stage 2. Temperature dependence of susceptibilities in two field directions [86W1].

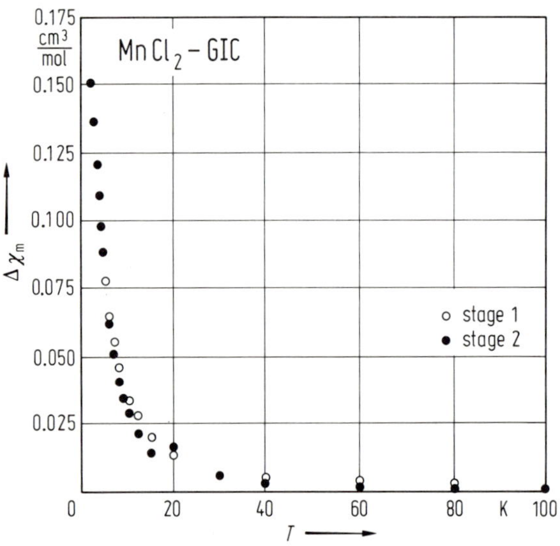

Fig. 2.68. $MnCl_2$–GIC. Temperature dependence of anisotropy $\Delta\chi (= \chi_\perp - \chi_\parallel)$ for stage 1 and 2 samples [86W1].

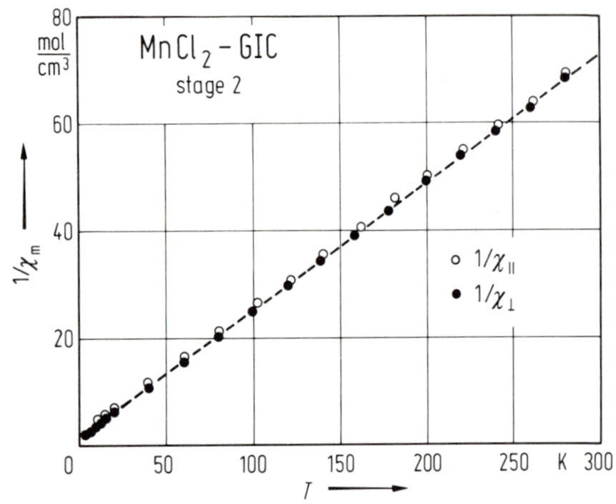

Fig. 2.69. $MnCl_2$–GIC, stage 2. Temperature dependence of inverse susceptibilities in two field directions [86W1].

▶

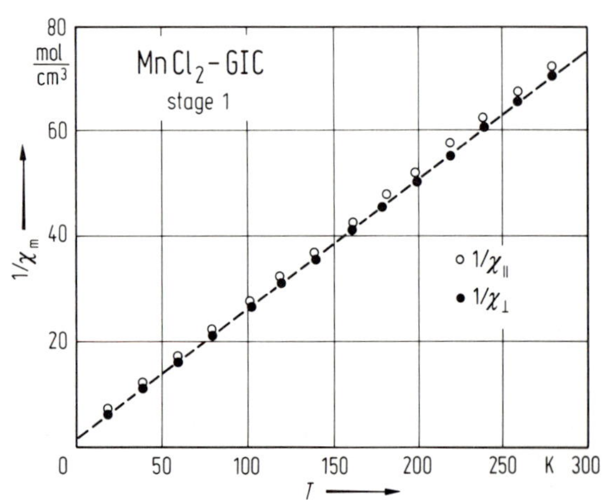

Fig. 2.70. $MnCl_2$–GIC, stage 1. Temperature dependence of inverse susceptibilities in two field directions [86W1].

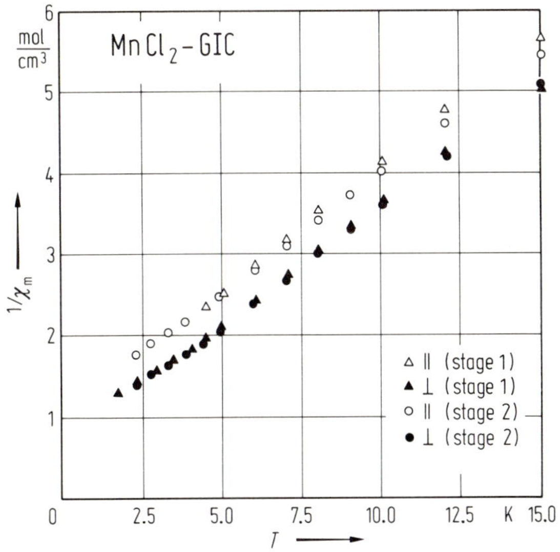

Fig. 2.71. $MnCl_2$–GIC. Temperature dependence of inverse susceptibilities in two field directions for stage 1 and 2 samples at low temperatures [86W1].

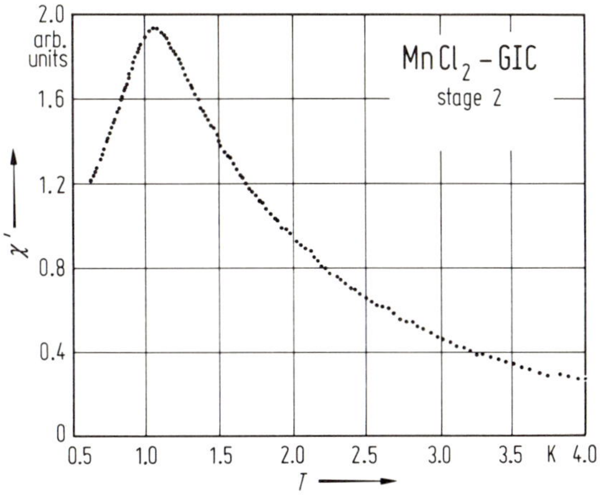

Fig. 2.72. $MnCl_2$–GIC, stage 2. Temperature dependence of AC susceptibility. $H = 0$, $h(\perp c) = 10$ mOe, $\nu = 10$ Hz [87M1].

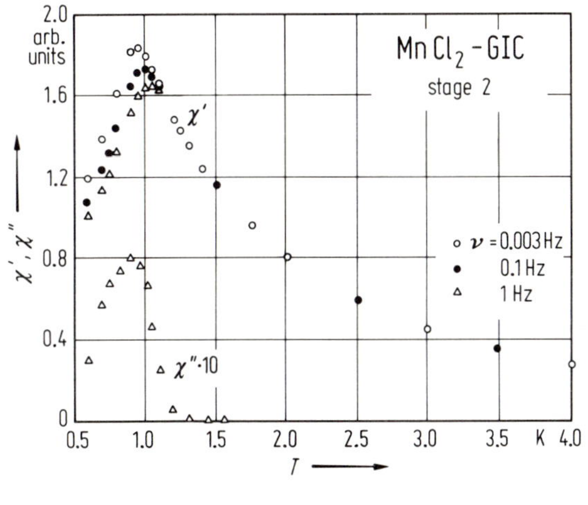

Fig. 2.73. $MnCl_2$–GIC, stage 2. Temperature dependence of AC susceptibilities χ' and χ'' at various frequencies. $H = 0$, $h(\perp c) = 20$ mOe [87M1].

For Fig. 2.74 see next page.

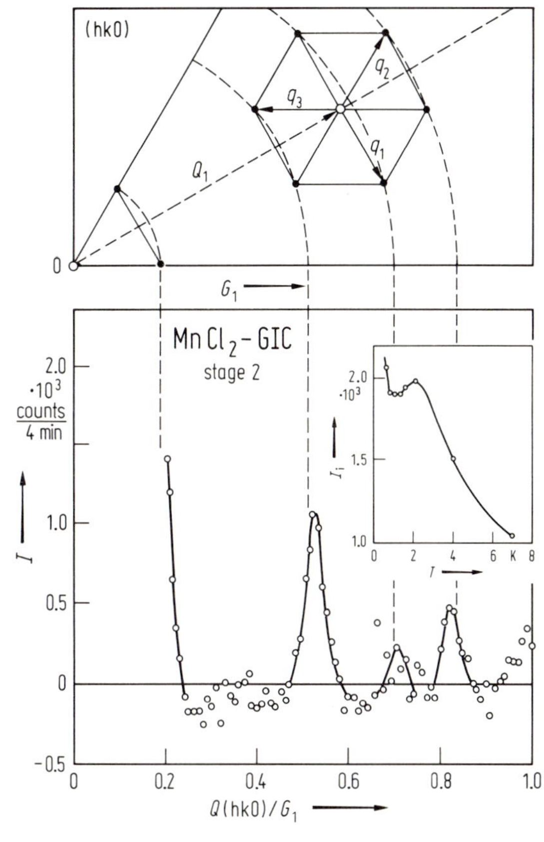

Fig. 2.75. $MnCl_2$–GIC, stage 2. Neutron scattering intensity. $T = 0.6$ K. The upper and the lower parts show the $(h, k, 0)$ reciprocal lattice plane and the magnetic scattering intensity along the $[h, k, 0]$ direction, respectively. The inset shows the temperature dependence of integrated magnetic Bragg intensity at $|\boldsymbol{Q}| = |\boldsymbol{Q}_1 - \boldsymbol{Q}_2| = |\boldsymbol{Q}_1 + \boldsymbol{Q}_3|$ with $l = 0$ [86S3].

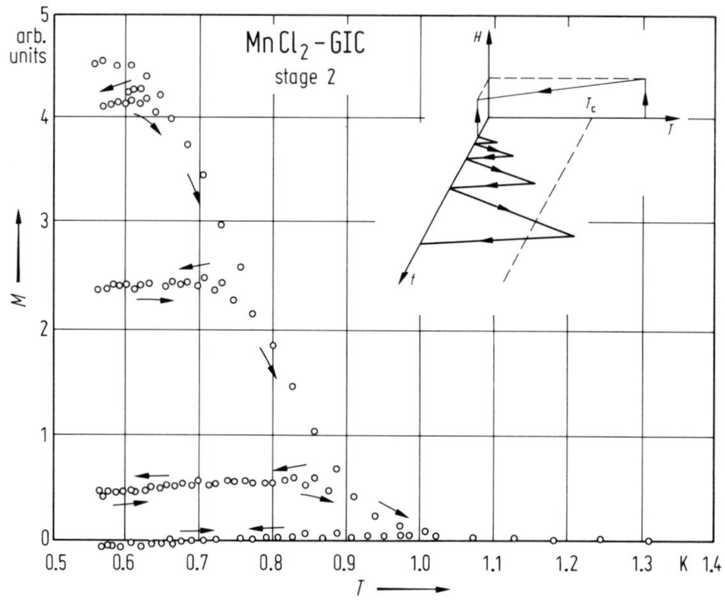

Fig. 2.74. $MnCl_2$–GIC, stage 2. Temperature dependence of thermoremanent magnetization in a series of heating and cooling processes shown in the inset. $H = 0$, $H_c = 76$ mOe [87M1].

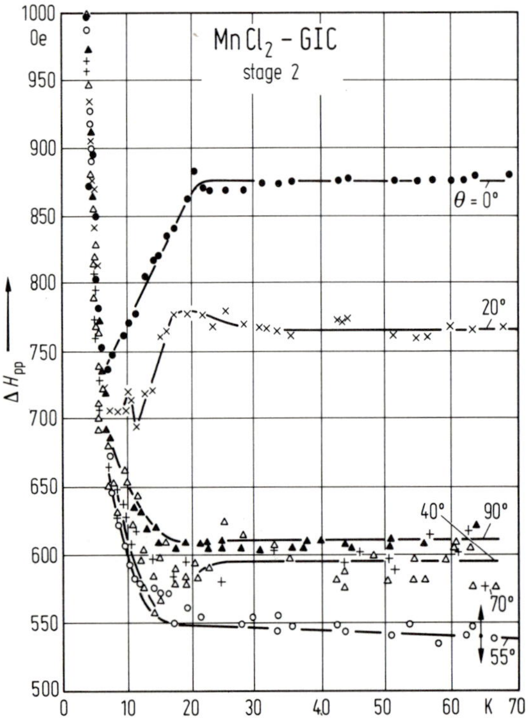

Fig. 2.76. $MnCl_2$–GIC, stage 2. Temperature dependence of EPR line widths for various external field directions. $\nu = 9.42$ GHz. Angles θ from the c-axis are indicated [86K2].

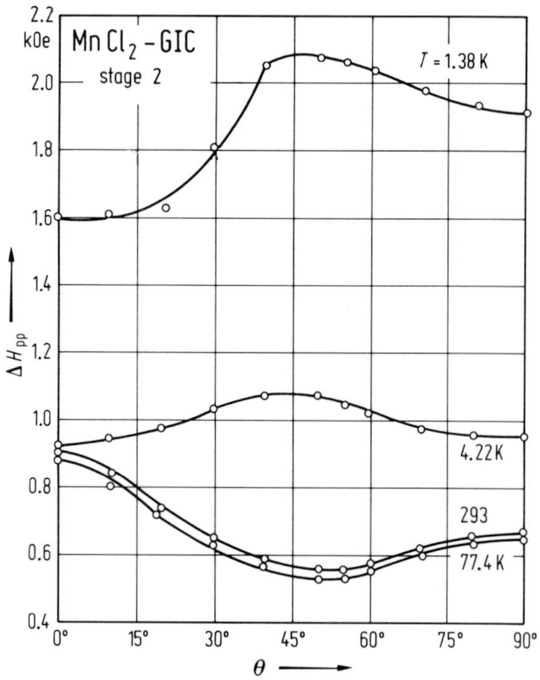

Fig. 2.77. $MnCl_2$–GIC, stage 2. Angular dependence of EPR line widths for various temperatures. $\nu = 9.42$ GHz. Measured temperature is indicated [86K2].

9.4.2.4.2 Stage n $CuCl_2$–GIC
Survey of properties

I. Lattice parameters

d = 9.3 Å for S1 [83H]
d = 12.7 Å for S2 [83H]
d = 12.72 Å for S2 [89Y2], [86R2]

II. Magnetic data

T_{cu} = 22 K for S1 [85N]
T_{cl} = 14 K for S1 [85N]
J/k = – 40 K for S1, S2, S3 [85N]
p_{eff} = 0.34 μ_B for S1, S2, S3 [85N]

III. Other physical data

STM: [90O]
DHvA: [89Y2]
ESR: [85K2]

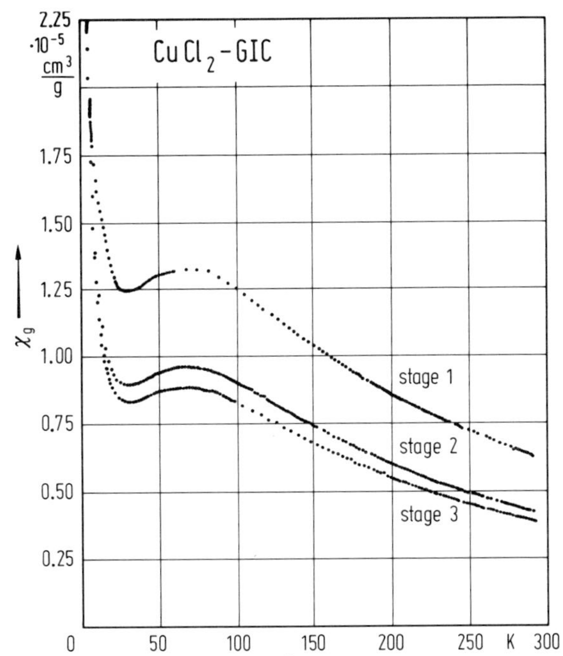

Fig. 2.78. $CuCl_2$–GIC. Temperature dependence of DC magnetic susceptibilities for stage 1, 2 and 3 samples. $\boldsymbol{H} \parallel c$ [85N].

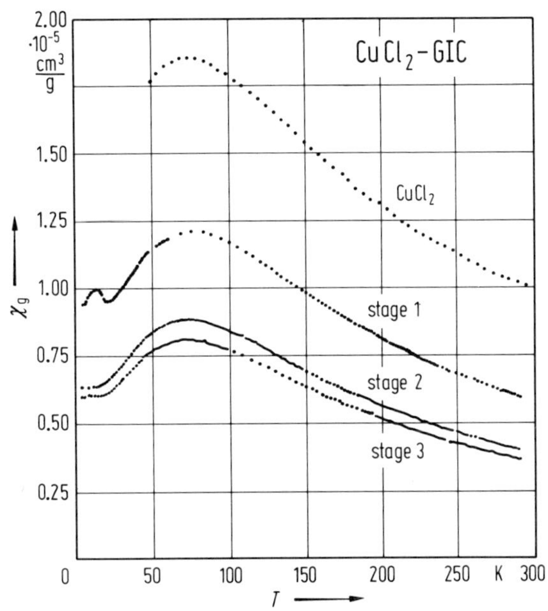

Fig. 2.79. $CuCl_2$–GIC. Temperature dependence of DC magnetic susceptibilities after subtraction of low temperature Curie contribution in Fig. 2.78 for stage 1, 2 and 3 samples. $\boldsymbol{H} \parallel c$. The data for pristine $CuCl_2$ from [78B] are also shown [85N].

9.4.2.5 MCl_3–GIC (M = Cr, Fe)

9.4.2.5.1 Stage n $CrCl_3$–GIC
Survey of the properties

I. Lattice parameters
$d = 16.0$ Å for S3 [90R2]

II. Magnetic data
$T_c = 11.4$ K ($M'_{3\omega}$) for S3 [90H]
$\Theta_\perp = 26$ K for S3 and 25 K for S1 [91C]
$\Theta_\parallel = 25$ K for S3 and 24 K for S1 [91C]
$g_\perp = 1.983$ for S3, 1.986 for S2 and 1.993 for S1 [91C]
$g_\parallel = 1.976$ for S3, 1.978 for S2 and 1.981 for S1 [91C]

III. Other physical data
ESR: Fig. 2.85–88 [91C]

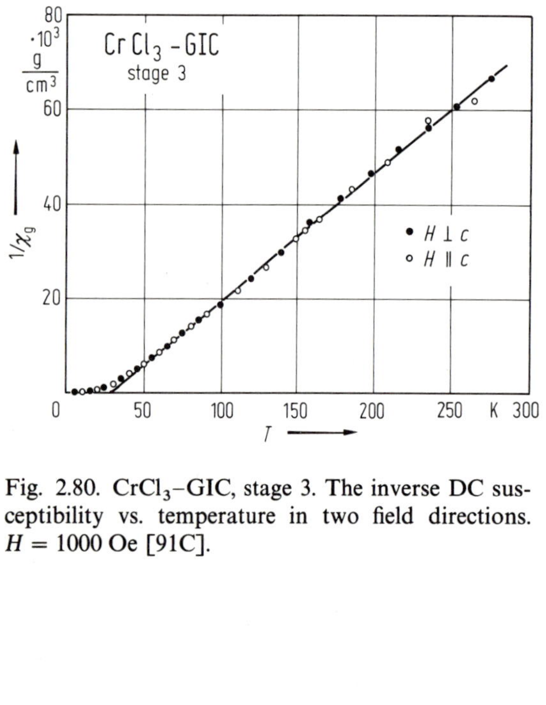

Fig. 2.80. $CrCl_3$–GIC, stage 3. The inverse DC susceptibility vs. temperature in two field directions. $H = 1000$ Oe [91C].

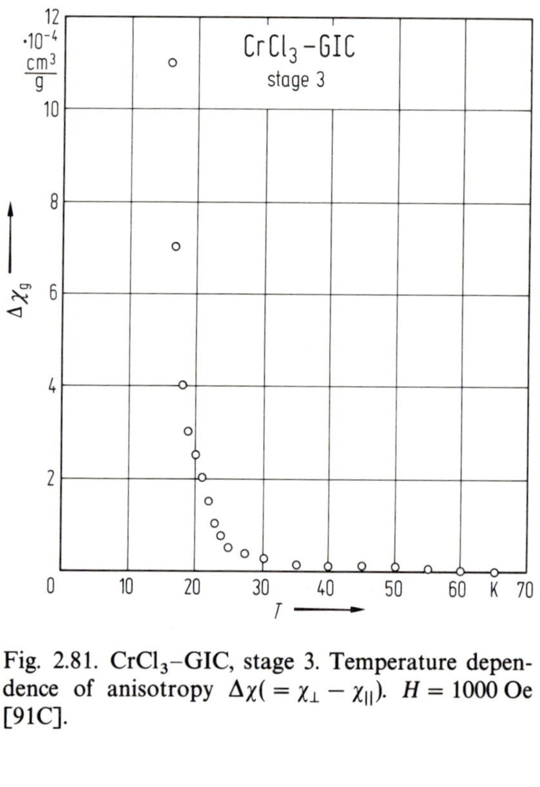

Fig. 2.81. $CrCl_3$–GIC, stage 3. Temperature dependence of anisotropy $\Delta\chi(=\chi_\perp - \chi_\parallel)$. $H = 1000$ Oe [91C].

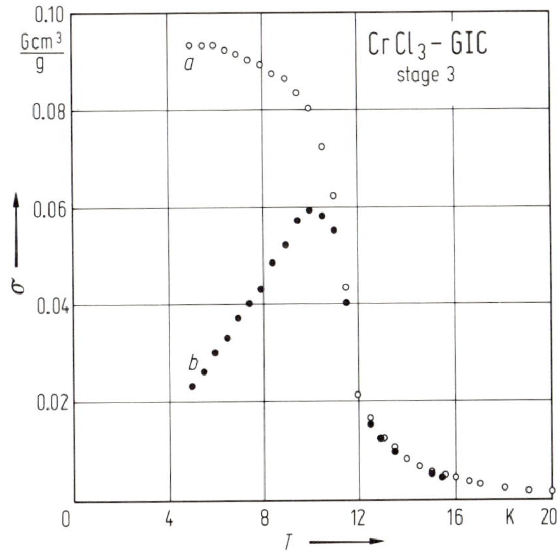

Fig. 2.82. $CrCl_3$–GIC, stage 3. Temperature dependence of FCM (curve *a*) and ZFCM (*b*). $H(\perp c) = 2$ Oe and $H_c(\perp c) = 2$ Oe [90R2].

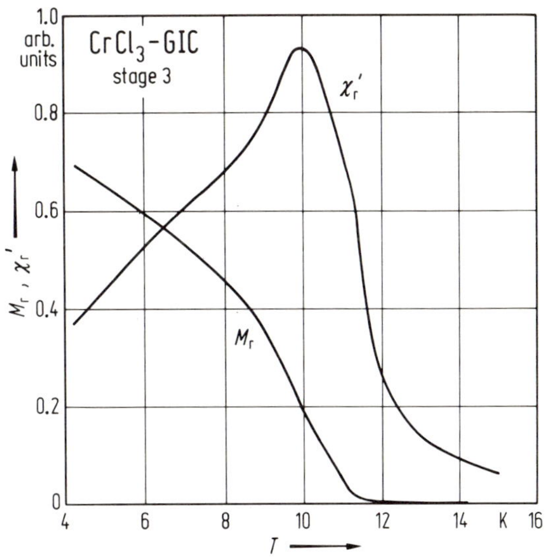

Fig. 2.83. $CrCl_3$–GIC, stage 3. Temperature dependence of TRM and AC susceptibility measured simultaneously. $H = 0$, $H_c = 0.5$ Oe, $h = 10$ mOe, $\nu = 100$ Hz [90H].

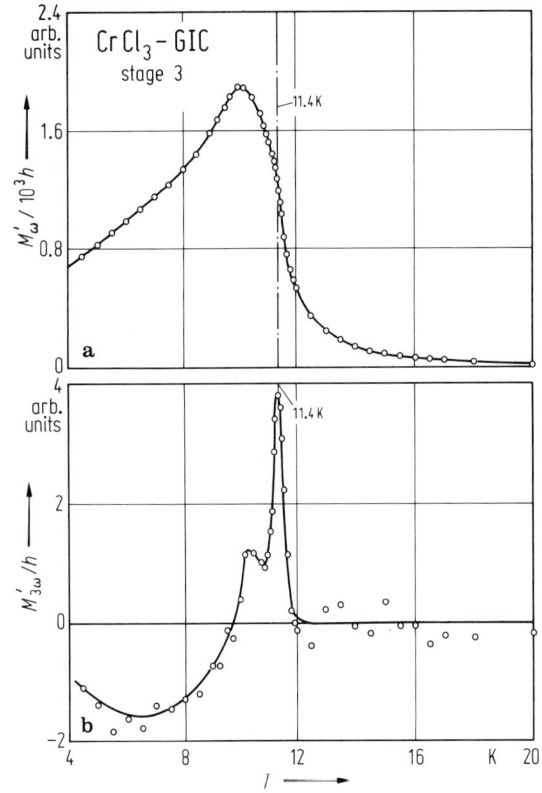

Fig. 2.84. $CrCl_3$–GIC, stage 3. Temperature dependence of (a) linear and (b) nonlinear magnetic responses to an AC field measured simultaneously at $h = 0.8$ Oe, $\nu = 0.1$ Hz [90H].

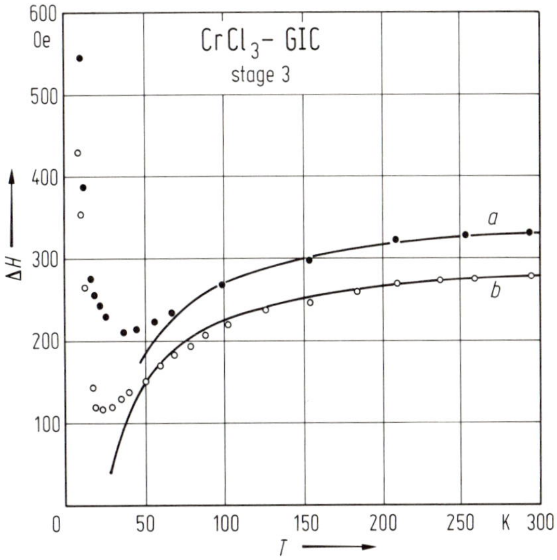

Fig. 2.85. $CrCl_3$–GIC, stage 3. Temperature dependence of EPR line width for different field directions, (curve *a*) $\boldsymbol{H} \parallel c$, (*b*) $\boldsymbol{H} \perp c$. The solid lines show the calculated curves. $\nu = 9.35$ GHz [91C].

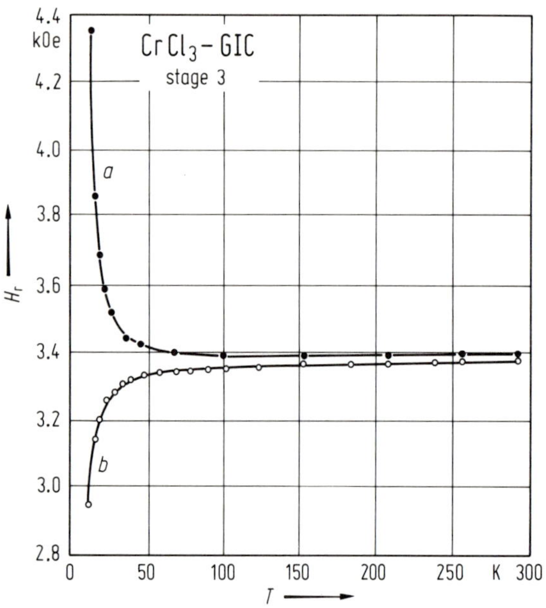

Fig. 2.86. $CrCl_3$–GIC, stage 3. Temperature dependence of EPR fields for different field directions, (curve *a*) $\boldsymbol{H} \parallel c$, (*b*) $\boldsymbol{H} \perp c$. $\nu = 9.35$ GHz [91C].

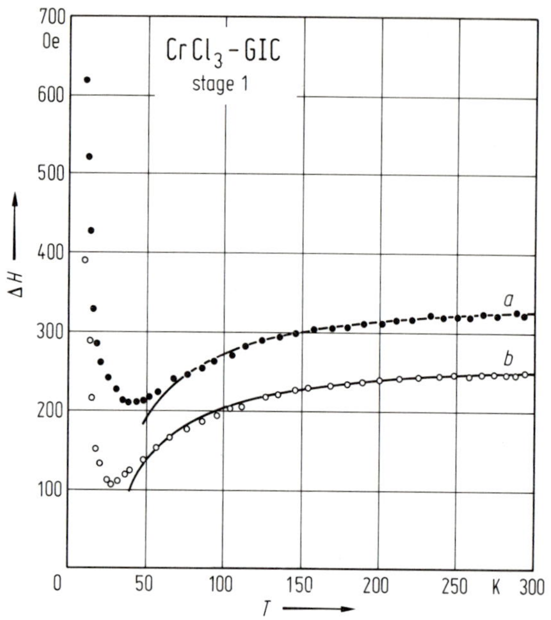

Fig. 2.87. $CrCl_3$–GIC, stage 1. Temperature dependence of EPR line width for different field directions, (curve *a*) $\boldsymbol{H} \parallel c$, (*b*) $\boldsymbol{H} \perp c$. The solid lines show the calculated curves. $\nu = 9.35$ GHz [91C].

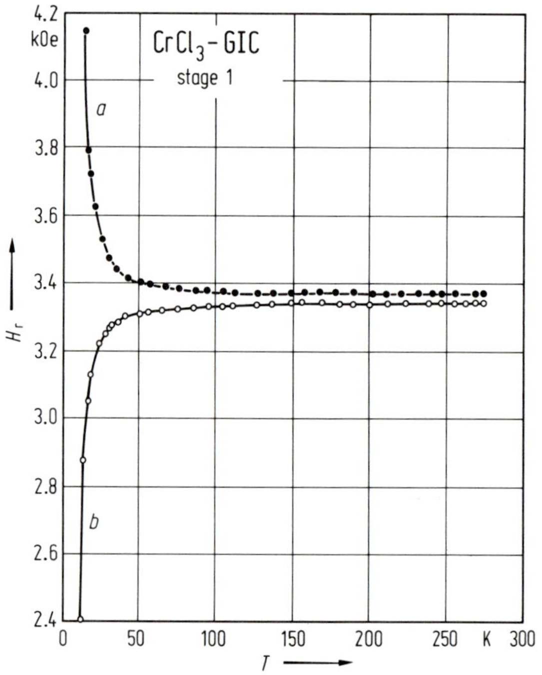

Fig. 2.88. $CrCl_3$–GIC, stage 1. Temperature dependence of EPR fields for different various field directions, (curve *a*) $\boldsymbol{H} \parallel c$, (*b*) $\boldsymbol{H} \perp c$. $\nu = 9.35$ GHz [91C].

9.4.2.5.2 Stage 1 $FeCl_3$–GIC
Survey of the properties

I. Lattice parameters
$d = 9.4$ Å [82M4]

II. Magnetic data
$T_c = 4.2$ K [82M1], [82M2]
$T_c = 4.3$ K [82M4]
$\Theta_\perp = -3.8$ K [82M4]

III. Other physical data
C_p: Fig. 2.93, 94 [82O], [86S4]
Electric resistivity: [89K]
Thermal conductivity: [88K]
Thermoelectric power: [88K]
Mössbauer: Fig. 2.89–92 [82M1], [83M2], [83M3], 3.6 [88R]
Susceptibility: Fig. 2.97 [87I2]

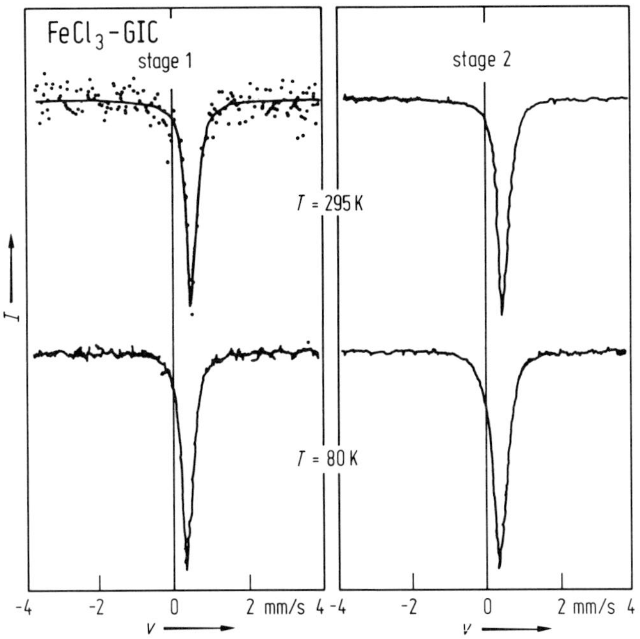

Fig. 2.89. $FeCl_3$–GIC. Mössbauer spectra for stage 1 and 2 samples. Experimental results are shown with the theoretical fits. Measured temperatures are indicated [82M1]. See also Fig. 3.6(a).

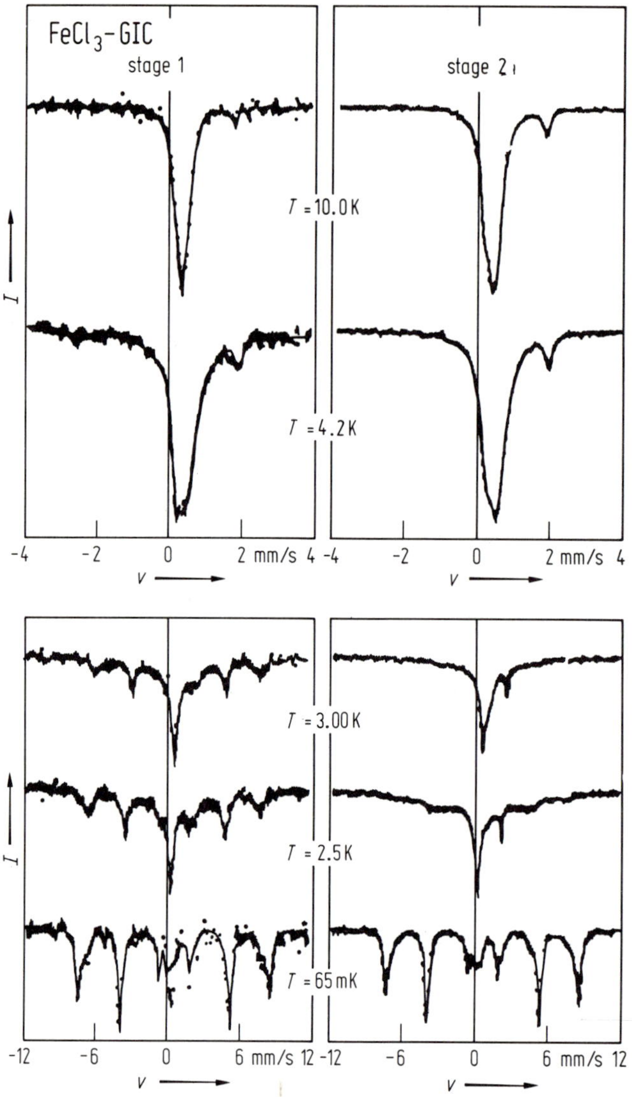

Fig. 2.90. $FeCl_3$–GIC. Mössbauer spectra for stage 1 and 2 samples at low temperatures. Experimental results are shown with the theoretical fits. Measured temperatures are indicated [82M2].

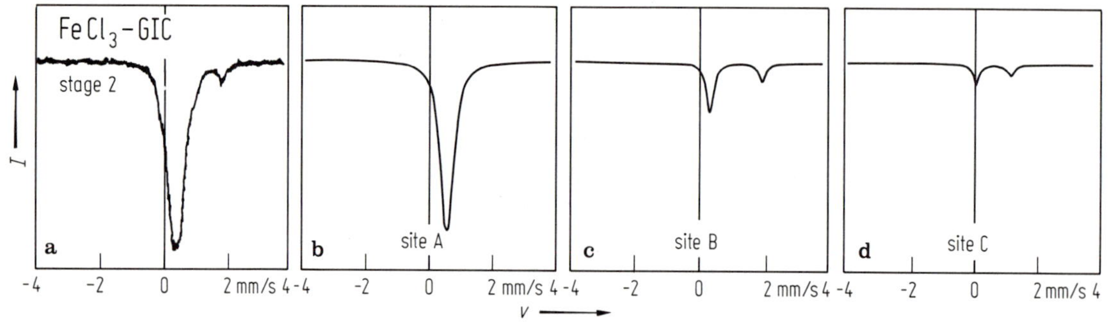

Fig. 2.91. $FeCl_3$–GIC, stage 2. Mössbauer spectra. Experimental result at $T = 10$ K with the theoretical fit (a) and the decomposition of the theoretical fit into three individual subspectra (b), (c) and (d) [83M3].

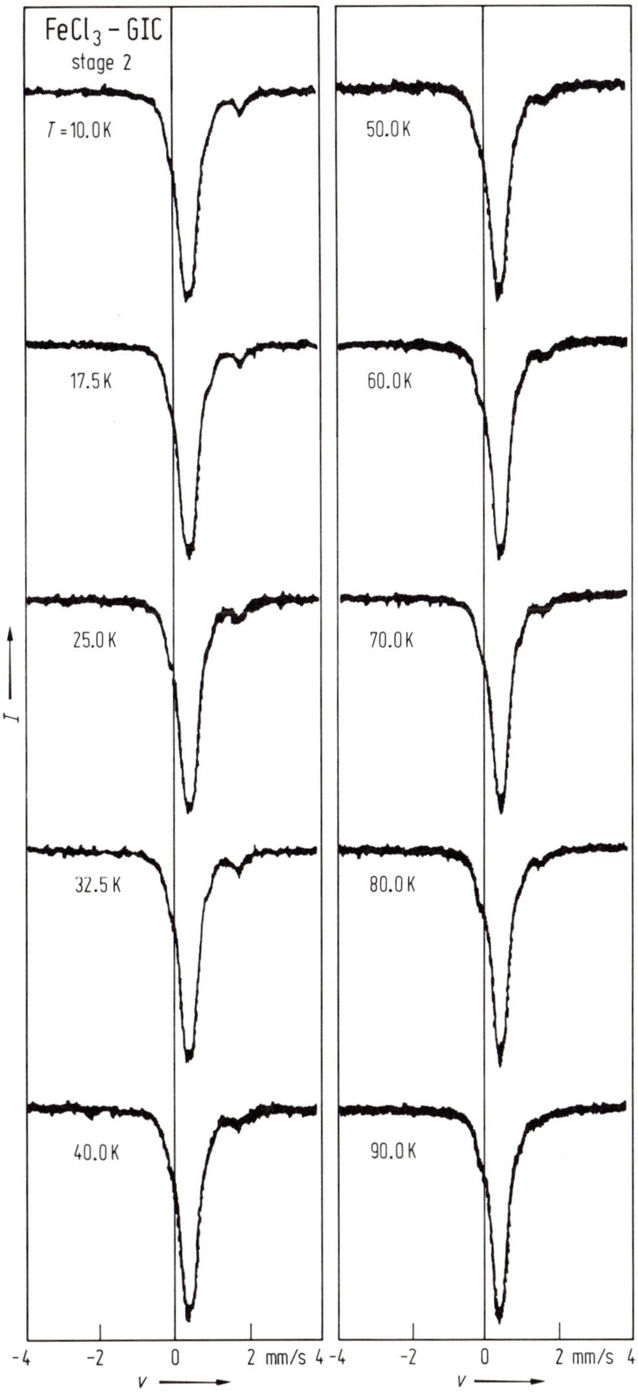

Fig. 2.92. $FeCl_3$–GIC, stage 2. Mössbauer spectra at various temperatures. Experimental results are shown with the theoretical fits to the spectra assuming the static model. Measured temperatures are indicated [83M3].

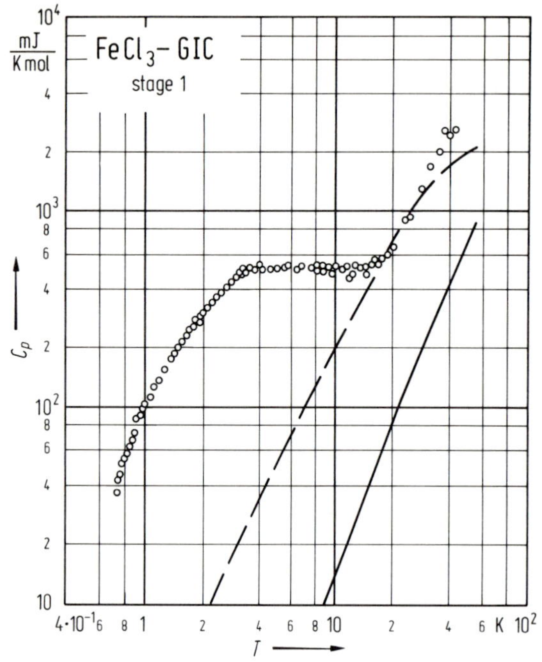

Fig. 2.93. $FeCl_3$–GIC, stage 1. Temperature dependence of heat capacity of $C_{6.2}FeCl_3$. Solid and dashed lines are the heat capacities of HOPG and $C_{17.8}AlCl_3$, respectively [82O].

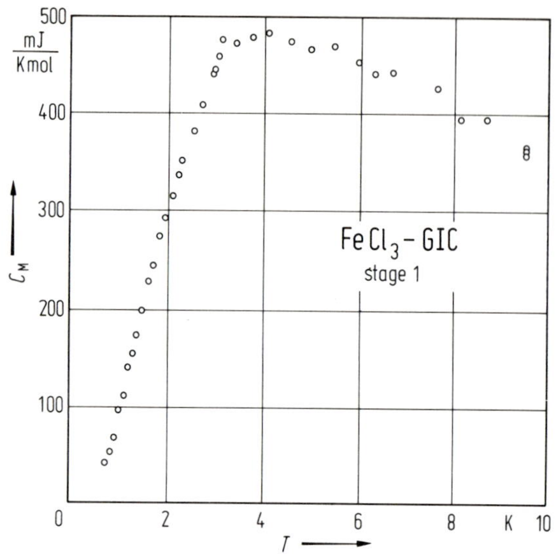

Fig. 2.94. $FeCl_3$–GIC, stage 1. Temperature dependence of magnetic heat capacity of $C_{6.2}FeCl_3$ [82O].

9.4.2.5.3 Stage n $FeCl_3$–GIC (n ≧ 2)
Survey of the properties

I. Lattice parameters

$d = 12.6$ Å for S2 [82M4]

II. Magnetic data

$T_c = 1.7$ K for S2 [83M2]
$T_c = 2.0$ K for S2 [82M1], [82M2]
$T_c = 1.3$ K for S2 [82M4]
$T_c = 1.7 \cdots 1.8$ K for S2–S9 [87I2]
$\Theta_\perp = -7.6$ K [82M4]

III. Other physical data

Thermal conductivity: [88K]
Thermoelectric power: [88K]
Electron microscopy: [80T]
C_p: Fig. 2.95, 96 [82O]
Magnetoresistance: [87I1]
SdH: [87I3]
Electric resistivity: Fig. 2.98 [88P]
Hall effect: [89A]
Mössbauer: Fig. 2.89, 90 [82M1], [82M2]

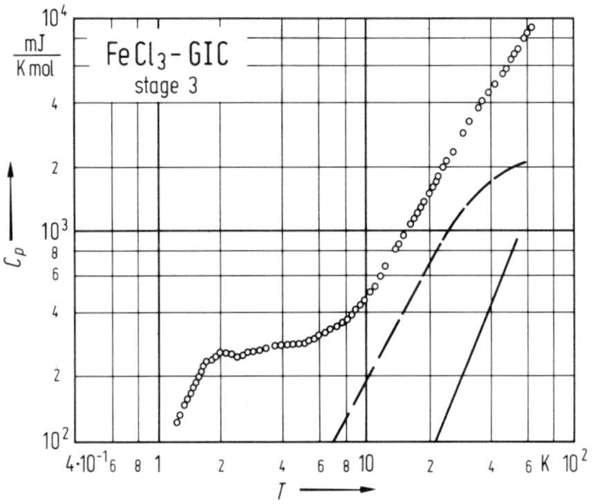

Fig. 2.95. $FeCl_3$–GIC, stage 3. Temperature dependence of heat capacity of $C_{17.8}FeCl_3$. Solid and dashed lines are the heat capacities of HOPG and $C_{17.8}AlCl_3$, respectively [82O].

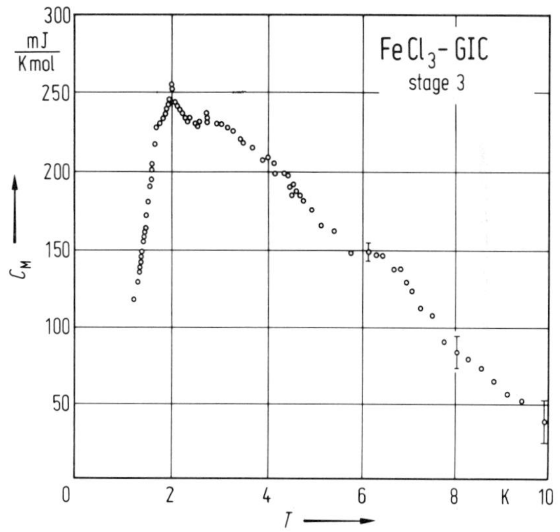

Fig. 2.96. $FeCl_3$–GIC, stage 3. Temperature dependence of magnetic heat capacity of $C_{17.8}FeCl_3$ [82O].

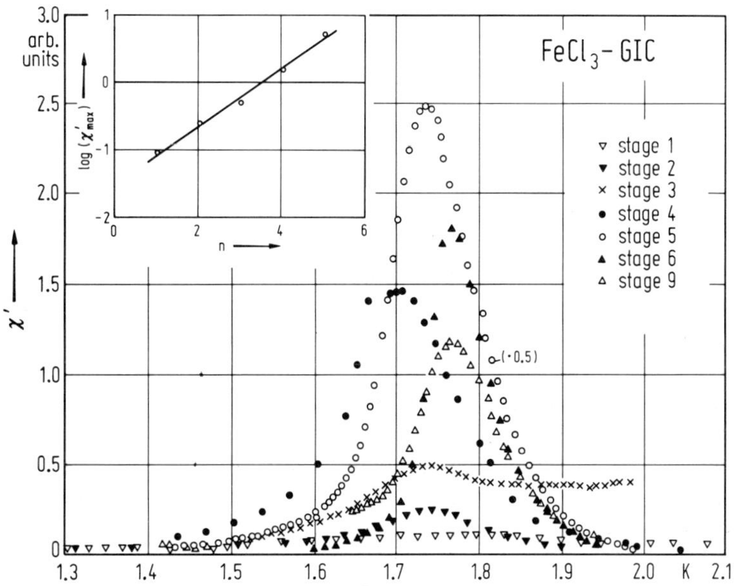

Fig. 2.97. $FeCl_3$–GIC. In-plane magnetic AC susceptibility vs. temperature for different stage samples. $H = 0$. $\nu = 397$ Hz. $h(\perp c) \leq 0.1$ Oe. The data for stage 5 is multiplied by 0.5. The inset shows the logarithm of the peak height as a function of stage index n [87I2].

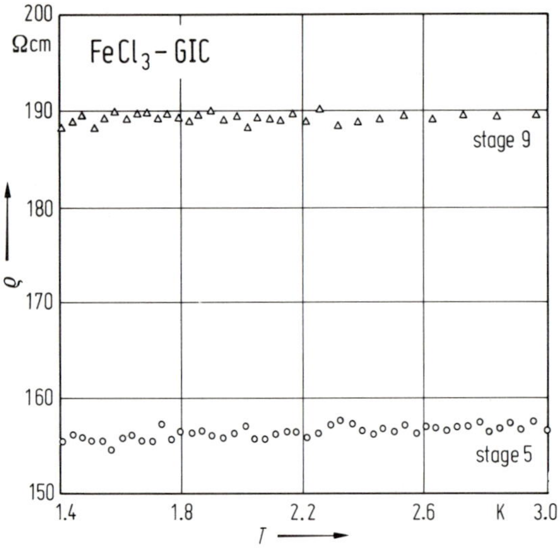

Fig. 2.98. $FeCl_3$–GIC. The *c*-axis resistivity vs. temperature in the vicinity of magnetic anomaly for stage 5 and 9 samples [86I].

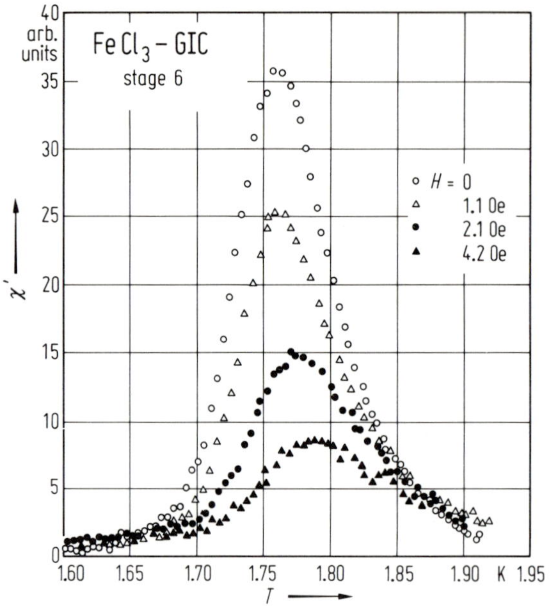

Fig. 2.99. $FeCl_3$–GIC, stage 6. In-plane magnetic AC susceptibility vs. temperature in various magnetic fields $\boldsymbol{H} \perp c$. $h(\perp c) \leq 0.1$ Oe, $\nu = 39$ Hz [87I2].

9.4.3 Bi-intercalation compounds

9.4.3.1 General survey

In this section, magnetic transitions in graphite bi-intercalation compounds (GBIC) are presented. Such a compound is usually synthesized by introducing an intercalant substance into the vacant spaces of an already prepared stage 2 or higher stage GIC with another different intercalant. Therefore, GBICs are essentially stage 1 compounds but different qualitatively from the singly intercalated ones mentioned already in the following points. Firstly, two different intercalant layers are piled up alternatively along the *c*-axis in GBICs. Secondly, two kinds of islands which are composed of two different intercalant species respectively, coexist in each intercalant layer.

In the case where the first intercalant is magnetic and the second one is nonmagnetic, magnetic ordering of such a GBIC will not qualitatively be different from the original stage 2 or higher stage magnetic GIC, except that the interlayer magnetic interaction could be reduced by elongation of the interlayer distance between the adjacent magnetic intercalants. The characteristics of magnetic ordering in these compounds are presented in 9.4.3.2.

In the case where the first and the second intercalants are both magnetic, magnetic ordering of such a GBIC may be quite different from both original magnetic GICs of stage 2 or higher stage, because of the inter-island couplings among two kinds of magnetic islands which are composed of two different intercalant species. The characteristics of magnetic ordering of such GBICs are presented in 9.4.3.3. Indeed, some of the ordering characteristics are complicated and peculiar.

In Table 3.1, the magnetic properties of GBICs are summarized and in Table 3.2, the references for the compounds are classified, according to the investigated ways.

Table 3.1. Magnetic properties

Compounds	*c*-axis repeat distance *d* [Å]	Transition temperature T_c [K]	Magnetic properties	References
$CoCl_2$–GIC (S2)	12.70[a])	6.8 (T_{cl})		[a]) 87W1
$CoCl_2$–$GaCl_3$ –GBIC (1:2)	> 12.7	6.1 (T_{cl})	$CoCl_2$–GIC (S2) -like	86R3
$CrCl_3$–GIC (S3)	16[b])	11[c])		[b]) 90R2, [c]) 90H
$CrCl_3$–$AlCl_3$ –GBIC (1:2)		13	$CrCl_3$–GIC (S3) -like	90R1
$CrCl_3$–$CdCl_2$ –GBIC (1:2)		13	$CrCl_3$–GIC (S3) -like	90R1
$CoCl_2$–GIC (S2)	12.7	9.1 (T_{cu}) 8.0 (T_{cl})		[d]) 87Y1 [e]) 82M4
$CoCl_2$–GIC (S1)	9.4[d])	9.9[f])		[f]) 85I
$CoCl_2$–$FeCl_3$ –GBIC (1:1)	18.6	8.8 (T_{cu}) 8.0 (T_{cl})	$CoCl_2$–GIC (S2) -like	84S2
$FeCl_3$–GIC (S1)	9.4[e])	4.3[e])		
$NiCl_2$–GIC (S2)	12.6	21 (T_{cu}) 18 (T_{cl})[g])		[g]) 83S2 [h]) 89H2
$NiCl_2$–GIC (S1)	9.37[h])	22.0[j])		[i]) 82M4 [j]) 89N1
$NiCl_2$–$FeCl_3$ –GBIC (1:1)	18.6	20.5	[$NiCl_2$–GIC (S2) + Super-para] -like	88R
$FeCl_3$–GIC (S1)	9.4[i])	4.3[i])		
$NiCl_2$–GIC (S2)	12.7	20.2 (T_{cu}) 18 (T_{cl})[k])		[k]) 83S2
$NiCl_2$–GIC (S1)	9.2			
$NiCl_2$–$CrCl_3$ –GBIC (2:1)	28	19.9 (T_{cu}) 11.3 (T_{cl})	[$NiCl_2$–GIC (S2) + $CrCl_3$–GIC (S3)] -like	90R2
$CrCl_3$–GIC (S3)	16.0	11.3		

The data without superscript are quoted from the reference without superscript in the same block.

Table 3.2. List of references.

Compounds	Crystal properties	Magnetic properties	MR, ND, Mössbauer	Others
$CoCl_2$–$GaCl_3$–GBIC		86R3 86S4		
$CoCl_2$–$AlCl_3$–GBIC				88K 89K
$CrCl_3$–$AlCl_3$–GBIC		90R1		
$CrCl_3$–$CdCl_2$–GBIC		90R1 91C	91C	
$CoCl_2$–$FeCl_3$–GBIC	84S2	84S2 86R1		
$NiCl_2$–$FeCl_3$–GBIC	88R	88R	88R	
$NiCl_2$–$CrCl_3$–GBIC	90R2	90R2		
$MnCl_2$–$CrCl_3$–GBIC		91C	91C	

9.4.3.2 M′ (magnetic) Cl_n–M″ (nonmagnetic) Cl_m–GBIC (M′ = Co, Cr; M″ = Al, Ga, Cd)

9.4.3.2.1 $CoCl_2$–$GaCl_3$–GBIC
Survey of the properties

II. Magnetic data

T_{c1} = 6.1 K (χ max) [86R3]

9.4.3.2.2 $CoCl_2$–$AlCl_3$–GBIC
Survey of the properties

III. Other physical data

Thermal conductivity: [88K]

Thermoelectric power: [88K], [89K]

9.4.3.2.3 $CrCl_3$–$AlCl_3$–GBIC
Survey of the properties

II. Magnetic data

T_c = 13 K [90R1]

$\Theta_\perp$ = 32 K [90R1]

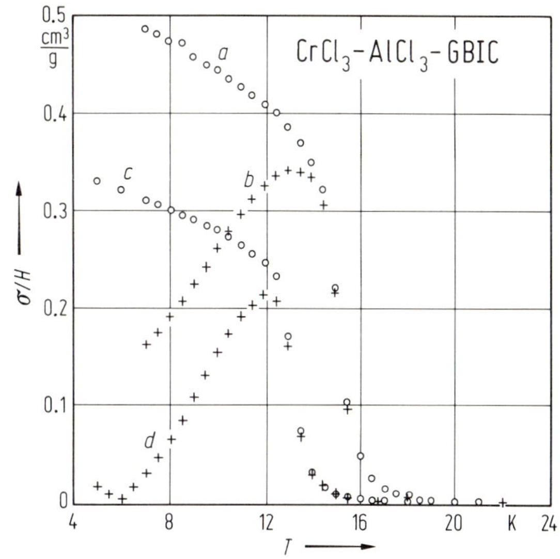

Fig. 3.1. $CrCl_3$–$AlCl_3$–GBIC. Temperature dependence of magnetization. (Curve *a*) and (*b*) are FCM and ZFCM at $H(\perp c) = 4.0$ Oe, (*c*) and (*d*) are FCM and ZFCM at $H(\perp c) = 2.0$ Oe. The temperature scale is shifted by 2 K for the higher field measurement [90R1].

9.4.3.2.4 $CrCl_3$–$CdCl_2$–GBIC
Survey of the properties

II. Magnetic data

$T_c = 13$ K [90R1]
$\Theta_\perp = 32$ K [90R1]
$\Theta_\perp = 22$ K [91C]
$\Theta_\parallel = 20$ K [91C]
$g_\perp = 1.992$ [91C]
$g_\parallel = 1.988$ [91C]

III. Other physical data

ESR: Fig. 3.3 [91C]

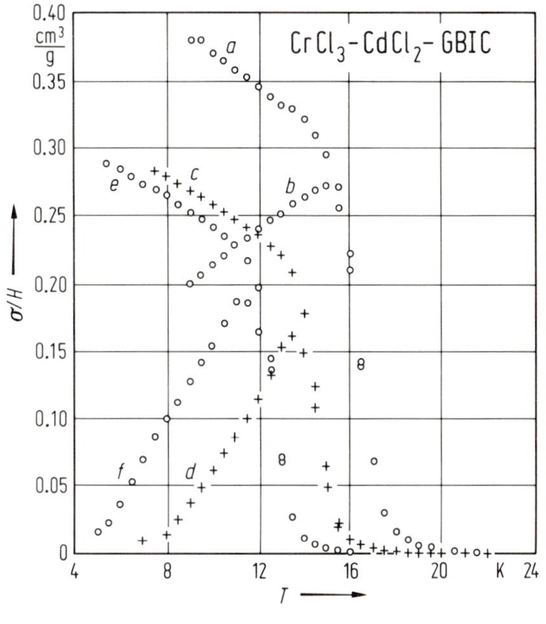

Fig. 3.2. $CrCl_3$–$CdCl_2$–GBIC. Temperature dependence of magnetization. (Curve *a*) and (*b*) are FCM and ZFCM at $H(\perp c) = 4$ Oe. (*c*) and (*d*), (*e*) and (*f*) are the corresponding quantities at $H(\perp c) = 2.0$ Oe and 1.0 Oe, respectively. The temperature scale is shifted by 2 K for the higher field measurements [90R1.]

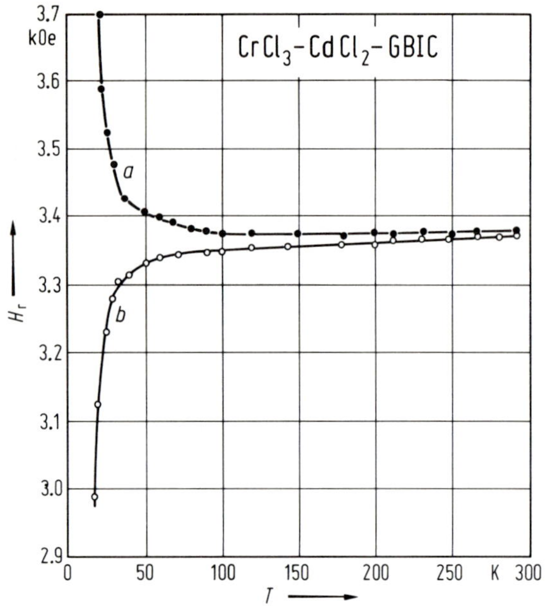

Fig. 3.3. $CrCl_3$–$CdCl_2$–GBIC. Temperature dependence of EPR fields for two field directions, $\boldsymbol{H}\perp c$ (curve *a*) and $\boldsymbol{H}\parallel c$ (*b*). $\nu = 9.35$ GHz [91C].

9.4.3.3 M' (magnetic) Cl_n–M'' (magnetic) Cl_m–GBIC (M', M'' = Cr, Mn, Fe, Co, Ni)

9.4.3.3.1 $CoCl_2$–$FeCl_3$–GBIC
Survey of the properties

I. Lattice parameters
$d = 18.60$ Å [84S2]

II. Magnetic data
$T_{cu} = 8.8$ K (χ kink) [84S2]
$T_{cl} = 8.0$ K (χ max) [84S2]

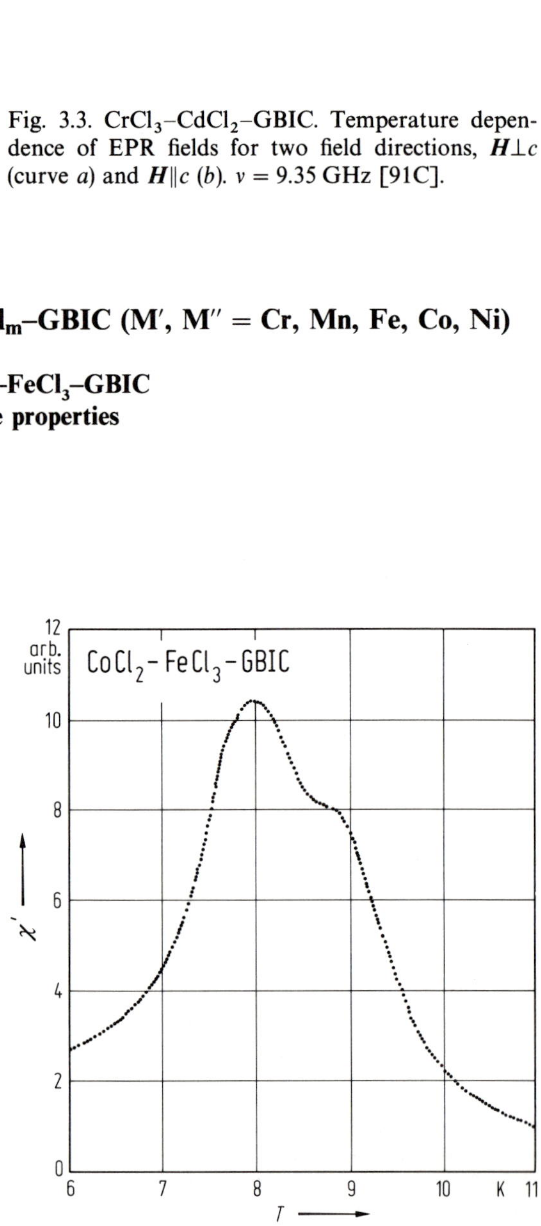

Fig. 3.4. $CoCl_2$–$FeCl_3$–GBIC. Temperature dependence of AC susceptibility. $H = 0$, $\boldsymbol{h}\perp c$ [84S2].

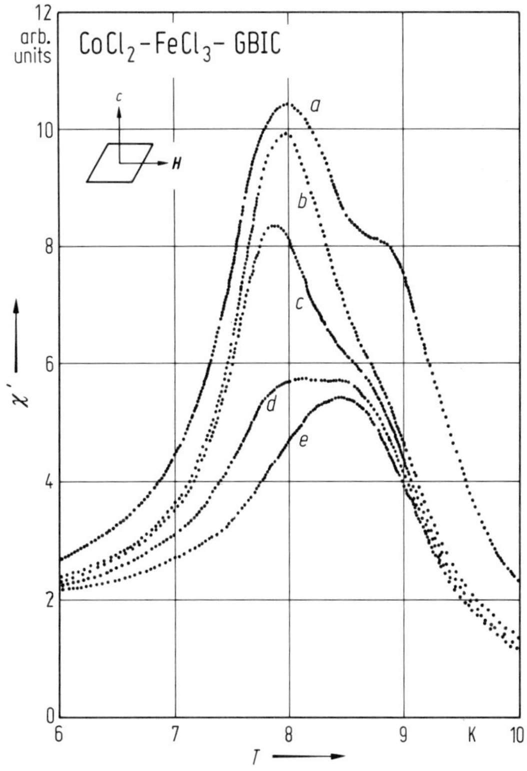

Fig. 3.5. $CoCl_2$–$FeCl_3$–GBIC. Temperature dependence of AC susceptibilities in various magnetic fields. $\boldsymbol{h} \perp c$, $H(\perp c) = 0$ (curve *a*), 14.3 Oe (*b*), 49.5 Oe (*c*), 60 Oe (*d*) and 75 Oe (*e*) [84S2].

9.4.3.3.2 $NiCl_2$–$FeCl_3$–GBIC
Survey of the properties

I. Lattice parameters
$d = 18.6$ Å [88R]

II. Magnetic data
$T_c = 20.5$ K (M_r) [88R]

III. Other physical data
Mössbauer spectra: Fig. 3.6 [88R]

For Fig. 3.6 see next page.

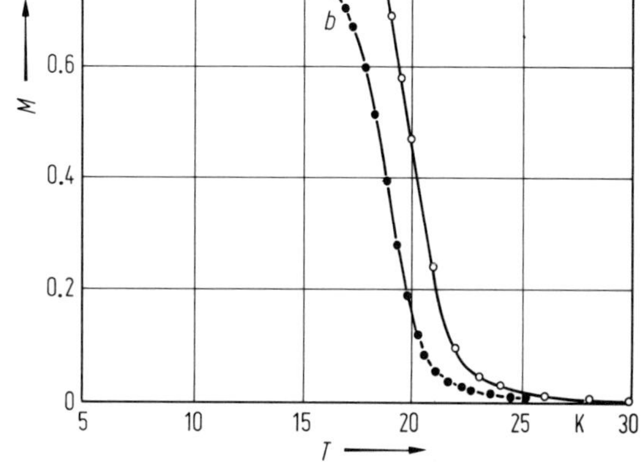

Fig. 3.7. $NiCl_2$–$FeCl_3$–GBIC. Temperature dependence of FCM for various samples. $H(\perp c) = 1$ Oe. (Curve *a*) Stage 2 $NiCl_2$–GIC. (*b*) $NiCl_2$–$FeCl_3$–GBIC; T2, T3 and T7 [88R].

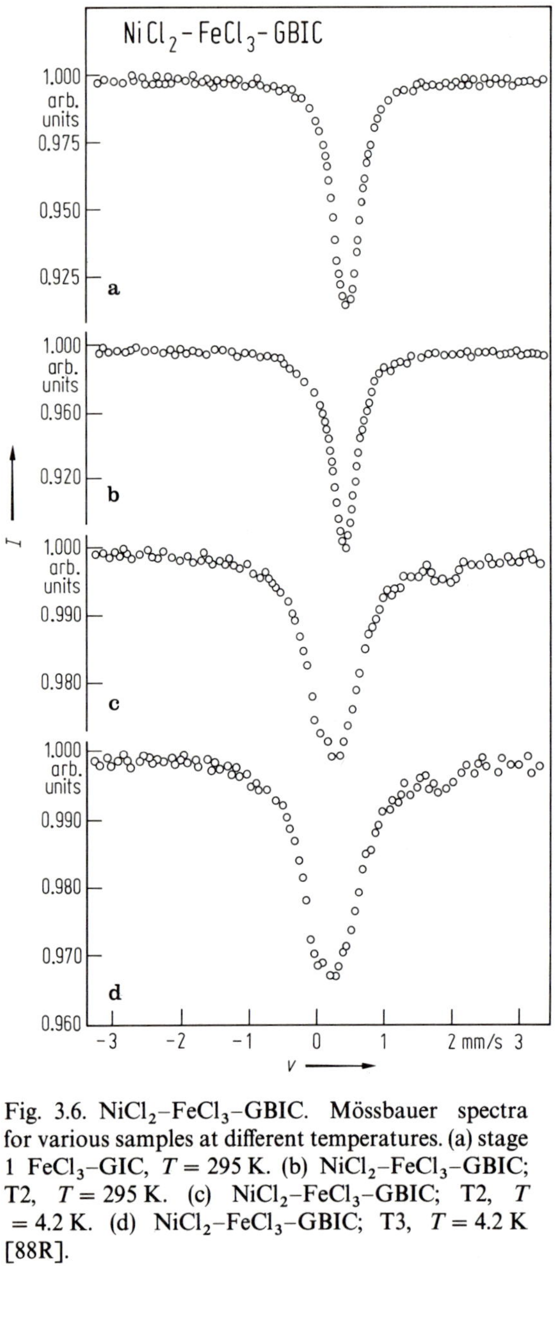

Fig. 3.6. $NiCl_2$–$FeCl_3$–GBIC. Mössbauer spectra for various samples at different temperatures. (a) stage 1 $FeCl_3$–GIC, $T = 295$ K. (b) $NiCl_2$–$FeCl_3$–GBIC; T2, $T = 295$ K. (c) $NiCl_2$–$FeCl_3$–GBIC; T2, $T = 4.2$ K. (d) $NiCl_2$–$FeCl_3$–GBIC; T3, $T = 4.2$ K [88R].

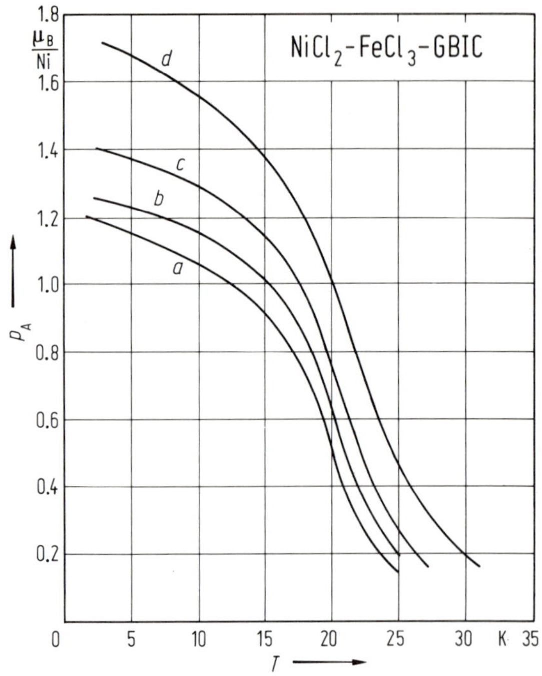

Fig. 3.8. $NiCl_2$–$FeCl_3$–GBIC. Temperature dependence of FCM per Ni atom for sample T3 of Fig. 3.6 and 3.7. $H(\perp c) = 250$ Oe (curve *a*), 350 Oe (*b*), 500 Oe (*c*) and 1000 Oe (*d*) [88R]. p_A: magnetic moment per atom.

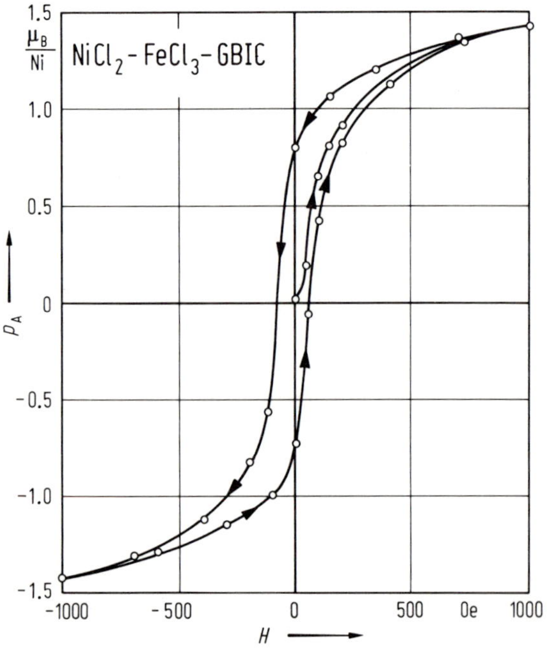

Fig. 3.9. $NiCl_2$–$FeCl_3$–GBIC. Hysteresis loop of magnetization per Ni atom for sample T2 of Fig. 3.6 and 3.7. $T = 1.8$ K. $H(\perp c)$ is swept in the sequence $0 \rightarrow 1000$ Oe $\rightarrow 0 \rightarrow -1000$ Oe $\rightarrow 0 \rightarrow 1000$ Oe [88R].

9.4.3.3.3 $NiCl_2$–$CrCl_3$–GBIC
Survey of the properties

I. Lattice parameters

d = 28 Å [90R2]

II. Magnetic data

T_{cu} = 19.9 K (M_{FC}) [90R2]

T_{cl} = 11.3 K (M_{FC}) [90R2]

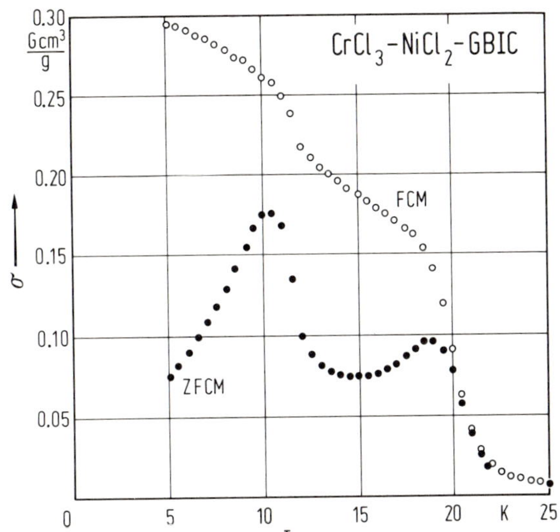

a

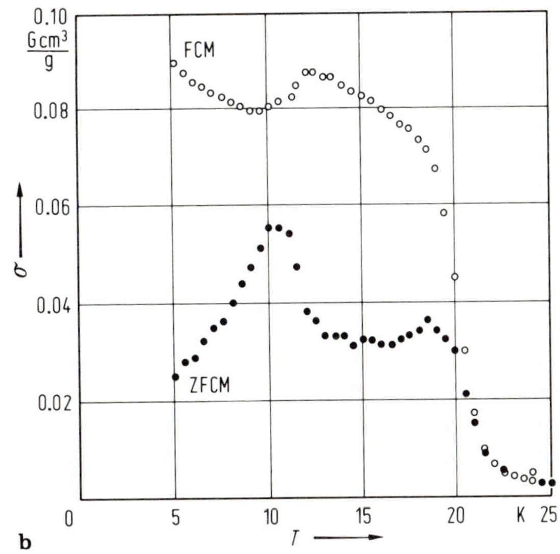

b

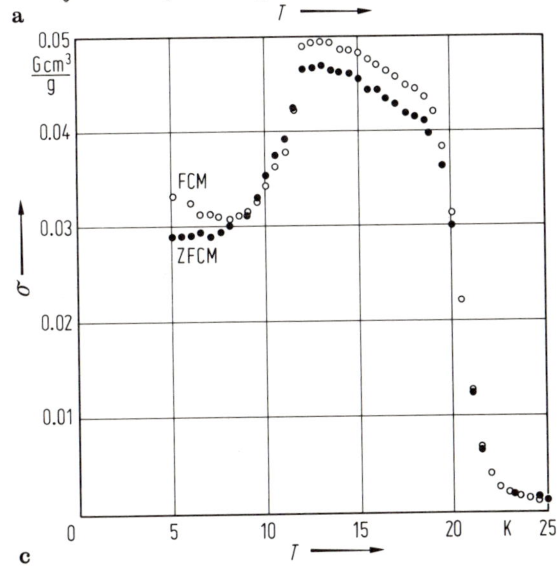

c

Fig. 3.10. $NiCl_2$–$CrCl_3$–GBIC. Temperature dependence of FCM and ZFCM for various applied fields. $H(\perp c)$ = 4 Oe (a), 2 Oe (b) and 1 Oe (c) [90R2].

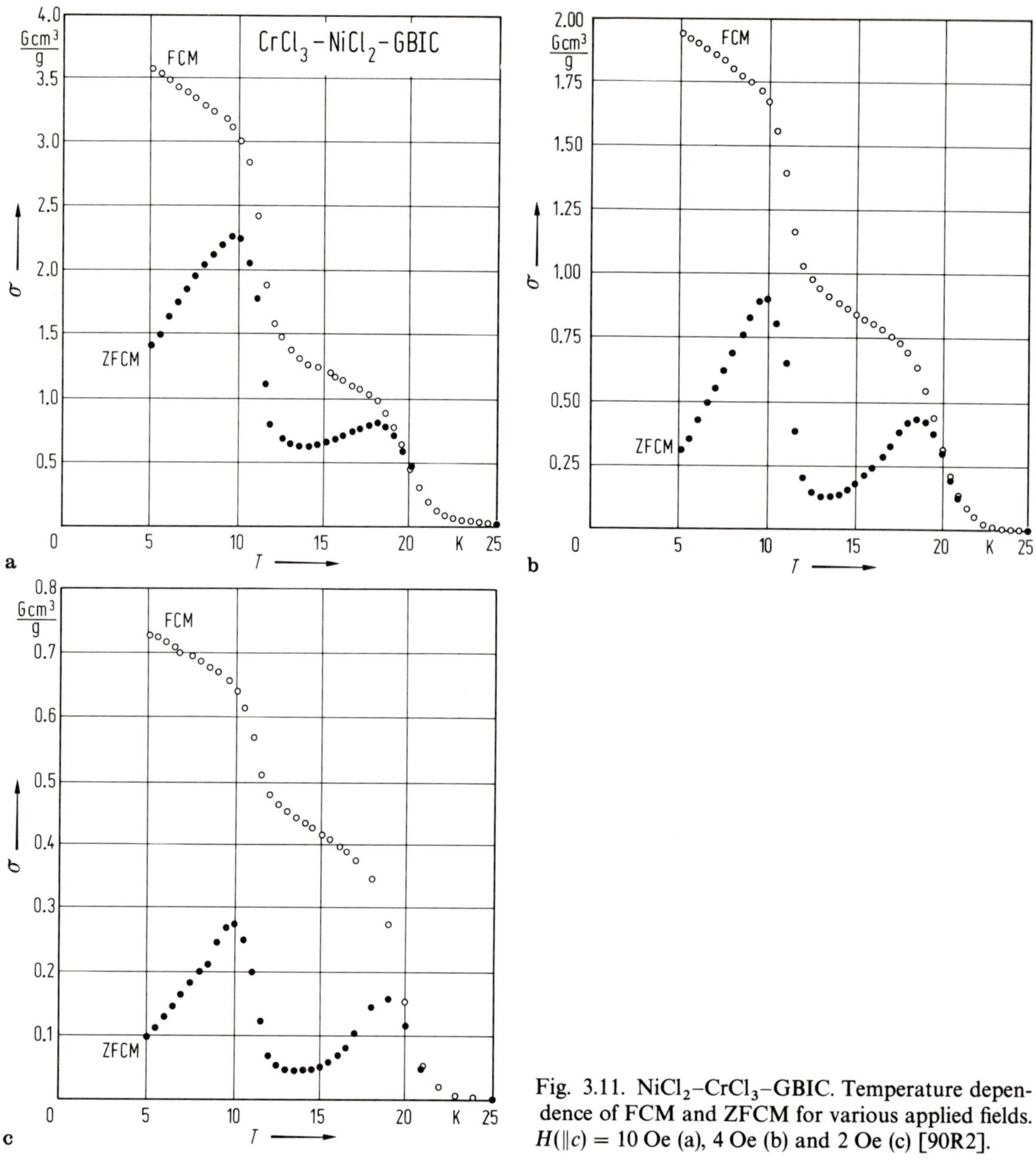

Fig. 3.11. $NiCl_2$–$CrCl_3$–GBIC. Temperature dependence of FCM and ZFCM for various applied fields. $H(\|c) = 10$ Oe (a), 4 Oe (b) and 2 Oe (c) [90R2].

9.4.3.3.4 $MnCl_2$–$CrCl_3$–GBIC
Survey of the properties

II. Magnetic data
$\Theta_\perp = 28$ K [91C]
$\Theta_\parallel = 27$ K [91C]
$g_\perp = 1.978$ [91C]
$g_\parallel = 1.974$ [91C]

III. Other physical data
ESR: [91C]

9.4.4 Mixed intercalation compounds

9.4.4.1 General survey

In this section, magnetic transitions of graphite mixed intercalation compounds are presented. The lattice structure of mixed intercalation compounds is qualitatively the same as that of singly intercalated GICs mentioned in 9.4.2.1. The essential difference is that each intercalant island in a mixed GIC is composed of a random mixture of two intercalant substances, e.g. $M'_{1-x}M''_xCl_2$, in this case. Thus, $M'_{1-x}M''_x$–GICs are taken as model systems of a 2D mixed magnet and several experimental studies have been carried out, intending to look after some new aspect of magnetic ordering in random systems.

The characteristics of magnetic ordering in mixed magnets are presented in 9.4.4.2 and those in diluted magnets are in 9.4.4.3. In Table 4.1, the references for the compounds are classified, according to the investigated ways.

Table 4.1. List of references.

Compounds	Crystal properties	Magnetic properties	MR, ND, Mössbauer	Others
$Co_{1-x}Ni_xCl_2$–GIC (S2)	89Y1 90Y 90S1	89Y1 90S1 90Y		
$Co_{1-x}Mn_xCl_2$–GIC (S2)	90S2 91S2	90S2 91S2		
$Co_{1-x}Fe_xCl_2$–GIC (S1, 2)	90S1	89S4 90S1 (S2)		89S4
$Co_{1-x}Al_xCl_2$–GIC (S2, 4, 5)	83S4	83S4		83S4
$Co_{1-x}Mg_xCl_2$–GIC (S1)	90N2	90N2 90N3		90N2

9.4.4.2 Mixed compounds

9.4.4.2.1 $Co_{1-x}Ni_xCl_2$–GIC
Survey of the properties

I. Lattice parameters

d: Fig. 4.1 and Table 2 in [90Y]

II. Magnetic data

$\Theta_{\perp}$: Fig. 4.5 and Table 3 in [90Y]

p_{eff}: Fig. 4.6 and Table 3 in [90Y]

$C_{\perp}$: Table 3 in [90Y]

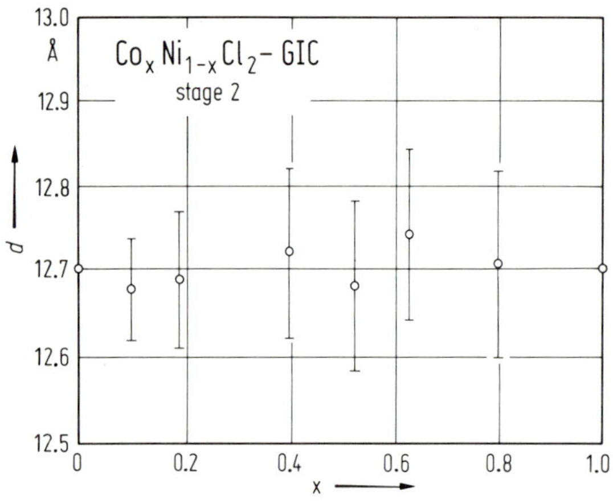

Fig. 4.1. $Co_xNi_{1-x}Cl_2$–GIC, stage 2. Intercalate *c*-axis repeat distance as a function of concentration x [90S1].

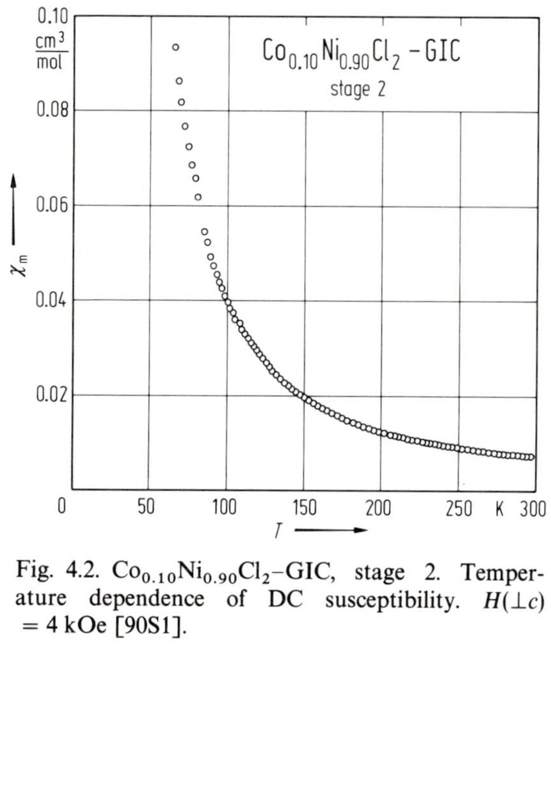

Fig. 4.2. $Co_{0.10}Ni_{0.90}Cl_2$–GIC, stage 2. Temperature dependence of DC susceptibility. $H(\perp c)$ = 4 kOe [90S1].

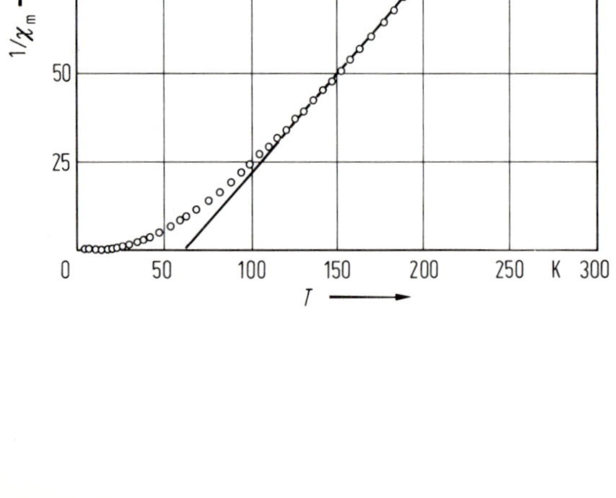

Fig. 4.3. $Co_{0.10}Ni_{0.90}Cl_2$–GIC, stage 2. Temperature dependence of inverse susceptibility. $H(\perp c)$ = 4 kOe. Solid line is the least square fit to Curie–Weiss law for 150 K < *T* < 300 K [90S1].

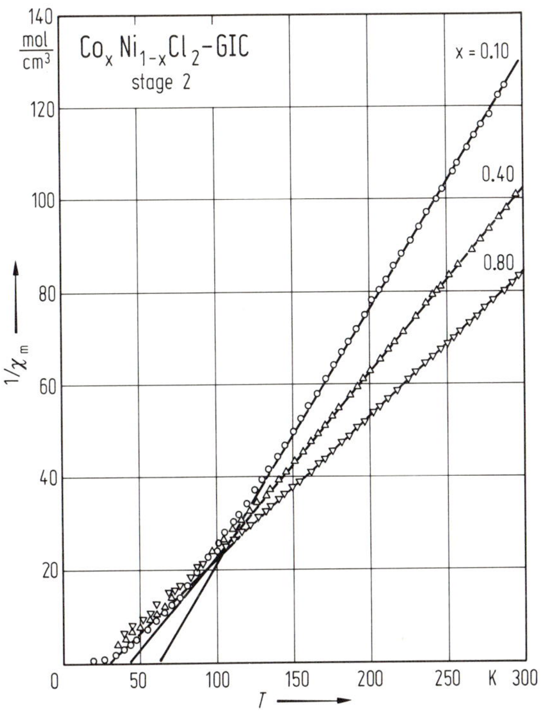

Fig. 4.4. $Co_xNi_{1-x}Cl_2$–GIC, stage 2. Temperature dependence of inverse susceptibilities for various concentrations. $H(\perp c) = 4$ kOe. Solid lines are the least square fits to Curie–Weiss law [90Y].

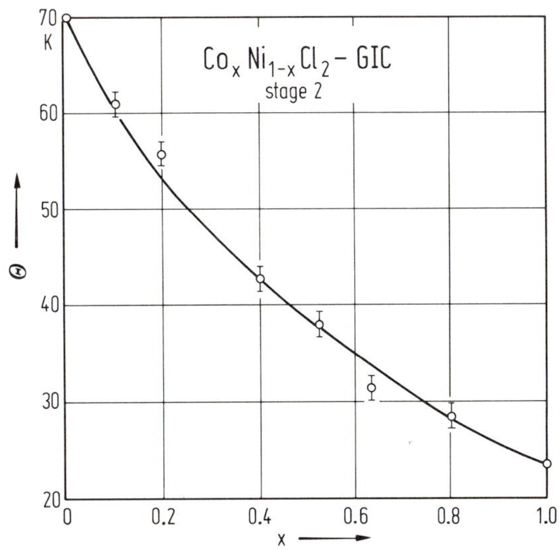

Fig. 4.5. $Co_xNi_{1-x}Cl_2$–GIC, stage 2. Curie–Weiss temperature vs. concentration x [89Y1].

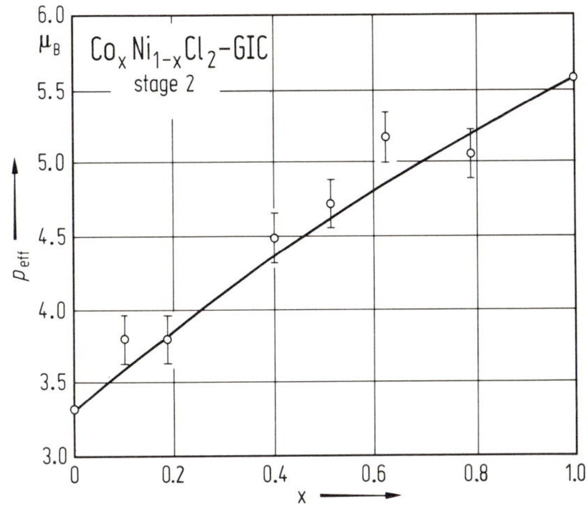

Fig. 4.6. $Co_xNi_{1-x}Cl_2$–GIC, stage 2. Average effective magnetic moment vs. concentration x [89Y1]. The solid line is a plot of Eq. (9) in [89Y1].

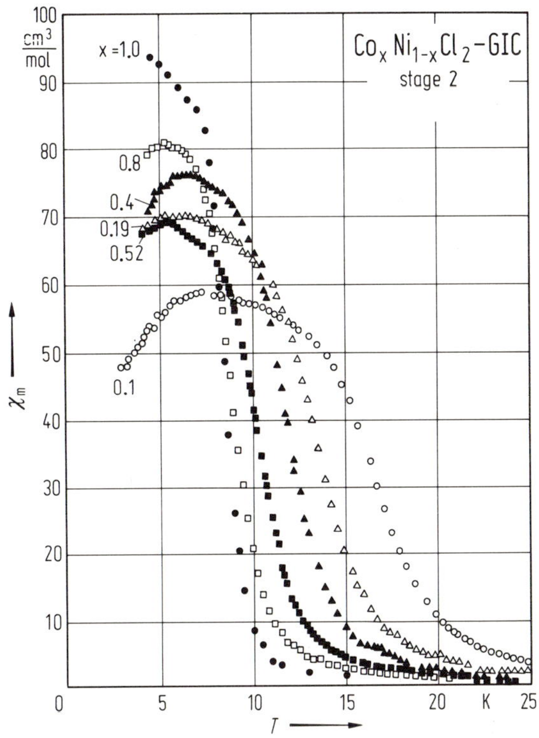

Fig. 4.7. $Co_xNi_{1-x}Cl_2$–GIC, stage 2. Temperature dependence of DC susceptibilities for different samples. $H(\perp c) = 100$ Oe. The value of x is indicated on each curve [90Y].

9.4.4.2.2 $Co_{1-x}Mn_xCl_2$–GIC
Survey of the properties

I. Lattice parameters

d: Fig. 4.8 and Table II in [91S2]

II. Magnetic data

$\Theta_\perp$: Fig. 4.10 and Table III in [91S2]

p_{eff}: Fig. 4.11 and Table III in [91S2]

$C_\perp$: Table III in [91S2]

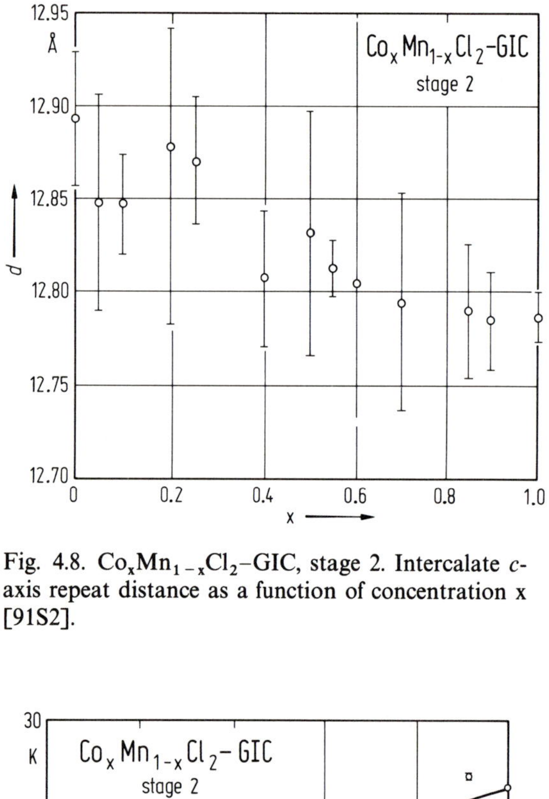

Fig. 4.8. $Co_xMn_{1-x}Cl_2$–GIC, stage 2. Intercalate *c*-axis repeat distance as a function of concentration x [91S2].

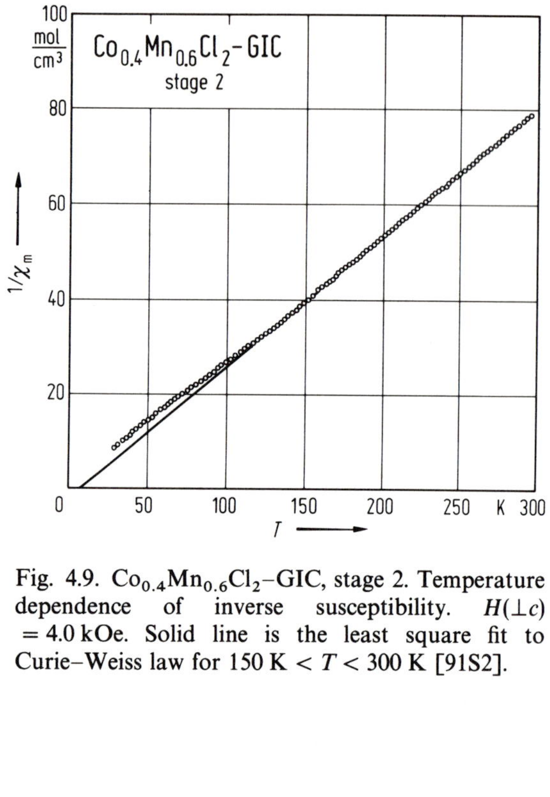

Fig. 4.9. $Co_{0.4}Mn_{0.6}Cl_2$–GIC, stage 2. Temperature dependence of inverse susceptibility. $H(\perp c)$ = 4.0 kOe. Solid line is the least square fit to Curie–Weiss law for 150 K < *T* < 300 K [91S2].

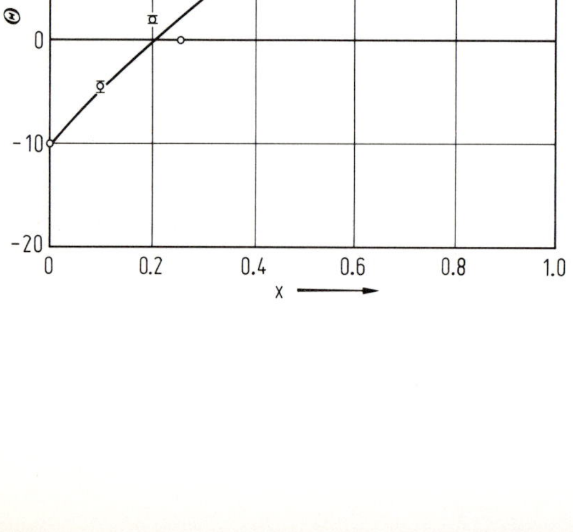

Fig. 4.10. $Co_xMn_{1-x}Cl_2$–GIC, stage 2. Curie–Weiss temperature vs. concentration x [91S2].

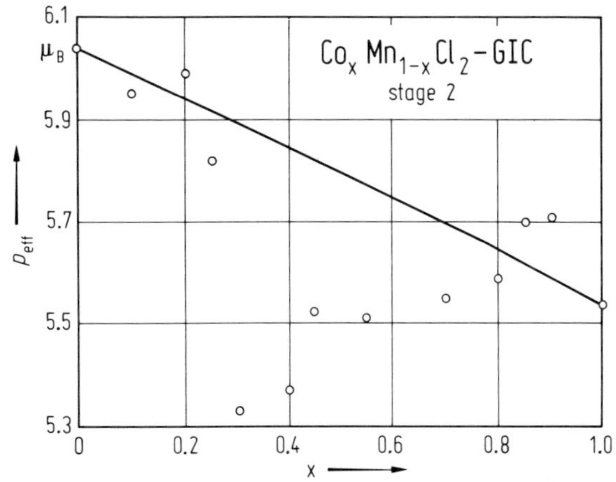

Fig. 4.11. $Co_xMn_{1-x}Cl_2$–GIC, stage 2. Average effective magnetic moment vs. concentration x [91S2].

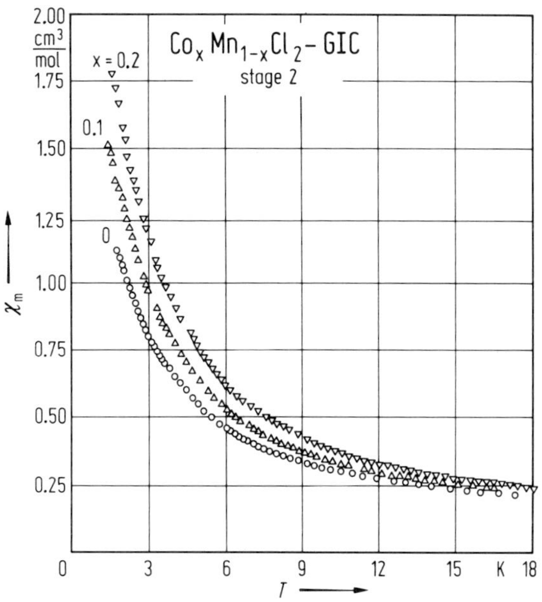

Fig. 4.12. $Co_xMn_{1-x}Cl_2$–GIC, stage 2. Temperature dependence of DC susceptibilities for $0 \le x \le 0.2$. $H(\perp c) = 100$ Oe [91S2].

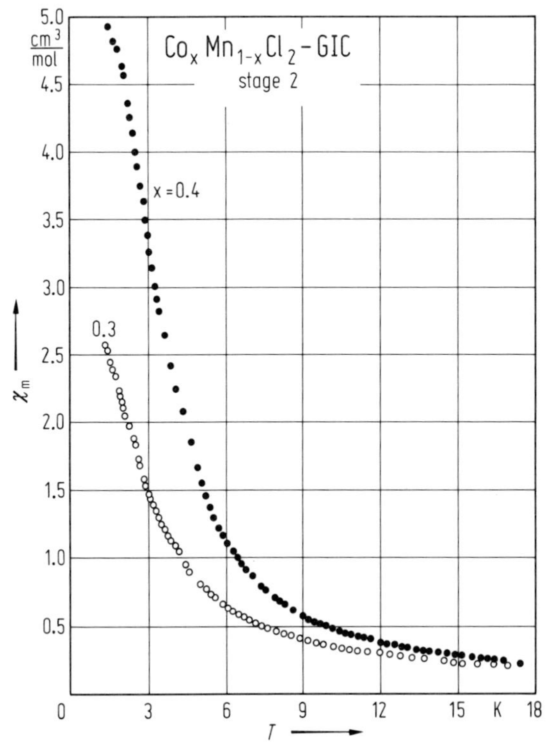

Fig. 4.13. $Co_xMn_{1-x}Cl_2$–GIC, stage 2. Temperature dependence of DC susceptibilities for x = 0.3 and 0.4. $H(\perp c) = 100$ Oe [91S2].

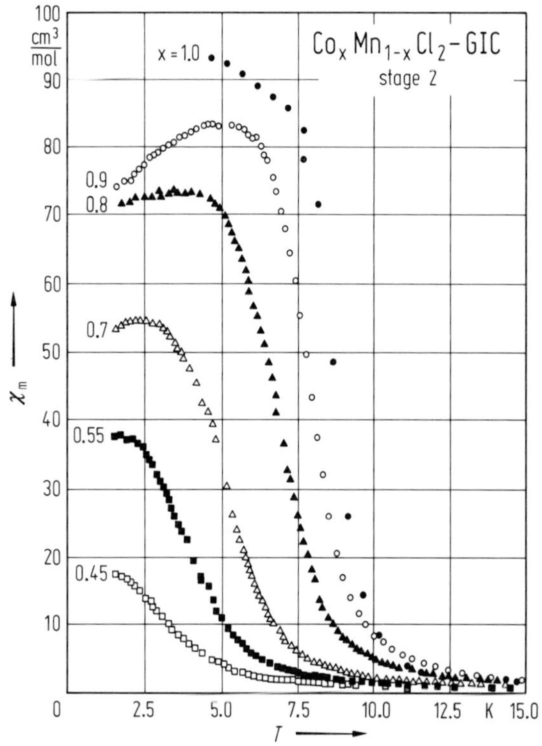

Fig. 4.14. $Co_xMn_{1-x}Cl_2$–GIC, stage 2. Temperature dependence of DC susceptibilities for $0.45 \le x \le 1.0$. $H(\perp c) = 100$ Oe [91S2].

9.4.4.2.3 $Co_{1-x}Fe_xCl_2$–GIC
Survey of the properties

I. Lattice parameters
d: Fig. 4.15 and Table III in [90S1]

II. Magnetic data
$\Theta_\perp$: Table III in [90S1]
p_{eff}: Table III in [90S1]

III. Other physical data
Electric resistivity: [89S4]

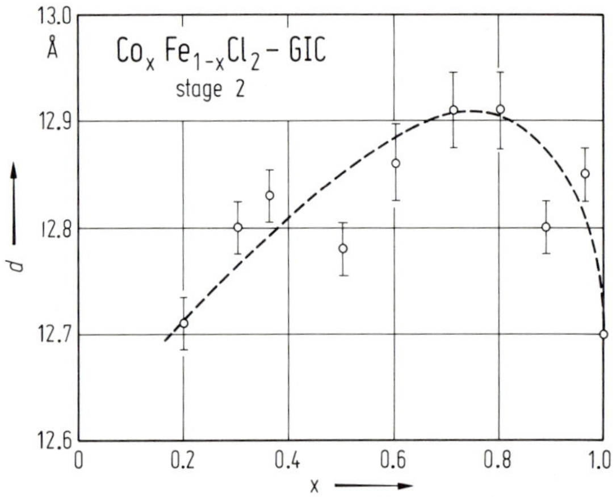

Fig. 4.15. $Co_xFe_{1-x}Cl_2$–GIC, stage 2. Intercalate *c*-axis repeat distance as a function of concentration x [90S1].

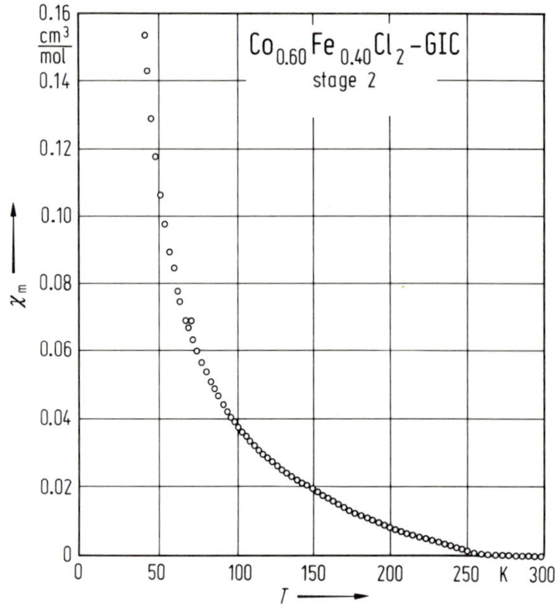

Fig. 4.16. $Co_{0.6}Fe_{0.4}Cl_2$–GIC, stage 2. Temperature dependence of DC susceptibility. $H(\perp c) = 4$ kOe [90S1].

9.4.4.3 Diluted compounds

9.4.4.3.1 $Co_{1-x}Al_xCl_2$–GIC
Survey of the properties

II. Magnetic data
T_c (C_M max): [83S4]
C_M: [83S4]

III. Other physical data
C_p: [83S4]
SdH frequency: [83S4]

9.4.4.3.2 $Co_{1-x}Mg_xCl_2$–GIC
Survey of the properties

I. Lattice parameters

d: Fig. 4.17 [90N2]

II. Magnetic data

$T_{\chi max}$(x): [90N2]
H_t(x): [90N2]
$H_t(T)$: [90N2]
$M(H)$ (high-field): Fig. 4.21 [90N2], 4.22 [90N3]
p_A: [90N3]
J_A: [90N3]

III. Other physical data

Electric resistivity: Fig. 4.23 [90N2]

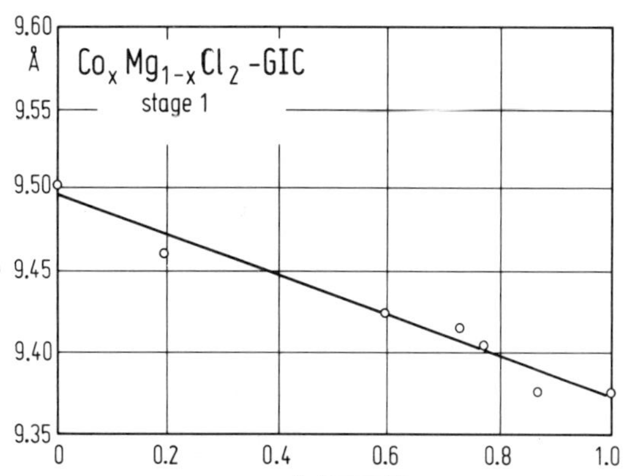

Fig. 4.17. $Co_xMg_{1-x}Cl_2$–GIC, stage 1. Intercalate *c*-axis repeat distance as a function of concentration x [90N2].

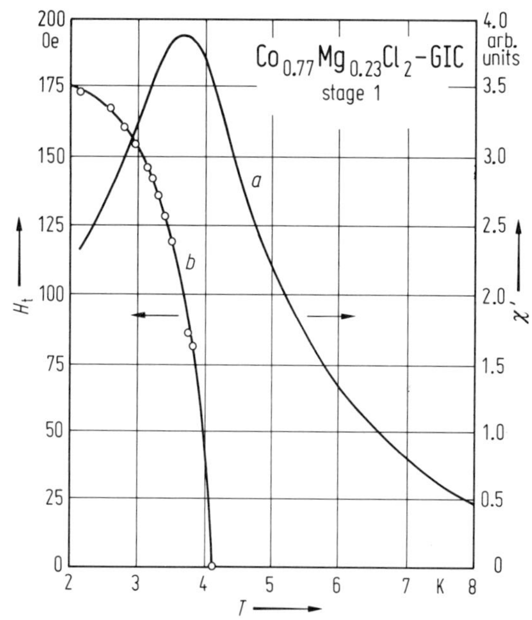

Fig. 4.18. $Co_{0.77}Mg_{0.23}Cl_2$–GIC, stage 1. Temperature dependence of in-plane AC susceptibility at $H = 0$ (curve *a*), and of transition field obtained from the field dependence of AC susceptibility in Fig. 4.19 (*b*) [90N2].

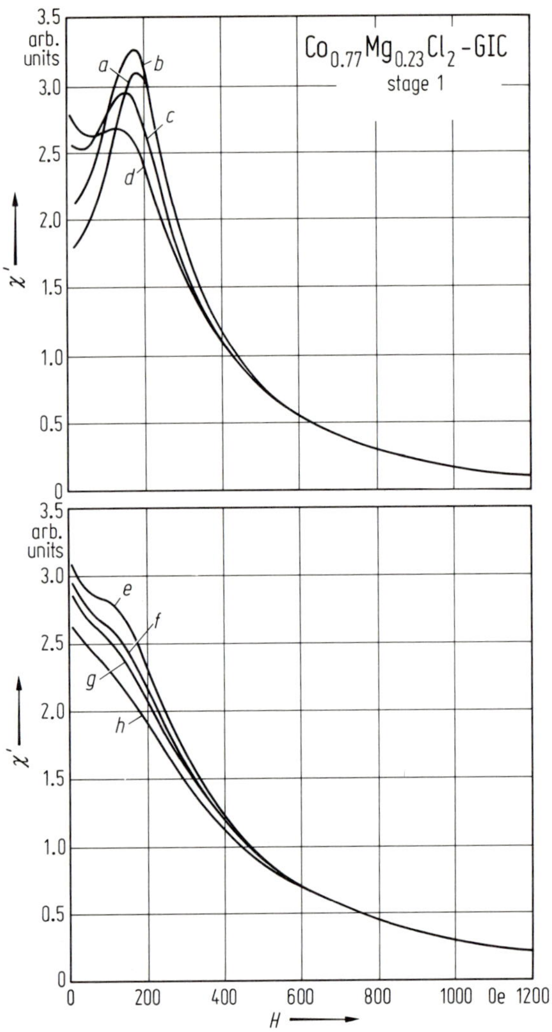

Fig. 4.19. $Co_{0.77}Mg_{0.23}Cl_2$–GIC, stage 1. Magnetic field dependence of in-plane AC susceptibilities for various temperatures. (Curve *a*) $T = 2.04$ K, (*b*) 2.62 K, (*c*) 3.14 K, (*d*) 3.50 K, (*e*) 3.59 K, (*f*) 3.75 K, (*g*) 3.88 K and (*h*) 4.14 K [90N2].

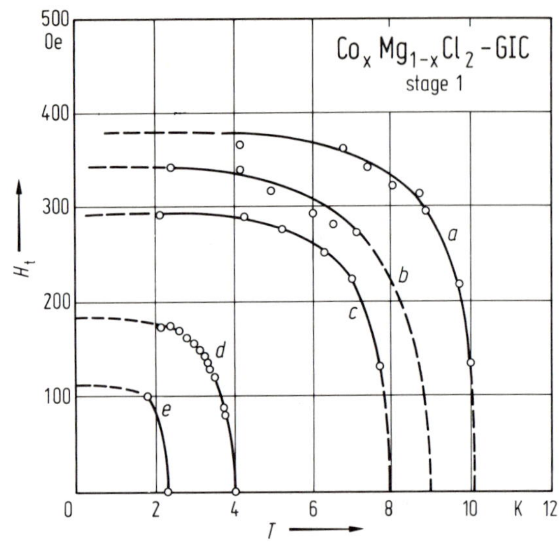

Fig. 4.20. $Co_xMg_{1-x}Cl_2$–GIC, stage 1. Temperature dependence of transition field H_t obtained from the field dependence of AC susceptibility for different samples. (Curve *a*) x = 1.00, (*b*) 0.92, (*c*) 0.88, (*d*) 0.77, (*e*) 0.70. Solid curves are guides to the eye [90N2].

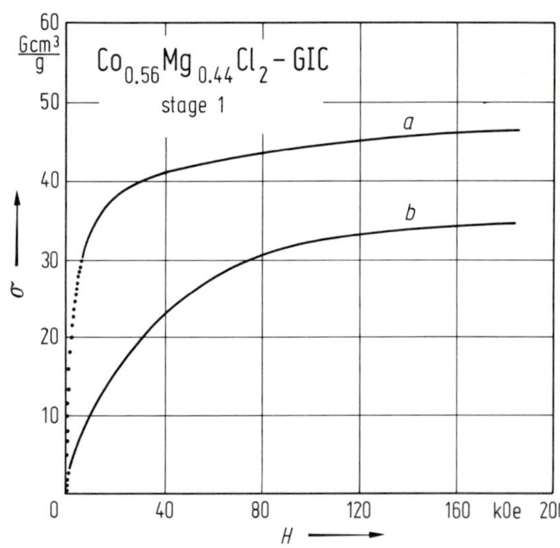

Fig. 4.21. $Co_{0.56}Mg_{0.44}Cl_2$–GIC, stage 1. The high field magnetizations for two field directions. $T = 1.3$ K. (Curve *a*) for $\boldsymbol{H} \perp c$ and (*b*) for $\boldsymbol{H} \parallel c$ [90N2].

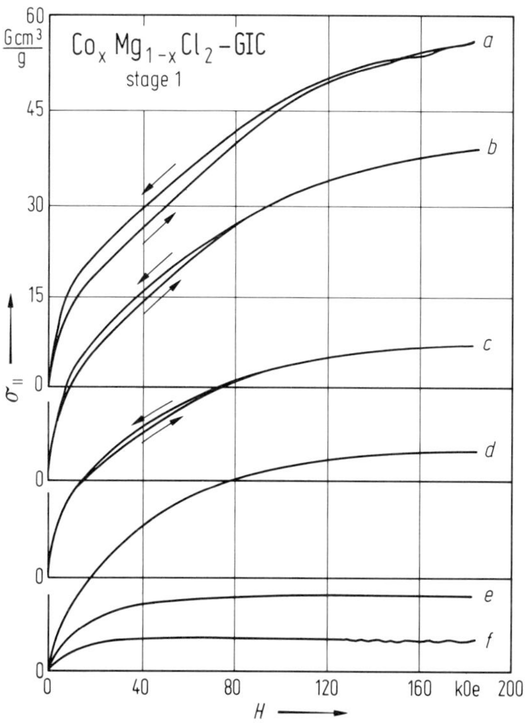

Fig. 4.22. $Co_xMg_{1-x}Cl_2$–GIC, stage 1. The virgin *c*-axis magnetization traces for various samples. (Curve *a*) x = 1.00, (*b*) 0.92, (*c*) 0.77 (*d*) 0.56, (*e*) 0.20 and (*f*) 0.09 [90N2].

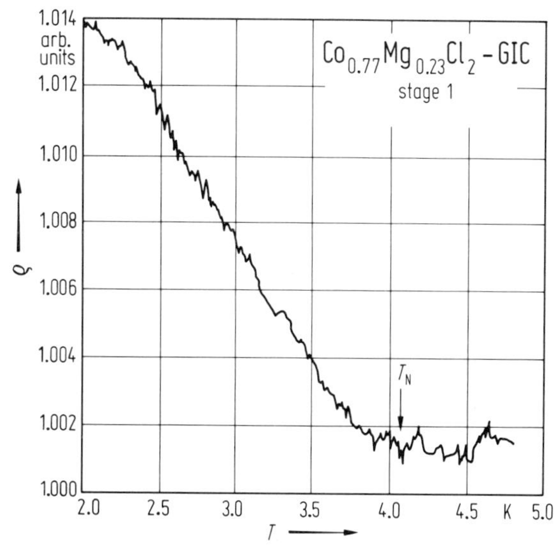

Fig. 4.23. $Co_{0.77}Mg_{0.23}Cl_2$–GIC, stage 1. Temperature dependence of the in-plane resistivity [90N2].

9.4.5 References for 9.4

44O Onsager, L.: Phys. Rev. **65** (1944) 117.
47W Wells, A.F.: J. Chem. Soc. (1947) 1870.
50H Husimi, K., Syoji, I.: Prog. Theor. Phys. **5** (1950) 177.
50W Wannier, G.M.: Phys. Rev. **79** (1950) 357.
55M Murray, R.B.: Phys. Rev. **100** (1955) 1071.
59K Koehler, W.C., Wilkinson, M.K., Cable, J.W., Wollan, E.O.: J. Phys. Radium **20** (1959) 180.
59W Wilkinson, M.K., Cable, J.W., Wollan, E.O., Koehler, W.C.: Phys. Rev. **113** (1959) 497.
62S Stout, J.W., Chisholm, R.C.: J. Chem. Phys. **36** (1962) 979.
65N Narath, A., Davis, L.: Phys. Rev. **137** (1965) 161.
66M Mermin, N.D., Wagner, H.: Phys. Rev. Lett. **17** (1966) 1133.
66S Stanley, H.E., Kaplan, T.A.: Phys. Rev. Lett. **17** (1966) 913.

73H Hutchings, M.T.: J. Phys. **C 6** (1973) 3143.
73K1 Kostryukova, M.O., Luk'yanova, L.V.: ZhETF Pis'ma. Red. **17** No. 2 **78–81** (1973) 54.
73K2 Kosterlitz, J.M., Thouless, D.J.: J. Phys. **C 6** (1973) 1181.
73S Stampfel, J.P., Oosterhuis, W.T., Window, B., Barros, F. de S.: Phys. Rev. **B 8** (1973) 4371.
74K1 Karimov, Yu.S.: Sov. Phys. JETP **38** (1974) 129.
74K2 Karimov, Yu.S.: Sov. Phys. JETP **39** (1974) 547.
74K3 Karimov, Yu.S., Novikov, Yu.N.: Sov. Phys. JETP Lett. **19** (1974) 159.
74K4 Kosterlitz, J.M.: J. Phys. **C 7** (1974) 1046.
75L Lindgård, P.A., Birgeneau, R.J., Als-Nielsen, J., Guggenheim, H.J.: J. Phys. **C 8** (1975) 1059.
76K Karimov, Yu.S.: Sov. Phys. JETP **41** (1976) 772.
77B Billerey, D., Terrier, C., Mainard, R., Meriel, P.: C.R. Acad. Sci. (Paris) **B 39** (1977) 495.
77J Jose, J.V., Kadanoff, L.P., Kirkpatrick, S., Nelson, D.R.: Phys. Rev. **B 16** (1977) 1217.
78B Billerey, D., Terrier, C., Mainard, R., Perrin, M., Hubsch, J.: Phys. Lett. **A 68** (1978) 275.
80E Escribe, C., Bouillot, J., Ziebeck, K.R.A.: J. Phys. **C 13** (1980) 4053.
80F Flandrois, S., Amiell, J., Masson, J.-M.: Phys. Lett. **A 80** (1980) 328.
80S1 Safran, S.A.: Synth. Met. **2** (1980) 1.
80S2 Shenker, S.H., Tobochnik, J.: Phys. Rev. **B 22** (1980) 4462.
80T Thomas, J.M., Millward, G.R., Schlögl, R.F., Boehm, H.P.: Mater. Res. Bull. **15** (1980) 671.
81D Dresselhaus, A.S., Dresselhaus, G.: Adv. Phys. **30** (1981) 139.
81F Flandrois, S., Masson, J.M., Rouillon, J.C., Gaultier, J., Hauw, C.: Synth. Met. **3** (1981) 1.
81S Suzuki, M., Ikeda, H.: J. Phys. **C 14** (1981) L923.
82B Baron, F., Flandrois, S., Hauw, C., Gaultier, J.: Solid State Commun. **42** (1982) 759.
82E Elahy, M., Nicolini, C., Dresselhaus, G., Zimmerman, G.O.: Solid State Commun. **41** (1982) 289.
82K Kuhlow, B.: Phys. Status Solidi (a) **72** (1982) 161.
82M1 Millman, S.E., Holmes, B.W., Zimmerman, G.O.: Solid State Commun. **43** (1982) 903.
82M2 Millman, S.E.: Solid State Commun. **44** (1982) 23.
82M3 Millman, S. E., Kirczenow, G.: Solid State Commun. **44** (1982) 1217.
82M4 Millman, S.E., Corson, M.R., Hoy, G.R.: Phys. Rev. **B 25** (1982) 6595.
82O Onn, D.G., Alexander, M.G., Ritsko, J.J., Flandrois, S.: J. Appl. Phys. **53** (1982) 2751.
83B Blatt, F.J., Zabala-Martinez, I., Heremans, J., Issi, J.-P., Shayegan, M., Dresselhaus, M.S.: Phys. Rev. **B 27** (1983) 2558.
83E1 Elahy, M., Dresselhaus, G.: Mater. Res. Soc. Symp. Proc. **20** (1983) 207.
83E2 Elahy, M., Shayegan, M., Szeto, K.Y., Dresselhaus, G.: Synth. Met. **8** (1983) 35.
83F Flandrois, S., Hewat, A.W., Hauw, C., Bragg, R.H.: Synth. Met. **7** (1983) 305.
83H Hauw, C., Gaultier, J., Flandrois, S., Gonzalez, O., Doringnac, D., Jagut, R.: Synth. Met. **7** (1983) 313.
83M1 Murakami, Y., Matsuura, M., Suzuki, M., Ikeda, H.: J. Magn. Magn. Mater. **31–34** (1983) 1171.
83M2 Millman, S.E., Zimmerman, G.O.: J. Phys. **C 16** (1983) L89.
83M3 Millman, S.E., Kirczenow, G.: Phys. Rev. **B 28** (1983) 5019.
83S1 Suematsu, H., Nishitani, R., Yoshizaki, R., Suzuki, M.: J. Phys. Soc. Jpn. **52** (1983) 3874.
83S2 Suzuki, M., Ikeda, H., Murakami, Y., Matsuura, M., Suematsu, H., Nishitani, R., Yoshizaki, R.: J. Magn. Magn. Mater. **31–34** (1983) 1173.
83S3 Shayegan, M., Salamanca-Riba, L., Heremans, J., Dresselhaus, G., Issi, J.-P.: Mater. Res. Soc. Symp. Proc. **20** (1983) 213.
83S4 Shayegan, M., Dresselhaus, M.S., Salamanca-Riba, L., Dresselhaus, G., Heremans, J., Issi, J.-P.: Phys. Rev. **B 28** (1983) 4799.
83S5 Suzuki, M., Ikeda, H., Endoh, Y.: Synth. Met. **8** (1983) 43.
83S6 Simon, Ch., Batallan, F., Rosenman, I., Schweizer, J., Lauter, H., Vangelisti, R.: Synth. Met. **8** (1983) 53.
84C Chen, S.T., Szeto, K.Y., Elahy, M., Dresselhaus, G.: J. Chim. Phys. **81** (1984) 863.

84E Elahy, M., Dresselhaus, G.: Phys. Rev. **B 30** (1984) 7225.
84G Gonzalez, O., Flandrois, S., Maaroufi, A., Amiell, J.: Solid State Commun. **51** (1984) 499.
84H Hawrylak, P., Subbaswamy, K.R.: Phys. Rev. Lett. **53** (1984) 2098.
84K1 Koga, K., Suzuki, M.: J. Phys. Soc. Jpn. **53** (1984) 786.
84K2 Kawamura, H., Miyashita, S.: J. Phys. Soc. Jpn. **53** (1984) 9.
84M Miyashita, S., Shiba, H.: J. Phys. Soc. Jpn. **53** (1984) 1145.
84S1 Simon, Ch., Batallan, F., Rosenman, I., Schweitzer, J., Lauter, H., Vangelisti, R.: J. Appl. Phys. **55** (1984) 2441.
84S2 Suzuki, M., Oguro, I., Jinzaki, Y.: J. Phys. **C 17** (1984) L575.
84S3 Suzuki, M., Koga, K., Zinzaki, Y.: J. Phys. Soc. Jpn. **53** (1984) 2745.
85D Dresselhaus, G., Chen, S.T., Szeto, K.Y.: Synth. Met. **12** (1985) 433.
85I Ikeda, H., Endoh, Y., Mitsuda, S.: J. Phys. Soc. Jpn. **54** (1985) 3232.
85K1 Kimishima, Y., Furukawa, A., Nagano, H., Chow, P., Wiesler, D., Zabel, H., Suzuki, M.: Synth. Met. **12** (1985) 455.
85K2 Koga, K., Suzuki, M., Yasuoka, H.: Synth. Met. **12** (1985) 467.
85K3 Kirczenow, J.: Synth. Met. **12** (1985) 143.
85M1 Matsuura, M., Murakami, Y., Takeda, K., Ikeda, H., Suzuki, M.: Synth. Met. **12** (1985) 427.
85M2 Murakami, Y., Matsuura, M., Kataoka, T.: Synth. Met. **12** (1985) 443.
85N Nishihara, H., Ogura, I., Suzuki, M., Koga, K., Yasuoka, H.: Technical Report of ISSP **A1539** (1985) 1.
85O Oguro, I., Suzuki, M., Yasuoka, H.: Synth. Met. **12** (1985) 449.
85R1 Rancourt, D.G.: J. Magn. Magn. Mater. **51** (1985) 133.
85R2 Rosenmann, I., Batallan, F., Simon, Ch., Ayache, C., Schweitzer, J., Lauter, H., Vangelisti, R.: Synth. Met. **12** (1985) 439.
85S1 Szeto, K.Y., Dresselhaus, G.: Phys. Rev. **B 32** (1985) 3142.
85S2 Szeto, K.Y., Dresselhaus, G.: Phys. Rev. **B 32** (1985) 3186.
85S3 Szeto, K.Y., Chen, S.T., Dresselhaus, G.: Phys. Rev. **B 32** (1985) 4628.
85T1 Takeda, K., Amaya, K., Murakami, Y.: J. Phys. Soc. Jpn. **54** (1985) 3091.
85T2 Tsuda, T., Yasuoka, H., Suzuki, M.: Synth. Met. **12** (1985) 461.
85Y Yazami, R., Chouteau, G., Touzain, Ph., Briggs, A.: J. Phys. (Paris) **46** (1985) 1961.
86A Ayache, C., Bonjour, E., Simon, Ch., Batallan, F., Rosenman, I., Vangelisti, R.: J. Magn. Magn. Mater. **54–57** (1986) 675.
86I Ibrahim, A.K., Zimmerman, G.O.: Phys. Rev. **B 34** (1986) 4224.
86K1 Kawabata, C., Bishop, A.R.: Z. Phys. B: Condens. Matter **65** (1986) 225.
86K2 Kimishima, Y., Furukawa, A., Suzuki, M., Nagano, H.: J. Phys. **C 19** (1986) L34.
86K3 Kawabata, C., Bishop, A.R.: Solid State Commun. **60** (1986) 167.
86L Laguës, M., Marchand, D., Frétigny, C.: Solid State Commun. **59** (1986) 583.
86M Matsuura, M.: Ann. Phys. (France) **11** No. 2S (1986) 117.
86R1 Rancourt, D.G., Hun, B., Flandrois, S.: Ann. Phys. (France) **11** (1986) No. 2S 107.
86R2 Rancourt, D.G., Meschi, C., Flandrois, S.: Phys. Rev. **B 33** (1986) 347.
86R3 Rosenman, I., Batallan, F., Simon, Ch., Hachim, L.: J. Phys. (Paris) **47** (1986) 1221.
86S1 Szeto, K.Y., Chen, S.T., Dresselhaus, G.: Phys. Rev. **B 33** (1986) 3453.
86S2 Szeto, K.Y., Dresselhaus, G.: J. Phys. **C 19** (1986) 2063.
86S3 Suzuki, M., Wiesler, D.G., Chow, P.C., Zabel, H.: J. Magn. Magn. Mater. **54–57** (1986) 1275.
86S4 Simon, Ch., Batallan, F., Rosenman, I., Furdin, G., Vangelisti, R., Lauter, H., Schweitzer, J., Ayache, C., Pepy, G.: Ann. Phys. (France) **11** No. 2S (1986) 143.
86W1 Wiesler, D.G., Suzuki, M., Chow, P.C., Zabel, H.: Phys. Rev. **B 34** (1986) 7951.
86W2 Wiesler, D.G., Suzuki, M., Zabel, H., Shapiro, S.M., Nicklow, R.M.: Physica **B 136** (1986) 22.
87C Chouteau, G., Yazami, R.: Europhys. Lett. **3** (2) (1987) 229.

87D Dresselhaus, G., Nicholls, J.T., Chen, S.T., Dresselhaus, M.S.: J. Appl. Phys. **61** (1987) 4398.
87I1 Ibrahim, A.K., Powers, R., Zimmerman, G.O., Tahar, M.: J. Appl. Phys. **61** (1987) 4382.
87I2 Ibrahim, A.K., Zimmerman, G.O.: Phys. Rev. **B 35** (1987) 1860.
87I3 Ibrahim, A.K., Powers, R., Zimmerman, G.O., Tahar, M.: Jpn. J. Appl. Phys. **26** (1987) 635.
87M1 Matsuura, M., Karaki, Y., Yonezawa, T., Suzuki, M.: Jpn. J. Appl. Phys. **26** (1987) 773.
87M2 Matsuura, M., Endoh, Y., Kataoka, T., Murakami, Y.: J. Phys. Soc. Jpn. **56** (1987) 2233.
87M3 Matsuura, M., Tanaka, N., Karaki, Y., Murakami, Y.: Jpn. J. Appl. Phys. **26** (1987) 797.
87W1 Wiesler, D.G., Suzuki, M., Zabel, H.: Phys. Rev. **B 36** (1987) 7051.
87W2 Wiesler, D.G., Zabel, H.: Phys. Rev. **B 36** (1987) 7303.
87Y1 Yazami, R., Choteau, G.: Synth. Met. **18** (1987) 543.
87Y2 Yazami, R.: Synth. Met. **20** (1987) 383.
87Z1 Zabel, H., Shapiro, S.M.: Phys. Rev. **B 36** (1987) 7292.
87Z2 Zimmerman, G.O., Ibrahim, A.K.: J. Appl. Phys. **61** (1987) 4416.
88C Chouteau, G., Yazami, R.: Synth. Met. **23** (1988) 243.
88K Kinany-Alaoui, M., Piraux, L., Issi, J.-P., Pernot, P., Vangelisti, R.: Solid State Commun. **68** (1988) 1065.
88L Laguës, M., Hao, X., Dresselhaus, M.S.: Phys. Rev. **B 38** (1988) 967.
88M1 Miura, K., Iye, Y., Nicholls, J.T., Dresselhaus, G.: Technical Report of ISSP **A2018** (1988) 1.
88M2 Murakami, Y., Matsuura, M.: J. Phys. Soc. Jpn. **57** (1988) 1056.
88M3 Matsuura, M., in: Dynamics of Ordering Processes in Condensed Matter, Komura, S., Furukawa, H., (eds.), New York: Plenum Pub., 1988, p. 349.
88N Nicholls, J.T., Shapira, Y., McNiff, jr., E.J., Dresselhaus, G.: Synth. Met. **23** (1988) 231.
88P Powers, R., Ibrahim, A.K., Zimmerman, G.O., Tahar, M.: Phys. Rev. **B 38** (1988) 680.
88R Rancourt, D.G., Hun, B., Flandrois, S.: Can. J. Phys. **66** (1988) 776.
88W1 Wiesler, D.G., Zabel, H.: J. Appl. Phys. **63** (1988) 3554.
88W2 Wiesler, D.G., Zabel, H., Suzuki, M.: Synth. Met. **23** (1988) 237.
88Z Zimmerman, G.O., Ibrahim, A.K.: Phys. Rev. **B 37** (1988) 2059.
89A Abou Aly, A.I., Awad, R., Ibrahim, I.H., Ibrahim, A.K., Zimmerman, G.O., Powers, R.E., Tahar, M.Z.: Synth. Met. **34** (1989) 479.
89F Flandrois, S., Amiell, J., Agricole, B., Stumpp, E., Ehrhardt, C., Schubert, P.: Synth. Met. **34** (1989) 531.
89H1 Hagiwara, M., Kawaguchi, T., Matsuura, M.: Springer Ser. Synergetics **43** (1989) 175.
89H2 Hafidi, M.E., Chouteau, G., Yazami, R.: Synth. Met. **34** (1989) 525.
89K Kinany-Alaoui, M., Piraux, L., Bayot, V., Issi, J-P., Pernot, P., Vangelisti, R.: Synth. Met. **34** (1989) 537.
89M Mertens, F.G., Bishop, A.R., Wysin, G.M.: Phys. Rev. **B 39** (1989) 591.
89N1 Nicholls, J.T., Speck, J.S., Dresselhaus, G.: Phys. Rev. **B 39** (1989) 10047.
89N2 Nicholls, J.T., Dresselhaus, G.: Synth. Met. **34** (1989) 519.
89R Rogerie, J., Simon, Ch., Rosenman, I., Schweizer, J., Vangelisti, R., Pernot, P., Perignon, A.: Synth. Met. **34** (1989) 513.
89S1 Sugihara, K., Yeh, N.-C., Dresselhaus, M.S.: Phys. Rev. **B 39** (1989) 4577.
89S2 Suzuki, M., Sampere, M., Koga, K.: Phys. Rev. **B 39** (1989) 6979.
89S3 Speck, J.S., Dresselhaus, M.S.: Synth. Met. **34** (1989) 211.
89S4 Sugihara, K., Shiozaki, I., Sampere, S.M., Suzuki, M.: Synth. Met. **34** (1989) 543.
89W1 Wiesler, D.G., Zabel, H., Shapiro, S.M.: Physica **B 156** & **157** (1989) 292.
89W2 Wiesler, D.G., Zabel, H., Shapiro, S.M.: Synth. Met. **34** (1989) 505.
89Y1 Yeh, M., Suzuki, M., Burr, C.R.: Phys. Rev. **B 40** (1989) 1422.
89Y2 Yoshida, Y.: Synth. Met. **34** (1989) 387.
89Y3 Yeh, N.-C., Sugihara, K., Dresselhaus, M.S., Dresselhaus, G.: Phys. Rev. **B 40** (1989) 622.
90H Hagiwara, M., Kanaboshi, A., Flandrois, S., Biensan, P., Matsuura, M.: J. Magn. Magn. Mater. **90** & **91** (1990) 277.

90M1 Meyer, C., Yazami, R., Chouteau, G.: J. Phys. (Paris) **51** (1990) 1239.
90M2 Matsuura, M., Hagiwara, M.: J. Phys. Soc. Jpn. **59** (1990) 3819.
90M3 Matsuura, M., Zabel, H.: J. Magn. Magn. Mater. **90** & **91** (1990) 260.
90M4 Matsuura, M., Hagiwara, M.: Tanso (in Japanese) **145** (1990) 283.
90N1 Nicholls, J.T., Murayama, C., Takahashi, H., Môri, N., Tamegai, T., Iye, Y., Dresselhaus, G.: Phys. Rev. **B 41** (1990) 4953.
90N2 Nicholls, J.T., Dresselhaus, G.: Phys. Rev. **B 41** (1990) 9744.
90N3 Nicholls, J.T., McNiff, jr., E.J., Dresselhaus, G.: Phys. Rev. **B 42** (1990) 5555.
90O Olk, C.H., Heremans, J., Dresselhaus, M.S., Speck, J.S., Nicholls, J.T.: Phys. Rev. **B 42** (1990) 7524.
90R1 Rancourt, D.G., Lamarche, G., Tume, P., Lalonde, A.E., Biensan, P., Flandrois, S.: Can. J. Phys. **68** (1990) 1134.
90R2 Rancourt, D.G., Flandrois, S., Biensan, P., Lamarche, G.: Can. J. Phys. **68** (1990) 1435.
90S1 Suzuki, M., Santodonato, L.J., Yeh, M., Sampere, S.M., Smith, A.V., Burr, C.R.: J. Mater. Res. **5** (1990) 422.
90S2 Suzuki, M., Li-Fen, T., Suzuki, I. Burr, C.R.: J. Appl. Phys. **67** (1990) 5749.
90Y Yeh, M., Suzuki, I., Suzuki, M., Burr, C.R.: J. Phys.: Condens. Matter **2** (1990) 9821.
91C Chehab, S., Biensan, P., Amiell, J., Flandrois, S.: J. Phys. I **1** (1991) 537.
91H Hafidi, M.E., Chouteau, G., Yazami, R.: Synth. Met. **40** (1991) 325.
91M Matsuura, M., Hagiwara, M., Zabel, H.: Memo. Fac. Engin. Des. Kyoto Inst. Technol. **39** (1991) 31.
91S1 Suzuki, M., Santodonato, L.J., Suzuki, I., White, jr. B.E., Cotts, E.J.: Phys. Rev. **B 43** (1991) 5805.
91S2 Suzuki, I., Suzuki, M., Tien, L.-F., Burr, C.R.: Phys. Rev. **B 43** (1991) 6393.

9.5 MM'F$_5$ and MM'F$_5 \cdot$nH$_2$O compounds (M, M' = 3d element or Al; n = 2 or 7)

9.5.1 List of symbols and abbreviations

Symbols

a, b, c	Å	lattice parameters
C_m	$cm^3\,K\,mol^{-1}$	molar Curie constant
H	kOe	magnetic field
H_c	kOe	coercive force
H_{hyp}	kOe	effective magnetic hyperfine field
$p, p_{x,y,z}$	μ_B	magnetic moment and its orthogonal components
p_{eff}	μ_B	effective magnetic moment
p_{FU}	μ_B	magnetic moment per formula unit
p_M	μ_B	magnetic moment per ion M
p_s	μ_B	spontaneous magnetic moment per formula unit
T	K	temperature
T_N	K	magnetic ordering temperature, Néel temperature
T_t	K	spin rotation temperature
v	$mm\,s^{-1}$	velocity of absorber in Mössbauer effect
δ	$mm\,s^{-1}$	isomer shift
Δ	$mm\,s^{-1}$	quadrupole splitting
ε	$mm\,s^{-1}$	quadrupole shift
θ	deg	angle
Θ_p	K	paramagnetic Curie temperature
μ_B	$J\,T^{-1}$	Bohr magneton
σ	$emu\,g^{-1} = G\,cm^3\,g^{-1}$	specific mass magnetization
χ_g	$cm^3\,g^{-1}$	magnetic mass susceptibility
χ_m	$cm^3\,mol^{-1}$	molar magnetic susceptibility

Abbreviations

FU	formula unit
M	metal
NMR	nuclear magnetic resonance
RT	room temperature

9.5.2 MM'F$_5$ compounds

An excellent survey of the properties of the compounds $M^{2+}M^{3+}F_5$ is given by Tressaud and Dance in 1977 [77T1, see also 85B1 and 85D1], since that time no further data have been given in literature. The compounds crystallize in three related structures. They have in common parallel strings of trans-corner-shared and more or less regular octahedra containing the M^{3+} ion. The larger M^{2+} ions are placed in between four parallel running strings and keep them together. They are thus the centre of deformed octahedra that share an edge with these strings. For crystallographic data see Table 1 and Figs. 1,2. The monoclinic Cr^{2+} containing structures, Table 1, may be derived from the orthorhombic structure by a cooperative Jahn–Teller distortion of the edge-shared Cr^{2+} octahedra. For magnetic configurations see Table 2 and for magnetic properties Table 3 and Figs. 3 ··· 9.

Table 1. MM'F$_5$ compounds. X-ray crystallographic data at RT. P and X powder sample or single crystal, respectively. z number of formula units per unit cell. Lattice parameters a, b and c. Angle β between the a and c axis. M^{2+}–F–M^{3+} bond angle between the M^{2+} and M^{3+} cation via the intermediate anion [85B1].

Compound	P or X	z	Symmetry Space group	a [Å]	b [Å]	c [Å]	β [deg]	M^{2+}–F–M^{3+} angle [deg]	Ref.
V_2F_5 [a])	P	2	tetragonal	4.77	—	3.20	—	—	68S1
Fe_2F_5 [a, b])	P	6	tetragonal	8.05	—	9.56	—	—	58B1
Cr_2F_5	X	4	monoclinic	7.773	7.540	7.440	124.25	158	64S1 [c])
			C 2/c	7.770(5)	7.540(5)	7.440(5)	124.25(1)		73T1
$CrAlF_5$	X	4	monoclinic C 2(c)	7.58(1)	7.46(1)	7.25(1)	123.7(2)	—	73T1
$CrTiF_5$	X	4	monoclinic C 2/c	7.98(1)	7.65(1)	7.70(1)	125.2(2)	—	73T1
$CrVF_5$	X	4	monoclinic C 2/c	7.91	7.60(1)	7.63(1)	125.0(2)	—	73T1
$MnCrF_5$	X	4	monoclinic C 2/c	8.586(5)	6.291(3)	7.381(4)	115.46(7)	150	77F1 [c])
Mn_2F_5	P	8	orthorhombic	15.44(2)	7.27(1)	6.17(1)	—	—	74T1
$MnAlF_5$	X	4	orthorhombic, Ama 2	9.54	9.85	3.58	—	—	69C1
									70R1
$MnAl_{1-x}Fe_xF_5$	P		orthorhombic, Ama 2	Fig. 2	Fig. 2	Fig. 2			73T2

a) Rutile related lattice constants.

b) This compound could not be obtained from a reaction between FeF_2 and FeF_3. This applies also for the case of the cobalt compounds [74T1].

c) In these references more detailed positional parameters for the ions are given.

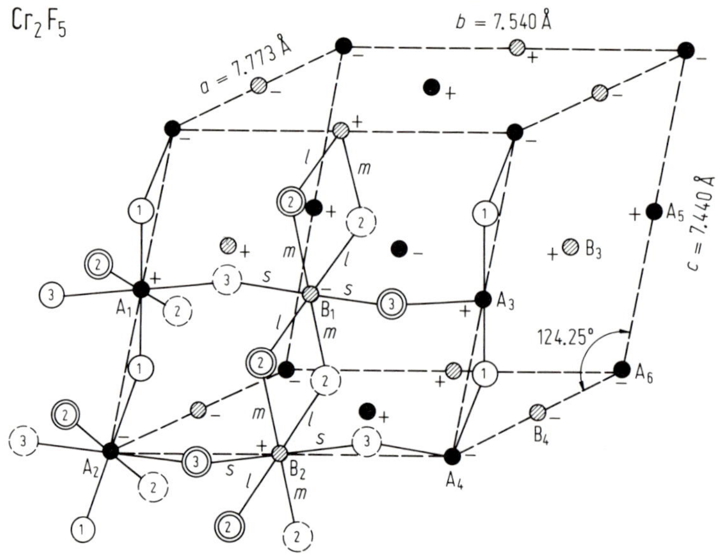

Fig. 1. Cr_2F_5. Monoclinic unit cell. Linked fluorine octahedra around two Cr^{3+} (A_1, A_2 full circles) and two Cr^{2+} (B_1, B_2 hatched circles) cations in the (100) face. The three categories of anions are F(1) in the (100) face (open circles), F(2) close to the (010) planes of cations and F(3) close to the (100) and (001) planes of cations. Of the last two categories of anions, those in front of the (100) face are shown as double circles and those behind this face as broken circles. The Cr^{3+} octahedra are nearly regular, Cr–F bond length 1.89 Å. The Cr^{2+} octahedra are Jahn–Teller distorted, $l = 2.57$ Å, $m = 2.01$ Å and $s = 1.95$ Å. The magnetic spin arrangement is indicated by plus and minus signs. Crystallographic data [64S1]. Magnetic data based on the Goodenough–Kanamori rules [66O1].

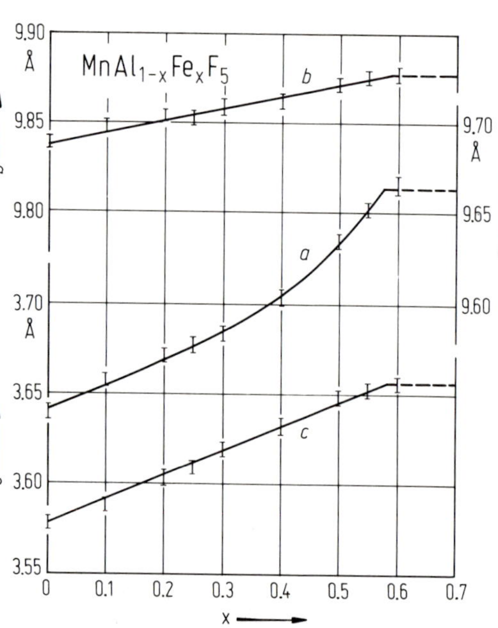

Fig. 2. $MnAl_{1-x}Fe_xF_5$. Lattice parameters a, b and c vs. composition parameter x [73T2].

Table 2. $MM'F_5$. Magnetic ordering types and configurations. Ordering temperatures derived from magnetic measurements and from NMR experiments. The extrapolated spontaneous magnetization p_s per molecule at 0 K.

Compound	Ordering type	Magnetic configuration	T_N magn. meas. [K]	T_N NMR [K]	p_s [μ_B]	Ref.
Cr_2F_5	antiferromagnetic	Ferromagnetic in (011) plane. Opposed magnetizations in alternate planes, Fig. 1, with weak antiferromagnetic canting from the *b* axis	40(3)			66O1 73T1
$CrVF_5$	ferrimagnetic	Strong ferromagnetic canting, Fig. 3	40(3)	0.86(10)		73T1
$CrTiF_5$	id.	id.	26(3)		1.78(10)	73T1
$CrAlF_5$	paramagnetic					73T1
Mn_2F_5	antiferromagnetic		54(3)			74T1
$MnAlF_5$	id.	$p_{Mn^{2+}} = 4.05\ \mu_B$ at 1.2 K, oriented close to the *b* axis. Magnetic lattice is double the crystal lattice in *b* and *c* direction.		2.35(10)		75W1
$MnAl_{1-x}Fe_xF_5$	ferrimagnetic		Fig. 4			77T1
$MnCrF_5$	antiferromagnetic	magnetic cell in $(a, 2b, c)$ with two orthogonal antiferromagnetic sublattices. $p_{Mn^{2+}} = 3.9\ \mu_B$ parallel to [010] $p_{Cr^{3+}} = 2.5\ \mu_B$ parallel to [001]. See also Fig. 5		6.0(5)		78F1

Table 3. $MM'F_5$. Paramagnetic properties. Θ_p paramagnetic Curie temperature, C_m molar Curie constant.

Compound	Θ_p [K]	C_m [$cm^3 K mol^{-1}$]	Fig.	Ref.
V_2F_5	− 78	—	—	68S1
Cr_2F_5	− 95(5)	4.98(3)	6	73T1
$CrVF_5$	− 51(5)	4.02(3)	6	73T1
$CrTiF_5$	− 78(5)	3.59(3)	6	73T1
$CrAlF_5$	− 3(2)	2.57(3)	6	73T1
Mn_2F_5	− 45(5)	7.29(5)	7	74T1
$MnAlF_5$	− 6	4.43	8	73T2
$MnAl_{0.9}Fe_{0.1}F_5$	− 20	4.55	—	73T2
$MnAl_{0.75}Fe_{0.25}F_5$	− 62	5.25	—	73T2
$MnAl_{0.6}Fe_{0.4}F_5$	− 122	5.55	—	73T2
$MnAl_{0.45}Fe_{0.55}F_5$	− 136	6.25	—	73T2
$MnCrF_5$	− 16	6.16	9	78F1

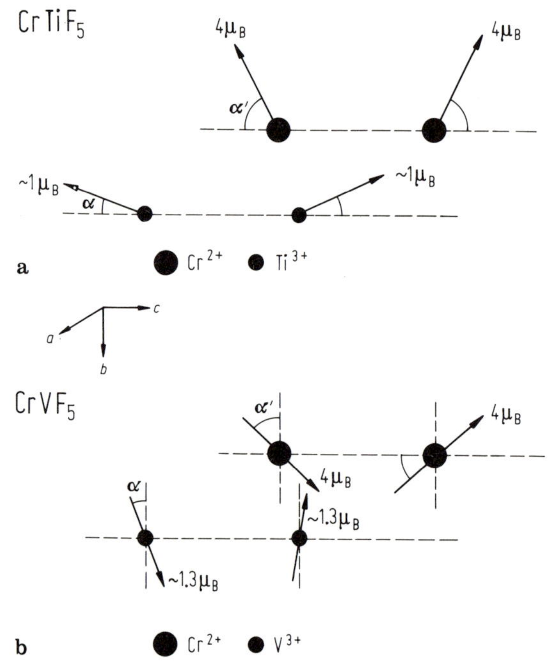

Fig. 3. (a) $CrTiF_5$ and (b) $CrVF_5$. Predicted ferrimagnetic spin configurations as derived from the structure and the rules of superexchange. The angles α and α' indicate expected magnetic moment directions with respect to main crystallographic axes [73T1].

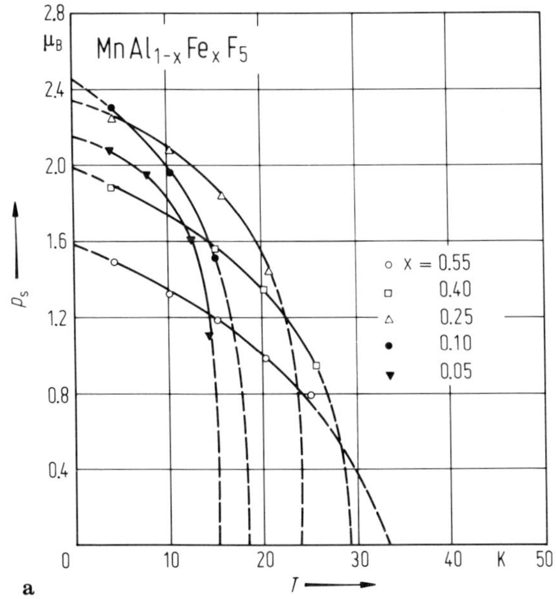

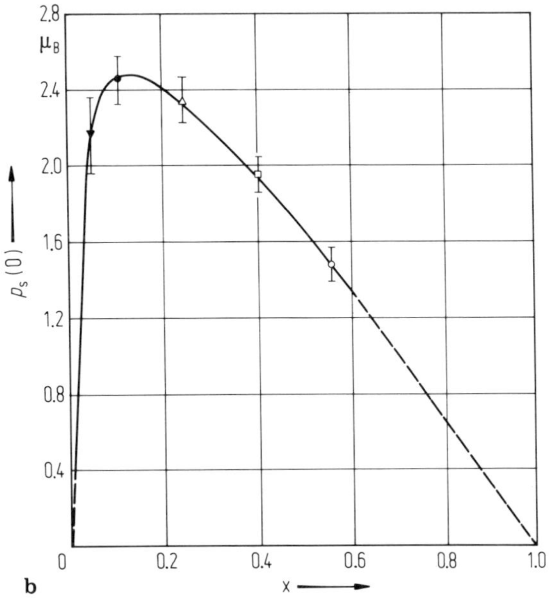

Fig. 4. $MnAl_{1-x}Fe_xF_5$. (a) Spontaneous magnetic moment per FU, p_s, vs. temperature T. (b) Extrapolated zero temperature spontaneous magnetic moment vs. composition parameter x [73T2, 77T1].

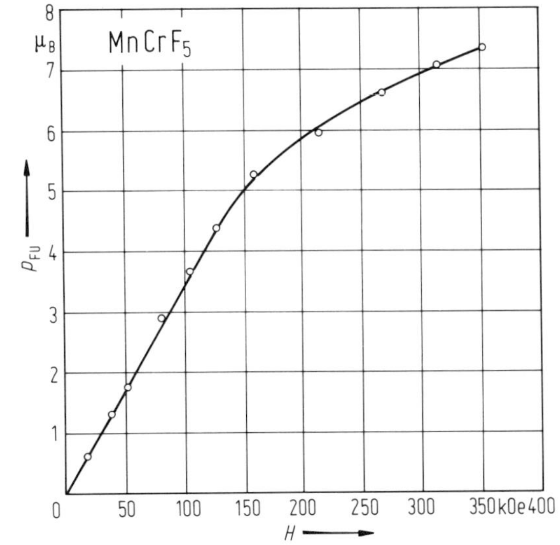

Fig. 5. $MnCrF_5$. Magnetic moment per formula unit p_{FU} vs. applied magnetic field H at 4.2 K [78F1].

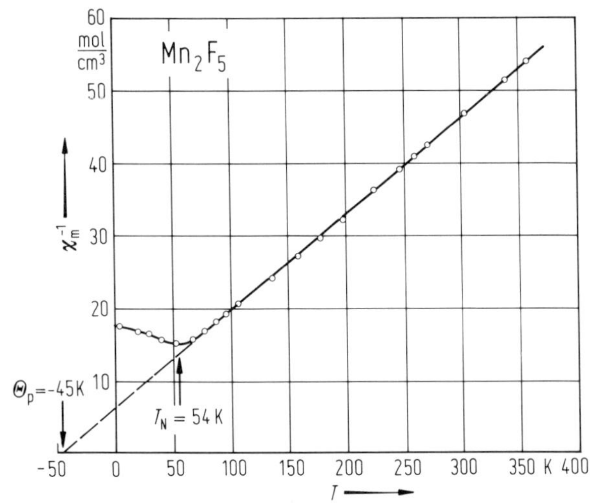

Fig. 7. Mn_2F_5. Inverse molar magnetic susceptibility χ_m^{-1} vs. temperature T [74T1].

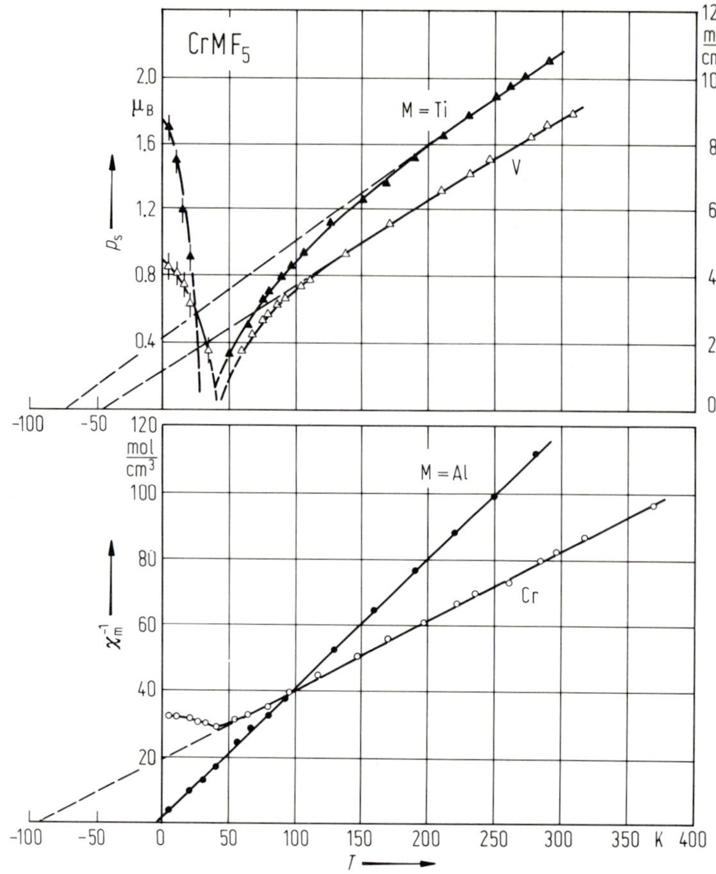

Fig. 6. $CrMF_5$, M = Ti, V, Cr, Al. Temperature dependence of reciprocal molar magnetic susceptibilities χ_m^{-1} and spontaneous magnetic moment p_s per FU [73T1].

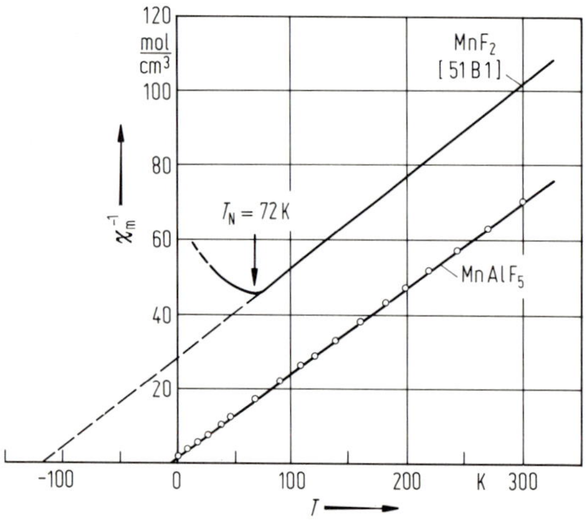

Fig. 8. $MnAlF_5$, MnF_2. Inverse molar magnetic susceptibility χ_m^{-1} vs. temperature T [73T2].

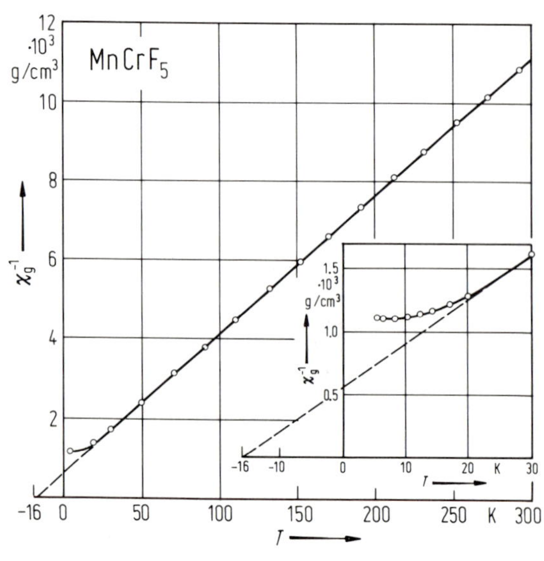

Fig. 9. $MnCrF_5$. Inverse magnetic mass susceptibility χ_g^{-1} vs. temperature T [78F1]. Inset shows χ_g^{-1} vs. T in the low-temperature range on an enlarged scale.

9.5.3 $MM'F_5 \cdot 2H_2O$ compounds

9.5.3.1 Crystal structure

In recent years the dihydrate compounds $MFeF_5 \cdot 2H_2O$, where M = Mn, Fe, Co, Ni, Zn or Al have been prepared and their magnetic properties investigated. The first chemically well-defined member of this group was the mixed valency ion compound $Fe^{2+}Fe^{3+}F_5 \cdot 2H_2O$ [74S1, 75C1]. The first crystal structure analysis is given by Hall et al. [77H1]. The compounds crystallize in the orthorhombic system, the unit cell contains four formula weights. The structure has been described as an inverse weberite. This should mean that the M^{2+} and the M^{3+} positions are interchanged as compared to the situation that exists in the weberite structure type of the compounds $Na_2M^{2+}M^{3+}F_7$, which are isostructural with the mineral weberite Na_2MgAlF_7. Moreover the H_2O molecules replace two trans F^- ions of the divalent metal coordination octahedra. The alkali cation sites □ remain empty, leading to the more informative notation of $\square_2M^{2+}M^{3+}F_5(H_2O)_2$ [86L1]. A perspective view of the structure is shown in Fig. 10. Two trans chains of corner sharing Fe^{3+} octahedra are oriented parallel to the [100] axis. They are interconnected by isolated trans-hydrated $M^{2+}F_4(H_2O)_2$ octahedra, which leads to hexagonal tungsten bronze type layers in the (011) plane, Fig. 11. In these planes the network of cations consists of triangles which lead to 1-D magnetism if M^{2+} is a diamagnetic ion and to frustrated 3-D magnetism for paramagnetic M^{2+} ions [86L3]. For unit cell dimensions see Table 4. A detailed survey of interatomic distances and bond angles is given in [87L2], [86L1] and [86L3] for M = Mn^{2+}, Fe^{2+} and Zn^{2+}, respectively.

Table 4. Unit cell dimensions *a*, *b*, *c* of the orthorhombic $MFeF_5 \cdot 2H_2O$ compounds.

Compound	*T* [K]	*a* [Å]	*b* [Å]	*c* [Å]	Space group	Measuring method	Ref.
$MnFeF_5 \cdot 2H_2O$	300	7.5635(2)	10.901(1)	6.7319(2)	Imma	MoK_α [a])	87L1
	50	7.475(1)	10.766(1)	6.594(1)	Imm2	neutron [b])	87L2
$Fe_2F_5 \cdot 2H_2O$	RT	7.489(7)	10.897(8)	6.671(6)	Imma	MoK_α [a])	77H1
	RT	7.477(1)	10.862(2)	6.652(1)	Imma	MoK_α [a, c])	86L2
	60	7.472(1)	10.928(2)	6.606(2)	Imma	neutron [b])	86L1
	30	7.469(1)	10.927(1)	6.602(1)	Imma	neutron [b])	86L1
	4.2	7.470(1)	10.930(1)	6.603(1)	I 2/b [d])	neutron [b])	86L1
$ZnFeF_5 \cdot 2H_2O$	300	7.475(1)	10.766(1)	6.594	Imma	CuK_α [a])	86L3
	1.5	7.451(1)	10.747	6.524	Imm2	neutron [b])	87L2

[a]) X-ray single crystal diffraction.
[b]) Neutron powder diffraction.
[c]) New refinement with the program PROMETHEUS [83Z1].
[d]) Below the magnetic transition temperature T_t = 26.5(5) K the structure becomes monoclinic [86L1].

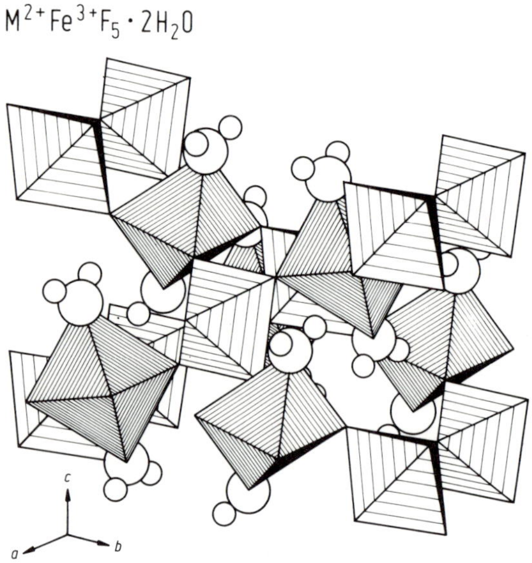

Fig. 10. $M^{2+}Fe^{3+}F_5 \cdot 2H_2O$. Perspective view of the orthorhombic crystal structure, build up out of two types of octahedra. The FeF_6 octahedra are thinly hatched, the $M^{2+}F_4 \cdot 2H_2O$ octahedra are strongly hatched. [87L1].

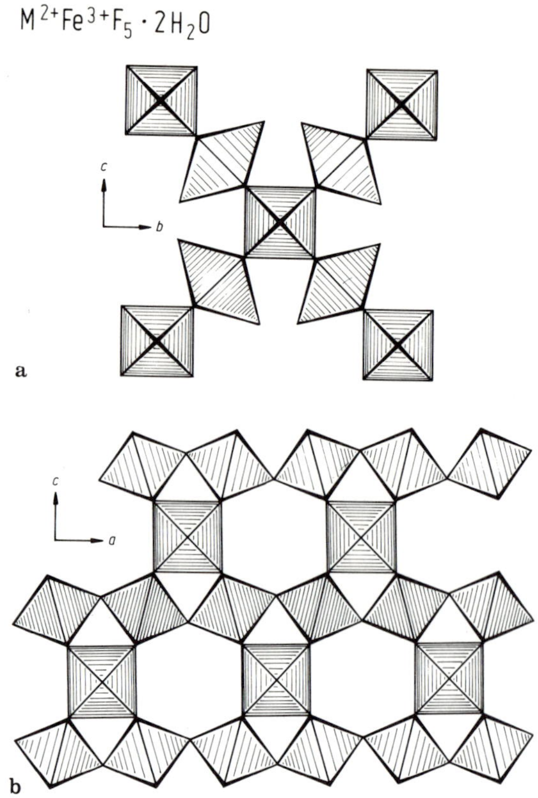

Fig. 11. $M^{2+}Fe^{3+}F_5 \cdot 2H_2O$. Diagrammatic representation of the structure viewed parallel to the *a* axis (a) and to the *b* axis (b). (a) squares represent columns of trans-vertex-sharing FeF_6 octahedra, the rhombi the cross-linking $MF_4 \cdot 2H_2O$ polyhedra. (b) Squares represent the cross-linking $MF_4 \cdot 2H_2O$ polyhedra, the rhombi the trans-vertex-sharing FeF_6 octahedra [77H1].

9.5.3.2 Magnetic structures

For the various types of magnetic structures and magnetic ordering temperatures in these compounds see Table 5 and Figs. 12 ··· 14. The situation for $Fe_2F_5 \cdot 2H_2O$ is complicated in so far that below its Néel point (48.6 K) an additional magnetic transition occurs at $T_t = 26.5(5)$K, corresponding to a rotation of the ferric spins, the ferrous spins remaining fixed along the local anisotropy axis, see Fig. 13a, b.

Table 5. $MFeF_5 \cdot 2H_2O$ compounds. Magnetic structures derived from powder neutron diffraction.

Compound	T [K]	Type of magnetism	Magnetic configuration	Spin orientation[b]	Magnitude of magnetic moments	Ref.
$MnFeF_5 \cdot 2H_2O$	1.5	3D frustrated ferrimagnetism, triangle configuration	Fig. 12	$\theta_{Mn,Fe} = 140.2°$; 102.8° $\theta_{Fe,Fe} = 115.7°$	Table 6	87L2
$Fe_2F_5 \cdot 2H_2O$[a])	30	3D frustrated ferrimagnetism, triangle configuration	Fig. 13b	Table 8 $p_{Fe^{2+}} \parallel (100)$ $p_{Fe^{3+}} \parallel [010]$	Table 7	86L1
	4.2	3D frustrated ferrimagnetism, triangle configuration	Fig. 13a	Table 8	Table 7	86L1
$ZnFeF_5 \cdot 2H_2O$	1.5	1D antiferro-magnetism	Fig. 14	[c])	$p_{Fe^{3+}} = 3.78(5)\,\mu_B$	87L2

[a]) Above and below the magnetic transition temperature $T_t = 26.5(5)$ K the magnetic and the nuclear cells are identical. This transition is a reorientation of the ferric moments, the ferrous moments remaining parallel to the local anisotropy axis, which is the principal axis of the electric field gradient at the Fe^{2+} site. Below T_t the Fe^{3+}–Fe^{3+} antiferromagnetic interaction becomes relatively more important and dominates at 4.2 K, where the magnetic structure becomes similar to that of $MnFeF_5 \cdot 2H_2O$ [76I1, 86L1].

[b]) θ denotes the angle between the magnetic moments of the two ions indicated.

[c]) Pure antiferromagnetism in the Fe^{3+} chains. The magnetic ferric moments are oriented perpendicular to these chains [87L2].

Table 6. $MnFeF_5 \cdot 2H_2O$. Refined values of the components of the ionic magnetic moment p_M in μ_B at $T = 1.5$ K, see also Fig. 12. The magnetic moment per FU derived from these results is $p_{FU} = 2.89\,\mu_B$ [87L2].

Ion	p_x	p_y	p_z	p_M
Fe^{3+}	1.83	—	± 2.90	3.43(9)
Mn^{2+}	− 4.59	0.86	± 1.60	4.93(11)

Table 7. $Fe_2F_5 \cdot 2H_2O$. Refined values of the components of the ionic magnetic moment p_M in μ_B at two temperatures T, see also Figs. 13a, b [86L1].

T [K]	Ion	p_x	p_y	p_z	p_M
30	Fe^{3+}	—	2.80(10)	—	2.80(10)
30	Fe^{2+}	—	− 3.52(8)	± 1.60(8)	3.86(16)
4.2	Fe^{3+}	− 0.7(2)	3.12(10)	± 3.73(8)	4.9(3)
4.2	Fe^{2+}	1.6(2)	− 3.4(1)	± 1.0(2)	3.80(3)

Table 8. $Fe_2F_5 \cdot 2H_2O$. Canting angles θ between the spins of the magnetic ions [86L1].

	T = 30 K	4.2 K
$\theta_{Fe^{3+},Fe^{3+}}$	0°	101.1°
$\theta_{Fe^{2+},Fe^{2+}}$	48.9°	46.6°
$\theta_{Fe^{3+},Fe^{2+}}$	155.6°	114.7°
		144.0°

Table 9. $MFeF_5 \cdot 2H_2O$. Magnetic ordering temperatures T_N.

Compound	T_N [K]	Measuring method	Ref.
$MnFeF_5 \cdot 2H_2O$	38.00(25)	magnetization	85J1
	39(1)	magnetization	86L3
	39.5(1)	Mössbauer	86L3
$Fe_2F_5 \cdot 2H_2O$	48.6	Mössbauer	76I1
	48.0	magnetization	77W1
	48.5(1)	magnetization	86L1
	48.50(25)	magnetization	78J1
$CoFeF_5 \cdot 2H_2O$	34.5(10)	Mössbauer	88G1
	27.00(25)	magnetization	79J1
$NiFeF_5 \cdot 2H_2O$	31(1)	Mössbauer	88G1
$ZnFeF_5 \cdot 2H_2O$	9(2)	Mössbauer	86L3

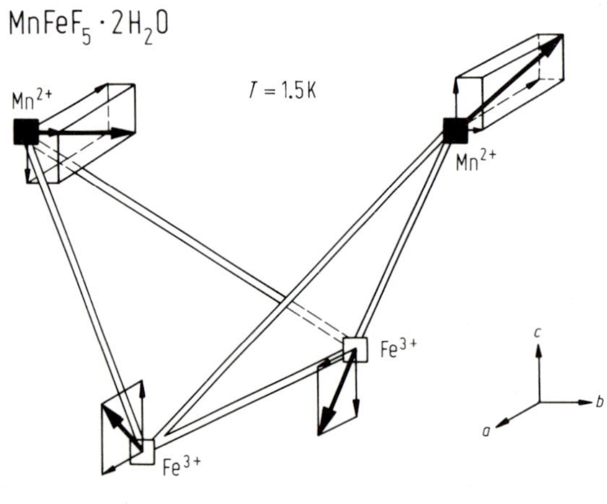

Fig. 12. $MnFeF_5 \cdot 2H_2O$, T = 1.5 K. The 3D frustrated ferrimagnetic spin configuration in the double cation triangle formed by the centres of two corner sharing Fe^{3+} octahedra and their adjacent M^{2+} octahedra shown in Fig. 10 [87L2].

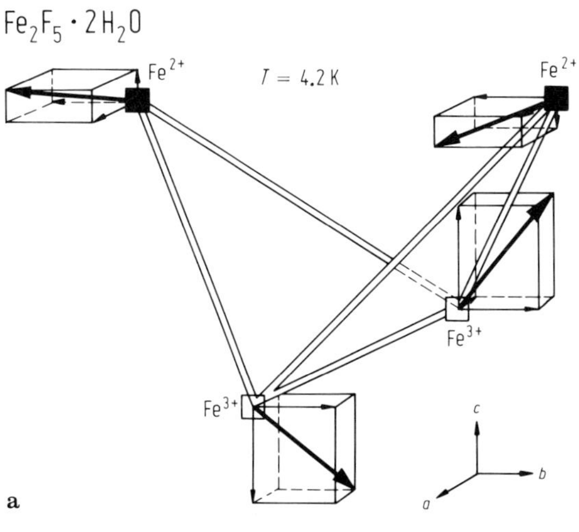

a

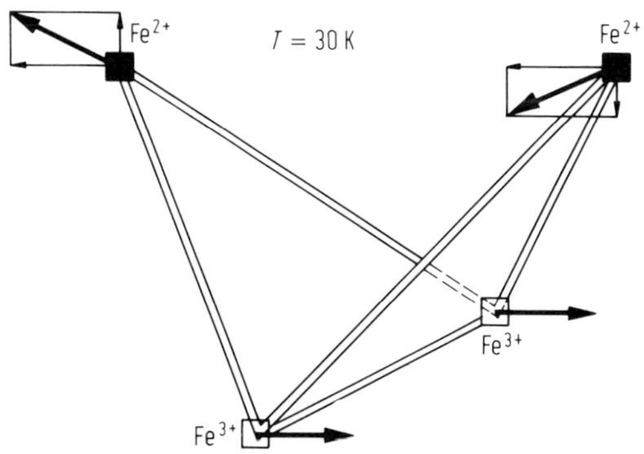

b

Fig. 13. $Fe_2F_5 \cdot 2H_2O$. The 3D frustrated ferrimagnetic spin configuration similar to Fig. 12 at two temperatures [87L2]. (a) $T = 4.2$ K, (b) $T = 30$ K.

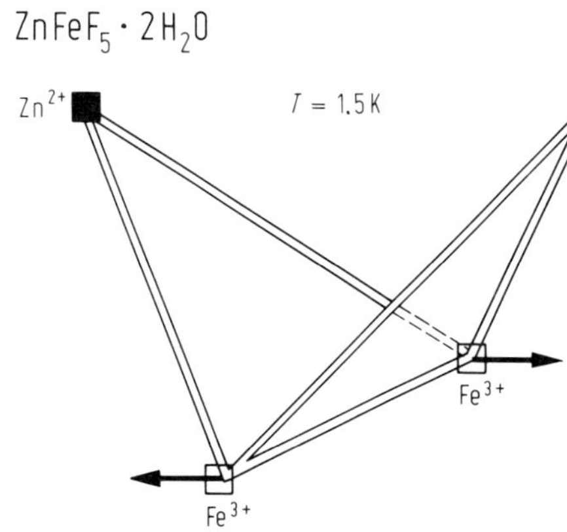

Fig. 14. $ZnFeF_5 \cdot 2H_2O$. The 1D antiferromagnetic spin configuration at $T = 1.5$ K [87L2].

9.5.3.3 Mössbauer data

An ^{57}Fe Mössbauer spectrum of $Fe_2F_5 \cdot 2H_2O$ is reproduced in Fig. 15. For various data see Table 10 and Figs. 15 ⋯ 20.

Table 10. $MFeF_5 \cdot 2H_2O$. ^{57}Fe Mössbauer data. Isomer shift δ; quadrupole splitting Δ and quadrupole shift 2ε of the first order Fe^{3+} Zeeman lines; quadrupole splitting Δ in the case of Fe^{2+} ions; magnetic effective hyperfine field H_{hyp}; magnetic ordering temperature T_N. For Mössbauer spectra in the presence of an applied magnetic field see [88G1, 90G1].

Compound	T	Fe^{3+} in FeF_6 octahedra			Fe^{2+} in FeF_4–$2H_2O$ octahedra			T_N	Ref.
	[K]	δ a) [mm s^{-1}]	Δ or 2ε [mm s^{-1}]	H_{hyp} [kOe]	δ a) [mm s^{-1}]	Δ [mm s^{-1}]	H_{hyp} [kOe]	[K]	
$MnFeF_5 \cdot 2H_2O$	293	0.46	0.64						87K1
	4 ⋯ RT	0.43 ⋯ 0.46	Fig. 16	Fig. 16				38.5(5)	87K1
	295	0.483	0.568						85J1
	40	0.556(4)	0.636(4)	—					
	4.2	0.56	0.26(2)	534(2) d)				39.5(1)	86L3
$Fe_2F_5 \cdot 2H_2O$	295	b) 0.25(2)	0.59(2)		b) 1.11(2)	2.50		48.6(5) e)	76I1
	50	0.35(2)	0.58(2)		1.25(2)	3.30			
	28	0.35	− 0.32	301	2.52	3.27	40.5		
	4.2	0.34	− 0.21	538(3)	2.52	3.29	40.7		
			Fig. 17	Fig. 17		Fig. 18	Fig. 18		
	295	c) 0.70	0.65		c) 1.60	2.44			77W1
	55	0.81	0.58		1.71	3.28			
	20			460					
	4.2			532					
	293	0.47	0.56(2)		1.32	2.57(2)			75S1, 2
	80	0.56	0.59(2)		1.45	3.31(2)			
$CoFeF_5 \cdot 2H_2O$	300	0.45(1)	0.63(2)						88G1
	77	0.58(1)	0.54(2)					34.5(10)	
	4.2	0.37(1)	0.15(2)	536(2)					
			Fig. 19	Fig. 20					
	297	0.430(10)	0.55(10)						78B1
	76	0.530(10)	0.553(10)						
	293	0.48(2)	0.55(2)						75S1
	80	0.57(2)	0.52(2)						
$NiFeF_5 \cdot 2H_2O$	300	0.45(1)	0.57(2)						88G1
	77	0.58(1)	0.53(1)					31(1)	
	4.2	0.55(1)	0.05(2)	508(3)					
			Fig. 19	Fig. 20					

$ZnFeF_5 \cdot 2H_2O$	293	0.46(2)	0.59(2)					75S1, 2
	80	0.55(2)	0.55(2)					
	297	0.432	0.587					78B1
	76	0.529	0.543					
	300	0.442(4)	0.599(4)				9(2)	86L3
	77	0.532(4)	0.553(4)					
	1.5	0.51(2)	−0.08(4)	463(2)				
$AlFeF_5 \cdot 2H_2O$[g])	297				1.31(3)	[f]) 2.62(3)		78B1
						1.89(3)		
	76				1.43(3)	3.16(3)		
						2.60(3)		
	293				1.35(2)	2.40(2)		75S1
	80				1.47(2)	3.01(2)		

[a]) Relative to metallic iron at RT, unless otherwise stated.
[b]) Relative to the source $^{57}Co/Cu$ [76I1].
[c]) Relative to a sodium nitroprusside standard [77W1].
[d]) Angle between H_{hyp} and the principal axis of the electric field gradient is 39° [76I1].
[e]) See footnote [a]) of Table 5.
[f]) The assumption of two different Fe^{2+} sites was necessary for an acceptable analysis of the spectrum [78B1].
[g]) See also Fig. 35.

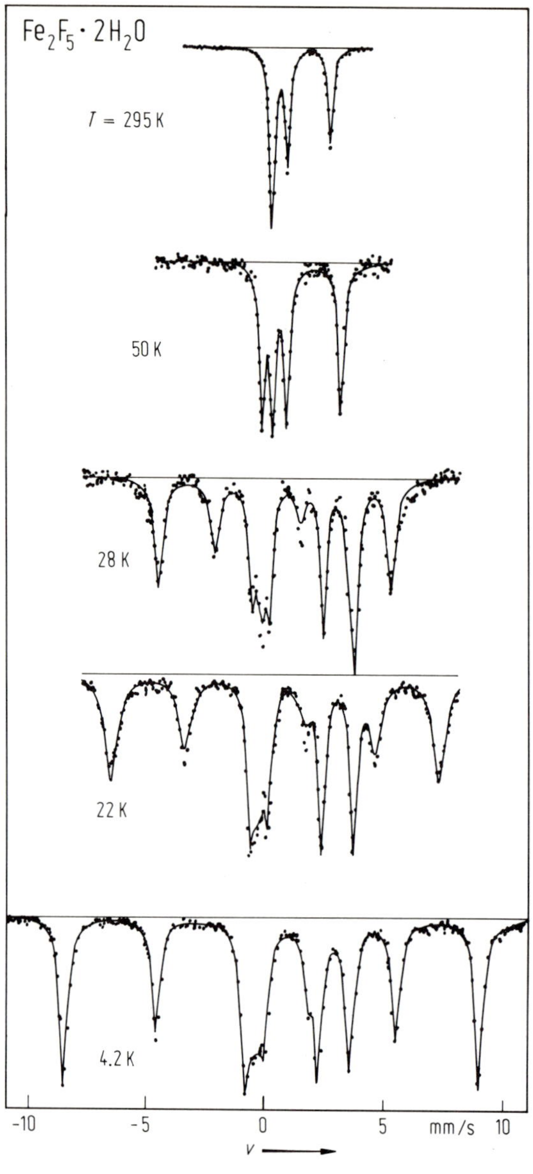

Fig. 15. $Fe_2F_5 \cdot 2H_2O$. ^{57}Fe Mössbauer spectra of a powder at various temperatures T [76I1].

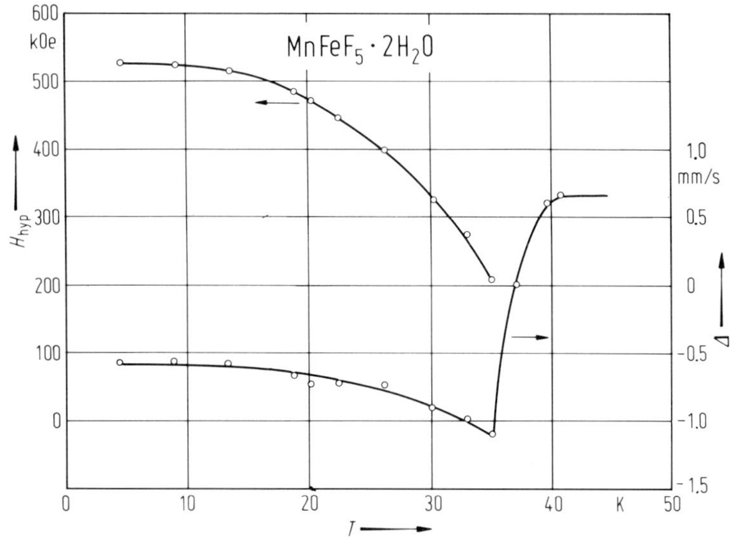

Fig. 16. MnFeF$_5$·2H$_2$O. Effective magnetic hyperfine field H_{hyp} and quadrupole splitting constant Δ as derived from ^{57}Fe Mössbauer experiments at various temperatures T [87K1].

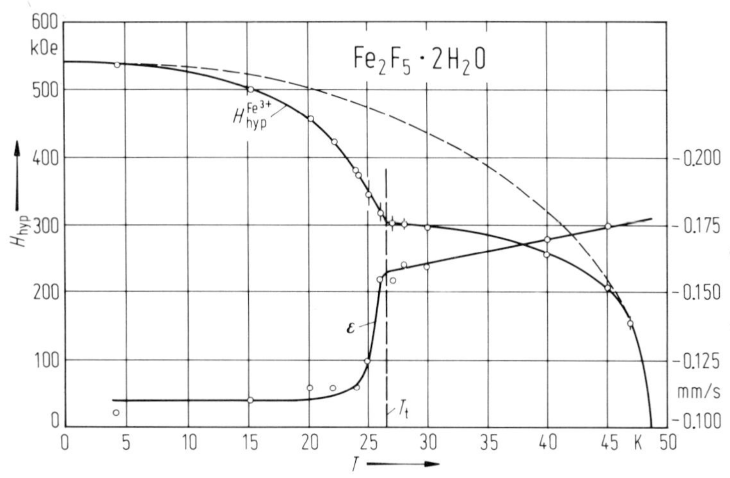

Fig. 17. Fe$_2$F$_5$·2H$_2$O. Temperature variation of the effective magnetic hyperfine field H_{hyp} of the Fe^{3+} ion and the quadrupole shift of the first order Fe^{3+} Zeeman line, as obtained from the Mössbauer spectra of Fig. 15. Dashed curve: Brillouin law for $S = 5/2$ [76I1].

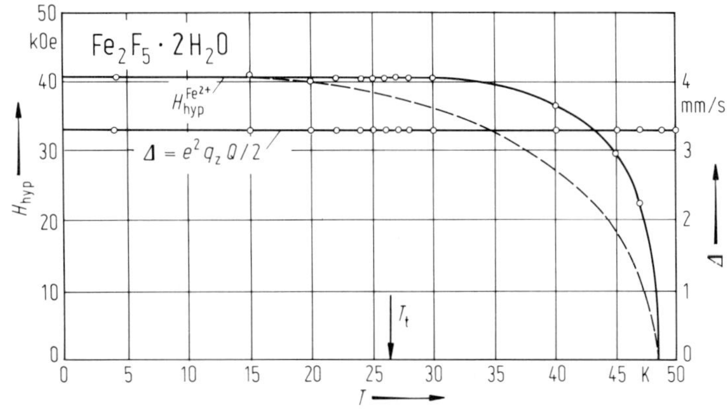

Fig. 18. Fe$_2$F$_5$·2H$_2$O. Similar to Fig. 17, but now for the Fe^{2+} ions. The quantity $\Delta = e^2 q_z Q/2$ is the main component of the quadrupole interaction. Dashed curve: Brillouin law for $S = 1/2$ [76I1].

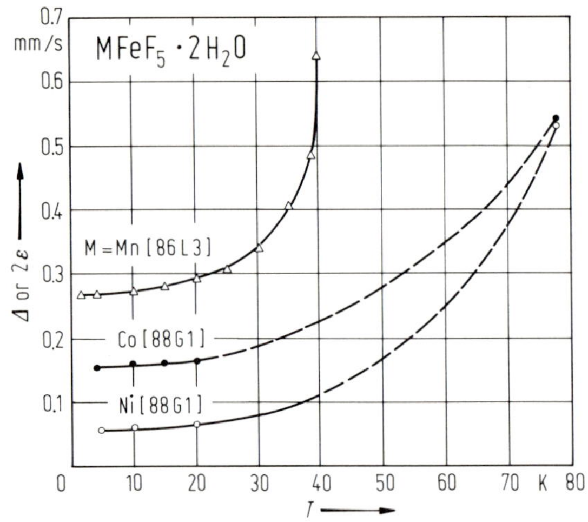

Fig. 19. MFeF$_5$·2H$_2$O, M = Mn, Co or Ni. Quadrupole splitting Δ or quadrupole shift 2ε of the first order Fe^{3+} Zeeman line found in ^{57}Fe Mössbauer spectroscopy as a function of temperature T [86L3, 88G1].

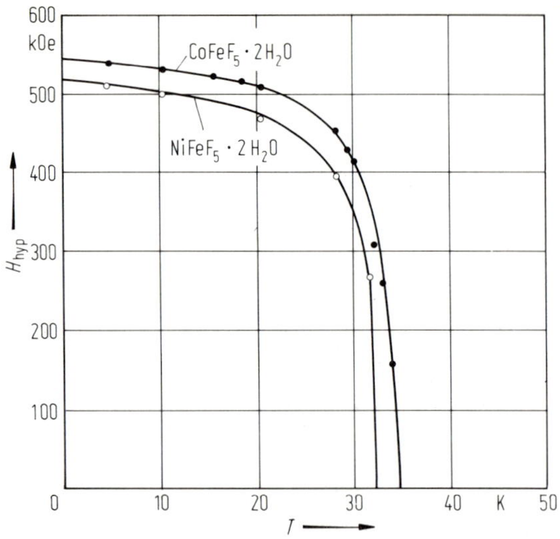

Fig. 20. CoFeF$_5$·2H$_2$O, NiFeF$_5$·2H$_2$O. Effective magnetic hyperfine field H_{hyp} at ferric ions vs. temperature T [88G1].

9.5.3.4 Magnetization curves

Table 11. MFeF$_5$·2H$_2$O. Spontaneous magnetic moment per formula unit, p_s, as derived from magnetization measurements at low temperature.

Compound	p_s [μ_B]	Ref.
MnFeF$_5$·2H$_2$O	2.50(3)	86L3
Fe$_2$F$_5$·2H$_2$O	1.2	77W1
	0.82	78J1
	0.85(2)	86L1

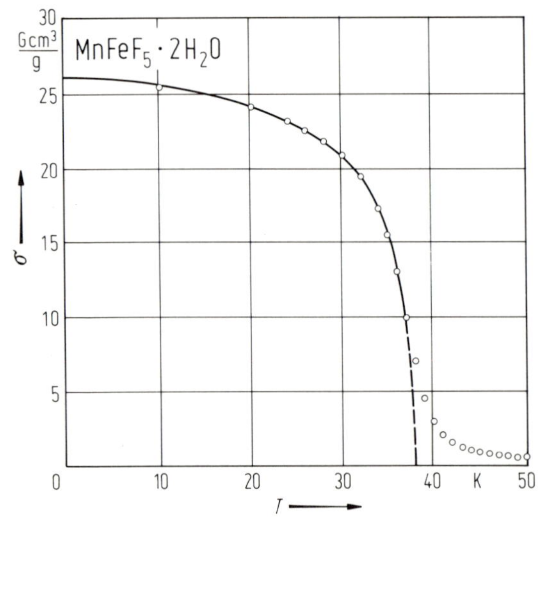

Fig. 21. MnFeF$_5$·2H$_2$O, polycrystalline powder. Mass magnetization σ in an external magnetic field of 1 kOe vs. temperature T. Extrapolated ordering temperature T_N = 38.00(25) K [85J1].

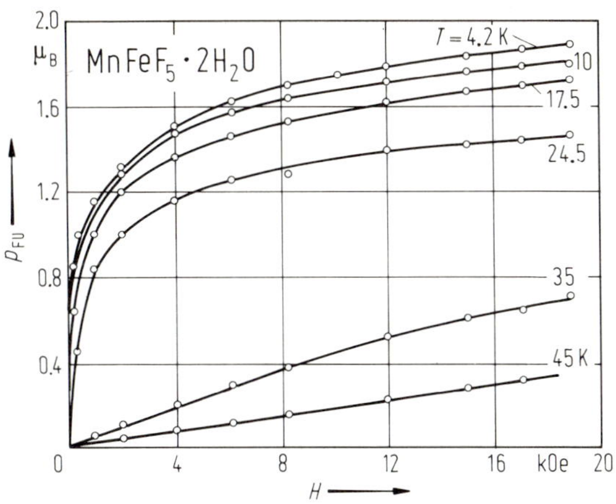

Fig. 22. $MnFeF_5 \cdot 2H_2O$, polycrystalline powder. Magnetic moment per FU at various temperatures T [86L3].

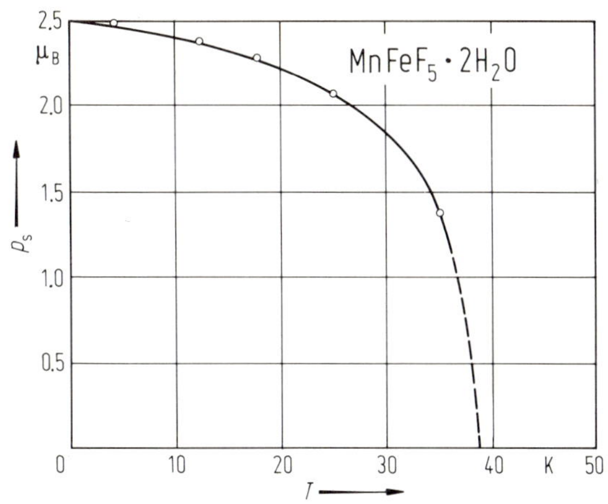

Fig. 23. $MnFeF_5 \cdot 2H_2O$. Spontaneous magnetic moment per FU as obtained after extrapolation of the curves of Fig. 22 according to a $H^{-3/2}$ law. The spontaneous magnetic moment per FU at $T = 0$ corresponds to 2.5 μ_B [86L3].

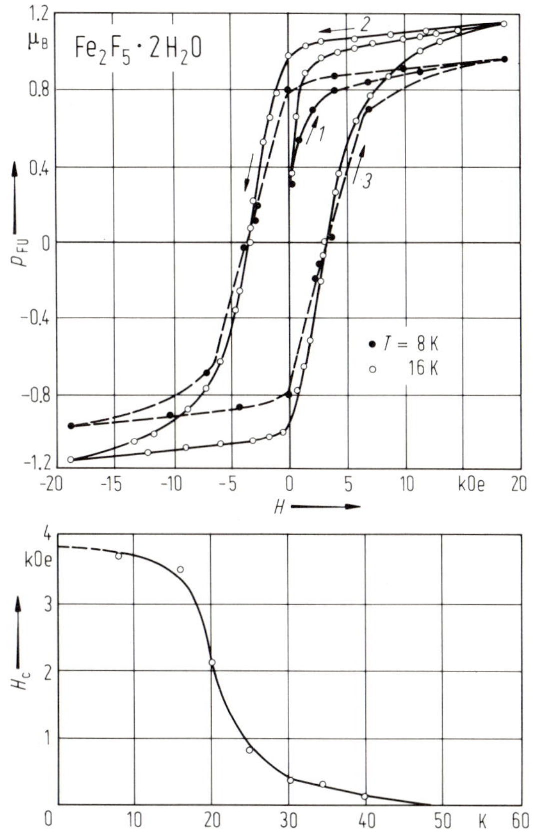

Fig. 25. $Fe_2F_5 \cdot 2H_2O$, polycrystalline powder. Magnetic moment per FU measured starting from the demagnetized state at $T = 8$ K and 16 K. Sequence of magnetization: 1) increase magnetic field to 19 kOe, 2) decrease to $-$ 19 kOe, followed by 3) an increase to + 19 kOe. Lower figure: coercive force H_c vs. T [86L1, see also 78J1].

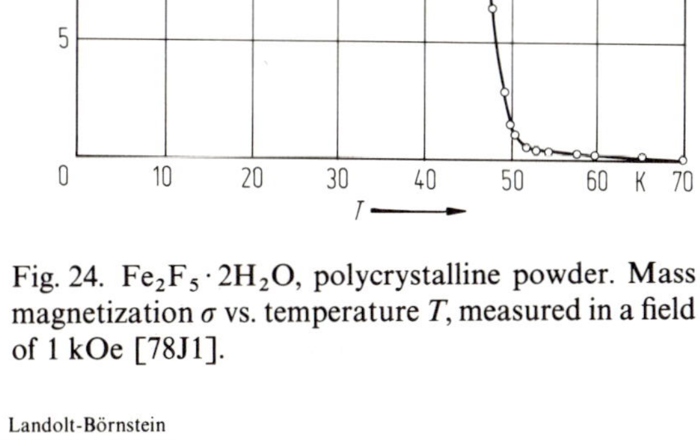

Fig. 24. $Fe_2F_5 \cdot 2H_2O$, polycrystalline powder. Mass magnetization σ vs. temperature T, measured in a field of 1 kOe [78J1].

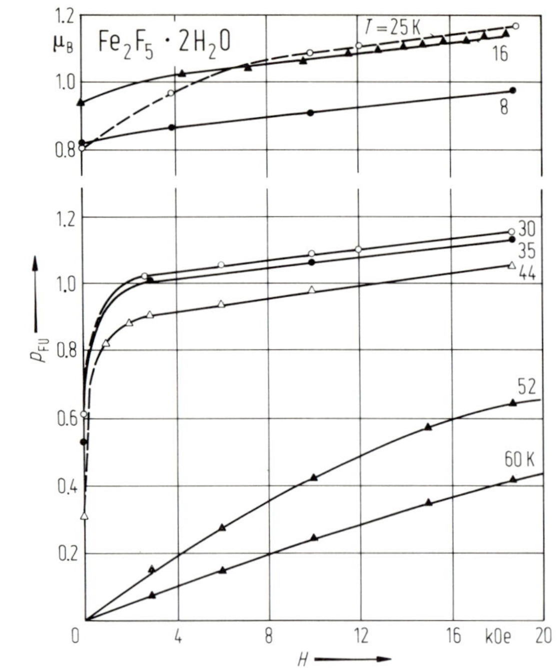

Fig. 26. $Fe_2F_5 \cdot 2H_2O$, polycrystalline powder. Magnetization curves at various temperatures T, each measured in a sequence of decreasing field strengths [86L1].

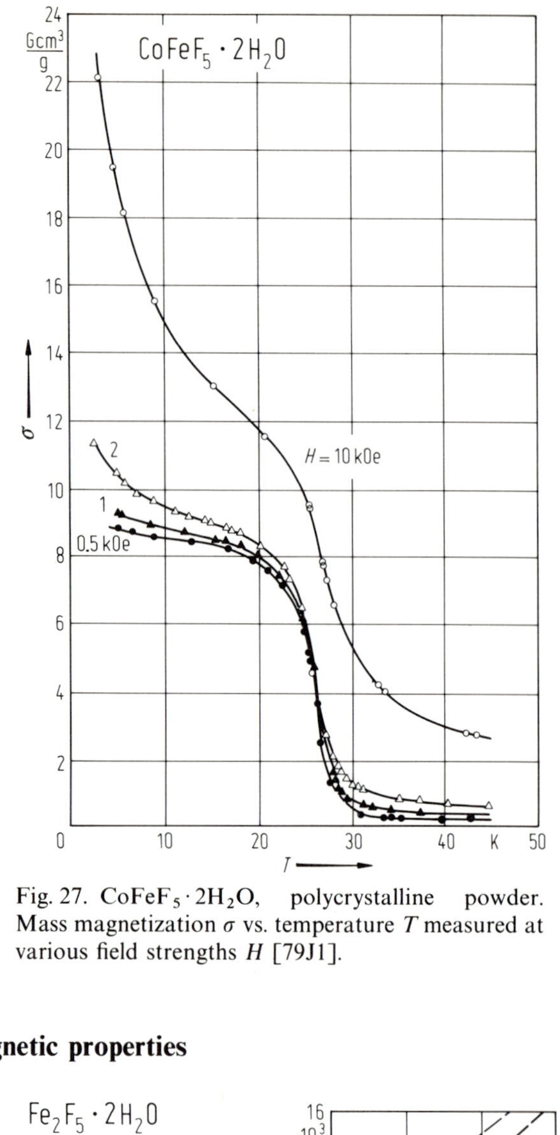

Fig. 27. $CoFeF_5 \cdot 2H_2O$, polycrystalline powder. Mass magnetization σ vs. temperature T measured at various field strengths H [79J1].

9.5.3.5 Paramagnetic properties

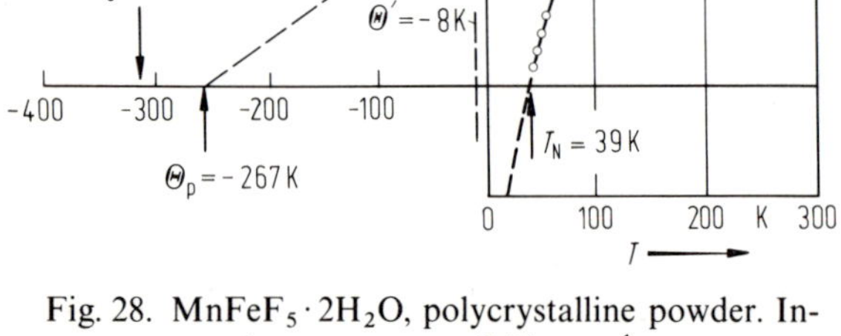

Fig. 28. $MnFeF_5 \cdot 2H_2O$, polycrystalline powder. Inverse magnetic mass susceptibility χ_g^{-1} vs. temperature T. For the paramagnetic parameters see Table 12 [86L3].

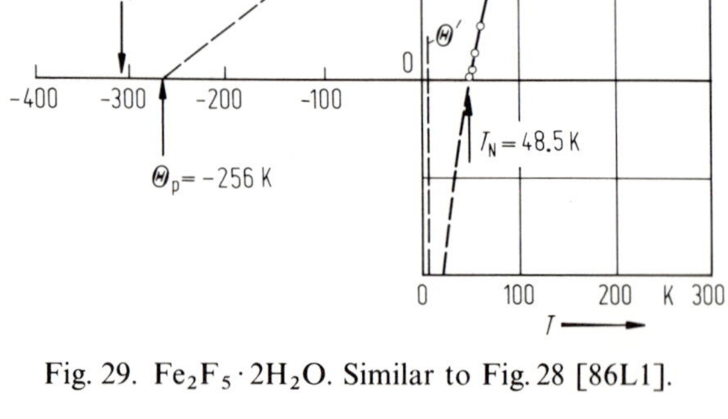

Fig. 29. $Fe_2F_5 \cdot 2H_2O$. Similar to Fig. 28 [86L1].

Table 12. $MFeF_5 \cdot 2H_2O$. Paramagnetic parameters derived from the temperature dependence of the molar magnetic susceptibility χ_m with a least-squares fitting of the experimental points to the equation a) $\chi_m^{-1} = (T - T_N)(T - T'_C)/C_m(T - \Theta')$, see Figs. 28, 29 or b) to a Curie–Weiss fit in the indicated temperature range.

Compound	Temperature range [K]	T_N [K]	Θ_p [K]	C_m [$cm^3K\,mol^{-1}$]	$p_{eff,exp}$ [μ_B]	$p_{eff,theor.}$ [μ_B]	T'_C [K]	Θ' [K]	Fig.	Ref.
$MnFeF_5 \cdot 2H_2O$	—	39(1)	− 267(5)	8.70(10)	8.34	8.37	− 314(3)	− 8.4(8)	28	86L3
$Fe_2F_5 \cdot 2H_2O$	94 ··· 300	48.50(25)	− 138	—	7.5(1)	7.68	—	—		79J1
	—	48.0	− 179	6.74	7.34	7.68	− 200.9	25.8		77W1
	—	48.5(1)	− 256.1(4)	6.70(4)	7.32(3)	7.68	− 308.8	6.43	29	86L1
$CoFeF_5 \cdot 2H_2O$	80 ··· 296	27.00(25)	− 56	—	6.8(1)	7.07	—	—	30	79J1
$ZnFeF_5 \cdot 2H_2O$[a])	200 ··· 300	—	− 182(5)	4.43(18)	5.95	5.92	—	—	31	86L3
$AlFeF_5 \cdot 2H_2O$	3 ··· 300	—	− 7.8	—	5.20(10)	4.90	—	—	32	79J1

[a]) Exchange integral $J/k = -11.8(2)$ K [86L3].

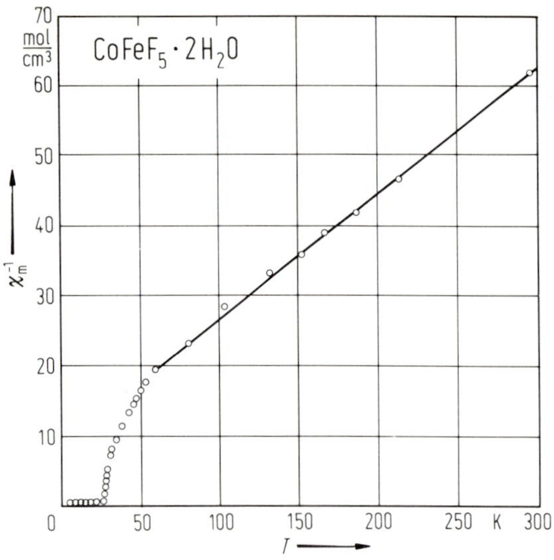

Fig. 30. CoFeF$_5$·2H$_2$O. Inverse magnetic molar susceptibility χ_m^{-1} vs. temperature T [79J1].

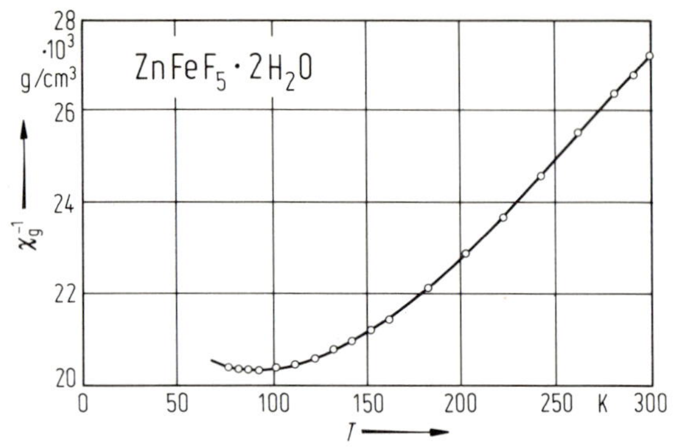

Fig. 31. ZnFeF$_5$·2H$_2$O. Inverse magnetic mass susceptibility χ_g^{-1} vs. temperature T [86L3].

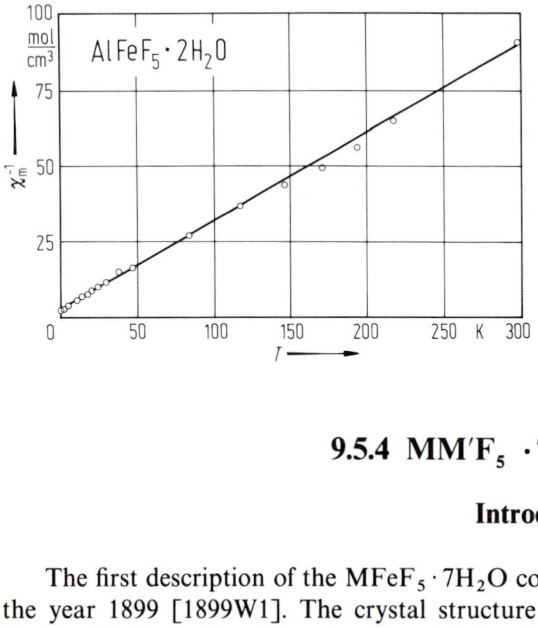

Fig. 32. AlFeF$_5$·2H$_2$O, polycrystalline powder. Inverse magnetic molar susceptibility χ_m^{-1} vs. temperature T. The straight line corresponds to a Curie–Weiss law with $p_{eff} = 5.20\ \mu_B$ and $\Theta_p = -7.8$ K [79J1].

9.5.4 MM'F$_5$ ·7H$_2$O compounds

Introduction

The first description of the MFeF$_5$·7H$_2$O compounds (M = Fe, Co, Ni, Zn or Al) dates back to the year 1899 [1899W1]. The crystal structure of the mixed-valency iron fluoride heptahydrate Fe$_2$F$_5$·7H$_2$O was first determined as triclinic with 5 formula units per unit cell by Gallagher [75G1] on a single crystal sample. The other compounds have an isomorphous structure, however some controversies still exist, see Table 13. The spectra obtained by means of ^{19}F NMR experiments indicate [75S1, 2] that all the fluorine atoms are bonded to the trivalent ions, the structure of these fluorides can be generally represented as $[M^{3+}F_5OH_2]\cdot[M'^{2+}(OH_2)_6]$. This means that each metal ion is in the centre of an octahedron consisting either of 5 fluorine and one oxygen or of six oxygen ions. These octahedra are connected by a three-dimensional array of hydrogen bonds. Thermal dehydration experiments are reported in [75C1, 75G1, 80B1, 80I1, 88N1]. For Mössbauer data see Table 14 and Figs. 33 ··· 35. There are indications in the Mössbauer spectra of Fe$_2$F$_5$·7H$_2$O that the ferric ions give rise to two different patterns pointing to the existence of at least two Fe^{3+} crystal chemical species [81E1]. For magnetic susceptibilities see Table 15 and Figs. 36 ··· 41.

Table 13. $MM'F_5 \cdot 7H_2O$. RT unit cell parameters obtained from powder X-ray spectra; a, b, c in [Å], α, β, γ in [deg].

Compound	X-ray	Symmetry	a	b	c	α	β	γ	Ref.
$CrCoF_5 \cdot 7H_2O$	Cu-K_α	triclinic[b])	6.533	8.907	10.446	106.02	123.14	83.25	77G1
$CrNiF_5 \cdot 7H_2O$	Cu-K_α	triclinic[b])	6.495	8.885	10.391	106.16	123.44	83.20	77G1
$Fe_2F_5 \cdot 7H_2O$	Fe-K_α	triclinic[a])	6.582	8.988	10.542	105.81	123.28	82.93	75G1
	Fe-K_α	tetragonal[c])	12.82	6.92					58B1
	Cu-K_α	tetragonal[c])	12.64(3)	7.05(3)					75S1, 2
$CoFeF_5 \cdot 7H_2O$	Fe-K_α	triclinic[b])	6.506	8.991	10.464	106.38	122.42	83.02	77G1
$NiFeF_5 \cdot 7H_2O$	Fe-K_α	triclinic[b])	6.506	8.942	10.433	106.30	123.11	83.02	77G1
$ZnFeF_5 \cdot 7H_2O$	Cu-K_α	tetragonal[c])	12.89(3)	6.94(3)					75S1, 2
$AlFeF_5 \cdot 7H_2O$	Cu-K_α	tetragonal[c])	12.71(3)	6.99(3)					75S1, 2
$AlCoF_5 \cdot 7H_2O$		monoclinic C2/m	10.917	13.863	6.525		100.3		80M1
$AlZnF_5 \cdot 7H_2O$		tetragonal[c])	12.75(3)	7.01(3)					75S1, 2
$NiMnF_5 \cdot 7H_2O$		monoclinic C2/m	11.089(5)	14.063(5)	6.375(5)		100.39		88N1
$CuMnF_5 \cdot 7H_2O$		id. $P2_1$ (c)	8.979(5)	18.237(5)	6.010(5)		96.40		88N1

[a]) The triclinic structure is derived from a single crystal X-ray analysis.
[b]) X-ray powder spectrum similar to that of $Fe_2F_5 \cdot 7H_2O$, therefore indicated as originating from a triclinic system.
[c]) X-ray powder spectrum indicated as originating from a tetragonal system.

Table 14. MM'$F_5 \cdot 7H_2O$ compounds. Temperature dependence of the isomer shift δ, the quadrupole splitting Δ and quadrupole shift 2ε above and below the magnetic ordering temperature T_N, respectively, and the magnetic hyperfine field H_{hyp}.

Compound	T [K]	Fe^{3+} in $FeF_5(OH_2)$ octahedra: δ[a]) [mm s^{-1}]	Δ or 2ε [mm s^{-1}]	H_{hyp} [kOe]	Fe^{2+} in $Fe(OH_2)_6$ octahedra: δ[a]) [mm s^{-1}]	Δ or 2ε [mm s^{-1}]	H_{hyp} [kOe]	T_N [K]	Ref.
$Fe_2F_5 \cdot 7H_2O$	300	0.22(1)	0.58(2)		1.03(1)	3.31(2)		3.30(5)	73I1[d])
(Fig. 33)	4.2	0.32(1)	0.60(2)		1.17(1)	3.48(2)			
	1.3[b])	0.38(2)	−0.51(1)	561(3)	1.16(2)	3.00(2)	146(3)[c])		
		0.32(2)	−0.08(1)	573(3)					
	297	0.393(10)	0.590(10)		1.215(10)	3.284(10)			78B1
	76	0.498(10)	0.643(10)		1.333(10)	3.527(10)			
	298[b])	0.462(3)	0.414(4)		1.228(1)	3.322(1)			81E1
		0.459(2)	0.817(3)						
$CoFeF_5 \cdot 7H_2O$	293	0.44(2)	0.44(2)						75S1
	80	0.53(2)	0.58(2)						
	297	0.399(10)	0.589(10)						78B1
	76	0.512(10)	0.612(10)						
$NiFeF_5 \cdot 7H_2O$	300[b])	0.33	0.34						77G1
(Fig. 34)		0.33	0.36						
	4.2[b])	0.43	0.41						
		0.43	0.83						
$ZnFeF_5 \cdot 7H_2O$	293	0.41(2)	0.47(2)						75S1, 2
	80	0.53(2)	0.52(2)						
	297	0.406(10)	0.589(10)						78B1
	76	0.498(10)	0.646(10)						
$AlFeF_5 \cdot 7H_2O$	293				1.24(2)	3.27(2)			75S1, 2
(Fig. 35)	80				1.38(2)	3.63(2)			
	297				1.217(10)	3.337(10)			78B1
	76				1.337(10)	3.672(10)			

a) Relative to metallic iron at RT.
b) The assumption of two different Fe^{3+} sites was necessary for an acceptable analysis of the spectra. Relative frequencies of the sites 1:4 in the case of $Fe_2F_5 \cdot 7H_2O$ and 1:1 for the $NiFeF_5 \cdot 7H_2O$ compound.
c) The hyperfine field makes an angle of 26(5)° with the the principal axis of the electric field gradient.
d) See also [75S1, 2] and [76W1].

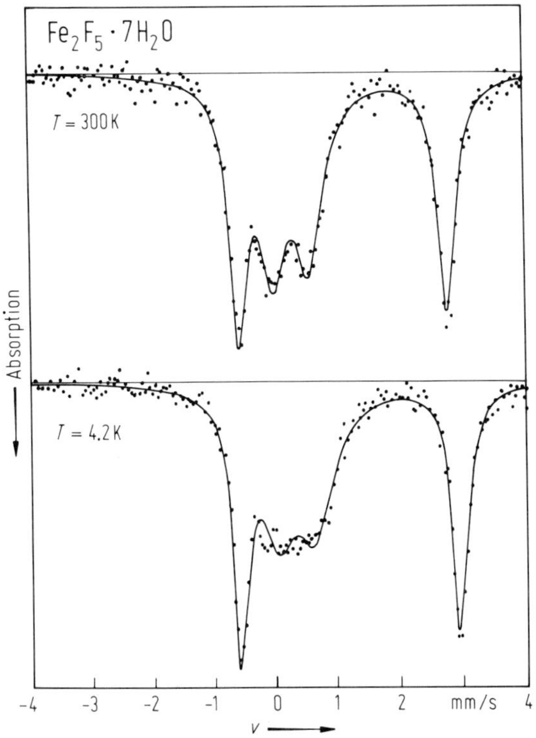

Fig. 33. $Fe_2F_5 \cdot 7H_2O$. ^{57}Fe Mössbauer spectra for a powder in the paramagnetic state at 300 K and 4.2 K [73I1].

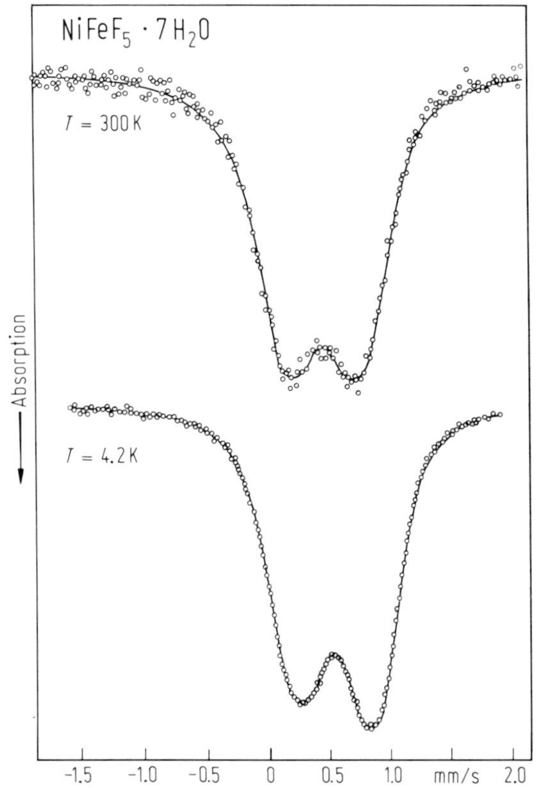

Fig. 34. $NiFeF_5 \cdot 7H_2O$. ^{57}Fe Mössbauer spectra for a powder in the paramagnetic state at 300 K and 4.2 K [77G1].

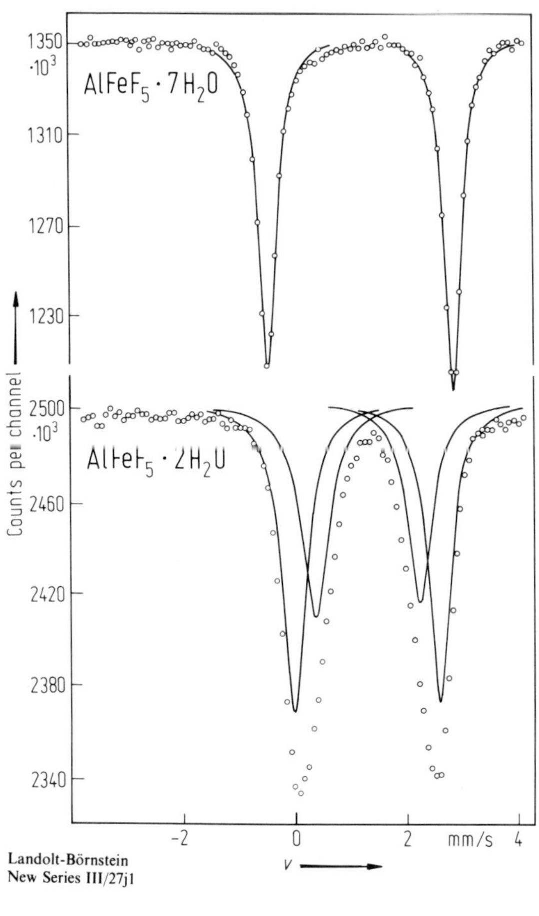

Fig. 35. $AlFeF_5 \cdot 7H_2O$ and $AlFeF_5 \cdot 2H_2O$. ^{57}Fe Mössbauer spectra for powders in the paramagnetic state at RT [78B1].

Table 15. $MM'F_5 \cdot 7H_2O$ compounds. Paramagnetic properties derived from the Curie–Weiss law: $\chi = C_m/(T - \Theta_p)$. Approximation of the temperature dependence of the susceptibility in the indicated temperature range. The ordering temperature is given as Néel temperature T_N.

Compound	Temperature region [K]	Θ_p [K]	C_m [$cm^3 K mol^{-1}$]	p_{eff} [μ_B] exp	p_{eff} [μ_B] spin-only value	T_N [K]	$\chi_m^{-1}(T)$ Fig.	Ref.
$V_2O_5 \cdot 7H_2O$	80 ⋯ 290	− 18		3.35				68S1
$Fe_2F_5 \cdot 7H_2O$	20 ⋯ RT	− 5.0(3)	7.94(1)	7.97	7.68			76W1
	4.2 ⋯ 60	− 6.9		7.70(6)	7.68	3.0(2)	36, 37	77J1
$CoFeF_5 \cdot 7H_2O$	3.5 ⋯ 55	− 5.2		7.50(6)	7.07	3.42(10)	36, 38	77J1
$NiFeF_5 \cdot 7H_2O$	10 ⋯ 50	− 4.35		6.69(8)	6.56	< 2.3	36, 39	77J1
$ZnFeF_5 \cdot 7H_2O$	2.5 ⋯ 50	− 1.05		5.94(5)	5.92	—	36, 40	77J1
$AlFeF_5 \cdot 7H_2O$	3.0 ⋯ 80	− 2.62		5.05(4)	4.90		40	77J1
$NiMnF_5 \cdot 7H_2O$	2.2 ⋯ 300	0(2)	4.32				41	88N1
$CuMnF_5 \cdot 7H_2O$	2.2 ⋯ 300	0(2)	3.51				41	88N1

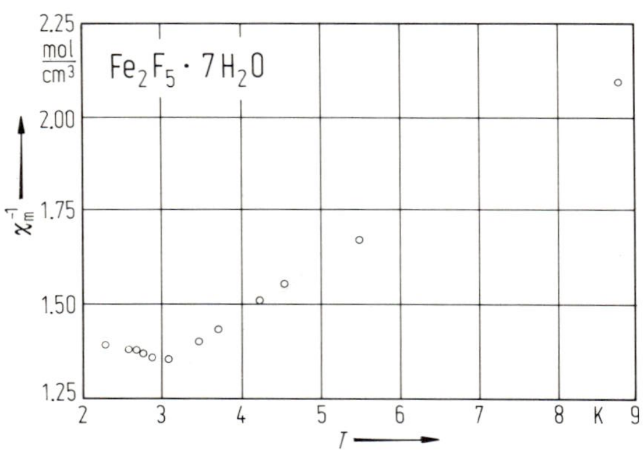

Fig. 37. $Fe_2F_5 \cdot 7H_2O$. Low temperature inverse molar magnetic susceptibility χ_m^{-1} vs. temperature T [77J1].

For Fig. 36 see next page.

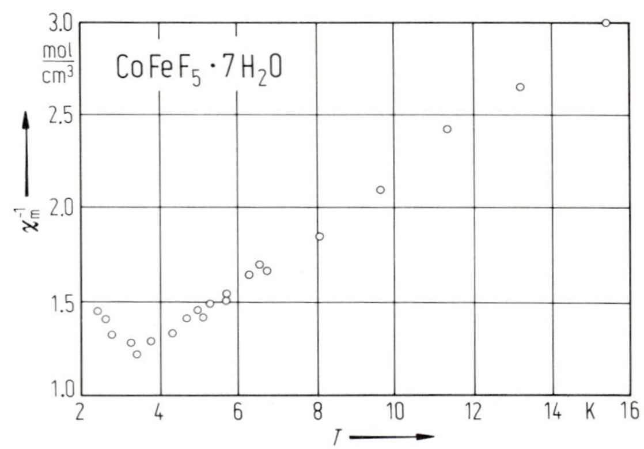

Fig. 38. $CoFeF_5 \cdot 7H_2O$. Low temperature inverse molar magnetic susceptibility χ_m^{-1} vs. temperature T [77J1].

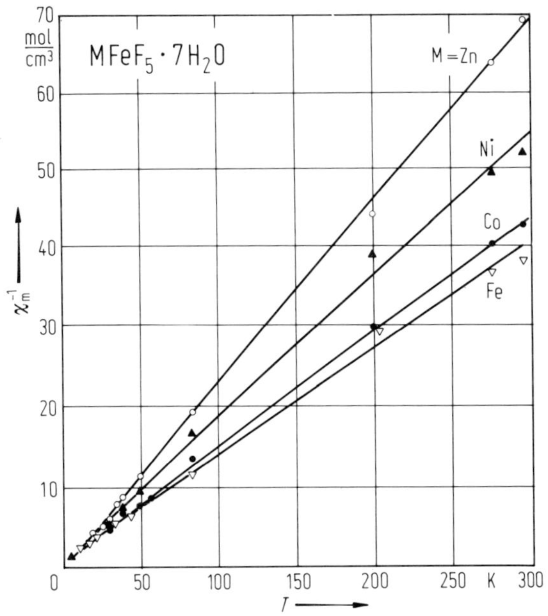

Fig. 36. MFeF$_5$·7H$_2$O, M = Fe, Co, Ni or Zn. Inverse molar magnetic susceptibility χ_m^{-1} vs. temperature T [77J1].

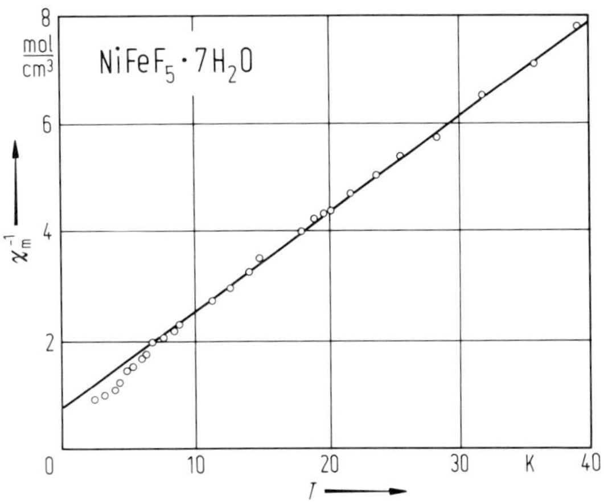

Fig. 39. NiFeF$_5$·7H$_2$O. Low temperature inverse molar magnetic susceptibility χ_m^{-1} vs. temperature T [77J1].

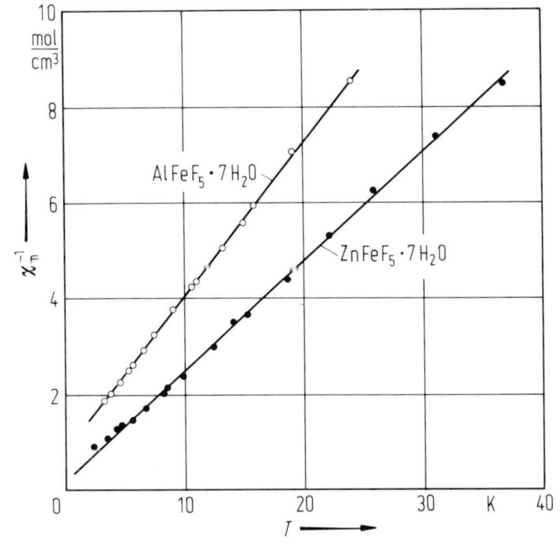

Fig. 40. ZnFeF$_5$·7H$_2$O, AlFeF$_5$·7H$_2$O. Low temperature inverse molar magnetic susceptibility χ_m^{-1} vs. temperature T [77J1].

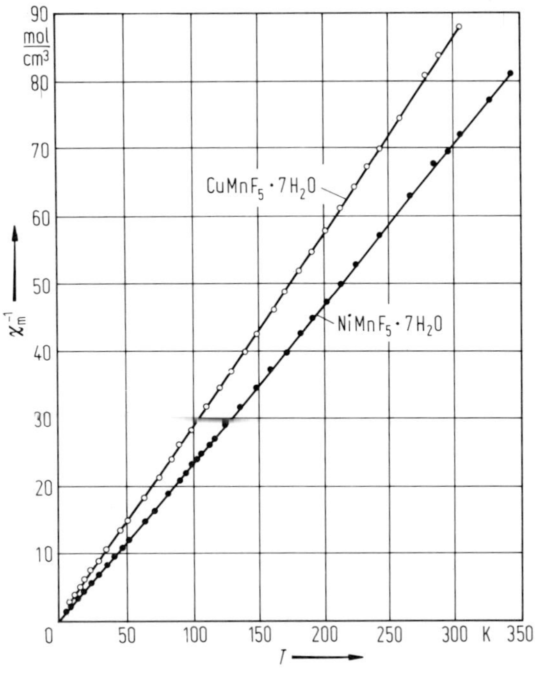

Fig. 41. NiMnF$_5$·7H$_2$O, CuMnF$_5$·7H$_2$O. Inverse molar magnetic susceptibility χ_m^{-1} vs. temperature T [88N1].

9.5.5 References for 9.5

1899W1 Weinland, R., Köppen, O.: Z. Anorg. Chem. **22** (1899) 266.
51B1 Bizette, H.: J. Phys. Radium **12** (1951) 161.
58B1 Brauer, G., Eichner, M.: Z. Anorg. Allg. Chem. **296** (1958) 13.
64S1 Steinfink, H., Burns, J.H.: Acta Crystallogr. **17** (1964) 823.
66O1 Osmond, W.P.: Proc. Phys. Soc. **87** (1966) 767.
68S1 Seifert, H.J., Loh, H.W., Jungnickel, K.: Z. Anorg. Allgem. Chem. **360** (1968) 62.
69C1 Cousseins, J-C., Erb, A., Freundlich, W.: C.R. Acad. Sci. (Paris) **C 268** (1969) 717.
70R1 Rimsky, A., Thoret, J., Freundlich, W.: C.R. Acad. Sci. (Paris) **C 270** (1970) 407.
73I1 Imbert, P., Machetau, Y., Varret, F.: J. Phys. (Paris) **34** (1973) 49.
73T1 Tressaud, A., Dance, J.M., Ravez, J., Portier, J., Hagenmuller, P.: Mater. Res. Bull. **8** (1973) 1467.
73T2 Tressaud, A., Parenteau, J.M., Dance, J.M., Portier, J., Hagenmuller, P.: Mater. Res. Bull. **8** (1973) 565.
74S1 Sakai, T., Tominaga, T.: Radioisotopes **23** (1974) 33.
74T1 Tressaud, A., Dance, J.M.: C.R. Acad. Sci. (Paris) **C 278** (1974) 463.
75C1 Charpin, P., Macheteau, Y.: C.R. Acad. Sci. (Paris) **C 280** (1975) 61.
75G1 Gallagher, K.J., Ottaway, M.R.: J. Chem. Soc. Dalton Trans. (1975) 978.
75S1 Sakai, T., Tominaga, T.: Bull. Chem. Soc. Jpn. **48** (1975) 3168.
75S2 Sakai, T., Tominaga, T.: Radiochem. Radioanal. Lett. **22** (1975) 11.
75W1 Wintenberger, M., Dance, J.M., Tressaud, A.: Solid State Commun. **17** (1975) 185.
76I1 Imbert, P., Jehanno, G., Macheteau, Y.: J. Phys. (Paris) **37** (1976) 969.
76W1 Walton, E.G., Corvan, P.J., Brown, D.B., Day, P.: Inorg. Chem. **15** (1976) 1737.
77F1 Férey, G., de Pape, R., Poulain, M., Grandjean, D., Hardy, A.: Acta Crystallogr. **B 33** (1977) 1409.
77G1 Gallagher, K.J., Ottaway, M.R.: J. Chem. Soc. Dalton Trans. **22** (1977) 2212.
77H1 Hall, W., Kim, S., Zubieta, J., Walton, E.G., Brown, D.B.: Inorg. Chem. **16** (1977) 1884.
77J1 Jones, E.R., Hendricks, M.E., Auel, T., Amma, E.L.: J. Chem. Phys. **66** (1977) 3252.
77T1 Tressaud, A., Dance, J.M.: Adv. Inorg. Chem. Radiochem. **20** (1977) 133.
77W1 Walton, E.G., Brown, D.B., Wong, H., Reiff, W.M.: Inorg. Chem. **16** (1977) 2425.
78B1 Balcerek, T.W., Cathey, L., Karraker, D.G.: J. Inorg. Nucl. Chem. **40** (1978) 773.
78F1 Férey, G., de Pape, R., Boucher, B.: Acta Crystallogr. **B 34** (1978) 1084.
78J1 Jones, E.R., Dawson, R.: J. Chem. Phys. **69** (1978) 3289.
79J1 Jones, E.R., Dawson, R.: J. Chem. Phys. **71** (1979) 202.
79J2 Jones, E.R., Dawson, R.: J. Chem. Phys. **70** (1979) 2586.
80B1 Brown, D.B., Walton, E.G., Dilts, J.A.: J. Chem. Soc. Dalton Trans. **6** (1980) 845.
80G1 Gallie, D., Vogel, R.H., Evans, B.J.: J. Phys. (Paris) Colloq. **41** (1980) C 1–289.
80I1 Ippolitov, E.G., Tripol'skaya, T.A.: Zh. Neorg. Khim. **25** (1980) 775; Russian J. Inorg. Chem. (English Transl.) **25** (1980) 425.
80M1 Massa, W.: J. Fluorine Chem. **16** (1980) 634.
81E1 Evans, B.J.: Inorg. Chem. **20** (1981) 504.
81F1 Férey, G., Leblanc, M., de Pape, R.: J. Solid State Chem. **40** (1981) 1.
83Z1 Zucker, U.H., Perenthaler, E., Kuhs, W.F., Bachmann, R., Schultz, H.: J. Appl. Crystallogr. **16** (1983) 358.
85B1 Babel, D., Tressaud, A.: Inorganic Solid Fluorides, Hagenmuller, P. (ed.), Academic Press. Inc. 1985, p. 77.
85D1 Dance, J.M., Tressaud, A.: Inorganic Solid Fluorides, Hagenmuller, P. (ed.), Academic Press. Inc. 1985, p. 371.
85J1 Jones, E.R., van Hine, C., Datta, T., Cathey, L., Karraker, D.G.: Inorg. Chem. **24** (1985) 3888.
85F1 Fischer, R.X.: J. Appl. Crystallogr. **18** (1985) 258.

86L1 Laligant, Y., Leblanc, M., Pannetier, J., Férey, G.: J. Phys. C.: Solid State Phys. **19** (1986) 1081.

86L2 Laligant, Y., Pannetier, J., Labbe, P., Férey, G.: J. Solid State. Chem. **62** (1986) 274.

86L3 Laligant, Y., Calage, Y., Torres-Tapia, E., Greneche, J.M., Varret, F., Férey, G.: J. Magn. Magn. Mater. **61** (1986) 283.

87K1 Karraker, D.G.: Inorg. Chem. **26** (1987) 3814.

87L1 Laligant, Y., Pannetier, J., Leblanc, M., Labbe, P., Heger, G., Férey, G.: Z. Kristallogr. **181** (1987) 1.

87L2 Laligant, Y., Pannetier, J., Férey, G.: J. Solid State Chem. **66** (1987) 242.

87S1 Swanepoel, S., Stander, C.M.: J. Phys. Chem. Solids **48** (1987) 275.

88G1 Greneche, J.M., Linares, J., Varret, F., Laligant, Y., Férey, G.: J. Magn. Magn. Mater. **73** (1988) 115.

88N1 Nunez, P., Tressaud, A., Darriet, J., Hagenmuller, P., Massa, W., Kummer, S., Babel, D.: J. Solid State Chem. **77** (1988) 240.

90G1 Greneche, J.M., Linares, J., Varret, F.: J. Phys: Condens. Matter **2** (1990) 4243.

MULTIPLE CRITERIA DECISION ANALYSIS: STATE OF THE ART SURVEYS

Recent titles in the
INTERNATIONAL SERIES IN OPERATIONS RESEARCH & MANAGEMENT SCIENCE
Frederick S. Hillier, Series Editor, *Stanford University*

Zhu, J. / *QUANTITATIVE MODELS FOR PERFORMANCE EVALUATION AND BENCHMARKING*
Ehrgott, M. & Gandibleux, X. / *MULTIPLE CRITERIA OPTIMIZATION: State of the Art Annotated Bibliographical Surveys*
Bienstock, D. / *Potential Function Methods for Approx. Solving Linear Programming Problems*
Matsatsinis, N.F. & Siskos, Y. / *INTELLIGENT SUPPORT SYSTEMS FOR MARKETING DECISIONS*
Alpern, S. & Gal, S. / *THE THEORY OF SEARCH GAMES AND RENDEZVOUS*
Hall, R.W./*HANDBOOK OF TRANSPORTATION SCIENCE - 2nd Ed.*
Glover, F. & Kochenberger, G.A. / *HANDBOOK OF METAHEURISTICS*
Graves, S.B. & Ringuest, J.L. / *MODELS AND METHODS FOR PROJECT SELECTION: Concepts from Management Science, Finance and Information Technology*
Hassin, R. & Haviv, M./ *TO QUEUE OR NOT TO QUEUE: Equilibrium Behavior in Queueing Systems*
Gershwin, S.B. et al/*ANALYSIS & MODELING OF MANUFACTURING SYSTEMS*
Maros, I./ *COMPUTATIONAL TECHNIQUES OF THE SIMPLEX METHOD*
Harrison, T., Lee, H. & Neale, J./ *THE PRACTICE OF SUPPLY CHAIN MANAGEMENT: Where Theory And Application Converge*
Shanthikumar, J.G., Yao, D. & Zijm, W.H./ *STOCHASTIC MODELING AND OPTIMIZATION OF MANUFACTURING SYSTEMS AND SUPPLY CHAINS*
Nabrzyski, J., Schopf, J.M., Węglarz, J./ *GRID RESOURCE MANAGEMENT: State of the Art and Future Trends*
Thissen, W.A.H. & Herder, P.M./ *CRITICAL INFRASTRUCTURES: State of the Art in Research and Application*
Carlsson, C., Fedrizzi, M., & Fullér, R./ *FUZZY LOGIC IN MANAGEMENT*
Soyer, R., Mazzuchi, T.A., & Singpurwalla, N.D./ *MATHEMATICAL RELIABILITY: An Expository Perspective*
Chakravarty, A.K. & Eliashberg, J./ *MANAGING BUSINESS INTERFACES: Marketing, Engineering, and Manufacturing Perspectives*
Talluri, K. & van Ryzin, G./ *THE THEORY AND PRACTICE OF REVENUE MANAGEMENT*
Kavadias, S. & Loch, C.H./*PROJECT SELECTION UNDER UNCERTAINTY: Dynamically Allocating Resources to Maximize Value*
Brandeau, M.L., Sainfort, F., Pierskalla, W.P./ *OPERATIONS RESEARCH AND HEALTH CARE: A Handbook of Methods and Applications*
Cooper, W.W., Seiford, L.M., Zhu, J./ *HANDBOOK OF DATA ENVELOPMENT ANALYSIS: Models and Methods*
Luenberger, D.G./ *LINEAR AND NONLINEAR PROGRAMMING, 2nd Ed.*
Sherbrooke, C.C./ *OPTIMAL INVENTORY MODELING OF SYSTEMS: Multi-Echelon Techniques, Second Edition*
Chu, S.-C., Leung, L.C., Hui, Y. V., Cheung, W./ *4th PARTY CYBER LOGISTICS FOR AIR CARGO*
Simchi-Levi, Wu, Shen/ *HANDBOOK OF QUANTITATIVE SUPPLY CHAIN ANALYSIS: Modeling in the E-Business Era*
Gass, S.I. & Assad, A.A./ *AN ANNOTATED TIMELINE OF OPERATIONS RESEARCH: An Informal History*
Greenberg, H.J./ *TUTORIALS ON EMERGING METHODOLOGIES AND APPLICATIONS IN OPERATIONS RESEARCH*
Weber, C./ *UNCERTAINTY IN THE ELECTRIC POWER INDUSTRY: Methods and Models for Decision Support*

**** A list of the early publications in the series is at the end of the book ****

MULTIPLE CRITERIA DECISION ANALYSIS: STATE OF THE ART SURVEYS

Edited by

JOSÉ FIGUEIRA
University of Coimbra

SALVATORE GRECO
University of Catania

MATTHIAS EHRGOTT
University of Auckland

Springer

Jose Figueira
University of Coimbra
Portugal

Salvatore Greco
University of Catania
Italy

Matthias Ehrgott
University of Auckland
New Zealand

Library of Congress Cataloging-in-Publication Data
A C.I.P. Catalogue record for this book is available
from the Library of Congress.

ISBN 0-387-23067-X e-ISBN 0-387-23081-5 Printed on acid-free paper.

Printed in the United States of America.

9 8 7 6 5 4 3 2 1 SPIN 11052333

springeronline.com

Contents

List of Figures

List of Tables

Introduction

José Figueira, Salvatore Greco, Matthias Ehrgott

1. Human Reflection about Decision

Decision has inspired reflection of many thinkers since the ancient times. The great philosophers Aristotle, Plato, and Thomas Aquinas, to mention only a few names, discussed the capacity of humans to decide and in some manners claimed that this possibility is what distinguishes humans from animals. To illustrate some important aspects of decision, let us briefly quote two important thinkers: Ignatius of Loyola (1491-1556) and Benjamin Franklin (1706-1790).

> To consider, reckoning up, how many advantages and utilities follow for me from holding the proposed office or benefice [...] , and, to consider likewise, on the contrary, the disadvantages and dangers which there are in having it. Doing the same in the second part, that is, looking at the advantages and utilities there are in not having it, and likewise, on the contrary, the disadvantages and dangers in not having the same. [...] After I have thus discussed and reckoned up on all sides about the thing proposed, to look where reason more inclines: and so, according to the greater inclination of reason, [...], deliberation should be made on the thing proposed.

This fragment from the "Spiritual Exercises" of St. Ignatius of Loyola [14] has been taken from a paper by Fortemps and Słowiński [12].

> London, Sept 19, 1772
>
> Dear Sir,
>
> In the affair of so much importance to you, wherein you ask my advice, I cannot, for want of sufficient premises, advise you what to determine, but if you please I will tell you how. [...], my way is to divide half a sheet of paper by a line into two columns; writing over the one Pro, and over the other Con. [...] When I have thus got them all together in one view, I endeavor to estimate their respective weights; and where I find two, one on each side, that seem equal, I strike them both out. If I find a reason pro equal to some two reasons con, I strike out the three. If I judge some two reasons con, equal to three reasons pro, I strike out the five; and thus proceeding I find at length where the balance lies; and if, after a day or two of

> further consideration, nothing new that is of importance occurs on either side, I come to a determination accordingly. [...] I have found great advantage from this kind of equation, and what might be called moral or prudential algebra. Wishing sincerely that you may determine for the best, I am ever, my dear friend, yours most affectionately.
>
> B. Franklin

This letter from Benjamin Franklin to Joseph Prestly has been taken from a paper by MacCrimmon [17].

What is interesting in the above two quotations is the fact that decision is strongly related to the comparison of different points of view, some in favour and some against a certain decision. This means that decision is intrinsically related to a plurality of points of view, which can roughly be defined as criteria. Contrary to this very natural observation, for many years the only way to state a decision problem was considered to be the definition of a single criterion, which amalgamates the multidimensional aspects of the decision situation into a single scale of measure. For example, even today the textbooks of Operations Research suggest to deal with a decision problem as follows: to first define an objective function, i.e., a single point of view like a comprehensive profit index (or a comprehensive cost index) representing the preferability (or dis-preferability) of the considered actions and then to maximize (minimize) this objective. This is a very reductive, and in some sense also unnatural, way to look at a decision problem. Thus, for at least thirty years, a new way to look at decision problems has more and more gained the attention of researchers and practitioners. This is the approach considered by Loyola and Franklin, i.e., the approach of explicitly taking into account the pros and the cons of a plurality of points of view, in other words the domain of Multiple Criteria Decision Analysis (MCDA). Therefore, MCDA intuition is closely related to the way humans have always been making decisions. Consequently, despite the diversity of MCDA approaches, methods and techniques, the basic ingredients of MCDA are very simple: a finite or infinite set of actions (alternatives, solutions, courses of action, ...), at least two criteria, and, obviously, at least one decision-maker (DM). Given these basic elements, MCDA is an activity which helps making decisions mainly in terms of choosing, ranking or sorting the actions.

2. Technical Reflection about Decision: MCDA Researchers before MCDA

Of course, not only philosophers reasoned about decision-making. Many important technical aspects of MCDA are linked to classic works in economics, in particular, welfare economics, utility theory and voting oriented social choice theory (see [28]). Aggregating the opinion or the preferences of voters or individuals of a community into collective or social preferences is quite similar a problem to devising comprehensive preferences of a decision-maker from a set of conflicting criteria in MCDA [7].

Despite the importance of Ramon Llull's (1232-1316) and Nicolaus Cusanus's (1401-1464) concerns about and interests in this very topic, the origins of voting systems are often attributed to Le Chevalier Jean-Charles de Borda (1733-1799) and Marie Jean Antoine Nicolas de Caritat (1743-1794), Le Marquis de Condorcet. However, Ramon Llull introduced the pairwise comparison concept before Condorcet [13], while Nicolaus Cusanus introduced the scoring method about three and a half centuries before Borda [27]. Furthermore, it should be noted that a letter from Pliny the Younger ($\approx$ AD 105) to Titus Aristo shows that he introduced the ternary approval voting strategy and was interested in voting systems a long time before Ramon Llull and Nicolaus Cusanus [18, Chapter 2]. Anyway, Borda's scoring method [4] has some similarities with current utility and value theories as has Condorcet's method [10] with the outranking approach of MCDA. In the same line of concerns, i.e., the aggregation of individual preferences into collective ones, Jeremy Bentham (1748-1832) introduced the utilitarian calculus to derive the total utility for the society from the aggregation of the personal interests of the individuals of a community [3]. Inspired by Bentham's works, Francis Ysidro Edgeworth (1845-1926), a utilitarian economist, was mainly concerned with the maximization of the utility of the different competing agents in economy. Edgeworth tried to find the competitive equilibrium points for the different agents. He proposed to draw indifference curves (lines of equal utility) for each agent and then derive the contract curve, a curve that corresponds to the notion of the Pareto or efficient set [21]. Not long afterwards, Vilfredo Federico Damaso Pareto (1848-1923) gave the following definition of ophelimity [utility] for the whole community [22]:

> We will say that the members of a collectivity enjoy maximum ophelimity in a certain position when it is impossible to find a way of moving from that position very slightly in such a manner that the ophelimity enjoyed by each of the individuals of that collectivity increases or decreases. That is to say, any small displacement in departing from that position necessarily has the effect of increasing the ophelimity which certain individuals enjoy, of being agreeable to some, and disagreeable to others.

From this definition it is easy to derive the concept of dominance, which today is one of the fundamental concepts in MCDA.

MCDA also benefits from the birth and development of game theory. Félix Edouard Justin Emile Borel (1871-1956) and John von Neumann (1903-1957) are considered the founders of game theory [5, 6, 20, 19]. Many concepts from this discipline had a strong impact on the development of MCDA.

The concept of efficient point was first introduced in 1951 by Tjalling Koopmans (1910-1985) in his paper "Analysis of production as an efficient combination of activities" [15]:

> A possible point in the commodity space is called efficient whenever an increase in one of its coordinates (the net output of one good) can be achieved only at the cost of a decrease in some other coordinate (the net output of a good).

In the same year (1951) Harold William Kuhn (born 1925) and Albert William Tucker (1905-1995) introduced the concept of vector maximum problem [16]. In the sixties, basic MCDA concepts were explicitly considered for the first time. As two examples we mention Charnes' and Cooper's works on goal programming [8] and the proposition of ELECTRE methods by Roy [23]. The seventies saw what is conventionally considered the "official" starting point of MCDA, the conference on "Multiple Criteria Decision Making" organised in 1972 by Cochrane and Zeleny at Columbia University in South Carolina [9]. Since then MCDA has seen a tremendous growth which continues today.

3. The Reasons for this Collection of State-of-the-Art Surveys

The idea of MCDA is so natural and attractive that thousands of articles and dozens of books have been devoted to the subject, with many scientific journals regularly publishing articles about MCDA. To propose a new collection of state-of-the-art surveys of MCDA in so rich a context may seem a rash enterprise. Indeed, some objections come to mind. There are many and good handbooks and reviews on the subject (to give an idea consider [1, 11, 25, 26, 29]). The main ideas are well established for some years and one may question the contributions this volume can provide. Moreover, the field is so large and comprises developments so heterogeneous that it is almost hopeless to think that an exhaustive vision of the research and practice of MCDA can be given.

We must confess that at the end of the work of editing this volume we agree with the above remarks. However, we believe that a new and comprehensive collection of state-of-the-art surveys on MCDA can be very useful. The main reasons which, despite our original resistance, brought us to propose this book are the following:

1. Many of the existing handbooks and reviews are not too recent. Since MCDA is a field which is developing very quickly this is an important reason.

2. Even though the field of research and application of MCDA is so large, there are some main central themes around which MCDA research and applications have been developed. Therefore our approach was to try to present the – at least in our opinion – most important of these ideas.

With reference to the first point, we can say that we observed many theoretical developments which changed MCDA over the last ten years. We tried to consider these changes as much as possible and in this perspective strong points of the book are the following:

1 It presents the most up-to-date discussions on well established methodologies and theories such as outranking based methods and MAUT.

2 The book also contains surveys of new, recently emerged fields such as conjoint measurement, fuzzy preferences, fuzzy integrals, rough sets and others.

Following these points we drafted a list of topics and asked well known researchers to present them. We encouraged the authors to cooperate with the aim to present different perspectives if topics had some overlap. We asked the authors to present a comprehensive presentation of the most important aspects of the field covered by their chapters, a simple yet concise style of exposition, and considerable space devoted to bibliography and survey of relevant literature. We also requested a sufficiently didactic presentation and a text that is useful for researchers in MCDA as well as for people interested in real life applications.

The importance of these requirements is related also to the specific way the MCDA community looks at its research field. It can be summarized in the observation that there is a very strong and vital link between theoretical and methodological developments on the one hand and real applications on the other hand. Thus, the validity of theoretical and methodological developments can only be measured in terms of the progress given to real world practice. Moreover, interest of MCDA to deal with concrete problems is related to the consideration of a sound theoretical basis which ensures the correct application of the methodologies taken into account.

In fact, not only the chapters of our book but rather all MCDA contributions should satisfy the requirements stated out above, because they should be not too "esoteric" and therefore understandable for students, theoretically well founded, and applicable to some advantage in reality.

4. A Guided Tour of the Book

Of course, this book can be read from the first to the last page. However, we think that this is not the only possibility and it may not even be the most interesting possibility. In the following we propose a guided tour of the book suggesting some reference points that are hopefully useful for the reader.

4.1 Part I: An Overview of MCDA Techniques Today

This part is important because MCDA is not just a collection of theories, methodologies, and techniques, but a specific perspective to deal with decision problems. Losing this perspective, even the most rigorous theoretical developments and applications of the most refined methodologies are at risk of being meaningless, because they miss an adequate consideration of the aims and of the role of MCDA. We share this conviction with most MCDA researchers. Bernard Roy discusses these "pre-theoretical" assumptions of MCDA and gives an overview of the field. Bernard Roy, besides giving many important theoretical contributions, engaged himself in thorough reflections on the meaning and the value of MCDA, proposing some basic key concepts that are accepted throughout the MCDA community.

4.2 Part II: Foundations of MCDA

This part of the book is related to a fundamental problem of MCDA, the representation of preferences. Classically, for example in economics, it is supposed that preference can be represented by a utility function assigning a numerical value to each action such that the more preferable an action, the larger its numerical value. Moreover, it is very often assumed that the comprehensive evaluation of an action can be seen as the sum of its numerical values for the considered criteria. Let us call this the classical model. It is very simple but not too realistic. Indeed, there is a lot of research studying under which conditions the classical model holds. These conditions are very often quite strict and it is not reasonable to assume that they are satisfied in all real world situations. Thus, other models relaxing the conditions underlying the classical model have been proposed. This is a very rich field of research, which is first of all important for those interested in the theoretical aspects of MCDA. However, it is also of interest to readers engaged in applications of MCDA. In fact, when we adopt a formal model it is necessary to know what conditions are supposed to be satisfied by the preferences of the DM. In the two chapters of this part problems related to the representations of preferences are discussed.

Meltem Öztürk, Alexis Tsoukiàs, and Philippe Vincke present a very exhaustive review of preference modelling, starting from classical results but arriving at the frontier of some challenging issues of scientific activity related to fuzzy logic and non-classical logic.

Denis Bouyssou and Marc Pirlot discuss the axiomatic basis of the different models to aggregate multiple criteria preferences. We believe that this chapter is very important for the future of MCDA. Initially, the emphasis of MCDA research was on proposal of new methods. But gradually the necessity to understand the basic conditions underlying each method and its specific axiomatization became more and more apparent. This is the first book on MCDA with so much space dedicated to the subject of foundations of MCDA.

4.3 Part III: Outranking Methods

In this part of the book the class of outranking based multiple criteria decision methods is presented. Given what is known about the decision-maker's preferences and given the quality of the performances of the actions and the nature of the problem, an outranking relation is a binary relation S defined on the set of potential actions A such that aSb if there are enough arguments to decide that a is at least as good as b, whereas there is no essential argument to refute that statement [24]. Methods which strictly apply this definition of outranking relation are the ELECTRE methods. They are very important in many respects, not least historically, since ELECTRE I was the first outranking method [2].

However, within the class of outranking methods we generally consider all methods which are based on pairwise comparison of actions. Thus, another class of very well known multiple criteria methods, PROMETHEE methods, are considered in this part of the book. Besides ELECTRE and PROMETHEE methods, many other interesting MCDA methods are based on the pairwise comparison of actions. José Figueira, Vincent Mousseau and Bernard Roy present the ELECTRE methods; Jean-Pierre Brans and Bertrand Mareschal present the PROMETHEE methods and Jean-Marc Martel and Benedetto Matarazzo review the rich literature of other outranking methods.

4.4 Part IV: Multiattribute Utility and Value Theories

In this part of the book we consider multiple attribute utility theory (MAUT). This MCDA approach tries to assign a utility value to each action. This utility is a real number representing the preferability of the considered action. Very often the utility is the sum of the marginal utilities that each criterion assigns to the considered action. Thus, this approach very often coincides with what we called the classical approach before. As we noted in commenting Part I, this approach is very simple at first glance. It is often applied in real life, e.g., every time we aggregate some indices by means of a weighted sum we are applying this approach. Despite its simplicity the approach presents some technical problems. The first are related to the axiomatic basis and to the construction of marginal utility functions (i.e., the utility functions relative to each single criterion), both in case of decision under certainty and uncertainty. These problems are considered by James Dyer in a comprehensive chapter about the fundamentals of this approach.

Yannis Siskos, Vangelis Grigoroudis and Nikolaos Matsatsinis present the very well known UTA methods, which on the basis of the philosophy of the aggregation-disaggregation approach and using linear programming, build a MAUT model that is as consistent as possible with the DM's preferences expressed in actual previous decisions or on a "training sample". The philosophy of aggregation-disaggregation can be summarized as follows: How is it possible to assess the decision-maker's preference model leading to exactly the same decision as the actual one or at least the most "similar" decision?

Thomas Saaty presents a very well known methodology to build utility functions, the AHP (Analytic Hierarchy Process) and its more recent extension, the ANP (Analytic Network Process). AHP is a theory of measurement that uses pairwise comparisons along with expert judgments to deal with the measurement of qualitative or intangible criteria. The ANP is a general theory of relative measurement used to derive composite priority ratio scales from individual ratio scales that represent relative measurements of the influence of elements that interact with respect to control criteria. The ANP captures the outcome of dependence and feedback within and between clusters of elements. Therefore AHP with its dependence assumptions on clusters and elements is a special case of the ANP.

Carlos Bana e Costa, Jean-Claude Vansnick, and Jean-Marie De Corte present another MCDA methodology based on the additive utility model. This methodology is MACBETH (Measuring Attractiveness by a Categorical Based Evaluation Technique). It is an MCDA approach that requires only qualitative judgements about differences of values of attractiveness of one action over another action to help an individual or a group to quantify the relative preferability of different actions. In simple words, the MACBETH approach tries to answer the following questions: How can we build an interval scale of preferences on a set of actions without forcing evaluators to produce direct numerical representations of their preferences? How can we coherently aggregate these qualitative evaluations using an additive utility model?

4.5 Part V: Non-Classical MCDA Approaches

Many approaches have been proposed in MCDA besides outranking methods and multiattribute utility theory. In this part of the book we try to collect information about some of the most interesting proposals. First, the question of uncertainty in MCDA is considered. Theo Stewart discusses risk and uncertainty in MCDA. It is necessary to distinguish between internal uncertainties (related to decision-maker values and judgements) and external uncertainties (related to imperfect knowledge concerning consequences of actions). The latter, corresponding to the most accepted interpretation of uncertainty in the specialized literature, has been considered in the chapter. Four broad approaches for dealing with external uncertainties are discussed. These are multiattribute utility theory and some extensions; stochastic dominance concepts, primarily in the context of pairwise comparisons of alternatives; the use of surrogate risk measures such as additional decision criteria; and the integration of MCDA and scenario planning.

The second consideration is the fuzzy set approach to MCDA. Most real world decision problems take place in a complex environment where conflicting systems of logic, uncertain and imprecise knowledge, and possibly vague preferences have to be considered. To face such complexity, preference modeling requires the use of specific tools, techniques, and concepts which allow the available information to be represented with the appropriate granularity. In this perspective, fuzzy set theory has received a lot of attention in MCDA for a long time. Patrick Meyer and Marc Roubens present the fuzzy set approach to MCDA for choice, ranking, and sorting problems. In this chapter, several MCDA approaches based on fuzzy evaluations are reviewed. The authors give details on a sorting procedure for the assignment of alternatives to graded classes when the available information is given by interacting points of view and a subset of prototypic alternatives whose assignment is given beforehand. A software dedicated to that approach (TOMASO) is briefly presented. Finally they recall the concepts of good and bad choices based on dominant and absorbent kernels in the valued digraph that corresponds to an ordinal valued outranking relation.

Salvatore Greco, Benedetto Matarazzo and Roman Słowiński present the decision rule approach to MCDA. This approach represents the preferences in terms of "if ..., then ..." decision rules such as, for example, "if the maximum speed of car x is at least 175 km/h and its price is at most $12000, then car x is comprehensively at least medium". This approach is related to rough set theory and to artificial intelligence. Its main advantages are the following. The DM gives information in the form of examples of decisions, which requires relatively low cognitive effort and which is quite natural. The decision model is also expressed in a very natural way by decision rules. This permits an absolute transparency of the methodology for the DM. Another interesting feature of the decision rule approach is its flexibility, since any decision model can be expressed in terms of decision rules and, even better, the decision rule model can be much more general than all other existing decision models used in MCDA.

Michel Grabisch and Christophe Labreuche present the fuzzy integral approach that is known in MCDA for the last two decades. In very simple words this methodology permits a flexible modeling of the importance of criteria. Indeed, fuzzy integrals are based on a capacity which assigns an importance to each subset of criteria and not only to each single criterion. Thus, the importance of a given set of criteria is not necessarily equal to the sum of the importance of the criteria from the considered subset. Consequently, if the importance of the whole subset of criteria is smaller than the sum of the importances of its individual criteria, then we observe a redundancy between criteria, which in some way represents overlapping points of view. On the other hand, if the importance of the whole subset of criteria is larger than the sum of the importances of its members, then we observe a synergy between criteria, the evaluations of which reinforce one another. On the basis of the importance of criteria measured by means of a capacity, the criteria are aggregated by means of specific fuzzy integrals, the most important of which are the Choquet integral (for cardinal evaluations) and the Sugeno integral (for ordinal evaluations).

Finally, Helen Moshkovich, Alexander Mechitov and David Olson present the verbal decision methods MCDA. This is a class of methods originated from the work of one of the MCDA pioneers, the late Oleg Larichev. The idea of verbal decision analysis is to build a decision model using mostly qualitative information expressed in terms of a language that is natural for the DM. Moreover, measurement of criteria and preference elicitation should be psychologically valid. The methods, besides being mathematically sound, should check the DM's consistency and provide transparent recommendations.

4.6 Part VI: Multiobjective Mathematical Programming

The classical formulation of an Operations Research model is based on the maximization or minimization of an objective function subject to some constraints. A very rich and powerful arsenal of methodologies and techniques has been developed and continues to be developed within Operations Research. However, it is very difficult to summarize all the points of view related to the desired results of the decision at hand in only one objective function. Thus, it seems natural to consider a very general formulation of decision problems where a set of objective functions representing different criteria have to be "optimized". To deal with these types of problems requires not only to generalize the methodologies developed for classical single objective optimization problems, but also to introduce new methodologies and techniques permitting to compare different objectives according to the preferences of the DM. In this part of the book we tried to give adequate space to these two sides of multiobjective programming problems.

Emphasis on the side of gathering information from the decision-maker and consequent preference representation is given in the first chapter of this part, in which Pekka Korhonen introduces the main concepts and basic ideas of interactive methods dealing with multiobjective programming problems. The basic observation is that, since the DM tries to "maximize" a set of criteria in conflict with each other and an increment of one criterion can only be reached by accepting a decrement of at one or more other criteria, we need to compare the advantages coming from increments with respect to some criteria with the disadvantages coming from corresponding decrements of other criteria. A utility or value function representing DM preferences would seem the most appropriate for this aim, but the key assumption in multiple objective programming is that this utility function is unknown. Therefore many methodologies have been proposed with the aim of developing a fruitful dialogue with the DM permitting, on the one hand, to provide the DM with relevant information about non-dominated solutions and, on the other hand, to obtain useful information about the preferences of the DM. This dialogue is generally assisted by specific software, very often employing graphical representations of the results. It permits to define a solution which the DM can accept as a good compromise.

In the next chapter, Matthias Ehrgott and Margaret Wiecek introduce mathematical methods to solve multiobjective programming (MOP) problems. In their survey, they present solution concepts of MOP, properties of efficient and nondominated sets, optimality conditions, solution techniques, approximation of efficient and nondominated sets, and specially-structured problems including linear and discrete MOPs as well as selected nonlinear MOPs. The contents of the chapter have been selected on the idea that the primary (although not necessarily the ultimate) goal of multiobjective programming is to seek solutions of MOPs and therefore a special attention was paid to methods suitable for finding these solutions. Since the ultimate goal of MOP problem is selection of a preferred solution, for which an adequate representation of DM preferences is necessary, this chapter is well complemented by the previous one.

Masahiro Inuiguchi deals with multiple objective programming problems with fuzzy coefficients. The introduction of fuzziness in multiple objective programming is due to the observation that in real world problems imprecise specifications of parameters fluctuating in certain ranges are very usual. For example, let us consider an activity for which the acceptable expense is 100 million dollars. However, the DM may accept the expense of 100.1 million dollars if the objective functions take much better values by this small violation of the constraint. Due to their specific nature, fuzzy multiobjective programming problems need an interpretation which leads to specific approaches to the problem. Since fuzzy programming has a relatively long history, many approaches related to different interpretations of the fuzzy MOP have been proposed. In this chapter the approach based on necessity and possibility is considered, as many of the approaches proposed in the specialized literature are of this type. The difference to other approaches often lies solely in the measures employed for the evaluation of a fuzzy event. Thus, describing the approaches based on possibility and necessity measures would be sufficient to acknowledge the essence of multiple objective programming problems with fuzzy coefficients.

Finally, this part is concluded by a chapter that deals with an area of Operations Research in which multiobjective programming has been used quite frequently. Stefan Nickel, Justo Puerto and Antonio Rodríguez-Chía present the multiple criteria approach to locational analysis. An important characteristic of location models is their intrinsic multiple criteria nature. In this context different criteria are related to one or several new facilities and depend on the distances of these facilities to the set of fixed or demand facilities. There are at least two natural ways of deriving the different criteria. First, a decision about a new facility to be located is typically a group decision and each decision maker will have his own preferences, which may be expressed by a corresponding criterion. Secondly, the functions may represent different evaluation criteria for the new facility to be located, like cost, reachability, risk, etc. The chapter provides a broad overview of the most representative multiple criteria location problems which have been divided into the three classes of continuous, network, and discrete problems.

4.7 Part VII: Applications

It is apparent that the validity and success of all the developments of MCDA research are measured by the number and quality of the decisions supported by MCDA methodologies. Applications in this case discriminate between results that are really interesting for MCDA and results that, even though beautiful and interesting for economics, mathematics, psychology, or other scientific fields, are not interesting for MCDA. The applications of MCDA in real world problems are very numerous and in very different fields. Therefore, it was clear from the outset that it would be impossible to cover all the fields of application of MCDA. We decided to select some of the most significant areas.

Jaap Spronk, Ralph Steuer and Constantin Zopounidis discuss the contributions of MCDA in finance. A very valuable feature of their chapter is the focus on justification of the multidimensional character of financial decisions and the use of different MCDA methodologies to support them. The presentation of the contributions of MCDA in finance permits to structure complex evaluation problems in a scientific context and in a transparent and flexible way, with the introduction of both quantitative (i.e., financial ratios) and qualitative criteria in the evaluation process.

Danae Diakoulaki, Carlos Henggeler Antunes and António Gomes Martins present applications of MCDA in energy planning problems. In modern technologically developed societies, decisions concerning energy planning must be made in complex and sometimes ill-structured contexts, characterized by technological evolution, changes in market structures, and new societal concerns. Decisions to be made by different agents (at utility companies, regulatory bodies, and governments) must take into account several aspects of evaluation such as technical, socio-economic, and environmental ones, at various levels of decision making (ranging from the operational to the strategic level) and with different time frames. Thus, energy planning problems inherently involve multiple, conflicting and incommensurate axes of evaluation. The chapter aims at examining to which extent the use of MCDA in energy planning applications has been influenced by those changes currently underway in the energy sector, in the overall socio-economic context, and in particular to which extent it is adapted to the new needs and structuring and modelling requirements.

João Clímaco and José Craveirinha present multiple criteria decision analysis in telecommunication network planning and design. Decision making processes in this field take place in an increasingly complex and turbulent environment involving multiple and potentially conflicting options. Telecommunication networks is an area where different socio-economic decisions involving communication issues have to be made, but it is also an area where technological issues are of paramount importance. This interaction between a complex socio-economic environment and the extremely fast development of new telecommunication technologies and services justifies the interest in using multiple criteria evaluation in decision making processes. The chapter presents a review of contributions in these areas, with particular emphasis on network modernisation planning and routing problems and outlines an agenda of current and future research trends and issues for MCDA in this area.

Finally, Giuseppe Munda addresses applications of MCDA in problems concerning sustainable development. Sustainable development is strongly related to environmental questions, i.e., sustainable development generalizes environmental management taking into account not only an ecological but also socio-economic, technical and ethical perspectives. Ecological problems were among the first to be dealt with by MCDA. Therefore, there is a strong tradition in this field and many interesting stimuli for MCDA research came from there. The extensive perspective of sustainable development is very significant because it improves the quality of decisions concerning the environment taking into account other criteria, which are not strictly environmental but which strongly interact with it. In making sustainability policies operational, basic questions to be answered are sustainability of what and whom? As a consequence, sustainability issues are characterised by a high degree of conflict. Therefore, in this context MCDA appears as an adequate approach.

4.8 Part VIII: MCDM Software

Application of an MCDA method requires such a considerable amount of computation that even the development of many MCDA methodologies without the use of a specialized software is hardly imaginable. While software is an even more important element in the application of MCDA methodologies, this does not mean that to have a good software is sufficient to apply an MCDA methodology correctly. Clearly, software is a tool and it should be used as a tool. Before using a software, it is necessary to have a sound knowledge of the adopted methodology and of the decision problem at hand.

After these remarks about cautious use of software, the problem is: What software is available for MCDA? Heinz Roland Weistroffer, Subhash Narula and Charles H. Smith present well known MCDA software packages. While there is certainly some MCDA software available that is not present in the chapter, it can help the reader. She may get suggestions of well known software, but also information about aspects to be taken into account when evaluating a software for adoption in an application.

5. Acknowledgment to the Referees

The editors are very grateful to Euro Beinat, Nabil Belacel, Denis Bouyssou, John Buchanan, João Clímaco, Danae Diakoulaki, Luís Dias, Michael Doumpos, Ernest Forman, Philippe Fortemps, Lorraine Gardiner, Christophe Gonzales, Michel Grabisch, Winfried Hallerbach, Raimo P. Hämäläinen, Carlos Henggeler Antunes, Masahiro Inuiguchi, Robin Keller, Pekka Korhonen, Masahiro Inuiguchi, Christophe Labreuche, Risto Lahdelma, Thierry Marchant, Benedetto Matarazzo, Manuel Matos, Nikolaos Matsatsinis, Kaisa Miettinen, Maria Franca Norese, Wlodzimierz Ogryczak, Patrice Perny, Jacques Pictet, Marc Pirlot, Jean-Charles Pomerol, Justo Puerto, Marc Roubens, Roman Słowiński, Jerzy Stefanowsky, Ralph Steuer, Theo Stewart, Christianne Tammer, Jean-Claude Vansnick, Luis Vargas, Philippe Vincke, Peter Wakker, Margaret Wiecek, Szymon Wilk who served as referees for this volume.

The editors would also like to express their gratitude to people who supported them with very valuable advice along all the preparation of the book: Bernard Roy, Denis Bouyssou, Benedetto Matarazzo, Roman Słowiński, Gary Folven and Frederick S. Hillier.

Acknowledgments

José Figueira was supported by the grant SFRH/BDP/6800/2001 (Fundação para a Ciência e Tecnologia, Portugal) and gratefully acknowledges DIMACS Research Center at Rutgers University and LAMSADE at University Paris-Dauphine for the welcome during his sabbatical leave and the short visits to the Catania University, Auckland University and London School of Economics. His research has partially benefited also from MONET research grant (POCTI/GES/37707) and the luso-french scientifique collaborations ICCTI/Embassy of France in Lisbon (500B4) and Program Pessoa 2004. Matthias Ehrgott was partially supported by University of Auckland grant 3602178/9275 and by the Deutsche Forschungsgemeinschaft grant Ka 477/27-1.

References

[1] C. Bana e Costa, editor. *Readings in Multiple Criteria Decision Aid.* Springer Verlag, Heidelberg, 1990.

[2] R. Benayoun, B. Roy, and B. Sussman. ELECTRE : Une méthode pour guider le choix en présence de points de vue multiples. Note de travail 49, SEMA-METRA International, Direction Scientifique, 1966.

[3] J. Bentham. *The Principles of Morals and Legislation.* Prometheus Books, New York, 1988.

[4] J.-Ch. Borda. Mémoire sur les elections au scrutin. *Histoire de l'Academie Royale des Sciences (Paris)*, Année MDCCLXXXI:657–665, 1784. [Translated by Alfred de Garzia: Mathematical Derivation of an Election System, *Isis*, 44(1-2), 42-51, 1953.].

[5] E. Borel. La théorie du jeu et les équations integrales à noyau symétrique gauche. *Comptes Rendus de l'Académie des Sciences (Paris)*, 173:1304–1308, 1921.

[6] E. Borel. *Traité du Calcul des Probabilités et de ses Applications*. Gauthier-Villars, Paris, 1938.

[7] D. Bouyssou, T. Marchant, M. Pirlot, P. Perny, A. Tsoukiàs, and Ph. Vincke. *Evaluation and Decision Model. A Critical Perspective*. Kluwer Academic Publishers, Dordrecht, 2000.

[8] A. Charnes and W.W. Cooper. *Management Models and Industrial Applications of Linear Programming*. John Wiley & Sons, New York, 1961.

[9] J.L. Cochrane and M. Zeleny. *Multiple Criteria Decision Making*. University of South Carolina Press, 1973.

[10] Marquis de Condorcet. *Essai sur l'Application de l'Analyse à la Probabilité des Décisions Rendues à la Pluralité des Voix*. L'Imprimerie Royale, Paris, 1785.

[11] G. Fandel, J. Spronk, and B. Matarazzo. *Multiple Criteria Decision Methods and Applications*. Springer-Verlag, Berlin, 1985.

[12] Ph. Fortemps and R. Slowinski. A graded quadrivalent logic for preference modelling: Loyola-like approach. *Fuzzy Optimization and Decision Making*, 1(1):93–111, 2002.

[13] G. Hagele and F. Pukelsheim. Llull's writtings on electoral systems. *Studia Lulliana*, 41:3–38, 2001.

[14] St. Ignatius of Loyola. *Spiritual Exercises*. 1548. No. 178-183.

[15] T. Koopmans. Analysis of production as an efficient combination of activities. In T. Koopmans, editor, *Activity Analysis of Production and Allocations*, volume 13 of *Cowles Comission Monograph*, pages 33–97. Jonh Wiley and Sons, New York, 1951.

[16] H. Kuhn and A. Tucker. Nonlinear programming. In *Proceedings of the Second Symposium on Mathematical Statistics and Probability*, pages 481–492. University of California Press, Berkeley CA, 1951.

[17] K.R. MacCrimmon. An overview of multiple objective decision making. In J.L. Cochrane and M. Zeleny, editors, *Multiple Criteria Decision Making*, pages 18–43. University of South Carolina Press, 1973.

[18] I. McLean and A. Urken, editors. *Classics of Social Choice*. The University of Michigan Press, Ann Arbor, 1995.

[19] J. Neumann. Zur theorie der gesellschaftsspiele. *Mathematische Annalen*, 100:295–320, 1928.

[20] J. Neumann and O. Morgenstern. *Theory of Games and Economic Bahavior*. Princeton University Press, Princeton, 1943.

[21] P. Newman, editor. *F.Y. Edgeworth's "Mathematical Psychics" and Further Papers on Political Economy*. Oxford University Press, Oxford, 2003.

[22] V. Pareto. *Manuale di Economia Politica*. Società Editrice Libraria, Milano, 1906. Translated into English by Ann S. Schwier, edited by Ann S. Schwier and Alfred N. Page (1971) Manual of Political Economy, Augustus M. Kelley, New York.

[23] B. Roy. Classement et choix en présence de point de vue multiples: Le méthode ELECTRE. *Revue Francaise d'Informatique et de Recherche Opérationnelle*, 8:57–75, 1968.

[24] B. Roy. Critéres multiple et modélisation des préf'erence: L'apport des relations de surclassement. *Revue d'Economie Politique*, 1:1–44, 1974.

[25] B. Roy. *Méthodologie Multicritère d'Aide à la Décision*. Economica, Paris, 1985.

[26] B. Roy and D. Bouyssou. *Aide Multicritère à la Décision: Méthodes et Cas*. Economica, Paris, 1993.

[27] P. Sigmund. *Nicholas of Cusa and Medieval Political Thought*. Harvard University Press, Cambridge, MA, 1963.

[28] W. Stadler. A survey of multicriteria optimization or the vector maximum problem, Part I: 1776-1960. *Journal of Optimization Theory and Applications*, 29(1):1–52, 1979.

[29] Ph. Vincke. *Multicriteria Decision-Aid*. John Wiley & Sons, Chichester, 1992.

I

AN OVERVIEW OF MCDA TECHNIQUES TODAY

Chapter 1

PARADIGMS AND CHALLENGES

Bernard Roy
LAMSADE
Université Paris-Dauphine
Place du Maréchal De Lattre de Tassigny, 75775 Paris Cedex 16
France
roy@lamsade.dauphine.fr

Abstract The purpose of this introductory part is to present an overall view of what MCDA is today. In Section 1, I will attempt to bring answers to questions such as: what is it reasonable to expect from MCDA? Why decision aiding is more often multicriteria than monocriterion? What are the main limitations to objectivity? Section 2 will be devoted to a presentation of the conceptual architecture that constitutes the main keys for analyzing and structuring problem situations. Decision aiding cannot and must not be envisaged jointly with a hypothesis of perfect knowledge. Different ways for apprehending the various sources of imperfect knowledge will be introduced in Section 3. A robustness analysis is necessary in most cases. The crucial question of how can we take into account all criteria comprehensively in order to compare potential actions to one another will be tackled in Section 4. In this introductory part, I will only present a general framework for positioning the main operational approaches that exist today. In Section 5, I will discuss some more philosophical aspects of MCDA. For providing some aid in a decision context, we have to choose among different paths which one seems to be the most appropriate, or how to combine some of them: the path of realism which leads to the quest for a discussion for discovering, the axiomatic path which is often associated with the quest of norms for prescribing, or the path of constructivism which goes hand in hand with the quest of working hypothesis for recommending.

Keywords: Multiple criteria decision aiding, imperfect knowledge, aggregation procedures.

1. What Are the Expectations that Multicriteria Decision Aiding (MCDA) Responds to?

The purpose of this introductory chapter is to present an overview of what MCDA is today. Since the 60s, this discipline has produced, and it still produces, a great number of theoretical as well as applied papers and books. The major part of them will be presented in the following chapters of this book. It is important at the outset to understand their specific contributions are in terms of enlarging the operations research field and, more generally, to bringing light to decision making contexts. That is why I shall begin this chapter by considering the three following questions: what is reasonable to expect from MCDA? Why is decision aiding is more often multicriteria than monocriterion? What are the main limitations to objectivity which must be taken into account? The next section will be devoted to a brief presentation of three basic concepts which can be viewed as initial and fundamental keys for analyzing and structuring problem situations. In practice, it is very important to draw attention to questions such as: what is the quality of the information which can be obtained? What is the meaning of the data which are available or can be elaborated? In Section 3, I shall examine how the existing models and procedures take into account various types of answers to such questions which refer to a given problem's real world context.

Another difficulty in an MCDA context comes from the fact that comparisons between potential actions must be made comprehensively, with respect to all criteria. Various aggregation techniques which will be described in detail throughout the successive chapters of this book have been proposed and used in order to overcome this kind of difficulty. In Section 4, I shall present a general framework for positioning the main operational approaches in which these aggregation techniques come into play. Some more general philosophical considerations will complete this introductory chapter.

1.1 What Is Reasonable to Expect from Decision Aiding (DA)?

Decision aiding can be defined (see [61]) as follows: Decision aiding is the activity of the person who, through the use of explicit but not necessarily completely formalized models, helps obtain elements of responses to the questions posed by a stakeholder in a decision process. These elements work towards clarifying the decision and usually towards recommending, or simply favoring, a behavior that will increase the consistency between the evolution of the process and this stakeholder's objectives and value system. In this definition, the word "recommending" is used to draw attention to the fact that both analyst and decision maker are aware that the decision maker is completely free to behave as he or she sees fit after the recommendation is made. This term is increasingly used in DA to replace "prescription". The latter is, in many cases, inappropriate (see [34, 58]) for designating what a team of analysts accompanying a decision making process might achieve.

Thus defined, DA aims at establishing, on recognized scientific bases, with reference to working hypotheses, formulations of propositions (elements of responses to questions, a presentation of satisfying solutions or possible compromises, ...) which are then submitted to the judgment of a decision maker and/or the various actors involved in the decision making process. According to the case, DA can thus reasonably contribute to:

- analyzing the decision making context by identifying the actors, the various possibilities of action, their consequences, the stakes, ...;
- organizing and/or structuring how the decision making process unfolds in order to increase coherence among, on the one hand, the values underlying the objectives and goals, and, on the other hand, the final decision arrived at;
- getting the actors to cooperate by proposing keys to a better mutual understanding and a framework favorable to debate;
- elaborating recommendations using results taken from models and computational procedures conceived of within the framework of a working hypothesis;
- participating in the final decision legitimization.

For a deeper understanding of the bases reviewed above, the reader can refer to [12, 13, 19, 20, 40, 48, 59, 68].

1.2 Why Is DA More Often Multicriteria than Monocriterion?

Even when DA is provided for a single decision maker, it is rare for her or him to have in mind a single clear criterion. Thus, when DA takes place in a multi-actor decision making process, it is even rarer for there to be *a priori* a single, well-defined criterion deemed acceptable by all actors to guide the process. This process is often not very rational. Each actor plays a more or less well defined role which gives priority to her or his own objectives and value system.

In both cases, it is necessary to take into consideration various points of view dealing with, for example, finance, human resources, environmental aspects, delays, security, quality, ethics,... By considering each pertinent point of view separately, independently from the others, it is generally possible to arrive at a clear and common elicitation of preferences regarding the single point of view considered. This naturally leads to associating a specific criterion to each pertinent point of view. Each of these criteria is used to evaluate any potential action on an appropriate qualitative or quantitative scale. In most cases, there is no obvious and acceptable arithmetic rule which can keep account of these heterogeneous scales by substituting a single scale based on a common unit for each of them (see Section 4 below).

Such a scale, bringing a common unit into play, must be introduced a priori when we want to avoid a multicriteria approach, i.e., when we prefer to choose what is called a *monocriterion* approach. This choice, in many decision making contexts, might:

- lead to wrongly neglecting certain aspects of realism;
- facilitate the setting up of equivalencies, the fictitious nature of which remains invisible;
- tend to present features of one particular value system as objective.

On the contrary, a multicriteria approach contributes to avoiding such dangers by:

- delimiting a broad spectrum of points of view likely to structure the decision process with regard to the actors involved;
- constructing a family of criteria which preserves, for each of them, without any fictitious conversion, the original concrete meaning of the corresponding evaluations;
- facilitating debate on the respective role (weight, veto, aspiration level, rejection level,...) that each criterion might be called upon to play during the decision aiding process.

Additional considerations about relative advantages on monocriterion and multicriteria approaches can be found in [10, 14, 21, 56, 61].

1.3 Can MCDA Be Always Totally Objective?

In many cases, those who claim to shed light objectively on a decision in fact take a stand – consciously or unconsciously – for an a priori position or for a prevailing hypothesis which they then seek to justify. Arguments for making a decision are thus put forward more in the spirit of advocacy than in that of an objective search (see [3, 32]).

In what follows, we will only consider situations in which DA is motivated by a strong desire for objectivity. Even in such situations, it is important to be sensitive to the existence of some fundamental limitations on objectivity. Their origins lie in the following facts:

a) The borderline between what is and what is not feasible is often fuzzy in real decision making contexts. Moreover, this borderline is frequently modified in the light of what is found through the study itself.

b) Even in cases for which DA is provided for a well-defined decision maker, his or her preferences very seldom seem well-shaped. In and among areas of firm convictions lie hazy zones of uncertainty, half held belief, or indeed conflicts and contradictions. Such sources of ambiguity or arbitrariness concerning preferences which are to be elicited and modeled are even more present when the decision maker (entity for whom or in the name of whom decision aiding is provided for) is a mythical person, or when decision aiding is provided in a multicriteria context. We have to admit, therefore, that the study itself contributes to eliminating questioning, solving conflicts, transforming contradictions and destabilizing certain convictions. Any interaction and questioning between the analyst and the decision maker, or any actors involved into the decision making process, may have some an unpredictable or imperceptible effect.

c) Many data (see Section 3 below) are imprecise, uncertain, or ill-defined. There is a real risk of making them say much more than they mean. Moreover, some of them only reflect features of a particular individual value system.

d) In general, it is impossible to say that a decision is a good one or a bad one by referring only to a mathematical model. Organizational, pedagogical, and/or cultural aspects of the whole decision process which lead to making a given decision also contribute to its quality and success.

Rather than dismissing or canceling the subjectivity which results from the limitations of objectivity described above, decision aiding must make an objective place for it. (For a pedagogical overview of MCDA approaches, see [58, 59, 64].)

2. Three Basic Concepts

From the beginning to the end of work in MCDA, three concepts usually play a fundamental role for analyzing and structuring the decision aiding process in close connection with the decision process itself. The presentation of these concepts in the three following sub-sections is obviously succinct. It nevertheless aims to draw attention to some important features.

2.1 Alternative, and More Generally, Potential Action

By *potential action*, we usually designate that which constitutes the object of the decision, or that which decision aiding is directed towards. The concept of *action* does not *a priori* incorporate any notion of feasibility, or possible implementation. An action is qualified as potential when it is deemed possible to implement it, or simply when it deserves some interest within the decision aiding process.

The concept of *alternative* corresponds to the particular case in which modeling is such that two distinct potential actions can in no way be conjointly put into operation. This mutual exclusion comes from a way of modeling which in a comprehensive way tackles that which is the object of the decision, or that towards which DA is directed. Many authors implicitly suppose that potential actions are, by definition, mutually exclusive. Nevertheless, such an hypothesis is in no way compulsory. In many real world decision aiding contexts, it can be more appropriate to adopt another way of modeling such that several potential actions can be implemented conjointly (see examples in [55, 61]).

In all cases, A will denote the set of potential actions considered at a given stage of the DA process. This set is not necessarily stable, i.e., it can evolve throughout the decision aiding process. Such an evolution may come from the study's environment, but also from the study itself. The study may shed light on some aspects of the problem, which could lead to revising some of the data and then, possibly, to modifying the borderline between what is and what is not feasible.

By a, we will denote any potential action or alternative. When the number of actions is finite ($|A| = m$) we shall let:

$$A = \{a_1, a_2, \ldots, a_m\}.$$

When modeling of actions can be done by referring to some variables $x_1, x_2, \ldots$ it is possible to write:

$$a = (x_1, x_2, \ldots).$$

In such cases, A is generally defined by a set of analytic constraints which characterize the borderline between what is feasible and is not feasible. Multi-objective mathematical programming constitutes an important particular case of this type of modeling (see [25] and Part VI).

In another type of modeling, the value of each variable x_i ($i = 1, 2, \ldots, n$) designates a possible score on an appropriate scale X_i built for evaluating actions according to a specified point of view or criterion. In such cases, A can be viewed as a subset of the Cartesian product $X = \prod_{i=1}^{n} X_i$. This type of modeling is commonly used in multiattribute utility theory (MAUT) (see Part IV). Let us observe that this type of modeling necessitates some precautions: since each potential action is identified with the n components of its evaluation, it loses all concrete identity; in particular, two actions having the same evaluations $x_1, \ldots, x_n$ are no longer distinguishable.

More details and illustrations of the concepts and ways of modeling presented above could be found in [61, Chapter 5], [84, Chapter 1], and [36].

2.2 Criterion and Family of Criteria

The reader, who is not yet familiar with some of the terms used in this subsection, will find more precise definitions in [64, Chapter 1, Appendix 1, Glossary].

Let us remember that a criterion g is a tool constructed for evaluating and comparing potential actions according to a point of view which must be (as far as it is possible) well-defined. This evaluation must take into account, for each action a, all the pertinent effects or attributes linked to the point of view considered. It is denoted by $g(a)$ and called the *performance* of a according to this criterion.

Frequently, $g(a)$ is a real number, but in all cases, it is necessary to define explicitly the set X_g of all the possible evaluations to which this criterion can lead. For allowing comparisons, it should be possible to define a complete order $<_g$ on X_g: $(<_g, X_g)$ is called the *scale* of criterion g. To be accepted by all stakeholders, a criterion should not bring into play, in a way which might be determinant, any aspects reflecting a value system that some of these stakeholders would find necessary to reject. This implies in particular that the direction to which the preferences increase along the scale (and more generally the complete order $<_g$) is not open to contest.

Elements $x \in X_g$ are called *degrees* or *scores* of the scale. Each degree can be characterized by a number, a verbal statement or a pictogram. When in order to compare two actions according to criterion g we compare the two degrees used for evaluating their respective performances, it is important to analyze the concrete meaning in terms of preferences covered by such degrees. This leads to distinguishing various types of scales:

a) *Purely ordinal scale*: Scale such that the gap between two degrees does not have a clear meaning in terms of difference preferences; this is the case with:

- a *verbal scale* when nothing allows us to state that the pairs of consecutive degrees reflect equal preference differences all along the scale;
- a *numerical scale* when nothing allows us to state that a given difference y between two degrees reflects an invariant preference difference when we move the pair of degrees considered along the scale.

This type of scale is often called a qualitative scale.

b) *Quantitative scale*: Numerical scale whose degrees are defined by referring to a clear, concrete defined quantity in a way that it gives meaning, on the one hand, to the absence of quantity (degree 0), and on the other hand, to the existence of a unit allowing us to interpret each degree as the addition of a given number (integer or fractional) of such units. In such conditions, the ratio between two degrees can receive a meaning which does not depend on the two particular degrees considered; this is another way of defining quantitative scales which are also called *cardinal* or *ratio scales*.

c) *Other types*: In MCDA, we do not always work with scales belonging to one of the above two extreme types (especially interval scales). The most interesting intermediate types are presented in [41, Section 2] and [64].

In MCDA, it is essential to know which type of scale we are working with to be sure of using its degrees in a meaningful way. According to the type of scales considered, certain kinds of reasoning and arithmetical operations are significant in terms of preference (see Chapter 2).

Moreover, the use of the degrees in a significant way must take the following fact into account: the difference between two degrees that are sufficiently close together may be non significant for justifying an indisputable preference in favor of one of the two actions. This stems from the procedure used to position the two actions on the scale considered. This procedure can appear insufficiently precise (with regard to the complexity of the reality in question), or insufficiently reliable (with regard to uncertainty concerning the future) for founding such an indisputable preference on such a small difference. I will come back to this subject in the next section.

In most cases, the first step of DA consists of building n criteria with $n > 1$ (see 1.2 above). They constitute what we call the *family F of criteria*. In order to be sure that F is able to play its role in the DA process correctly, i.e., in laying the foundations for convictions, communicating concerning the latter, debating and orienting the process towards the decision, and in contributing in some cases to legitimating this decision, it is necessary to verify that:

- what is apprehended by each criterion is sufficiently intelligible for each of the stakeholders;
- each criterion is perceived to be a relevant instrument for comparing potential actions along the scale which is associated with it without pre-judging their relative importance, which could vary considerably from one stakeholder to another;
- the n criteria considered all together satisfy some logical requirements (exhaustiveness, cohesiveness, and non redundancy) which insure coherence of the family (for more details, see [64, Chapter 1, Appendix 1, Glossary], [67]).

It is important to observe that none of the above requirements implies that the n criteria of F must be independent. The concept of independence is very complex, and if dependence is desirable, it is necessary to specify what type of independence is needed. Multicriteria analysis has led to important distinctions between structural independence, preferential independence, and utility independence (see [37, 54], [61, Chapter 10], and [67, Chapter 2]).

Additional developments concerning this basic concept of criterion can be found in [1, 2, 4, 5, 7].

2.3 Problematic as a Way in which DA May Be Envisaged

The word "problematic" is used here in the sense indicated by the heading. Other expressions such as "statement", "problem formulation" or "problem type" have been used as substitutes, but in my view, they are inappropriate and may lead to misunderstanding.

Let us underline first that DA must not be envisaged solely in the perspective of solving a problem of choice. In some cases, DA consists only of elaborating an appropriate set A of potential actions, building a suitable family F of criteria, and determining, for all or some $a \in A$, their performances sometimes completed by additional information (possible values for discrimination thresholds, aspiration and/or rejection levels, weights,...). For designating this manner of conceiving of DA's aim without seeking to elaborate any prescription, or recommendation, we use the term *description problematic* often coded $P.\delta$.

In MCDA, the word *problematic* refers to the way in which DA is envisaged. This means that the problematic deals with answers to questions such as the following: in what terms should we pose the problem?, what type of results should we try to obtain?, how does the analyst see himself fitting into the decision process to aid in arriving at these results?, what kind of procedure seems the most appropriate for guiding his investigation? In addition to $P.\delta$, three other reference problematics are currently used in practice. They can be briefly described as follows (for more details, see [52], [61, Chapter 6]):

- The *choice problematic* ($P.\alpha$): The aid is oriented towards and lies on a *selection* of a small number (as small as possible) of "good" actions in such a way that a single alternative may finally be chosen; this does not mean that the selection is necessarily oriented towards the determination of one or all the actions of A which can be regarded as optimum; the selection procedure can also, more modestly, be based on comparisons between actions so as to justify the elimination of the greatest number of them, the subset N of those actions which are selected (which can be viewed as a first choice) containing all the most satisfying actions, which remain non comparable between one another.

- The *sorting problematic* ($P.\beta$): The aid is oriented towards and lies on an *assignment* of each action to one category (judged the most appropriate) among those of a family of predefined categories; this family must be conceived on the basis of the diverse types of treatments, or judgments conceivable for the actions which motivate the sorting. For instance, a family of four categories can be based on a comprehensive appreciation leading to distinguishing between: actions for which implementation (i) is fully justified, (ii) could be advised after only minor modifications, (iii) can only be advised after major modifications, (iv) is unadvisable. Let us observe that categories are not necessarily ordered as it is the case in the above examples.

- The *ranking problematic* ($P.\gamma$): The aid is oriented towards and lies on a *complete* or *partial preorder* on A which can be regarded as an appropriate instrument for comparing actions between one another; this preorder is the result of a *classifying* procedure allowing us to put together in classes actions which can be judged as indifferent, and to rank these classes (some of them may remain non-comparable).

The four problematics described above are not the only possible ones (see [10, 11]). Whatever the problematic adopted, the result arrived at by treating a given set of data through a single procedure is (except under unusual conditions) not sufficient for founding a prescription or a recommendation (see Section 4 below).

3. How to Take Into Account Imperfect Knowledge?

DA cannot be correctly provided without trying to analyze and to take into account reasons and factors which can be responsible for contingency, arbitrariness, and ignorance in the way the problem is envisaged and procedures implemented. In addition to their subjective characteristics, these reasons and factors may take on various forms whose presence and/or importance greatly depends on the decision making context considered. Their presence comes essentially from three sources (for more details, see [16, 57]):

- *Source* α ($S.\alpha$): The imprecise, uncertain and, more generally, poorly understood or ill-defined nature of certain specific features or factual quantities or qualities present in the problem.

- *Source* β ($S.\beta$): the conditions for implementing the decision taken; these will be influenced by:

  - The state of the context at the time the decision is implemented if it is a once-and-for-all decision;

- The successive states of the context if the decision is sequential.

- *Source* γ ($S.\gamma$): the fuzzy or incomplete, sometimes unstable and easily influenced character of the system or systems of values to be taken into account; these values involve, in particular, the system and most often the systems of preferences which should prevail in order to evaluate the feasibility and relative interest of diverse potential actions, by considering the conditions envisaged for implementing these actions.

Considering the problem formulation, a study of these three sources must shed light on that which appears imprecise, uncertain, unstable or ill-defined. This can leads, for instance:

- starting from $S.\alpha$, to delimiting a domain of reasonable instantiation values for various data and parameters;

- from $S.\beta$, to building a set of scenarios describing different possible future contexts;

- from $S.\gamma$, to eliciting a set of weight vectors; for this purpose, it is important to remember that it makes no sense and is theoretically incorrect to specify measures of relative importance for the criteria without considering the nature of the overall evaluation model which will be used, i.e., without having defined the type of mathematical aggregation rules (see next section) which allow us to derive comprehensive preferences.

The DA process must clearly take into account all the results of this study. To do so, many approaches (formalisms, models, methods, ...) have been proposed. A panorama of such approaches can be found in Chapter 11, and in [39, 76]. These approaches rest upon various concepts, tools and theories; the main ones are:

- probability theory mainly used in MAUT (see Chapter 7), but also used in many other approaches, particularly for building criteria when uncertainty can be characterized by a probabilistic distribution;

- possibility theory [22, 24];

- multi-valued logic (see Chapter 3, [80]);

- concept of discrimination thresholds and quasi or pseudo-criterion (see Chapter 2, and for more details, [35, 70, 71, 72, 87]) mainly used in outranking methods (see Chapter 4).

- concept of fuzzy binary relations [18, 23, 27, 28, 44, 53];

- rough sets theory (see Chapter 13).

Whatever the formalism, the models, the methods used, it is generally indispensable to undertake a robustness analysis. According to which author uses it (*cf.* [66]), this term can cover different ways of proceeding. Nevertheless, the aim is always to distinguish the part of the results which can be firmly established in order to choose an appropriate method (*cf.* [85, 86]), to derive robust solutions (*cf.*[26, 33, 38, 50, 51]), or to formulate robust conclusions (*cf.* [62, 63, 78]). In [62, 63], the reader could find some comments on links and differences between robustness analysis and sensitivity analysis.

4. An Operational Point of View

As soon as more than one criterion comes into play, a crucial question arises: how can we take into account all criteria comprehensively in order to compare potential actions to one another? Let us consider two potential actions a and b together with their respective performances on the n criteria considered. More often, a will be better than b for some of the criteria, and b better than a for others. In such cases, in comparing a and b, on what basis can we found a *comprehensive judgment*, i.e., taking into account, in a comprehensive way, the n performances of a and the n performances of b. This problem is usually called the *aggregation problem*. In many of the chapters in this book, the reader will find a wide variety of solutions to this fundamental problem. In the present introductory chapter, I shall present only a general framework for positioning the main operational approaches provided for DA today (for more details on what the operational approach concept covers, see, [61, Chapter 11].

4.1 About Multicriteria Aggregation Procedures

The most frequently used decision aiding methods are based on mathematically explicit *multicriteria aggregation procedures* (MCAP). By definition, an MCAP is a procedure which, for any pair of potential actions, gives a clear answer to the aggregation problem. It brings into play:

i) various inter-criteria parameters such as weights, scaling constants, veto, aspiration levels, rejection levels,... which allow us to define the specific role that each criterion can play with respect to the others; some more technical parameters can also be present;

ii) A logic of aggregation: this logic should take into account:

- The possible types of dependence which we might want to bring into play concerning criteria,
- The conditions under which we accept or refuse compensation between "good" and "bad" performances.

In order to give a numerical value to inter-criteria parameters and more technical parameters, it is absolutely necessary to refer to the logic of aggregation of the MCAP considered. Outside of this logic, those parameters have no meaning.

For more details on the above considerations, see [15, 42, 60, 67, 69, 79].

Methods which are based on a mathematically explicit MCAP come under one of two types of operational approaches usually designated by the expressions approach based on a *synthesizing criterion* and approach based on a *synthesizing preference relational system.*

4.2 Approach Based on a Synthesizing Criterion

This approach is the most traditional. It can be characterized as follows: formal rules taking into account the n performances of any potential action $a \in A$ are defined so as to assign to a a well defined position (generally by means of a numerical value) on an appropriate scale.

The way the aggregation is addressed in this approach leads to defining a complete preorder on A. Most often, the formal rules consist of a mathematical formula which leads to an explicit definition of a unique criterion synthesizing the n criteria. This is the case with MAUT, SMART, TOPSIS, MACBETH, AHP, ... (see Chapters 7 – 10). The complete preorder on A can also be obtained by the use of a set of formal rules without any mathematically explicit expression of the synthesizing criterion, which remains implicit (see [6, 7]). In any case, this approach does not allow any incomparability.

Building a synthesizing criterion using such a multicriteria approach is not equivalent to a monocriterion approach. The dangers of the monocriterion approach have been presented above (see Section 1.2). Nevertheless, even if a multicriteria approach based on a synthesizing criterion contributes to reducing these dangers, it forces us to introduce a common scale (monetary scale, utility scale,...) on which performances of each of the n criteria have to be evaluated. Moreover, with this approach, imperfect knowledge (*cf.* Section 3 above) can be taken into account solely through probabilistic or fuzzy models.

4.3 The Operational Approach Based on a Synthesizing Preference Relational System

As is the first, this second operational approach is based on a mathematically explicit MCAP. A major difference with the preceding approach comes from the fact that here the MCAP does not work on each potential action a separately from the others, but it successively compares a to each of the other $b \in A$.

In other words, the aggregation problem is no longer addressed in terms of defining a complete preorder on A, it is now addressed in terms of pairwise comparisons so as to design a synthesizing preference relational system. Taking into account the n performances of a and the n performances of b, the role of the MCAP is to give an answer to the question: what is the preference relation which can be validated between a and b? Mathematical rules, which lead to answering this question, are based on:

- various inter-criteria parameters, as in the first approach; but also, unlike the first approach, on discrimination thresholds (see Section 3 above) and veto thresholds;
- a logic of aggregation which easily allows us to take into account (and this is much more difficult with the approach based on a synthesizing criterion), on one hand, some limitations to compensation, and on the other, no quantitative performances.

The synthesizing preference relational system can be reduced to a single binary relation, which can be crisp or fuzzy. But it can also bring into play more than one binary relation. In all instances, the advantages of this second type of MCAP relative to the first cause certain difficulties to arise when we consider the operational approach based on such an MCAP. These difficulties stem from the fact that:

- pairwise comparisons can cause some intransitivities to appear;
- incomparability can be the most appropriate conclusion for comparing certain pairs (a, b);
- consequently, a synthesizing relational preference system is not a tool which is immediately usable for elaborating a recommendation.

For these reasons, this second operational approach necessitates completing the MCAP by a second procedure called *exploitation procedure*. This procedure is conditioned by the problematic considered (see above 2.3).

This second operational approach has led to various methods, most of which are covered by the label of *outranking methods* . The second part of this book is devoted to them. Other works related to this approach are presented in Part V.

4.4 About Other Operational Approaches

All the operational approaches which are based on a mathematically explicit MCAP are not exactly in accordance with one of the two preceding approaches. Regarding this subject, the reader can refer to [8, 17, 31, 43, 75].

Finally, let us mention the existence of operational approaches which are not based on a mathematically explicit MCAP, when this procedure remains implicit. Such approaches often make use of interactivity. A formal procedure is then conceived for asking questions of the decision maker or some other actor.

This procedure leads to an ad hoc sequence of judgments and a progression by trial and error. These judgments have only a local meaning because they refer to the neighborhood of one or a very small number of actions. For more details on this kind of approach, see Chapter 16, [29, 45, 46], [67, Chapter 7], [77, 81, 82, 83, 84, 89, 90].

In any case, whatever the operational approach considered, there is a possible confusion which should be avoided. Except under very unusual conditions, the results arrived at by treating a set of data through any appropriate procedure should not be confused with a well founded scientific recommendation. Repeated calculations using different but equally realistic versions of the DA problem (sets of data, scenarios,...) are generally necessary to elaborate a recommendation on the basis of robust conclusions stemming from the multiple results thus obtained. The statement of the proposals which make up the recommendation should be submitted to the assessment and discernment of the decision maker and/or the actors involved in the DA process (see [34, 47, 63, 64, 65, 66, 78]).

5. Conclusion

The final objective of MCDA is, of course, to help managers to make "better" decisions. But what is the meaning of better? This meaning depends, in part, on the process by which the decision is made and implemented. This, combined with limitations on objectivity described above (see Section 1.3), shows that we cannot hope to prove scientifically, in a decision making context, that a given decision is the best. In other words, it is impossible to consider that in every situation there exists, somewhere, the right selection, the right assignment, the right ranking which could be considered and discovered or approximated independently of any procedure. This implies that the concepts, models and procedures presented in this book must not be viewed as being conceived from the perspective of discovering, with a better or a worst good approximation, a pre-existing truth which could be universally imposed. They have to be seen as keys capable of opening doors giving access to answers and/or expectations as described in Section 1.1.

Thus conceived, methodological decision aiding based upon appropriate concepts, models and procedures can play a significant and beneficial role helping us to make our way in the presence of ambiguity, uncertainty and an abundance of bifurcations in order to guide the decision making process.

To achieve this goal, three non exclusive paths can be envisaged:

- the path of realism which leads to the quest for a description for discovering;
- the axiomatic path which is often associated with the quest for norms for prescribing;

- the path of constructivism which goes hand in hand with the quest for a working hypothesis for recommending.

(for more details on each of these paths, see [58]). In a DA process, it is important, when following one or a combination of such paths, to shed light on:

- those aspects of reality which give meaning, value and order to facts;
- the influence exerted upon this reality by observing it, organizing it, provoking within it certain forms of debate, or even having certain tools placed there.

Personally, I consider that the path of realism can only play a role in producing certain descriptions of physical, institutional, socio-economic, financial or psychological systems which form the decision making context. Insofar as such descriptions are produced by other disciplines than DA strictly speaking, the contribution of DA comes essentially, in my opinion, from the constructivism path taken in conjunction with (observing certain precautions) the axiomatic path. Interesting developments and other points of view can be found in [9, 12, 30, 36, 40, 49, 73, 74, 88, 91, 92, 93].

Acknowledgments

I would like to express my most heartfelt thanks to Evelyn Perry and Dominique François for their invaluable help in putting together the final version of this chapter. Blurred nuances, obscure passages, possible misinterpretations – these are the risks we take when presenting a synthesis of rich and complex concepts, results and points of view in a language that is not our own. Such risks have been greatly reduced through a number of informative discussions with Evelyn Perry, discussions including a sentence-by-sentence examination of my first draft, typed from dictation by Dominique François, who with all her customary care and intelligence has followed the process resulting in the present version from beginning to end. The questions asked and suggestions given by Dominique François and Evelyn Perry have also proved extremely useful. Once again, I would like to thank them both for the quality of their collaboration.

References

[1] A. Aït Younes. *Problèmes liés à la construction d'un pseudo-critère: Développements théoriques et implémentation informatique*. PhD thesis, Université Paris-Dauphine, 2001.

[2] A. Aït Younes and B. Roy. Prise en compte d'une connaissance imparfaite à l'aide d'un pseudo-critère : Procédure interactive de construction. *Journal of Decision Systems*, 2004. To appear.

[3] J. Armstrong. Advocacy and objectivity in science. *Management Science*, 25(5):423–428, 1979.

[4] R. Azibi. *Construction des critères en aide à la décision: Aspects méthodologiques, techniques et pratiques.* PhD thesis, Université Paris-Dauphine, 2003.

[5] R. Azibi and D. Vanderpooten. Elaboration de critères agrégeant des conséquences dispersées: Deux exemples concrets. In A. Colorni, M. Paruccini, and R. Roy, editors, *A-MCD-A: Multiple Criteria Decision Aiding*, pages 13–30. Joint Research Center. The European Commission, Luxembourg, 2001.

[6] R. Azibi and D. Vanderpooten. Construction of rule-based assignment models. *European Journal of Operational Research*, 138(2):274–293, 2002.

[7] R. Azibi and D. Vanderpooten. Aggregation of dispersed consequences for constructing criteria: the evaluation of flood risk reduction strategies. *European Journal of Operational Research*, 44:397–411, 2003.

[8] C. Bana e Costa, editor. *Readings in Multiple Criteria Decision Aid.* Springer Verlag, Heidelberg, 1990.

[9] C. Bana e Costa and M. Pirlot. Thoughts on the future of multicriteria field: Basic convictions and outline for general methodology. In J. Clímaco, editor, *Multicriteria Analysis*, pages 562–568. Springer Verlag, Berlin, 1997.

[10] C. Bana e Costa, T. Stewart, and J.C. Vansnick. Multicriteria decision analysis: Some thoughts based on the tutorial and discussion sessions of ESIGMA meetings. *European Journal of Operational Research*, 99:28–37, 1997.

[11] C.A Bana e Costa. *Structuration, construction et exploitation d'un modèle multicritère d'aide à la décision.* PhD thesis, Instituto Superior Técnico, Lisboa, Portugal, 1992.

[12] M. Banville, M. Landry, J-M. Martel, and C. Boulaire. A stakeholder approach to MCDA. *Systems Research and Behavioral Science*, 15:15–32, 1998.

[13] V. Belton and J. Pictet. Talking about the practice of MCDA. In D. Bouyssou, E. Jacquet-Lagrèze, P. Perny, R. Slowinski, D. Vanderpooten, and Ph. Vincke, editors, *Aiding Decisions with Multiple Criteria: Essays in Honour of Bernard Roy*, pages 71–88. Kluwer Academic Publishers, 2001.

[14] P. Bertier and J. de Montgolfier. Comment choisir en tenant compte de points de vue non commensurables? *Analyse et Prévision*, XI:521–548, 1971.

[15] D. Bouyssou. Some remarks on the notion of compensation in MCDM. *European Journal of Operational Research*, 26:150–160, 1986.

[16] D. Bouyssou. Modelling inaccurate determination, uncertainty, imprecision using multiple criteria. In A. Lockett and G. Islei, editors, *Improving Decision Making in Organisations*, volume 335 of *Lecture Notes in Economics and Mathematical Systems*, pages 78–87. Springer Verlag, Berlin, 1989.

[17] A. Chipperfield, J. Whidborne, and P. Fleming. Evolutionary algorithms and simulated annealing for MCDM. In T. Gal, T. Stewart, and T. Hanne, editors, *Multicriteria Decision Making: Advances in MCDM Models, Algorithms, Theory and Applications*, pages 16.1–16.32. Kluwer Academic Publishers, Boston, 1999.

[18] P. Czyzak and R. Slowinski. Possibilistic construction of fuzzy outranking relation for multiple-criteria ranking. *Fuzzy Sets and Systems*, 81 (1):123–131, 1996.

[19] S. Damart. A note on the concept of knowledge management in decision aid activity. In *DSI-Age 2002 (Decision Making and Decision Support in the Internet Age 2002)*, Cork, Ireland, July 4-7, 2002.

[20] A. David. Decision-aid between tools and organisations. In D. Bouyssou, E. Jacquet-Lagrèze, P. Perny, R. Slowinski, D. Vanderpooten, and Ph. Vincke, editors, *Aiding Decisions with Multiple Criteria: Essays in Honour of Bernard Roy*, pages 45–69. Kluwer Academic Publishers, 2001.

[21] J. de Montgolfier and P. Bertier. *Approche Multicritère des Problèmes de Décision*. Editions Hommes et Techniques, Paris, 1978.

[22] D. Dubois. *Modèles mathématiques de l'imprécis et de l'incertain en vue d'applications aux techniques d'aide à la décision*. Mémoire d'habilitation, Université Paul Sabatier, Toulouse, 1983.

[23] D. Dubois, J. Marichal, H. Prade, and M. Roubens. The use of the discrete Sugeno integral in decision-making: A survey. In A. Colorni, M. Paruccini, and B. Roy, editors, *A-MCD-A: Multiple Criteria Decision Aiding*, pages 81–98. Joint Research Center. The European Commission, Luxembourg, 2001.

[24] D. Dubois and H. Prade. *Théorie des possibilités: Application à la représentation des connaissances en informatique*. Masson, Paris, 1985.

[25] M. Ehrgott and X. Gandibleux, editors. *Multiple Criteria Optimization. State of the Art Annotated Bibliographic Surveys*. Kluwer Academic Publishers, Dordrecht, 2002.

[26] L. Escudero. Robust decision making as a decision making aid under uncertainty. In S. Rios, editor, *Decision Theory and Decision Analysis: Trends and Challenges*, pages 127–138. Kluwer Academic Publishers, Dordrecht, 1994.

[27] H. Fargier and P. Perny. Modélisation des préférences par une règle de concordance généralisée. In A. Colorni, M. Paruccini, and B. Roy, editors, *A-MCD-A: Multiple Criteria Decision Aiding*, pages 99–116. Joint Research Center. The European Commission, Luxembourg, 2001.

[28] J. Fodor and M. Roubens. *Fuzzy Preference Modeling and Multicriteria Decision Support*. Kluwer Academic Publishers, Dordrecht, 1994.

[29] L. Gardiner and D. Vanderpooten. Interactive multiple criteria procedures: Some reflections. In J. Clímaco, editor, *Multicriteria Analysis*, pages 290–301. Springer Verlag, Berlin, 1997.

[30] J. Genard and M. Pirlot. Multi-criteria decision-aid in a philosophical perspective. In D. Bouyssou, E. Jacquet-Lagrèze, P. Perny, R. Slowinski, D. Vanderpooten, and Ph. Vincke, editors, *Aiding Decisions with Multiple Criteria: Essays in Honour of Bernard Roy*, pages 89–117. Kluwer Academic Publishers, Dordrecht, 2001.

[31] S. Greco, B. Matarazzo, and R. Slowinski. The use of rough sets and fuzzy sets in MCDM. In T. Gal, T. Stewart, and T. Hanne, editors, *Multicriteria Decision Making: Advances in MCDM Models, Algorithms, Theory, and Applications*, pages 14.1–14.59. Kluwer Academic Publishers, Dordrecht, 1999.

[32] Groupe de Reflexion sur l'Economie des Transports Urbains (GRETU). *Une étude économique a montré... Mythes et réalités sur les études de transports*. Editions Cujas, Paris, 1980.

[33] S. Gupta and J. Rosenhead. Robustness in sequential investment decisions. *Management Science*, 15(2):18–29, 1972.

[34] A. Hatchuel. Coopération et conception collective: Variété et crises des rapports de prescription. In G. De Terssac and E. Friedberg, editors, *Coopération et Conception*, pages 101–121. Octares Editions, 1996.

[35] E. Jacquet-Lagrèze. *Modélisation des préférences: Préordres, quasi-ordres et relations floues*. Thèse de doctorat de 3e Cycle, Université Paris V René Decartes, 1975.

[36] R. Keeney. Creativity in decision-making with value-focused thinking. *Sloan Management Review*, 35(4):33–41, 1994.

[37] R. Keeney and H. Raiffa. *Decisions with Multiple Objectives: Preferences and Value Tradeoffs*. John Wiley & Sons, New York, 1976.

[38] P. Kouvelis and G. Yu. *Robust Discrete Optimization and its Applications*. Kluwer Academic Publishers, Dordrecht, 1997.

[39] R. Kruse, E. Schwecke, and J. Heinsohn, editors. *Uncertainty and Vagueness in Knowledge Based Systems*. Springer Verlag, Berlin, 1991.

[40] M. Landry. A note on the concept of problem. *Organization Studies*, 16:315–343, 1995.

[41] J. Martel and B. Roy. Analyse de la signifiance de diverses procédures d'agrégation multicritère. Annales du LAMSADE 1, Université Paris-Dauphine, 2002.

[42] V. Mousseau. *Problèmes liés à l'évaluation de l'importance relative des critères en aide multicritère à la décision: Réflexions théoriques, expérimentations et implémentations informatiques*. PhD thesis, Université Paris-Dauphine, 1993.

[43] P. Perny and J.Ch. Pomerol. Use of artificial intelligence in MCDM. In T. Gal, T. Stewart, and T. Hanne, editors, *Multicriteria Decision Making: Advances in MCDM Models, Algorithms, Theory, and Applications*, pages 15:1–15:43. Kluwer Academic Publishers, Dordrecht, 1999.

[44] P. Perny and B. Roy. The use of fuzzy outranking relations in preference modelling. *Fuzzy Sets and Systems*, 49:33–53, 1992.

[45] P. Perny and D. Vanderpooten. An interactive multiobjective procedure for selecting medium-term countermeasures after nuclear accidents. *Journal of Multi-Criteria Decision Analysis*, 7(1):48–60, 1998.

[46] J. Pomerol and S. Barba-Romero. *Multicriterion Decision Making in Management*. Kluwer Academic Publishers, Dordrecht, 2000.

[47] J-C. Pomerol. La décision humaine: Reconnaissance plus raisonnement. In D. Bouyssou, D. Dubois, M. Pirlot, and H. Prade, editors, *Concepts et Méthodes pour l'Aide à la Décision*, Traité IC2, Information, Commande, Communication. Hermès, Paris, To appear 2004.

[48] F Rauschmayer. How to consider ethics in MCDA? In A. Colorni, M. Paruccini, and B. Roy, editors, *A-MCD-A: Multiple Criteria Decision Aiding*, pages 273–280. Joint Research Center. The European Commission, Luxembourg, 2001.

[49] J. Rosenhead. *Rational Analysis of a Problematic World*. John Wiley & Sons, New York, 1989. 2nd edition revised in 2001.

[50] J. Rosenhead. Robustness analysis: Keeping your options open. In J. Rosenhead and J. Mingers, editors, *Rational Analysis for a Problematic World Revisited: Problem Structuring Methods for Complexity, Uncertainty and Conflict*, pages 181–207. John Wiley & Sons, Chichester, 2001.

[51] J. Rosenhead. Robustness to the first degree. In J. Rosenhead and J. Mingers, editors, *Rational Analysis for a Problematic World Revisited: Problem Structuring Methods for Complexity, Uncertainty and Conflict*, pages 209–223. John Wiley & Sons, Chichester, 2001.

[52] B. Roy. Vers une méthodologie générale d'aide à la décision. *Revue METRA*, 14(3):459–497, 1975.

[53] B. Roy. Partial preference analysis and decision-aid: The fuzzy outranking relation concept. In D. Bell, R. Keeney, and H. Raiffa, editors, *Conflicting Objectives in Decisions*, pages 40–75. John Wiley & Sons, New York, 1977.

[54] B. Roy. Quelques remarques sur le concept d'indépendance dans l'aide à la décision multicritère. *Foundations of Control Engineering*, 8(3-4):183–191, 1983.

[55] B. Roy. *Méthodologie Multicritère d'aide à la Décision*. Economica, Paris, 1985.

[56] B. Roy. Des critères multiples en recherche opérationnelle: Pourquoi? In G. Rand, editor, *Operational Research '87*, pages 829–842. North-Holland, Amsterdam, 1988.

[57] B. Roy. Main sources of inaccurate determination, uncertainty and imprecision. *Mathematical and Computer Modelling*, 12(10-11):1245–1254, 1989.

[58] B. Roy. Decision science or decision aid science? *European Journal of Operational Research*, 66:184–203, 1993.

[59] B. Roy. On operational research and decision aid. *European Journal of Operational Research*, 73:23–26, 1994.

[60] B. Roy. Les logiques compensatoires et les autres. Note de Recherche du LAMSADE 16, Université Paris-Dauphine, France, 1996.

[61] B. Roy. *Multicriteria Methodology for Decision Aiding*, volume 12 of *Nonconvex Optimization and its Applications*. Kluwer Academic Publishers, Dordrecht, 1996.

[62] B. Roy. Une lacune en RO-AD: Les conclusions robustes. Cahier du LAMSADE 144, Université Paris-Dauphine, France, 1997.

[63] B. Roy. A missing link in OR-DA, robustness analysis. *Foundations of Computing and Decision Sciences*, 23(3):141–160, 1998.

[64] B. Roy. Decision-aiding today: What should we expect? In T. Gal, T. Stewart, and T. Hanne, editors, *Multicriteria Decision Making: Advances in MCDM Models, Algorithms, Theory, and Applications*, pages 1.1–1.35. Kluwer Academic Publishers, Boston, 1999.

[65] B. Roy. A propos de robustesse en recherche opérationnelle et aide à la décision. In J.Ch. Billaut, A. Moukrim, and E. Sanlaville, editors, *Flexibilité et Robustesse en Ordonnancement*. Hermès, Paris, To appear 2004.

[66] B. Roy. Robustesse de quoi et vis-à-vis de quoi mais aussi robustesse pourquoi en aide à la décision? In C. Henggeler-Antunes, J. Figueira, and J. Clímaco, editors, *Proceedings of 56th Meeting of the European Working Group Multiple Criteria Decision Aiding, Coimbra, Portugal, 3-5 October 2002*. CCRC, Coimbra, Portugal, To appear 2004.

[67] B. Roy and D. Bouyssou. *Aide Multicritère à la Décision: Méthodes et Cas*. Economica, Paris, 1993.

[68] B. Roy and S. Damart. L'analyse coûts-avantages, outil de concertation et de légitimation? *METROPOLIS*, 108-109:7–16, 2002.

[69] B. Roy and V. Mousseau. A theoretical framework for analysing the notion of relative importance of criteria. *Journal of Multi Criteria Decision Analysis*, 5:145–159, 1996.

[70] B. Roy and Ph. Vincke. Systèmes relationnels de préférences en présence de critères multiples avec seuils. *Cahiers du CERO*, 22(1):23–38, 1980.

[71] B. Roy and Ph. Vincke. Relational systems of preference with one or more pseudo-criteria: Some new concepts and results. *Management Science*, 30(11):1323–1335, 1984.

[72] B. Roy and Ph. Vincke. Pseudo-orders: Definition, properties and numerical representation. *Mathematical Social Sciences*, 14(3):263–274, 1987.

[73] B. Roy and Ph. Vincke. The case of the vanishing optimum revisited again. *Journal of Multi-Criteria Decision Analysis*, 7:351, 1998.

[74] A. Schärlig. The case of the vanishing optimum. *Journal of Multi-Criteria Decision Analysis*, 5:160–164, 1996.

[75] R. Slowinski, editor. *Intelligent Decision Support: Handbook of Applications and Advances of the Rough Sets Theory*. Kluwer Academic Publishers, Dordrecht, 1992.

[76] R. Slowinski and J. Teghem, editors. *Stochastic versus Fuzzy Approaches to Multiobjective Mathematical Programming under Uncertainty*. Kluwer Academic Publishers, Dordrecht, 1990.

[77] T. Stewart. Concepts of interactive programming. In T. Gal, T. Stewart, and T. Hanne, editors, *Multicriteria Decision Making: Advances in MCDM Models, Algorithms, Theory, and Applications*, pages 10.1–10.28. Kluwer Academic Publishers, Boston, 1999.

[78] T. Tanino. Sensitivity analysis in multicriteria decision making. In T. Gal, T. Stewart, and T. Hanne, editors, *Multicriteria Decision Making: Advances in MCDM Models, Algorithms, Theory, and Applications*, pages 7.1–7.29. Kluwer Academic Publishers, Boston, 1999.

[79] A. Tsoukiàs. *Sur la généralisation des concepts de concordance et discordance en aide multicritře à la décision*. Mémoire d'habilitation, Université Paris Dauphine, 1997. Appeared also as Document du LAMSADE 117.

[80] A. Tsoukiàs and Ph. Vincke. A new axiomatic foundation of partial comparability. *Theory and Decision*, 39:79–114, 1995.

[81] D. Vanderpooten. The interactive approach in MCDA: A technical framework and some basic conceptions. *Mathematical and Computer Modelling*, 12:1213–1220, 1989.

[82] D. Vanderpooten. The use of preference information in multiple criteria interactive procedures. In A.G Lockett and G. Islei, editors, *Improving Decision Making in Organisations*, pages 390–399. Springer Verlag, Berlin, 1989.

[83] D. Vanderpooten. Modelling in decision aiding. In D. Bouyssou, E. Jacquet-Lagrèze, P. Perny, R. Slowinski, D. Vanderpooten, and Ph. Vincke, editors, *Aiding Decisions with Multiple Criteria: Essays in Honour of Bernard Roy*, pages 195–210. Kluwer Academic Publishers, Dordrecht, 2001.

[84] Ph. Vincke. *Multicriteria Decision-Aid*. John Wiley & Sons, Chichester, 1992.

[85] Ph. Vincke. Robust and neutral methods for aggregating preferences into an outranking relation. *European Journal of Operational Research*, 112(2):405–412, 1999.

[86] Ph. Vincke. Robust solutions and methods in decision-aid. *Journal of Multi-Criteria Decision Analysis*, 8(3):181–187, 1999.

[87] Ph. Vincke. Preferences and numbers. In A. Colorni, M. Paruccini, and B. Roy, editors, *A-MCD-A: Multiple Criteria Decision Aiding*, pages 343–354. Joint Research Center. The European Commission, Luxembourg, 2001.

[88] A. Wierzbicki. On the role of intuition in decision making and some ways of multicriteria aid of intuition. *Journal of Multi-Criteria Decision Analysis*, 6(2):65–76, 1997.

[89] A. Wierzbicki. Reference point approaches. In T. Gal, T. Stewart, and T. Hanne, editors, *Multicriteria Decision Making: Advances in MCDM Models, Algorithms, Theory, and Applications*, pages 9.1–9.39. Kluwer Academic Publishers, Boston, 1999.

[90] A.P. Wierzbicki. The use of reference objectives in multiobjective optimisation. In Fandel G. and Gal T., editors, *MCDM Theory and Application*, volume 177 of *Lecture Notes in*

Economics and Mathematical Systems, pages 468–486, Hagen, 1980. Springer Verlag, Berlin.

[91] M. Zeleny. *Multiple Criteria Decision Making*. McGraw Hill, New York, 1982.

[92] M. Zeleny, editor. *MCDM: Past Decade and Future Trends, A Source Book of Multiple Criteria Decision Making*. JAI Press, London, 1984.

[93] S. Zionts. The case of the vanishing optimum revisited. *Journal of Multi-Criteria Decision Analysis*, 6:247, 1997.

II

FOUNDATIONS OF MCDA

Chapter 2

PREFERENCE MODELLING

Meltem Öztürk, Alexis Tsoukiàs
LAMSADE-CNRS, Université Paris Dauphine,
75775 Paris Cedex 16,
France
{ozturk,tsoukias}@lamsade.dauphine.fr

Philippe Vincke
Université Libre de Bruxelles
CP 210/1, Bld. du Triomphe, 1050 Bruxelles,
Belgium
pvincke@smg.ulb.ac.be

Abstract This chapter provides the reader with a presentation of preference modelling fundamental notions as well as some recent results in this field. Preference modelling is an inevitable step in a variety of fields: economy, sociology, psychology, mathematical programming, even medicine, archaeology, and obviously decision analysis. Our notation and some basic definitions, such as those of binary relation, properties and ordered sets, are presented at the beginning of the chapter. We start by discussing different reasons for constructing a model or preference. We then go through a number of issues that influence the construction of preference models. Different formalisations besides classical logic such as fuzzy sets and non-classical logics become necessary. We then present different types of preference structures reflecting the behavior of a decision-maker: classical, extended and valued ones. It is relevant to have a numerical representation of preferences: functional representations, value functions. The concepts of thresholds and minimal representation are also introduced in this section. In Section 8, we briefly explore the concept of deontic logic (logic of preference) and other formalisms associated with "compact representation of preferences" introduced for special purposes. We end the chapter with some concluding remarks.

Keywords: Preference modelling, decision aiding, uncertainty, fuzzy sets, non classical logic, ordered relations, binary relations.

1. Introduction

The purpose of this chapter is to present fundamental notions of preference modelling as well as some recent results in this field. Basic references on this issue can be considered: [4, 75, 78, 82, 110, 118, 161, 165, 167, 184, 189].

The chapter is organized as follows: The purpose for which formal models of preference and more generally of objects comparison are studied, is introduced in Section 2. In Section 3, we analyse the information used when such models are established and introduce different sources and types of uncertainty. Our notation and some basic definitions, such as those of binary relation, properties and ordered sets, are presented in Section 4. Besides classical logic, different formalisms can be used in order to establish a preference model, such as fuzzy sets and non-classical logics. These are discussed in Section 5. In Section 6, we then present different types of preference structures reflecting the behavior of a decision-maker: classical, extended and valued ones. It appears relevant to have a numerical representation of preferences: functional representations, value functions and intervals. These are discussed in Section 7. The concepts of thresholds and minimal representation are also introduced in this section. Finally, after briefly exploring the concept of deontic logic (logic of preference) and other related issued in Section 8, we end the chapter with some concluding remarks.

2. Purpose

Preference modelling is an inevitable step in a variety of fields. Scientists build models in order to better understand and to better represent a given situation; such models may also be used for more or less operational purposes (see [30]). It is often the case that it is necessary to compare objects in such models, basically in order to either establish if there is an order between the objects or to establish whether such objects are "near". Objects can be everything, from candidates to time intervals, from computer codes to medical patterns, from prospects (lotteries) to production systems. This is the reason why preference modelling is used in a great variety of fields such as economy [9, 10, 11, 50], sociology, psychology [37, 42, 45, 112, 111], political science [13, 179], artificial intelligence [65], computer science [82, 177, 189], temporal logic (see [5]) and the interval satisfiability problem [92, 150], mathematical programming [157, 158], electronic business, medicine and biology [22, 38, 108, 114, 138], archaeology [102], and obviously decision analysis.

In this chapter, we are going to focus on preference modelling for decision aiding purposes, although the results have a much wider validity.

Throughout this chapter, we consider the case of somebody (possibly a decision-maker) who tries to compare objects taking into account different points of view. We denote the set of alternatives A^1, to be labelled $a, b, c, \ldots$ and the set of points of view J, labelled $j = 1, 2, ..., m$. In this framework, a data $g_j(a)$ corresponds to the evaluation of the alternative a from the point of view $j \in J$.

As already mentioned, comparing two objects can be seen as looking for one of the two following possible situations:

- Object a is "before" object b, where "before" implies some kind of order between a and b, such an order referring either to a direct preference (a is preferred to b) or being induced from a measurement and its associated scale (a occurs before b, a is longer, bigger, more reliable, than b);

- Object a is "near" object b, where "near" can be considered either as indifference (object a or object b will do equally well for some purpose), or as a similarity, or again could be induced by a measurement (a occurs simultaneously with b, they have the same length, weight, reliability).

The two above-mentioned "attitudes" (see [142]) are not exclusive. They just stand to show what type of problems we focus on. From a decision aiding point of view we traditionally focus on the first situation. Ordering relations is the natural basis for solving ranking or choice problems. The second situation is traditionally associated with problems where the aim is to be able to put together objects sharing a common feature in order to form "homogeneous" classes or categories (a classification problem).

The first case we focus on is the ordering relation: given the set A, establishing how each element of A compares to each other element of A from a "preference" point of view enables to obtain an order which might be used to make either a choice on the set A (identify the best) or to rank the set A. Of course, we have to consider whether it is possible to establish such an ordering relation and of what type (certain, uncertain, strong, weak etc.) for all pairs of elements of A. We also have to establish what "not preference" represents (indifference, incomparability etc.). In the following sections (namely in Section 6), we are going to see that different options are available, leading to different so called preference structures.

In the second case we focus on the "nearness" relation since the issue here is to put together objects which ultimately are expected to be "near" (whatever the concept of "near" might represent). In such a case, there is also the problem how to consider objects which are "not near". Typical situations in this case include the problems of grouping, discriminating and assigning [98]. A further distinction in such problems concerns the fact that the categories with which the objects might be associated could already exist or not and the fact that such categories might be ordered or not. Putting objects into non pre-existing non ordered categories is the typical classification problem, conversely, assigning objects to pre-existing ordered categories is known as the "sorting" problem [149, 154, 220].

It should be noted that although preference relations have been naturally associated to ranking and choice problem statements, such a separation can be argued. For instance, there are sorting procedures (which can be seen as classification problems) that use preference relations instead of "nearness" ones [126, 136, 215]. The reason is the following: in order to establish that two objects belong to the same category we usually either try to check whether the two objects are "near" or whether they are near a "typical" object of the category (see for instance [154]). If, however, a category is described, not through its typical objects, but through its boundaries, then, in order to establish if an object belongs to such a category it might make sense to check whether such an object performs "better" than the "minimum", or "least" boundary of the category and that will introduce the use of a preference relation.

Recently Ngo The [142] claimed that decision aiding should not exclusively focus on preference relations, but also on "nearness relations", since quite often the problem statement to work with in a problem formulation is that of classification (on the existence of different problem statements and their meaning the reader is referred to [172, 173, 52, 204]).

3. Nature of Information

As already mentioned, the purpose of our analysis is to present the literature associated with objects comparison for either a preference or a nearness relation. Nevertheless, such an operation is not always as intuitive as it might appear. Building up a model from reality is always an abstraction (see [28]). This can always be affected by the presence of uncertainty due to our imperfect knowledge of the world, our limited capability of observation and/or discrimination, the inevitable errors occurring in any human activity etc. [170]. We call such an uncertainty exogenous. Besides, such an activity might generate uncertainty since it creates an approximation of reality, thus concealing some features of reality. We call this an endogenous uncertainty (see [191]).

As pointed out by Vincke [205] preference modelling can be seen as either the result of direct comparison (asking a decision-maker to compare two objects and to establish the relation between them) from which it might be possible to infer a numerical representation, or as the result of the induction of a preference relation from the knowledge of some "measures" associated to the compared objects.

In the first case, uncertainty can arise from the fact that the decision-maker might not be able to clearly state a preference relation for any pair of actions. We do not care why this may happen, we just consider the fact that the the decision-maker may reply when asked if "x is preferred to y": yes, no, I do not know, yes and no, I am not sure, it might be, it is more preference than indifference, but ... etc.. The problem in such cases is how to take such replies into account when defining a model of preferences.

In the second case, we may have different situations such as: incomplete information (missing values for some objects), uncertain information (the value of an object lies within an interval to which an uncertainty distribution might be associated, but the precise value is unknown), ambiguous information (contradictory statements about the present state of an object). The problem here is how to establish a preference model on the basis of such information and to what extent the uncertainty associated with the original information will be propagated to the model and how.

Such uncertainties can be handled through the use of various formalisms (see Section 5 of this chapter). Two basic approaches can be distinguished (see also [71]).

1 Handling uncertain information and statements. In such a case, we consider that the concepts used in order to model preferences are well-known and that we could possibly be able to establish a preference relation without any uncertainty, but we consider this difficult to do in the present situation with the available information. A typical example is the following: we know that x is preferred to y if the price of x is lower than the price of y, but we know very little about the prices of x and y. In such cases we might use an uncertainty distribution (classical probability, ill-known probabilities, possibility distributions, see [43, 70, 75, 107]) in order to associate a numerical uncertainty with each statement.

2 Handling ambiguous concepts and linguistic variables. With such a perspective we consider that sentences such as "x is preferred to y" are ill-defined, since the concept of preference itself is ill-defined, independently from the available information. A typical example is a sentence of the type: "the largest the difference of price between x and y is, the strongest the preference is". Here we might know the prices of x and y perfectly, but the concept of preference is defined through a continuous valuation. In such cases, we might use a multi-valued logic such that any preferential sentence obtains a truth value representing the "intensity of truth" of such a sentence. This should not be confused with the concept of "preference intensity", since such a concept is based on the idea of "measuring" preferences (as we do with temperature or with weight) and there is no "truth" dimension (see [117, 118, 164, 165]). On the other hand such a subtle theoretical distinction can be transparent in most practical cases since often happens that similar techniques are used under different approaches.

4. Notation and Basic Definitions

The notion of binary relation appears for the first time in De Morgan's study [51] and is defined as a set of ordered pairs in Peirce's works [151, 152, 153]. Some of the first work dedicated to the study of preference relations can be found in [72] and in [178] (more in general the concept of models of arbitrary relations will be introduced in [185, 186]). Throughout this chapter, we adopt Roubens' and Vincke's notation [167].

DEFINITION 1 (BINARY RELATION) *Let A be a finite set of elements $(a, b, c, ..., n)$, a binary relation R on the set A is a subset of the cartesian product $A \times A$, that is, a set of ordered pairs (a, b) such that a and b are in A: $R \subseteq A \times A$.*

For an ordered pair (a, b) which belongs to R, we indifferently use the notations:

$$(a, b) \in R \text{ or } aRb \text{ or } R(a, b).$$

Let R and T be two binary relations on the same set A. Some set operations are:

The Inclusion:	$R \subseteq T$	iff	$aRb \longrightarrow aTb$
The Union:	$a(R \cup T)b$	iff	aRb or(inclusive) aTb
The Intersection:	$a(R \cap T)b$	iff	aRb and aTb
The Relative Product:	$a(R.T)b$	iff	$\exists c \in A: \; aRc$ and cTb (aR^2b iff $aR.Rb$).

When such concepts apply we respectively denote (R^a), (R^s), $(\hat{R})$ the asymmetric, the symmetric and the complementary part of binary relation R:

$$\begin{aligned} aR^ab &\text{ iff } aRb \text{ and not}(bRa) \\ aR^sb &\text{ iff } aRb \text{ and } bRa \\ a\hat{R}b &\text{ iff } \text{not}(aRb) \text{ and not}(bRa). \end{aligned}$$

The complement (R^c), the converse (the dual)$(\overline{R})$ and the co-dual (R^{cd}) of R are respectively defined as follows:

$$\begin{aligned} aR^cb &\text{ iff } \text{not}(aRb) \\ a\overline{R}b &\text{ iff } bRa \\ aR^{cd}b &\text{ iff } \text{not}(bRa). \end{aligned}$$

The relation R is called

reflexive, if	$aRa, \ \forall a \in A$
irreflexive, if	$aR^c a, \ \forall a \in A$
symmetric, if,	$aRb \longrightarrow bRa, \ \forall a, b \in A$
antisymmetric, if	$(aRb, bRa) \longrightarrow a = b, \ \forall a, b \in A$
asymmetric, if	$aRb \longrightarrow bR^c a, \ \forall a, b \in A$
complete, if	$(aRb \text{ or } bRa), \ \forall a \neq b \in A$
strongly complete, if	$aRb \text{ or } bRa, \ \forall a, b \in A$
transitive, if,	$aRb, bRc) \longrightarrow aRc, \ \forall a, b, c \in A$
negatively transitive,	$(aR^c b, bR^c c) \longrightarrow aR^c c, \ \forall a, b, c \in A$
negatively transitive, if	$aRb \longrightarrow (aRc \text{ or } \ cRb), \ \forall a, b, c \in A$
semitransitive, if	$(aRb, bRc) \longrightarrow (aRd \text{ or } dRc), \ \forall a, b, c, d \in A$
Ferrers relation, if	$(aRb, cRd) \longrightarrow (aRd \text{ or } cRb), \ \forall a, b, c, d \in A$

The equivalence relation E associated with the relation R is a reflexive, symmetric and transitive relation, defined by:

$$aEb \text{ iff } \forall a \in A \begin{cases} aRc & \longleftrightarrow & bRc \\ cRa & \longleftrightarrow & cRb. \end{cases}$$

A binary relation R may be represented by a direct graph (A, R) where the nodes represent the elements of A, and the arcs, the relation R. Another way to represent a binary relation is to use a matrix M^R; the element M^R_{ab} of the matrix (the intersection of the line associated to a and the column associated to b) is 1 if aRb and 0 if $not(aRb)$.

EXAMPLE 1 *Let R be a binary relation defined on a set A, such that the set A and the relation R are defined as follows: $A = \{a, b, c, d\}$ and $R = \{(a, b), (b, d), (b, c), (c, a), (c, d), (d, b)\}$.*

The graphical and matrix representation of R are given in Figures 2.1 and 2.2.

5. Languages

Preference models are formal representations of comparisons of objects. As such they have to be established through the use of a formal and abstract language capturing both the structure of the world being described and the manipulations of it. It seems natural to consider formal logic as such a language. However, as already mentioned in the previous sections, the real world might be such that classical formal logic might appear too rigid to allow the definition of useful and expressive models. For this purpose, in this section, we introduce some further formalisms which extend the expressiveness of classical logic, while keeping most of its calculus properties.

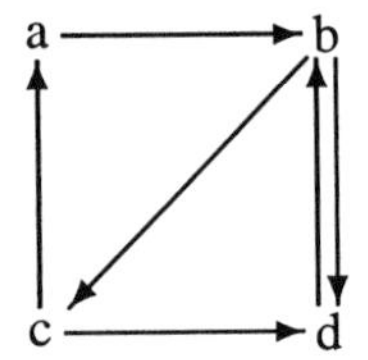

Figure 2.1. Graphical representation of R.

	a	b	c	d
a	0	1	0	0
b	0	0	1	1
c	1	0	0	1
d	0	1	0	0

Figure 2.2. Matrix representation of R.

5.1 Classical Logic

The interested reader can use two references: [128, 200] as introductory books to the use and the semantics of classical logic. All classic books mentioned in this chapter, implicitly or explicitly use classical logic, since binary relations are just sets and the calculus of sets is algebraically equivalent to truth calculus. Indeed the semantics of logical formulas as established by Tarski [185, 186], show the equivalence between membership of an element to a set and truth of the associate sentence.

Building a binary preference relation, a valuation of any proposition takes the values $\{0, 1\}$:

$$\begin{aligned}\mu(aRb) &= 1 \text{ iff } aRb \text{ is true}\\ \mu(aRb) &= 0 \text{ iff } aRb \text{ is false.}\end{aligned}$$

The reader will note that all notations introduced in the previous section are based on the above concept. He/she should also note that when we write "a preference relation P is a subset of $A \times A$", we introduce a formal structure where the universe of discourse is $A \times A$ and P is the model of the sentence

"x in relation P with y", that is, P is the set of all elements of $A \times A$ (ordered pairs of x and y) for which the sentence is true.

The above semantic can be in sharp contrast with decision analysis experience. For this purpose we will briefly introduce two more semantics: fuzzy sets and four-valued logic.

5.2 Fuzzy Sets

In this section, we provide a survey of basic notions of fuzzy set theory. We present definitions of connectives and several valued binary relation properties in order to be able to use this theory in the field of decision analysis. Basic references for this section include, [70, 85, 182, 219].

Fuzzy sets were first introduced by Zadeh [217, 218]. The concept and the associated logics were further developed by other researchers: [67, 93, 115, 116, 130, 131, 139, 144],

Fuzzy measures can be introduced for two different uses: either they can represent a concept imprecisely known (although well defined) or a concept which is vaguely perceived such as in the case of a linguistic variable. In the first case they represent possible values, while in the second they are better understood as a continuous truth valuation (in the interval $[0, 1]$). To be more precise:

- in the first case we associate a possibility distribution (an ordinal distribution of uncertainty) to classical logic formulas;
- in the second case we have a multi-valued logic where the semantics allow values in the entire interval $[0, 1]$.

A fuzzy set can be associated either with the set of alternatives considered in a decision aiding model (consider the case where objects are represented by fuzzy numbers) or with the preference relations. In decision analysis we may consider four possibilities[2]:

- Alternatives with crisp values and crisp preference relations
- Alternatives with crisp values and fuzzy preference relations
- Alternatives with fuzzy values and crisp preference relations (defuzzification , [124] with gravity center, [214] with means interval)
- Alternatives with fuzzy values and fuzzy preference relations (possibility graphs, [69] four fuzzy dominance index, [168]); but in this chapter we are going to focus on fuzzy preference relations.

In the following we introduce the definitions required for the rest of the chapter.

DEFINITION 2 (FUZZY SET) *A fuzzy set (or a fuzzy subset) F on a set Ω is defined by the result of an application:*

$$\mu_F : \Omega \longrightarrow [0,1]$$

where $\forall x \in \Omega$, $\mu(x)$ *is the membership degree of* x *to* F.

DEFINITION 3 (NEGATION) *A function* $n : [0,1] \longrightarrow [0,1]$ *is a negation if and only if it is non-increasing and:*

$$n(0) = 1 \text{ and } n(1) = 0.$$

If the negation n is strictly decreasing and continuous then it is called *strict*.

In the following we investigate the two basic classes of operators, the operators for the intersection (triangular norms called t-norms) and the union (triangular conorms called t-conorms or s-norms) of fuzzy sets:

DEFINITION 4 (T-NORM) *A function* $T : [0,1]^2 \longrightarrow [0,1]$ *is a triangular norm (t-norm), if and only if it satisfies the four conditions:*

Equivalence Condition: $T(1,x) = x \ \forall x \in [0,1]$

T *is commutative:* $T(x,y) = T(y,x) \ \forall x,y \in [0,1]$

T *is nondecreasing in both elements:* $T(x,y) \le T(u,v)$ *for all* $0 \le x \le u \le 1$ *and* $0 \le y \le v \le 1$

T *is associative:* $T(x,T(y,z)) = T(T(x,y),z) \ \forall x,y,z \in [0,1]$.

The function T defines a general class of intersection operators for fuzzy sets.

DEFINITION 5 (T-CONORM) *A function* $S : [0,1]^2 \longrightarrow [0,1]$ *is a t-conorm, if and only if it satisfies the four conditions:*

Equivalence Condition: $S(0,x) = x \ \forall x \in [0,1]$

S *is commutative:* $S(x,y) = S(y,x) \ \forall x,y \in [0,1]$

S *is nondecreasing in both elements:* $S(x,y) \le S(u,v)$ *for all* $0 \le x \le u \le 1$ *and* $0 \le y \le v \le 1$

S *is associative:* $S(x,S(y,z)) = S(S(x,y),z) \ \forall x,y,z \in [0,1]$.

T-norms and t-conorms are related by duality. For suitable negation operators[3] pairs of t-norms and t-conorms satisfy the generalisation of the De Morgan law:

DEFINITION 6 (DE MORGAN TRIPLETS) *Suppose that T is a t-norm, S is a t-conorm and n is a strict negation. $\langle T, S, n\rangle$ is a De Morgan triple if and only if:*

$$n(S(x,y)) = T(n(x), n(y)).$$

Such a definition extends De Morgan's law to the case of fuzzy sets. There exist different proposed De Morgan triplets: [60, 68, 90, 176, 210, 213, 216].

The more frequent t-norms and t-conorms are presented in Table 2.1.

Table 2.1. Principal t-norms and t-conorms.

Names	t-norms	t-conorms
Zadeh	$\min(x,y)$	$\max(x,y)$
probabilistic	$x * y$	$x + y - xy$
Lukasiewicz	$\max(x+y-1,0)$	$\min(x+y,1)$
Hamacher($\gamma > 0$)	$\frac{xy}{\gamma+(1-\gamma)(x+y-xy)}$	$\frac{x+y+xy-(1-\gamma)xy}{1-(1-\gamma)xy}$
Yager($p > 0$)	$\max(1-((1-x)^p+(1-y)^p)^{1/p},0)$	$\min((x^p+y^p)^{1/p},1)$
Weber($\lambda > -1$)	$\max((x+y-1+\lambda xy)/(1+\lambda),0))$	$\min(x+y+\lambda xy,1)$
drastic	x if $y=1$; y if $x=1$; 0 if not	x if $y=0$; y if $x=0$; 1 if not

We make use of De Morgan's triplet $\langle T, S, n\rangle$ in order to extend the definitions of the operators and properties introduced above in crisp cases. First, we give the definitions of operators of implication I_T and equivalence E_T:

$$\begin{aligned} I_T(x,y) &= \sup\{z \in [0,1] : T(x,z) \leq y\} \\ E_T(x,y) &= T(I_T(x,y), I_T(y,x)). \end{aligned}$$

Since preference modelling makes use of binary relations, we extend the definitions of binary relation properties to the valued case. For the sake of simplicity $\mu(R(x,y))$ will be denoted $R(x,y)$: a valued binary relation $R(x,y)$ is ($\forall a,b,c,d \in A$)

reflexive,	if $R(a,a) = 1$
irreflexive,	if $R(a,a) = 0$
symmetric,	if $R(a,b) = R(b,a)$
T-antisymmetric,	if $a \neq b \longrightarrow T(R(a,b), R(b,a)) = 0$
T-asymmetric,	if $T(R(a,b), R(b,a)) = 0$
S-complete,	if $a \neq b \longrightarrow S(R(a,b), R(b,a)) = 1$
S-strongly complete,	if $S(R(a,b), R(b,a))1$
T-transitive,	if $T(R(a,c), R(c,b)) \leq R(a,b)$
negatively S-transitive,	if $R(a,b) \leq S(R(a,c), R(c,b))$
T-S-semitransitive,	if $T(R(a,d), R(d,b)) \leq S(R(a,c), R(c,b))$
T-S-Ferrers relation,	if $T(R(a,b), R(c,d)) \leq S(R(a,d), R(c,b))$.

Different instances of De Morgan triplets will provide different definitions for each property.

The equivalence relation is one of the most-used relations in decision analysis and is defined in fuzzy set theory as follows:

DEFINITION 7 (EQUIVALENCE RELATION) *A function* $E : [0,1]^2 \rightarrow [0,1]$ *is an equivalence if and only if it satisfies:*

- $E(x,y) = E(y,x) \forall x,y \in [0,1]$
- $E(0,1) = E(1,0) = 0$
- $E(x,x) = 1 \forall x \in [0,1]$
- $x \leq x' \leq y' \leq y \longrightarrow E(x,y) \leq E(x',y')$.

In Section 6.3 and Chapter 12, some results obtained by the use of fuzzy set theory are represented.

5.3 Four-valued Logics

When we compare objects, it might be the case that it is not possible to establish precisely whether a certain relation holds or not. The problem is that such a hesitation can be due either to incomplete information (missing values, unknown replies, unwillingness to reply etc.) or to contradictory information (conflicting evaluation dimensions, conflicting reasons for and against the relation, inconsistent replies etc.). For instance, consider the query "is Anaxagoras intelligent?" If you know who Anaxagoras is you may reply "yes" (you came to know that he is a Greek philosopher) or "no" (you discover he is a dog). But if you know nothing you will reply "I do not know" due to your ignorance (on this particular issue). If on the other hand you came to know both that Anaxagoras is a philosopher and a dog you might again reply "I do not know", not due to ignorance, but to inconsistent information. Such different reasons for hesitation can be captured through four-valued logics allowing for different truth values for four above-mentioned cases. Such logics were first studied in [66] and introduced in the literature in [17] and [18]. Further literature on such logics can be found in [8, 23, 73, 84, 88, 113, 188, 192].

In the case of preference modelling, the use of such logics was first suggested in [190] and [54]. Such logics extend the semantics of classical logic through two hypotheses:

- the complement of a first order formula does not necessarily coincide with its negation;
- truth values are only partially ordered (in a bilattice), thus allowing the definition of a boolean algebra on the set of truth values.

The result is that using such logics, it is possible to formally characterise different states of hesitation when preferences are modelled (see [195, 196]. Furthermore, using sucha formalism, it becomes possible to generalise the concordance/discordance principle (used in several decision aiding methods) as shown in [193] and several characterisation problems can be solved (see for instance [197]). Recently (see [89, 159]) it has been suggested to use the extension of such logics for continuous valuations.

6. Preference Structures

DEFINITION 8 (PREFERENCE STRUCTURE) *A preference structure is a collection of binary relations defined on the set A and such that:*

- *for each couple a, b in A; at least one relation is satisfied*
- *for each couple a, b in A; if one relation is satisfied, another one cannot be satisfied.*

In other terms a preference structure defines a partition[4] of the set $A \times A$. In general it is recommended to have two other hypotheses with this definition (also denoted as fundamental relational system of preferences):

- Each preference relation in a preference structure is uniquely characterized by its properties (symmetry, transitivity, etc.).
- For each preference structure, there exists a unique relation from which the different relations composing the preference structure can be deduced. Any preference structure on the set A can thus be characterised by a unique binary relation R in the sense that the collection of the binary relations are be defined through the combinations of the epistemic states of this characteristic relation[5].

6.1 $\langle P, I \rangle$ Structures

The most traditional preference model considers that the decision-maker confronted with a pair of distinct elements of a set A, either:

- clearly prefers one element to the other, or
- feels indifferent about them.

The subset of ordered pairs (a, b) belonging to $A \times A$ such that the statement "a is preferred to b" is true, is called *preference relation* and is denoted by P.

The subset of pairs (a, b) belonging to $A \times A$ such that the statement "a and b are indifferent" is true, is called *indifference relation* and is denoted by I (I being considered the complement of $P \cup P^{-1}$ with respect to $A \times A$).

In the literature, there are two different ways of defining a specific preference structure:

- the first defines it by the properties of the binary relations of the relation set;
- the second uses the properties of the characteristic relation. In the rest of the section, we give definitions in both ways.

DEFINITION 9 ($\langle$ P, I $\rangle$ STRUCTURE) *A $\langle P, I \rangle$ structure on the set A is a pair $\langle P, I \rangle$ of relations on A such that:*

- *P is asymmetric,*
- *I is reflexive, symmetric.*

The characteristic relation R of a $\langle P, I \rangle$ structure can be defined as a combination of the relations P and I as:

$$aRb \text{ iff } a(P \cup I)b. \tag{2.1}$$

In this case P and I can be defined from R as follows:

$$aPb \quad \text{iff} \quad aRb \text{ and } bR^c a \tag{2.2}$$

$$aIb \quad \text{iff} \quad aRb \text{ and } bRa. \tag{2.3}$$

The construction of *orders* is of a particular interest, especially in decision analysis since they allow an easy operational use of such preference structures. We begin by representing the most elementary orders (weak order, complete order). To define such structures we add properties to the relations P and I (namely different forms of transitivity).

DEFINITION 10 (TOTAL ORDER) *Let R be a binary relation on the set A, R being a characteristic relation of $\langle P, I \rangle$, the following definitions are equivalent:*

i. R is a total order.

ii. R is reflexive, antisymmetric, complete and transitive.

iii. $\begin{cases} I = \{(a,a), \forall a \in A\} \\ P \text{ is transitive} \\ P \cup I \text{ is reflexive and complete.} \end{cases}$

iv. $\begin{cases} P \text{ is transitive} \\ PI \subset P \text{ (or equivalently } IP \subset P) \\ P \cup I \text{ is reflexive and complete.} \end{cases}$

With this relation, we have an indifference between any two objects only if they are identical. The total order structure consists of an arrangement of objects from the best one to the worst one without any ex aequo.

In the literature, one can find different terms associated with this structure: total order, complete order, simple order or linear order.

DEFINITION 11 (WEAK ORDER) *Let R be a binary relation on the set A, R being a characteristic relation of $\langle P, I\rangle$, the following definitions are equivalent:*

i. R is a weak order.

ii. R is reflexive, strongly complete and transitive.

iii. $\begin{cases} I \text{ is transitive} \\ P \text{ is transitive} \\ P \cup I \text{ is reflexive and complete.} \end{cases}$

This structure is also called complete preorder or total preorder. In this structure, indifference is an equivalence relation. The associated order is indeed a total order of the equivalence (indifference) classes of A.

The first two structures consider indifference as a transitive relation. This is empirically falsifiable. Literature studies on the intransitivity of indifference show this; undoubtedly the most famous is that of Luce [125], who gives the example of a cup of sweetened tea[6]. Before him, [9, 74, 91, 97] and [162] already suggested this phenomenon. For historical commentary on the subject, see [83]. Relaxing the property of transitivity of indifference results in two well-known structures: semi-orders and interval orders.

DEFINITION 12 (SEMI-ORDER) *Let R be a binary relation on the set A, R being a characteristic relation of $\langle P, I\rangle$, the following definitions are equivalent:*

i. R is a semi-order.

ii. R is reflexive, complete, Ferrers relation and semitransitive.

$$iii. \quad \begin{cases} P.I.P \subset P \\ P^2 \cap I^2 = \emptyset \\ P \cup I \textit{ is reflexive and complete.} \end{cases}$$

$$iv. \quad \begin{cases} P.I.P \subset P \\ P^2 I \subset P \textit{ (or equivalently } IP^2 \subset P) \\ P \cup I \textit{ is reflexive and complete.} \end{cases}$$

DEFINITION 13 (INTERVAL ORDER (IO)) *Let R be a binary relation on the set A, R being a characteristic relation of $\langle P, I\rangle$, the following definitions are equivalent:*

i. R is an interval order.

ii. R is reflexive, complete and Ferrers relation.

$$iii. \quad \begin{cases} P.I.P \subset P \\ P \cup I \textit{ is reflexive and complete.} \end{cases}$$

A detailed study of this structure can be found in [78, 132, 161]. It is easy to see that this structure generalizes all the structures previously introduced.

Can we relax transitivity of preference? Although it might appear counter-intuitive there is empirical evidence that such a situation can occur: [127, 199]. Similar work can be found in: [29, 31, 32, 33, 77, 79, 80, 206].

6.2 Extended Structures

The $\langle P, I\rangle$ structures presented in the previous section neither take into account all the decision-maker's attitudes, nor all possible situations. In the literature, there are two non exclusive ways to extend such structures:

- Introduction of several distinct preference relations representing (one or more) hesitation(s) between preference and indifference;

- Introduction of one or more situations of incomparability.

6.2.1 Several Preference Relations. One can wish to give more freedom to the decision-maker and allow more detailed preference models, introducing one or more intermediate relations between indifference and preference. Such relations might represent one or more zones of ambiguity and/or uncertainty where it is difficult to make a distinction between preference and indifference. Another way to interpret such "intermediate" relations is to consider them as different "degrees of preference intensity". From a technical point of view these structures are similar and we are not going to further discuss such semantics. We distinguish two cases: one where only one such intermediate relation is introduced (usually called weak preference and denoted by Q), and another where several such intermediate relations are introduced.

1 $\langle P, Q, I\rangle$ preference structures. In such structures we introduce one more preference relation, denoted by Q which is an asymmetric and irreflexive binary relation. The usual properties of preference structures hold. Usually such structures arise from the use of thresholds when objects with numerical values are compared or, equivalently, when objects whose values are intervals are compared. The reader who wants to have more information on thresholds can go to Section 7.1 where all definitions and representation theorems are given.

   $\langle P, Q, I\rangle$ preference structures have been generally discussed in [203]. Two cases are studied in the literature:

   - PQI interval orders and semi-orders (for their characterisation see [198]). The detection of such structures has been shown to be a polynomial problem (see [143]).
   - Double threshold orders (for their characterisation see [197, 203]) and more precisely pseudo-orders (see [174, 175]).

   One of the difficulties of such structures is that it is impossible to define P, Q and I from a single characteristic relation R as is the case for other conventional preference structures.

2 $\langle P_1, \cdots P_n\rangle$ preference structures. Practically, such structures generalise the previous situation where just one intermediate relation was considered. Again, such structures arise when multiple thresholds are used in order to compare numerical values of objects. The problem was first introduced in [47] and then extensively studied in [57, 59, 166], see also [2, 58, 135, 187]. Typically such structures concern the coherent representation of multiple interval orders. The particular case of multiple semi-orders was studied in [55].

6.2.2 Incomparability. In the classical preference structures presented in the previous section, the decision-maker is supposed to be able to compare the alternatives (we can have aPb, bPa or aIb). But certain situations, such as lack of information, uncertainty, ambiguity, multi-dimensional and conflicting preferences, can create incomparability between alternatives. Within this framework, the partial structures use a third symmetric and irreflexive relation J ($aJb \longleftrightarrow not(aPb), not(bPa), not(aIb), not(aQb), not(bQa)$), called incomparability, to deal with this kind of situation. To have a partial structure $\langle P, I, J\rangle$ or

$\langle P, Q, I, J \rangle$, we add to the definitions of the preceding structures (total order, weak order, semi-order, interval order and pseudo-order), the relation of incomparability ($J \neq \emptyset$); and we obtain respectively partial order, partial preorder (quasi-order), partial semi-order, partial interval order and partial pseudo-order [167].

DEFINITION 14 (PARTIAL ORDER) *Let R be a binary relation ($R = P \cup I$) on the set A, R being a characteristic relation of $\langle P, I, J \rangle$, the following definitions are equivalent:*

i. R is a partial order.

ii. R is reflexive, antisymmetric, transitive.

$$iii. \begin{cases} P \text{ is asymmetric, transitive} \\ I \text{ is reflexive, symmetric} \\ J \text{ is irreflexive and symmetric} \\ I = \{(a,a), \forall a \in A\}. \end{cases}$$

DEFINITION 15 (QUASI-ORDER) *Let R be a binary relation ($R = P \cup I$) on the set A, R being a characteristic relation of $\langle P, I, J \rangle$, the following definitions are equivalent:*

i. R is a quasi-order.

ii. R is reflexive, transitive.

$$iii. \begin{cases} P \text{ is asymmetric, transitive} \\ I \text{ is reflexive, symmetric and transitive} \\ J \text{ is irreflexive and symmetric} \\ (P.I \cup I.P) \subset P. \end{cases}$$

A fundamental result [72, 78] shows that every partial order (resp. partial preorder) on a finite set can be obtained as an intersection of a finite number of total orders (resp. total preorders, see [25]).

A further analysis of the concept of incomparability can be found in [195] and [196]. In these papers it is shown that the number of preference relations that can be introduced in a preference structure, so that it can be represented through a characteristic binary relation, depends on the semantics of the language used for modelling. In other terms, when classical logic is used in order to model preferences, no more than three different relations can be established (if one characteristic relation is used). The introduction of a four-valued logic allows to extend the number of independently defined relations to 10, thus introducing different types of incomparability (and hesitation) due to the different combination of positive and negative reasons (see [193]). It is therefore possible with such a language to consider an incomparability due to ignorance separately from one due to conflicting information.

6.3 Valued Structures

In this section, we present situations where preferences between objets are defined by a valued preference relation such that $\mu(R(a,b))$ represents either the intensity or the credibility of the preference of a over b[7] or the proportion of people who prefer a to b or the number of times that a is preferred to b. In this section, we make use of results cited in [85] and [155]. To simplify the notation, the valued relation $\mu(R(a,b))$ is denoted $R(a,b)$ in the rest of this section. We begin by giving a definition of a valued relation:

DEFINITION 16 (VALUED RELATION) *A valued relation R on the set A is a mapping from the cartesian product $A \times A$ onto a bounded subset of $\mathbb{R}$, often the interval [0,1].*

REMARK 1 *A valued relation can be interpreted as a family of crisp nested relations. With such an interpretation, each α-cut level of a fuzzy relation corresponds to a different crisp nested relation.*

In this section, we show some results obtained by the use of fuzzy set theory as a language which is capable to deal with uncertainty. The seminal paper by Orlovsky [147] can be considered as the first attempt to use fuzzy set theory in preference modelling. Roy in [169] will also make use of the concept of fuzzy relations in trying to establish the nature of a pseudo-order. In his paper Orlovsky defines the strict preference relation and the indifference relation with the use of Lukasiewicz and min t-norms. After him, a number of researchers were interested in the use of fuzzy sets in decision aiding, most of these works are published in the journal Fuzzy Sets and Systems.

In the following we give some definitions of fuzzy ordered sets. We derive the following definitions from the properties listed in Section 5.2:

DEFINITION 17 (FUZZY TOTAL ORDER) *A binary relation R on the set A, is a fuzzy total order iff R is antisymmetric, strongly complete and T-transitive.*

DEFINITION 18 (FUZZY WEAK ORDER) *A binary relation R on the set A is a fuzzy weak order iff R is strongly complete and transitive.*

DEFINITION 19 (FUZZY SEMI-ORDER) *A binary relation R on the set A is a fuzzy semi-order iff R is strongly complete, a Ferrers relation and semitransitive.*

DEFINITION 20 (FUZZY INTERVAL ORDER (IO)) *A binary relation R on the set A is a fuzzy interval order iff R is a strongly complete Ferrers relation.*

DEFINITION 21 (FUZZY PARTIAL ORDER) *A binary relation R on the set A is a fuzzy partial order iff: R is antisymmetric reflexive and T-transitive.*

DEFINITION 22 (FUZZY PARTIAL PREORDER) *A binary relation R on the set A is a fuzzy partial preorder iff R is reflexive and T-transitive.*

All the definitions above are given in terms of the characteristic relation R. The second step is to define valued preference relations (valued strict preference, valued indifference and valued incomparability) in terms of the characteristic relation [85, 86, 87, 148, 156]. For this, equations (2.1) – (2.3) are interpreted in terms of fuzzy logical operations:

$$P(a,b) = T[R(a,b), nR(b,a)] \quad (2.4)$$
$$I(a,b) = T[R(a,b), R(b,a)] \quad (2.5)$$
$$R(a,b) = S[P(a,b), I(a,b)]. \quad (2.6)$$

However, it is impossible to obtain a result satisfying these three equations using a De Morgan triplet. [6, 85] present this result as an impossibility theorem that proves the non-existence of a single, consistent many-valued logic as a logic of preference. A way to deal with this contradiction is to consider some axioms to define $\langle P, I, J\rangle$. In different papers [85, 86, 148], Fodor, Ovchinnikov, Roubens propose to define three general axioms that they call Independence of Irrelevant Alternatives (IA), Positive Association (PA), Symmetry (SY). With their axioms, the following propositions hold:

PROPOSITION 1 (FUZZY WEAK ORDER) *If $\langle P, I\rangle$ is a fuzzy weak order then*

- *P is a fuzzy strict partial order*
- *I is a fuzzy similarity relation (reflexive, symmetric, transitive).*

PROPOSITION 2 (FUZZY SEMI-ORDER) *If $\langle P, I\rangle$ is a fuzzy semi-order then*

- *P is a fuzzy strict partial order*
- *I is not transitive.*

PROPOSITION 3 (FUZZY INTERVAL ORDER (IO)) *If $\langle P, I\rangle$ is a fuzzy interval order then*

- *P is a fuzzy strict partial order*
- *I is not transitive.*

De Baets, Van de Walle and Kerre [48, 201, 202] define the valued preference relations without considering a characteristic relation:

P is $T-$ asymmetric $(P \cap_T P^{-1}) = \emptyset)$
I is reflexive and J is irreflexive $(I(a,a) = 1, (a,a) = 0 \forall a \in A)$
I and J are symmetric $(I = I^{-1}, J = J^{-1})$
$P \cap_T I = \emptyset, P \cap_T J = \emptyset, I \cap_T J = \emptyset$
$P \cup_T P^{-1} \cup_T I \cup_T = A \times A.$

With a continuous t-norm and without zero divisors, these properties are satisfied only in crisp case. To deal with this problem, we have to consider a continuous t-norm with zero divisor.

In multiple criteria decision aiding, we can make use of fuzzy sets in different ways. One of these helps to construct a valued preference relation from the crisp values of alternatives on each criteria. We cite the proposition of Perny and Roy [156] as an example here. They define a fuzzy outranking relation R from a real valued function θ defined on $\mathbb{R} \times \mathbb{R}$, such that $R(a,b)\theta(g(a), g(b))$ verifies the following conditions for all a, b in A:

$$\forall y \in X, \quad \theta(x,y) \text{ is a nondecreasing function of } x \tag{2.7}$$
$$\forall x \in X, \quad \theta(x,y) \text{ is a nonincreasing function of } y \tag{2.8}$$
$$\forall z \in X, \quad \theta(z,z) = 1. \tag{2.9}$$

The resulting relation R is a fuzzy semi-order (i.e. reflexive, complete, semi-transitive and Ferrers fuzzy relation). Roy (1978) proposed in Electre III to define the outranking relation R characterized by a function θ for each criterion as follows:

$$\theta(x,y) = \frac{p(x) - \min\{y-x, p(x)\}}{p(x) - \min\{y-x, q(x)\}},$$

where $p(x)$ and $q(x)$ are thresholds of the selected criteria.

We may work with alternatives representing some imprecision or ambiguity for a criterion. In this case, we make use of fuzzy sets to define the evaluation of the alternative related to the criterion. In the ordered pair $\{x, \mu_j^a\}$, μ_j^a represents the grade of membership of x for alternative a related to the criterion j. The fuzzy set μ is supposed to be normal $(sup_x(\mu_j^a) = 1)$ and convex $(\forall x, y, z \in \mathbb{R}$, $y \in [x,z]$, $\mu_j^a(y) \leq min\{\mu_j^a(x), \mu_j^a(z)\})$. The credibility of the preference of a over b is obtained from the comparison of the fuzzy intervals (normal, convex fuzzy sets) of a and b with some conditions:

- The method used should be sensitive to the specific range and shape of the grades of membership.
- The method should be independent of the irrelevant alternatives.
- The method should satisfy transitivity.

Fodor and Roubens [85] propose to use two procedures.

In the first one, the credibility of the preference of a over b for j is defined as the possibility that $a \geq b$:

$$\Pi_j(a \geq b) = \bigvee_{x \geq y} \left[\mu_j^a(x) \wedge \mu_j^b(y)\right] = \sup_{x \geq y} \left[min(\mu_j^a(x), \mu_j^b(y))\right]. \quad (2.10)$$

The credibility as defined by (2.10) is a fuzzy interval order (Π_j is reflexive, complete and a Ferrers relation) and

$$\min\left(\Pi_j(a,b), \Pi_j(b,a)\right) = \sup_x \min\left(\mu_j^a(x), \mu_j^b(x)\right).$$

In the case of a symmetrical fuzzy interval (μ^a), the parameters of the fuzzy interval can be defined in terms of the valuation $g_j(a)$ and thresholds $p(g_j(a)$ and $q(g_j(a)$. Some examples using trapezoidal fuzzy numbers can be found in the work of Fodor and Roubens.

The second procedure proposed by Fodor and Roubens makes use of the shapes of membership functions, satisfies the three axioms cited at the beginning of the section and gives the credibility of preference and indifference as follows:

$$\begin{aligned} P_j(a,b) &= R_j^d(a,b) = 1 - \Pi_j(b \geq a) = N_j(a > b) \quad &(2.11)\\ I_j(a,b) &= \min[\Pi_j(a \geq b), \Pi_j(b \geq a)]. \quad &(2.12) \end{aligned}$$

Where Π (the possibility degree) and N (the necessity degree) are two dual distributions of the possibility theory that are related to each other with the equality: $\Pi(A) = 1 - N(A)$ (see [71] for an axiomatic definition of the theory of possibility).

7. Domains and Numerical Representations

In this section we present a number of results concerning the numerical representation of the preference structures introduced in the previous section. This is an important operational problem. Given a set A and a set of preference relations holding between the elements of A, it is important to know whether such preferences fit a precise preference structure admitting a numerical representation. If this is the case, it is possible to replace the elements of A with their numerical values and then work with these. Otherwise, when to the set A is already associated a numerical representation (for instance a measure), it is important to test which preference structure should be applied in order to faithfully interpret the decision-maker's preferences [205].

7.1 Representation Theorems

THEOREM 1 (TOTAL ORDER) *Let $R = \langle P, I\rangle$ be a reflexive relation on a finite set A, the following definitions are equivalent:*

i. *R is a total order structure (see 10).*

ii. $\exists$ *g: $A \mapsto \mathbb{R}^+$ satisfying for all $a, b \in A$:* $\begin{cases} aPb \text{ iff } g(a) > g(b) \\ a \neq b \longrightarrow g(a) \neq g(b). \end{cases}$

iii. $\exists g : A \mapsto \mathbb{R}^+$ *satisfying for all $a, b \in A$:* $\begin{cases} aRb \text{ iff } g(a) > g(b) \\ a \neq b \longrightarrow g(a) \neq g(b). \end{cases}$

In the infinite not enumerable case, it can be impossible to find a numerical representation of a total order. For a detailed discussion on the subject, see [16]. The necessary and sufficient conditions to have a numerical representation for a total order are present in many works: [36, 49, 75, 118].

THEOREM 2 (WEAK ORDER) *Let $R = \langle P, I \rangle$ be a reflexive relation on a finite set A, the following definitions are equivalent:*

i. *R is a weak order structure (see 11).*

ii. $\exists g : A \mapsto \mathbb{R}^+$ *satisfying for all $a, b \in A$:* $\begin{cases} aPb & \text{iff} \quad g(a) > g(b) \\ aIb & \text{iff} \quad g(a) = g(b). \end{cases}$

iii. $\exists g : A \mapsto \mathbb{R}^+$ *satisfying for all $a, b \in A$: aRb iff $g(a) \geq g(b)$.*

REMARK 2 *Numerical representations of preference structures are not unique. All monotonic strictly increasing transformations of the function g can be interpreted as equivalent numerical representations*[8].

Intransitivity of indifference or the appearance of intermediate hesitation relations is due to the use of thresholds that can be constant or dependent on the value of the objects under comparison (in this case values of the threshold might obey further coherence conditions).

THEOREM 3 (SEMI-ORDER) *Let $R = \langle P, I \rangle$ be a binary relation on a finite set A, the following definitions are equivalent:*

1 *R is a semi-order structure (see 12).*

2 $\exists g : A \mapsto \mathbb{R}^+$ *and a constant $q \geq 0$ satisfying for all $a, b \in A$:*

$$\begin{cases} aPb & \text{iff} \quad g(a) > g(b) + q \\ aIb & \text{iff} \quad |g(a) - g(b| \leq q. \end{cases}$$

3 $\exists g : A \mapsto \mathbb{R}^+$ *and a constant $q \geq 0$ satisfying for all $a, b \in A$:*

$$aRb \text{ iff } g(a) \geq g(b) - q.$$

4 $\exists g : A \mapsto \mathbb{R}^+$ *and* $\exists q : \mathbb{R} \mapsto \mathbb{R}^+$ *satisfying for all* $a, b \in A$*:*

$$\begin{cases} aRb \text{ iff } g(a) \geq g(b) - q(g(b)) \\ (g(a) > g(b)) \longrightarrow (g(a) + q(g(a)) \geq g(b) + q(g(b))). \end{cases}$$

For the proofs of these theorems see [78, 119, 161, 178].

The threshold represents a quantity for which any difference smaller than this one is not significant for the preference relation. As we can see, the threshold is not necessarily constant, but if it is not, it must satisfy the inequality which defines a coherence condition.

Here too, the representation of a semi-order is not unique and all monotonic increasing transformations of g appear as admissible representations provided the condition that the function q also obeys the same transformation[9].

THEOREM 4 (PI INTERVAL ORDER) *Let* $R = \langle P, I \rangle$ *be a binary relation on a finite set A, the following definitions are equivalent:*

i. *R is an interval order structure (see 13).*

ii. $\exists g : A \mapsto \mathbb{R}^+$ *satisfying* $\forall a, b \in A$*:*

$$\begin{cases} aPb \quad \text{iff} \quad g(a) > g(b) + q(b) \\ aIb \quad \text{iff} \quad \begin{array}{l} g(a) \leq g(b) + q(b) \\ g(b) \leq g(a) + q(a). \end{array} \end{cases}$$

It should be noted that the main difference between an interval order and a semi-order is the existence of a coherence condition on the value of the threshold. One can further generalise the structure of interval order, by defining a threshold depending on both of the two alternatives. As a result, the asymmetric part appears without circuit: [1, 2, 3, 4, 53, 183]. For extensions on the use of thresholds see [81, 99, 134]. For the extension of the numerical representation of interval orders in the case A is infinite not denumerable see [36, 40, 76, 140, 146].

We can now see the representation theorems concerning preference structures allowing an intermediate preference relation (Q). Before that, let us mention that numerical representations with thresholds are equivalent to numerical representations of intervals. It is sufficient to note that associating a value $g(x)$ and a strictly positive value $q(g(x))$ to each element x of A is equivalent to associating two values: $l(x) = g(x)$ (representing the left extreme of an interval) and $r(x) = g(x) + q(g(x))$ (representing the right extreme of the interval to each x; obviously: $r(x) > l(x)$ always holds).

THEOREM 5 (PQI INTERVAL ORDERS) *Let* $R = \langle P, Q, I \rangle$ *be a relation on a finite set A, the following definitions are equivalent:*

i. *R is a PQI interval order.*

ii. *There exists a partial order L such that:*

1) $I = L \cup R \cup I_d$ *where* $I_d = \{(x, x), x \in A\}$ *and* $R = L^{-1}$*;*
2) $(P \cup Q \cup L).P \subset P$*;*
3) $P.(P \cup Q \cup R) \subset P$*;*
4) $(P \cup Q \cup L).Q \subset P \cup Q \cup L$*;*
5) $Q.(P \cup Q \cup R) \subset P \cup Q \cup R$.

iii. $\exists l, r : A \mapsto \mathbb{R}^+$ *satisfying:*

$$\begin{cases} r(a) \geq l(a) \\ aPb \text{ iff } l(a) > r(b) \\ aQb \text{ iff } r(a) > r(b) \geq l(a) \geq l(b) \\ aIb \text{ iff } r(a) \geq r(b) \geq l(a) \text{ or } r(b) \geq r(a) \geq l(a) \geq l(b). \end{cases}$$

For proofs, further theory on the numerical representation and algorithmic issues associated with such a structure see [141, 143, 198].

THEOREM 6 (DOUBLE THRESHOLD ORDER) *Let* $R = \langle P, Q, I \rangle$ *be a relation on a finite set A, the following definitions are equivalent:*

i. *R is a double Threshold Order (see [203]).*

ii. $$\begin{cases} Q.I.Q \subset Q \cup P \\ P.I.P \subset P \\ Q.I.P \subset P \\ P.Q^{-1}.P \subset P. \end{cases}$$

iii. $\exists g, q, p : A \mapsto \mathbb{R}^+$ *satisfying:*

$$\begin{cases} aPb \text{ iff } g(a) > g(b) + p(b)) \\ aQb \text{ iff } g(b) + p(b) \geq g(a) > g(b) + q(b) \\ aIb \text{ iff } g(b) + q(b) > g(a) > g(b) - q(a). \end{cases}$$

THEOREM 7 (PSEUDO-ORDER) *Let* $R = \langle P, Q, I \rangle$ *be a relation on a finite set A, the following definitions are equivalent:*

i. *R is a pseudo-order.*

ii. $$\begin{cases} R \text{ is a double threshold order} \\ \langle (P \cup Q), I \rangle \text{ is a semi-order} \\ \langle P, (Q \cup I \cup Q^{-1}) \rangle \text{is a semi-order} \\ P.I.Q \subset P. \end{cases}$$

$$\text{iii.}\quad \begin{cases} R \text{ is a double threshold order} \\ g(a) > g(b) \longleftrightarrow \quad g(a) + q(a) > g(b) + q(b) \\ \qquad\qquad\qquad\qquad\;\; g(a) + p(a) > g(b) + p(b). \end{cases}$$

A pseudo-order is a particular case of double threshold order, such that the thresholds fulfil a coherence condition. It should be noted however, that such a coherence is not sufficient in order to obtain two constant thresholds. This is due to different ways in which the two functions can be defined (see [59]). For the existence of multiple constant thresholds see [55].

For partial structures of preference, the functional representations admit the same formulas, but equivalences are replaced by implications. In the following, we present a numerical representation of a partial order and a quasi-order examples:

THEOREM 8 (PARTIAL ORDER) *If $\langle P, I, J\rangle$ presents a partial order structure,then $\exists$ g: $A \mapsto \mathbb{R}^+$ such that:*

$$aPb \quad \longrightarrow \quad g(a) > g(b).$$

THEOREM 9 (PARTIAL WEAK ORDER)*If $\langle P, I, J\rangle$ presents a partial weak order structure, then $\exists$ g: $A \mapsto \mathbb{R}^+$ such that:*

$$\begin{cases} aPb \longrightarrow g(a) > g(b) \\ aIb \longrightarrow g(a) = g(b). \end{cases}$$

The detection of the dimension of a partial order[10] is an NP-hard problem [57, 78].

REMARK 3 *In the preference modelling used in decision aiding, there exist two different approaches: In the first one, the evaluations of alternatives are known (they can be crisp or fuzzy) and we try to reach conclusions about the preferences between the alternatives. For the second one, the preferences between alternatives (pairwise comparison) are given by an expert (or by a group of experts), and we try to define an evaluation of the alternatives that can be useful. The first approach uses the inverse implication of the equivalences presented above (for example for a total order we have $g(a) > g(b) \longrightarrow aPb$); and the second one the other implication of it (for the same example, we have $aPb \longrightarrow g(a) > g(b)$).*

REMARK 4 *There is a body of research on the approximation of a preference structure by another one; here we cite some studies on the research of a total order with a minimum distance to a tournament (complete and antisymmetric relation): [14, 15, 24, 39, 106, 133, 181].*

7.2 Minimal Representation

In some decision aiding situations, the only available preferential information can be the kind of preference relation holding between each pair of alternatives. In such a case we can try to build a numerical representation of each alternative by choosing a particular functional representation of the ordered set in question and associating this with the known qualitative relations.

This section aims at studying some minimal or parsimonious representations of ordered sets, which can be helpful for this kind of situation. Particularly, given a countable set A and a preference relation $R \subseteq A \times A$, we are interested to find a numerical representation $\hat{f} \in \mathcal{F} = \{f : A \mapsto \mathbb{R},\ f \text{ homomorph to } R\}$, such that for all $x \in A$, $\hat{f}$ is minimal.

7.2.1 Total Order, Weak Order. The way to build a minimal representation for a total order or a weak order is obvious since the preference and the indifference relations are transitive: The idea is to minimize the value of the difference $g(a) - g(b)$ for all a, b in A. To do this we can define a unit $k = \min_{a,b \in A}(g(a) - g(b))$ and the minimal evaluation $m = \min_{a \in A}(g(a))$. The algorithm will be:

- Choose any value for k and m , e.g. $k = 1$, $m = 0$;
- Find the alternative i which is dominated by all the other alternatives j in A and evaluate it by $g(i) = m$;
- For all the alternatives l for which we have lIi, note $g(l) = g(i)$;
- Find the alternative i' which is dominated by all the alternatives j' in $A - \{i\}$ and evaluate it by $g(i') = m + k$;
- For all the alternatives l' for which we have $l'Ii'$, note $g(l') = g(i')$;
- Stop when all the alternatives are evaluated.

7.2.2 Semi-order. The first study on the minimal representation of semi-orders was done in [160] who proved its existence and proposed an algorithm to build it. One can find more information about this in [56, 129, 161] and [142]. Pirlot uses an equivalent definition of the semi-order which uses a second positive constant: *total semi-order*: A reflexive relation $R = (P, I)$ on a finite set A is a semi-order iff there exists a real function g, defined on A, a non negative constant q and a positive constant ε such that $\forall a, b \in A$:

$$aPb \quad \text{iff} \quad g(a) > g(b) + q + \varepsilon \tag{2.13}$$
$$aIb \quad \text{iff} \quad |g(a) - g(b)| \leq q. \tag{2.14}$$

Such a triple (g, q, ε) is called an $\varepsilon - representation$ of (P, I). Any representation (g, q), as in the definition of semi-order given in Section 6.1, yields an ε-representation where

$$\varepsilon = \min_{(a,b)\in P} (g(a) - g(b) - q).$$

Let (A, R) be an associated to the semi-order $R = (P, I)$, we denote $G(q, \varepsilon)$ the valued graph obtained by giving the value $(q + \varepsilon)$ to the arcs P and $(-q)$ to the arcs I.

THEOREM 10 *If $R = (P, I)$ is a semi-order on the finite set A, there exists an ε-representation with threshold q iff:*

$$\frac{q}{\varepsilon} \geq \alpha = \max_C \left\{ \frac{|C \cap P|}{|C \cap I| - |C \cap P|}, \; C \textit{ circuit of } (A, R) \right\}$$

where $|C \cap P|$ (resp. $|C \cap I|$), represents the number of arcs P (resp. I) in the circuit C of the graph (A, R).

An algorithm to find a numerical representation of a semi-order is as follows:

- Choose any value for ε, e.g. $\varepsilon = 1$;
- Choose a large enough value of $\frac{q}{\varepsilon}$, e.g. $\frac{q}{\varepsilon} = |P|$;
- Solve the maximal value path problem in the graph $G(q, \varepsilon)$ (e.g. by using the Bellman algorithm, see [122]).

Denote by $g_{q,\varepsilon}$, the solution of the maximal path problem in $G(q, \varepsilon)$; we have:

$$g_{q,\varepsilon} \leq g(a) \forall a \in A.$$

EXAMPLE 2 *We consider the example given by Pirlot and Vincke [161]: Let $S = (P, I)$ be a semiorder on $A = \{a, b, c\}$ defined by $P = \{(a, c)\}$. The inequality (2.13) gives the following equations:*

$$\begin{aligned} g(a) &\geq g(c) + q + \varepsilon \\ g(a) &\geq g(b) - q \\ g(b) &\geq g(a) - q \\ g(b) &\geq g(c) - q \\ g(c) &\geq g(b) - q. \end{aligned}$$

Figure 2.3 shows the graphical representation of this semiorder.

As the non-trivial circuit $C = \{(a, c), (c, b), (b, a)\}$ is $-q + \varepsilon$ ($-q + \varepsilon = (q + \varepsilon) + (-q) + (-q)$), necessary and sufficient condition for the existence of an ε-representation is $q \geq \varepsilon$.

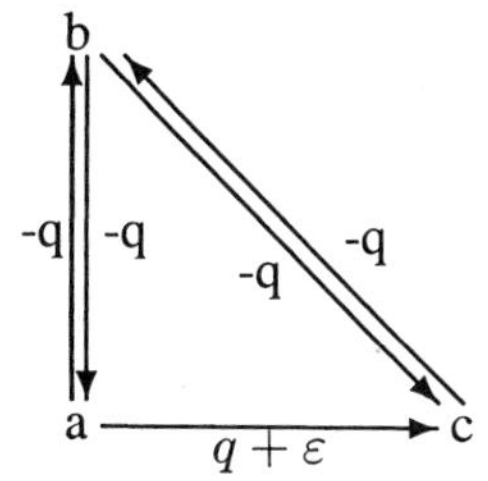

Figure 2.3. Graphical representation of the semiorder.

Table 2.2. Various ε-representations with ε=1.

		a	b	c
q=1	g_1=1	2	1	0
q=1	g_2=1	9.5	8.5	7.5
q=2.5	g_3=1	3.5	1	0
q=2.5	g_4=1	10.5	8.5	7
q=2.5	g_5=1	3.5	2.5	0

Table 2.2 provides an example of possible numerical representation of this semiorder.

DEFINITION 23 *A representation (g^*, q^*, ε) is minimal in the set of all non-negative ε-representations (g, q, ε) of a semiorder iff $\forall a \in A$ $g^*(a) \leq g(a)$.*

THEOREM 11 *The representation $(g_{q^*,\varepsilon}, q^*, \varepsilon)$ is minimal in the set of all ε-representations of a semiorder R.*

7.2.3 Interval Order. An interval can be represented by two real functions l and r on the finite set A which satisfy:

$$(\forall a \in A, l(a) \leq r(a))^{11}.$$

DEFINITION 24 *A reflexive relation R = $(P \cup I)$ on a finite set A is an interval order iff there exists a pair of functions l, r: A $\longrightarrow$ R^+ and a positive constant ε such that $\forall a, b \in A$*

$$\begin{cases} aPb & \textit{iff} \quad l(a) > r(b) + q + \varepsilon \\ aIb & \textit{iff} \quad l(a) \geq r(b) \quad \textit{and} \quad l(b) \geq r(a). \end{cases}$$

Such a triplet (l, r, ε) is called an ε-representation of the interval order $P \cup I$.

DEFINITION 25 *The ε-representation (l^*, r^*, ε) of the interval order $P \cup I$ is minimal iff for any other ε-representation (l, r, ε) we have, $\forall a \in A$,*

$$\begin{aligned} l^*(a) &\leq l(a) \\ r^*(a) &\leq r(a). \end{aligned}$$

THEOREM 12 *For any interval order $P \cup I$, there exists a minimal ε-representation (l^*, r^*, ε); the values of l^* and r^* are integral multiples of ε.*

8. Logic of Preferences

The increasing importance of preference modelling immediately interested people from other disciplines, particularly logicians and philosophers. The strict relation with deontic logic (see [7]) raised some questions such as:

- Does a general logic exist where any preferences can be represented and used?
- If yes, what is the language and what are the axioms?
- Is it possible, via this formalisation, to give a definition of bad or good as absolute values?

It is clear that this attempt had a clear positivist and normative objective: to define the one well-formed logic that people should follow when expressing preferences. The first work on the subject is the one by Halldén [95], but it is Von Wright's book [208] that tries to give the first axiomatisation of a logic of preferences. Inspired by this work some important contributions have been made [41, 42, 100, 101, 103, 163]. Influence of this idea can also be found in [109] and [164], but in related fields (statistics and value theory, respectively). The discussion apparently was concluded by Von Wright [209], but Huber [104, 105] continued on. Later on Halldin [96] and Widmeyer [211, 212] also worked on this.

The general idea can be presented as follows. At least two questions should be clarified: preferences among what? How should preferences be understood? Von Wright [208] argues that preferences can be distinguished as extrinsic and intrinsic. The first ones are derived as *a reason from a specific purpose*, while the second ones are *self-referential* to an actor expressing the preferences. In this sense intrinsic preferences are the expression of the actor's system of values of the actor. Moreover, preferences can be expressed for different things, the most general being (following Von Wright) *"states of affairs"*. That is, the expression "a is preferred to b" should be understood as the preference of a state (a world) where a occurs (whatever a represents: sentences, objects, relations etc.) over a state where b occurs. On the basis of Von Wright expressed a theory based on five axioms:

A^W1. $\forall x, y \;\; p(x, y) \;\; \longrightarrow \;\; \neg p(y, x)$

A^W2. $\forall x, y, z \;\; p(x, y) \;\wedge\; p(y, z) \;\; \longrightarrow \;\; p(x, z)$

A^W3. $p(a, b) \;\equiv\; p(a \wedge \neg b, \neg a \wedge b)$

A^W4. $p(a \vee b, c) \;\equiv\; p(a \wedge b \wedge \neg c, \neg a \wedge \neg b \wedge c) \;\wedge\; p(a \wedge \neg b \wedge \neg c, \neg a \wedge \neg b \wedge c) \;\wedge\; p(\neg a \wedge b \wedge \neg c, \neg a \wedge \neg b \wedge c)$

A^W5. $p(a, b) \;\equiv\; p(a \wedge c, b \wedge c) \;\wedge\; p(a \wedge \neg c, b \wedge \neg c)$.

The first two axioms are asymmetry and transitivity of the preference relation, while the following three axioms face the problem of combinations of states of affairs. The use of specific elements instead of the variables and quantifiers reflects the fact that von Wright considered the axioms not as logical ones, but as "reasoning principles". This distinction has important consequences on the calculus level. In the first two axioms, preference is considered as a binary relation (therefore the use of a predicate), in the three "principles", preference is a proposition. Von Wright does not make this distinction directly, considering the expression aPb ($p(a, b)$ in our notation) as a well-formed formulation of his logic. However, this does not change the problem since the first two axioms are referred to the binary relation and the others are not. The difference appears if one tries to introduce quantifications; in this case the three principles appear to be weak. The problem with this axiomatisation is that empirical observation of human behavior provides counterexamples of these axioms. Moreover, from a philosophical point of view (following the normative objective that this approach assumed), a logic of intrinsic preferences about general states of affairs should allow to define what is good (the always preferred?) and what is bad (the always not preferred?). But this axiomatization fails to enable such a definition.

Chisholm and Sosa [42] rejected axioms A^W3 to A^W5 and built an alternative axiomatization based on the concepts of *"good"* and *"intrinsically better"*. Their idea is to postulate the concept of good and to axiomatize preferences consequently. So a *good* state of affairs is one that is always preferred to its negation ($p(a, \neg a)$); Chisholm and Sosa, use this definition only for its operational potential as they argue that it does not capture the whole concept of "good"). In this case we have:

A^S1. $\forall x, y \;\; p(x, y) \;\; \longrightarrow \;\; \neg p(y, x)$

A^S2. $\forall x, y, z \;\; \neg p(x, y) \;\wedge\; \neg p(y, z) \;\; \longrightarrow \;\; \neg p(x, z)$

A^{S}3. $\forall x,y \;\; \neg p(x,\neg x) \;\wedge\; \neg p(\neg x,x) \;\wedge\; \neg p(y,\neg y) \;\wedge\; \neg p(\neg y,y) \;\longrightarrow\; \neg p(y,x) \;\wedge\; \neg p(x,y)$

A^{S}4. $\forall x,y \;\; p(x,y) \;\wedge\; \neg p(y,\neg y) \;\wedge\; \neg p(\neg y,y) \;\longrightarrow\; p(x,\neg x)$

A^{S}5. $\forall x,y \;\; p(y,\neg x) \;\wedge\; \neg p(y,\neg y) \;\wedge\; \neg p(\neg y,y) \;\longrightarrow\; p(x,\neg x)$.

Again in this axiomatisation there are counterexamples of the axioms. The assumption of the concept of good can be argued as it allows circularities in the definitions of preferences between combinations of states of affairs. This criticism lead Hansson [101] to consider only two fundamental, universally recognised axioms:

A^{H}1. $\forall x,y,z \;\; s(x,y) \;\wedge\; s(y,z) \;\longrightarrow\; s(x,z)$

A^{H}2. $\forall x,y \;\; s(x,y) \;\vee\; s(y,x)$,

where s is a "large preference relation" and two specific preference relations are defined, p (strict preference) and i (indifference):

D^{H}1. $\forall x,y \;\; p(x,y) \;\equiv\; s(x,y) \;\wedge\; \neg s(y,x)$

D^{H}2. $\forall x,y \;\; i(x,y) \;\equiv\; s(x,y) \;\wedge\; s(y,x)$.

He also introduces two more axioms, although he recognises their controversial nature:

A^{H}3. $\forall x,y,z \;\; s(x,y) \;\wedge\; s(x,z) \;\longrightarrow\; s(x,y \vee z)$

A^{H}4. $\forall x,y,z \;\; s(x,z) \;\wedge\; s(y,z) \;\longrightarrow\; s(x \vee y,z)$

Von Wright in his reply [209], trying to argue for his theory, introduced a more general frame to define intrinsic *"holistic"* preferences or as he called them *"ceteris paribus"* preferences. In this approach he considers a set S of states where the elements are the ones of A (n elements) and all the 2^n combinations of these elements. Given two states s and t (elementary or combinations of m states of S) you have i ($i = 2^{n-m}$) combinations C_i of the other states. You call an s-world any state that holds when s holds. A combination C_i of states is also a state so you can define it in the same way a C_i-world. Von Wright gives two definitions (strong and weak) of preference:

1 (strong): s is preferred to t under the circumstances C_i **iff** every C_i-world that is also an s-world and not a t-world is preferred to every C_i-world that is also t-world and not s-world.

2 (weak): s is preferred to t under the circumstances C_i **iff** some C_i-world that is an s-world is preferred to a C_i-world that is a t-world, but a C_i-world that is a t-world that is preferred to a C_i-world that is an s-world does not exist.

Now s is *"ceteris paribus"* preferred to t **iff** it is preferred under all C_i. We leave the discussion to the interested reader, but we point out that, with these definitions, it is difficult to axiomatize both transitivity and complete comparability unless they are assumed as necessary truths for "coherence" and "rationality" (see [209]).

It can be concluded that the philosophical discussion about preferences failed the objective to give a unifying frame of generalized preference relations that could hold for any kind of states, based on a well-defined axiomatization (for an interesting discussion see [137]). It is still difficult (if not impossible) to give a definition of good or bad in absolute terms based on reasoning about preferences and the properties of these relations are not unanimously accepted as axioms of preference modelling. For more recent advances in deontic logic see [145].

More recently, Von Wright's ideas and the discussion about "logical representation of preferences" attracted attention again. This is due to problems found in the field of Artificial Intelligence field due to essentially two reasons:

- the necessity to introduce some "preferential reasoning" (see [26, 27, 34, 62, 63, 64, 120, 123, 180]);
- the large dimension of the sets to which such a reasoning might apply, thus demanding a compact representation of preferences (see [19, 20, 21, 61, 121]).

9. Conclusion

We hope that this chapter on preference modelling, gave the non-specialist reader a general idea of the field by providing a list of the most important references of a very vast and technical literature. In this chapter, we have tried to present the necessary technical support for the reader to understand the following chapters. One can note that our survey does not interpret all the questions related to preference modelling. Let us mention some of them:

- How to get and validate preference information [12, 207];
- Relation between preference modelling and the problem of signifiance in measurement theory [165];
- Statistical analysis of preferential data [44, 94];

- Interrogations on the relations between preferences and the value system, and the nature of these values [37, 46, 194, 208].

Notes

1. We can use the word *action* instead of alternative.

2. Lets take an example: Imagine that we have to choose one car between two. We have to know the performance of each car in order to establish the relation of preference:

- In the first case, the performance of each car is known and noted between 1 and 10 ($p(car1) = 8$ and $p(car2) = 5$); the relation of preference is known too (car1 is preferred to car2: $car1Pcar2$ ($\mu(car1Pcar2) = 1$)).
- In the second case, the performance of each car is known and noted between 1 and 10 ($p(car1) = 8$ and $p(car2) = 7$); we are not sur about the preference relation that is why the relation of preference is fuzzy ($\mu(car1Pcar2) = 0.75$).
- In the third case, the performance of each car is fuzzy (in this case the performances of each car will be defined by fuzzy numbers ; in this case we can use triangular or trapezoidal fuzzy number to represent the performance); the relation of preference is crisp (car1 is preferred to car2: $car1Pcar2$ ($\mu(car1Pcar2) = 1$)).
- In the fourth case, the performance of each car is fuzzy (in this case the performances of each car will be defined by fuzzy numbers); the relation of preference is fuzzy too (($\mu(car1Pcar2) = 0.75$)).

3. A suitable one can be the complement operator defined: $n(\mu(x)) = 1 - \mu(x)$.

4. To have a partition of the set $A \times A$, the inverse of the asymmetric relation must be considered too.

5. While several authors prefer using both of them, there are others for which one is sufficient. For example Fishburn does not require the use of preference structures with a characteristic relation.

6. One can be indifferent between a cup of tea with n milligrams of sugar and one with $n+1$ milligrams of sugar, if one admits the transitivity of the indifference, after a certain step of transitivity, one will have the indifference between a cup of tea with n milligram of sugar and that with $n+N$ milligram of sugar with N large enough, even if there is a very great difference of taste between the two; which is contradictory with the concept of indifference.

7. This value can be given directly by the decision-maker or calculated by using different concepts, such values (indices) are widely used in many MCDA methods such as ELECTRE, PROMETHEE [171, 35].

8. The function g defines an ordinal scale for both structures.

9. But in this case the scale defined by g is more complex than an ordinal scale.

10. When it is a partial order of dimension 2, the detection can be made in polynomial time.

11. One can imagine that $l(a)$ represents the evaluation of the alternative a ($g(a)$) which is the left limit of the interval and $r(a)$ represents the value of ($g(a) + q(a)$) which is the right limit of the interval. One can remark that a semi-order is an interval order with a constant length.

References

[1] M. Abbas. Any complete preference structure without circuit admits an interval representation. *Theory and Decision*, 39:115–126, 1995.

[2] M. Abbas and Ph. Vincke. Preference structures and threshold models. *Journal of Multi-Criteria Decision Analysis*, 2:171–178, 1993.

[3] R. Agaev and F. Aleskerov. Interval choice: Classic and general cases. *Mathematical Social Sciences*, 26:249–272, 1993.

[4] F. Aleskerov and B. Monjardet. *Utility Maximization, Choice and Preference*. Springer Verlag, Heidelberg, 2002.

[5] J.F. Allen. Maintaining knowledge about temporal intervals. *Journal of the ACM*, 26:832–843, 1983.

[6] C. Alsina. On a family of connectives for fuzzy sets. *Fuzzy Sets and Systems*, 16:231–235, 1985.

[7] L. Åqvist. Deontic logic. In D. Gabbay and F. Guenther, editors, *Handbook of Philosophical Logic*, volume II, pages 605–714. D. Reidel, Dordrecht, 1986.

[8] O. Arieli and A. Avron. The value of the four values. *Artificial Intelligence*, 102:97–141, 1998.

[9] W.E. Armstrong. The determinateness of the utility function. *The Economic Journal*, 49:453–467, 1939.

[10] W.E. Armstrong. Uncertainty and utility function. *Economics Journal*, 58:1–10, 1948.

[11] W.E. Armstrong. A note on the theory of consumer's behavior. *Oxford Economics*, 2:119–122, 1950.

[12] C.A. Bana e Costa and J.C. Vansnick. Preference relation and MCDM. In T. Gal, T. Stewart, and T. Hanne, editors, *Multicriteria Decision Making: Advances in MCDM Models, Algorithms, Theory, and Applications*, volume 21 of *International Series in Operations Research & Management Science*, pages 4.1–4.23. Kluwer Academic Publishers, Dordrecht, 1999.

[13] J.-P. Barthélemy, C. Flament, and B. Monjardet. Ordered sets and social sciences. In I. Rankin, editor, *Ordered Sets*, pages 721–787. D. Reidel, Dordrecht, 1982.

[14] J.-P. Barthélémy, A. Guénoche, and O. Hudry. Median linear orders: Heuristics and a branch and bound algorithm. *European Journal of Operational Research*, 42(3):313–325, 1989.

[15] J.-P. Barthélémy and B. Monjardet. The median procedure in cluster analysis and social choice theory. *Mathematical Social Sciences*, 1:235–267, 1981.

[16] A.F. Beardon, J.C. Candeal, G. Herden, E. Induráin, and G.B. Mehta. The non-existence of a utility function and a structure of non-representable preference relations. *Journal of Mathematical Economics*, 37:17–38, 2002.

[17] N.D. Belnap. How computer should think. In G. Ryle, editor, *Proceedings of the Oxford International Symposium on Contemporary Aspects of Philosophy*, pages 30–56. Oriel Press, Stocksfield, 1976.

[18] N.D. Belnap. A useful four-valued logic. In G. Epstein and J. Dunn, editors, *Modern Uses of Multiple Valued Logics*, pages 8–37. D. Reidel, Dordrecht, 1977.

[19] S. Benferhat, D. Dubois, S. Kaci, and H. Prade. Bipolar representation and fusion of preferences in the possibilistic logic framework. In *Proceedings of the 8th International Conference on Knowledge Representation and Reasoning, KR'02*, pages 421–432. Morgan Kauffamn, San Francisco, 2002.

[20] S. Benferhat, D. Dubois, S. Kaci, and H. Prade. Possibilistic logic representation of preferences: Relating prioritized goals and satisfaction levels expressions. In *Proceedings of the 15th European Conference on Artificial Intelligence, ECAI 2002*, pages 685–689. ISO Press, Amsterdam, 2002.

[21] S. Benferhat and S. Kaci. A possibilistic logic handling of strong preferences. In *Proceedings 9th IFSA (International Fuzzy Systems Association) World Congress and 20th NAFIPS (North American Fuzzy Information Processing Society) International Conference*, volume 2, pages 962–967. IEEE Press, Piscataway, 2001.

[22] S. Benzer. The fine structure of the gene. *Scientific American*, 206(1):70–84, 1962.

[23] J.A. Bergstra, I. Bethke, and P. Rodenburg. A propositional logic with four values: True, false, divergent and meaningless. *Journal of Applied Non-Classical Logics*, 5:199–217, 1995.

[24] J.C. Bermond. Ordres à distance minimum d'un tournoi et graphes partiels sans circuits maximaux. *Mathématiques et Sciences Humaines*, 37:5–25, 1972.

[25] W. Bossert, Y. Sprumont, and K. Suzumura. Upper semicontinuous extensions of binary relations. *Journal of Mathematical Economics*, 37:231–246, 2002.

[26] C. Boutilier. Toward a logic for qualitative decision theory. In *Proceedings of the 4th International Conference on Knowledge Representation and Reasoning, KR'94*, pages 75–86. Morgan Kaufmann, San Francisco, 1994.

[27] C. Boutilier, R.I. Brafman, H.H. Hoos, and D. Poole. Reasoning with conditional ceteris paribus preference statements. In *Proceedings of the 15th Conference on Uncertainty in Artificial Intelligence, UAI'99*, pages 71–80. Morgan Kaufmann, San Francisco, 1999.

[28] D. Bouyssou. Modelling inaccurate determination, uncertainty, imprecision using multiple criteria. In A.G. Lockett and G. Islei, editors, *Improving Decision Making in Organisations*, volume 335 of *Lecture Notes in Economics and Mathematical Systems*, pages 78–87. Springer Verlag, Berlin, 1989.

[29] D. Bouyssou. Outranking relations: Do they have special properties? *Journal of Multi-Criteria Decision Analysis*, 5:99–111, 1996.

[30] D. Bouyssou, T. Marchant, M. Pirlot, P. Perny, A. Tsoukiàs, and Ph. Vincke. *Evaluation and Decision Models: A Critical Perspective*. Kluwer Academic Publishers, Dordrecht, 2000.

[31] D. Bouyssou and M. Pirlot. Conjoint measurement without additivity and transitivity. In N. Meskens and M. Roubens, editors, *Advances in Decision Analysis*, pages 13–29. Kluwer Academic Publishers, Dordrecht, 1999.

[32] D. Bouyssou and M. Pirlot. A characterization of strict concordance relations. In D. Bouyssou, E. Jacquet-Lagrèze, P. Perny, R. Słowiński, D. Vanderpooten, and Ph. Vince, editors, *Aiding Decisions with Multiple Criteria: Essays in Honour of Bernard Roy*, pages 121–145. Kluwer Academic Publishers, Dordrecht, 2002.

[33] D. Bouyssou and M. Pirlot. Non transitive decomposable conjoint measurement. *Journal of Mathematical Psychology*, 46:677–703, 2002.

[34] R.I. Brafman and N. Friedman. On decision-theoretic foundations for defaults. *Artificial Intelligence*, 133:1–33, 2001.

[35] J.P. Brans, B. Mareschal, and Ph. Vincke. Promethee: A new family of outranking methods in multicriteria analysis. In J.P. Brans, editor, *Operational Research, IFORS 84*, pages 477–490. North Holland, Amsterdam, 1984.

[36] D.S. Briges and G.B. Mehta. *Representations of Preference Orderings*. Springer Verlag, Berlin, 1995.

[37] J. Broome. *Weighting Goods*. Basil Blackwell, London, 1991.

[38] A.V. Carrano. Establishing the order to human chromosome-specific DNA fragments. In A. Woodhead and B. Barnhart, editors, *Biotechnology and the Human Genome*, pages 37–50. Plenum Press, New York, 1988.

[39] I. Charon-Fournier, A. Germa, and O. Hudry. Utilisation des scores dans des méthodes exactes déterminant les ordres médians des tournois. *Mathématiques, Informatique et Sciences Humaines*, 119:53–74, 1992.

[40] A. Chateauneuf. Continuous representation of a preference relation on a connected topological space. *Journal Mathematical Economics*, 16:139–146, 1987.

[41] R.M. Chisholm and E. Sosa. Intrinsic preferability and the problem of supererogation. *Synthese*, 16:321–331, 1966.

[42] R.M. Chisholm and E. Sosa. On the logic of intrinsically better. *American Philosophical Quarterly*, 3:244–249, 1966.

[43] M. Cohen and J.-Y.Jaffray. Rational behavior under complete ignorance. *Econometrica*, 48:1281–1299, 1980.

[44] C.H. Coombs. *A Theory of Data*. John Wiley & Sons, New York, 1964.

[45] C.H. Coombs and J.E.K. Smith. On the detection of structures in attitudes and developmental processes. *Psychological Reviews*, 80:337–351, 1973.

[46] T.A. Cowan and P.C. Fishburn. Foundations of preference. In *Essays in Honor of Werner Leinfellner*, pages 261–271. D. Reidel, Dordrecht, 1988.

[47] M.B. Cozzens and F.S. Roberts. Double semiorders and double indifference graphs. *SIAM Journal on Algebraic and Discrete Methods*, 3:566–583, 1982.

[48] B. De Baets, E. Kerre, and B. Van De Walle. Fuzzy preference structures and their characterization. *Journal of Fuzzy Mathematics*, 3:373, 1995.

[49] G. Debreu. Representation of a preference ordering by a numerical function. In R. Thrall, C.H. Coombs, and R. Davies, editors, *Decision Processes*, pages 159–175. John Wiley & Sons, New York, 1954.

[50] G. Debreu. *Theory of Value: An Axiomatic Analysis of Economic Equilibrium*. John Wiley & Sons, New York, 1959.

[51] A. De Morgan. On the symbols of logic I. *Transactions of the Cambridge Philosophical Society*, 9:78–127, 1864.

[52] L.C. Dias and A. Tsoukiàs. On the constructive and other approaches in decision aiding. In C.A. Henggeler Antunes and J. Figueira, editors, *Proceedings of the 57th Meeting of the EURO MCDA Working Group*, 2003. to appear.

[53] M.A. Diaye. Variable interval orders. *Mathematical Social Sciences*, 38:21–33, 1999.

[54] P. Doherty, D. Driankov, and A. Tsoukiàs. Partial logics and partial preferences. In *Proceedings of the International Conference on Economics, Managemant and Information Technology CEMIT'92, Tokyo*, pages 525–528, 1992.

[55] J.P. Doignon. Thresholds representation of multiple semiorders. *SIAM Journal on Algebraic and Discrete Methods*, 8:77–84, 1987.

[56] J.P. Doignon. Sur les représentations minimales des semiordres et des ordres d'intervalles. *Mathématiques et Sciences Humaines*, 101:49–59, 1988.

[57] J.P. Doignon, A. Ducamp, and J.C. Falmagne. On realizable biorders and the biorder dimension of a relation. *Journal of Mathematical Psychology*, 28:73–109, 1984.

[58] J.P. Doignon and J.C. Falmagne. Well graded families of relations. *Discrete Mathematics*, 173(1-3):35–44, 1997.

[59] J.P. Doignon, B. Monjardet, M. Roubens, and Ph. Vincke. Biorder families, valued relations and preference modelling. *Journal of Mathematical Psychology*, 30:435–480, 1986.

[60] J. Dombi. A general class of fuzzy operators, the De Morgan class of fuzzy operators and fuzziness measures induced by fuzzy operators. *Fuzzy Sets and Systems*, 8(2):149–163, 1982.

[61] C. Domshlak and R.I. Brafman. CP-nets - reasoning and consistency testing. In *Proceedings of the 8th International Conference on Knowledge Representation and Reasoning, KR'02*, pages 121–132. Morgan Kaufmann, San Francisco, 2002.

[62] J. Doyle. Constructive belief and rational representation. *Computational Intelligence*, 5:1–11, 1989.

[63] J. Doyle. Rationality and its roles in reasoning. In *Proceedings of the 8th National Conference on Artificial Intelligence (AAAI'90)*, pages 1093–1100. MIT Press, Cambridge, 1990.

[64] J. Doyle. Reasoned assumptions and rational psychology. *Fundamenta Informaticae*, 20:35–73, 1994.

[65] J. Doyle and M.P. Wellman. Modular utility representation for decision-theoretic planning. *Proceedings of the First International Conference on Artificial Intelligence Planning Systems (AIPS'92)*, pages 236–242, 1992.

[66] D. Dubarle. Essai sur la généralisation naturelle de la logique usuelle. *Mathématique, Informatique, Sciences Humaines*, 107:17–73, 1989. 1963 manuscript, published posthumously.

[67] D. Dubois and H. Prade. Operations on fuzzy numbers. *International Journal of Systems Sciences*, 9:613–626, 1978.

[68] D. Dubois and H. Prade. *Fuzzy Sets and Systems - Theory and Applications*. Academic Press, New York, 1980.

[69] D. Dubois and H. Prade. Ranking fuzzy numbers in the setting of possibility theory. *Information Sciences*, 30:183–224, 1983.

[70] D. Dubois and H. Prade. *Possibility Theory*. Plenum Press, New York, 1988.

[71] D. Dubois and H. Prade. Possibility theory, probability theory and multiple-valued logics: A clarification. *Annals of Mathematics and Artificial Intelligence*, 32:35–66, 2001.

[72] B. Dushnik and E.W. Miller. Partially ordered sets. *American Journal of Mathematics*, 63:600–610, 1941.

[73] F. Fages and P. Ruet. Combining explicit negation and negation by failure via belnap's logic. *Theoretical Computer Science*, 171:61–75, 1997.

[74] G.T. Fechner. *Elemente der Psychophysik*. Breitkof und Hartel, Leipzig, 1860. Reprinted Thoemmes Press, Bristol, 1999.

[75] P.C. Fishburn. *Utility Theory for Decision Making*. John Wiley & Sons, New York, 1970.

[76] P.C. Fishburn. Interval representations for interval orders and semiorders. *Journal of Mathematical Psychology*, 10:91–105, 1973.

[77] P.C. Fishburn. Nontransitive measurable utility. *Journal of Mathematical Psychology*, 26:31–67, 1982.

[78] P.C. Fishburn. *Interval Orders and Interval Graphs*. John Wiley & Sons, New York, 1985.

[79] P.C. Fishburn. Nontransitive additive conjoint measurement. *Journal of Mathematical Psychology*, 35:1–40, 1991.

[80] P.C. Fishburn. Nontransitive preferences in decision theory. *Journal of Risk and Uncertainty*, 4:113–134, 1991.

[81] P.C. Fishburn. Generalisations of semiorders: A review note. *Journal of Mathematical Psychology*, 41:357–366, 1997.

[82] P.C. Fishburn. Preference structures and their numerical presentations. *Theoretical Computer Science*, 217:359–389, 1999.

[83] P.C. Fishburn and B. Monjardet. Norbert Wiener on the theory of measurement (1914, 1915, 1921). *Journal of Mathematical Psychology*, 36:165–184, 1992.

[84] M.C. Fitting. Bilattices and the semantics of logic programming. *Journal of Logic Programming*, 11:91–116, 1991.

[85] J. Fodor and M. Roubens. *Fuzzy preference modelling and multicriteria decision support*. Kluwer Academic Publishers, Dordrecht, 1994.

[86] J.C. Fodor. An axiomatic approach to fuzzy preference modelling. *Fuzzy Sets and Systems*, 52:47–52, 1992.

[87] J.C. Fodor. Valued preference structures. *European Journal of Operational Research*, 79:277–286, 1994.

[88] J.M. Font and M. Moussavi. Note on a six valued extension of three valued logics. *Journal of Applied Non-Classical Logics*, 3:173–187, 1993.

[89] Ph. Fortemps and R. Słowiński. A graded quadrivalent logic for ordinal preference modelling: Loyola-like approach. *Fuzzy Optimization and Decision Making*, 1:93–111, 2002.

[90] M. Frank. On the simultaneous associativity of $f(x, y)$ and $x+y-f(x, y)$. *Aequationes Mathematicae*, 19:194–226, 1979.

[91] N. Georgescu-Roegen. The pure theory of consumer's behavior. *Quarterly Journal of Economics*, 50:545–593, 1936.

[92] M.C. Golumbic and R. Shamir. Complexity and algorithms for reasoning about time. A graph theoretic approach. *Journal of the ACM*, 40:1108–1133, 1993.

[93] J.A. Gougen. The logic of inexact concepts. *Synthese*, 19:325–373, 1969.

[94] P.E. Green, D.S. Tull, and G. Albaum. *Research for marketing decisions*. Prentice Hall, Englewood Cliffs, 1988. 2nd edition.

[95] S. Halldén. *On the Logic of Better*. Library of Theoria, Lund, 1957.

[96] C. Halldin. Preference and the cost of preferential choice. *Theory and Decision*, 21:35–63, 1986.

[97] E. Halphen. La notion de vraisemblance. *Publications de l'Institut de Statistique de l'Université de Paris*, 4:41–92, 1955.

[98] D.J. Hand. *Discrimination and Classification*. John Wiley & Sons, New York, 1981.

[99] E. Hansen. *Global Optimization Using Interval Analysis*. Marcel Dekker, New York, 1992.

[100] B. Hansson. Choice structures and preference relations. *Synthese*, 18:443–458, 1966.

[101] B. Hansson. Fundamental axioms for preference relations. *Synthese*, 18:423–442, 1966.

[102] F.R. Hodson, D.G. Kendall, and P. Tautu. *Mathematics in the Archaeological and Historial Sciences*. Edinburgh University Press, 1971.

[103] H.S. Houthakker. The logic of preference and choice. In A.T. Tymieniecka, editor, *Contributions to Logic and Methodology in Honour of J.M. Bochenski*, pages 193–207. North Holland, Amsterdam, 1965.

[104] O. Huber. An axiomatic system for multidimensional preferences. *Theory and Decision*, 5:161–184, 1974.

[105] O. Huber. Non transitive multidimensional preferences. *Theory and Decision*, 10:147–165, 1979.

[106] O. Hudry and F. Woirgard. Ordres médians et ordres de Slater des tournois. *Mathématiques, Informatique et Sciences Humaines*, 133:23–56, 1996.

[107] J.-Y. Jaffray and P.P. Wakker. Decision making with belief functions: Compatibility and incompatibility with the sure-thing principle. *Journal of Risk and Uncertainty*, 7:255–271, 1993.

[108] M. Janowitz, F.-J. LaPointe, F.R. McMorris, B. Mirkin, and F.S. Roberts, editors. *Bioconsensus*, volume 61 of *DIMACS Series*. American Mathematical Society, Providence, 2003.

[109] R.C. Jeffrey. *The Logic of Decision*. McGraw Hill, New York, 1965.

[110] J. Kacprzyk and M. Roubens. *Non Conventional Preference Relations in Decision Making*, volume 301 of *Lecture Notes in Economics and Mathematical Systems*. Springer Verlag, Berlin, 1988.

[111] D. Kahneman, P. Slovic, and A. Tversky. *Judgement under Uncertainty - Heuristics and Biases*. Cambridge University Press, Cambridge, 1981.

[112] D. Kahneman and A. Tversky. Prospect theory: An analysis of decision under risk. *Econometrica*, 47:263–291, 1979.

[113] Y. Kaluzhny and A.Y. Muravitsky. A knowledge representation based on the belnap's four valued logic. *Journal of Applied Non-Classical Logics*, 3:189–203, 1993.

[114] R. M Karp. Mapping the genome: Some combinatorial problems arising in molecular biology. In *Procceding 25th Symposium Theory of Computating (STOC)*, pages 278–285. ACM Press, New York, 1993.

[115] A. Kaufmann. *Theory of Fuzzy Sets*. Academic Press, New York, 1975.

[116] A. Kaufmann and M.M. Gupta. *Introduction to Fuzzy Arthmetic-Theory and Application*. Van Nostrand Reinhold, New York, 1985.

[117] R.L. Keeney and H. Raiffa. *Decisions with Multiple Objectives: Preferences and Value Tradeoffs*. John Wiley & Sons, New York, 1976.

[118] D.H. Krantz, R.D. Luce, P. Suppes, and A. Tversky. *Foundations of Measurement 1: Additive and Polynomial Rpresentations*. Academic Press, New York, 1971.

[119] D.H. Krantz, R.D. Luce, P. Suppes, and A. Tversky. *Foundations of Measurement 2: Geometrical, Threshold and Probabilistic Representations*. Academic Press, New York, 1989.

[120] S. Kraus, D. Lehmann, and M. Magidor. Nonmonotonic reasoning, preferential models and cumulative logics. *Artificial Intelligence*, 44:167–207, 1990.

[121] C. Lafage and J. Lang. Logical representation of preferences for group decision making. In *Proceedings of the 7th International Conference on Knowledge Representation and Reasoning (KR'2000)*, pages 457–468. Morgan Kaufmann, San Mateo, 2000.

[122] E. Lawler. *Combinatorial Optimization: Networks and Matroids*. Holt, Rinehart, Winston, New York, 1976.

[123] D. Lehmann. Nonmonotonic logics and semantics. *Journal of Logic and Computation*, 11:229–256, 2001.

[124] E. Li and R.J. Lee. Comparison of fuzzy numbers based on the probabiliy measure of fuzzy events. *Computers and Mathematics with Applications*, 15:887–896, 1988.

[125] R.D. Luce. Semiorders and a theory of utility discrimination. *Econometrica*, 24:178–191, 1956.

[126] R. Massaglia and A. Ostanello. N-tomic: A support system for multicriteria segmentation problems. In P. Korhonen, A. Lewandowski, and J. Wallenius, editors, *Multiple Criteria Decision Support*, volume 356 of *Lecture Notes in Economics and Mathematical Systems*, pages 167–174. Springer Verlag, Berlin, 1991.

[127] K.O. May. Intransitivity, utility and the aggregation of preference patterns. *Econometrica*, 22:1–13, 1954.

[128] E. Mendelson. *Introduction to Mathematical Logic*. Van Nostrand, Princeton, 1964.

[129] J. Mitas. Minimal representation of semiorders with intervals of same length. In V. Bouchitté and M. Morvan, editors, *Orders, algorithms and applications*, volume 831 of *Lecture Notes in Computer Science*, pages 162–175. Springer Verlag, Berlin, 1994.

[130] M. Mizumoto and K. Tanaka. The four operations of arithmetic on fuzzy numbers. *Systems-Computers-Controls*, 7(5):73–81, 1976.

[131] M. Mizumoto and K. Tanaka. Some properties of fuzzy sets of type 2. *Information and Control*, 31:312–340, 1976.

[132] B. Monjardet. Axiomatiques et propriétés des quasi ordres. *Mathématiques et Sciences Humaines*, 63:51–82, 1978.

[133] B. Monjardet. Relations à éloignement minimum des relations binaires: note bibliographique. *Mathématiques, Informatique et Sciences Humaines*, 67:115–122, 1979.

[134] R. Moore. *Interval Analysis*. Prentice-Hall, Englewood Cliffs, 1966.

[135] J.A. Moreno. A transitivity approach to preference relational systems. *European Journal of Operational Research*, 60(1):68–75, 1992.

[136] J. Moscarola and B. Roy. Procédure automatique d'examen de dossiers fondée sur une segmentation trichotomique en présence de critères multiples. *RAIRO Recherche Opérationnelle*, 11(2):145–173, 1977.

[137] J.D. Mullen. Does the logic of preference rest on a mistake? *Metaphilosophy*, 10:247–255, 1979.

[138] R. Nagaraja. Current approaches to long-range physical mapping of the human genome. In R. Anand, editor, *Techniques for the Analysis of Complex Genomes*, pages 1–18. Academic Press, London, 1992.

[139] S. Nahmias. Fuzzy variables. *Fuzzy Sets and Systems*, 1:97–110, 1976.

[140] Y. Nakamura. Real interval representations. *Journal of Mathematical Psychology*, 46:140–177, 2002.

[141] A. Ngo The and A. Tsoukiàs. Numerical representation of PQI interval orders. *Discrete Applied Mathematics*, To appear, 2003.

[142] A. Ngo The. *Structures de préférence non conventionnelles en aide à la décision*. PhD thesis, Université Paris Dauphine, Paris, 2002.

[143] A. Ngo The, A. Tsoukiàs, and Ph. Vincke. A polynomial time algorithm to detect PQI interval orders. *International Transactions in Operational Research*, 7:609–623, 2000.

[144] T.H. Nguyen, V. Kreinovitch, V. Nesterov, and M. Nakamura. On hardware support for interval computations and for soft computing: Theorems. *IEEE Transactions on Fuzzy Systems*, 5:108–127, 1978.

[145] D. Nute. *Defeasible Deontic Logic*. Kluwer Academic Publishers, Dordrecht, 1997.

[146] E. Oloriz, J.C. Candeal, and E. Induráin. Representability of interval orders. *Journal of Economic Theory*, 78:219–227, 1998.

[147] S.A. Orlovsky. Decision making with a fuzzy preference relation. *Fuzzy Sets and Systems*, 1:155–167, 1978.

[148] S.N. Ovchinnikov and M. Roubens. On strict preference relations. *Fuzzy Sets and Systems*, 43:319–326, 1991.

[149] Z. Pawlak and R. Słowiński. Decision analysis using rough sets. *International Transactions on Operational Research*, 1:107–114, 1994.

[150] I. Pe'er and R. Shamir. Satisfiability problems on intervals and unit intervals. *Theoretical Computer Science*, 175:349–372, 1997.

[151] C.S Peirce. On the algebra of logic. *American Journal of Mathematics*, 3:15–57, 1880.

[152] C.S. Peirce. On the logic of number. *American Journal of Mathematics*, 4:85–95, 1881.

[153] C.S. Peirce. A theory of probable inference. In C.S. Peirce, editor, *Studies in Logic: By Members of Johns Hopkins University*, pages 126–181. Little, Brown and Company, Boston, 1883.

[154] P. Perny. Multicriteria filtering methods based on concordance/non-discordance principles. *Annals of Operations Research*, 80:137–167, 1998.

[155] P. Perny and M. Roubens. Fuzzy preference modelling. In R. Słowiński, editor, *Fuzzy sets in decision analysis, operations research and statistics*, pages 3–30. Kluwer Academic Publishers, Dordrecht, 1998.

[156] P. Perny and B. Roy. The use of fuzzy outranking relations in preference modelling. *Fuzzy Sets and Systems*, 49:33–53, 1992.

[157] P. Perny and O. Spanjaard. Preference-based search in state space graphs. In *Proceedings of the Eighteenth National Conference on Artificial Intelligence (AAAI'2002)*, pages 751–756, Edmonton, Canada, July 2002. AAAI Press, Cambridge.

[158] P. Perny and O. Spanjaard. A preference-based approach to spanning trees and shortest paths problems. *European Journal of Operational Research*, to appear.

[159] P. Perny and A. Tsoukiàs. On the continuous extension of a four valued logic for preference modelling. In *Proceedings of the Information Processing and Management of Uncertainty in Knowledge-Based Systems (IPMU 1998) Conference, Paris*, pages 302–309, 1998.

[160] M. Pirlot. Minimal representation of a semiorder. *Theory and Decision*, 28:109–141, 1990.

[161] M. Pirlot and Ph. Vincke. *Semi Orders*. Kluwer Academic Publishers, Dordrecht, 1997.

[162] H. Poincaré. *La Valeur de la Science*. Flammarion, Paris, 1905.

[163] N. Rescher. Semantic foundations for the logic of preference. In N. Rescher, editor, *The Logic of Decision and Action*, pages 37–62. University of Pittsburgh, Pittsburgh, 1967.

[164] N. Rescher. *Introduction to Value Theory*. Prentice Hall, Englewood Cliffs, 1969.

[165] F.S. Roberts. *Measurement theory, with applications to Decision Making, Utility and the Social Sciences*. Addison-Wesley, Boston, 1979.

[166] M. Roubens and Ph. Vincke. On families of semiorders and interval orders imbedded in a valued structure of preference: A survey. *Information Sciences*, 34:187–198, 1984.

[167] M. Roubens and Ph. Vincke. *Preference Modelling*, volume 250 of *Lecture Notes in Economics and Mathematical Systems*. Springer Verlag, Berlin, 1985.

[168] M. Roubens and Ph. Vincke. Fuzzy possibility graphs and their application to ranking fuzzy numbers. In J. Kacprzyk and M. Roubens, editors, *Non-conventional Preference Relations in Decision Making*, volume 301 of *Lecture Notes in Economics and Mathematical Systems*, pages 119–128. Springer Verlang, Berlin, 1988.

[169] B. Roy. Partial preference analysis and decision aid: The fuzzy outranking relation concept. In D.E. Bell, R.L. Keeney, and H. Raiffa, editors, *Conflicting Objectives in Decisions*, pages 40–75. John Wiley & Sons, New York, 1977.

[170] B. Roy. Main sources of inaccurate determination, uncertainty and imprecision in decision models. In B. Munier and M. Shakun, editors, *Compromise, Negotiation and Group Decision*, pages 43–67. D. Reidel, Dordrecht, 1988.

[171] B. Roy. The outranking approach and the foundations of ELECTRE methods. *Theory and Decision*, 31:49–73, 1991.

[172] B. Roy. *Multicriteria Methodology for Decision Aiding*. Kluwer Academic Publishers, Dordrecht, 1996.

[173] B. Roy and D. Bouyssou. *Aide Multicritère à la Décision: Méthodes et Cas*. Economica, Paris, 1993.

[174] B. Roy and Ph. Vincke. Relational systems of preference with one or more pseudo-criteria: Some new concepts and results. *Management Science*, 30:1323–1335, 1984.

[175] B. Roy and Ph. Vincke. Pseudo-orders: Dfinition, properties and numerical representation. *Mathematical Social Sciences*, 14:263–274, 1987.

[176] B. Schweizer and A. Sklar. *Probabilistic Metric Spaces*. Elsevier Science, New York, 1983.

[177] D. Scott. Some ordered sets in computer science. In I. Rival, editor, *Ordered Sets*, pages 677–718. D. Reidel, Dordrecht, 1982.

[178] D. Scott and P. Suppes. Foundational aspects of theories of measurement. *Journal of Symbolic Logic*, 23:113–128, 1958.

[179] A.K. Sen. Social choice theory. In K.J. Arrow and M.D. Intriligator, editors, *Handbook of Mathematical Economics*, volume 3, pages 1073–1181. North-Holland, Amsterdam, 1986.

[180] Y. Shoham. Nonmonotonic logic: Meaning and utility. In *Proceedings of the 10th International Joint Conference in Artificial Intelligence, IJCAI87*, pages 388–393. Morgan Kaufmann, San Francisco, 1987.

[181] P. Slater. Inconsistencies in a schedule of paired comparisons. *Biometrika*, 48:303–312, 1961.

[182] R. Słowiński, editor. *Fuzzy sets in decision analysis, operations research and statistics*. Kluwer Academic Publishers, Dordrecht, 1998.

[183] B. Subiza. Numerical representation of acyclic preferences. *Journal of Mathematical Psychology*, 38:467–476, 1994.

[184] A.S. Tanguiane. *Aggregation and Representation of Preferences*. Springer Verlag, Berlin, 1991.

[185] A. Tarski. Contributions to the theory of models I, II. *Indagationes Mathematicae*, 16:572–588, 1954.

[186] A. Tarski. Contributions to the theory of models III. *Indagationes Mathematicae*, 17:56–64, 1955.

[187] L. Valadares Tavares. Generalized transitivity and preferences modelling: The concept of hyper-order. *European Journal of Operational Research*, 36:14–26, 1988.

[188] R. Thomason and J. Horty. Logics for inheritance theory. In M. Reinfrank, J. de Kleer, M.L. Ginsberg, and E. Sandewall, editors, *Non-Monotonic Reasoning*, volume 346 of *Lecture Notes in Artificial Intelligence*, pages 220–237. Springer Verlag, Berlin, 1987.

[189] W.T. Trotter. *Combinatorics and Partially Ordered Sets*. Johns Hopkins University Press, Baltimore, 1992.

[190] A. Tsoukiàs. Preference modelling as a reasoning process: A new way to face uncertainty in multiple criteria decision support systems. *European Journal of Operational Research*, 55:309–318, 1991.

[191] A. Tsoukiàs. Sur la généralisation des concepts de concordance et discordance en aide multicritère à la décision. Mémoire présenté pour l'obtention de l'habilitation à diriger des recherches, Université Paris Dauphine, 1997. Appeared also as Document du LAMSADE. n. 117.

[192] A. Tsoukiàs. A first-order, four valued, weakly paraconsistent logic and its relation to rough sets semantics. *Foundations of Computing and Decision Sciences*, 12:85–108, 2002.

[193] A. Tsoukiàs, P. Perny, and Ph. Vincke. From concordance/discordance to the modelling of positive and negative reasons in decision aiding. In D. Bouyssou, E. Jacquet-Lagrèze, P. Perny, R. Słowiński, D. Vanderpooten, and Ph. Vincke, editors, *Aiding Decisions with Multiple Criteria: Essays in Honour of Bernard Roy*, pages 147–174. Kluwer Academic Publishers, Dordrecht, 2002.

[194] A. Tsoukiàs and Ph. Vincke. A survey on non conventional preference modelling. *Ricerca Operativa*, 61:5–49, 1992.

[195] A. Tsoukiàs and Ph. Vincke. A new axiomatic foundation of partial comparability. *Theory and Decision*, 39:79–114, 1995.

[196] A. Tsoukiàs and Ph. Vincke. Extended preference structures in MCDA. In J. Climaco, editor, *Multicriteria Analysis*, pages 37–50. Springer Verlag, Berlin, 1997.

[197] A. Tsoukiàs and Ph. Vincke. Double threshold orders: A new axiomatization. *Journal of Multi-criteria Decision Analysis*, 7:285–301, 1998.

[198] A. Tsoukiàs and Ph. Vincke. A characterization of pqi interval orders. *Discrete Applied Mathematics*, 127(2):387–397, 2003.

[199] A. Tversky. Intransitivity of preferences. *Psychological Review*, 76:31–48, 1969.

[200] D. van Dalen. *Logic and Structure*. Springer Verlag, Berlin, 1983.

[201] B. Van De Walle, B. De Baets, and E. Kerre. Recent advances in fuzzy preference modelling. *Intelligent Systems and Soft Computing for Nuclear Science and Industry*, pages 98–104, 1996.

[202] B. Van De Walle, B. De Baets, and E. Kerre. Charaterizable fuzzy preference structures. *Annals of Operations Research*, 80:105–136, 1998.

[203] Ph. Vincke. P,Q,I preference structures. In J. Kacprzyk and M. Roubens, editors, *Non Conventional Preference Relations in Decision Making*, volume 301 of *Lecture Notes in Economics and Mathematical Systems*, pages 72–81. Springer Verlag, Berlin, 1988.

[204] Ph. Vincke. *Multicriteria Decision-Aid*. John Wiley & Sons, New York, 1992.

[205] Ph. Vincke. Preferences and numbers. In A. Colorni, M. Paruccini, and B. Roy, editors, *A-MCD-A - Aide Multi Critère à la Décision - Multiple Criteria Decision Aiding*, pages 343–354. The European Commission Joint Research Center, 2001.

[206] K. Vind. Independent preferences. *Journal of Mathematical Economics*, 20:119–135, 1991.

[207] D. von Winterfeldt and W. Edwards. *Decision Analysis and Behavorial Research*. Cambridge University Press, Cambridge, 1986.

[208] G.H. von Wright. *The logic of preference*. Edinburgh University Press, Edinburgh, 1963.

[209] G.H. von Wright. The logic of preference reconsidered. *Theory and Decision*, 3:140–169, 1972.

[210] S. Weber. A general concept of fuzzy connectives, negations and implications based on t-norms and t-conorms. *Fuzzy Sets and Systems*, 11(2):115–134, 1983.

[211] G.R. Widmeyer. Logic modeling with partially ordered preferences. *Decision Support Systems*, 4:87–95, 1988.

[212] G.R. Widmeyer. Reasoning with preferences and values. *Decision Support Systems*, 6:183–191, 1990.

[213] R.R. Yager. On the general class of fuzzy connectives. *Fuzzy Sets and Systems*, 4(3):235–242, 1980.

[214] R.R. Yager. A procedure for ordering fuzzy subsets of the unit interval. *Information Sciences*, 24:143–161, 1981.

[215] W. Yu. *Aide multicritère à la décision dans le cadre de la problématique du tri: méthodes et applications*. PhD thesis, LAMSADE, Université Paris Dauphine, Paris, 1992.

[216] Y.D. Yu. Triangular norms and tnf-sigma-algebras. *Fuzzy Sets and Systems*, 16(3):251–264, 1985.

[217] L.A. Zadeh. Fuzzy sets. *Information Control*, 8:338–353, 1965.

[218] L.A. Zadeh. The concept of a linguistic variable and its application to approximate reasoning. *Information Sciences*, 1:119–249, 1975.

[219] H.J. Zimmermann. *Fuzzy Set Theory and its applications*. Kluwer Academic Publishers, Dordrecht, 1985.

[220] C. Zopounidis and M. Doumpos. Multicriteria classification and sorting methods: A literature review. *European Journal of Operational Research*, 138:229–246, 2002.

Chapter 3

CONJOINT MEASUREMENT TOOLS FOR MCDM

A Brief Introduction

Denis Bouyssou
CNRS
LAMSADE, Université Paris Dauphine
F-75775 Paris Cedex 16
France
bouyssou@lamsade.dauphine.fr

Marc Pirlot
Faculté Polytechnique de Mons
9, rue de Houdain
B-7000 Mons
Belgium
marc.pirlot@fpms.ac.be

Abstract This paper offers a brief and nontechnical introduction to the use of conjoint measurement in multiple criteria decision making. The emphasis is on the, central, additive value function model. We outline its axiomatic foundations and present various possible assessment techniques to implement it. Some extensions of this model, e.g. nonadditive models or models tolerating intransitive preferences are then briefly reviewed.

Keywords: Conjoint Measurement, additive value function, preference modelling.

1. Introduction and Motivation

Conjoint measurement is a set of tools and results first developed in Economics [44] and Psychology [141] in the beginning of the '60s. Its, ambitious, aim is to provide measurement techniques that would be adapted to the needs of the Social Sciences in which, most often, multiple dimensions have to be taken into account.

Soon after its development, people working in decision analysis realized that the techniques of conjoint measurement could also be used as tools to structure preferences [51, 165]. This is the subject of this paper which offers a brief and nontechnical introduction to conjoint measurement models and their use in multiple criteria decision making. More detailed treatments may be found in [63, 79, 121, 135, 209]. Advanced references include [58, 129, 211].

1.1 Conjoint Measurement Models in Decision Theory

The starting point of most works in decision theory is a binary relation $\succsim$ on a set A of objects. This binary relation is usually interpreted as an "at least as good as" relation between alternative courses of action gathered in A.

Manipulating a binary relation can be quite cumbersome as soon as the set of objects is large. Therefore, it is not surprising that many works have looked for a *numerical representation* of the binary relation $\succsim$. The most obvious numerical representation amounts to associate a real number $V(a)$ to each object $a \in A$ in such a way that the comparison between these numbers faithfully reflects the original relation $\succsim$. This leads to defining a real-valued function V on A, such that:

$$a \succsim b \Leftrightarrow V(a) \geq V(b), \tag{3.1}$$

for all $a, b \in A$. When such a numerical representation is possible, one can use V instead of $\succsim$ and, e.g. apply classical optimization techniques to find the most preferred elements in A given $\succsim$. We shall call such a function V a *value function.*

It should be clear that not all binary relations $\succsim$ may be represented by a value function. Condition (3.1) imposes that $\succsim$ is complete (i.e. $a \succsim b$ or $b \succsim a$, for all $a, b \in A$) and transitive (i.e. $a \succsim b$ and $b \succsim c$ imply $a \succsim c$, for all $a, b, c \in A$). When A is finite or countably infinite, it is well-known [58, 129] that these two conditions are, in fact, not only necessary but also sufficient to build a value function satisfying (3.1).

REMARK 5 *The general case is more complex since* (3.1) *implies, for instance, that there must be "enough" real numbers to distinguish objects that have to be distinguished. The necessary and sufficient conditions for* (3.1) *can be found in [58, 129]. An advanced treatment is [13]. Sufficient conditions that are well-*

adapted to cases frequently encountered in Economics can be found in [42, 45]; see [34] for a synthesis.

It is vital to note that, when a value function satisfying (3.1) exists, it is by no means unique. Taking any increasing function ϕ on $\mathbb{R}$, it is clear that $\phi \circ V$ gives another acceptable value function. A moment of reflection will convince the reader that only such transformations are acceptable and that if V and U are two real-valued functions on A satisfying (3.1), they must be related by an increasing transformation. In other words, a value function in the sense of (3.1) defines an *ordinal scale*.

Ordinal scales, although useful, do not allow the use of sophisticated assessment procedures, i.e. of procedures that allow an analyst to assess the relation $\succsim$ through a structured dialogue with the decision-maker. This is because the knowledge that $V(a) \geq V(b)$ is strictly equivalent to the knowledge of $a \succsim b$ and no inference can be drawn from this assertion besides the use of transitivity.

It is therefore not surprising that much attention has been devoted to numerical representations leading to more constrained scales. Many possible avenues have been explored to do so. Among the most well-known, let us mention:

- the possibility to compare *probability distributions* on the set A [58, 207]. If it is required that, not only (3.1) holds but that the numbers attached to the objects should be such that their expected values reflect the comparison of probability distributions on the set of objects, a much more constrained numerical representation clearly obtains,
- the introduction of "preference difference" comparisons of the type: the difference between a and b is larger than the difference between c and d, see [44, 81, 123, 129, 159, 180, 199]. If it is required that, not only (3.1) holds, but that the differences between numbers also reflect the comparisons of preference differences, a more constrained numerical representation obtains.

When objects are evaluated according to several dimensions, i.e. when $\succsim$ is defined on a product set, new possibilities emerge to obtain numerical representations that would specialize (3.1). The purpose of conjoint measurement is to study such kinds of models.

There are many situations in decision theory which call for the study of binary relations defined on product sets.Among them let us mention:

- *Multiple criteria decision making* using a preference relation comparing alternatives evaluated on several attributes [16, 121, 162, 173, 209],
- *Decision under uncertainty* using a preference relation comparing alternatives evaluated on several states of nature [68, 107, 177, 184, 210, 211],

- *Consumer theory* manipulating preference relations for bundles of several goods [43],
- *Intertemporal decision making* using a preference relation between alternatives evaluated at several moments in time [121, 125, 126],
- *Inequality measurement* comparing distributions of wealth across several individuals [5, 17, 18, 217].

The purpose of this paper is to give an introduction to the main models of conjoint measurement useful in multiple criteria decision making. The results and concepts that are presented may however be of interest in all of the aforementioned areas of research.

REMARK 6 *Restricting ourselves to applications in multiple criteria decision making will not allow us to cover every aspect of conjoint measurement. Among the most important topics left aside, let us mention: the introduction of statistical elements in conjoint measurement models [54, 108] and the test of conjoint measurement models in experiments [135].*

Given a binary relation $\succsim$ on a product set $X = X_1 \times X_2 \times \cdots \times X_n$, the theory of conjoint measurement consists in finding conditions under which it is possible to build a convenient numerical representation of $\succsim$ and to study the uniqueness of this representation. The central model is the *additive value function* model in which:

$$x \succsim y \Leftrightarrow \sum_{i=1}^{n} v_i(x_i) \geq \sum_{i=1}^{n} v_i(y_i) \tag{3.2}$$

where v_i are real-valued functions, called *partial value functions*, on the sets X_i and it is understood that $x = (x_1, x_2, \ldots, x_n)$ and $y = (y_1, y_2, \ldots, y_n)$. Clearly if $\succsim$ has a representation in model (3.2), taking any common increasing transformation of the v_i will *not* lead to another representation in model (3.2).

Specializations of this model in the above-mentioned areas give several central models in decision theory:

- The Subjective Expected Utility model, in the case of decision-making under uncertainty,
- The discounted utility model for dynamic decision making,
- Inequality measures *à la* Atkinson/Sen in the area of social welfare.

The axiomatic analysis of this model is now quite firmly established [44, 129, 211]; this model forms the basis of many decision analysis techniques [79, 121, 209, 211]. This is studied in sections 3 and 4 after we introduce our main notation and definitions in section 2.

REMARK 7 *One possible objection to the study of model* (3.2) *is that the choice of an* additive *model seems arbitrary and restrictive. It should be observed here that the functions v_i will precisely be assessed so that additivity holds. Furthermore, the use of a simple model may be seen as an advantage in view of the limitations of the cognitive abilities of most human beings.*

It is also useful to notice that this model can be reformulated so as to make addition disappear. Indeed if there are partial value functions v_i such that (3.2) *holds, it is clear that $V = \sum_{i=1}^{n} v_i$ is a value function satisfying* (3.1). *Since V defines an ordinal scale, taking the exponential of V leads to another valid value function W. Clearly W has now a multiplicative form:*

$$x \succsim y \Leftrightarrow W(x) = \prod_{i=1}^{n} w_i(x_i) \geq W(y) = \prod_{i=1}^{n} w_i(y_i).$$

where $w_i(x_i) = e^{v_i(x_i)}$.

The reader is referred to chapter 7 for the study of situations in which V defines a scale that is more constrained than an ordinal scale, e.g. because it is supposed to reflect preference differences or because it allows to compute expected utilities. In such cases, the additive form (3.2) *is no more equivalent to the multiplicative form considered above.*

In section 5 we present a number of extensions of this model going from non-additive representations of transitive relations to model tolerating intransitive indifference and, finally, nonadditive representations of nontransitive relations.

REMARK 8 *In this paper, we shall restrict our attention to the case in which alternatives may be evaluated on the various attributes without risk or uncertainty. Excellent overviews of these cases may be found in [121, 209]; recent references include [142, 150].*

Before starting our study of conjoint measurement oriented towards MCDM, it is worth recalling that conjoint measurement aims at establishing measurement models in the Social Sciences. To many, the very notion of "measurement in the Social Sciences" may appear contradictory. It may therefore be useful to briefly consider how the notion of measurement can be modelled in Physics, an area in which the notion of "measurement" seems to arise quite naturally, and to explain how a "measurement model" may indeed be useful in order to structure preferences.

1.2 An Aside: Measuring Length

Physicists usually take measurement for granted and are not particularly concerned with the technical and philosophical issues it raises (at least when they work within the realm of Newtonian Physics). However, for a Social Scientist, these question are of utmost importance. It may thus help to have an idea of how things appear to work in Physics before tackling more delicate cases.

Suppose that you are on a desert island and that you want to "measure" the length of a collection of rigid straight rods. Note that we do not discuss here the "pre-theoretical" intuition that "length" is a property of these rods that can be measured, as opposed, say, to their softness or their beauty.

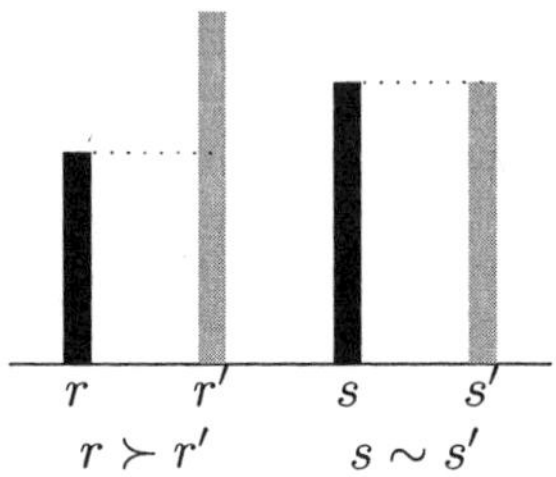

Figure 3.1. Comparing the length of two rods.

A first simple step in the construction of a measure of length is to place the two rods side by side in such a way that one of their extremities is at the same level (see Figure 3.1). Two things may happen: either the upper extremities of the two rods coincide or not. This seems to be the simplest way to devise an experimental procedure leading to the discovery of which rod "has more length" than the other. Technically, this leads to defining two binary relations $\succ$ and $\sim$ on the set of rods in the following way:

$r \succ r'$ when the extremity of r is higher than the extremity of r',

$r \sim r'$ when the extremities of r and r' are at the same level,

Clearly, if length is a quality of the rods that can be measured, it is expected that these pairwise comparisons are somehow consistent, e.g.,

- if $r \succ r'$ and $r' \succ r''$, it should follow that $r \succ r''$,
- if $r \sim r'$ and $r' \sim r''$, it should follow that $r \sim r''$,
- if $r \sim r'$ and $r' \succ r''$, it should follow that $r \succ r''$.

Although quite obvious, these consistency requirements are stringent. For instance, the second and the third conditions are likely to be violated if the experimental procedure involves some imprecision, e.g if two rods that slightly differ in length are nevertheless judged "equally long". They represent a form of *idealization* of what could be a perfect experimental procedure.

With the binary relations $\succ$ and $\sim$ at hand, we are still rather far from a full-blown measure of length. It is nevertheless possible to assign numbers to each of the rods in such a way that the comparison of these numbers reflects what has been obtained experimentally. When the consistency requirements mentioned above are satisfied, it is indeed generally possible to build a real-valued function Φ on the set of rods that would satisfy:

$$r \succ r' \Leftrightarrow \Phi(r) > \Phi(r') \text{ and}$$
$$r \sim r' \Leftrightarrow \Phi(r) = \Phi(r').$$

If the experiment is costly or difficult to perform, such a numerical assignment may indeed be useful because it summarizes, once for all, what has been obtained in experiments. Clearly there are many possible ways to assign numbers to rods in this way. Up to this point, they are equally good for our purposes. The reader will easily check that defining $\succsim$ as $\succ$ or $\sim$, the function Φ is noting else than a "value function" for length: any increasing transformation may therefore be applied to Φ.

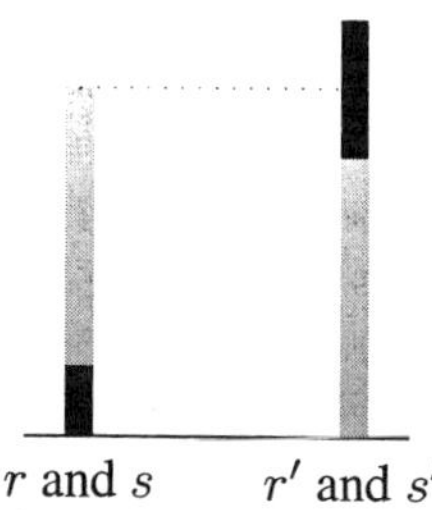

Figure 3.2. Comparing the length of composite rods.

The next major step towards the construction of a measure of length is the realization that it is possible to form new rods by simply placing two or more rods "in a row", i.e. you may *concatenate* rods. From the point of view of length, it seems obvious to expect this concatenation operation $\circ$ to be "commutative" ($r \circ s$ has the same length as $s \circ r$) and associative ($(r \circ s) \circ t$ has the same length as $r \circ (s \circ t)$).

You clearly want to be able to measure the length of these composite objects and you can always include them in our experimental procedure outlined above (see Figure 3.2). Ideally, you would like your numerical assignment Φ to be somehow compatible with the concatenation operation: knowing the numbers assigned to two rods, you want to be able to deduce the number assigned to their concatenation. The most obvious way to achieve that is to require that the numerical assignment of a composite object can be deduced by addition from the numerical assignments of the objects composing it, i.e. that

$$\Phi(r \circ r') = \Phi(r) + \Phi(r').$$

This clearly places many additional constraints on the results of your experiment. An obvious one is that $\succ$ and $\sim$ should be compatible with the concatenation operation $\circ$, e.g.

$$r \succ r' \text{ and } t \sim t' \text{ should lead to } r \circ t \succ r' \circ t'.$$

These new constraints may or may not be satisfied. When they are, the usefulness of the numerical assignment Φ is even more apparent: a simple arithmetic operation will allow to infer the result of an experiment involving composite objects.

Let us take a simple example. Suppose that you have five rods $r_1, r_2, \ldots, r_5$ and that, because space is limited, you can only concatenate at most two rods and that not all concatenations are possible. Let us suppose, for the moment, that you do not have much technology available so that you may only experiment using *different* rods. You may well collect the following information, using obvious notation exploiting the transitivity of $\succ$ which holds in this experiment,

$$r_1 \circ r_5 \succ r_3 \circ r_4 \succ r_1 \circ r_2 \succ r_5 \succ r_4 \succ r_3 \succ r_2 \succ r_1.$$

Your problem is then to find a numerical assignment Φ to rods such that using an addition operation, you can infer the numerical assignment of composite objects consistently with your observations. Let us consider the following three assignments:

	Φ	Φ'	Φ''
r_1	14	10	14
r_2	15	91	16
r_3	20	92	17
r_4	21	93	18
r_5	28	100	29

These three assignments are equally valid to reflect the comparisons of single rods. Only the first and the third allow to capture the comparisons of composite objects that were performed. Note that, going from Φ to Φ'' does not involve just changing the "unit of measurement": since $\Phi(r_1) = \Phi''(r_1)$ this would imply that $\Phi = \Phi''$, which is clearly false.

Such numerical assignments have limited usefulness. Indeed, it is tempting to use them to predict the result of comparisons that we have not been able to perform. But this turns out to be quite disappointing: using Φ you would conclude that $r_2 \circ r_3 \sim r_1 \circ r_4$ since $\Phi(r_2) + \Phi(r_3) = 15 + 20 = 35 = \Phi(r_1) + \Phi(r_4)$, but, using Φ'', you would conclude that $r_2 \circ r_3 \succ r_1 \circ r_4$ since $\Phi''(r_2) + \Phi''(r_3) = 16 + 17 = 33$ while $\Phi''(r_1) + \Phi''(r_4) = 14 + 18 = 32$.

Intuitively, "measuring" calls for some kind of a *standard* (e.g. the "Mètre-étalon" that can be found in the Bureau International des Poids et Mesures in Sèvres, near Paris). This implies choosing an appropriate "standard" rod *and* being able to prepare perfect copies of this standard rod (we say here "appropriate" because the choice of a standard should be made in accordance with the lengths of the objects to be measured: a tiny or a huge standard will not facilitate experiments). Let us call s_0 the standard rod. Let us suppose that you have been able to prepare a large number of perfect copies $s_1, s_2, \ldots$ of s_0. We therefore have:

$$s_0 \sim s_1, s_0 \sim s_2, s_0 \sim s_3, \ldots$$

Let us also agree that the length of s_0 is 1. This is your, arbitrary, unit of length. How can you use s_0 and its perfect copies so as to determine unambiguously the length of any other (simple or composite) object? Quite simply, you may prepare a "standard sequence of length n", $S(n) = s_1 \circ s_2 \circ \ldots \circ s_{n-1} \circ s_n$, i.e. a composite object that is made by concatenating n perfect copies of our standard rod s_0. The length of a standard sequence of length n is exactly n since we have concatenated n objects that are perfect copies of the standard rod of length 1. Take any rod r and let us compare r with several standard sequences of increasing length: $S(1), S(2), \ldots$

Two cases may arise. There may be a standard sequence $S(k)$ such that $r \sim S(k)$. In that case, we know that the number $\Phi(r)$ assigned to r must be exactly k. This is unlikely however. The most common situation is that we will find two consecutive standard sequences $S(k-1)$ and $S(k)$ such that $r \succ S(k-1)$ and $S(k) \succ r$ (see Figure 3.3). This means that $\Phi(r)$ must be such that $k-1 < \Phi(r) < k$. We seem to be in trouble here since, as before, $\Phi(r)$ is not exactly determined. How can you proceed? This depends on your technology for preparing perfect copies.

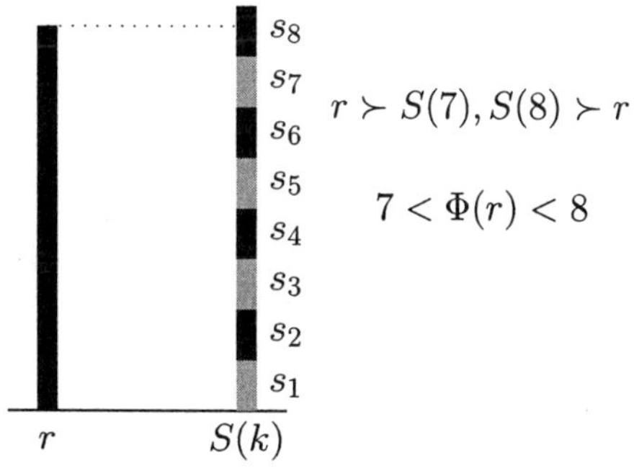

Figure 3.3. Using standard sequences.

Imagine that you are able to prepare perfect copies not only of the standard rod but also of any object. You may then prepare several copies $(r_1, r_2, \ldots)$ of the rod r. You can now compare a composite object made out of two perfect copies of r with your standard sequences $S(1), S(2), \ldots$ As before, you shall eventually arrive at locating $\Phi(r_1 \circ r_2) = 2\Phi(r)$ within an interval of width 1. This means that the interval of imprecision surrounding $\Phi(r)$ has been divided by two. Continuing this process, considering longer and longer sequences of perfect copies of r, you will keep on reducing the width of the interval containing $\Phi(r)$. This means that you can approximate $\Phi(r)$ with any given level of precision. Mathematically, a unique value for $\Phi(r)$ will be obtained using a simple argument.

Supposing that you are in position to prepare perfect copies of any object is a strong technological requirement. When this is not possible, there still exists a way out. Instead of preparing a perfect copy of r you may also try to increase the granularity of your standard sequence. This means building an object t that you would be able to replicate perfectly and such that concatenating t with one of its perfect replicas gives an object that has exactly the length of the standard object s_0, i.e. $\Phi(t) = 1/2$. Considering standard sequences based on t, you will be able to increase by a factor 2 the precision with which we measure the length of r. Repeating the process, i.e. subdividing t, will lead, as before, to a unique limiting value for $\Phi(r)$.

The mathematical machinery underlying the measurement process informally described above (called "extensive measurement") rests on the theory of ordered groups. It is beautifully described and illustrated in [129]. Although the underlying principles are simple, we may expect complications to occur e.g. when not all concatenations are feasible, when there is some level (say the velocity of light if we were to measure speed) that cannot be exceeded or when it comes to relate different measures. See [129, 140, 168] for a detailed treatment.

Clearly, this was an overly detailed and unnecessary complicated description of how length could be measured. Since our aim is to eventually deal with "measurement" in the Social Sciences, it may however be useful to keep the above process in mind. Its basic ingredients are the following:

- well-behaved relations $\succ$ and $\sim$ allowing to compare objects,
- a concatenation operation $\circ$ allowing to consider composite objects,
- consistency requirements linking $\succ$, $\sim$ and $\circ$,
- the ability to prepare perfect copies of some objects in order to build standard sequences.

Basically, conjoint measurement is a quite ingenious way to perform related measurement operations when no concatenation operation is available. This will however require that objects can be evaluated along several dimensions.

Before explaining how this might work, it is worth explaining the context in which such measurement might prove useful.

REMARK 9 *It is often asserted that "measurement is impossible in the Social Sciences" precisely because the Social Scientist has no way to define a concatenation operation. Indeed, it would seem hazardous to try to concatenate the intelligence of two subjects or the pain of two patients (see [56, 106]). Under certain conditions, the power of conjoint measurement will precisely be to provide a means to bypass this absence of readily available concatenation operation when the objects are evaluated on several dimensions.*

Let us remark that, even when there seems to be a concatenation operation readily available, it does not always fit the purposes of extensive measurement. Consider for instance an individual expressing preferences for the quantity of the two goods he consumes. The objects therefore take the well structured form of points in the positive orthant of $\mathbb{R}^2$. There seems to be an obvious concatenation operation here: $(x, y) \circ (z, w)$ might simply be taken to be $(x + y, z + w)$. However a fairly rational person, consuming pants and jackets, may indeed prefer $(3, 0)$ (3 pants and no jacket) to $(0, 3)$ (no pants and 3 jackets) but at the same time prefer $(3, 3)$ to $(6, 0)$. This implies that these preferences cannot be explained by a measure that would be additive with respect to the concatenation operation consisting in adding the quantities of the two goods consumed. Indeed $(3, 0) \succ (0, 3)$ implies $\Phi(3, 0) > \Phi(0, 3)$, which implies $\Phi(3, 0) + \Phi(3, 0) > \Phi(0, 3) + \Phi(3, 0)$. Additivity with respect to concatenation should then imply that $(3, 0) \circ (3, 0) \succ (0, 3) \circ (3, 0)$, that is $(6, 0) \succ (3, 3)$.

1.3 An Example: Even Swaps

The even swaps technique described and advocated in [120, 121, 165] is a simple way to deal with decision problems involving several attributes that does not have recourse to a formal representation of preferences, which will be the subject of conjoint measurement. Because this technique is simple and may be quite useful, we describe it below using the same example as in [120]. This will also allow to illustrate the type of problems that are dealt with in decision analysis applications of conjoint measurement.

EXAMPLE 3 (EVEN SWAPS TECHNIQUE) *A consultant considers renting a new office. Five different locations have been identified after a careful consideration of many possibilities, rejecting all those that do not meet a number of requirements.*

His feeling is that five distinct characteristics, we shall say five attributes, of the possible locations should enter into his decision: his daily commute time (expressed in minutes), the ease of access for his clients (expressed as the percentage of his present clients living close to the office), the level of services offered by the new office (expressed on an ad hoc scale with three levels: A (all facilities available), B (telephone and fax), C (no facilities)), the size of the office expressed in square feet, and the monthly cost expressed in dollars.

The evaluation of the five offices is given in Table 3.1. The consultant has

Table 3.1. Evaluation of the 5 offices on the 5 attributes.

	a	b	c	d	e
Commute	45	25	20	25	30
Clients	50	80	70	85	75
Services	A	B	C	A	C
Size	800	700	500	950	700
Cost	1850	1700	1500	1900	1750

well-defined preferences on each of these attributes, independently of what is happening on the other attributes. His preference increases with the level of access for his clients, the level of services of the office and its size. It decreases with commute time and cost. This gives a first easy way to compare alternatives through the use of dominance.

An alternative y is dominated by an alternative x if x is at least as good as y on all *attributes while being strictly better for at least one attribute. Clearly dominated alternatives are not candidate for the final choice and may, thus, be dropped from consideration. The reader will easily check that, on this example, alternative b dominates alternative e: e and b have similar size but b is less expensive, involves a shorter commute time, an easier access to clients and a better level of services. We may therefore forget about alternative e. This is the only case of "pure dominance" in our table. It is however easy to see that d is "close" to dominating a, the only difference in favor of a being on the cost attribute (50 $ per month). This is felt more than compensated by the differences in favor of d on all other attributes: commute time (20 minutes), client access (35 %) and size (150 sq. feet).*

Dropping all alternatives that are not candidate for choice, this initial investigation allows to reduce the problem to:

	b	c	d
Commute	*25*	*20*	*25*
Clients	*80*	*70*	*85*
Services	*B*	*C*	*A*
Size	*700*	*500*	*950*
Cost	*1700*	*1500*	*1900*

A natural way to proceed is then to assess tradeoffs. Observe that all alternatives but c have a common evaluation on commute time. We may therefore ask the consultant, starting with office c, what gain on client access would compensate a loss of 5 *minutes on commute time. We are looking for an alternative c' that would be evaluated as follows:*

	c	c'
Commute	*20*	**25**
Clients	*70*	$\mathbf{70+\delta}$
Services	*C*	*C*
Size	*500*	*500*
Cost	*1500*	*1500*

and judged indifferent to c. Although this is not an easy question, it is clearly crucial in order to structure preferences.

REMARK 10 *In this paper, we do not consider the possibility of lexicographic preferences, in which such tradeoffs do not occur, see [59, 60, 160]. Lexicographic preferences may also be combined with the possibility of "local" tradeoffs, see [22, 64, 136].*

REMARK 11 *Since tradeoffs questions may be difficult, it is wise to start with an attribute requiring few assessments (in the example, all alternatives but one have a common evaluation on commute time). Clearly this attribute should be traded against one with an underlying "continuous" structure (cost, in the example).*

Suppose that the answer is that for $\delta = 8$, it is reasonable to assume that c and c' would be indifferent. This means that the decision table can be reformulated as follows:

	b	c'	d
Commute	*25*	*25*	*25*
Clients	*80*	*78*	*85*
Services	*B*	*C*	*A*
Size	*700*	*500*	*950*
Cost	*1700*	*1500*	*1900*

It is then apparent that all alternatives have a similar evaluation on the first attribute which, therefore, is not useful to discriminate between alternatives and may be forgotten. The reduced decision table is as follows:

	b	c'	d
Clients	*80*	*78*	*85*
Services	*B*	*C*	*A*
Size	*700*	*500*	*950*
Cost	*1700*	*1500*	*1900*

There is no case of dominance in this reduced table. Therefore further simplification calls for the assessment of new tradeoffs. Using cost as the reference attribute, we then proceed to "neutralize" the service attribute. Starting with office c′, this means asking for the increase in monthly cost that the consultant would just be prepared to pay to go from level "C" of service to level "B". Suppose that this increase is roughly 250 $. This defines alternative c″. Similarly, starting with office d we ask for the reduction of cost that would exactly compensate a reduction of services from "A" to "B". Suppose that the answer is 100 $ a month, which defines alternative d′. The decision table is reshaped as:

	b	*c″*	*d′*
Clients	*80*	*78*	*85*
Services	*B*	**B**	**B**
Size	*700*	*500*	*950*
Cost	*1700*	***1750***	***1800***

We may forget about the second attribute which does not discriminate any more between alternatives. When this is done, it is apparent that c″ is dominated by b and can be suppressed. Therefore, the decision table at this stage looks like the following:

	b	*d′*
Clients	*80*	*85*
Size	*700*	*950*
Cost	*1700*	*1800*

Unfortunately, this table reveals no case of dominance. New tradeoffs have to be assessed. We may now ask, starting with office b, what additional cost the consultant would be ready to incur to increase its size by 250 square feet. Suppose that the rough answer is 250 $ a month, which defines b′. We are now facing the following table:

	b′	*d′*
Clients	*80*	*85*
Size	**950**	*950*
Cost	**1950**	*1800*

Attribute size may now be dropped from consideration. But, when this is done, it is clear that d′ dominates b′. Hence it seems obvious to recommend office d as the final choice.

The above process is simple and looks quite obvious. If this works, why be interested at all in "measurement" if the idea is to help someone to come up with a decision?

First observe that in the above example, the set of alternatives was relatively small. In many practical situations, the set of objects to compare is much larger than the set of alternatives in our example. Using the even swaps technique could then require a considerable number of difficult tradeoff questions. Furthermore, as the output of the technique is not a preference model but just the recommendation of an alternative in a given set, the appearance of new alternatives (e.g. because a new office is for rent) would require starting a new round of questions. This is likely to be highly frustrating. Finally, the informal even swaps technique may not be well adapted to the, many, situations, in which the decision under study takes place in a complex organizational environment. In such situations, having a formal model to be able to communicate and to convince is an invaluable asset. Such a model will furthermore allow to conduct extensive sensitivity analysis and, hence, to deal with imprecision both in the evaluations of the objects to compare and in the answers to difficult questions concerning tradeoffs.

This clearly leaves room for a more formal approach to structure preferences. But where can "measurement" be involved in the process? It should be observed that, beyond surface, there are many analogies between the even swaps process and the measurement of length considered above.

First, note that, in both cases, objects are compared using binary relations. In the measurement of length, the binary relation $\succ$ reads "is longer than". Here it reads "is preferred to". Similarly, the relation $\sim$ reading before "has equal length" now reads "is indifferent to". We supposed in the measurement of length process that $\succ$ and $\sim$ would nicely combine in experiments: if $r \succ r'$ and $r' \sim r''$ then we should observe that $r \succ r''$. Implicitly, a similar hypothesis was made in the even swaps technique. To realize that this is the case, it is worth summarizing the main steps of the argument.

We started with Table 3.1. Our overall recommendation was to rent office d. This means that we have reasons to believe that d is preferred to all other potential locations, i.e. $d \succ a$, $d \succ b$, $d \succ c$, and $d \succ e$. How did we arrive logically at such a conclusion?

Based on the initial table, using dominance and quasi-dominance, we concluded that b was preferable to e and that d was preferable to a. Using symbols, we have $b \succ e$ and $d \succ a$. After assessing some tradeoffs, we concluded, using dominance, that $b \succ c''$. But remember, c'' was built so as to be indifferent to c' and, in turn, c' was built so as to be indifferent to c. That is, we have $c'' \sim c'$ and $c' \sim c$. Later, we built an alternative d' that is indifferent to d ($d \sim d'$) and an alternative b' that is indifferent to b ($b \sim b'$). We then concluded, using dominance, that d' was preferable to b' ($d' \succ b'$). Hence, we know that:

$$
\begin{gathered}
d \succ a, b \succ e, \\
c'' \sim c', c' \sim c, b \succ c'', \\
d \sim d', b \sim b', d' \succ b'.
\end{gathered}
$$

Using the consistency rules linking $\succ$ and $\sim$ that we considered for the measurement of length, it is easy to see that the last line implies $d \succ b$. Since $b \succ e$, this implies $d \succ e$. It remains to show that $d \succ c$. But the second line leads to, combining $\succ$ and $\sim$, $b \succ c$. Therefore $d \succ b$ leads to $d \succ c$ and we are home. Hence, we have used the same properties for preference and indifference as the properties of "is longer than" and "has equal length" that we hypothesized in the measurement of length.

Second it should be observed that expressing tradeoffs leads, indirectly, to equating the "length" of "preference intervals" on different attributes. Indeed, remember how c' was constructed above: saying that c and c' are indifferent more or less amounts to saying that the interval $[25, 20]$ on commute time has exactly the same "length" as the interval $[70, 78]$ on client access. Consider an alternative f that would be identical to c except that it has a client access at 78%. We may again ask which increase in client access would compensate a loss of 5 minutes on commute time. In a tabular form we are now comparing the following two alternatives:

	f	f'
Commute	20	25
Clients	78	$78+\delta$
Services	C	C
Size	500	500
Cost	1500	1500

Suppose that the answer is that for $\delta = 10$, f and f' would be indifferent. This means that the interval $[25, 20]$ on commute time has exactly the same length as the interval $[78, 88]$ on client access. Now, we know that the preference intervals $[70, 78]$ and $[78, 88]$ have the same "length". Hence, tradeoffs provide a means to equate two preference intervals on the same attribute. This brings us quite close to the construction of standard sequences. This, we shall shortly do.

How does this information about the "length" of preference intervals relate to judgements of preference or indifference? Exactly as in the even swaps technique. You can use this measure of "length" modifying alternatives in such a way that they only differ on a single attribute and then use a simple dominance argument.

Conjoint measurement techniques may roughly be seen as a formalization of the even swaps technique that leads to building a numerical model of preferences much in the same way that we built a numerical model for length. This will require assessment procedures that will rest on the same principles as the standard sequence technique used for length. This process of "measuring preferences" is not an easy one. It will however lead to a numerical model of preference that will not only allow us to make a choice within a limited number of alternatives but that can serve as an input of computerized optimization algorithms that will be able to deal with much more complex cases.

2. Definitions and Notation

Before entering into the details of how conjoint measurement may work, a few definitions and notation will be needed.

2.1 Binary Relations

A *binary relation* $\succsim$ on a set A is a subset of $A \times A$. We write $a \succsim b$ instead of $(a, b) \in \succsim$. A binary relation $\succsim$ on A is said to be:

- *reflexive* if $[a \succsim a]$,
- *complete* if $[a \succsim b$ or $b \succsim a]$,
- *symmetric* if $[a \succsim b] \Rightarrow [b \succsim a]$,
- *asymmetric* if $[a \succsim b] \Rightarrow [Not[b \succsim a]]$,
- *transitive* if $[a \succsim b$ and $b \succsim c] \Rightarrow [a \succsim c]$,
- *negatively transitive* if $[Not[a \succsim b]$ and $Not[b \succsim c]] \Rightarrow Not[a \succsim c]$,

for all $a, b, c \in A$.

The *asymmetric* (resp. *symmetric*) part of $\succsim$ is the binary relation $\succ$ (resp. $\sim$) on A defined letting, for all $a, b \in A$, $a \succ b \Leftrightarrow [a \succsim b$ and $Not(b \succsim a)]$ (resp. $a \sim b \Leftrightarrow [a \succsim b$ and $b \succsim a]$). A similar convention will hold when $\succsim$ is subscripted and/or superscripted.

A *weak order* (resp. an *equivalence relation*) is a complete and transitive (resp. reflexive, symmetric and transitive) binary relation. For a detailed analysis of the use of binary relation as tools for preference modelling we refer to [4, 58, 66, 161, 167, 169]. The weak order model underlies the examples that were presented in the introduction. Indeed, the reader will easily prove the following.

PROPOSITION 4 *Let $\succsim$ be a weak order on A. Then:*

- *$\succ$ is transitive,*
- *$\succ$ is negatively transitive,*
- *$\sim$ is transitive,*

- $[a \succ b$ *and* $b \sim c] \Rightarrow a \succ c$,
- $[a \sim b$ *and* $b \succ c] \Rightarrow a \succ c$,

for all $a, b, c \in A$.

2.2 Binary Relations on Product Sets

In the sequel, we consider a set $X = \prod_{i=1}^{n} X_i$ with $n \geq 2$. Elements $x, y, z, \ldots$ of X will be interpreted as alternatives evaluated on a set $N = \{1, 2, \ldots, n\}$ of attributes. A typical binary relation on X is still denoted as $\succsim$, interpreted as an "at least as good as" preference relation between multi-attributed alternatives with $\sim$ interpreted as indifference and $\succ$ as strict preference.

For any nonempty subset J of the set of attributes N, we denote by X_J (resp. X_{-J}) the set $\prod_{i \in J} X_i$ (resp. $\prod_{i \notin J} X_i$). With customary abuse of notation, (x_J, y_{-J}) will denote the element $w \in X$ such that $w_i = x_i$ if $i \in J$ and $w_i = y_i$ otherwise. When $J = \{i\}$ we shall simply write X_{-i} and (x_i, y_{-i}).

REMARK 12 *Throughout this paper, we shall work with a binary relation defined on a product set. This setup conceals the important work that has to be done in practice to make it useful:*

- *the structuring of objectives [3, 15, 16, 117, 118, 119, 157, 163],*
- *the definition of adequate attributes to measure the attainment of objectives [80, 96, 116, 122, 173, 208, 216],*
- *the definition of an adequate family of attributes [24, 121, 173, 174, 209],*
- *the modelling of uncertainty, imprecision and inaccurate determination [23, 27, 121, 171].*

The importance of this "preliminary" work should not be forgotten in what follows.

2.3 Independence and Marginal Preferences

In conjoint measurement, one starts with a preference relation $\succsim$ on X. It is then of vital importance to investigate how this information makes it possible to define preference relations on attributes or subsets of attributes.

Let $J \subseteq N$ be a nonempty set of attributes. We define the *marginal relation* $\succsim_J$ induced on X_J by $\succsim$ letting, for all $x_J, y_J \in X_J$:

$$x_J \succsim_J y_J \Leftrightarrow (x_J, z_{-J}) \succsim (y_J, z_{-J}), \text{ for all } z_{-J} \in X_{-J},$$

with asymmetric (resp. symmetric) part $\succ_J$ (resp. $\sim_J$). When $J = \{i\}$, we often abuse notation and write $\succsim_i$ instead of $\succsim_{\{i\}}$. Note that if $\succsim$ is reflexive (resp.

transitive), the same will be true for $\succsim_J$. This is clearly not true for completeness however.

DEFINITION 26 (INDEPENDENCE) *Consider a binary relation $\succsim$ on a set $X = \prod_{i=1}^{n} X_i$ and let $J \subseteq N$ be a nonempty subset of attributes. We say that $\succsim$ is independent for J if, for all $x_J, y_J \in X_J$,*

$$[(x_J, z_{-J}) \succsim (y_J, z_{-J}), \textit{ for some } z_{-J} \in X_{-J}] \Rightarrow x_J \succsim_J y_J.$$

If $\succsim$ is independent for all nonempty subsets of N, we say that $\succsim$ is independent. *If $\succsim$ is independent for all subsets containing a single attribute, we say that $\succsim$ is* weakly independent.

In view of (3.2), it is clear that the additive value model will require that $\succsim$ is independent. This crucial condition says that common evaluations on some attributes do not influence preference. Whereas independence implies weak independence, it is well-know that the converse is not true [211].

REMARK 13 *Under certain conditions, e.g. when X is adequately "rich", it is not necessary to test that a weak order $\succsim$ is independent for J, for* all $J \subseteq N$ *in order to know that $\succsim$ is independent, see [21, 89, 121]. This is often useful in practice.*

REMARK 14 *Weak independence is referred to as "weak separability" in [211]; in section 5, we use "weak separability" (and "separability") with a different meaning.*

REMARK 15 *Independence, or at least weak independence, is an almost universally accepted hypothesis in multiple criteria decision making. It cannot be overemphasized that it is easy to find examples in which it is inadequate.*

If a meal is described by the two attributes, main course and wine, it is highly likely that most gourmets will violate independence, preferring red wine with beef and white wine with fish. Similarly, in a dynamic decision problem, a preference for variety will often lead to violating independence: you may prefer Pizza to Steak, but your preference for meals today (first attribute) and tomorrow (second attribute) may well be such that (Pizza, Steak) preferred to (Pizza, Pizza), while (Steak, Pizza) is preferred to (Steak, Steak).

Many authors [119, 173, 209] have argued that such failures of independence were almost always due to a poor structuring of attributes (e.g. in our choice of meal example above, preference for variety should be explicitly modelled).

When $\succsim$ is a weakly independent weak order, marginal preferences are well-behaved and combine so as to give meaning to the idea of dominance that we already encountered. The proof of the following is left to the reader as an easy exercise.

PROPOSITION 5 *Let $\succsim$ be a weakly independent weak order on $X = \prod_{i=1}^n X_i$. Then:*

- $\succsim_i$ *is a weak order on* X_i,
- $[x_i \succsim_i y_i$, *for all* $i \in N] \Rightarrow x \succsim y$,
- $[x_i \succsim_i y_i$, *for all* $i \in N$ *and* $x_j \succ_j y_j$ *for some* $j \in N] \Rightarrow x \succ y$,

for all $x, y \in X$.

3. The Additive Value Model in the "Rich" Case

The purpose of this section and the following is to present the conditions under which a preference relation on a product set may be represented by the additive value function model (3.2) and how such a model can be assessed. We begin here with the case that most closely resembles the measurement of length described in section 1.2.

3.1 Outline of Theory

When the structure of X is supposed to be "adequately rich", conjoint measurement is a quite clever adaptation of the process that we described in section 1.2 for the measurement of length. What will be measured here are the "length" of preference intervals on an attribute using a preference interval on another attribute as a standard.

3.1.1 The Case of Two Attributes. Consider first the two attribute case. Hence the relation $\succsim$ is defined on a set $X = X_1 \times X_2$. Clearly, in view of (3.2), we need to suppose that $\succsim$ is an *independent weak order*. Consider two levels $x_1^0, x_1^1 \in X_1$ on the first attribute such that $x_1^1 \succ_1 x_1^0$, i.e. x_1^1 is preferable to x_1^0. This makes sense because, we supposed that $\succsim$ is *independent*. Note also that we shall have to exclude the case in which all levels on the first attribute would be indifferent in order to be able to find such levels.

Choose any $x_2^0 \in X_2$. The, arbitrarily chosen, element $(x_1^0, x_2^0) \in X$ will be our "reference point". The basic idea is to use this reference point and the "unit" on the first attribute given by the reference preference interval $[x_1^0, x_1^1]$ to build a standard sequence on the preference intervals on the second attribute. Hence, we are looking for an element $x_2^1 \in X_2$ that would be such that:

$$(x_1^0, x_2^1) \sim (x_1^1, x_2^0). \tag{3.3}$$

Clearly this will require the structure of X_2 to be adequately "rich" so as to find the level $x_2^1 \in X_2$ such that the reference preference interval on the first attribute $[x_1^0, x_1^1]$ is exactly matched by a preference interval of the same "length" on the second attribute $[x_2^0, x_2^1]$. Technically, this calls for a solvability assumption or, more restrictively, for the supposition that X_2 has a (topological) structure that is close to that of an interval of $\mathbb{R}$ and that $\succsim$ is "somehow" continuous.

If such a level x_2^1 can be found, model (3.2) implies:

$$\begin{aligned} v_1(x_1^0) + v_2(x_2^1) &= v_1(x_1^1) + v_2(x_2^0) \text{ so that} \\ v_2(x_2^1) - v_2(x_2^0) &= v_1(x_1^1) - v_1(x_1^0). \end{aligned} \tag{3.4}$$

Let us fix the origin of measurement letting:

$$v_1(x_1^0) = v_2(x_2^0) = 0,$$

and our unit of measurement letting:

$$v_1(x_1^1) = 1 \text{ so that } v_1(x_1^1) - v_1(x_1^0) = 1.$$

Using (3.4), we therefore obtain $v_2(x_2^1) = 1$. We have therefore found an interval between levels on the second attribute ($[x_2^0, x_2^1]$) that exactly matches our reference interval on the first attribute ($[x_1^0, x_1^1]$). We may proceed to build our standard sequence on the second attribute (see Figure 3.4) asking for levels $x_2^2, x_2^3, \ldots$ such that:

$$\begin{aligned} (x_1^0, x_2^2) &\sim (x_1^1, x_2^1), \\ (x_1^0, x_2^3) &\sim (x_1^1, x_2^2), \\ &\ldots \\ (x_1^0, x_2^k) &\sim (x_1^1, x_2^{k-1}). \end{aligned}$$

As above, using (3.2) leads to:

$$\begin{aligned} v_2(x_2^2) - v_2(x_2^1) &= v_1(x_1^1) - v_1(x_1^0), \\ v_2(x_2^3) - v_2(x_2^2) &= v_1(x_1^1) - v_1(x_1^0), \\ &\ldots \\ v_2(x_2^k) - v_2(x_2^{k-1}) &= v_1(x_1^1) - v_1(x_1^0), \end{aligned}$$

so that:

$$v_2(x_2^2) = 2, v_2(x_2^3) = 3, \ldots, v_2(x_2^k) = k.$$

This process of building a standard sequence of the second attribute therefore leads to defining v_2 on a number of, carefully, selected elements of X_2.

Remember the standard sequence that we built for length in section 1.2. An implicit hypothesis was that the length of any rod could be exceeded by the length of a composite object obtained by concatenating a sufficient number of perfect copies of a standard rod. Such an hypothesis is called "Archimedean" since it mimics the property of the real numbers saying that for any positive

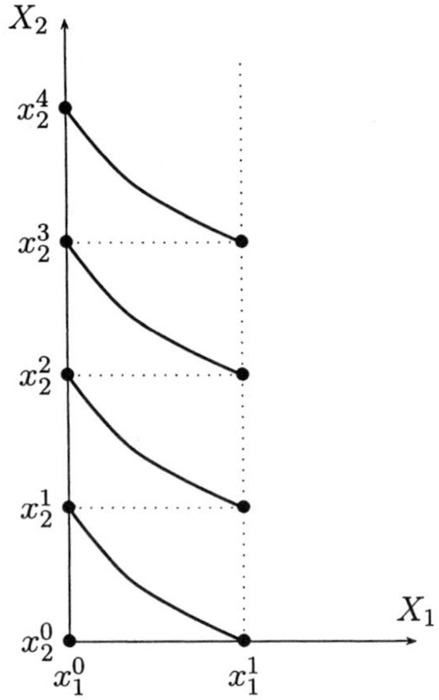

Figure 3.4. Building a standard sequence on X_2.

real numbers x, y it is true that $nx > y$ for some integer n, i.e. y, no matter how large, may always be exceeded by taking any x, no matter how small, and adding it with itself and repeating the operation a sufficient number of times. Clearly, we will need a similar hypothesis here. Failing it, there might exist a level $y_2 \in X_2$ that will never be "reached" by our standard sequence, i.e. such that $y_2 \succ_2 x_2^k$, for $k = 1, 2, \dots$. For measurement models in which this Archimedean condition is omitted, see [155, 193].

REMARK 16 *At this point a good exercise for the reader is to figure out how we may extend the standard sequence to cover levels of X_2 that are "below" the reference level x_2^0. This should not be difficult.*

Now that a standard sequence is built on the second attribute, we may use any part of it to build a standard sequence on the first attribute. This will require finding levels $x_1^2, x_1^3, \dots \in X_1$ such that (see Figure 3.5):

$$
\begin{aligned}
(x_1^2, x_2^0) &\sim (x_1^1, x_2^1), \\
(x_1^3, x_2^0) &\sim (x_1^2, x_2^1), \\
&\dots \\
(x_1^k, x_2^0) &\sim (x_1^{k-1}, x_2^1).
\end{aligned}
$$

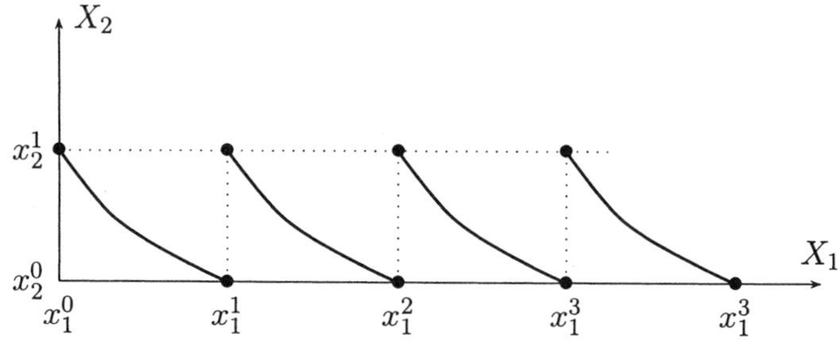

Figure 3.5. Building a standard sequence on X_1.

Using (3.2) leads to:

$$v_1(x_1^2) - v_1(x_1^1) = v_2(x_2^1) - v_2(x_2^0),$$
$$v_1(x_1^3) - v_1(x_1^2) = v_2(x_2^1) - v_2(x_2^0),$$
$$\dots$$
$$v_1(x_1^k) - v_1(x_1^{k-1}) = v_2(x_2^1) - v_2(x_2^0),$$

so that:

$$v_1(x_1^2) = 2, v_1(x_1^3) = 3, \dots, v_1(x_1^k) = k.$$

As was the case for the second attribute, the construction of such a sequence will require the structure of X_1 to be adequately rich, which calls for a solvability assumption. An Archimedean condition will also be needed in order to be sure that all levels of X_1 can be reached by the sequence.

We have defined a "grid" in X (see Figure 3.6) and we have $v_1(x_1^k) = k$ and $v_2(x_2^k) = k$ for all elements of this grid. Intuitively such numerical assignments seem to define an adequate additive value function on the grid. We have to prove that this intuition is correct. Let us first verify that, for all integers $\alpha, \beta, \gamma, \delta$:

$$\alpha + \beta = \gamma + \delta = \epsilon \Rightarrow (x_1^\alpha, x_2^\beta) \sim (x_1^\gamma, x_2^\delta). \tag{3.5}$$

When $\epsilon = 1$, (3.5) holds by construction because we have: $(x_1^0, x_2^1) \sim (x_1^1, x_2^0)$. When $\epsilon = 2$, we know that $(x_1^0, x_2^2) \sim (x_1^1, x_2^1)$ and $(x_1^2, x_2^0) \sim (x_1^1, x_2^1)$ and the claim is proved using the transitivity of $\sim$.

Consider the $\epsilon = 3$ case. We have $(x_1^0, x_2^3) \sim (x_1^1, x_2^2)$ and $(x_1^0, x_2^3) \sim (x_1^1, x_2^2)$. It remains to be shown that $(x_1^2, x_2^1) \sim (x_1^1, x_2^2)$ (see the dotted arc in Figure 3.6). This does not seem to follow from the previous conditions that we more or less explicitly used: transitivity, independence, "richness", Archimedean. Indeed, it does not. Hence, we have to suppose that: $(x_1^2, x_2^0) \sim (x_1^0, x_2^2)$

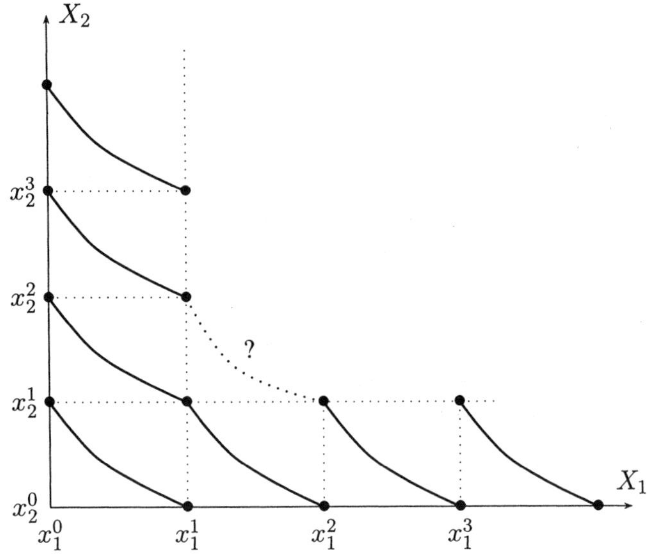

Figure 3.6. The grid.

and $(x_1^0, x_2^1) \sim (x_1^1, x_2^0)$ imply $(x_1^2, x_2^1) \sim (x_1^1, x_2^2)$. This condition, called the Thomsen condition, is clearly necessary for (3.2). The above reasoning easily extends to all points on the grid, using weak ordering, independence and the Thomsen condition. Hence, (3.5) holds on the grid.

It remains to show that:

$$\epsilon = \alpha + \beta > \epsilon' = \gamma + \delta \Rightarrow (x_1^\alpha, x_2^\beta) \succ (x_1^\gamma, x_2^\delta). \tag{3.6}$$

Using transitivity, it is sufficient to show that (3.6) holds when $\epsilon = \epsilon' + 1$. By construction, we know that $(x_1^1, x_2^0) \succ (x_1^0, x_2^0)$. Using independence this implies that $(x_1^1, x_2^k) \succ (x_1^0, x_2^k)$. Using (3.5) we have $(x_1^1, x_2^k) \sim (x_1^{k+1}, x_2^0)$ and $(x_1^0, x_2^k) \sim (x_1^k, x_2^0)$. Therefore we have $(x_1^{k+1}, x_2^0) \succ (x_1^k, x_2^0)$, the desired conclusion.

Hence, we have built an additive value function of a suitably chosen grid (see Figure 3.7). The logic of the assessment procedure is then to assess more and more points somehow considering more finely grained standard sequences. The two techniques evoked for length may be used here depending on the underlying structure of X. Going to the limit then unambiguously defines the functions v_1 and v_2. Clearly such v_1 and v_2 are intimately related. Once we have chosen an arbitrary reference point (x_1^0, x_2^0) and a level x_1^1 defining the unit of measurement, the process just described entirely defines v_1 and v_2. It follows that the only possible transformations that can be applied to v_1 and v_2

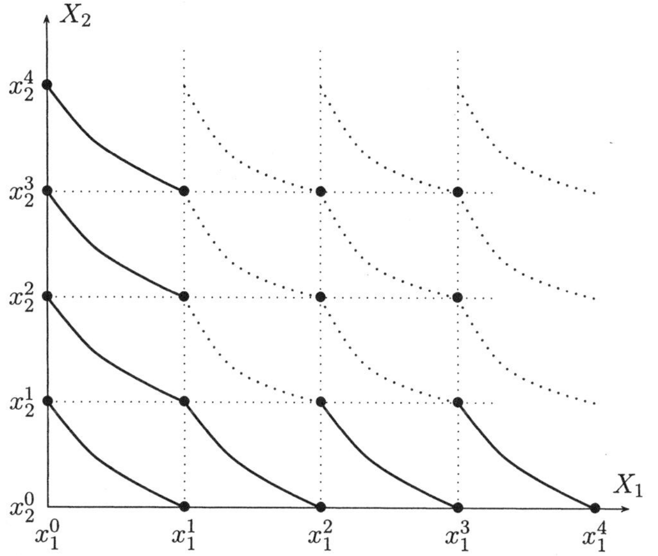

Figure 3.7. The entire grid.

is to multiply both by the same positive number α and to add to both a, possibly different, constant. This is usually summarized saying that v_1 and v_2 define interval scales with a common unit.

The above reasoning is a rough sketch of the proof of the existence of an additive value function when $n = 2$, as well as a sketch of how it could be assessed. Careful readers will want to refer to [58, 129, 211].

REMARK 17 *The measurement of length through standard sequences described above leads to a scale that is unique once the unit of measurement is chosen. At this point, a good exercise for the reader is to find an intuitive explanation to the fact that, when measuring the "length" of preference intervals, the origin of measurement becomes arbitrary. The analogy with the the measurement of duration on the one hand and dates, as given in a calendar, on the other hand should help.*

REMARK 18 *As was already the case with the even swaps technique, it is worth emphasizing that this assessment technique makes no use of the vague notion of the "importance" of the various attributes. The "importance" is captured here in the lengths of the preference intervals on the various attributes.*

A common but critical mistake is to confuse the additive value function model (3.2) *with a weighted average and to try to assess weights asking whether an attribute is "more important" than another. This makes no sense.*

3.1.2 The Case of More than Two Attributes. The good news is that the process is exactly the same when there are more than two attributes. With one surprise: the Thomsen condition is no more needed to prove that the standard sequences defined on each attribute lead to an adequate value function on the grid. A heuristic explanation of this strange result is that, when $n = 2$, there is no difference between independence and weak independence. This is no more true when $n \geq 3$ and assuming independence is much stronger than just assuming weak independence.

3.2 Statement of Results

We use below the "algebraic approach" [127, 129, 141]. A more restrictive approach using a topological structure on X is given in [44, 58, 211]. We formalize below the conditions informally introduced in the preceding section. The reader not interested in the precise statement of the results or, better, having already written down his own statement, may skip this section.

DEFINITION 27 (THOMSEN CONDITION) *Let $\succsim$ be a binary relation on a set $X = X_1 \times X_2$. It is said to satisfy the Thomsen condition if*

$$(x_1, x_2) \sim (y_1, y_2) \text{ and } (y_1, z_2) \sim (z_1, x_2) \Rightarrow (x_1, z_2) \sim (z_1, y_2),$$

for all $x_1, y_1, z_1 \in X_1$ and all $x_2, y_2, z_2 \in X_2$.

Figure 3.8 shows how the Thomsen condition uses two "indifference curves" (i.e. curves linking points that are indifferent) to place a constraint on a third one. This was needed above to prove that an additive value function existed on our grid. Remember that the Thomsen condition is only needed when $n = 2$; hence, we only stated it in this case.

DEFINITION 28 (STANDARD SEQUENCES) *A standard sequence on attribute $i \in N$ is a set $\{a_i^k : a_i^k \in X_i, k \in K\}$ where K is a set of consecutive integers (positive or negative, finite or infinite) such that there are $x_{-i}, y_{-i} \in X_{-i}$ satisfying $Not[x_{-i} \sim_{-i} y_{-i}]$ and $(a_i^k, x_{-i}) \sim (a_i^{k+1}, y_{-i})$, for all $k \in K$.*

A standard sequence on attribute $i \in N$ is said to be *strictly bounded* if there are $b_i, c_i \in X_i$ such that $b_i \succ_i a_i^k \succ_i c_i$, for all $k \in K$. It is then clear that, when model (3.2) holds, any strictly bounded standard sequence must be finite.

DEFINITION 29 (ARCHIMEDEAN) *For all $i \in N$, any strictly bounded standard sequence on $i \in N$ is finite.*

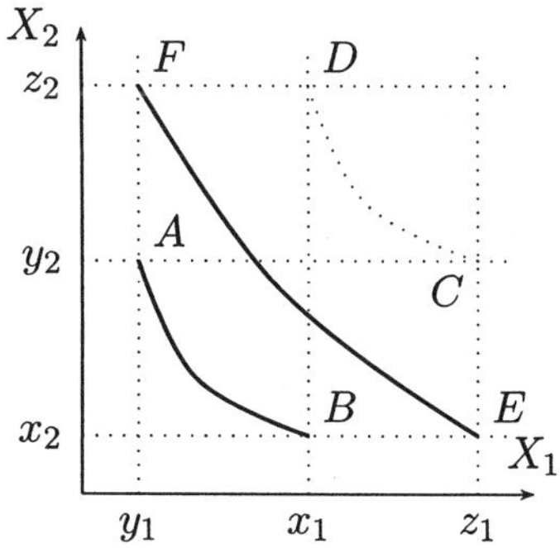

$$\left.\begin{array}{c} A \sim B \\ E \sim F \end{array}\right\} \Rightarrow C \sim D$$

Figure 3.8. The Thomsen condition.

The following condition rules out the case in which a standard sequence cannot be built because all levels are indifferent.

DEFINITION 30 (ESSENTIALITY) *Let $\succsim$ be a binary relation on a set $X = X_1 \times X_2 \times \cdots \times X_n$. Attribute $i \in N$ is said to be essential if $(x_i, a_{-i}) \succ (y_i, a_{-i})$, for some $x_i, y_i \in X_i$ and some $a_{-i} \in X_{-i}$.*

DEFINITION 31 (RESTRICTED SOLVABILITY) *Let $\succsim$ be a binary relation on a set $X = X_1 \times X_2 \times \cdots \times X_n$. Restricted solvability is said to hold with respect to attribute $i \in N$ if, for all $x \in X$, all $z_{-i} \in X_{-i}$ and all $a_i, b_i \in X_i$, $[(a_i, z_{-i}) \succsim x \succsim (b_i, z_{-i})] \Rightarrow [x \sim (c_i, z_{-i})$, for some $c_i \in X_i]$.*

REMARK 19 *Restricted solvability is illustrated in Figure 3.9 in the case where $n = 2$. It says that, given any $x \in X$, if it is possible find two levels $a_i, b_i \in X_i$ such that when combined with a certain level $z_{-i} \in X_{-i}$ on the other attributes, (a_i, z_{-i}) is preferred to x and x is preferred to (b_i, z_{-i}), it should be possible to find a level c_i, "in between" a_i and b_i, such that (c_i, z_{-i}) is exactly indifferent to x.*

A much stronger hypothesis is unrestricted solvability asserting that for all $x \in X$ and all $z_{-i} \in X_{-i}$, $x \sim (c_i, z_{-i})$, for some $c_i \in X_i$. Its use leads however to much simpler proofs [58, 86].

It is easy to imagine situations in which restricted solvability might hold while unrestricted solvability would fail. Suppose, e.g. that a firm has to choose between several investment projects, two attributes being the Net Present Value (NPV) of the projects and their impact on the image of the firm in the public.

Consider a project consisting in investing in the software market. It has a reasonable NPV and no adverse consequences on the image of the firm. Consider another project that could have dramatic consequences on the image of the firm, because it leads to investing the market of cocaine. Unrestricted solvability would require that by sufficiently increasing the NPV of the second project it would become indifferent to the more standard project of investing in the software market. This is not required by restricted solvability.

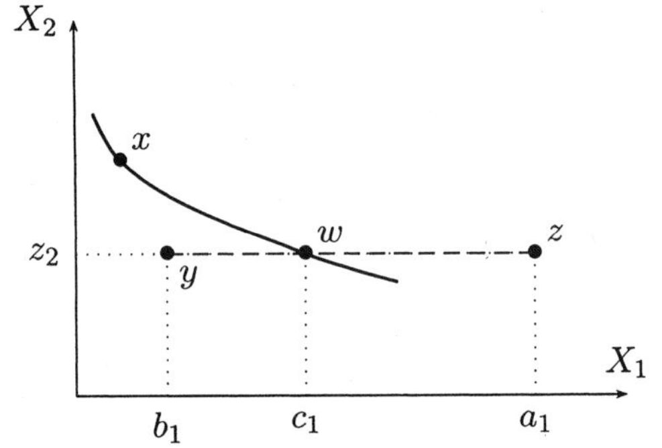

Figure 3.9. Restricted solvability on X_1.

We are now in position to state the central results concerning model (3.2). Proofs may be found in [129, 213].

THEOREM 1 (ADDITIVE VALUE FUNCTION WHEN $n = 2$) *Let $\succsim$ be a binary relation on a set $X = X_1 \times X_2$. If restricted solvability holds on all attributes and each attribute is essential then $\succsim$ has a representation in model (3.2) if and only if $\succsim$ is an independent weak order satisfying the Thomsen and the Archimedean conditions.*

Furthermore in this representation, v_1 and v_2 are interval scales with a common unit, i.e. if v_1, v_2 and w_1, w_2 are two pairs of functions satisfying (3.2), there are real numbers α, β_1, β_2 with $\alpha > 0$ such that, for all $x_1 \in X_1$ and all $x_2 \in X_2$

$$v_1(x_1) = \alpha w_1(x_1) + \beta_1 \text{ and } v_2(x_2) = \alpha w_2(x_2) + \beta_2.$$

When $n \geq 3$ and at least three attributes are essential, the above result simplifies in that the Thomsen condition can now be omitted.

THEOREM 2 (ADDITIVE VALUE FUNCTION WHEN $n \geq 3$) *Let $\succsim$ be a binary relation on a set $X = X_1 \times X_2 \times \ldots \times X_n$ with $n \geq 3$. If restricted solvability holds on all attributes and at least 3 attributes are essential then $\succsim$ has a representation in model (3.2) if and only if $\succsim$ is an independent weak order satisfying the Archimedean condition.*

Furthermore in this this representation $v_1, v_2, \ldots, v_n$ are interval scales with a common unit.

REMARK 20 *As mentioned in introduction, the additive value model is central to several fields in decision theory. It is therefore not surprising that much energy has been devoted to analyze variants and refinements of the above results. Among the most significant ones, let us mention:*

- *the study of cases in which solvability holds only on some or none of the attributes [75, 85, 86, 87, 88, 112, 113, 154],*
- *the study of the relation between the "algebraic approach" introduced above and the topological one used in [44], see e.g. [115, 124, 211, 213].*

The above results are only valid when X is the entire Cartesian product of the sets X_i. Results in which X is a subset of the whole Cartesian product $X_1 \times X_2 \times \ldots \times X_n$ are not easy to obtain, see [37, 181] (the situation is "easier" in the special case of homogeneous product sets, see [214, 215]).

3.3 Implementation: Standard Sequences and Beyond

We have already shown above how additive value functions can be assessed using the standard sequence technique. It is worth recalling here some of the characteristics of this assessment procedure:

- It requires the set X_i to be *rich* so that it is possible to find a preference interval on X_i that will exactly match a preference interval on another attribute. This excludes using such an assessment procedure when some of the sets X_i are discrete.
- It relies on *indifference* judgements which, a priori, are less firmly established than preference judgements.
- It relies on judgements concerning fictitious alternatives which, a priori, are harder to conceive than judgements concerning real alternatives.
- The various assessments are thoroughly intertwined and, e.g., an imprecision on the assessment of x_2^1, i.e. the endpoint of the first interval in the standard sequence on X_2 (see Figure 3.4) will propagate to many assessed values,

- The assessment of tradeoffs may be plagued with cognitive biases, see [46, 197].

The assessment procedure based on standard sequences is therefore rather demanding; this should be no surprise given the proximity between this form of measurement and extensive measurement illustrated above on the case of length. Hence, the assessment procedure based on standard sequences seems to be seldom used in the practice of decision analysis [121, 209]. The literature on the experimental assessment of additive value functions, see e.g. [197, 208, 216], suggests that this assessment is a difficult task that may be affected by several cognitive biases.

Many other simplified assessment procedures have been proposed that are less firmly grounded in theory. In many of them, the assessment of the partial value functions v_i relies on *direct* comparison of preference differences without recourse to an interval on another attribute used as a "meter stick". We refer to [50] for a theoretical analysis of these techniques. They are also studied in detail in 7 of this volume.

These procedures include:

- *direct rating* techniques in which values of v_i are directly assessed with reference to two arbitrarily chosen points [52, 53],

- procedures based on *bisection*, the decision-maker being asked to assess a point that is "half way" in terms of preference two reference points [209],

- procedures trying to build *standard sequences* on each attribute in terms of "preference differences" [129, ch. 4].

An excellent overview of these techniques may be found in [209].

4. The Additive Value Model in the "Finite" Case

4.1 Outline of Theory

In this section, we suppose that $\succsim$ is a binary relation on a finite set $X \subseteq X_1 \times X_2 \times \cdots \times X_n$ (contrary to the preceding section, dealing with subsets of product sets will raise no difficulty here). The finiteness hypothesis clearly invalidates the standard sequence mechanism used till now. On each attribute there will only be finitely many "preference intervals" and exact matches between preference intervals will only happen exceptionally, see [212].

Clearly, independence remains a necessary condition for model (3.2) as before. Given the absence of structure of the set X, it is unlikely that this condition is sufficient to ensure (3.2). The following example shows that this intuition is indeed correct.

EXAMPLE 4 *Let* $X = X_1 \times X_2$ *with* $X_1 = \{a, b, c\}$ *and* $X_2 = \{d, e, f\}$. *Consider the weak order on* X *such that, abusing notation in an obvious way,*

$$ad \succ bd \succ ae \succ af \succ be \succ cd \succ ce \succ bf \succ cf.$$

It is easy to check that $\succsim$ *is independent. Indeed, we may for instance check that:*

$$ad \succ bd \text{ and } ae \succ be \text{ and } af \succ bf,$$
$$ad \succ ae \text{ and } bd \succ be \text{ and } cd \succ ce.$$

This relation cannot however be represented in model (3.2) *since:*

$$\begin{aligned}
af \succ be &\Rightarrow v_1(a) + v_2(f) > v_1(b) + v_2(e),\\
be \succ cd &\Rightarrow v_1(b) + v_2(e) > v_1(c) + v_2(d),\\
ce \succ bf &\Rightarrow v_1(c) + v_2(e) > v_1(b) + v_2(f),\\
bd \succ ae &\Rightarrow v_1(b) + v_2(d) > v_1(a) + v_2(e).
\end{aligned}$$

Summing the first two inequalities leads to:

$$v_1(a) + v_2(f) > v_1(c) + v_2(d).$$

Summing the last two inequalities leads to:

$$v_1(c) + v_2(d) > v_1(a) + v_2(f),$$

a contradiction.

Note that, since no indifference is involved, the Thomsen condition is trivially satisfied. Although it is clearly necessary for model (3.2), *adding it to independence will therefore not solve the problem.*

The conditions allowing to build an additive value model in the finite case were investigated in [1, 2, 179]. Although the resulting conditions turn out to be complex, the underlying idea is quite simple. It amounts to finding conditions under which a system of linear inequalities has a solution.

Suppose that $x \succ y$. If model (3.2) holds, this implies that:

$$\sum_{i=1}^{n} v_i(x_i) > \sum_{i=1}^{n} v_i(y_i). \tag{3.7}$$

Similarly if $x \sim y$, we obtain:

$$\sum_{i=1}^{n} v_i(x_i) = \sum_{i=1}^{n} v_i(y_i). \tag{3.8}$$

The problem is then to find conditions on $\succsim$ such that the system of finitely many equalities and inequalities (3.7-3.8) has a solution. This is a classical problem in Linear Algebra [83].

DEFINITION 32 (RELATION E^m) *Let m be an integer ≥ 2. Let $x^1, x^2, \ldots, x^m, y^1, y^2, \ldots, y^m \in X$. We say that*

$$(x^1, x^2, \ldots, x^m) E^m (y^1, y^2, \ldots, y^m)$$

if, for all $i \in N$, $(x_i^1, x_i^2, \ldots, x_i^m)$ is a permutation of $(y_i^1, y_i^2, \ldots, y_i^m)$.

Suppose that $(x^1, x^2, \ldots, x^m) E^m (y^1, y^2, \ldots, y^m)$ then model (3.2) implies that

$$\sum_{j=1}^{m} \sum_{i=1}^{n} v_i(x_i^j) = \sum_{j=1}^{m} \sum_{i=1}^{n} v_i(y_i^j).$$

Therefore if $x^j \succsim y^j$ for $j = 1, 2, \ldots, m-1$, it cannot be true that $x^m \succ y^m$. This condition must hold for all $m = 2, 3, \ldots$.

DEFINITION 33 (CONDITION C^m) *Let m be an integer ≥ 2. We say that condition C^m holds if*

$$[x^j \succsim y^j \text{ for } j = 1, 2, \ldots, m-1] \Rightarrow Not[x^m \succ y^m]$$

for all $x^1, x^2, \ldots, x^m, y^1, y^2, \ldots, y^m \in X$ such that

$$(x^1, x^2, \ldots, x^m) E^m (y^1, y^2, \ldots, y^m).$$

REMARK 21 *It is not difficult to check that:*

- $C^{m+1} \Rightarrow C^m$,
- $C^2 \Rightarrow \succsim$ *is independent,*
- $C^3 \Rightarrow \succsim$ *is transitive.*

We already observed that C^m was implied by the existence of an additive representation. The main result for the finite case states that requiring that $\succsim$ is complete and that C^m holds for $m = 2, 3, \ldots$ is also sufficient. Proofs can be found in [58, 129].

THEOREM 3 *Let $\succsim$ be a binary relation on a finite set $X \subseteq X_1 \times X_2 \times \cdots \times X_n$. There are real-valued functions v_i on X_i such that* (3.2) *holds if and only if $\succsim$ is complete and satisfies C^m for $m = 2, 3, \ldots$.*

REMARK 22 *Contrary to the "rich" case considered in the preceding section, we have here necessary and sufficient conditions for the additive value model*

(3.2). *However, it is important to notice that the above result uses a denumerable scheme of conditions. It is shown in [180] that this denumerable scheme cannot be truncated: for all* $m \geq 2$, *there is a relation* $\succsim$ *on a finite set* X *such that* C^m *holds but violating* C^{m+1}. *This is studied in more detail in [139, 201, 218]. Therefore, no finite scheme of axioms is sufficient to characterize model* (3.2) *for all finite sets* X.

Given a finite set X of given cardinality, *it is well-known that the denumerable scheme of condition can be truncated. The precise relation between the cardinality of X and the number of conditions needed raises difficult combinatorial questions that are studied in [77, 78].*

REMARK 23 *It is clear that, if a relation* $\succsim$ *has a representation in model* (3.2) *with functions* v_i, *it also has a representation using functions* $v_i' = \alpha v_i + \beta_i$ *with* $\alpha > 0$. *Contrary to the rich case, the uniqueness of the functions* v_i *is more complex as shown by the following example.*

EXAMPLE 5 *Let* $X = X_1 \times X_2$ *with* $X_1 = \{a, b, c\}$ *and* $X_2 = \{d, e\}$. *Consider the weak order on* X *such that, abusing notation in an obvious way,*

$$ad \succ bd \succ ae \succ cd \succ be \succ ce.$$

This relation has a representation in model (3.2) *with*

$$v_1(a) = 3, v_1(b) = 1, v_1(c) = 0, v_2(d) = 3, v_2(e) = 0.5.$$

An equally valid representation would be given taking $v_1(b) = 2$. *Clearly this new representation cannot be deduced from the original one applying a positive affine transformation.*

REMARK 24 *Theorem 3 has been extended to the case of an arbitrary set* X *in [113, 112], see also [75, 81]. The resulting conditions are however quite complex. This explains why we spent time on this "rich" case in the preceding section.*

REMARK 25 *The use of a denumerable scheme of conditions in theorem 3 does not facilitate the interpretation and the test of conditions. However it should be noticed that, on a given set* X, *the test of the* C^m *conditions amounts to finding if a system of finitely many linear inequalities has a solution. It is well-known that Linear Programming techniques are quite efficient for such a task.*

4.2 Implementation: LP-based Assessment

We show how to use LP techniques in order to assess an additive value model (3.2), without supposing that the sets X_i are rich. For practical purposes, it is not restrictive to assume that we are only interested in assessing a model for a limited range on each X_i. We therefore assume that the sets X_i are bounded so that, using independence, there is a worst value x_{i*} and a most preferable value x_i^*. Using the uniqueness properties of model (3.2), we may always suppose, after an appropriate normalization, that:

$$v_1(x_{1*}) = v_2(x_{2*}) = \ldots = v_n(x_{n*}) = 0 \text{ and} \tag{3.9}$$

$$\sum_{i=1}^{n} v_i(x_i^*) = 1. \tag{3.10}$$

Two main cases arise (see Figures 3.10 and 3.11):

- attribute $i \in N$ is discrete so that the evaluation of any conceivable alternative on this attribute belongs to a finite set. We suppose that $X_i = \{x_{i*}, x_i^1, x_i^2, \ldots, x_i^{r_i}, x_i^*\}$. We therefore have to assess $r_i + 1$ values of v_i,

- the attribute $i \in N$ has an underlying continuous structure. It is hardly restrictive in practice to suppose that $X_i \subset \mathbb{R}$, so that the evaluation of an alternative on this attribute may take any value between x_{i*} and x_i^*. In this case, we may opt for the assessment of a piecewise linear approximation of v_i partitioning the set X_i in $r_i + 1$ intervals and supposing that v_i is linear on each of these intervals. Note that the approximation of v_i can be made more precise simply by increasing the number of these intervals.

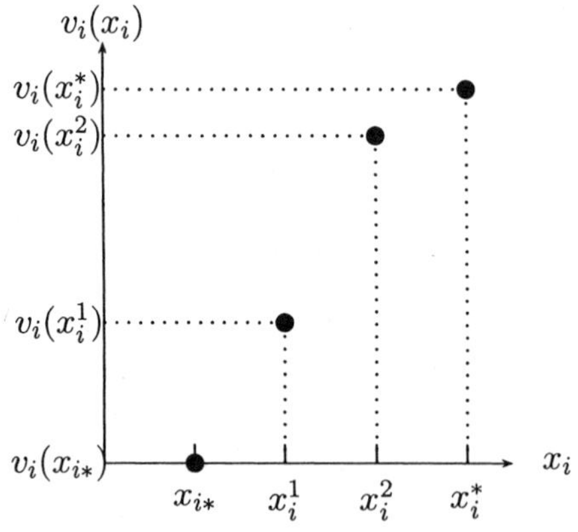

Figure 3.10. Value function when X_i is discrete.

With these conventions, the assessment of the model (3.2) amounts to giving a value to $\sum_{i=1}^{n}(r_i + 1)$ unknowns. Clearly any judgment of preference linking x and y translate into a *linear inequality* between these unknowns. Similarly

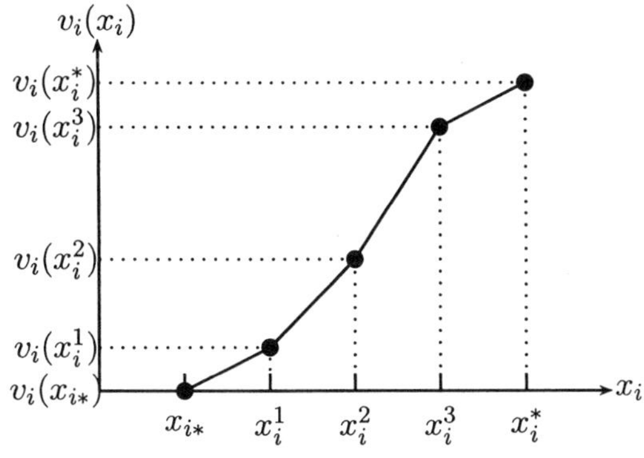

Figure 3.11. Value function when X_i is continuous.

any judgment of indifference linking x and y translate into a *linear equality*. Linear Programming (LP) offers a powerful tool for testing whether such a system has solutions. Therefore, an assessment procedure can be conceived on the following basis:

- obtain judgments in terms of preference or indifference linking several alternatives in X,
- convert these judgments into linear (in)equalities,
- test, using LP, whether this system has a solution.

If the system has no solution then one may either propose a solution that will be "as close as possible" from the information obtained, e.g. violating the minimum number of (in)equalities or suggest the reconsideration of certain judgements. If the system has a solution, one may explore the set of all solutions to this system since they are all candidates for the establishment of model (3.2). These various techniques depend on:

- the choice of the alternatives in X that are compared: they may be real or fictitious, they may differ on a different number of attributes,
- the way to deal with the inconsistency of the system and to eventually propose some judgments to be reconsidered,
- the way to explore the set of solutions of the system and to use this set as the basis for deriving a prescription.

Linear programming offers of simple and versatile technique to assess additive value functions. All restrictions generating linear constraints of the coefficient of the value function can easily be accommodated. This idea has been often exploited, see [16]. We present below two techniques using it. It should be noticed that rather different techniques have been proposed in the literature on Marketing [35, 103, 104, 114, 132].

4.2.1 UTA [111]. UTA ("UTilité Additive", i.e. additive utility in French) is one of the oldest techniques belonging to this family. It is supposed in UTA that there is a subset $Ref \subset X$ of reference alternatives that the decision-maker knows well either because he/she has experienced them or because they have received particular attention. The technique amounts to asking the DM to provide a weak order on Ref. Each preference or indifference relation contained in this weak order is then translated into a linear constraint:

- $x \sim y$ gives an equality $v(x) - v(y) = 0$ and
- $x \succ y$ gives an inequality $v(x) - v(y) > 0$,

where $v(x)$ and $v(y)$ can be expressed as a linear combination of the unknowns as remarked earlier. Strict inequalities are then translated into large inequalities as is usual in Linear Programming, i.e. $v(x) - v(y) > 0$ becomes $v(x) - v(y) \geq \epsilon$ where $\epsilon > 0$ is a very small positive number that should be chosen according to the precision of the arithmetics used by the LP package.

The test of the existence of a solution to the system of linear constraints is done via standard Goal Programming techniques [36] adding appropriate deviation variables. In UTA, each equation $v(x) - v(y) = 0$ is translated into an equation $v(x) - v(y) + \sigma_x^+ - \sigma_x^- + \sigma_y^+ - \sigma_y^- = 0$, where $\sigma_x^+, \sigma_x^-, \sigma_y^+$ and σ_y^- are nonnegative deviation variables. Similarly each inequality $v(x) - v(y) \geq \epsilon$ is written as $v(x) - v(y) + \sigma_x^+ - \sigma_x^- + \sigma_y^+ - \sigma_y^- \geq \epsilon$. It is clear that there will exist a solution to the original system of linear constraints if there is a solution of the LP in which all deviation variables are zero. This can easily be tested using the objective function

$$\text{Minimize } Z = \sum_{x \in Ref} \sigma_x^+ + \sigma_x^- \tag{3.11}$$

Two cases arise. If the optimal value of Z is 0, there is an additive value function that represents the preference information. It should be observed that, except in exceptional cases (e.g. if the preference information collected is identical to the preference information collected with the standard sequence technique), there are infinitely many such additive value functions (that are not related via a simple change of origin and of unit, since we already fixed them through normalization (3.9-3.10)). The one given as the "optimal" one by the LP does not have a special status since it is highly dependent upon the arbitrary choice of the objective function; instead of minimizing the sum of the deviation variables, we could have as well, and still preserving linearity, minimized the largest of these variables. The whole polyhedron of feasible solutions of the original (in)equalities corresponds to adequate additive value functions: we have a whole set $\mathcal{V}$ of additive value functions representing the information collected on the set of reference alternatives Ref.

The size of $\mathcal{V}$ is clearly dependent upon the choice of the alternatives in Ref. Using standard techniques in LP, several functions in $\mathcal{V}$ may be obtained, e.g. the ones maximizing or minimizing, within $\mathcal{V}$, $v_i(x_i^*)$ for each attribute [111]. It is often interesting to present them to the decision-maker in the pictorial form of Figures 3.10 and 3.11.

If the optimal value of Z is strictly greater than 0, there is no additive value function representing the preference information available. The solution given as optimal (note that it is not guaranteed that this solution leads to the minimum possible number of violations w.r.t. the information provided—this would require solving an integer linear programme) is, in general, highly dependent upon the choice of the objective function.

This absence of solution to the system might be due to several factors:

- the piecewise linear approximation of the v_i for the "continuous" attributes may be too rough. It is easy to test whether an increase in the number of linear pieces on some of these attributes may lead to a nonempty set of additive value functions.
- the information provided by the decision-maker may be of poor quality. It might then be interesting to present to the decision-maker one additive value function (e.g. one may present an average function after some post-optimality analysis) in the pictorial form of Figures 3.10 and 3.11 and to let him react to this information either by modifying his/her initial judgments or even by letting him/her react directly on the shape of the value functions. This is the solution implemented in the well-known PREFCALC system [109].
- the preference provided by the decision-maker might be inconsistent with the conditions implied by an additive value function. The system should then help locate these inconsistencies and allow the DM to think about them. Alternatively, since many alternative attribute descriptions are possible, it may be worth investigating whether a different definition of the various attributes may lead to a preference model consistent with model (3.2). Several examples of such analysis may be found in [119, 121, 209]

When the above techniques fail, the optimal solution of the LP, even if not compatible with the information provided, may still be considered as an adequate model. Again, since the objective function introduced above is somewhat arbitrary and it is recommended in [111] to perform a post-optimality analysis, e.g. considering additive value functions that are "close" to the optimal solution through the introduction of a linear constraint:

$$Z \leq Z^* + \delta,$$

where Z^* is the optimal value of the objective function of the original LP and δ is a "small" positive number. As above, the result of the analysis is a set $\mathcal{V}$ of additive value functions defined by a set of linear constraints. A representative sample of additive value functions within $\mathcal{V}$ may be obtained as above.

It should be noted that many possible variants of UTA can be conceived building on the following comments. They include:

- the addition of monotonicity properties of the v_i with respect to the underlying continuous attributes,
- the addition of constraints on the shape of the marginal value functions v_i, e.g. requiring them to be concave, convex or S-shaped,
- the addition of constraints linked to a possible indication of preference intensity for the elements of *Ref* given by the DM, e.g. the difference between x and y is larger than the difference between z and w.

For applications of UTA-like techniques, we refer to [38, 47, 48, 105, 110, 148, 185, 186, 187, 188, 189, 190, 192, 195, 196, 219, 221, 220, 223, 222]. Variants of the method are considered in [19, 20, 191]. This method is also studied in detail in Chapter 8 of this volume.

4.2.2 MACBETH [12]. It is easy to see that (3.9) and (3.10) may equivalently be written as:

$$x \succsim y \Leftrightarrow \sum_{i=1}^{n} k_i u_i(x_i) \geq \sum_{i=1}^{n} k_i u_i(y_i), \tag{3.12}$$

where

$$u_1(x_{1*}) = u_2(x_{2*}) = \ldots u_n(x_{n*}) = 0, \tag{3.13}$$

$$u_1(x_1^*) = u_2(x_2^*) = \ldots u_n(x_n^*) = 1 \text{ and} \tag{3.14}$$

$$\sum_{i=1}^{n} k_i = 1. \tag{3.15}$$

With such an expression of an additive value function, it is tempting to break down the assessment into two distinct parts: a value function u_i is assessed on each attribute and, then, scaling constants k_i are assessed taking the shape of the value functions u_i as given. This is the path followed in MACBETH.

REMARK 26 *Again, note that we are speaking here of k_i as* scaling constants *and not as* weights. *As already mentioned weights that would reflect the "importance" of attributes are irrelevant to assess the additive value function model. Notice that, under (3.12-3.15) the ordering of the scaling constant k_i is dependent upon the choice of x_{i*} and x_i^*. Increasing the width of the interval $[x_{i*}, x_i^*]$ will lead to increasing the value of the scaling constant k_i. The value k_i has, therefore, nothing to do with the "importance" of attribute i. This point is unfortunately too often forgotten when using a weighted average of some numerical attributes. In the latter model, changing the units in which the attributes are measured should imply changing the "weights" accordingly.*

The assessment procedure of the u_i is conceived in such a way as to avoid comparing alternatives differing on more than one attribute. In view of what was said before concerning the standard sequence technique, this is clearly an advantage of the technique. But can it be done? The trick here is that MACBETH asks for judgments related to the difference between the desirability of alternatives and not only judgments in terms of preference or indifference. Partial value functions u_i are approximated in a similar way than in UTA: for discrete attributes, each point on the function is assessed, for continuous ones, a piecewise linear approximation is used.

MACBETH asks the DM to compare pairs of levels on each attribute. If no difference is felt between these levels, they receive an identical partial value level. If a difference is felt between x_i^k and x_i^r, MACBETH asks for a judgment qualifying the strength of this difference. The method and the associated software propose three different semantical categories:

Categories	Description
$C1$	weak
$C2$	strong
$C3$	extreme

with the possibility of using intermediate categories, i.e. between null and weak, weak and strong, strong and extreme (giving a total of six distinct categories). This information is then converted into linear inequations using the natural interpretation that if the "difference" between the levels x_i^k and x_i^r has been judged larger than the "difference" between $x_i^{k'}$ and $x_i^{r'}$ then it should follow that $u_i(x_i^k) - u_i(x_i^r) > u_i(x_i^{k'}) - u_i(x_i^{r'})$. Technically the six distinct categories are delimited by thresholds that are used in the establishment of the constraints of the LP. The software associated to MACBETH offers the possibility to compare all pairs of levels on each attribute for a total of $(r_i + 1)r_i/2$ comparisons. Using standard Goal Programming techniques, as in UTA, the test of the compatibility of a partial value function with this information is performed via the solution of a LP. If there is a partial value function compatible with the information, a "central" function is proposed to the DM who has the possibility to modify it. If not, the results of the LP are exploited in such a way to propose modifications of the information that would make it consistent.

The assessment of the scaling constant k_i is done using similar principles. The DM is asked to compare the following $(n+2)$ alternatives by pairs:

$$\begin{gathered}(x_{1*}, x_{2*}, \ldots, x_{n*}),\\ (x_1^*, x_{2*}, \ldots, x_{n*}),\\ (x_{1*}, x_2^*, \ldots, x_{n*}),\\ \ldots\\ (x_{1*}, x_{2*}, \ldots, x_n^*) \text{ and}\\ (x_1^*, x_2^*, \ldots, x_n^*),\end{gathered}$$

placing each pair in a category of difference. This information immediately translates into a set of linear constraints on the k_i. These constraints are processed as before. It should be noticed that, once the partial value functions u_i are assessed, it is not necessary to use the levels x_{i*} and x_i^* to assess the k_i since they may well lead to alternatives that are too unrealistic. The authors of MACBETH suggest to replace x_{i*} by a "neutral" level which appears neither desirable nor undesirable and x_i^* by a "desirable" level that is judged satisfactory. Although this clearly impacts the quality of the dialogue with the DM, this has no consequence on the underlying technique used to process information.

We refer to [6, 7, 8, 9, 10, 11] for applications of the MACBETH technique. This method is also studied in detail in Chapter 10 of this volume.

5. Extensions

The additive value model (3.2) is the central model for the application of conjoint measurement techniques to decision analysis. In this section, we consider various extensions to this model.

5.1 Transitive Decomposable Models

The transitive decomposable model has been introduced in [129] as a natural generalization of model (3.2). It amounts to replacing the addition operation by a general function that is increasing in each of its arguments.

DEFINITION 34 (TRANSITIVE DECOMPOSABLE MODEL) *Let $\succsim$ be a binary relation on a set $X = \prod_{i=1}^n X_i$. The transitive decomposable model holds if, for all $i \in N$, there is a real-valued function v_i on X_i and a real-valued function g on $\prod_{i=1}^n v_i(X_i)$ that is increasing in all its arguments such that:*

$$x \succsim y \Leftrightarrow g(v_1(x_1), \ldots, v_n(x_n)) \geq g(v_1(y_1), \ldots, v_n(y_n)), \qquad (3.16)$$

for all $x, y \in X$.

An interesting point with this model is that it admits an intuitively appealing simple characterization. The basic axiom for characterizing the above transitive decomposable model is weak independence, which is clearly implied by (3.16). The following theorem is proved in [129, ch. 7].

THEOREM 4 *A preference relation $\succsim$ on a finite or countably infinite set X has a representation in the transitive decomposable model iff $\succsim$ is a weakly independent weak order.*

REMARK 27 *This result can be extended to sets of arbitrary cardinality adding a, necessary, condition implying that the weak order $\succsim$ has a numerical representation, see [42, 45].*

The weak point of such a model is that the function g is left unspecified so that the model will be difficult to assess. Furthermore, the uniqueness results for v_i and g are clearly much less powerful than what we obtained with model (3.2), see [129, ch. 7]. Therefore, practical applications of this model generally imply specifying the type of function g, possibly by verifying further conditions on the preference relation that impose that g belongs to some parameterized family of functions, e.g. some polynomial function of the v_i. This is studied in detail in [129, ch. 7] and [14, 82, 139, 138, 156, 166, 202]. Since such models have, to the best of our knowledge, never been used in decision analysis, we do not analyze them further.

The structure of the decomposable model however suggests that assessment techniques for this model could well come from Artificial Intelligence with its "rule induction" machinery. Indeed the function g in model (3.16) may also be seen as a set of "rules". We refer to [97, 98, 100, 101] for a thorough study of the potentiality of such an approach.

REMARK 28 *A simple extension of the decomposable model consists in simply asking for a function g that would be nondecreasing in each of its arguments. The following result is proved in [30] (see also [100]) (it can easily be extended to cover the case of an arbitrary set X, adding a, necessary, condition implying that $\succsim$ has a numerical representation).*

We say that $\succsim$ is weakly separable if, for all $i \in N$ and all $x_i, y_i \in X_i$, it is never true that $(x_i, z_{-i}) \succ (y_i, z_{-i})$ and $(y_i, w_{-i}) \succ (x_i, w_{-i})$, for some $z_{-i}, w_{-i} \in X_{-i}$. Clearly this is a weakening of weak independence since it tolerates to have at the same time $(x_i, z_{-i}) \succ (y_i, z_{-i})$ and $(x_i, w_{-i}) \sim (y_i, w_{-i})$.

THEOREM 5 *A preference relation $\succsim$ on a finite or countably infinite set X has a representation in the weak decomposable model:*

$$x \succsim y \Leftrightarrow g(u_1(x_1), \ldots, u_n(x_n)) \geq g(u_1(y_1), \ldots, u_n(y_n))$$

with g nondecreasing in all its arguments iff $\succsim$ is a weakly separable weak order.

A recent trend of research has tried to characterize special functional forms for g in the weakly decomposable model, such as max, min *or some more complex forms. The main references include [26, 100, 102, 182, 194].*

REMARK 29 *The use of "fuzzy integrals" as tools for aggregating criteria has recently attracted much attention [49, 90, 91, 93, 94, 95, 143, 145, 144, 146], the Choquet Integral and the Sugeno integral being among the most popular. It should be strongly emphasized that the very definition of these integrals requires to have at hand a weak order on $\cup_{i=1}^{n} X_i$, supposing w.l.o.g. that the sets X_i are disjoint. This is usually called a "commensurability hypothesis". Whereas this hypothesis is quite natural when dealing with an homogeneous Cartesian product, as in decision under uncertainty (see e.g. [211]), it is far less so in the area of multiple criteria decision making. A neat conjoint measurement analysis of such models and their associated assessment procedures is an open research question, see [92].*

5.2 Intransitive Indifference

Decomposable models form a large family of preferences though not large enough to encompass all cases that may be encountered when asking subjects to express preferences. A major restriction is that not all preferences may be assumed to be weak orders. The example of the sequence of cups of coffee, each differing from the previous one by an imperceptible quantity of sugar added [133], is famous; it leads to the notions of semiorder and interval order [4, 57, 66, 133, 161], in which indifference is not transitive, while strict preference is.

Ideally, taking intransitive indifference into account, we would want to arrive at a generalization of (3.2) in which:

$$x \sim y \Leftrightarrow |V(x) - V(y)| \leq \epsilon,$$
$$x \succ y \Leftrightarrow V(x) > V(y) + \epsilon,$$

where $\epsilon \geq 0$ and $V(x) = \sum_{i=1}^{n} v_i(x_i)$.

In the finite case, it is not difficult to extend the conditions presented in section 4 to cover such a case. Indeed, we are still looking here for the solution to a system of linear constraints. Although this seems to have never been done, it would not be difficult to adapt the LP-based assessment techniques to this case.

On the contrary, extending the standard sequence technique of section 3 is a formidable challenge. Indeed, remember that these techniques crucially rest on indifference judgments which lead to the determination of "perfect copies" of a given preference interval. As soon as indifference is not supposed to be transitive, "perfect copies" are not so perfect and much trouble is expected. We refer to [84, 128, 134, 161, 198] for a study of these models.

REMARK 30 *Even if the analysis of such models proves difficult, it should be noted that the semi-ordered version of the additive value model may be interpreted as having a "built-in" sensitivity analysis via the introduction of the threshold ϵ. Therefore, in practice, we may usefully view ϵ not as a parameter to be assessed but as a simple trick to avoid undue discrimination, because of the imprecision inevitably involved in our assessment procedures, between close alternatives*

REMARK 31 *Clearly the above model can be generalized to cope with a possibly non-constant threshold. The literature on the subject remains minimal however, see [161].*

5.3 Nontransitive Preferences

Many authors [147, 203] have argued that the reasonableness of supposing that strict preference is transitive is not so strong when it comes to comparing objects evaluated on several attributes. As soon as it is supposed that subjects may use an "ordinal" strategy for comparing objects, examples inspired from the well-known Condorcet paradox [176, 183] show that intransitivities will be difficult to avoid. Indeed it is possible to observe predictable intransitivities of strict preference in carefully controlled experiments [203]. There may therefore be a descriptive interest to studying such models. When it comes to decision analysis, intransitive preferences are often dismissed on two grounds:

- on a practical level, it is not easy to build a recommendation on the basis of a binary relation in which $\succ$ would not be transitive. Indeed, social choice theorists, facing a similar problem, have devoted much effort to devising what could be called reasonable procedures to deal with such preferences [41, 62, 130, 131, 149, 158, 178]. This literature does not lead, as was expected, to the emergence of a single suitable procedure in all situations.

- on a more conceptual level, many others have questioned the very rationality of such preferences using some version of the famous "money pump" argument [137, 164].

P. C. Fishburn has forcefully argued [73] that these arguments might not be as decisive as they appear at first sight. Furthermore some MCDM techniques make use of such intransitive models, most notably the so-called outranking methods [25, 172, 204, 205] and Part III in this volume. Besides the intellectual challenge, there might therefore be a real interest in studying such models.

A. Tversky [203] was one of the first to propose such a model generalizing (3.2), known as the *additive difference model*, in which:

$$x \succsim y \Leftrightarrow \sum_{i=1}^{n} \Phi_i(u_i(x_i) - u_i(y_i)) \geq 0 \tag{3.17}$$

where Φ_i are increasing and odd functions.

It is clear that (3.17) allows for intransitive $\succsim$ but implies its completeness. Clearly, (3.17) implies that $\succsim$ is independent. This allows to unambiguously define marginal preferences $\succsim_i$. Although model (3.17) can accommodate intransitive $\succsim$, a consequence of the increasingness of the Φ_i is that the marginal preference relations $\succsim_i$ are weak orders. This, in particular, excludes the possibility of any perception threshold on each attribute which would lead to an intransitive indifference relation on each attribute. Imposing that Φ_i are non-decreasing instead of being increasing allows for such a possibility. This gives rise to what is called the "weak additive difference model" in [22].

As suggested in [22, 70, 69, 72, 206], the subtractivity requirement in (3.17) can be relaxed. This leads to *nontransitive additive* conjoint measurement models in which:

$$x \succsim y \Leftrightarrow \sum_{i=1}^{n} p_i(x_i, y_i) \geq 0 \tag{3.18}$$

where the p_i are real-valued functions on X_i^2 and may have several additional properties (e.g. $p_i(x_i, x_i) = 0$, for all $i \in \{1, 2, \ldots, n\}$ and all $x_i \in X_i$).

This model is an obvious generalization of the (weak) additive difference model. It allows for intransitive and incomplete preference relations $\succsim$ as well as for intransitive and incomplete marginal preferences $\succsim_i$. An interesting specialization of (3.18) obtains when p_i are required to be *skew symmetric* i.e. such that $p_i(x_i, y_i) = -p_i(y_i, x_i)$. This skew symmetric nontransitive additive conjoint measurement model implies that $\succsim$ is complete and independent.

An excellent overview of these nontransitive models is [73]. Several axiom systems have been proposed to characterize them. P. C. Fishburn gave [70, 69, 72] axioms for the skew symmetric version of (3.18) both in the finite and the infinite case. Necessary and sufficient conditions for a nonstandard version of (3.18) are presented in [76]. [206] gives axioms for (3.18) with $p_i(x_i, x_i) = 0$ when $n \geq 4$. [22] gives necessary and sufficient conditions for (3.18) with and without skew symmetry in the denumerable case when $n = 2$.

The additive difference model (3.17) was axiomatized in [74] in the infinite case when $n \geq 3$ and [22] gives necessary and sufficient conditions for the weak additive difference model in the finite case when $n = 2$. Related studies of nontransitive models include [39, 64, 136, 153]. The implications of these models for decision-making under uncertainty were explored in [71] (for a different path to nontransitive models for decision making under risk and/or uncertainty, see [65, 67]).

It should be noticed that even the weakest form of these models, i.e. (3.18) without skew symmetry, involves an addition operation. Therefore it is unsurprising that the axiomatic analysis of these models share some common features with the additive value function model (3.2). Indeed, except in the special case in which $n = 2$, this case relating more to ordinal than to conjoint measurement (see [72]), the various axiom systems that have been proposed involve either:

- a denumerable set of cancellation conditions in the finite case or,
- a finite number of cancellation conditions together with unnecessary structural assumptions in the general case (these structural assumptions generally allow us to obtain nice uniqueness results for (3.18): the functions p_i are unique up to the multiplication by a common positive constant).

A different path to the analysis of nontransitive conjoint measurement models has recently been proposed in [30, 29, 31]. In order to get a feeling for these various models, it is useful to consider the various strategies that are likely to be implemented when comparing objects differing on several dimensions [40, 151, 152, 175, 200, 203].

Consider two alternatives x and y evaluated on a family of n attributes so that $x = (x_1, x_2, \ldots, x_n)$ and $y = (y_1, y_2, \ldots, y_n)$.

A first strategy that can be used in order to decide whether or not it can be said that "x is at least as good as y" consists in trying to measure the "worth" of each alternative on each attribute and then to combine these evaluations adequately. Giving up all idea of transitivity and completeness, this suggests a model in which:

$$x \succsim y \Leftrightarrow F(u_1(x_1), \ldots, u_n(x_n), u_1(y_1), \ldots, u_n(y_n)) \geq 0 \tag{3.19}$$

where u_i are real-valued functions on the X_i and F is a real-valued function on $\prod_{i=1}^{n} u_i(X_i)^2$. Additional properties on F, e.g. its nondecreasingness (resp. nonincreasingness) in its first (resp. last) n arguments, will give rise to a variety of models implementing this first strategy.

A second strategy relies on the idea of measuring "preference differences" separately on each attribute and then combining these (positive or negative) differences in order to know whether the aggregation of these differences leads to an advantage for x over y. More formally, this suggests a model in which:

$$x \succsim y \Leftrightarrow G(p_1(x_1, y_1), p_2(x_2, y_2), \ldots, p_n(x_n, y_n)) \geq 0 \tag{3.20}$$

where p_i are real-valued functions on X_i^2 and G is a real-valued function on $\prod_{i=1}^{n} p_i(X_i^2)$. Additional properties on G (e.g. its oddness or its nondecreasingness in each of its arguments) or on p_i (e.g. $p_i(x_i, x_i) = 0$ or $p_i(x_i, y_i) = -p_i(y_i, x_i)$) will give rise to a variety of models in line with the above strategy.

Of course these two strategies are not incompatible and one may well consider using the "worth" of each alternative on each attribute to measure "preference differences". This suggests a model in which:

$$x \succsim y \Leftrightarrow H(\phi_1(u_1(x_1), u_1(y_1)), \ldots, \phi_n(u_n(x_n), u_n(y_n))) \geq 0 \qquad (3.21)$$

where u_i are real-valued functions on X_i, ϕ_i are real-valued functions on $u_i(X_i)^2$ and H is a real-valued function on $\prod_{i=1}^{n} \phi_i(u_i(X_i)^2)$.

The use of general functional forms, instead of additive ones, greatly facilitate the axiomatic analysis of these models. It mainly relies on the study of various kinds of *traces* induced by the preference relation on coordinates and does not require a detailed analysis of tradeoffs between attributes.

The price to pay for such an extension of the scope of conjoint measurement is that the number of parameters that would be needed to assess such models is quite high. Furthermore, none of them is likely to possess any remarkable uniqueness properties. Therefore, although proofs are constructive, these results will not give direct hints on how to devise assessment procedures. The general idea here is to use numerical representations as guidelines to understand the consequences of a limited number of cancellation conditions, without imposing any transitivity or completeness requirement on the preference relation and any structural assumptions on the set of objects. Such models have proved useful to:

- understand the ordinal character of some aggregation models proposed in the literature [170, 172], known as the "outranking methods" (see Part III of this volume) as shown in [28],
- understand the links between aggregation models aiming at enriching a dominance relation and more traditional conjoint measurement approaches [30],
- to include in a classical conjoint measurement framework, noncompensatory preferences in the sense of [22, 33, 55, 60, 61] as shown in [28, 32, 99].

Acknowledgments

We thank two anonymous referees for their very helpful comments and suggestions. Part of this work was accomplished while Denis Bouyssou was visiting the *Service de Mathématique de la Gestion* at the *Université Libre de Bruxelles* (Brussels, Belgium). The warm hospitality of the *Service de Mathématique de la Gestion*, the support of the Belgian *Fonds National de la Recherche Scientistic* and the Brussels-Capital Region through a "Research in Brussels" action grant are gratefully acknowledged. Many Figures in this text were inspired by the ones drawn by Christophe Gonzales in [86].

References

[1] E. W. Adams. Elements of a theory of inexact measurement. *Philosophy of Science*, 32:205–228, 1965.

[2] E. W. Adams and R. F. Fagot. A model of riskless choice. *Behavioral Science*, 4:1–10, 1959.

[3] L. Adelman, P. J. Sticha, and M. L. Donnell. An experimental investigation of the relative effectiveness of two techniques for structuring multiattributed hierarchies. *Organizational Behavior and Human Decision Processes*, 37:188–196, 1986.

[4] F. Aleskerov and B. Monjardet. *Utility Maximization, Choice and Preference*. Springer Verlag, Heidelberg, 2002.

[5] A. B. Atkinson. On the measurement of inequality. *Journal of Economic Theory*, 2:244–263, 1970.

[6] C. A. Bana e Costa. The use of multi-criteria decision analysis to support the search for less conflicting policy options in a multi-actor context: Case study. *Journal of Multi-Criteria Decision Analysis*, 10(2):111–125, 2001.

[7] C. A. Bana e Costa, É. C. Corrêa, J.-M. De Corte, and J.-C. Vansnick. Facilitating bid evaluation in public call for tenders: A socio-technical approach. *Omega*, 30(3):227–242, April 2002.

[8] C. A. Bana e Costa, M. L. Costa-Lobo, I. A. J. Ramos, and J.-C. Vansnick. Multicriteria approach for strategic town planning: The case of Barcelos. In D. Bouyssou, É. Jacquet-Lagrèze, P. Perny, R. Słowiński, D. Vanderpooten, and Ph. Vincke, editors, *Aiding Decisions with Multiple Criteria: Essays in Honour of Bernard Roy*, pages 429–456. Kluwer Academic Publishers, Dordrecht, 2002.

[9] C. A. Bana e Costa, L. Ensslin, É. C. Corrêa, and J.-C. Vansnick. Decision support systems in action: Integrated application in a multicriteria decission aid process. *European Journal of Operational Research*, 113(2):315–335, March 1999.

[10] C. A. Bana e Costa, F. Nunes da Silva, and J.-C. Vansnick. Conflict dissolution in the public sector: A case-study. *European Journal of Operational Research*, 130(2):388–401, 2001.

[11] C. A. Bana e Costa and R. C. Oliveira. Assigning priorities for maintenance and repair and refurbishment in managing a municipal housing stock. *European Journal of Operational Research*, 138:380–91, 2002.

[12] C. A. Bana e Costa and J.-C. Vansnick. MACBETH – An interactive path towards the construction of cardinal value functions. *International Transactions in Operational Research*, 1:489–500, 1994.

[13] A. F. Beardon, J. C. Candeal, G. Herden, E. Induráin, and G. B. Mehta. The non-existence of a utility function and the structure of non-representatble preference relations. *Journal of Mathematical Economics*, 37:17–38, 2002.

[14] D. E. Bell. Multilinear representations for ordinal utility functions. *Journal of Mathematical Psychology*, 31:44–59, 1987.

[15] V. Belton, F. Ackermann, and I. Shepherd. Integrated support from problem structuring through alternative evaluation using COPE and V• I• S• A. *Journal of Multi-Criteria Decision Analysis*, 6:115–130, 1997.

[16] V. Belton and T. Stewart. *Multiple Criteria Decision Analysis: An Integrated Approach*. Kluwer Academic Publishers, Dordrecht, 2001.

[17] E. Ben-Porath and I. Gilboa. Linear measures, the Gini index and the income-equality tradeoff. *Journal of Economic Theory*, 64:443–467, 1994.

[18] E. Ben-Porath, I. Gilboa, and D. Schmeidler. On the measurement of inequality under uncertainty. *Journal of Economic Theory*, 75:194–204, 1997.

[19] M. Beuthe and G. Scannella. Applications comparées des méthodes d'analyse multicritère UTA. *RAIRO Recherche Opérationnelle / Operations Research*, 30(3):293–315, 1996.

[20] M. Beuthe and G. Scannella. Comparative analysis of UTA multicriteria methods. *European Journal of Operational Research*, 130(2):246–262, 2001.

[21] C. Blackorby, D. Primont, and R. Russell. *Duality, Separability, and Functional Structure: Theory and Economic Applications*. North-Holland, New York, 1978.

[22] D. Bouyssou. Some remarks on the notion of compensation in MCDM. *European Journal of Operational Research*, 26:150–160, 1986.

[23] D. Bouyssou. Modelling inaccurate determination, uncertainty, imprecision using multiple criteria. In A. G. Lockett and G. Islei, editors, *Improving Decision Making in Organisations*, pages 78–87. Springer Verlag, Heidelberg, 1989.

[24] D. Bouyssou. Builing criteria: A prerequisite for MCDA. In C. A. Bana e Costa, editor, *Readings in Multiple Criteria Decision Aid*, pages 58–81, Berlin, 1990. Springer Verlag.

[25] D. Bouyssou. Outranking methods. In C. A. Floudas and P. M. Pardalos, editors, *Encyclopedia of Optimization*, volume 4, pages 249–255. Kluwer Academic Publishers, Dordrecht, 2001.

[26] D. Bouyssou, S. Greco, B. Matarazzo, M. Pirlot, and R. Słowiński. Characterization of 'max', 'min' and 'order statistics' multicriteria aggregation functions. Communication to *IFORS'2002*, 8 – 12 July, 2002, Edinburgh, U.K., July 2002.

[27] D. Bouyssou, Th. Marchant, M. Pirlot, P. Perny, A. Tsoukiàs, and Ph. Vincke. *Evaluation and Decision Models: A Critical Perspective*. Kluwer Academic Publishers, Dordrecht, 2000.

[28] D. Bouyssou and M. Pirlot. A characterization of strict concordance relations. In D. Bouyssou, É. Jacquet-Lagrèze, P. Perny, R. Słowiński, D. Vanderpooten, and Ph. Vincke, editors, *Aiding Decisions with Multiple Criteria: Essays in Honour of Bernard Roy*, pages 121–145. Kluwer Academic Publishers, Dordrecht, 2002.

[29] D. Bouyssou and M. Pirlot. Nontransitive decomposable conjoint measurement. *Journal of Mathematical Psychology*, 46:677–703, 2002.

[30] D. Bouyssou and M. Pirlot. Preferences for multiattributed alternatives: Traces, dominance, and numerical representations. www.lamsade.dauphine.fr/~bouyssou, 2002. Working Paper, LAMSADE, Université Paris-Dauphine, submitted.

[31] D. Bouyssou and M. Pirlot. 'Additive difference' models without additivity and subtractivity. www.lamsade.dauphine.fr/~bouyssou, 2003. Working Paper, LAMSADE, Université Paris-Dauphine.

[32] D. Bouyssou and M. Pirlot. A characterization of concordance relations. www.lamsade.dauphine.fr/~bouyssou, 2003. Working Paper, LAMSADE, Université Paris-Dauphine.

[33] D. Bouyssou and J.-C. Vansnick. Noncompensatory and generalized noncompensatory preference structures. *Theory and Decision*, 21:251–266, 1986.

[34] D. S. Briges and G. B. Mehta. *Representations of Preference Orderings*. Springer Verlag, Berlin, 1995.

[35] J. D. Carroll and P. E. Green. Psychometric methods in marketing research. Part 1. Conjoint analysis. *Journal of Marketing Research*, 32:385–391, 1995.

[36] A. Charnes and W. W. Cooper. *Management Models and Industrial Applications of Linear Programming*. John Wiley & Sons, New York, 1961.

[37] A. Chateauneuf and P. P. Wakker. From local to global additive representation. *Journal of Mathematical Economics*, 22(6):523–545, 1993.

[38] J.-C. Cosset, Y. Siskos, and C. Zopounidis. The evaluation of country risk: A decision support approach. *Global Finance Journal*, 3:79–95, 1992.

[39] M. A. Croon. The axiomatization of additive difference models for preference judgements. In E. Degreef and G. van Buggenhaut, editors, *Trends in Mathematical Psychology*, pages 193–227, Amsterdam, 1984. North-Holland.

[40] V. Dahlstrand and H. Montgomery. Information search and evaluation processes in decision-making: A computer-based process tracking study. *Acta Psychologica*, 56:113–123, 1984.

[41] P. de Donder, M. Le Breton, and M. Truchon. Choosing from a weighted tournament. *Mathematical Social Sciences*, 40:85–109, 2000.

[42] G. Debreu. Representation of a preference ordering by a numerical function. In R. Thrall, C. H. Coombs, and R. Davies, editors, *Decision Processes*, pages 159–175, New York, 1954. John Wiley & Sons.

[43] G. Debreu. *Theory of Value: An Axiomatic Analysis of Economic Equilibrium*. John Wiley & Sons, New York, 1959.

[44] G. Debreu. Topological methods in cardinal utility theory. In K. J. Arrow, S. Karlin, and P. Suppes, editors, *Mathematical Methods in the Social Sciences*, pages 16–26. Stanford University Press, Stanford, 1960.

[45] G. Debreu. Continuity properties of Paretian utility. *International Economic Review*, 5(3):285–293, September 1964.

[46] Ph. Delquié. Inconsistent trade-offs between attributes: New evidence in preference assessment biases. *Management Science*, 39(11):1382–1395, 1993.

[47] A. I. Dimitras, C. Zopounidis, and C. Hurson. Assessing financial risks using a multicriteria sorting procedure: The case of country risk assessment. *Foundations of Computing and Decision Sciences*, 29:97–109, 2001.

[48] M. Doumpos, S. H. Zanakis, and C. Zopounidis. Multicriteria preference disaggregation for classification problems with an application to global investing risk. *Decision Sciences*, 32(2):333–385, 2001.

[49] D. Dubois, J.-L. Marichal, H. Prade, M. Roubens, and R. Sabbadin. The use of the discrete Sugeno integral in decision-making: A survey. *International Journal of Uncertainty, Fuzziness and Knowledge-Based Systems*, 9(5):539–561, 2001.

[50] J. S. Dyer and R. K. Sarin. Measurable multiattribute value functions. *Operations Research*, 27:810–822, 1979.

[51] W. Edwards. Social utilities. *Engineering Economist*, 6:119–129, 1971.

[52] W. Edwards. How to use multiattribute utility measurement for social decision making. *IEEE Transactions on Systems, Man and Cybernetics*, 7(5):326–340, 1977.

[53] W. Edwards and F. Hutton Barron. SMART and SMARTER: Improved simple methods for multiattribute utility measurement. *Organizational Behavior and Human Decision Processes*, 60:306–325, 1994.

[54] J.-C. Falmagne. On a class of probabilistic conjoint measurement models: Some diagnostic properties. *Journal of Mathematical Psychology*, 19(2):73–88, 1979.

[55] H. Fargier and P. Perny. Modélisation des préférences par une règle de concordance généralisée. In A. Colorni, M. Paruccini, and B. Roy, editors, A-MCD-A, *Aide Multicritère à la Décision/Multiple Criteria Decision Aid*, pages 99–115. European Commission, Joint Research Centre, 2001.

[56] Final report of the British Association for the Advancement of Science. 2:331–349, 1940.

[57] P. C. Fishburn. Intransitive indifference in preference theory: A survey. *Operations Research*, 18(2):207–228, 1970.

[58] P. C. Fishburn. *Utility Theory for Decision-Making*. John Wiley & Sons, New York, 1970.

[59] P. C. Fishburn. Lexicographic orders, utilities and decision rules: A survey. *Management Science*, 20(11):1442–1471, 1974.

[60] P. C. Fishburn. Axioms for lexicographic preferences. *Review of Economic Studies*, 42:415–419, 1975.

[61] P. C. Fishburn. Noncompensatory preferences. *Synthese*, 33:393–403, 1976.

[62] P. C. Fishburn. Condorcet social choice functions. *SIAM Journal on Applied Mathematics*, 33:469–489, 1977.

[63] P. C. Fishburn. A survey of multiattribute/multicriteria evaluation theories. In S. Zionts, editor, *Multicriteria Problem Solving*, pages 181–224. Springer Verlag, Berlin, 1978.

[64] P. C. Fishburn. Lexicographic additive differences. *Journal of Mathematical Psychology*, 21:191–218, 1980.

[65] P. C. Fishburn. Nontransitive measurable utility. *Journal of Mathematical Psychology*, 26:31–67, 1982.

[66] P. C. Fishburn. *Interval Orders and Intervals Graphs*. John Wiley & Sons, New York, 1985.

[67] P. C. Fishburn. *Nonlinear Preference and Utility Theory*. Johns Hopkins University Press, Baltimore, 1988.

[68] P. C. Fishburn. Normative theories of decision making under risk and under uncertainty. In *Nonconventional Preference Relations in Decision Making*, pages 1–21. Springer Verlag, Berlin, 1988.

[69] P. C. Fishburn. Additive non-transitive preferences. *Economic Letters*, 34:317–321, 1990.

[70] P. C. Fishburn. Continuous nontransitive additive conjoint measurement. *Mathematical Social Sciences*, 20:165–193, 1990.

[71] P. C. Fishburn. Skew symmetric additive utility with finite states. *Mathematical Social Sciences*, 19:103–115, 1990.

[72] P. C. Fishburn. Nontransitive additive conjoint measurement. *Journal of Mathematical Psychology*, 35:1–40, 1991.

[73] P. C. Fishburn. Nontransitive preferences in decision theory. *Journal of Risk and Uncertainty*, 4:113–134, 1991.

[74] P. C. Fishburn. Additive differences and simple preference comparisons. *Journal of Mathematical Psychology*, 36:21–31, 1992.

[75] P. C. Fishburn. A general axiomatization of additive measurement with applications. *Naval Research Logistics*, 39(6):741–755, 1992.

[76] P. C. Fishburn. On nonstandard nontransitive additive utility. *Journal of Economic Theory*, 56:426–433, 1992.

[77] P. C. Fishburn. Finite linear qualitative probability. *Journal of Mathematical Psychology*, 40:21–31, 1996.

[78] P. C. Fishburn. Cancellation conditions for multiattribute preferences on finite sets. In M. H. Karwan, J. Spronk, and J. Wallenius, editors, *Essays in Decision Making*, pages 157–167. Springer Verlag, Berlin, 1997.

[79] S. French. *Decision Theory – An Introduction to the Mathematics of Rationality*. Ellis Horwood, London, 1993.

[80] D. G. Fryback and R. L. Keeney. Constructing a complex judgmental model: An index of trauma severity. *Management Science*, 29:869–883., 1983.

[81] G. Furkhen and M. K. Richter. Additive utility. *Economic Theory*, 1:83–105, 1991.

[82] G. Furkhen and M. K. Richter. Polynomial utility. *Economic Theory*, 1:231–249, 1991.

[83] D. Gale. *The Theory of Linear Economic Models*. McGraw-Hill, New York, 1960.

[84] I. Gilboa and R. Lapson. Aggregation of semiorders: Intransitive indifference makes a difference. *Economic Theory*, 5:109–126, 1995.

[85] Ch. Gonzales. Additive utilities when some components are solvable and others not. *Journal of Mathematical Psychology*, 40:141–151, 1996.

[86] Ch. Gonzales. *Utilités additives: Existence et construction*. Thèse de doctorat, Université Paris 6, Paris, France, 1996.

[87] Ch. Gonzales. Two factor additive conjoint measurement with one solvable component. *Journal of Mathematical Psychology*, 44:285–309, 2000.

[88] Ch. Gonzales. Additive utility without restricted solvavibility on every component. *Journal of Mathematical Psychology*, 47:47–65, 2003.

[89] W. M. Gorman. The structure of utility functions. *Review of Economic Studies*, XXXV:367–390, 1968.

[90] M. Grabisch. The application of fuzzy integrals to multicriteria decision making. *European Journal of Operational Research*, 89:445–456, 1996.

[91] M. Grabisch. Fuzzy integral as a flexible and interpretable tool of aggregation. In *Aggregation and Fusion of Imperfect Information*, pages 51–72. Physica, Heidelberg, 1998.

[92] M. Grabisch, Ch. Labreuche, and J.-C. Vansnick. On the extension of pseudo-boolean functions for the aggregation of interacting criteria. *European Journal of Operational Research*, 148(1):28–47, 2003.

[93] M. Grabisch, S. A. Orlovski, and R. R. Yager. Fuzzy aggregation of numerical preferences. In R. Słowiński, editor, *Fuzzy Sets in Decision Analysis, Operations Research and Statistics*, pages 31–68. Kluwer Academic Publishers, Boston, MA, 1998.

[94] M. Grabisch and P. Perny. Agrégation multicritère. In B. Bouchon-Meunier and C. Marsala, editors, *Logique Floue, Principes, Aide à la Décision*, IC^2, pages 82–120. Hermès, 2002.

[95] M. Grabisch and M. Roubens. Application of the Choquet integral in multicriteria decision making. In M. Grabisch, T. Murofushi, and M. Sugeno, editors, *Fuzzy Measures and Integrals*, pages 348–374. Physica Verlag, Heidelberg, 2000.

[96] N. Grassin. Constructing criteria population for the comparison of different options of high voltage line routes. *European Journal of Operational Research*, 26:42–47, 1886.

[97] S. Greco, B. Matarazzo, and R. Słowiński. Rough approximation of a preference relation by dominance relations. *European Journal of Operational Research*, 117:63–83, 1999.

[98] S. Greco, B. Matarazzo, and R. Słowiński. The use of rough sets and fuzzy sets in MCDM. In T. Gal, T. Hanne, and T. Stewart, editors, *Multicriteria Decision Making, Advances in MCDM Models, Algorithms, Theory and Applications*, pages 14.1–14.59. KluwerAcademic Publishers, Dordrecht, 1999.

[99] S. Greco, B. Matarazzo, and R. Słowiński. Axiomatic basis of noncompensatoty preferences. Communication to *Foundations of Utility and Risk Theory, FUR X*, 30 May–2 June, Torino, Italy, 2001.

[100] S. Greco, B. Matarazzo, and R. Słowiński. Conjoint measurement and rough set approach for multicriteria sorting problems in presence of ordinal criteria. In A. Colorni, M. Paruccini, and B. Roy, editors, A-MCD-A, *Aide Multicritère à la Décision/Multiple Criteria Decision Aid*, pages 117–144. European Commission, Joint Research Centre, 2001.

[101] S. Greco, B. Matarazzo, and R. Słowiński. Rough sets theory for multicriteria decision analysis. *European Journal of Operational Research*, 129:1–47, 2001.

[102] S. Greco, B. Matarazzo, and R. Słowiński. Preference representation by means of conjoint measurement and decision rule model. In D. Bouyssou, É. Jacquet-Lagrèze, P. Perny, R. Sł owiński, D. Vanderpooten, and Ph. Vincke, editors, *Aiding Decisions with Multiple Criteria: Essays in Honour of Bernard Roy*, pages 263–313. Kluwer Academic Publishers, Dordrecht, 2002.

[103] P. E. Green and V. Srinivasan. Conjoint analysis in consumer research: Issues and outlook. *Journal of Consumer Research*, 5:103–152, 1978.

[104] P. E. Green, D. S. Tull, and G. Albaum. *Research for Marketing Decisions*. Englewood Cliffs, 1988.

[105] E. Grigoroudis, Y. Siskos, and O. Saurais. TELOS: A customer satisfaction evaluation software. *Computers and Operations Research*, 27(7–8):799–817, 2000.

[106] J. Guild. Part III of Quantitative estimation of sensory events. Interim report, British Academy for the Advancement of Science, 1936. pp. 296–328.

[107] F. Gul. Savage's theorem with a finite number of states. *Journal of Economic Theory*, 57:99–110, 1992.

[108] G. Iverson and J.-C. Falmagne. Statistical issues in measurement. *Mathematical Social Sciences*, 10:131–153, 1985.

[109] É. Jacquet-Lagrèze. Interactive assessment of preferences using holistic judgments. The PREFCALC system. In C. A. Bana e Costa, editor, *Readings in Multiple Criteria Decision Aid*, pages 335–350. Springer Verlag, Berlin, 1990.

[110] É. Jacquet-Lagrèze. An application of the UTA discriminant model for the evaluation of R&D projects. In P. Pardalos, Y. Siskos, and C. Zopounidis, editors, *Advances in Multicriteria Analysis*, pages 203–211. Kluwer Academic Publishers, Dordrecht, 1995.

[111] É. Jacquet-Lagrèze and J. Siskos. Assessing a set of additive utility functions for multicriteria decision making: The UTA method. *European Journal of Operational Research*, 10:151–164, 1982.

[112] J.-Y. Jaffray. *Existence, propriétés de continuité, additivité de fonctions d'utilité sur un espace partiellement ou totalement ordonné*. Thèse de doctorat d'état, Université Paris 6, Paris, France, 1974.

[113] J.-Y. Jaffray. On the extension of additive utilities to infinite sets. *Journal of Mathematical Psychology*, 11:431–452, 1974.

[114] R. M. Johnson. Trade-off analysis of consumer values. *Journal of Marketing Research*, 11:121–127, 1974.

[115] E. Karni and Z. Safra. The hexagon condition and additive representation for two dimensions: An algebraic approach. *Journal of Mathematical Psychology*, 42:393–399, 1998.

[116] R. L. Keeney. Measurement scales for quantifying attributes. *Behavioral Science*, 26:29–36, 1981.

[117] R. L. Keeney. Building models of values. *European Journal of Operational Resesearch*, 37(2):149–157, 1988.

[118] R. L. Keeney. Structuring objectives for problems of public interest. *Operations Research*, 36:396–405, 1988.

[119] R. L. Keeney. *Value-focused Thinking*. Harvard University Press, Cambridge, MA, 1992.

[120] R. L. Keeney, J. S. Hammond, and H. Raiffa. *Smart Choices: A Guide to Making Better Decisions*. Harvard University Press, Boston, 1999.

[121] R. L. Keeney and H. Raiffa. *Decisions with Multiple Objectives: Preferences and Value Tradeoffs*. John Wiley & Sons, New York, 1976.

[122] R. L. Keeney and G. A. Robillard. Assessing and evaluating environ-mental impacts at proposed nuclear power plant sites. *Journal of Environmental Economics and Management*, 4:153–166, 1977.

[123] V. Köbberling. Strenth of preference and cardinal utility. Working paper, University of Maastricht, Maastricht, June 2002.

[124] V. Köbberling. Comment on: Edi Karni & Zvi Safra (1998) The hexagon condition and additive representation for two dimensions: An algebraic approach. *Journal of Mathematical Psychology*, 47(3):370, 2003.

[125] T. C. Koopmans. Stationary ordinal utility and impatience. *Econometrica*, 28:287–309, 1960.

[126] T. C. Koopmans. Representation of prefernce orderings over time. In C. B. McGuire and R. Radner, editors, *Decision and Organization*, pages 57–100. Noth-Holland, Amsterdam, 1972.

[127] D. H. Krantz. Conjoint measurement: The Luce-Tukey axiomatization and some extensions. *Journal of Mathematical Psychology*, 1:248–277, 1964.

[128] D. H. Krantz. Extensive measurement in semiorders. *Philosophy of Science*, 34:348–362, 1967.

[129] D. H. Krantz, R. D. Luce, P. Suppes, and A. Tversky. *Foundations of Measurement,* Vol. 1: *Additive and Polynomial Representations*. Academic Press, New York, 1971.

[130] G. Laffond, J.-F. Laslier, and M. Le Breton. Condorcet choice correspondences: A set-theoretical comparison. *Mathematical Social Sciences*, 30:23–36, 1995.

[131] J.-F. Laslier. *Tournament Solutions and Majority Voting*. Springer Verlag, Berlin, 1997.

[132] J. Louvière. *Analyzing Decision Making: Metric Conjoint Analysis*. Sage, Park, CA, 1988.

[133] R. D. Luce. Semiorders and a theory of utility discrimination. *Econometrica*, 24:178–191, 1956.

[134] R. D. Luce. Three axiom systems for additive semiordered structures. *SIAM Journal of Applied Mathematics*, 25:41–53, 1973.

[135] R. D. Luce. Conjoint measurement: A brief survey. In David E. Bell, Ralph L. Keeney, and Howard Raiffa, editors, *Conflicting Objectives in Decisions*, pages 148–171. John Wiley & Sons, New York, 1977.

[136] R. D. Luce. Lexicographic tradeoff structures. *Theory and Decision*, 9:187–193, 1978.

[137] R. D. Luce. *Utility of Gains and Losses: Measurement-theoretical and Experimental Approaches*. Lawrence Erlbaum Publishers, Mahwah, New Jersey, 2000.

[138] R. D. Luce and M. Cohen. Factorizable automorphisms in solvable conjoint structures. I. *Journal of Pure and Applied Algebra*, 27(3):225–261, 1983.

[139] R. D. Luce, D. H. Krantz, P. Suppes, and A. Tversky. *Foundations of Measurement,* Vol. 3: *Representation, Axiomatisation and Invariance*. Academic Press, New York, 1990.

[140] R. D. Luce and A. A. J. Marley. Extensive measurement when concatenation is restricted. In S. Morgenbesser, P. Suppes, and M. G. White, editors, *Philosophy, Science and Method: Essays in Honor of Ernest Nagel*, pages 235–249. St. Martin's Press, New York, 1969.

[141] R. D. Luce and J. W. Tukey. Simultaneous conjoint measurement: A new type of fundamental measurement. *Journal of Mathematical Psychology*, 1:1–27, 1964.

[142] A. Maas and P. P. Wakker. Additive conjoint measurement for multiattribute utility. *Journal of Mathematical Psychology*, 38:86–101, 1994.

[143] J.-L. Marichal. An axiomatic approach of the discrete Choquet integral as a tool to aggregate interacting criteria. *IEEE Transactions on Fuzzy Systems*, 8, 2000.

[144] J.-L. Marichal. On Choquet and Sugeno integrals as aggregation functions. In M. Grabisch, T. Murofushi, and M. Sugeno, editors, *Fuzzy Measures and Integrals*, pages 247–272. Physica Verlag, Heidelberg, 2000.

[145] J.-L. Marichal. On Sugeno integralsas an aggregation function. *Fuzzy Sets and Systems*, 114, 2000.

[146] J.-L. Marichal and M. Roubens. Determination of weights of interacting criteria from a reference set. *European Journal of Operational Research*, 124:641–50, 2000.

[147] K. O. May. Intransitivity, utility and the aggregation of preference patterns. *Econometrica*, 22:1–13, 1954.

[148] G. Mihelis, E. Grigoroudis, Y. Siskos, Y. Politis, and Y. Malandrakis. Customer satisfaction measurement in the private bank sector. *European Journal of Operational Research*, 130(2):347–360, 2001.

[149] N. R. Miller. Graph theoretical approaches to the theory of voting. *American Journal of Political Science*, 21:769–803, 1977.

[150] J. Miyamoto and P. P. Wakker. Multiattribure utility theory without expected utility foundations. *Operations Research*, 44(2):313–326, 1996.

[151] H. Montgomery. A study of intransitive preferences using a think aloud procedure. In H. Jungerman and G. de Zeeuw, editors, *Decision-making and Change in Human Affairs*, pages 347–362. D. Reidel, Dordrecht, 1977.

[152] H. Montgomery and O. Svenson. On decision rules and information processing strategies for choice among multiattribute alternatives. *Scandinavian Journal of Psychology*, 17:283–291, 1976.

[153] Y. Nakamura. Lexicographic additivity for multi-attribute preferences on finite sets. *Theory and Decision*, 42:1–19, 1997.

[154] Y. Nakamura. Additive utility on densely ordered sets. *Journal of Mathematical Psychology*, 46:515–530, 2002.

[155] L. Narens. *Abstract Measurement Theory*. MIT press, Cambridge, MA, 1985.

[156] L. Narens and R. D. Luce. The algebra of measurement. *Journal of Pure and Applied Algebra*, 8(2):197–233, 1976.

[157] A. Ostanello. Action evaluation and action structuring – Different decision aid situations reviewed through two actual cases. In C. A. Bana e Costa, editor, *Readings in Multiple Criteria Decision Aid*, pages 36–57. Springer Verlag, Berlin, 1990.

[158] J. E. Peris and B. Subiza. Condorcet choice correspondences for weak tournaments. *Social Choice and Welfare*, 16:217–231, 1999.

[159] J. Pfanzagl. *Theory of Measurement*. Physica Verlag, Würzburg, 2nd revised edition edition, 1971.

[160] M. Pirlot and Ph. Vincke. Lexicographic aggregation of semiorders. *Journal of Multicriteria Decision Analysis*, 1:47–58, 1992.

[161] M. Pirlot and Ph. Vincke. *Semiorders. Properties, Representations, Applications*. Kluwer Academic Publishers, Dordrecht, 1997.

[162] J.-Ch. Pomerol and S. Barba-Romero. *Multicriterion Decision in Management, Principles and Practice*. Kluwer Academic Publsihers, Dordrecht, 2000.

[163] M. Pöyhönen, H. Vrolijk, and R. P. Hämäläinen. Behavioral and procedural consequences of structural variations in value trees. *European Journal of Operational Research*, 134(1):216–227, 2001.

[164] H. Raiffa. *Decision Analysis – Introductory Lectures on Choices under Uncertainty*. Addison-Wesley, Reading, MA, 1968.

[165] H. Raiffa. Preference for multi-attributed alternatives. RAND Memorandum, RM-5868-DOT/RC, December 1968.

[166] M. K. Richter. Rational choice and polynomial measurement theory. *Journal of Mathematical Psychology*, 12:99–113, 1975.

[167] F. S. Roberts. *Measurement Theory with Applications to Decision Making, Utility and the Social Sciences*. Addison-Wesley, Reading, 1979.

[168] F. S. Roberts and R. D. Luce. Axiomatic thermodynamics and extensive measurement. *Synthese*, 18:311–326, 1968.

[169] M. Roubens and Ph. Vincke. *Preference Modelling*. Springer Verlag, Berlin, 1985.

[170] B. Roy. Classement et choix en présence de points de vue multiples (la méthode ELECTRE). *RIRO, Revue d'Informatique et de Recherche Operationelle*, 2:57–75, 1968.

[171] B. Roy. Main sources of inaccurate determination, uncertainty and imprecision in decision models. *Mathematical and Computer Modelling*, 12(10–11):1245–1254, 1989.

[172] B. Roy. The outranking approach and the foundations of ELECTRE methods. *Theory and Decision*, 31:49–73, 1991.

[173] B. Roy. *Multicriteria Methodology for Decision Aiding*. Kluwer Academic Publishers, Dordrecht, 1996. Original version in French "*Méthodologie Multicritère d'Aide à la Décision*", Economica, Paris, 1985.

[174] B. Roy and D. Bouyssou. *Aide Multicritère à la Décision : Méthodes et Cas*. Economica, Paris, 1993.

[175] J. E. Russo and B. A. Dosher. Strategies for multiattribute binary choice. *Journal of Experimental Psychology: Learning, Memory and Cognition*, 9:676–696, 1983.

[176] D. G. Saari. Connecting and resolving Sen's and Arrow's theorems. *Social Choice and Welfare*, 15:239–261, 1998.

[177] L. J. Savage. *The Foundations of Statistics*. John Wiley & Sons, New York, 1954.

[178] T. Schwartz. *The Logic of Collective Choice*. Columbia University Press, 1986.

[179] D. Scott. Measurement structures and linear inequalities. *Journal of Mathematical Psychology*, 1:233–247, 1964.

[180] D. Scott and P. Suppes. Foundational aspects of theories of measurement. *Journal of Symbolic Logic*, 23:113–128, 1958.

[181] U. Segal. A sufficient condition for additively separable functions. *Journal of Mathematical Economics*, 23:295–303, 1994.

[182] U. Segal and J. Sobel. Min, Max and Sum. *Journal of Economic Theory*, 106:126–150, 2002.

[183] A. K. Sen. Social choice theory. In K. J. Arrow and M. D. Intriligator, editors, *Handbook of Mathematical Economics*, volume 3, pages 1073–1181. North-Holland, Amsterdam, 1986.

[184] L. Shapiro. Conditions for expected uility maximization. *Annals of Statistics*, 7:1288–1302, 1979.

[185] Y. Siskos. Evaluating a system of furniture retail outlets using an interactive ordinal regression model. *European Journal of Operational Research*, 23:179–193, 1986.

[186] Y. Siskos and D. K. Despotis. A DSS oriented method for multiobjective linear programming problems. *Decision Support Systems*, 5(1):47–56, 1989.

[187] Y. Siskos, D. K. Despotis, and M. Ghediri. Multiobjective modelling for regional agricultural planning: Case study in Tunisia. *European Journal of Operational Research*, 77(3):375–391, 1993.

[188] Y. Siskos, E. Grigoroudis, C. Zopounidis, and O. Saurais. Measuring customer satisfaction using a collective preference disaggregation model. *Journal of Global Optimization*, 12(2):175–195, 1998.

[189] Y. Siskos, N. F. Matsatsinis, and G. Baourakis. Multicriteria analysis in agricultural marketing: The case of French olive oil market. *European Journal of Operational Research*, 130(2):315–331, April 2001.

[190] Y. Siskos, A. Spyridakos, and D. Yannacopoulos. Using artificial intelligence and visual techniques into preference disaggregation analysis: The MIIDAS system. *European Journal of Operational Research*, 113(2):281–299, 1999.

[191] Y. Siskos and D. Yannacopoulos. UTASTAR, An ordinal regression method for building additive value functions. *Investigaçao Operacional*, 5(1):39–53, 1985.

[192] Y. Siskos and C. Zopounidis. The evaluation of venture capital investment activity: An interactive assessment. *European Journal of Operational Research*, 31(3):304–313, 1987.

[193] H. J. Skala. *Non-Archimedean Utility Theory*. Kluwer Academic Publishers, Dordrecht, 1975.

[194] J. Sounderpandian. Value functions when decision criteria are not totally substitutable. *Operations Research*, 39:592–600, 1991.

[195] A. Spyridakos, Y. Siskos, D. Yannacopoulos, and A. Skouris. Multicriteria job evaluation for large organizations. *European Journal of Operational Research*, 130(2):375–387, April 2001.

[196] T. Stewart. An interactive multiple objective linear programming method based on piecewise linear additive value functions. *IEEE Transactions on Systems, Man, and Cybernetics*, SMC-17(5):799–805, 1987.

[197] W. G. Stillwell, D. von Winterfeldt, and R. S. John. Comparing hierarchical and non-hierarchical weighting methods for eliciting multiattribute value models. *Management Science*, 33:442–50, 1987.

[198] P. Suppes, D. H. Krantz, R. D. Luce, and A. Tversky. *Foundations of Measurement,* Vol. 2: *Geometrical, Threshold, and Probabilistic Representations*. Academic Press, New York, 1989.

[199] P. Suppes and M. Winet. An axiomatization of utility based on the notion of utility difference. *Management Science*, 1:259–270, 1955.

[200] O. Svenson. Process description of decision making. *Organizational Behavior and Human Performance*, 23:86–112, 1979.

[201] R. J. Titiev. Measurement structures in classes that are not universally axiomatizable. *Journal of Mathematical Psychology*, 9:200–205, 1972.

[202] A. Tversky. A general theory of polynomial conjoint measurement. *Journal of Mathematical Psychology*, 4:1–20, 1967.

[203] A. Tversky. Intransitivity of preferences. *Psychological Review*, 76:31–48, 1969.

[204] Ph. Vincke. *Multi-criteria Decision Aid.* John Wiley & Sons, New York, 1992. Original version in French "*L'Aide Multicritère à la Décision*", Éditions de l'Université de Bruxelles-Éditions Ellipses, Brussels, 1989.

[205] Ph. Vincke. Outranking approach. In T. Gal, T. Stewart, and T. Hanne, editors, *Multicriteria Decision Making, Advances in MCDM Models, Algorithms, Theory and Applications*, pages 11.1–11.29. Kluwer Academic Publishers, Dordrcht, 1999.

[206] K. Vind. Independent preferences. *Journal of Mathematical Economics*, 20:119–135, 1991.

[207] J. von Neumann and O. Morgenstern. *Theory of Games and Economic Behavior*. John Wiley & Sons, New York, 2nd edition, 1947.

[208] R. von Nitzsch and M. Weber. The effect of attribute ranges on weights in multiattribute utility measurements. *Management Science*, 39(8):937–943, August 1993.

[209] D. von Winterfeldt and W. Edwards. *Decision Analysis and Behavioral Research*. Cambridge University Press, Cambridge, 1986.

[210] P. P. Wakker. Cardinal coordinate independence for expected utility. *Journal of Mathematical Psychology*, 28(1):110–117, 1984.

[211] P. P. Wakker. *Additive Representations of Preferences: A New Foundation of Decision Analysis*. Kluwer Academic Publishers, Dordrecht, 1989.

[212] P. P. Wakker. Additive representation for equally spaced structures. *Journal of Mathematical Psychology*, 35:260–266, 1991.

[213] P. P. Wakker. Additive representations of preferences, A new foundation of decision analysis; The algebraic approach. In J.-P. Doignon and J.-C. Falmagne, editors, *Mathematical Psychology: Current Developments*, pages 71–87. Springer Verlag, Berlin, 1991.

[214] P. P. Wakker. Additive representations on rank-ordered sets. I. The algebraic approach. *Journal of Mathematical Psychology*, 35(4):501–531, 1991.

[215] P. P. Wakker. Additive representations on rank-ordered sets. II. The topological approach. *Journal of Mathematical Economics*, 22(1):1–26, 1993.

[216] M. Weber, F. Eisenfuhr, and D. von Winterfeld. The effects of splitting attributes on weights in multiattribute utility measurement. *Management Science*, 34(4):431–445, 1988.

[217] J. A. Weymark. Generalized Gini inequality indices. *Mathematical Social Sciences*, 1:409–430, 1981.

[218] U. Wille. Linear measurement models: Axiomatizations and axiomatizability. *Journal of Mathematical Psychology*, 44:617–650, 2000.

[219] C. Zopounidis and M. Doumpos. Stock evaluation using a preference disaggregation methodology. *Decision Science*, 30:313–336, 1999.

[220] C. Zopounidis and M. Doumpos. Building additive utilities for multi-group hierarchical discrimination: The MHDIS method. *Optimization Methods & Software*, 14(3):219–240, 2000.

[221] C. Zopounidis and M. Doumpos. PREFDIS: A multicriteria decision support system for sorting decision problems. *Computers & Operations Research*, 27(7–8):779–797, June 2000.

[222] C. Zopounidis and M. Doumpos. Multicriteria preference disaggregation for classification problems with an application to global investing risk. *Decision Sciences*, 32(2):333–385, 2001.

[223] C. Zopounidis and M. Doumpos. A preference disaggregation decision support system for financial classification problems. *European Journal of Operational Research*, 130(2):402–413, April 2001.

III

OUTRANKING METHODS

Chapter 4

ELECTRE METHODS

José Figueira
Faculdade de Economia and INESC-Coimbra
Universidade de Coimbra
Av. Dias da Silva, 165, 3004-512 Coimbra
Portugal
figueira@fe.uc.pt

Vincent Mousseau, Bernard Roy
LAMSADE
Université Paris-Dauphine
Place du Maréchal De Lattre de Tassigny, 75775 Paris Cedex 16
France
{mousseau,roy}@lamsade.dauphine.fr

Abstract Over the last three decades a large body of research in the field of ELECTRE family methods appeared. This research has been conducted by several researchers mainly in Europe. The purpose of this chapter is to present a survey of the ELECTRE methods since their first appearance in mid-sixties, when ELECTRE I was proposed by Bernard Roy and his colleagues at SEMA consultancy company. The chapter is organized in five sections. The first section presents a brief history of ELECTRE methods. The second section is devoted to the main features of ELECTRE methods. The third section describes the different ELECTRE methods existing in the literature according to the three main problematics: choosing, ranking and sorting. The fourth section presents the recent developments and future issues on ELECTRE methods. Finally, the fifth section is devoted to the software and applications. An extensive and up-to-date bibliography is also provided in the end of this chapter.

Keywords: Multiple criteria decision aiding, Outranking approaches, ELECTRE methods.

1. Introduction: A Brief History

How far back in history should we go to discover the origins of ELECTRE methods? Some years ago B. Roy and D. Vanderpooten [119] published an article ("The European School of MCDA: Emergence, Basic Features and Current Works", *Journal of Multi-Criteria Decision Analysis*) on this very topic. This introduction is largely based on their paper, but additional material has been included to define the origins more precisely and to look more deeply into the history of ELECTRE methods. We have also benefited from an old, but nonetheless excellent, bibliography containing a lot of references collated by Y. Siskos, G. Wascher and H. Winkels [127]. The latter only covers the period 1966-1982, but contains many valuable references.

The origins of ELECTRE methods go back to 1965 at the European consultancy company SEMA, which is still active today. At that time, a research team from SEMA worked on a concrete, multiple criteria, real-world problem regarding decisions dealing with the development of new activities in firms. For "solving" this problem a general multiple criteria method, MARSAN (*Méthode d'Analyse, de Recherche, et de Sélection d'Activités Nouvelles*) was built. The analysts used a weighted-sum based technique included in the MARSAN method for the selection of the new activities [57]. When using the method the engineers from SEMA noticed serious drawbacks in the application of such a technique. B. Roy was thus consulted and soon tried to find a new method to overcome the limitations of MARSAN. The ELECTRE method for choosing the best action(s) from a given set of actions was thus devised in 1965, and was later referred to as ELECTRE I (electre one). In that same year (July, 1965) the new multiple criteria outranking method was presented for the first time at a conference (*les journées d'études sur les méthodes de calcul dans les sciences de l'homme*), in Rome (Italy). Nevertheless, the original ideas of ELECTRE methods were first merely published as a research report in 1966, the notorious *Note de Travail 49 de la SEMA* [10]. Shortly after its appearance, ELECTRE I was found to be successful when applied to a vast range of fields [18], but the method did not become widely known until 1968 when it was published in *RIRO, la Revue d'Informatique et de Recherche Opérationnelle* [89]. This article presents a comprehensive description of ELECTRE and the foundations of the outranking approach; the reader may also consult the graph theory book by B. Roy [90]. The method has since evolved and given rise to an "unofficial" version, ELECTRE Iv (electre one vee). This version took into account the notion of a veto threshold. A further version known as ELECTRE IS (electre one esse) appeared subsequently (see [117]) and was used for modelling situations in which the data was imperfect (see below). This is the current version of ELECTRE methods for choice problematic.

The acronym ELECTRE stands for [10, 95]: *ELimination Et Choix Traduisant la REalité* (ELimination and Choice Expressing the REality), and was cited for commercial reasons. At the time it seemed adequate and served well to promote the new tool. Nevertheless, the developments in ELECTRE methods over the last three decades, the way in which we consider the tool today and the methodological foundations of multiple criteria decision aiding have made the meaning of the acronym unsatisfactory.

An atypical ELECTRE method was also created to deal with the problem of highway layout in the *Ile de France* region; it was called the meaningful compensation method [11, 12, 25, 91, 109]. This approach was based on substitution rates. These rates were ill-defined (stakeholders views about their values strongly differed), it was only possible to fix a minimum and maximum value for each one. On such a basis a set of embedded fuzzy relations has been defined.

In the late sixties, a different real-world decision making situation arose in media planning, concerning the definition of an advertising plan. For such a purpose the question was: how to establish an adequate system of ranking for periodicals (magazines, newspapers, ...)? This led to the birth of ELECTRE II (electre two): a method for dealing with the problem of ranking actions from the best option to the worst [1, 43, 106, 107]. However, in a world where perfect knowledge is rare, imperfect knowledge only could be taken into account in ELECTRE methods through the use of probabilistic distributions and expected utility criterion. Clearly more work needed to be done. Research in this area was still in its initial stages. Another way to cope with uncertain, imprecision and ill-determination has been introduced, the threshold approach [19, 49, 50, 114]. For more details and a comprehensive treatment of this issue see [14, 96, 97]. Just a few years later a new method for ranking actions was devised: ELECTRE III (electre three), [93, 116]. The main new ideas introduced by this method were the use of pseudo-criteria (see [92]) and fuzzy binary outranking relations. Another ELECTRE method, known as ELECTRE IV (electre four), arose from a new real-world problem related to the Paris subway network [38, 45, 110, 111, 113]. It now became possible to rank actions without using the relative criteria importance coefficients; this is the only ELECTRE method which does not make use of such coefficients. In addition, the new method was equipped with an embedded outranking relations framework.

Methods created up to this point were particularly designed to help decision making in choosing and ranking actions. However, in the late seventies a new technique of sorting actions into predefined and ordered categories was proposed i.e. the trichotomy procedure [67, 68, 94]. This is a decision tree based approach. Several years later, in order to help decision making in a large banking company which faced to the problem of accepting or refusing credits requested by firms, a specific method, ELECTRE A, was devised and applied in

10 sectors of activity. This should have remained confidential. The most recent sorting method, ELECTRE TRI (electre tree), was greatly inspired by these earlier works. It removed everything they had of specific given their context of application. Indeed, this new method is, at the same time, both simpler and more general [141, 142].

ELECTRE methods are still evolving. Section 4 presents recent developments on the topic and avenues for future research.

2. Main Features of ELECTRE Methods

This section presents a set of key issues concerning ELECTRE methods: the context in which they are relevant, modelling with an outranking relation, their structure, the role of criteria, and how to account for imperfect knowledge.

2.1 In What Context Are ELECTRE Methods Relevant?

ELECTRE methods are relevant when facing decision situations with the following characteristics (see, [99, 109, 122]).

1. The decision-maker (DM) wants to include in the model at least three criteria. However, aggregation procedures are more adapted in situations when decision models include more than five criteria (up to twelve or thirteen).

And, at least one of the following situations must be verified.

2. Actions are evaluated (for at least one criterion) on an ordinal scale (see [84]) or on a weakly interval scale (see [63]). These scales are not suitable for the comparison of differences. Hence, it is difficult and/or artificial to define a coding that makes sense in terms of preference differences of the ratios $\frac{g_j(a)-g_j(b)}{g_j(c)-g_j(d)}$, where $g_j(x)$ is the evaluation of action x on criterion g_j.

3. A strong heterogeneity related with the nature of evaluations exists among criteria (e.g., duration, noise, distance, security, cultural sites, monuments, ...). This makes it difficult to aggregate all the criteria in a unique and common scale.

4. Compensation of the loss on a given criterion by a gain on another one may not be acceptable for the DM. Therefore, such situations require the use of noncompensatory aggregation procedures (see Chapter 1).

5. For at least one criterion the following holds true: small differences of evaluations are not significant in terms of preferences, while the accumulation of several small differences may become significant. This requires

the introduction of discrimination thresholds (indifference and preference) which leads to a preference structure with a comprehensive intransitive indifference binary relation (see Chapter 3).

2.2 Modelling Preferences Using an Outranking Relation

Preferences in ELECTRE methods are modelled by using binary outranking relations, S, whose meaning is "*at least as good as*". Considering two actions a and b, four situations may occur:

- aSb and not bSa, i.e., aPb (a is *strictly preferred* to b).
- bSa and not aSb, i.e., bPa (b is *strictly preferred* to a).
- aSb and bSa, i.e., aIb (a is *indifferent* to b).
- Not aSb and not bSa, i.e., aRb (a is *incomparable* to b).

ELECTRE methods build one or several (crisp, fuzzy or embedded) outranking relations.

Note that using outranking relations to model preferences introduces a new preference relation, R (incomparability). This relation is useful to account for situations in which the DM and/or the analyst are not able to compare two actions.

The construction of an outranking relation is based on two major concepts:

1 *Concordance*. For an outranking aSb to be validated, a *sufficient* majority of criteria should be in favor of this assertion.

2 *Non-discordance*. When the concordance condition holds, none of the criteria in the minority should oppose too strongly to the assertion aSb.

These two conditions must be fulfilled for validating the assertion aSb.

Given a binary relation on set A it is extremely helpful to build a graph $G = (V, U)$, where V is the set of vertices and U the set of arcs. For each action $a \in A$ we associate a vertex $i \in V$ and for each pair of actions $(a, b) \in A$ the arc (i, l) exists either if aPb or aIb. An action a outranks b if and only if the arc (i, l) exists. If there is no arc between vertices i and l, it means that a and b are incomparable; if two reversal arcs exist, there is an indifference between both a and b.

An outranking relation is not necessarily transitive. Preference intransitivities come from two different situations: Condorcet effect (see Chapter 2), and incomparabilities between actions. This requires an exploitation procedure to derive from such a relation results that fit the problematic (see Chapter 1).

2.3 Structure of ELECTRE Methods

ELECTRE methods comprise two main procedures: construction of one or several outranking relation(s) followed by an exploitation procedure.

The construction of one or several outranking relation(s) aims at comparing in a comprehensive way each pair of actions. The exploitation procedure is used to elaborate recommendations from the results obtained in the first phase. The nature of the recommendations depends on the problematic (choosing, ranking or sorting). Hence, each method is characterized by its construction and its exploitation procedures.

For more details the reader may consult the following references: [70, 98, 99, 109, 135, 138].

2.4 About the Relative Importance of Criteria

The relative role attached to criteria in ELECTRE methods is defined by two distinct sets of parameters: the importance coefficients and the veto thresholds.

The importance coefficients in ELECTRE methods refer to intrinsic "weights". For a given criterion the weight, w_j, reflects its voting power when it contributes to the majority which is in favor of an outranking. The weights do not depend neither on the ranges nor the encoding of the scales. Let us point out that these parameters can not be interpreted as substitution rates as in compensatory aggregation procedures AHP [120], MACBETH [7] and MAUT [55].

Veto thresholds express the power attributed to a given criterion to be against the assertion "a outranks b", when the difference of the evaluation between $g(b)$ and $g(a)$ is greater than this threshold. These thresholds can be constant along a scale or it can also vary.

A large quantity of works have been published on the topic of relative importance of criteria. The following list is not exhaustive: [35, 64, 69, 86, 87, 115, 116, 125, 136].

2.5 Discrimination Thresholds

To take into account the imperfect character of the evaluation of actions (see Chapter 1), ELECTRE methods make use of discrimination (indifference and preference) thresholds. This leads to a pseudo-criterion model on each criterion (see Chapter 2).

Discrimination thresholds account for the imperfect nature of the evaluations, and are used for modelling situations in which the difference between evaluations associated with two different actions on a given criterion may either:

- justify the preference in favor of one of the two actions (*preference threshold*, p_j);

- be compatible with indifference between the two actions (*indifference thresholds*, q_j).

- be interpreted as an hesitation between opting for a preference or an indifference between the two actions.

These thresholds can be constant or vary along the scale. When they are variable we must distinguish between *direct* (the evaluation of the best action is taken into account) and *inverse* (when they are computed by using the worst evaluation).

How to assign values to such thresholds? There are several techniques which can be used, some of them come directly from the definition of threshold and other ask for the concept of *dispersion threshold* (see Section 4.2).

A dispersion threshold allow us to take into account the concept of probable value and the notion of optimistic and pessimistic values. It translates the plausible difference, due to over or under-estimations, which affect the evaluation of a *consequence* or of a performance level.

It should be noticed that there are no true values for thresholds. Therefore, the values chosen to assign to the thresholds are the most convenient (the best adapted) for expressing the imperfect character of the knowledge.

For more details about thresholds see, [2, 17, 95, 100, 102, 103, 104, 109]

3. A Short Description of ELECTRE Methods

A comprehensive treatment of ELECTRE methods may be found in the books by B. Roy and D. Bouyssou [109] and Ph. Vincke [139]. Much of the theory developed on this field is presented in these books. This theory, however, was foreshadowed in earlier papers namely by B. Roy and his colleagues at SEMA and later at LAMSADE (some of these papers were cited in the introduction). The books [64, 95, 100, 122, 123] are also good references in the area. ELECTRE software manuals also contain much material both on theoretical and pedagogical issues [2, 43, 75, 117, 134, 142]. Finally, several other works deserve to be mentioned because they include information concerning ELECTRE methods: [5, 15, 16, 20, 37, 52, 79, 87, 125].

In what follows we will only summarize the elementary concepts underlying ELECTRE methods; details will be omitted. More sophisticated presentations can, however, be found in the references cited above.

Description of methods is presented in problematic and chronological order.

3.1 Choice Problematic

Let us remind the purpose of choice problematic before presenting methods. The objective of this problematic consists of aiding DMs in selecting a subset of actions, as small as possible, in such a way that a single action may finally be chosen.

The order in which methods will be presented permit us to understand the historical introduction of the two fundamental concepts in multiple criteria decision aiding, *veto thresholds* and *pseudo-criteria*.

3.1.1 ELECTRE I. The purpose underlying the description of this method is rather theoretical and pedagogical. The method does not have a significant practical interest, given the very nature of real-world applications, having usually a vast spectrum of quantitative and qualitative elementary consequences, leading to the construction of a contradictory and very heterogeneous set of criteria with both numerical and ordinal scales associated with them. In addition, a certain degree of imprecision, uncertainty or ill-determination is always attached to the knowledge collected from real-world problems.

The method is very simple and it should be applied only when all the criteria have been coded in numerical scales with identical ranges. In such a situation we can assert that an action "a outranks b" (that is, "*a is at least as good as* b") denoted by aSb, only when two conditions hold.

On the one hand, the *strength of the concordant coalition* must be powerful enough to support the above assertion. By strength of the concordant coalition, we mean the sum of the weights associated to the criteria forming that coalition. It can be defined by the following *concordance index* (assuming, for the sake of formulae simplicity, that $\sum_{j \in \mathcal{J}} w_j = 1$, where $\mathcal{J}$ is the set of the indices of the criteria):

$$c(aSb) = \sum_{\{j \,:\, g_j(a) \geq g_j(b)\}} w_j$$

(where $\{j \,:\, g_j(a) \geq g_j(b)\}$ is the set of indices for all the criteria belonging to the concordant coalition with the outranking relation aSb.)

In other words, the value of the concordance index must be greater than or equal to a given *concordance level*, s, whose value generally falls within the range $[0.5, 1 - \min_{j \in \mathcal{J}} w_j]$, i.e., $c(aSb) \geq s$.

On the other hand, no *discordance* against the assertion "a is at least as good as b" may occur. The discordance is measured by a *discordance level* defined as follows:

$$d(aSb) = \max_{\{j \,:\, g_j(a) < g_j(b)\}} \Big\{ g_j(b) - g_j(a) \Big\}$$

This level measures in some way the power of the discordant coalition, meaning that if its value surpasses a given level, v, the assertion is no longer valid. Discordant coalition exerts no power whenever $d(aSb) \leq v$.

Both concordance and discordance indices have to be computed for every pair of actions (a, b) in the set A, where $a \neq b$.

It is easy to see that such a computing procedure leads to a binary relation in comprehensive terms (taking into account the whole set of criteria) on the set A. Hence for each pair of actions (a, b), only one of the following situations may occur:

- aSb and not bSa, i.e., aPb (a is strictly preferred to b).
- bSa and not aSb, i.e., bPa (b is strictly preferred to a).
- aSb and bSa, i.e., aIb (a is indifferent to b).
- Not aSb and not bSa, i.e., aRb (a is incomparable to b).

This preference-indifference framework with the possibility to resort to incomparability, says nothing about how to select the best compromise action, or a subset of actions the DM will focus his attention on. In the construction procedure of ELECTRE I method only one outranking relation S is matter of fact.

The second procedure consists of exploiting this outranking relation in order to identify a small as possible subset of actions, from which the best compromise action could be selected. Such a subset, $\hat{A}$, may be determined with the help of the *graph kernel* concept, K_G. The justification of the use of this concept can be found in [109]. When the graph contains no direct cycles, there exists always a unique kernel; otherwise, the graph contains no kernels or several. But, let us point out that a graph G may contain direct cycles. If that is the case, a preprocessing step must take place where maximal direct cycles are reduced to singleton elements, forming thus a partition on A. Let $\bar{A}$ denote that partition. Each class on $\bar{A} = \{\bar{A}_1, \bar{A}_2, \ldots\}$ is now composed of a set of (considered) equivalent actions. It should be noticed that a new preference relation, $\succ$, is defined on $\bar{A}$:

$$\bar{A}_p \succ \bar{A}_q \;\Leftrightarrow\; \exists a \in \bar{A}_p \text{ and } \exists b \in \bar{A}_q \text{ such that } aSb \;\text{ for }\; \bar{A}_p \neq \bar{A}_q$$

In ELECTRE I all the actions which form a cycle are considered indifferent, which may be, criticized. ELECTRE IS was designed to mitigate this inconvenient (see Section 3.1.3).

3.1.2 ELECTRE Iv. The name ELECTRE Iv was an unofficial name created for designating ELECTRE I with veto threshold [64]. This method is equipped with a different but extremely useful tool. The new tool made possible for analysts and DMs to overcome the difficulties related to the heterogeneity of scales. Whichever the scales type, this method is always able to select the best compromise action or a subset of actions to be analyzed by DMs.

The new tool introduced was the *veto threshold*, v_j, that can be attributed to certain criteria g_j belonging to the family of criteria F. The concept of veto threshold is related in some way, to the definition of an upper bound beyond which the discordance about the assertion "a outranks b" can not surpass and allow an outranking. In practice, the idea of threshold is, however, quite different from the idea of the disconcordance level like in ELECTRE I. Indeed, while discordance level is related to the scale of criterion g_j in absolute terms for an action a from A, threshold veto is related to the preference differences between $g_j(a)$ and $g_j(b)$.

In terms of structure and formulae, little changes occur when moving from ELECTRE I to ELECTRE Iv. The only difference being the discordance condition, now called *no veto condition*, which may be stated as follows:

$$g_j(a) + v_j(g_j(a)) \geq g_j(b), \quad \forall \ j \in \mathcal{J}$$

To validate the assertion "a outranks b" it is necessary that, among the minority of criteria that are opposed to this assertion, none of them puts its veto.

ELECTRE Iv uses the same exploitation procedure as ELECTRE I.

But, this method is by no means complete; the problem of imperfect knowledge remains.

3.1.3 ELECTRE IS. How general an ELECTRE method can be when applied to choice decision-making problems? Is it possible to take into account simultaneously the heterogeneity of criteria scales, and imperfect knowledge about real-world decision-making situations? Previous theoretical research done on thresholds and semi-orders may, however, illuminate the issue of inaccurate data and permit to build a more general procedure, the so-called ELECTRE IS method.

The main novelty of ELECTRE IS is the use of pseudo-criteria instead of true-criteria. This method is an extension of the previous one aiming at taking into account a double objective: primarily the use of possible no nil indifference and preference thresholds for certain criteria belonging to F and, correlatively, a backing up (reinforcement) of the veto effect when the importance of the concordant coalition decreases. Both concordance and no veto conditions change. Let us present separately the formulae for each one of theses conditions.

- *Concordance condition*

  Let us start by building the following two indices sets:

  1 concerning the coalition of criteria in which aSb

  $$\mathcal{J}^S = \Big\{ j \in \mathcal{J} : \ g_j(a) + q_j\big(g_j(a)\big) \geq g_j(b) \Big\}$$

  2 concerning the coalition of criteria in which bQa

  $$\mathcal{J}^Q = \Big\{ j \in \mathcal{J} : \ g_j(a) + q_j\big(g_j(a)\big) < g_j(a) \leq g_j(b) + p_j\big(g_j(b)\big) \Big\}$$

The concordance condition will be:

$$c(aSb) = \sum_{j \in \mathcal{J}^S} w_j + \sum_{j \in \mathcal{J}^Q} \varphi_j w_j \geq s$$

where,

$$\varphi_j = \frac{g_j(a) + p_j\big(g_j(a)\big) - g_j(b)}{p_j\big(g_j(a)\big) - q_j\big(g_j(a)\big)}$$

the coefficient φ_j decreases linearly from 1 to 0, when g_j describes the range $[g_j(a) + q_j\big(g_j(a)\big),\ g_j(a) + p_j\big(g_j(a)\big)]$.

- *no veto condition*
The no veto condition can be stated as follows:

$$g_j(a) + v_j\big(g_j(a)\big) \geq g_j(b) + q_j\big(g_j(b)\big)\eta_j$$

where,

$$\eta_j = \frac{1 - c(aSb) - w_j}{1 - s - w_j}$$

In the exploitation procedure, actions belonging to a cycle are no longer considered as indifferent as in the previous versions of ELECTRE for choice problems. Now, we take into account the concept of degree of robustness of "a outranks b". It is a reinforcement of veto effect and allow us to build true classes of *ex æquo* (ties) and thus define an acycle graph over these classes. In such conditions there is always a single kernel.

3.2 Ranking Problematic

In ranking problematic we are concerned with the ranking of all the actions belonging to a given set of actions from the best to the worst, possibly with *ex æquo*. There are three different ELECTRE methods to deal with this problematic.

3.2.1 ELECTRE II. From an historical and pedagogical point of view it is interesting to present ELECTRE II. This method was the first of ELECTRE methods especially designed to deal with ranking problems.

Without going into further detail, it is important to point out that ELECTRE II was also the first method, to use a technique based on the construction of an embedded outranking relations sequence.

The construction procedure is very closed to ELECTRE Iv, in the sense that it is also a true-criteria based procedure. Hence, it is not surprising that the no veto condition remains the same. However, concordance condition is modified in order to take into account the notion of embedded outranking relations. There are two embedded relations: a *strong outranking* relation followed by a *weak outranking* relation. Both the strong and weak relations are built thanks to the definition of two concordance levels, $s^1 > s^2$, where $s^1, s^2 \in [0.5,\ 1 - min_{j \in \mathcal{J}} w_j]$. Now, the concordance condition with the assertion "a outranks b" can be defined as follows:

$$c(aSb) \geq s^r \text{ and } c(aSb) \geq c(bSa), \quad \text{for } r = 1, 2$$

The exploiting procedure is a four-step algorithm:

1 *Partioning the set* A. First, let us consider the relation S^1 over A. In a similar way like in ELECTRE I, this relation may define on A one or several cycles. If all the actions belonging to each maximal cycle are grouped together into a single class, a partition on A will be obtained. Let $\bar{A}$ denote this partition. When each class of $\bar{A}$ is not a singleton, the actions belonging to that class will be considered as *ex æquo*. For the purpose of comparison between elements of $\bar{A}$ a preference relation $\succ^1$ will be used. This relation has the same meaning as the relation $\succ$ for ELECTRE I.

2 *Building a complete pre-order* Z_1 *on* $\bar{A}$. After obtaining $\bar{A}$, the procedure identifies a subset B^1 of classes of $\bar{A}$ following the rule "no other is preferred to them" according to the relation $\succ^1$. After removing B^1 from $\bar{A}$ and applying the same rule to $\bar{A} \backslash B^1$, a subset B^2 will be found. The procedure iterates in the same way till define the final partition on $\bar{A}$, $\{B^1, B^2, \ldots\}$.

   Now, on the basis of S^1, we may define a rough version of the complete pre-order Z_1, while placing in the head of this pre-order and in an *ex æquo* position all classes of B^1, then those of B^2 and so forth. In order to define Z_1 in a more accurate way, we examine if it is possible to refine this pre-order on the basis of the relation S^2. This refinement consists of using the information that brings this less believable outranking to decide between the various classes of a subset B^p when it contains several classes. This refinement of the rough version is obtained while using S^2 to define over B^p a complete pre-order that takes place between B^{p-1} and B^{p+1}.

3 *Determining a complete pre-order* Z_2 *on* $\bar{A}$. The procedure to obtain this pre-order is quite similar to the above one; only two modifications are needed:

   - apply the rule "they are not preferred to any other" instead of "no other is preferred to them"; let $\{B^{1'}, B^{2'}, \ldots\}$ denote the partition thus obtained;

- define the rough version of the complete pre-order Z_2 by putting it in the queue of this pre-order, and in an *ex æquo* position all classes of $B^{1'}$, then those of $B^{2'}$ and so forth.

4 *Defining the partial pre-order* Z. The partial pre-order Z is an intersection of Z_1 and Z_2, $Z = Z_1 \cap Z_2$, and it is defined in the following way:

$$aZb \Leftrightarrow aZ_1b \text{ and } aZ_2b.$$

3.2.2 ELECTRE III. ELECTRE III was designed to improve ELECTRE II and thus deal with inaccurate, imprecise, uncertain or ill-determination of data. This purpose was actually achieved, and ELECTRE III was applied with success during the last two decades on a broad range of real-life applications.

In the current description of ELECTRE III we will omit several formulae details. The novelty of this method is the introduction of pseudo-criteria instead of true-criteria.

In ELECTRE III the outranking relation can be interpreted as a fuzzy relation. The construction of this relation requires the definition of a *credibility index*, which characterizes the credibility of the assertion "a outranks b", aSb; let $\rho(aSb)$ denote this index. It is defined by using both the concordance index (as determined in ELECTRE IS), $c(aSb)$, and a discordance index for each criterion g_j in F, that is, $d_j(aSb)$.

The discordance of a criterion g_j aims at taking into account the fact that this criterion is more or less discordant with the assertion aSb. The discordance index reaches its maximal value when criterion g_j puts its veto to the outranking relation; it is minimal when the criterion g_j is not discordant with that relation. To define the value of the discordance index on the intermediate zone, we simply admitted that this value grows in proportion to the difference $g_j(b) - g_j(a)$. This index can now be presented as follows:

$$d_j(aSb) = \begin{cases} 1 \text{ if } & g_j(b) > g_j(a) + v_j(g_j(a)) \\ 0 \text{ if } & g_j(b) \leq g_j(a) + p_j(g_j(a)) \\ \frac{g_j(b) - g_j(a) - p_j(g_j(a))}{v_j(g_j(a)) - p_j(g_j(a))}, & \text{otherwise} \end{cases}$$

The credibility index is defined as follows,

$$\rho(aSb) = c(aSb) \prod_{\{j \in \mathcal{J} \,:\, d_j(aSb) > c(aSb)\}} \frac{1 - d_j(aSb)}{1 - c(aSb)}$$

Notice that, when $d_j(aSb) = 1$, it implies that $\rho(aSb) = 0$, since $c(aSb) < 1$.

The definition of $\rho(aSb)$ is thus based on the following main ideas:

a) When there is no discordant criterion, the credibility of the outranking relation is equal to the comprehensive concordance index.

b) When a discordant criterion activates its veto power, the assertion is not credible at all, thus the index is null.

c) For the remaining situations in which the comprehensive concordance index is strictly lower than the discordance index on the discordant criterion, the credibility index becomes lower than the comprehensive concordance index, because of the opposition effect on this criterion

The index $\rho(aSb)$ corresponds to the index $c(aSb)$ weakened by possible veto effects.

In [71] a modification of the valued outranking relation used in the ELECTRE III and ELECTRE TRI was proposed. The modification requires the implementation of the discordance concept. Such a modification is shown to preserve the original discordance concept; the new outranking relation makes it easier to solve inference programs.

The exploitation procedure starts by deriving from the fuzzy relation two complete pre-orders as in ELECTRE II. A final partial pre-order Z is then built as the intersection of the two complete pre-orders, Z_1 and Z_2, which are obtained according to two variants of the same principle, both acting in an antagonistic way on the floating actions. The partial pre-order Z_1 is defined as a partition on the set A into q ordered classes, $\bar{B}_1, \ldots, \bar{B}_h, \ldots, \bar{B}_q$, where $\bar{B}_1$ is the head-class in Z_1. Each class $\bar{B}_h$ is composed of *ex æquo* elements according to Z_1. The complete pre-order Z_2 is determined in a similar way, where A is partitioned into u ordered classes, $\underline{B}_1, \ldots, \underline{B}_h, \ldots, \underline{B}_u$, $\underline{B}_u$ being the head-class. Each one of these classes is obtained as a final distilled of a distillation procedure.

The procedure designed to compute Z_1 starts (first distillation) by defining an initial set $D_0 = A$; it leads to the first final distilled $\bar{B}_1$. After getting $\bar{B}_h$, in the distillation $h+1$, the procedure sets $D_0 = A \backslash (\bar{B}_1 \cup \ldots \cup \bar{B}_h)$. According to Z_1, the actions in class $\bar{B}_h$ are, preferable to those of class $\bar{B}_{h+1}$; for this reason, distillations that lead to these classes will be called as descending (top-down).

The procedure leading to Z_2 is quite identic, but now the actions in $\bar{B}_{h+1}$ are preferred to those in class $\bar{B}_h$; these distillations will be called ascending (bottom-up).

The partial pre-order Z will be computed as the intersection of Z_1 and Z_2.

A complete pre-order is finally suggested taking into account the partial pre-orders and some additional considerations. The way the incomparabilities which remain in the pre-order are treated is nevertheless subject to criticism.

3.2.3 ELECTRE IV. In Section 2.4 we pointed out the difficulty to define the relative importance coefficients of criteria. However, in several circumstances we are not able, we do not want, or we do not know how to assign a value to those coefficients. It does not mean that we would be satisfied with the pre-order obtained, when applying ELECTRE III with the same value for all the coefficients w_j. Another approach we could take would be determining a pre-order, which takes into account all the pre-orders obtained from the application of several combinations of the weights. Obviously, this situation will be unmanageable.

ELECTRE IV is also a procedure based on the construction of a set of embedded outranking relations. There are five different relations, $S^1, \ldots, S^5$. The S^{r+1} relation ($r = 1, 2, 3, 4$) accepts an outranking in a less credible circumstances than the relation S^r. It means (while remaining on a merely ordinal basis) the assignment of a value ρ_r for the credibility index $\rho(aSb)$ to the assertion aSb. The chosen values must be such that $\rho_r > \rho_{r+1}$. Furthermore, the movement from one credibility value ρ_r to another ρ_{r+1} must be perceived as a considerable loss.

The ELECTRE IV exploiting procedure is the same as in ELECTRE III.

3.3 Sorting Problematic

A set of categories must be *a priori* defined. The definition of a category is based on the fact that all potential actions which are assigned to it will be considered further in the same way. In sorting problematic, each action is considered independently from the others in order to determine the categories to which it seems justified to assign it, by means of comparisons to profiles (bounds, limits), norms or references. Results are expressed using the absolute notion of "assigned" or "not assigned" to a category, "similar" or "not similar" to a reference profile, "adequate" or "not adequate" to some norms. The sorting problematic refers thus to absolute judgements. It consists of assigning each action to one of the pre-defined categories which are defined by norms or typical elements of the categories. The assignment of an action a to a specific category does not influence the category, to which another action b should be assigned.

3.3.1 ELECTRE TRI. ELECTRE TRI is designed to assign a set of actions, objects or items to categories. In ELECTRE TRI categories are ordered; let us assume from the worst (C_1) to the best (C_k). Each category must be characterized by a lower and an upper profile. Let $C = \{C_1, \ldots, C_h, \ldots, C_k\}$ denote the set of categories. The assignment of a given action a to a certain category C_h results from the comparison of a to the profiles defining the lower and upper limits of the categories; b_h being the upper limit of category C_h and the lower limit of category C_{h+1}, for all $h = 1, \ldots, k$. For a given category limit, b_h, this comparison rely on the credibility of the assertions aSb_h and b_hSa. This credibility (index) is defined as in ELECTRE III. In what follows, we will assume, without any loss of generality, that preferences increase with the value on each criterion.

After determining the credibility index, we should introduce a λ-*cutting level* of the fuzzy relation in order to obtain a crisp outranking relation. This level can be defined as the credibility index smallest value compatible with the assertion aSb_h.

Let $\succ$ denote the preference, I denote the indifference relation and R denote the incomparability binary relations.

The action a and the profile b_h may be related to each other as follows:

$a)$ aIb_h iff aSb_h and b_hSa

$b)$ $a \succ b_h$ iff aSb_h and not b_hSa

$c)$ $b_h \succ a$ iff not aSb_h and b_hSa

$d)$ aRb_h iff not aSb_h and not b_hSa

The objective of the exploitation procedure is to exploit the above binary relations. The role of this exploitation is to propose an assignment. This assignment can be grounded on two well-known logics.

1 The conjunctive logic in which an action can be assigned to a category when its evaluation on each criterion is at least as good as the lower limit which has been defined on the criterion to be in this category. The action is hence assigned to the highest category fulfilling this condition.

2 The disjunctive logic in which an action can be assigned to a category, if it has, on at least one criterion, an evaluation at least as good as the lower limit which has been defined on the criterion to be in this category. The action is hence assigned to the highest category fulfilling this condition.

With this disjunctive rule, the assignment of an action is generally higher than with the conjunctive rule. This is why the conjunctive rule is usually interpreted as pessimistic while the disjunctive rule is interpreted as optimistic. This interpretation (optimistic-pessimistic) can be permuted according to the semantic attached to the outranking relation.

When no incomparability occurs in the comparison of an action a to the limits of categories, a is assigned to the same category by both the optimistic and the pessimistic procedures. When a is assigned to different categories by the optimistic and pessimistic rules, a is incomparable to all "intermediate" limits within the highest and lowest assignment categories.

ELECTRE TRI is a generalization of the two above mentioned rules. The generalization is the following,

- in the conjunctive rule: replace, in the condition "*on each criterion*" by "*on a sufficient majority of criteria and in the absence of veto*"
- in the disjunctive rule: replace, the condition "*on at least one criterion*" by "*on a sufficient minority of criteria and in the absence of veto*"

The two procedures can be stated as follows,

1 *Pessimistic rule.* An action a will be assigned to the highest category C_h such that aSb_{h-1}.

   $a)$ Compare a successively with b_r, $r = k-1, k-2, \ldots, 0$.

   $b)$ The limit b_h is the first encountered profile such that aSb_h. Assign a to category C_{h+1}.

2 *Optimistic rule.* An action a will be assigned to the lowest category C_h such that $b_h \succ a$.

   $a)$ Compare a successively with b_r, $r = 1, 2, \ldots, k-1$.

   $b)$ The limit b_h is the first encountered profile such that $b_h \succ a$. Assign a to category C_h.

4. Recent Developments and Future Issues

Although, several decades past since the birth of the first ELECTRE method, research on ELECTRE family method stills active today. Some of the recent developments are shortly described in this Section.

4.1 Robustness Concerns

When dealing with real-world decision problems, DMs and analysts are often facing with several sources of imperfect knowledge regarding the available data. This leads to the assignment of arbitrary values to certain "variables". In addition, modelling activity frequently requires to choose between some technical options, introducing thus an additional source of arbitrariness to the problem. For these reasons, analysts hesitate when assigning values to the *preference parameters* (weights, thresholds, categories lower and upper limits, ...), and the *technical parameters* (discordance and concordance indices, $\lambda-$cutting level, ...) of ELECTRE methods.

In practice, it is frequent to define a *reference system* built from the assignment of *central values* to these two types of parameters. Then, an exploitation procedure should be applied in order to obtain outputs which are used to elaborate recommendations. But, what about the meaningfulness of such recommendations? They strongly depend on the set of central values attributed to the parameters. Should the analyst analyze the influence of a variation of each parameter, considered separately, on the results? And, then enumerate those parameters which provoke a strong impact on the results when their values vary from the central positions. This is a frequent way to proceed in classical operations research methods and it is called *sensitivity analysis* [32, 53, 79, 82]. But, this kind of analyzes has rather a theoretical interest than a practical one. Analysts are most often interested in building recommendations which remain acceptable for a large range of the parameters values. Such recommendations should be elaborated from what we call the *robust conclusions* (Chapter 1, [101, 105, 109]).

DEFINITION 35 *A conclusion, C^{r}, is said to be robust with respect to a domain, Ω, of possible values for the preference and technical parameters, if there is no a particular set of parameters, $\bar{\omega} \in \Omega$, which clearly invalidates the conclusion C^{r}.*

A *robustness concern* consists of all the possible ways that contribute to build synthetic recommendations based on the robust conclusions.

Possible ways to deal with robustness concerns in ELECTRE methods are illustrated, for example, in [26, 27, 29, 109, 116], Chapters 8, 9 and 10.

4.2 Elicitation of Parameter Values

Implementing ELECTRE methods requires to determine values (or intervals of variation) for the preference parameters.

DEFINITION 36 *A preference elicitation process proceeds through an interaction between DMs and analysts in which DMs express information about their preferences within a specific aggregation procedure.*

It is possible to distinguish among direct and indirect elicitation techniques.

4.2.1 Direct Elicitation Techniques. In direct elicitation procedures DMs should provide information directly on the values of the preference parameters. A major drawback of such techniques is that it is difficult to understand the precise meaning of the assertions of the DMs. This is why ELECTRE methods are usually implemented by using indirect elicitation procedures.

4.2.2 Indirect Elicitation Techniques. Indirect elicitation techniques do not require from DMs to provide answers to questions related to the values of the preference parameters. On the contrary, these techniques proceeds indirectly by posing questions whose answers can be interpreted through the aggregation procedure. Such techniques make use of the disaggregation paradigm [51, 60]. For instance, DIVAPIME [70] and SRF [35] elicitation techniques make it possible to determine the vector of the relative importance coefficients from pairwise comparisons of fictitious actions or a relative importance ranking of criteria.

Recent developments concerning elicitation techniques have been proposed for the ELECTRE TRI method. Inference procedures have been developed to elicit the parameters values from assignment examples, i.e., an assignment that is imposed by DMs on specific actions. It is possible to infer all the preference parameters simultaneously [74]; we will refer to such a case by complete inference. The induced mathematical programming model to be solved is, however, non-linear. Thus, its resolution is computationally difficult for real-world problems. In such cases, it is possible to infer a subset of parameters only (see Figure 1.1):

- Concordant coalition parameters: weights and $\lambda-$cutting level [72];
- Discordance related parameters: veto thresholds [28];
- Category limits [76].

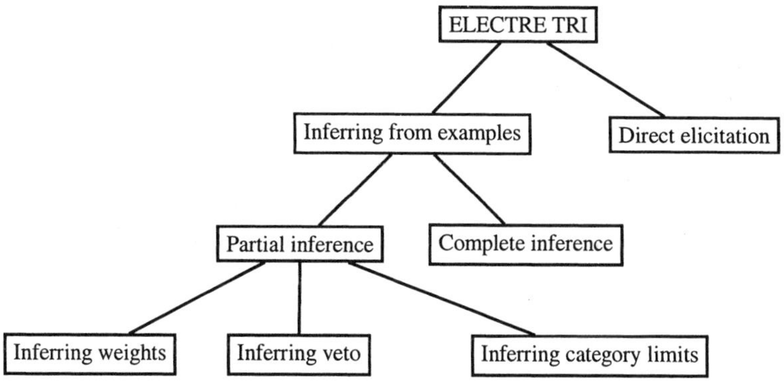

Figure 4.1. Inferring parameter values for ELECTRE TRI.

5. Software and Applications

The implementation of ELECTRE methods in real-world decision problems involving DMs requires software packages. Some of them are widely used in large firms and universities, in particular ELECTRE IS, ELECTRE III-IV, ELECTRE TRI and IRIS. Among the software available at LAMSADE are (`http://www.lamsade.dauphine.fr/english/ software.html`):

1 *ELECTRE IS* is a generalization of ELECTRE I. It is an implementation of ELECTRE IS described in Section 3.1. This software runs on a IBM-compatible computer on Windows 98 and higher.

2 *ELECTRE III-IV* is a software which implements ELECTRE III and ELECTRE IV methods described in Section 3.2. It runs on Windows 3.1, 95, 98, 2000, Millennium and XP.

3 *ELECTRE TRI* is a multiple criteria decision aiding tool designed to deal with sorting problems. This software implements ELECTRE TRI method described in Section 3.3. The ELECTRE TRI software versions 2.x were developed with the C++ programming language and runs on Microsoft Windows 3.1, 95, 98, Me, 2000, XP and NT. This software integrates, *ELECTRE TRI Assistant* which enables the user to define the weights indirectly, i.e., fixing the model parameters by giving some assignment examples (corresponding to desired assignments or past decisions). The weights are thus inferred through a certain form of regression. Hence, ELECTRE TRI Assistant reduces the cognitive effort required from the DM to elicit the preference parameters.

4 *IRIS*. Interactive Robustness analysis and parameters' Inference for multiple criteria Sorting problems. This DSS has been built to support the assignment of actions described by their evaluation on multiple criteria to a set of predefined ordered categories, using a variant of ELECTRE TRI. Rather than demanding precise values for the model's parameters, IRIS allows to enter constraints on these values, namely assignment examples that it tries to restore. When the constraints are compatible with multiple assignments for the actions, IRIS infers parameter values and allows to draw robust conclusions by indicating the range of assignments (for each action) that do not contradict any constraint. If it is not possible to fulfill all of the constraints, IRIS tells the user where is the source of inconsistency. It was developed with Delphi Borland and runs on Windows 98, Me, 2000, NT and XP.

5 *SFR* was designed to determine the relative importance coefficients for ELECTRE family methods. It is based on a very simple procedure (the pack of cards technique created by J. Simos) and try to assess these coefficients by questioning the DM in an indirect way. It was developed with the Delphi Borland 3.0 and runs on Windows 98, Me, 2000 and XP.

The software ELECTRE IS, III-IV, TRI and TRI Assistant were developed under a collaborative project between researchers from the Institute of Computing Science of the Technical University of Poznan (Poland) and LAMSADE, Université Paris-Dauphine (France), while IRIS and SRF result from a collaborative project between researchers from LAMSADE and the Faculty of Economics of the University of Coimbra / INESC-Coimbra (Portugal).

ELECTRE methods were successful applied in many areas.

1 *Agriculture and Forest Management*: [4, 31, 62, 118, 128, 130, 131]

2 *Energy*: [8, 9, 19, 39, 40, 54, 108, 126]

3 *Environment and Water Management*: [12, 40, 41, 44, 59, 78, 80, 85, 86, 118, 121, 124, 125, 131, 132, 77, 88, 58]

4 *Finance*: [3, 30, 46, 47, 48, 56, 61, 143, 144, 145, 146]

5 *Military*: [6, 36, 140]

6 *Project selection (call for tenders)*: [13, 21, 24, 65, 107, 137].

7 *Transportation*: [11, 12, 23, 38, 45, 73, 111, 112, 110, 114, 116]

8 *Varia*: [33, 34, 81, 83, 129, 107].

6. Conclusion

Since their first appearance, in 1965 (see [10]), ELECTRE methods, on one side, had a strong impact on the Operational Research community, mainly in Europe, and provoked the development of other outranking methods (see, for example, Chapters 5 and 6), as well as other complementary multiple criteria methodologies (see, for example, Chapters 8 and 9). Most importantly, the development of ELECTRE methods is strongly connected with the birth of the European Working Group of Multiple Criteria Decision Aiding (see `www.inescc.pt/~ewgmcda/`). On the other side, ELECTRE methods experienced a widespread and large use in real-world situations.

Despite their almost four decades of existence, research stills active in this field. We can also mention some of recent developments and avenues for future research: generalization of the concordance and non-discordance methods [133]; robustness analysis [26, 27, 29]; parameters elicitation techniques [74]; interaction between criteria [66, 42], multiple DMs and social interaction [22].

Acknowledgments

This work has benefited from the Luso-French grant n°500B4 (ICCTI/ Ambassade de France au Portugal). The first author was also supported by the grant SFRH/BDP/6800/2001 (Fundação para a Ciência e Tecnologia, Portugal) and gratefully acknowledges DIMACS Research Center at Rutgers University for the welcome during his sabbatical leave (August 2002 – May 2003). The authors are thankful to Dominique Champ-Brunet for the many valuable references she provided.

References

[1] R. Abgueguen. *La Sélection des Supports de Presse*. Robert Lafont, Paris, 1971.

[2] A. Aït Younes, R. Azibi, and B. Roy. ELECTRE IS : Manuel d'utilisation. Document du LAMSADE 118 and 118bis, Université Paris-Dauphine, 2000.

[3] A. Andenmatten. *Evaluation du Risque de Défaillance des Emetteurs d'Obligations, Une Approche par l'Aide Multicritère à la Décision.* Presses Polytechniques et Universitaires Romandes, Lausanne, Suisse, 1995.

[4] C. Arondel and P. Girardin. Sorting cropping systems on the basis of their impact on groundwater quality. *European Journal of Operational Research*, 127:467–482, 2000.

[5] C. Bana e Costa, editor. *Readings in Multiple Criteria Decision Aid.* Springer Verlag, Heidelberg, 1990.

[6] C. Bana e Costa and C. Dinis das Neves. Describing and formalizing the evaluation process of Portuguese navy officiers. In A. Lockett and G. Islei, editors, *Improving Decision Making in Organisations*, volume 335 of *Lecture Notes in Economics and Mathematical Systems*, pages 355–369. Springer Verlag, Berlin, 1989.

[7] C. Bana e Costa and J. Vansnick. MACBETH – An interactive path towards the construction of cardinal value functions. *International Transactions in Operational Research*, 1:489–500, 1994.

[8] O. Barda, J. Dupuis, and P. Lencioni. Multicriteria location of thermal power plants. *European Journal of Operational Research*, 45:332–346, 1990.

[9] M. Beccali, M. Cellura, and D. Ardente. Decision making in energy planning: The ELECTRE multicriteria analysis approach compared to a fuzzy-sets methodology. *Energy Conversion and Management*, 39 (16-18):1869–1881, 1998.

[10] R. Benayoun, B. Roy, and B. Sussman. ELECTRE : Une méthode pour guider le choix en présence de points de vue multiples. Note de travail 49, SEMA-METRA International, Direction Scientifique, 1966.

[11] P. Bertier, G. Gagey, J. de Montgolfier, and B. Roy. Choix de tracés autoroutiers en milieu suburbain : Faut-il vraiment endommager des sites urbains et/ou forestiers, lesquels? Communication aux Journées de Formation "Transports et Agglomérations", Nice, 1972.

[12] Y. Bétolaud and R. Février. Conservation des forêts suburbaines et passage des autoroutes. *Revue Forestière Française*, pages 197–200, 1973.

[13] P. Blondeau, M. Spérandio, and F. Allard. Multicriteria analysis of ventilation in summer period. *Building and Environment*, 37 (2):165–176, 2002.

[14] D. Bouyssou. Modelling inaccurate determination, uncertainty, imprecision using multiple criteria. In A. Lockett and G. Islei, editors, *Improving Decision Making in Organisations*, volume 335 of *Lecture Notes in Economics and Mathematical Systems*, pages 78–87. Springer Verlag, Berlin, 1989.

[15] D. Bouyssou, E. Jacquet-Lagrèze, P. Perny, R. Slowinski, D. Vanderpooten, and Ph. Vincke, editors. *Aiding Decisions with Multiple Criteria: Essays in Honour of Bernard Roy.* Kluwer Academic Publishers, Dordrecht, 2001.

[16] D. Bouyssou, T. Marchant, M. Pirlot, P. Perny, A. Tsoukiàs, and Ph. Vincke. *Evaluation and Decision Model. A Critical Perspective.* Kluwer Academic Publishers, Dordrecht, 2000.

[17] D. Bouyssou and B. Roy. La notion de seuils de discrimination en analyse multicritère. *INFOR*, 25(4):302–313, 1987.

[18] P. Buffet, J. Grémy, M. Marc, and B Sussmann. Peut-on choisir en tenant compte de critères multiples ? Une méthode ELECTRE et trois applications. *Revue METRA*, VI:283–316, 1967.

[19] A. Charpentier and E. Jacquet-Lagrèze. La promotion de l'électricité dans l'entreprise et l'utilisation des méthodes multicritères. In H. Thiriez and S. Zionts, editors, *Multiple Criteria Decision Making*, volume 130 of *Lecture Notes in Economics and Mathematical Systems*, pages 364–377. Springer Verlag, Berlin, 1976.

[20] A. Colorni, M. Paruccini, and B. Roy, editors. *A-MCD-A – Aide Multi-Critère à la Décision – Multiple Criteria Decision Aiding*. The European Commission Joint Research Center, Luxemburg, 2001.

[21] G. Colson. The OR's prize winner and the software ARGOS: How a multijudge and multicriteria ranking GDSS helps a jury to attribute a scientific award. *Computers & Operations Research*, 27 (7-8):741–755, 2000.

[22] S. Damart. *Aide à la décision et processus de décision collectifs: La contribution des outils d'aide à la décision aux démarches de concertation*. PhD thesis, Université Paris-Dauphine, Paris, 2003.

[23] S. Damart, V. Mousseau, and I. Sommerlatt. Du mode d'implication d'acteurs multiples dans le cadre de l'utilisation d'un modèle d'affectation multicritère : Analyse au regard d'une application à la tarification des transports publics. *INFOR*, 40(3):199–222, 2002.

[24] N. Danila. *Méthodologie d'aide à la décision dans le cas d'investissements fort dépendants*. PhD thesis, Université Paris-Dauphine, 1980.

[25] J. de Montgolfier and P. Bertier. *Approche Multicritère des Problèmes de Décision*. Editions Hommes et Techniques, Paris, 1978.

[26] L. Dias and J. Clímaco. On computing ELECTRE's credibility indices under partial information. *Journal of Multi-Criteria Decision Analysis*, 8(2):74–92, 1999.

[27] L. Dias and J. Clímaco. ELECTRE TRI for groups with imprecise information on parameter values. *Group Decision and Negotiation*, 9(5):355–377, 2000.

[28] L. Dias and V. Mousseau. Inferring ELECTRE's veto-related parameters from outranking examples. *European Journal of Operational Research*, to appear, 2004.

[29] L. Dias, V. Mousseau, J. Figueira, and J. Clímaco. An aggregation/disaggregation approach to obtain robust conclusions with ELECTRE TRI. *European Journal of Operational Research*, 138:332–348, 2002.

[30] A.I. Dimitras, C. Zopounidis, and C. Hurson. Assessing financial risks using a multicriteria sorting procedure: The case of country risk asessment. *Foundations of Computing and Decision Sciences*, 29:97–109, 2001.

[31] L. Duckstein and M. Gershon. Multicriterion analysis of a vegetation management problem using ELECTRE II. *Applied Mathematical Modelling*, 7 (4):254–261, 1983.

[32] J. Evans. Sensitivity analysis in decision theory. *Decision Sciences*, 15:239–247, 1984.

[33] D. Ferrand and J. Martel. Le choix multicritère des items d'une échelle de mesure. *Mathématiques et Sciences Humaines*, 89:35–59, 1985.

[34] J. Fichefet and J. Leclercq. A multicriteria microcomputer-assisted system for the identification of bacteria. *European Journal of Operational Research*, 24:12–22, 1986.

[35] J. Figueira and B. Roy. Determining the weights of criteria in the ELECTRE type methods with a revised Simos' procedure. *European Journal of Operational Research*, 139:317–326, 2002.

[36] V. Gabrel. *Méthodologie pour la planification de production de système d'observation de la terre par satellites : Aspects algorithmiques et multicritères*. PhD thesis, Université Paris-Dauphine, 1994.

[37] T. Gal, T. Stewart, and T. Hanne, editors. *Advances in MultCriteria Decision Making – MCDM Models, Algorithms, Theory, and Applications*. Kluwer Academic Publishers, Boston, 1999.

[38] L. Gargaillo. Réponse à l'article "Le plan d'extension du métro en banlieue parisienne, un cas type de l'analyse multicritère". *Les Cahiers Scientifiques de la Revue Transports*, 7:52–57, 1982.

[39] E. Georgopoulou, D. Lalas, and L. Papagiannakis. A multicriteria decision aid approach for energy planning problems: The case of renewable energy option. *European Journal of Operational Research*, 103 (1):38–54, 1997.

[40] E. Georgopoulou, Y. Sarafidis, S. Mirasgedis, S. Zaimi, and D. Lalas. A multiple criteria decision-aid approach in defining national priorities for greenhouse gases emissions reduction in the energy sector. *European Journal of Operational Research*, 146:199–215, 2003.

[41] M. Gershon, L. Duckstein, and R. McAniff. Multiobjective river basin planning with qualitative criteria. *Water Resources Research*, 18:193–202, 1982.

[42] S. Greco and J. Figueira. Dealing with interaction between bi-polar multiple criteria preferences in outranking methods. Research Report 11-2003, INESC-Coimbra, Portugal, 2003.

[43] J. Grolleau and J. Tergny. Manuel de référence du programme ELECTRE II. Document de travail 24, SEMA-METRA International, Direction Scientifique, 1971.

[44] J. Hokkanen and P. Salminen. Choosing a solid waste management system using multicriteria decision analysis. *European Journal of Operational Research*, 98 (1):19–36, 1997.

[45] J. Hugonnard and B. Roy. Le plan d'extension du métro en banlieue parisienne, un cas type d'application de l'analyse multicritère. *Les Cahiers Scientifiques de la Revue Transports*, 6:77–108, 1982.

[46] C. Hurson and C. Zopounidis. On the use of multicriteria decision aid methods to portfolio selection. *The Journal of Euro-Asian Management*, 1(2):69–94, 1995.

[47] C. Hurson and C Zopounidis. *Gestion de Portefeuille et Analyse Multicritère*. Economica, Paris, 1997.

[48] C. Hurson and C. Zopounidis. On the use of multicriteria decision aid methods to portfolio selection. In J. Clímaco, editor, *Multicriteria Analysis*, pages 496–507. Springer Verlag, Berlin, 1997.

[49] E. Jacquet-Lagrèze. Le problème de l'agrégation des préférences : Une classe de procédures à seuil. *Mathématiques et Sciences Humaines*, 43:29–37, 1973.

[50] E. Jacquet-Lagrèze. How we can use the notion of semi-orders to build outranking relations in multicriteria decision making. In E. Wendt and C. Vleck, editors, *Utility, Subjective Probability and Human Decision Making*, pages 87–112. Reidel, Dordrecht, 1975.

[51] E. Jacquet-Lagrèze and Y. Siskos. Assessing a set of additive utility functions for multicriteria decision making: The UTA method. *European Journal of Operational Research*, 10:151–164, 1982.

[52] E. Jacquet-Lagrèze and Y. Siskos, editors. *Méthode de Décision Multicritère*. Editions Hommes et Techniques, Paris, 1983.

[53] R. Kampke. Sensitivity analysis for assessing preferentially independent order relations. *Computers & Operations Research*, 23(12):1119–1130, 1996.

[54] A. Karagiannidis and N. Moussiopoulos. Application of ELECTRE III for the integrated management of municipal solid wastes in the greater Athens area. *European Journal of Operational Research*, 97 (3):439–449, 1997.

[55] R. Keeney and H. Raiffa. *Decisions with Multiple Objectives: Preferences and Value Tradeoffs*. John Wiley & Sons, New York, 1976.

[56] N. Khoury, J. Martel, and P. Yourougou. A multicriterion approach to country selection for global index founds. Document de travail FSA 91-13, Faculté des Sciences de l'Administration, Université Laval, Québec, Canada, 1991.

[57] R. Laffy. La méthode MARSAN pour la recherche de produits nouveaux. Presented at the ESOMAR congress, Copenhagen, September 1966.

[58] R. Lahdelma, P. Salminen, and J. Hokkanen. Locating a waste treatment facility by using stochastic multicriteria acceptability analysis with ordinal criteria. *European Journal of Operational Research*, 142:345–356, 2002.

[59] M. Mahmoud and L. Garcia. Comparison of different multicriteria evaluation methods for the Red Bluff diversion dam. *Environmental Modelling and Software*, 15 (5):471–478, 2000.

[60] J. March. Ambiguity and the engineering of choice. *Bell Journal of Economics*, 9:587–608, 1978.

[61] J. Martel, N. Khoury, and M. Bergeron. An application of a multicriteria approach to portfolio comparisons. *Journal of the Operational Research Society*, 39:617–628, 1988.

[62] J. Martel and A. Rousseau. Cadre de référence d'une démarche multicritère de gestion intégrée des ressources en milieu forestier. Document technique 93-11, Ministère de l'environment et de la faune, Québec, 1993.

[63] J. Martel and B. Roy. Analyse de la signifiance dc diverses procédures d'agrégation multicritère. Annales du LAMSADE 1, Université Paris-Dauphine, 2002.

[64] L. Maystre, J. Pictet, and J. Simos. *Les Méthodes Multicritères ELECTRE*. Presses Polytechniques et Universitaires Romandes, Lausanne, 1994.

[65] J. Michel. La sélection des projects du programme architecture nouvelle. In *Actes du congrès Association Française pour la Cybernétique Economique et Technique (AFCET): Aide à la Décision, Paris, May 16-17, 1974*, pages 111–121, 1974.

[66] L. Montano-Guzmán. *Fuzzy measures and integrals in the MCDA sorting problematic: Methodology and application to the diagnosis of firms*. PhD thesis, Université Libre de Bruxelles, Brussels, Belgium, 2002.

[67] J. Moscarola. *Aide à la décision en présence de critères multiples fondée sur une procédure trichotomique : Méthodologie et application*. Thèse de doctorat de 3e cycle, Université Paris V René Decartes, 1977.

[68] J. Moscarola and B. Roy. Procédure automatique d'examen de dossiers fondée sur une segmentation trichotomique en présence de critères multiples. *RAIRO Recherche Opérationnelle*, 11(2):145–173, 1977.

[69] V. Mousseau. *Problèmes liés à l'évaluation de l'importance relative des critères en aide multicritère à la décision : Réflexions théoriques, expérimentations et implémentations informatiques*. PhD thesis, Université Paris-Dauphine, 1993.

[70] V. Mousseau. Eliciting information concerning the relative importance of criteria. In P. Pardalos, Y. Siskos, and C. Zopounidis, editors, *Advances in Multicriteria Analysis*, Nonconvex Optimization and its Applications, pages 17–43. Kluwer Academic Publishers, Dordrecht, 1995.

[71] V. Mousseau and L. Dias. Valued outranking relations in ELECTRE providing manageable disaggregation procedures. *European Journal of Operational Research*, in press, 2004.

[72] V. Mousseau, J. Figueira, and J. Naux. Unsing assignment examples to infer weights for ELECTRE TRI method: Some experimental results. *European Journal of Operational Research*, 130(2):263–275, 2001.

[73] V. Mousseau, B. Roy, and I. Sommerlatt. Elaboration d'un outil d'aide à la décision en vue de l'évolution de la tarification des transports publics en Ile de France. *Journal of Decision Systems*, 9(2):289–315, 2000.

[74] V. Mousseau and R. Slowinski. Inferring an ELECTRE TRI model from assignment examples. *Journal of Global Optimization*, 12(2):157–174, 1998.

[75] V. Mousseau, R. Slowinski, and P. Zielniewicz. ELECTRE TRI 2.0a: Methodological guide and user's documentation. Document du LAMSADE 111, Université Paris-Dauphine, 1999.

[76] A. Ngo The and V. Mousseau. Using assignment examples to infer category limits for the ELECTRE TRI method. *Journal of Multi-Criteria Decision Analysis*, 11(1):29–43, 2002.

[77] M.F. Norese and S. Viale. A multi-profile sorting procedure in the public administration. *European Journal of Operational Research*, 138:365–379, 2002.

[78] E. Parent and P. Schnabele. Le choix d'un aménagement aquacole : Exemple d'utilisation de la méthode ELECTRE III et comparaison avec d'autres méthodes multicritères d'aide à la décision. Cahier du LAMSADE 47, Université Paris-Dauphine, 1988.

[79] J. Pomerol and S. Barba-Romero. *Multicriterion Decision Making in Management*. Kluwer Academic Publishers, Dordrecht, 2000.

[80] K. Raju and C. Pillai. Multicriterion decision making in river basin planning and development. *European Journal of Operational Research*, 112 (2):249–257, 1999.

[81] J. Richard. Aide à la décision stratégique en PME. In E. Jacquet-Lagrèze and Y. Siskos, editors, *Méthode de décision multicritère*, pages 119–143. Editions Hommes et Techniques, Paris, 1983.

[82] D. Rios Insua and S. French. A framework for sensitivity analysis in discrete multi-objective decision making. *European Journal of Operational Research*, 54:176–190, 1991.

[83] E. Roba, B. Sussmann, and M. Theys. *Les Méthodes de Choix Multicritère Appliquées à la Sélection du Personnel: Models of Manpower*. The English University Press, London, 1970.

[84] F. Roberts. *Measurement Theory, with Applications to Decision Making, Utility and the Social Sciences*. Addison-Wesley, New York, 1979.

[85] M. Rogers and M. Bruen. Choosing realistic values of indifference, preference and veto thresholds for use with environmental criteria within ELECTRE. *European Journal of Operational Research*, 107 (3):542–551, 1998.

[86] M. Rogers and M. Bruen. A new system for weighting environmental criteria for use within ELECTRE III. *European Journal of Operational Research*, 107 (3):552–563, 1998.

[87] M. Rogers, M. Bruen, and L. Maystre. *ELECTRE and Decision Support*. Kluwer Academic Publishers, London, 2000.

[88] J. Rohmer and J. Veret. Choix d'une stratégie commerciale dans la distribution succursaliste. In E. Jacquet-Lagrèze and Y. Siskos, editors, *Méthode de décision multicritère*, pages 91–118. Editions Hommes et Techniques, Paris, 1983.

[89] B. Roy. Classement et choix en présence de points de vue multiples (la méthode ELECTRE). *RIRO*, 8:57–75, 1968.

[90] B. Roy. *Algèbre Moderne et Théorie des Graphes Orientées vers les Sciences Economiques et Sociales*. Dunod, Paris, 1969 (volume 1) 1970 (volume 2).

[91] B. Roy. Critères multiples et modélisation des préférences : l'apport des relations de surclassement. *Revue d'Économie Politique*, 1:1–44, 1974.

[92] B. Roy. From optimization to multicriteria decision aid: Three main operational attitudes. In H. Thiriez and S. Zionts, editors, *Multiple Criteria Decision Making*, volume 130 of *Lecture Notes in Economics and Mathematical Systems*, pages 1–34. Springer Verlag, Berlin, 1976.

[93] B. Roy. ELECTRE III : Un algorithme de classements fondé sur une représentation floue des préférences en présence de critères multiples. *Cahiers du CERO*, 20(1):3–24, 1978.

[94] B. Roy. A multicriteria analysis for trichotomic segmentation problems. In P. Nijkamp and J. Spronk, editors, *Multiple Criteria Analysis: Operational Methods*, pages 245–257. Gower Press, Farnborough, 1981.

[95] B. Roy. *Méthodologie Multicritère d'aide à la Décision*. Economica, Paris, 1985.

[96] B. Roy. Main sources of inaccurate determination, uncertainty and imprecision in decision models. In B. Munier and M. Shakun, editors, *Compromise, Negotiation and Group Decision*, pages 43–62. Reidel Publishing Company, Dordrecht, 1988.

[97] B. Roy. Main sources of inaccurate determination, uncertainty and imprecision. *Mathematical and Computer Modelling*, 12(10/11):1245–1254, 1989.

[98] B. Roy. The outranking approach and the foundations of ELECTRE methods. In C.A. Bana e Costa, editor, *Readings in Multiple Criteria Decision Aid*, pages 155–183. Springer-Verlag, Berlin, 1990.

[99] B. Roy. The outranking approach and the foundations of ELECTRE methods. *Theory and Decision*, 31:49–73, 1991.

[100] B. Roy. *Multicriteria Methodology for Decision Aiding*, volume 12 of *Nonconvex Optimization and its Applications*. Kluwer Academic Publishers, Dordrecht, 1996.

[101] B. Roy. A missing link in OR-AD: Robustness analysis. *Foundations of Computing and Decision Sciences*, 23(3):141–160, 1998.

[102] B. Roy. Decision-aiding today: What should we expect? In T. Gal, T. Stewart, and T. Hanne, editors, *Advances in MultCriteria Decision Making – MCDM Models, Algorithms, Theory, and Applications*, pages 1.1–1.35. Kluwer Academic Publishers, Boston, 1999.

[103] B. Roy. A French-English decision aiding glossary. In *Newsletter of the EWG on MCDA (Supplement 1)*, volume Series 3(1). Faculty of Economics, INESC-Coimbra, Portugal, 2000.

[104] B. Roy. Optimisation et analyse multicritère. In C. Jessua, C. Labrousse, and D. Vitry, editors, *Dictionnaire des Sciences Economiques*, pages 640–643. Presses Universitaires de Frances, Paris, 2001.

[105] B. Roy. Robustesse de quoi et vis-à-vis de quoi mais aussi robustesse pourquoi en aide à la décision ? In C. Henggeler-Antunes, J. Figueira, and J. Clímaco, editors, *Proceedings*

of the 56th Meeting of the European Working Group Multiple Criteria Decision Aiding, Coimbra, Portugal, 3-5 October 2002. Commissão de Coordenação da Região Centro, Coimbra, Portugal, to appear 2004.

[106] B. Roy and P. Bertier. La méthode ELECTRE II. Note de travail 142, SEMA-METRA Metra International, 1971.

[107] B. Roy and P. Bertier. La méthode ELECTRE II – Une application au média-planning. In M. Ross, editor, *OR'72*, pages 291–302. North-Holland Publishing Company, Amsterdam, 1973.

[108] B. Roy and D. Bouyssou. Comparison of two decision-aid models applied to a nuclear plant siting example. *European Journal of Operational Research*, 25:200–215, 1986.

[109] B. Roy and D. Bouyssou. *Aide Multicritère à la Décision : Méthodes et Cas*. Economica, Paris, 1993.

[110] B. Roy and J. Hugonnard. Classement des prolongements de lignes de métro en banlieue parisienne (présentation d'une méthode multicritère originale). *Cahiers du CERO*, 24(2-3-4):153–171, 1982.

[111] B. Roy and J. Hugonnard. Le plan d'extension du métro en banlieue parisienne, un cas type de l'analyse multicritère. *Les Cahiers Scientifiques de la Revue Transports*, 6:77–108, 1982.

[112] B. Roy and J. Hugonnard. Ranking of suburban line extension projects on the Paris metro system by a multicriteria method. *Transportation Research*, 16A(4):301–312, 1982.

[113] B. Roy and J. Hugonnard. Réponse à Monsieur Gargaillo. *Les Cahiers Scientifiques de la Revue Transports*, 7:58–64, 1982.

[114] B. Roy, E. Jacquet-Lagrèze, and M. Blancher. Elaboration de critères permettant une intégration des divers aspects liés au temps dans l'aide à la décision en matière de transports. 1ère phase 1975, 2e phase 1977, Délégation Générale à la Recherche Scientifique et Technique DGRST-SAEI, Action Thématique Programmée ATP Socio-Economie des Transports, 1977.

[115] B. Roy and V. Mousseau. A theoretical framework for analysing the notion of relative importance of criteria. *Journal of Multi-Criteria Decision Analysis*, 5:145–159, 1996.

[116] B. Roy, M. Présent, and D. Silhol. A programming method for determining which Paris metro stations should be renovated. *European Journal of Operational Research*, 24:318–334, 1986.

[117] B. Roy and J. Skalka. ELECTRE IS : Aspects méthodologiques et guide d'utilisation. Document du LAMSADE 30, Université Paris Dauphine, 1984.

[118] B. Roy, R. Slowinski, and W. Treichel. Multicriteria programming of water resources supply systems for rural areas. *Water Resources Bulletin*, 28 (1):13–31, 1992.

[119] B. Roy and D. Vanderpooten. The European school of MCDA: Emergence, basic features and current works. *Journal of Multi-Criteria Decision Analysis*, 5(1):22–37, 1996.

[120] T. Saaty. *The Analytical Hierarchy Process*. McGraw-Hill, New York, 1980.

[121] P. Salminen, J. Hokkanen, and R. Lahdelma. Comparing multicriteria methods in the context of environmental problems. *European Journal of Operational Research*, 104 (3):485–496, 1998.

[122] A. Schärlig. *Décider sur Plusieurs Critères, Panorama de l'Aide à la Décision Multicritère*. Presses Polytechniques et Universitaires Romandes, Lausanne, 1985.

[123] A. Schärlig. *Pratiquer ELECTRE et PROMÉTHÉE : Un Complément à Décider sur Plusieurs Critères*. Presses Polytechniques et Universitaires Romandes, Lausanne, 1996.

[124] R. Silva Lopez. Une technique systématique de modélisation multicritère pour l'évaluation d'alternatives de lutte contre la pollution industrielle : Technologies propres ou curatives. Document du LAMSADE 60, Université Paris-Dauphine, 1990.

[125] J. Simos. *Gestion des Déchets Solides Urbains Genevois : Les Faits, le Traitement, l'Analyse*. Presses Polytechniques et Universitaires Romandes, Lausanne, 1990.

[126] Y. Siskos and Ph. Hubert. Multi-criteria analysis of the impacts of energy alternatives: A survey and a new comparative approach. *European Journal of Operational Research*, 13:278–299, 1983.

[127] Y. Siskos, G. Wascher, and H. Winkels. A bibliography on outranking approaches (1966-1982). Cahier du LAMSADE 31, Université Paris Dauphine, 1983.

[128] R. Srinivasa, L. Duckstein, and C. Arondel. Multicriterion analysis for sustainable water resources planning: A case study in Spain. *Water Resources Management*, 14:435–456, 2000.

[129] Y. Tanguy and N. Leclere. Une application des méthodes de choix multi-critères à l'étude de l'adéquation des personnels aux postes de travail. In *Actes du congrès Association Française pour la Cybernétique Economique et Technique (AFCET): Aide à la Décision, Paris, May 16-17, 1974*, pages 139–147, 1974.

[130] A. Tecle and L. Duckstein. A procedure for selecting MCDM techniques for forest resources management. In A. Goicoechea, L. Duckstein, and S. Zionts, editors, *Multiple Criteria Decision Making and Support at the Interface of Industry, Business and Finance*, pages 19–32. Springer Verlag, New York, 1992.

[131] A. Tecle, M. Fogel, and L. Duckstein. Multicriterion analysis of forest watershed management alternatives. *Water Resources Bulletin*, 24:1169–1178, 1988.

[132] J. Teng and G. Tzeng. Multicriteria evaluation for strategies of improving and controlling air quality in the super city: A case study of Taipei city. *Journal of Environmental Management*, 40 (3):213–229, 1994.

[133] A. Tsoukiàs, P. Perny, and Ph. Vincke. From concordance/discordance to the modelling of positive and negative reasons in decision aiding. In D. Bouyssou, E. Jacquet-Lagrèze, P. Perny, R. Słowiński, D. Vanderpooten, and Ph. Vincke, editors, *Aiding Decisions with Multiple Criteria: Essays in Honour of Bernard Roy*, pages 147–174. Kluwer Academic Publishers, Dordrecht, 2002.

[134] D. Vallée and P. Zielniewicz. ELECTRE III-IV, version 3.x, Aspects Méthodologiques (tome 1), Guide d'utilisation (tome 2). Document du LAMSADE 85 et 85bis, Université Paris Dauphine, 1994.

[135] D. Vanderpooten. The construction of prescriptions in outranking methods. In C. Bana e Costa, editor, *Readings in Multiple Criteria Decision Aid*, pages 184–216. Springer Verlag, Heidelberg, 1990.

[136] J. Vansnick. On the problem of weight in multiple criteria decision making (the non-compensatory approach). *European Journal of Operational Research*, 24:288–294, 1986.

[137] M. Ventura. Evry 1: Concours d'aménagement urbain. Cahier 31, Institut d'Amenagement et d'Urbanisme de la Région Parisienne (IAURP), Paris, 1973.

[138] Ph. Vincke. Exploitation of a crisp relation in a ranking problem. *Theory and Decision*, 32(3):221–240, 1992.

[139] Ph. Vincke. *Multicriteria Decision-Aid.* John Wiley & Sons, New York, 1992.

[140] D. Wijnmalen. A case-study in military decision-making with mission-oriented multiple objectives. In J. Clímaco, editor, *Multicriteria Analysis*, pages 549–560. Springer Verlag, Berlin, 1997.

[141] W. Yu. *Aide multicritère à la décision dans le cadre de la problématique du tri : Concepts, méthodes et applications.* PhD thesis, Université Paris-Dauphine, 1992.

[142] W. Yu. ELECTRE TRI : Aspects méthodologiques et manuel d'utilisation. Document du LAMSADE 74, Université Paris-Dauphine, 1992.

[143] C. Zopounidis. *Evaluation du Risque de Défaillance de l'Entreprise : Méthodes et Cas d'Application.* Economica, Paris, 1995.

[144] C. Zopounidis. *Operational Tools in the Management of Financial Risks.* Kluwer Academic Publishers, Dordrecht, 1998.

[145] C. Zopounidis and A Dimitras. *Multicriteria Decision Aid Methods for the Prediction of Business Failure.* Kluwer Academic Publishers, Dordrecht, 1998.

[146] C. Zopounidis and C. Hurson. Portfolio selection and multicriteria analysis. In C. Floudas and P. Pardalos, editors, *Encyclopedia of Optimization, Vol. IV*, pages 310–322. Kluwer Academic Publishers, Dordrecht, 2001.

Chapter 5

PROMETHEE METHODS

Jean-Pierre Brans
Centrum voor Statistiek en Operationeel Onderzoek
Vrije Universiteit Brussel
Pleinlaan 2, B-1050 Brussels
Belgium
jpbrans@vub.ac.be

Bertrand Mareschal
Service de Mathématiques de la Gestion
Université Libre de Bruxelles
Boulevard du Triomphe CP 210-01, B-1050 Brussels
Belgium
bmaresc@ulb.ac.be

Abstract This paper gives an overview of the PROMETHEE-GAIA methodology for MCDA. It starts with general comments on multicriteria problems, stressing that a multicriteria problem cannot be treated without additional information related to the preferences and the priorities of the decision-makers. The information requested by PROMETHEE and GAIA is particularly clear and easy to define for both decision-makers and analysts. It consists in a preference function associated to each criterion as well as weights describing their relative importance. The PROMETHEE I, the PROMETHEE II complete ranking, as well as the GAIA visual interactive module are then described and commented. The two next sections are devoted to the PROMETHEE VI sensitivity analysis procedure (human brain) and to the PROMETHEE V procedure for multiple selection of alternatives under constraints. An overview of the PROMETHEE GDSS procedure for group decision making is then given. Finally the DECISION LAB software implementation of the PROMETHEE-GAIA methodology is described using a numerical example.

Keywords: MCDA, outranking methods, PROMETHEE-GAIA, DECISION LAB.

1. History

The PROMETHEE I (partial ranking) and PROMETHEE II (complete ranking) were developed by J.P. Brans and presented for the first time in 1982 at a conference organised by R. Nadeau and M. Landry at the Université Laval, Québec, Canada (L'Ingéniérie de la Décision. Elaboration d'instruments d'Aide à la Décision). The same year several applications using this methodology were already treated by G. Davignon in the field of Heath care.

A few years later J.P. Brans and B. Mareschal developed PROMETHEE III (ranking based on intervals) and PROMETHEE IV (continuous case). The same authors proposed in 1988 the visual interactive module GAIA which is providing a marvellous graphical representation supporting the PROMETHEE methodology.

In 1992 and 1994, J.P. Brans and B. Mareschal further suggested two nice extensions: PROMETHEE V (MCDA including segmentation constraints) and PROMETHEE VI (representation of the human brain).

A considerable number of successful applications has been treated by the PROMETHEE methodology in various fields such as Banking, Industrial Location, Manpower planning, Water resources, Investments, Medicine, Chemistry, Health care, Tourism, Ethics in OR, Dynamic management, ... The success of the methodology is basically due to its mathematical properties and to its particular friendliness of use.

2. Multicriteria Problems

Let us consider the following multicriteria problem:

$$\max\{g_1(a), g_2(a), \ldots, g_j(a), \ldots, g_k(a) | a \in A\}, \tag{5.1}$$

where A is a finite set of possible alternatives $\{a_1, a_2, \ldots a_i, \ldots, a_n\}$ and $\{g_1(\cdot), g_2(\cdot), \ldots, g_j(\cdot), \ldots g_k(\cdot)\}$ a set of evaluation criteria. There is no objection to consider some criteria to be maximised and the others to be minimised. The expectation of the decision-maker is to identify an alternative optimising all the criteria.

Usually this is a *ill-posed mathematical* problem as there exists no alternative optimising all the criteria at the same time. However most (nearly all) human problems have a multicriteria nature. According to our various human aspirations, it makes no sense, and it is often not fair, to select a decision based on one evaluation criterion only. In most of cases at least technological, economical, environmental and social criteria should always be taken into account. Multicriteria problems are therefore extremely important and request an appropriate treatment.

The basic data of a multicriteria problem (5.1) consist of an evaluation table (Table 5.1).

Table 5.1. Evaluation table.

a	$g_1(\cdot)$	$g_2(\cdot)$	$\ldots$	$g_j(\cdot)$	$\ldots$	$g_k(\cdot)$
a_1	$g_1(a_1)$	$g_2(a_1)$	$\ldots$	$g_j(a_1)$	$\ldots$	$g_k(a_1)$
a_2	$g_1(a_2)$	$g_2(a_2)$	$\ldots$	$g_j(a_2)$	$\ldots$	$g_k(a_2)$
$\vdots$	$\vdots$	$\vdots$	$\ddots$	$\vdots$	$\ddots$	$\vdots$
a_i	$g_1(a_i)$	$g_2(a_i)$	$\ldots$	$g_j(a_i)$	$\ldots$	$g_k(a_i)$
$\vdots$	$\vdots$	$\vdots$	$\ddots$	$\vdots$	$\ddots$	$\vdots$
a_n	$g_1(a_n)$	$g_2(a_n)$	$\ldots$	$g_j(a_n)$	$\ldots$	$g_k(a_n)$

Let us consider as an example the problem of an individual purchasing a car. Of course the price is important and it should be minimised. However it is clear that in general individuals are not considering only the price. Not everybody is driving the cheapest car! Most people would like to drive a luxury or sports car at the price of an economy car. Indeed they consider many criteria such as price, reputation, comfort, speed, reliability, consumption, ... As there is no car optimising all the criteria at the same time, a *compromise* solution should be selected. Most decision problems have such a multicriteria nature.

The solution of a multicriteria problem depends not only on the basic data included in the evaluation table but also on the decision-maker himself. All individuals do not purchase the same car. There is no absolute best solution! The best compromise solution also depends on the individual *preferences* of each decision-maker, on the *"brain"* of each decision-maker.

Consequently, *additional information* representing these preferences is required to provide the decision maker with useful decision aid.

The natural dominance relation associated to a multicriteria problem of type (5.1) is defined as follows:
For each $(a, b) \in A$:

$$\begin{aligned}
&\begin{cases} \forall j : g_j(a) \geq g_j(b) \\ \exists k : g_k(a) > g_k(b) \end{cases} && \Longleftrightarrow \quad aPb, \\
&\quad \forall j : g_j(a) = g_j(b) && \Longleftrightarrow \quad aIb, \\
&\begin{cases} \exists s : g_s(a) > g_s(b) \\ \exists r : g_r(a) < g_r(b) \end{cases} && \Longleftrightarrow \quad aRb,
\end{aligned} \tag{5.2}$$

where P, I, and R respectively stand for *preference*, *indifference* and *incomparability*. This definition is quite obvious. An alternative is better than another if it is at least as good as the other on all criteria. If an alternative is better on a criterion s and the other one better on criterion r, it is impossible to decide which is the best one without additional information. Both alternatives are therefore incomparable!

Alternatives which are not dominated by any other are called *efficient solutions*. Given an evaluation table for a particular multicriteria problem, most of the alternatives (often all of them) are usually efficient. The dominance relation is very poor on P and I. When an alternative is better on one criterion, the other is often better on another criterion. Consequently incomparability holds for most pairwise comparisons, so that it is impossible to decide without additional information. This information can for example include:

- Trade-offs between the criteria;
- A value function aggregating all the criteria in a single function in order to obtain a mono-criterion problem for which an optimal solution exists;
- Weights giving the relative importance of the criteria;
- Preferences associated to each pairwise comparison within each criterion;
- Thresholds fixing preference limits;
- ...

Many multicriteria decision aid methods have been proposed. All these methods start from the same evaluation table, but they vary according to the additional information they request. The PROMETHEE methods require very clear additional information, that is easily obtained and understood by both decision-makers and analysts.

The purpose of all multicriteria methods is to enrich the dominance graph, i.e. to reduce the number of incomparabilities (R). When a utility function is built, the multicriteria problem is reduced to a single criterion problem for which an optimal solution exists. This seems exaggerated because it relies on quite strong assumptions (do we really make all our decisions based on a utility function defined somewhere in our brains?) and it completely transforms the structure of the decision problem. For this reason B. Roy proposed to build outranking relations including only realistic enrichments of the dominance relation (see [86] and [87]). In that case, not all the incomparabilities are withdrawn but the information is reliable. The PROMETHEE methods belong to the class of outranking methods.

In order to build an appropriate multicriteria method some requisites could be considered:

Requisite 1: The amplitude of the deviations between the evaluations of the alternatives within each criterion should be taken into account:

$$d_j(a,b) = g_j(a) - g_j(b). \tag{5.3}$$

This information can easily be calculated, but is not used in the efficiency theory. When these deviations are negligible the dominance relation can possibly be enriched.

Requisite 2: As the evaluations $g_j(a)$ of each criterion are expressed in their own units, *the scaling effects* should be completely eliminated. It is not acceptable to obtain conclusions depending on the scales in which the evaluations are expressed. Unfortunately not all multicriteria procedures are respecting this requisite!

Requisite 3: In the case of pairwise comparisons, an appropriate multicriteria method should provide the following information:

a is preferred to b;
a and b are indifferent;
a and b are incomparable.

The purpose is of course to reduce as much as possible the number of incomparabilities, but not when it is not realistic. Then the procedure may be considered as fair. When, for a particular procedure, all the incomparabilities are systematically withdrawn the provided information can be more disputable.

Requisite 4: Different multicriteria methods request different additional information and operate different calculation procedures so that the solutions they propose can be different. It is therefore important to develop methods being *understandable* by the decision-makers. "Black box" procedures should be avoided.

Requisite 5: An appropriate procedure should not include technical parameters having no significance for the decision-maker. Such parameters would again induce "Black box" effects.

Requisite 6: An appropriate method should provide information on the *conflicting nature* of the criteria.

Requisite 7: Most of the multicriteria methods are allocating weights of relative importance to the criteria. These weights reflects a major part of the *"brain"* of the decision-maker. It is not easy to fix them. Usually the decision-makers strongly hesitate. An appropriate method should offer *sensitivity tools* to test easily different sets of weights.

The PROMETHEE methods and the associated GAIA visual interactive module are taking all these requisites into account. On the other hand some mathematical properties that multicriteria problems possibly enjoy can also be considered. See for instance [95]. Such properties related to the PROMETHEE methods have been analysed by [7] in a particularly interesting paper.

The next sections describe the PROMETHEE I and II rankings, the GAIA methods, as well as the PROMETHEE V and VI extensions of the methodology. The PROMETHEE III and IV extensions are not discussed here. Additional information can be found in [17]. Several actual applications of the PROMETHEE methodology are also mentioned in the list of references.

3. The PROMETHEE Preference Modelling Information

The PROMETHEE methods were designed to treat multicriteria problems of type (5.1) and their associated evaluation table.

The additional information requested to run PROMETHEE is particularly clear and understandable by both the analysts and the decision-makers. It consists of:

- Information between the criteria;
- Information within each criterion.

3.1 Information between the Criteria

Table 5.2 should be completed, with the understanding that the set $\{w_j, j = 1, 2, \ldots, k\}$ represents weights of relative importance of the different criteria. These weights are non-negative numbers, independent from the measurement

Table 5.2. Weights of relative importance.

$g_1(\cdot)$	$g_2(\cdot)$	$\cdots$	$g_j(\cdot)$	$\cdots$	$g_k(\cdot)$
w_1	w_2	$\cdots$	w_j	$\cdots$	w_k

units of the criteria. The higher the weight, the more important the criterion. There is no objection to consider normed weights, so that:

$$\sum_{j=1}^{k} w_j = 1. \tag{5.4}$$

In the PROMETHEE software PROMCALC and DECISION LAB, the user is allowed to introduce arbitrary numbers for the weights, making it easier to express the relative importance of the criteria. These numbers are then divided by their sum so that the weights are normed automatically.

Assessing weights to the criteria is not straightforward. It involves the priorities and perceptions of the decision-maker. The selection of the weights is his *space of freedom*. PROMCALC and DECISION LAB include several sensitivity tools to experience different set of weights in order to help to fix them.

3.2 Information within the Criteria

PROMETHEE is not allocating an intrinsic absolute utility to each alternative, neither globally, nor on each criterion. We strongly believe that the decision-makers are not proceeding that way. The preference structure of PROMETHEE is based on *pairwise comparisons*. In this case the deviation between the evaluations of two alternatives on a particular criterion is considered. For small deviations, the decision-maker will allocate a small preference to the best alternative and even possibly no preference if he considers that this deviation is negligible. The larger the deviation, the larger the preference. There is no objection to consider that these preferences are real numbers varying between 0 and 1. This means that for each criterion the decision-maker has in mind a function

$$P_j(a,b) = F_j\left[d_j(a,b)\right] \quad \forall a,b \in A, \tag{5.5}$$

where:

$$d_j(a,b) = g_j(a) - g_j(b) \tag{5.6}$$

and for which:

$$0 \leq P_j(a,b) \leq 1. \tag{5.7}$$

In case of a criterion to be maximised, this function is giving the preference of a over b for observed deviations between their evaluations on criterion $g_j(\cdot)$. It should have the following shape (see Figure 5.1). The preferences equals 0 when the deviations are negative.

The following property holds:

$$P_j(a,b) > 0 \Rightarrow P_j(b,a) = 0. \tag{5.8}$$

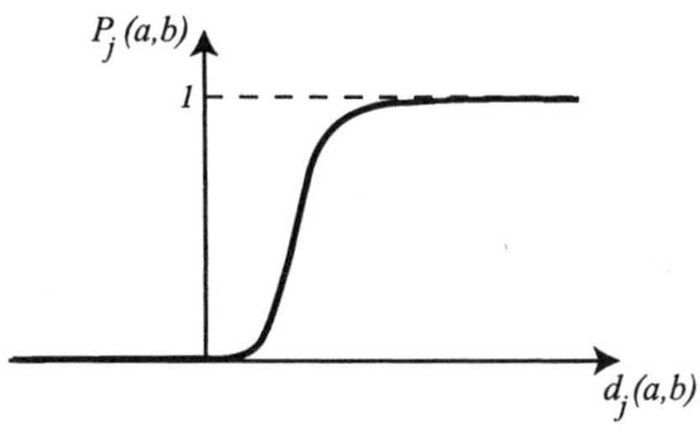

Figure 5.1. Preference function.

For criteria to be minimised, the preference function should be reversed or alternatively given by:

$$P_j(a,b) = F_j\left[-d_j(a,b)\right]. \tag{5.9}$$

We have called the pair $\{g_j(\cdot), P_j(a,b)\}$ the *generalised criterion* associated to criterion $g_j(\cdot)$. Such a generalised criterion has to be defined for each criterion. In order to facilitate the identification six types of particular preference functions have been proposed (see table 5.3).

Table 5.3. Types of generalised criteria ($P(d)$: Preference function).

Generalised criterion	Definition	Parameters to fix
Type 1: Usual Criterion 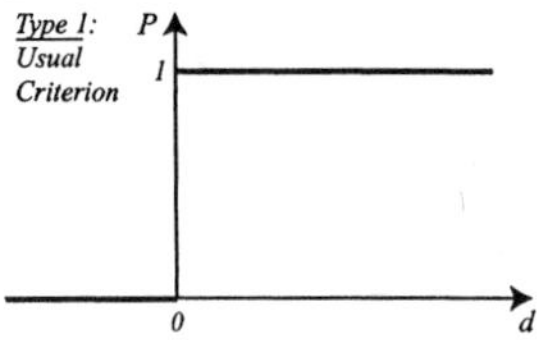	$P(d) = \begin{cases} 0 & d \leq 0 \\ 1 & d > 0 \end{cases}$	–
Type 2: U-shape Criterion 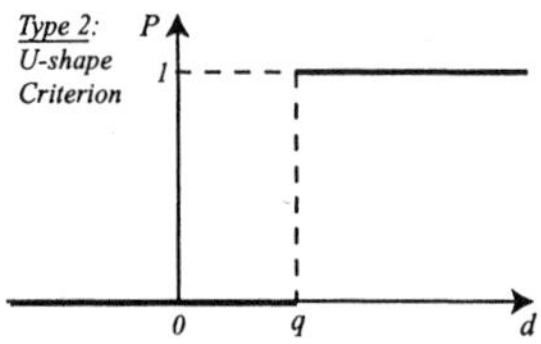	$P(d) = \begin{cases} 0 & d \leq q \\ 1 & d > q \end{cases}$	q
Type 3: V-shape Criterion 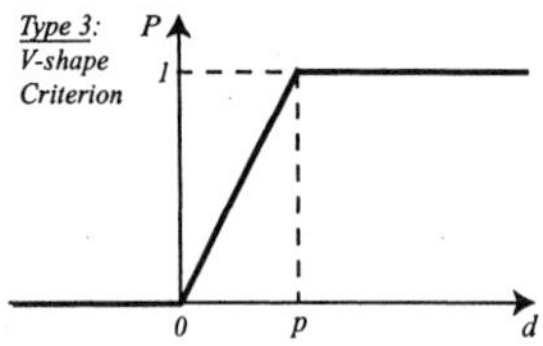	$P(d) = \begin{cases} 0 & d \leq 0 \\ \frac{d}{p} & 0 \leq d \leq p \\ 1 & d > p \end{cases}$	p
Type 4: Level Criterion 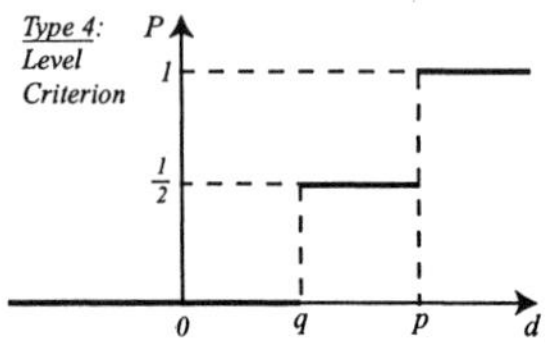	$P(d) = \begin{cases} 0 & d \leq q \\ \frac{1}{2} & q < d \leq p \\ 1 & d > p \end{cases}$	p, q
Type 5: V-shape with indifference Criterion 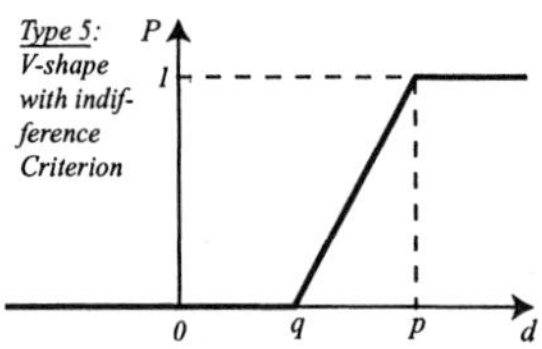	$P(d) = \begin{cases} 0 & d \leq q \\ \frac{d-q}{p-q} & q < d \leq p \\ 1 & d > p \end{cases}$	p, q
Type 6: Gaussian Criterion 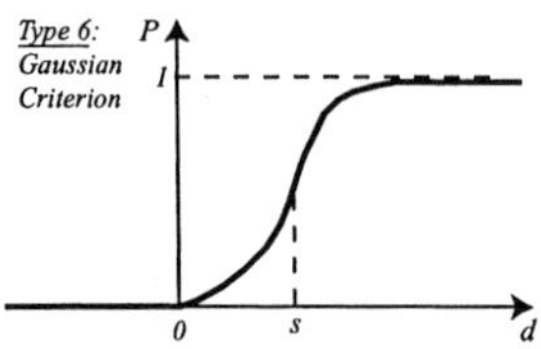	$P(d) = \begin{cases} 0 & d \leq 0 \\ 1 - e^{-\frac{d^2}{2s^2}} & d > 0 \end{cases}$	s

In each case 0, 1 or 2 parameters have to be defined, their significance is clear:

q is a threshold of indifference;
p is a threshold of strict preference;
s is an intermediate value between q and p.

The q indifference threshold is the largest deviation which is considered as negligible by the decision maker, while the p preference threshold is the smallest deviation which is considered as sufficient to generate a full preference.

The identification of a generalised criterion is then limited to the selection of the appropriate parameters. It is an easy task.

The PROMCALC and DECISION LAB software are proposing these six shapes only. As far as we know they have been satisfactory in most real-world applications. However there is no objection to consider additional generalised criteria.

In case of type 5 a threshold of indifference q and a threshold of strict preference p have to be selected.

In case of a Gaussian criterion (type 6) the preference function remains increasing for all deviations and has no discontinuities, neither in its shape, nor in its derivatives. A parameter s has to be selected, it defines the inflection point of the preference function. We then recommend to determine first a q and a p and to fix s in between. If s is close to q the preferences will be reinforced for small deviations, while close to p they will be softened.

As soon as the evaluation table $\{g_j(\cdot)\}$ is given, and the weights w_j and the generalised criteria $\{g_j(\cdot), P_j(a,b)\}$ are defined for $i = 1, 2, \ldots, n$; $j = 1, 2, \ldots, k$, the PROMETHEE procedure can be applied.

4. The PROMETHEE I and II Rankings

The PROMETHEE procedure is based on pairwise comparisons (cfr. [8]–[16], [59], [60]). Let us first define aggregated preference indices and outranking flows.

4.1 Aggregated Preference Indices

Let $a, b \in A$, and let:

$$\begin{cases} \pi(a,b) = \sum_{j=1}^{k} P_j(a,b)w_j, \\ \pi(b,a) = \sum_{j=1}^{k} P_j(b,a)w_j. \end{cases} \tag{5.10}$$

$\pi(a,b)$ is expressing with which degree a is preferred to b over all the criteria and $\pi(b,a)$ how b is preferred to a. In most of the cases there are criteria for which a is better than b, and criteria for which b is better than a, consequently $\pi(a,b)$ and $\pi(b,a)$ are usually positive. The following properties hold for all $(a,b) \in A$.

$$\begin{cases} \pi(a,a) = 0, \\ 0 \leq \pi(a,b) \leq 1, \\ 0 \leq \pi(b,a) \leq 1, \\ 0 \leq \pi(a,b) + \pi(b,a) \leq 1. \end{cases} \tag{5.11}$$

It is clear that:

$$\begin{cases} \pi(a,b) \sim 0 \text{ implies a weak global preference of } a \text{ over } b, \\ \pi(a,b) \sim 1 \text{ implies a strong global preference of } a \text{ over } b. \end{cases} \tag{5.12}$$

As soon as $\pi(a,b)$ and $\pi(b,a)$ are computed for each pair of alternatives of A, a complete valued outranking graph, including two arcs between each pair of nodes, is obtained (see Figure 5.2).

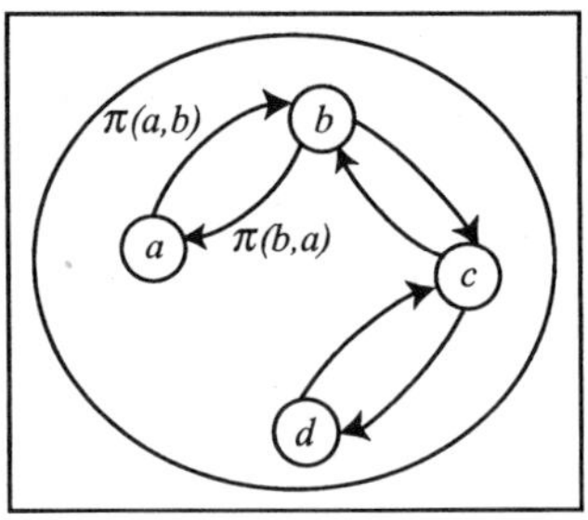

Figure 5.2. Valued outranking graph.

4.2 Outranking Flows

Each alternative a is facing $(n-1)$ other alternatives in A. Let us define the two following outranking flows:

- *the positive outranking flow:*

$$\phi^{+}(a) = \frac{1}{n-1} \sum_{x \in A} \pi(a,x), \tag{5.13}$$

- *the negative outranking flow:*

$$\phi^{-}(a) = \frac{1}{n-1} \sum_{x \in A} \pi(x,a). \tag{5.14}$$

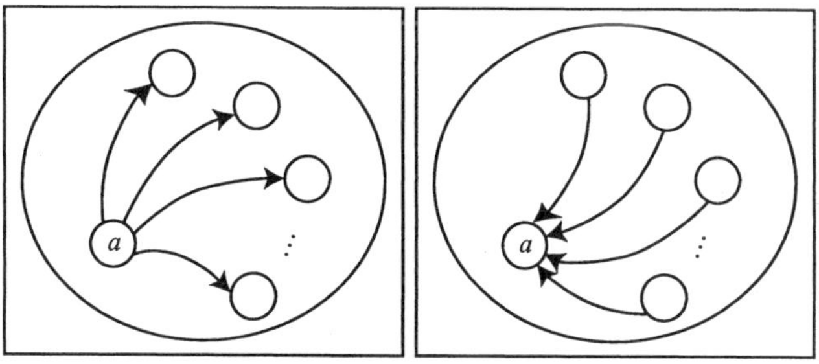

(a) The $\phi^+(a)$ outranking flow. (b) The $\phi^-(a)$ outranking flow.

Figure 5.3. The PROMETHEE outranking flows.

The positive outranking flow expresses how an alternative a is *outranking* all the others. It is its *power, its outranking character*. The higher $\phi^+(a)$, the better the alternative (see Figure 5.3(a)).

The negative outranking flow expresses how an alternative a is *outranked* by all the others. It is its *weakness, its outranked character*. The lower $\phi^-(a)$ the better the alternative (see Figure 5.3(b)).

4.3 The PROMETHEE I Partial Ranking

The PROMETHEE I partial ranking (P^I, I^I, R^I) is obtained from the positive and the negative outranking flows. Both flows do not usually induce the same rankings. PROMETHEE I is their intersection.

$$\begin{cases} aP^I b & \text{iff} \quad \begin{cases} \phi^+(a) > \phi^+(b) \text{ and } \phi^-(a) < \phi^-(b), \text{ or} \\ \phi^+(a) = \phi^+(b) \text{ and } \phi^-(a) < \phi^-(b), \text{ or} \\ \phi^+(a) > \phi^+(b) \text{ and } \phi^-(a) = \phi^-(b); \end{cases} \\ aI^I b & \text{iff} \quad \phi^+(a) = \phi^+(b) \text{ and } \phi^-(a) = \phi^-(b); \\ aR^I b & \text{iff} \quad \begin{cases} \phi^+(a) > \phi^+(b) \text{ and } \phi^-(a) > \phi^-(b), \text{ or} \\ \phi^+(a) < \phi^+(b) \text{ and } \phi^-(a) < \phi^-(b); \end{cases} \end{cases} \tag{5.15}$$

where P^I, I^I, R^I respectively stand for preference, indifference and incomparability.

When $aP^I b$, a higher power of a is associated to a lower weakness of a with regard to b. The information of both outranking flows is consistent and may therefore be considered as sure.

When $aI^I b$, both positive and negative flows are equal.

When $aR^I b$, a higher power of one alternative is associated to a lower weakness of the other. This often happens when a is good on a set of criteria on which b is weak and reversely b is good on some other criteria on which a is weak. In such a case the information provided by both flows is not consistent. It seems then reasonable to be careful and to consider both alternatives as incomparable. The PROMETHEE I ranking is prudent: it will not decide which action is best in such cases. It is up to the decision-maker to take his responsibility.

4.4 The PROMETHEE II Complete Ranking

PROMETHEE II consists of the (P^{II}, I^{II}) complete ranking. It is often the case that the decision-maker requests a complete ranking. The *net outranking flow* can then be considered.

$$\phi(a) = \phi^+(a) - \phi^-(a). \tag{5.16}$$

It is the balance between the positive and the negative outranking flows. The higher the net flow, the better the alternative, so that:

$$\begin{cases} aP^{II}b & \text{iff} \quad \phi(a) > \phi(b), \\ aI^{II}b & \text{iff} \quad \phi(a) = \phi(b). \end{cases} \tag{5.17}$$

When PROMETHEE II is considered, all the alternatives are comparable. No incomparabilities remain, but the resulting information can be more disputable because more information gets lost by considering the difference (5.16).

The following properties hold:

$$\begin{cases} -1 \leq \phi(a) \leq 1, \\ \sum\limits_{x \in A} \phi(a) = 0. \end{cases} \tag{5.18}$$

When $\phi(a) > 0$, a is more outranking all the alternatives on all the criteria, when $\phi(a) < 0$ it is more outranked.

In real-world applications, we recommend to both the analysts and the decision-makers to consider both PROMETHEE I and PROMETHEE II. The complete ranking is easy to use, but the analysis of the incomparabilities often helps to finalise a proper decision.

As the net flow $\phi(\cdot)$ provides a complete ranking, it may be compared with a utility function. One advantage of $\phi(\cdot)$ is that it is built on clear and simple preference information (weights and preferences functions) and that it does rely on comparative statements rather than absolute statements.

4.5 The Profiles of the Alternatives

According to the definition of the positive and the negative outranking flows (5.13) and (5.14) and of the aggregated indices (5.10), we have:

$$\phi(a) = \phi^+(a) - \phi^-(a) = \frac{1}{n-1} \sum_{j=1}^{k} \sum_{x \in A} \left[P_j(a,x) - P_j(x,a) \right] w_j. \tag{5.19}$$

Consequently,

$$\phi(a) = \sum_{j=1}^{k} \phi_j(a) w_j \tag{5.20}$$

if

$$\phi_j(a) = \frac{1}{n-1} \sum_{x \in A} [P_j(a,x) - P_j(x,a)] \,. \tag{5.21}$$

$\phi_j(a)$ is the single criterion net flow obtained when only criterion $g_j(\cdot)$ is considered (100% of the total weight is allocated to that criterion). It expresses how an alternative a is outranking ($\phi_j(a) > 0$) or outranked ($\phi_j(a) < 0$) by all the other alternatives on criterion $g_j(\cdot)$.

The profile of an alternative consists of the set of all the single criterion net flows: $\phi_j(a), j = 1, 2, \ldots, k$.

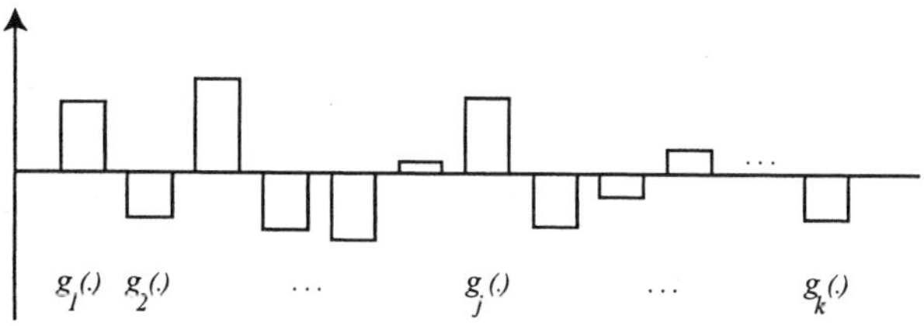

Figure 5.4. Profile of an alternative.

The profiles of the alternatives are particularly useful to appreciate their "*quality*" on the different criteria. It is extensively used by decision-makers to finalise their appreciation.

According to (5.20), we observe that the global net flow of an alternative is the scalar product between the vector of the weights and the profile vector of this alternative. This property will be extensively used when building up the GAIA plane.

5. The GAIA Visual Interactive Module

Let us first consider the matrix $M(n \times k)$ of the single criterion net flows of all the alternatives as defined in (5.21).

5.1 The GAIA Plane

The information included in matrix M is more extensive than the one in the evaluation table 5.1, because the degrees of preference given by the generalised criteria are taken into account in M. Moreover the $g_j(a_i)$ are expressed on their own scale, while the $\phi_j(a_i)$ are dimensionless. In addition, let us observe, that M is not depending on the weights of the criteria.

Table 5.4. Single criterion net flows.

	$\phi_1(\cdot)$	$\phi_2(\cdot)$	$\cdots$	$\phi_j(\cdot)$	$\cdots$	$\phi_k(\cdot)$
a_1	$\phi_1(a_1)$	$\phi_2(a_1)$	$\cdots$	$\phi_j(a_1)$	$\cdots$	$\phi_k(a_1)$
a_2	$\phi_1(a_2)$	$\phi_2(a_2)$	$\cdots$	$\phi_j(a_2)$	$\cdots$	$\phi_k(a_2)$
$\vdots$	$\vdots$	$\vdots$	$\ddots$	$\vdots$	$\ddots$	$\vdots$
a_i	$\phi_1(a_i)$	$\phi_2(a_i)$	$\cdots$	$\phi_j(a_i)$	$\cdots$	$\phi_k(a_i)$
$\vdots$	$\vdots$	$\vdots$	$\ddots$	$\vdots$	$\ddots$	$\vdots$
a_n	$\phi_1(a_n)$	$\phi_2(a_n)$	$\cdots$	$\phi_j(a_n)$	$\cdots$	$\phi_k(a_n)$

Consequently the set of the n alternatives can be represented as a cloud of n points in a k-dimensional space. According to (5.18) this cloud is centered at the origin. As the number of criteria is usually larger than two, it is impossible to obtain a clear view of the relative position of the points with regard to the criteria. We therefore project the information included in the k-dimensional space on a plane. Let us project not only the points representing the alternatives but also the unit vectors of the coordinate-axes representing the criteria. We then obtain:

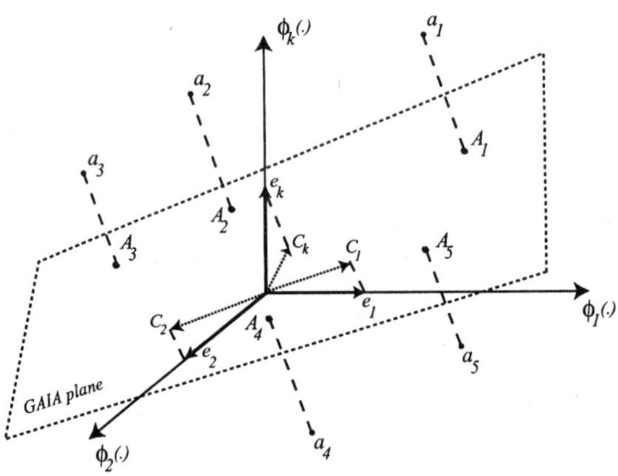

Figure 5.5. Projection on the GAIA plane.

The GAIA *plane* is the plane for which as much information as possible is preserved after projection. According to the *principal components analysis* technique it is defined by the two eigenvectors corresponding to the two largest eigenvalues of the covariance matrix $M'M$ of the single criterion net flows.

Of course some information get lost after projection. The GAIA plane is a *meta model* (a model of a model). Let δ be the quantity of information preserved.

In most applications we have treated so far δ was larger than 60% and in many cases larger than 80%. This means that the information provided by the GAIA plane is rather reliable. This information is quite rich, it helps to understand the structure of a multicriteria problem.

5.2 Graphical Display of the Alternatives and of the Criteria

Let $(A_1, A_2, \ldots, A_i, \ldots, A_n)$ be the projections of the n points representing the alternatives and let $(C_1, C_2, \ldots, C_j, \ldots, C_k)$ be the projections of the k unit vectors of the coordinates axes of $I\!R^k$ representing the criteria. We then obtain a GAIA plane of the following type:

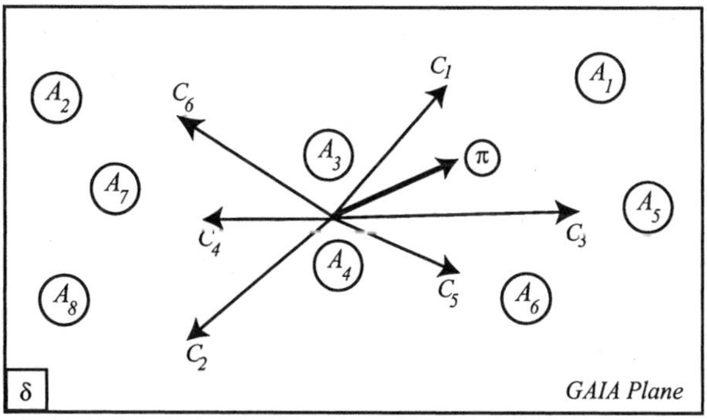

Figure 5.6. Alternatives and criteria in the GAIA plane.

Then the following properties hold (see [59] and [16]) provided that δ is sufficiently high:

P 1: The longer a criterion axis in the GAIA plane, the more discriminating this criterion.

P 2: Criteria expressing similar preferences are represented by axes oriented in approximatively the same direction.

P 3: Criteria expressing conflicting preferences are oriented in opposite directions.

P 4: Criteria that are not related to each others in terms of preferences are represented by orthogonal axes.

P 5: Similar alternatives are represented by points located close to each other.

P 6: Alternatives being good on a particular criterion are represented by points located in the direction of the corresponding criterion axis.

On the example of Figure 5.6, we observe:

- That the criteria $g_1(\cdot)$ and $g_3(\cdot)$ are expressing similar preferences and that the alternatives a_1 and a_5 are rather good on these criteria.
- That the criteria $g_6(\cdot)$ and $g_4(\cdot)$ are also expressing similar preferences and that the alternatives a_2, a_7, and a_8 are rather good on them.
- That the criteria $g_2(\cdot)$ and $g_5(\cdot)$ are rather independent
- That the criteria $g_1(\cdot)$ and $g_3(\cdot)$ are strongly conflicting with the criteria $g_4(\cdot)$ and $g_2(\cdot)$
- That the alternatives a_1, a_5 and a_6 are rather good on the criteria $g_1(\cdot)$, $g_3(\cdot)$ and $g_5(\cdot)$
- That the alternatives a_2, a_7 and a_8 are rather good on the criteria $g_6(\cdot)$, $g_4(\cdot)$ and $g_2(\cdot)$
- That the alternatives a_3 and a_4 are never good, never bad on all the criteria,
- ...

Although the GAIA plane includes only a percentage δ of the total information, it provides a powerful graphical visualisation tool for the analysis of a multicriteria problem. The discriminating power of the criteria, the conflicting aspects, as well as the "quality" of each alternative on the different criteria are becoming particularly clear.

5.3 The PROMETHEE Decision Stick. The PROMETHEE Decision Axis

Let us now introduce the impact of the weights in the GAIA plane. The vector of the weights is obviously also a vector of $\mathbb{R}^k$. According to (5.20), the PROMETHEE net flow of an alternative a_i is the scalar product between the vector of its single criterion net flows and the vector of the weights:

$$\begin{aligned} a_i : \quad & (\phi_1(a_i), \phi_2(a_i), \ldots, \phi_j(a_i), \ldots, \phi_k(a_i)), \\ w : \quad & (w_1, w_2, \ldots, w_j, \ldots, w_k). \end{aligned} \tag{5.22}$$

This also means that the PROMETHEE net flow of a_i is the projection of the vector of its single criterion net flows on w. Consequently, the relative positions of the projections of all the alternatives on w provides the PROMETHEE II ranking.

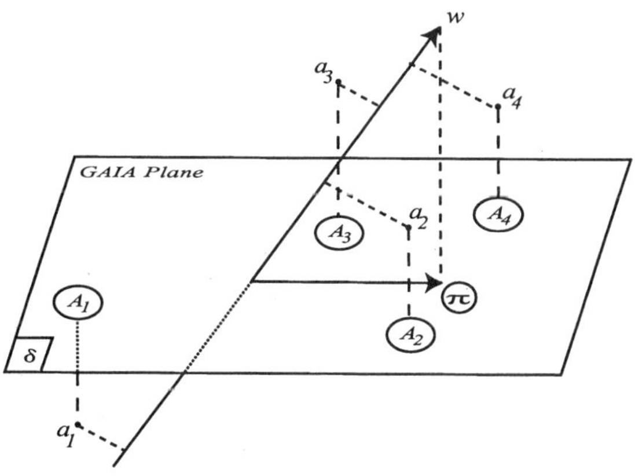

Figure 5.7. PROMETHEE II ranking. PROMETHEE decision axis and stick.

Clearly the vector w plays a crucial role. It can be represented in the GAIA plane by the projection of the unit vector of the weights. Let π be this projection, and let us call π the *PROMETHEE decision axis*.

On the example of Figure 5.7, the PROMETHEE ranking is: $a_4 \succ a_3 \succ a_2 \succ a_1$. A realistic view of this ranking is given in the GAIA plane although some inconsistencies due to the projection can possibly occur.

If all the weights are concentrated on one criterion, it is clear that the PROMETHEE decision axis will coincide with the axis of this criterion in the GAIA plane. Both axes are then the projection of a coordinate unit vector of $\mathbb{R}^k$. When the weights are distributed over all the criteria, the PROMETHEE decision axis appears as a weighted resultant of all the criterion axes $(C_1, C_2, \ldots, C_j, \ldots, C_k)$.

If π is long, the PROMETHEE decision axis has a strong decision power and the decision-maker is invited to select alternatives as far as possible in its direction.

If π is short, the PROMETHEE decision axis has no strong decision power. It means, according to the weights, that the criteria are strongly conflicting and that the selection of a good compromise is a hard problem.

When the weights are modified, the positions of the alternatives and of the criteria remain unchanged in the GAIA plane. The weight vector appears as a *decision stick* that the decision-maker can move according to his preferences in favour of particular criteria. When a sensitivity analysis is applied by modifying the weights, the PROMETHEE decision stick (w) and the PROMETHEE decision axis (π) are moving in such a way that the consequences for decision-making are easily observed in the GAIA plane (see Figure 5.8).

Decision-making for multicriteria problems appears, thanks to this methodology, as a piloting problem. Piloting the decision stick over the GAIA plane.

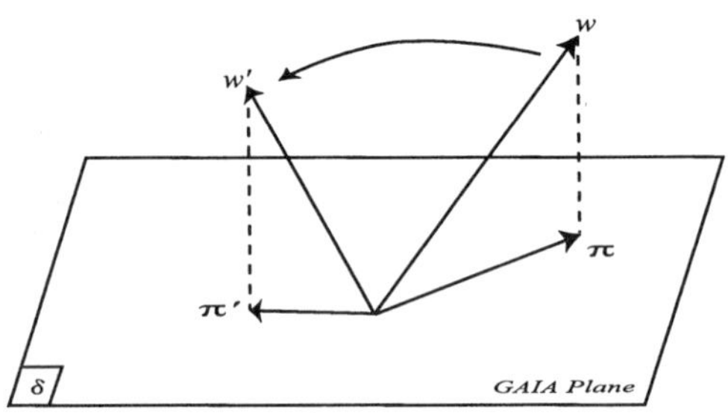

Figure 5.8. Piloting the PROMETHEE decision stick.

The PROMETHEE decision stick and the PROMETHEE decision axis provide a strong sensitivity analysis tool. Before finalising a decision we recommend to the decision-maker to simulate different weight distributions. In each case the situation can easily be appreciated in the GAIA plane, the recommended alternatives are located in the direction of the decision axis. As the alternatives and the criteria remain unchanged when the PROMETHEE decision stick is moving, the sensitivity analysis is particularly easy to manage. Piloting the decision stick is instantaneously operated by the PROMCALC and the DECISION LAB softwares. The process is displayed graphically so that the results are easy to appreciate.

6. The PROMETHEE VI Sensitivity Tool (The "Human Brain")

The PROMETHEE VI module provides the decision-maker with additional information on his own personal view of his multicriteria problem. It allows to appreciate whether the problem is *hard or soft* according to his personal opinion.

It is obvious that the distribution of the weights plays an important role in all multicriteria problems. As soon as the weights are fixed, a final ranking is proposed by PROMETHEE II. In most of the cases the decision-maker is hesitating to allocate immediately precise values of the weights. His hesitation is due to several factors such as *indetermination, imprecision, uncertainty, lack of control,* ... on the real-world situation.

However the decision-maker has usually in mind some order of magnitude on the weights, so that, despite his hesitations, he is able to give some intervals including their correct values. Let these intervals be:

$$w_j^- \leq w_j \leq w_j^+,\ j = 1, \ldots, k. \qquad (5.23)$$

Let us then consider the set of all the extreme points of the unit vectors associated to all allowable weights. This set is limiting an area on the unit hypersphere in $\mathbb{R}^k$. Let us project this area on the GAIA plane and let us call (HB) (*"Human Brain"*) the obtained projection. Obviously (HB) is the area including all the extreme points of the PROMETHEE decision axis (π) for all allowable weights.

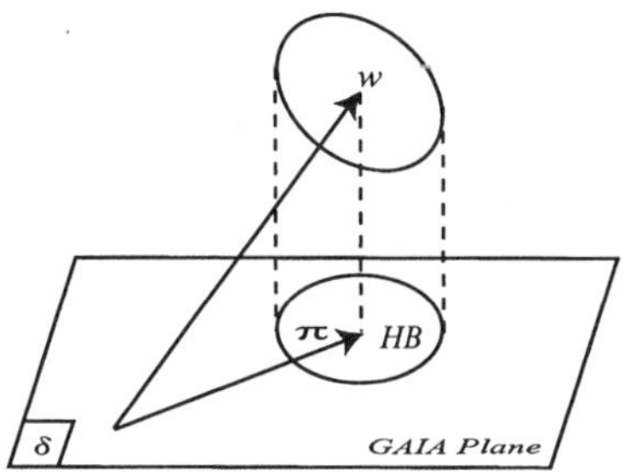

Figure 5.9. "Human Brain".

Two particular situations can occur:

S1: (HB) does not include the origin of the GAIA plane. In this case, when the weights are modified, the PROMETHEE decision axis (π) remains globally oriented in the same direction and all alternatives located in this direction are good. The multicriteria problem is rather easy to solve, it is a *soft* problem.

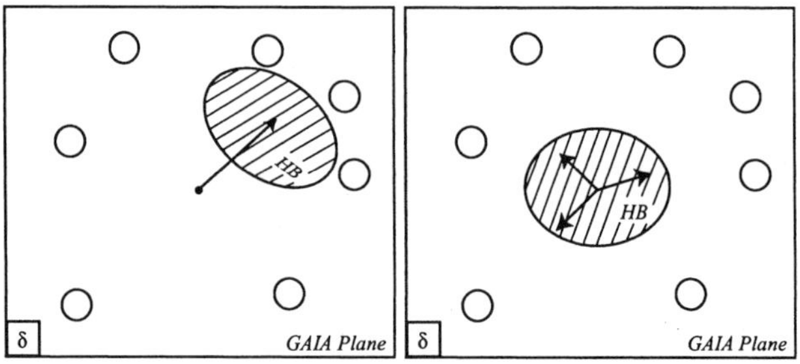

(a) Soft problem. (b) Hard problem.

Figure 5.10. Two types of decision problems.

S2: Reversely if (HB) is including the origin, the PROMETHEE decision axis (π) can take any orientation. In this case compromise solutions can be possibly obtained in all directions. It is then actually difficult to make a final decision. According to his preferences and his hesitations, the decision-maker is facing a *hard* problem.

In most of the practical applications treated so far, the problems appeared to be rather soft and not too hard. This means that most multicriteria problems offer at the same time good compromises and bad solutions. PROMETHEE allows to select the good ones.

7. PROMETHEE V: MCDA under Constraints

PROMETHEE I and II are appropriate to select one alternative. However in some applications a subset of alternatives must be identified, given a set of constraints. PROMETHEE V is extending the PROMETHEE methods to that particular case. (see [13]).

Let $\{a_i, i = 1, 2, \ldots, n\}$ be the set of possible alternatives and let us associate the following boolean variables to them:

$$x_i = \begin{cases} 1 & \text{if } a_i \text{ is selected,} \\ 0 & \text{if not.} \end{cases} \tag{5.24}$$

The PROMETHEE V procedure consists of the two following steps:

STEP 1: The multicriteria problem is first considered without constraints. The PROMETHEE II ranking is obtained for which the net flows $\{\phi(a_i), i = 1, 2, \ldots, n\}$ have been computed.

STEP 2: The following $\{0,1\}$ linear program is then considered in order to take into account the additional constraints.

$$\max\left\{\sum_{i=1}^{k}\phi(a_i)x_i\right\} \tag{5.25}$$

$$\sum_{i=1}^{n}\lambda_{p,i}x_i \sim \beta_p \quad p = 1, 2, \ldots, P \tag{5.26}$$

$$x_i \in \{0,1\} \quad i = 1, 2, \ldots, n, \tag{5.27}$$

where $\sim$ holds for $=$, $\geq$ or $\leq$. The coefficients of the objective function (5.25) are the net outranking flows. The higher the net flow, the better the alternative. The purpose of the $\{0,1\}$ linear program is to select alternatives collecting as much net flow as possible and taking the constraints into account.

The constraints (5.26) can include cardinality, budget, return, investment, marketing, ... constraints. They can be related to all the alternatives or possibly to some clusters.

After having solved the $\{0,1\}$ linear program, a subset of alternatives satisfying the constraints and providing as much net flow as possible is obtained. Classical 0-1 linear programming procedures may be used.

The PROMCALC software includes this PROMETHEE V procedure.

8. The PROMETHEE GDSS Procedure

The PROMETHEE Group Decision Support System has been developed to provide decision aid to a group of decision-makers $(DM_1), (DM_2), \ldots, (DM_r)$, $\ldots (DM_R)$ (see [54]). It has been designed to be used in a GDSS room including a PC, a printer and a video projector for the facilitator, and R working stations for the DM's. Each working station includes room for a DM (and possibly a collaborator), a PC and Tel/Fax so that the DM's can possibly consult their business base. All the PC's are connected to the facilitator through a local network.

There is no objection to use the procedure in the framework of teleconference or video conference systems. It this case the DM's are not gathering in a GDSS room, they directly talk together through the computer network.

One iteration of the PROMETHEE GDSS procedure consists in 11 steps grouped in three phases:

- PHASE I: Generation of alternatives and criteria
- PHASE II: Individual evaluation by each DM
- PHASE III: Global evaluation by the group

Feedback is possible after each iteration for conflict resolution until a final consensus is reached.

8.1 PHASE I: Generation of Alternatives and Criteria

STEP 1: First contact Facilitator — DM's
The facilitator meets the DM's together or individually in order to enrich his knowledge of the problem. Usually this step takes place in the business base of each DM prior to the GDSS room session.

STEP 2: Problem description in the GDSS room
The facilitator describes the computer infrastructure, the PROMETHEE methodology, and introduces the problem.

STEP 3: Generation of alternatives
It is a computer step. Each DM implements possible alternatives including their extended description. For instance strategies, investments, locations, production schemes, marketing actions, ... depending on the problem.

STEP 4: Stable set of alternatives
All the proposed alternatives are collected and displayed by the facilitator one by one on the video-screen, anonymously or not. An open discussion takes place, alternatives are canceled, new ones are proposed, combined ones are merged, until a stable set of n alternatives $(a_1, a_2, \ldots, a_i, \ldots, a_n)$ is reached. This brainstorming procedure is extremely useful, it often generates alternatives that were unforeseen at the beginning.

STEP 5: Comments on the alternatives
It is again a computer step. Each DM implements his comments on all the alternatives. All these comments are collected and displayed by the facilitator. Nothing gets lost. Complete minutes can be printed at any time.

STEP 6: Stable set of evaluation criteria
The same procedure as for the alternatives is applied to define a stable set of evaluation criteria $(g_1(\cdot), g_2(\cdot), \ldots, g_j(\cdot), \ldots g_k(\cdot))$. Computer and open discussion activities are alternating. At the end the frame of an evaluation table (Type Table 5.1) is obtained. This frame consists in a $(n \times k)$ matrix. This ends the first phase. Feedbacks are already possible to be sure a stable set of alternatives and criteria is reached.

8.2 PHASE II: Individual Evaluation by each DM

Let us suppose that each DM has a decision power given by a non-negative weight $(\omega_r, r = 1, 2, \ldots, R)$ so that:

$$\sum_{r=1}^{R} \omega_r = 1. \tag{5.28}$$

STEP 7: Individual evaluation tables

The evaluation table $(n \times k)$ has to be completed by each DM. Some evaluation values are introduced in advance by the facilitator if there is an objective agreement on them (prices, volumes, budgets, ...). If not each DM is allowed to introduce his own values.

All the DM's implement the same $(n \times k)$ matrix, if some of them are not interested in particular criteria, they can simply allocate a zero weight to these criteria.

STEP 8: Additional PROMETHEE information

Each DM develops his own PROMETHEE-GAIA analysis. Assistance is given by the facilitator to provide the PROMETHEE additional information on the weights and the generalised criteria.

STEP 9: Individual PROMETHEE-GAIA analysis

The PROMETHEE I and II rankings, the profiles of the alternatives and the GAIA plane as well as the net flow vector $\phi_r(\cdot)$ are instantaneously obtained, so that each DM gets his own clear view of the problem.

8.3 PHASE III: Global Evaluation by the Group

STEP 10: Display of the individual investigations

The rankings and the GAIA plane of each DM are collected and displayed by the facilitator so that the group of all DM'S is informed of the potential conflicts.

STEP 11: Global evaluation

The net flow vectors $\{\phi_r(\cdot), r = 1, 2, \ldots, R\}$ of all the DM's are collected by the facilitator and put in a $(n \times R)$ matrix. It is a rather small matrix which is easy to analyse. Each criterion of this matrix expresses the point of view of a particular DM.

Each of these criteria has a weight ω_r and an associated generalised criterion of Type 3 ($p = 2$) so that the preferences allocated to the deviations between the $\phi_i^r(\cdot)$ values will be proportional to these deviations.

A global PROMETHEE II ranking and the associated GAIA plane are then computed. As each criterion is representing a DM, the conflicts between them are clearly visualised in the GAIA plane. See for example Figure 5.11 where DM_3 is strongly in conflict with DM_1, DM_2 and DM_4.

The associated PROMETHEE decision axis (π) gives the direction in which to decide according to the weights allocated to the DM's.

If the conflicts are too sensitive the following feedbacks could be considered: Back to the weighting of the DM's. Back to the individual evaluations. Back to the set of criteria. Back to the set of alternatives. Back to the starting phase and to include an additional stakeholder ("DM") such as a social negotiator or a government mediator.

The whole procedure is summarised in the following scheme (Figure 5.12):

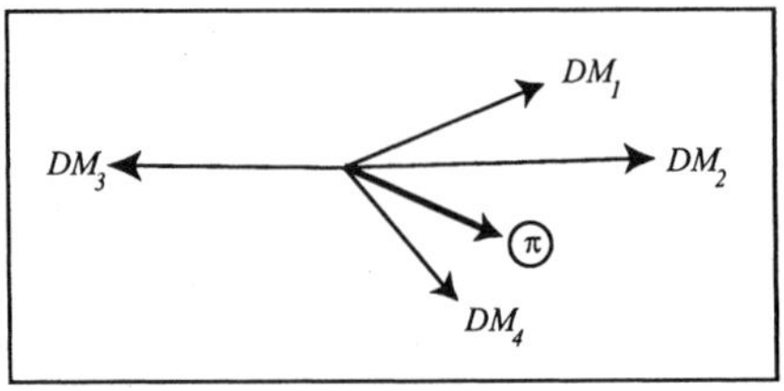

Figure 5.11. Conflict between DM's.

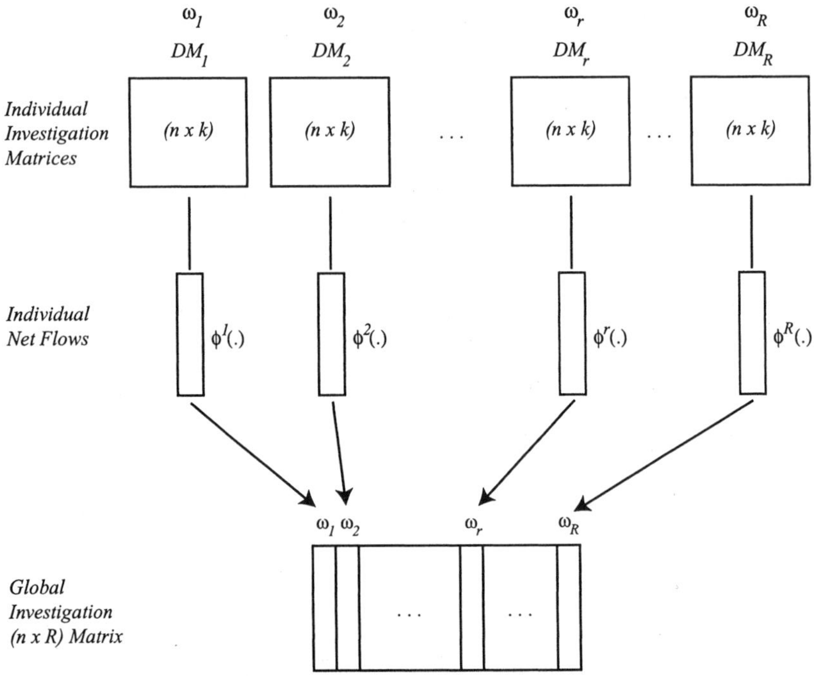

Figure 5.12. Overview PROMETHEE GDSS procedure.

9. The DECISION LAB Software

DECISION LAB is the current software implementation of the PROMETHEE and GAIA methods. It has been developed by the Canadian company Visual Decision, in cooperation with the authors. It replaces the PROMCALC software that the authors had previously developed.

DECISION LAB is a Windows application that uses a typical spreadsheet interface to manage the data of a multicriteria problem (Figure 5.13).

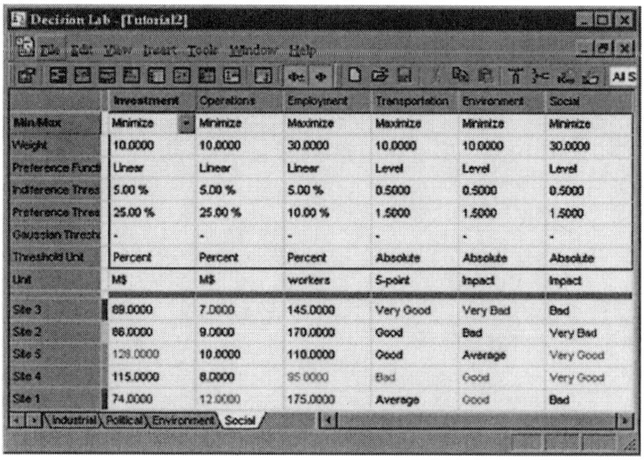

Figure 5.13. Main window.

All the data related to the PROMETHEE methods (evaluations, preference functions, weights, ...) can be easily defined and input by the user. Besides, DECISION LAB provides the user with additional features like the definition of qualitative criteria, the treatment of missing values in the multicriteria table or the definition of percentage (variable) thresholds in the preference functions. Categories of alternatives or criteria can also be defined to better identify sub-groups of related items and to facilitate the analysis of the decision problem.

All the PROMETHEE and GAIA computations take place in real-time and any data modification is immediately reflected in the output windows. The PROMETHEE rankings, action profiles and GAIA plane are displayed in separate windows and can easily be compared (Figure 5.14).

Several interactive tools and displays are available for facilitating extensive sensitivity and robustness analyses. It is possible to compute weight stability intervals for individual criteria or categories of criteria. The walking weights display (Figure 5.15) can be used to interactively modify the weights of the criteria and immediately see the impact of the modification on the PROMETHEE II complete ranking and on the position of the decision axis in the GAIA plane. This can particularly useful when the decision-maker has no clear idea of the appropriate weighting of the criteria and wants to explore his space of freedom.

The PROMETHEE GDSS procedure is also integrated in DECISION LAB through the definition of several scenarios for a same decision problem. Scenarios share the same lists of alternatives and criteria but can include different preference functions, different sets of weights and even different evaluations for some criteria. Each scenario can be analysed separately using PROMETHEE and GAIA. But it is also possible to aggregate all the scenarios and to generate

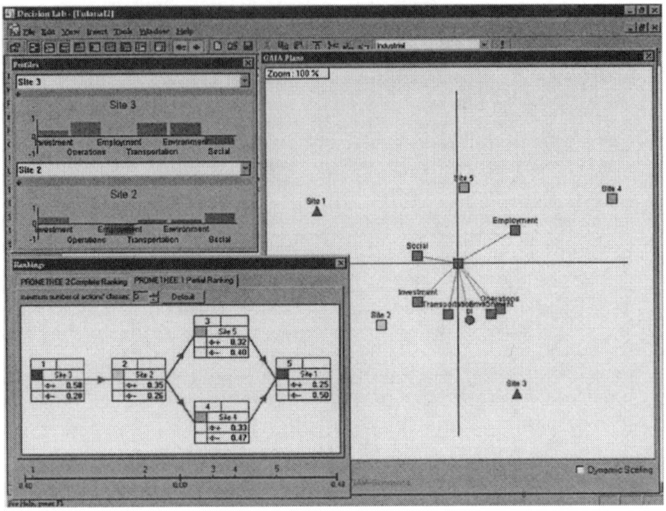

Figure 5.14. PROMETHEE rankings, action profiles, GAIA plane.

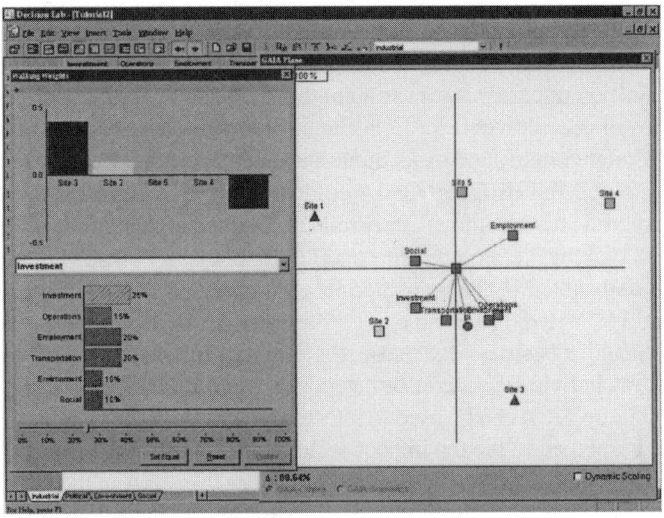

Figure 5.15. Walking weights.

the PROMETHEE group rankings as well as the group GAIA plane. Conflicts between decision-makers can easily be detected and analysed.

At the end of an analysis, the DECISION LAB report generator can produce tailor-made reports including the tables and graphics required by the user. The reports are in the html format so that they can easily be edited in a word processor or published on paper or on the web.

DECISION LAB can easily be interfaced with other programs like for instance databases. Its own interface can also be adapted to specific needs (special menus or displays, additional analysis modules, ...).

The next step in PROMETHEE software is a web-based implementation which is being developed under the Q-E-D name (Quantify-Evaluate-Decide). The Q-E-D demo web site will be launched during the spring 2003 at `http://www.q-e-d.be`.

Additional information on DECISION LAB can also be obtained on the following web sites: `http://www.idm-belgium.com` and `http://www.visualdecision.com`.

Acknowledgments

The authors wish to address their most sincere acknowledgements to their colleague and friend Professor Johan Springael for his valuable assistance and comments.

References

[1] M.F. Abutaleb and B. Mareschal. Water-resources planning in the Middle-East – Application of the Promethee-V multicriteria method. *European Journal of Operational Research*, 81(3):500–511, 1995.

[2] B. Al-Kloub and M.F. Abu-Taleb. Application of multicriteria decision aid to rank the Jordan-Yarmouk basin co-riparians according to the Helsinki and ILC rules. *Water International*, 23(3):164–173, 1998.

[3] D. Al-Rashdan, B. Al-Kloub, A. Dean, and T. Al-Shemmeri. Environmental impact assessment and ranking the environmental projects in Jordan. *European Journal of Operational Research*, 118(1):30–45, 1999.

[4] T. Al-Shemmeri, B. Al-Kloub, and A. Pearman. Model choice in multicriteria decision aid. *European Journal of Operational Research*, 97(3):550–560, 1997.

[5] Z. Babic and N. Plazibat. Ranking of enterprises based on multicriterial analysis. *International Journal of Production Economics*, 56(7):29–35, 1998.

[6] C. Bana e Costa. *Readings in Multiple Criteria Decision Aid*. Springer-Verlag, Berlin, Germany, 1990.

[7] D. Bouyssou and P. Perny. Ranking methods for valued preference relations – A characterization of a method based on leaving and entering flows. *European Journal of Operational Research*, 61(1-2):186–194, 1992.

[8] J.P. Brans. L'ingénièrie de la décision; Elaboration d'instruments d'aide à la décision. La méthode PROMETHEE. In R. Nadeau and M. Landry, editors, *L'aide à la décision: Nature, Instruments et Perspectives d'Avenir*, pages 183–213, Québec, Canada, 1982. Presses de l'Université Laval.

[9] J.P. Brans. The space of freedom of the decision maker modelling the human brain. *European Journal of Operational Research*, 92(3):593–602, 1996.

[10] J.P. Brans. Ethics and decision. *European Journal of Operational Research*, 136(2):340–352, 2002.

[11] J.P. Brans, C. Macharis, P.L. Kunsch, and M. Schwaninger. Combining multicriteria decision aid and system dynamics for the control of socio-economic processes. An iterative real-time procedure. *European Journal of Operational Research*, 109(2):428–441, 1998.

[12] J.P. Brans and B. Mareschal. BANK ADVISER: Un système interactif multicritère pour l'évaluation financière des entreprises à l'aide des méthodes PROMETHEE. *Actualité Economique*, 68(4), 1992.

[13] J.P. Brans and B. Mareschal. Promethee-V – MCDM problems with segmentation constraints. *INFOR*, 30(2):85–96, 1992.

[14] J.P. Brans and B. Mareschal. The Promcalc and Gaia decision-support system for multicriteria decision aid. *Decision Support Systems*, 12(4-5):297–310, 1994.

[15] J.P. Brans and B. Mareschal. The PROMETHEE VI procedure. How to differentiate hard from soft multicriteria problems. *Journal of Decision Systems*, 4:213–223, 1995.

[16] J.P. Brans and B. Mareschal. *PROMETHEE-GAIA. Une Méthodologie d'Aide à la Décision en Présence de Critères Multiples*. Ellipses, Paris, France, 2002.

[17] J.P. Brans, B. Mareschal, and Ph. Vincke. PROMETHEE: A new family of outranking methods in multicriteria analysis. In J.P. Brans, editor, *Operational Research '84*, pages 477–490. North-Holland, Amsterdam, 1984.

[18] J.P. Brans, B. Mareschal, and Ph. Vincke. How to select and how to rank projects: The PROMETHEE method. *European Journal of Operational Research*, 24(2):228–238, 1986.

[19] J.P. Brans and P. Mareschal. The PROMETHEE-GAIA decision support system for multicriteria investigations. *Investigation Operativa*, 4(2):107–117, 1994.

[20] J.P. Brans and P. Vincke. A preference ranking organisation method: The PROMETHEE method for MCDM. *Management Science*, 31(6):647–656, 1985.

[21] T. Briggs, P. Kunsch, and B. Mareschal. Nuclear waste management – An application of the multicriteria PROMRTHEE methods. *European Journal of Operational Research*, 44(1):1–10, 1990.

[22] C. Chareonsuk, N. Nagarur, and M.T. Tabucanon. A multicriteria approach to the selection of preventive maintenance intervals. *International Journal of Production Economics*, 49(1):55–64, 1997.

[23] G. Colson. The OR's prize winner and the software ARGOS: How a multijudge and multicriteria ranking GDSS helps a jury to attribute a scientific award. *Computers & Operations Research*, 27(7-8):741–755, 2000.

[24] G. Davignon and B. Mareschal. Specialization of hospital services in Quebec – An application of the PROMETHEE and GAIA methods. *Mathematical and Computer Modelling*, 12(10-11):1393–1400, 1989.

[25] I. De Leeneer and H. Pastijn. Selecting land mine detection strategies by means of outranking MCDM techniques. *European Journal of Operational Research*, 139(2):327–338, 2002.

[26] P. De Lit, P. Latinne, B. Rekiek, and Delchambre. A. Assembly planning with an ordering genetic algorithm. *International Journal of Production Research*, 39(16):3623–3640, 2001.

[27] D. Diakoulaki and N. Koumoutsos. Cardinal ranking of alternative actions – Extension of the PROMETHEE method. *European Journal of Operational Research*, 53(3):337–347, 1991.

[28] L.C. Dias, J.P. Costa, and J.L. Climaco. A parallel implementation of the PROMETHEE method. *European Journal of Operational Research*, 104(3):521–531, 1998.

[29] Ph. Du Bois, J.P. Brans, F. Cantraine, and B. Mareschal. MEDICIS: An expert system for computer-aided diagnosis using the PROMETHEE multicriteria method. *European Journal of Operational Research*, 39:284–292, 1989.

[30] H.A. Eiselt and G. Laporte. The use of domains in multicriteria decision-making. *European Journal of Operational Research*, 61(3):292–298, 1992.

[31] B. Espinasse, G. Picolet, and E. Chouraqui. Negotiation support systems: A multi-criteria and multi-agent approach. *European Journal of Operational Research*, 103(2):389–409, 1997.

[32] J. Esser. Complete, consistent and compatible preference relations. *OR Spektrum*, 23(2):183–201, 2001.

[33] M. Fendek. The multi-criteria and multi level evaluation of manufacturing capacities of industrial firms in the Slovak Republic. *Ekonomicky Casopis*, 43(10):746–763, 1995.

[34] G.M. Fernandez Barberis. New preference structures for multiple criteria decision making: Its extension to PROMETHEE methods. *Central European Journal for Operations Research and Economics*, 2(1):23–52, 1993.

[35] J. Geldermann and O. Rentz. Integrated technique assessment with imprecise information as a support for the identification of best available techniques (BAT). *OR Spektrum*, 23(1):137–157, 2001.

[36] J. Geldermann, T. Spengler, and O. Rentz. Fuzzy outranking for environmental assessment. Case study: iron and steel making industry. *Fuzzy Sets and Systems*, 115(1):45–65, 2001.

[37] E. Georgopoulou, Y. Sarafidis, and D. Diakoulaki. Design and implementation of a group DSS for sustaining renewable energies exploitation. *European Journal of Operational Research*, 109(2):483–500, 1998.

[38] M. Goumas and V. Lygerou. An extension of the PROMETHEE method for decision making in fuzzy environment: Ranking of alternative energy exploitation projects. *European Journal of Operational Research*, 123(3):606–613, 2000.

[39] M. Hababou and J.M. Martel. A multicriteria approach for selecting a portfolio manager. *INFOR*, 36(3):161–176, 1998.

[40] D.A. Haralambopoulos and H. Polatidis. Renewable energy projects: Structuring a multi-criteria group decision-making framework. *Renewable Energy*, 28(6):961–973, 2003.

[41] M.M.W.B. Hendriks, J.H. Deboer, A.K. Smilde, and D.A. Doornbos. Multicriteria decision-making. *Chemometrics and Intelligent Laboratory Systems*, 16(3):175–191, 1992.

[42] L. Hens, H. Pastijn, and W. Struys. Multicriteria analysis of the burden sharing in the European Community. *European Journal of Operational Research*, 59(2):248–261, 1992.

[43] J. Jablonsky. Promcalc, Gaia, Bankadviser – Software for methods of the class PROMETHEE. *Ekonomicko-Matematicky Obzor*, 27(2):188–190, 1991.

[44] A. Kangas, J. Kangas, and J. Pykäläinen. Outranking methods as tools in strategic natural resources planning. *Silva Fennica*, 35(2):215–227, 2001.

[45] J. Karkazis. Facilities location in a competitive environment – A PROMETHEE based multiple criteria analysis. *European Journal of Operational Research*, 42(3):294–304, 1989.

[46] H.R. Keller, D.L. Massart, and J.P. Brans. Multicriteria decision-making – A case-study. *Chemometrics and Intelligent Laboratory Systems*, 11(2):175–189, 1991.

[47] S. Kokot, G. King, H.R. Keller, and D.L. Massart. Application of chemometrics for the selection of microwave digestion procedures. *Analytica Chimica Acta*, 268(1):81–94, 1992.

[48] S. Kokot, G. King, H.R. Keller, and D.L. Massart. Microwave digestion – an analysis of procedures. *Analytica Chimica Acta*, 259(2):267–279, 1992.

[49] S. Kokot and T. D. Phuong. Elemental content of Vietnamese rice – Part 2. Multivariate data analysis. *Analyst*, 124(4):561–569, 1999.

[50] S. Kolli and H.R. Parsaei. Multicriteria analysis in the evaluation of advanced manufacturing technology using PROMETHEE. *Computers & Industrial Engineering*, 23(1-4):455–458, 1992.

[51] J.F. Le Teno and B. Mareschal. An interval version of PROMETHEE for the comparison of building products' design with ill-defined data on environmental quality. *European Journal of Operational Research*, 109(2):522–529, 1998.

[52] D. Lerche, R. Bruggemann, P. Sorensen, L. Carlsen, and O.J. Nielsen. A comparison of partial order technique with three methods of multi-criteria analysis for ranking of chemical substances. *Journal of Chemical Information and Computer Sciences*, 42(5):1086–1098, 2002.

[53] P.J. Lewi, J. Van Hoof, and P. Boey. Multicriteria decision-making using Pareto optimality and PROMETHEE preference ranking. *Chemometrics and Intelligent Laboratory Systems*, 16(2):139–144, 1992.

[54] C. Macharis, J. P. Brans, and B. Mareschal. The GDSS PROMETHEE procedure – A PROMETHEE-GAIA based procedure for group decision support. *Journal of Decision Systems*, 7:283–307, 1998.

[55] M. R. Mahmoud and L. A. Garcia. Comparison of different multicriteria evaluation methods for the Red Bluff diversion dam. *Environmental Modelling & Software*, 15(5):471–478, 2000.

[56] B. Mareschal. Stochastic multicriteria decision-making under uncertainty. *European Journal of Operational Research*, 26(1):58–64, 1986.

[57] B. Mareschal. Aide à la décision multicritère: Développements récents des méthodes PROMETHEE. *Cahiers du Centre d'Etudes en Recherche Operationelle*, 29:175–214, 1987.

[58] B. Mareschal. Weight stability intervals in multicriteria decision aid. *European Journal of Operational Research*, 33(1):54–64, 1988.

[59] B. Mareschal and J.P. Brans. Geometrical representations for MCDA. the GAIA module. *European Journal of Operational Research*, 34:69–77, 1988.

[60] B. Mareschal and J.P. Brans. BANK ADVISER. An industrial evaluation system. *European Journal of Operational Research*, 54:318–324, 1991.

[61] B. Mareschal and D. Mertens. BANKS: A multicriteria decision support system for financial evaluation in the international banking sector. *Journal of Decision Systems*, 1(2-3):175–189, 1992.

[62] B. Mareschal and D. Mertens. Evaluation financière par la méthode multicritère GAIA: Application au secteur de l'assurance. *Actualité Economique*, 68(4), 1992.

[63] J.M. Martel and B. Aouni. Incorporating the decision-makers preferences in the goal-programming model. *Journal of the Operational Research Society*, 41(12):1121–1132, 1990.

[64] J.M. Martel and B. Aouni. Multicriteria method for site selection – Example of an airport for New Quebec. *INFOR*, 30(2):97–117, 1992.

[65] N. J. Martin, B. St Onge, and J.P. Waaub. An integrated decision aid system for the development of Saint Charles River alluvial plain, Quebec, Canada. *International Journal of Environment and Pollution*, 12(2-3):264–279, 1999.

[66] K. Meier. Methods for decision making with cardinal numbers and additive aggregation. *Fuzzy Sets and Systems*, 88(2):135–159, 1997.

[67] N. Mladineo, I. Lozic, S. Stosic, D. Mlinaric, and T. Radica. An evaluation of multicriteria analysis for DSS in public-policy decision. *European Journal of Operational Research*, 61(1-2):219–229, 1992.

[68] N. Mladineo, J. Margeta, J.P. Brans, and B. Mareschal. Multicriteria ranking of alternative locations for small scale hydro plants. *European Journal of Operational Research*, 31:215–222, 1997.

[69] V. Mlynarovic. Complex evaluation and optimalization of financial structure of the firm. *Ekonomicky Casopis*, 43(10):734–745, 1995.

[70] V. Mlynarovic and E. Hozlar. PROMETHEE – a family of outranking methods in multi-criteria analyses. *Ekonomicko-Matematicky Obzor*, 25(4):435–452, 1989.

[71] Y.N. Ni, S.H. Chen, and S. Kokot. Spectrophotometric determination of metal ions in electroplating solutions in the presence of EDTA with the aid of multivariate calibration and artificial neural networks. *Analytica Chimica Acta*, 463(2):305–316, 2002.

[72] D.L. Olson. Comparison of three multicriteria methods to predict known outcomes. *European Journal of Operational Research*, 130(3):576–587, 2001.

[73] F. Ouellet and J.M. Martel. Multicriteria method for evaluation and selection of interdependent R-and-D projects. *Revue Canadienne des Sciences de l'Administration – Canadian Journal of Administrative Sciences*, 12(3):195–209, 1995.

[74] E. Ozelkan and L. Duckstein. Analysing water resources alternatives and handling criteria by multi criterion decision techniques. *Journal of Environmental Management*, 48(1):69–96, 1996.

[75] P.C. Pandey and A. Kengpol. Selection of an automated inspection system using multiattribute decision-analysis. *International Journal of Production Economics*, 39(3):289–298, 1995.

[76] H.R. Parsaei, M. Wilhelm, and S.S. Kolli. Application of outranking methods to economic and financial justification of CIM systems. *Computers & Industrial Engineering*, 25(1-4):357–360, 1993.

[77] M. Paruccini. *Applying Multiple Criteria Aid for Decision to Environmental Management.* Kluwer Academic Publishing, Dordrecht, Boston, 1994.

[78] I. Pavic and Z. Babic. The use of the PROMETHEE method in the location choice of a production system. *International Journal of Production Economics*, 23(1-3):165–174, 1991.

[79] K.S. Raju, L. Duckstein, and C. Arondel. Multicriterion analysis for sustainable water resources planning: A case study in Spain. *Water Resources Management*, 14(6):435–456, 2000.

[80] K.S. Raju and D.N. Kumar. Multicriterion decision making in irrigation planning. *Agricultural Systems*, 62(2):117–129, 1999.

[81] K.S. Raju and C.R. Pillai. Multicriterion decision making in performance evaluation of an irrigation system. *European Journal of Operational Research*, 112(3):479–488, 1999.

[82] K.S. Raju and C.R. Pillai. Multicriterion decision making in river basin planning and development. *European Journal of Operational Research*, 112(2):249–257, 1999.

[83] A. Raveh. Co-plot: A graphic display method for geometrical representations of MCDM. *European Journal of Operational Research*, 125(3):670–678, 2000.

[84] B. Rekiek, P. De Lit, and A. Delchambre. A multiple objective grouping genetic algorithm for assembly line design. *Journal of Intelligent Manufacturing*, 12(5-6):467–485, 2001.

[85] B. Rekiek, P. De Lit, F. Pellichero, T. L'Eglise, P. Fouda, E. Falkenauer, and A. Delchambre. Hybrid assembly line design and user's preferences. *International Journal of Production Research*, 40(5):1095–1111, 2002.

[86] B. Roy. *Méthodologie d'Aide à la Décision*. Economica, Paris, France, 1985.

[87] B. Roy and D. Bouyssou. *Aide Multicritère à la Décision. Méthodes et Cas*. Economica, Paris, France, 1993.

[88] P. Salminen, J. Hokkanen, and R. Lahdelma. Comparing multicriteria methods in the context of environmental problems. *European Journal of Operational Research*, 104(3):485–496, 1998.

[89] J. Sarkis. A comparative analysis of DEA as a discrete alternative multiple criteria decision tool. *European Journal of Operational Research*, 123(3):543–557, 2000.

[90] F. Ulengin, Y.I. Topcu, and S.Ö. Sahin. An integrated decision aid system for Bosphorus water-crossing problem. *European Journal of Operational Research*, 134(1):179–192, 2001.

[91] B. Urli. Multicriteria approach for resource reallocation at the Child and Youth Protection Centre of the Lower St.-Lawrence (CJEP-01). *Canadian Journal of Administrative Sciences – Revue Canadienne des Sciences de l'Administration*, 17(1):52–62, 2000.

[92] B. Urli and D. Beaudry. Multicriteria approach for allocation of financial resources in the area of health care. *RAIRO – Recherche Operationnelle/Operations Research*, 29(4):373–389, 1995.

[93] K. Vaillancourt and J.P. Waaub. Environmental site evaluation of waste management facilities embedded into EUGENE model: A multicriteria approach. *European Journal of Operational Research*, 139(2):436–448, 2002.

[94] G. Vanhuylenbroeck. The conflict-analysis method – Bridging the gap between ELECTRE, PROMETHEE and ORESTE. *European Journal of Operational Research*, 82(3):490–502, 1995.

[95] P. Vincke. *Multicriteria Decision Aid*. John Wiley & Sons, New York, 1989.

[96] D. Vuk, B.N. Kozelj, and N. Mladineo. Application of multicriterional analysis on the selection of the location for disposal of communal waste. *European Journal of Operational Research*, 55(2):211–217, 1991.

[97] W. Wolters and B. Mareschal. Novel types of sensitivity analysis for MCDM. *European Journal of Operational Research*, 81(2):281–290, 1995.

[98] X.Z. Xu. The SIR method: A superiority and inferiority ranking method for multiple criteria decision making. *European Journal of Operational Research*, 131(3):587–602, 2001.

[99] C. Zopounidis. Multicriteria decision aid in financial management. *European Journal of Operational Research*, 119(2):404–415, 1999.

Chapter 6

OTHER OUTRANKING APPROACHES

Jean-Marc Martel
Facultés des Sciences de l'Administration
University of Laval
Canada
jean-marc.martel@fsa.ulaval.ca

Benedetto Matarazzo
Department of Economics and Quantitative Methods
University of Catania
Corso Italia, 55, Catania
Italy
matarazz@unict.it

Abstract In this chapter, we shortly describe some outranking methods other than ELECTRE and PROMETHEE. All these methods (QUALIFLEX, REGIME, ORESTE, ARGUS, EVAMIX, TACTIC and MELCHIOR) propose definitions and computations of particular binary relations, more or less linked to the basic idea of the original ELECTRE methods. Beside them, we will also describe other outranking methods (MAPPAC, PRAGMA, IDRA and PACMAN) that have been developed in the framework of the Pairwise Criterion Comparison Approach (PCCA) methodology, whose peculiar feature is to split the binary relations construction phase in two steps: in the first one, each pair of actions is compared with respect to two criteria a time; in the second step, all these partial preference indices are aggregated in order to obtain the final binary relations. Finally, one outranking method for stochastic data (the Martel and Zaras' method) is presented, based on the use of stochastic dominance relations between each pair of alternatives.

Keywords: Multiple criteria decision analysis, outranking methods, pairwise criteria comparison approach.

1. Introduction

The outranking methods constitute one of the most fruitful approach in MCDA. They main feature is to compare all feasible alternatives or actions by pair building up some binary relations, crisp or fuzzy, and then to exploit in an appropriate way these relations in order to obtain final recommendations. In this approach, the ELECTRE family and PROMETHEE methods (see Chapters 4 and 5 in this book) are very well known and have been applied in a lot of real life problems. But beside them, there are also other outranking methods, interesting both from theoretical and operational points of view. All these methods propose definitions and computations of particular binary relations, more or less linked to the basic idea of the original ELECTRE methods, i.e. taking explicitly into account the reasons in favor and against an outranking relation (concordance-discordance analysis using appropriate veto thresholds). Some of these methods, moreover, present also a peculiar way to build up final recommendations, by exploiting the relations obtained in the previous step. In this chapter, we shortly describe some outranking methods other than ELECTRE and PROMETHEE. In Section 2 we present some outranking methods dealing with different kind of data (QUALIFLEX, REGIME, ORESTE, ARGUS, EVAMIX, TACTIC and MELCHIOR). Some of these methods are based on concordance-discordance analysis between the rankings of alternatives according to the considered criteria and the comprehensive ranking of them; others on direct comparison of each pair of alternatives, more or less strictly linked to the concordance-discordance analysis of ELECTRE type methods. In Section 3 some outranking methods (MAPPAC, PRAGMA, IDRA and PACMAN) are described. They have been developed in the framework of the Pairwise Criterion Comparison Approach (PCCA) methodology. Its peculiar feature is to split the binary relations construction phase in two steps: in the first one, each pair of actions is compared with respect to two criteria a time, among those considered in the problem, and partial preference indices are built up. In the second step, all these partial preference indices are aggregated in order to obtain the global indices and binary relations. An appropriated exploitation of these indices gives us the final recommendations. Finally, in Section 4 one outranking method for stochastic data (the Martel and Zaras' method) is presented. The main feature of this method is that the concordance-discordance analysis is based on the use of stochastic dominance relations on the set of feasible alternatives, comparing their cumulative distribution functions associated with each criterion. Some short conclusions are sketched in final Section.

2. Other Outranking Methods

The available information is not always of cardinal level; some times the evaluations of alternatives are ordinal scales, especially in social sciences. These evaluations may take the form of preorders. Several methods were been developed to aggregate this type of local evaluation in order to obtain a comprehensive comparison of alternatives. For example, we can mention Borda, Condorcet, Copeland, Blin, Bowmam and Colantoni, Kemeny and Snell, etc. (see [31]). Some methods that we will present in this Section drawn inspiration by some of them.

We present some outranking methods consistent with ordinal data, since they do not need to convert ordinal information to cardinal values, as it is the case, for example, in [15]. We will present some methods frequently mentioned in the literature on MCDA, where the general idea of outranking is globally implemented: QUALIFLEX, REGIME, ORESTE, ARGUS, EVAMIX, TACTIC and MELCHIOR, these methods are not too complex and do not introduce the mathematical programming within their algorithm as it is the case, for example, in [3]. We present also EVAMIX even if it was been developed for ordinal and cardinal evaluations.

2.1 QUALIFLEX

The starting point of QUALIFLEX [28, 27] was a generalization of Jacquet-Lagrèze's permutation method [8].

It is a metric procedure and it is based on the evaluation of all possible rankings (permutations) of alternatives under consideration. Its mechanism of aggregation is based on Kemeny and Snell's rule.

This method is based on the comparison among the comprehensive ranking of the alternatives and the evaluations of alternatives according to each criterion from family F (impact matrix). These evaluations are ordinal and take the form of preorders. For each permutation, one computes a concordance/discordance index for each couple of alternatives, that reflects the concordance and the discordance of their ranks and their evaluation preorders from the impact matrix. This index is firstly computed at the level of single criterion, after at a comprehensive level with respect to all possible rankings. One tries to identify the permutation that maximizes the value of this index, i.e. the permutation whose ranking best reflects (the best compromise between) the preorders according to each criterion from F and the multi-criteria evaluation table.

The information concerning the coefficients of relative importance (weights) of criteria may be explicitly known or expressed as a ranking (for example a preorder). In this case, [27] has show that one can circumscribe the exploration to extreme points (the vertex) of polyhedron formed by the feasible weights.

Given the set of alternatives A, the concordance/discordance index for each couple of alternatives $(a, b), a, b \in A$, at the level of preorder according to the criterion $g_j \in F$ and the ranking corresponding to the kth permutation is:

$$I_{jk}(a,b) = \begin{cases} 1 & \text{if there is concordance} \\ 0 & \text{if there is ex aequo} \\ -1 & \text{if there is discordance.} \end{cases}$$

There is concordance (discordance) if a and b are ranked (not ranked) in the same order within the two preorders, and *ex aequo* if they have the same rank. The concordance/discordance index between the pre-order according to the criterion g_j and the ranking corresponding to the kth permutation is:

$$I_{jk} = \sum_{a,b \in A} I_{jk}(a,b).$$

The comprehensive concordance/discordance index for the kth permutation is:

$$I_k = \sum_j \pi_j I_{jk}(a,b),$$

where π_j is the weight of criterion g_j, $j = 1, 2, \ldots, n$. The number of permutations k (Per_k) is $m!$ where $m = |A|$. The best compromise corresponds to the permutation that maximize I_k. If π_j are not explicitly known, but expressed by a ranking, then the best compromise is the permutation that:

$$\max_{P(\pi_j)} I_k,$$

where $P(\pi_j)$ is the set of feasible weights

EXAMPLE 6 *Given 3 alternatives $a_1, a_2, a_3 \in A$; 3 criteria g_1, g_2, g_3 and the evaluation table (see Table 6.1 where a rank number 1 indicates the best outcome, while a rank 3 is assigned to the worst outcome with respect to each criterion), there are 3! possible permutations:*

$$\begin{array}{rl} Per_1: & a_1 > a_2 > a_3 \\ Per_2: & a_2 > a_1 > a_3 \\ Per_3: & a_2 > a_3 > a_1 \\ Per_4: & a_3 > a_2 > a_1 \\ Per_5: & a_3 > a_1 > a_2 \\ Per_6: & a_1 > a_3 > a_2. \end{array}$$

One index is computed for each pair (g_j, Per_k), that, for our example, give a total of 18 concordance/discordance indices. For example for the pair (g_1, Per_1), we have for the criterion g_1: $a_1 > a_2$, $a_2 \approx a_3$, $a_1 > a_3$, and for

Table 6.1. Rank evaluation of alternatives (impact matrix).

		Criterion		
		g_1	g_2	g_3
Alternative	a_1	1	2	1
	a_2	2	1	3
	a_3	2	3	2

Table 6.2. The concordance/discordance indices.

		Criterion		
		g_1	g_2	g_3
Permutation	Per_1	2	1	1
	Per_2	0	3	-1
	Per_3	-2	1	-3
	Per_4	-2	-1	-3
	Per_5	0	-3	1
	Per_6	2	-1	3

the Per_1*:* $a_1 > a_2$*,* $a_1 > a_3$*,* $a_2 > a_3$*, that gives* $+1$ *for the couple* (a_1, a_2)*,* $+1$ *for the couple* (a_1, a_3) *and 0 for the couple* (a_2, a_3)*. Thus, the value of the index* I_{11} *is equal to 2.*

The concordance/discordance indices are given in the Table 6.2.

Concerning the weights, for example:

1 *If the three criteria have the same importance, i.e.* $\pi_j = \frac{1}{3}, j = 1, 2, 3$*, then we obtain that the maximum value of the index is* $\frac{4}{3}$ *for the permutations* Per_1 *and* Per_6*.*

2 *If we know that* $\pi_1 \geq \pi_2$*,* $\pi_2 \geq \pi_3$ *and* $\pi_j \geq 0$ *for all j, then* $\pi_3 = 1 - \pi_1 - \pi_2$ *(see Figure 6.1).*

Then, to obtain the permutation that maximizes the index I_k*, we must check for the three vertices* $(1, 0)$*,* $(\frac{1}{2}, \frac{1}{2})$ *and* $(\frac{1}{3}, \frac{1}{3})$*. The maximum value of the index is equal to 2 for the permutations* Per_1 *and* Per_6*, for the weights (1,0,0).*

The result of this method is a ranking of alternatives under consideration. QUALIFLEX is based on pairwise criterion comparison of alternatives, but no outranking relation is constructed. An important limitation of this method concerns the fact that the number of permutations increases tremendously with the number of alternatives. This problem may be solved. Ancot [1] formulated this problem as a particular case of Quadratic Assignment Problem; this algorithm is implemented in the software MICROQUALIFLEX.

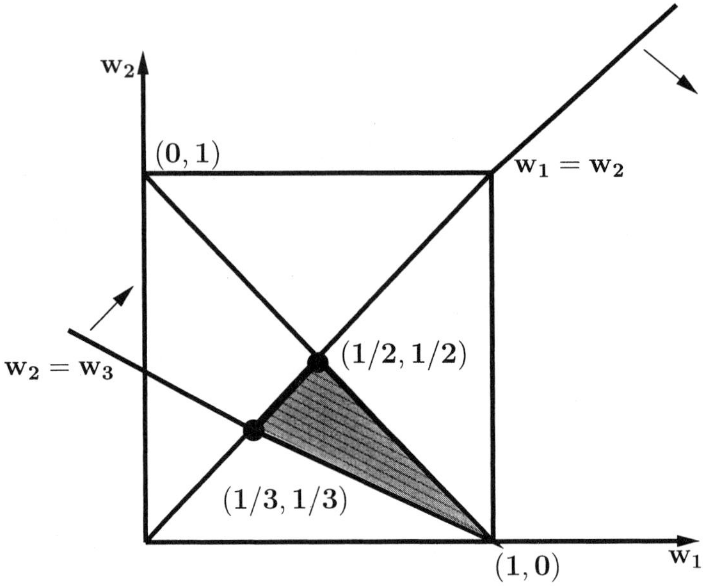

Figure 6.1. Set of feasible weights.

2.2 REGIME

The REGIME method [9, 10] can be viewed as an ordinal generalization of pairwise comparison methods such as concordance analysis. The starting point of this method is the concordance C_{il} defined in the following way:

$$C_{il} = \sum_{j \in \hat{C}_{il}} \pi_j,$$

where $\hat{C}_{il}$ is the concordance set, i.e. the set of criteria for which a_i is at least as good as a_l, a_i and $a_l \in A$ and π_j is the weight of criterion $g_j \in F$. The focus of this method is on the sign of $C_{il} - C_{li}$ for each pair of alternatives. If this sign is positive, alternative a_i is preferred to a_l; and the reverse if the sign is negative.

The first step of the REGIME method is the construction of the so-called regime matrix. The regime matrix is formed by pairwise comparison of alternatives in the multi-criteria evaluation table. Given a and $b \in A$, for every criterion we check whether a has a better rank than b, then on the corresponding place in the regime matrix the number $+1$ is noted, while if b is a better position than a, the number -1 is the result.

More explicitly, for each criterion $g_j, j = 1, 2, \cdots, n$, we can defined an indicator $c_{il,j}$ for each pair of alternatives (a_i, a_l).

$$c_{il,j} = \begin{cases} +1 & \text{if } r_{ij} < r_{lj} \\ 0 & \text{if } r_{ij} = r_{lj} \\ -1 & \text{if } r_{ij} > r_{lj}, \end{cases}$$

where r_{ij} (r_{lj}) is the rank of the alternative a_i (a_l) according to criterion g_j. When two alternatives are compared on all criteria, it is possible to form a vector

$$c_{il} = (c_{il,1}, \cdots, c_{il,j}, \cdots, c_{il,n})$$

that is called a regime and the regime matrix is formed of these regimes. These regimes will be used to determine rank order of alternatives.

The concordance index, in favor of the alternative a_i, is given by:

$$C_{il} = \sum_j \pi_j c_{il,j},$$

If the π_j are explicitly known, we can obtain a concordance matrix $\mathbf{C} = [C_{il}]$, with zero on the main diagonal (Table 6.3).

Table 6.3. Concordance matrix.

$$\begin{array}{c} \\ a_1 \\ \vdots \\ a_i \\ \vdots \\ a_m \end{array} \begin{array}{c} \begin{array}{ccccccccc} a_1 & \cdots & \cdots & \cdots & \cdots & \cdots & a_l & \cdots & \cdots & a_m \end{array} \\ \left[\begin{array}{ccccccccc} 0 & & & & & & \vdots & & \\ & & & & & & \vdots & & \\ \cdots & \cdots & \cdots & \cdots & \cdots & \cdots & C_{il} & \cdots & \cdots & \cdots \\ & & & & & & \vdots & & \\ & & & & & & \vdots & & 0 \end{array}\right] \end{array}$$

One half of this matrix can be ignored, since $C_{il} = -C_{li}$.

In general the available information concerning the weights is not explicit (not quantitative) and then it is not possible to compute the matrix $\mathbf{C}$. If the available information concerning the weights is ordinal, the sign of C_{il} may be determined with certainty only for some regimes [30]. For others regimes a such unambiguous result can not be obtained; such regime is called critical regime.

EXAMPLE 7 *We can illustrated this method on the basis of multi-criteria evaluation table with 3 alternatives and 4 criteria (Table 6.4, [10]).*

Table 6.4. Rank evaluation of alternatives (impact matrix).

		Criterion			
		g_1	g_2	g_3	g_4
Alternatives	a_1	3	1	1	2
	a_2	2	2	3	1
	a_3	1	3	2	3

Table 6.5. Regime matrix.

		Criterion			
		g_1	g_2	g_3	g_4
Comparison	(a_1,a_2)	-1	$+1$	$+1$	-1
	(a_1,a_3)	-1	$+1$	$+1$	$+1$
	(a_2,a_1)	$+1$	-1	-1	$+1$
	(a_2,a_3)	-1	$+1$	-1	$+1$
	(a_3,a_1)	$+1$	-1	-1	-1
	(a_3,a_2)	$+1$	-1	$+1$	-1

For this example, the regime matrix is presented in the Table 6.5.

If we make the hypothesis that $\pi_1 = \pi_2 = \pi_3 = \pi_4 = \frac{1}{4}$, we find $C_{12} = 0$, $C_{13} > 0$, $C_{21} = 0$, $C_{23} = 0$, $C_{31} < 0$ and $C_{32} = 0$. Thus a_1 is preferred to a_3, but we can not conclude between a_1 and a_2, a_2 and a_3. If we know for example that:

$$\pi_2 \geq \pi_4 \geq \pi_3 geq \pi_1, \sum_j \pi_j = 1 \text{ and } \pi_j \geq 0,$$

then we find that $C_{12} = -\pi_1 + \pi_2 + \pi_3 - \pi_4 \geq 0$ in all cases, which means that, on the basis of a pairwise comparison, a_1 is preferred to a_2. In a similar way it can be shown that, given the same information on the weights, a_1 is preferred to a_3, and that a_2 is preferred to a_3. Thus we arrive at a transitive rank order of alternatives.

It is not possible to arrive at such definite conclusions for all rankings of the weights. If we assume that:

$$\pi_1 \geq \pi_2 \geq \pi_3 \geq \pi_4, \sum_j \pi_j = 1 \text{ and } \pi_j \geq 0,$$

it is easy to see that from the first regime may result both positive and negative values of C_{il}'s. For example if π = (.40, .30, .25, .05), $C_{12} > 0$, whereas for π = (.45, .30, .15, .10), $C_{12} < 0$. Therefore, the corresponding regime is called a critical regime. The main idea of regime analysis is to circumvent these difficulties by partitioning the set of feasible weights so that for each region a final conclusion can be drawn about the sign of C_{il}.

Let the ordinal information available about the weights be:

$$\pi_1 \geq \pi_2 \geq \pi_3 \geq \pi_4, \sum_j \pi_j = 1 \text{ and } \pi_j \geq 0.$$

The set of weights satisfying this information will be denoted as T. We have to check, for all regimes c_{il}, if c_{il} may assume both positive and negative values, given that π is an element of T. The total number of regimes to be examined is $2^n = 2^4 = 16$. For our example, the number of critical regimes is equal to four:

$$\begin{array}{cccc} -1 & +1 & +1 & -1 \\ +1 & -1 & -1 & +1 \\ -1 & +1 & +1 & +1 \\ +1 & -1 & -1 & -1 \end{array}$$

The number of critical regimes is even, since we known that if c_{il} is a critical regime then $c_{li} = -c_{il}$ is critical. The subsets of T can be characterized by means of the structure of the critical regimes. The four critical regimes of our example give two critical equations:

$$\begin{aligned} f_1(\pi) = \pi_1 - \pi_2 - \pi_3 + \pi_4 &= 0 \\ f_2(\pi) = \pi_1 - \pi_2 - \pi_3 - \pi_4 &= 0. \end{aligned}$$

The following subsets of T can be distinguished by means of these equations:

$$\begin{aligned} T_1 &= T \cap \{\pi : f_1(\pi) > 0 \text{ and } f_2(\pi) > 0\}, \\ T_2 &= T \cap \{\pi : f_1(\pi) > 0 \text{ and } f_2(\pi) < 0\}, \\ T_3 &= T \cap \{\pi : f_1(\pi) < 0 \text{ and } f_2(\pi) < 0\}, \\ T_4 &= T \cap \{\pi : f_1(\pi) < 0 \text{ and } f_2(\pi) > 0\}. \end{aligned}$$

An examination of $T_1, \ldots, T_4$ reveals that T_4 is empty, so that ultimately three relevant subsets remain. The subsets T_1, T_2 and T_3 are convex polyhedra, as it is the case for the set T. The extreme points of these polyhedra can be determined graphically in the case of four criteria. The extreme points for T are:

$$\begin{array}{ll} A: & (1, 0, 0, 0) \\ B: & \left(\frac{1}{2}, \frac{1}{2}, 0, 0\right) \\ C: & \left(\frac{1}{3}, \frac{1}{3}, \frac{1}{3}, 0\right) \\ D: & \left(\frac{1}{4}, \frac{1}{4}, \frac{1}{4}, \frac{1}{4}\right) \end{array}$$

In addition to these four points, the extreme points

$$E: \quad \left(\frac{1}{2}, \frac{1}{4}, \frac{1}{4}, 0\right) \text{ and}$$
$$F: \quad \left(\frac{1}{2}, \frac{1}{6}, \frac{1}{6}, \frac{1}{6}\right)$$

are needed to characterize T_1, T_2 and T_3. The characterization of T_1, T_2 and T_3 by means of the extreme points are for T_1: A, B, E, F; for T_2: B, D, E, F and for T_3: B, C, D, E.

Once the partitioning of the weight set has been achieved, for each subset of T it is possible to indicate unambiguously the sign of C_{il} for each pair of alternatives. Let ν_{il} be defined as follows:

$$\begin{aligned} \nu_{il} &= +1 \text{ if } C_{il} > 0, \\ \nu_{il} &= -1 \text{ if } C_{il} < 0. \end{aligned}$$

Then a pairwise comparison matrix $\mathbf{V}$ can be constructed consisting of elements equal to +1 or -1, and zeros on the main diagonal. A final ranking of alternatives can be achieved on the basis of $\mathbf{V}$.

For example, take an interior point of subset T_1 (e.g. the centroid computed as the mean of the extreme points). Determine the sign of C_{il} for all regimes occurring in the regime matrix (Table 6.5). Thus we find for the pairwise comparison matrix $\mathbf{V_1}$:

$$\mathbf{V_1} = \begin{bmatrix} 0 & -1 & -1 \\ +1 & 0 & -1 \\ +1 & +1 & 0 \end{bmatrix}$$

On the basis of $\mathbf{V_1}$ we may conclude that a_3 is preferred to a_2 which in turn is preferred to a_1. For the two other subsets of weights we find:

$$\mathbf{V_2} = \begin{bmatrix} 0 & -1 & +1 \\ +1 & 0 & -1 \\ -1 & +1 & 0 \end{bmatrix} \qquad \mathbf{V_3} = \begin{bmatrix} 0 & +1 & +1 \\ -1 & 0 & -1 \\ -1 & +1 & 0 \end{bmatrix}.$$

The second pairwise comparison matrix does not give a definitive ranking of alternatives, but on the basis of $\mathbf{V_3}$ we may conclude that a_1 is preferred to a_3 which is again preferred to a_2.

The relative size of subsets T_1, T_2 and T_3 are not equal. If we assume that the weights are uniformly distributed in T, the relative size of the subsets of T can be interpreted as the probability that alternative a_i is preferred to a_l. Probabilities are aggregated to produce an overall ranking of alternatives. The relative sizes of the subsets can also be estimated using a random generator. This is recommended if there are seven criteria or more, since the number of subsets increases exponentially with the number of criteria [30].

The relevant subsets given an arbitrary number of criteria can be found in [10]. The REGIME method can be applied to mixed evaluations (ordinal and cardinal criteria) without losing the information contained in the quantitative evaluation. This requires a standardization of the quantitative evaluation. Israels and Keller [12] has been proposed a variant of REGIME method where the incomparability is accepted. The REGIME method is implemented in a system to support Decision on a finite set of alternatives: DEFINITE [13].

2.3 ORESTE

ORESTE (see [32, 33]) has been developed to deal with the situation where the alternatives are ranked according to each criterion and the criteria themselves are ranked according to their importance. In fact the ORESTE method can deal with the following multi-criteria problem. Let A be a finite set of alternatives a_i, $i = 1, 2, \cdots, m$. The consequences of the alternatives are analysed by a family F of n criteria. The relative importance of the criteria is given by a preference structure on the set of criteria F, which can be defined by a complete preorder S (the relation $S = I \cup P$ is strongly complete and transitive, the indifference I is symmetric and the preference P is asymmetric). For each criterion g_j, $j = 1, 2, \cdots, n$, we consider a preference structure on the set A, defined by a complete preorder. The objective of the method is to find a global preference structure on A which reflects the evaluation of alternatives on each criterion and the preference structure among the criteria.

The ORESTE method operates in three distinct phases:

First phase. Projection of the position-matrix.

Second phase. Ranking the projections.

Third phase. Aggregation of the global ranks.

We start from n complete preorders of the alternatives from A related to the n criteria, (for each alternative is given a rank with respect to each criterion). Also for each criterion is given a rank related to its position in the complete preorder among the criteria. The mean rank discussed by Besson [2] is used. For example, if the following preorder is given for the criteria $g_1 P g_2 I g_3 P g_4$, then $r_1 = 1, r_2 = r_3 = 2.5$ and $r_4 = 4$, where r_j is the Besson-rank of criterion g_j; idem for the alternatives, $r_j(a)$ is the average (Besson) rank of alternative a with respect to the criterion g_j. Given $\{r_j(a), r_j\}$, ORESTE tries to build a preference structure $O = \{I, P, R\}$ on A such as:

- $a_i P a_l$ if a_i is comprehensively preferred to $a_l (O_{il} = 1, O_{li} = 0)$,

- $a_i I a_l$if a_i is indifferent to $a_l (O_{il} = O_{li} = 1)$,
- $a_i R a_l$if a_i and a_l are comprehensively incomparable $(O_{il} = O_{li} = 0)$.

Projection. Considering an arbitrary origin **0**, a distance $d(0, a_j)$ is defined with the use of $\{r_j(a), r_j\}$ such that $d(0, a_j) < d(0, b_j)$ if aP_jb, where $a_j = g_j(a)$ is the evaluation of alternative a with respect to criterion g_j. When ties occur, an additional property is: if $g_j I g_k$ and $r_j(a) = r_k(b)$ then $d(0, a_j) = d(0, b_k)$. For the author, the "city-block" distance is adequate:

$$d(0, a_j) = \alpha r_j(a) + (1 - \alpha) r_j,$$

where α stands for a substitution rate $(0 < \alpha < 1)$. The projection may be performed in different ways [29, 33].

EXAMPLE 8 *Given the following example with 3 alternatives and 3 criteria (without ties). The complete preorders of alternatives are: aP_1bP_1c, bP_2cP_2a and cP_3aP_3b, and for the criteria: $g_1Pg_2Pg_3$. This example may be visualized by a position matrix (Table 6.6).*

Table 6.6. Position-matrix.

$$r_j(.) : \quad \begin{array}{c} \\ a \\ b \\ c \end{array} \begin{array}{c} \begin{array}{ccc} 1 & 2 & 3 \end{array} \\ \begin{bmatrix} 1 & 3 & 2 \\ 2 & 1 & 3 \\ 3 & 2 & 1 \end{bmatrix} \end{array}$$

Being $r_1 = 1$, $r_2 = 2$, $r_3 = 3$, the city-block distance for this example is given in Table 6.7.

Table 6.7. City-block.

$$d(0, a_j) : \quad \begin{array}{c} \\ a \\ b \\ c \end{array} \begin{array}{c} \begin{array}{ccc} 1 & 2 & 3 \end{array} \\ \begin{bmatrix} 1 & 2 + \alpha & 3 - \alpha \\ 1 + \alpha & 2 - \alpha & 3 \\ 1 + 2\alpha & 2 & 3 - 2\alpha \end{bmatrix} \end{array}$$

Ranking. Since it is the relative position of projections that is important and not the exact value of $d(0, a_j)$, the projections will be ranked. To rank the projections a mean rank $R(a_j)$ is assigned to a pair (a, g_j) such that $R(a_j) \leq R(b_k)$ if $d(0, a_j) \leq d(0, b_k)$. These ranks are called comprehensive ranks and are in the closed interval $(1, mn)$. For our example $R(a_1) < R(b_2)$ since $1 < 2 - \alpha$ $(0 < \alpha < 1)$.

Aggregation. For each alternative one computes the summation of their comprehensive ranks over the set of criteria. For an alternative a this yields the final aggregation

$$R(a) = \sum_j R(a_j).$$

For our example, if $\frac{1}{3} < \frac{\alpha}{2} < \frac{1}{2}$ we obtain:

$$\begin{array}{ccccccccccccc}
1 & < & 1+\alpha & < & 2-\alpha & < & 1+2\alpha & < & 2 & < & 3-2\alpha & < & 2+\alpha \\
R(a_1) & < & R(b_1) & < & R(b_2) & < & R(c_1) & < & R(c_2) & < & R(c_3) & < & R(a_2) \\
1 & & 2 & & 3 & & 4 & & 5 & & 6 & & 7 \\
 & < & 3-\alpha & < & 3 & & & & & & & & \\
 & < & R(a_3) & < & R(b_3) & & & & & & & & \\
 & & 8 & & 9 & & & & & & & &
\end{array}$$

$$R(a) = 16, R(b) = 14, R(c) = 15.$$

In the ORESTE method, the following index is also computed:

$$C(a, b) = \sum_{j:aP_jb} [R(b_j) - R(a_j)].$$

It is easily shown that $C(a, b) - C(b, a) = R(b) - R(a)$. Moreover, the maximum value of $R(b) - R(a)$ equals $n^2(m - 1)$.

For our example with $\frac{1}{3} < \alpha < \frac{1}{2}$, we obtain: $C(c, b) = 3$, $C(a, b) = 2$ and $C(a, c) = 3$. Thus, we may obtain the preference structure $O = \{I, P, R\}$ in such way that if $R(a) \leq R(b)$ then aIb or aPb or aRb, where β stands for an indifference level and γ for an incomparability level (see Figure 6.2).

For our example with $\frac{1}{3} < \alpha < \frac{1}{2}$, we have $\frac{C(c,b)}{R(c)-R(b)} = 3$, $\frac{C(a,b)}{R(a)-R(b)} = 1$ and $\frac{C(a,c)}{R(a)-R(c)} = 3$. Thus, if $\beta = \frac{1}{18} = \frac{1}{n^2(m-1)}$ and $1 \leq \gamma \leq 3$ we obtain bPa, aRc and cRb.

These thresholds are interpreted in [29]. When $\gamma = \infty$, the outranking relation is a semi-order which becomes a weak order if $\beta = 0$.

The global preference relation P built by ORESTE is transitive [29]. The axiom known as the Pareto principle or citizen's sovereignty holds if $\beta < \frac{1}{n(m-1)}$, but the axiom of independence of irrelevant alternatives is generally violated [33].

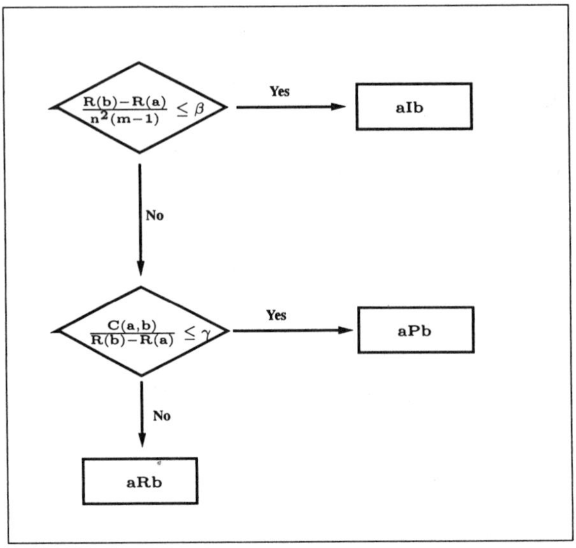

Figure 6.2. ORESTE flow chart.

2.4 ARGUS

The ARGUS method [14] uses qualitative values for representing the intensity of preference on an ordinal scale. They express this intensity of preference between two alternatives $a, b \in A$ by selecting one of the following qualitative relations: indifference, small, moderate, strong or very strong preference. All evaluation on the criteria are treated as evaluations on an ordinal scale, but the evaluations of each alternative with respect to each criterion can be quantitative (interval or ratio scale) or qualitative (ordinal scale).

The way of obtaining the required information from the decision maker (DM) to model his/her preference structure, depends on the scale of measurement of each criterion. If the scale is ordinal, we may use the following possible values: very poor, poor, average, good, very good. To model the preference structure of the DM on this criterion, the DM must indicate his/her preference for each pair of values. He must construct a preference matrix (Table 6.8).

In fact the DM must fill only the lower triangle of this matrix. The number of rows and columns of this matrix depends on the number of different values the ordinal criterion can have. The preference of the DM on an interval scale criterion will depend on $d = g_j(a) - g_j(b)$, while his/her preference on a ratio scale criterion will depend either on d only or on d, $g_j(a)$ and $g_j(b)$. For example, if his/her preference depends on d only, this means that only

Table 6.8. Preference matrix for a criterion with ordinal evaluation.

$g_i(b)$	very poor	poor	average	good	very good
$g_i(a)$ very poor	indiff.				
poor		indiff.			
average			indiff.		
good				indiff.	
very good					indiff.

the absolute difference determines his/her preference. The preference structure of the DM for an interval scale criterion can be modeled by determining for which absolute difference d the DM is indifferent, for which d he/she has a moderate preference, for which d he/she has a strong and for which d he has a very strong preference. For a ratio scale criterion, he/she can also consider the relative difference δ (see Table 6.9). We must indicate if the criterion must be MIN or MAX.

Table 6.9. Preference matrix for a criterion (Max) with evaluation on a ratio scale.

$g_i(a) \geq g_i(b) > 0$	$d = g_i(a) - g_i(b)$	$\delta = \frac{g_i(a)-g_i(b)}{g_i(b)}$
indifferent	$0 \leq d < d_1$	$0\% \leq \delta < \delta_1\%$
small preference	$d_1 \leq d < d_2$	$\delta_1\% \leq \delta < \delta_2\%$
moderate preference	$d_2 \leq d < d_3$	$\delta_2\% \leq \delta < \delta_3\%$
strong preference	$d_3 \leq d < d_4$	$\delta_3\% \leq \delta < \delta_4\%$
very strong preference	$d_4 \leq d$	$\delta_4\% \leq \delta$

The following ordinal scale may be used to reflect the importance of a criterion: not important, small, moderately, very and extremely important. The DM must indicate for each criterion, by selecting a value from this ordinal scale, how important one criterion is for him/her.

When the preference structure of the DM for each criterion is known as well as the importance of each criterion, the comparison of two alternatives a and b with respect to the criterion g_j leads to a two-dimensional table (Table 6.10).

In a cell, f_{st} stands for the number of criteria of a certain importance for which a certain preference between the alternatives a and b occurs, $\sum_s \sum_t f_{st} = n$.

In order to get one overall appreciation of the comparison between the alternatives a and b, the DM must rank all cells of Table 6.10 where $g_j(a) > g_j(b)$. A ranking in eight classes is proposed to DM. Through this ranking a one dimensional ordinal variable is created. In fact there is a combined preference with respect to difference on evaluations and importance of weights where $g_j(a) > g_j(b)$ and where $g_j(a) < g_j(b)$ (see Table 6.11).

Table 6.10. Preference importance table for g_j, a, b.

	criteria preference	not imp.	little imp.	moderate imp.	very imp.	extremly imp.	w_j
$g_j(a) > g_j(b)$	very strong	f_{11}	f_{12}	f_{13}	f_{14}	f_{15}	a
	strong	f_{21}	f_{22}	f_{23}	f_{24}	f_{25}	b
	moderate	f_{31}	f_{32}	f_{33}	f_{34}	f_{35}	$\vdots$
	small	f_{41}	f_{42}	f_{43}	f_{44}	f_{45}	$\vdots$
$g_j(a) = g_j(b)$	no	f_{51}	f_{52}	f_{53}	f_{54}	f_{55}	$\vdots$
$g_j(a) < g_j(b)$	small	f_{61}	f_{62}	f_{63}	f_{64}	f_{65}	$\vdots$
	moderate	f_{71}	f_{72}	f_{73}	f_{74}	f_{75}	$\vdots$
	strong	f_{81}	f_{82}	f_{83}	f_{84}	f_{85}	b
	very strong	f_{91}	f_{92}	f_{93}	f_{94}	f_{95}	a

Table 6.11. Combined preferences with weights variable.

	$g_j(a) > g_j(b)$	$g_j(a) < g_j(b)$
1	$u_1 = f_{15}$	$\nu_1 = f_{95}$
2	$u_2 = f_{14} + f_{25}$	$\nu_2 = f_{85} + f_{94}$
3	$u_3 = f_{13} + f_{24} + f_{45}$	$\nu_3 = f_{75} + f_{84} + f_{93}$
4	$u_4 = f_{12} + f_{23} + f_{34} + f_{45}$	$\nu_4 = f_{65} + f_{74} + f_{93} + f_{92}$
5	$u_5 = f_{11} + f_{22} + f_{33} + f_{44}$	$\nu_5 = f_{64} + f_{73} + f_{82} + f_{91}$
6	$u_6 = f_{21} + f_{32} + f_{43}$	$\nu_6 = f_{63} + f_{72} + f_{81}$
7	$u_7 = f_{31} + f_{42}$	$\nu_7 = f_{62} + f_{71}$
8	$u_8 = f_{41}$	$\nu_8 = f_{61}$

The decision maker can alter this ranking (by moving a cell from one class to another, by considering more or less classes) until it matches his/her personal conception. Based on those two variables an outranking (S), indifference (I) or incomparability (R) relation between two alternatives is constructed:

$$\text{if } \sum_{k=1}^{h} u_k = \sum_{k=1}^{h} v_k \text{ for all } h = 1, \ldots, 8, \text{ then } aIb;$$

$$\text{if } \sum_{k=1}^{h} u_k \geq \sum_{k=1}^{h} v_k \text{ for all } h = 1, \ldots, 8, \text{ then } aSb;$$

$$\text{if } \sum_{k=1}^{h} u_k \leq \sum_{k=1}^{h} v_k \text{ for all } h = 1, \ldots, 8, \text{ then } bSa.$$

in all other cases aRb.

According to the basic idea of outranking, if alternative a is much better than alternative b on one (or more) criteria and b is much better than a on other criteria, there can be discordance between alternative b and alternative a, and b will not outranking a. The DM must explicitely indicate for each criterion when there is discordance between two evaluations on that particular criterion. For an ordinal criterion he/she can indicate in the upper triangle of the preference matrix (Table 6.8) when discordance occurs. For an interval or ratio criterion, the DM must indicate from which difference (absolute or relative), between the evaluations of two alternatives on that criterion, there is discordance.

EXAMPLE 9 *We have 4 alternatives, 4 criteria and the evaluation table (Table 6.12). In this example, the criteria g_1, g_2, g_3 are ordinal scales, and criterion g_4 is a ratio scale to be minimized.*

Table 6.12. Evaluation of alternatives*.

	g_1	g_2	g_3	g_4
a_1	$\square$	$\oplus$	$-$	13
a_2	$\oplus$	$-$	$\square$	10
a_3	$\square$	$-$	$-$	17
a_4	$+$	$\square$	$\square$	17

* $\oplus$: *very good*; +: *good*; $\square$: *acceptable*; $-$: *moderate*.

The following dominance relation can be observed from the data: $a_4 D\, a_3$, so that after deleting a_3, the set of alternatives is $A = \{a_1, a_2, a_4\}$. It is necessary to make this pre-processing step.

The preference modeling of alternatives with respect to the criteria are given in Tables 6.13, 6.14, and 6.15.

Table 6.13. Criteria g_1 and g_3 (ordinal scales).

$g_2(b)$	$\ominus$	$-$	$\square$	$+$	$\oplus$
$g_2(a)$ $\ominus$	indifferent			discordance	discordance
$-$	moderate	indifferent			discordance
$\square$	strong	moderate	indifferent		
$+$	very strong	strong	moderate	indifferent	
$\oplus$	very strong	very strong	strong	small	indifferent

The preference structure of weights of the criteria is given in Table 6.16.

Suppose that the ranking in eight classes of the combined preference with weight of two alternatives presented in Table 6.11 is approved. Table 6.17 gives an example of a pairwise comparison between a_1 and a_4.

Table 6.14. Criterion g_2 (ordinal scale).

$g_6(b)$	$\ominus$	$-$	$\square$	$+$	$\oplus$
$g_6(a)$ $\ominus$	indifferent				discordance
$-$	small	indifferent			
$\square$	moderate	small	indifferent		
$+$	strong	moderate	small	indifferent	
$\oplus$	very strong	strong	moderate	small	indifferent

Table 6.15. Criterion g_4 (ratio scale MIN).

Preference (a above b)	$d = g_j(a) > g_j(b)$
Indifferent	$0 \leq d < 1$
small	$1 \leq d < 3.5$
moderate	$3.5 \leq d < 6$
strong	$6 \leq d < 9$
very strong	$9 \leq d < \infty$
discordance	$d < -\infty$

Table 6.16. Preference structure of weights.

Weight	
not important	
little important	g_1, g_3
moderately important	g_4
very important	g_2
extremely important	

Table 6.17. Pairwise comparison between a_1 and a_4.

	$g_j(a_1) > g_j(a_4)$	$g_j(a_1) < g_j(a_4)$
1	0	0
2	0	0
3	0	0
4	1	0
5	1	0
6	0	2
7	0	0
8	0	0

The pairwise comparison of all pair of alternatives from A permits to construct the following binary relations: a_1Sa_4, a_1Ra_2 *and* a_2Sa_4 *(see Figure 6.3).*

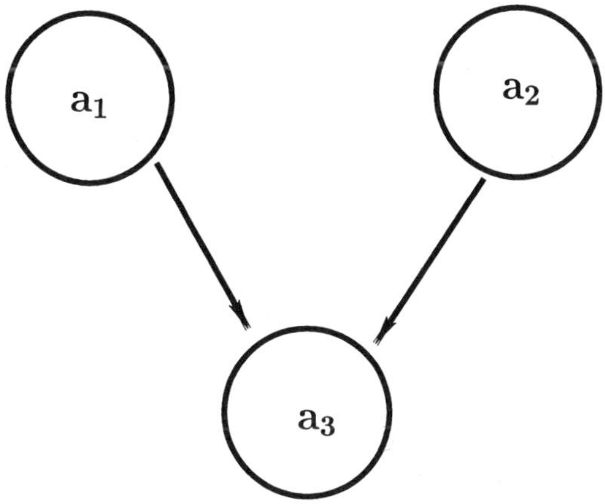

Figure 6.3. Outranking graph.

The ARGUS method demands a relatively great effort from the DM to model his/her preferences.

2.5 EVAMIX

The EVAMIX method [30, 39, 40] is a generalization of concordance analysis in the case of mixed information on the evaluation of alternatives on the judgment criteria. Thus a pairwise comparison is made for all pairs of alternatives to determine the so called concordance and discordance indices. The difference with standard concordance analysis is that separate indices are constructed for the qualitative and quantitative criteria. The comprehensive ranking of alternatives is the result of a combination of the concordance and discordance indices for the qualitative and quantitative criteria.

The set of criteria in the multi-criteria evaluation table is divided into a set of qualitative (ordinal) criteria O and a set of quantitative (cardinal) criteria C. It is assumed that the differences between alternatives can be expressed by means of two dominance measures: a dominance score $a_{ii'}$ for the ordinal criteria, and a dominance score $a_{ii'}$ for the cardinal criteria. These scores represent the degree to which alternative a_i dominates alternative $a_{i'}$. They have the following structure:

$$
\begin{aligned}
\alpha_{ii'} &= f(e_{ij}, e_{i'j}, \pi_j), \text{ for all } j \in O, \\
a_{ii'} &= g(e_{ij}, e_{i'j}, \pi_j), \text{ for all } j \in C,
\end{aligned}
$$

where e_{hj} represents the evaluation of alternative a_h on the criterion g_j and π_j the importance weight associated to this criterion. These scores can be defined as follows:

$$\alpha_{ii'} = \left[\sum_{j \in O} \{\pi_j sgn(e_{ij} - e_{i'j})\}^c \right]^{\frac{1}{c}},$$

where

$$\text{sgn}(e_{ij} - e_{i'j}) = \begin{cases} +1 & \text{if } e_{ij} > e_{i'j} \\ 0 & \text{if } e_{ij} = e_{i'j} \\ -1 & \text{if } e_{ij} < e_{i'j}. \end{cases}$$

The symbol c denotes an arbitrary scaling parameter, for which any positive odd value may be chosen, $c = 1, 3, 5, \ldots$. In a similar manner, the quantitative dominance measure can be made explicit:

$$a_{ii'} = \left[\sum_{j \in C} \{\pi_j (e_{ij} - e_{i'j})\}^c \right]^{\frac{1}{c}}.$$

In order to be consistent, the same value for the scaling parameter c should be used as in formula for $\alpha_{ii'}$. It is assumed that the quantitative employed evaluation e_{ij} have been standardized ($0 \leq e_{ij} \leq 1$). Evidently, all standardized scores should have the same direction, i.e., a 'higher' score should (for instance) imply a 'larger' preference. It should be noted that the rankings e_{ij} ($j \in O$) of the qualitative criteria also have to represent 'the higher, the better'. Since $\alpha_{ii'}$ and $a_{ii'}$ will have different measurement units, a standardization into the same unit is necessary. The standardized dominance measures can be written as:

$$\delta_{ii'} = h(\alpha_{ii'}) \text{ and } d_{ii'} = h(a_{ii'}),$$

where h represents a standardization function.

Let us assume that weights π_j have quantitative properties. The overall dominance measure $D_{ii'}$ for each pair of alternatives (a_i, $a_{i'}$) is:

$$D_{ii'} = \pi_o \delta_{ii'} + \pi_c d_{ii'},$$

where $\pi_o = \sum_{j \in O} \pi_j$ and $\pi_c = \sum_{j \in C} \pi_j$. This overall dominance score reflects the degree to which alternative a_i dominates alternative $a_{i'}$ for the given set of criteria and the weights. The last step is to determine an appraisal score s_i for each alternative. In general the measure $D_{ii'}$ may be considered as function k of the constituent appraisal scores, or:

$$D_{ii'} = k(s_i, s_{i'}).$$

This expression represents a well-known pairwise comparison problem. Depending on the way function k is made explicit, the appraisal scores can be calculated. The most important assumptions behind the EVAMIX method concern the definition of the various functions. It is shown in [40], that at least three different techniques can be distinguished which are based on different definitions of $\delta_{ii'}$, $d_{ii'}$ and $D_{ii'}$. The most straightforward standardization is probably the additive interval technique. The overall dominance measure $D_{ii'}$ is defined as:

$$D_{ii'} = \frac{s_i}{s_i + s_{i'}},$$

which implies that $D_{ii'} + D_{i'i} = 1$. To arrive at such overall dominance measures with this additivity characteristic, the following standardization is used:

$$\delta_{ii'} = \frac{(\alpha_{ii'} - \alpha^-)}{(\alpha^+ - \alpha^-)} \text{ and } d_{ii'} = \frac{(a_{ii'} - a^-)}{(a^+ - a^-)}$$

where α^- (α^+) is the lowest (highest) qualitative dominance score of any pair of alternatives $(a_i, a_{i'})$ and a^- (a^+) is the lowest (highest) quantitative dominance score of any pair of alternatives $(a_i, a_{i'})$. The resulting appraisal score is:

$$s_i = \left[\sum_{i'} \frac{D_{i'i}}{D_{ii'}} \right]^{-1}.$$

This expression means that the appraisal scores add up to unity, $\sum_i s_i = 1$.

In the previous elaboration, quantitative weights π_j, $j = 1, 2, \ldots, n$, were assumed. In some circumstances, only qualitative priority expressions can be given. If only ordinal information is given, at least two different approaches may be followed: an expected value approach (see [30, Appendix 4.I]) or a random weight approach. The random weight approach implies that quantitative weights are created by a random selection out an area defined by the extreme weight sets. These random weights γ_j, $j = 1, \cdots, n$, have to fulfill the following conditions:

1 for each $\gamma_j, \gamma_{j'}, \omega_j \leq \omega_{j'} \Rightarrow \gamma_j \geq \gamma_{j'}$,

2 $\sum_j \gamma_j = 1$,

where ω_j denotes a ranking number expressing a qualitative weight with "lower" means "better". For each set of metric weights γ_j, $j = 1, \cdots, n$, generated during one run of the random number generator, a set of appraisal scores can be determined. By repeating this procedure many times a frequency matrix can be constructed. Its element f_{ri} represents the number of times, alternative a_i was placed in the $r - th$ position in the final ranking. A probability matrix with element p_{ri} can be constructed, where:

$$p_{ri} = \frac{f_{ri}}{\sum_i f_{ri}}.$$

So, p_{ri} represents the probability that a_i will receive an $r - th$ position. We can make a comprehensive ranking of the alternatives in the following way:

$$\begin{aligned} a_i &= 1, \text{ if } p_{1i} \text{ is maximal}, \\ a_{i'} &= 2, \text{ if } p_{1i} + p_{2i'} \text{ is maximal and } i' \neq i, \\ a_{i''} &= 3, \text{if } p_{1i} + p_{2i'} + p_{3i''} \text{ is maximal and } i'' \neq i' \neq i, \end{aligned}$$

and so forth.

The EVAMIX method is based on important assumptions: 1) the definition of the various functions f, g, h and k; 2) the definition of the weights of the sets O and C and 3) the additive relationship of the overall dominance measure.

2.6 TACTIC

In the TACTIC method, proposed by Vansnick (see [37]), the family of criteria F is composed of true-criteria or quasi-criteria (criteria with an indifference threshold $q > 0$) $g_j, j = 1, \ldots, n$, and the preference structures correspondent are (P, I) or (P, I, R), where R is the incomparability relation, if no veto-threshold $v_j(\cdot), j \in \mathcal{J}$ is considered or at least one v_j is introduced respectively.

To each criterion $g_j \in F$ an importance weight $\lambda_j > 0$ is associated, as in the ELECTRE methods (see chapter 4 in this book). To model the preferences, the following subset of $\mathcal{J}$ is defined, $\forall a, b \in A, a \neq b$:

$$\mathcal{J}_T(a, b) = \{j \in | : g_j(a) > g_j(b) + q_j[g_j(b)]\},$$

where $q_j[g_j(b)]$ is the marginal indifference threshold as a function of the worst evaluation between $g_j(a)$ and $g_j(b)$, and therefore in this case we have aP_jb.

If in the set F only true criteria are considered, the statement aPb is true if and only if the following *concordance condition* is satisfied:

$$\sum_{j \in \mathcal{J}_T(a,b)} \lambda_j > \rho \sum_{j \in \mathcal{J}_T(b,a)} \lambda_j, \text{ i.e. } \frac{\sum_{j \in \mathcal{J}_T(a,b)} \lambda_j}{\sum_{j \in \mathcal{J}_T(b,a)} \lambda_j} > \rho \text{ if } \mathcal{J}_T(b, a) \neq \emptyset, \quad (6.1)$$

where the coefficient ρ is called required concordance level (usually, $1 \leq \rho \leq \frac{\sum_{j \in I} \lambda_j}{\min_{j \in I} \lambda_j} - 1$) and the two summations represent the absolute importance of the coalition of criteria in favor of a or b respectively.

If also some quasi-criterion is in the set F, in the preference structure (P, I, R) aPb is true if and only if both concordance condition 6.1 and the following *non-veto condition* are satisfied:

$$\forall j \in \mathcal{J}, g_j(a) + v_j[g_j(a)] \geq g_j(b), \tag{6.2}$$

where $v_j[g_j(a)]$ is the marginal veto threshold.

If the condition (6.2) is not satisfied by at least one criterion from F, we have aRb. On the other hand, we have aIb if and only if both pairs (a, b) and (b, a) do not satisfy condition (6.1) and no veto situation arises.

We remark that if $\rho = \rho^* = \frac{\sum_{j \in I} \lambda_j}{\min_{j \in I} \lambda_j} - 1$, the condition (6.1) is equivalent to the complete absence of criteria against the statement aPb, i.e. $\mathcal{J}_T(b, a) = \emptyset$ (and therefore in this case, (6.2) automatically holds). If $q_j = 0$ for each criterion g_j, the relation P is transitive for $\rho > \rho^*$. When ρ is decreasing from level ρ^*, we can have two types of intransitivity:

- aPb, bPc, aIc (or aRc),
- aPb, bPc, cPa.

If in equation (6.1) $\rho = 1$, we obtain the basic concordance-discordance procedure of Rochat type:

- for structures (P, I) (see [35]),

$$aPb \quad \text{iff} \quad \sum_{j \in \mathcal{J}_T(a,b)} \lambda_j > \sum_{j \in \mathcal{J}_T(b,a)} \lambda_j;$$

$$aIb \quad \text{iff} \quad \sum_{j \in \mathcal{J}_T(a,b)} \lambda_j = \sum_{j \in \mathcal{J}_T(b,a)} \lambda_j;$$

- for structures (P, I, R),

$$\begin{aligned}
aPb \quad \text{iff} \quad & \sum_{j \in \mathcal{J}_T(a,b)} \lambda_j > \sum_{j \in \mathcal{J}_T(b,a)} \lambda_j \text{ and} \\
& g_j(b) - g_j(a) \leq v_j[g_j(a)], \forall j \in \mathcal{J}; \\
aIb \quad \text{iff} \quad & \sum_{j \in \mathcal{J}_T(a,b)} \lambda_j = \sum_{j \in \mathcal{J}_T(b,a)} \lambda_j \text{ and} \\
& g_j(b) - g_j(a) \leq v_j[g_j(a)], \forall j \in \mathcal{J}, \text{ and} \\
& g_j(a) - g_j(b) \leq v_j[g_j(a)], \forall j \in \mathcal{J}. \\
aRb \quad \text{iff} \quad & \text{non}(aPb), \text{ non}(bPa) \text{ and non}(aIb).
\end{aligned}$$

The main difference between the ELECTRE I and TACTIC preference modeling is that the latter method is based on the binary relation aPb, while the former aims to build up the outranking relation aSb, $a, b \in A$. Moreover, the validation of the preference relation is now based on a sufficiently large ratio between the importance of criteria in favor and against the statement aPb. Roy and Bouyssou [35] show that this second difference is actually just a formal one. They also remark that, as a consequence of the peculiar characterization of the statement aPb, in TACTIC method is difficult to split indifference and incomparability situations. No particular exploitation procedure is suggested in TACTIC method.

2.7 MELCHIOR

In the MELCHIOR method [16] the basic information is a family F of pseudo-criteria, i.e. criteria g_j with an indifference threshold q_j and a preference threshold p_j ($p_j > q_j \geq 0$) such that, $\forall j \in \mathcal{J}$ and $\forall a, b \in A$:

- a is strictly preferred to b (aP_jb) with respect to g_j iff $g_j(a) > g_j(b) + p_j[(g_j(b)]$,
- a is weakly preferred to b (aQ_jb) with respect to g_j iff $g_j(b)+p_j[(g_j(b)] \geq g_j(a) > g_j(b) + q[(g_j(b)]$,
- a and b are indifferent (aI_jb) iff there is no strict or weak preference between them.

No importance weights are attached to criteria, but a binary relation M is defined on F such that g_iMg_j means "criterion g_i is as least as important as criterion g_j". In order to state the comprehensive outranking relation aSb, the Author proposes to "match" in a particular way the criteria in favor and the criteria against the latter relation (concordance analysis) and to verify that no discordance situation exists, i.e. no criterion g_j from F exists such that $g_j(b) > g_j(a)+v_j$, where v_j is a suitable veto-threshold for criterion g_j (absence of discordance). In this method, a criterion $g_j \in F$ is said to be in favor of the outranking relation aSb if one of the following situations is verified:

- aP_jb (marginal strict preference of a over b) (1st condition)
- aP_jb or aQ_jb (marginal strict or weak preference of a over b) (2nd condition)
- $g_j(a) > g_j(b)$ (3rd condition).

A criterion $g_j \in F$ is said to be against the outranking relation aSb if one of the following situations is verified:

- bP_ja (marginal strict preference of b over a) (1st condition)
- bP_ja or bQ_ja (marginal strict or weak preference of b over a) (2nd condition)
- $g_j(b) > g_j(a)$ (3rd condition).

The *concordance analysis* with respect to the outranking relation aSb, $a, b \in A$, is made by checking if the family of criteria G in favor of this relation "hides" the family of criteria H that are against relation aSb. These subsets of criteria are compared just using the binary relation M on F. A subset G of criteria is said to "hide" a subset H of criteria ($G, H \subset F$, $F \cap G = \emptyset$) if, for each criterion g_i from H, there exists a criterion g_j from G such that

- $g_j M g_i$ (1st condition) or

- $g_j M g_i$ or not($g_i M g_j$) (2nd condition),

where the same criterion g_j from G is allowed to hide at most one criterion from H.

By choosing two suitable combinations (see [16]) of the above conditions, the first stricter than the other, and verifying the concordance and the absence of discordance, a strong and a weak comprehensive outranking relation can be respectively built up. Then these relations are in turn exploited as in ELECTRE IV method (see chapter 4 in this book). We remark that the latter in fact coincides with MELCHIOR if the same importance is assigned to each criterion.

We finally observe that in both TACTIC and MELCHIOR methods no possibility of interaction among criteria (see Chapter 14 in this book) is taken into consideration, since the first one considers additive weights for the importance of each coalitions of criteria and the last one just "matches" one to one criteria in favor and against the comprehensive outranking relation aSb.

3. Pairwise Criterion Comparison Approach

In this approach, after the evaluations of potential alternatives with respect to different criteria, the phase of building up the outranking relations is split in two different steps, making comparisons at first level (partial aggregation) with respect to each subset of criteria $G_k \subset F$ ($|F| = m, G_k \neq \emptyset$, $|G| = k$, $k = 2, 3, \cdots, m-1$) and then aggregating at the second level these partial results (global aggregation).

With respect to weighting, this way of aggregating preferences allows to take into consideration the marginal *substitution rate* (trade-off) of each criterion from subset G_k at the first step and the *importance* of each coalition of criteria G_k at the second step, with the possibility to explicitly modeling the different meaning of these "weights" and the eventual *interaction* among criteria from each G_k (see chapter 14 in this book). Moreover, peculiar preference attitudes with respect to compensation, indifference and veto relations may be usefully introduced at each step of preference aggregation process; therefore, these particular options may be modelled at "local" and global level, when the partial and aggregated preferences indices respectively are built up.

For $k = 2$, (i.e. when two criteria a time are considered in the first phase of aggregation), we speak of Pairwise Criterion Comparison Approach (PCCA), that is therefore a methodology in which first all the feasible actions are compared with respect to pairs of criteria from F, and then all the partial information so obtained are suitably aggregated.

Given $a, b \in A$, in the Multiple Attribute Utility Theory (see chapter 7 in this book) the partial utility functions $u_i[g_i(a)]$, $i \in \mathcal{J}$, are aggregated in different ways to obtain the global utility $u(a)$ of each alternative and then the final recommendation.

In the outranking ELECTRE and PROMETHEE (see Chapter 4 in this book) families methods, from the evaluations of each action with respect to each criterion $g_i \in F$, some (crisp or fuzzy) marginal outranking or preference relations $\phi_i(a, b)$ are built up as elementary indices, or relations, with respect each criterion $i \in \mathcal{J}$ and each (ordered) pair of actions (a, b); then, using these marginal relations and other inter-criteria information, a comprehensive outranking relation or index $\phi(a, b)$ is obtained. In PCCA, in the first stage for each pair of actions (a, b) a fuzzy binary preference index $\delta_{ij}(a, b)$, $i, j \in \mathcal{J}$, is built up as elementary index taking into consideration two different criteria a time; then, by suitable aggregation of these partial indices, a global index $\delta(a, b)$ is obtained, expressing the comprehensive fuzzy preference of a over b.

As in all the other outranking methods, the exploitation of the indices expressing the comprehensive relation allows to obtain the recommendation for the decision problem at hand.

The main reasons that suggest this two levels aggregation procedure are the following:

- limited capacity of the human mind to compare a large number of elements at the same time, taking into consideration numerous and often conflicting evaluations simultaneously;
- limited ability of the DM for assessing a lot of parameters concerning subjective evaluations of general validity and considering all available information together.

Of course, this approach requires a larger number of computations and preference information, but:

- it actually helps in understanding and it supports the entire decision making process itself;
- it allows DM to use in an appropriate way all own preference information, requiring weaker coherence conditions, and to obtain further information about partial comparisons;

- it compares actions with respect to two criteria a time and, then it is easier to set appropriate parameters reflecting the partial comparison at hand;
- it offers greater flexibility in the preference modeling, allowing explicitly the representation of specific preference framework and information DM wants to use each time in the considered comparison;
- it allows useful extensions of some well-known basic concepts, like weighting, compensation, dominance, indifference, incomparability, etc.
- it actually allows to model interaction between each couple of criteria, possibly the most important and really workable in an effective way.

Therefore, in our opinion the PCCA satisfies the following principles, rel evant in any decision process, to build up realistic preference models and to obtain actual recommendations:

- *transparency*, making some light in any phase of the "black box" process (about the aggregation procedure in itself, the meaning of each parameter and index, their exploitation, etc.);
- *faithfulness*, respecting accurately the DM's preferences, without imposing too axiomatic constraints;
- *flexibility*, accepting and using any kind of information the DM wants and is able to give, neither more, nor less.

This means that DM will not be forced to "consistency" or "rationality". In other words, not too "external conditions" will be imposed to DM in expressing his/her preferences, but all actual information will be used. So, for example, not transitive trade-offs, $w_{i,k}$, (different from $w_{i,j} \cdot w_{j,k}$ where $w_{r,s}$ is the trade-off between criteria g_r and g_s), and or not complete importance weights (to some criterion no weight is associated) and also aggregated information (i.e., pooled importance weights, reflecting the interaction among criteria of each coalition) will be accepted as input.

Roughly speaking, the PCCA aggregation procedure can be applied to a lot of well-known compensatory or noncompensatory aggregation procedures resulting in binary preference indices. For each $j \in \mathcal{J}$, let $g_j \in F$ be an *interval scale* of measurement (i.e., unique up to a positive linear transformation) and $w_j, w_j \in \mathbb{R}^+$, be a suitable scale constant, called *trade-off* weight or *constant substitution rate*, reflecting (in a compensatory aggregation procedure) the increase on criterion value g_j necessary to compensate a unitary decrease on other criterion from F in terms of global preference. In other words, w_j is used to transform the scale g_j for normalizing and weighting the criteria values in order to compare units on different criterion scales, for each $g_j \in F$. Often this normalization is made introducing two parameters g_j^* and g_{*j}, $j \in \mathcal{J}$, $(g_{*j} < g_j^*)$, usually fixed *a priori* by DM according to the specific decision problem at hand and related with the discrimination power of the criterion scales. These parameters represent, in the DM's view, respectively two suitable "levels" on criterion g_j to normalize its evaluations of feasible actions. For example, g_{*j} and g_j^* can be respectively the "neutral" and the "excellent" level or the minimum and maximum value that can be assumed on criterion g_j; currently, $g_{*j} \leq \min\{g_j(x)\}$ and $g_j^* \geq \max\{g_j(x)\}$. Therefore we can write $w_j = \frac{t_j}{g_j^* - g_{*j}}$, where t_j represent the marginal weight ("importance") of criterion g_j after normalization of its scale.

Let consider the following subsets of $\mathcal{J}$:

$$\begin{aligned} \mathcal{J}_{a>b} &= \{j \in \mathcal{J} : g_j(a) > g_j(b)\}, \\ \mathcal{J}_{a=b} &= \{j \in \mathcal{J} : g_j(a) = g_j(b)\}, \\ \mathcal{J}_{a<b} &= \{j \in \mathcal{J} : g_j(a) < g_j(b)\}; \end{aligned}$$

In this way, each doubleton $\{a, b\} \subseteq A$ determines a partition of $\mathcal{J}$, (possible an improper one, since some of the three subsets may be empty), whose elements are the subsets of criteria for which there is preference of a over b, indifference of a and b, preference of b over a, respectively.

Moreover, let be

$$\mathcal{J}_{a \geq b} = \{j \in \mathcal{J} \colon g_j(a) \geq g_j(b)\},$$

i.e. the subset of criteria for which there is a weak preference of a over b.

Let us remember, for example, the following elementary indices:

$$\begin{aligned} m(a,b) &= |\mathcal{J}_{a>b}| \text{ (majority index)}, \\ \lambda(a,b) &= \sum_{(j \in \mathcal{J}_{a>b})} \lambda_j \text{ (Condorcet index)}, \end{aligned}$$

where $\lambda_j \in \mathbb{R}^+$ is the importance weight associated with criterion $g_j \in F$ and

$$w(a,b) = \sum_{(j \in \mathcal{J}_{a \geq b})} w_j \Delta_j(a,b) \text{ (weighted difference)},$$

where $\Delta_j(a,b) = g_j(a) - g_j(b)$ and all criteria are interval scales.

If we consider the subset of criteria $G = \{g_i, g_j\} \subseteq F$, indicating by f_{ij} any one of the above indices, computed with respect to G, it is possible to derive thence a new binary preference index $\delta_{ij}(a,b)$, defined as follows:

$$\delta_{ij}(a,b) = \begin{cases} \frac{f_{ij}(a,b)}{f_{ij}(a,b) + f_{ij}(b,a)} & \text{if } f_{ij}(a,b) + f_{ij}(b,a) > 0 \\ \frac{1}{2} & \text{if } f_{ij}(a,b) + f_{ij}(b,a) = 0 \end{cases} \tag{6.3}$$

The following properties hold, $\forall (a,b) \in A^2$:

$$\begin{aligned} 0 \le \delta_{ij} \le 1, \qquad & \delta_{ij}(a,b) + \delta_{ij}(b,a) = 1, \\ \delta_{ij}(a,b) &= 1 \Leftrightarrow a \text{ partially dominates } b, \\ \delta_{ij}(a,b) &= 0 \Leftrightarrow b \text{ partially dominates } a, \end{aligned}$$

both being partial dominance relations defined with respect to the considered couple of criteria $\{g_i, g_j\} \subseteq F$.

Therefore, the general index $\delta_{ij}(a,b)$, obtained by the PCCA partial aggregation procedure, indicates the credibility of the dominance of a over b with respect to criteria g_i and g_j.

Let now λ_j, $\lambda_j \in \mathbb{R}^+$ be the normalized weight used in a noncompensatory aggregation procedure, called *importance weight*, associated with criterion $g_j \in F$, indicating the intrinsic importance of each criterion, independently by its evaluation scale. Then, we can aggregate the partial indices $\delta_{ij}(a,b)$ computed with respect to all the pairs of different criteria g_i and g_j from F according to the PCCA logic, considering also the normalized *importance weight* λ_{ij} (i.e. $\sum_{i<j} \lambda_{ij} = 1$) *of the coalition* (couple) of criteria g_i and g_j, $i,j \in \mathcal{J}$.

We obtain the following aggregated index:

$$\delta(a,b) = \frac{1}{n-1} \sum_{ij(i<j)} \lambda_{ij}\delta_{ij}(a,b). \tag{6.4}$$

If there is no interaction between these criteria, additive weights can be used in equation (6.4), i.e. $\lambda_{ij} = \lambda_i + \lambda_j$. The following properties hold, $\forall (a,b) \in A^2$ (see Section 3.1):

$$\begin{aligned} 0 \le \delta(a,b) \le 1, \qquad & \delta(a,b) + \delta(b,a) = 1, \\ \delta(a,b) &= 1 \text{ if and only if } a \text{ strictly dominates } b, \\ \delta(a,b) &= 0 \text{ if and only if } b \text{ strictly dominates } a. \end{aligned}$$

Therefore, the particular meanings (credibility of dominance) of the partial and global indices $\delta_{ij}(a,b)$ and $\delta(a,b)$ respectively are results essentially linked to the *peculiar aggregation procedure* of PCCA and not to the specific bicriteria index considered each time.

In the framework of the PCCA methodology, different methods have been proposed: MAPPAC, PRAGMA, IDRA, PACMAN, each one with its own features to build up the correspondent outranking relations and indices.

3.1 MAPPAC

We recall that *a dominates b* (aDb), $a, b \in A$, with respect criteria from F if a is at least as good as b for the considered criteria and is strictly preferred to b

for at least one criterion:

$$aDb \Leftrightarrow g_i(a) \geq g_i(b), \forall g_i \in F \text{ and } \exists j \in \mathcal{J} : g_j(a) > g_j(b) .$$

We say that a *weakly dominates* b (aD_wb) if a is at least as good as b for all the criteria from F:

$$aD_wb \Leftrightarrow g_i(a) \geq g_i(b), \forall\, g_i \in F.$$

We say that a *strictly dominates* b (aD_sb) iff $g_i(a) \geq g_i(a)$, $\forall i \in F$, where at most only one equality is valid. The binary relation D_w is a partial pre-order (reflexive and transitive), while D (and D_s) is a partial order (irreflexive, asymmetric and transitive); the correspondent preference structures are partial order and strict partial order respectively. Of course, $D_s \subset D \subset D_w$, aDb, $bD_wc \Rightarrow aDc$ and $aD_wb, bDc \Rightarrow aDc, \forall a, b, c \in A$.

In PCCA, where a subset of criteria $G = \{g_i, g_j\} \subset F$, is considered at the first level of aggregation, we say that a partially dominates b ($aD_{ij}b$), if the relation of dominance is defined on G. We say that a is partially preferred or is partially indifferent to b ($aP_{ij}b_j$ and $aI_{ij}b$ respectively) if these relations hold with respect to the set of criteria $\{g_i, g_j\}$.

We observe that

$$\begin{aligned} aD_{ij}b &\Rightarrow aP_{ij}b, \\ \text{and } aD_{ij}b, \forall i, j \in \mathcal{J} &\Leftrightarrow aD_sb \Rightarrow aDb \Rightarrow aPb, \end{aligned}$$

if all criteria from F are true criteria.

In the MAPPAC method [25] the basic (or partial) indices $\pi_{ij}(a, b)$ can be interpreted as credibility indices of the partial dominance $aD_{ij}b$, indicating also the fuzzy degree of preference of a over b; the global index $\pi(a, b)$ can be interpreted as the credibility index of strict dominance aD_sb, i.e. as the fuzzy degree of comprehensive preference of a over b.

If all criteria from F are interval scales, recalling that $\Delta_j(a, b) = g_j(a) - g_j(b)$, for each $j \in \mathcal{J}$ and $a, b \in A$, w_j, is the trade-off weight and λ_j the (normalized) importance weight of criterion g_j, $j \in \mathcal{J}$, the axiomatic system of MAPPAC partial indices can be summarized as follows (see Table 6.18) for each $a, b \in A$:

- The basic indices $\pi_{ij}(a, b)$ are functions only of the signs of the differences in evaluations of a and b with respect to criteria g_i and g_j in case of concordant evaluations, i.e. iff $\Delta_i(a, b)\Delta_j(a, b) \geq 0$. In this case,

$$aD_{ij}b \Leftrightarrow \Delta_i(a, b) + \Delta_j(a, b) > 0$$

and

$$bD_{ij}a \Leftrightarrow \Delta_i(a, b) + \Delta_j(a, b) < 0$$

Table 6.18. Axiomatic system of MAPPAC basic indices.

$\pi_{ij}(a,b)$	Binary Relations	Signs of $\Delta_i(a,b) \cdot \Delta_j(a,b)$	Signs of $\Delta_i(a,b) + \Delta_j(a,b)$	Pair of Signs of $\Delta_i(a,b)$, $\Delta_j(a,b)$
$]0,1[$	$aP_{ij}b, bP_{ij}a,$ $aI_{ij}b$	< 0	$\gtreqless$	$(+,-),(-,+)$
$\frac{1}{2}$	$aI_{ij}b$	$= 0$	$= 0$	$(0,0)$
1	$aD_{ij}b$	≥ 0	> 0	$(+,+),(+,0),$ (0,+)
0	$bD_{ij}a$	≥ 0	< 0	$(-,-),(-,0),$ (0,-)

and then $\pi_{ij}(a,b) = 1$ and $\pi_{ij}(b,a) = 0$ in the first case, and $\pi_{ij}(a,b) = 0$ and $\pi_{ij}(b,a) = 1$ in the second case.

- The basic indices $\pi_{ij}(a,b)$ are functions of the values of the differences in evaluations of a and b with respect to criteria g_i and g_j and of trade-off weights w_i and w_j in case of discordant evaluations, i.e. iff $\Delta_i(a,b)\,\Delta_j(a,b) < 0$. In this case, the indices $\pi_{ij}(a,b)$ and $\pi_{ij}(b,a)$ will be of a compensatory type, lying in the interval $]0,1[$, and they will indicate the fuzzy degree of preference of a over b and of b over a respectively; if $w_i\Delta_i(a,b) + w_j\Delta_j(a,b) = 0$, $\pi_{ij}(a,b) = \pi_{ij}(b,a) = \frac{1}{2}$.
- The global indices $\pi(a,b)$ are functions of all the basic indices $\pi_{ij}(a,b)$ and of the importance weights λ_{ij} of all coalitions $\{g_i, g_j\}$ of criteria. If there is no interaction between criteria g_i and g_j, we have $\lambda_{ij} = \lambda_i + \lambda_j$. In case of strict dominance aD_sb or bD_sa, $\pi(a,b) = 1$ and $\pi(b,a) = 0$, or $\pi(a,b) = 0$ and $\pi(b,a) = 1$, respectively. Otherwise, $\pi(a,b)$ and $\pi(b,a)$ will lie in the interval $]0,1[$ and they will indicate the fuzzy degree of comprehensive preference of a over b and of b over a respectively.

Preference Indices. We recall that $w_j\Delta_j(a,b) = w_j(g_j(a) - g_j(b)), j \in \mathcal{J}$, $a,b \in A$, is the normalized weighted difference of evaluations of actions a and b with respect to criterion g_j.

If we assume $f_{ij}(a,b) = \sum_{h\in\{i,j\}\cap\mathcal{J}_{a\geq b}} w_h\Delta_h(a,b)$ in the equation (6.3) we obtain the partial index $\pi_{ij}(a,b)$ of MAPPAC, $a,b \in A$, $\{g_i, g_j\} \subset F$ ($|F| \geq 3$). This index can also be explicitly written as shown in Table 6.19.

It is invariant to the admissible transformation of any $g_j \in F$, i.e. all the affine transformations of the type $g_j'(\bullet) = \alpha g_j + \beta$, with $\alpha \in \mathbb{R}^+$ and $\beta \in \mathbb{R}$, being the criteria interval scales. It is the image of a valued binary relation, strictly complete, transitive and ipsodual (i.e. $\pi_{ij}(a,b) = 1 - \pi_{ij}(b,a)$), that constitutes a complete preorder on A, and it indicates the fuzzy partial preference intensity of a over b.

Table 6.19. Preference indices.

$\pi_{ij}(a,b)$	$\pi_{ij}(b,a)$	
1	0	if $g_i(a) > g_i(b)$ and $g_j(a) > g_j(b)$
0	1	if $g_i(a) < g_i(b)$ and $g_j(a) < g_j(b)$
0.5	0.5	if $g_i(a) = g_i(b)$ and $g_j(a) = g_j(b)$
$\frac{w_i(g_i(a)-g_i(b))}{w_i(g_i(a)-g_i(b))+w_j(g_j(b)-g_j(a))}$	$\frac{w_j(g_j(b)-g_j(a))}{w_i(g_j(a)-g_i(b))+w_j(g_j(b)-g_j(a))}$	if $g_i(a) > g_i(b)$ and $g_j(a) \leq g_j(b)$ $g_i(a) = g_i(b)$ and $g_j(a) < g_j(b)$
$\frac{w_j(g_j(a)-g_j(b))}{w_i(g_i(b)-g_i(a))+w_j(g_j(a)-g_j(b))}$	$\frac{w_i(g_i(b)-g_i(a))}{w_i(g_i(b)-g_i(a))+w_j(g_j(a)-g_j(b))}$	if $g_i(a) \leq g_i(b)$ and $g_j(a) > g_j(b)$ $g_i(a) < g_i(b)$ and $g_j(a) = g_j(b)$

The basic preference index $\pi_{ij}(a,b)$ may be immediately interpreted geometrically by considering the partial profiles of the actions a and b with respect to criteria g_i and g_j (see Fig. 6.4).

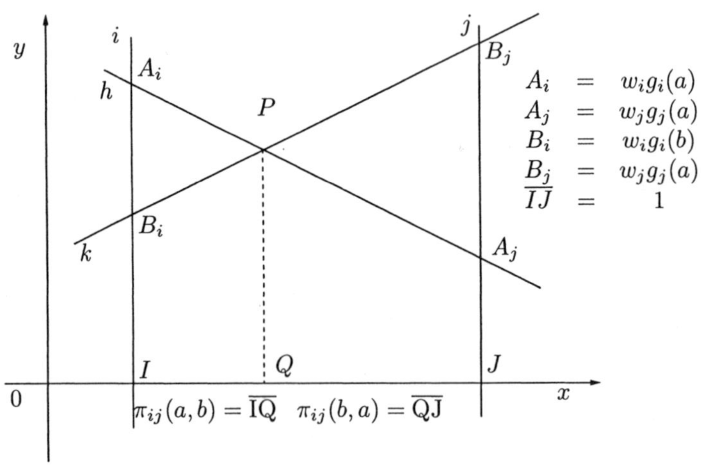

Figure 6.4. Geometrical interpretation of preferences indices.

Let us consider the following subsets of F:

$$\begin{aligned}
G^+(a,b) &= \{g_h \in F : \Delta_h(a,b) > 0\}; \quad |G^+(a,b)| = p,\\
G^=(a,b) &= \{g_h \in F : \Delta_h(a,b) = 0\}; \quad |G^=(a,b)| = o,\\
G^-(a,b) &= \{g_h \in F : \Delta_h(a,b) < 0\}; \quad |G^-(a,b)| = n,\\
D_1(a,b) &= \{(g_i,g_j) \in F^2, g_i \neq g_j : aD_{ij}b\},\\
D_0(a,b) &= \{(g_i,g_j) \in F^2, g_i \neq g_j : bD_{ij}a\}.
\end{aligned}$$

Of course, $G^+(a,b) \cup G^=(a,b) \cup G^-(a,b) = F$ and $|F| = m = p+o+n$, $(p,o,n \geq 0)$. Since (see [26])

$$\begin{aligned}
&\binom{m}{2} = \binom{p+o+n}{2} =\\
&\binom{p+o}{2} + \binom{p+n}{2} + \binom{o+n}{2} - \binom{p}{2} - \binom{o}{2} - \binom{n}{2} =\\
&\binom{p}{2} + \binom{o}{2} + \binom{n}{2} + po + pn + on,
\end{aligned}$$

we can split all the $\binom{m}{2}$ basic preference indices $\pi_{ij}(a,b)$ as follows:

$$\binom{m}{2} = |D_1(a,b)| + |D_o(a,b)| + \binom{o}{2+pn}.$$

Thus, $|D_1(a,b)| = \binom{m}{2}$ if and only if $p \geq m-1$ and $n = 0$ (i.e., $\Delta_h(a,b) \geq 0$ for each $h \in \mathcal{J}$, with at most only one equality), $|D_0(a,b)| = \binom{m}{2}$ if and only if $n \geq m-1$ and $p = 0$, $|D_1(a,b)| = |D_0(a,b)| = 0$ and $\pi_{ij}(a,b) = \frac{1}{2}$ for each $i,j \in \mathcal{J}$ if and only if $o = \frac{m}{2}$ (i.e., $\Delta_h(a,b) = 0$ for each $h \in \mathcal{J}$).

The *global preference index* $\pi(a,b)$ is the sum of all the $\binom{m}{2}$, $m > 2$, basic preference indices $\pi_{ij}(a,b)$, weighted each time by the normalized importance weights λ_{ij} of the considered couple of criteria g_i, g_j:

$$\pi(a,b) = \sum_{ij(i<j)} \pi_{ij}(a,b)\frac{\lambda_{ij}}{\Lambda},$$

where $\Lambda = \sum_{ij(i<j)} \lambda_{ij}$.

If there is no interaction between each couple of criteria, we have $\lambda_{ij} = \lambda_i + \lambda_j$, where λ_h is the normalized importance weight of criterion g_h, $h = 1, 2 \cdots m$, and therefore:

$$\pi(a,b) = \sum_{ij(i<j)} \pi_{ij}(a,b)\frac{\lambda_i + \lambda_j}{m-1}, \quad i,j \in \mathcal{J}, \left(\sum_{ij(i<j)} (\lambda_i + \lambda_j) = m-1\right) \tag{6.5}$$

Therefore, in this case we can write $\pi(a,b)$ as:

$$\pi(a,b) = \pi_{PP}(a,b) + \pi_{PO}(a,b) + \pi_{NN}(a,b) + \pi_{NO}(a,b) + \pi_{OO}(a,b) + \pi_{PN}(a,b), \quad (6.6)$$

where:

$$\begin{aligned}
\pi_{PP}(a,b) &= \frac{p-1}{m-1} \sum_{i \in G^+(a,b)} \lambda_i; \\
\pi_{PO}(a,b) &= \frac{1}{m-1}\left[p \sum_{i \in G^=(a,b)} \lambda_i + o \sum_{i \in G^+(a,b)} \lambda_i\right]; \\
\pi_{NN}(a,b) &= \pi_{NO}(a,b) = 0; \\
\pi_{OO}(a,b) &= \frac{1}{2}\frac{o-1}{m-1} \sum_{i \in G^=(a,b)} \lambda_i; \\
\pi_{PN}(a,b) &= \sum_{rs} \pi_{rs}(a,b) \frac{\lambda_r + \lambda_s}{m-1}, (g_r, g_s) \in G^+(a,b) \times G^-(a,b).
\end{aligned}$$

Let $S(a,b) = G^+(a,b) \cup G^=(a,b)$. We can write:

$$\pi_{D_1} = \pi_{PP}(a,b) + \pi_{PO}(a,b)$$

and, recalling equation (6.6),

$$\pi_S = \pi_{D_1}(a,b) + \pi_{OO}(a,b) = \frac{1}{m-1}\left[(p+o-1) \sum_{i \in S(a,b)} \lambda_i - \frac{1}{2}(o-1) \sum_{i \in G^=(a,b)} \lambda_i\right].$$

We observe that:

a) if $G^=(a,b) = \emptyset$ or $|G^=(a,b)| = 1, \pi_{D_1}(a,b) = \pi_S(a,b)$;

b) if $G^-(a,b) = \emptyset$, $\pi(a,b) = \pi_S(a,b)$;

c) the index $\pi_S(a,b)$ is a linear combination of the *crisp* concordance index $c(a,b)$ of the ELECTRE methods (see Chapter 4 in this book) and the opposite of semi-sum of the importance weights of criteria from set $G^=(a,b)$; their coefficients are respectively given by the ratios between the number of criteria belonging to the corresponding classes and the total number of criteria up to one unit (i.e., the number of *significant* criteria for a comparisons by means of pairs of criteria);

d) if $|S(a,b)| \geq 2$, $\frac{1}{2} \leq \pi_S(a,b) \leq c(a,b) \leq 1$, and $\pi_S(a,b) = c(a,b) = 1$ if and only if aD_Sb (but $c(a,b) = 1$ does not imply $\pi_S(a,b) = 1$);

e) $\pi_S(a,b) = 0$ if and only if $|S(a,b)| < 2$, and $\pi_S(a,b) = c(a,b) = 0$ if and only if $S(a,b) = \emptyset$ (but $\pi_S(a,b) = 0$ does not imply $c(a,b) = 0$);

f) the compensatory component $\pi_{PN}(a,b)$ of $\pi(a,b)$ (see equation (6.6)) may be methodologically linked to the MAUT approach, in particular to the weighted sum with constant marginal substitution rates (trade-off weights);

g) if the number o of the criteria g_h from F for which $g_h(a) = g_h(b)$ changes without modification in the sum of the relative importance weights of coalitions $G^+(a,b)$, $G^+(b,a)$ and $G^=(a,b)$, the value of the aggregate preference index $\pi(a,b)$ may vary, as a consequence of changing of its component $\pi_{PN}(a,b)$ value. More precisely:

- $G^+(a,b) = \emptyset$ and $G^-(a,b) \neq \emptyset \Rightarrow \pi(a,b)$ increases with o, i.e. $\Delta_o\pi(a.b) > 0$,
- $G^+(a,b) \neq \emptyset$ and $G^-(a,b) = \emptyset \Rightarrow \pi(a,b)$ decreases with o, i.e. $\Delta_o\pi(a.b) < 0$,
- $\lim_{o\to+\infty} \Delta_o\pi(a,b) = 0$,
- $\lim_{o\to+\infty} \pi(a,b) = \sum_{i\in G^+(a,b)} \lambda_i + \frac{1}{2}\sum_{i\in G^=(a,b)} \lambda_i$ (< 1 since $G^+(a,b) \subset F$),
- if the relative importance of $G^+(a,b)$ and $G^-(a,b)$ are equal, the relation aIb is stable with respect to o;
- $\forall o \geq 1, \frac{p}{p+n}\sum_{i\in S(a,b)} \lambda_i \geq \frac{1}{2} \Rightarrow aPb$ stable with respect to o,
- if there is a perfect compensation between the normalized weighted differences in evaluations of opposite signs (i.e. neutral behavior of $\pi_{PN}(a,b)$), $\Delta_o\pi(a,b) > 0\ [< 0] \Leftrightarrow n\sum_{i\in S(a,b)} \lambda_i - p\sum_{i\in S(a,b)} \lambda_i > 0\ [< 0]$, i.e. if and only if $n > [<] p$,
- the aggregate preference index $\pi(a,b)$ is an increasing function of p (i.e. $\Delta_p\pi(a,b) > 0$) if $\pi(a,b) < 1$,
- $\lim_{p\to+\infty} \Delta_p\pi(a,b) = 0$,
- $\lim_{p\to+\infty} \pi(a,b) = 1 - \frac{1}{2}\sum_{i\in G^-(a,b)} \lambda_i$.

Following the same principle of PCCA, it is possible to build up other partial and global preference indices, based on a logic of noncompensatory aggregation [24]. The common feature of all these indices is that they are based on bicriteria and global indices, measuring respectively the credibility of partial dominance and of strict dominance of a over b, $a, b \in A$. So, for example, if no 2-level interaction occurs among considered criteria, let us consider the following two aggregated indices:

$$\pi'(a,b) = \frac{1}{m-1}\left[(m-1)\sum_{i\in G^+(a,b)} \lambda_i - (p + \frac{o-1}{2})\sum_{i\in G^=(a,b)} \lambda_i\right],$$

$$\pi^*(a,b) = \frac{1}{m-1}\sum_{i,j;(i<j)}\left[\sum_{(i,j):\pi_{ij}(a,b)>0.5} (\lambda_i + \lambda_j) + \frac{1}{2}\sum_{(i,j):\pi_{ij}(a,b)=0.5} (\lambda_i + \lambda_j)\right].$$

We can observe that index $\pi^{'}(a, b)$ is totally noncompensatory and it is analogous to the concordance indices of ELECTRE I and II methods. On the other hand, index $\pi^*(a, b)$ is PCCA-totally noncompensatory (see [24]), depending on the *"coalition strength"* of the subsets (couples of criteria) of G^2 such that $aP_{ij}b$ or $aI_{ij}b$. Both these indices, like index $\pi(a, b)$, are also functions of p, n, o.

Taking into account the above properties and the peculiar features of the basic preference indices with respect to the dominance and compensation, MAPPAC and – more generally – PCCA may be considered as an *"intermediate"* MCDA methodology between the outranking (particularly ELECTRE) and MAUT methods.

Indifference Modelling. Since the evaluations of actions a and b with respect to the couple of criteria g_i, g_j from F are compared each time to build up index $\pi_{ij}(a, b)$, and recalling that $\Delta_i(a, b)\ \Delta_j(a, b) > 0$ means by definition active or passive partial dominance of a over b (and then $\pi_{ij}(a, b) = 1$ or 0 respectively), it is useful to confine the dominance relation only if well founded situations will occur. Therefore, in order to take into account the inevitable inaccuracies and approximations in the actions evaluations, and in order to prevent small differences between these evaluations from creating partial dominance relations or preference intensities close to the maximum or minimum values, it is advisable to introduce suitable indifference areas on the plane $Og_i(a)g_j(a)$ in the neighborhood of point $I = (g_i(a) = g_i(b), g_j(a) = g_j(b))$, see Fig. 6.5.

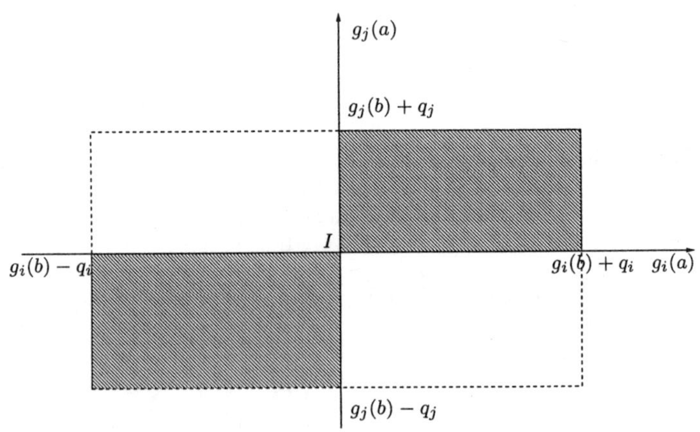

Figure 6.5. Indifference areas.

These areas may be defined in various way, as functions of correspondent indifference thresholds, one for each criterion considered (see [22]). The marginal indifference threshold for criterion g_j, denoted by q_j, is not negative and unique for every couple of distinct actions $a, b \in A$ ($q_j(a,b) = q_j(b,a) \geq 0, \forall a, b \in A$) and it is a function of the evaluations of these actions according to the criterion considered:

$$q_j(a,b) = \alpha_j + \beta_j \left| \frac{g_j(a) + g_j(b)}{2} \right|, g_j \in F, \alpha_j, \beta_j \geq 0. \quad (6.7)$$

The first parameter α_j is expressed in the same scale of values as the criterion g_j, and q_j is a linear function of the arithmetical mean of the evaluations of the considered actions, being β_j the constant of proportionality. Then, if $\beta_j = 0$ or $\alpha_j = 0$, equation (6.7) supplies constant indifference thresholds, in absolute or relative value respectively. It is therefore possible to define an indifference area IA_{ij} for each pair of actions $a, b \in A$ and criteria $g_i, g_j \in F$ as a function of the marginal indifference thresholds (6.7). This area may assume various shapes, for example:

- rectangular, if $\pi_{ij}(a,b) = \frac{1}{2}$ for $|g_i(a) - g_i(b)| \leq q_i(a,b)$ and $|g_j(a) - g_j(b)| \leq q_j(a,b)$ (see Fig. 6.6);

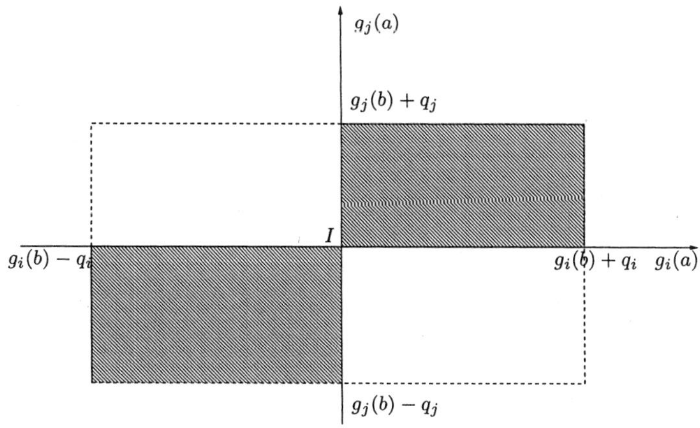

Figure 6.6. Indifference areas: rectangular.

- rhomboidal, if $\pi_{ij}(a,b) = \frac{1}{2}$ for $\frac{|g_i(a)-g_i(b)|}{q_i(a,b)} + \frac{|g_j(a)-g_j(b)|}{q_j(a,b)} \leq 1$ (see Fig. 6.7);
- elliptical, if $\pi_{ij}(a,b) = \frac{1}{2}$ for $\frac{(g_i(a)-g_i(b))^2}{q_i^2(a,b)} + \frac{(g_j(a)-g_j(b))^2}{q_j^2(a,b)} \leq 1$ (see Fig. 6.8).

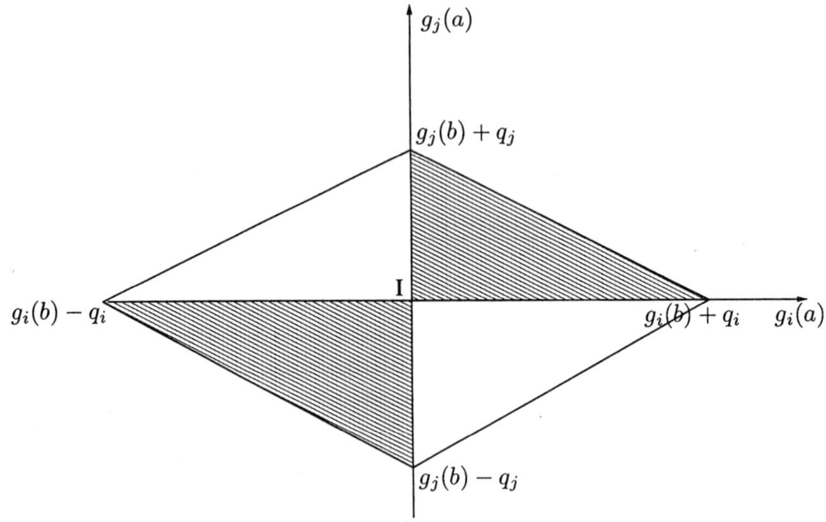

Figure 6.7. Indifference areas: rhomboidal.

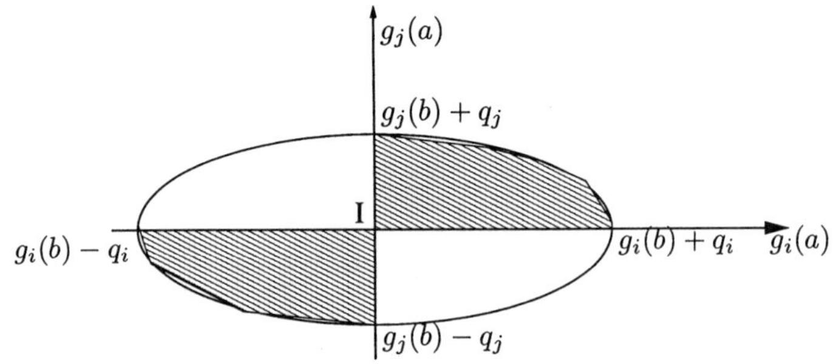

Figure 6.8. Indifference areas: elliptical.

It also possible to introduce semi-rectangular, semi-rhomboidal and semi-elliptical indifference areas, corresponding to the shadowed areas in Figures 6.6, 6.7, 6.8 respectively, with the specific aim of eliminating the effect of partial dominance only, adding each time the further conditions:

$$\begin{cases} g_i(b) & \leq & g_i(a) \\ g_j(b) & \leq & g_j(a) \end{cases} \text{ or } \begin{cases} g_i(a) & \leq & g_i(b) \\ g_j(a) & \leq & g_j(b). \end{cases}$$

Finally, it is also possible to consider mixed indifference areas, as a suitable combination of two or more of the cases considered above for each quadrant centered in point I. We can then modeling indifference in a flexible way, by setting different thresholds and/or shapes for each couple of criteria, according to the DM's preferential information.

Therefore, two separate indifference relations are obtained: strict indifference, denoted by $aI_{ij}b$, iff $\pi_{ij}(a,b) = \frac{1}{2}$ as a result of Table 6.19; large indifference, denoted by $aI^*_{ij}b$, iff a vector $\mathbf{q} \geq \mathbf{0}$ is introduced, $\mathbf{q} = [q_j(a,b)]$, $j \in \mathcal{J}$, and some of the corresponding above indifference area conditions are satisfied, and thus $\pi_{ij}(a,b) = \frac{1}{2}$ is assumed.

Note that I_{ij} is an equivalence relation, whereas the relation $aI^*_{ij}b$ is not necessarily transitive.

Preference Structures. Using the basic and global preference indices $\pi_{ij}(a,b)$ and $\pi(a,b)$, it is possible to immediately define the following correspondent binary relations of partial and comprehensive indifference and preference relations respectively, with the particular cases of dominance recalled above:

- Partial relations

$$\begin{aligned}
\pi_{ij}(a,b) = 0.5 &\Leftrightarrow aI_{ij}b, \\
0.5 < \pi_{ij}(a,b) \leq 1 &\Leftrightarrow aP_{ij}b, \\
(\pi_{ij}(a,b) = 1 &\Leftrightarrow aD_{ij}b), \\
0 \leq \pi_{ij}(a,b) < 0.5 &\Leftrightarrow bP_{ij}a \\
(\pi_{ij}(a,b) = 0 &\Leftrightarrow bD_{ij}a).
\end{aligned}$$

- Comprehensive relations

$$\begin{aligned}
\pi(a,b) = 0.5 &\Leftrightarrow aIb, \\
0.5 < \pi(a,b) \leq 1 &\Leftrightarrow aPb \quad (\pi(a,b) = 1 \Leftrightarrow aDb), \\
0 \leq \pi(a,b) < 0.5 &\Leftrightarrow bPa \quad (\pi(a,b) = 0 \Leftrightarrow bDa).
\end{aligned}$$

Both these structures constitute a complete preorder on A. We observe that, if no indifference areas are introduced, will be $\pi_{ij}(a,b) + \pi_{ij}(b,a) = 1$ for each $i, j \in \mathcal{J}$ and $(a,b) \in A^2$ and therefore also $\pi(a,b) + \pi(b,a) = 1$.

Of course, by means of the same indices, we can also build up some other particular complete valued preference structures. For example, we may consider the structure of semiorder, obtained by introducing a real parameter $\delta \in [1/2, 1]$, which emphasizes the partial or global indifference relations (see Figure 6.9).

In this case, the indifference relations are reflexive, symmetric and not transitive, while the preference relations are transitive, non reflexive and asymmetric. We note that if $\delta = \frac{1}{2}$ we obtain again a complete preorder with "punctual" indifference, i.e. only for $\pi(a,b) = \pi(b,a) = \frac{1}{2}$, while if $\delta = 1$, the binary preference relation is empty. Alternatively, by introducing two real parameters

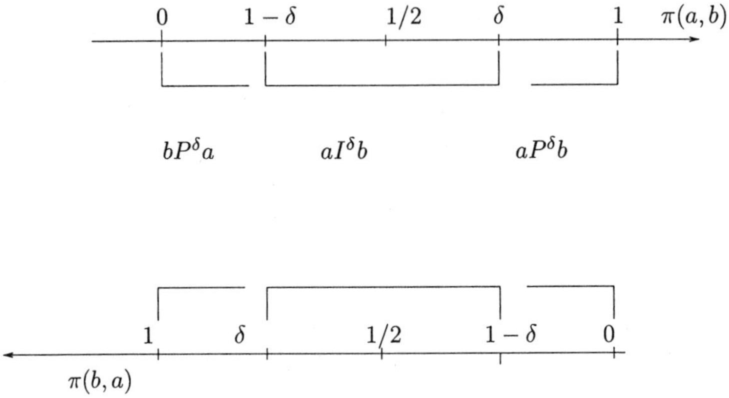

Figure 6.9. Aggregated semiorder structure.

δ and ϵ, $\frac{1}{2} \leq \delta < \epsilon \leq 1$, it is possible to build a complete two-valued preference structure, assuming that there are two preference intensity levels, represented by the preference relations P^{τ} (strict preference) and Q^{τ} (weak preference) (see Figure 6.10).

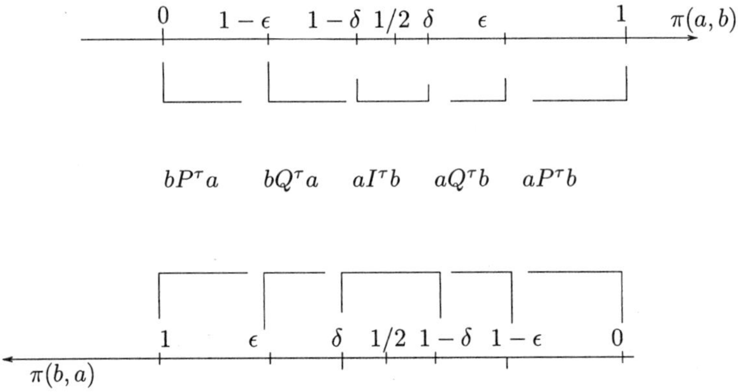

Figure 6.10. Aggregated pseudo-order structure.

In this case the relations of indifference and of weak preference are not transitive and the preference model presents the properties of the well-known pseudo-order structure (see [38]).

Conflict Analysis. Besides the concept of discordant criterion and veto threshold often used for building outranking relations, another interesting feature of PCCA approach is the possibility to consider a peculiar conflict analysis, taking into consideration the differences in evaluations of two actions with respect to each couple of criteria. The main aims of this analysis are the following:

- to explicitly define binary *incomparability relations* in presence of evaluations of two actions a and b in strong contrast on two criteria g_i and g_j, in the preference modeling phase (refusal to make a decision)
- to allow compensation only if differences in the conflicting evaluations are not too large; otherwise, to use non compensatory basic indices (functions only of importance weights), obtaining a *partially compensatory approach* (reduction of compensation) (see [24]).

These aims can be reached by defining a suitable *partial discordance index* $d_{ij}(a,b)$, $i,j \in \mathcal{J}$, $a,b \in A$, as a function of conflicting evaluations and entropy of information, and comparing this one with correspondent incomparability threshold r_{ij}, given by DM (see [22]). If we note by R_{ij} the partial incomparability relation with respect the couple of criteria g_i and g_j, we have:

$$d_{ij}(a,b) \geq r_{ij} \Leftrightarrow aR_{ij}b,\ (a,b) \in A^2, g_i, g_j \in F.$$

Then, considering all the possible couples of distinct criteria g_i, g_j from F, we have:

$$aRb \Leftrightarrow [aR_{ij}b \text{ for at least one couple } i,j \in \mathcal{J}].$$

This global incomparability relation R, symmetric but neither reflexive nor transitive, arise if at least one partial incomparability relation holds with respect to actions a and b.

The symmetric discordance index $d_{ij}(a,b)$, $i,j \in \mathcal{J}$, is defined as follows [21].

$$d_{ij} = |w_i\Delta_i(a,b) + w_j\Delta_j(b,a)|(1 - 2|\pi_{ij}(a,b) - 0.5|).$$

It lies in $[0, t_i + t_j]$ and reaches its maximum value only in case of maximum effective discordance of evaluations of a and b with respect to g_i and g_j (i.e. $g_i(a) = g_i^*$, $g_j(a) = g_{j*}$ and $g_i(b) = g_{i*}$, $g_j(b) = g_j^*$ or viceversa) and $t_i = t_j$ (equal normalized trade-off weights). Moreover, $d_{ij}(a,b) = 0$ if $\Delta_i(a,b) = \Delta_j(b,a) = 0$ or in case of partial dominance (evaluation concordance). Therefore, it is possible to set the incomparability thresholds r_{ij} according to the real preferential information of DM about the different level of compensation for each couple of criteria g_i and g_j:

$$r_{ij} = \begin{cases} = & 0 & \text{completely non compensatory approach} \\ \cong & 0 & \text{low compensation is allowed} \\ \cong & t_i + t_j & \text{high compensation is allowed} \\ = & t_i + t_j & \text{totally compensatory approach.} \end{cases}$$

The concepts introduced above therefore permit also a modelling by means of the four binary relations I,P,Q,R, defined on A, which are exhaustive and mutually exclusive and constitute a fundamental relational preference system.

Exploitation Phase. The results of the relational model in the form of fuzzy binary relations obtained can be presented in the form of suitable $\binom{m}{2}$ bicriteria $n \times n$ (i.e. $|A| \times |A|$) square matrices: $\mathbf{\Pi_{ij}} = [\pi_{ij}(a,b)]$, one for each couple of criteria g_i, g_j from F, containing the partial preference indices, and one aggregated matrix $\mathbf{\Pi} = [\pi(a,b)]$, with the comprehensive preference indices, $(a,b) \in A^2$.

The peculiar preference modeling flexibility of PCCA allows to respect accurately the DM's preference, without imposing too strong axiomatic constraints, and accepting and using any kind of information the DM is able to give. Therefore, DM is not forced to be *"consistent"*, *"rational"* or *"complete"*, but all information given by DM is accepted and used, neither more, nor less. Consequently, with respect to two criteria trade-offs w_{ij}, $i,j \in I$, it is possible to use as input not transitive (i.e. $w_{ij}w_{jk} = \frac{w_i}{w_j}\frac{w_j}{w_k} \neq \frac{w_i}{w_k} = w_{ik}$) or not complete (some w_{ij} not given by DM) trade-offs for some pairs of criteria (and therefore the component $\pi_{ij}(a,b)$ of index $\pi(a,b)$ correspondent to these criteria will be absent); and, with reference to importance weights λ_j, $j \in \mathcal{J}$, the DM may assign non additive weights λ_{ij} to some couple of criteria, modelling thus their interaction (i.e. weighting some index $\pi_{ij}(a,b)$ with a weight different from $\lambda_i + \lambda_j$). In all these cases, the aggregate index $\pi(a,b)$ will be computed taking into account the peculiar information actually used as input.

The indices of preference intensity contained in the aggregated matrix $\mathbf{\Pi}$ may, among other things, permit in the *exploitation phase* the building of specific partial or complete rankings of feasible actions as final prescription.

A first possible technique to build rankings can be based on the concept of qualification of a feasible action, introduced by Roy (see [34]). But, in order to take into consideration the most complete preference information given by the fuzzy relations, we can sum the global preference indices referred to each feasible action in comparison with others, obtaining its comprehensive preference index, aiming to build up the partition of A into S equivalence classes $C_1, C_2, \cdots, C_S$, $S \leq n$ (complete preorder), by means of a descending procedure (from the best action to the worst) or by an ascending procedure (from the worst to the best).

In either case, the peculiar feature of these techniques is that at every step they select the action(s) assigned to a certain position in the ranking considered and then repeat the procedure with respect to the subset of the remaining actions, eliminating at each iteration the action, selected in the preceding one. Here is a brief example of one of the possible techniques.

Computation of the *comprehensive preference index*, $a \in A$:

$$\sigma_{+}^{(1)}(a) = \sum_{b \in A \setminus \{a\}} \pi(a, b).$$

This will be:

$$0 \leq \sigma_{+}^{(1)}(a) \leq n - 1, \ \ \forall a \in A.$$

In particular we obtain:

$$\sigma_{+}^{(1)}(a) = n - 1 \ \text{ or } \ \sigma_{+}^{(1)}(a) = 0,$$

if and only if a strictly dominates, or is strictly dominated by, respectively, all the remaining feasible actions. We then select the action(s) with the highest index $\sigma_{+}^{(1)}$. This action, or these actions, will occupy the first place in the decreasing ranking, forming class C_1. Then, given $A^{(1)} = A \setminus C_1$, we repeat the procedure with reference to the actions from this new subset, obtaining the indices:

$$\sigma_{+}^{(2)}(a) = \sum_{b \in A^{(1)} \setminus \{a\}} \pi(a, b), \ a \in A^{(1)}.$$

This iteration will make it possible to form class C_2, and so on (*descending procedure*).

The *increasing solution* may be obtained by calculating for each action a the comprehensive index

$$\sigma_{-}^{(1)}(a) = \sum_{b \in A \setminus \{a\}} \pi(b, a),$$

and placing in the last class C_s the action(s) which present the highest value for this index. We then proceed with the calculation of the indices $\sigma_{-}^{(2)}(a)$ related to the subset $A \setminus C_s$ and so on.

This way to build the rankings is suggested in order to reduce the risk that an action dominating or dominated by one or more feasible actions may assume a discriminatory role over these. A dominated action has a distorting effect during the descending procedure, while a dominating action produces the same effect during the ascending procedure.

A useful geometrical interpretation on omometric axes of the complete pre-orders related to the actions considered each time in the k-th iteration may efficaciously express the different rankings with the corresponding comprehensive intensities of preference (see[22]). If the broken lines connecting the points representing the comprehensive preferences of each action at all different iterations prove to be more or less parallel, the relative comprehensive preferences tend to remain constant. On the other hand, if these broken lines intersect one another, the ranking will present inversion in terms of comprehensive preferences at the considered iterations.

Of course, in the building of all complete preorders it is possible to introduce suitable indifferent thresholds, to prevent small differences in the comprehensive indices considered at every iteration from assuming a discriminating role (see [22]).

The building of preorders allows also to solve the choice problem. But it is also possible to directly use the information about strict dominance (given by the comprehensive preference) indices to support DM in choice problem.

Let $P(a,b) = \max[\pi(a,b) - \pi(b,a), 0]$, that is $P(a,b) = T_L[\pi(a,b), 1 - \pi(b,a)]$, where $T_L[.,.]$ means Lukasiewicz t-norm. Choice is usually based on the following scoring functions:

- non domination degree

$$\mu_{ND+}(a,\pi) = \min_{x \in A}[1 - P(x,a)] = \min_{x \in A} P^d(a,x),$$

where $P^d(\bullet,\bullet)$ means "dual"of $P(\bullet,\bullet)$;

- non dominance degree

$$\mu_{ND-}(a,\pi) = \min_{x \in A}[1 - P(a,x)] = 1 - \max_{x \in A} P(a,x).$$

Let $A^{UND+} = \{a \in A : \mu_{ND+}(a,\pi) = 1\}$ (i.e. the subset of non-dominated actions from A) and $A^{UND-} = \{a \in A : \mu_{ND-}(\pi,a) = 1\}$(i.e. the subset of non-dominating actions from A). Clearly, best action(s) will belong to set A^{UND+} and worst action(s) to set A^{UND-}. We observe that, if relation $\pi(a,b)$ is transitive, A^{UND+} and A^{UND-} are non empty.

3.2 PRAGMA

The Preference RAnking Global frequencies in Multicriteria Analysis (PRAGMA) [23] method is based on the peculiar PCCA aggregation logic (that is firstly on pairwise comparisons by means of couples of distinct criteria, and then on the aggregation of these partial results), and use the same data input and preferential information of MAPPAC, of which it constitutes a useful complement and presents the same flexibility in preference modeling. Moreover, it instrumentally uses the MAPPAC basic preferences indices to compute its specific information to support DM in his/her decision problem at hand. From the methodological point of view, PRAGMA is neither a classical outranking neither a MAUT method. In fact, the output of this approach are not binary outranking relations or scores. But, following the aggregation procedure of PCCA, in the first and in the second phase partial and global ranking frequencies are respectively built, one for each feasible action, and these frequencies are then exploited to give DM a useful recommendation (partial or complete preorders are the final output).

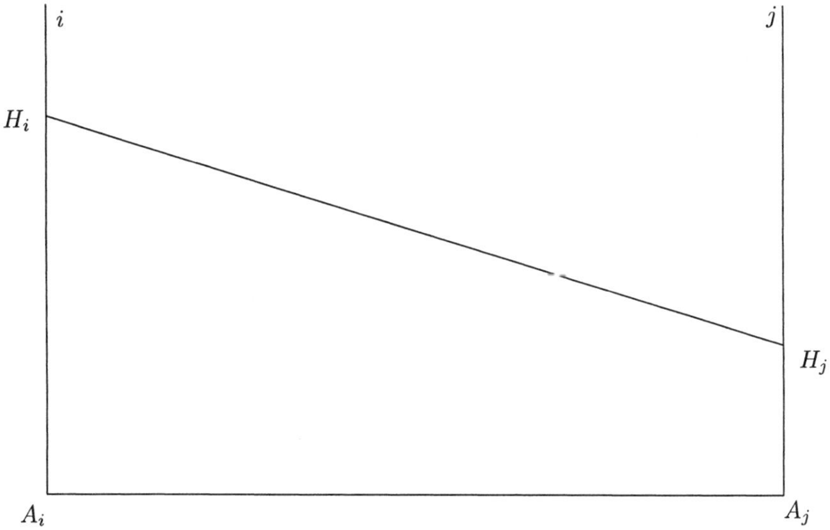

Figure 6.11. Partial profile of action a_h.

Partial and Global Frequencies. Let the segment H_iH_j (see Fig 6.11) be the *partial profile* of action $a_h \in A$, where the points H_i and H_j have as ordinates the weighted normalized evaluations of action a_h with respect to criteria g_i and g_j respectively, $g_i, g_j \in F$.

Considering all couples of criteria, it is possible to obtain $\binom{m}{2}$ distinct partial profiles of a_h and we call *global profile* of a_h the set of these $\binom{m}{2}$ partial profiles.

We define as *partial broken line-k*, or partial broken line of level k of a_h, $k = 1, 2 \cdots, n$ the set of consecutive segments of its partial profiles, to which correspond, for each point, $k - 1$ partial profiles (distinct or coinciding) of greater ordinate. If, for example, it is $A = \{a_r, a_s, a_t\}$, we obtain the partial profiles and partial broken lines represented in Figure 6.12.

We observe that the partial broken line-k, $k = 1, 2 \cdots, n$, coincides with the partial profiles of $a_h, a_h \in A$, if and only if a_h is partially dominated by d actions ($0 \leq d \leq k - 1$) and dominates the remaining ones and/or if p couples of actions from A ($0 \leq p \leq k - 1, d + p = k - 1$) exist such that, for each couple, their partial profiles come from opposite sides with respect to profile of a_h, and they intersect this profile at the same point.

Figure 6.12. Partial profiles and partial broken lines of a_r, a_s, a_t.

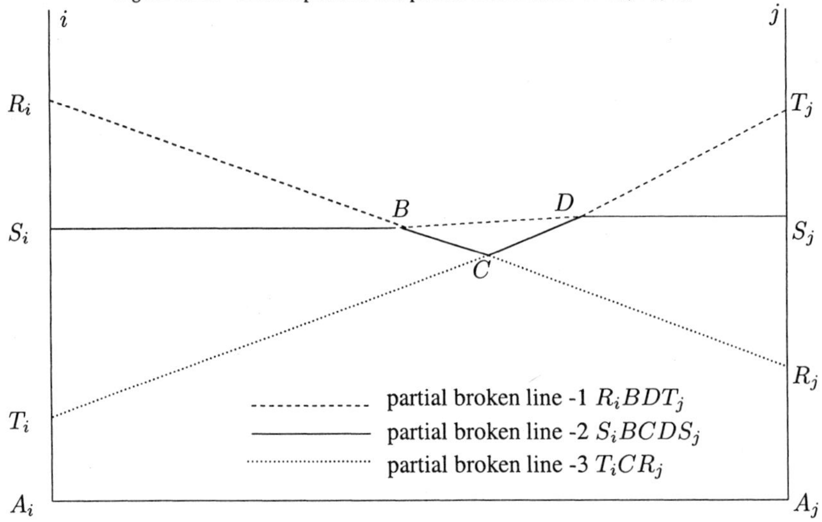

Further, we define as *global broken line-k* or global broken lines of level k ($k = 1, 2, \cdots, n$) the set of $\binom{m}{2}$ partial broken lines-k obtained by considering all the couples of distinct criteria $g_i, g_j \in G$. The global broken line-k coincides with the global profiles of a_h if and only if all the partial broken lines of level k, obtained by considering each of the $\binom{m}{2}$ couples of criteria, coincide with the corresponding partial profiles of a_h.

We define as the *partial frequency* of level k ($k = 1, 2 \cdots, n$) of a_h, with reference to the criteria g_i and g_j, the value of the orthogonal projection on the straight line A_iA_j (given $\overline{A_iA_j} = 1$) of the intersection of the partial profile of a_h with the corresponding partial broken line of level k. If we indicate this frequency as $f_{ij}^k(a_h)$, it will be $0 \leq f_{ij}^k(a_h) \leq 1$, for all $a_h \in A, k = 1, 2 \cdots, n$. Thus, for example, from the graphics in Figure 6.13.

$$\begin{aligned}
\overline{A_iA_j} &= 1; & \overline{A_iB} &= 0.3; & \overline{BC} &= 0.1;\\
\overline{CD} &= 0.2; & \overline{DA_j} &= 0.4; & &\\
f_{ij}^{(1)}(a) &= 0.3; & f_{ij}^{(2)}(a) &= 0.3; & f_{ij}^{(3)}(a) &= 0.6;\\
f^{(1)}{}_{ij}(b) &= 0.3; & f_{ij}^{(2)}(b) &= 0.7; & f_{ij}^{(3)}(b) &= 0\\
f_{ij}^{(1)}(c) &= 0.4; & f_{ij}^{(2)}(c) &= 0.2; & f_{ij}^{(3)}(c) &= 0.4
\end{aligned}$$

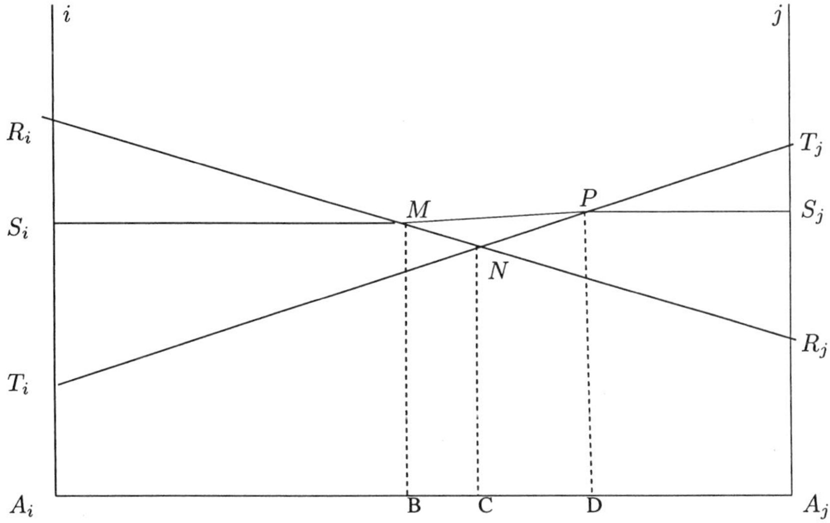

Figure 6.13. Partial frequencies of a_r, a_s, a_t.

The partial frequencies may be represented in matrix form, obtaining $\binom{m}{2}$ square $n \times n$ matrices $\mathbf{F_{ij}}$, which is the matrix of the partial ranking frequencies:

$$\mathbf{F_{ij}} = \left[f_{ij}^{(k)}(a_h) \right], a_h \in A;\ k = 1, 2 \cdots, n;\ i, j \in I. \tag{6.8}$$

The elements of the hth line of matrix (6.8) indicate in order the fractions of the interval unitary (A_i, A_j) for which the action a_h is in the kth position ($k = 1, 2 \cdots, n$), while the elements of the kth column of the same matrix indicate those fractions for which the kth position (in the partial preference ranking considered) is assigned to the actions $a_1, a_2, \cdots, a_n$, respectively. Obviously:

$$\sum_{k=1}^{n} f_{ij}^{(k)}(a_h) = 1,\ \forall a_h \in A \text{ and } \sum_{h=1}^{n} f_{ij}^{(k)}(a_h) = 1, k = 1, 2, \cdots n.$$

If $f_{ij}^{(k)}(a_h) \in \{0, 1\}$, for all $a_h \in A$ and $k = 1, 2 \cdots, n$, the partial profiles of all the actions will be non-coinciding, and there will be no inversions with respect to the preference relation in the two complete preference preorders with respect to the criteria g_i and g_j, i.e. all actions from A partially dominate one another.

If v ($v = 2, 3 \cdots, n$) partial profiles are coinciding, the corresponding partial broken lines-k must be built taking distinctly into account the coinciding profiles v times (see [23]).

Let us then define *global frequency of level* k, $(k = 1, 2 \cdots, n)$ of a_h as the weighted arithmetic mean of all the $\binom{m}{2}$ partial frequencies of level k of a_h, obtained by considering all the couples of distinct criteria g_i and g_j. Therefore, designating this frequency by $f^{(k)}(a_h)$, we obtain, if no interaction between criteria is considered (see Section 3.1):

$$f^{(k)}(a_h) = \sum_{(i<j)i,j} f_{ij}^{(k)}(a_h) \frac{\lambda_i + \lambda_j}{m-1},\ a_h \in A,\ k = 1, 2 \cdots, n.$$

The linear combination of the matrices (6.8) with weights $\frac{\lambda_i+\lambda_j}{m-1}$ will therefore give the square $n \times n$ matrix $\mathbf{F} = [f^k(a_h)]$ $(h = 1, 2, \cdots, n;\ k = 1, 2, \cdots, n)$, called the global ranking frequency matrix. Its generic element $f^k(a_h)$ indicates the relative frequency with which $a_h \in A$ is present in the kth position $(k = 1, 2, \cdots, n)$ in the particular ranking obtained by considering all the criteria $g_j \in F$ and the global profiles of all the feasible actions. It will therefore be:

$$\sum_{k=1}^{n} f^{(k)}(a_h) = 1, \forall a_h \in A \text{ and } \sum_{h=1}^{n} f^{(k)}(a_h) = 1, k = 1, 2, \cdots, n.$$

It is possible to calculate the partial frequencies $f_{ij}^{(k)}(a_h)$ by means of an algorithm which uses the indices $\pi_{ij}(a_h, a_k)$ of the MAPPAC method (see [23]). It is therefore possible to consider marginal indifference thresholds and suitable indifference areas also when the PRAGMA method is implemented. In other words, the indices $\pi_{ij}(a_h, a_k)$ here instrumentally introduced, may be calculated in advance by using all the techniques adopted with reference to the MAPPAC method (see Section 3.1).

Apart from these calculations, it is useful in any case to remember among others some particular features of the ranking frequencies obtained by the PRAGMA method:

1 The partial frequencies (and therefore also the global ones) of $a_h \in A$ are functions of the value of the normalized weighted differences between the evaluations of a_h and those of the remaining feasible actions with respect to the criteria considered. The values of these weighted differences may be overlooked only in the case of partial dominance (for partial frequencies) or strict dominance (for global frequencies), active or passive, of the action a_h.

2 If a_h partially dominates $n - k$ actions and it is partially dominated by the remaining $k - 1$ actions, $k = 1, 2 \cdots n$ the result is $f_{ij}^{(k)}(a_h) = 1$, whatever the values λ_i and λ_j.

3 If a_h strictly dominates $n-k$ actions and is strictly dominated by the remaining $k-1$ actions, $k=1,2\cdots,n$, the result is $f^{(k)}(a_h)=1$, whatever the values of the weights λ_j, $j \in \mathcal{J}$.

4 If $f^{(k)}(a_h)=1$, the action a_h occupies the kth position, $k=1,2\cdots,n$, in every monocriterion ranking and a_h is preceded and followed by the same subset of actions in these rankings.

Therefore, the information obtained by means of analysis of the global frequencies $f^{(k)}(a_h)$ is more complete and more accurate than that obtained from an examination of all the distinct monocriterion rankings of the feasible actions, or from a mixture of these.

Exploitation and Recommendation. In order to support DM in the decision problem at hand, it is often sufficient to analyze the elements of matrices $\mathbf{F}_{ij}$ and/or $\mathbf{F}$. For example, a straightforward reading of the global frequencies of matrix $\mathbf{F}$ could indicate which action(s) will finally be chosen. But the concise and accurate information regarding the frequencies of ranks each action may occupy can be extremely useful to build up final rankings.

If we want to obtain complete or partial rankings of the feasible actions in order to build up comprehensive evaluations and recommendations, it is possible, for example, to proceed in this way. Calculate for each action $a_h \in A$, the *accumulated frequencies of order* k, $k=1,2\cdots,n$, summing the first k elements of the hth row of matrix $\mathbf{F}$, that is:

$$F^{(1)}(a_h)=f^{(1)}(a_h) \text{ and } F^{(k)}(a_h)=\sum_{i-1}^{k} f^{(i)}(a_h), k=2,3\cdots,n$$

Then establish the order q ($q=1,2,\cdots,n-1$) of the frequencies which are considered relevant to the building of the ranking, that is indicate to what order q we intend to take into consideration the accumulated frequencies $F^k(a_h)$ for this purpose. The following comprehensive index is then built:

$$S^q(a_h)=\sum_{k=1}^{q}\alpha_k F^{(k)}(a_h),\ a_h \in A; 1 \geq \alpha_1 \geq \alpha_2 \geq \cdots \geq \alpha_q > 0. \quad (6.9)$$

This gives the measure of the "strength" with which a_h occupies the first q positions in the aggregated ranking. This in practice will be $1 \leq q \leq \frac{n}{2}$, which regards the first positions in the ranking; the coefficients α_k indicate the relative importance (not increasing with k) of accumulated frequence of order k. In the first class C_1 of the decreasing ranking will be placed the action(s) to which the maximum value of $S^{(q)}(a_h)$ corresponds. In order to avoid ex aequo rankings, we proceed by selecting whichever actions have obtained an equal value of $S^{(q)}$ on the basis of the values of the indices $S^{(q+1)}$ and, in the case of further equality, on those of the indices $S^{(q+2)}$ and so on. In this case ex aequo actions would be accepted only if their corresponding indices $S^{(i)}$ proved equal for $i = q, q+1, \cdots, n$. If, on the other hand, we desire to prevent small differences in the indices $S^{(q)}$ from having a discriminatory role in the building of the rankings, it is possible to consider global indifference thresholds (see [22]).

If we place $\alpha_k = 1$, for all k, in 6.9, we do not emphasize the greater importance of the global ranking frequencies of the first positions. On the other hand, if we accept $q = 1$, we take into account only the global frequencies of the first position for the purpose of building the rankings.

After building class C_1, with reference to the subset of the remaining actions $A^{(1)} = A - C_1$, we calculate again the partial, global and accumulated frequencies and the index 6.9, proceeding as above in order to build class C_2, and so on. We observe that at each iteration t the order q_t, on the basis of the which the index $S^{(q_t)}$ 6.9 is to be calculated, must be restated so that it is a non increasing whole number and, taking into account the number $|A^{(t)}|$ of actions of the evaluation set, so that at each iteration t the ratio $\frac{q_t}{|A^{(t)}|}$ is as near as possible to the ratio $\frac{q}{n}$ of the first iteration (see [23]). In general, the rankings obtained are a function of the value of the order q originally selected (see [22]).

If at each useful iteration $F^{(k)}(a_r) \geq F^{(k)}(a_s)$ for all $k = 1, 2, \ldots, n$ and $F^{(k)}(a_r) > F^{(k)}(a_s)$ for some k, or if $\sum_{k=1}^{s}[F^{(k)}(a_r) - F^{(k)}(a_s)] \geq 0$ for all $s = 1, 2, \cdots, n$ and $F^{(k)}(a_r) \neq F^{(k)}(a_s)$ for some k, it is possible to speak of first degree or second degree *frequency dominance*, respectively, of a_r over a_s. In both cases, if $\alpha_1 > \alpha_2 > \cdots > \alpha_q$, a_r will precede a_s in any of the rankings obtained, whatever value may be chosen for the other q.

Besides the partition of the actions of A into equivalence classes (complete preorder) obtained with the descending procedure (or procedure *from above*) described, it is also possible to build another complete preorder in the same way using the ascending procedure (or procedure *from below*), that is selecting the action(s) to be placed in the last, next to last, $\cdots$ and finally in the first equivalence class.

In conclusion, it is possible to build a final ranking (partial preorder) of the feasible actions, as the intersections of the two decreasing and increasing rankings obtained by means of two separate procedures described. Using the PRAGMA method for the building of rankings, it is possible not only to establish any implicit incomparability deriving from the inversion of preferences in the preorders obtained by means of the two separate procedures, but also in this case it is possible to consider an explicit incomparability, obtained if the relative tests give a positive result, during the preference modeling phase. Since, as we have said, the PRAGMA method makes instrumental use of the basic preference indices, it is possible to use once again the same discordance indices already introduced in the MAPPAC method (see Section 3.1).

Besides these, moreover, it is also possible to consider other analogous discordance indices peculiar to the PRAGMA method, that is using the partial and global ranking frequencies. Thus, for example, with respect to a couple or all criteria simultaneously, a strengthening of the ranking frequencies of an action a_h, respectively partial or global, corresponding to the first and last positions in the ranking, can reveal strongly discordant evaluations of a_h by means of those criteria. Therefore this kind of situation, suitably analyzed, could lead the DM to reconsider the nature of a_h, therefore, in the building phase of the rankings, this situation may lead to a rapid choice of a_h both in the descending and in the ascending procedure, resulting in situations of conflictuality and implicit incomparability.

Software. M&P (MAPPAC and PRAGMA) is a software to rank alternatives using the methods previously described. It presents a lot of options in order to be very flexible in the preference modeling, according to the PCCA philosophy. After loading or writing a file concerning the decisional problem at hand, in the Edit menu it is possible to set all the parameters required to compute the basic and global preference indices or ranking frequencies, i.e. trade-off and importance weights etc.. Some classical statistical analyses on the alternatives evaluations are also allowed (average values, standard deviations, correlations between criteria). The indifference areas can be performed in the Calculation menu. For each couple of criteria, suitable indifference thresholds and shapes can be defined. This option results in some non punctual indifference relations, that can also be seen on useful graphics, showing the indifference area and each pair of alternatives in the chosen plane Og_ig_j, g_i,$g_j \in F$. It is also possible to graphically represent the partial and global profiles and levels of the considered alternatives. Going to Solutions menu, after setting other optional parameters, we can firstly obtaining the (partial and global) preference matrices (MAPPAC) and frequencies matrices (PRAGMA); then, exploiting these data, the descending and ascending complete preorders and the final (partial) preorder (as their intersection) can be built up, respectively for MAPPAC and PRAGMA methods. On interesting geometrical interpretation on omometric axes of the complete preorders computation procedure, expresses with respect to each iteration the different rankings with the corresponding global preference intensities of the alternatives considered each time. This representation shows eventual inversion of preferences (as intersection of the corresponding straight lines) due to the presence of some strong dominance effect. Finally, it is possible to perform a suitable Conflict analysis among the alternatives, by setting the parameters needed to compute the bicriteria discordance indices and the incomparability relations, each time according to the corresponding compensation level established by the DM. The indifference and incomparability relations are also suitably presented in a geometrical way in the bicriteria planes Og_ig_j, for each $g_i, g_j \in F$, where the pairs of action are represented using different colours for different binary relation.

3.3 IDRA

A new MCDA methodology in the framework of PCCA was presented by Greco [7] in IDRA (Intercriteria Decision Rule Approach). Its main (and original) features are: to use mixed utility function (i.e. in the decision process both trade-off and importance intercriteria information are considered) and to allow bounded consistency, i.e. no hard constraint is imposed to the satisfaction of some axiomatic assumptions concerning intercriteria information obtained by DM. With respect to the last point, in a MCDA perspective two different kinds of coherence should be considered: the judgemental and the methodological. The first one concerns the intercriteria information supplied by DM and there is no room for technical judgement with respect to its internal coherence. The second one is related to the exploitation of intercriteria information in order to obtain the final recommendation and a coherence judgment based on some MCDA principles and axioms is allowed. Therefore, according to the judgemental coherence principle, within the IDRA method DM is allowed to give both trade-off and importance intercriteria information, without checking its not requested coherence.

Let $g_j\colon A \to \mathbb{R}$, $\forall j \in I$, an interval scale of measurement; a normalized value c_{hj} of $g_j(a_h)$, $a_h \in I$, can be obtained by introducing two suitable parameters $a(j)$, a minimum aspiration level, and $b(j)$, a maximum aspiration level, for each criterion $g_j \in F$, with $a(j) \leq \min g_j(x)$ and $b(j) \geq \max g_j(x)$ by defining

$$c_{hj} = \begin{cases} \frac{(g_j(a_h)-(a_j))}{b(j)-a(j)} & \text{if } a(j) < b(j), \\ 0 & \text{if } a(j) = b(j). \end{cases}$$

In IDRA, as above emphasized, the compensatory approach and the non-compensatory approach are complementary, rather than alternative, aggregation procedures, following the line coming out from some well known experiments carried out by Slovic [36] and others. The basic idea within IDRA [7] is that matching (i.e. comparing two actions by making the action that is superior on one criterion to be so inferior in the other one that the previous advantage is canceled) is not a decision problem: it is rather a questioning procedure for obtaining the intercriteria information called trade-off. On the contrary, choosing among equated (by matching) packing of actions is a typical decision problems, as ranking and sorting. Therefore, if this assumption is accepted, in each decision problem, like choice, there are two different types of intercriteria information: trade-off, which can be derived from a matching, and importance weights, linked to the intrinsic importance of each subset (also a singleton) of criteria from F.

As a consequence, there is only one utility function U^M, called mixed (see [7]), because both trade-off (α_j) and importance (λ_j) weights are considered, $j \in \mathcal{J}$; thus for each $a_h \in A$:

$$U^M(a_h) = \sum_{j=1}^{m} \lambda_j \alpha_j g_j(a_h).$$

The bounded consistency hypothesis

- for trade-off weights, $w_{ik}w_{kj} = w_{ij}, i, j, k \in \mathcal{J}$, where, in general, w_{pq} is the tradeoff between the criteria g_p and g_q;
- for importance-weights, given $G_1, G_2 \subset F$, if G_1 is more important than G_2, then $\sum_{g_i \in G_1} \alpha_j > \sum_{g_i \in G_2} \alpha_j$; if G_2 is more important than G_1, then $\sum_{g_i \in G_1} \alpha_j < \sum_{g_i \in G_2} \alpha_j$; if G_1 and G_2 are equally important, then $\sum_{g_i \in G_1} \alpha_j = \sum_{g_i \in G_2} \alpha_j$;

Very often these requirements are not satisfied by the answers given by the DM and the DM is said "incoherent". But, as remarked by Greco [7], most of these "inconsistencies" derive from the attempt to use information relative to partial comparisons (i.e. with respect to only some criteria from F) for global comparisons (i.e. where all the criteria from F are considered). In IDRA, the hypothesis of bounded consistency means that the information obtained from DM with respect to some criteria from F must be used only for comparisons with respect to the same criteria, according to the principle of judgemental coherence. Therefore, every above problem of intercriteria information consistency is "dissolved" in its origin. In IDRA the framework of PCCA is used to implement the bounded consistency hypothesis, considering therefore a couple of criteria at a time. We observe that, in particular, no requirement of completeness of the relations "more important than" and "equally important to" is assumed. As a consequence, for any couple of distinct criteria $g_i, g_j \in F$, one of the following intercriteria information can be obtained by the DM:

1 both the trade-off and the judgement about the relative importance of the criteria;

2 only the trade-off;

3 only the judgement about the relative importance of the criteria;

4 neither the trade-off nor the judgement about the relative importance of the criteria.

Using this information, a basic preference index $\pi^*_{ij}(a,b)$ can be suitably defined (see [7]). The index $\pi_{ij}: A \times A \to [0,1]$ is the image of a valued binary relation, complete and ipsodual, and constitutes a complete valued preference structure (complete preorder) on set A. The index $\pi^*_{ij}(a,b)$ can be interpreted as the probability that a is preferred to b, with respect to a mixed utility function in which the trade-off and importance weights are randomly chosen in the set of intercriteria information furnished by the DM. In IDRA, each piece of intercriteria information concerning the trade-off or the relative importance of criteria can be considered a "decision rule" (tradeoff-rule or importance-rule respectively), since it constitutes a basis for an argumentation about the preference between the potential actions. The DM is asked to give a non negative credibility-weight to each decision rule, according to his/her judgment about the relevance of the corresponding pairwise criterion comparisons in order to establish a global preference [7]. Therefore, from the sum of the basic indices $\pi^*_{ij}(a,b)$, with respect all the considered couple of criteria, weighted by the correspondent credibility-weights for the tradeoff-rule or the importance-rule, the aggregated index $\pi(a,b)$ is obtained, for each $a,b \in A$. These indices can be then exploited using the same procedure proposed for MAPPAC in order to obtain two complete preorders (decreasing and increasing solutions); the intersection of these two rankings gives the final ranking (partial preorder). The aggregated index of IDRA mainly differs from the analogous index of MAPPAC in this point: in MAPPAC all (i.e. with respect to each couple of criteria from F) basic indices are aggregated, while in IDRA only the elementary indices corresponding to couples of criteria about which the DM has given decision rules are aggregated (faithfulness principle). In IDRA there is a peculiar characteristic: distinction between:

1 intercriteria information which is not supplied by the DM (i.e. the DM does not says anything about the relative importance between g_i and g_j);

2 intercriteria information by which the DM expresses his/her incapacity to say what is the trade-off or the relative importance between g_i and g_j (i.e. the DM says that he/she is not able to give this information).

In IDRA, in case 1. the comparison with respect to criteria g_i and g_j plays no part; in case 2. the same comparison contributes to the aggregated index by means of considering the corresponding basic index calculated taking into account all the possible importance-weights as equally probable, according to the "principle of insufficient reason" (so called Laplace criterion in the case of decision making under uncertainty).

3.4 PACMAN

A new DM-oriented approach to the concept of compensation in multicriteria analysis was presented by Giarlotta [5, 6] in PACMAN (Passive and Active Compensability Multicriteria ANalysis). The main feature of this approach is that the notion of compensability is analyzed by taking into consideration two criteria at a time and distinguishing the compensating (or active) criterion from the compensated (or passive) one. Separating active and passive effects of compensation allows one to point out a possible asymmetry of the notion of compensability and to introduce a suitable valued binary relation of compensated preference.

The concept of compensation has been analyzed in many papers [35, 37, 38]. The literature on this topic is mainly concentrated on the study of decision methodologies, aggregation procedures and preference structures on the basis of this concept. Therefore definition and usage of compensation have essentiality been *method-oriented*, since this concept has been regarded as a theoretical device of classification.

On the contrary, the notion of compensation examined in PACMAN, namely *compensability*, is aimed at capturing the behavior of a decision maker towards the possibility to compensate among criteria. In our approach, intercriteria compensability remains somehow "the possibility that an advantage on one criterion can offset a disadvantage on another one", but as it is determined by a DM and not by a method. Therefore, being more or less compensatory is not regarded here as the characteristic of a multicriteria methodology or of an aggregation procedure. Instead, it is an intrinsic feature of a DM. In this sense, we speak of a *DM-oriented* usage of the concept of compensation.

There are three steps in PACMAN:

- *compensability analysis*, the procedure aimed at modeling intercriteria relations by means of compensability;
- evaluation of the degree of active and passive preference of an alternative over another one by the construction (at several levels of aggregation) of *binary indices*;
- determination of a binary relation of strict preference, weak preference, indifference or incomparability for each couple of alternatives, on the basis of two valued relations of *compensated preference*.

At each step of the procedure PACMAN requires a strict interaction between the actors of the decision process. Therefore, also this approach allows application of the principles of faithfulness (to the information provided by DM), transparency (at each stage of the procedure) and flexibility (in preference modelization).

Compensability Analysis. Let $g_j: A \rightarrow \mathbb{R}$ be an interval scale of measurement, representing the j-th criterion according to a non decreasing preference. For each $j \in \mathcal{J}$, let $\Delta_j: A \times A \rightarrow \mathbb{R}$ be the normalized difference function, defined by $\Delta_j(a,b) = (g_j(a) - g_j(b))/(\beta_j - \alpha_j)$, where α_j and β_j ($\alpha_j < \beta_j$) are respectively the minimum and the maximum value that can be assumed on $j \in \mathcal{J}$.

The aim of compensability analysis is to translate into numerical form the definition of bicriteria compensability for each pair of criteria.This is done by constructing, for each pair (i,j) of criteria, the compensatory function $CF_{i \rhd j}$ of i over j, which evaluates the compensating effect of a positive normalized difference on the passive criterion j.

Since a proper and complete estimation of the compensatory effect for every possible active and passive difference is too demanding in terms of amount and preciseness of the related information provided by the DM , we build $CF_{i \rhd j}$ as a *fuzzy* function. This function associates to any pair of normalized differences $(\Delta_i, \Delta_j) \in]0,1] \times [-1,0[$ a number belonging to $[0,1]$ the degree of confidence that the positive difference Δ_i totally compensates the negative differences Δ_j. Extending the function in frontier by continuity, we obtain a fuzzy compensatory function $CF_{i \rhd j}: [0,1] \times [-1,0] \rightarrow [0,1]$, which satisfies the following conditions:

Weak monotonicities

$$0 \leq \Delta_{i_1} \leq \Delta_{i_2} \leq 1 \text{ and } -1 \leq \Delta_j \leq 0 \quad \Rightarrow \quad CF_{i \rhd j}(\Delta_{i_1}, \Delta_j) \leq CF_{i \rhd j}(\Delta_{i_2}, \Delta_j),$$
$$0 \leq \Delta_i \leq 1 \text{ and } -1 \leq \Delta_{j_1} \leq \Delta_{j_2} \leq 0 \quad \Rightarrow \quad CF_{i \rhd j}(\Delta_i, \Delta_{j_1},) \leq CF_{i \rhd j}(\Delta_i, \Delta_{j_2})$$

Continuity $CF_{i \rhd j}$ is continuous everywhere on $[0,1] \times [-1,0]$.

The reason for a fuzzy modelling is to minimize the amount of information required from the DM, without losing too much in content. The two conditions stated above are very helpful in this sense. In fact, in order to assess a compensatory function, the DM is asked to determine just the zones where the *degree of confidence* expressed by $CF_{i \rhd j}$ is *maximmum* (usually equal to one) or *minimum* (usually equal to zero). Using monotonicity and continuity, it is possible to extend by linearization its definition to the whole domain $[0,1] \times [-1,0]$, without any further information. By definition, $CF_{i \rhd i} \equiv 0$ for each $i \in \mathcal{J}$.

The procedure for the construction of compensatory functions aims at simplifying the task for the DM in providing meaningful information. On the other hand, this procedure requires the DM to provide a large amount of information. In fact, according to the PCCA philosophy, we estimate intercriteria compensability for each couple of criteria. Moreover, we still distinguish their compensatory reaction within the couple, according to whether they effect or endure compensation. This results in the necessity of assessing a compensatory function for each *ordered* pair of distinct criteria.

However, the large amount of information required by PACMAN allows one to model the relationships between each couple of criteria in a rather faithful and flexible way, according to the PCCA philosophy. Usually, an important criterion is relevant both actively (i.e., contributing to preference) and passively (i.e., opposing to preference). Therefore for each criterion we can treat separately passive resistance and active contribution, concepts related to the notion of veto thresholds and preference thresholds respectively in the outranking approach [35]. For a detailed description of the procedure used to construct compensatory functions see [6].

Preference Modeling. In PACMAN preferences are modelled on the basis of compensability analysis. This is accomplished in steps (2) and (3) of the procedure.

(2): Let $g_j \in G^+(a,b)$, i.e., $\Delta_j(a,b) > 0$. The positive difference $\Delta_j(a,b)$ has a double effect.

- active, because it gives some contribution to the (possible) overall preference of a over b (accept this global preference);
- passive, because it states a resistance to the (possible) overall preference of b over a (reject this global preference).

Active contribution and passive resistance of a over b are evaluated for each $g_j \in G^+(a,b)$, computing the *partial indices* $\Pi_j^+(a,b)$ and $\Pi_j^-(a,b)$, respectively. Successively, active and passive effects are separately aggregated, thus obtaining an evaluation of the total strength of the arguments in favour of a preference of a over b, and of those against a preference of b over a, respectively. Numerically, this is done by computing the two binary *global indices* $\Pi^+(a,b)$ and $\Pi^-(a,b)$. Clearly, the same evaluations are done for the pair (b,a), first computing the partial indices $\Pi_j^+(b,a)$ and $\Pi_j^-(b,a)$, and then the global indices $\Pi^+(b,a)$ and $\Pi^-(b,a)$.

The final output of this stage is a pair of *global net indices* $\Pi(a,b)$ and $\Pi(b,a)$ for each couple of alternatives $a, b \in A$. These indices express the degree of *compensated preference* of a over b and b over a, respectively. The index $\Pi(a,b)$ is obtained from the values of the indices $\Pi^+(a,b)$ and $\Pi^-(b,a)$; similarly, the index $\Pi(a,b)$ is obtained from the values of the indices $\Pi^+(b,a)$ and $\Pi^-(a,b)$. A formalization of the whole procedure can be found in [5].

(3) The last step of PACMAN is the construction of a fundamental system of preferences (P, Q, I, R). The relation between the alternatives a and b is determined from the values of the two global net indices $\Pi(a,b)$ and $\Pi(b,a)$.

One of the main interesting features of PACMAN is that intercriteria compensability can be modelled with respect to the real scenarios, treating each pair of criteria in a peculiar way. Complexity and length of the related decision

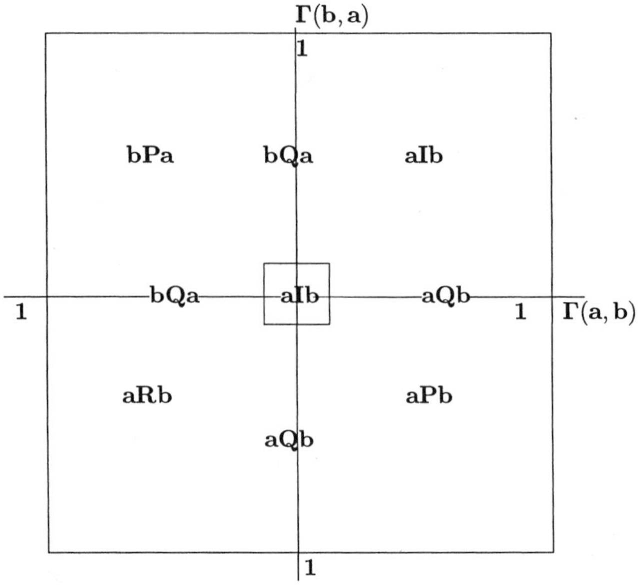

Figure 6.14. Determination of a relation between the two alternatives $a,b \in A$ on the basis of the values of global indices.

process is the price to pay for the attempt to satisfy the principles of faithfulness, transparency and flexibility.

4. One Outranking Method for Stochastic Data

It frequently happens that we have to treat a decision context in which the performance of the alternatives according to each criterion/attribute is subject to various forms of imperfection of the available data. The form of imperfection that interests us here concerns the uncertainty, in the sense of probability (statistic or stochastic data). For example, frequently the decision maker calls upon several experts in order to obtain judgements which then forms the basic data. Since each alternative is not necessarily evaluated at the same level of anticipated performance by all experts, each combination of 'alternative-criterion' leads to a distribution of expert's evaluation. This type of distributional evaluation is considered as stochastic data.

Even if the multi-criteria analysis with stochastic data has so far been treated nearly exclusively in the theory of the multi-attributes utility framework, the outranking synthesis approach can be constituted an appropriate alternative. Some multi-criteria aggregation procedures belonging to this second approach have been developed specially to treat stochastic data. For example, we can mention the works by [4, 19, 17, 18, 42, 20]. The majority of these methods construct outranking relations as in ELECTRE or PROMETHEE. In this chapter we have choose to present the Martel and Zaras' method that makes a link between the multi-attributes utility framework and the outranking approach.

4.1 Martel and Zaras' Method

We consider a multi-criteria problem which can be represented by the (A. A. E.) model (Alternatives, Attributes/Criteria, Evaluators). The elements of this model are as follows:

- $A = \{a_1, a_2, \ldots, a_m\}$ representing the set of all potential alternatives;
- $F = \{X_1, X_2, \ldots, X_n\}$ representing the set of attributes/criteria, an attribute X_j defined in the interval $[x_j^0, x_j^1]$ where x_j^0 is the worst value obtained with the attribute X_j and x_j^1 is the best value; $E = \{f_1, f_2, \cdots, f_n\}$ the set of evaluators, an evaluator $f_j(x_{ij})$ being a probability function associating to each alternative a_i a non-empty set of x_{ij} (a random variable) representing the evaluation of a_i relative to the attribute X_j.

In this method, it is assume known the distributional evaluation of the alternatives according to each attribute and the weight of the attributes.

These attributes (criteria) are defined such that a larger value is preferred to a small value and that the probability functions are known. It is also assume that the attribute set F obeys the additive independence condition. Huang, Kira and Vertinsky (see [11]) showed in the case of the probability independence and the additive multi-attributes utility function, that the necessary condition for the multi-attributes stochastic dominance is to verify stochastic dominance on the level of each attribute. In practice, the essential characteristic of a multi-attributes problem is that the attributes are conflicting. Consequently, the Multi-attributes Stochastic Dominance relation results poor and useless to the DM. It seems to be reasonable to weaken this unanimity condition and accept a majority attribute condition.

Thus, Martel and Zaras' method [20] use the stochastic dominance to compare the alternatives two by two, on each attribute. These comparisons are interpreted in terms of partial preferences. Next, the outranking approach is used for constructing outranking relations based on a concordance index and eventually on a discordance index. With this approach, a majority attribute condition (concordance test) replaces the unanimity condition of the classic dominance. Finally, these outranking relations are used in order to construct the prescription according to a specific problem statement.

Often, in order to conclude that alternative a_i is preferred or is at least as good as $a_{i'}$, with respect to the attribute X_j, it is unnecessary to make completely explicit all the decision-maker's partial preferences. In fact, it can be possible to conclude on the basis of stochastic dominance conditions of first, second and third order (i.e. FSD, SSD and TSD relations), for a class of concave utility functions with decreasing absolute risk aversion (i.e. DARA utility functions class). If the decision-maker's (partial) preference for each attribute X_j can be related by the utility function $U_j \in$ DARA, then his preference for the $F_j(x_{ij})$ distribution associated with alternative a_i for each attribute X_j will be:

$$g_j(F_j(x_{ij})) = \int_{x_j^0}^{x_j^1} U_j(x_{ij})\, dF_j(x_{ij}).$$

THEOREM 1 (HADAR AND RUSSEL, 1969) *: If $F_j(x_{ij})$ FSD $F_j(x_{i'j})$ or $F_j(x_{ij})$ SSD $F_j(x_{i'j})$ or $F_j(x_{ij})$ TSD $F_j(x_{i'j})$ and $F_j(x_{ij}) \geq F_j(x_{i'j})$, then $g_j(F_j(x_{ij})) \leq g_j(F_j(x_{i'j}))$ for all $U_j \in$ DARA, where $F_j(x_{ij})$ and $F_j(x_{i'j})$ represent cumulative distribution functions associated with a_i and $a_{i'}$ respectively.*

This theorem allows to conclude clearly that a_i is preferred to $a_{i'}$, with respect to the attribute X_j. We refer the reader to Zaras (see [41]) to review the concept of stochastic dominance.

In the MZ's model, two situations are identified; **clear situation** where the conditions imposed by the theorem are verified (SD=FSD ∪ SSDU ∪ TSD situations), and **unclear situation** where none of the three stochastic dominance is verified. The value of the concordance index can be decomposed into two parts:

Explicable concordance, that corresponds to cases in which the expression of the decision-maker's preferences is trivial or clear.

$$C_E(a_i, a_{i'}) = \sum_{j=1}^{n} \pi_j \delta_j^E(a_i, a_{i'}),$$

where

$$\delta_j^E(a_i, a_{i'}) = \begin{cases} 1 & \text{if } F_j(x_{ij})\ SD\ F_j(x_{i'j}) \\ 0 & \text{otherwise} \end{cases}$$

and π_j is the weight of attribute X_j, with $\pi_j \geq 0$ and $\sum_j^n \pi_j = 1$.

Non-Explicable concordance that corresponds to the potential value of the cases in which the expression of the decision-maker's preferences is unclear.

$$C_{NE}(a_i, a_{i'}) = \sum_{j=1}^{n} \pi_j \delta_j^{NE}(a_i, a_{i'}),$$

where

$$\delta_j^{NE}(a_i, a_{i'}) = \begin{cases} 1 & \text{if no } F_j(x_{ij})\ SD\ F_j(x_{i'j}) \text{ and} \\ & \text{no } F_j(x_{i'j})\ SD\ F_j(x_{ij}) \\ 0 & \text{otherwise.} \end{cases}$$

This second part of the concordance is only a potential value, as it is not certain that for each of these attribute $F_j(x_{ij})$ will be preferred to $F_j(x_{i'j})$.

In these cases, it may be useful to state a condition which tries to make explicit the decision-maker's value functions $U_j(x_{ij})$. If the condition

$$0 \le p - C_E(a_i, a_{i'}) \le C_{NE}(a_i, a_{i'}),$$

where $p \in [0.5, 1]$ is the concordance threshold, is fulfilled, then the explication of the unclear cases leads to a value of the concordance index such that the concordance test is satisfied for the proposition that "a_i globally outranks $a_{i'}$". The objective is to reduce as far as possible, without increasing the risk of erroneous conclusions, the number of time where the $U_j(x_{ij})$ functions must be to make explicit. It is notably in the case of unclear situation that [20] used the probabilistic dominance, as a complementary tool to the stochastic dominance, to build preference relationships.

A discordance index $D_j(a_i, a_i')$ for each attribute X_j may be eventually defined as the ratio between of the difference of the means of the distributions of a_i and a_i' to the range of the scale (if it is justified by the scale level of distributional evaluation):

$$D_j(a_i, a_{i'}) = \begin{cases} \frac{\mu(F_j(x_{i'j})) - \mu(F_j(x_{ij}))}{(x_i^1 - x_i^0)} & \text{if } F_j(x_{i'j})\ FSD_j\ F_j(x_{ij}) \\ 0 & \text{if } F_j(x_{i'j}) \text{ not } FSD_j F_j(x_{ij}). \end{cases}$$

The difference between the average values of two distributions gives a good indication of the difference in performance of the two compared alternatives. If this difference is large enough in relation to the range of the scale, and FSD is fulfilled on attribute X_j, then the chances are large that a_i is 'dominated' by $a_{i'}$. In this case, MZ assume a minimum level ν_j, called a veto threshold, of the discordance index $D_j(a_i, a_{i'}$ giving to a discordant attribute X_j the power of withdrawing all credibility that a_i globally outranks $a_{i'}$).

The discordance test is related to veto threshold ν_j for each attribute. The concordance and discordance relations for the potential alternatives from A are formulated in a classical manner:

$$\begin{array}{lcl} \text{For all } (a_i, a_{i'}) \in A \times A, (a_i, a_{i'}) \in C_p & \longleftrightarrow & C(a_i, a_{i'}) \ge p \\ \text{For all } (a_i, a_{i'}) \in A \times A, (a_i, a_{i'}) \in D_\nu & \longleftrightarrow & \exists j / D_j(a_i, a_{i'}) \ge \nu_j. \end{array}$$

The outranking relations result from the intersection between the concordance set and the complementary set of discordance set:

$$S(p, \nu_j) = C_p \cap \bar{D}_\nu = C_p \setminus D_\nu.$$

Therefore, like in ELECTRE I, we can conclude that a_i globally outranks $a_{i'}$ ($a_i S a_{i'}$) if and only if $C(a_i, a_{i'}) \ge p$ and $D_j(a_i, a_{i'}) < \nu_j$ for all j. If we have no $a_iSa_{i'}$ and no $a_{i'}Sa_i$, then a_i and $a_{i'}$ are incomparable, where S is a crisp outranking relation. On the basis on the level of overlapping of the compared distributions, Martel *et al.* [18] developed preference indices associated to the three types of stochastic dominance and constructed the valued outranking relations.

Depending on whether one is dealing with a choice or a ranking problematic, either the core of the graph of outranking relations is determined or the outranking relations are exploited as in ELECTRE II, for example.

EXAMPLE 10 *Given 6 alternatives a_1, a_2, a_3, a_4, a_5 and a_6, 4 attributes X_1, X_2, X_3 and X_4 and the stochastic dominance relation observed between each pair of alternatives according to each attribute (Table 6.20).*

Table 6.20. Table of observed stochastic dominances.

	X_1						X_2					
	a_1	a_2	a_3	a_4	a_5	a_6	a_1	a_2	a_3	a_4	a_5	a_6
a_1	×	—	—	—	TSD	—	×	FSD	FSD	FSD	FSD	FSD
a_2	FSD	×	—	—	FSD	—	—	×	FSD	FSD	FSD	FSD
a_3	FSD	FSD	×	SSD	FSD	FSD	—	—	×	FSD	FSD	FSD
a_4	FSD	FSD	—	×	SSD	FSD	—	—	—	×	FSD	—
a_5	—	—	—	—	×	—	—	—	—	—	×	—
a_6	FSD	FSD	—	—	FSD	×	—	—	—	—	FSD	×
	X_3						X_4					
	a_1	a_2	a_3	a_4	a_5	a_6	a_1	a_2	a_3	a_4	a_5	a_6
a_1	×	—	SSD	SSD	FSD	FSD	×	FSD	—*	FSD	FSD	SSD
a_2	FSD	×	SSD	SSD	FSD	FSD	—	×	FSD	FSD	FSD	FSD
a_3	—	—	×	—	FSD	FSD	—*	—	×	FSD	FSD	FSD
a_4	—	—	SSD	×	FSD	FSD	—	—	—	×	FSD	FSD
a_5	—	—	—	—	×	—	—	—	—	—	×	—
a_6	—	—	—	—	FSD	×	—	—	—	—	FSD	×

* no a_1SDa_3 and no a_3SDa_1 according to X_4.

It is assumed that the weights of the attributes are respectively .09, .55, .27 and .09. The explicable concordance indices was calculated and are presented in Table 6.21. The discordance indices are not considered in this example.

On the basis of the explicable concordance indices, we can build up the following outranking relations for a concordance threshold $p = .90$: a_1Sa_4, a_1Sa_5, a_1Sa_6; a_2Sa_3, a_2Sa_4, a_2Sa_5, a_2Sa_6; a_3Sa_5, a_3Sa_6; a_4Sa_5 and a_6Sa_5. It is possible to construct the following partial pre-order graph (Figure 6.15); within this graph, the transitivity is respected.

In the Table 6.20 we observe that the relation between a_1 and a_3 according to attribute X_4 is **unclear** *since no $F_4(x_{14})$ SD $F_4(x_{34})$ and no $F_4(x_{34})$ SD $F_4(x_{14})$. If the decision-maker can explicit $U_4(x)$ and if a_1 is preferred to a_3*

Table 6.21. Explicable concordances indices.

	a_1	a_2	a_3	a_4	a_5	a_6
a_1	×	0.64	0.82*	0.91	1	0.91
a_2	0.36	×	0.91	0.91	1	0.91
a_3	0.09*	0.09	×	0.73	1	1
a_4	0.09	0.09	0.27	×	1	0.45
a_5	0	0	0	0	×	0
a_6	0.09	0.09	0	0.55	1	×

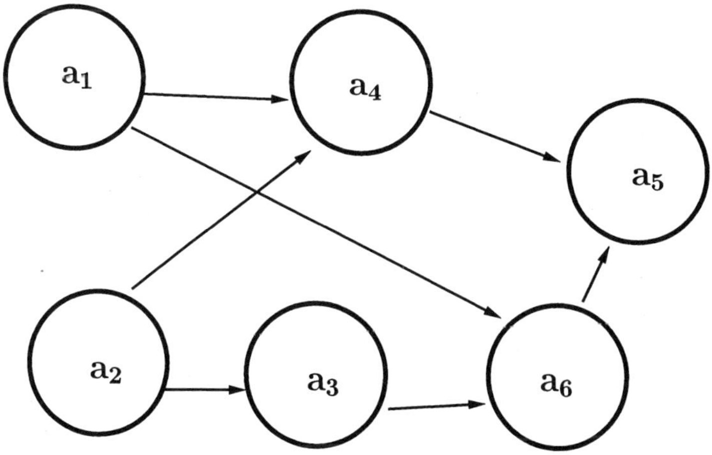

Figure 6.15. Partial preorder.

according to this attribute, then globally $a_1 S a_3$ *with a concordance thresholds* $p = .90$ *since* $C(a_1, a_3) = .91$ *(.82 in Table 6.21* $+.09$ *(the weight of* X_4*)).*

5. Conclusions

In this chapter some outranking methods different from ELECTRE and PROMETHEE family have been presented, able to manage different type of data (ordinal, cardinal and stochastic). Their description proved again the richness and flexibility of the outranking approach in preference modelling and in supporting DM in a lot of decisional problem at hand. Some properties of this approach are common to all the outranking methods, others are peculiar features of some of them. In the following we recall the main characteristics of the considered methods.

a) The input of these methods are alternative evaluations that can be given in the form of qualitative (ordinal scale), numerical non-quantitative (with the particular case of interval scales) or stochastic (probability distribution) data with respect to all considered criteria. Sometimes also some technical parameters should be supplied by DM as infracriterion information (indifference, preference, veto thresholds).

b) All these methods need as infracriterion information the importance weights in numerical terms. In some of them, just a particular order of criteria is explicitly requested, otherwise a random weight approach.

d) The outranking methods within the PCCA approach need the elicitation of both importance and trade-off weights, but the information concerning weights does not need to respect completeness (i.e. all pairwise trade-off and/or importance weights given) and transitivity with respect to trade off weights.

e) In their first step, all these methods (apart from PRAGMA) give as results some preference or outranking relations, crisp or fuzzy (preference relations and/or indices).

f) The preference structures associated with these methods is usually P, I, R, obtained at global level (comprehensive evaluation). In the PCCA approach is also possible to obtain the same binary relations with respect to each couple or pair (g_i, g_j) of considered criteria (P_{ij}, I_{ij}, R_{ij}).

g) Usually the final recommendation (complete or partial preorder) is obtained by the exploitation of the binary relations previously obtained. But in some ordinal method the complete final preorder is directly obtained as a result of the concordance-discordance analysis between different rankings.

Acknowledgments

I am grateful to Dr. Silvia Angilella for her kind cooperation and great help in the editing of this chapter.

References

[1] J. Ancot. *Micro-Qualiflex*. Kluwer Academic Publishers, Dordrecht, 1988.

[2] M. Besson. Rang moyen et agrégation de classements. *Revue Française d'Automatique, d'Informatique et de Recherche Operationnelle*, 1:37–58, 1975.

[3] W.D. Cook and M. Kress. A multiple criteria decision model with ordinal preference data. *European Journal of Operational Research*, 54:191–198, 1991.

[4] B.A. Dendrou, S.A. Dendrou, and E.N. Houtis. Multiobjective decisions analysis for engineering systems. *Computers & Operations Research*, 7:310–312, 1980.

[5] A. Giarlotta. Passive and active compensability multicriteria analysis (PACMAN). *Journal of Multi-Criteria Decision Analysis*, 7(4):204–216, 1998.

[6] A. Giarlotta. Multicriteria compensability analysis. *European Journal of Operational Research*, 133(1):190–209, 2001.

[7] S. Greco. A new PCCA method: IDRA. *European Journal of Operational Research*, 98(3):587–601, 1997.

[8] J.L. Guigou. *Analyse des données et choix à critères multiples*. Dunod, Paris, 1973.

[9] E. Hinloopen and P. Nijkamp. Regime-methode voor ordinal multicriteria-analyses. *Kwantitatieve Methoden*, 7(22):61–78, 1986.

[10] E. Hinloopen, P. Nijkamp, and P. Rietveld. Qualitative discrete multiple criteria choice models in regional planning. *Regional Science and Urban Economics*, 13:77–102, 1983.

[11] C.C. Huang, D. Kira, and I. Vertinsky. Stochastic dominance rules for multiattribute utility functios. *Review of Economic Studies*, 41:611–616, 1969.

[12] A.Z. Israels and W.J. Keller. Multicriteria analyse voor kwalitatieve data. *Kwantitatieve Methoden*, 7:49–74, 1986.

[13] R. Janssen. *Multiobjective Decision Support for Environmental Management*. Kluwer Academic Publishers, Dordrecht, 1992.

[14] W.S.M. De Keyser and P.H.M. Peeters. ARGUS – A new multiple criteria method based on the general idea of outranking. In M. Paruccini, editor, *Applying Multiple Criteria Aid for Decision to Environmental Management*, pages 263–278. Kluwer Academic Publishers, Dordrecht, 1994.

[15] M. Köksalan, M.H. Karwan, and S. Zionts. An approach for solving discret alternative multiple criteria problems involving ordinal criteria. *Naval Research Logistics*, 35:625–641, 1988.

[16] J.P. Leclercq. Propositions d'extensions de la notion de dominance en présence de relations d'ordre sur le pseudo-critéres: MELCHIOR. *Revue Belge de Recherche Operationnelle, de Statistique et d' Informatique*, 24(1):32–46, 1984.

[17] B. Mareschal. Stochastic multicriteria decision making and uncertainty. *European Journal of Operational Research*, 26:58–64, 1986.

[18] J.M. Martel, S. Azondékon, and K. Zaras. Preference relations in multicriterion analysis under risk. *JORBEL*, 31(3–4):55–83, 1992.

[19] J.M. Martel and G.R. D'Avignon. Projects ordering with multicriteria analysis. *European Journal of Operational Research*, 10:59–69, 1982.

[20] J.M. Martel and K. Zaras. Modeling preferences using stochastic and probabilistic dominances. In *International Conference on Methods and Applications of Multicriteria Decision Making*, pages 256–260. Facultés Universitaires Catholiques de Mons, Belgium, 1997.

[21] B. Matarazzo. Multicriteria analysis of preferences by means of pairwise actions and criterion comparisons (MAPPAC). *Applied Mathematics and Computation*, 18(2):119–141, 1986.

[22] B. Matarazzo. A more effective implementation of the MAPPAC and PRAGMA methods. *Foundations of Control Engineering*, 13(4):155–173, 1988.

[23] B. Matarazzo. Preference global frequencies in multicriterion analysis (PRAGMA). *European Journal of Operational Research*, 36(1):36–49, 1988.

[24] B. Matarazzo. PCCA and compensation. In P. Korhonen, A. Lewandoski, and J. Wallenius, editors, *Multicriteria Decision Support*, volume 356 of *Lecture Notes in Economics and Mathematical Systems*, pages 99–108. Springer Verlag, Berlin, 1989.

[25] B. Matarazzo. A pairwise criterion comparison approach: The MAPPAC and PRAGMA methods. In C. Bana e Costa, editor, *Readings in Multiple Criteria Decision Aid*, pages 253–273. Springer Verlag, Berlin, 1990.

[26] B. Matarazzo. MAPPAC as a compromise between outranking methods and MAUT. *European Journal of Operational Research*, 54:48–65, 1991.

[27] J.H.P. Paelinck. Qualitative multiple criteria analysis, environmental protection and multiregional development. *Papers of the Regional Science Association*, 36:59–74, 1976.

[28] J.H.P. Paelinck. Qualitative multiple criteria analysis: An airport location. *Environment and Planning*, 9:883–895, 1977.

[29] H. Pastijn and J. Leysen. Constructing an outranking relation with ORESTE. *Mathematical and Computer Modelling*, 12(10/11):1255–1268, 1989.

[30] P.Nijkamp, P. Rietveld, and H. Voogd. *Multicriteria Evaluation in Physical Planning*. North Holland, Amsterdam, 1990.

[31] J.-C. Pomerol and S. Barba-Romero. *Choix Multiple dans l'Entreprise*. Hermes, Paris, 1993.

[32] M. Roubens. Analyse et agrégation des préférences: Modélisation, ajustement et résumé de données relationnelles. *Revue Belge de Recherche Operationnelle, de Statistique et d' Informatique*, 20(2):36–67, 1980.

[33] M. Roubens. Preference relations on actions and criteria in multicriteria decision making. *European Journal of Operational Research*, 10:51–55, 1982.

[34] M. Roubens and Ph. Vincke. *Preference Modelling*. Springer Verlag, Berlin, 1985.

[35] B. Roy and D. Bouyssou. *Aide Multicritère à la Décision: Méthodes et Cas*. Economica, Paris, 1993.

[36] P. Slovic. Choice between equally-valued alternatives. *Journal of Experimental Economics: Human Perception Perfomance*, 1:280–287, 1975.

[37] J.C. Vansnick. On the problem of weighs in multiple criteria decision making (the noncompensatory approach). *European Journal of Operational Research*, 24:288–294, 1986.

[38] Ph. Vincke. *Multicriteria Decision Aid*. John Wiley & Sons, New York, 1992.

[39] H. Voogd. Multicriteria evaluation with mixed qualitative and quantitative data. *Environment and Planning B*, 9:221–236, 1982.

[40] H. Voogd. *Multicriteria Evaluation for Urban and Regional Planning*. Pion Ltd., London, 1983.

[41] K. Zaras. Dominance stochastique pour deux classes de fonctions d'utilité: concaves et convexes. *Recherche opérationnelle, Operations Research*, 23(1):57–65, 1989.

[42] K. Zaras and J.M. Martel. Multiattribute analysis based on stochastic dominance. In B. Munier and M. Machina, editors, *Models and Experiments in Risk and Rationality*, pages 225–248. Kluwer Academic Publishers, 1994.

IV

MULTIATTRIBUTE UTILITY AND VALUE THEORIES

Chapter 7

MAUT – MULTIATTRIBUTE UTILITY THEORY

James S. Dyer
Department of Management Science and Information Systems
The Graduate School of Business
University of Texas at Austin
Austin, TX 78712
USA
Jim.Dyer@bus.utexas.edu

Abstract In this chapter, we provide a review of multiattribute utility theory. We begin with a brief review of single-attribute preference theory, and we explore preference representations that measure a decision maker's strength of preference and her preferences for risky alternatives. We emphasize the distinction between these two cases, and then explore the implications for multiattribute preference models. We describe the multiattribute decision problem, and discuss the conditions that allow a multiattribute preference function to be decomposed into additive and multiplicative forms under conditions of certainty and risk. The relationships among these distinct types of multiattribute preference functions are then explored, and issues related to their assessment and applications are surveyed.

Keywords: Multiattribute utility theory, additive value functions, preference modeling.

1. Introduction

In this chapter, we provide a review of multiattribute utility theory. As we shall discuss, multiattribute preference theory would be a more general term for this topic that covers several multiattribute models of choice. These models are based on alternate sets of axioms that have implications for their assessment and use. We begin with a brief review of single-attribute preference theory, and explore preference representations that measure a decision maker's preferences on an ordinal scale, her strength of preference and her preferences for risky alternatives. We emphasize the distinctions among these cases, and then explore their implications for multiattribute preference theory.

In order to differentiate between theories for preference based on the notions of ordinal comparisons and strength of preference versus theories for risky choices, we will use the term value function to refer to the former and utility function to refer to the latter. This distinction was made by Keeney and Raiffa in 1976[1] and has been generally adopted in the literature. Further, we will use the term preference model or multiattribute preference model to include all of these cases.

We describe the multiattribute decision problem, and discuss the conditions that allow a multiattribute preference function to be decomposed into additive and multiplicative forms under conditions of certainty and risk. The relationships between multiattribute preference functions under conditions of certainty and risk are then explored, and issues related to their assessment and applications are surveyed.

There are several important points related to the field of multi-criteria decision analysis that we wish to make. First, multiattribute preference theory provides an axiomatic foundations for choices involving multiple criteria. As a result, one can examine these axioms and determine whether or not they are reasonable guides to rational behavior. Most applications of the methods of multi-criteria decision analysis are developed for individuals who are making decisions on behalf of others, either as managers of publicly held corporations or as government officials making decisions in the best interests of the public. In such cases, one should expect these decision makers to use decision-making strategies that can be justified based on a reasonable set of axioms, rather than some ad hoc approach to decision making that will violate one or more of these axioms.

Often arguments are made that decision makers do not always make decisions that are consistent with the rational axioms of decision theory. While this may be true as a descriptive statement for individual decision making, it is much more difficult to identify situations involving significant implications for other parties where a cavalier disregard for normative theories of choice can be defended.

Second, multiattribute utility theory can be based on different sets of axioms that are appropriate for use in different contexts. Specifically, the axioms that are appropriate for risky choice do not have to be satisfied in order to use multiattribute models of preference for cases that do not explicitly involve risk. Much of the work on multiobjective mathematical programming, for example, does not require the consideration of risk, and many applications of the Analytical Hierarchy Procedure (AHP) are also developed in the context of certainty, Saaty 1980 [35].

The broad popularity of the award-winning textbook on multiattribute utility theory by Keeney and Raiffa (1993) [27] emphasized the use of multiattribute preference models based on the theories of von Neumann and Morgenstern (1947) [41], which rely on axioms involving risk. As a result, this approach has become synonymous in the view of many scholars with multiattribute preference theory. However, this theory is not the appropriate one for decisions involving multiple objectives when risk is not a consideration.[2] Instead, the multiattribute preference theories for certainty are based on ordinal comparisons of alternatives or on estimates of the strength of preference between pairs of alternatives.

Third, many existing approaches to multi-criterion decision analysis can be viewed as special cases or approximations to multiattribute preference models. We shall make this case for the popular methods of goal programming and the AHP as examples. By viewing these seeming disparate methods from this unifying framework, it is possible to gain new insights into the methodologies, recognize ways that these approaches might be sharpened or improved, and provide a basis for evaluating whether their application will result in solutions that are justified by a normative theory.

2. Preference Representations Under Certainty and Under Risk

Preference theory studies the fundamental aspects of individual choice behavior, such as how to identify and quantify an individual's preferences over a set of alternatives, and how to construct appropriate preference representation functions for decision making. An important feature of preference theory is that it is based on rigorous axioms which characterize an individual's choice behavior. These preference axioms are essential for establishing preference representation functions, and provide the rationale for the quantitative analysis of preference.

The basic categories of preference studies can be divided into characterizations of preferences under conditions of certainty or risk and over alternatives described by a single attribute or by multiple attributes. In the following, we will begin with the introduction of basic preference relations and then discuss preference representation under certainty and under risk for the single attribute case. We shall refer to a preference representation function under certainty as a *value function*, and to a preference representation function under risk as a *utility function*.

Preference theory is primarily concerned with properties of a binary preference relation $\succ$ on a choice set X, where X could be a set of commodity bundles, decision alternatives, or monetary gambles. For example, we might present an individual with a pair of alternatives, say x and y (e.g., two cars) where $x, y \in X$ (e.g., the set of all cars), and ask how they compare (e.g., do you prefer x or y?). If the individual says that x is preferred to y, then we write $x \succ y$, where $\succ$ means strict preference. If the individual states that he or she is indifferent between x and y, then we represent this preference as $x \sim y$. Alternatively, we can define $\sim$ as the absence of strict preference; that is, not $x \succ y$ and not $y \succ x$. If it is not the case that $y \succ x$, then we write $x \succeq y$, where $\succeq$ represents a weak preference (or preference-indifference) relation. We can also define $\succeq$ as the union of strict preference $\succ$ and indifference $\sim$; that is, both $x \succ y$ and $x \sim y$.

Preference studies begin with some basic assumptions (or axioms) of individual choice behavior. First, it seems reasonable to assume that an individual can state her preference over a pair of alternatives without contradiction; that is, the individual does not strictly prefer x to y and y to x simultaneously. This leads to the following definition for *preference asymmetry*: preference is asymmetric if there is no pair x and y in X such that $x \succ y$ and $y \succ x$.

Asymmetry can be viewed as a criterion of preference consistency. Furthermore, if an individual makes the judgment that x is preferred to y, then he or she should be able to place any other alternative z somewhere on the ordinal scale determined by the following: either better than y, or worse than x, or both. Formally, we define *negative transitivity* by saying that preferences are negatively transitive if given $x \succ y$ in X and any third element z in X, it follows that either $x \succ z$ or $z \succ y$, or both.

If the preference relation $\succ$ is asymmetric and negatively transitive, then it is called a *weak order*. The weak order assumption implies some desirable properties of a preference ordering, and is a basic assumption in many preference studies. If the preference relation $\succ$ is a weak order, then the associated indifference and weak preference relationships are well behaved. The following statements summarize some of the properties of some of these relationships.

If strict preference $\succ$ is a weak order, then

1 strict preference $\succ$ is *transitive* (if $x \succ y$ and $y \succ z$, then $x \succ z$);

2 indifference $\sim$ is transitive, *reflexive* ($x \sim x$ for all x); and *symmetric* ($x \sim y$ implies $y \sim x$);

3 exactly one of $x \succ y$, $y \succ x$, $x \sim y$ holds for each pair x and y; and

4 weak preference $\succsim$ is transitive and *complete* (for a pair x and y, either $x \succsim y$ or $y \succsim x$).

Thus, an individual whose strict preference can be represented by a weak order can rank all alternatives considered in a unique order. Further discussions of the properties of binary preference relations are presented in Fishburn (1970) [15, Chapter 2], Kreps (1990) [31, Chapter 2], and by Bouyssou and Pirlot in Chapter 3 of this volume.

2.1 Preference Functions for Certainty (Value Functions)

If strict preference $\succ$ on X is a weak order and X is finite or denumerable, then there exists a numeric representation of preference, a real-valued function $\overset{\circ}{v}$ on X such that $x \succ y$ if and only if $\overset{\circ}{v}(x) > \overset{\circ}{v}(y)$, for all x and y in X (Fishburn, 1970 [15]). Since $\overset{\circ}{v}$ is a preference representation function under certainty, it is often called a *value function* (Keeney and Raiffa, 1993 [27]). This value function is said to be order-preserving since the numbers $\overset{\circ}{v}(x), \overset{\circ}{v}(y), \ldots$ ordered by $>$ are consistent with the order of $x, y, \ldots$ under $\succ$. Thus, any monotonic transformations of $\overset{\circ}{v}$ will also be order-preserving for this binary preference relation. Since such a function only rank orders different outcomes, there is no added meaning of the values of $\overset{\circ}{v}$ beyond the order that they imply.

Notice that we use the symbol "$\circ$" to indicate that $\overset{\circ}{v}$ is an ordinal function. While the notion of an ordinal value function is very important for economic and decision theories, such a function is seldom assessed in practice. For example, if we know that preferences are monotonically increasing for some real-valued attribute x (e.g., more is better), then $\overset{\circ}{v}(x) = x$ is valid ordinal preference function. Therefore, we may choose an objective function of maximizing profits or minimizing costs, and be comfortable assuming implicitly that these objective functions are order-preserving preference functions for a decision maker. However, the notion of an ordinal value function does become important when we speak of multiattribute value functions, as we shall discuss.

In order to replicate the preferences of a decision maker with less ambiguity, we may wish to consider a "strength of preference" notion that involves comparisons of preference differences between pairs of alternatives. To do so, we need more restrictive preference assumptions, including that of a weak order over preferences between exchanges of pairs of alternatives (Krantz et al., 1971 [30, Chapter 4]). We use the term *measurable value function* for a value function that may be used to order the differences in the strength of preference between pairs of alternatives or, more simply, the "preference differences" between the alternatives.

Once again, let X denote the set of all possible consequences in a decision situation, $w, x, y, z, w', x', y' \in X$; define X^* as a nonempty subset of $X \times X$, and $\succeq^*$ as a binary relation on X^*. We shall interpret $wx \succeq^* yz$ to mean that the strength of preference for w over x is greater than or equal to the strength of preference for y over z. The notation $wx \sim^* yz$ means both $wx \succeq^* yz$ and $yz \succeq^* wx$ and $wx \succ^* yz$ means not $yz \succeq^* wx$.

There are several alternative axiom systems for measurable value functions, including the topological results of Debreu (1960) [7] and the algebraic development by Scott and Suppes (1958) [39]. Some of these systems allow both "positive" and "negative" preference differences and are called algebraic difference structures. For example, the "degree of preference" for x over w would be "negative" if w is preferred to x. Our development is based on an axiom system presented by Krantz et al. (1971) [30, Definition 4.1] that does not allow negative differences; hence it is called a positive difference structure.

This set of axioms includes several technical assumptions that have no significant implications for behavior. However, a key axiom that does have an intuitive interpretation in terms of preferences is the following one: If $wx, xy, w'x', x'y' \in X^*$, $wx \succeq^* w'x'$, and $xy \succeq^* x'y'$, then $wy \succeq^* w'y'$. That is, if the difference in the strength of preference between w and x exceeds the difference between w' and x', and the difference in the strength of preference between x and y exceeds the difference between x' and y', then the difference in the strength of preference between w and y must exceed the difference between w' and y'. Some introspection should convince most readers that this would typically be true for preference comparisons of alternative pairs.

The axioms of Krantz et al. (1971) [30] imply that there exists a real-valued function v on X such that, for all $w, x, y, z \in X$, if w is preferred to x and y to z, then $wx \succeq^* yz$ if and only if

$$v(w) - v(x) \geq v(y) - v(z). \tag{7.1}$$

Further, v is unique up to a positive linear transformation, so it is a *cardinal function* (i.e., v provides an *interval scale of measurement*). That is, if v' also satisfies (7.1), then there are real numbers $a > 0$ and b such that $v'(x) = av(x) + b$ for all $x \in X$ (Krantz et al. [30, Theorem 4.1]).

We define the binary preference relation $\succeq$ on X from the binary relation $\succeq^*$ on X^* in the natural way by requiring $wx \succeq^* yx$ if and only if $w \succeq y$ for all $w, x, y \in X$. Then from (7.1) it is clear that $w \succeq y$ if and only if $v(w) \geq v(y)$. Thus, v is a value function on X and, by virtue of (7.1), it is a *measurable value function.*

The ideas of strength of preference and of measurable value functions are important concepts that are often used implicitly in the implementation of preference theories in practice. Intuitively, it may be useful to think of a measurable value function as the unique preference function in the case of certainty that reveals the marginal value of additional units of the underlying commodity. For example, we would expect that the measurable value function over wealth for most individuals would be concave, since the first million dollars would be "worth" more to the individual than the second million dollars, and so on. This notion would be consistent with the traditional assumption in economics of diminishing marginal returns to scale.

Further, the measurable value function can be assessed using questions for subjects that do not require choices among lotteries, which may be artificial distractions in cases where subjects are trying to choose among alternatives that do not require the consideration of risk. Examples of methods for assessing measurable value functions would include direct rating of alternatives on a cardinal scale, or direct comparisons of preference differences. For a detailed discussion of these approaches, see Farquhar, and Keller (1989) [13], von Winterfeldt and Edwards (1986) [42], and Kirkwood (1997)[29].

In addition, subjects can be asked to make ratio comparisons of preference differences. For example, they might be comparing automobiles relative to a "base case", say a Ford Taurus. Then, they could be asked to compare the improvement in acceleration offered by a BMW over a Taurus to the improvement offered by a Mercedes (relative to the same Taurus) in terms of a ratio. This ratio judgment could be captured and analyzed using the tools of the AHP, and this provides a link between measurable value functions and ratio judgments. This point has been made on numerous occasions, and is worth further exploration (e.g., see Kamenetzky, 1982 [26]; Dyer 1990 [8]; Salo and Hämäläinen 1997 [36]).

2.2 Preference Functions for Risky Choice (Utility Functions)

We turn to preference representation for risky options, where the risky options are defined as lotteries or gambles with outcomes that depend on the occurrence from a set of mutually exclusive and exhaustive events. For example, a lottery could be defined as the flip of a fair coin, with an outcome of $10 if heads occurs and an outcome of -$2 if tails occurs.

Perhaps the most significant contribution to this area of concern was the formalization of *expected utility theory* by von Neumann and Morgenstern (1947) [41]. This development has been refined by a number of researchers and is most commonly presented in terms of three basic axioms (Fishburn, 1970 [15]).

Let P be a convex set of simple probability distributions or lotteries $\{p, q, r, \ldots\}$ on a nonempty set X of outcomes. We shall use p, q and r to refer to probability distributions and random variables interchangeably. For lotteries p, q, r in P and all λ, $0 < \lambda < 1$, the expected utility axioms are:

1 (*Ordering*) $\succ$ is a weak order;

2 (*Independence*) If $p \succ q$ then $(\lambda p + (1 - \lambda)r) \succ (\lambda q + (1 - \lambda)r)$for all r in P;

3 (*Continuity*) If $p \succ q \succ r$ then there exist some $0 < \alpha < 1$ and $0 < \beta < 1$ such that $\alpha p + (1 - \alpha)r \succ q \succ \beta p + (1 - \beta)r$

The von Neumann-Morgenstern expected utility theory asserts that the above axioms hold if and only if there exists a real-valued function u such that for all p, q in P, $p \succ q$ if and only if

$$\sum_{x \in X} p(x)u(x) \geq \sum_{x \in X} q(x)u(x) \tag{7.2}$$

Moreover, such a u is unique up to a positive linear transformation.

The expected utility model can also be used to characterize an individual's risk attitude (Pratt, 1964 [34]; Keeney and Raiffa, 1993 [27, Chapter 4]). If an individual's utility function is concave, linear, or convex, then the individual is risk averse, risk neutral, or risk seeking, respectively.

The von Neumann-Morgenstern theory of risky choice presumes that the probabilities of the outcomes of lotteries are provided to the decision maker. Savage (1954) [38] extended the theory of risky choice to allow for the simultaneous determination of subjective probabilities for outcomes and for a utility function u defined over those outcomes. Deduced probabilities in Savage's model are personal or subjective probabilities. The model itself is a subjective expected utility representation.

The assessment of von Neumann-Morgenstern utility functions will almost always involve the introduction of risk in the form of simple lotteries. For a discussion of these assessment approaches, see Keeney and Raiffa (1993) [27, Chapter 4] or von Winterfeldt and Edwards (1986) [42].

As a normative theory, the expected utility model has played a major role in the prescriptive analysis of decision problems. However, for descriptive purposes, the assumptions of this theory have been challenged by empirical studies (Kahneman and Tversky, 1979 [25]). Some of these empirical studies demonstrate that subjects may choose alternatives that imply a violation of the independence axiom. One implication of the independence axiom is that the expected utility model is "linear in probabilities". For a discussion, see Fishburn and Wakker (1995) [18]. A number of contributions have been made by relaxing the independence axiom and developing some *nonlinear utility models* to accommodate actual decision behavior (Fishburn, 1988 [17] and Camerer, 1995 [3]).

2.3 Comment

Note that both the measurable value function $v(x)$ and the von Neumann and Morgenstern utility function $u(x)$ are cardinal measures, unique up to a positive linear transformation. However, the theory supporting the measurable function is based on axioms involving preferences differences, and it is assessed based on questions that rely on the idea of strength of preference. In contrast, the von Neumann and Morgenstern utility function is based on axioms involving lotteries, and it is assessed based on questions that typically involve lottery comparisons.

Suppose we find a subject and assess her measurable value function $v(x)$ and her utility function $u(x)$ over the same attribute (e.g., over monetary outcomes). Would these two functions be identical, except for measurement error? A quick reaction might be that they would be identical, since they are each unique representations of the subject's preferences, up to a positive linear transformation. However, that is not necessarily the case. Intuitively, a measurable value function $v(x)$ may be concave, indicating decreasing marginal value for the underlying attribute. However, a utility function $u(x)$ may be even more concave, since it will incorporate not only feelings regarding the marginal value of the attribute, but also it may incorporate psychological reactions to taking risks. Empirical tests of this observation are provided by Krzysztofowicz (1983) [32] and Keller (1985) [28] and generally support this intuition. This is an important point, and one that we will emphasize again in the context of multiattribute preference functions (see Ellsberg, 1954 [11]; Dyer and Sarin, 1982 [10]; Sarin, 1982 [37], and Jia and Dyer, 1996 [24]).

3. Ordinal Multiattribute Preference Functions for the Case of Certainty

A decision maker uses the appropriate preference function, $\mathring{v}(x)$ or $v(x)$ in the case of certainty or $u(x)$ in the case of risk, to choose among available alternatives. The major emphasis of the work on multiattribute utility theory has been on questions involving conditions for the decomposition of a preference function into simple polynomials, on methods for the assessment of these decomposed functions, and on methods for obtaining sufficient information regarding the multiattribute preference functions so that the evaluation can proceed without its explicit identification with full precision.

Suppose that the alternatives defined for single attribute preference functions are now considered to be vectors. That is, suppose that $X = \prod_{i=1}^{n} X_i$ where X_i represents the performance of an alternative on attribute i. We will be interested in conditions allowing the determination that $(x_1, \ldots, x_n) \succeq (y_1, \ldots, y_n)$ if and only if $\mathring{v}(x_1, \ldots, x_n) \geq \mathring{v}(y_1, \ldots, y_n)$ for example. Essentially, all that is required is the assumption that the decision maker's preferences are a weak order on the vectors of attribute values.

In some cases, methods for multiattribute optimization do not need any additional information regarding a multiattribute preference function, other than perhaps invoking concavity to allow maximization. Geoffrion, Dyer, and Feinberg (1972) [20] provide an example of an early approach to multiattribute optimization that does proceed with only local information regarding the implicit multiattribute preference function. Additional conditions are needed to decompose the multiattribute preference function into simple parts.

3.1 Preference Independence

The most common approach for evaluating multiattribute alternatives is to use an additive representation. For simplicity, we will assume that there exist a most preferred outcome x_i^* and a least preferred outcome x_i° on each attribute $i = 1$ to n. In the additive representation, a real value $\overset{\circ}{v}$ is assigned to each outcome $(x_1, \ldots, x_n)$ by

$$\overset{\circ}{v}(x_1, \ldots, x_n) = \sum_{i=1}^{n} \overset{\circ}{v}_i(x_i) \tag{7.3}$$

where the $\overset{\circ}{v}_i$ are single attribute value functions over X_i [3]. When it is convenient, we may choose the scaling $\overset{\circ}{v}_i(x_i^*) = 1, \overset{\circ}{v}_i(x_i^\circ) = 0$, and write $\overset{\circ}{v}(x_1, x_2, \ldots, x_n) = \sum_{i=1}^n \overset{\circ}{\lambda}_i \overset{\circ}{v}_i(x_i)$ where $\sum_{i=1}^n \overset{\circ}{\lambda}_i = 1$.

If our interest is in simply rank-ordering the available alternatives then the key condition for the additive form in (7.3) is *mutual preference independence.* Suppose that we let $I \subset \{1, 2, \ldots, n\}$ be a subset of the attribute indices, and define X_I as the subset of the attributes designated by the subscripts in I. Also, we let $\bar{X}_I$ represent the complementary subset of the n attributes. Then,

1 X_I is *preference independent* of $\bar{X}_I$ if $(w_I, \bar{w}_I) \succeq (x_I, \bar{w}_I)$ for any $w_I, x_I \in X_I$ and $\bar{w}_I \in \bar{X}_I$ implies $(w_I, \bar{x}_I) \succeq (x_I, \bar{x}_I)$ for all $\bar{x}_I \in \bar{X}_I$.

2 The attributes $X_1, \ldots, X_n$ are *mutually preference independent* if for every subset $I \subseteq \{1, \ldots, n\}$ the set X_I of these attributes is preference independent of $\bar{X}_I$.

When coupled with a solvability condition and some technical assumptions, mutual preference independence implies the existence of an *additive ordinal multiattribute value function* for $n \geq 3$ attributes. Furthermore, this additive ordinal value function is unique up to a positive linear transformation.

Attributes X_i and X_j are preference independent if the tradeoffs (substitution rates) between X_i and X_j are independent of all other attributes. Mutual preference independence requires that preference independence holds for all pairs X_i and X_j. Essentially, mutual preference independence implies that the indifference curves for any pair of attributes are unaffected by the fixed levels of the remaining attributes. Debreu (1959) [7], Luce and Tukey (1964) [33], and Gorman (1968) [21] provide axiom systems and analysis for the additive form (7.3).

An example may help to illustrate the idea of preference independence. Suppose that a subject is attempting to evaluate automobiles based on the three criteria of cost, horsepower, and appearance. Assume that the subject decides that her preferences between two automobiles differing in cost and horsepower but with identical values for appearance are as follows: ($24,000, 150 hp, ugly) $\succ$($25,000, 170 hp, ugly). If the level of appearance does not affect the subject's indifference curve between cost and horsepower, then she will also prefer ($24,000, 150 hp, beautiful) to ($25,000, 170 hp, beautiful), and will maintain the same preference relation for any common value of appearance.

As a practical matter, it is only necessary for preference independence to hold for the $n - 1$ pairs of criteria involving the first criterion and the other $n - 1$ criteria taken one at a time. See Keeney and Raiffa (1993) [27, Chapter 3] for a discussion.

In Chapter 3 of this volume, Bouyssou and Pirlot provide an excellent discussion of the additive ordinal value function which they present as the use of *conjoint measurement* for multiple criteria decision making. In our development, we use the terminology *ordinal additive value function* instead in order to contrast this preference representation with other additive and non-additive preference models. We also use the term *preference independence* rather than simply *independence* to distinguish this key assumption from other forms of independence conditions that are appropriate for multiple criteria decision making in different contexts.

3.2 Assessment Methodologies

The additive ordinal value function would seem to be an attractive choice for practical applications of multiattribute decision making. However, the resulting additive function is, in general, difficult to assess. The problem arises because the single attribute functions $\overset{\circ}{v}_i$ cannot be assessed using the methods appropriate for the single-attribute *measurable* value functions. Instead, these functions can only be assessed through protocols that require tradeoffs between two attributes throughout the process, and these protocols are therefore burdensome for the decision makers. Further, the resulting additive function will only have an ordinal interpretation, rather than providing a measure of the strength of preference.

Keeney and Raiffa (1993) [27, Chapter 3] illustrate two assessment procedures for ordinal additive value functions. However, an example may be helpful to emphasize that the resulting additive value function may only provide an ordinal ranking of alternatives, since this important point is also a subtle one.

Suppose that an analyst is attempting to assess a preference function from a decision maker on three attributes X, Y, and Z that are related in the mind of the decision maker in a multiplicative form; that is, the decision maker's true preferences are represented by the product xyz where x, y, and z are attribute values. Of course, the analyst is not aware of this multiplicative form, and is attempting to develop an appropriate preference representation from the decision maker based on a verbal assessment procedure. Further, suppose that there is no risk involved, so the analyst would like to consider the use of an additive ordinal multiattribute value function.

An example of a situation that might involve this type of a preference function would be the ranking of oil exploration opportunities based on estimates of their oil reserves. Suppose that the decision maker thinks that these reserves can be estimated by multiplying the area (x) of the structure containing oil by its depth (y) to obtain the volume of the structure, and then multiplying this volume by its rate of recovery per volumetric unit (z). In practice, this is a simplification of the approach actually used in many cases to estimate oil reserves.

This multiplication of the relevant parameters could be done explicitly in this case, but this example should suggest that such a true preference structure could occur naturally. For simplicity, and to avoid complications associated with units of measurement, we will assume that $X = Y = Z = [1, 10]$, which might occur if the analyst rescaled the actual units of measurement.

The analyst does not know the true underlying preference model of the decision maker, and so he might ask a series of questions to determine if mutual preference independence holds in this case. Consider alternative 1, with $x_1 = 2$, $y_1 = 3$, and $z_1 = 4$, versus alternative 2, with $x_2 = 4$, $y_2 = 2$, and $z_2 = 4$. The decision maker would be asked to compare (2,3,4) with (4,2,4), and would reply that she prefers alternative 2 (because $2 \times 3 \times 4 = 24$ and $4 \times 2 \times 4 = 32$, although these calculations are unknown to the analyst). It is easy to see that alternative 2 would remain preferred to alternative 1 for all common values of z_1 and z_2, so attributes X and Y are preference independent of Z. Likewise, a similar set of questions would reveal that X and Z are preference independent of Y, and Y and Z are preference independent of X, so these three attributes are mutually preference independent.

Therefore, the analyst concludes that the preferences of the decision maker can be represented by the ordinal additive multiattribute preference function

$$\mathring{v}(x, y, z) = \mathring{v}_x(x) + \mathring{v}_y(y) + \mathring{v}_z(z)$$

As we shall see, this is not a mistake even though the true preference function is multiplicative, and the assessment procedure will construct the correct ordinal additive function that will result in the same rank ordering of alternatives as the multiplicative preference function.

For this example only, we will abuse the notation and let subscripts of the attributes indicate the corresponding values of the single attribute functions. For examples, we will let x_0 indicate the value of attribute X such that $\mathring{v}_x(x_0) = 0$, and let y_1 indicate the value of attribute Y such that $\mathring{v}_y(y_1) = 1$, and so forth.

Suppose the analyst begins the assessment procedure by letting $x_0 = y_0 = z_0$ = 1, which is allowable given the fact that the function is unique up to a linear transformation. That is, the analyst scales $\overset{\circ}{v}_x$ so that $\overset{\circ}{v}_x(x_0) = \overset{\circ}{v}_x(1) = 0$, and similarly scales $\overset{\circ}{v}_y(1) = \overset{\circ}{v}_z(1) = 0$. Therefore, we would have

$$\overset{\circ}{v}(1,1,1) = \overset{\circ}{v}_x(1) + \overset{\circ}{v}_y(1) + \overset{\circ}{v}_z(1) = 0 + 0 + 0 = 0.$$

The analyst then arbitrarily selects $x_1 = 2$; that is, he sets $\overset{\circ}{v}_x(x_1) = \overset{\circ}{v}_x(2) = 1$, which is also allowable by virtue of the scaling convention. Finally, the analyst involves the decision maker, and asks her to specify a value y_1 so that she is indifferent between the alternative (2,1,1) and the alternative (1,y_1,1). Based on her true multiplicative preference model unknown to the analyst, if she is indifferent between (2,1,1) and (1,y_1,1) it must be the case that $2 \times 1 \times 1 = 1 \times y_1 \times 1$, so she responds y_1 = 2. Based on this response, the analyst sets $\overset{\circ}{v}_y(y_1) = \overset{\circ}{v}_y(2) = 1$.

This means that $\overset{\circ}{v}(2,1,1) = \overset{\circ}{v}_x(2) + \overset{\circ}{v}_y(1) + \overset{\circ}{v}_z(1) = 1 + 0 + 0 = 1$, and that $\overset{\circ}{v}(1,2,1) = \overset{\circ}{v}_x(1) + \overset{\circ}{v}_y(1) + \overset{\circ}{v}_z(1) = 0 + 1 + 0 = 1$, which verifies to the analyst that the additive representation indicates that the decision maker is indifferent between the alternatives (2,1,1) and (1,2,1). In addition, the analyst knows that $\overset{\circ}{v}(2,2,1) = \overset{\circ}{v}_x(2) + \overset{\circ}{v}_y(2) + \overset{\circ}{v}_z(1) = 1 + 1 + 0 = 2$.

Now, the analyst asks the decision maker to specify a value for x_2 so that she is indifferent between the alternatives (2,2,1) and (x_2,1,1). This response will determine the value of x_2 such that $\overset{\circ}{v}_x(x_2) = 2$, because indifference between these two alternatives will require $\overset{\circ}{v}(x_2,1,1) = \overset{\circ}{v}_x(x_2) + \overset{\circ}{v}_y(1) + \overset{\circ}{v}_z(1) = 2 + 0 + 0 = 2$ also.

Using her implicit multiplicative preference function for the alternative (2, 2, 1), she obtains $2 \times 2 \times 1 = 4$, and since indifference requires $x_2 \times 1 \times 1 = 4$, she would identify $x_2 = 4$, so $\overset{\circ}{v}_x(4) = 2$. The reader should confirm that similar questions would determine $\overset{\circ}{v}_y(4) = \overset{\circ}{v}_x(4) = 2$, and that $\overset{\circ}{v}_x(8) = 3$, and so forth. Continuing in this fashion, and using similar questions to develop the assessments of $\overset{\circ}{v}_y(y)$ and $\overset{\circ}{v}_z(z)$, the analyst would develop graphs that would indicate $\overset{\circ}{v}_x(x)$=ln x/ln 2, $\overset{\circ}{v}_y(y)$= ln y/ln 2, and $\overset{\circ}{v}_z(z)$= ln z/ln 2, so that the ordinal additive multiattribute value function would be given by the sum of the logs of the variables. Notice that this ordinal value function is an order preserving transformation of the true underlying preference representation of the decision maker, which was never revealed explicitly to the analyst.

As this example illustrates, the assessment procedure will determine an additive ordinal value function that may be an order preserving transformation of a true preference relation that is not additive. The log function provides an example of such a transformation for a multiplicative preference relation, but other non-additive relationships may also be transformed to order preserving additive value functions. See Krantz et al. (1971) [30] for a discussion of other such transformations.

This example also illustrates the fact that the assessment methods required for accurately capturing an additive ordinal multiattribute value function may be tedious, and will require tradeoffs involving two or more attributes. This same point is made by Bouyssou and Pirlot in Chapter 3 of this same volume. Thus, while this approach could be used in practice, it would be desirable to have simpler means of assessing the underlying preference functions. This can be accomplished if some additional preference conditions are satisfied, but the requirement of mutual preference independence will still be common to the preference models that are to follow.

4. Cardinal Multiattribute Preference Functions for the Case of Risk

When $X = \prod_{i=1}^{n} X_i$ in a von Neumann-Morgenstern utility model and the decision maker's preferences are consistent with some additional independence conditions, then $u(x_1, x_2, \ldots, x_n)$ can be decomposed into additive, multiplicative, and other well-structured forms that simplify assessment. In comparison with other sections, our coverage of this topic will be relatively brief since it is perhaps the most well known multiattribute preference model.

4.1 Utility Independence

An attribute X_i is said to be *utility independent* of its complementary attributes if preferences over lotteries with different levels of X_i do not depend on the fixed levels of the remaining attributes. Attributes $X_1, X_2, \ldots, X_n$ are mutually utility independent if all proper subsets of these attributes are utility independent of their complementary subsets. Further, it can be shown that if these same attributes are mutually preference independent, then they will also be mutually utility independent if any pair of the attributes is utility independent of its complementary attributes (Keeney and Raiffa, 1993 [27]).

Returning to the automobile selection example, suppose that a decision maker is considering using the attributes of cost, horsepower, and appearance as before, but there is some uncertainty regarding some new environmental laws that may impact the cost and the horsepower of a particular automobile. Further, assume that the decision maker prefers more horsepower to less, lower costs and more attractive automobiles. The current performance levels of one of the alternatives may be ($25,000, 170 hp, ugly), but if the legislation is passed a new device will have to be fitted that will increase cost and decrease horsepower to ($25,700, 150 hp, ugly). An alternative automobile might have possible outcomes of ($28,000,

200 hp, ugly) and ($29,000, 175 hp, ugly) depending on this same legislation, which the decision maker estimates will pass with probability 0.5.

Therefore, the decision maker may consider choices between lotteries such as the one shown in Figure 7.1. For example, the decision maker may prefer Auto 1 to Auto 2 because the risks associated with the cost and the horsepower for Auto 1 are more acceptable to her than the risks associated with the cost and horsepower of Auto 2. If the decision maker's choices for these lotteries do not depend on common values of the third attribute, then cost and horsepower are utility independent of appearance.

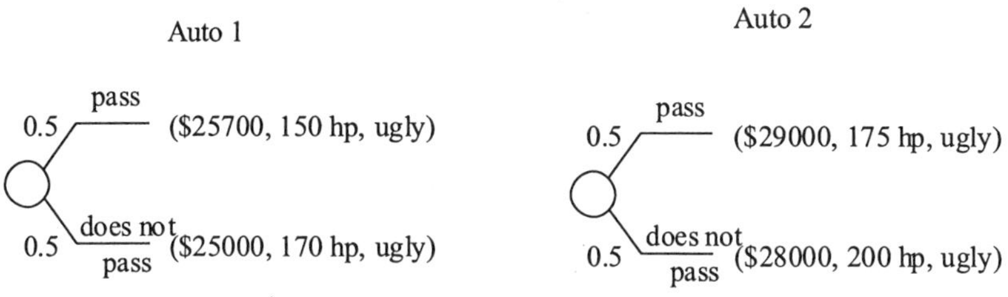

Figure 7.1. Choice between two lotteries.

A multiattribute utility function $u(x_1, x_2, \ldots, x_n)$ can have the multiplicative form

$$1 + ku(x_1, x_2, \ldots, x_n) = \prod_{i=1}^{n} [1 + kk_i u_i(x_i)] \tag{7.4}$$

if and only if the attributes $X_1, X_2, \ldots, X_n$ are mutually utility independent, where the u_i are single-attribute functions over X_i scaled from 0 to $1, 0 \leq k_i \leq 1$ are positive scaling constants, and k is an additional scaling constant. If the scaling constant k is determined to be 0 through the appropriate assessment procedure, then (7.4) reduces to the additive form

$$\sum_{i=1}^{n} k_i u_i(x_i) \tag{7.5}$$

where $\sum_{i=1}^{n} k_i = 1$.

4.2 Additive Independence

A majority of the applied work in multiattribute utility theory deals with the case when the von Neumann-Morgenstern utility function is decomposed into the additive form (7.5). Fishburn (1965) [14] has derived necessary and sufficient conditions for a utility function to be additive. The key condition for additivity is the marginality condition which states that the preferences for any lotteries

$p, q \in P$ should depend only on the marginal probabilities of the attribute values, and not on their joint probability distributions.

Returning to the automobile example once again, for additivity to hold, the decision maker must be indifferent between the two lotteries shown in Figure 7.2, and for all other permutations of the attribute values that maintain the same marginal probabilities for each.

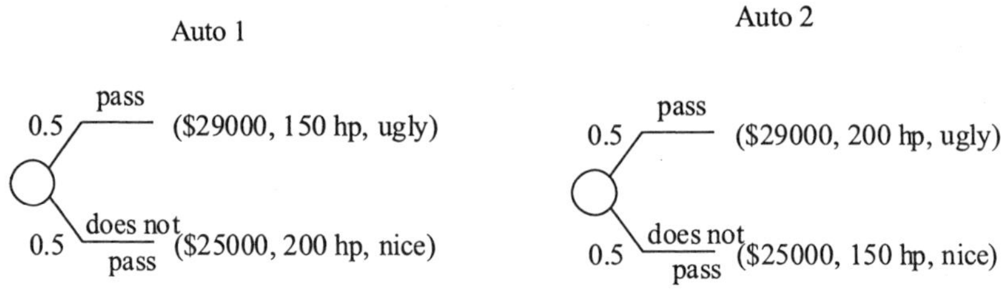

Figure 7.2. Additive Independence Criterion for Risk.

Notice that in either lottery, the marginal probability of receiving the most preferred outcome or the least preferred outcome on each attribute is identical (0.5). However, a decision maker may prefer the right-hand side lottery over the left-hand side lottery if the decision maker wishes to avoid a 0.5 chance of the poor outcome ($29,000, 150 hp, ugly) on all three attributes, or she may have the reverse preference if she is willing to accept some risk in order to have a chance at the best outcome on all three attributes. In either of the latter cases, utility independence may still be satisfied, and a multiplicative decomposition of the multiattribute utility function (7.4) may be appropriate.

Other independence conditions have been identified that lead to more complex non-additive decompositions of a multiattribute utility function. These general conditions are reviewed in Farquhar (1977) [12].

4.3 Assessment Methodologies

The assessment of the multiplicative or additive form implied by the condition of mutual utility independence is simplified by the fact that each of the single-attribute utility functions may be assessed independently (more accurately, while all of the other attributes are held constant at arbitrarily selected values), using the well-known utility function assessment techniques suitable for single attribute utility functions. In addition, the constants k_i and k can be assessed using n relatively simple tradeoff questions. See Keeney and Raiffa (1993) [27] or Kirkwood (1997) [29]for additional details and examples.

5. Measurable Multiattribute Preference Functions for the Case of Certainty

We have delayed the discussion of measurable multiattribute preference functions until after the review of multiattribute utility theory because the latter may be more familiar to the reader. If so, this transposition of a more natural order of presentation may be helpful in providing the opportunity to discuss similarities between these models of preference, and therefore to enhance an intuitive understanding of the relationships among some important concepts.

Again let X denote the set of all possible consequences in a particular decision problem. In the multiattribute problem $X = \prod_{i=1}^{n} X_i$ where X_i is the set of possible consequences for the ith attribute. In this section, we use the letters w, x, y, and z to indicate distinct elements of X. For example, $w \in X$ is represented by $(w_1, \ldots, w_n)$, where w_i is a level in the nonempty attribute set X_i for $i = 1, \ldots, n$. Once again, we let $I \subset \{1, 2, \ldots, n\}$ be a subset of the attribute indices, define X_I as the subset of the attributes designated by the subscripts in I, and let $\bar{X}_I$ represent the complementary subset of the n attributes. We may write $w = (w_I, \bar{w}_I)$ or use the notation $(w_i, \bar{w}_i)$ and $(x_i, \bar{w}_i)$ to denote two elements of X that differ only in the level of the ith attribute. Finally, we also assume that the preference relation $\succeq$ on X is a weak order.

Next we introduce the notation necessary to define preferences based on strength of preference between vector-valued outcomes. We let $X^* = \{wx : w, x \in X\}$ be a nonempty subset of $X \times X$, and $\succeq^*$ denote a weak order on X^*. Once again, we may interpret $wx \succeq^* yz$ to mean that the preference difference between w and x is greater than the preference difference between y and z.

It seems reasonable to assume a relationship between $\succeq$ on X and $\succeq^*$ on X^* as follows. Suppose the attributes $X_1, \ldots, X_n$ are mutually preference independent. These two orders are *difference consistent* if, for all $w_i, x_i \in X_i$, $(w_i, \bar{w}_i) \succeq (x_i, \bar{w}_i)$ if and only if $(w_i, \bar{w}_i)(x_i^\circ, \bar{w}_i) \succeq^* (x_i, \bar{w}_i)(x_i^\circ, \bar{w}_i)$ for some $x_i^\circ \in X_i$ and some $\bar{w}_i \in \bar{X}_i$, and for any $i \in \{1, \ldots, n\}$, and if $w \sim x$ then $wy \sim^* xy$ or $yw \sim^* yx$ or both for any $y \in X$. Loosely speaking, this means that if one multiattributed alternative is preferred to another differing only on the value of attribute X_i, then the preference difference between that alternative and some common reference alternative $(x_i^\circ, \bar{w}_i)$ will be larger than the difference between the alternative that is not preferred and this reference alternative.

5.1 Weak Difference Independence

In this section we identify a condition that we refer to as *weak difference independence*. This condition plays a role similar to the utility independence condition in multiattribute utility theory. We show how this condition can be exploited to obtain multiplicative and other nonadditive forms of the measurable multiattribute value function.

Specifically, the subset of attributes X_I is weak difference independent of $\bar{X}_I$ if, given any $w_I, x_I, y_I, z_I \in X_I$ and some $\bar{w}_I \in \bar{X}_I$ such that the subject's judgments regarding strength of preferences between pairs of multiattributed alternatives is as follows: $(w_I,\bar{w}_I)(x_I,\bar{w}_I) \succeq^* (y_I,\bar{w}_I)(z_I,\bar{w}_I)$ then the decision maker will also consider $(w_I,\bar{x}_I)(x_I,\bar{x}_I) \succeq^* (y_I,\bar{x}_I)(z_I,\bar{x}_I)$ for any $\bar{x}_I \in \bar{X}_I$. That is, the ordering of preference differences depends only on the values of the attributes X_I and not on the fixed values of $\bar{X}_I$.

The attributes are mutually weak difference independent if all proper subsets of these attributes are weak difference independent of their complementary subsets. Further, it can be shown that if these same attributes are mutually preference independent, then they will also be mutually weak difference independent if any pair of the attributes is weak difference independent of its complementary attributes (Dyer and Sarin, 1979 [9]).

Notice the similarity of the definition of weak difference independence to that of utility independence. In the latter case, preferences among lotteries depend only on the values of the attributes X_I and not on the fixed values of $\bar{X}_I$. In the case of certainty, the same notion applies to preference differences. Therefore, it should not be surprising that this condition leads to a decomposition of a measurable value function that is identical to the one implied by utility independence for utility functions.

This intuition may be formalized as follows. A measurable multiattribute value function $v(x_1, x_2, \ldots, x_n)$ on X can have the multiplicative form

$$1 + \lambda v(x) = \prod_{i=1}^{n} [1 + \lambda\lambda_i v_i(x_i)] \tag{7.6}$$

if and only if $X_1, \ldots, X_n$ are mutually weak difference independent, where v_i is a single-attribute measurable value function over X_i scaled from 0 to 1, the λ_i are positive scaling constants, and λ is an additional scaling constant. If the scaling constant λ is determined to be 0 through the appropriate assessment procedure, then (7.4) reduces to the additive form

$$v(x) = \sum_{i=1}^{n} \lambda_i v_i(x_i) \tag{7.7}$$

where $\sum_{i=1}^{n} \lambda_i = 1$. Therefore, we obtain either an additive or a multiplicative measurable preference function that is based on notions of strength of preference.

5.2 Difference Independence

Finally, we are interested in the conditions that are required to ensure the existence of an additive multiattribute measurable value function. Recall that mutual preference independence guarantees the existence of an additive preference function for the case of certainty that will provide an ordinal ranking of alternatives, but it may not capture the underlying strength of preference of the decision maker. Further, the appropriate assessment technique will require tradeoffs that simultaneously consider two or more attributes as illustrated in Section 3.2.

Recall the example from Section 3.2 where the decision maker's true preferences were represented by the product of the attributes. If we were to ask the decision maker to express her preferences for the first attribute while holding the other attributes constant at some given values, she would respond in such a way that we would obtain a linear function for each attribute, rather than the correct logarithmic form. We would like to exclude this case, and be assured that the preference function that also measures strength of preference is additive.

Perhaps this point is worth some elaboration. Recall that the true preferences of the hypothetical decision maker introduced in Section 3.2 were consistent with the multiplicative representation xyz. Suppose we set $y = z = 1$, and ask the decision maker to consider the importance of changes in the attribute x while holding these other attribute values constant. Considering the alternatives (1,1,1), (3,1,1), and (5,1,1), she would indicate that the preference difference between (3,1,1) and (1,1,1) would be the same as the preference difference between (5,1,1) and (3,1,1). This is because her true preference relation gives $1 \times 1 \times 1 = 1$, $3 \times 1 \times 1 = 3$, $5 \times 1 \times 1 = 5$, and the preference difference between (3,1,1) and (1,1,1) is 3 - 1 = 2, which is also the preference difference between (5,1,1) and (3,1,1). If the analyst is not aware of the fact that this assessment approach cannot be used when only preference independence is satisfied, he might mistakenly conclude that $\overset{\circ}{v}(x, y, z) = x + y + z$ rather than the appropriate logarithmic transformation that we obtained earlier in Section 3.2.

The required condition for additivity that also provides a measurable preference function is called *difference independence* . The attribute X_i is difference independent of $\bar{X}_i$ if, for all $w_i, x_i \in X_i$ such that $(w_i, \bar{w}_i) \succeq (x_i, \bar{w}_i)$ for some $\bar{w}_i \in \bar{X}_i$, $(w_i, \bar{w}_i)(x_i, \bar{w}_i) \sim^* (w_i, \bar{x}_i)(x_i, \bar{x}_i)$ for any $\bar{x}_i \in \bar{X}_i$. Intuitively, the preference difference between two multiattributed alternatives differing only on one attribute does not depend on the common values of the other attributes.

The attributes are mutually difference independent if all proper subsets of these attributes are difference independent of their complementary subsets. Again, it can be shown that if these same attributes are mutually preference independent, then they will also be mutually difference independent if X_1 is difference independent of $\bar{X}_1$ (Dyer and Sarin, 1979[9]). For the case of $n \geq 3$, mutual difference independence along with some additional structural and technical conditions[4] ensure that if $wx, yz \in X^*$, then $wx \succeq^* yz$ if and only if

$$\sum_{i=1}^{n} \lambda_i v_i(w_i) - \sum_{i=1}^{n} \lambda_i v_i(x_i) \geq \sum_{i=1}^{n} \lambda_i v_i(y_i) - \sum_{i=1}^{n} \lambda_i v_i(z_i) \qquad (7.8)$$

and $x \succeq y$ if and only if

$$\sum_{i=1}^{n} \lambda_i v_i(x_i) \geq \sum_{i=1}^{n} \lambda_i v_i(y_i), \qquad (7.9)$$

where v_i is a single-attribute measurable value function over X_i scaled from 0 to 1, and $\sum_{i=1}^{n} \lambda_i = 1$. Further, if $v_i', i = 1, \ldots, n$ are n other functions with the same properties, then there exist constants $\alpha > 0, \beta_1, \ldots, \beta_n$ such that $v_i' = \alpha v_i + \beta_i, i = 1, \ldots, n$.

Result (7.9) is well known and follows immediately from the assumption that the attributes are mutually preference independent (Section 3.1). The significant result is (7.8), which means that $v(x_1, x_2, \ldots, x_n) = \sum_{i=1}^{n} \lambda_i v_i(x_i)$ also provides difference measurement on X. Note that this latter result is obtained based on the observation that any arbitrarily selected attribute is difference independent of its complementary attributes.

5.3 Assessment Methodologies

Because the notion of a measurable multiattribute value function may not be familiar to many readers, we will briefly consider methods for the assessment of them. Further details and examples are provided by von Winterfeldt and Edwards (1986) [42], and by Kirkwood (1997) [29].

5.3.1 Verification of the Independence Conditions. The first issue to be considered is the verification of the independence conditions. Since methods for verifying mutual preference independence are discussed in Keeney and Raiffa (1993) [27], we focus on the independence conditions involving preference differences .

Difference consistency is so intuitively appealing that it could simply be assumed to hold in most practical applications. The following procedure could be used to verify difference independence. We determine $w_1, x_1 \in X_1$ such that $(w_1, \bar{w}_1) \succeq (x_1, \bar{w}_1)$ for some $\bar{w}_1 \in \bar{X}_1$. We then ask the decision maker to imagine that she is in situation 1: She already has $(x_1, \bar{w}_1)$ and she can exchange it for $(w_1, \bar{w}_1)$. Next, we arbitrarily choose $\bar{x}_1 \in \bar{X}_1$ and ask her to imagine situation 2: She already has $(x_1, \bar{x}_1)$, and she can exchange it for $(w_1, \bar{x}_1)$. Would she prefer to make the exchange in situation 1 or in situation 2, or is she indifferent between the two exchanges? If she is indifferent between the two exchanges for several different values of $w_1, x_1 \in X_1$ and $\bar{w}_1, \bar{x}_1 \in \bar{X}_1$ then we can conclude that X_1 is difference independent of $\bar{X}_1$.

For example, suppose we ask the decision maker to consider exchanging a car described by ($ 25,000, 150 hp, ugly) for a car described by ($ 25,000, 180 hp, ugly). Next, we ask her to consider exchanging ($ 35,000, 150 hp, nice) for ($ 35,000, 180 hp, nice). Would the opportunity to exchange a car with 150 hp for one with 180 hp be more important to the decision maker when the cost and appearance are $25,000 and ugly, or when they are $ 35,000 and nice? If the common values of these two attributes do not affect her judgments of the importance of these exchanges, then horsepower would be difference independent of cost and appearance.

Before using this procedure, we must ensure that the decision maker understands that we are asking her to focus on the exchange rather than on the final outcomes. For example, if she states that she prefers an exchange of $1,000,000 for $1,000,001 to an exchange of $5 for $500, then she undoubtedly is not focusing on the substitution of one outcome for another, but she is focusing instead on the final outcome. Thus, some training may be required before this approach to verification of difference independence is attempted.

To verify weak difference independence, partition X into X_I and $\bar{X}_I$, and choose $w_I, x_I, y_I, z_I \in X_I$ and $\bar{w}_I \in \bar{X}_I$ so that $(w_I, \bar{w}_I) \succ (x_I, \bar{w}_I)$, $(y_I, \bar{w}_I) \succ (z_I, \bar{w}_I)$ and the exchange of $(x_I, \bar{w}_I)$ for $(w_I, \bar{w}_I)$ is preferred to the exchange of $(z_I, \bar{w}_I)$ for $(y_I, \bar{w}_I)$. Then pick another value $\bar{x}_I \in \bar{X}_I$ and ask if the decision maker still prefers the exchange of $(x_I, \bar{x}_I)$ for $(w_I, \bar{x}_I)$ to the exchange of $(z_I, \bar{x}_I)$ for $(y_I, \bar{x}_I)$. This must be true if the subset X_I is weakly difference independent of $\bar{X}_I$. If the decision maker's response is affirmative, we repeat the question for other quadruples of consequences from X_I with the values of the criteria in $\bar{X}_I$ fixed at different levels. Continuing in this manner and asking the decision maker to verbally rationalize her responses, the analyst can either verify that X_I is weakly difference independent of $\bar{X}_I$ or discover that the condition does not hold.

Note that for the multiplicative measurable value function, it would only be necessary to verify weak difference independence for the special case of $I = \{i, j\}$, where i and j indicate the subscripts of an arbitrarily chosen pair of alternatives. This is true so long as the attributes are mutually preference independent.

For example, suppose we establish that the decision maker would prefer the exchange of the car ($25,000, 150 hp, ugly) for the car ($27,000, 200 hp, ugly) to the exchange of the car ($24,000, 130 hp, ugly) for the car ($25,000, 150 hp, ugly). If this preference for the first exchange over the second exchange does not depend on the common value of appearance, and if it also holds true for all other combinations of the values of cost and horsepower, then cost and horsepower are weak difference independent of appearance.

5.3.2 Assessment of the Measurable Value Functions. If difference independence or weak difference independence holds, each conditional measurable value function v_i can be assessed while holding $\bar{x}_i$ constant at any arbitrary value (generally at $\bar{x}_i^\circ$). With the additive value function that does not provide difference measurement, this strategy cannot be used as illustrated above. As a result, any of the approaches for assessing a single attribute measurable value function referenced in Section 2.1 may be used, including the direct rating of attribute values on an arbitrary scale (e.g., from 0 to 100), or direct estimates of preference differences.

If the measurable value function is additive, the scaling constants may be assessed using the same trade-off approach suggested for estimating the scaling constants for the additive ordinal value function (Keeney and Raiffa, 1993 [27, Chapter 3]). In this volume (Chapter 10), Bana e Costa, De Corte, and Vansnick discuss the use of MACBETH to assess a preference scale measuring preference differences based on qualitative judgments about "the difference of attractiveness" between two alternatives. For a discussion of other approaches to the assessment of the scaling constants for the additive and multiplicative cases, see also Dyer and Sarin (1979) [9].

Measurable multiattribute value functions may also be assessed using the ratio judgments and tools provided by the AHP methodology, and used as a basis for relating the AHP to formal preference theories that are widely accepted by economists and decision analysts. This point has been made by several authors, notably Kamenetsky (1982) [26] and Dyer (1990) [8].

Perhaps the best discussion of this important point is provided by Salo and Hämäläinen (1997) [36]. As they observe, once a suitable range of performance $[x_i^\circ, x_i^*]$ has been defined for each attribute, the additive measurable value function representation may be scaled so that the values $v(x^\circ) = v(x_1^\circ, \ldots, x_n^\circ) = 0$ and $v(x^*) = v(x_1^*, \ldots, x_n^*) = 1$ are assigned to the worst and best conceivable consequences, respectively. By also normalizing the component value functions onto the [0,1] range, the additive representation can be written as

$$\begin{aligned} v(x) &= \textstyle\sum_{i=1}^n v_i(x_i) = \sum_{i=1}^n [v_i(x_i) - v_i(x_i^\circ)] &&= \\ &= \textstyle\sum_{i=1}^n [v_i(x_i^*) - v_i(x_i^\circ)] \frac{v_i(x_i) - v_i(x_i^\circ)}{v_i(x_i^*) - v_i(x_i^\circ)} &&= \textstyle\sum_{i=1}^n w_i s_i(x_i), \end{aligned}$$

where $s_i(x_i) = v_i(x_i) - v_i(x_i^\circ) / v_i(x_i^*) - v_i(x_i^\circ) \in [0, 1]$ is the normalized score of x on the ith attribute and $w_i = v(x_i^*) - v(x_i^\circ)$ is the scaling constant or weight of the ith attribute.

A careful evaluation of this representation leads Salo and Hämäläinen to the conclusion that pair wise comparisons in ratio estimation should be interpreted in terms of ratios of value differences between pairs of underlying alternatives. This, in turn, provides the link between traditional models of preference theory and the AHP, and reveals that the latter can be an alternative assessment technique for measurable multiattribute value functions (with some simple adjustments for normalization and scaling).

5.4 Goal Programming and Measurable Multiattribute Value Functions

Goal programming was originally proposed by Charnes, Cooper, and Ferguson (1955) [6] as an ingenious approach to developing a scheme for executive compensation. As noted by Charnes and Cooper (1977) [5] in a review of the field, this approach to multiple objective optimization did not receive significant attention until the mid-1960's. However, during the past forty years, we have witnessed a flood of professional articles and books (e.g. Ijiri (1965) [23], Ignizio (1986) [22], Trzaskalik and Michnik (2002) [40]) dealing with applications of this methodology.

This discussion is limited to the use of goal programming as a methodology for solving problems with multiple, compensatory objectives. That is, we do not address problems that do not allow tradeoffs among the objectives. These non-compensatory models involve the use of the non-Archimedian, or "preemptive priority", weights. An analysis of these models would be based on the theory of lexicographic orders, summarized by Fishburn (1974)[16]. The conditions that would justify the use of a non-compensatory model are very strict, and are unlikely to be met in a significant number of real-world applications.

5.4.1 Goal Programming as an Approximation to Multiattribute Preferences. Let us begin with a simple example. Suppose a manager has identified a problem that can be formulated as a traditional mathematical programming problem with one exception – there are two criterion functions, $f_1(x)$ and $f_2(x)$ where $x \in X$ is an n-vector of controllable and uncontrollable variables, and the non-empty feasible set X is defined by a set of constraints. For simplicity, and without loss of generality, we assume that our choice of X ensures $0 \leq f_i(x) \leq 1,\ i = 1, 2$.

To use goal programming, we ask the manager if she has any "goals" in mind for the criteria. She replies that she would be happy if $f_1(\cdot)$ were at least as large as b_1, but she does not feel strongly about increasing $f_1(\cdot)$ beyond b_1. However, she would like for $f_2(\cdot)$ to be somewhere *between* b_{2L} and b_{2U}. Finally, we ask her to assign "weights" of relative importance to the deviations of $f_1(\cdot)$ from b_1, and of $f_2(\cdot)$ from b_{2L} and b_{2U}, respectively. After some thought, she responds with the weights w_1,w_2 and w_3.

Now, we can immediately write down this problem as follows:

$$\begin{aligned}
\min_{x\in X} w_1 y_1^- + w_2 y_2^- + w_3 y_3^+ & \\
\text{subject to} f_1(x) - y_1^+ + y_1^- &= b_1 \\
f_2(x) - y_2^+ + y_2^- &= b_{2L} \\
f_2(x) - y_3^+ + y_3^- &= b_{2U} \\
y_i^+, y_i^- &\geq 0, \quad i = 1,2,3.
\end{aligned} \tag{GP}$$

Notice that (GP) includes a "one-sided" formulation with respect to $f_1(\cdot)$, and a "goal interval" formulation with respect to $f_2(\cdot)$.

Let us pause a moment to reflect on this formulation. First, notice that $y_1^- = b_1 + y_1^+ - f_1(x)$. Suppose we introduce the relationship $f_1(x) - y_{12}^+ = 0$ as a new constraint for (GP). Since b_1 is a constant, minimizing $w_1 y_1^-$ is obviously equivalent to minimizing $w_1\left(y_1^+ - y_{12}^+\right)$.

Similarly, if we introduce the constraint $f_2(x) - y_{22}^+ = 0$, minimizing $w_2 y_2^-$ is equivalent to minimizing $w_2\left(y_2^+ - y_{22}^+\right)$, and minimizing $w_3 y_3^+$ is equivalent to minimizing $w_3\left(y_3^- + y_{22}^+ - b_{2U}\right)$. The constant b_{2U} is maintained in the last expression in order to facilitate a graphical portrayal of the objective function as we shall see. Combining the results and re-writing (GP) as a maximization problem, we have the equivalent problem statement:

$$\max_{x\in X} w_1\left(y_{12}^+ - y_1^+\right) + w_2\left(y_{22}^+ - y_2^+\right) + w_3\left(y_{22}^+ + y_3^- - b_{2U}\right)$$

$$\begin{aligned}
\text{subject to } f_1(x) - y_1^+ + y_1^- &= b_1 \\
f_2(x) - y_2^+ + y_2^- &= b_{2L} \\
f_2(x) - y_3^+ + y_3^- &= b_{2U} \\
f_1(x) - y_{12}^+ &= 0 \\
f_2(x) - y_{22}^+ &= 0 \\
y_i^+, y_i^- &\geq 0 \quad i = 1,2,3 \\
y_{12}^+, y_{22}^+ &\geq 0,
\end{aligned} \tag{VA}$$

where the objective function may be interpreted as the sum of two piecewise linear functions (e.g. see Charnes and Cooper, 1961 [4, pp. 351-355])

Figures 7.3 and 7.4 illustrate these two piecewise linear functions. Recall that piecewise linear transformations are commonly used to transform additive separable nonlinear programming problems into linear programming problems. The lines labeled $v_1(\cdot)$ and $v_2(\cdot)$ in Figures 7.3 and 7.4 respectively suggest nonlinear preference functions that *might* be approximated by the bold piecewise linear functions.

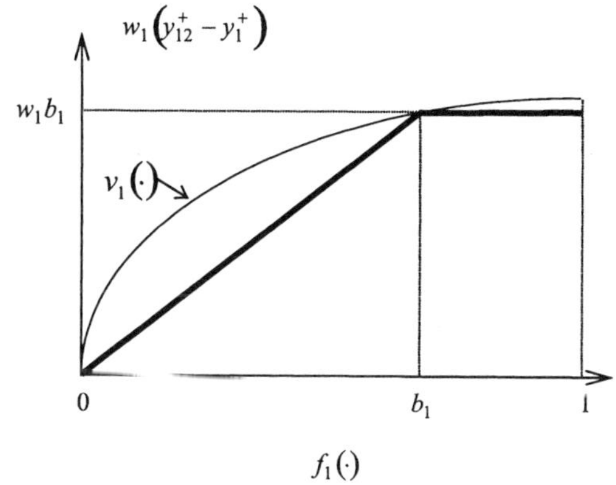

Figure 7.3. Piecewise linear approximation of $v_1(\cdot)$.

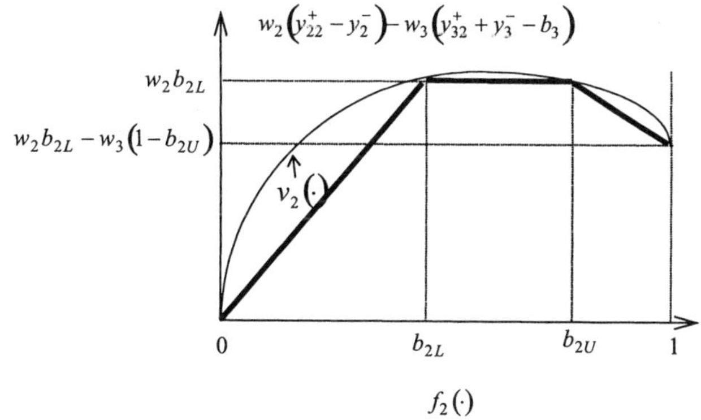

Figure 7.4. Piecewise linear approximation of $v_2(\cdot)$.

Thus since (VA) is equivalent to (GP), and (VA) may be viewed as a piecewise linear approximation to an additive separable nonlinear objective function, we are led to the conclusion that (GP) is an implicit approximation to the problem:

$$\max_{x \in X} v_1(f_1(x)) + v_2(f_2(x)). \tag{V}$$

And how do we interpret (V)? Since the choice of goals and goal intervals in (GP) reflect the decision maker's preferences and no uncertainty is involved in the decision, $v_1(\cdot)$ and $v_2(\cdot)$ are measurable functions, and their sum, $v(f_1(\cdot), f_2(\cdot)) = v_1(f_1(\cdot)) + v_2(f_2(\cdot))$, is an additive separable measurable value function.

Goal programming is generally applied to problems where risk is not explicitly involved in the formulation. Therefore, the additive utility function theory developed for risky choice is not relevant for these applications. Likewise, the ordinal additive theories are not operational here because they require a simultaneous conjoint scaling of the separable terms. Goal programming applications generally allow the selection of each goal or goal interval independent of consideration of the values of the other criteria. This practice implies the existence of a measurable additive utility function under certainty.

This point has been made recently by Bordley and Kirkwood (2004) [2] in a general discussion of the relationship between goals and multiattribute preference models. This perspective provides some insights regarding the nature of goal programming, as well as some challenges. For example, how should the piecewise linear approximations to the nonlinear value functions be selected in order to minimize error? Geoffrion (1977) [19] provides some useful guidelines for choosing "goals" or "goal intervals" for each criterion so that the piecewise linear approximation to the implicit utility function provides the best fit.

One important implication of this point of view is that goal programming should not be considered an *ad hoc*, heuristic approach to solving multiple objective problems. Rather, the approach is based on a set of implicit, well-understood assumptions from multiattribute preference theory. Goal programming formulations should be either criticized or justified on the basis of these assumptions.

6. The Relationships Among the Multiattribute Preference Functions

The necessary conditions for the additive and multiplicative measurable value functions and risky utility functions, notably mutual preference independence, are also necessary and sufficient for the ordinal additive value function that does not provide difference measurement. Therefore, it is natural to investigate the relationships among them. The following choice of scaling will be imposed. For $f = \overset{\circ}{v}, v$, or u, f is normalized by $f(x_1^*, \ldots, x_n^*) = 1$ and $f(x_1^\circ, \ldots, x_n^\circ) = 0$ and $f_i(x_i)$ is a conditional function on X_i scaled by $f_i(x_i^*) = 1$ and $f_i(x_i^\circ) = 0$. Finally, $\overset{\circ}{\lambda}, \lambda$ and k will be used as scaling constants for the ordinal and measurable value functions and the utility function, respectively.

6.1 The Additive Functions

The relationships among the alternative developments of the additive forms of real-valued functions on X follow immediately from their respective uniqueness properties. This may be summarized as follows. Assume $n \geq 3$ and $X_1, \ldots, X_n$ are mutually preference independent. Then

1 if $X_1, \ldots, X_n$ are difference consistent and X_1 is difference independent of $\bar{X}_1$ then $\overset{\circ}{v} = v$;

2 if there exists a utility function u on X and if preferences over lotteries on $X_1, \ldots, X_n$ depend only on their marginal probability distributions and not on their joint probability distributions, then $\overset{\circ}{v} = u$.

3 if both 1 and 2 are satisfied , $\overset{\circ}{v} = v = u$.

Note the implication of this result. In order for $\overset{\circ}{v} = v = u$ for a single decision maker, she must have preferences simultaneously consistent with mutual preference independence, difference independence, and additive independence for risky alternatives. Mutual preference independence will hold in all cases, but it may be the case that difference independence and/or additive independence for risky alternatives will not hold. Further, difference independence may hold for the preferences of a decision maker, implying that an additive measurable value function would provide a valid representation of her preferences, but additive independence for risky alternatives may not be satisfied, implying that an additive utility function would not be a valid representation of her preferences in decision scenarios involving risk.

6.2 The Multiplicative Functions

Throughout this section we assume that the following conditions are satisfied:

1 There are $n \geq$ 3 attributes, and $X_1, \ldots, X_n$ are mutually preference independent;

2 There exists a measurable value function v on X and X_1 is weak difference independent of $\bar{X}_1$, and

3 There exists a utility function u on X and X_1 is utility independent of $\bar{X}_1$.

Suppose we have assessed the additive value function $\overset{\circ}{v}$ and wish to obtain either v or u.Then the following relationships will hold (Dyer and Sarin (1979) [9, Theorem 5]). Either

1 $\overset{\circ}{v}(x) = v(x)$ and $\overset{\circ}{v}_i(x_i) = v_i(x_i), i = 1, \ldots, n$, or

2 $\overset{\circ}{v}(x)\ln(1+\lambda) = \ln[1+\lambda v(x)]$ and $\overset{\circ}{\lambda}_i\,\overset{\circ}{v}_i(x_i)\ln(1+\lambda) = \ln[1+\lambda\lambda_i v_i(x_i)], i=1,\ldots,n$.

Either

1 $\overset{\circ}{v}(x) = u(x)$ and $\overset{\circ}{v}_i(x_i) = u_i(x_i), i = 1,\ldots,n$, or,

2 $\overset{\circ}{v}(x)\ln(1+k) = \ln[1+ku(x)]$ and $\overset{\circ}{\lambda}_i\,\overset{\circ}{v}_i(x_i)\ln(1+k) = \ln[1+kk_iu_i(x_i)], i = 1,\ldots,n$.

These relationships may be used to simplify the assessment of multiattribute preference functions. For example, suppose we define x_i' as the *equal difference point* for attribute X_i if $(x_i', \bar{x}_i)(x_i^\circ, \bar{x}_i) \sim^* (x_i^*, \bar{x}_i)(x_i', \bar{x}_i)$ for any $\bar{x}_i \in \bar{X}_i$. Notice that $v_i(x_i') = 1/2$ because of our choice of scaling. Given $\overset{\circ}{v}$, the assessment of x_i' for any attribute X_i is enough to completely specify v, because if $v_i(x_i') = 1/2$ for some $i \in \{1,\ldots,n\}$ then $v = \overset{\circ}{v}$. Otherwise, $1+(1+\lambda)^{\overset{\circ}{\lambda}_i} = 2(1+\lambda)^{\overset{\circ}{\lambda}_i\,\overset{\circ}{v}_i(x_i)}$.

Finally, to derive u from $\overset{\circ}{v}$, find x_i'' for some attribute X_i such that the decision maker is indifferent between x_i'' and an equal chance lottery between x_i^* and x_i° with the other criteria held fixed. A parallel result to the above relationship between ordinal and measurable value functions holds. Specifically, if $\overset{\circ}{v}_i(x_i'') = 1/2$ for some $i \in \{1,\ldots,n\}$, then $u = \overset{\circ}{v}$. Otherwise, $1+(1+k)^{\overset{\circ}{\lambda}_i} = 2(1+k)^{\overset{\circ}{\lambda}_i\,\overset{\circ}{v}_i(x_i)}$.

These results can also be used to derive v after u has been assessed, or vice versa. For example, suppose $u$has been assessed using appropriate procedures. To obtain v, we find the equal difference point x_i' for some criterion X_i. The second result above is used to obtain $\overset{\circ}{\lambda}_i$ and $\overset{\circ}{v}_i$ for each criterion, and we can obtain v. In a similar manner, u can be obtained from v after assessing x_i'' for some criterion X_i.

Since the AHP can be interpreted as ratios of preference differences, this relationship also allows the results from assessments based on the AHP to be suitably transformed into multiattribute utility functions appropriate for use in risky situations. This completes the circle required to synthesize ordinal multiattribute value functions, measurable multiattribute value functions, multiattribute utility functions, and multiattribute functions based on ratio judgments. As a result, the analyst is justified in choosing among a variety of assessment tools, and making the appropriate adjustments in order to calibrate the results into a coherent and theoretically sound representation of preferences.

7. Concluding Remarks

In this chapter, we have presented an informal discussion of "mutiattribute utility theory". In fact, this discussion has emphasized that there is no single version of multiattribute utility theory that is relevant to multicriteria decision analysis. Instead, there are three distinct theories of multiattribute preference functions that may be used to represent a decision maker's preferences.

The ordinal additive multiattribute preference model requires the assumption of mutual preference independence, and is appropriate for use in the case of certainty. Most of the applications and methods of multicriteria decision analysis are presented in the context of certainty, and so this would seem to be an appealing theory to use for framing these approaches. However, as we have emphasized, the ordinal additive multiattribute preference model requires assessment techniques that are cumbersome in practice, and that force the decision maker to make explicit tradeoffs between two or more criteria in the assessment of the value functions defined on the individual criteria.

The measurable value functions also require the assumption of mutual preference independence, along with the stronger assumptions of weak difference independence or difference independence in order to obtain convenient decompositions of the model that are easy to assess. The assessment of these preference models is relatively easy, and they can be interpreted intuitively as providing a measure of strength of preference. In addition, the ratio judgments of the AHP can be interpreted as ratios of preference differences based on this theory, linking the AHP methodology to traditional models of preference accepted in the decision analysis and economics literatures.

Finally, multiattribute utility theory is an elegant and useful model of preference suitable for applications involving risky choice. The brilliant work of Keeney and Raiffa (1993) [27] has made this theory synonymous to many scholars with multiple criterion decision making, and the ordinal and measurable theories are often overlooked or ignored as a result. In fact, these latter approaches may provide more attractive and appropriate theories for many applications of multicriteria decision analysis.

Notes

1. The classic book *Decisions with Multiple Objectives* by R. L. Keeney and H. Raiffa was originally published by Wiley in 1976. The Cambridge University Press version was published in 1993 [27].

2. "The important addition since 1976 concerns value functions that address strength of preference between pairs of consequences (see Dyer and Sarin, 1979 [9]; Bell and Raiffa, 1988 [1])." A quote from the Preface to the Cambridge University Press Edition, R. L. Keeney and H. Raiffa, *Decisions with Multiple Objectives*, Cambridge University Press, 1993 [27].

3. Note that the $\overset{o}{v}_i$ are called partial value functions by Bouyssou and Pirlot in Chapter 3 of this volume.

4. Specifically, we assume restricted solvability from below, an Archimedian property, at least three attributes are essential, and that the attributes are bounded from below. If $n = 2$, we assume that the two attributes are preferentially independent of one another and that the Thomsen condition is satisfied (see Krantz et al. (1971) [30] and the discussion by Bouyssou and Pirlot in this volume).

References

[1] D. Bell and H. Raiffa. Risky choice revisited. In D.E. Bell, H. Raiffa, and A. Tversky, editors, *Decision Making: Descriptive, Normative, and Prescriptive Interactions*. Cambridge University Press, New York, 1988.

[2] R. Bordley and C. Kirkwood. Multiattribute preference analysis with performance targets. *Operations Research*, in press, 2004.

[3] C. Camerer. Individual decision making. In J.H. Kagel and A.E. Roth, editors, *Handbook of Experimental Economics*. Princeton University Press, Princeton, NJ, 1995.

[4] A. Charnes and W.W. Cooper. *Management Models and Industrial Applications of Linear Programming*. John Wiley & Sons, New York, 1961.

[5] A. Charnes and W.W. Cooper. Goal programming and multiple objective optimizations. *European Journal of Operational Research*, 1(1):37–54, 1977.

[6] A. Charnes, W.W. Cooper, and R.O. Ferguson. Optimal estimation of executive compensation by linear programming. *Management Science*, 1(2):138–151, 1955.

[7] G. Debreu. Topological methods in cardinal utility theory. In K.J. Arrow, S. Karlin, and P. Suppes, editors, *Mathematical Methods in the Social Sciences 1959*. Stanford University Press, Stanford, 1960.

[8] J.S. Dyer. Remarks on the analytic hierarchy process. *Management Science*, 36:249–258, 1990.

[9] J.S. Dyer and R.A. Sarin. Measurable multiattribute value functions. *Operations Research*, 22:810–822, 1979.

[10] J.S. Dyer and R.K. Sarin. Relative risk aversion. *Management Science*, 28:875–886, 1982.

[11] D. Ellsberg. Classic and current notions of "measurable utility". *Econometrica*, 64:528–556, 1954.

[12] Peter H. Farquhar. A survey of multiattribute utility theory and applications. In M.K. Starr and M. Zeleny, editors, *Multiple Criteria Decision Making*, volume 6 of *TIMS Studies in the Management Sciences*, pages 59–90. North-Holland, Amsterdam, 1977.

[13] P.H. Farquhar and L.R. Keller. Preference intensity measurement. *Annals of Operations Research*, 19:205–217, 1989.

[14] P.C. Fishburn. Independence in utility with whole product sets. *Operations Research*, 13:28–45, 1965.

[15] P.C. Fishburn. *Utility Theory for Decision Making*. John Wiley & Sons, New York, 1970.

[16] P.C. Fishburn. Lexicographic orders, utilitites, and decision rules. *Management Science*, 20:1442–1471, 1974.

[17] P.C. Fishburn. *Nonlinear Preference and Utility Theory*. Johns Hopkins University Press, Baltimore MD, 1988.

[18] P.C. Fishburn and P. Wakker. The invention of the independence condition for preferences. *Management Science*, 41(7):1130–1144, 1995.

[19] A. Geoffrion. Objective function approximations in mathematical programming. *Mathematical Programming*, 13(1):23–37, 1977.

[20] A. Geoffrion, J.S. Dyer, and A. Feinberg. An interactive approach for multi-criterion optimization with an application to the operation of an academic department. *Management Science*, 19:357–368, 1972.

[21] W.M. Gorman. The structure of utility functions. *Review of Economic Studies*, 35:367–390, 1968.

[22] J.P. Ignizio. *Introduction to Linear Goal Programming*. Sage Publications, 1986.

[23] J. Ijiri. *Management Goals and Accounting for Control*. American Elsevier, New York, 1965.

[24] J. Jia and J.S. Dyer. A standard measure of risk and risk-value models. *Management Science*, 42(12):1691–1705, 1996.

[25] D. Kahneman and A. Tversky. Prospect theory: An analysis of decisions under risk. *Econometrica*, 47:276–287, 1979.

[26] R.D. Kamenetzky. The relationship between the analytic hierarchy process and the additive value function. *Decision Sciences*, 13:702–713, 1982.

[27] R.L. Keeney and H. Raiffa. *Decision with Multiple Objectives: Preference and Value Tradeoffs*. Cambridge University Press, New York, 1993.

[28] L.R. Keller. An empirical test of relative risk aversion. *IEEE Transactions on Man, Systems, and Cybernetics*, 15:476–482, 1985.

[29] C.W. Kirkwood. *Strategic Decision Making*. Duxbury Press, Belmont, CA, 1996.

[30] D.H. Krantz, R.D. Luce, P. Suppes, and A. Tversky. *Foundations of Measurement. Volume I: Additive and Polynomial Representations*. Academic Press, New York, 1971.

[31] D.M. Kreps. *A Course in Microeconomic Theory*. Princeton University Press, Princeton, NJ, 1990.

[32] R. Krzysztofowicz. Strength of preference and risk attitude in utility measurement. *Organizational Behavior and Human Performance*, 31:88–113, 1983.

[33] R.D. Luce and J.W. Tukey. Simultaneous conjoint measurement: A new type of fundamental measurement. *Journal of Mathematical Psychology*, 1:1–27, 1964.

[34] J.W. Pratt. Risk aversion in the small and in the large. *Econometrica*, 32:122–136, 1964.

[35] T.L. Saaty. *The Analytic Hierarchy Process*. McGraw-Hill, NewYork, 1980.

[36] A.A. Salo and R.P. Hämäläinen. On the measurement of preferences in the analytic hierarchy process. *Journal of Multi-Criterion Decision Analysis*, 6:309–319, 1997.

[37] R.K. Sarin. Strength of preference and risky choice. *Operations Research*, 30:982–997, 1982.

[38] L.J. Savage. *The Foundations of Statistics*. John Wiley & Sons, New York, 1954.

[39] D. Scott and P. Suppes. Foundational aspects of theories of measurement. *Journal of Symbolic Logic*, 23:113–128, 1958.

[40] T. Trzaskalik and J. Michnik, editors. *Multiple Objective and Goal Programming: Recent Developments*. Physica Verlag, Heidelberg, 2002.

[41] J. von Neumann and O. Morgenstern. *Theory of Games and Economic Behavior*. Princeton University Press, Princeton, NJ, 1947.

[42] D. von Winterfeldt and W. Edwards. *Decision Analysis and Behavioral Research*. Cambridge University Press, New York, 1986.

Chapter 8

UTA METHODS

Yannis Siskos
University of Piraeus
Department of Informatics
80 Karaoli & Dimitriou Str.
18534 Piraeus, Greece
ysiskos@unipi.gr

Evangelos Grigoroudis, Nikolaos F. Matsatsinis
Technical University of Crete
Decision Support Systems Laboratory
University Campus, Kounoupidiana
73100 Chania,– Greece
{vangelis,nikos}@ergasya.tuc.gr

Abstract UTA methods refer to the philosophy of assessing a set of value or utility functions, assuming the axiomatic basis of MAUT and adopting the preference disaggregation principle. UTA methodology uses linear programming techniques in order to optimally infer additive value/utility functions, so that these functions are as consistent as possible with the global decision-maker's preferences (inference principle). The main objective of this chapter is to analytically present the UTA method and its variants and to summarize the progress made in this field. The historical background and the philosophy of the aggregation-disaggregation approach are firstly given. The detailed presentation of the basic UTA algorithm is presented, including discussion on the stability and sensitivity analyses. Several variants of the UTA method, which incorporate different forms of optimality criteria, are also discussed. The implementation of the UTA methods is illustrated by a general overview of UTA-based DSSs, as well as real-world decision-making applications. Finally, several potential future research developments are discussed.

Keywords: UTA methods, preference disaggregation, ordinal regression, additive utility, multicriteria analysis.

1. Introduction

1.1 General Philosophy

In decision-making involving multiple criteria, the basic problem stated by analysts and Decision-Makers (DMs) concerns the way that the final decision should be made. In many cases, however, this problem is posed in the opposite way: assuming that the decision is given, how is it possible to find the rational basis for the decision being made? Or equivalently, how is it possible to assess the DM's preference model leading to exactly the same decision as the actual one or at least the most "similar" decision? The philosophy of preference disaggregation in multicriteria analysis is to assess/infer preference models from given preferential structures and to address decision-aiding activities through operational models within the aforementioned framework.

Under the term "multicriteria analysis" two basic approaches have been developed involving:

1. a set of methods or models enabling the aggregation of multiple evaluation criteria to choose one or more actions from a set A, and
2. an activity of decision-aid to a well-defined DM (individual, organization, etc.).

In both cases, the set A of potential actions (or objectives, alternatives, decisions) is analyzed in terms of multiple criteria in order to model all the possible impacts, consequences or attributes related to the set A.

Roy (1985) [76] outlines a general modeling methodology of decision-making problems, which includes four modeling steps starting with the definition of the set A and ending with the activity of decision-aid, as follows:

- *Level 1*: Object of the decision, including the definition of the set of potential actions A and the determination of a problem statement on A.
- *Level 2*: Modeling of a consistent family of criteria assuming that these criteria are non-decreasing value functions, exhaustive and non-redundant.
- *Level 3*: Development of a global preference model, to aggregate the marginal preferences on the criteria.
- *Level 4*: Decision-aid or decision support, based on the results of level 3 and the problem statement of level 1.

In level 1, Roy (1985) [76] distinguishes four reference problem statements, each of which does not necessarily preclude the others. These problem statements can be employed separately, or in a complementary way, in all phases of the decision-making process. The four problem statement are the following:

- *Problem statement* α: Choosing one action from *A* (choice).
- *Problem statement* β: Sorting the actions into predefined and preference ordered categories.
- *Problem statement* γ: Ranking the actions from the best one to the worst one (ranking).
- *Problem statement* δ: Describing the actions in terms of their performances on the criteria (description).

In level 2, the modeling process must conclude with a consistent family of criteria $\{g_1, g_2, ..., g_n\}$. Each criterion is a non-decreasing real valued function defined on *A*, as follows:

$$g_i : A \rightarrow [g_{i^*}, g_i^*] \subset \Re / a \rightarrow g(a) \in \Re, \tag{8.1}$$

where $[g_{i^*}, g_i^*]$ is the criterion evaluation scale, g_{i^*} and g_i^* are the worst and the best level of the i-th criterion respectively, $g_i(a)$ is the evaluation or performance of action a on the i-th criterion and $\mathbf{g}(a)$ is the vector of performances of action a on the n criteria.

From the above definitions, the following preferential situations can be determined:

$$\begin{cases} g_i(a) > g_i(b) \Leftrightarrow a \succ b & (a \text{ is preferred to } b) \\ g_i(a) = g_i(b) \Leftrightarrow a \sim b & (a \text{ is indifferent to } b). \end{cases} \tag{8.2}$$

So, having a weak-order preference structure on a set of actions, the problem is to adjust additive value or utility functions based on multiple criteria, in such a way that the resulting structure would be as consistent as possible with the initial structure. This principle underlies the disaggregation-aggregation approach presented in the next section.

This chapter is devoted to UTA methods, which are regression based approaches that have been developed as an alternative to multiattribute utility theory (MAUT). UTA methods not only adopt the aggregation-disaggregation principles, but they may also be considered as the main initiatives and the most representative examples of preference disaggregation theory. Another, more recent example of the preference disaggregation theory is the dominance-based rough set approach (DRSA) leading to decision rule preference model via inductive learning (see Chapter 13 of this book).

1.2 The Disaggregation-aggregation Paradigm

In the traditional aggregation paradigm, the criteria aggregation model is known a priori, while the global preference is unknown. On the contrary, the philosophy of disaggregation involves the inference of preference models from given global preferences (Figure 8.1).

The disaggregation-aggregation approach ([43, 81, 98, 97]) aims at analyzing the behavior and the cognitive style of the DM. Special iterative interactive procedures are used, where the components of the problem and the DM's global judgment policy are analyzed and then they are aggregated into a value system (Figure 8.2). The goal of this approach is to aid the DM to improve his/her knowledge about the decision situation and his/her way of preferring that entails a consistent decision to be achieved.

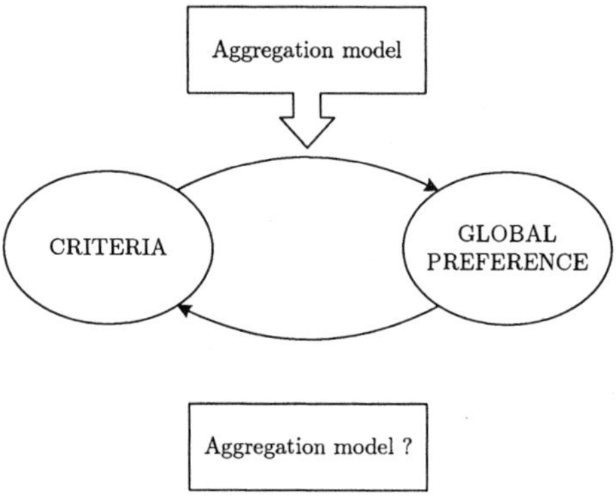

Figure 8.1. The aggregation and disaggregation paradigms in MCDA [44].

In order to use global preference given data, Jacquet-Lagrèze and Siskos (2001) [44] note that the clarification of the DM's global preference necessitates the use of a set of reference actions A_R. Usually, this set could be:

1 a set of past decision alternatives (A_R: past actions),

2 a subset of decision actions, especially when A is large ($A_R \subset A$),

3 a set of fictitious actions, consisting of performances on the criteria, which can be easily judged by the DM to perform global comparisons (A_R: fictitious actions).

In each of the above cases, the DM is asked to externalize and/or confirm his/her global preferences on the set A_R taking into account the performances of the reference actions on all criteria.

1.3 Historical Background

The history of the disaggregation principle in multidimensional/ multicriteria analyses begins with the use of goal programming techniques, a special form

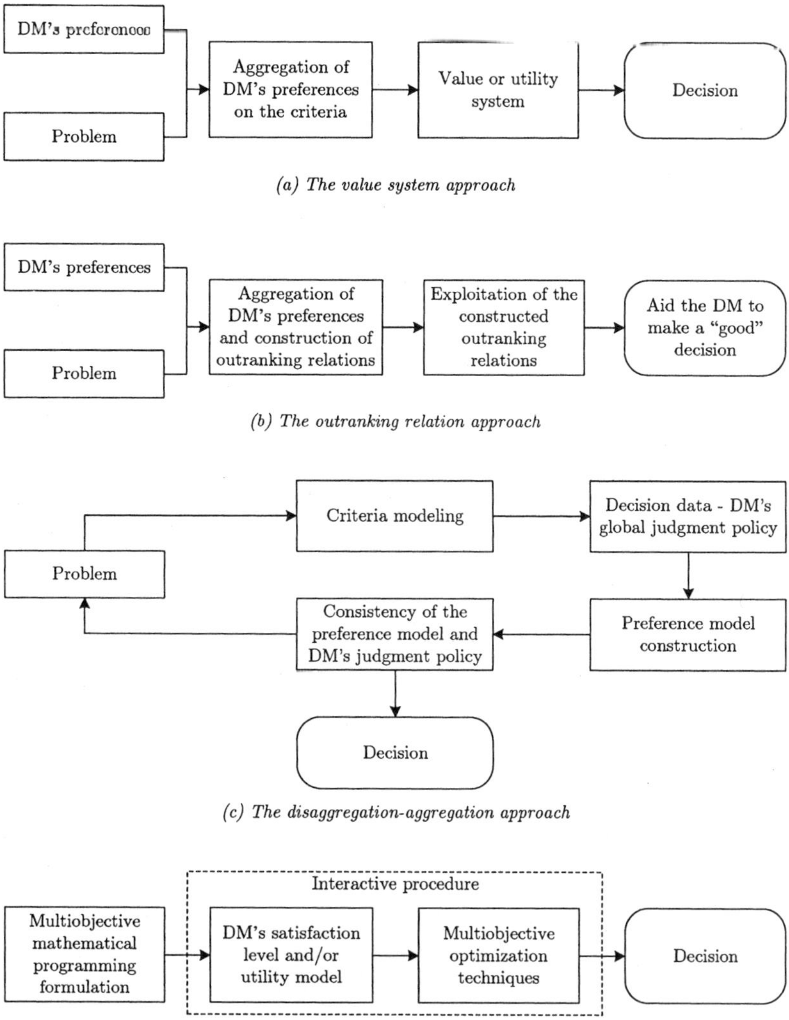

(a) The value system approach

(b) The outranking relation approach

(c) The disaggregation-aggregation approach

(d) The multiobjective optimization approach

Figure 8.2. The disaggregation-aggregation approach [96].

of linear programming structure, in assessing/inferring preference/aggregation models or in developing linear or nonlinear multidimensional regression analyses [83].

Charnes, Cooper, and Ferguson (1955) [14] proposed a linear model of optimal estimation of executive compensation by analyzing or disaggregating pairwise comparisons and given measures (salaries); the model was estimated so that it could be as consistent as possible with the data from the goal programming point of view.

Karst (1958) [46] minimized the sum of absolute deviations via goal programming in linear regression with one variable, while Wagner (1959) [108] generalized Karst's model in the multiple regression case. Later Kelley (1958) [49] proposed a similar model to minimize the Tchebycheff's criterion in linear regression.

Srinivasan and Shocker (1973) [104] outlined the ORDREG ordinal regression model to assess a linear value function by disaggregating pairwise judgments. Freed and Glover (1981) [26] proposed goal programming models to infer the weights of linear value functions in the frame of discriminant analysis (problem statement β).

The research on handling ordinal criteria began with the studies of Young, De Leeuw, and Takane (1976) [109], and Jacquet-Lagrèze and Siskos (1978) [42]. The latter research refers to the presentation of the UTA method in the "Cahiers du LAMSADE" series and indicates the actual initiation of the development of disaggregation methods. Both research teams faced the same problem: to infer additive value functions by disaggregating a ranking of reference alternatives. Young (1976) [109] proposed alternating least squares techniques, without ensuring, however, that the additive value function is optimally consistent with the given ranking. In the case of the UTA method, optimality is ensured through linear programming techniques.

2. The UTA Method

2.1 Principles and Notation

The UTA (UTilitès Additives) method proposed by Jacquet-Lagrèze and Siskos (1982) [43] aims at inferring one or more additive value functions from a given ranking on a reference set A_R. The method uses special linear programming techniques to assess these functions so that the ranking(s) obtained through these functions on A_R is (are) as consistent as possible with the given one.

The criteria aggregation model in UTA is assumed to be an additive value function of the following form [43]:

$$u(\mathbf{g}) = \sum_{i=1}^{n} p_i u_i(g_i) \tag{8.3}$$

subject to normalization constraints:

$$\begin{cases} \sum_{i=1}^{n} p_i = 1 \\ u_i(g_{i^*}) = 0, \quad u_i(g_i^*) = 1, \quad \forall i = 1, 2, \cdots, n; \end{cases} \tag{8.4}$$

where $u_i, i = 1, 2, \cdots, n$ are non decreasing real valued functions, named marginal value or utility functions, which are normalized between 0 and 1, and p_i is the weight of u_i (Figure 8.3)

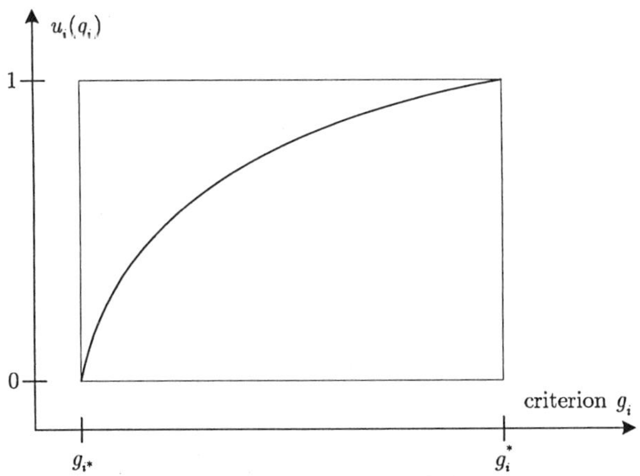

Figure 8.3. The normalized marginal value function.

Both the marginal and the global value functions have the monotonicity property of the true criterion. For instance, in the case of the global value function the following properties hold:

$$\begin{cases} u[\mathbf{g}(a)] > u[\mathbf{g}(b)] \Leftrightarrow a \succ b \quad \text{(preference)} \\ u[\mathbf{g}(a)] = u[\mathbf{g}(b)] \Leftrightarrow a \sim b \quad \text{(indifference)} \end{cases} \tag{8.5}$$

The UTA method infers an unweighted form of the additive value function, equivalent to the form defined from relations (8.3) and (8.4), as follows:

$$u(\mathbf{g}) = \sum_{i=1}^{n} u_i(g_i) \tag{8.6}$$

subject to normalization constraints:

$$\begin{cases} \sum_{i=1}^{n} u_i(g_i^*) = 1 \\ u_i(g_{i*}) = 0, \qquad \forall i = 1, 2, \cdots, n. \end{cases} \tag{8.7}$$

Of course, the existence of such a preference model assumes the preferential independence of the criteria for the DM [48], while other conditions for additivity have been proposed by Fishburn (1966, 1967) [25].

2.2 Development of the UTA Method

On the basis of the additive model (8.6)–(8.7) and taking into account the preference conditions (8.5), the value of each alternative $a \in A_R$ may be written as:

$$u'[\mathbf{g}(a)] = \sum_{i=1}^{n} u_i[g_i(a)] + \sigma(a) \; \forall a \in A_R, \tag{8.8}$$

where $\sigma(a)$ is a potential error relative to $u'[\mathbf{g}(a)]$.

Moreover, in order to estimate the corresponding marginal value functions in a piecewise linear form, Jacquet-Lagrèze and Siskos (1982) [43] propose the use of linear interpolation. For each criterion, the interval $[g_{i*}, g_i^*]$ is cut into $(\alpha_i - 1)$ equal intervals, and thus the end points g_i^j are given by the formula:

$$g_i^j = g_{i*} + \frac{j-1}{\alpha_i - 1}(g_i^* - g_{i*}) \; \forall j = 1, 2, \cdots, \alpha_i. \tag{8.9}$$

The marginal value of an action a is approximated by a linear interpolation, and thus, for $g_i(a) \in [g_i^j - g_i^{j+1}]$

$$u_i[g_i(a)] = u_i(g_i^j) + \frac{g_i(a) - g_i^j}{g_i^{j+1} - g_i^j}[u_i(g_i^{j+1}) - u_i(g_i^j)] \tag{8.10}$$

The set of reference actions $A_R = \{a_1, a_2, \cdots, a_m\}$ is also "rearranged" in such a way that a_1 is the head of the ranking (best action) and a_m its tail (worst action). Since the ranking has the form of a weak order R, for each pair of consecutive actions (a_k, a_{k+1}) it holds either $a_k \succ a_{k+1}$ (preference) or $a_k \sim a_{k+1}$ (indifference). Thus, if

$$\Delta(a_k, a_{k+1}) = u'[\mathbf{g}(a_k)] - u'[\mathbf{g}(a_{k+1})] \tag{8.11}$$

then one of the following holds:

$$\begin{cases} \Delta(a_k, a_{k+1}) \geq \delta & \text{iff } a_k \succ a_{k+1} \\ \Delta(a_k, a_{k+1}) = 0 & \text{iff } a_k \sim a_{k+1}, \end{cases} \tag{8.12}$$

where δ is a small positive number so as to discriminate significantly two successive equivalence classes of R.

Taking into account the hypothesis on monotonicity of preferences, the marginal values $u_i(g_i)$ must satisfy the set of the following constraints:

$$u_i(g_i^{j+1}) - u_i(g_i^j) \geq s_i \; \forall j = 1, 2, \cdots, \alpha_i - 1, \; i = 1, 2, \cdots, n \tag{8.13}$$

with $s_i \geq 0$ being indifference thresholds defined on each criterion g_i. Jacquet-Lagrèze and Siskos (1982) [43] urge that it is not necessary to use these thresholds in the UTA model ($s_i = 0$), but they can be useful in order to avoid phenomena such as $u_i(g_i^{j+1}) = u_i(g_i^j)$ when $g_i^{j+1} \succ g_i^j$.

The marginal value functions are finally estimated by means of the following Linear Program (LP) with (8.6), (8.7), (8.12), (8.13) as constraints and with an objective function depending on the $\sigma(a)$ and indicating the amount of total deviation:

$$\begin{cases} [\min]F = \sum_{a \in A_R} \sigma(a) \\ \text{subject to} \\ \left.\begin{array}{l} \Delta(a_k, a_{k+1}) \geq \delta \quad \text{if } a_k \succ a_{k+1} \\ \Delta(a_k, a_{k+1}) = 0 \quad \text{if } a_k \sim a_{k+1} \end{array}\right\} \quad \forall k \\ u_i(g_i^{j+1}) - u_i(g_i^j) \geq 0 \qquad \forall i \text{ and } j \\ \sum_{i=1}^{n} u_i(g_i^*) = 1 \\ u_i(g_{i*}) = 0, u_i(g_i^j) \geq 0, \sigma(a) \geq 0 \qquad \forall a \in A_R, \forall i \text{ and } j. \end{cases} \tag{8.14}$$

The stability analysis of the results provided by LP (8.14) is considered as a post-optimality analysis problem. As Jacquet-Lagrèze and Siskos (1982) [43] note, if the optimum $F^* = 0$, the polyhedron of admissible solutions for $u_i(g_i)$ is not empty and many value functions lead to a perfect representation of the weak order R. Even when the optimal value F^* is strictly positive, other solutions, less good for F, can improve other satisfactory criteria, like Kendall's τ.

As shown in Figure 8.4, the post-optimal solutions space is defined by the polyhedron:

$$\begin{cases} F \leq F^* + k(F^*) \\ \text{all the constraints of LP (8.14)}, \end{cases} \tag{8.15}$$

where $k(F^*)$ is a positive threshold which is a small proportion of F^*.

The algorithms which could be used to explore the polyhedron (8.15) are branch and bound methods, like reverse simplex method [107], or techniques

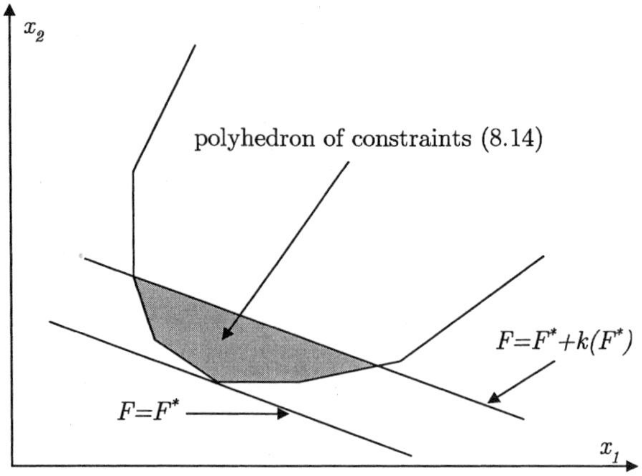

Figure 8.4. Post-optimality analysis [43].

dealing with the notion of the labyrinth in graph theory, such as Tarry's method [13], or the method of [54]. Jacquet-Lagrèze and Siskos (1982) [43], in the original UTA method, propose the partial exploration of polyhedron (8.15) by solving the following LPs:

$$\begin{cases} [\min] u_i(g_i^*) \text{ and } [\max] u_i(g_i^*) \\ \text{in} \\ \text{polyhedron (8.15)} \end{cases} \quad \forall i = 1, 2, \cdots, n. \tag{8.16}$$

The average of the previous LPs may be considered as the final solution of the problem. In case of instability, a large variation of the provided solutions appears, and this average solution is less representative. In any case, the solutions of the above LPs give the internal variation of the weight of all criteria g_i, and consequently give an idea of the importance of these criteria in the DM's preference system.

2.3 The UTASTAR Algorithm

The UTASTAR method proposed by Siskos and Yannacopoulos (1985) [98] is an improved version of the original UTA model presented in the previous section. In the original version of UTA [43], for each packed action $a \in A_R$, a single error $\sigma(a)$ is introduced to be minimized. This error function is not sufficient to minimize completely the dispersion of points all around the monotone curve of Figure 8.5. The problem is posed by points situated on the right of the curve, from which it would be suitable to subtract an amount of value/utility and not increase the values/utilities of the others.

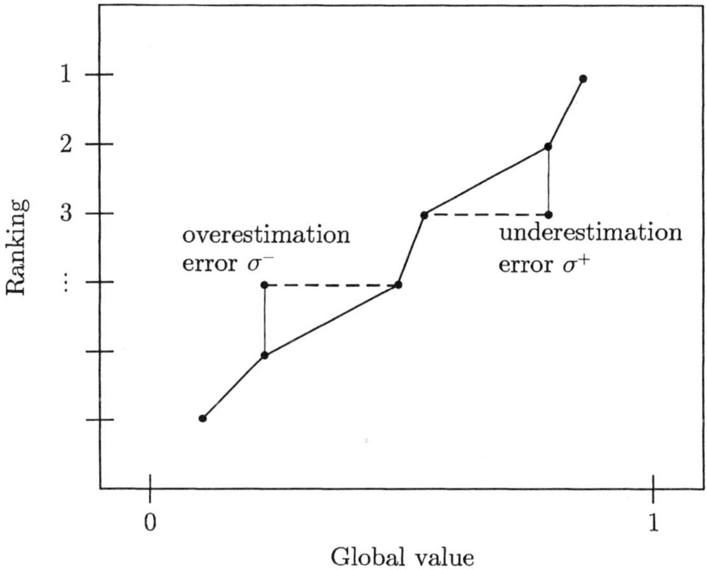

Figure 8.5. Ordinal regression curve (ranking versus global value).

In UTASTAR method, Siskos and Yannacopoulos (1985) [98] introduced a double positive error function, so that formula (8.8) becomes:

$$u'[\mathbf{g}(a)] = \sum_{i=1}^{n} u_i[g_i(a)] - \sigma^+(a) + \sigma^-(a) \; \forall \, a \in A_R, \qquad (8.17)$$

where σ^+ and σ^- are the overestimation and the underestimation error respectively.

Moreover, another important modification concerns the monotonicity constraints of the criteria, which are taken into account through the transformations of the variables:

$$w_{ij} = u_i(g_i^{j+1}) - u_i(g_i^j) \;\; \geq \;\; 0 \qquad (8.18)$$
$$\forall \, i = 1, 2, \cdots, n \text{ and } j = 1, 2, \cdots, \alpha_i - 1$$

and thus, the monotonicity conditions (8.13) can be replaced by the non-negative constraints for the variables w_{ij} (for $s_i = 0$).

Consequently, the UTASTAR algorithm may be summarized in the following steps:

Step 1:

Express the global value of reference actions $u[\mathbf{g}(a_k)], k = 1, 2, \cdots, m$, first in terms of marginal values $u_i(g_i)$, and then in terms of variables w_{ij} according to the formula (8.18), by means of the following expressions:

$$\begin{cases} u_i(g_i^1) = 0 & \forall i = 1, 2, \cdots, n \\ u_i(g_i^j) = \sum_{t=1}^{j-1} w_{it} & \forall i = 1, 2, \cdots, n \text{ and } j = 2, 3, \cdots, \alpha_i - 1. \end{cases} \tag{8.19}$$

Step 2:

Introduce two error functions σ^+ and σ^- on A_R by writing for each pair of consecutive actions in the ranking the analytic expressions:

$$\begin{aligned} \Delta(a_k, a_{k+1}) &= u[\mathbf{g}(a_k)] - \sigma^+(a_k) + \sigma^-(a_k) \\ &\quad - u[\mathbf{g}(a_{k+1})] + \sigma^+(a_{k+1}) - \sigma^-(a_{k+1}). \end{aligned} \tag{8.20}$$

Step 3:

Solve the LP:

$$\begin{cases} [min] z = \sum_{k=1}^{m} [\sigma^+(a_k) + \sigma^-(a_k)] \\ \text{subject to} \\ \left.\begin{array}{ll} \Delta(a_k, a_{k+1}) \geq \delta & \text{if } a_k \succ a_{k+1} \\ \Delta(a_k, a_{k+1}) = 0 & \text{if } a_k \sim a_{k+1} \end{array}\right\} \quad \forall k \\ \sum_{i=1}^{n} \sum_{j=1}^{\alpha_i - 1} w_{ij} = 1 \\ w_{ij} \geq 0, \sigma^+(a_k) \geq 0, \sigma^-(a_k) \geq 0 \qquad \forall i, j, \text{and } k \end{cases} \tag{8.21}$$

with δ being a small positive number.

Step 4:

Test the existence of multiple or near optimal solutions of the LP (8.21) (stability analysis); in case of non uniqueness, find the mean additive value function of those (near) optimal solutions which maximize the objective functions:

$$u_i(g_i^*) = \sum_{j=1}^{\alpha_i - 1} w_{ij} \ \forall i = 1, 2, \cdots, n \tag{8.22}$$

on the polyhedron of the constraints of the LP (8.21) bounded by the new constraint:

$$\sum_{k=1}^{m} [\sigma^+(\alpha_k) + \sigma^-(\alpha_k)] \leq z^* + \varepsilon, \tag{8.23}$$

where z^* is the optimal value of the LP in Step 3 and ε is a very small positive number.

A comparison analysis between UTA and UTASTAR algorithms is presented by Siskos and Yannacopoulos (1985) [98] through a variety of experimental data. UTASTAR method has provided better results concerning a number of comparison indicators, like:

1 The number of the necessary simplex iterations for arriving at the optimal solution.

2 The Kendall's τ between the initial weak order and the one produced by the estimated model.

3 The minimized criterion z (sum of errors) taken as the indicator of dispersion of the observations.

2.4 A Numerical Example

The implementation of the UTASTAR algorithm is illustrated by a practical example presented by Siskos and Yannacopoulos (1985) [98]. The problem concerns a DM who wishes to analyze the choice of transportation means during the peak hours (home-work place). Suppose that the DM is interested only in the following three criteria:

1 price (in monetary units),

2 time of journey (in minutes), and

3 comfort (possibility to have a seat).

The evaluation in terms of the previous criteria is presented in Table 8.1, where it should be noted that the following qualitative scale has been used for the comfort criterion: 0 (no chance of seating), + (little chance of seating) ++ (great chance of finding a seating place), and +++ (seat assured). Also, the last column of Table 8.1 shows the DM's ranking with respect to the five alternative means of transportation.

Table 8.1. Criteria values and ranking of the DM.

Means of transportation	Price (mu)	Time (min)	Comfort	Ranking of the DM
RER	3	10	+	1
METRO (1)	4	20	++	2
METRO (2)	2	20	0	2
BUS	6	40	0	3
TAXI	30	30	+++	4

The first step of UTASTAR, as presented in the previous section, consists of making explicit the utilities of the five alternatives. For this reason the following scales have been chosen:

$$\begin{aligned}
[g_{1*}, g_1^*] &= [30, 16, 2] \\
[g_{2*}, g_2^*] &= [40, 30, 20, 10] \\
[g_{3*}, g_3^*] &= [0, +, ++, +++] \,.
\end{aligned}$$

Using linear interpolation for the criterion according to formula (8.10), the value of each alternative may be written as:

$$\begin{aligned}
u[\text{g(RER)}] &= 0.07u_1(16) + 0.93u_1(2) + u_2(10) + u_3(+) \\
u[\text{g(METRO1)}] &= 0.14u_1(16) + 0.86u_1(2) + u_2(20) + u_3(++) \\
u[\text{g(METRO2)}] &= u_1(2) + u_2(20) + u_3(0) \\
&= u_1(2) + u_2(20) \\
u[\text{g(BUS)}] &= 0.29u_1(16) + 0.71u_1(2) + u_2(40) + u_3(0) \\
&= 0.29u_1(16) + 0.71u_1(2) \\
u[\text{g(TAXI)}] &= u_1(30) + u_2(30) + u_3(+++) \\
&= u_2(30) + u_3(+++),
\end{aligned}$$

where the following normalization conditions for the marginal value functions have been used: $u_1(30) = u_2(40) = u_3(0) = 0$.

Also, according to formula (8.19), the global value of the alternatives may be expressed in terms of the variables w_{ij}:

$$\begin{aligned}
u[\text{g(RER)}] &= w_{11} + 0.93w_{12} + w_{21} + w_{22} + w_{23} + w_{31} \\
u[\text{g(METRO1)}] &= w_{11} + 0.86w_{12} + w_{21} + w_{22} + w_{31} + w_{32} \\
u[\text{g(METRO2}] &= w_{11} + w_{12} + w_{21} + w_{22} \\
u[\text{g(BUS)}] &= w_{11} + 0.71w_{12} \\
u[\text{g(TAXI)}] &= w_{21} + w_{31} + w_{32} + w_{33}.
\end{aligned}$$

According to the second step of the UTASTAR algorithm, the following expressions are written, for each pair of consecutive actions in the ranking:

$$\begin{aligned}
\Delta(\text{RER,METRO1}) &= 0.07w_{12} + w_{23} - w_{32} \\
&\quad -\sigma^+_{RER} + \sigma^-_{RER} + \sigma^+_{METRO1} - \sigma^-_{METRO1} \\
\Delta(\text{METRO1,METRO2}) &= -0.14w_{12} + w_{31} + w_{32} \\
&\quad -\sigma^+_{METRO1} + \sigma^-_{METRO1} + \sigma^+_{METRO2} - \\
&\quad \sigma^-_{METRO2} \\
\Delta(\text{METRO2,BUS}) &= 0.29w_{12} + w_{21} + w_{22} \\
&\quad -\sigma^+_{METRO2} + \sigma^-_{METRO2} + \sigma^+_{BUS} - \sigma^-_{BUS} \\
\Delta(\text{BUS,TAXI}) &= w_{11} + 0.71w_{12} - w_{21} - w_{31} - w_{32} - w_{33} \\
&\quad -\sigma^+_{BUS} + \sigma^-_{BUS} + \sigma^+_{TAXI} - \sigma^-_{TAXI}.
\end{aligned}$$

Based on the aforementioned expression, an LP according to (8.21) is formulated, with $\delta = 0.05$. An optimal solution is: $w_{11} = 0.5$, $w_{21} = 0.05$,$w_{23} = 0.05$, $w_{33} = 0.4$ with $[min]z = z^* = 0$. This solution corresponds to the marginal value functions presented in Table 8.2 and produces a ranking which is consistent with the DM's initial weak order.

Table 8.2. Marginal value functions (initial solution).

Price	Time	Comfort
$u_1(30) = 0.000$	$u_2(40) = 0.000$	$u_3(0) = 0.000$
$u_1(16) = 0.500$	$u_2(30) = 0.050$	$u_3(+) = 0.000$
$u_1(2) = 0.500$	$u_2(20) = 0.050$	$u_3(++) = 0.000$
$u_2(10) = 0.100$	$u_3(+++) = 0.400$	

It should be emphasized that this solution is not unique. Through post-optimality analysis (Step 4), the UTASTAR algorithm searches for multiple optimal solutions, or more generally, for near optimal solutions corresponding to error values between z^* and $z^* + \epsilon$. For this reason, the error objective should be transformed to a constraint of the type (8.23).

In the presented numerical example, the initial LP has multiple optimal solutions, since $z^* = 0$. Thus, in the post-optimality analysis step, the algorithm searches for more characteristic solutions, which maximize the expressions (8.22), i.e. the weights of each criterion. Furthermore, in this particular case we have:

$$z^* = 0 \Longleftrightarrow \sigma^+(a_k) = \sigma^-(a_k) = 0 \; \forall \; k$$

so the error variables may be excluded from the LPs of the post-optimality analysis. Table 8.3 presents the formulation of the LP that has to be solved during this step.

Table 8.3. Linear programming formulation (post-optimality analysis).

w_{11}	w_{12}	w_{21}	w_{22}	w_{23}	w_{31}	w_{32}	w_{33}	RHS
0	0.07	0	0	1	0	-1	0	≥ 0.05
0	-0.14	0	0	0	1	1	0	$= 0$
0	0.29	1	1	0	0	0	0	≥ 0.05
1	0.71	-1	0	0	-1	-1	-1	≥ 0.05
1	1	1	1	1	1	1	1	$= 1$
1	1	0	0	0	0	0	0	$[max]u_1(g_1^*)$
0	0	1	1	1	0	0	0	$[max]u_2(g_2^*)$
0	0	0	0	0	1	1	1	$[max]u_3(g_3^*)$

The solutions obtained during post-optimality analysis are presented in Table 8.4. The average of these three solutions is also calculated in the last row of Table 8.4. This centroid is taken as a unique utility function, provided that it is considered as a more representative solution of this particular problem.

Table 8.4. Post-optimality analysis and final solution.

	w_{11}	w_{12}	w_{21}	w_{22}	w_{23}	w_{31}	w_{32}	w_{33}
$[max]u_1(g_1^*)$	0.7625	0.175	0	0	0.0375	0.025	0	0
$[max]u_2(g_2^*)$	0.05	0	0	0.05	0.9	0	0	0
$[max]u_3(g_3^*)$	0.3562	0.175	0	0	0.0375	0.025	0	0.4063
Average	0.3896	0.1167	0	0.0167	0.3250	0.0167	0	0.1354

This final solution corresponds to the marginal value functions presented in Table 8.5. Also, the utilities for each alternative are calculated as follows:

$$\begin{aligned} u[\mathbf{g}(\text{RER})] &= 0.856 \\ u[\mathbf{g}(\text{METRO1})] &= 0.523 \\ u[\mathbf{g}(\text{METRO2})] &= 0.523 \\ u[\mathbf{g}(\text{BUS})] &= 0.473 \\ u[\mathbf{g}(\text{TAXI})] &= 0.152, \end{aligned}$$

where it is obvious that these values are consistent with the DM's weak order.

These marginal utilities may be normalized by dividing every value $u_i(g_i^j)$ by $u_i(g_i^*)$. In this case the additive utility can be written as:

$$u_{(\mathbf{g})} = 0.506u_1(g_1) + 0.342u_2(g_2) + 0.152u_3(g_3),$$

where the normalized marginal value functions are presented in Figure 8.6.

Table 8.5. Marginal value functions (final solution).

Price	Time	Comfort
$u_1(30) = 0.000$	$u_2(40) = 0.000$	$u_3(0) = 0.000$
$u_1(16) = 0.390$	$u_2(30) = 0.000$	$u_3(+) = 0.017$
$u_1(2) = 0.506$	$u_2(20) = 0.017$	$u_3(++) = 0.017$
$u_2(10) = 0.342$	$u_3(+++) = 0.152$	

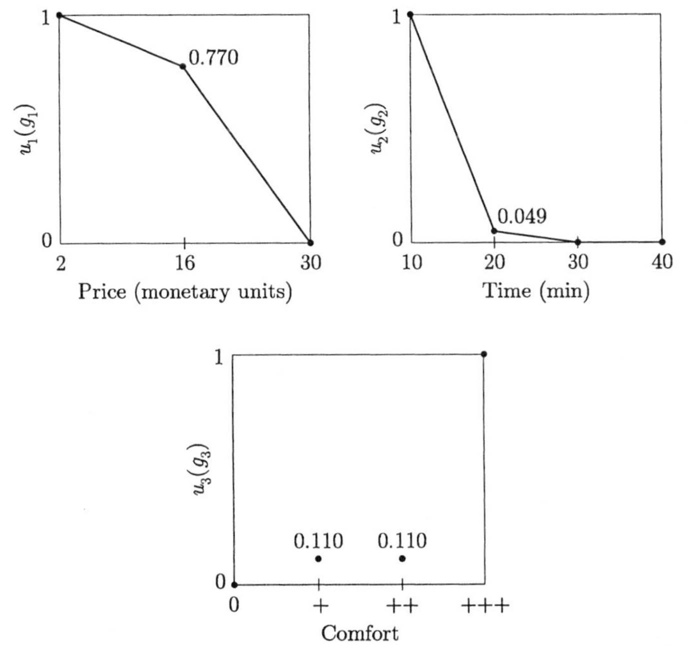

Figure 8.6. Normalized marginal value functions.

3. Variants of the UTA Method

3.1 Alternative Optimality Criteria

Several variants of the UTA method have been developed, incorporating different forms of global preference or different forms of optimality criteria used in the linear programming formulation.

An extension of the UTA methods, where $u[\mathrm{g}(a)]$ is inferred from pairwise comparisons is proposed by Jacquet-Lagrèze and Siskos (1982) [43]. This subjective preference obtained by pairwise judgments is most often not transitive, and thus, the modified model may be written as in the following LP:

$$\begin{cases} [min]F = \sum_{(a,b):a \succ b} \lambda_{ab} z_{ab} + \sum_{(a,b):a \sim b} \lambda_{ab} z_{ba} \\ \text{subject to} \\ \sum_{i=1}^{n} \{u_i[g_i(a)] - u_i[g_i(b)]\} + z_{ab} \geq 0 & \text{if } a \succ b \\ \sum_{i=1}^{n} \{u_i[g_i(a)] - u_i[g_i(b)]\} + z_{ab} - z_{ba} = 0 & \text{if } a \sim b \; (\Rightarrow b \sim a) \\ u_i(g_i^{j+1}) - u_i(g_i^j) \geq s_i & \forall i, j \\ \sum_{i=1}^{n} u_i(g_i^*) = 1 \\ u_i(g_{i*}) = 0, u_i(g_i^j) \geq 0, & \forall i, j \\ z_{ab} \geq 0 & \forall (a,b) \in R, \end{cases} \quad (8.24)$$

λ_{ab} being a non-negative weight reflecting a degree of confidence in the judgment between a and b.

An alternative optimality criterion would be to minimize the number of violated pairs of an order R provided by the DM in ranking R' given by the model, which is equivalent to maximize Kendall's τ between the two rankings.

This extension is given by the mixed integer LP (8.25), where $\gamma_{ab} = 0$ if $u[\mathbf{g}(a)] - u[\mathbf{g}(b)] \geq \delta$ for a pair $(a, b) \in R$ and the judgment is respected, otherwise $\gamma_{ab} = 1$ and the judgment is violated. Thus, the objective function in this LP represents the number of violated pairs in the overall preference aggregated by $u(\mathbf{g})$.

$$\begin{cases} [min]F = \sum_{(a,b) \in R} \gamma_{ab} \Leftrightarrow [max]\tau(R, R') \\ \text{subject to} \\ \sum_{i=1}^{n} \{u_i[g_i(a)] - u_i[g_i(b)]\} + M \cdot \gamma_{ab} \geq \delta & \forall (a,b) \in R \\ u_i(g_i^{j+1}) - u_i(g_i^j) \geq s_i & \forall i, j \\ \sum_{i=1}^{n} u_i(g_i^*) = 1 \\ u_i(g_{i*}) = 0, u_i(g_i^j) \geq 0 & \forall i, j \\ \gamma_{ab} = 0 \text{ or } 1 & \forall (a,b) \in R, \end{cases} \quad (8.25)$$

where M is a large number. Beuthe and Scannella (2001) [11] propose to handle separately the preference and indifference judgments, and modify the previous

LP using the constraints:

$$\begin{cases} \sum_{i=1}^{n}\{u_i[g_i(a)] - u_i[g_i(b)]\} + M \cdot \gamma_{ab} \geq \delta & \text{if } a \succ b \\ \left. \begin{array}{l} \sum_{i=1}^{n}\{u_i[g_i(a)] - u_i[g_i(b)]\} + M \cdot \gamma_{ab} \geq 0 \\ \sum_{i=1}^{n}\{u_i[g_i(a)] - u_i[g_i(b)]\} + M \cdot \gamma_{ba} \geq 0 \end{array} \right\} & \text{if } a \sim b. \end{cases} \tag{8.26}$$

The assumption of monotonicity of preferences, in the context of separable value functions, means that the marginal values are monotonic functions of the criteria. This assumption, although widely used, is sometimes not applicable to real-world situations. One way to deal with non-monotonic preferences is to divide the range of the criteria into intervals, so that the preferences are monotonic in each interval, and then treat each interval separately [48]. In the same spirit, Despotis and Zopounidis (1993) [18] present a variation of the UTASTAR method for the assessment of non-monotonic marginal value functions. In this model, the range if each criterion is divided into two intervals (see also Figure 8.7):

$$\begin{cases} G_i^1 = \{g_{i*} = g_i^1, g_i^2, \cdots, g_i^{p_i} = d_i\} \\ G_i^2 = \{d_i = g_i^{p_i}, g_i^{p_i+1}, \cdots, g_i^{p_i+q_i} = g_i^*\}, \end{cases} \tag{8.27}$$

where d_i is the most desirable value of g_i, and the parameters p_i and q_i are determined according to the dispersion of the input data; of course it holds that $p_i + q_i = \alpha_i$. In this approach, the main modification concerns the assessment of the decision variables w_{ij} of the LP (8.21). Hence, formula (8.19) becomes:

$$u_i(g_i^j) = \begin{cases} \sum_{t=1}^{j-1} w_{it} & \text{if } 1 < j \leq p_i \\ \sum_{t=1}^{p_i-1} w_{it} - \sum_{t=p_i}^{j-1} w_{it} & \text{if } p_i < j \leq \alpha_i \end{cases} \tag{8.28}$$

without considering the conditions $u_i(g_i^1) = 0$.

Another extension of the UTA methods refers to the intensity of the DM's preferences, similar to the context proposed by Srinivasan and Shocker (1973) [104]. In this case, a series of constraints may be added during the LP formulation. For example, if the preference of alternative a over alternative b is stronger than the preference of b over c, then the following condition may be written:

$$[u'[\mathbf{g}(a)] - u'[\mathbf{g}(b)]] - [u'[\mathbf{g}(b)] - u'[\mathbf{g}(c)]] \geq \phi, \tag{8.29}$$

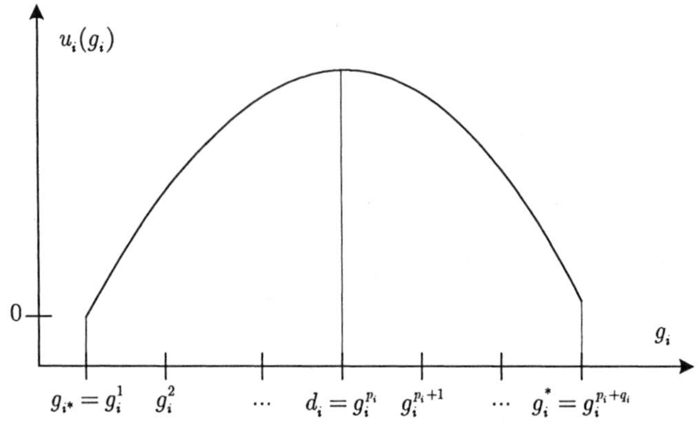

Figure 8.7. A non-monotonic partial utility function [18].

where $\phi > 0$ is a measure of preference intensity and $u'(\mathbf{g})$ is given by formula (8.8). Thus, using formula (8.11), the following constraint should be added in LP (8.14):

$$\Delta(a,b) - \Delta(b,c) \geq \phi. \tag{8.30}$$

In general, if the DM wishes to expand these preferences to the whole set of alternatives, a minimum number of $m-2$ constraints of type (8.33) is required.

Despotis and Zopounidis (1993) [18] consider the case where the DM ranks the alternatives using an explicit overall index I. Thus, formula (8.12) may be replaced by the following condition:

$$\Delta(a_k, a_{k+1}) = I_k - I_{k+1} \; \forall k = 1, 2, \cdots, m-1. \tag{8.31}$$

Besides the succession of the alternatives in the preference ranking, these constraints state that the difference of global value of any successive alternatives in the ranking should be consistent with the difference of their evaluation on the ratio scale.

In the same context, Oral and Kettani (1989) [71] (1989) propose the optimization of lexicographic criteria without discretisation of criteria scales G_i, where a ratio scale is used in order to express intensity of preferences.

Other variants of the UTA method concerning different forms of global preference are mainly focused on:

- additional properties of the assessed value functions, like concavity [18];
- construction of fuzzy outranking relations based on multiple value functions provided by UTA's post-optimality analysis [82].

The dimensions of the aforementioned UTA models affect the computational complexity of the formulated LPs. In most cases, as noted by Jacquet-Lagrèze and Siskos (1982) [43], it is preferable to solve the dual LP due to the structure of these LPs. Table 8.6 summarizes the size of all LPs presented in the previous sections, where $|P|$ and $|I|$ denote the number of preference and indifference relations respectively, considering all possible pairwise comparisons in R. Also, it should be noted that LP (8.25) has $m(m-1)/2$ binary variables.

Table 8.6. LP size of UTA models.

LP Model	Constraints	Variables
LP (8.14)	$m+\sum_{i=1}^{n}(\alpha_i-1)$	$m+\sum_{i=1}^{n}(\alpha_i-1)$
LP (8.21)	m	$2m+\sum_{i=1}^{n}(\alpha_i-1)$
LP (8.24)	$1+[m(m-1)/2]+\sum_{i=1}^{n}(\alpha_i-1)$	$\|P\|+\|I\|+\sum_{i=1}^{n}(\alpha_i-1)$
LP (8.25)	$1+[m(m-1)/2]+\sum_{i=1}^{n}(\alpha_i-1)$	$[m(m-1)/2]+\sum_{i=1}^{n}(\alpha_i-1)$

3.2 Meta-UTA Techniques

Other techniques, named meta-UTA, aimed at the improvement of the value function with respect to near optimality analysis or to its exploitation for decision support.

Despotis and Yannacopoulos (1990) [17] propose to minimize the dispersion of errors (Tchebycheff criterion) within the UTASTAR's Step 4 (see Section 2.3). In case of a strictly positive error z^*, the aim is to investigate the existence of near optimal solutions of the LP (8.21) which give rankings R' such that $\tau(R', R) > \tau(R^*, R)$, with R^* being the ranking corresponding to the optimal value functions. The experience with the model (cf. [16]) confirms that apart from the total error z^*, it is also the dispersion of the individual errors that is crucial for $\tau(R^*, R)$. Therefore, in the proposed post-optimality analysis, the difference between the maximum (σ_{max}) and the minimum error is minimized. As far as the individual errors are non-negative, this requirement can be satisfied by minimizing the maximum individual error (the L_∞ norm) according to the following LP:

$$
\left\{
\begin{array}{l}
[min]\sigma_{max} \\
\text{subject to} \\
\text{all the constraints of LP (8.21)} \\
\sum_{k=1}^{m}[\sigma^{+}(a_k)+\sigma^{-}(a_k)] \leq z^{*}+\varepsilon \\
\left.\begin{array}{l}
\sigma_{max}-\sigma^{+}(a_k) \geq 0 \\
\sigma_{max}-\sigma^{-}(a_k) \geq 0
\end{array}\right\} \qquad \forall k \\
\sigma_{max} \geq 0.
\end{array}
\right. \tag{8.32}
$$

With the incorporation of the model (8.32) in UTASTAR, the value function assessment process becomes a lexicographic optimization process. That is, the final solution is obtained by minimizing successively the L_1 and the L_∞ norms.

Another approach concerning meta-UTA techniques refers to the UTAMP models. Beuthe and Scannella (1996, 2001) [9, 11] note that the values given to parameters s and δ in the UTA and UTASTAR methods, respectively, influence the results as well as the predictive quality of the models. Hence, in the framework of the research by Srinivasan and Shocker (1973) [104], they look for optimal values of s and/or δ in the case of positive errors $(z^* > 0)$, as well as when UTA gives a sum of error equal to zero $(z^* = 0)$.

In the post-optimality analysis step of UTASTAR (see Section 2.3), UTAMP1 model maximizes δ, which is the minimum difference between the global value of two consecutive reference actions. The name of the model denotes that, on the basis of UTA, maximizing δ leads to better identification for the relations of preference between actions.

Beuthe and Scannella (1996) [9] have also proposed to maximize the sum $(\delta + s)$ in order to stress not only the differences of utilities between actions, but also the differences between values at successive bounds. This more general approach was named UTAMP2. Note that s corresponds to the minimum of marginal value step w_{ij} in the UTASTAR algorithm. Although the simple addition of these parameters is legitimate since both of them are defined in the same value units, Beuthe and Scannella (2001) [11] note that a weighted sum formula may also be considered.

3.3 Stochastic UTA Method

Within the framework of multicriteria decision-aid under uncertainty, Siskos (1983) [83] developed a specific version of UTA (Stochastic UTA), in which the aggregation model to infer from a reference ranking is an additive utility function of the form:

$$u(\mathbf{d}^a) = \sum_{i=1}^{n} \sum_{j=1}^{\alpha_i} d_i^a(g_i^j) u_i(g_i^j) \tag{8.33}$$

subject to normalization constraints (8.7), where d_i^a is the distributional evaluation of action a on the i-th criterion, $d_i^a(g_i^j)$ is the probability that the performance of action a on the i-th criterion is g_i^j , $u_i(g_i^j)$ is the marginal value of the performance g_i^j, $\mathbf{d}^a$ is the vector of distributional evaluations of action a, and $u(\mathbf{d}^a)$and is the global utility of action a (see also Figure 8.8).

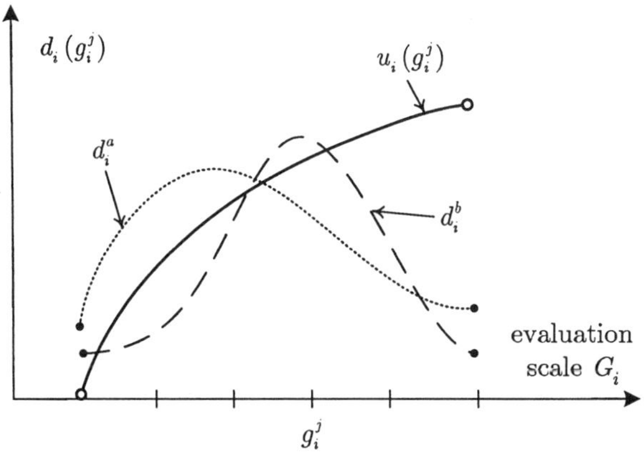

Figure 8.8. Distributional evaluation and marginal value function.

This global utility is of the von Neumann-Morgenstern form (cf. [47]), in the case of discrete g_i, where:

$$\sum_{j=1}^{\alpha_i} d_i^a(g_i^j) = 1. \tag{8.34}$$

Of course, the additive utility function (8.33) has the same properties as the value function:

$$\begin{cases} u(\mathbf{d}^a) > u(\mathbf{d}^b) \Leftrightarrow a \succ b & \text{(preference)}, \\ u(\mathbf{d}^a) = u(\mathbf{d}^b) \Leftrightarrow a \sim b & \text{(indifference)}. \end{cases} \tag{8.35}$$

Similarly to the cases of UTA and UTASTAR described in sections 2.2-2.3, the stochastic UTA method disaggregates a ranking of reference actions [87]. The algorithmic procedure could be expressed in the following way:

Step 1:
Express the global expected utilities of reference actions $u(d^{\alpha_k}), k = 1, 2, \cdots,$ m, in terms of variables:

$$w_{ij} = u_i(g_i^{j+1}) - u_i(g_i^j) \geq 0. \tag{8.36}$$

Step 2:
Introduce two error functions σ^+ and σ^- by writing the following expressions for each pair of consecutive actions in the ranking:

$$\begin{aligned} \Delta(a_k, a_{k+1}) &= u(d^{a_k}) - \sigma^+(a_k) + \sigma^-(a_k) \\ &\quad -u(d^{a_{k+1}}) + \sigma^+(a_{k+1}) - \sigma^-(a_{k+1}). \end{aligned} \tag{8.37}$$

Step 3:
Solve the LP (8.21) by using formulae (8.36) and (8.37).
Step 4:
Test the existence of multiple or near optimal solutions.

Of course, the ideas employed in all variants of the UTA method are also applicable in the same way in the case of the stochastic UTA.

3.4 UTA-type Sorting Methods

The extension of the UTA method in the case of a discriminant analysis model was firstly discussed by Jacquet-Lagrèze and Siskos (1982) [43]. The aim is to infer u from assignment examples in the context of problem statement β (cf. [76]). In the presence of two classes, if the model is without errors, the following inequalities must hold:

$$\begin{cases} a \in A_1 \Leftrightarrow u[\mathrm{g}(a)] \geq u_0 \\ a \in A_2 \Leftrightarrow u[\mathrm{g}(a)] < u_0 \end{cases} \tag{8.38}$$

with u_0 being the level of acceptance/rejection, which must be found in order to distinguish the set of accepted actions called A_1 and the set of rejected actions called A_2.

Introducing the error variables $\sigma(a)$, $a \in A_R$, the objective is to minimize the sum of deviations from the threshold u_0 for the ill classified actions (see

Figure 8.9). Hence, $u(\mathbf{g})$ can be estimated by means of the LP:

$$\left\{\begin{array}{ll} [\min]F = \sum_{a\in A_R} \sigma(a) & \\ \text{subject to} & \\ \sum_{i=1}^{n} u_i[g_i(a)] - u_0 + \sigma(a) \geq 0 & \forall\, a \in A_1 \\ \sum_{i=1}^{n} u_i[g_i(a)] - u_0 - \sigma(a) \leq 0 & \forall\, a \in A_2 \\ u_i(g_i^{j+1}) - u_i(g_i^j) \geq s_i & \forall i \text{ and } j \\ \sum_{i=1}^{n} u_i(g_i^*) = 1 & \\ u_i(g_{i*}) = 0, u_0 \geq 0, u_i(g_i^j) \geq 0, \sigma(a) \geq 0 & \forall\, a \in A_R, \forall\, i \text{ and } j. \end{array}\right. \tag{8.39}$$

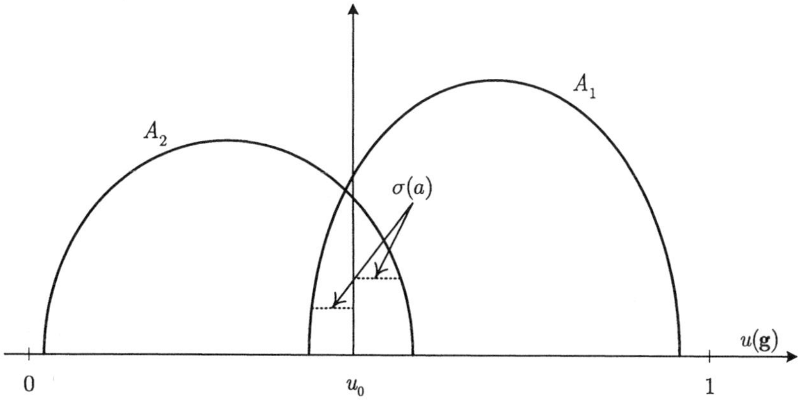

Figure 8.9. Distribution of the actions A_1 and A_2 on $u(\mathbf{g})$ [43].

In the general case, the DM's evaluation is expressed in terms of a classification of the reference alternatives into homogenous ordinal groups $A_1 \succ A_2 \succ \cdots \succ A_q$ (i.e. group A_1 includes the most preferred alternatives, whereas group A_q includes the least preferred ones). Within this context, the assessed additive value model will be consistent with the DM's global judgment, if the following conditions are satisfied:

$$\left\{\begin{array}{ll} u[\mathbf{g}(a)] \geq u_1 & \forall\, a \in A_1 \\ u_l \leq u[\mathbf{g}(a)] < u_{l-1} & \forall\, a \in A_l (l = 2, 3, \cdots, q-1) \\ u[\mathbf{g}(a)] < u_{q-1} & \forall\, a \in A_q, \end{array}\right. \tag{8.40}$$

where $u_1 > u_2 > \cdots > u_{q-1}$ are thresholds defined in the global value scale $[0, 1]$ to discriminate the groups, and u_l is the lower bound of group A_l.

This approach is named UTADIS method (UTilits Additives DIScriminantes) and is presented by Devaud et al. (1980) [19], Jacquet-Lagrèze (1995) [39], Zopounidis and Doumpos (1997, 2001) [113, 119], Doumpos and Zopounidis (2002) [23]. Similarly to the UTASTAR method, two error variables are employed in the UTADIS method to measure the differences between the model's results and the predefined classification of the reference alternatives. The additive value model is developed to minimize these errors using a linear programming formulation of type (8.39). In this case, the two types of errors are defined as follows:

1 $\sigma_k^+ = \max\{0, u_l - u[\mathbf{g}(a_k)]\}$ $\forall a_k \in A_l$ $(l = 1, 2, \cdots, q-1)$ represents the error associated with the violation of the lower bound u_l of a group A_l by an alternative $a_k \in A_l$,

2 $\sigma_k^- = \max\{0, u[\mathbf{g}(a_k) - u_{l-1}]\}$ $\forall a_k \in A_l(l = 2, 3, \cdots, q)$ represents the error associated with the violation of the upper bound u_{l-1} of a group A_l by an alternative $a_k \in A_l$.

Recently, several new variants of the original UTADIS method have been proposed (UTADIS I, II, III) to consider different optimality criteria during the development of the additive value classification model ([113, 119, 23]. The UTADIS I method considers both the minimization of the classification errors, as well as the maximization of the distances of the correctly classified alternatives from the value thresholds. The objective in the UTADIS II method is to minimize the number of misclassified alternatives, whereas UTADIS III combines the minimization of the misclassified alternatives with the maximization of the distances of the correctly classified alternatives from the value thresholds.

In the same context, Zopounidis and Doumpos (2000) [116] proposed the MHDIS method (Multi-group Hierarchical DIScrimination) extending the preference disaggregation analysis framework of the UTADIS method in complex sorting/classification problems involving multiple-groups. MHDIS addresses sorting problems through a hierarchical (sequential) procedure starting by discriminating group A_1 from all the other groups $\{A_2, A_3, \cdots, A_q\}$, and then proceeding to the discrimination between the alternatives belonging to the other groups. At each stage of this sequential/hierarchical process, two additive value functions are developed for the classification of the alternatives. Assuming that the classification of the alternatives should be made into q ordered classes , $A_1 \succ A_2 \succ \cdots \succ A_q$, $2(q-1)$ additive value functions are developed. These value functions have the following additive form:

$$\begin{cases} u_l(\mathbf{g}) = \sum_{i=1}^{n} u_{li}(g_i) \\ u_{\sim l}(\mathbf{g}) = \sum_{i=1}^{n} u_{\sim li}(g_i) \end{cases} \tag{8.41}$$

where u_l measures the value for the DM of a decision to assign an alternative into group A_l, whereas the $u_{\sim l}$ corresponds to the classification into the set of groups $A_{\sim l} = \{A_{l+1}, A_{l+2}, \cdots, A_q\}$ and both functions are normalized in the interval $[0, 1]$.

The rules used to perform the classification of the alternatives have the following form:

$$\begin{cases} \text{if } u_1(a_k) > u_{\sim 1}(a_k) \text{ then } a_k \in A_1 \\ \quad \text{else if } u_2(a_k) > u_{\sim 2}(a_k) \text{ then } a_k \in A_2 \\ \quad \cdots\cdots\cdots\cdots\cdots\cdots\cdots\cdots\cdots\cdots \\ \qquad \text{else if } u_{q-1}(a_k) > u_{\sim (q-1)}(a_k) \text{ then } a_k \in A_{q-1} \\ \qquad\quad \text{else } a_k \in A_q. \end{cases} \tag{8.42}$$

The development of all value functions in the MHDIS method is performed through the solution of three mathematical programming problems at each stage l of the discrimination process $l = 1, 2, \cdots, q-1$. Initially, an LP is solved to minimize the magnitude of the classification errors (in distance terms similarly to the UTADIS approach). Then, a mixed-integer LP is solved to minimize the total number of misclassifications among the misclassifications that occur after the solution of the initial LP, while retaining the correct classifications. Finally, a second LP is solved to maximize the clarity of the classification obtained from the solutions of the previous LPs.

3.5 Other Variants and Extensions

In all previous approaches, the value function was built in a one-step process by formulating an LP that requires only the DM's global preferences. In some cases, however, it would be more appropriate to build such a function from a two-step questioning process, by dissociating the construction of the marginal value functions and the assessment of their respective scaling constants.

In the first step, the various marginal value functions are built outside the UTA algorithm. These functions may be facilitated, for instance, by proposing specific parametrical marginal value functions to the DM and asking him/her to choose the one that matches his/her preferences on that specific criterion. Those functions should be normalized according to (8.4) conditions. Generally, the approaches applied in this construction step are:

a) techniques based on MAUT theory and described by Keeney and Raiffa (1976) [48] and Klein et al. (1985) [51],

b) the MACBETH method [3, 4, 2] and Chapter 10 in this book,

c) the Quasi-UTA method by Beuthe et al. (2000) [8], that uses "recursive exponential" marginal value functions, and

d) the MIIDAS system (see Section 4) that combines artificial intelligence and visual procedures in order to extract the DM's preferences [95].

In the second step, after the assessment of these value functions, the DM is asked to give a global ranking of alternatives in a similar way as in the basic UTA method. From this information, the problem may be formulated via an LP, in order to assess only the weighting factors p_i of the criteria (scaling constants of criteria). Through this approach, initially named UTA II model [81], formula (8.11) becomes:

$$\begin{aligned}\Delta(a_k, a_{k+1}) &= \sum_{i=1}^{n} p_i\{u_i[g_i(a_k)] - u_i[g_i(a_{k+1})]\} \\ &\quad -\sigma^+(a_k) + \sigma^-(a_k) + \sigma^+(a_{k+1}) - \sigma^-(a_{k+1})\end{aligned} \tag{8.43}$$

and the LP (8.14) is modified as follows:

$$\begin{cases} [min]F = \sum_{k=1}^{m} [\sigma^+(a_k) + \sigma^-(a_k)] \\ \text{subject to} \\ \quad \left.\begin{array}{ll} \Delta(a_k, a_{k+1}) \geq \delta & \text{if } a_k \succ a_{k+1} \\ \Delta(a_k, a_{k+1}) = 0 & \text{if } a_k \sim a_{k+1} \end{array}\right\} \quad \forall k \\ \sum_{i=1}^{n} p_i = 1 \\ p_i \geq 0, \sigma^+(a_k) \geq 0, \sigma^-(a_k) \geq 0 \qquad \forall\, i, k. \end{cases} \tag{8.44}$$

The main principles of the UTA methods are also applicable in the specific field of multiobjective optimization, mainly in the field of linear programming with multiple objective functions. For instance, in the classical methods of Geoffrion et al. (1972) [27] and Zionts and Wallenius (1976) [111], the weights of the linear combinations of the objectives are inferred locally from trade-offs or pairwise judgments given by the DM at each iteration of the methods. Thus, these methods exploit in a direct way the DM's value functions and seek the best compromise solution through successive maximization of these assessed value functions.

Stewart (1987) [106] proposed a procedure of pruning the decision alternatives using the UTA method. In this approach a sequence of alternatives is presented to the DM, who places each new presented alternative in rank order relative to the earlier alternatives evaluated. This ranking of elements in a subset of the decision space is used to eliminate other alternatives from further consideration. In the same context, Jacquet-Lagrèze et al. (1987) [40] developed a disaggregation method, similar to UTA, to assess a whole value function of multiple objectives for linear programming systems. This methodology enables to find compromise solutions and is mainly based on the following steps:

1 Generation of a limited subset of feasible efficient solutions as representative as possible of the efficient set.

2 Assessment of an additive value function using PREFCALC system (see Section 4).

3 Optimization of the additive value function on the original set of feasible alternatives.

Finally, Siskos and Despotis (1989) [80], in the context of UTA-based approaches in multiobjective optimization problems, proposed the ADELAIS method. This approach refers to an interactive method that uses UTA iteratively, in order to optimize an additive value function within the feasible region defined on the basis of the satisfaction levels and determined in each iteration.

3.6 Other Disaggregation Methods

The main principles of the aggregation-disaggregation approach may be combined with outranking relation methods. The most important efforts concern the problem of determining the values of several parameters when using these methods. The set of these parameters is used to construct a preference model with which the DM accepts as a working hypothesis in the decision-aid study. In several real-world applications the assumption that the DM is able to give explicitly the values of each parameter is not realistic.

In this framework, the ELECCALC system has been developed [50], which estimates indirectly the parameters of the ELECTRE II method. The process is based on the DM's responses to questions of the system regarding his/her global preferences.

Furthermore, concerning problem statement β, several approaches consist in inferring the parameters of ELECTRE TRI through holistic information on DM's judgments. These approaches aim at substituting assignment examples for direct elicitation of the model parameters. Usually, the values of these parameters are inferred through a regression-type analysis on assignment examples.

Mousseau and Slowinski (1998) [68] propose an interactive aggregation-disaggregation approach that infers ELECTRE TRI parameters simultaneously starting from assignment examples. In this approach, the determination of the parameters' values (except the veto thresholds) that best restore the assignment examples is formulated through a non-linear optimization program.

Several efforts have tried to overcome the limitations of the aforementioned approach (computational difficulty, estimation of the veto threshold):

a) Mousseau et al. (2000) [67, 69] consider the subproblem of the determination of the weights only, assuming that the thresholds and category limits have been fixed. This leads to formulate an LP (rather than non-linear in the global inference model). Through experimental analysis, they show that this approach is able to infer weights that restore in a stable way the assignment examples and it is also able to identify possible inconsistencies in these assignment examples.

b) Doumpos and Zopounidis (2000) [24] use linear programming formulations in order to estimate all the parameters of the outranking relation classification model. However, in this approach, the parameters are estimated sequentially rather than through a global inference process. Thus, the proposed methodology does not specify the optimal parameters of the outranking relation (i.e. the ones that lead to a global minimum of the classification error). The results of this approach ("reasonable" specification of the parameters) serve rather as a basis for a thorough decision-aid process.

The problem of robustness and sensitivity analysis, through the extension of the previous research efforts is discussed by Dias et al. (2002) [21]. They consider the case where the DM can not provide exact values for the parameters of the ELECTRE TRI method, due to uncertain, imprecise or inaccurately determined information, as well as from lack of consensus among them. The proposed methodology combines the following approaches:

1 The first approach infers the value of parameters from assignment examples provided by the DM, as an elicitation aid.

2 The second approach considers a set of constraints on the parameter values reflecting the imprecise information that the DM is able to provide.

In the context of UTA-based ordinal regression analysis (cf. [84]), the MUSA method has been developed in order to measure and analyze customer satisfaction [92, 32]. The method is used for the assessment of a set of marginal satisfaction functions in such a way that the global satisfaction criterion becomes as consistent as possible with customer's judgments. Thus, the main objective of the method is the aggregation of individual judgments into a collective value function.

The MUSA method assesses global and partial satisfaction functions Y^* and X_i^* respectively, given customers' ordinal judgments Y and X_i (for the *i*-th criterion). The ordinal regression analysis equation has the following form:

$$\hat{Y}^* = \sum_{i=1}^{n} b_i X_i^* - \sigma^+ + \sigma^- \tag{8.45}$$

where $\hat{Y}^*$ is the estimation of the global value function Y^*, n is the number of criteria, b_i is a positive weight of the i-th criterion, σ^+ and σ^- are the overestimation and the underestimation errors, respectively, and the value functions Y^* and X_i^* are normalized in the interval [0,100]. In the MUSA method the notation of ordinal regression analysis is adopted, where a criterion g_i is considered as a monotone variable X_i and a value function is denoted as X_i^*.

Similarly to the UTASTAR algorithm, the following transformation equations are used:

$$\begin{cases} z_m = y^{*m+1} - y^{*m} & \text{for } m = 1, 2, \cdots, \alpha - 1 \\ w_{ik} = b_i x_i^{*k+1} & \text{for } k = 1, 2, \cdots, \alpha_i - 1 \text{ and } i = 1, 2, \cdots, n \end{cases} \tag{8.46}$$

where y^{*m} is the value of the y^m satisfaction level, x_i^{*k} is the value of the x_i^* satisfaction level, and α and α_i are the number global and partial satisfaction levels.

According to the previous definitions and assumptions, the MUSA estimation model can be written in an LP formulation, as follows:

$$\begin{cases} [min]F = \sum_{j=1}^{m} \sigma_j^+ + \sigma_j^- \\ \text{subject to} \\ \sum_{i=1}^{n} \sum_{k=1}^{x_i^j - 1} w_{ik} - \sum_{m=1}^{y^j - 1} z_m - \sigma_j^+ + \sigma_j^- \quad \text{for } j = 1, 2, \cdots, M \\ \sum_{m=1}^{\alpha - 1} z_m = 100 \\ \sum_{i=1}^{n} \sum_{k=1}^{\alpha_i - 1} w_{ik} = 100 \\ z_m, w_{ik}, \sigma_j^+, \sigma_j^- \forall m, i, j, k, \end{cases} \tag{8.47}$$

where M is the size of the customer sample, and x_i^j and y^j are the *j*-th level on which variables X_i and Y are estimated (i.e. global and partial satisfaction judgments of the *j*-th customer). The MUSA method includes also a post-optimality analysis stage, similarly to Step 4 of the UTASTAR algorithm.

An analytical development of the method and the provided results is given by Grigoroudis and Siskos (2002) [32], while the presentation of the MUSA DSS can be found in [35] and [33].

The problem of building non-additive utility functions may also be considered in the context of aggregation-disaggregation approach. A characteristic case refers to positive interaction (synergy) or negative interaction among criteria (redundancy). Two or more criteria are synergic (redundant) when their joint weight is more (less) than the sum of the weights given to the criteria considered singularly.

In order to represent interaction among criteria, some specific formulations of the utility functions expressed in terms of fuzzy integrals have been proposed [70, 29, 56]. In this context, Angilella et al. (2003) [1] propose a methodology that allows the inclusion of additional information such as an interaction among criteria. The method aims at searching a utility function representing the DM's preferences, while the resulting functional form is a specific fuzzy integral (Choquet integral). As a result, the obtained weights may be interpreted as the "importance" of coalitions of criteria, exploiting the potential interaction between criteria. The method can also provide the marginal utility functions relative to each one of the considered criteria, evaluated on a common scale, as a consequence of the implemented methodology.

The general scheme of the disaggregation philosophy is also employed in other approaches, including rough sets [73, 101, 22, 110], machine learning [74], and neural networks [53, 105]. All these approaches are used to infer some form of decision model (a set of decision rules or a network) from given decision results involving assignment examples, ordinal or measurable judgments.

4. Applications and UTA-based DSS

The methods presented in the previous sections adopt the aggregation-disaggregation approach. This approach constitutes a basis for the interaction between the analyst and the DM, which includes:

- the consistency between the assessed preference model and the a priori preferences of the DM,
- the assessed values (values, weights, utilities, $\cdots$ ...), and
- the overall evaluation of potential actions (extrapolation output).

A general interaction scheme for this decision support process is given in Figure 8.10.

Several decision support systems (DSSs), based on the UTA model and its variants, have been developed on the basis of disaggregation methods. These systems include:

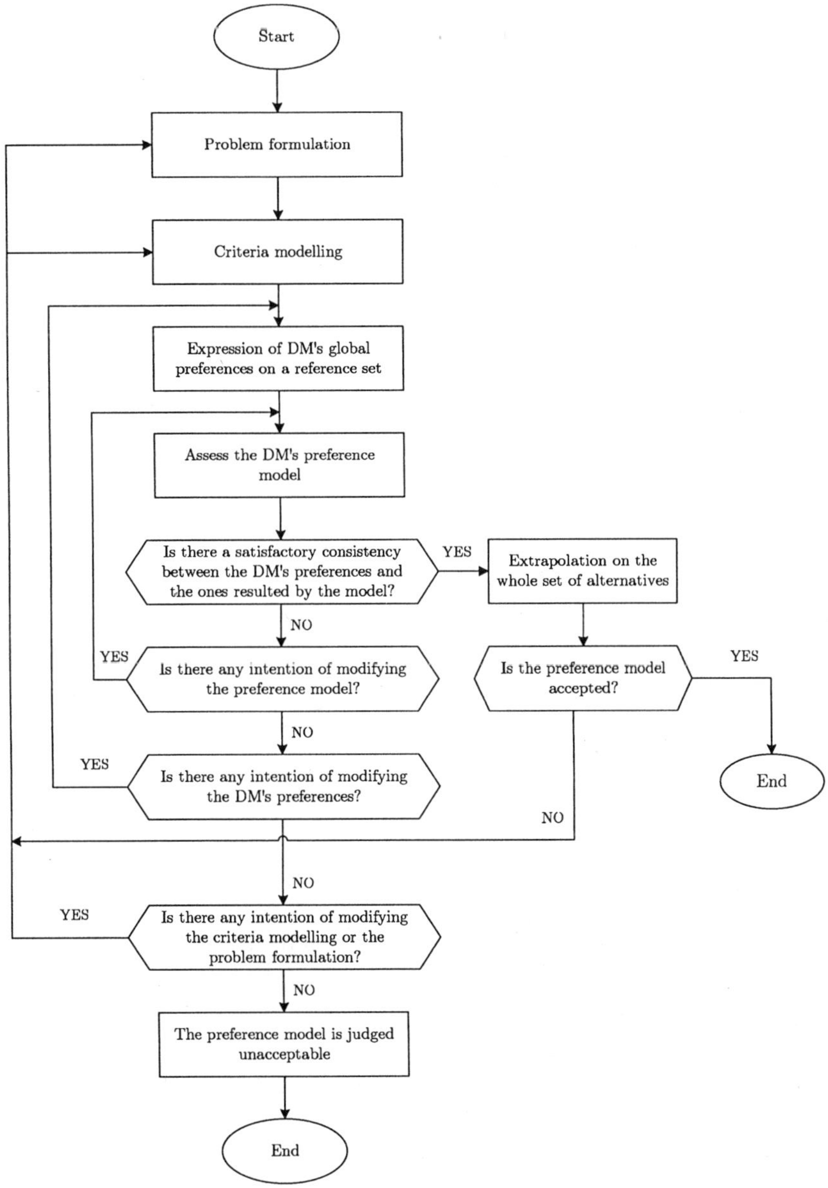

Figure 8.10. Simplified decision support process based on disaggregation approach [44].

a) The PREFCALC system [38] is a DSS for interactive assessment of preferences using holistic judgments. The interactive process includes the classical aggregation phase where the DM is asked to estimate directly the parameters of the model (i.e. weights, trade-offs, etc.), as well as the disaggregation phase where the DM is asked to express his/her holistic judgments (i.e. global preference order on a subset of the alternatives) enabling an indirect estimation of the parameters of the model.

b) MINORA (Multicriteria Interactive Ordinal Regression Analysis) is a multicriteria interactive DSS with a wide spectrum of supported decision making situations [97]. The core of the system is based on the UTASTAR method and it uses special interaction techniques in order to guide the DM to reach a consistent preference system.

c) MIIDAS (Multicriteria Interactive Intelligence Decision Aiding System) is an interactive DSS that implements the extended UTA II method [95]. In the first step of the decision-aid process, the system assess the DM's value functions, while in the next step, the system estimates the DM's preference model from his/her global preferences on a reference set of alternative actions. The system uses Artificial Intelligence and Visual techniques in order to improve the user interface and the interactive process with the DM.

d) The UTA PLUS software [52] `http://www.lamsade.dauphine.fr/english/software.html#uta+` is an implementation of the UTA method, which allows the user to modify interactively the marginal value functions within limits set from a sensitivity analysis of the formulated ordinal regression problem. During all these modifications, a friendly graphical interface helps the DM to reach an accepted preference model.

e) MUSTARD (Multicriteria Utility-based Stochastic Aid for Ranking Decisions) is an interactive DSS developed by Beuthe and Scannella (1999) [10], which incorporates several variants of the UTA method. The system provides several visual tools in order to structure the DM's preferences to a specific problem (see also [86]). The interactive process with the DM contains the following main steps: problem structuring, preference questionnaire, optimization solver-parameter computing, final results (full rankings and graphs).

UTA methods have also been used in several works for conflict resolution in multi-actor decision situations [41, 12, 62]. In the same context, the MEDIATOR system was developed [45, 78, 79], which is a negotiation support system based on Evolutionary Systems Design (ESD) and database-centered implementation. ESD visualizes negotiations as a collective process of searching for designing a mutually acceptable solution. Participants are seen as playing a dynamical difference game in which a coalition of players is formed, if it can achieve a set of agreed upon goals. In MEDIATOR, negotiations are supported by consensus seeking through exchange of information and, where consensus is incomplete, by compromise. It assists in consensus seeking by aiding the players to build a group joint problem representation of the negotiations-in effect, joint mappings from control space to goal space (and through marginal utility functions) to utility space. Individual marginal utility functions are estimated by applying the UTA method. Players can arrive to a common coalition utility function through exchange of information and negotiation until players' marginal utility functions are identical. In addition to exchanging information and negotiating to expand targets, players can consider the use of axioms to contract the feasible region. In the area of intelligent multicriteria DSSs, the MARKEX system has been proposed in [93, 63, 65]. The system includes the UTASTAR algorithm and is used for the new product development process. It acts as a consultant for marketers, providing visual support to enhance understanding and to overcome lack of expertise. The data bases of the system are the results of consumer surveys, as well as financial information of the enterprises involved in the decision-making process. The system's model base encompasses statistical analysis, preference analysis, and consumer choice models. Figure 8.11 presents a general methodological flowchart of the system. Also, MARKEX incorporates partial knowledge bases to support DMs in different stages of the product development process. The system incorporates threepartial expert systems, functioning independently of each other. These expert systems use the following knowledge bases for the:

- selection of data analysis method,
- selection of brand choice model, and
- evaluation of the financial status of enterprises.

Furthermore, an intelligent web-based DSS, named DIMITRA, has been developed by Matsatsinis and Siskos [64]. The system is a consumer survey-based DSS, focusing on the decision-aid process for agricultural product development. Besides the implementation of the UTASTAR method in the preference analysis module, the DIMITRA system comprises several statistical analysis tools and consumer choice models. The system provides visual support to the DM (agricultural cooperatives, agribusiness firms, etc.) for several complex tasks, such as:

- evaluation of current and potential market shares,
- determination of the appropriate communication and penetration strategies, based on consumer attitudes and beliefs,

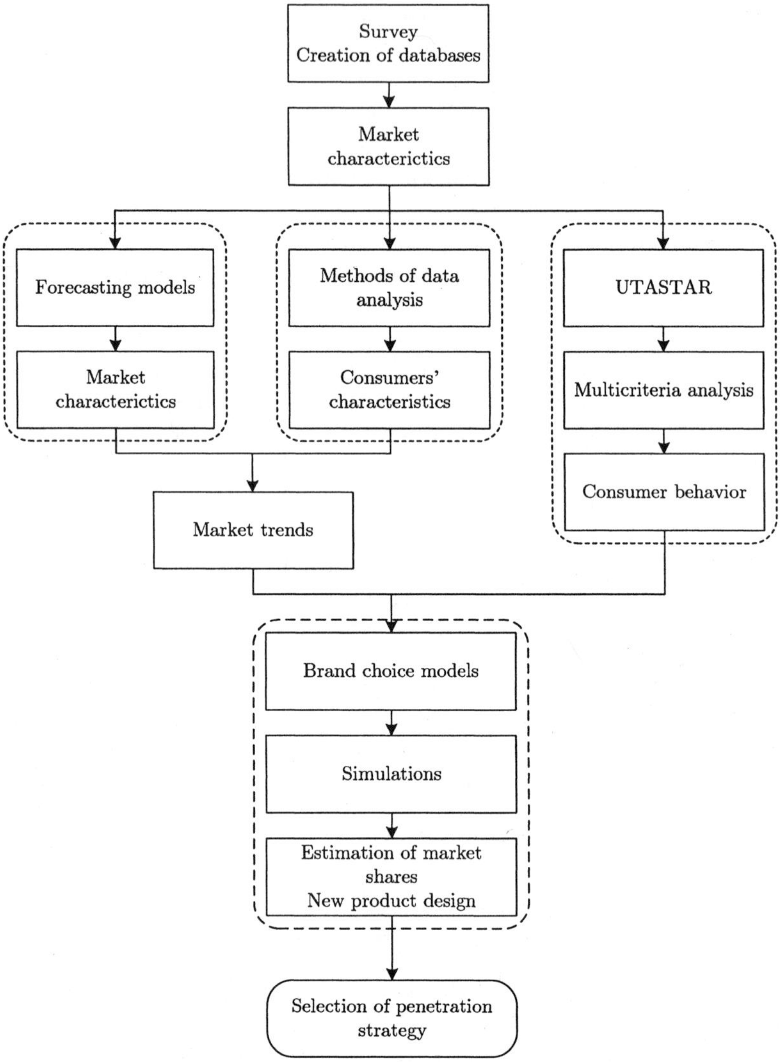

Figure 8.11. Methodological flowchart of MARKEX [63].

- adjustment of the production according to product's demand, and
- detection of the most promising markets.

In the same context, new research efforts have combined UTA-based DSSs with intelligent agents' technology. In general, the proposed methodologies engage the UTA models in a multi-agent architecture in order to assess the DM's preference system. These research efforts include mainly the following:

a) An intelligent agent-based DSS, focusing on the determination of product penetration strategies has been developed by Matsatsinis et al. (1999, 2000, 2004) [59, 60, 61]. The system implements an original consumer-based methodology, in which intelligent agents operate in a functional and a structural level, simultaneously. Task, information and interface agents are included in the functional level in order to coordinate, collect necessary information and communicate with the DM. Likewise, the structural level includes elementary agents based on a generic reusable architecture and complex agents which aim to the development of a dynamical agent organization in a recursive way.

b) A multi-agent architecture is proposed by Manouselis and Matsatsinis (2001) [55] for modeling electronic consumer's behavior. The implementation of the system refers to electronic marketplaces and incorporates a step-by-step methodology for intelligent systems analysis and design, used in the particular decision-aid process. The system develops consumer behavioral models for the purchasing and negotiation process adopting additional operational research tools and techniques. The presented application refers to the case of Internet radio.

c) The AgentAllocator system [58] implements the UTA II method in the task allocation problem. These problems are very common to any multi-agent system in the context of Artificial Intelligence. The system is an intelligent agent DSS, which allows the DM to model his/her preferences in order to reach and employ the optimal allocation plan.

The need to combine data and knowledge in order to solve complex and ill-structured decision problems is a major concern in the modern marketing-management science. Matsatsinis (2002) [57] has proposed a DSS thatimplements the UTASTAR algorithm along with rule-induction data mining techniques. The main aim of the system is to derive and apply a set of rules that relate the global and the marginal value functions. A comparison between the original and the rule-based global values is used in the validity and stability analysis of the proposed methodology.

Furthermore, in the area of financial management, a variety of UTA-based DSSs has been developed, including mainly the following systems:

a) The FINEVA system [122] is a multicriteria knowledge-based DSS developed for the assessment of corporate performance and viability. The system implements multivariate statistical techniques (e.g. principal components analysis), expert systems technology, and the UTASTAR method to provide integrated support in evaluating the corporate performance.

b) The FINCLAS system [114] is a multicriteria DSS developed to study financial decision-making problems in which a classification (sorting) of the alternatives is required. The present form of the system is devoted to corporate credit risk assessment, and it can be used to develop classification models to assign a set of firms into predefined credit risk classes. The analysis performed by the system is based on the family of the UTADIS methods.

c) The INVESTOR system [117] is developed to study problems related to portfolio selection and management. The system implements the UTADIS method, as well as goal programming techniques to support portfolio managers and investors in their daily practice.

d) The PREFDIS system [118] is a multicriteria DSS developed to address classification problems. The system implements a series of preference disaggregation analysis techniques, namely the family of the UTADIS methods, in order to develop an additive utility function to be used for classification purposes.

Finally, as presented in Section 3.5, Siskos and Despotis (1989) [80] have developed the ADELAIS system, which is designed to decision-aid in multiobjective linear programming (MOLP) problems.

Over the past two decades UTA-based methods have been applied in several real-world decision-making problems from the fields of financial management, marketing, environmental management, as well as human resources management, as presented in Table 8.7. These applications have provided insight on the applicability of preference disaggregation analysis in addressing real-world decision problems and its efficiency.

5. Concluding Remarks and Future Research

The UTA methods presented in this chapter belong to the family of ordinal regression analysis models aiming to assess a value system as a model of the preferences of the DM. This assessment is implemented through an aggregation-disaggregation process. With this process the analyst is able to infer an analytical model of preferences, which is as consistent as possible with the DM' preferences. The acceptance of such a preference model is accomplished through a repetitive interaction between the model and the DM. This approach contributes towards an alternative reasoning for decision-aid (see Figure 8.2).

Future research regarding UTA methods aims to explore further the potentials of the preference disaggregation philosophy within the context of multicriteria decision-aid. Jacquet-Lagrèze and Siskos (2001) [44] propose that potential research developments may be focused on:

Table 8.7. Indicative applications of the UTA methods.

Field	Scope	Reference
Financial management	Venture capital evaluation	[99]
	Portfolio selection and management	[37, 120]
	Business failure prediction	[112, 115]
	Business financing	[100, 122, 114]
	Country risk assessment	[15, 72, 121]
Marketing	Marketing of new products	[102]
	Marketing of agricultural products	[5, 93, 7, 60, 65, 94, 64]
	Consumer behavior	[89, 90, 6, 55, 57]
	Customer satisfaction	[30, 66, 91, 88, 31, 34, 77]
	Sales strategy problems	[75, 85]
Management (general)	Project evaluation	[39, 8]
	Environmental management	[87, 36, 20]
	Job evaluation	[103, 28]

a) the inference of more sophisticated aggregation models by disaggregation, and

b) the experimental evaluation of disaggregation procedures.

Finally, it would be interesting to explore the relationship of aggregation and disaggregation procedures in terms of similarities and/or dissimilarities regarding the evaluation results obtained by both approaches [44]. This will enable the identification of the reasons and the conditions under which aggregation and disaggregation procedures will lead to different or the same results.

References

[1] S. Angilella, S. Greco, F. Lamantia, and B. Matarazzo. Assessing non-additive utility for multicriteria decision aid. *European Journal of Operational Research*, in press, 2003.

[2] C.A. Bana e Costa, F. Nunes Da Silva, and J.C. Vansnick. Conflict dissolution in the public sector: A case-study. *European Journal of Operational Research*, 130(2):388–401, 2001.

[3] C.A. Bana e Costa and J.C. Vansnick. MACBETH: An interactive path towards the construction of cardinal value functions. *International Transactions in Operations Research*, 1(4):489–500, 1994.

[4] C.A. Bana e Costa and J.C. Vansnick. Applications of the MACBETH approach in the framework of an additive aggregation model. *Journal of Multi-Criteria Decision Analysis*, 6(2):107–114, 1997.

[5] G. Baourakis, N.F. Matsatsinis, and Y. Siskos. Agricultural product design and development. In J. Janssen and C.H. Skiadas, editors, *Applied stochastic models and data analysis*, pages 1108–1128. World Scientific, Singapore, 1993.

[6] G. Baourakis, N.F. Matsatsinis, and Y. Siskos. Consumer behavioural analysis using multicriteria methods. In J. Janssen, C.H. Skiadas, and C. Zopounidis, editors, *Advances in stochastic modelling and data analysis*, pages 328–338. Kluwer Academic Publishers, Dordrecht, 1995.

[7] G. Baourakis, N.F. Matsatsinis, and Y. Siskos. Agricultural product development using multidimensional and multicriteria analyses: The case of wine. *European Journal of Operational Research*, 94(2):321–334, 1996.

[8] M. Beuthe, L. Eeckhoudt, and G. Scanella. A practical multicriteria methodology for assessing risky public investments. *Socio-Economic Planning Sciences*, 34(2):121–139, 2000.

[9] M. Beuthe and G. Scannella. Applications comparées des méthodes d'analyse multicritère UTA. *RAIRO Recherche Opérationelle*, 30(3):293–315, 1996.

[10] M. Beuthe and G. Scannella. *MUSTARD User's Guide*. GTM, Facultés Universitaires Catholiques de Mons (FUCaM), Mons, 1999.

[11] M. Beuthe and G. Scannella. Comparative analysis of UTA multicriteria methods. *European Journal of Operational Research*, 130(2):246–262, 2001.

[12] T.X. Bui. *Co-oP: A Group Decision Support System for Cooperative Multiple Criteria Group Decision Making*, volume 290 of *Lecture Notes in Computer Science*. Springer-Verlag, Berlin, 1987.

[13] A. Charnes and W. Cooper. *Management Models and Industrial Applications of Linear Programming Vol. 1*. John Wiley & Sons, New York, 1961.

[14] A. Charnes, W. Cooper, and R.O. Ferguson. Optimal estimation of executive compensation by linear programming. *Management Science*, 1(2):138–151, 1955.

[15] J.C. Cosset, Y. Siskos, and C. Zopounidis. Evaluating country risk: A decision support approach. *Global Finance Journal*, 3(1):79–95, 1992.

[16] D.K. Despotis and D. Yannacopoulos. Méthode d'estimation d'utilités additives concaves en programmation linéare multiobjectifs. *RAIRO Recherche Opérationelle*, 24:331–349, 1990.

[17] D.K. Despotis, D. Yannacopoulos, and C. Zopounidis. A review of the uta multicriteria method and some improvements. *Foundations of Computing and Decision Sciences*, 15(2):63–76, 1990.

[18] D.K. Despotis and C. Zopounidis. Building additive utilities in the presence of non-monotonic preference. In P.M. Pardalos, Y. Siskos, and C. Zopounidis, editors, *Advances in Multicriteria Analysis*, pages 101–114. Kluwer Academic Publisher, Dordrecht, 1993.

[19] J.M. Devaud, G. Groussaud, and E. Jacquet-Lagrèze. UTADIS: Une méthode de construction de fonctions d'utilité additives rendant compte de jugements globaux. European Working Group on Multicriteria Decision Aid, Bochum, 1980.

[20] D. Diakoulaki, C. Zopounidis, G. Mavrotas, and M. Doumpos. The use of a preference disaggregation method in energy analysis and policy making. *Energy – The International Journal*, 24(2):157–166, 1999.

[21] L. Dias, V. Mousseau, J. Figueira, and J. Climaco. An aggregation/disaggregation approach to obtain robust conclusions with electre tri. *European Journal of Operational Research*, 138(2):332–348, 2002.

[22] A.I. Dimitras, R. Slowinski, R. Susmaga, and C. Zopounidis. Business failure prediction using rough sets. *European Journal of Operational Research*, 114(2):263–280, 1999.

[23] M. Doumpos and C. Zopounidis. *Multicriteria Decision Aid Classification Methods*. Kluwer Academic Publishers, Dordrecht, 2002.

[24] M. Doumpos and C. Zopounidis. On the development of an outranking relation for ordinal classification problems: An experimental investigation of a new methodology. *Optimization Methods and Software*, 17(2):293–317, 2002.

[25] P. Fishburn. A note on recent developments in additive utility theories for multiple factors situations. *Operations Research*, 14:1143–1148, 1966.

[26] N. Freed and G. Glover. Simple but powerful goal programming models for discriminant problems. *European Journal of Operational Research*, 7:44–60, 1981.

[27] A.M. Geoffrion, J.S. Dyer, and A. Feinberg. An interactive approach for multi-criterion optimization, with an application to the operation of an academic department. *Management Science*, 19(4):357–368, 1972.

[28] M.C. González-Araya, L.A.D. Rangel, M.P.E. Lins, and L.F.A.M. Gomes. Building the additive utility functions for CAD-UFRJ evaluation staff criteria. *Annals of Operations Research*, 116(1–4):271–288, 2002.

[29] M. Grabisch. The application of fuzzy integrals in multicriteria decision making. *European Journal of Operational Research*, 89(3):445–456, 1996.

[30] E. Grigoroudis, J. Malandrakis, J. Politis, and Y. Siskos. Customer satisfaction measurement: An application to the Greek shipping sector. In D.K. Despotis and C. Zopounidis, editors, *Integrating Technology & Human Decisions: Global Bridges into the 21st Century. Proceedings of the 5th Decision Sciences Institute International Conference, July 4-7 1999, Athens, Greece*, volume 2, pages 1363–1365, 1999.

[31] E. Grigoroudis, Y. Politis, and Y. Siskos. Satisfaction benchmarking and customer classification: An application to the branches of a banking organization. *International Transactions in Operational Research*, 9(5):599–618, 2002.

[32] E. Grigoroudis and Y. Siskos. Preference disaggregation for measuring and analysing customer satisfaction: The MUSA method. *European Journal of Operational Research*, 143(1):148–170, 2002.

[33] E. Grigoroudis and Y. Siskos. MUSA: A decision support system for evaluating and analysing customer satisfaction. In K. Margaritis and I. Pitas, editors, *Proceedings of the 9th Panhellenic Conference in Informatics, Thessaloniki, Greece*, pages 113–127, 2003.

[34] E. Grigoroudis and Y. Siskos. A survey of customer satisfaction barometers: Results from the transportation-communications sector. *European Journal of Operational Research*, 152(2):334–353, 2003.

[35] E. Grigoroudis, Y. Siskos, and O. Saurais. TELOS: A customer satisfaction evaluation software. *Computers and Operations Research*, 27(7–8):799–817, 2000.

[36] I. Hatzinakos, D. Yannacopoulos, C. Faltsetas, and C. Ziourkas. Application of the MINORA decision support system to the evaluation of landslide favourability in Greece. *European Journal of Operational Research*, 50(1):60–75, 1991.

[37] Ch. Hurson and C. Zopounidis. *Gestion de Portefeuille et Analyse Multicritère*. Economica, Paris, 1997.

[38] E. Jacquet-Lagrèze. Interactive assessment of preferences using holistic judgments: The PREFCALC system. In C.A. Bana e Costa, editor, *Readings in Multiple Criteria Decision Aid*, pages 225–250. Springer-Verlag, Berlin, 1990.

[39] E. Jacquet-Lagrèze. An application of the UTA discriminant model for the evaluation of R&D projects. In P.M. Pardalos, Y. Siskos, and C. Zopounidis, editors, *Advances in Multicriteria Analysis*, pages 203–211. Kluwer Academic Publishers, Dordrecht, 1995.

[40] E. Jacquet-Lagrèze, R. Meziani, and R. Slowinski. MOLP with an interactive assessment of a piecewise linear utility function. *European Journal of Operational Research*, 31(3):350–357, 1987.

[41] E. Jacquet-Lagrèze and M.F. Shakun. Decision support systems for semistructured buying decisions. *European Journal of Operational Research*, 16:48–56, 1984.

[42] E. Jacquet-Lagrèze and J. Siskos. Une méthode de construction de fonctions d' utilité additives explicatives d' une prèfèrence globale. Technical Report Cahiers du LAMSADE 16, Université de Paris-Dauphine, Paris, 1978.

[43] E. Jacquet-Lagrèze and Y. Siskos. Assessing a set of additive utility functions for multicriteria decision making: The UTA method. *European Journal of Operational Research*, 10(2):151–164, 1982.

[44] E. Jacquet-Lagrèze and Y. Siskos. Preference disaggregation: 20 years of MCDA experience. *European Journal of Operational Research*, 130(2):233–245, 2001.

[45] M. Jarke, M.T. Jelassi, and M.F. Shakun. MEDIATOR: Toward a negotiation support system. *European Journal of Operational Research*, 31(3):314–334, 1987.

[46] O.J. Karst. Linear curve fitting using least deviations. *Journal of the American Statistical Association*, 53:118–132, 1958.

[47] R.L. Keeney. *Siting Energy Facilities*. Academic Press, New York, 1980.

[48] R.L. Keeney and H. Raiffa. *Decisions with Multiple Objectives: Preferences and Value Tradeoffs*. John Wiley & Sons, New York, 1976.

[49] J.E. Kelley. An application of linear programming to curve fitting. *Journal of Industrial and Applied Mathematics*, 6(1):15–22, 1958.

[50] L.N. Kiss, J.M. Martel, and R. Nadeau. ELECCALC – An interactive software for modelling the decision maker's preferences. *Decision Support Systems*, 12(4–5):311–326, 1994.

[51] G. Klein, H. Moskowitz, S. Mahesh, and A. Ravindran. Assessment of multiattribute measurable value and utility functions via mathematical programming. *Decision Sciences*, 16(3):309–324, 1985.

[52] M. Kostkowski and R. Slowinski. UTA+ application (v. 1.20) – User's manual. Technical Report Documents du LAMSADE 95, Université de Paris-Dauphine, Paris, 1996.

[53] B. Malakooti and Y.Q. Zhou. Feedforward artificial neural networks for solving discrete multiple criteria decision making problems. *Management Science*, 40(11):1542–1561, 1994.

[54] M. Manas and J. Nedoma. Finding all vertices of a convex polyhedron. *Numerical Mathematics*, 12:226–229, 1968.

[55] N. Manouselis and N.F. Matsatsinis. Introducing a multi-agent, multi-criteria methodology for modeling electronic consumer's behavior: The case of internet radio. In M. Klush and F. Zambonelli, editors, *Cooperative Information Agents V*, volume 2182 of *Lecture Notes in Artificial Intelligence*, pages 190–195. Springer-Verlag, Berlin, 2001.

[56] J.L. Marichal and M. Roubens. Determination of weights of interactive criteria from a reference set. *European Journal of Operational Research*, 124(3):641–650, 2000.

[57] N.F. Matsatsinis. New agricultural product development using data mining techniques and multicriteria methods. In A. Sideridis, editor, *Proceedings of the 1st Hellenic Association of Information and Communication Technology in Agriculture, Food and Environment (HAICTA 2002), June 6-7 2002, Athens, Greece*, 2002.

[58] N.F. Matsatsinis and P. Delias. AgentAllocator: An agent-based multi-criteria decision support system for task allocation. In V. Marik, D. McFarlane, and P. Valckenaers, editors, *Holonic and Multiagent Systems for Manufacturing*, volume 2744 of *Lecture Notes in Artificial Intelligence*, pages 225–235. Springer-Verlag, Berlin, 2003.

[59] N.F. Matsatsinis, P. Moraitis, V. Psomatakis, and N. Spanoudakis. Intelligent software agents for products penetration strategy selection. In F.G. Garijo and M. Boman, editors, *Proceedings of the 9th European Workshop on Modelling Autonomous Agents in a Multi Agent World (MAAMAW99), June 30 - July 2 1999, Valencia, Spain*, 1999.

[60] N.F. Matsatsinis, P. Moraitis, V. Psomatakis, and N. Spanoudakis. Multi-agent architecture for agricultural products development. In G. Schiefer, R. Helbig, and U. Rickert, editors, *Proceedings of the 2nd European Conference of the European Federation for Information Technology in Agriculture, Food and the Environment (EFITA 2000), September 27-30 2000, Bonn, Germany*, pages 187–196, 2000.

[61] N.F. Matsatsinis, P. Moraitis, V. Psomatakis, and N. Spanoudakis. An agent-based system for products penetration strategy selection. *Applied Artificial Intelligence*, 17(10):901–925, 2003.

[62] N.F. Matsatsinis and A. Samaras. MCDA and preference disaggregation in group decision support systems. *European Journal of Operational Research*, 130(2):414–429, 2001.

[63] N.F. Matsatsinis and Y. Siskos. MARKEX: An intelligent decision support system for product development decision. *European Journal of Operational Research*, 113(2):363–354, 1999.

[64] N.F. Matsatsinis and Y. Siskos. DIMITRA: An intelligent decision support system for agricultural products development decisions. In F. Sevila, editor, *Proceedings of the 3rd European Conference of the European Federation for Information Technology in Agriculture, Food and the Environment (EFITA 2001), June 18-20, Montpellier, France*, 2001.

[65] N.F. Matsatsinis and Y. Siskos. *Intelligent Support Systems for Marketing Decision.* Kluwer Academic Publishers, Dordrecht, 2003.

[66] G. Mihelis, E. Grigoroudis, Y. Siskos, Y. Politis, and Y. Malandrakis. Customer satisfaction measurement in the private bank sector. *European Journal of Operational Research*, 130(2):347–360, 2001.

[67] V. Mousseau, J. Figueira, and J.Ph. Naux. Using assignment examples to infer weights for ELECTRE-TRI method: Some experimental results. *European Journal of Operational Research*, 130(2):263–275, 2000.

[68] V. Mousseau and R. Slowinski. Inferring an ELECTRE-TRI model from assignment examples. *Journal of Global Optimization*, 12(2):157–174, 1998.

[69] V. Mousseau, R. Slowinski, and P. Zielniewicz. A user-oriented implementation of the ELECTRE-TRI method integrating preference elicitation support. *Computers and Operations Research*, 27(7–8):757–777, 2000.

[70] T. Murofushi and M. Sugeno. An interpretation of fuzzy measure and the Choquet integral as an integral withrespect to a fuzzy measure. *Fuzzy Sets and Systems*, 29(2):201–227, 1989.

[71] M. Oral and O. Kettani. Modelling the process of multiattribute choice. *Journal of the Operational Research Society*, 40(3):281–291, 1989.

[72] M. Oral, O. Kettani, J.C. Cosset, and M. Daouas. An estimation model for country risk rating. *International Journal of Forecasting*, 8(4):583–593, 1992.

[73] Z. Pawlak. Rough sets. *International Journal of Information and Computer Sciences*, 11(5):341–356, 1982.

[74] J.R. Quinlan. Induction of decision trees. *Machine Learning*, 1(1):81–106, 1986.

[75] J.L. Richard. Aide à la décision stratégique en P.M.E. In E. Jacquet-Lagrèze and Y. Siskos, editors, *Méthode de Décision Multicritère*, pages 119–142. Hommes et Techniques, Paris, 1983.

[76] B. Roy. *Méthodologie Multicritère d'Aide à la Décision*. Economica, Paris, 1985.

[77] E. Sandalidou, E. Grigoroudis, and Y. Siskos. Organic and conventional olive oil consumers: A comparative analysis using customer satisfaction evaluation approach. In G. Baourakis, editor, *Marketing Trends for Organic Food in the Advent of the 21st Century*. World Scientific, 2003. to appear.

[78] M.F. Shakun. *Evolutionary Systems Design: Policymaking Under Complexity and Group Decision Support Systems*. Holden-Day, San Francisco, CA, 1988.

[79] M.F. Shakun. Airline buyout: Evolutionary systems design and problem restructuring in group decision and negotiation. *Management Science*, 37(10):1291–1303, 1991.

[80] J. Siskos and D.K. Despotis. A DSS oriented method for multiobjective linear programming problems. *Decision Support Systems*, 5(1):47–55, 1989.

[81] Y. Siskos. Comment modéliser les préférences au moyen de fonctions d'utilité additives. *RAIRO Recherche Opérationelle*, 14:53–82, 1980.

[82] Y. Siskos. A way to deal with fuzzy preferences in multicriteria decision problems. *European Journal of Operational Research*, 10(3):314–324, 1982.

[83] Y. Siskos. Analyse de systèmes de décision multicritère en univers aléatoire. *Foundations of Control Engineering*, 10(3–4):193–212, 1983.

[84] Y. Siskos. Analyses de regression et programmation lineaire. *Revue de Statistique Appliquée*, 23(2):41–55, 1985.

[85] Y. Siskos. Evaluating a system of furniture retail outlets using an interactive ordinal regression method. *European Journal of Operational Research*, 23:179–193, 1986.

[86] Y. Siskos. MUSTARD: Multicriteria utility-based stochastic aid for ranking decisions. *Journal of Behavioral Decision Making*, 15(5):461–465, 2002.

[87] Y. Siskos and N. Assimakopoulos. Multicriteria highway planning: A case study. *Mathematical and Computer Modelling*, 12(10-11):1401–1410, 1989.

[88] Y. Siskos and E. Grigoroudis. Measuring customer satisfaction for various services using multicriteria analysis. In D. Bouyssou, E. Jacquet-Lagrèze, P. Perny, R. Sllowinski, D. Vanderpooten, and P. Vincke, editors, *Aiding Decisions with Multiple Criteria: Essays in Honor of Bernard Roy*, pages 457–482. Kluwer Academic Publishers, Dordrecht, 2002.

[89] Y. Siskos, E. Grigoroudis, N.F. Matsatsinis, and G. Baourakis. Preference disaggregation analysis in agricultural product consumer behaviour. In P.M. Pardalos, Y. Siskos, and C. Zopounidis, editors, *Advances in Multicriteria Analysis*, pages 185–202. Kluwer Academic Publishers, Dordrecht, 1995.

[90] Y. Siskos, E. Grigoroudis, N.F. Matsatsinis, G. Baourakis, and F. Neguez. Comparative behavioural analysis of european olive oil consumer. In J. Janssen, C.H. Skiadas, and C. Zopounidis, editors, *Advances in Stochastic Modelling and Data Analysis*, pages 293–310. Kluwer Academic Publishers, Dordrecht, 1995.

[91] Y. Siskos, E. Grigoroudis, Y. Politis, and Y. Malandrakis. Customer satisfaction evaluation: Some real experiences. In A. Colorni, M. Paruccini, and B. Roy, editors, *A-MCD-A: Multiple Criteria Decision Aiding*, pages 297–314. European Commission Joint Research Centre, Ispra, Italy, 2001.

[92] Y. Siskos, E. Grigoroudis, C. Zopounidis, and O. Saurais. Measuring customer satisfaction using a collective preference disaggregation model. *Journal of Global Optimization*, 12(2):175–795, 1998.

[93] Y. Siskos and N.F. Matsatsinis. A DSS for market analysis and new product design. *Journal of Decision Systems*, 2(1):35–60, 1993.

[94] Y. Siskos, N.F. Matsatsinis, and G. Baourakis. Multicriteria analysis in agricultural marketing: The case of French olive oil market. *European Journal of Operational Research*, 130(2).315–331, 2001.

[95] Y. Siskos, A. Spiridakos, and D. Yannacopoulos. Using artificial intelligence and visual techniques into preference disaggregation analysis: The MIIDAS system. *European Journal of Operational Research*, 113(2):236–246, 1999.

[96] Y. Siskos and A. Spyridakos. Intelligent multicriteria decision support: Overview and perspectives. *European Journal of Operational Research*, 113(2):236–246, 1999.

[97] Y. Siskos, A. Spyridakos, and D. Yannacopoulos. MINORA: A multicriteria decision aiding system for discrete alternatives. *Journal of Information Science and Technology*, 2(2):136–149, 1993.

[98] Y. Siskos and D. Yannacopoulos. UTASTAR: An ordinal regression method for building additive value functions. *Investigação Operacional*, 5(1):39–53, 1985.

[99] Y. Siskos and C. Zopounidis. The evaluation criteria of the venture capital investment activity: An interactive assessment. *European Journal of Operational Research*, 31(3):304–313, 1987.

[100] Y. Siskos, C. Zopounidis, and A. Pouliezos. An integrated DSS for financing firms by an industrial development bank in Greece. *Decision Support Systems*, 112(2):151–168, 1994.

[101] R. Slowinski. Rough set approach to decision analysis. *AI Expert Magazine*, 10(3):18–25, 1995.

[102] P. Spiliopoulos. *Analyse et Simulation du Marché pour le Lancement d'un Nouveau Produit Réalisation d'un SIAD*. PhD thesis, Université de Paris-Dauphine, Paris, 1987. Thèse de 3e cycle.

[103] A. Spyridakos, Y. Siskos, D. Yannakopoulos, and A. Skouris. Multicriteria job evaluation for large organisations. *European Journal of Operational Research*, 130(2):375–387, 2000.

[104] V. Srinivasan and A.D. Shocker. Linear programming techniques for multidimensional analysis of preferences. *Psychometrika*, 38(3):337–396, 1973.

[105] A. Stam, M. Sun, and M. Haines. Artificial neural network representations for hierarchical preference structures. *Computers and Operations Research*, 23(12):1191–1201, 1996.

[106] T.J. Stewart. Pruning of decision alternatives in multiple criteria decision making, based on the UTA method for estimating utilities. *European Journal of Operational Research*, 28(1):79–88, 1987.

[107] C. Van de Panne. *Methods for Linear and Quadratic Programming*. North-Holland Publishing Company, Amsterdam, 1975.

[108] H.M. Wagner. Linear programming techniques for regression analysis. *Journal of the American Statistical Association*, 54:206–212, 1959.

[109] F.W. Young, J. De Leeuw, and Y. Takane. Regression with qualitative and quantitative variables: An alternating least squares method with optimal scaling features. *Phychometrika*, 41(4):505–529, 1976.

[110] K. Zaras. Rough approximation of a preference relation by a multi-attribute stochastic dominance for deterministic and stochastic evaluation problems. *European Journal of Operational Research*, 130(2):305–314, 2000.

[111] S. Zionts and J. Wallenius. An interactive programming method for solving the multiple criteria problem. *Management Science*, 22(6):652–663, 1976.

[112] C. Zopounidis. A multicriteria decision-making methodology for the evaluation of the risk of failure and an application. *Foundations of Control Engineering*, 12(1):45–67, 1987.

[113] C. Zopounidis and M. Doumpos. A multicriteria decision aid methodology for the assessment of country risk. *European Research on Management and Business Economics*, 3(3):13–33, 1997.

[114] C. Zopounidis and M. Doumpos. Developing a multicriteria decision support system for financial classification problems: The FINCLAS system. *Optimization Methods and Software*, 8((3-4)):277–304, 1998.

[115] C. Zopounidis and M. Doumpos. Business failure prediction using UTADIS multicriteria analysis. *Journal of the Operational Research Society*, 50(11):1138–1148, 1999.

[116] C. Zopounidis and M. Doumpos. Building additive utilities for multi–group hierarchical discrimination: The MHDIS method. *Optimization Methods and Software*, 14(3):219–240, 2000.

[117] C. Zopounidis and M. Doumpos. INVESTOR: A decision support system based on multiple criteria for portfolio selection and composition. In A. Colorni, M. Paruccini, and B. Roy, editors, *A-MCD-A: Multiple Criteria Decision Aiding*, pages 371–381. European Commission Joint Research Centre, Ispra, Italy, 2000.

[118] C. Zopounidis and M. Doumpos. PREFDIS: A multicriteria decision support system for sorting decision problems. *Computers and Operations Research*, 27((7-8)):779–797, 2000.

[119] C. Zopounidis and M. Doumpos. A preference disaggregation decision support system for financial classification problems. *European Journal of Operation Research*, 130(2):402–413, 2001.

[120] C. Zopounidis, M. Doumpos, and S.H. Zanakis. Stock evaluation using a preference disaggregation methodology. *Decision Sciences*, 30(2):313–336, 1999.

[121] C. Zopounidis, Ch. Hurson, and M. Doumpos. *Risque-Pays: Evaluation des Aspects Economiques, Sociaux et Politiques*. Economica, Paris, 2000.

[122] C. Zopounidis, N.F. Matsatsinis, and M. Doumpos. Developing a multicriteria knowledge-based decision support system for the assessment of corporate performance and viability: The FINEVA system. *Fuzzy Economic Review*, 1(2):35–53, 1996.

Chapter 9

THE ANALYTIC HIERARCHY AND ANALYTIC NETWORK PROCESSES FOR THE MEASUREMENT OF INTANGIBLE CRITERIA AND FOR DECISION-MAKING

Thomas L. Saaty
Katz Graduate School of Business
University of Pittsburgh
USA
saaty@katz.pitt.edu

Abstract The Analytic Hierarchy Process (AHP) and its generalization to dependence and feedback, the Analytic Network Process (ANP), are theories of relative measurement of intangible criteria. With this approach to relative measurement, a scale of priorities is derived from pairwise comparison measurements only after the elements to be measured are known. The ability to do pairwise comparisons is our biological heritage and we need it to cope with a world where everything is relative and constantly changing. In traditional measurement one has a scale that one applies to measure any element that comes along that has the property the scale is for, and elements are measured one by one, not by comparing them with each other. In the AHP paired comparisons are made with judgments using numerical values taken from the AHP absolute fundamental scale of 1-9. A scale of relative values is derived from all these paired comparisons and it also belongs to an absolute scale that is invariant under the identity transformation like the system of real numbers. The AHP/ANP is useful for making multicriteria decisions involving benefits, opportunities, costs and risks. The ideas are developed in stages and illustrated with examples of real life decisions. The subject is transparent and despite some mathematics, it is easy to understand why it is done the way it is along the lines discussed here.

Keywords: Analytic Hierarchy Process, decision-making, prioritization, negative priorities, rating, benefits, opportunities, costs, risks.

1. Introduction

The purpose of decision-making is to help people make decisions according to their own understanding. They would then feel that they really made the decision themselves justified completely according to their individual or group values, beliefs, and convictions even as one tries to make them understand these better. Because decision-making is the most frequent activity of all people all the time, the techniques used today to help people make better decisions should probably remain closer to the biology and psychology of people than to the techniques conceived and circulated at a certain time and that are likely to become obsolete, as all knowledge does, even though decisions go on and on forever. This suggests that methods offered to help make better decisions should be closer to being descriptive and considerably transparent. They should also be able to capture standards and describe decisions made normatively. Natural science, like decision-making, is mostly descriptive and predictive to help us cope intelligently with a complex world.

Not long ago, people believed that the human mind is an unreliable instrument for performing measurement and that the only meaningful measurement is obtained on a physical scale like the meter and the kilogram invented by clever people who care about precision and objective truth. They did not think how the measurements came to have meaning for people and that this meaning depends on people's purpose each time they obtain a reading on that scale. In the winter, ice may be a source of discomfort but an ice drink in the summer can be a refreshing source of comfort. A number has no meaning except that assigned to it by someone. We may all agree on the numerical value of a reading on a physical scale, but not on what exactly that number means to each of us in practical terms. We tend to parrot abstractions that define a number but often forget that numbers are meant to serve some need that is inevitably subjective, which is ultimately more important for our survival. Thus it is our subjective values that are essential for interpreting the readings obtained through measurement. This interpretation depends on what one has in mind at the time and different people may interpret the same reading differently for the same situation depending on their goal. The reading may be called objective, but the interpretation is predominantly subjective. In this sense subjectivity is important, because without it objectivity has no intrinsic meaning. If the mind of an expert can produce measurement close to what we obtain through measuring instruments, then it has greater power than instruments to deal with a complexity for which we have no way to measure. What we have to do is examine the possibility and validity of this assumption as critically as we can. It turns out that when we have knowledge and experience, our brains are very good measuring instruments. That does not mean that we should discard what we use in science that enhances our understanding, but rather we should use it to support and strengthen what we do directly with our minds.

The subject of this chapter is the Analytic Hierarchy Process (AHP), the original theory of prioritization that derives relative scales of absolute numbers known as priorities from judgments expressed numerically on an absolute fundamental scale. It is also about a more general approach to decisions that is a generalization of hierarchies to networks with dependence and feedback, the Analytic Network Process (ANP). Both the AHP and ANP are descriptive approaches to decision-making. The AHP/ANP evolved out of my experience at the Arms Control and Disarmament Agency (ACDA) in the Department of State during the Kennedy and Johnson years. ACDA negotiated arms agreements with the Soviets in Geneva. I was invited to join ACDA, I think because of work I had done for the military using Operations Research mathematics. I published on it and wrote the first book on mathematical methods of operations research. At ACDA I supervised a team of foremost internationally known scientists, economists and game theorists (including three people who later won the Nobel Prize in economics: Debreu, Harsanyi and Selten) who advised ACDA on arms tradeoffs, but we had some insurmountable difficulties in making lucid and usable recommendations to our highly intelligent and experienced negotiators who were guided by strong intuition deriving from long practice.

The basic problem is that we need to quantify intangibles of which there is nearly an infinite number and we can only do it by making comparisons in relative terms. Even if everything were measurable, we would still need to compare the different types of measurements on the different scales and determine how important they are to us to make tradeoffs among them and reach a final answer. If we use tangibles and their measurements we would need to reduce them to a common relative frame of reference and then weight and combine them along with intangibles. Combining priorities of measurable quantities with those of non-measurable qualities needs ratio or even the stronger absolute scales, because we can then multiply and add the outcomes particularly when there is interdependence among all the elements involved in a decision.

The AHP is a theory of relative measurement on absolute scales of both tangible and intangible criteria based both on the judgment of knowledgeable and expert people and on existing measurements and statistics needed to make a decision. How to measure intangibles is the main concern of the mathematics of the AHP. The AHP has been mostly applied to multi-objective, multi-criteria and multiparty decisions because decision-making has this diversity. To make tradeoffs among the many intangible objectives and criteria, the judgments that are usually made in qualitative terms are expressed numerically. To do this, rather than simply assign a score out of a person's memory that is hard to justify, one must make reciprocal pairwise comparisons in a carefully designed scientific way. In the end, we must fit our entire world experience into our system of priorities if we are going to understand it. The AHP is based on four axioms: (1) reciprocal judgments, (2) homogeneous elements, (3) hierarchic or feedback dependent structure, and (4) rank order expectations. The synthesis of the AHP combines multidimensional scales of measurement into a single "unidimensional" scale of priorities. Decisions are determined by a single number for the best outcome or by a vector of priorities that gives a proportionate ordering of the different possible outcomes to which one can then allocate resources in an optimal way subject to both tangible and intangible constraints. We can also combine the judgments obtained from a group when several people are involved in a decision. It is known that with the reciprocal condition, the geometric mean is a necessary condition for combining individual judgments and that, contrary to the impossibility of combining individual judgments into a social welfare function when ordinals are used subject to certain conditions, with absolute judgments it is possible to construct with the AHP such a social welfare function that satisfies these conditions [8].

It is not idiosyncratic for one to believe that making a decision is more complex than just listing all the factors, good and bad, that one can think of and then plunge into numerical manipulations that surface a best outcome according to some plausible way of analysis. Nor is it less idiosyncratic to confine the analysis of decisions to risk and use risk aversion as a way to justify how to make a good choice. For every decision there are positive and negative factors to consider, usually interpreted psychologically in the form of benefits (gains) and opportunities (potential gains), and costs (losses) and risks (potential losses). How to evaluate a decision according to these merits (demerits) and how to combine them into a single overall answer is not easy to do and is something that leaders in business and government do qualitatively with the help of advisors to satisfy the broad goals that they serve. Multicriteria decision-making needs to provide meaningful quantitative assistance on this important, complex, and inevitable concern with its many intangibles.

2. Pairwise Comparisons; Inconsistency and the Principal Eigenvector

The psychologist Arthur Blumenthal writes in his book *The Process of Cognition*, Prentice-Hall, Inc., Englewood Cliffs, New Jersey, 1977, that there are two types of judgment: "Comparative judgment which is the identification of some relation between two stimuli both present to the observer, and absolute judgment which involves the relation between a single stimulus and some information held in short term memory about some former comparison stimuli or about some previously experienced measurement scale using which the observer rates the single stimulus."

Comparative or relative judgment is made on pairs of elements to ensure accuracy. In paired comparisons, the smaller or lesser element is used as the unit, and the larger or greater element is estimated as a multiple of that unit with respect to the common property or criterion for which the comparisons are made. In this sense, measurement with many pairwise comparisons is made more scientifically than by assigning numbers more or less arbitrarily through guessing. What is really the scale to which such numbers belong so they can be operated on arithmetically in a legitimate way? For example, one cannot simply add numbers that belong to an ordinal or an interval scale. Because our brains are limited in size and the firings of their neurons are limited in intensity, it is clear that there is a limit to their ability to compare the very small with the very large. It is precisely for this reason that pairwise comparisons are made on elements or alternatives that are close or homogeneous and the more separated they are, the more need there is to put them in different groups and link these groups with a common element from one group to an adjacent group of slightly greater or slightly smaller elements. One can then compare the elements in each homogeneous group and then combine them through appropriate use of the measurement of the elements (pivots) that are common to consecutive groups.

We learn from making paired comparisons in the AHP that if A is 5 times larger in size than B and B is 3 times larger in size than C, then A is 15 times larger in size than C and thus we say that A dominates C 15 times. That is different from A having 5 dollars more than B and B having 3 dollars more than C implies that A has 8 dollars more than C. Defining intensity along the arcs of a graph and raising the resulting matrix of comparisons to powers measures the first kind of dominance precisely and never the second. It has definite meaning and as we shall see, because of the inconsistency inherent in making judgments , in the limit it is measured uniquely by the principal eigenvector. There is a useful connection between what we do with dominance priorities in the AHP and what is done with transition probabilities both of which use matrix algebra to find their answers. Transitions between states are multiplied and added. To compose the priorities of the alternatives of a decision with respect to different criteria, it is also necessary that the priorities of the alternatives with respect to each criterion be multiplied by the priority of that criterion and then added over all the criteria.

Paired comparisons deal with comparative judgment. However, in conformity with Blumenthal's observation above, the AHP also provides a way to rate alternatives one at a time to deal with absolute judgment. In absolute judgment the criteria are first prioritized through comparisons and then for each criterion one creates a scale of relative intensities possibly of widely ranging orders of magnitude. The priorities of these intensities are again appropriately derived through paired comparisons with respect to their criterion, and in the end the alternatives are rated one at a time by assigning each one an intensity level for each criterion, then weighting by the priorities of the criteria and adding to obtain their overall rating priority [2]. Thus rating only applies to alternatives taken one at a time and relies on standards (good or poor) in the memory of the decision maker to rate the alternatives. It is useful when the number of alternatives is large and we want to standardize our treatment of them. When alternatives are fundamentally new, different and not fully understood, paired comparisons are essential because there are no familiar and widely accepted standards on which they can be rated.

To derive priorities for criteria or attributes we either think of a need to be satisfied, or of a property of alternatives that we already have. In either case when there are several criteria we need to establish their priorities to select the best alternative that meets all the requirements.

Assume that one is given n stones, $A_1, \cdots, A_n$, with known weights $w_1, \cdots, w_n$, respectively, and suppose that a matrix of pairwise ratios is formed whose rows give the ratios of the weights of each stone with respect to all others. We have:

$$\begin{array}{c} A_1 \\ \vdots \\ A_n \end{array} \begin{array}{c} \begin{array}{ccc} A_1 & \cdots & A_n \end{array} \\ \begin{bmatrix} w_1/w_1 & \cdots & w_1/w_n \\ \vdots & & \vdots \\ w_n/w_1 & \cdots & w_n/w_n \end{bmatrix} \end{array}.$$

To recover the vector $w = (w_1, ..., w_n)$ we introduce the system of equations:

$$Aw = \begin{bmatrix} w_1/w_1 & \cdots & w_1/w_n \\ \vdots & & \vdots \\ w_n/w_1 & \cdots & w_n/w_n \end{bmatrix} \begin{bmatrix} w_1 \\ \vdots \\ w_n \end{bmatrix} = n \begin{bmatrix} w_1 \\ \vdots \\ w_n \end{bmatrix} = nw,$$

where A has been multiplied on the right by the vector of weights w. The result of this multiplication is nw. To recover the scale from the matrix of ratios, one must solve the problem $Aw = nw$ or $(A - nI)w = 0$. This is a system of homogeneous linear equations. It has a nontrivial solution if and only if the determinant of $A - nI$ vanishes, that is, n is an eigenvalue of A. Now A has unit rank since every row is a constant multiple of the first row. As a result, all its eigenvalues except one are zero. The sum of the eigenvalues of a matrix is equal to its trace, the sum of its diagonal elements, and in this case the trace of A is equal to n. Thus n is an eigenvalue of A, and one has a nontrivial solution. The solution consists of positive entries and is unique to within a multiplicative constant.

To make w unique, we can normalize its entries by dividing by their sum. Thus, given the comparison matrix, we can recover the scale. In this case, the solution is any column of A normalized. Notice that in A the reciprocal property $a_{ji} = 1/a_{ij}$ holds; thus, also $a_{ii} = 1$. Another property of A is that it is consistent: its entries satisfy the condition $a_{jk} = a_{ik}/a_{ij}$. The entire matrix can be constructed from a set of n elements that form a chain across the rows and columns of A.

In the general case, the precise value of w_i/w_j cannot be given, but instead only an estimate of it as a judgment. For the moment, consider an estimate of these values by an expert whose judgments are small perturbations of the coefficients w_i/w_j. This implies small perturbations of the eigenvalues.

Let us for generality call $A_1, \ldots, A_n$ stimuli instead of stones. The quantified judgments on pairs of stimuli A_i, A_j, are represented by an n-by-n matrix $A' = (a_{ij})$, $ij = 1, 2, ..., n$. The entries a_{ij} are defined by the following entry rules.

Rule 1. If $a_{ij} = a$, then $a_{ji} = 1/a, a > 0$.

Rule 2. If A_i is judged to be of equal relative intensity to A_j then $a_{ij} = 1$, $a_{ji} = 1$; in particular, $a_{ii} = 1$ for all i.

Thus the matrix A' has the form:

$$A' = \begin{pmatrix} 1 & a_{12} & \cdots & a_{1n} \\ 1/a_{12} & 1 & \cdots & a_{2n} \\ \vdots & \vdots & \vdots & \vdots \\ 1/a_{1n} & 1/a_{2n} & \cdots & 1 \end{pmatrix}.$$

Having recorded the quantified judgments on pairs of stimuli (A_i, A_j) as numerical entries a_{ij} in the matrix , the problem now is to assign to the n stimuli $A_1, A_2, \ldots, A_n$ a set of numerical weights that would "reflect the recorded judgments." In order to do that, the vaguely formulated problem must first be transformed into a precise mathematical one. This essential, and apparently harmless, step is the most crucial one in any problem that requires the representation of a real life situation in terms of an abstract mathematical structure. It is particularly crucial in the present problem where the representation involves a number of transitions that are not immediately discernible. It appears, therefore, desirable in the present problem to identify the major steps in the process of representation and to make each step as explicit as possible to enable the potential user to form his own judgment as to the *meaning and value* of the method in relation to *his* problem and *his* goal.

Why we must solve the principal eigenvalue problem in general has a simple justification based on the idea of dominance among the elements represented by the coefficients of the matrix. Dominance between two elements is obtained as the normalized sum of path intensities defined by the numerical judgments assigned to the arcs along a path. The overall dominance of an element is the sum of the entries in its row given by Ae, $e = (1, \ldots, 1)$ when A is consistent because then $A^k = n^{k-1}A$. When A is inconsistent, we must consider paths of dominance of all lengths between the two points. All the paths of a given length k are obtained by raising the matrix to the power k. According to Cesaro summability, the limit of the average or Cesaro sum $\lim_{N\to\infty} 1/N \sum_{k=0}^{N} A^k e$ that represents the average of all order dominance vectors up to N, is the same as the limit of the sequence of the powers of the matrix i.e. $(\lim_{k\to\infty} A^k)e$. Now we know from Perron theory that the sequence A^k converges to a matrix all whose columns are identical and are proportional to the principal right eigenvector of A. Thus $(\lim_{k\to\infty} A^k)e$ is also proportional to the principal right eigenvector of A.

Without the theory of Perron, the proof (not given here but known in eigenvalue theory) of how to go from $Aw = nw$ to $Aw = \lambda_{\max} w$, is related to small perturbation theory and the amount of inconsistency one allows. A modicum of inconsistency is necessary to change our mind about old relations when we learn new things.

Another way to prove the necessity of the principal eigenvector is based on the need for the invariance of priorities. No matter what method we use to derive the weights , by using them to weight and add the entries in each row to determine the dominance of the element represented in that row, we must get these priorities back as proportional to the expression $\sum_{j=1}^{n} a_{ij}w_j$ $i = 1, 2 \cdots n$, that is, we must solve $\sum_{j=1}^{n} a_{ij}w_j = cw_i, i = 1, 2 \cdots n$, because in the end they can be normalized. Otherwise $\sum_{j=1}^{n} a_{ij}w_j, i = 1, \ldots, n$ would yield another set of different weights and they in turn can be used to form new expressions $\sum_{j=1}^{n} a_{ij}w_j$, $i = 1, 2 \ldots n$, and so on ad infinitum violating the need to have priorities that are invariant, unless in any case we solve the principal eigenvalue problem.

Our general problem takes the form:

$$A'w = \begin{bmatrix} 1 & a_{12} & \cdots & a_{1n} \\ 1/a_{12} & 1 & \cdots & a_{2n} \\ \vdots & \vdots & \vdots & \vdots \\ 1/a_{1n} & 1/a_{2n} & \cdots & 1 \end{bmatrix} \begin{bmatrix} w_1 \\ w_2 \\ \vdots \\ w_n \end{bmatrix} = cw.$$

We now show that the perturbed eigenvalue from the consistent case is the principal eigenvalue of A'. Our argument involves both left and right eigenvectors of A'. Two vectors $x = (x_1, \cdots, x_n)$, $y = (y_1, \cdots, y_n)$ are orthogonal if their scalar product $x_1y_1 + \cdots x_ny_n$ is equal to zero. It is known that any left eigenvector of a matrix corresponding to an eigenvalue is orthogonal to any right eigenvector corresponding to a different eigenvalue. This property is known as bi-orthogonality using which we can prove:

THEOREM 1 *For a given positive matrix A, the only positive vector w and only positive constant c that satisfy $Aw = cw$, is a vector w that is a positive multiple of the principal eigenvector of A, and the only such c is the principal eigenvalue of A.*

Thus we see that both requirements of dominance and invariance lead us to the principal right eigenvector. The problem now is how good is the estimate of w. Notice that if w is obtained by solving this problem, the matrix whose entries are w_i/w_j is a consistent matrix. It is a consistent estimate of the matrix A'. The matrix itself need not be consistent. In fact, the entries of A' need not even be transitive; that is, A_1 may be preferred to A_2 and A_2 to A_3 but A_3 may be preferred to A_1. What we would like is a measure of the error due to inconsistency. It turns out that A' is consistent if and only if $\lambda_{\max} = n$ and that we always have $\lambda_{\max} \geq n$ when we solve the system of equations $Aw = \lambda_{\max} w$ for a non-negative reciprocal matrix A to obtain the priorities.

Thus the story is very different if the judgments are inconsistent, and as we said before, we need to allow inconsistent judgments for good reasons. In sports, team A beats team B, team B beats team C, but team C beats team A. How would we admit such an occurrence in our attempt to explain the real world if we do not allow inconsistency? So far we have legislated inconsistency, which is natural in making judgments, by assuming axiomatically that it should not exist particularly with regard to transitivity!

The priorities that we seek are concerned with the order to be captured from dominance judgments involving all order transitivity. Thus the problem of deriving unique priorities in decision-making by solving the principal eigenvalue problem of A' belongs to the field of mathematics known as *order topology*. In general priorities are not obtainable directly by the many methods of *metric topology* involving minimization of a metric such as the method of least squares (LSM) which determines a priority vector by minimizing the Frobenius norm of the difference between A and a positive rank one reciprocal matrix $[y_i/y_j]$:

$$\min_{y>0} \sum_{i,j=1}^{n} \left(a_{ij} - \frac{y_i}{y_j} \right)^2 \tag{9.1}$$

and the method of logarithmic least squares (LLSM) which determines a vector by minimizing the Frobenius norm of $[\log(a_{ij}x_j/x_i)]$:

$$\min_{x>0} \sum_{i,j=1}^{n} \left[\log a_{ij} - \log \left(\frac{x_i}{x_j} \right) \right]^2. \tag{9.2}$$

Metric methods not only ignore transitivity, but also yield a variety of different answers thus violating the overall justification of the need for a single unique set of priorities. There is however a connection between order and optimization.

Solving the principal eigenvalue problem to obtain priorities is equivalent to the two problems of optimization that follow: Find w_i, $i = 1, \cdots, n$ which

1 maximize $\frac{1}{n}\sum_{i=1}^{n}\sum_{j=1}^{n} a_{ij}w_j/w_i$, or, in the simpler linear optimization setting,

2 maximize $\sum_{j=1}^{n} w_j \sum_{i=1}^{n} a_{ij}$, obtained by multiplying the sum of each column j by its corresponding w_j and summing over j, subject to $\sum_{i=1}^{n} w_i = 1$.

3. Stimulus Response and the Fundamental Scale

What numbers should we use when we only have qualitative judgments to express our understanding in making pairwise comparisons of elements that are close or homogeneous? We note that to be able to perceive and sense objects in the environment our brains miniaturize them within our system of neurons so that we have a proportional relationship between what we perceive and what is out there. Without proportionality we cannot coordinate our thinking with our actions with the accuracy needed to control the environment. Proportionality with respect to a single stimulus requires that our response to a proportionately amplified or attenuated stimulus we receive from a source should be proportional to what our response would be to the original value of that stimulus. If $w(s)$ is our response to a stimulus of magnitude s, then the foregoing gives rise to the functional equation $w(as) = b\,w(s)$. This equation can also be obtained as the necessary condition for solving the Fredholm equation of the second kind:

$$\int_a^b K(s,t)w(t)dt = \lambda_{max}w(s)$$

obtained as the continuous generalization of the discrete formulation $Aw = \lambda_{max}w(s)$. The solution of this functional equation in the real domain is given by

$$w(S) = Ce^{\log b \frac{\log s}{\log a}} P\left(\frac{\log s}{\log a}\right),$$

where P is a periodic function of period 1 and $P(0) = 1$. One of the simplest such examples with $u = \log a / \log a$ is $P(u) = \cos(u/2\pi)$ for which $P(0) = 1$ and from which the logarithmic law of response to stimuli can be obtained as a first order approximation as:

$$v(u) = C_1 e^{-\beta u} P(u) \approx C_2 \log s + C_3$$

$\log ab = -\beta, \beta > 0$. The expression on the right is the well-known Weber-Fechner law of logarithmic response $M = a \log s + b$, $a \neq 0$ to a stimulus of magnitude s. It belongs to an interval scale. The larger the stimulus, the larger a change in it is needed for that change to be detectable. The ratio of successive just noticeable differences (the well-known "jnd" in psychology) is equal to the ratio of their corresponding successive stimuli values. Proportionality is maintained. Thus, starting with a stimulus s_0 successive magnitudes of the new stimuli take the form:

$$\begin{aligned}
s_1 &= s_0 + \Delta s_0 &&= s_0 + \tfrac{\Delta s_0}{s_0} &&= s_0(1+r) \\
s_2 &= s_1 + \Delta s_1 &&= s_0(1+r)^2 &&\equiv s_0\alpha^2 \\
&\vdots \\
s_n &= s_{n-1}\alpha &&= s_0\alpha^n (n = 0, 1, 2, \ldots).
\end{aligned}$$

We consider the responses to these stimuli to be measured on a ratio scale $(b = 0)$. A typical response has the form $M_i = a \log \alpha^i$, $i = 1, \cdots, n$, or one after another they have the form:

$$M_1 = a \log \alpha, M_2 = 2a \log \alpha, \cdots M_n = na \log \alpha.$$

We take the ratios $M_i/M_1, i = 1, \cdots, n$ of these responses in which the first is the smallest and serves as the unit of comparison, thus obtaining the integer values $1, 2, \cdots, n$ of the fundamental scale of the AHP.

A person may not be schooled in the use of numbers but still have feelings, judgment and understanding that enable him or her to make accurate comparisons (equal, moderate, strong, very strong and extreme and compromises between these intensities). Such judgments can be applied successfully to compare stimuli that are not too disparate but homogeneous in magnitude. By homogeneous we mean that they fall within specified bounds. The foregoing may be summarized to represent the fundamental scale for paired comparisons shown in Table 9.1.

We know now that a judgment or comparison is the numerical representation of a relationship between two elements that share a common parent. We also know that the set of all such judgments can be represented in a square matrix in which the set of elements is compared with itself. Each judgment represents the dominance of an element in the column on the left over an element in the row on top. It reflects the answers to two questions: which of the two elements is more important with respect to a higher level criterion, and how strongly, using the 1-9 scale shown in Table 9.1 for the element on the left over the element at the top of the matrix. If the element on the left is less important than that on the top of the matrix, we enter the reciprocal value in the corresponding position in the matrix. It is important to note that the lesser element is always used as the unit and the greater one is estimated as a multiple of that unit. From all the paired comparisons we calculate the priorities and exhibit them on the right of the matrix. For a set of n elements in a matrix one needs $n(n-1)/2$

Table 9.1. The fundamental scale of absolute numbers.

Intensity of Importance	Definition	Explanation
1	Equal importance	Two activities contribute equally to the objective
2	Weak	
3	Moderate importance	Experience and judgment slightly favor one activity over another
4	Moderate plus	
5	Strong importance	Experience and judgment strongly favor one activity over another
6	Strong plus	
7	Very strong or demonstrated importance	An activity is favored very strongly over another; its dominance demonstrated in practice
8	Very, very strong	
9	Extreme importance	The evidence favoring one activity over another is of the highest possible order of affirmation
Reciprocals of above	If activity i has one of the above nonzero numbers assigned to it when compared with activity j, then j has the reciprocal value when compared with i	A reasonable assumption
Rationals	Ratios arising from the scale	If consistency were to be forced by obtaining n numerical values to span the matrix

comparisons because there are n 1's on the diagonal for comparing elements with themselves and of the remaining judgments, half are reciprocals. Thus we have $(n^2 - n)/2$ judgments. In some problems one may elicit only the minimum of $n - 1$ judgments.

In a judgment matrix A, instead of assigning two numbers w_i and w_j (that generally we do not know), as one does with tangibles, and forming the ratio w_i/w_j we assign a single number drawn from the fundamental scale of absolute numbers shown in Table 9.1 to represent the ratio $(w_i/w_j)/1$. It is a nearest integer approximation to the ratio $(w_i/w_j)/1$. The ratio of two numbers from a ratio scale (invariant under multiplication by a positive constant) is an absolute number (invariant under the identity transformation). The derived scale will reveal what w_i and w_i are.

This is a central fact about the relative measurement approach. It needs a fundamental scale to express numerically the relative dominance relationship.

If one wishes to use actual measurements or use fractional values for judgments one of course can. In the end one needs to justify with care what one does.

REMARK 32 *The reciprocal property plays an important role in combining the judgments of several individuals to obtain a judgment for a group. Judgments must be combined so that the reciprocal of the synthesized judgments must be equal to the syntheses of the reciprocals of these judgments. It has been proved that the geometric mean is the unique way to do that. If the individuals are experts, they my not wish to combine their judgments but only their final outcome from a hierarchy. In that case one takes the geometric mean of the final outcomes. If the individuals have different priorities of importance their judgments (final outcomes) are raised to the power of their priorities and then the geometric mean is formed [2].*

3.1 Validation Example

Here is an example (one of many) which shows that the scale works well on homogeneous elements of a real life problem. A matrix of paired comparison judgments is used to estimate relative drink consumption in the United States as shown in Table 9.2. To make the comparisons, the types of drinks are listed on the left and at the top, and judgment is made as to how strongly the consumption of a drink on the left dominates that of a drink at the top. For example, when coffee on the left is compared with wine at the top, it is thought that it is consumed extremely more and a 9 is entered in the first row and second column position. A $1/9$ is automatically entered in the second row and first column position. If the consumption of a drink on the left does not dominate that of a drink at the top, the reciprocal value is entered. For example in comparing coffee and water in the first row and eighth column position, water is consumed more than coffee slightly and a $1/2$ is entered. Correspondingly, a value of 2 is entered in the eighth row and first column position. At the bottom of Table 9.2, we see that the derived values and the actual values are close.

3.2 Clustering and Homogeneity; Using Pivots to Extend the Scale from 1-9 to 1-∞

Most real life decisions are not widely separated in ranges of criteria (one or two) because what is important to individuals or to groups to corporations and finally to governments needs to meet their most essential requirements. Note that the priorities in two adjacent categories would be sufficiently different, one being an order of magnitude smaller than the other, that in the synthesis, the priorities of the elements in the smaller set would ordinarily have little effect on the decision.

We note that our ability to make accurate comparisons of widely disparate objects on a common property is limited. We cannot compare with any reliability the very small with the very large. However, we can do it in stages by comparing objects of relatively close magnitudes and gradually increase their sizes until we reach the desired object of large size (see example later). In this process, we can think of comparing several close or homogeneous objects for which we obtain a scale of relative values, and then again pairwise compare the next set of larger objects that includes for example the largest object from the previous already compared collection, and then derive a scale for this second set. We then divide all the measurements in the second set by the value of the common object and multiply all the resulting values by the weight of the common element in the first set, thus rendering the two sets to be measurable on the same scale and so on to a third collection of the objects using a common object from the second set.

Table 9.2. Which drink is consumed more in the U.S.? An example of estimation using judgments.

Drink Consumption in th U.S.	Coffee	Wine	Tea	Beer	Sodas	Milk	Water
Coffee	1	9	5	2	1	1	1/2
Wine	1/9	1	1/3	1/9	1/9	1/9	1/9
Tea	1/5	2	1	1/3	1/4	1/3	1/9
Beer	1/2	9	3	1	1/2	1	1/3
Sodas	1	9	4	2	1	2	1/2
Milk	1	9	3	1	1/2	1	1/3
Water	2	9	9	3	2	3	1

The derived scale based on the judgments in the matrix is:

Coffee	Wine	Tea	Beer	Sodas	Milk	Water
.177	.019	.042	.116	.190	.129	.327

with a consistency index of 0.022.

The actual consumption (from statistical sources) is:

Coffee	Wine	Tea	Beer	Sodas	Milk	Water
.180	.010	.040	.120	.180	.140	.330

In Figure 9.1 a cherry tomato is eventually and indirectly compared with a large watermelon by first comparing it with a small tomato and a lime, the lime is then used again in a second cluster with a grapefruit and a honey dew where we then divide by the weight of the lime and then multiply by its weight in the first cluster, and then use the honey dew again in a third cluster and so on. In the end we have a comparison of the cherry tomato with the large watermelon and would accordingly extended the scale from 1-9 to 1-721.

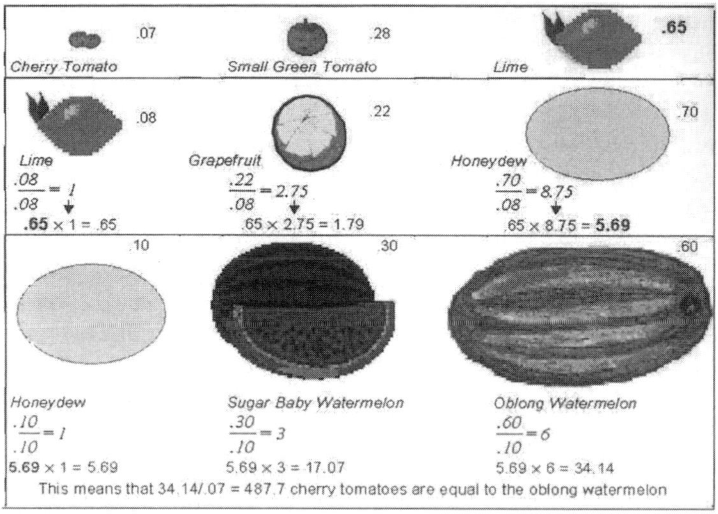

Figure 9.1. Comparisons according to volume.

4. Hospice Decision

Westmoreland County Hospital in Western Pennsylvania, like hospitals in many other counties around the United States, has been concerned with the costs of the facilities and manpower involved in taking care of terminally ill patients. Normally these patients do not need as much medical attention as do other patients. Those who best utilize the limited resources in a hospital are patients who require the medical attention of its specialists and advanced technology equipment, whose utilization depends on the demand of patients admitted into the hospital. The terminally ill need medical attention only episodically. Most of the time, such patients need psychological support. Such support is best given by the patient's family, whose members are able to supply the love and care the patients most need. For the mental health of the patient, home therapy is a benefit. From the medical standpoint, especially during a crisis, the hospital provides a greater benefit. Most patients need the help of medical professionals only during a crisis. Some will also need equipment and surgery. The planning association of the hospital wanted to develop alternatives and to choose the best one considering various criteria from the standpoint of the patient, the hospital, the community, and society at large.

In this problem, we need to consider the costs and benefits of the decision. Costs include economic costs and all sorts of intangibles, such as inconvenience and pain. Such disbenefits are not directly related to benefits as their mathematical inverses, because patients infinitely prefer the benefits of good health to these intangible disbenefits. To study the problem, one needs to deal with benefits and with costs separately.

I met with representatives of the planning association for several hours to decide on the best alternative. To make a decision by considering benefits and costs, one must first answer the question: In this problem, do the benefits justify the costs? If they do, then either the benefits are so much more important than the costs that the decision is based simply on benefits, or the two are so close in value that both the benefits and the costs should be considered. Then we use two hierarchies for the purpose and make the choice by forming the ratio from them of the benefits priority/costs priority for each alternative. One asks which is most beneficial in the benefits hierarchy (Figure 9.2) and which is most costly in the costs hierarchy (Figure 9.3).

If the benefits do not justify the costs, the costs alone determine the best alternative, which is the least costly. In this example, we decided that both benefits and costs had to be considered in separate hierarchies. In a risk problem, a third hierarchy is used to determine the most desired alternative with respect to all three: benefits, costs, and risks. In this problem, we assumed risk to be the same for all contingencies.

The planning association thought the concepts of benefits and costs were too general to enable it to make a decision. Thus, the planners and I further subdivided each (benefits and costs) into detailed subcriteria to enable the group to develop alternatives and to evaluate the finer distinctions the members perceived between the three alternatives. The alternatives were to care for terminally ill patients at the hospital, at home, or partly at the hospital and partly at home.

The two hierarchies are fairly clear and straightforward in their description. They descend from the more general criteria in the second level to secondary subcriteria in the third level and then to tertiary subcriteria in the fourth level on to the alternatives at the bottom or fifth level. At the general criteria level, each of the hierarchies, benefits or costs, involved three major interests. The decision should benefit the recipient, the institution, and society, and their relative importance is the prime determinant as to which outcome is more likely to be preferred. We located these three elements on the second level of the benefits hierarchy. As the decision would benefit each party differently and the importance of the benefits to each recipient affects the outcome, the group thought that it was important to specify the types of benefit for the recipient and the institution. Recipients want physical, psycho-social and economic benefits, while the

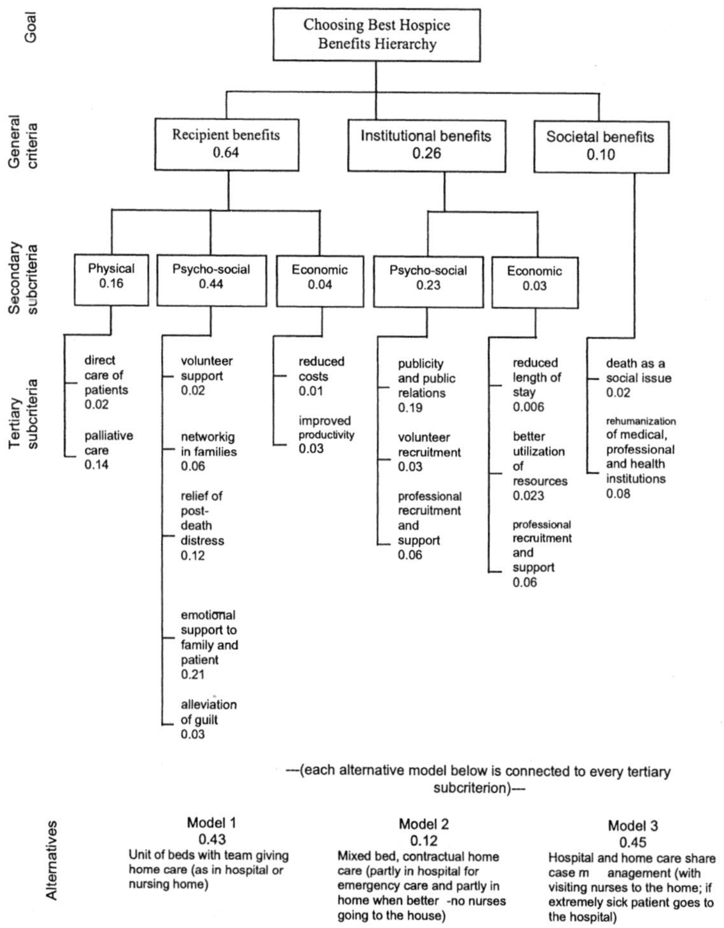

Figure 9.2. To choose the best hospice plan, one constructs a hierarchy modeling the benefits to the patient, to the institution, and to society. This is the benefits hierarchy of two separate hierarchies.

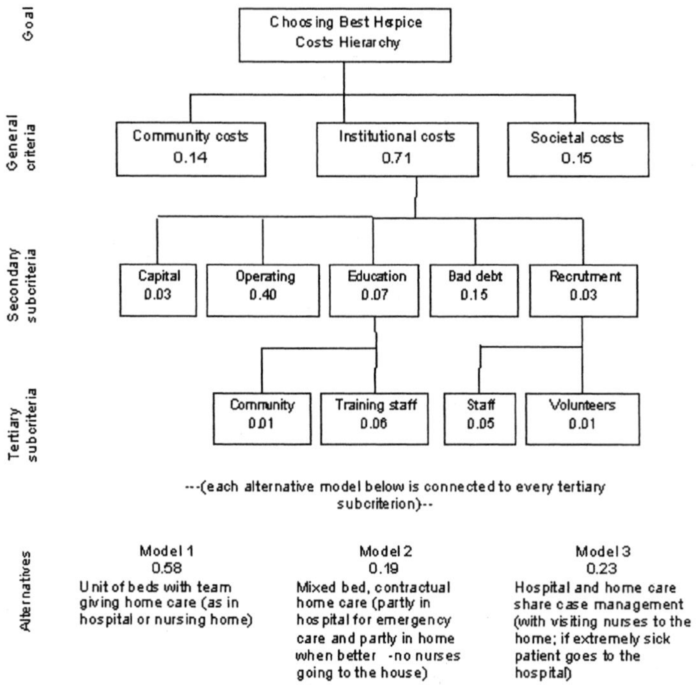

Figure 9.3. To choose the best hospice plan, one constructs a hierarchy modeling the community, institutional, and societal costs. This is the costs hierarchy of two separate hierarchies.

institution wants only psycho-social and economic benefits. We located these benefits in the third level of the hierarchy. Each of these in turn needed further decomposition into specific items in terms of which the alternatives could be evaluated. For example, while the recipient measures economic benefits in terms of reduced costs and improved productivity, the institution needed the more specific measurements of reduced length of stay, better utilization of resources, and increased financial support from the community. There was no reason to decompose the societal benefits into a third level subcriteria, hence societal benefits connects directly to the fourth level. The group considered three models for the alternatives, and they are at the bottom (or fifth level in this case) of the hierarchy: in Model 1, the hospital provided full care to the patients; in Model 2, the family cares for the patient at home, and the hospital provides only emergency treatment (no nurses go to the house); and in Model 3, the hospital and the home share patient care (with visiting nurses going to the home).

In the costs hierarchy there were also three major interests in the second level that would incur costs or pains: community, institution, and society. In this decision the costs incurred by the patient were not included as a separate factor. Patient and family could be thought of as part of the community. We thought decomposition was necessary only for institutional costs. We included five such costs in the third level: capital costs, operating costs, education costs, bad debt costs, and recruitment costs. Educational costs apply to educating the community and training the staff. Recruitment costs apply to staff and volunteers. Since both the costs hierarchy and the benefits hierarchy concern the same decision, they both have the same alternatives in their bottom levels, even though the costs hierarchy has fewer levels.

As usual with the AHP, in both the costs and the benefits models, we compared the criteria and subcriteria according to their relative importance with respect to the parent element in the adjacent upper level. For example, in the first matrix of comparisons of the three benefits criteria with respect to the goal of choosing the best hospice alternative, recipient benefits are moderately more important than institutional benefits and are assigned the absolute number 3 in the (1, 2) or first-row second-column position. Three signifies three times more. The reciprocal value is automatically entered in the (2, 1) position, where institutional benefits on the left are compared with recipient benefits at the top. Similarly a 5, corresponding to strong dominance or importance, is assigned to recipient benefits over social benefits in the (1, 3) position, and a 3, corresponding to moderate dominance, is assigned to institutional benefits over social benefits in the (2, 3) position with corresponding reciprocals in the transpose positions of the matrix.

REMARK 33 *In order to give the reader familiarity with the AHP without too much theory, we have delayed discussion of the measurement of the inconsistency and random inconsistency and of the ratio C.R. of the inconsistency of a given matrix and the corresponding random inconsistency to a later section. However, we have indicated the C.R. corresponding to each matrix immediately under that matrix.*

Judgments in a matrix may not be consistent. In eliciting judgments, one makes redundant comparisons to improve the validity of the answer, given that respondents may be uncertain or may make poor judgments in comparing some of the elements. Redundancy gives rise to multiple comparisons of an element with other elements and hence to numerical inconsistencies. For example, where we compare recipient benefits with institutional benefits and with societal benefits, we have the respective judgments 3 and 5. Now if $x = 3y$ and $x = 5z$ then $3y = 5z$ or $y = 5/3z$. If the judges were consistent, institutional benefits would be assigned the value 5/3 instead of the 3 given in the matrix. Thus the judgments are inconsistent. In fact, we are not sure which judgments are the

Table 9.3. The entries in this matrix respond to the question: Which criterion is more important with respect to choosing the best hospice alternative and how strongly?

Choosing best hospice	Recipient benefits	Institutional benefits	Social benefits	Priorities
Recipient benefits	1	3	5	.64
Institutional benefits	1/3	1	3	.26
Social benefits	1/5	1/3	1	.11
		C.R. = .033		

accurate ones and which are the cause of the inconsistency. Inconsistency is inherent in the judgment process. Inconsistency may be considered a tolerable error in measurement only when it is of a lower order of magnitude (10 %) than the actual measurement itself; otherwise the inconsistency would bias the result by a sizable error comparable to or exceeding the actual measurement itself.

When the judgments are inconsistent, the decision-maker may not know where the greatest inconsistency is. The AHP can show one by one in sequential order which judgments are the most inconsistent, and suggests the value that best improves consistency. However, this recommendation may not necessarily lead to a more accurate set of priorities that correspond to some underlying preference of the decision-maker. Greater consistency does not imply greater accuracy and one should go about improving consistency (if one can, given the available knowledge) by making slight changes compatible with one's understanding. If one cannot reach an acceptable level of consistency, one should gather more information or reexamine the framework of the hierarchy. For a 3-by-3 matrix this ratio should be about 5 %, for a 4-by-4 matrix about 8 %, and for larger matrices, about 10 %.

The process is repeated in all the matrices by asking the appropriate dominance or importance question. For example, for the matrix comparing the subcriteria of the parent criterion institutional benefits (Table 9.4), psycho-social benefits are regarded as very strongly more important than economic benefits, and 7 is entered in the (1, 2) position and 1/7 in the (2, 1) position.

In comparing the three models for patient care, we asked members of the planning association which model they preferred with respect to each of the covering or parent secondary criteria in level 3 or with respect to the tertiary criteria in level 4. For example, for the subcriterion direct care (located on the left-most branch in the benefits hierarchy), we obtained a matrix of paired comparisons (Table 9.5) in which Model 1 is preferred over Models 2 and 3 by

Table 9.4. The entries in this matrix respond to the question: Which subcriterion yields the greater benefit with respect to institutional benefits and how strongly?

Institutional benefits	Psycho-social	Economic	Priorities
Psycho-social	1	7	.875
Economic	1/7	1	.125
	C.R. = .000		

5 and 3 respectively and Model 3 is preferred by 3 over Model 2. The group first made all the comparisons using semantic terms for the fundamental scale and then translated them to the corresponding numbers.

Table 9.5. The entries in this matrix respond to the question: Which model yields the greater benefit with respect to direct care and how strongly?

Direct care of patient	Model I	Model II	Model III	Priorities
Model I unit team	1	5	3	.64
Model II mixed/home care	1/5	1	1/3	.10
Model III case management	1/3	3	1	.26
		C.R. = .033		

For the costs hierarchy, I again illustrate with three matrices. First the group compared the three major cost criteria and provided judgments in response to the question: which criterion is a more important determinant of the cost of a hospice model (Table 9.6)?

Table 9.6. The entries in this matrix respond to the question: Which criterion is a greater determinant of cost with respect to the care method and how strongly?

Choosing best hospice (costs)	Community	Institutional	Societal	Priorites
Community costs	1	1/5	1	.14
Institutional costs	5	1	5	.71
Societal costs	1	1/5	1	.14
		C.R. = .000		

The group then compared the subcriteria under institutional costs and obtained the importance matrix shown in Table 9.7.

Table 9.7. The entries in this matrix respond to the question: Which criterion incurs greater institutional costs and how strongly?

Institutional costs	Capital	Operating	Education	Bad debt	Recruit-ment	Priorities
Capital	1	1/7	1/4	1/7	1	.05
Operating	7	1	9	4	5	.57
Education	4	1/9	1	1/2	1	.01
Bad debt	7	1/4	2	1	3	.21
Recruitment	1	1/5	1	1/3	1	.07
			C.R. = .08			

Finally, we compared the three models to find out which incurs the highest cost for each criterion or subcriterion. Table 9.8 shows the results of comparing them with respect to the costs of recruiting staff.

Table 9.8. The entries in this matrix respond to the question: Which model incurs greater cost with respect to institutional costs for recruiting staff and how strongly?

Institutional costs for recruiting staff	Model I	Model II	Model III	Priorities
Model I unit team	1	5	3	.64
Model II mixed/home care	1/5	1	1/3	.10
Model III case management	1/3	3	1	.26
		C.R. = .000		

As shown in Table 9.9 we divided the benefits priorities by the costs priorities for each alternative to obtain the best alternative, Model 3, that with the largest value for the ratio.

Table 9.9 shows two ways or modes of synthesizing the local priorities of the alternatives using the global priorities of their parent criteria: The distributive mode and the ideal mode. In the distributive mode, the weights of the alternatives sum to one. It is used when there is dependence among the alternatives and a unit priority is distributed among them. The ideal mode is used to obtain the single best alternative regardless of what other alternatives there are. In the ideal mode, the local priorities of the alternatives under each criterion are divided by the largest value among them. This is done for each criterion; for each criterion one alternative becomes an ideal with value one. In both modes, the local priorities are weighted by the global priorities of the parent criteria and synthesized and the benefit-to-cost ratios formed. In Table 9.9 we rounded off the numbers to two decimal places. Unfortunately, that causes substantial difference from the actual results obtained in the AHP calculations. We request that the reader accept this as an illustration.

Table 9.9. Synthesis (P=Priorities, M=Model).

		Distributive Mode			Ideal Mode		
Benefits	P	M 1	M 2	M 3	M 1	M 2	M 3
Direct Care of Patient	.02	0.64	0.10	0.26	1.00	0.16	0.41
Palliative Care	.14	0.64	0.10	0.26	1.00	0.16	0.41
Volunteer Support	.02	0.09	0.17	0.74	0.12	0.23	1.00
Networking in Families	.06	0.46	0.22	0.32	1.00	0.48	0.70
Relief of Post Death Stress	.12	0.30	0.08	0.62	0.48	0.13	1.00
Emotional Support of Family and Patient	.21	0.30	0.08	0.62	0.48	0.13	1.00
Alleviation of Guilt	.03	0.30	0.08	0.62	0.48	0.13	1.00
Reduced Economic Costs for Patient	.01	0.12	0.65	0.23	0.18	1.00	0.35
Improved Productivity	.03	0.12	0.27	0.61	0.20	0.44	1.00
Publicity and Public Relations	.19	0.63	0.08	0.29	1.00	0.13	0.46
Volunteer Recruitment	.03	0.64	0.10	0.26	1.00	0.16	0.41
Professional Recruitment and Support	.06	0.65	0.23	0.12	1.00	0.35	0.18
Reduced Length of Stay	.01	0.26	0.10	0.64	0.41	0.41	1.00
Better Utilization of Resources	.02	0.09	0.22	0.69	0.13	0.13	1.00
Increased Monetary Support	.06	0.73	0.08	0.19	1.00	1.00	0.26
Death as a Social Issue	.02	0.20	0.20	0.60	0.33	0.33	1. 00
Rehumanization of Institutions	.08	0.24	0.14	0.62	0.39	0.23	1.00
Synthesis (taken from original AHP without approximations)		0.428	0.121	0.451	0.424	0.123	0.453

When the criteria priorities do not depend on the values of the alternatives with regard to those criteria, we need to derive their priorities by comparing them pairwise with each other with respect to higher-level criteria or goal. It is a process of trading off one unit of one criterion against a unit of another, an ideal alternative from one against an ideal alternative from another. To determine

Table 9.9 (continued)
Synthesis (P=Priorities, M=Model).

		Distributive Mode			Ideal Mode		
Costs	P	M 1	M 2	M 3	M 1	M 2	M 3
Community Costs	.14	0.33	0.33	0.33	1.00	1.00	1.00
Institutional Capital Costs	.03	0.76	0.09	0.15	1.00	0.12	0.20
Institutional Operating Costs	.40	0.73	0.08	0.19	1.00	0.11	0.26
Institutional Costs for Educating the Community	.01	0.65	0.24	0.11	1.00	0.37	0.17
Institutional Costs for Training Staff	.06	0.56	0.32	0.12	1.00	0.57	0.21
Institutional Bad Debt	.15	0.60	0.20	0.20	1.00	0.33	0.33
Institutional Costs of Recruiting Staff	.05	0.66	0.17	0.17	1.00	0.26	0.26
Institutional Costs of Recruiting Volunteers	.01	0.60	0.20	0.20	1.00	0.33	0.33
Societal Costs	.15	0.33	0.33	0.33	1.00	1.00	1.00
Synthesis (taken from original AHP without approximations)		0.583	0.192	0.224	0.523	0.229	0.249
Benefit/Cost Ratio		0.734	0.630	2.013	0.811	0.537	1.819

the ideal, the alternatives are divided by the largest value among them for each criterion. In that case, the process of weighting and adding assigns each of the remaining alternatives a value that is proportionate to the value 1 given to the highest rated alternative. In this way the alternatives are weighted by the priorities of the criteria and summed to obtain the weights of the alternatives. This is the ideal mode of the AHP.

The distributive mode is essential for synthesizing the weights of alternatives with respect to tangible criteria with the same scale of measurement into a single criterion for that scale and then they are treated as intangibles and compared pairwise and combined with other intangibles with the ideal mode. The dominant mode of synthesis in the AHP where the criteria are independent from the alternatives is the ideal mode. The standard mode for synthesizing in the ANP where criteria depend on alternatives and also alternatives may depend on other alternatives is the distributive mode.

In this case, both modes lead to the same outcome for hospice, which is Model 3. As we shall see below, we need both modes to deal with the effect of adding (or deleting) alternatives on an already ranked set. The priorities of the alternatives in the benefits hierarchy belong to an absolute scale of relative numbers and the priorities of the alternatives in the costs hierarchy also belong to another absolute scale of relative numbers. These two relative scales cannot be arbitrarily combined. Later we provide another way to combine them. In this exercise they were assumed to be commensurate and were combined in the traditional way by forming benefit to cost ratios. To derive the answer we divide the benefits priority of each alternative by its costs priority. We then choose the alternative with the largest of these ratios.

Model 3 has the largest benefit to cost ratio in both the distributive and ideal modes, and the hospital selected it for treating terminal patients. This need not always be the case. In this case, there is dependence of the personnel resources allocated to the three models because some of these resources would be shifted based on the decision. Therefore the distributive mode is the appropriate method of synthesis. If the alternatives were sufficiently distinct with no dependence in their definition, the ideal mode would be the way to synthesize.

I also performed marginal analysis to determine where the hospital should allocate additional resources for the greatest marginal return. To perform marginal analysis, I first ordered the alternatives by increasing cost priorities and then formed the benefit-to-cost ratios corresponding to the smallest cost, followed by the ratios of the differences of successive benefits to differences in costs. If this difference in benefits is negative, the new alternative is dropped from consideration and the process continued. The alternative with the largest ratio is then chosen. For the costs and corresponding benefits from the synthesis rows in Table 9.9 one obtains:

- Benefits: .12, .45, .43;
- Costs: .20, .21, .59;
- Ratios: $.12/.20 = .60$, $(.45-.12)/(.21-.20) = 33$, $(.43-.45)/(.59-.21) = -0.051$.

The third alternative is not a contender for resources because its marginal return is negative. The second alternative is the best. In fact, in addition to adopting the third model, the hospital management chose the second model of hospice care for further development.

5. Rating Alternatives One at a Time in the AHP – Absolute Measurement

The AHP has a second way to derive priorities known as absolute measurement. It involves making paired comparisons but the criteria just above the alternatives, known as the covering criteria, are assigned intensities that vary in number and type. For example they can simply be: high, medium and low; or they can be:

excellent, very good, good, average, poor and very poor; or for experience: more than 15 years, between 10 and 15, between 5 and 10 and less than 5 and so on. These intensities themselves are also compared pairwise to obtain their priorities as to importance, and they are then put in ideal form by dividing by the largest value. Finally each alternative is assigned an intensity, along with its accompanying priority, for each criterion. This process of assigning intensities is called rating the alternatives. The priority of each intensity is weighted by the priority of its criterion and summed over the weighted intensities for each alternative to obtain that alternative's final rating that also belongs to a ratio scale. It is often necessary to have categories of ratings for alternatives that are widely disparate so that one can rate the alternatives correctly. Ratings are useful when standards are established with which the alternatives must comply. They are also useful when the number of alternatives n is very large to perform pairwise comparisons on them for each criterion. In this case if the number of criteria is c, the number of rating operations in rating the alternatives is cn, whereas doing all the pairwise judgments involves $cn(n-1)/2$ comparisons. Here is an example of absolute measurement.

5.1 Evaluating Employees for Salary Raises

Employees are evaluated for raises. The criteria are Dependability, Education, Experience, and Quality. Each criterion is subdivided into intensities, standards, or discrimination categories as shown in Figure 9.4. Priorities are set for the criteria by comparing them in pairs. The intensities are then pairwise compared according to importance with respect to their parent criterion (example as in Table 9.10). Their priorities are often divided by the largest intensity for each criterion (second column of priorities in Figure 9.4) particularly useful in preserving the ranks of the alternatives from the addition or deletion of other alternatives. Finally, each individual is rated in Table 9.11 by assigning the intensity rating that applies to him or her under each criterion and adding. The score is obtained by weighting the intensities by the priority of their criteria and then summing over the criteria to derive a total score for each individual. This approach can be used whenever it is possible to set priorities for intensities of the criteria, which is usually possible when sufficient experience with a given operation has been accumulated. The raises can be made in proportion to the normalized values on the right.

One needs to choose the intensities widely enough by putting them in different order-of- magnitude categories in which the elements can be compared with the fundamental scale, and then combine the categories with pivots as in the cherry with watermelon example. Any alternative can be appropriately rated and receives its correct final value no matter how large or how small. When rating widely contrasting alternatives and the rating of an alternative is exceed-

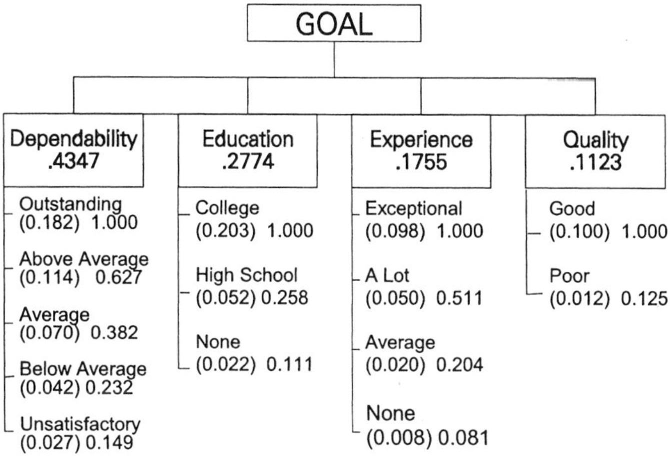

Figure 9.4. Employee evaluation hierarchy.

Table 9.10. Ranking intensities.

	Out-standing	Above Average	Average	Below Average	Unsatis-factory	Priorities
Outstanding	1.0	2.0	3.0	4.0	5.0	0.419
Above Average	1/2	1.0	2.0	3.0	4.0	0.263
Average	1/3	1/2	1.0	2.0	3.0	0.160
Below Average	1/4	1/3	1/2	1.0	2.0	0.097
Unsatisfactory	1/5	1/4	1/3	1/2	1.0	0.062
Inconsistency Ratio = 0.015						

ingly small with respect to a certain criterion, a zero value can be assigned to that alternative.

In ratings, adding new alternatives has no effect on the rank of existing alternatives. In paired comparisons the alternatives depend on each other and a new alternative can affect the relative ranks of existing alternatives. Using the ideal mode each time a new alternative is added prevents rank reversal with respect to irrelevant alternatives. However, if it is done only the first time and new alternatives are only compared with the first ideal so their values go above that ideal (more than one when necessary) there can be no rank reversal. It is clear that when alternatives are independent they can be rated one at a time and there would be no rank reversal. But even with independence, how many

Table 9.11. Ranking alternatives.

	Dependability	Education .2774	Experience .1775	Quality .1123	Total	Norm
1. Adams, V.	Outstand.	College	Excep.	Good	1.000	0.245
2. Becker, L.	Average	College	Average	Good	0.592	0.145
3. Hayat, F.	Average	College	A Lot	Good	0.645	0.158
4. Kesselman, S.	Above Av.	HighSch.	None	Poor	0.373	0.091
5. O'Shea, K.	Average	College	Average	Poor	0.493	0.121
6. Petres, T.	Average	College	None	Good	0.570	0.140
7. Tobias, K.	Average	None	A Lot	Poor	0.407	0.100

other alternatives of the same kind (sometimes also of a different kind) there are, can affect their rank. However, the number of alternatives cannot be used as a criterion for rating because it implies dependence of an alternative on how many others there are and a fortiori on their presence.

6. Paired Comparisons Imply Dependence

In most multicriteria decision problems the criteria are assumed independent of the alternatives and the alternatives independent of other alternatives. Paired comparisons imply dependence of a different kind. The common understanding is that when alternatives depend on each other it is according to their function like the electric industry depending on the coal industry for its output. In paired comparisons, the importance assigned to an alternative depends on what other alternatives it is compared with and how many there are. This is dependence not according to function but according to structure. This dependence happens even when the alternatives may be independent of each other according to function. Independence means that the rank of an alternative does not depend on what other alternatives there are and how many of them there may be. The situation with pairwise comparisons is that it automatically implies structural dependence. When a new alternative is added or an old one deleted the ranks of the other alternatives relative to each other may change. However one can preserve rank from adding new but irrelevant alternatives by creating an ideal alternative each time alternatives are added or deleted, or preserve it from any new alternative by simply idealizing the first time but never after and only comparing new alternatives with the first ideal and allowing the priority value of the new alternative to exceed one. Rating alternatives one at a time with appropriate and exhaustive orders of intensities for each criterion always preserves rank from structural effects, but is not always the best way to prioritize alternatives that may depend on the number and quality of other alternatives.

As we increase the number of copies of an alternative, it often loses (or conversely increases) its importance. For example, if gold, which is important, were to increase in quantity to fill the universe, it could lose its importance. No new criterion is added and no judgment is changed but only the quantity of gold. Relative measurementmeasurement,relative takes quantity into consideration. We often need to consider this kind of dependence known as structural dependence. When we add more alternatives, the ranks among old ones may change and what was preferred to another now because of the presence of new ones may no longer be preferred to the other. Another example is that of a company that sells cars A and B. Car B is better than car A but it costs more to make. It is more desirable all around for people to buy car B but they buy A because it is cheaper. The company advertises that it is going to make car C that is similar to B but much more expensive. People are now observed more and more to buy car B. The company never makes car C. This is a real life example from marketing. However, in some decision problems we may want to treat by fiat the alternatives of a decision as completely independent both in property and in number and quality and want to preserve the ranks of existing alternatives when new ones are added or old ones deleted. The AHP allows for both these possibilities. Actually, change in rank in the presence of relevant alternatives is a fact of our world. It is also a fact that when the number of irrelevant alternatives is very large, they can cause rank to change. Viruses are irrelevant in most decisions but they can eventually cause the death of all decision makers and make mockery of the decisions they thought were so important. In essence reality is much more interdependent than we have allowed for in our limited ways of thinking. Admittedly there are times when we wish to preserve rank no matter what the situation may be. We need to allow for both in our decision theories and not take the simple way out by always assuming independence.

7. When is a Positive Reciprocal Matrix Consistent?

In light of the foregoing, for the validity of the vector of priorities to describe response, we need greater redundancy and therefore also a large number of comparisons. Because of the reciprocal relation, in all we need $n(n-1)/2$ comparisons. An expert may provide $(n-1)$ comparisons to fill one row or a spanning tree from which the matrix is consistent and the priorities are easily obtained. Let us relate the psychological idea of the consistency of judgments and its measurement to a central concept in matrix theory and also to the size of our channel capacity to process information. Let $A = [a_{ij}]$ be an n-by-n positive reciprocal matrix, so all $a_{ii} = 1$ and $a_{ij} = 1/a_{ji}$ for all $i, j = 1, \cdots, n$. Let $w = [w_i]$ be the principal right eigenvector of A, let $D = diag(w_1, ..., w_n)$ be the n-by-n diagonal matrix whose main diagonal entries are the entries of w, and set $E = D^{-1}AD = [a_{ij}w_j/w_i] = [\varepsilon_{ij}]$. Then E is similar to A and is a positive reciprocal matrix since $\varepsilon_{ji} = (a_{ji}w_i/w_j) = (a_{ij}w_j/w_i)^{-1} = 1/\varepsilon_{ij}$. Moreover, all the row sums of E are equal to the principal eigenvalue of A:

$$\sum_{j=1}^{n} \varepsilon_{ij} = \sum_{j} \frac{a_{ij}w_j}{w_i} = \frac{[Aw]_i}{w_i} = \frac{\lambda_{max}w_i}{w_i} = \lambda_{\max}.$$

The computation

$$\begin{aligned} n\lambda_{\max} &= \sum_{i=1}^{n}(\sum_{j=1}^{n} \varepsilon_{ij}) = \sum_{i=1}^{n} \varepsilon_{ii} + \sum_{i,j=1\ i\neq j}^{n} (\varepsilon_{ij} + \varepsilon_{ji}) \\ &= n + \sum_{i,j=1\ i\neq j}^{n} (\varepsilon_{ij} + \varepsilon_{ij}^{-1}) \\ &\geq n + (n^2 - n) = n^2 \end{aligned}$$

reveals that $\lambda_{\max} \geq n$. Moreover, since $(x+1)/x \geq 2$ for all $x > 0$, with equality if and only if $x = 1$, we see that $\lambda_{\max} = n$ if and only if all $\varepsilon_{ij} = 1$, which is equivalent to having all $a_{ij} = w_i/w_j$.

The foregoing arguments show that a positive reciprocal matrix A has $\lambda_{\max} \geq n$, with equality if and only if A is consistent. When A is consistent we have $A^k = n^{k-1}A$. As our measure of deviation of A from consistency, we choose the *consistency index*

$$\mu = \frac{\lambda_{max} - n}{n-1}.$$

We have seen that $\mu \geq 0$ and $\mu = 0$ if and only if A is consistent. We can say that as $\mu \to 0$, $a_{ij} \to w_i/w_j$, or $\varepsilon_{ij} = a_{ij}(w_j/w_i) \to 1$. These two desirable properties explain the term "n" in the numerator of μ; what about the term "$n-1$" in the denominator? Since trace $(A) = n$ is the sum of all the eigenvalues of A, if we denote the eigenvalues of A that are different from $\lambda_{\max}$ by $\lambda_2, \cdots, \lambda_{n-1}$, we see that $n = \lambda_{\max} + \sum_{i=2}^{n} \lambda_i$, so $n - \lambda_{\max} = \sum_{i=2}^{n} \lambda_i$ and $\mu = -(1/(n-1)) \sum_{i=2}^{n} \lambda_i$ is the average of the non-principal eigenvalues of A.

Table 9.12. Random index.

Order	1	2	3	4	5	6	7	8	9	10
R.I.	0	0	0.52	0.89	1.11	1.25	1.35	1.40	1.45	1.49

In order to get some feel for what the consistency index might be telling us about a positive n-by-n reciprocal matrix A, consider the following simulation: choose the entries of A above the main diagonal at random from the 17 values

$\{1/9, 1/8, \cdots, 1, 2, \cdots, 8, 9\}$. Then fill in the entries of A below the diagonal by taking reciprocals. Put ones down the main diagonal and compute the consistency index. Do this many thousands of times and take the average, which we call the random index. Table 9.12 shows the values obtained from one set of such simulations and also their first order differences, for matrices of size $1, 2, \cdots, 10$.

A plot of the first two rows of Table 9.12 shows the asymptotic nature of random inconsistency. We also have shown that one should not compare more than about seven elements because increase in inconsistency is so small that it becomes difficult to perceive the ensuing small changes in the judgments needed to improve consistency [7]. In passing we note that there are several algorithms to change judgment to improve consistency, the best known among them is the gradient method of Patrick Harker [1, 3].

For a given positive reciprocal matrix $A = [a_{ij}]$ and a given pair of distinct indices $k > l$, define $A(t) = [a_{ij}(t)]$ by $a_{kl}(t) \equiv a_{kl} + t$, $a_{lk}(t) \equiv (a_{kl} + t)^{-1}$, and $a_{ij}(t) \equiv a_{ij}$ for all $i \neq k$, $j \neq l$, so $A(0) = A$. Let $\lambda_{\max}(t)$ denote the Perron eigenvalue of $A(t)$ for all t in a neighborhood of $t = 0$ that is small enough to ensure that all entries of the reciprocal matrix $A(t)$ are positive there. Finally, let $v = [v_i]$ be the unique positive eigenvector of the positive matrix A^T that is normalized so that $v^T w = 1$. Then a classical perturbation formula tells us that

$$\left.\frac{d\lambda_{\max}(t)}{da_{ji}}\right|_{t=0} = \frac{v^T A'(0)w}{v^T w} = v^T A'(0)w = v_k w_l - \frac{1}{a_{kl}^2} v_l w_k.$$

We conclude that

$$\frac{\partial \lambda_{\max}}{\partial a_{ij}} = v_i w_j - a_{ij}^2 v_j w_i, \quad i, j = 1, \cdots, n.$$

Because we are operating within the set of positive reciprocal matrices, $\partial\lambda_{\max}/\partial a_{ji} = -\partial\lambda_{\max}/\partial a_{ij}$ for all i and j. Thus, to identify an entry of A whose adjustment within the class of reciprocal matrices would result in the largest rate of change in $\lambda_{\max}$ we should examine the $n(n-1)/2$ values $\{v_i w_j - a_{ji}^2 v_j w_i\}, i > j$ and select (any) one of largest absolute value.

8. In the Analytic Hierarchy Process Additive Composition is Necessary

Sometimes people have assigned criteria different weights when they are measured in the same unit. Others have used different ways of synthesis than multiplying and adding. An example should clarify what we must do. Synthesis in the AHP involves weighting the priorities of elements compared with respect to an element in the next higher level, called a parent element, by the priority

Table 9.13. Calculating returns arithmetically.

Alternatives	Criterion C_1 Unnormalized weight = 1.0	Criterion C_2 Unnormalized weight = 1.0	Weighted Sum Unnormalized	Normalized or Relative values
A_1	200	150	350	350/1300=.269
A_2	300	50	350	350/1300=.269
A_3	500	100	600	600/1300=.462
Column totals	1000	300	1300	1

of that element and adding over all such parents for each element in the lower level. Consider the example of two criteria C_1 and C_2 and three alternatives A_1, A_2 and A_3 measured in the same scale such as dollars. If the criteria are each assigned the value 1, then the weighting and adding process produces the correct dollar value as in Table 9.13.

However, it does not give the correct outcome if the weights of the criteria are normalized, with each criterion having a weight of .5. Once the criteria are given in relative terms, so must the alternatives also be given in relative terms. A criterion that measures values in pennies cannot be as important as another measured in thousands of dollars. In this case, the only meaningful importance of a criterion is the ratio of the total money for the alternatives under it to the total money for the alternatives under both criteria. By using these weights for the criteria, rather than .5 and .5, one obtains the correct final relative values for the alternatives.

What is the relative importance of each criterion? Normalization indicates relative importance. Relative values require that criteria be examined as to their relative importance with respect to each other. What is the relative importance of a criterion, or what numbers should the criteria be assigned that reflect their relative importance? Weighting each criterion by the proportion of the resource under it, as shown in Table 9.14, and multiplying and adding as in the additive synthesis of the AHP, we get the same correct answer. For criterion C_1 we have (200+300+500)/[(200+300+500)+ (150+50+100)]=1000/1300 and for criterion C_2 we have $(150+50+100)/[(200+300+500)+(150+50+100)] = 300/1300$. Here the criteria are automatically in normalized form, and their weights sum to one. We see that when the criteria are normalized, the alternatives must also be normalized to get the right answer. For example, if we look in Table 9.13 we have 350/1300 for the priority of alternative A_1. Now if we simply weight and add the values for alternative A_1 in Table 9.14 we get for its final value $(200/1000)(1000/1300) + (150/300)(300/1300) = 350/1300$. It is clear that if the priorities of the alternatives are not normalized one does not get the desired outcome.

Table 9.14. Normalized criteria weights and normalized alternative weights from measurements in the same scale (additive synthesis).

Alternatives	Criterion C_1 Normalized weight = 1000/1300=0.7692	Criterion C_2 Normalized weight = 300/1300=0.2308	Weighted Sum
A_1	200/1000	150/300	350/1300 = .2692
A_2	300/1000	50/300	350/1300 = .2692
A_3	500/1000	100/300	600/1300 = .4615

We have seen in this example that in order to obtain the correct final relative values for the alternatives when measurements on a measurement scale are given, it is essential that the priorities of the criteria be derived from the priorities of the alternatives. Thus when the criteria depend on the alternatives we need to normalize the values of the alternatives to obtain the final result. This procedure is known as the distributive mode of the AHP. It is also used in case of functional dependence of the alternatives on the alternatives and of the criteria on the alternatives. The AHP is a special case of the Analytic Network Process. The dominant mode of synthesis in the ANP with all its interdependencies is the distributive mode. The ANP automatically assigns the criteria the correct weights, if one only uses the normalized values of the alternatives under each criterion and also the normalized values for each alternative under all the criteria without any special attention to weighting the criteria.

9. Benefits, Opportunities, Costs and Risks

In many decision problems four kinds of concerns or merits are considered: benefits, opportunities, costs and risks, which we abbreviate as BOCR. The first two are advantageous and hence are positive and the second two are disadvantageous and are therefore negative [5, 6]. Later we show how to determine the relative importance of each of the BOCR.

There are two ways to combine BOCR priorities. The first is the traditional one (used by economists) in which one does not need the relative importance of the BOCR by simply forming their ratio BO/CR for each alternative obtained from a separate hierarchy for each of the four BOCR merits and selecting that alternative with the largest ratio. It is known as the ratio outcome. The second derives corresponding normalized weights b, o, c, and r obtained respectively by rating the best alternative (one at a time) for each of the BOCR with respect to strategic criteria illustrated with an example later. One then forms for the four values of each alternative the expression

$$bB + oO - cC - rR.$$

The first way is a tradeoff between a unit of BO against a unit of CR, a unit of the desirable against a unit of the undesirable. It may be advisable, for example, that if the costs are considered to be negligibly smaller than the benefits to use only the benefits for the best alternative of a decision and not form the ratio and vice versa. The second way simply subtracts the sum of the weighted undesirables from the sum of the weighted desirables to give the total gain or loss. It can give rise to negative priorities and when applied to measurements in dollars, for example, where the weights b, o, c, and r are the same, gives back the correct answer. We have seen examples in which numbers or differences of numbers are made so small that one faces the classical problem of dividing by zero or comparing things whose measurements are near zero.

Two other formulas have been considered and set aside. They are $bB + oO + c/C + r/R$, and $bB + oO + c(1 - C) + r(1 - R)$. The first with only $bB + c/C$ makes the benefits determine the outcome when the cost is very high, which is counter intuitive. The second is always positive and is equal to $bB + oO - cC - rR + (c + r)$, and adds a constant to the subtractive formula $bB + oO - cC - rR$.

Note that there is no advantage in using the weights b, o, c and r in the formula BO/CR because we would be multiplying the result for each alternative by the same constant bo/cr. Because all values lie between zero and one, we have from the series expansions of the exponential and logarithmic functions the approximation:

$$\begin{aligned}\frac{bBoO}{cCrR} &= \exp(\log \mathrm{bB} + \log \mathrm{oO} - \log \mathrm{cC} - \log \mathrm{rR}) \\ &= 1 + (\log bB + \log oO - \log cC - \log rR) + \ldots \\ &\approx 1 + (bB - 1) + (oO - 1) - (cC - 1) - (rR - 1) \\ &= 1 + bB + oO - cC - rR.\end{aligned}$$

Because one is added to the overall value of each alternative we can eliminate it. The approximate result is that the ratio formula is similar to the total formula with equal weights assumed for the B, O, C, R.

10. On the Admission of China to the World Trade Organization (WTO)

This section was taken from an analysis done in 2000 carried out before the US Congress acted favorably on China joining the WTO and was hand-delivered to many of the members of the committee including its Chairperson. Since 1986, China had been attempting to join the multilateral trade system, the General Agreement on Tariffs and Trade (GATT) and, its successor, the World Trade Organization (WTO). According to the rules of the 135-member nations of WTO, a candidate member must reach a trade agreement with any existing member country that wishes to trade with it. By the time this analysis was done, China signed bilateral agreements with 30 countries – including the US (November 1999) out of 37 members that had requested a trade deal with it [5].

As part of its negotiation deal with the US, China asked the US to remove its annual review of China's Normal Trade Relations (NTR) status, until 1998 called Most Favored Nation (MFN) status. In March 2000, President Clinton sent a bill to Congress requesting a Permanent Normal Trade Relations (PNTR) status for China. The analysis was done and copies sent to leaders and some members in both houses of Congress before the House of Representatives voted on the bill, May 24, 2000. The decision by the US Congress on China's trade-relations status will have an influence on US interests, in both direct and indirect ways. Direct impacts include changes in economic, security and political relations between the two countries as the trade deal is actualized. Indirect impacts will occur when China becomes a WTO member and adheres to WTO rules and principles. China has said that it would join the WTO only if the US gives it Permanent Normal Trade Relations status.

It is likely that Congress will consider four options. The least likely is that the US will deny China both PNTR and annual extension of NTR status. The other three options are:

1 **Passage of a clean PNTR bill**: Congress grants China Permanent Normal Trade Relations status with no conditions attached. This option would allow implementation of the November 1999 WTO trade deal between China and the Clinton administration. China would also carry out other WTO principles and trade conditions.

2 **Amendment of the current NTR status bill**: This option would give China the same trade position as other countries and disassociate trade from other issues. As a supplement, a separate bill may be enacted to address other matters, such as human rights, labor rights, and environmental issues.

3 **Annual extension of NTR status**: Congress extends China's Normal Trade Relations status for one more year, and, thus, maintains the status quo.

The conclusion of the study is that the best alternative is granting China PNTR status. China now has that status.

Our analysis involves four steps. First, we prioritize the criteria in each of the benefits, costs, opportunities and risks hierarchies with respect to the goal. Figure 9.5 shows the resulting prioritization of these criteria. The alternatives and their priorities are shown under each criterion both in the distributive and in the ideal modes. The ideal priorities of the alternatives were used appropriately to synthesize their final values beneath each hierarchy.

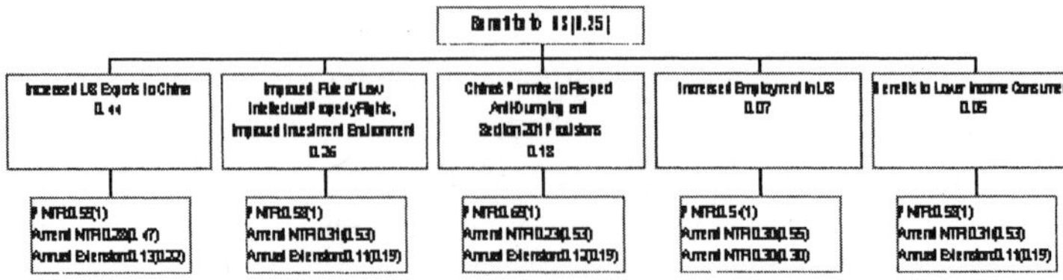

Benefits Synthesis (Ideal): PNTR 1.00; Amend NTR 0.51; Annual Extension 0.21

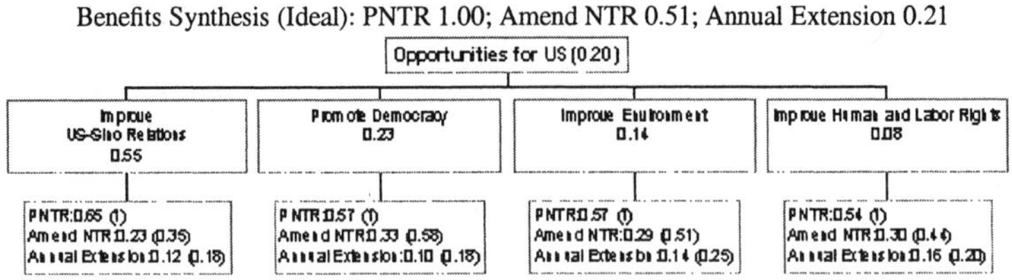

Opportunities Synthesis (Ideal): PNTR 1.00; Amend NTR 0.43; Annual Extension 0.13

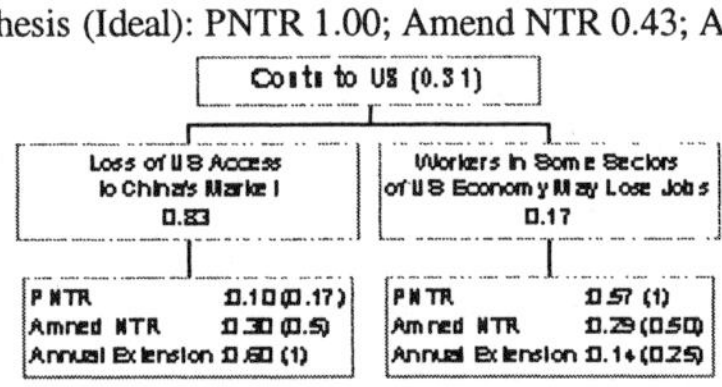

Costs Synthesis (more costly, Ideal): PNTR 0.31; Amend NTR 0.50; Annual Extension 0.87

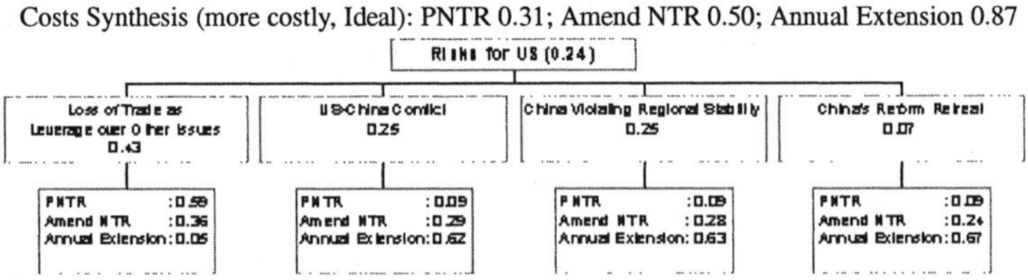

Risks Synthesis (more risky, Ideal): PNTR 0.54; Amend NTR 0.53; Annual Extension 0.58

Figure 9.5. Hierarchies for rating benefits, costs, opportunities, and risks.

The priorities shown in Figure 9.5 were derived from judgments that compared the elements involved in pairs. For readers to estimate the original pairwise judgments (not shown here) one forms the ratio of the corresponding two priorities shown, leave them as they are, or take the closest whole number, or its reciprocal if it is less than 1.0.

The idealized values are shown in parentheses after the original distributive priorities obtained from the eigenvector. The ideal values are obtained by dividing each of the distributive priorities by the largest one among them. For the

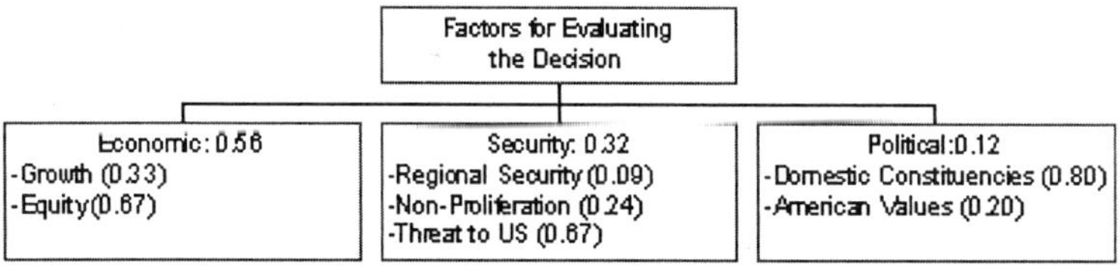

Figure 9.6. Prioritizing the strategic criteria to be used in rating the BOCR.

Costs and Risks structures, the question is framed as to which is the *most* costly or risky alternative. That is, the most costly alternative ends up with the highest priority.

Table 9.15. Priority Ratings for the Merits: Benefits, Costs, Opportunities, and Risks. Intensities: Very High (0.42), High (0.26), Medium (0.16), Low (0.1), Very Low (0.06).

		Benefits	Opportunities	Costs	Risks
Economic (0.56)	Growth (0.19)	High	Medium	Very Low	Very Low
	Equity (0.37)	Medium	Low	High	Low
Security (0.32)	Regional (0.03) (0.03)	Low	Medium	Medium	High
	Non-Proliferation (0.08)	Medium	High	Medium	High
	Threat to US (0.21)	High	High	Very High	Very High
Political (0.12)	Constituencies (0.1)	High	Medium	Very High	High
	American Values (0.02)	Very Low	Low	Low	Medium
Priorities		0.25	0.20	0.31	0.24

It is likely that, in a particular decision, the benefits, costs, opportunities and risks (BOCR) are not equally important, so we must also prioritize them. This is shown in Table 9.15. The priorities for the economic, security and political factors themselves were established as shown in Figure 9.6 and used to rate the importance of the top ideal alternative for each of the benefits, costs, opportunities and risks from Table 9.15. Finally, we used the priorities of the latter to combine the synthesized priorities of the alternatives in the four hierarchies, using both formulas BO/CR and $bB + oO - cC - rR$ to obtain their final ranking, as shown in Table 9.11.

How to derive the priority shown next to the goal of each of the four hierarchies in Figure 9.5 is outlined in Table 9.15. We rated each of the four merits: benefits, costs, opportunities and risks of the dominant PNTR alternative, as it happens to be in this case, in terms of intensities for each assessment criterion. The intensities, Very High, High, Medium, Low, and Very Low were themselves prioritized in the usual pairwise comparison matrix to determine their priorities. We then assigned the appropriate intensity for each merit on all assessment criteria. The outcome is as found in the bottom row of Table 9.15.

We are now able to obtain the overall priorities of the three major decision alternatives, given in the last two columns of Table 9.16. We see in bold that PNTR is the dominant alternative either way we synthesize as in the last two columns.

Table 9.16. Four methods of synthesizing BOCR using the ideal mode.

Alternatives	Benefits (0.25)	Opportunities (0.20)	Costs (0.31)	Risks (0.24)	BO/CR	$bB + oO$ $-cC - rR$
PNTR	1	1	0.31	0.54	**5.97**	**0.22**
Amend NTR	0.51	0.43	0.50	0.53	0.83	-0.07
Annual Exten.	0.21	0.13	0.87	0.58	0.05	-0.33

We have laid the basic foundation with hierarchies for what we need to deal with networks involving interdependencies. Let us now turn to that subject.

11. The Analytic Network Process (ANP)

To simplify and deal with complexity, people who work in decision-making use mostly very simple hierarchic structures consisting of a goal, criteria, and alternatives. Yet, not only are decisions obtained from a simple hierarchy of three levels different from those obtained from a multilevel hierarchy, but also decisions obtained from a network can be significantly different from those obtained from a multilevel hierarchy. We cannot collapse complexity artificially into a simplistic structure of two levels, criteria and alternatives, and hope to capture the outcome of interactions in the form of highly condensed judgments that correctly reflect all that goes on in the world. For 30 years we have worked with people to decompose these judgments through more elaborate structures to organize our reasoning and calculations in sophisticated but simple ways to serve our understanding of the complexity around us. Experience indicates that it is not very difficult to do this although it takes more time and effort, but not too much more. We have consulted and lectured on this subject in many countries: extensively in the US, in Brazil, Chile, the Czech Republic, Turkey, Poland, Indonesia, Switzerland, and soon in England and in China. There seems to be worldwide interest in decisions with dependence and feedback. My book on this subject has been translated to two languages. *Indeed, we must use feedback networks to arrive at the kind of decisions needed to cope with the future.*

Many decision problems cannot be structured hierarchically because they involve the interaction and dependence of higher-level elements in a hierarchy on lower-level elements. Not only does the importance of the criteria determine the importance of the alternatives as in a hierarchy, but also the importance of the alternatives themselves determines the importance of the criteria. Two elephants chosen for work should have powerful trunks. One of them is slightly stronger but has only one ear. Strength alone would lead one to choose the strong but less attractive elephant unless the criteria of strength and attractiveness are evaluated in terms of the elephants, and strength receives a smaller value, and appearance a larger value because both elephants are strong. Feedback also enables us to factor the future into the present to determine what we have to do to attain a desired future. The Analytic Network Process is a generalization of the Analytic Hierarchy Process. The basic structures are networks. Priorities are established in the same way they are in the AHP using pairwise comparisons and judgments.

The feedback structure does not have the top-to-bottom form of a hierarchy but looks more like a network, with cycles connecting its components of elements, which we can no longer call levels, and with loops that connect a component to itself (see Figure 9.7). It also has sources and sinks. A **source** node is an origin of paths of influence (importance) and never a destination of such paths. A **sink** node is a destination of paths of influence and never an origin of such paths. A full network can include source nodes; intermediate nodes that fall on paths from source nodes, lie on cycles,cycle or fall on paths to sink nodes; and finally sink nodes. Some networks can contain only source and sink nodes. Still others can include only source and cycle nodes or cycle and sink nodes or only cycle nodes. A decision problem involving feedback arises often in practice. It can take on the form of any of the networks just described. The challenge is to determine the priorities of the elements in the network and in particular the alternatives of the decision and to justify the validity of the outcome. Because feedback involves cycles, and cycling is an infinite process, the operations needed to derive the priorities become more demanding than is familiar with hierarchies.

To obtain the overall dependence of elements such as the criteria, one proceeds as follows: Construct a zero-one matrix of criteria against criteria using the number one to signify dependence of one criterion on another, and zero otherwise. A criterion need not depend on itself as an industry, for example, may not use its own output. For each column of this matrix, construct a pairwise comparison matrix only for the dependent criteria, derive an eigenvector, and augment it with zeros for the excluded criteria. If a column is all zeros, then assign a zero vector to represent the priorities. The question in the comparison would be: For a given criterion, which of two criteria depends more on that criterion with respect to the goal or with respect to a higher-order controlling criterion?

In Figure 9.7, a view is shown of a hierarchy and a network. A hierarchy is comprised of a goal, levels of elements and connections between the elements. These connections go only to elements in lower levels. A network has clusters of elements, with the elements being connected to elements in another cluster (outer dependence) or the same cluster (inner dependence). A hierarchy is a special case of a network with connections going only in one direction. In a view of a hierarchy, such as that shown in Figure 9.7, the levels in the hierarchy correspond to clusters in a network. One example of inner dependence in a component consisting of a father mother and baby is whom does the baby depend on more for its survival, its mother or itself. The baby depends more on its mother than on itself. Again suppose one makes advertising by newspaper and by television. It is clear that the two influence each other because the newspaper writers watch television and need to make their message unique in some way, and vice versa. If we think about it carefully everything can be seen to influence everything including itself according to many criteria. The world is far more interdependent than we know how to deal with using our existing ways of thinking and acting. We know it but how to deal with it. The ANP appears to be a plausible logical way to deal with dependence.

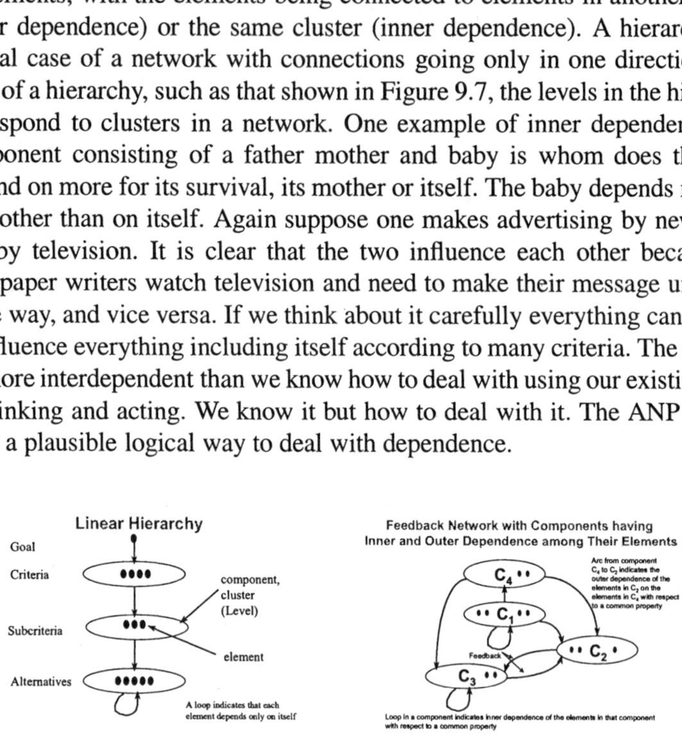

Figure 9.7. How a hierarchy compares to a network.

The priorities derived from pairwise comparison matrices are entered as parts of the columns of a supermatrix. The supermatrix represents the influence priority of an element on the left of the matrix on an element at the top of the matrix. A supermatrix along with an example of one of its general entry matrices is shown in Figure 9.8. The component C_i in the supermatrix includes all the priority vectors derived for nodes that are "parent" nodes in the C_i cluster. Figure 9.9 gives the supermatrix of a hierarchy along with the kth power that yields the principle of hierarchic composition in its $(k, 1)$ position.

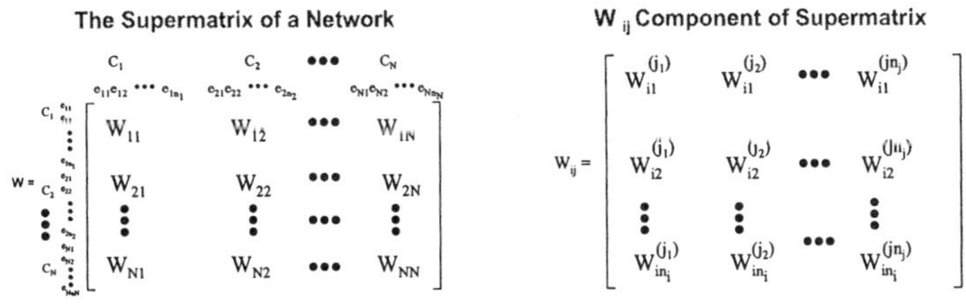

Figure 9.8. The supermatrix of a network and detail of a component in it.

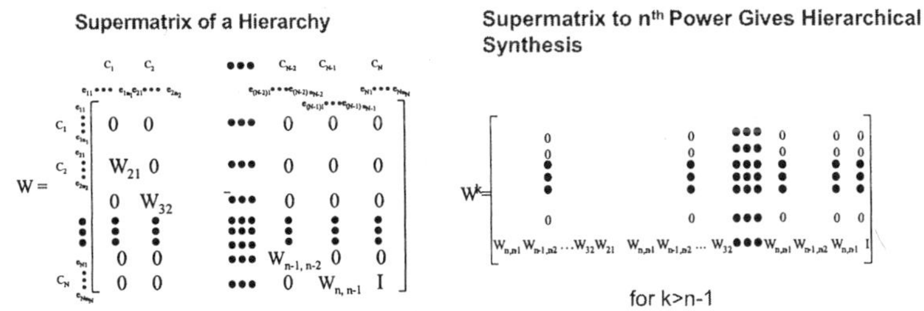

Figure 9.9. The supermatrix of a hierarchy with the resulting limit matrix corresponding to hierarchical composition.

Hierarchic composition yields multilinear forms that are of course nonlinear and have the form

$$\sum_{i_1,\cdots,i_p} x_1^{i_1} x_2^{i_2} \cdots x_p^{i_p}$$

where i_j indicates the jth level of the hierarchy and the x_j is the priority of an element in that level. The richer the structure of a hierarchy in breadth and depth, the more elaborate are the multilinear forms derived from it. There seems to be a good opportunity to investigate the relationship obtained by composition to covariant tensors and their algebraic properties. Powers of a variable allow for the possibility that the variable is repeated in the composition. Multilinear forms are related to polynomials and these by the Stone-Weierstrass theorem can be used to approximate arbitrarily close to continuous functions. Such functions may be assumed to underlie the representations of complex events in a decision. In this manner, mathematics and the apparent complicated use of numbers in decision-making can be related in a way that one can understand.

More concretely we have the covariant tensor

$$w_i^h = \sum_{i_2,\cdots,i_{h-1}=1}^{N_{h-1},\cdots,N_1} w_{i_1,i_2}^{h-1} \cdots w_{i_{h-2},i_{h-1}}^{2} w_{i_{h-1}}^{1} \qquad i_1 \equiv i$$

for the priority of the ith element in the hth level of the hierarchy. The composite vector W^h for the entire hth level is represented by the vector with covariant tensorial components. Similarly, the left eigenvector approach to a hierarchy gives rise to a vector with contravariant tensor components.

The classical problem of relating space (geometry) and time to subjective thought can perhaps be examined by showing that the functions of mathematical analysis (and hence also the laws of physics) are derivable as truncated series from the above tensors by composition in an appropriate hierarchy. The foregoing is reminiscent of the theorem in dimensional analysis that any physical variable is proportional to the product of powers of primary variables.

Multilinear forms are obviously nonlinear and are a powerful building stone to go from linearity to non-linearity through the use of complex structures (hierarchies and networks) and enable us to deal with the world according to our deepest ways of understanding and judgment.

In the ANP we look for steady state priorities from a limit supermatrix. To obtain the limit we must raise the matrix to powers. The reason for that is that to capture overall influence (dominance) one must consider all transitivities of different length. These are each represented by the corresponding power of the supermatrix. For each such matrix, the influence of an element on all others is obtained by taking the sum of its corresponding row. If we do that for all the elements, we obtain a vector of influence from that matrix. The sum of all such vectors gives the overall influence. Cesaro summability tells us that it is sufficient to obtain the outcome from the limiting power of the supermatrix.

The outcome of the ANP is nonlinear and rather complex. We know, from a theorem due to J.J. Sylvester that when the multiplicity of each eigenvalue of a matrix W is equal to one that an entire function $f(x)$ (power series expansion of $f(x)$ converges for all finite values of x) with x replaced by W, is given by

$$\begin{aligned} f(W) &= \sum_{i=1}^{n} f(\lambda_i) Z(\lambda_i), \\ Z(\lambda_i) &= \frac{\prod_{j\neq i}(\lambda_j I - A)}{\prod_{j\neq i}(\lambda_j - \lambda_i)}, \end{aligned}$$

$$\sum_{i=1}^{n} Z(\lambda_i) = 1, \quad Z(\lambda_i)Z(\lambda_j) = 0, \quad Z^2(\lambda_i) = Z(\lambda_i),$$

where I and 0 are the identity and null matrices respectively.

A similar expression is also available when some or all of the eigenvalues have multiplicities greater than one. We can easily see that if, as we need in our case, $f(W) = W^k$, then $f(\lambda_i) = \lambda_i^k$ and as $k \to \infty$ the only terms that give a finite nonzero value are those for which the modulus of λ_i is equal to one.

The fact that W is stochastic ensures this because

$$\max \sum_{j=1}^{n} a_{ij} \geq \sum_{j=1}^{n} a_{ij} \frac{w_j}{w_i} = \lambda_{\max} \text{ for } \max w_i$$

$$\min \sum_{j=1}^{n} a_{ij} \leq \sum_{j=1}^{n} a_{ij} \frac{w_j}{w_i} = \lambda_{\max} \text{ for } \min w_i$$

Thus for a row stochastic matrix we have

$$1 = \min \sum_{j=1}^{n} a_{ij} \leq \lambda_{\max} \leq \max \sum_{j=1}^{n} a_{ij} = 1,$$

and $\lambda_{\max} = 1$. See this author's 2001 book on the ANP [4], and also the manual for the ANP software [2]. Here are two examples that illustrate the validity of the supermatrix as a general framework for prioritization. The first as a generalization of hierarchies that gives back hierarchic answers, and the second as a method of computation and synthesis that carries the burden of computation with the user mostly providing judgments.

11.1 The Classic AHP School Example as an ANP Model

We show in Figures 9.10a and 9.10b below the hierarchy, and its corresponding supermatrix, and its limit supermatrix to obtain the priorities of three schools involved in a decision to choose one for the author's son. They are precisely what one obtains by hierarchic composition using the AHP. Figure 9.10a shows the priorities of the criteria with respect to the goal and those of the alternatives with respect to each criterion. There is an identity submatrix for the alternatives with respect to the alternatives in the lower right hand part of the matrix, because each alternative depends on itself. The level of alternatives in a hierarchy is a sink cluster of nodes that absorbs priorities but does not pass them on. This calls for using an identity submatrix for them in the supermatrix. The last three entries of column one of Figure 9.10b give the overall priorities of the alternatives with respect to the goal.

11.2 Criteria Weights Automatically Derived from Supermatrix

Let us revisit the example we gave earlier in Table 9.13 of three alternatives and two criteria measured in the same unit. We use interdependence to determine

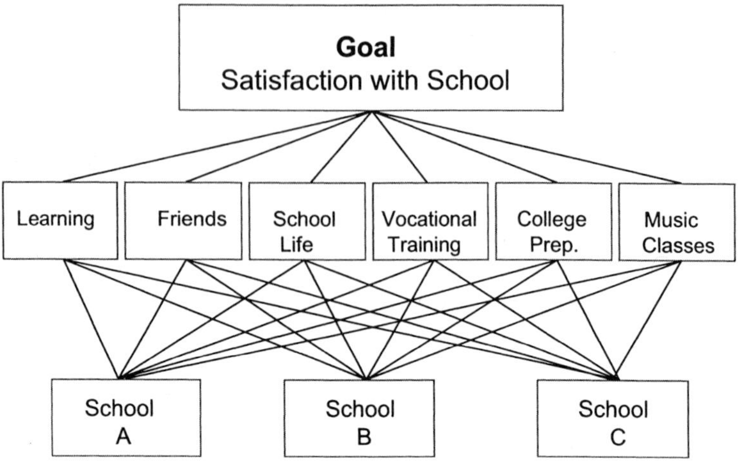

Figure 9.10a. School choice hierarchy composition.

The School Hierarchy as Supermatrix

	Goal	Learning	Friends	School life	Vocational training	College preparation	Music classes	A	B	C
Goal	0	0	0	0	0	0	0	0	0	0
Learning	0.32	0	0	0	0	0	0	0	0	0
Friends	0.14	0	0	0	0	0	0	0	0	0
School life	0.03	0	0	0	0	0	0	0	0	0
Vocational training	0.13	0	0	0	0	0	0	0	0	0
College preparation	0.24	0	0	0	0	0	0	0	0	0
Music classes	0.14	0	0	0	0	0	0	0	0	0
Alternative A	0	0.16	0.33	0.45	0.77	0.25	0.69	1	0	0
Alternative B	0	0.59	0.33	0.09	0.06	0.50	0.09	0	1	0
Alternative C	0	0.25	0.34	0.46	0.17	0.25	0.22	0	0	1

Limiting Supermatrix & Hierarchic Composition

	Goal	Learning	Friends	School life	Vocational training	College preparation	Music classes	A	B	C
Goal	0	0	0	0	0	0	0	0	0	0
Learning	0	0	0	0	0	0	0	0	0	0
Friends	0	0	0	0	0	0	0	0	0	0
School life	0	0	0	0	0	0	0	0	0	0
Vocational training	0	0	0	0	0	0	0	0	0	0
College preparation	0	0	0	0	0	0	0	0	0	0
Music classes	0	0	0	0	0	0	0	0	0	0
Alternative A	0.3676	0.16	0.33	0.45	0.77	0.25	0.69	1	0	0
Alternative B	0.3781	0.59	0.33	0.09	0.06	0.50	0.09	0	1	0
Alternative C	0.2543	0.25	0.34	0.46	0.17	0.25	0.22	0	0	1

Figure 9.10b. Supermatrix of school choice hierarchy gives same results as hierarchic composition.

what overall weight the criteria should have without computing the relative sum of the alternatives under each criterion to the total. Since we are dealing with tangibles we normalize each column to obtain the priorities for the alternatives under each criterion. We also normalize each row to obtain the priorities of the criteria with respect to each alternative. We enter these in a supermatrix as

Table 9.17. The supermatrix.

		Alternatives			Criteria	
		A_1	A_2	A_3	C_1	C_2
Alternatives	A_1	0.000	0.000	0.000	0.200	0.500
	A_2	0.000	0.000	0.000	0.300	0.167
	A_3	0.000	0.000	0.000	0.500	0.333
Criteria	C_1	0.571	0.857	0.833	0.000	0.000
	C_2	0.429	0.143	0.167	0.000	0.000

Table 9.18. The limit supermatrix.

		Alternatives			Criteria	
		A_1	A_2	A_3	C_1	C_2
Alternatives	A_1	0.135	0.135	0.135	0.135	0.135
	A_2	0.135	0.135	0.135	0.135	0.135
	A_3	0.231	0.231	0.231	0.231	0.231
Criteria	C_1	0.385	0.385	0.385	0.385	0.385
	C_2	0.115	0.115	0.115	0.115	0.115

shown in Table 9.17; there is no need to weight the supermatrix because it is already column stochastic, so we can raise it to limiting powers right away and obtain the limit supermatrix in Table 9.18 in which, in this case it turns out that, all the columns are identical.

12. Two Examples of Estimating Market Share – The ANP with a Single Benefits Control Criterion

A market share estimation model is structured as a network of clusters and nodes. The object is to determine the relative market share of competitors in a particular business, or endeavor, by considering what affects market share in that business and introducing them as clusters, nodes and influence links in a network. No actual statistics are used in these examples, but only judgments by experts about relative influence. The decision alternatives are the competitors and the synthesized results are their relative dominance. The relative dominance results can then be compared against some outside measure such as dollars. If dollar income is the measure being used, the incomes of the competitors must be normalized to get it in terms of relative market share.

The clusters might include customers, service, economics, advertising, and quality of goods. The customers cluster might then include nodes for the age groups of the people that buy from the business: teenagers, 20-33 year olds,

34-55 year olds, 55-70 year olds, and over 70. The advertising cluster might include newspapers, TV, Radio, and Fliers. After all the nodes are created start by picking a node and linking it to the other nodes in the model that influence it. The "children" nodes will then be pairwise compared with respect to that node as a "parent" node. An arrow will automatically appear going from the cluster the parent node is in to the cluster with its children nodes. When a node is linked to nodes in its own cluster, the arrow becomes a loop on that cluster and we say there is inner dependence.

The linked nodes in a given cluster are pairwise compared for their influence on the node they are linked from (the parent node) to determine the priority of their influence on the parent node. Comparisons are made as to which is more important to the parent node in capturing "market share". These priorities are then entered in the supermatrix.

The clusters are also pairwise compared to establish their importance with respect to each cluster they are linked from, and the resulting matrix of numbers is used to weight the components of the original unweighted supermatrix to give the weighted supermatrix. This matrix is then raised to powers until it converges to give the limit supermatrix. The relative values for the companies are obtained from the columns of the limit supermatrix that in this case, with the help of Cesaro summability, are reduced in the software to be all the same. Normalizing these numbers yields the relative market share.

If comparison data in terms of sales in dollars, or number of members, or some other known measures are available, one can use their relative values to validate the outcome. The AHP/ANP has a compatibility metric to determine how close the ANP result is to the known measure. It involves taking the Hadamard product of the matrix of ratios of the ANP outcome and the transform of the matrix of ratios of the actual outcome summing all the coefficients and dividing by n^2. The requirement is that the value should be close to 1 and certainly not much more than 1.1.

We will give two examples of market share estimation showing details of the process in the first example and showing only the models and results in the second.

12.1 Example 1. Estimating the Relative Market Share of Walmart, Kmart and Target

The network for the ANP model shown in Figure 9.11 describes quite well the influences that determine the market share of these companies. We will not use space in this chapter to describe the clusters and their nodes in greater detail.

12.1.1 The Unweighted Supermatrix. The unweighted supermatrix is constructed from the priorities derived from the different pairwise comparisons.

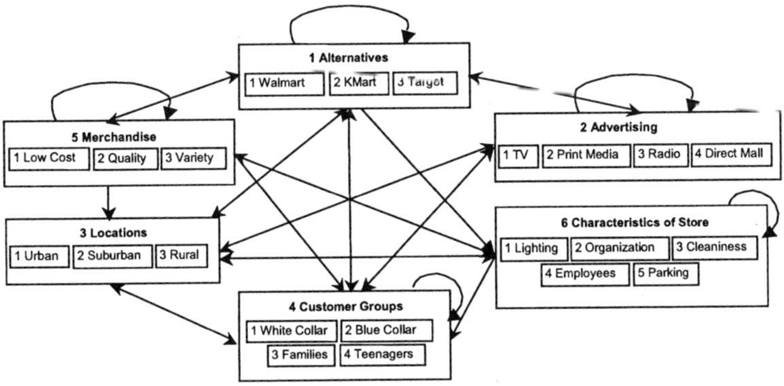

Figure 9.11. The clusters and nodes of a model to estimate the relative market share of Walmart, Kmart and Target.

The nodes, grouped by the clusters they belong to, are the labels of the rows and columns of the supermatrix. The column for a node a contains the priorities of the nodes that have been pairwise compared with respect to a. The supermatrix for the network in Figure 9.11 is shown in Table 9.19. In Tables 9.19 – 9.21 the following abbreviations have been used:

- Al - Alternatives, WM - Walmart, KM - KMart, Ta - Target;
- Ad - Advertising, TV, PM - Print Media, Ra - Radio, DM - Direct Mail;
- Lo - Location, Ur - Urban, Su - Suburban, Ru - Rural;
- CG - Custommer Groups, WC - White Collar, BC - Blue Collar, Fa - Families, Te - Teenagers;
- Me - Merchandise, LC - Low Cost, Qu - Quality, Va - Variety;
- CS - Characteristics of Store, Li - Lighting, Or - Organization, Cl - Cleanliness, Em - Employees, Pa - Parking.

Table 9.19. The unweighted supermatrix.

		1 Al			2 Ad				3 Lo		
		1 WM	2 KM	3 Ta	1 TV	2 PM	3 Ra	4 DM	1 Ur	2 Su	3 Ru
1 Al	1 WM	0.000	0.833	0.833	0.687	0.540	0.634	0.661	0.614	0.652	0.683
	2 KM	0.750	0.000	0.167	0.186	0.297	0.174	0.208	0.268	0.235	0.200
	3 Ta	0.250	0.167	0.000	0.127	0.163	0.192	0.131	0.117	0.113	0.117
2 Ad	1 TV	0.553	0.176	0.188	0.000	0.000	0.000	0.000	0.288	0.543	0.558
	2 PM	0.202	0.349	0.428	0.750	0.000	0.800	0.000	0.381	0.231	0.175
	3 Ra	0.062	0.056	0.055	0.000	0.000	0.000	0.000	0.059	0.053	0.048
	4 DM	0.183	0.420	0.330	0.250	0.000	0.200	0.000	0.273	0.173	0.219
3 Lo	1 Ur	0.114	0.084	0.086	0.443	0.126	0.080	0.099	0.000	0.000	0.000
	2 Su	0.405	0.444	0.628	0.387	0.416	0.609	0.537	0.000	0.000	0.000
	3 Ru	0.481	0.472	0.285	0.169	0.458	0.311	0.364	0.000	0.000	0.000
4 CG	1 WC	0.141	0.114	0.208	0.165	0.155	0.116	0.120	0.078	0.198	0.092
	2 BC	0.217	0.214	0.117	0.165	0.155	0.198	0.203	0.223	0.116	0.224
	3 Fa	0.579	0.623	0.620	0.621	0.646	0.641	0.635	0.656	0.641	0.645
	4 Te	0.063	0.049	0.055	0.048	0.043	0.045	0.041	0.043	0.045	0.038
5 Me	1 LC	0.362	0.333	0.168	0.000	0.000	0.000	0.000	0.000	0.000	0.000
	2 Qu	0.261	0.140	0.484	0.000	0.000	0.000	0.000	0.000	0.000	0.000
	3 Va	0.377	0.528	0.349	0.000	0.000	0.000	0.000	0.000	0.000	0.000
6 CS	1 Li	0.000	0.000	0.000	0.000	0.000	0.000	0.000	0.000	0.000	0.000
	2 Or	0.000	0.000	0.000	0.000	0.000	0.000	0.000	0.000	0.000	0.000
	3 Cl	0.000	0.000	0.000	0.000	0.000	0.000	0.000	0.000	0.000	0.000
	4 Em	0.000	0.000	0.000	0.000	0.000	0.000	0.000	0.000	0.000	0.000
	5 Pa	0.000	0.000	0.000	0.000	0.000	0.000	0.000	0.000	0.000	0.000

Table 9.19 (continued)
The unweighted supermatrix.

		4 CG				5 Me			6 CS				
		1 WC	2 BC	3 Fa	4 Te	1 LC	2 Qu	3 Va	1 Li	2 Or	3 Cl	4 Em	5 Pa
1 Al	1 WM	0.637	0.661	0.630	0.691	0.661	0.614	0.648	0.667	0.655	0.570	0.644	0.558
	2 KM	0.105	0.208	0.218	0.149	0.208	0.117	0.122	0.111	0.095	0.097	0.085	0.122
	3 Ta	0.258	0.131	0.151	0.160	0.131	0.268	0.230	0.222	0.250	0.333	0.271	0.320
2 Ad	1 TV	0.323	0.510	0.508	0.634	0.000	0.000	0.000	0.000	0.000	0.000	0.000	0.000
	2 PM	0.214	0.221	0.270	0.170	0.000	0.000	0.000	0.000	0.000	0.000	0.000	0.000
	3 Ra	0.059	0.063	0.049	0.096	0.000	0.000	0.000	0.000	0.000	0.000	0.000	0.000
	4 DM	0.404	0.206	0.173	0.100	0.000	0.000	0.000	0.000	0.000	0.000	0.000	0.000
3 Lo	1 Ur	0.167	0.094	0.096	0.109	0.268	0.105	0.094	0.100	0.091	0.091	0.111	0.067
	2 Su	0.833	0.280	0.308	0.309	0.117	0.605	0.627	0.433	0.455	0.455	0.444	0.293
	3 Ru	0.000	0.627	0.596	0.582	0.614	0.291	0.280	0.466	0.455	0.455	0.444	0.641
4 CG	1 WC	0.000	0.000	0.279	0.085	0.051	0.222	0.165	0.383	0.187	0.242	0.165	0.000
	2 BC	0.000	0.000	0.649	0.177	0.112	0.159	0.165	0.383	0.187	0.208	0.165	0.000
	3 Fa	0.857	0.857	0.000	0.737	0.618	0.566	0.621	0.185	0.583	0.494	0.621	0.000
	4 Te	0.143	0.143	0.072	0.000	0.219	0.053	0.048	0.048	0.043	0.056	0.048	0.000
5 Me	1 LC	0.000	0.000	0.000	0.000	0.000	0.800	0.800	0.000	0.000	0.000	0.000	0.000
	2 Quality	0.000	0.000	0.000	0.000	0.750	0.000	0.200	0.000	0.000	0.000	0.000	0.000
	3 Variety	0.000	0.000	0.000	0.000	0.250	0.200	0.000	0.000	1.000	0.000	0.000	0.000
6 CS	1 Li	0.000	0.000	0.000	0.000	0.000	0.000	0.000	0.000	0.169	0.121	0.000	0.250
	2 Or	0.000	0.000	0.000	0.000	0.000	0.000	0.000	0.251	0.000	0.575	0.200	0.750
	3 Cl	0.000	0.000	0.000	0.000	0.000	0.000	0.000	0.673	0.469	0.000	0.800	0.000
	4 Em	0.000	0.000	0.000	0.000	0.000	0.000	0.000	0.000	0.308	0.304	0.000	0.000
	5 Pa	0.000	0.000	0.000	0.000	0.000	0.000	0.000	0.075	0.055	0.000	0.000	0.000

Table 9.20. The cluster matrix.

	1. Al	2. Ad	3. Li	4. CG	5. Me	6. CG
1. Alternatives (Al)	0.137	0.174	0.094	0.057	0.049	0.037
2. Advertising (Ad)	0.091	0.220	0.280	0.234	0.000	0.000
3. Locations (Li)	0.276	0.176	0.000	0.169	0.102	0.112
4. Customer Groups (CG)	0.054	0.429	0.627	0.540	0.252	0.441
5. Merchandise (Me)	0.442	0.000	0.000	0.000	0.596	0.316
6. Characteristics of Store (CS)	0.000	0.000	0.000	0.000	0.000	0.094

12.1.2 The Cluster Matrix. The cluster themselves must be compared to establish their relative importance and use it to weight the supermatrix to make it column stochastic. A cluster impacts another cluster when it is linked from it, that is, when at least one node in the source cluster is linked to nodes in the target cluster. The clusters linked from the source cluster are pairwise compared for the importance of their impact on it with respect to market share, resulting in the column of priorities for that cluster in the cluster matrix. The process is repeated for each cluster in the network to obtain the matrix shown in Table 9.20. An interpretation of the priorities in the first column is that Merchandise (0.442) and Locations (0.276) have the most impact on Alternatives, the three competitors.

12.1.3 The Weighted Supermatrix. The weighted supermatrix shown in Table 9.21 is obtained by multiplying each entry in a block of the component at the top of the supermatrix by the priority of influence of the component on the left from the cluster matrix in Table 9.20. For example, the first entry, 0.137, in Table 9.20 is used to multiply each of the nine entries in the block (Alternatives, Alternatives) in the unweighted supermatrix shown in Table 9.19. This gives the entries for the (Alternatives, Alternatives) component in the weighted supermatrix of Table 9.21. Each column in the weighted supermatrix has a sum of 1, and thus the matrix is stochastic.

Table 9.21. The weighted supermatrix.

		1 Al			2 Ad				3 Lo		
		1 WM	2 KM	3 Ta	1 TV	2 PM	3 Ra	4 DM	1 Ur	2 Su	3 Ru
1 Al	1 WM	0.000	0.114	0.114	0.120	0.121	0.110	0.148	0.058	0.061	0.064
	2 KM	0.103	0.000	0.023	0.033	0.066	0.030	0.047	0.025	0.022	0.019
	3 Ta	0.034	0.023	0.000	0.022	0.037	0.033	0.029	0.011	0.011	0.011
2 Ad	1 TV	0.050	0.016	0.017	0.000	0.000	0.000	0.000	0.080	0.152	0.156
	2 PM	0.018	0.032	0.039	0.165	0.000	0.176	0.000	0.106	0.064	0.049
	3 Ra	0.006	0.005	0.005	0.000	0.000	0.000	0.000	0.016	0.015	0.014
	4 DM	0.017	0.038	0.030	0.055	0.000	0.044	0.000	0.076	0.048	0.061
3 Lo	1 Ur	0.031	0.023	0.024	0.078	0.028	0.014	0.022	0.000	0.000	0.000
	2 Su	0.112	0.123	0.174	0.068	0.094	0.107	0.121	0.000	0.000	0.000
	3 Ru	0.133	0.130	0.079	0.030	0.103	0.055	0.082	0.000	0.000	0.000
4 CG	1 WC	0.008	0.006	0.011	0.071	0.086	0.050	0.066	0.049	0.124	0.058
	2 BC	0.012	0.011	0.006	0.071	0.086	0.085	0.112	0.140	0.073	0.141
	3 Fa	0.031	0.033	0.033	0.267	0.356	0.275	0.350	0.411	0.402	0.404
	4 Te	0.003	0.003	0.003	0.021	0.024	0.019	0.023	0.027	0.028	0.024
5 Me	1 LC	0.160	0.147	0.074	0.000	0.000	0.000	0.000	0.000	0.000	0.000
	2 Qu	0.115	0.062	0.214	0.000	0.000	0.000	0.000	0.000	0.000	0.000
	3 Va	0.166	0.233	0.154	0.000	0.000	0.000	0.000	0.000	0.000	0.000
6 CS	1 Li	0.000	0.000	0.000	0.000	0.000	0.000	0.000	0.000	0.000	0.000
	2 Or	0.000	0.000	0.000	0.000	0.000	0.000	0.000	0.000	0.000	0.000
	3 Cl	0.000	0.000	0.000	0.000	0.000	0.000	0.000	0.000	0.000	0.000
	4 Em	0.000	0.000	0.000	0.000	0.000	0.000	0.000	0.000	0.000	0.000
	5 Pa	0.000	0.000	0.000	0.000	0.000	0.000	0.000	0.000	0.000	0.000

Table 9.21 (continued)
The weighted supermatrix.

		4 CG				5 Me			6 CS				
		1 WC	2 BC	3 Fa	4 Te	1 LC	2 Qu	3 Va	1 Li	2 Or	3 Cl	4 Em	5 Pa
1 Al	1 WM	0.036	0.038	0.036	0.040	0.033	0.030	0.032	0.036	0.024	0.031	0.035	0.086
	2 KM	0.006	0.012	0.012	0.009	0.010	0.006	0.006	0.006	0.004	0.005	0.005	0.019
	3 Ta	0.015	0.007	0.009	0.009	0.006	0.013	0.011	0.012	0.009	0.018	0.015	0.049
2 Ad	1 TV	0.076	0.119	0.119	0.148	0.000	0.000	0.000	0.000	0.000	0.000	0.000	0.000
	2 PM	0.050	0.052	0.063	0.040	0.000	0.000	0.000	0.000	0.000	0.000	0.000	0.000
	3 Ra	0.014	0.015	0.012	0.023	0.000	0.000	0.000	0.000	0.000	0.000	0.000	0.000
	4 DM	0.095	0.048	0.040	0.023	0.000	0.000	0.000	0.000	0.000	0.000	0.000	0.000
3 Lo	1 Ur	0.028	0.016	0.016	0.018	0.027	0.011	0.010	0.016	0.010	0.015	0.018	0.031
	2 Su	0.141	0.047	0.052	0.052	0.012	0.062	0.064	0.071	0.051	0.074	0.073	0.135
	3 Ru	0.000	0.106	0.101	0.098	0.063	0.030	0.029	0.076	0.051	0.074	0.073	0.295
4 CG	1 WC	0.000	0.000	0.151	0.046	0.013	0.056	0.042	0.247	0.082	0.156	0.107	0.000
	2 BC	0.000	0.000	0.350	0.096	0.028	0.040	0.042	0.247	0.082	0.134	0.107	0.000
	3 Fa	0.463	0.463	0.000	0.398	0.156	0.143	0.157	0.119	0.257	0.318	0.400	0.000
	4 Te	0.077	0.077	0.039	0.000	0.055	0.013	0.012	0.031	0.019	0.036	0.031	0.000
5 Me	1 LC	0.000	0.000	0.000	0.000	0.000	0.477	0.477	0.000	0.000	0.000	0.000	0.000
	2 Qu	0.000	0.000	0.000	0.000	0.447	0.000	0.119	0.000	0.000	0.000	0.000	0.000
	3 Va	0.000	0.000	0.000	0.000	0.149	0.119	0.000	0.000	0.316	0.000	0.000	0.000
6 CS	1 Li	0.000	0.000	0.000	0.000	0.000	0.000	0.000	0.000	0.016	0.017	0.000	0.097
	2 Or	0.000	0.000	0.000	0.000	0.000	0.000	0.000	0.035	0.000	0.079	0.027	0.290
	3 Cl	0.000	0.000	0.000	0.000	0.000	0.000	0.000	0.092	0.044	0.000	0.110	0.000
	4 Em	0.000	0.000	0.000	0.000	0.000	0.000	0.000	0.000	0.029	0.042	0.000	0.000
	5 Pa	0.000	0.000	0.000	0.000	0.000	0.000	0.000	0.010	0.005	0.000	0.000	0.000

The limit supermatrix is not shown here to save space. It is obtained from the weighted supermatrix by raising it to powers until it converges so that all columns are identical. From the top part of the first column of the limit supermatrix we get the priorities we seek and normalize. We show what they are in Table 9.22.

Table 9.22. The synthesized results for the alternatives.

Alternatives	Values from Limit Supermatrix	Actual Values July13, 1998	Normalized Values from Supermatrix	Actual Market Share as Dollar Sales Normalized
Walmart	0.057	58 billion $	0.599	54.8
KMart	0.024	27.5 billion $	0.248	25.9
Target	0.015	20.3 billion $	0.254	19.2

12.1.4 Synthesized Results from the Limit Supermatrix. The relative market share of the alternatives Walmart, Kmart and Target from the limit supermatrix are: 0.057, 0.024 and 0.015. When normalized they are 0.599, 0.248 and 0.154.

The relative market share values obtained from the model were compared with the actual sales values by computing the compatibility index. The Compatibility Index, illustrated in the next example, is used to determine how close two sets of numbers from a ratio scale or an absolute scale are to each other. We form the matrix of ratios of each set and multiply element-wise one matrix by the transpose of the other (the Hadamard product), add all the entries of the resulting matrix and divide the outcome by n^2, where n is the order of the matrix which is the number of entries in each vector. The outcome should not exceed the value of 1.1. In this example the result is equal to 1.016 and falls below 1.1 and therefore is an acceptable outcome.

12.2 Example 2: US Athletic Footwear Market in 2000

My student Maria Lagasca has studied the US Athletic Footwear market. That market has seen tremendous growth over the years. Not only are these products used for specific athletic purposes but also they have been used as casual wear because of its ability to provide comfort and agility to consumers. Interest in the industry has grown to a large extent because of advances in research and development for durable yet comfortable materials. The industry is also considered as one of the heaviest advertisers based on a study made last year along with other industries such as apparel, beer/wine/liquor, computers and electronics. The study illustrated in Figure 9.12 aims at estimating the market share using the ANP with the aid of SuperDecisions software. The estimates are then compared against the actual market share of various manufacturers in the year 2000. As the industry is fragmented (with many players holding fewer shares of the market), the other manufacturers have been lumped under the "Others" category as they are considered as homogeneous given the factors used in the analysis.

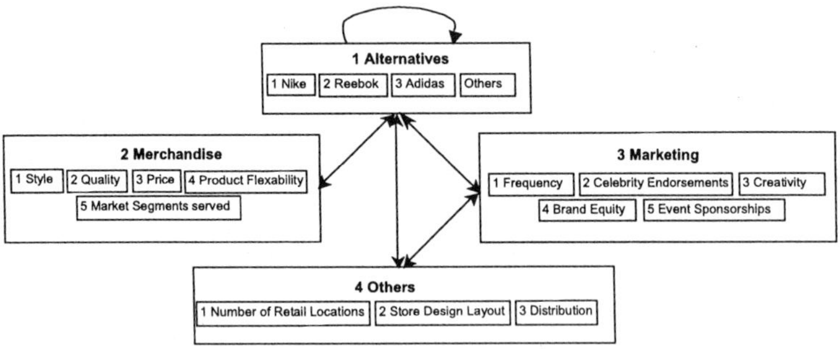

Figure 9.12. The clusters and nodes of a model to estimate the relative market share of footware.

12.2.1 Clusters and Elements (Nodes).

1 Alternatives (brands competing against each other in the market)

(a) Nike - Nike as an alternative brand for athletic footwear.

(b) Reebok - Reebok as an alternative brand for athletic footwear.

(c) Adidas - Adidas as an alternative brand for athletic footwear.

(d) Others - Other alternative brands (And1, Skechers, New Balance, Timberland, etc) for athletic footwear.

2 Merchandise (affects each brand and each brand affects the type of merchandising strategy)

(a) Style – the ability of a manufacturer to immediately respond to customers tastes and needs or create demand by introducing new products to the market.

(b) Quality – Quality includes the reliability / durability of products including the ability to withstand pressure and frequent use.

(c) Price – defined as value for money.

(d) Product Flexibility – Ability of the product to substitute for other footwear, i.e. Running shoes can be used for casual wear and other purposes.

(e) Market Segments Served – Ability of the manufacturer to cover various target segments through their different product lines, i.e. men, women, children, basketball players, soccer players, etc.

3 Marketing (Marketing affects each of the brands and each brand affects the type of marketing strategy)

(a) Frequency – frequency of advertising regardless of media.

(b) Celebrity Endorsements – endorsement by a well-known popular sports celebrity.

(c) Creativity – Creativity of marketing advertisements regardless of length

(d) Brand Equity – Ability to create brand awareness and recognition among various segments of the market.

(e) Event Sponsorships-a marketing tool to advertise and create awareness for brand.

4 Others (Other factors affect the brand and each brand affects the type of strategy for these factors; also the Marketing strategy affects these factors)

(a) Number of retail locations – The number and the coverage of retail locations across the United States.

(b) Store design and layout – includes placement and effective layout of merchandise vis-à-vis competitors.

(c) Distribution – shelf space and coverage of merchandise across the United States. Includes relationships with distributors and even with own distribution chain.

Comparisons were done based on information gathered for each individual manufacturer. Advertising was determined due to factors such as each manufacturer's relative selling, general, and administrative expenses from their annual reports. Advertisements (mostly in print) were also viewed and use of celebrity endorsements in the same period were also assessed relative to each brand to measure creativity as well as frequency. Brand equity was measured on more intuitive terms i.e. Nike's Swoosh logo is considered as one of the most recognized logos and brands, which gave them an advantage over the other brands.

Other factors, such as the number of retail locations, were assessed by counting the total number of such locations (from individual websites). Store layout and distribution information were gathered from the websites as well to assess the relative effectiveness of each factor. For instance, Reebok and Adidas, have fewer individual stores than Nike (Factory outlets and Niketown) and tend to be distributed in department stores or sporting goods stores facing more competition from other brands because of less exclusivity.

In terms of merchandise, prices are relatively the same for all brands although some like Adidas and Reebok may seem to be higher than other brands because of quality. Nike and other athletic footwear products tend to be more flexible in terms of how consumers use the products i.e. their basketball shoes are often substituted for casual wear and running shoes, which leads to a broader target segment. Also, Nike and the other brands seem to serve broader market segments specifically women and children. Their line extensions, e.g. Michael Jordan for men have been extended to children.

As more and more people substitute athletic footwear for everyday use, Nike and the other brands seem to be stronger in catering to this need thereby leading to more market share

Table 9.23 gives the actual and the estimated market share for each brand. They are surprisingly close. This example was done as a take home exercise. In this case the compatibility index obtained from the study is 1.001428, which is very small. We would be glad to provide the interested reader with at least a dozen such market share examples often worked out in class in about one hour without prior preparation or looking at numbers. They all have such close outcomes, because students, interested in the example, provided the judgments.

We now look at full blown decisions with their BOCR. First we give an outline of the steps recommended in applying the ANP.

13. Outline of the Steps of the ANP

1 Describe the decision problem in detail including its objectives, criteria and subcriteria, actors and their objectives and the possible outcomes of that decision. Give details of influences that determine how that decision may come out.

2 Determine the control criteria and subcriteria in the four control hierarchies one each for the benefits, opportunities, costs and risks of that decision and obtain their priorities from paired comparisons matrices. If a control criterion or subcriterion has a global priority of 3% or less, you may consider carefully eliminating it from further consideration. The software automatically deals only with those criteria or subcriteria that have subnets under them. For benefits and opportunities, ask what gives the most benefits or presents the greatest opportunity to influence fulfillment of that control criterion. For costs and risks, ask what incurs the most cost or faces the greatest risk. Sometimes (very rarely), the comparisons are made simply in terms of benefits, opportunities, costs, and risks in the aggregate without using control criteria and subcriteria.

Table 9.23. Footwear actual statistics and model results along with the compatibility index.

Alternatives	A1	A2	A3	A4	
Actual Market Share	39.200	15.100	10.900	34.800	
Estimated Market Share from ANP Model	40.670	15.040	11.330	32.970	
Pairwise Comparison Matrix from Actual Market Share Data					
	A1	A2	A3	A4	
A1	1	2.596026	3.59633	1.12643678	
A2	0.385204082	1	1.385321	0.43390805	
A3	0.278061224	0.721854	1	0.31321839	
A4	0.887755102	2.304636	3.192661	1	
Transpose of Comparison Matrix from Estimated Market Share					
	A1	A2	A3	A4	
A1	1	0.369806	0.278584	0.81067126	
A2	2.70412234	1	0.753324	2.19215426	
A3	3.589585172	1.327449	1	2.90997352	
A4	1.233545648	0.456172	0.343646	1	
Result of Hadamard (Element-wise) Multiplication of Previous Two Matrices					
	A1	A2	A3	A4	Row Sums
A1	1	0.960026	1.001879	0.91316992	3.875075
A2	1.041638963	1	1.043596	0.95119337	4.036429
A3	0.998124448	0.958225	1	0.91145722	3.867807
A4	1.095086442	1.051311	1.097144	1	4.243542
				SUM =	16.02285

Number of Alternatives: $n = 4$
Compatibility Index = (SUM/n^2) = 1.001428

3 Determine the most general network of clusters (or components) and their elements that applies to all the control criteria. To better organize the development of the model as well as you can, number and arrange the clusters and their elements in a convenient way (perhaps in a column). Use the identical label to represent the same cluster and the same elements for all the control criteria.

4 For each control criterion or subcriterion, determine the clusters of the general feedback system with their elements and connect them according to their outer and inner dependence influences. An arrow is drawn from a cluster to any cluster whose elements influence it.

5 Determine the approach you want to follow in the analysis of each cluster or element, influencing (the preferred approach) other clusters and elements with respect to a criterion, or being influenced by other clusters

and elements. The sense (being influenced or influencing) must apply to all the criteria for the four control hierarchies for the entire decision.

6 For each control criterion, construct the supermatrix by laying out the clusters in the order they are numbered and all the elements in each cluster both vertically on the left and horizontally at the top. Enter in the appropriate position the priorities derived from the paired comparisons as subcolumns of the corresponding column of the supermatrix.

7 Perform paired comparisons on the elements within the clusters themselves according to their influence on each element in another cluster they are connected to (outer dependence) or on elements in their own cluster (inner dependence). In making comparisons, you must always have a criterion in mind. Comparisons of elements according to which element influences a given element more and how strongly more than another element it is compared with are made with a control criterion or subcriterion of the control hierarchy in mind.

8 Perform paired comparisons on the clusters as they influence each cluster to which they are connected with respect to the given control criterion. The derived weights are used to weight the elements of the corresponding column blocks of the supermatrix. Assign a zero when there is no influence. Thus obtain the weighted column stochastic supermatrix.

9 Compute the limit priorities of the stochastic supermatrix according to whether it is irreducible (primitive or imprimitive [cyclic]) or it is reducible with one being a simple or a multiple root and whether the system is cyclic or not. Two kinds of outcomes are possible. In the first all the columns of the matrix are identical and each gives the relative priorities of the elements from which the priorities of the elements in each cluster are normalized to one. In the second the limit cycles in blocks and the different limits are summed and averaged and again normalized to one for each cluster. Although the priority vectors are entered in the supermatrix in normalized form, the limit priorities are put in idealized form because the control criteria do not depend on the alternatives.

10 Synthesize the limiting priorities by weighting each idealized limit vector by the weight of its control criterion and adding the resulting vectors for each of the four merits: Benefits (B), Opportunities (O), Costs (C) and Risks (R). There are now four vectors, one for each of the four merits. An answer involving ratio values of the merits is obtained by forming the ratio BO/CR for each alternative from the four vectors. The alternative with the largest ratio is chosen for some decisions. Companies and individuals with limited resources often prefer this type of synthesis.

11 Determine strategic criteria and their priorities to rate the top ranked (ideal) alternative for each of the four merits one at a time. The synthesized ideals for all the control criteria under each merit may result in an ideal whose priority is less than one for that merit. Only an alternative that is ideal for all the control criteria under a merit receives the value one after synthesis for that merit. Normalize the four ratings thus obtained and use them to calculate the overall synthesis of the four vectors. For each alternative, subtract the sum of the weighted costs and risks from the sum of the weighted benefits and opportunities.

12 Perform sensitivity analysis on the final outcome. Sensitivity analysis is concerned with "what if" kind of question to see if the final answer is stable to changes in the inputs whether judgments or priorities. Of special interest is to see if these changes change the order of the alternatives. How significant the change is can be measured with the Compatibility Index of the original outcome and each new outcome.

14. Complex Decisions with Dependence and Feedback

With the China example for hierarchies and with the market share examples it is now easier to deal with complex decisions involving networks. For each of the four BOCR merits we have criteria (and subcriteria where relevant) called control criteria that are prioritized under that merit through paired comparisons. For each of the control criteria we create a network of influences with respect to that control criterion as we did in the market share examples. We obtain the ideal outcome ranking for each control criterion and then synthesize these outcomes by weighting by the importance of the control criteria for each merit. We then rate the top alternative under each merit to obtain the weights b,o,c and r for the BOCR and use them to synthesize and obtain the final weights for the alternatives using the two formulas BO/CR and more importantly, $bB + oO - cC - rR$. Let us sketch out an example using as little space as possible.

14.1 The National Missile Defense (NMD) Example

Not long ago, the United States government faced the crucial decision of whether or not to commit itself to the deployment of a National Missile Defense (NMD) system. Many experts in politics, the military, and academia had expressed different views regarding this decision. The most important rationale behind supporters of the NMD system was protecting the U.S. from potential threats said to come from countries such as North Korea, Iran and Iraq. According to the Central Intelligence Agency, North Korea's Taepo Dong long-range missile tests were successful, and it has been developing a second generation capable of reaching the U.S. Iran also tested its medium-range missile Shahab-3 in July 2000. Opponents expressed doubts about the technical feasibility, high costs (estimated at $60 billion), political damage, possible arms race, and the exacerbation of foreign relations. The idea for the deployment of a ballistic missile defense system has been around since the late 1960s but the current plan for NMD originated with President Reagan's Strategic Defense Initiative (SDI) in the 1980s. SDI investigated technologies for destroying incoming missiles. The controversies surrounding the project were intensified with the National Missile Defense Act of 1996, introduced by Senator Sam Nunn (D-GA) in June 25, 1996. The bill required Congress to make a decision on whether the U.S. should deploy the NMD system by 2000. The bill also targeted the end of 2003 as the time for the U.S. to be capable of deploying NMD.

The ANP was applied to analyze this decision. It was done in the usual three steps of the ANP process: 1) the BOCR merits and their control criteria and subcriteria prioritized with respect to each merit, 2) the network of influence for each control criterion from which priorities for the alternatives are derived as in the market share examples and then synthesized using the weights of the control criteria for each merit and finally, 3) the use of strategic criteria as in Figure 9.13 to rate the merits one at a time as in Table 9.24 through their top alternative and use the resulting normalized ratings as priorities to weight and combine the priorities of each alternative with respect to the four merits to get the final answer.

On February 21, 2002 this author gave a half-day presentation on the subject to the National Defense University in Washington. In December 2002, President George W. Bush and his advisors decided to build the NMD. This study may have had no influence on the decision but still two years earlier (September 2000) it had arrived at the same decision produced by this analysis. The alternatives we considered for this analysis are: Deploy NMD, Global defense, R&D, Termination of the NMD program. Complete analysis of this example is given in the author's book on the ANP published in 2001. There were 23 criteria under the BOCR merits, including economic, terrorism, technological progress and everything else people were thinking about as important to develop or not to develop the NMD. After prioritization they were reduced to 9 control criteria for all four merits. Each criterion was treated in a very similar way to the single market share (essentially economic benefits) examples. Table 9.25 gives the final outcome. Here we see that the two formulas give the same outcome to deploy as the best alternative. The conclusion of this analysis is that pursuing the deployment of NMD is the best alternative. Sensitivity analysis indicates that the final ranks of the alternatives might change, but such change requires making extreme assumptions on the priorities of BOCR and of their corresponding control criteria.

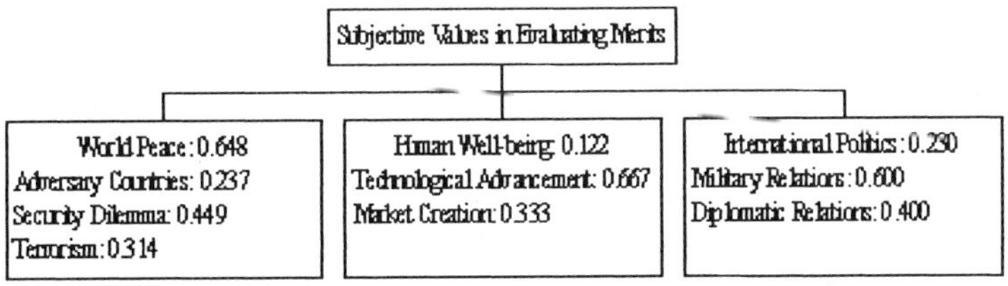

Figure 9.13. Hierarchy for rating benefits, opportunities, costs and risks.

Table 9.24. Priority ratings for the merits: Benefits, opportunities, costs and risks. Very High (0.419), High (0.263), Medium (0.160), Low (0.097), Very Low (0.061)

		Benefits	Opportunities	Costs	Risks
World Peace	Adversary Countries	Very High	Medium	High	Very Low
	Security Dilemma	Very Low	Very Low	Very High	Very Low
	Terrorism	Medium	Very Low	High	High
Human Well-Being	Technological Advancement	High	High	Low	Very Low
	Market Creation	Medium	High	Very Low	Very Low
International Politics	Military Relations	High	High	Medium	Very Low
	Diplomatic Relations	Low	Low	Low	Very High
Priorities		b=0.264	o=0.185	c=0.361	r=0.190

15. Conclusions

Numerous other examples along with the software Super Decisions for the ANP can be obtained from www.superdecisions.com. We hope that the reader now has a good idea as to how to use the AHP/ANP in making a complex decision. The AHP and ANP have found application in practice by many companies and governments. My book Decision Making for Leaders is now in nearly 10 languages. Another recent policy study was done regarding whether the US should go to war with Iraq directly or through the UN done in September 2002. The analysis found that the US should go with the UN with priority more than double those of going alone or of going with a coalition. There is also the ongoing Middle East conflict. An ANP analysis showed that the best option is

Table 9.25. Overall syntheses of the alternatives.

Alternatives	BO/CR (from unweighted columnns in Table 9.24)	Normalized	$bB + oO - cC - rR$ (from weighted columns in Table 9.24	Unitized*
Deploy	2.504	0.493	0.111	1.891
Global	1.921	0.379	0.059	1.000
R&D	0.560	0.110	-0.108	-1.830
Terminate	0.090	0.018	-0.278	-4.736

*Unitized means to divide each number in the column by the number with the smallest absolute value (it is recommended that one not unitize when such a number is close to zero).

for Israel and the US to help the Palestinians both set up a state and in particular achieve a viable economy. My forthcoming book The Encyclicon has about 100 summarized examples of applications of the ANP. A list of more than a thousand references until the early 1990's on the AHP appears in reference [3].

Acknowledgments

I am grateful to my colleague Professor Dr. Klaus Dellmann for his careful reading and suggestions to improve the paper and to my student Yeonmin Cho for her untiring efforts in coauthoring with me the case study about China and the WTO, and to my friend Ania Greda from Krakow Poland for her great help in making this manuscript more attractive for the reader. My thanks also go to my friend Kirti Peniwati for her valuable suggestions regarding the NMD example.

References

[1] P.T. Harker. Derivatives of the Perron root of a positive reciprocal matrix: With applications to the Analytic Hierarchy Process. *Appl. Math. Comput.*, 22:217–232, 1987.

[2] R.W. Saaty. *Decision Making in Complex Environments: The Analytic Network Process (ANP) for Dependence and Feedback; A Manual for the ANP Software SuperDecisions.* Creative Decisions Foundation, 4922 Ellsworth Avenue, Pittsburgh, PA 15213, 2002.

[3] T.L. Saaty. *Fundamentals of the Analytic Hierarchy Process.* RWS Publications, 4922 Ellsworth Avenue, Pittsburgh, PA 15413, 2000.

[4] T.L. Saaty. *The Analytic Network Process.* RWS Publications, 4922 Ellsworth Avenue, Pittsburgh, PA 15213, 2001.

[5] T.L. Saaty and Y. Cho. The decision by the US Congress on China's trade status: A multicriteria analysis. *Socio-Economic Planning Sciences*, 35(6):243–252, 2001.

[6] T.L. Saaty and M.S. Ozdemir. Negative priorities in the analytic hierarchy process. *Mathematical and Computer Modelling*, 37:1063–1075, 2003.

[7] T.L. Saaty and M.S. Ozdemir. Why the magic number seven plus or minus two. *Mathematical and Computer Modelling*, 38:233–244, 2003.

[8] T.L. Saaty and L.G. Vargas. The possibility of group choice: Pairwise comparisons and merging functions. *To appear*, 2004.

Chapter 10

ON THE MATHEMATICAL FOUNDATIONS OF MACBETH

Carlos A. Bana e Costa
Centre for Management Studies of Instituto Superior Técnico
Technical University of Lisbon
Av. Rovisco Pais, 1049-001 Lisbon, Portugal
and
Department of Operational Research, London School of Economics
Houghton Street, London WC2A 2AE, U.K.
carlosbana@netcabo.pt,c.bana@lse.ac.uk

Jean-Marie De Corte, Jean-Claude Vansnick
Centre de Recherche Warocqué, Université de Mons-Hainaut
Place du Parc, 20, 7000 Mons
Belgium
{Jean-Marie.DeCorte,Jean-Claude.Vansnick}@umh.ac.be

Abstract MACBETH (Measuring Attractiveness by a Categorical Based Evaluation Technique) is a multicriteria decision analysis approach that requires only qualitative judgements about differences of value to help an individual or a group quantify the relative attractiveness of options. This chapter presents an up-to-date survey of the mathematical foundations of MACBETH. Reference is also made to real-world applications and an extensive bibliography, spanning back to the early 1990's, is provided.

Keywords: MACBETH, questioning procedure, qualitative judgements, judgmental inconsistency, cardinal value measurement, interaction.

1. Introduction

Let X (with $\#X = n \geq 2$) be a finite set of elements (alternatives, choice options, courses of action) that an individual or a group, J, wants to compare in terms of their relative attractiveness (desirability, value).

Ordinal value scales (defined on X) are quantitative representations of preferences that reflect, numerically, the order of attractiveness of the elements of X for J. The construction of an ordinal value scale is a straightforward process, provided that J is able to rank the elements of X by order of attractiveness – either directly or through pairwise comparisons of the elements to determine their relative attractiveness. Once the ranking is defined, one needs only to assign a real number $v(x)$ to each element x of X, in such a way that:

1 $v(x) = v(y)$ if and only if J judges the elements x and y to be equally attractive;

2 $v(x) > v(y)$ if and only if J judges x to be more attractive than y.

The problem, however, is that, in a multiple criteria decision analysis, conclusions based on a additive value model may be quantitatively meaningless, because "to be quantitatively meaningful a statement should be unaffected by admissible transformations of all the quantities involved." [53, p. 91]. A necessary condition is that each value scale should be unique up to a positive affine transformation (an interval scale), as it is with a value difference scale. A value difference scale (defined on X) is a quantitative representation of preferences that is used to reflect, not only the order of attractiveness of the elements of X for J, but also the differences of their relative attractiveness, or in other words, the strength of J's preferences for one element over another. Unfortunately, the construction of an interval value scale is usually a difficult task.

Both numerical and non-numerical techniques have been proposed and used to build a value difference scale (hereafter, simply called a value scale) – see [51] for a survey. Examples of numerical techniques are direct rating and difference methods – see descriptions in [61, 62] and [41]. They require J to be able to produce, either directly or indirectly, numerical representations of his or her strengths of preferences, which is not a natural cognitive task. Non-numerical techniques, such as the bisection method (also described by the same authors), are based on indifference judgements, forcing J to compare his or her strengths of preferences between two pairs of elements of X, therefore involving at least three different elements in each judgement. This requires J to perform an intensive cognitive task and is prone to be substantively meaningless – "substantive meaningfulness (...) requires that the qualitative relations (...) being modelled should be unambiguously understood by the decision maker." [53, p. 91].

The aforementioned difficulties inspired the development of MACBETH "**M**easuring **A**ttractiveness by a **C**ategorical **B**ased **E**valuation **T**ec**h**nique".

The original research on the MACBETH approach was carried out in the early 1990's – see [2, 29] and [35] – as a response to the following question:

How can a value scale be built on X, both in a qualitatively and quantitatively meaningful way, without forcing J to produce direct numerical representations of preferences and involving only two elements of X for each judgement required from J?

Using MACBETH, J is asked to provide preferential information about two elements of X at a time, firstly by giving a judgement as to their relative attractiveness (ordinal judgement) and secondly, if the two elements are not deemed to be equally attractive, by expressing a qualitative judgement about the difference of attractiveness between the most attractive of the two elements and the other. Moreover, to ease the judgemental process, six semantic categories of difference of attractiveness, "very weak", "weak", "moderate", "strong", "very strong" or "extreme", or a succession of these (in case hesitation or disagreement arises) are offered to J as possible answers. This is somewhat in line with similar ideas previously proposed by Saaty [59] in a ratio measurement framework, or by Freeling [52] and Belton [40] in difference value measurement. By pairwise comparing the elements of X a matrix of qualitative judgements is filled in, with either only a few pairs of elements, or with all of them (in which case $n \cdot (n-1)/2$ comparisons would be made by J).

A brief review of the previous research on MACBETH is offered in Section 2, together with the evolution of its software's development. It shows that, on a technical level, MACBETH has evolved through the course of theoretical research and also through its extension to the multicriteria value measurement framework in numerous practical applications (see Section 10). Its essential characteristics, however, have never changed.

Sections 3 through 9 of this chapter present an up-to-date survey of the mathematical foundations of MACBETH. Section 3 describes the two MACBETH modes of questioning mentioned above (both involving only two elements at a time) used to acquire preferential information from J, as well as the types of information that can be deduced from each of them. The subsequent sections are devoted to an up-to-date rigorous survey of the mathematical foundations of MACBETH. Section 4 addresses the numerical representation of those different types of information. These numerical representations are only possible if J's responses satisfy certain rational working hypotheses. Section 5 deals with the "consistency / inconsistency" of the preferential information gathered from J and Section 6 explores the practical problem of testing the consistency of preferential information. How should an inconsistency be dealt with? The answer to this question is the subject of Section 7. Sections 8 and 9 present what MACBETH proposes to J once the preference information provided by J is consistent. Finally, Section 10 lists several real-world applications of multicriteria value analysis in which the MACBETH approach was used.

This chapter will use the following notation:

- J is an evaluator, either a individual or group.
- X (with $\#X = n \geq 2$) is a finite set of elements (alternatives, choice options, courses of action) that J wants to compare in terms of their relative attractiveness (desirability, value).
- $\Delta att(x, y)$ is the "difference of attractiveness between x and y for J", where x and y are elements of X such that x is more attractive than y for J.
- $\Delta att(x, y) \succ \Delta att(z, w)$ means that $\Delta att(x, y)$ is greater than Δatt (z, w).
- ϕ is an empty set.
- $\mathbb{R}$ is the set of real numbers.
- $\mathbb{R}_+ = \{x \in \mathbb{R} \mid x \geq 0\}$.
- $\mathbb{R}^* = \mathbb{R} \setminus \{0\}$.
- $\mathbb{R}_+^* = \mathbb{R}_+ \setminus \{0\}$.
- $\mathbb{Z}$ is the set of integer numbers.
- $\mathbb{N}$ is the set of non-negative integer numbers.
- $\mathbb{N}^* = \mathbb{N} \setminus \{0\}$.
- $\mathbb{N}_{s,t} = \{s, s+1, \ldots, t\} = \{x \in \mathbb{N} \mid s \leq x \leq t\}$ where $s, t \in \mathbb{N}$, and $s < t$.
- The transpose of a matrix A will be denoted by ${}^tA$.

2. Previous Research and Software Evolution

In order to build an interval (value) scale based on the qualitative judgements of difference of attractiveness formulated by J, it is necessary that the six MACBETH categories "very weak", "weak", "moderate", "strong", "very strong" or "extreme" be represented by non-overlapping (disjoint) intervals of real numbers. The basic idea underlying the initial development of MACBETH was that the limits of these intervals should not be arbitrarily fixed a priori, but determined simultaneously with numerical value scores for the elements of X. Research was then conducted on how to test for the existence of such intervals and how to propose numerical values for the elements of X and for the limits of the intervals – see [2, Chapter IV]. This gave rise to the formulation of a chain of four linear programs – see [31, 29, 30] and [32] – that, implemented in GAMS, were used in the first real-world applications of MACBETH as a decision aiding tool to derive value scores and criteria weights in the framework of an additive aggregation model – see [42, 43, 35] and [37]. Theoretical research conducted at the same time, and first presented in 1994 at the 11th International Conference on MCDM, demonstrated the equivalence of the approach by constant thresholds and the approach by measurement conditions – see [36].

The first MACBETH software was developed in 1994. In it, the objective function used in the GAMS implementation to determine a value scale was modified, on the basis of a simple principle – see [37] and [38] – that makes it possible, for simple cases, to determine the scale "by hand" [34]. However, complete procedures to address and manage all cases of inconsistency were not available at that time. Therefore, the software offered its users the possibility of obtaining a compromise scale in the case of inconsistency. This initial software was used in several real world applications – see, for example, [19, 21, 23, 24, 32, 39] and [48]. However, it had several important limitations:

1 The determination of suggestions was still heuristic and did not guarantee the minimal number of changes necessary to achieve consistency;

2 It was not possible for the evaluator to hesitate between several semantic categories when expressing judgements. It, therefore, did not enable one to facilitate the management of group judgemental disagreements;

3 It forced the evaluator to first provide all of the judgements before it could run any procedure. Consequently, judgemental inconsistency could only be detected for a full matrix of judgements. As a result, suggestions of changes to resolve inconsistency could only then be discussed, a restriction that did not lend itself to good interaction.

Subsequent theoretical research was therefore concentrated on resolving these problems. Results reported in [46] and [56], allowing inconsistencies to be dealt with in a mathematically sound manner, were the turning point in the search for a more interactive formulation. Indeed, it was then possible to implement a procedure that automatically detects "inconsistency", even for an incomplete matrix of judgements, in a new software called M-MACBETH – see `www.m-macbeth.com` and [16] – which has been used to produce some of the figures in this paper. The objective of abandoning the suggestion of a compromise scale could also finally be achieved, since the origin of the inconsistency could now be found (detection of elementary incompatible systems) and explained to J. M-MACBETH finds the minimal number of necessary changes and, for any number of changes not greater than five, suggests all of the possible ways in which the inconsistency can be resolved. Furthermore, it is able to provide suggestions of multiple category changes, where a "k categories change" is considered to be equivalent to k "1 category changes".

Real-world applications in the specific context of bid evaluation (see references in Section 10) inspired research regarding the concepts of "robustness" [46] and sensitivity [9], the results of which were then included in the software, together with the possibility of addressing potential imprecision (uncertainty) associated with impacts of options, incorporating reference levels for one criterion at any time, and graphically representing comparisons of options on any two groups of criteria. These issues are out of the scope of the present chapter and they are not also included in the version of the software, limited to scoring and weighting, embedded into the HIVIEW3 software in 2003 – see [45] and `www.catalyze.co.uk`.

3. Types of Preferential Information

3.1 Type 1 Information

Type 1 information refers to preferential information obtained from J by means of Questioning Procedure 1.

Let x and y be two different elements of X.

Questioning Procedure 1 *A first question (Q1) is asked of J:*

Q1: Is one of the two elements more attractive than the other?
J's response (R1) can be: "Yes", or "No", or "I don't know".
If R1 = "Yes", a second question (Q2) is asked:
Q2: Which of the two elements is the most attractive?

The responses to Questioning Procedure 1 for several pairs of elements of X enable the construction of three binary relations on X:

$$P = \{(x,y) \in X \times X : x \text{ is more attractive than } y\}$$

$$I = \{(x,y) \in X \times X : x \text{ is not more attractive than } y \text{ and } y \text{ is not more attractive than } x, \text{ or } x = y\}$$

$$? = \{(x,y) \in X \times X : x \text{ and } y \text{ are not comparable in terms of their attractiveness}\}.$$

P is asymmetric, I is reflexive and symmetric, and ? is irreflexive and symmetric. Note that $? = X \times X \setminus (I \cup P \cup P^{-1})$, with $P^{-1} = \{(x,y) \in X \times X \mid yPx\}$.

DEFINITION 37 *Type 1 information about X is a structure $\{P, I, ?\}$ where P, I and ? are disjoint relations on X, P is asymmetric, I is reflexive and symmetric, and $? = X \times X \setminus (I \cup P \cup P^{-1})$.*

3.2 Type 1+2 Information

Suppose that type 1 information $\{P, I, ?\}$ about X is available.

Questioning Procedure 2 *The following question (Q3) is asked, for all $(x, y) \subset P$:*

Q3: How do you judge the difference of attractiveness between x and y?

J's response (R3) would be provided in the form "d_s" (where $d_1, d_2, \ldots, d_Q$ ($Q \in \mathbb{N} \setminus \{0,1\}$) are semantic categories of difference of attractiveness defined so that, if $i < j$, the difference of attractiveness d_i is weaker than the difference of attractiveness d_j) or in the more general form (possibility of hesitation) "d_s to d_t", with $s \leq t$ (the response "I don't know" is assimilated to the response "d_1 to d_Q").

REMARK 34 *When $Q = 6$ and d_1 = very weak, d_2 = weak, d_3 = moderate, d_4 = strong, d_5 = very strong, d_6 = extreme, Questioning Procedure 2 is the mode of interaction used in the MACBETH approach and its M-MACBETH software.*

R3 responses give rise to relations C_{st} $(s, t \in \mathbb{N}, 1 \leq s \leq t \leq Q)$ where $C_{st} = \{(x, y) \in P \mid \Delta_{att}(x, y)$ is "d_s to d_t"$\}$. They enable the construction of an asymmetric relation on $P : \{((x, y), (z, w)) \in P \times P \mid \exists\, i, j, s, t \in \mathbb{N}$ with $1 \leq i \leq j < s \leq t \leq Q, (x, y) \in C_{st}, (z, w) \in C_{ij}\}$. Hereafter, C_{ss} will simply be referred to as C_s.

DEFINITION 38 *Type 1+2 information about X is a structure $\{P, I, ?, P^e\}$ where $\{P, I, ?\}$ is type 1 information about X and P^e is an asymmetric relation on P, the meaning of which is "$(x, y) P^e (z, w)$ when $\Delta_{att}(x, y) \succ \Delta_{att}(z, w)$".*

4. Numerical Representation of the Preferential Information

4.1 Type 1 Scale

Suppose that type 1 information $\{P, I, ?\}$ about X is available.

DEFINITION 39 *A type 1 scale on X relative to $\{P, I\}$ is a function $\mu : X \to \mathbb{R}$ satisfying Condition 1.*

CONDITION 1 $\forall\, x, y \in X$, $[xPy \Rightarrow \mu(x) > \mu(y)]$ *and* $[xIy \Rightarrow \mu(x) = \mu(y)]$.

Let $Sc_1(X, P, I) = \{\mu : X \to \mathbb{R} \mid \mu$ is a type 1 scale on X relative to $\{P, I\}\}$. When X, P and I are well determined, $Sc_1(X, P, I)$ will be noted Sc_1.

When $? = \phi$ and $Sc_1(X, P, I) \neq \phi$, each element of $Sc_1(X, P, I)$ is an ordinal scale on X.

4.2 Type 1+2 Scale

Suppose that type 1+2 information $\{P, I, ?, P^e\}$ about X is available.

DEFINITION 40 *A type 1+2 scale on X relative to $\{P, I, ?, P^e\}$ is a function $\mu : X \to \mathbb{R}$ satisfying Condition 1 and Condition 2.*

CONDITION 2 $\forall\, x, y, z, w \in X$, $[(x, y)P^e(z, w) \Rightarrow \mu(x) - \mu(y) > \mu(z) - \mu(w)]$.

Let $Sc_{1+2}(X, P, I, P^e) = \{\mu : X \to \mathbb{R} \mid \mu$ is a type 1+2 scale on X relative to $\{P, I, P^e\}\}$. When X, P, I and P^e are well determined, $Sc_{1+2}(X, P, I, P^e)$ will be noted Sc_{1+2}.

5. Consistency – Inconsistency

DEFINITION 41 *Type 1 information $\{P, I, ?\}$ about X is*

- *consistent when $Sc_1(X, P, I) \neq \phi$*
- *inconsistent when $Sc_1(X, P, I) = \phi$.*

DEFINITION 42 *Type 1+2 information $\{P, I, ?, P^e\}$ about X is*

- *consistent when $Sc_{1+2}(X, P, I, P^e) \neq \phi$*
- *inconsistent when $Sc_{1+2}(X, P, I, P^e) = \phi$.*

When $Sc_{1+2}(X, P, I, P^e) = \phi$, one can have $Sc_1(X, P, I) = \phi$ or $Sc_1(X, P, I) \neq \phi$. In the first case, the message "no ranking" will appear in M-MACBETH; it occurs namely when J declares, in regards to elements x, y and z of X, that $[xIy, yIz$ and $xPz]$ or $[xPy, yPz$ and $zPx]$. In the second case, the message "inconsistent judgement" will appear in M-MACBETH.

Although this is the only difference between the types of inconsistency introduced in M-MACBETH, it is interesting to mention, from a theoretical perspective, that one could further distinguish two sub-types of inconsistency (sub-type a and sub-type b) when $Sc_{1+2}(X, P, I, P^e) = \phi$ and $Sc_1(X, P, I) \neq \phi$.

Sub-type a inconsistency arises when there is a conflict between type 1 information and P^e that makes the simultaneous satisfaction of conditions 1 and 2 impossible. These kinds of conflicts are found essentially in four types of situations; namely when $x, y, z \in X$ exist such that

$$\begin{aligned} & [\, xPy, yPz, xPz \text{ and } (y, z)P^e(x, z)\,] \\ \text{or} \quad & [\, xPy, yPz, xPz \text{ and } (x, y)P^e(x, z)\,] \\ \text{or} \quad & [\, xIy, yPz, xPz \text{ and } (x, z)P^e(y, z)\,] \\ \text{or} \quad & [\, xIy, zPy, zPx \text{ and } (z, x)P^e(z, y)\,]. \end{aligned}$$

Sub-type b inconsistency arises when there is no conflict between type 1 information and P^e but at least one conflict exists inside P^e that makes satisfying Condition 2 impossible. An example of this type of conflict is (see Figure 10.1):

$$xPy, xPw, yPz, wPz, xPz, yPw$$
$$(x, y) \in C_1, (y, z) \in C_2$$
$$(x, w) \in C_3, (w, z) \in C_2.$$

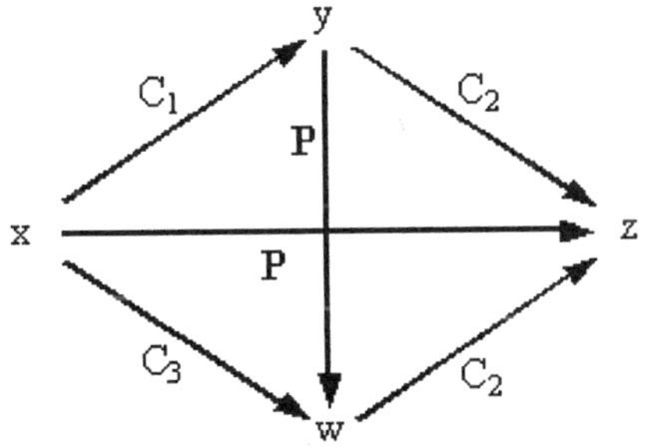

Figure 10.1. Example of sub-type b inconsistency.

In such a case, Condition 2 cannot be respected, because one should have

$$\begin{cases} \mu(x) - \mu(w) > \mu(y) - \mu(z) & (1) \\ \mu(w) - \mu(z) > \mu(x) - \mu(y) & (2) \end{cases}$$

which is impossible.

On the other hand, it is easily shown that the following two systems are compatible, that is, there is no conflict between type 1 information and P^e:

$$\begin{cases} \mu(x) - \mu(w) > \mu(y) - \mu(z) \\ \mu(x) - \mu(y) > 0 \\ \mu(x) - \mu(w) > 0 \\ \mu(x) - \mu(z) > 0 \\ \mu(y) - \mu(z) > 0 \\ \mu(w) - \mu(z) > 0 \\ \mu(y) - \mu(w) > 0 \end{cases} \qquad \begin{cases} \mu(w) - \mu(z) > \mu(x) - \mu(y) \\ \mu(x) - \mu(y) > 0 \\ \mu(x) - \mu(w) > 0 \\ \mu(x) - \mu(z) > 0 \\ \mu(y) - \mu(z) > 0 \\ \mu(w) - \mu(z) > 0 \\ \mu(y) - \mu(w) > 0 \end{cases}$$

For a detailed study of inconsistency, see [46].

6. Consistency Test for Preferential Information

6.1 Testing Procedures

Suppose that $X = \{a_1, a_2, \ldots, a_n\}$.

During the interactive questioning process conducted with J, each time that a new judgement is obtained, the consistency of all the responses already provided is tested. This consistency test begins with a pre-test aimed at detecting the (potential) presence of cycles within the relation P and, if no such cycle exists, making a permutation of the elements of X in such a way that, in the matrix of judgements, all of the cells P or C_{ij} will be located above the main diagonal.

When there is no cycle in P, the consistency of type 1 information $\{P, I, ?\}$ is tested as follows:

- If $? \neq \phi$, a linear program named LP-test$_1$ is used;
- if $? = \phi$, rather than linear programming, a method named DIR-test$_1$ is used, which has the advantage of being easily associated with a very simple visualization of an eventual ranking within the matrix of judgements.

When $\{P, I, ?\}$ is consistent, the consistency of type 1+2 information $\{P, I, ?, P^e\}$ is tested with the help of a linear program named LPσ-test$_{1+2}$.

6.2 Pre-test of the Preferential Information

The pre-test of the preferential information is based on Property 1. (Evident because $\#X$ is finite).

PROPERTY 1 *Let* $X^* \subset X$; *if* $\forall\, x \in X^*, \exists\, y \in X^*$ *such that* xPy, *then* $\exists\, x_1, x_2, \ldots, x_p \in X^*$ *such that* $x_1Px_2P \ldots Px_pPx_1$ *(cycle).*

The pre-test consists of seeking a permutation $\varphi : \mathbb{N}_{1,n} \rightarrow \mathbb{N}_{1,n}$ such that

$$\forall\, i, j \in \mathbb{N}_{1,n},\ [\, i > j \Rightarrow a_{\varphi(i)}(notP)a_{\varphi(j)}\,].$$

The permutation of the elements of X is made by the algorithm PRETEST, that detects cycles within P and sorts the elements(s) of X.

PRETEST:

1 $s \leftarrow n$;

2 among $a_1, a_2, \ldots, a_s$ find a_i which is not preferred over any other:
if a_i exists, go to 3.;
if not, return FALSE ($Sc_1 = \phi$, according to Property 1); finish.

3 permute a_i and a_s;

4 $s \leftarrow s - 1$:
if $s = 1$, return TRUE; finish.
If not, go to 2.

6.3 Consistency Test for Type 1 Information

Suppose that PRETEST detected no cycle within P and that the elements of X were renumbered as follows (to avoid the introduction of a permutation in the notation):

$$\forall\, i,j \in \mathbb{N}_{1,n},\ [\, i > j \Rightarrow a_i (notP) a_j \,].$$

6.3.1 Consistency Test for Incomplete ($? \neq \phi$) Type 1 Information.

Consider the linear program LP-test$_1$ with variables $x_1, x_2, \ldots, x_n$:

$$\begin{array}{ll} \min x_1 & \\ \text{subject to} & \\ x_i - x_j \geq d_{\min} & \forall\,(a_i,a_j) \in P \\ x_i - x_j = 0 & \forall\,(a_i,a_j) \in I \text{ with } i \neq j \\ x_i \geq 0 & \forall\, i \in \mathbb{N}_{1,n} \end{array}$$

where $d_{\min}$ is a positive constant, and the variables $x_1, x_2, \ldots, x_n$ represent the numbers $\mu(a_1), \mu(a_2), \ldots, \mu(a_n)$ that should satisfy Condition 1 so that μ is a type 1 scale.

The objective function min x_1 of LP-test$_1$ is obviously arbitrary. It is trivial that $Sc_1 \neq \phi \Leftrightarrow$ LP-test$_1$ is feasible.

6.3.2 Consistency Test for Complete ($? = \phi$) Type 1 Information.

When $? = \phi$ and the elements of X have been renumbered (after the application of PRETEST), another simple test (DIR-test$_1$) allows one to verify if $P \cup I$ is a complete preorder on X. DIR-test$_1$ is based on Proposition 1 (Proved in [46]).

PROPOSITION 6 *If* $[\, \forall\, i,j \in \mathbb{N}_{1,n}$ *with* $i < j,\ (a_i,a_j) \in P \cup I \,]$ *then* $P \cup I$ *is a complete preorder on* X *if and only if* $\forall\, i,j \in \mathbb{N}_{1,n}$ *with* $i < j : \left[a_i P a_j \Rightarrow \begin{cases} \forall\, s \leq i, \forall\, t \geq j, a_s P a_t \\ \exists\, s : i \leq s \leq j-1 \textit{ and } a_s P a_{s+1} \end{cases} \right]$.

Proposition 1 means that when the "P cases" of the matrix of judgements forms a "staircase", a ranking exists such that each step of the "staircase" rests, at least partly, on the principal diagonal of the matrix.

6.4 Consistency Test for Type 1+2 Information

It would be possible to test the consistency of type 1+2 information with a linear program based on Conditions 1 and 2. However, the more efficient linear program LP-test$_{1+2}$, which includes "thresholds conditions" equivalent to Conditions 1 and 2, is used instead. LP-test$_{1+2}$ is based on Lemma 1 (Proved in [46]).

LEMMA 1 *Let* $\mu : X \to \mathbb{R}$. μ *satisfies Conditions 1 and 2 if and only if there exist* Q *"thresholds"* $0 < \sigma_1 < \sigma_2 < \ldots < \sigma_Q$ *that satisfy Conditions 3, 4 and 5.*

CONDITION 3 $\forall\, (x, y) \in I,\ \mu(x) = \mu(y)$.

CONDITION 4 $\forall\, i, j \in \mathbb{N}_{1,Q}$ *with* $i \le j$, $\forall\, (x, y) \in C_{ij},\ \sigma_i < \mu(x) - \mu(y)$.

CONDITION 5 $\forall\, i, j \in \mathbb{N}_{1,Q-1}$ *with* $i \le j$, $\forall\, (x, y) \in C_{ij},\ \mu(x) - \mu(y) < \sigma_{j+1}$.

Program LP-test$_{1+2}$ has variables $x_1 (= \mu(a_1)), \ldots, x_n (= \mu(a_n)), \sigma_1, \ldots, \sigma_Q$:

$$
\begin{array}{ll}
\min x_1 & \\
\text{subject to} & \\
x_p - x_r = 0 & \forall\, (a_p, a_r) \in I \text{ with } p < r \\
\sigma_j + d_{\text{min}} \le x_p - x_r & \forall\, i, j \in \mathbb{N}_{1,Q} \text{ with } i \le j,\ \forall\, (a_p, a_r) \in C_{ij} \\
x_p - x_r \le \sigma_{j+1} - d_{\text{min}} & \forall\, i, j \in \mathbb{N}_{1,Q-1} \text{ with } i \le j,\ \forall\, (a_p, a_r) \in C_{ij} \\
d_{\text{min}} \le \sigma_1 & \\
\sigma_{i-1} + d_{\text{min}} \le \sigma_i & \forall\, i \in \mathbb{N}_{2,Q} \\
x_i \ge 0 & \forall\, i \in \mathbb{N}_{1,n} \\
\sigma_i \ge 0 & \forall\, i \in \mathbb{N}_{1,Q}
\end{array}
$$

Taking into account Lemma 1, it is trivial that $Sc_{1+2} \ne \phi$ if and only if the linear program LP-test$_{1+2}$, which is based on Conditions 3, 4 and 5, is feasible.

7. Dealing with Inconsistency

When a type 1+2 information $\{P, I, ?, P^e\}$ about X is inconsistent, it is convenient to be able to show J systems of constraints that render his or her judgements inconsistent and modifications of these judgements that would render LPσ-test$_{1+2}$ feasible.

7.1 Systems of Incompatible Constraints

Suppose that LP-test$_{1+2}$ is not feasible or, in other words, that the following system is incompatible (variables $x_1 (= \mu(a_1)), \ldots, x_n (= \mu(a_n)), \sigma_1, \ldots, \sigma_Q$ nonnegative):

$$
\left\{
\begin{array}{lll}
x_p - x_r = 0 & \forall\, (a_p, a_r) \in I \text{ with } p < r & \text{(t1)} \\
\sigma_i < x_p - x_r & \forall\, i, j \in \mathbb{N}_{1,Q} \text{ with } i \le j,\ \forall\, (a_p, a_r) \in C_{ij} & \text{(t2)} \\
x_p - x_r < \sigma_{j+1} & \forall\, i, j \in \mathbb{N}_{1,Q-1} \text{ with } i \le j,\ \forall\, (a_p, a_r) \in C_{ij} & \text{(t3)} \\
0 < \sigma_1 & & \text{(t4)} \\
\sigma_{i-1} < \sigma_i & \forall\, i \in \mathbb{N}_{2,Q} & \text{(t5)}
\end{array}
\right.
$$

Conventions:

- $\mathbb{R}^{m \times n}$ is the set of the real matrices with m lines and n columns.
- Matrix $M \in \mathbb{R}^{m \times n}$ is "non-zero" ($M \neq 0$) if at least one of its elements is not null.
- Matrix $M \in \mathbb{R}^{m \times n}$ is positive or null ($M \geq 0$) if all of its elements are positive or null.

The system of incompatible constraints can be written in the matrix format as follows:

$$\begin{cases} C \cdot Z > 0 & \text{(by grouping constraints (t2))} \\ D \cdot Z > 0 & \text{(by grouping constraints (t3))} \\ E \cdot Z > 0 & \text{(by grouping constraints (t4) and (t5))} \\ B \cdot Z = 0 & \text{(by grouping constraints (t1))} \end{cases}$$

where

$$Z = \begin{pmatrix} x_1 \\ x_2 \\ \vdots \\ x_n \\ \sigma_1 \\ \sigma_2 \\ \vdots \\ \sigma_Q \end{pmatrix}$$

$C \in \mathbb{R}^{p_1 \times (n+Q)}$(where p_1 is the number of constraints (t2))
$D \in \mathbb{R}^{p_2 \times (n+Q)}$(where p_2 is the number of constraints (t3))
$E \in \mathbb{R}^{p_3 \times (n+Q)}$(where p_3 is the number of constraints (t4) and (t5))
$B \in \mathbb{R}^{r \times (n+Q)}$(where r is the number of constraints (t1))

Note: if $r = 0$, one could consider that $B = 0 \in \mathbb{R}^{1 \times (n+Q)}$ without losing generality.

Let A be the matrix $\begin{bmatrix} C \\ D \\ E \end{bmatrix} \in \mathbb{R}^{p \times (n+Q)} \quad (p = p_1 + p_2 + p_3)$. The system of incompatible constraints can be written more simply as

$$S \begin{cases} A \cdot Z > 0 & \text{(by grouping constraints (t2), (t3), (t4) and (t5))} \\ B \cdot Z = 0 & \text{(by grouping constraints (t1)).} \end{cases}$$

In order to detect incompatibilities between the constraints (t1), (t2), (t3), (t4) and (t5) and propose eventual corrections, we apply Proposition 2 (Proved in [46]), which is a corollary of Mangasarian's [55] version of the *Theorem of the Alternative*.

PROPOSITION 7 *The system* $S \quad \{A \cdot Z > 0; B \cdot Z = 0\}$ *admits a solution* $Z \in \mathbb{R}^{(n+Q)\times 1}$ *or there exists* $Y \in \mathbb{R}^{p\times 1}, V, W \in \mathbb{R}^{r\times 1}$ *with* $Y \neq 0, Y \geq 0, V \geq 0, W \geq 0$ *such that* ${}^tA \cdot Y + {}^tB \cdot (V - W) = 0$ *and* $\forall i \in \mathbb{N}_{1,r}$, $V_i \cdot W_i = 0$ *but never both.*

The interest of Proposition 2 is that vectors Y, V and W have positive or null components, thus making it compatible with linear programming (see Sections 7.3 and 7.4)

7.2 Example 1

Suppose that $X = \{a_1, a_2, a_3, a_4\}$ and that J has formulated the following judgements:

- $P = \{(a_1, a_2), (a_1, a_3), (a_2, a_3), (a_3, a_4)\}$
- $(a_1, a_2) \in C_1$, $(a_1, a_3) \in C_4$, $(a_2, a_3) \in C_2$, $(a_3, a_4) \in C_2$.

Suppose that J also judges that $a_2 P a_4$ and that $(a_2, a_4) \in C_3$. LP-test$_1$ is feasible: the judgements are compatible with a ranking. LP-test$_{1+2}$ is not feasible: the software informs J that his or her judgements are "inconsistent".

Suppose now that J confirms his or her judgements. One must then have:

$$
\begin{array}{llllll}
\sigma_1 < x_1 - x_2 & (1) & x_1 - x_2 < \sigma_2 & (2) & 0 < \sigma_1 & (11) \\
\sigma_2 < x_2 - x_3 & (3) & x_2 - x_3 < \sigma_3 & (4) & \sigma_1 < \sigma_2 & (12) \\
\sigma_2 < x_3 - x_4 & (5) & x_3 - x_4 < \sigma_3 & (6) & \sigma_2 < \sigma_3 & (13) \\
\sigma_3 < x_2 - x_4 & (7) & x_2 - x_4 < \sigma_4 & (8) & \sigma_3 < \sigma_4 & (14) \\
\sigma_4 < x_1 - x_3 & (9) & x_1 - x_3 < \sigma_5 & (10) & \sigma_4 < \sigma_5 & (15) \\
 & & & & \sigma_5 < \sigma_6 & (16)
\end{array}
$$

or, in matrix format (which one can denote as $A \cdot Z > 0$):

$$\begin{pmatrix} 1 & -1 & 0 & 0 & -1 & 0 & 0 & 0 & 0 & 0 \\ 1 & 1 & 0 & 0 & 0 & 1 & 0 & 0 & 0 & 0 \\ 0 & 1 & -1 & 0 & 0 & -1 & 0 & 0 & 0 & 0 \\ 0 & -1 & 1 & 0 & 0 & 0 & 1 & 0 & 0 & 0 \\ 0 & 0 & 1 & -1 & 0 & -1 & 0 & 0 & 0 & 0 \\ 0 & 0 & -1 & 1 & 0 & 0 & 1 & 0 & 0 & 0 \\ 0 & 1 & 0 & -1 & 0 & 0 & -1 & 0 & 0 & 0 \\ 0 & -1 & 0 & 1 & 0 & 0 & 0 & 1 & 0 & 0 \\ 1 & 0 & -1 & 0 & 0 & 0 & 0 & -1 & 0 & 0 \\ -1 & 0 & 1 & 0 & 0 & 0 & 0 & 0 & 1 & 0 \\ 0 & 0 & 0 & 0 & 1 & 0 & 0 & 0 & 0 & 0 \\ 0 & 0 & 0 & 0 & -1 & 1 & 0 & 0 & 0 & 0 \\ 0 & 0 & 0 & 0 & 0 & -1 & 1 & 0 & 0 & 0 \\ 0 & 0 & 0 & 0 & 0 & 0 & -1 & 1 & 0 & 0 \\ 0 & 0 & 0 & 0 & 0 & 0 & 0 & -1 & 1 & 0 \\ 0 & 0 & 0 & 0 & 0 & 0 & 0 & 0 & -1 & 1 \end{pmatrix} \cdot \begin{pmatrix} x_1 \\ x_2 \\ x_3 \\ x_4 \\ \sigma_1 \\ \sigma_2 \\ \sigma_3 \\ \sigma_4 \\ \sigma_5 \\ \sigma_6 \end{pmatrix} > 0$$

Since it is known, according to Proposition 2, that the system has no solution, there necessarily exists $Y \in \mathbb{R}^{16\times 1}(Y \neq 0, Y \geq 0)$ such that ${}^tA \cdot Y = 0$. Thus, positive or null (but not all null) real numbers $y_1, y_2, \ldots, y_{16}$ exist such that $\sum_{i=1}^{16} y_i \cdot Col_i = 0$ (where Col_i is the column i of the matrix ${}^tA$).

In this simple example, one can see that it is enough to make $y_2 = y_5 = y_8 = y_9 = 1$ and $y_1 = y_3 = y_4 = y_6 = y_7 = y_{10} = y_{11} = y_{12} = y_{13} = y_{14} = y_{15} = y_{16} = 0$:

$$1 \cdot \underbrace{\begin{pmatrix} -1 \\ 1 \\ 0 \\ 0 \\ 0 \\ 1 \\ 0 \\ 0 \\ 0 \\ 0 \end{pmatrix}}_{Col_2} +1 \cdot \underbrace{\begin{pmatrix} 0 \\ 0 \\ 1 \\ -1 \\ 0 \\ -1 \\ 0 \\ 0 \\ 0 \\ 0 \end{pmatrix}}_{Col_5} +1 \cdot \underbrace{\begin{pmatrix} 0 \\ -1 \\ 0 \\ 1 \\ 0 \\ 0 \\ 0 \\ 1 \\ 0 \\ 0 \end{pmatrix}}_{Col_8} +1 \cdot \underbrace{\begin{pmatrix} 1 \\ 0 \\ -1 \\ 0 \\ 0 \\ 0 \\ 0 \\ -1 \\ 0 \\ 0 \end{pmatrix}}_{Col_9} = \begin{pmatrix} 0 \\ 0 \\ 0 \\ 0 \\ 0 \\ 0 \\ 0 \\ 0 \\ 0 \\ 0 \end{pmatrix}$$

These four vectors correspond to the four constraints (2), (5), (8) and (9) above:

$$\left.\begin{array}{ll} \sigma_4 > x_2 - x_4 & (8) \\ x_1 - x_3 > \sigma_4 & (9) \end{array}\right\} \Rightarrow x_1 - x_3 > x_2 - x_4 \quad (*)$$

$$\left.\begin{array}{ll}\sigma_2 > x_1 - x_2 & (2)\\ x_3 - x_4 > \sigma_2 & (5)\end{array}\right\} \Rightarrow x_3 - x_4 > x_1 - x_2 \quad (**)$$

$(*)$ and $(**)$ bring to the contradiction $x_1 - x_4 > x_1 - x_4$. The incompatibility between $(*)$ and $(**)$ is presented in M-MACBETH as shown in Figure 10.2.

Diff.	Couples		Couples	Diff.
strong	a1 - a3	>	a2 - a4	moderate
weak	a3 - a4	>	a1 - a2	very weak

Figure 10.2. Example of incompatibility between $(*)$ and $(**)$.

Note that the problem disappears if

$$\begin{array}{lll} & (a_1, a_3) \in C_3 \text{ instead of } C_4 & ((*) \text{ disappears})\\ \text{or} & (a_2, a_4) \in C_4 \text{ instead of } C_3 & ((*) \text{ disappears})\\ \text{or} & (a_3, a_4) \in C_1 \text{ instead of } C_2 & ((**) \text{ disappears})\\ \text{or} & (a_1, a_2) \in C_2 \text{ instead of } C_1 & ((**) \text{ disappears}).\end{array}$$

Note also that the inconsistency would not be eliminated for any modification of the judgement "$(a_2, a_3) \in C_2$".

If J confirms the judgement "$(a_2, a_4) \in C_3$", M-MACBETH calculates the different possibilities (four in example 1) that J can follow to make his or her judgements consistent with a "minimal" number of changes of category (one in Example 1). (We will specify in Section 7.4 the meaning of this notion).

In M-MACBETH, the "suggestions" of changes are presented (graphically) in the matrix of judgements. They are:

- to replace the judgement $(a_1, a_3) \in C_4$ with the judgement $(a_1, a_3) \in C_3$
- or to replace the judgement $(a_2, a_4) \in C_3$ with the judgement $(a_2, a_4) \in C_4$
- or to replace the judgement $(a_3, a_4) \in C_2$ with the judgement $(a_3, a_4) \in C_1$
- or to replace the judgement $(a_1, a_2) \in C_1$ with the judgement $(a_1, a_2) \in C_2$.

7.3 Identifying Constraints which Cause Inconsistency

Let us detail the various stages of our search for "suggestions". The first step consists of determining the constraints (t1), (t2) and (t3) which are "the origin of the incompatibilities" present in the system

$$S\left\{\begin{array}{l} A \cdot Z > 0\\ B \cdot Z = 0\end{array}\right. \quad \text{(see Section 7.1)}$$

We consider that a constraint is "at the origin of an incompatibility" when it is part of a system S' that

- is a "sub-system" of S,
- is incompatible,
- does not contain any incompatible "sub-system".

Mathematically, this idea can be represented by Definition 7.

DEFINITION 43 *An incompatible elementary system (SEI) is a system*

$$S' \left\{ \begin{array}{l} A' \cdot Z > 0 \\ B' \cdot Z = 0 \end{array} \right.$$

such that

1 $A' \in \mathbb{R}^{p' \times (n+Q)}$ *is a sub-matrix of A, and* $B' \in \mathbb{R}^{r' \times (n+Q)}$ *is a sub-matrix of B;*

2 S' *is incompatible;*

3 *If* $\left\{ \begin{array}{l} A'' \in \mathbb{R}^{p'' \times (n+Q)} \text{ is a sub-matrix of } A', \\ B'' \in \mathbb{R}^{r'' \times (n+Q)} \text{ is a sub-matrix of } B', \\ p'' + r'' < p' + r' \end{array} \right.$ *then* $\left\{ \begin{array}{l} A'' \cdot Z > 0 \\ B'' \cdot Z = 0 \end{array} \right.$ *is compatible.*

However, our goal is not to determine all the SEI that could be extracted from the constraints using LPσ-test$_{1+2}$. We just want to find all of the judgements of the type $(a_s, a_t) \in C_{ij}$ that "generate" an incompatibility. In Section 7.4.3, we will explain how we use these judgements.

We know that an inconsistency occurs when the system

$$S \left\{ \begin{array}{l} A \cdot Z > 0 \\ B \cdot Z = 0 \end{array} \right.$$

is incompatible; that is, $\exists Y \in \mathbb{R}^p$ and $V, W \in \mathbb{R}^r$ such that

$$\left\{ \begin{array}{l} {}^tA \cdot Y +^t B \cdot (V - W) = 0 \\ Y \geq 0,\ V \geq 0,\ W \geq 0 \\ \forall\, i \in \mathbb{N}_{1,r},\ V_i \cdot W_i = 0 \\ \exists\ i_0 \in \mathbb{N}_{1,p} \text{ such that } Y_{i_0} \neq 0 \end{array} \right.$$

In such a case, if $i_0 \leq p_1 + p_2$, where p_1 is the number of constraints (t2) and p_2 is the number of constraints (t3) (see Section 7.1), a constraint of the type $x_s - x_t < \sigma_j$ or $x_s - x_t > \sigma_j$ will correspond to S.

Consider, then, the system (with $i \leq p_1 + p_2$):

$$\text{Syst-}Y_i \begin{cases} {}^tA \cdot Y + {}^tB \cdot (V - W) = 0 \\ Y_i = 1 \end{cases}$$

If Syst-Y_i is compatible, for one of its solutions it corresponds to a system of incompatible constraints (t1), (t2), (t3), (t4) and (t5) where at least one constraint (that which corresponds to $Y_i = 1$) is of the type $x_s - x_t < \sigma_j$ or $x_s - x_t > \sigma_j$ and is part of a SEI. If Syst-Y_i is incompatible, the constraint that corresponds to Y_i is not part of any SEI.

To find all of the constraints (t2) and (t3) which are part of a SEI, it is sufficient to study the compatibility of all of the systems Syst-Y_i, for $i = 1, 2, \ldots, p_1 + p_2$.

We will proceed in a similar way, using the systems Syst-V_i and Syst-W_i, to find all of the constraints (t1) which are part of a SEI:

$$\text{Syst-}V_i \begin{cases} {}^tA \cdot Y + {}^tB \cdot (V - W) = 0 \\ W_i = 0 \\ V_i = 1 \end{cases}$$

and

$$\text{Syst-}W_i \begin{cases} {}^tA \cdot Y + {}^tB \cdot (V - W) = 0 \\ V_i = 0 \\ W_i = 1 \end{cases}$$

It is not necessary to examine all of the systems Syst-Y_i, Syst-V_i and Syst-W_i:

- If Syst-Y_i is compatible and has the solution Y, V, W, then
  - $\forall\, j > i$ such that $Y_j \neq 0$, Syst-Y_i is compatible;
  - $\forall\, j \in \mathbb{N}_{1,r}$ such that $V_j \neq 0$, Syst-V_i is compatible;
  - $\forall\, j \in \mathbb{N}_{1,r}$ such that $W_j \neq 0$, Syst-W_i is compatible.
- If Syst-V_i is compatible and has the solution Y, V, W, then
  - $\forall\, j > i$ such that $V_j \neq 0$, Syst-V_i is compatible;
  - $\forall\, j \in \mathbb{N}_{1,r}$ such that $W_j \neq 0$, Syst-W_i is compatible.
- If Syst-W_i is compatible and has the solution Y, V, W, then
  - $\forall\, j > i$ such that $W_j \neq 0$, Syst-W_i is compatible.

It is for this reason that a "witness-vector" $T \in \mathbb{N}^{p_1+p_2+2\cdot r}$ must be used, initially null, updated as follows:

- For any solution Y, V, W of a system Syst-Y_i, Syst-V_i or Syst-W_i do
  - $\forall\, j \in \mathbb{N}_{1,p_1+p_2}, \ [\, Y_j \neq 0 \Rightarrow T_j = 1\,]$

- $\forall\, j \in \mathbb{N}_{1,r}$, $[\, V_j \neq 0 \Rightarrow T_{p_1+p_2+j} = 1\,]$
- and $[\, W_j \neq 0 \Rightarrow T_{p_1+p_2+r+j} = 1\,]$.

To find the interesting pairs, the compatibility of at most p_1+p_2+2r systems should be studied. The general algorithm to seek equations (t1) and inequalities (t2) and (t3) that are part of a SEI is the following:

- $T = (0, 0, \ldots, 0)$
- for $i = 1, 2, \ldots, p_1 + p_2$ do:
  - $T_i = 0$,
  - then if Syst-Y_i compatible and Y, V, W solution of Syst-Y_i

    then update T
- for $i = 1, 2, \ldots, r$ do:
  - if $T_{p_1+p_2+i} = 0$,
  - then if Syst-V_i compatible and Y, V, W solution of Syst-V_i

    then update T
- for $i = 1, 2, \ldots, r$ do:
  - if $T_{p_1+p_2+r+i} = 0$,
  - then if Syst-W_i compatible and Y, V, W solution of Syst-W_i

    then update T.

In this way one obtains the set of all of the equations and inequalities that make up the SEI.

7.4 Augmentation – Reduction in a Judgement with p Categories

7.4.1 Preliminaries. Notation:

- Judgement $(x, y) \in C_{ij}$ will be represented by element (x, y, i, j) of $X \times X \times \mathbb{N}_{1,Q} \times \mathbb{N}_{1,Q}$.
- Judgement $(x, y) \in I$ will be represented by element $(x, y, 0, 0)$ of $X \times X \times \mathbb{N} \times \mathbb{N}$.

DEFINITION 44 *A reduction in judgement (s, t, i, j) with p categories $(1 \leq p \leq Q + i)$ is the replacement of this judgement*

- *by the judgement $(s, t, i - p, i - p)$ if $i \geq p$*

- *by the judgement* $(t, s, p - i, p - i)$ *if* $i < p$.

DEFINITION 45 *An augmentation of the judgement* (s, t, i, j) *with* p *categories* $(1 \leq p \leq Q - j)$ *is the replacement of this judgement by the judgement* $(s, t, j + p, j + p)$.

DEFINITION 46 *A change of judgement* (s, t, i, j) *with* p *categories is an augmentation or a reduction of the judgement with* p *categories.*

Comment: It is evident that one obtains the same final judgement as a result of "1 reduction of a judgement with p categories" or the "p successive reductions of a category of 1 judgement".

Convention: A "change in judgement (s, t, i, j) with p categories" will be represented by $(s, t, i, j, p) \in X \times X \times \mathbb{N}_{1,Q} \times \mathbb{N}_{1,Q} \times \mathbb{Z}$ (augmentation if $p > 0$, reduction if $p < 0$).

7.4.2 Exploitation of the Constraints of SEI. Let us recall from 7.3 that

- if $T_i > 0$, it has a corresponding constraint (t2) or (t3) or (t1) that is part of an SEI;
- if $T_i = 0$, it has no corresponding constraint that is part of an SEI.

These variables, then, provide us with an indication as to the future "modification" to be made to the judgements associated with these constraints. Indeed, suppose that $T_i > 0$:

a) if $1 \leq i \leq p_1$, a constraint $\sigma_u < x_s - x_t$ which is part of an SEI corresponds to variable T_i; if a change in its judgement $(s, t, \ldots, \ldots)$ can help to eliminate the SEI, it ensures that it will be a "reduction" (evident).

b) if $p_1 + 1 \leq i \leq p_1 + p_2$, a constraint $x_s - x_t < \sigma_u$ which is part of an SEI corresponds to variable T_i; if a change in its judgement $(s, t, \ldots, \ldots)$ can help to eliminate the SEI, it ensures that it will be an "augmentation" (evident).

c) if $p_1 + p_2 + 1 \leq i \leq p_1 + p_2 + r$, a constraint $x_s - x_t = 0$ which is part of an SEI corresponds to variable T_i; if a change in its judgement $(s, t, 0, 0)$ can help to eliminate the SEI, it ensures that it will be a "reduction".

d) if $p_1 + p_2 + r + 1 \leq i \leq p_1 + p_2 + 2r$, a constraint $x_s - x_t = 0$ which is part of an SEI corresponds to variable T_i; if a change in its judgement $(s, t, 0, 0)$ can help to eliminate the SEI, it ensures that it will be an "augmentation" (proof similar to that of c).

Proof of c):

Being $h = i - (p_1 + p_2)$, one knows (by the definition of T_i) that $\exists\, Y \in \mathbb{R}^p$, $\exists\, V, W \in \mathbb{R}^r$ with $Y \geq 0,\ V \geq 0,\ W \geq 0,\ Y \neq 0,\ V_h \neq 0$ and $W_h = 0$ such that ${}^t(A') \cdot Y +^t (B') \cdot (V - W) = 0$ or, if one notes $LineB_j$ the jth line of B',

$$ {}^t(A') \cdot Y + {}^tLineB_h \cdot V_h + \sum_{\substack{j=1 \\ j \neq h}}^{r} {}^tLineB_j \cdot V_j - \sum_{\substack{j=1 \\ j \neq h}}^{r} {}^tLineB_j \cdot W_j = 0 $$

(because $W_h = 0$).

The corresponding SEI $\left\{ \begin{array}{l} A' \cdot Z > 0 \\ B' \cdot Z = 0 \end{array} \right.$ can be written $\left\{ \begin{array}{l} A' \cdot Z > 0 \\ x_s - x_t = 0 \\ B'' \cdot Z = 0, \end{array} \right.$ where

$$ B'' = \begin{bmatrix} LineB_1 \\ \vdots \\ LineB_{h-1} \\ LineB_{h+1} \\ \vdots \\ LineB_r \end{bmatrix} \quad \text{(the matrix } B' \text{ without line } LineB_h\text{).} $$

If one considers an "augmentation" of judgement $(s, t, 0, 0)$, the constraint $x_s - x_t = 0$ would be replaced by the constraint $x_s - x_t > 0$. The new system $\left\{ \begin{array}{l} A' \cdot Z > 0 \\ x_s - x_t > 0 \\ B'' \cdot Z = 0 \end{array} \right.$ can be written $\left\{ \begin{array}{l} A'' \cdot Z > 0 \\ B'' \cdot Z = 0, \end{array} \right.$ where $A'' = \begin{bmatrix} A' \\ LineB_h \end{bmatrix}$ (the matrix A' "augmented" with line $LineB_h$).

The system is still incompatible; indeed, if one poses

- $Y' = (Y_1, Y_2, \ldots, Y_p, V_h) \in \mathbb{N}^{p+1}$
- $V' = (V_1, \ldots, V_{h-1}, V_{h+1}, \ldots, V_r) \in \mathbb{N}^{r-1}$
- $W' = (W_1, \ldots, W_{h-1}, W_{h+1}, \ldots, W_r) \in \mathbb{N}^{r-1}$.

$$ {}^t(A') \cdot Y + {}^tLineB_h \cdot V_h + \sum_{\substack{j=1 \\ j \neq h}}^{r} {}^tLineB_j \cdot V_j - \sum_{\substack{j=1 \\ j \neq h}}^{r} {}^tLineB_j \cdot W_j = 0 $$

can be written: ${}^t(A'') \cdot Y' +^t (B'') \cdot (V' - W') = 0$, where $Y' \neq 0$ (since $Y \neq 0$), which proves the incompatibility of the system.

Each "suggestion" of a potential change ($T_i = 1$) of a judgement $(s, t, \ldots, \ldots)$ can thus be stored in a vector S of $\mathbb{N}^4$ where

$$S_1 = s$$
$$S_2 = t$$
$$S_3 = \begin{cases} 1 & \text{if } \exists\, i \in \mathbb{N}_{1,p_1} \cup \mathbb{N}_{p_1+p_2+1,p_1+p_2+r} \text{ such that } T_i = 1 \\ & \quad \text{(reduction)} \\ 0 & \text{otherwise} \end{cases}$$
$$S_4 = \begin{cases} 1 & \text{if } \exists\, i \in \mathbb{N}_{p_1+1,p_1+p_2} \cup \mathbb{N}_{p_1+p_2+r+1,p_1+p_2+2r} \text{ such that} \\ & T_i = 1 \ \text{(augmentation)} \\ 0 & \text{otherwise} \end{cases}$$

We will denote by $PreSugg$ the set of these "pre-suggestions". In the case of example 1 (see Section 7.3) one has

$$PreSugg = \{(a_1, a_3, 1, 0), (a_3, a_4, 1, 0), (a_1, a_2, 0, 1), (a_2, a_4, 0, 1)\}.$$

7.4.3 Search for Suggestions.

DEFINITION 47 *Changing judgements by m categories is any set $Modif_m$ of the form $Modif_m = \{(s_1, t_1, i_1, j_1, p_1), (s_2, t_2, i_2, j_2, p_2), \ldots, (s_u, t_u, i_u, j_u, p_u) \,|\forall v \in \mathbb{N}_{1,u}, (s_v, t_v, i_v, j_v, p_v)$ is a change of judgement (s_v, t_v, i_v, j_v) with p_v categories$\}$ such that $\sum_{v=1}^{u} |p_v| = m$*

Within Example 1, $\{(a_1, a_2, 1, 1, 2), (a_3, a_4, 2, 2, -1)\}$ is a "change of judgements with 3 categories", which consists of

- to replace the judgement $(a_1, a_2) \in C_1$ with the judgement $(a_1, a_2) \in C_3$ (augmentation of 2 categories)
- to replace the judgement $(a_3, a_4) \in C_2$ with the judgement $(a_3, a_4) \in C_1$ (reduction of 1 category)

Notation: the set of "judgement changes with m categories" which renders the judgements consistent will be denoted by $Sugg_m$.

Within Example 1,

- $\{(a_1, a_2, 1, 1, 2), (a_3, a_4, 2, 2, -1)\} \in Sugg_3$
- $\{(a_1, a_3, 4, 4, -1)\}$, $\{(a_3, a_4, 2, 2, -1)\}$, $\{(a_1, a_2, 1, 1, 1)\}$ and $\{(a_2, a_4, 3, 3, 1)\} \in Sugg_1$,

these are the 4 changes suggested in Section 7.3.

Once the PreSugg group is determined, the third step is to:

- determine the "minimum number of changes" (some possibly successive) necessary to render the judgements consistent;
- determine all of the combinations of such "minimal" changes.

More rigorously, this means

- find $m_0 = \min\{m \in \mathbb{N}^* | Sugg_m \neq \emptyset\}$
- clarify $Sugg_m$

In Example 1, we have already seen that $m_0 = 1$ (since $Sugg_1 \neq \emptyset$).

We will proceed as follows for all cases of inconsistency (see Figure 10.3).

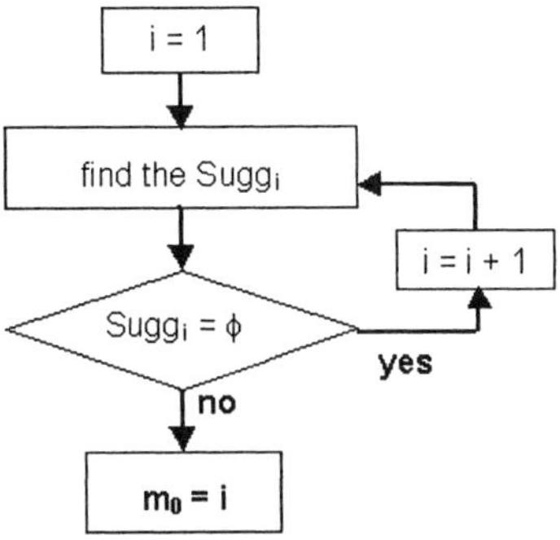

Figure 10.3. Procedure for all cases of inconsistency.

At each step i,

- the set of all "judgement changes of i categories", built on the basis of element PreSugg are considered;
- for each of the elements in this group:
  - carry out the modifications included in the selected item;
  - test the consistency of the new matrix of judgements; if it is consistent, store the element in $Sugg_i$;
  - restore the matrix to the initial judgements.

It is worth mentioning that we consider the possibility of changing a judgement by several categories.

This algorithm is always convergent since one can always give consistent judgements in a finite number of changes.

We emphasize that in practice, the cases of inconsistency that require more than 2 "changes of 1 category" are almost non-existent. The main reason being that any change in judgement that generates an inconsistency is immediately announced to J, who must then confirm or cancel his or her judgement.

This procedure allows one to avoid

- coarse errors of distraction (by cancelling the judgement);
- the "accumulation" of inconsistencies since, if J confirms his or her judgement, suggestions of changes that will eliminate the inconsistency are made.

7.5 Example 2

Suppose that $X = \{a_1, a_2, a_3, a_4\}$ and that J has formulated the following consistent judgements:

- $P = \{(a_1, a_2), (a_1, a_3), (a_2, a_3), (a_3, a_4)\}$
- $(a_1, a_2) \in C_1, (a_1, a_3) \in C_4, (a_2, a_3) \in C_2, (a_3, a_4) \in C_3$

Suppose that J adds that $a_2 P a_4$ and that $(a_2, a_4) \in C_3$: M-MACBETH informs J that his or her judgements are "inconsistent".

If J confirms the judgement $(a_2, a_4) \in C_3$, M-MACBETH will display the message: "Inconsistent judgements: MACBETH has found 6 ways to render the judgements matrix consistent with 2 category changes."

This time, it will be necessary to make at least 2 "changes of 1 category" to render the judgements consistent; there are 6 distinct combinations of such changes. Each of these 6 suggestions is presented graphically (see Figure 10.4) within the table of judgements, accompanied by SEI which, moreover, shows why the suggestions made eliminate this incompatibility: Figure 10.4 presents the first of six suggestions.

8. The MACBETH Scale

8.1 Definition of the MACBETH Scale

Suppose that $Sc_{1+2} \neq \phi$ and $a_1(P \cup I)a_2 \dots a_{n-1}(P \cup I)a_n$. The linear program LP-MACBETH with variables $x_1, \dots, x_n, \sigma_1, \dots, \sigma_Q$ is therefore feasible:

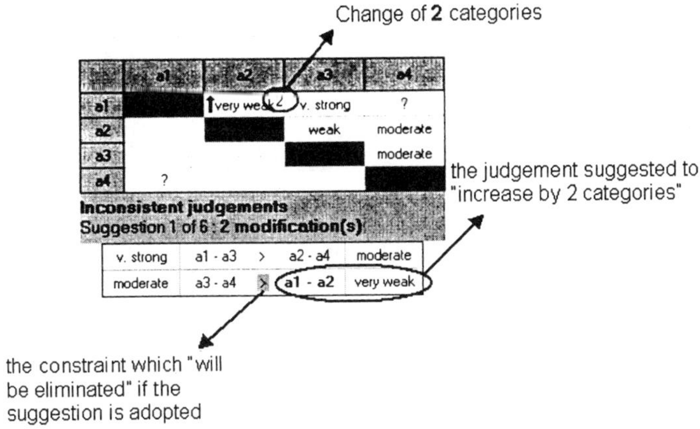

Figure 10.4. Suggestion of change to resolve inconsistency.

$$
\begin{array}{lll}
\min x_1 & & \\
\text{subject to} & & \\
x_p - x_r = 0 & \forall\, (a_p, a_r) \in I \text{ with } p < r & \text{(t1)} \\
\sigma_i + \frac{1}{2} \leq x_p - x_r & \forall\, i, j \in N_{1,Q} \text{ with } i \leq j, \forall (a_p, a_r) \in C_{ij} & \text{(t2')} \\
x_p - x_r \leq \sigma_{j+1} - \frac{1}{2} & \forall\, i, j \in N_{1,Q-1} \text{ with } i \leq j, \forall (a_p, a_r) \in C_{ij} & \text{(t3')} \\
\sigma_1 = \frac{1}{2} & & \text{(t4')} \\
\sigma_{i-1} + 1 \leq \sigma_i & \forall\, i \in N_{2,Q} & \text{(t5')} \\
x_i \geq 0 & \forall\, i \in N_{1,n} & \\
\sigma_i \geq 0 & \forall\, i \in N_{1,Q} &
\end{array}
$$

DEFINITION 48 *Any function* $EchMac : X \rightarrow \mathbb{R}$ *such that* $\forall\, i \in \mathbb{N}_{1,n}$, $EchMac(a_i) = x_i^*$ *– where* $(x_1^*, \ldots, x_n^*)$ *is an optimal solution of LP-MACBETH – is called a basic MACBETH scale.*

DEFINITION 49 $\forall\, a \in \mathbb{R}_+^*, \forall\, b \in \mathbb{R}$ *with* $(a, b) \neq (1, 0)$, $a \cdot EchMac + b$ *is a transformed MACBETH scale.*

8.2 Discussing the Uniqueness of the Basic MACBETH Scale

Nothing guarantees that a LP-MACBETH optimal solution is unique. For example, consider the matrix of judgements and the basic MACBETH scale shown is Figure 10.5.

One can verify that, $\forall\, x \in [6, 7]$, $(8, x, 5, 2, 1, 0)$ is still an optimal solution of LP-MACBETH. Thus, a basic MACBETH scale is not necessarily unique.

	a1	a2	a3	a4	a5	a6
a1	no	very weak	weak	moderate	moderate	strong
a2		no	very weak	moderate	moderate	moderate
a3			no	weak	moderate	moderate
a4				no	very weak	very weak
a5					no	very weak
a6						no

	Macbeth basic
a1	8.00
a2	6.50
a3	5.00
a4	2.00
a5	1.00
a6	0.00

Figure 10.5. Matrix of judgements and basic MACBETH scale.

As long as the MACBETH scale is interpreted as a technical aid whose purpose is to provide the foundation for a discussion with J, this does not constitute a true problem. However, we have observed that in practice decision makers often adopt the MACBETH scale as the final scale. It is, therefore, convenient to guarantee the uniqueness of the MACBETH scale. This is obtained technically, as follows (where S_{mac} is the group of the constraints of LP-MACBETH):

Step 1) solution of LP-MACBETH
$\rightarrow$ optimal solution $x_1, x_2, \ldots, x_n$
$\rightarrow \mu(a_1) = x_1, \mu(a_n) = x_n = 0$ (remark: $\mu(a_1)$ is unique)

Step 2) for $i = 2$ to $n - 1$

to solve $\max x_i$ under $\begin{cases} S_{mac} \\ x_1 = \mu(a_1), \ldots, x_{i-1} = \mu(a_{i-1}) \end{cases}$
$\rightarrow$ optimal solution $x_1, x_2, \ldots, x_n$
$\rightarrow xmax = x_i$

to solve $\min x_i$ under $\begin{cases} S_{mac} \\ x_1 = \mu(a_1), \ldots, x_{i-1} = \mu(a_{i-1}) \end{cases}$
$\rightarrow$ optimal solution $x_1, x_2, \ldots, x_n$
$\rightarrow xmin = x_i$

$$\mu(a_i) = \frac{xmin + xmax}{2}$$

Thus,

- to calculate $\mu(a_2)$, the variable x_1 is "fixed" to the value $\mu(a_1)$, the minimum and maximum values of x_2 are calculated and the average of the two results is taken as the value of $\mu(a_2)$;
- to calculate $\mu(a_3)$, the variable x_1 is "fixed" to the value of $\mu(a_1)$, the variable x_2 is "fixed" to the value of $\mu(a_2)$, the minimum and maximum values of x_3 are calculated and the average of the two values is taken as the value of $\mu(a_3)$;

- etc.

This method guarantees that $\mu(a_1), \mu(a_2), \ldots \mu(a_n)$ are unique for a given preferential information $\{P, I, ? = \phi, P^e\}$. It permits us to speak of "the" basic MACBETH scale, instead of "one" MACBETH scale.

8.3 Presentation of the MACBETH Scale

The MACBETH scale that corresponds to $\{P, I, ? = \phi, P^e\}$ consistent information is represented in two ways in M-MACBETH: a table and a "thermometer". In the example in Figure 10.6, the transformed MACBETH scale represented in the thermometer was obtained by imposing the values of the elements d and c as 100 and 0 respectively.

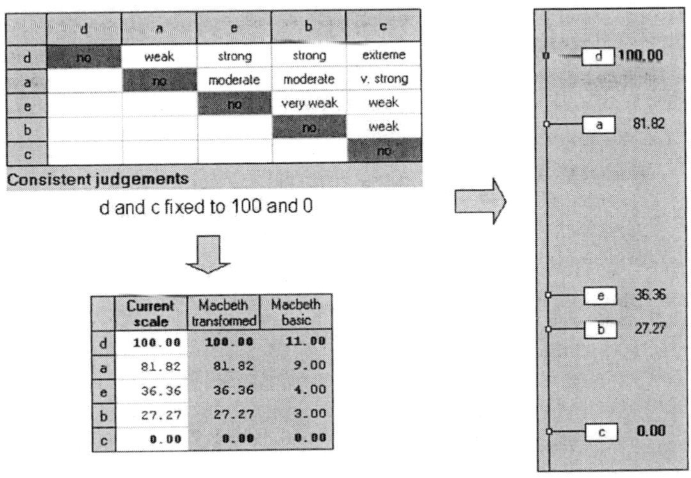

Figure 10.6. Representations of the MACBETH scale.

Even though the values attributed to c and d are fixed, in general an infinite number of scales that satisfy Conditions 1 and 2 exist. It is, thus, necessary to allow J to, should he or she want to, modify the values suggested. This is the subject of the next section.

9. Discussion About a Scale

Suppose that, in the example in Figure 10.6, J considers that the element a is badly positioned when compared to elements c and d and therefore J wants to redefine the value of a. It is then interesting to show J the limits within which the value of a can vary without violating the preferential information provided by J. Let us suppose in this section that we have a type 1+2 information about X which is consistent and that $? = \phi$.

Let μ_0 be a particular scale of Sc_{1+2}, L and H be two fixed elements of X with HPL (H more attractive than L) and a be an element of X (not indifferent to L and not indifferent to H) that J would like to have repositioned.

Let

- $Sc_{(\mu_0,H,L)} = \{\mu \in Sc_{1+2} \mid \mu(H) = \mu_0(H) \text{ and } \mu(L) = \mu_0(L)\}$ (scales for which values associated with H and L have been fixed)
- $Sc_{(\mu_0,\hat{a})} = \{\mu \in Sc_{1+2} \mid \forall\, y \in X$ with y not indifferent to a: $\mu(y) = \mu_0(y)\}$ (scales for which the values of all of the elements of X except a and its eventual equals have been fixed).

We call *free interval* associated to interval a :

$$\left]\inf_{\mu\in Sc_{(\mu_0,H,L)}} \mu(a), \sup_{\mu\in Sc_{(\mu_0,H,L)}} \mu(a)\right[$$

We call *dependent interval* associated to interval a :

$$\left]\inf_{\mu\in Sc_{(\mu_0,\hat{a})}} \mu(a), \sup_{\mu\in Sc_{(\mu_0,\hat{a})}} \mu(a)\right[$$

In the example in Figure 10.6, if one selects a, two intervals are presented to J (see Figure 10.7) which should be interpreted as follows:

$$\forall\, \mu \in Sc_{1+2}\,,\ [\,\mu(c) = 0, \mu(d) = 100\,] \Rightarrow 66.69 \leq \mu(a) \leq 99.98.$$

$$\forall\, \mu \in Sc_{1+2}\,,\ [\,\mu(c) = 0, \mu(d) = 100, \mu(e) = 36.36, \mu(b) = 27.27\,]$$
$$\Rightarrow 72.74 \leq \mu(a) \leq 90.9.$$

The closed intervals (in the example $[66.69, 99.98]$ and $[72.74, 90.9]$) that have been chosen to present to J are not the precise free and dependent intervals associated to a (which, by definition, are open); however, by taking a precision of 0.01 into account, they can be regarded as the "greatest" closed intervals included in the free and dependent intervals.

M-MACBETH permits the movement of element a with the mouse but, obviously, only inside of the dependent interval associated to a.

If J wants to give element a a value that is outside of the dependent interval (but still inside the free interval), the software points out that the values of the other elements must be modified. If J confirms the new value of a, a new MACBETH scale is calculated, taking into account the additional constraint that fix the new value of a.

The ("closed") free interval is calculated by integer linear programming. The ("closed") dependent interval could be also calculated in the same manner. However, M-MACBETH computes it by "direct" calculation formulas which make the determination of these intervals extremely fast – for details, see [46].

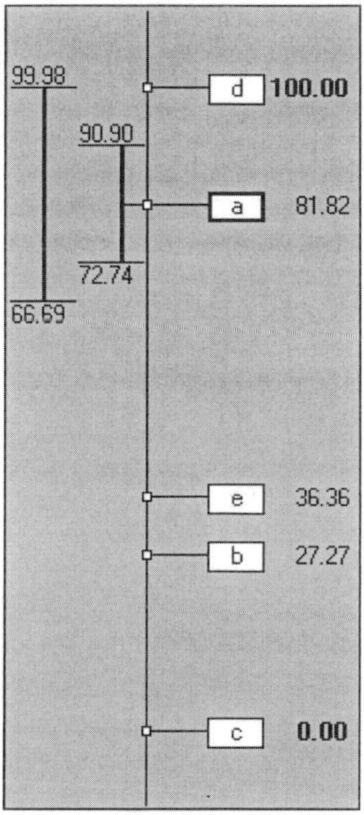

Figure 10.7. "Greatest" closed intervals included in the free and dependent intervals.

10. MACBETH and MCDA

The MACBETH approach and the M-MACBETH software have been used to derive preference scales or value functions and scaling constants in many public and private applications of multicriteria additive value analysis, some of them reported in the literature:

- Evaluation of bids in international public calls for tenders and contractors' choice – see [3, 5, 10, 11, 12, 19, 20, 21, 33] and [58].
- Management of European structural programs – see [37, 42] and [43].

- Public policy analysis, prioritization of projects, resources allocation and conflict management – see [4, 22, 23, 24, 25, 26, 33] and [60].
- Suppliers performance evaluation – see [7] and [57].
- Credit scoring – see [8].
- Strategic town planning – see [14] and [15].
- Environmental management and evaluation of flood control measures – see [1, 6] and [17].
- Portfolio management – see [27].
- Airport management – see [39].
- Human resources evaluation and management – see [50, 47, 48] and [54].
- Total Quality Management – see [11].
- Firms' competitiveness, resource allocation and risk management – see [13] and [49].
- Location of military facilities – see [28].
- Applications in the telecommunications sector – see [18] and [44].

It is worth noting that in all these applications MACBETH was applied in a constructive framework of multicriteria additive aggregation, whose theoretical foundations are reviewed in James Dyer's chapter in this book (Chapter 7).

Acknowledgments

The authors thank Manuel Pedro Chagas and two anonymous referees for their valuable comments on an earlier version of this chapter.

References

[1] P. Antão da Silva, C.A. Bana e Costa, and F.N. Correia. Avaliação multicritério das incidências ambientais de medidas de controlo de cheias: Aplicação à bacia hidrográfica da Ribeira do Livramento. In *Trabalhos Técnicos do 1° Congresso sobre Aproveitamento e Gestão de Recursos Hídricos em Países de Idioma Português*, pages 210–217, Rio de Janeiro, 2000. Associação Brasileira de Engenharia Sanitária e Ambiental. CD-ROM edition.

[2] C.A Bana e Costa. *Structuration, Construction et Exploitation d'un Modèle Multicritère d'Aide à la Décision.* PhD thesis, Technical University of Lisbon, Lisbon, 1992.

[3] C.A. Bana e Costa. O modelo de apoio à avaliação de propostas nos concursos do Metro do Porto. *FER XXI*, XVIII:111–115, 1999.

[4] C.A. Bana e Costa. The use of multicriteria decision analysis to support the search for less conflicting policy options in a multi-actor context: Case-study. *Journal of Multi-Criteria Decision Analysis*, 10(2):111–125, 2001.

[5] C.A. Bana e Costa. Issues in facilitating bid evaluation in public call for tenders. In F. Khosrowshahi, editor, *Proceedings of the 3rd International Conference on Decision Making in Urban and Civil Engineering*, pages 703–709, London, 2002. SOAS. CD-ROM edition.

[6] C.A. Bana e Costa, P. Antão da Silva, and F.N. Correia. Multicriteria evaluation of flood control measures: The case of Ribeira do Livramento. *Water Resources Management*, to appear, 2004.

[7] C.A. Bana e Costa, G. Appa, M.P. Chagas, F.C. Ferreira, J.O. Soares, R.C. Souza, and A. Zanini. A methodology to estimate the relative operational performance of electricity distributors. Technical report, PUC-Rio, Pontifícia Universidade Católica do Rio de Janeiro, 2002.

[8] C.A. Bana e Costa, L. Barroso, and J.O. Soares. Qualitative modelling of credit scoring: A case study in banking. *Journal of European Research Studies*, V(1-2):37–51, 2002.

[9] C.A. Bana e Costa and M.P. Chagas. A career choice problem: An example of how to use MACBETH to build a quantitative value model based on qualitative value judgements. *European Journal of Operational Research*, 153:323–331, 2004.

[10] C.A. Bana e Costa and E.C. Corrêa. Framing public calls for tenders through a multicriteria approach. In D.K. Despotis and C. Zopounidis, editors, *Proceedings of the 5th International Conference of the Decision Sciences Institute*, volume II, pages 1353–1355, Athens, 1999. Athens University of Business and Economics.

[11] C.A. Bana e Costa and E.C. Corrêa. Construction of a total quality index using a multicriteria approach: The case of Lisbon Gas Company. Research paper 10/2000, CEG-IST, Technical University of Lisbon, 2000.

[12] C.A. Bana e Costa, E.C. Corrêa, J.M. De Corte, and J.C. Vansnick. Facilitating bid evaluation in public call for tenders: A socio-technical approach. *OMEGA: The International Journal of Management Science*, 30(3):227–242, 2002.

[13] C.A. Bana e Costa, E.C. Corrêa, L. Ensslin, and J.C. Vansnick. Decision support systems in action: Integrated application in a multicriteria decision aid process. *European Journal of Operational Research*, 113(2):315–335, 1999.

[14] C.A. Bana e Costa, M.L. Costa-Lobo, and I.A. Ramos. Contributo da metodologia multicritério na elaboração do Plano Estratégico de Barcelos. *Sociedade e Território*, 24:102–115, 1997.

[15] C.A. Bana e Costa, M.L. da Costa-Lobo, I.A. Ramos, and J.C. Vansnick. Multicriteria approach for strategic town planning: The case of Barcelos. In D. Bouyssou, E. Jacquet-Lagrèze, P. Perny, R. Słowiński, D. Vanderpooten, and Ph. Vincke, editors, *Aiding Decisions with Multiple Criteria. Essays in Honor of Bernard Roy*, pages 429–456. Kluwer Academic Publishers, Boston, 2002.

[16] C.A. Bana e Costa, J.M. De Corte, and J.C. Vansnick. MACBETH. Working paper 03.56, London School of Economics, 2003.

[17] C.A. Bana e Costa, L. Ensslin, and E. Holz. Avaliação ambiental em situação de conflito de interesses – Um estudo de caso. In *IX SILUBESA, Simpósio Luso-Brasileiro de Engenharia Sanitária e Ambiental*, pages 1914–1923. Associação Brasileira de Engenharia Sanitária e Ambiental and Associação Portuguesa de Recursos Hídricos, Porto Seguro (BA), 2000. CD-ROM edition.

[18] C.A. Bana e Costa, L. Ensslin, and I.J. Zanella. A real-world MCDA application in cellular telephony systems. In T.J. Stewart and R.C. van den Honert, editors, *Trends in Multicriteria Decision Making*, volume 465 of *Lecture Notes in Economics and Mathematical Systems*, pages 412–423. Springer–Verlag, Berlin, 1998.

[19] C.A. Bana e Costa and J.A. Ferreira. Concursos de obras públicas: Metodologia de avaliação de propostas. Aplicação a uma grande empreitada de infraestruturas de transportes. In *Actas do Congresso 94, Engenharia Portuguesa na Viragem do Século*, pages 47–w/9, Lisbon, 1994. Ordem dos Engenheiros.

[20] C.A. Bana e Costa, J.A. Ferreira, and E.C. Corrêa. Metodologia multicritério de apoio à avaliação de propostas em concursos públicos. In C.H. Antunes and L.V. Tavares, editors, *Casos de Aplicação da Investigação Operacional*, pages 336–363. McGraw-Hill Portugal, Lisbon, 2000.

[21] C.A. Bana e Costa, J.A. Ferreira, and J.C. Vansnick. Avaliação multicritério de propostas: O caso de uma nova linha do Metropolitano de Lisboa. *Revista de Transporte e Tecnologia*, VII(14):31–65, 1995.

[22] C.A. Bana e Costa and R.C. Oliveira. Assigning priorities for maintenance, repair and refurbishment in managing a municipal housing stock. *European Journal of Operational Research*, 138(2):380–391, 2002.

[23] C.A. Bana e Costa, R.C. Oliveira, and M.L. Alvarez. Concepção de um sistema multicritério de definição de prioridades de intervenção/conservação do património habitacional da Câmara Municipal de Lisboa. In *Comunicações ao 2º Encontro sobre Conservação e Reabilitação de Edifícios*, volume II, pages 925–934. LNEC, Laboratório Nacional de Engenharia Civil, Lisbon, 1994.

[24] C.A. Bana e Costa, R.C. Oliveira, and M.L. Alvarez. SMAI: Sistema multicritério de avaliação de intervenções num parque habitacional municipal. In A.A. Bezelga and A.B. Leitão, editors, *A Avaliação do Imobiliário*, pages 109–122. Associação Portuguesa dos Avaliadores de Engenharia, Lisbon, 1997.

[25] C.A. Bana e Costa and F.N. Silva. Concepção de uma 'boa' alternativa de ligação ferroviária ao porto de Lisboa: Uma aplicação da metodologia multicritério de apoio à decisão e à negociação. *Investigação Operacional*, 14(2):115–131, 1994.

[26] C.A. Bana e Costa, F.N. Silva, and J.C. Vansnick. Conflict dissolution in the public sector: A case-study. *European Journal of Operational Research*, 130(2):388–401, 2001.

[27] C.A. Bana e Costa and J.O. Soares. A multicriteria model for portfolio management. *European Journal of Finance*, 10(3):to appear, 2004.

[28] C.A. Bana e Costa and J.F. Thomaz. Locating centres of information and recruitment of volunteers for the Portuguese armed forces: A decision-analysis case-study. In *Proceedings of the 42nd Annual Conference of the International Military Testing Association*, pages 173–180. International Military Testing Association, Edinburgh, 2000,. CD-ROM edition.

[29] C.A. Bana e Costa and J.C. Vansnick. Sur la quantification des jugements de valeur: L'approche MACBETH. Cahiers du LAMSADE, 117, Université Paris-Dauphine, Paris, 1993.

[30] C.A. Bana e Costa and J.C. Vansnick. MACBETH – An interactive path towards the construction of cardinal value functions. *International Transactions in Operational Research*, 1(4):489–500, 1994.

[31] C.A. Bana e Costa and J.C. Vansnick. General overview of the MACBETH approach. In P.M. Pardalos, Y. Siskos, and C. Zopounidis, editors, *Advances in Multicriteria Analysis*, pages 93–100. Kluwer Academic Publishers, Dordrecht, 1995.

[32] C.A. Bana e Costa and J.C. Vansnick. Uma nova abordagem ao problema de construção de uma função de valor cardinal: MACBETH. *Investigação Operacional*, 15(1):15–35, 1995.

[33] C.A. Bana e Costa and J.C. Vansnick. Applications of the MACBETH approach in the framework of an additive aggregation model. *Journal of Multi-Criteria Decision Analysis*, 6(2):107–114, 1997.

[34] C.A. Bana e Costa and J.C. Vansnick. L'approche MACBETH : Note technique de presentation. Research paper 779, CESUR, Technical University of Lisbon, 1997.

[35] C.A. Bana e Costa and J.C. Vansnick. The MACBETH approach: Basic ideas. In *Proceedings of the International Conference on Methods and Applications of Multicriteria Decision Making*, pages 86–88. FUCAM, Facultés Universitaires Catholiques de Mons, 1997.

[36] C.A. Bana e Costa and J.C. Vansnick. A theoretical framework for measuring attractiveness by a categorical based evaluation technique (MACBETH). In J. Clímaco, editor, *Multicriteria Analysis*, pages 15–24. Springer–Verlag, Berlin, 1997.

[37] C.A. Bana e Costa and J.C. Vansnick. The MACBETH approach: Basic ideas, software and an application. In N. Meskens and M. Roubens, editors, *Advances in Decision Analysis*, pages 131–157. Kluwer Academic Publishers, Dordrecht, 1999.

[38] C.A. Bana e Costa and J.C. Vansnick. Cardinal value measurement with MACBETH. In S.H. Zanakis, G. Doukidis, and C. Zopounidis, editors, *Decision Making: Recent Developments and Worldwide Applications*, pages 317–329. Kluwer Academic Publishers, Dordrecht, 2000.

[39] C.A. Bana e Costa and A.C. Vitorino. Construção multicritério da 'matriz de custos' de um modelo de afectação para gestão de stands de aeronaves. *Investigação Operacional*, 17(2):179–199, 1997.

[40] V. Belton. *A Comparative Study of Methods for Multiple Criteria Decision Aiding*. PhD thesis, University of Cambridge, Cambridge, U.K., 1986.

[41] V. Belton and T.J. Stewart. *Multiple Criteria Decision Analysis: An Integrated Approach*. Kluwer Academic Publishers, Boston, 2002.

[42] C3E, editor. *Applying the Multi-criteria Method to the Evaluation of Structural Programmes*, volume 4 of *MEANS Handbook*. European Commission, DG XVI/02, Brussels, 1995.

[43] C3E, editor. *Evaluation Pilote Multicritère du Hainaut*. Rapport d'Expérience. European Commission, DG XVI/02, Brussels, 1995.

[44] F.J.P. Cardoso, L.F.A.M. Gomes, and M. Sanglard Júnior. Administração das operações de telecomunicação: Uma análise de decisão. *Revista Portuguesa e Brasileira de Gestão*, 2(2):91–103, 2003.

[45] Catalyze Ltd. *HIVIEW3 Starter Guide*. Catalyze Ltd, London, 2003.

[46] J.M. De Corte. *Un Logiciel d'Exploitation d'Informations Préférentielles pour l'Aide à la Décision. Bases Mathématiques et Algorithmiques*. PhD thesis, University of Mons-Hainaut, Mons, 2002.

[47] L. Ensslin, A. Dutra, and S. Ensslin. MCDA: A constructivist approach to the management of human resources at a governmental agency. *International Transactions in Operational Research*, 7(1):79–100, 2000.

[48] L. Ensslin and S. Ensslin. Elaboração de um modelo construtivista para identificação de oportunidades de aperfeiçoamento de docentes do EPS-UFSC. *Revista Produto & Produção*, 2(3):143–149, 1998.

[49] L. Ensslin, G.N. Montibeller, and M.V.A Lima. Constructing and implementing a DSS to help evaluate perceived risk of accounts receivable. In Y.Y. Haimes and R.E. Steuer, editors, *Research and Practice in Multiple Criteria Decision Making*, volume 487 of *Lecture Notes in Economics and Mathematical Systems*, pages 245–259. Springer–Verlag, Berlin, 2000.

[50] L. Ensslin, G.N. Montibeller, and S. Noronha. *Apoio à Decisão: Metodologias para Estruturação de Problemas e Avaliação Multicritério de Alternativas*. Editora Insular, Florianópolis (SC), 2001.

[51] P.H. Farquhar and L.R. Keller. Preference intensity measurement. In P.C. Fishburn and I.H. LaValle, editors, *Choice under Uncertainty*, volume 19 of *Annals of Operations Research*, pages 205–217. 1989.

[52] A.N.S. Freeling. *Belief and Decision Aiding*. PhD thesis, University of Cambridge, U.K., 1983.

[53] S. French. *Decision Theory: An Introduction to the Mathematics of Rationality*. Ellis Horwood Limited, Chichester, 1988.

[54] M.A.L Leitão, J.F. Thomaz, F.V. Freire, and C.A. Bana e Costa. Using decision conferencing and process consultation to facilitate the structuring of a performance appraisal model in the Portuguese army. Research paper 5/2002, CEG-IST, Technical University of Lisbon, 2002.

[55] O.L. Mangasarian. *Nonlinear Programming*. McGraw-Hill, New York, 1969.

[56] C. Mousset and J.C. Vansnick. About the representation of a precardinal information. Technical report, Université de Mons-Hainaut, 2002.

[57] R.C. Oliveira and J.C. Lourenço. A multicriteria model for assigning new orders to service suppliers. *European Journal of Operational Research*, 139(2):390–399, 2002.

[58] J.L. Porto. Analyse multicritère dans le cadre des appels d'offres pour la construction de travaux publics et privés : Le cas du Métro de Porto au Portugal. *Newsletter of the European Working Group 'Multiple Criteria Decision Aiding'*, Serie II (15):1–2, 1999.

[59] T.L. Saaty. *The Analytic Hierarchy Process*. McGraw-Hill, New York, 1980.

[60] J.S.V. Silva and C.A. Bana e Costa. Especificação de uma ferramenta de apoio à decisão para gestão pública em regiões semi-áridas. Research paper 23/2001, CEG-IST, Technical University of Lisbon, 2001.

[61] D. von Winterfeldt and W. Edwards. *Decision Analysis and Behavioral Research*. Cambridge University Press, Cambridge, 1986.

[62] S.R. Watson and D.M. Buede. *Decision Synthesis: The Principles and Practice of Decision Analysis*. Cambridge University Press, Cambridge, 1987.

V

NON-CLASSICAL MCDA APPROACHES

Chapter 11

DEALING WITH UNCERTAINTIES IN MCDA

Theodor J Stewart
Department of Statistical Sciences
University of Cape Town
Rondebosch 7701
South Africa
tjstew@stats.uct.ac.za

Abstract Many MCDA models are based on essentially deterministic evaluations of the consequences of each action in terms of each criterion, possibly subjecting final results and recommendations to a degree of sensitivity analysis. In many situations, such an approach may be justified when the primary source of complexity in decision making relates to the multicriteria nature of the problem rather than to the stochastic nature of individual consequences. Nevertheless, situations do arise, especially in strategic planning problems, when risks and uncertainties are as critical as the issue of conflicting management goals. In such situations, more formal modelling of these uncertainties become necessary.

In this paper, we start by reviewing the meaning and origin of risk and uncertainty. We recognize both internal uncertainties (related to decision maker values and judgements) and external uncertainties (related to imperfect knowledge concerning consequences of action), but for this paper focus on the latter. Four broad approaches to dealing with external uncertainties are discussed. These are multiattribute utility theory and some extensions; stochastic dominance concepts, primarily in the context of pairwise comparisons of alternatives; the use of surrogate risk measures as additional decision criteria; and the integration of MCDA and scenario planning. To a large extent, the concepts carry through to all schools of MCDA. A number of potential areas for research are identified, while some suggestions for practice are included in the final section.

Keywords: Multicriteria analysis, multiobjective programming, uncertainty, risk, utility theory.

1. What is Uncertainty?

The term uncertainty can have many different meanings. The Chambers Dictionary (1998 edition) defines "uncertain" as not definitely known or decided; subject to doubt or question. Klir and Folger [30] quote six different definitions for "uncertainty" from Webster's Dictionary. In the context of practical applications in multicriteria decision analysis, however, the definition given by Zimmermann [59] would appear to be particularly appropriate. With minor editing, this is as follows:

> Uncertainty implies that in a certain situation a person does not possess the information which quantitatively and qualitatively is appropriate to describe, prescribe or predict deterministically and numerically a system, its behaviour or other characteristics.

At a most fundamental level, uncertainty relates to a state of the human mind, i.e. lack of complete knowledge about something. Many writers also use the term "risk", although the definition of the term varies widely. Some earlier work tended to apply the term "risk" to situations in which probabilities on outcomes are (to a large extent) known objectively (cf. Goicoechea et al. [16], p. 389, and Millet and Wedley [37] for some reference to this view). More recently, the concept of risk has come to refer primarily to the desirability or otherwise of uncertain outcomes, in addition to simple lack of knowledge. Thus, for example, Fishburn [13] refers to risk as "a chance of something bad happening", and in fact separates uncertainty (alternatives with several possible outcome values) from the fundamental concept of risk as a bad outcome. Sarin and Weber [45] state that "judgements about riskiness depend on both the probability and the *magnitude of adverse effects*" (my emphasis), while Jia and Dyer [25] also discuss the psychological aspects of establishing a preference order on risks.

For the most part in this chapter, we shall make use of the value-neutral term "uncertainty", referring to "risk" only when direct preference orderings of the uncertainty *per se* are relevant (for example, in Section 4). It is interesting to note in passing that while the thrust of the present discussion is to give consideration to the effects of uncertainty on MCDA, there has also been work on applying multicriteria concepts to the measurement of risk for other purposes, as for example in credit risk assessment (Dimitras et al. [12], who make use of a rough sets approach).

A number of authors (e.g. French [14], Zimmermann [59]) have attempted to categorize types or sources of uncertainty in the context of decision making. French [14], for example, identifies no less than 10 different sources of uncertainty which may arise in model building for decision aid, which he classifies into three groups referring broadly to uncertainties in the modelling (or problem structuring) process, in the use of models for exploring trends and options, and in interpreting results. The common theme underlying such categorizations, as well as those of other authors, such as Friend [15] and Levary and Wan [34], is the need at very least distinguish between *internal uncertainty*, relating to the process of problem structuring and analysis, and *external uncertainty*, regarding the nature of the environment and thereby the consequences of a particular course of action which may be outside of the control of the decision maker. Let us briefly examine each of these broad categories of uncertainty.

Internal uncertainty. This refers to both the structure of the model adopted and the judgmental inputs required by those models, and can take on many forms, some of which are resolvable and others which are not. Resolvable uncertainties relate to imprecision or ambiguity of meaning – for example, what exactly may be meant by a criterion such as "quality of life"? Less easily resolvable problems may arise when different stakeholders generate different sets of criteria which are not easily reconciled; or perceive alternatives in such different ways that they differ fundamentally on how they contribute to the same criterion.

Imprecisions in human judgments, whether these relate to specifications of preferences or values (for example importance weights in many models), or to assessments of consequences of actions, have under certain circumstances been modelled by fuzzy set (see, for example, Chapters 4 and 5 of Klir and Folger [30]) and related approaches (such as the use of rough sets as described by Greco et al. [20, 19, 21]. From the point of view of practical decision aid, such models of imprecision add complexity to an already complex process, and the result may often be a loss of transparency to the decision maker, contrary to the ethos of MCDA. For this reason, the view espoused here is that internal uncertainties should ideally be resolved as far as is possible by better structuring of the problem (cf. Belton and Stewart [6], Chapter 3) and/or by appropriate sensitivity and robustness analysis where not resolvable.

External uncertainty. This refers to lack of knowledge about the consequences of a particular choice. Friend [15] and French [14] both recognize a further distinction between uncertainty about the environment and uncertainty about related decision areas, as described below.

- *Uncertainty about the environment* represents concern about issues outside the control of the decision maker. Such uncertainty may be a consequence of a lack of understanding or knowledge (in this sense it is similar to uncertainty about related decision areas) or it may derive from the randomness inherent in processes (for example the chance of equipment failure, or the level of the stock market). For example, the success of an investment in new production facilities may rest on the size of the potential market, which may depend in part on the price at which the good will be sold, which itself depends on factors such as the cost of raw materials and labour costs. A decision about whether or not to invest in the new facilities must take all of these factors into account. This kind of uncertainty may be best handled by responses of a technical nature such as market research, or forecasting.

- *Uncertainty about related decision areas* reflects concern about how the decision under consideration relates to other, interconnected decisions. For example, suppose a company which supplies components to computer manufacturers is looking to invest in a management information system. They would like their system to be able to communicate directly with that of their principal customers; however, at least one of these customers may be planning to install a new system in the near future. This customer's decision could preclude certain of the options open to the supplier and would certainly have an impact on the attractiveness of options. The appropriate response to uncertainty of this kind may be to expand the decision area to incorporate interconnected decisions, or possibly to collaborate or negotiate with other decision makers.

Under many circumstances, both internal and external uncertainties can be treated in much the same manner, for example by appropriate sensitivity analyses *post hoc*. In other words, the approach might be to make use of a crisp deterministic MCDA methodology, and to subject the results and conclusions to extensive sensitivity studies. Indeed, we would assert that such sensitivity studies should routinely be part of any MCDA application.

Where uncertainties are of sufficient magnitude and importance to be modelled explicitly as part of the MCDA methodology, however, the modelling approaches for internal and external uncertainties may often become qualitatively different in nature. It seems, therefore, that the treatment of the two types of uncertainty should preferably be discussed in separate papers or chapters. In order to provide focus for the present paper, our attention will be focussed primarily on consideration of the *external uncertainties* as defined above. Without in any way minimizing the importance of dealing with internal uncertainties, our choice of the problem of external uncertainties as the theme for this chapter is in part due to the present author's practical experience, which suggests that it is the external uncertainties which are often of sufficient magnitude and importance to require more explicit modelling.

Admittedly, the boundary between external uncertainty and imprecision is, well, fuzzy! To this extent, at least some of the material in this chapter may well be appropriate to internal uncertainties as well, while some methods formulated to deal with human imprecision might equally well be useful in dealing with external uncertainties. We leave it to the reader to decide where this may be true. We do not attempt here a comprehensive review of literature related primarily to internal uncertainties, but the interested reader may wish to consult some of the following references:

- Fuzzy set approaches: Klir and Folger [30]; Chang et al. [9, 8]; Yeh et al. [57]; (Some of these do partially relate to external uncertainties as well.)

- Rough set approaches: Greco et al. [20, 19, 21, 22].

- Identifying potentially optimal solutions amongst uncertainty ranges: Cook and Kress [10]; Lahdelma and Salminen [32]; Lahdelma et al. [33].

Our approach will also be pragmatic, motivated by practical needs of real-world decision analysis. In particular, the fundamental philosophical point of departure is a belief in the over-riding need for *transparency* in any MCDA: it is vitally and critically important that any approaches to MCDA are fully understandable to all participants in the process. Elegant mathematical models which are inaccessible to such participants are of very little practical value.

Within the context of the opening discussion, let us now define a notational framework within which to consider MCDA under uncertainty (primarily "external uncertainty" as defined earlier). Let X be the set of actions or decision alternatives. When there is no uncertainty about the outcomes, there exists a one-to-one correspondence between elements of X and consequences in terms of the criteria, and X may written as the product space $\prod_{i=1}^{n} X_i$, where X_i is the set of evaluations with respect to criterion i. In other words, any $x \in X$ may be viewed as an n-dimensional vector with elements $x_i \in X_i$, where x_i represents the evaluation of x with respect to the criterion i.

Under uncertainty, however, the one-to-one correspondence between actions and evaluations or consequences breaks down. It may be possible to postulate or to conceptualize an ultimate set of consequences $Z_1(x), Z_2(x), \ldots, Z_n(x)$ corresponding to each of the criteria, but at decision time there will still exist many possible values for each $Z_i(x)$. For ease of notation, we shall use $\mathbf{Z}(x)$ to indicate the vector of $Z_i(x)$ values.

In some cases, it may be possible and useful to structure $Z_i(x)$ (or $\mathbf{Z}(x)$) in the form $Z_i(x, \xi)$ (or $\mathbf{Z}(x, \xi)$), where $\xi \in \Xi$ fully characterizes the external conditions, sometimes termed the "states of nature", and Ξ represents the set of all possible states of nature. The assumption is then that once ξ (the state of nature) is established or revealed, then the consequences in terms of each criterion will also be known. We observe, however, that even Ξ might not be fully known or understood at decision time, and that Ξ could possibly depend upon the action x (although, for ease of notation, we shall not show this explicitly).

The question to be addressed in this chapter is that of constructing some form of (possibly partial) preference ordering on X, when the consequences are incompletely known or understood in the sense described in the previous paragraph.

As indicated earlier, one approach may be initially to ignore the uncertainty, and to conduct the analysis on the basis of a nominal set of consequences $x_1, x_2, \ldots, x_n$ chosen to be representative of the possible $Z_i(x)$, followed by extensive sensitivity analysis which takes into account the range of uncertainty in each $Z_i(x)$. Under many circumstances this may be adequate. Care needs to be exercised in undertaking sensitivity analyses, however, as simple "one-at-a-time" variations in unknown parameter values may fail to identify effects of higher order interactions. Some of the complications inherent in undertaking properly validated sensitivity analyses, and suggestions as to how these may be addressed, are discussed by Rios Insua [41], Parnell et al. [39] and Saltelli et al. [44]. In the remainder of this chapter, our focus will be on situations in which the ranges of uncertainty are simply too large to be handled purely by such sensitivity analysis.

In Section 2 we discuss the use of probability models to represent the uncertainties, emphasizing particularly the comprehensively axiomatized approach of multiattribute utility theory. The potential for relaxing the needs to specify complete utility functions are addressed in Section 3, which leads naturally to the use of pairwise comparison models for MCDA. In many practical situations, decision maker preferences for various types of risk (magnitude and impact of the uncertainties) may be modelled by defining explicit risk-avoidance criteria, and these are discussed in Section 4. Finally, links between MCDA and scenario planning for dealing with uncertainties are presented in Section 5, before concluding with some general implications for practice.

2. Probabilistic Models and Expected Utility

The most thoroughly axiomatized mathematical treatment of uncertainty is that of probability theory. The application of probability concepts would require the specification of a (multivariate) probability distribution on $\mathbf{Z}(x)$ for each action x, so that in effect the decision requires a comparison of probability distributions (sometimes called "lotteries" in this context). Let $¶^x(\mathbf{z})$ denote the probability distribution function on $\mathbf{Z}(x)$, i.e.:

$$¶^x(\mathbf{z}) = \Pr[Z_1(x) \leq z_1, Z_2(x) \leq z_2, \ldots, Z_n(x) \leq z_n].$$

Define $P_i^x(z_i)$ as the corresponding marginal probability distribution function for $Z_i(x)$.

Where uncertainties are structured in terms of "states of nature", the probability distributions may be defined on the ξ (rather than on the $\mathbf{Z}(x)$ directly). In some situations, the probability distribution on ξ may be independent of the action which would make the application of probability models much more tractable, but this will not necessarily always be the case.

A possibility at this stage is to construct a deterministic MCDA model based only on expectations, and to subject the results to some form of (possibly interactive) sensitivity analysis guided by the broader distributional properties. Examples of this are in the PROTRADE method described by Goicoechea et al. [16], Chapter 7, dealing with an interactive method for multiobjective mathematical programming problems, and in the stochastic extensions to outranking proposed by Mareschal [35].

Simple expectation models do not, however, take full account of the ranges of outcome which may occur. Multiattribute utility theory (MAUT) extends the concept of expectation to include explicit modelling of risk preferences, i.e. of the magnitudes of dispersion that may occur. MAUT is discussed by Dyer in Chapter 7 of this volume, and also more comprehensively in the now classic text of Keeney and Raiffa [27] and by von Winterfeldt and Edwards [53]. In essence, MAUT seeks to construct a "*utility function*" $U(\mathbf{Z})$, such that for any two actions x and y in X, $x \succsim y$ if and only if $\mathrm{E}[U(\mathbf{Z}(x))] \geq \mathrm{E}[U(\mathbf{Z}(y))]$, where expectations are taken with respect to the probability distributions on $\mathbf{Z}(x)$ and on $\mathbf{Z}(y)$ respectively.

Practically, the construction of the global utility function $U(\mathbf{Z})$ starts with the construction of partial or marginal utility functions individually for each attribute, say $u_i(Z_i)$, satisfying the expected utility hypothesis for variations in Z_i only. The axioms underlying the existence of such marginal utility functions and the methods for their construction are well-known from univariate decision analysis (see, for example, Chapter 7, or Goodwin and Wright[17], Chapter 5). It is well-established that these axioms are not descriptively valid, in the sense that decision makers do systematically violate them (see, for example, the various paradoxes described by Kahnemann and Tversky [26], or in the text of Bazerman [4]). Attempts have been made to extend the utility models to account for observed behaviour (see, for example, Miyamoto and Wakker [38] for a review of such extensions in the multicriteria context). Nevertheless, as we have argued elsewhere (e.g., Belton and Stewart [6], Section 4.3.1), descriptive failures do not lessen the value of the simpler axiomatically based theory of MAUT as a coherent discipline within which to construct preferences in a simple, transparent and yet defensible manner.

The real challenge relates to the aggregation of the $u_i(Z_i)$ into a $U(\mathbf{Z})$ still satisfying the expected utility hypothesis for the multivariate outcomes. The two simplest forms of aggregation are the *additive* and *multiplicative*, which we shall now briefly review (although a full description can be found in Chapter 7).

Additive aggregation. In this case, we define:

$$U(\mathbf{Z}) = \sum_{i=1}^{n} k_i u_i(Z_i). \tag{11.1}$$

This model is only justifiable if the criteria are *additive independent*, i.e. if preferences between the multivariate lotteries depend only on the marginal probability distributions. That this is not an entirely trivial assumption may be seen by considering two-dimensional lotteries ($n = 2$) in which there are only two possible outcomes on each criterion, denoted by z_i^0 and z_i^1 for $i = 1, 2$. Suppose that $z_i^1 \succ z_i^0$. Then without loss of generality, the partial utility functions can be standardized such that $u_1(z_1^0) = u_2(z_2^0) = 0$ and $u_1(z_1^1) = u_2(z_2^1) = 1$. Consider then a choice between two lotteries defined as follows:

- The lottery giving equal chances on $(z_1^0\ ;\ z_2^0)$ and $(z_1^1\ ;\ z_2^1)$; and
- The lottery giving equal chances on $(z_1^0\ ;\ z_2^1)$ and $(z_1^1\ ;\ z_2^0)$.

We note that both lotteries give the same marginal distributions on each Z_i, i.e. equal chances on each of z_i^0 and on z_i^1 for each i. It is easily verified that with additive aggregation defined by (11.1), both of these lotteries yield an expected utility of $(k_1 + k_2)/2$. The additive model thus suggests that the decision maker should always be indifferent between these two lotteries. There seems, however, to be no compelling axiomatic reason for forcing indifference between the above two options. Where there is some measure of compensation between the criteria (in the sense that good performance on one can compensate for poorer outcomes on the other), the second option may be preferred as it ensures that one always gets some benefit (a form of multivariate risk aversion). On the other hand, if there is need to ensure equity between the criteria (if they represent benefits to conflicting social groups, for example), then the first lottery (in which loss or gain is always shared equally) may be preferred.

Multiplicative aggregation. Now we define $U(\mathbf{Z})$ such that:

$$1 + kU(\mathbf{Z}) = \prod_{i=1}^{n} [1 + kk_i u_i(Z_i)] \tag{11.2}$$

where the multivariate risk aversion k parameter satisfies:

$$1 + k = \prod_{i=1}^{n} [1 + kk_i] \,. \tag{11.3}$$

Use of the multiplicative model requires that the condition of *mutual utility independence* be satisfied. A subset of criteria, say $C \subset \{1, 2, \ldots, n\}$ is set to be utility independent of its complement $\bar{C} = \{1, 2, \ldots, n\} \setminus C$, if preferences for lotteries involving only Z_i for $i \in C$ *for fixed values* of Z_i for $i \in \bar{C}$ are independent of these fixed values. The criteria are said to be mutually utility independent if every subset of the criteria is utility independent of its complement.

In principle, however, there are no good reasons why criteria *should necessarily be* mutually utility independent, and in fact it can be difficult in practice to verify that the condition holds. Good problem structuring for MCDA would seek to ensure preferential independence of some form between criteria (for example, such that trade-offs between pairs of criteria are independent of outcomes on other criteria), but mutual utility independence is a much stronger assumption and a more elusive concept than this.

Models based on weaker preference assumptions have been developed, such as the multilinear model given by:

$$U(\mathbf{Z}) = \sum_{i=1}^{n} k_i u_i(Z_i) + \sum_{i=1}^{n} \sum_{i<j\leq n} k_{ij} u_i(Z_i) u_j(Z_j) + \ldots + k_{12\ldots n} u_1(Z_1) u_2(Z_2) \ldots u_n(Z_n). \quad (11.4)$$

The large number of parameters which have to fitted to decision maker preferences is prohibitive in most real world applications. Even the multiplicative model is far from trivial to apply in practice. Its construction involves the following steps:

- *Assessment of the partial utilities* $u_i(Z_i)$ by standard single attribute lottery procedures.
- *Parameter estimation:* The multiplicative model includes $n + 1$ parameters which have in principle to be estimated. In the light of (11.3), however, only n independent parameters need estimation. Estimates thus require at least n preference statements concerning hypothetical choices to be made by the decision maker. Some of these can be based on deterministic trade-off assessments, but at least one of the hypothetical choices must involve consideration of preferences between multivariate lotteries.

In exploring the literature, it is difficult to find many reported applications even of the multiplicative model, let alone the multilinear model. Some of the practical complications of properly implementing these models are illustrated by Rosqvist [42] and Yilmaz [58].

Such difficulties of implementation raise the question as to how sensitive the results of analysis may be to the use of the additive model (11.1) instead of the more theoretically justifiable aggregation models given by (11.2) or (11.4). We have seen earlier that situations can be constructed in which the additive model may generate misleading results. But how serious is this in practice? Construction of the additive model requires much less demanding inputs from the decision maker, and it may be that the resultant robustness or stability of the model will compensate for biases introduced by use of the simpler model. In Stewart [46] a number of simulation studies are reported in which the effects are studied of using the additive aggregation model when "true preferences" follow a multiplicative aggregation model. Details may be found in the cited reference, but in essence it appeared that the errors introduced by using the additive model were generally extremely small for realistic ranges of problem settings. The errors were in any case substantially smaller than those introduced by incorrect modelling of the partial utility functions (such as by over-linearization of the partial functions which appears to be a frequent but erroneous simplification). Related work (Stewart [47]) has also demonstrated that more fundamental violations of preferential independence may also introduce substantial errors.

Concerns about the validity of the axiomatic foundations of utility theory have led other writers to formulate alternative models to circumvent these. Miyamoto and Wakker [38] review generalizations to utility theory, while others (e.g. Beynon et al. [7] and Yang [56]) relax the demands of probability theory by invoking concepts from Dempster-Shafer theory of evidence. Unfortunately, these generalizations tend often to make the models even more complex and thus less transparent to decision makers, further aggravating difficulties of implementation.

Our overall conclusion is thus that in the practical application of expected utility theory to decision making under uncertainty, the use of the additive aggregation model is likely to be more than adequate in the vast majority of settings. The imprecisions and uncertainties involved in constructing the partial utilities, which need in any case to be addressed by careful sensitivity analysis, are likely to far outweigh any distinctions between the additive and multiplicative models. In fact, given that marginal utility functions based on preferences between hypothetical lotteries may generally not differ markedly from deterministic value functions based on relative strengths of preference (e.g. von Winterfeldt and Edwards [53], Chapter 10), we conjecture that even the first step of the model construction could be based on the latter (e.g. by use of the SMART methodology, von Winterfeldt and Edwards [53], Section 8.2).

3. Pairwise Comparisons

As indicated in the previous section, the requirements of fitting a complete utility function can be extremely demanding both for the decision maker (in providing the necessary judgemental inputs) and for the analysts (in identifying complete multivariate distributions). We have seen how the assumption of a simple additive model may substantially reduce these demands without serious penalty in many practical situations. Nevertheless, other attempts at avoiding the construction of the full utility model have been made.

Even for single criterion models, the construction and validation of the complete utility model may be seen as too burdensome. Quite early work recognized, however, that it may often not be necessary to construct the full utility function in order to confirm whether one alternative is preferred to another. The conclusions may be derived from the concepts of *stochastic dominance* introduced by Hadar and Russell [23], and extended (to include third order stochastic dominance) by Whitmore [55].

For purposes of defining stochastic dominance, suppose for the moment that there is only one criterion which we shall denote by $Z(x)$ (i.e. unsubscripted). Then let $P^x(z)$ be the (univariate) probability distribution function of $Z(x)$, i.e.: $P^x(z) = \Pr[Z(x) \leq z]$. With some abuse of notation, we shall use P^x (without argument) to denote the probability distribution described by the function $P^x(z)$. Suppose also that values for $Z(x)$ are bounded between z^L and z^U.

Three degrees of stochastic dominance may then be defined as follows.

First degree stochastic dominance (FSD): P^x stochastically dominates P^y in the *first degree* if and only $P^x(z) \leq P^y(z)$ for all $z \in [z^L, z^U]$ (Hadar and Russell [23]).

Second degree stochastic dominance (SSD): P^x stochastically dominates P^y in the *second degree* if and only:

$$\int_{z^L}^{\zeta} P^x(z)dz \leq \int_{z^L}^{\zeta} P^y(z)dz$$

for all $\zeta \in [z^L, z^U]$ (Hadar and Russell [23]).

Third degree stochastic dominance (TSD): P^x stochastically dominates P^y in the *third degree* if and only $\mathrm{E}[Z(x)] \geq \mathrm{E}[Z(y)]$ and:

$$\int_{z^L}^{\eta} \int_{z^L}^{\zeta} P^x(z)dzd\zeta \leq \int_{z^L}^{\eta} \int_{z^L}^{z} P^y(z)dzd\zeta$$

for all $\eta \in [z^L, z^U]$ (Whitmore [55]).

In this single-criterion case, the standard axioms of expected utility theory imply the existence of a utility function $u(z)$ such that $x \succ y$ if and only if:

$$\int_{z^L}^{z^U} u(z)dP^x(z > \int_{z^L}^{z^U} u(z)dP^y(z).$$

Without having explicitly to identify the utility function, however, considerations of stochastic dominance allow us to conclude the following (Bawa [3]):

1 If P^x stochastically dominates P^y in the first degree ($P^x\ FSD\ P^y$), then $x \succ y$ provided that $u(z)$ is an increasing function of z (which can be generally be assumed to be true in practical problems).

2 If P^x SSD P^y, then $x \succ y$ provided that $u(z)$ is a concave increasing function of z (i.e. the decision maker is risk averse).

3 If P^x TSD P^y, then $x \succ y$ provided that $u(z)$ is a concave increasing function of z with positive third derivative (corresponding to a risk averse decision maker exhibiting decreasing absolute risk aversion).

The potential importance of the above results lies in the claim that has been made that in practice some form of stochastic dominance may hold between many pairs of probability distributions. In other words, we may often be able to make pairwise comparisons between alternatives according to a particular criterion on the basis of stochastic dominance considerations, without needing to establish the partial value function for comparison of lotteries. In fact, we may often argue that FSD provides a strict pairwise preference, while SSD and TSD provide weaker forms of pairwise preference. Only in the absence of any stochastic dominance would we be unable to determine a preference without obtaining much stronger preference information from the decision maker.

The existence of pairwise preferences at the level of a single criterion under uncertainty suggests that some form of outranking approach may be appropriate to aggregation across multiple criteria under uncertainty. D'Avignon and Vincke [11] did in fact propose an outranking approach to dealing with uncertainty, in which they started by comparing univariate probability distributions for each criterion in order to obtain "preference indices" measuring degree of preference for one lottery over another in terms of one criterion, which were then aggregated according to an outranking philosophy. Their preference indices may not be easily interpretable by many decision makers however, and perhaps with this problem in mind, Martel and Zaras [36] (but see also Azondékon and Martel [1]) suggested an alternative outranking approach in which preferences according to individual criteria were established as far as possible by stochastic dominance considerations.

Martel and Zaras found it useful to introduce two forms of concordance index, which they term "explicable" and "non-explicable". For the "explicable" concordance, x is judged at least as good as y according to criterion i if P_i^x stochastically dominates P_i^y at first, second or third degrees. This is quite a strong assumption, as it implies decreasing absolute risk aversion. The "non-explicable" concordance arises if neither of P_i^x or P_i^y stochastically dominates the other. The authors concede that in this case it is not certain that x is at least as good as y, but they do combine the two indices under certain conditions. The discordance when comparing x to y is only non-zero in their model if P_i^y FSD P_i^x.

Although some of the implementation details are not clear from the paper, the method of Martel and Zaras does appear to offer potential as an approach to dealing with uncertainty in MCDA using quite minimal preference information from the decision maker. This might at least be valuable for a first-pass screening of alternatives. Two problems may, however, limit wide applicability:

- Strong independence assumptions are implicitly made: The approach is based entirely on the marginal distributions of the elements of $\mathbf{Z}(x)$. This would only be valid if these elements (i.e. the criteria) were stochastically independent, or if the decision maker's preferences were additively independent in the sense of Keeney and Raiffa [27]. Either assumption would need to be carefully justified.

- Strong risk aversion assumptions are made: As indicated above, the method as proposed bases concordance measures on risk aversion and on decreasing absolute risk aversion. Especially the latter assumption may not always be easy to verify. The method can be weakened by basing concordance either only on FSD or on FSD and SSD, but this may not generate such useful results.

There is clear scope for further research aimed at addressing the above problems.

4. Risk Measures as Surrogate Criteria

In this and the next sections, we move to more pragmatic approaches to dealing with uncertainty in the multicriteria context.

One obvious modelling approach is to view avoidance of risks as decision criteria in their own right. For example, the standard Markowitz portfolio theory (cf. Jia and Dyer [25]) represents a risky single-criterion objective (monetary reward) in terms of what are effectively two non-stochastic measures, namely expectation and standard deviation of returns. In this sense a single criterion decision problem under uncertainty is structured as a deterministic bi-criterion decision problem. The extension to risk components for each of number of fundamental criteria is obvious (see, for example, Millet and Wedley [37], p. 104, in the context of AHP).

There has, in fact, been a considerable literature on the topic of measuring risk for purposes of decision analysis, much of it motivated by the descriptive failures of expected utility theory. Papers by Sarin and Weber [45], and by Jia and Dyer [25] contain many useful references. This literature is virtually entirely devoted to the single criterion case (typically financial returns), but it is worth recalling some of the key results with a view to extending the approaches to the multicriteria case.

The common theme has been that of developing axiomatic foundations for representation of psychological perceptions of risk (including consideration of importance and impact in addition to simple uncertainty), often based on some form of utility model. For example, Bell [5] considers situations in which, if a decision maker switches from preferring one (typically more risky) lottery to another as his/her wealth increases, then he/she never switches back to preference for the first as wealth further increases. This he terms the "one-switch" rule for risk preferences, and demonstrates that if the decision maker is decreasingly risk averse, obeys the one switch rule, and approaches risk neutrality as total wealth tends to infinity, then the utility as a function of wealth w must take on the form $w - be^{-cw}$ for some positive parameters b and c. Taking expectations results in an additive aggregation of two criteria, namely:

- The expectation of wealth (to be maximized); and
- The expectation of be^{-cw} (to be minimized), which can be viewed as a measure of risk.

Sarin and Weber [45] and Jia and Dyer [25] provide arguments for general moments of the distribution of returns (including but not restricted to variance) and/or expectations of terms such as be^{-cw}, as measures of risk. While these may be useful as descriptive measures of risk behaviour, from the point of view of practical decision aid it is doubtful whether the decision maker would be able to interpret anything but variance (or standard deviation) for purposes of providing necessary preference information (to establish tradeoffs, relative weights, goals, etc.).

Limited empirical and simulation work which we have undertaken in the context of fisheries management (Stewart [48]) suggested that perceptions of risk of fishery collapse might be modelled better by probabilities of achieving one or more goals (in that case, periods of time before a collapse of the fishery). One advantage of such measures is that they might be much more easily interpreted by decision makers for purposes of expressing preferences or value judgements.

Given the modelling success in representing preferences under uncertainty by simple additive models of expected return and one or more risk measures, there seems to be no reason why such results should not be extended to the general multicriteria problem under uncertainty. In other words, each criterion (not necessarily financial) for which there exists substantial uncertainties might be restructured in terms of two separate criteria, viz. expected return and risk. Many of the above results produce an axiomatic justification for an additive aggregation of expected return and risk, so that these sub-criteria would be preferentially independent under the same axiomatic assumptions.

In spite of how obvious such multicriteria extensions might be, there seems to be little reference in the literature to explicit multicriteria modelling of returns and risks. It is this author's experience, however, that various risk-avoidance criteria arise almost naturally during the structuring phase of decision modelling, so that in practice risk avoidance criteria may in fact be more common than is apparent from the literature.

Some of the few explicit references to multicriteria modelling in terms of a risk-return decomposition appear in the context of goal programming. For example, Ballestero [2] expresses a stochastic multicriteria problem in terms of goals on combinations of risks and returns which are then solved by goal programming, but he does not separate out the risk and return components which may have led to a simpler model structure. Korhonen [31] develops a multicriteria model for financial management, in which a number of different financial performance measures are used as criteria, some of which have a risk interpretation. Details of the solution procedure are not given, but the formulation clearly lends itself to a goal programming structure.

A somewhat earlier paper by Keown and Taylor [28] describes an integer goal programming model for capital budgeting, which can be viewed (together with the STRANGE method of Teghem et al. [50]) as an extension of chance-constrained stochastic programming (see Ruszczyński and Shapiro [43] for a broad introduction to stochastic programming). Keown and Taylor define goals in terms of desired probability levels, which may generically be expressed in the form:

$$\Pr\left[g(Z) \leq \beta\right] \geq \alpha$$

where $g(Z)$ is some performance function based on the unknown attribute values, β the desired level of performance, and α a desired probability of achieving such performance. By using normal approximations, however, Keown and Taylor reduce the probability goal to one expressed in terms of a combination of mean and standard deviation which is subsequently treated in a standard goal programming manner. This suggests opportunity for research into investigation of generalized goal programming models which deal directly with deviations from both the desired performance levels (b, above) and the desired probability levels (α, above).

Some work on fuzzy multiobjective programming (e.g. Chang et al. [9] and Chang and Wang [8]) can be viewed in a similar manner, in the sense that a degree of anticipated level of goal achievement, measured in a fuzzy membership sense, may be interpreted as a risk measure.

More generally, the structuring of MCDA problems under uncertainty in terms of expected value and risk sub-criteria for each main criterion does have the advantage of being relatively simple and transparent to users. Such an approach appears to be easily integrated into any of the main MCDA methodologies, namely value measurement, outranking and goal programming/reference point methods. As indicated earlier, however, a decidedly open research question relates to the manner in which risk is most appropriately measured *for this purpose*.

A further practical issue is the extent to which the necessary independence properties can be be verified. In other words, to what extent can "risk" on one criterion be measured and assessed without taking into consideration ranges of uncertainties on the other criteria. Once again, this offers much scope for further research.

5. Scenario Planning and MCDA

Scenario planningscenario planning (van der Heijden [52]) was developed as a technique for facilitating the process of identifying uncertain and uncontrollable factors which may impact on the consequences of decisions in the strategic management context. Scenario analysis has been widely accepted as an important component of strategic planning, and it is thus somewhat surprising how little appears to have been written concerning links between MCDA and scenario planning. A discussion of the link between scenario planning and decision making is provided by Harries [24], but does not place this in an MCDA framework.

Scenario planning may be described as a process of organizational learning, distinguished by an emphasis on the explicit and ongoing consideration of multiple futures. The scenarios themselves are constructed as stories which describe the current and plausible, but challenging, future states of the organizational environment. They provide alternative perspectives that will challenge an organization in viewing the future and in evaluating its strategies and action plans. The primary goal of scenario planning is in the first instance to provide a structured "conversation" to sensitize decision makers to external and uncontrollable uncertainties, and to develop a shared understanding of such uncertainties. The approach is, however, naturally extended to the more analytical process of designing, evaluating and selecting courses of action on the basis of robustness to these uncertainties, which suggests close parallels with MCDA (as discussed, for example, by Goodwin and Wright [18]). We shall explore these parallels shortly.

Scenarios are meant to represent fairly extreme futures than can still be viewed as plausible. As to what constitutes sufficiently "extreme" would depend on the facilitator, as in a very real sense, there will always be a possible future more extreme (and thus with greater potential impact on the consequences of decisions) than any which is incorporated into formal scenarios.

Van der Heijden suggests five principles which should guide scenario construction:

- At least two scenarios are required to reflect uncertainty, but more than four has proved (in his experience) to be impractical;

- Each scenario must be plausible, meaning that it can be seen to evolve in a logical manner from the past and present;

- Each scenario must be internally consistent;

- Scenarios must be relevant to the client's concerns and they must provide a useful, comprehensive and challenging framework against which the client can develop and test strategies and action plans;

- The scenarios must produce a novel perspective on the issues of concern to the client.

Once scenarios are constructed, they may be used to explore and to evaluate alternative strategies for the organization. Most proponents of scenario planning seem to avoid formal evaluation and analysis procedures, preferring to leave the selection of strategy to informed judgement. For example, van der Heijden [52] (pp. 232–235) rejects "traditional rationalistic decision analysis" as an approach which seeks to find a "right answer". This, however, represents are rather limited and technocratic view of decision analysis, contrary to the constructive and learning view espoused by most in the MCDA field. The constructivist perspective is discussed at a number of places by Belton and Stewart [6] (see particularly Chapters 3, 4 and 11), where it is argued that the underlying axioms are not meant to suggest a "right answer", but to provide a coherent discipline within which to construct preferences and strategies. Within such a view, the aims of scenario planning and MCDA share many commonalities, suggesting the potential for substantial synergies in seeking to integrate MCDA and scenario planning. On the one hand, MCDA can enrich the evaluation process in scenario planning, while the scenario planning approach can contribute to deeper understanding of the effects of external uncertainties in MCDA.

Various authors have hinted at the concept of scenarios in MCDA. These include, for example, Klein et al. [29], although this is largely in the context of a two state stochastic programming model; Watkins et al. [54], also in a stochastic programming context; Millet and Wedley [37], Section 3, who refer to "states of nature"; Urli and Nadeau [51] in the context of multiple objective linear programming. These authors do not refer directly to the philosophical basis of scenario planning, however, and in many senses the models are structured to suggest that the scenarios or states of nature constitute a complete sample space (see later).

Pomerol [40] is one of the few to discuss scenario planning in the context of decision theory or decision analysis, but without substantive link to MCDA. He does however warn (page 199) of the danger that what might appear to be a robust choice of action (perhaps through unstructured and unsupported use of scenarios) may in fact be an illusion resulting from the fact that some events have simply been ignored. Such a danger suggests another perspective on the potential for two-way synergistic advantage between scenario planning and formal decision analysis: not only may scenario planning provide a means of dealing with uncertainties in MCDA, but decision analysis might contribute to avoiding of illusions of robustness or control in decision making. In the latter context, MCDA might contribute to the choice of scenarios as well as to the formal analysis of alternative courses of action.

Preliminary suggestions for such integration of scenario planning and MCDA is made on pages 312–315 of Belton and Stewart [6], which extended an earlier discussion in Chapter 14 of of Goodwin and Wright [17]. In the remainder of this section, we seek to explore these potentialities in greater detail. For this purpose, suppose that a set of scenarios $Y = \{y_1, \ldots, y_s\}$ have been selected for purposes of evaluating alternatives. Let us then define $z_i(x, y_r)$ (expressed by a lower case letter to emphasize that this is no longer viewed as a random variable) as the consequence of action x in terms of criterion i, under the conditions defined by scenario y_r. As before, $\mathbf{z}(x, y_r)$ will represent the corresponding vector of consequences.

Standard assumptions of MCDA imply that it should be possible for each individual criterion, to obtain at least partial preference orderings on any given set of specific (deterministic) consequences, independently of any other criteria, whether or not these outcomes refer to real or hypothetical alternatives. This observation forms the basis of a scenario-based approach to MCDA under uncertainty.

A direct MAUT approach would presumably still strive to establish a preference ordering of the alternatives in terms of an "expected" utility defined by:

$$\sum_{r=1}^{s} p_r U(\mathbf{z}(x, y_r))$$

where p_r represent the "probability" associated with scenario y_r. There is, however, an immediate theoretical problem concerning the definition and interpretation of p_r. The set of scenarios Y does not constitute a complete probability space. More importantly, each element of this set, y_r, cannot in general be expected to represent the same hypervolume in probability space, so that even a relative probability density (or "likelihood") at the point in probability space represented by y_r cannot be used as a surrogate for p_r. Thus both the practical and theoretical questions regarding the assessment of the p_r remain fundamentally unanswered, and alternative procedures need to be defined.

It will simplify further discussion (and often the implementation) of the models to be discussed if now restrict consideration to the case in which the space of alternatives is also discrete, i.e. the alternatives belong to the set $A = \{a_1, \ldots, a_m\}$. With some abuse of notation we shall then use z_i^{kr} to denote the performance level of alternative a_k in terms of criterion i under the conditions of scenario y_r. The vector $\mathbf{z}^{kr}$ will be interpreted in a similar manner.

In searching for an appropriate and broadly applicable theoretical basis for modelling preferences in this context, two approaches immediately suggest themselves as an extension of the approach discussed by Goodwin and Wright [17], Chapter 14:

Model A: Apply a standard MCDA approach, to construct a preference model (ordinal or cardinal) across all ms possible outcomes (combinations of alternatives and scenarios) given by the performance level vectors $\mathbf{z}^{kr}$. This process involves aggregation across the original n criteria. Goodwin and Wright [17] adopt this model, making use of an n-dimensional additive value function to generate preference values for each $\mathbf{z}^{kr}$. Other MCDA approaches may equally well be employed, however, such as outranking (to generate a classification into preference classes) or goal programming (to measure achievements in terms of distance from a goal or reference level). An $m \times s$ table can then be constructed, giving for each alternative an aggregate measure of performance or goal satisfaction under each scenario. A second evaluation is then required to select the alternative which is "best" in some sense across all scenarios.

Model B: Treat each of the ns criterion-scenario combinations as *metacriteria* (much as in Teghem et al. [50]), and apply some form of MCDA to the problem of comparing m alternatives in terms of the $n \times s$ metacriteria.

Let us now explore the above two possibilities in somewhat greater detail.

5.1 Model A

Here the first step is to evaluate the $m \times s$ distinct "outcomes" in terms of the n criteria by some form of MCDA process, to provide an aggregate comparative evaluation of each outcome. As indicated above, Goodwin and Wright [17] suggested such an approach, and applied a simple value measurement model (SMART) to this step. In other words, the approach adopted was as follows:

1 A value function $v_i(z_i)$ was constructed for each criterion, standardized (e.g. to a 0–100 scale) over an appropriate range of performance levels covering at the least the ms outcomes.

2 Swing weights w_i were assessed by considering the ranges of outcomes used to standardize the scale for each criterion.

3 An overall value for each outcome was computed as:

$$V(\mathbf{z}^{kr}) = \sum_{i=1}^{n} w_i v_i(z_i^{kr}).$$

As an alternative to the value measurement suggested by Goodwin and Wright, the analyst might:

- Use an outranking method to construct a valued pairwise preference relation as done by Mareschal [35], or a (perhaps partial) preference ordering of the full set of ms outcomes; or

- Apply a goal programming method obtain an aggregate distance measure between each of the ms outcomes $\mathbf{z}^{kr}$ and a pre-specified set of goals for each of the n criteria.

Whichever methodology of MCDA is applied, the result will be some numerical scoring, say ζ_{kr} indicating a level of performance or goal satisfaction achieved by each alternative k under the conditions of each scenario y_r. The ζ_{kr} scores can be represented in a two-dimensional matrix, to give a form of "pay-off" table. The places the problem into a framework which can be viewed either as a standard monocriterion decision problem under uncertainty, or as an MCDA problem with aggregate performances under each scenario playing the role of "criteria". The final step is to select the alternative i which is robust against the uncertainties (according to the first view), or which best satisfies these "criteria" (according to the second view).

Goodwin and Wright leave this second phase selection problem to direct holistic judgement, and this does indeed seem to be consistent with the usual scenario planning philosophy. Nevertheless, if a value function approach is adopted and properly implemented in the first step, then the ζ_{kr} values should constitute an interval preference scale. It should then be permissible to construct an additive aggregation of the form $\sum_{r=1}^{s} \omega_r \zeta_{kr}$, where the ω_r represent relative weights on the scenarios. It may be difficult to elicit appropriate values for the scenario weights, however, as these may not be intuitively self-evident. Certainly, as we have indicated earlier, an assumption that ω_r should be equated to a "probability" for scenario y_r cannot really be supported. Some form of "swing-weighting" approach would perhaps be more justifiable.

An alternative approach may be to adopt a "max-min" strategy, i.e. to select alternative k which maximizes the worst aggregate performance given by $\min_{r=1}^{s} \zeta_{kr}$. This could plausibly be construed as the most robust solution, but is unsatisfying from an MCDA perspective, as no consideration is given to possibilities of trade-offs between performances under different scenarios. For example, if one alternative is very good under all but one scenario, but marginally worst on the remaining scenario, should it summarily be rejected? The second level MCDA problem thus poses some challenging questions to the MCDA research community.

5.2 Model B

In this model, the approach is of a standard MCDA form, treating all $n \times s$ combinations of criteria and scenarios as "metacriteria" (where each represents the desire of the decision maker to achieve satisfactory performance according to a particular criterion under a particular scenario).

At the outset, this formulation fits neatly into the MCDA framework, as the operational requirement of being able to compare alternatives in terms of each criterion without reference to performance on other criteria, will typically be satisfied if true for the original criteria. Even the stricter preferential independence condition of additive value function models may be expected to apply if satisfied for the original criteria. This would follow, provided that tradeoffs between outcomes under two scenarios do not depend on how well the alternative performs under other scenarios. On *prima facie* grounds it is difficult to conceive of situations in which such independence would not apply.

The process would then follow standard MCDA procedures, and any of the well-known MCDA methodologies (e.g. value measurement, goal programming or outranking) should in principle all be applicable (not necessarily equally easily or transparently, however). An important point distinguishing model B from model A, is that preference structures across criteria would be allowed to differ across scenarios, in the sense that (a) relative tradeoffs between criteria (importance weights) and (b) intensities of preference for different increments in performance on any one criterion may differ from scenario to scenario. It is an open question as to whether such changes may or should be expected.

Perhaps the most critical question would relate to importance weights placed on each of the metacriteria, as required in some or other sense by most MCDA methods. In principle, we require a relative weight, say w_{ir} to be placed on each metacriterion. There seems to be no difficulty in principle in establishing ratios $w_{ir}/w_{\ell r}$ for any pair of criteria under the assumption of the same scenario r. This would correspond exactly to standard MCDA considerations (e.g. swing weights for value functions). If decision makers can also express the relative importance of changes in performance level for the same criterion under different scenarios, by considering the question as to whether the same range of outcomes on the criterion would have a more or less important impact on the final decision under one scenario than another, this would generate estimates of $w_{ir}/w_{i\rho}$ for the pair of scenarios. From the two sets of ratios, it would be possible to infer relative weights for all combinations. In fact, by repeating the $w_{ir}/w_{i\rho}$ assessments for two or more criteria, some evaluation of the consistency of the estimates would also be possible.

5.3 The Way Forward

A formal integration of MCDA and scenario planning would thus appear to offer substantial potential benefits, and anecdotal evidence suggests that something along this line is done from time to time. At the present time, however, a completely integrated procedure would require answers to the following research questions:

- Do preference structures tend to change from scenario to scenario? If so, then this might better be handled by Model B.
- Are particular MCDA methods more appropriate to one model or the other?
- How many scenarios are needed for effective application of MCDA?
- How should these scenarios be constructed? Should the primary emphasis be on plausibility of the scenarios (as in standard scenario planning) or on achieving representivity of ranges of variation that can occur?
- How should weights be assessed?

At time of writing, a series of simulation studies are under way (based broadly on the approach described by Stewart [47, 49]) to address some of the above questions, especially those related to the number and selection of scenarios. Definitive results are not yet available, but early indications are extremely encouraging, in the sense that good results can be obtained with as few as 3–5 scenarios.

6. Implications for Practice

It should be evident from the preceding discussion that there still remains considerable scope for research into the treatment of substantive external uncertainties within an MCDA framework. It is hoped that such research will lead to ever-improved methodologies. Nevertheless, for the practitioner, certain guidelines can be given at the present time. These may be summarized as follows.

1 For those working within a value or utility function framework, the expectation of a simple additive value function can generate quite useful insights for the decision maker, *provided that* due attention is given to the shape (changing marginal values) of the function (cf. Stewart [46]). On the other hand, complete multiplicative or multilinear multiattribute utility functions may be difficult to implement correctly.

2 With any MCDA approach, there is value and some theoretical justification in decomposing those criteria for which there is substantial uncertainty regarding outcomes, into two subcriteria of expected value and a risk measure respectively. An open question remains as to whether variance or standard deviation (which are conventionally used in this context) are the most appropriate risk measures for all problem types.

3 The integration of MCDA and scenario planning is relatively easy to apply in at least two different ways, and may be particularly transparent to many decision makers. Once again, there do remain some open questions, especially as regards the number of scenarios to be used and the means by which they are constructed or selected.

References

[1] S. H. Azondékon and J.-M. Martel. "Value" of additional information in multicriterion analysis under uncertainty. *European Journal of Operational Research*, 117:45–62, 1999.

[2] E. Ballestero. Stochastic goal programming: A mean-variance approach. *European Journal of Operational Research*, 131:476–481, 2001.

[3] V. S. Bawa. Optimal rules for ordering uncertain prospects. *Journal of Financial Economics*, 2:95—121, 1975.

[4] M. H. Bazerman. *Judgment in Managerial Decision Making*. John Wiley & Sons, New York, fifth edition, 2002.

[5] D. E. Bell. One-switch utility functions and a measure of risk. *Management Science*, 34:1416–1424, 1988.

[6] V. Belton and T. J. Stewart. *Multiple Criteria Decision Analysis: An Integrated Approach*. Kluwer Academic Publishers, Boston, 2002.

[7] M. Beynon, B. Curry, and P. Morgan. The Dempster-Shafer theory of evidence: an alternative approach to multicriteria decision modelling. *OMEGA: International Journal of Management Science*, 28:37–50, 2000.

[8] N.-B. Chang and S. Wang. A fuzzy goal programming approach for the optimal planning of metropolitan solid waste management systems. *European Journal of Operational Research*, 99:303–321, 1997.

[9] N.-B. Chang, C. Wen, and Y. Chen. A fuzzy multi-objective programming approach for optimal management of the reservoir watershed. *European Journal of Operational Research*, 99:289–302, 1997.

[10] W. D. Cook and M. A. Kress. Multiple criteria decision model with ordinal preference data. *European Journal of Operational Research*, 54:191–198, 1991.

[11] G. R. D'Avignon and P. Vincke. An outranking method under uncertainty. *European Journal of Operational Research*, 36:311–321, 1988.

[12] A. Dimitras, R. Slowinski, R. Susmaga, and C. Zopounidis. Business failure prediction using rough sets. *European Journal of Operational Research*, 114:263–280, 1999.

[13] P. C. Fishburn. Foundations of risk measurement. I. Risk as probable loss. *Management Science*, 30:396–406, 1984.

[14] S. French. Uncertainty and imprecision: modelling and analysis. *Journal of the Operational Research Society*, 46:70–79, 1995.

[15] J. Friend. The strategic choice approach. In J. Rosenhead and J. Mingers, editors, *Rational Analysis for a Problematic World Revisited*, pages 115–149. John Wiley & Sons, Chichester, second edition, 2001.

[16] A. Goicoechea, D. R. Hansen, and L. Duckstein. *Multiobjective Decision Analysis with Engineering and Business Applications*. John Wiley & Sons, New York, 1982.

[17] P. Goodwin and G. Wright. *Decision Analysis for Management Judgement*. John Wiley & Sons, Chichester, second edition, 1997.

[18] P. Goodwin and G. Wright. Enhancing strategy evaluation in scenario planning: a role for decision analysis. *Journal of Management Studies*, 38:1–16, 2001.

[19] S. Greco, B. Matarazzo, and R. Slowinski. Rough approximation of a preference relation by dominance relations. *European Journal of Operational Research*, 117:63–83, 1999.

[20] S. Greco, B. Matarazzo, and R. Slowinski. The use of rough sets and fuzzy sets in MCDM. In T. Gal, T. J. Stewart, and T. Hanne, editors, *Multicriteria Decision Making: Advances in MCDM Models, Algorithms, Theory, and Applications*, chapter 14. Kluwer Academic Publishers, Boston, 1999.

[21] S. Greco, B. Matarazzo, and R. Slowinski. Rough sets theory for multicriteria decision analysis. *European Journal of Operational Research*, 129:1–47, 2001.

[22] S. Greco, B. Matarazzo, and R. Slowinski. Rough sets methodology for sorting problems in presence of multiple attributes and criteria. *European Journal of Operational Research*, 138:247–259, 2002.

[23] J. Hadar and W. R. Russell. Rules for ordering uncertain prospects. *The American Economic Review*, 59:25–34, 1969.

[24] C. Harries. Correspondence to what? Coherence to what? What is good scenario-based decision making. *Technological Forecasting & Social Change*, 70:797–817, 2003.

[25] J. Jia and J. S. Dyer. A standard measure of risk and risk-value models. *Management Science*, 42:1691–1705, 1996.

[26] D. Kahneman and A. Tversky. Prospect theory: An analysis of decision under risk. *Econometrica*, 47:263–291, 1979.

[27] R. L. Keeney and H. Raiffa. *Decisions with Multiple Objectives*. John Wiley & Sons, New York, 1976.

[28] A. J. Keown and B. W. Taylor III. A chance-constrained integer goal programming model for capital budgeting in the production area. *Journal of the Operational Research Society*, 31:579–589, 1980.

[29] G. Klein, H. Moskowitz, and A. Ravindran. Interactive multiobjective optimization under uncertainty. *Management Science*, 36:58–75, 1990.

[30] G. J. Klir and T. A. Folger. *Fuzzy Sets, Uncertainty and Information*. Prentice Hall, Englewood Cliffs, New Jersey, 1988.

[31] A. Korhonen. Strategic financial management in a multinational financial conglomerate: A multiple goal stochastic programming approach. *European Journal of Operational Research*, 128:418–434, 2001.

[32] R. Lahdelma and P. Salminen. Pseudo-criteria versus linear utility function in stochastic multi-criteria acceptability analysis. *European Journal of Operational Research*, 141:454—-469, 2002.

[33] R. Lahdelma, P. Salminen, and J. Hokkanen. Locating a waste treatment facility by using stochastic multicriteria acceptability analysis with ordinal criteria. *European Journal of Operational Research*, 142:345—-356, 2002.

[34] R. R. Levary and K. Wan. A simulation approach for handling uncertainty in the analytic hierarchy process. *European Journal of Operational Research*, 106:116–122, 1998.

[35] B. Mareschal. Stochastic multicriteria decision making and uncertainty. *European Journal of Operational Research*, 26:58–64, 1986.

[36] J.-M. Martel and K. Zaras. Stochastic dominance in multicriterion analysis under risk. *Theory and Decision*, 39:31–49, 1995.

[37] I. Millet and W. C. Wedley. Modelling risk and uncertainty with the Analytic Hierarchy Process. *Journal of Multi-Criteria Decision Analysis*, 11:97–107, 2002.

[38] J. M. Miyamoto and P. Wakker. Multiattribute utility theory without expected utility foundations. *Operations Research*, 44:313–326, 1996.

[39] G. S. Parnell, J. A. Jackson, R. C. Burk, L. J. Lehmkuhld, and J. A. Engelbrecht Jr. R&D concept decision analysis: using alternate futures for sensitivity analysis. *Journal of Multi-Criteria Decision Analysis*, 8:119—127, 1999.

[40] J.-C. Pomerol. Scenario development and practical decision making under uncertainty. *Decision Support Systems*, 31:197–204, 2001.

[41] D. Rios Insua. *Sensitivity Analysis in Multi-Objective Decision Making*, volume 347 of *Lecture Notes in Economics and Mathematical Systems*. Springer, Berlin, 1990.

[42] T. Rosqvist. Simulation and multi-attribute utility modelling of life cycle profit. *Journal of Multi-Criteria Decision Analysis*, 10:205–218, 2001.

[43] A. Ruszczyński and A. Shapiro, editors. *Stochastic Programming*. Handbooks in Operations Research and Management Science, Volume 10. Elsevier, Amsterdam, 2003.

[44] A. Saltelli, A. S. Tarantola, and K. Chan. A role for sensitivity analysis in presenting the results from MCDA studies to decision makers. *Journal of Multi-Criteria Decision Analysis*, 8:139—145, 1999.

[45] R. K. Sarin and M. Weber. Risk-value models. *European Journal of Operational Research*, 70:135–149, 1993.

[46] T. J. Stewart. Simplified approaches for multi-criteria decision making under uncertainty. *Journal of Multi-Criteria Decision Analysis*, 4:246–258, 1995.

[47] T. J. Stewart. Robustness of additive value function methods in MCDM. *Journal of Multi-Criteria Decision Analysis*, 5:301–309, 1996.

[48] T. J. Stewart. Measurements of risk in fisheries management. *ORiON*, 14:1–15, 1998.

[49] T. J. Stewart. Evaluation and refinement of aspiration-based methods in MCDM. *European Journal of Operational Research*, 113:643–652, 1999.

[50] J. Teghem Jr., D. Dufrane, and M. Thauvoye. STRANGE: An interactive method for multi-objective linear programming under uncertainty. *European Journal of Operational Research*, 26:65–82, 1986.

[51] B. Urli and R. Nadeau. PROMISE/scenarios: An interactive method for multiobjective stochastic linear programming under partial uncertainty. *European Journal of Operational Research*, 155:361–372, 2004.

[52] K. Van der Heijden. *Scenarios: The Art of Strategic Conversation*. John Wiley & Sons, Chichester, 1996.

[53] D. von Winterfeldt and W. Edwards. *Decision Analysis and Behavioral Research*. Cambridge University Press, Cambridge, 1986.

[54] D. W. Watkins Jr., D. C. McKinney, L. S. Lasdon, S. S. Nielsen, and Q. W. Martin. A scenario-based stochastic programming model for water supplies from the highland lakes. *International Transactions in Operational Research*, 7:211–230, 2000.

[55] G. A. Whitmore. Third order stochastic dominance. *The American Economic Review*, 60:457–459, 1970.

[56] J.-B. Yang. Rule and utility based evidential reasoning approach for multiattribute decision analysis under uncertainties. *European Journal of Operational Research*, 131:31–61, 2001.

[57] C. Yeh, H. Deng, and H. Pan. Multi-criteria analysis for dredger dispatching under uncertainty. *Journal of the Operational Research Society*, 50:35–43, 1999.

[58] M. R. Yilmaz. An information-expectation framework for decisions under uncertainty. *Journal of Multi-Criteria Decision Analysis*, 1:65–80, 1992.

[59] H. Zimmermann. An application-oriented view of modeling uncertainty. *European Journal of Operational Research*, 122:190–198, 2000.

Chapter 12

CHOICE, RANKING AND SORTING IN FUZZY MULTIPLE CRITERIA DECISION AID

Patrick Meyer, Marc Roubens
Department of Mathematics
University of Liège
Grande Traverse, 12
4000 Liège
Belgium
patrick.meyer@internet.lu, m.roubens@ulg.ac.be

Abstract In this chapter we survey several approaches to derive a recommendation from some preference models for multiple criteria decision aid. Depending on the specificities of the decision problem, the recommendation can be a selection of the best alternatives, a ranking of these alternatives or a sorting. We detail a sorting procedure for the assignment of alternatives to graded classes when the available information is given by interacting points of view and a subset of prototypic alternatives whose assignment is given beforehand. A software dedicated to that approach (TOMASO) is briefly presented. Finally we define the concepts of good and bad choices based on dominant and absorbant kernels in the valued digraph that corresponds to an ordinal valued outranking relation.

Keywords: Aggregation with fuzzy environment, fuzzy choice, ordinal ordered sorting, choquet integral, TOMASO.

1. Introduction

Let $A = \{\ldots, x, y, \ldots\}$ be a finite set of potential alternatives, and $\mathcal{J}$ be a set of n points of view. The Multiple Criteria Decision Problem can often be formulated as comparing and/or discriminating between the alternatives on the basis of several points of view.

As clearly stated by B.Roy in his book on Multicriteria Methodology [27] , Multiple Criteria Decision Aid is an activity that creates models to provide the decision maker (DM) with guidelines with respect to his decision problem. Three basic problems are usually put forward:

- the *choice* problem that aims to select a subset of potential alternatives, as restricted as possible, containing the "satisfactory" actions,
- the *sorting* problem that corresponds to the assignment of each alternative into pre-defined categories. These categories correspond to a set M of classes. If M is just a set of labels we talk about a classification problem. If the labels of M can be ordered, we are dealing with an ordered sorting,
- the *ordering* problem that aims at ranking the alternatives by decreasing order of preference. The prescription may be given in terms of a partial or a complete order.

A first step in the Decision Aiding Process consists in the evaluation of the alternatives on each of the points of view and is possibly followed by the definition of a valued preference relation R_j on A for each dimension $j \in \mathcal{J}$.

A second step consists in either determining a global ranking on the alternatives, a sorting into different classes, or a choice function which results in a subset of alternatives of A. Two different procedures can be used: the pre-ranking methods and the pre-aggregation methods.

The *pre-ranking methods* first determine a score $S(x, R_j) = S_j(x)$ for each alternative $x \in A$ and each point of view $j \in \mathcal{J}$. An aggregation rule M_v then transforms those partial scores into a global score $S(x; R_1, \ldots, R_n)$, where v represents weights linked to the points of view. v is either a vector $(v(1), \ldots, v(n))$ or a monotone set function $v : 2^{\mathcal{J}} \to [0, 1]$ fulfilling $v(\emptyset) = 0$ and $v(\mathcal{J}) = 1$. This procedure will be used in the TOMASO method which deals with ordinal data and interacting points of view. An ordered sorting is obtained and all alternatives are comparable.

The *pre-aggregation methods* first determine a global binary relation R on A using an aggregation rule $M_v : R = M_v(R_1, \ldots, R_n)$. Comparisons of partial evaluations are performed dimension by dimension and their results are then aggregated. Usually this relation is constructed so as to reflect the majoritarian preference among the set of points of view. This approach allows a fine and flexible description of preferences without forcing arbitrarily alternatives to be comparable and allows to take into account not only concordance between pairs of alternatives but also discordance. A global score $S'(x, R)$ transforms the global information on each pair of alternatives into a global rating related to each alternative. However a global partial order on the alternatives might be obtained if top-down or bottom-up procedures are considered (as the combination of in and out-flows in PROMETHEE [3] and the intersection of direct and inverse complete preorders in ELECTRE II [28]).

This chapter is built around three main subjects. First of all, a general description of the different ways to deal with a multiple criteria decision problem is proposed. In Section 2 we describe the different types of data one may encounter. Section 3 presents the concepts of valued preference relation and outranking relation. Section 4 describes the two possibilities for aggregation: pre-aggregation and pre-scoring. Section 5 deals with the particular multiple criteria decision aiding problematic called the sorting. This is done in view of Section 6. There we focus on a particular sorting procedure called TOMASO. It is a multiple criteria sorting procedure for the assignment of alternatives to ordered classes based on a pre-ranking method. The alternatives are evaluated on different interacting points of view using performance levels (scores). The objective is to aggregate these partial evaluations by the Choquet integral. The basic technique we present is due to Roubens [23]. An evolution to this method is explicited, in case the basic procedure has no solution. The fuzzy measures associated to the Choquet integral can be learnt from a subset of alternatives (called prototypes) which are assigned beforehand to the classes by the DM. This leads in a first stage to solving a linear constraint satisfaction problem whose unknown variables are the coefficients of the fuzzy measure. If a fuzzy measure is found, the boundaries of the classes are calculated, and the alternatives are classified. If no solution is found to this problem, an alternate way is suggested, which can lead to ambiguous assignments of the prototypes.

Both results can be analysed by means of the importance indexes and the interaction indexes of the assessed fuzzy measure. These two parameters give the following indications on the fuzzy measure:

- the importance indexes make it possible to appraise the overall importance of each point of view and each combination of points of view;
- the interaction indexes measure the extent to which the points of view interact (positively or negatively).

Finally, in Section 7, we focus on a choice procedure for the selection of a set of "good" alternatives that includes a fuzzy approach based on a pre-aggregation method. It can be considered as a substitute to the ELECTRE IS [17, 30] method or a complement to its prescriptions. The chapter finishes on some conclusions and perspectives.

2. The Data Set

Without any loss of generality, we will suppose hereafter that the higher an evaluation of an alternative on a point of view, the better the alternative is in the eyes of the decision maker.

For each point of view $j \in \mathcal{J}$, the evaluation related to each alternative is possibly given under one of the following forms:

- An *ordinal value* g_j defined on a s_j-point performance scale, that is a totally ordered set $X_j := \{g_1^j \prec_j \ldots \prec_j g_{s_j}^j\}$. It usually corresponds to linguistic ordered data.

- A *fuzzy ordinal value*, i.e. a membership function $\mu_j(u) \in [0,1]$, $\forall u \in X_j$. The degree of membership can be interpreted as the degree of compatibility of the evaluation with u. The fuzzy set is supposed to be normal ($\sup_u \mu_j(u) = 1$) and convex ($\forall u, v, w \in X_j, v \in [u,w], \mu_j(v) \leq \min\{\mu_j(u), \mu_j(w)\}$).

- A *cardinal value* g_j that associates the alternative with a real number indicating its performance. This is the most conventional way of building a preference model and in that case we are talking about a true-criterion.

- A *fuzzy interval*, i.e. a membership function $\mu_j(u) \in [0,1], \forall u \in \mathbb{R}$ that is supposed to be normal and convex. Every λ-cut is a closed interval $I_j^\lambda = \{u : \mu_j(u) \geq \lambda\}$.

  A particular example of a fuzzy interval corresponds to a *trapezoidal fuzzy number* defined by the parameters $(g_j^-, g_j^+, \sigma_j^-, \sigma_j^+)$:

$$\mu_j(u) = \begin{cases} 1 - \frac{g_j^- - u}{\sigma_j^-} & \text{if } g_j^- - \sigma_j^- \leq u \leq g_j^- \\ 1 & \text{if } g_j^- \leq u \leq g_j^+ \\ 1 - \frac{u - g_j^+}{\sigma_j^+} & \text{if } g_j^+ \leq u \leq g_j^+ + \sigma_j^+ \end{cases}$$

  This may correspond to imprecise information on the evaluation of a given alternative: it lies possibly in the support ($g_j^- - \sigma_j^- \leq u \leq g_j^+ + \sigma_j^+$) and belongs certainly to the kernel ($g_j^- \leq u \leq g_j^+$).

  A symmetric trapezoidal fuzzy number is such that $\sigma_j^- = \sigma_j^+$ and may translate the indifferences and preferences that might exist between values that are assessed to an alternative. In that situation we call

$$\begin{cases} g_j^+ - g_j^- = q_j \text{ (indifference threshold)} \\ g_j^+ + \sigma_j^+ - (g_j^- - \sigma_j^-) = p_j \text{ (preference threshold)} \\ \frac{g_j^+ + g_j^-}{2} = g_j \end{cases}$$

These definitions are interesting as a help to understand the concepts of indifference and preference thresholds. All the values between $g_j - \frac{1}{2}q_j$ and $g_j + \frac{1}{2}q_j$ are considered as indifferent. Values greater than $g_j + \frac{1}{2}p_j$ are better than g_j and those lower than $g_j - \frac{1}{2}p_j$ are worse than g_j. Even in the case of complete and precise information, a small positive difference does not always justify the preference.

3. Valued Preference Relation and Outranking Relation

Now that we have described the different possible evaluations in Section 2 the goal of this section is to recall the concepts of valued preference relation and outranking relation. We define the degree to which an alternative x is not worse than y for point of view j. Let $R_j(x, y)$ be this degree, for each ordered pair (x, y) of alternatives. We use the same notations as in Section 2 for the different possible evaluations.

Similarly to the different possibilities described in Section 2, the degree $R_j(x, y)$ has different definitions and properties:

- For an *ordinal* or *cardinal value* g_j:

$$R_j(x,y) = \begin{cases} 1 & \text{if } g_j(x) \geq g_j(y) \\ 0 & \text{otherwise.} \end{cases}$$

This crisp binary relation is a linear quasiorder.

- For a *fuzzy ordinal value*, $R_j(x, y)$ defines the degree of the preference of x over y and is considered as the possibility that x is not worse than y:

$$\begin{aligned} R_j(x,y) = \Pi_j(x \geq y) &= \max\nolimits_{u \geq v} \min(\mu_j^x(u), \mu_j^x(v)), u, v \in X_j \\ &= \max\nolimits_{u} \min(\mu_j^x(u), \mu_j^y(u)), u \in X_j. \end{aligned}$$

Π_j is a valued binary relation such that $\max(\Pi_j(x,y), \Pi_j(y,x)) = 1, \forall x, y \in A$. Roubens and Vincke [24] have proved that Π_j is a fuzzy interval order and every λ-cut is a crisp interval order.

- For *fuzzy intervals*, $R_j(x, y)$ is also defined as the possibility that x is not worse than y:

$$R_j(x,y) = \Pi_j(x \geq y) = \max_{u \geq v} \min(\mu_j^x(u), \mu_j^y(v)), u, v \in \mathbb{R}.$$

If the kernel of μ_j^x is located to the right of the kernel of μ_j^y, then $\Pi_j(x \geq y) = 1$ and $\Pi_j(y \geq x)$ equals the height of the intersection of μ_j^x and μ_j^y, $h_j(x, y)$ (see Figure 12.1). This valued binary relation presents the same properties as the fuzzy ordinal value.

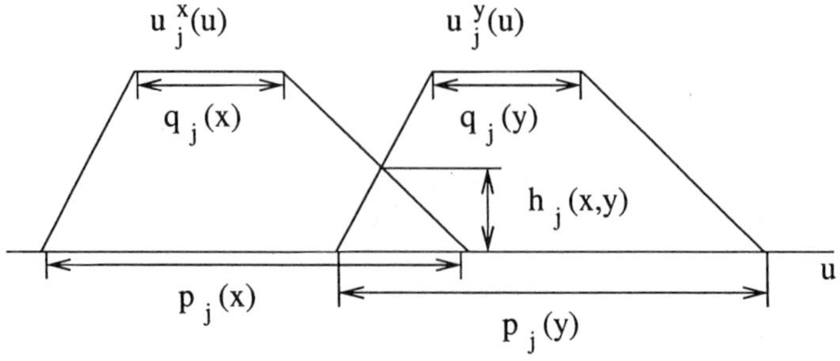

Figure 12.1. Comparing two fuzzy intervals.

Starting from the credibility of the preference of x over y it is possible to define [6, 7]:

- the degree of strict preference of x over y as the necessity that x is strictly better than y:

$$P_j(x,y) = 1 - \Pi_j(y \geq x) = 1 - R_j(y,x)$$

- the degree of indifference between x and y as:

$$I_j(x,y) = \min(R_j(x,y), R_j(y,x)).$$

- For of a *symmetric trapezoidal number*:

$$R_j(x,y) = \frac{\frac{p_j(x)+p_j(y)}{2} - \min\{g_j(y) - g_j(x), \frac{p_j(x)+p_j(y)}{2}\}}{\frac{p_j(x)+p_j(y)}{2} - \min\{g_j(y) - g_j(x), \frac{q_j(x)+q_j(y)}{2}\}},$$

where $\frac{0}{0}$ should be taken as 0.

If p_j and q_j are linear functions of g_j, then R_j is a fuzzy semiorder and every λ-cut is a crisp semiorder [6, 7].

If $p_j = q_j$, then we obtain a crisp interval order. Let us define $g'_j(x) = g_j(x) - \frac{q_j(x)}{2}$. We then have:

- for the strict preference:

$$\begin{aligned} xP_jy &\iff R_j(x,y) = 1, R_j(y,x) = 0 \\ &\iff g'_j(x) - g'_j(y) > q_j(y) \end{aligned}$$

- for the indifference:

$$\begin{aligned} xI_jy &\iff R_j(x,y) = R_j(y,x) = 1 \\ &\iff |g'_j(x) - g'_j(y)| \leq \min(q_j(x), q_j(y)) \end{aligned}$$

An extra condition of local consistency should be added [27]:

$$g'_j(x) > g'_j(y) \Rightarrow g'_j(x) + q_j(x) > g'_j(y) + q_j(y)$$

The criterion function g'_j and the threshold function q_j define a semi-criterion and the structure (P_j, I_j) is a semiorder.

The classical procedures ELECTRE III [26] and PROMETHEE [3] are using this approach based on the intersection of fuzzy sets.

According to Perny [22], the degree of preference of x over y may be considered in very general terms as

$$R_j(x,y) = f_j[g_j(x), \mathcal{N} g_j(y)]$$

where f_j is a non-decreasing function of both arguments, $\mathcal{N}$ is a strong negation and $R_j(x,x) = 1$. Perny proved that such a valued preference relation is a fuzzy semiorder and every λ-cut constitutes a crisp semiorder [22]. As a particular case, we have the concordance index defined by Roy [26]:

$$R_j(x,y) = \min\left\{1, \max\left\{0, \frac{g_j(x) - g_j(y) + p_j(g_j(x))}{p_j(g_j(x)) - q_j(g_j(x))}\right\}\right\}$$

where p_j and q_j are non-decreasing functions of g_j and correspond respectively to a preference threshold and an indifference threshold. For consistency reasons, $p_j(g_j(x)) \geq q_j(g_j(x))$. The concordance index R_j is meaningful (i.e. is invariant under admissible transformations of g_j) if g_j is defined on an interval scale (admissible transformations are $h_j = r \cdot g_j + s, r > 0$). q_j corresponds to a constant or a proportion of g_j and p_j is expressed as a proportion of g_j [26].

Similarly, according to Perny, we may also define a degree of discredit as

$$D_j(x,y) = h_j[g_j(y), \mathcal{N} g_j(x)]$$

where h_j is a non decreasing function of both arguments, $D_j(x,x) = 0$ and $\min\{R_j(x,y), D_j(x,y)\} = 0$. Under these conditions, D_j is a fuzzy partial order and every λ-cut represents a crisp partial order. As previously, we can consider the particular case of the discordance index defined by Roy [26]:

$$D_j(x,y) = \min\left\{1, \max\left\{0, \frac{g_j(y) - g_j(x) - p_j(g_j(x))}{v_j(g_j(x)) - p_j(g_j(x))}\right\}\right\},$$

where v_j corresponds to a veto threshold which expresses the existence of a discordant point of view that prohibits to accept the idea that x is globally preferred to y.

4. Aggregation Procedures

4.1 Pre-aggregation Methods

Let us first consider the methods that propose to merge the marginal information about each pair of alternatives (x, y) in terms of concordance (and possibly discordance) indexes into a global relation that expresses the overall importance of the consensus on the fact that "x is globally not worse than y".

Roy [27] introduces an outranking relation $\mathcal{O}(x, y)$ that corresponds to the "agreement versus discordance" measure linked to the proposition that x is globaly not worse than y. It indicates the importance of the coalition of the points of view that agree with the proposition by taking also into account the discordance.

In general, if v_j represents the relative importance of each point of view j, $(j \in \mathcal{J}, |\mathcal{J}| = n)$, we may consider two aggregation operators M_R and M_D such that

$$R = M_R(R_1, \dots, R_n; v_1, \dots, v_n)$$
$$D = M_D(D_1, \dots, D_n; v_1, \dots, v_n).$$

M_R is a monotonic function of the first arguments such that $M_R(0, \dots, 0; v_1, \dots, v_n) = 0$ and $M_R(1, \dots, 1; v_1, \dots, v_n) = 1$. M_D is a monotonic function of the first arguments that should satisfy:

$$(\exists j, j \in \mathcal{J} : D_j(x, y) = 1) \Rightarrow D(x, y) = 1$$

stating that if at least one point of view is totally discordant with the proposition that x is not worse than y, the global discordance should be maximal for that specific pair of alternatives.

We could consider the following approach:

- for R the compensative idempotent operator (weighted sum)

$$R(x, y) = \sum_{j \in \mathcal{J}} v_j R_j(x, y), \sum_{j \in \mathcal{J}} v_j = 1$$

- for $1 - D$ the non-discordance index (geometric mean)

$$1 - D(x, y) = \Pi_{j \in \mathcal{J}} (1 - D_j(x, y))^{v_j}.$$

R measures the overall importance of the agreement and D allows to give a bad rating as soon as one important partial evaluation of the discordance is achieved.

Finally the outranking relation is obtained as a combination of concordant and discordant aspects as:

$$\mathcal{O}(x, y) = R(x, y) \cdot [1 - D(x, y)].$$

Roy on his side considered in ELECTRE III:

- for R the compensative weighted sum operator
- for the outranking degree

$$\begin{aligned}\mathcal{O}(x,y) &= R(x,y) \text{ if } D_j(x,y) \leq R(x,y) \text{ for all } j \in \mathcal{J}, \\ &= R(x,y)\Pi_{j\in\mathcal{J}(x,y)} \tfrac{1-D_j(x,y)}{1-R(x,y)} \text{ otherwise,}\end{aligned}$$

where $\mathcal{J}(x,y)$ corresponds to the subset of points of view for which $D_j(x,y) > R_j(x,y)$.

In this case, the outranking degree is thus equal to the concordance index if no point of view is discordant, or if no veto is used and is lowered if the level of discordance $(1 - D_j)$ increases above a threshold value.

Most of the existing proposals linked to pre-aggregation methods simply merge the marginal information related to the agreement on the proposal that x is globally not worse than y. They are thus directly linked to the concordance measures $R_j(x,y)$. The subjectivity of the decision maker with respect to the importance of each of the points of view can be used in different ways to obtain a global compromise. We consider here three of these approaches.

- the weighted sum (good items compensate bad ones with respect to different points of view):

$$R = \sum_{j\in\mathcal{J}} v_j R_j, \sum_{j\in\mathcal{J}} v_j = 1$$

- the weighted minimum (the outranking value is high if the partial evaluations are favorable on each of the points of view)

$$R = \min_{j\in\mathcal{J}} \max(1 - v_j, R_j), \max_{j\in\mathcal{J}} v_j = 1$$

- the weighted maximum (the outranking value is high if at least one of the points of view presents a good evaluation)

$$R = \max_{j\in\mathcal{J}} \min(v_j, R_j), \max_{j\in\mathcal{J}} v_j = 1.$$

Weighted maximum and minimum can be interpreted as weighted medians (see [5]). The interested reader can refer to [6, 11] and [7] for a more elaborate list of aggregators.

In the case of a *choice* problem the outranking relations $\mathcal{O}$ (initially with crisp outranking relations and later with λ-cuts of the valued outranking relations) were exploited by Roy using the kernel concept (internally stable and dominating subset of A) in ELECTRE I [17, 25] and later in ELECTRE IS [17, 30]. The idea is to track the maximal circuits to transform them into indifference cliques or suppress these circuits by eliminating the less credible outrankings.

Another approach was proposed by Orlovski [22]. He considers the fuzzy set of non dominated elements over A as

$$\mathrm{ND}(x) = \min_{y \in A \setminus \{x\}} \neg P(y, x), \quad \forall x \in A,$$

where $P(y,x)$ corresponds to the degree of strict preference associated to $R(y,x)$ (see [6, 7, 20]). The rational choice corresponds to these alternatives giving the maximal value of ND:

$$A^{\mathrm{ND}} := \{x \in A : \max_{x \in A} \mathrm{ND}(x)\}.$$

Under certain sufficient conditions (transitivity of R) this subset corresponds to maximal values equal to one; such good alternatives are called *unfuzzy non dominated alternatives* (UND-alternatives) and the corresponding rational choice is

$$A^{\mathrm{UND}} := \{x \in A : P(y, x) = 0 \quad \forall y \in A\}.$$

In Section 7 we consider the case where the valued relations R are ordinal values defined on a discrete finite set L (L-valued binary relations) and we determine the choice set as a kernel with a maximum degree of credibility.

In the case of a *sorting* problem the outranking relations $\mathcal{O}$ are used in procedures where a decision tree is used or by filtering as in ELECTRE TRI [17, 29]. These procedures use a cutting procedure that transforms the fuzzy outranking relations into a sequence of crisp and nested outranking relations.

In the case of an *ordering* problem the outranking relations $\mathcal{O}$ are used to construct two complete pre-orders, one arising from a ascending distillation procedure and another constructed from a descending distillation procedure. Another prescription consists in the intersection of the two previous pre-orders. These exploitation procedures are described in ELECTRE III [17, 26].

4.2 Pre-scoring Methods

In this type of approach, the implicit assumption that there exists a complete and transitive comparability of the alternatives is made. The most typical example of such methods corresponds to an ordering or a sorting that is based on the weighted sum of some partial scores $S_i(x)$ ($x \in A$). The additive representation of the utilities (expressed in terms of the partial scores) however implies preferential independance of the utilities.

One way to avoid this independance condition is to use the Choquet integral [4] as an aggregator.

Let us consider an alternative x which is described by its partial scores vector $S(x) = (S_1(x), \ldots, S_n(x))$. The Choquet integral of x is then defined by:

$$\mathcal{C}_v(S(x)) := \sum_{i=1}^{n} S_{(i)}(x)[v(A_{(i)}) - v(A_{(i+1)})]$$

where v represents a fuzzy measure on $\mathcal{J}$, that is a monotone set function $v : 2^{\mathcal{J}} \to [0, 1]$ fulfilling $v(\emptyset) = 0$ and $v(\mathcal{J}) = 1$. This fuzzy measure merely expresses the importance of each subset of points of view. The parentheses used for indexes represent a permutation on $\mathcal{J}$ such that

$$S_{(1)}(x) \leq \ldots \leq S_{(n)}(x),$$

and $A_{(i)}$ represents the subset $\{(i), \ldots, (n)\}$.

We note that for additive measures ($v(S \cup T) = v(S) + v(T)$, whenever $S \cap T = \emptyset$) the Choquet integral coincides with the usual discrete Lebesgue integral and the set function v is simply determined by the importance of each point of view: $v(1), \ldots, v(n)$. In this particular case

$$\mathcal{C}_v(S(x)) = \sum_{i=1}^{n} v(i) S_i(x) \ \ (x \in A),$$

which is the natural extension of the Borda score as defined in voting theory if alternatives play the role of candidates and points of view represent voters.

If points of view cannot be considered as being independent, the importance of the combinations $S \subseteq \mathcal{J}$, namely $v(S)$ has to be taken into account.

Some combinations of points of view might present a positive interaction or *synergy*. Although the importance of some points of view, members of a combination S, might be low, the importance of a pair, a triple, ..., might be substantially larger and $v(S) > \sum_{i \in S} v(i)$.

In other situations, points of view might exhibit negative interaction or *redundancy*. The union of some points of view do not have much impact on the decision and for such combinations S, $v(S) < \sum_{i \in S} v(i)$.

The Choquet integral presents standard properties for aggregation [13, 15, 35]: it is continuous, non decreasing, located between min and max.

The major advantage linked to the use of the Choquet integral derives from the large number of parameters ($2^n - 2$) associated with a fuzzy measure. On the other hand, this flexibility can also be considered as a serious drawback when assessing real values to the importance of all possible combinations. We will come back to this important question in Section 6.

Let us present an equivalent definition of the Choquet integral. Let v be a fuzzy measure on $\mathcal{J}$. The Möbius transform of v is a set function $m : 2^{\mathcal{J}} \to \mathbb{R}$ defined by

$$m(S) = \sum_{T \subseteq S} (-1)^{|S|-|T|} v(T) \ \ (S \subseteq \mathcal{J}).$$

This transformation is invertible and thus constitutes an equivalent form of a fuzzy measure and v can be recovered from m by using

$$v(S) = \sum_{T \subseteq S} m(T) \ (S \subseteq N).$$

This transformation can be used to redefine the Choquet integral without reordering the partial scores:

$$\mathcal{C}_v(S(x)) = \sum_{T \subseteq \mathcal{J}} m(T) \bigwedge_{i \in T} S_i(x).$$

A fuzzy measure is k-additive [8] if its Möbius transform m satisfies $m(S) = 0$ for S such that $|S| > k$ and there exists at least one subest S such that $|S| = k$ and $m(S) \neq \emptyset$. Thus, k-additive fuzzy measures can be represented by at most $\sum_{j=1}^{k} \binom{n}{j}$ coefficients.

For a k-additive fuzzy measure,

$$\mathcal{C}_v(S(x)) = \sum_{\substack{T \subseteq \mathcal{J} \\ |T| \leq k}} m(T) \bigwedge_{j \in T} S_j(x).$$

In order to assure boundary and monotonicity conditions imposed on v, the Möbius transform of a k-additive fuzzy measure must satisfy:

$$\begin{cases} m(\emptyset) = 0, \ \sum\limits_{\substack{T \subseteq \mathcal{J} \\ |T| \leq k}} m(T) = 1 \\ \sum\limits_{\substack{T : i \in T \subseteq S \\ |T| \leq k}} m(T) \geq 0, \quad \forall S \subseteq \mathcal{J}, \ \forall j \in S \end{cases}$$

In Section 6 we present a sorting method using the Choquet integral and based on supervised learning. But first let us introduce some general considerations on the problematic of sorting alternatives.

5. The Sorting Problem

Let A be a set of q potential alternatives which are to be assigned to disjoint ordered classes. Let $F = \{g_1, \ldots, g_n\}$ be a set of points of view. For each index of point of view $j \in \mathcal{J} = \{1, \ldots, n\}$, the alternatives are evaluated according to a s_j-point ordinal performance scale represented by a totally ordered set

$$X_j := \{g_1^j \prec_j \ldots \prec_j g_{s_j}^j\}.$$

Therefore, an alternative $x \in A$ can be identified with its corresponding profile

$$(x_1, \ldots, x_n) \in \prod_{j=1}^{n} X_j =: X,$$

where for any $j \in \mathcal{J}$, x_j is the partial evaluation of x on point of view j.

Let us consider a partition of X into m nonempty increasingly ordered classes $\{Cl_t\}_{t=1}^m$. This means that for any $r, s \in \{1, \ldots, m\}$, with $r > s$, the elements of Cl_r are considered as better than the elements of Cl_s.

The sorting problem we are dealing with consists in partitioning the alternatives of A into the classes $\{Cl_t\}_{t=1}^m$.

In Greco *et al.* [10], a very general theorem states that, under a simple condition of monotonicity, a discriminant function can be found which strictly separates the classes $\{Cl_t\}_{t=1}^m$ by thresholds. In Roubens [23] a restriction to the class of n-place Choquet integrals and normalised scores as criteria functions is made. Hereafter we present the sorting procedure derived from this particular case.

6. The TOMASO Method

The TOMASO method (**T**echnique for **O**rdinal **M**ultiattribute **S**orting and **O**rdering) is mainly based on two techniques (which can lead to the same results under certain conditions). The original method has first been described in [23]. In the following Subsection, we present its basics. In Subsection 6.2 we show how it is possible to deal with a larger set of problems.

6.1 The Classical Way

The different stages of the original TOMASO are listed below:

1 Modification of the criteria evaluations into scores;

2 Use of a Choquet integral as a discriminant function;

3 Assessment of fuzzy measures by questionning the DM and by solving a linear constraint satisfaction problem;

4 Calculation of the borders of the classes and assignment of the alternatives to the classes;

5 Analysis of the results (interaction, importance, leave one out, visualisation).

In this Section we roughly present these different elements.

One of the most difficult tasks is to modify the original ordinal evaluations of the alternatives on the criteria into some "scores" which can be aggregated by means of a Choquet integral. For example, two ordinal scales X_j and X_k can have a distinct number of evaluation levels and very different intrinsic meanings. The transformations of the scales should take into account these possible characteristics in order to obtain comparable evaluations. Two natural possibilities appear: the scores are built on basis of the data which are to be analysed or the scores are constructed completely out of the context of the problem. In the first case, the scores are solely based on the information which is contained in the set of alternatives which are considered. In the second case, the scales $X_j, j \in \mathcal{J}$ are modified in a general way, without taking into account the particular structure of the analysed set of data. At the present stage of our research, we suggest three possible alternatives for the building of these evaluation scores. In each of the cases, the DM must be aware of the consequences of his choice. Therefore, a deep analysis of the problem is important for its complete understanding.

First of all, in case the problem can be resumed to the set A of potential alternatives and if the DM is a single person, then one possible way to build the scores is to consider pairwise comparisons of the alternatives on each of the points of view. For each point of view $j \in \mathcal{J}$, the order on X_j ($\preceq_j$) can be caracterised by a valuation $R_j : A \times A \to \{0,1\}$ defined by $R_j(x,y) = 1$ if $x_j \succeq y_j$, 0 otherwise. Starting from this valuation we define a *partial net score* $S_j : A \to \mathbb{R}$ by

$$S_j(x) := \sum_{y \in A} [R_j(x,y) - R_j(y,x)] \quad (x \in A, j \in \mathcal{J}). \tag{12.1}$$

The interpretation of the integer $S_j(x)$ is natural: it represents the number of times that x is preferred to any other alternative of A minus the number of times that any other alternative of A is preferred to x for point of view j. One can show that the partial net scores identify the corresponding partial evaluations. We furthermore normalise these scores so that they range in the unit interval. The highest partial net score which can be obtained corresponds to the following general case:

- one single alternative $x_{\max} \in A$ has $\text{ord}_j(x_{\max}) = i$;
- no alternative $x \in A$ has $\text{ord}_j(x) > i$;
- the remaining alternatives $x \in A \backslash \{x_{\max}\}$ have $\text{ord}_j(x) < i$.

Therefore, the highest possible partial net score is $S_{j,\max}(x) = q - 1$. Similarly, the lowest possible partial net score is $S_{j,\min}(x) = -(q-1)$. We can therefore write the normalised partial net scores S_j^N as follows:

$$S_j^N(x) := \frac{S_j(x) + (q-1)}{2(q-1)} \in [0,1] \quad \forall j \in \mathcal{J}, \forall x \in A. \tag{12.2}$$

On contrary of the original ordinal partial evaluations, the partial net scores (and the normalised partial net scores) are commensurable. During the whole chapter we will use the notation $S^N(x) := (S_1^N(x), \ldots, S_n^N(x))$.

Two important questions now arise: how can this choice be motivated, and how can it be interpreted? First of all, the DM must understand that the selection of the set of potential alternatives A will have an influence on the final result. Therefore this choice must be made with much care. Then, his way of thinking must be a comparison of the alternatives on each of the points of view. Let us consider a short example which clearly illustrates this way of obtaining the scores. Suppose that we have to deal with a sorting problem with two qualitative ordinal criteria on a set of cars. The first point of view C_1 expresses the degree of comfort of the alternatives and is evaluated on a 3-point ordinal scale $X_1 = \{\text{Bad} \prec_1 \text{Medium} \prec_1 \text{Good}\}$. The second one C_2 expresses the fuel consumption of the cars on a 3-point ordinal scale $X_2 = \{\text{High} \prec_2 \text{Normal} \prec_2 \text{Low}\}$. The set of potential alternatives consists in 6 cars. The DM is aware that the results will depend on these 6 alternatives, but he considers that they have been chosen in a right way (for example, they are the only possible cars that he can afford with his tight budget). One can then assume that the absolute value of an alternative on a point of view is not informative, unless considered in relation with the other elements of A. We summarise this short example in Table 12.1. It shows the distribution of the alternatives among the different evaluation levels of the two points of view.

Table 12.1. Number of alternatives per evaluation level.

C_1		C_2	
Good	4	Low	2
Medium	1	Normal	2
Bad	1	High	2

If one reasons according to the comparison philosophy, it appears clearly that it is less exceptional to be "Good" than to be "Low". In fact, there are many good cars, but fewer cars with a low fuel consumption. Similarly, being "Bad" is worse than being "High". This means that having a high fuel consumption is less exceptional than being an uncomfortable car. The scores, as defined earlier, reflect these properties. They are given in Table 12.2.

Table 12.2. Score of each of the evaluation levels.

C_1		C_2	
Good	7/10	Low	9/10
Medium	2/10	Normal	5/10
Bad	0/10	High	1/10

This example shows that it is not senseless to modelise the DM's way of thinking by these scores. Three conditions should be satisfied: the decisions must be taken by a single DM, the set of potential alternatives must be chosen carefully and the DM should evaluate the alternatives by comparisons.

Secondly, let us consider the cases where multiple DMs intervene or where the decisions are not taken according to the previously described comparison philosophy. Here, the scoring functions are built "out of the context". This means that the values given to each of the evaluation levels of the ordinal scales don't depend on the set A. If the DM cannot help us with the building of such scores, we can approximate these discrete utility functions by the following formula:

$$S_j^N(x) := \frac{\text{ord}_j(x) - 1}{s_j - 1} \in [0,1] \quad \forall j \in \mathcal{J}, \forall x \in A.,$$

where $\text{ord}_j : A \to \{1, \ldots, s_j\}$ is a mapping defined by $\text{ord}_j(x) = r \iff x_j = g_r^j$. $S_j^N(x)$ does not represent a real utility and probably does not correspond to the utility the DM has in mind. We therefore continue to call it a score.

Finally, we would like to point out a particular situation, where the DM considers that any possible alternative which can be built out of the evaluation scales is a potential alternative. In this case, A equals the set of all possible alternatives which can be built from the sets $X_j, j \in \mathcal{J}$, i.e. $A = \prod_{j=1}^n X_i$. The partial net score formula (12.1) then becomes

$$S_j(x) = q(\frac{2\text{ord}_j(x) - 1}{s_j} - 1) \quad (x \in X, j \in \mathcal{J}). \tag{12.3}$$

These partial net scores are normalised according to the formula (12.2).

We now come to the crucial part of the aggregation of the normalised partial net scores of a given alternative x by means of a Choquet integral [4]. The advantage of this aggregator is mainly that it allows to deal with interacting (depending) points of view. According to what has been said in Section 4:

$$\mathcal{C}_v(S^N(x)) := \sum_{j=1}^{n} S_{(j)}^N(x)[v(A_{(j)}) - v(A_{(j+1)})],$$

where v is a fuzzy measure on $\mathcal{J}$; that is a monotone set function $v : 2^{\mathcal{J}} \to [0,1]$ fulfilling $v(\emptyset) = 0$ and $v(\mathcal{J}) = 1$. The parentheses used for indexes stand for a permutation on $\mathcal{J}$ such that

$$S_{(1)}^N(x) \leq \ldots \leq S_{(n)}^N(x),$$

and for any $j \in \mathcal{J}$, $A_{(j)}$ represents the subset $\{(j), \ldots, (n)\}$. The characterisation of the Choquet integral by Marichal [12, 13] clearly justifies the way the partial scores are aggregated. points of view.

The next step of this method is to assess the fuzzy measures in order to classify the alternatives of A. One can easily understand that it is impossible to ask the DM to give values for the $2^n - 2$ free parameters of the fuzzy measure v. Practically, the assessment of the fuzzy measures is done by asking the DM to provide a set of prototypes $P \subseteq A$ and their assignments to the given classes; that is a partition of P into prototypic classes $\{P_t\}_{t=1}^m$ where $P_t := P \cap Cl_t$ for $t \in \{1, \ldots, m\}$. The values of the fuzzy measure are then derived from this information as described hereafter.

We would like the Choquet integral to strictly separate the classes Cl_t. Therefore, the following necessary condition is imposed

$$\mathcal{C}_v(S^N(x)) - \mathcal{C}_v(S^N(x')) \geq \varepsilon \tag{12.4}$$

for each ordered pair $(x, x') \in P_t \times P_{t-1}$ and each $t \in \{2, \ldots, m\}$, where ε is a given strictly positive thershold.

Due to the increasing monotonicity of the Choquet integral, the number of separation constraints (12.4) can be reduced significantly. Thus, it is enough to consider *border elements* of the classes. To formalise this concept, we first define a dominance relation D (partial order) on X by

$$xDy \iff x_j \succsim_j y_j, \text{ for all } j \in \mathcal{J}.$$

As *upper border* of the prototypic class P_t we use the set of non-dominated alternatives of P_t defined by

$$ND_t := \{x \in P_t \text{ such that } \nexists x' \in P_t \setminus \{x\} : x'Dx\}.$$

Similarly, the *lower border* of the prototypic class is given by the set of non-dominating alternatives of P_t which is defined by

$$Nd_t := \{x \in P_t \text{ such that } \nexists x' \in P_t \setminus \{x\} : xDx'\}.$$

The separation conditions restricted to the prototypes of the subsets $ND_t \cup Nd_t$, $t \in \{1, \ldots, m\}$ put together with the monotonicity constraints on the fuzzy measure, form a linear program [16] whose unknowns are the capacities $v(S)$, $S \subset \mathcal{J}$ and where ε is a non-negative variable to be maximised in order to deliver well separated classes.

We use the principle of parsimony for the resolution of this problem. If there exists a k-additive fuzzy measure v^*, k being kept as low as possible, then we determine the boundaries of the classes as follows:

- lower boundary of Cl_t: $z(t) := \min_{x \in Nd_t} \mathcal{C}_{v^*}(S^N(x))$;
- upper boundary of Cl_t: $Z(t) := \max_{x \in ND_t} \mathcal{C}_{v^*}(S^N(x))$.

At this point, any alternative $x \in A$ can be classified in the following way:

- x is assigned to class Cl_t if $z_t \leq \mathcal{C}_{v^*}(S^N(x)) \leq Z_t$;
- x is assigned to class $Cl_t \cup Cl_{t-1}$ if $Z_{t-1} < \mathcal{C}_{v^*}(S^N(x)) < z_t$.

A final step of the classical TOMASO method concerns the evaluation of the results and the interpretation of the behavior of the Choquet integral. The meaning of the values $v(T)$ is not clear to the DM. They don't immediatly indicate the global importance of the points of view, nor their degree of interaction. It is possible to derive some indexes from the fuzzy measure which are helpful to interpret its behavior. Among them, the TOMASO method proposes to have a closer look at the importance indexes [32] and the interaction indexes [19]. We present the calculation of these indexes in Section 6.3.

6.2 An Alternate Way

It may happen that the linear program described in Subsection 6.1 has no solution. This occurs when the prototypic elements violate the axioms that are imposed to produce a discriminant function of Choquet type [13, 35], in particular the triple cancellation axiom.

In such a case, and in order to present a solution to the DM, we suggest to find a fuzzy measure by solving the following quadratic program

$$\min \sum_{x \in \cup_{t \in \{1,\ldots,m\}} \{ND_t \cup ND_t\}} [\mathcal{C}_v(S^N(x)) - y(x)]^2,$$

where the unkowns are

- the capacities $v(S)$ which determine the fuzzy measure;
- some global evaluations $y(x)$ for each $x \in \cup_{t \in \{1,\ldots,m\}} \{ND_t \cup Nd_t\}$.

The capacities $v(S)$ are constrained by the monotonicity conditions (as previously shown in Section 6.1). The global evaluations $y(x)$ must verify the classification imposed by the DM. In other words, for every ordered pair $(x, x') \in Nd_t \times ND_{t-1}$, $t \in \{2, \ldots, m\}$ the condition $y(x) - y(x') \geq \varepsilon'$, $\varepsilon' > 0$ must be satisfied.

Intuitively, for a given alternative $x \in A$, its Choquet integral $\mathcal{C}_v(S^N(x))$ should be as close as possible to the global evaluation $y(x)$, without being constrained by monotonicity conditions which might violate the triple cancellation axiom for example. On the other hand, the evaluation $y(x)$ is constrained by the these conditions derived from the original classification given by the DM on the prototypes.

Unlike the method described in Section 6, in this case, ε' plays the role of a parameter, which needs to be fixed by the DM. As previously, we use the principle of parsimony when searching for a solution (keep k as low as possible;

at worst k equals the number of points of view). A correct choice of ε' remains one of the main challenges of our future research. It is clear that ε' has to be chosen in $]0, 1/n[$.

As in the classical method, the next step is to determine the structure of the classes. We determine an assignment for every alternative of X in terms of intervals of contiguous classes on the basis of the information provided by the Choquet integrals related to the prototypes of $P \subseteq A$.

First of all, let us suppose that $S^N(x^-) := (0, \dots, 0)$ is classified to the worst class, Cl_1 and that $S^N(x^+) := (1, \dots, 1)$ is classified to the best class, Cl_m.

To each assignment $I(x)$ correspond a lower class label $\underline{l}(x)$ and an upper class label $\overline{l}(x)$, $\underline{l}, \overline{l} \in \mathcal{J}$. We say that the alternative $x \in X$ is *precisely assigned* to $Cl_{l(x)}$ if for the assignment $I(x)$ we have $\underline{l}(x) = \overline{l}(x) =: l(x)$. Else, the alternative x is said to be *ambiguously assigned* to the interval of labels $I(x) = [\underline{l}(x), \overline{l}(x)]$. The *degree of the assignment* corresponds to the number of contiguous classes contained in $I(x)$, $d(x) = \overline{l}(x) - \underline{l}(x) + 1$.

The assignments are done according to the procedure described hereafter. Starting from the prototypes $x \in P$, their Choquet integrals $\mathcal{C}_v(S^N(x))$ and their original classification label $Cl(x)$ (according to the DM's choice), we define for every $u \in [0, 1]$,

$$m(u) = \max_{x \in P: \mathcal{C}_v(S^N(x)) \leq u} Cl(x), \text{ and}$$

$$M(u) = \min_{x \in P: \mathcal{C}_v(S^N(x)) \geq u} Cl(x).$$

m (resp. M) is a right (resp. left) continuous stepwise function of argument u with values belonging to the discrete finite set $\mathcal{J}$.

We now define for each $u \in [0, 1]$ an interval of contiguous classes $I(u) = [\underline{l}(u), \overline{l}(u)]$ where

$$\underline{l}(u) = \min\{m(u), M(u), \}$$
$$\overline{l}(u) = \max\{m(u), M(u)\}.$$

Obviously $\underline{l}(u) \leq \overline{l}(u)$ and due to monotonicity of m and M we have: $\underline{l}(u) \leq \underline{l}(v), \overline{l}(u) \leq \overline{l}(v), \forall u, v \in [0, 1]$ with $u \leq v$.

The interval $[0, 1]$ is partitioned into (closed, semi-open or open) intervals $I_s, s = 1, \dots, S$, and each of those intervals of $[0, 1]$ receives an assignment of the type $[\underline{l}(s), \overline{l}(s)]$ (or semi-open or open) in such a way that: if $u, v \in [0, 1], u \leq v$ and if u is assigned to $I_r := [\underline{l}(r), \overline{l}(r)]$ and v is assigned to $I_{r'} := [\underline{l}(r'), \overline{l}(r')]$ then $\underline{l}(r) \leq \underline{l}(r')$ and $\overline{l}(r) \leq \overline{l}(r')$.

Moreover if $u = \mathcal{C}_v(S^N(x)), x \in P$ then $\underline{l}(u) \leq Cl(x) \leq \overline{l}(u)$. This means that each prototype is *correctly classified*, possibly with ambiguity if $d(x) \geq 1$.

The assignment of a prototype a to the intervals of classes leads now to two scenarios:

- a is assigned to a single class (interval of length 0) which corresponds to the original class decided by the DM
- a is assigned to an interval of classes and the original class decided by the DM belongs to this interval.

The quality of a model (classifier) depends on different ratios. A good model has the following *natural* properties:

- a simple model according to parsimony (low k);
- a high number of precise assignments of the elements of P;
- a low number of ambiguous assignments of the elements of P (and the lower the degree of the assignment, the better the model)

For a given ε', the DM has to select a model (k) which seems the best compromise to him in terms of the previously described assignments. The simplest additive model ($k = 1$) can in certain situations be this ideal compromise between simplicity and quality. But in more complex problems, k has to be increased in order to obtain a satisfying number of precisely assigned prototypes.

The next Section briefly presents some indexes (importance, interaction) which give indications on the behaviour of the fuzzy measure.

6.3 Behavioral Analysis of Aggregation

Now that we have a sorting model for assigning alternatives to classes (based on the linear program or the quadratic program), an important question arises: How can we interpret the behavior of the Choquet integral or that of its associated fuzzy measure? Of course the meaning of the values $v(T)$ is not always clear for the DM. These values do not give immediately the global importance of the points of view, nor the degree of interaction among them.

In fact, from a given fuzzy measure, it is possible to derive some indexes or parameters that will enable us to interpret the behavior of the fuzzy measure. These indexes constitute a kind of *id card* of the fuzzy measure. In this Section, we present two types of indexes: importance and interaction. Other indexes, such as tolerance and dispersion, were proposed and studied by Marichal [12, 14].

6.3.1 Importance Indexes. The overall importance of a point of view $j \in \mathcal{J}$ in a decision problem is not solely determined by the value of $v(\{j\})$, but also by all $v(T)$ such that $j \in T$. Indeed, we may have $v(\{j\}) = 0$, suggesting a priori that element j is unimportant, but it may happen that for many subsets $T \subseteq \mathcal{J}$, $v(T \cup \{j\})$ is much greater than $v(T)$, suggesting that j is actually an important element in the decision.

Shapley [32] proposed in 1953 a definition of a coefficient of importance, based on a set of reasonable axioms. The *importance index* or *Shapley value* of point of view j with respect to v is defined by:

$$\phi(v, \{j\}) := \sum_{T \subseteq \mathcal{J} \setminus \{j\}} \frac{(n - |T| - 1)!\,|T|!}{n!} [v(T \cup \{j\}) - v(T)]. \quad (12.5)$$

This index is a fundamental concept in game theory and it expresses a power index. It can be interpreted as a weighted average value of the marginal contribution $v(T \cup \{j\}) - v(T)$ of element j alone in all combinations. To make this clearer, we rewrite the index as follows:

$$\phi(v, \{j\}) = \frac{1}{n} \sum_{t=0}^{n-1} \frac{1}{\binom{n-1}{t}} \sum_{\substack{T \subseteq \mathcal{J} \setminus \{j\} \\ |T| = t}} [v(T \cup \{j\}) - v(T)].$$

Thus, the average value of $v(T \cup \{j\}) - v(T)$ is computed first over the subsets of same size t and then over all the possible sizes. Consequently, the subsets containing about $n/2$ points of view are the less important in the average, since they are numerous and a same point of view j is very often involved into them.

The use of the Shapley value in multicriteria decision making was proposed in 1992 by Murofushi [19]. It is worth noting that a basic property of the Shapley value is

$$\sum_{j=1}^{n} \phi(v, \{j\}) = 1.$$

Note also that, when v is additive, we clearly have $v(T \cup \{j\}) - v(T) = v(\{j\})$ for all $j \in \mathcal{J}$ and all $T \subseteq \mathcal{J} \setminus \{j\}$, and hence

$$\phi(v, \{j\}) = v(\{j\}), \quad j \in \mathcal{J}. \quad (12.6)$$

If v is non-additive then some points of view are dependent and (12.6) generally does not hold anymore. This shows that it is useful to search for a coefficient of overall importance for each point of view.

6.3.2 Interaction Indexes. A further interesting concept is that of *interaction* among points of view. We have seen that when the fuzzy measure is not additive then some points of view interact. Of course, it would be interesting to appraise the degree of interaction among any subset of points of view.

Consider first a pair $\{i, j\} \subseteq \mathcal{J}$ of points of view. It may happen that $v(\{i\})$ and $v(\{j\})$ are small and at the same time $v(\{i, j\})$ is large. The Shapley index $\phi(v, \{j\})$ merely measures the average contribution that point of view j brings to all possible combinations, but it gives no information on the phenomena of interaction existing among points of view.

Clearly, if the marginal contribution of j to every combination of points of view that contains i is greater (resp. less) than the marginal contribution of j to the same combination when i is excluded, the expression

$$[v(T \cup \{i,j\}) - v(T \cup \{i\})] - [v(T \cup \{j\}) - v(T)]$$

is positive (resp. negative) for any $T \subseteq \mathcal{J} \setminus \{i,j\}$. We then say that i and j positively (resp. negatively) interact.

This latter expression is called the *marginal interaction* between i and j, conditioned to the presence of elements of the combination $T \subseteq \mathcal{J} \setminus \{i,j\}$. Now, an interaction index for $\{i,j\}$ is given by an average value of this marginal interaction. Murofushi and Soneda [19] proposed in 1993 to calculate this average value as for the Shapley value. Setting

$$(\Delta_{ij}\, v)(T) := v(T \cup \{i,j\}) - v(T \cup \{i\}) - v(T \cup \{j\}) + v(T),$$

the *interaction index* of points of view i and j related to v is then defined by

$$I(v, \{i,j\}) := \sum_{T \subseteq N \setminus \{i,j\}} \frac{(n - |T| - 2)!\,|T|!}{(n-1)!}\,(\Delta_{ij}\, v)(T). \tag{12.7}$$

It should be mentioned that, historically, the interaction index (12.7) was first introduced in 1972 by Owen (see Eq. (28) in [21]) in game theory to express a degree of complementarity or competitiveness between elements i and j.

6.4 Interpretation of the Behaviour of the Fuzzy Measure

In this Section we briefly show the main advantage to use a Choquet integral rather than the weighted sum as a discriminant function. We therefore take the simple case of two points of view, which can be represented in a plane. Figure 12.2 presents 5 possible ranges of values for the weights v and the corresponding structures of the limits of the classes. One can see that the main difference between the classical weighted sum and the Choquet integral is the greater flexibility of the borders of the classes. The Choquet integral creates piecewise linear borders, which allows to build more precise classes. The different possibilities are summarised by the following list.

- I: $v(1) + v(2) < v(12)$: synergy
- II: $v(1) + v(2) > v(12)$: redundancy
- III: $v(1) + v(2) = v(12) = 1$: additivity
- IV: $v(1) = v(2) = 0$: limit case; maximal synergy
- V: $v(1) = v(2) = 1$: limit case; maximal redundancy

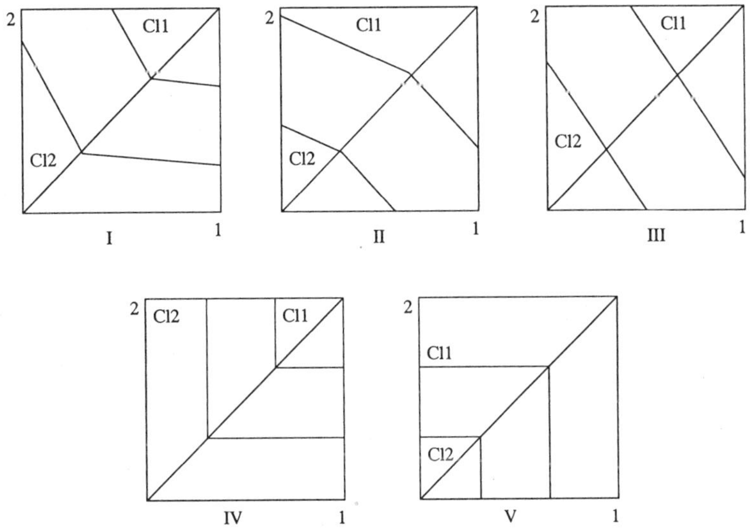

Figure 12.2. Interpretation of the discriminant functions.

In [9] the authors give an interpretation to the first two cases. In case of synergy, although the importance of a single criterion for the decision is rather low, the importance of the pair is large. The criteria are said to be *complementary*. In case of redundance, or negative synergy, the union of criteria does not bring much, and the importance of the pair might be roughly the same as the importance of a single criterion.

The limit case (IV) occurs for maximal synergy. In that case, the Choquet integral corresponds to the aggregation by the min function. Maximal redundancy occurs for case (V), where the Choquet integral is the max function.

In case the number of points of view is larger than two, it becomes quite hard to represent the problem. Nevertheless, the previous short example helps to understand how the borders of the classes are built in such more general examples.

6.5 The Software TOMASO

In this short part of the chapter we briefly present the key characteristics of the software TOMASO. It can be downloaded on http://patrickmeyer.tripod.com. It is an implementation of the algorithms which were presented previously. Its name stands for "Tool for Ordinal MultiAttribute Sorting and Ordering". It is written in Visual Basic and uses two external solvers: a free linear program solver (lp_solve 3.0, ftp://ftp.ics. ele.tue.nl/pub/lp_solve/, released under the

LGPL license), and a non free quadratic program solver (bpmpd, free trial version at http://www. sztaki.hu/ meszaros/bpmpd/).

It is still under development and many improvements are added on a regular basis. The general steps of the software are outlined hereafter:

- Loading of the ordinal data;
- Choice of a scoring method according to the problem's specificities and calculation of the normalised partial net scores;
- Definition of the prototypes by the DM;
- Search for a fuzzy measure (either by the linear program, or the quadratic program);
- Analysis of the results (classes, Shapley indexes, interaction indexes, accuracies, ...).

A detailed description of the software can be obtained from the authors.

6.6 Testing the Method on Two Problems

In this Section, we apply the previously presented method on two particular problems. The following part describes briefly the two problems. Then they are analysed by the TOMASO method.

6.6.1 Description of the Problems.

The Students Problem This small example clearly illustrates the procedure when no solution can be found to the linear program. We consider a set of 8 students evaluated on 2 courses (C1, C2). For each matter, the evaluation scale has 10 ordered qualitative levels (1-10). In total, this makes 100 possible different ratings. Besides, for each student, the DM has given a global evaluation on a 6-levelled qualitative ordinal scale (the classes): (very good (6)> good (5)> above average (4)> below average (3)> bad (2)> very bad(1)). A summary of the problem is given in Table 12.3.

The Noise Annoyance Problem This real-life example concerns noise annoyance caused by different sources. Details on these data can be found in [2] and [33]. It was obtained by a survey performed on 2661 persons (alternatives). They were asked to give an estimation of their annoyance level (not at all annoyed (n) $\preceq$ slightly annoyed (s) $\preceq$ moderately annoyed (m) $\preceq$ very annoyed (v) $\preceq$ extremely annoyed(e)) on 21 different potential noise sources (points of view) (noise annoyance caused by road traffic, by rail traffic, ...). This ordering

Table 12.3. Profiles of the students.

student	profile	class
A	(7,5)	2
B	(6,6)	1
C	(7,7)	3
D	(6,8)	4
A'	(10,7)	6
B'	(8,8)	5
C'	(10,5)	3
D'	(8,6)	4

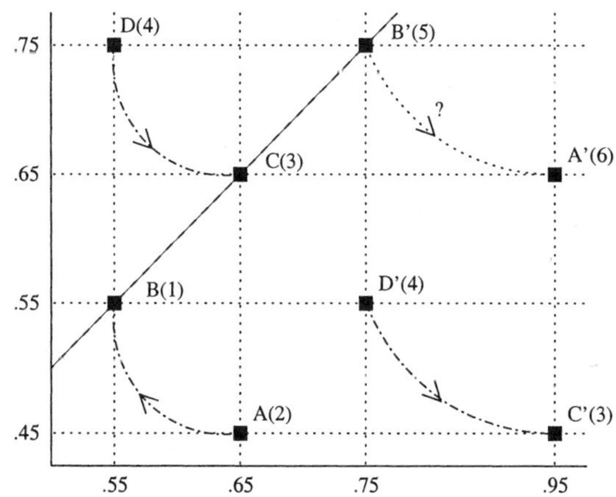

Figure 12.3. Representation of the students problem.

and the exact wording of the questions is in accordance with international standards. Besides, the questionned persons had to give an overall noise annoyance level on the same scale.

The original dataset contains 2661 alternatives and 21 points of view. But unfortunately, its structure is not proper for the TOMASO method as it contains a lot of inconsistencies (aDb but a is in a worse class than b). For the purpose of this chapter, we restrict ourselves to a consistent subset of 155 alternatives and 6 points of view (road traffic (cars, busses,...), air traffic, truck loading and unloading, factories, dance halls, agricultural equipment).

The goal is therefore to find a Choquet integral as a discriminant function which can reproduce the overall noise annoyance level by using the separate noise annoyances as an input.

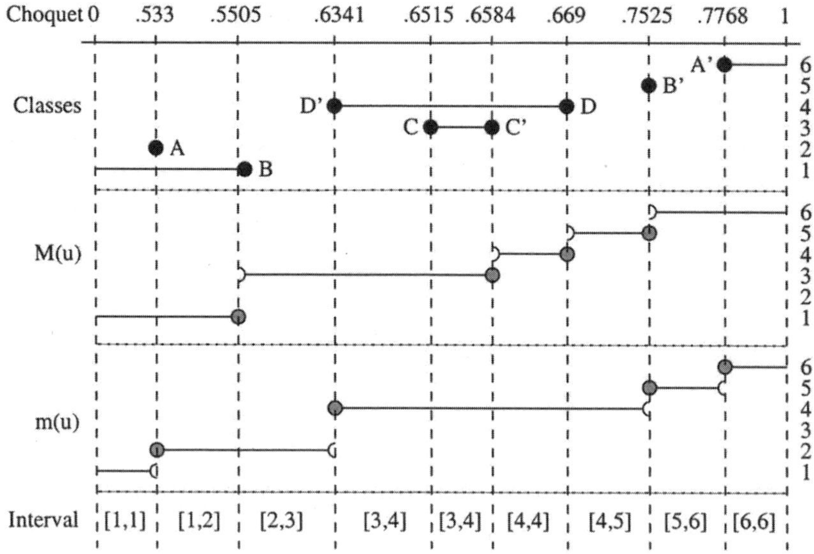

Figure 12.4. Classes for $\varepsilon = 0.1, k = 1$.

6.6.2 Solving the Students Problem. For this problem, the scales on the points of view are quite rich, but we only have a few alternatives. We therefore suppose that information can be extracted from X. This means that any possible student which can be built out of the evaluation scales on C_1 and C_2 is a potential alternative. A representation of the 2-dimensional problem is given on Figure 12.3. It helps to understand why the linear program has no solution. If the triple cancellation property [35] is violated, there exists no Choquet integral which satisfies the constraints imposed by the classification of the prototypes. If triple cancellation was verified in this example, we would have:

$$\left\{ \begin{array}{c} \textsc{Class}(B) \leq \textsc{Class}(A) \\ \& \\ \textsc{Class}(D) \geq \textsc{Class}(C) \\ \& \\ \textsc{Class}(D') \geq \textsc{Class}(C') \end{array} \right\} \Rightarrow \textsc{Class}(B') \geq \textsc{Class}(A'),$$

where $\textsc{Class}(X)$ stands for the index of the class to which X belongs (the higher the better). But in this particular example, we clearly have $\textsc{Class}(D') \leq \textsc{Class}(C')$. Therefore, no solution can be found to the linear program. In other words, this problem cannot be described by the classical TOMASO method by means of a Choquet integral. We therefore go over to the method based on the quadratic probgram.

In this case, a solution can be found for different values of ε and k. For $k = 1$, the best solution is found with $\varepsilon' = 0.01$. 3 out of 8 alternatives are precisely assigned to their classes. The other 5 elements are ambiguously assigned with a degree 2.

Figure 12.4 explains how the assignment described in 6.2 to the intervals of classes works for this particular simplest model ($k = 1$). The original classes, which are shown by means of the 8 prototypes look somewhat chaotic. Both functions $m(u)$ and $M(u)$ help to build the final ordered intervals of classes. It shows that the alternatives A', B' and D are assigned precisely.

A better, but more complex model can be found for $k = 2$. A good solution can be found for $\varepsilon' = .10$. 6 out of 8 alternatives are assigned precisely, whereas 2 out of 8 students (C' and D') are assigned ambiguously with degree 2. This shows the better performance of a Choquet integral over the weighted sum aggregator. A representation of the solution in this case ($k = 2, \varepsilon = .10$) is given on Figure 12.5.

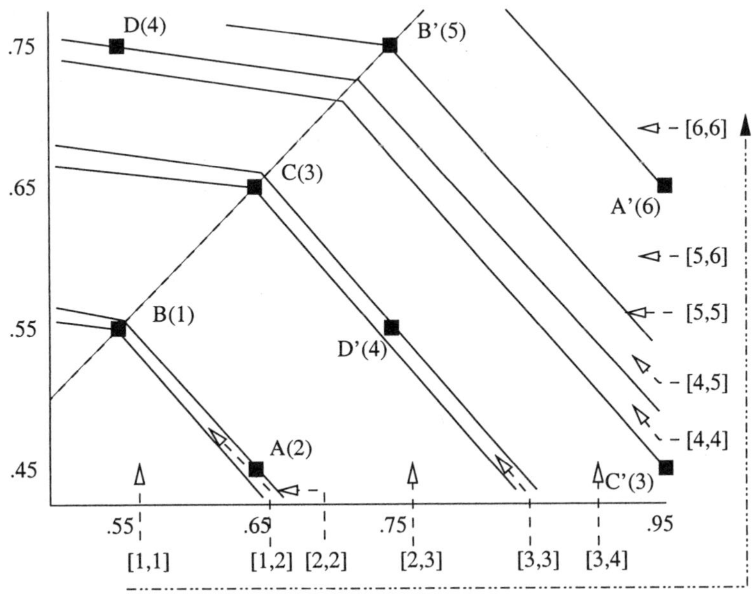

Figure 12.5. Classes for $\varepsilon = 0.1, k = 1$.

We observe that the borders of the classes are piecewise linear, and that this allows to cope with a larger set of problems. We can also observe the overlapping zone between classes 3 and 4, which induce the ambiguous assignments of C' and D'.

To conclude this example, in Table 12.4 we present the importance indexes for this example, in both models. We can see that the values are quite identical; the order on the importance of the criteria is clearly respected.

Table 12.4. Importance indexes for the students problem.

Model	Course 1	Course 2
$k = 1$	.414	.586
$k = 2$	.325	.675

6.6.3 Solving the Noise Annoyance Problem. Let us get back to the second problem described in 6.6.1. The data set of 155 prototypes violates the triple cancellation axiom. Therefore, as no solution can be found to the linear program, we switch to the resolution of the quadratic program.

This problem is not adapted for the comparison philosophy for the scores. This is clearly not a decision problem with a single DM. In fact, each of the 155 decisions has been taken by a separate person. One of these persons could not compare his profile to the other ones, before giving a global noise annoyance level. Furthermore, as we cannot ask these 155 people to give us hints on the shape of the discrete utility functions linked to the evaluation scales, we have no other option than considering formula (12.2).

Let us first start with the best possible solution that we can find, for $k = 6$, which means a non-additive fuzzy measure. Quite similar solutions exist for values of ε' between 0.05 and 0.18. We chose an average value of $\varepsilon' = 0.1$.

Table 12.5. Global accuracy.

$\varepsilon' = 0.1, k = 6$	
precise	ambiguous $d = 2$
46.45%	53.55%

We can also analyse the accuracy of this discriminant function for each class separately. Table 12.6 shows its performance for each of the 5 ordered classes.

We can see that the class *very annoyed* is nearly unpredictable in a precise way with this discriminant function. This is due to the fact that it overlaps strongly with the classes *extremely annoyed* and *moderately annoyed*. This phenomenon can be observed on Figure 12.6. It represents the assignments to the 5 ordered

Table 12.6. Per class accuracy: Precise assignments.

		$\varepsilon' = 0.1, k - 6$		
extremely	very	moderately	slightly	not at all
45.83%	2.33%	66.67%	69.05%	71.43%

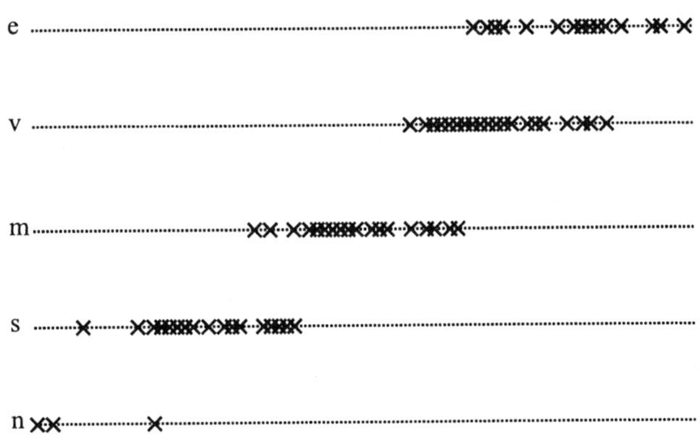

Figure 12.6. Visual representation of the classes, $k = 6$, $\varepsilon' = 0.1$.

classes. A cross can represent more than one alternative (if they have equal Choquet integrals).

The assignments of the alternatives to the classes according to their Choquet integral is shown in Table 12.7.

Table 12.7. Assignments of the elements to intervals of classes.

$\varepsilon' = 0.1, k = 6$	
Interval of Choquets	Interval of classes
$[0, 0.0375[$	$[1, 1]$
$[0.0375, 0.1011]$	$[1, 2]$
$]0.1011, 0.1903[$	$[2, 2]$
$[0.1903, 0.2182]$	$[2, 3]$
$]0.2182, 0.3139[$	$[3, 3]$
$[0.3139, 0.3642[$	$[3, 4]$
$[0.3642, 0.3754]$	$[4, 4]$
$]0.3754, 0.4565]$	$[4, 5]$
$]0.4565, 1]$	$[5, 5]$

The importance indexes of the 6 criteria are given in Table 12.8.

Table 12.8. Importance indexes.

		$\varepsilon' = 0.1, k = 6$			
street	air	truck	factory	dancing	agriculture
.166	.158	.164	.0938	.272	.146

Taking into account our scoring method, we can state that the noise annoyance caused by dancing halls is the most important, followed by street noises, truck loading, air traffic disturbances, agricultural annoyances and finally factory noises.

One should note that the global results of Table 12.5 are not too bad. They should not be misinterpreted: there are no erroneous classifications, but only ambiguous ones. In order to obtain a classification of the prototypes into single classes, we suggest to use a l-nearest neighbourhood algorithm to force an assignment. Each prototype $a \in P$ and its Choquet integral is presented to the remaining set $P \backslash \{a\}$ of elements and their Choquet integral. The l closest neighbours of a in terms of the Choquet integral are then selected. Among these l elements, we search for the original class (as decides by the DM) which appears most often. In case of identity, the class is chosen randomly among the equally present classes. a is then assigned to this majority class. A global accuracy and a weighted accuracy are then computed. The global accuracy is simply the ratio of correctly assigned alternatives over the total number of alternatives. The weighted accuracy is the average of the separate accuracies of each class.

Figure 12.7 shows these accuracies for different values of l (let us notice here that the axis for the separate class accuracies is on the left side of the figure, and the axis for both the global and the weighted accuracies is on the right side of the figure).

Let us make a few observations. Classes 1 (not at all annoyed) and 5 (extremely annoyed) only contain a few alternatives. A consequence is that if we select l too high, no alternative will be assigned to one of these two classes. As a consequence, the weighted accuracy strongly depends on the right choice of l. On Figure 12.7 one can see this influence for both of these classes. As an example, above a value of $l = 11$, the accuracy of class 1 is equal to 0. The choice of l remains a critical one on which the resulting accuracies strongly depend, see Table 12.9

These results can be compared to those obtained by the methods described in [33, 34] and [2]. On this same data set of 155 alternatives and 6 points of view, with a genetic optimization of a Choquet integral with a possibility measure (1-maxitive [18]), their performance is 76.77% for the global accuracy and 80.57% for the weighted accuracy. Globally, the results are comparable.

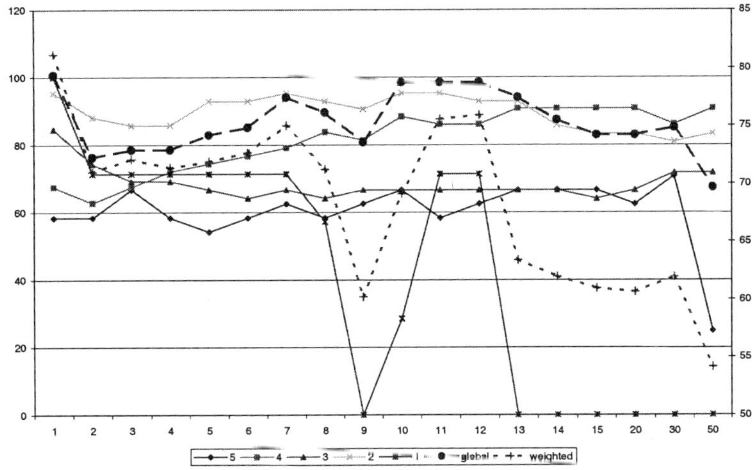

Figure 12.7. Results for the l-nearest neighbour algorithm, $\varepsilon' = 0.1$, $k = 6$.

Table 12.9. Global and weighted accuracy in %.

	$\varepsilon' = 0.1, k = 6$	
l	Global	Weighted
1	79.35	81.12
2	72.25	71.00
3	72.90	72.09
4	72.90	71.36
5	74.19	71.91
6	74.84	72.69
7	77.42	74.98
8	76.13	71.23
9	73.55	60.21
10	78.71	69.10
11	78.71	75.54
12	78.71	75.90
average	75.81	72.26

But unfortunately, we have worse results on the weighted accuracy. This is due to the nonuniform distribution of the alternatives among the 5 ordered classes, which is a big disadvantage for the use of the l-nearest neighbour method for the forced classification.

We would like to point out that very similar results (accuracies, shapley indexes) can be obtained with a 3-additive fuzzy measure ($k = 3$). For $k < 3$, the discriminating power of the Choquet integral is quite low for this particular example. In particular, the number of precisely assigned alternatives (before the l-NN procedure) becomes very low. This means that the classes overlap a lot.

7. The Choice Problem

In this Section we consider a way to select a subset of alternatives to consider as a good choice.

Consider a binary relation R whose credibility is evaluated as follows:

$R(x,y) = C_v[R_1(x,y),\ldots,R_i(x,y),\ldots,R_n(x,y)] \in [0,1]$, for all $x,y \in A$. In the sequel we will only use the ordering of $R(x,y)$ and not their cardinality and we will obtain a L-valued binary relation R (see [1]).

For all $x,y \in A$, $R(x,y)$ belongs to a finite set $L = \{c_0 = 0, c_1, \ldots, c_m = 0.5, \ldots, c_{2m} = 1$ that constitutes a $(2m+1)$-element chain $c_0 \prec \ldots \prec c_{2m}$. $R(x,y)$ may be understood as the credibility that "x is at least as good as y". The set L is built using the values of R taking into consideration an antitone unary contradiction operator $\neg$ such that $\neg c_l = c_{2m-l}$ for $l = 0,\ldots,2m$.

If $R(x,y)$ is one of the elements of L, then automatically, $\neg R(x,y)$ belongs to L. We call such a relation an L-valued binary relation.

We denote $L^{\succ m} := \{c_{m+1},\ldots,c_{2m}\}$ and $L^{\prec m} := \{c_0,\ldots,c_{m-1}\}$.

If $R(x,y) \in L^{\succ m}$ we say that the proposition "$(x,y) \in R$" is L-true. If however $R(x,y) \in L^{\prec m}$, we say that the proposition is L-false. If $R(x,y) = c_m$, the median level (a fix point of the negation operator), then the proposition "$(x,y) \in R$" is L-undetermined.

In the classical case, where R is a crisp binary relation we define a digraph $G(A,R)$ with vertex set A and arc family R. A choice in $G(A,R)$ is a non-empty set Y of A.

A (dominant) kernel is a choice that is stable in G, i.e. $\forall x \neq y \in Y, (x,y) \notin R$ and dominant, i.e. $\forall x \notin Y, \exists y \in Y$ such that $(x,y) \in R$.

We now denote $G^L = G^L(A,R)$ a digraph with vertices set A and a valued arc family that corresponds to the L-valued binary relation R.

We define the level of stability qualification of subset Y of X as

$$\Delta^{\text{sta}}(Y) = \begin{cases} c_{2m} \text{ if } Y & \text{is a singleton,} \\ \min_{y\neq x} \max_{x\neq y}\{\neg R(x,y)\} & \text{otherwise;} \end{cases}$$

and the level of dominance qualification of Y as

$$\Delta^{\text{dom}}(Y) = \begin{cases} c_{2m} & \text{if } Y = A, \\ \min_{x\notin Y} \max_{y\in Y} R(x,y) & \text{otherwise.} \end{cases}$$

Y is considered to be an L-good choice, i.e. L-stable and L-dominant, if $\Delta^{\text{sta}}(Y) \in L^{\succ m}$ and $\Delta^{\text{dom}}(Y) \in L^{\succ m}$. Its qualification corresponds to $Q^{\text{good}}(L) = \min\{\Delta^{\text{sta}}(Y), \Delta^{\text{dom}}(Y)\}$.

We denote $C^{\text{good}}(G^L)$ the possibly empty set of L-good choices in G^L.

The determination of this set in an NP-complete problem even if, following a result of Kitainik [11], we do not have to enumerate the elements of the power set of A, but only have to consider the kernels of the corresponding crisp strict median-level cut relation $R^{\succ m}$ associated to R, i.e. $(x, y) \in R^{\succ m}$ if $R(x, y) \in L^{\succ m}$.

As the kernel in $G(X, R^{\succ m})$ is by definition a stable and dominant crisp subset of A, we consider the possibly empty set of kernels of $G^{\succ m} = G^{\succ m}(A, R^{\succ m})$ which we denote $C^{\text{good}}(G^{\succ m})$.

Kitiainik proved that

$$C^{\text{good}}(G^L) \subseteq C^{\text{good}}(G^{\succ m}).$$

The determination of crisp kernels has been extensively described in the literature (see, for example [31]) and the definition of $C^{\text{good}}(G^L)$ is reduced to the enumeration of the elements of $C^{\text{good}}(G^{\succ m})$ and the calculation of their qualification.

The decision maker might also be interested in bad choices. These choices correspond to absorbent kernels with a qualification greater than c_m. In the classical Boolean framework (see [31]) an (absorbent) kernel is a choice that is stable and absorbent, i.e. $\forall x \notin Y, \exists y \in Y$ such that $(x, y) \in R$. As $(x, y) \in R$ is equivalent to $(y, x) \in R^t$, where matrix R^t represents the transpose of matrix R, all the results obtained for dominant kernels can be immediately transposed for absorbent kernels and definitions like Δ^{bad} and Q^{bad} are obviously and straightforwardly obtained from Δ^{good} and Q^{good}.

Indeed the level of absorbance qualification of Y is defined as

$$\Delta^{\text{abs}}(Y) = \begin{cases} c_{2m} & \text{if } Y = A, \\ \min_{x \notin Y} \max_{y \in Y} R(y, x) & \text{otherwise.} \end{cases}$$

In order to determine a unique rational choice (if any), we first compute dominant kernels in G^L (see [1, 31]) and determine their qualification as not being bad choices, i.e. $\neg Q^{\text{bad}}(Y)$ where $Q^{\text{bad}}(Y) = \min(\Delta^{\text{sta}}(Y), \Delta^{\text{abs}}(Y))$.

The selection is based on

$$\max Q^{\text{good}}(Y).$$

If more than one candidate remain, other discriminant functions may be added as minimal absorbancy, lowest cardinality, ...

8. Conclusion

In this chapter we have presented a few approaches to multiple criteria decision aiding. In particular, we have focussed on fuzzy methods for choice, sorting and ordering. We have also described in details the sorting procedure TOMASO which can deal with interacting criteria. Some tests on examples (theoretical and real-life) have shown the interestingness of this method. Further investigations have to be done on the validation of the models. We intend to implement a cross-validation procedure to make stability tests on the data and the method.

References

[1] R. Bisdorff and M. Roubens. On defining and computing fuzzy kernels and l-valued simple graphs. In D. Ruan, P. D'Hondt, P. Govaerts, and E. Kerre, editors, *Intelligent Systems and Soft Computing for Nuclear Science and Industry*, pages 113–122. World Scientific, Singapore, 1996.

[2] D. Botteldooren and A. Verkeyn. Fuzzy models for accumulation of reported community noise annoyance from combined sources. *Journal of the Acoustical Society of America*, 112:496–1508, 2002.

[3] J.P. Brans and P. Vincke. A preference ranking organization method. *Management Science*, 31:647–656, 1984.

[4] G. Choquet. Theory of capacities. *Annales de l'Institut Fourier*, 5:131–295, 1953.

[5] D. Dubois and H. Prade. Weighted minimum and maximum in fuzzy set theory. *Information Science*, 39:205–210, 1986.

[6] J. Fodor, S.A. Orlovski, P. Perny, and M. Roubens. The use of fuzzy preference models in multiple criteria: Choice, ranking and sorting. In D. Dubois and H. Prade, editors, *Operations Research and Statistics*, volume 5 of *Handbooks of Fuzzy Sets*, pages 69–101. Kluwer Academic Publishers, Dordrecht, Boston, London, 1998. Chapter 3.

[7] J. Fodor, P. Perny, and M. Roubens. Decision making and optimization. In E. Ruspini, P. Bonissone, and W. Pedrycz, editors, *Handbook of Fuzzy Computation*, pages F.5.1:1–14. Institute of Physics Publication and Oxford University Press, Bristol, 1998. Chapter 5.1.

[8] M. Grabisch. k-order additive discrete fuzzy measure and their representation. *Fuzzy Sets and Systems*, 92:167–189, 1997.

[9] M. Grabisch and M. Roubens. Application of the Choquet integral in multicriteria decision making. In M. Grabisch, T. Murofushi, and M. Sugeno, editors, *Fuzzy Measures and Integrals*, pages 348–374. Physica Verlag, Heidelberg, 2000.

[10] S. Greco, B. Matarazzo, and R. Slowinski. Conjoint measurement and rough set approach for multicriteria sorting problems in presence of ordinal criteria. In A. Colorni, M. Paruccini, and B. Roy, editors, *A-MCD-A: Aide Multicritère à la Décision (Multiple Criteria Decision Aiding)*, pages 17–144. Joint Research Centre, Ispra, 2001. European Commission Report, EUR 19808 EN.

[11] L. Kitainik. *Fuzzy Decision Procedures with Binary Relations. Towards a Unified Theory*. Kluwer Academic Publishers, Dordrecht, 1993.

[12] J.-L. Marichal. *Aggregation Operators for Multicriteria Decision Aid*. PhD thesis, Institute of Mathematics, University of Liège, Liège, Belgium, 1998.

[13] J.-L. Marichal. An axiomatic approach of the discrete Choquet integral as a tool to aggregate interacting criteria. *IEEE Transactions on Fuzzy Systems*, 8:800–807, 2000.

[14] J.-L. Marichal. Behavioral analysis of aggregation in multicriteria decision aid. In J. Fodor, B. De Baets, and P. Perny, editors, *Preferences and Decisions under Incomplete Knowledge*, volume 51 of *Studies in Fuzziness and Soft Computing*, pages 153–178. Physica Verlag, Heidelberg, 2000.

[15] J.-L. Marichal. Aggregation of interacting criteria by means of the discrete Choquet integral. In T. Calvo, G. Mayor, and R. Mesiar, editors, *Aggregation Operators: New Trends and Applications*, volume 97 of *Studies in Fuzziness and Soft Computing*, pages 224–244. Physica Verlag, Heidelberg, 2002.

[16] J.-L. Marichal and M. Roubens. On a sorting procedure in the presence of qualitative interacting points of view. In J. Chojean and J. Leski, editors, *Fuzzy Sets and their Applications*, pages 217–230. Silesian University Press, Gliwice, 2001.

[17] L.Y. Maystre, J. Pictet, and J. Simos. *Méthodes Multicritères ELECTRE*. Presses polytechniques et universitaires romandes, Lausanne, 1994.

[18] R. Mesiar. Generalizations of k-order additive discrete fuzzy measures. *Fuzzy Sets and Systems*, 102:423–428, 1999.

[19] T. Murofushi and S. Soneda. Techniques for reading fuzzy measures (III): Interaction index. In *9th Fuzzy System Symposium, May 19-21 1993, Sapporo, Japan*, pages 693–696, 1993. In Japanese.

[20] S.V. Ovchinnikov and M. Roubens. On strict preference relations. *Fuzzy Sets and Systems*, 43:319–326, 1991.

[21] G. Owen. Multilinear extensions of games. *Management Science*, 18:64–79, 1972.

[22] P. Perny. *Modélisation, Agrégation et Exploitation de Préférences Floues dans une Problématique de Rangement: Bases Axiomatiques, Procédures et Logiciel*. PhD thesis, Université Paris-Dauphine, 1992. LAFORIA document 9.

[23] M. Roubens. Ordinal multiattribute sorting and ordering in the presence of interacting points of view. In D. Bouyssou, E. Jacquet-Lagrèze, P. Perny, R. Slowinski, Vanderpooten D., and P. Vincke, editors, *Aiding Decisions with Multiple Criteria: Essays in Honour of Bernard Roy*, pages 229–246. Kluwer Academic Publishers, Dordrecht, 2001.

[24] M. Roubens and P. Vincke. Fuzzy preferences in an optimisation perspective. In J. Kacprzyk and S.A. Orlovski, editors, *Optimization Models Using Fuzzy Sets and Possibility Theory*, pages 77–90. D. Reidel, Dordrecht, Boston, Lancaster, Tokyo, 1987.

[25] B. Roy. Classement et choix en présence de points de vue multiples (la méthode ELECTRE). *Revue d'Information et de Recherche Opérationnelle*, 8:57–75, 1968.

[26] B. Roy. ELECTRE III: Un algorithme de classement fondé sur une représentation floue des préférences en présence de critères multiples. *Cahiers du Centre d'Etudes en Recherche Opérationnelle*, 20(1):3–24, 1978.

[27] B. Roy. *Multicriteria Methodology for Decision Aiding*. Kluwer Academic Publishers, Dordrecht, 1996.

[28] B. Roy and P. Bertier. La méthode ELECTRE II. une application au media planning. In M. Ross, editor, *Operational Research 72*, pages 291–302. North Holland Publishing Co., 1973.

[29] B. Roy and D. Bouyssou. Aide à la décision fondée sur une PAMC de type ELECTRE. Technical Report 69, Université Paris-Dauphine, 1991. Document du LAMSADE.

[30] B. Roy and J.-M. Skalka. ELECTRE IS – Aspects méthodologiques et guide d'utilisation. Technical Report 30, Université Paris-Dauphine, 1985. Document du LAMSADE.

[31] G. Schmidt and T. Ströhlein. *Relations and Graphs – Discrete Mathematics for Computer Scientists*. Springer Verlag, Berlin, Heidelberg, New York, 1999.

[32] L.S. Shapley. A value for n-person games. In H.W. Kuhn and A.W. Tucker, editors, *Contributions to the Theory of Games, Vol. II*, volume 28 of *Annals of Mathematics Studies*, pages 307–317. Princeton University Press, Princeton, NJ, 1953.

[33] A. Verkeyn, D. Botteldooren, B. De Baets, and G. De Tre. Modeling annoyance aggregation with Choquet integrals. In B. De Baets, editor, *Proceedings of Eurofuse Workshop on Information Systems, September 23-25 2002, Varenna, Italy*, pages 259 – 264, 2002.

[34] A. Verkeyn, D. Botteldooren, D. De Baets, and G. de Tre. Sugeno integrals for the modelling of noise annoyance aggregation. In T. Bilgic and et al, editors, *IFSA 2003*, volume 2715 of *Lecture Notes in Artificial Intelligence*, pages 277–284. Springer Verlag, Berlin, 2003.

[35] P. Wakker. *Additive Representations of Preferences: A new Foundation of Decision Analysis*. Kluwer Academic Publishers, Dordrecht, Boston, London, 1989.

Chapter 13

DECISION RULE APPROACH

Salvatore Greco, Benedetto Matarazzo
Faculty of Economics, University of Catania
Corso Italia 55, 95129 Catania
Italy
{salgreco,matarazzo}@unict.it

Roman Słowiński
Institute of Computing Science, Poznań University of Technology, 60-965 Poznań and
Systems Research Institute, Polish Academy of Sciences, 01-447 Warsaw
Poland
Roman.Slowinski@cs.put.poznan.pl

Abstract We present the methodology of Multiple-Criteria Decision Aiding (MCDA) based on preference modelling in terms of "*if…, then …*" decision rules. The basic assumption of the decision rule approach is that the decision maker (DM) accepts to give preferential information in terms of examples of decisions and looks for simple rules justifying her decisions. An important advantage of this approach is the possibility of handling inconsistencies in the preferential information, resulting from hesitations of the DM. The proposed methodology is based on the elementary, natural and rational principle of dominance. It says that if action x is at least as good as action y on each criterion from a considered family, then x is also comprehensively at least as good as y. The set of decision rules constituting the preference model is induced from the preferential information using a knowledge discovery technique properly modified, so as to handle the dominance principle. The mathematical basis of the decision rule approach to MCDA is the Dominance-based Rough Set Approach (DRSA) developed by the authors. We present some basic applications of this approach, along with didactic examples whose aim is to show in an easy way how DRSA can be used in various contexts of MCDA.

Keywords: Dominance, rough sets, decision rules, multiple criteria classification, choice and ranking.

1. Introduction

Multiple-criteria decision support aims at givinge the decision maker (DM) a recommendation [51] in terms of the best actions (choice), or of the assignment of actions to pre-defined and preference-ordered classes (classification, called also sorting), or of the ranking of actions from the best to the worst (ranking). None of these recommendations can be elaborated before the DM provides some preferential information suitable to the preference model assumed.

There are two major preference models used until now in multiple-criteria decision analysis: Multi-Attribute Utility Theory (MAUT; see [40] and Chapter 7) and the outranking approach (see Chapter 4 and [50]). These models require specific preferential information more or less explicitly related to their parameters. For example, the DM is often asked for pairwise comparisons of actions from which one can assess the substitution rates for a MAUT model or the importance weights for an outranking model (see [5], [44]). This kind of preferential information seems to be close to the natural reasoning of the DM. She is typically more confident exercising her decisions than explaining them. The transformation of this information into MAUT or outranking models seems, however, less natural. According to Slovic [53], people make decisions by searching for *rules* which provide good justification of their choices. So, after getting the preferential information in terms of exemplary decisions, it would be natural to build the preference model in terms of "*if*..., *then* ..." rules. Examples of such rules are the following:

- "*if* maximum speed of car x is at least 175 km/h *and* its price is at most \$12000, *then* car x is comprehensively at least medium",
- "*if* car x is at least weakly preferred to car y with respect to acceleration *and* the price of car x is no more than slightly worse than that of car y, *then* car x is at least as good as car y".

The rules induced from exemplary decisions represent a preferential attitude of the DM and enable her understanding of the reasons of her preference. The acceptance of the rules by the DM justifies, in turn, their use for decision support. This is concordant with the principle of posterior rationality by March [43] and with aggregation-disaggregation logic by Jacquet-Lagrèze [39].

The set of decision rules accepted by the DM can be applied to a set of potential actions in order to obtain specific preference relations. From the exploitation of these relations, a suitable recommendation can be obtained to support the DM in decision problem at hand.

So, the preference model in the form of decision rules induced from examples fulfils both representation and recommendation tasks (see [51]).

The induction of rules from examples is a typical approach of artificial intelligence. This explains our interest in rough set theory [46, 47, 49, 54] which proved to be a useful tool for analysis of vague description of decision situations [48, 55]. The aim of rough set analysis is the explanation of the dependence between the values of some decision attributes, playing the role of "dependent variables", by means of the values of other condition attributes, playing the role of "independent variables". For example, in a diagnostic context, data about the presence of some diseases are given by decision attributes, while data about symptoms are given by condition attributes. An important advantage of the rough set approach is that it can deal with partly inconsistent examples, i.e. objects indiscernible by condition attributes but discernible by decision attributes (for example, cases where the presence of different diseases is associated with the presence of the same symptoms). Moreover, it provides useful information about the role of particular attributes and their subsets, and prepares the ground for representation of knowledge hidden in the data by means of "*if..., then ...*" decision rules relating values of some condition attributes with values of decision attributes (for example "if symptom A and B are present, then there is disease X").

For a long time, however, the use of the rough set approach and, in general, of data mining techniques, has been restricted to classification problems where the preference order of evaluations is not considered. Typical examples of such problems come from medical diagnostics. In this context, symptom A is not better or worse than symptom B, or disease X is not preferable to disease Y. It is thus sufficient to consider A as different from B, and X as different from Y. There are, however, situations, where discernibility is not sufficient to handle all relevant information. Consider, for example, two firms, α and β, evaluated for assessment of bankruptcy risk by a set of criteria including the "debt ratio" (total debt/total assets). If firm α has a low value of the debt ratio while firm β has a high value of the debt ratio, then, within data mining and classical rough set theory, α is different (discernible) from β with respect to the considered attribute (debt ratio). However, from the viewpoint of preference analysis and, say, bankruptcy risk evaluation, the debt ratio of α is not simply different from the debt ratio of β but, clearly, the former is better than the latter.

The basic principle of the classical rough set approach and data mining techniques is the following *indiscernibility principle*: if x is indiscernible with y, i.e. x has the same characteristics as y, then x should belong to the same class as y; if not, there is an inconsistency between x and y. According to this principle, if a patient has symptom A and disease X while another patient has symptom B and disease Y, there is not any inconsistency and one can draw a simple conclusion that symptom A is associated with disease X, while symptom B is associated with disease Y.

In multiple-criteria decision analysis, indiscernibility principle is not sufficient to convey all important semantics of the available information. Consider again the above two firms: α having a low value of debt ratio and β having a high value of the debt ratio. Suppose that evaluations of these firms on other attributes (profitability indices, quality of managers, market competitive situation, etc.) are equal. Suppose, moreover, that firm α has been assigned by a DM to a class of higher risk than firm β. According to the indiscernibility principle, one can say that α and β are discernible, and it follows that low debt ratio is associated with high risk while high debt ratio is associated with low risk. This is contradictory, of course. The reason is that, within multiple-criteria decision analysis, the indiscernibility principle has to be substituted by the following *dominance principle*: if x dominates y, i.e. x is at least as good as y with respect to all considered criteria, then x should belong to a class not worse than the class of y; if not, there is an inconsistency between x and y. Applying the dominance principle to α and β, one can state an inconsistency between their debt ratio and the risk of their bankruptcy, which leads to a paradoxical conclusion that the lower the debt ratio the higher the risk of bankruptcy.

For this reason, Greco, Matarazzo and Slowinski ([17, 20, 22, 23, 26, 33, 57]; for an elementary introduction see [25]) have proposed an extension of rough set theory based on the dominance principle, which permits to deal with MCDA problems. This innovation is mainly based on substitution of the indiscernibility relation by a dominance relation in the rough approximation of decision classes. An important consequence of this fact is a possibility of inferring from exemplary decisions the preference model in terms of decision rules being logical statements of the type "*if..., then...*". The separation of certain and doubtful knowledge about the DM's preferences is done by distinction of different kinds of decision rules, depending whether they are induced from examples consistent with the dominance principle or from examples inconsistent with the dominance principle. The latter rules are very important because they represent situations of hesitation in the DM's expression of preferences.

This is to say that, using a properly modified technique of artificial intelligence, one can construct a preference model in form of decision rules, by induction from exemplary decisions. Such a preference model has some interesting properties: it is expressed in a natural language without using any complex analytical formulation, its interpretation is immediate and does not depend on technical parameters, often it uses only a subset of the considered attributes in each rule, and, finally, it can represent situations of hesitation, typical for a real expression of preferences. The rule preference model is therefore a new approach to MCDA, which becomes a strong alternative for MAUT and outranking approaches.

In this chapter, we present the Dominance-based Rough Set Approach to multiple-criteria decision problems (DRSA), starting by multiple-criteria classification problems, and then going through decision under uncertainty, hierarchical decision making, classification problems with partially missing information, problems with imprecise information modelled by fuzzy sets, until multiple-criteria choice and ranking problems.

2. Dominance-based Rough Set Approach (DRSA) to Multiple-criteria Classification

2.1 Data Table

For the sake of a didactic exposition, we introduce the main ideas of DRSA (for a detailed exposition see [35] through a very simple example. Let us suppose that students of a technical college are evaluated taking into account their marks in Mathematics, Physics and Literature. Let us suppose that one is interested in finding some general rules for a comprehensive evaluation of the students. These general rules can be inferred from some previous examples of decisions, i.e. comprehensive evaluations of students made in the past. Let us consider the examples presented in Table 13.1.

Table 13.1. Data table presenting examples of comprehensive evaluations of students.

Student	Mathematics	Physics	Literature	Comprehensive evaluation
S1	*Good*	*Medium*	*Bad*	*Bad*
S2	*Medium*	*Medium*	*Bad*	*Medium*
S3	*Medium*	*Medium*	*Medium*	*Medium*
S4	*Good*	*Good*	*Medium*	*Good*
S5	*Good*	*Medium*	*Good*	*Good*
S6	*Good*	*Good*	*Good*	*Good*
S7	*Bad*	*Bad*	*Bad*	*Bad*
S8	*Bad*	*Bad*	*Medium*	*Bad*

Each application of the DRSA is based on a data table having the form of Table 13.1. In general, a data table can be described as follows. Each row of the table corresponds to an *object*. In multiple-criteria decision analysis we can also speak of an *action*. In the considered example, the objects (actions) are students. Each column of the table corresponds to an *attribute*, i.e. to a different type of information. In our example, the attributes are: Mathematics, Physics, Literature, Comprehensive evaluation. Each cell of this table indicates an *evaluation* (quantitative or qualitative) of the object placed in the corresponding row by means of the attribute in the corresponding column. In the above example, the evaluation is a mark of the considered student in a given course (Mathematics, Physics, Literature) or in the Comprehensive evaluation.

Therefore, formally, a *data table* is the 4-tuple $\boldsymbol{S} =< U, Q, V, f >$, where U is a finite set of *objects* (universe), $Q = \{q_1, q_2, ..., q_m\}$ is a finite set of *attributes*, V_q is the domain of attribute q, $V = \bigcup_{q \in Q} V_q$ and $f : U \times Q \rightarrow V$

is a total function such that $f(x,q) \in V_q$ for each $q \in Q, x \in U$, called *information function.*

In our example, U = {S1, S2, S3, S4, S5, S6, S7, S8}, Q={Mathematics, Physics, Literature, Comprehensive evaluation}, $V_{Mathematics} = V_{Physics} = V_{Literature} = V_{Comprehensive\_evaluation} = V = \{Bad, Medium, Good\}$, the information function $f : U \times Q \rightarrow V$ can be rebuild from Table 13.1 such that, for example, f(S1, Math.) = *Good*, f(S1, Phys.) = *Medium*, f(S1, Lit.) = *Bad* and so on.

Let us remark that the domain of each attribute is monotonically ordered according to preference. Such an attribute is called *criterion.* As one can see in Table 13.1, *Good* is better than *Medium* and *Medium* is better than *Bad*. In general, this fact can be formally expressed as follows. Let $\succeq_q$ be a *weak preference* relation on U with reference to criterion $q \in Q$, such that $x \succeq_q y$ means "x is at least as good as y with respect to criterion q". Suppose that $\succeq_q$ is a complete preorder, i.e. a strongly complete (which means that for each $x, y \in U$, at least one of $x \succeq_q y$ and $y \succeq_q x$ is verified, and thus x and y are always comparable with respect to criterion q) and transitive binary relation. In the following we shall denote by $\succ_q$ the asymmetric part of $\succeq_q$ and by $\sim_q$, its symmetric part. The meaning of $x \succ_q y$ is "x is preferred to y with respect to criterion q" and the meaning of $x \sim_q y$ is "x is indifferent to y with respect to criterion q". For example, in Table 13.1, we see that, with respect to Mathematics, S1, being *Good*, is preferred to S2, being *Medium*, which is denoted by S1$\succ_{Mathematics}$S2. Analogously, with respect to Physics, S1, being *Medium*, is indifferent to S2, being also *Medium*, which is denoted by S1$\sim_{Physics}$S2.

The attributes from Q are divided in two sets, C and D with $C \neq \emptyset$, $D \neq \emptyset$, $C \cap D = \emptyset$ and $C \cup D = Q$. The attributes from C are called condition attributes, while the attributes from D are called decision attributes. This distinction is made with the aim of explaining the evaluations on D using the evaluations on C. Very often $D = \{d\}$ and then its evaluations are considered in terms of a classification of objects from U. The case in which there are more than one decision attribute, i.e. $D = \{d_1, \ldots, d_m\}$, is also very interesting because it is related to decisions with multiple decision makers, $DM_1, \ldots, DM_m$, expressing different classifications corresponding to particular decision attributes, i.e. d_1 represents classification of $DM_1, \ldots, d_m$ represents classification of DM_m. For the sake of simplicity, in the following, we shall consider the case of a single decision attribute, i.e. $D = \{d\}$. More formally, let $\boldsymbol{Cl} = \{Cl_t, t \in \{1, ..., n\}\}$, be a classification of U, such that each $x \in U$ belongs to one and only one class $Cl_t \in \boldsymbol{Cl}$. In the above example, the set of condition attributes is C={Mathematics, Physics, Literature} and the decision attribute is d=Comprehensive evaluation. In this case, the aim is to explain evaluation on d, using evaluations on C. Consequently, the classification is referred to the comprehensive evaluation, ***Cl***$=\{Cl_1, Cl_2, Cl_3\}$, $Cl_1 = \{$*Bad* students$\}$ = $\{$S1, S7, S8$\}$, $Cl_2 = \{$*Medium* students$\}$ = $\{$S2, S3$\}$ and $Cl_3 = \{$*Good* students$\}$ = $\{$S4, S5, S6$\}$.

We assume that, for all $r, s \in \{1, ..., n\}$, such that $r > s$, each element of Cl_r is preferred to each element of Cl_s. More formally, if $\succeq$ is a comprehensive weak preference relation on U, i.e. $x \succeq y$ means: "x is at least as good as y" for any $x, y \in U$, then it is supposed that

$$[x \in Cl_r, y \in Cl_s, r > s] \Rightarrow x \succ y,$$

where $x \succ y$ means $x \succeq y$ and *not* $y \succeq x$.

In our example, each element of Cl_2 is preferred to each element of Cl_1 (each *Medium* student is better than each *Bad* student) and each element of Cl_3 is preferred to each element of Cl_2 (each *Good* student is better than each *Medium* student).

2.2 Dominance Principle

A natural question with respect to Table 13.1 arises: what classification patterns can be induced from the data table? They represent *knowledge* which may be useful for explanation of a policy of comprehensive evaluation and for prediction of future decisions. In this sense, it is a preference model of a decision maker (DM) who made the comprehensive evaluations and provided the exemplary decisions. Knowledge discovery from Table 13.1 will respect the following *dominance principle*: given $x, y \in U$, if x is at least as good as y with respect to all criteria from a subset $P \subseteq C$, then x should have a comprehensive evaluation at least as good as y. If this is not the case, the reasons for that may be as follows:

1) some aspects relevant to the comprehensive evaluation are ignored, i.e. some significant criteria are missing in subset P, or

2) given the evaluations of students on criteria from P, the DM hesitates with the comprehensive evaluation.

The following two examples drawn from Table 13.1 explain reason 1) and reason 2).

Example 1, relative to reason 1). Let us consider students S1 and S3 with respect to their evaluations on Mathematics and Physics. Remark that student S1 is not worse than S3, in both Mathematics and Physics, however, S1 is comprehensively evaluated as *Bad*, while S3 is comprehensively evaluated as *Medium*. This contradicts the dominance principle with respect to Mathematics and Physics. This inconsistency with the dominance principle is solved by taking into account the Literature, which gives advantage to S3 (whose evaluation is *Medium*) over S1 (whose evaluation is *Bad*). Thus, considering Mathematics,

Physics and Literature, S1 does not dominate S3, i.e. it is no more true that S1 has an evaluation at least as good as S3 on all considered criteria (Mathematics, Physics and Literature). In consequence, after including the Literature in the set of criteria, the dominance principle is no more contradicted. In other words, the evaluation on Literature is necessary to avoid contradiction with the dominance principle while giving comprehensive evaluation to S1 and S3.

Example 2, relative to reason 2). Let us consider students S1 and S2 with respect to evaluations on Mathematics, Physics and Literature. Student S1 is not worse than S2 with respect to all the considered criteria, however, S1 is comprehensively evaluated as *Bad* while S2 is comprehensively evaluated as *Medium*. This contradicts the dominance principle. This inconsistency with the dominance principle cannot be solved by taking into account one criterion more, because all the available information has been used. In consequence, given all the available information, the comprehensive evaluation of S1 and S2 contradicts the dominance principle. This contradiction may be interpreted as a hesitation of the DM.

DRSA permits to detect all the inconsistencies with the dominance principle following from hesitations, but this is not the sole interesting feature. The main advantage of DRSA is its capacity of discovering certain and doubtful knowledge from the data table, that is a preference model which has also its certain and doubtful part; the certain part is inferred from decision examples consistent with the dominance principle, while the doubtful part is inferred from decision examples inconsistent with the dominance principle. The preference model is useful for both explanation of past decisions and recommendation for new decisions.

2.3 Decision Rules

To have a first idea of the multiple-criteria decision analysis performed with DRSA, let us take into account the following example relative to Table 13.1. Consider student S3 with comprehensive evaluation *Medium* and the set of students classified as *at least Medium*, that is *Medium* or *Good*. Taking into account all three criteria – Mathematics, Physics and Literature – the comprehensive evaluation of S3 is not inconsistent with the dominance principle. Indeed, in Table 13.1 there is no other student dominated by S3 and having a better comprehensive evaluation. Remark, however, that less than three criteria are sufficient to ensure the consistency. In fact, the evaluations on Mathematics and Literature are sufficient for the comprehensive evaluation of S3 consistent with the dominance principle. Further reduction of criteria (to Mathematics or Literature only) makes the comprehensive evaluation of S3 inconsistent. For example, considering only Mathematics, we can see that S1 dominates S3, but it has a worse comprehensive evaluation. Therefore, {Mathematics, Literature} is a minimal set of criteria ensuring the consistent evaluation of S3. In other words, one can induce from Table 13.1 a minimal conclusion that each student having not worse evaluations than S3, has also a not worse comprehensive evaluation; this conclusion creates the following decision rule:

ρ : "*if* Mathematics $\geq$ *Medium and* Literature$\geq$*Medium*, *then* the comprehensive evaluation is at least *Medium*
(that is *Medium* or *Good*)".

It is interesting to remark that this decision rule, possibly useful as an element of a preference model, is a result of search of a boundary line between consistency and inconsistency with the dominance principle. In general, we can say that the preference model, and all the decision analysis using DRSA, can be seen as a search of this boundary line between consistency and inconsistency.

2.4 Rough Approximations

Let us continue the presentation of the most important concepts relative to DRSA.

As it was told before, the considered objects are evaluated by criteria from set C from one side, and by the comprehensive decision d from the other side. Using the dominance relation with respect to d, we can define unions of classes relative to a particular dominated or dominating class – these unions of classes are called *upward* and *downward unions of classes*, defined, respectively, as:

$$Cl_t^{\geq} = \bigcup_{s\geq t} Cl_s, \quad Cl_t^{\leq} = \bigcup_{s\leq t} Cl_s.$$

Observe $Cl_1^{\geq} = Cl_n^{\leq} = U$, $Cl_n^{\geq} = Cl_n$ and $Cl_1^{\leq} = Cl_1$.

In the above example, $Cl_1^{\geq} = Cl_1 \cup Cl_2 \cup Cl_3$ = {students comprehensively *Bad*, *Medium* or *Good*} = {all students in Table 13.1 }, $Cl_2^{\geq}$ = $Cl_2 \cup Cl_3$ = {students comprehensively *Medium* or *Good*}={S2, S3, S4, S5, S6} and $Cl_3^{\geq}$ = Cl_3 = {students comprehensively *Good*}={S4, S5, S6}.

On the other hand, using the dominance relation with respect to criteria from set C, we can define sets of objects dominatin or dominated by a particular object. It is said that object x *P-dominates* object y with respect to $P \subseteq C$ (denotation xD_Py) if $x \succeq_q y$ for all $q \in P$. For example, in Table 13.1, S1 dominates S3 with respect to P = {Mathematics, Physics} because S1 $\succeq_{Mathematics}$ S3 and S1 $\succeq_{Physics}$ S3. Since the intersection of complete preorders is a partial preorder and $\succeq_q$ is a complete preorder for each $q \in P$, and $D_P = \bigcap_{q\in P} \succeq_q$, then the dominance relation D_P is a partial preorder. Given $P \subseteq C$ and $x \in U$, let

$$D_P^+(x) = \{y \in U : yD_Px\},$$
$$D_P^-(x) = \{y \in U : xD_Py\}$$

represent, so-called, *P-dominating set* and *P-dominated set* with respect to x, respectively. For example in Table 13.1, for P = {Mathematics, Physics},

$$D^{+}_{\{\text{Mathematics,Physics}\}}(S1) = \{S1, S4, S5, S6\},$$
$$D^{-}_{\{\text{Mathematics,Physics}\}}(S1) = \{S1, S2, S3, S5, S7, S8\}.$$

Given a set of criteria $P \subseteq C$, an object $x \in U$ creates an *inconsistency with the dominance principle* with respect to the upward union of classes $Cl_t^{\geq}$, $t = 2, \ldots, n$, if one of the following conditions holds:

(i) x belongs to class Cl_t or better but it is P-dominated by an object y belonging to a class worse than Cl_t, i.e. $x \in Cl_t^{\geq}$ but $D_P^{+}(x) \cap Cl_{t-1}^{\leq} \neq \emptyset$ (for example, considering P = {Mathematics, Physics} and the upward union of classes composed of students comprehensively evaluated as "at least *Medium*", student x=S3 creates an inconsistency with the dominance principle: in fact x=S3 belongs to the class of *Medium* students, but there is another student y = S1, P-dominating x and belonging to the class of *Bad* students, that is to a worse class),

(ii) x belongs to a worse class than Cl_t but it P-dominates an object y belonging to class Cl_t or better, i.e. $x \notin Cl_t^{\geq}$ but $D_P^{-}(x) \cap Cl_t^{\geq} \neq \emptyset$ (consider the example from point (i), but taking x = S1 and y = S3).

If for a given set of criteria $P \subseteq C$, the assignment of $x \in U$ to $Cl_t^{\geq}$, $t = 2, \ldots, n$, creates an inconsistency with the dominance principle, we say that x belongs to $Cl_t^{\geq}$ *with some ambiguity*. Thus, x belongs to $Cl_t^{\geq}$ *without any ambiguity* with respect to $P \subseteq C$, if $x \in Cl_t^{\geq}$ and there is no inconsistency with the dominance principle. This means that all objects P-dominating x belong to $Cl_t^{\geq}$, i.e. $D_P^{+}(x) \subseteq Cl_t^{\geq}$.

Furthermore, x *possibly belongs to* $Cl_t^{\geq}$ with respect to $P \subseteq C$ if one of the following conditions holds:

- according to decision attribute d, x belongs to $Cl_t^{\geq}$ (in the example from point (i) above, this is the case of x=S3 which, taking into account criteria from P = {Mathematics, Physics}, could belong to $Cl_2^{\geq}$, i.e. to the set of student comprehensively evaluated as "at least *Medium*"),

- according to decision attribute d, x does not belong to $Cl_t^{\geq}$ but it is inconsistent in the sense of the dominance principle with an object y belonging to $Cl_t^{\geq}$ (in the example from point (i) above, this is the case of x=S1 which, taking into account criteria from P = {Mathematics, Physics}, could belong to $Cl_2^{\geq}$, even if S1 is comprehensively evaluated as *Bad*).

In terms of ambiguity, x possibly belongs to $Cl_t^{\geq}$ with respect to $P \subseteq C$, if x belongs to $Cl_t^{\geq}$ with or without any ambiguity. Due to reflexivity of the dominance relation D_P, conditions (i) and (ii) can be summarized as follows: x *possibly belongs* to class Cl_t or better, with respect to $P \subseteq C$, if among the objects P-dominated by x there is an object y belonging to class Cl_t or better, i.e. $D_P^-(x) \cap Cl_t^{\geq} \neq \emptyset$.

In DRSA, the sets to be approximated are upward and downward unions of classes and the items (granules of knowledge) used for this approximation are P-dominating and P-dominated sets.

The *P-lower* and the *P-upper approximation*of upward union $Cl_t^{\geq}$, $t \in \{1, \dots, n\}$, with respect to $P \subseteq C$ (denotation $\underline{P}(Cl_t^{\geq})$ and $\overline{P}(Cl_t^{\geq})$, respectively), are defined as:

$$\begin{aligned}
\underline{P}(Cl_t^{\geq}) &= \{x \in U : D_P^+(x) \subseteq Cl_t^{\geq}\}, \\
\overline{P}(Cl_t^{\geq}) &= \bigcup_{x \in Cl_t^{\geq}} D_P^+(x) = \{x \in U : D_P^-(x) \cap Cl_t^{\geq} \neq \emptyset\}.
\end{aligned}$$

Let us comment the above definitions. The P-lower approximation of an upward union $Cl_t^{\geq}$, $\underline{P}(Cl_t^{\geq})$, is composed of all objects x from the universe such that all objects y having at least the same evaluations on all the considered criteria from P also belong to class Cl_t or better. Thus, one can say that if an object y has at least as good evaluations on criteria from P as object x belonging to $\underline{P}(Cl_t^{\geq})$, then, certainly, y belongs to class Cl_t or better. Therefore, taking into account all decision examples from the considered data table, one can conclude that the evaluations on criteria from $P \subseteq C$ of an object x belonging to $\underline{P}(Cl_t^{\geq})$ create a *partial profile* (partial, because $P \subseteq C$), such that for an object y it is sufficient to dominate this partial profile in order to belong to class Cl_t or better. This is the case of above decision rule ρ, using the partial profile built on criteria from $P = \{$Mathematics, Literature$\}$; this profile corresponds to x = S3, which belongs to $\underline{P}(Cl_2^{\geq})$. The assignment of a decision rule is true for all objects from the considered data table, but it can also be used by induction for objects that are not in U. Indeed, it is rather natural to admit such a working hypothesis that, if for a new object z, its evaluations on criteria from P are not worse than the evaluations of x, then z should be assigned to class Cl_t or better. This is because we can consider the data table as a record of experience of the DM. Thus, if according to the experience of the DM – i.e. according to the data table at hand – all objects having evaluations on criteria from P not worse than x are assigned to class Cl_t or better, then the simplest classification strategy following from the previous experience is to assign any other object z, having evaluations on criteria from P not worse than x, to class Cl_t or better.

Let us come back to Table 13.1. Given P=$\{$Mathematics, Physics$\}$, $\underline{P}(Cl_3^{\geq})$ = $\underline{P}(Cl_2^{\geq})$ = $\{$S4, S6$\}$ and $\underline{P}(Cl_1^{\geq})$ = $\{$all the students$\}$. Precisely, given the information provided by evaluations on Mathematics and Physics, S4 and S6 belong to $\underline{P}(Cl_3^{\geq})$, the P-lower approximation of the union of (at least) *Good* students, because there is no other student having at least the same evaluations on Mathematics and Physics and belonging to a class comprehensively worse. This is not the case of student S5, even if she belongs to the class of students comprehensively evaluated as *Good*. Indeed, S5 does not belong to P-lower approximation of (at least) *Good* students because there is another student, S1, having not worse evaluations (in this case, exactly the same) on Mathematics and Physics and, nevertheless, not belonging to the class of (at least) *Good* students. The fact that S4 and S6 belong to $\underline{P}(Cl_3^{\geq})$ permits to conclude that, according to the information given by Table 13.1, an evaluation (at least) *Good* in Mathematics and (at least) *Good* in Physics is enough to assign a student to the union of (at least) *Good* students.

The P-upper approximation of an upward union $Cl_t^{\geq}$, $\overline{P}(Cl_t^{\geq})$, is composed of all objects x from the universe which, in comparison with an object y belonging to union $Cl_t^{\geq}$, have at least the same evaluations on all the considered criteria from P. In other words, the P-upper approximation of an upward union $Cl_t^{\geq}$, $\overline{P}(Cl_t^{\geq})$, is composed of all objects x from the universe, whose evaluations on criteria from P are not worse than evaluations of at least one other object y belonging to class Cl_t or better. Thus, one can say that, if an object z has not worse evaluations on criteria from P than an object x belonging to $\overline{P}(Cl_t^{\geq})$, then z *possibly* belongs to class Cl_t or better. Therefore, taking into account all decision examples from the considered data table, one can conclude that the evaluations of an object x belonging to $\overline{P}(Cl_t^{\geq})$, on criteria from P, create a *partial profile*, such that an object z dominating this profile possibly belongs to class Cl_t or better. This conclusion is true for all objects from the considered data table but it can also be used by induction for objects that are not in U. Indeed, it is again natural to admit such a working hypothesis that, if for a new object z, its evaluations on criteria from P are not worse than the evaluations of x, then z could be assigned to class Cl_t or better.

Coming back to our example, one can see that, given P = {Mathematics, Literature}, $\overline{P}(Cl_3^{\geq})$ = {S4, S5, S6}, $\overline{P}(Cl_2^{\geq})$ = {S1, S2, S3, S4, S5, S6} and $\overline{P}(Cl_1^{\geq})$ = {all the students}. Precisely, given the information provided by the comprehensive evaluation, S1 is not (at least) *Medium* student, i.e. S1 does not belong to union $Cl_2^{\geq}$. However, S1 belongs to $\overline{P}(Cl_2^{\geq})$, the upper approximation of the union of at least *Medium* students, because there is another student, S2, having at most the same evaluations on Mathematics and Literature and belonging to the set of students comprehensively evaluated as at least *Medium*.

Analogously, the P-lower and the P-upper approximation of downward union $Cl_t^{\leq}$, $t \in \{1, \ldots, n\}$, with respect to $P \subseteq C$ (denotation $\underline{P}(Cl_t^{\leq})$ and

$\overline{P}(Cl_t^{\leq})$, respectively), are defined as:

$$\underline{P}(Cl_t^{\leq}) = \{x \in U : D_P^-(x) \subseteq Cl_t^{\leq}\},$$
$$\overline{P}(Cl_t^{\leq}) = \bigcup_{x \in Cl_t^{\leq}} D_P^-(x) = \{x \in U : D_P^+(x) \cap Cl_t^{\leq} \neq \emptyset\}.$$

The P-lower and P-upper approximation of $Cl_t^{\leq}$ have analogous interpretation of the P-lower and P-upper approximation of $Cl_t^{\geq}$.

2.5 Properties of Rough Approximations

The P-lower and P-upper approximations defined as above satisfy the following properties for all $t \in \{1, \ldots, n\}$ and for any $P \subseteq C$:

$$\underline{P}(Cl_t^{\geq}) \subseteq Cl_t^{\geq} \subseteq \overline{P}(Cl_t^{\geq}), \quad \underline{P}(Cl_t^{\leq}) \subseteq Cl_t^{\leq} \subseteq \overline{P}(Cl_t^{\leq}).$$

This property means that all the objects belonging to $Cl_t^{\geq}$ without any ambiguity belong also to $Cl_t^{\geq}$, and all the objects from $Cl_t^{\geq}$ are among the objects that belong to $Cl_t^{\geq}$ with some possible ambiguity. With respect to Table 13.1, one can see that, given $P = C = \{$Mathematics, Physics, Literature$\}$ and the union of at least *Medium* students, $Cl_2^{\geq}$, we have:

$$\underline{P}(Cl_2^{\geq}) = \{S3, S4, S5, S6\},$$
$$Cl_2^{\geq} = \{S2, S3, S4, S5, S6\},$$
$$\overline{P}(Cl_2^{\geq}) = \{S1, S2, S3, S4, S5, S6\}.$$

Let us observe that all objects belonging to $\underline{P}(Cl_2^{\geq})$ are from $Cl_2^{\geq}$. However, S2 from $Cl_2^{\geq}$ does not belong to $\underline{P}(Cl_2^{\geq})$ because comprehensive evaluation of S2 is inconsistent with comprehensive evaluation of S1 (classified as *Bad*) in the sense of the dominance principle. This inconsistency creates an ambiguity because S1 dominates S2 on criteria from P and, nevertheless, S1 has been classified worse than S2. Let us also observe that all objects belonging to $Cl_2^{\geq}$, belong also to $\overline{P}(Cl_2^{\geq})$. However, S1 from $\overline{P}(Cl_2^{\geq})$ does not belong to $Cl_2^{\geq}$ because, basing on the available information, it only possibly belongs to $Cl_2^{\geq}$ due to the above ambiguity with S2.

The above examples point out that the differences between $Cl_2^{\geq}$ and $\underline{P}(Cl_2^{\geq})$ from one side, and between $\overline{P}(Cl_2^{\geq})$ and $Cl_2^{\geq}$ from the other side, are related to the inconsistency (or ambiguity) of information. This observation can be generalized: the set difference between upper and lower approximation is composed of objects, whose assignment to the considered upward union $Cl_t^{\geq}$ or downward union $Cl_t^{\leq}$ is ambiguous, that is inconsistent, with the dominance principle. This justifies the following definition. The *P-boundaries* (P-doubtful regions) of $Cl_t^{\geq}$ and $Cl_t^{\leq}$ are defined as:

$$Bn_P(Cl_t^{\geq}) = \overline{P}(Cl_t^{\geq}) - \underline{P}(Cl_t^{\geq}), \qquad Bn_P(Cl_t^{\leq}) = \overline{P}(Cl_t^{\leq}) - \underline{P}(Cl_t^{\leq}).$$

In the above example, we have: $Bn_P(Cl_2^{\geq}) = \overline{P}(Cl_2^{\geq}) - \underline{P}(Cl_2^{\geq})$ = {S1, S2}; indeed, the ambiguity of S1 and S2 with respect to $Cl_2^{\geq}$ was already explained.

The P-lower and P-upper approximations satisfy the following specific complementarity properties:

$$\begin{aligned}
\underline{P}(Cl_t^{\geq}) &= U - \overline{P}(Cl_{t-1}^{\leq}),\ t = 2, \ldots, n, \\
\underline{P}(Cl_t^{\leq}) &= U - \overline{P}(Cl_{t+1}^{\geq}),\ t = 1, \ldots, n-1, \\
\overline{P}(Cl_t^{\geq}) &= U - \underline{P}(Cl_{t-1}^{\leq}),\ t = 2, \ldots, n, \\
\overline{P}(Cl_t^{\leq}) &= U - \underline{P}(Cl_{t+1}^{\geq}),\ t = 1, \ldots, n-1.
\end{aligned}$$

The first expression above has the following interpretation: if object x belongs without any ambiguity to class Cl_t or better, it is impossible that it could belong, even with some ambiguity, to class Cl_{t-1} or worse, i.e. $\underline{P}(Cl_t^{\geq}) = U - \overline{P}(Cl_{t-1}^{\leq})$. Let us consider the set of at least *Medium* students, i.e. $Cl_2^{\geq}$, and the set of (at most) *Bad* students, i.e. $Cl_1^{\leq}$. The above complementarity property means that a student is at least *Medium* without any ambiguity if and only if, basing on available information, it is impossible that she could be comprehensively evaluated as *Bad*, even with some ambiguity. According to this definition, if we consider $P = C = \{$ Mathematics, Physics, Literature $\}$, we have: $\underline{P}(Cl_2^{\geq})$ = {S3, S4, S5, S6} and $\overline{P}(Cl_1^{\leq})$ = {S1, S2, S7, S8}, thus, $\underline{P}(Cl_2^{\geq}) = U - \overline{P}(Cl_1^{\leq})$.

Due to the complementarity property, $Bn_P(Cl_t^{\geq}) = Bn_P(Cl_{t-1}^{\leq})$, for $t = 2, ..., n$, which means that if x belongs with ambiguity to class Cl_t or better, it also belongs with ambiguity to class Cl_{t-1} or worse. In our example, $Bn_P(Cl_2^{\geq})$ = $Bn_P(Cl_1^{\leq})$={S1, S2}. In simple words, this can be expressed as follows: the students, whose assignment to at least *Medium* class is ambiguous, are the same as students, whose assignment to at most *Bad* class is also ambiguous.

A very important property related to the value of information is the following monotonicity of rough approximations with respect to the considered set of attributes: given $R \subseteq P \subseteq C$,

$$\underline{R}(Cl_t^{\geq}) \subseteq \underline{P}(Cl_t^{\geq}), \quad \overline{R}(Cl_t^{\geq}) \supseteq \overline{P}(Cl_t^{\geq})$$

$$\underline{R}(Cl_t^{\leq}) \subseteq \underline{P}(Cl_t^{\leq}), \quad \overline{R}(Cl_t^{\leq}) \supseteq \overline{P}(Cl_t^{\leq})$$

$$Bn_R(Cl_t^{\geq}) \supseteq Bn_P(Cl_t^{\geq}), \quad Bn_R(Cl_t^{\leq}) \supseteq Bn_P(Cl_t^{\leq}).$$

This property has the following interpretation. When the considered information is augmented, then the ambiguity decreases or, at least, does not increase. This means that, if object x is ambiguous with respect to a set of criteria R, then with respect to another set of criteria $P \supseteq R$ the same object x may become non-ambiguous, because the new information conveyed by criteria from $P - R$ may remove this ambiguity. Let us consider the assignment of student S5, in our example, to the union $Cl_3^{\geq}$ of (at least) *Good* students. If P={Physics}, then S5 is ambiguous and does not belong to $\underline{P}(Cl_3^{\geq})$. Indeed, S5 creates an ambiguity with students S1, S2 and S3 that are evaluated at least as good as S5 with respect to Physics and, nevertheless, their comprehensive evaluation is worse than S5. If we augment the available information by evaluation on Mathematics, considering therefore P={Physics, Mathematics}, then S5 is still ambiguous and does not belong to $\underline{P}(Cl_3^{\geq})$. In this case, however, the set of students ambiguous with S5 is reduced to S1 only. Finally, if P={Physics, Mathematics, Literature}, then S5 is no more ambiguous in comparison with other students and thus S5 belongs to $\underline{P}(Cl_3^{\geq})$.

2.6 Quality of Approximation, Reducts and Core

The ratio

$$\gamma_P(\mathbf{Cl}) = \frac{|U - (\bigcup_{t\in\{2,\ldots,n\}} Bn_P(Cl_t^{\geq}))|}{|U|} = \frac{|U - (\bigcup_{t\in\{1,\ldots,n-1\}} Bn_P(Cl_t^{\leq}))|}{|U|}$$

defines the *quality of approximation of the classification* **Cl** by means of criteria from set $P \subseteq C$, or, briefly, *quality of classification*, where $|\cdot|$ means cardinality of a set. This ratio expresses the proportion of all $P-$correctly classified objects, i.e. all the non-ambiguous objects, to all the objects in the data table. Let us calculate this ratio for Table 13.1; taking P={Mathematics, Physics}, we have

$$\begin{aligned}
\underline{P}(Cl_2^{\geq}) &= \{S4, S6\},\\
\underline{P}(Cl_3^{\geq}) &= \{S4, S6\},\\
\overline{P}(Cl_2^{\geq}) &= \{S1, S2, S3, S4, S5, S6\},\\
\overline{P}(Cl_3^{\geq}) &= \{S1, S4, S5, S6\},\\
\underline{P}(Cl_1^{\leq}) &= \{S7, S8\},\\
\underline{P}(Cl_2^{\leq}) &= \{S2, S3, S7, S8\},\\
\overline{P}(Cl_1^{\leq}) &= \{S1, S2, S3, S5, S7, S8\},\\
\overline{P}(Cl_2^{\leq}) &= \{S1, S2, S3, S5, S7, S8\}.
\end{aligned}$$

This means that

$$\begin{aligned} Bn_P(Cl_2^{\geq}) &= \{S1, S2, S3, S5\}, \\ Bn_P(Cl_3^{\geq}) &= \{S1, S5\}, \\ Bn_P(Cl_1^{\leq}) &= \{S1, S2, S3, S5\}, \\ Bn_P(Cl_2^{\leq}) &= \{S1, S5\}. \end{aligned}$$

Thus, the quality of classification with respect to ***Cl*** and criteria from set P is

$$\begin{aligned} \gamma_P(\boldsymbol{Cl}) &= \frac{|U - (Bn_P(Cl_2^{\geq}) \cup Bn_P(Cl_3^{\geq}))|}{|U|} = \\ &= \frac{|U - (Bn_P(Cl_1^{\leq}) \cup Bn_P(Cl_2^{\leq}))|}{|U|} = \frac{|\{S4, S6, S7, S8\}|}{|U|} = \frac{4}{8}. \end{aligned}$$

Due to the above monotonicity property, for all $R, P \subseteq C$ the following implication is true

$$R \subseteq P \Rightarrow \gamma_R(\boldsymbol{Cl}) \leq \gamma_P(\boldsymbol{Cl}).$$

This property is illustrated for Table 13.1 by the results of calculation presented in Table 13.2.

Table 13.2. Quality of classification and Shapley value for classification ***Cl*** and set of criteria P.

Set of criteria P	Ambiguous objects	Non-ambiguous objects	Quality of classification	Shapley value
{Mathematics}	S1, S2, S3 S4, S5, S6	S7, S8	0.25	0.167
{Physics}	S1, S2, S3, S5	S4, S6, S7, S8	0.5	0.292
{Literature}	S1, S2, S3 S4, S7, S8	S5, S6	0.25	0.292
{Mathematics, Physics}	S1, S2, S3, S5	S4, S6, S7, S8	0.5	-0.375
{Mathematics, Literature}	S1, S2	S3, S4, S5 S6, S7, S8	0.75	0.125
{Physics, Literature}	S1, S2	S3, S4, S5 S6, S7, S8	0.75	-0.125
{Mathematics, Physics, Literature}	S1, S2	S3, S4, S5 S6, S7, S8	0.75	0.125

Every minimal subset of criteria $P \subseteq C$ such that $\gamma_P(\boldsymbol{Cl}) = \gamma_C(\boldsymbol{Cl})$ is called a *reduct* of C with respect to ***Cl*** and is denoted by RED_{Cl}(P). $\gamma_P(\boldsymbol{Cl}) = \gamma_C(\boldsymbol{Cl})$

means that, if P is a reduct, then no object which is non-ambiguous with respect to C, is ambiguous with respect to P. In other words, reducing the information from the set of all criteria C to the subset P, no new ambiguity arises. The condition $\gamma_P(\mathbf{Cl}) = \gamma_C(\mathbf{Cl})$ is not sufficient for declaring P a reduct. The other important condition in the definition of reduct is the minimality. Supposing that P is a reduct, minimality means that, for any $q \in P$, $\gamma_{P-\{q\}}(\mathbf{Cl}) < \gamma_C(\mathbf{Cl})$. Therefore, the reducts are all the subsets $P \subseteq C$ which keep the same number of ambiguous objects as C, and such that removing any criterion from P one creates new ambiguous objects.

Looking at the results presented in Table 13.2, one can conclude that in our example there are two reducts: $RED^1_{\mathbf{Cl}}$ = {Mathematics, Literature} and $RED^2_{\mathbf{Cl}}$ = {Physics, Literature}.

A data table may have more than one reduct. The intersection of all the reducts is known as the *core*, denoted by $\text{CORE}_{\mathbf{Cl}}$. In our example, the core is

$$\text{CORE}_{\mathbf{Cl}} = RED^1_{\mathbf{Cl}} \cap RED^2_{\mathbf{Cl}} = \{\text{Mathematics, Literature}\} \cap \{\text{Physics, Literature}\} = \{\text{Literature}\}.$$

The criteria from the core are indispensable for keeping the quality of classification at the level attained for set C. Other criteria from different reducts are *exchangeable*, in the sense that they can substitute each other and their joint presence is not necessary to keep the quality of classification at the level attained for set C. The criteria which do not appear in any reduct are *superfluous* and they have no influence on the quality of approximation of the classification. In our example, the criterion of Literature is indispensable because, for all $P \subseteq C$ such that Literature does not belong to P, we have $\gamma_P(\mathbf{Cl}) < \gamma_C(\mathbf{Cl})$ = 0.75. This means that removing a core criterion from C creates new ambiguous objects. For the monotonicity of rough approximations with respect to the considered set of attributes, these new ambiguous objects will be present in all the subsets of criteria P which do not include the core criterion. One can see in Table 13.2 that S3 and S5 are ambiguous objects for all the subsets of criteria not including Literature. This is not the case for other criteria belonging to the reducts. In our example, Mathematics and Physics are exchangeable, so it is sufficient that one of them stays with Literature in order to keep the number of ambiguous objects unchanged. This means that the information supplied by the core criteria cannot be substituted by the information supplied by other criteria.

2.7 Importance and Interaction Among Criteria

In [15, 30], the information about the quality of classification with respect to all subsets of the considered set of criteria was analysed in view of finding the relative importance and the interaction among criteria. The main idea is based on observation that the quality of classification with respect to all subsets of criteria is a fuzzy measure with the property of Choquet capacity [3]. Such a measure can be used to calculate some specific indices introduced in cooperative game theory (for example the Shapley value [52] and in the fuzzy measure theory [8, 45]; see also Chapter 14). Using the quality of classification from Table 13.2, the Shapley value indicating the importance of particular criteria is equal to 0.167 for Mathematics and to 0.292 for Physics and Literature. Therefore, Physics and Literature are quite more important than Mathematics. The Shapley interaction index for pairs of criteria is equal, respectively, to -0.375 for Mathematics and Physics, 0.125 for Mathematics and Literature, and -0.125 for Physics and Literature. It follows that there is a redundancy of information between Mathematics and Physics, and between Physics and Literature, while there is a synergy of information between Mathematics and Literature.

Such a type of analysis can be conducted also on the decision rules in order to determine the importance of each condition and the interaction among different conditions in the considered rules [13]. Let us consider again the rule

ρ : "*if* Mathematics$\geq$*Medium and* Literature$\geq$*Medium*, *then* the comprehensive evaluation is at least *Medium*".

Consider now the two following rules having only one of the two conditions with the same conclusion:

ρ': "*if* Mathematics$\geq$*Medium*, *then* the comprehensive evaluation is at least *Medium*",

ρ": "*if* Literature$\geq$*Medium*, *then* the comprehensive evaluation is at least *Medium*".

In Table 13.1, rule ρ is always verified, while rule ρ' is verified in 5 on 6 cases (the one counterexample is student S1) and rule ρ" is verified in 4 on 5 cases (the one counterexample is student S8). Thus, the credibility of rule ρ is 1, while the credibility of rules ρ' and ρ" is $\frac{5}{6}$ and $\frac{4}{5}$, respectively. This means that in rule ρ the importance of condition "Mathematics$\geq$*Medium*" is $\frac{5}{6}$ and the importance of condition "Literature$\geq$*Medium*" is $\frac{4}{5}$. Moreover, there is a negative interaction between the two conditions which can be measured as the difference between the credibility of rule ρ on one side and the sum of the credibility of rules ρ' and ρ" on the other side, that is $1 - \frac{5}{6} - \frac{4}{5} \approx -0.63$. This means that the general tendency is such that students at least *Medium*, who are at least *Medium* in Mathematics, are also at least *Medium* in Literature. Observe that the sign of interaction among criteria can be different at the global level of the whole decision table and at the local level of a specific decision rule. In the case considered, Mathematics and Literature present a positive synergy at the global level, measured by the Shapley value equal to 0.125 (see Table 13.2), while at the local level of the decision rule ρ, there is a negative interaction between conditions concerning the same criteria (-0.63). Let us remark that the core of this analysis is the same as the analysis of the importance and interaction among criteria based on the quality of classification, that is the Shapley value and the Shapley interaction indices.

3. Variable-Consistency Dominance-Based Rough Set Approach (VC-DRSA)

The definitions of rough approximations introduced in Section 2 are based on a strict application of the dominance principle. However, when defining non-ambiguous objects, it is reasonable to accept a limited proportion of negative examples, particularly for large data tables. Such extended version of DRSA is called Variable-Consistency DRSA model (VC-DRSA) [12]. It is presented below.

For any $P \subseteq C$, we say that $x \in U$ belongs to $Cl_t^{\geq}$ *without any ambiguity at consistency level* $l \in (0, 1]$, if $x \in Cl_t^{\geq}$ and at least $l*100\%$ of all objects $y \in U$ dominating x with respect to P also belong to $Cl_t^{\geq}$, i.e.

$$\frac{\left|D_P^+(x) \cap Cl_t^{\geq}\right|}{\left|D_P^+(x)\right|} \geq l.$$

For example, student S3 in Table 13.1 does not belong to P-lower approximation of the union of at least *Medium* students, where P={Math-ematics, Physics}, because there is student S1 who is at least as good as S3 both in Mathematics and Physics but comprehensively evaluated as Bad. Anyway, if we fix $l \leq \frac{5}{6}$, then S3 belongs to P-lower approximation of the union of at least *Medium* students, because there are no more counterexamples than $(1-l)*100\%$ of all students being not worse than S3 on Mathematics and Physics.

The level l is called *consistency level* because it controls the degree of consistency between objects qualified as belonging to $Cl_t^{\geq}$ without any ambiguity. In other words, if $l < 1$, then at most $(1-l)*100\%$ of all objects $y \in U$ dominating x with respect to P do not belong to $Cl_t^{\geq}$ and thus contradict the inclusion of x in $Cl_t^{\geq}$.

Analogously, for any $P \subseteq C$ we say that $x \in U$ belongs to $Cl_t^{\leq}$ *without any ambiguity at consistency level* $l \in (0, 1]$, if $x \in Cl_t^{\leq}$ and at least $l*100\%$ of all the objects $y \in U$ dominated by x with respect to P also belong to $Cl_t^{\leq}$, i.e.

$$\frac{\left|D_P^-(x) \cap Cl_t^{\leq}\right|}{\left|D_P^-(x)\right|} \geq l.$$

For example, student S1 in Table 13.1 does not belong to P-lower approximation of the union of at most *Bad* students, where P = {Physics, Literature}, because there is student S2 who is at most as good as S1 both on Physics and

Literature, but comprehensively evaluated as Medium. Anyway, if we fix $l \leq \frac{2}{3}$, then S3 belongs to P-lower approximation of the union of at most *Bad* students, because there are no more counterexamples than $(1 - l) * 100\%$ of all students being not better than S1 on Physics and Literature.

The concept of non-ambiguous objects at some consistency level l leads naturally to the definition of P-lower approximations of the unions of classes $Cl_t^{\geq}$ and $Cl_t^{\leq}$, respectively:

$$\underline{P}^l(Cl_t^{\geq}) = \left\{ x \in Cl_t^{\geq} : \frac{\left|D_P^+(x) \cap Cl_t^{\geq}\right|}{\left|D_P^+(x)\right|} \geq l \right\}$$

$$\underline{P}^l(Cl_t^{\leq}) = \left\{ x \in Cl_t^{\leq} : \frac{\left|D_P^-(x) \cap Cl_t^{\leq}\right|}{\left|D_P^-(x)\right|} \geq l \right\}.$$

Given $P \subseteq C$ and consistency level l, we can define the *P-upper approximations* of $Cl_t^{\geq}$ and $Cl_t^{\leq}$, denoted by $\overline{P}^l(Cl_t^{\geq})$ and $\overline{P}^l(Cl_t^{\leq})$, respectively, by complementation of $\underline{P}^l(Cl_{t-1}^{\leq})$ and $\underline{P}^l(Cl_{t+1}^{\geq})$ with respect to U:

$$\overline{P}^l(Cl_t^{\geq}) = U - \underline{P}^l(Cl_{t-1}^{\leq}),$$
$$\overline{P}^l(Cl_t^{\leq}) = U - \underline{P}^l(Cl_{t+1}^{\geq}).$$

$\overline{P}^l(Cl_t^{\geq})$ can be interpreted as a set of all the objects belonging to $Cl_t^{\geq}$, *possibly ambiguous* at consistency level l. Analogously, $\overline{P}^l(Cl_t^{\leq})$ can be interpreted as a set of all the objects belonging to $Cl_t^{\leq}$, *possibly ambiguous* at consistency level l. The *P-boundaries* (P-doubtful regions) of $Cl_t^{\geq}$ and $Cl_t^{\leq}$ at consistency level l are defined as:

$$Bn_P^l(Cl_t^{\geq}) = \overline{P}^l(Cl_t^{\geq}) - \underline{P}^l(Cl_t^{\geq}),$$
$$Bn_P^l(Cl_t^{\leq}) = \overline{P}^l(Cl_t^{\leq}) - \underline{P}^l(Cl_t^{\leq}), \quad t = 1, ..., n.$$

The *variable consistency* model of the dominance-based rough set approach provides some degree of flexibility in assigning objects to lower and upper approximations of the unions of decision classes. The following property can be easily proved: for $0 < l' < l \leq 1$ and $t = 2, ..., n$,

$$\underline{P}^l(Cl_t^{\geq}) \subseteq \underline{P}^{l'}(Cl_t^{\geq}) \text{ and } \overline{P}^{l'}(Cl_t^{\geq}) \subseteq \overline{P}^l(Cl_t^{\geq}).$$

The *variable consistency* model is inspired by the *variable precision* model proposed by Ziarko [63, 64] within the classical, indiscernibility-based rough set approach.

4. Induction of Decision Rules from Rough Approximations of Upward and Downward Unions of Decision Classes

4.1 A Syntax of Decision Rules Involving Dominance with Respect to Partial Profiles

The end result of DRSA is a representation of the information contained in the considered data table in terms of simple "*if..., then...*" decision rules. Considering Table 13.1, one can induce, for example, the following decision rules (within parentheses there are symbols of students supporting the corresponding rule):

Rule 1): "*if* the evaluations in Physics *and* Literature are at least *Med-ium, then* the student is comprehensively at least *Medium*" (S3, S4, S5, S6)

Rule 2): "*if* the evaluation in Physics is at most *Medium and* the evaluation in Literature is at most *Bad, then* the student is comprehensively at most *Medium*" (S1, S2, S7)

or

Rule 3): "*if* the evaluation in Physics is at least *Medium and* the evaluation in Literature is at most *Bad, then* the student is comprehensively *Bad* or *Medium* (due to ambiguity of information)" (S1, S2).

In fact, the decision rules are not induced directly from the data table but from lower and upper approximations of upward and downward unions of decision classes. For a given upward or downward union of classes, $Cl_t^{\geq}$ or $Cl_s^{\leq}$, the decision rules induced under a hypothesis that objects belonging to $\underline{P}(Cl_t^{\geq})$ or $\underline{P}(Cl_s^{\leq})$ are *positive* (i.e. must be covered by the induced rules) and all the others *negative* (i.e. must not be covered by the induced rules), suggest an assignment to "class Cl_t or better", or to "class Cl_s or worse", respectively. For example, Rule 1) is based on the observation that student S3 belongs to $\underline{P}(Cl_2^{\geq})$, while Rule 2) is based on the observation that S1 belongs to $\underline{P}(Cl_2^{\leq})$, where P={Physics, Literature}.

On the other hand, the decision rules induced under a hypothesis that, for $s < t$, objects belonging to the intersection $\overline{P}(Cl_s^{\leq}) \cap \overline{P}(Cl_t^{\geq})$ are *positive* and all the others *negative*, are suggesting an assignment to some classes between Cl_s and Cl_t. For example, Rule 3) is based on the observation that students S1 and S2 belong to $\overline{P}(Cl_1^{\leq}) \cap \overline{P}(Cl_2^{\geq})$.

Generally speaking, in case of preference-ordered data it is meaningful to consider the following five types of decision rules:

1) *certain* $D_{\geq}$*-decision rules*, providing lower profile descriptions for objects belonging to union $Cl_t^{\geq}$ without ambiguity: "*if* $x_{q1} \succeq_{q1} r_{q1}$ *and*

$x_{q2} \succeq_{q2} r_{q2}$ *and* $\ldots x_{qp} \succeq_{qp} r_{qp}$, *then* $x \in Cl_t^{\geq}$", where for each $w_q, z_q \in V_q$, "$w_q \succeq_q z_q$ "means "w_q is at least as good as z_q"; this is the case of Rule 1) which can be re- written as

if $x_{Physics} \succeq_{Physics}$ *Medium and* $x_{Literature} \succeq_{Literature}$ *Medium, then* x belongs to $Cl_2^{\geq}$;

2) *possible* $D_{\geq}$*-decision rules*, providing lower profile descriptions for objects belonging to union $Cl_t^{\geq}$ with or without any ambiguity: "*if* $x_{q1} \succeq_{q1} r_{q1}$ *and* $x_{q2} \succeq_{q2} r_{q2}$ *and* $\ldots x_{qp} \succeq_{qp} r_{qp}$, *then* x possibly belongs to $Cl_t^{\geq}$"; this is the case of the following

   Rule 4): "*if* the evaluation in Physics is at least *Medium, then* the student could be comprehensively at least *Medium*" (S1, S2, S3, S4, S5, S6).

   Let us remark that the conclusion of Rule 4), "the student could be comprehensively at least *Medium*" should be read as "it is not completely certain that the student is *Bad*, so it is possible that she is at least *Medium*". Rule 4) can also be re-written as

if $x_{Physics} \succeq_{Physics}$ *Medium, then* x possibly belongs to $Cl_2^{\geq}$;

3) *certain* $D_{\leq}$*-decision rules*, providing upper profile descriptions for objects belonging to union $Cl_t^{\leq}$ without ambiguity: "*if* $x_{q1} \preceq_{q1} r_{q1}$ *and* $x_{q2} \preceq_{q2} r_{q2}$ *and* $\ldots$ $x_{qp} \preceq_{qp} r_{qp}$, *then* $x \in Cl_t^{\leq}$, where for each $w_q, z_q \in V_q$, "$w_q \preceq_q z_q$" means "w_q is at most as good as z_q"; this is the case of Rule 2) which can be re- written as

if $x_{Physics} \preceq_{Physics}$*Medium and* $x_{Literature} \preceq_{Literature}$ *Bad, then* x belongs to $Cl_2^{\leq}$;

4) *possible* $D_{\leq}$*-decision rules*, providing upper profile descriptions for objects belonging to union $Cl_t^{\leq}$ with or without any ambiguity: "*if* $x_{q1} \preceq_{q1} r_{q1}$ *and* $x_{q2} \preceq_{q2} r_{q2}$ *and* $\ldots$ $x_{qp} \preceq_{qp} r_{qp}$, *then* x possibly belongs to $Cl_t^{\leq}$", this is the case of the following

   Rule 5): "*if* the evaluation in Literature is at most *Bad, then* the student could be comprehensively at most *Medium*" (S1, S2, S7).

   Let us remark that the conclusion of Rule 5) "the student could be comprehensively at most *Medium*" should be read as "it is not completely certain that the student is *Good*, so it is possible that she is at most *Medium*". Rule 5 can also be re-written as

if $x_{Literature} \preceq_{Literature}$*Bad, then* x possibly belongs to $Cl_2^{\leq}$;

5) *approximate $D_{\geq\leq}$-decision rules*, providing simultaneously lower and upper profile descriptions for objects belonging to classes $Cl_s \cup Cl_{s+1} \cup \ldots \cup Cl_t$, without possibility of discerning to which class: "*if* $x_{q1} \succeq_{q1} r_{q1}$ *and* ... $x_{qk} \succeq_{qk} r_{qk}$ *and* $x_{qk+1} \preceq_{qk+1} r_{qk+1}$ *and* ... $x_{qp} \preceq_{qp} r_{qp}$, *then* $x \in Cl_s \cup Cl_{s+1} \cup \ldots \cup Cl_t$"; this is the case of Rule 3) which can be re-written as

if $x_{Physics} \succeq_{Physics}$ *Medium and* $x_{Literature} \preceq_{Literature}$ *Bad, then* x belongs to Cl_1 or Cl_2.

In the left hand side of a $D_{\geq\leq}$-decision rule we can have "$x_q \succeq_q r_q$" and "$x_q \preceq_q r'_q$", where $r_q \leq r'_q$, for the same $q \in C$. Moreover, if $r_q = r'_q$, the two conditions boil down to "$x_q \sim_q r_q$", where for each $w_q, z_q \in V_q$, "$w_q \sim_q z_q$" means "w_q is indifferent to z_q".

The rules of type 1) and 3) represent *certain knowledge* induced from the data table, while the rules of type 2), 4) represent *possible knowledge*, and rules of type 5) represent *doubtful knowledge.*

The rules of type 1) and 3) are *exact*, if they do not cover negative examples, and they are *probabilistic* otherwise. In the latter case, each rule is characterized by a *confidence ratio*, representing the probability that an object matching the left hand side (LHS) of the rule matches also its right hand side (RHS). Probabilistic rules are concordant with the VC-DRSA model presented above. To give an example, consider the following probabilistic $D_{\geq}$-decision rules and $D_{\leq}$-decision rules obtained from Table 13.1 (within parentheses, the symbols of students supporting the corresponding rule but not concordant with its RHS are underlined):

Rule 6): "*if* the evaluation in Mathematics is at least *Good*, *then* the student is comprehensively at least *Good* in 75% of cases (*confidence*)", (S1 [underlined], S4, S5, S6),

Rule 7): "*if* the evaluation in Physics is at most *Medium*, *then* the student is at most *Medium* in 83.3% of cases (*confidence*)", (S1, S2, S3, S5 [underlined], S7, S8).

Let us remark that the probabilistic decision rules are very useful when large data tables are considered. In large data tables, an ambiguity typically exists and prevents finding some very strong patterns because the certain decision rules are contradicted by the ambiguous examples. Probabilistic decision rules, permitting a limited number of counterexamples, may represent these strong patterns.

Since a decision rule is a kind of implication, by a *minimal* rule we understand such an implication that there is no other implication with the antecedent (the LHS of the rule) of at least the same weakness (in other words, a rule using a subset of its elementary conditions and/or weaker elementary conditions) and the consequent (the RHS of the rule) of at least the same strength (in other words, a $D_{\geq}$ or a $D_{\leq}$-decision rule assigning objects to the same union or sub-union of classes, or a $D_{\geq\leq}$-it decision rule assigning objects to the same or larger set of classes). Consider, for example, the following decision rules which both are true for objects from Table 13.1:

Rule A): "*if* the evaluation in Mathematics, Physics and Literature is at least *Good, then* the student is comprehensively at least *Medium*" (S6)

Rule B): "*if* the evaluation in Mathematics is at least *Good* and the evaluation in Literature is at least *Medium, then* the student is comprehensively at least *Good*" (S4, S5, S6).

Comparison of decision rules A) and B) shows that rule A) is not minimal indeed:

(i) rule B) has weaker conditions than rule A) because rule A) has conditions on all three criteria, while rule B) has conditions on Mathematics and Literature only; moreover, rule A) has a stronger requirement than rule B) with respect to Literature (at least *Good* instead of at least *Medium*), while the requirement with respect to Mathematics is not weaker (both rules require an evaluation at least *Good*);

(ii) rule B) has a stronger conclusion than rule A) because "comprehensively at least *Good*" is more precise than "comprehensively at least *Medium*".

A set of decision rules is *complete* if it is able to cover all objects from the data table in such a way that consistent objects are re-classified to their original classes and inconsistent objects are classified to clusters of classes referring to this inconsistency. An example of a complete set of decision rules induced from Table 13.1 is given below (between parentheses there are symbols of students supporting the considered rule):

Rule α): "*if* the evaluation in Mathematics *and* Physics is at most *Bad, then* the student is comprehensively at most *Bad*", (S7, S8);

Rule β): "*if* the evaluation in Physics *and* Literature is at most *Medium, then* the student is comprehensively at most *Medium*", (S1, S2, S3, S7, S8);

Rule γ): "*if* the evaluation in Mathematics *and* Physics is at most *Medium, then* the student is comprehensively at most *Medium*", (S2, S3, S7, S8);

Rule δ): "*if* the evaluation in Physics *and* Literature is at least *Medium, then* the student is comprehensively at least *Medium*", (S3, S4, S5, S6);

Rule ε): "*if* the evaluation in Physics is at least *Good and* the evaluation in Literature is at least *Medium, then* the student is comprehensively at least *Good*", (S4, S6);

Rule ζ): "*if* the evaluation in Physics is at least *Medium and* the evaluation in Literature is at least *Good, then* the student is comprehensively at least *Good*", (S5, S6);

Rule η): "*if* the evaluation in Physics is at least *Medium* and the evaluation in Literature is at most *Bad, then* the student is comprehensively *Bad* or *Medium* (due to ambiguity of information)", (S1, S2).

4.2 Different Strategies of Decision Rule Induction

We call *minimal* each set of decision rules that is complete and non-redundant, i.e. exclusion of any rule from this set makes it non-complete. Remark that our set of decision rules, α) to η), presented at the end of point 4.1 is not minimal. Indeed, one can remove rule γ) and the remaining set of rules is still complete and minimal; elimination of any other rule does not permit a proper reclassification of at least one student from Table 1.

One of three induction strategies can be adopted to obtain a set of decision rules [60]:

- *minimal* description, i.e. generation of a minimal set of rules,
- *exhaustive* description, i.e. generation of all rules for a given data table,
- *characteristic* description, i.e. generation of a set of "strong" rules covering relatively many objects each, however, all together not necessarily all objects from U.

Let us also remark that, contrary to traditional rule induction in machine learning, within DRSA the domains of the considered criteria need not to be discretized, because the syntax of dominance-based rules makes them much less specific than the traditional rules with elementary conditions of the type "attribute=value". This is particularly true for the minimal description strategy of induction because, in the case of exhaustive description, the number of all rules may also augment exponentially with the number of different evaluations on particular criteria. Specific algorithms for induction of decision rules consistent with the dominance principle have been proposed in [11, 14, 62].

4.3 Application of Decision Rules

A set of decision rule can be seen as a preference model and used to support future decisions. Let us suppose that two new students, S9 and S10, not considered in above Table 13.1, are to be evaluated comprehensively. Evaluations of these students in Mathematics, Physics and Literature are given in Table 13.3.

Table 13.3. Evaluations of new students.

Student	Mathematics	Physics	Literature
S9	*Medium*	*Good*	*Good*
S10	*Bad*	*Bad*	*Good*

Using decision rules proposed within DRSA, different types of preference models can be considered. In general, one can consider the following models:

1) preference model composed of $D_{\geq}$-decision rules only,

2) preference model composed of $D_{\leq}$-decision rules only,

3) preference model composed of $D_{\geq}$-decision rules, $D_{\leq}$-decision rules and $D_{\geq\leq}$-decision rules.

In case of model 1), when applying $D_{\geq}$-it decision rules to object x, it is possible that x either matches LHS of at least one decision rule or does not match LHS of any decision rule. Let us consider a preference model composed of three $D_{\geq}$-decision rules: rule δ), rule ε) and rule ζ) presented at the end of point 4.1. One can see that S9 matches the LHS of all three rules while S10 does not match the LHS of any of the three rules.

According to rule δ), S9 is comprehensively at least *Medium*. According to rule ε) and rule ζ), S9 is comprehensively at least *Good*. Thus, it is reasonable to assign S9 to the class of *Good* students. In general, in the case of at least one matching of $D_{\geq}$-decision rules, it is reasonable to conclude that x belongs to class Cl_t, being the lowest class of the upward union $Cl_t^{\geq}$, where $Cl_t^{\geq}$ is the upward union resulting from intersection of all RHS of rules matching x. Precisely, if x matches LHS of rules ρ_1, ρ_2,...,ρ_u, whose RHS are $x \in Cl_{t1}^{\geq}$, $x \in Cl_{t2}^{\geq}$,..., $x \in Cl_{tu}^{\geq}$, respectively, then x is assigned to class Cl_t, where $Cl_t^{\geq} = \cap_{i=1}^{u} Cl_{ti}^{\geq}$ or, equivalently, $t = \max\{t1, t2, \ldots, tu\}$.

Since S10 does not match any $D_{\geq}$-decision rules, decision rule among rule δ), rule ε) and rule ζ), it is reasonable to conclude that S10 is neither at least *Medium* nor at least *Good*. Therefore, S10 is classified as comprehensively *Bad*. The idea behind is that the induced $D_{\geq}$-decision rules are considered as arguments for assignment of new objects to classes Cl_t, where $t > 1$. Therefore, if there is no $D_{\geq}$-decision rules matching a new object x, there is no argument to assign x to Cl_t with $t > 1$; it remains to conclude that x belongs to Cl_1, i.e. to the worst class. In general, in the case of no matching of $D_{\geq}$-decision rules, it is concluded that x belongs to Cl_1, i.e. to the worst class, since no rule with RHS suggesting a better classification of x is matching this object.

Now, let us consider model 2) composed of $D_{\leq}$-decision rules only. Let us assume that it is composed of three rules: rule α), rule β) and rule γ) from 4.1. One can see that S9 does not match the LHS of any rule while S10 matches the LHS of rule α) and rule γ). Since S9 does not match any decision rule from the considered preference model, it is reasonable to conclude that S9 is neither at most *Medium* nor at most *Bad*. Therefore, S9 is classified as comprehensively *Good*. In general, in the case of no matching of $D_{\leq}$-decision rules, it is concluded that x belongs to the best class Cl_n because no rule with RHS suggesting a worse classification of x is matching this object.

According to rule α), S10 is comprehensively at most *Bad*, while, according to rule γ), S10 is comprehensively at most *Medium*. Thus, it is reasonable to assign S10 to the class of *Bad* students. In general, in the case of at least one matching of $D_{\leq}$-decision rules, it is reasonable to conclude that x belongs to class Cl_z, being the highest class of the downward union $Cl_z^{\leq}$ resulting from intersection of all RHS of rules matching x. Precisely, if x matches the LHS of rules ρ_1, ρ_2,...,ρ_v, whose RHS are $x \in Cl_{t1}^{\leq}$, $x \in Cl_{t2}^{\leq}$,..., $x \in Cl_{tv}^{\leq}$, respectively, then x is assigned to class Cl_z, where $Cl_z^{\leq} = \cap_{i=1}^{v} Cl_{ti}^{\leq}$ or, equivalently, $z = \min\{t1, t2, \ldots, tv\}$.

In model 3), $D_{\geq}$-decision rules, $D_{\leq}$-decision rules and $D_{\geq\leq}$-decision rules are used. For example, let us consider a preference model composed of five rules: rule α), rule γ), rule δ), rule ζ) and rule η) from 4.1, and suppose that two new students, S11 and S12, are to be evaluated comprehensively. Evaluations of these students in Mathematics, Physics and Literature are given in Table 13.4.

Table 13.4. Evaluations of new students.

Student	Mathematics	Physics	Literature
S11	*Bad*	*Medium*	*Good*
S12	*Bad*	*Medium*	*Bad*

S11 matches the LHS of $D_{\leq}$-decision rule γ) and of two $D_{\geq}$-decision rules, δ) and ζ). Thus, on the basis of rule γ), S11 is at most *Medium*, while according to rule δ), S11 is at least *Medium*, and according to rule ζ), S11 is at least *Good*. In other words, rule γ) suggests that S11 is comprehensively at most *Medium*, while rules δ) and ζ) suggest that S11 is comprehensively at least *Good*. This means that there is an ambiguity in the comprehensive evaluation of S11 by rule γ) from one side, and rules δ) and ζ) from the other side. In this situation the classes *Medium* and *Good* fix the range of the ambiguous classification and it is reasonable to conclude that student S11 is comprehensively *Medium* or *Good*. In general, for this kind of preference model, the final assignment of an object x matching both, $D_{\geq}$-decision rules $\rho_1^{\geq}, \rho_2^{\geq}, \ldots, \rho_u^{\geq}$, whose RHS are $x \in Cl_{t1}^{\geq}$, $x \in Cl_{t2}^{\geq}, \ldots, x \in Cl_{tu}^{\geq}$, and $D_{\leq}$-decision rules $\rho_1^{\leq}, \rho_2^{\leq}, \ldots, \rho_v^{\leq}$, whose RHS are

$x \in Cl_{z1}^{\leq}, x \in Cl_{z2}^{\leq}, \ldots, x \in Cl_{zv}^{\leq}$, is made to the union of all classes between Cl_t and Cl_z, i.e. to $Cl_t \cup Cl_{t+1} \cup \ldots Cl_z$ if $t \leq z$, or to $Cl_z \cup Cl_{z+1} \cup \ldots Cl_t$ if $t \geq z$, such that $t = \max\{t1, t2, \ldots, tu\}$ and $z = \min\{z1, z2, \ldots, zv\}$. Remark that if only $D_{\geq}$-decision rules or only $D_{\leq}$-decision rules are matching x, then the above union boils down to a single class, Cl_t or Cl_z, respectively.

S12 matches the LHS of $D_{\leq}$-decision rule $\gamma)$ and of $D_{\geq\leq}$-decision rule $\eta)$. Thus, on the basis of rule $\gamma)$, S12 is at most *Medium*, while according to rule $\eta)$, S12 is *Bad* or *Medium*, without possibility of discerning to which one of the two classes it must be assigned. In this situation, it is reasonable to conclude that student S12 is comprehensively *Bad* or *Medium*. In general, for this kind of preference model, the final assignment of an object x is made as follows. Let us suppose that x matches, on one hand, $D_{\geq}$-decision rules $\rho_1^{\geq}$, $\rho_2^{\geq}, \ldots, \rho_u^{\geq}$, whose RHS are $x \in Cl_{t1}^{\geq}, x \in Cl_{t2}^{\geq}, \ldots, x \in Cl_{tu}^{\geq}$, and $D_{\leq}$-decision rules $\rho_1^{\leq}$, $\rho_2^{\leq}, \ldots, \rho_v^{\leq}$, whose RHS are $x \in Cl_{z1}^{\leq}$, $x \in Cl_{z2}^{\leq}, \ldots, x \in Cl_{zv}^{\leq}$, and, on the other hand, $D_{\geq\leq}$-decision rules $\rho_1^{\geq\leq}$, $\rho_2^{\geq\leq}, \ldots, \rho_w^{\geq\leq}$, whose RHS are $x \in Cl_{a1}^{\geq} \cap Cl_{b1}^{\leq}$, $x \in Cl_{a2}^{\geq} \cap Cl_{b2}^{\leq}, \ldots, x \in Cl_{aw}^{\geq} \cap Cl_{bw}^{\leq}$, $ai \leq bi$ for all $i = 1, 2, \ldots, w$. Then, let $t = max\{t1, t2, \ldots, tu\}$, $z = min\{z1, z2, \ldots, zv\}$, $k = min\{a1, a2, \ldots, aw\}$ and $h = max\{b1, b2, \ldots, bw\}$. Now, define A and B as follows:

$$
\begin{aligned}
A &= \begin{cases} Cl_t \cup Cl_{t+1} \cup \ldots Cl_z & \text{if } t \leq z \\ Cl_z \cup Cl_{z+1} \cup \ldots Cl_t & \text{if} t > z \end{cases}, \\
B &= Cl_k \cup Cl_{k+1} \cup \ldots Cl_h.
\end{aligned}
$$

Finally, x is assigned to $A \cup B$.

Recently a new classification procedure for dominance-based probabilistic decision rules coming from VC-DRSA model has been proposed in [1].

4.4 Decision Trees – An Alternative to Decision Rules

The dominance-based rough approximations can also serve to induce *decision trees* representing knowledge discovered from preference-ordered data. Several forms of decision trees, useful for representation of classification patterns, have been proposed by Giove, Greco, Matarazzo and Slowinski [7]. One of these trees, representing knowledge discovered from Table 13.1, is presented in Figure 13.1.

The decision tree presented in Figure 13.1 can be interpreted as follows. The root (node 1) of the tree is a test node. The test formulates the following question with respect to all the students: "is the evaluation in Mathematics at least *Medium*?". The root has two child nodes (node 2 and node 3). The right child node (node 2) concerns the students who passed the test, i.e. all the students being at least *Medium* in Mathematics (S1, S2, S3, S4, S5, S6), while the left child node (node 3) concerns the students who did not pass the test, i.e.

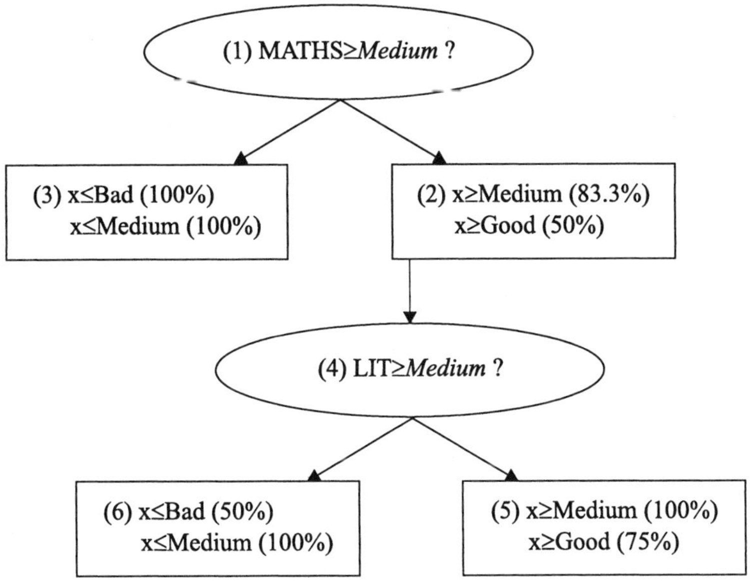

Figure 13.1. Decision tree representing knowledge included from Table 13.1.

all the students being worse than *Medium* in Mathematics (S7, S8). According to node 2, 83.3% of students being at least *Medium* in Mathematics are comprehensively at least *Medium* (S2, S3, S4, S5, S6); moreover, 50% of students being at least *Medium* in Mathematics are comprehensively at least *Good* (S4, S5, S6). According to node 3, 100% of students being worse than *Medium* in Mathematics are comprehensively at most *Bad* (S7, S8). Node 3 says also that the students having an evaluation worse than *Medium* in Mathematics are comprehensively at most *Medium* in 100%. Let us observe that the information that students are at most *Medium* in 100% of cases is redundant with respect to the information that the same students are at most *Bad* in 100% of cases, because a student at most *Bad* is of course also at most *Medium*. Node 4 is the second test node. The test formulates the following question with respect to the students who passed the first test: "is the evaluation in Literature at least *Medium*?". The right node 5 concerns all the students being at least *Medium* in Mathematics and Literature (S3, S4, S5, S6); 100% of these students are comprehensively at least *Medium* (S3, S4, S5, S6) and 75% of them are at least *Good* (S4, S5, S6). The left node 6, in turn, concerns all the students being at least *Medium* in Mathematics and worse than *Medium* in Literature (S1, S2); 50% of these students are comprehensively at most *Bad* (S1) and 100% of them are at most *Medium* (S1, S2).

Let us show how decision tree classifies new objects. Consider the above decision tree and students S9 and S10 from Table 13.3. As evaluation of S9 in Mathematics is *Medium*, according to the test node from the root of the tree, we can conclude that S9 is comprehensively at least *Medium* with credibility = 83.3% and at least *Good* with credibility = 50%. Furthermore, as evaluation of S9 in Literature is *Good*, according to the second test node, the student is for sure comprehensively at least *Medium* (credibility = 100%) and at least *Good* with credibility = 75%. With respect to student S10, the decision tree says that she is for sure comprehensively *Bad* (credibility = 100%). Let us remark the great transparency of the classification decision provided by the decision tree: indeed it explains in detail how the comprehensive evaluation is reached and what is the impact of particular elementary conditions on the confidence of classification.

5. Extensions of DRSA

5.1 DRSA with Joint Consideration of Dominance, Indiscernibility and Similarity Relations

Very often in data tables describing realistic decision problems, there are data referring to a preference order (criteria) and data not referring to any specific preference order (attributes).

The following example illustrates the point. In Table 13.5, six companies are described by means of four attributes:

- A_1, capacity of management,
- A_2, number of employees,
- A_3, localization,
- A_4, company profit or loss.

Table 13.5. Information table of the illustrative example.

Warehouse	Attributes			
	A_1	A_2	A_3	A_4
C1	high	700	A	profit
C2	high	420	A	loss
C3	medium	500	B	profit
C4	medium	555	B	loss
C5	low	400	A	loss
C6	low	100	B	loss

The objective is to induce decision rules explaining profit or loss on the basis of attributes A_1, A_2 and A_3. Let us observe that

- attribute A_1 is a criterion, because the evaluation with respect to the capacity of management is preferentially ordered (high is better than medium, and medium is better than low);

- attribute A_2 is a quantitative attribute, because the values of the number of employees are not preferentially ordered (neither the high number of employees is in general better than the small number, nor the inverse); for quantitative attributes it is reasonable to use a similarity relation, which, in general, is a binary relation, only reflexive and neither transitive nor symmetric; for example, with respect to the data from Table 13.5, similarity between companies can be defined as follows: company a is similar to company b with respect to the attribute "number of employees" if

$$\frac{|number\ of\ employees\ of\ a\ - number\ of\ employees\ of\ b|}{number\ of\ employees\ of\ b} \leq 10\%;$$

  Let us remark that C3 is similar to C4 because $\frac{|500-555|}{555} \leq 10\%$, while C4 is not similar to C3 because $\frac{|555-500|}{500} > 10\%$. This shows how similarity relation may not satisfy symmetry. Let us suppose now that there is another company, C7, having 530 employees. Then, C4 is similar to C7 because $\frac{|555-530|}{530} \leq 10\%$ and C7 is similar to C3 because $\frac{|530-500|}{500} \leq 10\%$. However, we have already verified that C4 is not similar to C3. This shows how similarity may not satisfy transitivity;

- attribute A_3 is a qualitative attribute, because there is no preference order between different types of localization: two companies are indiscernible with respect to localization if they have the same localization;

- decision classes defined by attribute A_4 are preferentially ordered (obviously, profit is better than loss).

Let us remark that indiscernibility is the typical binary relation considered within the classical rough set approach (CRSA) while an extension of rough sets to the similarity relation has been proposed by Slowinski and Vanderpooten [58, 59] (for a fuzzy extension of this approach see [16, 29]). Greco, Matarazzo and Slowinski [18, 21] proposed an extension of DRSA to deal with data table like Table 13.5, where preference, indiscernibility and similarity are to be considered jointly. Applying this approach to Table 13.5, several decision rules can be induced; the following set of decision rules covers all the examples (within parentheses there are symbols of companies supporting the corresponding decision rule):

Rule 1): "*if* capacity of management is medium, *then* the company makes profit or loss", (C3, C4),

Rule 2): "*if* capacity of management is (at least) high *and* the number of employees is similar to 700, *then* the company makes profit", (C1),

Rule 3): "*if* capacity of management is (at most) low, *then* the company makes loss", (C5, C6),

Rule 4): "if the number of employees is similar to 420, then the company makes loss", (C2, C5).

5.2 DRSA and Interval Orders

In the previous sections we considered precise evaluations of objects on particular criteria and precise assignment of each object to one class. In practice, however, due to imprecise measurement, random variation of some parameters, unstable perception or incomplete definition of decision classes and preference scales of criteria, the evaluations and/or assignment may not be univocal. This was not the case in our example considered above; however, it is realistic to ask how DRSA should change in order to handle Table 13.1 augmented by students S13, S14 and S15 presented in Table 13.6.

Table 13.6. Students with interval evaluations.

Student	Mathematics	Physics	Literature	Comprehensive evaluation
S13	*Medium-Good*	*Medium*	*Bad-Medium*	*Bad*
S14	*Medium*	*Good*	*Medium*	*Medium-Good*
S15	*Medium-Good*	*Medium-Good*	*Medium*	*Medium-Good*

This adaptation of DRSA has been considered in [4]. Its basic idea consists in approximation of an interval order of the comprehensive evaluation by means of interval orders on particular criteria; the key concept of this approximation is a specially defined dominance relation.

5.3 Fuzzy DRSA – Rough Approximations by Means of Fuzzy Dominance Relations

The concept of dominance can be refined by introducing gradedness through the use of fuzzy sets in the sense of semantics expressing preferences for pairs of objects (for a detailed presentation of fuzzy preferences see [6]; see also

Chapter 2). The gradedness introduced by the use of fuzzy sets refines the classic crisp preference structures. The idea is the following. Let us consider the problem of classifying some enterprises according to their profitability. Let us suppose that the DM decides that the enterprises should be classified according to their ROI (Return On Investment). More precisely, she considers that enterprise x with ROI not smaller than 12% should be assigned to the class of profitable enterprises $Cl_2^{\geq}$ and, otherwise, it should be assigned to the class of non profitable enterprises $Cl_1^{\leq}$. Now, consider enterprise a with ROI equal to 12% and enterprise b with ROI equal to 11.9%. The difference between the ROI of a and b is very small, however, it is enough to make a radically different assignment of these two enterprises. This example shows that it would be more reasonable to consider a smooth transition from $Cl_1^{\leq}$ to $Cl_2^{\geq}$. Such a transition can be controlled by a graded credibility $Cl_2^{\geq}(x)$, telling to what degree enterprise x belongs to $Cl_2^{\geq}$, defined as follows:

$$Cl_2^{\geq}(x) = \begin{cases} 0 & \text{if ROI}(x) < 10\% \\ (ROI(x) - 10)/2 & \text{if } 10\% \leq ROI(x) < 12\% \\ 1 & \text{if } ROI(x) \geq 12\% \end{cases}$$

The correlative credibility that enterprise x belongs to $Cl_1^{\leq}$ can be defined as: $Cl_1^{\leq}(x)$=1–$Cl_2^{\geq}(x)$.

According to the above definition, we get $Cl_2^{\geq}(a)$=1 and $Cl_2^{\geq}(b)$=0.95, which means that a is for sure a profitable enterprise while b is profitable with a credibility of 95%. Thus, the small difference between ROI(a) and ROI(b) does not lead to radically different classification of the two enterprises with respect to profitability.

The above reasoning about a smooth transition from truth to falsity of an inclusion relation can be applied to the dominance relation considered in the rough approximations. In Section 3, the dominance relation xD_Py has been declared true if evaluations of object x on all criteria from set P are not worse than those of object y. Continuing our example of classification with respect to profitability, let us consider among criteria the percentage growth of the sales, denoted by GS. Let us also define the weak preference relation $\succeq_{GS}$ with respect GS as

$$x \succeq_{GS} y \Leftrightarrow GS(x) \geq GS(y). \tag{13.1}$$

(13.1) can be read as "enterprise x is at least as good as y with respect to GS if and only if GS of x is greater than or equal to GS of y". Considering two enterprises, a and b, one can remark that if the difference between $GS(a)$ and $GS(b)$ is small, for example $GS(a)$=12% and $GS(b)$=12.1%, then definition (i) is too restrictive; in this situation it is hard to say that enterprise b is definitely better than enterprise a. It is thus realistic to assume an indifference threshold $q > 0$ on criterion GS, so that the following definition would replace (13.1)

$$x \succeq_{GS} y \Leftrightarrow GS(x) \geq GS(y) - q. \qquad (13.2)$$

(13.2) can be read as "enterprise x is at least as good as y with respect to GS if and only if GS of x is greater than or equal to GS of y decreased by an indifference threshold q". For example, if q=1%, then enterprise a is indifferent to enterprise b. Definition (13.2) is not yet completely satisfactory. Let us consider enterprises c and d such that $GS(c) = 13\%$ and $GS(d) = 13.1\%$. Using (13.2) with q=1% we have to conclude that $a \succeq_{GS} c$ while *non* $a \succeq_{GS} d$. This is counterintuitive because the difference between $GS(c)$ and $GS(d)$ is very small and one would expect a similar result of comparison of c and d with a. Therefore, the following reformulation of the definition of weak preference $\succeq_{GS}$ seems reasonable:

$$\text{“}x \text{ is at least as good as } y \text{ with a credibility } \succeq_{GS}(x, y)\text{''} \qquad (13.3)$$

where

$$\succeq_{GS}(x,y) = \begin{cases} 0 & \text{if } GS(x) < GS(y) - p \\ \frac{p-(GS(y)-GS(x))}{p-q} & \text{if } -p \leq GS(x) - GS(y) < -q \\ 1 & \text{if } GS(x) \geq GS(y) - q \end{cases}$$

and p is a preference threshold such that $p > q$.

(13.3) has the following interpretation:

- it is completely true ($\succeq_{GS}(x, y) = 1$) that enterprise x is at least as good as y with respect to GS under the same condition as (ii);
- it is completely false ($\succeq_{GS}(x, y) = 0$) that enterprise x is at least as good as y with respect to GS when $GS(x)$ is smaller than $GS(y)$ by at least p;
- between the two extremes ($0 < \succeq_{GS}(x, y) < 1$), the credibility that enterprise x is at least as good as y with respect to GS increases linearly with the opposite of the difference $GS(y)$–$GS(x)$.

Applying (13.3) with q=1% and p=2% to enterprises a, c, and d described above, one gets $\succeq_{GS}(a, c) = 1$ and $\succeq_{GS}(a, d) = 0.9$. Thus, considering comparison of c and d with a on GS, the small difference between $GS(c)$ and $GS(d)$ does not give as radically different results as before.

In [23, 27, 28] and in [9], DRSA was extended by using in two different ways *fuzzy dominance relation*. These extensions of the rough approximation into the fuzzy context maintain the same desirable properties of the crisp rough approximation of preference-ordered decision classes. These generalizations follow the traditional line of using fuzzy logical connectives in definitions of lower and upper approximation. In fact, there is no rule for the choice of the "right" connective, so this choice is always arbitrary to a certain extent. For this reason, in [10], a new fuzzy rough approximation was proposed. It avoids the use of fuzzy connectives, such as T-norm, T-conorm and fuzzy implication, which extend "and", "or" and "if..., then..." operators within fuzzy logic (see, for example, [6]), but at the price of introducing a certain degree of subjectivity related to the choice of one or another of their functional form. The proposed approach solves this problem because it is based on the ordinal properties of fuzzy membership functions only.

5.4 DRSA with Missing Values – Multiple-Criteria Classification Problem with Missing Values

In practical applications, the data table is often incomplete because some data are missing. For example, let us consider the profiles of students presented in Table 13.7, where "*" means that the considered evaluation is missing (for example students S16 and S17 have not yet passed the examination in Literature and Physics, respectively).

Table 13.7. Example of missing values in the evaluation of students.

Student	Mathematics	Physics	Literature	Comprehensive evaluation
S16	*Medium*	*Bad*	$*$	*Bad*
S17	*Medium*	$*$	*Good*	*Good*
S18	*Medium*	*Medium*	*Medium*	*Medium*

An extension of DRSA enabling the analysis of incomplete data tables has been proposed in [19] and [24]. In this extension it is assumed that the dominance relation between two objects is a directional statement, where a subject object is compared to a referent object having no missing values on considered criteria. With respect to Table 13.7, one can say that, taking into account all the criteria,

a) subject S17 dominates referent S18: in fact, referent S18 has no missing value;

b) it is unknown if subject S18 dominates referent S17: in fact, referent S17 has no evaluation in Physics.

From a) we can derive the following decision rule: "*if* a student is at least *Medium* in Mathematics, Physics and Literature, *then* the student is comprehensively at least *Medium*". This rule can be simplified into one of the following rules: "if a student is at least *Medium* in Physics, *then* the student is comprehensively at least *Medium*" or "if a student is at least *Medium* in Literature, *then* the student is comprehensively at least *Medium*".

The advantage of this approach is that the rules induced from the rough approximations defined according to the extended dominance relation are *robust*, i.e. each rule is supported by at least one object with no missing value on the criteria represented in the condition part of the rule. To better understand this feature, let us compare the above approach with another approach suggested to deal with missing values [41, 42]. In the latter it is proposed to substitute an object having a missing value by a set of objects obtained by putting all possible evaluations in the place of the missing value. Thus, from Table 13.7 one would obtain the following Table 13.8.

Table 13.8. Substitution of missing values in the evaluation of students.

Student	Mathematics	Physics	Literature	Comprehensive evaluation
S16A	*Medium*	*Bad*	*Bad*	*Bad*
S16B	*Medium*	*Bad*	*Medium*	*Bad*
S16C	*Medium*	*Bad*	*Good*	*Bad*
S17A	*Medium*	*Bad*	*Good*	*Good*
S17B	*Medium*	*Medium*	*Good*	*Good*
S17C	*Medium*	*Good*	*Good*	*Good*
S18	*Medium*	*Medium*	*Medium*	*Medium*

From Table 13.8, one can induce the rule: "*if* a student is at least *Medium* in Physics *and* at least *Good* in Literature, *then* the student is comprehensively at least *Good*". However, this rule is not robust because in the original Table 13.7, no student has such a profile.

DRSA extended to deal with missing values maintains all good characteristics of the dominance-based rough set approach and boils down to the latter when there are no missing values. This approach can also be used to deal with decision table in which dominance, similarity and indiscernibility must be considered jointly with respect to criteria and attributes.

5.5 DRSA for Decision under Uncertainty

In [32] we opened a new avenue for applications of the rough set concept to analysis of preference-ordered data. We considered the classical problem of decision under uncertainty extending DRSA by using *stochastic dominance*. In a risky context, an act A stochastically dominates an act B if, for all possible levels k of gain or loss, the probability of obtaining an outcome at least as good as k with A is not smaller than with B. In this context we have an ambiguity if an act A stochastically dominates an act B, but, nevertheless, B has a comprehensive evaluation better than A. On this basis, it is possible to restate all the concepts of DRSA and adapt this approach to preference analysis under risk and uncertainty. We considered the case of traditional additive probability distribution over the set of future states of the world; however, the model is rich enough to handle non-additive probability distributions and even qualitative ordinal distributions. The rough set approach gives a representation of DM's preferences under uncertainty in terms of "*if..., then...*" decision rules induced from rough approximations of sets of exemplary decisions (preference-ordered classification of acts described in terms of outcomes in uncertain states of the world). This extension is interesting with respect to MCDA from two different viewpoints:

1) each decision under uncertainty can be viewed as a multiple-criteria decision, where the criteria are the outcomes in different states of the world;

2) DRSA adapted to decision under uncertainty can be applied to deal with multiple-criteria decision under uncertainty, i.e. decision problem where in each future state of the world the outcomes are expressed in terms of a set of criteria (see Chapter 7 and Chapter 11).

5.6 DRSA for Hierarchical Structure of Attributes and Criteria

In many real life situations, the process of decision-making is decomposable into sub-problems; this decomposition may either follow from a natural hierarchical structure of the evaluation or from a need of simplification of a complex decision problem. These situations are referred to *hierarchical decision problems*. The structure of a hierarchical decision problem has the form of a *tree* whose *nodes* are attributes and criteria describing objects. An example structure of a hierarchical classification problem is shown in Figure 13.2. The cars are sorted into three classes: acceptable, hardly acceptable and non-acceptable, on the basis of three criteria (Price, Max speed, Fuel consumption) and two regular attributes (Colour and Country of production); one of criteria – Fuel consumption – is further composed of four sub-criteria.

In [4], hierarchical decision problems are considered where the decision is made in a finite number of steps due to hierarchical structure of regular attributes and criteria. The proposed methodology is based on decision rule preference model induced from examples of hierarchical decisions made by the DM on a set of reference objects. To deal with inconsistencies appearing in decision examples, DRSA has been adapted to hierarchical classification problems. In these problems, the main difficulty consists in *propagation* of inconsistencies along the tree, i.e. taking into account at each node of the tree the inconsistent

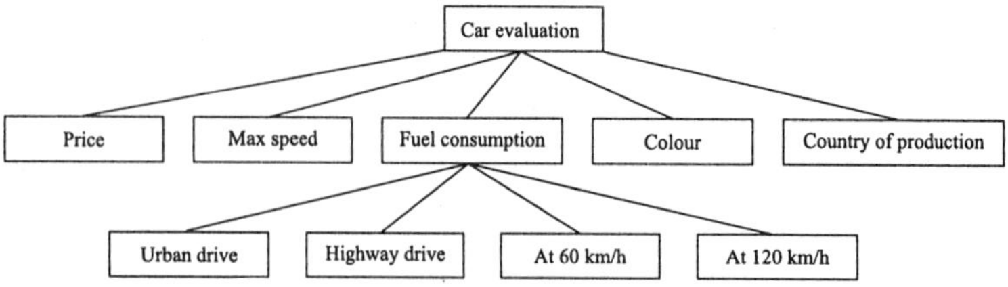

Figure 13.2. The hierarchy of attributes and criteria for a car classification problem.

information coming from lower level nodes. In the proposed methodology, the inconsistencies are propagated from the bottom to the top of the tree in the form of *subsets* of possible attribute values instead of single values. In the case of hierarchical criteria, these subsets are *intervals* of possible criterion values. Subsets of possible values may also appear in leafs of the tree, i.e. in evaluations of objects by the lowest-level attributes and criteria. To deal with multiple values of attributes in description of objects the classical rough set approach has been adapted adequately. Interval evaluations of objects on particular criteria can be handled by DRSA extended to interval orders (see point 5.2).

6. DRSA for Multiple-criteria Choice and Ranking

DRSA can also be applied to multiple-criteria choice and ranking problems. However, there is a basic difference between classification problems from one side and choice and ranking from the other side. To give a didactic example, consider a set of companies A for evaluation of a risk of failure, taking into account the debt ratio criterion. To assess the risk of failure of company x, we will not compare the debt ratio of x with the debt ratio of all the other companies from A. The comparison will be made with respect to a fixed risk threshold on the debt ratio criterion. Indeed, the debt ratio of x can be the highest of all companies from A and, nevertheless, x can be classified as a low risk company if its debt ratio is below the fixed risk threshold. Consider, in turn, the situation, in which we must choose the lowest risk company from A or we want to rank the companies from A from the less risky to the most risky one. In this situation, the comparison of the debt ratio of x with a fixed risk threshold is not useful and, instead, a pairwise comparison of the debt ratio of x with the debt ratio of all other companies in A is relevant for the choice or ranking. Thus, in general, while classification is based on absolute evaluation of objects (e.g. comparison of the debt ratio with the fixed risk threshold), choice and ranking refer to relative evaluation, by means of pairwise comparisons of objects (e.g. comparisons of the debt ratio of pairs of companies).

The necessity of pairwise comparisons of objects in multiple-criteria choice and ranking problems requires some further extensions of DRSA. Simply speaking, in this context we are interested in the approximation of a binary relation, corresponding to a comprehensive preference relation, using other binary relations, corresponding to marginal preference relations on particular criteria, for pairs of objects. In the above example, we would approximate the binary relation "from the viewpoint of the risk of failure, company x is comprehensively preferred to company y" using binary relations on the debt ratio criterion, like "the debt ratio of x is *much better* than that of y" or "the debt ratio of x is *weakly better* than that of y", and so on.

Technically, the modification of DRSA necessary to approach the problems of choice and ranking are twofold:

1) *pairwise comparison table* (PCT) is considered instead of the simple data table [20]: PCT is a decision table whose rows represent pairs of objects for which multiple-criteria evaluations and a comprehensive preference relation are known;

2) *dominance principle* is considered for *pairwise comparisons* instead of simple objects: if object x is preferred to y at least as strongly as w is preferred to z on all the considered criteria, then x must be comprehensively preferred to y at least as strongly as w is comprehensively preferred to z.

The application of DRSA to the choice or ranking problems proceeds as follows. First, the DM gives some examples of pairwise comparisons with respect to some reference objects, for example a complete ranking from the best to the worst of a limited number of objects – well known to the DM. From this set of examples, a preference model in terms of "*if* ..., *then*..." decision rules is induced. These rules are applied to a larger set of objects. A proper exploitation of the results so obtained gives a final recommendation for the decision problem at hand. Below, we present more formally and in greater detail this methodology.

6.1 Pairwise Comparison Table (PCT) as a Preferential Information and a Learning Sample

Let A be the set of objects for the decision problem at hand. Let us also consider a set of reference objects $B \subseteq A$ on which the DM is expressing her preferences by pairwise comparisons. Let us represent the comprehensive preference by a function $P : A \times A \to \boldsymbol{R}$. In general, for each $x, y \in A$,

- if $P(x, y) > 0$, then $P(x, y)$ can be interpreted as a degree to which x is evaluated better than y,
- if $P(x, y) < 0$, then $P(x, y)$ can be interpreted as a degree to which x is evaluated worse than y,

- if $P(x,y) = 0$, then x is evaluated equivalent to y.

The semantic value of preference P can be different. We remember two possible interpretations:

(a) $P(x,y)$ represents a *degree of outranking* of x over y, i.e. $P(x,y)$ is the credibility of the proposition "x is at least as good as y";

(b) $P(x,y)$ represents a *degree of net preference* of x over y, i.e. $P(x,y)$ is the strength with which x is preferred to y.

In case (a), $P(x,y)$ measures the strength of arguments in favor of x and against y, while $P(y,x)$ measures the arguments in favor of y and against x. Thus, there is no relation between values of $P(x,y)$ and $P(y,x)$. In case (b), $P(x,y)$ synthesizes arguments in favor of x and against y together with arguments in favor of y and against x. $P(y,x)$ has a symmetric interpretation and the relation $P(x,y) = -P(y,x)$ is expected.

Let us suppose that objects from set A are evaluated by a consistent family of n criteria $g_i : A \to \boldsymbol{R}$, $i = 1,2,\dots,n$, such that, for each object $x \in A$, $g_i(x)$ represents the evaluation of x with respect to criterion g_i. Using the terms of the rough set approach, the family of criteria constitutes the set C of condition attributes. With respect to each criterion $g_i \in C$ one can consider a particular preference function $P_i : \boldsymbol{R} \times \boldsymbol{R} \to \boldsymbol{R}$, such that for all $x, y \in A$, $P_i[g_i(x), g_i(y)]$ for criterion g_i has an interpretation analogous to comprehensive preference relation $P(x,y)$, i.e.

- if $P_i[g_i(x), g_i(y)] > 0$, then $P_i[g_i(x), g_i(y)]$ is a degree to which x is better than y on criterion g_i,
- if $P_i[g_i(x), g_i(y)] < 0$, then $P_i[g_i(x), g_i(y)]$ is a degree to which x is worse than y on criterion g_i,
- if $P_i[g_i(x), g_i(y)] = 0$, then x is equivalent to y on criterion g_i.

Let us suppose that the DM expresses her preferences with respect to pairs (x,y) from $E \subseteq B \times B, |E| = m$. These preferences are represented in an $m \times (n+1)$ Pairwise Comparison Table $\boldsymbol{S}_{PCT}$. The m rows correspond to the pairs from E. For each $(x,y) \in E$ in the corresponding row, the first n columns include information about preferences $P_i[g_i(x), g_i(y)]$ on particular criteria from set C, while the last, $(n+1) - th$ column represents the comprehensive preference $P(x,y)$.

6.2 Multigraded Dominance

Given subset $P \subseteq C$ ($P \neq \emptyset$) of criteria and pairs of objects $(x,y), (w,z) \in A \times A$, the pair (x,y) is said to *P-dominate* the pair (w,z) (denotation

$(x, y)D_P(w, z))$, if $P_i[g_i(x), g_i(y)] > P_i[g_i(w), g_i(z)]$ for all $g_i \in P$, i.e. if x is preferred to y at least as strongly as w is preferred to z with respect to each criterion $g_i \subset P$. Let us remark that the dominance relation D_P is a partial preorder on $A \times A$; as, in general, it involves different grades of preference on particular criteria, it is called *multigraded* dominance relation.

Given $P \subseteq C$ and $(x, y) \in E$, we define:

- a set of pairs of objects *P-dominating* (x, y), called *P-dominating set*, $D_P^+(x, y) = \{(w, z) \in E : (w, z)D_P(x, y)\}$,
- a set of pairs of objects *P-dominated* by (x, y), called *P-dominated set*, $D_P^-(x, y) = \{(w, z) \in E : (x, y)D_P(w, z)\}$.

The P-dominating sets and the P-dominated sets defined on E for considered pairs of reference objects from E are "granules of knowledge" that can be used to express *P-lower* and *P-upper approximations* of set $B_k^{\geq} = \{(x, y) \in E : P(x, y) \geq k\}$, corresponding to comprehensive preference of degree *at least* k, and set $B_k^{\leq} = \{(x, y) \in E : P(x, y) \leq k\}$, corresponding to comprehensive preference of degree *at most* k, respectively:

$$\begin{aligned}
\underline{P}(B_k^{\geq}) &= \{(x, y) \in E : D_P^+(x, y) \subseteq B_k^{\geq}\},\\
\overline{P}(B_k^{\geq}) &= \bigcup_{(x,y)\in B_k^{\geq}} D_P^+(x, y) = \{(x, y) \in E : D_P^-(x, y) \cap B_k^{\geq} \neq \emptyset\},\\
\underline{P}(B_k^{\leq}) &= \{(x, y) \in E : D_P^-(x, y) \subseteq B_k^{\leq}\},\\
\overline{P}(B_k^{\leq}) &= \bigcup_{(x,y)\in B_k^{\geq}} D_P^-(x, y) = \{(x, y) \in E : D_P^+(x, y) \cap B_k^{\geq} \neq \emptyset\}.
\end{aligned}$$

The set difference between P-lower and P-upper approximations of sets $B_k^{\geq}$ and $B_k^{\leq}$ contains all the ambiguous pairs (x, y):

$$Bn_P(B_k^{\geq}) = \overline{P}(B_k^{\geq}) - \underline{P}(B_k^{\geq}),\ Bn_P(B_k^{\leq}) = \overline{P}(B_k^{\leq}) - \underline{P}(B_k^{\leq}),$$

The above rough approximations of $B_k^{\geq}$ and $B_k^{\leq}$ satisfy properties analogous to the rough approximations of upward and downward unions of classes $Cl_t^{\geq}$ and $Cl_t^{\leq}$; precisely, these are:

- inclusion: $\underline{P}(B_k^{\geq}) \subseteq B_k^{\geq} \subseteq \overline{P}(B_k^{\geq})$, $\underline{P}(B_k^{\leq}) \subseteq B_k^{\leq} \subseteq \overline{P}(B_k^{\leq})$,
- complementarity: $\underline{P}(B_k^{\geq}) = E - \overline{P}(B_k^{\leq})$, $\overline{P}(B_k^{\geq}) = E - \underline{P}(B_k^{\leq})$, $\underline{P}(B_k^{\leq}) = E - \overline{P}(B_k^{\geq})$, $\overline{P}(B_k^{\leq}) = E - \underline{P}(B_k^{\geq})$, where $B_k^{>} = E - B_k^{\leq}$

and $B_k^{<} = E - B_k^{\geq}$ and the rough approximation of $B_k^{>}$ and $B_k^{<}$ are analogous to those of $B_k^{\geq}$ and $B_k^{\leq}$, for example, $\underline{P}(B_k^{>}) = \{(x,y) \in E : D_P^{+}(x,y) \subseteq B_k^{>}\}$,

- monotonicity: for each $R, P \subseteq C$, such that $R \subseteq P$, $\underline{R}(B_k^{\geq}) \subseteq \underline{P}(B_k^{\geq})$, $\overline{R}(B_k^{\geq}) \supseteq \overline{P}(B_k^{\geq})$, $\underline{R}(B_k^{\leq}) \subseteq \underline{P}(B_k^{\leq})$, $\overline{R}(B_k^{\leq}) \supseteq \overline{P}(B_k^{\leq})$.

The concepts of the quality of approximation, reducts and core can be extended also to the approximation of the comprehensive preference relation by multigraded dominance relations. In particular, the coefficient

$$\gamma_P = \frac{\left| E - \bigcup_k Bn_P\left(B_k^{\geq}\right)\right|}{|E|} = \frac{\left| E - \bigcup_k Bn_P\left(B_k^{\leq}\right)\right|}{|E|}.$$

defines the *quality of approximation of comprehensive preference* $P(x,y)$ by criteria from $P \subseteq C$. It expresses the ratio of all pairs $(x,y) \in E$ whose degree of preference of x over y is correctly assessed using set P of criteria, to all the pairs of objects contained in E. Each minimal subset $P \subseteq C$, such that $\gamma_P = \gamma_C$, is called *reduct* of C (denoted by $RED_{\mathbf{S}_{PCT}}$). Let us remark that $\boldsymbol{S}_{PCT}$ can have more than one reduct. The intersection of all reducts is called the *core* (denoted by $CORE_{\mathbf{S}_{PCT}}$).

It is also possible to use the Variable Consistency Model on $\boldsymbol{S}_{PCT}$ [56] relaxing the definitions of P-lower approximations of graded comprehensive preference relations represented by sets $B_k^{\geq}$ and $B_k^{\leq}$, such that some pairs in P-dominated or P-dominating sets belong to the opposite relation but at least $l{*}100\%$ of pairs belong to the correct one. Then, the definition of P-lower approximations of $B_k^{\geq}$ and $B_k^{\leq}$ with respect to set $P \subseteq C$ of criteria boils down to:

$$\underline{P}^{l}\left(B_k^{\geq}\right) = \left\{(x,y) \in E : \frac{\left|D_P^{+}(x,y)) \cap B_k^{\geq}\right|}{\left|D_P^{+}(x,y)\right|} \geq l\right\},$$

$$\underline{P}^{l}\left(B_k^{\leq}\right) = \left\{(x,y) \in E : \frac{\left|D_P^{-}(x,y)) \cap B_k^{\leq}\right|}{\left|D_P^{-}(x,y)\right|} \geq l\right\},$$

6.3 Induction of Decision Rules from Rough Approximations of Graded Preference Relations

Using the rough approximations of sets $B_k^{\geq}$ and $B_k^{\leq}$, that is rough approximations of comprehensive preference relation $P(x,y)$ of degree at least or at most k, respectively, it is possible to induce a generalized description of the preferential information contained in a given $\boldsymbol{S}_{PCT}$ in terms of decision rules with a special syntax. We are considering decision rules of the following types:

1) $D_{\geq}$-*decision rules*:
   if $P_{i1}[g_{i1}(x), g_{i1}(y)] \geq k_{i1}$ and ... $P_{ir}[g_{ir}(x), g_{ir}(y)] \geq k_{ir}$, then $P(x, y) \geq k$

   where $\{g_{i1}, ..., g_{ir}\} \subseteq C$; for example: "*if* car x is much better than y with respect to maximum speed *and* at least weakly better with respect to acceleration, *then* x is comprehensively better than y"; these rules are supported by pairs of objects from the P-lower approximation of sets $B_k^{\geq}$ only;

2) $D_{\leq}$-*decision rules*:
   if $P_{i1}[g_{i1}(x), g_{i1}(y)] \leq k_{i1}$ and ... $P_{ir}[g_{ir}(x), g_{ir}(y)] \leq k_{ir}$, then $P(x, y) \leq k$

   where $\{g_{i1}, ..., g_{ir}\} \subseteq C$; for example: "*if* car x is much worse than y with respect to price *and* weakly worse with respect to comfort, *then* x is comprehensively worse than y"; these rules are supported by pairs of objects from the P-lower approximation of sets $B_k^{\leq}$ only;

3) $D_{\geq\leq}$-*decision rules*:
   if $P_{i1}[g_{i1}(x), g_{i1}(y)] \geq k_{i1}$ and ... $P_{ir}[g_{ir}(x), g_{ir}(y)] \geq k_{ir}$ and $P_{j1}[g_{j1}(x), g_{j1}(y)] \leq k_{j1}$ and ... $P_{js}[g_{js}(x), g_{js}(y)] \leq k_{js}$, then $h \leq P(x, y) \leq k$

   where $\{g_{i1}, ..., g_{ir}\}, \{g_{j1}, ..., g_{js}\} \subseteq C$; for example: "*if* car x is much worse than y with respect to price *and* much better with respect to comfort, *then* x is indifferent or better than y, and there is not enough information to distinguish between the two situations"; these rules are supported by pairs of objects from the intersection of the P-upper approximation of sets $B_k^{\geq}$ and $B_h^{\leq}$ $(h < k)$ only.

6.4 Use of Decision Rules for Decision Support

The decision rules induced from rough approximations of sets $B_k^{\geq}$ and $B_k^{\leq}$ for a given $\boldsymbol{S}_{PCT}$, describe the comprehensive preference relations $P(x, y)$ either exactly ($D_{\geq}$ and $D_{\leq}$-decision rules) or approximately ($D_{\geq\leq}$-decision rules). A set of these rules covering all pairs of $\boldsymbol{S}_{PCT}$ represent a preference model of the DM who gave the pairwise comparison of reference objects. Application of these decision rules on a new subset $M \subseteq A$ of objects induces a specific preference structure on M.

For simplicity, in the following we consider the case where $P(x, y)$ is interpreted as outranking and assumes two values only: $P(x, y) = 1$, which means that x is at least as good as y, and $P(x, y) = -1$, which means that x is *not* at least as good as y. In the following $P(x, y) = 1$ will be denoted by xSy and $P(x, y) = -1$ will be denoted by xS^cy.

In fact, any pair of objects $(u, v) \in M \times M$ can match the decision rules in one of four ways:

- at least one $D_{\geq}$-decision rule and neither $D_{\leq}$ nor $D_{\geq\leq}$-decision rules,
- at least one $D_{\leq}$-decision rule and neither $D_{\geq}$ nor $D_{\geq\leq}$-decision rules,
- at least one $D_{\geq}$-decision rule and at least one $D_{\leq}$-decision rule, or at least $D_{\geq\leq}$-decision rules,
- no decision rule.

These four ways correspond to the following four situations of outranking, respectively:

- uSv and *not* uS^cv, that is *true outranking* (denoted by uS^Tv),
- uS^cv and *not* uSv, that is *false outranking* (denoted by uS^Fv),
- uSv and uS^cv, that is *contradictory outranking* (denoted by uS^Kv),
- not uSv and not uS^cv, that is *unknown outranking* (denoted by uS^Uv).

The four above situations, which together constitute the so-called *four-valued outranking* [37], have been introduced to underline the presence and absence of *positive* and *negative* reasons for the outranking. Moreover, they make it possible to distinguish contradictory situations from unknown ones.

A final *recommendation* (choice or ranking) can be obtained upon a suitable exploitation of this structure, i.e. of the presence and the absence of outranking S and S^c on M. A possible exploitation procedure consists in calculating a specific score, called Net Flow Score, for each object $x \in M$:

$$S_{nf}(x) = S^{++}(x) - S^{+-}(x) + S^{-+}(x) - S^{--}(x),$$

where

$$\begin{aligned}
S^{++}(x) &= \text{card}(\{y \in M\text{: there is at least one decision rule which affirms } xSy\}),\\
S^{+-}(x) &= \text{card}(\{y \in M\text{: there is at least one decision rule which affirms } ySx\}),\\
S^{-+}(x) &= \text{card}(\{y \in M\text{: there is at least one decision rule which affirms } yS^cx\}),\\
S^{--}(x) &= \text{card}(\{y \in M\text{: there is at least one decision rule which affirms } xS^cy\}).
\end{aligned}$$

The recommendation in ranking problems consists of the total preorder determined by $S_{nf}(x)$ on M; in choice problems, it consists of the object(s) $x^* \in M$, such that $S_{nf}(x^*) = \max_{x \in M}\{S_{nf}(x)\}$.

The above procedure has been characterized with reference to a number of desirable properties in [37].

6.5 Illustrative Example

Let us suppose that a company managing a chain of warehouses wants to buy some new warehouses. To choose the best proposals or to rank them all, the managers of the company decide to analyze first the characteristics of eight warehouses already owned by the company (reference objects). This analysis should give some indications for the choice and ranking of the new proposals. Eight warehouses belonging to the company have been evaluated by three following criteria: capacity of the sales staff (g_1), perceived quality of goods (g_2) and high traffic location (g_3). The domains (scales) of these attributes are presently composed of three preference-ordered echelons: $V_1 = V_2 = V_3 = \{$*sufficient, medium, good*$\}$. The decision attribute (d) indicates the profitability of warehouses, expressed by the Return On Equity (ROE) ratio (in %). Table 13.9 presents a decision table with the considered reference objects.

Table 13.9. Decision table with reference objects.

Warehouse	g_1	g_2	g_3	d (ROE %)
1	*good*	*medium*	*good*	10.35
2	*good*	*sufficient*	*good*	4.58
3	*medium*	*medium*	*good*	5.15
4	*sufficient*	*medium*	*medium*	-5
5	*sufficient*	*medium*	*medium*	2.42
6	*sufficient*	*sufficient*	*good*	2.98
7	*good*	*medium*	*good*	15
8	*good*	*sufficient*	*good*	-1.55

With respect to the set of criteria $C = \{g_1, g_2, g_3\}$, the following numerical representation is used for criterion g_i, $i = 1, 2, 3$: $g_i(x) = 1$ if x is *sufficient*, $g_i(x) = 2$ if x is *medium*, $g_i(x) = 3$ if x is *good*.

The degree of preferences with respect to pairs of actions are defined as $P_i[(x), g_i(y)] = g_i(x) - g_i(y), i = 1, 2, 3$, and they are coded as follows:

$P_i[g_i(x), g_i(y)] = -2 \Leftrightarrow P_i[g_i(y), g_i(x)] = 2$, which means that "$x$ is worse than y" and "y is better than x",

$P_i[g_i(x), g_i(y)] = -1 \Leftrightarrow P_i[g_i(y), g_i(x)] = 1$, which means that "$x$ is weakly worse than y" or "y is weakly better than x",

$P_i[g_i(x), g_i(y)] = P_i[g_i(y), g_i(x)] = 0$, which means that "$x$ is equivalent to y".

Using the decision attribute, the comprehensive outranking relation was build as follows: warehouse x is at least as good as warehouse y with respect to profitability $(P(x,y) = 1 \Leftrightarrow xSy)$ if

$$\text{ROE}(x) \geq \text{ROE}(y) - 2\%.$$

Otherwise, i.e. if ROE(x) <ROE(y) − 2%, warehouse x is *not* at least as good as warehouse y with respect to profitability $(P(x,y) = -1 \Leftrightarrow xS^c y)$.

The pairwise comparisons of reference objects result in $\boldsymbol{S}_{PCT}$. In Table 13.10, there is a small fragment of $\boldsymbol{S}_{PCT}$.

Table 13.10. A fragment of $\boldsymbol{S}_{PCT}$.

Pair(x,y) of warehouses	$P_1[g_1(x), g_1(y)]$	$P_2[g_2(x), g_2(y)]$	$P_3[g_3(x), g_3(y)]$	S or S^c
(1,2)	0	1	0	S
(4,7)	-2	0	-1	S^c
(6,3)	-1	-1	0	S^c
...	...	...	...	...

The rough set analysis of the $\boldsymbol{S}_{PCT}$ leads to conclusion that the set of decision examples on reference objects is inconsistent. The quality of approximation of S and S^c by all criteria from set C is equal to 0.44. Moreover, $RED_{S_{PCT}} = CORE_{S_{PCT}} = \{g_1, g_2, g_3\}$; this means that no criterion is superfluous.

The C-lower approximations and the C-upper approximations of S and S^c, obtained by means of multigraded dominance relations, are as follows:

$$\begin{aligned}\underline{C}(S) &= \{(1,2),(1,4),(1,5),(1,6),(1,8),(3,2),(3,4),(3,5),(3,6),\\ &\quad (3,8),(7,2),(7,4),(7,5),(7,6),(7,8)\},\\ \underline{C}(S^c) &= \{(2,1),(2,7),(4,1),(4,3),(4,7),(5,1),(5,3),(5,7),(6,1),\\ &\quad (6,3),(6,7),(8,1),(8,7)\}.\end{aligned}$$

All the remaining 36 pairs of reference objects belong to the *C-boundaries* of S an S^c, i.e. $Bn_C(S) = Bn_C(S^c)$.

The following minimal $D_{\geq}$-decision rules and $D_{\leq}$-decision rules can be induced from lower approximations of S and S^c, respectively (within parentheses there are the pairs of objects supporting the corresponding rules):

- if $P_1[g_1(x), g_1(y)] \geq 1$ and $P_2[g_2(x), g_2(y)] \geq 1$, then xSy; $((1,6),(3,6),(7,6))$,

- if $P_2[g_2(x), g_2(y)] \geq 1$ and $P_3[g_3(x), g_3(y)] \geq 0$, then xSy; $((1,2),(1,6),(1,8),(3,2),(3,6),(3,8),(7,2),(7,6),(7,8))$,
- if $P_2[g_2(x), g_2(y)] \geq 0$ and $P_3[g_3(x), g_3(y)] \geq 1$, then xSy; $((1,4),(1,5),(3,4),(3,5),(7,4),(7,5))$,
- if $P_1[g_1(x), g_1(y)] \leq -1$ *and* $P_2[g_2(x), g_2(y)] \leq -1$, then xS^cy; $((6,1),(6,3),(6,7))$,
- if $P_2[g_2(x), g_2(y)] \leq 0$ and $P_3[g_3(x), g_3(y)] \leq -1$, then xS^cy; $((4,1),(4,3),(4,7),(5,1),(5,3),(5,7))$,
- if $P_1[g_1(x), g_1(y)] \leq 0$ and $P_2[g_2(x), g_2(y)] \leq -1$, and $P_3[g_3(x), g_3(y)] \leq 0$ then xS^cy; $((2,1),(2,7),(6,1),(6,3),(6,7),(8,1),(8,7))$.

Moreover, it was possible to induce five minimal $D_{\geq\leq}$-decision rules from the boundary of approximation of S and S^c:

- if $P_2[g_2(x), g_2(y)] \leq 0$ and $P_2[g_2(x), g_2(y)] \geq 0$ (i.e. $P_2[g_2(x), g_2(y)] = 0$) and $P_3[g_3(x), g_3(y)] \leq 0$ and $P_3[g_3(x), g_3(y)] \geq 0$ (i.e. $P_3[g_3$ $((1,1),(1,3),(1,7),(2,2),(2,6),(2,8),(3,1),(3,3),(3,7),(4,4),(4,5),(5,4),(5,5),(6,2),(6,6),(6,8),(7,1),(7,3),(7,7),(8,2),(8,6),(8,8))$,
- if $P_2[g_2(x), g_2(y)] \leq 1$ and $P_3[g_3(x), g_3(y)] \geq 1$, then xSy or xS^cy; $((2,4),(2,5),(6,4),(6,5),(8,4),(8,5))$,
- if $P_2[g_2(x), g_2(y)] \geq 1$ and $P_3[g_3(x), g_3(y)] \leq -1$, then xSy or xS^cy; $((4,2),(4,6),(4,8),(5,2),(5,6),(5,8))$,
- if $P_1[g_1(x), g_1(y)] \geq 1$ and $P_2[g_2(x), g_2(y)] \leq 0$, and $P_3[g_3(x), g_3(y)] \leq 0$ then xSy or xS^cy; $((1,3),(2,3),(2,6),(7,3),(8,3),(8,6))$,
- if $P_1[g_1(x), g_1(y)] \geq 1$ and $P_2[g_2(x), g_2(y)] \leq -1$, then xSy *or* xS^cy; $((2,3),(2,4),(2,5),(8,3),(8,4),(8,5))$.

Using all above decision rules and the Net Flow Score exploitation procedure on ten other warehouses proposed for sale, the managers obtained the result presented in Table 13.11. The dominance-based rough set approach gives a clear recommendation:

- for the **choice problem** it suggests to **select warehouse 2' and 6'**, having maximum score (11),

- for the **ranking problem** it suggests the **ranking** presented in the last column of Table 13.11, as follows:

$$(2',6') \rightarrow (8') \rightarrow (9') \rightarrow (1') \rightarrow (4') \rightarrow (5') \rightarrow (3') \rightarrow (7',10')$$

Table 13.11. Ranking of warehouses for sale by decision rules and the Net Flow Score procedure.

Warehouse	g_1	g_2	g_3	Net Flow Score	Ranking
1'	*good*	*sufficient*	*medium*	1	5
2'	*sufficient*	*good*	*good*	11	1
3'	*sufficient*	*medium*	*sufficient*	-8	8
4'	*sufficient*	*good*	*sufficient*	0	6
5'	*sufficient*	*sufficient*	*medium*	-4	7
6'	*sufficient*	*good*	*good*	11	1
7'	*medium*	*sufficient*	*sufficient*	-11	9
8'	*medium*	*medium*	*medium*	7	3
9'	*medium*	*good*	*sufficient*	4	4
10'	*medium*	*sufficient*	*sufficient*	-11	9

6.6 Fuzzy Preferences

Let us consider the case where the preferences $P_i[g_i(x), g_i(y)]$ with respect to each criterion $g_i \in C$, as well as the comprehensive preference $P(x, y)$, can assume values from a finite set. For example, given $x, y \in A$, the preferences $P_i[g_i(x), g_i(y)]$ and $P(x, y)$ can assume the following qualitatively ordinal values: x is much better than y, x is better than y, x is equivalent to y, x is worse than y, x is much worse than y. Let us suppose, moreover, that each possible value of $P_i[g_i(x), g_i(y)]$ and $P(x, y)$ is fuzzy in the sense that it is true at some level of credibility between 0 and 100%, e.g. "x is better than y on criterion g_i with credibility 75%", or "x is comprehensively worse than y with credibility 80%". Greco, Matarazzo and Slowinski [23] proved that the fuzzy comprehensive preference $P(x, y)$ can be approximated by means of fuzzy preferences $P_i[g_i(x), g_i(y)]$ after translating the dominance-based rough approximations of S_{PCT} defined for the crisp case, by means of fuzzy operators.

6.7 Preferences without Degree of Preferences

The values of $P_i[g_i(x), g_i(y)]$ considered in the dominance-based rough approximation of S_{PCT} represent a degree (strength) of preference. It is possible, however, that in some cases, the concept of degree of preference with respect to some criteria is meaningless for a DM. In these cases, there does not exist a function $P_i[g_i(x), g_i(y)]$ expressing how much x is better than y with respect to criterion g_i and, on the contrary, we can directly deal with values $g_i(x)$ and $g_i(y)$ only. For example let us consider the car decision problem and four cars x, y, w, z with the maximum speed of 210 km/h, 180 km/h, 150 km/h and 140 km/h, respectively. Even if the concept of degree of preference is meaningless, it is possible to say that with respect to the maximum speed, x is preferred to z at least as much as y is preferred to w. On the basis of this observation, Greco, Matarazzo and Slowinski [23] proved that comprehensive preference $P(x, y)$ can be approximated by means of criteria with only ordinal scales, for which the concept of degree of preference is meaningless. An example of decision rules obtained in this situation is the following:

> "*if* car x has a maximum speed of at least 180km/h while car y has a maximum speed of at most 140km/h *and* the comfort of car x is at least good while the comfort of car y is at most medium, *then* car x is at least as good as car y".

7. Conclusions

In this chapter, we made a synthesis of the contribution of the dominance-based rough set theory, called DRSA, to MCDA.

Some remarks relative to comparison of DRSA with other MCDA methodologies will be useful to fully appreciate the decision rule approach:

1) for multiple-criteria sorting problems, a set of DRSA decision rules is equivalent to a general utility function, simply increasing with respect to each criterion, with a set of thresholds corresponding to frontiers between preference-ordered decision classes [31, 33]; more generally, for multiple-criteria decision problems, using a utility function is equivalent to adopt a set of DRSA decision rules; these rules have a specific syntax when the utility function assumes specific formulations (for example an associative operator) [36];

2) for multiple-criteria choice and ranking problems, a set of DRSA decision rules is equivalent to a general conjoint measurement model (see Chapter 3, non-additive and non-transitive proposed by Bouyssou and Pirlot ([2]; see also [34]);

3) decision rules obtained by DRSA are more general than Sugeno integral ([61]; see also Chapter 14), being the most general max-min ordinal aggregator; in fact, Sugeno integral is equivalent to a set of single-graded decision rules where evaluations with respect to conditions and conclusion of a rule are of the same degree, for example,

ρ_1 : "*if* Mathematics$\geq$*Medium and* Literature$\geq$*Medium*, *then* the comprehensive evaluation is at least *Medium*", is a single-grade decision rule, while

ρ_2 : "*if* Physics$\geq$*Good and* Literature$\geq$*Medium*, *then* the comprehensive evaluation is at least *Good*"

is not a single-graded decision rule; this means that if in the set of decision rules there is at least one rule which is not single graded, then the DM's preferences cannot be represented by the Sugeno integral; for example, if DM's preferences are represented by a set of decision rules containing rule ρ_2, then these preferences cannot be represented by the Sugeno integral [31, 36];

4) preferences modelled by outranking methods from ELECTRE family can be represented by a set of specific DRSA decision rules based on PCT [34, 38].

The main features of DRSA can be summarized as follows:

- preferential information necessary to deal with a multiple-cri-teria decision problem is asked to the DM in terms of exemplary decisions,
- rough set analysis of preferential information supplies some useful elements of knowledge about the decision situation; these are: the relevance of attributes and/or criteria, information about their interaction (from quality of approximation and its analysis using fuzzy measures theory), minimal subsets of attributes or criteria (reducts) conveying the relevant knowledge contained in the exemplary decisions, the set of the non-reducible attributes or criteria (core),
- preference model induced from the preferential information is expressed in a natural and comprehensible language of "*if* ..., *then* ..." decision rules,
- heterogeneous information (qualitative and quantitative, preference-ordered or not, crisp and fuzzy evaluations, and ordinal and cardinal scales of preferences, with a hierarchical structure and with missing values) can be processed within DRSA, while classical MCDA methods consider only quantitative ordered evaluations with rare exceptions,
- decision rule preference model resulting from the rough set approach is more general than all existing models of conjoint measurement due to its capacity of handling inconsistent preferences,
- proposed methodology is based on elementary concepts and mathematical tools (sets and set operations, binary relations), without recourse to

any algebraic or analytical structures; the main idea is very natural and the key concept of dominance relation is even objective.

There is no doubt that the use of the decision rule model and the capacity of handling inconsistent preferential information with DRSA opened a fascinating research field to MCDA and moved it towards artificial intelligence and knowledge discovery.

Acknowledgments

The first two authors wish to acknowledge financial support from the Italian Ministry of Education, University and Scientific Research MIUR. The research of the second author has been supported by the State Committee for Scientific Research.

References

[1] J. Blaszczynski, S. Greco, and R. Slowinski. Multi-criteria classification using decision rules. *European Journal of Operational Research*, to appear, 2004.

[2] D. Bouyssou and M. Pirlot. Nontransitive decomposable conjoint measurement. *Journal of Mathematical Psychology*, 46:677–703, 2002.

[3] G. Choquet. Theory of capacities. *Annales de l'Institut Fourier*, 5:131–295, 1953.

[4] K. Dembczynski, S. Greco, and R. Slowinski. Methodology of rough-set-based classification and sorting with hierarchical structure of attributes and criteria. *Control & Cybernetics*, 31:891–920, 2002.

[5] P.C. Fishburn. Methods for estimating additive utilities. *Management Science*, 13:435–453, 1967.

[6] J. Fodor and M. Roubens. *Fuzzy Preference Modelling and Multicriteria Decision Support*. Kluwer, Dordrecht, 1994.

[7] S. Giove, S. Greco, B. Matarazzo, and R. Slowinski. Variable consistency monotonic decision trees. In J.J. Alpigini, J.F. Peters, A. Skowron, and N. Zhong, editors, *Rough Sets and Current Trends in Computing*, pages 247–254, Berlin, 2002. Springer-Verlag.

[8] M. Grabisch. k-order additive discrete fuzzy measures and their representation. *Fuzzy Sets and Systems*, 92:167–189, 1997.

[9] S. Greco, M. Inuiguchi, and R. Slowinski. Dominance-based rough set approach using possibility and necessity measures. In J.J. Alpigini, J.F. Peters, A. Skowron, and N. Zhong, editors, *Rough Sets and Current Trends in Computing*, pages 85–92, Berlin, 2002. Springer-Verlag.

[10] S. Greco, M. Inuiguchi, and R. Slowinski. A new proposal for fuzzy rough approximations and gradual decision rule representation. In D. Dubois, J. Grzymala-Busse, M. Inuiguchi, and L. Polkowski, editors, *Rough Fuzzy and Fuzzy Rough Sets*. Springer-Verlag, Berlin, 2004.

[11] S. Greco, Matarazzo, R. Slowinski, and J. Stefanowski. An algorithm for induction of decision rules consistent with dominance principle. In W.Ziarko and Y.Yao, editors, *Rough Sets and Current Trends in Computing*, pages 304–313, Berlin, 2001. Springer-Verlag.

[12] S. Greco, Matarazzo, R. Slowinski, and J. Stefanowski. Variable consistency model of dominance-based rough set approach. In W.Ziarko and Y.Yao, editors, *Rough Sets and Current Trends in Computing*, pages 170–181, Berlin, 2001. Springer-Verlag.

[13] S. Greco, Matarazzo, R. Slowinski, and J. Stefanowski. Importance and interaction of conditions in decision rules. In J.J. Alpigini, J. F. Peters, A. Skowron, and N. Zhong, editors, *Rough Sets and Current Trends in Computing*, pages 255–262, Berlin, 2002. Springer-Verlag.

[14] S. Greco, Matarazzo, R. Slowinski, and J. Stefanowski. Mining association rules in preference-ordered data. In M.-S. Hacid, Z.W. Ras, D.A. Zighed, and Y. Kodratoffa, editors, *Foundations of Intelligent Systems*, pages 442–450, Berlin, 2002. Springer-Verlag.

[15] S. Greco, B. Matarazzo, and R. Slowinski. Fuzzy measures technique for rough set analysis. In *Proc. 6th European Congress on Intelligent Techniques & Soft Computing*, 1998.

[16] S. Greco, B. Matarazzo, and R. Slowinski. Fuzzy similarity relation as a basis for rough approximations. In L.Polkowski and A.Skowron, editors, *Rough Sets and Current Trends in Computing*, pages 283–289, Berlin, 1998. Springer-Verlag.

[17] S. Greco, B. Matarazzo, and R. Slowinski. A new rough set approach to evaluation of bankruptcy risk. In C.Zopounidis, editor, *Rough Fuzzy and Fuzzy Rough Sets*, pages 121–136. Kluwer, Dordrecht, 1998.

[18] S. Greco, B. Matarazzo, and R. Slowinski. A new rough set approach to multicriteria and multiattribute classification. In L.Polkowski and A.Skowron, editors, *Rough Sets and Current Trends in Computing*, pages 60–67, Berlin, 1998. Springer-Verlag.

[19] S. Greco, B. Matarazzo, and R. Slowinski. Handling missing values in rough set analysis of multi-attribute and multi-criteria decision problems. In N. Zhong, A. Skowron, and S. Ohsuga, editors, *New Directions in Rough Sets, Data Mining and Granular-Soft Computing*, pages 146–157, Berlin, 1999. Springer-Verlag.

[20] S. Greco, B. Matarazzo, and R. Slowinski. Rough approximation of a preference relation by dominance relations. *European Journal of Operational Research*, 117:63–83, 1999.

[21] S. Greco, B. Matarazzo, and R. Slowinski. Rough sets methodology for sorting problems in presence of multiple attributes and criteria. *European Journal of Operational Research*, 138:247–259, 1999.

[22] S. Greco, B. Matarazzo, and R. Slowinski. Rough sets theory for multicriteria decision analysis. *European Journal of Operational Research*, 129:1–47, 1999.

[23] S. Greco, B. Matarazzo, and R. Slowinski. The use of rough sets and fuzzy sets in MCDM. In T.Gal, T.Stewart, and T.Hanne, editors, *Advances in Multiple Criteria Decision Making*, pages 14.1–14.59. Kluwer, Dordrecht, 1999.

[24] S. Greco, B. Matarazzo, and R. Slowinski. Dealing with missing data in rough set analysis of multi-attribute and multi-criteria decision problems. In S.H. Zanakis, G. Doukidis, and C. Zopounidis, editors, *Decision Making: Recent Developments and Worldwide Applications*, pages 295–316. Kluwer, Dordrecht, 2000.

[25] S. Greco, B. Matarazzo, and R. Slowinski. Decision rules. In *Encyclopedia of Management, 4th edition*, pages 178–183. Gale Group, Farmington Hills, MI, 2000.

[26] S. Greco, B. Matarazzo, and R. Slowinski. Extension of the rough set approach to multicriteria decision support. *INFOR*, 38:161–196, 2000.

[27] S. Greco, B. Matarazzo, and R. Slowinski. Fuzzy dominance-based rough set approach. In F. Masulli, R. Parenti, and G. Pasi, editors, *Advances in Fuzzy Systems and Intelligent Technologies*, pages 56–66. Shaker Publishing, Maastricht (NL), 2000.

[28] S. Greco, B. Matarazzo, and R. Slowinski. Fuzzy extension of the rough set approach to multicriteria and multiattribute sorting. In J. Fodor, B. De Baets, and P. Perny, editors, *Preferences and Decisions under Incomplete Knowledge*, pages 131–151. Physica-Verlag, Heidelberg, 2000.

[29] S. Greco, B. Matarazzo, and R. Slowinski. Rough set processing of vague information using fuzzy similarity relations. In C.S. Calude and G. Paun, editors, *Finite Versus Infinite – Contributions to an Eternal Dilemma*, pages 149–173. Springer-Verlag, London, 2000.

[30] S. Greco, B. Matarazzo, and R. Slowinski. Assessment of a value of information using rough sets and fuzzy measures. In J. Chojcan and J. Leski, editors, *Fuzzy Sets and their Applications*, pages 185–193, Gliwice, 2001. Silesian University of Technology Press.

[31] S. Greco, B. Matarazzo, and R. Slowinski. Conjoint measurement and rough set approach for multicriteria sorting problems in presence of ordinal criteria. In A. Colorni, M. Paruccini, and B. Roy, editors, *A-MCD-A: Aide Multi Critère à la Décision – Multiple Criteria Decision Aiding*, pages 117–144, Ispra, 2001. European Commission Report, EUR 19808 EN.

[32] S. Greco, B. Matarazzo, and R. Slowinski. Rough set approach to decisions under risk. In W.Ziarko and Y.Yao, editors, *Rough Sets and Current Trends in Computing*, pages 160–169, Berlin, 2001. Springer-Verlag.

[33] S. Greco, B. Matarazzo, and R. Slowinski. Multicriteria classification, chapter 16.1.9. In W. Kloesgen and J. Zytkow, editors, *Handbook of Data Mining and Knowledge Discovery*, pages 318–328. Oxford University Press, New York, 2002.

[34] S. Greco, B. Matarazzo, and R. Slowinski. Preference representation by means of conjoint measurement and decision rule model. In D. Bouyssou, E. Jacquet-Lagrèze, P. Perny, R. Slowinski, D. Van-derpooten, and Ph. Vincke, editors, *Aiding Decisions with Multiple Criteria – Essays in Honor of Bernard Roy*, pages 263–313. Kluwer, Boston, 2002.

[35] S. Greco, B. Matarazzo, and R. Slowinski. Rough approximation by dominance relations. *International Journal of Intelligent Systems*, 17:153–171, 2002.

[36] S. Greco, B. Matarazzo, and R. Slowinski. Axiomatic characterization of a general utility function and its particular cases in terms of conjoint measurement and rough-set decision rules. *European Journal of Operational Research*, to appear, 2004.

[37] S. Greco, B. Matarazzo, R. Slowinski, and A. Tsoukias. Exploitation of a rough approximation of the outranking relation in multicriteria choice and ranking. In T.J.Stewart and R.C. van den Honert, editors, *Trends in Multicriteria Decision Making*, pages 45–60. Springer-Verlag, Berlin, 1998.

[38] S. Greco, B. Predki, and R. Slowinski. Searching for an equivalence between decision rules and concordance-discordance preference model in multicriteria choice problems. *Control and Cybernetics*, 31:921–935, 2002.

[39] E. Jacquet-Lagrèze. *Systèmes de décision et acteurs multiples - Contribution à une théorie de l'action pour les sciences des organisations*. Thèse de doctorat d'état, Université de Paris-Dauphine, Paris, France, 1981.

[40] R.L. Keeney and H. Raiffa. *Decision with Multiple Objectives - Preferences and value Tradeoffs*. Wiley, New York, 1976.

[41] M. Kryszkiewicz. Rough set approach to incomplete information systems. *Information Sciences*, 112:39–49, 1998.

[42] M. Kryszkiewicz. Rules in incomplete information systems. *Information Sciences*, 113:271–292, 1999.

[43] J. G. March. Bounded rationality, ambiguity, and the engineering of choice. In D. E. Bell, H. Raiffa, and A. Tversky, editors, *Decision Making, Descriptive, Normative and Prescriptive Interactions*, pages 33–58. Cambridge University Press, New York, 1988.

[44] V. Mousseau. *Problèmes liés à l'évaluation de l'importance en aide multicritère à la décision: Réflexions théoriques et expérimentations*. Thèse de doctorat d'état, Université de Paris-Dauphine, Paris, France, 1993.

[45] T. Murofushi and S. Soneda. Techniques for reading fuzzy measures (iii): interaction index. In *Proc. 9th Fuzzy Systems Symposium, Sapporo, Japan, May 1993*, pages 693–696, 2002.

[46] Z. Pawlak. Rough Sets. *International Journal of Information &Computer Sciences*, 11:341–356, 1982.

[47] Z. Pawlak. *Rough Sets. Theoretical Aspects of Reasoning about Data*. Kluwer, Dordrecht, 1991.

[48] Z. Pawlak and R. Slowinski. Rough set approach to multi-attribute decision analysis. *European Journal of Operational Research*, 72:443–459, 1994.

[49] L. Polkowski. *Rough Sets: Mathematical Foundations*. Physica-Verlag, Heidelberg, 2002.

[50] B. Roy. The outranking approach and the foundation of ELECTRE methods. *Theory and Decision*, 31:49–73, 1991.

[51] B. Roy. Decision science or decision aid science? *European Journal of Operational Research*, 66:184–203, 1993.

[52] L. S. Shapley. A value for n-person games. In H.W. Kuhn and A.W. Tucker, editors, *Contributions to the Theory of Games II*, pages 307–317. Princeton University Press, Princeton, 1953.

[53] P. Slovic. Choice between equally-valued alternatives. *Journal of Experimental Psychology: Human Perception Performance*, 1:280–287, 1975.

[54] R. Slowinski. *Intelligent Decision Support. Handbook of Applications and Advances of the Rough Sets Theory*. Kluwer, Dordrecht, 1992.

[55] R. Slowinski. Rough set learning of preferential attitude in multi-criteria decision making. In J.Komorowski and Z.Ras, editors, *Methodologies for Intelligent Systems*, pages 642–651. Springer-Verlag, Berlin, 1993.

[56] R. Slowinski, S. Greco, and B. Matarazzo. Mining decision-rule preference model from rough approximation of preference relation. In *Proc. 26th IEEE Annual Int. Conference on Computer Software & Applications (COMPSAC 2002)*, pages 1129–1134, Oxford, 2002.

[57] R. Slowinski, S. Greco, and B. Matarazzo. Rough set analysis of preference-ordered data. In J.J. Alpigini, J.F. Peters, A. Skowron, and N. Zhong, editors, *Rough Sets and Current Trends in Computing*, pages 44–59, Berlin, 2002. Springer-Verlag.

[58] R. Slowinski and D. Vanderpooten. Similarity relation as a basis for rough approximations. In P.P. Wang, editor, *Advances in Machine Intelligence & Soft-Computing, vol.IV*, pages 17–33. Duke University Press, Durham, NC, 1997.

[59] R. Slowinski and D. Vanderpooten. A generalized definition of rough approximations based on similarity. *IEEE Transactions on Data and Knowledge Engineering*, 12:331–336, 2000.

[60] J. Stefanowski. On rough set based approaches to induction of decision rules. In L.Polkowski and A. Skowron, editors, *Rough Sets in Data Mining and Knowledge Discovery. Vol.1*, pages 500–529. Physica-Verlag, Heidelberg, 1998.

[61] M. Sugeno. *Theory of fuzzy integrals and its applications*. Doctoral thesis, Tokyo Institute of Technology, Tokyo, Japan, 1974.

[62] R. Susmaga, R Slowinski, S. Greco, and B. Matarazzo. Generation of reducts and rules in multi-attribute and multi-criteria classification. *Control and Cybernetics*, 29:969–988, 2000.

[63] W. Ziarko. Variable precision rough sets model. *Journal of Computer and Systems Sciences*, 1993:39–59, 1993.

[64] W. Ziarko. Rough sets as a methodology for data mining. In L.Polkowski and A.Skowron, editors, *Rough Sets in Knowledge Discovery. Vol. 1*, pages 554–576. Physica-Verlag, Heidelberg, 1998.

Chapter 14

FUZZY MEASURES AND INTEGRALS IN MCDA

Michel Grabisch
Université Paris I – Panthéon-Sorbonne
LIP6, 8 rue du Capitaine Scott
75015 Paris, France
michel.grabisch@lip6.fr

Christophe Labreuche
Thales Research & Technology
Domaine de Corbeville
91404 Orsay Cedex, France
christophe.labreuche@thalesgroup.com

Abstract This chapter aims at a unified presentation of various methods of MCDA based on fuzzy measures (capacity) and fuzzy integrals, essentially the Choquet and Sugeno integral. A first section sets the position of the problem of multicriteria decision making, and describes the various possible scales of measurement (cardinal unipolar and bipolar, and ordinal). Then a whole section is devoted to each case in detail: after introducing necessary concepts, the methodology is described, and the problem of the practical identification of fuzzy measures is given. The important concept of interaction between criteria, central in this chapter, is explained in detail. It is shown how it leads to k-additive fuzzy measures. The case of bipolar scales leads to the general model based on bi-capacities, encompassing usual models based on capacities. A general definition of interaction for bipolar scales is introduced. The case of ordinal scales leads to the use of Sugeno integral, and its symmetrized version when one considers symmetric ordinal scales. A practical methodology for the identification of fuzzy measures in this context is given.

Keywords: Choquet integral, fuzzy measure, interaction, bi-capacities.

1. Introduction

MultiCriteria Decision Aid (MCDA) aims at modeling the preferences of a Decision Maker (DM) over alternatives described by several points of view, which are denoted by $X_1, \ldots, X_n$. An alternative is characterized by a value w.r.t. each point of view and is thus identified with a point in the Cartesian product X of the points of view: $X = X_1 \times \cdots \times X_n$. We denote by $N := \{1, \ldots, n\}$ the index set of points of view. The preference relation of the DM over alternatives is denoted by $\succeq$. For $x, y \in X$, "$x \succeq y$" means that the DM prefers alternative x to y.

The main concern in practice is to come up with the knowledge of $\succeq$ on $X \times X$ from a relatively small amount of questions asked to the DM on $\succeq$. The information provided by the DM can be composed of examples of comparisons between alternatives, which gives $\succeq$ on a subset of $X \times X$, as well as more qualitative judgments, whose modelling is more complex, and depends on the kind of representation of $\succeq$ we choose. In general, we look for a *numerical representation* [43] $u : X \to \mathbb{R}$ such that:

$$\forall x, y \in X \ , \ x \succeq y \Leftrightarrow u(x) \geq u(y). \tag{14.1}$$

It is classical to write u in the following way [42]:

$$u(x) = F\left(u_1(x_1), \ldots, u_n(x_n)\right) \ \forall x \in X, \tag{14.2}$$

where the u_i's : $X_i \to \mathbb{R}$ are called the *utility functions* and $F : \mathbb{R}^n \to \mathbb{R}$ is an *aggregation function*. A result by Krantz *et al.* gives the axioms that characterize the representation of $\succeq$ by (14.2) [43]. As it will be detailed in Section 2.1, the *weak separability* axiom is the key axiom that justifies the construction of utility functions, that is partial preference relations over the points of view, from the overall preference relation $\succeq$. A criterion is defined as a preference relation $\succeq_i$ over one point of view X_i. Thus a criterion is the association of one point of view X_i with its related utility function u_i.

In practice, we restrict ourself to a family $\mathcal{F}$ of aggregation functions (parameterized by some coefficients). The justification of the use of a special family is based on an axiomatic approach. The axioms that characterize the family should be in accordance with the problem in consideration and the behaviour of the decision maker. The DM has then to provide the needed information to set the parameters of the model. The more restrictive the family is, the less representative it is, but the less information the DM shall give.

The most classical functions used to aggregate the criteria are the weighted sums $F(u_1, \ldots, u_n) = \sum_{i=1}^n \alpha_i \, u_i$. As an aggregation operator, they are characterized by an independence axiom [42, 73]. This property implies some limitations in the way the weighted sum can model typical decision behaviours.

To make this more precise, let us consider the example of two criteria having the same importance, an example which we will consider in more details in Section 3.5. We are interested in the following four alternatives: x is bad in both criteria, y is bad in the first criterion but good at the second one, z is good in the first criterion but bad in the second one, and t is good in both. Clearly $x \prec t$ and the DM is equally satisfied by y and z since the two criteria have the same importance. However, the comparison of y, z with x and t leads to several cases. First, the DM may say that $x \sim y \sim z \prec t$, where $\sim$ means indifference. This depicts a DM who is *intolerant*, since both criteria have to be satisfied in order to get a satisfactory alternative. In the opposite way, the DM may think that $x \prec y \sim z \sim t$, which depicts a *tolerant* DM, since only one criterion has to be satisfactory in order to get a satisfactory alternative. Finally, we may have all intermediate cases, where $x \prec y \sim z \prec t$. An important fact is that, due to additivity, the weighted sum is unable to distinguish among all these cases, in particular, all decision behaviours related to tolerance or intolerance are missed. These phenomena are called *interaction* between criteria. They encompass also other phenomena such as *veto*. We will show in this chapter that the notions of capacity and fuzzy integrals enable to model previous phenomena.

The construction of the utility functions and the determination of the parameters of the aggregation function are often carried out in two separate steps. The utility functions are generally set up first, that is without the knowledge of the precise aggregator F within $\mathcal{F}$. However, the utility functions have no intrinsic meaning to the DM and shall be determined from questions regarding only the overall preference relation $\succeq$. It is not assumed that the DM can isolate attributes and give information directly on u_i. This point is generally not considered in the literature. The main reason is probably that due to the use of a weighted sum as an aggregation function, the independence assumption (preferential or cardinal independence) makes it possible in some sense to separate each attribute and thus construct the utility functions directly. This becomes far more complicated when this assumption is removed. Besides, these approaches are not relevant from a theoretical standpoint. To our knowledge, the only approach that addresses this problem with the use of a weighted sum is the so-called MACBETH approach designed by Bana e Costa and Vansnick [2, 1, 3]. A generalization of this approach to more complex aggregation operators has been proposed by the authors [33]. These approaches are considered in this chapter.

The determination of the utility function is not concerned only with measurement considerations. The main difficulty is to ensure commensurateness between criteria. Commensurateness means that one shall be able to compare any element of one point of view with any element of any other point of view. This is inter-criteria comparability:

For $x_i \in X_i$ and $x_j \in X_j$, we have $u_i(x_i) \geq u_j(x_j)$ iff x_i is considered at least as good as x_j by the DM.

Commensurateness implies the existence of a preference relation over $\bigcup_{i=1}^{n} X_i$. This assumption, considered by Modave *et al.* [56], is very strong. Taking a simple example involving two criteria (for instance consumption and maximal speed), this amounts to know whether the DM prefers a consumption of 5 liters/100km to a maximum speed of 200 km/h. This does not generally make sense to the DM, so that he or she is not generally able to make this comparison directly.

In Sections 3 and 4 we push the previous method one step further by considering on top of intra-criteria information some natural inter-criteria information to determine the aggregation functions as well. We will show that the requirements induced by measurement considerations naturally imply the use of fuzzy integrals as aggregation operators. In Section 5, we deal with the case of ordinal information. It will be seen that this induces difficulties, so that the previous construction no more applies.

2. Measurement Theoretic Foundations

As explained in the introduction, we focus on a model called *decomposable* given by Eq. (14.2), involving an aggregation function $F : \mathbb{R}^n \longrightarrow \mathbb{R}$, and utility functions $u_i : X_i \longrightarrow \mathbb{R}$, $i = 1, \ldots, n$.

In this section we will give some considerations coming from measurement theory as well as more practical considerations coming from the MACBETH approach around this kind of model. This will help us in giving a firm theoretical basis to our construction.

2.1 Basic Notions of Measurement, Scales

This section is based on [43, 64], to which the reader is referred for more details.

The fundamental aim of measurement theory is to build homomorphisms f between a relational structure $\mathcal{A}$ coming from observation, and a relational structure $\mathcal{B}$ based on real numbers (or more generally, some totally ordered set). Doing so, we get a numerical *representation* of our observation. A *scale* (of measurement) is the triplet $(\mathcal{A}, \mathcal{B}, f)$. If no ambiguity occurs, f alone denotes the scale.

A simple example is when $\mathcal{A} = (A, \succeq)$, where $\succeq$ is a binary relation expressing e.g. the preference of the DM on some set A, and $\mathcal{B}$ is simply $(\mathbb{R}, \geq)$. As usual, $\sim$ and $\succ$ denote respectively the symmetric and asymmetric parts of $\succeq$, and $A/\sim$ is the set of equivalence classes of $\sim$ (when defined). This measurement problem is called *ordinal measurement*. The homomorphism satisfies the following condition

(Ord[A]) $a \succeq b$ iff $f(a) \geq f(b), \quad \forall a, b \in A$.

Obviously, f is not unique since any strictly increasing transform $\phi \circ f$ of f is also a homomorphism. Generally speaking, the set of functions $\phi : \mathbb{R} \longrightarrow \mathbb{R}$ such that $\phi \circ f$ remains a homomorphism is called the *set of admissible transformations*.

Types of scale are defined by their set of admissible transformations. The most common ones are:

- *ordinal scales*, where the set of admissible transformations are all strictly increasing functions. Examples: scale of hardness, of earthquakes intensity.
- *interval scales*, where all $\phi(t) = \alpha t + \beta$, $\alpha > 0$ are admissible (positive affine transformations). Example: temperature in Celsius.
- *ratio scales*, where the admissible transformations are of the form $\phi(t) = \alpha t$, $\alpha > 0$. Examples: temperature in Kelvin, mass.

Thus, our condition **(Ord[A])** defines an ordinal scale. The conditions under which such a f exists are well known. A necessary condition is that $\succeq$ is a weak order (reflexive, complete, transitive). A second condition (and then both are necessary and sufficient) is that $A/\sim$ contains a countable order-dense subset (this is known as the Birkhoff-Milgram theorem, we do not enter further into details).

An ordinal scale is rather poor, and does not really permit to handle numbers, since usual arithmetic operations are not invariant under admissible transformations. It would be better to build an interval scale in the above sense. This is related to the *difference measurement* problem: in this case, $\mathcal{A} = (A, \succeq^*)$, where $\succeq^*$ is a quaternary relation. The meaning of $ab \succeq^* st$ is the following: the difference of intensity (e.g. of preference) between a and b is larger than the difference of intensity between s and t. Then, the homomorphism f should satisfy:

$$ab \succeq^* st \Leftrightarrow f(a) - f(b) \geq f(s) - f(t). \tag{14.3}$$

It is shown that under several conditions on $\mathcal{A}$, such a function f exists, and that it defines an interval scale. Thus the ratio $\frac{f(a)-f(b)}{f(s)-f(t)}$ is meaningful (invariant under any admissible transformation).

Based on this remark, we express the interval scale condition under a form which is suitable for our purpose.

(Inter[A]). $\forall a, b, s, t \in A$ such that $a \succ b$ and $s \succ t$, we have

$$\frac{f(a) - f(b)}{f(s) - f(t)} =: k(a, b, s, t)\ ,\quad k(a, b, s, t) \in \mathbb{R}_+$$

if and only if the difference of satisfaction degree that the DM feels between a and b is $k(a, b, s, t)$ times as large as the difference of satisfaction between s and t.

The conditions of existence of f amounts to verify the following condition.

(C-Inter[A]). $\forall a, b, s, t, u, v \in A$ such that $a \succ b$, $s \succ t$ and $u \succ v$,

$$k(a,b,s,t) \times k(s,t,u,v) = k(a,b,u,v).$$

We end this section by addressing the case where A is a product space, as for $X = X_1 \times \cdots \times X_n$. Conditions for an ordinal representation by $u : X \longrightarrow \mathbb{R}$ are given by the Birkhoff-Milgram theorem. However, we are interested in a decomposable form of u (see (14.2). If F is one-to-one in each place, then necessarily $\sim$ satisfies *substitutability*:

$$(x_i, z_{-i}) \sim (y_i, z_{-i}) \Leftrightarrow (x_i, z'_{-i}) \sim (y_i, z'_{-i}), \quad \forall x, y, z, z' \in X. \qquad (14.4)$$

Notation $z = (x_A, y_{-A})$ means that z is defined by $z_i = x_i$ if $i \in A$, else $z_i = y_i$ (hence, $-A$ stands for $N \setminus A$). This property implies the existence of equivalence relations $\sim_i$ on each X_i. If F is strictly increasing, then $\sim$ has to be replaced by $\succeq$ in (14.4) (this is called *weak separability*), and relations $\succeq_i$ are obtained on each X_i.

Reciprocally, substitutability (or weak separability) and the conditions of the Birkhoff-Milgram theorem lead to an *ordinal* representation: hence, u is unique up to a strictly increasing function.

This result remains of theoretical interest, since not verifiable in practice, and moreover, it does not lead to an interval scale. The MACBETH methodology will serve as a basis for such a construction, whose essence is briefly addressed below. Before that, some words on unipolar and bipolar scales are in order.

2.2 Bipolar and Unipolar Scales

Let us view scales under a different point of view. Let $(A, \succeq)$ be a relational system, and f a scale, which is supposed to be numerical, without loss of generality. It may exists in A a particular element or level e, called *neutral level*, such that if $a \succ e$, then a is considered as "good", while if $e \succ a$, then a is considered as "bad" for the DM. We may choose for convenience f such that $f(e) = 0$.

Such a neutral level exists whenever relation $\succeq$ corresponds to two opposite notions of common language. For example, this is the case when $\succeq$ means "more attractive than", "better than", etc., whose pairs of opposite notions are respectively "attractiveness/repulsiveness", and "good/bad". By contrast, relations as "more prioritary than", "more allowed than", "belongs more to category C than" do not clearly exhibit a neutral level.

A scale is said to be *bipolar* if A contains such a neutral level. A *unipolar scale* has no neutral level, but has a least level, i.e. an element or level a_0 in A such that $a \succeq a_0$ for all $a \in A$. We may for convenience choose f so that $f(a_0) = 0$.

A scale has a greatest element if there exists an element or level $a_1 \in A$ such that $a_1 \succeq a$, for all $a \in A$. We say that a unipolar scale is *bounded* if it has a greatest level. A bipolar scale is bounded if it has a least and a greatest level (since there is an inherent symmetry in bipolar scales, the existence of a greatest level implies the existence of a least level).

Taking our previous examples, the relations "more attractive than", "better than", "more prioritary than" may not be bounded, while "more allowed than" and "belongs more to category C than" are clearly bounded, the greatest levels being respectively "fully authorized" and "fully belongs to C".

Typically, f maps on $\mathbb{R}$ (resp. $\mathbb{R}_+$) when the scale is unbounded bipolar (resp. unipolar). In the case of bounded scales, f maps respectively to a closed interval centered on 0, and an interval such as $[0, \beta]$.

It is convenient to denote by $\mathbf{0}$ the neutral level of a bipolar scale, or the least level of a unipolar scale. We may also use $\mathbf{1}$ to denote the greatest level when it exists, and $-\mathbf{1}$ for the least level of a bipolar scale.

When the scale is unbounded, it may be convenient to introduce another particular level, called the *satisfactory level*, and denoted by $\mathbf{1}$. This level is considered as *good and completely satisfactory* if the DM could obtain it, even if more attractive elements could exist in A (due to unboundness). The existence of such a level has been the main argument of H. Simon in his theory of *satisficing bounded rationality* [69], and a fundamental assumption in the MACBETH methodology, as described in the next section. For convenience, we may fix $f(\mathbf{1}) = 1$. If in addition the scale is bipolar, the same considerations lead to a level denoted $-\mathbf{1}$ (unsatisfactory level).

Finally, let us remark that there is no direct relation between unipolar/bipolar scales and the types of scales given in Section 2.1 (interval, ratio, etc.). For example, the temperature scales are clearly unipolar with a least level (at least in the physical sense), but may be of the ratio type (in Kelvin) or of the interval type (in Celsius, Fahrenheit). However, the neutral level of a bipolar scale clearly plays the role of the zero in a ratio scale, since it cannot be shifted.

2.3 Construction of the Measurement Scales and Absolute References Levels

The MACBETH methodology [2, 1, 3], described in Chapter 9, permits to build interval scales from a questionary. We limit ourselves here to necessary notions.

We consider A a finite set on which the decision maker is able to express some preference (the finiteness assumption is necessary for the method. If A is infinite, then a finite subset $\tilde{A}$ of representative objects should be chosen). The decision maker is asked for any pair $(a, b) \in A^2$:

1 Is a more attractive than b?

2 If yes, is the difference of attractivity between a and b *very weak, weak, moderate, strong, very strong*, or *extreme*?

The first question concerns ordinal measurement: we are looking for a function $f : A \longrightarrow \mathbb{R}$ satisfying condition **(Ord[A])**. The second question is related to difference measurement. The six ordered categories *very weak,...,extreme* define a quaternary relation on A, as defined in Section 2.1. MACBETH is able to test in a simple way if f as in (14.3) exists, and if yes, produces such a function, unique up to a positive affine transformation. In summary, we get an interval scale satisfying conditions **(Inter[A])** and **(C-Inter[A])**.

As explained in Section 2.2, we may have a unipolar or a bipolar scale, in which case a $\mathbf{0}$ level exists. It is convenient to choose f such that $f(\mathbf{0}) = 0$. If several sets $A_1, \ldots, A_n$ are involved, then commensurability between the scales $f_1, \ldots, f_n$ may be required, as it will be seen later.

We say that scales f_i, f_j are *commensurate* if $f_i(a_i) = f_j(a_j)$ means that the DM has the same intensity of attractiveness (or satisfaction, etc.) for a_i and a_j. A set of scales is commensurate if any pair is commensurate. Under the assumption that all f_i's are interval scales, it is sufficient to find two levels on each A_i, $i = 1, \ldots, n$ for which the DM feels an equal satisfaction for all i (they are in a sense *absolute levels*), and to impose equality of the scales for those levels.

Obviously, the levels $\mathbf{0}_i$ of each A_i have an identical absolute meaning, provided the A_i's are either all bipolar or all unipolar, but not mixed. We fix $f_i(\mathbf{0}_i) = 0$, $i = 1, \ldots, n$.

The second absolute levels could be the levels $\mathbf{1}_i$ (satisfactory levels in case of unbounded scales, and greatest elements otherwise). As suggested in Section 2.2, we may fix $f_i(\mathbf{1}_i) = 1$, $i = 1, \ldots, n$.

The same considerations apply to the absolute levels $-\mathbf{1}_i$.

To conclude this section, let us stress the fact that the underlying assumptions on which MACBETH (and hence, the method presented here) is based is that the DM is able to deliver information concerning difference measurement, and that the DM is able to exhibit on A two elements or levels with an absolute meaning, denoted $\mathbf{0}$ and $\mathbf{1}$, the precise meaning of them being dependent on the type of scale. We adopt throughout the paper the convention that

$$f(\mathbf{0}) = 0, \quad f(\mathbf{1}) = 1. \tag{14.5}$$

3. Unipolar Scales

We address in this section the construction of our model in the case of unipolar scales. As explained in Section 2, we have on each X_i two absolute levels $\mathbf{0}_i$ and $\mathbf{1}_i$ given by the DM.

3.1 Notion of Interaction – A Motivating Example

To introduce more precisely the idea of interaction and show some flaws of the weighted sum, let us give an example. The director of a university decides on students who are applying for graduate studies in management where some prerequisites from school are required. Students are indeed evaluated according to mathematics (M), statistics (S) and language skills (L). All the marks with respect to the scores are given on the same scale from 0 to 20. These three criteria serve as a basis for a preselection of the candidates. The best candidates have then an interview with a jury of members of the university to assess their motivation in studying in management. The applicants have generally speaking a strong scientific background so that mathematics and statistics have a big importance to the director. However, he does not wish to favor too much students that have a scientific profile with some flaws in languages. Besides, mathematics and statistics are in some sense *redundant*, since, usually, students good at mathematics are also good at statistics. As a consequence, for students good in mathematics, the director prefers a student good at languages to one good at statistics. Consider the following student A

	mathematics (M)	statistics (S)	languages (L)
student A	16	13	7

Student A is highly penalized by his performance in languages. Henceforth, the director would prefer a student (with the same mark in mathematics) that is a little bit better in languages even if the student would be a little bit worse in statistics. This means that the director prefers the following student to A

	mathematics (M)	statistics (S)	languages (L)
student B	16	11	9

We have thus

$$A \prec B \tag{14.6}$$

Consider now a student that has a weakness in mathematics. In this case, since the applicants are supposed to have strong scientific skills, a student good in statistics is now preferred to one good in languages. Consider the following two students

	mathematics (M)	statistics (S)	languages (L)
student C	6	13	7
student D	6	11	9

Following above arguments, C is preferred to D even though C has poor language skills.

$$C \succ D \tag{14.7}$$

Satisfying (14.6) and (14.7) at the same time leads to the following requirement

$$F(16,13,7) > F(16,11,9) \quad \text{and} \quad F(6,13,7) < F(6,11,9).$$

No weighted sum can model such preferences since (14.6) implies that languages is more important than statistics whereas (14.7) tells exactly the contrary. There is an inversion of preferences between (14.6) and (14.7) in the sense that the relative importance of languages compared to statistics depends on the satisfaction level in mathematics. This behaviour is a typical example of *interaction* between criteria.

3.2 Capacities and Choquet Integral

The natural generalization of giving weights on criteria is to assign weights on *coalitions* (i.e. groups, subsets) of criteria. This can be achieved by introducing particular functions on $\mathcal{P}(N)$, called fuzzy measures or capacities. We recall that $N := \{1, \ldots, n\}$ is the index set of criteria.

A *fuzzy measure* [70] or *capacity* [5] is a set function $\mu : 2^N \rightarrow \mathbb{R}$ satisfying

(FM$_a$) $A \subset B \Rightarrow \mu(A) \leq \mu(B)$,

(FM$_b$) $\mu(\emptyset) = 0$,

(FM$_c$) $\mu(N) = 1$.

Property **(FM$_a$)** is called *monotonicity* of the capacity. In MCDA, $\mu(A)$ is interpreted as the overall assessment of the binary alternative $(1_A, 0_{-A})$. A set function satisfying only **(FM$_b$)** is called a *game* or a *non-monotonic fuzzy measure*.

The *conjugate* μ^* of a capacity μ is defined by $\mu^*(S) = \mu(N) - \mu(N \setminus S)$. The capacity is said to be *additive* if $\mu(A \cup B) = \mu(A) + \mu(B)$, whenever $A \cap B = \emptyset$, while it is said to be *symmetric*capacity,symmetric if $\mu(A)$ depends only on $|A|$.

Let $a := (a_1, \ldots, a_n) \in \mathbb{R}^n_+$. The Choquet integral [5] of a w.r.t. a capacity μ has the following expression :

$$\mathcal{C}_\mu(a) = a_{\tau(1)}\, \mu(N) + \sum_{i=2}^{n} \left(a_{\tau(i)} - a_{\tau(i-1)}\right) \mu\left(\{\tau(i), \cdots, \tau(n)\}\right), \tag{14.8}$$

where τ is a permutation on N such that $a_{\tau(1)} \leq a_{\tau(2)} \leq \cdots \leq a_{\tau(n)}$. Note that the Choquet integral is also well-defined w.r.t. set functions which are games.

When the capacity is additive, the Choquet integral reduces to a weighted sum.

We say that $a, b \in \mathbb{R}^n_+$ are *comonotone* if $a_i < a_j \Rightarrow b_i \leq b_j$ for any $i, j \in N$. In other words, a, b are comonotone if they belong to $\Gamma_\tau := \{a \in \mathbb{R}^n_+ \mid a_{\tau(1)} \leq a_{\tau(2)} \leq \cdots \leq a_{\tau(n)}\}$ for the same permutation τ. Thus, it is clear from (14.8) that for comonotone a, b we have $\mathcal{C}_\mu(a+b) = \mathcal{C}_\mu(a)+\mathcal{C}_\mu(b)$. This property, called *comonotonic additivity*, is characteristic of the Choquet integral, as shown by Schmeidler [66].

For other properties and characterizations of the Choquet integral, we refer the reader to survey papers [7, 49, 61].

Taking F as the Choquet integral, let us see whether it exists some capacity μ such that $\mathcal{C}_\mu$ is able to model relation (14.6) and (14.7). The modeling of (14.6) implies that $2\mu(\mathrm{M}, \mathrm{S}) > \mu(\mathrm{M}) + 1$, while (14.7) gives $2\mu(\mathrm{S}) > \mu(\mathrm{S}, \mathrm{L})$. There is no contradiction between previous two inequalities, hence the Choquet integral can model the preferences of the DM.

3.3 General Method for Building Utility Functions

Let us describe now a general method to construct the utility functions u_i without the prior knowledge of F [33, 46]. The utility functions shall be determined through questions regarding elements of X. Following the MACBETH approach [2, 1, 3], the subset $X\rfloor_i$ (for $i \in N$) of X will serve as a basis for the determination of u_i:

$$X\rfloor_i = \{(x_i, \mathbf{0}_{-i}) \ , \ x_i \in X_i\}.$$

We apply the MACBETH methodology to each set $X\rfloor_i$, which amounts to satisfy conditions **(Ord[$X\rfloor_i$])**, **(Inter[$X\rfloor_i$])**, **(C-Inter[$X\rfloor_i$])**. This gives the numerical representation $u_{X\rfloor_i}$ of $X\rfloor_i$. It is uniquely determined if (14.5) is applied. Since $\mathbf{0}_i$ is a least level of X_i, the utility function u_i is non-negative. Besides, it satisfies (14.5).

For $(x_i, \mathbf{0}_{-i}) \in X\rfloor_i$, one has by (14.2) and (14.5), since $u_{X\rfloor_i}(x_i, \mathbf{0}_{-i})$ corresponds to the overall utility of the act $(x_i, \mathbf{0}_{-i})$:

$$u_{X\rfloor_i}(x_i, \mathbf{0}_{-i}) = F\left(u_i(x_i), u_{-i}(\mathbf{0}_{-i})\right) = F\left(u_i(x_i), 0_{-i}\right).$$

Assume that the family $\mathcal{F}$ of aggregation functions satisfies

$$\exists \alpha_i \in \mathbb{R}^*_+ \ , \quad F\left(a_i, 0_{-i}\right) = \alpha_i \, a_i \ \text{ for all } a_i \in \mathbb{R}_+. \tag{14.9}$$

Since $u_{X\rfloor_i}(\mathbf{1}_i, \mathbf{0}_{-i}) = F\left(1_i, 0_{-i}\right) = \alpha_i$, we get for any $x_i \in X_i$:

$$u_i(x_i) = \frac{F\left(u_i(x_i), 0_{-i}\right)}{F\left(1_i, 0_{-i}\right)} = \frac{u_{X\rfloor_i}(x_i, \mathbf{0}_{-i})}{u_{X\rfloor_i}(\mathbf{1}_i, \mathbf{0}_{-i})}. \tag{14.10}$$

This shows that if all aggregation functions belonging to $\mathcal{F}$ satisfy (14.9) then u_i can be determined by (14.10) from cardinal information related to $X\rfloor_i$.

Note that we do not need to assume weak separability, thanks to (14.9).

Considering the case of the Choquet integral, it is easy to see that whenever $\mu(\{i\}) > 0$ for any $i \in N$, condition (14.9) is fulfilled so that the utility functions can be constructed with $\mathcal{F}$ being equal to the Choquet integral w.r.t. capacities satisfying previous condition.

3.4 Justification of the Use of the Choquet Integral

We adopt here a slightly different approach than the one described in the introduction. We show that if we consider natural information that allow the modeling of interaction between criteria on top of information regarding $X\rfloor_i$, the Choquet integral comes up as a natural aggregation function. The justification of the use of the Choquet integral does not come from a pure axiomatic approach but rather from some reasonable information asked to the DM.

3.4.1 Required Information. As said in Section 3.3, each utility function u_i is built from the set $X\rfloor_i$, which requires the satisfaction of conditions **(Ord[**$X\rfloor_i$**])**, **(Inter[**$X\rfloor_i$**])**, and **(C-Inter[**$X\rfloor_i$**])**, and is uniquely determined by (14.5).

Now that we have described intra-criterion information, let us give the inter-criteria information, that is data needed for the gathering of all criteria. The information regarding the aggregation of the criteria can be limited to alternatives whose scores on criteria are either $\mathbf{0}_i$ or $\mathbf{1}_i$. In order to be able to model subtle interaction phenomena, all combinations of $\mathbf{0}_i$ and $\mathbf{1}_i$ must be considered. This leads to defining the following set:

$$X\rceil_{\{0,1\}} := \{(\mathbf{1}_A, \mathbf{0}_{-A})\ ,\ A \subset N\}\,,$$

called the set of *binary alternatives*. The application of the MACBETH methodology leads to the interval scale $u_{X\rceil_{\{0,1\}}}$, which requires the satisfaction of conditions **(Ord[**$X\rceil_{\{0,1\}}$**])**, **(Inter[**$X\rceil_{\{0,1\}}$**])**, and **(C-Inter [**$X\rceil_{\{0,1\}}$**])**. Applying (14.5) to this scale, it becomes uniquely determined:

$$u_{X\rceil_{\{0,1\}}}(\mathbf{0}_N) = 0, \quad u_{X\rceil_{\{0,1\}}}(\mathbf{1}_N) = 1. \tag{14.11}$$

The second condition in (14.11) says that an alternative which is completely satisfactory on each criteria should be completely satisfactory, and similarly for the first condition.

3.4.2 Measurement Conditions. $u_{X\rceil_{\{0,1\}}}$ represents the importance that the DM gives to the coalition A in the DM process for any $A \subset N$. It depicts the way criteria are aggregated. It leads to the definition of a capacity μ defined by $\mu(A) := u_{X\rceil_{\{0,1\}}}(\mathbf{1}_A, \mathbf{0}_{-A})$. Consequently, it is natural to write u as follows:

$$u(x) = F_\mu(u_1(x_1), \ldots, u_n(x_n)), \tag{14.12}$$

where F_μ is the aggregation operator. F_μ depends on μ in a way that is not known for the moment.

The u_i's correspond to interval scales, whose admissible transformations are the positive affine transformations (see Section 2.1). Hence, one could change all u_i's in $\alpha u_i + \beta$, for any $\alpha > 0$ and $\beta \in \mathbb{R}$, without any change in the model. On the other hand, $\mu(A)$ corresponds in fact to the difference of the satisfaction degrees between the alternatives $(\mathbf{1}_A, \mathbf{0}_{-A})$ and $\mathbf{0}_N$. Applying this to $A = \emptyset$, the value $\mu(\emptyset)$ shall always be equal to zero, whatever the interval scale attached to $X\rceil_{\{0,1\}}$ may be. Henceforth, μ corresponds to a ratio scale, and can be replaced by $\gamma\mu$, with $\gamma \in \mathbb{R}_+$, since these are the admissible transformations for ratio scales. Hence one shall have [46]:

(Meas-Inter) The preference relation $\succeq$ and the ratio $\frac{u(x)-u(y)}{u(z)-u(t)}$ for $x, y, z, t \in X\rfloor_i$ (for all $i \in N$) and for $x, y, z, t \in X\rceil_{\{0,1\}}$ shall not be changed if all the u_i's are changed into $\alpha u_i + \beta$ with $\alpha > 0$ and $\beta \in \mathbb{R}$, and μ is changed into $\gamma\mu$ with $\gamma \in \mathbb{R}_+$.

From **(Ord[$X\rfloor_i$])**, **(Inter[$X\rfloor_i$])**, **(C-Inter[$X\rfloor_i$])**, **(Ord[$X\rceil_{\{0,1\}}$])**, (14.5), **(Inter[$X\rceil_{\{0,1\}}$])**, **(C-Inter[$X\rceil_{\{0,1\}}$])** (14.11) and **(Meas-Inter)** it can be shown that [46, Lemma 2]

$$\frac{F_\mu((\alpha+\beta)_A, \beta_{-A}) - F_\mu((\alpha+\beta)_B, \beta_{-B})}{F_\mu((\alpha+\beta)_C, \beta_{-C}) - F_\mu((\alpha+\beta)_D, \beta_{-D})} = \frac{\mu(A) - \mu(B)}{\mu(C) - \mu(D)}.$$

Taking this with $B = D = \emptyset$ and $C = N$, we get

$$\frac{F_\mu((\alpha+\beta)_A, \beta_{-A}) - F_\mu(\beta_N)}{F_\mu((\alpha+\beta)_N) - F_\mu(\beta_N)} = \mu(A).$$

Since F_μ acts on commensurate scales and returns a value on the same scale, it is natural to assume that F_μ satisfies idempotency [15]

$$F_\mu(\beta, \ldots, \beta) = \beta \ , \ \forall \beta \in \mathbb{R}.$$

Plugging this into previous relation one gets

$$F_\mu((\alpha+\beta)_A, \beta_{-A}) = \alpha\, \mu(A) + \beta.$$

This equality with $\alpha = 1$ and $\beta = 0$ gives

Properly Weighted (PW): If μ satisfies conditions **(FM$_b$)** and **(FM$_c$)**, then $F_\mu\left(1_A, 0_{-A}\right) = \mu(A), \forall A \subset N$.

Previous relation together with **(PW)** gives

Stability for the admissible Positive Linear transformations (weak SPL): If μ satisfies conditions **(FM$_b$)** and **(FM$_c$)**, then for all $A \subset N$, $\alpha > 0$, and $\beta \in \mathbb{R}$,

$$F_\mu\left((\alpha+\beta)_A, \beta_{-A}\right) = \alpha F_\mu\left(1_A, 0_{-A}\right) + \beta$$

Since F_μ aggregates satisfaction scales, it is natural to assume that $x \mapsto F_\mu(x)$ is increasing. Hence F_μ shall satisfy the following axiom.

Increasingness (In): If μ satisfies conditions **(FM$_a$)** and **(FM$_b$)**, then $\forall x, x' \in \mathbb{R}^n$,

$$x_i \leq x'_i \, \forall i \in N \Rightarrow F_\mu(x) \leq F_\mu(x')$$

Measurement considerations yield linearity of the mapping $\mu \mapsto F_\mu(x)$ [46]. Hence F_μ shall satisfy to the following axiom.

Linearity w.r.t. the Measure (LM): If μ satisfies condition **(FM$_b$)**, then for all $x \in \mathbb{R}^n$ and $\gamma, \delta \in \mathbb{R}$,

$$F_{\gamma\mu+\delta\mu'}(x) = \gamma F_\mu(x) + \delta F_{\mu'}(x).$$

The following result can be shown.

THEOREM 1 (THEOREM 1 IN [46]) *F_μ satisfies* **(LM)**, **(In)**, **(PW)** *and* **(weak SPL)** *if and only if $F_\mu \equiv \mathcal{C}_\mu$ in $\mathbb{R}^n$.*

We have seen that the measurement conditions we have on u_i and $u_{X]\{0,1\}}$ lead naturally to axioms **(LM)**, **(In)**, **(PW)** and **(weak SPL)**. There is only one aggregation function that satisfies these axioms, namely the Choquet integral w.r.t. μ. So the cardinal information we work with leads naturally to the use of the Choquet integral.

Let us remark that Theorem 1 is a weak version of an axiomatic characterization obtained by Marichal [49].

3.5 Shapley Value and Interaction Index

By construction, the capacity μ expresses the score of binary alternatives. Since there are 2^n such alternatives, it may be difficult to analyse or explain the behaviour of the decision maker through the values taken by μ.

A first question of interest is: "What is the importance of a given criterion for the decision?". We may say that a criterion i is important if whenever added to some coalition A of criteria, the score of $(\mathbf{1}_{A\cup i}, \mathbf{0}_{-(A\cup i)})$ is significantly larger than the score of $(\mathbf{1}_A, \mathbf{0}_{-A})$. Hence, an importance index should compute an average value Δ_i of the quantity $\mu(A \cup i) - \mu(A)$ for all $A \subset N \setminus i$.

A second requirement is that the sum of importance indices for all criteria should be a constant, say 1. Lastly, the importance index should not depend on the numbering of the criteria. Strangely enough, these three requirements plus a linearity assumption, which imposes that the average Δ_i is a weighted arithmetic mean, suffices to determine uniquely the importance index, known as the *Shapley importance index* [67]

$$\phi^{\mu}(i) := \sum_{K \subset N \setminus i} \frac{(n-k-1)!k!}{n!} [\mu(K \cup i) - \mu(K)] \tag{14.13}$$

with $k := |K|$. We omit the superscript if no ambiguity occurs. The *Shapley value* is the vector $(\phi(1), \ldots, \phi(n))$. As said above, we have $\sum_{i=1}^{n} \phi(i) = \mu(N) = 1$. Another fundamental property is that $\phi(i) = \mu(\{i\})$ if μ is additive.

We have shown by an example in Section 3.1 that interaction may occur among criteria, and that the Choquet integral was able to deal with situations where interaction occurs. We define this notion more precisely. Let us consider for simplicity 2 criteria and the following alternatives (see Figure 14.1):

- $x = (\mathbf{0}_1, \mathbf{0}_2)$
- $y = (\mathbf{1}_1, \mathbf{0}_2)$
- $z = (\mathbf{0}_1, \mathbf{1}_2)$
- $t = (\mathbf{1}_1, \mathbf{1}_2)$

Clearly, t is more attractive than x, but preferences over other pairs may depend on the decision maker. Due to monotonicity ($\mathbf{FM}_a$), we can range from the two extremal following situations (recall that $\mu(\{1,2\}) = 1$ and $\mu(\emptyset) = 0$):

extremal situation 1 (lower bound): we put $\mu(\{1\}) = \mu(\{2\}) = 0$, which is equivalent to the preferences $x \sim y \sim z$ (Figure 14.1, left) (strictly speaking, $\mu(\{i\})$ cannot attain the value 0: see Section 3.3). This means that for the DM, both criteria have to be satisfactory in order to get a satisfactory alternative, the satisfaction of only one criterion being useless. We say that the criteria are *complementary*.

extremal situation 2 (upper bound): we put $\mu(\{1\}) = \mu(\{2\}) = 1$, which is equivalent to the preferences $y \sim z \sim t$ (Figure 14.1, middle). This means that for the DM, the satisfaction of one of the two criteria is sufficient to have a satisfactory alternative, satisfying both being useless. We say that the criteria are *substitutive*.

Clearly, in these two situations, the criteria are not independent, in the sense that the satisfaction of one of them acts on the usefulness of the other in order

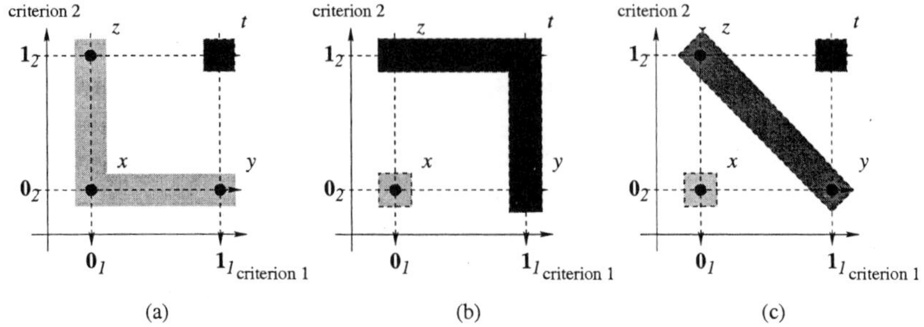

Figure 14.1. Different cases of interaction.

to get a satisfactory object (necessary in the first case, useless in the second). We say that there is some *interaction* between the criteria.

A situation without interaction is such that the satisfaction of each criterion brings its own contribution to the overall satisfaction, hence:

$$\mu(\{1,2\}) = \mu(\{1\}) + \mu(\{2\}) \tag{14.14}$$

(additivity) (see Figure 14.1, right). In the first situation, $\mu(\{1,2\}) > \mu(\{1\}) + \mu(\{2\})$, while the reverse inequality holds in the second situation. This suggests that the interaction I_{12} between criteria 1 and 2 should be defined as :

$$I_{12}^{\mu} := \mu(\{1,2\}) - \mu(\{1\}) - \mu(\{2\}) + \mu(\emptyset). \tag{14.15}$$

This is simply the difference between binary alternatives on the diagonal (where there is strict dominance) and on the anti-diagonal (where no dominance relation exists). The interaction is positive when criteria are complementary, while it is negative when they are substitutive. This is consistent with intuition considering that when criteria are complementary, they have no value by themselves, but put together they become important for the DM.

In the case of more than 2 criteria, the definition of *interaction index* follows the same idea as with the Shapley index, in the sense that all coalitions of N have to be taken into account. The following definition has been first proposed by Murofushi and Soneda [60], for a pair of criteria i, j:

$$I_{ij}^{\mu} := \sum_{K \subset N \setminus \{i,j\}} \frac{(n-k-2)!k!}{(n-1)!} \big[\mu(K \cup \{i,j\}) - \mu(K \cup \{i\}) - \mu(K \cup \{j\}) + \mu(K)\big], \tag{14.16}$$

The definition of this index has been extended to any coalition $\emptyset \neq A \subset N$ of criteria by Grabisch [19]:

$$I^{\mu}(A) := \sum_{K \subset N \setminus A} \frac{(n-k-|A|)!k!}{(n-|A|+1)!} \sum_{L \subset A} (-1)^{|A|-|L|} \mu(K \cup L), \forall A \subset N, A \neq \emptyset. \quad (14.17)$$

We have $I_{ij} = I(\{i,j\})$. When $A = \{i\}$, $I(\{i\})$ coincides with the Shapley index $\phi(i)$. It is easy to see that when the fuzzy measure is additive, we have $I(A) = 0$ for all A such that $|A| > 1$. Also $I_{ij} > 0$ (resp. $< 0, = 0$) for complementary (resp. substitutive, non-interactive) criteria.

The definition can be extended to the case $A = \emptyset$, by just putting $\sum_{L \subset A} (-1)^{|A|-|L|} \mu(K \cup L) = \mu(K)$. Hence I defines a set function $I : \mathcal{P}(N) \longrightarrow \mathbb{R}$. Properties of this set function has been studied and related to the Möbius transform [8, 34]. In particular, it is possible to recover μ if I is given for each $A \subset N$, which means that the interaction index can be viewed as a particular *transform* of a fuzzy measure, which is invertible, as the Möbius transform. Also, I has been characterized axiomatically by Grabisch and Roubens [37], in a way similar to the Shapley index.

Another important property is that the interaction index can be obtained recursively from the Shapley importance index, by considering sub-problems with less criteria [37]. For I^{μ}_{ij}, the relation writes:

$$I^{\mu}_{ij} = \phi^{\mu^{[ij]}}([ij]) - \phi^{\mu_{N \setminus i}}(j) - \phi^{\mu_{N \setminus j}}(i), \quad (14.18)$$

where $[ij]$ stands for an artificial criterion (i and j taken together), $\mu^{[ij]} : \mathcal{P}((N \setminus \{i,j\}) \cup \{[ij]\}) \longrightarrow [0,1]$, with $\mu^{[ij]}(A) := \mu(A \cup \{i,j\})$ if $A \ni [ij]$, and $\mu(A)$ else, and $\mu_{N \setminus i}$ is the restriction of μ to $N \setminus i$.

3.6 k-additive Measures

Although we have shown that our construction is able to model in a clear way interaction, this has to be paid by an exponential complexity, since the number of binary alternatives is 2^n. There exists a way to cope with complexity by defining sub-families of fuzzy measures, which require less than 2^n coefficients to be defined. The first such family which has been defined is the one of *decomposable measures* [11, 75], which includes the well-known class of λ-measures proposed by Sugeno [70]. These fuzzy measures are defined by a kind of density function, and thus need only $n-1$ coefficients. However, they have a very limited ability to represent interaction since e.g. I_{ij} has the same sign for all i, j.

A second family is given by the concept of k-additive measure, which is detailed in this section.

DEFINITION 50 *[19] Let* $k \in \{1, \ldots, n-1\}$. *A fuzzy measure* μ *is said to be k-additive if* $I(A) = 0$ *whenever* $|A| > k$, *and there exists some* $A \subset N$ *with* $|A| = k$ *such that* $I(A) \neq 0$.

From the properties of interaction cited in Section 3.5, a 1-additive measure is simply an additive measure, hence the name. Also, since μ is completely determined by the values of I on $\mathcal{P}(N)$, a k-additive measure is determined by $1+n+\binom{n}{2}+\ldots+\binom{n}{k}$ parameters, among which 2 are not free.

The 2-additive measure, which needs only $\frac{n(n+1)}{2}-1$ parameters, permits to model interaction between pair of criteria, which is in general sufficient in practice (it is in fact fairly difficult to have a clear understanding of interaction among more than 2 criteria).

The Choquet integral can be expressed using I instead of μ in a very instructive way when the measure is 2-additive[18]:

$$\mathcal{C}_\mu(a_1,\ldots,a_n) = \sum_{I_{ij}>0}(a_i \wedge a_j)I_{ij} + \sum_{I_{ij}<0}(a_i \vee a_j)|I_{ij}| + \sum_{i=1}^{n} a_i(\phi_i - \frac{1}{2}\sum_{j\neq i}|I_{ij}|), \quad \forall a \in [0,1]^n, \quad (14.19)$$

for all $(a_1,\ldots,a_n) \in \mathbb{R}^n_+$, with the property that $\phi_i - \frac{1}{2}\sum_{j\neq i}|I_{ij}| \geq 0$ for all i. It can be seen that the Choquet integral for 2-additive measures is the sum of a conjunctive, a disjunctive and an additive part, corresponding respectively to positive interaction indices, negative interaction indices, and the Shapley value. Equation (14.19) shows clearly the disjunctive and conjunctive effects of negative and positive interaction between criteria, which has been explained in Section 3.5. It is important to notice that, due to the normalization $\sum_{i=1}^n \phi_i = 1$, (14.19) is a convex combination of disjunctions, conjunctions, and a linear part. Hence, as illustrated in [21] in a graphical way, the Choquet integral is the convex closure of all conjunctions and all disjunctions of pair of criteria, and of all dictators (single criteria).

Before ending this section, we mention a third family of fuzzy measures introduced by Miranda and Grabisch, the *p-symmetric fuzzy measures* [55]. The idea is to generalize symmetric fuzzy measures (see Section 3.2), by considering a partition $\{A_1,\ldots,A_p\}$ of N into subsets of indifference: taking elements in $A_1,\ldots,A_p$, the value of μ does not depend on the particular elements which are chosen in each A_i, but only on their number. Hence a symmetric measure corresponds to a 1-symmetric measure (i.e. the partition is N itself). The number of parameters needed to define a p-symmetric measure is $\prod_{i=1}^p(|A_i|+1)-2$.

3.7 Identification of Capacities

We assume here that the utility functions u_i are known. Their construction is carried out with the help of cardinal data on the sets $X\rfloor_i$ (See Section 3.4). So, we focus in this section on the determination of the capacity.

In Section 3.4, we proposed to determine the aggregation function with the help of cardinal information related to binary alternatives. The main advantage of this method is that by **(PW)** each alternative is associated to one term of the capacity. However, this way is not considered in practice because of the following two reasons. The first one is that it may not be natural for a DM to give his preferences on the prototypical alternatives $(\mathbf{1}_A, \mathbf{0}_{-A})$. The second one is that it forces the DM to construct a ratio scale over 2^n alternatives using the MACBETH approach. This requires roughly $4^n/2$ questions to be asked to the DM. This is too much in practice.

The first idea is to replace $X\rceil_{\{0,1\}}$ by a set of more intuitive alternatives. The DM provides a set of learning examples x^1, ..., x^p in X. As for $X\rceil_{\{0,1\}}$, we want a numerical representation of these learning examples. In order to obtain a unique interval scale, the two prototypical alternatives $\mathbf{0}_N$ and $\mathbf{1}_N$ are added to the learning examples. Let

$$X\rceil := \{x^1, \ldots, x^p\} \bigcup \{\mathbf{0}_N, \mathbf{1}_N\}.$$

An interval scale $u_{X\rceil}$ representing the preference on $X\rceil$ can be obtained using the MACBETH methodology, if conditions **(Ord[$X\rceil$])**, **(Inter[$X\rceil$])**, **(C-Inter[$X\rceil$])** are satisfied. The application of (14.5) makes the scale unique, putting 0 for $\mathbf{0}_N$ and 1 for $\mathbf{1}_N$. One wishes to determine the capacity μ solution to the following set of equations:

$$\forall i \in \{1, \ldots, p\} \ , \ \mathcal{C}_\mu\left(u_1(x_1^i), \ldots, u_n(x_n^i)\right) = u_{X\rceil}\left(x^i\right). \qquad (14.20)$$

Unfortunately, no solution may exist or there may be more than one solution. In these cases, in order to get an approximate solution, previous problem is written as a minimization problem [16, 36] in which the unknowns are the parameters of the capacity:

Minimize

$$\sum_{i=1}^{p} \left|\mathcal{C}_\mu\left(u_1(x_1^i), \ldots, u_n(x_n^i)\right) - u_{X\rceil}\left(x^i\right)\right|^2$$

under the constraints **(FM$_a$)**, **(FM$_b$)** and **(FM$_c$)**.

It can be shown that the above problem is a quadratic minimization problem under linear constraints [17, 36]. Thanks to **(FM$_b$)** and **(FM$_c$)**, there are $2^n - 2$ unknowns. Moreover, there are $n(2^{n-1} - 1)$ monotony constraints [17]. There is generally not a unique solution to this problem [53]. Experiments on real data have shown some drawbacks of this method.

- if there is too few data, the solution is of course not unique, and the solution proposed by quadratic optimization libraries may be counterintuitive, because many coefficients are near 0 or 1.

- as n grows up, the dimensions of vectors and matrices grows exponentially, so does the memory required and the computation time. $n = 8$ is already a large value, and $n = 10$ is nearly infeasible.

For these reasons, some authors have looked for more heuristic methods, as Ishii and Sugeno [41] and Mori and Murofushi [58]. Based on this last one, Grabisch has proposed an optimization algorithm [16], which although suboptimal, gives better results than previous attempts. The basic idea is that, in the absence of any information, the most non-arbitrary (least specific) way of aggregation is the arithmetic mean, thus a Choquet integral with respect to an additive equidistributed fuzzy measure. Any input of information tends to move away the fuzzy measure from this equilibrium point. This means that, in case of few data, coefficients of the fuzzy measure which are not concerned with the data are kept as near as possible to the equilibrium point, in order to ensure monotonicity.

Experiments done in classification problems show the good performance of the algorithm, even better than the optimal method when n is large. Especially, the memory and computation time required are much smaller than for the quadratic program, and it is possible to treat problems with $n = 16$.

The DM may not be able to give cardinal information on alternatives. So, the second idea is to use a set of examples of comparisons between alternatives provided by the DM. In other words, the DM gives two sets of alternatives x^1, ..., x^p and $y^1, \ldots, y^p$ in X such that $x^1 \succ y^1, \ldots, x^p \succ y^p$. One looks then for a fuzzy measure that is consistent with previous relations and thus that satisfies

$$\forall i \in \{1,\ldots,p\}\,,\; \mathcal{C}_\mu\left(u_1(x_1^i),\ldots,u_n(x_n^i)\right) > \mathcal{C}_\mu\left(u_1(y_1^i),\ldots,u_n(y_n^i)\right). \tag{14.21}$$

Most of the time, there is a huge number of solutions. In order to reduce the solution space, additional constraints must be added. As remarked by Marichal and Roubens in [52], when the DM states that $x^i \succ y^i$, he or she generally means that x^i is significantly preferred to y^i. If the overall utilities of the two alternatives x^i and y^i are almost the same, it will probably not represent the DM's intention. Henceforth, among all solutions to (14.21), one should prefer the ones with the highest margin. This leaded Marichal and Roubens to introduce a positive coefficient ϵ in the right-hand side of (14.21), and to maximize ϵ:

Maximize ϵ
under the constraints **(FM$_a$)**, **(FM$_b$)**, **(FM$_c$)**, $\epsilon \geq 0$ and for all $i \in \{1,\ldots,p\}$

$$\mathcal{C}_\mu\left(u_1(x_1^i),\ldots,u_n(x_n^i)\right) \geq \mathcal{C}_\mu\left(u_1(y_1^i),\ldots,u_n(y_n^i)\right) + \epsilon$$

This is a linear programming problem. It is a simplified version of a linear method proposed by Marichal and Roubens [52].

Other learning methods have been tried, principally using genetic algorithms (see in particular Wang [74], Kwon and Sugeno [44], and Grabisch [23]).

4. Bipolar Scales

We address now the construction of the model in the case of bipolar scales. As explained in Section 2, we have on each X_i one neutral level $\mathbf{0}_i$ and another absolute level $\mathbf{1}_i$ given by the DM.

4.1 A Motivating Example

Let us go a little deeper in the example described in Section 3.1. We have seen in Section 3.1 that for students good in mathematics, the director prefers someone good at languages to one good at statistics. In other words, when the mark with respect to mathematics is good, the director thinks that languages is more important than statistics. This leads to the following rule

(R1): For a student good at mathematics (M), L is more important than S.

The comparison between students A and B in Section 3.1 are governed by this rule. Let us consider now another set of students. Consider the following students E and F

	mathematics (M)	statistics (S)	languages (L)
student E	14	16	7
student F	14	15	8

According to rule **(R1)**, the director prefers student F to E

$$E \prec F \tag{14.22}$$

As justified in Section 3.1, when the score w.r.t. mathematics is bad, a student good in statistics is now preferred to one good in languages. More precisely, we have the following statement

(R2): For a student bad in mathematics M, S is more important than L.

Consider the following two students

	mathematics (M)	statistics (S)	languages (L)
student G	9	16	7
student H	9	15	8

Following rule **(R2)**, G is preferred to H even though G is very bad in languages.

$$G \succ H \tag{14.23}$$

Relations (14.22) and (14.23) look similar to (14.6) and (14.7). However, we will see that they exhibit a weakness of the Choquet integral. Let us indeed try to model (14.22) and (14.23) with the help of the Choquet integral. We have $\mathcal{C}_\mu(E) = 7+7\mu(\{\mathrm{M},\mathrm{S}\})+2\mu(\{\mathrm{S}\})$ and $\mathcal{C}_\mu(F) = 8+6\mu(\{\mathrm{M},\mathrm{S}\})+\mu(\{\mathrm{S}\})$. This shows that (14.22) is equivalent to

$$\mu(\{\mathrm{M},\mathrm{S}\}) + \mu(\{\mathrm{S}\}) < 1.$$

Similarly, relation (14.23) is equivalent to $\mu(\{\mathrm{M},\mathrm{S}\})+\mu(\{\mathrm{S}\}) > 1$, which contradicts previous inequality. Hence, the Choquet integral cannot model (14.22) and (14.23).

It is no surprise that the Choquet integral cannot model both **(R1)** and **(R2)**. This is due to the fact that the Choquet integral satisfies comonotonic additivity (see Section 3.2). In our example, the marks of the four students E, F, G and H are ranked in the same way: languages is the worst score, mathematics is the second best score, and statistics is the best score. Those four students are comonotonic. The Choquet integral is able to model rules of the following type :

(R1'): If M is the best satisfied criteria, L is more important than S.

(R2'): If M is the worst satisfied criteria, S is more important than L.

On the other hand, rules **(R1)** and **(R2)** make a reference to absolute values (good/bad in mathematics). The Choquet integral does not allow to model this type of property. The Choquet integral fails to represent the expertise that makes an explicit reference to an absolute value. This happens quite often in applications.

Let us study the meaning of the reference point used in rules **(R1)** and **(R2)**. In our example, the satisfaction level is either rather good (good in mathematics) or rather bad (bad in mathematics). This makes an implicit reference to a neutral level that is neither good nor bad. This suggests to construct criteria on ratio scales. In such scales, the zero element is the neutral element. It has an absolute meaning and cannot be shifted. Values above this level are attractive (good) whereas values below the zero level are repulsive (bad).

4.2 The Symmetric Choquet Integral and Cumulative Prospect Theory

4.2.1 Definitions. Let $f : N \longrightarrow \mathbb{R}$ be a real-valued function, and let us denote by $f^+(i) := f(i) \vee 0$, $\forall i \in N$, and $f^- := (-f)^+$ the positive and negative parts of f.

The *symmetric Choquet integral* [6] (also called the *Šipoš integral* [72]) of f w.r.t. μ is defined by:

$$\check{\mathcal{C}}_\mu(f) := \mathcal{C}_\mu(f^+) - \mathcal{C}_\mu(f^-).$$

This differs from the usual definition of Choquet integral for real-valued functions, sometimes called *asymmetric Choquet integral* [6], which is

$$\mathcal{C}_\mu(f) := \mathcal{C}_\mu(f^+) - \mathcal{C}_{\overline{\mu}}(f^-).$$

The Cumulative Prospect Theory model [71] generalizes these definitions, by considering different capacities for the positive and negative parts of the integrand.

$$\mathrm{CPT}_{\mu_1,\mu_2}(f) := \mathcal{C}_{\mu_1}(f^+) - \mathcal{C}_{\mu_2}(f^-).$$

4.2.2 Application to the Example. Let us go back to the example of Section 4.1. In this example, value 10 for the marks seems to be the appropriate neutral value. Hence, in order to transform the regular marks given in the interval $[0, 20]$ to a ratio scale, it is enough to subtract 10 to each mark yielding the mark 10 to the zero level. This gives :

	mathematics (M)	statistics (S)	languages (L)
student E'	4	6	-3
student F'	4	5	-2
student G'	-1	6	-3
student H'	-1	5	-2

Modeling our example with the Šipoš integral, a straightforward calculation shows that (14.22) is equivalent to $\mu(\{S\}) < \mu(\{L\})$ whereas relation (14.23) is equivalent to $\mu(\{S\}) > \mu(\{L\})$, which contradicts previous inequality. Henceforth, the Šipoš integral is not able to model both (14.22) and (14.23).

Trying now the representation of our example with the CPT model, it is easy to see that (14.22) is equivalent to $\mu_1(\{S\}) < \mu_2(\{L\})$, and relation (14.23) is equivalent to $\mu_1(\{S\}) > \mu_2(\{L\})$. Henceforth, the CPT model too fails to model both (14.22) and (14.23).

4.3 Bi-capacities and the Corresponding Integral

The Choquet, Šipoš and CPT models are limited by the fact that they are constructed on the notion of capacity. The idea is thus to generalize the notion of capacity. Such generalizations have first been introduced in the context of game theory. The concept of ternary voting games has recently been defined by D. Felsenthal and M. Machover as a generalization of binary voting games [14]. Binary voting games model the result of a vote when some voters are in favor of the bill and the other voters are against [68]. The main limitation of such games is that they cannot represent decision rules in which *abstention* is an alternative option to the usual *yes* and *no* opinions. This leaded D. Felsenthal and M. Machover to introduce *ternary voting games* [14]. These voting games can be represented by a function v with two arguments, one for the *yes* voters and the other one for the *no* voters. This concept of ternary voting game has been generalized by J.M. Bilbao *et al.* in [4], yielding the definition of *bi-cooperative game*. Let

$$\mathcal{Q}(N) = \{(A, B) \in \mathcal{P}(N) \times \mathcal{P}(N) \mid A \cap B = \emptyset\}.$$

A bi-cooperative game is a function $\nu : \mathcal{Q}(N) \to \mathbb{R}$ satisfying $\nu(\emptyset, \emptyset) = 0$. In the context of game theory, the first argument A in $\nu(A, B)$ is called the *defender* part, and the second argument B in $\nu(A, B)$ is called the *defeater* part.

This generalization has recently been rediscovered independently by the authors in the context of MCDA [29, 48]. A *bi-capacity* is a function $\nu : \mathcal{Q}(N) \to \mathbb{R}$ satisfying

(BFM$_a$) $A \subset A' \Rightarrow \nu(A, B) \leq \nu(A', B)$,

(BFM$_b$) $B \subset B' \Rightarrow \nu(A, B) \geq \nu(A, B')$,

(BFM$_c$) $\nu(\emptyset, \emptyset) = 0$,

(BFM$_d$) $\nu(N, \emptyset) = 1$, $\nu(\emptyset, N) = -1$

Conditions **(BFM$_a$)** and **(BFM$_b$)** together define *monotonic* bi-capacities. Bi-capacities are special cases of bi-cooperative games. In MCDA, $\nu(A, B)$ is interpreted as the overall assessment of the *ternary alternative* $\left(1_A, -1_B, 0_{-(A \cup B)}\right)$. Thanks to that interpretation, the first argument A in $\nu(A, B)$ is called the *positive* part, and the second argument B in $\nu(A, B)$ is called the *negative* part.

The *conjugate* or *dual* ν^* of a bi-capacity ν can be defined by $\nu^*(S, T) = -\nu(T, S)$ for all $(S, T) \in \mathcal{Q}(N)$ [45, 47]. In the context of Game Theory, it means that the defenders and the defeaters are switched, and the abstentionists are untouched. This definition of dual bi-capacity coincides with that proposed in [14] for ternary voting games.

A bi-capacity ν is of the *CPT type* if it can be written $\nu(A, B) = \mu_1(A) - \mu_2(B)$, for all $(A, B) \in \mathcal{Q}(N)$, where μ_1, μ_2 are capacities. If $\mu_1 = \mu_2$, we say that the bi-capacity is *symmetric*. If μ_1 and μ_2 are additive, then ν is said to be *additive*.

A similar concept has also been introduced by S. Greco *et al.* leading to the concept of *bipolar capacity* [39]. A bipolar capacity is a function $\zeta : \mathcal{Q}(N) \to [0, 1] \times [0, 1]$ with $\zeta(A, B) =: (\zeta^+(A, B), \zeta^-(A, B))$ such that

- If $A \supset A'$ and $B \subset B'$ then $\zeta^+(A, B) \geq \zeta^+(A', B')$ and $\zeta^-(A, B) \leq \zeta^-(A', B')$.

- $\zeta^-(A, \emptyset) = 0$, $\zeta^+(\emptyset, A) = 0$ for any $A \subset N$.

- $\zeta(N, \emptyset) = (1, 0)$ and $\zeta(\emptyset, N) = (0, 1)$.

$\zeta^+(A, B)$ can be interpreted as the importance of coalition A of criteria in the presence of B for the positive part. $\zeta^-(A, B)$ can be interpreted as the importance of coalition B of criteria in the presence of A for the negative part.

The Choquet integral w.r.t. a bi-capacity ν proposed in [29] is now given. For any $a \in \mathbb{R}^n$,

$$\mathcal{BC}_\nu(a) := \mathcal{C}_{\mu_{N^+}}(|a|)$$

where $\mu_{N^+}(C) := \nu(C \cap N^+, C \cap N^-)$, $N^+ = \{i \in N \mid a_i \geq 0\}$, $N^- := N \setminus N^+$, and $|a|$ stands for $(|a_1|, \ldots, |a_n|)$. Note that μ_{N^+} is a non-monotonic capacity.

The Choquet integral w.r.t. a bipolar capacity can also be defined [39]. For $a \in \mathbb{R}^n$, let τ be a permutation on N such that

$$\left|a_{\tau(1)}\right| \leq \cdots \leq \left|a_{\tau(n)}\right|. \tag{14.24}$$

Let

$$A_i^+ := \left\{\tau(j)\,,\; j \in \{i, \ldots, n\} \text{ such that } a_{\tau(j)} \geq 0\right\}$$
$$A_i^- := \left\{\tau(j)\,,\; j \in \{i, \ldots, n\} \text{ such that } a_{\tau(j)} < 0\right\}$$

and

$$C^+(a;\zeta) = \textstyle\sum_{i \in N} \left(a^+_{\tau(i)} - a^+_{\tau(i-1)}\right) \zeta^+\left(A_i^+, A_i^-\right)$$
$$C^-(a;\zeta) = \textstyle\sum_{i \in N} \left(a^-_{\tau(i)} - a^-_{\tau(i-1)}\right) \zeta^-\left(A_i^+, A_i^-\right)$$

where $a_{\tau(0)} := 0$ and for $a \in \mathbb{R}$ we set $a^+ = \max(a, 0)$ and $a^- = (-a)^+$. Finally the Choquet integral w.r.t. ζ is defined by

$$C(a;\zeta) := C^+(a;\zeta) - C^-(a;\zeta).$$

For $a \in \mathbb{R}^n$ for which several permutations τ satisfy (14.24), it is easy to see that the previous expression depends on the choice of the permutation. This is not the case of the usual Choquet integral or the Choquet integral w.r.t. a bi-capacity. Enforcing that the results are the same for all permutations satisfying (14.24), we obtain the following constraints on the bipolar capacity:

$$\forall (A, B) \in \mathcal{Q}(N)\,,\;\; \zeta^+(A, B) - \zeta^-(\emptyset, B) = \zeta^+(A, \emptyset) - \zeta^-(A, B).$$

It can be shown then that the bipolar capacity ζ reduces exactly to a bi-capacity ν defined by

$$\nu(A, B) := \zeta^+(A, B) - \zeta^-(\emptyset, B).$$

One has indeed $\zeta^+(A, B) = \nu(A, B) - \nu(\emptyset, B)$ and $\zeta^-(A, B) = \nu(A, \emptyset) - \nu(A, B)$. Moreover, it can be shown that the Choquet integral w.r.t. ζ is equal to $\mathcal{BC}_\nu$. As a consequence, the concept of bipolar capacity reduces to bi-capacities when the Choquet integral is used. For this reason, we will consider only bi-capacities from now on. Note however that the concept of bipolar capacities has some interests in itself for other domains than MCDA.

The concept of bi-capacities is now applied to the example of Section 4.2.2. Let us try to model (14.22) and (14.23) with the extension of the Choquet integral to bi-capacities. We have $\mathcal{BC}_\nu(4,6,-3) = \mathcal{C}_{\mu_{N^+}}(4,6,3) = 3\mu(\{\mathrm{M},\mathrm{S},\mathrm{L}\}) + \mu(\{\mathrm{M},\mathrm{S}\}) + 2\mu(\{\mathrm{S}\}) = 3\nu(\{\mathrm{M},\mathrm{S}\},\{\mathrm{L}\}) + \nu(\{\mathrm{M},\mathrm{S}\},\emptyset) + 2\nu(\{\mathrm{S}\},\emptyset)$ and $\mathcal{BC}_\nu(4,5,-2) = 2\nu(\{\mathrm{M},\mathrm{S}\},\{\mathrm{L}\}) + 2\nu(\{\mathrm{M},\mathrm{S}\},\emptyset) + \nu(\{\mathrm{S}\},\emptyset)$. Hence (14.22) is equivalent to

$$\nu(\{\mathrm{M},\mathrm{S}\},\emptyset) - \nu(\{\mathrm{M},\mathrm{S}\},\{\mathrm{L}\}) > \nu(\{\mathrm{S}\},\emptyset)$$

Similarly, relation (14.23) is equivalent to

$$\nu(\{\mathrm{S}\},\{\mathrm{L}\}) > 0.$$

There is no contradiction between these two inequalities. Henceforth, $\mathcal{BC}_\nu$ is able to model the example. This aggregation operator models the expertise that makes an explicit reference to an absolute value.

Before ending this section, we would like to stress that bi-capacities cannot account for *all* decision behaviours involving bipolar scales. To illustrate this, let us change the scores of E' and F' as follows.

	mathematics (M)	statistics (S)	languages (L)
student E''	2	6	-4
student F''	2	5	-3

It is easy to check that maintining $E'' \prec F''$ is equivalent to

$$\nu(\{\mathrm{S}\},\{\mathrm{L}\}) < 0,$$

a contradiction with $G' \succ H'$. The fact is that with E'', F'', the score on mathematics is now too weak with respect to the score on languages. Hence E'' should be preferred to F'' since the latter one is better in statistics.

4.4 General Method for Building Utility Functions

Let us now describe a general method to construct the utility functions u_i without the prior knowledge of F. It is possible to extend the method described in Section 3.3 in a straightforward way. Due to the existence of a neutral level, utility functions can now take positive and negative values. Hence assumption (14.9) is replaced by the following one:

$$\exists \alpha_i \in \mathbb{R}_+^* \ , \quad F(a_i, 0_{-i}) = \alpha_i \, a_i \ \text{ for all } a_i \in \mathbb{R}. \qquad (14.25)$$

Then the utility function can be derived from (14.10). It has been shown in [33] that the Šipoš integral satisfies (14.25). However, this condition is too restrictive since the usual Choquet does not fulfill it [33]. As a consequence, we are looking for a more general method.

Since the neutral level has a central position, the idea is to process separately elements which are "above" the neutral level (attractive part), and "below" it (repulsive part). Doing so, we may avoid difficulties due to some asymmetry between attractive and repulsive parts [29, 48]. The positive part of the utility function of X_i will be based on the two absolute levels $\mathbf{0}_i$ and $\mathbf{1}_i$, while the negative part is based on the absolute levels $\mathbf{0}_i$ and $-\mathbf{1}_i$, as defined in Section 2.3.

Generalizing (14.5), we set

$$u_i(\mathbf{0}_i) = 0\ ,\ u_i(\mathbf{1}_i) = 1 \ \text{ and } \ u_i(-\mathbf{1}_i) = -1. \tag{14.26}$$

The two values 1 and -1 are opposite to express the symmetry between $\mathbf{1}_i$ and $-\mathbf{1}_i$.

The construction of the positive and negative parts of the utility function u_i is performed through the MACBETH methodology from the following two sets $X\rfloor_i^+$ and $X\rfloor_i^-$:

$$X\rfloor_i^\pm = \left\{(x_i, \mathbf{0}_{-i})\ ,\ x_i \in X_i^\pm\right\},$$

where $X_i^+ = \{x_i \in X_i\ ,\ (x_i, \mathbf{0}_{-i}) \succeq \mathbf{0}_N\}$ and $X_i^- = \{x_i \in X_i\ ,\ (x_i, \mathbf{0}_{-i}) \preceq \mathbf{0}_N\}$. Interval scales $u_{X\rfloor_i^+}, u_{X\rfloor_i^-}$ are obtained for $i = 1, \ldots, n$, provided that conditions **(Ord[$X\rfloor_i^+$])**, **(Inter[$X\rfloor_i^+$])**, **(C-Inter [$X\rfloor_i^+$])**, **(Ord [$X\rfloor_i^-$])**, **(Inter[$X\rfloor_i^-$])**, and **(C-Inter[$X\rfloor_i^-$])** are satisfied for $i = 1, \ldots, n$. Now the scales are uniquely determined if one applies (14.5) to all positive scales, and the symmetric condition

$$u_{X\rfloor_i^-}(\mathbf{0}_N) = 0 \ \text{ and } \ u_{X\rfloor_i^-}(-\mathbf{1}_i, \mathbf{0}_{-i}) = -1. \tag{14.27}$$

to all negative scales. Like for interval scales, one has for $x_i \in X_i^\pm$

$$u_{X\rfloor_i^\pm}(x_i, \mathbf{0}_{-i}) = F\left(u_i(x_i), 0_{-i}\right).$$

The assumption on the family $\mathcal{F}$ becomes

$$\exists \alpha_i^\pm \in \mathbb{R}_+^*\ ,\quad F\left(a_i, 0_{-i}\right) = \alpha_i^\pm\, a_i \ \text{ for all } a_i \in \mathbb{R}_\pm. \tag{14.28}$$

Hence by (14.26), one has for any $x_i \in X_i^\pm$:

$$u_i(x_i) = \frac{F\left(u_i(x_i), 0_{-i}\right)}{F\left(\pm 1_i, 0_{-i}\right)} = \frac{u_{X\rfloor_i^\pm}(x_i, \mathbf{0}_{-i})}{u_{X\rfloor_i^\pm}(\pm\mathbf{1}_i, \mathbf{0}_{-i})}. \tag{14.29}$$

Hence, under assumption (14.28), the positive and negative parts of the utility functions can be constructed in two separate steps by (14.29) from cardinal information related to $X\rfloor_i^{\pm}$.

It can be shown that the Choquet integral, Šipoš integral, the CPT model and the generalized Choquet integral fulfills (14.28).

4.5 Justification of the Use of the Generalized Choquet Integral

4.5.1 Required Information. For any $i \in N$, the utility function u_i is built from $u_{X\rfloor_i^+}$ and $u_{X\rfloor_i^-}$ like in Section 4.4.

Inter-criteria information is a generalization of the set $X\rceil_{\{0,1\}}$. The three reference levels $-\mathbf{1}_i$, $\mathbf{0}_i$ and $\mathbf{1}_i$ are now used to build the set of *ternary alternatives*:

$$X\rceil_{\{-1,0,1\}} := \left\{ \left(\mathbf{1}_A, -\mathbf{1}_B, \mathbf{0}_{-(A\cup B)}\right) \ , \ (A,B) \in \mathcal{Q}(N) \right\}.$$

Let $u_{X\rceil_{\{-1,0,1\}}}$ be a numerical representation of $X\rceil_{\{-1,0,1\}}$. In the previous set, three special points can be exhibited: $\mathbf{1}_N$, $\mathbf{0}_N$ and $-\mathbf{1}_N$. Thanks to commensurateness between the $\mathbf{1}_i$ levels, between the $\mathbf{0}_i$ levels and between the $-\mathbf{1}_i$ levels, it is natural to set

$$u_{X\rceil_{\{-1,0,1\}}}(-\mathbf{1}_N) = -1 \ , \ u_{X\rceil_{\{-1,0,1\}}}(\mathbf{0}_N) = 0 \ \text{ and } \ u_{X\rceil_{\{-1,0,1\}}}(\mathbf{1}_N) = 1. \qquad (14.30)$$

Relation $u_{X\rceil_{\{-1,0,1\}}}(\mathbf{1}_N) = 1$ means that the alternative which is satisfactory on all attributes is also satisfactory. Relation $u_{X\rceil_{\{-1,0,1\}}}(\mathbf{0}_N) = 0$ means that the alternative which is neutral on all attributes is also neutral. Finally, relation $u_{X\rceil_{\{-1,0,1\}}}(-\mathbf{1}_N) = -1$ means that the alternative which is unsatisfactory on all attributes is also unsatisfactory. Since there are only two degrees of freedom in a scale of difference, one of these three points must be removed for the practical construction of the scale. We decide to remove the act $-\mathbf{1}_i$. Let $X\rceil^*_{\{-1,0,1\}} := X\rceil_{\{-1,0,1\}} \setminus \{-\mathbf{1}_N\}$.

The numerical representation $u_{X\rceil^*_{\{-1,0,1\}}}$ on $X\rceil^*_{\{-1,0,1\}}$ is ensured by **(Ord[$X\rceil^*_{\{-1,0,1\}}$])**, **(Inter[$X\rceil^*_{\{-1,0,1\}}$])**, **(C-Inter[$X\rceil^*_{\{-1,0,1\}}$])** and the last two conditions in (14.30). $u_{X\rceil^*_{\{-1,0,1\}}}$ is uniquely determined by previous requirements. In summary

$$u_{X\rceil_{\{-1,0,1\}}}\left(\mathbf{1}_A, -\mathbf{1}_B, \mathbf{0}_{-(A\cup B)}\right) = \begin{cases} u_{X\rceil^*_{\{-1,0,1\}}}\left(\mathbf{1}_A, -\mathbf{1}_B, \mathbf{0}_{-(A\cup B)}\right) & \text{if } (A,B) \neq (\emptyset, N) \\ -1 \ \text{otherwise} & \end{cases}$$

4.5.2 Measurement Conditions.

$u_{X\rceil_{\{-1,0,1\}}}$ can be described by a bi-capacity ν defined by: $\nu(A,B) := u_{X\rceil_{\{0,1\}}}(\mathbf{1}_A, -\mathbf{1}_B, \mathbf{0}_{-(A\cup B)})$. Consequently, it is natural to write u as follows:

$$u(x) = F_\nu\left(u_1(x_1), \ldots, u_n(x_n)\right), \qquad (14.31)$$

where F_ν is the aggregation function.

We introduce the following axioms.

(Bi-LM): For any bi-capacities ν, ν' on $\mathcal{Q}(N)$ satisfying **(BFM$_c$)**, for all $x \in \mathbb{R}^n$ and $\gamma, \delta \in \mathbb{R}$,

$$F_{\gamma\nu+\delta\nu'}(x) = \gamma F_\nu(x) + \delta F_{\nu'}(x)$$

(Bi-In): For any bi-capacity ν on $\mathcal{Q}(N)$ satisfying **(BFM$_a$)**, **(BFM$_b$)** and **(BFM$_c$)**, $\forall x, x' \in \mathbb{R}^n$,

$$x_i \leq x'_i, \ \forall i \subset N \ \rightarrow F_\nu(x) \leq F_\nu(x')$$

(Bi-PW): For any bi-capacity ν satisfying **(BFM$_a$)**, **(BFM$_b$)**, and **(BFM$_c$)**, F_ν $(1_A, -1_{A'}, 0_{-A\cup A'}) = \nu(A, A'), \forall (A, A') \in \mathcal{Q}(N)$.

(Bi-weak SPL$^+$): For any bi-capacity v on $\mathcal{Q}(N)$ satisfying **(BFM$_a$)**, **(BFM$_b$)**, **(BFM$_c$)**, for all $A, C \subset N$, $\alpha > 0$, and $\beta \geq 0$,

$$F_\nu\left((\alpha+\beta)_A, \beta_{-A}\right) = \alpha F_v\left(1_A, 0_{-A}\right) + \beta v(N, \emptyset).$$

These axioms are basically deduced from the measurement conditions on $u_{X\rfloor_i^\pm}$ and ν. This is done exactly as in Section 3.4.2 [29, 48].

For $A \subset N$, consider the following application $\Pi_A : \mathbb{R}^n \to \mathbb{R}^n$ defined by $(\Pi_A(x))_i = x_i$ if $i \in A$ and $-x_i$ otherwise. By **(Bi-PW)**, $\nu(B, B')$ corresponds to the point $(1_B, -1_{B'}, 0_{(B\cup B')})$. Define $\Pi_A \circ \nu(B, B')$ as the term of the bi-capacity associated to the point $\Pi_A(1_B, -1_{B'}, 0_{-B\cup B'}) = (1_{(B\cap A)\cup(B'\setminus A)}, -1_{(B\setminus A)\cup(B'\cap A)}, 0_{-B\cup B'})$. Hence we set

$$\Pi_A \circ \nu(B, B') := \nu\left((B\cap A)\cup(B'\setminus A), (B\setminus A)\cup(B'\cap A)\right).$$

By symmetry arguments, it is reasonable to have $F_{\Pi_A\circ\nu}(\Pi_A(x))$ being equal to $F_\nu(x)$.

(Bi-Sym): For any $\nu : \mathcal{Q}(N) \to \mathbb{R}$ satisfying **(BFM$_c$)**, we have for all $A \subset N$

$$F_\nu(x) = F_{\Pi_A\circ\nu}\left(\Pi_A(x)\right).$$

We have the following result.

THEOREM 2 (THEOREM 1 IN [48]) $\{F_\nu\}_\nu$ *satisfies* **(Bi-LM)**,**(Bi-In)**, **(Bi-PW)**, **(Bi-weak SPL$^+$)** *and* **(Bi-Sym)** *if and only if for any* $\nu : \mathcal{Q}(N) \to \mathbb{R}$ *satisfying* **(BFM$_a$)**, **(BFM$_b$)**, **(BFM$_c$)** *and* **(BFM$_d$)**, *and for any* $a \in \mathbb{R}^n$,

$$F_\nu(a) = \mathcal{BC}_\nu(a).$$

The measurement conditions we have on u_i and $u_X\rceil_{\{-1,0,1\}}$ lead to axioms **(Bi-LM)**, **(Bi-In)**, **(Bi-PW)**, **(Bi-weak SPL$^+$)** and **(Bi-Sym)**. The Choquet integral w.r.t a bi-capacity ν is the only aggregation operator satisfying the previous set of axioms. So the generalized Choquet integral comes up very naturally when one works with information related to a bi-capacity.

4.6 Shapley Value, Interaction Index and k-additive Bi-capacities

As for capacities, due to the complexity of the model, involving 3^n coefficients, it is necessary to be able to analyze a bi-capacity in terms of decision behaviour, namely importance of criteria and interaction among them.

We address first the importance index. Keeping the same rationale than for capacities, we may say that a criterion i is important if whenever it is added to a coalition of satisfied criteria, or dropped from a coalition of unsatisfied criteria, there is a significant improvement. In terms of the bi-capacity, it means that the importance index should be an average of the quantities $\nu(A\cup i, B) - \nu(A, B)$ and $\nu(A, B) - \nu(A, B\cup i)$ over all $(A, B) \in \mathcal{Q}(N\setminus i)$. Summing up these two expressions gives $\nu(A\cup i, B) - \nu(A, B\cup i)$, where the term where i is a criterion with neutral value has disappeared. We choose here to take as basis of the importance index this last expression, making the assumption that the importance index of i should not depend on situations where i is neutral (an alternative way taken by Felsenthal and Machover [14] is to keep separate the two expressions above in the average; see a detailed discussion of this issue in [47]).

As for capacities, under a linearity assumption, it suffices to impose a symmetry condition (the result should not depend on the numbering of criteria) and a normalization condition (the sum of importance indices over all criteria is constant) to determine uniquely the importance index, we call by analogy the *Shapley importance index for bi-capacities*, which writes

$$\phi^{\nu}(i) = \sum_{K\subset N\setminus\{i\}} \frac{(n-k-1)!k!}{n!}\left[\nu(K\cup\{i\}, N\setminus(K\cup\{i\})) - \nu(K, N\setminus K)\right].$$

The expression is very similar to the original Shapley index (see (14.13)). Observe that only vertices of $\mathcal{Q}(N)$ (i.e. elements of the form (A, A^c)) are used. We have given in [45, 47] an axiomatization of this Shapley index in the spirit of the original axiomatization of Shapley.

The normalization property writes $\sum_{i=1}^{n}\phi(i) = \nu(N,\emptyset) - \nu(\emptyset, N) = 2$. If ν is of the CPT type with $\nu(A, B) := \mu_1(A) - \mu_2(B)$, then $\phi^{\nu}(i) = \phi^{\mu_1}(i) + \phi^{\mu_2}(i)$.

Let us turn to the notion of interaction. As for the case of bi-capacities, we may define an interaction index $I(A)$, $A\subset N$, obtained recursively from the

Shapley importance index for bi-capacities, as with Eq. (14.18) [29]. However, due to bipolarity, it seems more natural to distinguish criteria which are satisfied from those which are not. Denoting A, B the coalitions of satisfactory and unsatisfactory criteria, we are led to an interaction index with 2 arguments $I_{A,B}$ (this is called *bi-interaction* in [29]). Let us explain this in the case of $n = 2$, following the same argument than for capacities (see Section 3.5). Due to bipolarity, we have now 9 ternary alternatives, as given on Figure 14.2. In

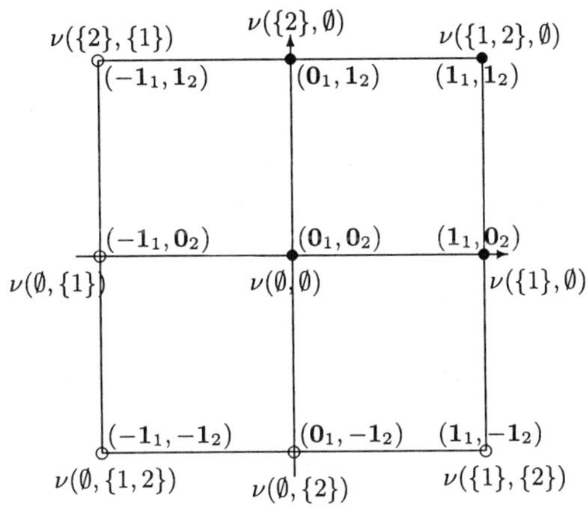

Figure 14.2. Ternary alternatives for $n = 2$.

each subsquare of $[-1, 1]^2$, it suffices to apply the classical interaction index for capacities, i.e. Formula (14.14). This gives, using our notation:

$$\begin{aligned} I_{\{1,2\},\emptyset} &:= \nu(\{1,2\},\emptyset) - \nu(\{1\},\emptyset) - \nu(\{2\},\emptyset) + \nu(\emptyset,\emptyset) \\ I_{\emptyset,\{1,2\}} &:= \nu(\emptyset,\emptyset) - \nu(\emptyset,\{1\}) - \nu(\emptyset,\{2\}) + \nu(\emptyset,\{1,2\}) \\ I_{1,2} &:= \nu(\{1\},\emptyset) - \nu(\emptyset,\emptyset) - \nu(\{1\},\{2\}) + \nu(\emptyset,\{2\}) \\ I_{2,1} &:= \nu(\{2\},\emptyset) - \nu(\{2\},\{1\}) - \nu(\emptyset,\emptyset) + \nu(\emptyset,\{1\}). \end{aligned} \tag{14.32}$$

Based on this principle, the general formula is the following

$$I^\nu(A,B) = \sum_{K\subset N\setminus(A\cup B)} \frac{(n-a-b-k)!k!}{(n-a-b+1)!} \Delta_{A,B}\nu(K, N\setminus(A\cup K)),$$

with $\Delta_{A,B}\nu(S,T) := \sum_{K\subset A, L\subset B}(-1)^{(a-k)+(b-l)}\nu(S\cup K, T\setminus(K\cup L))$. It is easy to check that our previous Shapley index writes

$$\phi(i) = I_{i,\emptyset} + I_{\emptyset,i}$$

suggesting that the Shapley index too could be divided into an index for satisfied criteria, and one for unsatisfied criteria.

If ν is of CPT type with $\nu(S,T) := \mu_1(S) - \mu_2(T)$, the interaction is expressed by:

(i) $I^{\nu}_{S,T} = 0$ unless $S = \emptyset$ or $T = \emptyset$.

(ii) denoting I^{μ_i} the interaction index of capacity μ_i, we have:

$$I^{\nu}_{S,\emptyset} = I^{\mu_1}(S), \quad \forall \emptyset \neq S \subseteq N$$
$$I^{\nu}_{\emptyset,T} = I^{\overline{\mu_2}}(T), \quad \forall T \subseteq N.$$

Property (i) clearly expresses the fact that for a CPT model, there is no interaction between the positive part and the negative part. Property (ii) explains the relation between the interaction for bi-capacities and for capacities.

Since the complexity of bi-capacities is of order 3^n, the necessity to have simplified models is yet more crucial than with capacities. The concept of k-additive bi-capacities can be defined in a way similar to the case of capacities. We refer the reader to [28] for the reasons underlying the definition hereafter.

DEFINITION 51 *A bi-capacity is said to be* k-additive *for some* k *in* $\{1, \ldots, n-1\}$ *if the interaction index is such that* $I_{A,B} = 0$ *whenever* $|B| < n - k$, *and there exists* (A, B) *with* $|B| = n - k$ *such that* $I_{A,B} \neq 0$.

As for capacities, a bi-capacity is completely determined by the values of I on $\mathcal{Q}(N)$, hence a k-additive bi-capacity is determined by $1+2\binom{n}{n-1}+2^2\binom{n}{n-2}+\cdots+2^k\binom{n}{n-k}$ coefficients, among which three are not free. Again, the case of 2-additive bi-capacities seems of particular interest, the number of coefficients being $2n^2 - 3$.

The expression of the Choquet integral for 2-additive bi-capacities is however complex (see [32]). This is not surprising since the expression contains as particular case the one of the symmetric Choquet integral [30], which is already complex compared to (14.19).

The concept of p-symmetry, as well as decomposable bi-capacities, has also been generalized to bi-capacities [28, 54].

4.7 Identification of Bi-capacities

For $v \in \mathbb{R}^n$ fixed, the mapping $\nu \mapsto \mathcal{BC}_\nu(v)$ is linear. Henceforth, the methods described in Section 3.7 for the determination of a capacity can be extended with no change to the case of bi-capacities. In particular, this enables the determination of ν with a quadratic method from a set of alternatives with the associated scores, and with a linear method from a set of comparisons between alternatives. The constraints on the bi-capacity are composed of conditions **(BFM$_a$)**, **(BFM$_b$)**, **(BFM$_c$)** and **(BFM$_d$)**.

However, we are faced here to another difficulty. A bi-capacity contains 3^n unknowns which makes its determination quite delicate. As an example, with 5 criteria, a capacity has $2^5 = 32$ coefficients whereas a bi capacity holds $3^5 = 243$ coefficients. Ten well-chosen learning examples are generally enough to determine a capacity with 5 criteria. It would require maybe 80 learning examples to determine a bi-capacity with 5 criteria. This is obviously beyond what a human being could stand.

The way out to this problem is to reduce the complexity of the model. The first idea is to restrict to sub-classes of bi-capacities, such as the $k-$additive bi-capacities described above. For instance, there are $2\,n^2 - 3 = 47$ unknowns for a $2-$additive bi-capacity with 5 criteria. Other approaches are also possible.

5. Ordinal Scales

5.1 Introduction

So far, we have supposed that the quantities we deal with (score, utilities, ...) are defined on some numerical scale, either an interval or a ratio scale, let us say a *cardinal scalescale,cardinal scale*. In practical applications, most of the time it is not possible to have directly cardinal information, but merely ordinal information. The MACBETH methodology we presented in Section 2.3 is a well-founded means to produce cardinal information from ordinal information. In some situations, this method may not apply, the decision maker being not able to give the required amount of information or being not consistent. In such a case, there is nothing left but to use the ordinal information as such, coping with the poor structure behind ordinal scales. We try in this section to define a framework and build tools as close as possible to those existing in the cardinal case, although many difficulties arise. All problems are not solved in this domain, we will present a state of the art, indicating main difficulties.

In the sequel, ordinal scales are denoted by L or similar, and are supposed to be finite totally ordered sets, with top and bottom denoted $\mathbb{1}$ and $\mathbb{O}$.

Since ordinal scales forbid the use of usual arithmetic operations (see Section 2.1), minimum ($\wedge$) and maximum ($\vee$) become the main operations. Hence, decision models are more or less limited to combinations of these operations. We call *Boolean polynomials* expressions $P(a_1, \ldots, a_n)$ involving n variables and coefficients valued in L, linked by $\wedge$ or $\vee$ in an arbitrary combination of parentheses, e.g. $((\alpha\wedge a_1)\vee(a_2\wedge(\beta\vee a_3)))\wedge a_4$. An important result by Marichal [51] says that the *Sugeno integral* w.r.t. a capacity coincides with the class of Boolean polynomials such that $P(\mathbb{O}, \mathbb{O}, \ldots, \mathbb{O}) = \mathbb{O}$, $P(\mathbb{1}, \mathbb{1}, \ldots, \mathbb{1}) = \mathbb{1}$, and P is non-decreasing w.r.t. each variable. Since these conditions are natural in decision making, this shows that the Sugeno integral plays a central role when scales are ordinal, and the whole section is devoted to it.

Before entering into details, we wish to underline the fact that however, this is not the only way to deal with ordinal information. Roubens has proposed a methodology based on the Choquet integral (which has far better properties than the Sugeno integral, as we will show), where scores of an alternative on criteria are related to the number of times this alternative is better or worse than the others on the same criteria (see Chapter 11 by Roubens in this book, and [65]).

Let us begin by pinpointing fundamental difficulties linked to the ordinal context.

- **finiteness of scales:** sticking to a decomposable model of the type (14.2), the function F is now defined from L^n to L. Clearly it is impossible that F be strictly increasing due to the finiteness of L. A solution may be to map F on L', with $|L'| \geq |L|^n$. Anyway, most measurement theoretic results are based on a solvability condition and Archimedean axioms, which cannot hold on a finite set.

- **ordinal nature:** the Sugeno integral, even defined as a function from $\mathbb{R}^n$ to $\mathbb{R}$, can never be strictly increasing, and large domains of indifference exist. Hence, the decomposable model cannot satisfy weak separability (see Section 2.1). Specifically, Marichal [51] has shown that the Sugeno integral satisfies weak separability if and only if there is a dictator criterion. However, any Sugeno integral induces a preference relation $\succeq$ which satisfies *directional weak separability*, defined by:

$$(x_i, z_{-i}) \succ (y_i, z_{-i}) \Rightarrow (x_i, z'_{-i}) \succeq (y_i, z'_{-i}), \quad \forall x, y, z, z' \in X.$$

  This weaker condition ensures that no preference reversal occurs.

- **construction of utility functions u_i:** since on ordinal scales arithmetical operations are not permitted, the method described in Sections 3.3 and 4.4 cannot be applied directly. The ordinal counterpart of the multiplication being the minimum operator ($\wedge$), Equation (14.9) becomes:

$$F(a_i, \mathbb{O}_{-i}) = \alpha_i \wedge a_i.$$

  The term α_i acts as a saturation level, hiding all utilities a_i larger than α_i. Hence relation (14.9) cannot be satisfied and the previous method cannot be applied to build the utility functions.

  To our knowledge, there is no method that enables the construction of utility functions in an ordinal framework. However, Greco *et al.* [40] have shown from a theoretical standpoint that this is possible (see Section 5.2). As a consequence, to avoid this problem most of works done in this area suppose that the attributes are defined on a common scale L, although this is not in general a realistic assumption.

5.2 Making Decision with the Sugeno Integral

We consider a capacity μ on N taking its value in L, with $\mu(\emptyset) = \mathbb{O}$ and $\mu(N) = \mathbb{1}$. Let $a := (a_1, \ldots, a_n)$ be a vector of scores in L^n. The *Sugeno integral* of a w.r.t. μ is defined by [70]:

$$\mathcal{S}_\mu(a) := \bigvee_{i=1}^{n} [a_{\tau(i)} \wedge \mu(A_{\tau(i)})], \tag{14.33}$$

where τ is a permutation on N so that $a_{\tau(1)} \leq a_{\tau(2)} \leq \cdots \leq a_{\tau(n)}$, and $A_{\tau(i)} := \{\tau(i), \ldots, \tau(n)\}$. One can notice the similarity with the Choquet integral. Taking $L = [0, 1]$, Choquet and Sugeno integrals coincide when either the capacity or the integrand is 0-1 valued, specifically:

$$\mathcal{S}_\mu(1_A, 0_{-A}) = \mu(A) = \mathcal{C}_\mu(1_A, 0_{-A}), \quad \forall A \subset N$$
$$\mathcal{S}_\mu(a) = \mathcal{C}_\mu(a) \quad \forall a \in [0,1]^n \text{ iff } \mu(A) \in \{0, 1\} \quad \forall A \subset N.$$

We refer the reader to survey papers [10, 61] and to [50, 51] for properties of the Sugeno integral, especially in a decision making perspective. We mention that in the context of decision under uncertainty, an axiomatic construction similar to the one of Savage has been done by Dubois *et al.* [12, 13].

We cite here an interesting result by Greco *et al.* [40], giving a very simple characterization of the Sugeno integral in MCDA. Assuming finiteness of X (or $X/\sim$ contains a countable order-dense subset), they have shown that the preference relation $\succeq$ on X is representable by a Sugeno integral (i.e. there exist utility functions $u_i : X_i \longrightarrow [0, 1]$ and a capacity μ such that $x \succeq y$ iff $\mathcal{S}_\mu(u_1(x_1), \ldots, u_n(x_n)) \geq \mathcal{S}_\mu(u_1(y_1), \ldots, u_n(y_n))$) iff $\succeq$ is a weak order and satisfies

$$\big[(x_i, a_{-i}) \succeq w \text{ and } (y_i, b_{-i}) \succeq t\big] \Rightarrow \big[(z_i, a_{-i}) \succeq w \text{ or } (x_i, b_{-i}) \succeq t\big]$$

for $i = 1, \ldots, n$ and $x, y, z, a, b \in X$.

As said in the introduction, making decision with the Sugeno integral has some drawbacks, which are clearly put into light with the following results [50, 59]. Let $\succeq$ be a weak order (complete, reflexive, transitive) on $[0, 1]^n$, and for $a, b \in [0, 1]^n$, denote $a \geq b$ if $a_i \geq b_i$ for all $i \in N$, and $a > b$ if $a \geq b$ and $a_i > b_i$ for some $i \in N$, and $a \gg b$ if $a_i > b_i$ for all $i \in N$. We say that $\succeq$ satisfies *monotonicity* if $a \geq b$ implies $a \succeq b$, the *strong Pareto condition* if $a > b$ implies $a \succ b$, and the *weak Pareto condition* if $a \gg b$ implies $a \succ b$. Then the following holds.

PROPOSITION 8 *Let μ be a capacity on N, and $\succeq_\mu$ the weak order induced by the Sugeno integral $\mathcal{S}_\mu$.*

(i) $\succeq_\mu$ always satisfies monotonicity.

(ii) $\succeq_\mu$ *satisfies the weak Pareto condition iff* μ *is 0-1 valued.*

(iii) $\succeq_\mu$ *never satisfies the strong Pareto condition.*

Note that the Choquet integral always satisfies the weak Pareto condition, and the strong one iff μ is strictly monotone.

Since arithmetic operations cannot be used with ordinal scales, our definitions of importance and interaction indices cannot work, and alternatives must be sought. Grabisch [20] has proposed definitions which more or less keep mathematical properties of the original Shapley value and interaction index. However, these indices, especially the interaction index, do not seem to convey the meaning they are supposed to have.

5.3 Symmetric Ordinal Scales and the Symmetric Sugeno Integral

This section introduces *bipolar ordinal scalesscale,bipolar ordinal scale*, i.e. ordinal scales with a central neutral level, and a symmetry around it, and is based on [22, 24, 25]. The aim is to have a structure similar to cardinal bipolar scales, so as to build a counterpart of the CPT model, using a Sugeno integral for the "positive" part (above the neutral level), and another one for the "negative" part (below the neutral level):

$$\mathrm{OCPT}_{\mu_1,\mu_2}(a) := \mathcal{S}_{\mu_1}(a^+) \ominus \mathcal{S}_{\mu_2}(a^-)$$

("O" stands for "ordinal") where $a^+ := a \vee 0$, $a^- := (-a)^+$, and $\ominus$ is a suitable difference operator. We will show that this task is not easy.

Let us call L^+ some ordinal scale, and define $L := L^+ \cup L^-$, where L^- is a reversed copy of L^+, i.e. for any $a, b \in L^+$, we have $a \leq b$ iff $-b \leq -a$, where $-a, -b$ are the copies of a, b in L^-. We want to endow L with operations $\varovee, \varowedge$, respectively called *symmetric maximum and minimum* satisfying (among possible other conditions):

(C1) $\varovee, \varowedge$ coincide with $\vee, \wedge$ respectively on L^+

(C2) $-a$ is the symmetric of a, i.e. $a \varovee (-a) = \mathbb{O}$.

Hence we may extend to L what exists on L^+ (e.g. the Sugeno integral), and a difference operation could be defined. The problem is that conditions (**C1**) and (**C2**) imply that $\varovee$ would be non-associative in general. Take $\mathbb{O} < a < b$ and consider the expression $(-b) \varovee b \varovee a$. Depending on the place of parentheses, the result differs since $((-b) \varovee b) \varovee a = \mathbb{O} \varovee a = a$, but $(-b) \varovee (b \varovee a) = (-b) \varovee b = \mathbb{O}$.

It can be shown that the best solution (i.e. associative on the largest domain) for $\varovee$ is given by:

$$a \varovee b := \begin{cases} -(|a| \vee |b|) & \text{if } b \neq -a \text{ and } |a| \vee |b| = -a \text{ or } = -b \\ \mathbb{O} & \text{if } b = -a \\ |a| \vee |b| & \text{else.} \end{cases} \tag{14.34}$$

Except for the case $b = -a$, $a \varovee b$ equals the absolutely larger one of the two elements a and b.

The extension of $\wedge$, viewed as the counterpart of multiplication, is simply done on the principle that the rule of sign should hold: $-(a \varowedge b) = (-a) \varowedge b$, $\forall a, b \in L$. It leads to an associative operator, defined by:

$$a \varowedge b := \begin{cases} -(|a| \wedge |b|) & \text{if sign } a \neq \text{sign } b \\ |a| \wedge |b| & \text{else.} \end{cases} \tag{14.35}$$

Based on these definitions, the OCPT model writes:

$$\mathrm{OCPT}_{\mu_1,\mu_2}(a) := \mathcal{S}_{\mu_1}(a^+) \varovee (-\mathcal{S}_{\mu_2}(a^-)).$$

When $\mu_1 = \mu_2 =: \mu$, we get the *symmetric Sugeno integral*, denoted $\check{\mathcal{S}}_\mu$.

Going a step further, it is possible to define the Sugeno integral w.r.t. bi-capacities, following the same way as with the Choquet integral. One can show that, defining $\mathcal{BS}_\nu(a) := \mathcal{S}_{\mu_{N^+}}(|a|)$, with same notations as in Section 4.3 and replacing in the definition of Sugeno integral $\vee, \wedge$ by $\varovee, \varowedge$, the expression is [31] (see also Greco *et al.* [39] for a similar definition):

$$\mathcal{S}_\nu(a) = \langle \bigvarovee_{i=1}^{n} \Big[|a_{\tau(i)}| \varowedge \nu(A_{\tau(i)} \cap N^+, A_{\tau(i)} \cap N^-) \Big] \rangle, \tag{14.36}$$

where τ is a permutation on N so that $|a_{\tau(1)}| \leq \cdots \leq |a_{\tau(n)}|$, $N^+ := \{i \in N \mid a_i \geq \mathbb{O}\}$, $N^- := N \setminus N^+$, and the expression $\langle \bigvarovee_{i=1}^{n} b_i \rangle$ is a shorthand for $(\bigvarovee_{i=1}^{n} b_i^+) \varovee (-\bigvarovee_{i=1}^{n} b_i^-)$. It can be shown that if ν is of the CPT type, one recovers the OCPT model.

Lastly, we mention Denneberg and Grabisch, who have proposed a general formulation of the Sugeno integral on arbitrary bipolar spaces [9].

5.4 Building a Model from Preferences

The previous sections have shown many difficulties underlying the construction. We try in this section to build a model from preferences, in a spirit close to the one of Sections 3 and 4, and based on the symmetric model [26]. We assume the existence on each attribute X_i of a neutral element $\mathbf{0}_i$ and a greatest element $\mathbf{1}_i$

in the sense that $(\mathbf{1}_i, \mathbf{0}_{-i}) \succeq (x_i, \mathbf{0}_{-i})$, for all $x_i \in X_i$. We suppose in addition that $(\mathbf{1}_i, x_{-i}) \succeq (\mathbf{0}_i, x_{-i})$ for each x_{-i}. We consider as in Section 4.4 the sets $X_i^+ := \{x_i \in X_i \mid (x_i, \mathbf{0}_{-i}) \succeq \mathbf{0}_N\}$ and $X_i^- := \{x_i \in X_i \mid (x_i, \mathbf{0}_{-i}) \preceq \mathbf{0}_N\}$.

Our aim is to represent the preference of the DM on X by a (symmetric) Sugeno integral with respect to some capacity μ, that is:

$$a \succeq b \text{ iff } \check{S}_\mu(u_1(a_1), \dots, u_n(a_n)) \geq \check{S}_\mu(u_1(b_1), \dots, u_n(b_n)) \qquad (14.37)$$

where $u_i : X_i \longrightarrow L, i = 1, \dots, n$ are commensurable utility functions, defined on some scale L, which we will build. As explained in Section 5.1, one cannot build separately the utility functions and the capacity. In our approach, we need to determine the capacity first.

The determination of the capacity is done through the set of binary alternatives $(\mathbf{1}_A, \mathbf{0}_{-A})$, denoted as before $X\rceil_{\{0,1\}}$. We suppose that $\succeq$ restricted to this set is reflexive, transitive, and complete, and in addition that it satisfies monotonicity in the following sense: if $A \subset B$, then $(\mathbf{1}_A, \mathbf{0}_{-A}) \preceq (\mathbf{1}_B, \mathbf{0}_{-B})$.

Let us denote by m the number of equivalence classes of $\sim$ on $X\rceil_{\{0,1\}}$. From this, we build the ordinal scale $L^+ = \{e_0, \dots, e_{m-1}\}$, with $e_0 < e_1 < \cdots < e_{m-1}$, assigning to each equivalence class a degree of the scale, which reflects the rank of the equivalence class. Then, due to monotonicity:

- e_0, denoted $\mathbb{O}$, corresponds to $(\mathbf{0}_1, \dots, \mathbf{0}_n) = \mathbf{0}_N$.
- e_{m-1}, denoted $\mathbb{1}$, corresponds to $(\mathbf{1}_1, \dots, \mathbf{1}_n) = \mathbf{1}_N$.

We define $\mu(A) := u(\mathbf{1}_A, \mathbf{0}_{A^c})$, where $u : X\rceil_{\{0,1\}} \longrightarrow L^+$ assigns to each binary alternative the value on L^+ of its equivalence class. By monotonicity, μ is a capacity on L^+.

We turn to the identification of the utility functions. The approach is related to the one proposed by Marichal [51]. u_i should be a representation of the preference of the DM among alternatives in $X\rfloor_i := \{(x_i, \mathbf{0}_{-i}), x_i \in X_i\}$, i.e.

$$u_i(x_i) \geq u_i(y_i) \text{ iff } (x_i, \mathbf{0}_{-i}) \succeq (y_i, \mathbf{0}_{-i}),$$

supposing that $\succeq$ is a weak order when restricted to each $X\rfloor_i$. In order to ensure commensurability, we impose

$$u_i(\mathbf{1}_i) = \mathbb{1}, \quad u_i(\mathbf{0}_i) = \mathbb{O}, \quad \forall i = 1, \dots, n.$$

We suppose to be in the bipolar case (otherwise we just need L^+ and an ordinary Sugeno integral), hence we build the symmetrized scale $L = L^+ \cup L^- = \{e_{-m+1}, \dots, e_{-1}, e_0, e_1, \dots, e_{m-1}\}$, which we equip with $\varovee, \varowedge$. We denote naturally e_{-m+1} by $-\mathbb{1}$. From now on, all u_i's are from X_i to L.

We first try to determine $u_i(x_i)$ for all $x_i \in X_i^+$. Suppose the DM assigns $(x_i, \mathbf{0}_{-i})$ to e_k (more exactly, the DM thinks that $(x_i, \mathbf{0}_{-i})$ is indifferent with any alternative from the equivalence class assigned to e_k). Then, from (14.37) and the definition of the Sugeno integral, we necessarily have

$$e_k = u_i(x_i) \wedge \mu(i) \leq \mu(i).$$

We have two possible cases.

- if $e_k = \mu(i)$, then $u_i(x_i) \geq e_k = \mu(i)$
- if $e_k < \mu(i)$, then $u_i(x_i) = e_k$.

Suppose the DM assigns $(x_i, \mathbf{1}_{-i})$ to e_l. Then, from the representation condition by the Sugeno integral, we should have

$$e_l = u_i(x_i) \vee \mu(N \setminus i) \geq \mu(N \setminus i),$$

with again two possible cases.

- if $e_l = \mu(N \setminus i)$, then $u_i(x_i) \leq e_l = \mu(N \setminus i)$
- if $e_l > \mu(N \setminus i)$, then $u_i(x_i) = e_l$.

By a repeated application of the assumption $(\mathbf{1}_i, x_{-i}) \succeq (\mathbf{0}_i, x_{-i})$, we deduce that $e_l \geq e_k$. Equality means that $(x_i, \mathbf{0}_{-i}) \sim (x_i, \mathbf{1}_{-i})$, which can be interpreted as a dictatorship of attribute X_i. Combining the above and supposing $e_k < e_l$, three cases can happen:

- **Case 1:** $e_k = \mu(i)$, $e_l > \mu(N \setminus i)$, which entails

$$\begin{cases} (x_i, \mathbf{0}_{-i}) & \sim (\mathbf{1}_i, \mathbf{0}_{-i}) \\ (x_i, \mathbf{1}_{-i}) & \succ (\mathbf{0}_i, \mathbf{1}_{-i}) \end{cases}$$

  This could be interpreted as x_i "close to" $\mathbf{1}_i$, and in this case $u_i(x_i) = e_l$.

- **Case 2:** $e_l = \mu(N \setminus i)$, $e_k < \mu(i)$, which entails

$$\begin{cases} (x_i, \mathbf{1}_{-i}) & \sim (\mathbf{0}_i, \mathbf{1}_{-i}) \\ (x_i, \mathbf{0}_{-i}) & \prec (\mathbf{1}_i, \mathbf{0}_{-i}) \end{cases}$$

  This could be interpreted as x_i "close to" $\mathbf{0}_i$, and in this case $u_i(x_i) = e_k$.

- **Case 3:** $e_k = \mu(i)$, $e_l = \mu(N \setminus i)$, which entails

$$\begin{cases} (x_i, \mathbf{0}_{-i}) & \sim (\mathbf{1}_i, \mathbf{0}_{-i}) \\ (x_i, \mathbf{1}_{-i}) & \sim (\mathbf{0}_i, \mathbf{1}_{-i}) \end{cases}$$

  This causes the indetermination of $u_i(x_i)$ since $e_k \leq u_i(x_i) \leq e_l$.

The last case corresponds to $e_k < \mu(i)$ and $e_l > \mu(N \setminus i)$, which implies $e_k = e_l$, a case we have eliminated since it corresponds to a dictatorship of X_i.

The same procedure can applied to "negative" values $x_i \in X_i^-$. Let us assume that the DM assigns $(x_i, \mathbf{0}_{-i})$ to e_{-k}. Then, by the symmetric Sugeno integral, one should satisfy

$$e_{-k} = u_i(x_i) \oslash \mu(i).$$

Then, if $e_{-k} = -\mu(i)$, we have $u_i(x_i) \leq e_{-k}$, and if $e_{-k} > -\mu(i)$, we get $u_i(x_i) = e_{-k}$.

Now we suppose that the DM assigns to $(x_i, -\mathbf{1}_{-i})$ the value e_{-l}. We find that

$$e_{-l} = u_i(x_i) \varovee (-\mu(N \setminus i))$$

Then, if $e_{-l} = -\mu(N \setminus i)$, we have $u_i(x_i) \geq e_{-l}$, and if $e_{-l} < -\mu(N \setminus i)$, then $u_i(x_i) = e_{-l}$.

As before, we have three cases.

- **Case 1:** $e_{-k} = -\mu(i)$, $e_{-l} < -\mu(N \setminus i)$. Then $u_i(x_i) = e_{-l}$.
- **Case 2:** $e_{-l} = -\mu(N \setminus i)$, $e_{-k} > -\mu(i)$. Then $u_i(x_i) = e_{-k}$.
- **Case 3:** $e_{-k} = -\mu(i)$, $e_{-l} = -\mu(N \setminus i)$. Then $u_i(x_i) \in [e_{-l}, e_{-k}]$.

The above methodology can be easily extended to have a representation by an OCPT model, the case of the bipolar Sugeno integral being more tricky. Remark that the procedure may leave some indetermination for the utility functions, hence several solutions are possible. Also, the set of equivalence classes can be enriched if necessary when utility functions are built, e.g. if the DM thinks that some alternative $(x_i, \mathbf{0}_{-i})$ is strictly between two consecutive equivalence classes.

5.5 Identification of Capacities

In situations where utility functions are known, the problem of the identification of capacities when the model is a Sugeno integral (or OCPT, bipolar Sugeno integral) in an ordinal contex, or even when $L = [0, 1]$ or $[-1, 1]$, appears to be rather different from the case of the Choquet integral. The main reason is that we are not able to write the identification problem as a minimization problem *stricto sensu* (see Section 3.7), since the notion of difference between values, hence of error, is not defined in a way which is suitable on an ordinal scale, to say nothing about "squared errors" and "average values".

Even if we take L as a real interval, which permits to define a squared error criterion as for the Choquet integral, the minimization problem obtained is not easy to solve, since it involves non-linear, non-differentiable operations $\vee, \wedge, \varovee, \oslash$. In such cases, only meta-heuristic methods can be used, as genetic algorithms, simulated annealing, etc. There exist some works in this direction, although most of the time used for the Choquet integral, which is questionable [23, 74].

What can be done without error criterion to minimize? The second option, also used for the Choquet integral (see Section 3.7), is to find capacities which enable the representation of the preference of the DM over a set of alternatives of interest by the Sugeno integral (or OCPT,...). A detailed study of this problem has been done by Rico *et al.* [63] for the Sugeno integral. We mention also the work of Greco *et al.* based on decision rules, which can be found in Chapter 12 of this book (see also [38]). We give a short description of the work by Rico *et al.*

Since utility functions are assumed to be known and commensurable, defined on some scale L (supposed to be unipolar here), the preference relation $\succeq$ of the DM is expressed directly on L^n. We call $\mathcal{A} \subset L^n$ the set of alternatives of interest. We distinguish two levels of representation.

- the *strong representation*, where the capacity μ must satisfy $\mathcal{S}_\mu(a) \geq \mathcal{S}_\mu(b)$ if and only if $a \succeq b$.
- the *weak representation*, where we merely forbid a reversal: $a \succ b$ implies $\mathcal{S}_\mu(a) \geq \mathcal{S}_\mu(b)$.

We can guess by properties of the Sugeno integral (see. e.g. weak separability vs. directional weak separability) that the weak representation is more appropriate.

Let us suppose that the alternatives in O can be put into p equivalence classes $[a^1], \dots, [a^p]$ by $\sim$, assuming $a^1 \prec \cdots \prec a^p$. The strong representation problem amounts to find p values $\alpha_1 < \alpha_2 < \cdots < \alpha_p$ in L such that there exists a capacity μ satisfying $\mathcal{S}_\mu(a) = \alpha_i$, for all $a \in [a^i]$, $i = 1, \dots, p$. For the weak representation problem, it suffices to find $p - 1$ numbers $0 =: \alpha_0 \leq \alpha_1 \leq \alpha_2 \leq \cdots \leq \alpha_p := 1$ in L such that there exists a capacity μ satisfying $\alpha_{i-1} \leq \mathcal{S}_\mu(a) \leq \alpha_i$, for all $a \in [a^i]$, $i = 1, \dots, p$.

The set of capacities such that $\mathcal{S}_\mu(a) = \alpha$ is non-empty iff $a_{(n)} < \alpha$ or $a_{(1)} > \alpha$, and is the interval $[\check{\mu}^{a,\alpha}, \hat{\mu}^{a,\alpha}]$, where for all $A \neq \emptyset, N$

$$\hat{\mu}^{a,\alpha}(A) := \begin{cases} \alpha & \text{if } \; A \subset A_{\left(i^{>}_{a,\alpha}\right)} \\ \mathbb{1} & \text{otherwise} \end{cases}$$

$$\check{\mu}^{a,\alpha}(A) := \begin{cases} \alpha & \text{if } \; A_{\left(i^{\geq}_{a,\alpha}\right)} \subset A \\ \mathbb{O} & \text{otherwise} \end{cases}$$

with $i^{\geq}_{a,\alpha} \in N$ such that $a_{\left(i^{\geq}_{a,\alpha}-1\right)} < \alpha \leq a_{\left(i^{\geq}_{a,\alpha}\right)}$, and $i^{>}_{a,\alpha} \in N$ such that $a_{\left(i^{>}_{a,\alpha}-1\right)} \leq \alpha < a_{\left(i^{>}_{a,\alpha}\right)}$. The set of solutions for the strong representation is then the intersection of all these intervals for all α_i.

The set of capacities solution of the weak representation problem is empty iff $\exists i$ such that $a_{(1)} > \alpha_i$ for some $a \in [a^i]$ or $\exists i$ such that $b_{(n)} < \alpha_i$ for some $b \in [a^{i+1}]$, and otherwise is the interval $[\check{\mu}, \hat{\mu}]$, with

$$\check{\mu}(A) = \bigvee_{i=1}^{p-1} \bigvee_{a \in [a^{i+1}]} \check{\mu}^{a,\alpha_i}(A), \quad \hat{\mu}(A) = \bigwedge_{i=1}^{p-1} \bigwedge_{a \in [a^i]} \hat{\mu}^{a,\alpha_i}(A).$$

6. Concluding Remarks

This chapter has tried to give a unified presentation of MCDA methods based on fuzzy integrals. It has shown that the concepts of capacity and bi-capacity naturally arise as overall utility of binary and ternary alternatives, and that the Choquet integral appears to be the unique solution for aggregating criteria, under a set of natural axioms.

This methodology has been applied in various fields of MCDA from a long time, particularly in subjective evaluation, and seems to receive more and more attention. Following the pionnering works of Sugeno [70], many researchers in the eighties in Japan have applied in practical problems the Sugeno integral, for example to opinion poll [62], and later the Choquet integral (see a summary of main works in [36]). More recent applications can be found in [23, 35], see also [27, 57].

References

[1] C.A. Bana e Costa and J.C. Vansnick. Applications of the MACBETH approach in the framework of an additive aggregation model. *Journal of Multi-Criteria Decision Analysis*, 6:107–114, 1997.

[2] C.A. Bana e Costa and J.C. Vansnick. A theoretical framework for Measuring Attractiveness by a Categorical Based Evaluation TecHnique (MACBETH). In J. Climaco, editor, *Multicriteria Analysis. Proceedings of the XIth International Conference on MCDM*, pages 15–24. Springer Verlag, Berlin, 1997.

[3] C.A. Bana e Costa and J.C. Vansnick. The MACBETH approach: Basic ideas, software and an application. In N. Meskens and M. Roubens, editors, *Advances in Decision Analysis*, pages 131–157. Kluwer Academic Publishers, Dordrecht, 1999.

[4] J.M. Bilbao, J.R. Fernandez, A. Jiménez Losada, and E. Lebrón. Bicooperative games. In J.M. Bilbao, editor, *Cooperative games on combinatorial structures*, pages 23–26. Kluwer Academic Publishers, Dordrecht, 2000.

[5] G. Choquet. Theory of capacities. *Annales de l'Institut Fourier*, 5:131–295, 1953.

[6] D. Denneberg. *Non-Additive Measure and Integral*. Kluwer Academic Publishers, Dordrecht, 1994.

[7] D. Denneberg. Non-additive measure and integral, basic concepts and their role for applications. In M. Grabisch, T. Murofushi, and M. Sugeno, editors, *Fuzzy Measures and Integrals – Theory and Applications*, pages 42–69. Physica Verlag, Heidelberg, 2000.

[8] D. Denneberg and M. Grabisch. Interaction transform of set functions over a finite set. *Information Sciences*, 121:149–170, 1999.

[9] D. Denneberg and M. Grabisch. Measure and integral with purely ordinal scales. *Journal of Mathematical Psychology*, to appear.

[10] D. Dubois, J.-L. Marichal, H. Prade, M. Roubens, and R. Sabbadin. The use of the discrete Sugeno integral in decision making: A survey. *International Journal of Uncertainty, Fuzziness and Knowledge-Based Systems*, 9(5):539–561, 2001.

[11] D. Dubois and H. Prade. A class of fuzzy measures based on triangular norms. *International Journal of General Systems*, 8:43–61, 1982.

[12] D. Dubois, H. Prade, and R. Sabbadin. Qualitative decision theory with Sugeno integrals. In M. Grabisch, T. Murofushi, and M. Sugeno, editors, *Fuzzy Measures and Integrals – Theory and Applications*, pages 314–332. Physica Verlag, Heidelberg, 2000.

[13] D. Dubois, H. Prade, and R. Sabbadin. Decision-theoretic foundations of qualitative possibility theory. *European Journal of Operational Research*, 128:459–478, 2001.

[14] D. Felsenthal and M. Machover. Ternary voting games. *International Journal of Game Theory*, 26:335–351, 1997.

[15] J.C. Fodor and M. Roubens. *Fuzzy Preference Modelling and Multi-Criteria Decision Aid.* Kluwer Academic Publishers, Dordrecht, 1994.

[16] M. Grabisch. A new algorithm for identifying fuzzy measures and its application to pattern recognition. In *International Joint Conference of the 4th IEEE International Conference on Fuzzy Systems and the 2nd International Fuzzy Engineering Symposium*, pages 145–150, Yokohama, Japan, March 1995.

[17] M. Grabisch. The application of fuzzy integrals in multicriteria decision making. *European Journal of Operational Research*, 89:445–456, 1996.

[18] M. Grabisch. Alternative representations of discrete fuzzy measures for decision making. *International Journal of Uncertainty, Fuzziness, and Knowledge Based Systems*, 5:587–607, 1997.

[19] M. Grabisch. k-order additive discrete fuzzy measures and their representation. *Fuzzy Sets and Systems*, 92:167–189, 1997.

[20] M. Grabisch. On the representation of k-decomposable measures. In *7th International Fuzzy Systems Association World Congress (IFSA'97), Vol. I*, pages 478–483, Prague, Czech Republic, June 1997.

[21] M. Grabisch. A graphical interpretation of the Choquet integral. *IEEE Transactions on Fuzzy Systems*, 8:627–631, 2000.

[22] M. Grabisch. The Möbius function on symmetric ordered structures and its application to capacities and integrals. In *9th Internationl Conference on Information Processing and Management of Uncertainty in Knowledge-Based systems (IPMU'2002)*, pages 755–762, Annecy, France, July 2002. ESIA, Université de Savoie.

[23] M. Grabisch. Modelling data by the Choquet integral. In V. Torra, editor, *Information Fusion in Data Mining*, pages 135–148. Physica Verlag, Heidelberg, 2003.

[24] M. Grabisch. The symmetric Sugeno integral. *Fuzzy Sets and Systems*, 139:473–490, 2003.

[25] M. Grabisch. The Möbius function on symmetric ordered structures and its application to capacities on finite sets. *Discrete Mathematics*, submitted.

[26] M. Grabisch, S. Dia, and Ch. Labreuche. A multicriteria decision making framework in ordinal context based on Sugeno integral. In *Joint 9th International Fuzzy Systems Association World Congress and 20th North American Fuzzy Information Processing Society International Conference (IFSA/NAFIPS'2001)*, Vancouver, Canada, July 2001.

[27] M. Grabisch, J. Duchêne, F. Lino, and P. Perny. Subjective evaluation of discomfort in sitting position. *Fuzzy Optimization and Decision Making*, 1(3):287–312, 2002.

[28] M. Grabisch and Ch. Labreuche. Bi-capacities. In *Joint International Conference on Soft Computing and Intelligent Systems and 3d International Symposium on Advanced Intelligent Systems*, Tsukuba, Japan, October 2002.

[29] M. Grabisch and Ch. Labreuche. Bi-capacities for decision making on bipolar scales. In B. De Baets, J. Fodor, and G. Pasi, editors, *EURO Working Group on Fussy Sets Workshop on Informations Systems (EUROFUSE'2002)*, pages 185–190, Varenna, Italy, September 2002.

[30] M. Grabisch and Ch. Labreuche. The symmetric and asymmetric Choquet integrals on finite spaces for decision making. *Statistical Papers*, 43:37–52, 2002.

[31] M. Grabisch and Ch. Labreuche. Bi-capacities. Technical Report 2003/002, Laboratoire d'Informatique de Paris 6, Université Pierre et Marie Curie, http://www.lip6.fr/reports/lip6.2003.002.html, 2003.

[32] M. Grabisch and Ch. Labreuche. The Choquet integral for 2-additive bi-capacities. In M. Wagenknecht and R. Hampel, editors, *3rd International Conference of the European Society for Fuzzy Logic and Technology (EUSFLAT 2003)*, pages 300–303, Zittau, Germany, September 2003.

[33] M. Grabisch, Ch. Labreuche, and J.C. Vansnick. On the extension of pseudo-Boolean functions for the aggregation of interacting bipolar criteria. *European Journal of Operational Research*, 148:28–47, 2003.

[34] M. Grabisch, J.L. Marichal, and M. Roubens. Equivalent representations of a set function with applications to game theory and multicriteria decision making. *Mathematics of Operations Research*, 25(2):157–178, 2000.

[35] M. Grabisch, T. Murofushi, and M. Sugeno, editors. *Fuzzy Measures and Integrals. Theory and Applications*. Studies in Fuzziness and Soft Computing. Physica Verlag, Heidelberg, 2000.

[36] M. Grabisch, H.T. Nguyen, and E.A. Walker. *Fundamentals of Uncertainty Calculi, with Applications to Fuzzy Inference*. Kluwer Academic Publishers, Dordrecht, 1995.

[37] M. Grabisch and M. Roubens. An axiomatic approach to the concept of interaction among players in cooperative games. *International Journal of Game Theory*, 28:547–565, 1999.

[38] S. Greco, B. Matarazzo, and R. Słowinski. Conjoint measurement and rough set approach for multicriteria sorting problems in presence of ordinal criteria. In A. Colorni, M. Paruccini, and B. Roy, editors, *A-MCD-A Aide Multicritère à la Décision – Multiple Criteria Decision Aiding*, pages 117–144. The European Commission Joint Reserch Center, Ispra, 2001.

[39] S. Greco, B. Matarazzo, and R. Słowinski. Bipolar Sugeno and Choquet integrals. In B. De Baets, J. Fodor, and G. Pasi, editors, *EURO Working Group on Fuzzy Sets Workshop on Informations Systems (EUROFUSE'2002)*, pages 191–196, Varenna, Italy, September 2002.

[40] S. Greco, B. Matarazzo, and R. Słowinski. The axiomatic basis of Sugeno integral and associative aggregation operator. In *4th International Workshop on Preferences and Decisions*, pages 33–38, Trento, Italy, September 2003.

[41] K. Ishii and M. Sugeno. A model of human evaluation process using fuzzy measure. *International Journal of Man-Machine Studies*, 22:19–38, 1985.

[42] R.L. Keeney and H. Raiffa. *Decision with Multiple Objectives*. John Wiley & Sons, New York, 1976.

[43] D.H. Krantz, R.D. Luce, P. Suppes, and A. Tversky. *Foundations of Measurement. Volume 1: Additive and Polynomial Representations*. Academic Press, New York, 1971.

[44] S.H. Kwon and M. Sugeno. A hierarchical subjective evaluation model using non-monotonic fuzzy measures and the Choquet integral. In M. Grabisch, T. Murofushi, and M. Sugeno, editors, *Fuzzy Measures and Integrals – Theory and Applications*, pages 375–391. Physica Verlag, Heidelberg, 2000.

[45] Ch. Labreuche and M. Grabisch. Bi-cooperative games and their importance and interaction indices. In R. Bisdorff, editor, *14th Mini-EURO Conference on Human Centered Processes (HCP'2003)*, pages 287–291, Luxembourg, May 2003.

[46] Ch. Labreuche and M. Grabisch. The Choquet integral for the aggregation of interval scales in multicriteria decision making. *Fuzzy Sets and Systems*, 137:11–26, 2003.

[47] Ch. Labreuche and M. Grabisch. The Shapley value and the interaction indices for bi-cooperative games. Technical Report 2003/001, Laboratoire d'Informatique de Paris 6, Université Pierre et Marie Curie, http://www.lip6.fr/reports/lip6.2003.001.html, 2003.

[48] Ch. Labreuche and M. Grabisch. Generalized Choquet-like aggregation functions for handling ratio scales. *European Journal of Operational Research*, submitted.

[49] J.L. Marichal. An axiomatic approach of the discrete Choquet integral as a tool to aggregate interacting criteria. *IEEE Transactions on Fuzzy Systems*, 8(6):800–807, 2000.

[50] J.L. Marichal. On Sugeno integral as an aggregation function. *Fuzzy Sets and Systems*, 114:347–365, 2000.

[51] J.L. Marichal. An axiomatic approach of the discrete Sugeno integral as a tool to aggregate interacting criteria in a qualitative framework. *IEEE Transactions on Fuzzy Systems*, 9(1):164–172, 2001.

[52] J.L. Marichal and M. Roubens. Dependence between criteria and multiple criteria decision aid. In *2nd International Workshop on Preferences and Decisions*, pages 69–75, Trento, Italy, 1998.

[53] P. Miranda and M. Grabisch. Optimization issues for fuzzy measures. In *7th International Conference on Information Processing and Management of Uncertainty in Knowledge-Based Systems (IPMU'98)*, pages 1204–1211, Paris, France, July 1998. Editions EDK, Paris.

[54] P. Miranda and M. Grabisch. p-symmetric bi-capacities. In *International Summer School on Aggregation Operators and their Applications*, pages 123–129, Alcala, Spain, July 2003.

[55] P. Miranda, M. Grabisch, and P. Gil. p-symmetric fuzzy measures. *International Journal of Uncertainty, Fuzziness, and Knowledge-Based Systems*, 10 (Suppl.):105–123, 2002.

[56] F. Modave and M. Grabisch. Preference representation by a Choquet integral: Commensurability hypothesis. In *7th International Conference on Information Processing and Management of Uncertainty in Knowledge-Based Systems (IPMU'98)*, pages 164–171, Paris, France, July 1998. Editions EDK, Paris.

[57] J. Montmain, A. Akharraz, and G. Mauris. Knowledge management as a support for collective decision-making and argumentation processes. In *9th International Conference on Information Processing and Management of Uncertainty in Knowledge-Based systems (IPMU'2002)*, pages 91–98, Annecy, France, July 2002. ESIA, Université de Savoie.

[58] T. Mori and T. Murofushi. An analysis of evaluation model using fuzzy measure and the Choquet integral. In *5th Fuzzy System Symposium*, pages 207–212, Kobe, Japan, 1989. Japan Society for Fuzzy Sets and Systems. In Japanese.

[59] T. Murofushi. Lexicographic use of Sugeno integrals and monotonicity conditions. *IEEE Transactions on Fuzzy Systems*, 9(6):783–794, 2001.

[60] T. Murofushi and S. Soneda. Techniques for reading fuzzy measures (III): Interaction index. In Japan Society for Fuzzy Sets and Systems, editors, *9th Fuzzy System Symposium*, pages 693–696, Sapporo, Japan, May 1993. In Japanese.

[61] T. Murofushi and M. Sugeno. Fuzzy measures and fuzzy integrals. In M. Grabisch, T. Murofushi, and M. Sugeno, editors, *Fuzzy Measures and Integrals – Theory and Applications*, pages 3–41. Physica Verlag, Heidelberg, 2000.

[62] T. Onisawa, M. Sugeno, Y. Nishiwaki, H. Kawai, and Y. Harima. Fuzzy measure analysis of public attitude towards the use of nuclear energy. *Fuzzy Sets & Systems*, 20:259–289, 1986.

[63] A. Rico, M. Grabisch, Ch. Labreuche, and A. Chateauneuf. Preference modelling on totally ordered sets by the Sugeno integral. *Discrete Applied Mathematics*, to appear.

[64] F.S. Roberts. *Measurement Theory*. Addison-Wesley, Reading MA, 1979.

[65] M. Roubens. Ordinal multiattribute sorting and ordering in the presence of interacting points of view. In D. Bouyssou, E. Jacquet-Lagrèze, P. Perny, R. Slowinsky, D. Vanderpooten, and Ph. Vincke, editors, *Aiding Decisions with Multiple Criteria: Essays in Honor of Bernard Roy*, pages 229–246. Kluwer Academic Publishers, Dordrecht, 2001.

[66] D. Schmeidler. Integral representation without additivity. *Proceedings of the American Mathematical Society*, 97(2):255–261, 1986.

[67] L.S. Shapley. A value for n-person games. In H.W. Kuhn and A.W. Tucker, editors, *Contributions to the Theory of Games, Vol. II*, volume 28 of *Annals of Mathematics Studies*, pages 307–317. Princeton University Press, Princeton, 1953.

[68] L.S. Shapley. Simple games: An outline of the descriptive theory. *Behavioral Science*, 7:59–66, 1962.

[69] H. Simon. Rational choice and the structure of the environment. *Psychological Review*, 63(2):129–138, 1956.

[70] M. Sugeno. *Theory of Fuzzy Integrals and its Applications*. PhD thesis, Tokyo Institute of Technology, 1974.

[71] A. Tversky and D. Kahneman. Advances in prospect theory: Cumulative representation of uncertainty. *Journal of Risk and Uncertainty*, 5:297–323, 1992.

[72] J. Šipoš. Integral with respect to a pre-measure. *Mathematica Slovaca*, 29:141–155, 1979.

[73] P. Wakker. *Additive Representations of Preferences*. Kluwer Academic Publishers, Dordrecht, 1989.

[74] Z. Wang, K.S. Leung, and J. Wang. A genetic algorithm for determining nonadditive set functions in information fusion. *Fuzzy Sets and Systems*, 102:462–469, 1999.

[75] S. Weber. $\perp$-decomposable measures and integrals for archimedean t-conorms $\perp$. *Journal of Mathematical Analysis and Applications*, 101:114–138, 1984.

Chapter 15

VERBAL DECISION ANALYSIS

Helen Moshkovich, Alexander Mechitov
College of Business
The University of Montevallo
Montevallo, AL 35115
USA
MoshHM,Mechitov@montevallo.edu

David Olson
Department of Management
University of Nebraska, Lincoln
Lincoln, NE 68588-0491
USA
dolson3@unl.edu

Abstract Verbal Decision Analysis is a new methodological approach for the construction of decisions methods with multiple criteria. The approach is based on cognitive psychology, applied mathematics, and computer science. Problems of eliciting exact quantitative estimations from the decision makers may be overcome by using preferential information from the decision makers in the ordinal form (e.g., "more preferable", "less preferable",...). This type of judgments is known to be much more stable and consistent. Ways of how to obtain and use ordinal judgments for multicriteria alternatives' evaluation are discussed. Decision methods ZAPROS, and ORCLASS based on the approach are briefly described.

Keywords: Decision analysis, multiple criteria, ordinal judgments, preference elicitation, ZAPROS, ORCLASS.

1. Features of Unstructured Decision Problems

According to Simon [48] decision problems may be divided into three main groups: 1) well-structured problems, 2) ill-structured problems, and 3) unstructured problems.

Well-structured problems are problems where the essential dependencies between parameters are known and may be expressed in a formal way. Problems of this class are being rather successfully solved by operations management methods.

Ill-structured or mixed problems have both qualitative and quantitative elements, but unknown and undefined problem elements tend to dominate these tasks. Problems in this class are rather diversified and methods from different areas may be used to work with them including "cost-benefit" analysis, as well as multicriteria decision making and multicriteria decision aids.

Unstructured problems are the problems with mostly qualitative parameters with no objective model for their aggregation. We can see examples of such tasks in policy making and strategic planning in different fields, as well as in personal decisions. These problems are in the area of multicriteria decision aids but require some special considerations in the methods used.

Larichev and Moshkovich [33, 34] proposed the following list of general features for the unstructured problems:

- the problems in this class are unique in the sense that each problem is new to the decision maker and has characteristics not previously experienced;
- parameters (criteria) in these problems are mostly qualitative in nature, most often formulated in a natural language;
- in many cases evaluations of alternatives against these parameters may be obtained only from experts (or the decision maker him/her self);
- an overall evaluation of alternatives' quality may be obtained only through subjective preferences of the decision maker.

Human judgment is the basic source of information in unstructured problems. Being interested in the result, the decision maker would like to control the whole process, including selection of experts and formation of the decision rule(s). Verbal Decision Analysis (VDA) was proposed as a framework for the unstructured problems [34].

2. Main Principles of Verbal Decision Analysis

The role of decision making methods applied to unstructured problems should be to help the decision maker to structure the problem (form a set of alternatives and elaborate a set of relevant criteria) and work out a consistent policy for evaluating/comparing multicriteria alternatives.

As human judgment is the central source of information in unstructured problems, the proposed methods should consider the constraints of the human information processing system as well as the psychological validity of input data in decision analysis. This requires that the methods should: 1) use language for problem description that is natural to the decision maker; 2) implement psychologically valid measurement of criteria and psychologically valid preference elicitation procedures; 3) incorporate means for consistency check of the decision maker's information; 4) be "transparent" to the decision maker and provide explanations of the result.

Verbal Decision Analysis is oriented on construction of a set of methods for different types of decision tasks within the stated framework.

2.1 Natural Language of a Problem Description

Verbal Decision Analysis tries to structure a decision problem by using the natural language commonly used by a decision maker and other parties participating in the decision process [26]. The goal of problem structuring is to define alternatives and the primary criteria to be used for evaluation.

In unstructured practical decision tasks most decisions involve qualitative criteria with no natural numerical equivalents [28, 34].

People are known to be poor at estimating and comparing objects that are close in value. It is reasonable for qualitative as well as for originally quantitatively measured criteria to have scales with several distinct levels, possibly differentiated in words and examples [17, 20, 52]. For example, experts were found to have much closer estimates of applicants over separate criteria using scales with a small number of verbal estimates than when using a 1 to 10 quality scale [40].

Verbal descriptions over criteria scale levels instead of numerical values, not only allow the decision maker to be more confident in his(her) own evaluations, but also should lead to information from experts that is more stable. Therefore, Verbal Decision Analysis uses scales with verbal descriptions of criteria levels for unstructured problems.

2.2 Psychological Basis for Decision Rules Elaboration

The measurements discussed in the previous section may be referred to as primary measurements. These primary measurements structure the problem to allow construction of a decision rule for overall evaluation and/or comparison of alternatives. Construction of the decision rule for unstructured problems includes elicitation of the decision maker's preferences as there are almost no objective dependencies between decision criteria.

The complexity involved in eliciting preference information from human subjects has been widely recognized. The process of eliciting necessary information for such decisions is one of the major challenges facing the field [19, 27, 28, 49].

The limitations in human ability to evaluate and to compare multiattribute options can lead to inconsistencies in human judgments [45, 51] or to application of simplified rules that do not consider essential aspects of the options under consideration [32, 38, 43].

It is important to understand what input information is reliable. Larichev [28] attempted to collect and classify all elementary operations in information processing used in normative decision-making. Twenty-three operations were defined and analyzed from the perspective of their complexity for human subjects. The study concluded that quantitative evaluation and comparison of different objects was much more difficult for subjects than conducting the same operations through qualitative ordinal expression of preference.

The following operations were found admissible on the basis of the known research results [34]:

- rank ordering of criteria importance;
- qualitative comparison of attribute values for one criterion or two criteria;
- qualitative evaluation of probabilities.

Some other operations are expected to be admissible although not enough research has been obtained to date to be sure of admissibility.

Qualitative judgments are preferable for the majority of operations. Therefore, Verbal Decision Analysis uses ordinal (cardinal) judgments as compared to interval data.

2.3 Theoretical Basis for Decision Rules Elaboration

Ordinal comparisons are always the first practical step in preference elicitation procedures in multicriteria analysis. Rather often, scaling procedures follow this step (resulting in quantitative values for all elements of the model). There are ways to analyze the decision on the basis of ordinal judgments, sometimes leading to the preferred decision without resort to numbers [4, 22, 23, 34]. Possible types of available ordinal preference information can be grouped as follows:

- rank ordering of separate levels upon criterion scales (ordinal scales);
- rank ordering of criteria upon their importance;
- pairwise comparison of real alternatives;

- ordinal tradeoffs: pairwise comparison of hypothetical alternatives differing in estimates of only two criteria.

Ordinal Scales are used in the rule of dominance (Pareto Principle). This rule states that one alternative is more preferable than another if it has criterion levels that are not less preferable on all attributes and is more preferable on at least one. This rule does not utilize criterion importance and is not necessarily connected with an additive form of a value function but it requires preferential independence of each separate criterion from all other criteria.

Rank Ordering of Criteria upon Importance does not provide any decision rule by itself. In combination with ordinal scales and lexicographical criterion ranking, the rule for selection of the best alternative may be as follows: first select alternatives with the best possible level upon the most important criterion. From the resulting subset select alternatives with the best possible level upon the next important criterion and so on. This rule is based on the assumption that in the criterion ranking one attribute is more important than all the other attributes, which follow it in the ranking. This preemptive rule does not necessarily imply the additive value function, but has the obvious drawback of its non-compensatory nature, and is theoretically unpopular.

Pairwise Comparison of real alternatives may be directly used in some methods (see, e.g. [24]). In general this information by itself will lead to the solution (if you compare all pairs of alternatives then you can construct a complete rank order of alternatives). But the whole area of multicriteria decision analysis has evolved from the notion that this task is too difficult for the decision maker. This approach is mostly used in multicriteria mathematical programming (in which there is not a finite number of alternatives for consideration). Still this information is considered to be highly unstable [28, 51].

Ordinal Tradeoffs [33] exploit the idea of tradeoffs widely used in decision analysis for deriving criterion weights, but is carried out in a verbal (ordinal) form for each pair of criteria and for all possible criterion levels. To find the tradeoff we have to ask the decision maker to consider two criteria and choose which he(she) prefers to sacrifice to some lower level of attainment. When levels are changed from the best to the worst attribute level, this corresponds to the questions in the "swing" procedure for criterion weights [11, 52], but does not require quantitative estimation of the preference.

The use of such tradeoffs is valid if there is preferential independence of pairs of criteria from all other criteria. Two of these preference elicitation methods provide the safest basis for preference identification: ordinal criterion scales and ordinal tradeoffs.

2.4 Consistency Check of Decision Maker's Information

Valid implementation of both ordinal criterion scales and ordinal tradeoffs requires preferential independence of one or two criteria (for all practical purposes if there is pairwise criterion independence, there exists an additive value function and it is reasonable to conclude that any group of criteria is independent from the rest – see [53]). In addition, in many practical cases the decision rule would require transitivity of preferences. It is necessary to check for these conditions for the method to be valid.

The use of preferential independence conditions stems from the desire to construct an efficient decision rule from relatively weak information about the decision maker's preferences. On the other hand complete checking for this condition will require an exhaustive number of comparisons. Therefore it is reasonable [33, 34] to carry out a partial check of the independence condition over pairs of alternatives. First all necessary tradeoff comparisons are carried out with all criterion levels except those being considered held at their most preferable level. Then, the same tradeoffs are carried out with all other criteria held at their least preferable level. If preferences are the same in both cases, those two criteria are considered to be preferentially independent from all other criteria.

This check is considered to be profound as the change in criterion levels is the most drastic (from the best to the worst) and stability of preferences under those conditions is good evidence of independence.

In case of dependency Verbal Decision Analysis recommends trying to reformulate the problem: group some criteria if they seem to be dependent, or decompose some criteria if their dependence seems to have a root in some essential characteristic combining several others that should be considered separately (see [34] for more details).

To be able to check for consistency of the information elicited (for ordinal information in the form of transitivity of preferences), Verbal Decision Analysis applies "closed procedures" where subsequent questions can be used to check information over all previous questions. For instance, if we ask the decision maker to compare A and B, then B and C, it's a good idea to ask the decision maker to compare A and C as well. If A is preferred to B, B is preferred to C, and A is preferred to C, then everything is consistent. If C is preferred to A, the preferences are intransitive. Within our approach, transitivity of preferences is assumed, so the decision maker is asked to reconsider comparisons from which intransitivity arises.

2.5 Explanation of the Analysis Result

The last but not the least requirement for Verbal Decision Analysis is to demonstrate the results of the analysis to the decision maker in a way that connects the problem structure and the elicited information with the resulting recommended alternative or alternatives.

It should be possible for the decision maker to see how information provided by him(her) lead to the result obtained. This is a necessary condition for the decision maker to rely on the result and to have the necessary information for re-analysis in case the result does not seem plausible. Methods based on Verbal Decision Analysis principles provide the ability to give explanations due to their logical and valid elicitation and their use of qualitative information.

In the next two sections methods based on these principles are presented for two important decision problems: rank ordering of multicriteria alternatives and ordinal classification/sorting [9] of multicriteria alternatives.

3. Decision Methods for Multicriteria Alternatives Ranking

The problems of ranking alternatives evaluated against a set of criteria are wide spread in real life. There are many decision aiding methods oriented on the solution of these problems [21, 34, 44, 46].

Within the Verbal Decision Analysis framework, we consider an unstructured problem where there is a large number of alternatives with mostly qualitative characteristics evaluated by human experts. The task is to elaborate a subjective decision rule able to establish at least a partial order on the set of alternatives.

Alternatives are evaluated against a set of criteria with *verbal* formulations of quality grades along their scales and as the number of alternatives is large enough the idea is to construct a decision rule in the criteria space and then use it on any set of real alternatives.

A good example of such a problem is selection of applicants for an interview for a faculty position [40]. A variant of a set of criteria with simple ordinal scales for evaluation of an applicant for a position in Management Information Systems is presented in Table 15.1

Method ZAPROS was proposed to deal with this type of problem and was based on the VDA principles. The ideas of ZAPROS started to be developed in 80s by a group of Russian scientists under the leadership of Larichev. The first publication in English presenting fully developed version of earlier ideas appeared in a 1995 issue of European Journal of Operational Research [33].

The method is based on the implementation of ordinal verbal scales and ordinal tradeoffs on the scales of criterion pairs near two reference situations. The goal is the construction of the Joint Ordinal Scale for all criteria. The name ZAPROS is the abbreviation of Russian words: Closed Procedures near Reference Situations.

Let us look more closely at the method and its enhancement during recent years.

Table 15.1. Criteria for applicant evaluation.

Criteria	Scale
A. Ability to teach our students	A1. Above average
	A2. Average
	A3. Below average
B. Ability to teach SA&D and DBMS	B1. Above average
	B2. Average
	B3. Below average
C. Evaluation of completed research and scholarship	C1. Above average
	C2. Average
	C3. Below average
D. Potential in publications	D1. Above average
	D2. Average
	D3. Below average
E. Potential leadership in research	E1. Above average
	E2. Average
	E3. Below average
F. Match of research interests	F1. Above average
	F2. Average
	F3. Below average

3.1 Problem Formulation

Formal presentation of the problem under consideration is as follows:

Given:

1 There is a set of n criteria for evaluation of alternatives.

2 X_i is a finite set of possible verbal values on the scale of criterion $i = 1, ..., n$, where $|X_i| = n_i$.

3 $X = \prod_{i=1}^{n} X_i$ is a set of all possible vectors in the space of n criteria.

4 $A = \{a_1, ..., a_i, ..., a_m\} \subsetneq X$ is a subset of vectors from X describing real alternatives.

Required: to rank order alternatives from the set A on the basis of the decision-maker's preferences.

We will use the following notations for relationships between alternatives:

- $\succeq_i$ is the weak preference relationship with respect to criterion i: for $a, b \in A, a \succeq_i b$ means a is at least as good as b with respect to criterion i ;

- $\succ_i$ is the strict preference relationship with respect to criterion i: $a, b \in A, a \succ_i b$ iff $a \succeq_i b$ and not $b \succeq_i a$;

- $\sim_i$ is the indifference relationship with respect to criterion i: $a, b \in A$, $a \sim_i b$ iff $a \succeq_i b$ and $b \succeq_i a$;
- $\succeq$ is the weak preference relationship: for $a, b \in A$, $a \succeq b$ means a is at least as good as b;
- $\succ$ is the strict preference relationship: $a, b \in A$, $a \succ b$ iff $a \succeq b$ and not $b \succeq a$;
- $\sim$ is the indifference relationship with respect to criterion i: $a, b \in A$, $a \sim b$ iff $a \succeq b$ and $b \succeq a$.

3.2 Formation and Implementation of the Joint Ordinal Scale

The first step in any decision analysis is to form the set of alternatives, form the set of criteria, and to evaluate alternatives against criteria. As we have decided to use only ordinal judgments for comparison of alternatives, the first step in this direction is to elaborate ordinal scales for attributes.

Formally, ordering criterion values along one criterion scale requires the decision maker to select the preferred alternative out of two hypothetical vectors from X differing in values with respect to one criterion (with all other values being at the same level).

This information allows formation of a strict preference relation $\succ_i$ for each criterion $i = 1, ..., n$.

Ordinal scales allow pairwise comparison of real alternatives according to the rule of *dominance*.

DEFINITION 52 *Alternative a is not less preferable than alternative b, if for each criterion i alternative a is not less preferable than alternative b ($a \succeq_i b$ for $i = 1, ..., n$).*

The next level of preference elicitation is based on comparison in an *ordinal form* of combinations of values with respect to two criteria.

To carry out such a task we need to ask a decision maker questions of the kind: "what do you prefer: to have this (better) level with respect to criterion i and that (inferior) level with respect to criterion j, or this (better) level for criterion j and that (inferior) level for criterion i if all other criteria are at the same level?"

Possible responses in this case are: more preferable, less preferable or equally preferable [33].

The decision-maker may be asked to make these "ordinal tradeoffs" for each pair of criteria and for each pair of possible values in their scales.

The same information may be obtained with far fewer questions by comparing two hypothetical vectors from X differing in values with respect to two criteria (with all other values being at the same level). Still the number of the comparisons for all possible combinations of criterion values may be quite large.

ZAPROS [33, 34] uses only part of this information for the construction of the Joint Ordinal Scale (JOS). The decision-maker is asked to compare pairs of hypothetical vectors from $Y \subset X$, each vector with the *best possible values* for all criteria but one. The number of these vectors is not large $|Y| = \sum_{i=1}^{n}(n_i - 1) + 1$.

The goal is to construct a complete rank ordering of all vectors from Y on the basis of the decision maker's preferences. An example of a possible preference elicitation question is presented in Table 15.2.

Table 15.2. Comparison of hypothetical alternatives.

Criteria	Alternative 1		Alternative 2	
A. Ability to teach our students	Above average	A1	Above Average	A1
B. Ability to teach SA&D and DBMS	Above average	B1	Above Average	B1
C. Evaluation of completed research and scholarship	Above average	C1	Above Average	C1
D. Potential in publications	*Average*	*D2*	*Above Average*	*D1*
E. Potential leadership in research	Above average	E1	Above Average	E1
F. Match of research interests	*Above average*	*F1*	*Below Average*	*F3*

Possible Answers:
1. Alt.1 is more preferable than Alt.2
2. Alt.1 and Alt.2 are equally preferable
3. Alt.1 is less preferable than Alt.2

DEFINITION 53 *Joint Ordinal Scale (JOS) is a complete rank order of vectors from Y, where Y is a subset of vectors from X with all the best values but one. Complete rank order means that for each $x, y \subseteq Y$ $x \succ y$ or $y \succ x$ or $x \sim y$.*

If the comparisons do not violate transitivity of preferences, we are able to construct a complete rank order of the vectors from Y on the basis of this information, forming the Joint Ordinal Scale. An example of the JOS for the applicants' problem is presented in Table 15.3 with the JOS rank for the most preferred vector marked as 1.

Construction of the Joint Ordinal Scale provides a simple rule for comparison of multiattribute alternatives. The correctness of rule 54 in case of pairwise preferential independence of criteria was proven in [33]. The crucial difference between the rule of dominance and this rule is that we are able now to compare criterion values with respect to *different* criteria.

Table 15.3. An Example of a joint ordinal scale.

Equal Criterion Values	Rank in JOS	Corresponding vector(s)
A1,B1,C1,D1,E1,F1	1	A1B1C1D1E1F1
C2, E2	2	A1B1C2D1E1F1 A1B1C1D1E2F1
A2, D2, F2	3	A2B1C1D1E1F1 A1B1C1D2E1F1 A1B1C1D1E1F2
B2	4	A1B2C1D1E1F1
B3, E3, F3	5	A1B3C1D1E1F1 A1B1C1D1E3F1 A1B1C1D1E1F3
A3, C3, D3	6	A3B1C1D1E1F1 A1B1C3D1E1F1 A1B1C1D3E1F1

DEFINITION 54 *Alternative a is not less preferable than alternative b, if for each criterion value of a there may be found a not more preferable unique criterion value of alternative b.*

There is an easy way to implement this rule, introduced and proven correct in [39]. Let us substitute a criterion value in each alternative by the corresponding rank in the Joint Ordinal Scale ($JOS(a)$). Then rearrange them in the ascending order (from the most preferred to the least preferred one), so that

$$JOS_1(a) \leq JOS_2(a) \leq ... \leq JOS_n(a)$$

.

Then the following rule for comparison of two alternatives may be presented.

DEFINITION 55 *Alternative a is not less preferable than alternative b if for each $i = 1, ..., n$ $JOS_i(a) \leq JOS_i(b)$.*

Let us use our Joint Ordinal Scale presented in Table 15.3 to compare the following two applicants, incomparable on the basis of the dominance rule: a=(A1,B2,C1,D1,E1,F2) and b=(A1,B1,C1,D2,E2,F1)

If we substitute each criterion value in alternatives a and b with corresponding rank from the JOS and rearrange them in an ascending order, we will obtain the following two vectors, which can be easily compared: JOS(a)= (1,1,1,1,3,4) and

JOS(b)=(1,1,1,1,2,3). It is clear now that alternative b is preferred to alternative a.

ZAPROS suggests using Joint Ordinal Scale for pairwise comparison of alternatives from A, thus constructing a partial order on this set.

The construction and implementation of Joint Ordinal Scale, as stated above, is based on two assumptions: transitivity of the decision maker's preferences and preferential independence of pairs of criteria (the last condition leads to an additive value function in the decision maker's preferences [33, 34]). This is the basis for the correctness of rule 55.

For the decision making method to be valid within the paradigm of Verbal Decision Analysis it should provide means for verification of underlying assumptions. ZAPROS provides these means as follows.

3.3 Verification of the Structure of the Decision Maker's Preferences

When comparing vectors from Y (for JOS construction) the decision maker can give contradictory responses. In the problem under consideration these responses may be determined as violations of transitivity in the constructed preference relation.

Possible responses of the decision maker in comparison of hypothetical vectors y_i and y_j from Y (see Table 15.2) reflect the binary relation of strict preference ($\succ$) or indifference ($\sim$) between these two alternatives. The following conditions should be met as a result of the decision maker's responses:

if $y_i \succ y_j$ and $y_j \succ$ or $\sim y_k$ then $y_i \succ y_k$

if $y_i \sim y_j$ and $y_j \sim y_k$ then $y_i \sim y_k$

if $y_i \sim y_j$ and $y_j \succ y_k$ then $y_i \succ y_k$.

These conditions are checked in the process of preference elicitation, the intransitive pairs are presented to the decision maker for reconsideration.

The procedure for transitivity verification is described in details in [33, 34], is implemented in a corresponding computerized system and was used in a number of different tasks [34, 37, 40].

The next assumption necessary to check is the pairwise preferential independence of criteria.

DEFINITION 56 *Criteria i and j are* preferentially independent *from the other criteria, if preference between vectors with equal values with respect to all criteria but i and j, does not depend on the values of equal components.*

As it is impossible to carry out preference elicitation for all possible combinations of equal values, it was proposed to check preferential independence for pairs of criteria near two very different "reference situations". One variant is based on all the best values for equal components (used in the construction of JOS). The second with the worst possible values for equal components.

If the decision maker's preferences among criterion values are the same when elicited using these two different points, then it is assumed the criteria are preferentially independent.

Although this check is not comprehensive, the preferential stability when using essentially different criterion values as the "reference" point suggests it would hold with the intermediate levels as well [34].

3.4 Contemporary Modifications of ZAPROS

The general direction in enhancing method ZAPROS in recent years [29, 39] was concentrated on the efforts to ensure higher level of compatibility among real alternatives. To achieve that in both publications it was proposed to carry out more ordinal tradeoffs.

For construction of the Joint Ordinal Scale (see section 3.2), only a relatively small number of comparisons are carried out, limited to vectors with all the best criterion values but one.

In general, the decision-maker may be asked to compare any two hypothetical vectors from X differing in values with respect to two criteria (with all other values being at the same level).

Larichev [29] proposed just that in a method called ZAPROS III . The method requires comparing all criterion values for all pairs of criteria and using this information for comparison of real alternatives.

As the number of such comparisons may be quite large, it is reasonable to use this approach for relatively small problems (small number of criteria and small number of possible criterion values with relatively large number of real alternatives).

In [39] the authors proposed to use additional comparisons only after applying Joint Ordinal Scale for comparison of real alternatives. The goal is to elicit information necessary to compare alternatives left incomparable only if there is such a need for making the decision. The process is iterative (as needed), that's why it was named STEP-ZAPROS. The authors carried out simulations to evaluate effectiveness of the procedure and the number of additional comparisons carried out by the decision maker for different problem sizes.

Let's look briefly at each of these new methods.

3.4.1 ZAPROS III. ZAPROS III introduces a notion of *Quality Variation (QV)* which is the result of changing one value on the scale of one criterion (e.g., from *Average ability to teach our students* to *Below Average* level).

The decision maker is to compare all possible QVs for each pair of criteria with the assumption that all other criterion values are at the same level. The number of QVs for each scale is $n_i(n_i - 1)/2$, where n_i is the number of values on the criterion scale. In addition, the decision maker is to compare some QVs on the same scale (e.g., QV from Above Average to Average compared to QV from Average to Below Average).

Once all comparisons for two criteria are carried out all QVs for them are rank ordered forming the Joint Scale for Quality Variation (JSQV). For example, let's assume that the JSQV for the first two criteria in applicants' evaluation example, are as follows (we will use A1,A2 to show changing value from A1 to A2) :

$$A1, A2 \succ B1, B2 \succ A1, A3 \succ B1, B3 \succ A2, A3 \succ B2, B3.$$

It is proposed to carry out these comparisons at two reference situations (as in ZAPROS): with all the best and all the worst values with respect to other criteria. If the comparisons provide the same JSQV, these criteria are considered to be preferentially independent.

Those rankings are carried out for all pairs of criteria and can be used to construct a Joint Scale of Criteria Variations (JSCV).

Let's look at a simple example for three criteria: a = (A3, B1, C2) and b = (A2, B2, C1). Suppose JSQVs for criteria A & B, B & C, and A & C are as follows:

$$A1, A2 \succ B1, B2 \succ A1, A3 \succ B1, B3 \succ A2, A3 \succ B2, B3.$$

$$C1, C2 \succ B1, B2 \succ B1, B3 \succ C1, C3 \succ B2, B3 \succ C2, C3.$$

$$C1, C2 \succ A1, A2 \succ A1, A3 \succ C1, C3 \succ A2, A3 \succ C2, C3.$$

If we combine all this information together the JSCV is:

$$C1, C2 \succ A1, A2 \succ B1, B2 \succ A1, A3 \succ B1, B3 \succ C1, C3$$

$$C1, C3 \succ A2, A3 \succ B2, B3 \succ C2, C3.$$

If in this process violations of transitivity of preferences are discovered, they are presented to the decision maker, and resolved.

Each QV for each criterion gets a rank (e.g., C1,C2 has rank 1, A1,A2 has rank 2, and so on). This rank can be used to compare alternatives. In ZAPROS III [29] it is proposed to present each real alternative as a combination of JSCV ranks. It is not possible, e.g., in alternative a=(A3,B1,C2) it is not clear if A3 should be presented as A1,A3 or A2,A3. In ZAPROS we have only information on A1,A2 and A1,A3. We do not have information on A2,A3 and so there is no question about the rank to use. With JSCV we need to differentiate these two cases. To overcome this, ranks describing two alternatives at the same time should be used.

We can rewrite vectors a and b as follows. Criterion A: the change is from A2 to A3, so we change A3 to rank 7 of A2,A3 in the JSCV and A2 to rank 0. Criterion B: change is from B1 to B2, so we change B2 to rank 3 and B1 to rank 0. Criterion C: change is from C1 to C2, so we change C2 to rank 1 and C1 for rank 0. As a result alternative a is presented as (7,0,1) or (0,1,7) and alternative

b is (0,3,0)or (0,0,3). Vector (0,0,3) dominates vector (0,1,7), so alternative b is preferred to alternative a.

Although the amount of additional information on the decision maker's preferences is rather large, there still may be incomparable alternatives. In ZAPROS III it is proposed to sequentially select non-dominated nuclei (analogous to ZAPROS [33]). Alternatives from the first nucleus are assigned rank 1. An alternative has a rank r if it is dominated by an alternative ranked r-1 and itself dominates alternative ranked r+1. As a result some alternatives can have a "fuzzy" rank (e.g., 5-7).

3.4.2 STEP-ZAPROS. This approach views the general application of ordinal preferences for comparison of real alternatives as a three-step procedure:

1 use rule of dominance to compare real alternatives on the basis of ordinal scales. If required decision accuracy is obtained, stop here

2 construct Joint Ordinal Scale and use it to compare real alternatives. If required decision accuracy is obtained, stop here

3 use additional ordinal tradeoffs to compare real alternatives as necessary. Use restructuring procedures if the necessary accuracy is not achieved.

Additional comparisons are carried out only when necessary and only the necessary comparisons are carried out. Thus, the procedure is oriented on efficient acquisition of necessary information.

When comparing real alternatives using Joint Ordinal Scale, alternatives are presented through JOS ranks: JOS(a) and JOS(b) (see section 3.2). If alternatives a and b have been left incomparable it means we have at least two ranks such that $JOS_i(a) < JOS_i(b)$ while $JOS_j(a) > JOS_j(b)$. These ranks represent some criterion values in JOS.

The idea is to form two vectors from X different in values with respect to only two criteria (with all the best values with respect to all other criteria). Different criterion values represent the "contradicting" ranks in JOS(a) and JOS(b).

Let our JOS(a)= (1,1,1,2,3,3) and JOS(b)= (1,1,1,1,1,5). They are incomparable according to JOS as rank 5 is less preferable than rank 2 or 3. If, for example, rank 5 is more preferable than ranks 3 and 3 *together*, then alternative b would be preferable to alternative a.

Rank 3 is presented in the JOS (see Table 15.3) by corresponding criterion values A2, D2, and F2. Rank 5 corresponds to criterion values B3, E3, and F3. It allows formation of the following vectors, representing combination of ranks (3,3) and (1,5) and differing in only two criterion values: (A1,B1,C1,D2,E1,F2) and (A1,B1,C1,D1, E1,F3). Comparison of these two vectors will compare D1,D2 with F2,F3 (see ZAPROS III).

If the second vector is preferred to the first one then alternative b is preferred to alternative a. If not, they may be left incomparable.

As the comparison of such specially formed vectors reflects comparison of pairs of ranks in the Joint Ordinal Scale, it is referred to as Paired Joint Ordinal Scale (PJOS) and allows the following rule for comparison of real alternatives:

DEFINITION 57 *Alternative a is not less preferable than alternative b if for each pair of criterion values (a_i, a_j) of alternative a there exists a pair of values (b_k, b_l) of alternative b such that PJOS$(a_i, a_j) \leq$ PJOS(b_k, b_l).*

The proof of the correctness of the rule in case of an additive value function is given in [39].

Preferential independence of criteria is checked while constructing the Joint Ordinal Scale (see section 3.2). Transitivity of preferences at the third step is checked only partially in the process of comparisons (as we have previous information on preferences among some of pairs of JOS ranks). It is technically possible to carry out auxiliary comparisons (as in ZAPROS) to ensure transitive closure. It can be applied as necessary at the discretion of the consultant.

To demonstrate the potential of these three steps, simulation results were presented in [39]. Partial information for different problem sizes in presented in Table 15.4.

Table 15.4. Effectiveness of STEP-ZAPROS.

Parameters								
Number of criteria	5	5	5	5	7	7	7	7
Number of criterion values	3	3	5	5	3	3	5	5
Number of alternatives	30	50	30	50	30	50	30	50
% of compared alternatives	76	76	73	74	63	64	56	59
Additional comparisons	14	17	63	96	21	30	86	141

Data show that 1) the number of real alternatives does not influence the efficiency of the procedure very much; 2) the number of criteria to some extent influences overall comparability of alternatives; 3) the number of criterion values has a *crucial* influence on the number of additional comparisons carried out in the third step.

Overall the data show that method ZAPROS is most efficient for tasks where number of criteria is relatively small and number of alternatives for comparison is relatively large.

4. Decision Methods for Multicriteria Alternatives' Classification

Along with multicriteria choice/ranking problems, people may face multicriteria classification problems [9]. Rather a large number of classification tasks in business applications may be viewed as tasks with classes which reflect the levels of the same property. Evaluating creditworthiness of clients is rather often measured on an ordinal level as, e.g., "excellent", "good", "acceptable", or "poor" [6]. Articles submitted to the journals in the majority of cases are divided into four groups: "accepted", "accepted with minor revisions", "may be accepted after revision and additional review", "rejected" [34]. Applicants for a job are divided into accepted and rejected, but sometimes there may be also a pool of applicants left for further analysis as they may be accepted in some circumstances [5, 50].

Multicriteria problems with ordinal criterion scales and ordinal decision classes were named problems of *ordinal classification* (ORCLASS). As with method ZAPROS the ideas of ORCLASS were developed in 80s by a group of Russian scientists under the leadership of Larichev. Journal publications in English appeared only in mid 90s [41, 1].

4.1 Problem Formulation

Formal presentation of the problem under consideration is close to the one in section 3.1 as we use criteria scales with finite set of verbal values and analyze the criterion space. Thus items 1-4 are the same while item 5 and what is required in the problem differ.

Given:

1 There is a set of n criteria for evaluation of alternatives.

2 X_i is a finite set of possible verbal values on the scale of criterion $i = 1, ..., n$, where $|X_i| = n_i$.

3 $X = \prod_{i=1}^{n} X_i$ is a set of all possible vectors in the space of n criteria.

4 $A = \{a_1, ..., a_i, ..., a_m\} \subseteq X$ is a subset of vectors from X describing real alternatives

5 $C = \{C_1, ..., C_i, ..., C_k\}$ is a set of decision classes.

Required: distribute alternatives from A among decision classes C on the basis of the decision-maker's preferences.

For example, the applicants' problem presented in Table 15.1 may be viewed as a classification problem if we need to divide all applicants into three classes: 1) accepted for an interview, 2) left for further consideration, 3) rejected.

We will use the same notation for preferences as in section 3.1. In addition, notation C(a) means class for alternative a, e.g., C(a)=C_2 means alternative a belongs to the second class.

4.2 An Ordinal Classification Approach

As in ZAPROS the VDA framework assumes ordinal criterion scales establishing a *dominance* relationship among vectors from X (see definition 52). In ordinal classification there is an *ordinal relationship among decision classes* as well. This means that alternatives from class C_1 are preferred to alternatives in class C_2 and so on. The least preferable alternatives are presented in class C_k. As a result alternatives with "better" qualities (criterion values) should be placed in a "better" class.

These ordinal qualities allow formation of an effective decision maker's preference elicitation approach [30, 25, 41, 34, 42, 3, 1, 2].

The decision maker is presented with vectors from X and asked directly to define an appropriate decision class. The cognitive validity of this form of preference elicitation was thoroughly investigated and found admissible [32, 36].

It is possible to present the decision maker with all possible vectors from X to construct a universal classification rule in the criterion space. However, it is impractical even for relatively small problem sizes. The ordinal nature of criterion scales and decision classes allows formulation of a strict preference relation: if vector x is placed in a better class than vector y, then vector x is more preferable than vector y.

DEFINITION 58 *For any vectors $x, y \in X$ where C(x)=C_i and C(y)=C_j if $i < j$ then $x \succ y$.*

As a result we can formulate a condition for a non-contradictory classification of vectors x and y: if vector x dominates vector y and is placed into i-th class, then vector y should be placed into a class not more preferable than the i-th class.

DEFINITION 59 *For any vectors $x, y \in X$ if y is dominated by x ($x \succ y$) and C(x)=C_i, then C(y)=C_j where $j \geq i$.*

Using this quality we can introduce a notion of *expansion by dominance* [25].

DEFINITION 60 *If vector $x \in X$ is assigned class C_i by a decision maker, then for all $y \in X$ such that $x \succ y$ possible classes are C_j where $j \geq i$. For all $y \in X$ such that $y \succ x$ possible classes are C_j where $j \leq i$.*

Each classification of a vector from X by a decision maker limits possible classes for all dominating it and dominated by it vectors from X. When the number of admissible classes for the vector becomes equal to one, we have a unique class assigned to a vector.

Using expansion by dominance we can obtain classification for some vectors from X not presented to the decision maker (there are some results [25, 41, 34, 42] showing that between 50 and 75% of vectors may be classified indirectly using this rule).

In addition, there is a simple way to discover possible errors in the decision maker's classifications: if an assigned class is outside the admissible range, there is a contradiction in the ordinal classification. Contradictory classifications may be presented to the decision maker for reconsideration.

For more details on the procedure see [41, 34].

The efficiency of the *indirect* classification of vectors from set X depends on the vectors presented to the decision maker as well as on the class assigned [41, 34]. Ideally, we would like to present the decision maker with as few questions as possible and still be able to construct a complete classification of vectors from set X. Different heuristic approaches were proposed to deal with this problem, based on the desire to find the most "informative" vectors to be presented to the decision maker for classification.

4.3 Class Boundaries and Effectiveness of Preference Elicitation

Ordinal classification allows not only a convenient method of preference elicitation, but also an efficient way to present the final classification of set X.

Let assume we have a classification of set X into classes C. We will view C_i as a subset of vectors from X, assigned to the i-th class.

Two special groups of vectors may be differentiated among them: *lower border* of the class LB_i and the *upper border* UB_i. Upper border includes all *non-dominated* vectors in the class, while lower border includes all *non-dominating* vectors in this class.

These two borders accurately represent the i-th class: we can classify any other vector as belonging to class C_i if its criterion values are between values of vectors from LB_i and UB_i.

Let us look at vector C(x)=C_i which is not in the upper or lower border of the class. It means there is a vector $y \in UB_i$ for which $y \succ x$, thus $C(y) \leq C(x)$. Analogously there is object $z \in LB_i$ for which $x \succ z$. Thus $C(x) \geq C(z)$. But $C(y) = C(z) = C_i$. This leads to $C(x) = C_i$.

Borders summarize classification rules. If we know classification of vectors in the class borders only, it would be enough to classify any vector from set X [41, 34, 42]. That is why, heuristic methods are oriented on finding potential "border vectors" for presentation to the decision maker.

The first approach was based on the maximum "informativeness" of unclassified vectors [34]. Each class was presented by its "center" (average of criterion values of vectors already in the class). For each unclassified vector x for each its admissible class "similarity" measure $p_i(x)$ was calculated (it evaluated how probable that class was for that vector). Also, for each admissible class the number of indirectly classified vectors $g_i(x)$ if x is assigned class C_i was evaluated.

Informativeness $F(x)$ for vector x was calculated as $F(x) = \sum p_i(x)\, g_i(x)$ for all admissible classes. The vector with the largest informativeness value was selected for classification by the decision maker. After that the expansion by dominance was carried out and informativeness of all vectors was recalculated.

Simulations showed high effectiveness of the procedure with only 5 to 15% of all vectors from X necessary to be classified by the decision maker [34]. The drawback of the approach is its high computational complexity.

Another approach was proposed in [42]. It is based on a maxmin principle. For each unclassified vector the minimum number of indirectly classified vectors in case of admissible classes is defined and the vector with the maximum number is selected for classification by the decision maker. The computational complexity of the approach is a bit lower than in the previous case.

A new algorithm called CYCLE was presented in [25]. The idea is to construct "chains" of vectors between vectors x and y which belong to different classes. The "chain" is constructed sequentially by changing one criterion value in vector x by one level until we obtain criterion values of vector y. Then the most "informative" vector is searched only in the chain, thus essentially lowering the computational complexity of the algorithm.

The effectiveness of the approach was compared to the algorithms of monotone function decoding and appeared much more effective for smaller problems and simpler borders while being somewhat less effective in more complicated cases.

The computation complexity of CYCLE is not stated in this work and there is a question of how we select x and y for the "chain" construction (in the beginning we have only two classified vectors: with the best criterion values and with the worst criterion values, so the chain contains all other vectors from X), but the direction seems promising.

5. Place of Verbal Decision Analysis in MCDA

The decision maker is the central figure in decision making based on multiple criteria. Elicitation of the decision makers' preferences should take into account peculiarities of human behavior in the decision processes. This is the goal of Verbal Decision Analysis.

Like outranking methods (e.g. ELECTRE, PROMETHEE) VDA provides outranking relationships among multicriteria alternatives. At the same time, VDA is designed to elicit a sound preference relationship that can be applied to future cases while outranking methods are intended to compare a given set of alternatives. VDA is more oriented on tasks with rather large number of alternatives while number of criteria is usually relatively small. Outranking methods deal mostly with reverse cases.

VDA bases its outranking on axiomatic relationships, to include direct assessment, dominance, transitivity, and preferential independence. Outranking methods use weights as well as other parameters, which serve an operational purpose but also introduce heuristics and possible intransitivity of preferences. VDA is based on the same principles as multiattribute utility theory (MAUT), but is oriented on using the verbal form of preference elicitation and on evaluation of alternative decisions without resort to numbers. That is why we consider that it is oriented on the same tasks as MAUT and will be compared in a more detail to this approach to multicriteria decision making.

5.1 Multi Attribute Utility Theory and Verbal Decision Analysis Methods

The central part of MAUT is in deriving numeric scores for criterion values and relative criterion weights which are combined in an overall evaluation of an alternative's value.

There are a number of methods and procedures for eliciting criterion weights and scores. Some of these methods are based on sound theory, while others use simplified heuristic approaches.

Experiments show that different techniques may lead to different weights [7, 47], but in modelling situations varying criterion weights often does not change the result thus leading to the conclusion that equal weights work sufficiently well [10, 12]. However, the situation may not be the same for real decision tasks when differences between alternatives are small. Slight differences in weights can lead to reversals in the ranking of alternatives [35, 37, 54].

Two approaches (MAUT and VDA) were applied to the same decision making problems [15, 26, 31]. Positive and negative features of each approach were analyzed, the circumstances under which one or the other would be favored were examined.

Three groups of criteria for comparison were considered: methodological, institutional and personal [15, 26].

Methodological criteria characterize an approach from the following perspectives:

- measurements of alternatives with respect to criteria;
- consideration of alternatives;

- complexity reduction;
- quality of output;
- cognitive burden.

Measurements. VDA uses verbal scales, while MAUT is oriented on obtaining numerical values.

People use verbal communication much more readily than quantitative communication. Words are perceived as more flexible and less precise, and therefore seem better suited to describe vague opinions. Erev and Cohen stated that "forcing people to give numerical expressions for vague situations where they can only distinguish between a few levels of probability may result in misleading assessments" [13].

But there are positive factors in utilization of quantitative information: people attach a degree of precision, authority and confidence to numerical statements that they do not ordinarily associate with verbal statements, and it is possible to use quantitative methods of information processing.

The experiments made over many years by Prof. T. Wallsten and his colleagues demonstrated no essential differences in the accuracy of evaluations [8, 13], but there was essential difference in the number of preference reversals. The frequency of reversals was significantly decreased when using the verbal mode [16].

The two methods differ considerably in whether they *force consideration of alternatives.* If the best alternative is not found by using "verbal" comparisons, VDA seeks to form another alternative that has not previously been considered (generating new knowledge) by acknowledging the fact that there is no best alternative among presented. VDA assumes that if it is not possible to find better alternative on an ordinal level, there is either no satisfactory alternative or alternatives are too close in quality to differentiate between them.

The numerical approach does not force thorough consideration of alternatives, as it is capable to evaluate even very small differences among alternatives. It is always possible to find the best alternative in this case. The question is if the result is reliable enough.

Complexity. VDA diminishes complexity of judgments required from the decision maker as it concentrates only on *essential* differences. The MAUT method requires very exact (numerical) comparisons of differences among criteria and/or alternatives in majority of cases.

Quality of output. MAUT provides overall utility value for each alternative. This makes it possible to not only identify the best alternative but also to define the difference in utility between alternatives. This means that the output of MAUT methods is rich enough to give the decision-maker the basis for detailed evaluation and comparison of any set of alternatives.

VDA attempts to construct a binary relation between alternatives which may lead to incomparable alternatives, but assures that comparisons are based on sound information elicitation.

Cognitive burden. A goal of all decision methods is reducing the confusing effect of ambiguity in preferences. Methods deal with this phenomenon in very different ways. VDA alters ambiguity and corresponding compensations into levels (rather than exact numbers).

MAUT attempts to estimate the exact amount of uncertainty. The payoff is that the analysis can derive a single estimate of uncertainty to go with the single estimate of utility.

Institutional criteria include: the ease of using the approach within organizations, and consequences of cultural differences.

Both MAUT and VDA can be considered improvements over confounding cost-benefit analysis based upon data with little hope of shared acceptance. Achieving greater clarity does, to some extent, provide improved communication within organizations. However, the information upon which MAUT develops utility is of suspect reliability.

The VDA approach uses more direct communication and active groups are used to assign the verbal quality grades on criteria scales. The VDA approach does not require the decision-maker or expert to have previous knowledge in decision methods. On the other hand, MAUT findings can be presented graphically and provide sensitivity analysis because of its numerical basis.

Some cultural differences may influence the applicability of different approaches. Americans tend to use numerical evaluations more often than in some other countries (e.g., Russia). American analysts are usually required "to put a price tag on goods not traded in any market place" [14]. That is not always the case in Europe.

Personal criteria include: the educational level required of decision-makers to use methods; and how the professional habits of analysts influence the selection of an approach.

The practical experience and intellectual ability of the decision-maker are presuppositions for the utilization of any analytical technique. MAUT requires more detailed trade-off balancing, calling for deeper ability to compare pairs of criteria performances. VDA is designed to focus on more general concepts.

Training in decision analysis helps decision-makers to understand and accept the MAUT approach. VDA methods do not require any special knowledge in decision analysis on the part of the decision-maker. The VDA approach is especially useful when a decision is made under new circumstances or in conditions of high ambiguity.

Comparison: The MAUT approach has a strong mathematical basis. MAUT provides a strong justification of the type of utility function used for aggregation of single-attribute utilities over criteria. Different kinds of independence conditions can be assumed [21]. In the case of criteria dependence, a nonlinear form quite different from the simple additive linear model is available. The involvement of the decision-maker is needed to elaborate a utility function. But after this is accomplished, it is possible to compare many alternatives. Should a new alternative appear, no additional decision-maker efforts are needed. Possible inaccuracy in the measurements could be compensated for by sensitivity analysis.

Conversely, the questions posed to decision-makers have no psychological justification. Some questions could be very difficult for humans to completely understand. Decision-makers require special training or orientation in order for MAUT methods to be used. Possible human errors in evaluating model parameters are not considered. Sensitivity analysis is recommended to evaluate stability of the result.

Verbal Decision Analysis has both psychological and mathematical basis. In all stages of the method natural language is used to describe concepts and information gathered relating to preference. Preferential criteria independence is checked. If criteria are dependent, we may try to transform the verbal description of a problem to obtain independence [34]. For example, sometimes criteria (or their scales) may be too detailed (not necessary information) or too general (not possible to differentiate). In these cases introducing two or three more detailed criteria instead of one too general for evaluation or collapsing a couple of criteria into one on a more general level may lead to preferential independence. In addition VDA has special procedures for the identification of contradictions in the information provided by the decision-maker.

Conversely, there are some cases when incomparability (due to lack of reliable information) does not guarantee identification of one best alternative. There may be more than one alternative ranked at the best level. The decision rule might not be decisive enough in cases when a decision must be reached quickly. There is no guarantee that experts could find a better alternative after formulation of directions for improvements of existing alternatives.

5.2 Practical Value of the Verbal Decision Analysis Approach

VDA has positive features of:

- Using psychologically valid preference input;
- Providing checks for input consistency;
- It is based on mathematically sound rules.

It was used in a number of applications for different types of decision problems. ZAPROS (and its variations) was used in R&D planning [33, 34], applicants' selection [40], job selection [35, 37], and pipeline selection [15, 26, 31].

R&D planning problem was connected with a state agency financing different research projects. Number of applications for funding was around several thousand each year, approximately 70% of them were awarded required (or reduced) funding. The decisions were to be made rather quickly after the deadline for applications (couple of months). To be able to cope with this level of complexity, it was decided to construct a decision rule in the criterion space and apply it to alternatives' descriptions against the criteria which were obtained through experts. ZAPROS was used to construct Joint Ordinal Scale in the criterion space which was used to form ordered groups of alternatives (for sequential distribution of funds). The number of criteria ranged from 5 to 7 for different subgroups of projects.

The task of applicants selection was implemented in one of the American universities where there could be more than 100 applicants for a faculty position. Six criteria with three level (verbal) scales were used to construct the Joint Ordinal Scale to be used to select s subset of better applicants for further analysis and an interview. The department chair was the decision maker in this case.

Pipeline selection was a somewhat different type of problem where there were relatively small number of very complicated alternatives: possible routes for a new gas pipeline. Modified variant of ZAPROS was used to elicit preferences from the decision maker in this case and use it to analyze the quality of presented alternatives. All alternatives were found out to be not good enough for implementation. The analysis was directed towards "redefining" the problem (through more detailed and/or less detailed criteria) and formation of a new "adjusted" alternative acceptable for the authorities.

The ordinal classification approach was used for R&D planning and journals' evaluation, as well as for job selection [34, 1]. In addition, this approach was found to be very useful in the area of knowledge base construction for expert systems.

Ordinal classification can be rather easily applied to nominal classification tasks if the decision maker (expert) is asked to evaluate the "level of appropriateness" of each nominal class for the presented vector from the criterion space [42]. Quite a number of applications were in the area of medical diagnostics [30, 42]. Ideas of ordinal classification were also implemented within the framework of case-base reasoning [3, 2] and data mining [18]. Transformation of initially nominal classification problems into problems with ordinal classes and ordinal scales enabled more effective procedures for data analysis.

6. Conclusion

MCDA is an applied science. The primary goal of research in MCDA is to develop tools to help people to make more reasonable decisions. In many cases the development of such tools requires combination of knowledge derived from such areas as applied mathematics, cognitive psychology, and organizational behavior. Verbal Decision Analysis is an example of such a combination. It is based on valid mathematical principles, takes into account peculiarities of human information processing system, and places the decision process within the organizational environment of the decision making.

Acknowledgments

We would like to recognize that Prof. Oleg Larichev of the Russian Academy of Sciences was the founder of Verbal Decision Analysis paradigm. His untimely death prevented Prof. Larichev from writing this chapter. The authors were fortunate to work with him over many years and have endeavored to present shared ideas and approaches he was instrumental in developing.

References

[1] Mechitov A.I., Moshkovich H.M., and Olson D.L. Problems of decision rules elicitation in a classification task. *Decision Support Systems*, 12:115–126, 1994.

[2] Mechitov A.I., Moshkovich H.M., Olson D.L., and Killingsworth B. Knowledge acquisition tool for case-based reasoning systems. *Expert Systems with Applications*, 9(2):201–212, 1995.

[3] Mechitov A.I., Moshkovich H.M., Bradley J.H., and Schellenberger R.E. An ordinal model for case-based reasoning in a classification task. *International Journal of Expert Systems: Research and Applications*, 9(2):225–242, 1996.

[4] F.H. Barron and B.E. Barrett. Decision quality using ranked attribute weights. *Management Science*, 42:1515–1523, 1996.

[5] A. Ben-David. Automated generation of symbolic multiattribute ordinal knowledge-based dsss: Methodology and applications. *Decision Sciences*, 23:1357–1372, 1992.

[6] A. Ben David, L.A. Sterling, and Y.H. Pao. Learning and classification of monotonic ordinal concepts. *Computational Intelligence*, 5(1):45–49, 1989.

[7] K. Borcherding, T. Eppel, and D. von Winterfeldt. Comparison of weighting judgments in multiattribute utility measurement. *Management Science*, 27:1603–1619, 1991.

[8] D. Budescu and T. Wallsten. *Processing Linguistic Probabilities: General Principles and Empirical Evidence*, volume 32 of *The Psychology of Learning and Motivation*. Academic Press, 1995.

[9] Zopounidis C. and Doumpos M. Multicriteria classification and sorting methods: A literature review. *European Journal of Operational Research*, 138:229–246, 2002.

[10] R.M. Dawes and B. Corrigan. Linear models in decision making. *Psychological Bulletin*, 81:95–106, 1974.

[11] W. Edwards and F.H. Barron. SMARTS and SMARTER: Improved simple methods for multiattribute utility measurement. *Organizational Behavior and Human Decision Processes*, 60:306–325, 1994.

[12] H. Einhorn and R. Hogarth. Unit weighting schemes for decision making. *Organizational Behavior and Human Decision Processes*, 13:171–192, 1975.

[13] I. Erev and B. Cohen. Verbal versus numerical probabilities: Efficiency, biases, and the preference paradox. *Organizational Behavior and Human Decision Processes*, 5:1–18, 1990.

[14] B. Fischoff. The real world: What good is it? *Organizational Behavior and Human Decision Processes*, 65:232–248, 1996.

[15] N.E. Flanders, R.V. Brown, Y. Andre'eva, and O. Larichev. Justifying public decisions in Arctic oil and gas development: American and Russian approaches. *Arctic*, 51(3):262–279, 1998.

[16] C. Gonzalez-Vallejo and T. Wallsten. The effects of communication mode on preference reversal and decision quality. *Journal of Experimental Psychology: Learning, Memory and Cognition*, 18:855–864, 1992.

[17] P. Goodwin and G. Wright. *Decision Analysis for Management Judgments*. John Wiley & Sons, New York, 1991.

[18] Moshkovich H.M., Mechitov A.I., and Olson D. Rule induction in data mining: Effect of ordinal scales. *Expert Systems with Applications Journal*, 22:303–311, 2002.

[19] D. Kahneman, P. Slovic, and A. Tversky, editors. *Judgment under Uncertainty: Heuristics and Biases*. Cambridge University Press, Cambridge, 1982.

[20] R. Keeney. *Value-Focused Thinking*. Harvard University Press, Cambrisge MA, 1992.

[21] R.L. Keeney and H. Raiffa. *Decisions with Multiple Objectives: Preferences and Value Tradeoffs*. John Wiley & Sons, New York, 1976.

[22] C.W. Kirkwood and J.L. Corner. The effectiveness of partial information about attribute weights for ranking alternatives in multiattribute decision making. *Organizational Behavior and Human Decision Processes*, 54:456–476, 1993.

[23] C.W. Kirkwood and R.K. Sarin. Ranking with partial information: A method and an application. *Operations Research*, 33:38–48, 1985.

[24] P. Korhonen. A visual reference direction approach to solving discrete multiple criteria problems. *European Journal of Operational Research*, 34(2):152–159, 1988.

[25] O. Larichev, A. Asanov, and Y. Naryzhny. Effectiveness evaluation of expert classification methods. *European Journal of Operational Research*, 138:260–273, 2002.

[26] O. Larichev and R. Brown. Numerical and verbal decision analysis: Comparison on practical cases. *Journal of Multi-Criteria Decision Analysis*, 9:263–273, 2000.

[27] O.I. Larichev. Psychological validation of decision methods. *Journal of Applied Systems Analysis*, 11:37–46, 1984.

[28] O.I. Larichev. Cognitive validity in design of decision-aiding techniques. *Journal of Multi-Criteria Decision Analysis*, 1(3):127–138, 1992.

[29] O.I. Larichev. Ranking multicriteria alternatives: The method ZAPROS III. *European Journal of Operational Research*, 131:550–558, 2001.

[30] O.I. Larichev. Close imitation of expert knowledge: The problem and methods. *International Journal of Information Technology and Decision Making*, 1(1):27–42, 2002.

[31] O.I. Larichev, R. Brown, E. Andre'eva, and N. Flanders. Categorical decision analysis for environmental management: A Siberian gas distributing case. In J.-P. Caverni, M. Bar-Hillel, F.H. Barron, and H. Jungermann, editors, *Contribution to Decision Making*, pages 255–286. North-Holland, Amsterdam, 1995.

[32] O.I. Larichev and H.M. Moshkovich. Limits to decision-making ability in direct multiattribute alternative evaluation. *Organizational Behavior and Human Decision Processes*, 42:217–233, 1988.

[33] O.I. Larichev and H.M. Moshkovich. ZAPROS-LM – A method and system for ordering multiattribute alternatives. *European Journal of Operational Research*, 82:503–521, 1995.

[34] O.I. Larichev and H.M. Moshkovich. *Verbal Decision Analysis for Unstructured Problems*. Kluwer Academic Publishers, Boston, 1997.

[35] O.I. Larichev, H.M. Moshkovich, Mechitov A.I., and D.L. Olson. Experiments comparing qualitative approaches to rank ordering of multiattribute alternatives. *Journal of Multi-Criteria Decision Analysis*, 2(1):5–26, 1993.

[36] O.L. Larichev, H.M. Moshkovich, and S.B. Rebrik. Systematic research into human behavior in multiattribute object classification problems. *Acta Psychologica*, 68:171–182, 1998.

[37] O.L. Larichev, D.L. Olson, Moshkovich H.M., and A.I. Mechitov. Numeric vs. cardinal measurements in multiattribute decision making: How exact is enough? *Organizational Behavior and Human Decision Processes*, 64:9–21, 1995.

[38] H. Montgomery. A study of intransitive preferences using a think aloud procedure. In H. Jungerman and G. de Zeeuw, editors, *Decision Making and Change in Human Affairs*. Reidel, Dordrecht, 1977.

[39] H.M. Moshkovich, A. Mechitov, and D.L. Olson. Ordinal judgments for comparison of multiattribute alternatives. *European Journal of Operational Research*, 137:625–641, 2002.

[40] H.M. Moshkovich, R. Schellenberger, and D.L. Olson. Data influences the result more than preferences: Some lessons from implementation of multiattribute techniques in a real decision task. *Decision Support Systems*, 22:73–84, 1998.

[41] Larichev O.I. and Moshkovich H.M. An approach to ordinal classification problems. *International Transactions of Operational Research*, 1(3):375–385, 1994.

[42] Larichev O.I., Moshkovich H.M., Furems E.M., Mechitov A.I., and Morgoev V.K. *Knowledge Acquisition for the Construction of Full and Contradiction Free Knowledge Bases*. IecProGamma, Groningen, The Netherlands, 1991.

[43] J.W. Payne. Task complexity and contingent processing in decision making: An information search and protocol analysis. *Organizational Behavior and Human Processes*, 16:366–387, 1976.

[44] B. Roy. *Multicriteria Methodology for Decision Aiding*. Kluwer Academic Publishers, Dordrecht, 1996.

[45] J.E Russo and L.D. Rosen. An eye fixation analysis of multiattribute choice. *Memory Cognition*, 3:267–276, 1975.

[46] T.L. Saaty. *The Analytic Hierarchy Process*. McGraw-Hill, New York, 1980.

[47] P.J.H Shoemaker and C.C. Waid. An experimental comparison of different approaches to determining weights in additive utility models. *Management Science*, 28(2):182–196, 1982.

[48] H. Simon. *The New Science of Manaqement Decision*. Harper and Row, New York, 1960.

[49] P. Slovic, B. Fishhoff, and S. Lichtenstein. Behavioral decision theory. *Annual Psychological Review*, 28:1–39, 1977.

[50] R. Slowinski. Rough set approach to decision analysis. *AI Expert*, 9(2):19–25, 1995.

[51] A. Tversky. Intransitivity of preferences. *Psychological Review*, 76:31–48, 1969.

[52] D. von Winterfeldt and W. Edwards. *Decision Analysis and Behavioral Research*. Cambridge University Press, Cambridge, 1986.

[53] D. von Winterfeldt and G.W. Fischer. Multiattribute utility theory: Models and assessment procedures. In D. Wendt and C. Vlek, editors, *Utility, Probability and Human Decision Making*. Reidel, Dordrecht, 1975.

[54] H. Wainer. Estimating coefficients in linear models: It don't make no nevermind. *Psychological Bulletin*, 85:213–217, 1976.

VI

MULTIOBJECTIVE MATHEMATICAL PROGRAMMING

Chapter 16

INTERACTIVE METHODS

Pekka Korhonen
Helsinki School of Economics
Department of Economics and Management Science
Runeberginkatu 14–16, 00100 Helsinki
Finland
Pekka.Korhonen@hkkk.fi

Abstract We provide an introduction to the use of interactive methods in multiple objective programming. We focus on discussing the principles to implement those methods. Our purpose is not to review existing procedures, but some examples are picked to illustrate the main ideas behind those procedures. Furthermore, we discuss two available software systems developed to implement interactive methods.

Keywords: Decision making, multiple objective, multiple criteria, interactive, behavioral.

1. Introduction

In most decision and planning situations a *Decision Maker* (DM) will have to consider multiple criteria implicitly or explicitly. For a minor problem, the DM may not even recognize the presence of multiple criteria in his/her evaluation. However, in major decision and planning situations it is important that the DM recognizes all relevant criteria and evaluate the decision (or planning) alternatives using them.

The term *Multiple Criteria Decision Making* (MCDM) refers to decision and planning problems involving multiple (generally conflicting) criteria. For an MCDM-problem it is typical that no unique solution for the problem exists. The solution is determined by the preferences of the DM. *Solving a Multiple Criteria Problem* means that a DM will choose one "*reasonable*" alternative which pleases him/her most. The word "reasonable" is defined more precisely by using the terms *efficient* or *nondominated.*

To find a solution for MCDM-problems requires the intervention of a DM. The main idea is simple: the system generates reasonable alternatives, and the DM will make choices. Those choices are used to lead the algorithm to generate more alternatives until the DM will reach the solution that pleases him/her most. Helping DMs to deal with multiple criteria decision and planning problems has been the subject to intensive studies since the beginning of the 1970's (see, e.g., [7, 15, 41]), but many theoretical concepts were defined much earlier (see, e.g., [24, 40]). In the 1970's, the research focused on the theory of multiple objective mathematical programming and the development of procedures and algorithms for solving such problems. Many ideas originated from the theory of mathematical programming. The algorithms were programmed for mainframe computers and were used mainly for illustrative purposes. The systems were often of a prototypical nature and lacked user-friendly interfaces. Examples of such early prototypical systems were [13] and [51].

During the 1980's, a clear shift towards multiple criteria decision support occurred. Accordingly, more research has focused on the user interface, on the behavioural aspects of decision-making, and on supporting the entire decision-making process from problem structuring to solution implementation. Several *Multiple Criteria Decision Making Support Systems* (MCDSS) were developed where a graphical presentation was an essential part of the system (see, e.g., [5, 6, 11, 23, 26, 28, 35, 36]). There were also published some articles where the behavioural realism of the MCDSSs was critically evaluated (see, e.g., [29, 32, 34, 48]).

In practice, MCDM-problems are not often so well-structured that they can be considered just as a choice problem. Before a decision problem is ready to be "*solved*", the following questions require a lot of preliminary work: How to structure the problem? How to find essential criteria? How to handle uncertainty? These questions are by no means outside the interest area of MCDM-researchers.

In this article we take a narrower perspective and focus on an essential supporting problem in Multiple Criteria Decision Making: How to assist a DM to find the "best" solution from among a set of available "reasonable" alternatives, when the alternatives are evaluated by using several criteria? The criteria are assumed to be given and alternatives are assumed to be defined explicitly or implicitly by means of a mathematical model.

The article consists of seven sections. In Section 2, we give a brief introduction to some basic definitions, and in Section 3, we consider the main principles to implement interactive methods. How to generate nondominated alternatives are considered in Section 4. The properties of Multiple Criteria Decision Support Systems (MCDSS) are discussed in Section 5. Section 6 considers stopping conditions, and in Section 7, we represent two examples of interactive systems: VIG and VIMDA. Concluding remarks are given in Section 8.

2. Basic Definitions and Some Theory

Multiple Objective Programming (MOP) problem in a so-called *criterion space* can be characterized as a vector maximization problem as follows:

$$\max \{\mathbf{q} = (q_1, q_2, ..., q_k)\} \quad \text{such that } \mathbf{q} \in Q, \tag{16.1}$$

where set $Q \subset \mathbb{R}^k$ is a so-called *feasible region* in a k-dimensional criterion space $\mathbb{R}^k$. The set Q is of special interest. Most considerations in multiple objective programming are made in a criterion space.

Set Q may be convex or nonconvex, bounded or unbounded, precisely known or unknown, consist of a finite or infinite number of alternatives, etc. When Q consists of a finite number of elements which are explicitly known in the beginning of the solution process, we have the class of problems which may be called *(Multiple Criteria) Evaluation Problems.* Sometimes those problems are referred to as *Discrete Multiple Criteria Problems* or *Selection Problems* (for a survey, see, e.g., [39]).

When the number of alternatives in Q is nondenumerable, the alternatives are usually defined using a mathematical model formulation, and the problem is called continuous. In this case we say that the alternatives are only implicitly known. This kind of problem is referred as a *Multiple Criteria Design Problem* or a *Continuous Multiple Criteria Problem*(for a survey, see, e.g., [46, 38]).

In this case, the set Q is specified by means of decision variables as usually done in single optimization problems:

$$\max\{\mathbf{q} = \mathbf{f}(\mathbf{x})(f_1(\mathbf{x}), ..., f_k(\mathbf{x}))\} \qquad (16.2)$$
$$\text{s.t. } \mathbf{x} \in X,$$

where $X \subset \mathbb{R}^n$ is a *feasible set* and $\mathbf{f} : \mathbb{R}^n \to \mathbb{R}^k$. The space $\mathbb{R}^n$ is called *variable space*. The functions f_i, $i = 1, 2, \ldots, k$ are *objective functions*. The feasible region Q can now be written as $Q = \{\mathbf{q} \mid \mathbf{q} = \mathbf{f}(\mathbf{x}),\ \mathbf{x} \in X\}$.

Conceptually the multiple objective mathematical programming problem may be regarded as a *value (or utility) function* maximization program:

$$\max\{v(\mathbf{q})\} \qquad (16.3)$$
$$\text{s.t. } \mathbf{q} \in Q,$$

where v is a strictly increasing in each argument and real-valued, and defined at least in the feasible region Q. It is mapping the feasible region into a one-dimensional *value space*. Function v specifies the DM's preference structure over the feasible region. However, the key assumption in multiple objective programming is that v is **unknown**. Even if the value function v is not explicitly known, the solution methods are heavily dependent on the additional assumptions we make about its form and stability. Those assumptions vary from very strict assumptions to "no assumptions". We may also assume that v is existing, but not stable during the solution process.

Regardless of the assumptions concerning v, possible solutions to our problem are the alternatives that can be the solutions of (16.1) for some value function $v : Q \to \mathbb{R}$. Those solutions are called *efficient* or *nondominated* depending on the space where the alternatives are considered. The term nondominated is used in criterion space and efficient in variable space. (Some researchers use the term efficient to refer to efficient and nondominated solutions without making any difference.). Any choice from among the set of nondominated (efficient) solutions is possible, unless we have no additional information about the DM's preference structure.

Nondominated solutions are defined as follows:

DEFINITION 61 *In (16.1),* $\mathbf{q}^* \in Q$ *is* nondominated *iff there does not exist another* $\mathbf{q} \in Q$ *such that* $\mathbf{q} \geq \mathbf{q}^*$ *and* $\mathbf{q} \neq \mathbf{q}^*$ *(i.e.* $q_i \geq q_i^*$ *for all* $i = 1, 2, \ldots, k$ *and* $q_i \neq q_i^*$ *for some* i*).*

DEFINITION 62 *In (16.1),* $\mathbf{q}^* \in Q$ *is* weakly nondominated iff there does not exist another $\mathbf{q} \in Q$ such that $\mathbf{q} > \mathbf{q}^*$ (i.e. $q_i > q_i^*$, $i = 1, 2, \ldots, k$).

Correspondingly, efficient solutions are defined as follows:

DEFINITION 63 *In (16.2),* $\mathbf{x}^* \in X$ *is* efficient *iff there does not exist another* $\mathbf{x} \in X$ *such that* $\mathbf{q} = \mathbf{f}(\mathbf{x}) \geq \mathbf{q}^* = \mathbf{f}(\mathbf{x}^*)$ *and* $\mathbf{q} = \mathbf{f}(\mathbf{x}) \neq \mathbf{q}^* = \mathbf{f}(\mathbf{x}^*)$.

DEFINITION 64 *In (16.2),* $\mathbf{x}^* \in X$ *is* weakly efficient *iff there does not exist another* $\mathbf{x} \in X$ *such that* $\mathbf{q} = \mathbf{f}(\mathbf{x}) > \mathbf{q}^*\mathbf{f}(\mathbf{x}^*)$.

The set of all nondominated (efficient) solutions is called the nondominated (efficient) frontier. The final ("best") solution $\mathbf{q} \in Q$ of the problem (16.1) is called the *Most Preferred Solution* (MPS). It is the solution preferred by the DM to all other solutions. At the conceptual level, we may think that it is the solution maximizing an (unknown) value function in problem (16.3). How to find the maximum of a function we do not know is a key problem in multiple objective programming.

In the following, we use the term MCDM to refer to a multiple criteria problem generally without emphasizing a mathematical model and the term MOP to emphasize the mathematical formulation of the problem. In case the mathematical model is linear, we use the term multiple objective linear programming (MOLP).

3. Principles for Implementing Interactive Methods

As we defined in the previous section, in MCDM we assume that the DM tries to find a solution preferred to all other solutions. We do not need to consider dominated solutions, because for each dominated solution there exists at least one (nondominated) solution which is better at least on one criterion and worse on no criterion. Although we are able to reduce the number of alternatives in this way, it is not realistic to assume that the DM is able to compare all nondominated solutions simultaneously, and name the best one. That's why the above characterization of the MPS is not very operational.

Another way is to approach the problem through the value function. We can distinguish the following principles, which most MCDM approaches use implicitly or explicitly in solving MCDM problems. The following considerations are based on Korhonen et al. [30] where you can find more details:

1 Assume the existence of a value function v, and assess it explicitly.

2 Assume the existence of a stable value function v, but do not attempt to assess it explicitly. Make assumptions of its general functional.

3 Assume the existence of a value function v, but do not assume it stable. Let it change with time, and assume that the DM's final choice is based on its specific form.

4 Do not assume the existence of a value function v.

Those principles lead to the following general approaches to solving MCDM problems.

Approach 1: Prior Articulation of Preferences The value function v is explicitly constructed by means of preference information received from a DM. The multiple attribute utility theory (MAUT) ([22] and 7 in this book) provides a classical example of this approach. Using Keeney-Raiffa type of interaction, the following steps can be identified:

a) assumptions are made about the underlying value function, in particular, its functional form;

b) the parameters of the value function are assessed using elaborate interview techniques;

c) the internal consistency of the DM's responses is controlled;

d) the value function is used to determine a value score for each alternative. Those scores are used to determine the MPS or just to rank alternatives.

In other words, once an explicit value function has been assessed, the function determines a "value" or a "score" for each alternative making it possible to rank (a finite number of) the decision alternatives or to find the alternative having an optimal value. Currently, the popular AHP (the Analytic Hierarchy Process) developed by Saaty [42] is also based on those principles. In the AHP, the value function is assumed to be linear. MACBETH by Bana e Costa and Vansnick, [3] and Chapter 10 in this book, is another example of the approach, in which a cardinal value function is constructed in an interactive manner. The construction of the value function is based on preference difference measurement.

The interaction between the system and the DM is needed in step b. If inconsistency is revealed in step c, in the MAUT, the DM is asked to revise his/her preference information; in the AHP (s)he is informed about inconsistency, but not required to remove it. One of the basic assumptions in the AHP is that a DM cannot be always fully consistent.

Actually, this approach is widely used in practical decision making. A weighted average (sum) may be the most common way to aggregate several criteria. The value function is thus implicitly assumed to be linear. However, the users are seldom aware of many implicit assumptions they have made concerning the meaning of weights, the dependence of criteria, the functional form of the value function, etc.

A classical example of the use of weights to aggregate preferences is goal programming (see, e.g., [16, 17]). Since Charnes and Cooper [9, 10] developed goal programming, it has been a very popular method to deal with multiple criteria. In Archimedean goal programming, the weighted-sums of the deviational variables are used to find the solutions that best satisfy the goals. The deviational variables measure overachievement and underachievement from the target or threshold levels.

For other quite recent methods based on the use of the weights, see, e.g., [1, 8].

Approach 2: Interactive Articulation of Preferences

i) Based on an Implicit Value Function. The value function is neither assumed to be known nor tried to be estimated explicitly. DM's responses to specific questions are used to guide the solution process towards an "optimal" or "most preferred" solution. The DM is assumed to behave according to some specific underlying value function which is known only of its functional form. Classical examples are the methods developed by Geoffrion et al. [15], Zionts and Wallenius [51], and Steuer [45].

Geoffrion et al. presented their approach in a general framework, where objective functions were assumed to be differentiable and concave, and the feasible set X is compact and convex. The value function was assumed concave increasing. The idea was adopted from a well-known Frank–Wolfe algorithm. In the GDF-method [15], the direction of improvement was chosen by estimating the gradient of a concave value function on the basis of marginal rates of substitution, which a DM evaluated. Zionts and Wallenius [51] developed a method for MOLP-problems. In their method, the value function was also assumed to be linear. (In 1983, the authors extended the method to deal with concave value functions [52].) The weights of the value function were determined on the basis of the preference evaluation of the DM. The DM compared two alternatives at a time, and expressed his/her preference over them. Based on the idea that a value function has a higher value at a more preferred alternative, the method generated a sequence of inequalities, which finally specified the most preferred (extreme point) solution – in theory. In practice, the responses of the DM often lead to conflicting information. As a solution to this problem, the authors proposed the ignoring of the oldest responses. It is quite plausible to assume that the DM may learn during the search process, and will be more competent to give more precise information about his/her "true" preferences.

The Interval Criterion Weights/Vector-Maximum Approach by Steuer [45] is based on the idea to restrict the region where the optimal solution of the value function may lie. The DM's preferences are used to reduce the possible weights of the objective functions. Steuer uses the term "criterion cone reduction method" to describe the class of the methods his method belongs to.

The following steps typically characterize this approach:

1 assumptions are made of the functional form of the underlying value function;

2 an initial solution is provided for evaluation;

3 the DM is asked to give preference information which is used to update knowledge about the underlying value function;

4 an improved solution or improved solutions are generated;

5 iteration is stopped when an "optimal" solution has been identified.

ii) Based on No Stable Value Function. These approaches are typically based on the idea to generate nondominated solutions for the DM's evaluation without making any specific assumptions concerning the value function. The DM is free to make a search on the efficient frontier and stop at any time (s)he likes. For instance, change of mind and learning is allowed, but no explicit assumptions are made about those behavioral aspects. A quite common approach is to let the DM freely express aspiration levels for the objectives, and to let the system to show feasible solutions to him/her. The previous responses are used to "guess" more preferred solutions. This projection is usually accomplished via minimizing so called achievement scalarizing functions [49, 47].

Typically the following steps are included:

- present the DM with an efficient solution and provide him/her with as much information as possible about the nondominated region, in particular in the "neighborhood" of the current solution;
- ask the DM to provide preference information in the form of aspiration levels, weights, etc.;
- use the responses to generate a single nondominated solution or a set of nondominated solutions for the DM's evaluation;
- iterate until the DM stops, or some specified termination criteria for the search have been satisfied.

A typical example of the method, where no assumptions about the value function are made until the DM likes to stop, is the method by Korhonen and Laakso [28]. The DM is free to make a search on the efficient frontier, but at the moment (s)he likes to stop, the DM is helped to evaluate whether (s)he has found the most preferred solution or not. At this moment, specific assumptions about the functional form of the value function are needed.

Another example of the method which helps the DM to search the efficient frontier is Light Beam Search (LBS) approach by Jaszkiewicz and Slowinski [18]. LBS enables the DM to analyze multiple objective decision problems by presenting samples of non-dominated points. The DM can control the search by either modifying the aspiration and reservation points, or by shifting the current point to a selected better point from its neighborhood. Michalowski and Szapiro [37] have also developed an interactive method which is based on the use of the aspiration and reference points.

Approach 3: Posterior Articulation of Preferences This approach tries to find a good approximation to a nondominated frontier. The choice problem does not play a significant role. The main idea is to provide information to the DM about possible solutions. The presentation and visualization of the frontier is emphasized. A classical example is the ADBASE system by Steuer [46]. ADBASE finds all nondominated extreme point solutions for an MOLP-problem. The original idea was to approximate a nondominated frontier by means of nondominated extreme point solutions. Unfortunately, the number of nondominated extreme points can become large even in problems of reasonable size. Nowadays, the approach has become popular in the problems where the functional forms of the objective functions are too complex for traditional optimization methods. Genetic algortithms are widely used for those problems, see, e.g., [12]. These approaches seem to work fine in case of two objectives.

4. Generating Nondominated Solutions

Despite many variations among different methods of generating nondominated solutions, the ultimate principle is the same in all methods: a single objective optimization problem is solved to generate a new solution or solutions. The objective function of this single objective problem may be called a *scalarizing function* according to Wierzbicki [49]. It has typically the original objectives and a set of parameters as its arguments. The form of the scalarizing function as well as what parameters are used depends on the assumptions made concerning the DM's preference structure and behavior.

Two classes of parameters are widely used in multiple objective optimization: 1) *weighting coefficients for objective functions* and 2) *reference/aspiration/reservation levels for objective function values*. Based on those parameters, there exist several ways to specify a scalarizing function. An important requirement is that this function completely characterizes the set of nondominated solutions: "*for each parameter value, all solution vectors are nondominated, and for each nondominated criterion vector, there is at least one parameter value, which produces that specific criterion vector as a solution*" (see, for theoretical considerations, e.g., [50]).

4.1 A Linear Scalarizing Function

A classic method to generate nondominated solutions is to use the weighted sum of objective functions, i.e. to use the following linear scalarizing function:

$$\max\{\lambda'\mathbf{q} \mid \mathbf{q} \in Q\}. \qquad (16.4)$$

If $\lambda > \mathbf{0}$, then the solution $\mathbf{q}$ of (16.4) is nondominated, but if we allow that $\lambda \geq \mathbf{0}$, then the solution is weakly nondominated (see, e.g., [46, pp. 215 and 221]). Using the parameter set $\Lambda = \{\lambda \mid \lambda > \mathbf{0}\}$ in the weighted-sums linear program we can completely characterize the efficient set provided the constraint set is convex. However, Λ is an open set, which causes difficulties in a mathematical optimization problem. If we use $\mathrm{cl}(\Lambda) = \{\lambda \mid \lambda \geq \mathbf{0}\}$ instead, the nondominance of $\mathbf{q}$ cannot be guaranteed. It is surely weakly-efficient, and not necessarily efficient. When the weighted-sums are used to specify a scalarizing function in MOLP problems, the optimal solution corresponding to nonextreme points of Q is never unique. The set of optimal solutions always consists of at least one extreme point, or the solution is unbounded. In early methods, a common feature was to operate with weight vectors $\lambda \in \mathbb{R}^k$, limiting considerations to nondominated extreme points (see, e.g., [51]).

4.2 A Chebyshev-type Scalarizing Function

Currently, most solution methods are based on the use of a so-called Chebyshev-type scalarizing function first proposed by Wierzbicki [49]. We will refer to this function by the term *achievement (scalarizing) function.* The achievement (scalarizing) function projects any given (feasible or infeasible) point $\mathbf{g} \in \mathbb{R}^k$ onto the set of nondominated solutions. Point $\mathbf{g}$ is called a *reference point*, and its components represent the desired values of the objective functions. These values are called *aspiration levels.*

The simplest form of achievement function is:

$$s(\mathbf{g}, \mathbf{q}, \mathbf{w}) = \max\left\{\frac{g_k - q_k}{w_k} \mid k \in K = \{1, 2, \ldots, k\}\right\}, \tag{16.5}$$

where $\mathbf{w} > \mathbf{0} \in \mathbb{R}^k$ is a (given) vector of weights, $\mathbf{g} \in \mathbb{R}^k$, and $\mathbf{q} \in Q = \{\mathbf{f}(\mathbf{x}) \mid \mathbf{x} \in X\}$. By minimizing $s(\mathbf{g}, \mathbf{q}, \mathbf{w})$ subject to $\mathbf{q} \in Q$, we find a weakly nondominated solution vector $\mathbf{q}^*$ (see, e.g., [49, 50]). However, if the solution is unique for the problem, then $\mathbf{q}^*$ is nondominated. If $\mathbf{g} \in \mathbb{R}^k$ is feasible, then $\mathbf{q}^* \in Q$, $\mathbf{q}^* \geq \mathbf{g}$. To guarantee that only nondominated (instead of weakly nondominated) solutions will be generated, more complicated forms for the achievement function have to be used, for example,

$$s(\mathbf{g}, \mathbf{q}, \mathbf{w}, \rho) = \max_{k \in K}\left\{\frac{g_k - q_k}{w_k} + \rho \sum_{i=1}^{k}(g_i - q_i)\right\}, \tag{16.6}$$

where $\rho > 0$. In practice, we cannot operate with a definition "any positive value". We have to use a pre-specified value for ρ. Another way is to use a lexicographic formulation (see, e.g., [46, pp. 292–296]).

The applying of the scalarizing function (16.6) is easy, because given $\mathbf{g} \in \mathbb{R}^k$, the minimum of $s(\mathbf{g}, \mathbf{v}, \mathbf{w}, \rho)$ is found by solving the following problem:

$$\min\left\{\varepsilon + \rho\sum_{i=1}^{k}(g_i - q_i)\right\} \tag{16.7}$$

$$\text{s.t.} \begin{cases} \mathbf{q} \in Q, \\ \varepsilon \geq (g_i - q_i)/w_i, \quad i = 1, 2, ..., k. \end{cases}$$

The problem (16.7) can be further written as:

$$\min\left\{\varepsilon + \rho\sum_{i=1}^{k}(g_i - q_i)\right\} \tag{16.8}$$

$$\text{s.t.} \begin{cases} \mathbf{q} \in Q, \\ \mathbf{q} + \varepsilon\mathbf{w} - \mathbf{z} = \mathbf{g}, \\ \mathbf{z} \geq \mathbf{0}. \end{cases}$$

To illustrate the use of the achievement scalarizing function, consider a two-criteria problem with a feaxsible region having four extreme points $\{(0,0), (0, 3), (2,3), (8,0)\}$, as shown in Figure 16.1. In case, Q is a polyhedron, the model (16.8) is an LP-model.

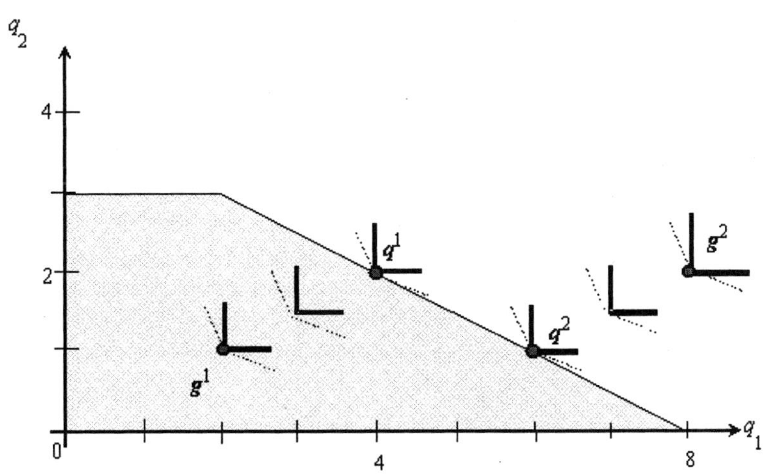

Figure 16.1. Illustrating the projection of a feasible and an infeasible aspiration level point onto the nondominated surface.

In Figure 16.1, the thick solid lines describe the indifference curves when $\rho = 0$ in the achievement scalarizing function. The thin dotted lines stand for the case $\rho > 0$. Note that the line from (2,3) to (8,0) is nondominated and the line from (0,3) to (2,3) is weakly-nondominated, but dominated. Let us assume that the DM first specifies a feasible aspiration level point $\mathbf{g}^1 = (2,1)$. Using a weight vector $\mathbf{w} = [2,1]^T$, the minimum value of the achievement scalarizing function (-1) is reached at point $\mathbf{q}^1(4,2)$ (cf. Figure 16.1). Correspondingly, if an aspiration level point is infeasible, say $\mathbf{g}^2 = (8,2)$, then the minimum of the achievement scalarizing function (+1) is reached at point $\mathbf{q}^2 = (6,1)$. When a feasible point dominates an aspiration level point, then the value of the achievement scalarizing function is always negative; otherwise it is nonnegative. It is zero, if an aspiration level point is weakly nondominated.

As Figure 16.1 illustrates, by varying aspiration levels different nondominated solutions are generated. Any nondominated point is a possible (unique) solution. Instead, a linear scalarizing function has not such property. Depending on the (strictly positive) weights used, the unique solutions are either point (2,3) or point (8,0). In case the ratio of the components of the weight vector is $\lambda_2/\lambda_1 = 2$, all nondominated solutions have the same value.

5. Solving Multiple Objective Problems

The MCDM always assumes the intervention of a DM at some stage in the solution process. The preference information can be gathered in advance, during the search process, or afterwards. In that sense, all MCDM-methods are interactive. However, generally the term "*interactive*" is used to refer to the support systems, where the dialogue step with the DM and the computation step are iterated until the final solution is reached. We will use this narrower interpretation in discussion on interactive systems. The following steps typically appear in any interactive system:

Step 1: Initial Solution(s)

Step 2: Evaluation

Step 3: Solution(s) Generation

Step 4: Termination?

One or several solutions are generated in Step 1 and displayed to the DM. The DM considers the solution(s) and provides preference information in Step 2. Based on that information, new solutions are generated in Step 3, and they are evaluated in Step 2. Steps 2 and 3 are repeated until the DM is willing to consider termination. The DM may simply stop the search or use termination conditions to help him/her to make a final decision.

The term "*support*" is also used in connection with MCDM in a broad perspective to refer to all research associated with the relationship between the problem and the decision-maker. For instance, the following questions may require support: How to structure the problem? How to find essential criteria? How to handle uncertainty? However, in this paper we have an essential supporting problem in Multiple Criteria Decision Making: How to assist a DM to find the "best" solution from among a set of available alternatives?

5.1 Properties of a Multiple Criteria Decision Support System

There is no single criterion for evaluating multiple criteria decision support systems. Several relevant criteria can be introduced:

1 the system recognizes and generates nondominated solutions;

2 the system helps the DM feel convinced that the final solution is the most preferred one, or at least close enough to that one;

3 the system helps the DM to have a "holistic" view over the nondominated frontier;

4 the system does not require too much time from the DM to find the final solution;

5 the communication between the DM and a system is not too complicated;

6 the system provides reliable information about alternatives available;

7 the system provides a possibility to evaluate optimality conditions.

Provided that the problem is correctly specified, the final solution of a rational DM is always nondominated. Therefore it is important that the system is able to recognize and generate nondominated solutions. The system can operate with dominated solutions during the search process for behavioral reasons, but it has to lead the DM finally to a nondominated solution. The Geoffrion–Dyer–Feinberg method [15] is a typical method, where the DM is making a search on the line passing through the feasible region, but finally (s)he has the possibility to reach the efficient frontier. Another example is the method by Arbel and Korhonen [2], in which the search is started from the nadir criterion values (worst criterion values over the nondominated set). To start from the worst solution enables the DM to proceed with a win-win strategy until the nondominated frontier is achieved. (S)he may change the search direction, but does not need to worsen any criterion value for gaining more on some other criterion value.

No system can provide a DM with a capability to compare all alternatives simultaneously. However, a good system can provide a holistic view over the alternatives and assists the DM to feel convinced that his/her final choice is best or at least close to the best solution. The user interface plays an important role in that aspect.

A good system does not waste the time of the DM, and the communication language is easy. Irrelevant questions are very boring to the DM. It is good to increase the intelligence of the system, but it is important to remember that the DM wants to keep the control of the system in his/her own hands. There are several ways to implement "discussion" between a system and the DM. For instance, Shin and Ravindran [44] list the following eight typical interaction styles:

- Binary pairwise comparisons – the DM must compare two-dimensional vectors at each iteration (see, e.g., [21]);
- Pairwise comparisons – the DM must compare a pair of p-dimensional vectors and specify a preference (see, e.g., [51]);
- Vector comparisons – the DM must compare a set of p-dimensional vectors and specify the best, the worst or the order of preference (see, e.g., [45]);
- Precise local tradeoff ratios – the DM must specify precise values of local tradeoff ratios at a given point (see, e.g., [15]).
- Interval local tradeoff ratios – the DM must specify an interval for each local tradeoff ratio (see, e.g., [43]).
- Comparative tradeoff ratios – the DM must specify his/her preference for a given tradeoff ratio (see, e.g., [20]).
- Index specification and value tradeoff – the DM must list the indices of objectives to be improved or sacrificed, and specify the amount [7];
- Aspiration levels (reference points) – the DM must specify or adjust the values of the objectives which indicate his/her optimistic wish concerning the outcomes of the objectives (see, e.g., [49]).

There are several ways to implement those principles. The comparison information can be given to the DM in the numeric or visual form. "One picture speaks more than one thousand words" is very true, when the communication aspects are considered. Graphics can often be used to illustrate the effects of the DM's choices much more effective than using numbers.

A very important aspect is that the system provides reliable information about the alternatives available. For instance, if the system always produces the same nondominated solution for the DM's evaluation, (s)he is misled to believe that the solution is the only possible choice.

The optimality checking can also be implemented in interactive systems, but in that case, the DM has to be willing to accept some assumptions concerning the functional form of the value function. Those optimality conditions are considered in more details in the next section.

5.2 The Role of Interface

Mathematical programming models are developed by mathematicians who are mainly interested in mathematical aspects – not necessarily practical or behavioral aspects. Consequently, many systems make irrelevant assumptions about the behavior of a DM. A typical example is how "importance" is considered in models. When the DM says that one criterion is more important than another one, a standard interpretation is that the more important criterion has to have a higher weight than the less important criterion. However, in most cases it is not a correct interpretation. The DM may be interested to reach a specific (reservation) level for a criterion value. That's why (s)he may experience that the criterion is important. Obviously, in some problems importance means the amount of attention the DM is going to pay to the criterion value at a certain moment. For instance, in a house buying problem, the DM may say that price is not important to him or her. The reason may simply be that the price of all the houses under consideration does not vary very much. That's why price does not require his/her primary attention. To discuss "importance" was an example demonstrating how easily behavioral features may cause insidious pitfalls for the decision systems.

Because we do not know very well how the brain of a DM is working, it is important that we try to avoid unrealistic assumptions in our systems. From a behavioral point of view, a critical point is, when the system uses and interprets preference information received from the DM. The question the system asks the DM may be very clear and simple like "Do you prefer this to that?", but the problem arises, when the system interprets this information. The purpose of the system is to produce alternatives which are probably closer to the most preferred solution than the previous ones. If we do not make any assumptions concerning the value function, there are no guidelines to generate a better solution. If the linear value function is a realistic assumption, then the problem is easy provided that DM is able to give consistent preference information. Even more general assumptions help the system to generate more preferable solutions. But whether those assumptions meet the requirements of behavioral realism is a difficult question with no definite answer.

A way to solve the above problems is to use a free search type of approach like VIG [25], the DM may move on the nondominated frontier, until (s)he is satisfied. In the approach, the search process is always in his/her own hands. We can avoid misinterpretation, but the approach is not without problems. The premature stopping may happen, because people experience sacrifices in criterion values more strongly than gains. This kind of behavior can be explained by prospect theory [19].

Currently, a very common way is to use graphics to illustrate nondominated solutions. In case of two criteria, the approach is very powerful, because all solutions (nondominated frontier) can be presented in one picture. Into some extent, the idea is also working in three dimensions, but the picture is not necessarily so illustrative than in two dimensions. In more than three dimensions, it is not possible generally to visualize the whole nondominated set. However, to illustrate single solution vectors graphically is helpful. A brief look at using various graphical techniques in the context of multiple criteria decision making is given in [38, pp. 239–249].

6. Final Solution

The speed of (infinite) convergence and a stopping rule are important in mathematical programming algorithms. However, it is not the infinite convergence that matters in interactive procedures, because all interactive algorithms converge in a finite number of steps – actually in few steps. Thus, the convergence in interactive methods has to be considered from the behavioral point of view. We may speak about *behavioral convergence*. In mathematical convergence, the main interest is in the general convergence. In the behavioral convergence, for instance, the initial rate of convergence is more interesting as pointed out already by Geoffrion, Dyer, and Feinberg [15]. However, the initial rate of convergence is not the only aspect we have to take into account in the behavioral convergence. Human beings are making errors in evaluations, not being able to give precise information, etc. In addition, the function we optimize is not usually known, and it may even change during the search process. Instead of studying the convergence process of an interactive method, it is often more useful to study the termination situations.

To test the "optimality" of the final solution in interactive methods, we may adopt ideas from mathematical programming with some exceptions. In mathematical programming, generally used optimality testing is based on the Karush–Kuhn–Tucker conditions (see, e.g., [4]). Those conditions are based on the use of the gradients of the objective functions and those of the functions that define the constraints. The gradients of the active constraints define a cone. Loosely defining the main idea of the Karush–Kuhn–Tucker conditions, we may say that when the gradient vector of the objective function is in the before mentioned cone, the optimality conditions are satisfied.

However, this kind of the optimality conditions cannot be used in interactive methods. It is not realistic to assume that the DM could be able to compare the gradient vectors of implicitly defined value functions. Instead, the DM is able to compare feasible directions. A typical idea in interactive methods is based on that idea: if there exists no direction of improvement for the DM, the solution is considered the most preferred one. If no assumptions are made about the value function, the idea is purely heuristic and based on the philosophy: "Because no better solution can be found, let's stop!" Provided that there is no feasible direction of improvement, the final solution is at least locally most preferred. A critical point is whether the DM has really been able to "see" all feasible directions. If we make assumptions about the functional form of the value function, we may reduce the number of the directions we have to consider to be sure that no direction of improvement exists. Zionts and Wallenius [51] assumed that the set Q is a polyhedron and the value function is linear. Based on this assumption it is sufficient to study only all adjacent efficient tradeoffs to prove the optimality of the current extreme point solution. Korhonen and Laakso [28] introduced general optimality conditions for a pseudoncave value function. In case the set Q is a polyhedron, generally it is needed only k (= the number of objectives) directions to check the optimality of the current solution.

7. Examples of Software Systems: VIG and VIMDA

7.1 VIG

Today, many interactive systems use an aspiration level projection principle in generating nondominated solutions for the DM's evaluation. The projection is performed using Chebyshev-type achievement scalarizing functions as explained above. In Section 4.2, we described how these functions can be controlled by varying the aspiration levels $\mathbf{g} \in \mathbb{R}^k$. Those functions may also be controlled by varying the weight vector $\mathbf{w} \in \mathbb{R}^k$ (keeping aspiration levels fixed). Instead of using aspiration levels, some algorithms ask the DM to specify the reservation levels for the criteria (see, e.g., [37]).

An achievement scalarizing function projects one aspiration (reservation) level point at a time onto the nondominated frontier. By parametrizing the function, it is possible to project a whole direction onto the nondominated frontier as originally proposed by Korhonen and Laakso [28]. The vector to be projected is called a *Reference Direction Vector* and the method *Reference Direction Method*, correspondingly. When a direction is projected onto the nondominated frontier, a curve traversing across the nondominated frontier is obtained. Then an interactive line search is performed along this curve. The idea enables the DM to make a continuous search on the nondominated frontier. The corresponding mathematical model is a simple modification from the original model (16.8) developed for projecting a single point:

$$\min\left\{\varepsilon+\rho\sum_{i=1}^{k}(g_i-q_i)\right\} \tag{16.9}$$

$$\text{s.t.}\begin{cases}\mathbf{q}\in Q,\\ \mathbf{q}+\varepsilon\mathbf{w}-\mathbf{z}=\mathbf{g}+t\mathbf{r},\\ \mathbf{z}\geq\mathbf{0},\end{cases}$$

where $t : 0 \to \infty$ and $\mathbf{r} \in \mathbb{R}^k$ is a reference direction. In the original approach, a reference direction was specified as a vector starting from the current solution and passing through the aspiration levels. The DM was asked to give aspiration levels for criteria. In this way, the DM can specify where (s)he would like to go, and the system lets him/her start to move in the desired direction. At any time, (s)he is free to change the search direction by specifying a new aspiration level vector. The interpretation of the DM's desire is required in the projection of the aspiration level vector, i.e. how the weight vector $\mathbf{w}$ is chosen. However, a slight misinterpretation is not drastic, because the DM can immediately re-specify new aspiration levels for the criteria, if (s)he does not like the search direction. The idea of the reference direction approach can be applied in any multiple objective programming problem – in principle, but in practice only in the multiple objective linear programming problems, where the objective functions are linear and the constraints set is a polyhedron. For multiple objective linear problems, to generate a nondominated curve is an easy task by using parametric programming.

The original method has been further developed in many directions. First, Korhonen and Wallenius [31] improved upon the original procedure by making the specification of a reference direction dynamic. The dynamic version was called *Pareto Race*. In Pareto Race, the DM can freely move in any direction on the nondominated frontier (s)he likes, and no restrictive assumptions concerning the DM's behavior are made. Furthermore, the objectives and constraints are presented in a uniform manner. Thus, their role can also be changed during the search process. The whole software package consisting of Pareto Race is called VIG.

In Pareto Race, a reference direction $\mathbf{r}$ is determined by the system on the basis of preference information received from the DM. By pressing number keys corresponding to the ordinal numbers of the objectives, the DM expresses which objectives (s)he would like to improve and how strongly. In this way (s)he implicitly specifies a reference direction. Figure 16.2 shows the Pareto Race interface for the search, embedded in the VIG software [25].

In Pareto Race, the user sees the objective function values on a display in numeric form and as bar graphs, as he/she travels along the nondominated frontier. The keyboard controls include an accelerator, gears, brakes, and a steering mechanism. The search on the nondominated frontier is like driving a car. The DM can, e.g., increase/decrease the speed, make a turn and brake at any moment he/she likes.

To implement those features, Pareto Race uses certain control mechanisms, which are controlled by the following keys:

(SPACE) BAR: An "Accelerator"
Proceed in the current direction at constant speed.

F1: "Gears (Backward)"
Increase speed in the backward direction.

F2: "Gears (Forward)"
Increase speed in the forward direction.

F3: "Fix"
Use the current value of objective i as the worst acceptable value.

F4: "Relax"
Relax the "bound" determined with key F3.

F5: "Brakes"
Reduce speed.

F10: "Exit"

num: "Turn"
Change the direction of motion by increasing the component of the reference direction corresponding to the goal's ordinal number $i \in [1, k]$ pressed by DM.

An example of the Pareto Race screen is given in Figure 16.2. The screen is associated with the numerical example described in the next section.

Pareto Race does not specify restrictive behavioral assumptions for a DM. (S)he is free to make a search on the nondominated surface, until he/she believes that the solution found is his/her most preferred one.

Pareto Race is only suitable for solving moderate size problems consisting of less than 10 objectives, and few hundreds of rows and variables. When the size of the problem becomes large, computing time makes the interactive mode inconvenient. To solve large-scale problems, Korhonen, Wallenius and Zionts [33] proposed a method based on Pareto Race. An interactive local free search is first performed to find the most preferred direction. Based on the direction, a nondominated curve can be generated in a batch mode if desired.

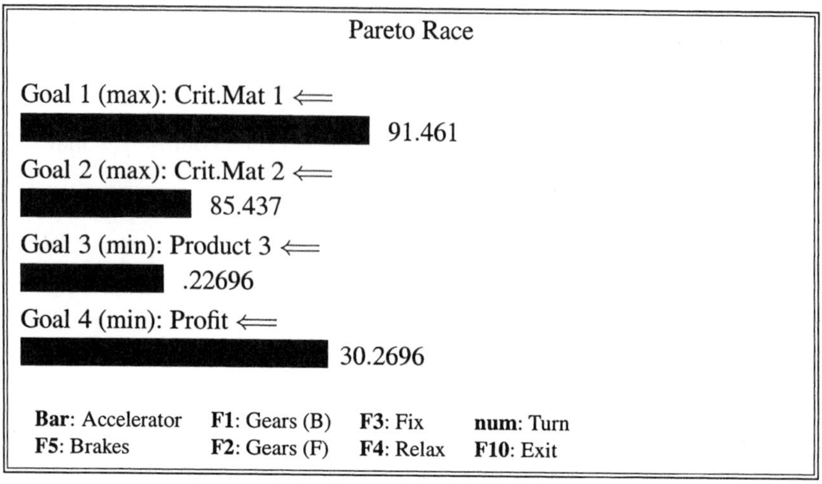

Figure 16.2. An example of the Pareto Race screen.

7.2 VIMDA

When the MCDM-problem is an (Multiple Criteria) Evaluation Problem, i.e. Q consists of a finite number of elements which are explicitly known in the beginning of the solution process, the Pareto Race type of approach is not a feasible method, because the nondominated frontier is not "smooth". However, the reference direction method is a valid approach for generating nondominated alternatives for the DM's evaluation. Instead of an efficient curve, in the evaluation problem the path is composed of a set of nondominated points which are displayed to the DM. Those points can be shown to the DM for instance by using the visual representation like that in Figure 16.3.

The current alternative is shown on the left hand margin. Each line is standing for one criterion. The vertical position of the line gives information about the value of the criterion. The ideal value of each criterion is the highest position on the screen and the worst value is the lowest position, respectively. The alternatives are evenly distributed horizontally. The vertical lines point to the positions of alternatives. The thick vertical line refers to the alternative for which the numerical values are shown on the top of the screen. The DM can control the position of the thick line by using the cursor "left" and "right". The consecutive alternatives have been connected with lines for improving illustration. By using lines, it is easier to the DM to observe how the criterion values change when moved from alternative to next.

The DM is asked to choose the most preferred alternative from the screen by moving the cursor to point to such an alternative. The chosen (most preferred)

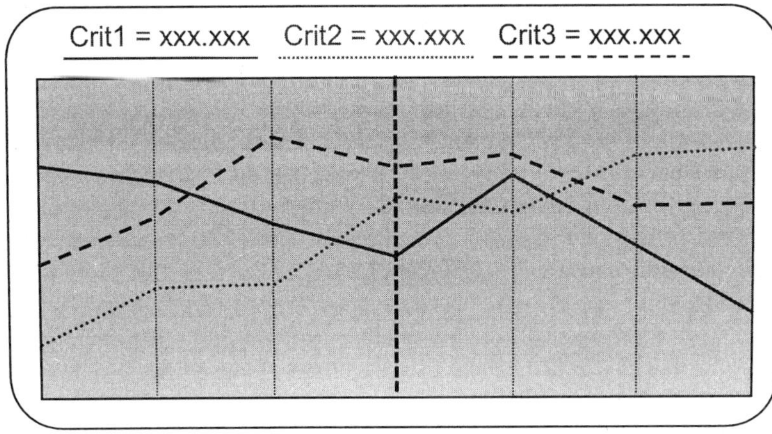

Figure 16.3. The screen of the computer-graphics interface in VIMDA.

alternative becomes the current solution for the next iteration. Next, the user is asked to reconsider his/her aspiration levels, etc.

The process stops when the DM is satisfied with the solution. Actually, mathematics behind the approach is quite complex, because to project a direction to the set of finite points, requires to solve a non-convex parametrization problem (see, e.g., [27]). However, mathematics can be fully hidden from the DM without losing anything in understandability of the method. The alternatives displayed on the screen reflect his/her desire to change the values of the criterion. The nondominance concept is also easy to explain to him/her.

8. Concluding Remarks

An interactive approach is a very natural way to solve multiple criteria decision making problems, because a DM is always an essential part of the solution process. The system generates "reasonable" (nondominated) solutions, and the DM provides preference information to the system and make choices. The DM's responses are used to generate new potential most preferred alternatives.

In interactive methods, there are few critical points which require careful consideration:

- How preference information is gathered;
- How information is used to generate new alternatives for the DM's evaluation; and
- How the system generates alternatives.

If the information gathering process is too complicated, the DM is unable to give reliable information and the system all the time "interprets" his/her responses erroneously. It is important that the way how the system deals with the DM's responses is based on behavioral realism. Unrealistic interpretation of preference information produces to the DM for evaluation the alternatives which are not consistent with his/her wishes. On the other hand, the system cannot restrict too much the choices.

It is also important that the system enables the DM to search the whole nondominated frontier, not only a part of that.

There a lot of interactive methods developed for MOP problems. To choose the best interactive method is a multiple criteria decision problem. Gardiner and Steuer [14] have tried to help the users in their choice problems by proposing a unified interactive multiple-objective programming framework by putting together the best ideas developed in the MCDM-community.

Obviously, in the future we have to focus on the issues which improve our understanding on the behavior of human being. The lack of behavioral realism in the MCDM systems may be one reason that the systems (with some exceptions like the AHP [42]) are not much used in practice, even if there is a lot of potential need for them.

References

[1] C.H. Antunes and J.N. Clìmaco. Sensitivity analysis in MCDM using the weight space. *Operations Research Letters*, 12:187–196, 1992.

[2] A. Arbel and P.J. Korhonen. Using objective values to start multiple objective linear programming algorithms. *European Journal of Operational Research*, 128:587–596, 2001.

[3] C.A. Bana e Costa and J.C. Vansnick. MACBETH – An interactive path towards the construction of cardinal value functions. *International Transactions in Operational Research*, 1:387–500, 1994.

[4] M.S. Bazaraa, J.J. Jarvis, and H.D. Sherali. *Linear Programming and Network Flows*. John Wiley & Sons, Chichester, 1990.

[5] V. Belton and M. Elder. Visual interactive modeling. *European Journal of Operational Research*, 54:273–273, 1991.

[6] V. Belton and S.P. Vickers. V.I.S.A. – VIM for MCDM. In G. Lockett and G. Islei, editors, *Improving Decisions Making in Organisations*, volume 335 of *Lecture Notes in Economics and Mathematical Systems*, pages 287–304. Springer Verlag, Berlin, 1988.

[7] R. Benayoun, J. de Montgolfier, J. Tergny, and O. Larichev. Linear programming with multiple objective functions: Step method (Stem). *Mathematical Programming*, 1:366–375, 1971.

[8] A.R. Borges and C.H. Antunes. A weight space-based approach to fuzzy multiple-objective linear programming. *Decision Support Systems*, 34:427–443, 2003.

[9] A. Charnes and W.W. Cooper. *Management Models and Industrial Applications of Linear Programming (Appendix B), Vol. I*. John Wiley & Sons, Chichester, 1961.

[10] A. Charnes, W.W. Cooper, and R.O. Ferguson. Optimal estimation of executive compensation by linear programming. *Management Science*, 1:138–151, 1955.

[11] J.N. Clìmaco and C.H. Antunes. Implementation of a user-friendly software package – A guided tour of TRIMAP. *Mathematical and Computer Modelling*, 12:1299–1309, 1989.

[12] C. Coello, D. Van Veldhuizen, and G. Lamont. *Evolutionary Algorithms for Solving Multi-Objective Problems*, volume 5 of *Genetic Algorithms and Evolutionary Computation*. Kluwer Academic Publishers, Boston, 2002.

[13] J. Dyer. A time-sharing computer program for the solution of the multiple criteria problem. *Management Science*, 19:1379–1383, 1973.

[14] L.R. Gardiner and R.E. Steuer. Unified interactive multiple-objective programming – An open-architecture for accommodating new procedures. *Journal of the Operational Research Society*, 45:1456–1466, 1994.

[15] A. Geoffrion, J. Dyer, and A. Feinberg. An interactive approach for multi-criterion optimization, with an application to the operation of an academic department. *Management Science*, 19:357–368, 1972.

[16] J.P. Ignizio. *Goal Programming and Its Extensions*. D.C. Heath, Lexington, MA, 1976.

[17] J.P. Ignizio, editor. *Computers & Operations Research*, volume 10. 1983. Special Issue on Generalized Goal Programming.

[18] A. Jaszkiewicz and R. Slowinski. The light beam search approach – An overview of methodology and applications. *European Journal of Operational Research*, 113:300–314, 1999.

[19] D. Kahneman and A. Tversky. Prospect theory: An analysis of decisions under risk. *Econometrica*, 47:262–291, 1979.

[20] I. Kaliszewski and W. Michalowski. Efficient solutions and bounds on tradeoffs. *Journal of Optimization Theory and Applications*, 94:381–394, 1997.

[21] E.K. Karasakal and M. Koksalan. A simulated annealing approach to bicriteria scheduling problems on a single machine. *Journal of Heuristics*, 6:311–327, 2000.

[22] R. Keeney and H. Raiffa. *Decisions with Multiple Objectives*. John Wiley & Sons, New York, 1976.

[23] R. Klimberg. GRADS – A new graphical display system for visualizing multiple criteria solutions. *Computers & Operations Research*, 19:707–711, 1992.

[24] T. Koopmans. Analysis of production as an efficient combination of activities. In T. Koopmans, editor, *Activity Analysis of Production and Allocation*, volume 13 of *Cowles Commission Monograph*, pages 33–97. John Wiley & Sons, New York, 1951.

[25] P.J. Korhonen. VIG – A visual interactive support system for multiple criteria decision making. *Belgian Journal of Operations Research, Statistics and Computer Science*, 27:3–15, 1987.

[26] P.J. Korhonen. Using harmonious houses for visual pairwise comparison of multiple criteria alternatives. *Decision Support Systems*, 7:47–54, 1991.

[27] P.J. Korhonen and J. Karaivanova. An algorithm for projecting a reference direction onto the nondominated set of given points. *IEEE Transactions on Systems, Man, and Cybernetics – Part A: Systems and Humans*, 29(5):429–435, 1999.

[28] P.J. Korhonen and J. Laakso. A visual interactive method for solving the multiple criteria problem. *European Journal of Operational Research*, 24:277–287, 1986.

[29] P.J. Korhonen, H. Moskowitz, and J. Wallenius. Choice behavior in interactive multiple criteria decision-making. *Annals of Operations Research*, 23:161–179, 1990.

[30] P.J. Korhonen, H. Moskowitz, and J. Wallenius. Multiple criteria decision support – A review. *European Journal of Operational Research*, 63:61–375, 1992.

[31] P.J. Korhonen and J. Wallenius. A Pareto race. *Naval Research Logistics*, 35:615–623, 1988.

[32] P.J. Korhonen and J. Wallenius. Observations regarding choice behavior in interactive multiple criteria decision-making environments: An experimental investigation. In A. Lewandowski and I. Stanchev, editors, *Methodology and Software for Interactive Decision Support*, pages 163–170. Springer Verlag, Berlin, 1989. Proceedings of the Workshop sponsored by IIASA.

[33] P.J. Korhonen, J. Wallenius, and S. Zionts. A computer graphics-based decision support system for multiple objective linear programming. *European Journal of Operational Research*, 60:280–286, 1992.

[34] O.I. Larichev and H.M. Moshkovich. Limits to decision-making ability in direct multiattribute alternative evaluation. *Organizational Behavior And Human Decision Processes*, 42:217–233, 1988.

[35] A. Lewandowski and M. Granat. Dynamic BIPLOT as the interaction interface for aspiration based decision support systems. In P.J. Korhonen, A. Lewandowski, and J. Wallenius, editors, *Multiple Criteria Decision Support*, pages 229–241. Springer-Verlag, Heidelberg, 1991. Proceedings of the Workshop organized by IIASA and sponsored by IIASA.

[36] B. Mareschal and J.P. Brans. Geometrical representations for MCDA. *European Journal of Operational Research*, 34:69–77, 1988.

[37] W. Michalowski and T. Szapiro. A bi-reference procedure for interactive multiple criteria programming. *Operations Research*, 40:247–258, 1992.

[38] K. Miettinen. *Nonlinear Multiobjective Optimization*. Kluwer Academic Publishers, Boston, 1999.

[39] D. Olson. *Decision Aids for Selection Problems*. Springer Series in Operations Research. Springer Verlag, New York, 1996.

[40] V. Pareto. *Manuel d'Economie Politique*. Marcel Giard, Paris, 1909.

[41] B. Roy. Problems and methods with multiple objective functions. *Mathematical Programming*, 1:239–266, 1972.

[42] T. Saaty. *The Analytic Hierarchy Process*. McGraw-Hill, New York, 1980.

[43] A.A. Salo and R.P. Hämäläinen. Preference assessment by imprecise ratio statements. *Operations Research*, 40:1053–1061, 1992.

[44] W.S. Shin and A. Ravindran. Interactive multiple objective optimization: Survey I – Continuous case. *Computers & Operations Research*, 18:97–114, 1991.

[45] R.E. Steuer. An interactive multiple objective linear programming procedure. *TIMS Studies in the Management Sciences*, 6:225–239, 1977.

[46] R.E. Steuer. *Multiple Criteria Optimization: Theory, Computation, and Application*. John Wiley & Sons, New York, 1986.

[47] R.E. Steuer and E.-U. Choo. An interactive weighted Tchebycheff procedure for multiple objective programming. *Mathematical Programming*, 26:326–344, 1983.

[48] M. Weber and K. Borcherding. Behavioral-influences on weight judgments in multiattribute decision-making. *European Journal of Operational Research*, 67:1–12, 1993.

[49] A. Wierzbicki. The use of reference objectives in multiobjective optimization. In G. Fandel and T. Gal, editors, *Multiple Objective Decision Making, Theory and Application*, volume 177 of *Lecture Notes in Economics and Mathematical Systems*. Springer Verlag, Berlin, 1980.

[50] A. Wierzbicki. On the completeness and constructiveness of parametric characterizations to vector optimization problems. *OR Spektrum*, 8:73–87, 1986.

[51] S. Zionts and J. Wallenius. An interactive programming method for solving the multiple criteria problem. *Management Science*, 22:652–663, 1976.

[52] S. Zionts and J. Wallenius. An interactive multiple objective linear programming method for a class of underlying nonlinear utility functions. *Management Science*, 29:519–529, 1983.

Chapter 17

MULTIOBJECTIVE PROGRAMMING

Matthias Ehrgott
Department of Engineering Science
The University of Auckland
Private Bag 92019, Auckland
New Zealand
m.ehrgott@auckland.ac.nz, wmalgor@clemson.edu

Margaret M. Wiecek
Department of Mathematical Sciences
Clemson University
Clemson, SC 29634-0975
USA
wmalgor@clemson.edu

Abstract We present our view of the state of the art in multiobjective programming. After an introduction we formulate the multiobjective program (MOP) and define the most important solution concepts. We then summarize the properties of efficient and nondominated sets. In Section 4 optimality conditions are reviewed. The main part of the chapter consists of Sections 5 and 6 that deal with solution techniques for MOPs and approximation of efficient and nondominated sets. In Section 7 we discuss specially-structured problems including linear and discrete MOPs as well as selected nonlinear MOPs. In Section 8 we present our perspective on future research directions.

Keywords: Multiobjective programming, efficient solution, nondominated solution, scalarization, approximation.

1. Introduction

Multiobjective programming is a part of mathematical programming dealing with decision problems characterized by multiple and conflicting objective functions that are to be optimized over a feasible set of decisions. Such problems, referred to as multiobjective programs (MOPs), are commonly encountered in many areas of human activity including engineering, management, and others. Throughout the chapter we understand multiobjective programming as pertaining to situations where feasible alternatives are available implicitly, through constraints in the form of mathematical functions. An optimization problem (typically a mathematical program) has to be solved to explicitly find the alternatives. Decision problems with multiple criteria and explicitly available alternatives are treated within multicriteria decision analysis (MCDA). This view constitutes the difference between multiobjective programming and multicriteria decision analysis (MCDA) which complement each other within multicriteria decision making (MCDM).

In the last fifty years, a great deal of theoretical, methodological and applied studies have been undertaken in the area of multiobjective programming. This chapter presents a review of the theory and methodology of MOPs in finite dimensions. The content of the review is based on the understanding that the primary (although not necessarily the ultimate) goal of multiobjective programming is to seek solutions of MOPs. Consequently, methods suitable for finding these solutions are considered the most fundamental tools for dealing with MOPs and therefore given special attention. The selection of a preferred solution of the MOP performed by the decision maker can be considered the ultimate goal of MCDM. However, the modelling of decision maker preferences is outside the scope of this chapter and belongs to the domain of MCDA.

In Sections 2, 3, and 4 we review theoretical foundations of multiobjective programming. In Section 2 we define MOPs and relevant solution concepts. Sections 3 and 4 contain a summary of properties of the solution sets and conditions for efficiency, respectively. The subsequent sections focus on methodological aspects of multiobjective programming. In Sections 5 and 6, numerous methods for generating individual elements or subsets of the solution sets are collected. In these sections we present scalarization, nonscalarizing and approximation methods. Specially structured problems, including linear, combinatorial and nonlinear MOPs, are discussed in Section 7. The chapter is concluded in Section 8 with our view of current and future research directions.

We point out that the results are not always presented chronologically but rather with respect to the order implied by the content of this chapter and with respect to their level of generality.

The following notation is used. Let $\mathbb{R}^p$ be the Euclidean vector space and $y, y' \in \mathbb{R}^p$.

- $y < y'$ denotes $y_k < y'_k$ for all $k = 1, \ldots, p$. $y \leqq y'$ denotes $y_k \leq y'_k$ for all $k = 1, \ldots, p$. $y \leq y'$ denotes $y \leqq y'$ but $y \neq y'$.
- Let $\mathbb{R}^p_{\geqq} := \{y \in \mathbb{R}^p : y \geqq 0\}$. The sets $\mathbb{R}^p_{\geq}, \mathbb{R}^p_{>}$ are defined accordingly.

For a subset $S \subset \mathbb{R}^n$ we use $\operatorname{bd} S, \operatorname{int} S, \operatorname{ri} S$, and $\operatorname{cl} S$ to denote the boundary, interior, relative interior, and closure of S. Furthermore $\operatorname{cone} S$ and $\operatorname{conv} S$ denote the conical and convex hulls of S.

2. Problem Formulation and Solution Concepts

Let $\mathbb{R}^n$ and $\mathbb{R}^p$ be Euclidean vector spaces referred to as the decision space and the objective space. Let $X \subset \mathbb{R}^n$ be a feasible set and let f be a vector-valued objective function $f : \mathbb{R}^n \to \mathbb{R}^p$ composed of p real-valued objective functions, $f = (f_1, \ldots, f_p)$, where $f_k : \mathbb{R}^n \to \mathbb{R}$ for $k = 1, \ldots, p$. A multiobjective program (MOP) is given by

$$\begin{aligned} \min \quad & (f_1(x), \ldots, f_p(x)) \\ & \text{subject to } x \in X. \end{aligned} \tag{17.1}$$

Throughout this chapter we refer to problem (17.1) as the MOP. When $p = 2$ the problem is referred to as the biobjective program (BOP). We usually assume that the set X is given implicitly in the form of constraints, i.e., $X := \{x \in S \subseteq \mathbb{R}^n : g_j(x) \leq 0, j = 1, \ldots, l; h_j(x) = 0, j = 1, \ldots, m\}$. A feasible solution $x \in X$ is evaluated by p objective functions producing the outcome $f(x)$. We define the set of all attainable outcomes or criterion vectors for all feasible solutions in the objective space, $Y := f(X) \subset \mathbb{R}^p$. Occasionally, we will deal with a special case of the MOP with the feasible set defined by $X' := \{x \in S' \subseteq \mathbb{R}^n : g_j(x) \leq 0, j = 1, \ldots, l\}$. This MOP with inequality constraints will be referred to as the MOP'.

The symbol "min" in the MOP is generally understood as finding optimal or preferred outcomes in Y and their pre-images in X, where the preference between the outcomes results from a binary relation $\mathcal{R}$ defined on Y. Let $y^1 \succ y^2$ denote that an outcome y^1 is preferred to an outcome y^2. $y^1 \succeq y^2$ denotes preference of y^1 over y^2 or indifference between y^1 and y^2. Given a binary relation $\mathcal{R}$, we say that an outcome y^1 is preferred (or indifferent) to an outcome y^2 with respect to this relation if and only if y^1 is in relation with y^2, i.e., $y^1 \succeq_{\mathcal{R}} y^2$ if and only if $y^1 \mathcal{R} y^2$.

To derive a definition for a class of preferences between outcomes, we consider cones. A set $\mathcal{C} \subset \mathbb{R}^p$ is a cone, if $\alpha y \in \mathcal{C}$ whenever $y \in \mathcal{C}$ and $0 \leq \alpha \in \mathbb{R}$. Given a cone $\mathcal{C}$, we say that an outcome y^1 dominates (is preferred to) an outcome y^2 with respect to this cone, $y^1 \succ_{\mathcal{C}} y^2$, if and only if $y^2 - y^1 \in \mathcal{C} \setminus \{0\}$, or equivalently there exists a direction $d \in \mathcal{C}, d \neq 0 : y^2 = y^1 + d$. Then $y^1 \succeq_{\mathcal{C}} y^2$ if $y^1 \succ_{\mathcal{C}} y^2$ or $y^1 = y^2$.

Given a cone $\mathcal{C}$, we can also define a relation $\mathcal{R}_\mathcal{C}$ on $\mathbb{R}^p$ by $y^1 \mathcal{R}_\mathcal{C} y^2$ if and only if $y^2 - y^1 \in \mathcal{C}$. This relation is compatible with addition and scalar multiplication (i.e., $y^1 \mathcal{R} y^2$ implies $(y^1+z)\mathcal{R}(y^2+z)$ for all $z \in \mathbb{R}^p$ and $y^1 \mathcal{R} y^2$ implies $\alpha y^1 \mathcal{R} \alpha y^2$ for all $0 < \alpha \in \mathbb{R}$). Conversely, given a relation $\mathcal{R}$ on $\mathbb{R}^p$ we can define a set $\mathcal{C}_\mathcal{R}$ as $\mathcal{C}_\mathcal{R} := \{d \in \mathbb{R}^p : d = y^2 - y^1 \text{ and } y^1 \mathcal{R} y^2\}$, see Ehrgott [72]. If $\mathcal{R}$ is compatible with scalar multiplication and $\{0\} \in \mathcal{C}_\mathcal{R}$ then $\mathcal{C}_\mathcal{R}$ is a cone; (otherwise $\mathcal{C}_\mathcal{R} \cup \{0\}$ is a cone). If $\mathcal{R}$ is compatible with addition then $d \in \mathcal{C}_\mathcal{R}$ and $d = y^2 - y^1$ imply that $y^1 \mathcal{R} y^2$.

Theorem 1 provides relationships between binary relations and cones. Note that a reflexive, antisymmetric, and transitive binary relation is a partial order on $\mathbb{R}^p$.

THEOREM 1

1 *Let $\mathcal{R}$ be a binary relation on $\mathbb{R}^p$ which is compatible with addition. Then $0 \in \mathcal{C}_\mathcal{R}$ if and only if $\mathcal{R}$ is reflexive; $\mathcal{C}_\mathcal{R}$ is pointed (i.e., $\mathcal{C} \cap (-\mathcal{C}) = \{0\}$) if and only if $\mathcal{R}$ is antisymmetric; $\mathcal{C}_\mathcal{R}$ is convex if and only if $\mathcal{R}$ is transitive.*

2 *Let $\mathcal{C}$ be a cone. Then $\mathcal{R}_\mathcal{C}$ is reflexive if and only if $0 \in \mathcal{C}$; $\mathcal{R}_\mathcal{C}$ is antisymmetric if and only if $\mathcal{C}$ is pointed; $\mathcal{R}_\mathcal{C}$ is transitive if and only if $\mathcal{C}$ is convex.*

Thus some binary relations and cones are equivalent concepts, and we can define a notion of nondominated solutions for MOPs (Yu [232]).

DEFINITION 65 *Let $\mathcal{C} \subset \mathbb{R}^p$ be a cone and $Y \subset \mathbb{R}^p$. Then $y \in Y$ is called a nondominated outcome of the MOP if*

- *there does not exist $y^1 \in Y$ and $d \in \mathcal{C}, d \neq 0 : y = y^1 + d$, or equivalently,*
- $(y - \mathcal{C}) \cap Y = \{y\}$.

We shall denote the set of all nondominated outcomes of the MOP by $N(X, f, \mathcal{C})$ or $N(Y, \mathcal{C})$. One typically assumes that the cone $\mathcal{C}$ is proper (i.e., $\{0\} \neq \mathcal{C} \neq \mathbb{R}^p$) and pointed. The pre-images of the nondominated outcomes are called efficient solutions and are denoted by $E(X, f, \mathcal{C})$. We also define weakly nondominated solutions in the objective space the pre-images of which in the decision space are called weakly efficient.

DEFINITION 66 *Let $\mathcal{C} \subset \mathbb{R}^p$ be a cone and $Y \subset \mathbb{R}^p$. Then $y \in Y$ is called a weakly nondominated outcome of the MOP if $(y - \operatorname{int} \mathcal{C}) \cap Y = \emptyset$.*

We state some basic properties of nondominated sets, see Sawaragi et al. [180].

THEOREM 2 *Let Y, Y_1, Y_2 be subsets of $\mathbb{R}^p$, C, C_1 and C_2 be cones in $\mathbb{R}^p$.*

- *If $C \neq \emptyset$ and $0 \in C$ then $N(Y,C) \supset N(Y+C,C)$. If, additionally, C is pointed and convex then the inclusion becomes an equality.*
- $N(Y,C) \subset \text{bd}(Y)$.
- $N(Y_1+Y_2,C) \subset N(Y_1,C)+N(Y_2,C)$.
- $N(\alpha Y,C) = \alpha N(Y,C)$ *for* $0 < \alpha \in \mathbb{R}$.
- *If* $C_1 \subset C_2$ *then* $N(Y,C_2) \subset N(Y,C_1)$.
- $N(Y,C_1 \cup C_2) = N(Y,C_1) \cap N(Y,C_2)$.

Specific results have been obtained when the cone C is convex and polyhedral.

THEOREM 3 *Let C be a convex polyhedral cone represented by $\{d \subset \mathbb{R}^p : Ld \geqq 0\}$ where L is a $q \times p$ matrix. Then*

1 *[219]* $E(X,f,C) \subseteq E(X,Lf,\mathbb{R}^q_{\geqq})$.

2 *[119]* $L[N(Y,C)] \subseteq N(L[Y],\mathbb{R}^q_{\geqq})$.

What constitutes a nondominated solution of the MOP depends on the definition of the preference between outcomes which is modeled as a binary relation or a cone in $\mathbb{R}^p$. The most commonly used preference is based on the Pareto relation according to which $y^1 \succ_{\text{Pareto}} y^2$ if and only if $y^1_k \leq y^2_k$ for $k = 1,\dots,p$ with a strict inequality for at least one index k. The cone $C = \mathbb{R}^p_{\geqq}$ is equivalent to the Pareto preference and referred to as the Pareto cone. For the set $N(X,f,\mathbb{R}^p_{\geqq})$ we simply write Y_N. The corresponding solutions in the decision space with two other variations of the Pareto relation are summarized in Definition 67.

DEFINITION 67 *Consider the MOP. A point $x \in X$ is called*

1 *a weakly efficient solution if there is no $x' \in X$ such that $f(x') < f(x)$;*

2 *an efficient solution if there is no $x' \in X$ such that $f(x') \leq f(x)$;*

3 *a strictly efficient solution if there is no $x' \in X$, $x' \neq x$, such that $f(x') \leqq f(x)$.*

We shall denote the weakly efficient solutions, efficient solutions, and strictly efficient solutions by X_{wE}, X_E, X_{sE}, respectively, and shall call their images weak Pareto points and Pareto points, respectively. The latter are denoted by Y_{wN}, Y_N. Note that strictly efficient solutions correspond to unique efficient solutions, and therefore they do not have a counterpart in the objective space.

For many of the solution approaches presented in Section 5, statements of the form "If $\hat{x} \in X$ is a unique optimal solution of the approach, then $\hat{x} \in X$ is an efficient solution" are presented. Uniqueness of a solution actually implies that $\hat{x}$ is strictly efficient for the MOP.

All the classes of solutions defined above are global solutions. However, we also define local solutions of the MOP. A point $x \in X$ is called a locally efficient solution of the MOP if there exists a neighborhood $N(x)$ such that there is no $x' \in N(x) \cap X$ such that $f(x') \leq f(x)$. Similarly, all other classes of local solutions in the decision space and the objective space can be defined. In this chapter, all solutions of optimization problems are global unless stated otherwise.

Additionally, the following authors define properly efficient solutions: Kuhn and Tucker [147], Klinger [138], Geoffrion [103], Borwein [33], Benson [18], Wierzbicki [227] and Henig [117]. Borwein and Zhuang define super efficient solutions [35, 36].

DEFINITION 68 *A point $\hat{x} \in X$ is called a properly efficient solution of the MOP' in the sense of Kuhn-Tucker if $\hat{x} \in X_E$ and if there does not exist a $d \in \mathbb{R}^n$ such that $\nabla f_k(\hat{x})^T d \leq 0$ for all $k = 1, \ldots, p$ with a strict inequality for some k and $\nabla g_j(\hat{x})^T d \leq 0$ for all $j \in I(\hat{x}) = \{j : g_j(\hat{x}) = 0\}$.*

DEFINITION 69 *A point $\hat{x} \in X$ is called a properly efficient solution of the MOP in the sense of Geoffrion if $\hat{x} \in X_E$ and if there exists $M > 0$ such that for each $k = 1, \ldots, p$ and each $x \in X$ satisfying $f_k(x) < f_k(\hat{x})$ there exists an $l \neq k$ with $f_l(x) > f_l(\hat{x})$ and $(f_k(\hat{x}) - f_k(x))/(f_l(x) - f_l(\hat{x})) \leq M$.*

The sets of all properly efficient solutions and properly nondominated outcomes (in the sense of Geoffrion) are denoted by X_{pE} and Y_{pN}, respectively.

Approximate efficient solutions are defined by Loridan [153] in the following way.

DEFINITION 70 *Let $\epsilon \in \mathbb{R}^p_{\geqq}$. A point $\hat{x} \in X$ is called an ϵ-efficient solution of the MOP if there is no $x' \in X$ such that $f(x') \leq f(\hat{x}) - \epsilon$.*

Other types of approximate efficient solutions are defined by White [222]. Similarly, weakly ϵ-efficient solutions and strictly ϵ-efficient solutions and their images can be defined.

Let $y^I_k := \min\{f_k(x) : x \in X\}$ be the (global) minimum of $f_k(x), k = 1, \ldots, p$. The point $y^I \in \mathbb{R}^p, y^I = (y^I_1, \ldots, y^I_p)$ is called the ideal point for the MOP.

The point y^U where $y^U_k := \min\{f_k(x) : x \in X\} - \epsilon_k, k = 1, \ldots, p$, where the components of $\epsilon = (\epsilon_1, \ldots, \epsilon_p) \in \mathbb{R}^p$ are small positive numbers, is called a utopia point for the MOP.

Furthermore, the point y^N with $y_k^N := \max\{f_k(x) : x \in X_E\}$ is called the nadir point for the MOP. For each $x \in X_E$ it holds: $y^U < y^I \leq f(x) \leq y^N$. We shall assume that $y^I < y^N$ for the MOP.

3. Properties of the Solution Sets

In this section we discuss properties of the nondominated and efficient sets including existence, stability, convexity, and connectedness of the solution sets $N(Y, C)$ and $E(X, f, C)$ of the MOP. Here we assume that C is a pointed, closed, convex cone. We first consider existence of nondominated points and efficient solutions.

THEOREM 4 *[34] Let $Y \neq \emptyset$ and suppose there exists a $y^0 \in Y$ such that $Y^0 = (y^0 - C) \cap Y$ is compact. Then $N(Y, C) \neq \emptyset$.*

An earlier result by Corley requires Y to be C-semicompact, i.e., every open cover $\{Y \setminus (y^\tau - \mathrm{cl}\, C) : y^\tau \in Y,\ \tau \in T\}$ of Y, where T is an index set, has a finite subcover.

THEOREM 5 *[54] If $Y \neq \emptyset$ and Y is C-semicompact then $N(Y, C) \neq \emptyset$.*

COROLLARY 1 *[114] If $Y \neq \emptyset$ and Y is C-compact (i.e., $(y - C) \cap Y$ is compact for all $y \in Y$) then $N(Y, C) \neq \emptyset$.*

Sawaragi et al. [180] also give necessary and sufficient conditions for $N(Y, C)$ to be nonempty for the case of a nonempty, closed, convex set Y. Essentially, the existence of nondominated points can be guaranteed under some compactness assumption. Consistently, the existence of efficient solutions can be guaranteed under appropriate continuity assumptions on the objective functions and compactness assumptions on X. The function $f : \mathbb{R}^n \to \mathbb{R}^p$ is said to be C-semicontinuous if the preimage of $y - \mathrm{cl}\, C$ is a closed subset of $\mathbb{R}^n$ for all y in $\mathbb{R}^p$.

THEOREM 6 *Let $\emptyset \neq X \subset \mathbb{R}^n$ be a compact set and assume $f : \mathbb{R}^n \to \mathbb{R}^p$ is C-semicontinuous. Then $E(X, f, C) \neq \emptyset$.*

A review of existence results for nondominated and efficient sets is provided by Sonntag and Zălinescu [196].

Stability of MOPs is studied, among others, in the following context. Let $y \in Y$ be a feasible solution. If y is dominated then there exists a $y' \in Y, y' \neq y$, such that $y' \succ_C y$. The question arises whether y' is nondominated. In this case, $N(Y, C)$ is called externally stable. Note that the external stability condition can also be written as $Y \subset N(Y, C) + C$.

THEOREM 7 *[180] Let C be a pointed, closed, convex cone and let $Y \neq \emptyset$ a C-compact set. Then $N(Y, C)$ is externally stable.*

In addition, Sawaragi et al. [180] prove that the necessary and sufficient conditions for existence of nondominated points for nonempty, closed, convex sets Y are also necessary and sufficient for external stability of $N(Y, C)$ in that case.

We now state some relationships between the various nondominated and efficient sets. From Definitions 67 and 69 it is clear that $X_{pE} \subseteq X_E \subseteq X_{wE}$ and $X_{sE} \subseteq X_E$, and therefore $Y_{pN} \subseteq Y_N \subseteq Y_{wN}$. Again, for convex sets a stronger result holds. Hartley [114] proves that if Y is C-closed ($Y + C$ is closed) and C-convex ($Y + C$ is convex) then $Y_{pN} \subseteq Y_N \subseteq \operatorname{cl} Y_{pN}$ and that equality holds if Y is polyhedral. The C-convexity condition on Y is satisfied if, e.g., the set X is convex and the objective functions f_k are convex. Therefore it makes sense to define the convex MOP.

If all the objective functions $f_k, k = 1, \ldots, p$ of the MOP are convex and the feasible set X is convex then the problem is called convex MOP. The outcome set Y of the convex MOP is $\mathbb{R}^p_{\geqq}$-convex, i.e., $Y + \mathbb{R}^p_{\geqq}$ is a convex set.

The last property of the efficient and the nondominated sets that we discuss is connectedness.

THEOREM 8 *[218] Assume that $f_1, \ldots, f_p$ are continuous and that X satisfies one of the following conditions.*

1. *$X \subset \mathbb{R}^n$ is a compact, convex set.*
2. *X is a closed, convex set and for all $y \in Y, X(y) := \{x \in X : f(x) \leqq y\}$ is compact.*

Then the following statements hold:

1. *If $f_k : \mathbb{R}^n \to \mathbb{R}$ are quasiconvex on X for $k = 1, \ldots, p$ then X_{wE} is connected.*
2. *If $f_k : \mathbb{R}^n \to \mathbb{R}$ are strongly quasiconvex on X for $k = 1, \ldots, p$ then X_E is connected.*

Warburton [218] also gives examples showing that X_E may be disconnected if X is compact and convex but f_k only quasiconvex, and that X_E (respectively X_{wE}) may be disconnected if $X(y)$ is not compact for some y.

Because convex functions are continuous and the image of a connected set under a continuous function is connected, it follows immediately that the sets Y_{wN} and Y_N are connected under the assumptions stated in Theorem 8, if the objective functions $f_k, k = 1, \ldots, p$ are continuous. However, connectedness of Y_N can also be proved under more general assumptions.

THEOREM 9 *[160] Let C be a closed, convex, nonempty cone that does not contain a nontrivial subspace of $\mathbb{R}^p$ and let Y be a closed, convex, and C-compact set. Then $N(Y, C)$ is connected.*

In the remaining sections we will mainly be considering nondominance and efficiency with respect to the Pareto cone, i.e., $C = \mathbb{R}^p_{\geqq}$. Thus, throughout the rest of the chapter, efficiency is meant for the MOP (17.1) in the sense of Definition 67.

4. Conditions for Efficiency

Conditions for efficiency are powerful theoretical tools for determining whether a feasible point is efficient. Denote the set of indices of active inequality constraints at $\hat{x} \in X$ by $I(\hat{x}) = \{j \in \{1, \ldots, l\} : g_j(\hat{x}) = 0\}$.

4.1 First Order Conditions

Assume that the objective functions $f_k, k = 1, \ldots, p$, and the constraint functions $g_j, j = 1, \ldots, l;\ h_j, j = 1, \ldots, m$ of the MOP are continuously differentiable.

THEOREM 10 *Fritz-John necessary conditions for efficiency [57]. If $\hat{x}$ is efficient then there exist vectors $w \in \mathbb{R}^p_{\geqq}$, $u \in \mathbb{R}^l_{\geqq}$, and $v \in \mathbb{R}^m$, $(w, u, v) \neq 0$ such that*

$$\begin{aligned} \sum_{k=1}^{p} w_k \nabla f_k(\hat{x}) + \sum_{j=1}^{l} u_j \nabla g_j(\hat{x}) + \sum_{j=1}^{m} v_j \nabla h_j(\hat{x}) &= 0 \\ u_j g_j(\hat{x}) &= 0 \textit{ for all } j - 1, \ldots, l. \end{aligned}$$

Kuhn-Tucker type conditions for efficiency have also been studied for MOPs'.

DEFINITION 71 *The MOP' is said to satisfy the Kuhn-Tucker constraint qualification at $\hat{x} \in X$ if for any $d \in \mathbb{R}^n$ such that $\nabla g_j(\hat{x})^T d \leq 0$ for all $j \in I(\hat{x})$, there exist a continuously differentiable function $a : [0,1] \to \mathbb{R}^n$ and a real scalar $\alpha > 0$ such that $a(0) = \hat{x}, g(a(\beta)) \leqq 0$ for all $\beta \in [0,1]$ and $a'(0) = \alpha d$.*

THEOREM 11 *Kuhn-Tucker necessary conditions for efficiency [158]. Let the Kuhn-Tucker constraint qualification hold at $\hat{x} \in X$. If $\hat{x}$ is efficient for the MOP' then there exist vectors $w \in \mathbb{R}^p_{\geq}$ and $u \in \mathbb{R}^l_{\geqq}$ such that*

$$\begin{aligned} \sum_{k=1}^{p} w_k \nabla f_k(\hat{x}) + \sum_{j=1}^{l} u_j \nabla g_j(\hat{x}) &= 0 \\ u_j g_j(\hat{x}) &= 0 \textit{ for all } j = 1, \ldots, l. \end{aligned}$$

THEOREM 12 *Kuhn-Tucker necessary conditions for proper efficiency [147, 180]. If $\hat{x} \in X$ is properly efficient (in the sense of Kuhn-Tucker) for the MOP'* *then there exist vectors* $w \in \mathbb{R}^p_>$ *and* $u \in \mathbb{R}^l_{\geqq}$ *such that*

$$\begin{aligned} \sum_{k=1}^{p} w_k \nabla f_k(\hat{x}) + \sum_{j=1}^{l} u_j \nabla g_j(\hat{x}) &= 0 \\ u_j g_j(\hat{x}) &= 0 \textit{ for all } j = 1, \ldots, l. \end{aligned}$$

If the Kuhn-Tucker constraint qualification is satisfied at a point $\hat{x} \in X$, then the condition in Theorem 12 is also necessary for $\hat{x}$ to be properly efficient in the sense of Geoffrion for the MOP', as shown by Sawaragi et al. [180].

THEOREM 13 *Kuhn-Tucker sufficient conditions for proper efficiency [147, 158, 180]. Let the MOP' be convex and let* $\hat{x} \in X$. *If there exist vectors* $w \in \mathbb{R}^p_>$ *and* $u \in \mathbb{R}^l_{\geqq}$ *such that*

$$\begin{aligned} \sum_{k=1}^{p} w_k \nabla f_k(\hat{x}) + \sum_{j=1}^{l} u_j \nabla g_j(\hat{x}) &= 0 \\ u_j g_j(\hat{x}) &= 0 \textit{ for all } j = 1, \ldots, l \end{aligned}$$

then $\hat{x}$ *is properly efficient in the sense of Kuhn-Tucker for the MOP'.*

4.2 Second Order Conditions

Various types of second-order conditions for efficiency have been developed. For this type of conditions it is usually assumed that the objective functions $f_k, k = 1, \ldots, p$ and the constraint functions $g_j, j = 1, \ldots, l; h_j, j = 1, \ldots, m$ of the MOP are twice continuously differentiable.

Several necessary and sufficient second-order conditions for the MOP are developed by Wang [216]. Cambini et al. [39] establish second order conditions for MOPs with general convex cones while Cambini [41] develops second order conditions for MOPs with the Pareto cone. Aghezzaf [5] and Aghezzaf and Hachimi [6] develop second-order necessary conditions for the MOP'. Recent works include Bolintinéanu and El Maghri [32], Bigi and Castellani [28], and Jiménez and Novo [127].

5. Generation of the Solution Sets

There are two general approaches to generate solution sets of MOPs, scalarization methods and nonscalarizing methods. These approaches convert the MOP into a single objective program (SOP), a sequence of SOPs, or another MOP. Under some assumptions solution sets of these new programs yield solutions of the original problem. Scalarization methods explicitly employ a scalarizing function to accomplish the conversion while nonscalarizing methods use other means. Solving the SOP typically yields one solution of the MOP so that a repetitive solution scheme is needed to obtain a subset of solutions of the MOP.

5.1 Scalarization Methods

The traditional approach to solving MOPs is by scalarization which involves formulating an MOP-related SOP by means of a real-valued scalarizing function typically being a function of the objective functions of the MOP, auxiliary scalar or vector variables, and/or scalar or vector parameters. Sometimes the feasible set of the MOP is additionally restricted by new constraint functions related to the objective functions of the MOP and/or the new variables introduced.

In this section we review the most well-known scalarization techniques and list related results on the generation of various classes of solutions of the MOP.

5.1.1 The Weighted Sum Approach. In the weighted sum approach a weighted sum of the objective functions is minimized:

$$\begin{aligned} \min \quad & \sum_{k=1}^{p} \lambda_k f_k(x) \\ \text{subject to } x \quad & \in \quad X, \end{aligned} \tag{17.2}$$

where $\lambda \in \mathbb{R}^p_{\geq}$.

THEOREM 14 *[103]*

1 *Let* $\lambda \in \mathbb{R}^p_{\geq}$. *If* $\hat{x} \in X$ *is an optimal solution of problem (17.2) then* $\hat{x} \in X_{wE}$. *If* $\hat{x} \in X$ *is a unique optimal solution of problem (17.2) then* $\hat{x} \in X_E$.

2 *Let* $\lambda \in \mathbb{R}^p_{>}$. *If* $\hat{x} \in X$ *is an optimal solution of problem (17.2) then* $\hat{x} \in X_{pE}$.

3 *Let the MOP be convex. A point* $\hat{x} \in X$ *is an optimal solution of problem (17.2) for some* $\lambda \in \mathbb{R}^p_{>}$ *if and only if* $\hat{x} \in X_{pE}$.

5.1.2 The Weighted t-th Power Approach. In the weighted t-th power approach a weighted sum of the objective functions taken to the power of t is minimized:

$$\begin{aligned} \min \quad & \sum_{k=1}^{p} \lambda_k (f_k(x))^t \\ \text{subject to } x \quad & \in \quad X, \end{aligned} \tag{17.3}$$

where $\lambda \in \mathbb{R}^p_{>}$ and $t > 0$.

THEOREM 15 *[223] Let* $\lambda \in \mathbb{R}^p_>$.

1 *For all* $t > 0$, *if* $\hat{x} \in X$ *is an optimal solution of problem (17.3) then* $\hat{x} \in X_E$.

2 *If a point* $\hat{x} \in X$ *is efficient then there exists a* $\hat{t} > 0$ *such that for every* $t \geq \hat{t}$ $\hat{x}$ *is an optimal solution of problem (17.3).*

Under certain conditions, applying the t-th power to the objective functions of nonconvex MOPs may convexify the set $Y_N + \mathbb{R}^p_{\geqq}$ so that the weighting approach can be successfully applied to generate efficient solutions of these MOPs, Li [149].

5.1.3 The Weighted Quadratic Approach. In the weighted quadratic approach a quadratic function of the objective functions is minimized:

$$\begin{aligned} \min \quad & f(x)^T Q f(x) + q^T f(x) \\ & \text{subject to } x \in X, \end{aligned} \tag{17.4}$$

where Q is a $p \times p$ matrix and q is a vector in $\mathbb{R}^p$.

THEOREM 16 *[211] Under conditions of quadratic Lagrangian duality, if* $\hat{x} \in X$ *is efficient then there exist a symmetric* $p \times p$ *matrix* Q *and a vector* $q \in \mathbb{R}^p$ *such that* $\hat{x}$ *is an optimal solution of problem (17.4).*

5.1.4 The Guddat et al. Approach. Let x^0 be an arbitrary feasible point for the MOP. Consider the following problem:

$$\begin{aligned} \min \quad & \sum_{k=1}^{p} \lambda_k f_k(x) \\ \text{subject to} \quad & f_k(x) \leq f_k(x^0), \; k = 1, \ldots, p \\ & x \in X \end{aligned} \tag{17.5}$$

where $\lambda \in \mathbb{R}^p_{\geqq}$.

THEOREM 17 *[107] Let* $\lambda \in \mathbb{R}^p_>$. *A point* $x^0 \in X$ *is an optimal solution of problem (17.5) if and only if* $x^0 \in X_E$.

In [107], this result is also generalized for scalarizations in the form of problem (17.5) with an objective function being strictly increasing on $\mathbb{R}^p$ (cf. Definition 73).

5.1.5 The ε-constraint Approach. In the ε-constraint method one objective function is retained as a scalar-valued objective while all the other objective functions generate new constraints. The k-th ε-constraint problem is formulated as:

$$\begin{array}{rrcl} \min & f_k(x) & & \\ \text{subject to} & f_i(x) & \leq & \varepsilon_i,\ i=1,\ldots,p;\ i\neq k \\ & x & \in & X. \end{array} \tag{17.6}$$

Let $\varepsilon_{-k} \in \mathbb{R}^{p-1}, \varepsilon_{-k} = (\varepsilon_1, \ldots, \varepsilon_{k-1}, \varepsilon_{k+1}, \ldots, \varepsilon_p)$. Let the set $\Psi = \{\varepsilon \in \mathbb{R}^p$: Problem (17.6) is feasible for $\varepsilon_{-k} = (\varepsilon_1, \ldots, \varepsilon_{k-1}, \varepsilon_{k+1}, \ldots, \varepsilon_p)$ for all $k = 1, \ldots, p\}$.

THEOREM 18 *[43]*

1 *If, for some $k, k \in \{1, \ldots, p\}$, there exists $\varepsilon_{-k} \in \mathbb{R}^{p-1}$ such that $\hat{x}$ is an optimal solution of problem (17.6) then $\hat{x} \in X_{wE}$.*

2 *If, for some $k, k \in \{1, \ldots, p\}$, there exists $\varepsilon_{-k} \in \mathbb{R}^{p-1}$ such that $\hat{x}$ is a unique optimal solution of problem (17.6) then $\hat{x} \in X_E$.*

3 *A point $\hat{x} \in X$ is efficient if and only if there exists $\varepsilon \in \Psi$ such that $\hat{x}$ is an optimal solution of problem (17.6) for every $k = 1, \ldots, p$ and with $f_i(\hat{x}) = \varepsilon_i, i = 1, \ldots, p, i \neq k$.*

The method of proper equality constraints is a modification of the ε-constraint method in which the constraints with the right-hand side parameters ε_i are equalities (Lin [151]).

5.1.6 The Elastic Constraint Approach. The ε-constraint approach has numerical disadvantages when applied to problems with a specific structure, in particular discrete multiobjective problems, see Ehrgott and Ryan [83]. The elastic constraint approach tries to overcome those difficulties using the following scalarization:

$$\begin{array}{rrcl} \min & f_k(x) + \sum\limits_{i\neq k} \phi_i s_i & & \\ \text{subject to} & f_i(x) + l_i - s_i & = & \varepsilon_i,\ i=1,\ldots,p;\ i\neq k \\ & s, l & \geqq & 0 \\ & x & \in & X. \end{array} \tag{17.7}$$

Here, l_i and s_i are, respectively, slack and surplus variables for the inequality constraints in problem (17.6). Thus problem (17.7) is always feasible. Furthermore, since $s_i = 0$ is desired, penalty terms are added to the objective function in (17.7) with penalty parameters ϕ_i if $s_i > 0$.

THEOREM 19 *[83]*

1 *If* $(\hat{x}, \hat{l}, \hat{s})$ *is an optimal solution of problem (17.7) then* $\hat{x} \in X_{wE}$.

2 *If* $\hat{x} \in X_{pE}$ *then there exists an* $\varepsilon \in \mathbb{R}^{p-1}$, *a* $p \in \mathbb{R}^{p-1}$ *and* $\hat{l}, \hat{s} \in \mathbb{R}^{p-1}$ *such that* $(\hat{x}, \hat{l}, \hat{s})$ *is an optimal solution of problem (17.7).*

It is worth noting that the method contains the ε-constraint method (let $p_i = \infty, i \neq k$) and the weighted sum method (let $\varepsilon_i \leq y_i^I, i \neq k$) as special cases.

5.1.7 The Benson Approach. Benson [17] introduces an auxiliary vector variable $l \in \mathbb{R}^p$ and uses a known feasible point x^0 in the following scalarization:

$$\begin{array}{rcl} \max \displaystyle\sum_{k=1}^{p} l_k & & \\ \text{subject to } f_k(x) + l_k & = & f_k(x^0),\ k = 1, \ldots, p \\ l & \geqq & 0 \\ x & \in & X. \end{array} \tag{17.8}$$

Not only can this approach find an efficient solution but it can also check whether the available point x^0 is efficient.

THEOREM 20 *[17]*

1 *The point* $x^0 \in X$ *is efficient if and only if the optimal objective value of problem (17.8) is equal to zero.*

2 *If* $(\hat{x}, \hat{l})$ *is an optimal solution of problem (17.8) with a positive optimal objective value then* $\hat{x} \in X_E$.

3 *Let the MOP be convex. If no finite optimal objective value of problem (17.8) exists then* $X_{pE} = \emptyset$.

Earlier, Charnes and Cooper [44] have proposed problem (17.8) and proved part 1 of Theorem 20.

5.1.8 Reference Point Approaches. The family of reference point approaches includes a variety of methods in which a feasible or infeasible reference point in the objective space is used. A reference point in the objective space, $r \in \mathbb{R}^p$, is typically a vector of satisfactory or desirable criterion values referred to as aspiration levels, $r_k, k = 1, \ldots, p$. However, it may also be a vector representing a currently available outcome or a worst outcome.

Distance-function-based approaches. These methods employ a distance function, typically based on a norm, to measure the distance between a utopia (or ideal) point and the points in the Pareto set. Let $d : \mathbb{R}^p \times \mathbb{R}^p \to \mathbb{R}_{\geq}$ denote a distance function. The generic problem is formulated as:

$$\begin{aligned} \min \quad & d(f(x), r) \\ & \text{subject to } x \in X, \end{aligned} \tag{17.9}$$

where $r \in \mathbb{R}^p$ is a reference point.

Under suitable assumptions satisfied by the distance function, problem (17.9) yields efficient solutions which in this case are often called compromise solutions. Let d be a distance function derived from a norm, i.e., $d(y^1, y^2) = ||y^1 - y^2||$ for some norm $|| \cdot || : \mathbb{R}^p \to \mathbb{R}_{\geq}$.

DEFINITION 72 *1 A norm $|| \cdot || : \mathbb{R}^p \to \mathbb{R}_{\geq}$ is called monotonically increasing if $||y^1|| \leq ||y^2||$ holds for all $y^1, y^2 \in \mathbb{R}^p$ with $|y^1_k| \leq |y^2_k|$, $k = 1, \ldots, p$ and $||y^1|| < ||y^2||$ holds if $|y^1_k| < |y^2_k|$, $k = 1, \ldots, p$.*

2 A norm $|| \cdot ||$ is called strictly monotonically increasing, if $||y^1|| < ||y^2||$ holds for all $y^1, y^2 \in \mathbb{R}^p$ with $|y^1_k| \leq |y^2_k|$, $k = 1, \ldots, p$ and $|y^1_k| \neq |y^2_k|$ for some k.

THEOREM 21 *[72]*

1 Let $|| \cdot || : \mathbb{R}^p \to \mathbb{R}_{\geq}$ be monotonically increasing and assume $r = y^I$.

- *If $\hat{x}$ is an optimal solution of problem (17.9) then $\hat{x} \in X_{wE}$.*
- *If $\hat{x}$ is a unique optimal solution of problem (17.9) then $\hat{x} \in X_E$.*

2 Let $|| \cdot ||$ be strictly monotonically increasing and assume $r = y^I$. If $\hat{x}$ is an optimal solution of problem (17.9) then $\hat{x} \in X_E$.

The norms studied in the literature include the family of weighted l_p-norms for $1 \leq p \leq \infty$ (Yu [231]; Zeleny [235]; Bowman [37]), a family of norms proposed by Gearhart [101], composite norms (Bardossy et al. [11]; Jeyakumar and Yano [126]) and oblique norms (Schandl et al. [185]).

Among the l_p-norms, the weighted l_∞-norm (also known as the Chebyshev or Tchebycheff norm) has been extensively studied. Since it produces all weakly efficient solutions of convex and nonconvex MOPs, it has been modified to ensure that efficient rather than weakly efficient solutions are found. The modified norms include the augmented l_∞-norm (Steuer and Choo [203]; Steuer [201]) and the modified l_∞-norm (Kaliszewski [131]).

Scalarizations based on more general distance functions such as gauges have also been considered and proved to generate weakly efficient or properly efficient solutions for convex and nonconvex MOPs (Klamroth et al. [136]). Since these approaches implicitly use not only the utopia point but also gauge-related directions, they are discussed later.

The achievement function approach. A certain class of real-valued functions $s_r : \mathbb{R}^p \to \mathbb{R}$, referred to as achievement functions, is used to scalarize the MOP. The scalarized problem is given by

$$\begin{aligned} \min \quad & s_r(f(x)) \\ \text{subject to } x \quad & \in \quad X. \end{aligned} \tag{17.10}$$

Similar to distance functions discussed above, certain properties of achievement functions guarantee that problem (17.10) yields (weakly) efficient solutions.

DEFINITION 73 *An achievement function $s_r : \mathbb{R}^p \to \mathbb{R}$ is said to be*

1 *increasing if for $y^1, y^2 \in \mathbb{R}^p$, $y^1 \leqq y^2$ then $s_r(y^1) \leq s_r(y^2)$,*

2 *strictly increasing if for $y^1, y^2 \in \mathbb{R}^p$, $y^1 < y^2$ then $s_r(y^1) < s_r(y^2)$,*

3 *strongly increasing if for $y^1, y^2 \in \mathbb{R}^p$, $y^1 \leq y^2$ then $s_r(y^1) < s_r(y^2)$.*

THEOREM 22 *[228, 229]*

1 *Let an achievement function s_r be increasing. If $\hat{x} \in X$ is a unique optimal solution of problem (17.10) then $\hat{x} \in X_E$.*

2 *Let an achievement function s_r be strictly increasing. If $\hat{x} \in X$ is an optimal solution of problem (17.10) then $\hat{x} \in X_{wE}$.*

3 *Let an achievement function s_r be strongly increasing. If $\hat{x} \in X$ is an optimal solution of problem (17.10) then $\hat{x} \in X_E$.*

Among many achievement functions satisfying the above properties we mention the strictly increasing function

$$s_r(y) \quad = \quad \max_{k=1,\ldots,p} \{\lambda_k(y_k - r_k)\}$$

and the strongly increasing functions

$$\begin{aligned} s_r(y) &= \max_{k=1,\ldots,p} \{\lambda_k(y_k - r_k)\} + \rho_1 \sum_{k=1}^{p} \lambda_k(y_k - r_k) \\ s_r(y) &= -\|y - r\|^2 + \rho_2 \|(y - r)_+\|^2, \end{aligned}$$

where $r \in \mathbb{R}^p$, $\lambda \in \mathbb{R}^p_>$ is a vector of positive weights, $\rho_1 > 0$ and sufficiently small, $\rho_2 > 1$ is a penalty parameter, and $(y - r)_+$ is a vector with components $\max\{0, y_k - r_k\}$ (Wierzbicki [228, 229]).

The weighted geometric mean approach. Consider the weighted geometric mean of the differences between the nadir point y^N and the objective functions with the weights in the exponents

$$\begin{array}{rrcl} \max & \Pi_{k=1}^{p}(y_k^N - f_k(x))^{\lambda_k} & & \\ & \text{subject to } f_k(x) & \leq & y_k^N, \; k = 1, \ldots, p \\ & x & \in & X, \end{array} \tag{17.11}$$

where $\lambda \in \mathbb{R}^p_>$. Let the MOP be convex. According to Lootsma et al. [152], an optimal solution of problem (17.11) is efficient.

Goal programming. In (GP) one is interested in achieving a desirable goal or target established for the objective functions of the MOP. The vector of these goals produces a reference point in the objective space and therefore goal programming can be viewed as a variation of the reference point approaches. Let $r \in \mathbb{R}^p$ be a goal. The general formulation of GP is

$$\begin{array}{rcl} \min & a(\delta^-, \delta^+) & \\ \text{subject to } f_k(x) + \delta_k^- - \delta_k^+ & = & r_k, \; k = 1, \ldots, p \\ x & \in & X, \end{array} \tag{17.12}$$

where $\delta^-, \delta^+ \in \mathbb{R}^p$ are variables representing negative and positive deviations from the goal r, and $a(\delta^-, \delta^+)$ is an achievement function.

A real-valued achievement function, $a : \mathbb{R}^p \times \mathbb{R}^p \to \mathbb{R}$, is typically defined as the weighted sum of deviations (weighted or non-preemptive GP) or the maximum deviation from among the weighted deviations (l_∞-GP). Solving problem (17.12) with this function results in minimizing all deviations simultaneously. A vector-valued achievement function, $a : \mathbb{R}^p \times \mathbb{R}^p \to \mathbb{R}^L$, is associated with L priority levels to which the objective functions are assigned, and results in a new GP-related MOP typically solved with the lexicographic approach based on the priority ranking (see Section 5.2.1). In this case the deviations are minimized sequentially and the technique is known as lexicographic GP. Jones and Tamiz [129] provide a recent bibliography of GP while Romero [176] presents a general achievement function including the weighted, maximum and lexicographic functions as special cases.

Whether an optimal solution of problem (17.12) is efficient for the MOP depends on the achievement function a and the goal r. In fact, when solving problem (17.12) non-Pareto criterion vectors $f(\hat{x})$ are quite common. Tamiz and Jones [208] develop tests for efficiency and methods to restore efficiency in case problem (17.12) produces a solution $\hat{x}$ that is not efficient.

Dependent on the type of goals (i.e., whether over- and/or underachievement of r_k is penalized, or only values outside a certain interval are penalized) one can define a subset of the objective space $G := \{y \in \mathbb{R}^p : y_k \geq r_k, k \in K_1;\ y_k \leq r_k, k \in K_2;\ y_k = r_k, k \in K_3;\ y_k \in [r_k^l, r_k^u], k \in K_4\}$, where $K_1 \cup K_2 \cup K_3 \cup K_4 = \{1, \ldots, p\}$, and interpret goal programming as finding a feasible solution x that is in or close to G (Steuer [201]). In this way, G can be understood as a reference set for the MOP.

The reference set approach. The concept of a reference point has been generalized by some authors. Michalowski and Szapiro [157] use two reference points to search the Pareto set of multiobjective linear programs. Skulimowski [191] studies the notion of a reference set. Simple examples of reference sets include the sets $r - C$ or $(r^1 - C) \cap (r^2 + C)$ for $r^2 \succ_C r^1$, where r, r^1, r^2 are reference points and C is a closed, convex, and pointed cone. Under suitable conditions, a solution to the MOP obtained by means of minimizing the distance from a reference set is nondominated. Cases in which a reference set can be reduced to a reference point are also examined in [191].

5.1.9 Direction-based Approaches. This group of scalarizing approaches employs a reference point $r \in \mathbb{R}^p$, a direction in the objective space along which a search is performed, and a real variable α measuring the progress along the direction.

The Roy approach. Perhaps the first approach of that kind has been proposed by Roy [177, p. 242] which (slightly reformulated) can be written as

$$\begin{array}{rcl} \max & \alpha & \\ \text{subject to } f_k(x) + \alpha e & \leq & r_k,\ k = 1, \ldots, p \\ x & \in & X, \end{array} \tag{17.13}$$

where $e \in \mathbb{R}^p$ is a vector of ones and determines the fixed direction of search. Depending on the choice of the reference point r the approach finds a (weakly) efficient solution.

The goal-attainment approach. Given a (feasible or infeasible) goal vector r and a direction $d \leq 0$ along which the search is performed the goal-attainment approach is formulated as

$$\begin{array}{rcl} \max & \alpha & \\ \text{subject to } f_k(x) - \alpha d_k & \leq & r_k, k = 1, \ldots, p \\ x & \in & X, \end{array} \tag{17.14}$$

and produces a weakly efficient point (Gembicki and Haimes [102]) .

The Pascoletti and Serafini approach. This is a more general approach with an unrestricted search direction $d \in \mathbb{R}^p$ and an auxiliary vector variable $l \in \mathbb{R}^p$:

$$\begin{aligned} \max \quad & \alpha \\ \text{subject to } f_k(x) - \alpha d_k + l_k &= r_k, \ k = 1, \ldots, p \\ l &\geqq 0 \\ x &\in X. \end{aligned} \tag{17.15}$$

THEOREM 23 *[168]*

1 If $(\hat{\alpha}, \hat{x}, \hat{l})$ is a finite optimal solution of problem (17.15) then $\hat{x} \in X_{wE}$.

2 If $(\hat{\alpha}, \hat{x}, \hat{l})$ is a unique finite optimal solution of problem (17.15) then $\hat{x} \in X_E$.

The reference direction approach. Independently of [102], Korhonen and Wallenius [139] propose an approach analogous to the goal-attainment method and refer to it as a generalized goal programming model. With the inclusion of nonnegative slack variables they arrive at problem (17.15) with a feasible reference point $r \in Y$ and a search direction $d \leq 0$. Additionally, to move on the Pareto set they parametrize the reference point r or the search direction d, and obtain:

$$\begin{aligned} \max \quad & \alpha \\ \text{subject to } f_k(x) - \alpha(d_k + \alpha_1 \Delta d_k) + l_k &= (r_k + \alpha_2 \Delta r_k), \ k = 1, \ldots, p \\ l &\geqq 0 \\ x &\in X, \end{aligned} \tag{17.16}$$

where $\Delta d = [\Delta d_1, \ldots, \Delta d_p]$ and $\Delta r = [\Delta r_1, \ldots, \Delta r_p]$ are auxiliary vectors used for the parametrization with the parameters $\alpha_1 \geq 0$ and $\alpha_2 \geq 0$. The vector Δr is called the reference direction and problem (17.16) is called the reference direction approach (see also [133, 140]).

The modified Pascoletti and Serafini approach. Since a solution to problem (17.15) may not be finite, the following modification has been developed:

$$\begin{aligned} \text{lex max} \quad & (\alpha, \|l\|_p) \\ \text{subject to } f_k(x) - \alpha d_k + l_k &= r_k, k = 1, \ldots, p \\ l &\geqq 0 \\ x &\in X. \end{aligned} \tag{17.17}$$

THEOREM 24 *[186] Let $r \in Y + \mathbb{R}^p_{\geqq}$, $d \in \mathbb{R}^p \setminus \mathbb{R}^p_{\geqq}$ and $1 \leq p \leq \infty$. Then problem (17.17) has a finite optimal solution $(\hat{\alpha}, \hat{x}, \hat{l})$ where $\hat{x} \in X_E$.*

The normal boundary intersection approach. The approach by Das and Dennis [59] is motivated by the interest in obtaining an evenly distributed set of Pareto solutions. Let y^I be the ideal point and let Φ be the $p \times p$ matrix the k-th column of which is given by $f(x^k) - y^I$, where x^k is a global minimizer of the objective function $f_k, k = 1, \ldots, p$. The set of points in $\mathbb{R}^p$ that are convex combinations of $f(x^k) - y^I$, i.e., $\{\Phi\lambda : \lambda \in \mathbb{R}^p_{\geq}, \sum_{k=1}^p \lambda_k = 1\}$, is referred to as the convex hull of the individual global minima of the objective functions (CHIM). Let a search direction be given by the unit normal, denoted by ξ, to the CHIM pointing toward the origin. Consider the following SOP:

$$\begin{aligned} \max \quad & \alpha \\ \text{subject to } f(x) - \alpha\xi - \Phi\lambda &= y^I \\ x &\in X. \end{aligned} \tag{17.18}$$

When the approach is iteratively applied to convex MOPs with evenly distributed coefficients λ_k of the convex combination, the authors claim and demonstrate by examples, that an evenly distributed set of Pareto solutions is produced. A scalarization using the CHIM and a direction normal to it is developed by Ismail-Yahaya and Messac [124].

5.1.10 Gauge-based Approaches. Assume without loss of generality that $0 \in Y + \mathbb{R}^p_{\geqq}$. Let B be a polytope in $\mathbb{R}^p$ containing the origin in its interior and let $y \in \mathbb{R}^p$. The polyhedral gauge $\gamma : \mathbb{R}^p \to \mathbb{R}$ of y is defined as $\gamma(y) = \min\{\alpha \geq 0 : y \in \alpha B\}$. The vectors defined by the extreme points of the unit ball B of γ are called fundamental vectors and are denoted by ν^i.

Gauges are used to measure the distance in the objective space either in the interior or the exterior of the outcome set. The former leads to the inner scalarization while the latter is related to the outer scalarization.

The inner gauge-based approach. Consider the gauge problem

$$\begin{aligned} \max \quad & \gamma(y) \\ \text{subject to } y &= f(x) \\ y &\in Y \cap (-\mathbb{R}^p_{\geqq}). \end{aligned} \tag{17.19}$$

THEOREM 25 *[136] Let the set Y be strictly* int $\mathbb{R}^p_{\geqq}$*-convex, i.e.,* $Y + \text{int}\,\mathbb{R}^p_{\geqq}$ *is strictly convex. If $\hat{x}$ is an optimal solution of problem (17.19) then $\hat{x} \in X_{pE}$.*

The outer gauge-based approach. Let B be the unit ball of a polyhedral gauge γ such that the fundamental vectors $\nu^1, \ldots, \nu^t$ of $B \cap (-\mathbb{R}^p_{\geqq})$ satisfy

$Y \cap (-\mathbb{R}^p_{\geqq}) \subseteq \{y \leqq 0 : y \geqq \sum_{i=1}^t \lambda_i \nu^i, \sum_{i=1}^t \lambda_i = 1, \lambda_i \geq 0\}$ and consider the problem

$$\begin{array}{rrcl} & \max & \alpha & \\ \text{subject to} & \alpha\nu^i - f(x) & \geqq & 0,\ i = 1, \ldots, t \\ & \alpha & \geq & 0 \\ & x & \in & X. \end{array} \tag{17.20}$$

THEOREM 26 *[136] Let the set* Y *be strictly* int $\mathbb{R}^p_{\geqq}$*-convex. If* $(\hat{\alpha}, \hat{x})$ *is an optimal solution of problem (17.20) then* $\hat{x} \in X_{pE}$.

A combination of each of the gauge-based approaches together with the weighted l_∞ norm method results in two other approaches generating weakly efficient solutions of nonconvex MOPs (Klamroth et al. [136]).

5.1.11 Composite and Other Approaches. In order to achieve certain properties of scalarizations, some authors develop composite approaches involving a combination of methods. The hybrid method is composed of the weighted sum and the ϵ-constraint approaches (Wendell and Li [220]). More generally, an increasing scalarizing function can be combined with constraints on some or all objective functions (Soland [193]). Dual approaches include the ε-constraint approach coupled with Lagrangian duality (Chankong and Haimes [43]) or generalized Lagrangian duality (TenHuisen and Wiecek [209, 210]). The method of exact penalty functions is based on the weighted sum approach equipped with exact penalty terms (Bernau [26]).

A very general scalarization method using continuous functionals has been proposed by Gerth and Weidner [104] and Ester and Tröltzsch [87]. (Weakly, properly) nondominated points of (nonconvex) MOPs are characterized through the existence of functionals with certain properties.

5.2 Nonscalarizing Approaches

In contrast to scalarizing approaches discussed in Section 5.1, nonscalarizing methods do not explicitly use a scalarizing function but rather rely on other optimality concepts or auxiliary sets. In effect, there are usually strong links to efficiency. In the following sections we summarize the most important approaches.

5.2.1 The Lexicographic Approach. The lexicographic approach assumes a ranking of the objective functions according to their importance. Let π be a permutation of $\{1, \ldots, p\}$ and assume that $f_{\pi(k)}$ is more important than $f_{\pi(k+1)}$, $k = 1, \ldots, p-1$. Let $f^\pi : \mathbb{R}^n \to \mathbb{R}^p$ be $(f_{\pi(1)}, \ldots, f_{\pi(p)})$.

We denote $y^1 \succeq_{lex} y^2$ if $y^1 = y^2$ or there is some $k, 1 \leq k \leq p$, such that $y_i^1 = y_i^2, i = 1, \ldots, k-1$ and $y_k^1 < y_k^2$. The lexicographic problem is formulated as

$$\begin{aligned} \text{lex min} \quad & f^\pi(x) \\ \text{subject to } x \quad & \in \quad X. \end{aligned} \tag{17.21}$$

This problem is solved as follows. Let $X_0^\pi = X$ and define recursively $X_k^\pi := \{\hat{x} \in X_{k-1}^\pi : f_{\pi(k)}(\hat{x}) = \min f_{\pi(k)}(x), x \in X_{k-1}^\pi\}$ for $k = 1, \ldots, p$.

THEOREM 27 *Let π be a permutation of $\{1, \ldots, p\}$.*

1 *If $X_k^\pi = \{\hat{x}\}$ is a singleton then $\hat{x}$ is an optimal solution of problem (17.21) and $\hat{x} \in X_E$.*

2 *All elements of X_p^π are optimal solutions of problem (17.21) and $X_p^\pi \subset X_E$.*

Note that the inclusion $\bigcup_\pi X_p^\pi \subset X_E$ is usually strict.

5.2.2 The Max-ordering Approach. The max-ordering approach does only consider the objective function f_k which has the highest (worst) value. The preference relation of the max-ordering approach is $y^1 \succeq_{MO} y^2$ if $\max_{k=1,\ldots,p} y_k^1 \leq \max_{k=1,\ldots,p} y_k^2$. This preference relation does not define a constant cone as described in Section 2. The max-ordering problem is formulated as

$$\begin{aligned} \min \max_{k=1,\ldots,p} \quad & f_k(x) \\ \text{subject to } x \quad & \in \quad X. \end{aligned} \tag{17.22}$$

An optimal solution of problem (17.22) is weakly efficient. Furthermore, if $X_E \neq \emptyset$ and problem (17.22) has an optimal solution then there exists an optimal solution of problem (17.22) which is efficient, and consequently a unique optimal solution of problem (17.22) is efficient.

It is possible to include a weight vector $\lambda \in \mathbb{R}^p_{\geq}$ so that the weighted max-ordering problem becomes

$$\begin{aligned} \min \max_{k=1,\ldots,p} \quad & \lambda_k f_k(x) \\ \text{subject to } x \quad & \in \quad X. \end{aligned} \tag{17.23}$$

THEOREM 28 *[145] Let $Y \subset \mathbb{R}^p_{>}$.*

1 *Let $\lambda \in \mathbb{R}^p_{\geq}$. If $\hat{x} \in X$ is an optimal solution of problem (17.23) then $\hat{x} \in X_{wE}$.*

2 *If $\hat{x} \in X_{wE}$ there exists a $\lambda \in \mathbb{R}^p_>$ such that $\hat{x}$ is an optimal solution of problem (17.23)*

A result similar to Theorem 28 can be proved for efficient solutions when an additional SOP to minimize $\sum_{k=1}^p f_k(x)$ subject to $\lambda_k f_k(x) \leq y^*$, where y^* is the optimal value of problem (17.23), is solved.

The max-ordering approach plays an important role in robust optimization, where each objective function f_k is interpreted as the objective for making a decision in a scenario k, see Kouvelis and Yu [145].

5.2.3 The Lexicographic Max-ordering Approach. The idea of the max-ordering approach can be extended to consider the second worst, third worst, etc., objective. That means, for $x \in X$ one reorders the components of $f(x)$ in nonincreasing order. For $y \in \mathbb{R}^p$ let $\Theta(y) = (\theta_1(y), \ldots, \theta_p(y))$ be a permutation of y such that $\theta_1(y) \geq \ldots \geq \theta_p(y)$. The lexicographic max-ordering approach then seeks to lexicographically minimize $\Theta(f(x))$ over the feasible set X. This corresponds to seeking preferred outcomes according to the lexicographic max-ordering relation $y^1 \succeq_{lex-MO} y^2$ if $\theta(y^1) \succeq_{lex} \theta(y^2)$.

It is easy to see that an optimal solution of the lexicographic max-ordering problem is also an optimal solution of the max-ordering problem and efficient for the MOP, which strengthens the corresponding result for the max-ordering approach. When a weight vector $\lambda \in \mathbb{R}^p_>$ is introduced (again assuming that $Y \subset \mathbb{R}^p_>$), the problem becomes

$$\begin{aligned} \text{lex min} \quad & (\lambda_1\theta_1(f(x)), \ldots, \lambda_p\theta_p(f(x))) \\ & \qquad\qquad \text{subject to } x \in X. \end{aligned} \tag{17.24}$$

THEOREM 29 *[71] A point $\hat{x} \in X$ is efficient if and only if there exists a $\lambda \in \mathbb{R}^p_>$ such that $\hat{x}$ is an optimal solution of (17.24).*

For problems with a special structure further results can be obtained. For convex problems, Behringer [16] shows that problem (17.24) can be solved through solving a sequence of max-ordering problems. A solution of problem (17.24) is also known as a nucleolar solution, Marchi and Oviedo [155]. Ehrgott and Skriver [84] give an algorithm for the bicriteria discrete case and Sankaran [179] provides one for the convex case.

5.2.4 The Equitability Approach. Kostreva and Ogryczak [142] introduce the concept of equitability in multiobjective programming. This concept is applicable if the objective functions are anonymous and measured on a common scale (or normalized to a common scale).

While the Pareto preference assumes a binary relation between outcome vectors that is reflexive, transitive and strictly monotone ($y - \epsilon e_i$ is preferred to

y for $\epsilon > 0$ and $k = 1, \ldots, p$, where $e_i \in \mathbb{R}^p$ is the i-th unit vector), equitability makes two further assumptions. It is assumed that there is indifference between y and y' if there is a permutation π such that $y' = (y_{\pi(1)}, \ldots, y_{\pi(p)})$. The second assumption is the principle of transfers: Let $y \in \mathbb{R}^p$ and $y_{k'} > y_{k''}$, then $y - \epsilon e_{k'} + \epsilon e_{k''}$ is preferred to y for $0 < \epsilon < y_{k'} - y_{k''}$. A preference relation $\succeq_{equi}$ satisfying these assumptions is not derived from a cone as described in Section 2.

In order to obtain equitably efficient solutions of the MOP one proceeds as follows. For $y \in \mathbb{R}^p$ let $\Theta(y) = (\theta_1(y), \ldots, \theta_p(y))$ be as in Section 5.2.3. Next, using this vector of ordered outcomes define the cumulative ordered outcome vector $\bar{\Theta}(y) = (\bar{\theta}_1(y), \ldots, \bar{\theta}_p(y))$, where

$$\bar{\theta}_k(y) = \sum_{i=1}^{k} \theta_i(y).$$

The equitability preference can then be defined by $y^1 \succeq_{equi} y^2$ if and only if $\bar{\Theta}(y^1) \succeq_{\mathbb{R}^p_{\geqq}} \bar{\Theta}(y^2)$. An equitable MOP can be written as

$$\begin{aligned} \min \quad & (\bar{\theta}_1(f(x)), \ldots, \bar{\theta}_p(f(x))) \\ & \text{subject to } x \in X. \end{aligned} \tag{17.25}$$

The relationship between equitably efficient solutions and efficient solutions is provided by the following theorem.

THEOREM 30 *An efficient solution of problem (17.25) is an equitably efficient solution of the MOP.*

To generate equitably efficient solutions, scalarization using strictly convex functions can be applied. Kostreva and Ogryczak [142] show that any optimal solution of $\min_{x\in X} \sum_{k=1}^{p} s(f_k(x))$, where $s : \mathbb{R} \to \mathbb{R}$ is strictly convex and increasing, is equitably efficient. Ogryczak applies the equitability approach to solve portfolio optimization problems [166] and location problems [165].

5.2.5 The Balance and Level Set Approaches. Galperin [97] introduces the balance space approach. The approach is based on sets of points with a bounded deviation from global minima of the individual objective functions. For $k = 1, \ldots, p$ define the sets

$$X_k(\eta) := \{x \in X : f_k(x) - y_k^I \leq \eta_k\}. \tag{17.26}$$

Then $\eta \in \mathbb{R}^p$ is called a balance point if the intersection $\cap_{k=1}^{p} X_k(\eta) \neq \emptyset$ but for any $\eta', \eta' \leq \eta$, the intersection $\cap_{k=1}^{p} X_k(\eta') = \emptyset$. The set of all balance points for the MOP is called the balance set denoted by Υ. Galperin and Wiecek [98] demonstrate the applicability of this approach on example problems. Ehrgott et al. [81] show that the balance set is equal to the Pareto set translated by the ideal point: $\Upsilon = Y_N - y^I$.

It is possible to require all η_k to be equal to one another in (17.26). The smallest $\eta \in \mathbb{R}$ such that the intersection $\cap_{k=1}^{p} X_k(\eta) \neq \emptyset$ is then called the balance number which, according to Ehrgott and Galperin [74], can be found by solving the problem

$$\begin{aligned} \min \max_{k=1,\dots,p} \quad & (f_k(x) - y_k^I) \\ \text{subject to } x \quad & \in \quad X. \end{aligned} \tag{17.27}$$

The definition of balance number can be extended to include weights, i.e., $\eta_k = \eta\lambda_k$ in (17.26). For positive weights $\lambda \in \mathbb{R}^P_>$ the smallest η with the nonempty intersection of $X_k(\eta)$ (called the apportioned balance number $\eta(\lambda)$) can again be computed via a min-max problem. Ehrgott [73] compares the sets Υ and $\{\eta(\lambda)\lambda : \sum_{k=1}^{p} \lambda_k = 1, \lambda \in \mathbb{R}^p_>\}$ and gives conditions for them being equal.

The balance space approach is closely related to the level set approach. The level set of objective function f_k with respect to $\bar{x} \in X$ is

$$L^k_{\leq}(\bar{x}) := \{x \in X : f_k(x) \leq f_k(\bar{x})\},$$

the strict level set is

$$L^k_{<}(\bar{x}) := \{x \in X : f_k(x) < f_k(\bar{x})\},$$

and the level curve islevel curve

$$L^k_{=}(x) := \{x \in X : f_k(x) = f_k(\bar{x})\}.$$

Therefore the sets $X_k(\eta)$ used in the balance space approach are level sets of f_k with respect to levels $y_k^I + \eta_k$. The main result on level sets is the following theorem.

THEOREM 31 *[81]*

1 A point $\hat{x} \in X$ is weakly efficient if and only if $\cap_{k=1}^{p} L^k_{<}(\hat{x}) = \emptyset$.

2 A point $\hat{x} \in X$ is efficient if and only if $\cap_{k=1}^{p} L^k_{\leq}(\hat{x}) = \cap_{k=1}^{p} L^k_{=}(\hat{x})$.

3 A point $\hat{x} \in X$ is strictly efficient if and only if $\cap_{k=1}^{p} L^k_{\leq}(\hat{x}) = \{\hat{x}\}$.

This geometric characterization of efficiency can be exploited when dealing with problems that have a geometric structure, e.g., location problems as in Ehrgott et al. [82].

5.2.6 The ϵ-Efficiency Approach. Let $\{\epsilon^\tau\}$ be a sequence in $\mathbb{R}^p$ with $\lim_{\tau \to \infty} \epsilon^\tau = 0$. Lemaire [148] introduces a notion of a sequence of auxiliary MOPs converging to the original MOP and studies properties of the sequence of ϵ^τ-efficient sets of these problems with respect to the efficient set X_E of the original problem. In particular, he shows that every weakly efficient point of the MOP can be obtained as a limit of a sequence of ϵ^τ-efficient points of the auxiliary problems.

6. Approximation of the Pareto Set

It is of interest to design methods for obtaining a complete description of the Pareto and efficient sets since solving MOPs is understood as finding these sets. An exact description might be available analytically as a closed-form formula, numerically as a set of points, or in a mixed form as a parameterized set of points.

Unfortunately, for a majority of MOPs it is not easy to obtain an exact description of the Pareto set that typically includes a very large or infinite number of points. Even if it is theoretically possible to find these points exactly, this is computationally challenging and expensive and therefore usually abandoned. For some other problems finding elements of the Pareto set is even impossible due to numerical complexity of resulting optimization problems.

Since the exact solution set is very often not attainable, an approximated description of this set becomes an appealing alternative. Approximating approaches have been developed for the following purposes: to represent the Pareto set when this set is numerically available (linear or convex MOPs); to approximate the Pareto set when some but not all Pareto points are numerically available (nonlinear MOPs); and to approximate the Pareto set when Pareto points are not numerically available (discrete MOPs).

For any MOP, the approximation requires less effort and often may be accurate enough to play the role of the solution set. Additionally, if the approximation represents this set in a simplified, structured, and understandable way, it may effectively support the decision-maker. An important aspect of the approximating approaches includes the approximation quality and a measure for evaluating it.

Approximation approaches employ an iterative method to produce points or objects approximating the Pareto set. Some approaches are exact and based on algorithms equipped with theoretical proofs for correctness and optimality while some other approaches are heuristic and often theoretically unsupported.

6.1 Exact Approaches

A majority of exact approaches employ a scalarization technique as an integrated component of the resulting approximating algorithm. The scalarization is used to generate Pareto points that either become the final approximation or are used to construct other approximating objects such as polyhedral sets and functions, curves, and rectangles. For a review of exact approximation techniques we refer to Ruzika and Wiecek [178].

6.1.1 Point-wise approximation. Point-wise approximation approaches produce a set of Pareto points obtained by means of a scalarization method of choice. Approaches for continuous biobjective programs (BOPs) are proposed by Helbig [115] and Jahn and Merkel [125] and for discrete BOPs by Schandl et al. [184].

Various methods have been developed for MOPs. Approximation by means of a finite set of elements is studied by Nefedov [162]with special attention given to convergence of the approximating set to the Pareto set. Helbig [115] presents an approach to approximate the nondominated set of general MOPs with convex cones. A global shooting procedure to find a representation of the Pareto set for problems with compact outcome sets is proposed by Benson and Sayin [24] while an approach producing representative subsets of the Pareto set for linear MOPs is introduced by Sayin [182]. A method based on an interior point algorithm for convex quadratic MOPs is given by Fliege and Heseler [90]. A target-level method using an infinite set of reference points and the l_∞-norm is proposed by Churkina [47].

6.1.2 Piece-wise linear approximation. Approaches producing piece-wise linear approximations first generate Pareto points using a scalarization method and then construct approximating polyhedral sets or functions.

For BOPs, the generated Pareto points are connected with line segments. For convex BOPs, Cohon [51] and Poliščuk [175] develop similar inner approximations while Cohon et al. [52], Fruhwirth et al. [93], and Yang and Goh [230] propose sandwich approximations composed of inner and outer approximations.

Approaches for linear MOPs are proposed by Voinalovich [215], whose method yields a system of linear inequalities as an outer approximation, by Solanki et al. [195], who extends the sandwich approach of Cohon et al. [52], and by Benson [23], whose algorithm produces the Pareto set exactly.

Chernykh [45] approximates the set $Y + \mathbb{R}^p_{\geqq}$ for a convex set Y in the form of a system of linear inequalities. A cone-based approach for general MOPs is proposed by Kaliszewski [132]. Kostreva et al. [144] develop a method using simplices, which is applicable to MOPs with discontinuous objective functions and/or a disconnected feasible set. A brief outline of an approach based on the normal-boundary intersection technique is offered by Das [58].

Schandl et al. [186] and Klamroth et al. [136] base their approaches on polyhedral distance functions that are constructed successively during the execution of the algorithm and utilize both to evaluate the quality of the approximation and to generate additional Pareto points. Norms and gauges are used for convex MOPs while nonconvex functions are used for nonconvex MOPs. The approximation itself is used to define a problem-dependent distance function and is independent of objective function scaling. In Klamroth et al. [136], inner and outer approximation approaches are proposed for convex and nonconvex MOPs and in all cases the approximation improves where it is currently the worst, a unique property among the approximation approaches.

6.1.3 Nonlinear approximation. This group of methods includes approaches producing quadratic, cubic, and other approximations for BOPs. Wiecek et al. [225] use piece-wise quadratic approximating functions while Payne et al. [171] and Polak [174] use cubic functions to interpolate the Pareto set. Other structures used for approximation purposes include rectangles (Payne and Polak [170] and Payne [169]) and the hyper-ellipse (Li et al. [150]). Each of these nonlinear functions may provide a closed-form approximating formula.

6.2 Heuristic Approaches

For MOPs Pareto solutions of which are difficult or impossible to obtain due to the computational effort involved, one resorts to heuristic approaches for approximating the Pareto set. This is often done for large-scale discrete and nonconvex continuous MOPs. Both problem types share the feature that due to nonconvexity there exist Pareto solutions that cannot be generated by the very popular weighted sum approach. An in-depth review of heuristic methods for multiobjective combinatorial optimization can be found in [80]

6.2.1 Parameter Space Investigation. Statnikov and Matusov describe this method in [198, 199]. They assume that the MOP is given by a set S defined by box constraints $\underline{x_i} \leq x_i \leq \overline{x_i}, i = 1, \ldots, n$, functional constraints $l_j \leq g_j(x) \leq u_j, j = 1, \ldots, m$, and objective functions f_k. The method includes an interactive part to elicit reservation levels $\bar{f}_k$ above which objective values are unacceptable, i.e., $f_k(x) \leq \bar{f}_k, k = 1, \ldots, p$.

The method consists in selecting N trial points $x^1, \ldots, x^N$, evaluating $f_k(x^i)$, $k = 1, \ldots, p$; $i = 1, \ldots, N$, and sorting the values for each criterion nondecreasingly. Having considered the sorted values $f_k(x^i)$ the decision maker is then asked to specify $\bar{f}_k$. If the trial set does not contain a point that is below all reservation levels, these are changed or more trial points are selected. Finally, dominated solutions are removed from the trial set.

The most important step is the choice of trial points, which is based on sequences of uniformly distributed points such as P_τ and LP_τ nets (see [198] for details). Matusov and Statnikov [199] show that the set $\{y \in Y : y \leq \bar{f}_k\}$ and the set of Pareto points contained therein can be approximated to any given accuracy, when enough trial points are chosen. Sobol' and Levitan [192] find an estimate of the approximation error.

6.2.2 Classic Heuristics. Heuristic methods are mainly developed for discrete optimization problems. Heuristic schemes such as the greedy algorithm or local search can be applied in a multiobjective context. In single objective optimization, the greedy algorithm constructs a feasible solution by extending a partial solution in a way that deteriorates the objective function least. Local search algorithms iteratively explore a neighborhood of a feasible solution and choose the best solution in the neighborhood. Although multiobjective versions of greedy and local search algorithms can be formulated (by extending "deteriorates" and "best" to the Pareto concept), these schemes are mostly applied to scalarized MOPs using a set of values for the parameters of the scalarizing function, e.g., a set of weight vectors λ in the weighted sum scalarization.

Serafini [188] presents a greedy algorithm based on topological orders and shows that all efficient bases of a multiobjective matroid problem can be found using appropriate topological orders. Corley [55] as well as Lind et al. [7] use a local search based on an exchange neighborhood for multiobjective spanning tree problems.

Very few results are available on the quality of the approximation of the Pareto set for multiobjective discrete problems. Warburton [217] gives an ε-approximation algorithm for the Pareto set of the multiobjective shortest path problem and Erlebach et al. [86] develop a fully polynomial time approximation scheme for the multiobjective knapsack problem.

When heuristics are combined with lower bounds, it is possible to extend the concept of lower and upper bounds on optimal objective values to the multiobjective case. Ehrgott and Gandibleux [76] pursue that approach.

6.2.3 Local-search-based Metaheuristics. Extensions of simulated annealing and tabu search methods to multiobjective programming are most often found for discrete MOPs.

MOSA by Ulungu [212] is a prototype of a multiobjective simulated annealing method. For a set of weighting vectors λ a simulated annealing procedure is performed on the problem scalarized with the weighted sum method. A starting solution x is randomly chosen, a solution x' in some neighborhood of x is selected and compared with x. If $f(x') \leq f(x)$ or $\sum_{k=1}^{p} \lambda_k f_k(x') < \sum_{k=1}^{p} \lambda_k f_k(x)$, x' is accepted as a better solution, otherwise it is accepted with some probability that decreases in the course of the iterations. The result is a set of potentially efficient solutions in direction λ. After the procedures for all λ are completed, the corresponding sets of potentially efficient solutions are merged. Variations and applications of simulated annealing for MOPs can be found in Czyzak and Jaszkiewicz [56], Engrand [85], Nam and Park [161], Parks and

Suppapitnarm [167] and Serafini [189]. See Ehrgott and Gandibleux [79] for more details.

Multiobjective tabu search methods are also based on neighborhood search principles. Gandibleux's MOTS [99] starts from an initial solution x and chooses new solutions in a neighborhood of the current solution based on a scalarization using a weighted distance from a utopia point y^U. In order to overcome local optima, some solutions in the neighborhood are declared "tabu". The tabu status depends on the iterations performed so far. A set of weights is chosen prior to the start of the algorithm, and for each weight a single objective tabu search procedure is performed. At the end of the algorithm the sets of potentially efficient solutions are merged as in the simulated annealing case. Other tabu search implementations are presented by Ben Abdelaziz et al. [1], Baykasoglu et al. [12], Hansen [113] and Sun [205]. These methods have been applied to a number of problems, see again Ehrgott and Gandibleux [79] for more details.

6.2.4 Population-based Metaheuristics. While local search based metaheuristics typically work with one solution at a time, population based methods maintain a whole set of solutions (the population) and try to evolve the population towards the Pareto set. Many different techniques, described in the literature as evolutionary and genetic algorithms, have been developed to evaluate the fitness of individual solutions in a multiobjective context and guarantee enough diversity to achieve a uniform distribution of solutions over the whole Pareto set. Research on this topic was initiated by Schaffer's vector evaluated genetic algorithm (VEGA, Schaffer [183]). Important references in this area include Fonseca and Fleming [91], Horn et al. [118], Murata and Ishibuchi [159], Srinivas and Deb [197], and Ziztler and Thiele [238], the survey papers by Coello [48, 49], Fonseca and Fleming [92], Jones et al. [128], and the books by Deb [64] and Coello [50].

6.2.5 Other Heuristic Approaches. Other heuristic approaches have been proposed in recent years. These include artificial neural networks (Malakooti et al. [154], Sun et al. [206, 207]), a greedy randomized adaptive search procedure (Gandibleux et al. [100]), ant colony systems (Dörner et al. [66], Gravel et al. [106], Iredi et al. [120], Shelokar et al. [190]), and a scatter search (Beausoleil [15]).

7. Specially Structured Problems

7.1 Multiobjective Linear Programming

A linear MOP is the following problem:

$$\begin{aligned} \min \quad & Cx & & \\ \text{subject to } Ax & = & b \\ x & \geqq & 0, \end{aligned} \tag{17.28}$$

where C is a $p \times n$ objective function matrix, A is an $l \times n$ constraint matrix, and $b \in \mathbb{R}^l$. It is usually assumed that the rows of A are linearly independent. For ease of exposition we shall assume that X is bounded, therefore compact. In consequence Y is also a compact polyhedron. Throughout this section we do not discuss issues arising from degeneracy which is important for Simplex type MOLP algorithms. Degeneracy is addressed in some of the papers referred to in Section 7.1.1.

In this section, we review methods for finding efficient solutions of multiobjective programs of type (17.28) referred to as multiobjective linear programs (MOLPs). Because problem (17.28) is a special case of a convex MOP, all efficient solutions can be found by the weighted sum scalarization. For $\lambda \in \mathbb{R}^p_{\geq}$, let LP($\lambda$) denote the linear program $\min\{\lambda^T Cx : Ax = b, x \geqq 0\}$.

THEOREM 32 *[123] $\hat{x} \in X_E$ if and only if $\hat{x}$ is an optimal solution of LP(λ) for some $\lambda \in \mathbb{R}^p_{>}$.*

In view of Geoffrion's Theorem 14, Theorem 32 implies that $X_E = X_{pE}$ for MOLPs. The polyhedral structure of X and Y allows a more thorough investigation of the efficient and Pareto sets (Fruhwirth and Mekelburg [94] present a detailed analysis of the structure of Y_N for the case of $p = 3$ criteria). Let F be a face of X. If $F \subset X_E$, it is called an efficient face. F is called a maximal efficient face if it is an efficient face and for all faces F' such that $F \subset F'$, F' is not an efficient face.

THEOREM 33

1. *$X_E = X$ if and only if there exists $x^0 \in \operatorname{ri} X$ such that $x^0 \in X_E$. Otherwise $X_E \subset \operatorname{bd} X$ and $X_E = \cup_{j \in J} F_j$, where F_j are maximal efficient faces and J is a finite index set.*
2. *Let F be a face of X. F is an efficient face if and only if there exists $\hat{x} \in \operatorname{ri} F$ such that $\hat{x} \in X_E$.*
3. *Let F be a face of X and $F = \operatorname{conv}\{x^1, \ldots, x^q\}$. Then $F \subset X_E$ if and only if $\{x^1, \ldots, x^q\} \in X_E$.*
4. *For each maximal efficient face F_i there is a subset $\Lambda_i \subset \{\lambda \in \mathbb{R}^p_{>} : \sum_{k=1}^p \lambda_k = 1\}$ such that all $x \in F_i$ are optimal for LP(λ) for all $\lambda \in \Lambda_i$.*

The decomposition of the weight space indicated in the last point of Theorem 33 can be further elaborated. Let $\Lambda := \{\lambda \in \mathbb{R}^P_> : \sum_{i=1}^p \lambda_i = 1\}$ denote the set of weights. Theorem 32 suggests a decomposition of Λ into subsets, such that for each λ in a subset LP(λ) has the same optimal solutions. Such a partition can be attempted with respect to efficient bases (see page 698) of the MOLP or with respect to extreme points of X_E or Y_N. Some of the simplex based algorithms mentioned below use such decompositions. The main results for weight set decomposition with respect to extreme points of Y_N are summarized in the following theorem. We assume that Y is of dimension p.

THEOREM 34 *[25] Let $\{y^1, \ldots, y^q\}$ be the Pareto extreme points of Y and $\Lambda(y) = \{\lambda \in \mathbb{R}^p : \lambda^T y \leq \lambda^T y' \text{ for all } y' \in Y\}$. Then the following statements hold.*

- $\Lambda \subset \cup_{i=1}^q [\Lambda(y^i) \cap \Lambda]$.
- $\operatorname{int} \Lambda(y^i) \neq \emptyset$, $\Lambda \cap \operatorname{int} \Lambda(y^i) = \operatorname{int}(\Lambda \cap \Lambda(y^i)) \neq \emptyset$.
- *If $\lambda \in \Lambda \cap \operatorname{int} \Lambda(y^i)$ then y^i is a unique optimal solution of the problem* $\max\{\lambda^T y : y \in Y\}$.
- $[\Lambda \cap \operatorname{int} \Lambda(y^i)] \cap [\Lambda \cap \Lambda(y^j)] = \emptyset$ *and therefore* $\Lambda \cap \Lambda(y^i) \neq \Lambda \cap \Lambda(y^j)$ *when* $i \neq j$.
- *If F is a proper face of Y and $\bar{y}, y^* \in \operatorname{ri} F$ then $\Lambda(\bar{y}) = \Lambda(y^*)$.*

Due to Theorem 32, any MOLP can in principle be solved using parametric linear programming. However, simplex, interior point, and objective-space methods have also been developed to deal with MOLPs directly.

7.1.1 Multicriteria Simplex Methods. Some notation is first needed to explain multicriteria simplex algorithms.

- An extreme point (zero dimensional face) of X that is efficient, is called efficient extreme point.
- A basis B of (17.28) (an index set of l linearly independent columns of A) is called efficient basis if there exists a $\lambda \in \mathbb{R}^p_>$ such that B is an optimal basis of LP(λ).
- Let B be a basis and $N := \{1, \ldots, n\} \backslash B$. Let C_B and C_N be the columns of C indexed by B and N, respectively. A_B, A_N, x_B and x_N are defined accordingly. Then $\bar{C} = C - C_B A_B^{-1} A$ is called the reduced cost matrix with respect to B. $\bar{C}_B$ and $\bar{C}_N$ are defined analogously to C_B and C_N.
- Let B be an efficient basis. A variable x_j is called efficient nonbasic variable if there exists a $\lambda \in \mathbb{R}^p_>$ such that $\lambda^T \bar{C}_N \geqq 0$ and $\lambda^T \bar{c}^j = 0$, where $\bar{c}^j$ is the jth column of $\bar{C}$, $j \in N$.

- Let B be an efficient basis and x_j be an efficient nonbasic variable. A feasible pivot from B with x_j as entering variable is called an efficient pivot.

If B is a basis then (x_B, x_N) with $x_N = 0$ and $x_B = A_B^{-1}b$ is a basic solution, and it is a basic feasible solution if additionally $x_B \geq 0$. A basic feasible solution is an extreme point of X.
THEOREM 35

1 *If $X_E \neq \emptyset$ then there exists an efficient extreme point.*

2 *Let B be an efficient basis. Then $(x_B, 0) \in X_E$.*

3 *Let $x \in X$ be an efficient extreme point. Then there exists an efficient basis B such that x is a basic feasible solution for B.*

4 *Let B be an efficient basis and x_j be an efficient nonbasic variable. Then any efficient pivot leads to an efficient basis.*

Multicriteria Simplex algorithms proceed in three phases. First, an auxiliary single objective LP $\min\{e^T z : Ax + z = b; x, z \geqq 0\}$ is solved to check feasibility (assuming, without loss of generality, $b \geqq 0$). $X \neq \emptyset$ if and only if the optimal value of this LP is 0. If $X \neq \emptyset$, in Phase 2 an initial efficient extreme point is found or the algorithm stops with the conclusion that $X_E = \emptyset$. Finally in Phase 3, efficient pivots are performed to explore all efficient extreme points or efficient bases.

THEOREM 36 *Tests for efficient nonbasic variables. Let B be an efficient basis and $\bar{C}$ the reduced cost matrix with respect to B. Let $j \in N$. Let e be a vector of all ones and let I be an identity matrix of appropriate dimension.*

1 *[89] Consider the LP* $\min\{e^T z : -\bar{C}_N y + \bar{c}^j \delta + Iz = 0; y, \delta, z \geqq 0\}$. *$x_j$ is an efficient nonbasic variable if and only if the optimal objective value of this LP is zero.*

2 *[122] Consider a subset $J \subseteq N$ and the LP* $\min\{e^T z : -\bar{C}_N y + \bar{C}_J \delta + Iz = e; y, \delta, v \geqq 0\}$. *Each variable $x_j, j \in J$, is efficient if and only if this LP has an optimal solution.*

3 *[70] Variable x_j is efficient if and only if there is a solution $(\hat{z}, \hat{y})$ of the linear system $\bar{C}_N^T(z + e) - y = 0; z, y \geqq 0$ with $\hat{y}_j = 0$.*

4 *[237] Variable x_j is efficient if and only if the LP* $\{\min 0^T z : \sum_{i \neq j} \bar{C}_{ki} z_i \geq \bar{C}_{kj}, k = 1, \ldots, p; z \geqq 0\}$ *is infeasible.*

THEOREM 37 *Finding an efficient (extreme) point.*

1 *[69] Let $x^0 \in X$ and consider the LP $\{\max e^T z : Cx - Iz = Cx^0; Ax = b; x, z \geqq 0\}$. If $(\hat{x}, \hat{z})$ is an optimal solution of this LP then $\hat{x} \in X_E$. If the LP is unbounded $X_E = \emptyset$.*

2 *[19] Let $x^0 \in X$. If the LP $\{\min -z^T Cx^0 + u^T b : z^T C - u^T A + w^T = -e^T C; w, z \geqq 0\}$ has no optimal solution then $X_E = \emptyset$. Otherwise let $(\hat{z}, \hat{u}, \hat{w})$ be an optimal solution. Then an optimal extreme point of the LP $\{\max(\hat{z} + e)^T Cx : x \in X\}$ is an efficient extreme point of the MOLP.*

Algorithms based on the simplex method are proposed by Armand and Malivert [9, 10], Evans and Steuer [89], Ecker and Kouada [68, 70], Isermann [122], Gal [95], Philip [172, 173], Schönfeld [187], Strijbosch [204], Yu and Zeleny [233, 234], and Zeleny [236]. Some numerical experiments for an implementation of Steuer's [201] multicriteria simplex method (called ADBASE [202]) on randomly generated problems are available in [200, 226].

In order to determine the whole efficient set, it is necessary to find subsets of efficient extreme points, the convex hulls of which determine maximal efficient faces. This process can be considered an additional phase of the multicriteria simplex method. Approaches that follow this strategy can be considered bottom up, as they build efficient faces starting from faces of dimension 0 (extreme points).

In [181] Sayin proposes a top-down approach instead. Consider an MOLP, where X is given in the form $X = \{x \in \mathbb{R}^n : \bar{A}x \leqq \bar{b}\}$, i.e., the nonnegativity constraints are included in $\bar{A}$. Let $M = \{1, \ldots, n + l\}$ denote the set of indices of constraints and $\mathcal{M} := \{J : J \subseteq M\}$. Then each $J \in \mathcal{M}$ represents a face $F(J)$ of X, $|J_1| \leq |J_2|$ implies $\dim F(J_1) \geq \dim F(J_2)$, and $J_1 \subseteq J_2$ implies $F(J_2) \subseteq F(J_1)$. Sayin solves the LP (17.29) to check whether or not a face is efficient. For $J \in \mathcal{M}$ let $\bar{A}^J$ and $\bar{b}^J$ denote submatrices (subvectors) of $\bar{A}$ and $\bar{b}$ containing only rows with indices in J.

$$\begin{array}{rrcl} \max & e^T Cx - e^T Cy & & \\ & \text{subject to } \bar{A}x & \leqq & \bar{b} \\ & \bar{A}y & \leqq & \bar{b} \\ & \bar{A}^J y & = & \bar{b}^J \\ & -Cx + Cy & \leqq & 0. \end{array} \tag{17.29}$$

THEOREM 38 *[181]*

1 *$F(J)$ is a proper face of X if and only if the LP (17.29) is feasible.*

2 *$F(J)$ is an efficient proper face of X if and only if the optimal objective value of the LP (17.29) is 0.*

In conjunction with 1 of Theorem 33 it is now shown that there are subsets $\mathcal{E}^s$ of $\mathcal{M}^s := \{J \in \mathcal{M} : |J| = s\}$ such that $X_E = \cup_{s=0}^{n+l} \cup_{J \in \mathcal{E}^s} F(J)$. Sayin's algorithm checks whether $J = \emptyset$, i.e., whether $X = X_E$, and then proceeds to larger sets J, i.e., faces of smaller and smaller dimension. In this process, supersets of sets already checked can be eliminated (as they correspond to subsets of faces already classified as efficient or nonefficient).

7.1.2 Interior Point Methods. Interior point methods are not easy to adapt for MOLPs since they construct a sequence of points that converges to a single point on the boundary of X. Thus most interior point methods proposed in the literature are of an interactive nature, see, e.g., Arbel [8] and references therein.

Abhyankar et al. [2] propose a method of centers of polytopes to find a nondominated point of Y. Assume that $Y = \{y : Gy \geqq H\}$, where $H \in \mathbb{R}^p$, is bounded and full dimensional and given in the form $Y = \{G_s^T x \geq H_s, s = 1, \ldots, q_1\}$, where G_s denotes the s-th row of G. Let $\mathcal{C}$ be a polyhedral cone given in the form $\mathcal{C} = \{d \in \mathbb{R}^p : Ld \geqq 0\} = \{d : L_s^T d \geq 0, s = 1, \ldots, q_2\}$, where L_s denotes the s-th row of L. Starting from $y^0 \in \operatorname{int} Y$, a sequence of points $\{y^\tau\} \subseteq \operatorname{int} Y$ is constructed so that $y^\tau \to \hat{y} \in N(Y, \mathcal{C})$. In this sequence, $y^{\tau+1}$ is the center of a polytope $Y^\tau := (y^\tau - \mathcal{C}) \cap Y$. Therefore $y^{\tau+1}$ can be determined as a unique maximizer of the potential function

$$f_{Y^\tau}(y) := \sum_{s=1}^{q_1} \ln(G_s^T y - H_s) + \sum_{s=1}^{q_2} (L_s^T y^\tau - L_s^T y)$$

by solving

$$\sum_{s=1}^{q_1} \frac{G_s^T}{G_s^T y - H_s} - \sum_{s=1}^{q_2} \frac{L_s^T}{L_s^T y^s - L_s^T y} = 0.$$

THEOREM 39 *[2] Every subsequence of $\{y^\tau\}$ converges to some point $\hat{y} \in N(Y, \mathcal{C})$.*

Abhyankar et al. [2] also construct an approximation of a portion of the nondominated faces of Y by constructing a sequence of algebraic surfaces (ellipsoids) $\{\tilde{Y}^\tau\}$ that approaches a part of $N(Y, \mathcal{C})$. In this way, they are able to parameterize the nondominated set.

To obtain a description of X_E they consider the polar cone of $\mathcal{C}$, $\mathcal{C}^* := \{d \in \mathbb{R}^n : d = \sum_{i=1}^q : \lambda_i v^i : \lambda_i \geq 0\}$ and the cone of increasing directions $\mathcal{C}^> = \{d \in \mathbb{R}^n : (CV)d \geq 0\}$, where C is the objective function matrix, and $CV = (Cv^1, \ldots, Cv^q)$ and $v^i, \ldots, v^q$ are the generators of $\mathcal{C}^*$. Then the methodology above can be applied to X with cone $\mathcal{C}^>$.

7.1.3 Objective Space Methods. Objective space methods for solving MOLPs are based on the assumption that the dimension of the objective space is typically smaller than the dimension of the decision space and therefore the number of Pareto extreme points of the set $Y := \{y \in \mathbb{R}^p : y = Cx$ for some $x \in X\} = C[X]$ to examine should be smaller than the number of efficient extreme points of the set X.

Dauer and Liu [62] propose a simplex-like method performed at those extreme points of X that correspond to the extreme points of Y. Let $\bar{C} = [\bar{C}_B, \bar{C}_N]$ be the reduced cost matrix associated with an extreme point $x \in X$. Define the cone spanned by the columns of $\bar{C}_N$ as cone $\bar{C}_N := \{d \in \mathbb{R}^p : d = \sum_{j \in N} \lambda_j \bar{c}^j, \lambda_j \geq 0\}$. A frame of cone $\bar{C}_N$ is a minimal collection of vectors selected from among the columns of $\bar{C}_N$ that determine this cone.

THEOREM 40 *[62] Let $y = Cx$ be an extreme point of Y and let $\bar{C}$ be the reduced cost matrix associated with the extreme point $x \in X$. Let E^j be the edge of X determined by a column $\bar{c}^j$ in $\bar{C}_N$. The image of E^j under C is contained in an edge of Y if and only if $\bar{c}^j$ is in a frame of the* cone $\bar{C}_N$.

Dauer and Saleh [63] develop an algebraic representation of $Y = \{y \in \mathbb{R}^p : Gy \geqq H\}$ and propose an algorithm to construct this set. Additionally, they develop an algebraic representation of a polyhedral set $\tilde{Y}$ that has the same Pareto structure as that of Y and that has no extreme points that are not Pareto. In this way, any method designed for finding all vertices of convex polyhedral sets becomes suitable to find the set of Pareto extreme points of Y. However, they also propose an algorithm using a single objective linear program in $\mathbb{R}^{p+1}$, the set of optimal basic solutions of which corresponds to the set of extreme points of $\tilde{Y}$ (and to the set of Pareto extreme points of Y). Dauer [60] presents an improved version of the algorithm of Dauer and Liu [62] to generate the set of all Pareto extreme points and edges of Y. He gives special attention to degenerate extreme points and the collapsing effect that reduces the number of extreme points of X that are necessary to analyze in order to fully determine the structure of Y. Almost a parallel effort to represent the set Y in terms of a set of inequalities is undertaken by White [224].

More recently, Dauer and Gallagher [61] have developed algorithm MEF for determining high-dimensional maximal Pareto faces of Y. Algorithm MEF requires as input a nonredundant system of linear inequalities representing Y (or $\tilde{Y}$).

THEOREM 41 *[61]*

1 *If F_X is a maximal efficient face of X then $C[F_X] := \{y \in \mathbb{R}^p : y = Cx$ for some $y \in F_X\}$ is a maximal Pareto face of Y.*

2 *If F_Y is a maximal Pareto face of Y then $C^{-1}[F_Y] \cap X := \{x \in \mathbb{R}^n : Cx \in F_Y\}$ is a maximal efficient face of X.*

THEOREM 42 *[61] Let* $Y = \{y \in \mathbb{R}^p : Gy \geqq H\}$ *and* G_k *be the k-th row of the matrix G. Let* F *be a nonempty face of* Y *and let* $I(F)$ *denote the set of indices of the active constraints defining* F *(i.e.,* $I(F) = \{i : G_i y = H_i$ *for all* $y \in F\}$*). Then*

1. $F \subseteq Y_N$ *if and only if* $\sum_{j \in I(F)} \lambda_j G_j > 0$ *for some collection of scalars* $\lambda_j \geq 0, j \in I(F)$.

2. F *is a maximal Pareto face of dimension* $p - |I(F)|$ *if and only if*

   (a) *there exist scalars* $\lambda_j \geq 0$, $j \in I(F)$, *such that* $\sum_{j \in I(F)} \lambda_j G_j > 0$ *and*

   (b) $\sum_{j \in I(F)} \mu_j G_j \not> 0$ *for every* $J \subsetneqq I(F)$ *and every collection of scalars* $\mu_j \geq 0, j \in J$.

The work of Benson follows on the work of Dauer et al. He describes the outcome set-based algorithm in [23]. Its purpose is to generate all efficient extreme points of Y_N. Let $\bar{Y} := \{y \in \mathbb{R}^p : Cx \leqq y \leqq \hat{y}\}$, where $\hat{y}$ is such that $y_k < \hat{y}_k, k = 1, \ldots, p$ for all $y \in Y = C[X]$. Then $\bar{Y}_N = Y_N$. The method starts by finding a simplex S^0 containing $\bar{Y}$ and its vertices. S^0 is given by $S^0 := \{y \in \mathbb{R}^p : y \leq \hat{y}, \beta \leq e^T y\}$, where $\beta = \min\{e^T y : y \in \bar{Y}\}$. Given S^τ, an extreme point y^τ of S^τ is chosen that is not contained in $\bar{Y}$. Then w^τ is chosen as a unique point on the boundary of $\bar{Y}$ on the line segment connecting an interior point y^0 of Y with y^τ. Now, $S^{\tau+1}$ can be obtained by

$$S^{\tau+1} = S^\tau \cap \{y \in \mathbb{R}^p : (u^\tau)^T y \leq b^T v^\tau\}$$

where $u^\tau, v^\tau \in \mathbb{R}^p$ are such that $F^\tau = \{y \in \bar{Y} : (u^\tau)^T y = b^T v^\tau\}$ is a face of Y containing w^τ. The computation of the extreme points of $S^{\tau+1}$ completes the iteration.

THEOREM 43 *[23] The outcome set-based algorithm is finite and at termination* $S^K = \bar{Y}$, *where* K *is the number of iterations.*

All Pareto extreme points of Y are found by eliminating from the extreme points of $\bar{Y}$ those for which $y_k = \hat{y}_k$ for some k. Benson [21] shows that also all weak Pareto extreme points of Y are found. In [22], Benson combines the algorithm with a simplicial partitioning technique, which makes the computation of the extreme points of $S^{\tau+1}$ more efficient.

7.2 Discrete MOPs

In discrete MOPs some or all variables are restricted to a discrete set of values. Most often, these variables are allowed to take only integer values yielding multiobjective integer programs (MOIP) or values 0 and 1 resulting in multi-objective zero-one programs (MOZOP). The discrete nature of these problems implies they are non-convex in general.

Bitran [29, 30] solves the MOLP with zero-one variables by analyzing the efficient set of the unconstrained problem $\max\{Cx : x \in \{0,1\}^n\}$ and using directions of preference v with $Cv \geq 0$, where $v_i \in \{0, 1, -1\}, i = 1, \ldots, n$. Bitran and Rivera [31] develop a branch and bound algorithm for the same problem. The branch and bound method by Kiziltan and Yucaoglu [135] for this problem uses multiple upper bound vectors which can be seen as local ideal points for the subproblem considered at each node of the branch and bound tree. Deckro and Winkofsky [65] use a similar technique in another implicit enumeration scheme.

Burkard et al. [38] introduce a scalarization for MOZOPs (with nonlinear objective functions)

$$\begin{aligned} \min \quad & \lambda_1 f_1(x) + \lambda_2 (f_2(x))^{t_2} + \ldots + \lambda_p (f_p(x))^{t_p} \\ & \text{subject to } x \in X. \end{aligned} \tag{17.30}$$

THEOREM 44 *[38] There exist a* $\hat{t}_2 \geq 0$ *and functions* $b_k : \mathbb{R}_{\geq} \to \mathbb{R}_{\geq}, k = 1, \ldots, p$, *such that for all* $t_2 \geq \hat{t}_2$ *and* $t_k \geq b_k(t_{k-1})$ X_E *is the set of optimal solutions of problem (17.30) solved for all* $\lambda \in \mathbb{R}^p_{>}, \sum_{k=1}^p \lambda_k = 1$.

White [221] considers a special MOZOP with nonlinear objective functions but with only one (nonlinear) constraint $g(x) = 0$ which is assumed to be non-negative on $\{0,1\}^n$. He proposes a branch and bound method and characterizes X_E as the efficient set of an unconstrained problem obtained by Lagrangian relaxation.

White [223] proposes a variation of the weighted sum approach for MOPs with a finite feasible set X not necessarily consisting of integer vectors only. This approach is similar to that of Burkard et al. [38] but White uses a common power t for all objective functions.

The more general problem of MOIPs has been treated by some authors. Biobjective problems have first been studied. Chalmet et al. [42] consider the problem $\min\{(f_1(x), f_2(x)) : x \in X, \text{ integer}\}$ where f_1 and f_2 are supposed to be integer valued on X. They use the weighted sum scalarization but impose additional constraints that exclude previously generated efficient solutions from the problem. Thus, the scalarized problem is of the form $\min\{\lambda_1 f_1(x) + \lambda_2 f_2(x) : x \in X, f_k(x) \geq \varepsilon_k + 1\}$ where ε_k are objective values of previously determined efficient solutions.

Some authors focus on biobjective linear programs with integer variables. Kaliszewski [130] uses a number theoretic approach to enumerate efficient solutions of such programs with equality constraints. Solanki [194] applies the

l_∞-norm to scalarize the problems and find their efficient solutions. Algorithms using the same norm as well as a quadratic function as an auxiliary objective are presented by Neumayer and Schweigert [163].

Eswaran et al. [88] propose the use of the l_∞-norm in the scalarization $\min\{\max_{k=1,2}\{\lambda_k|f_k(x) - y_k^I| : x \in X\}\}$ to solve $\max\{(f_1(x), f_2(x)) : g_j(x) \leq 0, x \text{ integer}\}$, i.e., a nonlinear integer problem with two objective functions. Their algorithm achieves a decomposition of the weight interval $\Lambda = [0,1]$ into sets $\Lambda(y^i)$ such that y^i is an optimal solution of $\min\{z : z \geq \lambda(y_1^I - y_1), z \geq (1-\lambda)(y_2^I - y_2), y \in Y\}$ for all $\lambda \in \Lambda(y^i)$.

We finally mention articles on MOIPs with linear constraints. A similar approach to that of Chalmet et al. [42] is taken by Klein and Hannan [137] to study MOLPs with integer variables. They also use additional constraints enforced by efficiency of solutions. They consider the scalarization $\min\{c^s x : Ax \geq b, x \geq 0, x \text{ integer}, \cap_{i=1}^q \cup_{k=1,k\neq s}^p c^k x \leq c^k x^i - \varepsilon_k\}$, where $\varepsilon_k \geq 1$ and integer, $x^1, \ldots, x^q$ are the efficient solutions determined so far, and c^k is the k-th row of C. Villarreal and Karwan [213, 214] discuss multiobjective extensions of dynamic programming to solve MOLPs with integer and bounded variables. They use sets of upper and lower bounds to limit the search space.

Cooper and Farhangian [53] consider a problem with nonlinear but separable objectives $\max\{\sum_{i=1}^n f_{ki}(x_i), k = 1, \ldots, p : Ax \leq b, x \text{ integer}\}$, where all $f_{ki} : \mathbb{R} \to \mathbb{R}$ are nondecreasing functions and the coefficients of the constraints are nonnegative. They use the ε-constraint scalarization and change the right-hand-side values of the constraints on $f_2(x), \ldots, f_p(x)$ to compute some part of X_E.

On a more theoretical note, Helbing [116] proves some relationships between MOIPs and group theory that can be used to find efficient solutions.

The general methods considered so far proved to be unable to solve MOIPs of practically relevant size. In the 1990s interest began to switch from general MOIPs to problems with particular combinatorial structures. Multiple objective combinatorial optimization (MOCO) problems may still be formulated as problems with integer variables, but they also have an underlying combinatorial structure (often graph theoretical objects like trees, paths, etc.) that can be exploited to design more efficient techniques for their solution. This area of research has seen enormous growth since 1990. Because recent exhaustive surveys are available, we do not go into any detail but refer to Ehrgott and Gandibleux [75, 77].

7.3 Nonlinear MOPs

In this section we review results on some classes of MOPs with nonlinear objective functions including piecewise linear, quadratic and polynomial, and fractional functions.

MOPs with piecewise linear objective functions are not much studied. Achary [3] develops a simplex-type method to enumerate all efficient solutions of a biobjective transportation problem. Nickel and Wiecek [164] propose an approach in which the task of finding efficient solutions of the original problem is replaced by tasks of finding efficient solutions of simpler subproblems starting with subsets of the highest dimension.

MOPs with quadratic functions have been of interest to many authors. Unconstrained quadratic MOPs with strictly convex objective functions are analyzed by Beato-Moreno et al. [13]. They obtain an explicit characterization of the efficient set for the biobjective case and show that the p-objective case can be reduced to the $(p-1)$-objective case. The set of weakly efficient solutions of quadratic MOPs with convex objective functions is examined by Beato-Moreno et al. [14]. A method to produce an analytic description of the efficient set of linearly constrained convex quadratic MOPs is proposed by Goh and Yang [105]. Some researchers relate quadratic MOPs to linear complementarity problems. Kostreva and Wiecek [143] demonstrate and Isac et al. [121] exploit equivalence between the linear complementarity problem and a nonconvex quadratic MOP with linear constraints. Generalized linear complementarity problems are related to MOPs by Ebiefung [67] and Bhatia and Gupta [27]. MOPs with polynomial objective and constraint functions are studied by Kostreva et al. [143] who use the Benson approach to develop a method for finding efficient solutions of those problems. The resulting SOP is solved with a homotopy continuation method. Korhonen and Yu [140] also use a linear complementarity approach to MOPs with one quadratic and several linear objective functions.

Multiobjective fractional problems (MOFPs), objective functions of which are fractional functions, have been extensively studied and this review covers only a small part of available articles. A survey on biobjective problems in this class has recently been given by Cambini et al. [40]. If numerators and denominators of the objective functions of MOFPs are affine functions, the problems are referred to as multiobjective linear fractional programs (MOLFPs). Kornbluth and Steuer [141] and Benson [20] develop a simplex-based procedure to find weakly efficient vertices of MOLFPs. Gupta [111] relates efficient points of these problems to efficient points of an MOLP and to efficient points of a number of biobjective linear programs. Connectedness of the weakly efficient set of MOLFPs is examined by Choo and Atkins [46]. Scalarizations have recently been applied by Metev and Gueorguieva [156] to generate weakly efficient solutions of MOLFPs. An algorithm to find all efficient solutions of MOLFPs with zero-one variables is proposed by Gupta [110] while MOLFPs with integer variables are examined by Gupta and Malhotra [112]. Conditions for efficiency for MOLFPs with convex constraints are developed by Gulati and Islam [108, 109]. If numerators and denominators of the objective functions are nonlinear functions, the problems are referred to as multiobjective nonlinear fractional programs. For these problems, conditions for the existence of efficient solutions are developed by Kaul and Lyall [134] and Fritz-John and Kuhn-Tucker type conditions for efficiency are proposed by Gulati and Ahmad [4].

8. Current and Future Research Directions

The rapid development of optimization techniques and computational power over the last decade has made it possible to solve MOPs of practically relevant size in a reasonable time. At the same time we observe an increasing awareness of decision makers that it is necessary to incorporate multiple objectives in decision processes. Thus in the future we can expect to see more real world applications of multiobjective programming.

We would like to mention two such applications. Küfer et al. [146] describe an MOLP formulation of the radiation therapy planning problem. These models can have thousands of variables, tens of thousands of constraints, nevertheless an approximation of (a part of) the efficient set can be computed effectively. Ehrgott and Ryan [83] solve bicriteria set partitioning problems with a few hundred constraints and many thousands of variables for an application in airline crew scheduling.

From this perspective we believe that the following are valuable directions of future research.

- *Theoretical studies into the structure of MOCO problems.* As the structure of these discrete MOPs is better understood, more effective solution techniques can be developed.

- *Research on evolutionary methods and metaheuristics.* The advent of metaheuristics has provided the MOP community with algorithmic schemes that are relatively easily adaptable to many special problems. At present, evolutionary techniques constitute probably the most successful approach for solving MOPs in practice and we expect this trend to continue.

- *Theoretical and methodological studies motivated by applications.* There is no "one size fits all" methodology for MOPs. A method that works well in theory can fail in practice, one that works well on some problem may not be suitable for another one. So MOP methodology will increasingly be studied in problem contexts.

- *Applications in new areas.* Disciplines such as astronomy in sciences, quality control and design in engineering, and medicine will provide new opportunities for challenging applications of multiobjective programming.

- *Integration of multiobjective programming with multicriteria decision analysis (MCDA).* In the current multicriteria decision making (MCDM) methodology, multiobjective programming methods and MCDA methods are often seen as two ends of a spectrum. However, current applications indicate that both paradigms are needed in order for MCDM to succeed. In many, if not all, applications (e.g., the two mentioned above) there is an objective stage, where multiobjective programming techniques are appropriate, and there is a subjective stage, where human judgment modeled within MCDA comes into play. At this stage the formal mathematical approach is likely to be less effective and human factors-oriented strategies are needed to guide the decision maker.

9. Conclusions

In this chapter we summarized the state of the art in multiobjective programming. Our main attention was devoted to optimality concepts, optimality conditions, solution techniques and approximations of the Pareto and efficient set. We recognize that the content of this chapter is subjective, as we excluded many facets of the subject such as duality, relations between parametric and multiobjective optimization, other stability results, variational inequalities, generalized convexity, nonsmoothness, Arrow-Barankin-Blackwell theorems, results for more general problems in vector spaces, other special classes of problems, etc. The topics of this chapter as well as other related topics have been discussed in other sources such as Ehrgott and Gandibleux [78], Gal et al. [96], and some other chapters in this book.

Acknowledgments

The work of M. Ehrgott was partially supported by University of Auckland grant 3602178/9275 and by the Deutsche Forschungsgemeinschaft grant Ka 477/27-1. The work of M. M. Wiecek was done while on sabbatical leave at the Department of Engineering Science, University of Auckland, Auckland, New Zealand, and at the Department of Mathematics, University of Kaiserslautern, Kaiserslautern, GermanyA; supported by the American Association for the Advancement of Science and Deutsche Forschungsgemeinschaft grant Ka 477/24-1.

References

[1] F. Ben Abdelaziz, J. Chaouachi, and S. Krichen. A hybrid heuristic for multiobjective knapsack problems. In S. Voss, S. Martello, I. Osman, and C. Roucairol, editors, *Meta-Heuristics: Advances and Trends in Local Search Paradigms for Optimization*, pages 205–212. Kluwer Academic Publishers, Dordrecht, 1999.

[2] S.S. Abhyankar, T.L. Morin, and T. Trafalis. Efficient faces of polytopes: Interior point algorithms, parametrization of algebraic varieties, and multiple objective optimization. *Contemporary Mathematics*, 114:319–341, 1990.

[3] K.K. Achary. A direct simplex type method for determining all efficient solution pairs of time-cost transportation problem. *Econ. Comput. Econ. Cybern. Stud. Res.*, 24(2/3):139–150, 1989.

[4] B. Aghezzaf. Efficiency and duality in multiobjective fractional programming. *Opsearch*, 32(1):31–43, 1995.

[5] B. Aghezzaf. Second-order necessary conditions of the Kuhn-Tucker type in multiobjective programming problems. *Control and Cybernetics*, 28(2):213–224, 1999.

[6] B. Aghezzaf and M. Hachimi. Second order optimality conditions in multiobjective optimization problems. *Journal of Optimization Theory and Applications*, 102(1):37–50, 1999.

[7] K.A. Andersen, K. Jørnsten, and M. Lind. On bicriterion minimal spanning trees: An approximation. *Computers and Operations Research*, 23:1171–1182, 1996.

[8] A. Arbel. An interior multiobjective primal-dual linear programming algorithm based on approximated gradients and efficient anchoring points. *Computers and Operations Research*, 24(4):353–365, 1997.

[9] P. Armand. Finding all maximal efficient faces in multiobjective linear programming. *Mathematical Programming*, 61:357–375, 1993.

[10] P. Armand and C. Malivert. Determination of the efficient set in multiobjective linear programming. *Journal of Optimization Theory and Applications*, 70(3):467–489, 1991.

[11] A. Bárdossy, I. Bogárdi, and L. Duckstein. Composite programming as an extension of compromise programming. In P. Serafini, editor, *Mathematics of Multi Objective Optimization*, pages 375–408. Springer Verlag, Wien, 1985.

[12] A. Baykasoglu, S. Owen, and N. Gindy. A taboo search based approach to find the Pareto optimal set in multiple objective optimisation. *Journal of Engineering Optimization*, 31:731–748, 1999.

[13] A. Beato-Moreno, P. Luque-Calvo, R. Osuna-Gómez, and A. Rufián-Lizana. Finding efficient points in multiobjective quadratic programming with strictly convex objective functions. In R. Caballero, F. Ruiz, and R.E. Steuer, editors, *Advances in Multiple Objective and Goal Programming*, volume 455 of *Lecture Notes in Economics and Mathematical Systems*, pages 365–373. Springer Verlag, Berlin, 1997.

[14] A. Beato-Moreno, P. Ruiz-Canales, P.-L. Luque-Calvo, and R. Blanquero-Bravo. Multiobjective quadratic problem: Characterization of the efficient points. In J.P. Crouzeix, J.E. Martinez-Legaz, and M. Volle, editors, *Generalized Convexity, Generalized Monotonicity: Recent Results*, pages 425–438. Kluwer Academic Publishers, Dordrecht, 1998.

[15] R. Beausoleil. Multiple criteria scatter search. In *MIC'2001 - 4th Metaheuristics International Conference*, pages 539–543, 2001. Porto, Portugal - July 16-20, 2001.

[16] F.A. Behringer. Lexicographic quasiconcave multiobjective programming. *Zeitschrift für Operations Research*, 21:103–116, 1977.

[17] H.P. Benson. Existence of efficient solutions for vector maximization problems. *Journal of Optimization Theory and Applications*, 26(4):569–580, 1978.

[18] H.P. Benson. An improved definition of proper efficiency for vector maximization with respect to cones. *Journal of Optimization Theory and Applications*, 71:232–241, 1979.

[19] H.P. Benson. Finding an initial efficient extreme point for a linear multiple objective program. *Journal of the Operational Research Society*, 32:495–498, 1981.

[20] H.P. Benson. Finding certain weakly-efficient vertices in multiple objective linear fractional programming. *Management Science*, 31(2):240–245, 1985.

[21] H.P. Benson. Further analysis of an outcome set-based algorithm for multiple-objective linear programming. *Journal of Optimization Theory and Applications*, 97(1):1–10, 1998.

[22] H.P. Benson. Hybrid approach for solving multiple-objective linear programs in outcome space. *Journal of Optimization Theory and Applications*, 98(1):17–35, 1998.

[23] H.P. Benson. An outer approximation algorithm for generating all efficient extreme points in the outcome set of a multiple objective linear programming problem. *Journal of Global Optimization*, 13:1–24, 1998.

[24] H.P. Benson and S. Sayin. Towards finding global representations of the efficient set in multiple objective mathematical programming. *Naval Research Logistics*, 44:47–67, 1997.

[25] H.P. Benson and E. Sun. Outcome space partition of the weight set in multiobjective linear programming. *Journal of Optimization Theory and Applications*, 105(1):17–36, 2000.

[26] H. Bernau. Use of exact penalty functions to determine efficient decisions. *European Journal of Operational Research*, 49:348–355, 1990.

[27] D. Bhatia and P. Gupta. Generalized linear complementarity problem and multiobjective programming problem. *Optimization*, 46(2):199–214, 1999.

[28] G. Bigi and M. Castellani. Second order optimality conditions for differentiable multiobjective programming. *RAIRO Operations Research*, 34:411–426, 2000.

[29] G.R. Bitran. Linear multiple objective programs with zero-one variables. *Mathematical Programming*, 13:121–139, 1977.

[30] G.R. Bitran. Theory and algorithms for linear multiple objective programs with zero-one variables. *Mathematical Programming*, 17:362–390, 1979.

[31] G.R. Bitran and J.M. Rivera. A combined approach to solve binary multicriteria problems. *Naval Research Logistics Quarterly*, 29:181–201, 1982.

[32] S. Bolintinéanu and M. El Maghri. Second-order efficiency conditions and sensitivity of efficient points. *Journal of Optimization Theory and Applications*, 98(3):569–592, 1998.

[33] J. Borwein. Proper efficient points for maximization with respect to cones. *SIAM Journal on Control and Optimization*, 15(1):57–63, 1977.

[34] J.M. Borwein. On the existence of Pareto efficient points. *Mathematics of Operations Research*, 8(1):64–73, 1983.

[35] J.M. Borwein and D.M. Zhuang. Super efficiency in convex vector optimization. *Zeitschrift für Operations Research*, 35(3):175–184, 1991.

[36] J.M. Borwein and D.M. Zhuang. Super efficiency in vector optimization. *Transactions of the American Mathematical Society*, 338(1):105–122, 1993.

[37] V.J. Bowman. On the relationship of the Tchebycheff norm and the efficient frontier of multiple-criteria objectives. In H. Thieriez, editor, *Multiple Criteria Decision Making*, volume 130 of *Lecture Notes in Economics and Mathematical Systems*. Springer Verlag, Berlin, 1976.

[38] R.E. Burkard, H. Keiding, J. Krarup, and P.M. Pruzan. A relationship between optimality and efficiency in multiple criteria 0-1 programming problems. *Computers and Operations Research*, 8:241–247, 1981.

[39] A. Cambini, L. Martein, and R. Cambini. A new approach to second order optimality conditions in vector optimization. In R. Caballero, F. Ruiz, and R.E. Steuer, editors, *Advances in Multiple Objective and Goal Programming*, volume 455 of *Lecture Notes in Economics and Mathematical Systems*, pages 219–227. Springer Verlag, Berlin, 1997.

[40] A. Cambini, L. Martein, and I.M. Stancu-Minasian. A survey on bicriteria fractional problems. *Mathematic Reports (Bucureşti.)*, 2(52)(2):127–162, 2000.

[41] R. Cambini. Second order optimality conditions in multiobjective programming. *Optimization*, 44(2):139–160, 1998.

[42] L.G. Chalmet, L. Lemonidis, and D.J. Elzinga. An algorithm for the bi-criterion integer programming problem. *European Journal of Operational Research*, 25:292–300, 1986.

[43] V. Chankong and Y.Y. Haimes. *Multiobjective Decision Making Theory and Methodology*. Elsevier Science, New York, 1983.

[44] A. Charnes and W. Cooper. *Management Models and Industrial Applications of Linear Programming*. John Wiley and Sons, New York, 1961.

[45] O. Chernykh. Approximating the Pareto-hull of a convex set by polyhedral sets. *Computational Mathematics and Mathematical Physics*, 35:1033–1039, 1995.

[46] E.U. Choo and A.D. Atkins. Connectedness in multiple linear fractional programming. *Management Science*, 29(2):250–255, 1983.

[47] S. Yu. Churkina. Search for efficient solutions of multi-criterion problems by target-level method. *Computational Mathematics and Modeling*, 13(2):208–214, 2002.

[48] C.A. Coello. A comprehensive survey of evolutionary-based multiobjective optimization techniques. *Knowledge and Information Systems*, 1(3):269–308, 1999.

[49] C.A. Coello. An updated survey of GA-based multiobjective optimization techniques. *ACM Computing Surveys*, 32(2):109–143, 2000.

[50] C.A. Coello Coello, D.A. Van Veldhuizen, and G.B. Lamont. *Evolutionary Algorithms for Solving Multi-Objective Problems*. Kluwer Academic Publishers, Boston, MA, 2002.

[51] J.L. Cohon. *Multiobjective Programming and Planning*. Academic Press, New York, 1978.

[52] J.L.. Cohon, R.L. Church, and D.P. Sheer. Generating multiobjective tradeoffs: An algorithm for bicriterion problems. *Water Resources Research*, 15:1001–1010, 1979.

[53] M.W. Cooper and K. Farhangian. Multi-criteria optimization for nonlinear integer-variable problems. *Large Scale Systems*, 9:73–78, 1985.

[54] H.W. Corley. An existence result for maximization with respect to cones. *Journal of Optimization Theory and Applications*, 31(2):277–281, 1980.

[55] H.W. Corley. Efficient spanning trees. *Journal of Optimization Theory and Applications*, 45(3):481–485, 1985.

[56] P. Czyzak and A. Jaszkiewicz. Pareto simulated annealing – A metaheuristic technique for multiple objective combinatorial optimization. *Journal of Multi-Criteria Decision Analysis*, 7(1):34–47, 1998.

[57] N.O. Da Cuhna and E. Polak. Constrained minimization under vector-valued criteria in finite dimensional spaces. *Journal of Mathematical Analysis and Applications*, 19(1):103–124, 1967.

[58] I. Das. An improved technique for choosing parameters for Pareto surface generation using normal-boundary intersection. In *Short Paper Proceedings of the Third World Congress of Structural and Multidisciplinary Optimization*, volume 2, pages 411–413, 1999.

[59] I. Das and J.E. Dennis. Normal-boundary intersection: A new method for generating the Pareto surface in nonlinear multicriteria optimization problems. *SIAM Journal of Optimization*, 8(3):631–657, 1998.

[60] J.P. Dauer. On degeneracy and collapsing in the construction of the set of objective values in a multiple objective linear program. *Annals of Operations Research*, 46/47(1-4):279–292, 1993.

[61] J.P. Dauer and R.J. Gallagher. A combined constraint-space, objective-space approach for determining high-dimensional maximal efficient faces of multiple objective linear programs. *European Journal of Operational Research*, 88:368–381, 1990.

[62] J.P. Dauer and Y.H. Liu. Solving multiple objective linear programs in objective space. *European Journal of Operational Research*, 46:350–357, 1990.

[63] J.P. Dauer and O.A. Saleh. Constructing the set of efficient objective values in multiple objective linear programs. *European Journal of Operational Research*, 46:358–365, 1990.

[64] K. Deb. *Multi-Objective Optimization using Evolutionary Algorithms*. Wiley-Interscience Series in Systems and Optimization. John Wiley & Sons, Chichester, 2001.

[65] R.F. Deckro and E.P. Winkofsky. Solving zero-one multiple objective programs through implicit enumeration. *European Journal of Operational Research*, 12:362–274, 1983.

[66] K. Dörner, W.J. Gutjahr, R.F. Hartl, C. Strauß, and C. Stummer. Pareto ant colony optimization: A metaheuristic approach to multiobjective portfolio selection. *Annals of Operations Research*, 2003, in print.

[67] A. Ebiefung. Nonlinear mappings associated with the generalized linear complementarity problem. *Mathematical Programming*, 69(2):255–268, 1995.

[68] J.G. Ecker, N.S. Hegner, and I.A. Kouada. Generating all maximal efficient faces for multiple objective linear programs. *Journal of Optimization Theory and Applications*, 30(3):353–381, 1980.

[69] J.G. Ecker and I.A. Kouada. Finding efficient points for linear multiple objective programs. *Mathematical Programming*, 8:375–377, 1975.

[70] J.G. Ecker and I.A. Kouada. Finding all efficient extreme points for multiple objective linear programs. *Mathematical Programming*, 14(2):249–261, 1978.

[71] M. Ehrgott. Discrete decision problems, multiple criteria optimization classes and lexicographic max-ordering. In T.J. Stewart and R.C. van den Honert, editors, *Trends in Multicriteria Decision Making*, volume 465 of *Lecture Notes in Economics and Mathematical Systems*, pages 31–44. Springer Verlag, Berlin, 1998.

[72] M. Ehrgott. *Multicriteria Optimization*, volume 491 of *Lecture Notes in Economics and Mathematical Systems*. Springer Verlag, Berlin, 2000.

[73] M. Ehrgott. The balance space approach to multicriteria decision making - Involving the decision maker. *Computers & Mathematics with Applications*, 44(7):909–923, 2002.

[74] M. Ehrgott and E.A. Galperin. Min-max formulation of the balance number in multiobjective global optimization. *Computers & Mathematics with Applications*, 44(7):899–907, 2002.

[75] M. Ehrgott and X. Gandibleux. A survey and annotated bibliography of multiobjective combinatorial optimization. *OR Spektrum*, 22:425–460, 2000.

[76] M. Ehrgott and X. Gandibleux. Bounds and bound sets for biobjective combinatorial optimization problems. In M. Köksalan and S. Zionts, editors, *Multiple Criteria Decision Making in the New Millenium*, volume 507 of *Lecture Notes in Economics and Mathematical Systems*, pages 242–253. Springer Verlag, Berlin, 2001.

[77] M. Ehrgott and X. Gandibleux. Multiobjective combinatorial optimization. In *Multiple Criteria Optimization – State of the Art Annotated Bibliographic Surveys*, pages 369–444. Kluwer Academic Publishers, Boston, MA, 2002.

[78] M. Ehrgott and X. Gandibleux, editors. *Multiple Criteria Optimization – State of the Art Annotated Bibliographic Surveys*, volume 52 of *Kluwer's International Series in Operations Research and Management Science*. Kluwer Academic Publishers, Boston, MA, 2002.

[79] M. Ehrgott and X. Gandibleux. Multiple objective combinatorial optimization – A tutorial. In T. Tanino, T. Tanaka, and M. Inuiguchi, editors, *Multi-Objective Programming and Goal-Programming*, pages 3–18. Springer Verlag, Berlin, 2003.

[80] M. Ehrgott and X. Gandibleux. Approximative solution methods for multiobjective combinatorial optimization. *TOP*, 12(1):1–88, 2004.

[81] M. Ehrgott, H.W. Hamacher, K. Klamroth, S. Nickel, A. Schöbel, and M.M. Wiecek. A note on the equivalence of balance points and Pareto solutions in multiple-objective programming. *Journal of Optimization Theory and Applications*, 92(1):209–212, 1997.

[82] M. Ehrgott, S. Nickel, and H.W. Hamacher. Geometric methods to solve max-ordering location problems. *Discrete Applied Mathematics*, 93:3–20, 1999.

[83] M. Ehrgott and D.M. Ryan. Constructing robust crew schedules with bicriteria optimization. *Journal of Multi-Criteria Decision Analysis*, 11(3):139–150, 2002.

[84] M. Ehrgott and A.J.V. Skriver. Solving biobjective combinatorial max-ordering problems by ranking methods and a two-phases approach. *European Journal of Operational Research*, 147:657–664, 2003.

[85] P. Engrand. A multi-objective approach based on simulated annealing and its application to nuclear fuel management. In *Proceedings of the 5th ASME/SFEN/JSME International Conference on Nuclear Engineering. ICONE 5, Nice, France 1997*, pages 416–423. American Society of Mechanical Engineers, New York, NY, 1997.

[86] T. Erlebach, H. Kellerer, and U. Pferschy. Approximating multi-objective knapsack problems. In F. Dehne, J.R. Sack, and R. Tamassia, editors, *Algorithms and Data Structures. 7th International Workshop, Providence, RI, August 8-10, 2001*, volume 2125 of *Lecture Notes in Computer Science*, pages 210–221. Springer Verlag, Berlin, 2001.

[87] J. Ester and F. Tröltzsch. On generalized notions of efficiency in multicriteria decision making. *System Analysis Modelling Simulation*, 3:153–161, 1986.

[88] P.K. Eswaran, A. Ravindran, and H. Moskowitz. Algorithms for nonlinear integer bicriterion programs. *Journal of Optimization Theory and Applications*, 63(2):261–279, 1989.

[89] J.P. Evans and R.E. Steuer. A revised simplex method for linear multiple objective programs. *Mathematical Programming*, 5:375–377, 1973.

[90] J. Fliege and A. Heseler. Constructing approximations to the efficient set of convex quadratic multiobjective problems. Technical Report Ergebnisberichte Angewandte Mathematik No. 211, Universität Dortmund, Fachbereich Mathematik, 2002.

[91] C.M. Fonseca and P.J. Fleming. Genetic algorithms for multiobjective optimization: Formulation, discussion and generalization. In S. Forrest, editor, *Proceedings of the Fifth International Conference on Genetic Algorithms, San Mateo, California, 1993. University of Illinois at Urbana-Champaign*, pages 416–423. Morgan Kaufman, San Francisco, CA, 1993.

[92] C.M. Fonseca and P.J. Fleming. An overview of evolutionary algorithms in multiobjective optimization. *Evolutionary Computation*, 3(1):1–16, 1995.

[93] B. Fruhwirth, R.E. Burkard, and G. Rote. Approximation of convex curves with application to the bicriterial minimum cost flow problem. *European Journal of Operational Research*, 42:326–388, 1989.

[94] B. Fruhwirth and K. Mekelburg. On the efficient point set of tricriteria linear programs. *European Journal of Operational Research*, 72:192–199, 1994.

[95] T. Gal. A general method for determining the set of all efficient solutions to a linear vectormaximum problem. *European Journal of Operational Research*, 1:307–322, 1977.

[96] T. Gal, T. Stewart, and T. Hanne, editors. *Multicriteria Decision Making – Advances in MCDM Models, Algorithms, Theory, and Applications*, volume 21 of *Kluwer's International Series in Operations Research and Management Science*. Kluwer Academic Publishers, Norwell, MA, 2002.

[97] E.A. Galperin. Nonscalarized multiobjective global optimization. *Journal of Optimization Theory and Applications*, 75(1):69–85, 1992.

[98] E.A. Galperin and M. M. Wiecek. Retrieval and use of the balance set in multiobjective global optimization. *Computers & Mathematics with Applications*, 37(4/5):111–123, 1999.

[99] X. Gandibleux, N. Mezdaoui, and A. Fréville. A tabu search procedure to solve multiobjective combinatorial optimization problems. In R. Caballero, F. Ruiz, and R. Steuer, editors, *Advances in Multiple Objective and Goal Programming*, volume 455 of *Lecture Notes in Economics and Mathematical Systems*, pages 291–300. Springer Verlag, Berlin, 1997.

[100] X. Gandibleux, D. Vancoppenolle, and D. Tuyttens. A first making use of GRASP for solving MOCO problems. Technical report, University of Valenciennes, France, 1998. Paper presented at MCDM 14, June 8-12 1998, Charlottesville, VA.

[101] W.B. Gearhart. Compromise solutions and estimation of the noninferior set. *Journal of Optimization Theory and Applications*, 28:29–47, 1979.

[102] F.W. Gembicki and Y.Y. Haimes. Approach to performance and multiobjective sensitivity optimization: The goal attainment method. *IEEE Transactions on Automatic Control*, 20:769–771, 1975.

[103] A.M. Geoffrion. Proper efficiency and the theory of vector maximization. *Journal of Mathematical Analysis and Applications*, 22:618–630, 1968.

[104] C. Gert and P. Weidner. Nonconvex separation theorems and some applications in vector optimization. *Journal of Optimization Theory and Applications*, 67:297–320, 1990.

[105] C.J. Goh and X.Q. Yang. Analytic efficient solution set for multi-criteria quadratic programs. *European Journal of Operational Research*, 92(1):166–181, 1996.

[106] M. Gravel, W.L. Price, and C. Gagné. Scheduling continuous casting of aluminium using a multiple-objective ant colony optimization metaheuristic. *European Journal of Operational Research*, 143(1):218–229, 2002.

[107] J. Guddat, F. Guerra Vasquez, K. Tammer, and K. Wendler. *Multiobjective and Stochastic Optimization Based on Parametric Optimization*. Akademie-Verlag, Berlin, 1985.

[108] T.R. Gulati and M.A. Islam. Proper efficiency in a linear fractional vector maximum problem with generalized convex constraints. *European Journal of Operational Research*, 36(3):339–345, 1988.

[109] T.R. Gulati and M.A. Islam. Efficiency in linear fractional vector maximization problem with nonlinear constraints. *Optimization*, 20(4):477–482, 1989.

[110] B. Gupta. Finding the set of all efficient solutions for the linear fractional multiobjective program with zero-one variables. *Opsearch*, 18(4):204–214, 1982.

[111] B. Gupta. Programming with multi-objective linear fractional functionals. *Acta Ciencia Indica. Mathematics*, 9(1-4):195–201, 1983.

[112] R. Gupta and R. Malhotra. Multi-criteria integer linear fractional programming problem. *Optimization*, 35(4):373–389, 1995.

[113] M.P. Hansen. Tabu search for multiobjective combinatorial optimization: TAMOCO. *Control and Cybernetics*, 29(3):799–818, 2000.

[114] R. Hartley. On cone-efficiency, cone-convexity and cone-compactness. *SIAM Journal on Applied Mathematics*, 34(2):211–222, 1978.

[115] S. Helbig. On a constructive approximation of the efficient outcomes in bicriterion vector optimization. *Journal of Global Optimization*, 5:35–48, 1994.

[116] A. Helbing. Relations between integer linear vector optimization and multicriteria problems on groups and graphs. *Optimization*, 38(1):61–71, 1996.

[117] M. Henig. Proper efficiency with respect to cones. *Journal of Optimization Theory and Applications*, 36(3):387–407, 1982.

[118] J. Horn, N. Nafpliotis, and D.E. Goldberg. A niched Pareto genetic algorithm for multiobjective optimization. In *Proceedings of the First IEEE Conference on Evolutionary Computation, IEEE World Congress on Computational Intelligence, Orlando, FL, 29 June - 1 July 1994*, volume 1, pages 82–87. IEEE Service Center, Piscataway, NJ, 1994.

[119] B.J. Hunt and M.M. Wiecek. Cones to aid decision making in multicriteria programming. In T. Tanino, T. Tanaka, and M. Inuiguchi, editors, *Multi-Objective Programming and Goal Programming*, pages 153–158. Springer Verlag, Berlin, 2003.

[120] S. Iredi, D. Merkle, and M. Middendorf. Bi-criterion optimization with multi colony ant algorithms. In E. Zitzler, K. Deb, L. Thiele, C.A. Coello Coello, and D. Corne, editors, *First International Conference on Evolutionary Multi-Criterion Optimization*, volume 1993 of *Lecture Notes in Computer Science*, pages 359–372. Springer Verlag, Berlin, 2001.

[121] G. Isac, M.M. Kostreva, and M.M. Wiecek. Multiple-objective approximation of feasible but unsolvable linear complementarity problems. *Journal of Optimization Theory and Applications*, 86(2):389–405, 1995.

[122] H. Isermann. The enumeration of the set of all efficient solutions for a linear multiple objective program. *Operational Research Quarterly*, 28(3):711–725, 1977.

[123] H. Isermann. Linear lexicographic optimization. *OR Spektrum*, 4:223–228, 1982.

[124] A. Ismail-Yahaya and A. Messac. Effective generation of the pareto frontier: the normalized normal constraint method. In *43rd AIAA/ASME/ASCE/AHS Structures, Structural Dynamics, and Materials Conference, Denver, CO, April 2002*, volume AIAA 2002-1232, 2002.

[125] J. Jahn and A Merkel. Reference point approximation method for the solution of bicriterial nonlinear optimization problems. *Journal of Optimization Theory and Applications*, 74:87–103, 1992.

[126] V. Jeyakumar and X.Q. Yang. Convex composite multi-objective nonsmooth programming. *Mathematical Programming*, 59(3):325–343, 1993.

[127] B. Jimenez and V. Novo. First and second order sufficient conditions for strict minimality in multiobjective programming. *Numerical Functional Analysis and Optimization*, 23(3/4):303–322, 2002.

[128] D. Jones, S.K. Mirrazavi, and M. Tamiz. Multi-objective meta-heuristics: An overview of the current state-of-the-art. *European Journal of Operational Research*, 137(1):1–9, 2002.

[129] D.F. Jones and M. Tamiz. Goal programming in the period 1900-2000. In M. Ehrgott and X. Gandibleux, editors, *Multiple Criteria Optimization: State of the Art Annotated Bibliographic Surveys*, pages 129–170. Kluwer Academic Publishers, Boston, MA, 2002.

[130] I. Kaliszewski. Enumeration of efficient solutions for bicriteria integer programs by number theoretic approach. *Control and Cybernetics*, 12:39–54, 1983.

[131] I. Kaliszewski. A modified weighted Tchebycheff metric for multiple objective programming. *Computers and Operations Research*, 14(4):315–323, 1987.

[132] I. Kaliszewski. *Quantitative Pareto Analysis by Cone Separation Technique*. Kluwer Academic Publishers, Dordrecht, 1994.

[133] J. Karaivanova, P. Korhonen, S. Narula, J. Wallenius, and V. Vassilev. A reference direction approach to multiple objective integer linear programming. *European Journal of Operational Research*, 81:176–187, 1995.

[134] R.N. Kaul and V. Lyall. A note on nonlinear fractional vector maximization. *Opsearch*, 26(2):108–121, 1989.

[135] G. Kiziltan and E. Yucaoglu. An algorithm for multiobjective zero-one linear programming. *Management Science*, 29(12):1444–1453, 1983.

[136] K. Klamroth, J. Tind, and M.M. Wiecek. Unbiased approximation in multicriteria optimization. *Mathematical Methods of Operations Research*, 36(3):413–437, 2002.

[137] D. Klein and E. Hannan. An algorithm for the multiple objective integer linear programming problem. *European Journal of Operational Research*, 9:378–385, 1982.

[138] A. Klinger. Improper solutions of the vector maximum problem. *Operations Research*, 15(3):570–572, 1967.

[139] P. Korhonen and J. Wallenius. A Pareto race. *Naval Research Logistics*, 35:615–623, 1988.

[140] P. Korhonen and G.Y. Yu. A reference direction approach to multiple objective quadratic-linear programming. *European Journal of Operational Research*, 102(3):601–610, 1997.

[141] J.S.H. Kornbluth and R.E. Steuer. On computing the set of all weakly efficient vertices in multiple objective linear fractional programming. In G. Fandel and T. Gal, editors, *Multiple Criteria Decision Making Theory and Application*, Lecture Notes in Economics and Mathematical Systems, pages 189–202. Springer Verlag, Berlin, 1980.

[142] M.M. Kostreva and W. Ogryczak. Linear optimization with multiple equitable criteria. *RAIRO Operations Research*, 33:275–297, 1999.

[143] M.M. Kostreva and M.M. Wiecek. Linear complementarity problems and multiple objective programming. *Mathematical Programming*, 60:349–359, 1993.

[144] M.M. Kostreva, Q. Zheng, and D. Zhuang. A method for approximating solutions of multicriteria nonlinear optimization problems. *Optimization Methods and Software*, 5:209–226, 1995.

[145] P. Kouvelis and G. Yu. *Robust Discrete Optimization and Its Applications*, volume 14 of *Nonconvex Optimization and its Applications*. Kluwer Academic Publishers, Dordrecht, 1997.

[146] K.H. Küfer, A. Scherrer, M. Monz, F. Alonso, H. Trinkaus, T. Bortfeld, and C. Thieke. Intensity-modulated radiotherapy – A large scale multi-criteria programming problem. *OR Spectrum*, 25(2):223–249, 2003.

[147] H.W. Kuhn and A.W. Tucker. Nonlinear programming. In J. Neyman, editor, *Proceedings of the Second Berkeley Symposium on Mathematical Statistics and Probability*, pages 481–492. University of California Press, Berkeley, California, 1951.

[148] B. Lemaire. Approximation in multiobjective optimization. *Journal of Global Optimization*, 2:117–132, 1992.

[149] D. Li. Convexification of a noninferior frontier. *Journal of Optimization Theory and Applications*, 88(1):177–196, 1996.

[150] Y. Li, G.M. Fadel, and M.M. Wiecek. Minimum effort approximation of the Pareto space of convex bi-criteria problems. *Optimization and Engineering*, 4:231–261, 2003.

[151] J.G. Lin. Multiple-objective problems: Pareto-optimal solutions by methods of proper equality constraints. *IEEE Transactions on Automatic Control*, 21(5):641–650, 1976.

[152] F.A. Lootsma, T.W. Athan, and P.Y. Papalambros. Controlling the search for a compromise solution in multi-objective optimization. *Enginering Optimization*, 25(1):65–81, 1995.

[153] P. Loridan. ε-solutions in vector minimization problems. *Journal of Optimization Theory and Applications*, 43:265–276, 1984.

[154] B. Malakooti, J. Wang, and E.C. Tandler. A sensor-based accelerated approach for multi-attribute machinability and tool life evaluation. *International Journal of Production Research*, 28:2373, 1990.

[155] E. Marchi and J.A. Oviedo. Lexicographic optimality in the multiple objective linear programming: The nucleolar solution. *European Journal of Operational Research*, 57:355–359, 1992.

[156] B. Metev and D. Gueorguieva. A simple method for obtaining weakly efficient points in multiobjective linear fractional prgramming problems. *European Journal of Operational Research*, 126(2):386–390, 2000.

[157] W. Michalowski and T. Szapiro. A bi-reference procedure for interactive multiple criteria programming. *OR*, 40(2):247–258, 1992.

[158] K. Miettinen. *Nonlinear Multiobjective Optimization*, volume 12 of *International Series in Operations Research and Management Science*. Kluwer Academic Publishers, Dordrecht, 1999.

[159] T. Murata and H. Ishibuchi. MOGA: Multi-objective genetic algorithms. In *Proceedings of the 2nd IEEE International Conference on Evolutionary Computing, Perth, Australia*, pages 289–294. IEEE Service Center, Piscataway, NJ, 1995.

[160] P.H. Naccache. Connectedness of the set of nondominated outcomes in multicriteria optimization. *Journal of Optimization Theory and Applications*, 25:459–467, 1978.

[161] D. Nam and C.H. Park. Multiobjective simulated annealing: A comparative study to evolutionary algorithms. *International Journal of Fuzzy Systems*, 2(2):87–97, 2000.

[162] V.N. Nefedov. On the approximation of the Pareto set. *USSR Computational Mathematics and Mathematical Physics*, 24(4):19–28, 1984.

[163] P. Neumayer and D. Schweigert. Three algorithms for bicriteria integer linear programs. *OR Spectrum*, 16:267–276, 1994.

[164] S. Nickel and M.M. Wiecek. Multiple objective programming with piecewise linear functions. *Journal of Multi-Criteria Decision Analysis*, 8:322–332, 1999.

[165] W. Ogryczak. Inequality measures and equitable approaches to location problems. *European Journal of Operational Research*, 122:374–391, 2000.

[166] W. Ogryczak. Multiple criteria linear programming model for portfolio selection. *Annals of Operations Research*, 97:143–162, 2000.

[167] G. Parks and A. Suppapitnarm. Multiobjective optimization of PWR reload core designs using simulated annealing. In *Proceedings of the International Conference on Mathematics and Computation, Reactor Physics and Environmental Analysis in Nuclear Applications. Madrid, Spain, September 1999*, volume 2, pages 1435–1444, 1999.

[168] A. Pascoletti and P. Serafini. Scalarizing vector optimization problems. *Journal of Optimization Theory and Applications*, 42(4):499–524, 1984.

[169] A.N. Payne. Efficient approximate representation of bi-objective tradeoff sets. *Journal of the Franklin Institute*, 330:1219–1233, 1993.

[170] A.N. Payne and E. Polak. An interactive rectangle elimination method for biobjective decision making. *IEEE Transactions on Automatic Control*, 25(3):421–432, 1980.

[171] H.J. Payne, E. Polak, D.C. Collins, and W. S. Meisel. An algorithm for bicriteria optimization based on the sensitivity function. *IEEE Transactions on Automatic Control*, 20(4):546–648, 1975.

[172] J. Philip. Algorithms for the vector maximization problem. *Mathematical Programming*, 2:207–229, 1972.

[173] J. Philip. Vector maximization at a degenerate vertex. *Mathematical Programming*, 13:357–359, 1977.

[174] E. Polak. On the approximation of solutions to multiple criteria decision making problems. In M. Zeleny, editor, *Multiple Criteria Decision Making*, volume 123 of *Lecture Notes in Economics and Mathematical Systems*, pages 271–282. Springer, 1976.

[175] L.I. Poliščuk. Piece-wise linear approximation of the Pareto boundary of convex two-criteria problems (Russian). In *Models and Methods of Investigating Economic Systems*, pages 108–116. Nauka Sibirsk. Otdel., Novosibirsk, 1979.

[176] C. Romero. A general structure of achievement function for a goal programming model. *European Journal of Operational Research*, 153:675–686, 2004.

[177] B. Roy. Problems and methods with multiple objective functions. *Mathematical Programming*, 1:239–266, 1971.

[178] S. Ruzika and M. M. Wiecek. A survey of approximation methods in multiobjective programming. Technical Report 90, Department of Mathematics, University of Kaiserslautern, 2003.

[179] J.K. Sankaran. On a variant of lexicographic multi-objective programming. *European Journal of Operational Research*, 107:669–674, 1998.

[180] Y. Sawaragi, H. Nakayama, and T. Tanino. *Theory of Multiobjective Optimization*. Academic Press, Orlando, 1985.

[181] S. Sayin. An algorithm based on facial decomposition for finding the efficient set in multiple objective linear programming. *Operations Research Letters*, 19:87–94, 1996.

[182] S. Sayin. Measuring the quality of discrete representations of efficient sets in multiple objective mathematical programming. *Mathematical Programming*, 87:543–560, 2000.

[183] J.D. Schaffer. *Multiple Objective Optimization with Vector Evaluated Genetic Algorithms*. PhD thesis, Vanderbilt University, Nashville, TN, 1984.

[184] B. Schandl, K. Klamroth, and M.M. Wiecek. Norm-based approximation in bicriteria programming. *Computational Optimization and Applications*, 20:23–42, 2001.

[185] B. Schandl, K. Klamroth, and M.M. Wiecek. Introducing oblique norms into multiple criteria programming. *Journal of Global Optimization*, 23:81–97, 2002.

[186] B. Schandl, K. Klamroth, and M.M. Wiecek. Norm-based approximation in multicriteria programming. *Computers and Mathematics with Applications*, 44(7):925–942, 2002.

[187] K.P. Schönfeld. *Effizienz und Dualität in der Aktivitätsanlyse (German)*. PhD thesis, Freie Universität Berlin, Germany, 1964.

[188] P. Serafini. Some considerations about computational complexity for multi objective combinatorial problems. In J. Jahn and W. Krabs, editors, *Recent advances and historical development of vector optimization*, volume 294 of *Lecture Notes in Economics and Mathematical Systems*, pages 222–232. Springer Verlag, Berlin, 1986.

[189] P. Serafini. Simulated annealing for multiobjective optimization problems. In *Proceedings of the 10th International Conference on Multiple Criteria Decision Making, Taipei-Taiwan*, volume I, pages 87–96, 1992.

[190] P.S. Shelokar, S. Adhikari, R. Vakil, V.K. Jayaraman, and B.D. Kulkarni. Multiobjective ant algorithm for continuous function optimization: Combination of strength Pareto fitness assignment and thermo-dynamic clustering. *Foundations of Computing and Decision Sciences Journal*, 25(4):213–230, 2000.

[191] A.M. Skulimowski. *Decision Support Systems Based on Reference Sets*, volume 40 of *Dissertations Monographies*. Wydawnictwa AGH, Krakow, 1996.

[192] I.M. Sobol' and Y.L. Levitan. Error estimates for the crude approximation of the trade-off curve. In G. Fandel and T. Gal, editors, *Multiple Criteria Decision Making*, volume 448 of *Lecture Notes in Economics and Mathematical Systems*, pages 83–92. Springer, Berlin, 1997.

[193] R.M. Soland. Multicriteria optimization: A general characterization of efficient solutions. *Decision Sciences*, 10:26–38, 1979.

[194] R. Solanki. Generating the noninferior set in mixed integer biobjective linear programs: an application to a location problem. *Computers and Operations Research*, 18:1–15, 1991.

[195] R.S. Solanki, P.A. Appino, and J.L. Cohon. Approximating the noninferior set in multiobjective linear programming problems. *European Journal of Operational Research*, 68:356–373, 1993.

[196] Y. Sonntag and C. Zălinescu. Comparison of existence results for efficient points. *JOTA*, 105(1):161–188, 2000.

[197] N. Srinivas and K. Deb. Multiobjective optimization using non-dominated sorting in genetic algorithms. *Evolutionary Computation*, 2(3):221–248, 1994.

[198] R.B. Statnikov and J. Matusov. Use of p_τ-nets for the approximation of the Edgeworth-Pareto set in multicriteria optimization. *Journal of Optimization Theory and Applications*, 91(3):543–560, 1996.

[199] R.B. Statnikov and J.B. Matusov. *Multicriteria Optimization and Engineering*. Chapman & Hall, New York, 1995.

[200] R. Steuer. Random problem generation and the computation of efficient extreme points in multiple objective linear programming. *Computational Optimization and Applications*, 3:333–347, 1994.

[201] R.E. Steuer. *Multiple Criteria Optimization: Theory, Computation and Application*. John Wiley & Sons, New York, NY, 1985.

[202] R.E. Steuer. ADBASE: A multiple objective linear programming solver. Technical report, Terry College of Business, University of Georgia, Athens, GA, 30602 USA, 2000.

[203] R.E. Steuer and E.U. Choo. An interactive weighted Tchebycheff procedure for multiple objective programming. *Mathematical Programming*, 26:326–344, 1983.

[204] L.W.G. Strijbosch, A.G.M. Van Doorne, and W.J. Selen. A simplified MOLP algorithm: The MOLP-S procedure. *Computers and Operations Research*, 18(8):709–716, 1991.

[205] M. Sun. Applying tabu search to multiple objective combinatorial optimization problems. In *Proceedings of the 1997 DSI Annual Meeting, San Diego, California*, volume 2, pages 945–947. Decision Sciences Institute, Atlanta, GA, 1997.

[206] M. Sun, A. Stam, and R. Steuer. Solving multiple objective programming problems using feed-forward artificial neural networks: The interactive FFANN procedure. *Management Science*, 42(6):835–849, 1996.

[207] M. Sun, A. Stam, and R. Steuer. Interactive multiple objective programming using Tchebycheff programs and artificial neural networks. *Computers and Operations Research*, 27:601–620, 2000.

[208] M. Tamiz and D. Jones. Goal programming and Pareto efficiency. *Journal of Information & Optimization Sciences*, 17(2):291–307, 1996.

[209] M.L. TenHuisen and M.M. Wiecek. Vector optimization and generalized Lagrangian duality. *Annals of Operations Research*, 51:15–32, 1994.

[210] M.L. TenHuisen and M.M. Wiecek. Efficiency and solution approaches to bicriteria nonconvex programs. *Journal of Global Optimization*, 11(3):225–251, 1997.

[211] J. Tind and M. M. Wiecek. Augmented Lagrangian and Tchebycheff approaches in multiple objective programming. *Journal of Global Optimization*, 14:251–266, 1999.

[212] E.L. Ulungu. *Optimisation combinatoire multicritère: Détermination de l'ensemble des solutions efficaces et méthodes interactives*. PhD thesis, Faculté des Sciences, Université de Mons-Hainaut. Mons, Belgium, 1993.

[213] B. Villarreal and M.H. Karwan. Multicriteria integer programming: A (hybrid) dynamic programming recursive approach. *Mathematical Programming*, 21:204–223, 1981.

[214] B. Villarreal and M.H. Karwan. Multicrietria dynamic programming with an application to the integer case. *Journal of Optimization Theory and Applications*, 38(1):43–69, 1982.

[215] V. Voinalovich. External approximation to the Pareto set in criterion space for multicriterion linear programming tasks. *Cybernetics and Computing Technology*, 1:135–142, 1984.

[216] S. Wang. Second-order necessary and sufficient conditions in multiobjective programming. *Numerical Functional Analysis and Optimization*, 12(1-2):237–252, 1991.

[217] A. Warburton. Aproximation of Pareto optima in multiple-objective shortest-path problems. *Operations Research*, 35(1):70 –79, 1987.

[218] A.R. Warburton. Quasiconcave vector maximization: Connectedness of the sets of Pareto-optimal and weak Pareto-optimal alternatives. *Journal of Optimization Theory and Applications*, 40:537–557, 1983.

[219] P. Weidner. Complete efficiency and interdependencies between objective functions in vector optimization. *Zeitschrift für Operations Research*, 34:91–115, 1990.

[220] R.E. Wendell and D.N. Lee. Efficiency in multiple objective optimization problems. *Mathematical Programming*, 12(3):406–414, 1977.

[221] D.J. White. A branch and bound method for multiobjective Boolean problems. *European Journal of Operational Research*, 15(1):126–130, 1984.

[222] D.J. White. Epsilon efficiency. *Journal of Optimization Theory and Applications*, 49(2):319–337, 1986.

[223] D.J. White. Weighting factor extensions for finite multiple objective vector minimization problems. *European Journal of Operational Research*, 36:256–265, 1988.

[224] D.J. White. A characterisation of the feasible set of objective function vectors in linear multiple objective problems. *European Journal of Operational Research*, 52:361–366, 1991.

[225] M.M. Wiecek, W. Chen, and J. Zhang. Piecewise quadratic approximation of the non-dominated set for bi-criteria programs. *Journal of Multi-Criteria Decision Analysis*, 10:35–47, 2001.

[226] M.M. Wiecek and H. Zhang. A parallel algorithm for multiple objective linear programs. *Computational Optimization and Applications*, 8(1):41–56, 1997.

[227] A.P. Wierzbicki. A methodological guide to multiobjective optimization. In K. Iracki, K. Malanowski, and S. Walukiewicz, editors, *Optimization Techniques, Part 1*, volume 22 of *Lecture Notes in Control and Information Sciences*, pages 99–123. Springer, Berlin, 1980.

[228] A.P. Wierzbicki. A methodological approach to comparing parametric characterizations of efficient solutions. In G. Fandel, M. Grauer, A. Kurzhanski, and A.P. Wierzbicki, editors, *Large-scale modelling and interactive decision analysis*, volume 273 of *Lecture Notes in Economics and Mathematical Systems*, pages 27–45. Springer Verlag, Berlin, 1986.

[229] A.P. Wierzbicki. On the completeness and constructiveness of parametric characterizations to vector optimization problems. *OR Spectrum*, 8(2):73–87, 1986.

[230] X.Q. Yang and C.J. Goh. A method for convex curve approximation. *European Journal of Operational Research*, 97:205–212, 1997.

[231] P.L. Yu. A class of solutions for group decision making. *Management Science*, 19(8):936–946, 1973.

[232] P.L. Yu. Cone convexity, cone extreme points and nondominated solutions in decision problems with multiobjectives. *Journal of Optimization Theory and Applications*, 14:319–377, 1974.

[233] P.L. Yu and M. Zeleny. The set of all nondominated solutions in linear cases and a multicriteria simplex method. *Journal of Mathematical Analysis and Applications*, 49:430–468, 1975.

[234] P.L. Yu and M. Zeleny. Linear multiparametric programming by multicriteria simplex method. *Management Science*, 23(2):159–170, 1976.

[235] M. Zeleny. Compromise programming. In J.L. Cochrane and M. Zeleny, editors, *Multiple Criteria Decision Making*, pages 262–301. University of South Carolina Press, Columbia, SC, 1973.

[236] M. Zeleny. *Linear Multiobjective Programming*, volume 95 of *Lecture Notes in Economics and Mathematical Systems*. Springer-Verlag, Berlin, 1974.

[237] S. Zionts and J. Wallenius. Identifying efficient vectors: Some theory and computational results. *Operations Research*, 28(3):785–793, 1980.

[238] E. Zitzler and L. Thiele. An evolutionary algorithm for multiobjective optimization: The strength Pareto approach. Technical Report 43, Computer Engineering and Communication Networks Lab (TIK), Swiss Federal Institute of Technology (ETH), Zürich, Switzerland, 1998.

Chapter 18

MULTIPLE OBJECTIVE LINEAR PROGRAMMING WITH FUZZY COEFFICIENTS

Masahiro Inuiguchi
Department of Systems Innovation
Graduate School of Engineering Science, Osaka University
1-3, Machikaneyama, Toyonaka, Osaka 560-8531
Japan
inuiguti@sys.es.osaka-u.ac.jp

Abstract In this paper, we treat multiple objective programming problems with fuzzy coefficients. We introduce the approaches based on possibility and necessity measures. Our aim in this paper is to describe the treatments of the problem rather than the solution method for the problem. We describe the modality constrained programming approach, the modality goal programming approach and modal efficiency approach. In the first approach, we discuss treatments of fuzziness in the programming problems. The extensions of a fuzzy relation to the relation between fuzzy numbers are developed in order to treat generalized constraints. In the second approach, we show that two kinds of differences between a fuzzy objective function value and a fuzzy target are conceivable under the fuzziness. We describe the distinction of their applications in programming problems. In the third approach, we describe how the efficiency can be extended to multiple objective programming problems with fuzzy coefficients. Necessary and sufficient conditions for a feasible solution to satisfy the extended efficiency are discussed. Finally some concluding remarks are given.

Keywords: Multiple objective programming, fuzzy coefficient, fuzzy relation, possibility measure, necessity measure.

1. Introduction

In multiple objective programming problems, parameters such as coefficients and right-hand side values of constraints are assumed to be known as real numbers. However, in real world problems, we may face cases where the expert knowledge is not so certain as to specify the parameters as real numbers and cases where parameters fluctuate in certain ranges. For example, demands of products, future profit and parameters depends on human ability such as 'manual processing time' are hard to estimate their values as real numbers. To cope with such uncertainties, stochastic programming approaches were proposed [42, 43]. In stochastic programming approaches, we should estimate proper probability distributions of parameters. However, the estimation is not always a simple task because of the following reasons: (1) historical data of some parameters cannot be obtained easily especially when we face a new uncertain variable, and (2) subjective probabilities cannot be specified easily when many parameters exist. Moreover, even if we succeeded to estimate the probability distribution from historical data, there is no guarantee that the current parameters obey the distribution actually.

On the other hand, it is often that we can estimate the possible ranges of the uncertain parameters. In such cases, it is conceivable that we can represent the possible ranges by fuzzy sets so that we formulate the problems as multiple objective programming problems with fuzzy coefficients. From this point of view, many approaches to the problems have been proposed. Since we treat the uncertainty as well as multiplicity of objectives, we should discuss not only the solution procedure but also the treatment of the problem.

In this paper, we introduce the approaches to multiple objective programming problems with fuzzy coefficients based on possibility and necessity measures. Since fuzzy programming has a relatively long history, a lot of approaches have been proposed. However, many of them can be regarded as approaches based on possibility and necessity measures (see [16, 21]). Many other approaches are not very different because the difference is often only in the employed measures for the evaluation of a fuzzy event. Thus, describing the approaches based on possibility and necessity measures would be sufficient to know the ideas of treatments of multiple objective programming problems with fuzzy coefficients. Since this book is devoted to multiple criteria decision making, we describe mainly the treatments of the problem rather than solution algorithms for the problem. References [3, 22, 28, 29, 31, 37, 38, 39, 41, 42] are good books and papers to know various approaches as well as solution procedures.

We describe three approaches, the modality constrained programming approach [16, 18], the modality goal programming approach [20, 24] and modal efficiency approach [1, 23]. In the first approach, we describe the treatments of fuzziness in the problems. Namely, we show how we transform ill-posed programming problems with fuzzy coefficients to well-posed conventional programming problems. Here we also discuss how a fuzzy relation between elements is extended to fuzzy relations between fuzzy sets. These extensions would be useful for various decision making problems with fuzzy parameters. Moreover, we describe the relations with some other approaches. In the second approach, modality goal programming approach, we show that we can define two kinds of differences between a fuzzy objective function value and a fuzzy target due to the fuzziness involved in the problem. We also describe the distinction of their applications. In the third approach, we discuss how the efficiency can be extended to problems with fuzzy coefficients. Necessary and sufficient conditions for a feasible solution to satisfy the extended efficiency are described.

This paper is organized as follows. First we describe the problem statement. We introduce a generalized multiple objective programming problems with fuzzy coefficients. In next two sections, we describe modality constrained programming and modality goal programming approaches. Then we describe extended efficient solutions. Finally, some concluding remarks are given.

2. Problem Statement and Approaches

Since the aim of this paper is to describe the models to treat fuzziness in multiple objective programming problems rather than solution algorithms for the problems, we restrict ourselves into multiple objective linear programming problems with fuzzy coefficients.

Multiple objective linear programming (MOLP) problems can be written as

$$\begin{array}{ll} \text{minimize} & (\boldsymbol{c}_1^{\mathrm{T}}\boldsymbol{x}, \boldsymbol{c}_2^{\mathrm{T}}\boldsymbol{x}, \ldots, \boldsymbol{c}_p^{\mathrm{T}}\boldsymbol{x})^{\mathrm{T}}, \\ \text{subject to} & \boldsymbol{a}_i^{\mathrm{T}}\boldsymbol{x} \leq b_i,\ i = 1, 2, \ldots, m_1, \\ & \boldsymbol{a}_i^{\mathrm{T}}\boldsymbol{x} = b_i,\ i = m_1 + 1, \ldots, m, \\ & \boldsymbol{x} \geq \boldsymbol{0}, \end{array} \tag{18.1}$$

where $\boldsymbol{c}_k = (c_{k1}, c_{k2}, \ldots, c_{kn})^{\mathrm{T}}$, $k = 1, 2, \ldots, p$ and $\boldsymbol{a}_i = (a_{i1}, a_{i2}, \ldots, a_{in})^{\mathrm{T}}$, $i = 1, 2, \ldots, m$ are constant vectors and b_i, $i = 1, 2, \ldots, m$ constants. $\boldsymbol{x} = (x_1, x_2 \ldots, x_n)^{\mathrm{T}}$ is the decision variable vector. In multiple objective programming problems, efficient solutions are considered as reasonable solutions (see [11]). An efficient solution is a feasible solution $\bar{\boldsymbol{x}}$ to which there is no feasible solution $\boldsymbol{x}$ such that $\boldsymbol{c}_k^{\mathrm{T}}\boldsymbol{x} \leq \boldsymbol{c}_k^{\mathrm{T}}\bar{\boldsymbol{x}}$, $k = 1, 2, \ldots, p$ with at least one strict inequality.

In real world applications, the constraints are not always extremely strict. For example, let $\boldsymbol{a}_1^{\mathrm{T}}\boldsymbol{x}$ show the required expense for the activity $\boldsymbol{x}$ and let b_1 be the acceptable expense say \$ 100 million. We may accept the expense of \$ 100.1 million if the objective functions take much better values by this small violation of the constraint. Such constraints are called soft constraints while the conventional strict constraints are called hard constraints.

From this point of view, we may relax the constraints by replace the inequality relation $\leq$ and the equality relation $=$ with a fuzzy inequality relation $\lesssim_i$ and a fuzzy equality relation $\simeq_i$. Then the problem becomes

$$\begin{array}{ll} \text{minimize} & (\boldsymbol{c}_1^{\mathrm{T}}\boldsymbol{x}, \boldsymbol{c}_2^{\mathrm{T}}\boldsymbol{x}, \ldots, \boldsymbol{c}_p^{\mathrm{T}}\boldsymbol{x})^{\mathrm{T}}, \\ \text{subject to} & \boldsymbol{a}_i^{\mathrm{T}}\boldsymbol{x} \lesssim_i b_i,\ i = 1, 2, \ldots, m_1 \\ & \boldsymbol{a}_i^{\mathrm{T}}\boldsymbol{x} \simeq_i b_i,\ i = m_1 + 1, \ldots, m, \\ & \boldsymbol{x} \geq \boldsymbol{0}, \end{array} \tag{18.2}$$

where $r_1 \lesssim_i r_2$ and $r_1 \simeq_i r_2$ may have a linguistic expressions 'r_1 is approximately smaller than r_2' and 'r_1 is approximately equal to r_2'. The subscript i in $\lesssim_i$ (resp. $\simeq_i$) shows that a fuzzy inequality (resp. equality) relation can be different by the constraint. The fuzzy inequality relations and the fuzzy equality relations are fuzzy relations, a fuzzy set on $\mathbf{R} \times \mathbf{R}$ (see [12] for details of fuzzy relations).

Such fuzzy inequality and equality relations can be modeled by the following equations:

$$\mu_{\lesssim_i}(r_1, r_2) = \nu_i(r_1 - r_2), \tag{18.3}$$

$$\mu_{\simeq_i}(r_1, r_2) = \eta_i(r_1 - r_2), \tag{18.4}$$

where $\nu_i : \mathbf{R} \to [0, 1]$ and $\eta_i : \mathbf{R} \to [0, 1]$ are upper semi-continuous non-increasing and upper semi-continuous quasi-concave functions such that $\nu_i(0) = \eta_i(0) = 1$. Functions ν_i and η_i are illustrated in Figure 18.1. The membership function values show the degrees of the satisfaction. Note that if $\nu_i(r) = 0$ for all $r > 0$ and $\eta_i(r) = 0$ for all $r \neq 0$ then fuzzy inequality and equality relations $\lesssim_i$ and $\simeq_i$ degenerate to usual inequality and equality relations $\leq$ and $=$, respectively.

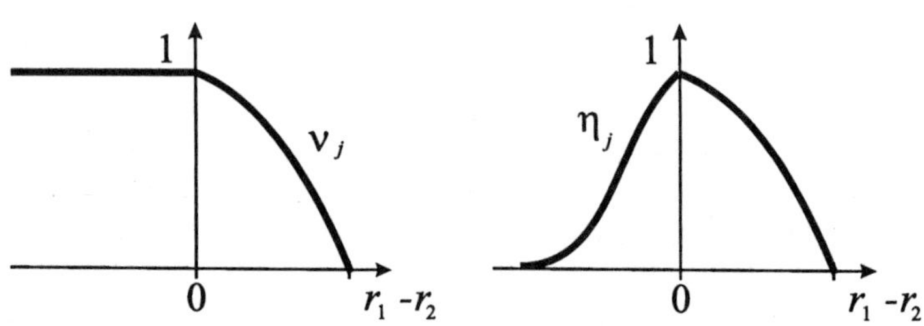

(a) Function ν_j. (b) Function η_j.

Figure 18.1. Fuzzy inequality and equality relations.

This fuzzification can be found in the beginning of fuzzy mathematical programming problems [46, 48, 49] called flexible programming problems. In flexible programming, a fuzzy goal G_i is defined as a fuzzy set of satisfactory left-hand side values. The fuzzy goal can be seen as a combination of a fuzzy inequality or equality relation and a right-hand value, i.e.,

$$\mu_{G_i}(r) = \mu_{\lesssim_i}(r, b_i), \quad \text{or} \quad \mu_{G_i}(r) = \mu_{\simeq_i}(r, b_i), \tag{18.5}$$

where μ_{G_i} is a membership function of a fuzzy goal G_i. The ideal solution of a flexible programming problem maximizes all membership functions μ_{G_i}, $i = 1, 2, \dots, m$ at the same time. However usually there is no such a solution. To obtain a compromise solution by solving a mathematical programming problem, aggregation operators of all membership functions are discussed (see [7, 49, 50]). Recently, the idea of flexible programming is introduced to constraint satisfaction problems [4, 5]. We will not discuss these topics in this paper because we will concentrate on the treatments of MOLP problems with fuzzy coefficients.

In Problems (18.1) and (18.2), coefficients $\boldsymbol{c}_k$, $\boldsymbol{a}_i$ and right-hand side values b_i are assumed to be known as real numbers. However, we may face problems whose coefficients and right-hand side values are not known exactly but roughly. We assume that the expert can represent the ranges of those values as fuzzy sets. Then Problem (18.2) becomes

$$\begin{array}{ll} \text{minimize} & (\tilde{\boldsymbol{c}}_1^{\mathrm{T}}\boldsymbol{x}, \tilde{\boldsymbol{c}}_2^{\mathrm{T}}\boldsymbol{x}, \dots, \tilde{\boldsymbol{c}}_p^{\mathrm{T}}\boldsymbol{x})^{\mathrm{T}}, \\ \text{subject to} & \tilde{\boldsymbol{a}}_i^{\mathrm{T}}\boldsymbol{x} \lesssim_i \tilde{b}_i, \ i = 1, 2, \dots, m_1 \\ & \tilde{\boldsymbol{a}}_i^{\mathrm{T}}\boldsymbol{x} \simeq_i \tilde{b}_i, \ i = m_1 + 1, \dots, m, \\ & \boldsymbol{x} \geq \boldsymbol{0}, \end{array} \tag{18.6}$$

where $\tilde{\boldsymbol{c}}_k = (\tilde{c}_{k1}, \tilde{c}_{k2}, \dots, \tilde{c}_{kn})^{\mathrm{T}}$, $k = 1, 2, \dots, p$, $\tilde{\boldsymbol{a}}_i = (\tilde{a}_{i1}, \tilde{a}_{i2}, \dots, \tilde{a}_{in})^{\mathrm{T}}$, $i = 1, 2, \dots, m$ are fuzzy coefficient vectors. $\tilde{c}_{kj}$, $\tilde{a}_{ij}$ and $\tilde{b}_i$, $i = 1, 2, \dots, m$, $j = 1, 2, \dots, n$, $k = 1, 2, \dots, p$ are fuzzy numbers. A fuzzy number $\tilde{a}$ is a fuzzy set on a real line whose membership function $\mu_{\tilde{a}} : \mathbf{R} \to [0, 1]$ satisfies (see, for example [8]):

(i) $\tilde{a}$ is normal, i.e., there exists $r \in \mathbf{R}$ such that $\mu_{\tilde{a}}(r) = 1$.

(ii) $\mu_{\tilde{a}}$ is upper semi-continuous, i.e., the h-level set $[\tilde{a}]_h = \{r \in \mathbf{R} \mid \mu_{\tilde{a}}(r) \geq h\}$ is a closed set for any $h \in (0, 1]$.

(iii) $\tilde{a}$ is a convex fuzzy set. Namely, $\mu_{\tilde{a}}$ is a quasi-concave function, i.e., for any $r_1, r_2 \in \mathbf{R}$, for any $\lambda \in [0, 1]$, $\mu_{\tilde{a}}(\lambda r_1 + (1 - \lambda) r_2) \geq \min(\mu_{\tilde{a}}(r_1), \mu_{\tilde{a}}(r_2))$. In other words, h-level set $[\tilde{a}]_h$ is a convex set for any $h \in (0, 1]$.

(iv) $\tilde{a}$ is bounded, i.e., $\lim_{r \to +\infty} \mu_{\tilde{a}}(r) = \lim_{r \to -\infty} \mu_{\tilde{a}}(r) = 0$. In other words, the h-level set $[\tilde{a}]_h$ is bounded for any $h \in (0, 1]$.

From (ii) to (iv), an h-level set $[\tilde{a}]_h$ is a bounded closed interval for any $h \in (0, 1]$ when $\tilde{a}$ is a fuzzy number. L-R fuzzy numbers are often used in literature. An

L-R fuzzy number $\tilde{a}$ is a fuzzy number defined by the following membership function:

$$\mu_{\tilde{a}}(r) = \begin{cases} L\left(\dfrac{a^{\mathrm{L}} - r}{\alpha}\right), & \text{if } r \le a^{\mathrm{L}} \text{ and } \alpha > 0, \\ 1, & \text{if } r \in [a^{\mathrm{L}}, a^{\mathrm{R}}], \\ R\left(\dfrac{r - a^{\mathrm{R}}}{\beta}\right), & \text{if } r \ge a^{\mathrm{R}} \text{ and } \beta > 0, \\ 0, & \text{otherwise,} \end{cases} \tag{18.7}$$

where L and $R : [0, +\infty) \to [0, 1]$ are reference functions, i.e., $L(0) = R(0) = 1$, $\lim_{r\to+\infty} L(r) = \lim_{r\to+\infty} R(r) = 0$ and L and R are upper semi-continuous non-increasing functions. α and β are assumed to be non-negative.

An example of L-R fuzzy number $\tilde{a}$ is illustrated in Figure 18.2. As shown in Figure 18.2, a^{L} and a^{R} are lower and upper bounds of the core of $\tilde{a}$, i.e., $\mathrm{Core}(\tilde{a}) = \{r \mid \mu_{\tilde{a}}(r) = 1\}$. α and β show the left and right spreads of $\tilde{a}$. Functions L and R specify the left and right shapes. Using those parameters and functions, fuzzy number $\tilde{a}$ is represented as $\tilde{a} = (a^{\mathrm{L}}, a^{\mathrm{R}}, \alpha, \beta)_{LR}$.

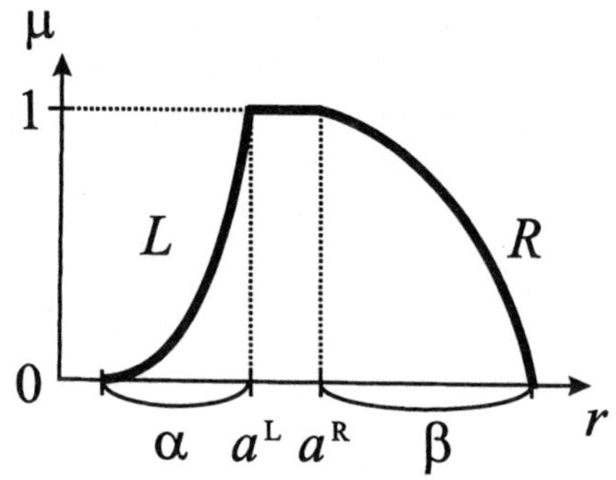

Figure 18.2. L-R fuzzy number $\tilde{a} = (a^{\mathrm{L}}, a^{\mathrm{R}}, \alpha, \beta)_{LR}$.

A membership degree $\mu_{\tilde{a}}(r)$ of fuzzy coefficient $\tilde{a}$ shows the possibility degree of an event 'the coefficient value is r' while a membership degree $\mu_{\lesssim_i}(r_1, r_2)$ of a fuzzy inequality relation $\lesssim_i$ shows the satisfaction degree of a fuzzy inequality '$r_1 \lesssim_i r_2$'.

In Problem (18.6), we should calculate fuzzy linear function values $\tilde{\boldsymbol{c}}_k^{\mathrm{T}}\boldsymbol{x}$ and $\tilde{\boldsymbol{a}}_k^{\mathrm{T}}\boldsymbol{x}$. Those function values can be fuzzy quantities since the coefficients are fuzzy numbers. The extension principle [8] defines the fuzzy quantity of function values of fuzzy numbers. Let $g : \mathbf{R}^q \to \mathbf{R}$ be a function. A function value of $(\tilde{a}_1, \tilde{a}_2, \ldots, \tilde{a}_q)$, i.e., $g(\tilde{a}_1, \tilde{a}_2, \ldots, \tilde{a}_q)$ is a fuzzy quantity $\tilde{Y}$ with the following membership function:

$$\mu_{\tilde{Y}}(y) = \begin{cases} \sup_{r:g(r)=y} \min\left(\mu_{\tilde{a}_1}(r_1), \mu_{\tilde{a}_2}(r_2), \ldots, \mu_{\tilde{a}_q}(r_q)\right), \\ \qquad \text{if } \exists \boldsymbol{r} = (r_1, r_2, \ldots, r_q); g(\boldsymbol{r}) = y, \\ 0, \quad \text{otherwise.} \end{cases} \tag{18.8}$$

Since $\tilde{\boldsymbol{a}}$ is a vector of fuzzy numbers $\tilde{a}_i$ whose h-level set is a bounded closed interval for any $h \in (0,1]$, we have the following equation (see [8]) when g is a continuous function;

$$[\tilde{Y}]_h = g([\tilde{\boldsymbol{a}}]_h), \ \forall h \in (0,1], \tag{18.9}$$

where $[\tilde{\boldsymbol{a}}]_h = ([\tilde{a}_1]_h, [\tilde{a}_2]_h, \ldots, [\tilde{a}_q]_h)$. Note that $[\tilde{a}_j]_h$ is a closed interval since $\tilde{a}_j$ is a fuzzy number. (18.9) implies that h-level set of function value $\tilde{Y}$ can be obtained by interval calculations. Moreover, since g is continuous, from (18.9), we know that $[\tilde{Y}]_h$ is also a closed interval and $[\tilde{Y}]_1 \neq \emptyset$. Therefore, $\tilde{Y}$ is also a fuzzy number.

Let $g(\boldsymbol{r}) = \boldsymbol{r}^{\mathrm{T}}\boldsymbol{x}$, where we define $\boldsymbol{r} = (r_1, r_2, \ldots, r_n)^{\mathrm{T}}$. We obtain the fuzzy linear function value $\tilde{\boldsymbol{c}}_k^{\mathrm{T}}\boldsymbol{x}$ as a fuzzy number $g(\tilde{\boldsymbol{c}}_k)$. For $\boldsymbol{x} \geq \boldsymbol{0}$, we have

$$[\tilde{\boldsymbol{c}}_k^{\mathrm{T}}\boldsymbol{x}]_h = \left[\sum_{j=1}^{n} c_{kj}^{\mathrm{L}}(h)x_j, \sum_{j=1}^{n} c_{kj}^{\mathrm{R}}(h)x_j\right], \ \forall h \in (0,1], \tag{18.10}$$

where $c_{kj}^{\mathrm{L}}(h)$ and $c_{kj}^{\mathrm{R}}(h)$ are lower and upper bounds of h-level set $[\tilde{c}_{kj}]_h$, i.e., $c_{kj}^{\mathrm{L}}(h) = \inf[\tilde{c}_{kj}]_h$ and $c_{kj}^{\mathrm{R}}(h) = \sup[\tilde{c}_{kj}]_h$. Note that when $\tilde{c}_{kj}$ is an L-R fuzzy number $(c_{kj}^{\mathrm{L}}, c_{kj}^{\mathrm{R}}, \gamma_{kj}^{\mathrm{L}}, \gamma_{kj}^{\mathrm{R}})_{L_{kj}^c R_{kj}^c}$, we have

$$c_{kj}^{\mathrm{L}}(h) = c_{kj}^{\mathrm{L}} - \gamma_{kj}^{\mathrm{L}} L_{kj}^{c(-1)}(h), \quad c_{kj}^{\mathrm{R}}(h) = c_{kj}^{\mathrm{R}} + \gamma_{kj}^{\mathrm{R}} R_{kj}^{c(-1)}(h), \tag{18.11}$$

where $L_{kj}^{c(-1)}$ and $R_{kj}^{c(-1)}$ are pseudo-inverse functions of L_{kj}^c and R_{kj}^c defined by $L_{kj}^{c(-1)}(h) = \sup\{r \mid L_{kj}^c(r) \geq h\}$ and $R_{kj}^{c(-1)}(h) = \sup\{r \mid R_{kj}^c(r) \geq h\}$.

In Problem (18.6), each objective function value $\tilde{\boldsymbol{c}}_k^{\mathrm{T}}\boldsymbol{x}$ is obtained as a fuzzy number and left- and right-hand side values of a constraint, $\tilde{\boldsymbol{a}}_i^{\mathrm{T}}\boldsymbol{x}$ and $\tilde{b}_i$ are also fuzzy numbers. Minimizing a fuzzy number $\tilde{\boldsymbol{c}}_k^{\mathrm{T}}\boldsymbol{x}$ cannot be clearly understood. Moreover, since fuzzy relations $\lesssim_i$ and $\simeq_i$ are defined between real numbers, the meaning of relations $\lesssim_i$ and $\simeq_i$ between fuzzy numbers $\tilde{\boldsymbol{a}}_i^{\mathrm{T}}\boldsymbol{x}$ and $\tilde{b}_i$ are not defined. Therefore, Problem (18.6) is an ill-posed problem. We should introduce an interpretation of Problem (18.6) so that we can transform the problem to a well-posed problem.

Many interpretations were proposed. However, most of them were proposed in special cases of Problem (18.6). Namely, many of them treat Problem (18.6) where fuzzy inequality and equality relations are conventional inequality and equality relations. Many other of them treat Problem (18.6) with crisp right-hand side values in the constraints, i.e., problems with fuzzy coefficients and goals. Therefore Problem (18.6) is a generalized problem.

In this paper, we concentrate the approaches to MOLP problems based on possibility and necessity measures. Many approaches can be included as special cases of the approaches based on possibility and necessity measures. Moreover many other approaches can be obtained by the replacement of possibility and necessity measures with other measures. Thus, the concentration on approaches based on possibility and necessity measures may be sufficient to know the essence of the treatments of MOLP problems with fuzzy coefficients.

Possibility and necessity measures of a fuzzy set $\tilde{B}$ under a fuzzy set $\tilde{A}$ are defined as follows (see [10]):

$$\Pi_{\tilde{A}}(\tilde{B}) = \sup_r T(\mu_{\tilde{A}}(r), \mu_{\tilde{B}}(r)), \tag{18.12}$$

$$N_{\tilde{A}}(\tilde{B}) = \inf_r I(\mu_{\tilde{A}}(r), \mu_{\tilde{B}}(r)), \tag{18.13}$$

where $T : [0,1] \times [0,1] \to [0,1]$ and $I : [0,1] \times [0,1] \to [0,1]$ are conjunction and implication functions such that $T(0,0) = T(0,1) = T(1,0) = I(1,0) = 0$ and $T(1,1) = I(0,0) = I(0,1) = I(1,1) = 1$. $\mu_{\tilde{A}}$ and $\mu_{\tilde{B}}$ are membership functions. In this paper, we restrict ourselves into the case where $T(u,v) = \min(u,v)$ and $I(u,v) = \max(1-u,v)$ (Dienes implication), i.e.,

$$\Pi_{\tilde{A}}(\tilde{B}) = \sup_r \min(\mu_{\tilde{A}}(r), \mu_{\tilde{B}}(r)), \tag{18.14}$$

$$N_{\tilde{A}}(\tilde{B}) = \inf_r \max(1 - \mu_{\tilde{A}}(r), \mu_{\tilde{B}}(r)). \tag{18.15}$$

Those possibility and necessity measures are dep'cted in Figure 18.3.

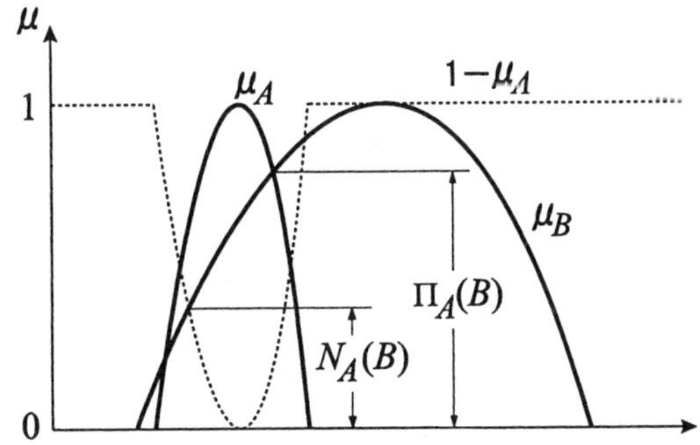

Figure 18.3. Possibility and necessity measures.

By this restriction, we cannot include several conventional approaches to programming problems with fuzzy coefficients such as approaches based on fuzzy max [42] and robust programming approaches [41] in the modality constrained programming approach we describe later. However, those approaches can be included in the approach by using other implication functions (see Section 3.6 and, for details, [21]).

When we assume that fuzzy sets $\tilde{A}$ and $\tilde{B} \subseteq \mathbf{R}^q$ are bounded and they have upper semi-continuous membership functions, for any $h \in (0, 1]$, we have

$$\Pi_{\tilde{A}}(\tilde{B}) \geq h \quad \Leftrightarrow \quad [\tilde{A}]_h \cap [\tilde{B}]_h \neq \emptyset, \tag{18.16}$$

$$N_{\tilde{A}}(\tilde{B}) \geq h \quad \Leftrightarrow \quad (\tilde{A})_{1-h} \subseteq [\tilde{B}]_h, \tag{18.17}$$

where $\tilde{A}$ is said to be bounded when $[\tilde{A}]_h$ is bounded for any $h \in (0, 1]$. $(\tilde{A})_{1-h}$ is a strong $(1-h)$-level set of $\tilde{A}$ defined by $(\tilde{A})_{1-h} = \{r \mid \mu_{\tilde{A}}(r) > 1-h\}$. (18.16) and (18.17) play an important role to reduce Problem (18.6) to a conventional programming problem.

Despite we restricted ourselves into approaches based on possibility and necessity measures, three different approaches have been proposed. They are (a) modality constrained programming approach as a counterpart of chance constrained programming approach (see [43]), (b) modality goal programming approach as the extension of goal programming approach (see [15]) and (c) modal efficiency approach as the direct extension of efficient solutions.

3. Modality Constrained Programming Approach

3.1 Fuzzy Inequality and Equality Relations between Two Fuzzy Numbers

In order to formulate Problem (18.6) as a conventional programming problem, we should discuss the treatment of constraints and treatment of objective functions. In this subsection, we discuss the treatment of constraints. The constraints of Problem (18.6) are fuzzy inequality and equality relations $\precsim_i$ and $\simeq_i$ between two fuzzy numbers $\tilde{\boldsymbol{a}}_i^{\mathrm{T}}\boldsymbol{x}$ and $\tilde{b}_i$. Fuzzy inequality and equality relations are defined as a fuzzy set of $\mathbf{R}^2$, i.e., a fuzzy relation between real numbers. We should extend these relations to the relations between fuzzy numbers.

We describe mainly the extension of fuzzy equality relation. However the way of the extension for fuzzy inequality relation is the same. The difference is only in the properties of the extended relations.

An approach [2] to the extension is based on fuzzy goal G_i with linear membership function in flexible programming, i.e.,

$$\mu_{G_i}(r) = \max\left(0, \min\left(1, \frac{b_i + \kappa^+ - r}{\kappa^+}, \frac{r - b_i + \kappa^-}{\kappa^-}\right)\right), \tag{18.18}$$

where $\kappa^+, \kappa^- > 0$ show excess and shortage tolerances. Given a satisfaction degree h_i, we often treat the constraint $\boldsymbol{a}_i^{\mathrm{T}}\boldsymbol{x} \simeq_i b_i$ with fuzzy goal G_i by

$$\mu_{G_i}(\boldsymbol{a}_i^{\mathrm{T}}\boldsymbol{x}) \geq h_i \;\Leftrightarrow\; \boldsymbol{a}_i^{\mathrm{T}}\boldsymbol{x} \leq b_i + (1 - h_i)\kappa^+ \text{ and } \boldsymbol{a}_i^{\mathrm{T}}\boldsymbol{x} \geq b_i - (1 - h_i)\kappa^-. \tag{18.19}$$

From this result, it is conceivable to treat $\tilde{\boldsymbol{a}}_i^{\mathrm{T}}\boldsymbol{x} \simeq_i \tilde{b}_i$ as follows (see [2]).

$$\tilde{\boldsymbol{a}}_i^{\mathrm{T}}\boldsymbol{x} \leq \tilde{b}_i + (1 - h_i)\tilde{\kappa}^+ \text{ and } \tilde{\boldsymbol{a}}_i^{\mathrm{T}}\boldsymbol{x} \geq \tilde{b}_i - (1 - h_i)\tilde{\kappa}^-, \tag{18.20}$$

where $h_i \in (0, 1]$ is a predetermined value and $\tilde{\kappa}^+, \tilde{\kappa}^-$ are fuzzy numbers whose membership functions $\mu_{\tilde{\kappa}^+}$ and $\mu_{\tilde{\kappa}^-}$ satisfy $\mu_{\tilde{\kappa}^+}(r) = \mu_{\tilde{\kappa}^-}(r) = 0$ for all $r < 0$. Namely, by this way, the fuzzy equality relation between fuzzy numbers $\tilde{\boldsymbol{a}}_i^{\mathrm{T}}\boldsymbol{x}$ and $\tilde{b}_i$ is reduced to two inequality relations between fuzzy numbers. Thus, we can introduce any treatment of the inequality relation between fuzzy numbers to (18.19). This approach is successful but the meaning of the fuzziness in fuzzy tolerances $\tilde{\kappa}^+$ and $\tilde{\kappa}^-$ may be controversial.

Apart from the approach described above, we discuss a way to extend a fuzzy relation between real numbers to the relations between fuzzy numbers based on possibility and necessity measures.

Let $\widetilde{AE}_i^+(b)$ be a fuzzy set of real numbers approximately equal b in the sense of fuzzy equality relation $\simeq_i$. The fuzzy set $\widetilde{AE}_i^+(b)$ can be characterized by the following membership function:

$$\mu_{\widetilde{AE}_i^+(b)}(r) = \mu_{\simeq_i}(r, b). \tag{18.21}$$

Similarly, $\widetilde{AE}_i^-(a)$, a fuzzy set of real numbers to which a is approximately equal in the sense of fuzzy equality relation $\widetilde{AE}_i$ is characterized by the following membership function:

$$\mu_{\widetilde{AE}_i^-(a)}(r) = \mu_{\simeq_i}(a, r). \tag{18.22}$$

Using possibility and necessity measures, we obtain possibility and necessity degrees of the event that a fuzzy number $\tilde{a}$ approximately equals a real number b as follows:

$$\Pi_{\tilde{a}}(\widetilde{AE}_i^+(b)) = \sup_r \min\left(\mu_{\tilde{a}}(r), \mu_{\widetilde{AE}_i^+(b)}(r)\right), \tag{18.23}$$

$$N_{\tilde{a}}(\widetilde{AE}_i^+(b)) = \inf_r \max\left(1 - \mu_{\tilde{a}}(r), \mu_{\widetilde{AE}_i^+(b)}(r)\right). \tag{18.24}$$

$\Pi_{\tilde{a}}(\widetilde{AE}_i^+(b))$ can be also interpreted as a degree to what extent $\tilde{a}$ is possibly approximately equal to b. Similarly, $N_{\tilde{a}}(\widetilde{AE}_i^+(b))$ shows a degree to what extent $\tilde{a}$ is necessarily approximately equal to b. Namely, $\Pi_{\tilde{a}}(\widetilde{AE}_i^+(b))$ and $N_{\tilde{a}}(\widetilde{AE}_i^+(b))$ can be seen as membership degrees of fuzzy relations between a fuzzy number $\tilde{a}$ and a real number b. From this point of view, we may define fuzzy relations $\simeq_i^{\Pi^{\mathrm{F}}}$ and $\simeq_i^{N^{\mathrm{F}}}$ between a fuzzy number $\tilde{a}$ and a real number b by the following membership functions:

$$\mu_{\simeq_i^{\Pi^{\mathrm{F}}}}(\tilde{a}, b) = \Pi_{\tilde{a}}(\widetilde{AE}_i^+(b)), \qquad \mu_{\simeq_i^{N^{\mathrm{F}}}}(\tilde{a}, b) = N_{\tilde{a}}(\widetilde{AE}_i^+(b)), \tag{18.25}$$

In the same way, we can define possibility and necessity degrees of the event that a fuzzy number $\tilde{b}$ is an object to which a is approximately equal as follows:

$$\Pi_{\tilde{b}}(\widetilde{AE}_i^-(a)) = \sup_r \min\left(\mu_{\tilde{b}}(r), \mu_{\widetilde{AE}_i^-(a)}(r)\right), \tag{18.26}$$

$$N_{\tilde{b}}(\widetilde{AE}_i^-(a)) = \inf_r \max\left(1 - \mu_{\tilde{b}}(r), \mu_{\widetilde{AE}_i^-(a)}(r)\right). \tag{18.27}$$

Therefore, we may define fuzzy relations $\simeq_i^{\Pi^{\mathrm{L}}}$ and $\simeq_i^{N^{\mathrm{L}}}$ between a real number a and a fuzzy number $\tilde{b}$ by the following membership functions:

$$\mu_{\simeq_i^{\Pi^{\mathrm{L}}}}(a, \tilde{b}) = \Pi_{\tilde{b}}(\widetilde{AE}_i^-(a)), \qquad \mu_{\simeq_i^{N^{\mathrm{L}}}}(a, \tilde{b}) = N_{\tilde{b}}(\widetilde{AE}_i^-(a)), \tag{18.28}$$

$\mu_{\simeq_i^{\Pi^{\mathrm{L}}}}(a, \tilde{b})$ shows a degree to what extent a fuzzy number $\tilde{b}$ is possibly a number to which a is approximately equal. $\mu_{\simeq_i^{N^{\mathrm{L}}}}(a, \tilde{b})$ shows a degree to what extent a fuzzy number $\tilde{b}$ is necessarily a number to which a is approximately equal.

Now we have obtained fuzzy equality relations between a fuzzy number and a real number. Using those fuzzy relations, we can obtain fuzzy inequality relations between fuzzy numbers. Fuzzy sets $\widetilde{AE}_i^{\Pi^{\mathrm{L}}+}(\tilde{b})$ and $\widetilde{AE}_i^{N^{\mathrm{L}}+}(\tilde{b})$ of real numbers which are possibly and necessarily approximately equal to a fuzzy set $\tilde{b}$ can be defined by the following membership functions, respectively:

$$\mu_{\widetilde{AE}_i^{\Pi^{\mathrm{L}}+}(\tilde{b})}(r) = \mu_{\simeq_i^{\Pi^{\mathrm{L}}}}(r, \tilde{b}), \qquad \mu_{\widetilde{AE}_i^{N^{\mathrm{L}}+}(\tilde{b})}(r) = \mu_{\simeq_i^{\Pi^{\mathrm{L}}}}(r, \tilde{b}). \tag{18.29}$$

Similarly, fuzzy sets $\widetilde{AE}_i^{\Pi^{\rm F}-}(\tilde{a})$ and $\widetilde{AE}_i^{N^{\rm F}-}(\tilde{a})$ of real numbers to which a fuzzy set $\tilde{a}$ is possibly and necessarily approximately equal can be defined by the following membership functions, respectively:

$$\mu_{\widetilde{AE}_i^{\Pi^{\rm F}-}(\tilde{a})}(r) = \mu_{\simeq_i^{\Pi^{\rm F}}}(\tilde{a}, r), \qquad \mu_{\widetilde{AE}_i^{N^{\rm F}-}(\tilde{a})}(r) = \mu_{\simeq_i^{\Pi^{\rm L}}}(\tilde{a}, r). \tag{18.30}$$

Using four fuzzy sets (18.29) and (18.30), we can define fuzzy inequality relations between fuzzy numbers based on possibility and necessity measures as follows:

1 Fuzzy relation $\simeq_i^{\Pi^{\rm F}\Pi^{\rm L}}$: A membership function value $\mu_{\simeq_i^{\Pi^{\rm F}\Pi^{\rm L}}}(\tilde{a}, \tilde{b})$ shows a possibility degree of an event that $\tilde{a}$ is possibly approximately equal to $\tilde{b}$. We define

$$\mu_{\simeq_i^{\Pi^{\rm F}\Pi^{\rm L}}}(\tilde{a}, \tilde{b}) = \sup_r \min\left(\mu_{\tilde{a}}(r), \mu_{\widetilde{AE}_i^{\Pi^{\rm L}+}(\tilde{b})}(r)\right). \tag{18.31}$$

2 Fuzzy relation $\simeq_i^{N^{\rm F}\Pi^{\rm L}}$: A membership function value $\mu_{\simeq_i^{N^{\rm F}\Pi^{\rm L}}}(\tilde{a}, \tilde{b})$ shows a necessity degree of an event that $\tilde{a}$ is possibly approximately equal to $\tilde{b}$. We define

$$\mu_{\simeq_i^{N^{\rm F}\Pi^{\rm L}}}(\tilde{a}, \tilde{b}) = \inf_r \max\left(1 - \mu_{\tilde{a}}(r), \mu_{\widetilde{AE}_i^{\Pi^{\rm L}+}(\tilde{b})}(r)\right). \tag{18.32}$$

3 Fuzzy relation $\simeq_i^{\Pi^{\rm F}N^{\rm L}}$: A membership function value $\mu_{\simeq_i^{\Pi^{\rm F}N^{\rm L}}}(\tilde{a}, \tilde{b})$ shows a possibility degree of an event that $\tilde{a}$ is necessarily approximately equal to $\tilde{b}$. We define

$$\mu_{\simeq_i^{\Pi^{\rm F}N^{\rm L}}}(\tilde{a}, \tilde{b}) = \sup_r \min\left(\mu_{\tilde{a}}(r), \mu_{\widetilde{AE}_i^{N^{\rm L}+}(\tilde{b})}(r)\right). \tag{18.33}$$

4 Fuzzy relation $\simeq_i^{N^{\rm F}N^{\rm L}}$: A membership function value $\mu_{\simeq_i^{N^{\rm F}N^{\rm L}}}(\tilde{a}, \tilde{b})$ shows a necessity degree of an event that $\tilde{a}$ is necessarily approximately equal to $\tilde{b}$. We define

$$\mu_{\simeq_i^{N^{\rm F}N^{\rm L}}}(\tilde{a}, \tilde{b}) = \inf_r \max\left(1 - \mu_{\tilde{a}}(r), \mu_{\widetilde{AE}_i^{N^{\rm L}+}(\tilde{b})}(r)\right). \tag{18.34}$$

5 Fuzzy relation $\simeq_i^{\Pi^{\rm L}\Pi^{\rm F}}$: A membership function value $\mu_{\simeq_i^{\Pi^{\rm L}\Pi^{\rm F}}}(\tilde{a}, \tilde{b})$ shows a possibility degree of an event that $\tilde{b}$ is possibly a number to which $\tilde{a}$ is approximately equal. We define

$$\mu_{\simeq_i^{\Pi^{\rm L}\Pi^{\rm F}}}(\tilde{a}, \tilde{b}) = \sup_r \min\left(\mu_{\tilde{b}}(r), \mu_{\widetilde{AE}_i^{\Pi^{\rm F}-}(\tilde{a})}(r)\right). \tag{18.35}$$

6 Fuzzy relation $\simeq_i^{N^{\rm L}\Pi^{\rm F}}$: A membership function value $\mu_{\simeq_i^{N^{\rm L}\Pi^{\rm F}}}(\tilde{a},\tilde{b})$ shows a necessity degree of an event that $\tilde{b}$ is possibly a number to which $\tilde{a}$ is approximately equal . We define

$$\mu_{\simeq_i^{N^{\rm L}\Pi^{\rm F}}}(\tilde{a},\tilde{b}) = \inf_r \max\left(1-\mu_{\tilde{b}}(r), \mu_{\widetilde{AE}_i^{\Pi^{\rm F}-}(\tilde{a})}(r)\right). \qquad (18.36)$$

7 Fuzzy relation $\simeq_i^{\Pi^{\rm L}N^{\rm F}}$: A membership function value $\mu_{\simeq_i^{\Pi^{\rm L}N^{\rm F}}}(\tilde{a},\tilde{b})$ shows a possibility degree of an event that $\tilde{b}$ is necessarily a number to which $\tilde{a}$ is approximately equal. We define

$$\mu_{\simeq_i^{\Pi^{\rm L}N^{\rm F}}}(a,\tilde{b}) = \sup_r \min\left(\mu_{\tilde{b}}(r), \mu_{\widetilde{AE}_i^{N^{\rm F}-}(\tilde{a})}(r)\right). \qquad (18.37)$$

8 Fuzzy relation $\simeq_i^{N^{\rm L}N^{\rm F}}$: A membership function value $\mu_{\simeq_i^{N^{\rm L}N^{\rm F}}}(\tilde{a},\tilde{b})$ shows a necessity degree of an event that $\tilde{b}$ is necessarily a number to which $\tilde{a}$ is approximately equal. We define

$$\mu_{\simeq_i^{N^{\rm L}N^{\rm F}}}(\tilde{a},\tilde{b}) = \inf_r \max\left(1-\mu_{\tilde{b}}(r), \mu_{\widetilde{AE}_i^{N^{\rm F}-}(\tilde{a})}(r)\right). \qquad (18.38)$$

We can prove

$$\simeq_i^{\Pi^{\rm F}\Pi^{\rm L}} = \simeq_i^{\Pi^{\rm L}\Pi^{\rm F}}, \qquad \simeq_i^{N^{\rm F}N^{\rm L}} = \simeq_i^{N^{\rm L}N^{\rm F}}. \qquad (18.39)$$

Thus, we have six kinds of extended fuzzy relations between fuzzy numbers. The difference among six kinds of extended fuzzy relations can be depicted in Figure 18.4. In Figure 18.4, we consider a case when $\simeq_i$ is a tolerance relation, $\mu_{\simeq_i}(r_1, r_2)$ takes 1 if $|r_1 - r_2| \le \varepsilon$ and 0 otherwise, and $\tilde{a}$ and $\tilde{b}$ are intervals. Since $\tilde{a}$ and $\tilde{b}$ are intervals, we use notation $r_1 \in \tilde{a}$ and $r_2 \in \tilde{b}$ in order to represent that r_1 is included in interval $\tilde{a}$ and r_2 is included in interval $\tilde{b}$, respectively. In this case, the membership values of extended fuzzy relations take 0 or 1. Therefore each extended fuzzy relation corresponds to a logical statement as shown in Figure 18.4.

Among them, we have the following relations (see [17]):

$$\begin{aligned}\mu_{\simeq_i^{\Pi^{\rm F}\Pi^{\rm L}}}(\tilde{a},\tilde{b}) &\ge \mu_{\simeq_i^{N^{\rm F}\Pi^{\rm L}}}(\tilde{a},\tilde{b}) \\ &\ge \mu_{\simeq_i^{\Pi^{\rm L}N^{\rm F}}}(\tilde{a},\tilde{b}) \ge \mu_{\simeq_i^{N^{\rm F}N^{\rm L}}}(\tilde{a},\tilde{b}), \qquad (18.40)\\ \mu_{\simeq_i^{\Pi^{\rm F}\Pi^{\rm L}}}(\tilde{a},\tilde{b}) &\ge \mu_{\simeq_i^{N^{\rm L}\Pi^{\rm F}}}(\tilde{a},\tilde{b}) \\ &\ge \mu_{\simeq_i^{\Pi^{\rm F}N^{\rm L}}}(\tilde{a},\tilde{b}) \ge \mu_{\simeq_i^{N^{\rm F}N^{\rm L}}}(\tilde{a},\tilde{b}). \qquad (18.41)\end{aligned}$$

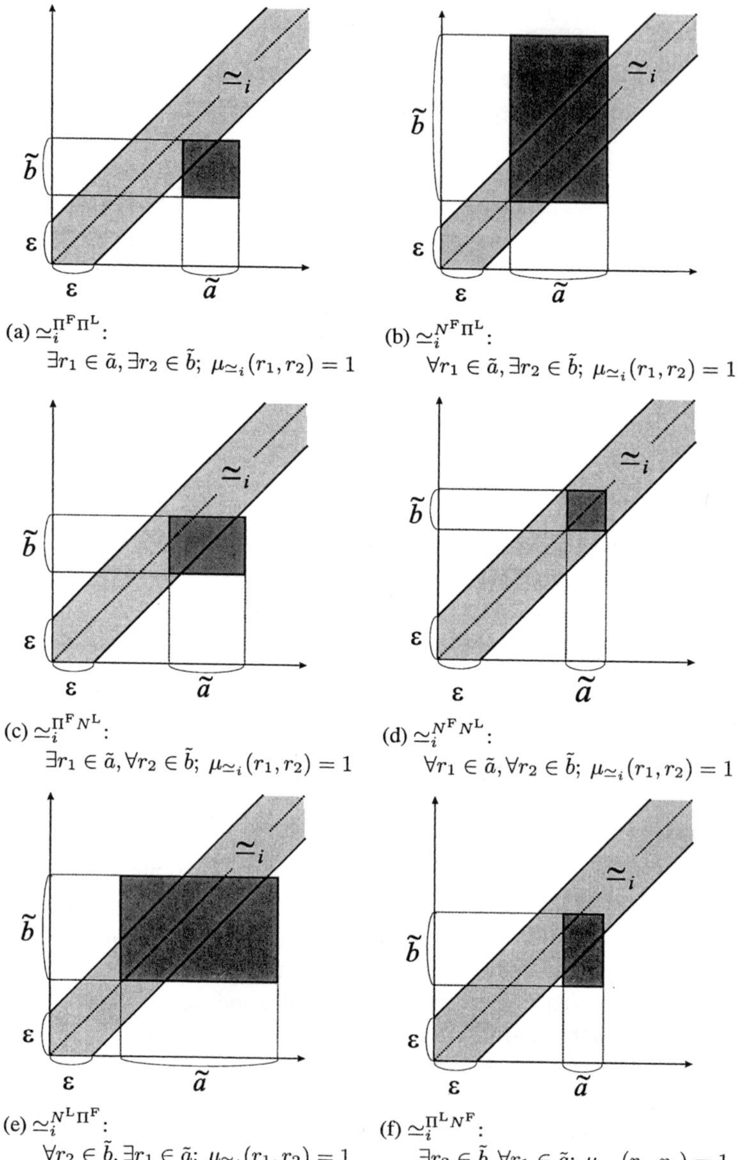

(a) $\simeq_i^{\Pi^{\mathrm{F}}\Pi^{\mathrm{L}}}$: $\exists r_1 \in \tilde{a}, \exists r_2 \in \tilde{b};\ \mu_{\simeq_i}(r_1, r_2) = 1$

(b) $\simeq_i^{N^{\mathrm{F}}\Pi^{\mathrm{L}}}$: $\forall r_1 \in \tilde{a}, \exists r_2 \in \tilde{b};\ \mu_{\simeq_i}(r_1, r_2) = 1$

(c) $\simeq_i^{\Pi^{\mathrm{F}}N^{\mathrm{L}}}$: $\exists r_1 \in \tilde{a}, \forall r_2 \in \tilde{b};\ \mu_{\simeq_i}(r_1, r_2) = 1$

(d) $\simeq_i^{N^{\mathrm{F}}N^{\mathrm{L}}}$: $\forall r_1 \in \tilde{a}, \forall r_2 \in \tilde{b};\ \mu_{\simeq_i}(r_1, r_2) = 1$

(e) $\simeq_i^{N^{\mathrm{L}}\Pi^{\mathrm{F}}}$: $\forall r_2 \in \tilde{b}, \exists r_1 \in \tilde{a};\ \mu_{\simeq_i}(r_1, r_2) = 1$

(f) $\simeq_i^{\Pi^{\mathrm{L}}N^{\mathrm{F}}}$: $\exists r_2 \in \tilde{b}, \forall r_1 \in \tilde{a};\ \mu_{\simeq_i}(r_1, r_2) = 1$

Figure 18.4. Differences among six extended fuzzy relations.

From those relations, $\simeq_i^{\Pi^{\mathrm{F}}\Pi^{\mathrm{L}}}$ is the weakest so that the relation holds easily. On the other hand, $\simeq_i^{N^{\mathrm{F}}N^{\mathrm{L}}}$ is the strongest. $\simeq_i^{N^{\mathrm{F}}\Pi^{\mathrm{L}}}$ and $\simeq_i^{N^{\mathrm{L}}\Pi^{\mathrm{F}}}$ are the second weakest. $\simeq_i^{\Pi^L N^F}$ and $\simeq_i^{\Pi^F N^L}$ are second strongest. Two extremes $\simeq_i^{\Pi^F \Pi^L}$ and $\simeq_i^{N^F N^L}$ can be regarded as possibility and necessity extensions of the relation and often used in many approaches. In order to moderate them so that we can express the decision maker's manifold attitude to the uncertainty, we use other four extensions. As shown in Figure 18.4, $\simeq_i^{N^F \Pi^L}$ and $\simeq_i^{\Pi^L N^F}$ are averse to the uncertainty of $\tilde{a}$, the left-hand side of the relation while $\simeq_i^{N^L \Pi^F}$ and $\simeq_i^{\Pi^F N^L}$ are averse to the uncertainty $\tilde{b}$, the righ-hand side of the relation.

We can apply the extension methods (a)–(h) to any fuzzy relation. Thus, we can extend a fuzzy inequality $\lesssim_i$ defined by (18.3). In this case, we have $\lesssim_i^{N^F \Pi^L} = \lesssim_i^{\Pi^L N^F}$ and $\lesssim_i^{N^L \Pi^F} = \lesssim_i^{\Pi^F N^L}$ other than (18.39)–(18.41) (see [17]). Other properties of the extended fuzzy relations are investigated in [17].

3.2 Treatment of Constraints

Using extended fuzzy relations, we can define feasibility degrees of each constraint $\tilde{\boldsymbol{a}}_i^{\mathrm{T}} \boldsymbol{x} \lesssim_i \tilde{b}_i$ as

$$
\begin{aligned}
\mathrm{VWF}\left(\tilde{\boldsymbol{a}}_i^{\mathrm{T}} \boldsymbol{x} \lesssim_i \tilde{b}_i\right) &= \mu_{\lesssim_i^{\Pi^F \Pi^L}}(\tilde{\boldsymbol{a}}_i^{\mathrm{T}} \boldsymbol{x}, \tilde{b}_i), && (18.42)\\
\mathrm{MF1}\left(\tilde{\boldsymbol{a}}_i^{\mathrm{T}} \boldsymbol{x} \lesssim_i \tilde{b}_i\right) &= \mu_{\lesssim_i^{N^F \Pi^L}}(\tilde{\boldsymbol{a}}_i^{\mathrm{T}} \boldsymbol{x}, \tilde{b}_i), && (18.43)\\
\mathrm{MF2}\left(\tilde{\boldsymbol{a}}_i^{\mathrm{T}} \boldsymbol{x} \lesssim_i \tilde{b}_i\right) &= \mu_{\lesssim_i^{N^L \Pi^F}}(\tilde{\boldsymbol{a}}_i^{\mathrm{T}} \boldsymbol{x}, \tilde{b}_i), && (18.44)\\
\mathrm{VSF}\left(\tilde{\boldsymbol{a}}_i^{\mathrm{T}} \boldsymbol{x} \lesssim_i \tilde{b}_i\right) &= \mu_{\lesssim_i^{N^F N^L}}(\tilde{\boldsymbol{a}}_i^{\mathrm{T}} \boldsymbol{x}, \tilde{b}_i), && (18.45)
\end{aligned}
$$

where VWF, MF and VSF stand for 'very weak feasibility', 'medium feasibility' and 'very strong feasibility', respectively. Similarly, we can define feasibility degrees of each constraint $\tilde{\boldsymbol{a}}_i^{\mathrm{T}} \boldsymbol{x} \simeq_i \tilde{b}_i$ as

$$
\begin{aligned}
\mathrm{VWF}\left(\tilde{\boldsymbol{a}}_i^{\mathrm{T}} \boldsymbol{x} \simeq_i \tilde{b}_i\right) &= \mu_{\simeq_i^{\Pi^F \Pi^L}}(\tilde{\boldsymbol{a}}_i^{\mathrm{T}} \boldsymbol{x}, \tilde{b}_i), && (18.46)\\
\mathrm{WF1}\left(\tilde{\boldsymbol{a}}_i^{\mathrm{T}} \boldsymbol{x} \simeq_i \tilde{b}_i\right) &= \mu_{\simeq_i^{N^F \Pi^L}}(\tilde{\boldsymbol{a}}_i^{\mathrm{T}} \boldsymbol{x}, \tilde{b}_i), && (18.47)\\
\mathrm{WF2}\left(\tilde{\boldsymbol{a}}_i^{\mathrm{T}} \boldsymbol{x} \simeq_i \tilde{b}_i\right) &= \mu_{\simeq_i^{N^L \Pi^F}}(\tilde{\boldsymbol{a}}_i^{\mathrm{T}} \boldsymbol{x}, \tilde{b}_i), && (18.48)\\
\mathrm{SF1}\left(\tilde{\boldsymbol{a}}_i^{\mathrm{T}} \boldsymbol{x} \simeq_i \tilde{b}_i\right) &= \mu_{\simeq_i^{\Pi^L N^F}}(\tilde{\boldsymbol{a}}_i^{\mathrm{T}} \boldsymbol{x}, \tilde{b}_i), && (18.49)\\
\mathrm{SF2}\left(\tilde{\boldsymbol{a}}_i^{\mathrm{T}} \boldsymbol{x} \simeq_i \tilde{b}_i\right) &= \mu_{\simeq_i^{\Pi^F N^L}}(\tilde{\boldsymbol{a}}_i^{\mathrm{T}} \boldsymbol{x}, \tilde{b}_i), && (18.50)\\
\mathrm{VSF}\left(\tilde{\boldsymbol{a}}_i^{\mathrm{T}} \boldsymbol{x} \simeq_i \tilde{b}_i\right) &= \mu_{\simeq_i^{N^F N^L}}(\tilde{\boldsymbol{a}}_i^{\mathrm{T}} \boldsymbol{x}, \tilde{b}_i), && (18.51)
\end{aligned}
$$

where WF and SF stand for 'weak feasibility' and 'strong feasibility'. Because of $\simeq_i^{N^F \Pi^L} \neq \simeq_i^{\Pi^L N^F}$ and $\simeq_i^{\Pi^F N^L} \neq \simeq_i^{N^L \Pi^F}$, medium feasibilities in (18.45) are replaced with weak and strong feasibilities.

Since Problem (18.6) includes uncertainty, the number of solutions which satisfy constraints for all possible realizations of uncertain parameters are extremely small or zero. Therefore, it is not practical to render the solution feasible for all possible realizations. We assume that the decision maker can afford to accept the infeasibility risk so that he/she specify the minimally required degrees of feasibility indices, VWF, WF1, WF2, MF1, MF2, SF1, SF2 and VSF. Let $h_i^1, h_i^2, h_i^3, h_i^4, h_i^5$ and $h_i^6 \in [0, 1]$ be specified degrees of feasibility indices to a constraint $\tilde{\boldsymbol{a}}_i^{\mathrm{T}}\boldsymbol{x} \simeq_i \tilde{b}_i$. The fuzzy equality constraint is treated as

$$\begin{aligned} &\mathrm{VWF}\left(\tilde{\boldsymbol{a}}_i^{\mathrm{T}}\boldsymbol{x} \simeq_i \tilde{b}_i\right) \geq h_i^1, \quad \mathrm{WF1}\left(\tilde{\boldsymbol{a}}_i^{\mathrm{T}}\boldsymbol{x} \simeq_i \tilde{b}_i\right) \geq h_i^2, \\ &\mathrm{WF2}\left(\tilde{\boldsymbol{a}}_i^{\mathrm{T}}\boldsymbol{x} \simeq_i \tilde{b}_i\right) \geq h_i^3, \quad \mathrm{SF1}\left(\tilde{\boldsymbol{a}}_i^{\mathrm{T}}\boldsymbol{x} \simeq_i \tilde{b}_i\right) \geq h_i^4, \qquad i = 1, 2, \ldots, m, \\ &\mathrm{SF2}\left(\tilde{\boldsymbol{a}}_i^{\mathrm{T}}\boldsymbol{x} \simeq_i \tilde{b}_i\right) \geq h_i^5, \quad \mathrm{VSF}\left(\tilde{\boldsymbol{a}}_i^{\mathrm{T}}\boldsymbol{x} \simeq_i \tilde{b}_i\right) \geq h_i^6, \end{aligned} \tag{18.52}$$

where we assume $h_i^1 \geq h_i^2 \geq h_i^4 \geq h_i^6$ and $h_i^1 \geq h_i^3 \geq h_i^5 \geq h_i^6$ because of (18.40) and (18.41). If $h_i^l = 0$ then the corresponding constraint in (18.52) is discarded.

The selection of indices for each constraint from VWF, WF1, WF2, MF1, MF2, SF1, SF2 and VSF is based on Figure 18.5. Namely, if the constraint should be satisfied with certainty, VSF should be adopted. If the constraint satisfaction is just a decision maker's wish, VWF can be adopted. Moreover if the constraint satisfaction is strongly required, SF1 and SF2 can be used at the same time. If the constraint satisfaction is desirable, WF1 or WF2 will be suitable. In such a way, the indices are selected according to the required assurance. In Figure 18.5, the indices located in upper place are more suitable for constraints with stronger assurance while the indices in lower place are more suitable for constraints with weaker assurance. The assurance of the constraints can be also controlled by the specifications of degrees h_i^j.

Similarly, a fuzzy inequality constraint $\tilde{\boldsymbol{a}}_i^{\mathrm{T}}\boldsymbol{x} \lesssim_i \tilde{b}_i$ is treated as

$$\begin{aligned} &\mathrm{VWF}\left(\tilde{\boldsymbol{a}}_i^{\mathrm{T}}\boldsymbol{x} \lesssim_i \tilde{b}_i\right) \geq h_i^1, \quad \mathrm{MF1}\left(\tilde{\boldsymbol{a}}_i^{\mathrm{T}}\boldsymbol{x} \lesssim_i \tilde{b}_i\right) \geq h_i^2, \\ &\mathrm{MF2}\left(\tilde{\boldsymbol{a}}_i^{\mathrm{T}}\boldsymbol{x} \lesssim_i \tilde{b}_i\right) \geq h_i^3, \quad \mathrm{VSF}\left(\tilde{\boldsymbol{a}}_i^{\mathrm{T}}\boldsymbol{x} \simeq_i \tilde{b}_i\right) \geq h_i^4, \end{aligned} \qquad i = 1, 2, \ldots, m, \tag{18.53}$$

where we assume $1 \geq h_i^1 \geq h_i^2 \geq h_i^4 \geq 0$ and $1 \geq h_i^1 \geq h_i^3 \geq h_i^4 \geq 0$ are given by the decision maker.

When $\lesssim_i$ and $\simeq_i$ are defined by (18.3) and (18.4), from $\boldsymbol{x} \geq \boldsymbol{0}$ it is known that constraints (18.52) and (18.53) are reduced to linear constraints (see [18]).

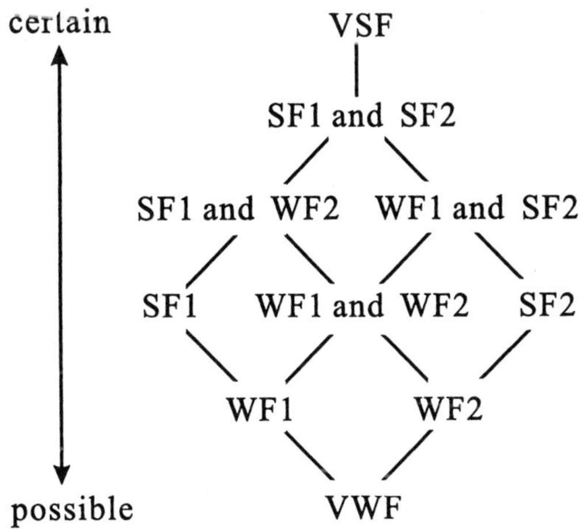

Figure 18.5. Relations among VWF, WF1, WF2, MF1, MF2, SF1, SF2 and VSF.

For example, (18.53) is reduced to

$$\begin{aligned}
&\sum_{j=1}^{n} a_{ij}^{\mathrm{L}}(h_i^1)x_j \le b_i^{\mathrm{R}}(h_i^1) + \nu_i^{(-1)}(h_i^1),\\
&\sum_{j=1}^{n} \bar{a}_{ij}^{\mathrm{R}}(1-h_i^2)x_j \le b_i^{\mathrm{R}}(h_i^2) + \nu_j^{(-1)}(h_i^2),\\
&\sum_{j=1}^{n} a_{ij}^{\mathrm{L}}(h_i^4)x_j \le \bar{b}_i^{\mathrm{L}}(1-h_i^4) + \nu_i^{(-1)}(h_i^4),\\
&\sum_{j=1}^{n} \bar{a}_{ij}^{\mathrm{R}}(1-h_i^6)x_j \le \bar{b}_i^{\mathrm{L}}(1-h_i^6) + \nu_i^{(-1)}(h_j^6),
\end{aligned} \tag{18.54}$$

where the j-th component $\tilde{a}_{ij}$ of $\tilde{\boldsymbol{a}}_i$ and $\tilde{b}_i$ are L-R fuzzy numbers $(a_{ij}^{\mathrm{L}}, a_{ij}^{\mathrm{R}}, \alpha_{ij}^{\mathrm{L}}, \alpha_{ij}^{\mathrm{R}})_{L_{ij}^a R_{ij}^a}$ and $(b_i^{\mathrm{L}}, b_i^{\mathrm{R}}, \beta_i^{\mathrm{L}}, \beta_i^{\mathrm{R}})_{L_i^b R_i^b}$, respectively. We define $a_{ij}^{\mathrm{L}}(h) = a_{ij}^{\mathrm{L}} - \alpha_{ij}^{\mathrm{L}} L_{ij}^{a(-1)}(h)$, $\bar{a}_{ij}^{\mathrm{R}}(1-h) = a_{ij}^{\mathrm{R}} + \alpha_{ij}^{\mathrm{R}} \bar{R}_{ij}^{a(-1)}(1-h)$, $b_i^{\mathrm{R}}(h) = b_i^{\mathrm{R}} + \beta_i^{\mathrm{R}} R_i^{b(-1)}(h)$, $\bar{b}_i^{\mathrm{L}}(1-h) = b_i^{\mathrm{L}} - \beta_i^{\mathrm{L}} \bar{L}_i^{b(-1)}(1-h)$, $L_{ij}^{a(-1)}(h) = \sup_r\{r \mid L_{ij}^a(r) \ge h\}$, $\bar{R}_{ij}^{a(-1)}(1-h) = \sup_r\{r \mid R_{ij}^a(r) > 1-h\}$, $R_i^{b(-1)}(h) = \sup_r\{r \mid R_i^b(r) \ge h\}$ and $\bar{L}_i^{b(-1)}(1-h) = \sup_r\{r \mid L_i^b(r) > 1-h\}$.

The selection of indices for each constraint from VWF, MF1, MF2 and VSF is based on Figure 18.6. Namely, if the conclusive constraint satisfaction is required, VSF should be selected. If the constraint satisfaction is just a decision maker's wish, VWF can be used. Moreover, MF1 and MF2 are useful for the moderate requirement.

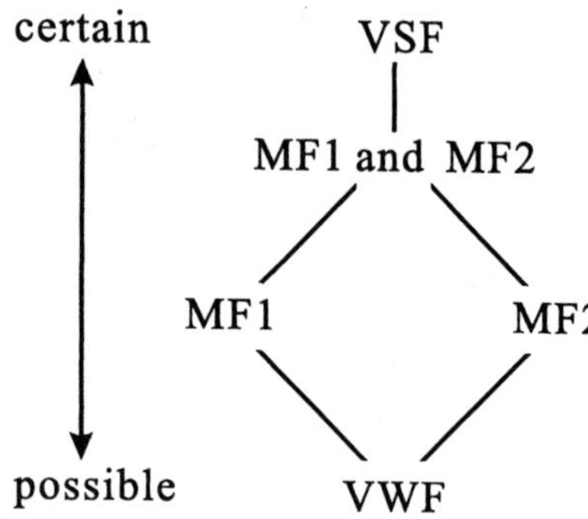

Figure 18.6. Relations among VWF, MF1, MF2 and VSF.

3.3 Fractile Optimization Models

We describe the treatments of objective functions. In the first model, we assume that the decision maker is interested in the value z such that we can expect the possibility or necessity degree that the objective function value is not larger than z is at least a given degree $\bar{h} \in (0, 1]$. Such value z is called $\bar{h}$-possibility or necessity fractile. Namely, $\bar{h}^{\Pi}$-possibility fractile is a value z^{Π} satisfying

$$\mathrm{POS}(\tilde{\boldsymbol{c}}_k^{\mathrm{T}}\boldsymbol{x} \le z^{\Pi}) = \Pi_{\tilde{\boldsymbol{c}}_k^{\mathrm{T}}\boldsymbol{x}}((-\infty, z^{\Pi}]) = \sup_{r \le z^{\Pi}} \mu_{\tilde{\boldsymbol{c}}_k^{\mathrm{T}}\boldsymbol{x}}(r) \ge \bar{h}^{\Pi}. \tag{18.55}$$

Similarly, $\bar{h}^N$-necessity fractile is a value z^N satisfying

$$\mathrm{NES}(\tilde{\boldsymbol{c}}_k^{\mathrm{T}}\boldsymbol{x} \le z^{N}) = N_{\tilde{\boldsymbol{c}}_k^{\mathrm{T}}\boldsymbol{x}}((-\infty, z^{N}]) = \inf_{r > z^{N}} (1 - \mu_{\tilde{\boldsymbol{c}}_k^{\mathrm{T}}\boldsymbol{x}}(r)) \ge \bar{h}^{N}. \tag{18.56}$$

Using possibility and necessity fractiles, the objective functions of Problem (18.6) are treated as

$$\begin{array}{ll} \text{minimize} & (z_1^{\Pi}, z_2^{\Pi}, \ldots, z_p^{\Pi}, z_1^{N}, z_2^{N}, \ldots, z_p^{N})^{\mathrm{T}}, \\ \text{subject to} & \mathrm{POS}(\tilde{\boldsymbol{c}}_k^{\mathrm{T}}\boldsymbol{x} \le z_k^{\Pi}) \ge \bar{h}_k^{\Pi},\ k = 1, 2, \ldots, p, \\ & \mathrm{NES}(\tilde{\boldsymbol{c}}_k^{\mathrm{T}}\boldsymbol{x} \le z_k^{N}) \ge \bar{h}_k^{N},\ k = 1, 2, \ldots, p, \end{array} \tag{18.57}$$

where $\bar{h}_k^{\Pi}$, $\bar{h}_k^{N}$, $k = 1, 2, \ldots, p$ are predetermined degrees by the decision maker. When the decision maker is not interested in $z_{k_1}^{\Pi}$ and $z_{k_2}^{N}$, we discard the corresponding constraints $\mathrm{POS}(\tilde{\boldsymbol{c}}_{k_1}^{\mathrm{T}}\boldsymbol{x} \le z_{k_1}^{\Pi}) \ge \bar{h}_{k_1}^{\Pi}$ and $\mathrm{NES}(\tilde{\boldsymbol{c}}_{k_2}^{\mathrm{T}}\boldsymbol{x} \le z_{k_2}^{N}) \ge \bar{h}_{k_2}^{N}$. z_k^{Π} such that $\mathrm{POS}(\tilde{\boldsymbol{c}}_k^{\mathrm{T}}\boldsymbol{x} \le z_k^{\Pi}) \ge \bar{h}_k^{\Pi}$ is used when the decision maker wants to manage the possible minimum value of the objective function. On the other hand, z_k^N such that $\mathrm{NES}(\tilde{\boldsymbol{c}}_k^{\mathrm{T}}\boldsymbol{x} \lesssim z_k^N) \geq \bar{h}_k^N$ is adopted when the decision is interested in managing the possible worst value of the objective function.

In this model, p objective functions in Problem (18.6) are transformed to $2p$ objective functions with $2p$ constraints. (18.57) is reduced to the following $2p$ linear objective functions:

$$\text{minimize} \quad \left(\sum_{j=1}^{n} c_{1j}^{\mathrm{L}}(\bar{h}_1^{\Pi})x_j, \ldots, \sum_{j=1}^{n} c_{pj}^{\mathrm{L}}(\bar{h}_p^{\Pi})x_j, \right. \\ \left. \sum_{j=1}^{n} \bar{c}_{1j}^{\mathrm{R}}(1-\bar{h}_1^{N})x_j, \ldots, \sum_{j=1}^{n} \bar{c}_{pj}^{\mathrm{R}}(1-\bar{h}_p^{N})x_j \right), \tag{18.58}$$

defining $c_{kj}^{\mathrm{L}}(h) = c_{kj}^{\mathrm{L}} - \gamma_{kj}^{\mathrm{L}} L_{kj}^{c(-1)}(h)$ and $\bar{c}_{kj}^{\mathrm{R}}(h) = c_{kj}^{\mathrm{R}} + \gamma_{kj}^{\mathrm{R}} \bar{R}_{kj}^{c(-1)}(h)$.

Together with the treatment of constraints, this model transforms Problem (18.6) to a conventional multiple objective programming problem with $2p$ objective functions. Thus, we can apply the concept of efficient solutions to the reduced problem.

We can apply any multiple objective programming technique to obtain an efficient solution. For example, let us consider Problem (18.6) without fuzzy equality constraints, i.e., $m_1 = m$ and fuzzy inequality relations defined by (18.3). We obtain reduced constraints (18.54) and reduced objective functions (18.58). Thus, we have a multiple objective linear programming problem. We can apply a multicriteria simplex method [47, 44] to enumerate all efficient solutions. Moreover, when we apply a weighting method with a weighting vector $\boldsymbol{w} = (w_1^{\Pi}, \ldots, w_p^{\Pi}, w_1^{N}, \ldots, w_p^{N})^{\mathrm{T}} > \mathbf{0}$ such that $\sum_{i=1}^{p} w_i^{\Pi} + \sum_{i=1}^{p} w_i^{N} = 1$, the reduced problem becomes the following linear programming problem:

$$\begin{aligned}
\text{minimize} \quad & \sum_{k=1}^{p} w_i^{\Pi} \sum_{j=1}^{n} c_{kj}^{\mathrm{L}}(\bar{h}_k^{\Pi}) x_j + \sum_{k=1}^{p} w_i^{N} \sum_{j=1}^{n} \bar{c}_{kj}^{\mathrm{R}}(1 - \bar{h}_k^{N}) x_j, \\
\text{subject to} \quad & \sum_{j=1}^{n} a_{ij}^{\mathrm{L}}(h_i^1) x_j \leq b_i^{\mathrm{R}}(h_i^1) + \nu_i^{(-1)}(h_i^1), \\
& \sum_{j=1}^{n} \bar{a}_{ij}^{\mathrm{R}}(1 - h_i^2) x_j \leq b_i^{\mathrm{R}}(h_i^2) + \nu_i^{(-1)}(h_i^2), \\
& \sum_{j=1}^{n} a_{ij}^{\mathrm{L}}(h_i^3) x_j \leq \bar{b}_i^{\mathrm{L}}(1 - h_i^3) + \nu_i^{(-1)}(h_i^3), \\
& \sum_{j=1}^{n} \bar{a}_{ij}^{\mathrm{R}}(1 - h_i^4) x_j \leq \bar{b}_i^{\mathrm{L}}(1 - h_i^4) + \nu_i^{(-1)}(h_i^4), \\
& \qquad\qquad i = 1, 2, \ldots, m, \\
& x_j \geq 0, \ j = 1, 2, \ldots, n.
\end{aligned} \tag{18.59}$$

By this problem, we obtain an efficient solution to the problem composed of (18.54) and (18.58).

3.4 Modality Optimization Models

A modality optimization model is a dual model of a fractile optimization model. In this model, we assume that the decision maker specifies the possibility and necessity aspiration levels $\bar{z}_k^{\Pi}$ and $\bar{z}_k^{N}$, $k = 1, 2, \ldots, p$ with respect to p objective function values. While the possibility aspiration level $\bar{z}_k^{\Pi}$ is the objective function value the decision maker would like to keep a chance to achieve, the necessity aspiration level $\bar{z}_k^{N}$ is the objective function value the decision maker would like to achieve certainly. Then we maximize the possibility degree $\mathrm{POS}(\tilde{\boldsymbol{c}}_k^{\mathrm{T}} \boldsymbol{x} \leq \bar{z}_k^{\Pi})$ and necessity degrees $\mathrm{NES}(\tilde{\boldsymbol{c}}_k^{\mathrm{T}} \boldsymbol{x} \leq \bar{z}_k^{\Pi})$.Namely, the objective functions of Problem (18.6) are treated as

$$\begin{aligned}
\text{maximize} \quad & \big(\mathrm{POS}(\tilde{\boldsymbol{c}}_1^{\mathrm{T}} \boldsymbol{x} \leq \bar{z}_1^{\Pi}), \ldots, \mathrm{POS}(\tilde{\boldsymbol{c}}_p^{\mathrm{T}} \boldsymbol{x} \leq \bar{z}_p^{\Pi}), \\
& \mathrm{NES}(\tilde{\boldsymbol{c}}_1^{\mathrm{T}} \boldsymbol{x} \leq \bar{z}_1^{N}), \ldots, \mathrm{NES}(\tilde{\boldsymbol{c}}_p^{\mathrm{T}} \boldsymbol{x} \leq \bar{z}_p^{N})\big) .
\end{aligned} \tag{18.60}$$

When the decision maker is not interested in maximizing some of the POS $(\tilde{\boldsymbol{c}}_k^{\mathrm{T}} \boldsymbol{x} \leq \bar{z}_k^{\Pi})$ and $\mathrm{NES}(\tilde{\boldsymbol{c}}_k^{\mathrm{T}} \boldsymbol{x} \leq \bar{z}_k^{N})$, $k = 1, 2, \ldots, p$, we discard them in Problem (18.60). While $\mathrm{POS}(\tilde{\boldsymbol{c}}_k^{\mathrm{T}} \boldsymbol{x} \leq \bar{z}_k^{\Pi})$ is used when the decision maker seeks the possibility of objective function value not larger than $\bar{z}_k^{\Pi}$, $\mathrm{NES}(\tilde{\boldsymbol{c}}_k^{\mathrm{T}} \boldsymbol{x} \leq \bar{z}_k^{N})$ is used when the decision maker maximizes the safety (certainty) that the objective function value not larger than $\bar{z}_k^{N}$. In this model, p original objective functions of Problem (18.6) are reduced to at most $2p$ objective functions. Together with the treatment of constraints, Problem (18.6) is reduced to a multiple objective programming problem. Thus, an efficient solution to the reduced multiple objective programming problem is regarded as a good solution.

Problem (18.60) is reduced to

$$\begin{aligned} \text{maximize} \quad & (h_1^{\Pi}, h_2^{\Pi}, \ldots, h_p^{\Pi}, h_1^{N}, h_2^{N}, \ldots, h_p^{N}), \\ \text{subject to} \quad & \sum_{j=1}^{n} c_{kj}^{\mathrm{L}}(h_k^{\Pi})x_j \le \bar{z}_k^{\Pi},\ k = 1, 2, \ldots, p, \\ & \sum_{j=1}^{n} \bar{c}_{kj}^{\mathrm{R}}(1 - h_k^{N})x_j \le \bar{z}_k^{N},\ k = 1, 2, \ldots, p. \end{aligned} \tag{18.61}$$

If $L_{kj} = L_i$ and $R_{kj} = R_k$, $j = 1, 2, \ldots, n$, $k = 1, 2, \ldots, p$ and we can assume that $\sum_{j=1}^{n} \gamma_{kj}^{\mathrm{L}} x_j > 0$ and $\sum_{j=1}^{n} \gamma_{kj}^{\mathrm{R}} x_j > 0$, $k = 1, 2, \ldots, p$ at the optimal solution, then, from (18.11), Problem (18.61) is further reduced to

$$\text{minimize} \left(\frac{\sum_{j=1}^{n} c_{1j}^{\mathrm{L}} x_j - \bar{z}_1^{\Pi}}{\sum_{j=1}^{n} \gamma_{1j}^{\mathrm{L}} x_j}, \ldots, \frac{\sum_{j=1}^{n} c_{pj}^{\mathrm{L}} x_j - \bar{z}_p^{\Pi}}{\sum_{j=1}^{n} \gamma_{pj}^{\mathrm{L}} x_j}, \frac{\bar{z}_i^{N} - \sum_{j=1}^{n} \bar{c}_{1j}^{\mathrm{R}} x_j}{\sum_{j=1}^{n} \gamma_{1j}^{\mathrm{R}} x_j}, \ldots, \frac{\bar{z}_p^{N} - \sum_{j=1}^{n} c_{pj}^{\mathrm{R}} x_j}{\sum_{j=1}^{n} \gamma_{pj}^{\mathrm{R}} x_j} \right). \tag{18.62}$$

Namely, we have multiple linear fractional functions. Therefore, since constraints of Problem (18.6) are reduced to linear constraints such as (18.54), Problem (18.6) is reduced to a multiple objective linear fractional programming problem (see, for example, [44]).

3.5 Numerical Examples

In order to illustrate a modality constrained programming approach, let us consider the following bi-objective linear programming problem with fuzzy coefficients;

$$\begin{aligned} \text{minimize} \quad & (\tilde{c}_{11}x_1 + \tilde{c}_{12}x_2,\ \tilde{c}_{21}x_1 + \tilde{c}_{22}x_2), \\ \text{subject to} \quad & \tilde{a}_{i1}x_1 + \tilde{a}_{i2}x_2 \lesssim_i b_i,\ i = 1, 2, 3, \\ & x_1 \ge 0,\ x_2 \ge 0, \end{aligned} \tag{18.63}$$

where $\tilde{c}_{kj}$, $\tilde{a}_{ij}$ and $\tilde{b}_i$, $i = 1, 2, 3$, $j = 1, 2$, $k = 1, 2$ are L-R fuzzy numbers, i.e., $\tilde{c}_{11} = (-4, -4, 0.5, 0.5)_{LL}$, $\tilde{c}_{12} = (-5, -5, 1, 1)_{LL}$, $\tilde{c}_{21} = (-10, -10, 1,$

$1)_{LL}$, $\tilde{c}_{22} = (-2,-2,1,1)_{LL}$, $\tilde{a}_{11} = (2.5,2.5,0.7,0.7)_{LL}$, $\tilde{a}_{12} = (5,5,0.5,0.5)_{LL}$, $\tilde{a}_{21} = (5,5,0.9,0.9)_{LL}$, $\tilde{a}_{22} = (6,6,1.2,1.2)_{LL}$, $\tilde{a}_{31} = (3,3,0.6,0.6)_{LL}$, $\tilde{a}_{32} = (2,2,0.4,0.4)_{LL}$, $\tilde{b}_1 = (330,330,10,10)_{LL}$ and $\tilde{b}_2 = (440,440,5,5)_{LL}$. Reference function L is defined by $L(r) = \max(1-r,0)$. Fuzzy inequalities $\lesssim_i$, $i = 1,2,3$ are defined by the following membership functions;

$$\mu_{\lesssim_i}(r_1,r_2) = \begin{cases} 1, & \text{if } r_1 \le r_2, \\ x\ 1 + \dfrac{r_2 - r_1}{\gamma_i}, & \text{if } r_1 \le r_2 + \gamma_i, \\ 0, & \text{otherwise,} \end{cases} \tag{18.64}$$

with $\gamma_1 = 10$, $\gamma_2 = 15$ and $\gamma_3 = 8$.

In application of a modality optimization model, we may obtain

$$\begin{array}{ll} \text{maximize} & (\text{POS}(\tilde{c}_{11}x_1 + \tilde{c}_{12}x_2 \le -390),\ \text{NES}(\tilde{c}_{21}x_1 + \tilde{c}_{22}x_2 \le -480)) \\ \text{subject to} & \text{MF2}(\tilde{a}_{11}x_1 + \tilde{a}_{12}x_2 \lesssim_1 b_1) \ge 0.5, \\ & \text{MF1}(\tilde{a}_{21}x_1 + \tilde{a}_{22}x_2 \lesssim_2 b_2) \ge 0.5, \\ & \text{VSF}(\tilde{a}_{31}x_1 + \tilde{a}_{32}x_2 \lesssim_3 b_3) \ge 0.5, \\ & x_1 \ge 0,\ x_2 \ge 0. \end{array} \tag{18.65}$$

This problem is reduced to bi-objective linear fractional programming problem,

$$\begin{array}{ll} \text{maximize} & \left(\dfrac{4.5x_1 + 6x_2 - 390}{0.5x_1 + x_2}, \dfrac{9x_1 + x_2 - 480}{x_1 + x_2} \right) \\ \text{subject to} & 2.85x_1 + 5.25x_2 \le 340, \\ & 4.55x_1 + 5.4x_2 \le 445, \\ & 3.3x_1 + 2.2x_2 \le 233, \\ & x_1 \ge 0,\ x_2 \ge 0. \end{array} \tag{18.66}$$

Solving this problem, we find that all solutions on the line segment between $(x_1.x_2)^{\mathrm{T}} = (35.7490, 52.2855)$ and $(x_1,x_2)^{\mathrm{T}} = (70.6061, 0)^{\mathrm{T}}$ are efficient.

On the other hand, in application of fractile optimization model, we may obtain

$$\begin{array}{ll} \text{minimize} & (z_1^{\Pi},\ z_2^{N}) \\ \text{subject to} & \text{POS}(\tilde{c}_{11}x_1 + \tilde{c}_{12}x_2 \le z_1^{\Pi}) \ge 0.5, \\ & \text{NES}(\tilde{c}_{21}x_1 + \tilde{c}_{22}x_2 \le z_2^{N}) \ge 0.5, \\ & \text{MF2}(\tilde{a}_{11}x_1 + \tilde{a}_{12}x_2 \lesssim_1 b_1) \ge 0.5, \\ & \text{MF1}(\tilde{a}_{21}x_1 + \tilde{a}_{22}x_2 \lesssim_2 b_2) \ge 0.5, \\ & \text{VSF}(\tilde{a}_{31}x_1 + \tilde{a}_{32}x_2 \lesssim_3 b_3) \ge 0.5, \\ & x_1 \ge 0,\ x_2 \ge 0. \end{array} \tag{18.67}$$

This problem is reduced to bi-objective linear fractional programming problem,

$$\begin{array}{ll} \text{minimize} & (-4.25x_1 - 5.5x_2,\ -9.5x_1 - 1.5x_2) \\ \text{subject to} & 2.85x_1 + 5.25x_2 \leq 340, \\ & 4.55x_1 + 5.4x_2 \leq 445, \\ & 3.3x_1 + 2.2x_2 \leq 233, \\ & x_1 \geq 0,\ x_2 \geq 0. \end{array} \tag{18.68}$$

Solving this problem, we find that all solutions on the line segment between $(x_1.x_2)^{\mathrm{T}} = (35.7490, 52.2855)$ and $(x_1, x_2)^{\mathrm{T}} = (70.6061, 0)^{\mathrm{T}}$ are efficient.

3.6 Relations to Other Approaches

This subsection is devoted to a brief discussion about the relations between approaches described above and the other major approaches.

3.6.1 Robust Programming. In the robust programming [33, 42], one of the most traditional approach to treat linear programming problems with fuzzy coefficients, the following set-inclusive constraint is treated;

$$\tilde{\boldsymbol{a}}_i^{\mathrm{T}} \boldsymbol{x} \subseteq G_i. \tag{18.69}$$

The set-inclusion bewteen fuzzy sets is defined as $\mu_{\tilde{\boldsymbol{a}}_i^{\mathrm{T}}\boldsymbol{x}}(r) \leq \mu_{G_i}(r)$, for all $r \in \mathbf{R}$. This constraint means that we would like to control the distribution of left-hand side values within a given fuzzy goal G_i.

Define a fuzzy equality relation $\simeq_i$ and a real number b_i so as to satisfy $\mu_{G_i}(r) = \mu_{\lesssim_i}(r, b_i)$. The set-inclusive constraint (18.69) can be treated by a modality constraint,

$$\mathrm{WF1}\left(\tilde{\boldsymbol{a}}_i^{\mathrm{T}} \boldsymbol{x} \simeq_i \tilde{b}_i\right) \geq 1 \tag{18.70}$$

with adoption of reciprocal Gödel implication $I^{\mathrm{r-G}}$ instead of Dienes implication in definition of the necessity measure, where reciprocal Gödel implication is defined by $I^{\mathrm{r-G}}(u, v) = 1$ if $u \leq v$ and $1 - u$ otherwise. Therefore, a set-inclusive constraint can be treated in the framework of modality constrained programming problem by adoption of reciprocal Gödel implication. By generalizing the constraint (18.70) to

$$\mathrm{WF1}\left(\tilde{\boldsymbol{a}}_i^{\mathrm{T}} \boldsymbol{x} \simeq_i \tilde{b}_i\right) \geq h_i, \tag{18.71}$$

we can treat a weak set-inclusive constraint which requires the degree of set-inclusion more than $h_i \in (0, 1]$ (see Figure 18.7).

3.6.2 Fuzzy Max. In order to treat inequalities between fuzzy numbers $\tilde{a}$ and $\tilde{b}$, a fuzzy max $\widetilde{\max}(\tilde{a}, \tilde{b})$ is often treated. A fuzzy max $\widetilde{\max}(\tilde{a}, \tilde{b})$ is

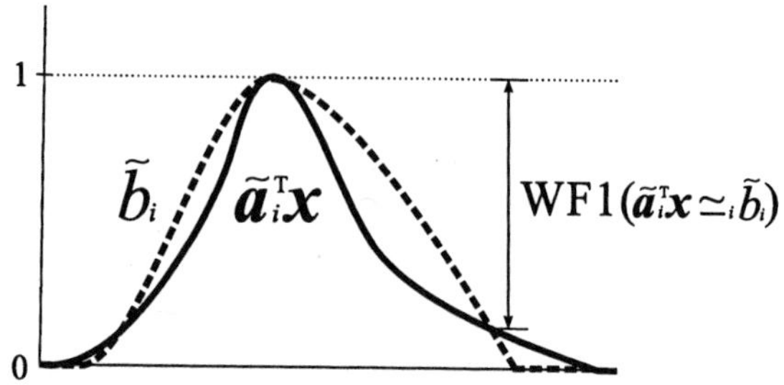

Figure 18.7. $\mathrm{WF1}(\tilde{\boldsymbol{a}}_i^{\mathrm{T}}\boldsymbol{x} \simeq_i \tilde{b}_i)$.

defined by the following membership function based on the extension principle (see Figure 18.8);

$$\mu_{\widetilde{\max}(\tilde{a},\tilde{b})}(r) = \sup_{r=\max(r_1,r_2)} \min\left(\mu_{\tilde{a}}(r_1), \mu_{\tilde{b}}(r_2)\right). \tag{18.72}$$

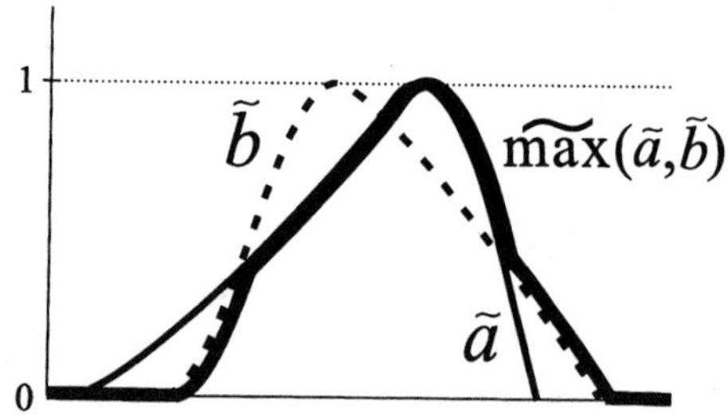

Figure 18.8. Fuzzy max $\widetilde{\max}(\tilde{a}, \tilde{b})$.

Using the fuzzy max, an inequality constraint with fuzzy coefficient, $\tilde{\boldsymbol{a}}_i^{\mathrm{T}}\boldsymbol{x} \le \tilde{b}_i$, is treated as (see [45])

$$\widetilde{\max}(\tilde{\boldsymbol{a}}_i^{\mathrm{T}}\boldsymbol{x}, \tilde{b}_i) = \tilde{b}_i. \tag{18.73}$$

By a discussion independent of the fuzzy max, this treatment has been proposed also by Ramík and Řimánek [35].

By the adoption of reciprocal Gödel implication instead of Dienes implication in definition of the necessity measure, (18.73) is equivalent to the following modality constraints;

$$\mathrm{MF1}\left(\tilde{\boldsymbol{a}}_i^{\mathrm{T}}\boldsymbol{x} \lesssim_i \tilde{b}_i\right) \ge 1, \qquad \mathrm{MF2}\left(\tilde{\boldsymbol{a}}_i^{\mathrm{T}}\boldsymbol{x} \lesssim_i \tilde{b}_i\right) \ge 1, \tag{18.74}$$

where fuzzy inequality $\lesssim_i$ is defined by (18.3) with ν_i satisfying $\nu_i(r) = 0$ for all $r > 0$, i.e., $\lesssim_i$ degenerates to the usual inequality relation $\le$.

Generalizing (18.74), we have

$$\mathrm{MF1}\left(\tilde{\boldsymbol{a}}_i^{\mathrm{T}}\boldsymbol{x} \lesssim_i \tilde{b}_i\right) \geq h_i, \qquad \mathrm{MF2}\left(\tilde{\boldsymbol{a}}_i^{\mathrm{T}}\boldsymbol{x} \lesssim_i \tilde{b}_i\right) \geq h_i. \tag{18.75}$$

This treatment is equivalent to a treatment of inequality relation with fuzzy coefficients by Tanaka et al. [45].

3.6.3 Imprecise Probabilistic Information. In the possibilistic interpretation of fuzzy numbers, possibility and necessity measures are often related to an intersection point of two membership functions. They are not related to area of regions defined by membership functions.

By the correspondence between possibility theory and Dempster-Shafer theory of evidence, fuzzy numbers can be seen as imprecise probability distributions (see [9]). Under this interpretation, lower and upper expected values $E_*(\tilde{a})$ and $E^*(\tilde{a})$ of a fuzzy number $\tilde{a}$ are defined by

$$E_*(\tilde{a}) = \int_0^1 \inf[\tilde{a}]_h dh, \qquad E^*(\tilde{a}) = \int_0^1 \sup[\tilde{a}]_h dh. \tag{18.76}$$

Therefore, a real number $\mathcal{E}(\tilde{a}) = (E_*(\tilde{a}) + E^*(\tilde{a}))/2$ can be regarded as a representative number of $\tilde{a}$. Indeed, Parra et al. [34] and Maleki [32] applied $\mathcal{E}(\tilde{a})$ to linear programming problems with fuzzy coefficients. Fortemps and Teghem [14] successfully showed the relation of the area compensation method [13] for comparing two fuzzy numbers with lower and upper expectations. Because of the limited space, we do not discuss the area compensation method deeply but the equivalent approach using lower and upper expectations.

In order to compare fuzzy numbers $\tilde{a}$ and $\tilde{b}$, consider an index $\mathcal{I}(\tilde{a} \leq \tilde{b})$ by

$$\mathcal{I}(\tilde{a} \leq \tilde{b}) = \frac{E^*(\tilde{b}) - E_*(\tilde{a})}{E^*(\tilde{a}) - E_*(\tilde{a}) + E^*(\tilde{b}) - E_*(\tilde{b})}. \tag{18.77}$$

This index is originally defined based on the area compensation meth-od but Fortemps and Teghem [14] gave the representation (18.77) using lower and upper expectations. Regarding fuzzy sets $\tilde{a}$ and $\tilde{b}$ as intervals $[E_*(\tilde{a}), E^*(\tilde{a})]$ and $[E_*(\tilde{b}), E^*(\tilde{b})]$, index $\mathcal{I}(\tilde{a} \leq \tilde{b})$ corresponds to the index proposed by Ishibuchi and Tanaka [25] for comparison between intervals. In this sense, approaches based on index $\mathcal{I}$ identify fuzzy numbers $\tilde{a}$ and $\tilde{b}$ with intervals $[E_*(\tilde{a}), E^*(\tilde{a})]$ and $[E_*(\tilde{b}), E^*(\tilde{b})]$, respectively.

Moreover, index $\mathcal{I}$ relates to VWF and VSF. Using lower and upper expectations $E_*(\tilde{a})$, $E^*(\tilde{a})$ of a fuzzy number $\tilde{a}$, let us define a symmetric triangular fuzzy number (STFN) $\hat{a}$ by the following membership function (see

Figure 18.9):

$$\mu_{\hat{a}}(r) = \begin{cases} \dfrac{2r - (3E_*(\tilde{a}) - E^*(\tilde{a}))}{2(E^*(\tilde{a}) - E_*(\tilde{a}))}, \\ \quad \text{if } 2r \in [3E_*(\tilde{a}) - E^*(\tilde{a}), E_*(\tilde{a}) + E^*(\tilde{a})], \\ \dfrac{(3E^*(\tilde{a}) - E_*(\tilde{a})) - 2r}{2(E^*(\tilde{a}) - E_*(\tilde{a}))}, \\ \quad \text{if } 2r \in [E_*(\tilde{a}) + E^*(\tilde{a}), 3E^*(\tilde{a}) - E_*(\tilde{a})], \\ 0, \quad \text{otherwise}. \end{cases} \tag{18.78}$$

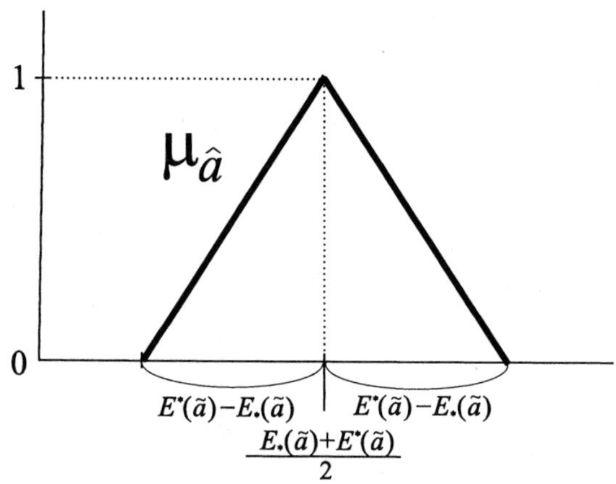

Figure 18.9. Symmetric triangular fuzzy number $\hat{a}$ defined by $E_*(\tilde{a})$ and $E^*(\tilde{a})$.

Obviously, we have $E_*(\hat{a}) = E_*(\tilde{a})$ and $E^*(\hat{a}) = E^*(\tilde{a})$. Therefore $\hat{a}$ can be seen as an STFN approximation of $\tilde{a}$. Let $\hat{b}$ an STFN approximation of a fuzzy set $\tilde{b}$. Then we have the following relations;

$$\text{when VWF}(\hat{a} \leq \hat{b}) \geq 0.5, \quad \text{VWF}(\hat{a} \leq \hat{b}) = \min(1, \mathcal{I}(\tilde{a} \leq \tilde{b}) + 0.5), \tag{18.79}$$

$$\text{when VSF}(\hat{a} \leq \hat{b}) \leq 0.5, \quad \text{VSF}(\hat{a} \leq \hat{b}) = \max(0, \mathcal{I}(\tilde{a} \leq \tilde{b}) - 0.5), \tag{18.80}$$

where we define VWF$(\hat{a} \leq \hat{b})$ and VSF$(\hat{a} \leq \hat{b})$ by (18.42) and (18.45) replacing $\tilde{\boldsymbol{a}}_i^{\mathrm{T}}\boldsymbol{x}$, $\precsim_i$ and $\tilde{b}_i$ with $\hat{a}$, $\leq$ and $\hat{b}$, respectively. From (18.79) and (18.80), index $\mathcal{I}(\tilde{a} \leq \tilde{b})$ can be expressed by VWF$(\hat{a} \leq \hat{b})$ and VSF$(\hat{a} \leq \hat{b})$ as

$$\mathcal{I}(\tilde{a} \leq \tilde{b}) = \max(0, \text{VWF}(\hat{a} \leq \hat{b}) - 0.5) + \min(0.5, \text{VSF}(\hat{a} \leq \hat{b})). \tag{18.81}$$

Fortemps and Teghem [14] proved that

$$\mathcal{I}(\tilde{a} \leq \tilde{b}) \geq 0.5 \text{ if and only if } \mathcal{E}(\tilde{a}) \leq \mathcal{E}(\tilde{b}). \tag{18.82}$$

Using $VWF(\hat{a} \leq \hat{b})$, this equivalence can be expressed as

$$VWF(\hat{a} \leq \hat{b}) \geq 1 \text{ if and only if } \mathcal{E}(\tilde{a}) \leq \mathcal{E}(\tilde{b}). \tag{18.83}$$

Fortemps and Teghem proposed to use the index to treat a constraint with fuzzy coefficients, $\tilde{\boldsymbol{a}}_i^{\mathrm{T}}\boldsymbol{x} \leq \tilde{b}_i$ by

$$\mathcal{I}\left(\tilde{\boldsymbol{a}}_i^{\mathrm{T}}\boldsymbol{x} \leq \tilde{b}_i\right) \geq p_i. \tag{18.84}$$

If we apply the extension principle for calculation of fuzzy quantity $\tilde{\boldsymbol{a}}_i^{\mathrm{T}}\boldsymbol{x}$ then (18.84) is reduced to a linear constraint. Moreover when $p_i = 0.5$, (18.84) is reduced to

$$\mathcal{E}(\tilde{\boldsymbol{a}}_i^{\mathrm{T}}\boldsymbol{x}) \leq \mathcal{E}(\tilde{b}_i), \tag{18.85}$$

which is treated by Maleki et al. [32].

Using STFN approximations $\hat{\boldsymbol{a}}_i$ and $\hat{b}_i$ of $\tilde{\boldsymbol{a}}_i$ and $\tilde{b}_i$, we have

$$\mathcal{I}\left(\tilde{\boldsymbol{a}}_i^{\mathrm{T}}\boldsymbol{x} \leq \tilde{b}_i\right) \geq p_i \Leftrightarrow \begin{cases} \mathrm{VWF}\left(\hat{\boldsymbol{a}}_i^{\mathrm{T}}\boldsymbol{x} \leq \hat{b}_i\right) \geq p_i + 0.5, & \text{if } p_i \leq 0.5, \\ \mathrm{VSF}\left(\hat{\boldsymbol{a}}_i^{\mathrm{T}}\boldsymbol{x} \leq \hat{b}_i\right) \geq p_i - 0.5, & \text{if } p_i > 0.5, \end{cases} \tag{18.86}$$

where $\hat{\boldsymbol{a}}_i$ is a vector of $\hat{a}_{ij}$, $j = 1, 2, \ldots, n$ which are STFN approximations of components $\tilde{a}_{ij}$, $j = 1, 2, \ldots, n$ of fuzzy number vector $\tilde{\boldsymbol{a}}_i$.

As shown above, the approach by Fortemps and Teghem [14] can be represented as a modality constrained programming model with respect to MOLP problems with STFN approximations of fuzzy coefficients.

4. Modality Goal Programming

In modality constrained programming approaches, we discussed the way to treat the fuzziness in a given problem so that we reduce it to a conventional multiple objective programming problem. In modality optimization models, we use a target value to each fuzzy objective function and maximize the possibility and necessity degrees that fuzzy objective function value is not greater than the target value. But we did not consider the deviation of a fuzzy objective function value from the given target value. Such a deviation is often considered in goal programming approach.

In this section, we discuss the deviation of a fuzzy objective function value from a target value given as a fuzzy number.

4.1 Two Kinds of Differences

The deviation is closely related to a difference between two fuzzy numbers. We discuss the difference between two fuzzy numbers $\tilde{a}$ and $\tilde{b}$.

By the extension principle (18.8), we obtain the difference $\tilde{b} - \tilde{a}$ with a membership function,

$$\mu_{\tilde{b}-\tilde{a}}(y) = \sup_{y=r_2-r_1} \min(\mu_{\tilde{a}}(r_1), \mu_{\tilde{b}}(r_2)), \tag{18.87}$$

where $\mu_{\tilde{b}-\tilde{a}}$, $\mu_{\tilde{a}}$ and $\mu_{\tilde{b}}$ are membership functions of $\tilde{b}-\tilde{a}$, $\tilde{a}$ and $\tilde{b}$, respectively. Of course this is a difference between two fuzzy numbers but another difference can be also defined by an alternative way.

Now let us discuss the following equation with an unknown fuzzy number $\tilde{w}$,

$$\tilde{a} + \tilde{w} = \tilde{b}. \tag{18.88}$$

This problem was treated in relation with fuzzy relational equation by Sanchez [40] and also by Dubois and Prade [6]. The problem of finding $\tilde{w}$ in (18.88) is known as deconvolution and has been treated extensively also by Kaufmann and Gupta [26]. The sum $\tilde{a}+\tilde{w}$ is defined by the extension principle (18.8). The difference $\tilde{b} - \tilde{a}$ defined by (18.87) is not a solution of (18.88). The solution $\tilde{w}$ does not always exist but only if there exists $r \in \mathbf{R}$ such that (see [6])

$$\mu_{\tilde{a}}(y - r) \leq \mu_{\tilde{b}}(y), \ \forall y \in \mathbf{R}. \tag{18.89}$$

Let $\tilde{a}$ and $\tilde{b}$ be symmetrical L-L fuzzy numbers $(a, a, \alpha, \alpha)_{LL}$ and $(b, b, \beta, \beta)_{LL}$, respectively. Then the greatest solution $\tilde{w}^*$ in the sense of inclusion relation is simply obtained as $\tilde{w}^* = (b - a, b - a, \beta - \alpha, \beta - \alpha)_{LL}$ when $\beta \geq \alpha$.

The solution of (18.88) is the difference between right- and left-hand side values when all fuzzy numbers are degenerated to real numbers. Taking this fact into consideration, the solution $\tilde{w}^* = (b - a, b - a, \beta - \alpha, \beta - \alpha)_{LL}$ can be seen as the difference between $\tilde{b}$ and $\tilde{a}$.

Moreover, let us consider another equation with an unknown fuzzy number $\tilde{u}$,

$$\tilde{b} - \tilde{u} = \tilde{a}, \tag{18.90}$$

where $\tilde{b} - \tilde{u}$ is defined by the extension principle (18.8). The difference $\tilde{b} - \tilde{a}$ defined by (18.87) is not a solution of (18.90), again. Let $\tilde{a} = (a, a, \alpha, \alpha)_{LL}$ and $\tilde{b} = (b, b, \beta, \beta)_{LL}$. When $\alpha \geq \beta$, the greatest solution $\tilde{u}^*$ in the sense of inclusion relation is obtained as $\tilde{u}^* = (b - a, b - a, \alpha - \beta, \alpha - \beta)_{LL}$. In the same consideration, this can be also seen as a difference between $\tilde{a}$ and $\tilde{b}$.

However, $\tilde{w}^*$ and $\tilde{u}^*$ are conditionally defined even when $\tilde{a}$ and $\tilde{b}$ are symmetric L-L fuzzy numbers. Their conditions are complementary each other. Then we combine two equations (18.88) and (18.90) with restriction $\tilde{w} = 0$ or $\tilde{u} = 0$. We obtain an equation with unknown fuzzy numbers $\tilde{w}$ and $\tilde{u}$,

$$\tilde{a} + \tilde{w} = \tilde{b} - \tilde{u}, \ \tilde{w} = 0 \text{ or } \tilde{u} = 0. \tag{18.91}$$

When $\tilde{a} = (a, a, \alpha, \alpha)_{LL}$ and $\tilde{b} = (b, b, \beta, \beta)_{LL}$, the greatest solution in the sense of inclusion is obtained as $\tilde{w}^* = (b - a, b - a, \beta - \alpha, \beta - \alpha)_{LL}$, $\tilde{u}^* = 0$ if $\beta \geq \alpha$, and $\tilde{w}^* = 0$, $\tilde{u}^* = (b - a, b - a, \alpha - \beta, \alpha - \beta)_{LL}$ otherwise. If $\tilde{w}^* = \tilde{u}^* = 0$ is the greatest solution, $\tilde{a}$ coincides with $\tilde{b}$ and vice versa. Therefore $\tilde{w}^* + \tilde{u}^*$ shows how much $\tilde{a}$ and $\tilde{b}$ do not coincide. Therefore, we may define a difference of $\tilde{a}$ from $\tilde{b}$ as $\tilde{b} \setminus \tilde{a} = \tilde{w}^* + \tilde{u}^*$.

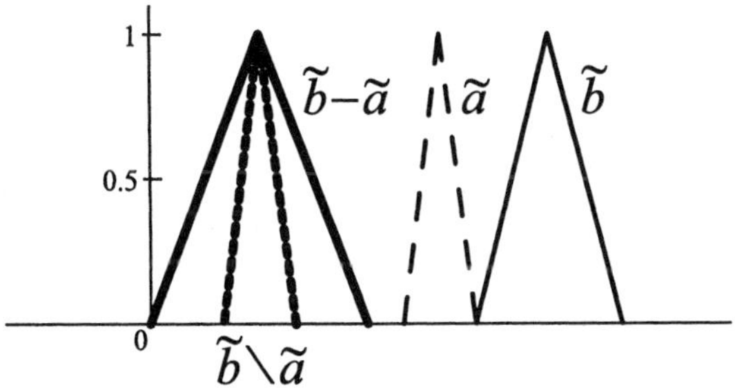

Figure 18.10. Two differences between fuzzy numbers $\tilde{a}$ and $\tilde{b}$.

Note that the difference $\tilde{b} - \tilde{a}$ defined by the extension principle is not the same as $\tilde{w}^* + \tilde{u}^*$. The former is a symmetrical L-L fuzzy number $(b - a, b - a, \alpha + \beta, \alpha + \beta)_{LL}$ while the latter is a symmetrical L-L fuzzy number $(b - a, b - a, |\alpha - \beta|, |\alpha - \beta|)_{LL}$. The two differences $\tilde{b} - \tilde{a}$ and $\tilde{b} \setminus \tilde{a}$ are depicted in Figure 18.10. The center $b - a$ is the same but the width is different. The difference $\tilde{b} - \tilde{a}$ shows the possible range of the difference between two uncertain numbers expressed by fuzzy numbers $\tilde{a}$ and $\tilde{b}$ while $\tilde{b} \setminus \tilde{a}$ shows how much $\tilde{a}$ and $\tilde{b}$ do not coincide. We call $\tilde{b} - \tilde{a}$ a value difference between $\tilde{a}$ and $\tilde{b}$ while $\tilde{b} \setminus \tilde{a}$ a distribution difference between $\tilde{a}$ and $\tilde{b}$. Moreover $\tilde{b} - \tilde{a}$ can be defined for any pairs of fuzzy numbers $\tilde{a}$ and $\tilde{b}$ while $\tilde{b} \setminus \tilde{a}$ cannot always be defined. We define $\tilde{b} \setminus \tilde{a}$ when $\tilde{a}$ and $\tilde{b}$ are symmetric L-L fuzzy numbers.

4.2 Applications to the Objective Function Treatments

In real world applications, we may have two types of fuzzy targets to a fuzzy objective function: one is a target value which is unknown and the other is a target distribution by which the decision maker would like to control the distribution of uncertain objective function values.

To the first type, the difference $\widetilde{dif}_k(\boldsymbol{x})$ between a fuzzy objective function value $\tilde{\boldsymbol{c}}_k^{\mathrm{T}}\boldsymbol{x}$ and a given fuzzy target value $\tilde{t}_k$,

$$\widetilde{dif}_k(\boldsymbol{x}) = \tilde{t}_k - \tilde{\boldsymbol{c}}_k^{\mathrm{T}}\boldsymbol{x}. \tag{18.92}$$

To the second type, the difference $\widetilde{Dif}_k(\boldsymbol{x})$ between a fuzzy objective function value $\tilde{\boldsymbol{c}}_k^{\mathrm{T}}\boldsymbol{x}$ and a given fuzzy target distribution $\tilde{d}_k$ is defined by

$$\widetilde{Dif}_k(\boldsymbol{x}) = \tilde{w}_k + \tilde{u}_k,\ \tilde{\boldsymbol{c}}_k^{\mathrm{T}}\boldsymbol{x} + \tilde{w}_k = \tilde{d}_k - \tilde{u}_k,\ \tilde{w}_k = 0 \text{ or } \tilde{u}_k = 0, \quad (18.93)$$

where we assume that $\tilde{c}_{kj}$, $\tilde{d}_k$, $\tilde{w}_k$ and $\tilde{u}_k$ are symmetrical L-L fuzzy numbers $(c_{kj}, c_{kj}, \gamma_{kj}, \gamma_{kj})_{LL}$, $(d_k, d_k, \delta_k, \delta_k)_{LL}$, $(w_k, w_k, \omega_k, \omega_k)_{LL}$ and $(u_k, u_k, v_k, v_k)_{LL}$, respectively. The sum of $\tilde{w}_k$ and $\tilde{u}_k$ is the greatest solution of the second equation of (18.93) in the sense of inclusion.

A guide to utilize those two types of differences in the real world problem is as follows. When the decision maker would like to make the realization of the objective function value closed to a target value, $\widetilde{dif}_k(\boldsymbol{x})$ should be adopted. On the other hand, when the decision maker would like to control the distribution of the objective function values closed to a target distribution, $\widetilde{Dif}_k(\boldsymbol{x})$ should be adopted.

Now, we briefly describe the approaches to MOLP problems with fuzzy targets based on two types of differences described above. In either type, we will have p new fuzzy objective functions,

$$\text{optimize } (\widetilde{dif}_1(\boldsymbol{x}), \widetilde{dif}_2(\boldsymbol{x}), \ldots, \widetilde{dif}_p(\boldsymbol{x}))^{\mathrm{T}} \quad (18.94)$$

or

$$\text{optimize } (\widetilde{Dif}_1(\boldsymbol{x}), \widetilde{Dif}_2(\boldsymbol{x}), \ldots, \widetilde{Dif}_p(\boldsymbol{x}))^{\mathrm{T}} \quad (18.95)$$

To each deviation we specify a fuzzy goal $\tilde{Z}_k$ with linguistic expression 'approximately zero'. Then objective functions (18.94) can be treated as

$$\text{maximize } (N_{G_1}(\widetilde{dif}_1(\boldsymbol{x})), N_{G_2}(\widetilde{dif}_2(\boldsymbol{x})), \ldots, N_{G_p}(\widetilde{dif}_p(\boldsymbol{x}))). \quad (18.96)$$

We may also have model (18.96) with necessity measures $N_{G_k}(\widetilde{dif}_k(\boldsymbol{x}))$ replaced with possibility measures $\Pi_{G_k}(\widetilde{dif}_k(\boldsymbol{x}))$. Now we have p objective functions. A solution can be obtained by applying a multiple objective programming technique to (18.96).

On the other hand, as in goal programming, we can also consider deviations of fuzzy objective function values from fuzzy targets by taking the absolute values of differences. The absolute value $|\tilde{a}|$ of a fuzzy number $\tilde{a}$ is also defined by the extension principle (18.8). Namely we have

$$\mu_{|\tilde{a}|}(r) = \begin{cases} \max(\mu_{\tilde{a}}(r), \mu_{\tilde{a}}(-r)), & \text{if } r \geq 0, \\ 0, & \text{if } r < 0, \end{cases} \quad (18.97)$$

where $\mu_{|\tilde{a}|}$ is the membership function of $\tilde{a}$.

Giving a regret function $R : \mathbf{R}_+^p \to \mathbf{R}$, we will have a single objective function with respect to (18.94),

$$\text{minimize } R(|\widetilde{dif}_1(\boldsymbol{x})|, |\widetilde{dif}_2(\boldsymbol{x})|, \ldots, |\widetilde{dif}_p(\boldsymbol{x})|). \quad (18.98)$$

Similarly, with respect to (18.95), we obtain

$$\text{minimize}\ \ R(|\widetilde{Dif}_1(\boldsymbol{x})|, |\widetilde{Dif}_2(\boldsymbol{x})|, \ldots, |\widetilde{Dif}_p(\boldsymbol{x})|). \tag{18.99}$$

(18.98) and (18.99) are single fuzzy objective function so that we can apply the treatments of objective functions described in the previous section.

For example, let R be defined by

$$R(r_1, r_2, \ldots, r_p) = \sum_{k=1}^{p} \lambda_k r_k, \tag{18.100}$$

where $\lambda_k > 0$ are weights for the commensurability and the importance. We may apply a necessity fractile model to (18.99),

$$\begin{aligned} &\text{minimize} \quad z, \\ &\text{subject to} \quad \mathrm{NES}\left(\sum_{k=1}^{p} \lambda_k |\widetilde{Dif}_k(\boldsymbol{x})| \leq z\right) \geq \bar{h}^N, \end{aligned} \tag{18.101}$$

where $\bar{h}^N \in (0, 1]$ is the necessity level given by the decision maker.

(18.101) is reduced to the following problem (see [24]):

$$\begin{aligned} &\text{minimize} \quad \sum_{k=1}^{p} \lambda_k e_k, \\ &\text{subject to} \quad \sum_{j=1}^{n} c_{kj} x_j - L^{(-1)}(1 - \bar{h}^N) \sum_{k=1}^{n} \gamma_{kj} |x_{kj}| \\ &\qquad\qquad + e_k^{\mathrm{L}-} - e_k^{\mathrm{L}+} = d_k - L^{(-1)}(1 - \bar{h}^N)\delta_k,\ k = 1, 2, \ldots, p, \\ &\qquad \sum_{j=1}^{n} c_{kj} x_j + L^{(-1)}(1 - \bar{h}^N) \sum_{j=1}^{n} \gamma_{kj} |x_j| \\ &\qquad\qquad + e_k^{\mathrm{R}-} - e_k^{\mathrm{R}+} = d_k + L^{(-1)}(1 - \bar{h}^N)\delta_k,\ k = 1, 2, \ldots, p, \\ &\qquad e_k^{\mathrm{L}-} + e_k^{\mathrm{L}+} \leq e_k,\ e_k^{\mathrm{R}-} + e_k^{\mathrm{R}+} \leq e_k,\ k = 1, 2, \ldots, p, \\ &\qquad e_k^{\mathrm{L}-},\ e_k^{\mathrm{L}+},\ e_k^{\mathrm{R}-},\ e_k^{\mathrm{R}+} \geq 0,\ k = 1, 2, \ldots, p. \end{aligned} \tag{18.102}$$

As shown in this example, the linearity can be preserved even when we use the difference $\tilde{\boldsymbol{c}}_k^{\mathrm{T}} \boldsymbol{x} \setminus \tilde{d}_k$. However, we cannot preserve the linearity if we adopt a possibility measure instead of a necessity measure in (18.101).

As a similar approach to minimizing the deviation $|\tilde{\boldsymbol{c}}_k^{\mathrm{T}} \boldsymbol{x} \setminus \tilde{d}_k|$, we may minimize Hausdorf-like distance between fuzzy numbers $\tilde{\boldsymbol{c}}_k^{\mathrm{T}} \boldsymbol{x}$ and $\tilde{d}_k$. An approach using Hausdorf distance between h-level sets of $\tilde{\boldsymbol{c}}_k^{\mathrm{T}} \boldsymbol{x}$ and $\tilde{d}_k$ is discussed in [27]. Moreover, if we use index $\mathcal{I}$ instead of NES and if we define $\bar{h}^N = 0.5$ then (18.101) is the problem treated by Parra [34].

5. Modal Efficiency Approach

5.1 Problem Statement and Efficiency

In the previous two approaches, we discussed the reduction of a given MOLP problem with fuzzy coefficients to a conventional multiple objective programming problem. The solution obtained in this approach would be efficient to the reduced conventional multiple objective programming problem. However, we do not know how the solution is reasonable from the point of the original MOLP problem with fuzzy coefficients. To evaluate the reasonability, we should discuss the efficiency and the feasibility of a solution in view of the original MOLP problem with fuzzy coefficients. Such topics have been discussed in [19, 23, 30]. In [19], various non-dominated solutions and efficient solutions are defined and the relationships among them are discussed. In this section, we describe possible and necessary efficient solutions since they are simple and basic in the approaches

In this section, we treat the following MOLP problems with a fuzzy objective coefficient matrix:

$$\begin{array}{ll} \text{minimize} & \tilde{C}\boldsymbol{x}, \\ \text{subject to} & A\boldsymbol{x} \leq \boldsymbol{b}, \end{array} \tag{18.103}$$

where $\tilde{C}$ is a $p \times n$ matrix whose (k, j)-component is a fuzzy number $\tilde{c}_{kj}$. A is an $m \times n$ constant matrix. Namely, no fuzziness is involved in the constraints. Since the efficiency and feasibility are discussed independently, we do not loose the generality for the discussion of the efficiency by this simplification. The feasibility have been already discussed in the treatments of constraints in the modality constrained programming approach so that we may apply those when fuzziness is included in the constraints in order to obtain (18.103). For the sake of simplicity, let F be the feasible solution set and then Problem (18.103) is simply written as $\min\{\tilde{C}\boldsymbol{x} \mid \boldsymbol{x} \in F\}$.

When each $\tilde{c}_{kj}$ is a closed interval, Problem (18.103) degenerates to an MOLP problem with an interval objective coefficients treated in [1]. Since $\tilde{C}$ shows the range of possible coefficient matrices with membership degrees, it can be seen as a fuzzy set with a membership function $\mu_{\tilde{C}}$. Thus, when $\tilde{c}_{kj}$ is a closed interval, $\tilde{C}$ is a set of matrices Q. Therefore, in the interval case, we can write $Q \in \tilde{C}$ if and only if $\mu_{\tilde{C}}(Q) = 1$.

Let $EF(Q)$ be a set of efficient solutions of an MOLP problem with a $p \times n$ matrix Q, $\min\{Q\boldsymbol{x} \mid \boldsymbol{x} \in F\}$. Namely, we have

$$EF(Q) = \{\boldsymbol{x} \in F \mid \not\exists \bar{\boldsymbol{x}} \in F;\ Q\bar{\boldsymbol{x}} \leq Q\boldsymbol{x} \text{ and } Q\bar{\boldsymbol{x}} \neq Q\boldsymbol{x}\}. \tag{18.104}$$

Let $K(Q) = \{\boldsymbol{r} \in \mathbf{R}^n \mid Q\boldsymbol{r} \leq \mathbf{0}\}$. Then we have $\boldsymbol{x} \in EF(Q) \Leftrightarrow (\{\boldsymbol{x}\} + K(Q)) \cap F = \{\boldsymbol{x}\}$.

Moreover, $\boldsymbol{x} \in EF(Q)$ is confirmed by the existence of $(\boldsymbol{y}^{\mathrm{T}}, \boldsymbol{z}^{\mathrm{T}})^{\mathrm{T}}$ such that

$$A^{\mathrm{T}}\boldsymbol{y} = -Q^{\mathrm{T}}(\mathbf{1}+\boldsymbol{z}),\ (A\boldsymbol{x}-\boldsymbol{b})^{\mathrm{T}}\boldsymbol{y},\ \boldsymbol{y} \geq \mathbf{0},\ \boldsymbol{z} \geq \mathbf{0}, \tag{18.105}$$

where $\mathbf{1} = (1, 1, \ldots, 1)^{\mathrm{T}}$. Given $\boldsymbol{x} \in F$, A and $\boldsymbol{b}$ can be represented as

$$A = \begin{pmatrix} A^0 \\ A^- \end{pmatrix}, \qquad \boldsymbol{b} = \begin{pmatrix} \boldsymbol{b}^0 \\ \boldsymbol{b}^- \end{pmatrix} \tag{18.106}$$

such that $A^0\boldsymbol{x} - \boldsymbol{b}^0 = \mathbf{0}$ and $A^-\boldsymbol{x} - \boldsymbol{b}^- < \mathbf{0}$ by a suitable permutation. From (18.105), $\boldsymbol{x} \in EF(Q)$ is confirmed by the existence of $(\boldsymbol{y}^{0\mathrm{T}}, \boldsymbol{z}^{\mathrm{T}})^{\mathrm{T}}$ such that

$$A^{0\mathrm{T}}\boldsymbol{y}^0 = -Q^{\mathrm{T}}(\mathbf{1}+\boldsymbol{z}),\ \boldsymbol{y}^0 \geq \mathbf{0},\ \boldsymbol{z} \geq \mathbf{0}. \tag{18.107}$$

When A^0 is empty, we define $A^{0\mathrm{T}}\boldsymbol{y}^0 = \mathbf{0}$.

5.2 Interval Case

When each $\tilde{c}_{kj}$ is a closed interval $[c_{kj}^{\mathrm{L}}, c_{kj}^{\mathrm{R}}]$, we can define possibly efficient solution set ΠS and necessarily efficient solution set NS by

$$NS = \bigcap_{Q \in \tilde{C}} EF(Q), \qquad \Pi S = \bigcup_{Q \in \tilde{C}} EF(Q). \tag{18.108}$$

An element of NS is a solution efficient for any $Q \in \tilde{C}$ and called a necessarily efficient solution. On the other had, an element of ΠS is a solution efficient for at least one $Q \in \tilde{C}$ and called a possibly efficient solution. While the possible efficiency shows the minimum rationality, the necessary efficiency shows the ideality.

Define $\Pi K(\tilde{C}) = \bigcup_{Q \in \tilde{C}} K(Q)$. Then we have

$$\boldsymbol{x} \in NS \Leftrightarrow (\{\boldsymbol{x}\} + \Pi K(\tilde{C})) \cap F = \{\boldsymbol{x}\}. \tag{18.109}$$

Let C^{L} and C^{R} be $p \times n$ matrices whose (k, j)-components are c_{kj}^{L} and c_{kj}^{R}. Let $M = \{Q \mid Q_{\cdot j} = C_{\cdot j}^{\mathrm{L}} \text{ or } Q_{\cdot j} = C_{\cdot j}^{\mathrm{R}},\ j = 1, 2, \ldots, n\}$, where $Q_{\cdot j}$, $C_{\cdot j}^{\mathrm{L}}$ and $C_{\cdot j}^{\mathrm{R}}$ are the j-th columns of Q, C^{L} and C^{R}, respectively. It is shown that $\Pi K(\tilde{C}) = \bigcup_{Q \in M} K(Q)$ (see [1]). Then we have

$$\boldsymbol{x} \in NS \Leftrightarrow \left(\{\boldsymbol{x}\} + \bigcup_{Q \in M} K(Q)\right) \cap F = \{\boldsymbol{x}\}. \tag{18.110}$$

Moreover, this implies $NS = \bigcap_{Q \in M} EF(Q)$. Therefore, we do not need to consider infinitely many $Q \in \tilde{C}$ but only 2^n matrices $Q \in M$.

Given a feasible solution $\boldsymbol{x} \in F$, $\boldsymbol{x} \in NS$ is confirmed by the existence of a solution $(\boldsymbol{y}^{0\mathrm{T}}, \boldsymbol{z}^{\mathrm{T}})^{\mathrm{T}}$ in (18.107) for all $Q \in M$. An implicit enumeration method was proposed to test the necessary efficiency of a feasible basic solution in [1].

On the other hand, given $\boldsymbol{x} \in F$, $\boldsymbol{x} \in \Pi S$ is confirmed by the existence of a solution $(\boldsymbol{y}^{0\mathrm{T}}, \boldsymbol{z}^{\mathrm{T}})^{\mathrm{T}}$ of (18.107) for a $Q \in \tilde{C}$. Since we have $\tilde{C} = \{Q \mid C^{\mathrm{L}} \leq Q \leq C^{\mathrm{R}}\}$ and $\boldsymbol{z} \geq \boldsymbol{0}$, such $(\boldsymbol{y}^{0\mathrm{T}}, \boldsymbol{z}^{\mathrm{T}})^{\mathrm{T}}$ exists if and only if it satisfies

$$-C^{\mathrm{R}^{\mathrm{T}}}(\boldsymbol{1}+\boldsymbol{z}) \leq A^{0\mathrm{T}}\boldsymbol{y}^0 \leq -C^{\mathrm{L}^{\mathrm{T}}}(\boldsymbol{1}+\boldsymbol{z}),\ \boldsymbol{y}^0 \geq \boldsymbol{0},\ \boldsymbol{z} \geq \boldsymbol{0}. \qquad (18.111)$$

(18.111) is a system of linear inequalities. Thus, possibility efficiency of $\boldsymbol{x} \in F$ is easily confirmed.

5.3 Fuzzy Coefficient Case

Let $P(\boldsymbol{x}) = \{Q \mid \nexists \bar{\boldsymbol{x}} \in F;\ Q\bar{\boldsymbol{x}} \leq Q\boldsymbol{x} \text{ and } Q\bar{\boldsymbol{x}} \neq Q\boldsymbol{x}\}$. Then necessarily and possibly efficient solution sets are defined as fuzzy sets by the following membership functions (see [23]): for $\boldsymbol{x} \in F$,

$$\mu_{NS}(\boldsymbol{x}) = N_{\tilde{C}}(P(\boldsymbol{x})) = \inf_Q\{1 - \mu_{\tilde{C}}(Q) \mid Q \notin P(\boldsymbol{x})\}, \qquad (18.112)$$

$$\mu_{\Pi S}(\boldsymbol{x}) = \Pi_{\tilde{C}}(P(\boldsymbol{x})) = \sup_Q\{\mu_{\tilde{C}}(Q) \mid Q \in P(\boldsymbol{x})\}, \qquad (18.113)$$

and for $\boldsymbol{x} \notin F$, $\mu_{NS}(\boldsymbol{x}) = \mu_{\Pi S}(\boldsymbol{x}) = 0$. In interval case, we only discuss whether $\boldsymbol{x}$ is a necessarily (or possibly) efficient solution or not. In fuzzy coefficient case, each feasible solution can take a degree of necessary (or possible) efficiency. As in interval coefficient case, while the possible efficiency relates to the minimum rationality, the necessary efficiency relates to the ideality. In fuzzy coefficient case, we can discuss the degrees of possible and necessary efficiencies.

From the property of necessity and possibility measures, we have

$$\mu_{NS}(\boldsymbol{x}) = \sup_h\{h \mid [\tilde{C}]_{1-h} \subseteq P(\boldsymbol{x})\}, \qquad (18.114)$$

$$\mu_{\Pi S}(\boldsymbol{x}) = \sup_h\{h \mid [\tilde{C}]_h \cap P(\boldsymbol{x}) \neq \emptyset\}. \qquad (18.115)$$

Therefore, $\mu_{NS}(\boldsymbol{x})$ and $\mu_{\Pi S}(\boldsymbol{x})$ are the upper bounds of $h \in (0,1]$ satisfying $[\tilde{C}]_{1-h} \subseteq P(\boldsymbol{x})$ and $[\tilde{C}]_h \cap P(\boldsymbol{x}) \neq \emptyset$, respectively. Since $[\tilde{C}]_{1-h} \subseteq P(\boldsymbol{x})$ is equivalent to $\boldsymbol{x} \in \cap_{Q \in [\tilde{C}]_{1-h}} EF(Q)$ and $[\tilde{C}]_h \cap P(\boldsymbol{x}) \neq \emptyset$ is equivalent $\boldsymbol{x} \in \cup_{Q \in [\tilde{C}]_h} EF(Q)$, we can obtain $\mu_{NS}(\boldsymbol{x})$ and $\mu_{\Pi S}(\boldsymbol{x})$ by varying h together with testing necessary efficiency with respect to interval coefficient matrix $[\tilde{C}]_{1-h}$ and possible efficiency with respect to interval coefficient matrix $[\tilde{C}]_h$, respectively.

6. Concluding Remarks

We described approaches to multiple objective programming problems with fuzzy coefficients based on possibility and necessity measures. In the modality constrained programming approach, we discussed treatments of the fuzziness involved in the problems. The extension methods of a fuzzy relation to the relation between fuzzy numbers are reviewed in order to treat generalized constraints. In the modality goal programming approach, we emphasized that two kinds of differences between a fuzzy objective function value and a fuzzy target are conceivable. We described a guide for applications of those differences. In the modal efficiency approach, we extended the efficiency to the case of multiple objective programming problems. We discussed some necessary and sufficient conditions for a feasible solution to satisfy the extended efficiencies. In those approaches, (1) by the combination of possibility and necessity measures, we have a lot of indices for evaluating the solution, and thus (2) by a suitable management of possibility and necessity measures, we may reflect decision maker's attitude to the uncertainty to the model.

As demonstrated in this paper, a concept in multiple objective programming problems diverges into several extensions depending on how we treat the fuzziness. We obtained many approaches even if we restrict ourselves only the approaches based on possibility and necessity measures. Using other measures, we will obtain more approaches. For example, expected values of fuzzy numbers are used in [14, 34] and the extension principle by Yager's parameterized t-norm is used in [36].

Despite that a lot of models were proposed in multiple objective programming problems with fuzzy coefficients, there are still a lot of issues to be discussed. For example, multiple objective nonlinear programming problems with fuzzy coefficients, the problems with interrelated uncertain parameters, extended efficiency tests and so on. The author hopes that the approaches will be further developed and applied to real world problems.

References

[1] G.R. Bitran. Linear multiple objective problems with interval coefficients. *Management Science*, 26:694–706, 1980.

[2] M. Delgado, J.L. Verdegay, and M.A. Vila. A general model for fuzzy linear programming. *Fuzzy Sets and Systems*, 29:21–29, 1989.

[3] D. Dubois. Linear programming with fuzzy data. In J.C. Bezdek, editor, *Analysis of Fuzzy Information, Vol. III: Application in Engineering and Science*, pages 241–263. CRC Press, Boca Raton, FL, 1987.

[4] D. Dubois, H. Fargier, and H. Prade. Possibility theory in constraint satisfaction problems: Handling priority, preference and uncertainty. *Applied Intelligence*, 6:287–309, 1996.

[5] D. Dubois and P. Fortemps. Computing improved optimal solutions to max-min flexible constraint satisfaction problems. *European Journal of Operational Research*, 118:95–126, 1999.

[6] D. Dubois and H. Prade. Inverse operations for fuzzy numbers. In *Proceedings of IFAC Fuzzy Information, Marseille, France*, pages 399–404, 1983.

[7] D. Dubois and H. Prade. A review of fuzzy set aggregation connectives. *Information Sciences*, 36:85–121, 1985.

[8] D. Dubois and H. Prade. Fuzzy numbers: An overview. In J.C. Bezdek, editor, *Analysis of Fuzzy Information, Vol.I: Mathematics and Logic*, pages 3–39. CRC Press, Boca Raton, FL, 1987.

[9] D. Dubois and H. Prade. The mean value of a fuzzy number. *Fuzzy Sets and Systems*, 24:279–300, 1987.

[10] D. Dubois and H. Prade. *Possibility Theory: An Approach to Computerized Processing of Uncertainty*. Plenum Press, New York, 1988.

[11] M. Ehrgott and X. Gandibleux, editors. *Multiple Criteria Optimization: State of the Art Annotated Bibliographic Surveys*. Kluwer Academic Publishers, Boston, 2002.

[12] J. Fodor and M. Roubens. *Fuzzy Preference Modeling and Multicriteria Decision Making*. Kluwer Academic Publishers, Dordrecht, 1994.

[13] P. Fortemps and M. Roubens. Ranking and defuzzification methods based on area compensation. *Fuzzy Sets and Systems*, 82:319–330, 1996.

[14] P. Fortemps and J. Teghem. Multi-objective fuzzy linear programming: The MOFAC method. In B. DeBaets J. Fodor and P. Perny, editors, *Preferences and Decisions under Incomplete Knowledge*, pages 11–31. Physica-Verlag, Heidelberg, 2000.

[15] J.P. Ignizio. *Linear Programming in Single- & Multiple-Objective Systems*. Prentice-Hall Inc., New Jersey, 1982.

[16] M. Inuiguchi, H. Ichihashi, and Y. Kume. Relationships between modality constrained programming problems and various fuzzy mathematical programming problems. *Fuzzy Sets and Systems*, 49:243–259, 1992.

[17] M. Inuiguchi, H. Ichihashi, and Y. Kume. Some properties of extended fuzzy preference relations using modalities. *Information Sciences*, 61:187–209, 1992.

[18] M. Inuiguchi, H. Ichihashi, and Y. Kume. Modality constrained programming problems: A unified approach to fuzzy mathematical programming problems in the setting of possibility theory. *Information Sciences*, 67:93–126, 1993.

[19] M. Inuiguchi and Y. Kume. Solution concepts for fuzzy multiple objective programming problems. *Japanese Journal of Fuzzy Theory and Systems*, 2:1–22, 1990.

[20] M. Inuiguchi and Y. Kume. Goal programming problems with interval coefficients and target intervals. *European Journal of Operational Research*, 52:345–360, 1991.

[21] M. Inuiguchi and Y. Kume. Modality constrained programming problems introduced gödel implication and various fuzzy mathematical programming problems (in Japanese). *Transactions of the Institute of Systems, Control, and Information Engineers*, 4:382–392, 1991.

[22] M. Inuiguchi and J. Ramík. Possibilistic linear programming: A brief review of fuzzy mathematical programming and a comparison with stochastic programming in portfolio selection problem. *Fuzzy Sets and Systems*, 111:3–28, 2000.

[23] M. Inuiguchi and M. Sakawa. Possible and necessary efficiency in possibilistic multiobjective linear programming problems and possible efficiency test. *Fuzzy Sets and Systems*, 78:231–241, 1996.

[24] M. Inuiguchi and T. Tanino. Modality goal programming models with interactive fuzzy numbers. In *Proceedings of he Eighth International Fuzzy Systems Association World Congress*, volume 1, pages 519–523, 1999.

[25] H. Ishibuchi and H. Tanaka. Formulation and analysis of linear programming problem with interval coefficients (in Japanese). *Journal of the Japan Industrial Management Association*, 40:320–329, 1989.

[26] A. Kaufmann and M.M. Gupta. *Fuzzy Mathematical Models in Engineering and Management Science*. Elsevier, Amsterdam, 1988.

[27] H. Kuwano. On the fuzzy multi-objective linear programming problem: Goal programming approach. *Fuzzy Sets and Systems*, 82:57–64, 1996.

[28] Y.-J. Lai and C.L. Hwang. *Fuzzy Mathematical Programming*. Springer-Verlag, Berlin, 1992.

[29] B. Liu. *Theory and Practice of Uncertain Programming*. Physica-Verlag, Heidelberg, 2002.

[30] M.K. Luhandjula. Multiple objective programming problems with possibilistic coefficients. *Fuzzy Sets and Systems*, 21:135–145, 1987.

[31] M.K. Luhandjula. Fuzzy optimization: An appraisal. *Fuzzy Sets and Systems*, 30:257–282, 1989.

[32] H.R. Maleki, M. Tata, and M. Mashinichi. Linear programming with fuzzy variables. *Fuzzy Sets and Systems*, 109:21–33, 2000.

[33] C.V. Negoita. The current interest in fuzzy optimization. *Fuzzy Sets and Systems*, 6:261–269, 1981.

[34] M.A. Parra, A.B. Terol, and M.V.R Uria. Solution of a possibilistic multiobjective linear programming problem. *European Journal of Operational Research*, 119:338–344, 1999.

[35] J. Ramík and J. Římánek. Inequality relation between fuzzy numbers and its use in fuzzy optimization. *Fuzzy Sets and Systems*, 16:123–138, 1985.

[36] H. Rommelfanger and T. Keresztfalvi. Multicriteria fuzzy optimization based on Yager's parameterized t-norm. *Foundation of Computing and Decision Sciences*, 16:99–110, 1991.

[37] H. Rommelfanger and R. Słowiński. Fuzzy linear programming with single or multiple objective functions. In R. Słowiński, editor, *Fuzzy Sets in Decision Analysis, Operations Research and Statistics*. Kluwer Academic Publishers, Boston, 1998.

[38] M. Sakawa. *Fuzzy Sets and Interactive Multiobjective Optimization*. Plenum Press, New York, 1993.

[39] M. Sakawa. Fuzzy multiobjective and multilevel programming. In M. Ehrgott and X. Gandibleux, editors, *Multiple Criteria Optimization: State of the Art Annotated Bibliographic Surveys*, pages 171–226. Kluwer Academic Publishers, Boston, 2002.

[40] E. Sanchez. Solution of fuzzy equations with extended operations. *Fuzzy Sets and Systems*, 12:237–248, 1984.

[41] R. Słowiński. Fuzzy multi-objective linear programming. *Belgian Journal of Operations Research, Statistics and Computer Science*, 37:83–98, 1997.

[42] R. Słowiński and J. Teghem, editors. *Stochastic versus Fuzzy Approaches to Multiobjective Mathematical Programming under Uncertainty*. Kluwer Academic Publishers, Dordrecht, 1990.

[43] I.M. Stancu-Minasian. *Stochastic Programming with Multiple Objective Functions*. D. Reidel Publishing Company, Dordrecht, 1984.

[44] R.E. Steuer. *Multiple Criteria Optimization: Theory, Computation, and Application*. John Wiley & Sons, New York, 1986.

[45] H. Tanaka, H. Ichihashi, and K. Asai. A formulation of fuzzy linear programming problem based on comparison of fuzzy numbers. *Control and Cybernetics*, 13:185–194, 1984.

[46] H. Tanaka, T. Okuda, and K. Asai. On fuzzy mathematical programming. *Journal of Cybernetics*, 3:37–46, 1974.

[47] M. Zeleny. *Linear Multiobjective Programming*. Springer-Verlag, Berlin, 1974.

[48] H.-J. Zimmermann. Description and optimization of fuzzy systems. *International Journal of General Systems*, 2:209–215, 1976.

[49] H.-J. Zimmermann. Fuzzy programming and linear programming with several objective functions. *Fuzzy Sets and Systems*, 1:45–55, 1978.

[50] H.-J. Zimmermann. Applications of fuzzy sets theory to mathematical programming. *Information Sciences*, 36:29–58, 1985.

Chapter 19

MCDM LOCATION PROBLEMS

Stefan Nickel
Fakultät für Rechts- und Wirtschaftswissenschaften, Universität des Saarlands
66041 Saarbrücken, Germany
and
Fraunhofer Insitute for Industrial Mathematics (ITWM)
67663 Kaiserslautern, Germany
s.nickel@wiwi.uni-sb.de

Justo Puerto
Facultad de Matemáticas
Universidad de Sevilla
41012 Seville, Spain
puerto@us.es

Antonio M. Rodríguez-Chía
Facultad de Ciencias
Universidad de Cádiz
11510 Puerto Real (Cádiz), Spain
antonio.rodriguezchia@uca.es

Abstract In this chapter, we provide a broad overview of the most representative multicriteria location problems as well as of the most relevant achievements in this field, indicating the relationship between them whenever possible. We consider a large number of references which have been classified in three sections depending on the type of decision space where the analyzed models are stated. Therefore, we distinguish between continuous, network, and discrete multicriteria location problems.

Keywords: Locational Analysis, multicriteria location problems, point-objective location problems, multiobjective location problems.

1. Introduction

Locational Analysis has become a very active field of research in the last decades. Since the seminal papers by Hakimi [68] and Witzgall [156] in the early sixties the number of researchers and publications have grown and Locational Analysis has become very popular among both practitioners and academia. The interested reader can find excellent surveys in the literature providing reviews of location references (among others [15, 42, 44, 63, 64, 93, 123]).

In this chapter, we present a survey of the most representative multicriteria location problems. Our goal is to give a broad overview of the different models and resolution procedures used in this field as well as to indicate how they relate to one another. Although we have not been exhaustive, we have tried to cover the most fruitful lines of research of Multicriteria Locational Analysis. Our hope is that this chapter will provide the readers with a helpful tool: the location analysts may complete their knowledge about the state of the art of their research field while the rest of the readers can find a comprehensive overview to introduce them to the main streams of this area.

Since our study focuses on multicriteria location problems, we proceed by giving a general formulation of this type of problems. To do so, we consider a family of possibly conflicting objective functions $F_i(\cdot)$ with $i \in I$. These functions represent different criteria to locate one or several new facilities and depend on the distances from these facilities to the set of fixed or demand facilities, usually called A. There are at least two natural ways of deriving the different F_i. First, a decision about a new facility to be located is typically a group decision and each decision maker i will have his own preferences, which my be expressed by F_i. Secondly, the functions F_i may represent different quality criteria for the new facility to be located, like cost, reachability, risk, etc. The general formulation is given by

$$\underset{X_p \subseteq S \subseteq X,\, |X_p|=p}{v-\min} \left(F_i(X_p, A)\right)_{i \in I}, \tag{19.1}$$

where $v - \min$ stands for vector minimization, X is the decision space, X_p is the finite set of service facilities, $|X_p|$ its cardinality, and S is the feasible region (see [51] for a survey of multicriteria optimization). The reader may note that the classical median and center problem in the literature of Locational Analysis are just particular aggregation procedures of the multiple criteria formulation in (19.1).

Problem (19.1) is a valid formulation for the general multifacility multicriteria problem. Nevertheless, a majority of the results published in the literature refers to the single facility case, $p = 1$. Therefore, in this chapter, the results will be generally referred to single facility models, unless the multifacility character is stated explicitly.

Formulation (19.1) corresponds to a multicriteria problem. Therefore, it is common to propose as solution concept different sets of feasible points that correspond to different levels of exigency regarding the ordering relationship between vectors. The most classical solution sets used in the literature are included in the following definition (for more details see for instance [50]).

DEFINITION 74 *Let X_p and X_p^* denote sets with cardinality p.*

i) *$X_p^* \subseteq S$ is a weakly efficient solution for Problem (19.1) if there is no $X_p \subseteq S$ such that $F_i(X_p, A) < F_i(X_p^*, A)$ for all $i \in I$.*

ii) *$X_p^* \subseteq S$ is an efficient or Pareto solution for Problem (19.1) if there is no $X_p \subseteq S$ such that $F_i(X_p, A) \leq F_i(X_p^*, A)$ for all $i \in I$ and it holds that $F_{i_o}(X_p, A) < F_{i_o}(X_p^*, A)$ for some $i_o \in I$.*

iii) *$X_p^* \subseteq S$ is a strictly efficient solution for Problem (19.1) if there is no $X_p^* \neq X_p \subseteq S$ such that $F_i(X_p, A) \leq F_i(X_p^*, A)$ for all $i \in I$.*

iv) *$X_p^* \subseteq S$ is a properly efficient solution for Problem (19.1) if it is an efficient solution and if there is a number $M > 0$ such that for all $i \in I$ and $X_p \subseteq S$ satisfying $F_i(X_p, A) \leq F_i(X_p^*, A)$ there exist $i_o \in I$ such that $F_{i_o}(X_p^*, A) < F_{i_o}(X_p, A)$ and moreover*

$$\frac{F_i(X_p^*, A) - F_i(X_p, A)}{F_{i_o}(X_p, A) - F_{i_o}(X_p^*, A)} \leq M.$$

Specific choices of solutions among the solution sets defined above have been suggested in the literature of Location Analysis. In the following definition we recall two of them that will be used later.

DEFINITION 75 *Let X_p and X_p^* denote sets with cardinality p.*

i) *$X_p^* \subseteq S$ is a lexicographic solution (or lex-optimal) if there exists a permutation π of the set I such that*

$$(F_{\pi(1)}(X_p^*, A), \dots, F_{\pi(|I|)}(X_p^*, A)) \leq_{\text{lex}} (F_{\pi(1)}(X_p, A), \dots, F_{\pi(|I|)}(X_p, A))$$

for all $X_p \subseteq S$, where $|I|$ is the cardinality of the set I and

$$z \leq_{\text{lex}} \bar{z} \quad :\Leftrightarrow \quad z = \bar{z} \text{ or } z_{i0} < \bar{z}_{i0} \text{ for } i_0 := \min\{i \in I \ : \ z_i \neq \bar{z}_i\}.$$

ii) *$X_p^* \subseteq S$ is a max-ordering solution if*

$$\max_{i \in I} F_i(X_p^*, A) \leq \max_{i \in I} F_i(X_p, A) \quad \text{for all } X_p \subseteq S.$$

Our chapter is organized in five sections. After the introduction we present the standard models of location theory and describe their inherent multicriteria nature. The third, fourth and fifth sections are devoted to analyze the main models and results of continuous, network and discrete multicriteria location analysis, respectively. The chapter ends with the list of references cited in the text.

2. Location Problems

In order to establish a classification of the different problems of Locational Analysis, it is assumed that they can be divided into three branches: continuous, network and discrete location problems. Within each of these branches a further distinction can be made with respect to the number of new facilities, the distances used and pecularities such as forbidden regions. For more advanced classification schemes the reader is referred to [73].

An important characteristic of location models is their intrinsic multicriteria behavior. In any location problem with attractive criteria, every user wants to have the service as close as possible. Therefore, this behavior gives rise to a trade-off among users that leads to a multicritera formulation:

$$v-\min_{X_p\subseteq X,\,|X_p|=p} \Big(d(a,X_p)\Big)_{a\in A}, \tag{19.2}$$

being X the decision space, X_p the finite set of service facilities, $|X_p|$ its cardinality, $A \subset X$ the set of demand facilities, $\bar{d} : X \times X \longrightarrow \mathbb{R}_+$ the function used to measure the distances and

$$d(a,X_p) = \min_{x\in X_p} \bar{d}(a,x).$$

The formulation in (19.2) represents a new general trend in Operations Research. Considering more than one objective reflects better the actual world where usually several objectives, some of them in conflict, must be considered to model a problem. The reader can find excellent arguments justifying the multicriteria character of Locational Analysis and a detailed presentation of several aspects in [39].

In location theory many criteria have been used to locate one or several new facilities. However, median and center problems have attracted special attention of researchers for many years. The median problem has received different names in these years, as for instance, Fermat-Weber, Weber, Steiner or minisum problem, among others (see for instance [154] or Chapter 1 in [44]). This model uses as criterion to locate a new service the minimization of the average distances to all the users. The formulation of this problem, with the notation of Problem (19.2), is given by

$$\min_{X_p\subseteq X,\,|X_p|=p} \sum_{a\in A} w_a d(a,X_p), \tag{19.3}$$

where w_a is the weight associated to a. The median objective function is used in real world situations to locate a new facility minimizing the transportation costs. In a practical setting, the demand facilities represent customers or demands, the new facilities denote the unknown location of the servers, and the weighted distances are cost components associated with the interactions of flows between each new facility and its customers. For this model, we can find many applications cited in the literature involving communication network design, distribution centers, location and routing of robots, or the optimal location of utility and manufacturing plants, among others.

The center problem, which is also called the minimax problem or the problem of minimizing the eccentricity, uses as criterion to locate the new facilities the minimization of the largest distance supported by the users (see for instance [55]). Therefore, with the conventions of Problem (19.2), the center problem can be stated as

$$\min_{X_p \subseteq X,\, |X_p|=p} \max_{a \in A} u_a d(a, X_p), \tag{19.4}$$

where u_a is the weight associated to a. The minimax models may correspond to the social oriented notion of justice.

The median and center are the most frequently used criteria to locate new facilities. However, many real-world situations cannot be exactly modelled with one of these criteria. Indeed, since the median approach is based on averaging, it often provides solutions in which remote low-population density areas are discriminated in terms of accessibility. In the same sense, the center approach provides solutions where there may exist high population density areas with central locations, which have not been taken into account when locating the new facilities. However, to locate, for instance, a fire station, one goal may be to locate the station as close as possible to the farthest potential customer, while another goal would be to locate the station as close as possible to a majority of customers. Therefore, a possible approach to study this kind of situations is the cent-dian problem problem which consists of minimizing the convex combination of median and center objective functions, i.e.,

$$\min_{X_p \subseteq X,\, |X_p|=p} \lambda \sum_{a \in A} w_a d(a, X_p) + (1-\lambda) \max_{a \in A} u_a d(a, X_p) \tag{19.5}$$

with $\lambda \in [0, 1]$.

Notice that depending on the choice of λ, we are considering criteria more similar to the median objective function or to the center one, i.e., for λ close to 1 we are giving more importance to the averaged distances while for λ close to 0 we are giving more importance to the largest distance. Once more, we find in the cent-dian problem the intrinsic multicriteria nature of location problems. The analysis of the optimal solutions for varying λ coincides with the trade-off

Table 19.1. Summary of references.

	Continuous	*Networks*	*Discrete*
Surveys & Textbooks	[15, 39, 42, 44, 50, 51, 63, 64, 68, 93, 123, 154, 156]		
Point-Objective	[2, 26, 27, 28] [34, 37, 41, 43] [47, 48, 49, 61] [62, 79, 84, 86] [95, 101, 117] [118, 121, 122] [132, 140, 149] [150, 151, 152] [153, 155] *Constrained case:* [20, 22, 29] [105, 130, 131] *Majority rules:* [6, 19, 23, 24, 46]	[80]	[32]
Bicriteria Median & Center	[3, 21, 65] [81, 115, 129]	[69, 70, 71] [77, 78, 110] [119, 120, 136] *Extensive facility:* [4, 92, 100, 137]	[18]
Bicriteria Other	[10, 11, 88, 107]	[38, 60, 83] [87, 128, 139]	[104, 135]
Semiobnoxious	[14, 16, 25, 30] [31, 116, 134]	[75, 134]	[57]
Multicriteria Problems	[17, 33, 36, 35] [52, 59, 76, 106] [108, 109, 124] [125, 126] *Best aproximation:* [45, 54, 82, 138] [142, 143, 144] [145, 146, 147] [148] *Equity measurement:* [99, 113]	[74, 94, 127]	[13, 58, 90, 91] [111, 112, 114] [141, 157, 158] *Applications:* [1, 5, 7, 12] [53, 66, 67] [85, 89, 133]

analysis between the minisum and minimax solutions: the multicriteria analysis of the problem.

Although, many other different models have been considered in the literature, we have only described some of them because our purpose is to provide the reader a general overview that illustrates the use of different criteria to locate new facilities. For readers interested in location software a possibility is the public domain software LoLA (Library of Location Algorithms) [72]. LoLA contains several algorithms for multicriteria location problems in the plane and on networks.

We have summarized in Table 19.1 the references reviewed in this chapter.

3. Continuous Multicriteria Location Problems

In this section we give an annotated exposition of the literature on continuous multicriteria location problems. Before dealing with the references of this area, we recall the concept of a gauge which is a general function used to measure distances in continuous models. A gauge is a function defined with respect to a compact, convex set B containing the origin in its interior as:

$$\gamma_B(x) := \inf\{\, r > 0 \,:\, x \in rB \,\}.$$

For instance, when B is the unit disk (ball) centered at 0, we have that $\gamma_B(\cdot) = \|\cdot\|_2$ (the Euclidean norm) or when B is a square of side two and centered at 0, we have that $\gamma_B(\cdot) = \|\cdot\|_\infty$ (the Tchebychev norm). We say that γ_B is: 1) a polyhedral or block gauge if B is a polytope, 2) a strictly convex gauge if B is a strictly convex set, 3) a norm if B is symmetric with respect to 0 and 4) a round norm if B is in addition a smooth set. Moreover, we denote by $co(A)$ to the convex hull of the set A, by $\bar{A}$ its topological closure and by $\mathrm{ri}(A)$ its relative interior.

The models analyzed in this section are organized in three subsections. The first one is devoted to study the point-objective location problem. The second subsection analyzes continuous bicriteria location problems and the last one considers multicriteria problems with more than two objective functions.

3.1 Point-objective Location Problems

The problem of locating one or several facilities to serve a certain number of demand facilities depends strongly on the criteria used to place such services. In order to obtain a general approach to this problem independently of the criterion, and having in mind that each demand facility wants to have the service as close as possible, the location problem is stated as follows:

$$\underset{X_p \subseteq S \subseteq X,\, |X_p| = p}{v-\min} \left(\min_{x \in X_p} \gamma_a(x - a) \right)_{a \in A} \tag{19.6}$$

where S is the feasible region, A is the set of demand facilities and $\gamma_a(\cdot)$ is the gauge associated to the demand facility a.

In location theory, Problem (19.6) is called Point-Objective location problem. This problem may be considered the first multicriteria model in location theory. The demand facilities may be communities that have to be served by some other facilities (fire houses, schools, hospitals, etc.) which have to be as close as possible. The distance to each demand facility a is measured by its corresponding gauge $\gamma_a(\cdot)$.

Because of the multiple objective nature of this problem, we are interested in the solution sets introduced in Definition 74. The final location is usually chosen from these sets in conjunction with other non-quantifiable criteria that the decision maker may have.

In this case, the different sets of efficient solutions of Definition 74 correspond to different level of exigency regarding the proximity to each demand facility. For Problem (19.6), the weakly efficient, efficient, strictly efficient and properly efficient sets are denoted by $WE(A,S)$, $E(A,S)$, $SE(A,S)$ and $PE(A,S)$ respectively. In the unconstrained case, i.e., $S = X$, these sets are denoted by $WE(A)$, $E(A)$, $SE(A)$ and $PE(A)$ respectively.

It is worth noting that the different solution sets, in addition of being considered as solution of the point-objective location problem, can be regarded as a global sensitivity analysis onto the weights of the solution set of the median problems with the same demand set. Hence, the first references that we overview do no state properly the formulation of the point-objective problem but the parametric analysis of weighted minisum problems. This fact is due to the scalarization results that establishes the relationship between the solution sets of a multicriteria problem and the set of minimizers of the weighted sum of their corresponding functions. In particular, if we denote by $M_\lambda(A)$, $\lambda = (\lambda_1, \dots, \lambda_{|A|}) \in \mathbb{R}^{|A|}$, the set of minimizers of the function $\sum_{a\in A} \lambda_a \gamma_a(x-a)$, we have (see [65] for the second statement)

$$x^* \in WE(A) \quad \text{if and only if} \quad \exists\, \lambda \in \mathbb{R}_+^{|A|} \text{ such that } x^* \in M_\lambda(A)$$
$$x^* \in PE(A) \quad \text{if and only if} \quad \exists\, \lambda \in \text{int}(\mathbb{R}_+^{|A|}) \text{ such that } x^* \in M_\lambda(A).$$

In two dimensional space [49, 151] prove that $co(A) \cap M_\lambda(A) \neq \emptyset$, which implies that there exists at least one weakly efficient solution in the convex hull of A ($WE(A) \cap co(A) \neq \emptyset$). If a single block norm is used, [140] obtains that $E(A) \cap co(A) \cap IP \cap M_\lambda(A) \neq \emptyset$, where IP is the set of intersection points defined by the fundamental directions of the unit ball associated to the block norm starting at each demand point. In the case of mixed l_p−norms (different norms associated to each demand point), [79] obtains that the octogonal hull of the demand points has nonempty intersection with $M_\lambda(A)$. Later, [122] shows that this result fails for general norms as soon as the dimension of the space is at least three. [121] obtains that $E(A) = co(A)$ for: 1) any norm on the line, 2) any round norm on the plane and 3) any norm derived from inner product in spaces with finite dimension greater than two. [43] obtains that the smallest set which includes at least one point of $M_\lambda(A)$ is $\text{int}(\text{co}(A)) \cup A$ for l_p-norms with $1 < p < \infty$ and for $p = 1$ or ∞, this set is $SE(A) \cap IP$.

Compared with location problems in the plane, location problems on the sphere have received little attention in the literature. However, to model situations where the distances between the demand facilities and their corresponding servers are too long, it is necessary to take into account the spherical surface of the Earth. [2] extends the results of [151] to location problems on the surface of a sphere. In fact, they show that we can restrict ourselves to the spherically convex hull of the demand points to search a solution of the single facility median problem on the sphere if the demand points are not located entirely on a great circle arc. In addition, [41] obtains that if the demand points are located on a great circle arc, then the optimal solution is in this arc and some demand point is optimal.

For the multifacility case in a two dimensional space (with interaction), whatever the norm is, it holds that the optimal locations for all the new facilities can be found in $WE(A)$, [101], and they belong to the convex hull of the existing facilities when l_p-norms ($1 < p < +\infty$) are used, [62, 86].

In addition to the weighted sum approach, an alternative procedure to deal with the point-objective location problem and, in general, with a multicriteria problem is the $\varepsilon-$constrained method. Probably, this is among the most well-known techniques to solve multicriteria problems and it consists of the minimization of one of the original objective functions while the others are transformed to constraints, representing security or satisfaction levels that must be fulfilled by these criteria. For Problem (19.1), [84] studies properties of the optimal solutions for this kind of constrained problems involving generalizations of the median objective function.

Concerning the relationship between the different solution sets previously defined, we have that in general it holds that $SE(A,S) \subseteq E(A,S) \subseteq WE(A,S)$. In what follows we analyze these relationships for the unconstrained case and latter we will study the constrained one. In the case of a single gauge, that is, $\gamma_a = \gamma\ \forall a \in A$, it holds that $WE(A) = E(A) = SE(A) = \overline{co}(A)$ when γ is: 1) a round norm, [140]; 2) generated by a scalar product, [48]; or 3) strictly convex norm in a two dimensional real space with A being a bounded set, [48]. Besides, when γ is a strictly convex norm in a general real space, [48] proves that $WE(A) = E(A) = SE(A) = SE(\bar{A})$. In addition, when A is finite in $\mathbb{R}^n$, [95] proves that $WE(A) = \cup_{B \subseteq A} E(B)$. In the case of the Euclidean norm, [155] obtains that $E(A) = E(co(A)) = co(A)$ and for the l_1-norm that $E(A) = E(co(A))$.

Problem (19.6) also has some limit properties under particular hypotheses. In particular, when $X = \mathbb{R}^2$ and $\gamma_a = \gamma_n \; \forall a$ where $\{\gamma_n\}_{n \in \mathbb{N}}$ is a sequence of block norms approaching a round norm, we have that $WE_n(A)$, $E_n(A)$ and $SE_n(A)$ (the corresponding sets using γ_n) approach the convex hull of the demand facilities, [140].

Concerning topological properties, if A is bounded then $WE(A)$, $E(A)$ and $SE(A)$ are bounded. Moreover, $WE(A) = WE(\bar{A})$ and $E(A) = E(\bar{A})$ but not necessarily $SE(A) = SE(\bar{A})$. If A is compact then $SE(A)$ is closed. It holds that $WE(A)$, $SE(A)$ and $E(A)$ are weakly compact when X is an infinite dimensional, reflexive and strictly convex normed space with A being a compact set, [48].

Concerning geometrical characterizations, [48] gives a description of WE (A), $E(A)$ and $SE(A)$ for A being compact, using recession cones in any arbitrary normed space. If we impose further A to be finite, in the rectilinear case the set of efficient solutions is a region enclosed by a boundary defined by horizontal and vertical lines throug each demand point, [37, 152, 153]. In $\mathbb{R}^n$ and mixed gauges (different gauges associated to each demand point) the set $WE(A)$ is characterized as the region enclosed by $\cup_{B \subseteq A, |B| \leq n} WE(B)$, [149], a similar result for the planar case is obtained in [132]. In finite dimension and mixed l_p-norms, [26, 27] show that the efficient set is a subset of the octogonal hull defined by the demand points (this result was also proved by [79] using a different methodology). In addition, they prove that under certain conditions these sets coincide. [28] obtains a similar result to those above for problems with polyhedral mixed norms. In fact, they propose a procedure to obtain a set containing the efficient solution set with certain properties of minimality, called pseudoefficient set.

Apart from the unconstrained case ($S = X$), there are also some results for the constrained models. The following references correspond to those models where the location decisions are restricted to a given set S. We assume in the following unless stated otherwise that S is convex and closed. This model is usually called constrained point-objective location problem. In this case, it holds that $WE(A, S) = E(A, S) = \text{proj}_S(\overline{co}(A))$, being A compact and $\text{proj}_S(\cdot)$ the projection operator using $\gamma(\cdot)$ onto S, whenever γ is: 1) strictly convex in a two dimensional space, [29]; 2) the Euclidean norm in $\mathbb{R}^n$ and A finite, [20]; or 3) generated by a scalar product in a two dimensional space, [105]. We can also find geometrical characterizations of constrained solution sets, using recession cones, in [105]; and a theoretical characterization of $WE(A, S)$ using the convex hull of subdifferentials in [29]. In addition, $PE(A, S) = A \cap S \cap \text{proj}_S(\text{ri}(co(A)))$ when a Euclidean norm in $\mathbb{R}^n$ is used and A is finite, [20]. The case where S is not necessarily convex but can be decomposed into a finite number of polyhedra was studied by [22]. On the plane with mixed gauges (different gauges associated to each demand point) one can find a complete geometrical description of the weakly efficient and efficient solution sets in [130, 131].

In addition to the theoretical results already presented there also exist several algorithms to compute some of these sets. In general the problem is very difficult and in many cases of enumerative nature. When total polyhedrality is given through block norms and linear constraints, the problem reduces to a multicriteria linear problem. Notice that even in this very easy case the general problem is already NP-hard. However, there are some particular cases where efficient algorithms exist. The set of efficient solutions using l_1-norm in two dimensional spaces was obtained by [152] with an algorithm based on generating the boundary of the set of efficient solutions. [34] presents a simple row algorithm based entirely on a geometrical analysis, that constructs all efficient solutions with complexity $O(|A| \log |A|)$. They also prove that no alternative algorithm can be of a lower order. [150] considers a simple schematic algorithm for characterizing the efficient solution set for the one-infinity norm. [117, 118] propose a polynomial algorithm for the case of polyhedral norms in the plane. Finally, for the case of polyhedral norms in $\mathbb{R}^n$, [47] presents a general method for determining, in a finite number of steps, the set of all efficient solutions. Besides, he states a geometrical characterization of properly efficient points which later is proved by [61] that only works on dimension one and two.

A different line of research is concerned with the use of majority rules in Locational Analysis. The relationship between Simpson decisions (those preferred by a majority of voters) and Pareto solutions is well-known among the researchers in voting theory. The application of these concepts to Locational Analysis was first given by [6, 46] for problems without locational constraints and later extended by [23, 24] to the constrained case. It is worth noting that this line of research offers interesting open problems, some of them already solved in [19].

3.2 Bicriteria Problems

For many practical situations it is sufficient to deal with two criteria. This allows to obtain a better knowledge of different solution sets and their properties. Most of the references dealing with this type of problems consider the median and the center or some modification of them as objective functions. Using the notation of Problem (19.2), this type of problems can be formulated as

$$\underset{X_p \subseteq S \subseteq X, \, |X_p|=p}{v-\min} \left(\sum_{a \in A} w_a d(a, X_p), \max_{a \in A} u_a d(a, X_p) \right), \tag{19.7}$$

where w_a and u_a are the weights associated with the demand facility a by the median and center criteria, respectively. Therefore, the first part of this subsection is devoted to analyze this kind of problems and the last part considers other bicriteria location problems.

In order to study Problem (19.7), we first notice that these two functions are convex, so by [65] the properly efficient solution set coincides with the set of minimizers of the convex combination of these two criteria, that is, the cent-dian problem, see (19.5). [21] proposes an axiomatic characterization of this criterion, which leads to an interpretation of the parameter $\lambda \in \mathbb{R}$ as a marginal rate of compensation.

The first reference we found that considers the bicriteria problem with median and center objectives is [81]. In this paper, the access cost of users is defined by a non-decreasing, continuous function of the distances which are measured by l_p-norms. For determining the set of efficient points, it provides a simple and practical approach based on the Big Square-Small Square method. Later, [3] shows that all the efficient solutions for Problem (19.7) can be obtained by solving constrained problems. These problems consist of minimizing the weighted sum of the distances so that the minimax function satisfies a varying upper bound. This result has the advantage that solving a constrained median problem is simpler than solving directly a cent-dian problem. In the plane, [115] studies the unweighted case with squared Euclidean distances and proposes a polynomial time algorithm to find the set of Pareto optimal locations based on the use of the farthest point Voronoi diagram.

The bicriteria problem with median and center objective functions in the presence of forbidden regions was considered in [129]. They use a direct search procedure based on Hooke and Jeeves algorithm to solve the rectangular norm planar location problem with forbidden regions, which is interesting because of its simplicity and versatility. A bicriteria location problem with a line barrier is considered in [88]. Their solution approach is based on [74].

The multifacility planar case (with interaction), using l_1-norm, is studied by [10]. They present a fuzzy goal programming model for locating new facilities in a region bounded by a convex polygon. Later, [11] proposes an interactive method to solve the problem above. In order to obtain a satisfactory solution for the Decision Maker (D-M), this procedure requires the D-M to know how much he/she can concede from the most satisfactory fuzzy goals at each current solution to improve the degree of satisfaction of other objectives.

Now that we have analyzed the bicriteria problems with median and center objective functions, we will start with the second part of this section where we study bicriteria problems where at least one the objective functions is none of them. [107] considers the bicriteria 2-Facility median problem using l_1-norm with interaction in $\mathbb{R}^d$ and gives a polynomial algorithm for determining all efficient locations. This algorithm is based on a discretization of the original continuous problem using geometrical and combinatorial arguments.

In a variety of practical settings the new facility to be located cannot be classified as being either purely desirable or obnoxious. These facilities falling somewhere between these two extremes are called semidesirable. As an illustrative example, consider the problem of locating a new chemical factory. For public safety, air pollution, and other reasons, such a facility should not be situated too close to populations centers. On the other hand, there are decreasing marginal benefits from locating the factory further away, because transportation cost for the users is steadily increasing. [25] presents a critical overview of the mathematical methods commonly used in semi-obnoxious facility location.

A common method to solve this kind of problems is to consider a bicriteria problem where each one of these objectives represents an attractive and a repulsive criterion respectively. [16] considers a bicriteria semidesirable location problem where the objective functions are the median criterion and the minimization of the weighted sum of Euclidean distances raised to a negative power. To solve the problem, they develop a heuristic method based on the computation of a trajectory determined by combining the first order necessary condition with the truncated Taylor series of the convex combination of these two criteria. Notice that this trajectory may not represent the complete set of efficient solutions. [30] considers a semi-obnoxious location problem where the objectives are the transportation and environmental costs. Since the usual solution set has, in general, infinite cardinality, they propose as solution a finite feasible set representing the best compromise solutions using the concept of α-dominance. Other applications of global optimization techniques to semi-obnoxious bicriteria location problems can be found in [31, 14]. On the other hand, [116] considers a semidesirable location problem using a bicriteria model with the center and anti-center (minimax) objective functions. He presents a geometrical characterization of the efficient set as well as the trade-off curve and develops a polynomial time algorithm for finding them. Finally, [134] considers planar bicriteria semi-obnoxious location problems where the importance of the obnoxious criterion with respect to the cost objective is not determined in advance.

3.3 Multicriteria Problems

As it was announced, the third subsection is devoted to the general case of multicriteria location problems where more than two objective functions are considered.

We start by mentioning the paper by [17]. It presents an axiomatic foundation of objective functions employed in multicriteria location theory that allows to characterize single objective reductions of these multicriteria problems. This procedure also simplifies the search of the Decision-Maker for suitable objective functions on the basis of desirable properties.

The first references that we consider in this section deal with the multicriteria median problem, that is, a multicriteria location problem where each of the involved objective functions is of the median type. This problem can be considered as the first actual multicriteria problem (more than two criteria not being distance functions) in continuous location. The formulation of this problem is as follows:

$$v - \min_{x \in X} \left(\sum_{a \in A} w_a^i \gamma(x-a) \right)_{i \in I},$$

where w_a^i is the weight associated to the demand facility a by the ith criterion. [33] studies this problem when the l_1-norm is used and develops a graphic-type algorithm that generates the set of all efficient solutions. [76] extends the analysis of this problem for the case of l_p-norms. In [106] the multicriteria median problem with polyhedral gauges is investigated. In addition, both papers also deal with the multicriteria center problem, that is, all the involved objective functions are of the center type. This problem can be formulated as

$$v - \min_{x \in X} \left(\max_{a \in A} u_a^i \gamma(x-a) \right)_{i \in I},$$

where u_a^i is the weight associated to the demand facility a by the ith criterion. For these two problems [76] analyzes the set of lexicographic locations, the set of Pareto locations and the set of max-ordering locations. A relationship between these three sets is established; and they develop efficient algorithms to compute the lexicographic location set for these two kinds of problems. Moreover, using the convex hull of their optimal solutions, they give a geometrical description of the set of efficient and properly efficient solutions for the case of median objectives with squared Euclidean norm. Finally, they develop an algorithm, based on a combinatorial approach, to compute efficient solution sets for the multicriteria median problem with l_1- and $l_\infty-$norm.

The multicriteria median problem with a general norm is studied in [125] which introduces the null vector condition for characterizing the set of properly efficient solutions. This condition is based on the computation of the cone generated by the subdifferentials of the functions considered in the multicriteria problem. They also analyze the relationship between the set of properly efficient solutions of this problem and the set of properly efficient solutions of the point-objective location problem defined by the demand points of the considered median objective functions. In the polyhedral case, they develop an algorithm to compute the set of efficient solutions with polynomial complexity. For the case of only one strict norm and assuming that the demand points are not collinear, [124] proves that the set of efficient solutions can be obtained as the limit of the set of weakly efficient solutions with a polyhedral gauge converging to the original strict norm.

A mixed version of the multicriteria median and center problem is analyzed by [52] which considers a multicriteria problem where all the objectives are either median or center ones. In particular, they characterize the set of max ordering locations using the lexicographic and Pareto location sets. Three different strategies are proposed to find efficiently this set based on: 1) a direct approach, 2) the decision space approach, and 3) the objective space approach. Finally, they introduce the lexicographic max-ordering locations as a further specialization of max-ordering locations, which can be found efficiently.

One of the most general approaches to locate new facilities is the so called ordered median problem (see [108, 126]). Indeed, this criterion includes as particular instances the median, the center and the cent-dian problems among others. The multicriteria version of this problem with polyhedral gauges is studied in [109]. In this paper, the authors give geometrical characterizations of the set of efficient solutions and a polynomial time algorithm to compute it.

An alternative multicriteria location problem where neither center nor median objective functions are used, is proposed by [59]. In this paper, the authors consider the multicriteria minmax regret which combines the robustness approach using the minmax regret criterion together with Pareto-optimality. Its formulation is as follows:

$$\min_{x\in\mathbb{R}^2}\max_{w\in W}\sum_{a\in A} w_a\|x-a\|_2^2-\sum_{a\in A} w_a\|x(w)-a\|_2^2,$$

where $W \subseteq \mathbb{R}^{|A|}$ and $x(w)$ is the optimal solution of $\min_{x\in\mathbb{R}^2}\sum_{a\in A} w_a \|x-a\|_2^2$. For the bicriteria case, the set of efficient locations is characterized as a particular set of line segments. Using this result the authors also give an algorithm for the general multicriteria case based on the solutions of bicriteria problems.

The important issue of equity measurement in Locational Analysis has also been modeled as a multiobjective problem. The interested reader can find a good review and a framework for this problem in [99]. A more recent approach as a multiobjective problem is given in [113]. For further details on this subject the reader is referred to [8, 9, 56, 98, 102, 103].

Another fruitful area of research in this field deals with the so called vectorial best approximation location problem. We are given two real linear spaces X, Y and a convex cone $C \subset Y$. We also consider a vectorial norm $|||\cdot|||$ being a mapping from X into C that satisfies for $x, \bar{x} \in X$ and $\lambda \in \mathbb{R}$:

i) $|||x||| = 0_Y$ (the null element in Y) if and only if $x = 0_X$,

ii) $|||\lambda x||| = |\lambda| |||x|||$,

iii) $|||x||| + |||\bar{x}||| - |||x+\bar{x}||| \in C$ ($|||x+\bar{x}||| \leq_C |||x||| + |||\bar{x}|||$).

The vectorial best approximation problem is

$$v - \min_{u \in V} |||x - u|||,$$

where $V \subset X$ is the feasible (constrained) set.

It is clear that most of the problems considered in the above sections fall into this very general formulation. In particular, the reader can check that considering, $X = (\mathbb{R}^n)^{|A|}$, $Y = \mathbb{R}^{|A|}$, $C = \mathbb{R}_+^{|A|}$, $|||z||| = (\|z^a\|_a)_{a \in A}$ being $z = (z^a)_{a \in A}$, $z^a \in \mathbb{R}^n$, $a \in A$; and $V = \{u \in X \,:\, u^a = u^b \quad \forall a, b \in A\}$, we get the so called point-objective problem.

Early references in the literature stating the relationships between multicriteria location problems and this general vectorial best approximation optimization problem can be found in [45, 142, 143, 144, 145].

The interested reader will find very important results scanning this line of research. However, they are scattered in journals that are hardly considered by location analysts (locators).

In [45] several topological properties as well as a geometrical description of the set of vectorial best approximants is given. On the other hand, [144, 145] emphasize more the conditions of the Kolmogorov type that characterize weakly and properly efficient solutions of the vectorial approximation problem.

Another topic pursued by the authors in this field is the use of general duality results characterizing the different notions of efficiency. In [142, 143, 148] the reader can find characterizations using duality under different hypotheses, with their corresponding applications to multicriteria location problems.

We also want to recall the concept of ε-optimality in vectorial approximation location problems. Without entering the details of this concept, we would like to mention at least that powerful results are known. The results are based on a generalization of Ekeland's variational principle (see [54]) for vector approximation problems (see [138]) that has been later applied to get results in approximating efficient solution sets [82, 147, 146]. The interested reader can find all the details in the references above and those cited therein.

We finish this section by mentioning a different multicriteria location problem. The goal is to find efficient designs (shapes) for a given area provided that disutilities for the users are known. [35, 36] study this problem and give necessary and sufficient conditions for a design to be efficient.

4. Multicriteria Network Location Problems

In a general network location problem, one or several facilities are to be placed in a graph optimizing a function of the distances between these facilities and the set of demand facilities located in the graph. Therefore the main difference with respect to the continuous problem is that the decision space is a network. This fact provides many intrinsic peculiarities both in the theoretical and practical point of view. In particular, this kind of models adapt better to some specific real world situations, as for instance road networks, power lines, etc. This justifies that many efforts have been devoted to improve the performance of facility systems to deal with network location problems. These problems can be classified depending on the graph structure (general graph, trees, etc.), the type of objective function (center, median, etc.) or the number of objective functions considered (single criterion or multicriteria problem).

We start introducing some basic notation to understand the formulation of this type of problems. Let $N = (G, l)$ denote a network with underlying graph $G = (V, E)$, where the node set is V (demand points) and the edge set is E.Therefore, we write the edge that joins the nodes v and v' as $[v, v']$.

The *length* of an edge $e \in E$ is denoted by $l(e) = l(v, v')$ and it represents the cost of going once through the edge to satisfy the demand of one user. By $\bar{d}(v, v')$, we denote the length of the shortest path between v and v' measured by l.

A point x on an edge $e = [v, v']$ is determined by a value t, $0 \leq t \leq l(e)$, which represents the length of the proportion of the edge between x and v, the point x is then denoted by $x = p(e, t) = p([v, v'], t)$. Hence, for instance in the case of an undirected graph, the distance from this point to another node v_k is:

$$\bar{d}(x, v'') := \bar{d}(v'', x) := \min\{\bar{d}(v'', v) + t\ ,\ \bar{d}(v'', v') + l(e) - t\}.$$

Notice that the function $d(\cdot, v)$ for any $v \in V$ is concave over each edge of the graph, in fact, it is a concave piecewise linear function. Besides, if the graph is a tree, this function is convex over paths what implies that the sum of the distances from x to each node is a convex function over each path of the tree. This property allows to apply results of convex analysis to the resolution of location problems stated on a tree graph.

The set of all the points of a network (G, l) is denoted by $P(G)$. It should be noted that this set also contains the node set. Therefore, in order to locate p service facilities, we have to consider the distance from a node to a set of p points, $X_p \subseteq P(G)$, as

$$d(v, X_p) = \min_{x \in X_p} \bar{d}(v, x).$$

In order to present the references considering networks multicriteria location problems, analogously to the continuous case, we have divided this section in two subsections. In the first one, we consider the case of two criteria, i.e., bicriteria problems and in the second one, we deal with the general case where more than two objective functions are used.

4.1 Bicriteria Problems

In this subsection we analyze bicriteria location problems on networks. The most popular models are those with median and center objective functions. Similar to the continuous case this problem is

$$\underset{X_p \subseteq P(G),\, |X_p|=p}{v-\min} \left(\sum_{v\in V} w_v d(v, X_p), \max_{v\in V} u_v d(v, X_p) \right), \tag{19.8}$$

where w_v and u_v are the weights associated to the demand facility v by the median and center criteria, respectively. We start with the analysis of these problems.

A first method to handle this bicriteria problem is to transform it into a single objective problem via scalarization. This can be mathematically expressed by the minimization of different single objective functions as: cent-dian, generalized center or medi-center, among others. The cent-dian objective function, already defined in (19.5), was introduced by [69], who coined the term cent-dian for the point of a graph that minimizes the convex combination of the center and median functions. On the second hand, the generalized center objective, introduced and studied by [78], minimizes the difference between the center and the median functions,

$$\min_{X_p \subseteq P(G),\, |X_p|=p} \left\{ \max_{v\in V} u_v d(v, X_p) - \sum_{v\in V} w_v d(v, X_p) \right\}.$$

This criterion allows to deal with distributional justice considerations in the access to the facilities and corresponds to an aggregation procedure of semi-obnoxious location problems (center, anti-median).

Finally, the medi-center problem, considered in [71], minimizes one criterion subject to a restriction on the value of the other:

$$\begin{aligned} \min \quad & \max_{v\in V} u_v d(v, X_p) \\ \text{s.t.} \quad & \sum_{v\in V} w_v d(v, X_p) \le \mu \\ & X_p \subseteq P(G), \;\; |X_p| = p \end{aligned} \tag{19.9}$$

or

$$\begin{aligned} \min \quad & \sum_{v\in V} w_v d(v, X_p) \\ \text{s.t.} \quad & \max_{v\in V} u_v d(v, X_p) \le \mu', \\ & X_p \subseteq P(G), \;\; |X_p| = p \end{aligned} \tag{19.10}$$

where μ and μ' are upper bounds to the median and center objectives respectively.

The first approach that we consider in order to deal with a bicriteria location problem with center and median objective functions, is the parametric analysis of the cent-dian problem. This parametric analysis is very informative in a general network, however it does not provide a complete characterization of the efficient solution set. It is due to the non-convexity of these objective functions (recall that distances are concave). For the particular case oftree networks and due to the convexity properties of this case, this parametric analysis gives the whole set of efficient solutions, similarly to the continuous problems.

In any case, the solution set of the cent-dian problem for any parameter defining the convex combination of the center and median objectives is included in the set of efficient solutions of the bicriteria problem defined by these two criteria. Thus, its characterization continues to be interesting from multicriteria point of view. [69] shows that the cent-dian of a tree has the attractive property of being located either at the center of a tree or at a vertex on the path connecting the center and a median. Unfortunately, a cent-dian of a general graph does not satisfy this property in general. [70] proposes a procedure based on the computation of an upper bound to identify a cent-dian of an undirected graph; and traces its location as it moves from a graph median to its center as the weight of the latter objective is increased and of the former is decreased.

Since a one-to-one correspondence between cent-dian or generalized center solutions and efficient solutions does not exist, [110] analyzes a different solution concept for this bicriteria problem that provides some compromise between them. This is the Tchebychev cent-dian solution which is the set of lexicographic solutions of a bicriteria problem with the cent-dian and the weighted Tchebychev norm of the center and median criteria. This new solution concept allows to identify all Pareto locations on any network by means of a parametric analysis. Besides, he proposes an algorithm to generate the set of Tchebychev cent-dian solutions.

The models above only consider the case of locating one facility, however we can find situations where more than one facility is required. Hence, we analyze the p-faciltiy case, where the goal is to locate p points on the network so that the demand of the given facilities is covered by the closest new facility,

$$\min_{X_p \subseteq P(G),\, |X_p|=p} \left(\lambda \sum_{v \in V} w_v d(v, X_p) + (1-\lambda) \max_{v \in V} u_v d(v, X_p) \right).$$

[119] studies the unweighted p-facility cent-dian network location problem. They give a finite dominating set and also provide a solution method that solves this problem based on an exhaustive search in the set of all combinations of p points within this finite dominating set. For the case, $p = 2$, [120] provides a

different algorithm based on an exhaustive search that solves the problem with complexity $O(|V|^2)$.

[136] considers the weighted p-cent-dian problem on tree networks. The authors identify a set of points of polynomial size which is guaranteed to contain an optimal solution. Then, they exploit some convexity properties to develop a $O(|V| \log |V|)$ time algorithm that solves this problem.

The generalized center problem is introduced in [78], which includes an algorithm to solve this problem. Moreover, this paper provides a complete characterization of the cent-dian problem in the case of a tree. For the case of a general network, the authors present a new algorithm to find the set of cent-dian solutions which is conceptually much simpler than that developed by [70], although with the same computational complexity. This algorithm is based on the computation of the lower envelope of the bottleneck points and local minima of the median and maximum distance objective functions on the image space.

The third approach that we are looking at to study Problem (19.8) considers the medi-center problems, see (19.9) and (19.10). [71] analyzes this bicriteria location problem on general undirected graphs using the cent-dian problem and two medi-center problems. Between these two medi-center problems, a duality relation is stated, where solving one problem is equivalent to solving the other one when the upper bounds defining the constraints correspond to each other in a definite way. [71] presents a procedure for the identification of all efficient solutions based on solving only one of the two constrained problems. Finally, the author shows that the cent-dian problem is in some sense a special case of these medi-center problems since its solutions correspond to the extreme points of the solution set of a medi-center problem when the upper bounds in the constraint vary. [77] considers a medi-center problem, which minimizes the average travel time subject to the constraint that no individual response will be more than a determined number of time units long. Efficient algorithms are developed for locating a single facility on a tree. This efficiency is again due to the fundamental convexity characteristic for the distance measures on trees.

Most of the models dealing with network location problems use points to represent the facilities to be located. However, there are circumstances where these facilities cannot be modelled by points on a network, as for instance the problem of locating railroad lines, highways, transit routes, pipelines, etc. In order to solve these situations, some models have been developed where the goal is to locate an extensive facility (see [100] for a survey of this type of problems). In particular, one can find some papers in the literature considering multicriteria problems with path or tree shaped facilities.

Locating a path using the cent-dian criterion can be formulated as:

$$\begin{array}{ll} \min & \lambda \sum_{a \in A} w_a d(a,P) + (1-\lambda) \max_{a \in A} u_a d(a,P) \\ \text{s.t.} & d(v,P) = \min_{x \in P} d(v,x) \\ & l(P) \leq L \\ & P \subseteq P(G) \quad \text{being a path.} \end{array}$$

where $l(P)$ is the length of the path P and L its upper bound. The case of a tree shaped facility can be formulated in a similar way by considering this shape instead of a path. For the case of a path, [92] gives a complete characterization of the cent-dian function for tree graphs. To solve this problem, [4] proposes an efficient algorithm based on dynamic programming. For the case of a subtree, [137] presents an algorithm to find an optimal solution based on two facts: 1) that the point solution for the cent-dian problem belongs to an optimal subtree; and 2) the characterization of a finite set of breakpoints of the considered objective function. Its overall complexity is $O(|V| \log |V|)$.

After the analysis of the references considering bicriteria location problems with median and center objective functions we will study other bicriteria models in the second part of this section.

[139] considers a biobjective multifacilily minimax location problem on a tree network, which involves as objectives the maximum of the weighted distances between specified pairs of new and existing facilitics, and the maximum of the weighted distances between specified pairs of new facilities. They develop an algorithm for constructing the efficient frontier and also provide a general result which gives necessary and sufficient conditions for a location vector to be efficient.

The problem of determining the absolute center of a network with two objective functions was studied by [128]. The authors consider a bicriteria problem where the objective functions are two center criteria using independent lengths on each edge. The problem is solved by a polynomial time algorithm based on [87].

The minimization of the superior section in a graph consists of finding the path, such that, the edge with the longest length is minimum. Applications can be found for instance in transportation of hazardous materials, where the weight associated to each edge is the risk of accident on that edge. [60] considers the bicriteria location problem of locating a path on a tree with respect to the minimization of the eccentricity or farthest distance and the superior section. They propose an algorithm that obtains all the efficient paths with complexity $O(|V|^3)$ based on two results: 1) on paths, the superior section function is a maximum function over the edge lengths and it may use a progressive reduction of the original tree and 2) there exist linear time algorithms to find path centers on trees. Moreover, they propose modifications of this algorithm that can be applied to a variety of bicriteria path problems on trees where one of the objective functions is the superior section.

The balance criterion is an equity objective function defined as the difference of the distance from the service facility to the farthest and to the nearest demand point. This model is induced by the situation when e.g. according to a designed schedule or by some equity reason, the distances to the facility are to be as balanced as possible. [83] studies a bicriteria location problem with the center and balance criterion. The set of efficient solutions for this problem is generated by minimizing a constrained problem, namely the center objective function subject to an upper bound on the balance objective.

At the beginning of this section we have considered the cent-dian objective function as an approach to deal with bicriteria location problems with center and median objectives. However, [38] considers two cent-dian objective functions in a bicriteria location problem on a network, where one function minimizes the distance and the other one minimizes the cost. The efficient solutions of this problem are derived by a polynomial algorithm based on computational geometry.

4.2 Multicriteria Problems

Considering more than two objective functions implies that several methods very useful in bicriteria problems, as for instance those based on projections onto the image space in order to find the efficient solution set, are useless. Hence, different techniques are needed to deal with these problems.

The point-objective location problem in networks is considered in [80]. The authors give a polynomial time algorithm for the set of efficient points on a general network and a linear time algorithm for the problem on trees.

In the case of tree networks, [94] considers a more general multicriteria location problem where each objective function is a continuous convex function constrained to a compact set. He characterizes the set of efficient solutions as a subtree delimited by the optimal solutions of each criteria. Besides, he provides a procedure for determining such a set. Finally, extensions to the case of non-convex feasible regions are analyzed.

The single facility multicriteria median problem on networks can be formulated as follows:

$$v-\min_{x\in P(G)}\left(\sum_{v\in V} w_v^i \bar{d}(v,x)\right)_{i\in I},$$

where w_v^i is the weight associated to the demand facility v by the ith criterion. Due to the non-convexity of this problem searching the efficient solutions is not restricted to a specific part of the network but rather it should be extended to all its edges, [127]. They develop a polynomial time algorithm to determine the efficient solution set. The procedure, first, determines the distance function for each objective with their corresponding breakpoints and then removes edges according to a simple rule.

Also for this problem, [74] develops a polynomial time algorithm to find the lexicographic and efficient solutions. The complexity of the algorithm is considerably improved for the case of tree graphs. The analysis in this case is based heavily on the partition of the objective function into subedges. For the case of lexicographical solutions, it reduces the search over a finite set of vectors. For the Pareto locations case, a procedure based on two stages is developed. In the first stage, the set of efficient points on an edge is obtained, while in the second, these points are tested for global domination.

In multicriteria network location analysis, we can also find models considering the location of a semi-obnoxious facility. [75] presents different models using criteria of median type with positive and negative weights. To solve these problems, they propose efficient algorithms based on the methodology used by [74]. These results are extended to models with maximin and minimax objectives. Recently, also ε-approximated solutions to the semi-obnoxious location problem have been discussed in [134].

5. Multicriteria Discrete Location Problems

Discrete location models consider the problem of determining where to locate one or several facilities within a finite set of given potential places to cover the demand of a region. Therefore, the mathematical formulations of these models mainly rely on (mixed) integer programming. Thus, one of the most classical models in this area, namely the $p-$median can be formulated as:

$$\begin{array}{ll} \min & \sum_{x \in PL \subseteq X} \sum_{a \in A} w_a d(a,x) y_{a,x} \\ \text{s.t.} & \sum_{x \in PL} y_{a,x} = 1 \quad \forall a \in A \\ & z_x - y_{a,x} \geq 0 \quad \forall x \in PL \\ & \sum_{x \in PL} z_x = p \\ & z_x, y_{a,x} \in \{0,1\} \quad \forall a \in A \text{ and } \forall x \in PL, \end{array}$$

where PL represents the finite set of potential locations for the service facilities. It should be mentioned the narrow relationship between discrete and networks location problems, especially when the latter is restricted to the vertex locations, [96, 97]. Multicriteria discrete location problems add to the above models the consideration of several criteria to be optimized simultaneously. The reader can find an introduction to these models in Chapter 8 of [40]. Depending on the different criteria used to locate the new facilities we can find a large variety of models in this field of location theory. This fact makes a difference between continuous or network multicriteria location problems and multicriteria discrete ones. In the former, most of the papers deal with specific problems (median, center, cent-dian,...) and focus on theoretical results. In the latter, the effort is put more on applications than on methodological results. As a consequence our presentation in this section does not classify the papers by areas. To review this material, we mainly follow a chronological scheme combined with a description of the most used techniques.

A possible approach to solve this type of problems are the interactive algorithms. These procedures help the D-M to explore and analyze his/her preferences in conjunction with an exploration of the set of feasible solutions. In other words, it combines what is desirable with a consideration of what is possible. [133] elaborates a specific interactive algorithm for a multicriteria location problem involving public facilities. Besides they give arguments showing that practical problems involving the location of public facilities are really multicriteria problems. [114] considers an interactive procedure which is an extension of the classical reference point approach to solve various multicriteria transshipment problems with facility location. In this new approach, the decision maker forms his/her requirements in terms of aspiration and reservation levels, i.e., he/she specifies acceptable and required values for the given objectives. [112] develops an interactive process that generates the solutions belonging to the symmetrically efficient set which is applied to discrete location problems. Notice that symmetric efficiency is a new solution concept based on the principle of impartiality, i.e., on the assumption that any permutation of the achievement vector is equally good as the original achievement vector.

A second approach to deal with multicriteria discrete location problems is goal programming. It is a very valuable tool, since it gives the D-M the opportunity to include many aspects of problems that usually are not included by other methodologies (e.g. quality of life, compliance with states laws, etc.). In what follows, we present four references that have used this procedure to solve multicriteria discrete location problems. [7] considers the location and size of day nurseries within a town by means of a multicriteria discrete model and its solution consists of finding a compromise among three conflicting objectives which represent educational needs, accessibility and budget considerations. [90] applies a branch and bound integer goal programming approach to a multicriteria location-allocation problem. [5] describes a model for evaluating and determining locations of fire stations. The model considers multiple objectives that incorporate both travel times and travel distances from stations to demand sites. [66] considers the problem of locating disposal or treatment centres and routing hazardous wastes through an underlying transportation network. The considered objectives are: minimization of total operating cost, minimization of total perceived risk, minimization of maximum individual risk and minimization of maximum individual disutility. In order to solve the problem, the author shows how monotonically increasing penalty functions can be used to obtain more satisfactory solutions. Location of waste disposals of several materials have also been addressed using other multicriteria techniques as in [1, 67, 89].

The third resolution method that we analyze are the enumeration procedures. [13] develops an implicit enumeration algorithm to determine the set of efficient points in zero-one multiple criteria problems. The algorithm is specialized on the solution of a particular class of facility location problems. The procedure is complemented with the use of the utility function of the decision maker to identify a subset of efficient candidates for the final selection. [57] applies this resolution method to a multicriteria model for locating one or more undesirable facilities to service a region. The objectives are to minimize the total cost of the facilities located, the total opposition to the facilities, and the maximun disutility imposed on any individual. Opposition and disutility are assumed to be nonlinearly decreasing functions of distances, and increasing functions of facilities size.

The point-objective location problem has also been considered in the discrete case. [32] studies the discrete version of this model with rectilinear distances and develops an enumerative algorithm that checks efficiency for each one of the candidate sites.

There also exist results that establish the relationship between the efficient solution set of a bicriteria (median-center) problem and the solution set of a single criterion problem resulting from the combination of both objective functions. In particular, the parametric analysis of the cent-dian problem only gives a subset of the efficient solutions of the considered problem. However, a modification of the cent-dian problem allows to obtain a criterion whose parametric analysis provides the whole set of efficient solutions, [18]. In addition, this paper suggests a solution procedure for the cent-dian problem.

[157, 158] employ five newly developed multiple attribute decision making methods for different versions of the manufacturing plant site selection problem. They consider the single plant strategy with qualitative and quantitative data and cover the multiplant strategy with budget constraints and relocation strategies.

An algorithm for generating an approximate representation of the efficient solutions in biobjective problems which are modeled as mixed integer linear programs is developed by [135]. A geometrical measure of the error is given to assure that the deviation of the approximation from the exact solution set is within a maximum allowable error. The author illustrates the algorithm with a biobjective model which seeks to locate p facilities in a set of potential facility sites to maximize the objectives of single coverage and multiple coverage over a set of demand points.

[91] proposes a facility site selection algorithm. Since in facility site selection it is common to find imprecise assessments of alternatives versus criteria as well as weighting factors, the conventional quantitative approaches may not be applicable. The paper suggests the application of the hierarchical structure analysis to aggregate the decision maker's linguistic assessments about weighting factors

and the suitability of facility sites. This procedure allows the decision-makers to obtain the final ranking of the alternatives automatically.

A general approach to consider multicriteria problems is to apply weights to the criteria to obtain overall scores for the purposes of simplifying the comparison. DEA (Data Envelopment Analysis) is an interesting and non-subjective method for obtaining weights. [141] presents an application of DEA called "profiling" in order to assist in the choice of a location for a particular facility when various criteria are considered. This application provides much greater discrimination than conventional DEA which greatly eases the site selection process.

The classical uncapacitated facility location has also been analyzed from a multicriteria point of view. In particular, [104] considers a biobjective model for this problem where one objective is to maximize the net profit and the other to maximize the profitability of the investment. To solve the problem, they develop a heuristic procedure to generate the efficient solutions which has computational advantages over existing methods. On the other hand, [58] presents the multicriteria version of this problem (where each objective represents a different scenario) and develops two approaches to obtain the set of efficient solutions based on the decomposition of the problem into two nested subproblems and the use of multicriteria dynamic programming.

[111] develops the concept of the lexicographic minimax solution (lexicographic center) being a refinement of the standard minimax approach to location problems. It is shown that the lexicographic minimax approach complies with both the Pareto-optimality (efficient) principle (crucial in multiple criteria optimization) and the principle of transfers (essential for equity measures) whereas the standard approach may violate both these principles. Computational algorithms are developed for the lexicographic minimax solution of discrete location problems.

An application of multicriteria discrete location analysis consists of locating regional service offices in the expanded operating territories of a large property and liability insurer. These offices serve as first line administrative centers for sales support and claims processing. For solving this real situation, [12] proposes a zero-one linear multicriteria programming formulation where the criteria and constraints of the model reflect investment and operating cost, budget considerations and a measure of the service level provided. The reader can also find another application of multiobjective integer programming to spatial decision for housing mobility planning in [85]. In addition, [53] gives an analysis of a part of the distribution system of the company BASF AG, which involves the construction of warehouses at various locations. The authors evaluate 14 different scenarios and each one of these scenarios is evaluated with the minimal cost solution obtained through linear programming and the resulting average delivery time at this particular solution. It is illustrated that a bicriteria analysis is certainly superior to a decision based on the cost or the service criterion alone.

6. Conclusions

We have shown in this chapter that location problems are multicriteria by their own nature. Location decisions are typically group decisions and different quality criteria have to be taken into account. The three main areas of location problems have been reviewed: continuous, network and discrete location problems. When looking at the references discussed, one can easily see that still many interesting open problems remain. In the continuous as well as the network cases multifacility problems are not adequately treated yet. Also the development of efficient algorithms is still in an early stage.

Moreover, the location of new facilities conditioned to the existence of other facilities that have already been located (conditional problems) have attracted the attention of researchers in Locational Analysis. Thus, this kind of problems opens a future avenue of research in the multicriteria case. Although some references have dealt with nonconvex problems, they only consider particular situations. The study of general models is another open line of research. Nonconvexities in the objective function may be modelled by the ordered median function that has been proven to be very useful in different problems of Locational Analysis.

For discrete problems a more systematic treatment of the different problem types is missing. For all three areas there is nearly no software available. Summing up, we can conclude that although an amazing number of publications dealing with multicriteria location problems is around, a lot of work is still waiting for the research community.

Acknowledgments

The research of the second and third authors is partially supported by Spanish research grants BMF2001-2378 and BMF2001-4028.

References

[1] M.F. Abu-Taleb. Application of multicriteria analysis to the design of wastewater treatment in a nationally protected area. *Environmental Engineering and Policy*, 2:37–46, 2000.

[2] A.A. Aly, D.C. Kay, and D.W. Litwhiler. Location dominance on spherical surfaces. *Operations Research*, 27(5):972–981, 1979.

[3] A.A. Aly and B. Rahali. Analysis of a bicriteria location model. *Naval Research Logistics Quarterly*, 37:937–944, 1990.

[4] I. Averbakh and O. Berman. Algorithms for path medi-centers of a tree. *Computers and Operations Research*, 26:1395–1409, 1999.

[5] M.A. Badri, A.K. Mortagy, and C.A. Alsayed. A multiobjective model for locating fire stations. *European Journal of Operational Research*, 110:243–260, 1998.

[6] M. Baronti and E. Casini. Simpson points in normed spaces. *Rivista di Matematica della Universita di Parma*, 5(5):103–107, 1996.

[7] M.A. Benito-Alonso and P. Devaux. Location and size of day nurseries- a multiple goal approach. *European Journal of Operational Research*, 6:195–198, 1981.

[8] O. Berman. Mean-variance location problems. *Transportation Science*, 24(4):287–293, 1990.

[9] O. Berman and E.H. Kaplan. Equity maximizing facility location schemes. *Transportation Science*, 24(2):137–144, 1990.

[10] U. Bhattacharya, J.R. Rao, and R.N. Tiwari. Bi-criteria multi facility location problem in fuzzy environment. *Fuzzy Sets and Systems*, 56:145–153, 1993.

[11] U. Bhattacharya and R. N. Tiwari. Multi-objective multiple facility location problem: A fuzzy interactive approach. *Journal of Fuzzy Mathematics*, 4:483–490, 1996.

[12] G.R. Bitran and K.D. Lawrence. Location service offices: a multicriteria approach. *OMEGA*, 8(2):201–206, 1980.

[13] G.R. Bitran and J.M. Rivera. A combined approach to solve binary multicriteria problems. *Naval Research Logistics Quarterly*, 29(2):181–200, 1982.

[14] R. Blanquero and E. Carrizosa. A d.c. biobjective location model. *Journal of Global Optimization*, 23:139–154, 2002.

[15] M.L. Brandeau and S.S. Chiu. An overview of representative problems in location research. *Management Science*, 35(6):645–674, 1989.

[16] J. Brimberg and H. Juel. A bicriteria model for locating a semi-desirable facility in the plane. *European Journal of Operational Research*, 106:144–151, 1998.

[17] H.U. Buhl. Axiomatic considerations in multi-objective location theory. *European Journal of Operational Research*, 37:363–367, 1988.

[18] R.E. Burkard, J. Krarup, and P.M. Pruzan. Efficiency and optimality in minisum-minimax 0-1 programming problems. *Journal of the Operational Research Society*, 33:137–151, 1982.

[19] C.M. Campos and J.A. Moreno. Relaxation of the condorcet and simpson conditions in voting location. *European Journal of Operational Research*, 145:673–683, 2003.

[20] E. Carrizosa, E. Conde, F.R. Fernández, and J. Puerto. Efficiency in euclidean constrained location problems. *Operations Research Letters*, 14:291–295, 1993.

[21] E. Carrizosa, E. Conde, F.R. Fernández, and J. Puerto. An axiomatic approach to the cent-dian criterion. *Location Science*, 2:165–171, 1994.

[22] E. Carrizosa, E. Conde, M. Muñoz-Márquez, and J. Puerto. Planar point-objective location problems with nonconvex constraints: A geometrical construction. *Journal of Global Optimization*, 6:77–86, 1995.

[23] E. Carrizosa, E. Conde, M. Muñoz-Márquez, and J. Puerto. Simpson points in planar problems with locational constraints. The polyhedral-gauge case. *Mathematics of Operations Research*, 22:291–300, 1997.

[24] E. Carrizosa, E. Conde, M. Muñoz-Márquez, and J. Puerto. Simpson points in planar problems with locational constraints. The round-norm case. *Mathematics of Operations Research*, 22:276–290, 1997.

[25] E. Carrizosa, E. Conde, and M.D. Romero-Morales. Location of a semiobnoxious facility. A biobjective approach. In R. Caballero, F. Ruiz, and R.E. Steuer, editors, *Advances in Multiple Objective and Goal Programming*, volume 455 of *Lecture Notes in Economics and Mathematical Systems*, pages 338–346. Springer Verlag, Berlin, 1997.

[26] E. Carrizosa and F.R. Fernández. The efficient set in location problems with mixed norms. *Trabajos en Investigación Operativa*, 6:61–69, 1991.

[27] E. Carrizosa and F.R. Fernández. A polygonal upper bound for the efficient set for single-facility location problems with mixed norms. *TOP*, 1(1):107–116, 1993.

[28] E. Carrizosa, F.R. Fernández, and J. Puerto. Determination of a pseudoefficient set for single-location problem with polyhedral mixed norms. In F. Orban-Ferauge and J.P. Rasson, editors, *Proceedings of the Meeting V of the EURO Working Group on Locational Analysis*, pages 27–39, 1990.

[29] E. Carrizosa and F. Plastria. A characterization of efficient points in constrained location problems with regional demand. *Operations Research Letters*, 19:129–134, 1996.

[30] E. Carrizosa and F. Plastria. Location of semi-obnoxious facilities. *Studies in Locational Analysis*, 12:1–27, 1999.

[31] E. Carrizosa and F. Plastria. Dominators for multiple-objective quasiconvex maximization problems. *Journal of Global Optimization*, 18:35–58, 2000.

[32] L. G. Chalmet and S. Lawphongpanich. Efficient solutions for point-objective discrete facility location problems. In *Organizations: Multiple Agents with Multiple Criteria*, volume 190 of *Lecture Notes in Economics and Mathematical Systems*, pages 56–71. Springer Verlag, Berlin, 1981.

[33] L.G. Chalmet. *Efficiency in Minisum Rectilinear Distance Location Problems*, pages 431–445. North Holland, Amsterdam, 1983.

[34] L.G. Chalmet, R.L. Francis, and A. Kolen. Finding efficient solutions for rectilinear distance location problems efficiently. *European Journal of Operational Research*, 6:117–124, 1981.

[35] L.G. Chalmet, R.L. Francis, and J.F. Lawrance. Efficiency in integral facility design problems. *Journal of Optimization Theory and Applications*, 32(2):135–149, 1980.

[36] L.G. Chalmet, R.L. Francis, and J.F. Lawrance. On characterizing supremum and l_p−efficient facility designs. *Journal of Optimization Theory and Applications*, 35(1):129–141, 1981.

[37] D. Chhajed and V. Chandru. Hulls and efficient sets for the rectilinear norm. *ORSA Journal of Computing*, 7:78–83, 1995.

[38] M. Colebrook, M.T. Ramos, J. Sicilia, and R.M. Ramos. Efficient points in the biobjective cent-dian problem. *Studies in Locational Analysis*, 15:1–16, 2000.

[39] J. Current, H. Min, and D. Schilling. Multiobjective analysis of facility location decisions. *European Journal of Operational Research*, 49:295–307, 1990.

[40] M.S. Daskin. *Network and Discrete Location. Models, Algorithms and Applications.* Wiley-Interscience Series in Discrete Mathematics and Optimization. Wiley & Sons, Chichester, 1995.

[41] Z. Drezner. On location dominance on spherical surfaces. *Operations Research*, 29(6):1218–1219, 1981.

[42] Z. Drezner, editor. *Facility Location. A Survey of Applications and Methods.* Springer Series in Operations Research. Springer Verlag, New York, 1995.

[43] Z. Drezner and A.J. Goldman. On the set of optimal points to the Weber problem. *Transportation Science*, 25:3–8, 1991.

[44] Z. Drezner and H.W. Hamacher, editors. *Facility Location. Aplications and Theory*. Springer, New York, 2002.

[45] R. Durier. Meilleure approximation en norme vectorielle et théorie de la localisation. *Mathematical Modelling and Numerical Analysis – Modélisation Mathématique et Analyse Numérique*, 21:605–626, 1987.

[46] R. Durier. Continuous location theory under majority rule. *Mathematics of Operations Research*, 14(2):258–274, 1989.

[47] R. Durier. On pareto optima, the fermat-weber problem and polyhedral gauges. *Mathematical Programming*, 47:65–79, 1990.

[48] R. Durier and C. Michelot. Sets of efficient points in a normed space. *Journal of Mathematical Analysis and Applications*, 117:506–528, 1986.

[49] R. Durier and C. Michelot. On the set of optimal points to the weber problem: Further results. *Transportation Science*, 28:141–149, 1994.

[50] M. Ehrgott. *Multicriteria Optimization*, volume 491 of *Lecture Notes in Economics and Mathematical Systems*. Springer Verlag, Berlin, 2000.

[51] M. Ehrgott and X. Gandibleaux, editors. *Multiple Criteria Optimization. State of the Art Annotated Bibliographic Surveys*, volume 52 of *International Series in Operations Research and Management Science*. Kluwer Academic Publishers, Boston, 2002.

[52] M. Ehrgott, H.W. Hamacher, and S. Nickel. Geometric methods to solve max-ordering location problems. *Discrete Applied Mathematics*, 93:3–20, 1999.

[53] M. Ehrgott and A. Rau. Bicriteria costs versus service analysis of a distribution network – A case study. *Journal of Multi-Criteria Decision Analysis*, 8:256–267, 1999.

[54] I. Ekeland. On the variational principle. *Journal of Mathematical Analysis and Applications*, 47:324–353, 1974.

[55] J. Elzinga and D.W. Hearn. Geometrical solutions for some minimax location problems. *Transportation Science*, 6:379–394, 1972.

[56] E. Erkut. Inequality measures for location problems. *Location Science*, 1(3):199–217, 1993.

[57] E. Erkut and S. Neuman. A multiobjective model for locating undesirable facilities. *Annals of Operations Research*, 40:209–227, 1992.

[58] E. Fernández and J. Puerto. Multiobjective solution of the uncapacitated plant location problem. *European Journal of Operational Research*, 145(3):509–529, 2003.

[59] F.R. Fernández, S. Nickel, J. Puerto, and A.M. Rodríguez-Chía. Robustness in the Pareto-solutions for the multi-criteria minisum location problem. *Journal of Multi-Criteria Decision Analysis*, 10:191–203, 2001.

[60] P. Fernández, B. Pelegrín, and J. Fernández. Location of paths on trees with minimal eccentricity and superior section. *TOP*, 6:223–246, 1998.

[61] J. Fliege. A note on "On Pareto optima, the Fermat-Weber problem, and polyhedral gauges". *Mathematical Programming*, 84:435–438, 1999.

[62] R.L. Francis and A.V. Cabot. Properties of a multifacility location problem involving euclidean distances. *Naval Research Logistics Quartely*, 19:335–353, 1972.

[63] R.L. Francis, L.F. McGinnis, and J.A. White. Locational analysis. *European Journal of Operational Research*, 12(3):220–252, 1983.

[64] R.L. Francis, L.F. McGinnis, and J.A. White. *Facility Layout and Location: An Analytical Approach*. Prentice Hall, Englewood Cliffs, 1992.

[65] A.M. Geoffrion. Proper efficiency and the theory of vector maximization. *Journal of Mathematical Analysis and Applications*, 22:618–630, 1968.

[66] I. Giannikos. A multiobjective programming model for locating treatment sites and routing hazardous wastes. *European Journal of Operational Research*, 104:333–342, 1998.

[67] P. Haastrup, V. Maniezzo, M. Mattarelli, F. Mazzeo, I. Mendes, and M. Paruccini. A decision support system for urban waste management. *European Journal of Operational Research*, 109:330–341, 1998.

[68] S.L. Hakimi. Optimum locations of switching centers and the absolute centers and medians of a graph. *Operations Research*, 12:450–459, 1964.

[69] J. Halpern. The location of a centdian convex combination on a undirected tree. *Journal of Regional Science*, 16:237–245, 1976.

[70] J. Halpern. Finding minimal center-median convex combination (cent dian) of a graph. *Management Science*, 16:534–544, 1978.

[71] J. Halpern. Duality in the cent-dian of a graph. *Operations Research*, 28:722–735, 1980.

[72] H. W. Hamacher, H. Hennes, J. Kalcsics, and S. Nickel. LoLA – Library of Location Algorithms, Version 2.1, 2003. http://www.itwm.fhg.de/ opt/projects/lola.

[73] H. W. Hamacher and S. Nickel. Classification of location models. *Location Science*, 6:229–242, 1998.

[74] H.W. Hamacher, M. Labbé, and S. Nickel. Multicriteria network location problems with sum objectives. *Networks*, 33:79–92, 1999.

[75] H.W. Hamacher, M. Labbé, S. Nickel, and A.J.V. Skriver. Multicriteria semi-obnoxious network location problems (MSNLP) with sum and center objectives. *Annals of Operations Research*, 110:33–53, 2002.

[76] H.W. Hamacher and S. Nickel. Multicriteria planar location problems. *European Journal of Operational Research*, 94:66–86, 1996.

[77] G.Y. Handler. Medi-centers of a tree. *Transportation Science*, 19:246–260, 1985.

[78] P. Hansen, M. Labbé, and J.F. Thisse. From the median to the generalized center. *RAIRO Recherche Operationelle*, 25:73–86, 1991.

[79] P. Hansen, J. Perreur, and J.F. Thisse. Location theory, dominance, and convexity: Some further results. *Operations Research*, 28:1241–1250, 1980.

[80] P. Hansen, J.-F. Thisse, and R.E. Wendell. Efficient points on a network. *Networks*, 16(4):357–368, 1986.

[81] P. Hansen and J.F. Thisse. The generalized Weber-Rawls problem. In J.L. Brans, editor, *Operational Research '81 (Hamburg, 1981)*, pages 569–577. North-Holland, Amsterdam, 1981.

[82] E.C. Henkel and C. Tammer. ε-variational inequalities for vector approximation problems. *Optimization*, 38:11–21, 1996.

[83] O. Hudec and K. Zimmermann. Biobjective center-balance graph location model. *Optimization*, 45:107–115, 1999.

[84] A.P. Hurter, M.K. Shaeffer, and R.E. Wendell. Solutions of constrained location problems. *Management Science*, 22:51–56, 1975.

[85] M.P. Johnson. A spatial decision support system prototype for housing mobility program planning. *Journal of Geographical Systems*, 3:49–67, 2001.

[86] H. Juel and R. Love. Hull properties in location problems. *European Journal of Operational Research*, 12:262–265, 1983.

[87] O. Kariv and S.L. Hakimi. An algorithm approach to network location problems I. The p-centers. *SIAM Journal of Applied Mathematics*, 37:513–538, 1979.

[88] K. Klamroth and M.M. Wiecek. A bi-objective median location problem with a line barrier. *Operations Research*, 50:670–679, 2002.

[89] R. Lahdelma, P. Salminen, and J. Hokkanen. Locating a waste treatment facility by using stochastic multicriteria acceptability analysis with ordinal criteria. *European Journal of Operational Research*, 142:345–356, 2002.

[90] S.M. Lee, G.I. Green, and C.S. Kim. A multiple criteria model for the location-allocation problem. *Computers & Operations Research*, 8:1–8, 1981.

[91] G.S. Liang and M.J.J. Wang. A fuzzy multi-criteria decision-making method for facility site selection. *International Journal of Product Resources*, 11:2313–2330, 1991.

[92] M.C. López-Mozos and J.A. Mesa. Location of cent-dian path in tree graphs. In J.A. Moreno, editor, *Proceeding of Meeting VI of the EURO Workking Group on Location Analysis*, number 34 in Serie Informes, pages 135–144. Secretariado de Publicaciones de la Universidad de la Laguna, 1992.

[93] R.F. Love, J.G. Morris, and G.O. Wesolowsky. *Facilities Location. Models and Methods.* North Holland Publishing Company, 1988.

[94] T.J. Lowe. Efficient solutions in multiobjective tree network location problems. *Transportation Science*, 12:298–316, 1978.

[95] T.J. Lowe, J.F. Thisse, J.E. Ward, and R.E. Wendell. On efficient solutions to multiple objective mathematical programs. *Management Science*, 30:1346–1349, 1984.

[96] J. Malczewki and W. Ogryczak. The multiple criteria location problem: 1. A generalized network model and the set of efficient solutions. *Environment and Planning*, A27:1931–1960, 1995.

[97] J. Malczewki and W. Ogryczak. The multiple criteria location problem: 2. Preference-based techniques and interactive decision support. *Environment and Planning*, A28:69–98, 1996.

[98] M.B. Mandell. Modelling effectiveness-equity trade-offs in public service delivery systems. *Management Science*, 37:467–482, 1991.

[99] M.T. Marsh and D.A. Schilling. Equity measurement in facility location analysis: A review and framework. *European Journal of Operational Research*, 74(1):1–17, 1994.

[100] J.A. Mesa and T.B. Boffey. A review of extensive facility location in networks. *European Journal of Operational Research*, 95:592–603, 1996.

[101] C. Michelot. Localization in multifacility location theory. *European Journal of Operational Research*, 31:177–184, 1987.

[102] R.L. Morrill and J. Symons. Efficiency and equity aspects of optimum location. *Geographical Analysis*, 9:215–225, 1977.

[103] G.F. Mulligan. Equity measures and facility location. *Papers in Regional Science*, 70:345–365, 1991.

[104] Y.S. Myung, H.G. Kim, and D.W. Tcha. A bi-objective uncapacitated facility location problem. *European Journal of Operational Research*, 100:608–616, 1997.

[105] M. Ndiaye and C. Michelot. Efficiency in constrained continuous location. *European Journal of Operational Research*, 104:288–298, 1998.

[106] S. Nickel. *Discretization of Planar Location Problems*. Shaker Verlag, Aachen, 1995

[107] S. Nickel. Bicriteria and restricted 2-facility Weber problems. *Mathematical Methods of Operations Research*, 45:167–195, 1997.

[108] S. Nickel and J. Puerto. A unified approach to network location problems. *Networks*, 34:283–290, 1999.

[109] S. Nickel, J. Puerto, A.M. Rodríguez-Chía, and A. Weissler. General continuous multicriteria location problems. Technical report, University of Kaiserslautern, Department of Mathematics, 1997.

[110] W. Ogryczak. On cent-dians of general networks. *Location Science*, 5:15–28, 1997.

[111] W. Ogryczak. On the lexicographic minimax approach to location problems. *European Journal of Operational Research*, 100:566–585, 1997.

[112] W. Ogryczak. On the distribution approach to location problems. *Computers & Industrial Engineering*, 37:595–612, 1999.

[113] W. Ogryczak. Inequality measures and equitable approaches to location problems. *European Journal of Operational Research*, 122(2):374–391, 2000.

[114] W. Ogryczak, K. Studzinski, and K. Zorychta. A solver for the multiobjective transshioment problem with facility location. *European Journal of Operational Research*, 43:53–64, 1989.

[115] Y. Ohsawa. A geometrical solution for quadratic bicriteria location models. *European Journal of Operational Research*, 114:380–388, 1999.

[116] Y. Ohsawa. Bicriteria euclidean location associated with maximin and minimax criteria. *Naval Research Logistics Quarterly*, 47:581–592, 2000.

[117] B. Pelegrín and F. R. Fernández. Determination of efficient points in multiple-objective location problems. *Naval Research Logistics*, 35:697–705, 1988.

[118] B. Pelegrín and F. R. Fernández. Determination of efficient solutions for point-objective locational decision problems. *Annals of Operations Research*, 18:93–102, 1989.

[119] D. Pérez-Brito, J.A. Moreno-Pérez, and I. Rodríguez-Martín. Finite dominating set for the p-facility cent-dian network locating problem. *Studies in Locational Analysis*, 11:27–40, 1997.

[120] D. Pérez-Brito, J.A. Moreno-Pérez, and I. Rodríguez-Martín. The 2-facility centdian network problem. *Location Science*, 6:369–381, 1998.

[121] F. Plastria. *Continuous Location Problems and Cutting Plane Algorithms*. PhD thesis, Vrije Universiteit Brussel, 1983.

[122] F. Plastria. Localization in single facility location. *European Journal of Operational Research*, 18:215–219, 1984.

[123] J. Puerto. *Lecturas en Teoría de Localización*. Universidad de Sevilla. Secretariado de Publicaciones, 1996.

[124] J. Puerto and F.R. Fernández. A convergent approximation scheme for efficient sets of the multi-criteria weber location problem. *TOP*, 6:195–202, 1998.

[125] J. Puerto and F.R. Fernández. Multi-criteria minisum facility location problems. *Journal of Multi-Criteria Decision Analysis*, 8:268–280, 1999.

[126] J. Puerto and F.R. Fernández. Geometrical properties of the symmetrical single facility location problem. *Journal of Nonlinear and Convex Analysis*, 1(3):321–342, 2000.

[127] R.M. Ramos, M.T. Ramos, M. Colebrook, and J. Sicilia. Locating a facility on a network with multiple median-type objectives. *Annals of Operations Research*, 86:221–235, 1999.

[128] R.M. Ramos, J. Sicilia, and M.T. Ramos. The biobjective absolute center problem. *TOP*, 5(2):187–199, 1997.

[129] K. Ravindranath, P. Vrat, and N. Singh. Bicriteria single facility rectilinear location problems in the presence of a single forbidden region. *Operations Research*, 22:1–16, 1985.

[130] A.M. Rodríguez-Chía. *Advances on the Continuous Single Facility Location Problem*. PhD thesis, Universidad de Sevilla, 1998.

[131] A.M. Rodríguez-Chía and J. Puerto. Geometrical description of the weakly efficient solution set for constrained multicriteria location problems. Technical report, Facultad de Matematicas, Universidad de Sevilla, 2002.

[132] A.M. Rodríguez-Chía and J. Puerto. Geometrical description of the weakly efficient solution set for multicriteria location problem. *Annals of Operations Research*, 111:179–194, 2002.

[133] G.T. Ross and R.M. Soland. A multicriteria approach to the location of public facilities. *European Journal of Operational Research*, 4:307–321, 1980.

[134] A.J.V. Skriver and K.A. Andersen. The bicriterion semi-obnoxious location (BSL) problem solved by an ε-approximation. *European Journal of Operational Research*, 146:517–528, 2003.

[135] R. Solanki. Generating the noninferior set in mixed integer biobjective linear programs: An application to a location problem. *Computers & Operations Research*, 18:1–15, 1991.

[136] A. Tamir, D. Pérez-Brito, and J.A. Moreno-Pérez. A polynomial algorithm for the p-centdian problem on a tree. *Networks*, 32:255–262, 1998.

[137] A. Tamir, J. Puerto, and D. Pérez-Brito. The centdian subtree on tree networks. *Discrete Applied Mathematics*, 118:263–278, 2002.

[138] C. Tammer. A generalization of ekeland's variational principle. *Optimization*, 25:129–141, 1992.

[139] B.C. Tansel, R.L. Francis, and T.J. Lowe. A biobjective multifacility minimax location problem on a tree network. *Transportation Science*, 16(4):407–429, 1982.

[140] J.F. Thisse, J. E. Ward, and R. E. Wendell. Some properties of location problems with block and round norms. *Operations Research*, 32:1309–1327, 1984.

[141] C. Tofallis. Multi-criteria site selection using d.e.a. profiling. *Studies in Locational Analysis*, 11:211–218, 1997.

[142] G. Wanka. Duality in vectorial control approximation problems with inequality restrictions. *Optimization*, 22(5):755–764, 1991.

[143] G. Wanka. On duality in the vectorial control-approximation problem. *ZOR – Methods and Models of Operations Reseach*, 35:309–320, 1991.

[144] G. Wanka. Kolmogorov-conditions for vectorial approximation problems. *OR Spektrum*, 16:53–58, 1994.

[145] G. Wanka. Properly efficient solutions for vectorial norm approximation. *OR Spektrum*, 16:261–265, 1994.

[146] G. Wanka. Characterization of approximately efficient solutions to multiobjective location problems using ekeland's variational principle. *Studies in Locational Analysis*, 10:163–176, 1996.

[147] G. Wanka. ε-optimality to approximation in partially ordered spaces. *Optimization*, 38:1–10, 1996.

[148] G. Wanka. Multiobjective control approximation problems: Duality and optimality. *Journal of Optimization Theory and Applications*, 105(2):457–475, 2000.

[149] J. Ward. Structure of efficient sets for convex objectives. *Mathematics of Operations Research*, 14:249–257, 1989.

[150] J.E. Ward and R.E. Wendell. Characterizing efficient points in location problems under the one-infinity norm. In J.-F. Thisse and H.G. Zoller, editors, *Location Analysis of Public Facilities*, pages 413–429. North Holland, Amsterdam, 1983.

[151] R.E. Wendell and A.P. Hurter. Location theory, dominance, and convexity. *Operations Research*, 21:314–320, 1973.

[152] R.E. Wendell, A.P. Hurter, and T.J. Lowe. Efficient points in location problems. *AIIE Transactions*, 9:238–246, 1977.

[153] R.E. Wendell and D.N. Lee. Efficiency in multiple objective optimization problems. *Mathematical Programming*, 12:406–414, 1977.

[154] G.O. Wesolowsky. The weber problem: History and perspectives. *Location Science*, 1(1):5–23, 1993.

[155] D.J. White. *Optimality and Efficiency*. John Wiley & Sons, Chichester, 1982.

[156] C. Witzgall. Optimal location of a central facility: Mathematical models and concepts. *National Bureau of Standards Report 8388*, 1965.

[157] K. Yoon and C.L. Hwang. Manufacturing plant location analysis by multiple attribute decision making: Part I - Single-plant strategy. *International Journal of Production Research*, 23(2):345–359, 1985.

[158] K. Yoon and C.L. Hwang. Manufacturing plant location analysis by multiple attribute decision making: Part II - Multi-plant strategy and plant relocation. *International Journal of Production Research*, 23(2):361–370, 1985.

VII

APPLICATIONS

Chapter 20

MULTICRITERIA DECISION AID/ ANALYSIS IN FINANCE

Jaap Spronk

Erasmus University Rotterdam, Department of Finance and Investment
P.O.Box 1738, 3000 DR Rotterdam, The Netherlands
spronk@few.eur.nl

Ralph E. Steuer

University of Georgia, Department of Banking and Finance, Terry College of Business, Athens, Georgia 30602-6253
USA
rsteuer@uga.edu

Constantin Zopounidis

Technical University of Crete, Department of Production Engineering and Management, Financial Engineering Laboratory, University Campus, 73100 Chania, Greece
kostas@dpem.tuc.gr

Abstract Over the past decades the complexity of financial decisions has increased rapidly, thus highlighting the importance of developing and implementing sophisticated and efficient quantitative analysis techniques for supporting and aiding financial decision making. Multicriteria decision aid (MCDA), an advanced branch of operations research, provides financial decision makers and analysts with a wide range of methodologies well-suited for the complexity of modern financial decision making. The aim of this chapter is to provide an in-depth presentation of the contributions of MCDA in finance focusing on the methods used, applications, computation, and directions for future research.

Keywords: Multicriteria decision aid, finance, portfolio theory, multiple criteria optimization, outranking relations, preference disaggregation analysis.

1. Introduction

Over the past decades, the globalization of financial markets, the intensification of competition among organizations, and the rapid social and technological changes that have taken place have only led to increasing uncertainty and instability in the business and financial environment. Within this more recent context, both the importance of financial decision making and the complexity of the process by which financial decision making is carried out have increased. This is clearly evident by the variety and volume of new financial products and services that have appeared on the scene.

In this new era of financial reality, researchers and practitioners acknowledge the requirement to address financial decision-making problems through integrated and realistic approaches utilizing sophisticated analytical techniques. In this way, the connections between financial theory, the tools of operations research, and mathematical modelling have become more entwined. Techniques from the fields of optimization, forecasting, decision support systems, MCDA, fuzzy logic, stochastic processes, simulation, etc. are now commonly considered valuable tools for financial decision making.

The use of mathematics and operations research in finance got its start in the 1950s with the introduction of Markowitz's portfolio theory [81, 83]. Since then, in addition to portfolio selection and management, operations research has contributed to financial decision making problems in other areas including venture capital investments, bankruptcy prediction, financial planning, corporate mergers and acquisitions, country risk assessment, etc. These contributions are not limited to academic research; they are now often found in daily practice.

Within the field of operations research, MCDA has evolved over the last three decades into one of its pillar disciplines. The development of MCDA is based upon the common finding that a sole objective, goal, criterion, or point of view is rarely used to make real-world decisions. In response, MCDA is devoted to the development of appropriate methodologies to support and aid decision makers across ranges of situations in which multiple conflicting decision factors (objectives, goals, criteria, etc.) are to be considered simultaneously.

The methodological framework of MCDA is well-suited to the growing complexities encountered in financial decision making. While there have been in finance MCDA stirrings going back twenty to thirty years, the topic of MCDA, as can be seen from the bulk of the references, really hasn't come into its own until recently. As for early stirrings, we have, for example, Bhaskar [11] in which microeconomic theory was criticized for largely pursuing a single criterion approache arguing that things like profit maximization are too naive to meet the evolving decision-making demands in many financial areas. Also, in another paper [12], the unavoidable presence of multiple objectives in capital budgeting was noted and the necessity for developing ways to deal with the unique challenges posed by multiple criteria was stressed. It is upon what has taken place since these early roots, and on what are today promising directions in MCDA in finance, that this contribution is focused.

Such observations and findings have motivated researchers to explore the potentials of MCDA in addressing financial decision-making problems. The objective of this chapter is to provide a state-of-the-art comprehensive review of the research made up to date on this issue. Section 2 presents discussions to justify the presence of MCDA in financial decision making. Section 3, focuses on MCDA in resource allocation problems (continuous problems) as in the field of portfolio management. Section 4, presents the contribution of MCDA methodologies in supporting financial decisions that require the evaluation of a discrete set of alternatives (firms, countries, stocks, investment projects, etc.). Finally, Section 5 concludes the chapter and discusses possible future research directions on the implementation of multicriteria analysis in financial institutions and firms.

2. Financial Decision Making

Financial-economic decision problems come in great variety. Individuals are involved in decisions concerning their future pensions, the financing of their homes, and investments in mutual funds. Firms, financial institutions, and advisors are involved in cross-country mergers, complicated swap contracts, and mortgage-backed securities, to name just a few.

Despite the variety, such decisions have much in common. Maybe "money" comes first to mind, but there are typically other factors that suggest that financial-economic problems should most appropriately be treated as multiple criteria decision problems in general: multiple actors, multiple policy constraints, and multiple sources of risk (see e.g., Spronk & Hallerbach [115], and Hallerbach & Spronk [49, 50], Martel & Zopounidis [86], Zopounidis [135], and Steuer & Na [120]).

Two other common elements in financial decisions are that their outcomes are distributed over time and uncertainty, and thus involve risk. A further factor is that most decisions are made consciously, with a clear and constant drive to make "good", "better" or even "optimal" decisions. In this drive to improve on financial decisions, we stumble across an area of tension between decision making in practice on the one hand and the potential contributions of finance theory and decision tools on the other. Although the bulk of financial theory is of a descriptive nature, thus focusing on the "average" or "representative" decision maker, we observe a large willingness to apply financial theory in actual decision-making. At the same time, knowledge about decision tools that can be applied in a specific decision situation, is limited. Clearly, there is need of a framework that can provide guidance in applying financial theory, decision tools, and common sense to solving financial problems.

2.1 Issues, Concepts, and Principles

> Finance is a sub field of economics distinguished by both its focus and its methodology. The primary focus of finance is the workings of the capital markets and the supply and the pricing of capital assets. The methodology of finance is the use of close substitutes to price financial contracts and instruments. This methodology is applied to value instruments whose characteristics extend across time and whose payoffs depend upon the resolution of uncertainty. (Ross [101], p. 1)

The field of finance is concerned with decisions with respect to the efficient allocation of scarce capital resources over competing alternatives. The allocation is efficient when the alternative with the highest value is chosen. Current value is viewed as the (present) value of claims on future cash flows. Hence we can say that financial decisions involve the valuation of future, and hence uncertain or "risky," cash flow streams. Cash flow stream X is valued by comparing it with cash flow streams $\{A,...,Z\}$ that are traded on financial markets. When a traded cash flow stream Y has been identified that is a substitute for X, then their values must be the same. After all, when introducing X to the market, it cannot be distinguished value-wise from Y. Accepting the efficient market hypothesis (stipulating that all available information is fully and immediately incorporated in market prices), the market price of Y equals the value of Y, and hence the value of X. This explains the crucial role of financial markets.

The valuation of future cash flow streams is a key issue in finance. The process of valuation must be preceded by evaluation: without analyzing the characteristics of a cash flow stream, no potential substitute can be identified. Since it is uncertain what the future will bring, the analysis of the risk characteristics will be predominant. Moreover, as time passes, the current value must be protected against influences that may erode its value. This in turn implies the need for risk management. There are basically three areas of financial decisions:

1 **Capital budgeting**: to what portfolio of real investment projects should a firm commit its capital? The central issues here are how to evaluate investment opportunities, how to distinguish profitable from non-profitable projects and how to choose between competing projects.

2 **Corporate financing**: this encompasses the capital structure policy and dividend policy and addresses questions as: how should the firm finance its activities? What securities should the firm issue or what financial contracts should the firm engage in? What part of the firm's earnings should be paid as cash dividends and what part reinvested in the firm? How should the firm's solvency and liquidity be maintained?

3 **Financial investment**: this is the mirror image of the previous decision area and involves choosing a portfolio of financial securities with the objective to change the consumption pattern over time.

In each of these decision areas the financial key issues of valuation, risk analysis and risk management, and performance evaluation can be recognized, and from the above several financial concepts emerge: financial markets, efficient allocation and market value. In approaching the financial decision areas, some financial principles or maxims are formulated. The first is self-interested behavior: economic subjects are driven by *non-satiation* ("greed"). This ensures the goal of value maximization. Prices are based on financial markets, and under the efficient market hypothesis, prices of securities coincide with their value. Value has time and risk dimensions. With regard to the former, *time preference* is assumed (a dollar today is preferred to a dollar tomorrow). With respect to the latter, *risk aversion* is assumed (a safe dollar is preferred to a risky dollar). Overall risk may be reduced by *diversification*: combining risky assets or cash flow streams may be beneficial. In one way or another, the trade-off between expected return and risk that is imposed by market participants on the evaluation of risky ventures will translate into a *risk-return trade-off* that is offered by investment opportunities in the market.

Since value has time and risk aspects, the question arises about what mechanisms can be invoked to incorporate these dimensions in the valuation process. There are basically two mechanisms. The first is the arbitrage mechanism. Value is derived from the presumption that there do not exist arbitrage opportunities. This no-arbitrage condition excludes sure profits at no cost and implies that perfect substitutes have the same value. This is the *law of one price*, one of the very few laws in financial economics. It is a strong mechanism, requiring very few assumptions on market subjects, only non-satiation. Examples of valuation models built on no-arbitrage are the Arbitrage Pricing Theory for primary financial assets and the Option Pricing Theory for derivative securities. The second is the equilibrium mechanism. In this case value is derived from the market clearing condition that demand equals supply. The latter mechanism is much weaker than the former: the exclusion of arbitrage opportunities is a necessary but by no means a sufficient condition for market equilibrium. In addition to non-satiation also assumptions must be made regarding the risk attitudes of all market participants. Examples of equilibrium-based models are the Capital Asset Pricing Model and its variants. Below we discuss the differences between the two valuation approaches in more detail. It suffices to remark that it is still a big step from the principles to solving actual decision problems.

2.2 Focus of Financial Research

An alternative, albeit almost circular, definition of finance is provided by Jarrow [63], p.1.

> Finance theory (...) includes those models most often associated with financial economics. (...) [*A*] practical definition of financial economics is found in those topics that appear with some regularity in such publications as *Journal of Finance, Journal of Financial and Quantitative Analysis, Journal of Financial Economics,* and *Journal of Banking and Finance*.

Browsing through back volumes of these journals and comparing them to the more recent ones reveals a blatant development in nature and focus. In early days of finance, the papers were descriptive in a narrative way and in the main focused on financial instruments and institutions. Finance as a decision science emerged in the early 1950s, when Markowitz [80, 81] studied the portfolio selection decision and launched what now is known as "modern portfolio theory." In the 1960s and the early 1970s, many financial economic decision problems were approached by operational research techniques; see for example Ashford, Berry & Dyson [3] and McInnes & Carleton [87] for an overview. However, since then, this type of research has became more and more absorbed by the operations research community and in their journals.

But what direction did finance take? Over the last 25 years mathematical models have replaced the verbal models and finance has founded itself firmly in a neo-classical micro-economic tradition. Over this period we observe a shift to research that is descriptive in a sophisticated econometrical way and that focuses on the statistical characteristics of (mainly well-developed) financial markets where a host of financial instruments is traded. Bollerslev [15], p. 41, aptly describes this shift as follows.

> A cursory look at the traditional econometrics journals (...) severely underestimates the scope of the field [of financial econometrics], as many of the important econometric advances are now also published in the premier finance journals – the *Journal of Finance*, the *Journal of Financial Economics*, and the *Review of Financial Studies* – as well as a host of other empirically oriented finance journals.

The host of reported research addresses the behavior of financial market prices. The study of the pricing of primary securities is interesting for its own right, but it is also relevant for the pricing of derivative securities. Indeed, the description of the pricing of primary assets and the development of tools for pricing derivative assets mark the success story of modern finance.

The body of descriptive finance theory has grown enormously. According to modern definitions of the field of finance, the descriptive nature is even predominant.

> The core of finance theory is the study of the behavior of economic agents in allocating and deploying their resources, both spatially and across time, in an uncertain environment. (Merton [89], p. 7)

Compared to Ross' [101] definition cited earlier, the focus is purely positive. The question arises to what extent the insights gained from descriptive finance – how sophisticated they may be from a mathematical, statistical or econometric point of view – can serve as guidelines for financial decisions in practice. Almost thirty years ago, in the preface of their book *The Theory of Finance*, Eugene Fama and Merton Miller defended their omission of detailed examples, purporting to show how to apply the theory to real-world decision problems, as

> (...) a reflection of our belief that the potential contribution of the theory of finance to the decision-making process, although substantial, is still essentially indirect. The theory can often help expose the inconsistencies in existing procedures; it can help keep the really critical questions from getting lost in the inevitable maze of technical detail; and it can help prevent the too easy, unthinking acceptance of either the old clichés or new fads. But the theory of finance has not yet been brought, and perhaps never will be, to the cookbook stage. (Fama & Miller [41], p. viii)

Careful inspection of current finance texts reveals that in this respect not much has changed. However, pure finance theory and foolproof financial recipes are two extremes of a continuum. The latter cookbook stage will never be achieved, of course, and in all realism and wisdom this alchemic goal should not be sought for. But what we dearly miss is an extensive body of research that bridges the apparent gap between the extremes: research that shows how to solve real-world financial decision problems without violating insights offered by pure finance theory on the one hand and without neglecting the peculiarities of the specific decision problem on the other.

On another matter, the role of assumptions in modelling is to simplify the real world in order to make it tractable. In this respect the art of modelling is to make assumptions where they most contribute to the model's tractability and at the same time detract from the realism of the model as little as possible. The considerations in this trade-off are fundamentally different for positive (descriptive) models on the one hand and conditional-normative models on the other. In the next section we elaborate further on the distinctions between the two types of modelling as concerns the role of assumptions.

2.3 Descriptive vs. Conditional-Normative Modelling

In a positive or descriptive model simplified assumptions are made in order to obtain a testable implication of the model. The validity of the model is evaluated according to the inability to reject the model's implications at some level of significance. So validity is of an empirical nature, solely judged by the implications of the model. Consider the example of an equilibrium asset-pricing model. As a starting point, assumptions are made with respect to the preferences of an imaginary investor and the risk-return characteristics of the investment opportunities. These assumptions are sufficiently strong to allow solving the portfolio optimization problem. Next a homogeneity condition is imposed: all investors in the market possess the same information and share the same expectations. This allows focusing on "a representative investor". Finally the equilibrium market clearing condition is imposed: all available assets (supply) must be incorporated in the portfolio of the representative investor (demand). The first order conditions of portfolio optimality then stipulate the trade-off between risk and expected return that is required by the investor. Because of the market clearing, the assets offer the same trade-off. Hence a market-wide relationship between risk and return is established and this relationship is the object of empirical testing. As long as the pricing relationship is not falsified the model is accepted, irrespective of whether the necessary assumptions are realistic or not. When the model is falsified, deduction may help to amend the assumptions where after the same procedure is followed. This hypothetic-deductive cycle ends when the model is no longer falsified by the empirical data at hand.

In a conditional-normative model, simplifying assumptions are also made in order to obtain a tractable model. These assumptions relate to the preferences of the decision maker and to the representation of the set of choice alternatives. The object of the conditional-normative modelling is not to infer a testable implication but to obtain a decision rule. This derived decision rule is valid and can normatively be applied conditional on the fact that the decision maker satisfies the underlying assumptions; cf. Keynes [70].

In order to support decisions in finance, obviously both the preferences of the decision maker and the characteristics of the choice alternatives should be understood and related to each other. Unfortunately, the host of financial-economic modelling is of a positive nature and focuses on the "average" decision maker instead of addressing the particular (typically non-average) decision maker. The assumptions underlying financial theory at best describe "average individuals" and "average decision situations" and hence are not suited to describe specific individual decision problems. The assumptions made to simplify the decision situation often completely redefine the particular problem at hand. The real world is replaced by an over-simplified model-world. As a consequence, not the initial problem is solved but a synthesized and redefined problem that is not even recognized by the decision maker himself. The over-simplified model becomes a Procrustes bed for the financial decision maker who seeks advice.

For example, it is assumed that a decision maker has complete information and that this information can be molded into easily manipulated probability distributions. Even worse, positive knowledge and descriptive theories that by definition reflect the outcomes of decisions made by some representative decision maker are used to prescribe what actions to take in a specific decision situation. For example, equilibrium asset pricing theories predict the effects of decisions and actions of many individuals on the formation of prices in financial markets. Under the homogeneity condition the collection of investors is reduced to the representative investor. When the pricing implications of the model are simply used to guide actual investment behavior, then the decision maker is forced into the straitjacket of this representative investor.

Unfortunately we observe that conditional-normative financial modelling is only regarded as a starting point for descriptive modelling and is not pursued for its own sake. After almost twenty years, Hastie's [52] lament has not lost its poignancy.

> In American business today, particularly in the field of finance, what is needed are approximate answers to the precise problem rather than precise answers to the approximate problem.

Apart from the positive modelling of financial markets as described above, there is one other field in finance in which the achievements of applied modelling are apparent: option pricing theory, the set of models that enable the pricing of derivative securities and all kinds of contingent claims. Indeed, the option pricing formulas developed by Black & Scholes [13] and Merton [88] mark a huge success in the history of financial modelling. Contingent claims analysis made a flying start, and

> .... when judged by its ability to explain the empirical data, option pricing theory is the most successful theory not only in finance, but in all of economics. (Ross [101], p. 24)
>
> Given a theory that works so well, the best empirical work will be to use it as a tool rather than to test it. (Ross [101], p. 23)

Indeed, modern-day derivatives trading would be unthinkable without the decision support of an impressive coherent toolbox for analyzing the risk characteristics of derivatives and for pricing them in a consistent way. Compared to this framework, the models and theories developed and tested for primary assets look pale. What is the reason for the success of derivatives research?

For an explanation we turn to the principal tool used in option pricing theory: no-arbitrage valuation. By definition derivative securities derive their value from primary underlying assets. Under some mild assumptions, a dynamic trading strategy can be designed in which the derivative security is exactly replicated with a portfolio of the primary security and risk-free bonds. Under the no-arbitrage condition, the current value of the derivative security and the replicating portfolio should be identical. Looking from another perspective, a suitably chosen hedge combination of the derivative and the underlying security produces a risk-free position. On this position the risk-free rate must be earned, otherwise there exist arbitrage opportunities. Since the position is risk free, risk attitudes and risk aversion do not enter the story. Therefore a derivative security will have the same value in a market environment with risk neutral investors as in a market with risk averse investors. This in turn implies that a derivative can be priced under the assumption that investors are risk neutral.

As a consequence, no assumptions are required on preferences (other than non-satiation), utility functions, the degree of risk aversion, and risk premia. Thus, option pricing theory can escape from the burden of modelling of preference structures. Instead, research attention shifts to analyzing price dynamics on financial markets. An additional reason for the success in derivatives research is that the analytical and mathematical techniques are similar to those used in the physical sciences (see for example Derman [25]).

Of course, even in derivatives modelling some assumptions are required. This introduces model risk. When the functional relationships stipulated in the model are wrong, or when relevant input parameters of the model are incorrectly estimated, the model produces the wrong value and the wrong risk profile of the derivative. To an increasing degree, financial institutions are aware that great losses can be incurred because of model risk. Especially in risk management and derivatives trading model risk is a hot item (see Derman [24]). This spurred Merton to ventilate this warning.

> At times, the mathematics of the models become too interesting and we lose sight of the models' ultimate purpose. The mathematics of the models is precise, but the models are not, being only approximations to the complex, real world. Their accuracy as a useful approximation to that world varies considerably across time and place. The practitioner should therefore apply the models only tentatively, assessing their limitations carefully in each application. (Merton [89], p. 14)

Ironically this quote was taken just after the very successful launch of Long Term Capital Management (LTCM), the hedge fund of which Merton and Myron Scholes were the founding partners. In 1998, LTCM collapsed and model risk played a very important role in this debacle.

Summarizing we draw the conclusion that successful applied financial modelling does exist, and blossoms in the field of derivatives. Here also the validity of the assumptions is crucial, this in contrast to positive modelling. However, in the field of derivatives with replicating strategies and arbitrage-based valuation, the concept of "absence of risk" is well defined and no preference assumptions are needed in the modelling process. For modelling decisions regarding the underlying primary assets, in contrast, assumptions on the decision maker's preferences and on the "risk" attached to the outcomes of the choice alternatives are indispensable. For these types of financial problems, the host of simplifying assumptions that are made in the descriptive modelling framework invalidate the use of the model in a specific decision situation. Thus we face the following challenge: how can we retain the conceptual foundation of the financial-economic framework and still provide sound advice that can be applied in multifarious practice? As a first step we will sketch the relationship between decision sciences and financial decision-making.

2.4 Decision Support for Financial Decisions

Over the last fifty years or so, the financial discipline has shown continuously rapid and profound changes, both in theory and in practice. Many disciplines have been affected by globalization, deregulation, privatization, computerization, and communication technologies. Hardly any field has been influenced as much as finance. After the mainly institutional and even somewhat *ad hoc* approaches before the fifties, Markowitz [80, 81] has opened new avenues by formalizing and quantifying the concept of "risk". In the decades that followed, a lot of attention was paid to the functioning of financial markets and the pricing of financial assets including options. The year 1973 gave birth to the first official market in options (CBOE) and to crucial option pricing formulas that have become famous quite fast (Black-Scholes and Cox-Ross-Rubinstein, see Hull [53]), both in theory and practice. At that time, financial decision problems were structured by (a) listing a number of mutually exclusive decision alternatives, (b) describing them by their (estimated) future cash flows, including an estimation of their stochastic variation and later on including the effect of optional decisions, and (c) valuing them by using the market models describing financial markets.

In the seventies, eighties and nineties, the financial world saw enormous growth in derivative products, both in terms of variety and in terms of market volumes. Financial institutions have learned to work with complex financial products. Academia has contributed by developing many pricing models, notably for derivatives. Also, one can say that financial theory has been rewritten in the light of contingent claims ("optional decisions") and will soon be further reshaped by giving more attention to game elements in financial decisions. The rapid development of the use of complex financial products has certainly not been without accidents. This has led regulators to demand more precise evaluations and the reporting of financial positions (cf. e.g., the emergence of the Value-at-Risk concept, see Jorion [68]).

In addition to the analysis of financial risk, the structured management of financial risk has come to the forefront. In their textbook, Bodie & Merton [14] describe the threefold tasks of the financial discipline as Valuation, Risk Management, and Optimization. We would like to amend the threefold tasks of financial management to Valuation, Risk Management, and Decision Making. The reason is that financial decision problems often have to be solved in dynamic environments where information is not always complete, different stakeholders with possibly conflicting goals and constraints play a role and clear-cut optimization problems cannot always be obtained (and solved).

At the same time, many efforts from the decision-making disciplines are misdirected. For instance, some approaches fail to give room for the inherent complexity of the decision procedure given the decision maker's specific context. Other approaches concentrate on the beauties of a particular decision method without doing full justice to the peculiarities of the decision context. Aside from being partial in this respect, useful principles and insights offered by financial-economic theory are often not integrated in the decision modelling. It is therefore no surprise that one can observe in practice unstructured *ad hoc* approaches as well as complex approaches that severely restrict the decision process.

2.5 Relevance of MCDA for Financial Decisions

The central issue in financial economics is the efficient allocation of scarce capital and resources over alternative uses. The allocation (and redistribution) of capital takes place on financial markets and is termed "efficient" when market value is maximized. Just as water will flow to the lowest point, capital will flow to uses that offer the highest return. Therefore it seems that the criterion for guiding financial decisions is one-dimensional: maximize market value or maximize future return.

From a financial-economic perspective, the goal of the firm, for example, is very much single objective. Management should maximize the firm's contribution to the financial wealth of its shareholders. Also the shareholders are considered to be myopic. Their only objective is to maximize their single-dimensional financial wealth. The link between the shareholders and the firm is footed in law. Shareholders are the owners of the firm. They possess the property rights of the firm and are thus entitled to decide what the firm should aim for, which according to homogeneity is supposed to be the same for all shareholders, i.e., maximize the firm's contribution to the financial wealth of the shareholders. The firm can accomplish this by engaging in investment projects with positive net present value. This is the neo-classical view on the role of the firm and on the relationship between the firm and its shareholders in a capitalist society. Figure 20.1 depicts a simplified graphical representation of this line of thought.

It is important to note that this position is embedded in a much larger framework of stylized thinking in among others economics (general equilibrium framework) and law (property rights theory and limited liability of shareholders). Until today, this view is seen as an ideal by many; see for example Jensen [64]. Presently, however, the societal impact of the firm and its governance structure is a growing topic of debate. Here we will show that also in finance there are many roads leading to Rome, or rather to the designation MCDA. Whether one belongs to the camp of Jensen or to the camp of those advocating socially responsible entrepreneurship, one has to deal with multiple criteria.

There is a series of situations in which the firm chooses (or has to take account of) a multiplicity of objectives and (policy) constraints. An overview of these situations is depicted in Figure 20.2. One issue is who decides on the objective(s)

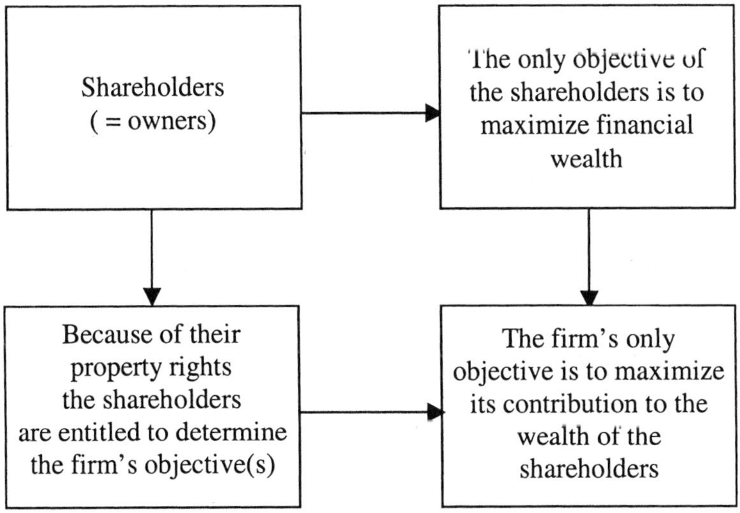

Figure 20.1. The neo-classical view on the objective of the firm.

of the firm. If there is a multiplicity of parties who may decide what the firm is aiming for, one generally encounters a multitude of goals, constraints and considerations that – more often than not – will be at least partially conflictive. A clear example is the conflicting objectives arising from agency problems (Jensen & Meckling, [65]). This means that many decision problems include multiple criteria and multiple actors (viz. group decision making, negotiation theory, see Box 3 in Figure 20.2). Sometimes, all those who decide on what the firm should aim for agree upon exactly the same objective(s). In fact, this is what neo-classical financial theory assumes when adopting shareholder value maximization (Box 1 in Figure 20.2). In practice, there are many firms that explicitly strive for a multiplicity of goals, which naturally leads to decision problems with multiple criteria (Box 2 in Figure 20.2).

However, although these firms do explicitly state to take account of multiple objectives, there are still very few of these firms that make use of tools provided by the MCDA literature. In most cases firms maximize one objective subject to (policy) constraints on the other objectives. As such there is nothing wrong with such a procedure as long as the location of these policy constraints is chosen correctly. In practice, however, one often observes that there is no discussion at all about the location of the policy constraints. Moreover, there is often no idea about the trade-offs between the location of the various constraints and the objective function that is maximized. In our opinion, multiple criteria decision

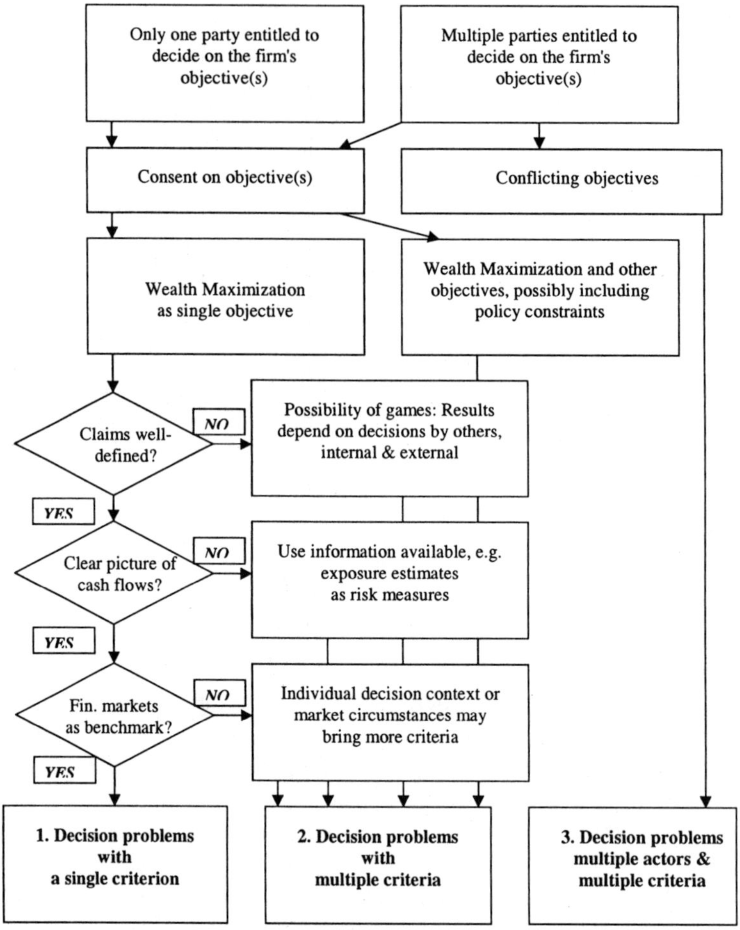

Figure 20.2. Situations leading to MCDA in the firm.

methodologies may help decision makers to gain better insights in the trade-offs they are confronted with.

Now let us get back to the case in which the owner(s) / shareholders do have only one objective in mind: wealth maximization. Although this is by definition the most prominent candidate for single criteria decision-making, we will argue that even in this case there are many circumstances in which the formulation as a multiple criteria decision problem is opportune.

In order to contribute maximally to the wealth of its shareholders, an individual firm should maximize the value of its shares. The value of these shares is determined on the financial markets by the forces of demand and supply. Shares represent claims on the future residual cash flows of the firm (and also on a usually very limited right on corporate control). In the view of the financial markets, the value of such a claim is determined relative to the claims of other firms that are traded on these markets. The financial markets' perception of the quality of these cash flow claims is crucial for the valuation of the shares. Translated to the management of the individual firm, the aim is not only to maximize the quality of the future residual cash flows of the firm but also to properly communicate all news about these cash flows to the financial markets. Only by the disclosure of such information can informational asymmetries be resolved and the fair market value of a cash flow claim be determined. In evaluating the possible consequences of its decision alternatives, management should estimate the effects on the uncertain (future) cash flows followed by an estimation of the financial markets' valuation of these effects. Then (and only then) the decision rule of management is very simple: choose the decision alternative that generates the highest estimated market value.

The first problem that might arise while following the above prescription is that residual claims cannot always be defined because of "gaming effects" (see Figure 20.2, Box 2). In other words, the future cash flows of the firm do not only depend on the present and future decisions of the firm's management, but also on the present and future decisions of other parties. An obvious example is the situation of oligopolistic markets in which the decisions of the competitors may strongly influence each other. Similar situations may arise with other external stakeholders such as powerful clients, powerful suppliers, and powerful financiers. Games may also arise within the firm, for instance between management and certain key production factors. The problem with game situations is that their effect on a firm's future cash flows caused by other parties involved cannot be treated in the form of simple constraints or as cost factors in cash flow calculations. MCDA may help to solve this problem by formulating multi-dimensional profiles of the consequences of the firm's decision alternatives. In these profiles, the effects on parties other than the firm are also included. These multi-dimensional profiles are the keys to open the complete MCDA toolbox.

A second problem in dealing with the single-objective wealth maximization problems is that the quality of information concerning the firm's future cash flows under different decision alternatives is far from complete. In addition, the available information may be biased or flawed. One way to approach the incomplete information problem is suggested by Spronk & Hallerbach [115]. In their multi-factorial approach, different sources of uncertainty should be identified after which the exposures of the cash flows to these risk sources are estimated. The estimated exposures can next be included in a multi-criteria decision method. In the case that the available information is not conclusive, different "views" on the future cash flows may develop. Next each of these views can be adopted as representing a different dimension of the decision problem. The resulting multi-dimensional decision problem can then be handled by using MCDA (see Figure 20.2, Box 2).

A third potential problem in wealth maximization is that the financial markets do not always provide relevant pricing signals to evaluate the wealth effects of the firm's decisions, for example, because of market inefficiencies. This means that the firm may want to include attributes in addition to the market's signals in order to measure the riskiness and wealth effects of its decisions.

2.6 A Multicriteria Framework for Financial Decision

In our view it, is the role of financial modelling to support financial decision making, as described in Hallerbach and Spronk [51], to build pointed models that take into account the peculiarities of the precise problem. The goal here is to bridge the gaps between decision-making disciplines, the discipline of financial economics, and the need for adequate decision support.

2.6.1 Principles. This framework is built on the principle that assumptions should be made where they help the modelling process the most and hurt the particular decision problem the least.[1] We call this the *Principle of Low Fat Modelling*. When addressing a decision situation, make use of all available information, but do not make unrealistic assumptions with respect to the availability of information. Do not make unrealistic assumptions that disqualify the decision context at hand. There should be ample room to incorporate idiosyncrasies of the decision context within the problem formulation, thus recognizing that the actual (non-average) decision maker is often very different from the "representative" decision maker. The preferences of the decision maker may not be explicitly available and may not even be known in detail by the decision maker himself. The uncertainty a decision maker faces with respect to the potential outcomes of his decisions may not be readily represented by means of a tractable statistical distribution. In many real-life cases, uncertainty can only be described in imprecise terms and available information is far from complete. And when the preferences of the decision maker are confronted with the characteristics of the decision alternatives, the conditional-normative nature of derived decision rules and advice should be accepted.

A second principle underlying our framework is the *Principle of Eclecticism*. One should borrow all the concepts and insights from modern financial theory that help to make better financial decisions. Financial theory can provide rich descriptions of uncertainty and risk. Examples are the multi-factor representation of risk in which the risk attached to the choice alternatives is conditioned on underlying factors such as the contingent claims approach in which the decision outcomes are conditioned on the opportunity to adjust or revise decisions in the future or game theory in which the outcomes are also conditioned on potential (conflicting) decisions made by other parties. But it is not the availability of theoretical insights that determines their application; it depends on the specific decision context at hand.

By restricting one thinking to a prechoosen set of problem characteristics, there is obviously more "to be seen" but at the same time it is possible to make observation errors, and maybe more worrisome, the problem and its context may be changing over time. This calls for the *Principle of Permanent Learning*, which stresses the process nature of decision making in which both the representation of the problem and the problem itself can change over time. Therefore, there is a permanent need to critically evaluate the problem formulation, the decisions made and their performance. Obviously, decision making and performance evaluation are two key elements in the decision-making process. As argued in Spronk and Vermeulen [116], performance evaluation of decisions should be structured such that the original idiosyncrasies of the problem (i.e., at the time the decision is made) are fully taken into account at the moment of evaluation, (i.e., *ex post*). By doing so, one increases the chance of learning from errors and misspecifications in the past.

2.6.2 Allocation Decisions. Financial decisions are allocation decisions, in which both time and uncertainty (and thus risk) play a crucial role. In order to support decisions in finance, both the preferences of the decision maker and the characteristics of the choice alternatives should be adequately understood and related to each other. A distinction can be made between "pure" financial decisions in which cash flows and market values steer the decision and "mixed" financial decisions in which other criteria are also considered. In financial theory, financial decisions are considered to be pure. In practice, most decisions are mixed. Hallerbach & Spronk [50] show that many financial decisions are mixed and thus should be treated as multiple criteria decision problems.

The solution of pure financial decisions requires the analysis, valuation, and management of risky cash flow streams and risky assets. The solution of mixed financial problems involves, in addition, the analysis of other effects. This implies that, in order to describe the effects of mixed decisions, multi-dimensional impact profiles should be used (cf. Spronk & Hallerbach [115]). The use of multi-dimensional impact profiles naturally opens the door to MCDA. Another distinction that can be made is between the financial decisions of individuals on one hand the financial decisions of companies and institutions on the other. The reason for the distinction results from the different ways in which decision makers steer the solutions. Individual decisions are guided by individual preferences (e.g., as described by utility functions), whereas the decisions of corporations and institutions are often guided by some aggregate objective (e.g., maximization of market value).

2.6.3 Uncertainty and Risk[2]. In each of the types of financial decisions just described, the effects are distributed over future time periods and are uncertain. In order to evaluate these possible effects, available information should be used to develop a "picture" of these effects and their likelihood. In some settings there is complete information but more often information is incomplete. In our framework, we use multi-dimensional risk profiles for modelling uncertainty and risk. This is another reason why multicriteria decision analysis is opportune when solving financial decision problems. *Two questions* play a crucial role:

1. Where does the uncertainty stem from or, in other words, what are the sources of risk?

2. When and how can this uncertainty be changed?

The answer to the first question leads to the decomposition of uncertainty. This involves attributing the inherent risk (potential variability in the outcomes) to the variability in several underlying state variables or factors. We can thus view the outcomes as being *generated* by the factors. Conversely, the stochastic outcomes are *conditioned* on these factors. The degree in which fluctuations in the factors propagate into fluctuations in the outcomes can be measured by response coefficients. These sensitivity coefficients can then be interpreted as exposures to the underlying risk factors and together they constitute the multi-dimensional risk profile of a decision alternative.

The answer to the second question leads to three prototypes of decision problems:

(1) The decision maker makes and implements a final decision and waits for its outcome. This outcome will depend on the evolution of external factors, beyond the decision maker's control.

(2) The decision maker makes and implements a decision and observes the evolution of external factors (which are still beyond the decision maker's control). However, depending on the value of these factors, the decision maker may make and implement additional decisions. For example, a decision maker may decide to produce some amount of a new and spectacular software package and then, depending on market reaction, he may decide to stop, decrease, or increase production.

(3) As in (2), but the decision maker is not the sole player and thus has to take account of the potential impact of decisions made by others sometime in the future (where the other(s) are of course confronted with a similar type of decision problem). The interaction between the various players in the field gives rise to dynamic game situations.

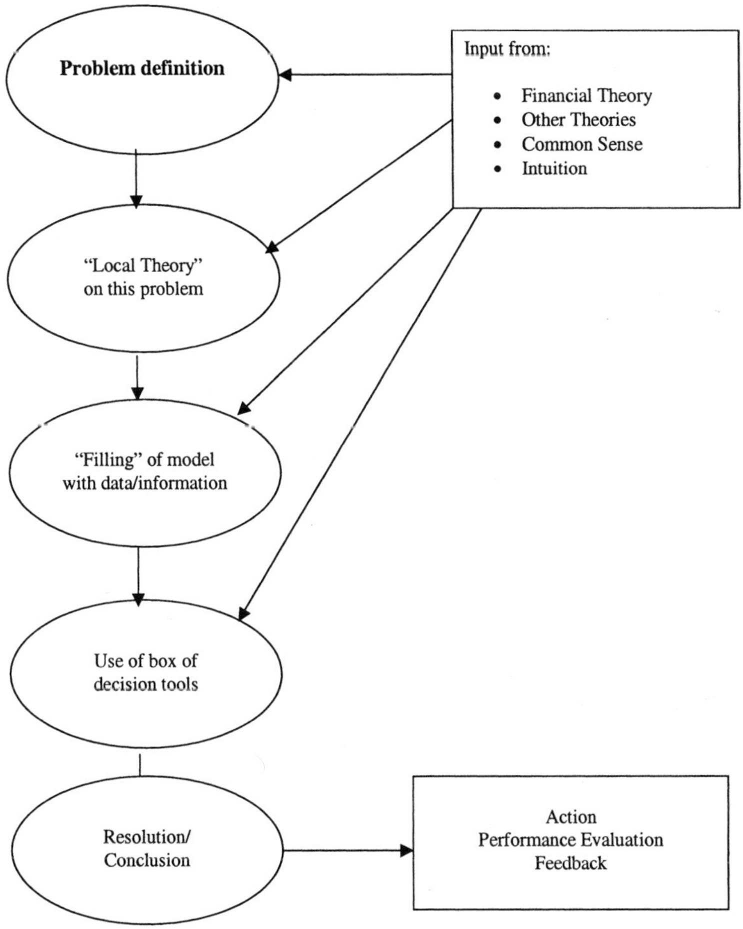

Figure 20.3. A bird's-eye view of the framework.

2.6.4 A Bird's-eye View of the Framework. In Figure 20.3, a bird's-eye view of the framework is presented. The framework integrates several elements in a process-oriented approach towards financial decisions. The left side of Figure 20.3 represents the elements that lead to decisions, represented by the Resolution/Conclusion box at the lower left hand side. As mentioned above, performance evaluation (shown at the lower right hand side of the figure) is an integral part of the decision-making process. However, in this article we do not pay further attention to performance evaluation or to the feedback leading from performance evaluation to other elements of the decision-making process.

Financial decision problems will often be put as allocation problems. At this stage, it is important to determine whether the problem is a mixed or pure financial problem. Also, one should know who decides and which objectives are to be served by the decisions.

In the next step, the problem is defined more precisely. Many factors play a role here. For instance, the degree of upfront structure in the problem definition, the similarity with other problems, time and commitment from the decision makers, availability of time, similarity to problems known in theory and so on. In this stage, the insights from financial theory often have to be supplemented (or even amended) by insights from other disciplines and by the discipline of common sense. The problem formulation can thus be seen as a theoretical description (we use the label "local theory") of the problem.

After the problem formulation, data have to be collected, evaluated and sometimes transformed into estimates. These data are then used as inputs for the formalization of the problem description. The structure of the problem, together with the quality and availability of the data determines what tools can be used and in which way. As explained above, the use of multi-dimensional impact profiles almost naturally leads to the use of multicriteria decision analysis.

2.6.5 The Framework and Modern Financial Theory. In our framework we try to borrow all concepts and insights from modern financial theory that help to make better financial decisions. Financial theory provides rich and powerful tools for describing uncertainty and risk. Examples are the multi-factor representation of risk, which leads to multi-dimensional impact profiles that can be integrated within multicriteria decision analysis. A very important contribution of financial theory is the contingent claims approach in which the decision outcomes are conditioned on the opportunity to adjust or revise decisions in the future. This comes together with financial markets where contingent claims are being traded in volume. This brings us to the role of financial markets as instruments to trade risks, to redistribute risks, and even to decrease or eliminate risk. We believe and hope that contingent claims thinking will also be used in other domains than finance. In the first place because of what it adds when describing decision problems. Secondly, new markets may emerge in which also non-financial risks can be handled in a better way.

In addition to helping to better describe decision problems, financial theory provides a number of crucial insights. The most obvious (which is clearly not limited to financial economics) is probably the concept of "best alternative opportunity" thinking. Whenever making an evaluation of decision alternatives, one should take into account that the decision maker may have alternative opportunities (often but not exclusively provided by markets), the best of which sets a benchmark for the evaluation of the decision alternatives considered.

Other concepts are the efficient market hypothesis and the no-arbitrage condition. These point both to the fact that in competitive environments, it is not obvious that one can outsmart all the others. So if you find ways to make easy money, you should at least try to answer the question why you have been so lucky and how the environment will react.

3. MCDA in Portfolio Decision-Making Theory

We now turn our attention to the area of finance known as portfolio theory. In portfolio theory, we study the attributes of collections of securities called portfolios and how investors process these attributes in order to determine the securities that are ultimately selected to form a portfolio.

At the core of portfolio theory is the portfolio selection problem. Formulated as an optimization problem, this is a problem that has been studied extensively. However, the problem that has been the subject of so much study for over fifty years is only two-dimensional, able to address only the two criteria of risk (as measured by standard deviation) and return. To more realistically model the problem and be better prepared for the future which will only be more complicated, we now discuss the issues involved in generalizing portfolio selection to include criteria beyond standard deviation and return. In this way, MCDA in the form of multiple criteria optimization enters the picture. The word "multiple" of course means two or more, but in this paper we will generally use it to mean more than two. We now explore the possibilities of multiple objectives in portfolio selection and discuss the effects of recognizing multiple criteria on the traditional assumption and practice of portfolio selection in finance.

In this portion of the paper, we are organized as follows. In Subsection 3.1 we describe the risk-return portfolio selection problem in finance. In Subsection 3.2 we show how the problem, although with only two objectives, can be recast in a multiple criteria optimization framework. In Subsection 3.3 we discuss two popular variants of the portfolio selection model, the short-sales permitted and short-sales prohibited models, and in Subsection 3.4 we discuss the bullet-shaped feasible regions that so often accompany portfolio optimization problems. In Subsection 3.5 we review some of key assumptions of portfolio analysis and discuss the sensitivity of the nondominated set to changes in various factors. With the sensitivity of the nondominated set indicating the presence of additional criteria beyond risk and return, an expanded multiple criteria portfolio optimization formulation is proposed in Subsection 3.6. With the nondominated frontier transformed into a nondominated surface, Subsection 3.7 reports on the idea that the "modern portfolio analysis" of today is probably best seen in the large as the projection onto the risk-return plane of the real multiple criteria portfolio selection problem in higher dimensional space.

In Subsection 3.8 we comment on some future research directions in MCDA in portfolio analysis.

3.1 Portfolio Selection Problem

In finance, due to Markowitz [81, 82, 83], we have the canonical portfolio selection problem as follows. Assume

(a) n securities
(b) an initial sum of money to be invested
(c) the beginning of a holding period
(d) end of a holding period.

Let $x_1, \dots, x_n$ denote *investment proportion* weights. These are the proportions of the initial sum to be invested at the beginning of the holding period in the n securities to form a portfolio. Also, for each $i \in \{1, \dots, n\}$, let r_i be the random variable for the percent return realized on security i between the beginning of the holding period and the end of the holding period. Then r_p, the random variable for the percent return realized on the portfolio between the beginning of the holding period and the end of the holding period, is given by

$$r_p = \sum_{i=1}^{n} r_i x_i$$

Unfortunately, r_p is not deterministic because it is based on upon the r_i. Thus it is not possible to know at the beginning of the holding period the value to be achieved by r_p at the end of the holding period. However, it is assumed that at the beginning of the holding period we have in our possession all expected values $E\{r_i\}$, predicted variances σ_{ii}, and predicted covariances σ_{ij} for the n securities.

Since r_p is not deterministic and an investor would presumably wish to protect against low values of r_p from in fact turning out to be the case, the approach considered prudent in portfolio selection is to seek a solution (that is, of investment proportion weights) that produces both a high expected value of r_p and a low predicted standard deviation value of r_p. Using the $E\{r_i\}$, σ_{ii} and σ_{ij}, the expected value of r_p is given by

$$E\{r_p\} = \sum_{i=1}^{n} E\{r_i\} x_i \tag{20.1}$$

and the predicted standard deviation of r_p is given by

$$\sigma\{r_p\} = \sqrt{\sum_{i=1}^{n} \sum_{j=1}^{n} x_i \sigma_{ij} x_j} \tag{20.2}$$

As for constraints, there is always the "sum-to-one" constraint

$$\sum_{i=1}^{n} x_i = 1 \tag{20.3}$$

Depending on the model being built, there may be constraints in addition to the above and restrictions on the variables such as

$$\ell_i \geq x_i \geq \mu_i \text{ for all } i \tag{20.4}$$

The way (20.1)-(20.4) is solved is as follows. First compute the set of all of the model's "nondominated" combinations of expected return and standard deviation. Then, after examining the set, which is portrayed graphically in the form of a curved line, the investor selects the nondominated combination that he or she feels strikes the best balance between expected return $E\{r_p\}$ and predicted standard deviation $\sigma\{r_p\}$.

With $E\{r_p\}$ to be maximized and $\sigma\{r_p\}$ to be minimized, (20.1)-(20.4) is a multiple objective program. Although the power of multiple criteria optimization is not necessary with two-objective programs (because they can be addressed with single criterion techniques), the theory of multiple criteria optimization is necessary when wishing to generalize portfolio selection, as we do, to take into account additional criteria.

3.2 Multiple Criteria Optimization

In operations research, there is the multiple criteria optimization problem. In its formulation, apart from having more than one objective, it looks like any other mathematical programming problem, but its solution is more involved. To handle both maximization and minimization objectives, we have

$$\begin{aligned} \max \textit{ or } \min \{f_1(\mathbf{x}) = z_1\} \\ \vdots \\ \max \textit{ or } \min \{f_k(\mathbf{x}) = z_k\} \\ \text{s.t.} \qquad \mathbf{x} \in S \end{aligned} \tag{MC}$$

where $\mathbf{x} \in R^n$, k is the number of objectives, and the z_i are *criterion values.* In multiple criteria optimization there are two feasible regions. One is $S \subset R^n$ in *decision space* and the other is $Z \subset R^k$ in *criterion space.* Let $\mathbf{z} \in R^k$. Then *criterion vector* $\mathbf{z} \in Z$ if and only if there exists an $\mathbf{x} \in S$ such that $\mathbf{z} = (f_1(\mathbf{x}), \ldots, f_k(\mathbf{x}))$. In this way, Z is the set of all *images* of the $\mathbf{x} \in S$

Criterion vectors in Z are either *nondominated* or *dominated*, and points in S are either *efficient* or *inefficient*. Letting $J^+ = \{\, i \mid f_i(\mathbf{x}) \text{ is to be maximized}\}$ and $J^- = \{j \mid f_j(\mathbf{x}) \text{ is to be minimized}\}$, we have

DEFINITION 76 *Assume formulation* (MC). *Then* $\bar{\mathbf{z}} \in Z$ *is a* nondominated *criterion vector if and only if there does not exist another* $\mathbf{z} \in Z$ *such that* **(i)** $z_i \geq \bar{z}_i$ *for all* $i \in J^+$, *and* $z_j \leq \bar{z}_j$ *for all* $j \in J^-$, *and* **(ii)** $z_i > \bar{z}_i$ *or* $z_j < \bar{z}_j$ *for at least one* $i \in J^+$ *or* $j \in J^-$. *Otherwise,* $\bar{\mathbf{z}} \in Z$ *is* dominated.

The set of all nondominated criterion vectors is designated N and is called the *nondominated set.*

DEFINITION 77 *Let* $\bar{\mathbf{x}} \in S$. *Then* $\bar{\mathbf{x}}$ *is* efficient *in* (MC) *if and only if its* image *criterion vector* $\bar{\mathbf{z}} = (f_1(\bar{\mathbf{x}}), \ldots, f_k(\bar{\mathbf{x}}))$ *is nondominated, that is, if and only if* $\bar{\mathbf{z}} \in N$. *Otherwise,* $\bar{\mathbf{x}}$ *is* inefficient.

The set of all efficient points is designated E and is called the *efficient set*. Note the distinction that is be made with regard to terminology. While nondominance is, of course, a criterion space concept, in multiple criteria optimization, efficiency is only a decision space concept.

To define optimality in a multiple criteria optimization problem, let $U : R^k \rightarrow R$ be the decision maker's utility function. Then, any $\mathbf{z}^o \in Z$ that maximizes U over Z is an *optimal criterion vector*, and any $\mathbf{x}^o \in S$ such that $(f_1(\mathbf{x}^o), \ldots, f_k(\mathbf{x}^o)) = \mathbf{z}^o$ is an *optimal solution*. We are interested in the efficient and nondominated sets because if U is such that *more-is-always-better-than-less* for each z_i, $i \in J^+$, and *less-is-always-better-than-more* for each z_j, $j \in J^-$, then any $\mathbf{z}^o$ optimal criterion vector is such that $\mathbf{z}^o \in N$, and any feasible *inverse image* $\mathbf{x}^o$ is such that $\mathbf{x}^o \in E$. The significance of this is that to find an optimal criterion vector $\mathbf{z}^o$, it is only necessary to find a best point in N. Since N is normally a portion of the surface of Z, this is much better than having to search all of Z. After a $\mathbf{z}^o$ has been found, it is only necessary to obtain an $\mathbf{x}^o \in S$ inverse image to know what to implement to achieve the k simultaneous performances indicated by the values in $\mathbf{z}^o$.

Unfortunately, although N is a portion of the surface of feasible Z in criterion space, locating the best solution in N, when $k > 2$, is often a non-trivial task because of the size of N. As a result, a large part of the field of multiple criteria optimization is concerned with procedures, mostly interactive, for searching N for an optimal or *near-optimal* solution, where a near-optimal solution is close enough to being optimal to terminate the decision process.

Thus, in this framework, the portfolio selection problem of (20.1)-(20.4) now appears as the two-objective multiple criteria optimization problem

$$\begin{aligned}
\min \quad & \left\{ \sum_{i=1}^{n} \sum_{j=1}^{n} x_i \sigma_{ij} x_j = z_1 \right\} \\
\max \quad & \left\{ \sum_{i=1}^{n} E\{r_i\} x_i = z_2 \right\} \\
\text{s.t.} \quad & \sum_{i=1}^{n} x_i = 1 \\
& \ell_i \geq x_i \geq \mu_i \text{ for all } i
\end{aligned} \tag{MC-Orig}$$

In (MC-Orig), z_1 is predicted variance and z_2 is expected return. But why variance instead of standard deviation? Whereas standard deviation is more intuitive, mathematical programming formulations as in (MC-Orig) most often have the risk objective expressed in terms of variance since quadratic routines are typically employed in a workhorse software capacity when analyzing the problem. Square roots can always be taken manually later.

3.3 Two Model Variants

Two model variants of (20.1)-(20.4) have evolved as classics. One is the *unrestricted* model

$$\begin{aligned}
\min \quad & \left\{ \sum_{i=1}^{n} \sum_{j=1}^{n} x_i \sigma_{ij} x_j = z_1 \right\} \\
\max \quad & \left\{ \sum_{i=1}^{n} E\{r_i\} x_i = z_2 \right\} \\
\text{s.t.} \quad & \sum_{i=1}^{n} x_i = 1 \\
& \text{all } x_i \text{ unrestricted}
\end{aligned} \tag{MC-Unrestr}$$

meaning that no constraints beyond the sum-to-one constraint are allowed in the model. The other is the *variable-restricted* model

$$\begin{aligned} \min & \left\{ \sum_{i=1}^{n}\sum_{j=1}^{n} x_i \sigma_{ij} x_j = z_1 \right\} \\ \max & \left\{ \sum_{i=1}^{n} E\{r_i\} x_i = z_2 \right\} \\ \text{s.t.} & \sum_{i=1}^{n} x_i = 1 \\ & \ell_i \geq x_i \geq \mu_i \text{ for all } i \end{aligned} \qquad \text{(MC-Bounds)}$$

in which lower and upper bounds exist on the weights. The significant aspect of the unrestricted model is that there are no limits on the negativities of the weights, meaning that *unlimited* short selling is permitted. To illustrate the short selling of a security, let $x_3 = -.2$. This would say the following to an investor. Borrow a position in security 3 to the extent of 20% of the initial sum to be invested. Then immediately sell it to generate extra cash. Now with the 120% of the initial sum, invest it as dictated by the other x_i weights to complete the portfolio.

The unrestricted model is the clear favorite in teaching and academic research. In research, the unrestricted model has long provided fertile ground for academics because of its beautiful mathematical properties. For example, as long as the variance/covariance matrix

$$\mathbf{V} = \begin{bmatrix} \sigma_{11} & \sigma_{12} & \cdots & \sigma_{1n} \\ \sigma_{21} & \sigma_{22} & & \\ \vdots & & & \vdots \\ \sigma_{n1} & & \cdots & \sigma_{nn} \end{bmatrix}$$

is nonsingular, every imaginable piece of information about the model appears to be derivable in closed form (see for example Roll [100], pp. 158–165). In teaching, this is an advantage because via this model portfolio selection can be taught without having to have mathematical programming included in the curriculum (which it hardly ever is any more in finance, even in Ph.D. programs).

One the other hand, the variable-restricted model is the clear favorite in practice. For instance, in the US, short selling is prohibited by law in the $6 trillion mutual fund business. It is also prohibited in the management of pension assets. And even in hedge funds where short selling is part of their strategy, it is all but impossible to imagine any situation in which there wouldn't be limits. The question is, when trying to locate an optimal solution, how much difference might there be in the locations of the nondominated frontiers of the two models, and might any differences be cause for concern?

3.4 Bullet-Shaped Feasible Regions

When looking through the portfolio chapters of almost any university investments text, it would be hard to miss seeing graphs of bullet-shaped regions, often with dots in them, as in Figure 20.4 (*top*). When unbounded (which in almost all books they are), and with standard deviation on the horizontal and expected return $E\{r_p\}$ on the vertical, these are all graphs of the feasible region Z of (MC-Unrestr) in criterion space. The dots within Z typically signify the criterion vectors $(\sigma\{r_i\}, E\{r_i\})$ of individual securities.

In contrast, the feasible region Z of an (MC-Bounds) formulation is as in Figure 20.4 (*bottom*). As a subset of the Z of its (MC-Unrestr) counterpart, the Z of an (MC-Bounds) formulation is easily recognizable by its "scalloped" rightmost boundary.

To see why a feasible region Z of (MC-Unrestr) is continuous, bullet-shaped, and unbounded, let us first consider the two securities A and B in Figure 20.5. The unbounded line sweeping through A and B, which is actually a hyperbola, is the set of criterion vectors of all two-stock portfolios resulting from all linear combinations of A and B whose weights sum to one. In detail, all points on the hyperbola strictly between A and B correspond to weights $x_a > 0$ and $x_b > 0$; all points on the hyperbola above and to the right of A correspond to weights $x_a > 1$ and $x_b < 0$; and all points on the hyperbola below and to the right of B correspond to weights $x_a < 0$ and $x_b > 1$. The degree of "bow" toward the vertical axis of the hyperbola is a function of the correlation coefficient ρ_{be} between A and B. This is seen by looking at the components of the $(\sigma\{r_{ab}\}, E\{r_{ab}\})$ criterion vector of any two-stock portfolio which are given by

$$\sigma\{r_{ab}\} = \sqrt{\sigma_{aa}x_a^2 + 2\rho_{ab}\sigma_a\sigma_b x_a x_b + \sigma_{bb}x_b^2} \tag{20.5}$$

and

$$E\{r_{ab}\} = E\{r_a\}x_a + E\{r_b\}x_b \tag{20.6}$$

in which $\sigma_a = \sqrt{\sigma_{aa}}$ and $\sigma_b = \sqrt{\sigma_{bb}}$. Whereas $E\{r_{ab}\}$ is linear, the positive value of $\sigma\{r_{ab}\}$ decreases nonlinearly as a function of ρ_{be} as its value goes from 1 to -1, and hence the bowing effect.

Through B and C there is another hyperbola. And since through any point on the hyperbola through A and B and any point on the hyperbola through B and C there is another hyperbola (not shown), and so forth, the feasible region fills in and takes on its bullet shape whose boundary, in the case of (MC-Unrestr), is a (single) hyperbola as well.

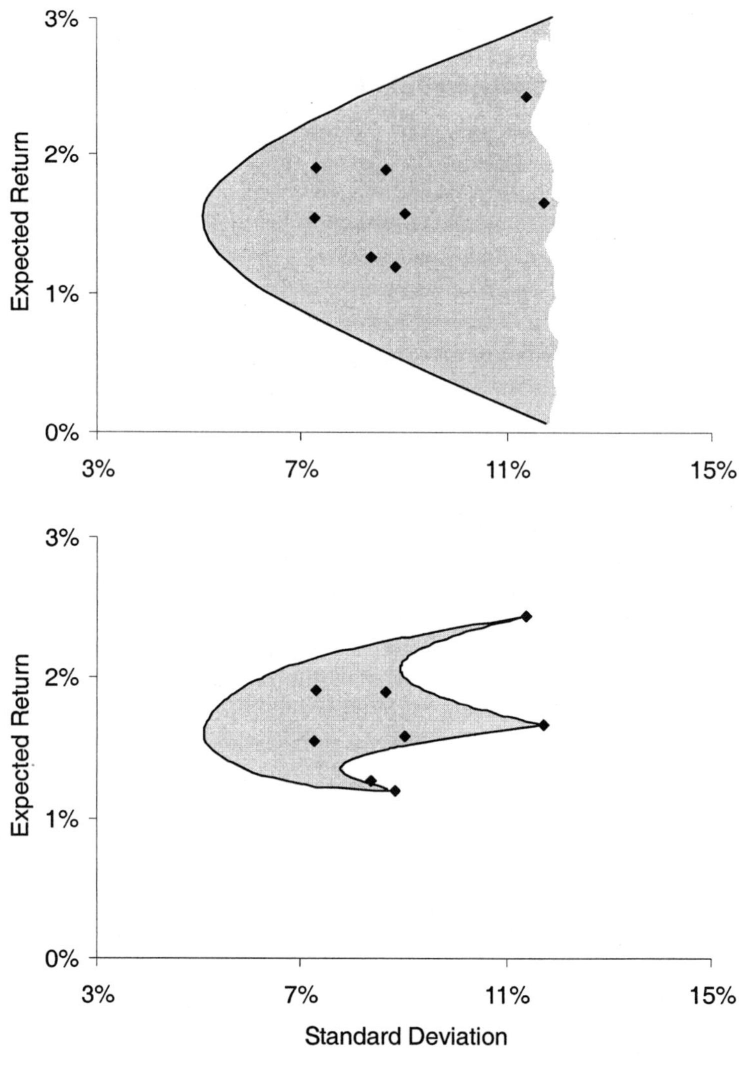

Figure 20.4. Feasible regions Z of (MC-Unrestr) and (MC-Bounds) for the same eight securities.

With regard to the feasible region Z of (MC-Bounds), the hyperbolic lines through the criterion vectors of any two financial products are not unbounded. In each case, they end in each direction at some point because of the bounds on the variables. While still filling in to create a bullet-shaped Z, the leftmost boundary, instead of being formed by a single hyperbola, is in general piecewise hyperbolic. The rightmost boundary, instead of being unbounded takes on it trademark "scalloped" effect.

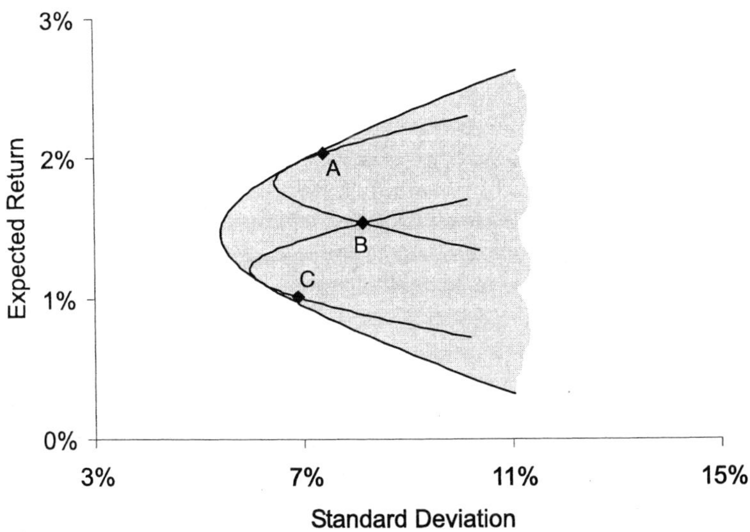

Figure 20.5. Continuous, bullet-shaped, and unbounded feasible region Z created by securities A, B and C.

Because predicted standard deviation is to be minimized and expected return is to be maximized, we look to the "northwest" of Z for the nondominated set. This causes the nondominated set to be the upper portion of the leftmost boundary (the portion that is non-negatively sloped). In finance, they call this the "efficient frontier." However, this causes a terminological conflict with the distinction indicated earlier about efficiency/inefficiency being reserved for points in decision space and nondominated/dominated being reserved for vectors in criterion space. Rather than the efficient frontier, we will refer to it as the "nondominated frontier," not only because this is consistent with the terminology of multiple criteria optimization discussed earlier, but because nondominated is the more intuitive term in criterion space.

3.5 Assumptions and Nondominated Sensitivities

The assumptions surrounding the use of model (MC-Unrestr) and model (MC-Bounds), and theories based upon them, in finance are largely as follows.

- **(a)** There are many investors, each small, none of which can affect prices.
- **(b)** There are no taxes.
- **(c)** There are no transactions costs.
- **(d)** Each investor's asset universe is all publicly traded securities.
- **(e)** All investors are rational mean-variance optimizers.
- **(f)** All investors have utility functions whose indifference curves are convex-to-the-origin.
- **(g)** All investors share the same expected returns, predicted variances, and predicted covariances about the future. This is called *homogeneous expectations*.
- **(h)** All investors have the same single holding period.
- **(i)** Each security is infinitely divisible.

We now discuss the sensitivity of the nondominated frontier to factors that have implications about the appropriateness of this set of the assumptions. Sensitivity is measured by noting what happens to the nondominated frontier as the parameter associated with a given factor changes. We start by looking at the sensitivity of the nondominated frontier to changes in an upper bound common to all investment proportion weights. Then we discuss the likely sensitivities of the nondominated frontier to changes in other things such a portfolio dividend requirement, a social responsibility attribute to be possessed by a portfolio, and other matters of concern. The computer work required for testing such sensitivities is outlined in the following 7-step algorithmic procedure.

1. Start the construction of what we recognize in multiple criteria optimization as an e-constraint program by converting the expected return objective in (MC-Unrestr) and (MC-Bounds) to a $\geq$ constraint with right-hand side ρ.
2. Install in the e-constraint program whatever constraints are required to accommodate the factor parameter to be varied.
3. Set the factor parameter to its starting value.
4. Set ρ to its starting value.
5. Solve the e-constraint program and take the square root of the outputted variance to form the nondominated point $(\sigma\{r_p\}, \rho)$.
6. If ρ has reached its ending value, go to Step 7. Otherwise, increment ρ and go to Step 5.
7. Connect on a graph all of the nondominated points obtained from the current value of the factor parameter to achieve a display of the nondominated frontier of this factor parameter value. If the factor parameter has

reached its ending value, stop. Otherwise increment the factor parameter and go to Step 4.

To illustrate with regard to the testing of the sensitivity of the nondominated frontier to changes in the common upper bound on the x_i, we form the *e*-constraint program

$$\min \left\{ \sum_{i=1}^{n} \sum_{j=1}^{n} x_i \sigma_{ij} x_j = \sigma^2\{r_p\} \right\} \qquad \text{(Exp-1)}$$

$$\text{s.t.} \quad \sum_{i=1}^{n} E\{r_i\} x_i \geq \rho$$

$$\sum_{i=1}^{n} x_i = 1$$

$$\ell \leq x_i \leq \mu \quad i \in \{1, \ldots, n\}$$

in which $n = 20$; $\ell = -.05$ in all runs to permit mild short selling; and μ is set in turn to $1.00, .15, .10$ to generate three frontiers. Running for 25 different ρ values (experimenter's choice) for each μ-value, the three nondominated frontiers of Figure 20.6 result. The topmost frontier is for $\mu = 1.00$, the middle frontier is for $\mu = .15$, and the bottommost frontier is for $\mu = .10$.

As seen in Figure 20.6, the nondominated frontier undergoes major changes as we step through the three values of μ. Hence there is considerable sensitivity to the value of μ. Since, in the spirit of diversification, investors would presumably prefer smaller values of μ to larger values as long as portfolio performance is not seriously deteriorated in other aspects, we can see that an examination of the tradeoffs among risk, return, and μ are involved before a final decision can be made. Since an investor would probably have no way of knowing in advance his or her optimal value of μ without reference to its effects on risk and return, we have demonstrated that μ should probably be considered a criterion to be minimized.

Using the same 7-step algorithmic procedure, other experiments (results not shown) could be conducted. For example, if we wished to test the sensitivity of the nondominated frontier to changes in a expected portfolio dividend yield

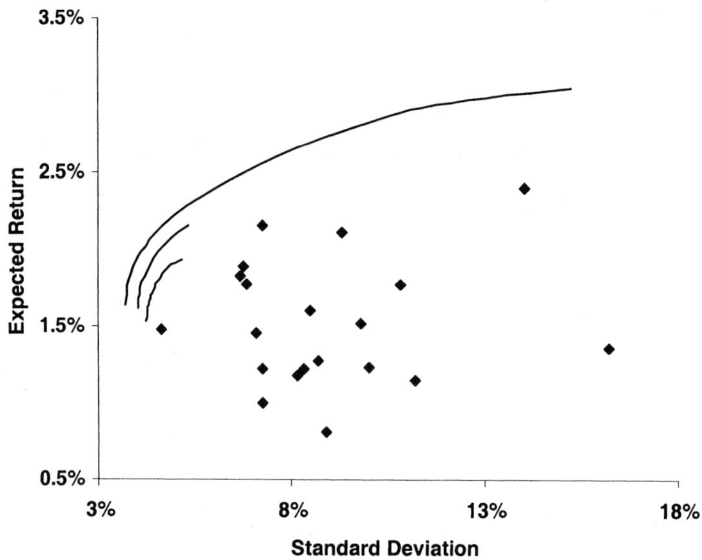

Figure 20.6. Nondominated frontiers as a function of changes in the value of upper bound parameter μ.

requirement, we would form the following e-constraint program

$$\min \left\{ \sum_{i=1}^{n} \sum_{j=1}^{n} x_i \sigma_{ij} x_j = \sigma^2\{r_p\} \right\} \quad \text{(Exp-2)}$$

$$\begin{aligned} \text{s.t.} \quad & \sum_{i=1}^{100} E\{r_i\} x_i \geq \rho \\ & \sum_{i=1}^{n} E\{d_i\} x_i \geq \delta \\ & \sum_{i=1}^{n} x_i = 1 \\ & \ell \leq x_i \leq \mu \quad i \in \{1, \ldots, n\} \end{aligned}$$

in which d_i is the random variable for the dividend yield realized on security i between the beginning and end of the holding period and δ is the expected portfolio dividend yield requirement. A similar type of formulation could be set up for social responsibility.

For both dividends and social responsibility we can probably expect to see nondominated frontier sensitivities along the lines of that for μ. If this is indeed the case, this would signal that dividends and social responsibility should probably be added to the list of criteria as well. With μ, we now see how it is easy to have more criteria than two in investing. Whereas the assumptions at the beginning of the subsection assume a two-criterion world, we are led to see new things by virtue of these experiments. One is that the assumption about risk and return being the only criteria is certainly under seige. Another is that, in the company of μ, dividends, and social responsibility, the last of which can be highly subjective, *individualism* should be given more play. By individualism, no investor's criteria, opinions, or assessments need conform to those of another. In direct conflict with the assumption about homogeneous expectations – which nobody believes in anyway – at the security level, individualism allows an investor to have differing opinions about any security's expected return, variance, covariance with any other security, liquidity, dividend outlook, social responsibility score, and so forth. At the portfolio level, for example, individualism allows investors to possess different lists of criteria, have differing objective functions for even the same criteria, work from different asset universes, and enforce different attitudes about the nature of allowable short selling or the number of securities to be tolerated in a portfolio. Therefore, in contrast to current theory, with different lists of criteria, different objective functions, and different sets of constraints, all investors would not face the same feasible region with the same nondominated set. Each would have his or her own portfolio problem with its own optimal solution. The benefit of this enlarged outlook would be that portfolio theory would then not only have to focus on explaining equilibrium solutions, but on customized solutions as well.

3.6 Expanded Formulations and New Assumptions

Generally, in multiple criteria, we recognize a constraint from an objective as follows. If when modelling as a constraint we realize that we can not easily fix a right-hand side value without knowing how other output measures turn out, then we are probably looking at an objective. With individualism, an investor could easily be looking at the expanded multiple criteria optimization problem as follows

$$
\begin{aligned}
&\min \{f_1(\mathbf{x}) = \text{risk}\} &&\text{(MC-Expand)}\\
&\max \{f_2(\mathbf{x}) = \text{return}\}\\
&\max \{f_3(\mathbf{x}) = \text{dividends}\}\\
&\min \{f_4(\mathbf{x}) = \text{maximum investment proportion weight}\}\\
&\max \{f_5(\mathbf{x}) = \text{social responsibility}\}\\
&\min \{f_6(\mathbf{x}) = \text{number of securities in portfolio}\}\\
&\min \{f_7(\mathbf{x}) = \text{short selling}\}\\
&\text{s.t.} \qquad \mathbf{x} \in S
\end{aligned}
$$

With regard to the number of securities in a portfolio, this can easily be a criterion. For individuals or mutual funds, every extra security is a paperwork headache and a distraction. Not only is resource time required to monitor the security, but there are also the monthly statements to absorb and file, annual reports to decide whether to read or not, proxy matters (shareholder proposals, mergers, name changes, spin offs, etc.) to be evaluated and voted upon, tax consequences to be dealt with, and the like. For most investors, they would just as soon wish to minimize much of the hassle. Also, as reflected by the last objective in (MC-Expand), investors open to the idea of short selling might nevertheless wish to minimize it if possible.

Updating to take a new look at portfolio selection, the following is proposed as a more appropriate set of assumptions with which to approach the study of portfolio theory when multiple criteria and individualism are to be taken into account.

- **(a)** There are many investors, each small, none of which can affect prices.
- **(b)** There are no taxes.
- **(c)** There are no transactions costs.
- **(d)** An investor's asset universe can be any subset of all publicly traded securities, even for large investors usually not more than a few hundred.
- **(e)** Investors may possess any mix of up to about six or eight objectives.
- **(f)** All investors have utility functions whose indifference curves are convex-to-the-origin.
- **(g)** Heterogeneity of expectations is the rule. That is, investors can be expected to have widely different forecasts about any security attribute including expected returns, predicted variances, and predicted covariances, expected dividends, and so forth.
- **(h)** Short selling is allowed but to only some limited extent.

The first three assumptions remain the same as they are nice to retain in that they establish benchmarks against which some of the world's imperfections can be measured. The assumption about convex-to-the-origin utility function contours is also retained as we see no compelling difficulty with it at the present time, but all the rest have either been modified or deleted.

3.7 Nondominated Surfaces

If multiple criteria exist in portfolio selection, then the nondominated set of current-day finance that exists as a frontier in R^2 is a *surface* in R^n. What evidence might there be to support this? In current-day finance there is the "market portfolio". By theory, the market portfolio contains every security in proportion to its market capitalization (number of shares times price), is somewhere in the midst of the nondominated frontier, and is supposed to be everyone's optimal portfolio when not including the risk-free asset. Since the market portfolio is impractical, indices like the S&P 500 are used as surrogates. But empirically, the surrogates, which should be essentially as desirable as the market portfolio, have always been found to be deep below the nondominated frontier, in fact so deep below that this cannot be explained by chance variation. Whereas this is an anomaly in conventional risk-return finance, this is exactly what we would expect in multiple criteria finance.

To take a glimpse at the logic as to why this is what we would expect, consider the following. In a risk-return portfolio problem, let us assume that the feasible region Z is the ellipse in Figure 20.7. Here, the nondominated frontier is the portion of the boundary of the ellipse in the second quadrant positioned at the center of the ellipse. Similarly, in a q-criterion portfolio problem (with $q-2$ objectives beyond risk and return), let us assume that the feasible region is an *ellipsoid* in q-space. Here, the nondominated surface is the portion of the surface of the ellipsoid in an orthant positioned at the center of the ellipsoid. Now assume that the market portfolio, which by theory is nondominated, is in the middle of the nondominated set. If this is the case, then the market portfolio would be at $\mathbf{z}^2$ on the ellipse. However, if (i) there is a third objective, (ii) the feasible region is ellipsoidal in three-space, and (iii) the market portfolio is in the middle of the nondominated surface in R^3, then the market portfolio would *project* onto risk-return space at $\mathbf{z}^3$. If (i) there is a fourth objective, (ii) the feasible region is ellipsoidal in four-space, and (iii) the market portfolio is in the middle of the nondominated surface in R^4, then the market portfolio would project onto risk-return space at $\mathbf{z}^4$. With five objectives under the same conditions, then the market portfolio would project onto risk-return space at $\mathbf{z}^5$, and so forth, becoming deeper and deeper.

Consequently, it may not be unreasonable to conjecture that what is, to use a term from Elton and Gruber [39], the "modern portfolio theory" of today is only a first-order approximation, a projection onto the risk-return plane, of the real multiple criteria problem from higher dimensional criterion space.

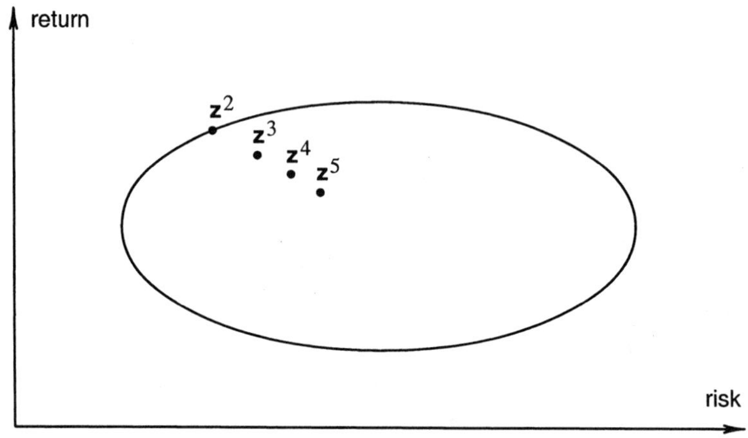

Figure 20.7. An ellipsoidal feasible region projected onto two-dimensional risk-return space.

3.8 Further Research in MCDA in Portfolio Analysis

In addition to further study into multiple criteria and individualism in investing, we also find intriguing for future research the area of special variable treatments in portfolio optimization. By special variable treatments, we mean conditions on the variables such as the following.

- **a.** No fewer than a given number of securities, and no more than a given number of securities, can be in a portfolio (either long or short).
- **b.** No more than a given number of securities can be sold short.
- **c.** If a stock is in a portfolio, then its weight must be in market cap proportion to the weights of all other stocks in the portfolio.
- **d.** No more than a given proportion of a portfolio can be involved in stocks sold short.
- **e.** Some or all of the x_i are semi-continuous. That is, an x_i is either zero or in a given interval $[a, b], a > 0$.
- **f.** No more than a given number of stocks may have a given upper bound. For instance, at most one stock (but which one is not known beforehand) may constitute as much as 25% of a portfolio, with all other stocks having an upper bound of 5%.

While some of these can be modelled with auxiliary 0-1 variables, others may only be amenable to local search algorithms as in Gandibleux, Caballero and Molina [42]. Having at one's disposal well-researched methods for dealing with such special variable treatments would extend the power at our new look at portfolio analysis when focusing on customized portfolio solutions. Another area of interest is the use of mean absolute deviation (MAD), which can be modelled linearly, as the risk measure in place of variance (standard deviation). Finally, it may be that multiple criteria and behavioral finance (see for example Shefrin [107]) reinforce one another as both area see much more going on in investing than the traditional.

References to some of the older classical papers on portfolio selection and to papers by authors who have most recently been taking a new look at portfolio theory along the lines discussed here are in the bibliography.

4. MCDA in Discrete Financial Decision-Making Problems

Several decision-making problems, including financial decision-making problems, require the evaluation of a finite set of alternatives $A = \{a_1, a_2, \ldots, a_m\}$, which may include firms, investment projects, stocks, credit applications, etc. These types of problems are referred to as "discrete" problems. The outcome of the evaluation process may have different forms, which are referred to as "problematics" [104]: (1) problematic α: Choosing one alternative, (2) problematic β: Sorting the alternatives in well defined groups defined in a preference order, (3) problematic γ: Ranking the alternatives from the best to the worst, and (4) problematic δ: Describing the alternatives in terms of their performance on the criteria. The selection of an investment project is a typical example of a financial decision-making problem where problematic α (choice) is applicable. The prediction of business failure is an example of problematic β (classification of firms as healthy or failed), the comparative evaluation and ranking of stocks according to their financial and stock market performance is an example of problematic γ, whereas the description of the financial characteristics of a set of firms is a good example of problematic δ.

The selection of one of these problematics depends solely on the objective of the analysis and the decision-making context. In each case, the evaluation process involves the aggregation of all the pertinent decision factors $F = (g_1, g_2, \ldots, g_n)$, which are referred to as "evaluation criteria" or simply "criteria". Formally, a criterion g_j is a non-decreasing real-valued function that describes an aspect of the global performance of the alternatives and defines how the alternatives are compared to each other, as follows:

$$\begin{aligned} g_{ij} > g_{kj} &\Leftrightarrow a_i \succ a_k \quad (a_i \text{ is preferred to } a_k) \\ g_{ij} = g_{kj} &\Leftrightarrow a_i \sim a_k \quad (a_i \text{ is indifferent to } a_k) \end{aligned}$$

where g_{ij} denotes the performance of alternative a_i on criterion g_j.

The aggregation of all criteria into an overall evaluation index can be performed in many different ways depending on the form of the criteria aggregation model. Within the MCDA field one can distinguish three main forms of aggregation models: (1) outranking relations (relational form), (2) utility functions (functional form), (3) decision rules (symbolic form). In all cases, the aggregation model is developed so as to respect the decision maker's judgment policy. To ensure that this purpose is achieved some information on the preferential system of the decision maker must be specified, such as the criteria weights. The required preferential information can be specified either through direct procedures in which a decision analyst elicits it directly from the decision maker, or through indirect procedures in which the decision maker provides examples of the decisions that he takes and the decision analyst analyzes them to determine the required preferential parameters which are most consistent with the decision maker's global evaluations. The latter approach is known in the MCDA field as "preference disaggregation analysis" [60].

The subsequent subsections in this portion of the paper present several MCDA discrete evaluation approaches which are suitable for addressing financial decision-making problems. The presentation is organized in terms of the criteria aggregation model employed by each approach (outranking relations, utility functions, decision rules).

4.1 Outranking Relations

The foundations of the outranking relations theory have been set by Bernard Roy during the late 1960s through the development of the ELECTRE family of methods (**EL**imination **E**t **C**hoix **T**raduisant la **RE**alité; [102]). Since then, they have been widely used by MCDA researchers, but mostly in Europe and Canada.

An outranking relation is a binary relation that enables the decision maker to assess the strength of the outranking character of an alternative a_i over an alternative a_k. This strength increases if there are enough arguments (coalition of the criteria) to confirm that a_i is at least as good as a_k, while there is no strong evidence to refuse this statement.

Outranking relations techniques operate into two stages. The first stage involves the development of an outranking relation among the considered alternatives, while the second stage involves the exploitation of the developed outranking relation to choose the best alternatives (problematic α), to sort them into homogenous groups (problematic β), or to rank them from the most to the least preferred ones (problematic γ).

Some of the most widely known outranking relations methods include the family of the ELECTRE methods [103] and the family of the PROMETHEE methods [16]. These methods are briefly discussed below. A detailed presentation of all outranking methods can be found in the books of Roy and Bouyssou [105] and Vincke [123].

ELECTRE Methods. The family of ELECTRE methods was initially introduced by Roy [102], through the development of the ELECTRE I method, the first method to employ the outranking relation concept. Since then several extensions have been proposed, including ELECTRE II, III, IV, IS and TRI [103]. These methods address different types of problems, including choice (ELECTRE I, IS), ranking (ELECTRE II, III, IV) and sorting/classification (ELECTRE TRI).

Given a set of alternatives $A = \{a_1, a_2, \dots a_m\}$ any of the above ELECTRE methods can be employed depending on the objective of the analysis (choice, ranking, sorting/classification). Despite their differences, all the ELECTRE methods are based on the identification of the strength of affirmations of the form $Q =$ "alternative a_i is at least as good as alternative a_k". The specification of this strength requires the consideration of the arguments that support the affirmation Q as well as the consideration of the arguments that are against it. The strength of the arguments that support Q is analyzed through the "concordance test". The measure used to assess this strength is the global concordance index $C(a_i, a_k) \in [0, 1]$. The closer is C to unity, the higher is the strength of the arguments that support the affirmation Q. The concordance index is estimated as the weighted average of partial concordance indices defined for each criterion:

$$C(a_i, a_k) = \sum_{j=1}^{n} w_j c_j(g_{ij} - g_{kj})$$

where w_j is the weight of criterion g_j ($\sum w_j = 1, w_j \geq 0$) and $c_j(g_{ij} - g_{kj})$ is the partial concordance index defined as a function of the difference $g_{ij} - g_{kj}$ between the performance of a_i and a_k on criterion g_j. The partial concordance index measures the strength of the affirmation Q' = "a_i is at least as good as a_k on the basis of criterion g_j". The partial index is normalized in the interval $[0, 1]$, with values close to 1 indicating that Q' is true and values close to 0 indicating that Q' is false.

Except for assessing the strength of the arguments that support the affirmation Q, the strength of the arguments against Q is also assessed. This is performed through the "discordance test", which leads to the calculation of the discordance index $D_j(g_{ij} - g_{kj})$ for each criterion g_j. Conceptually, the discordance index $D_j(g_{ij} - g_{kj})$ measures the strength of the indications against the affirmation Q'. The higher is the discordance index the more significant is the opposition of the criterion on the validity of the affirmation Q. If the strength of this opposition for criterion g_j is above a critical level (veto threshold), then the criterion vetoes the validity of the affirmation Q irrespective of the performance of the considered pair of alternatives (a_i, a_k) on the other criteria.

Once the concordance and discordance tests are performed, their results (concordance index C, discordance indices D_j) are combined to construct the final outranking relation. The way that this combination is performed, as well as the way that the results are employed to choose, rank, or sort the alternatives depends on the specific ELECTRE method that is used. Details on these issues can be found in the works of Roy [103, 104] as well as in the book of Roy and Bouyssou [105].

PROMETHEE Methods. The development of the PROMETHEE family of methods (**P**reference **R**anking **O**rganization **METH**od of **E**nrichment **E**valuations) began in the mid 1980s with the work of Brans and Vincke [16] on the PROMETHEE I and II methods.

The PROMETHEE method leads to the development of an outranking relation that can be used either to choose the best alternatives (PROMETHEE I) or to rank the alternatives from the most preferred to the least preferred (PROMETHEE II). For a given set of alternatives A, the evaluation process in PROMETHEE involves the pairwise comparisons (a_i, a_k) to determine the preference index $\pi(a_i, a_k)$ measuring the degree of preference for a_i over a_k, as follows:

$$\pi(a_i, a_k) = \sum_{j=1}^{n} w_j P_j(g_{ij} - g_{kj}) \in [0, 1]$$

The preference index is similar to the global concordance index of the ELECTRE methods. The higher is the preference index (closer to unity) the higher is the strength of the preference for a_i over a_k. The calculation of the preference index depends on the specification of the criteria weights w_j ($\sum w_j = 1$, $w_j \geq 0$) and the preference functions P_j for each criterion g_j. The preference functions are increasing functions of the difference $g_{ij} - g_{kj}$ between the performances of a_i and a_k on criterion g_j. The preference functions are normalized between 0 and 1. The case $P_j(a_i, a_k) \approx 1$ indicates a strong preference for a_i over a_k in terms of the criterion g_j, whereas the case $P_j(a_i, a_k) \approx 0$ indicates weak preference. Generally, the preference functions may have different forms, depending on the judgment policy of the decision maker. Brans and Vincke [16] proposed six specific forms (generalized criteria) which seem sufficient in practice.

The results of the comparisons made for all pairs of alternatives (a_i, a_k) are organized in a graph (value outranking graph). The nodes of the graph represent the alternatives under consideration, whereas the arcs between nodes a_i and a_k represent the preference of alternative a_i over a_k (if the direction of the arc is $a_i \rightarrow a_k$) or the opposite (if the direction of the arc is $a_k \rightarrow a_i$). Each arc is associated with a flow representing the preference index $\pi(a_i, a_k)$. The sum of all flows leaving a node a_i is called the leaving flow $\phi^+(a_i)$. The leaving flow provides a measure of the outranking character of alternative a_i over all the other alternatives in A. In a similar way, the sum of all flows entering a node a_i is called the entering flow $\phi^-(a_i)$. The entering flow measures the outranked character of alternative a_i compared to all the other alternatives in A.

On the basis of these flows the heuristic procedures of PROMETHEE I and II are employed to choose the best alternatives (PROMETHEE I) or to rank the alternatives from the most preferred to the least preferred (PROMETHEE II). The choice of the best alternatives in the PROMETHEE I method involves the definition of the preference (P), indifference (I) and incomparability (R) relations of the basis of the leaving and entering flows of the outranking graph [16]. The ranking of the alternatives in the PROMETHEE II method is based on the difference between the leaving and the entering flow $\phi(a_i) = \phi^+(a_i) - \phi^-(a_i)$, which provides the net flow for a node (alternative) a_i. The net flow constitutes the overall evaluation index of the performance of the alternatives. The most preferred alternatives are the ones with the higher net flows, whereas the alternatives with the lower net flows are considered as the least preferred ones.

4.2 Utility Functions-Based Approaches

The multiattribute utility theory (MAUT; [69]) extends the traditional utility theory to the multi-dimensional case. Even from the early stages of the MCDA field, the strong theoretical foundations of the MAUT framework have been among the cornerstones of the development of MCDA and its practical implementation. The objective of MAUT is to model and represent the decision maker's preferential system into a utility/value function $U(a_i)$. The utility function is defined on the criteria space, such that:

$$U(a_i) > U(a_k) \Leftrightarrow a_i \succ a_k \quad (a_i \text{ is preferred to } a_k) \tag{20.7}$$

$$U(a_i) = U(a_k) \Leftrightarrow a_i \sim a_k \quad (a_i \text{ is indifferent to } a_k) \tag{20.8}$$

The most commonly used form of utility function is the additive one:

$$U(a_i) = p_1 u_1(g_{i1}) + p_2 u_2(g_{i2}) + \ldots + p_n u_n(g_{in}) \tag{20.9}$$

where, $u_1, u_2, \ldots, u_n$ are the marginal utility functions corresponding the evaluation criteria. Each marginal utility function $u_j(g_j)$ defines the utility/value of the alternatives for each individual criterion g_j. The constants $p_1, p_2, \ldots, p_n$ represent the criteria trade-offs that the decision maker is willing to take. These constants are often considered to represent the weights of the criteria and they are defined such that they sum-up to unity.

A detailed description of the methodological framework underlying MAUT and its applications is presented in the book of Keeney and Raiffa [69].

Generally, the process for developing an additive utility function is based on the cooperation between the decision analyst and the decision maker. This process involves the specification of the criteria trade-offs and the form of the marginal utility functions. The specification of these parameters is performed through interactive procedures, such as the midpoint value technique [69]. The realization of such interactive procedures is often facilitated by the use of multicriteria decision support systems, such as the MACBETH system [7].

However, the implementation of such interactive procedures in practice can be cumbersome, mainly because it is rather time consuming and it depends on the willingness of the decision maker to provide the required information and the ability of the decision analyst to elicit it efficiently. The preference disaggregation approach of MCDA (PDA; [60]) provides the methodological framework to cope with this problem. PDA refers to the analysis (disaggregation) of the global preferences (judgement policy) of the decision maker in order to identify the criteria aggregation model that underlies the preference result (ranking or classification/sorting). Similarly to MAUT, preference disaggregation analysis uses common utility decomposition forms to model the decision maker's preferences. Nevertheless, instead of employing a direct procedure for estimating the global utility model (MAUT), preference disaggregation analysis uses regression-based techniques (indirect estimation procedure). More specifically, in PDA the parameters of the utility decomposition model are estimated through the analysis of the decision maker's overall preference on some reference alternatives A', which may include either examples of past decisions or a small subset of the alternatives under consideration. The decision maker is asked to provide a ranking or a classification of the reference alternatives according to his decision policy (global preferences). Then, using regression-based techniques the global preference model is estimated so that the decision maker's global evaluation is reproduced as consistently as possible by the model. A comprehensive bibliography on preference disaggregation methods can be found in Jacquet-Lagrèze and Siskos [60, 61].

PDA methods are particularly useful in addressing financial decision-making problems [136]. The repetitive character of financial decisions and the requirement for real-time decision support are two features of financial decisions which are consistent with the PDA framework. Thus, several PDA methods have been extensively used in addressing financial decision problems, mainly in cases where a ranking or sorting/classification of the alternatives is required. The following subsections provide a brief description of some representative PDA methods which have been used in financial problems.

UTA Method. The UTA method (**UT**ilités **A**dditives; [59]) is an ordinal regression method developed to address ranking problems. The objective of the method is to develop an additive utility function which is as consistent as possible with the decision maker's judgment policy. The input to the method involves a set of reference alternatives A'. For each reference alternative the decision maker is asked to provide his global evaluation so as to form a total pre-order of the alternatives in A': $a_1 \succ a_2 \succ \ldots \succ a_m$ (the indifference relation "$\sim$" as defined in (20.8) can also be used in the pre-order). The developed utility model is assumed to be consistent with the decision maker's judgment policy if it is able to reproduce the given pre-order of the reference alternatives as consistently as possible. In that regard, the utility model should be developed so that: $U(a_1) > U(a_2) > \ldots > U(a_m)$.

In developing the utility model to meet this requirement, there are two types of possible errors which may occur [110]): (1) the under-estimation error when the developed model assigns a reference alternative to a lower (better) rank than the one specified in the given pre-order (the alternative is under-estimated by the decision maker), and (2) the over estimation error when the developed model assigns a reference alternative to a higher (worse) rank than the one specified in the given pre-order (the alternative is over-estimated by the decision maker). The objective of the model development process is to minimize the sum of these errors. This is performed through linear programming techniques [59].

UTADIS Method. The UTADIS method (**UT**ilités **A**dditives **DIS**criminantes; [37, 58]) is a variant of the UTA method, developed for sorting/classification problems. Similarly to the UTA method, the decision maker is asked to provide his global evaluation on a set of reference alternatives A'. In this case, however, the decision maker is not asked to rank the alternatives is A'. Instead, he classifies the reference alternatives into homogenous groups $C_1, C_2, \ldots, C_q$ defined in an ordinal way, such that $C_1 \succ C_2 \succ \cdots \succ C_q$ (i.e., group C_1 includes the most preferred alternatives, whereas group C_q includes the least preferred ones). Within this context, the developed additive utility model will be consistent with the decision maker's global judgment if the following conditions are satisfied:

$$\left.\begin{array}{ll} U(a_i) \geq t_1, & \forall\, a_i \in C_1 \\ t_1 > U(a_i) \geq t_2, & \forall\, a_i \in C_2 \\ \cdots\cdots\cdots\cdots & \cdots\cdots \\ U(a_i) < t_{q-1}, & \forall\, a_i \in C_q \end{array}\right\} \tag{20.10}$$

where $t_1 > t_2 > \ldots > t_{q-1}$ are thresholds defined in the global utility scale $[0, 1]$ to discriminate the groups (each t_k is the lower bound of group C_k). Similarly, to the UTA method, the under-estimation and over-estimation errors are also used in the UTADIS method to measure the differences between the model's results and the predefined classification of the reference alternatives. In this case, the two types of errors are defined as follows: (1) the under-estimation error $\sigma_i^+ = \max\{0, t_k - U(a_i)\}, \forall\, a_i \in C_k,\ k = 1, 2, \ldots, q-1$, (2) the over-estimation error $\sigma_i^- = \max\{0, U(a_i) - t_{k-1}\}, \forall\, a_i \in C_k,\ k = 2, 3, \ldots, q$.

The additive utility model is developed to minimize these errors using a linear programming formulation [37].

Recently, several new variants of the original UTADIS method have been proposed (UTADIS I, II, III) to consider different optimality criteria during the development of the additive utility classification model [37, 139].

MHDIS Method. The MHDIS method (**M**ulti-group **H**ie-rarchical **DIS**crimination [143]) extends the PDA framework of the UTADIS method in complex sorting / classification problems involving multiple groups (of course the method is also applicable in the simple two-group case). As the name of the method implies, MHDIS addresses sorting problems through a hierarchical procedure, during which the groups are distinguished progressively, starting by discriminating group C_1 (most preferred alternatives) from all the other groups $\{C_2, C_3, \ldots, C_q\}$, and then proceeding to the discrimination between the alternatives belonging to the other groups. At each stage of this sequential/hierarchical process two additive utility functions are developed for the classification of the alternatives. Assuming that the classification of the alternatives should be made into q ordered classes $C_1 \succ C_2 \succ \cdots \succ C_q$, $2(q-1)$ additive utility functions are developed. These utility functions have the following additive form:

$$U_k(a_i) = \sum_{j=1}^{n} u_{kj}(g_{ij}), \quad U_{\sim k}(a_i) = \sum_{j=1}^{n} u_{\sim kj}(g_{ij}) \tag{20.11}$$

Both functions are defined between 0 and 1. The function U_k measures the utility for the decision maker of a decision to assign an alternative into group C_k, whereas the second function $U_{\sim k}$ corresponds to the classification into the set of groups $C_{\sim k} = \{C_{k+1}, C_{k+2}, \ldots, C_q\}$. The rules used to perform the classification of the alternatives are the following:

$$\left.\begin{array}{l}
\text{If } U_1(a_i) > U_{\sim 1}(a_i) \text{ then } a_i \in C_1 \\
\quad \text{Else if } U_2(a_i) > U_{\sim 2}(a_i) \text{ then } a_i \in C_2 \\
\quad \cdots\cdots\cdots\cdots\cdots\cdots \\
\qquad \text{Else if } U_{q-1}(a_i) > U_{\sim(q-1)}(a_i) \text{ then } a_i \in C_{q-1} \\
\qquad\quad \text{Else } a_i \in C_q
\end{array}\right\} \tag{20.12}$$

Except for the hierarchical classification framework, the MHDIS method has another special feature that distinguishes it from other MCDA sorting methods as well as from other linear programming classification approaches [119]. This involves the optimization framework used to develop the optimal sorting model (additive utility functions). In particular, during model development in the MHDIS method, three mathematical programming problems are solved. At each stage k of the hierarchical discrimination process ($k = 1, 2, \ldots, q-1$), two linear and one mixed-integer programming problems are solved to estimate the "optimal" pair of utility functions, where the term "optimal" refers both to the total number of misclassifications as well as to the clarity of the distinction between the groups. Initially, a linear programming problem (LP1) is solved to minimize the magnitude of the classification errors (in distance terms). Then, a mixed-integer programming problem (MIP) is solved to minimize the total number of misclassifications among the misclassifications that occur after the solution of LP1, while retaining the correct classifications. Finally, a second linear programming problem is solved to maximize the clarity of the classification obtained from the solutions of LP1 and MIP. A detailed description of the model optimization process in the MHDIS method can be found in Zopounidis and Doumpos [143].

4.3 Decision Rule Models: Rough Set Theory

Pawlak [95] introduced the rough set theory as a tool to describe dependencies between attributes, to evaluate the significance of attributes and to deal with inconsistent data. Generally, the rough set approach is a very useful tool in the study of sorting and classification problems, regarding the assignment of a set of alternatives into pre-specified groups. Recently, however, there have been several advances in this field to allow the application of the rough set theory to choice and ranking problems as well [43].

The rough set philosophy is founded on the assumption that with every alternative some information (data, knowledge) is associated. This information involves two types of attributes: condition and decision attributes. Condition attributes are those used to describe the characteristics of the alternatives (e.g., criteria), whereas the decision attributes define a partition of the alternatives into groups. Alternatives that have the same description in terms of the condition attributes are considered to be indiscernible. The indiscernibility relation constitutes the main mathematical basis of the rough set theory. Any set of all indiscernible alternatives is called an elementary set and forms a basic granule of knowledge about the universe. Any set of alternatives being a union of some elementary sets is referred to as crisp (precise) otherwise it is a rough set (imprecise, vague). A rough set can be approximated by a pair of crisp sets, called the lower and the upper approximation. The lower approximation includes the alternatives that certainly belong to the set and the upper approximation includes the alternatives that possibly belong to the set.

On the basis of these approximations, the first major capability that the rough set theory provides is to reduce the available information, so as to retain only what is absolutely necessary for the description and classification of the alternatives. This is achieved by discovering subsets of attributes, which provide the same quality of classification as the whole set of attributes. Such subsets of attributes are called reducts. Generally, the reducts are more than one. In such a case the intersection of all reducts is called the core. The core is the collection of the most relevant attributes, which cannot be excluded from the analysis without reducing the quality of the obtained description (classification).

The subsequent steps of the analysis involve the development of a set of "IF ... THEN ..." rules for the classification of the alternatives. The developed rules can be consistent if they include only one decision in their conclusion part, or approximate if their conclusion involves a disjunction of elementary decisions. Approximate rules are consequences of an approximate description of decision classes in terms of blocks of alternatives (granules) indiscernible by condition attributes. Such a situation indicates that using the available knowledge, one is unable to decide whether some alternatives belong to a given group (decision class) or not.

This traditional framework of the rough set theory, has been recently extended towards the development of a new preference modelling framework within the MCDA field [46, 45]. The main novelty of the recently developed rough set approach concerns the possibility of handling criteria, i.e., attributes with preference ordered domains, and preference ordered groups. Within this context the rough approximations of groups are defined according to the dominance relation, instead of the indiscernibility relation used in the basic rough sets approach. The decision rules derived from these approximations constitute a preference model.

4.4 Applications in Financial Decisions

MCDA discrete evaluation methods are well suited for the study of several financial decision-making problems. The diversified nature of the factors (evaluation criteria) that affect financial decisions, the complexity of the financial, business and economic environments, the subjective nature of many financial decisions, are only some of the features of financial decisions which are in accordance with the MCDA modelling framework. On the basis of these remarks this section reviews the up-to-date applications of MCDA discrete evaluation methods in several major financial decisions.

Bankruptcy and Credit Risk Assessment. The assessment of bankruptcy and credit risk have been major research fields in finance for the last three decades. Bankruptcy risk is derived by the failure of a firm to meet its debt obligations to its creditors, thus leading the firm either to liquidation (discontinuity of the firm's operations) or to a reorganization program [138]. The concept of credit risk is similar to that of bankruptcy risk, in the sense that in both cases the likelihood that a debtor (firm, organization or individual) will not be able to meet its debt obligations to its creditors, is a key issue in the analysis. Credit risk assessment decisions, however, are not simply based on the estimation of this likelihood; furthermore, they take into account the opportunity cost that arises when a good client (firm or individual) is denied credit. In both cases, the most common approach used to address bankruptcy and credit risk assessment problems is to develop appropriate models that sort/classify the firms or the individuals into predefined groups (problematic β), e.g., classification of firms as bankrupt/non-bankrupt, or as high credit risk firms/low credit risk firms. Statistical and econometric techniques (discriminant analysis, logit and probit analysis, etc.) have dominated this field for several decades, but recently new methodologies have attracted the interest of researchers and practitioners including several MCDA methods [28, 138]. A representative list of the MCDA evaluation approaches applied in bankruptcy and credit risk assessment is presented in Table 20.1.

Table 20.1. Applications of MCDA approaches in bankruptcy and credit risk assessment.

Approaches	*Methods*	*Studies*
Multiattribute utility theory	AHP	[57, 117, 118]
Outranking relations	ELECTRE	[9, 29, 71]
	Other methods	[2, 130]
Preference disaggregation	UTA	[131, 134]
	UTADIS	[140, 141, 142]
	MHDIS	[31, 35]
	Other methods	[33, 47, 117]
Rough set theory		[27, 44, 112, 114]

Portfolio Selection and Management. Portfolio selection and management involves the construction of a portfolio of securities (stocks, bonds, treasury bills, mutual funds, etc.) that maximizes the investor's utility. This problem can be realized as a two stage process [55, 56]: (1) the evaluation of the available securities to select the ones that best meet the investor's preferences, (2) specification of the amount of capital to be invested in each of the securities selected in the first stage. The implementation of these two stages in the traditional portfolio theory is based on the mean-variance approach developed by Markowitz [81, 83]. Recently, however, the multi-dimensional nature of the problem has been emphasized by researchers in finance [62], as well as by MCDA researchers [115, 127, 128]. Within this multi-dimensional context, MCDA discrete evaluation methods provide significant support in evaluating securities according to the investor's policy. Studies conducted on this topic have focused on the modelling and representation of the investor's policy, goals and objectives in a mathematical model. The model aggregates all the pertinent factors describing the performance of the securities and provides their overall evaluation. The securities with the higher overall evaluation are selected for portfolio construction purposes in a latter stage of the analysis. Table 20.2 summarizes several studies involving the application of MCDA evaluations methods in portfolio selection and management.

Table 20.2. Applications of MCDA approaches in portfolio selection and management.

Approaches	*Methods*	*Studies*
Multiattribute utility theory	AHP	[106]
	MACBETH	[5, 6]
	Other methods	[19, 30, 40, 66, 98]
Outranking relations	ELECTRE	[54, 56, 55, 73, 84, 85, 121]
	PROMETHEE	[48, 72, 85]
Preference disaggregation	UTA	[55, 56, 133, 146]
	UTADIS	[144, 145]
	MHDIS	[38]
Rough set theory		[67]

Corporate Performance Evaluation. The evaluation of the performance of corporate entities and organizations is an important activity for their management and shareholders as well as for investors and policy makers. Such an evaluation provides the management and the shareholders with a tool to assess the strength and weakness of the firm as well as its competitive advantages over its competitors, thus providing guidance on the choice of the measures that need to be taken to overcome the existing problems. Investors (institutional and individual) are interested in the assessment of corporate performance for guidance to their investment decisions, while policy makers may use such an assessment to identify the existing problems in the business environment and take measures that will ensure a sustainable economic growth and social stability. The performance of a firm or an organization is clearly multi-dimensional, since it is affected by a variety of factors of different nature, such as: (1) financial factors indicating the financial position of the firm/organization, (3) strategic factors of qualitative nature that define the internal operation of the firm and its relation to the market (organization, management, market trend, etc. [131], (2) economic factors that define the economic and business environment. The aggregation of all these factors into a global evaluation index is a subjective process that depends on the decision maker's values and judgment policy. These findings are in accordance with the MCDA paradigm, thus leading several operational researchers to the investigation of the capabilities that MCDA methods provide in supporting decision maker's in making decisions regarding the evaluation of corporate performance. An indicative list of studies on this topic is given in Table 20.3.

Table 20.3. Applications of MCDA approaches in the assessment of corporate performance.

Approaches	*Methods*	*Studies*
Multiattribute utility theory	AHP	[4, 75]
	Other methods	[26, 126]
Outranking relations	ELECTRE	[20]
	PROMETHEE	[4, 8, 20, 76, 77, 78, 79, 94, 129]
Preference disaggregation	UTA	[111, 137, 147]
	UTADIS	[90, 124]

Investment Appraisal. In most cases the choice of investment projects is an important strategic decision for every firm, public or private, large or small. Therefore, the process of an investment decision should be conveniently modelled. In general, the investment decision process consists of four main stages: perception, formulation, evaluation and choice. The financial theory intervenes only in the stages of evaluation and choice based on traditional financial criteria such as the payback period, the accounting rate of return, the net present value, the internal rate of return, the index of profitability, the discounted payback method, etc. [18]. This approach, however, entails some shortcomings such as the difficulty in aggregating the conflicting results of each criterion and the elimination of important qualitative variables from the analysis [135]. MCDA, on the other hand, contributes in a very original way to the investment decision process, supporting all stages of the investment process. Concerning the stages of perception and formulation, MCDA contributes to the identification of possible actions (investment opportunities) and to the definition of a set of potential actions (possible variants, each variant constituting an investment project in competition with others). Concerning the stages of evaluation and choice, MCDA supports the introduction in the analysis of both quantitative and qualitative criteria. Criteria such as the urgency of the project, the coherence of the objectives of the projects with those of the general policy of the firm, the social and environmental aspects should be taken into careful consideration. Therefore, MCDA contributes through the identification of the best investment projects according to the problematic chosen, the satisfactory resolution of the conflicts between the criteria, the determination of the relative importance of the criteria in the decision-making process, and the revealing of the investors' preferences and system of values. These attractive features have been the main motivation for the use of MCDA methods in investment appraisal in several real-world cases. A representative list of studies in presented in Table 20.4.

Other Financial Decision Problems. Except for the above financial decision-making problems, discrete MCDA evaluation methods are also applicable in several other fields of finance. Table 20.5 list some additional applications of

Table 20.4. Applications of MCDA approaches in investment appraisal.

Approaches	*Methods*	*Studies*
Multiattribute utility theory	AHP	[74]
	Other methods	[40, 96]
utranking relations	ELECTRE	[17, 23]
	PROMETHEE	[97, 125]
	ORESTE	[23]
Preference disaggregation	UTA	[10, 108]
	UTADIS	[58]

MCDA methods in other financial problems, including venture capital, country risk assessment and the prediction of corporate mergers and acquisitions. In venture capital investment decisions, MCDA methods are used both as tools to evaluate the firms that seek venture capital financing, as well as analysis tools to identify the factors that drive such financing decisions. In country risk assessment, MCDA methods are used to developed models that aggregate the appropriate economic, financial and socio-political factors, to support the evaluation of the creditworthiness and the future prospects of the countries. Finally, in corporate mergers and acquisitions MCDA methods are used to assess the likelihood that a firm will be merged or acquired on the basis of financial information (financial ratios) and strategic factors.

Table 20.5. Applications of MCDA approaches in other financial decision-making problems.

Topic	*Methodology*	*Studies*
Venture capital	Conjoint analysis	[91, 99]
	UTA	[109, 132]
Country risk	MAUT	[122]
	UTA	[1, 22]
	UTADIS	[1, 34, 139]
	MHDIS	[32, 34, 36]
	Other methods	[21, 92, 93]
Mergers and acquisitions	Rough sets	[113]

5. Conclusions and Future Perspectives

This chapter discussed the contribution of MCDA in financial decision-making problems, focusing on the justification of the multi-dimensional character of financial decisions and the use of different MCDA methodologies to support them.

Overall, the main advantages that the MCDA paradigm provides in financial decision making, could be summarized in the following aspects [135]: (1) the possibility of structuring complex evaluation problems, (2) the introduction of both quantitative (i.e. financial ratios) and qualitative criteria in the evaluation process, (3) the transparency in the evaluation, allowing good argumentation in financial decisions, and (4) the introduction of sophisticated, flexible and realistic scientific methods in the financial decision-making process.

In conclusion, MCDA methods seem to have a promising future in the field of financial management, because they offer a highly methodological and realistic framework to decision problems. Nevertheless, their success in practice depends heavily on the development of computerized multicriteria decision support systems. Financial institutions as well as firms acknowledge the multi-dimensional nature of financial decision problems [12]. Nevertheless, they often use optimization or statistical approaches to address their financial problems, since optimization and statistical software packages are easily available in relatively low cost, even though many of these software packages are not specifically designed for financial decision-making problems. Consequently, the use of MCDA methods to support real time financial decision making, calls upon the development of integrated user-friendly multicriteria decision support systems that will be specifically designed to address financial problems. Examples of such systems are the CGX system [118], the BANKS system [78], the BANKADVISER system [76], the INVEX system [125], the FINEVA system [147], the FINCLAS system [140], the INVESTOR system [144], etc. The development and promotion of such systems is a key issue in the successful application of MCDA methods in finance.

Notes

1. The underlying assumptions must be validated and the effectiveness and efficiency of the actions taken must be evaluated systematically. The latter calls for a sophisticated performance evaluation process that explicitly acknowledges the role of learning.
2. This section draws heavily on a part of Hallerbach and Spronk [49].

Acknowledgments

The authors wish to thank Winfried Hallerbach by allowing to draw in this text on work co-authored by him (see our list of references).

References

[1] Th. Anastassiou and C. Zopounidis. Country risk assessment: A multicriteria analysis approach. *Journal of Euro-Asian Management*, 3(1):51–73, 1997.

[2] A. Andenmatten. *Evaluation du Risque de Défaillance des Emetteurs d'Obligations: Une Approche par l'Aide Multicritère à la Décision.* Presses Polytechniques et Universitaires Romandes, Lausanne, 1995.

[3] R. W. Ashford, R. H. Berry, and R. G. Dyson. Operational research and financial management. *European Journal of Operational Research*, 36(2):143–152, 1988.

[4] Z. Babic and N. Plazibat. Ranking of enterprises based on multicriteria analysis. *International Journal of Production Economics*, 56-57:29–35, 1998.

[5] C. A. Bana e Costa and J. O. Soares. Multicriteria approaches for portfolio selection: An overview. *Review of Financial Markets*, 4(1):19–26, 2001.

[6] C. A. Bana e Costa and J. O. Soares. A multicriteria model for portfolio management. Technical Report LSE OR 01.43, London School of Economics, 2001.

[7] C. A. Bana e Costa and J. C. Vansnick. MACBETH: An interactive path towards the construction of cardinal value functions. *International Transactions in Operational Research*, 1:489–500, 1994.

[8] G. Baourakis, M. Doumpos, N. Kalogeras, and C. Zopounidis. Multicriteria analysis and assessment of financial viability of agri-businesses: The case of marketing co-operatives and juice producing companies. *Agribusiness*, 18(4):543–558, 2002.

[9] M. Bergeron, J. M. Martel, and P. Twarabimenye. The evaluation of corporate loan applications based on the MCDA. *Journal of Euro-Asian Management*, 2(2):16–46, 1996.

[10] M. Beuthe, L. Eeckhoudt, and G. Scannella. A practical multicriteria methodology for assessing risk public investments. *Socio-Economic Planning Sciences*, 34:121–139, 2000.

[11] K. Bhaskar. A multiple objective approach to capital budgeting. *Accounting and Business Research*, 9:25–46, 1979.

[12] K. Bhaskar and P. McNamee. Multiple objectives in accounting and finance. *Journal of Business Finance and Accounting*, 10(4):595–621, 1983.

[13] F. Black and M. Scholes. The pricing of options and corporate liabilities. *Journal of Political Economy*, 81(3):637–659, 1973.

[14] Z. Bodie and R. C. Merton. *Finance*. Prentice-Hall, Upper Saddle River, New Jersey, 2000.

[15] T. Bollerslev. Financial econometrics: Past developments and future challenges. *Journal of Econometrics*, 100:41–51, 2001.

[16] J. P. Brans and Ph. Vincke. A preference ranking organization method. *Management Science*, 31(6):647–656, 1985.

[17] J. Buchanan, Ph. Sheppard, and D. Vanderpooten. Project ranking using Electre III. Technical Report 1999–01, Department of Management Systems, University of Waikato, 1999.

[18] B. Colasse. *Gestion Financière de l'Entreprise*. Presses Universitaires de France, Paris, 1993.

[19] G. Colson and Ch. DeBruyn. An integrated multiobjective portfolio management system. *Mathematical and Computer Modelling*, 12(10-11):1359–1381, 1989.

[20] G. Colson and M. Mbangala. Evaluation multicritère d'entreprises publiques du rail. *FINECO*, 8(1):45–72, 1998.

[21] W. D. Cook and K. J. Hebner. A multicriteria approach to country risk evaluation: With an example employing Japanese data. *International Review of Economics and Finance*, 2(4):327–438, 1993.

[22] J. C. Cosset, Y. Siskos, and C. Zopounidis. Evaluating country risk: A decision support approach. *Global Finance Journal*, 3(1):79–95, 1992.

[23] N. Danila. *Méthodologie d' Aide à la Décision Dans le Cas d' Investissements Fort Dépendants*. Thèse de Doctorat de 3e Cycle, UER Sciences des Organisations, Université de Paris–Dauphine, 1980.

[24] E. Derman. Model risk. *Risk*, 9(5):34–37, 1996.

[25] E. Derman. Valuing models and modelling value. *The Journal of Portfolio Management*, 22(3):106–114, 1996.

[26] D. Diakoulaki, G. Mavrotas, and L. Papagyannakis. A multicriteria approach for evaluating the performance of industrial firms. *Omega*, 20(4):467–474, 1992.

[27] A. I. Dimitras, R. Slowinski, R. Susmaga, and C. Zopounidis. Business failure prediction using rough sets. *European Journal of Operational Research*, 114(2):263–280, 1999.

[28] A. I. Dimitras, S.H. Zanakis, and C. Zopounidis. A survey of business failures with an emphasis on prediction methods and industrial applications. *European Journal of Operational Research*, 90:487–513, 1996.

[29] A. I. Dimitras, C. Zopounidis, and C. Hurson. A multicriteria decision aid method for the assessment of business failure risk. *Foundations of Computing and Decision Sciences*, 20(2):99–112, 1995.

[30] C. Dominiak. Portfolio selection using the idea of reference solution. In G. Fandel and T. Gal, editors, *Lecture Notes in Economics and Mathematical Systems,* vol. 448, pages 593–602. Springer-Verlag, Berlin, 1997.

[31] M. Doumpos, K. Kosmidou, G. Baourakis, and C. Zopounidis. Credit risk assessment using a multicriteria hierarchical discrimination approach: A comparative analysis. *European Journal of Operational Research*, 138(2):392–412, 2002.

[32] M. Doumpos, K. Pentaraki, C. Zopounidis, and C. Agorastos. Assessing country risk using a multi-group discrimination method: A comparative analysis. *Managerial Finance*, 27:16–34, 2001.

[33] M. Doumpos, M. Spanos, and C. Zopounidis. On the use of goal programming techniques in the assessment of financial risks. *Journal of Euro-Asian Management*, 5(1):83–100, 1999.

[34] M. Doumpos, S. H. Zanakis, and C. Zopounidis. Multicriteria preference disaggregation for classification problems with an application to global investing risk. *Decision Sciences*, 32(2):333–385, 2001.

[35] M. Doumpos and C. Zopounidis. A multicriteria discrimination method for the prediction of financial distress: The case of Greece. *Multinational Finance Journal*, 3(2):71–101, 1999.

[36] M. Doumpos and C. Zopounidis. Assessing financial risks using a multicriteria sorting procedure: The case of country risk assessment. *Omega*, 29(1):97–109, 2000.

[37] M. Doumpos and C. Zopounidis. *Multicriteria Decision Aid Classification Methods*. Kluwer Academic Publishers, Dordrecht, 2002.

[38] M. Doumpos, C. Zopounidis, and P. M. Pardalos. Multicriteria sorting methodology: Application to financial decision problems. *Parallel Algorithms and Applications*, 15(1-2):113–129, 2000.

[39] E. J. Elton and M. J. Gruber. *Modern Portfolio Theory and Investment Analysis*. John Wiley, New York, 5th edition, 1995.

[40] Y. Evrard and R. Zisswiller. Une analyse des décisions d'investissement fondée sur les modèles de choix multi-attributs. *Finance*, 3(1):51–68, 1982.

[41] E. F. Fama and M. H. Miller. *The Theory of Finance*. Dryden Press, Hinsdale Ill., 1972.

[42] X. Gandibleux, R. Caballero, and J. Molina. MOAMP: A multiobjective programming metaheuristic using an adaptive memory procedure. Technical report, University of Valenciennes, Valenciennes, France, 2003.

[43] S. Greco, B. Matarazzo, and R. Slowinski. Rough set approach to multi-attribute choice and ranking problems. In G. Fandel and T. Gal, editors, *Lecture Notes in Economics and Mathematical Systems,* vol. 448, pages 318–329. Springer-Verlag, Berlin, 1997.

[44] S. Greco, B. Matarazzo, and R. Slowinski. A new rough set approach to evaluation of bankruptcy risk. In C. Zopounidis, editor, *Operational Tools in the Management of Financial Risks*, pages 121–136. Kluwer Academic Publishers, Dordrecht, 1998.

[45] S. Greco, B. Matarazzo, and R. Slowinski. The use of rough sets and fuzzy sets in MCDM. In T. Gal, T. Hanne, and T. Stewart, editors, *Advances in Multiple Criteria Decision Making*, pages 14.1–14.59. Kluwer Academic Publishers, Dordrecht, 1999.

[46] S. Greco, B. Matarazzo, and R. Slowinski. Extension of the rough set approach to multicriteria decision support. *INFOR*, 38(3):161–196, 2000.

[47] Y. P. Gupta, R. P. Rao, and P. K. Bagghi. Linear goal programming as an alternative to multivariate discriminant analysis: A note. *Journal of Business Finance and Accounting*, 17(4):593–598, 1990.

[48] M. Hababou and J. M. Martel. A multicriteria approach for selecting a portfolio manager. *INFOR*, 36(3):161–176, 1998.

[49] H. W. Hallerbach and J. Spronk. A multicriteria framework for risk analysis. In Y. Y. Haimes and R. E. Steuer, editors, *Lecture Notes in Economics and Mathematical Systems,* vol. 487, pages 272–283. Springer-Verlag, Berlin, 2000.

[50] W. G. Hallerbach and J. Spronk. A multidimensional framework for financial-economic decisions. *Journal of Multi-Criteria Decision Analysis*, 11(3):111–124, 2002.

[51] W. G. Hallerbach and J. Spronk. The relevance of MCDM for financial decisions. *Journal of Multi-Criteria Decision Analysis*, 11(4-5):187–195, 2002.

[52] K. L. Hastie. A perspective on management education. *Financial Management*, 11(4):55–62, 1982.

[53] J. C. Hull. *Options, Futures and Other Derivatives*. Prentice-Hall, Upper Saddle River, New Jersey, 2003.

[54] Ch. Hurson and N. Ricci. Multicriteria decision making and portfolio management with arbitrage pricing theory. In C. Zopounidis, editor, *Operational Tools in The Management of Financial Risks*, pages 31–55. Kluwer Academic Publishers, Dordrecht, 1998.

[55] Ch. Hurson and C. Zopounidis. On the use of multi-criteria decision aid methods to portfolio selection. *Journal of Euro-Asian Management*, 1(2):69–94, 1995.

[56] Ch. Hurson and C. Zopounidis. *Gestion de Portefeuille et Analyse Multicritère*. Economica, Paris, 1997.

[57] J. Jablonsky. Multicriteria evaluation of clients in financial houses. *Central European Journal of Operations Research and Economics*, 3(2):257–264, 1993.

[58] E. Jacquet-Lagrèze. An application of the UTA discriminant model for the evaluation of R & D projects. In P.M. Pardalos, Y. Siskos, and C. Zopounidis, editors, *Advances in Multicriteria Analysis*, pages 203–211. Kluwer Academic Publishers, Dordrecht, 1995.

[59] E. Jacquet-Lagrèze and Y. Siskos. Assessing a set of additive utility functions for multicriteria decision making: The UTA method. *European Journal of Operational Research*, 10(2):151–164, 1982.

[60] E. Jacquet-Lagrèze and Y. Siskos. *Méthodes de Décision Multicritère*. Editions Hommes et Techniques, Paris, 1983.

[61] E. Jacquet-Lagrèze and Y. Siskos. Preference disaggregation: Twenty years of MCDA experience. *European Journal of Operational Research*, 130(2):233–245, 2001.

[62] B. Jacquillat. Les modèles d'evaluation et de sélection des valeurs mobilières: Panorama des recherches américaines. *Analyse Financière*, 11(4e trim):68–88, 1972.

[63] R. A. Jarrow. *Finance Theory*. Prentice-Hall, Upper Saddle River, New Jersey, 1988.

[64] M. C. Jensen. Value maximization, stakeholder theory, and the corporate objective function. *Journal of Applied Corporate Finance*, 14(3):8–21, 2001.

[65] M. C. Jensen and W. H. Meckling. Theory of the firm: Managerial behavior, agency costs and ownership structure. *Journal of Financial Economics*, 3:305–360, 1976.

[66] V. Jog, I. Kaliszewski, and W. Michalowski. Using attribute trade-off information in investment. *Journal Multi-Criteria Decision Analysis*, 8(4):189–199, 1999.

[67] V. Jog, W. Michalowski, R. Slowinski, and R. Susmaga. The rough sets analysis and the neural networks classifier: A hybrid approach to predicting stocks' performance. In D.K. Despotis and C. Zopounidis, editors, *Integrating Technology and Human Decisions: Bridging into the 21st Century, Proceedings of the 5th International Meeting of the Decision Sciences Institute*, volume II, pages 1386–1388. New Technologies Editions, Athens, Greece, 1999.

[68] Ph. Jorion. *Value at Risk: The Benchmark for Controlling Market Risk*. McGraw-Hill, Chicago, 2001.

[69] R. L. Keeney and H. Raiffa. *Decisions with Multiple Objectives: Preferences and Value Trade-offs*. Cambridge University Press, Cambridge, 1993.

[70] J. N. Keynes. *The Scope and Method of Political Economy*. Macmillan, London, 1891.

[71] J. Khalil, J. M. Martel, and P. Jutras. A multicriterion system for credit risk rating. *Gestion 2000: Belgian Management Magazine*, 15(1):125–146, 2000.

[72] N. T. Khoury and J. M. Martel. The relationship between risk-return characteristics of mutual funds and their size. *Finance*, 11(2):67–82, 1990.

[73] N. T. Khoury, J. M. Martel, and M. Veilleux. Méthode multicritère de sélection de portefeuilles indiciels internationaux. *L'Actualité Economique, Revue d'Analyse Economique*, 69(1):171–190, 1993.

[74] H. Kivijärvi and M. Tuominen. A decision support system for semistructured strategic decisions: A multi-tool method for evaluating intangible investments. *Revue des Systèmes de Décision*, 1:353–376, 1992.

[75] H. Lee, W. Kwak, and I. Han. Developing a business performance evaluation system: An analytic hierarchical model. *Engineering Economist*, 30(4):343–357, 1995.

[76] B. Mareschal and J. P. Brans. BANKADVISER: An industrial evaluation system. *European Journal of Operational Research*, 54:318–324, 1991.

[77] B. Mareschal and D. Mertens. Evaluation financière par la méthode multicritère GAIA: Application au Secteur Bancaire Belge. *Revue de la Banque*, 6:317–329, 1990.

[78] B. Mareschal and D. Mertens. BANKS: A multicriteria PROMETHEE-based decision support system for the evaluation of the international banking sector. *Revue des Systèmes de Décision*, 1(2-3):175–189, 1992.

[79] B. Mareschal and D. Mertens. Evaluation multicritère par la méthode multicritère GAIA: Application au secteur de l'assurance en belgique. *L'Actualité Economique, Revue d'Analyse Economique*, 69(1):206, 1993.

[80] H. Markowitz. Theories of uncertainty and financial behavior. *Econometrica*, 19:325–326, 1950.

[81] H. Markowitz. Portfolio selection. *Journal of Finance*, 7(1):77–91, 1952.

[82] H. Markowitz. The optimization of a quadratic function subject to linear constraints. *Naval Research Logistics Quarterly*, 3:111–133, 1956.

[83] H. Markowitz. *Portfolio Selection: Efficient Diversification in Investments*. John Wiley, New York, 1959.

[84] J. M. Martel, N. T. Khoury, and M. Bergeron. An application of a multicriteria approach to portfolio comparisons. *Journal of the Operational Research Society*, 39(7):617–628, 1988.

[85] J. M. Martel, N. T. Khoury, and B. M'Zali. Comparaison performancetaille des fonds mutuels par une analyse multicritère. *L' Actualité Economique, Revue d'Analyse Economique*, 67(3):306–324, 1991.

[86] J. M. Martel and C. Zopounidis. Critères multiples et décisions financiers. *FINECO*, 8:1–12, 1998.

[87] J. M. McInnes and W. J. Carleton. Theory, models and implementation in financial management. *Management Science*, 28(9):957–978, 1982.

[88] R. C. Merton. Theory of rational option pricing. *Bell Journal of Economics*, 4(Spring):141–183, 1973.

[89] R. C. Merton. Influence of mathematical models in finance on practice: past, present, and future. *Financial Practice and Education*, 5(1):7–15, 1995.

[90] M. Michalopoulos, C. Zopounidis, and M. Doumpos. Evaluation des succursales bancaires è l'aide d'une méthode multicritère. *FINECO*, 8(2):123–136, 1998.

[91] D. Muzyka, S. Birley, and B. Leleux. Trade-offs in the investment decisions of European venture capitalists. *Journal of Business Venturing*, 11(4):273–287, 1996.

[92] M. Oral and H. Chabchoub. An estimation model for replicating the rankings of the world competitiveness report. *International Journal of Forecasting*, 13:527–537, 1997.

[93] M. Oral, O. Kettani, J. C. Cosset, and M. Daouas. An estimation model for country risk rating. *International Journal of Forecasting*, 8:583–593, 1992.

[94] P. M. Pardalos, M. Michalopoulos, and C. Zopounidis. On the use of multicriteria methods for the evaluation of insurance companies in Greece. In C. Zopounidis, editor, *New Operational Approaches for Financial Modelling*, pages 271–283. Physica-Verlag, Heidelberg, 1997.

[95] Z. Pawlak. Rough sets. *International Journal of Information and Computer Sciences*, 11:341–356, 1982.

[96] A. D. Pearman, P. J. Mackie, A. D. May, and D. Simon. The use of multi-criteria techniques to rank highway investment proposals. In A.G. Lockett and G. Islei, editors, *Lecture Notes in Economics and Mathematical Systems,* vol. 335, pages 157–165. Springer Verlag, Berlin, 1989.

[97] Z. Ribarovic and N. Mladineo. Application of multicriterional analysis to the ranking and evaluation of the investment programmes in the ready mixed concrete industry. *Engineering Costs and Production Economics*, 12:367–374, 1987.

[98] S. Rios-Garcia and S. Rios-Insua. The portfolio problem with multiattributes and multiple criteria. In P. Hansen, editor, *Lecture Notes in Economics and Mathematical Systems*, vol. 209, pages 317–325. Springer-Verlag, Berlin, 1983.

[99] H. Riquelme and T. Rickards. Hybrid conjoint analysis: An estimation probe in new venture decisions. *Journal of Business Venturing*, 7:505–518, 1992.

[100] R. Roll. A critique of the asset pricing theory's tests. *Journal of Financial Economics*, 4:129–176, 1977.

[101] S. A. Ross. *The New Palgrave Finance*, chapter Finance, pages 1–34. MacMillan, Hong Kong, 1989.

[102] B. Roy. Classement et choix en présence de points de vue multiples: La méthode ELECTRE. *Revue Francaise d'Informatique et de Recherche Operationnelle*, 8:57–75, 1968.

[103] B. Roy. The outranking approach and the foundations of ELECTRE methods. *Theory and Decision*, 31(1):49–73, 1991.

[104] B. Roy. *Multicriteria Methodology for Decision Aiding*. Kluwer Academic Publishers, Dordrecht, 1996.

[105] B. Roy and D. Bouyssou. *Aide Multicritère à la Décision: Méthodes et Cas*. Economica, Paris, 1993.

[106] T. L. Saaty, P. C. Rogers, and R. Pell. Portfolio selection through hierarchies. *Journal of Portfolio Management*, pages 16–21, Spring 1980.

[107] H. Shefrin. *Beyond Greed and Fear: Understanding Behavioral Finance and the Psychology of Investing*. Oxford University Press, New York, 2002.

[108] J. Siskos and N. Assimakopoulos. Multicriteria highway planning: A case study. *Mathematical and Computer Modelling*, 12(10-11):1401–1410, 1989.

[109] J. Siskos and C. Zopounidis. The evaluation criteria of the venture capital investment activity: An interactive assessment. *European Journal of Operational Research*, 31(3):304–313, 1987.

[110] Y. Siskos and D. Yannacopoulos. UTASTAR: An ordinal regression method for building additive value functions. *Investigação Operacional*, 5(1):39–53, 1985.

[111] Y. Siskos, C. Zopounidis, and A. Pouliezos. An integrated DSS for financing firms by an industrial development bank in Greece. *Decision Support Systems*, 12:151–168, 1994.

[112] R. Slowinski and C. Zopounidis. Application of the rough set approach to evaluation of bankruptcy risk. *International Journal of Intelligent Systems in Accounting Finance and Management*, 4:27–41, 1995.

[113] R. Slowinski, C. Zopounidis, and A. I. Dimitras. Prediction of company acquisition in Greece by means of the rough set approach. *European Journal of Operational Research*, 100:1–15, 1997.

[114] R. Slowinski, C. Zopounidis, A. I. Dimitras, and R. Susmaga. Rough set predictor of business failure. In R. R. Ribeiro, R. R. Yager, H.J. Zimmermann, and J. Kacprzyk, editors, *Soft Computing in Financial Engineering*, pages 402–424. Physica-Verlag, Heidelberg, 1999.

[115] J. Spronk and W. G. Hallerbach. Financial modelling: Where to go? With an illustration for portfolio management. *European Journal of Operational Research*, 99(1):113–125, 1997.

[116] J. Spronk and E. M. Vermeulen. Comparative performance evaluation under uncertainity. *European Journal of Operational Research*, to appear.

[117] V. Srinivasan and Y. H. Kim. Credit granting: A comparative analysis of classification procedures. *Journal of Finance*, XLII(3):665–683, 1987.

[118] V. Srinivasan and B. Ruparel. CGX: An expert support system for credit granting. *European Journal of Operational Research*, 45:293–308, 1990.

[119] A. Stam. Nontraditional approaches to statistical classification: Some perspectives on l_p-norm methods. *Annals of Operations Research*, 74:1–36, 1997.

[120] R. E. Steuer and P. Na. Multiple criteria decision making combined with finance: A categorized bibliography. *European Journal of Operational Research*, 150(3):496–515, 2003.

[121] A. Szala. *L' aide à la décision en gestion de portefeuille*. Diplôme Supérieur de Recherches Appliquées, Université de Paris–Dauphine, 1990.

[122] J. C. S. Tang and C. G. Espinal. A model to assess country risk. *Omega*, 17(4):363–367, 1989.

[123] Ph. Vincke. *Multicriteria Decision Aid*. John Wiley, New York, 1992.

[124] F. Voulgaris, M. Doumpos, and C. Zopounidis. On the evaluation of Greek industrial SMEs' performance via multicriteria analysis of financial ratios. *Small Business Economics*, 15:127–136, 2000.

[125] S. Vranes, M. Stanojevic, V. Stevanovic, and M. Lucin. INVEX: Investment advisory expert system. *Expert Systems*, 13(2):105–119, 1996.

[126] C. H. Yeh, H. Deng, and Y. H. Chang. Fuzzy multicriteria analysis for performance evaluation of bus companies. *European Journal of Operational Research*, 126(3):459–473, 2000.

[127] M. Zeleny. Multidimensional measure of risk: The prospect ranking vector. In S. Zionts, editor, *Lecture Notes in Economics and Mathematical Systems,* vol. 155, pages 529–548. Springer Verlag, Heidelberg, 1977.

[128] M. Zeleny. *Multiple Criteria Decision Making*. McGraw-Hill, New York, 1982.

[129] R. Zmitri, J. M. Martel, and Y. Dumas. Un indice multicritère de santé financière pour les succursales bancaires. *FINECO*, 8(2):107–121, 1998.

[130] M. Zollinger. L'analyse multicritère et le risque de crédit aux entreprises. *Revue Française de Gestion*, pages 56–66, Janvier-Février 1982.

[131] C. Zopounidis. A multicriteria decision making methodology for the evaluation of the risk of failure and an application. *Foundations of Control Engineering*, 12(1):45–67, 1987.

[132] C. Zopounidis. *La Gestion du Capital-Risque*. Economica, Paris, 1990.

[133] C. Zopounidis. On the use of the MINORA decision aiding system to portfolio selection and management. *Journal of Information Science and Technology*, 2(2):150–156, 1993.

[134] C. Zopounidis. *Evaluation du Risque de Défaillance de l' Entreprise: Méthodes et Cas d'Application*. Economica, Paris, 1995.

[135] C. Zopounidis. Multicriteria decision aid in financial management. *European Journal of Operational Research*, 119(2):404–415, 1999.

[136] C. Zopounidis. Preference disaggregation approach: Basic features, examples from financial decision making. In C.A. Floudas and P.M. Pardalos, editors, *Encyclopedia of Optimization,* vol. IV, pages 344–357. Kluwer Academic Publishers, Dordrecht, 2001.

[137] C. Zopounidis, D. K. Despotis, and E. Stavropoulou. Multiattribute evaluation of Greek banking performance. *Applied Stochastic Models and Data Analysis*, 11(1):97–107, 1995.

[138] C. Zopounidis and A. I. Dimitras. *Multicriteria Decision Aid Methods for the Prediction of Business Failure*. Kluwer Academic Publishers, Dordrecht, 1998.

[139] C. Zopounidis and M. Doumpos. A multicriteria decision aid methodology for the assessment of country risk. *European Research on Management and Business Economics*, 3(3):13–33, 1997.

[140] C. Zopounidis and M. Doumpos. Developing a multicriteria decision support system for financial classification problems: The FINCLAS system. *Optimization Methods and Software*, 8:277–304, 1998.

[141] C. Zopounidis and M. Doumpos. Business failure prediction using UTADIS multicriteria analysis. *Journal of the Operational Research Society*, 50(11):1138–1148, 1999.

[142] C. Zopounidis and M. Doumpos. A multicriteria decision aid methodology for sorting decision problems: The case of financial distress. *Computational Economics*, 14(3):197–218, 1999.

[143] C. Zopounidis and M. Doumpos. Building additive utilities for multi-group hierarchical discrimination: The MHDIS method. *Optimization Methods and Software*, 14(3):219–240, 2000.

[144] C. Zopounidis and M. Doumpos. INVESTOR: A decision support system based on multiple criteria for portfolio selection and composition. In A. Colorni, M. Paruccini, and B. Roy, editors, *A-MCD-A (Aide Multi Critere a la Decision - Multiple Criteria Decision Aiding)*, pages 371–381. European Commission Joint Research Centre, Brussels, 2000.

[145] C. Zopounidis, M. Doumpos, and S. H. Zanakis. Stock evaluation using a preference disaggregation methodology. *Decision Sciences*, 30(2):313–336, 1999.

[146] C. Zopounidis, M. Godefroid, and Ch. Hurson. Designing a multicriteria decision support system for portfolio selection and management. In J. Janssen, C.H. Skiadas, and C. Zopounidis, editors, *Advances in Stochastic Modelling and Data Analysis*, pages 261–292. Kluwer Academic Publishers, Dordrecht, 1995.

[147] C. Zopounidis, N. F. Matsatsinis, and M. Doumpos. Developing a multicriteria knowledge-based decision support system for the assessment of corporate performance and viability: The FINEVA system. *Fuzzy Economic Review*, 1(2):35–53, 1996.

Chapter 21

MCDA AND ENERGY PLANNING

Danae Diakoulaki
National Technical University of Athens
Department of Chemical Engineering
Labortaory of Industrial and Energy Economics
Zografou Campus, 15780 Athens
Greece
diak@chemeng.ntua.gr

Carlos Henggeler Antunes, António Gomes Martins
University of Coimbra and INESC Coimbra
Department of Electrical Engineering and Computers
Polo II, Pinhal de Marrocos, 3030 Coimbra
Portugal
cantunes,amartins@inescc.pt

Abstract The growing environmental awareness and the apparent conflict between economic and environmental objectives was the main impetus that pushed energy planners during the early eighties towards the use of MCDA methods. Thereafter, the rapid changes and the increasing complexity of the energy market gave rise to further methodological developments. Although the energy market restructuring and ongoing liberalization seemed to restrict the purpose for centralized energy decisions, they added new dimensions in energy planning. Increasing competition along with the prerequisite for sustainability have broadened the energy application field by bringing out new challenges for the development of integrated multicriteria and multi-stakeholders approaches also taking uncertainty into consideration. This paper aims at illustrating the evolution of MCDA approaches, in the context of the emerging problems faced by energy planners and other stakeholders involved in energy-related decision situations, one of the most active and exciting areas of application of MCDA models and methods.

Keywords: Multicriteria, multiobjective, energy planning, electricity.

1. Introduction

Energy is central to achieving the interrelated goals of modern societies: to meet human needs for heating, cooling, lighting, mobility and for running a large diversity of appliances, as well as to supply power and heat to production systems. Until the outbreak of the energy crisis, meeting these needs was a routine problem whose solution was principally a matter of money and technology availability. At these times, per capita energy consumption was a safe index of a nation's prosperity, while energy planning was aiming at supplying the energy required at the right time and in the least costly way.

The last 30 years have seen radical changes in the world's energy scene and in the mentality of energy planners. The first most dramatic occurrence was the energy crisis of the '70s. The sharp increase of energy prices disclosed all the hidden constraints behind the simple-minded perception of plentiful, affordable and cheap energy. At almost the same time, environmental considerations reflecting either the concern for the depletion of conventional energy resources or the need to cope with the ongoing environmental degradation, imposed a reconsideration of values and a shift towards new technological solutions. As a result, conventional energy technologies, although still dominant in the energy system, were increasingly disputed on environmental grounds. Cost, although still being the market's driving force, was no longer enough to reflect the society's multiple, incommensurate and often conflicting concerns.

In this context, energy planners came across unprecedented dilemmas that were no longer solvable with traditional tools. They had to look at a much broader spectrum of options and analyze all their multiple facets with respect to a much wider range of evaluation criteria under conditions of a higher uncertainty. Furthermore, they had to take into account the diverse, not clearly articulated preferences of all involved groups of interest. Typical questions illustrating their task in this new context were:

- Which type of energy resource or conversion technology to use?
- How to combine different energy sources and technologies in order to meet present and future energy needs?
- Where to locate new energy conversion or transmission facilities?

The energy sector has thus been a fertile ground for the emergence of several problems which are intrinsically of multiple criteria nature. Researchers and practitioners have responded to the challenges with ever-increasing sophisticated problem formulations, models and adequate methods to tackle the diversity of operational and planning problems arising in the energy sector. Moreover, multiple criteria decision aid (MCDA) models and methods have revealed an effective contribution to the successful resolution of several problems and provided the foundations for sound decision support.

The first historical applications of MCDA in energy planning, in the late 70's – early '80s, proved the strengths of the methodology and its capacity to be adapted in many different decision configurations. Just to mention a few, [99] made a multicriteria comparison of alternative power generation technologies, multiobjective linear programming models were developed in [62] and [98] in order to establish strategies in the power sector and in a regional energy system, respectively, whereas [63] and [48] proposed a multicriteria approach for identifying the most suitable location of energy facilities. Later on, the rapid developments in the field of multicriteria modelling resulted in an exponential increase in the number of real-world applications, exploiting in many different ways MCDA approaches to problem structuring, problem solving and decision making. Among these applications, the share of energy planning problems is steadily growing, while the range of questions to be answered is considerably widening. In former literature surveys on the use of decision analysis in energy and environmental modelling, a rising use of MCDA methods was detected along with an increased concern for adequately taking into account the uncertainty inherent to relevant decision situations [50, 56]. In two recent volumes of the Annals of Operational Research devoted to "OR Models for Energy Policy, Planning and Management", a great variety of energy planning problems are solved by means of the MCDA methodology [4]. [40] recently made a survey of the use of multicriteria decision making in the design of coordinated energy and environmental policies recommending the implementation of several MCDA methods in an integrated assessment framework. This rising interest is closely connected with the latest changes in the economic, social and natural environment, as they became more visible during the '90s.

- The ongoing trend for market deregulation. For several decades, energy supply systems were regarded as being necessarily large to secure economies of scale and state-owned to protect public interests. State-owned or privately-owned monopolies, in this case under severe franchising regulations, have dominated the electric power sector. Presently, a strong move towards liberalization is underway encompassing generation and sale, but even also transmission and local distribution. The trend for energy market deregulation was principally grounded on the wide recognition of the several market distortions and economic inefficiencies associated with the operation of energy monopolies of the public or private sector. In addition, it has been assisted by the ongoing globalization of the economy and the technological improvements facilitating the small-scale production of electricity. Since the moment when the possibility was granted to economic agents to have access to electric networks in order to celebrate electricity trade contracts, the formerly valid paradigm of natural monopoly fell. Provided that the network management part of the electricity business remains a regulated monopoly, generation, wholesale trading and retail trading may, in principle, be fully competitive. Energy market liberalization is widening the playing field and thus is expected to stimulate forces of competition to achieve a better allocation of resources than administrative processes do [83]. However, as long as fair-playing rules are not agreed and side-effects of energy production are not sufficiently reflected in the market mechanisms there are fears that competition may undermine public interests. Hence, although decentralization of energy systems may shrink the traditional opportunities for large-scale energy planning, new decision situations emerge in an attempt to combine economic efficiency and social interests. However, it must be noticed that this is an ongoing process and a large diversity of electricity systems remains, ranging from a more traditional vertically integrated structure to a totally unbundled structure with different entities in each branch of activity (generation, transmission, distribution).

- The exigency for sustainable development. The long identified need to secure a balance between economic, environmental and social goals, having found its most comprehensive phrasing in the United Nations' report (the so-called Brundtland report), has subsequently influenced several international conventions and EU policy documents. Particular attention is paid to the alarming threat of climate change, as well as to the severe impacts of atmospheric pollution on human health and natural ecosystems. Thus, the international community is forced to seek for common routes to jointly cope with these difficult and highly uncertain problems despite their differing interests, responsibilities and capabilities. Since the energy sector is the main contributor to the emissions of greenhouse gases and atmospheric pollutants, energy planning should explicitly take account of the long-term and large-scale effects of energy choices. In this sense, sustainable development policies appear as an essential complement of energy market deregulation.

In this context, energy planning has to be positioned in a global, intergenerational and interdisciplinary perspective. It becomes clear that in view of this highly complex and poorly understood reality, decision aid needs are correspondingly greater and different in nature. First and foremost, it is essential to elucidate the kind of the dilemmas faced along with the unstructured values of the multiple actors involved or affected by the outcome of the decision. According to [15], today it is more than ever necessary to make balanced decisions, that is, to incorporate Rationality, Subjectivity and Ethics in the decision making process. MCDA methodologies have an important role to play in providing a harmonious combination of quantitative and qualitative approaches, and in creating decision platforms allowing for rationality to be merged with subjective judgements and ethical concerns.

In the three decades of its existence, MCDA had generated an enormous amount of papers and reports devoted to problems and applications in the energy sector. It is therefore impossible to make an exhaustive review of all this literature. This paper aims at examining to which extent the use of MCDA in energy planning applications has been influenced by those changes currently underway in the energy sector and in the overall socio-economic context and, in particular, to which extent it is adapted to the new needs and the ensuing structuring and modelling requirements. The analysis performed is mostly confined to the literature of the last decade and in particular to publications in the most known OR and Energy journals. Although efforts have been made to include a large number of relevant articles, the review is by no means exhaustive since our main goal was to identify general trends and approaches, rather than to proceed to a detailed review.

The analysis distinguishes between the two broad multicriteria methodologies, namely the multiobjective programming (MOP) models (where alternatives are implicitly defined by a set of constraints) and the models dealing with discrete alternative options (where, in general, alternatives are explicitly known a-priori). The former are the natural evolution of the monocriterion optimization techniques, traditionally used for ensuring the supply of the required quantity of energy at the right time, generally using a monetary indicator as the objective function. Section 2 gives an overview of the new extended range of applications of MOP models by focusing on the modelling novelties developed to effectively handle the increased complexity of the planning process. Models dealing with discrete alternative options, although apparently assuming a simpler mathematical formulation, are suitable to deal with a larger variety of problems encountered in energy planning. Their main strength is their capability to help structuring these problems that are often vaguely defined and provide a deeper insight into their various components. In Section 3, relevant problems found in the literature are classified in broad categories and analyzed by giving emphasis to the structuring process and to the modelling techniques used to derive the DM's preferences and arrive at the most preferred solution. Section 4 summarizes the main evolutionary features of MCDA applications in the field of energy planning and decision or policy making.

2. Multiobjective Programming Models for Energy Planning

This section is devoted to multiobjective programming models to provide decision support in a wide range of energy planning problems. Therefore, it is mainly focused on the characterization of the problems, also mentioning the main aspects of evaluation, operationalized through objective functions, and the categories of constraints implicitly defining feasible action plans. A reference is also made to the type of models (linear, non-linear, integer, mixed integer) and methods used to compute solutions.

The actual real-world nature of applications reported in the literature is generally unclear. Most papers focus on the components of problem formulation, model building and algorithms developed, lacking the necessary details to assess the issues dealing with real-world application even when this is mentioned. It must be remarked that, even when they are assumed as academic case-studies, which possess nevertheless an important role to play not just as valuable experimentation frameworks but also as *evangelization* tools to show the potential benefits which can be harvested from an MCDA approach in complex decision situations, results cannot generally be reproduced due to the lack of data.

Planning problems in power systems can be broadly categorized according to the time frame under analysis and the decisions to be made. A common distinction is between long-term/strategic, operational and short-term problems (see Table 21.1).

Table 21.1. Broad categories of planning problems in power systems.

Planning	*Typical time frame*	*Examples of decisions made*
Long-term/Strategic	Several years – decades	Generation expansion planning Transmission facility expansion
Operational	Months – years	Generation scheduling Transmission scheduling VAR planning DSM planning
Short-term	Hours – days – weeks	Unit commitment Power flow

2.1 Capacity Expansion Planning

The planning of the expansion of power generation capacity is abundantly reported in the literature as inherently involving multiple, conflicting and incommensurate objectives, since it is now widely recognized that multiple competing objectives are generally pursued besides strictly economic ones, such as those reflecting social and environmental concerns.

In the power generation expansion planning problem the aim is identifying the power to be installed throughout a planning period (number and type of generating units, being the type associated with the primary energy source and the energy conversion technology) and output (energy to be produced by new and already installed units). The objective functions generally considered include the minimization of the total expansion cost (or production costs only) in the planning horizon, the minimization of pollutant emissions (SO2, CO2, NOx), the minimization of a surrogate for environmental impacts (an economic indicator obtained by monetizing the pollutant emissions, a tons-equivalent indicator or an aggregate dimensionless indicator), the maximization of the reliability/safety of the supply system, the minimization of radioactive wastes produced, the minimization of the external dependence of the country, the minimization of a risk/damage potential indicator [5, 20, 28, 62, 69, 77, 80, 89, 101, 105, 114].

The constraints generally refer to capacity limitations, minimum load requirements, demand satisfaction (including a reserve margin), resource availability, technology restrictions (by technical or political reasons, such as the amount of nuclear power allowed to be installed), domestic fuel quotas, energy security (to guarantee a certain diversification of the energy supply), bounds for committed power, budgetary limitations, operational availability of generating units, rate of growth of the addition of new capacity. Some studies consider pollutant emissions as constraints (generally expressing the regulatory framework of the country) rather than objective functions. In this case, models resort to aggregate indicators (for instance, penalizing the installed capacity and the energy output) for assessing environmental impacts as objective functions, considering the pollutant emissions in physical quantities (tons) as constraints.

Some of these models are multiple objective linear programming (MOLP) models, which do not enable the consideration of the discrete requirements of the candidate units for power generation expansion. Therefore, MOLP models ask for a post-processing phase of discretization of the continuous solutions by taking into account the actual modular capacities of the available expansion units. This issue is fully taken into account in multiple objective mixed integer linear programming (MOMILP) models.

Besides relating the decision variables to the supply options, some models also consider demand-side management (DSM) options, under the broad perspective of integrated resource planning (IRP). Hobbs and Horn [52] present a non-linear programming model whose objective functions are utility costs, environmental emissions, net value as measured by consumer surplus, and regional employment, to assess their significance in an IRP framework. The integration of DSM in the planning process is modelled as an equivalent-generating group with some associated constraints of operational nature in [77]. [21] integrates DSM in a multiobjective model for electric utility planning (with an application to an Indian utility), considering as objective functions the annual system cost, CO2 emissions and loss-of-load expectation (LOLE) of the generating system. Constraints are associated with coal production and transportation capacities, power output limits for thermal plants, availability of generating capacity, hydro-energy restrictions, electricity demand supply, coal demand supply balance, and gas availability. A compromise programming approach based on the minimization of a distance to the ideal solution is used to compute solutions.

DSM options are characterized as supply-side resources, by considering the categories: dispatchable technologies and lighting (energy conservation programs and efficient lighting), thermal generating unit (direct load control programs), limited energy plant (co-generation), pump storage (load shifting).

The algorithmic approaches used by the studies referred to above to tackle the power generation expansion problem range from goal programming (which is the most commonly known and used model explicitly considering multiple axes of evaluation and may be regarded as the "bridge" between single and multiple objective programming) to generating methods (to characterize exhaustively the nondominated solution set) to interactive methods. These are aimed at assisting decision makers/planners in selecting a final solution, or a set of nondominated solutions for further screening, through a feedback process including a phase of computation of a compromise solution and a dialogue phase in which the DM's input on the last computed solution is used to modify the scalarizing function to be used in the next computation phase.

[69] uses a compromise approach based on the L_1 and L_∞ metrics (and the AHP for preferential weights elicitation) applied to an electricity planning exercise in Spain. An extension of this work based on goal programming allows for the integration of individual cardinal preferences provided by several social groups (regulator, academics, electricity utility and environmentalists) for the different criteria [70]. In [80] the authors developed a branch-and-bound algorithm modified for the multiple objective case, capable of computing the whole set of nondominated solutions. The TRIMAP interactive approach is used in [28] and [77], while [105] developed the STRANGE method, which is an extension of STEM, thus using a minimax approach, including stochastic aspects by means of scenarios.

Uncertainties regarding the coefficients of the mathematical models are generally dealt with a-priori (that is, embedded in the models) by considering fuzzy sets or stochastic programming, or a-posteriori namely by performing sensitivity analysis studies of selected compromise solutions.

Having in mind the changes currently underway in the energy sector, namely regarding the shift towards the liberalization of the electricity market, most of these models (mostly developed much earlier than that trend emerged) require that the electricity market in the target region or country is organized in such a way that an important part of the load is being supplied in a franchise environment. And, of course, that generation capacity expansion is mostly centrally planned. However, this corresponds still to a very large number of cases around the world, either because electricity market has not been liberalized or the transformations have kept some fundamental characteristics of the traditional market organization.

The supply options considered in those models are almost exclusively thermoelectric ones. In the scientific literature references can also be found to hydroelectric generation planning [2, 73]. In fact, the analysis of hydroelectric systems and the optimization of reservoir operation and management is one of the oldest applications of MCDA [30, 42, 100, 112]. The conflicting nature of these problems derive from competing utility operators on the same basin (scheduling of reservoirs) and eventually competing energy and non-energy uses (flood protection and control, agriculture irrigation, industrial and domestic water supply, navigation, dilution of pollutants and heated effluents, recreation, ecological sustainability and protection of species, etc.).

[74] addresses the problems of short-term scheduling of the power system (the objective functions to be maximized are the total energy output of two utilities operating on the river Iguaçu, in Brazil) and the sharing of hydroelectric resources with a diversity of users (the objective functions to be maximized are the energy output and several other non-energy uses). In hydroelectric generation planning problems, constraints generally refer to water flows conservation, bounds on reservoir contents, discharges and spills, and limitations for acceptable deviations regarding energy output. The objective functions considered in the MOLP model proposed in [2] are: flood protection, dam safety and operational stability, ability to supply consumers with water, recreation, power generation, and environmental impact. The constraints refer to storage continuity equations, limits on releases due to ecological and flood control requirements, storage limits, bounds on power generation, transit releases and water requirements. The solution procedure is the weighted Tchebycheff method [102] which is based on a Tchebycheff metric and a contracted gradient cone approach. [7] presents an interactive fuzzy satisfying method based on evolutionary programming for short-term hydrothermal scheduling considering cost and emission objectives. The DM intervenes in the solution search process by updating the reference membership values until a satisfying solution is obtained. This approach deals with a multi-reservoir cascaded hydroelectric system with nonlinear relationships between water discharge, net head and power generation, also considering water transport delays between reservoirs.

2.2 Transmission and Distribution Network Planning

Transmission planning involves determining the location, the size and the time frame of the installation of new circuit additions to supply the forecasted load throughout the planning period in a way to balance economic, environmental and technical objectives subject to operating constraints given existing network configuration and generation units. Line routing problems may have associated aspects of evaluation such as population exposure to electromagnetic fields, potential damages to ecosystems, visual impact, etc.

The distribution network planning problem involves deciding the construction and/or reinforcement of facilities (substations) and branches to meet demand and satisfy operational and technical constraints (such as thermal limits and maximum voltage drop) while optimizing objectives of monetary and technical nature.

[27] presents a power system transmission network planning problem which is formulated as a multiobjective mathematical optimization problem, considering investment cost, reliability and environmental impact as objective functions. A genetic algorithm approach is developed to compute possible planning schemes. This step is followed by a fuzzy decision analysis method to select a final solution. A multiobjective fuzzy model for distribution network planning using a meta-heuristic simulated annealing approach to sample the nondominated solution set is developed in [88]. The fuzzy objective functions are the investment cost, the operating cost and the non-supplied energy. Constraints refer to the relationships of fuzzy flows and fuzzy injections, branch limits and maximum voltage drop. [93] addresses a distribution design problem considering as objective functions the line construction cost and the deviation of bus voltages to a target voltage. This is a nonlinear programming problem in which line locations and bus voltages are the decision variables. The solution computed by using STEM and goal programming are compared. [12] considers the minimization of the substation and feeder costs and the interruption costs as linear objective functions. Constraints refer to radiality, load, power flow and interruption duration.

2.3 Reactive Power Compensation Planning

The reactive power compensation planning (also referred to as VAR planning) problem considering multiple aspects of evaluation has recently deserved a great attention in the scientific literature. The installation of shunt capacitors in electrical distribution networks can effectively reduce energy and peak power losses, while improving quality of service particularly promoting a better voltage profile. Economic and operational benefits depend mainly on the number, location and sizes of new reactive power sources (capacitors) to be installed. This is intrinsically a non-linear problem with binary and continuous variables, although several studies use some forms of linearization and/or other simplifications in order to keep the problem manageable. Most recent methodological approaches to deal with this problem are based on meta-heuristics (simulated annealing and tabu search) as well as evolutionary strategies/genetic algorithms.

The objective functions considered are generally related with economical and technical aspects. A MOLP model is presented in [113] with an economical objective function (related with transmission losses and costs of reactive power compensation) and a technical objective function (related with the optimization of reactive power aimed at improving the quality in the distribution network). [25] considers three objective functions: economical operating condition of the system, the system security margin, and the voltage deviation, and an interactive simulated annealing algorithm based on the e-constraint method is used. [53] also uses simulated annealing in a model with four objective functions: operating efficiency (loss reduction), cost (investment, installation, and operation and maintenance costs), quality of service (voltage profile), and system security (line overloads due to excessive power flow). [24] extends the previous works by means of a weighted-norm approach. [108] uses a successive application of fuzzy linear programming to compute solutions in face of two objective functions: minimizing active energy losses and maximizing the voltage stability margin. A nondifferentiable model considering costs (investment plus monetized real power losses), security margin and the sum of deviations from the ideal voltage at buses as objective functions is developed by [22]. The DM is asked to specify fuzzy goals for the objective functions through an elicitation process of the corresponding membership functions. The reference membership values are updated interactively based on the current solution in each iteration. The same objective functions have already been considered in [23] to determine the candidate weak buses for installing new VAR sources. Weak buses are firstly identified by using a voltage collapse proximity indicator. A goal-attainment approach, based on simulated annealing, is then used to iterate on the current solution according to the preferences expressed by the DM regarding his/her goals for the objective functions.

2.4 Unit Commitment and Dispatch Problems

Load dispatch in electric energy systems involves the determination of a generation schedule (allocation of generation to the different units) that minimizes total cost and also addresses other objectives, namely environmental ones, satisfying several categories of constraints, mainly of operational nature, such as line overloading, bus voltage profile, deviations from standard values, etc. These problems are generally very complex and algorithmic approaches based on meta-heuristics have been predominantly used to tackle them in the last years.

[31] considers the total cost of generation and atmospheric hazardous emissions as objective functions, subject to power balance and capacity constraints. The authors use a multiobjective stochastic search technique which is a combination of genetic algorithms and simulated annealing. In [32] the aim is to optimize the scheduling of the real and reactive power generation, subject to constraints such as real power balance, line overload prevention, upper and lower bounds on generator output power, environmental impacts, upper and lower bounds on bus voltage. The algorithmic approach is based on linear programming with bounded variables. [111] proposes a model with economic, environmental and security objectives, which is then reduced to a bi-objective

problem by combining the economic and environmental objectives in a single monetized objective function. A simulated annealing algorithm operating on a weighted-sum function is then used for computing solutions and analyzing tradeoffs. [55] presents a fuzzy satisfaction maximizing approach for a model considering the minimization of fuel cost and environmental impact of NOx emissions. The distance to the ideal solution is used to obtain dispatch solutions and the corresponding tradeoff between fuel cost and emissions, given a fuzzy utility membership function mapped between the nadir and ideal points.

[17] considers cost and emission objectives in a thermal power dispatch problem to allocate the electricity demand among the committed generating units, subject to physical and technological constraints. A "best compromise" solution is obtained by searching for the optimal weighting pattern (the one that attains the maximum satisfaction level of the objective membership functions) using a genetic algorithm. [8] presents an interactive fuzzy satisfying weighting method to decide the generation schedule considering explicitly statistical uncertainties in the systems production cost data, pollutant emission data and system load demand. The objectives are the operating cost, NOx emissions and risk due to the variance of active and reactive power generation mismatch. Hooke-Jeeves and evolutionary search techniques are used to generate the "best" solution in the framework of an interactive approach.

[82] present an optimal load flow model with three objectives: cost of generation, system transmission loss and pollution. Solutions to the power system operation problem are obtained by minimizing the Euclidean distance to the ideal point.

An evolutionary programming approach is presented in [106] to solve the economical operation of a co-generation system under emission constraints. The objective functions are the minimization of cost and multi-emissions. The cost model includes fuels cost and tie-line energy. The emissions considered are CO2, SOx, and NOx, which are derived as a function of fuel enthalpy. The constraints include fuel mix, operational constraints, and emissions. The steam output, fuel mix, and power generations are computed by considering the time-of-use dispatch between cogeneration systems and utility companies.

[1] formulates the environmental/economic power dispatch problem as a nonlinear constrained model. A NSGA (nondominated sorting genetic algorithm) approach is used to generate a well-distributed nondominated frontier. Fuzzy sets are the used to extract a "best compromise" solution from the trade-off curve.

The implications of a deregulated market on unit commitment models is investigated in [51]. [19] develops a weighted mixed integer goal programming model to deal with the problem of converting an energy schedule into a power schedule, respecting the reserve schedule as well as technical constraints, and considering goals related with the energy schedule of a unit, the total energy scheduled for the company, the positive reserve schedule for a unit, the total cost for the company, and the smoothness of power changes. The constraints are related to the linearization of energy costs, limits for power generation and reserve, reserve limits intervals, ramp rates, and security limits. The model has been applied to real-sized problems of the Spanish electricity market.

2.5 Load Management

Electric utilities have used demand-side resources by changing the regular working cycles of loads through the implementation of appropriate power curtailment actions with the main goals of obtaining operational benefits (such as increasing load factor, reducing peak power demand, improving reliability or reducing loss) and costs reduction. Recently, this kind of programmes has raised further attention mainly due to the economic interests related with the volatility and spikes of wholesale electricity prices and also because of reliability concerns (transmission congestion and generation shortfalls). In the context of a progressively deregulated market, these programmes, which include direct load control (allowing to shed remote customer loads unilaterally), interruptible power and voluntary load shedding, have revealed to be attractive for a retailer dealing with volatile wholesale prices and fixed, over a certain time period, retail prices. Typical targets for these actions are loads which deliver energy services whose quality is not substantially affected by supply interruptions of short duration (in the residential sector, examples are forms of thermal storage such as electric water heaters and air conditioners). The aim is to select adequate load shedding actions to be implemented over sets of loads, considering a broad set of load management objectives and being useful for the different possible players in the power market.

A distribution utility (which owns and also manages the distribution grid) is generally interested in decreasing peak demand at, for example, sub-station and transformer stations levels due to capacity constraints, reliability concerns, or efficiency improvement through loss reduction. Power demand reduction may also be desirable due to costs associated with a specific demand level, where profits may substantially decrease because average wholesale prices are much higher than retail prices in a certain period of time. Peak reduction enables both the distribution utility and the retailer to have a better capability of continuously exploring the differences between purchasing and selling prices. The minimization of the peak power demand is considered in [11, 29, 38, 58, 68, 90, 110].

Profits are generally influenced by the amount of electricity sold and the time of day/season/year. In the presence of demand and wholesale price forecasts, the distribution utility/retailer can design adequate load shedding actions in order to maximize profits once retail prices are fixed. The maximization of the utility/retailer profits is addressed in [38, 58].

The electricity service provided by loads under control is changed, possibly postponed or even not provided at all, when load management actions are implemented. These changes can eventually cause some discomfort to customers that must be minimized, so that those actions become also attractive from the customers' point of view (with eventual reduction in their electricity bill) and/or at least not decrease their willingness to accept them. The discomfort caused to customers resulting from the implementation of power curtailment actions is incorporated in the models proposed in [11, 38, 58, 90]. This objective can be considered as a surrogate for energy service quality.

The minimization of costs is considered in [11, 29, 76, 110]. The impacts on reliability or on spinning reserve is taken into account in [68]. Consumer bill reduction (in the cases where tariff rates change during the day), as well as energy storage by consumers during the off-peak/low-cost periods is considered in [90]. The minimization of the loss factor is addressed in [38] and the maximization of the security margin (in the event of a generation shortage) is considered in [76].

Constraints of these models (typically involving continuous and binary variables) include technical and economical aspects. In some models, some of the objective functions presented above are included as hard or soft constraints (by establishing thresholds whose violation is taken into account by using a penalty function).

2.6 Energy-economy Planning Models

Multiple objective models are also being used in the study of the interactions between the economy (at national or regional levels), the energy sector and the corresponding impacts on the environment. In general, these models are developed based on data and inter-relationships emerging from input-output analysis. The analytical framework of input-output analysis enables to model the interactions between the whole economy and the energy sector, thus identifying the energy required for the provision of goods and services in an economy and also quantifying the corresponding pollutant emissions.

[54] used the NISE algorithm in conjunction with the inter-industrial input-output model to study the tradeoffs between the Gross Domestic Product (GDP) and energy consumption in Taiwan. [26] presents a model considering as objective functions the maximization of economic growth, the minimization of environmental pollution and the minimization of energy consumption. Compromise solutions – composition of sector outputs in the Chungbuk (Korea) economy – are obtained by using an interactive method. The employment, pollution and energy consumption multipliers are calculated from the Chungbuk multi-region input-output model. The impact multipliers are then combined with decision variables to form the objective functions of the MOLP model.

[85] proposes an economy-energy-environment planning model whose objective functions are private consumption, employment level, CO2 emissions and the self-production of electricity. Constraints refer to balance of payments, gross-added value, production capacity, bounds on exports and imports, public deficit, storage capacity and security stocks for hydrocarbons. Solutions to this MOLP model were obtained by using the STEM method. An interactive approach to tackle uncertainty and imprecision associated with the coefficients of this type of models is presented in [13], where some of the coefficients are triangular fuzzy numbers. Interactive techniques are used to perform the decomposition of the parametric (weight) diagram into indifference solutions corresponding to basic nondominated solutions. Three objective functions are considered which enables to graphically display the decomposition of the parametric diagram: energy imports, self-production of electricity and CO2 emissions. The model presented in [85] has been extended by constructing an adjusted input-output table suited to energy-environment analysis which considers an economical sphere and an environmental sphere and six main sectors. The objective functions considered in the model are: minimization of acidification potential, maximization of self-power generation, maximization of employment, maximization of GDP, and minimization of energy imports [86]. The nondominated solutions are computed by using a min-max scalarizing function associated with displaced reference points.

Due to the steady increase of population living in urban areas as well as the energy consumption in residential and commercial sectors, attention is being paid to models that can analyze energy systems in urban areas, in which costs, energy conservation and environmental impacts, among other aspects, are at stake. [107] develops a multiple objective model considering cost, primary energy consumption and CO2 emissions as objective functions, which are evaluated by using the results of simulating the operation of energy systems (such as co-generation, solar, electric turbo refrigerator with heat accumulation) for each of the major types of buildings in urban areas in Japan. [14] uses a reference energy system to map the flow of intermediate forms of energy from supply-side to demand nodes at the end use level in four major economic sectors (domestic, transport, industries, and services and commercial) in the framework of sustainable energy-environment management in an urban area (Delhi, India). A goal programming model is developed including as goals energy demand, energy budget, emissions, vehicle-utilization capacity, power supply capacity/system efficiency. Constraints are related with regional availability of energy sources for different sectoral end uses. The weights assigned to goals are elicited by using the AHP.

2.7 Energy Markets

The trend towards market deregulation possesses a major impact on transactions and electricity trade, namely whenever environmental aspects must be taken into account.

In [84] the authors analyze the tradeoff relationships between the different goals of power system operation and the influence of social policies, such as environmental impact minimization, upon deregulated electricity trade. An optimization procedure to reach a coordinated solution between different objectives is presented based on fuzzy interactive multiobjective optimization. Numerical examples are demonstrated on an IEEE 30 bus system. From the simulation, it has been found that the additional goals may reduce the volume of free trade of electricity, but the fuzzy multiobjective optimization can reach a good balance between conflicting goals.

A stochastic short-term planning model for supporting decisions of small energy suppliers (price takers) is presented in [60]. The technical constraints lead to a mixed integer linear programming problem. The uncertainty associated with market prices is modelled by developing a set of scenarios with assigned probabilities. The performance measures considered in the multiple criteria model for modelling the generator attitude towards risk are the mean return, the mean loss, the mean semi-deviation below the mean return, the worst return realization and the conditional value-at-risk. Solutions are computed by using an interactive approach based on aspiration/reservation levels and achievement scalarizing functions.

3. Energy Planning Decisions with Discrete Alternatives

The changes brought in the energy market and in the priorities of energy planners and policy makers have revealed a multiplicity of new tasks aiming at the choice, ranking or sorting of discrete options by taking into consideration different points of view. These tasks underpin the relevance of MCDA methods and may occur in a wide range of decision contexts.

There are different possible approaches to look at energy planning problems dealing with discrete alternatives in a more systematic way. The approach adopted herein firstly focuses on the subjects treated in relation with the main target of the decisions addressed. Subsequently, problem structuring and modelling issues will be highlighted in order to indicate:

- Common methodological foundations that are present in apparently very different decision situations,
- Different routes to approach more or less similar energy planning problems.

The intention is first to recognize the type of decision problems raised in the new conditions of the energy market, and then to examine the way MCDA is adapted to these conditions and the type of assistance it provides to effectively cope with the dilemmas faced.

3.1 Problem Classification

The majority of MCDA applications with discrete alternatives in the energy sector focuses on complex one-off decisions of strategic importance. Similarly to MOP problems, the electricity sector appears as a vast source of inspiration. However, there is a clear difference between the two methodologies regarding both decision context and means used. MOP formulations represent in great detail the real system's structure in order to establish the utility's medium- to long-term plans. Despite this fact, modelling limitations generally restrict the range of impacts that can be integrated in the evaluation procedure. On the contrary, formulations with discrete alternatives (in general, explicitly known a-priori) are very often trying to handle ill-structured problems while being able to look in more detail at all the different aspects that should be taken into account in order to identify strategic and policy directions to guide future actions. Furthermore, there is a great variety of decision problems that can be dealt with this kind of modelling approaches. More specifically, the major questions to be answered in power planning are:

- Which energy sources and energy technologies are the most appropriate for electricity generation?
- Which is the appropriate capacity plan coupling together various conversion technologies?
- Which is the appropriate pricing and how to proceed to the licensing of new facilities within a deregulated electricity market?
- Where should electricity facilities (power plants or transmission lines) be located?
- How to cope with unexpected problems occurring during the system's operation?

Besides the electricity sector, there is also a great variety of decision contexts arising at either the supply or the demand side of the whole energy sector. In addition to the above stated questions, further concerns that may be encountered include:

- Which is the most appropriate energy policy for the future?
- How to develop and exploit new energy resources?

- How to select among competing projects for increasing the production and/or the efficiency and/or the environmental performance of energy systems?

On the basis of the concerns described above and by taking into account the desired type of outcome, the publications examined are classified into the following major groups.

3.1.1 Group A: Comparative Evaluation of Power Generation Technologies. The problems of this category fall in their vast majority into the ranking typology as defined in [95]. The aim is to prioritize the available technological options, while the -often not explicitly stated- intention is to establish development plans and accordingly direct policy instruments. However, it is hardly visible how the obtained rankings will be translated into operational action plans or policy priorities. The considered alternatives are in all cases defined at the outset and reflect the technological progress and the prevailing concerns of utility planners in the period and area of the application.

Thus, in earlier publications the predominant dilemma was mostly to compare nuclear energy with its conventional competitors and sometimes also with emerging renewable technologies [99]. The work developed by Hämäläinen and his co-authors is characteristic of this research line, in which MCDA methods are used to clarify opposing views in the debate about the social benefits of the nuclear option if compared with coal fired power plants and a decentralized alternative based on conservation and small electricity units [43, 44, 45]. A common feature in these papers is that the authors' main preoccupation is to provide a rigid framework to handle this type of problems rather than to find a global answer. Problem structuring results from the investigation of the most critical issues with the implicit or explicit involvement of stakeholders, while particular attention is paid to the treatment of uncertainties. The dilemma between nuclear energy and fossil fuels is found also in a more recent publication [41] which assumes a complete lack of quantitative data in order to justify resorting to fuzzy decision analysis.

In the late nineties, the risks associated with nuclear energy and the serious concerns about climate change have shifted the interest of utilities and other stakeholders towards renewable energy sources. A ranking procedure based on the consideration of an apparently exhaustive set of evaluation criteria is followed in [9] in order to select the most promising set among alternative renewable technologies for establishing an action plan for the Sardinia region. The increasing emphasis on renewable power generation technologies is also reported in [3, 59, 75], where a small set of predefined alternatives are compared to each other in order to derive priorities complying with broad policy objectives.

3.1.2 Group B: Selection among Alternative Energy Plans and Policies.

Typical problems in this category are the choice dilemmas faced by energy planners or regulators at the national, regional or local level seeking to identify the most desired one among alternative scenarios for the future. Similar decision contexts are comparable with the selection among the set of efficient solutions determined by MOP formulations. In the electricity sector, relevant strategic decisions concern the choice among alternative strategies at a country or regional level by taking into account a number of scenarios for the likely evolution of external conditions [71, 87, 109]. Similarly, scenarios that have been formulated according to the specific characteristics of the autonomous electricity system of the island of Crete are evaluated with respect to a large set of sustainability criteria in [34]. Finally, in [61] policy options are perceived as composed of a set of consistent actions and the intention was to develop a framework to help utility managers to formulate alternatives, to express preferences and to find out the policy option securing the most satisfactory achievement of their strategic goals. The proposed structuring and analytical procedure has been illustrated with a case study referring to the selection of the most appropriate electricity pricing policy.

Besides the electricity sector, the same reasoning is followed for the selection among discrete policies and action plans for the whole energy system. [65] and [57] address the broader topic of energy policy at the national level by structuring and formalizing the whole decision procedure with the active involvement of several interest groups. In [35] alternative plans for the development of renewable energies in Greece are constructed and evaluated by a group of stakeholders in order to identify a compromise solution giving place to the widest possible consensus.

At a much smaller scale, choice problems are faced in the selection of the best option among a number of mutually exclusive alternative plans exploiting to a different extent and in a different way a certain energy resource, especially a locally confined renewable energy resource. Three papers found in the literature, authored by different researchers in different time periods, are all concerned with identifying the best exploitation plan of different geothermal fields in Greece [18, 39, 47]. Another relevant application is described in [46]. Here, the choice is among alternative biomass crops for use in electricity generation and/or in the transport sector, while the evaluation process is taking into account the economic and ecological impacts generated through each alternative's entire lifecycle.

3.1.3 Group C: Sorting Out a Subset of Candidate Energy Projects.

A quite different problem typology is developed in cases where the number of projects is large and the desired outcome is the identification of the most attractive subset of alternatives. In the absence of any further constraint or complementary relation between alternatives, the problem can be treated with a MCDA ranking procedure designed as to single out the best projects according to the decision maker(s)' preferences. In [81] and [49] the aim is to find out the best exploitation of the possibilities offered by the multiplicity of demand side management (DSM) options within the broader context of Integrated Resource Planning of electricity systems. In [37] the aim is to sort out a limited number of both supply and demand side energy projects among a large number of proposals submitted by various bodies and organizations for the restructuring of the Armenian energy sector. [78] has developed a fuzzy filtering method based on the degree of acceptance for reducing the initial large set of efficient solutions in the power distribution problem, in particular a set of alternative plans for expanding or reinforcing the network. Finally, in another interesting application, the Greek programme for the mitigation of greenhouse gases has resulted through the classification of a large number of DSM and supply options into groups of different attractiveness and then by elaborating the obtained results for scheduling their implementation [36].

A further subdivision in this group refers to decision contexts in which budget constraints are imposed and/or complementary or competitive relations between the candidate proposals should be taken into account. Thus, a portfolio approach has to be followed in order to single out the subset of projects that complies with the decision maker(s)'preference system and satisfies the imposed limitations. In [79], the problem faced by a regulatory authority is to license a number of independent electricity producers (wind farms) that are competing for a limited land area and a restricted grid's capacity, securing, at the same time, the competition rules of the free electricity market. A similar decision situation is faced by a regulatory authority trying to select among a large number of applications concerning gas transmission facilities [104], while in [72, 103], the task was the allocation of a certain budget to various technological areas for Energy R&D projects. The combinatorial nature of these problems and the need to consider a large number of evaluation criteria imposed the combination of integer programming models with MCDA approaches intended to assign an overall score to each alternative. In [103] the problem is formulated as a multiobjective non-linear knapsack problem tackled by a heuristic algorithm.

3.1.4 Group D: Siting and Dispatching Decisions in the Electricity Sector.

A further decision situation following the choice typology is the classical location problem. In the electricity sector major concern for the siting of facilities is concentrated on thermal or nuclear power plants and on transmission lines. In [6] the aim was to identify the most suitable among a large number of potential sites to locate thermal power plants in three coastal regions in Algeria, while [91] developed a stochastic MCDA approach for siting two new nuclear power plants in the Netherlands. Finally, an MCDA approach was applied in an environmental impact assessment problem related with the pathway of an electric transmission line [94].

Besides the medium-to long-term planning problems, MCDA is also exploited for assisting dispatchers in the routine operation of electricity systems. [33] developed a multicriteria DSS with the intention to support dispatching decisions in electricity generation in front of abnormal situations, i.e. various disturbances threatening the system's stability and security.

3.2 Problem Structuring and Model Building

The problem structuring phase is the starting point of any MCDA application. It is often quoted that "a well structured problem is a problem half solved". Although this statement does not seem to be always verified in the literature, an increasing number of applications is devoting particular attention to identifying the key concerns that should be encompassed in the model development phase and in better understanding the conditions in which the desired solution will be implemented. This trend is quite clear in energy planning applications mainly because of the high complexity of energy relevant issues and of the long-term and often irreversible nature of related decisions.

Generally speaking, a more systematic structuring procedure is followed in real life applications that are called to meet a specific energy planning problem for which a serious concern has already been expressed by the corresponding authority and/or in which the involvement of stakeholders is more pronounced. In such cases, all relevant aspects are thoroughly examined and it is their combined consideration which leads the decision procedure to an effective and legitimate outcome. On the contrary, in some publications dealing, in an abstract way, with technologies prioritization, the structuring effort is minimized and multicriteria models are mostly used as algorithms capable of synthesizing multiple impacts associated with each alternative into an overall assessment that could be exploited at a later -not explicitly defined- stage of the planning procedure.

In order to look more thoroughly at the problem structuring approach, we proceed to a detailed analysis of the examined publications following the CAUSE checklist (Criteria, Alternatives, Uncertainties, Stakeholders, Environment) suggested by [10]. Tables 21.2 – 21.5 present in a symbolic way the emphasis given in each publication to the key-aspects of the structuring process, as described in the following paragraphs.[1]

3.2.1 Criteria and Alternatives. Criteria and alternatives constitute the two major poles of any decision situation, around which decision maker(s) and other stakeholders try to elucidate their judgement and select the course of action that better fits to their value system. Considered from the two structuring perspectives, namely alternatives-focused and value-focused thinking, the former is clearly the dominant one in the type of problems encountered in energy

Table 21.2. Key aspects in problem structuring in papers of Group A.

Reference	*Criteria*	*Alternatives*	*Uncertainties*	*Stakeholders*	*Environment*
[99]	+	+	i	i	+
[45]	+		i	i	+
[44]	+		i	d	+
[43]	+		i	d	+
[41]			i		
[9]	+	+	i	i	
[59]	+	+		i	
[3]		+			+
[75]	+		i		

Table 21.3. Key aspects in problem structuring in papers of Group B.

Reference	*Criteria*	*Alternatives*	*Uncertainties*	*Stakeholders*	*Environment*
[71]	+		i,e	d	+
[109]		+	i,e	i	+
[87]		+	i,e	d	+
[34]	+	+	i,e	i	+
[61]	+	+	i,e	d	+
[65]	+	+	i	d	+
[57]	+			d	+
[35]	+	+	i,e	d	+
[18]	+	+	i,e	i	+
[39]	+	+	i,e		
[47]		+	e	d	
[46]	+		i,e		+

Table 21.4. Key aspects in problem structuring in papers of Group C.

Reference	*Criteria*	*Alternatives*	*Uncertainties*	*Stakeholders*	*Environment*
[81]	+		e	D	
[49]	+	+	i,e	d	+
[37]	+		i	d	
[78]		+	i,e	i	
[36]	+		i,e	i	+
[79]			i	i	+
[104]	+			d	+
[72]	+	+	i	d	+
[103]	+		i	D	+

Table 21.5. Key aspects in problem structuring in papers of Group D.

Reference	*Criteria*	*Alternatives*	*Uncertainties*	*Stakeholders*	*Environment*
[6]	+	+	i	i	+
[91]	+	+	i	d	+
[94]	+		i	d	+
[33]	+	+	i	D	+

planning. This should not necessarily be interpreted as an underestimation of the potential contribution of hard thinking about values, but rather as a consequence of the more or less constrained context in which relevant problems emerge. In fact, in most energy planning problems the available alternatives are defined at the outset or imposed by the overall decision framework. Blank cells in the Criteria columns of Tables 21.2 – 21.5 simply indicate that no particular attention is paid to the selection of criteria, and/or that the selected ones are not adequately justified in the reported publications. Similarly, blank cells in the Alternatives column usually denote an a-priori defined set of alternatives, or a set for which not sufficient justification is provided in the text.

Papers in Group A mostly deal with an a-priori defined set of alternative technologies that, in some cases, are evaluated with respect to a rather unprocessed set of criteria. However, there are examples in which evaluation criteria result from an in depth investigation of fundamental objectives and their analysis into natural or constructed attributes [9, 43, 44, 45, 99]. It is worth mentioning that in these applications stakeholders are recognized as a key problem component and an attempt is made to directly or indirectly include their concerns in the structuring phase.

Group B consists of papers that are enlightening the stimulating force of value-focused thinking in problem structuring. The rather unstructured decision situations are gradually shaped around some key-questions leaving enough space for creative thinking and for inventing potential courses of actions. The generation of alternatives (policy options, action plans or exploitation plans of renewable energies) goes along with a detailed analysis of the differing points of view in direct or indirect consultation with stakeholders.

The problems treated in Group C are also structured around a given set of alternatives, while in many cases additional constraints are imposed to decision makers. Thus, the essence of the decision procedure is shifted to the modelling phase and to the selection of a consistent set of criteria that are adequately reflecting the stakeholders' value system.

Finally, in Group D, alternatives may be defined at the outset or not, depending on the specific problem faced and the extent to which a pre-decision stage has been reached. However, especially in siting decisions, the elucidation of points of view of all actors involved is considered as a crucial step for moving towards a satisfactory solution. Operational problems that may often – though not on a regular basis – occur are analyzed in [33] by means of existing knowledge bases in order to identify, each time, a set of likely alternative solutions along with the appropriate evaluation criteria.

Regarding the type of criteria used, practically all examined papers are including criteria reflecting environmental values as a result of the increasing recognition of the close links between energy and environmental planning. Depending on the depth of the analysis performed, the environmental dimension may simply be included in the assessment procedure as an unstructured qualitative criterion [3, 41, 47] or may be broken down in order to develop a whole value tree and detect most environmental attributes or impacts. Moreover, in some publications, the main goal is to provide a framework for the identification, measurement, scaling and weighting of environmental criteria [94, 104]. In addition to environmental criteria, most of the examined problems are structured by integrating the four main value axes shown in Figure 21.1, along which individual technologies or development plans and scenarios for the future are evaluated.

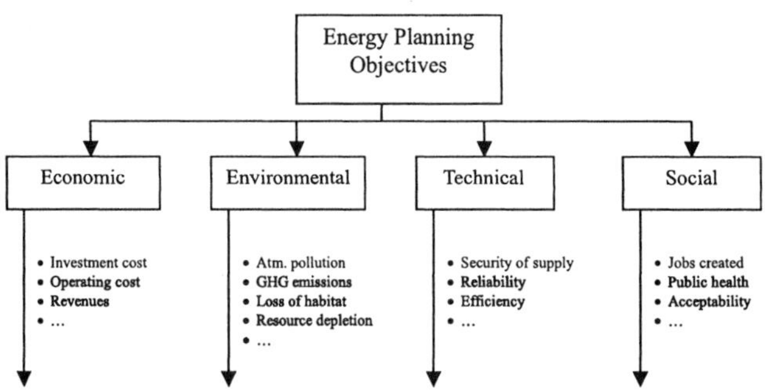

Figure 21.1. Typical hierarchical structure of criteria used in energy planning.

The selection of the particular set of criteria is clearly depending upon the particular type of the problem under consideration, the stakeholders' interests and/or the analyst's abilities and preoccupations. Data availability often imposes very severe limitations to the range of criteria included in the analysis and in the way they are measured, thus affecting the reliability of the obtained results. In any case, problems of redundancy or judgemental dependence are not apparent, while efforts are made to balance between completeness and conciseness, especially in publications which give particular emphasis to the deployment of values.

3.2.2 Internal and External Uncertainties. Uncertainty is a key characteristic of the real world that arises from the continuously increasing complexity of systems and the variability of parameters. In face of this intricate world, it is usually very difficult for the decision makers to capture all the complex phenomena, to get through all the necessary information and, last but not least, to express their value judgements.

Energy planning is by its very nature an intricate task concerned with complex technological systems interacting in multiple ways -not all being thoroughly investigated and understood- with the economic, natural and social environment. Furthermore, energy planning is by definition targeting at a more or less distant future, for which forecasts for certain aspects are very difficult to make, because of missing information, of the stochastic nature of the variables concerned (e.g. inflows into a hydro reservoir) or of a lack of human experience regarding some phenomena (e.g. greenhouse effect). To this purpose, a number of powerful techniques are implemented in energy planning applications, ranging from simple scenarios and sensitivity analysis to more sophisticated approaches based on the exploitation of fuzzy sets, stochastic methods, etc.

We shall examine the type of uncertainties identified in the reported literature and the way they are treated by distinguishing between internal and external uncertainties as defined in [10].

Internal uncertainties are related with either problem structuring issues or the elicitation of values. The former is mostly experienced by the analysts themselves and is not clearly discernible in the short text of a publication, though the insufficient structuring of some of them regarding the selection of alternatives and criteria gives an indication of relevant uncertainty problems.

The other source of internal uncertainties is present in practically all MCDA applications. However, as shown in Tables 21.2 – 21.5, it is often the case that uncertainties of this type are to a great extent overlooked. This gap may be attributed to the practical limitations of a written report summarizing the experience of a MCDA application, since in several papers the emphasis is given to other structuring or modelling issues, while authors refer to other works in which they look in more detail to preference modelling [81]. It should also be noted that a representative group of stakeholders secures that a significant part of relevant uncertainties is captured in their varying assessments of performances and the richer preferential information provided [34, 35, 37, 65, 104]. Finally, the use of outranking methods, which allow for sufficiently modelling the imprecision of data and the hesitations of decision makers, minimizes the need for treating uncertainties explicitly in a post-evaluation phase.

Among the techniques used for handling this type of uncertainty the most commonly used are:

- Construction of scenarios and sensitivity analysis: This is the most widely used approach in order to cope principally with the uncertainty characterizing the assignment of factors of relative importance to the evaluation criteria. The analysts' intention is first to capture different points of view reflecting the major stakeholders' value system, in case it was not possible to directly involve them into the decision making procedure and also to test the robustness of the obtained results [18, 36, 43, 45, 79, 99, 109]. Besides weighting factors, the uncertainty characterizing the definition of other preferential parameters such as thresholds, or of the criteria scores or even the relevance of particular criteria or alternatives is treated by means of scenarios assuming changes in the initial values or in the assumptions made during the problem structuring phase [34, 35, 44, 46].

- Qualitative scales: This technique is often used in the case of criteria that are not possible to be assessed in quantitative terms or the information available is not adequate to estimate cardinal values [9, 18, 34, 35, 57, 61, 79]. Qualitative or categorical scales are mostly used in combination with outranking methods because of the flexibility offered by the implied pairwise comparisons.

- Stochastic approaches: Although relevant techniques are more appropriate in handling uncertainties related with external conditions, in [91] randomly generated probabilities distributions are used to estimate weights and criteria scores for which only ordinal information is available. Similarly, the stochastic multicriteria acceptability analysis (SMAA) is also appropriate for energy planning problems, although most of its applications refer to environmental management [66, 67]. [87] is using composite utility variances in order to take account of both imprecise information and inconsistent subjective judgements in the evaluation of power expansion plans.

- Fuzzy sets: There is an extensive use of the fuzzy set theory in MCDA for the treatment of uncertainties. Besides its use in the development of outranking relations, it is possible to transform interval assessments with an indication of the most plausible value into triangular fuzzy numbers [39, 99] or to use linguistic terms in estimating performances or weights of criteria that are subsequently transformed in triangular or trapezoidal fuzzy numbers [9, 41]. Fuzzy acceptance levels along with fuzzy attribute values are used in [78] in a filtering algorithm targeting at the reduction of a large set of candidate actions.

- Parallel implementation of weighting and scaling methods: The use of different weighting methods is proposed in [49] in order to effectively cope with the uncertainties and inconsistencies inherent to the elicitation of weights. With a similar reasoning different scaling, weighting and aggregation techniques are implemented in [65, 71, 72]. In addition to dealing with relevant uncertainties, the multi-weighting or scaling approach is acting as a learning procedure for the participating stakeholders. It namely offers a better understanding of the problem's particular aspects, strengthens the stakeholders' ability to distinguish the most important issues and assists in building confidence in the decision made.

External uncertainties refer to the limited knowledge about the magnitude and evolution of some important parameters referring to the general economic, social or natural environment and which are outside the control of the decision makers. Although this type of uncertainty is typical in energy planning because of the multiple ambiguous or varying parameters affecting the energy market, the concern for handling this type of uncertainty is rather limited. The techniques used for this purpose are the following:

- Construction of scenarios: As in the case of internal uncertainties, scenarios are the preferred means to handle uncertainties regarding external conditions [18, 34, 35, 46, 49, 71]. Besides the ease of construction and limited computational requirements, scenarios offer a better insight into the problem's particularities and allow for the identification of the real threats to the success of the decision to be made.

- Stochastic approaches: The basic principles of decision theory are used to estimate expected utilities according to the probabilities estimated for a number of scenarios of external conditions [109].

- Construction of criteria: In this case, the uncertainty associated with specific aspects of the outside environment is expressed in the form of an extra attribute against which alternative options are valued usually through a qualitative scale. A common criterion of this type is the risk criterion [39, 47, 81], while in [36, 61] the criterion of applicability or implementability is suggested to take account of the different obstacles that may hinder or retard the realization of the selected projects or plans.

3.2.3 Stakeholders Involvement. Energy planning is all the more often concentrating the strong interest of the general public and of several authorities and non-governmental organizations. This is mainly due to the growing environmental concern about the serious impacts associated with energy production and use. In addition, the complex technological aspects, the high capital cost and the long lifetime of the necessary investments make the consultation of relevant experts and competent authorities an essential element of the decision process. Therefore, several actors are by definition involved in the planning procedure and others are simply wishing to actively participate in decisions they feel they may affect their own welfare or the environment's overall stability.

In fact, MCDA applications in energy planning are in their vast majority characterized by a notable involvement of a usually large and interdisciplinary group of stakeholders. This trend is clearly observable in the literature, although their direct participation is not always achieved. Several reasons are reported, while others are more difficult to be admitted. Among the latter, maybe the most important is that such a participatory process of sharing concerns, exchanging ideas, and accepting compromises is still not very common in the public or private sector. In addition, it is a costly and time consuming process. Therefore, in some publications no hint is made on any form of stakeholders' participation, although some form of consultation with experts may have taken place at an earlier stage of the analysis. In another group of publications this involvement is reduced to an informal and thus not binding process which takes one of the two following forms:

- Participation in a pre-decision stage in order to define the range of the alternatives to be considered and/or to identify major points of view and other key-concerns that should be taken into account [6, 9, 18, 45, 59, 79, 99, 109]. In these publications, there is either an explicit mention of this kind of involvement or an attempt to reflect the stakeholders' points of view through the elaboration of different sets of weights or different scenarios for the development of external conditions.

- Bilateral contacts with individual stakeholders in order to get a richer understanding of the problem at hand, and for directly extracting the relative importance to be granted to the criteria [34, 36].

Although, in these cases, the potential synergies of working together and interacting in the generation of ideas are lost, such informal involvement is of great value to capture the essence of the problems to be tackled.

In the remaining papers the participation of stakeholders appears as a crucial component of the whole decision process. We distinguish the following major types of contribution:

- In all these applications, stakeholders have actively participated in the elaboration of the criteria set and the assignment of weights. Their contribution starts with the identification of the fundamental objectives that should guide the decision process, extends to their breakdown into lower level attributes and ends up with retaining those criteria that are judged to be the most relevant for the problem to be tackled. In some cases, they furthermore contribute to the measurement and scaling of the defined criteria [43, 49, 57, 61, 65, 71, 72, 94].

- In some applications, stakeholders take part in the establishment of alternatives, especially if they refer to constructed scenarios or action plans

[35, 61], while in others they express their opinion about the options that should not be retained for evaluation in the final set [6, 37, 91].

- In only a few applications [35, 37, 49, 61] the involvement of stakeholders is extended in all major stages of the decision process. In particular, in [49] their participation is described in detail for all 9-steps in which the authors split the decision procedure. It is worth mentioning that it is exactly these publications which emphasize the significance of reaching consensus and suggest specific techniques to measure the disagreement between stakeholders and to achieve its resolution.

3.2.4 Consideration of the External Environment. This final aspect of problem structuring is closely but not exclusively connected with the extent the uncertainty related with external conditions is taken into account in problem structuring. It may happen that the analysts have thoroughly examined the various parameters of the economic, social and natural environment, along with technical constraints in building the alternatives and in defining the evaluation criteria, without considering the effect of changes in these parameters on the performance of the alternatives. As shown in Tables 21.2 – 21.5, in their vast majority the publications examined are founded on a sufficiently reliable representation of the environment in which the decision to be made will be implemented. It is worth emphasizing that the absence of a positive sign does not necessarily mean that the external environment has been totally overlooked, but simply that relevant hints are not included in the paper.

3.3 Method Selection and Model Development

The selection of the basic MCDA method and its adaptation to the problem's particularities is influenced by several parameters, of which theoretical grounds and scientific concerns play an important although not always the dominant role. As shown in Figure 21.2, the two major schools of thought, namely the Multi-Attribute Utility (or Value) Theory and the Outranking Approach, are represented in an approximately equal basis and, depending on the overall structuring procedure, they are providing valuable results.

3.3.1 Value and Utility Theory Approaches. This family of multicriteria methods -often referred to as the monocriterion synthesis approach- were the most widely used in earlier publications. Its theoretical foundation provides a consolidated framework for the deployment of the decision context, whereas the base hypothesis that the overall utility or value attributable to each alternative can be derived by aggregating partial utilities or values seems familiar and comprehensible by the decision makers [64, 92].

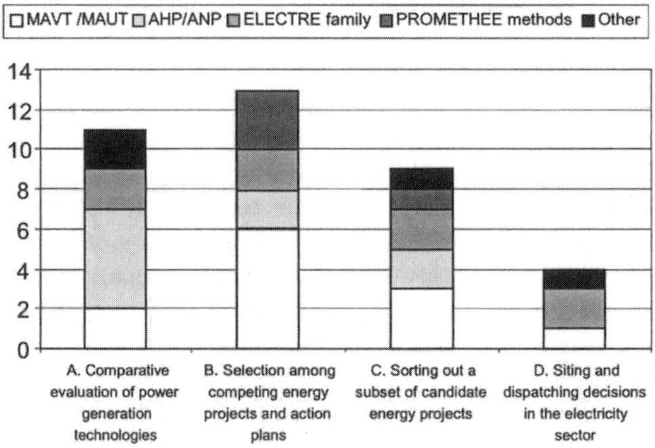

Figure 21.2. MCDA methods in energy planning applications.

The papers belonging to this broad category include various modelling approaches, such as the simple weighted average approach using a calibrated normalization procedure [46, 61, 65], the aggregation of fuzzy scores with fuzzy weights [9], the aspiration-led approach using a modified utility function expressing the degree of achievement or underachievement of the set aspiration levels [33, 81], as well as the exploitation of AHP for facilitating the process of eliciting partial utility functions and weights [87]. In [43, 49, 57] particular emphasis is given to the assessment of value functions and the elicitation of weights through an exemplary participatory learning process. Finally, AHP is by far the most widely used approach in this particular application field, since almost half of the papers classified in this group are relying on some form of the Saaty's analytical procedure [96, 97]. In most cases its use was grounded on the attractive feature of merging the problem decomposition process with the process of weights elicitation. Moreover, the information asked by the decision makers is easier to provide and therefore decision makers feel more comfortable than with other more demanding methods [43]. However, its ease of use turns, in some cases, into its greatest drawback, in that it simply serves as a convenient algorithm to solve the problem at hand without focusing on the essence of the problem itself and learning from a more tedious procedure of preference elicitation. There are applications avoiding this risk by paying particular attention to the extraction of the stakeholders' preferences and extensively using consistency checks [61, 71, 72].

Nevertheless, in decision situations where the ambiguity related with human judgements and/or with the imprecise information available cannot be overlooked, the results are considered as too precise to build the necessary confidence [94]. In these cases, decision makers feel more comfortable with partial rankings indicating incomparabilities and forcing to revisit the problem's elements and to get more precise preferential information. Furthermore, the total compensatory approach assumed behind the aggregation of partial values is often disputed, especially as far as environmental criteria are counterbalanced with economic or technical ones.

3.3.2 Outranking Approaches. Outranking approaches have known a remarkably rapid development and an extensive use in several application fields. Among these fields, energy and environmental planning have a prominent place, mainly because the imprecision associated with the measurement and evaluation of environmental parameters calls for modelling approaches giving more freedom to the decision makers to express their hesitations.

The ELECTRE family of methods developed by Roy and his collaborators at the LAMSADE Laboratory of the Paris Dauphine University [95] presents the higher frequency of use in the set of publications examined, with ELECTRE III being the most commonly used. ELECTRE-TRI is used in [36, 79] in problems involving as a first step the classification of the examined alternatives in ordered groups of preference. The PROMETHEE method developed by [16] at the Free University of Brussels is the other most widely applied outranking method in all kind of applications, among which energy planning problems. The capability of producing complete rankings is actually a significant reason justifying the use of PROMETHEE, together with all the other prominent characteristics of the outranking approaches [39, 47]. In [35] the PROMETHEE modelling approach is judged as simpler and more transparent by the involved stakeholders. A merging of ELECTRE III and PROMETHEE is proposed in [37] in order to draw advantage of their respective unique characteristics, namely the use of veto thresholds and the capacity to come out with complete rankings. Finally, fuzzy outranking approaches are proposed in [41] in order to cope with the complete lack of quantitative information.

The use of outranking approaches is very advantageous in applications where the stakeholders' involvement is considered as an essential element of the decision making process [6, 34, 35, 37, 47]. In these cases, indifference and preference thresholds are more convenient to capture a great part of the stakeholders' ambiguity, while the inter-criterion preferential information needed is provided in the form of factors of relative importance through significantly less demanding modelling approaches. However, it is exactly the rather arbitrary way in which thresholds are defined that is the most controversial aspect of outranking approaches.

4. Conclusions

The energy sector is of outstanding importance for the satisfaction of societal needs, providing directly or indirectly the fundamental requirements for almost all the activities involving Human beings, ranging from well-being and comfort needs to transportation and production systems. However, it is now widely recognized that most crucial environmental problems derive from energy demand to sustain Human needs and economic growth. The largest source of atmospheric pollution is fossil fuel combustion, on which current energy production and use patterns heavily rely. On the other hand, new requirements of reliability, quality of service and security of supply are at stake, namely having in mind the trend towards the liberalization of the electricity market. Therefore, in modern technologically developed societies, decisions concerning energy planning must be made in complex and sometimes ill-structured contexts characterized by technological evolution, changes in market structures and new societal concerns. Decisions to be made by different agents (at utilities, regulatory bodies and governments) must take into account several aspects of evaluation such as technical, socio-economic, and environmental ones, at various levels of decision making (ranging from the operational to the strategic level) and with different time frames.

Thus, energy planning problems inherently involve multiple, conflicting and incommensurate axes of evaluation. Models capturing these intrinsic characteristics of those problems not just become more realistic but also contribute to support reflection and creativity in face of a larger universe of potential solutions since a prominent solution no longer exists. MCDA models and methods thus enable decision makers to grasp the inherent conflicts and trade-offs among the distinct aspects of evaluation and to rationalize the comparison among different alternative solutions.

The approach developed in this paper to MCDA and energy planning distinguishes between multiobjective programming models and models dealing with discrete alternative options. An overview of the application of MOP models to an extended range of problems has been presented focusing on the sets of objective functions and constraints as well as the methods used. Regarding models dealing with discrete alternative options, relevant problems are classified and analyzed, emphasizing the structuring process and the modelling techniques used to derive the DM's preferences as well as the methods to obtain a recommendation.

It is shown that in both model categories, the decision context is increasingly more complex, parameters and attributes are often uncertain and imprecise, so that decision makers experience more difficulties in problem structuring, model building and in entrusting the provided solution. Therefore, a clear trend towards the effective treatment of uncertainty in implementing MCDA in energy planning is already discernible and is expected to be further enhanced in the future.

The energy sector presently is and will remain one of the most active and exciting areas of application of MCDA models and methods, providing new and challenging problems. The improvement of the implementation rate of MCDA studies should be a crucial concern of MCDA researchers and practitioners, namely having in mind the conceptual and operational validation of the use of MCDA techniques in real-world problems with actual decision makers.

Notes

1. In column Uncert. "i" and "e" denote internal and external factors of uncertainty, respectively, while in column Stakehold. "i" and "d" denote indirect or direct involvement of stakeholders, respectively, and "D" stands for DSS.

Acknowledgments

The research of C. H. Antunes and A. G. Martins has been partially supported by FCT under grant POCT/ESE/38422/2001.

References

[1] M. Abido. A novel multiobjective evolutionary algorithm for environmental/economic power dispatch. *Electric Power Systems Research*, 65(1):71–81, 2003.

[2] P. Agrell, B. Lence, and A. Stam. An interactive multicriteria decision model for multipurpose reservoir management: The Shellmouth reservoir. *Journal of Multi-Criteria Decision Analysis*, 7:61–86, 1998.

[3] B. Akash, O. Al-Jayyousi, and M. Mohsen. Multicriteria analysis of nonconventional energy technologies for water desalination in Jordan. *Desalination*, 114:1–12, 1997.

[4] C.H. Antunes and A.G. Martins, editors. *OR Models for Energy Policy, Planning and Management*. Annals of Operations Research, 120/121, 2003.

[5] C.H. Antunes, A.G. Martins, and I. Brito. A multiple objective mixed integer linear programming model for power generation expansion planning. *Energy*, 29(4):613–627, 2004.

[6] O. Barda, J. Dupuis, and P. Lencioni. Multicriteria location of thermal power plants. *European Journal of Operational Research*, 45:332–346, 1990.

[7] M. Basu. An interactive fuzzy satisfying method based on evolutionary programming technique for multiobjective short-term hydrothermal scheduling. *Electric Power Systems Research*, 69(2-3):277–285, 2004.

[8] S. Bath, J. Dhillon, and D.P. Kothari. Fuzzy satisfying stochastic multi-objective generation scheduling by weightage pattern search methods. *Electric Power Systems Research*, 69(2-3):311–320, 2004.

[9] M. Beccali, M. Cellura, and D Ardente. Decision making in energy planning: The ELECTRE multicriteria analysis approach compared to a fuzzy-sets methodology. *Energy Conversion Management*, 39:1869–1881, 1998.

[10] V. Belton and T. Stewart. *Multiple Criteria Decision Analysis: An Integrated Approach*. Kluwer Academic Publishers, Dordrecht, 2002.

[11] K. Bhattacharyya and M. Crow. A fuzzy logic based approach to direct load control. *IEEE Transactions on Power Systems*, 11(2):708–714, 1996.

[12] S. Bhowmick, S.K. Goswami, and P.K. Bhattacherjee. An LP-based multiobjective distribution system planning using goal attainment method. *Electric Power Components and Systems*, 29(5):479–490, 2001.

[13] A. Borges and C. Antunes. A fuzzy multiple objective decision support model for energy-economy planning. *European Journal of Operational Research*, 45(2):304–316, 2003.

[14] R. Bose and G. Anandalingam. Sustainable urban energy-environment management with multiple objectives. *Energy*, 4:305–318, 1996.

[15] J. Brans. Ethics and decision. *European Journal of Operational Research*, 136:340–352, 2002.

[16] J. Brans and P. Vincke. A preference ranking organization method: The PROMETHEE method for multiple criteria decision-making. *Management Science*, 31:647–656, 1985.

[17] Y. Brar, J.S. Dhillona, and D. Kotharib. Multiobjective load dispatch by fuzzy logic based searching weightage pattern. *Electric Power Systems Research*, 63(2):149–160, 2002.

[18] P. Capros, S. Papathanassiou, and J. Samouilidis. Multicriteria analysis of energy supply decisions in an uncertain future. *Omega*, 16:107–115, 1988.

[19] E. Centeno, B. Vitoriano, A. Campos, A. Muñoz, J Villar, and E. Sánchez. A goal programming model for rescheduling of generation power in deregulated markets. *Annals of Operations Research*, 120:45–57, 2003.

[20] D. Chattopadhyay. Cost versus emissions minimization in thermal electricity generation: A case of SO_2 emissions reduction from the Indian power system. *International Journal of Energy Research*, 19:721–727, 1985.

[21] D. Chattopadhyay, R. Banerjee, and J. Parikh. Integrating demand side options in electric utility planning: A multiobjective approach. *IEEE Transactions on Power Systems*, 10(2):657–663, 1995.

[22] C. Chen. An interactive fuzzy-norm satisfying method for multiobjective reactive power sources. *IEEE Transactions on Power Systems*, 15(3):1154–1160, 2000.

[23] Y. Chen. Weak bus-oriented optimal multiobjective VAR planning. *IEEE Transactions on Power Systems*, 11(4):1885–1890, 1996.

[24] Y. Chen. Weighted-norm approach for multiobjective VAR planning. *IEE Proceedings C, Generation, Transmission and Distribution*, 145(4):369–374, 1998.

[25] Y. Chen and C. Liu. Optimal multiobjective VAR planning using an interactive satisfying method. *IEEE Transactions on Power Systems*, 10(2):664–670, 1995.

[26] C. Cho. The economic-energy-environmental policy problem: An application of the interactive multiobjective decision method for Chungbuk Province. *Journal of Environmental Management*, 56(2):119–131, 1999.

[27] T. Chung, K. Li, G. Chen, J. Xie, and G. Tang. Multiobjective transmission network planning by a hybrid GA approach with fuzzy decision analysis. *International Journal of Electrical Power & Energy Systems*, 25(3):173–258, 2003.

[28] J. Clímaco, C. Antunes, A. Martins, and A. Almeida. A multiple objective linear programming model for power generation expansion planning. *International Journal of Energy Research*, 19:419–432, 1995.

[29] A. Cohen and C. Wang. An optimization method for load management scheduling. *IEEE Transactions on Power Systems*, 3(2):612–618, 1988.

[30] J. Cohon. *Multiobjective Programming and Planning*. Academic Press, New York, 1978.

[31] B. Das and Patvardhan. New multiobjective stochastic search technique for economic load dispatch. *IEE Proceedings C, Generation, Transmission, and Distribution*, 145(6):747–752, 1998.

[32] A. Farag, S. Al-Baiyat, and T. Cheng. Economic load dispatch multiobjective optimization procedures using linear programming techniques. *IEEE Transactions on Power Systems*, 10(2):731–738, 1995.

[33] X. Gandibleux. Interactive multicriteria procedure exploiting a knowledge-based module to select electricity production alternatives: The CASTART system. *European Journal of Operational Research*, 113:355–373, 1999.

[34] E. Georgopoulou, D. Lalas, and L. Papayannakis. A multicriteria decision aid approach for energy planning problems: The case of renewable energy option. *European Journal of Operational Research*, 103:38–54, 1997.

[35] E. Georgopoulou, Y. Sarafidis, and D. Diakoulaki. Design and implementation of a group DSS for sustaining renewable energies exploitation. *European Journal of Operational Research*, 109:483–500, 1998.

[36] E. Georgopoulou, Y. Sarafidis, S. Mirasgedis, S. Zaimi, and D. Lalas. A multiple criteria decision aid approach in defining national priorities for greenhouse gases emissions reduction in the energy sector. *European Journal of Operational Research*, 146:199–215, 2003.

[37] Y. Goletsis, J. Psarras, and J. Samouilidis. Project ranking in the Armenian energy sector using a multicriteria method for groups. *Annals of Operations Research*, 120:135–157, 2003.

[38] A. Gomes, C.H. Antunes, and A. Martins. A multiple objective evolutionary approach for the design and selection of load control strategies. *IEEE Transactions on Power Systems*, 19(2), 2004.

[39] M. Goumas and V. Lygerou. An extension of the PROMETHEE method for decision making in fuzzy environment: Ranking of alternative energy exploitation projects. *European Journal of Operational Research*, 123:606–613, 2000.

[40] L.A. Greening and S. Bernow. Design of coordinated energy and environmental policies: Use of multi-criteria decision-making. *Energy Policy*, 32:721–735, 2004.

[41] Z. Güngör and F. Arikan. A fuzzy outranking method in energy policy planning. *Fuzzy Sets and Systems*, 114:115–122, 2000.

[42] Y. Haimes, W. Hall, and H. Freedman. *Multiobjective Optimization in Water Resource Systems*. Elsevier, New York, 1975.

[43] R. Hämäaläainen and R. Karjaläinen. Decision support for risk analysis in energy policy. *European Journal of Operational Research*, 56:172–183, 1992.

[44] R. Hämäläinen. A decision aid in the public debate on nuclear power. *European Journal of Operational Research*, 48:66–76, 1990.

[45] R. Hämäläinen and T. Seppäläinen. The Analytic Network Process in energy policy planning. *Socio-Economic Planning Science*, 20:399–405, 1986.

[46] M. Hanegraaf, E. Biewinga, and G. Van der Bijl. Assessing the ecological and economic sustainability of energy crops. *Biomass and Bioenergy*, 15:345–355, 1998.

[47] D. Haralampopoulos and H. Polatidis. Renewable energy projects: Structuring a multi-criteria group decision-making framework. *Renewable Energy*, 28:961–973, 2003.

[48] B. Hobbs. A comparison of weighting methods in power plant siting. *Decision Sciences*, 11:725–737, 1980.

[49] B. Hobbs and G. Horn. Building public confidence in energy planning: A multimethod MCDM approach to demand-side planning at BC Gas. *Energy Policy*, 25:357–375, 1997.

[50] B. Hobbs and P. Meier. *Energy Decisions and the Environment: A Guide to the Use of Multicriteria Methods*. Kluwer Academic Publishers, Dordrecht, 2000.

[51] B. Hobbs, M. Rothkopf, R. O'Neill, and H. Chao, editors. *The Next Generation of Electric Power Unit Commitment Models*. Kluwer Academic Publishers, Dordrecht, 2001.

[52] D. Hoog and B. Hobbs. An integrated resource planning model considering customer value, emissions, and regional economic impacts. *Energy*, 18(11):1153–1160, 1993.

[53] Y. Hsiao, H. Hsiao, C. Liu, and Y. Chen. A computer package for optimal multiobjective VAR planning in large scale power systems. *IEEE Transactions on Power Systems*, 9(2):668–676, 1994.

[54] G. Hsu, P. Leung, and C. Ching. A multiobjective programming and inter-industry model for energy-economic planning in Taiwan. *Energy Systems and Policy*, 11:185–204, 1987.

[55] C. Huang, H. Yang, and C. Huang. Bi-objective power dispatch using fuzzy satisfaction maximizing decision approach. *IEEE Transactions on Power Systems*, 12(4):1715–1721, 1997.

[56] J. Huang, K. Poh, and B. Ang. Decision analysis in energy and environmental modelling. *Energy*, 20:843–855, 1995.

[57] M. Jones, C. Hope, and R. Hughes. A multi-attribute value model for the study of UK energy policy. *Journal of the Operational Research Society*, 41:919–929, 1990.

[58] H. Jorge, C. Antunes, and A. Martins. A multiple objective decision support model for the selection of remote load control strategies. *IEEE Transactions on Power Systems*, 15(2):865–872, 2000.

[59] M. Kablan. Prioritization of decentralized electricity options available for rural areas in Jordan. *Energy Conversion Management*, 38:1515–1521, 1997.

[60] M. Kaleta, W. Ogryczak, E. Toczylowski, and I. Zoltowska. On multiple criteria decision support for suppliers on the competitive electric power market. *Annals of Operations Research*, 121:79–104, 2003.

[61] R. Karni, P. Feigin, and A. Breiner. Multicriterion issues in energy policymaking. *European Journal of Operational Research*, 56:30–40, 1992.

[62] I. Kavrakoglu and G. Kiziltan. Multiobjective strategies in power systems planning. *European Journal of Operational Research*, 12:159–170, 1983.

[63] R.L. Keeney. *Siting Energy Facilities*. John Wiley & Sons, New York, 1980.

[64] R.L. Keeney and H. Raiffa. *Decisions with Multiple Objectives*. John Wiley & Sons, Chichester, 1976.

[65] R.L. Keeney, D. von Winterfeldt, and T. Eppel. Eliciting public values for complex policy decisions. *Management Science*, 36(9):1011–1030, 1990.

[66] R. Lahdelma, J. Hokkanen, and P. Salminen. SMAA – Stochastic multiobjective acceptability analysis. *European Journal of Operational Research*, 106:137–143, 1998.

[67] R. Lahdelma, K. Miettinen, and P. Salminen. Ordinal criteria in stochastic multicriteria acceptability analysis. *European Journal of Operational Research*, 147:117–127, 2003.

[68] F. Lee and A. Breipohl. Operational cost savings of direct load control. *IEEE Transactions on Power Applications and Systems*, 103(5):988–992, 1984.

[69] P. Linares and C. Romero. A multiple criteria decision making approach for electricity planning in Spain: Economic versus environmental objectives. *Journal of the Operational Research Society*, 51:736–743, 2000.

[70] P. Linares and C. Romero. Aggregation of preferences in an environmental economics context: A goal-programming approach. *Omega*, 30:89–95, 2002.

[71] F. Lootsma, P. Boonekamp, R. Cooke, and F. Van Oostvoorn. Choice of a long-term strategy for the national electricity supply via scenario analysis and multicriteria analysis. *European Journal of Operational Research*, 48:189–203, 1990.

[72] F. Lootsma, J. Meisner, and F. Schellemans. Multicriteria decision analysis as an aid to the strategic planning of energy R&D. *European Journal of Operational Research*, 25:216–234, 1986.

[73] C. Lyra, R. Castro, and L. Ferreira. Assessing decisions on multiple uses of water and hydroelectric facilities. *International Transactions in Operational Research*, 3(3-4):281–292, 1996.

[74] C. Lyra and L. Ferreira. A multiobjective approach to the short-term scheduling of a hydroelectric power system. *IEEE Transactions on Power Systems*, 10(4):1750–1754, 1995.

[75] R. Mamlook, B. Akash, and S. Nijmeh. Fuzzy sets programming to perform evaluation of solar systems in Jordan. *Energy Conversion Management*, 42:1717–1726, 2001.

[76] D. Manjure and E. Makram. Optimal load curtailment as a bi-criteria program. *Electric Power Systems Research*, 66(2):155–161, 2003.

[77] A. Martins, D. Coelho, C. Antunes, and J. Clímaco. A multiple objective linear programming approach to power generation planning with demand- side management. *International Transactions in Operational Research*, 3(3-4):305–317, 1996.

[78] M.A. Matos. A fuzzy filtering method applied to power distribution planning. *Fuzzy Sets and Systems*, 102:53–58, 1999.

[79] G. Mavrotas, D. Diakoulaki, and P. Capros. Combined MCDA-IP approach for project selection in the electricity market. *Annals of Operations Research*, 120:159–170, 2003.

[80] G. Mavrotas, D. Diakoulaki, and L. Papayannakis. An energy planning approach based on mixed $0-1$ multiple objective linear programming. *International Transactions in Operational Research*, 6:231–244, 1999.

[81] D. Mills, L. Vlacic, and I. Lowe. Improving electricity planning: Use of a multicriteria decision making model. *International Transactions in Operational Research*, 3:293–304, 1996.

[82] U. Nangia, N. Jain, and L. Wadhwa. Multiobjective optimal load flow based on ideal distance minimization in 3D space. *International Journal of Electrical Power & Energy Systems*, pages 847–855, 2001.

[83] United Nations. *World Energy Assessment: Energy and the Challenge of Sustainability*. United Nations Development Programme, New York, 2000.

[84] T. Niimura and T. Nakashima. Multiobjective tradeoff analysis of deregulated electricity transactions. *International Journal of Electrical Power & Energy Systems*, 25(3):179–185, 2003.

[85] C. Oliveira and C. Antunes. An input-output model for decision support in energy-economy planning – A multiobjective interactive approach. *Journal of Systems Analysis Modelling Simulation*, 42(5):769–790, 2002.

[86] C. Oliveira and C. Antunes. A multiple objective model to deal with economy-energy-environment interactions. *European Journal of Operational Research*, 153:370–385, 2004.

[87] J. Pan and S. Rahman. Multiattribute utility analysis with imprecise information: an enhanced decision support technique for the evaluation of electric generation expansion strategies. *Electric Power Systems Research*, 46:101–109, 1998.

[88] M. Ponce de Leão and M. Matos. Multicriteria distribution network planning using simulated annealing. *International Transactions in Operational Research*, 6:377–391, 1999.

[89] M. Quaddus and T. Goh. Electric power generation expansion planning with multiple objectives. *Applied Energy*, 19:301–319, 1985.

[90] B. Rautenbach and I. Lane. The multiobjective controller: A novel approach to domestic hot water load control. *IEEE Transactions on Power Systems*, 11(4):1832–1837, 1996.

[91] P. Rietveld and H. Ouwersloot. Ordinal data in multicriteria decision making: A stochastic dominance approach to siting nuclear power plants. *European Journal of Operational Research*, 56:249–262, 1992.

[92] F.S. Roberts. *Measurement Theory with Applications to Decisionmaking, Utility and the Social Sciences*. Addison-Wesley, London, 1979.

[93] I. Rosado and R. Adams. Multiobjective planning of the optimal voltage profile in electric power distribution systems. *COMPEL*, 10(2):115–128, 1991.

[94] A. Rousseau and J.M. Martel. Environmental assessment of an electric transmission line project: A MCDA method. In M. Paruccini, editor, *Applying Multiple Criteria Aid for Decisions to Environmental Management*, pages 163–185. Kluwer Academic Publishers, Dordrecht, 1994.

[95] B. Roy. *Multicriteria Methodology for Decision Aiding*. Kluwer Academic Publishers, Dordrecht, 1996.

[96] T.L. Saaty. *The Analytic Hierarchy Process*. McGraw-Hill, New York, 1980.

[97] T.L. Saaty. How to make a decision: The Analytic Hierarchy Process. *European Journal of Operational Research*, 48:9–26, 1990.

[98] V. Schulz and H. Stehfest. Regional energy supply optimization with multiple objectives. *European Journal of Operational Research*, 17:302–312, 1984.

[99] J. Siskos and Ph. Hubert. Multicriteria analysis of the impacts of energy alternatives: A survey and a new comparative approach. *European Journal of Operational Research*, 13:278–299, 1983.

[100] R. Slowinski. A multicriteria fuzzy linear programming for water supply system development planning. *Fuzzy Sets and Systems*, 19:217–237, 1986.

[101] D. Soloveitchick, N. Ben Aderet, M. Grimman, and A. Lotov. Multiobjective optimization and marginal pollution abatement cost in the electricity sector – an Israeli case study. *European Journal of Operational Research*, 140:571–583, 2002.

[102] R. Steuer and E. Choo. An interactive weighted Tchebycheff procedure for multiple objective programming. *Mathematical Programming*, 26:326–344, 1983.

[103] T. Stewart. A multicriteria decision support system for R&D project selection. *Journal of the Operational Research Society*, 42:17–26, 1991.

[104] W. Stewart and E. Horowitz. Environmental factor weighting at the federal energy regulatory commission. *Socio-Economic Planning Science*, 25:123–132, 1991.

[105] J. Teghem and P. Kunsch. Application of multiobjective stochastic linear programming to power systems planning. *Engineering Costs and Production Economics*, 9:83–89, 1985.

[106] M Tsay. Applying the multiobjective approach for operation strategy of cogeneration systems under environmental constraints. *International Journal of Electrical Power & Energy Systems*, 25(3):219–226, 2003.

[107] K. Tsuji and N. Chiba. Multiobjective evaluation of appropriate energy systems for urban areas. *International Journal of Global Energy Issues*, 11(1-4):171–177, 1998.

[108] B. Venkatesh, G. Sadasivam, and M. Khan. A new optimal reactive power scheduling method for loss minimization and voltage stability margin maximization using successive multiobjective fuzzy LP technique. *IEEE Transactions on Power Systems*, 15(2):844–851, 2000.

[109] N. Voropai and E. Ivanova. Multicriteria decision analysis techniques in electric power system expansion planning. *Electrical Power and Energy Systems*, 24:71–78, 2002.

[110] D. Wei and N. Chen. Air conditioner direct load control by multi-pass dynamic programming. *IEEE Transactions on Power Systems*, 10(1):307–313, 1995.

[111] K. Wong, B. Fan, C. Chang, and A. Liew. Multiobjective generation dispatch using bicriterion global optimization. *IEEE Transactions on Power Systems*, 10(4):1813–1819, 1995.

[112] W. Yeh and L. Becker. Multiobjective analysis of multireservoir operations. *Water Resources Research*, 18(5):1326–1336, 1982.

[113] M. Yehia, R. Ramadan, Z. El-Tawil, and K. Tarhini. An integrated technico-economical methodology for solving reactive power compensation problem. *IEEE Transactions on Power Systems*, 13(1):54–59, 1988.

[114] S. Zionts and D. Deshpande. Energy planning using a multiple criteria decision method. In P. Nijkamp and J. Spronk, editors, *Multiple Criteria Analysis: Operational Methods*, pages 153–162. Gower Press, London, 1981.

Chapter 22

MULTICRITERIA ANALYSIS IN TELECOMMUNICATION NETWORK PLANNING AND DESIGN – PROBLEMS AND ISSUES

João Clímaco
Faculty of Economics
The University of Coimbra and INESC – Coimbra
Portugal
jclimaco@inescc.pt

José Craveirinha
Department of Electrical Engineering Sience and Computers
Faculty of Science and Technology
The University of Coimbra and INESC – Coimbra
Portugal
jcrav@deec.uc.pt

Abstract The interaction between a complex socio-economic environment and the extremely fast pace of development of new telecommunication technologies and services justifies the interest in using multicriteria evaluation in decision making processes associated with several phases of network planning and design. Based on an overview of current and foreseen evolutions in telecommunication network technologies and services we begin by identifying and discussing challenges and issues concerning the use of multicriteria analysis (M.A.) in telecommunication network planning and design problems. Next we present a review of contributions in these areas, with particular emphasis on network modernisation planning and routing problems. We will also outline an agenda of current and future research trends and issues in this application area of multicriteria modelling.

Keywords: Telecommunication planning and design, multicriteria analysis.

1. Motivation

Telecommunication systems and network technologies and the associated services have been and are in a process of very rapid evolution. Major changes in telecommunication system technologies and service offerings are currently underway. The evolution of telecommunication networks is a process of paramount importance not only because of the large investments required but also due to its significant impacts on the economic activities and on the society as a whole. The development of these networks gives rise to a variety of complex multidimensional problems. Therefore, the interaction between a complex socio-economic environment and the extremely fast pace of development of new telecommunication technologies and services justifies the interest in using multicriteria evaluation in decision making processes associated with several phases of network planning and design. In the present work a state of the art review on this subject is done.

In the second section of this study an overview of current and foreseen evolutions in telecommunication network technologies and services is presented. Section 3 discusses general issues concerning the use of multicriteria analysis in telecommunication network planning and design. Section 4 is dedicated to a comprehensive discussion of applications of multicriteria analysis in telecommunication network planning and design problems. In the first part (Section 4.1) of this section strategic modernization planning applications are studied. Section 4.2 is dedicated to routing models followed by Section 4.3 that deals with operational planning problems. Section 4.4 is related to studies that present multicriteria evaluation approaches focusing on socio-economic evolutions associated with specific telecommunication issues. Of course it should be noted that there is no sharp frontier between the Sections 4.1 and 4.4.

It must be remarked that we decided to describe in more detail, as compared to other models, an interactive linear programming approach dedicated to a strategic planning problem and a bicriterion Quality of Service (QoS) routing model because, in our opinion, they typify well cases in which the use of multicriteria (in this chapter used as synonymous with multiple objective) mathematical programming models is justified. Applications of multiattribute approaches are just outlined because, in technical terms, they are not much different from their application in other areas and for lack of space for a thorough discussion.

2. Overview of Current Evolutions in Telecommunication Networks and Services

To give a better understanding of the decisive impact of network evolutions on the emergence of a significant number of new sets of problems of network planning and design involving multiple objectives and constraints, we now present an overview of the major trends and factors underlying current and future developments.

Firstly, from a historical perspective, it can be said that major telecommunication network evolutions have been centred on and around two major modes of information transfer: circuit switching (typical of classical telephone networks) and packet switching (typical of the Internet). When a call is generated in circuit switching the network routing mechanisms seek to find an available path (with the required bandwidth) from origin to destination. When that path (usually designated as route) is found then it is seized (in terms of the corresponding resources needed for each call) for the duration of the call; if no path is found in the required conditions the call is lost. In packet switching, the information to be transmitted is divided into packets (carrying the information about their origin and destination) of variable size that are routed through an available path and may suffer delays in the intermediate nodes. The various packets don't necessarily travel along the same path and can be reassembled at the destination node in order to regenerate the original packet sequence. These basic functionalities, including the possibility of establishing connection-oriented data communications and the interconnection of equipment of multiple vendors, were made possible by the emergence of the TCP/IP (Transmission Connection Protocol/Internet Protocol) protocol suite. This enabled the very rapid expansion of the Internet in the 80s, strongly accelerated in the 90s through the release of the basic Web technologies by the European Laboratory CERN, in 93. The public telephone networks rapidly evolved from the 80s through the development of ISDNs (Integrated Services Digital Networks) enabling the convergence of different types of services (namely telephone, facsimile, data and video services) on the same network by recurring to standardised equipment and functionalities. The extremely rapid expansion of the demand for data services and for new and more bandwidth "greedy" services, soon required the development of technologies enabling the implementation of the concept of broadband ISDNs (B-ISDNs). In the early to mid 90s, a new information transfer technology, ATM (Asynchronous Transfer Mode) became the most popular technology to implement the B-ISDN concept.

At the level of the transport infrastructure (underlying transmission networks) these trends were supported and have stimulated the development of optical networks capable of making the most of the large bandwidths associated with the very low wavelengths that may be carried by optical fibres. In particular WDM (Wavelength Division Multiplexing), enabling the simultaneous transport of several high capacity signals in each fibre, by assigning each signal to a different wavelength, permitted to take further advantage of the very large economies of scale provided by optical networks. Also several evolutions in digital radio communication technologies enabled a very rapid expansion of mobile networks with an increasing demand for mobile data services including Internet access.

In the last decade telecommunication networks have been subject to an extremely rapid evolution that is the result of the combination of two major forces: *traffic growth* and a very fast pace of *technological advances*.

Traffic growth is both quantitative and qualitative, i.e. it involves both the increase in traffic volumes in response to broad socio-economic developments and also the demand for new more bandwidth demanding services as these become available through technological evolution or are simply perceived as desirable by groups of customers. In this respect it should be stressed the extremely rapid increase in Internet traffic that has occurred in very recent years (2000-2001) attaining annual rates of 60-80 % (*apud* El-Sayed and Jaffe [36]). At the same time the increase of the number of subscribers of broadband services and wireless networks attained average rates of 60% and 25%, respectively.

It should be stressed the strong interactions between those two driving forces (traffic growth and technological advances) and socio-economic factors. A relevant example is the fact that the explosive growth of Internet enabled the rapid development of the so-called electronic commerce as an increasingly important business practice, with very strong impact on economy and society as a whole. The impacts of telecommunication network developments in the structure, management and organisational culture of the companies in association with the present day globalization are also obvious. On the other hand the needs of electronic commerce in terms of its basic functionalities, namely communication with the customers, processing environment, service management and transaction capabilities, foster developments in terms of improvements of the technological platforms. Also the expansion of that commerce is associated with the increase in traffic volumes and the demand for increased bandwidth at the lowest possible cost. Overall it can be said that there is a strong correlation between the technological development and expansion of telecommunication networks and economic and social evolutions. Also, at the market level, the steady transition from regulatory monopolies to liberalization leads to fierce competition among operators and service providers both at the level of national networks and local access networks. All these evolutions are multifaceted and prone to conflicts and contradictions, an example being the tensions between the recent drive for big mergers and acquisitions between operators and the antitrust policies of the regulatory bodies (Federal Trade Commission and Federal Communications Commission in the US and the EU Competition Directorate). Needless to say that there are strong social interactions associated with the development of new network technologies simultaneously in terms of strictly human interactions, in terms of the relationships between humans and all types of organisations and in terms of the intra and inter-organisational relations. The detailed analysis of these trends and interactions is naturally out of the scope of this study.

In a simplified manner it can be said that the factors mentioned above, favoured the development of technologies and network architectures capable of satisfying increasing traffic volumes and more sophisticated services, at the lowest possible cost (per basic information unit that can be carried with a certain QoS satisfaction degree).

Concerning network technologies a fast migration/integration of the technologies developed in the 90s, towards new technologies, very powerful in terms of transmission capacity, traffic carrying efficiency and integrative capacity, is foreseen in the near future (see El-Sayed and Jaffe [36], Banerjee et al. [16]) as will be discussed later. To analyse this trend we consider, as starting point, the most important installed technologies, namely the TCP/IP architecture (basis of todays Internet), ATM (Ashynchrous Tranfer Mode, the dominant information transfer technology in present broadband integrated service networks) and SONET (Synchronous Optical Network corresponding in Europe to the ITU standard SDH-Synchronous Digital Hierarchy) – a high-speed optical transport technology with signal rates of 51.84 Mb/s to 2.48832 Gb/s (corresponding to Optical Carrier signals OC-1 to OC-48).

The following great trends in future telecommunication network technological evolutions can be foreseen:

- The convergence of Internet wired transport infrastructure towards an intelligent optical network;
- The evolution of 3G (third generation) wireless networks in the direction of an all IP converged network;
- The increasing relevance of multidimensional QoS (Quality of Service) issues in the new technological platforms.

Each of these trends is now briefly analysed.

2.1 Convergence of IP over Optical Network

The explosive growth in data traffic resulted essentially from the extremely rapid increase in IP traffic and the emergence of large number of VPNs (Virtual Private Networks) in many countries. This, in association with the proliferation of the demand for IP based multimedia applications, has led to the necessity of developing technological solutions enabling to carry very large amounts of traffic at low costs. The mentioned traffic increase occurred mainly since the mid 90's and made the data traffic surpass the voice traffic at the turn of the millennium, a tendency likely to prevail for the next years.

On the other hand recent optical network technologies namely WDM (Wavelength Division Multiplexing) and DWDM (Dense WDM) and OXCs (Optical Cross-Connects) have paved the way for the development of a transport infrastructure with extremely high bandwidth capacities – up to the order of Terabits

(Tb) per second ($1Tb = 10^{12}b$). WDM is a technology that enables the simultaneous transmission of various optical signals, using different wavelengths in the same fibre. In the case of DWDM more than eight wavelengths can be used in a given fiber thereby enabling a significant increase in the traffic capacity (even without introducing new fibres) and a reduction in the transmission cost per bit. OXCs on the other hand have multiple ports and can switch an optical wavelength channel from an input to an output port and will also enable full wavelength transfer from an input to an output port. Hence important features of these emerging networks are the significant flexibility in the management of transmission resources and the provision of new optical services and, in general, increased operational efficiencies with respect to previous networks which, in essence, were just interconnected transmission systems with large transmission capacity. All these developments led to the concept of *intelligent optical network*. These features lead to the capability of a flexible mechanism of establishment of *lightpaths*, that is alternating sequences of OXCs and optical channels from an originating OXC to a destination OXC, normally associated with a secondary lightpath or backup route defined by specific service protection mechanisms.

At the same time a new Internet technology has emerged and developed in recent years: MPLS (Multiprotocol Label Switching). Basic features of MPLS are a single mechanism for forwarding packets based on "label swapping" (utilizing fixed length labels) and the flexibility to form "forward equivalent classes" (FEC) composed of packets carried over the same LSPs (Label Switched Paths), enabling the implementation of connection-oriented services from origin to destination. This is accomplished through LSRs (Label Switched Routers) which forward the data, step by step, in the network by using the label carried in the data units. In MPLS the control functions are separated from the data forwarding functions, enabling the introduction of advanced and flexible techniques aimed at achieving the largest operational efficiencies. Associated with these and other features of MPLS is the possibility of implementing sophisticated mechanisms of traffic engineering which until now were only possible in classical ISDN circuit-switched networks or ATM networks. The underlying objective of such mechanisms is to achieve the best performance from the traffic point of view and to optimize network resource utilization hence aiming at some form of maximal operational efficiency. For example, it is undesirable that subsets of network resources be congested or overutilized while, at the same time, other subsets which could be used by feasible LSPs, are underutilized. A central objective of traffic engineering is therefore the efficient management of bandwidth resources in order to optimize key QoS parameters such as packet loss probability, average delay, peak to peak delay variation or maximum packet transfer delay. Another important feature of MPLS is to allow existing transport technologies, namely ATM, FR (Frame Relay) and Ethernet to interoperate and coexist with IP based networks. This guarantees a non-disruptive technological evolution, great flexibility for network planners and significant cost reductions when articulating equipments and networks (with those "legacy" technologies) with MPLS based networks. MPLS also allows the easy deployment of multiservice applications enabling the satisfaction of the increasing needs of service providers and users hence contributing to economical growth. A detailed description of MPLS at an introductory level can be seen in Harnedy [47]. Note that an extension of MPLS, designated as GMPLS (Generalized Multiprotocol Label Switching), is being developed. It enables the utilization of the same label swapping technique and control plane functionalities with additional types of switching technologies, namely optical switching technologies.

All these developments combined with the dramatic increase in traffic volumes point to an evolution of the Internet transport infrastructure onto an *intelligent optical core network*. In this future network architecture high-speed Internet routers will be interconnected through intelligent optical networks capable of dynamically establishing switched lightpaths. This trend will enable, according to Rajagopalem et al. [84] and Barnerjee et al. [16], a rapid evolution of todays network architectures based on IP over ATM over SONET over DWDM towards an architecture based on IP (with GMPLS) directly over DWDM (with full optical switching). According to this evolutionary model ATM and SONET/SDH (at a later stage) would be rapidly surpassed to give rise to the IP over all-optical network paradigm, although we foresee, together with other authors, that these new technologies will coexist, for economical reasons, with the legacy technologies for a significant number of years.

2.2 Evolution of 3G Wireless Network towards an All IP Network

According to recent forecasts [18] wireless traffic will continue to increase at a significant rate: from 1.3 billion subscribers expected at the end of 2003 to 1.8 billion in 2007, corresponding to almost an average 10% per year. Similar to the trend in wired networks, an increase in the relative weight of data traffic with respect to voice traffic is expected to occur in wireless networks. With present wireless technology the average data throughput per user during busy hours is expected to be in the range of 30-40 Kb/s. In emerging networks, with 3G technology, (corresponding to the UMTS -Universal Mobile Telecommunication System – Releases 4 and 5) these rates can be expected to increase to a range of 100-150 Kb/s therefore allowing Internet access, image transfer and data VPN applications. 3G will therefore allow higher throughputs, new services and higher spectral efficiency. This was made possible by the developments in receiver/transmitter-air interface technologies, radio transmission techniques, mobile terminals and new protocols for subscriber services and QoS management mechanisms. The planned evolution of 3G, designated as 3G+, will enable the achievement of average 600-700 Kb/s throughputs and peak data rates up to 2.4 Mb/s, cf El-Sayed and Jaffe [36], paving the way for the mobile multiservice Internet. A whole range of new services such as roadside assistance, truck fleet management, information services and financial transfer services (or M-commerce) will be made available. Eventually the whole traffic from the network base stations will be IP-based, giving rise to the full integration of the wireless and wired Internets, into an *all-IP converged network.* According to this model legacy circuit-switched voice traffic will be converted into packet traffic at legacy mobile switching centres.

Concerning wireless technologies significant developments have also occurred in fixed broadband wireless access systems (BWA), associated with the rapid growth in the demand for IP and broadband access by residential and small business customers. This local access technology appears to offer some advantages in many situations by avoiding distance limitations and the cost of DSL (Digital Subscriber Line) and cable, enabling the use of data rates of 5-10 Mb/s on the downlink and 0.5-2 Mb/s on the uplink (Bölcskei et al. [22]). Typical BWA services include Internet access, multi-line voice, audio and video.

In the future 4G (fourth generation) mobile technological platform it is expected a convergence of broadband wireless access and broadband wireless mobile.

2.3 Increasing Relevance of QoS Issues in the New Technological Platforms

The simplicity of the Internet Protocol (IP) that provides the basic end-to-end data delivery service in the existing Internet, based on a "best-effort" service concept, lacks a mechanism capable of guaranteeing the multiple QoS requirements of new type of applications, namely multimedia applications. This leads to the introduction of new functionalities in the next generation Internet, namely the Integrated Service (IntServ) and the Differentiated Service (DiffServ) mechanisms, providing certain QoS guarantees concerning the transport of traffic of different types (for an overview see Manniatis et al. [67]). These mechanisms will support a variety of services ranging from voice over IP, video, teleconferencing to audio/video download and data-base queries by guaranteeing the appropriate treatment of the QoS parameters relevant to each class of packet traffic flow. Also the MPLS technology contains QoS mechanisms that enable different QoS parameter levels to be guaranted on separate label switched paths (or network "tunnels") as well as functions of network load balancing (through traffic engineering operations) and fast rerouting under failure. All these developments pave the way to a new, high performance multiservice Internet corresponding to the concept of *QoS-based packet network* proposed in El-Sayed and Jaffe [36].

On the other hand the UMTS platform provides mechanisms of QoS support for 3G wireless networks. These mechanisms are based on a QoS architecture that uses four traffic classes intended for different types of applications where each class corresponds to applications with similar statistical behaviour and similar QoS requirements. The mapping among the traffic classes of new generation Internet and 3G is investigated in Manniatis et al. [67] in order to permit the interoperability between the QoS mechanisms of the two types of networks. This and other developments will create the technical conditions for the full interoperability of these networks and its convergence towards an all IP network, as discussed above.

All these innovations and technological trends put in evidence the increasing relevance of the issues related to the definition and assessment of multidimensional QoS parameters and the associated network control mechanisms. These issues are reflected in the type and nature of many new problems of network planning and design, namely concerning routing methods and the choice of alternative network architectures. The inclusion of multiple, eventually conflicting objectives and various types of technical and socio-economic constraints, in the OR models associated with such problems lays the ground for the potential advantage of the introduction of multicriteria analysis methods. In fact, concerning the *type of problems* that need to be addressed, the demand for new services, the rapid traffic growth and the extremely rapid technological evolution lead to the multiplication of new types of problems of network planning and design (as it will become clear in the next sections). In many problems there is potential advantage in explicitly considering several criteria. With respect to the *nature* of many of such new problems, it is important to address explicitly the multidimensional character of the problems, together with the consideration of relevant technical and socio-economic constraints. This necessity becomes more apparent if one takes into account the increasing importance of the QoS issues (of a multidimensional nature) related to the development of new services and the rapid evolution of the technological platforms. Finally, the importance of the inclusion of negotiation processes involving various decision agents (in complex cases the customer, the end service provider and the network operator) and the uncertainty associated with many objective functions and constraint parameters, in various decision problems makes clear the interest in considering multicriteria analysis approaches in this context.

3. Multicriteria Analysis in Telecommunication Network Planning and Design

From the last section, it is clear that decision making processes related to telecommunication networks take place in an increasingly complex and turbulent environment characterised by a fast pace of technological evolution, drastic changes in available services, market structures and societal expectations, involving multiple and potentially conflicting options. This is obviously an area where different socio-economic decisions involving communication issues have to be made. But it is also an area where technological issues are of paramount importance as it is recognized, for instance, by Nurminen [77] : "...The network engineering process starts with a set of requirements or planning goals. Typical requirements deal with issues like functionality, cost, reliability, maintainability, and expandability. Often there are case specific additional requirements such as location of the maintenance personnel, access to the sites, company policies, etc. In practice the requirements are often obscure...". Nurminen, who has collaborated in the development of mathematical network planning models with Nokia, recognises the limitations of monocriterion models. However, he emphasizes the difficulties in the tuning of parameters in mathematical programming models and draws attention to the fact that this aspect becomes more difficult to tackle when multiple objective formulations are used, since the procedures of preference aggregation by the decision maker(s) imply, in general, the definition of specific parameters, such as, for example, the fixation of some kind of "weights". This difficulty does not justify less interest in multicriteria modelling but must be taken into account.

In many situations the mathematical models for decision support in this area become more realistic if different evaluation aspects are explicitly considered by building a consistent set of criteria (or objectives) rather than aggregating them *a priori* in a single economic indicator. In fact, multicriteria models explicitly address different concerns that are at stake so that decision makers may grasp the conflicting nature of the criteria and the compromises to be made in order to identify satisfactory solutions. In a context involving multiple and conflicting criteria, the concept of optimal solution gives place to the concept of nondominated solutions set that is feasible solutions for which no improvement in any criterion is possible without sacrificing at least one of the other criteria. In general, multicriteria approaches look for the identification of one or more nondominated, or approximately nondominated, satisfactory solutions. Of course, the choice of the approach or method to aggregate the preferences is also multicriteria in nature. Beyond the problem mentioned above, concerning the fixation of parameters, it must be taken into account whether or not there is a possibility of using interactive procedures specially in relation to the speed of the calculation. In fact, the procedure can not be interactive if the calculations in each interaction are too slow. In many telecommunication network decision problems no more than a few seconds (sometimes less) are available for finding the solution to be implemented. In such situations too interactive procedures cannot be applied. As we will see later on, when presenting a concrete example in Section 4.1, the simplicity of the questions the decision maker has to answer, in the phase of preference aggregation, is crucial. Cognitive as well as technical aspects are involved that may compromise, in many cases, the quality of the selected solutions.

Another aspect, in which there are compromises to be made, concerns the type of implementations to be executed with respect to monocriterion problems that have to be solved in a multicriteria approach. This question is not exclusive of multicriteria models but it is more critical in this case than in monocriterion models, since the programs with the monocriterion implementations have to run several times. Let us examine what is at stake. In many situations the mathematical programming models to be used have a network structure. In many cases there are very efficient specific algorithms for their solution, sometimes exact resolution procedures, sometimes heuristics. It should be noted, in this respect, the remarkable development of methaeuristics in recent years. The question is that the very rapid development of modern telecommunication networks makes it advisable, in many situations, the use of generic algorithms. These are often less efficient but more robust concerning its applicability when there are technological shifts, in order to avoid heavy implementation overheads for each specific new case.

It is also important to discuss in broad terms which multicriteria model is most adequate in each situation. Up to now we have talked about mathematical programming models that may be linear, nonlinear and additionally may have, or not, a special structure. On the other hand other types of models that we will designate as multiattribute models have been developed. While multicriteria mathematical programming models assume the set of feasible alternatives is defined implicitly through the introduction of constraints, in multiattribute models a finite and small set of alternatives is explicitly defined, which are analysed taking into account multiple criteria. This type of models allows a more detailed evaluation of the considered alternatives, without computational explosion, but in most situations it implies a very reductive point of view when considering telecommunication planning and design. In fact, the explicit definition of a small set of global alternatives is a hard task and not realistic in many cases. As we will see later on, in some circumstances the complementary utilisation of both types of models can be advisable. It is out of the scope of this paper to describe details of the approaches that are available for analysis of multicriteria models since it is a matter of study in other chapters of this book. Just a few words concerning multiattribute models. There is the so called American School where, to support the evaluation of a discrete set of alternatives, a multiattribute utility function, linear or not, depending on the approaches (see Keeney and Raiffa [52]) is built. Regarding the construction of value functions in telecommunication management see Keeney [51]. The Analytical Hierarchy Process (AHP) can be viewed as a special branch of the American School where a hierarchy of interrelated decision levels is identified (Saaty [90, 91, 92]). On the other hand, the so-called French School is based on the introduction of partial orders, i.e. outranking relations. No complete comparability of alternatives and transitivity are obtained. As an example of the French School approaches we can refer to ELECTRE methods (Roy and Bouyssou [89]). Depending on the situation the intention is to select the most preferred alternative, to rank the alternatives or to classify the alternatives in groups. In general outranking methods are less demanding than the American School approaches, namely in terms of fixing parameters. However, the results are less conclusive regarding the aggregation of the decision maker preferences.

Concerning the approaches dedicated to multicriteria mathematical programming models, attention should be paid to the dimension of the real problems to deal with and, many times, the necessity of a rapid execution, for the reasons discussed above. In this respect one should emphasize, from the bibliography concerning telecommunication applications (see Section 4): the use of interactive approaches dedicated to multicriteria linear programming models, the use of methaheuristics for analysing integer and mixed-integer programming models, and the use of approaches based on the resolution of shortest and k-shortest path problems. It should be noted that network multicriteria shortest path models are the only multicriteria mathematical programs for which sufficiently rapid exact algorithms are available, either to generate the whole or part of the nondominated solution set or to study the problem in an interactive manner.

Last but not the least, the uncertainties in various instances of the models are also a key issue in telecommunication planning and design. The uncertainty associated with the representation of traffic flows offered to the network is of major importance in many models. Such representation is a twofold task concerning: the use of adequate stochastic models (these are often approximations) for representing the traffic flows as required by the model and the obtainment of estimates of the probabilistic parameters that are needed in the stochastic sub-models. The uncertainties and/or imprecisions associated with other parameters of the OR model of different origins, from data collection to preference aggregation modelling (see Bouyssou [24]) are also a relevant issue in this context.

As it is well known multicriteria approaches allow the identification of the set of criteria related to the stable part of the decision makers preferences, leaving to later analysis further aggregation of their preferences. In many situations, the output of the multicriteria analysis is not a solution but some satisfactory solutions according to the model used. Therefore, an *a posteriori* analysis studying in more detail (namely, taking into account characteristics not included in the model) these solutions may be advisable. Furthermore, in some situations (as, for instance, in strategic planning) the analysis may not lead to a prescription but just to a clarification of the decision situation. This attitude towards dealing with the problems may help to reduce the gap between models and real world problems.

In the following sections of this paper a review and discussion of works using multicriteria models published in the context of planning and design of telecommunication networks, is presented. A discussion of future trends in these areas will also be outlined. Special attention will be paid to the section concerning routing models (Section 4.2) since, as it was seen in the previous section, this is an area that raises great challenges having in mind the introduction of new technologies and services, of a multidimensional character. In fact, beyond costs various dimensions associated with QoS are involved. A historical perspective about the way in which various dimensions were treated in different models and proposals to consider explicitly more than one criterion in situations of static routing and of dynamic routing, will be presented. In this context, and from a methodological point of view, exact algorithms for the calculation of shortest paths in monocriterion and multicriteria situations as well as heuristics, were used.

A reference to studies on strategic planning of the evolution of telecommunication networks, using multicriteria linear programming models and interactive methodologies of analysis (Section 4.1) will be made. A model that intends to evaluate the introduction of new basic services in the local access network, in face of some of the remarkable technological developments previously discussed, will be underlined. An expansion planning model, concerning the cellular phone system in a Brasilian state, based on a multiattribute approach, is also briefly outlined. Next, reference to several studies focusing on problems which may be grouped in the area of operational planning (Section 4.3) will be made. In particular we discuss: a link frequency assignment problem, a power management policy problem in wireless communication, an internet catches placement problem, a hub location problem dedicated to rural area telecommunication networks taking advantage of new technologies, a frequency allocation problem in mobile telephone networks, and a power management policy problem in a wireless communication system. Very different models were used in these applications, however all of them belong to the category of multicriteria mathematical programming.

Finally (in Section 4.4), some socio-economic application models related to telecommunication issues are reported, namely several strategic studies concerning electronic commerce decisions and a study of quality concerning the provided telecommunication services. In all these situations multiattribute models were used. Furthermore, some studies concerning the complementarity/substitution between travelling and telecommuting are referred to, namely studies where multicriteria network equilibrium modelling is proposed.

4. Review and Discussion of Applications of MA to Telecommunication Network Planning

4.1 Strategic Modernization Planning

Telecommunication networks have been subject to continuing and extremely rapid technical innovations and to permanently evolving modes of communication. In parallel, there is a significant increase in the demand for new services. It becomes more and more attractive for the telecommunication operating companies to offer the customers new ranges of new services, in order to take economic advantages of the new technology platforms and to respond to customer needs. Strategic planning is focused on the development and evaluation of scenarios of qualitative and quantitative network growth over a medium/long term period having in mind traffic increase, introduction of new technologies and services and the company economical objectives. This is a type of problem which involves a multiplicity of factors, some of which cannot be directly represented by an economic indicator.

In general, most network planning models try to express different aspects of these complex problems in currency units in order to encompass them in a unique economic objective function. These telecommunication network planning models fail to capture explicitly the different and conflicting aspects arising in evaluating network modernization policies. Multicriteria models taking explicitly into account (many times incommensurable) economic, technological and social aspects enable the decision makers to grasp the conflicting nature of the objectives and the compromises to be made in order to select a satisfactory solution.

In "A Multiple Objective Linear Programming Approach to the Modernization of Telecommunications Networks" Antunes et al. [6], in "On Multicriteria Decision Making Approaches Concerning Strategic New Telecommunication Planning" Antunes et al. [7], and in "A Multiple Criteria Model for New Telecommunication Service Planning" Antunes et al. [9], the authors propose a new multicriteria linear programming approach dedicated to the evaluation of the modernization planning of telecommunication networks.

Several trends are evident in recent rapid changes in telecommunication networks and services, which may be described in terms of functional types of networks, the services offered and the underlying basic technologies. The evolution and growth of these networks and services poses difficult problems of forecasting, planning and decision making. This stems from technological factors (namely the possibility of using alternative technologies for certain types of services and the difficulties in terms of standardization) and socio-economic factors (the difficulty in foreseeing the associated economic constraints and potential benefits). In addition, the development of these networks gives rise to a variety of options and conflicts involving the government and operator policies. For example, policy-makers must decide whether (and up to which extent) the potential economic and social benefits associated with these new networks justify public support of their extensive capital costs.

It is also clear that telecommunications, both at national and international levels have important impacts regarding the economic growth, the apparent reduction of geographical distances, social welfare and political options.

As referred to above a multicriteria linear programming model has been developed by Antunes et al. [6] to address an important strategic modernization problem: the planning of the evolution of subscriber lines in terms of classes of service offers and basic technologies. An extension of this model, which seems of practical interest, was done in Antunes et al. [7]. It concerns the possibility of evaluating the modernization plans in terms of particular regional environments.

The original model (Antunes et al. [6]) is based on a state transition diagram (in Figure 22.1) the nodes of which characterize a subscriber line in terms of service offers and supporting technologies, considering both the transition of lines to a more sophisticated state as well as the installation of new lines directly in any state. The planning period is discretized in years $j = 1, ..., J$, where J is the horizon of the planning period. Installation of new lines or upgrading of existing ones may take place for $j \geq 1$.

The state transition diagram depicts the transitions permitted among the states in consecutive years of the planning period. The transitions which are not permitted, express the irreversibility of service enhancement (once it is upgraded it will not be downgraded) and of digitization (once a facility is digitized it can not be replaced by analog equipment in the future).

The service offerings are $s \in S = \{P, E, R\}$ where $P \equiv$ traditional telephone service; $E \equiv$ enhanced service providing a narrow band data channel in addition to voice; $R \equiv$ wide-band integrated digital service suitable for voice, data and video communications.

The underlying basic supporting technologies considered are $t \in T = \{A, D\}$, where $A \equiv$ analog; $D \equiv$ digital. A subscriber line may be in any of the following states $n \in N_O = \{O, PA, EA, PD, ED, RD\}(N = \{N_O\}\backslash\{O\})$ where $O \equiv$ line not yet installed; $PA \equiv$ traditional telephone service / analog technology; $EA \equiv$ enhanced service / digital technology; $RD \equiv$ wide-band service / digital technology. Let $A = [a_{mn}]$ be a boolean transition matrix where $a_{mn} = 1$, if the state n can be reached from state m (m to n is a valid transition), $a_{mn} = 0$, otherwise.

The following auxiliary sets are defined: $A_n = \{i \in N_O : a_{in} = 1\}$, the set of all states from which it is possible to reach the state n and $B_m = \{i \in N_O : a_{mi} = 1\}$ the set of all states which can be reached from the state m.

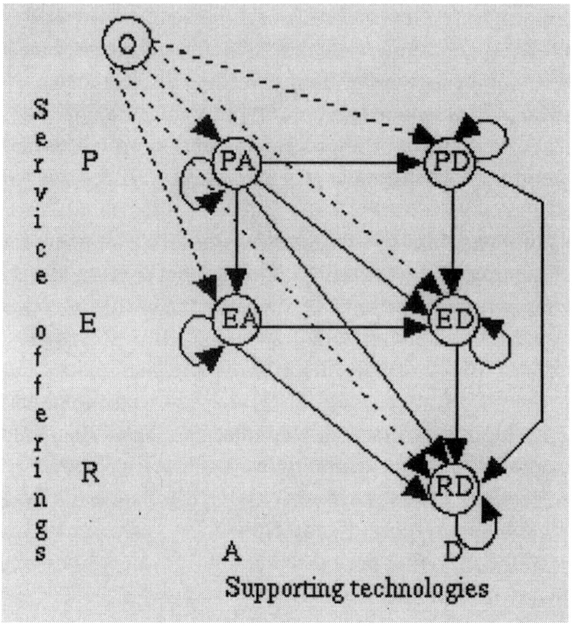

Figure 22.1. State transition diagram.

The decision variables consist of:

x^j_{mn} = number of lines making a transition from state m to state n in year j;

y^j_m = number of lines which are in state m at the end of year j.

Note that these two types of decision variables are obviously related:

$$y^j_n = \sum_{i \in A_n} x^j_{in} = \sum_{m \in B_n} x^{j+1}_{nm}.$$

Five cash flows are defined concerning capital costs associated with the transition of a line from state m to state n, salvage value after dismantling a line, annual operational and maintenance charges, annual revenue of a line at year m and final value of a line at the end of the planning period. From these cash-flows an objective function (to be maximized) quantifying the NPV (net present value) of network modernization is defined:

$$f_1 = \sum_{j=1}^{J} \sum_{n \in N} \sum_{m \in A_n} c_{mn}^j x_{mn}^j + \sum_{j=1}^{J} \sum_{m \in N} c_m^j y_m^j,$$

where c_{mn}^j is the present value of the salvage value obtained with a transition of a line from state m to state n in year j subtracted by the investment cost associated with the transition of a line from state m to state n in the same year and c_m^j is the present value of annual revenue associated with a line in state m (in year j) subtracted by the annual operational and maintenance charges and the final value of a line in state m.

An external dependence function (to be minimised) quantifies the imported components associated with the investment costs and operational and maintenance charges:

$$f_2 = \sum_{j=1}^{J} \sum_{n \in N} \sum_{m \in A_n} d_{mn}^j x_{mn}^j + \sum_{j=1}^{J} \sum_{m \in N} d_m^j y_m^j,$$

where d_{mn}^j is the present value of the fraction of the investment cost associated with the external/imported component of the investment cost corresponding to the transition of a line from state m to state n, in year j and d_m^j has similar meaning for the maintenance and operational charges of a line in state m in year j.

The "quality of service" is understood in this model as the "degree of modernization" associated with the "desirability" of new services (E and R). This quality of service quantifies the number of lines supporting new services weighted by a desirability factor, given by:

$$\begin{aligned} w_E^j &= \text{"weight" of the existence of a line offering service } E \text{ in year } j, \\ w_R^j &= \text{"weight" of the existence of a line offering service } R \text{ in year } j. \end{aligned}$$

This leads to the following objective function, to be maximized:

$$f_3 = \sum_{j=1}^{J} \sum_{m \in M(E)} w_E^j y_m^j + \sum_{j=1}^{J} \sum_{m \in M(R)} w_R^j y_m^j.$$

where $M(s)$ is the set of all the states which include service s and other less advanced services.

Finally, the model considers four main categories of constraints: upper bound on the cost and charges, degree of current satisfaction of the estimated demand, degree of penetration of the supporting technologies and continuity (line conservation) constraints. The policy of the telecommunication operator may also be reflected in the model through the inclusion of techno-economic constraints imposing upper bounds on the number of new lines of each technology to be installed at each year of the planning period.

Examples of application of this model using various sets of data may also be seen in Antunes et al. [6, 9].

It must be stressed that since this was a seminal work in multiple objective modelling of strategic modernization planning of telecommunication networks, the analysed model is naturally incomplete, subject to updates and modifications and its practical utilization would certainly require additional information from telecommunication operators and major network equipment suppliers. This information – which we think to be difficult to gather and which has a high degree of uncertainty, having in mind the very rapid changes in technical, economic and social factors – would enable network planners and managers to meet new challenges and opportunities associated with concrete scenarios of network evolution. In fact, by modifying the state transition diagram (namely through the consideration of new nodes and arcs) or by including new objectives and/or constraints, or changing those in the model, other aspects, which might require consideration by the decision makers, may be easily incorporated in the model without jeopardizing its basic philosophy. So, this multicriteria model is sufficiently flexible, namely enabling the incorporation of new evaluating criteria, which might become important in the assessment of network modernization strategies in new contexts.

In Antunes et al. [6] the interactive multicriteria analysis is based on the TRIMAP approach by Clímaco and Antunes [29]. TRIMAP is an interactive calculation tool the aim of which is to aid the DM in the progressive and selective learning of the set of nondominated solutions. It combines three main components: decomposition of the weighting space, introduction of constraints on the objective function space and introduction of constraints on the weighting space. One important innovative feature of TRIMAP is to enable the translation of additional constraints on the objective function values into the weighting space. This means the elimination in the weighting space of the areas for which the optimisation of weighted sums of the objective functions (where the weigts correspond to points in those areas) leads to efficient solutions which do not satisfy the additional constraints. The weighting space is used in TRIMAP mainly as a valuable means for collecting and presenting the information. In TRIMAP phases of computation alternate with phases of dialogue with the DM, this mainly in terms of the objective function values, allowing a progressive and selective learning of the nondominated solutions. In each computation phase a scalar problem consisting of a weighted sum of the objective functions is solved with the main purpose of performing a progressive filling of the weighting space. In each step the DM will be called to decide whether or not the study of solutions corresponding to not yet searched regions of the weighting space is of interest. In this way it is intended the prevention from the exhaustive search in regions with close objective function values, a situation found very often in real case studies. The underlying principle is to narrow progressively the scope of the search, using the knowledge accumulated in the previous interactions. The interactive process only ends when the DM considers to have gathered "sufficient knowledge" about the set of nondominated solutions, which enables him/her to make a decision. This method uses an interface that offers the DM a flexible and user-friendly human-computer interaction the use of which is easy and intuitive and enhances his/her capabilities of information processing and decision making.

The experience of the authors of [6] with implementations and applications of different interactive multicriteria linear programming methods led to the conclusion that no single method is better than all the others in all circumstances (Clímaco and Antunes [29]). This methodological posture led to the development of a flexible integrated computer package (Antunes et al. [7]): a method base which seeks to take advantage of the combination of different types of interactive multicriteria linear programming methods. The basic principle of this integrated model is "to support interactively the decision maker in the progressive narrowing of the scope of the search, using the knowledge accumulated in the previous interactions. As more knowledge about the problem is gathered in each interaction, the preference system of the DM progressively evolves, thus making the DM to reflect upon his/her previously stated indications, or even to revise his/her preferences"(*op. cit.* Antunes et al. [9], p. 343). It is assumed that in the process the DM, beyond gathering knowledge, will gain new insights into the problem under analysis, which may be used for specifying new preferences and search directions. The main goal of the method base is therefore to support the DM in the task of exploring the problem and expressing his/her preferences by enabling the DM to reinforce or weaken his/her current convictions at each step. The DM is considered a central and active element of this method base: the stopping criterion is the DM's "satisfaction" and not the verification of a convergence condition on any implicit utility function. The main purpose was to create a flexible decision aid tool capable of respecting the underlying characteristics of the methods and facilitating their combination by guaranteeing a consistent transfer of usable information. This computer package is called TOMMIX (Antunes et al. [5]) and integrates the STEM method, the Zionts Wallenius method, TRIMAP, Interval Criterion Weights method, and Pareto Race. In Antunes et al. [9] the application of this package to the problem of modernization planning of telecommunication networks, introduced above, is exemplified and discussed.

In "Flexible MOLP Approach to the Modernization of Telecommunication Networks Incorporating Sensitivity Analysis", Antunes et al. [10], the flexibility of the proposed approach is enlarged by showing the way in which sensitivity analysis can be associated with the model. Interactive sensitivity analysis techniques concerning changes in the coefficients of the three objective functions and the right hand side of the constraints, as well as the possibility of introducing new constraints, are proposed and discussed.

Finally, we must be refer to the extension of TOMMIX to more than three objective functions, developped in the package SOMMIX by Clímaco et al. [30]. This package can be of great interest in telecommunication strategic planning in those cases where the explicit consideration of more than three objective functions is advisable.

Later, in "Planning the Evolution to Broadband Access Networks: A Multicriteria Approach", Antunes et al. [11], the authors extended the type of analysis mentioned above to new strategic telecommunication planning problems, namely regarding the evolution paths towards the deployment of technologies capable of providing broadband services in a residential and small business setting.

The emergence of new services based on broadband access technologies is recognized as an essential driver to generate additional revenues and support a long-term growth and the financial strength of the operators. Several factors possessing many inter-related influences, are involved namely the rapid pace of technical innovations, the development of multiple modes of communication and the changing market structures (even in local access networks). Therefore, the model described above has been extended as an attempt to exploit new avenues for studying the evolution policies towards broadband services (Antunes et al. [11]). This model is based on an extended state transition table that considers as states of the system the feasible combinations of service categories and technology architectures for the access network. The set of service categories (S) consists of:

POTS – plain old telephone service, inherently a narrow-band symmetric service;

ES – enhanced services, as such the ones presently offered by narrow-band basic rate ISDN;

ASB – asymmetric switched broadband services, capable of providing at least 2 Mb/s downstream and 16 Kb/s upstream;

SSB – symmetric switched broadband services, capable of providing at least 2 Mb/s bi-directional;

CATV – broadcast (distributive) broadband services typically non-switched, such as cable TV;

AS – switched broadband advanced services, generally asymmetric.

The set of technology architectures (T) for access alternatives consists of:

- copper pairs;
- enhanced copper pairs (namely ADSL – asymmetric digital subscriber line);
- Hybrid Fibre Coaxial (HFC);
- Fibre To The Curb (FTTC).

These service categories provide distinct service applications, such as POTS, videotext, data transfer, video telephony, internet access, desktop multimedia, distance learning, video on demand, shopping/home ordering systems, interactive video games, telecommuting (at different levels), enhanced pay-per-view and broadcast TV. The relationships between these application services and the service categories in the diagram can be established in various ways with different degrees of plausibility and/or technical feasibility. For instance, video telephone belongs to the category SSB which can be supported by HFC or FTTC architectures.

The objective functions considered in the extended model are:

- the minimization of the net present value of the total evolution cost;
- the maximization of the near-term service capability; the maximization of the compatibility with the embedded base of subscriber equipment.

Three main categories of constraints have been considered:

- upper bounds on cost and charges;
- degree of satisfaction of the estimated demand;
- degree of penetration of the supporting technologies.

As it is said in Antunes et al. [11], the proposed approach required a great effort of data collection regarding the construction of the coefficients in the objective functions and constraints. Hence the reliability of the analysis results is clearly questionable taking into account all types of uncertainties and imprecisions associated with estimates of the demand for services, investment, operational and maintenance cost and so on, as previously mentioned.

The study of approaches and methods suitable for tackling the inherent uncertainty and imprecision of the input information required by this and other types of planning models, such as interval programming, stochastic programming and fuzzy programming approaches, is a quite relevant research issue. A certainly difficult, but decisive question, is trying to identify which approaches are best suited for a specific model of a particular problem, in a given decision environment. Naturally these questions and challenges are common to most of the problematic areas discussed in this study.

In any case we think the discussed multiple objective mathematical programming approach is of interest when trying to grasp certain compromises to be made and to discover trends in this type of problem, which can be helpful to network operators to make decisions concerning the upgrade and expansion of access networks. The experiments displayed in Antunes et al. [11], were carried out in the framework of an outline study more concerned with showing the usefulness of the multiple objective model rather than putting forward "prescriptive" conclusions. It would be required to perform more experimentation with updated and more accurate data, in particular involving sensitivity and robustness analysis on the model parameters and assumptions. Furthermore, in many cases, this type of studies could be complemented, at a lower level of analysis, with the screening of distinct alternatives to aid making some "intermediate" decisions. Again multiple evaluation aspects are involved. A possible approach to be developed would be to consider an impact matrix stating the level of performance of each potential course of action in terms of the evaluation criteria considered in this context, leading to a discrete alternative multiattribute decision model. This could be tackled by using several methods proposed in the scientific literature. An example of such an approach is the possible consideration of the choice between HFC or FTTC architectures using as evaluation criteria (among other significant possibilities): support for full service installation strategy, installation first cost, operational savings, fitness to the embedded plant, and evolutionary potential.

Finally, in Bana e Costa et al. [14] the authors deal with a real world multicriteria decision aiding problem regarding the strategic study of the expansion of cellular telephony systems. The original problem concerns the determination of the municipalities of a Brasilian State in which a given mobile operator intends to expand its network. Economic (including budget limitations, costs, return of investments) as well as a significant number of technical factors (such as ease of installation and QoS parameters) are considered in the model attributes or criteria.

The authors pay particular attention to the phase of structuring the problem (see Rosenhead [88]), i.e. the identification of the decision problem under study hence enabling to build a multiattribute model. Cognitive maps, imported from psychology, were used in this task of organizing and synthesizing the points of view of the various actors. The outcome of this procedure provides the adequate information for supporting the construction of a consistent family of criteria. Although the integration of structuring methods with multicriteria evaluation approaches, following, for instance, the lines defended in Belton et al. [19], is an important practical issue it is beyond the scope of this paper. The analysis of the obtained multiattribute model is carried out using an additive value function approach to evaluate the alternatives. In order to build the criteria and to assess the scaling constants (weights), the methodology MACBETH in Bana e Costa and Vansnick [15] was used.

4.2 Routing Models

4.2.1 Background Concepts. Routing is a key functionality in any communication network and has a decisive impact on network performance (in terms of traffic carried and supplied grade of service for end-to-end connections) and cost. Routing is essentially concerned with the definition and selection of a path or set of paths from an originating node to a terminating node (assuming the functional network topology is represented by a graph), seeking to optimise certain objective(s) and satisfy certain technical constraints. The routing problems have different natures and multiple formulations, depending fundamentally on the mode of information transfer, the type of service(s) associated with the routed "calls", the level of representation of the network (typically two levels are considered: the physical or transmission network and the logical or functional network), and the features of the routing paradigm (for example whether it is static or time varying according to traffic fluctuations or network conditions). The term "call" is here taken in its broadest sense, as an end-to-end service request with certain requirements that must be met by the path (or route) along which that call is routed. Examples are a telephone call, a video call, a data packet stream or a wavelength assignment (in an optical network). In the broader context of the planning and design activities routing is a fundamental network functionality that may be considered as an integral part of the network operational planning decision process, strongly related to other planning instances, namely network structure design (involving topological design and facility capacity calculation) and traffic network management. At a lower level of the network functionalities routing is intimately related to the entities usually designated as routing protocols that actually implement the routing in a real network. These are critically interrelated with the technological requirements. Two examples, for the Internet, are the OSPF (Open Shortest Path First) protocol and the BGP (Border Gateway Protocol). These aspects and interdependencies are a decisive factor in the formulation of the routing problems from the OR perspective. An overview of some of these issues and possible modelling and resolution approaches can be seen in Mahey and Ribeiro [66].

When formulating routing problems it is useful to model networks as *teletraffic networks* the specification of which includes the following elements: a graph (V,L) defining the network topology where the nodes (in V) may represent switches, exchanges (groups of switches interconnected in a certain manner) or routers, and the edges (or links in L) represent transmission facilities with a certain capacity; the capacities of the arcs that are expressed in terms of bandwidth (in bit/s) or equivalent number of certain basic transmission channels (for example in multiples of 64 kb/s channels); the node-to-node traffic flows that may be modelled in general as marked point processes cf Cox [33] (e.g. a marked Poisson process), which enable a representation of the call instants of arrivals, call durations and associated bandwidth requirements in the links; the routing principle(s) used i.e. the basic features of the network routing function (for example, whether it is static or dynamic and a prescribed maximal number of links per path). Here we define *routing method* as a particular specification of certain routing principle(s), including, as key element, the algorithm or set of rules which are used to perform the path computation and path selection for every traffic flow at a given time, having in mind to optimise the adopted routing metric(s) and satisfy certain constraints (associated with the underlying routing principle(s) and possible additional constraints reflecting bounds on relevant metrics or requirement(s) inherent to the method).

It must be emphasized that the specification of the objective(s) and constraint(s) depends strongly on the nature of the network and services (in various technical instances) and on the rationale of the routing method.

4.2.2 Multicriteria Routing Approaches. The extremely rapid pace of technological evolution and the increase in the demand for new communication services lead to the necessity of multiservice network functionalities dealing with multiple, heterogeneous QoS dimensions. This trend (discussed in Section 2) led to a new routing paradigm in telecommunication networks designated as *QoS routing* . This type of routing involves the selection of a chain of network resources along a feasible path satisfying certain requirements (dependent on traffic features associated with service types) and seeking to optimise some relevant metric(s) such as delay, cost, number of edges of a path and loss probability. Therefore, in this context, routing algorithms need to consider distinct metrics, Lee et al. [62].

In commonly used approaches the path calculation problem is formulated as a shortest path problem with a single objective function, corresponding either to a single metric or to a function encompassing different metrics, while QoS requirements are incorporated into these models by means of additional constraints. This is the usually proposed approach for QoS routing problems, generally designated as *constrained-based QoS routing*. This type of routing problems is particularly relevant in the new Internet technologies, namely MPLS, as explained in Section 2, and in some ATM routing protocols.

A well known approach in multicriteria model analysis consists of transforming the objective functions into constraints, excepting one of them which is then optimised. In adequate conditions the obtained solution will necessarily be nondominated in the original multicriteria model. Furthermore, by varying the right hand-side of the constraints it is possible to obtain different nondominated solutions (see Steuer [96]). In this sense constrained-based QoS routing models can be envisaged as a first tentative approach to multicriteria analysis. On the other hand, the necessity of determining the solution to be implemented in the network in a very short time (usually a few seconds or even less, depending on several factors) makes the most common approach the development of heuristics that include classical algorithms for shortest path computations.

Chen and Nahrstedt [28] present an overview of the majority of QoS routing procedures up to 1998. Also Zee [107] presents a report on the state of art on QoS routing up to 1999. Kuipers et al. [59, 60] provide a comprehensive review on constrained-based routing. The authors of these reviews recognise that QoS routing requires that multiple parameters are related to current network state and have to be frequently updated and the corresponding information has to be distributed throughout the network. Hence the creation of routing protocols capable of efficiently computing the required paths and processing and distributing that dynamically varying information, is still an open issue that needs further investigation. In these circumstances they opted for presenting a review of methods dedicated to this type of problem where network state is temporarily static. In the same study several exact algorithms and heuristics dedicated to the multiple-constrained path (MCP), to the multiple-constrained optimal path (MCOP) and to the restricted shortest path (RSP) problems, are discussed. In the MCP problem it is just intended to obtain path(s) which satisfy constraints on all metrics while in MCOP and RSP (this is a particular case of the former with one constraint alone) problems there is an objective function to be optimised.

Our bibliography includes several models on variants of QoS routing problems and various resolution procedures. These are: Hassin [48], Guo and Matta [45], Reeves and Salama [85] (focusing on procedures for the RSP problem); De Neve and Van Mieghem [76] (dealing with the MCP problem through a heuristic with tunable accuracy, based on a k-shortest path algorithm) and Van Mieghem et al. [71] (proposing an algorithm for dealing with the MCP and the MCOP problems, also based on a k-shortest path algorithm); Iwata et al. [49] (proposing a heuristic for the MCP problem based on the calculation of shortest paths and presenting an application to ATM networks) Chen and Nahrstedt [27] (proposing two heuristics for the MCP problem, based on the Dijkstra and Bellman-Ford algorithms); Yuan [104] (presenting two heuristics for the MCP problem); Korkmaz and Krunz [58, 57] (proposing a heuristic, based on a modified versions of Dijkstra algorithm for the MCOP problem); Liu and Ramakrishnam [63] (developping an exact algorithm for finding k-shortest paths satisfying multiple constraints); Aneja and Nair [4] (dealing with the constrained shortest path problem); Goel et al. [43] (proposing a heuristic for a specific RSP problem where one seeks to find a least-cost path from a given node to all destination nodes, satisfying a delay constraint for each path) Blokh and Gutin [21] and

Handler and Zang [46] (proposing procedures for the constrained shortest path problem using Lagrangean-based linear algorithms); Yuan and Liu [105] and Yuan [104], (both dealing with heuristics for the multiconstrained problems).

Also other papers dealing with specific application models involving problems of this type are included in the bibliography, namely: Banerjee et al. [16] (presenting an overview of application models for MPLS networks); De Neve and Van Mieghem [75] (focusing on an application to ATM networks); Ergun et al. [40], Fortz and Thorup [41] (dealing with apllications to Internet routing); Guerin and Orda (describing possible extensions to path selection algorithms which enable the incorporation of mechanisms of path reservation in advance) [44] ; Ma and Steenkisk [64, 65] (showing applications to routing protocols for traffic with bandwidth guarantees in integrated services networks); Pornavalai et al. [80] (dealing with applications to routing problems in integrated services packet networks); Kuipers and Mieghem [59] (presenting a QoS routing procedure for a constrained "multicast" path problem which involves the simultaneous selection of paths from a source node to multiple destination nodes); Wang and Crowcroft [100] (making an analysis of various formulations and mathematical properties of the MCP problem with respect to the metrics most relevant to QoS routing).

Special attention should be drawn to some cases where the concerns which lead to this type of approaches, are relevant to multicriteria analysis. Widyono [101] proposes an exact restricted shortest path (RSP) algorithm designated as constrained Bellman-Ford (CBF). This enables, for example, the calculation of successive shortest paths between pairs of nodes for different values of the right hand-side constraint on the delay, hence obtaining nondominated solutions. That paper proposes an exact algorithm dedicated to the RSP problem. The bicriterion nature of this proposal is clear and we could put in evidence that the bicriteria shortest path problem approach in Clímaco and Martins [32] could perform a similar study in a more efficient manner.

Consider now approaches based on Lagrangean decomposition, where, for example, one intends to calculate the minimal cost path subject to a delay constraint. The costs and delays on the links are combined linearly and thence the shortest path, regarding the obtained objective function, is calculated. Kuipers et al. [59] recognise that a key issue in such approaches is the way in which the appropriate multipliers are determined when delay and cost are combined, since this obviously conditions the solution that is obtained. It is a question of the same type that arises in the definition of weights when in multicriteria analysis one intends to optimise a weighted sum of objective functions. Note that in bicriterion shortest path problems, there may exist unsupported nondominated solutions. In the example above nothing guarantees that the obtained solution is optimal for the original RSP problem. Approaches where one seeks to close the gap between the optimal solution and the solution obtained from a linear combination by using k-shortest path algorithms are referred to in Kuipers et al. [59]. Also approaches for calculating unsupported nondominated solutions based on k-shortest path algorithms can be developed.

Finally we would like to draw attention to the fact that the principles underlying the bicritrerion approach that is described next (based on a specific k-shortest path algorithm and on the introduction of "soft constraints") have clear relations with the principles underlying Jaffe's algorithm [50] dedicated to the MCP problem and to other algorithms that seek to overcome some of the limitations of Jaffe's algorithm.

Other multidimensional approaches, where there is an *a priori* articulation of preferences in the path selection, taking as basis bandwidth, delay and hop count, are mentioned in Kuipers et al. [59]. Relevant examples are the widest-shortest and the shortest-widest path approaches. Examples of such approaches can be seen in: Ma and Steckiste [64, 65], Orda [78], Wang and Crowcroft [100] (in this case it is intended to calculate the shortest path in terms of delay, with maximal minimal arc bandwidth; note that the minimal bandwidth of all arcs of the path is usually known as *bottleneck bandwidth*), Van Mieghem et al. [71] and Oueslti-Boulahia and Oubagha [79] (presenting a heuristic approach based on a utility function, as an alternative to the widest-shortest path model for routing "elastic traffic flows" in the Internet).

Several QoS routing related path computation problems are treated in Sobrinho [94] in a unified form (including connectivity, shortest path, widest path, most-reliable path, widest-shortest path and most-reliable shortest path problems) by using an algebra of weights (hence treating the aggregation of preferences in an articulated manner). This approach also enables a specific requirement of the routing procedure implementation in the Internet (designated as 'hop by hop' routing) to be taken into account. As an application of this approach a variant of the Dijkstra algorithm, which guarantees the satisfaction of that requirement, is constructed.

There is yet a different type of multicriteria model which deserves a reference. In many types of telecommunication networks there is a mechanism, closely associated with the routing function, that is usually designated as *admission control.* This mechanism involves a decision on whether or not each call is accepted, as a function of certain call characteristics (eg. associated type of service, tariff system and QoS requirements) and, possibly, network working conditions (this is typical of dynamic routing methods that include admission control mechanisms). The underlying objective of this mechanism is to maximise the operator revenue while satisfying the QoS guarantees for every customer class. In B-ISDN and in broadband multimedia networks in general, this is a relevant issue, since the supplied QoS guarantees are directly related to the obtained revenue, via the tariff (or "charging") system (a comprehensive analysis and discussion on charging models for multiservice networks is in Songhurst [95]). Brown et al. [26] address an admission control problem in broadband multiservice networks, modelled as a specific semi-Markov decision process that might be considered as a first tentative stochastic multicriteria approach. In this approach the objective is to maximize the total revenue rate of ongoing calls while satisfying the QoS guarantees of all carried calls. The solution approach is based on a reinforcement learning technique. The solutions are compared with simple heuristic admission control solutions, by using a simulation model for a test communication system with two types of traffic sources. In the tested examples the model application enabled a 30% improvement in the average revenue.

Let us now consider the cases where the modelling is more explicitly multicriteria. We think there are potential advantages in considering many routing problems in modern telecommunication networks explicitly as multiple criteria problems. This type of modelling is potentially advantageous although one cannot ignore that, in the majority of situations, the solution to be implemented has to be obtained in a short time that may range from a fraction of a second to a few seconds. This practical limitation implies the impossibility of using interactive methods in most cases, hence leading to the necessity of implementing automatic path calculation procedures. The exception is in static routing problems or in some form of periodic dynamic routing models where the input parameters are estimated in advance (for example, node to node traffic intensities in different hours), cases in which an interactive procedure could be used to select the routes (for every node pair) to be memorised in routing tables assigned to every node. This explains the predominance of methods where there is an *a priori* articulation of preferences. It should be noted that, even in these cases, there are advantages of explicit multicriteria modelling hence rendering the mechanisms of preference aggregation transparent. In this manner, several aspects, namely cost and QoS parameters such as blocking probability, delay or bandwidth, can be addressed explicitly by the mathematical models, some as objective functions and the remainder as constraints, seeking to reflect in a more realistic way the underlying engineering problem.

On the other hand, as it will be seen in the cases that we are describing next in more detail, it is possible to conciliate the automatic path calculation and selection with some flexibility in the form of preference aggregation. This enables the grasp of the compromises among different objectives, taking into account certain QoS requirements, by treating the comparison among distinct routing possibilities in the context of a certain routing principle, in a consistent manner.

Following this methodological framework an explicit multiple objective routing model for telecommunication networks was firstly (as far as we know) presented by Antunes et al. [8]. In this approach a static routing problem (that is a routing problem where the objective functions coefficients are constant values) is formulated as a bi-objective shortest path problem. The model can be adapted to different metrics associated with different types of services. An algorithmic approach was developed to deal with this problem which computes nondominated paths based on the optimization of weighted sums of the two objective functions, using a very efficient k-shortest path algorithm in Martins and Pascoal [68]. QoS requirements are represented in the model through 'soft constraints' (that is constraints not directly incorporated in the mathematical formulation) in terms of 'acceptable' and 'requested' values for each of those metrics. Note that since the routing problem is modelled as a multiple objective shortest path problem without side constraints, no metrics other than the ones considered as objectives are represented in this model. This limitation could be surpassed, but then it is necessary to check whether each new calculated path respects the side constraints. Because of the importance of this contribution in the present context we will review its basic aspects.

The starting point of this approach is the formulation of the static routing problem as an M-objective shortest path problem where each metric $m = 1, ..., M$ is associated with an objective function to be minimized. Hence let us consider a network $G = (N, A)$, consisting of a set N of nodes and a set A of arcs ($A \subset N \times N$). Each arc a_{ij} connecting nodes i and j ($i, j \in N$) is assigned M real values c_{ij}^m which denote the cost per unit flow on that arc for metric m. A path p from an origin node $s \in N$ to a destination node $t \in N$ is a sequence of arcs: $p = \{a_{si}, a_{ij}, \ldots, a_{1t}\}$. Then the routing calculation problem can be formulated as follows:

$$\begin{aligned}
\min z^m &= \sum_{a_{ij} \in A} c_{ij}^m x_{ij} \\
& \quad m = 1, \ldots, M \\
\text{s.t.} \sum_{j \in N} x_{sj} &= 1 \\
\sum_{j \in N} x_{ij} - \sum_{q \in N} x_{jq} &= 1 \quad \forall j \in N \setminus \{s, t\} \qquad \text{(Problem } P1\text{)} \\
\sum_{j \in N} x_{jt} &= 1 \\
x_{ij} &\in \{0, 1\} \quad \forall a_{ij} \in A
\end{aligned}$$

where c_{ij}^m represents the cost of using arc a_{ij} with respect to metric m.

Routing metrics generally considered are delay, cost, hop-count, blocking (or loss) probability, error rate, and bandwidth. The aggregation function to compute the value v_m^p of path p depends on the type of the metric m:

- The metric is *additive* if

$$v_p^m = \sum_{a_{ij} \in p} c_{ij}^m; \tag{22.1}$$

- the metric is *multiplicative* if

$$v_p^m = \prod_{a_{ij} \in p} c_{ij}^m; \tag{22.2}$$

- the metric is *concave* if

$$v_p^m = \min_{a_{ij} \in p} c_{ij}^m. \tag{22.3}$$

The blocking (or loss) probability (and error rate), is calculated assuming that this metric follows the aggregation function:

$$v_p^m = 1 - \prod_{a_{ij} \in p} (1 - B_{ij}), \tag{22.4}$$

where B_{ij} is blocking probability on arc a_{ij}.

Delay, hop-count and cost follow the additive aggregation function. Path bandwidth (and throughput) is computed by using the concave aggregation rule. The loss probability (and error rate) metric can be transformed into an additive metric (and hence comply with the shortest path approach requirements) by defining:

$$c_{ij}^m = -\log(1 - B_{ij}) \tag{22.5}$$

hence transforming the minimisation of v_p^m into the minimisation of the sum of the corresponding c_{ij}^m.

The solution approach proposed in Antunes et al. [8] is inspired by the one presented in Rodrigues et al. [87], in the framework of a procedure enabling to search interactively nondominated supported and unsupported shortest paths in the bicriteria case. It should be stressed that the node-to-node routing plans are supposed to run in an automatic manner, in the framework of a routing control network mechanism. The procedure satisfies this requirement by integrating the use of a k-shortest paths algorithm in Martins et al. [68], (likewise in Rodrigues et al. [87]) and new devices called soft constraints. In resume, in this approach a specialised automatic algorithmic was developed which takes into account the specific aspects of a routing problem in a multiservice environment to obtain nondominated solutions. Note that, updating the thresholds regarding the soft constraints related to QoS requirements, according to the evolution of the network state, is a very simple and clear procedure in operational terms. We will come back to this point while discussing the routing model by Craveirinha et al. [34].

The main features of the approach [8] are:

i) to enable QoS requirements to be expressed as additional *soft* constraints on the objective function values in terms of *requested* and *acceptable thresholds* for each metric;

ii) the addition of this type of soft constraints defines priority regions, in which nondominated solutions are searched for according to the underlying QoS thresholds;

iii) the auxiliary objective function which is used to search for nondominated solutions is a weighted sum of the original objective functions, where the weights are calculated, for instance, from the optimal solutions for each objective function and the requested metric values (for example see constant cost line passing through Ac and Ad in Figure 22.2);

iv) the nondominated solutions (including those in the interior of the convex hull of the feasible solution set) are computed by means of an extremely efficient k-shortest path algorithm proposed in Martins et al. [68], designated as MPS algorithm.

To understand the main features of this algorithmic approach (see [8]) an illustrative example of its working is presented, based on the priority regions case in Figure 22.2, considering as metrics cost and delay.

First the vertex solutions, which optimise each objective function, are computed, by solving two shortest path problems using Dijkstra's algorithm. This yields information regarding the value range of each objective function over the nondominated solution set. Quality of service (QoS) requirements for each of those metrics are specified by means of the thresholds requested value (aspiration level) and acceptable value (reservation level). The addition of this type of *soft* constraints (that is, constraints not directly incorporated into the mathematical formulation) defines priority regions, in which nondominated solutions are searched. Region A is a first priority region where both requested values are satisfied. Regions B1 and B2 are second priority regions where only one of the requested values is met and the acceptable value for the other metric is also guaranteed. A further distinction can be made between these second priority regions by establishing a preference order on the objective functions. For instance, stating that cost is more important than delay, would give preference to region B1. Region C is a third priority (or fourth if B1 and B2 have different priorities) region in which only acceptable values for both metrics are fulfilled. For the example in Figure 22.2 the first solution found within (first priority) region A (solution 3) is selected. Note that any solution in the first priority region dominates any solution in region C. Of course, in other situations, solutions within

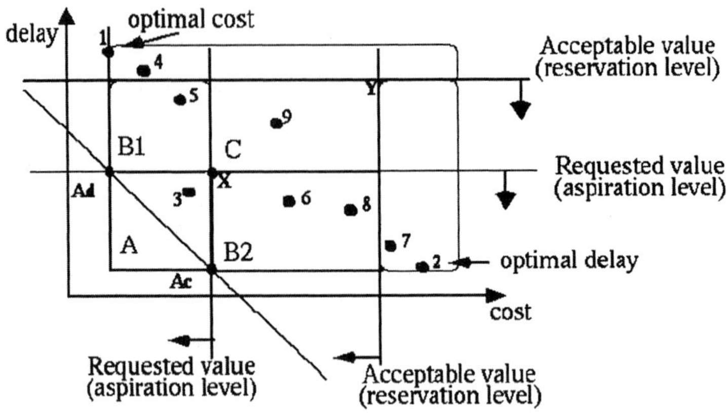

Figure 22.2. Example of priority regions.

second priority regions B1 and B2 could be found first. These solutions should be stored but not reported until the 1st priority region is entirely searched (i.e. when the constant cost line of the objective function used in the k-shortest path problem passes through point X). If there are no nondominated solutions within region A, the search proceeds to 2nd priority regions. The previously computed solutions in regions B1 and B2, if any, are now reconsidered. In the example, solutions 5, 6 and 8 are found within second priority regions. In general, it is (again) possible to obtain solutions in the third priority region (C) before all second priority regions (B1, B2) are searched. Again these solutions are stored and reported only when regions B are completely searched without finding any nondominated solutions within them. If the algorithm proceeds to this point it means that no paths exist satisfying at least one of the requested QoS values (aspiration levels) and only acceptable values (reservation levels) can be met. Beyond point Y even acceptable values for QoS requirements cannot be met. In this case a possible relaxation in the acceptable value thresholds would have to be considered. In fact nondominated solutions may possibly exist outside the priority regions (such as solution 4 and 7), which could be used as "last chance" routes.

The use of the capability of this type of model incorporating soft constraints is strongly dependent on the application environment, in terms of network technological constraints (with repercussion on the teletraffic network model) and capabilities, as well as on QoS requirements, types of traffic flows and characteristics of provided services. For example, in conventional NB-ISDN, only constraints concerning "acceptable" levels of QoS need to be considered, which should follow standard ITU (International Telecommunication Union) recommendations. On the other hand, in ATM networks where traffic sources of quite different nature and a multiplicity of requirements may occur, the connection oriented approach allows the user to indicate the communication needs during the connection set-up phase and the network may tailor the transfer properties of the connection to specific user needs. This gives rise, in particular, to the concept of traffic contract with its inherent flexibility in terms of resource management. In this framework both types of (soft) constraints, concerning "acceptable" and "requested" values become significant. In this context, it must be noted that the (possible) occurrence of nondominated paths which lead to a better value than the one "requested" by the user raises questions regarding their admissibility as outcomes of the algorithm, since they correspond to an over-utilization (albeit temporary) of network resources. This type of questions, which does not bring any further algorithmic or computational complexity to the proposed approach, nevertheless requires further analysis, which will be necessarily dependent on the network features. So, an important point put forward in this paper is to draw attention to the potential advantage of the application of multicriteria analysis to routing problems in multiservice networks and to provide an efficient algorithmic approach for resolving the problem with the consideration of relevant "soft" constraints.

On the other hand the advantages of using a *dynamic routing* principle in telecommunication networks are well known. The essential feature of dynamic routing is the dependence of routing decisions on measurable network parameters (e.g.number of channels occupied in a link, proportion of lost calls, packet delays, estimated traffic offered), or events (e.g. whether a call is successful or not) hence reflecting, in one way or another, the network working conditions. This implies that selected end-to-end routes vary rapidly in time, seeking to take advantage of the evolving working conditions of the network, with the aim of achieving, at any given time period, the best possible value(s) of some network performance criterion (or criteria). The impact of dynamic routing in network performance is particularly relevant in situations with highly variable traffic intensities, overload and failure conditions by enabling an effective response of the routing system to adverse network working states. The dynamic routing methods with this adaptive nature are usually designated as *adaptive routing* methods. A comprehensive review on dynamic routing is found in Ash [12] where the advantages of dynamic routing methods concerning network performance and cost are clearly shown.

Having in mind to explore the potential advantages of a multiple objective routing principle of the type analysed above (Antunes et al. [8]) and the inherent benefits of dynamic routing, Craveirinha et al. [34] proposes and describes the essential features of a multiple objective dynamic routing method (designated as MODR) of periodic type where the selected node-to-node routes for all traffic flows change periodically as a function of estimates of certain network QoS related parameters, obtained from measurements in the network. In its initial formulation, for circuit-switched networks, MODR uses a principle of alternative routing, that is any call of traffic flow f from node i to node j may attempt the routes (corresponding to loopless paths from i to j in the network graph) $r^1(f), r^2(f), \ldots, r^O(f)$, in this order. The first of these paths with at least one free capacity unit (usually designated as channel or 'circuit', corresponding to the minimal arc capacity necessary to carry a call of flow f) in every arc and satisfying other possible requirements of the routing method, is the one which will be used by the call. If none of those O routes satisfies this condition the call is lost, and the associated probability is designated as the (marginal) blocking probability for flow f or call congestion. The traffic flows were modelled as independent Poisson processes. In alternative dynamic routing methods the ordered route sets that may be used by calls of any traffic flow may vary in time in order to adapt the routing patterns to network conditions in order to obtain the possible "best" network performance, under a certain criterion (or criteria). In general these methods, when correctly designed, are the most efficient routing methods that may be used in this type of networks. MODR uses two metrics for path calculation purposes: blocking probability and implied costs, which define a specific form of a bi-objective shortest path problem of type (P1) with $c^1_{ij} = c_{ij}$ (implied cost associated with arc a_{ij}) and c^2_{ij} given by the transformation (22.5) for blocking probabilities. The implied cost c_{ij} associated with the acceptance of a call in link a_{ij} is an important concept, due to Kelly [53], for modelling routing problems in loss networks (that is in networks where calls are subject to a non-null blocking probability). It can be defined as the expected value (taking into account the revenue associated with the carried calls) of the increase in calls lost on all routes of all traffic flows which use arc a_{ij}, resulting from the acceptance of a call in this link. The method uses $O = 2$: the first attempted route ($r^1(f)$) is the direct arc from i to j whenever it exists; the second choice route (alternative route, $r^2(f)$) has a maximum number D of links and is obtained from a modified version of the algorithmic approach in Antunes et al. [8], reviewed above. This new version of the algorithm (designated as MMRA), adapted to MODR, has the following essential features: it enables nondominated paths to be selected, in the higher priority regions of the objective function space; the priority region boundaries associated with soft constraints (required and acceptable values of the two metrics) are calculated as function of periodic updates of the cost coefficients. In this model, in some situations dominated solutions calculated in the first priority region(s) may be interesting for selection, leading to a change in the original procedure (for details see Craveirinha et al. [34]). Examples in Craveirinha et al. [34] and Martins et al. [70], illustrating the application of this bi-objective model to a fully-meshed circuit-switched network with telephone type traffic show that path implied cost and blocking probability may be conflicting objectives in many practical network conditions, especially in cases of global or local traffic overload.

An instability problem in the path calculation model presented in the previous paper [34], when that model is used directly to obtain the set of routes for every node to node traffic flow, is put in evidence in Martins et al. [70]. This instability is expressed by the fact that the paths calculated by the algorithm MMRA for all traffic flows, in each path updating period, tend to oscillate among a few sets of solutions. A preliminary analytical model showed that solution sets may be obtained by MMRA which lead to poor network performance from the point of view of two global network performance criteria, network mean blocking probability B_m (that is the mean blocking probability for a call offered to the network) and maximal node-to-node blocking probability, B_M. It is also shown that the minimisation of the implied cost of the paths (z^1) tends to minimise B_m while the minimisation of the blocking probabilities of the paths (z^2) tends to minimise B_M. That instability problem is a new "bi-objective" case of a known instability in single objective adaptive routing models, of particular relevance in packet-switched data networks (see e.g. Bertsekas and Gallager [20]). The phenomenon analysed in Martins et al. [70] results from the interdependencies between the coefficients of the objective function z^1 and z^2, from the discrete nature of the problem and from the interdependencies between those coefficients and the set of paths calculated by the algorithm in previous iterations. To overcome this instability problem associated with the great complexity of the routing model, the main requirements of a heuristic procedure enabling the selection of "good" compromise solutions (set of routes for all traffic flows in every path up-dating period) from the point of view of the two mentioned global network performance criteria, are put forward in Martins et al. [70]. Note that even using a single-objective formulation of the adaptive alternative routing problem is NP-complete in the strong sense (also in the degenerated case where $O = 1$, i.e. no alternative route is provided), which is an indication of computational intractability even for near-optimal solutions.

A complete analytical model for the network routing problem in Martins et al. [70], is presented in Martins et al. [69] enabling the mentioned instability problem to be explicit and calculate the two global network performance values, for given traffic intensities and link capacities through the resolution of a system of non-linear teletraffic equations. This leads to a bi-objective dynamic alternative routing problem, formulated at the network level. A heuristic for solving this problem was developed in the report Martins et al. [69], enabling the calculation of good compromise solutions with respect to B_m and B_M, at every path up-dating period (heuristic for synchronous path selection), hence overcoming the mentioned instability problem. To show the effectiveness of the proposed approach, results from the MODR method (using this heuristic) are compared, for some test networks, with a reference dynamic routing method (RTNR developed by AT&T – see Ash [12]), by recurring to a discrete event simulation platform.

A multiple objective routing model for a stochastic network representing a large processing facility is proposed in Kerbache and Smith [54]. The nodes of the network correspond to finite capacity queues of different types (eg. M/G/1/m, GI/G/1/m) and include the possibility of reattempts and the arrival processes from the source nodes are renewal processes. The functions to be optimised are the average sojourn times for all customer types and the total routing costs and are often conflicting. It should be noted, as mentioned by the authors, that this type of model, although having originally a formulation for manufacturing facilities, could be adapted to telecommunication networks, namely packet switched networks. The proposed mathematical formulation is a multiple objective multi-commodity integer programming problem with extra constraints. A heuristic is developed for solving the problem, based on the calculation of k-shortest paths, enabling an approximation to the nondominated solution set, to be found.

Another important type of networks where multicriteria routing models have been proposed is multi-service networks supporting multimedia applications. The utilisation of a QoS routing principle as mentioned above, involves the selection of paths satisfying multiple constraints of a technical nature which seek to optimise some relevant metric(s). A multiple objective model for this type of routing problem intended for application to networks supporting multimedia applications, namely video services, was presented in Pornavalai et al. [81]. The objective functions to be minimised are the number of links of the path (usually designated as hop-count), z^1, and a cost z^2 that is obtained by considering that the cost of a call using a link a_{ij} is the inverse of its available bandwidth, b_{ij}. The first objective function is intended to minimise the number of resources used by a call while the second seeks to minimise the impact of the acceptance of a call by choosing 'least loaded' paths. As for the constraints, they are expressed by bounds on the minimal available bandwidth (bound BW_M), on the delay – sum of the delays d_{ij} on the links of the path – (bound DM_M), and on the delay jitter (bound J_M). This corresponds to the formulation of a bi-objective constrained shortest path problem ($P1_C$) obtained by adding to the classical bicriteria shortest path formulation, the three constraints:

$$\min_{a_{ij} \in p} \{b_{ij}\} \geq BW_M \quad (22.6)$$

$$\sum_{a_{ij} \in p} d_{ij} \leq DM_M \quad \text{(constraints in Problem } P1_C\text{)} \quad (22.7)$$

$$\sum_{a_{ij} \in p} J_{ij} \leq J_M \quad (22.8)$$

where J_{ij} is the delay jitter on the link a_{ij} of path p. The objective function coefficients are $c_{ij}^1 = 1$, $c_{ij}^2 = \frac{1}{b_{ij}}$ and the constraint coefficients d_{ij} and J_{ij} are calculated from stochastic models representing the queuing and jitter mechanisms associated with the link transmission functions, for each type of traffic flow. In some applications, such as video traffic in an ATM network using a specific queuing mechanism it is possible to transform the constraint (1.8) into a constraint on the number of links of the path. In Pornavalai et al. [81] the solution approach to this problem is a heuristic based on the Dijsktra shortest path (SP) algorithm. The heuristic is rule-based and has two phases: route metric selection (i.e. selection of the objective function that it seeks to optimise in each iteration) and route composition rule (where SPs from the origin to intermediate nodes in terms of one metric are concatenated with SPs from those nodes to the destination). For each selected routing metric and composition rule if the SP or the composed path do not satisfy the constraints the heuristc will retry a new route metric and/or new composition rules until a feasible route is found or all routes are exhausted. In spite of its capability in supplying feasible solutions in short times (in networks with hundreds of nodes and average node degree of 4) it doesn't guarantee that the obtained solutions are nondominated.

This type of routing problem was tackled in Clímaco et al. [31] by using an exact algorithmic approach for calculating the whole set of nondominated paths of problem $P1_C$. This approach is based on the bi-objective shortest path algorithm in Clímaco and Martins [32] and on the MPS algorithm in Martins et al. [68]. In this approach it was necessary to adapt a ranking algorithm for generating the set of nondominated paths. It might be expected that a labelling algorithm would be a better approach. However it was shown by the authors Clímaco and Martins [32] that the ranking algorithmic approach has better performance as a result of explicit consideration of the constraints in the bi-objective problem. This approach was applied to a problem of video traffic routing on ATM networks, by constructing random networks and networks based on the US inter-city spatial topology. In this particular application study it was shown that, although the used objective functions were not strongly conflicting, there was a significant number of problems with 2, 3 and 4 nondominated solutions. Also the algorithm proposed in Clímaco et al. [31], enabled the calculation of the whole set of nondominated solutions in networks with up to 3000 nodes and average degree of 4, in short processing times and modest memory requirements, up to certain bounds on the acceptable delay. This makes it attractive in many realistic problems.

A specific new routing problem in MPLS networks concerning "book ahead guaranteed services" (or BAG in short), modelled as a multicriteria decision problem, is approached in Thirumalasetty and Medhi [98]. This problem is focussed on the calculation ahead of time (with respect to the instant of generation of the actual call) of two paths, at the request of a user, with certain QoS guarantees. For example, the user may request from the network administrator, through a web-page sign-up, his/her access, at a future time, to the use of a supercomputer, with bandwidth and survivability guarantees in the event of failures. A pair of arc-disjoint paths (the first for the actual connection and the second to be activated in the event of failures) satisfying certain bandwidth constraints have to be calculated. The considered objectives are: to maximize the residual capacity in the network for other type of services (designated as "best effort services", such as e-mail or www), to minimise the routing costs of the BAG traffic, to minimize a penalty associated with the rejection of BAG service requests, and the maximization of revenue from accepted BAG demand. The proposed problem resolution is based on the aggregation of the four objective functions and uses a heuristic to solve the resulting integer-linear programming problem.

A multiobjective formulation for a QoS "off-line" routing problem in telecommunication networks is presented in Knowles et al. [56]. This model considers the three following objectives: minimization of the routing cost, minimization of the total positive deviations from a certain target utilisation of the links and minimization of over-utilisation of the links, expressed in terms of available landwidths. This approach uses an evolutionary algorithm to obtain approximate nondominated solutions. In Resende and Ribeiro [86] a bi-objective model for a private virtual circuit routing problem in the Internet, is described. The objectives are the minimization of the propagation delay (suffered by the packets along a path) and the minimization of a load balancing function which depends on the capacity and the load of each link of the route.

A multiobjective formulation for a QoS "off-line" routing problem in MPLS networks is developed in Erbas and Erbas [38]. This model considers three objectives: to minimise the route cost, to minimise a load balancing function similar to the one in Resende and Ribeiro [86] and to minimise the total number of LSPs (Label Switched Paths) assigned to all connection requests. The obtained formulation is a three-objective mixed-integer problem. An evolutionary algorithm is developed as resolution approach to this problem and its results are compared (in terms of running time and "quality" of the obtained approximate nondominated set) with the ones from an exact algorithm, in a specific case study. Related works focusing on the same type of routing models using evolutionary algorithmic approaches, are reported in Erbas and Mattar [39] and Erbas [37].

A multiple objective routing model for B-ISDN (based on ATM), using a fuzzy optimisation approach, was presented in Aboellla and Douligeris [1]. The fuzzy programming model is focused on maximizing the minimum membership function of all traffic class delays (corresponding to different service types) and the minimum membership function of the link utilization factor of all network links. The efficiency and applicability of the approach are studied, under different network load conditions, by calculating several performance measures and comparing their values with the ones obtained from single objective models. The authors discuss and recommend a hybrid resolution approach that combines the "generalised network model" that has been successfully applied to large zero-one integer programming problems (Glover and Mulvey [42]) with the fuzzy programming technique.

The paper by Anandalingam and Nam [3] proposes a game theoretic approach to deal with a dynamic alternative routing problem in international circuit-switched networks, considering the cooperative and non-cooperative case. In the non-cooperative case it is assumed that each player (corresponding to a given country involved in the network routing design) selects a routing strategy which optimises his/her payoff given the strategies chosen by the others and he equally assumes that the other players will attempt to use strategies which optimise their payoffs, where the payoff objectives of each player are expressed in terms of the minimization of the cost of adding more links (with the required capacities) in his/her own part of the global network. This problem is modelled as a bi-level integer linear programming problem characterised by a player who works as "leader" and makes the initial decision (by minimising his/her own cost function) and then the other players, or "followers", seek to minimise their own cost function given the leader decision; the leader has to pay a certain fraction of the link costs of a part of the jointly owned network. Several application examples, where approximate solutions to the model are obtained from the branch-and-bound algorithm by Bard and Moore [17], are discussed. The major conclusions stress the great cost savings in global networks obtained from the dynamic routing solutions (an example is presented for a network interconnecting the US, Japan and Hong-Kong), for all the involved "players", both in the cooperation and in the non-cooperation cases. This is a result of enabling the distribution of the peak traffic loads of one country to the idle parts of the routes in other countries by making the most of the different time zones of the countries involved.

A multiple-objective approach, based on genetic algorithms, is proposed in the report Zhu [109] for dealing with a specific routing problem in WDM optical networks. The problem is a particular version of the problem usually designated "route and wavelength assignment" (RWA in short) and is modelled as a three objective integer linear programming problem and the resolution approach is a genetic algorithm using a Pareto ranking technique. RWA refers to a type of routing problem that has become very important in optical networks, especially with the emergence of OXCs (optical cross-connectors) mentioned in Section 2.1, and is focused on the calculation of lightpaths (fixed bandwidth connection between two nodes via a succession of optical fibres). It can be decomposed in two inter-related sub-problems. Given an optical network, the arcs of which correspond to bundles of optical fibres each one with a number of available wavelengths, and the demand for node-to-node optical connections, the first sub-problem, or "routing problem", involves the determination of the path along which the connection should be established. The second sub-problem involves the assignment of wavelengths for every connection, on each arc of the selected path. RWA has multiple formulations depending on the nature of the traffic offered (optical connections), objectives (for example: to maximize the number of established connections for a fixed number of available wavelengths or to minimise the number of required wavelengths for a given set of requests) and technical constraints. An overview of the technical motivation and basic concepts is found in Assi et al. [13] and a review of resolution approaches for the RWA problem can be seen in Zang et al. [106].

Akkaren and Nurminen [2] present a study on the evolution of the routing algorithms that are used in a sequence of releases of a telecommunication network planning tool. Typical situations are adressed namely the evolution from a simple stage where the aim is to find shortest paths to more complex routing tasks shuch as procedures focusing on the finding of protected routes. In this problem the routing protection scheme involves the calculation of a primary and a secondary route in order to make sure that a connection continues operational in the event of failures. The usual goal in this context is to find two arc-disjoint paths for each traffic flow. The authors develop a routing optimisation model involving a trade-off between route length and disjointness. A thorough discussion on the problem of selection of algorithms taking into account their evolution capability is also put forward.

4.3 Operational Plannning

Operational planning covers a wide area of planning activities focused on the short term network design such as location, interconnection and dimensioning of transmission equipments and other facilities such as switching units, routers or traffic concentrators. In a few specific problems of this type there have been some proposals of multicriteria modelling. Next we review some papers in this area.

Wierzbicki [102] presents a multicriteria modelling approach for a problem concerning the placement of Internet caches. The underlying generic technical objective is to increase the network efficiency and the goal is to minimize the overall flow or the average packet delay. The problem of general cache location is formulated as a MILP (Multicriteria Integer Linear Programming Problem) and is reformulated using a reference point approach. Also the sensitivity of the model solutions, to simplifications of the problem are studied. Simple greedy heuristic resolution approaches are tested for some medium size network topologies.

Tiourine et al. [99] proposes search algorithms for the problem of link frequency assignment, which has great relevance for wireless networks, satellite communications, television and radio broadcast networks. The model proposed in this paper is in some sense a bi-objective combinatorial model. In fact it proposes a lexicographic sequence of two objective functions. The principal objective consists in minimising interference and a secondary objective is the minimisation of the used radio spectrum. When optimising the latter objective it is assumed that a zero value of interference was obtained when solving the former optimisation problem. This study was included in the CALMA (Combinatorial Algorithms for Military Applications) project, part of the long term European Cooperation Programme on Defence. Some local search approaches were considered such as tabu search, simulated annealling and variable-depth search, paying particular attention to the development of problem specific neighbourhood functions, as well as to the presentation and discussion of computational experiences.

The application of reinforcement learning methods to a problem related to packet wireless communication channels is proposed in Brown [25]. The addressed problem involves the search for a satisfactory power management policy considering simultaneously two criteria: trying to maximize the radio communication revenue and to decrease the battery usage. This problem is modelled as Markov Decision Process, where the generated traffic is modelled as the traffic from an ON/OFF source and rewards are assigned to packets carried in each direction (between the mobile and the base station); other technical elements of the communication system are also incorporated in the model in a simplified manner. This problem can be approached as a stochastic shortest path problem, introducing some simplifications that enable the reduction of the dimension and complexity of the state space. Encouraging results were obtained from a simulation study with the model, enabling power saves from 50% to 80% to be obtained in several cases.

4.4 Models Studying Interactions between Telecommunication Evolution and Socio-economic Issues

Some socio-economic application models related to telecommunication issues are reported here.

As far as we know the use of multiattribute models in telecommunication planning and design is still very limited at this moment. However, in some applications we foresee great potential for future applications. In "A Telecommunications Quality Study Using the Analytic Hierarchy Process" Douligeris and Pereira [35] show the way in which a customer can use AHP (Analytic Hierarchy Process) to choose a telecommunication company and/or particular services that are the best for satisfying his/her needs in terms of quality of service or to decide between two telecommunication service providers. For instance, in Kim [55] a survey on the evaluation of intranet functions using AHP, is presented. Also Raisinghani [83], Raisinghani and Schakade [82] studied multicriteria approaches for supporting strategic decisions on electronic-commerce (e-commerce), based on AHP and ANP (Analytical Network Process). In the approach based on ANP the evaluation methodology seeks an integration of Internet "domain strategies" (such as virtual communication, information, distribution and transaction spaces) and business strategies. ANP is considered as a generalization of the AHP decision methodology, where hierarchies are replaced by networks enabling the modelling of feedback loops (see Saaty [93]). The authors also discuss the possible advantages of this methodology, as a multicriteria decision analysis modelling approach, in the context of e-commerce.

Another application of AHP to socio-economic problems deals with the vendor selection of a telecommunication system and is reported in Tam and Tummala [97]. The developped model takes into account a double conflict related to multiple criteria and multiple decision makers. The authors emphasize the feasibility of this application of AHP and its potential capability to reduce the time taken to select a vendor.

Lee et al. [61] develop a specific multicriteria decision support mathematical programming model for dealing with the definition of a "hub-structure", that is the selection of a number of "nucleus cities" in the context of a rural network planning process. The model enables the consideration of several socio-economic criteria, namely economic measures of the effectiveness of telecommunication via hubs, economic activity, population, budget, health care and transportation means. The approach is a "compromise programming" technique (see Yu [103] and Zeleny [108]), a method of analysing the goal setting processes of decision makers, and uses a so called "regret function" to be minimised, i.e. a function of a number of goals set by the decision makers, aggregated through normalized weights. The authors also propose an interactive methodology for applying the model in a real decision support process and present an application example for the State of Nebraska.

In Nagurney et al. [74] a multicriteria network equilibrium modelling framework is developed for supporting decision making processes concerning the choice between physical transportation and the use of a telecommunication network. This type of modelling enables the prediction of the number of decision makers who will choose either one or the other option for a given set of criteria. This approach is applied to two problems: telecommuting versus commuting (i.e. physical transportation of the DM) and teleshopping versus shopping *in loco*. The same authors had previously addressed (in Nagurney et al. [73]) a particular problem of choice of teleshopping versus shopping by the same type of approach using a number of criteria, such as cost, time, security, or transaction safety. Also related to this topic Mokhtarian and Salomon [72] analyse and compare two approaches for dealing with the constraints in the previously mentioned type of decision problem, namely incorporating the constraints into a utility function, or using them to define the choice set. This study also addresses the importance of behavioural models in the forecasting of telecommuting adoption.

Keeney [51] discusses the issues concerning how to build a value model in the context of decision processes in telecommunication management. Special attention is paid to the identification and structuring of objectives both in qualitative and quantitative terms, including the use of utility functions.

5. Future Trends

Now we will seek to give an outline of possible research trends in some areas of network planning and design, where challenges and opportunities for multicriteria analysis may arise. For simplifying this presentation, of a prospective nature, we will take as basis, application areas (or sub-areas) identified in the previous section, although one must be aware that new problem areas are likely to emerge where multicriteria analysis may play a significant role in relation with some decision problems.

Concerning *strategic modernization planning* the following points can be explored.

I. The study and development of new types of models (concerning new planning problems and different decision processes) and of new variants of models previously presented (namely regarding objectives and constraints) is a natural trend. This having in mind the effects on the planning processes of the great turbulence of the socio-economic environment and the rapid market changes in interaction with an extremely fast technological evolution, as previously mentioned. Regarding the problem and modelling frameworks it can be said, in general, that economic, social and technological factors not only directly condition their form but also influence the perception of the decision makers *vis a vis* the problems and the associated models, namely concerning the relative significance and importance of criteria or constraints.

II. A preliminary level of decision analysis for screening distinct alternatives, seems worth considering. A particular case is the modernization planning of the access networks, associated with the trend for the introduction of broadband services (requiring in many premises optical fibre directly to the customer), a type of problem in which different technological architectures can be used. This level of analysis might be concerned with the evaluation, under different performance criteria (for example, based on up-grade cost, operator revenue, response to estimated demand and user satisfaction in different technical instances) of various technologies and associated architectures available to the operator in a given market scenario.

III. Furthermore, mathematical programming approaches can be used to help with the identification of more detailed multiattribute models, enabling a deeper analysis of the problem under study. It must be remarked that we believe in the complementary use of both types of approaches. Last but not least, we emphasise a point referred to in Section 3. In fact, modelling uncertainty requires particular attention in the future.

Other telecommunication applications with strong socio-economic implications deserve further investment in multicriteria modelling, in order to enable a more realistic evaluation of their impacts. As an example, we can refer to e-commerce and e-learning.

Routing is clearly an area where there is still a significant number of issues and challenges that can possibly be approached through multicriteria analysis, namely having in mind the multitude of variants of the basic QoS routing problems, the multiple ways in which they can be formulated, and the intrinsic nature of those problems. They involve multiple, often conflicting, criteria and requirements associated with different QoS metrics. This multiplicity of problems and possible formulations are decisively influenced by a rapidly changing network technological environment where the problems arise. Some particular trends can be pointed out in this area, according to the following points:

I. The investigation of multicriteria models for dealing with specific routing problems in multiservice networks, associated with different types of services and involving multiple QoS performance objectives and constraints. Typical application environments for possible development of such models are QoS routing in the Internet in the framework of the 'integrated service', or 'differentiated service' models, or in the context of MPLS 'constrained-based' routing, mentioned above.

II. Regarding the next generation of optical networks, often designated as intelligent optical networks, a particular type of routing problems, the route and wavelength assignment (RWA) problems (briefly characterised in Section 4.2) arise, some of which might be tackled through multicriteria approaches. Note that survivability requirements (imposing for example the provision of two paths for each optical connection, such that the second path is only activated in the event of failure of the first path) usually have to be considered in association with the RWA processes. These further complicate the problem formulations, hence adding a further dimension (reliability) to the problem analysis and solution evaluation.

III. There are other routing specific areas, already identified in the literature, where multicriteria analysis seems a promising approach. A first example, already mentioned Anandalingam and Nam [3] concerns routing problems in international networks, where several decision makers are involved (corresponding, in the example, to different national operators) and conflicting objectives may arise, for example in terms of the cost functions associated with the different decision makers. In the context of this and similar types of problems on network design and management, group decision and negotiation together with multicriteria analysis can play a very important role. Another example concerns problems of QoS negotiation involving several decision makers, namely customers of certain services and network operators, centred on the negotiation of supplied QoS levels and associated tariffs.

Operational planning certainly involves a vast number of problems some of which have already been treated, using multicriteria analysis models, as in the studies referred to in Section 4.3. It is foreseeable that in the future some other problems in this area will be prone to treatment in a multicriteria framework, especially having in mind the very rapid and multifaceted technological evolutions previously identified and their interactions with complex and fast changing economic and social trends. An example of such research challenges concerns cell partitioning and frequency allocation problems in the context of the very complex planning process of mobile cellular networks. Bourjolly et al. [23] presents an overview of the application of OR-based decision support tools in this area. In particular the authors draw attention to the fact that cell partitioning (a decision process the goal of which is to enable repeated use of the available frequencies hence increasing the network capacity) addresses two conflicting issues, namely covered area and capacity (involving, in essence, a choice between a smaller number of larger cells versus a larger number of smaller cells). As for the frequency allocation problem, it involves assigning to each cell a certain number of radio frequencies, according to some "optimality" criteria and satisfying various technical constraints. In this type of problem several objective functions can be considered such as those discussed by those authors Bourjolly et al. [23](namely the number of frequencies used, the frequency span and two types of signal interference, all to be minimised).

Acknowledgments

This chapter was partially supported by the project MONET under the FCT research grant POCTI/GES/30707/2001 and by the project "Models and Algorithms for Tackling Uncertainty in Decision Support Systems" under the FCT research grant POSI/SRI/37346/2001.

References

[1] E. Aboellela and C. Douligeris. Fuzzy generalyzed network approach for solving an optimization model for routing in B-ISDN. *Telecommunication Systems*, 12:237–263, 1999.

[2] J. Akkanen and J.K. Nurminen. Case study of the evolution of routing algorithms in a network plannig tool. *The Journal of Systems and Software*, 58(3):181–198, 2001.

[3] G. Anandalingam and K. Nam. Conflict and cooperation in designing internatioal telecommunication networks. *Journal of the Operational Research Society*, 48:600–611, 1997.

[4] Y. Aneja and K. Nair. The constrained shortest path problem. *Naval Research Logistics Quarterly*, 25:549–555, 1978.

[5] C.H. Antunes, M.J. Alves, A.L. Silva, and J. Clímaco. An integrated MOLP method base package – A guided tour of TOMMIX. *Computers and Operations Research*, 19(7):609–625, 1992.

[6] C.H. Antunes, J. Clímaco, and J. Craveirinha. A multiple objective linear programming approach to the modernization of telecommunication networks, in teletraffic and data-traffic – socioeconomic aspects. In A. Jensen and B. Iversen, editors, *Proceedings of the 13th International Teletraffic Congress – ITC 13, Copenhagen*, volume 17 of *North-Holland Studies in Telecommunications*, pages 369–374. North-Holland, Amsterdam, 1991.

[7] C.H. Antunes, J. Clímaco, and J. Craveirinha. On multicriteria decision making approaches concerning strategic new telecommunication planning. In T. Matsuda, J. Lesourne, and Y. Takahar, editors, *Proceedings of the International Conference on Economics Management and Information Technology*, volume 92, pages 95–98, Tokyo, 1992. The Japanese Society of Management Information.

[8] C.H. Antunes, J. Clímaco, J. Craveirinha, and C. Barrico. Multiple objective routing in integrated communication networks. In D. Smith and P. Key, editors, *Teletraffic Engineering in a Competitive World. Proceedings of the 16th International Teletraffic Congress – ITC 16, Edinburgh*, pages 1291–1300. Elsevier, Amsterdam, 1999.

[9] C.H. Antunes, J. Craveirinha, and J. Clímaco. A multiple criteria model for new telecommunication service planning. *European Journal of Operational Research*, 71(3):341–352, 1993.

[10] C.H. Antunes, J. Craveirinha, and J. Clímaco. A flexible MOLP approach to the modernization of telecommunication networks incorporating sensitivity analysis. In J. Labetoulle and J.W. Roberts, editors, *Proceedings of the 14th International Teletraffic Congress on Advances in Teletraffic Science and Engineering – ITC 14*, pages 1425–1434, Elsevier, Amsterdam, 1994.

[11] C.H. Antunes, J. Craveirinha, and J. Clímaco. Planning the evolution to broadband access networks – A multicriteria approach. *European Journal of Operational Research*, 109(2):530–540, 1998.

[12] G.R. Ash. *Dynamic Routing in Telecommunication Networks*. McGraw-Hill, New York, 1998.

[13] C. Assi, A. Shami, M.A. Ali, R. Kurtz, and D. Guo. Optical networking and real-time provisioning: An integrated vision for the next-generation internet. *IEEE Networks*, 15(4):36–45, 2001.

[14] C. Bana e Costa, L. Ehsslim, and I.J. Zanella. A real world MCDA application in cellular telephony systems. In T. Stewart and R.C. van der Honert, editors, *Trends in Multicriteria Decision Making*, volume 465 of *Lecture Notes in Economics and Mathematical Systems*, pages 412–426. Springer Verlag, Berlin, 1998.

[15] C. Bana e Costa and J.C. Vansnick. MACBETH – An interactive path towards the construction of cardinal value functions. *International Transactions in Operations Research*, 1(4):489–500, 1994.

[16] A. Banerjee, J. Drake, J.P. Lang, B. Turner, K. Kompella, and Y. Rekhter. Generalized multiprotocol label switching: An overview of routing and management enhancements. *IEEE Communications Magazine*, 39(1):144–149, 2001.

[17] J.F. Bard and J.J. Moore. A branch-and-bound algorithm for the bilevel linear programming problem. *SIAM Journal on Science and Statistical Computing*, 11:281–292, 1990.

[18] Baskerville Strategic Research. Global mobile forecasts to 2010. 3rd edition, April 2002.

[19] V. Belton, F. Ackermann, and I. Shepherd. Integrated support from problem structuring through to alternative evaluation using cope and visa. *Journal of Multi-Criteria Decision Analysis*, 6:115–130, 1997.

[20] D. Bertsekas and R. Gallager. *Data Networks*. Prentice-Hall, Englewood Cliffs, 2001.

[21] D. Blokh and G. Gutin. An approximate algorithm for combinatorial optimization problems with two parameters. *Australasian Journal of Combinatorics*, 14:157–164, 1996.

[22] H. Bölcskei, A.J. Paulraj, K.V.S. Hari, and R.U. Nabar. Fixed broadband wireless access: State of the art, challenges, and future directions. *IEEE Communications Magazine*, 39(1):100–107, 2001.

[23] J. Bourjolly, L. Déjoie, K.O. Dioume, and M. Lominy. Frequency allocation in cellular phone networks: An OR success story. *OR/MS Today*, 28(2):41–44, 2001.

[24] D. Bouyssou. Building criteria: A prerequisite for MCDA. In C.A. Bana e Costa, editor, *Readings in Multiple Criteria Decision Aid*, pages 58–80. Springer Verlag, Berlin, 1990.

[25] T.X. Brown. Low power wireless communication via reinforcement learning. In S.A. Solla, T.K. Leen, and K.-R. Müller, editors, *Advances in Neural Information Processing Systems*, volume 12, pages 893–899. MIT Press, Cambridge, MA, 2000.

[26] T.X. Brown, H. Tong, and S. Singh. Optimizing admission control while ensuring quality of service in multimedia networks via reinforcement learning. *Neural Information Processing Systems*, 1999.

[27] S. Chen and K. Nahrstedt. On finding multi-constrained paths. In *Proceedings of International Communications Conference '98 IEEE*, pages 874–879, New York, 1998. IEEE Communications Society.

[28] S. Chen and K. Nahrstedt. An overview of quality of service routing for next-generation high-speed networks: Problems and solutions. *IEEE Networks*, 12(6), 1998.

[29] J. Clímaco and C.H. Antunes. Implementation of a user friendly software package – A guide tour of TRIMAP. *Mathematical and Computer Modelling*, 10/11(12):1299–1309, 1989.

[30] J. Clímaco, C.H. Antunes, and M.J. Alves. From Trimap to Sommix – Building effective interactive MOLP computational tools in multicriteria decision making. In Fandel and Gal, editors, *Lecture Notes in Economics and Mathematical Systems*, volume 448, pages 285–296. Springer Verlag, Berlin, 1997.

[31] J. Clímaco, J. Craveirinha, and M. Pascoal. A bicriterion approach for routing problems in multimedia networks. *Networks*, 41(4):206–220, 2003.

[32] J. Clímaco and E.Q.V. Martins. A bicriterion shortest path algorithm. *European Journal of Operational Research*, 11:399–404, 1982.

[33] D.R. Cox and V. Ishem. *Point Processes*. Chapman and Hall, Englewood Cliffs, 1980.

[34] J. Craveirinha, L. Martins, T. Gomes, C.H. Antunes, and J. Clímaco. A new multiple objective dynamic routing method using implied cost. *Journal of Telecommunications and Information Technologies*, 3/2003:51–59, 2003. Special issue edited by the Institute of Telecommunications, Warsaw.

[35] C. Douligeris and I. Pereira. A telecommunications quality study using the analytic hierarchy process. *IEEE Journal on Select Areas in Communications*, 12(2):241–250, 1994.

[36] M. El-Sayed and J. Jaffe. A view of telecommunications network evolution. *IEEE Communications Magazine*, 40(12):74–81, 2002.

[37] S.C. Erbas. Utilizing evolutionary algorithms for multiobjective problems in traffic engineering. In W. Ben-Amur and A. Petrowski, editors, *Proceedings of the International Networks Optimisation Conference INOC 2003, Evry, October 2003*, pages 207–212. Institut National des Telecommunications, Evry, France, 2003.

[38] S.C. Erbas and C. Erbas. A multiobjective off-line routing model for MPLS networks. In J. Charzinski, R. Lehnert, and P. Tran-Gia, editors, *Proceedings of the 18th International Teletraffic Congress – ITC 18, Berlin*, pages 471–480. Elsevier, Amsterdam, 2003.

[39] S.C. Erbas and R. Mathar. An off-line traffic engineering model for MPLS networks. In B. Bakshi and B. Stiller, editors, *Proceedings of IEEE 27th Annual Conference on Local Computer Networks*, pages 166–174, Tampa, Florida, 2002. IEEE Computer Society.

[40] F. Ergun, R. Sinha, and L. Zhang. QoS routing with performance-dependent cost. In M. Sidi, I. Katzela, and Y. Shavitt, editors, *Proceedings of IEEE INFOCOM 2000, The Conference on Computer Communications, 19th Annual Meeting of the Computer and Communications Societies, Tel Aviv, March 2000*, volume 1, pages 137–146. IEEE Computer and Communications Societies, 2000.

[41] B. Fortz and M. Thorup. Internet traffic engineering by optimizing OSPF weights. In M. Sidi, I. Katzela, and Y. Shavitt, editors, *Proceedings of IEEE INFOCOM 2000, The Conference on Computer Communications, 19th Annual Meeting of the Computer and Communications Societies, Tel Aviv, March 2000*, volume 2, pages 519–528. IEEE Computer and Communications Societies, 2000.

[42] F. Glover and J.M. Mulvey. Equivalence of the 0-1 integer programming problem to discrete generalized and pure networks. *Operations Research*, 28:829–836, 2004.

[43] A. Goel, K.G. Ramakrishnan, D. Kataria, and D. Logothetis. Efficient computation of delay-sensitive routes from one source to all destinations. In B. Sengupta, F. Bauer, and D. Cavendish, editors, *Proceedings of IEEE INFOCOM 2001, The Conference on Computer Communications, 20th Annual Meeting of the Computer and Communications Societies, Anchorage, April 2001*, volume 2, pages 854–858. IEEE Computer and Communications Societies, April 2001.

[44] R. Guerin and A. Orda. Networks with advanced reservations: The routing perspective. In M. Sidi, I. Katzela, and Y. Shavitt, editors, *Proceedings of IEEE INFOCOM 2000, The Conference on Computer Communications, 19th Annual Meeting of the Computer and Communications Societies, Tel Aviv, March 2000*, volume 1, pages 26–30. IEEE Computer and Communications Societies, 2000.

[45] L. Guo and I. Matta. Search space reduction in qos routing. In *Proceedings of the 19th IEEE International Conference on Distributed Computing Systems*, volume 3, pages 142–149. IEEE Computer Society, 1999.

[46] G.Y. Handler and I. Zang. A dual algorithm for the constrained shortest path problem. *Networks*, 10:293–310, 1980.

[47] S. Harnedy. *The MPLS Primer – An Introduction to Multiprotocol Label Switching*. Prentice Hall, Englewood Cliffs, 2002.

[48] R. Hassin. Approximation schemes for the restriced shortest path problem. *Mathematics of Operations Research*, 17(1):36–42, 1992.

[49] A. Iwata, R. Izmailov, D.S. Lee, B. Sengupta, G. Ramamurthy, and H. Suzuki. ATM routing algorithms with multiple QoS requirements for multimedia internetworking. *IEICE Transactions and Communications E79-B*, 8:999–1006, 1996.

[50] J.M. Jaffe. Algorithms for finding paths multiple constraints. *Networks*, 14:95–116, 2004.

[51] R. Keeney. Modelling values for telecommunications management. *IEEE Transactions on Engineering Management*, 48(3):370–379, 2001.

[52] R.L. Keeney and H. Raiffa. *Decisions with Multiple Objectives Preferences and Value Tradeoffs*. Cambridge University Press, Cambridge, 1993.

[53] F. Kelly. Routing in circuit-switched networks: Optimization, shadow prices and decentralisation. *Applied Probability*, 20:112–144, 1988.

[54] L. Kerbache and J.M. Smith. Multi-objective routing within large scale facilities using open finite queueing network. *European Journal of Operational Research*, 121:105–123, 2000.

[55] J. Kim. A survey study on the relative importance of intranet functions: Using the analytic hierarchy, 1997. Paper presented at the International Decision Sciences Institute Meeting, Sydney, July 1997.

[56] J. Knowles, M. Oates, and D. Corne. Advanced muli-objecive evolutionary algorithms applied to two problems in telecommunications. *British Telecom Technology Journal*, 18(4):51–65, 2000.

[57] T. Korkmaz and M. Krunz. Multi-constrained optimal path selection. In B. Sengupta, F. Bauer, and D. Cavendish, editors, *Proceedings of IEEE INFOCOM 2001, The Conference on Computer Communications, 20th Annual Meeting of the Computer and Communications Societies, Anchorage, April 2001*. IEEE Computer and Communications Societies, 2001. www.ieee-infocom.org/2001/program.html.

[58] T. Korkmaz and M. Krunz. A randomized algorithm for finding a path subject to multiple QoS requirements. *Computer Networks*, 36:251–268, 2001.

[59] F. Kuipers, T. Korkmaz, M. Krunz, and P. Van Mieghem. A review of constraint-based routing algorithms. Technical report, Delft University of Technology, 2002.

[60] F. Kuipers, P. Van Mieghem, T. Korkmaz, and M. Krunz. An overview of constraint-based path selection algorithms for routing. *IEEE Communications Magazine*, 40(12):50–55, 2002.

[61] H. Lee, Y. Shi, S.M. Nazem, S. Kang, and M. Sohn. Multicriteria hub decision making for rural area telecommunication networks. *European Journal of Operational Research*, 133(3):483–495, 2001.

[62] W.C. Lee, M.G. Hluchyj, and P.A. Humblet. Routing subject to quality of service constraints in integrated communication networks. *IEEE Networks*, 9(4):46–55, 1995.

[63] G. Liu and K.G. Ramakrishnan. A* prune: An algorithm for finding k shortest paths subject to multiple constraints. In B. Sengupta, F. Bauer, and D. Cavendish, editors, *Proceedings of IEEE INFOCOM 2001, The Conference on Computer Communications, 20th Annual Meeting of the Computer and Communications Societies, Anchorage, April 2001*, pages 743–749. IEEE Computer and Communications Societies, 2001.

[64] Q. Ma and P. Steenkiste. On path selection for traffic with bandwidth guarantees. In M. Ammar and U. Shankar, editors, *Proceedings of IEEE International Conference on Network Protocols ICNP '97, Atlanta, October 1997*, pages 191–202. IEEE Computer and Communications Societies, 1997.

[65] Q. Ma and P. Steenkiste. Routing traffic with quality-of-service guarantees in integrated services networks. In I. Lesley and D. McAuley, editors, *Proceedings of 8th International Workshop on Network and Operating System Support for Digital Audio and Video (NOSSDAV'98), July 1998*, July 1998.

[66] P. Mahey and C.C. Ribeiro. Modeling modern multimedia traffic – Optimal routing for multi-service communication networks. *OR/MS TODAY*, 29(3):40–43, 2002.

[67] S. Maniatis, E.G. Nikolouzou, and I.S. Venieris. QoS issues in the converged 3G wireless and wired networks. *IEEE Communications Magazine*, 40(8):44–53, 2002.

[68] E.V. Martins, M. Pascoal, and J. Santos. Deviation algorithms for ranking shortest paths. *International Journal of Foundations of Computer Science*, 10(3):247–263, 1999.

[69] L. Martins, J. Craveirinha, J. Clímaco, and T. Gomes. Modelling and performance analysis of a new multiple objective dynamic routing method for multiexchange networks. Technical Report 11, ISSN 1645-2631, INESC-Coimbra, 2002.

[70] L. Martins, J. Craveirinha, J. Clímaco, and T. Gomes. Implementation and performance of a new multiple objective dynamic routing method for multiexchange networks. *Journal of Telecommunications and Information Technologies*, 3/2003:60–66, 2003. Special issue edited by the Institute of Telecommunications, Warsaw.

[71] P. Van Mieghem, H. De Neve, and F.A. Kuipers. Hop-by-hop quality of service routing. *Computer Networks*, 37(3-4):407–423, 2001.

[72] P.L. Mokhtarian and I. Salomon. Modeling the choice of telecommuting 3: Identifying the choice set and estimating binary choice models for technology-based alternatives. *Environment and Planning*, A, 28:1877–1894, 1996.

[73] A. Nagurney, J. Dong, and P.L. Mokhtarian. Teleshopping versus shopping: A multicriteria network equilibrium framework. *Mathematical and Computer Modeling*, 34:783–798, 2001.

[74] A. Nagurney, J. Dong, and P.L. Mokhtarian. Multicriteria network equilibrium modeling with variable weights for decision-making in the information age, with applications to telecommuting and teleshopping. *Journal of Economic Dynamics and Control*, 26(9-10):1629–1650, 2002.

[75] H. De Neve and P. Van Mieghem. A multiple quality of service routing algorithm for PNNI. *Proceedings of the IEEE ATM Workshop, Fairfax, May26-29, 1998*, pages 324–329, 1998.

[76] H. De Neve and P. Van Mieghem. TAMCRA, a tunable accuracy multiple constraints routing algorithm. *Computer Communications*, 33:667–679, 2000.

[77] J. Nurminen. Models and algorithms for network planning tool – Pratical experiences. Technical Report E14, Helsinki University of Technology, 2003.

[78] A. Orda. Routing with end-to-end QoS guarantees in broadband networks. *IEEE/ACM Transactions on Networking*, 7(3):365–374, 1999.

[79] S. Oueslti-Boulahia and E. Oubagha. An approach to routing elastic flows. In D. Smith and P. Key, editors, *Teletraffic Engineering in a Competitive World. Proceedings of the 16th International Teletraffic Congress – ITC 16, Edinburgh*, pages 1311–1320. Elsevier Science, Amsterdam, 1999.

[80] C. Pornavalai, G. Chakraborty, and N. Shiratori. QoS based routing algorithm in integrated services packet networks. In R. Miller and K. Calvert, editors, *IEEE International Conference on Network Protocols (ICNP'97)*, pages 167–174, Atlanta, Georgia, 1997. IEEE.

[81] C. Pornavalai, G. Chakraborty, and N. Shiratori. Routing with multiple QoS requirements for supporting multimedia applications. *Telecommunication Systems*, 9:357–373, 1998.

[82] M.S. Raisinghani. A balanced analytic approach to strategie electronic commerce decision: A framework of the evaluation method. In W.V. Grembergen, editor, *Information Technology Evaluation Methods and Management*, pages 185–197. Idea Group Publishing, 2001.

[83] M.S. Raisinghani and L. Schakade. Strategic evaluation of electronic commerce technologies. In N. Callaos, M. Bennamony, and J. Aguilar, editors, *Proceedings of the World Multiconference on Systemics, Cybernetics and Informations, Caracas, July 1997*, volume 2, pages 361–365, 1997.

[84] B. Rajagopalen, D. Pendarakis, D. Saha, R.S. Ramamoorthy, and K. Bala. IP over optical netwoks: Architectural aspects. *IEEE Communications Magazine*, 38(9):94–102, 2004.

[85] D.S. Reeves and H.F. Salama. A distributed algorithm for delay-constrained unicast routing. *IEEE/ACM Transactions on Networking*, 8(2):239–250, April 2000.

[86] M. Resende and C. Ribeiro. A GRASP with path-relinking for private virtual circuit routing. *Networks*, 41(3):104–114, 2003.

[87] J.C. Rodrigues, J.C.N. Clímaco, and J.R. Current. An interactive bi-objective shortest path approach: Searching for unsupported nondominated solutions. *Computers & Operations Research*, 26:789–798, 1999.

[88] J. Rosenhead. *Rational Analysis for a Problematic World – Problem Structuring Methods for Complexity and Conflict*. John Wiley & Sons, Chichester, 1999.

[89] B. Roy and D. Bouyssou. *Aide Multicritère à la Décision: Méthodes et Cas*. Economica, Paris, 1993.

[90] T.L. Saaty. *The Analytic Hiearchy Process*. McGraw-Hill, New York, 1980.

[91] T.L. Saaty. Highlights and critical point in the theory and application of the analytic hierarchy process. *European Journal of Operational Research*, 74:426–447, 1994.

[92] T.L. Saaty. How to make a decision: The analytic hierarchy process. *Interfaces*, 24(6):19–43, 1994.

[93] T.L. Saaty. *Decision Making with Dependence and Feedback The Analytic Network Process*. RWS Publications, Pittsburgh, 1996.

[94] J. Sobrinho. Algebra and algorithms for qos path computation and hop-by-hop routing in the internet. In B. Sengupta, F. Bauer, and D. Cavendish, editors, *Proceedings of IEEE INFOCOM 2001, The Conference on Computer Communications, 20th Annual Meeting of the Computer and Communications Societies, Anchorage, April 2001*. IEEE Compute and Communications Societies, 2001.

[95] D.J. Songhurst. *Charging Communication Networks – From Theory to Practice*. Elsevier, Amsterdam, 1999.

[96] R. Steuer. *Multiple Criteria Optimization: Theory Computation and Application*. John Wiley & Sons, New York, 1986.

[97] M. Tam and V.M.R. Tummala. An application of the AHP in vendor selection of a telecommunication system. *Omega – The International Journal of Management Science*, 29:171–182, 2001.

[98] S.R. Thirumalasetty and D. Medhi. MPLS traffic engineering for survivable book-ahead guaranteed services. Technical report in computer science and telecommunications, University of Missouri Kansas City, July 2001. www.estp.umkc.edu/public/papers/-dmedhi/tm-bag-te.pdf.

[99] S.R. Tiourine, C.A.J. Hurkens, and J.K. Lenstra. Local search algorithms for the radio link frequency assignment problem. *Telecommunication Systems*, 13:293–314, 2000.

[100] Z. Wang and J. Crowcroft. Quality-of-service routing for supporting multimedia applications. *IEEE Journal on Select Areas in Communications*, 14(7):1228–1234, 1996.

[101] R. Widyono. The design and evaluation of routing algorithms for real-time channels. Technical Report 94-024, University of California at Berkeley & International Computer Science Institute, June 1994.

[102] A. Wierzbicki. Multicriteria models for the placement of internet cache. In *MCDM Winter Conference (16th MCDM World Conference), Semmering*, Austria, February 2002.

[103] P.L. Yu. *Multiple-Criteria Optimization: Techniques and Extensions*. Plenum Press, New York, 1985.

[104] X. Yuan. Heuristic algorithms for multiconstrained qualiy-of-service routing. *IEEE/ACM Transactions on Networking*, 10(2), April 2002.

[105] X. Yuan and X. Liu. Heuristic algorithms for multi-constrained quality of service routing. In B. Sengupta, F. Bauer, and D. Cavendish, editors, *Proceedings of IEEE INFOCOM 2001, The Conference on Computer Communications, 20th Annual Meeting of the Computer and Communications Societies, Anchorage, April 2001*. IEEE Computer and Commxunications Societies, 2001.

[106] H. Zang, J. Jue, and B. Mukherjee. A review of routing and wavelength assignment approaches for wavelength-routed optical wdm networks. *Optical Networks Magazine*, 14(1):47–60, 2000.

[107] M.V. Zee. Quality of service routing – State of the art report. http://searchpdf.adobe.com/proxies/0/9/25/62.html, 1999. Report of Work Unit 4 of the EU "Internet Next Geenartion Project"; Ericsson 1/0362-FCP NB 102 88, 1999.

[108] M. Zeleny. A concept of compromise solutions and the method of the displaced ideal. *Computers and Operations Research*, 1:479–496, 1974.

[109] B. Zhu. Routing and wavelengh assignment in all-optical networks by multi-objective genetic algorithms. Technical report, University of California, Berkeley, November 2000.

Chapter 23

MULTIPLE CRITERIA DECISION ANALYSIS AND SUSTAINABLE DEVELOPMENT

Giuseppe Munda
Universitat Autonoma de Barcelona
Department of Economics and Economic History, Edifici B
and
Institute for Environmental Sciences and Technologies
08193, Bellaterra, Barcelona, Spain
giuseppe.munda@uab.es

Abstract Sustainable development is a multidimensional concept, including socio-economic, ecological, technical and ethical perspectives. In making sustainability policies operational, basic questions to be answered are sustainability of what and whom? As a consequence, sustainability issues are characterised by a high degree of conflict. The main objective of this Chapter is to show that multiple-criteria decision analysis is an adequate approach for dealing with sustainability conflicts at both micro and macro levels of analysis. To achieve this objective, lessons, learned from both theoretical arguments and empirical experience, are reviewed. Guidelines of "good practice" are suggested too.

Keywords: Sustainable development, economics, complex systems, incommensurability, social choice, social multi-criteria evaluation.

1. The Concept of Sustainable Development

In the eighties, the awareness of actual and potential conflicts between economic growth and the environment led to the concept of "*sustainable development*". Since then, all governments have declared, and still claim, their willingness to pursue economic growth under the flag of sustainable development although often development and sustainability are contradictory terms. The concept of sustainable development has wide appeal, partly because it does not set economic growth and environmental preservation in sharp opposition. Rather, sustainable development carries the ideal of a harmonisation or *simultaneous realisation* of economic growth and environmental concerns. For example Barbier [6, p.103] writes that sustainable development implies: "*to maximise simultaneously*[1] *the biological system goals (genetic diversity, resilience, biological productivity), economic system goals (satisfaction of basic needs, enhancement of equity, increasing useful goods and services), and social system goals (cultural diversity, institutional sustainability, social justice, participation)*". This definition correctly points out that sustainable development is a multidimensional concept, but as our everyday life teaches us, it is generally impossible to maximise different objectives at the same time, and as formalised by multi-criteria decision analysis, compromise solutions must be found.

Let us try to clarify some fundamental points of the concept of "*sustainable development*". In economics by "development" is meant "*the set of changes in the economical, social, institutional and political structure needed to implement the transition from a pre-capitalistic economy based on agriculture, to an industrial capitalistic economy*" [15]. Such a definition of development has two main characteristics:

- The changes needed are not only quantitative (like the growth of gross domestic product), but qualitative too (social, institutional and political).
- There is only a possible model of development, i.e. the one of western industrialised countries. This implies that the concept of development is viewed as a process of cultural fusion toward the best knowledge, the best set of values, the best organisation and the best set of technologies.

Adding the term "sustainable" to the "set of changes" (the first point) means adding an ethical dimension to development. The issue of distributional equity, both within the same generation (intra-generational equity, e.g. the North-South conflict) and between different generations (inter-generational equity) becomes crucial [61]. Going further, a legitimate question could be raised [2]: sustainable development of what and whom? Norgaard [71, p. 11] writes: "*consumers want consumption sustained, workers want jobs sustained. Capitalists and socialists have their 'isms', while aristocrats and technocrats have their 'cracies'*".

Martinez-Alier and O'Connor [44] have proposed the concept of ecological distribution to synthesise sustainability conflicts. The concept of *ecological distribution* refers to the social, spatial, and temporal asymmetries or inequalities in the use by humans of environmental resources and services. Thus, the territorial asymmetries between SO_2 emissions and the burdens of acid rain are an example of *spatial ecological distribution.* The inter-generational inequalities between the benefits of nuclear energy and the burdens of radioactive waste are an example of *temporal ecological distribution.* In the USA, "*environmental racism*", meaning locating polluting industries or toxic waste disposal sites in areas where poor people live, is an example of *social ecological distribution.* We can then conclude that sustainability management and planning is essentially a conflict analysis.

The second characteristic of the term "*development*" refers to the western industrialized production system as symbol of any successful development process. However, serious environmental problem may stem from this vision. For example, according to actual social values in western countries, to have a car per two/three persons could be considered a reasonable objective in less developed countries. This would imply a number of cars ten times greater than the existent one, with possible consequences on global warming, reserves of petroleum, loss of agricultural land and noise. The contradiction between the terms "development" and "sustainable" may not be reconcilable unless other models of development are considered.

This is proposed by the so-called co-evolutionary paradigm. According to this view of social evolution, borrowed from biology [25], there is a constant and active interaction of the organisms with their environment. Organisms are not simply the results but they are also the causes of their own environments [37, 71]. Economic development can be viewed as a process of adaptation to a changing environment while itself being a source of environmental change. In real world societies, "*people survive to a large extent as members of groups. Group success depends on culture: the system of values, beliefs, artefacts, and art forms which sustain social organisation and rationalise action. Values and beliefs which fit the ecosystem survive and multiply; less fit ones eventually disappear. And thus cultural traits are selected much like genetic traits. At the same time, cultural values and beliefs influence how people interact with their ecosystem and apply selective pressure on species. Not only have people and their environment coevolved, but social systems and environmental systems have coevolved*" [71, p. 41]. From the co-evolutionary paradigm the following lessons can be learned:

1 A priori, different models of co-evolution are possible, and then no unique optimal development path exists.

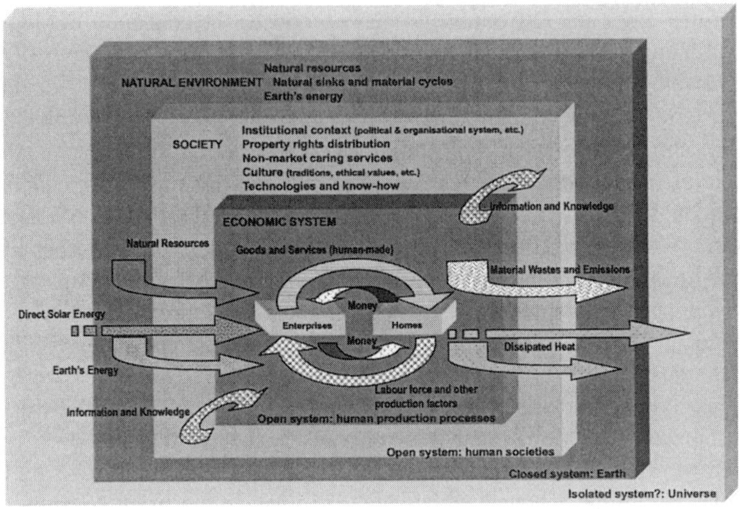

Figure 23.1. A systemic vision of sustainability issues.

2 The spatial dimension is a key feature of sustainable development; as a consequence the respect of cultural diversity is of a fundamental importance. In environmental management local knowledge and expertise (being the result of a long co-evolutionary process) sometimes are more useful than experts' opinions. Social participation is then essential for successful sustainability policies.

Taking sustainability seriously into account creates a need for the inclusion of the physical appraisal of the environmental impacts on the socio-economic system too. As shown in Figure 23.1, systemic approaches to sustainability issues consider the relationships between three systems: the *economic system*, the human system and the natural system [74]. The economic system includes the economic activities of humans, such as production, exchange and consumption. Given the scarcity phenomenon, such a system is efficiency oriented. The *human system* comprises all activities of human beings on our planet. It includes the spheres of biological human elements, of inspiration, of aesthetics, of social conflict, and of morality which constitute the frame of human life. Since it is clear that the economic system does not constitute the entire human system, one may assume that the economic system is a subsystem of the human system. Finally, the *natural system* includes both the human system and the economic system.

The previous discussion can be synthesised by using the philosophical concept of weak comparability [53, 54, 73]. *Weak comparability* implies *incom-*

mensurability i.e. there is an irreducible value conflict when deciding what common comparative term should be used to represent a real-world system and eventually implement sustainability policies. It is possible to further distinguish the concepts of social incommensurability and technical incommensurability [64].

Social incommensurability refers to the existence of a multiplicity of legitimate values in society, and to deal with it, there is a need to consider the public participation issue. Any social decision problem is characterised by conflicts between competing values and interests and different groups and communities that represent them. In sustainability policies, biodiversity goals, landscape objectives, the direct services of different environments as resources and sinks, the historical and cultural meanings that places have for communities, the recreational options environments provide are a source of conflict. Choosing any particular operational definition for *value* and its corresponding valuation technique involves making a decision about what is important and real. Distributional issues play a central role. Any policy option always implies winners and losers, thus it is important to check if a policy option seems preferable just because some dimensions (e.g. the environmental) or some social groups (e.g. the lower income groups) are not taken into account.

As a tool for conflict management, multi-criteria evaluation has demonstrated its usefulness in many sustainability policy and management problems (see e.g. [10, 11, 40, 45, 46, 59, 66, 70, 79, 81]). The main point of force is the fact that the use of various evaluation criteria has a direct translation in terms of plurality of values used in the evaluation exercise. From this point of view, multiple-criteria decision analysis can be considered as a tool for implementing political democracy.

When dealing with sustainability issues neither an economic reductionism nor an ecological one is possible. Since in general, economic sustainability has an ecological cost and ecological sustainability has an economic cost, an integrative framework such as multi-criteria evaluation is needed for tackling sustainability issues properly. *Technical incommensurability* comes from the multidimensional nature of sustainability issues. One should note that the construction of a descriptive model of a real-world system depends on very strong assumptions about

1 the *purpose* of this construction, e.g. to evaluate the sustainability of a given city,

2 the *scale* of analysis, e.g. a block inside a city, the administrative unit constituting a Commune or the whole metropolitan area and

3 the set of dimensions, objectives and criteria used for the evaluation process.

A reductionist approach for building a descriptive model can be defined as the use of just *one measurable indicator* (e.g. the monetary city product per person), *one dimension* (e.g. economic), *one scale of analysis* (e.g. the Commune), *one objective* (e.g. the maximisation of economic efficiency) and *one time horizon*. If one wants to avoid reductionism, there is a clear need to take into account incommensurable dimensions using different scientific languages coming from different legitimate representations of the same system [36]. This is what Neurath [69] called the need for an *"orchestration of sciences"*.

The use of a multi-criteria framework is a very efficient tool to implement a multi/inter-disciplinary approach. When experts involved have various backgrounds in the beginning, the communication process is always very difficult; however it is astonishing to realize that when a multi-criterion framework is used, *immediately a common language is created*. This virtue of multi-criterion approaches has been corroborated in a great number of real-world case studies tackled by means of a variety of methods (see e.g., Beinat [9] who mainly uses MAUT approaches; Janssen [45] who builds on the DEFINITE software; Maystre et al. [55] building on ELECTRE methods; Moreno-Jimenez et al. [56] using AHP; Espelta et al. [26] by means of NAIADE; Stewart and Joubert [89] who use SMART). In terms of inter-disciplinarity, the issue is to find agreement on the set of criteria to be used; in terms of multi-disciplinarity, the issue is to propose and compute an appropriate criterion score. The efficiency of the interaction process can greatly increase and its effectiveness too[2].

In the rest of this Chapter, I will first analyse the role of multi-criteria decision analysis at a macroeconomic level, in particular with reference to the problem of construction and aggregation of sustainability assessment indicators and indexes. Then, I will discuss the use of multi-criteria techniques at a project level, for sustainability management and planning. At both levels, particular emphasis will be put on issues such as the role of problem structuring, the quality of the social process and the meaning of mathematical properties.

2. Measuring Sustainability: The Issue of Sustainability Assessment Indexes

From an economic point of view, traditionally Gross Domestic Product (GDP) has been considered as the best performance indicator for measuring national economy and welfare. But if resource depletion and degradation are factored into economic trends, what emerges is a radically different picture from the one depicted by conventional methods. In environmental terms, the GDP measure is plainly defective because:

1 no account is taken of environmental destruction or degradation;

2 natural resources as such are valued at zero;

3 repair and remedial expenditure such as pollution abatement measures, health care, etc., are counted as positive contribution to GDP inasmuch as they involve expenditures of economic goods and services.

The purpose of "green accounting" is to provide information on the sustainability of the economy but there is no settled doctrine on how to combine different and sometimes contradictory indicators and indexes in a way immediately useful for policy (in the sense that GDP or other macroeconomic statistics have been useful for policy) [30, 31]. The expression "*Taking nature into account*" (much used both in the UN system and in the European Union) hides the tension between money valuation, and appraisal through physical indicators and indexes (which themselves might show contradictory trends). So far, the elementary question of whether the European economy is moving towards sustainability or away from sustainability cannot be answered with consensus on the indicators and the integrative framework to be used (see e.g. [7, 16, 28, 42, 43, 61, 68, 75]).

A point of scientific controversy present in the contemporary debate is on the use of monetary or physical indexes. Examples of monetary indexes are [19] ISEW (Index of Sustainable Economic Welfare), [76] Weak Sustainability Index, the so-called El Serafy approach [102]. Examples of physical indexes are HANPP (Human Appropriation of Net Primary Production [97], the Ecological Footprint [99], MIPS (Material Input Per unit of Service) [87].

Although these approaches may look different, they all have some common characteristics:

1 The subcomponents needed for the building the aggregate index are *ad hoc*. No clear justification is given why e.g. diet enters in the computation of the ecological footprint and the generation of waste does not.

2 All the indexes are based on the assumptions that a common measurement rod needs to be established for aggregation purposes (money, energy, space, and so on). This creates the need of making very strong assumptions on conversion coefficients to be used and on compensability allowed (i.e. till which point better economical performances may cause environmental destruction or social exclusion?). The mathematical aggregation convention behind an index thus needs an explicit and well thought formulation.

3 The policy objective is often not clear. Inter-country or inter-city comparisons are a different policy objective than managing a particular country or city sustainability. Moreover, aggregate indexes are somewhat confusing, if one wishes to derive policy suggestions. For example, by looking at ISEW, we could know that indeed a country has a worst sustainability performance than the one pictured by standard GDP, but so what? ISEW

being so aggregated does not supply any clear information of the cause of this bad performance and thus is useless for policy-making (while conventional GDP is at least giving clear information on the economic performance). The same applies to the ecological footprint, which sometimes can even give misleading policy suggestions (giving that diet is used, a more energy intensive agriculture might reduce the ecological footprint of e.g. a city, but in reality its environmental performance would be much worst!) or to the weak sustainability index (which is nothing but the classical golden rule of growth theory, where environmental physical destruction is never considered – above all if it is externalised outside the national borders).

4 All these approaches belong to the more general family of composite indicators (see Table 23.1), and as a consequence, the assumptions used for their construction are common to them all.

Let's discuss this fourth point more in depth. Composite indicators[3] are very common in fields such as economic and business statistics and a variety of policy domains such as industrial competitiveness, sustainable development, globalisation and innovation. The proliferation of this kind of indicators is a clear symptom of their political importance and operational relevance in decision-making. From a mathematical point of view, a composite indicator is a weighted linear aggregation rule applied to a set of variables (in a multi-criteria terminology variables can be considered criterion scores). A typical composite indicator, I, is built as follows [72, p. 5]:

$$I = \sum_{i=1}^{N} w_i x_i \tag{23.1}$$

where x_i is a normalised variable and w_i a weight attached to x_i, with $\sum_{i=1}^{N} w_i = 1$ and $0 \leq w_i \leq 1, i = 1, \ldots, N$. The main technical (i.e., without considering how variables have been selected) steps needed for its construction are two:

1 standardisation of the variables to allow comparison without scale effect,

2 weighted summation of these variables.

The standardisation step is a very delicate one. Main sources of a somewhat arbitrary assessment here are:

- *Normalisation technique* used for the different measurement units dealt with.
- *Scale adjustment* used, for example population or GDP of each country considered.

- *Common measurement* unit used (money, energy, space and so on).

Let's first discuss the issue of linear aggregation of the variables chosen. As it is well known, the aggregation of several variables implies taking a position on the fundamental issue of compensability. The use of weights with intensity of preference originates compensatory aggregation conventions and gives the meaning of trade-offs to the weights. On the contrary, the use of weights with ordinal variable scores originates non-compensatory aggregation procedures and gives the weights the meaning of importance coefficients [77, 80, 95] (see also Chapters 4 and 7 of this book).

Now the question arises: in their standard use weights in composite indicators are trade-offs or importance coefficients? "*Variables which are aggregated in a composite indicator have first to be weighted – all variables may be given equal weights or they may be given differing weights which reflect the significance, reliability or other characteristics of the underlying data. The weights given to different variables heavily influence the outcomes of the composite indicator. The rank of a country on a given scale can easily change with alternative weighting systems. ... Greater weight should be given to components which are considered to be more significant in the context of the particular composite indicator*". [72, p. 10]. The concept of a weight used by OECD can be then classified as symmetrical importance, that is "*... if we have two non-equal numbers to construct a vector in R2, then it is preferable to place the greatest number in the position corresponding to the most important criterion.*" [77, p. 241].

Clearly, the mathematical convention underlying the additive aggregation model is a completely compensatory one. This means that in the weighted summation case, the substitution rates are equal to the weights of the variables up to a multiplicative coefficient. As a consequence, the estimation of weights is equivalent to that of substitution rates: the questions to be asked are in terms of "*gain with respect to one variable allowing to compensate loss with respect to another*" and NOT in terms of "*symmetrical importance*" of variables [14]. As a consequence in composite indicators, a theoretical inconsistency exists between the way weights are actually used and what their real theoretical meaning is[4].

It is obvious that the aggregation convention used for composite indicators deal with the classical conflictual situation tackled in multi-criteria evaluation. Thus, the use of a multi-criterion framework for composite indicators in general and for sustainability indexes in particular is relevant and desirable [27, 31, 62, 93]. However, as made clear in this book, the so-called "*multi-criterion problem*" can be solved by means of a variety of mathematical approaches, all of them correct. This situation is due to Arrow's impossibility theorem [3], which proves that it is impossible to develop a "*perfect*" multi-criterion aggregation convention. This implies that it is desirable to have mathematical algorithms that may be recommended on some theoretical and empirical grounds. To deal with this problem, two main approaches can be distinguished.

Table 23.1. Example of composite indicators. (Source: OECD; JRC, 2002, cited in [72, p. 4])

Area	*Name of Composite Indicator*
Economy	Composite of Leading Indicators (OECD) OECD International Regulation database (OECD) Economic Sentiment Indicator (EC) Internal Market Index (EC) Business Climate Indicator (EC)
Environment	Environmental Sustainability Index (World Economic Forum) Wellbeing Index (Prescott-Allen) Sustainable Development Index (UN) Synthetic Environmental Indices (Isla M.) Eco-Indicator 99 (Pre Consultants) Concern about Environmental Problems (Parker) Index of Environmental Friendliness (Puolamaa) Environmental Policy Performance Index (Adriaanse)
Globalisation	Global Competitiveness Report (World Economic Forum) Transnationality Index (UNCTAD) Globalisation Index (A.T. Kearny) Globalisation Index (World Markets Research Centre)
Society	Human Development Index (UN) Overall Health Attainment (WHO) National Health Care Systems Performance (King's Fund) Relative Intensity of Regional Problems (EC) Employment Index (Storrie and Bjurek)
Innovation / Technology	Summary Innovation Index (EC) Networked Readiness Index (CID) National Innovation Capacity Index (Porter and Stern) Investment in Knowledge-Based Economy (EC) Performance in Knowledge-Based Economy (EC) Technology Achievement Index (UN) General Indicator of Science and Technology (NISTEP) Information and Communications Technologies Index (Fagerberg) Success of Software Process Improvement (Emam)

1 The attempt of looking for a complete set of formal axioms that can be attributed to a specific method (e.g., [4, 96]).

2 The attempt to check under which specific circumstances each method could be more useful than others, i.e. the search of the right method for the right problem (e.g., see [39] for a general approach, and [86] for a discussion in the context of environmental problems).

Next Section gives an example of the first approach in the framework of sustainability composite indicators. Section 6 will deal with the second approach in the framework of multi-criteria evaluation of sustainability policies at a micro-level.

3. A Defensible Axiomatic Setting for Sustainability Composite Indicators

As discussed in the previous Section, in the framework of sustainability composite indicators there is a need for a theoretical guarantee that weights are used with the meaning of " "*symmetrical importance*". As a consequence, complete compensability should be avoided. This implies that *variables have to be used with an ordinal meaning*. This is not a problem since no loss of information is implied [4]. Moreover, given that often the measurement of variables is rough, it seems even desirable to use indicator scores with an ordinal meaning. As it is well-known in social choice literature, *desirable ranking procedures using ordinal information are always of a Condorcet type* [4, 57]. A problem inherent to this family of algorithm is the presence of cycles. The probability $\pi(N, M)$ of obtain a cycle with N countries (regions, cities, etc.) and M individual indicators increases with N as well as the number of indicators. With many countries and individual indicators, cycles occur with an extremely high frequency. *As a consequence, the ranking procedure used has to deal with the cycle issue properly.*

Let's then discuss the cycle issue. A cycle breaking rule normally needs some arbitrary choice such as to delete the cycle with the lowest support. Now the question is: Is it possible to tackle the cycle issue in a more general way? Condorcet himself was aware of the problem of cycles in his approach; he built examples to explain it and he got close to find a consistent rule able to rank any number of alternatives when cycles are present. However, attempts to fully understand this part of Condorcet's voting theory came to a conclusions like "*...the general rules for the case of any number of candidates as given by Condorcet are stated so briefly as to be hardly intelligible ... and as no examples are given it is quite hopeless to find out what Condorcet meant*" (E.J. Nanson as quoted in [12, p. 175]. Or "*The obscurity and self-contradiction are without any parallel, so far as our experience of mathematical works extends ... no amount of examples can convey an adequate impression of the evils*" [91, p. 352] as cited in [100, p. 1234].

Attempts of clarifying, fully understanding and axiomatizing Condorcet's approach for solving cycles have been mainly done by Kemeny [49], who made the first intelligible description of the Condorcet approach, and by Young and Levenglick [101], who gave the clearest exposition and a complete axiomatisation. For this reason we can call this approach the Condorcet-Kemeny-Young-Levenglick (henceforth C-K-Y-L) ranking procedure.

Arrow and Raynaud [4, p. 77] also arrive at the conclusion that the highest feasible ambition for an aggregation algorithm building a multi-criterion ranking is to be Condorcet. These authors discard what they call the Kemeny's method, on the grounds that preference reversal phenomena may occur inside this approach [4, p. 96]. However, although the so-called Arrow-Raynaud's method does not present rank reversal, it is not applicable if cycles exist. Since in the context where composite indicators are built, cycles are very probable to occur, here the only solution is to choose the C-K-Y-L ranking procedure, thus accepting that rank reversals might appear[5]. The acceptance of rank reversals phenomena implies that the famous axiom of independence of irrelevant alternatives of Arrow's theorem is not respected. Anyway, Young [100, p. 1241] claims that the C-K-Y-L ranking procedure is the "*only plausible ranking procedure that is locally stable*". Where local stability means that the ranking of alternatives does not change if only an interval of the full ranking is considered.

The adaptation of C-K-Y-L ranking procedure to the case of composite indicators is very simple. The maximum likelihood ranking of countries (regions, cities, etc.) is the ranking supported by the maximum number of individual indicators for each pair-wise comparison, summed over all pairs of countries considered.

Formally, a simple ranking algorithm of sustainability composite indicators, based on these concepts, can be the following (for more details and formal proofs see [65]).

Given a set of individual indicators $G = \{g_m\}, m = 1, 2, ..., M$, and a finite set $A = \{a_n\}, n = 1, 2, ..., N$ of countries (cities or regions), let's assume that the evaluation of each country an with respect to an individual indicator g_m (i.e. the indicator score or variable) is based on an *interval* or *ratio* scale of measurement. For simplicity of exposition, let's assume that a higher value of an individual indicator is preferred to a lower one (the higher, the better), that is:

$$\begin{cases} a_j P a_k \Leftrightarrow g_m(a_j) > g_m(a_k) \\ a_j I a_k \Leftrightarrow g_m(a_j) = g_m(a_k) \end{cases} \tag{23.2}$$

where, P and I indicate a preference and an indifference relation respectively, both fulfilling the transitive property.

Let's also assume the existence of a set of individual indicator weights derived as importance coefficients. The mathematical problem to be dealt with is then how to use this available information to rank in a complete pre-order (i.e. without any incomparability relation) all the countries from the best to the worst one.

The mathematical aggregation convention can be divided into two main steps:

1 Pair-wise comparison of countries according to the whole set of individual indicators used.

2 Ranking of countries in a complete pre-order.

For carrying out the pair-wise comparison of countries the following axiomatic system is needed (adapted from [4, p. 81-82]).

Axiom 1: Diversity. Each individual indicator is a total order on the finite set A of countries to be ranked, and there is no restriction on the individual indicators; they can be any total order on A.

Axiom 2: Symmetry. Since individual indicators have incommensurable scales, the only preference information they provide is the ordinal pair-wise preferences they contain[6].

Axiom 3: Positive Responsiveness. The degree of preference between two countries a and b is a strictly increasing function of the number and weights of individual indicators that rank a before b[7].

Thanks to these three axioms a $N \times N$ matrix, E, called *outranking matrix* [4, 83] can be built. Any generic element of E: e_{jk}, $j \neq k$ is the result of the pair-wise comparison, according to all the M individual indicators, between countries j and k. Such a global pair-wise comparison is obtained by means of equation (23.3).

$$e_{jk} = \sum_{m=1}^{M} \left(w_m(P_{jk}) + \frac{1}{2} w_m(I_{jk}) \right) \tag{23.3}$$

where $w_m(P_{jk})$ and $w_m(I_{jk})$ are the weights of individual indicators presenting a preference and an indifference relation respectively. It clearly holds

$$e_{jk} + e_{kj} = 1 \tag{23.4}$$

All the $N(N-1)$ pair-wise comparisons compose the outranking matrix E. Call R the set of all $N!$ possible complete rankings of alternatives, $R = \{rs\}$, $s = 1, 2, ..., N!$. For each r_s, compute the corresponding score φ_s as the summation of e_{jk} over all the

$$\binom{N}{2}$$

pairs j, k of alternatives, i.e.

$$\varphi_s = \sum e_{jk} \tag{23.5}$$

where $j \neq k$, $s = 1, 2, \ldots, N!$ and $e_{jk} \in r_s$.

The final ranking (r^*) is the one which maximises equation (23.6), which is:

$$r^* \Leftrightarrow \varphi_* = \max \sum e_{jk} \tag{23.6}$$

where $e_{jk} \in R$.

Other formal properties of the C-K-Y-L ranking procedure are the following [101]:

- *Neutrality*: it does not depend on the name of any country, all countries are equally treated.
- *Unanimity* (sometimes called *Pareto Optimality*): if all individual indicators prefer country a to country b than b should not be chosen.
- *Monotonicity*: if country a is chosen in any pair-wise comparison and only the individual indicator scores (i.e. the variables) of a are improved, then a should be still the winning country.
- *Reinforcement*: if the set A of countries is ranked by 2 subsets G_1 and G_2 of the individual indicator set G, such that the ranking is the same for both G_1 and G_2, then $G_1 \cup G_2 = G$ should still supply the same ranking. This general consistency requirement is very important in the framework of composite indicators, since one may wish to apply the individual indicators belonging to each single dimension first and then pool them in the general model (see [62] for an example).

At this point a question arises: does the application of a formally correct mathematical aggregation procedure always guarantee the quality of the results obtained? This problem is tackled in the next section.

4. Warning! Not Always Rankings Have to Be Trusted ...

Let's now take into consideration an illustrative example regarding 4 cities, 2 belonging to highly industrialized Countries (Amsterdam and New York) and 2 belonging to transitional economies (Budapest and Moscow) [63]. The indicators used are typical of the literature on urban sustainability (see e.g. [8] or the Urban Indicator Programme). The profiles (i.e. the score of each city according to each indicator) of these 4 cities are the ones described in Figure 23.2.

Several techniques can be used to standardise variables [72, 85]. However, although each normalisation technique entails different absolute values, the ranking provided remains constant. In our example, the "*distance from the best and worst performers*" technique is applied, where positioning is in relation to the global maximum and minimum and the index takes values between 0 (laggard) and 100 (leader):

Matrix type Impact Case Study

Criteria \ Alternatives	Budapest	Moscow	Amsterdam	New York
Houses owned (%)	50.5	40.2	2.2	10.3
Residential density (pers. /hectare)	123.3	225.2	152.1	72
Use of private car (%)	31.1	10	60	32.5
Mean travel time to work (minutes)	40	62	22	36.5
Solid waste generated per capita (t/year)	0.2	0.29	0.4	0.61
City product per person (US$/year)	4750	5100	28251	30952
Income disparity (Q5/Q1)	9.19	7.61	5.25	14.81
Households below poverty line (%)	36.6	15	20.5	16.3
Crime rate per 1000 (theft)	39.4	4.3	144.05	56.7

Figure 23.2. Impact matrix for the 4 chosen cities according to the selected indicators.

$$100 \times \left(\frac{\text{actual value} - \text{minimum value}}{\text{maximum value} - \text{minimum value}}\right) \qquad (23.7)$$

By applying equation (23.7) to the values contained in Figure 23.2, the results presented in Table 23.2 are obtained. By applying equation (23.1) to the values contained in Table 23.2, the following results are obtained:

Budapest = 512.986

Moscow = 533.373

Amsterdam = 463.169

New York = 492.052

Thus the final ranking presents Amsterdam in the bottom position (worst than all the other cities considered), Moscow is in the top position, Budapest ranks second and New York ranks third. As a first reaction one might think that these somewhat surprising results are due to the use of the linear aggregation rule. Let's then apply the algorithm illustrated from equation (23.2) to equation (23.6) to the impact matrix shown in Figure 23.2. The outranking matrix E is the one shown in Table 23.3.

The 24 possible rankings and the corresponding φ_s scores are the shown in Table 23.4, where A is Budapest, B is Moscow, C is Amsterdam and D is New York.

Also in this case Moscow is clearly in the top position. New York is surely better than Amsterdam. The position of Budapest with respect to both New York and Amsterdam is not well defined.

Let's look at Figure 23.2 again. The 9 indicators used seem reasonable; they indeed belong to three dimensions, i.e. economical, social and environmental,

Table 23.2. Normalised impact matrix.

100	78.674	0	16.770
66.515	0	47.72	100
57.8	100	0	55
55	0	100	63.75
100	78.05	51.22	0
0	1.335	89.691	100
58.787	75.314	100	0
0	100	74.538	93.982
74.884	100	0	62.505

Table 23.3. Outranking matrix of the 4 cities according to the 9 indicators.

	Budapest	Moscow	Amsterdam	New York
Budapest	0	4	4	5
Moscow	5	0	5	6
Amsterdam	5	4	0	3
New York	4	3	6	0

Table 23.4. Possible ranking.

B	A	D	C	31	C	B	D	A	27
B	D	C	A	31	D	B	A	C	27
A	B	D	C	30	D	C	B	A	27
B	D	A	C	30	A	C	B	D	26
B	C	A	D	29	A	D	C	B	26
B	A	C	D	28	D	A	B	C	26
B	C	D	A	28	D	C	A	B	26
C	B	A	D	28	D	A	C	B	25
D	B	C	A	28	C	A	D	B	24
A	B	C	D	27	C	D	B	A	24
A	D	B	C	27	A	C	D	B	23
C	A	B	D	27	C	D	A	B	23

considered essential in any sustainability assessment. Let's then try to understand to which dimension each single indicator belongs. Roughly the following classification may be made:

Economic dimension

1. City product per person

Environmental dimension

2. Use of private car
3. Solid waste generated per capita

Social dimension

4. Houses owned
5. Residential density
6. Mean travel time to work
7. Income disparity
8. Households below poverty line
9. Crime rate

Clearly the social dimension is receiving implicitly a much bigger weight than any other dimension (considering that 6 indicators over 9 belong to this dimension). A reasonable decision might be to consider the three dimensions equally important. This would imply to give the same weight to each dimension considered and finally to split this weight among the indicators. That is, each dimension has a weight of 0.333; then the economic indicator has a weight of 0.333, the 2 environmental indicators have a weight of 0.1666 each, and each one of the 6 social indicators receives a weight equal to 0.0555. As one can see, if dimensions are considered, weighting indicators by means of importance coefficients is crucial.

Let's now see if this weighting exercise provokes any change in the final ranking. The new outranking matrix is the one presented in Table 23.5.

Table 23.5. Weighted outranking matrix.

	Budapest	Moscow	Amsterdam	New York
Budapest	0.0	0.3	0.4	0.4
Moscow	0.7	0.0	0.5	0.6
Amsterdam	0.6	0.5	0.0	0.3
New York	0.6	0.4	0.7	0.0

The 24 possible rankings and the new corresponding scores φ_s are shown in Table 23.6 (where A is Budapest, B is Moscow, C is Amsterdam and D is New York).

As one can see, Moscow is still on the top position, but this time Budapest is on the bottom one. New York scores again better than Amsterdam.

Table 23.6. Possible ranking with new scores.

B	D	C	A	3,6	B	C	A	D	2,9
D	B	C	A	3,5	C	B	A	D	2,9
D	C	B	A	3,5	A	B	D	C	2,9
B	D	A	C	3,5	B	A	C	D	2,8
D	B	A	C	3,4	A	D	B	C	2,8
B	A	D	C	3,3	A	D	C	B	2,8
B	C	D	A	3,2	C	D	A	B	2,7
C	B	D	A	3,2	C	A	B	D	2,6
D	C	A	B	3,2	C	A	D	B	2,5
C	D	B	A	3,1	A	B	C	D	2,5
D	A	B	C	3,1	A	C	B	D	2,5
D	A	C	B	3,1	A	C	D	B	2,4

Concluding, we can state that an advantage of this algorithm is to highlight the fact that rankings are not always robust, even if no parameter is changed. This type of lack of robustness is completely ignored by the linear aggregation rule. Moreover, the use of weights as importance coefficients can change the problem modelling significantly. However one has to note that the improvement of the mathematical aggregation procedure does not change the results spectacularly. The structuring process, and in this case above all, the input information used for the indicator scores determine clearly the ranking. *Garbage in, garbage out* phenomena are almost impossible to avoid.

At this point a general question needs to be answered: *From where are multi-criteria results coming from and what they mean?* The results obtained depend on:

1 *quality of the information available* (in our case for example the data concerning Amsterdam on the use of private cars and on criminality are suspiciously high, while criminality in Moscow or residential density in New York are suspiciously low),

2 *indicators chosen* (i.e. which representation of reality we are using, e.g. whose interests we are taken into account),

3 *direction of each indicator* (i.e. the bigger the better or vice versa, e.g. in our example, it has been used the principle that house owners should be maximized, but this could be quite disputable and culturally dependent),

4 *relative importance of these indicators* (indicated by the weighting factor attached),

5 *ranking method used.*

All these uncertainties have to be taken into account when we state that an evaluation is made. Points from 1 to 4 clearly concern the way a given assessment exercise is structured; this implies that the quality of the aggregation convention is an important step to guarantee consistency between the assumptions used and the ranking obtained; but *the overall quality of a multi-criteria study depends crucially on the way this mathematical model is embedded in the social, political and technical structuring process*. This is the reason why in multi-criteria decision aid (MCDA) it is claimed that what is really important is the "decision process" and not the final solution [82, 83].

However, while it is clear what this means in terms of single-person decisions, how can we deal with the issue of a social process? To answer this question will be the aim of the next Section.

5. The Issue of the "Quality of the Social Decision Processes"

In empirical evaluations of public projects and public provided goods, multi-criteria decision analysis seems to be an adequate policy tool since it allows taking into account a wide range of assessment criteria (e.g. environmental impact, distributional equity, and so on) and not simply profit maximisation, as a private economic agent would do. However, the management of a policy process involves many layers and kinds of decisions, and requires the construction of a *dialogue process* among many stakeholders, individual and collective, formal and informal, local and not.

In general, these concerns have not been considered very relevant by scientific research in the past (where the basic implicit assumption was that time was an infinite resource). On the other hand, the new nature of the policy problems faced in this third millennium (e.g., the mad cow, genetic modified organisms,...), implies that very often when using science for policy-making, long term consequences may exist and scientists and policy-makers are confronting issues where, "*facts are uncertain, values in dispute, stakes high and decisions urgent*" [33, 34]. In this case, scientists cannot provide any useful input without interacting with the rest of society and the rest of the society cannot perform any sound decision making without interacting with the scientists. That is, the question on "*how to improve the quality of a social decision process*" must be put, quite quickly, on the agenda of "scientists", "decision makers" and indeed the whole society.

An outcome of this discussion is that the political and social framework must find a place in multi-criteria decision analysis. An effective policy exercise should consider not merely the measurable and contrastable dimensions of the simple parts of the system, that even if complicated may be technically simulated (technical incommensurability). To be realistic it should also deal with the higher dimensions of the system. Those dimensions in which power relations, hidden interests, social participation, cultural constraints, and other "soft" values, become relevant, and unavoidable variables that heavily, but not deterministically, affect the possible outcomes of the strategies to be adopted (social incommensurability).

At this point in the discussion, one question arises, who is making the decisions? Some critics of multi-criteria evaluation say that *in principle*, in cost-benefit analysis, votes expressed on the market by the whole population can be taken into account (of course with the condition that the distribution of income is accepted as a means to allocate votes)[8]. On the contrary, multi-criteria evaluation can be based on the priorities and preferences of some decision-makers only (we could say that the way these decision-makers have reached their position is accepted as a way to allocate the right to express these priorities). This criticism may be correct if a "*technocratic approach*" is taken, where the analyst constructs the problem relying only upon experts' inputs (by experts meaning those who know the "technicalities" of a given problem).

For the formation of contemporary public policies, it is hard to imagine any viable alternative to *extended peer communities* [18, 23, 30, 33, 34, 38, 47]. They are already being created, in increasing numbers, either when the authorities cannot see a way forward, or know that without a broad base of consensus, no policies can succeed. They all have one important element in common: they assess the quality of policy proposals, including the scientific and technical component. And their verdicts all have some degree of moral force and hence political influence. Here the quality is not merely in the verification, but also in the *creation*; as local people can imagine solutions and reformulate problems in ways that the accredited experts, with the best will in the world, do not find natural.

This need of incorporating the general public into the policy processes has been more and more recognized by the multi-criteria community. Science for policy implies a responsibility of the scientists towards the whole society and not just towards a mythical decision-maker. The classical schematised relationship decision-maker/analyst is indeed embedded in a social framework, which is of a crucial importance in the case of sustainability management and planning. Banville et al. [5] offer a very well structured and convincing argumentation on the need to extend Roy's concept of Multiple Criteria decision Aid by incorporating the notion of stakeholder (extension called "*Participative Multi-criteria Evaluation*" (PMCE) or "*Stakeholder Multi-Criteria Decision Aid*" (SMCDA)).

However, in my opinion, participation is a *necessary* condition but not a sufficient one, since the scientific team cannot simply accept uncritically the inputs of a participatory process. The main justifications of this statement are the following:

1 In a focus group, powerful stakeholders may influence deeply all the others.

2 Some stakeholders might not desire or be able to participate, but ethically the scientific team should not ignore them.

3 The notion of stakeholders does only recognise relevant organised groups; this is the reason why the term "*social actor*" seems preferable to me.

4 Focus groups are never meant to be a representative sample of population. As a consequence, they can be a useful instrument to improve the knowledge of the scientific team of the institutional and social dimensions of the problem at hand, but never a way for deriving consistent conclusions on social preferences.

5 Since decision-makers search for legitimacy[9] of the decisions taken, it is extremely important that public participation or scientific studies do not become instruments of political de-responsibility. The deontological principles of the scientific team and policy-makers are essential for assuring the quality of the evaluation process. Social participation does not imply that scientists and decision-makers have no responsibility of policy actions defended and eventually taken.

Synthesising these arguments we can say that a participatory policy process can always be conditioned by heavy value judgements such as, have all the social actors the same importance (i.e. weight)? Should a socially desirable ranking be obtained on the grounds of the majority principle? Should some veto power be conceded to the minorities? Are income distribution effects important? And so on.

One of the most interesting research directions in the field of public economics is the attempt to introduce political constraints, interest groups and collusion effects explicitly (see e.g. [51]). In this context, *transparency* becomes an essential feature of public policy processes [90]. *Social Multi-Criteria Evaluation* (SMCE) has been explicitly designed to enhance transparency; the main idea being that results of an evaluation exercise depends on the way a given policy problem is *represented* and thus the assumptions used, the interests and values considered have to be made clear [64].

A clear example of these considerations can be found in the determination of criterion weights. Can we have an elicitation of weights from all the social actors involved to be used in the evaluation process? As we know in society there are different legitimate values and points of view. This creates social pressure for taking into account various policy dimensions, e.g. economic, social and environmental[10]. These dimensions are then translated by analysts into objectives and criteria. At this point a question arises who should attach criterion weights and how? To answer this question we have to accept a basic assumption: to attach weights to different criteria implies to give weights to different groups in society. This assumption has the following main consequences:

1 In social decision processes, weights cannot be derived as inputs coming from participatory techniques. This is *technically* very difficult (e.g., which elicitation method has to be used? Which statistical index is a good synthesis of the results obtained? Do average values of weights have meaning at all?), *pragmatically* not desirable (since strong conflicts among the various social actors are very probable to occur) and even *ethically* unacceptable (at least if a Kantian position is taken). A *plurality of ethical principles* seems the only consistent way to derive weights in a social framework.

2 Ethical judgements are unavoidable components of the evaluation exercise. These judgements always influence heavily the results. Let's imagine the extreme case where a development project in the Amazon forest will affect an indigenous community with little contact with other civilizations yet. Would it be ethically more correct to invite them in a focus group... or ethically compulsory to take into account the consequences of the project for their survival? As a consequence, transparency on the assumptions used is essential.

3 Weights in SMCE are clearly meaningful only as *importance coefficients* and not as trade-off (since different ethical positions leads to different ideas on criterion importance). This also implies that the aggregation conventions used should be non-compensatory mathematical algorithms. Non-compensability implies that minorities represented by criteria with smaller weights can still be very influent. This is for example clear in the use of the discordance index in the ELECTRE methods [82, 83].

4 *Sensitivity and robustness analysis* have a complete different meaning with respect to the case of single person and technical decisions. In fact in the case of SMCE, weights derive only from a few clear cut ethical positions. This means that sensitivity or robustness analysis have to check the consequences on the final ranking of only these positions and not of all the possible combinations of weights. Sensitivity and robustness analysis are then a way to improve transparency[11].

In a social multi-criteria evaluation framework, the pitfalls of the technocratic approach can be overtaken by applying different *methods of sociological research* (see Figure 23.3). For example, "*institutional analysis*", performed mainly on historical, legislative and administrative documents, can provide a map of the relevant social actors. By means of focus groups it is possible to

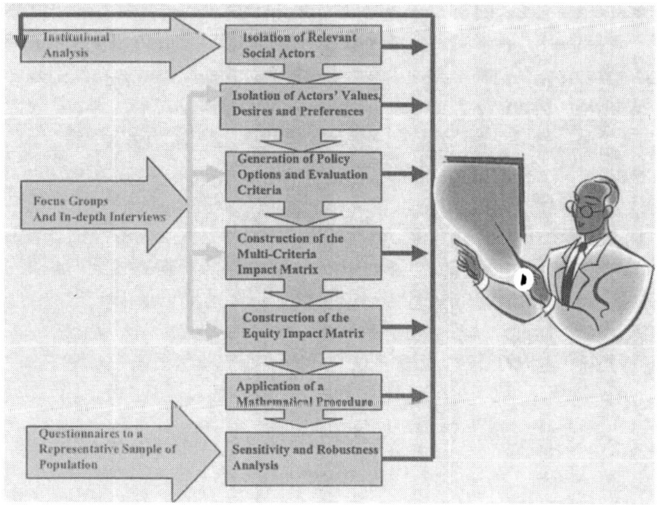

Figure 23.3. The ideal problem structuring in SMCE.

have an idea of people's desires and it is then possible to develop a set of policy options. Main limitations of the focus group technique are that they are not supposed to be a representative sample of the population and that sometimes people are not willing to participate or to state publicly what they really think (above all in small towns and villages). For this reason anonymous questionnaires and personal interviews are an essential part of the participatory process (for practical examples see e.g., [22, 47]).

The selection of evaluation criteria has to be also based on what it is learned through the participation process. However, at this stage a problem generally arises: the evaluation criteria should come directly from the public participation process or they should be "translated" by the research team? I think that the rough material collected during interviews and focus groups could be used as a source of inspiration but the technical formulation of criteria having properties such as "non-redundancy", "legibility" and so on (see [13]) is a clear job of the researchers. Of course in this step, subjectivity is unavoidable, for this reason a widespread information campaign on the assumptions and conclusions of the study including local people, regional and national authorities, international scientists and even children at school is, in my opinion, highly recommendable.

Finally one has to note that policy evaluation is not a one-shot activity. On the contrary, it takes place as a *learning process* which is usually highly dynamic, so that judgements regarding the political relevance of items, alternatives or impacts may present sudden changes, hence requiring a policy analysis to be flexible and adaptive in nature. This is the reason why evaluation processes have a *cyclic nature*. By this is meant the possible adaptation of elements of the evaluation process due to continuous feedback loops among the various steps and consultations among the actors involved.

At this stage a question arises: which is the role of mathematical aggregation procedures in a social evaluation process of sustainability policies? In this framework, of course mathematical aggregation conventions play an important role, i.e. to assure that the rankings obtained are *consistent* with the information and the assumptions used along the structuring process. Next Section then discusses the technical properties considered desirable for a multi-criteria algorithm to assure such a consistency.

6. The Issue of Consistency in Multi-Criteria Evaluation of Sustainability Policies

An issue, that makes multi-criterion aggregation conventions intrinsically complex, is the fact they are *formal, descriptive and normative* models simultaneously [58]. As a consequence, the properties of an approach have to be evaluated at least in the light of these three dimensions. Musgrave [67] in the framework of the debate on the maximisation assumption in microeconomics, made a very useful classification of the assumptions used in economic theory. He makes a distinction among *negligibility assumptions, domain assumptions and heuristic assumptions*. The first type is required to simplify and focus on the essence of the phenomena studied. The second type of assumptions is needed when applying a theory to specify the domain of applicability. The third type is needed either when a theory cannot be directly tested or when the essential assumptions give rise to such a complex model that successive approximation is required. One might see this last type of assumptions as the sake of learning about limits to the relationship between understandable implications and complexity.

In this Section, by using these categories, I try to isolate some main properties that may be considered desirable for a discrete multi-criteria method in the framework of sustainability policies. Of course in another framework, e.g. stock exchange investments, these properties can easily be irrelevant or even undesirable.

When an economic/environmental integration has to be dealt with, a fundamental issue is the one of *compensability*. As we already saw, compensability refers to the existence of trade-offs, i.e. the possibility of offsetting a disadvantage on some criteria by a sufficiently large advantage on another criterion, whereas smaller advantages would not do the same. Thus a preference relation is non-compensatory if no trade-off occurs and is compensatory otherwise. The use of weights with intensity of preference originates compensatory multi-criteria methods and gives the meaning of trade-offs to the weights. On the contrary, the use of weights with ordinal criterion scores originates non-compensatory aggregation procedures and gives the weights the meaning of importance coefficients.

Mathematical compensability plays an important role in the implementation of the so-called "*weak and strong sustainability concepts*". Weak sustainability has been theorised mainly by those economists who have a quite optimistic view of technological progress and economic growth. They generally recognise that even if the production technologies of an economy can potentially yield increases in output commensurate with increases in inputs, overall output will be constrained by limited supplies of resources (growth theory with exhaustible resources). But these limits can be overcome by technological progress: if the rate of technological progress is high enough to offset the decline in the per capita quantity of natural resource services available, output per worker can rise indefinitely. A stronger statement is the following: *even in the absence of any technological progress exhaustible resources do not pose a fundamental problem if reproducible man-made capital is sufficiently substitutable for natural resources* [20]. Pearce and Atkinson [76] state that an economy is sustainable, if it saves more than the combined depreciation of natural and man-made capital. "*We can pass on less environment so long as we offset this loss by increasing the stock of roads and machinery, or other man-made (physical) capital. Alternatively, we can have fewer roads and factories so long as we compensate by having more wetlands or mixed woodlands or more education*" [92, p. 56].

From an ecological perspective, the expansion of the economic subsystem is limited by the size of the overall finite global ecosystem, by its dependence on the life support sustained by intricate ecological connections which are more easily disrupted as the scale of the economic subsystem grows relative to the overall system. This calls for a different concept of sustainability, that of *strong sustainability*, according to which certain sorts of natural capital are deemed critical and not readily substitutable by man-made capital [7]. Human expansion, with the associated exploitation and disposal of waste and pollutants, not only affects the natural environment as such, but also the level and composition of environmentally produced goods and services required to sustain society. Thus, the economic subsystem will be limited by the impacts of its own actions on the environment [29].

Unlimited growth cannot take place in a physically limited planet. Technology is, obviously, an important tool for a development truly sustainable but should not be mystified. The scale of human activities has a maximum expansion possibility defined either by the *regenerative or absorptive capacity of the ecosystem*. Strong sustainability implies that certain sorts of natural capital are deemed critical and not readily substitutable by man-made capital; it is clear that if one wants to operationalize strong sustainability, there is a clear need to use non-compensatory multi-criterion algorithms. Another argument in favour of non-compensatory algorithm is given by the desirability, in the framework of social decisions, that criterion weights can be attached in the form of importance coefficients and not as trade-offs. Clear examples of non-compensatory methods are the ELECTRE methods (see Chapter 4 of this book and [82, 83]) and the Condorcet type algorithm described in Section 3 of this Chapter.

Another important desirable property is the possibility of dealing with mixed criterion scores. It has been argued that the presence of qualitative information in evaluation problems concerning socio-economic and physical planning is a rule, rather than an exception [70]. Thus, the idea of technical incommensurability implies that there is a clear need for methods that are able to take into account information of a "mixed" type (both qualitative and quantitative criterion scores). For simplicity, I refer to *qualitative information* as information measured on a nominal or ordinal scale, and to *quantitative information* as information measured on an interval or ratio scale. Examples of multi-criteria methods able to deal with mixed criterion scores are REGIME [41] and EVAMIX [98].

Moreover, ideally, this information should be precise, certain, exhaustive and unequivocal. But in reality, it is often necessary to use information which does not have those characteristics so that one has to face the uncertainty of a stochastic and/or fuzzy nature present in the data.

If it is impossible to establish exactly the future state of the system studied, a stochastic uncertainty exists, this type of uncertainty is well known in decision theory and economics, where it is called "*decisions under risk*". Applications of this concept in a multi-criteria framework can be found in [21, 52, 78] among others.

Another framing of uncertainty, called *fuzzy uncertainty*, focuses on the ambiguity of information in the sense that the uncertainty does not concern the occurrence of an event but the event itself, which cannot be described unambiguously. This situation is very common in human systems. These systems are complex systems characterised by subjectivity, incompleteness and imprecision. Zadeh [103] writes: "*as the complexity of a system increases, our ability to make a precise and yet significant statement about its behaviour diminishes until a threshold is reached beyond which precision and significance (or relevance) become almost mutually exclusive characteristics*" (incompatibility principle). Fuzzy set theory is a mathematical theory for modelling situations, in which traditional modelling languages which are dichotomous in character and unambiguous in their description cannot be used. For a survey of multi-criteria approaches able to deal with fuzzy uncertainty see Part IV of this book and [59]. In conclusion, *multi-criteria methods able to tackle consistently the widest types of mixed information and different sources of uncertainty should be considered as desirable ones.*

Another desirable property for mathematical aggregation procedures in the framework of sustainability decisions is *simplicity*, i.e. the use of a few parameters as possible. While in the context of multi-criteria decision aid, parameters helping the decision-maker to elicitate her/his preferences are desirable, in a social context there is the risk that their presence increases arbitrariness and reduces transparency. I think that in this second context the only exogenous parameters desirable are weights and, if absolutely necessary, indifference and preference thresholds.

Finally, in a policy framework, *to have a ranking of all the different courses of actions is better than to select just one alternative.* This mainly because in this way social compromises are easier (the second or the third alternative in the ranking may minimise opposition much more than the first one). Technically speaking this implies that multi-criteria methods able to deal with the ranking decision problem formulation have to be preferred and that dominated alternatives cannot be excluded a priori.

Concluding, we can summarise a set of desirable properties for choosing an appropriate method for dealing with sustainability decision problems, as follows.

<u>Descriptive domain assumptions:</u>

- Mixed information on criterion scores should be tackled in the form of ordinal, crisp, stochastic and fuzzy criterion scores.

<u>Normative domain assumptions:</u>

- Simplicity is desirable and means the use of as less ad hoc parameters as possible.
- The most useful result for policy-making is a complete ranking of alternatives.
- Weights are meaningful only as importance coefficients and not as trade-offs.
- Complete compensability is not desirable.

<u>Heuristic descriptive assumptions:</u>

- When not all intensities of preference are meaningful, indifference and preference thresholds are useful exogenous parameters.
- Dominated alternatives have to be considered.

Finally one should note that these selection properties can be applied only to methods who achieve a set of minimum formal requirements, the main important being the following.

<u>Formal domain assumptions:</u>

- Unanimity.
- Monotonicity.
- Neutrality.

Negligibility formal assumptions:

- Anonymity.

7. Conclusion

When science is used for policy making, an appropriate management of decisions implies including the multiplicity of participants and perspectives. This also implies the impossibility of reducing all dimensions to a single unity of measure. "*The issue is not whether it is only the marketplace that can determine value, for economists have long debated other means of valuation; our concern is with the assumption that in any dialogue, all valuations or 'numeraires' should be reducible to a single one-dimension standard*" [34, p. 198]. It is noteworthy that this call for citizen participation and transpareny, when science is used for policy making, is more and more supported institutionally inside the European Union, where perhaps the most significant examples are the White Paper on Governance and the Directive on Strategic Environmental Impact Assessment.

Multi-criteria evaluation supplies a powerful framework for the implementation of the incommensurability principle. In fact it accomplishes the goals of being *inter/multi-disciplinary* (with respect to the research team), *participatory* (with respect to the local community) and *transparent* (since all criteria are presented in their original form without any transformations in money, energy or whatever common measurement rod). As a consequence multi-criteria evaluation looks as an adequate assessment framework for (micro and macro) sustainability policies.

However, one should remember that we are in a second best world. A useful analogy here is with Flatland, the classic Victorian science fiction and social parody [1]. There, the inhabitants of spaces with more dimensions had a richer awareness of themselves, and also could see beyond and through the consciousness of the simpler creatures inhabiting fewer dimensions. At this stage it is not unfair to reveal the dénouement of the story, namely that the Sphere of three-dimensional space showed himself to be just another Flatlander at heart, when he angrily refused to accept the reality of higher dimensions of being.

Notes

1. Emphasis added to the original
2. Here I refer to the idea of orchestration of sciences as a combination of multi/inter-disciplinarity. Multi-disciplinarity: each expert takes her/his part. Inter-disciplinarity: methodological choices are discussed across the disciplines.

3. Composite indicators are indeed synthetic indexes, thus the two terms can be considered synonymous; here I use the term composite indicator since is the standard one in OECD terminology [72].

4. One should note that this inconsistency is present in the majority of the environmental impact assessment studies too. In fact it is a common practice to aggregate environmental impact indicators by means of a linear rule and to attach weights to them according to the relative importance idea. Moreover, the use of a linear aggregation procedure implies that among the different ecosystem aspects there are not phenomena of synergy or conflict. This appears to be quite an unrealistic assumption for environmental impact assessment studies [32]. For example, "*laboratory experiments made clear that the combined impact of the acidifying substances* SO_2, NO_X, NH_3 *and* O_3 *on plant growth is substantially more severe that the (linear) addition of the impacts of each of these substances alone would be*." [24].

5. Anyway a Condorcet consistent rule always presents smaller probabilities of the occurrence of a rank reversal in comparison with any Borda consistent rule. This is again a strong argument in favour of a Condorcet's approach in this framework.

6. In our case, this axiom is needed since the intensity of preference of individual indicators is not considered to be useful preference information given that compensability has to be avoided and weights have to be symmetrical importance coefficients. Moreover, thanks to this axiom, a normalisation step is not needed. This reduces the sources of uncertainty and imprecise assessment.

7. In social choice terms then the *anonymity* property (i.e. equal treatment of all individual indicators) is broken. Indeed, given that full *decisiveness* yields to dictatorship, Arrow's impossibility theorem forces us to make a trade-off between decisiveness (an alternative has to be chosen or a ranking has to be made) and anonymity. In our case the loss of anonymity in favour of decisiveness is even a positive property. In general, it is essential that no individual indicator weight is more than 50% of the total weight; otherwise the aggregation procedure would become lexicographic in nature, and the indicator would become a dictator in Arrow's term.

8. One should note that indeed cost-benefit analysis can be easily criticised both from the distributive and environmental points of view (see e.g., [60, 88]). However I prefer not to deal with this issue here.

9. On the issue of legitimacy see also [84].

10. By *dimension*, here I mean the highest hierarchical level of analysis which indicates the scope of objectives and criteria.

11. On this point I disagree with Kleijnen [50], who claims that "*modellers should try to develop robust models*", in the sense that models should not be very sensitive to modellers' assumptions. Some ethical positions might be very different and thus lead to different rankings of the policy options. What is essential in a social framework is then transparency on these assumptions.

References

[1] E.A. Abbott. *Flatland, a Romance of Many Dimensions*. Little, Brown & Co, Boston, 1935.

[2] T.F.H. Allen, J.A. Tainter, and T.W. Hoekstra. *Supply-Side Sustainability*. Columbia University Press, New York, 2002.

[3] K.J. Arrow. *Social Choice and Individual Values*. John Wiley & Sons, New York, second edition, 1963.

[4] K.J. Arrow and H. Raynaud. *Social Choice and Multicriterion Decision Making*. M.I.T. Press, Cambridge, second edition, 1963.

[5] C. Banville, M. Landry, J.M. Martel, and C. Boulaire. A stakeholder approach to MCDA. *Systems Research and Behavioral Science*, 15:15–32, 1998.

[6] E.B. Barbier. The concept of sustainable economic development. *Environmental Conservation*, 14(2):101–110, 1987.

[7] E.B. Barbier and A. Markandya. The conditions for achieving environmentally sustainable growth. *European Economic Review*, 34:659–669, 1990.

[8] G. Barbiroli. Direct and indirect indicators of sustainable development. In *Proceedings International Conference on Environmental Pollution, Barcelona*, pages 489–496, 1993.

[9] E. Beinat. *Value Functions for Environmental Management*. Kluwer Academic Publishers, Dordrecht, 1997.

[10] E. Beinat and P. Nijkamp, editors. *Multicriteria Evaluation in Land-Use Management: Methodologies and Case Studies*. Kluwer Academic Publishers, Dordrecht, 1998.

[11] M.L. Bell, B. Hobbs, E.M. Elliott, H. Ellis, and Z. Robinson. An evaluation of multi-criteria methods in integrated assessment of climate policy. *Journal of Multi-Criteria Decision Analysis*, 10(5):229–256, 2001.

[12] D. Black. *The Theory of Committees and Elections*. Cambridge University Press, Cambridge, 1958.

[13] D. Bouyssou. Building criteria: A prerequisite for MCDA. In C.A. Bana e Costa, editor, *Readings in Multiple Criteria Decision Aid*, pages 58–80. Springer Verlag, Berlin, 1990.

[14] D. Bouyssou, T. Marchant, P. Perny, M. Pirlot, A. Tsoukiàs, and Ph. Vincke. *Evaluation and Decision Models: A Critical Perspective*. Kluwer Academic Publishers, Dordrecht, 2000.

[15] M. Bresso. *Per Un' Economia Ecologica*. La Nuova Italia Scientifica, Rome, 1993.

[16] G. Chichilnisky. An axiomatic approach to sustainable development. *Social Choice and Welfare*, 13(2):219–248, 1996.

[17] Marquis de Condorcet. *Essai sur l'Application de l'Analyse à la Probabilité des Décisions Rendues à la Probabilité des Voix*. De l'Imprimerie Royale, Paris, 1785.

[18] S. Corral-Quintana, B. De Marchi, S. Funtowicz, G. Gallopín, Â. Guimarães Pereira, and B. Maltoni. The visions project at the jrc. Technical Report EUR 19926 EN, European Commission Joint Research Centre, Institute for the Protection and Safety of the Citizens, 2001. Available at http://alba.jrc.it/visions.

[19] H.E. Daly and J.J. Cobb. *For the Common Good: Redirecting the Economy Toward Community, the Environment and a Sustainable Future*. Beacon Press, Boston, 1989.

[20] P. Dasgupta and D. Heal. *Economic Theory and Exhaustible Resources*. Cambridge University Press, Cambridge, 1979.

[21] G. D'Avignon and Ph. Vincke. An outranking method under uncertainty. *European Journal of Operational Research*, 36:311–321, 1988.

[22] B. De Marchi, S.O. Funtowicz, S. Lo Cascio, and G. Munda. Combining participative and institutional approaches with multi-criteria evaluation. an empirical study for water issue in troina, sicily. *Ecological Economics*, 34(2):267–282, 2000.

[23] B. De Marchi and J. Ravetz. Participatory approaches to environmental policy, concerted action EVE. Policy research brief 10, European Commission, DGXII, 2001.

[24] F.J. Dietz and J. van der Straaten. Rethinking environmental economics: missing links between economic theory and environmental policy. *Journal of Economic Issues*, XXVI(1):27–51, 1992.

[25] P. Ehrlich and P. Raven. Butterflies and plants: A study in co-evolution. *Evolution*, 18:586–608, 1964.

[26] J.M. Espelta, J. Retana, and A. Habrouk. An economic and ecological multi-criteria evaluation of reforestation methods to recover burned Pinus Negra forests in NE Spain. *Forest Ecology and Management*, 180:185–198, 2003.

[27] F. Falconi. *Economía y Desarrollo Sostenible*. FLACSO, Quito, 2002.

[28] S. Faucheux and M. O'Connor, editors. *Valuation for Sustainable Development: Methods and Policy Indicators*. Edward Elgar, Cheltenham, 1998.

[29] C. Folke. Socio-economic dependence on the life-supporting environment. In C. Folke and T. Kaberger, editors, *Linking the Natural Environment and the Economy: Essays from the Eco-Eco Group*, pages 77–94. Kluwer Academic Publishers, Dordrecht, 1991.

[30] S. Funtowicz, J. Martinez-Alier, G. Munda, and J. Ravetz. Information tools for environmental policy under conditions of complexity. Environmental Issues Series 9, European Environmental Agency, Experts' Corner, 1999.

[31] S.O. Funtowicz, J. Martinez-Alier, G. Munda, and J. Ravetz. Multicriteria-based environmental policy. In H. Abaza and A. Baranzini, editors, *Implementing Sustainable Development*, pages 53–77. UNEP/Edward Elgar, Cheltenham, 2002.

[32] S.O. Funtowicz, G. Munda, and M. Paruccini. The aggregation of environmental data using multicriteria methods. *Environmetrics*, 1(4):353–368, 1990.

[33] S.O. Funtowicz and J.R. Ravetz. A new scientific methodology for global environmental issues. In R. Costanza, editor, *Ecological Economics*, pages 137–152. Columbia, New York, 1991.

[34] S.O. Funtowicz and J.R. Ravetz. The worth of a songbird: Ecological economics as a post-normal science. *Ecological Economics*, 10:197–207, 1994.

[35] M. Giampietro and K. Mayumi. Multiple-scale integrated assessment of societal metabolism: Integrating biophysical and economic representations across scales. *Population and Environment*, 22(2):155–210, 2000.

[36] M. Giampietro and K. Mayumi. Multiple-scale integrated assessment of societal metabolism: Introducing the approach. *Population and Environment*, 22(2):109–154, 2000.

[37] J.M. Gowdy. *Coevolutionary Economics: The Economy, Society, and the Environment.* Kluwer Academic Publishers, Boston, 1994.

[38] J.M. Gowdy and S. O'Hara. *Economic Theory for Environmentalists.* Saint Lucie Press, St Lucie County, 1996.

[39] A. Guitouni and J.M. Martel. Tentative guidelines to help choosing an appropriate MCDA method. *European Journal of Operational Research*, 109:501–521, 1998.

[40] K. Hayashi. Multicriteria analysis for agriculture resource management: A critical survey and future perspectives. *European Journal of Operational Research*, 122:486–500, 2000.

[41] E. Hinloopen and P. Nijkamp. Qualitative multiple criteria choice analysis, the dominant regime method. *Quality and Quantity*, 24:37–56, 1990.

[42] R. Horwarth and R.B. Norgaard. Intergenerational resource rights, efficiency and social optimality. *Land Economics*, 66:1–11, 1990.

[43] R. Horwarth and R.B. Norgaard. Environmental valuation under sustainable development. *American Economic Review Papers and Proceedings*, 80:473–477, 1992.

[44] Martinez-Alier J. and M. O'Connor. Ecological and economic distribution conflicts. In R. Costanza, O. Segura, and J. Martinez-Alier, editors, *Getting down to Earth: Practical Applications of Ecological Economics*. Island Press/ISEE, Washington D.C., 1996.

[45] R. Janssen. *Multiobjective decision support for environmental management.* Kluwer Academic Publishers, Dordrecht, 1992.

[46] R. Janssen and G. Munda. Multi-criteria methods for quantitative, qualitative and fuzzy evaluation problem. In J. van den Bergh, editor, *Handbook of Environmental and Resource Economics*, pages 837–852. Edward Elgar, Cheltenham, 1999.

[47] B. Kasemir, M. Gardner, J. Jäger, and Jaeger C., editors. *Public Participation in Sustainability Science*. Cambridge University Press, Cambridge, 2003.

[48] R. Keeney and H. Raiffa. *Decision with Multiple Objectives: Preferences and Value Trade-offs*. John Wiley & Sons, New York, 1976.

[49] J. Kemeny. Mathematics without numbers. *Daedalus*, 88:571–591, 1959.

[50] J.P.C. Kleijnen. Ethical issues in modelling: Some reflections. *European Journal of Operational Research*, 130:223–230, 2001.

[51] J.J. Laffont. *Incentives and Political Economy*. Oxford University Press, Oxford, 2000.

[52] J. Martel and K. Zaras. Stochastic dominance in multicriteria analysis under risk. *Theory and Decision*, 35(1):31–49, 1995.

[53] J. Martinez-Alier, G. Munda, and J. O'Neill. Weak comparability of values as a foundation for ecological economics. *Ecological Economics*, 26:277–286, 1998.

[54] J. Martinez-Alier, G. Munda, and J. O'Neill. Theories and methods in ecological economics: A tentative classification. In C.J. Cleveland, Stern D.J., and Costanza R., editors, *The Economics of Nature and the Nature of Economics*, pages 34–56. Edward Elgar, Cheltenham, 2001.

[55] L. Maystre, J. Pictet, and Simos J. *Méthodes Multicritères ELECTRE – Description, Conseils Pratiques et Cas d'Application à la Gestion Environnementale*. Presses Polytechniques et Universitaires Romandes, Lausanne, 1994.

[56] J. M. Moreno-Jiménez, J. Aguaron, T. Escobar, and A. Turon. Multicriteria procedural rationality on SISDEMA. *European Journal of Operational Research*, 119(2):388–403, 1999.

[57] H. Moulin. *Axioms of co-operative decision making*. Econometric Society Monographs. Cambridge University Press, Cambridge, 1988.

[58] G. Munda. Multiple-criteria decision aid: Some epistemological considerations. *Journal of Multi-Criteria Decision Analysis*, 2:41–55, 1993.

[59] G. Munda. *Multicriteria Evaluation in a Fuzzy Environment – Theory and Applications in Ecological Economics*. Contributions to Economics Series. Physica-Verlag, Heidelberg, 1995.

[60] G. Munda. Cost-benefit analysis in integrated environmental assessment: Some methodological issues. *Ecological Economics*, 19(2):157–168, 1996.

[61] G. Munda. Environmental economics, ecological economics and the concept of sustainable development. *Environmental Values*, 6(2):213–233, 1997.

[62] G. Munda. Multicriteria evaluation as a multidimensional approach to welfare measurement. In J. van den Bergh and J. van der Straaten, editors, *Economy and Ecosystems in Change: Analytical and Historical Approaches*, pages 96–115. Edward Elgar, Cheltenham, 1997.

[63] G. Munda. Measuring sustainability: A multi-criterion framework. *Environment, Development and Sustainability*, 2004. Forthcoming.

[64] G. Munda. Social multi-criteria evaluation (SMCE): Methodological foundations and operational consequences. *European Journal of Operational Research*, 158(3):662–677, 2004.

[65] G. Munda and M. Nardo. Mathematical modelling of composite indicators for ranking countries. In *Proceedings of the First OECD/JRC*

Workshop on Composite Indicators of Country Performance, JRC, Ispra, 2003. Available at http://webfarm.jrc.cec.eu.int/uasa/events/oecd_12may03/Background/.

[66] G. Munda, M. Paruccini, and G. Rossi. Multicriteria evaluation methods in renewable resource management: The case of integrated water management under drought conditions. In E. Beinat and P. Nijkamp, editors, *Multicriteria Evaluation in Land-Use Management: Methodologies and Case Studies*, pages 79–94. Kluwer Academic Publishers, Dordrecht, 1998.

[67] A. Musgrave. Unreal assumptions in economic theory: The F-twist untwisted. *Kyklos*, 34:377–387, 1981.

[68] I. Musu and D. Siniscalco. *National Accounts and the Environment*. Kluwer Academic Publishers, Dordrecht, 1996.

[69] O. Neurath. *Empiricism and Sociology*. Reidel, Dordrecht, 1973.

[70] P. Nijkamp, P. Rietveld, and H. Voogd. *Multicriteria Evaluation in Physical Planning*. North-Holland, Amsterdam, 1990.

[71] R.B. Norgaard. *Development Betrayed*. Routledge, London, 1994.

[72] OECD. Composite indicators of country performance: A critical assessment. DST/IND(2003)5, OECD, Paris, 2003.

[73] J. O'Neill. *Ecology, Policy and Politics*. Routledge, London, 1993.

[74] R. Passet. *L'Economique et le Vivant*. Payot, Paris, 1979.

[75] D. Pearce, G. Hamilton, and G.D. Atkinson. Measuring sustainable development: Progress on indicators. *Environment and Development Economics*, 1:85–101, 1996.

[76] D.W. Pearce and G.D. Atkinson. Capital theory and the measurement of sustainable development: An indicator of "weak" sustainability. *Ecological Economics*, 8:103–108, 1993.

[77] V.V. Podinovskii. Criteria importance theory. *Mathematical Social Sciences*, 27:237–252, 1994.

[78] P. Rietveld. Using ordinal information in decision making under uncertainty. *Systems Analysis, Modeling, Simulation*, 6:659–672, 1989.

[79] L. Ringius, T. Asbjørn, and B. Holtsmark. Can multi-criteria rules fairly distribute climate burdens? OECD results from three burden sharing rules. *Energy Policy*, 26(10):777–793, 1998.

[80] F.S. Roberts. *Measurement Theory with Applications to Decision Making, Utility and the Social Sciences*. Addison-Wesley, London, 1979.

[81] C. Romero and T. Rehman. *Multiple Criteria Analysis for Agricultural Decisions*. Elsevier, Amsterdam, 1989.

[82] B. Roy. *Méthodologie Multicritère d'Aide à la Décision*. Economica, Paris, 1985.

[83] B. Roy. *Multicriteria Methodology for Decision Analysis*. Kluwer Academic Publishers, Dordrecht, 1996.

[84] B. Roy and S. Damart. L'analyse coûts-avantages, outil de concertation et de légitimation? *Metropolis*, 108/109:7–16, 2002.

[85] M. Saisana and S. Tarantola. State-of-the-art report on current methodologies and practives for composite indicator development. EUR 20408 EN, European Commission, JRC, Ispra, Italy, 2002.

[86] P. Salminen, J. Hokkanen, and R. Lahdelma. Comparing multicriteria methods in the context of environmental problems. *European Journal of Operational Research*, 104:485–496, 1998.

[87] F. Schmidt-Bleek. *Wieviel Umwelt braucht der Mensch? MIPS Das Mass für Ökologisches Wirtschaften*. Birkhäuser, Berlin, 1994.

[88] C. Spash and N. Hanley. Preferences, information, and biodiversity preservation. *Ecological Economics*, 12:191–208, 1995.

[89] T.J. Stewart and A. Joubert. Conflicts between conservation goals and land use for exotic forest plantations in South Africa. In E. Beinat and P. Nijkamp, editors, *Multicriteria Evaluation in Land-Use Management: Methodologies and Case Studies*, pages 17–31. Kluwer Academic Publishers, Dordrecht, 1998.

[90] J. E. Stiglitz. New perspectives on public finance: Recent achievements and future challenges. *Journal of Public Economics*, 86:341–360, 2002.

[91] I. Todhunter. *A history of mathematical theory of probability*. Chelsea, New York, 1949.

[92] R.K. Turner, D.W. Pearce, and I. Bateman. *Environmental Economics: An Elementary Introduction*. Harvester Wheatsheaf, London, 1994.

[93] B. Ülengin, F. Ülengin, and Ü. Güvenç. A multidimensional approach to urban quality of life: The case of Istanbul. *European Journal of Operational Research*, 130:361–374, 2001.

[94] Urban indicators programme. http://www.urbanobservatory.org/indicators/database.

[95] Ph. Vincke. *Multicriteria Decision Aid*. John Wiley & Sons, New York, 1992.

[96] Ph. Vincke. Recent progresses in multicriteria decision-aid. *Rivista di Matematica per le Scienze Economiche e Sociali*, 2:21–32, 1994.

[97] P. Vitousek, P. Ehrlich, A. Ehrlich, and P. Matson. Human appropriation of the products of photosynthesis. *Bioscience*, 6:21–32, 1986.

[98] H. Voogd. *Multicriteria Evaluation for Urban and Regional Planning*. Pion, London, 1983.

[99] M. Wackernagel and W. E. Rees. *Our Ecological Footprint: Reducing Human Impact on the Earth*. New Society Publishers, Gabriola Island, BC and Philadelphia, PA, 1995.

[100] H.P. Young. Condorcet's theory of voting. *American Political Science Review*, 82(4):1231–1244, 1988.

[101] H.P. Young and A. Levenglick. A consistent extension of Condorcet's election principle. *SIAM Journal of Applied Mathematics*, 35:285–300, 1978.

[102] J.A. Yusuf, S. El Serafy, and E. Lutz, editors. *Environmental Accounting for Sustainable Development, A UNEP World Bank Symposium, Washington D.C.* The World Bank, Washington, 1989.

[103] L.A. Zadeh. Fuzzy sets. *Information and Control*, 8:338–353, 1965.

VIII

MCDM SOFTWARE

Chapter 24

MULTIPLE CRITERIA DECISION SUPPORT SOFTWARE

H. Roland Weistroffer, Charles H. Smith, Subhash C. Narula
School of Business
Virginia Commonwealth University
Box 844000, Richmond, Virginia 23284-4000
USA
{hrweistr,chsmith,snarula}@vcu.edu

Abstract We present an overview of the current state of multiple criteria decision-making (MCDM) decision support software. Many approaches have been proposed in the literature to solve multiple criteria decision-making problems, and there is an abundance of software that implements these approaches. Much of the software is still quasi-experimental, developed by academic researchers to test specific algorithms or to solve a specific problem on an ad hoc basis.

Keywords: DSS, MCDSS, software packages.

1. Introduction

It is well known that multiple criteria decision models do not possess a mathematically well-defined optimal solution; therefore the decision maker (DM) has to find a satisfactory (desirable, acceptable) compromise solution from among many non-dominated (efficient) solutions. Unless the utility function of the DM is known a priori and explicitly, interactive solution techniques are imperative to identify the most preferred solution or a manageable set of desirable compromise solutions.

Today a wide variety of software has been developed specifically to support multiple criteria decision-making. Many general software tools, such as linear programming packages and electronic spreadsheets that do not implement specific MCDM techniques, can also be used to analyze multiple criteria problems. MCDM software covers various stages of the decision making process, from problem exploration and structuring to discovering the DM's preferences and the most preferred compromise solution. Our primary objective in this paper is to report on the state of commercially or otherwise readily available multiple criteria decision support software.

In the next section we present an overview of multiple criteria decision support software. The software is organized into seven areas based on the type of problem to which the software is applied. These areas are qualitative problem structuring, general multiple attribute decision making, general multiple objective decision making, multiple criteria sorting problems, portfolio analysis, group decision support, and some application specific software. Within each section the software is listed in alphabetical order. Table 1 provides a list of the software described in each area. The software review is followed by a few concluding remarks.

2. Software Overview

Decision support software can assist DMs at various stages of structuring and solving decision problems. These stages can include problem exploration and formulation, decomposition, and preference and trade-off judgments. Many of the general commercially available decision aids have been included in the biennial decision support systems reviews in OR/MS Today [14, 15, 16, 17, 85, 86]. However, several other approaches that have been suggested in the literature have only been implemented on an ad hoc basis, to solve a specific problem situation, or as experimental software to demonstrate the salient points of the proposed methodology. While most software developed by academics is available free of charge, or for a nominal fee, commercial packages sell for hundreds or even thousands of dollars (though some give educational discounts). Most have their own websites and sophisticated marketing literature.

In assessing MCDM software, it is important to consider not only the technology (i.e., computer hardware and software, and MCDM methodology) aspects, but also the role of the DM in the interactive process, and the user-friendliness of the human-computer interface. The MCDM programs developed in the 1970s were mainly oriented towards the study of multiple objective mathematical programming problems [33]. These early systems were primarily developed for academic purposes. They were implemented on mainframe computers, with no documentation available. They also did not have any visual representation capabilities, mainly due to the limited capabilities of computer technology at that time. There are encouraging signs that some psychological and behavioral research is being integrated into MCDM theory and practice [87]. Korhonen et al. [62] note that during the 1980's, emphasis shifted away from the mathematical aspects of multiple objective programming towards providing decision support to the DM. Most modern MCDM software tools are designed for the Windows platform and provide graphical interfaces to assist in visualizing the effects of changes to problem parameters. An increasing number of packages are also available for interactive use via the Internet. In our coverage we have tried to focus on more recently developed software, though we have also included older software that appears to be continually maintained. We also note that information for several commercial and other computer programs is not available in a uniform format. This is reflected in the write-ups for the various decision support systems.

2.1 Qualitative Problem Structuring

Software in this category addresses the early stages of the decision making process: exploring and formulating the decision problem.

2.1.1 Decision Explorer. `http://www.banxia.com/`. Decision Explorer is oriented to organize and map qualitative information for complex, ill-structured problems [128]. The fundamental method employed is the causal mapping technique. The aim is to identify useful courses of action by the relationships established between variables as a cognitive map is built. Decision Explorer can facilitate group discussion and understanding by means of its visual development of problem issues. The software includes analytical tools that assist in evaluating the similarities and differences of sets and in developing and analyzing clusters of information about the problem. The website provides a tutorial, case study, demonstration downloads, and a bibliography of material related to the software or the cognitive mapping method.

Table 24.1. Software by problem type.

Area	*Software*
Problem Structuring	Decision Explorer
Multiple Attribute DM	Criterium Decision Plus, DAM, Decision Lab ELECCALC, ELECTRE IS, ELECTRE III-IV Equity, Expert Choice, HIVIEW Logical Decisions, MACBETH, M&P MacModel, MIIDAS, MINORA MUSTARD, NAIADE, OnBalance PREFCALC, PRIAM, PRIME Decisions REMBRANDT, RGDB, SANNA TOPSIS, UTA Plus, VIMDA VIP Analysis, VISA, WebHIPRE WINPRE
Multiple Objective DM	ADBASE, Feasible Goals Method Feasible Set in Criterion Space, MOMHLib++ MultiGen, SOLVEX, TOMMIX TRIMAP, VIG, WWW-NIMBUS Multistat Optimizer
Sorting Problems	ELECTRE TRI, IRIS, PREFDIS PROAFTN, TOMASO
Portfolio Analysis	HiPriority
Group Decision Support	AGAP, ARGOS, CTLite GMCR, Joint Gains, MEDIATOR SCDAS, WINGDSS
Application Specific Software	ACADEA, AgentAllocator, AutoMan BANKADVISOR, CASTART, CGX DIDASN++, DIMITRA Electrical Power Districting, ESY, FINCLAS FINEVA, INVEX, MARKEX MEDICS, MOIRA, SANEX Skills Evaluator, Steel Mill Scheduling TELOS, Water Quality Plan

2.2 General Multiple Attribute Decision Making (MADM)

MCDM problems can be roughly divided into two main groups, viz. multiple attribute decision-making (MADM) and multiple objective decision-making (MODM) problems. In the MADM problems, the decision-maker must choose from among a finite number of available alternatives characterized by a set of multiple attributes. Software in this category is designed to deal with any type of decision problem where one has to choose among a finite set of decision alternatives characterized by a set of attributes.

2.2.1 Criterium Decision Plus. `http://www.infoharvest.com`. Criterium Decision Plus 3.0 (CDP), (reviewed by Haerer in [45]), provides users a choice between a simple multiattribute rating technique and AHP. The primary strengths of CDP include, among others, the immediate graphical feedback from what-if analysis and the support of value of information analyses. Haerer also reports that CDP has been used live on the Internet and has supported decision-making via video conferencing. Users have the option of choosing non-linear value functions. Performance scores can be entered into a table or in a rating window that provides choices among numerical, graphic and verbal representations. Uncertainties can be accommodated through a choice of distributions or by a customized distribution. An earlier version of this software was evaluated in [130].

2.2.2 DAM. DAM (Decision Analysis Module) [101] was originally designed as a module in more complex software used to analyze electric system expansion scenarios. DAM utilizes imprecise information about the trade-offs in the form of ranges. The principal decision analysis options supported by the software include the testing of potential optimality, the identification of outperformed and not outperformed alternatives, and visual sensitivity analysis. To solve the linear programs arising from different analysis options, a fairly straightforward version of the simplex method is used.

2.2.3 Decision Lab. `http://www.visualdecision.com`. Known as PROMCALC in a previous version, Decision Lab 2000 is an interactive decision support system [39] based on the outranking methods PROMETHEE [12, 13] and GAIA [11]. Sensitivity analyses are generated by using techniques of walking weights, intervals of stability, and the graphical axis of decision displayed by the GAIA method. The software is now suitable for group decision support, providing profiles of actions and multi-scenario comparisons. The methodology used here requires fewer comparisons from the decision maker than the AHP method; it permits the user to define his own measurement scale.

The reference [11] describes the method, the decision support system, and also gives an illustrative example.

2.2.4 ELECCALC. Utilizing a user-friendly graphical interface, a decision maker can globally express preferences about a few reference alternatives, and then the method can specify initial values for parameters of ELECTRE II [107]. A disaggregation-aggregation procedure like that in PREFCALC [50] is used [57].

2.2.5 ELECTRE IS. `http://www.lamsade.dauphine.fr/english/software.html#elis`. ELECTRE IS is a generalization of the ELECTRE I method [105, 107], which enables the use of pseudo-criteria (criteria with thresholds). Given a finite set of alternatives evaluated on a consistent family of criteria, ELECTRE IS supports the user in the process of selecting one alternative or a subset of alternatives. The method consists of two parts: construction of one crisp outranking for modeling the decision-maker's preferences, and exploitation of the graph corresponding to this relation. The subset searched is the kernel of the graph. Software implementing ELECTRE IS is available from LAMSADE at the Université Paris-Dauphine. ELECTRE methods are also discussed in Chapter 4 of this volume.

2.2.6 ELECTRE III-IV. `http://www.lamsade.dauphine.fr/english/software.html#el34`. ELECTRE III starts with a finite set of actions evaluated on a consistent family of pseudo-criteria and aggregates these partial preferences into a fuzzy outranking relation [106, 107]. ELECTRE IV builds several non-fuzzy outranking relations when criteria cannot be weighted. Two complete preorders are then obtained through a "distillation" procedure, either from the fuzzy outranking relation of ELECTRE III, or from the non-fuzzy outranking relations provided by ELECTRE IV. The intersection of these preorders indicates the most reliable part of the global preference. The ELECTRE III-IV software is available from LAMSADE at the Université Paris-Dauphine. ELECTRE methods are also discussed in Chapter 4 of this volume.

2.2.7 Equity. `http://enterprise-lse.co.uk`. Equity is a multicriteria decision analysis (MCDA) tool that can be used to obtain better value-for-money in allocating scarce resources. It is highly adaptive and can be used to address a variety of problems. In stage 1, an outline of the model is constructed. In stage 2, each option is scored against a set of defined criteria. In stage 3, the decision maker must make a value judgment on the relative importance of different aspects of the model. In stage 4, the model is analyzed and recommendations are presented in stage 5. A 30-day evaluation version of the program is available to download.

2.2.8 Expert Choice. `http://www.expertchoice.com`. Expert Choice (reviewed in [37]) has been closely identified with AHP, and the software employs AHP as its core methodology. The latest versions emphasize group decision support and an easy-to-use interface. The software will accept judgments from multiple stakeholders using wireless keypads or the Internet. It has the capability to weight team members and evaluate outcomes based on team member demographics. The company website also states that Expert Choice offers a "freestyle, interactive technique for building a model that simulates the flow of ideas, and helps decision-makers organize the objectives of their decision into theme clusters." Graphs for sensitivity analysis are provided. The company claims that Expert Choice has over 50,000 users, including many large corporations and government agencies. Expert Choice is one of the packages evaluated in [130].

2.2.9 HIVIEW. `http://www.enterprise-lse.co.uk`. HIVIEW is a multicriteria decision analysis (MCDA) tool that can be used to support decisions among mutually exclusive options. It is highly adaptive and can be used to address a variety of problem areas. There are five main stages for modeling in HIVIEW. A model is constructed as a tree structure in stage 1. In stage 2, each action option is scored against the criteria set out in the tree structure. In stage 3, the decision maker must make a value judgment on the relative importance of different aspects of the model. The model is analyzed in stage 4, and recommendations are presented in stage 5. An evaluation version of HIVIEW is available to download.

2.2.10 Logical Decisions. `http://www.logicaldecisions.com`. Logical Decisions for structuring and analyzing multiple attribute decision analysis problems has been commercially available for several years. It is currently offered in both single and group user versions. The user interface is considered a significant attraction, with a graphical, point and click way to adjust weights. Historically associated with multiattribute utility theory, according to the website Logical Decisions offers five methods for assessing weights, "ranging from the easy-to-use 'Smarter' method, to the sophisticated 'tradeoff' method, to the popular 'analytic hierarchy process'." The results can be displayed in various ways, and one can compare pairs of alternatives to see their major differences. Interactive graphical sensitivity analysis displays are available. Logical Decisions is one of the packages evaluated in [130].

2.2.11 MACBETH. `http://www.umh.ac.be/~smq`. MACBETH – Measuring Attractiveness by a Categorical Based Evaluation Technique (Chapter 10 of this volume, and Bana e Costa and Chagas [5]) uses semantic judgments about the differences in attractiveness of several stimuli to help a decision maker quantify the relative attractiveness of each stimulus. It employs an initial, iterative, questioning procedure that compares two elements at a time, requesting only a qualitative preference judgment. MACBETH automatically verifies the consistency of the judgments and generates a representative numerical scale. Similarly, MACBETH generates weighting scales for the decision criteria, and also provides sensitivity analysis.

2.2.12 MacModel. `http://www.civil.ist.utl.pt/~lavt/ software.html`. MacModel is decision tree based software for multicriteria problems, developed at the Instituto Superior Técnico in Lisbon, Portugal [120].

2.2.13 M&P. M&P (MAPPAC and PRAGMA) implements the MAPPAC [80] and PRAGMA [81] outranking methods also described in Chapter 6 of this volume. M&P offers multiple options for preference modeling, such as specifying trade-off and importance weights, and normalization levels. Some classical statistical analyses on the evaluations of alternatives are also allowed (average values, standard deviations, correlations between criteria). For each pair of criteria, suitable indifference thresholds and shapes can be defined. It is also possible to graphically represent the partial and global profiles and levels of the alternatives.

2.2.14 MIIDAS. The Multicriteria Interactive Intelligence Decision Aiding System (MIIDAS) [115] is based on the UTA II method (see Chapter 8 of this volume). In UTA II, the assessment of the DM's additive utility model is carried out in a two step procedure: in the first step the DM expresses preferences, and in the second step the system estimates weighting factors of the decision criteria using special linear programming techniques. MIIDAS uses artificial intelligence, visual procedures, and data analysis techniques to improve the user interface and the interactive character of the system.

2.2.15 MINORA. MINORA (Multicriteria Interactive Ordinal Regression) [113] is an interactive DSS based on the UTA method [51]. The interaction takes the form of an analysis of inconsistencies between the decision maker's rankings and those derived from utility measures. The method stops when an acceptable compromise is determined. The result is an additive utility function which is used to rank the set of alternatives.

2.2.16 MUSTARD. The software MUSTARD [9] implements variants of the UTA [51] and the Quasi-UTA models [10]. It offers the basic deterministic UTA model of disaggregation, as well as its first programmed stochastic version. In both cases, the software proceeds stepwise and interactively helping the decision maker to formulate the problem and state preferences between projects;

in the stochastic case, the decision maker is also helped to build the criteria distributions.

2.2.17 NAIADE. `http://alba.jrc.it/ulysses/voyage-home/naiade/naisoft.htm`. NAIADE (Novel Approach to Imprecise Assessment and Decision Environments) [94] is a discrete multicriteria method [92] which provides an impact or evaluation matrix that may include either crisp, stochastic, or fuzzy measurements of the performance of an alternative with respect to an evaluation criterion. A peculiarity of NAIADE is the use of conflict analysis procedures integrated with the multicriteria results. NAIADE can give rankings of the alternatives with respect to the evaluation criteria (leading to a technical compromise solution), indications of the distance of the positions of the various interest groups (possibly leading to convergence of interests or to coalition formation), and rankings of the alternatives with respect to the actors' impacts or preferences (leading to a social compromise solution).

2.2.18 OnBalance. `http://www.krysalis.co.uk`. OnBalance is based on a simple weighting approach: each decision option is scored against each decision criterion, and each decision criterion is given a weight. The package then computes an overall weight for each option. Multiple hierarchies, called trees in the package, using different weights, can be created to allow for different perspectives. Thus the approach appears to be similar to AHP, but no indication is given as to how the overall weights are calculated. The package is designed to be easy to use by anyone, without much technical understanding required.

2.2.19 PREFCALC. PREFCALC [50] is an earlier implementation of the UTA method [51]. A more recent implementation of the UTA method is the UTA Plus system described separately in this chapter and also in Chapter 8 of this volume.

2.2.20 PRIAM. PRIAM (PRogramme utilisant l'Intelligence Artificielle en Multicritère) [66] takes an unstructured interactive approach to finding the most desirable alternative. The decision maker is required to make only a small number of pairwise comparisons and is not committed to an irrevocable path by the choices made on previous comparisons.

2.2.21 PRIME Decisions. `http://www.hut.fi/Units/SAL/Downloadables/`. PRIME Decisions [44] emphasizes its ability to use incomplete preference information. It relies on the PRIME method that uses interval valued ratio statements of preference. These lead to linear constraints for a series of linear programming problems. Solving the linear programs leads to dominance structures. There is an "elicitation tour" to guide the decision maker. The software is downloadable for academic use. Because of the large number of linear programs that must be solved, the approach is best suited to problems with relatively few nondominated alternatives.

2.2.22 REMBRANDT. The REMBRANDT (Ratio Estimation in Magnitudes or deci-Bells to Rate Alternatives which are Non-DominaTed) system [70, 71] requires that decision-makers make pairwise comparisons both between decision criteria to determine their relative importance and between alternatives under each criterion. Results are aggregated leading to a final impact score for each alternative, permitting a ranking of the alternatives. The REMBRANDT system was developed to overcome perceived flaws in AHP. The approaches will appear identical to the users because the same inputs are required, but some of the technical aspects are different. For example, direct ratings are on a logarithmic scale and weights are determined by use of the geometric mean, which avoids potential rank reversal. A performance comparison between the REMBRANDT system and AHP is reported in [97]. An adaptation of the system for application in negotiation is found in [122].

2.2.23 RGDB. `http://www.ccas.ru/mmes/mmeda/RGDB/index.htm`. RGDB (Reasonable Goals for Database) is a tool that supports the selection of preferable items (say, goods and services) from large lists using a simple graphic interface. The application server is based on the Reasonable Goals Method [52] and is implemented in Java. The prototype application server was developed by the Department of Mathematical Methods for Economic Decision Analysis (MMEDA) of the Russian Academy of Sciences.

2.2.24 SANNA. `http://nb.vse.cz/~jablon/sanna.htm`. SANNA [49] is an add-in application of MS Excel. It is freeware that enables solving multicriteria problems using several methods (WSA, TOPSIS, ELECTRE I, PROMETHEE II and MAPPAC). SANNA can solve problems up to 100 alternatives and 50 criteria.

2.2.25 TOPSIS. The Technique for Order Preference by Similarity to Ideal Solution (TOPSIS) proposed in [47] is based on the idea that the most preferred alternative should be the shortest distance from the ideal solution and the longest distance from the negative ideal solution. Recent modifications have extended the method to a fuzzy environment [18] and to inter-company performance comparisons using an entropy measure to provide objective weights of criteria importance [27].

2.2.26 UTA Plus. `http://www.lamsade.dauphine.fr/english/software.html#uta+`. UTA Plus is the latest Windows implementation of the UTA method, originally proposed in 1982 [51, 64]. The method can be used to solve multicriteria choice and ranking problems on a finite set of alternatives. It constructs an additive utility function from a weak preference order defined by the user on a subset of reference alternatives. The construction, based on a principle of ordinal regression, requires solving a small LP-problem. The software proposes marginal utility functions in piecewise linear form based on the given weak order, and then allows the user to interactively modify the marginal utility functions, helped by a graphical user interface. UTA methods are also discussed in Chapter 8 of this volume.

2.2.27 VIMDA. `http://www.numplan.fi/vimda/vimdaeng.htm`. VIMDA is a visual multiple-criteria DSS for MADM problems [59, 61]. VIMDA is also described in Chapter 16 of this volume, and is one of the packages evaluated in [130].

2.2.28 VIP Analysis. `http://www4.fe.uc.pt/lmcdias/english/vipa.htm`. VIP (Variable Interdependent Parameter) Analysis was introduced recently in [31]. While the approach uses a basic additive value function, it permits the decision maker to provide imprecise information for the parameters of criteria importance. In the authors' words, they propose "a methodology of analysis based on the progressive reduction of the number of alternatives, introducing a concept of tolerance that lets the decision makers use some of the approaches in a more flexible manner." Several output options exist depending on the size of the problem and the nature of the input data. For example, among the output displayed is the maximum regret for each alternative. The software is available free from the authors through the website, and an online manual is also available at the site.

2.2.29 V.I.S.A. `http://www.simul8.com/products/visa.htm`. This software's name stands for Visual Interactive Sensitivity Analysis, and the approach is described in [7]. Applying a linear multiattribute value function, it has been offered in a Windows version since 1994, emphasizing a friendly graphical interface for adjusting the criteria hierarchy and other components of the model. For example, one can interactively provide input of weights and scores using bar charts, thermometer scales, or numerical input. The weights and scores can be adjusted by dragging the computer mouse, and the effects can be seen immediately on several output windows. A variety of user licenses are available including a version oriented towards group usage. VISA is one of the packages evaluated in [130].

2.2.30 Web-HIPRE. `http://www.hipre.hut.fi/`. Web-HIPRE is Internet accessible software based on AHP and value functions [93]. The web feature permits information about the alternatives or criteria to be shared by a scattered group. The local use variant is called HIPRE 3+. It permits the user to customize the attribute scale and to combine approaches such as AHP and value functions in a single model.

2.2.31 WINPRE. `http://www.hut.fi/Units/ SAL/Downloadables/`. WINPRE [109] is another software tool available from the Systems Analysis Laboratory in Finland, the group that also offers PRIME Decisions and Web-HIPRE described earlier. WINPRE relies on a methodology called PAIRS (Preference Assessment by Imprecise Ratio Statements) that permits the decision maker to state a range of numbers to indicate preferences among alternatives. These result in linear constraints that lead to a feasible region for each criterion that is consistent with the decision maker's judgments. The software is available free for academic use.

2.3 General Multiple Objective Decision Making (MODM)

In the MODM models, the criteria are expressed in the form of mathematical objective functions that are to be optimized. The argument vectors of the objective functions are decision variables that can usually take on an infinite number of values. The MODM models may involve linear or nonlinear objective functions and constraints, and may have continuous or integer decision variables.

2.3.1 ADBASE. ADBASE, originally written in FORTRAN, implements multiobjective linear programming (MOLP) methods to enumerate extreme points and unbounded efficient edges [119]. ADBASE is maintained at the Terry College of Business, University of Georgia, Athens, Georgia, USA.

2.3.2 Feasible Goals Method (FGM). `http://www.ccas.ru/mmes/mmeda`. The concept of the Feasible Goals Method is to explore possible results of all feasible decisions [76]. In the FGM software package, the objective information on the decision situation is displayed in graphical form as various decision maps. An efficiency frontier displays an objective (criterion) tradeoff among two criteria. By changing one efficiency frontier for another, the user can see how the increment (or decrement) of the value of the third criterion influences the efficiency frontier. Decision maps are provided by the Interactive Decision Maps (IDM) technique [72, 74, 77].

2.3.3 Feasible Set in Criterion Space (FSCS). `http://www.ccas.ru/mmes/mmeda`. The FSCS software allows visualization of the feasible set in the criterion space for nonlinear problems [73]. The decision maker obtains a general orientation in the criterion space that may help him or her access the limits of what is possible in terms of the criteria. In the case of more than two criteria, visualization is based on approximating the feasible set in the criterion space by simple figures and subsequent on-line displays of the approximations using two-criterion slices. Visualization of the FSCS can be incorporated into various multicriteria methods. The software was coded in the form of an add-in tool for MS Excel. It consists of four subsystems. The first one helps to formulate a nonlinear model using MS Excel. The second one helps to specify criteria and approximation parameters. The covering base is constructed in the form of a table in the third subsystem. The last subsystem helps visualize the approximation and select a preferred goal.

2.3.4 MOMHLib++. `http://www-idss.cs.put.poznan.pl/~jaszkiewicz/MOMHLib/` MOMHLib++ (Multiple Objective MetaHeuristics Library in C++) is a library of C++ classes that implements a number of multiple objective metaheuristics. The library includes Pareto simulated annealing [25], multiple objective simulated annealing [110, 121], the Pareto memetic algorithm [55], multiple objective genetic local search [48, 54], multiple objective multiple start local search, non-dominated sorting genetic algorithm and controlled NSGA II [26, 116], and the Strength Pareto Evolutionary Algorithm [131]. Each method is implemented in a separate template class that utilizes a number of common library classes. The template classes are instantiated with classes corresponding to solutions of a given problem. In order to adapt one or more of the multiple objective metaheuristics to a given problem one has to implement a new class corresponding to the problem's solution by deriving from the library class TMOMHSolution. A detailed design pattern described in the documentation of MOMHLib++ illustrates the suggested way of adapting the library to a given problem. The library is implemented in standard C++.

2.3.5 MultiGen. MultiGen [89] contains both an optimization system and a heuristic genetic algorithm solver. It can be used for solving linear and nonlinear multiple objective programming models and also large integer problems. It is not a tool for the casual decision maker because several parameters must be set to guide the genetic algorithm's search process; however, an interactive environment permits the user to more easily change these parameters. A published study shows that the system is capable of finding the optimal solution based on decision maker preferences for models with up to 100 constraints and 200 variables.

2.3.6 Multistat Optimizer. `http://www.multistat.com`. Multistat Optimizer is based on a method of visualization for models by projection (VMPF). The VMPF method [99] differs from other multiple objective methods by working with a multidimensional dataset using visualization techniques.

2.3.7 SOLVEX. `http://www.ccas.ru/pma/product.htm`. SOLVEX is an integrated application package intended for solving nonlinear constrained optimization problems, multi-variable global optimization problems, and nonlinear multicriteria problems [102]. It uses convolution (including goal programming) and direct approximation algorithms for the multicriteria problem solving part.

2.3.8 TRIMAP. TRIMAP [19] is an interactive approach that explores the Pareto optimal set for three-criterion linear programming models. The aim is to aid the decision maker in eliminating parts of the Pareto optimal solution set that are judged to be of less value. The limitation to three objectives permits graphical displays that facilitate the decision maker's information processing. The procedure does not converge to a particular solution, but the decision maker can stop the process when sufficient information has been learned about the Pareto optimal solutions.

2.3.9 TOMMIX. This is an interactive package [4] designed to be a flexible tool for a decision maker. The software incorporates various methods of multiple objective optimization including STEM, Zionts-Wallenius, interval criterion weights [118], Pareto Race [63], and TRIMAP [19]. Designed for three-objective problems, TOMMIX has an emphasis on graphics and the decision maker interface.

2.3.10 VIG. `http://www.numplan.fi/vig/vigeng.htm`. VIG, a visual, dynamic, and interactive DSS for MODM problems, can handle linear programming constraint matrices with 96 columns and 100 rows, of which 10 rows may constitute the objective functions [58, 60, 61]. This software, also described in Chapter 16 of this volume, is based on the Pareto Race technique [63].

2.3.11 WWW-NIMBUS. `http://nimbus.mit.jyu.fi/`. WWW-NIMBUS [88] has been designed to solve differentiable and non-differentiable multi- and single objective optimization problems subject to nonlinear and linear constraints with bounds on the variables. It can also accommodate integer variables. WWW-NIMBUS can be accessed via the Internet and is free for academic use.

WWW-NIMBUS implements the classification-based NIMBUS me- thod. The problems are stated as minimization problems. Therefore, it is assumed that the user prefers less to more for the objective function values. At each iteration, the decision maker divides the objective functions into five mutually exclusive classes and provides the desirable changes. The proximal bundle method and genetic algorithms are used as underlying solvers.

2.4 Multiple Criteria Sorting Problems

Software in this category is designed to sort decision alternatives into predefined groups or classes.

2.4.1 ELECTRE TRI. `http://www.lamsade.dauphine.fr/english/software.html#TRI`. ELECTRE TRI [29, 90, 91] sorts alternatives by using reference alternatives and outranking relations. Two procedures (pessimistic and optimistic) are provided to deal with situations in which specific alternatives are incomparable with some reference alternatives. The ELECTRE TRI software, written for Windows in C++, was developed jointly by LAMSADE at the University of Paris-Dauphine, France, and the Institute of Computer Science at Poznan University of Technology, Poland.

2.4.2 IRIS. `http://www4.fe.uc.pt/lmcdias/iris.htm`. IRIS (Interactive Robustness analysis and parameters' Inference for multicriteria Sorting problems) is a DSS for sorting a set of actions (alternatives, projects, candidates) into predefined ordered categories, according to their evaluations (performances) on multiple criteria [30]. Application examples would be sorting funding requests according to merit categories, such as "Very good", "Good", "Fair", "Not eligible", or sorting loan applicants into categories such as "Accept", "Require more collateral", "Reject". IRIS uses a pessimistic concordance-only variant of the ELECTRE TRI method [29]. Rather than demanding precise values for the ELECTRE TRI parameters, IRIS allows one to enter constraints on these values. It adds a module to identify the source of inconsistency among the constraints when it is not possible to respect all of them at the same time, according to a method described in [20]. On the other hand, if the constraints are compatible with multiple assignments for the actions, IRIS allows drawing robust conclusions by indicating the range of assignments (for each action) that do not contradict any constraint.

2.4.3 PREFDIS. PREFDIS [133] is based on a preference disaggregation approach. Different sorting techniques are available, and the system provides a graphical user interface. It has been used in several applications, especially in financial management. These applications have included portfolio selection and management, country risk assessment, and the evaluation of bank branches.

2.4.4 PROAFTN. PROAFTN is a fuzzy multicriteria classification method belonging to the class of supervised learning algorithms; it enables the determination of fuzzy indifference relations by generalizing the indices (concordance and discordance) used in the ELECTRE III method. The fuzzy belonging degree of the alternatives is assigned to the categories. A clinical application of the proposed method in the cytopathological diagnosis of acute leukemia is presented in [6].

2.4.5 TOMASO. `http://cassandra.ro.math.ulg.ac.be/`. TOMASO (Tool for Ordinal Multiattribute Sorting and Ordering) is freeware written in Visual Basic for sorting in the presence of qualitative interacting points of view [79]. The underlying methodology is described in [104]. TOMASO is described in more detail in Chapter 12 of this volume.

2.5 Portfolio Analysis

Software in this category deals with problems where a set, or portfolio, of alternatives is required, rather than the best single alternative.

2.5.1 HiPriority. `http://www.krysalis.co.uk`. HiPriority is designed to find best portfolio solutions, i.e. each solution is a set of alternatives subject to resource constraints. Weights are assigned to criteria and alternatives, and the software allows specifying dependencies between alternatives, as well as specifying mutually exclusive alternatives. To visualize benefit/cost ratios, the package creates simple value trees of cost elements together with their corresponding benefits, where cost is defined as any scarce resource. Miniature graphical views of the models are used as navigational tools.

2.6 Group Decision Support

Software in this category is specifically designed to deal with the situation of multiple decision makers. However, several of the other packages described in this chapter also claim to be able to handle multiple decision maker situations.

2.6.1 AGAP. AGAP (Aid to Groups for Analysis and evaluation of Projects) is a distributed group decision support system allowing multiple decision makers to cooperate in the evaluation and selection of investment projects [22]. AGAP supports both synchronous and asynchronous usage, providing decision support at individual, inter-personal, and collective levels. For individual multi-criteria evaluation, AGAP offers additive and multiplicative utility functions, as well as the PROMOTHEE I and II methods [12]. For the sorting of projects, AGAP incorporates ELECTRE TRI, described separately in this chapter.

2.6.2 ARGOS. ARGOS is a software tool described in [21], which focuses on facilitating a small group in ranking projects or candidates using the outranking methods. It is illustrated in the reference by applying the methodology to a jury with the task of evaluating a group of candidates for a scientific award. ARGOS is run in two phases: a multicriteria phase and a multijudge phase. The first phase uses the outranking methods to determine the ranking of candidates for individual judges. In the second phase ARGOS uses several functions of social choice to arrive at the winning candidate.

2.6.3 CTLite. `http://www.CTLite.com`. ClearThinking Lite (CTLite) is an internet based, collaborative, multi-criteria decision modeling environment for evaluating and ranking alternatives along parameter sets. It uses a hierarchical or network approach, where criteria or attributes are established within "communities" of decision makers and weighted by the decision makers. The decision makers also score the decision alternatives with respect to each attribute. Multiple "communities" of decision makers are accommodated by adding another level to the hierarchy. CTLite is an end-to-end XML application built on an Oracle 8i database using Oracle XML Developer Kit and related components.

2.6.4 GMCR. The decision support system GMCR (Graph Model for Conflict Resolution) [46] can model strategic decisions, forecast compromise solutions, and assist in assessing the political, economic, environmental, and social viability of alternative scenarios to resolving conflicts. The software is based on the graph model for conflict resolution [36].

2.6.5 Joint Gains. `http://www.jointgains.hut.fi/mid.html`. Joint Gains is negotiation support software based on the method of improving directions [34]. In this method, joint gains are searched starting from an initial point, such as a previously reached agreement. Each iteration in the mediation process tries to find a jointly preferred alternative to the current one. An improving direction and a most preferred alternative in that direction is obtained from the participants by pair wise comparison questions. Joint Gains uses an algorithm based on optimization theory and the golden section method in identifying the most preferred direction. Alternatives are represented to the participants in the form of decision variable values and criteria function values. This software is web-based.

2.6.6 MEDIATOR. MEDIATOR is a negotiation support system (NSS) based on evolutionary systems design (ESD) and database-centered implementation [53, 111]. It supports negotiations by consensus seeking through exchange of information and, where consensus is incomplete, by compromise. The negotiation problem is shown graphically in three spaces as a mapping from control space to goal space and (through marginal utility functions) to utility space. Within each of these spaces the negotiation process is characterized by adaptive change, i.e., mappings of group target and feasible sets by which these sets are redefined in seeking a solution characterized by a single-point intersection between them. Each player employs private and shared database views, using his/her own micro-computer decision support system enhanced with a communications manager to interact with the MEDIATOR DSS.

2.6.7 SCDAS. (Selection Committee Decision Analysis and Support) This tool [67] is designed to support groups that have a common goal and need to work cooperatively to select a best alternative. It aids in identifying aspiration levels, assessing disagreements, aggregating the assessments of individual group members, etc.

2.6.8 WINGDSS. WINGDSS [24] is a group decision support system for multiple attribute problems. WINGDSS provides a final score for every alternative and thus a complete ranking. Voting powers are assigned to each decision maker for each criterion. Preference weights are given directly by the users. Sensitivity analysis permits studying the effect of the variations of parameters such as individual preferences, voting powers, and scores.

2.7 Some Application Specific Software

Here we reference some decision support software packages that have been developed for very specific applications. A large number of such packages exist and have been published in several journals and technical reports. We do not claim that our list is even close to being complete.

2.7.1 ACADEA. ACADEA is a multi-criteria decision support system for the performance review of individual faculty in a university [1]. The system considers the aggregate performance of an academic department using the result of individual faculty member evaluations. Criteria are established in the areas of research output, teaching output, external service, internal service and cost. Incorporating the approach of data envelopment analysis, the system can be used as an academic policy aid.

2.7.2 AgentAllocator. This is an agent-based multi-criteria DSS for task allocation [82].

2.7.3 AutoMan. `http://www.ntis.gov`. AutoMan [125] is an implementation of AHP [108] designed to support decisions about automated manufacturing investments. It is one of the packages evaluated in [98].

2.7.4 BANKADVISOR. Focused on industrial clients, this decision support tool [78] assists financial analysts in making decisions, such as offering loans and setting their terms. This DSS uses financial data from balance sheets and income statements. The multicriteria part is based on the PROMETHEE [13] method.

2.7.5 CASTART. CASTART is an interactive multicriteria package for selecting electricity production alternatives [38].

2.7.6 CGX. This is an expert system [117] designed to support credit granting decisions in non-financial firms. In addition to an inference engine, it uses AHP [108] to link credit evaluation and credit limit determination.

2.7.7 DIDASN++. DIDASN++ is an interactive, multi-criteria based, system for modeling engineering applications [42, 129]. It is a modular and more modern version, written in C++, of the older program DIDASN, originally written in Pascal [65].

2.7.8 DIMITRA. DIMITRA is a DSS for agricultural products development decisions [84].

2.7.9 Electrical Power Districting DSS. This DSS allows decision makers to partition a power grid into economically viable units as might be required under deregulation [8]. Criteria include measures of revenue balance among districts and the geographical compactness of districts. A genetic algorithm was used as the search engine for Pareto optimal solutions. Decision makers can use the DSS to explore non-Pareto optimal alternatives based on judgment applied to the less structured aspects of the problem.

2.7.10 ESY. ESY (Evaluation SYstem) [100] helps decision makers make more rational decisions and promote consistency in their decision making throughout all phases of a nuclear emergency. There are different requirements at each phase. For example, during the early phase, the decision makers are under pressure to take a decision in a short period of time whereas during the middle phases, the decision makers have more time to balance the costs and benefits of the protective actions. The ESY provides decision support not only in the evaluation of the strategies, but also in the formulation and appraisal of the decision problem. The authors also mention several other decision support systems, ranging from rule-based systems to those using multi-attribute value and utility theory, which evaluate strategies in nuclear emergencies.

2.7.11 FINCLAS. `http://www.dpem.tuc.gr/fel/`. The Financial Classification (FINCLAS) multicriteria decision support system [132, 134] incorporates financial modeling tools, along with preference disaggregation methods that lead to the development of additive utility models for the classification of the alternatives into predefined classes.

2.7.12 FINEVA. FINEVA is a knowledge based multi-criteria DSS for the assessment of corporate performance and viability [135].

2.7.13 INVEX. INVEX (Investment Advisory expert system) [124] combines several methods to aid business decision makers in selecting capital investment projects. The part that uses a multicriteria method relies on an extension of the PROMETHEE [13] approach. Several static and dynamic measures can be used, e.g., mean net present value of the investment, coefficient of variation for the return on investment, etc. Knowledge from experts and risk assessment methods are also employed in this system that the authors describe as a "multiparadigm" method.

2.7.14 MARKEX. Market Expert (MARKEX) [83, 112] provides decision support for various stages in the product development process. The system's model base encompasses statistical analysis, multicriteria analysis, and consumer choice models.

2.7.15 MEDICS. This is a knowledge-based system [32] to aid in medical diagnosis by distinguishing among possible diseases. It includes a final PROMETHEE [13] multicriteria analysis to improve results.

2.7.16 MOIRA. MOIRA is a DSS for selecting remedial strategies to restore water systems after accidental introduction of radioactive substances [103]. It includes an evaluation module based on a multi-attribute value model to rank alternatives and a module to perform multi-parametric sensitivity analyses with respect to both weights and values.

2.7.17 SANEX. `http://www.iees.ch/EcoEng001/EcoEng001_R3.html`. SANEX is a non-commercial computer program to support planners in assessing the suitability of sanitation systems (e.g. latrines, septic tanks, and sewerage) [68]. It uses socio-cultural, financial and technical criteria in connection with multicriterion decision analysis techniques [69]. SANEX was developed at the Advanced Wastewater Management Centre (AWMC) at the University of Queensland, Australia.

2.7.18 Skills Evaluator. `http://www.astrolavos.tuc.gr/contents/skills_evaluator.htm`. Skills Evaluator (SE) is a DSS for the evaluation of an individual's information technology qualifications and skills [3]. SE models the qualitative criteria that make up the problem, and uses a multicriteria approach of evaluation, based on aggregation-disaggregation procedures. It produces input data for the ELECTRE-TRI method described separately in this chapter, which can then be used to classify the individual based on his or her qualifications.

2.7.19 Steel Hot Rolling Mill Scheduling DSS. This multi-criteria decision support system [23] provides semi-automatic schedules using a variety of bespoke local and tabu search [40] heuristics.

2.7.20 TELOS. TELOS is marketing research software for evaluating customer satisfaction. Its features and capabilities are described in [43]. This reference also reviews other customer satisfaction-oriented software and the methodology on which TELOS is based, described as a multicriteria preference disaggregation method using ordinal regression. The main objective is the aggregation of individual preferences into a collective value function based on the notion that a customer's global satisfaction depends on a set of criteria representing the product's appeal. The individuals completing a questionnaire provide both a global indicator of satisfaction and judgments concerning individual attributes. The model develops marginal satisfaction functions to render the global satisfaction criterion as consistent as possible with customer's judgments on the individual criteria. An advantage of TELOS is that it provides for qualitative customer preference inputs.

2.7.21 Water Quality Planning DSS. `http://www.ccas.ru/mmes/mmeda/papers/vodhoz.htm`. The Water Quality Planning DSS [75] is based on the Feasible Goals Method [76], which provides experts and decision makers with objective trade-off curves among cost and pollution criteria. The information improves their understanding of the problem and helps to identify wastewater treatment strategies which provide reasonable balance between cost and pollution.

3. Concluding Remarks

A number of authors have provided overviews of the available MCDM methodology and software. One of the earliest overviews was a user-oriented listing of MCDM methods by Despontin, Moscarola, and Spronk [28]. Evans [35] gave an overview of techniques for multiobjective programs, and Korhonen, Moskowitz, and Wallenius [62] published a review of multiple criteria decision support. Aksoy, Butler, and Minor [2] provided a comprehensive overview of comparative studies in interactive multiple objective mathematical programming, and Weistroffer and Narula [126] reported on the state of the art of MCDM software in 1997. Siskos and Spyridakos [114] also presented a survey of multicriteria decision support systems in 1999. Various attempts at assessing and evaluating MCDM techniques and software have also been reported. Most of these involve the testing of a newly developed technique by comparing an experimental implementation with one or two other approaches. Olson [95] provided a review of empirical studies conducted between 1973 and 1990, and also recently reported a comparison of three multicriteria methods, SMART, PROMETHEE, and a centroid method [96]. Weistroffer, Narula, and Kim [127] did an exploratory study in which they compare four commercially advertised MCDM software packages. Goicochea and Li [41] also conducted an experimental evaluation of four MCDM packages. Zapatero, Smith, and Weistroffer [130] undertook an extensive comparative evaluation of five MCDM packages and compared their effectiveness to that of a simple spreadsheet package. Ossadnik and Lange [98] used AHP to evaluate three packages implementing AHP. Any evaluation of MCDM software, of course, tests specific implementations of MCDM techniques, and it is not easy to separate the software features from the characteristics of the methodology. One problem with experimental evaluations is that more commercially oriented software packages have new releases almost every year, rendering any comparative results quickly obsolete.

Though a large variety of MCDM methods has been proposed in the literature, dealing with all aspects of the decision problem, a large majority of commercially marketed packages deal primarily with MADM problem models and focus on the comparison of alternatives and the identification of the most acceptable solution. Furthermore, these packages tend to use the simpler algorithmic approaches, whereas the MCDM literature is full of sophisticated and complex solution approaches to MCDM problems. In our list of software packages, in Section 2, we included commercially available packages as well as software packages developed at academic or research institutions. These latter packages, more likely, implement newer and more sophisticated methodology. Only a few of the commercial packages handle MODM problems, though many methods for dealing with these problems exist.

Many approaches that have been suggested in the literature have only been implemented on an ad hoc basis, to solve a specific problem situation, or as experimental software to demonstrate the salient features of the proposed underlying methodology. One example of such software is the interactive reference direction algorithm for multiobjective convex nonlinear integer programming problems of Vassilev, Narula, and Gouljashki [123]. A paper by Karaivanova and Narula [56] presents an overview of multiobjective integer programming methods, many of which have been implemented as experimental software only. We did not include these packages in our listing of software in Section 2.

An area where there is also an apparent need for more or better software is the portfolio selection problem, i.e., the situation where a set of solutions is required, rather than the best compromise solution. Only one of the available packages listed in Section 2 is specifically designed for portfolio selection although others can be used to assist in addressing this problem.

We also note that several chapters in this volume describe specific MCDM software.

References

[1] P.J. Agrell and R.E. Steuer. ACADEA – A decision support system for faculty performance reviews. *Journal of Multi-Criteria Decision Analysis*, 9(5):191–204, 2000.

[2] Y. Aksoy, T.W. Butler, and E.D. Minor. Comparative studies in interactive multiple objective mathematical programming. *European Journal of Operational Research*, 89:408–422, 1996.

[3] G. Anestis, E. Grigoroudis, E. Krasadaki, N.F. Matsatsinis, and Y. Siskos. Skills Evaluator: A multicriteria decision support system for the evaluation of qualifications and skills in information technology. *Operational Research: An International Journal*, page To appear, 2004.

[4] C.H. Antunes, M.J. Alves, A.L. Silva, and J. Clímaco. An integrated MOLP method based package – A guided tour of TOMMIX. *Computers & Operations Research*, 1(4):609–625, 1992.

[5] C. Bana e Costa and M.P. Chagas. A career choice problem: An example of how to use MACBETH to build a quantitative value model based on qualitative value judgments. *European Journal of Operational Research*, 153(2):323–331, 2004.

[6] N. Belacel. Multicriteria assignment method PROAFTN: Methodology and medical application. *European Journal of Operational Research*, 125(1):175–183, 2000.

[7] V. Belton and T.J. Stewart. *Multiple Criteria Decision Analysis*. Kluwer Academic Publishers, Dordrecht, 2002.

[8] P.K. Bergey, C.T. Ragsdale, and M. Hoskote. A decision support system for the electrical power districting problem. *Decision Support Systems*, 36:1–17, 2003.

[9] M. Beuthe and G. Scannella. *MUSTARD User's Guide*. Facultés Universitaires Catholiques de Mons (FUCaM), Mons, 1999.

[10] M. Beuthe and G. Scannella. Comparative analysis of UTA multicriteria methods. *European Journal of Operational Research*, 130(2):246–262, 2001.

[11] J.P. Brans and B. Mareschal. The PROMCALC&GAIA decision support system for multicriteria decision aid. *Decision Support Systems*, 12:297–310, 1994.

[12] J.P. Brans, B. Mareschal, and P. Vincke. How to select and how to rank projects – The PROMETHEE method. *European Journal of Operational Research*, 24:228–238, 1986.

[13] J.P. Brans and P. Vincke. A preference ranking organisation method: The PROMETHEE method for MCDM. *Management Science*, 31(6):647–656, 1985.

[14] D. Buede. Aiding insight. *OR/MS Today*, 20(2):52–60, 1993.

[15] D. Buede. Aiding insight II. *OR/MS Today*, 21(3):62–71, 1994.

[16] D. Buede. Aiding insight III. *OR/MS Today*, 23(4):73–79, 1996.

[17] D. Buede. Aiding insight IV. *OR/MS Today*, 25(4):56–64, 1998.

[18] C.T. Chen. Extensions of the TOPSIS for group decision-making under fuzzy environment. *Fuzzy Sets and Systems*, 114(1):1–9, 2000.

[19] J. Clímaco and C.H. Antunes. Implementation of a user friendly software package – A guided tour of TRIMAP. *Mathematical and Computer Modelling*, 12(10-11), 1989.

[20] J. Clímaco, L. Dias, J. Figueira, C. Gomes da Silva, and V. Mousseau. Resolving inconsistencies among constraints on the parameters of an MCDA model. *European Journal of Operational Research*, 145(2):287–303, 2003.

[21] G. Colson. The OR's prize winner and the software ARGOS: How a multijudge and multicriteria ranking GDSS helps a jury to attribute a scientific award. *Computers & Operations Research*, 27:741–755, 2000.

[22] J.P. Costa, P. Melo, P. Godinho, and L.C. Dias. The AGAP system: A GDSS for project analysis and evaluation. *European Journal of Operational Research*, 145:287–303, 2003.

[23] P. Cowling. A flexible decision support system for steel hot rolling mill scheduling. *Computers and Industrial Engineering*, 45:307–321, 2003.

[24] P. Csaki, T. Rapcsak, P. Turchanyi, and M. Vermes. R and D for group decision aid in hungary by WINGDSS, a Microsoft Windows based group decision support system. *Decision Support Systems*, 14:205–217, 1995.

[25] P. Czyzak and A. Jaszkiewicz. Pareto simulated annealing – A metaheuristic technique for multiple-objective combinatorial optimization. *Journal of Multi-Criteria Decision Analysis*, 7:34–37, 1998.

[26] K. Deb and T. Goel. Controlled elitist non-dominated sorting genetic algorithms for better convergence. In *Evolutionary Multi-Crietriaon Optimization. EMO 2001, March 7 –9, 2001, Zürich. Proceedings*, volume 1993 of *Lecture Notes in Computer Science*. Springer Verlag, Berlin, 2001.

[27] H. Deng, C.H. Yeh, and R.J. Willis. Inter-company comparison using modified TOPSIS with objective weights. *Computers & Operations Research*, 27(10):963–973, 2000.

[28] M. Despontin, J. Moscarola, and J. Spronk. A user-oriented listing of multiple criteria decision methods. *Revue Belge de Statistique, d'Informatique et de Recherche Opérationelle*, 23(4):3–110, 1983.

[29] L. Dias and J. Clímaco. ELECTRE-TRI for groups with imprecise information on parameter values. *Group Decision and Negotiation*, 9(5):355–377, 2000.

[30] L. Dias, V. Mousseau, J. Figueira, J. Clímaco, and C.G. Silva. IRIS 1.0 software. *Newsletter of the European Working Group "Multicriteria Aid for Decisions"*, 3(5):4–6, 2002.

[31] L.C. Dias and J.N. Clímaco. Additive aggregation with variable independent parameters: The VIP Analysis software. *Journal of the Operational Research Society*, 51(9):1070–1082, 2000.

[32] P. Du Bois, J.P. Brans, F. Cantraine, and B. Mareschal. MEDICS: An expert system for computer-aided diagnosis using the PROMETHEE multicriteria method. *European Journal of Operational Research*, 39:284–292, 1989.

[33] J. Dyer. A time-sharing computer program for the solution of the multiple criteria problem. *Management Science*, 19:1379–1383, 1973.

[34] H. Ehtamo, M. Verkama, and R.P. Hämäläinen. How to select fair improving directions in a negotiation model over continuous issues. *IEEE Transactions on Systems, Man and Cybernetics – Part C: Applications and Reviews*, 29(1):26–33, 1999.

[35] G.W. Evans. An overview of techniques for multiobjective mathematical programs. *Management Science*, 30:1268–1282, 1984.

[36] L. Fang, K.W. Hipel, and D.M. Kilgour. *Interactive Decision Making: The Graph Model for Conflict Resolution*. John Wiley & Sons, New York, 1993.

[37] A. Fernandez. Software review: Expert choice. *OR/MS Today*, 23:80–83, 1996.

[38] X. Gandibleux. Interactive multicriteria procedure exploiting a knowledge-based module to select electricity production alternatives: The CASTART system. *European Journal of Operational Research*, 113(2):355–373, 1999.

[39] J. Geldermann and K. Zhang. Software review: Decision Lab 2000. *Journal of Multi-Criteria Decision Analysis*, 10:317–323, 2001.

[40] A.F. Glover, E. Taillard, and D. de Werra. A user's guide to tabu search. *Annals of Operations Research*, 41(1-4):3–28, 1993.

[41] A. Goicochea and F. Li. Does it matter which MCDM method you use? – A look at the experimental evaluation of four MCDM decision support systems. In *Proceedings of Tenth International Conferene on Multiple Criteria Decision Making, Taipei, Taiwan*, volume III, pages 143–165, 1992.

[42] J. Granat, T. Kreglewski, J. Paczyeski, and A. Stachurski. IAC-DIDAS-N++ modular modeling and optimization system. Technical report, Institute of Automatic Control, Warsaw University of Technology, Warsaw, Poland, 1994.

[43] E. Grigoroudis, Y. Siskos, and O. Saurais. TELOS: A customer satisfaction evaluation software. *Computers & Operations Research*, 27:799–817, 2000.

[44] J. Gustafsson, A. Salo, and T. Gustafsson. PRIME Decisions: An interactive tool for value tree analysis. In M. Köksalan and S. Zionts, editors, *Multiple Criteria Decision Making in the New Millennium*, volume 507 of *Lecture Notes in Economics and Mathematical Systems*, pages 165–176. Springer Verlag, Berlin, 2001.

[45] W. Haerer. Software review: Criterium Decision Plus 3.0. *OR/MS Today*, 27(1), 2000.

[46] K.W. Hipel, D.M. Kilgour, F. Liping, and X.J. Peng. The decision support system GMCR in environmental conflict management. *Applied Mathematics and Computation*, 83(2-3):117–152, 1997.

[47] C.L. Hwang and K. Yoon. *Multiple Attribute Decision Making: Methods and Applications*. Springer Verlag, New York, 1981.

[48] H. Ishibuchi and T. Murata. Multi-objective genetic local search algorithm and its application to flowshop scheduling. *IEEE Transactions on Systems, Man and Cybernetics*, 28(3):392–403, 1998.

[49] J. Jablonsky. Multicriteria evaluation of alternatives in spreadsheets. In L. Lenart, L. Zadnik Stirn, and S. Drobne, editors, *Proceedings of SOR '01 Conference, Preddvor, Slovenia*. Slovenian Society Infromatika, 2001.

[50] E. Jacquet-Lagrèze. Interactive assessment of preferences using holistic judgments: The PREFCALC system. In C. A. Bana e Costa, editor, *Readings in Multiple Criteria Decision Aid*, pages 225–250. Springer Verlag, Berlin, 1990.

[51] E. Jacquet-Lagrèze and J. Siskos. Assessing a set of additive utility functions for multicriteria decision-making: The UTA method. *European Journal of Operational Research*, 10:151–164, 1982.

[52] P. Jankowski, A. Lotov, and D. Gusev. Multiple criteria trade-off approach to spatial decision making. In J.-C. Thill, editor, *Spatial Multicriteria Decision Making and Analysis: A Geographical Information Sciences Approach*, pages 127–148. Ashgate, Brookfield, VT, 1999.

[53] M. Jarke, M. T. Jelassi, and M. F. Shakun. MEDIATOR: Toward a negotiation support system. *European Journal of Operational Research*, 31(3):314–334, 1987.

[54] A. Jaszkiewicz. Genetic local search for multiple objective combinatorial optimization. Technical Report RA-014/98, Institute of Computing Science, Poznan University of Technology, 1998.

[55] A. Jaszkiewicz. A comparative study of multiple-objective metaheuristics on the bi-objective set covering problem and the pareto memetic algorithm. Technical Report RA-003/01, Institute of Computing Science, Poznan University of Technology, 2001.

[56] J. N. Karaivanova and S. C. Narula. The current state of multiple objective integer programming methods. *International Journal of Management and Systems*, 14(1):17–46, 1998.

[57] L. N. Kiss, J.-M. Martel, and R. Nadeau. ELECCALC an interactive software for modelling the decision maker's preferences. *Decision Support Systems*, 12(1994):311–326, 1994.

[58] P. Korhonen. VIG – A visual interactive support system for multiple criteria decision making. *Belgian Journal of Operations Research, Statistics and Computer Science*, 27(1):3–15, 1987.

[59] P. Korhonen. A visual reference direction approach to solving discrete multiple criteria problems. *European Journal of Operational Research*, 34(2):152–159, 1988.

[60] P. Korhonen. The multiobjective linear programming decision support system VIG and its applications. In C. Bana e Costa, editor, *Readings in Multiple Criteria Decision Aid*, pages 471–491. Springer Verlag, Berlin, 1990.

[61] P. Korhonen. Two decision support systems for continuous and discrete multiple criteria decision making: VIG and VIMDA. In A. Lewandowski, P. Serafini, and M.G. Speranza, editors, *Methodology, Implementation and Applications of Decision Support Systems*, volume 320 of *CISM Courses and Lectures, International Centre for Mechanical Sciences*. Springer Verlag, Berlin, 1991.

[62] P. Korhonen, H. Moskowitz, and J. Wallenius. Multiple criteria decision support: A review. *European Journal of Operational Research*, 63:361–375, 1992.

[63] P. Korhonen and J. Wallenius. A Pareto race. *Naval Research Logistics*, 35(6):615–623, 1988.

[64] M. Kostkowski and R. Slowinski. UTA+ application (v.1.20) – user's manual. Document de LAMSADE 95, Université Paris-Dauphine, 1996.

[65] T. Kreglewski, J. Paczyeski, J. Granat, and A. P. Wierzbicki. IAC-DIDAS-N: A dynamic interactive decision support system for multicriteria analysis of nonlinear models with a nonlinear model generator supporting model analysis. Technical Report WP-88-112, International Institute for Applied Systems Analysis, Laxenburg, Austria, 1988.

[66] P. Levine and J. C. Pomerol. PRIAM, an interactive program for choosing among multiple attribute alternatives. *European Journal of Operational Research*, 25:272–280, 1986.

[67] A. Lewandowski. SCDAS – Decision support system for group decision making: Decision theoretic framework. *Decision Support Systems*, 5:403–423, 1989.

[68] T. Loetscher. A simple expert system for evaluating sanitation systems in developing countries. *EcoEng Newsletter, International Ecological Engineering Society*, 2(1):http://www.iees.ch/EcoEng001/EcoEng001_R3.html, 2000.

[69] T. Loetscher and J. Keller. A multicriterion model for the evaluation of sanitation systems in developing countries. *Socio-Economic Planning Sciences*, to appear.

[70] F.A. Lootsma. The REMBRANDT system for multi-criteria decision analysis via pairwise comparisons or direct rating. Technical Report 92-05, Faculty of Technical Mathematics and Informatics, Delft University of Technology, Delft, Netherlands, 1992.

[71] F.A. Lootsma, T.C.A. Mensch, and F.A. Vos. Multi-criteria analysis and budget reallocation in long-term research planning. *European Journal of Operational Research*, 47:293–305, 1990.

[72] A. Lotov, O. Chernykh, V. Bushenkov, H. Wallenius, and J. Wallenius. Interactive decision maps, with an example illustrating ocean waste management decisions. In T.J. Stewart and R.C. van den Honert, editors, *Trends in Multicriteria Decision Making*, volume 465 of *Lecture Notes in Economics and Mathematical Systems*, pages 313–323. Springer Verlag, Berlin, 1998.

[73] A. Lotov, G. Kamenev, and V. Berezkin. Software for visualization of the feasible set in criterion space in nonlinear MCDA problems. *Newsletter of the European Working Group "Multicriteria Aid for Decisions"*, 3(4):8–9, 2001.

[74] A.V. Lotov, V.A. Bushenkov, A.V. Chernov, D.V. Gusev, and G.K. Kamenev. Internet, GIS and interactive decision maps. *Journal of Geographical Information and Decision Analysis*, 1(2):118–149, 1997.

[75] A.V. Lotov, V.A. Bushenkov, and O.L. Chernykh. Multi-criteria DSS for river water quality planning. *Microcomputers in Civil Engineering*, 12(1), 1997.

[76] A.V. Lotov, V.A. Bushenkov, and G.K. Kamenev. *Feasible Goals Method – Search for Smart Decisions*. Computing Centre RAS, Moscow, 2001.

[77] A.V. Lotov, V.A. Bushenkov, and G.K. Kamenev. *Interactive Decision Maps*, volume 89 of *Applied Optimization*. Kluwer Academic Publishers, 2004.

[78] B. Mareschal and J.P. Brans. BANKADVISER: An industrial evaluation system. *European Journal of Operational Research*, 54:318–324, 1991.

[79] J.-L. Marichal, P. Meyer, and M. Roubens. On a sorting procedure in the presence of qualitative interacting points of view. Technical report, University of Liege, 2002.

[80] B. Matarazzo. Multicriterion analysis of preferences by means of pairwise actions and criterion comparisons (MAPPAC). *Applied Mathematics and Computation*, 18(2):119–141, 1986.

[81] B. Matarazzo. Preference ranking global frequencies in multicriterion analysis (PRAGMA). *European Journal of Operational Research*, 36(1):36–49, 1988.

[82] N.F. Matsatsinis and P. Delias. AgentAllocator: An agent-based multi-criteria decision support system for task allocation. In V. Marik, D. McFarlane, and P. Valckenaers, editors, *Holonic and Multi-Agent Systems for Manufacturing*, volume 2744 of *Lecture Notes in Computer Science*, pages 225–235. Springer Verlag, 2003.

[83] N.F. Matsatsinis and Y. Siskos. MARKEX: An intelligent decision support system for product development decisions. *European Journal of Operational Research*, 113:336–354, 1999.

[84] N.F. Matsatsinis and Y. Siskos. DIMITRA: An intelligent decision support system for agricultural products development decisions. In *Proceedings of 3rd European Conference of the European Federation for Information Technology in Agriculture, Food and the Environment (EFITA 2001), Montpellier, France*, 2001.

[85] D.T. Maxwell. Aiding insight V. *OR/MS Today*, 27(5):28–35, 2000.

[86] D.T. Maxwell. Decision analysis: Aiding insight VI. *OR/MS Today*, 29(3):44–51, 2002.

[87] K. McKean. Decisions, decisions. *Discover Magazine*, pages 22–31, June 1985.

[88] K. Miettinen and M. M. Mäkelä. Interactive multiobjective optimization system WWW-NIMBUS on the Internet. *Computers & Operations Research*, 27(7-8):709–723, 2000.

[89] S. K. Mirrazavi, D. F. Jones, and M. Tamiz. MultiGen: An integrated multiple-objective solution system. *Decision Support Systems*, 36(177-187), 2003.

[90] V. Mousseau, R. Slowinski, and P. Zielniewicz. Presentation of the ELECTRE Tri 2.0a software. *Newsletter of the European Working Group "Multicriteria Aid for Decisions"*, 3(2), 2000.

[91] V. Mousseau, R. Slowinski, and P. Zielniewicz. A user-oriented implementation of the ELECTRE-TRI method integrating preference elicitation support. *Computers & Operations Research*, 27(7-8):757–777, 2000.

[92] G. Munda. *Multicriteria Evaluation in a Fuzzy Environment*. Physica Verlag, Heidelberg, 1995.

[93] J. Mustajoki and R.P. Hämäläinen. Web-HIPRE: Global decision support by value tree and AHP analysis. *INFOR*, 38(3):208–220, 2000.

[94] *NAIADE Manual & Tutorial*. Joint Research Center of the European Commission, Ispra, Italy, 1996.

[95] D. L. Olson. Review of empirical studies in multiobjective mathematical programming: Subject reflection of nonlinear utility and learning. *Decision Sciences*, 23:1–20, 1992.

[96] D. L. Olson. Comparison of three multicriteria methods to predict known outcomes. *European Journal of Operational Research*, 130(3):576–587, 2001.

[97] D.L. Olson, G. Fliedner, and K. Currie. Comparison of the REMBRANDT system with Analytic Hierarchy Process. *European Journal of Operational Research*, 82:522–541, 1995.

[98] W. Ossadnik and O. Lange. AHP-based evaluation of AHP-software. *European Journal of Operational Research*, 118(578-588), 1999.

[99] M. Papadrakakis, N.D. Lagaros, and V. Sevastyanov. A new visualization method of multidimensional numerical models for structural optimization. In K.J. Bathe, editor, *Proceedings of Second MIT Conference on Computational Fluid and Solid Mechanics*. Elsevier, Amsterdam, 2003.

[100] K.N. Papamichail and S. French. Decision support in nuclear emergencies. *Journal of Hazardous Material*, 71:321–342, 2000.

[101] V.V. Podinovski. A DSS for multiple criteria decision analysis with imprecisely specified trade-offs. *European Journal of Operational Research*, 113:261–270, 1999.

[102] M.A. Potapov and P.N. Kabanov. SOLVEX – System for solving nonlinear, global and multicriteria problems. In J. Dolézal and J. Fidler, editors, *Proceedings of 3rd IFIP WG-7.6 Working Conference on Optimization-Based Computer-Aided Modelling and Design, Prague, Czech Republic*, pages 343–347. Institute of Information Theory and Automation, Academy of Sciences of the Czech Republic, Prague, 1994.

[103] D. Rios Insua, E. Gallego, A. Mateos, and S. Rios-Insua. MOIRA: A decision support system for decision making on aquatic ecosystems contaminated by radioactive fallout. *Annals of Operations Research*, 95(1-4):341–364, 2000.

[104] M. Roubens. Ordinal multiattribute sorting and ordering in the presence of interacting points of view. In D. Bouyssou, E. Jacquet-Lagrèze, P. Perny, R. Slowinski, D. Vanderpooten, and P. Vincke, editors, *Aiding Decisions with Multiple Criteria: Essays in Honour of Bernard Roy*, pages 229–246. Kluwer Academic Publishers, Dordrecht, 2001.

[105] B. Roy. Classement et choix en présence de points de vue multiples (la méthode ELECTRE). *RIRO*, 8:57–75, 1968.

[106] B. Roy. ELECTRE III: Un algorithme de classements fondé sur une représentation floue des préférences en présence de critères multiples. *Cahiers du Centre d'Etudes de Recherche Opérationnelle*, 20(1):3–24, 1978.

[107] B. Roy. The outranking approach and the foundations of ELECTRE methods. *Theory and Decision*, 31:49–73, 1991.

[108] T.L. Saaty. *The Analytic Hierarchy Process*. McGraw-Hill, New York, 1980.

[109] A Salo and R.P. Hämäläinen. Preference programming through approximate ratio comparisons. *European Journal of Operational Research*, 82:458–475, 1995.

[110] P. Serafini. Simulated annealing for multiple objective optimization problems. In G.H. Tzeng, H.F. Wang, V. P. Wen, and P.L. Yu, editors, *Multiple Criteria Decision Making. Expand and Enrich the Domains of Thinking and Application*, pages 283–292. Springer Verlag, 1994.

[111] M.F. Shakun. Airline buyout: Evolutionary systems design and problem restructuring in group decision and negotiation. *Management Science*, 37(10):1291–1303, 1991.

[112] J. Siskos and N.F. Matsatsinis. A DSS for market analysis and new product design. *Journal of Decision Systems*, 2(1):35–60, 1993.

[113] J. Siskos, A. Spyridakos, and D. Yannacopoulos. MINORA: A multicriteria decision aiding system for discrete alternatives. *Journal of Information Science and Technology*, 2(2):136–149, 1993.

[114] Y. Siskos and A. Spyridakos. Intelligent multicriteria decision support: Overview and perspectives. *European Journal of Operational Research*, 113:236–246, 1999.

[115] Y. Siskos, A. Spyridakos, and D. Yannacopoulos. Using artificial intelligence and visual techniques into preference disaggregation analysis: The MIIDAS system. *European Journal of Operational Research*, 113(2):281–299, 1999.

[116] N. Srinivas and Deb. K. Multiple objective optimization using nondominated sorting in genetic algorithms. *Evolutionary Computation*, 2(2):221–248, 1994.

[117] V. Srinivasan and B. Ruparel. CGX: An expert support system for credit granting. *European Journal of Operational Research*, 45:293–308, 1990.

[118] R.E. Steuer. *Multiple Criteria Optimization: Theory, Computation, and Application*. John Wiley & Sons, New York, 1986.

[119] R.E. Steuer. The ADBASE multiple objective linear programming package. In J. Gu, G. Chen, Q. Wei, and S. Wang, editors, *Multiple Criteria Decision Making*, pages 1–6. SCI-TECH, Windsor, England, 1995.

[120] L.V. Tavares. A review of major paradigms and models for the design of civil engineering systems. *European Journal of Operational Research*, 119(1):1–13, 1999.

[121] E.L. Ulungu, J. Teghem, P. Fortemps, and D. Tuyttens. MOSA method: A tool for solving multiobjective combinatorial optimization problems. *Journal of Multi-Criteria Decision Analysis*, 8:221–236, 1999.

[122] R. van den Honert and F. Lootsma. Assessing the quality of negotiated proposals using the REMBRANDT system. *European Journal of Operational Research*, 120(1):162–173, 2000.

[123] V.S. Vassilev, S.C. Narula, and V.J. Gouljashki. An interactive reference direction algorithm for solving multiobjective convex nonlinear integer programming problems. *International Transactions in Operational Research*, 8:1–14, 2001.

[124] S. Vrane, M. Stanojevic, V. Stevanovic, and M. Lucin. INVEX: Investment advisory expert system. *Expert Systems*, 13(2):105–119, 1996.

[125] S.F. Weber and B.C. Lippiatt. AutoMan 2.0: Decision support software for automated manufacturing investments. Technical report nistir 4543, National Institute of Standards and Technology, 1991.

[126] H.R. Weistroffer and S.C. Narula. The state of multiple criteria decision support software. *Annals of Operations Research*, 72:299–313, 1997.

[127] H.R. Weistroffer, S.C. Narula, and J.W. Kim. A comparative evaluation of MADM based DSSs. In *Proceedings of Tenth International Conference on Multiple Criteria Decision Making, Taipei, Taiwan*, volume II, pages 195–204, 1992.

[128] M.M. White and S.W. Palocsay. Software review: Decision Explorer. *OR/MS Today*, 26(5), 1999.

[129] A.P. Wierzbicki and J. Granat. Multi-objective modeling for engineering applications: DIDASN++ system. *European Journal of Operational Research*, 113:374–389, 1999.

[130] E.G. Zapatero, C.H. Smith, and H.R. Weistroffer. Evaluating multiple-attribute decision support systems. *Journal of Multi-Criteria Decision Analysis*, 6:201–214, 1997.

[131] E. Zitzler and L. Thiele. Multiobjective evolutionary algorithms: A comparative case study and the strength pareto evolutionary algorithm. *IEEE Transactions on Evolutionary Computation*, 3(4):257–271, 1999.

[132] C. Zopounidis and M. Doumpos. Developing a multicriteria decision support system for financial classification problems: The FINCLAS system. *Optimization Methods and Software*, 8(3-4):277–304, 1998.

[133] C. Zopounidis and M. Doumpos. PREFDIS: A multicriteria decision support system for sorting decision problems. *Computers & Operations Research*, 27(7-8):779–797, 2000.

[134] C. Zopounidis and M. Doumpos. A preference disaggregation decision support system for financial classification problems. *European Journal of Operational Research*, 130:402–413, 2001.

[135] C. Zopounidis, N.F. Matsatsinis, and M. Doumpos. Developing a multicriteria knowledge-based decision support system for the assessment of corporate performance and viability: The FINEVA system. *Fuzzy Economic Review*, 1(2):35–53, 1996.

Contributing Authors

Carlos A. Bana e Costa is Full Professor of Decision and Information at the Technical University of Lisbon "Instituto Superior Técnico" (IST), Department of Engineering and Management, Centennial Professor of Operational Research at the London School of Economics and Political Science, and president of the Center of Management Studies of IST. His primary interests are in the fields of management and decision sciences, namely multicriteria decision analysis and decision conferencing, and he has published widely in these areas. He is co-author of the MACBETH approach. He is also a decision-aid consultant of many private and public organizations in Portugal, Brazil and other countries, following the socio-technical facilitation perspective shared by the members of the International Decision Conferencing Forum.

Denis Bouyssou holds a doctorate in Operations Research. He is presently senior researcher at the Centre National de Recherche Scientifique (CNRS). His research interests include decision theory, social choice theory, and multiple criteria decision making.

Jean-Pierre Brans received his Ph.D. in Mathematics at the ULB/VUB University of Brussels (1966). He has been Professor in these Universities since 1964 and dean of the VUB-Solvay Business school (1975-78). He has held visiting professorships at the universities of Kinshasa, Constantine, Aix-en-Provence, ENSTA (Paris), AIT (Bangkok), KUL (Leuven), and Lulea (Sweden). He has taught courses taught on statistical analysis, OR, MCDA, mathematical programming, and system dynamics. Dr. Brans has been president of the Belgian OR society (1975-78), president of EURO (1983-84), vice-president of IFORS (1977-80 and 1989-92), initiator of the EURO K Conferences and organisor of the first one in Brussels (1975), initiator of the EURO Summer Institutes and organisor of the two first ones, initiator of the MINI EURO Conferences and organisor of the 1st, the 7th and the 12th ones. He has presented over 100 invited lectures all over the world and published over 100 papers in international scientific journals. He is the initiator of the PROMETHEE-GAIA Methodology and has written one book on this subject with B. Mareschal. He received the EURO Gold Medal 1994 and Doctor Honoris Causa from Copenhagen (2000) and Fribourg (2002). He has been elected Professor of the Year by the students of the VUB Brussels (2002).

João C. N. Clímaco is Full Professor at the School of Economics, University of Coimbra, and researcher at the Systems Engineering and Computers Institute INESC Coimbra. Currently he is president of the Scientific Committee of INESC Coimbra. He obtained an M.Sc. in Control Systems from Imperial College, University of London, and the Diploma of Membership of the Imperial College of Science and Technology (1978), a Ph.D. (1982) in Electrical Engineering from the University of Coimbra, and the Aggregation (1989) from the University of Coimbra. His current interests of research are in multicriteria decision making, group decision and negotiation, decision support systems, and telecommunication planning, design, and management. He is author (or co-author) of more than seventy papers published in refereed international scientific journals and about thirty papers (or chapters) published in thematic scientific books. He belongs to the editorial board of the "Group Decision and Negotiation Journal" and of "Investigação Operacional". He was vice-president of APDIO and of ALIO. Currently he belongs to the international executive committee of the International Society on MCDM.

José Manuel F. Craveirinha is Full Professor in Telecommunications at the Department of Electrical Engineering Science of the Faculty of Sciences and Technology of the University of Coimbra, Portugal, since 1997. He obtained the following degrees: Undergraduate Diploma in Electrical Engineering Science (E.E.S.) – Telecommunications & Electronics at IST, Lisbon Technical University (1975); M.Sc. (1981) and Ph.D. (1984) in E.E.S. at the University of Essex (UK) and Doctor of Science in E.E.S – Telecommunications at the University of Coimbra (1996). Previous positions were: Associate Professor, Assistant Professor, and Assistant Lecturer at FCUT, Coimbra University, and telecommunications R&D engineer at CET-Portugal Telecom. He has coordinated a research group in teletraffic theory & network planning at INESC-Coimbra R&D institute since 1986 and was director of this institute 1994-99. He is author/co-author of more than 95 scientific and technical publications in modelling of teletraffic, reliability analysis and planning of telecommunication networks. His main present interests are in traffic modelling and routing in Internet and multiple objective routing in multiservice networks.

Jean-Marie De Corte is Assistant Professor at the University of Mons-Hainaut and member of the Warocqué Research Center. He received his Ph.D. in Mathematics in 2002. His current research interest is in the field of multiple criteria decision aid, more specifically on development of algorithms for "minimal solving" of inconsistencies and for several robustness analyses in the framework of the MACBETH approach. He is also the designer of the MACBETH software. His work has been published in international journals such as Omega.

Danae Diakoulaki received her Ph.D. degree in Engineering (Operations Management) at the National Technical University of Athens (NTUA) in 1988. She is currently an Associate Professor at the Chemical Engineering Department of NTUA and heads the area of energy and environmental management at the Laboratory of Industrial and Energy Economics. Her research activities concern mainly the use of MCDA methods in energy and environmental planning as well as the exploitation of energy externalities in energy policy making. She has participated in several national and European Commission research projects and published numerous articles in refereed international journals.

James S. Dyer occupies the Fondren Centennial Chair in Business in the College of Business Administration, University of Texas at Austin. Dr. Dyer's research and teaching interests include risk management and capital budgeting, and he has published extensively on these subjects in various journals, including Management Science and Operations Research. He is the former Chair of the Decision Analysis Society of the Operations Research Society of America (now INFORMS). He is the Area Editor for the field of decision analysis for the journal Operations Research. In 2002 he received the Ramsey Award for outstanding career achievements from the Decision Analysis Society. Dr. Dyer has consulted with a number of companies regarding the application of decision and risk analysis tools to a variety of practical problems, including Amoco, Texaco, Unocal, ENI, and the Department of Energy.

Matthias Ehrgott is Associate Professor of Operations Research in the Department of Engineering Science, The University of Auckland, New Zealand. He obtained his Diploma, Ph.D. and Dr. habil. degrees in Mathematics from the University of Kaiserslautern, Germany, in 1992, 1997, and 2001, respectively. From 1997 to 2000 he was Assistant Professor at the University of Kaiserslautern. He has been invited professor at Université de Valenciennes, Copenhagen University and Mercator Visiting Profesor at the University of Kaiserslautern. Dr. Ehrgott is the author or editor of six books, and has written more than 30 refereed articles on multiobjective and combinatorial optimization that have been published in international journals or proceedings volumes. He is a member of the international executive committee of the MCDM Society, member of the council of the Operations Research Society of New Zealand, editor of OR Spectrum and associate editor of Asia Pacific Journal of Operational Research. He received the Wiley Prize for best applied paper in MCDA at the International Conference on MCDM in 2002.

José Figueira is an Associate Professor at the School of Economics of the University of Coimbra, Portugal, and researcher at INESC-Coimbra and LAMSADE, Paris-Dauphine University, France. He obtained his Ph.D. (1996) in Operations Research from the University of Paris-Dauphine. He has been invited researcher London School of Economics, Rutgers University, Auckland

University, University of Catania, University of Talca, Free University of Brussels, Clemson University, University of Georgia, University of Valenciennes and Carnegie Mellon University. His current research interests are in decision analysis, integer programming, network flows and multiple criteria decision aiding. His works have been published in international journals such as European Journal of Operational Research, Computers & Operations Research, Journal of the Operational Research Society, Journal of Mathematical Modelling and Algorithms, European Business Review. He is the current editor of the newsletter of the European Working Group on MCDA.

Michel Grabisch received his Graduate Engineer Diploma in 1979 and his Ph.D. degree in signal processing in 1982, both from Ecole Nationale des Ingénieurs Electriciens de Grenoble (ENSIEG). From 1984 to 1993 he worked at Thomson-Sintra Activités Sous-Marines and from 1993 to 2000 at Central Research Laboratory of Thomson-CSF. He spent two years from 1989 to 1991 at Tokyo Institute of Technology, Japan, and participated in the LIFE (Laboratory for International Fuzzy Engineering Research) project. From 2000 to 2002 he was Associate Professor at Université Pierre et Marie Curie, Paris. Since 2002, he is Professor of Computer Science at Université Panthéon-Sorbonne, Paris. He is area editor of IEEE Transactions on Fuzzy Systems, Fuzzy Optimization and Decision, and belongs to the editorial board of Fuzzy Sets and Systems. His interests are fuzzy measure and capacity theory, decision making, fuzzy sets and possibility theory as well as discrete mathematics.

Salvatore Greco is Full Professor at the Faculty of Economics of Catania University since 2001. His main research interests are in the field of multicriteria decision aid, in the application of the rough set approach to decision analysis, in the axiomatic foundation of multicriteria methodology and in the fuzzy integral approach to MCDA. In these fields he cooperates with many researchers of different countries. He received the Best Theoretical Paper Award by the Decision Sciences Institute (Athens, 1999). Together with Benedetto Matarazzo, he organized the VIIth International Summer School on MCDA (Catania, 2000). He is author of many articles published in international journals and specialized books. He has been invited professor at Poznan Technical University and at the University of Paris Dauphine. He has been invited speaker in international conferences. He is referee of the most relevant journals in the field of decision analysis.

Evangelos Grigoroudis received his M.Sc. and Ph.D. degrees in Decision Sciences and Operations Research from the Technical University of Crete (Greece) in 1996 and 1999, respectively. He is Lecturer on Management of Quality Processes at the Department of Production Engineering and Management, Technical University of Crete. During 1999-2002 he was Adjunct Professor at the Department of Political Sciences, University of Crete. He is member of the International Society on Multiple Criteria Decision Making, the EURO Working Groups on Multicriteria Aid for Decisions and Financial Modelling, the American Society for Quality, the Hellenic Operational Research Society, and the Hellenic Institute of Production and Operations Management. He is the author of a book on the measurement of service quality and a large number of research reports and papers in scientific journals and conference proceedings. His research interests include operational research, multicriteria decision analysis, and management and control of quality.

Carlos Henggeler-Antunes received his Ph.D. degree in Electrical Engineering (Optimization and Systems Theory) from the University of Coimbra in 1992. Presently, he is an Associate Professor at the Department of Electrical Engineering and Computers, University of Coimbra, and director of the R&D Institute INESC Coimbra. His research interests include multiple objective programming, management of uncertainty in decision support models, energy-environment models, and energy policy and planning. He was secretary (1993-95) and vice-president (1995-99) of APDIO (Portuguese OR Society). He was chairman of the Programme Committee of the 9th Congress of APDIO (2000) and the 15th Mini EURO Conference on "Managing Uncertainty in Decision Support Models" (2004). His most recent works have been published in the European Journal of Operational Research, Decision Support Systems, Energy, IEEE Transactions in Power Systems.

Masahiro Inuiguchi received a D.E. degree in Industrial Engineering at Osaka Prefecture University in 1991. He worked as a Research Associate at Osaka Prefecture University and an Associate Professor at Hiroshima University and later at Osaka University. Presently, he is a Full Professor at the Department of Systems Innovation, Graduate School of Engineering Science, Osaka University. He is an editor of two books and has written more than 100 refereed journal and proceedings articles. He is an area editor of Fuzzy Optimization and Decision Making and a member of editorial committee of Fuzzy Sets and Systems. He received the Best Paper Award by Japan Society for Fuzzy Theory and Systems in 1997. He is interested in possibility theory, fuzzy mathematical programming, rough sets, Dempster-Shafer's theory of evidence, and approximate reasoning.

Pekka J. Korhonen has been Professor of Statistics at the Helsinki School of Economics, Finland since 1988. He received a Ph.D. in Applied Mathematics from the University of Helsinki. His research interests are MCDM, productivity/efficiency analysis, and computational statistics. Over 60 refereed journal articles have appeared in journals like Journal of the Operational Research Society, Management Science, European Journal of Operational Research, Naval Research Logistics, Operations Research, etc. He is member of the editorial boards of the journals Theory and Decision, Group Decision and Negotiation, and Journal of Productivity Analysis. He was the President of International Society on MCDM during 1996-2000, and is currently the member of the international executive committee. In the ISI Web of Science, there are about 750 citations to his scientific work. The George Cantor-Prize was awarded to him by the International Society on MCDM in 1994. He is an honorary chairman of the Finnish Operations Research Society.

Christophe Labreuche received a Graduate Engineer Diploma from Ecole Centrale de Lyon and a Master in Numerical Analysis, both in 1993. He received a Ph.D. degree in Applied Mathematics in 1997 from Université de Paris IX Dauphine. His first publications were in the areas of partial differential equations and more specifically scattering and inverse scattering. He was working at University of Delaware (USA) during the 1995/1996 academic year. From 1997 to 1998 he was working at the research lab of Thales in numerical analysis. Then he joined the advanced software department to work on multicriteria decision aid. On top of conducting industrial applications on MCDA and developing a software on MCDA, he works in the areas of fuzzy logic as well as fuzzy measure theory, fuzzy integrals and their application in MCDA. His field of interest includes also the representation of uncertainty and vagueness, decision theory and the modeling of expert knowledge.

Bertrand Mareschal received his Ph.D. in Mathematics from the ULB Free University of Brussels (1989). He has been Professor at the ULB since 1993. He has been part-time professor at the UMH (Université de Mons-Hainaut, Belgium), HEC-Liège (Belgium), EDHEC (Lille, France) and Université de Lille-3 (France) as well as visiting professor at Ecole des Mines de Nancy (France). He taught courses in statistics, operations research, decision aid, computer science and mathematics. He is chairman of IDM (Innovative Decision for Management – www.idm-belgium.com) and iitiator of the QED multicriteria decision aid web project (www.qed-solutions.com).

Jean-Marc Martel received B.Sc. and M.Sc. degrees in Mathematics at Laval University in 1963 and 1965 and a Ph.D. degree in Applied Economics (Quantitative Methods) at Leuven University in 1975. Since 1965 he has been working at the Faculty of Business Adminstration of Laval University, from 1965 to 1976 as Associate Professor and from 1976 to 2000 as Full Professor in the Department of Operations and Decision Systems. Since 2001 he is Emeritus Professor at Laval University and since 2002 member of Royal Society of Canada He has been invited professor at several universities and research centers. He has been organiser of FRANCORO III (Québec, 2001), MOPGP'98, the 48th meeting of the European Working Group on MCDA and the Fourth International Summer School on MCDA (Québec, 1991). His main research interests are multi-criteria decision aid under uncertainty, Bayesian analysis, information value, group decision and participative processes.

António Gomes Martins received his Ph.D. degree in Electrical Engineering from the University of Coimbra in 1985. He is presently Full Professor at the Department of Electrical Engineering at this University, where he is responsible for a R&D group on efficient use of energy resources. Since 1999 he is leading the Institute of Systems Engineering and Computers of Coimbra. His current research interests are energy planning, load modelling, energy market transformation.

Benedetto Matarazzo is Full Professor at the Faculty of Economics of Catania University. He has been committee member of scientific societies of operational research. He has been organiser, member of the programme committee, and invited speaker in many scientific conferences. He is member of the editorial boards of the European Journal of Operational Research, Journal of Multi-Criteria Decision Analysis, Foundations of Computing and Decision Sciences. He has been chairman of the programme committee of EURO XVI (Brussels, 1998). His research is in the fields of MCDA and rough sets. He has been invited professor at and co-operates with several European universities. He received the Best Theoretical Paper Award by the Decision Sciences Institute (Athens, 1999). He is member of the Organising Committee of the International Summer School on MCDA, of which he organized the first (Catania, 1983) and the VIIth (Catania, 2000) editions.

Nikolaos F. Matsatsinis is Associate Professor of Information and Decision Support Systems at the Department of Production Engineering and Management, Technical University of Crete (Greece). He received his B.A. in Physics from Aristotle University of Thessaloniki (Greece) and his Ph.D. in Intelligence Decision Support Systems from Technical University of Crete (Greece) in 1980 and 1995, respectively. He is the author or co-author of five books and over 35 articles in international scientific journals and books. He teaches the following courses: Decision Support Systems, Knowledge Engineering, Electronic Commerce, Advanced Issues in Information and Decision Systems (postgraduate) and Distributed Artificial Intelligence and Multi-Agent Systems (postgraduate). His research interests fall into the areas of decision support systems, artificial intelligent and multi-agent systems, e-business, e-marketing, multicriteria decision analysis, and group decision support systems.

Alexander Mechitov received his Ph.D. degree in Management Information Systems from the Institute of Systems Analysis of the Russian Academy of Sciences (Moscow) in 1988. He is presently a Professor of MIS in the Michael E. Stephens College of Business, University of Montevallo. He has co-authored two books and has published over 40 refereed journal articles on multicriteria decision theory and applications. His research interests include multicriteria decision making, behavioral decision making, risk analysis, and expert systems.

Patrick Meyer is currently working as a Researcher at the University of Luxembourg. He received a Master in Mathematics in 2003 at the Faculté Polytechnique of Mons in Belgium. At present, he is working on his Ph.D. thesis. His main research interests are multiple criteria decision aiding and data mining. He has worked on projects involving financial portfolio management (CELOFA) and analysis of large amounts of financial data from stocks (MIKADO). He has contributed to the development of the TOMASO method for ordinal multiple criteria sorting and has implemented the tool in a software package.

Helen Moshkovich received her Ph.D. degree in Management Information Systems/Management Science from the Institute of Systems Analysis of the Russian Academy of Sciences (Moscow) in 1984. She is presently a Professor of MIS/Quantitative Methods in the Michael E. Stephens College of Business, University of Montevallo. Before that, she was with the Russian Academy of Sciences for 22 years. She has co-authored four books and has published over 50 refereed journal articles on multicriteria decision theory and applications. Her research interests include multicriteria decision making, behavioral decision making, decision support systems, and data mining.

Vincent Mousseau is Assistant Professor at the University of Paris Dauphine and member of the LAMSADE research laboratory. He received his Ph.D. in Computer Science/Operations Research in 1993. His reseach interest lies in the field of multiple criteria decision Aid and more specifically on preference modeling and elicitation, experimental analysis of decision behavior, and implementation of multiple criteria methodologies in real world decision problems. His work has been published in international journals such as EJOR, Journal of Global Optimization, and JMCDA.

Giuseppe Munda got a "Laurea" degree in Economics from the University of Catania, Italy (1988). He made his Ph.D. studies with the Free University of Amsterdam, where he got a Ph.D. in Economics and Econometrics (1993). At the moment he is tenured Professor of Economics of Natural Resources and Multi-Criteria Decision Analysis at Universitat Autonoma de Barcelona. Previously he has worked at the Joint Research Centre of the European Commission (Ispra site). He has been visiting lecturer at Université Panthéon-Sorbonne, University of Naples Federico II, Centre d'Economie et d'Ethique pour l' Environnement et le Développement (C3ED), University of Pisa and various universities and research centres in South-America. He has also been consultant for the Inter-American Development Bank, the European Commission, DG XII, and for the European Environment Agency. He has published one book and about 40 articles and book chapters.

Subhash C. Narula is a Professor of Management Science and Statistics in the Department of Management, School of Business, Virginia Commonwealth University, Richmond, Virginia, USA. He received his Ph.D. in Industrial and

Management Engineering from the University of Iowa. Prior to his current position, he held faculty positions at the State University of New York at Buffalo, Buffalo, New York, Rensselaer Polytechnic Institute, Troy, New York. During 1992-93, he was Chair of Optimization at the Linkoping Institute of Technology, Linkoping, Sweden. He is an elected Fellow of the American Statistical Association and the American Society for Quality. He is a recipient of the Constantine Porphyrogenetus International Association Award and was awarded the Distinguished Scholar Award of VCU in 2000. He has published papers in the leading journals of statistics, operations research and management science, contributed papers in national and international proceedings, and chapters to books.

Stefan Nickel is Full Professor (Chair) of Operations Research and Logistics at the Saarland University. He is also head of the optimization department of the Fraunhofer Institute for Industrial Mathematics. Stefan Nickel received his Diploma, his Ph.D. and his Habilitation in Mathematics at the University of Kaiserslautern, in 1992, 1995, and 1999. From 1992 he worked with the Department of Mathematics, University of Kaiserslautern as a Research Associate. From 1995 to 1999, he was an Assistant Professor at the Department of Mathematics at the University of Kaiserslautern. He is interested in location theory, combinatorial optimization, real world problems and computational geometry. Stefan Nickel is associate editor of Operations Research Letters and member of the editorial board of Computers & Operations Research.

David L. Olson is the James and H.K. Stuart Professor of MIS at the University of Nebraska. He has published research in over 60 refereed journal articles, primarily on the topic of multiple objective decision-making. He has authored three books, including Decision Aids for Selection Problems, and coauthored six others. He is a member of the Association for Information Systems, the Decision Sciences Institute, the Institute for Operations Research and Management Sciences, and the Multiple Criteria Decision Making Society. He was with Texas A&M University from 1981 through 2001 where he held a Lowry Mays endowed professorship.

Meltem Öztürk received her B.E. degree in Industrial Engineering at the University of Galatasaray (Turkey) in 2000 and her DEA at the Université Paris Dauphine in 2001. Since 2001 she is a Research Assistant and a Ph.D. Candidate in Computer Sciences at LAMSADE, Université Paris Dauphine. Her research interest lies in the field of multiple criteria decision aid, non classical logics (fuzzy set theory, Belnap logics etc.), and more specifically their use in preference modeling and aggregation methods.

Marc Pirlot is Professor of Mathematics and Operations Research at the Institute of Engineering, Mons, Belgium. He obtained his Ph.D. in Mathematics from the University of Mons in 1981. His main research interests are multicriteria decision analysis and metaheuristics for combinatorial optimizaztion. He is the co-author or editor of four books and of more than 50 refereed papers in international journals or conference proceedings. He has been president of the Operations Research Society of Belgium and is currently associate editor of the Journal of Multi-Criteria Decision Analysis and of Mathématiques et Sciences Humaines.

Justo Puerto is Full Professor (Chair) of Statistics and Operations Research at the University of Seville, where he has taught different subjects in the undergraduate and graduate programs of Mathematics, Statistics, Biology, Computer Science, and Chemistry among others. Besides, he has been visiting professor at several universities: NorthWestern University (USA), Kaiserslautern and Chemnitz (Germany), Bologna (Italy), Hirosaki (Japan), etc. Justo Puerto received his Ph.D. in Mathematics at the University of Seville in 1990. He is interested in location theory, game theory, combinatorial and classical optimization, computational geometry and mathematical education. He has published over 60 papers in professional journals. Justo Puerto has a large experience in the coordination of R&D projects and currently is associate editor of TOP, the Spanish journal of Operations Research.

Antonio Rodríguez-Chía received his Ph.D. degree in Mathematics (Statistics and Operations Research) from the Faculty of Mathematics at Sevilla University in 1998. From 1994 to 1998, he worked in Department of Mathematics at Cadiz University as a Research Associate. Since 1998 he is Full Professor of the Department of Statistics and Operations Research in the Sciences Faculty at Cadiz University. He is interested in location theory and game theory.

Marc Roubens has been Professor of Statistics and Operations Research at the Faculté Polytechnique de Mons from 1971 to 1989 and is currently Scientific Advisor of that institution. He has been with the Department of Mathematics, University of Liège, where he was Professor from 1986 to 2002. His primary interests have been in several areas of fuzzy sets theory (preference modelling, clustering and control), operations research (multiple criteria decision aid) and statistics (time series analysis , data mining) He is co-author of a book with Philippe Vincke on preference modeling and another book with Janos Fodor on fuzzy preference modelling and multiple criteria decision support. Professor Roubens has been on the editorial boards of two international journals (Fuzzy Sets and Systems, European Journal of Operations Research) and is Associate Editor of the International Journal of Approximate Reasoning. He has been president of the Belgian Operations Research Society and president of the European Chapter of the International Fuzzy Systems Association.

Bernard Roy is Emeritus Professor at Université Paris-Dauphine. He is the founder and, since 1999, honorary director of LAMSADE, a research group centered on the theme of decision aiding. Since 1980, he his scientific adviser of the Paris city transport authority. He is Graduate of the Institute of Statistics of Paris University (1957) and Doctor dés Sciences Mathématiques of Faculty of Paris (1961). After an extensive consulting experience at SEMA-METRA, he joined the Université Paris-Dauphine in 1972 and created LAMSADE. In 1975 he founded the EURO Working Group "Multiple Criteria Decision Aiding" which invariably held two annual meetings since then. He is Doctor Honoris Causa from several universities. He received the EURO Gold Medal (the highest distinction granted by EURO, the Association of European Operational Research Societies) in 1992 and the MCDM Gold Medal granted by the International MCDM Society in 1995. He his the author of several books and hundreds of research papers.

Thomas Saaty holds the Chair of University Professor, Katz Graduate School of Business, University of Pittsburgh. He has a Ph.D. in Mathematics from Yale University. Previously he was Professor at the Wharton School, University of Pennsylvania, prior to which he was involved in research at the Arms Control and Disarmament Agency, the State Department, in Washington, on nuclear arms reduction negotiations with the Soviets in Geneva. His current research is in decision-making, planning, conflict resolution and neural synthesis. He developed the Analytic Hierarchy Process (AHP) and its generalization to feedback, the Analytic Network Process (ANP) (co-developed Expert Choice for AHP and SuperDecisions for ANP) to deal with decision-making, weapons tradeoffs, and resource allocation. He has written more than 300 articles and 33 books on mathematics, operations research and decision-making. His latest books are The Brain: Unraveling the Mystery of How It Works and Creative Thinking, Problem Solving & Decision Making. He has consulted for many corporations and governments.

Yannis Siskos received his Doctorat d'Etat (1984) in Management Science from the University of Paris-Dauphine (France) and his DEA (1977) and Doctorat 3e Cycle (1979) in Computer Science and Operational Research from the University Pierre et Marie Curie (France). He is presently Professor of Decision Science at the Department of Informatics, University of Piraeus. During 1984-2001 he was Professor at the Technical University of Crete and founding director of the Decision Support Systems Laboratory. He has been visiting professor at the University of Paris-Dauphine (France), the University of Aix-Marseille II (France), the University of Laval (Canada), and the University of Cyprus. He is presently President of the Hellenic Operational Research Society and his research interests fall into the areas of multicriteria decision support systems, service quality, and mathematical programming. He is the author of several books and over sixty articles in international scientific journals.

Roman Slowinski is Professor and Head of the Laboratory of Intelligent Decision Support Systems within the Institute of Computing Science, Poznan University of Technology, Poland. He received a M.Sc. in Computer Science, Ph.D. in Operations Research, and Habilitation in Computing Science from the Poznan University of Technology in 1974, 1977, and 1981, respectively. He has been Professor on European Chair at the University of Paris Dauphine and invited professor at the Swiss Federal Institute of Technology in Lausanne and at the University of Catania. His research concerns operational research and artificial intelligence, including multiple-criteria decision analysis, preference modelling, knowledge-based decision support in medicine, technology and economics, project scheduling, and rough set theory approach to knowledge and data engineering. He is laureate of the EURO Gold Medal (1991) and Doctor Honoris Causa of Polytechnic Faculty of Mons (2000) and University of Paris Dauphine (2001). Since 1999 he is co-editor-in-chief of the European Journal of Operational Research.

Charles H. Smith received an M.B.A. from the College of William and Mary (Williamsburg, VA) in 1981. Previously he received his Ph.D. in Mathematics from the University of Maryland in 1975. Since 1982 he has taught in the School of Business at Virginia Commonwealth University, where he is currently Associate Professor in the Department of Management. His research publications have been in the areas of applied decision analysis and operations management in addition to multiple criteria decision making.

Jaap Spronk is Full Professor of Finance & Investment (since 1982, tenure), Vice-Dean and Director of Bachelor & Master Programs in Economics, Erasmus University, Rotterdam. He obtained a Ph.D. in Economics (Interactive Multiple Goal Programming for Financial Planning, 1980) at Erasmus University. His main current research interests include: financial risk management, performance evaluation, financial modelling, financial management and strategy. The main areas of application are the following: professional investment, banking, transportation, aviation in particular, government and state agencies. Scientific board functions include: founder (1986) and chairman of the EURO Working Group on Financial Modelling, member of the executive committee of the Special Interest Group on MCDM (1980-1992), president (1991-1992) of EURO (Association of European Operational Research Societies). He has been initiator (1983) of the First International Summer School on Multiple Criteria Decision Methods, Applications and Software. His honours include the Gold Medal of the International Society of Multiple Criteria Decision Making (2002).

Ralph E. Steuer is the Charles S. Sanford, Sr. Chair of Business in the Terry College of Business, University of Georgia, USA. His degrees are an Sc.B. from Brown University, an M.B.A. from Cornell University, and a Ph.D. from the University of North Carolina. Dr. Steuer is the author of "Multiple Criteria Optimization: Theory, Computation and Application," the ADBASE multiple objective linear programming package, and over 85 scientific publications. He is a co-founder of the International Society on Multiple Criteria Decision Making and was editor of the Society's newsletter, MCDM WorldScan, for seventeen years. Dr. Steuer's interests are in efficient sets and surfaces, models with multiple criteria, portfolio theory in finance, and interactive procedures in multiple criteria optimization.

Theodor J Stewart has been Professor of Statistical Sciences at the University of Cape Town since 1984, with main responsibility for operational research and decision analysis. He has published widely in these areas, including a recent book co-authored with Valerie Belton of the University of Strathclyde, on Multiple Criteria Decision Analysis. Professor Stewart has been involved in many applications of these topics in private and public sectors, linked primarily to strategic planning and natural resources management. Recent work has included water resources management, project prioritization problems, and strategic bundling of assets for the electricity industry. He is on the editorial boards of four international journals, including a period as guest editor of a number of issues of International Transactions in Operational Research. He is vice president (at large) of the International Federation of Operational Research Societies, and is president elect of the International Society of Multiple Criteria Decision Making.

Alexis Tsoukiàs is a Senior Researcher of CNRS within the LAMSADE, Université Paris Dauphine. He holds a Ph.D. in Systems and Computer Science Engineering from Politecnico di Torino, Italy. He is author or editor of three books and has published more than 50 papers in journals and contributed volumes on several fields including multiple criteria decision aiding, preference modelling, formal logic, artificial intelligence etc.. He has served on different editorial boards and at different positions to several European working groups and OR societies. Presently he is president elect of EURO, the European association of OR societies within IFORS.

Jean-Claude Vansnick is Professor at the Warocqué Faculty of Management of the University of Mons-Hainaut, Belgium and member of the Warocqué Research Center. He received, in 1973, his Ph.D. in Mathematics from the Free University of Brussels and, in 1974, the Royal Academy of Belgium award (Section Mathematical Sciences). Since then, he has extended his areas of interest to measurement theory and decision aid. His paper "Strength of Preference. Theoretical and Practical Aspects" was selected as National Contribution of Belgium for the Tenth Triennial Conference of the International Federation of Operational Research Societies (IFORS 84). He is member of groups on multicriteria analysis and was guest professor at several International Summer Schools on Multicriteria Decision Aid. He is co-author of the MACBETH approach.

Philippe Vincke is Full Professor and the director of the Service de Mathématiques de la Gestion, and vice-rector of the Université Libre de Bruxelles. His research interests are preference modelling, aggregation, axiomatics of decision-aid methods and applications of Operations Research and Decision Aid (through industrial collaborations). He was president of EURO (European Association of Operational Research) in 2001 and 2002 and is currently vice-president of IFORS (International Federation of Operational Research Societies). He is author or co-author of five books and about 100 papers in international journals.

H. Roland Weistroffer received his M.A. degree in Mathematics from Duke University in 1973, and his Doctor of Science degree in applied mathematics from the Free University Berlin in 1976. He taught computer science at the University of Natal in Durban, South Africa, from 1978 to 1979, and was a Chief Research Officer in operations research at the Centre for Scientific and Industrial Research in Pretoria, South Africa, from 1980 to 1983. Since then he has been at Virginia Commonwealth University in Richmond, Virginia. He is currently an Associate Professor in the Information Systems Department in the School of Business. His research interests are in decision support systems, multiple criteria decision making, and object oriented modeling.

Margaret M. Wiecek is Professor in the Department of Mathematical Sciences at Clemson University. She obtained a M.S. degree in Electrical Engineering and a Ph.D. degree in Systems Engineering from the University of Mining and Metallurgy in Krakow. Her research area includes theory, methodology, and applications of mathematical programming with special interest in multi-criteria optimization and decision-making. She has been the Sofia Kovalevskaia Visiting Professor and Mercator Visiting Professor at the University of Kaiserslautern and a visiting professor at the University of Copenhagen. She has published over sixty research articles. In the United States, her research has been funded by the National Automotive Research Center, the National Institute of Science and Technology, the National Science Foundation, and the Office of Naval Research. She is a member of the Institute for Operations Research and Management Sciences, the Mathematical Programming Society, and the Multiple Criteria Decision Making Society.

Constantin Zopounidis is Professor of Financial Management and Operations Research and Chairman of the Department of Production Engineering and Management, Technical University of Crete, Greece. He holds a Doctorat D'Etat in Management Science and a D.E.A. in Financial Management, both from the University of Paris-IX Dauphine, France. Prof. Zopounidis' research interests include multiple criteria decision making, financial engineering and financial risk management. His work has been published in such journals as Decision

Sciences, European Journal of Operational Research, Decision Support Systems, The Journal of the Operational Research Society, Expert Systems with Applications, Global Finance Journal, International Journal of Intelligent Systems in Accounting, Finance and Management and Computational Economics. He edited or co-edited more than 20 books on financial management and multicriteria decision aid.

Index

Early Titles in the
INTERNATIONAL SERIES IN OPERATIONS RESEARCH & MANAGEMENT SCIENCE
Frederick S. Hillier, Series Editor, *Stanford University*

Saigal/ *A MODERN APPROACH TO LINEAR PROGRAMMING*
Nagurney/ *PROJECTED DYNAMICAL SYSTEMS & VARIATIONAL INEQUALITIES WITH APPLICATIONS*
Padberg & Rijal/ *LOCATION, SCHEDULING, DESIGN AND INTEGER PROGRAMMING*
Vanderbei/ *LINEAR PROGRAMMING*
Jaiswal/ *MILITARY OPERATIONS RESEARCH*
Gal & Greenberg/ *ADVANCES IN SENSITIVITY ANALYSIS & PARAMETRIC PROGRAMMING*
Prabhu/ *FOUNDATIONS OF QUEUEING THEORY*
Fang, Rajasekera & Tsao/ *ENTROPY OPTIMIZATION & MATHEMATICAL PROGRAMMING*
Yu/ *OR IN THE AIRLINE INDUSTRY*
Ho & Tang/ *PRODUCT VARIETY MANAGEMENT*
El-Taha & Stidham/ *SAMPLE-PATH ANALYSIS OF QUEUEING SYSTEMS*
Miettinen/ *NONLINEAR MULTIOBJECTIVE OPTIMIZATION*
Chao & Huntington/ *DESIGNING COMPETITIVE ELECTRICITY MARKETS*
Weglarz/ *PROJECT SCHEDULING: RECENT TRENDS & RESULTS*
Sahin & Polatoglu/ *QUALITY, WARRANTY AND PREVENTIVE MAINTENANCE*
Tavares/ *ADVANCES MODELS FOR PROJECT MANAGEMENT*
Tayur, Ganeshan & Magazine/ *QUANTITATIVE MODELS FOR SUPPLY CHAIN MANAGEMENT*
Weyant, J./ *ENERGY AND ENVIRONMENTAL POLICY MODELING*
Shanthikumar, J.G. & Sumita, U./ *APPLIED PROBABILITY AND STOCHASTIC PROCESSES*
Liu, B. & Esogbue, A.O./ *DECISION CRITERIA AND OPTIMAL INVENTORY PROCESSES*
Gal, T., Stewart, T.J., Hanne, T. / *MULTICRITERIA DECISION MAKING: Advances in MCDM Models, Algorithms, Theory, and Applications*
Fox, B.L. / *STRATEGIES FOR QUASI-MONTE CARLO*
Hall, R.W. / *HANDBOOK OF TRANSPORTATION SCIENCE*
Grassman, W.K. / *COMPUTATIONAL PROBABILITY*
Pomerol, J-C. & Barba-Romero, S. / *MULTICRITERION DECISION IN MANAGEMENT*
Axsäter, S. / *INVENTORY CONTROL*
Wolkowicz, H., Saigal, R., & Vandenberghe, L. / *HANDBOOK OF SEMI-DEFINITE PROGRAMMING: Theory, Algorithms, and Applications*
Hobbs, B.F. & Meier, P. / *ENERGY DECISIONS AND THE ENVIRONMENT: A Guide to the Use of Multicriteria Methods*
Dar-El, E. / *HUMAN LEARNING: From Learning Curves to Learning Organizations*
Armstrong, J.S. / *PRINCIPLES OF FORECASTING: A Handbook for Researchers and Practitioners*
Balsamo, S., Personé, V., & Onvural, R./ *ANALYSIS OF QUEUEING NETWORKS WITH BLOCKING*
Bouyssou, D. et al. / *EVALUATION AND DECISION MODELS: A Critical Perspective*
Hanne, T. / *INTELLIGENT STRATEGIES FOR META MULTIPLE CRITERIA DECISION MAKING*
Saaty, T. & Vargas, L. / *MODELS, METHODS, CONCEPTS and APPLICATIONS OF THE ANALYTIC HIERARCHY PROCESS*
Chatterjee, K. & Samuelson, W. / *GAME THEORY AND BUSINESS APPLICATIONS*
Hobbs, B. et al. / *THE NEXT GENERATION OF ELECTRIC POWER UNIT COMMITMENT MODELS*
Vanderbei, R.J. / *LINEAR PROGRAMMING: Foundations and Extensions, 2nd Ed.*
Kimms, A. / *MATHEMATICAL PROGRAMMING AND FINANCIAL OBJECTIVES FOR SCHEDULING PROJECTS*
Baptiste, P., Le Pape, C. & Nuijten, W. / *CONSTRAINT-BASED SCHEDULING*
Feinberg, E. & Shwartz, A. / *HANDBOOK OF MARKOV DECISION PROCESSES: Methods and Applications*
Ramík, J. & Vlach, M. / *GENERALIZED CONCAVITY IN FUZZY OPTIMIZATION AND DECISION ANALYSIS*
Song, J. & Yao, D. / *SUPPLY CHAIN STRUCTURES: Coordination, Information and Optimization*
Kozan, E. & Ohuchi, A. / *OPERATIONS RESEARCH/ MANAGEMENT SCIENCE AT WORK*
Bouyssou et al. / *AIDING DECISIONS WITH MULTIPLE CRITERIA: Essays in Honor of Bernard Roy*

Early Titles in the
INTERNATIONAL SERIES IN OPERATIONS RESEARCH & MANAGEMENT SCIENCE
(Continued)

Cox, Louis Anthony, Jr. / *RISK ANALYSIS: Foundations, Models and Methods*
Dror, M., L'Ecuyer, P. & Szidarovszky, F. / *MODELING UNCERTAINTY: An Examination of Stochastic Theory, Methods, and Applications*
Dokuchaev, N. / *DYNAMIC PORTFOLIO STRATEGIES: Quantitative Methods and Empirical Rules for Incomplete Information*
Sarker, R., Mohammadian, M. & Yao, X. / *EVOLUTIONARY OPTIMIZATION*
Demeulemeester, R. & Herroelen, W. / *PROJECT SCHEDULING: A Research Handbook*
Gazis, D.C. / *TRAFFIC THEORY*

*** A list of the more recent publications in the series is at the front of the book ***